ASTRONOMY AND ASTROPHYSICS ABSTRACTS

A Publication of the Astronomisches Rechen-Institut Heidelberg
Member of the Abstracting Board
of the International Council of Scientific Unions
Astronomy and Astrophysics Abstracts is Prepared
Under the Auspices of the International Astronomical Union

Volume 31
Literature 1982, Part 1

Edited by
S. Böhme W. Fricke H. Hefele I. Heinrich W. Hofmann
D. Krahn V. R. Matas L. D. Schmadel G. Zech

Springer-Verlag Berlin Heidelberg GmbH

Astronomisches Rechen-Institut Heidelberg
Director: Professor Dr. Walter Fricke

Astronomy and Astrophysics Abstracts
Editors-in-Chief: Inge Heinrich, Dr. Lutz D. Schmadel

ISBN 978-3-662-12336-2   ISBN 978-3-662-12334-8 (eBook)
DOI 10.1007/978-3-662-12334-8

Preface

Astronomy and Astrophysics Abstracts, which has appeared in semi-annual volumes since 1969, is devoted to the recording, summarizing and indexing of astronomical publications throughout the world. It is prepared under the auspices of the International Astronomical Union (according to a resolution adopted at the 14th General Assembly in 1970).

Astronomy and Astrophysics Abstracts aims to present a comprehensive documentation of literature in all fields of astronomy and astrophysics. Every effort will be made to ensure that the average time interval between the date of receipt of the original literature and publication of the abstracts will not exceed eight months. This time interval is near to that achieved by monthly abstracting journals, compared to which our system of accumulating abstracts for about six months offers the advantage of greater convenience for the user.

Volume 31 contains literature published in 1982 and received before July 15, 1982; some older literature which was received late and which is not recorded in earlier volumes is also included.

We acknowledge with thanks contributions to this volume by Dr. J. Bouška, Prague, who surveyed journals and publications in Czech and supplied us with abstracts in English.

We express our warmest thanks again to Ms. Helga Ballmann, Ms. Mona El-Choura, Ms. Monika Kohl, Ms. Sylvia Matyssek and Ms. Angelika Meßmer for typing the text of this volume on IBM 72 Composers, for compiling the pages from abstract slips in a perfect form for offset reproduction, and for punching material for the author index and for the subject index, which finally were printed with a TN chain on a 1403 IBM high-speed printer. Finally, we have to thank Mr. Claus Leitherer, Mr. Uwe Reichert and Mr. Roland Zanella who supported our task by careful proofreading.

Heidelberg, September 1982

Siegfried Böhme   Dietlinde Krahn
Walter Fricke   Vladimir R. Matas
Herbert Hefele   Lutz D. Schmadel
Inge Heinrich   Gert Zech
Wilfried Hofmann

Contents

X   Contents

Introduction

Astronomical bibliographies

Astronomy and Astrophysics Abstracts begins documentation and abstracting from the year 1969. For information on astronomical literature before this date consultation of one of the following bibliographies is suggested:

(1) J. J. de Lalande, Bibliographie Astronomique, Paris 1803 (this work covers the time from 480 B. C. to the year 1803, VIII + 966 pages).

(2) J. C. Houzeau, A. Lancaster, Bibliographie générale de l'astronomie, Volume I (in two parts), Bruxelles 1887, 1889, Volume II, Bruxelles 1882. The complete title of Volume II is "Bibliographie générale de l'astronomie ou catalogue méthodique des ouvrages, des mémoires et des observations astronomiques, publiés depuis l'origine de l'imprimerie jusqu'en 1880". A new edition of these volumes was prepared by D. W. Dewhirst in 1964.

(3) Bibliography of Astronomy, 1881 - 1898. The literature of this period was recorded on standard slips by the Observatoire Royal de Belgique. From the material (some 52,000 items) a microfilm version was produced by University Microfilms Limited, Tylers Green, High Wycombe, Buckinghamshire, England, in 1970.

(4) Astronomischer Jahresbericht, 1899 gegründet von Walter Wislicenus, herausgegeben vom Astronomischen Rechen–Institut in Heidelberg (formerly in Berlin), Verlag W. de Gruyter, Berlin. For the period from 1899 to 1968 sixty-eight volumes were published, each of which, in general, covers the literature of one year.

(5) Bulletin Signalétique – Section 120: Astronomie, Physique Spatiale, Géophysique. Published by Centre de Documentation du Centre National de la Recherche Scientifique, Paris. This publication is a continuation of "Bibliographie Mensuelle de l'Astronomie" founded in 1933 by the Société Astronomique de France. The publication is continued.

(6) Referativnyj Zhurnal. Founded in 1953 and published by Vsesoyuznyj Institut Nauchnoj i Tekhnicheskoj Informatsii, Akademiya Nauk, Moskva. The publication is continued.

Concept of Astronomy and Astrophysics Abstracts

This abstracting service aims to present a comprehensive documentation of the literature in all fields of astronomy and astrophysics and their border fields. It appears in semi-annual volumes. Two of these volumes cover the literature of one calendar year. The half-yearly period of issue is regarded as an optimal period for summarizing papers into subject categories and for the presentation of abstracts as quickly as possible after the publication of the original literature. The recording summarizing and indexing of astronomical publications of the year 1982 received from January 1982 to July 1982 are subjects of **Volume 31**. It also records a number of papers issued before 1982 but received within this period.

The main characteristics of the concept of Astronomy and Astrophysics Abstracts may be summarized as follows:

(1) The subdivision of astronomy and its border fields into subject categories is facilitated by the fact that the astronomical objects appear to be particularly well suited for the formation of categories. It may be assumed that such subdivisions can be maintained for a long period. Experience shows, however, that progress in research might imply minor changes in the classification scheme.

(2) Each paper has been classified into one of 108 numbered subject categories and given a serial number within the category. In this way each item is numbered by six figures: the first three indicate the number of the category, the following three the serial number within the category. Reference to an abstract in Volume 1 is indicated by "01" before the number of the category; for example: 01.074.028, denotes Volume 1, category 074, abstract 028.

A paper might be classified into more than one category. In this case, its abstract is placed only in one category, whereas in the other categories only cross references are given. These are listed at the end of each category.

(3) Authors' abstracts are used whenever possible. Popular articles are not abstracted.

(4) If possible, titles of papers and abstracts are given in English. A special reference is made to titles which we have not taken in the original language.

Transliteration scheme for the Russian alphabet

The transliteration of the Russian alphabet in use in Astronomy and Astrophysics Abstracts is presented here.

А	а	a		П	п	p
Б	б	b		Р	р	r
В	в	v		С	с	s
Г	г	g		Т	т	t
Д	д	d		У	у	u
Е	е	e		Ф	ф	f
Ё	ё	e		Х	х	kh
Ж	ж	zh		Ц	ц	ts
З	з	z		Ч	ч	ch
И	и	i		Ш	ш	sh
Й	й	j		Ш	ш	shch
К	к	k		Ы	ы	y
Л	л	l		Ь	ь	'
М	м	m		Э	э	eh
Н	н	n		Ю	ю	yu
О	о	o		Я	я	ya

This transliteration was recommended by the Abstracting Board of the International Council of Scientific Unions in 1969. It corresponds essentially to the transliteration proposed by the Academy of Sciences, Moscow, which is used by the Referativnyj Zhurnal. In this case the letters can be read and printed by usual data processing machines.

If the names of Russian authors in the literature are transliterated in a different scheme, we present the names as they are given in the references cited and in addition in brackets according to our transliteration table.

Sources of information

The majority of sources of information for this volume is given in section **001 Periodicals** and in section **008 Observa-**

tories, Institutes. Section 001 records 656 periodicals indicating full titles and publishers. It may be noted that the titles of the periodicals are given in the original languages, and that Russian titles have been transliterated applying the transliteration scheme given above. Section 008 records 103 periodicals; these are publication series of observatories and astronomical institutes. Titles of the periodicals have been given following the recommendations of the "International List of Periodical Title Word Abbreviations" and its additions (see also **Abbreviations**, p. 10). In most cases they permit recognition of the full title without recourse to the key in section 001.

If other abstracting journals have been consulted in order to examine the degree of completeness of our service, we cite these papers and give reference to the abstracting service.

Author index and subject index

The subject category and the serial number have been used as a reference both in the author index and the subject index. These references are more precise than page references. They offer considerable advantages in indexing by means of data processing machines, and are more convenient for the user.

The author index of this volume contains 9740 names. A complete reference comprises six figures, three for the subject category and three for the serial number within the category. In the case of more than one reference to abstracts in one category, the number of the category is given only once and not repeated in the immediately following references. The total number of papers (some do not give names of authors) recorded in this volume amounts to 8436.

We consider the subject index as an approximation to an optimal index covering all fields of astronomy and astrophysics and their border fields. The assigning of one or more key words to a paper is, undoubtedly, a difficult task. Some journals have started giving key words together with the titles of papers. These key words are chosen by the authors themselves. Starting with Volume 18, the subject index was enlarged to a certain extent in order to provide a thesaurus of astronomical and astrophysical terms. This is done not only for the users' convenience, but also with the intention to propose the use of special key words to authors and publishers.

While each volume is scheduled to contain an author index and a subject index, the magnetic tapes containing the index information will be used to produce separate index volumes (authors and subjects) at intervals of five years.

The sorting program for the author and subject indexes is based on the IBM SORT/MERGE Program. This program sorts blank before hyphen (–) and before letters. Apostrophes are ignored by a special routine. The computations and printing were carried out on an IBM 360/44.

The two most common and widely used classification systems in astronomy and astrophysics are given by Class 9 of the International Classification System for Physics, published by the International Council of Scientific Unions Abstracting Board (Second edition 1978. ICSU-AB, 17 Rue Mirabeau, 75017 Paris, France, ISSN 0305-9618), and the Astronomy and Astrophysics Abstracts classification. In order to facilitate literature searches, we introduce a concordance relation between these two very different systems. This solution is only a unilateral one. Starting from the fourth hierarchical level of the ICSU-AB system, the appropriate Astronomy and Astrophysics Abstracts chapter numbers are listed. This cannot imply an identical content of the respective chapters in both systems. In many cases there is only a rather partial concordance, and therefore the Astronomy and Astrophysics Abstracts numbers are enclosed in parentheses. Due to the fact that our service only aims to present a comprehensive documentation of the literature in all fields of astronomy and astrophysics, only a certain part of Class 9 of the ISCU-AB scheme is covered.

The users are requested to inform us on spelling errors within the author and subject indexes in order to assist us in eliminating mistakes in future cumulative indexes.

Concordance Relation

between the ICSU-AB International Classification System for Physics
and the *Astronomy and Astrophysics Abstracts* Classification Scheme

ICSU-AB International Classification System for Physics	Astronomy and Astrophysics Abstracts Classification Scheme
0 General	
01.10 Announcements, news, and **organizational activities**	
01.10.C	006 (010)
01.10.F	011
01.10.H	013 (010)
01.30 Physics literature and publications	
01.30.B	012 (014)
01.30.C	012
01.30.E	003
01.30.K	002 (003)
01.30.M	003
01.30.P	003
01.30.R	014
01.30.T	002
01.40 Education	014
01.50 Educational aids	014
01.60 Biographical, historical, and personal notes	004 (005, 006, 007)
01.65 History of science	004
01.90 Other topics of general interest	015

ICSU-AB International Classification System for Physics	Astronomy and Astrophysics Abstracts Classification Scheme

9 Geophysics, Astronomy, and Astrophysics

91.10 Geodesy and gravity

91.10.B	046
91.10.N	044 (045)
91.10.Q	081

91.25 Geomagnetism and paleomagnetism; geoelectricity	084

91.35 Earth's interior structure and properties	081

91.90 Other topics in solid Earth physics	081

92.60 Meteorology	082

92.65 Atmospheric optics	082

94.10 Physics of the neutral atmosphere

94.10.B	082
94.10.D	082
94.10.F	082
94.10.G	082 (063)
94.10.H	082
94.10.L	082
94.10.N	106
94.10.Q	082
94.10.S	084

94.20 Physics of the ionosphere

94.20.B	083
94.20.D	083
94.20.M	083 (084)
94.20.P	083 (084)
94.20.W	083 (062)
94.20.Y	083 (084)

94.30 Physics of the magnetosphere

94.30.C	084
94.30.D	084

ICSU-AB International Classification System for Physics	Astronomy and Astrophysics Abstracts Classification Scheme
94.30.E	084
94.30.F	084 (062)
94.30.G	084 (062)
94.30.H	084
94.30.L	084
94.30.M	084
94.30.S	084
94.30.V	084 (074)
94.30.W	084 (078, 143)

94.40   Cosmic rays

94.40.C	143
94.40.E	143 (078, 106)
94.40.H	078
94.40.K	143 (085)
94.40.L	143 (078)
94.40.V	105 (143)

94.60   Interplanetary space

94.60.D	074
94.60.F	074
94.60.G	074 (062)
94.60.K	106
94.60.M	106
94.60.Q	074 (091, 094.0)
94.60.R	106 (062)

94.80   Aerospace facilities and techniques,
         space research

94.80.P	053 (051)
94.80.R	054 (051)
94.80.W	032.5

95.10   Fundamental astronomy

95.10.C	042 (043, 052)
95.10.E	042 (052)
95.10.G	041 (079, 095, 096)
95.10.J	041

ICSU-AB International Classification System for Physics	Astronomy and Astrophysics Abstracts Classification Scheme
95.30 Fundamental aspects of astrophysics	
95.30.C	061 (022)
95.30.E	022
95.30.G	022 (061)
95.30.J	063
95.30.L	062
95.30.Q	062
95.30.S	066.0 (162)
95.45 Observatories	008 (009)
95.55 Astronomical instruments	
95.55.B	032.0
95.55.C	032.0 (031.0)
95.55.E	032.0
95.55.J	033
95.55.L	032.5
95.65 Auxiliary and recording instruments	034
95.70 Other instrumentation and techniques (including clocks, frequency standards, etc.)	035 (031.5, 034, 036)
95.75 Techniques of observation and reduction	
95.75.D	031.5
95.75.F	031.5
95.75.H	031.5
95.75.K	031.5
95.75.M	031.5 (021)
95.75.P	021
95.80 Catalogues, atlases, etc.	002 (047)
96.10 General, solar nebula, and cosmogony	107 (091)
96.20 Moon	
96.20.B	094.0
96.20.D	094.0 (094.5)
96.20.J	094.0

ICSU-AB International Classification System for Physics	Astronomy and Astrophysics Abstracts Classification Scheme

96.30   Planets and satellites (excluding the moon)

96.30.D	092
96.30.E	093
96.30.G	097
96.30.H	098
96.30.K	099
96.30.M	100
96.30.T	101

96.50   Other objects in the planetary system

96.50.D	106
96.50.G	102 (103)
96.50.K	104
96.50.M	105

96.60   Solar physics

96.60.C	080 (075)
96.60.F	071 (080)
96.60.K	080
96.60.M	071
96.60.N	073 (074)
96.60.P	074
96.60.Q	072
96.60.R	073 (076, 077)
96.60.S	073
96.60.V	078 (074)

97.10   Stellar characteristics and properties

97.10.B	131
97.10.C	065 (061)
97.10.E	064 (063)
97.10.F	112 (064)
97.10.H	064 (112)
97.10.K	116 (065)
97.10.L	116 (065)
97.10.N	115
97.10.Q	115
97.10.R	115 (113, 114).
97.10.T	114
97.10.V	111
97.10.W	111

ICSU-AB **International Classification System for Physics**	Astronomy and Astrophysics Abstracts **Classification Scheme**

97.20   Normal stars (by class): general or individual

| 97.20.D | 121 |
| 97.20.R | 126 |

97.30   Variable and peculiar stars (including novae)

97.30.E	114
97.30.F	116
97.30.G	122 (123)
97.30.J	122 (123)
97.30.K	122 (123)
97.30.N	122 (123)
97.30.Q	124 (122, 123)
97.30.S	122 (123)

**97.60   Late stages of stellar evolution
(including black holes)**

97.60.B	125
97.60.G	141.5
97.60.J	066.5
97.60.L	066.0
97.60.S	066.0 (065)

**97.80   Binary and multiple stars (including
extrasolar planetary systems)**

97.80.D	118 (002)
97.80.F	120
97.80.H	119
97.80.J	142.0 (117)
97.80.K	118
97.80.M	118

| **98.10   Stellar dynamics** | 151 |

98.20   Stellar clusters and associations

98.20.C	152
98.20.E	153
98.20.H	154

ICSU-AB International Classification System for Physics	Astronomy and Astrophysics Abstracts Classification Scheme

98.40   Interstellar matter and nebulae

98.40.B	131
98.40.C	131
98.40.F	132
98.40.H	132 (134)
98.40.J	133 (112)
98.40.K	134 (131)
98.40.M	134
98.40.N	125

98.50   The Galaxy; extragalactic objects and systems

98.50.C	158
98.50.E	151 (158)
98.50.H	158 (162)
98.50.K	160
98.50.L	155 (156, 157)
98.50.M	160 (159)
98.50.R	158
98.50.T	161

98.70   Other objects and background radiations of unknown origin or distances

98.70.D	141.0
98.70.J	141.0
98.70.L	133
98.70.Q	142.0 (142.5)
98.70.S	143
98.70.V	066.0 (142.0, 142.5, 162)

98.80   Cosmology

98.80.B	162
98.80.D	162 (066.0)
98.80.F	061 (162)

Abbreviations

Abbreviations used in *Astronomy and Astrophysics Abstracts* are primarily based on the 'International List of Periodical Title Word Abbreviations', prepared for the UNISIST/ICSU-AB Working Group on Bibliographic Descriptions (1970).

A.A.B.	Associazione Astrofili Bolognesi	Atmos.	Atmosf–, Atmosph–
Aarg.	Aargang	BAA	British Astronomical Association
AAS	American Astronomical Society	Bayer.	Bayerisch–
AAVSO	American Association of Variable Star Observers	Beitr.	Beitrag, Beiträge
		Beob.	Beobacht–
Abh.	Abhandlung–	Ber.	Bericht–
Abstr.	Abstract–	Bibl.	Bibliot–
Abt.	Abteilung	Bibliogr.	Bibliograf–, Bibliograph–
Acad.	Academi–, Academy	BIH	Bureau International de l'Heure (Paris)
Accad.	Accademi–	Bimest.	Bimestr–
Act.	Active, Activit–	Bl.	Blatt, Blätter
Adm.	Administr–	Bol.	Boletin
Adv.	Advanc–	Boll.	Bolletino
Aehron.	Aehronomi–	Bul.	Buleten–, Buletin–, Bulten
Aeron.	Aeronom–	Bull.	Bulletin–, Bullettino
Aeronaut.	Aeronauti–	Bur.	Bureau–
Aerosp.	Aerospace	Byul.	Byuleten–, Byuletin–
AG	Astronomische Gesellschaft	Byull.	Byulleten–
AIAA	American Institute of Aeronautics and Astronautics	C.R.	Comptes Rendus
		Cah.	Cahier–
AJB	Astronomischer Jahresbericht	Calif.	California
Akad.	Akadem–	Cas.	Casopis
Ala.	Alabama	Cent.	Center–, Central, Centrale, Centrally, Centre
Alm.	Almanac–, Almanak–		
Amat.	Amateur–	Cercet.	Cercetary
An.	Anais, Anale–, Anali–, Anals	Chem.	Chemi–
Anal.	Analis–, Analit–, Analys–, Analyt–	Chim.	Chimi–
Angew.	Angewandt–	Chron.	Chronic–, Chronik, Chronique
Ann.	Annaes, Annal–	Chronom.	Chronometr–
Annu.	Annu–	Cie.	Compagnie
Anst.	Anstalt	Cienc.	Ciencia–
Anu.	Anual–, Anuar–	Cient.	Cientific–
Anz.	Anzeiger	Circ.	Circolar–, Circolo, Circolaire–, Circular–, Circulo
Appl.	Applied		
Arb.	Arbeit	Cirk.	Cirkulaer–
Arch.	Archiv–	Cl.	Clasa, Classe–
Årg.	Årgang	Co.	Companies, Company
Ariz.	Arizona	Coll.	College
Ark.	Arkiv–	Collect.	Collect–
Arkh.	Arkhiv–	Colloq.	Colloqui–
Artif.	Artifici–	Colo.	Colorado
ASA	Astronomical Society of Australia	Comet.	Cometary
Asoc.	Asocia–	Commentat.	Commentat–
ASP	Astronomical Society of the Pacific	Commun.	Communica–
ASSA	Astronomical Society of Southern Africa	Comput.	Computation, Computer–, Computing
Assem.	Assembl–	Comun.	Comunica–
Assoc.	Associ–	Conf.	Conferen–
Assoz.	Assozi–	Congr.	Congres–
Astrofis.	Astrofisic–	Conn.	Connecticut
Astrofiz.	Astrofizi–	Contract.	Contract–
Astrometr.	Astrometr–	Contrib.	Contribu–
Astron.	Astronom–	Cosm.	Cosmic–
Astronaut.	Astronauti–, Astronauty–	Cosmochim.	Cosmochimi–
Astrophys.	Astrophys–	COSPAR	Committee on Space Research
ASV	Astronomical Society of Victoria	Crystallogr.	Crystallograph–
ASWA	Astronomical Society of Western Australia	CSIRO	Commonwealth Scientific and Industrial Research Organization
At.	Atom–	Cult.	Cultur–, Cultuur

Curr.	Current
D.C.	District of Columbia
DDR	Deutsche Demokratische Republik
Del.	Delaware
Dep.	Departament, Département, Department
Dev.	Development−, Développement−
Diss.	Disserta−
Div.	Divis−
Doc.	Document−
Dok.	Dokument−
Dokl.	Doklad−
Ehksp.	Ehksperiment−
Eidg.	Eidgenössisch−
Eksp.	Eksperiment−
Electron.	Electroni−
Eng.	Engineer−
Environ.	Environment−
Equip.	Equipement, Equipment
Ergeb.	Ergebnis−
ESA	European Space Agency
ESO	European Southern Observatory
ESRO	European Space Research Organization
Eval.	Evaluation−
Exp.	Experiment−
Extraterr.	Extraterrestr−
F. R. Germany	Federal Republic of Germany
Fac.	Facolt−, Faculd−, Facult−
Fak.	Fakult
Fasc.	Fascicul−
Fenn.	Fenni−
Fis.	Fisic−, Fisik−
Fiz.	Fizic−, Fizik−, Fizyk−
Fla.	Florida
Fluid.	Fluidi−
Fond.	Fondation−, Fondazione
Fortschr.	Fortschritt−
Fotogr.	Fotograf−
Found.	Foundation−
Freq.	Frequen−
Fundam.	Fundamenta−
Fys.	Fysik−, Fysisch, Fysisk−
Fyz.	Fyzik−
G.	Giornale
Ga.	Georgia
Gaz.	Gazeta, Gazette
Gazz.	Gazzetta
Gen.	General
Geochem.	Geochem−
Geochim.	Geochim−
Geod.	Geodaes−, Geodaet−, Geodes−, Geodet−, Geodez−
Geofis.	Geofis−
Geofiz.	Geofiz−
Geofys.	Geofys−
Geogr.	Geograf−, Geograph−
Geokhim.	Geokhim−
Geol.	Geolog−, Geolosk−
Geomagn.	Geomagneti−
Geophys.	Geophys−
Ges.	Gesellschaft
Gesch.	Geschichte
Gl.	Glavno−
Glas.	Glasnik
Gos.	Gosudarst−
Gov.	Government−
Grenzgeb.	Grenzgebiet−
GSFC	Goddard Space Flight Center
H. M.	Her Majesty's, His Majesty's
Handb.	Handbook, Handbuch
Her.	Herald−
Hist.	History
Hochsch.	Hochschule

Hoegsk.	Hoegskol−
HR-diagram	Hertzsprung-Russell diagram
Hydrogr.	Hydrograf−, Hydrograph−
IAF	International Astronautical Federation
IAU	International Astronomical Union
IBM	International Business Machines Corporation
ICSU	International Council of Scientific Unions
ICSU-AB	International Council of Scientific Unions− Abstracting Board
IEEE	Institute of Electrical and Electronics Engineers
Ill.	Illinois
Inc.	Incorporated
Ind.	Industr−
Inf.	Informat−, Informaz−, Informe−
Ing.	Ingenieur
INIS	International Nuclear Information System
INSPEC	International Information Services for the Physics and Engineering Communities
Inst.	Institut−, Instytut−
Instn.	Institution
Instrum.	Instrument−
Int.	Internationa−, Internazional−
Inter.	Intérieur−, Interior
Interplanet.	Interplanetary
Intez.	Intezet−
Invest.	Investiga−
Ionos.	Ionosfer−, Ionospher−
Iskusstv.	Iskusstvenn−
Issled.	Issledovan−
Ist.	Istitut−
Izd.	Izdatel−
Izv.	Izvesti−
J.	Joernaal−, Jornal−, Journal−
Jaarb.	Jaarboek−
Jahrb.	Jahrbuch, Jahrbücher
Jahresber.	Jahresbericht−
Jahresschr.	Jahresschrift
Jahrg.	Jahrgang
JPL	Jet Propulsion Laboratory
K.	Königlich−, Koninklijk−, Kunglig−
Kans.	Kansas
Kartogr.	Kartograf−
Kernforsch.	Kernforschung
Kernphys.	Kernphysik−
Khem.	Khemyi−
Khim.	Khimi−
Kim.	Kimija−, Kimya
Kl.	Klass−
Kolloq.	Kolloquium−
Komet.	Kometnyj
Komm.	Kommission−
Konf.	Konfer−
Kongr.	Kongress
Kosm.	Kosmich−
Kosmog.	Kosmogon−
Kozp.	Kozponti
KPNO	Kitt Peak National Observatory
Kut.	Kutato
Ky.	Kentucky
La.	Louisiana
Lab.	Laborato−
Lett.	Letter−, Lettra, Lettre
Libr.	Librair−, Librar−
Mag.	Magasin, Magazin−
Magn.	Magneti−, Magnitn−
Mass.	Massachusetts
Mat.	Matemaat−, Matemat−
Mater.	Material−
Math.	Mathemat−
Md.	Maryland

Meas.	Measur–
Mec.	Mecani–
Mech.	Mechani–
Medd.	Meddelande–, Meddelelse
Meded.	Mededeeling–, Mededeling–
Mekh.	Mekhani–
Mem.	Memento–, Memoir–, Memori–, Memory–, Memuary
Memo.	Memorand–
Mens.	Mensile, Mensual–, Mensuel–
Messtech.	Messtechni–
Meteorol.	Meteorolog–
Mich.	Michigan
Micromec.	Micromecaniq–
Miner.	Mineral, Minerale–, Minerali–
Mineral.	Mineralog–
Minn.	Minnesota
Miss.	Mississippi
MIT	Massachusetts Institute of Technology
Mitt.	Mitteilung–
Mo.	Missouri
Mod.	Modern–
Mol.	Molecul–, Molekul–
Mon.	Monat, Monatlich–, Month–
Monogr.	Monograph–
Mont.	Montana
MPI	Max-Planck-Institut
Mus.	Museum
N. C.	North Carolina
N. D.	North Dakota
N. H.	New Hampshire
N. J.	New Jersey
N. M.	New Mexico
N. Y.	New York
Nablyud.	Nablyudeni–
Nac.	Nacion––
Nachr.	Nachricht–
NASA	National Aeronautics and Space Administration
Nat.	Natur–
Natl.	National–
Naturforsch.	Naturforsch–
Naturwiss.	Naturwissenschaft–
Natuurkd.	Natuurkunde
Nauchn.	Nauchny–
Nauk.	Nauka, Naukite, Naukov–, Naukow–
Naut.	Nautic–
Nav.	Naval–
Navig.	Navigat–
Naz.	Nazion–
Nebr.	Nebraska
Nev.	Nevada
Newsl.	Newsletter–
Not.	Notationes, Notic–, Notise–, Notizi–
Nouv.	Nouveau–, Nouvell–
Nov.	Novoe
Nucl.	Nucléaire–, Nuclear–, Nucl–
Nukl.	Nukle–
Numer.	Numeri–
O-va	Obshchestva
O-vo	Obshchestvo
Obs.	Observ–
Obz.	Obzor–
Okla.	Oklahoma
Opt.	Optic–, Optik–, Optique
Oreg.	Oregon
Oss.	Osserva–
Pa.	Pennsylvania
Paleontol.	Paleontolog–
Pap.	Paper–, Papier
Part.	Particle
Perem.	Peremenn–

Period.	Periodi–
Petrol.	Petrolog–
Philos.	Philosoph–
Photogr.	Photograf–, Photograph–
Photogramm.	Photogrammetr–
Photom.	Photometr–
Phys.	Physic–, Physik–, Physique–, Physisch–
Pict.	Picture–
Planet.	Planetary
Pr.	Prac–
Prelim.	Prelimin–
Prepr.	Preprint
Prib.	Pribor–
Prikl.	Prikladnoj
Prir.	Prirodn–
Prirodoved.	Prirodoved–
Probl.	Problem–
Proc.	Proceedings
Prod.	Prodott–, Produc–, Produkt,
Prog.	Progres–
Propag.	Propagation
Prov.	Provinc–, Provints–, Provinz–
Pubbl.	Pubblicazion–
Publ.	Publicac–, Publicas–, Publicat–, Publikas–, Publikat–
Q.	Quarterly
Quant.	Quantit–
R.	Royal
R. I.	Rhode Island
Radiat.	Radiati–
Radioact.	Radioactiv–, Radioaktiv–
Radioisot.	Radioisotop–
Rap.	Raport–
Rapp.	Rapport–
RAS	Royal Astronomical Society
Rec.	Record–
Rech.	Recherche–
Ref.	Referat–, Reference–, Referieren
Relat.	Related, Relation–
Relativ.	Relativit–
Rend.	Rendicont–
Rep.	Report–
Repr.	Reprint–
Repub.	Republi–
Res.	Research–
Result.	Resultad–, Resultat–
Rev.	Review–, Revisio, Revista, Revue–
Rezul't.	Rezul'tat–
Ric.	Ricerca, Ricerche
Riv.	Rivist–
Rundsch.	Rundschau
S. C.	South Carolina
S. D.	South Dakota
SAF	Société Astronomique de France
SAI	Società Astronomica Italiana
Samml.	Sammlung–
SAO	Smithsonian Astrophysical Observatory
SAS	Société Astronomique de Suisse
Satell.	Satellite
Sb.	Sbornik–
Schr.	Schrift–
Schriftenr.	Schriftenreihe
Sci.	Scienc–, Scient–, Scienz–
Scr.	Scripta, Scritt–
Secc.	Seccion–
Sect.	Secti–
Sekc.	Sekci–, Sekcj–
Sekt.	Sektion–, Sektor–
Sekts.	Sektsi–
Sel.	Seleccion–, Select–, Selek–, Selezione
Selsk.	Selskab–, Selskap–
Semin.	Séminair–, Seminar–

Sep.	Separat–
Ser.	Seria–, Serie–, Seriya
Serv.	Servic–, Serviz–
Sess.	Sessi–
Signal.	Signalétique–
Simp.	Simpoz–
Sitzungsber.	Sitzungsbericht–
Skr.	Skrift–
Soc.	Sociedad–, Societ–
Sol.	Solar
Soln.	Solnechn–
Sonderdr.	Sonderdruck–
Soobshch.	Soobshchen–
South.	Southern
Spacecr.	Spacecraft
Spat.	Spatial–
Spec.	Special–
Spectrosc.	Spectroscop–
Spectrosk.	Spectroskop–
Spets.	Spetsial–
Spez.	Spezial–, Speziell–
SSR	Sovetskaya Sotsialisticheskaya Respublika
SSSR	Soyuz Sovetskikh Sotsialisticheskikh Respublik
St.	Saint–, Sankt–, Sant–
–St.	–Straße, Street
Stand.	Standard–, Standart–
Sternw.	Sternwarte–
Stiint.	Stiintific–
Stn.	Station, Stazione
Stud.	Studia, Studie–, Studii
Supl.	Suplement–, Supliment–
Suppl.	Supplement–
Surv.	Survey–
Symp.	Sympos–, Sympoz–
Syst.	System–
Sz.	Szemle
Teach.	Teacher–, Teaching
Tec.	Tecni–
Tech.	Techni–
Technol.	Technolog–
Tecnol.	Tecnolog–
Teh.	Tehnic–, Tehnika, Tehnisk–
Tehnol.	Tehnolog–, Tehnolosk–
Tek.	Tekni–
Tekh.	Tekhni–
Tekhnol.	Tekhnolog–
Teknol.	Teknolog–
Telesc.	Telescop–
Telev.	Television–
Tenn.	Tennessee
Teor.	Teoret–, Teori–
Terr.	Terrestr–
Test.	Testing
Tex.	Texas
TH	Technische Hochschule
Theor.	Theoret–, Theori–
Tidschr.	Tidschrift–

Tidskr.	Tidskrift–
Tidsskr.	Tidsskrift–
Top.	Topic–
Tr.	Trudy
Trans.	Transactions, Transazione
Tsentr.	Tsentral–
Tsirk.	Tsirkulyar–
TU	Technical University
Uch.	Uchen–
Uchebn.	Uchebn–
UK	United Kingdom
Umsch.	Umschau
UN	United Nations
Univ.	Universidad–, Universit–, Univerzitet–
US	United States
USA	United States of America
USSR	Union of Soviet Socialist Republics
Va.	Virginia
Var.	Various
Ver.	Verein–, Verenig–
Veränderl.	Veränderlich–
Verh.	Verhandl–
Vermess.	Vermessung–
Vermessungswes.	Vermessungswesen
Veröff.	Veröffentlich–
Vesn.	Vesnik
Vestn.	Vestnik
Vetensk.	Vetenskap–
Vidensk.	Videnskab–, Videnskap
Vierteljahresschr.	Vierteljahresschrift–
Vierteljahrsschr.	Vierteljahrsschrift–
VLB	Very Long Baseline
Volcanol.	Volcanolog–
Vopr.	Vopros–
Vortr.	Vorträge
Vses.	Vsesoyuzn–
Vt.	Vermont
Vyp.	Vypusk–
Vyssh.	Vyssh–
Vyzk.	Vyzkum–
W. Va.	West Virginia
Wash.	Washington
West.	Western
Wet.	Wetenschap–, Wetenskap–
Wis.	Wisconsin
Wiss.	Wissenschaft–
Wyo.	Wyoming
Yad.	Yadern–
Z.	Zeitschrift–
ZA	Zero Age
ZAED	Zentralstelle für Atomkernenergie-Dokumentation
Zap.	Zapisk–, Zapyisk–
Zaved.	Zaveden–
Zent.	Zentral
Zentralbl.	Zentralblatt
Zesz.	Zeszyt
Zh.	Zhurnal–
Zirk.	Zirkular–

Periodicals, Proceedings, Books, Activities

001 Periodicals

A. A. O. Newsl.
Anglo-Australian Observatory Newsletter. Published by the Anglo-Australian Observatory, PO Box 296, Epping, NSW 2121, Australia.

AAS Photo-Bull.
AAS (American Astronomical Society) Photo-Bulletin. Published by the Working Group on Photographic Materials. Produced by Eastman Kodak Co., Rochester, N.Y.

AAVSO Bull.
American Association of Variable Star Observers Bulletin. 187 Concord Avenue, Cambridge, Mass., 02138, U.S.A.

Abh. Hamburger Sternw.
Abhandlungen aus der Hamburger Sternwarte. Hamburg-Bergedorf. ISSN 0374-1583.

Acad. R. Belgique, Bull. Cl. Sci.
Académie Royale de Belgique, Bulletin de la Classe des Sciences (Koninklijke Academie van België, Mededelingen van de Klasse der Wetenschappen). 5e Série, Palais des Académies, Bruxelles.

Acta Acust.
Acta Acustica. Science Press, Beijing, People's Republic of China. Subscription address: Guozi Shudian, P. O. Box 399, Beijing.

Acta Astron.
Acta Astronomica. An international quarterly journal. Publisher: Polska Akademia Nauk, Komitet Astronomii (Polish Academy of Sciences, Committee of Astronomy), Warszawa – Wrocław.

Acta Astron. Sinica
Acta Astronomica Sinica. Published by Purple Mountain Observatory, Academia Sinica, Nanking, China.

Acta Astronaut.
Acta Astronautica. Journal of the International Academy of Astronautics. Publisher: Pergamon Press Inc., Elmsford, New York, U.S.A.; Pergamon Press Ltd., Oxford, England.

Acta Astrophys. Sinica
Acta Astrophysica Sinica. Editorial Board "Acta Astrophysica Sinica", P. O. Box 399, Beijing, China.

Acta Cosmologica
Acta Cosmologica. Published by Obserwatorium Astronomiczne Uniwersytetu Jagiellońskiego, Kraków, Poland.

Acta Fac. Rerum Nat. Univ. Comenianae Phys.
Acta Facultatis Rerum Naturalium Universitatis Comenianae, Physica. Ústredná knižnica PFUK, 886 11 Bratislava, Ul. 29 Augusta č.5, Czechoslovakia.

Acta Geod. Geophys.
Acta Geodaetica et Geophysica. Edited by the Institute of Geodesy and Geophysics, Academia Sinica. Published by Science Press, Beijing, Cao Yan Men Nei Street No. 137, China.

Acta Geod. Geophys. Montan.
Acta Geodaetica, Geophysica et Montanistica. Akademiai Kiado, 1054 Budapest, Alkotmany utca 21, Hungary.

Acta Geophys. Polonica
Acta Geophysica Polonica. ARS Polona-Ruch, 00-068 Warszawa, Krakowskie Przedmiescie 7, P.O. Box 1001, Poland.

Acta Geophys. Sinica
Acta Geophysica Sinica. Chinese Academy of Sciences, Department of Geophysical Research. Published by Science Press, Peking, People's Republic of China.

Acta Mech. Sinica
Acta Mechanica Sinica, Science Press, Peking, People's Republic of China. Subscription address: Guozi Shudian, P.O. Box 399, Peking.

Acta Phys. Acad. Sci. Hungaricae
Acta Physica Academiae Scientiarum Hungaricae. Kultura, Hungarian Trading Co., H-1389 Budapest 62, P.O. Box 149, Hungary.

Acta Phys. Austriaca
Acta Physica Austriaca. Springer-Verlag, A-1011 Wien, Molkerbastei 5, Postfach 367, Austria.

Acta Phys. Polonica B
Acta Physica Polonica B. ARS Polona-Ruch, Warszawa 1, P.O. Box 154, Poland.

Acta Phys. Sinica
Acta Physica Sinica. Chinese Academy of Sciences, Institute of Physics, Peking, People's Republic of China. [English translation in: Chinese J. Phys. (*USA*)].

Acta Phys. Slovaca
Acta Physica Slovaca. VEDA Publishing House of the Slovak Academy of Sciences, 895 30 Bratislava, Klemensova 27, Czechoslovakia.

Acta Polytech. III
Acta Polytechnica. Series III. Elektrotechnická fakulta ČVUT v Praze, Technická ul. 2, Praha 6-Dejvice, Czechoslovakia.

Acta Sci. Nat. Univ. Pekinensis
Acta Scientiarum Naturalium Universitatis Pekinensis. Peking, People's Republic of China.

Acta Sci. Nat. Univ. Sunyatseni
Acta Scientiarum Naturalium Universitatis Sunyatseni

(Zhongshandaxue Xuebao). Canton Post Office, Canton, People's Republic of China.

Acta Tech. Acad. Sci. Hungaricae
Acta Technica Academiae Scientiarum Hungaricae. Akademiai Kiado, Budapest 1363, P. O. Box 24, Hungary.

Acta Tech. CSAV
Acta Technica Československá akademie věd. Academia,. Publishing House of the Czechoslovak Academy of Sciences, Vodickova 40, 112 29 Praha 1, Czechoslovakia. John Benjamins N.V., Periodical Trade, Warmoesstraat 54, Amsterdam, Netherlands.

Acta Univ. Carolinae Math. Phys.
Acta Universitatis Carolinae, Mathematica et Physica. Administrace: Matematicko-fyzikální fakulta University Karlovy, Praha.

Adv. Phys.
Advances in Physics. Taylor & Francis Ltd., 10–14 Macklin Street London, WC2B 5NF, England.

Adv. Space Res.
Advances in Space Research. The official journal of the Committee on Space Research (COSPAR). Pergamon Press, Oxford – New York – Toronto – Sydney – Paris – Frankfurt. ISSN 0273-1177.

Aeronaut. Astronaut.
L'Aeronautique et l'Astronautique. Editions Air et Cosmos, 6 Rue Anatole de la Forge, 75017 Paris, France.

AIAA J.
AIAA Journal. A Publication of the American Institute of Aeronautics and Astronautics devoted to Aerospace Research and Development. Published by the American Institute of Aeronautics and Astronautics, New York, N.Y.

AIP Conf. Proc.
AIP Conference Proceedings. American Institute of Physics, 335 East 45th Street, New York, N.Y. 10017, USA.

Alta Freq.
Alta Frequenza.Ufficio Centrale AEI-CEI, Viale Monza 259, 20126 Milano, Italy.

American J. Phys.
American Journal of Physics. Published for the American Association of Physics Teachers by the American Institute of Physics, 335 East 45th Street, New York, N. Y. 10017, USA.

American Mineral.
American Mineralogist. Mineralogical Society of America, 2000 Florida Avenue, N. W., Washington, DC 20009, USA.

American Sci.
American Scientist. Society of Sigma XI, 345 Whitney Avenue, New Haven, CT 06510, USA.

An. Acad. Brasil. Cienc.
Anais da Academia Brasileira de Ciencias. Caixa Postal 229, ZC-00 Rio de Janeiro gb, Brazil.

An. Fis.
Anales de Física. Real Sociedad Española de Física y Química (Facultad de Ciencias), Ciudad Universitaria, Madrid-3, Spain.

Ann. Acad. Sci. Fennicae, Ser. A. VI
Annales Academiae Scientiarum Fennicae, Series A VI (Physica). Snellmaninkato 9-11, 00170 Helsinki-17, Finnland.

Ann. Geofis.
Annali di Geofisica. Istituto Nazionale di Geofisica, Citta Universitaria, Via Ruggero Bonghi 11/B, 00184 Roma, Italy.

Ann. Géophys.
Annales de Géophysique. Service des Publications du CNRS, 15 Quai Anatole-France, 75700 Paris, France.

Ann. Inst. Henri Poincaré A
Annales de l'Institut Henri Poincaré, Section A (Physique Théorique). 11 Rue Pierre-Curie, Paris 5, France.

Ann. Israel Phys. Soc.
Annals of the Israel Physical Society. c/o Department of Physics, Bar-Ilan University, Ramat-Gan, Israel. Adam Hilger Ltd., Techno House, Recliffe Way, Bristol BS1 6NX England.

Ann. Nucl. Energy
Annals of Nuclear Energy. Pergamon Press Ltd., Headington Hill Hall Oxford, OX3 0BW, England.

Ann. Physics
Annals of Physics. Academic Press Inc., 111 Fifth Avenue, New York, NY 10003, USA.

Ann. Physik
Annalen der Physik. 7. Folge. Publisher: Johann Ambrosius Barth, Salomonstr. 18B, Leipzig 701, German Democratic Republic.

Ann. Physique
Annales de Physique. Publisher: Masson et Cie., 120 Boulevard Saint-Germain, Paris 6, France.

Ann. Sci.
Annals of Science. Taylor & Francis Ltd., 10-14 Macklin Street, London, WC2B 5NF, England.

Ann. Shanghai Obs., Acad. Sinica
Annuals of Shanghai Observatory, Academia Sinica. Published by Shanghai Scientific and Technical Publishers, Shanghai, Rei Jing Er Street, No. 450.

Ann. Soc. Sci. Bruxelles I
Annales de la Société Scientifique de Bruxelles. Série I: Sciences Mathématiques, Astronomiques et Physiques. Rue de Bruxelles 61, B 5000 Namur, Belgium.

Ann. Télécommun.
Annales des Télécommunications. Centre National d'Études des Télécommunications, 38 rue du Général Leclerc, 92 Issy-les-Moulineaux, France.

Ann. Tokyo Astron. Obs., Second Ser.
Annals of the Tokyo Astronomical Observatory, Second Series. University of Tokyo, Mitaka, Tokyo, Japan. ISSN 0082-4704.

Annu. Rep. Astron. Inst. Greece
Annual Reports of the Astronomical Institutes of Greece. Published by the Greek National Committee for Astronomy. Academy of Athens, Research Center for Astronomy and Applied Mathematics.

Annu. Rev. Astron. Astrophys.
Annual Review of Astronomy and Astrophysics. Annual Reviews Inc., 4139 El Camino Way, Palo Alto, Calif. 94306, USA. ISSN 0066-4146.

Annu. Rev. Earth Planet. Sci.
Annual Review of Earth and Planetary Sciences. Annual Reviews Inc., 4139 El Camino Way, Palo Alto, Calif. 94306, USA. ISSN 0084-6597.

Annu. Univ. Sofia Fac. Phys.
Annuaire de l'Université de Sofia Faculté de Physique, Sofiya, Bulgaria.

Antenna
L'Antenna. Via Monte Generoso 6/a, 20155 Milano, Italy.

Anz. Österreich. Akad. Wiss. Math.-Naturwiss. Kl.
Anzeiger. Österreichische Akademie der Wissenschaften. Mathematisch-Naturwissenschaftliche Klasse. Publisher: Springer-Verlag, Wien.

APL Tech. Dig.
APL Technical Digest. Applied Physics Laboratory, The John Hopkins University, 8621 Georgia Avenue, Silver Spring, MD 20910, USA.

Appl. Opt.
Applied Optics. A monthly publication of the Optical Society of America. Published for the Optical Society of America by the American Institute of Physics, 335 East 45th Street, New York, NY 10017, USA.

Appl. Phys.
Applied Physics. Springer-Verlag, Heidelberger Platz 3, D-1000 Berlin 33, F. R. Germany.

Appl. Phys. Lett.
Applied Physics Letters. American Institute of Physics, 335 East 45th Street, New York, N.Y. 10017, USA.

Appl. Spectrosc.
Applied Spectroscopy. 428 East Preston Street, Baltimore, MD 21202, USA.

Appl. Spectrosc. Rev.
Applied Spectroscopy Reviews. Marcel Dekker Inc., 95 Madison Avenue, New York, NY 10016, USA.

Arch. Hist. Exact Sci.
Archive for History of Exact Sciences. Springer-Verlag, Berlin - Heidelberg - New York. ISSN 0003 - 9519.

Arch. Mech.
Archives of Mechanics (Archiwum Mechaniki Stosowanej). Polish Scientific Publishers, Swietokrzyska 21, Warszawa, Poland.

Arch. Sci.
Archives des Sciences, éditées par la Société de Physique et d'Histoire Naturelle de Genève. Publisher: Imprimerie Kundig, Genève. Subscription address: Librairie Payot, Genève.

Archaeoastronomy (England)
Archaeoastronomy. Supplement to Journal for the History of Astronomy. Published by Science History Publications Ltd, Halfpenny Furze, Mill Lane, Chalfont St Giles, Bucks, England, HP8 4NR. ISSN 0142-7253.

Archaeoastronomy (U.S.A.)
Archaeoastronomy. The Bulletin of the Center for Archaeoastronomy, Space Sciences Building, University of Maryland, College Park, MD 20742 (301) 454-4460. ISSN 0190-9940.

Archaeometry
Archaeometry. Cambridge University Press, P.O. Box 92, London, NW1 2DB, England.

Ark. Fys. Semin. Trondheim
Arkiv for det Fysiske Seminar i Trondheim. c/o Institutt for Teoretisk Fysikk, Universitetet i Trondheim, NTH, N-7034 Trondheim, Norway.

Ark. Mat.
Arkiv för Matematik. Published by Institut Mittag-Leffler, Auravägen 17, S-182 62 Djursholm, Sweden.

Artif. Satell.
Artificial Satellites. Publication of Polish Scientific Institutions. Polish Academy of Sciences, National Committee of Geophysics and Geodesy, National Committee for Space Research, Warsaw. Space Research Centre, Pałac Kultury i Nauki 2301, 00-901 Warszawa, Poland.

Astrofiz. Issled. Izv. Spets. Astrofiz. Obs.
Astrofizicheskie Issledovaniya. Izvestiya Spetsial'noj Astrofizicheskoj Observatorii. Akademiya Nauk SSSR. Publishers: Izdatel'stvo "Nauka", Leningradskoe Otdelenie, Leningrad.

Astrofizika
Astrofizika. Izdatel'stvo Akademii Nauk Armyanskoj SSR, Erevan. [An English translation is published in "Astrophysics"].

Astrometr. Astrofiz.
Astrometriya i Astrofizika. Respublikanskij Mezhvedomstvennyj Sbornik. Akademiya Nauk Ukrainskoj SSR, Glavnaya Astronomicheskaya Observatoriya. Naukova Dumka, Kiev.

Astron. Astrophys.
Astronomy and Astrophysics. A European Journal. Published by Springer-Verlag, Berlin–Heidelberg–New York.

Astron. Astrophys., Suppl. Ser.
Astronomy and Astrophysics. Supplement Series. A European Journal. Published by les Editions de Physique, Orsay, France, on behalf of the Board of Directors of the European Southern Observatory. ISSN 0365-0138.

Astron. Circ.
Astronomical Circular. Compiled by the editor section of Acta Astronomica Sinica, Purple Mountain Observatory, Nanking, China. Edited by the Chinese Astronomical Society.

Astron. Data Cent. Bull.
Astronomical Data Center Bulletin. National Space Science Data Center/World Data Center A for Rockets and Satellites. National Aeronautics and Space Administration, Goddard Space Flight Center, Greenbelt, Maryland 20771, USA.

Astron. Her.
Astronomical Herald. Astronomical Society of Japan, Tokyo Astronomical Observatory, Oosawa Mitaka, Tokyo, Japan.

Astron. J.
The Astronomical Journal. Published for the American Astronomical Society by the American Institute of Physics, New York, N.Y. Editorial Office: Department of Astronomy, Columbia University, New York, N.Y.

Astron. Mitt. Eidg. Sternw. Zürich
Astronomische Mitteilungen der Eidgenössischen Sternwarte Zürich, Switzerland.

Astron. Nachr.
Astronomische Nachrichten. Publisher: Akademie-Verlag, Berlin.

Astron. Pap.
Astronomical Papers prepared for the use of the American Ephemeris and Nautical Almanac. Published by the Nautical Almanac Office, U.S. Naval Observatory by direction of the Secretary of the Navy and under the authority of Congress. U.S. Government Printing Office, Washington, D.C.

Astron. Q.
The Astronomy Quarterly. Pachart Publishing House, P.O. Box 35549, Tucson, Ariz. 85740. ISSN 0364-9229.

Astron. Rep.
The Astronomical Reports. Polish Amateur Astronomical Society. Polskie Towarzystwo Miłośników Astronomii, Kraków, Poland.

Astron. Schule
Astronomie in der Schule. Zeitschrift für die Hand des Astronomielehrers. Herausgegeben vom Verlag Volk und Wissen, Berlin. Redaktion: Sternwarte Bautzen.

Astron. Soc. Western Australia, Circ.
The Astronomical Society of Western Australia, Circular.

Astron. Tidsskr.
Astronomisk Tidsskrift. Edited by Astronomisk Selskab, København; Norsk Astronomisk Selskap, Oslo; Svenska Astronomiska Sällskapet, Stockholm. Printed by John Griegs Boktrykkeri, Bergen.

Astron. Tsirk.
Astronomicheskij Tsirkulyar, izdavaemyj Byuro Astronomicheskikh Soobshchenij Akademii Nauk SSSR. Tipografiya Astrosoveta AN SSSR, Moskva.

Astron. Vestn.
Astronomicheskij Vestnik. Publishers: Izdatel'stvo "Nauka", Moskva.

Astron. Zh.
Astronomicheskij Zhurnal. Akademiya Nauk SSSR. Publishers: Izdatel'stvo "Nauka", Moskva. [An English translation is published in "Soviet Astronomy"].

Astronomia
Astronomia. Periodico trimestrale dell'Unione Astrofili Italiani.

Astronomie
L'Astronomie et Bulletin de la Société Astronomique de France. Société Astronomique de France, Paris.

Astronomy
Astronomy. AstroMedia Corp., 757 North Broadway, Suite 204, Milwaukee, WI 53202, USA.

Astrophys. J.
The Astrophysical Journal. Published for the American Astronomical Society by the University of Chicago Press, Chicago, Illinois.

Astrophys. J., Lett.
The Astrophysical Journal. Letters to the Editors. Published for the American Astronomical Society by the University of Chicago Press, Chicago, Illinois.

Astrophys. J., Suppl. Ser.
The Astrophysical Journal. Supplement Series. Published for the American Astronomical Society by the University of Chicago Press, Chicago, Illinois.

Astrophys. Lett.
Astrophysical Letters. Published by NASA–Goddard Space Flight Center. Gordon and Breach Science Publishers Ltd., New York–London–Paris.

Astrophys. Space Sci.
Astrophysics and Space Science. An International Journal of Cosmic Physics. Published by D. Reidel Publishing Company, Dordrecht, Holland.

Astrophysics
Astrophysics. A cover-to-cover translation of Astrofizika (USSR). Consultants Bureau, New York, N.Y.

Atmos. Environ.
Atmospheric Environment. Pergamon Press Ltd., Headington Hill Hall, Oxford, OX3 OBW, England.

Atomkernenerg. Kerntech.
Atomkernenergie Kerntechnik. Verlag Karl Thiemig AG, Pilgersheimerstrasse 38, Postfach 900740, D-8000 München 90, F.R. Germany.

Atti Accad. Naz. Lincei, Mem. Ser. Ottava
Atti della Accademia Nazionale dei Lincei. Serie Ottava. Memorie. Classe di Scienze fisiche, matematiche e naturali. Sezione I: Matematica, Meccanica, Astronomia, Geodesia e Geofisica. Published by Accademia Nazionale dei Lincei, Roma.

Atti Accad. Naz. Lincei, Rend. Ser. Ottava
Atti della Accademia Nazionale dei Lincei. Serie Ottava. Rendiconti. Classe di Scienze fisiche, matematiche e naturali. Published by Accademia Nazionale dei Lincei, Roma.

Atti Accad. Sci. Torino I
Atti della Accademia delle Scienze di Torino. I. Classe di Scienze Fisiche, Mathematiche e Naturali. Via Accademia delle Scienze 6, Via Maria Vittoria 3, Torino (208), Italy.

Atti Fond. Giorgio Ronchi
Atti della Fondazione Giorgio Ronchi. Largo Enrico Fermi 1, 50125 Arcetri-Firenze, Italy.

Australian J. Phys.
Australian Journal of Physics. Published by the Commonwealth Scientific and Industrial Research Organization, 372 Albert Street, East Melbourne, Victoria 3002, Australia.

Australian J. Phys., Astrophys. Suppl.
Australian Journal of Physics, Astrophysical Supplement. Published by Commonwealth Scientific and Industrial Research Organization, 372 Albert Street, East Melbourne, Victoria 3002, Australia.

Autom. Strum.
Automazione e Strumentazione. Associazione Nazionale Italiana per l'Automazione, Via Le Premuda 2, 21029 Milano, Italy.

B. I. H., Paris, Circ.
Bureau International de l'Heure, B.I.H., Paris, Circulars. 61, Avenue de l'Observatoire, 75014-Paris.

BAV Rundbrief
BAV Rundbrief. Mitteilungsblatt der Berliner Arbeitsgemeinschaft für Veränderliche Sterne. Editor: BAV Berliner Arbeitsgemeinschaft für Veränderliche Sterne eV., Berlin.

BBSAG Bull.
Bedeckungsveränderlichen Beobachter der Schweizerischen Astronomischen Gesellschaft, [Swiss Astronomical Society's Eclipsing Variable Observers], Bulletin. To be obtained from R. Diethelm, Winterthur, Switzerland.

Biul. Obs. Astron. Uniw. M. Kopernika Toruniu
Biuletyn Obserwatorium Astronomicznego Uniwersytetu M. Kopernika w Toruniu.

Blick Weltall
Blick in das Weltall. Monatsprogramm und Mitteilungen für Sternfreunde. Archenhold-Sternwarte, Berlin-Treptow.

Bol. Acad. Cienc. Fis. Mat. Nat.
Boletin de la Academia de Ciencias Fisicas Matematicas y Naturales. Printed by Italgrafica, S.R. L. Republica de Venezuela.

Bol. Asoc. Argentina Astron.
Boletin de la Asociación Argentina de Astronomía, La Plata, Argentina. ISSN 0571-3285.

Bol. Astron.
Boletin Astronômico. Observatório do Capricórnio, Prefeitura Municipal de Campinas–SP, Brazil.

Bol. Astron. Obs. Madrid
Boletín Astronómico del Observatorio de Madrid. Instituto Geografico Nacional, General Ibáñez de Ibero, 3. Madrid 3. Spain.

Bol. Inst. Tonantzintla
Boletin del Instituto de Tonantzintla. Instituto Nacional de Astrofisica, Optica y Electronica, Apartados Postales Nos. 216 y 51, Puebla, Pue, Mexico.

Bol. Obs. Ebro
Boletín del Observatorio del Ebro, Tortosa. Printed by Cooperativa Gráfica Dertosense, Tortosa.

Boll. Geod. Sci. Affini
Bolletino di Geodesia e Scienze Affini. Pubblicazione dell'Istituto Geografico Militare, Firenze.

Boundary-Layer Meteorol.
Boundary-Layer Meteorology. D. Reidel Publishing Co., P.O. Box 17, Dordrecht, Netherlands.

British Astron. Assoc. Circ.
British Astronomical Association, Circular. Editorial Office: S. W. Milbourn, Brookhill Road, Copthorne Bank, Crawley, West Sussex, RH10 3QJ.

British J. Philos. Sci.
British Journal for the Philosophy of Science. Cambridge

University Press, Bentley House, 200 Euston Road, London, NW1 2DB, England.

British J. Photogr.
British Journal of Photography. Henry Greenwood & Co., 24 Wellington Street, London, WC2E 7DH, England.

Bul. Inst. Politeh. 'Gheorghe Gheorghiu-Dej' Bucuresti.
Buletinul Institutului Politehnic 'Gheorghe Gheorghiu-Dej' Bucuresti.Calea Grivitei 132, Bucuresti, Rumania. Journal split into three series, Bul. Inst. Politeh. 'Gheorghe Gheorghiu-Dej' Bucuresti Ser. Chim. – Metal., Ser. Electroteh. and Ser. Mec. (Rumania).

Bulgarian J. Phys.
Bulgarian Journal of Physics. Bulgarian Academy of Sciences, Faculty of Physics, 5 Anton Ivanov Blvd., 1126 Sofia, Bulgaria.

Bull. Acad. Polonaise Sci., Ser. Sci. Tech.
Bulletin de l'Académie Polonaise des Sciences. Série des Sciences Techniques. 00-901 Warszawa, Palac Kultury i Nauki, P. O. Box 20, Poland.

Bull. AFOEV
Bulletin de l'Association Française des Observateurs d'Etoiles Variables. Rédaction et publication: E. Schweitzer, "La Moineaudière", 16, rue de Plobsheim, 67100 Strasbourg–Neudorf, France.

Bull. American Astron. Soc.
Bulletin of the American Astronomical Society. Published for the American Astronomical Society by the American Institute of Physics, 335 East 45th Street, New York, N.Y. 10017, USA.

Bull. Astron. Inst. Czechoslovakia
Bulletin of the Astronomical Institutes of Czechoslovakia. Published under the auspices of the Czechoslovak Academy of Sciences by Academia, Praha. Editor: Astronomical Institute of the Czechoslovak Academy of Sciences, Praha.

Bull. Astron., Obs. R. Belgique
Bulletin Astronomique, Observatoire Royal de Belgique. (Astronomisch Bulletin, Koninklijke Sterrenwacht van België).

Bull. Astron. Soc. India
Bulletin of the Astronomical Society of India. Published by S. K. Trehan (Editor) on behalf of the Astronomical Society of India, Osmania University, Hyderabad.

Bull. Géod.
Bulletin Géodésique. The Journal of the International Association of Geodesy. Publié par le Bureau Central de l'Association Internationale de Géodésie, 39 Rue Gay-Lussac, 75005 Paris, France.

Bull. Inf. Cent. Données Stellaires
Bulletin d'Information du Centre de Données Stellaires. Compiled at Observatoire de Strasbourg, 11, rue de l'Université, 67000-Strasbourg, France.

Bull. Inst. Space Aeronaut. Sci. Univ. Tokyo A
Bulletin of the Institute of Space and Aeronautical Science, University of Tokyo A. Tokyo, Japan.

Bull. Inst. Space Aeronaut. Sci., Univ. Tokyo B
Bulletin of the Institute of Space and Aeronautical Science, University of Tokyo B. Tokyo, Japan.

Bull. Obs. Astron. Belgrade
Bulletin de l'Observatoire Astronomique de Belgrade.
Editor: Observatoire Astronomique de Belgrade. Printed
by Naucna delo, Belgrade.

Bull. Res. Inst. Sci. Meas. Tôhoku Univ.
Bulletin of the Research Institute for Scientific Measurements, Tôhoku University, Sendai, Japan.

Bull. Sci. Yougoslavie
Bulletin Scientifique. Conseil des Academies des Sciences
et des Arts de la RSF de Yougoslavie. Section A: Sciences
Naturelles, Techniques et Médicales. Rédaction et Administration: Opatička ul. 18/II, Zagreb, Yougoslavie.

Bull. Signal.
Bulletin Signalétique. Section 120: Astronomie, physique
spatiale, geophysique. Centre Nationale de la Recherche
Scientifique, Informascience, Centre de Documentation
Scientifique et Technique, 26, rue Boyer, 75971 Paris.
ISSN 0007-5337.

Bull. Soc. R. Sci. Liège
Bulletin de la Société Royale des Sciences de Liège.
L'Université, 15 Avenue des Tilleurs, Liège, Belgium.

Bull. Tokyo Gakugei Univ., Ser. IV
Bulletin of Tokyo Gakugei University. Series IV (Mathe-
matics and Natural Sciences) 4-1-1 Nukui-kita-machi,
Koganei, Tokyo, Japan.

Bull. Yamagata Univ. (Nat. Sci.)
Bulletin of the Yamagata University (Natural Science).
Yamagata, Japan.

Byull. Abastumanskaya Astrofiz. Obs.
Abastumanskaya Astrofizicheskaya Observatoriya, Gora
Kanobili. Byulleten'. Akademiya Nauk Gruzinskoj SSR.
Publishers: Izdatel'stvo "Metsniereba", Tbilisi.

Byull. Inst. Astrofiz.
Byulleten' Instituta Astrofiziki, Akademiya Nauk
Tadzhikskoj SSR. Izdatel'stvo Donish, Dushanbe.

Byull. Inst. Teor. Astron.
Byulleten' Instituta Teoreticheskoj Astronomii. Izdatel'-
stvo Nauka, Leningradskoe Otdelenie, Leningrad.

C. R. Acad. Sci. Paris
Comptes Rendus des Séances de l'Académie des Sciences.
Série II: mécanique, physique, chimie, sciences de
l'univers, sciences de la terre. Publiés par MM. les
Secrétaire perpétuels. Gauthier-Villars, C.D.R., Centrale
des Revues, B.P. No. 119, 93104 Montreuil Cedex,
France. ISSN 0567-6541.

Canadian J. Earth Sci.
Canadian Journal of Earth Sciences. National Research
Council of Canada, Ottawa KIA OR6, Canada.

Canadian J. Phys.
Canadian Journal of Physics. Published by the National
Research Council of Canada, Ottawa. Printed in Canada
by the University of Toronto Press, Toronto, Ont.

Carter Obs., Astron. Bull.
Carter Observatory, Astronomical Bulletin. Carter Observatory, P.O. Box 2909, Wellington 1, New Zealand.

Celestial Mech.
Celestial Mechanics. An International Journal of Space

Dynamics. Publishers: D. Reidel Publishing Company,
Dordrecht, Holland.

Cent. Astrophys. Prepr. Ser.
Center for Astrophysics, Preprint Series. Harvard College
Observatory, Smithsonian Astrophysical Observatory.
Center for Astrophysics, 60 Garden St., Cambridge,
Mass. 02138.

Centaurus
Centaurus. International magazine of the history of
mathematics, science, and technology. Munksgaard,
Copenhagen.

Ceskoslovensky Cas. Fyz., A.
Československý časopis pro fyziku. Sekce A. Academia
Publishing House of the Czechoslovak Academy of
Sciences, Vodičkova 40, 112 29 Praha 1, Czechoslovakia.

Chinese Astron.
Chinese Astronomy. A cover-to-cover translation of
Acta Astron. Sinica and Stud. Astron. Sinica. Published
by Pergamon Press, Headington Hill Hall, Oxford,
OX3 0BW, England – Maxwell House, Fairview Park,
Elmsford, N.Y. 10523, USA.

Chinese J. Phys.
Chinese Journal of Physics. Physical Society of the
Republic of China, Physics Department, National Taiwan
University, Taipei, Taiwan, China.

Chinese Phys.
Chinese Physics. American Institute of Physics. 335 East
45th Street, New York, NY 10017, USA. Contains selected translations from current issues of major Chinese
physics and astronomy journals.

Ciel
Le Ciel. Bulletin de la Société Astronomique de Liège.
Éditeur responsable: Françoise Rameau, 19 - 21, rue
des genêts, 4310 St. Nicolas.

Ciel Terre
Ciel et Terre. Bulletin de la Société Royale Belge
d'Astronomie, de Météorologie et de Physique du Globe.
Administration: Avenue Circulaire 3, B-1180 Bruxelles.

Circ. Czechoslovak Obs., Time and Latitude
Circular of the Czechoslovak Observatories, Time and
Latitude. Czechoslovak Academy of Sciences, Astro-
nomical Institute, Prague, Czechoslovakia.

Circ. Inf.
Circulaire d'Information. Union Astronomique Interna-
tionale. Commission des Etoiles Doubles. Address:
Observatoire de Meudon, Meudon, France.

Circ. Stn. Astron. Int. Latitudine, Carloforte-Cagliari
Circolari della Stazione Astronomica Internazionale di
Latitudine, Carloforte-Cagliari. Serie A printed by Tipo-
Offset "3T", Cagliari. Serie B printed by Multi Copy,
Milano.

Circ. Time and Latitude Serv.
Circular Time and Latitude Service. Polish Academy of
Sciences, Astronomical Latitude Observatory, Borowiec,
Poland.

CODATA Bull.
CODATA Bulletin. Committee on Data for Science and
Technology of the International Council of Scientific
Unions. CODATA Secretariat, 51 Boulevard de Mont-

morency, 75016 Paris, France. Pergamon Press, Oxford–
New York–Toronto–Sydney–Paris–Frankfurt. ISSN
0366-757 X.

Coelum
Coelum. Periodico bimestrale per la Divulgazione dell'
Astronomia. Editor: Osservatorio Astronomico Univer-
sitario di Bologna.

Comments Astrophys.
Comments on Astrophysics. A Journal of Critical Dis-
cussion of the Current Literature. Comments on Modern
Physics: Part C. Publishers: Gordon and Breach, Science
Publishers Ltd., 42 William IV Street, London WC2,
England.

Comments At. Mol. Phys.
Comments on Atomic and Molecular Physics. Gordon
& Breach Science Publishers Ltd., 41 and 42 William IV
Street, London, WC2, England.

Comments Nucl. Part. Phys.
Comments on Nuclear and Particle Physics. Gordon &
Breach Science Publishers Ltd., 41 and 42 William IV
Street, London, WC2, England.

Comments Plasma Phys. Controlled Fusion
Comments on Plasma Physics and Controlled Fusion.
Gordon & Breach Science Publishers Ltd., 41 and 42
William IV Street, London, WC2, England.

Commun. Fac. Sci. Univ. Ankara
Communications de la Faculté des Sciences de l'Université
d'Ankara, Série A_2: Physique, Série A_3: Astronomie.

Commun. Konkoly Obs.
Communications from the Konkoly Observatory of the
Hungarian Academy of Sciences, Budapest. HU
ISSN 0324-2234.

Commun. Math. Phys.
Communications in Mathematical Physics, Springer-
Verlag, Postfach 105280, 6900 Heidelberg 1, F.R.
Germany.

Comput. Geosci.
Computers & Geosciences. Pergamon Press Ltd.,
Headington Hill Hall, Oxford OX3 0BW, England.

Comput. Phys. Commun.
Computer Physics Communications. North-Holland
Publishing Co., P. O. Box 211 Amsterdam, Netherlands.

Comun. Obs. Astron. Univ. Coimbra
Comunicações do Observatório Astronomico da Universidade de Coimbra, Portugal.

Contemp. Phys.
Contemporary Physics.Taylor and Francis Ltd., 10 - 14
Macklin Street, London, WC2B 5NF, England.

Contrib. Astron. Obs. Skalnaté Pleso.
Contributions of the Astronomical Observatory Skalnaté
Pleso. VEDA, vydavateľstvo Slovenskej akadémie vied,
Bratislava, Czechoslovakia.

Contrib. Atmos. Phys.
Contributions to Atmospheric Physics – Beiträge zur
Physik der Atmosphäre. Publisher: Friedrich Vieweg &
Sohn, Braunschweig.

Contrib. Bosscha Obs. Lembang
Contributions from the Bosscha Observatory Lembang.
Bandung Institute of Technology, Department of
Science.

Czechoslovak J. Phys. B
Czechoslovak Journal of Physics, Section B. Czechoslovak
Academy of Science, Akademia, Vodičkova 40, 112 29
Praha 1, Czechoslovakia.

Data Rep. Hydrogr. Obs., Ser. Astron. Geod., Tokyo
Data Report of Hydrographic Observations. Series of
Astronomy and Geodesy. Maritime Safety Agency,
Hydrographic Department Tsukiji-5, Chuo-ku, Tokyo,
104 Japan.

Debrecen Heliophys. Obs. Hungarian Acad. Sci., Repr.
Debrecen Heliophysical Observatory of the Hungarian
Academy of Sciences, Reprints.

Deutsche Geod. Komm. Bayerisch. Akad. Wiss.
Deutsche Geodätische Kommission bei der Bayerischen
Akademie der Wissenschaften. Reihe A: Höhere Geo-
däsie; Reihe B: Angewandte Geodäsie; Reihe C: Disser-
tationen; Reihe D: Tafelwerke; Reihe E: Geschichte und
Entwicklung der Geodäsie. Published by Verlag der
Bayerischen Akademie der Wissenschaften, München.

Dimensions NBS
Dimensions NBS. U.S. Department of Commerce,
Washington, DC 20234, USA.

Dokl. Akad. Nauk SSSR
Doklady Akademii Nauk SSSR. Seriya Matematika,
Fizika. Publishers: Izdatel'stvo "Nauka", Moskva.

Dokl. Bolg. Akad. Nauk
Doklady Bolgarskoj Akademii Nauk. Sofiya, Bulgaria.

Dudley Obs. Rep.
Dudley Observatory Reports. Dudley Observatory,
Albany, N.Y., USA.

Dunsink Obs. Publ.
Dunsink Observatory Publications. The Observatory of
the School of Cosmic Physics, Dublin Institute for Advanced Studies, Dublin.

Earth Planet. Sci. Lett.
Earth and Planetary Science Letters. A Letter Journal
devoted to the Development in Time of the Earth and
Planetary System. Publisher: North-Holland Publishing
Company, Amsterdam, Netherlands.

Electro-Opt. Syst. Des.
Electro-Optical Systems Design. Milton S. Kiver Publications Inc., 222 West Adams, Chicago, IL 60606, USA.

Electron. Lett.
Electronics Letters. Institution of Electrical Engineers,
Savoy Place, London, WC2R 0BL, England.

Electronics
Electronics. McGraw-Hill Publishing Co., 1221 Avenue of
the Americas, New York, N.Y. 10020, USA.

Elektroteh. Vestn.
Elektrotehniski Vestnik. Editorial address: YU-61001
Ljubljana, P. O. Box 92 - II, Trzaska 25, Yugoslavia.

Endeavour New Ser.
Endeavour New Series. Pergamon Press Ltd., Headington Hill Hall, Oxford, OX3 0BW, England.

EOS Trans. American Geophys. Union
EOS Transactions of the American Geophysical Union. 1707 L Street, N.W., Washington, DC 20036, USA.

ESA Bull.
ESA Bulletin. Editorial Office: ESA Scientific and Technical Publications Branch, ESTEC, Noordwijk, The Netherlands. ISSN 0376-4265.

ESA IUE Newsl.
ESA IUE Newsletter. Published by The ESA IUE Observatory, Villafranca Satellite Tracking Station, Apartado 54065, Madrid, Spain.

ESA J.
ESA Journal. Editorial Office: ESA Scientific and Technical Publications Branch, ESTEC, Noordwijk, The Netherlands. ISSN 0379-2285.

ESO Sci. Prepr.
European Southern Observatory, Scientific Preprints. Published by European Southern Observatory, Karl-Schwarzschild-Straße 2, D-8046 Garching bei München.

ESO Tech. Rep.
European Southern Observatory, Technical Report. Published by European Southern Observatory, Karl-Schwarzschild-Straße 2, D-8046 Garching bei München.

European J. Phys.
European Journal of Physics. Institute of Physics, 47 Belgrave Square, London SW1X 8QX.

Europhys. News
Europhysics News. European Physical Society, P. O. Box 69, CH-1213 Petit-Lancy 2, Switzerland.

Exp. Tech. Phys.
Experimentelle Technik der Physik, VEB Deutscher Verlag der Wissenschaften, Traubenstrasse 10, 108 Berlin 8, German Democratic Republic.

Feingerätetechnik
Feingerätetechnik. VEB Verlag Technik, Oranienburger Strasse 13/14, 1020 Berlin, DDR.

Feinwerktech. Messtech.
F & M. Feinwerktechnik und Messtechnik. Fusion of "Feinwerktechnik" and "Messtechnik" (formerly Zeitschrift für Instrumentenkunde) beginning with Jahrgang 82, No. 5 (1974). Publishers: Karl Hanser Verlag, Kolbergerstr. 22, D-8000 München 80. F. R.Germany.

Fis. Tecnol.
Fisica e Tecnologia. Societa Italiana di Fisica, Via Loderingo Degli Andalo 2, 40124 Bologna, Italy.

Fiz. Sz.
Fizikai Szemle. Kiadja a Lapkiado Vallalat, Budapest VII, Lenin korut 9–11, Hungary.

Fizika
Fizika. 'Mladost' Export-Import, Zagreb, Ilica 30, Yugoslavia.

Folia Fac. Sci. Nat. Univ. Purkynianae Brunensis Phys.
Folia Facultatis Scientiarum Naturalium Universitatis Purkynianae Brunensis, Physica. University J. E. Purkyne, 61137 Brno-Kotlarska 2, Czechoslovakia.

Fortschr. Phys.
Fortschritte der Physik. Akademie-Verlag. DDR-108 Berlin, Leipzigerstrasse 3 - 4, Germany.

Found. Phys.
Foundations of Physics. Plenum Publishing Co., 8 Scrubs Lane, Harlesden, London, NW10 6SE, England.

Fra Fys. Verden
Fra Fysikkens Verden. Fysisk Institutt, Universitetet i Trondheim, Norges Laererhogskole, 7000 Trondheim, Norway.

Fujitsu Sci. Tech. J.
Fujitsu Scientific and Technical Journal. Fujitsu Ltd., 1015 Kamikodanaka, Nakahara-ku, Kawasaki 211, Kanagawa, Japan.

Fundam. Cosmic Phys.
Fundamentals of Cosmic Physics. Gordon and Breach Science Publishers Ltd., New York–London–Paris.

Funkschau
Funkschau. Francis-Verlag, 8 München 37, Postfach 37 01 20, Karlstrasse 37, Germany.

Fys. Tidsskr.
Fysisk Tidsskrift. Subscription address: Jul. Gjellerups Boghandel, Solvgade 87, 1307 Kobenhavn, Denmark.

G. A.A.B.
Giornale dell'A.A.B. Notiziario trimestrale delle attività culturali e scientifiche della Associazione Astrofili Bolognesi, Bologna, Italy.

G. Astron.
Giornale di Astronomia, Pubblicazione della Società Astronomica Italiana. Printed by Tipolitografia Lodigraf S.p.A. Lodi (MI).

Gen. Relativ. Gravitation
General Relativity and Gravitation. Published under the auspices of the International Committee on General Relativity and Gravitation GRG. Publishing Office: Plenum Publishing Corporation, 233 Spring Street, New York, N. Y. 10013, USA.

Geochim. Cosmochim. Acta
Geochimica et Cosmochimica Acta. Journal of the Geochemical Society. Publishing House: Pergamon Press, Ltd., Oxford.

Geod. Geophys. Veröff., Reihe III
Geodätische und Geophysikalische Veröffentlichungen. Reihe III: Physik der festen Erde. Herausgegeben vom Nationalkomitee für Geodäsie und Geophysik bei der Akademie der Wissenschaften der Deutschen Demokratischen Republik.

Geod. Kartogr.
Geodezja i Kartografia. Komitet Geodezji Polskiej Akademii Nauk. Publisher: Państwowe Wydawnictwo Naukowe, Warszawa.

Geol. Rundsch.
Geologische Rundschau. Ferdinand Enke Verlag, Stuttgart.

Geomagn. Aehron.
Geomagnetizm i Aehronomiya. Akademiya Nauk SSSR. Izdatel'stvo "Nauka", Moskva [An English translation is published in "Geomagnetism and Aeronomy", American Geophysical Union, Washington, D.C.].

Geomagn. Ser. Earth Phys. Branch
Geomagnetic Series, Earth Physics Branch. Energy, Mines and Resources Canada, 1 Observatory Crescent, Ottawa K1A OE4, Canada.

Geophys. Astrophys. Fluid Dyn.
Geophysical and Astrophysical Fluid Dynamics. Gordon and Breach Science Publishers Ltd., 41/42 William IV Street, London, WC2, England.

Geophys. J. R. Astron. Soc.
The Geophysical Journal of the Royal Astronomical Society. Published for the Royal Astronomical Society by Blackwell Scientific Publications, Oxford–Edinburgh. American Office of the Geophys. J., US Geological Survey, Stop 967, Box 25046, Federal Center, Denver, Colorado 80225, USA.

Geophys. Res. Lett.
Geophysical Research Letters. Published monthly by the American Geophysical Union, Washington, D.C., U.S.A.

Geophys. Surv.
Geophysical Surveys. D. Reidel Publishing Co., P.O. Box 17, Dordrecht, Netherlands.

Geophysics
Geophysics. Society of Exploration Geophysicists, P.O. Box 3098, Tulsa, OK 74101, USA.

GEOS
GEOS. Department of Energy, Mines and Resources, Ottawa, Canada.

GEOS Circ.
GEOS (Groupe: Etude et Observation Stellaire and Gruppo Europeo di Osservazione Stellare) Circulars, Series: RR (RR Lyrae type variables), SR (red variables), EB (eclipsing binaries), SA (small-amplitude variables). Published by A. Figer, GEOS, 12 rue Bezout, 75014 Paris, France.

Gerlands Beitr. Geophys.
Gerlands Beiträge zur Geophysik. Publisher: Akademische Verlagsgesellschaft Geest & Portig K.-G., Leipzig.

Glasnik Mat.
Glasnik Matematicki. Published by the Society of Mathematicians and Physicists of the S. R. of Croatia. Publisher: Drustvo Matematicara i Fizicara S. R. Hrvatske, Zagreb.

Heavens
The Heavens. The Oriental Astronomical Association, Otsu-shi, Shiga-ken, Japan. In Japanese.

Helvetica Phys. Acta
Helvetica Physica Acta. Schweizerische Physikalische Gesellschaft. Publisher: Birkhäuser Verlag, Elisabethenstrasse 19, CH-4000 Basel 10, Switzerland.

HHI Sol. Data
HHI Solar Data. Heinrich-Hertz-Institut, Solare Beobachtungsergebnisse. Akademie der Wissenschaften der DDR, Zentralinstitut für Solar-Terrestrische Physik (Heinrich-Hertz-Institut), DDR-1199 Berlin-Adlershof.

HHI-STP-Rep.
HHI Solar-Terrestrische Physik Reports. Heinrich-Hertz-Institut. Akademie der Wissenschaften der DDR, Zentralinstitut für Solar-Terrestrische Physik (Heinrich-Hertz-Institut), DDR-1199 Berlin-Adlershof.

Hvar Obs. Bull.
Hvar Observatory Bulletin. Faculty of Geodesy. 41000 Zagreb, Kačićeva 26, Yugoslavia. ISSN 0351-2651.

I. A. P. P. Commun.
International Amateur-Professional Photoelectric Photometry, Communication. R. M. Genet, Fairborn Observatory, 1247 Folk Road, Fairborn, Ohio 45324, USA.

I.U.A.A. Bull.
I.U.A.A. Bulletin. International Union of Amateur Astronomers, Contributions. I.U.A.A. c/o Achille Leani, via Bertesi 15, 26100 Cremona, Italy.

IAU Circ.
International Astronomical Union, Circular. Central Bureau for Astronomical Telegrams, Smithsonian Astrophysical Observatory, Cambridge, Mass.

IAU Inf. Bull.
International Astronomical Union Information Bulletin, Published by D. Reidel Publishing Company, P. O. Box 17, 3300 AA Dordrecht, Holland.

Icarus
Icarus. International Journal of Solar System Studies. Publisher: Academic Press, New York – London.

IEE J. Microwave Opt. Acoust.
IEE Journal on Microwave, Optics and Acoustics. Institution of Electrical Engineers, Publishing Department, P.O. Box 8, Southgate House, Stevenage, Herts. SG1 1HQ, England.

IEEE Spectrum
IEEE Spectrum. Published monthly by the Institute of Electrical and Electronics Engineers, 345 East 47th Street. New York, N.Y. 10017, USA.

IEEE Trans. Aerosp. Electron. Syst.
IEEE Transactions on Aerospace and Electronic Systems. Published by the Institute of Electrical and Electronics Engineers, 345 East 47th Street, New York, N.Y. 10017, USA.

IEEE Trans. Antennas Propag.
IEEE Transactions on Antennas and Propagation. Published by the Institute of Electrical and Electronics Engineers, 345 East 47th Street, New York, N.Y. 10017, USA.

IEEE Trans. Commun.
IEEE Transactions on Communications. Institute of Electrical and Electronics Engineers, 345 East 47th Street, New York, NY 10017, USA.

IEEE Trans. Consum. Electron.
IEEE Transactions on Consumer Electronics. Institute of Electrical and Electronics Engineers, 345 East 47th Street, New York, N. Y. 10017, USA.

IEEE Trans. Electromagn. Compat.
IEEE Transactions on Electromagnetic Compatibility. Institute of Electrical and Electronics Engineers, 345 East 47th Street, New York, NY 10017, USA.

IEEE Trans. Electron Devices
IEEE Transactions on Electron Devices. Published by the Institute of Electrical and Electronics Engineers, 345 East 47th Street, New York, N. Y. 10017, USA.

IEEE Trans. Geosci. Electron.
IEEE Transactions on Geoscience Electronics. Published by the Institute of Electrical and Electronics Engineers, 345 East 47th Street, New York, N.Y. 10017, USA.

IEEE Trans. Instrum. Meas.
IEEE Transactions on Instrumentation and Measurement. Published by the Institute of Electrical and Electronics Engineers, 345 East 47th Street, New York, N.Y. 10017, USA.

IEEE Trans. Magn.
IEEE Transactions on Magnetics. Institute of Electrical and Electronics Engineers, 345 East 47th Street, New York, N. Y. 10017, USA.

IEEE Trans. Microwave Theory Tech.
IEEE Transactions on Microwave Theory and Techniques. Published by the Institute of Electrical and Electronics Engineers, 345 East 47th Street, New York, N.Y. 10017, USA.

IEEE Trans. Nucl. Sci.
IEEE Transactions on Nuclear Science. Institute of Electrical and Electronics Engineers, 345 East 47th Street, New York, N.Y. 10017, USA.

IEEE Trans. Plasma Sci.
IEEE Transactions on Plasma Science. Institute of Electrical and Electronics Engineers, 345 East 47th Street, New York, NY 10017, USA.

Indian J. Hist. Sci.
Indian Journal of History of Science. Published and printed by Indian National Science Academy, Bahadur Shah Zafar Marg, New Delhi 110002, at Mudranika, 13-A Bepin Pal Road, Calcutta 700026.

Indian J. Phys. Part B
Indian Journal of Physics Part B. Indian Association for the Cultivation of Science, 2 & 3 Raja Subodh Chandra Mallik Road, Calcutta 700032, India.

Indian J. Pure Appl. Math.
Indian Journal of Pure and Applied Mathematics. National Institute of Sciences India, Bahadur Shah Zafar Marg, New Delhi 1, India.

Indian J. Pure Appl. Phys.
Indian Journal of Pure and Applied Physics. Council of Scientific and Industrial Research, Hillside Road, New Delhi 110012, India.

Indian J. Radio Space Phys.
Indian Journal of Radio & Space Physics. Council of Scientific & Industrial Research. Editorial address: Publications & Information Directorate, Hillside Road, New Delhi 110012, India.

Indian J. Theor. Phys.
Indian Journal of Theoretical Physics. Institute of Theoretical Physics, Bognan Kutir, 4 - 1 Mohan Bagan Lane, Calcutta 700004, India

Inf. Bull. Variable Stars
Commission 27 of the I.A.U. Information Bulletin on Variable Stars. Konkoly Observatory, Budapest.

Informeto Astron. Obs. Univ. Turku
Informeto Astronomia Observatorio Universitato de Turku, Finnlando.

Infrared Phys.
Infrared Physics. An International Research Journal. Publisher: Pergamon Press Ltd., Oxford, England.

Inst. Theor. Astrophys., Blindern–Oslo, Rep.
Institute of Theoretical Astrophysics, Blindern–Oslo, Report. Universitetsforlagets trykningssentral, Oslo.

Int. Comet Q.
The International Comet Quarterly, Physics Department, Appalachian State University, Boone, NC 28608.

Int. J. Electron.
International Journal of Electronics. Taylor and Francis Ltd., 10–14 Macklin Street, London, WC2B 5BF, England.

Int. J. Eng. Sci.
International Journal of Engineering Science. Pergamon Press Ltd., Headington Hill Hall, Oxford, OX3 0BW, England.

Int. J. Gen. Syst.
International Journal of General Systems. Gordon & Breach Science Publishers Ltd., 42 William IV Street, London WC2N 4DE, England.

Int. J. Heat Mass Transfer
International Journal of Heat and Mass Transfer, Pergamon Press Ltd., Headington Hill Hall, Oxford, OX3 0BW, England.

Int. J. Mass Spectrom. Ion Phys.
International Journal of Mass Spectrometry and Ion Physics. Elsevier Scientific Publishing Co., P.O. Box 211 Amsterdam, Netherlands.

Int. J. Theor. Phys.
International Journal of Theoretical Physics. Plenum Publishing Co. Ltd., Davis House, 8 Scrubs Lane, London, NW10 6SE, England.

Interdisciplinary Sci. Rev.
Interdisciplinary Science Reviews. Heyden & Son Ltd., Spectrum House, Alderton Crescent, London NW4 3XX, England.

Irish Astron. J.
The Irish Astronomical Journal. A Quarterly Publication under the auspices of the Observatories of Armagh and Dunsink. Armagh Observatory, Northern Ireland.

ISIS
ISIS. An international review devoted to the history of science and its cultural influences. Publication and Editorial Office, Department of History and Sociology of Science, University of Pennsylvania, Philadelphia 19104.

Istanbul Üniv. Fen Fak. Mec., Ser. C
Istanbul Üniversitesi Fen Fakultesi Mecmuasi, Serie C (Astronomie, Physique, Chimie). Istanbul, Turkey.

Izv. Akad. Nauk Armyansk. SSR
Izvestiya Akademii Nauk Armyanskoj SSR. Fizika. Publisher: Izdatel'stvo AN Armyanskoj SSR, Erevan.

Izv. Astron. Ehngel'gardt. Obs.
Izvestiya Astronomicheskoj Ehngel'gardtovskoj Observatorii. Izdatel'stvo Kazanskogo Universiteta, Kazan.

Izv. Glav. Astron. Obs. Pulkovo
Izvestiya Glavnoj Astronomicheskoj Observatorii v Pulkove. Akademiya Nauk SSSR. Izdanie Glavnoj astronomicheskoj observatorii v Pulkove, Leningrad.

Izv. Krymskoj Astrofiz. Obs.
Izvestiya Krymskoj Astrofizicheskoj Observatorii. Akademiya Nauk SSR. Publishers: Izdatel'stvo "Nauka", Moskva.

J. Acoust. Soc. America
Journal of the Acoustical Society of America. Published for the Acoustical Society of America by the American Institute of Physics, 335 East 45th Street, New York, NY 10017, USA

J. American Assoc. Variable Star Obs.
The Journal of the American Association of Variable Star Observers. Published by The American Association of Variable Star Observers, 187 Concord Avenue, Cambridge, Mass. 02138, USA.

J. Appl. Meteorol.
Journal of Applied Meteorology. American Meteorological Society, 45 Beacon Street, Boston, MA 02108, USA.

J. Appl. Photogr. Eng.
Journal of Applied Photographic Engineering. Society of Photographic Scientists and Engineers, Suite 204, 1330 Massachusetts Avenue, N.W. Washington, D.C. 20005, USA.

J. Appl. Phys.
Journal of Applied Physics. American Institute of Physics, 335 East 45th Street, New York, NY 10017, USA.

J. Astron. Soc. Egypt
Journal of the Astronomical Society of Egypt. Published by Astronomical Society of Egypt, Astronomy Department, Faculty of Sciences, Cairo University, Egypt.

J. Astron. Soc. Western Australia
The Journal of the Astronomical Society of Western Australia. Edited by the Astronomical Society of Western Australia, Perth, W. A.

J. Astronaut. Sci.
Journal of the Astronautical Sciences. American Astronautical Society, 6060 Duke Street, Alexandria, VA 22304, USA.

J. Astrophys. Astron.
Journal of Astrophysics and Astronomy. Published by Indian Academy of Sciences, Post Box No. 8005, Bangalore 560080, India.

J. Atmos. Sci.
Journal of the Atmospheric Sciences. American Meteorological Society, 45 Beacon Street, Boston, MA 02108, USA.

J. Atmos. Terr. Phys.
Journal of Atmospheric and Terrestrial Physics. Pergamon Press Ltd., Oxford, England.

J. British Astron. Assoc.
Journal of the British Astronomical Association.

Burlington House, Piccadilly, London, W1V ONL, England.

J. British Interplanet. Soc.
Journal of the British Interplanetary Society. British Interplanetary Society, 12 Bessborough Gardens, London, SW1V 2JJ, England.

J. Colloid Interface Sci.
Journal of Colloid and Interface Science. Academic Press Inc., 111 Fifth Avenue, New York, N.Y. 10003, USA.

J. Comput. Phys.
Journal of Computational Physics. Academic Press Inc., 111 Fifth Avenue, New York, NY 10003, USA.

J. Fluid Mech.
Journal of Fluid Mechanics. Cambridge University Press, Bentley House, 200 Euston Road, London, NW1 2DB, England.

J. Geomagn. Geoelectr.
Journal of Geomagnetism and Geoelectricity. Society of Terrestrial Magnetism and Electricity of Japan, Geophysical Institute, Tokyo University, Tokyo 113, Japan.

J. Geophys.
Journal of Geophysics / Zeitschrift für Geophysik. Springer Verlag, D-6900 Heidelberg 1, Postfach 105280, F. R. Germany.

J. Geophys. Res.
Journal of Geophysical Research. Published by American Geophysical Union, 1909 K Street, N.W. Washington D.C. First section: Space physics; Second section: Physics and chemistry of the solid earth, planetology, geodesy; Third section: Oceans and atmospheres.

J. Guid. Control
Journal of Guidance and Control. American Institute of Aeronautics and Astronautics, 1290 Avenue of the Americas, New York, NY 10019, USA.

J. Hist. Astron.
Journal for the History of Astronomy. Published by Science History Publications Ltd., Halfpenny Furze, Mill Lane, Chalfont St Giles, Buckinghamshire, England.

J. Inorg. Nucl. Chem.
Journal of Inorganic and Nuclear Chemistry. Pergamon Press Ltd., Headington Hill Hall, Oxford OX3 0BW, England.

J. Inst. Math. Appl.
Journal of the Institute of Mathematics and its Applications. Academic Press Inc. (London) Ltd., 24 - 28 Oval Road, London NW1 7DX, England.

J. Korean Astron. Soc.
The Journal of the Korean Astronomical Society. Published by the Korean Astronomical Society.

J. Magn. Magn. Mater.
Journal of Magnetism and Magnetic Materials. North-Holland Publishing Co., P.O. Box 211, Amsterdam, Netherlands.

J. Math. Phys.
Journal of Mathematical Physics. American Institute of Physics, 335 East 45th Street, New York, N.Y. 10017, USA.

J. Meteor Res.
The Journal of Meteor Research. Space News Publishing Co., P. O. Box 66521, Baton Rouge, LA 70896. ISSN 0277-6057.

J. Mol. Spectrosc.
Journal of Molecular Spectroscopy. Academic Press Inc., 111 Fifth Avenue, New York, NY 10003, USA.

J. Nanjing Univ.
Journal of Nanjing University. Nanjing Daxue Xuebao. (Natural Science Edition). Nanking University, Nanking, China.

J. Navig.
The Journal of Navigation. The Royal Institute of Navigation at the Royal Geographical Society, Kensington Gore, London, SW7 2AT. Scottish Academic Press Ltd., 33 Montgomery Street, Edinburgh EH7 5JX.

J. Opt. (*France*)
Journal of Optics. Masson Editeur, 120 Boulevard Saint-Germain, 75280 Paris Cedex 06, France.

J. Opt. (*India*)
Journal of Optics. Optical Society of India, Department of Applied Physics, University of Calcutta, 92 Acharya Prafulla Chandra Road, Calcutta-9, India.

J. Opt. Soc. America
Journal of the Optical Society of America. American Institute of Physics, 335 East 45th Street, New York, N.Y. 10017, USA.

J. Phys. A
Journal of Physics A, (Mathematical, Nuclear and General). Institute of Physics, 47 Belgrave Square, London, SW1X 8QX, England.

J. Phys. B
Journal of Physics B, (Atomic and Molecular Physics). Institute of Physics, 47 Belgrave Square, London, SW1X 8QX, England.

J. Phys. Chem. Ref. Data
Journal of Physical and Chemical Reference Data. American Chemical Society, 1155 Sixteenth Street, N.W., Washington, DC 20036, USA.

J. Phys. Colloq.
Journal de Physique Colloque. Société Française de Physique, 87 bis Avenue du Général Leclerc, 75014 Paris, France.

J. Phys. E
Journal of Physics E, (Scientific Instruments). Formerly: J. Sci. Instrum. (GB). Institute of Physics, 47 Belgrave Square, London, SW1X 8QX, England.

J. Phys. F
Journal of Physics F, (Metal Physics). Institute of Physics, 47 Belgrave Square, London, SW1X 8QX, England.

J. Phys. G
Journal of Physics G, (Nuclear Physics). Institute of Physics, 47 Belgrave Square, London, SW1X 8QX, England.

J. Phys. Soc. Japan
Journal of the Physical Society of Japan. Room 211, Kikai Shinko Building, Shiba Koen, Minato-ku, Tokyo 105, Japan.

J. Physique
Journal de Physique. Z. I. de Courtaboeuf, B. P. 112, 91402 Orsay, France.

J. Plasma Phys.
Journal of Plasma Physics. Cambridge University Press, Bentley House, 200 Euston Road, London, NW1 2DB, England.

J. Proc. R. Soc. New South Wales
Journal and Proceedings of the Royal Society of New South Wales. Science Centre, 35 Clarence Street, Sydney, N.S.W. 2000, Australia

J. Quant. Spectrosc. Radiat. Transfer
Journal of Quantitative Spectroscopy & Radiative Transfer. Pergamon Press Ltd., Headington Hill Hall, Oxford, OX3 OBW, England.

J. R. Astron. Soc. Canada
The Journal of the Royal Astronomical Society of Canada, devoted to the advancement of astronomy and allied sciences. The Royal Astronomical Society of Canada, 124 Merten Street, Toronto, Ontario, Canada.

J. Radio Res. Lab.
Journal of the Radio Research Laboratories. Chief Planning Section, Radio Research Laboratories, Ministry of Posts & Telecommunications, Nukui-Kitamachi, Konganei-shi, Tokyo 184, Japan.

J. Res. Natl. Bur. Stand. B
Journal of Research of the National Bureau of Standards. Section B (Mathematics and Mathematical Physics). US Government Printing Office, Division of Public Documents, Washington, DC 20402, USA.

J. Sci. Ind. Res.
Journal of Scientific and Industrial Research. Sales & Distribution Office, Publications & Information Directorate, Hillside Road, New Delhi 110012, India.

J. Sci. Res. Banaras Hindu Univ.
Journal of Scientific Research of the Banaras Hindu University. PO-Banaras Hindu University, India.

J. Spacecr. Rockets
Journal of Spacecraft and Rockets. American Institute of Aeronautics and Astronautics, 1290 Avenue of the Americas, New York, N. Y. 10019, USA.

J. Spectrosc. Soc. Japan
Journal of the Spectroscopical Society of Japan. (Bunkyo Kenkyu). 2-15-1, Nakai, Shinjuku-ku, Tokyo 161, Japan.

J. Toyo Univ., Gen. Educ., Nat. Sci.
Journal of the Toyo University, General Education, Natural Science. Published by The Toyo University, 28, Hakusan 5-chôme, Bunkyo-ku, Tokyo, Japan.

Japanese J. Appl. Phys.
Japanese Journal of Applied Physics. Publication Office, 2nd Toya Kaiji Building, 24-8 Shinbashi, Minato-ku, Tokyo 105, Japan.

Jenaer Rundsch. (Jena Rev.)
Jenaer Rundschau (Jena Review). Publisher: VEB Verlag Technik, Berlin, German Democratic Republic.

JETP Lett.
JETP Letters. A translation of JETP Pis'ma v Redaktsiyu

of the Academy of Sciences in the USSR. American
Institute of Physics, 335 East 45th Street, New York,
NY 10017, USA.

Kexue Tongbao
Kexue Tongbao. Academia Sinica, Peking, People's
Republic of China [English translation in: Kexue
Tongbao (Scientia)(USA)].

Kodaikanal Obs. Bull.
Kodaikanal Observatory Bulletins, Series A, Indian
Institute of Astrophysics, Bangalore, India.

Komet. Tsirk.
Kometnyj Tsirkulyar. Gruppa po Issledovaniyu Komet
Astrosoveta i Mezhduvedomstvennyj Geofizicheskij
Komitet Akademii Nauk SSSR. Kievskij Universitet im.
T. G. Shevchenko.

Komety i Meteory
Komety i Meteory. Akademiya Nauk Tadzhikskoj SSR.
Astronomicheskij Sovet Akademii Nauk SSSR. Publish-
ers: Izdatel'stvo "Donish", Dushanbe.

Kosm. Issled.
Kosmicheskie Issledovaniya. Akademiya Nauk SSSR.
Publishers: Izdatel'stvo "Nauka", Moskva [An English
translation is published as "Cosmic Research", Consultants Bureau, New York, N. Y.].

Kozmos
Kozmos. Popular Astronomical Journal of the Slovak
Central Observatory in Hurbanovo. Publisher: Slovenská
ústredná hvezdáren v Hurbanove.

Lett. Math. Phys.
Letters in Mathematical Physics. D. Reidel Publishing
Co., P.O. Box 17, Dordrecht, Netherlands.

Lett. Nuovo Cimento
Lettere al Nuovo Cimento. Società Italiana di Fisica.
Editrice Compositori viale XII Giugno 1, 40124 Bologna,
Italy.

L'Universo
L'Universo. Rivista dell'Istituto Geografico Militare.
Direzione, Redazione e Amministrazione: Istituto Geo-
grafico Militare, Firenze.

Mada
Mada (Science). Published by The Weizmann Science
Press of Israel, Jerusalem.

Magn. Polya Soln. Pyaten
Magnitnye Polya Solnechnykh Pyaten. (Supplements to
Solnechnye Dannye. Byulleten' (*Solar Data*)). Publishers:
Izdatel'stvo "Nauka", Leningrad.

Magy. Geofiz.
Magyar Geofizika. Lapkiado Vallalat, 1073 Budapest,
Lenin korut 9 - 11, Hungary

Mat.-Fys. Medd.
Matematisk-fysiske Meddelelser. Published by Det
Kongelige Danske Videnskabernes Selskab. Sold by
Munksgaard, Nørre Søgade 35, DK-1370 Copenhagen K.,
Denmark. ISSN 0023-3323.

Math. Intelligencer
The Mathematical Intelligencer. Springer-Verlag, Berlin—
Heidelberg—New York, 175 Fifth Avenue, New York,
NY 10010, USA.

Math. Proc. Cambridge Philos. Soc.
Mathematical Proceedings of the Cambridge Philosophi-
cal Society. Formerly: Proceedings of the Cambridge,
Philosophical Society (Mathematical and Physical
Sciences). Cambridge University Press, Bentley House,
200 Euston Road, London, NW1 2DB, England.

Mem. Astron. Soc. India
Memoirs of the Astronomical Society of India. Edited
and published by M.S. Vardya, Tata Institute of Funda-
mental Research, Bombay 400005 on behalf of the
Astronomical Society of India, Osmania University,
Hyderabad 500007.

Mem. Fac. Eng. Kyoto Univ.
Memoirs of the Faculty of Engineering, Kyoto University, Kyoto, Japan.

Mem. Fac. Eng. Osaka City Univ.
Memoirs of the Faculty of Engineering, Osaka City
University. 459 Sugimoto-cho, Sumi Yoshi-kum, Osaka,
Japan.

Mem. Fac. Sci. Kyoto Univ.
Memoirs of the Faculty of Science, Kyoto University.
Series of Physics, Astrophysics, Geophysics, and Chemis-
try. Printed by Yamashiro Printing Publishing Co. Ltd.,
Kamigyo, Kyoto.

Mem. Japan Astron. Study Assoc.
Memoirs of the Japan Astronomical Study Association.
c/o National Science Museum, Ueno Park, Taito-ku,
Tokyo, Japan.

Mem. Soc. Astron. Italiana
Memorie della Società Astronomica Italiana. Presso
Laboratorio di Astrofisica Spaziale, Castella Postale 67,
00044 Frascati, Italy.

Mercury
Mercury. The Journal of the Astronomical Society of the
Pacific. Published by the Astronomical Society of the
Pacific, 1290 24th Avenue, San Francisco, California
94122, USA. (415) 661 - 8660.

Messenger
The Messenger — El Mensajero. Published by European
Southern Observatory, Karl-Schwarzschild-Straße 2,
D-8046 Garching bei München.

Meteoritics
Meteoritics. The Journal of the Meteoritical Society. Pub-
lished quarterly by The Meteoritical Society and Arizona
State University Bureau of Publications. Editorial address:
Center for Meteorite Studies, The Arizona State Universi-
ty, Tempe, Ariz. 85281, USA.

Meteoritika
Akademiya Nauk SSSR. Komitet po Meteoritam. Publishers: Izdatel'stvo "Nauka", Moskva.

Meteorol. Rundsch.
Meteorologische Rundschau. Springer-Verlag, D-1000
Berlin 33, Heidelberger Platz 3, Germany.

Metrologia
Metrologia. Springer-Verlag, Heidelberger Platz 3,
D-1000 Berlin 33, F. R. Germany.

Microwave J.
Microwave Journal. To be obtained from 610 Washington
Street, Dedham Plaza, Dedham, Massachusetts, U.S.A.

Minor Planet Bull.
The Minor Planet Bulletin. Bulletin of the Minor Planets Section of the Association of Lunar and Planetary Observers. Editorial Office: R. G. Hodgson, Dordt College, Sioux Center, Iowa, U.S.A.

Minor Planet Circ., (M. P. C.)
The Minor Planet Circulars/Minor Planets and Comets. Edited under the supervision of B. G. Marsden. Published by Minor Planet Center, Smithsonian Astrophysical Observatory, Cambridge, Mass. 02138, USA.

Mitt. Astron. Ges.
Mitteilungen der Astronomischen Gesellschaft, Hamburg. Available from Astron. Instit. Univ. Bochum, Postfach 10 21 48, D-4630 Bochum. ISSN 0172–5483.

Mitt. Inst. Theor. Geod. Univ. Bonn
Mitteilungen aus dem Institut für Theoretische Geodäsie der Universität Bonn, Nußallee 17, 5300 Bonn 1, F. R. Germany.

Mitt. Karl-Schwarzschild-Obs. Tautenburg
Mitteilungen des Karl-Schwarzschild-Observatoriums Tautenburg der Deutschen Akademie der Wissenschaften der DDR. Zentralinstitut für Astrophysik.

Mitt. Satell.-Beobachtungsstn. Zimmerwald
Mitteilungen der Satelliten-Beobachtungsstation Zimmerwald. Hausdruckerei Institut für Exakte Wissenschaften, Bern, Switzerland.

Mitt. Sternw. Sonneberg
Akademie der Wissenschaften der DDR, Zentralinstitut für Astrophysik, Sternwarte Sonneberg. Mitteilungen der Sternwarte zu Sonneberg.

Mitt. Veränderl. Sterne (MVS)
Mitteilungen über Veränderliche Sterne. Herausgegeben von der Sternwarte Sonneberg der Akademie der Wissenschaften der DDR, Sonneberg, German Democratic Republic.

Mod. Geol.
Modern Geology. Gordon & Breach Science Publishers Ltd., 41 and 42 William IV Street, London WC2, England.

Mon. Not. R. Astron. Soc.
Monthly Notices of the Royal Astronomical Society. Published for the Royal Astronomical Society by Blackwell Scientific Publications, Oxford – London – Edinburgh – Melbourne.

Mon. Notes Astron. Soc. South. Africa
Monthly Notes of the Astronomical Society of Southern Africa. Published by the Astronomical Society of Southern Africa, S. A. Astronomical Observatory, Cape Province, South Africa.

Mon. Notes Int. Polar Motion Serv.
Monthly Notes of the International Polar Motion Service. Published by the Central Bureau, International Latitude Observatory of Mizusawa, Mizusawa-shi, Iwate-ken, Japan.

Moon Planets
The Moon and the Planets. An International Journal of Comparative Planetology. Publisher: D. Reidel Publishing Company, Dordrecht, Holland – Boston, USA. Formerly: Moon.

Nablyud. Iskusstv. Nebesn. Tel
Nablyudeniya Isskusstvennykh Nebesnykh Tel. Published by Astronomicheskij Sovet Akademii Nauk SSSR, Moskva.

Nachr. Akad. Wiss. Göttingen II
Nachrichten der Akademie der Wissenschaften in Göttingen. II. Mathematisch-Physikalische Klasse. Vandenhoeck & Ruprecht, Göttingen.

Nachr. Karten-, Vermessungswesen
Nachrichten aus dem Karten- und Vermessungswesen. Editor: Institut für Angewandte Geodäsie (Abt. II des Deutschen Geodätischen Forschungsinstituts). Published by Verlag des Instituts für Angewandte Geodäsie, Frankfurt a. M.

Nachr. Olbers-Ges. Bremen
Nachrichten der Olbers-Gesellschaft Bremen. Werderstraße 73, Bremen.

NASA Conf. Publ.
NASA Conference Publication. National Aeronautics and Space Administration. Scientific and Technical Information Branch, Washington, D.C. For sale by the National Technical Information Service, Springfield, Virginia 22161.

NASA Contract. Rep.
NASA Contractor Report. National Aeronautics and Space Administration, Washington, D.C. For sale by the National Technical Information Service, Springfield, Virginia 22161.

NASA Ref. Publ.
NASA Reference Publication. National Aeronautics and Space Administration. Scientific and Technical Information Office. Washington, D.C. 20546. For sale by the National Technical Information Service, Springfield, Virginia 22161.

NASA Tech Briefs
NASA Tech Briefs. NASA Technology Utilization Program, Technology Transfer Division, P. O. Box 8757, Baltimore/Washington International Airport, MD 21240, USA.

NASA Tech. Memo.
NASA Technical Memorandum. National Aeronautics and Space Administration, Washington, D.C. For sale by the National Technical Information Service, Springfield, Virginia 22161.

NASA Tech. Note
NASA Technical Note. National Aeronautics and Space Administration, Washington, D.C. For sale by the National Technical Information Service, Springfield, Virginia 22161.

NASA Tech. Pap.
NASA Technical Paper. National Aeronautics and Space Administration, Washington, D.C. For sale by the National Technical Information Service, Springfield, Virginia 22161.

Natl. Geogr.
National Geographic. Official Journal of the National Geographic Society, Washington, D.C. 17th and M Sts. N.W., Washington, D.C. 20036.

Nature
Nature. Editorial and Publishing Offices: Macmillan

Journals Limited, 4 Little Essex Street, London WC2R 3LF, 711 National Press Building, Washington, D.C. 20045.

Naturwissenschaften
Die Naturwissenschaften. Publisher: Springer-Verlag, Berlin – Heidelberg – New York.

Nauchn. Inf.
Nauchnye Informatsii. Astronomicheskij Sovet Akademii Nauk SSSR, Moskva.

Naučna Misao
Naučna Misao. Društvo za Unaprediivanje i Širenje Nauke, Zagreb, Babonićeva 54, Yugoslavia. Scientific Idea. Society for Promotion and Propagation of Science.

Navigation *(France)*
Navigation. Institut Française de Navigation, 3 avenue Octave-Greard, Paris 7, France.

Navigation *(USA)*
Navigation. Journal of the Institute of Navigation, Institute of Navigation, Suite 832, 815 15th Street, N. W., Washington, DC 20005, USA.

NBS Monogr.
National Bureau of Standards Monograph. U.S. Government Printing Office, Washington, D.C. 20402.

Nederlands Tijdschr. Natuurkd. A
Nederlands Tijdschrift voor Natuurkunde, Publisher: Martinus Nijhoff, Lange Voorhout 9, Den Haag, Netherlands.

New Scientist
New Scientist. New Science Publications, 128 Long Acre, London, WC2E 9QH, England.

New Zealand J. Sci.
New Zealand Journal of Science. Department of Scientific and Industrial Research, Private Bag, Wellington, New Zealand.

News Lett. Astron. Soc. N.Y.
News Letter of the Astronomical Society of New York. A. G. D. Philip (Editor). Astronomical Society of New York, Dudley Observatory, 69 Union Avenue, Schenectady, New York 12308.

NSSDC/WDC-A-R&S
National Space Science Data Center/World Data Center A for Rockets and Satellites. National Aeronautics and Space Administration, Goddard Space Flight Center, Greenbelt, Maryland 20771, USA.

Nucl. Instrum. Methods Phys. Res.
Nuclear Instruments and Methods in Physics Research. Formerly: Nucl. Instrum. Methods. North-Holland Publishing Co., P. O. Box 211, 1000 AE Amsterdam, Netherlands.

Nucl. Phys. A
Nuclear Physics, Volume A. North-Holland Publishing Co., P. O. Box 211, Amsterdam, Netherlands.

Nucl. Tracks Methods Instrum. Appl.
Nuclear Tracks, Methods, Instruments and Application. Formerly: Nucl. Track Detect. (GB) Pergamon Press Ltd., Headington Hill Hall, Oxford OX3 0BW, England.

Numer. Math.
Numerische Mathematik. Springer-Verlag, Berlin–Heidelberg–New York.

Nuovo Cimento A
Il Nuovo Cimento A. Società Italiana di Fisica, Editrice Compositori, viale XII Giugno 1, 40124 Bologna, Italy.

Nuovo Cimento B
Il Nuovo Cimento B. Società Italiana di Fisica, Editrice Compositori, viale XII Giugno 1, 40124 Bologna, Italy.

Nuovo Cimento C
Il Nuovo Cimento C. Società Italiana di Fisica, Editrice Compositori, viale XII Giugno 1, 40124 Bologna, Italy.

Obs. Artif. Earth Satell.
Observations of Artificial Satellites of the Earth (Nablyudeniya Iskusstvennykh Sputnikov Zemli). Magyar Tudományos Akadémia Csillagvizsgáló Intézete. Budapest.

Obs. Astron. Antares, Contrib. Cient.
Universidade Estadual de Feira de Santana, Observatório Astronômico Antares, Contribuição Cientifica. Feira de Santana, Brazil.

Obs. Astrophys. Lab., Univ. Helsinki.Rep.
Observatory and Astrophysics Laboratory, University of Helsinki. Report. Helsinki, Finland.

Obs. Lyon Prepr.
Observatoire de Lyon, Preprints, Observatoire de Lyon, 69230 St. Genis Laval, France.

Observatory
The Observatory. A Review of Astronomy. Publishers: The Editors of 'The Observatory', Royal Greenwich Observatory, Herstmonceux Castle, Hailsham, Sussex, England, BN27 1RP.

Occas. Rep. R. Obs. Edinburgh
Occasional Reports of the Royal Observatory, Edinburgh, Blackford Hill, Edinburgh EH9 3HJ, Scotland. ISSN 0309-099X.

Occultation Newsl.
Occultation Newsletter. Published by the International Occultation Timing Association (I.O.T.A.). 6 N 106 White Oak Lane, St. Charles, Ill. 60174, USA.

Österreich. Z. Vermessungswes. Photogramm.
Österreichische Zeitschrift für Vermessungswesen und Photogrammetrie. Editor and Publisher: Österreichischer Verein für Vermessungswesen und Photogrammetrie, Wien, Austria.

Opt. Acta
Optica Acta. Taylor and Francis Ltd., 10 - 14 Macklin Street, London, WC2B 5NF, England.

Opt. Commun.
Optics Communications. North-Holland Publishing Co., P.O. Box 211, Amsterdam, Netherlands.

Opt. Eng.
Optical Engineering. Society of Photo-Optical Instrumentation Engineers, 337 Tejon Place, Palos Verdes Estates, CA 90274, USA.

Opt. Lett.
Optics Letters. A publication of the Optical Society of

America. American Institute of Physics, 335 East 45th Street, New York, N. Y. 10017, USA.

Optik
Optik. Zeitschrift für das gesamte Gebiet der Licht- und Elektronenoptik. Publishers: Wissenschaftliche Verlagsgesellschaft mbH, Postfach 40, D-7000 Stuttgart, F. R. Germany.

Origins of Life
Origins of Life (Formerly Space Life Sciences). An International Journal. Publisher: D. Reidel Publishing Company, Dordrecht, Holland.

Orion
Orion. Zeitschrift der Schweizerischen Astronomischen Gesellschaft (SAG). Revue de la Société Astronomique de Suisse (SAS). Printed by A. Schudel & Co. AG, 4125 Riehen, Switzerland.

Orione
Orione. Rivista Trimestrale di Divulgazione Astronomica. Via Roma 6, 10025 Pino Torinese (Torino).

Oss. Astrofis. Catania, Pubbl.
Osservatorio Astrofisico di Catania, Pubblicazione. Printed by Scuola Salesiana del Libro, Catania.

Oss. Mem. Oss. Astrofis. Arcetri
Osservazioni e Memorie dell'Osservatorio Astrofísico di Arcetri. Università Degli Studi di Firenze, Firenze, Italy.

Oyo Buturi
Oyo Buturi. Japan Society of Applied Physics, Room No. 209-2, Kikai-Shinko Building, 21 Shiba-Koen Minato-ku, Tokyo, Japan.

Perem. Zvezdy
Peremennye Zvezdy, Byulleten', izdavaemyj Astronomicheskim Sovetom Akademii Nauk SSSR. Published by Astronomicheskij Sovet Akademii Nauk SSSR, Moskva.

Perem. Zvezdy, Prilozhenie
Peremennye Zvezdy, Prilozhenie (The Variable Stars, Supplement). Astronomicheskij Sovet Akademii Nauk SSSR, Moskva.

Philos. Mag. A
Philosophical Magazine A. (Physics of Condensed Matter. Defects and Mechanical Properties) Taylor & Francis Ltd., 10 - 14 Macklin Street, London, WC2B 5NF, England.

Philos. Trans. R. Soc. London, Ser. A
Philosophical Transactions of the Royal Society of London. Series A, Mathematical and Physical Sciences. Carlton House Terrace, London, SW1Y 5AG, England.

Photogr. J. Sun
Photographic Journal of the Sun. Supplement to Monthly Bulletin, Solar Phenomena. Osservatorio Astronomico di Roma.

Photogr. Sci. Eng.
Photographic Science and Engineering. Society of Photographic Scientists and Engineers, Suite 204, 1330 Massachusetts Avenue N.W., Washington, DC 20005, USA.

Photogramm. Eng. Remote Sensing
Photogrammetric Engineering and Remote Sensing. Formerly: Photogramm. Eng. American Society of Photogrammetry, 105 North Virginia Avenue, Falls Church, VA 22046, USA.

Photonics Spectra
Photonics Spectra. The Magazine of Optical/Electro-Optical/Laser Technology. Published by the Optical Publishing Co., Inc., 59 Bartlett Ave., P. O. Box 1146, Pittsfield, Mass. 01201, USA.

Photorin
Photorin, Mitteilungen der Lichtenberg-Gesellschaft e.V.

Phys. Abstr.
Physics Abstracts. Science Abstracts, Series A. An INSPEC Publication, published by The Institution of Electrical Engineers in Association with the Institute of Electrical and Electronics Engineers Inc. Printed by Pindar & Son Ltd., Scarborough, N. Yorkshire, England.

Phys. Bl.
Physikalische Blätter. Physik-Verlag GmbH, Pappelallee 3, Postfach 1260/1280, D-6940 Weinheim, F. R. Germany.

Phys. Briefs
Physics Briefs. Physikalische Berichte. Edited by Deutsche Physikalische Gesellschaft and Fachinformationszentrum Energie, Physik, Mathematik in cooperation with American Institute of Physics. Published by Physik Verlag GmbH, Postfach 1260/1280, D-6940 Weinheim, F. R. Germany. ISSN 0170-7434.

Phys. Chem. Earth
Physics and Chemistry of the Earth. Pergamon Press Ltd., Headington Hill Hall, Oxford OX3 0BW, England.

Phys. Chem. Miner.
Physics and Chemistry of Minerals. Springer-Verlag, D-1000 Berlin 33, Heidelberger Platz 3, Germany.

Phys. Earth Planet. Inter.
Physics of the Earth and Planetary Interiors. A journal devoted to observational and experimental studies of the Earth and Planetary interiors and their theoretical interpretation by the physical sciences. Publisher: North-Holland Publishing Company, Amsterdam, Netherlands.

Phys. Energ. Fortis Phys. Nucl.
Physica Energiae Fortis et Physica Nuclearis. Science Press, Peking, People's Republic of China. Subscription address: Guozo Shudian, P. O. Box 399, Peking.

Phys. Fluids
The Physics of Fluids. Published by the American Institute of Physics, 335 East 45th Street, New York, NY 10017, USA.

Phys. Lett.
Physics Letters. Volumes A and B. Publisher: North-Holland Publishing Company, Amsterdam.

Phys. Rep.
Physics Reports. North-Holland Publishing Co., P. O. Box 211, Amsterdam, Netherlands.

Phys. Rev. A
Physical Review A, General Physics. Published for the American Physical Society by the American Institute of Physics, 335 East 45th Street, New York, N. Y. 10017, USA.

Phys. Rev. B
Physical Review B, Solid State. Published for the Ameri-

can Physical Society by the American Institute of Physics, 335 East 45th Street, New York, NY 10017, USA.

Phys. Rev. C
Physical Review C, Nuclear Physics. Published for the American Physical Society by the American Institute of Physics, 335 East 45th Street, New York, NY 10017, USA.

Phys. Rev. D
Physical Review D, Particles and Fields. Published for the American Physical Society by the American Institute of Physics, 335 East 45th Street, New York, NY 10017, USA.

Phys. Rev. Lett.
Physical Review Letters. Published for the American Physical Society by the American Institute of Physics, 335 East 45th Street, New York, NY 10017, USA.

Phys. Scr.
Physica Scripta. (Formerly Arkiv för Fysik). Published by the Royal Swedish Academy of Sciences, S-104 05 Stockholm 50, Sweden.

Phys. Teach.
Physics Teacher. American Institute of Physics, 335 East 45th Street, New York, NY 10017, USA.

Phys. Technol.
Physics in Technology. Institute of Physics, 47 Belgrave Square, London SW1X 8QX, England.

Phys. Today
Physics Today. Published by the American Institute of Physics, 335 East 45th Street, New York, NY 10017, USA.

Phys. unserer Zeit
Physik in unserer Zeit. Verlag Chemie GmbH, Pappelallee 3, Postfach 1260/1280, D-6940 Weinheim, F. R. Germany.

Physica B, C
Physica B & C. Subscription address: North-Holland Publishing Co., P.O. Box 211, Amsterdam, Netherlands.

Physis
Physis. Rivista Internazionale di Storia della Scienza.

Pis'ma Astron. Zh.
Pis'ma v Astronomicheskij Zhurnal. Akademiya Nauk SSSR. Publishers: Izdatel'stvo 'Nauka', Moskva. Translation in Soviet Astron. Lett.

Planet. Space Sci.
Planetary and Space Science. Pergamon Press Ltd., Headington Hill Hall, Oxford, OX3 0BW, England.

Postępy Astron.
Postępy Astronomii. Czasopismo Poświecone Upowszechnianiu Wiedzy Astronomicznej. Polskie Towarzystwo Astronomiczne, Warszawa. Printed in Poland by Pánstwowe Wydawnictwo Naukowe, Lódź.

Postępy Fiz.
Postępy Fizyki. Polskie Towarzystwo Fizykczne, 00-681 Warszawa, ul. Hoza 69, Poland.

Pramãna
Pramãna. Indian Academy of Sciences, Bangalore 560006, India.

Pré-publ. Inst. Astrophys. Paris
Pré-publication Institut d'Astrophysique de Paris, 98bis, Boulevard Arago, 75014 Paris.

Priroda
Priroda. Publishers: Izdatel'stvo "Nauka", Moskva.

Probl. Kosm. Fiz.
Problemy Kosmichskoj Fiziki. Mezhvedomstvennyj Nauchnoj Sbornik. Izdatel'skoe Obedinenie Vishcha Shkola. Izdatel'stvo pri Kievskom Universitete, Kiev.

Proc. Astron. Soc. Australia
Proceedings of the Astronomical Society of Australia. Published for the Society by Sydney University Press, Sydney.

Proc. IEEE
Proceedings of the IEEE. Published by the Institute of Electrical and Electronics Engineers, 345 East 47th Street, New York, NY 10017, USA.

Proc. Indian Acad. Sci. Earth Planet. Sci.
Proceedings of the Indian Academy of Sciences, Earth and Planetary Sciences. Bangalore 560 006, India.

Proc. Indian Natl. Sci. Acad., Part A.
Proceedings of the Indian National Science Academy, Part A, Bahadur Shah Zafar Marg, New Delhi 1, India.

Proc. Int. Latitude Obs. Mizusawa
Proceedings of the International Latitude Observatory of Mizusawa. Published by the International Latitude Observatory of Mizusawa, Japan.

Proc. Japan Acad., Ser. B
Proceedings of the Japan Academy. Series B, Physical and Biological Sciences, Ueno Park, Tokyo 110, Japan.

Proc. K. Nederlandse Akad. Wet. B
Koninklijke Nederlandse Akademie van Wetenschappen. Proceedings. Series B, Physical Sciences. Publisher: North-Holland Publishing Company, Amsterdam, Netherlands.

Proc. Natl. Acad. Sci. India, Sect. A
Proceedings of the National Academy of Sciences of India. Section A (Physical Sciences). Lajpatrai Road, Allahabad-2, India

Proc. Natl. Acad. Sci. U.S.A.
Proceedings of the National Academy of Sciences of the United States of America. National Academy of Sciences, 2101 Constitution Avenue, Washington, DC 20418, USA.

Proc. R. Soc. London, Ser. A
Proceedings of the Royal Society of London. Series A: Mathematical and Physical Sciences. Published by the Royal Society, 6 Carlton House Terrace, London, SW1Y 5AG, England.

Proc. Res. Inst. Atmos. Nagoya Univ.
Proceedings of the Research Institute of Atmospherics Nagoya University. Nagoya University, 13 Honohara, 3 Chrome, Toyokawa 442, Japan.

Proc. Soc. Photo-Opt. Instrum. Eng.
Proceedings of the Society of Photo-Optical Instrumentation Engineers, Bellingham, WA 98225, USA.

Prog. Part. Nucl. Phys.
Progress in Particle and Nuclear Physics. Pergamon

Press Ltd., Headington Hill Hall, Oxford OX3 0BW, England.

Prog. Theor. Phys.
Progress of Theoretical Physics. Published for the Research Institute for Fundamental Physics and the Physical Society of Japan. Publication Office: Progress of Theoretical Physics, Yukawa Hall, Kyoto University, 606 Kyoto, Japan.

Prog. Theor. Phys. Suppl.
Supplement of the Progress of Theoretical Physics. Published for the Research Institute for Fundamental Physics and The Physical Society of Japan. Publication Office: Progress of Theoretical Physics. Yukawa Hall, Kyoto University, 606 Kyoto, Japan.

PTB Mitt.
PTB Mitteilungen. Forschen + Prüfen. Fachorgan für Wirtschaft und Wissenschaft. Amts- und Mitteilungsblatt der Physikalisch-Technischen Bundesanstalt. Braunschweig−Berlin. Deutscher Eichverlag, Postfach 3367, D-3300 Braunschweig, F.R. Germany.

Publ. Astron. Soc. Japan
Publications of the Astronomical Society of Japan. Published by the Astronomical Society of Japan. Office of the Society: Tokyo Astronomical Observatory, Mitaka, Tokyo. Agent: Maruzen Co. Ltd. (Export Department), Nihonbashi, Tokyo, Japan.

Publ. Astron. Soc. Pacific
Publications of the Astronomical Society of the Pacific. Published by the Astronomical Society of the Pacific, 1290 24th Avenue, San Francisco, California 94122, USA. (415) 661 - 8660.

Publ. Beijing Astron. Obs.
Publications of the Beijing Astronomical Observatory. Beijing Astronomical Observatory, Academia Sinica, China.

Publ. Bosscha Obs.
Publications of the Bosscha Observatory. Bandung Institute of Technology, Department of Science. Lembang, Indonesia.

Publ. Debrecen Heliophys. Obs.
Publications of Debrecen Heliophysical Observatory of the Hungarian Academy of Sciences.

Publ. Dep. Astron., Univ. Beograd
Publications of the Department of Astronomy, University of Beograd, Faculty of Sciences (Publications de la Chaire d'Astronomie, Université de Beograd, Faculté des Sciences) Beograd. YU ISSN 0350-3283.

Publ. Dep. Astron. Univ. Cape Town
Publications of the Department of Astronomy of the University of Cape Town.

Publ. Dep. Geod. Astron., Univ. Thessaloniki
Publications of the Department of Geodetic Astronomy, University of Thessaloniki.

Publ. Dominion Astrophys. Obs.
Publications of the Dominion Astrophysical Observatory, Victoria, B. C. National Research Council of Canada.

Publ. Eidg. Sternw. Zürich
Publikationen der Eidgenössischen Sternwarte Zürich. Schulthess Polygraphischer Verlag, Zürich.

Publ. Inst. Geophys., Polish Acad. Sci.
Publications of the Institute of Geophysics, Polish Academy of Sciences. Państwowe Wydawnictwo Naukowe, Warszawa-Łódź. ISBN 83-01-02210-8. ISSN 0138-0214.

Publ. Inst. R. Meteorol. Belgique A
Publications, Institut Royal Meteorologique de Belgique. Serie A. 3 Avenue Circulaire, Uccle-Bruxelles 1180, Belgium.

Publ. Int. Latitude Obs. Mizusawa
Publications of the International Latitude Observatory of Mizusawa. Published by the International Latitude Observatory of Mizusawa, Japan.

Publ. Korean Natl. Astron. Obs.
Publications of the Korean National Astronomical Observatory. Published by the Korean National Astronomical Observatory, Seoul, Korea.

Publ. Obs. Astron. Beograd
Publications de l'Observatoire Astronomique de Beograd. Editeur: Observatoire Astronomique de Belgrade, 11050 Beograd, Volgina 7, Yougoslavie.

Publ. R. Obs. Edinburgh
Publications of the Royal Observatory, Edinburgh. Published by The Royal Observatory, Edinburgh, Scotland.

Publ. Shaanxi Astron. Obs.
Publications of the Shaanxi Astronomical Observatory. Academia Sinica, P. O. Box 18, Lintong, near Xian, China.

Publ. United States Naval Obs.
Publications of the United States Naval Observatory. Department of the Navy, U.S. Naval Observatory, Washington. U.S. Government Printing Office, Washington, D.C.

Publ. Variable Star Sect.,R. Astron. Soc. New Zealand
Publications of the Variable Star Section, Royal Astronomical Society of New Zealand. Director: F. M. Bateson, Greerton, Tauranga, New Zealand.

Publ. Warner Swasey Obs.
Publications of the Warner and Swasey Observatory, Case Western Reserve University, Cleveland, Ohio 44106.

Q. Appl. Math.
Quarterly of Applied Mathematics. American Mathematical Society, P. O. Box 1571, Providence, RI 02901, USA.

Q. Bull. Sol. Act.
International Astronomical Union, Quarterly Bulletin on Solar Activity. Published by the Tokyo Astronomical Observatory. Beginning with No. 197; formerly Zürich.

Q. J. R. Astron. Soc.
Quarterly Journal of the Royal Astronomical Society. Burlington House, London, W1V ONL, England.

R. Greenwich Obs., Time Latitude Serv.
Royal Greenwich Observatory, Time and Latitude Service Royal Greenwich Observatory, Herstmonceux Castle, Hailsham, East Sussex BN27 1RP, England.

R Muscae
R Muscae. Revista de Estrellas Variables. Published by Departamento de Astronomía, Instituto Copérnico,

Buenos Aires, Argentina. Editorial address:
Montevideo 724, 1° "4", (1019), Buenos Aires,
Argentina.

R. Obs. Ann.
Royal Observatory Annals, Royal Greenwich Observatory, Herstmonceux Castle, Hailsham, East Sussex,
BN 27 1RP, England.

Radiotekh. Ehlektron.
Radiotekhnika i Ehlektronika. Moskva TSP-3, Pr. Karl
Marx 18, USSR.

Rech. Aerosp.
Recherche Aerospatiale. Office National d'Études et de
Recherches Aerospatiales, 29 Avenue de la Division
Leclerc, 92320-Chatillon, France.

Recherche
Recherche, 4 Place de l'Odéon, Paris 6, France.

Ref. Zh., 51. Astron.
Referativnyi Zhurnal, 51. Astronomiya. Vsesoyuznyj
Institut Nauchnoj i Tekhnicheskoj Informatsii. Moskva.

Ref. Zh., 52. Geod. Aehrosemka
Referativnyj Zhurnal, 52. Geodeziya i Aehrosemka.
Vsesoyuznyj Institut Nauchnoj i Tekhnicheskoj Informatsii. Moskva.

Ref. Zh., 62. Issled. kosm. prostranstva
Referativnyj Zhurnal, 62. Issledovanie Kosmicheskogo
Prostranstva. Vsesoyuznyj Institut Nauchnoj i Tekhnicheskoj Informatsii. Moskva.

Rep. Finnish Geod. Inst.
Reports of the Finnish Geodetic Institute. Suomen
Geodeettisen Laitoksen Tiedonantoja. Helsinki, Finland.

Rep. Math. Phys.
Reports on Mathematical Physics, Pergamon Press Ltd.,
Headington Hill Hall, Oxford OX3 0BW, England.
Subscription address: ARS Polona-Ruch Foreign Trade
Enterprise, Krakowskie Przedmiescie 7, 00-068
Warszawa.

Rep. Obs. Lund
Reports from the Observatory of Lund. Lunds
Universitet, Institutionen för Astronomie, S-22224 Lund,
Sweden. ISSN 0349-4217.

Rep. Prog. Phys.
Reports on Progress in Physics. Published by the Institute
of Physics, 47 Belgrave Square, London, SW1X 8QX,
England.

Rep. Ser. Dep. Phys. Sci., Univ. Turku
Report Series, Department of Physical Sciences,
Institute of Astronomy, University of Turku, SF-20500
Turku 50, Finland.

Rep. Univ. Electro-Commun.
Reports of the University of Electro-Communications.
University of Electro-Communications, 1-5-1 Chofugaoka,
Chofu-shi, Tokyo, Japan.

Res. Lab. Electron. Onsala Space Obs.
Research Laboratory of Electronics and Onsala Space
Observatory, Chalmers University of Technology,
Gothenburg, Sweden. Research Report.

Rev. Astron.
Revista Astronomica. Organo de la Asociación Argentina
Amigos de la Astronomia, Avenida Patricias Argentinas
550, Buenos Aires 5, Argentina.

Rev. Brasil. Fis.
Revista Brasileira de Fisica. Sociedade Brasileira de Fisica,
Cx. Postal 20553, Sao Paolo SP, Brazil.

Rev. Geophys. Space Phys.
Reviews of Geophysics and Space Physics (formerly Reviews of Geophysics). Published by the American Geophysical Union, 1909 K Street, N.W., Washington,
DC 20006, USA.

Rev. Hist. Sci.
Revue d'Histoire des Sciences. Revue trimestrielle publiée
avec le concours du C.N.R.S. Centre International de
Synthèse, Section d'Histoire des Sciences.

Rev. Mexicana Astron. Astrofis.
Revista Mexicana de Astronomia y Astrofisica. Dirección:
Instituto de Astronomia, Universidad Nacional Autónoma de México, Apartado Postal 70-264, Mexico 20,
D. F., Mexico.

Rev. Mexicana Fis.
Revista Mexicana de Fisica. Sociedad Mexicana de Fisica,
Apartado Postal No. 20-364, Mexico 20, D. F. Mexico.

Rev. Mod. Phys.
Reviews of Modern Physics. Published for the American
Physical Society by the American Institute of Physics,
335 East 35th Street, New York, NY 10017, USA.

Rev. Quest. Sci.
Revue des Questions Scientifiques. Publiée par la Société
Scientifiques de Bruxelles, 61, rue de Bruxelles,
5000 Namur, Belgique.

Rev. Radio Res. Lab.
Review of the Radio Research Laboratories. Ministry of
Posts & Telecommunications, Nukui-Kitamachi,
Konganei-shi, Tokyo 184, Japan.

Rev. Roumaine Phys.
Revue Roumaine de Physique. Academie Republicii
Populare Romine, Boite Postale 134 - 135, Bucuresti,
Rumania.

Rev. Sci. Instrum.
Review of Scientific Instruments. American Institute of
Physics, 335 East 45th Street, New York, NY 10017,
USA.

Rezul't. Nablyud. Iskusstv. Sputnikov Zemli
Rezul'taty Nablyudenij Iskusstvennykh Sputnikov
Zemli. Published by Astronomicheskij Sovet Akademii
Nauk SSSR, Ryazanskij Gosudarstvennyj Pedagogicheskij
Institut, Ryazan'.

Ric. Astron.
Ricerche Astronomiche. Specola Vaticana, Città del
Vaticano.

Ric. Spettrosc.
Ricerche Spettroscopiche. Specola Vaticana, Città del
Vaticano.

Říše hvězd
Říše hvězd. Czech popular astronomical journal.
Publisher: Panorama, Praha.

Riv. Nuovo Cimento
La Rivisty del Nuovo Cimento. Società Italiana di Fisica. Editrice Compositori, viale XII Giugno 1, Bologna 40124, Italy.

Sci. American
Scientific American. 415 Madison Avenue, New York, NY 10017, USA.

Sci. Atmos. Sinica
Scientia Atmospherica Sinica. Science Press, Peking. Subscription address: Guozi Shudian, P.O. Box 399, Peking, Peoples's Republic of China.

Sci. Dimension
Science Dimension. National Research Council of Canada Ottawa K1A 0R6, Canada.

Sci. Pap. Inst. Phys. Chem. Res.
Scientific Papers of the Institute of Physical and Chemical Research. Rikagaku Kenkyusho, Wako-shi, Saitama 351, Japan.

Sci. Prog.
Science Progress. Blackwell Scientific Publications, Oxford, England.

Sci. Rep. Tôhoku Univ., Ser. 1.
The Science Reports of the Tôhoku University. First Series (Physics, Chemistry, Astronomy). Published by the Faculty of Science, Tôhoku University, Sendai, Japan. ISSN 0040-8778.

Sci. Rep. Tôhoku Univ., Ser. 8.
The Science Reports of the Tôhoku University. Eighth Series (Physics and Astronomy). Published by the Faculty of Science, Tôhoku University, Sendai, Japan. ISSN 0388-5607.

Sci. Rev.
Scienca Revuo. Prof. B. Popovic, Ognjena Price 80, Beograd, Yugoslavia.

Sci. Sinica, Ser. A
Scientia Sinica, Series A (Mathematical, Physical, Astronomical and Technical Sciences). Published by Science Press, Beijing, China. Distributed by Scientific and Technical Books Service Ltd., P.O. Box 197, London WC2N 4DE, England.

Science
Science. American Association for the Advancement of Science, 1515 Massachusetts Avenue, N. W., Washington, D. C. 20005, USA.

Scr. Fac, Sci. Nat. Univ. Purkynianae Brunensis Phys.
Scripta Facultatis Scientiarum Naturalium Universitatis Purkynianae Brunensis, Physica. University J. E. Purkyne, 61137 Brno-Kotlarská 2, Czechoslovakia.

Shaanxi Astron. Obs. Repr.
Shaanxi Astronomical Observatory Reprints. Academia Sinica, P. O. Box 18, Lintong, near Xian, China.

SIAM J. Appl. Math.
SIAM Journal on Applied Mathematics. Society for Industrial and Applied Mathematics, 33 South 17th Street, Philadelphia, PA 19103, USA.

Sitzungsber. Akad. Wiss. DDR
Sitzungsberichte der Akademie der Wissenschaften der DDR. Mathematik-Naturwissenschaften-Technik. Akademie-Verlag, Berlin.

Sitzungsber. Bayerische Akad. Wiss.
Bayerische Akademie der Wissenschaften. Mathematisch-Naturwissenschaftliche Klasse. Sitzungsberichte. Publisher: Verlag der Bayerischen Akademie der Wissenschaften, München.

Sitzungsber. Heidelberger Akad. Wiss.
Sitzungsberichte der Heidelberger Akademie der Wissenschaften. Mathematisch-Naturwissenschaftliche Klasse. Publisher: Springer-Verlag, Heidelberg.

Sitzungsber. Österreich. Akad. Wiss.
Sitzungsberichte. Österreichische Akademie der Wissenschaften. Mathematisch-Naturwissenschaftliche Klasse. Abteilung II: Mathematik, Astronomie, Meteorologie und Technik. Publisher: Springer-Verlag, Wien.

Sky Telesc.
Sky and Telescope. Published by Sky Publishing Corporation, 49-50-51 Bay State Road, Cambridge, Mass. 02138, USA.

Smithsonian Astrophys. Obs., Spec. Rep.
Smithsonian Astrophysical Observatory, Special Report. Available from the Publications Division, Distribution Section, Smithsonian Astrophysical Observatory, Cambridge, Mass. 02138.

Smithsonian Contrib. Astrophys.
Smithsonian Contributions to Astrophysics. Smithsonian Institution Astrophysical Observatory, Cambridge, Mass. Printed by Smithsonian Institution Press, City of Washington. For sale by the Superintendent of Documents, U.S. Government Printing Office, Washington, D.C.

Sol. Energy
Solar Energy, Pergamon Press, Maxwell House, Fairview Park, Elmsford, NY 10523, USA.

Sol. Phenom.
Solar Phenomena. Osservatorio Astronomico di Roma.

Sol. Phys.
Solar Physics. A Journal for Solar Research and the Study of Solar Terrestrial Physics. Publisher: D. Reidel Publishing Company, Dordrecht, Holland.

Soln. Dannye, Byull.
Solnechnye Dannye. Byulleten'. (*Solar Data*). Publishers: Izdatel'stvo "Nauka", Leningradskoe Otdelenie, Leningrad.

Sonne
Sonne. Mitteilungsblatt der Amateursonnenbeobachter. Peter Völker, c/o Wilhelm-Foerster-Sternwarte, Munsterdamm 90, 1000 Berlin 41.

Soobshch. Byurakan. Obs.
Soobshcheniya Byurakanskoj Observatorii. Akademiya Nauk Armyanskoj SSR, Erevan.

Soobshch. Gos. Astron. Inst. Shternberg
Soobshcheniya Gosudarstvennogo Astronomicheskogo Instituta im. P. K. Shternberga. Publishers: Izdatel'stvo Moskovskogo Universiteta, Moskva.

Soobshch. Spets. Astrofiz. Obs.
Soobshcheniya Spetsial'noj Astrofizicheskoj Observatorii.

Izdanie Spetsial'noj Astrofizicheskoj Observatorii AN SSSR.

South African Astron. Obs. Circ.
South African Astronomical Observatory, Circulars. S.A. Astronomical Observatory, Observatory, Cape.

South African J. Phys.
South African Journal of Physics (Suid-Afrikaanse Tydskrif vir Fisika). Bureau for Scientific Publications, P.O. Box 1758, Pretoria 0001, South Africa.

South. Stars
Southern Stars. The Journal of the Royal Astronomical Society of New Zealand (Inc.). Address of the Society: P.O. Box 3181, Wellington C1, New Zealand.

Soviet Astron.
Soviet Astronomy. A translation of Astronomicheskij Zhurnal (Astronomical Journal). Published by the American Institute of Physics, New York, N.Y.

Soviet Astron. Lett.
Soviet Astronomy Letters. A translation of "Pis'ma v Astronomicheskij Zhurnal". Published by the American Institute of Physics.

Space Educ.
Space Education. Supplement to Spaceflight. Published by the British Interplanetary Society, 27/29 South Lambeth Road, London SW8 1SZ, England.

Space Sci. Rev.
Space Science Reviews. Publishers: D. Reidel Publishing Company, Dordrecht, Holland.

Spaceflight
Spaceflight. Published by the British Interplanetary Society. Printed by Unwin Brothers Ltd., at the Gresham Press, Old Woking, England.

Spaceworld
Spaceworld. Palmer Publications Inc., Amherst, WI 54406, USA.

Speculations Sci. Technol.
Speculations in Science and Technology. Elsevier Sequoia, S.A., P. O. Box 851, 1001 Lausanne 1, Switzerland.

Sterne
Die Sterne. Zeitschrift für alle Gebiete der Himmelskunde. Johann Ambrosius Barth, Leipzig, German Democratic Republic.

Sterne Weltraum
Sterne und Weltraum. Astronomische Monatsschrift. Publisher: Verlag Sterne und Weltraum Dr. Vehrenberg, Düsseldorf, F. R. Germany.

Sternenbote
Sternenbote. Monatsschrift für Österreichs Amateurastronomen. Publisher: Astronomisches Büro. Hermann Mucke, Wien, Austria.

Stockholms Obs. Ann.
Stockholms Observatorium Annaler. Printed by Almquist & Wiksell, Stockholm, Sweden.

Stockholms Obs. Rep.
Stockholms Observatorium, Saltsjöbaden, Sweden, Report.

Strolling Astron.
The Strolling Astronomer. The Journal of The Association of Lunar and Planetary Observers, Publication Office: The Strolling Astronomer, Box 3 AZ, University Park, New Mexico, USA.

Stud. Astron. Sinica
Studia Astronomica Sinica. Published by the Purple Mountain Observatory, Academia Sinica, Nanking, People's Republic of China.

Stud. Cercet. Fiz.
Studii si Cercetari de Fizica. Academia Republicii Populare Romine. P. O. Box 134-5, Calca Victoriei 126, Bucuresti, Rumania.

Stud. Geophys. Geod.
Studia geophysica et geodaetica. Published for the Geophysical Institute of the Czechoslovak Academy of Sciences by Academia, Praha.

Stud. Hist. Philos. Sci.
Studies in History and Philosophy of Science. Pergamon Press Ltd., Headington Hill Hall, Oxford OX3 0BW, England.

Stud. Soc. Sci. Torunensis
Studia Societatis Scientiarum Torunensis, Toruń — Polonia. Sectio F (Astronomia).

Stud. Univ. Babeş-Bolyai
Studia Universitatis Babeş-Bolyai. Series Mathematica-Physica. Publishers: Intreprinderea Poligrafica, Cluj.

Surv. High Energy Phys.
Surveys in High Energy Physics. Harwood Academic Publishers GmbH, P. O. Box 786, Cooper Station, New York, NY 10003, USA.

Syst. Int.
Systems International. IPC Electrical-Electronic Press Ltd., Dorset House, Stamford Street, London SE1 9LU, England.

Tartu Astrofüüs. Obs. Publ.
W. Struve nimelise Tartu Astrofüüsika Observatooriumi, Publikatsioonid. Eesti NSV Teaduste Akadeemia, Tartu.

Tartu Astrofüüs. Obs. Teated
Tartu Astrofüüsika Observatoorium Teated. Eesti NSV Teaduste Akadeemia W. Struve nim. Tartu Astrofüüsika Observatoorium, Tartu.

Telecommun. J.
Telecommunication Journal.(English Edition). International Telecommunications Union.Place des Nations, 1211 Genève 20, Switzerland.

Tellus
Tellus, a bi-monthly Journal of Geophysics. Svenska Geofysiska Foreningen, Arrhenius laboratoriet, Fack, S - 104 05 Stockholm, Sweden.

Time Freq. Serv. Bull.
Time and Frequency Services Bulletin. Shaanxi Astronomical Observatory, Chinese Academy of Sciences, Lintong, Xian, China.

Tôhoku Geophys. J. Sci. Rep. Tôhoku Univ. Fifth Ser.
Tôhoku Geophysical Journal, Science Reports of the Tôhoku University, Fifth Series. Formerly: Sci. Rep.

Tôhoku Univ. Fifth Ser. Geophys. Faculty of Science,
Tôhoku University, Sendai 980, Japan.

Tokyo Astron. Bull., Second Ser.
Tokyo Astronomical Observatory, Japan. Tokyo Astronomical Bulletin, Second Series.

Tokyo Astron. Obs. Rep.
University of Tokyo, Tokyo Astronomical Observatory,
Japan. Report. ISSN 0374-4639.

Tokyo Astron. Obs., Time and Latitude Bull.
Tokyo Astronomical Observatory, Time and Latitude
Bulletins. Mitaka, Tokyo, Japan.

Tr. Astrofiz. Inst. Alma-Ata
Trudy Astrofizicheskogo Instituta, Alma-Ata. Akademiya
Nauk Kazakhskoj SSR. Publishers: Izdatel'stvo "Nauka"
Kazakhskoj SSR, Alma-Ata.

Tr. Astron. Obs., *Leningrad*
Uchenye Zapiski Gosudarstvennogo Universiteta im.
A. A. Zhdanova, Seriya matematicheskikh nauk = Trudy
Astronomicheskoj Observatorii. Izdatel'stvo Leningradskogo Universiteta, Leningrad.

Tr. Glav. Astron. Obs. Pulkovo
Trudy Glavnoj Astronomicheskoj Observatorii v Pulkove.
Akademiya Nauk SSR. Izdanie Glavnoj astronomicheskoj observatorii v Pulkove, Leningrad.

Tr. Inst. Teor. Astron., *Leningrad*
Trudy Instituta Teoreticheskoj Astronomii. Akademiya
Nauk SSSR. Publishers: Izdatel'stvo "Nauka", Leningrad.

Tr. Kazan. Gorod. Astron. Obs.
Trudy Kazanskoj Gorodskoj Astronomicheskoj Observatorii. Izdatel'stvo Kazanskogo Universiteta, Kazan.

Tr. Tashkent. Astron. Obs.
Trudy Tashkentskoj Astronomicheskoj Observatorii.
Akademiya Nauk Uzbekskoj SSR. Publishers: Izdatel'stvo "FAN" Uzbekskoj SSR, Tashkent.

Trans. American Nucl. Soc.
Transactions of the American Nuclear Society. American
Nuclear Society, 555 North Kensington Avenue, La
Grange Park, IL 60525, USA.

Trans. Astron. Obs. Yale Univ.
Transactions of the Astronomical Observatory of Yale
University. Published by the Observatory, New Haven.

Trans. IAU
Transactions of the International Astronomical Union.
Published and distributed for the IAU (UAI) by D. Reidel Publishing Company, Dordrecht, Holland − Boston,
U.S.A.

Tsirk. Astron. Inst. Tashkent
Tsirkulyar Astronomicheskogo Instituta. Akademiya
Nauk Uzbekskoj SSR. Izdatel'stvo "FAN" Uzbekskoj
SSR, Tashkent.

Tsirk. Astron. Obs. L'vov
Tsirkulyar. Astronomicheskaya Observatoriya. L'vovskij
Ordena Lenina Gosudarstvennyj Universitet imeni Ivana
Franko. Publisher: Izdatel'stvo L'vovskogo Universiteta,
L'vov.

UKIRT Rep.
United Kingdom Infrared Telescope Report. ISSN 0260-
9983.

Umschau
Umschau in Wissenschaft und Technik. Umschau Verlag,
Stuttgarter Str. 18 - 24, D-6000 Frankfurt/M., F. R.
Germany.

United States Naval Obs., Circ.
United States Naval Observatory, Circular. U.S. Naval
Observatory, Washington, D.C. 20390.

Univ. Chile, Dep. Astron., Publ.
Universidad de Chile, Facultad de Ciencias Fisicas y
Matematicas, Departamento de Astronomía, Publicaciones. Observatorio Astronómico Nacional, Cerro
Calán, Santiago de Chile.

Urania Barcelona
Urania. Revista de Astronomía y Ciencias Afines. Órgano
de la Sociedad Astronómica de España y América, Barcelona; Unión Nacional de Astronomía y Ciencias Afines,
Madrid, Spain.

Urania Kraków
Urania. Miesięcznik Polskiego Towarzystwa Miłośnikōw
Astronomii, Kraków. Publisher: Z. N. im Ossolińskich,
Kraków, Poland.

Vac News
Vac News. Indian Vacuum Society, c/o Technical Physics
Division, Bhabha Atomic Research Centre, Trombay,
Bombay 400 085, India.

Vasiona
Vasiona. Revue d'Astronomie et d'Astronautique. Bulletin de la Société Astronomique "R. Bosković", Beograd.

Vatican Obs. Publ.
Vatican Observatory Publications, Specola Vaticana,
Città del Vaticano.

Veröff. Astron. Rechen-Inst. Heidelberg
Veröffentlichungen des Astronomischen Rechen-Instituts
Heidelberg. Verlag G. Braun, Karlsruhe, F.R. Germany.

**Veröff. Bayer. Komm. Int. Erdmessung, Bayer. Akad. Wiss.,
Astron.-Geod. Arb.**
Veröffentlichungen der Bayerischen Kommission für die
Internationale Erdmessung der Bayerischen Akademie der
Wissenschaften. Astronomisch-Geodätische Arbeiten.
Published by Verlag der Bayerischen Akademie der
Wissenschaften, München, F. R. Germany. ISSN 0340-
7691, ISBN 3-7696-9782-0.

Veröff. Remeis-Sternw. Bamberg
Veröffentlichungen der Remeis-Sternwarte Bamberg,
Astronomisches Institut der Universität Erlangen-Nürnberg.

Veröff. Sternw. Sonneberg
Akademie der Wissenschaften der DDR, Zentralinstitut
für Astrophysik, Veröffentlichungen der Sternwarte in
Sonneberg. Publisher: Akademie-Verlag, Berlin, German
Democratic Republic.

Veröff. Zentralinst. Phys. Erde
Akademie der Wissenschaften der DDR, Forschungsbereich Geo- und Kosmowissenschaften. Veröffentlichungen des Zentralinstituts für Physik der Erde, Potsdam,
German Democratic Republic.

Vesmír
Vesmír. Přírodovědecký časopis Čs. akademie věd. Publisher: Academia, Praha.

Vestn. Khar'kov. Univ.
Vestnik Khar'kovskogo Universiteta. Seriya Astronomicheskaya. Publishers: Izdatel'stvo Khar'kovskogo Universiteta, Khar'kov.

Vestn. Kiev. Univ.
Vestnik Kievskogo Universiteta. Seriya Astronomii. Publishers: Izdatel'stvo Kievskogo Universiteta, Kiev.

Vignana Bharathi
Vignana Bharathi. Department of Publications & Extension, Bangalore University, Bangalore 560056, India.

Vistas Astron.
Vistas in Astronomy. An international review journal. Pergamon Press, Oxford – New York – Braunschweig.

Weather
Weather. James Glaisher House, Grenville Place, Bracknell, Berks RG12 1BX, England.

Wiss. Z. Friedrich-Schiller-Univ. Jena
Wissenschaftliche Zeitschrift der Friedrich-Schiller-Universität. Jena. Mathematisch-Naturwissenschaftliche Reihe; Edited by the Rektor der Friedrich-Schiller-Universität Jena, Am Anger 24, Jena, German Democratic Republic.

Wiss. Z. Humboldt-Univ. Berlin
Wissenschaftliche Zeitschrift der Humboldt-Universität zu Berlin. Mathematisch-Naturwissenschaftliche Reihe. Edited by the Rektor der Humboldt-Universität Berlin, Unter den Linden 6, 108 Berlin, German Democratic Republic.

Wiss. Z. Tech. Univ. Dresden
Wissenschaftliche Zeitschrift der Technischen Universität Dresden. Mommsenstraße 13, Dresden A227, Germany.

Wuli
Wuli. Science Press, Peking, People's Republic of China. Subscription address: Guozi Shudian, P.O. Box 399, Peking.

Yamamoto Circ.
Yamamoto Circular. Published by the Yamamoto Observatory, Kamitanakami-Kiryutyo, Otu, Siga-Ken, [520-21] Japan.

Z. angew. Math. Mech.
Zeitschrift für angewandte Mathematik und Mechanik.

Akademie-Verlag GmbH, 108 Berlin, Leipziger Strasse 3–4, German Democratic Republic.

Z. angew. Math. Phys.
Zeitschrift für angewandte Mathematik und Physik. Verlag Birkhauser, Postfach 4000, Basel 24, Switzerland.

Z. Naturforsch.
Zeitschrift für Naturforschung. Teil A. A Europhysics Journal. Physik–Physikalische Chemie–Kosmophysik. Verlag der Zeitschrift für Naturforschung, Tübingen. P. O. Box 2645, D–7400 Tübingen, F. R. Germany.

Z. Phys. A
Zeitschrift für Physik A. Atoms and Nuclei. Springer-Verlag, Berlin–Heidelberg–New York.

Z. Phys. B
Zeitschrift für Physik B. Condensed Matter and Quanta. Springer-Verlag, Berlin–Heidelberg–New York.

Z. Phys. C
Zeitschrift für Physik C. Particles and Fields. Springer-Verlag, P.O. Box 105280, D-6900 Heidelberg 1, Germany.

Z. Vermessungswes.
Zeitschrift für Vermessungswesen. Verlag Konrad Wittwer, 7000 Stuttgart 1, Nordbahnhofstrasse 16, Postfach 147, F. R. Germany.

Zeiss Inf.
Zeiss Information. Carl Zeiss, Oberkochen. F. R. Germany.

Zemlya Vselennaya
Zemlya i Vselennaya. Astronomiya, Geofizika. Issledovaniya Kosmicheskogo Prostranstva, Nauchno-Populyarnyj Zhurnal Akademii Nauk SSSR. Publishers: Izdatel'stvo "Nauka", Moskva.

Zenit
Populair wetenschappelijk maandblad over sterrenkunde/weerkunde/ruimtevaart/ruimte-onderzoek/aanverwante wetenschappen en technieken. Bureau: Stichting De Koepel, Utrecht.

Zentralbl. Math. Grenzgeb. – Math. Abstr.
Zentralblatt für Mathematik und ihre Grenzgebiete – Mathematics Abstracts. Publisher: Springer-Verlag, Berlin–Heidelberg–New York.

Zvaigžņota Debess
Latvijas PSR Zinātņu Akadēmijas Radioastrofizikas Observatorijas Populārzinatnisks Gadalaiku Izdevums. Izdevnieciba "Zinātne", Riga.

Journals abstracted completely.

A selected number of periodicals listed in category 001 are central to the subject scope of *Astronomy and Astrophysics Abstracts*. Depending on their relevance, almost all papers of the journals listed below are abstracted in our service.

AAVSO Bull.
Acta Astron.
Acta Astron. Sinica
Acta Astrophys. Sinica
Acta Cosmologica
Ann. Shanghai Obs., Acad. Sinica
Ann. Tokyo Astron. Obs.
Annu. Rep. Astron. Inst. Greece
Annu. Rev. Astron. Astrophys.
Astrofiz. Issled. Izv. Spets. Astrofiz. Obs.
Astrofizika
Astrometr. Astrofiz.
Astron. Astrophys.
Astron. Astrophys., Suppl. Ser.
Astron. J.
Astron. Nachr.
Astron. Pap.
Astron. Q.
Astron. Tidsskr.
Astron. Tsirk.
Astron. Vestn.
Astron. Zh.
Astronomia
Astronomie
Astrophys. J.
Astrophys. J., Lett.
Astrophys. J., Suppl. Ser.
Astrophys. Lett.
Astrophys. Space Sci.
Australian J. Phys., Astrophys. Suppl.
BAV Rundbrief
BBSAG Bull.
Bol. Astron. Obs. Madrid
Bol. Inst. Tonantzintla
British Astron. Assoc. Circ.
Bull. AFOEV
Bull. American Astron. Soc.
Bull. Astron. Inst. Czechoslovakia
Bull. Astron., Obs. R. Belgique
Bull. Astron. Soc. India
Bull. Inf. Cent. Données Stellaires
Bull. Obs. Astron. Belgrade
Byull. Abastumanskaya Astrofiz. Obs.
Byull. Inst. Astrofiz.
Byull. Inst. Teor. Astron.
Carter Obs., Astron. Bull.
Celestial Mech.
Circ. Inf.
Circ. Stn. Astron. Int. Latitudine, Carloforte-Cagliari
Coelum
Comments Astrophys.
Comun. Obs. Astron. Univ. Coimbra
Dudley Obs. Rep.
Dunsink Obs. Publ.
ESO Sci. Prepr.
ESO Tech. Rep.
G. Astron.
IAU Circ.
Icarus
Inf. Bull. Variable Stars
Inst. Theor. Astrophys., Blindern—Oslo, Rep.
Irish Astron. J.
Izv. Astron. Ehngel'gardt. Obs.
Izv. Glav. Astron. Obs. Pulkovo
Izv. Krymskoj Astrofiz. Obs.

J. American Assoc. Variable Star Obs.
J. British Astron. Assoc.
J. Hist. Astron.
J. R. Astron. Soc. Canada
Komet. Tsirk.
Komety i Meteory
Mem. Soc. Astron. Italiana
Mercury
Messenger
Meteoritics
Meteoritika
Minor Planet Bull.
Minor Planet Circ.
Mitt. Astron. Ges.
Mitt. Sternw. Sonneberg
Mitt. Veränderl. Sterne (MVS)
Mon. Not. R. Astron. Soc.
Mon. Notes Astron. Soc. South Africa
Mon. Notes Int. Polar Motion Serv.
Moon Planets
Nablyud. Iskusstv. Nebesn. Tel
Nauchn. Inf.
News Lett. Astron. Soc. N.Y.
Obs. Astrophys. Lab., Univ. Helsinki. Rep.
Observatory
Occas. Rep. R. Obs. Edinburgh
Occultation Newsl.
Orion
Oss. Astrofis. Catania, Pubbl.
Oss. Mem. Oss. Astrofis. Arcetri
Perem. Zvezdy
Perem. Zvezdy, Prilozhenie
Pis'ma Astron. Zh.
Postępy Astron.
Probl. Kosm. Fiz.
Proc. Astron. Soc. Australia
Proc. Int. Latitude Obs. Mizusawa
Publ. Astron. Soc. Japan
Publ. Astron. Soc. Pacific
Publ. Dominion Astrophys. Obs.
Publ. Eidg. Sternw. Zürich
Publ. Int. Latitude Obs. Mizusawa
Publ. R. Obs. Edinburgh
Publ. United States Naval Obs.
Publ. Variable Star Sect. R. Astron. Soc. New Zealand
Q. Bull. Sol. Act.
Q. J. R. Astron. Soc.
R. Obs. Ann.
Rep. Obs. Lund
Rev. Mexicana Astron. Astrofis.
Rezul't. Nablyud. Iskusstv. Sputnikov Zemli
Sky Telesc.
Smithsonian Astrophys. Obs., Spec. Rep.
Smithsonian Contrib. Astrophys.
Sol. Phys.
Soln. Dannye, Byull.
Soobshch. Byurakan. Obs.
Soobshch. Gos. Astron. Inst. Shternberg
Soobshch. Spets. Astrofiz. Obs.
South African Astron. Obs. Circ.
South. Stars
Space Sci. Instrum.
Space Sci. Rev.
Sterne
Sterne Weltraum

Stockholms Obs. Ann.
Stockholms Obs. Rep.
Strolling Astron.
Tartu Astrofüüs. Obs. Publ.
Tartu Astrofüüs. Obs. Teated
Tokyo Astron. Bull.
Tokyo Astron. Obs. Rep.
Tr. Astrofiz. Inst. Alma-Ata
Tr. Astron. Obs., *Leningrad*
Tr. Glav. Astron. Obs. Pulkovo
Tr. Inst. Teor. Astron., *Leningrad*
Tr. Kazan. Gorod. Astron. Obs.
Tr. Tashkent. Astron. Obs.

Trans. IAU
Tsirk. Astron. Inst. Tashkent
Tsirk. Astron. Obs. L'vov
United States Naval Obs., Circ.
Urania Barcelona
Urania Kraków
Vatican Obs. Publ.
Veröff. Astron. Rechen-Inst. Heidelberg
Veröff. Sternw. Sonneberg
Vestn. Khar'kov. Univ.
Vestn. Kiev. Univ.
Vistas Astron.
Yamamoto Circ.

002 Bibliographical Publications, Catalogues, Atlases

002.001 Une revue mensuelle née il y a cent ans.
R. Sagot.
Astronomie, Vol. 96, 109 - 122 (1982).

002.002 On projections for stellar charts.
A. A. Mikhajlov.
Pis'ma Astron. Zh., Tom 8, 60 - 62 (1982). In Russian. English translation in Soviet Astron. Lett., Vol. 8.
It is shown that an equidistant polar projection on a tangent plane has smaller scale errors than an equatorial projection on a secant cylinder with the same extent in declination.

002.003 Catalogue of measurements in the DDO photoelectric photometric system (magnetic tape).
G. Meylan.
Astron. Astrophys., Suppl. Ser., Vol. 47, 483 - 484 (1982).
This catalogue on magnetic tape contains the data in the DDO photoelectric photometric system published up to 1980. It concerns 3406 measurements of 2911 stars.

002.004 Morphological classification (revised RS) of Abell clusters in $d \leqslant 4$ and an analysis of observed correlations. M. F. Struble, H. J. Rood.
Astron. J., Vol. 87, 7 - 46 (1982).
The authors present a catalog of revised Rood-Sastry (RS) types for the 276 Abell clusters in distance class $\leqslant 4$ obtained from the Palomar Observatory Sky Survey plate collection. Also included are the Hubble morphological types of the three brightest galaxies and the separations between them. Separations between binaries (which are counted as single galaxies) among the three brightest galaxies are also given; notation of other binaries are also listed. Abell counts (N_A), Bautz-Morgan (BM) type, redshifts, and line-of-sight velocity dispersions (σ_{vis}) are also tabulated. A possible progression of cluster evolution is suggested.

002.005 Some important publications on early Chinese astronomy from China and Japan, 1978 - 1980.
N. Sivin.
Archaeoastronomy (U.S.A.), Vol. 4, No. 1, p. 26 - 31 (1981).

002.006 KSS compiled catalogue of reference stars in the declination zones from −5° to +90°.
V. S. Borovskikh.
Kazan. inzh.-stroit. inst. Kazan', 1981. 18 pp. In Russian. Abstr. in Ref. zh., 51. Astron., 2.51.85 (1982).

002.007 Viking Orbiter Stereo Imaging Catalog: Second Edition.
K. R. Blasius, A. V. Vetrone, M. D. Martín.
Reports of Planetary Geology Program − 1981, (see 003.001), p. 523 - 524 (1981). − Abstract.

002.008 Archival storage of digital data on videotapes and videodisks.
R. E. Arvidson, L. K. Bolef, R. Lewis.
Reports of Planetary Geology Program − 1981, (see 003.001), p. 525 (1981). − Abstract.

002.009 Annotations on astrophysical papers published in the journal "Radiofizika", Tom XXIII, Nos. 1 - 12, 1980.
Astron. Zh., Tom 59, 202 - 203 (1982). In Russian. English translation in Soviet Astron., Vol. 26, No. 1.

002.010 Étoiles variables et documentation automatique.
J. Gunther.
Bull. AFOEV, No. 19, p. 1 - 2 (1982).

002.011 SS 433 − Bibliographie.
Bull. AFOEV, No. 19, p. 12 - 14 (1982).

002.012 Catalogue of white dwarfs.
A. G. Agayev, O. H. Guseinov (O. Kh. Gusejnov), H. I. Novruzova (Kh. I. Novruzova).
Astrophys. Space Sci., Vol. 81, 5 - 84 (1982).
This catalogue contains all presently available data (spectroscopic, photometric, kinematic) for 488 white dwarfs.

002.013 Second catalogue of X-ray sources.
P. R. Amnuel (Amnuehl'), O. H. Guseinov (O. Kh. Gusejnov), Sh. Yu. Rakhamimov.
Astrophys. Space Sci., Vol. 82, 3 - 103 (1982).
A catalogue of X-ray sources containing 677 objects known as of 1979 September has been compiled. Optical and radio counterparts are suggested. The number of X-ray sources of each type in the Galaxy is estimated.

002.014 Catalogue of declinations of 100 equatorial stars of A. A. Mikhajlov's list. N. F. Minyajlo.
Astrometr. Astrofiz., Vyp. (No.) 43, p. 84 - 88 (1981). In Russian.
100 equatorial stars listed by A. A. Mikhajlov were observed with the Wanschaff vertical circle of the Main Astronomical Observatory of the Ukrainian Academy of Sciences (Goloseevo near Kiev). The purpose of the work was not only to obtain a catalogue of declinations of these stars, but also to estimate the effect of corrections to the limb divisions on the results.

002.015 Catalogue of the positions of red variables.
B. V. Novopashennyj, M. Yu. Volyanskaya.
Astrometr. Astrofiz., Vyp. (No.) 44, p. 52 - 60 (1981). In Russian.
A catalogue of the positions of 217 red variable stars derived from differential meridian observations is given.

002.016 Choice of 315 stars in 87 areas with extragalactic radio sources for meridian observations in the FK4 system. P. F. Lazorenko.
Astrometr. Astrofiz., Vyp. (No.) 46, p. 73 - 80 (1982). In Russian.
A list of AGK3 and SAO stars in areas with extragalactic radio sources with declinations from −44° to + 90° is recommended for meridian observations in the FK4 system. Such observations will enable the positions of optical counterparts of radio sources to be obtained by means of photographic methods directly in the system of the fundamental catalogue.

002.017 Supplementary catalogue of the coordinates of 90 stars of the Moscow zenith zone and estimate of its accuracy. N. B. Frolova, E. I. Pampushnaya, S. P. Ushakov.
Dvizhenie iskusstv. i estestv. nebes. tel. Sverdlovsk, 1981, p. 144 - 152. In Russian. − Abstr. in Ref. zh., 51. Astron., 4.51.116 (1982).

002.018 Catálogo de elementos químicos identificados en estrellas anómalas.
H. Levato, E. B. de Hernández.
Bol. Asoc. Argentina Astron., No. 20 - 24, p. 471 (1981). Abstract.

002.019 Cuarto catálogo círculo meridiano San Juan (FK4-SUR). R. A. Carestia, W. Castro.
Bol. Asoc. Argentina Astron., No. 26, p. 9 - 26 (1981).

002.020 **Segundo catálogo círculo meridiano San Juan (FKSZ).** R. A. Carestia, M. Gallego.
Bol. Asoc. Argentina Astron., No. 26, p. 27 - 51 (1981).

002.021 **On supplement 1 to the second edition of the Catalogue of Star Clusters and Associations.**
B. Balazs.
Bull. Inf. Cent. Données Stellaires, No. 22, p. 51 - 52 (1982).

002.022 **Le catalogue général des données photométriques.** B. Hauck.
Bull. Inf. Cent. Données Stellaires, No. 22, p. 67 - 69 (1982).

002.023 **The Hyderabad (S) and Cordoba zones of the astrographic catalogue: a preliminary investigation.**
D. Herald.
Bull. Inf. Cent. Données Stellaires, No. 22, p. 79 - 86 (1982).

002.024 **A sample test of the reliability of the Bibliographic Star Index.** S. Nishimura.
Bull. Inf. Cent. Données Stellaires, No. 22, p. 90 - 95 (1982).

002.025 **Fichier cinématique des étoiles du HD et du HDE.** M. O. Mennessier, A. Gomez, M. Creze, D. Morin.
Bull. Inf. Cent. Données Stellaires, No. 22, p. 98 - 104 (1982).
A compilation of the most reliable kinematic data for stars of the HD and HD Extension Catalogues is available at the Centre de Données Stellaires. It also includes positions, and spectroscopic and photometric data.

002.026 **Stars named after astronomer's names.** F. Spite, R. Lahmek.
Bull. Inf. Cent. Données Stellaires, No. 22, p. 105 - 106 (1982).

002.027 **Compléments au catalogue de J. Stock, W. Osborn, M. Ibanez.** M. Duflot.
Bull. Inf. Cent. Données Stellaires, No. 22, p. 107 - 111 (1982).

002.028 **Errors or omissions in star-identifications in the General Catalogue of Trigonometric Stellar Parallaxes.** D. Hoffleit.
Bull. Inf. Cent. Données Stellaires, No. 22, p. 112 - 117 (1982).

002.029 **Some errata in the Fifth General Catalogue of MK Spectral Classifications.**
D. Hoffleit, P. Wlasuk.
Bull. Inf. Cent. Données Stellaires, No. 22, p. 118 (1982).

002.030 **Une banque de données pour l'étude des phénomènes de transfert radiatif dans les atmosphères planétaires: la banque "GEISA".**
A. Chedin, N. Husson, N. A. Scott.
Bull. Inf. Cent. Données Stellaires, No. 22, p. 121 - 124 (1982).

002.031 **The Harvard College Observatory plate collection.** M. H. Liller.
Bull. Inf. Cent. Données Stellaires, No. 22, p. 125 - 126 (1982).

002.032 **New and revised catalogues available from the Astronomical Data Center (ADC) at NASA/ Goddard Space Flight Center.**
W. H. Warren, Jr., T. A. Nagy, J. M. Mead, R. S. Hill.
Bull. Inf. Cent. Données Stellaires, No. 22, p. 127 - 130 (1982).

002.033 **New catalogues of stellar data.** W. Buscombe.
Bull. Inf. Cent. Données Stellaires, No. 22, p. 131 (1982).

002.034 **Catalogue of star clusters and associations. Supplement 1, Vols. I - III.**
J. Ruprecht, B. Balazs, R. E. White.
Bull. Inf. Cent. Données Stellaires, No. 22, p. 132 (1982).

002.035 **Catalogs recently published, to be published or in preparation. List XIV.** IAU Commission 45 Working Group on Spectroscopic and Photometric Data.
Bull. Inf. Cent. Données Stellaires, No. 22, p. 133 - 135 (1982).

002.036 **Star catalogs available at the C. D. S.: Corrections.**
Bull. Inf. Cent. Données Stellaires, No. 22, p. 136 - 137 (1982).

002.037 **Catalogs currently available on microfiche.**
Bull. Inf. Cent. Données Stellaires, No. 22, p. 138 - 144 (1982).

002.038 **A preliminary catalogue of 366 stars observed with the photoelectric astrolabe GD II No. 2.**
L. Z. Lu, D. J. Luo.
Special issue on astronomical geodynamics, (see 012.028), p. 176 - 181 (1981). In Chinese and English − Abstract.

002.039 **Introductory talk: bibliography of Be stars.** P. Koubský.
Be stars, (see 012.030), p. 259 - 260 (1982).

002.040 **A catalogue of Be stars.** M. Jaschek, D. Egret.
Be stars, (see 012.030), p. 261 - 263 (1982).

002.041 **A catalogue of Hα observations.** J. R. Ducati.
Be stars, (see 012.030), p. 265 - 268 (1982).

002.042 **Supplement to the Catalogue of RR Lyrae-type stars.**
V. G. Derevyagin, B. N. Firmanyuk, V. P. Tsesevich.
Astron. Tsirk., No. 1166, p. 7 - 8 (1981). In Russian.

002.043 **List of dissertations at the Sternberg Astronomical Institute in 1980.** L. N. Bondarenko.
Astron. Tsirk., No. 1170, p. 6 - 8 (1981). In Russian.

002.044 **Creation of a supernova database.** D. Branch, D. Clark, B. Patchett, R. Wood.
Inf. Bull. Variable Stars, No. 2079 (1982).

002.045 **Catalogue of Cometary Orbits.**
IAU Circ., No. 3701 (1982).

002.046 **Index of star names: BBSAG Bulletins 1 through 60.**
BBSAG Bull., No. 60, p. 5 - 6 (1982).

002.047 **Southern Astrographic Catalog data added to USNO C-catalog for extended-coverage total occultation predictions.** D. W. Dunham.
Occultation Newsl., Vol. 2, 222 - 224 (1982).

002.048 **Bibliographie der Schriften von Karl Friedrich Zöllner, 1834 - 1882.** J. Hamel.
Veröff. Archenhold-Sternw., Berlin-Treptow, Nr. 10, 30 pp. (1982).

002.049 **A simple source catalogue of galaxies south of declination $-17^1/_2°$ that have been observed spectroscopically. First Version (1981).**
P. J. K. Dobbie, A. P. Fairall.
Publ. Dep. Astron. Univ. Cape Town, No. 4, 27 pp. (1981). ISBN 0-7992-0459-5.
This catalogue represents an attempt to identify galaxies, in the fields of the ESO/SRC sky surveys, that have been observed spectroscopically. It is highly simplified; only R.A., Decl. and reference code are given (i.e. no catalogue designations and no radial velocities are provided here − but may be traced in the references). It is also very incomplete in that a large body of unpublished and unreduced data apparently exists.

002.050   A combined catalogue of astronomical sources.
J. M. Mead, T. A. Nagy, R. S. Hill.
Bull. American Astron. Soc., Vol. 13, 838 (1981). – Abstract.

002.051   New updated and corrected machine-readable
editions of a catalog of SAO-HD cross identifications and the Smithsonian Astrophysical Observatory Star
Catalog.   W. H. Warren, Jr., N. G. Roman.
Bull. American Astron. Soc., Vol. 13, 838 (1981). – Abstract.

002.052   Revised S201 Catalog of Far-Ultraviolet Objects.
H. M. Heckathorn, T. L. Page, G. R. Carruthers.
Bull. American Astron. Soc., Vol. 13, 857 (1981). – Abstract.

002.053   A search for "unidentified" S201 stellar objects.
C. B. Opal, H. M. Heckathorn.
Bull. American Astron. Soc., Vol. 13, 857 (1981). – Abstract.

002.054   Catalogue of radial velocities from Herstmonceux
and Kottamia 1964 - 1971.
R. Woolley, M. J. Penston, G. A. Harding, W. L. Martin,
J. E. Sinclair, C. M. Haslam, S. Aslan, A. Savage, K. Aly,
A. S. Asaad.
R. Obs. Ann., No. 14, 122 pp. (1981).
The authors report radial velocities for 1246 stars which
have been obtained from over 3700 spectra taken with
spectrographs on the Isaac Newton 98-inch, Yapp 36-inch
and 30-inch coudé telescopes at Herstmonceux and the
74-inch telescope at Kottamia, Egypt.

002.055   Catalogue of the archives of the Royal Observatory,
Edinburgh, 1764 - 1937.   M. F. I. Smyth.
Published by Royal Observatory, Edinburgh, 55 pp. (1981).
ISBN 0-902553-24-0.

002.056   The ESO Quick Blue Survey and ESO (B) Atlas.
R. M. West, H.-E. Schuster.
ESO Sci. Prepr. No. 188, 23 pp. (1982). – Submitted to
Astron. Astrophys. Suppl. Ser.

002.057   The catalog of 1111 northern PZT stars compiled
from meridian circle catalogs in the 1950s, (NPZT$_{58}$).
H. Yasuda, N. Miyauchi.
Ann. Tokyo Astron. Obs., Second Ser., Vol. 18, 339 - 366
(1982).
A catalog has been formed for the 1111 northern PZT
stars on the FK4 system by compiling the five meridian circle
catalogs made in the last half of the 1950s. The systematic
differences of those meridian circle catalogs from the FK4
have been investigated. The present catalog shall be used for
the establishment of a system of proper motions of northern
PZT stars.

002.058   Northern PZT stars catalog (NPZT$_{74}$).
H. Yasuda, K. Hurukawa, H. Hara.
Ann. Tokyo Astron. Obs., Second Ser., Vol. 18, 367 - 427
(1982).
The internationally cooperative observations of the stars
in the northern PZT stars list had been made with ten
meridian circles over the world in the 1970s, and a catalog of
their final positions, northern PZT stars catalog, was formed
on the FK4 system. The catalog contains 1719 stars for the
mean epoch 1974 with the average mean errors of
±0^{s}0033 sec δ and ±0$''$079 in right ascension and declination
respectively.

002.059   General catalog of stars observed with astrolabes.
Y.-x. Zhu, L.-z. Lu, D.-j. Luo, Z.-f. Ling.
Publ. Beijing Astron. Obs., No. 1, p. 1 - 30 (1981). In
Chinese.
Using the same method of reductions, 6 preliminary
catalogs observed with Danjon astrolabes and the photo-

electric astrolabes have been made at Shanghai Observatory,
Beijing Observatory, Shaanxi Observatory and Wuchang Time
Service. From these preliminary catalogs, the corrections for
the positions given by the fundamental catalog FK4 are
combined in a uniform system. The General Catalog consists
of 606 stars (466 FK4 stars, 130 FK4 Supp stars and 10 GC
stars) with declinations between +2°56' and +67°48'. The
mean epoch of the General Catalog is 1973.6. The method of
reductions of the catalog will be presented in another paper.

002.060   Preliminary catalogue of the photoelectric astrolabe
(PACP I).   L.-z. Lu, D.-j. Luo, Z.-z. Wang.
Publ. Beijing Astron. Obs., No. 1, p. 31 - 43 (1981). In
Chinese.

002.061   A catalogue of photoelectric magnitudes and colours
of visual double and multiple systems.
Å. Wallenquist.
Acta Universitatis Upsaliensis, Nova Acta Regiae Societatis
Scientiarum Upsaliensis, Ser. V:A. Vol. 4, p. 1 - 68 (1981) =
Uppsala Astron. Obs. Rep. No. 22.

002.062   An atlas of southern and equatorial dwarf novae.
N. Vogt, F. M. Bateson.
Astron. Astrophys., Suppl. Ser., Vol. 48, 383 - 407 (1982).
A total of 117 cataclysmic variables (δ < +30°), most of
them dwarf novae, is identified on reproductions of Sky
Survey prints. The sample is nearly complete for dwarf novae
brighter than 13^{m}5 at maximum. The identification of 78
variables is confirmed by means of photoelectric photometry,
spectroscopy or outburst photographs. 10 stars, which
previously were classified as dwarf novae, turned out not to be
cataclysmic variables.

002.063   Radial velocities from objective-prism plates in the
direction of the Large Magellanic Cloud.
C. Fehrenbach, M. Duflot.
Astron. Astrophys., Suppl. Ser., Vol. 48, 409 - 442 (1982).
In French.
Catalogue of 711 stars, members of the Large Magellanic
Cloud (LMC), known as members by their radial velocities
(RV). RV of 418 of these stars. RV of 1127 of galactic stars
in the direction of the LMC. Discussion of the precision of the
measures. Dispersion of the RV in different fields.

002.064   Geneva photometric boxes. 0. Announcement of
the catalogue (microfiches and magnetic tape).
B. Nicolet.
Astron. Astrophys., Suppl. Ser., Vol. 48, 485 - 490 (1982).
The catalogue of the Geneva photometric boxes (or
Golay's boxes) is announced. It is available on magnetic tape
at the Centre de Données Stellaires in Strasbourg and on
microfiches on request to the author.

002.065   Meridian observations made in Brorfelde
(Copenhagen University Observatory) 1969–1975.
Positions of 6427 stars brighter than 11.00 vis.mag.
L. Helmer, H. J. Fogh Olsen.
Astron. Astrophys., Suppl. Ser., Vol. 49, 13 - 60 (1982).
This catalogue presents positions for 6427 stars, observed
with the 7$''$ transit circle. It contains positions for 1586 NPZT
stars from the Northern Photographic Zenith Tube program
proposed by Yasuda (Yasuda et al., 1971), and selected stars
from various lists mainly faint GC stars and double stars. The
declinations range from −10° to the North Pole. The internal
mean error for a single observation is $\epsilon_{\alpha} \cos \delta = 0$^{s}0129 and
$\epsilon_{\delta} = 0$$''$233 and the positions are reduced relative to FK4. The
positions of all the FK4 stars used in the least squares solution
are also given in the catalogue.

002.066 **Homogeneous catalogue of red and infrared magnitudes in the photoelectric photometric system of Kron (magnetic tape).** G. Jasniewicz.
Astron. Astrophys., Suppl. Ser., Vol. 49, 99 - 100 (1982).
This catalogue on magnetic tape contains homogeneous data in the photoelectric photometric system of Kron and Smith. It concerns 5704 stars measured up to the end of 1981.

002.067 **Sky Atlas 2000.0 von Wil Tirion.**
E. Laager.
Orion, 40. Jahrg., 98 - 102 (1982).

002.068 **Dumbbell galaxies and precessing radio jets.**
A. Wirth, L. Smarr, J. S. Gallagher.
Astron. J., Vol. 87, 602 - 615 (1982).
The authors have compiled a catalog of ~100 close double elliptical galaxies which are radio sources. Many of these are dumbbell galaxies, i.e., two nearly equally bright elliptical galaxies in a common envelope. It is argued that the close proximity of the galaxies (~10 - 30 kpc) will dynamically affect the radio jet through gravitational interactions with its confining gas cloud.

002.069 **The second COS-B catalogue of high-energy γ-ray sources.** W. Hermsen.
Philos. Trans. R. Soc. London, Ser. A, Vol. 301, (see 012.043), 519 - 521 (1981). — Same as 30.002.048.

002.070 **The Texas data base.** P. D. Hemenway.
Abh. Hamburger Sternw., Band 10, 131 - 138 (1982).

002.071 **Candidate radio sources for a radio/optical reference catalog.** A. Witzel. K. J. Johnston.
Abh. Hamburger Sternw., Band 10, 151 - 164 (1982).

002.072 **Development of a radio-astrometric catalog by means of very long baseline interferometry observations.** J. L. Fanselow, O. J. Sovers, J. B. Thomas, F. R. Bletzacker, T. J. Kearns, E. J. Cohen, G. H. Purcell, Jr., D. H. Rogstad, L. J. Skjerve, L. E. Young.
Abh. Hamburger Sternw., Band 10, 171 (1982). — Abstract.

002.073 **A catalogue of extragalactic radio sources having flux densities greater than 1 Jy at 5 GHz.**
H. Kühr, A. Witzel, I. I. K. Pauliny-Toth, U. Nauber.
Abh. Hamburger Sternw., Band 10, 172 (1982). — Abstract.

002.074 **On the derivation of a catalogue of radio source positions from interferometric observations.**
H. G. Walter.
Abh. Hamburger Sternw., Band 10, 174 (1982). — Abstract.

002.075 **Catalogue of highest energy cosmic rays. Giant extensive air showers. No. 1. Volcano Ranch, Haverah Park.** M. Wada.
Published by World Data Center C2 for Cosmic Rays, Institute of Physical and Chemical Research, Itabashi, Tokyo, Japan. 99 pp. (1980).

002.076 **Charts for southern variables. Series No. 14.**
F. M. Bateson, M. Morel, B. Sumner, R. Winnett.
Published by Astronomical Research Ltd., P. O. Box 3093, Greerton, Tauranga, New Zealand. 3 pp. + charts 601 - 649 (1982).

002.077 **The Bright Star Catalogue. Fourth** revised edition. (Containing data compiled through 1979).
D. Hoffleit, with the collaboration of C. Jaschek.
Published by Yale University Observatory, New Haven, Conn., U. S. A. 15 + 472 pp. Price $ 35.00 (1982).

002.078 **MK spectral classifications. Fifth General Catalogue. Including standard stars and members of Magellanic Clouds.**
Compiled by W. Buscombe.
Published by Dearborn Observatory, Evanston, Ill. 60201, U. S. A. 243 pp. Price $ 15.00 (1981). ISBN 0-939160-03-X.

002.079 **Viking Orbiter Stereo Imaging Catalog.**
K. R. Blasius, A. V. Vetrone, B. H. Lewis, M. D. Martin.
NASA Contract. Rep., NASA CR-3501, 399 pp. (1981).
The extremely long missions of the two Viking Orbiter spacecraft produced a wealth of photos of surface features. Many of these photos can be used to form stereo images allowing the earth-bound student of Mars to examine a subject in 3-D. This catalog is a technical guide to the use of stereo coverage within the complex Viking imaging data set. This second edition of the catalog supersedes the first edition published in June 1980 as NASA CR-3277.

002.080 **The Viking Mosaic Catalog. Volume 1, 2.**
N. Evans.
NASA Contract. Rep., NASA CR-3496, 2456 pp. (1982).
This two-volume Catalog is a collection of more than 500 mosaics prepared from Viking Orbiter images. Accompanying each mosaic is a footprint plot, which identifies by location, picture number, and order number, each frame in the mosaic. Corner coordinates and pertinent imaging information are also included. A short text provides the camera characteristics, image format, and data processing information necessary for using the mosaic plates as a research aide. Procedures for ordering mosaic enlargements and individual images are also provided.

002.081 **Documentation for the machine-readable version of the 0.2-Å resolution far-ultraviolet stellar spectra measured with Copernicus.**
W. T. Sheridan, W. H. Warren, Jr.
NSSDC/WDC-A-R&S, 81-12, 10 pp. (1981).

002.082 **Documentation for the machine-readable version of OAO 2 filter photometry of 531 stars of diverse types.** W. H. Warren, Jr., W. T. Sheridan.
NSSDC/WDC-A-R&S, 82-02, 18 pp. (1982).

002.083 **Documentation for the machine-readable version of the Catalogue of Stars within 25 parsecs of the Sun.**
W. H. Warren, Jr.
NSSDC/WDC-A-R&S, 82-03, 17 pp. (1982).

002.084 **Documentation for the machine-readable version of the Lick Jupiter-Voyager Reference Star Catalogue.**
W. H. Warren, Jr.
NSSDC/WDC-A-R&S, 82-04, 11 pp. (1982).

002.085 **Documentation for the machine-readable version of the Lick Saturn-Voyager Reference Star Catalogue.**
W. H. Warren, Jr.
NSSDC/WDC-A-R&S, 82-05, 11 pp. (1982).

002.086 **Documentation for the machine-readable AGK3-BD and BD-AGK3 cross-index catalogues.**
W. H. Warren, Jr.
NSSDC/WDC-A-R&S, 82-06, 13 pp. (1982).

002.087 **Documentation for the machine-readable version of A Deep Objective-Prism Survey for Large Magellanic Cloud Members.** W. H. Warren, Jr.
NSSDC/WDC-A-R&S, 82-07, 14 pp. (1982).

002.088 **Documentation for the machine-readable version of A Catalogue of Homogeneous Photometry of Bright**

Stars on the DDO System.   W. H. Warren, Jr.
NSSDC/WDC-A-R&S, 82-08, 12 pp. (1982).

002.089   Documentation for the machine-readable version of
the Vilnius Photometric Catalogue 1980.
W. H. Warren, Jr.
NSSDC/WDC-A-R&S, 82-09, 23 pp. (1982).

002.090   Documentation for the machine-readable version of
the First Santiago-Pulkovo Fundamental Stars
Catalogue (SPF 1 Catalogue).   W. H. Warren, Jr.
NSSDC/WDC-A-R&S, 82-10, 13 pp. (1982).

002.091   Documentation for the machine-readable version of
The Absolute Calibration of Stellar Spectrophotom-
etry.   W. H. Warren, Jr.
NSSDC/WDC-A-R&S, 82-11, 10 pp. (1982).

002.092   Documentation for the machine-readable version of
Luminous Stars in the Northern Milky Way.
W. H. Warren, Jr.
NSSDC/WDC-A-R&S, 82-12, 17 pp. (1982).

002.093   Documentation for the machine-readable version of
Photometric Data for the Nearby Stars.
W. H. Warren, Jr.
NSSDC/WDC-A-R&S, 82-13, 18 pp. (1982).

002.094   Documentation for the machine-readable version of
Faint Blue Objects at High Galactic Latitude.
W. H. Warren, Jr.
NSSDC/WDC-A-R&S, 82-14, 12 pp. (1982).

002.095   Documentation for the machine-readable version of
the Catalogue of Individual UBV and uvbyβ
Observations in the Region of the Orion OB 1 Association.
W. H. Warren, Jr.
NSSDC/WDC-A-R&S, 82-16, 15 pp. (1982).

002.096   A bibliography of observations of molecular clouds
in galaxies.   L. J. Rickard.
Extragalactic molecules, (see 012.048), p. 211 - 214 (1982).

002.097   Report of IAU Commission 5: Documentation and
astronomical data (Documentation et données
astronomiques).   W. D. Heintz.
Trans. IAU, Vol. XVIIIA, (see 003.013), p. 15 - 18 (1982).

002.098   The ESO/Uppsala Survey of the ESO (B) Atlas.
A. Lauberts.
Published by European Southern Observatory, Karl-
Schwarzschild-Str. 2, D-8046 Garching bei München. 503 pp.
(1982). ISBN 3-923524-13-7.
   A systematic search for certain objects (NGC + IC
galaxies, all galaxies with a diameter larger than about 1.0
arcmin, all disturbed galaxies, all star clusters in the Budapest
Catalogue, and all listed planetary nebulae) has been carried
out by means of the ESO (B) Atlas, covering the southern
sky from −90 to −17.5 deg. A total of 18,438 objects is listed;
of these, about 60% for the first time. Magnitudes and radial
velocities are also given for a total of 2,102 galaxies.

002.099   Astronomy and Astrophysics Abstracts. Vol. 30.
Literature 1981, Part 2.
S. Böhme, W. Fricke, I. Heinrich, W. Hofmann, D. Krahn,
V. R. Matas, D. Rosa, L. D. Schmadel, G. Zech (Editors).
Published for Astronomisches Rechen-Institut by Springer-
Verlag, Berlin−Heidelberg−New York. 10 + 792 pp. Price
DM 148.00; US $ 66.00 [Subscription price DM 118.40;
US $ 52.80] (1982). ISBN 3-540-11721-0, ISBN 0-387-
11721-0.

002.100   Guide to the presentation of astronomical data.
G. A. Wilkins.
CODATA Bull., No. 46, p. 1 - 10 (1982).
   This guide contains general recommendations to editors,
referees and authors on the reporting of numerical data obtained from astronomical observations. The recommendations
are intended to facilitate the use and evaluation of the
reported data, which are usually space and time dependent.
They cover the description of observational procedures, the
treatment of data derived from them, and the presentation of
the final, numerical results.

002.101   Atlas cosmico.   J. de la Herran, J. M. Van Hoof.
Consejo Nacional de Ciencia y Tecnologia,
Insurgentes Sur 1677, Mexico 20, D. F. 34 pp. Price $ 10.00
(1981). ISBN 968-823-095-2. − Review in Sky Telesc. Vol. 64,
47 (1982).

002.102   Galaxien.
T. Ferris, with a preface by G. A. Tammann.
Translated from the American by A. Ehlers.
Birkhäuser Verlag, Basel − Boston − Stuttgart. 6 + 183 pp.
Price DM 128.00 (1981). ISBN 3-7643-1250-5. − Review in
Sternenbote, 25. Jahrg., 87 - 88 (1982).
   Contents: Die Milchstraße: Ein Spiralnebel von innen
gesehen. Die Lokale Gruppe von Galaxien. Form und Vielfalt
der Galaxien. Wechselwirkende Galaxien. Galaxienhaufen.
Galaxien und das Weltall.

002.103   Sky catalogue 2000.0. Vol. 1: stars to magnitude 8.0.
A. Hirshfeld, R. W. Sinnott (Editors).
Sky Publishing Corp., Cambridge, Mass., Cambridge University
Press, New York. 604 pp. Price $ 44.95 cloth, $ 29.95 paper
(1982). ISBN 0-933 346-35-2 cloth, ISBN 0-933 346-34-4
paper. − Reviews in Sky Telesc., Vol. 64, 44; 1982
(C. S. Morris); Strolling Astron., Vol. 29, 128; 1982
(J. R. Smith).

002.104   Atlas of the Andromeda galaxy.   P. W. Hodge.
University of Washington Press, Seattle, Wash.
98105. 79 pp. Price $ 50.00 (1981). ISBN 0-295-95795-6.
Review in Sky Telesc., Vol. 63, 374 (1982).

002.105   A dictionary of astronomy.
V. Illingworth (Editor).
Pan Books, London. 6 + 378 pp. Price £ 2.95 (1981).
Reviews in J. British Astron. Assoc., Vol. 92, 102 (1982);
J. British Astron. Assoc., Vol. 92, 151; 1982 (C. Ronan);
Spaceflight, Vol. 24, 96 (1981).

002.106   Handbook for astronomical societies, 1981.
B. Jones (Editor).
Federation of Astronomical Societies, 'Alcyone', 28 High
House Ave., Bradford, West Yorkshire. 97 pp. Price £ 2.30
paperback (1981). − Reviews in J. British Astron. Assoc.,
Vol. 92, 102; 1982 (C. A. Ronan); Sky Telesc., Vol. 63, 375
(1982).

002.107   The RAE table of earth satellites, 1957 - 1980.
D. G. King-Hele, J. A. Pilkington, H. Hiller,
D. M. C. Walker (Compilers).
Facts on File, Inc., 460 Park Ave. S., New York, Macmillan
Reference Books, London. 655 pp. Price $ 75.00, £ 30.00
(1981/82). ISBN 0-87196-599-2. − Reviews in Sky Telesc.,
Vol. 63, 268 (1982); Spaceflight, Vol. 24, 142 (1982).

002.108   A bibliography of astronomy 1970 - 1979.
R. A. Seal, S. S. Martin.
Libraries Unlimited, Inc., Littleton, Colo. 407 pp. Price
£ 22.50 (1982). ISBN 0-87287-280-7.

002.109 Dictionary of astronomy, space and atmospheric
         phenomena.   D. F. Tver.
Van Nostrand Reinhold. 281 pp. Price £ 11.00 (1982).
Review in J. British Interplanet. Soc., Vol. 35, 336 (1982).

002.110 BAA Star Charts.
         British Astronomical Association. Price £ 2.00.
Review in British Astron. Assoc. Circ., No. 619 (1981).

002.111 Catalogue of Star Clusters and Associations. Sup-
         plement 1.
J. Ruprecht, B. Balázs, R. E. White.
Part A: Introduction. Part B1: New data for open clusters.
Part B2: New data for associations, globular clusters and extra-
galactic objects.
Akadémiai Kiadó, Publishing House of the Hungarian Academy
of Sciences, Budapest. 732 pp. Price $ 65.00 (1981).
ISBN 963-05-2472-4.
    Supplement 1 to the Catalogue of Star Clusters and
Associations contains the supplementary data up to the end of
1973 on open and globular star clusters as well as on associa-
tion both in our Galaxy and in extragalactic objects.

         Astronomical activity in the period 1600 to 1880.
See Abstr. 004.090.

         On the Belgrade Catalogue of NPZT stars.
See Abstr. 041.026.

         Some indices of sunspot groups in the 11-year cycle
No. 20.   See Abstr. 072.020.

         Catalogue of observations of occultations of stars by
the Moon for the years 1623 to 1942 and solar eclipses for
the years 1621 to 1806.   See Abstr. 096.013.

         Catalogue of orbits of unnumbered minor planets.
See Abstr. 098.108.

         Catalogue of discoveries and identifications of minor
planets.   See Abstr. 098.109.

         Charting the moons of Saturn – II.
See Abstr. 100.001.

         Investigation of catalogues of absolute proper
motions of stars compiled in accordance with the KSZ plan.
See Abstr. 111.009.

         Proper motion survey with the forty-eight inch
Schmidt telescope. LV. First supplement to the NLTT
catalogue.   See Abstr. 111.030.

         Proper motion survey with the 48-inch Schmidt
telescope. LIX. A catalogue of 929 possible candidates for
Hyades membership.   See Abstr. 111.033.

         Comparison of spectrophotometric and photometric
data of stars.   See Abstr. 113.022.

         The intrinsic colours of stars in the ultraviolet.
See Abstr. 113.038.

         Fundamental data for southern stars. (Seventh list).
See Abstr. 113.072.

         A catalog of red stars near L1454.
See Abstr. 113.075.

         Ground-based photometric data.
See Abstr. 113.094.

         Errors in the HD identification in "Spectral Survey
of the Southern Milky Way III" by Loden, L. O., Loden, K.,
Nordström, B., Sundman, A.   See Abstr. 114.113.

         A list of stars with large expected angular diameters.
See Abstr. 115.003.

         A catalogue of variable and visual double stars.
See Abstr. 118.032.

         A 408 MHz all-sky continuum survey. II. The atlas
of contour maps.   See Abstr. 141.036.

         Faint blue objects at high galactic latitude.
III. Palomar Schmidt field centered on Selected Area 28.
See Abstr. 141.290.

         The positions, structures, and polarizations of
404 compact radio sources.   See Abstr. 141.306.

         Far infrared survey of extended molecular cloud
H II region complexes along the galactic plane.
See Abstr. 156.007.

         The angular correlation function at one minute of
arc as a function of magnitude.   See Abstr. 158.233.

         A catalog of infrared magnitudes and H I velocity
widths for nearby galaxies.   See Abstr. 158.271.

         Gravitational mechanics of systems of galaxies.
I. Corrections for errors in redshifts.
See Abstr. 158.291.

003 Books

003.001 Reports of Planetary Geology Program – 1981.
Compiled by H. E. Holt, with a foreword by
J. M. Boyce.
NASA Tech. Memo., NASA TM 84211, 20 + 561 pp. (1981).
The individual contributions within the subject scope of
Astronomy and Astrophysics Abstracts are included in their
corresponding categories – see abstracts 002.007, 002.008,
011.010, 013.011, 013.012, 022.046 - 022.067, 031.522,
031.523, 063.017 - 063.019, 081.016 - 081.033, 091.011 -
091.026, 092.003, 093.016, 093.017, 094.005- 094.007,
094.506 - 094.511, 097.003 - 097.079, 098.017, 099.016 -
099.037, 100.028 - 100.032, 102.009, 105.021, 107.004,
107.005.

003.002 Physics reviews. Vol. 3.
I. M. Khalatnikov (Editor).
Harwood Academic Publishers, Chur, Switzerland. 10 + 593 pp.
(1981). ISBN 3-7186-0068-4. – Review in Phys. Abstr.,
Vol. 85, Abstr. 28617 (1982).

003.003 B stars with and without emission lines.
Part I: A. Underhill (Editor), Part II: V. Doazan
(Editor), with contributions by J. Rountree Lesh,
M. L. Aizenman, R. N. Thomas.
Monograph Series on Nonthermal Phenomena in Stellar
Atmospheres. Centre National de la Recherche Scientifique,
Paris, France; National Aeronautics and Space Administration,
Scientific and Technical Information Branch, Washington,
D.C. NASA SP-456, 57 + 485 pp. (1982). For sale by the
National Technical Information Service, Springfield, Virginia
22161.
This book is divided into two parts: Part I is about
B stars for which emission lines do not occur on the main
sequence, only among the supergiants. Part II is about those
B stars for which the presence of emission in Hα is considered
to be a significant factor in delineating atmospheric structure.

003.004 Progress in optics. Vol. 19. E. Wolf (Editor).
North-Holland Publishing Company, Amsterdam,
Netherlands. 16 + 393 pp. (1981). ISBN 0-444-85444-4.
Review in Phys. Abstr., Vol. 85, Abstr. 48410 (1982). – See
Abstr. 082.046.

003.005 Solar system plasmas and fields.
J. Lemaire, M. J. Rycroft (Editors).
Adv. Space Res., Vol. 2, No. 1, 6 + 87 pp. (1982). ISBN
0-08-029125-2. – The individual contributions are included
in their corresponding subject categories – see abstracts
074.050, 084.062 - 084.069, 085.017, 085.018, 091.046,
099.071, 106.022.

003.006 The Mars reference atmosphere.
A. Kliore (Editor).
Adv. Space Res., Vol. 2, No. 2, 5 + 107 pp. (1982). ISBN
0-08-029126-0. – The individual contributions are included
in their corresponding subject categories – see abstracts
097.097 - 097.105.

003.007 Annual Review of Nuclear and Particle Science.
Vol. 31. J. D. Jackson, H. E. Gove,
R. F. Schwitters (Editors).
Annual Reviews Inc., 4139 El Camino Way, Palo Alto, Calif.
94306, USA 9 + 455 pp. Price $ 23.50 (1981). ISBN
0-8243-1531-6. – Review in Phys. Abstr., Vol. 85, Abstr.
53436 (1982).

003.008 Detrimental activities in space.
K. Rawer (Editor).

Adv. Space Res., Vol. 2, No. 3, 7 + 161 pp. (1982). ISBN
0-08-029694-7. – The individual contributions within the
subject scope of Astronomy and Astrophysics Abstracts are
included in their corresponding categories – see abstracts
013.046 - 013.048, 031.618 - 031.623.

003.009 Annual Review of Earth and Planetary Sciences.
Volume 10.
G. W. Wetherill, A. L. Albee, F. G. Stehli (Editors).
Annual Reviews Inc., 4139 El Camino Way, Palo Alto, Calif.
94306. 10 + 544 pp. Price $ 22.00 (1982). ISBN 0-8243-
2010-7. – The individual contributions within the subject
scope of Astronomy and Astrophysics Abstracts are included
in their corresponding categories – see abstracts 091.061,
091.062, 103.509, 107.031.

003.010 Researches on stellar physics.
Collection of papers of the Leningrad State Ped.
Inst., Leningrad. 102 pp. (1981). In Russian. – Review in Ref.
zh., 51. Astron., 5.51.59 (1982). – See abstracts 022.188,
022.189, 062.120, 062.121, 153.046.

003.011 Advances in planetary geology.
A. Woronow (Editor), with a foreword by J. M. Boyce.
NASA Tech. Memo., NASA TM-84412, 5 + 718 pp. (1981).
The individual contributions (dissertations) are included in
their corresponding subject categories – see abstracts
097.118, 098.016, 099.092.

003.012 Compendium in astronomy. A volume dedicated to
Professor John Xanthakis on the occasion of
completing twenty-five years of scientific activities as fellow
of the National Academy of Athens.
E. G. Mariolopoulos, P. S. Theocaris, L. N. Mavridis (Editors).
D. Reidel Publishing Company, Dordrecht, Holland–Boston,
U. S. A.–London, England. 16 + 464 pp. Price Dfl. 120.00,
US $ 49.50 (1982). ISBN 90-277-1373-1. – The individual
contributions are included in their corresponding subject
categories – see abstracts 004.089, 004.090, 005.013,
015.039, 015.040, 022.198 - 022.200, 031.039, 051.030,
061.057, 062.129, 064.101, 066.149, 072.101 - 072.103,
073.151, 075.043, 077.065, 085.037, 085.038, 094.063,
097.119, 098.106, 099.093, 102.054, 106.040, 114.223,
119.134, 122.248, 122.249, 131.266, 131.267, 151.095,
158.337, 159.039.

003.013 Reports on Astronomy. Transactions of the International Astronomical Union, Volume XVIIIA.
P. A. Wayman (Editor).
D. Reidel Publishing Company, Dordrecht, Holland – Boston,
U. S. A. – London, England. 8 + 669 pp. Price Dfl. 155.00,
$ 67.50 (1982). ISBN 90-277-1423-1. – The reports of the
various IAU commissions are included in their corresponding
subject categories – see abstracts 002.097, 004.091, 013.057,
013.058, 014.016, 022.201, 032.048, 041.056, 041.057,
042.131, 044.063, 044.064, 047.028, 051.031, 064.102,
065.072, 072.104, 080.109, 082.086, 082.087, 091.073,
091.074, 098.107, 102.055, 104.039, 106.041, 111.034,
113.097, 114.224, 114.225, 117.146, 118.039, 122.250,
131.278, 141.322, 142.112, 153.053, 155.056, 158.358,
162.226.

003.014 Initiation à l'astronomie. A. Acker.
3rd edition. Editions Masson. 153 pp. Price BF 555
(1982). – Review in Ciel Terre, Vol. 98, 119; 1982 *(J. Sauval)*.

003.015 Spatial distribution of galactic cosmic ray density
and flux. M. V. Alaniya, L. I. Dorman.

Metsniereba, Tbilisi. 119 pp. (1981). In Russian. – Review in Ref. zh., 51. Astron., 4.51.50 (1982).

003.016 Introduction to planetary physics.
Yu. V. Aleksandrov.
Vishcha shkola, Kiev. 303 pp. (1982). In Russian. – Review in Ref. zh., 51. Astron., 5.51.55 (1982).

003.017 Structure and evolutionary history of the solar system.   H. Alfvén, G. Arrhenius.
Translated from the English edition. Naukova dumka, Kiev. 331 pp. (1981). In Russian. – Review in Ref. zh., 51. Astron., 5.51.54 (1982).

003.018 Comets and corpuscular radiation of the sun.
D. A. Andrienko, V. N. Vashchenko.
Nauka, Moskva. 164 pp. (1981). In Russian. – Review in Ref. zh., 51. Astron., 4.51.45 (1982).

003.019 Aujourd'hui l'univers.   J. Audouze.
Editions Pierre Belfond, 216 Blvd. Saint-Germain, 75007 Paris, France. 350 pp. Price FF 79.00, BF 600, $ 12.95 (1981).   ISBN 2-7144-1423-0. – Reviews in Ciel Terre, Vol. 98, 117; 1982 *(J. Sauval)*; Sky Telesc., Vol. 64, 47 (1982).

003.020 The channels of Mars.   V. R. Baker.
Adam Hilger Ltd., Bristol, England. 13 + 198 pp. Price £ 22.50 (1982). ISBN 0-85274-467-6. – Published in North and South America by University of Texas Press, P. O. Box 7819, Austin, Texas 78 712, USA. Price $ 39.95. ISBN 0-292-71068-2. – Reviews in Nature, Vol. 298, 694; 1982 *(L. Wilson)*; Sky Telesc., Vol. 64, 46 (1982).
Contents: Telescopes, canals, and space probes. The geomorphology of Mars. Channel types, distribution, ages, and proposed origins. Patterns and networks of Martian valleys. Ice and the Martian surface. The outflow channels. The channeled scabland: an earth analog. Catastrophic flood processes. Mars: a water planet?

003.021 The telescope.   L. Bell.
Dover. 287 pp. Price $ 5.50. ISBN 0-486-24151-3. Available from Sky Publishing Corp. – Review in Sky Telesc., Vol. 63, 268 (1982).

003.022 Coding the universe.
A. Beller, R. Jensen, P. Welch.
London Mathematical Society Lecture Note Series, 47. Cambridge University Press, New York. 8 + 354 pp. Price $ 34.50 (1982). – From Science Vol. 216, 1104 (1982).

003.023 Contact with the stars. The search for extraterrestrial life.   R. Breuer, translated from the German by C. Payne-Gaposchkin and M. Lowery.
W. H. Freeman, Oxford – San Francisco. 10 + 292 pp. Price £ 11.95 (1982). ISBN 0-7167-1355-1 hbk.
Contents: From molecule to man. Interstellar exchange of information. The conquest of space.

003.024 Einsteins Universum.   N. Calder.
Translated from the English edition by W. Knapp. Umschau Verlag, DM 29.80 (1980). – Review in Sterne Weltraum, 21. Jahrg., 49; 1982 *(E. Bettwieser)*.

003.025 Kometen kommer!   N. Calder.
Brombergs Bokförlag, Uppsala. 202 pp. Price Sv. kr. 69.00 (1981). – Review in Astron. Tidsskr., Årg. 15, 47 (1982).

003.026 The surface of Mars.   M. H. Carr.
Yale Planetary Exploration Series. Yale University Press, New Haven – London. 12 + 232 pp. Price $ 45.00, £ 31.50 (1981).   ISBN 0-300-02750-8. – Reviews in Orion,

40. Jahrg., 110 - 111; 1982 *(A. Tarnutzer)*; Sky Telesc., Vol. 63, 375 (1982); Science, Vol. 216, 882 - 883; 1982 *(W. K. Hartmann)*; Spaceflight, Vol. 24, 287 (1982).

003.027 Variability of the general magnetic field of the sun as a source of cosmic ray modulation.
A. N. Charakhch'yan, T. N. Charakhch'yan.
Fiz. inst. AN SSSR, Moskva. 70 pp. (1981). In Russian. From Ref. zh., 51. Astron., 3.51.40 (1982).

003.028 The evolving Earth.   L. R. M. Cocks (Editor).
Cambridge University Press, England. 7 + 264 pp. (1981). ISBN 0-521-28229-2. – Review in Phys. Abstr., Vol. 85, Abstr. 23388 (1982).

003.029 Revealing the Universe. Prediction and proof in astronomy.   J. Cornell, A. P. Lightman (Editors).
MIT Press, Cambridge, Mass. 16 + 246 pp. Price $ 17.50, £ 12.25 (1982). ISBN 0-262-03080-2. – Review in Sky Telesc., Vol. 63, 583 (1982).

003.030 Life itself: its origin and nature.   F. Crick.
Simon and Schuster. 192 pp. Price $ 12.95 (1981). ISBN 0-671-25562-0. Available from Sky Publishing Corp. Review in Sky Telesc., Vol. 63, 267 (1982).

003.031 Enigmi del cosmo.   V. Croce.
Ed. Mediterranee, Roma. 272 pp. Price L 10,500 (1981). – Review in Orione, Vol. 3, 33; 1982 *(S. Baroni)*.

003.032 The view from planet earth.   V. Cronin.
Collins, 14 St. James's Place, London SW1A 1PS, England, 348 pp. Price £ 12.50 (1981). ISBN 0-00-211397-X. Reviews in J. British Astron. Assoc., Vol. 92, 151; 1982 *(P. Moore)*; Sky Telesc., Vol. 63, 374 (1982).

003.033 The edge of infinity. Where the universe came from and how it will end.   P. Davies.
Simon and Schuster, New York. 10 + 196 pp. Price $ 13.50 (1982). – From Science, Vol. 216, 1104 (1982).

003.034 Astronomie.   P. de La Cotardière (Editor).
Librairie Larousse, Paris, 326 pp. Price FF 255.00, $ 45.00 (1981). ISBN 2-03-505201-7. – Review in Sky Telesc., Vol. 63, 584 (1982).

003.035 Plurality of worlds.   S. J. Dick.
Cambridge University Press, 246 pp. Price $ 34.50, £ 19.00 (1982). ISBN 0-521-24308-4. – Reviews in J. British Interplanet. Soc., Vol. 35, 336 (1982); Sky Telesc., Vol. 63, 583 (1982).

003.036 Universe.   D. Dixon.
Houghton Mifflin Co., Boston. 240 pp. Price $ 35.00 (1981). ISBN 0-395-31290-6. – Review in Sky Telesc., Vol. 63, 579 - 580; 1982 *(J. Lomberg)*.

003.037 Meteorites. A petrologic-chemical synthesis.
R. T. Dodd.
Cambridge University Press, New York, USA – Cambridge, England. 11 + 368 pp. Price $ 69.50 (1981). ISBN 0-521-22570-1. – Review in Phys. Abstr., Vol. 85, Abstr. 48412 (1982).

003.038 Cosmic radiation. Historical review.
I. V. Dorman.
Nauka, Moskva. 192 pp. (1981). In Russian. – Review in Ref. zh., 62. Issled. kosm. prostranstva, 3.62.67 (1982).

003.039 Evolution of the earth.   R. H. Dott, R. L. Batten.
3rd edition. McGraw-Hill. 573 pp. Price $ 31.50,

£ 17.50 (1981). ISBN 0-07-017625-6. – Review in Nature, Vol. 295, 463; 1982 *(K. Bell).*

003.040 Between Sputnik and the Shuttle: new perspectives on American astronautics. F. C. Durant III (Editor).
AAS History Series, Vol. 3, American Astronautical Society, California. 334 pp. Price $ 40.00 hbk, $ 30.00 pbk (1981). ISBN 0-87703-145-2 hbk; ISBN 0-87703-149-5 pbk. – From Nature, Vol. 295, 444 (1982).

003.041 Doppelgalaxien. W. Eichendorf, M. Reinhardt.
Forschungsbericht des Landes Nordrhein-Westfalen, Nr. 3079, Fachgruppe Physik/Chemie/Biologie.
Westdeutscher Verlag GmbH, Opladen. 4 + 218 pp. Price DM 37.00 (1981). ISBN 3-531-03079-5.
Contents: Einführung. Zur Definition einer Doppelgalaxie. Eine statistisch vollständige Stichprobe von Doppelgalaxien auf der Nordhemisphäre. Messungen und Reduktion. Massenbestimmung und Dynamik von Doppelgalaxien. Statistische Auswertung der Messungen. Ergebnisse und Diskussion. Ein Katalog von Doppelgalaxien.

003.042 Die Sonne – Die Erforschung des kosmischen Feuers. J. W. Ekrutt.
Geo-Buch, Hamburg, 320 pp. Price DM 78.00 (1981). From Umschau, 82. Jahrg., 35 (1982).

003.043 Fotografia astronomica. W. Ferreri.
Second edition. Il Castello - Collane Tecniche, via C. Ravizza, 16 - 20149 Milano. 240 pp. Price L. 18 000 (1982).

003.044 Gamma ray astrophysics. New insight into the universe. C. E. Fichtel, J. I. Trombka.
NASA SP-453. Scientific and Technical Information Branch, National Aeronautics and Space Administration, Washington, DC. 12 + 401 pp. (1981). Available from the Superintendent of Documents U. S. Government Printing Office, Washington, DC 20402.
Contents: Introduction: Gamma ray astronomy in perspective. The solar system: Gamma ray observations of the solar system. Planets, comets, and asteroids. Solar observations. The Galaxy: The interstellar medium and galactic structure. Compact objects. Extragalactic radiation: Galaxies. Cosmology. Diffuse radiation. Instrumentation: Gamma ray interaction processes. Detectors for energies less than 10 MeV. Detectors for energies greater than 10 MeV. Analysis of observational spectra. The future: Prospects for gamma ray astronomy.

003.045 Les étoiles et les curiosités du ciel. C. Flammarion.
C. Marpon, E. Flammarion, éditeurs en 1882. Avec une Postface de J.-C. Pecker. Réédité par Flammarion Éditeur. 22 + 792 pp. (1981). – Review in Astronomie, Vol. 96, 270; 1982 *(S. Débarbat).*

003.046 Our turbulent sun. K. Frazier.
Prentice-Hall, Englewood Cliffs, N. J. 12 + 198 pp. Price $ 16.95 cloth, $ 7.95 paper (1982). – From Science, Vol. 216, 50 (1982).

003.047 Cosmology, physics, and philosophy. B. Gal-Or, with forewords by K. Popper and A. Cottrell.
Springer Verlag, New York. 15 + 522 pp. (1981). ISBN 0-387-90581-2.
Contents: Preliminary concepts: From terrestrial gravitational structures to black holes and neutrinos in astrophysics. From "conservation" in classical physics to solitons in particle physics. From general relativity and relativistic cosmology to gauge theories. From physics to philosophical crossroads and back: The arrows of time. The crisis in quantum physics. From Physics to cosmological crossroads and back: Cosmology, physics and philosophy. Black holes and the unification of asymmetries. Beyond present knowledge: Havayism – the science of the whole. Critique of Western thought: A few historical remarks on time, mind and symmetry. The philosophy of time & change: some historical notions. Structuralism and the divided American thought. Policy and publicity: a critique. Thought-provoking and thought-depressing quotations. Critique of Western methodology.

003.048 Astronomie nuove (a cura di). R. Gallino.
Cooperativa Libraria Universitaria, Torino. 253 pp. Price L. 2 400 (1980). – Review in G. Astron., Vol. 7, 351 - 352; 1981 *(M. Rigutti).*

003.049 Einführung in die Astronomie. R.-H. Giese.
Wissenschaftliche Buchgesellschaft, Darmstadt. 400 pp. Price DM 55.00 (1981). – Review in Sterne Weltraum, 21. Jahrg., 225; 1982 *(A. M. Quetsch).*

003.050 Hemelmechanica: het tweelichamenprobleem. E. Goffin.
Uitgave Volkssterrenwacht Urania en Vereniging voor Sterrenkunde (België) in de serie Monografieën over Astronomie en Astrofysica, Vol. 7. Price BF 200.00. Available from Volkssterrenwacht Urania, Mattheesenstr. 62, B-2540 België. Review in Zenit, 9. Jaarg., 90; 1982 *(G. P. Können).*

003.051 The Orion complex: a case study of interstellar matter. C. Goudis.
Astrophysics and space science library, Vol. 90.
D. Reidel Publishing Company, Dordrecht, Holland – Boston, U. S. A. – London, England. 14 + 311 pp. Price Dfl. 140.00, $ 59.50 (1982). ISBN 90-277-1298-0.
Contents: Large scale view of the Orion region. The H II regions M42 and M43. The Orion complex (M42/OMC 1, M43/OMC 2). Empirical models of the Orion complex. NGC 2024 and the associated molecular complex. Appendix I: Radiative transfer. Appendix II: Physical parameters of an H II region. Appendix III: Physical parameters of the dust associated with an H II region. Appendix IV: Physical parameters of a molecular cloud.

003.052 Rainbows, halos and glories. R. Greenler.
Cambridge University Press, Cambridge – London – New York – New Rochelle – Melbourne – Sydney. 10 + 195 pp. Price $ 24.95 (1980). – Review in J. Opt. Soc. America, Vol. 72, 821; 1982 *(R. A. R. Tricker).*

003.053 The Jupiter effect reconsidered. J. R. Gribbin, S. H. Plagemann.
Revision of 'The Jupiter effect' (1974). Vintage (Random House), New York. 16 + 184 pp. Price $ 3.95 (1982). – From Science, Vol. 216, 908 (1982).

003.054 The night sky: the science and anthropology of the stars and planets. R. Grossinger.
Sierra Club, San Francisco, 28 + 484 pp. Price $ 16.95 (1981). From Phys. Today, Vol. 35, No. 1, p. 77 (1982).

003.055 Planetarium: window to the universe. C. F. Hagar.
Carl Zeiss, Oberkochen, FRG. 193 pp. Price DM 45.50, $ 18.50 (1980). ISBN 3-87517-005-9. – Review in Sky Telesc., Vol. 63, 583 - 584 (1982).

003.056 Photoelectric photometry of variable stars. D. S. Hall, R. M. Genet.
IAPPP, 1247 Folk Rd., Fairborn, Ohio 45324. 272 pp. Price $ 17.95 (1982). – Review in Sky Telesc., Vol. 64, 47 (1982).

003.057 **Regiomontanus-Studien.**   G. Hamann (Editor).
Österreichische Akademie der Wissenschaften,
Philosophisch-historische Klasse, Sitzungsberichte, Vol.
CCCLXIV. Verlag der Österreichischen Akademie der Wissenschaften, Vienna. 448 pp. + 44 plates (1980). − Review in
J. Hist. Astron., Vol. 13, 66 - 67; 1982 *(M. Shank)*.

003.058 **Geometric theory of diffraction.**
R. C. Hansen (Editor).
IEEE Press Selected Reprint Series. John Wiley & Sons,
New York. 7 + 406 pp. Price $ 40.95 (1981). − Review in
J. Opt. Soc. America, Vol. 72, 171; 1982 *(E. W. Marchand)*.

003.059 **Extraterrestrials: where are they?**
M. H. Hart, B. Zuckerman (Editors).
Pergamon Press. 182 pp. Price $ 22.50 cloth, $ 9.50 paper
(1982). ISBN 0-08-026342 cloth, ISBN 0-08-026341 paper.
Review in Sky Telesc., Vol. 63, 584 (1982).

003.060 **Astronomy: the cosmic journey.**
W. K. Hartmann.
Second edition. Wadsworth Publishing Co., 10 Davis Dr.,
Belmont, Calif. 94002. 22 + 530 pp. Price 24.95 (1982).
ISBN 0-534-01005-9. − Review in Sky Telesc., Vol. 63,
481 (1982).

003.061 **Extragalactic adventure: our strange universe.**
J. Heidmann.
Translated from the French edition by M. Schaeffer,
A. Boesgaard. Cambridge University Press. 8 + 174 pp. Price
$ 19.95, £ 12.50 cloth, $ 7.95, £ 4.95 paper (1982). ISBN
0-521-23571-5 cloth, ISBN 0-521-28045-1 paper. − Reviews
in Sky Telesc., Vol. 63, 375 (1982); Spaceflight, Vol. 24, 287
(1982).

003.062 **Encountering the universe.**   M. Heller.
Translated from the Polish by A. Potocki, edited by
G. W. Collins II.
The Astronomy Quarterly Library, Vol. 2. Pachart, Tucson,
Ariz., 8 + 110 pp. Price $ 9.95 (1982). ISBN 0-912918-07-1.
Review in Sky Telesc., Vol. 64, 47 (1982).

003.063 **Astrofysica voor calculators.**   P. Hellings.
Monografieën van de Volkssterrenwacht Urania/
Vereniging voor Sterrenkunde, België. 139 pp. Price BF 250.00
Review in Zenith, 9. Jaarg., 215; 1982 *(T. Dethier)*.

003.064 **Landolt-Börnstein.** Numerical data and functional
relationships in science and technology. New Series.
Group VI: Astronomy, astrophysics and space research.
Volume 2: Astronomy and astrophysics. Extension and supplement to Volume 1. Subvolume b: Stars and star clusters.
K.- H. Hellwege (Editor-in-chief), K. Schaifers, H. H. Voigt
(Editors).
Springer-Verlag, Berlin−Heidelberg−New York. 14 + 456 pp.
Price DM 750.00, $ 348.80 (1982). ISBN 3-540-10976-5
(Berlin, Heidelberg, New York), ISBN 0-387-10976-5 (New
York, Heidelberg, Berlin).
Contents: Physical parameters of the stars
(T. Schmidt-Kaler). Magnitudes and colors *(E. Lamla)*. Physics
of stellar atmospheres *(B. Baschek, M. Scholz)*. Stellar structure and evolution *(E. Meyer-Hofmeister)*. Variable stars
(H. W. Duerbeck, W. C. Seitter). Peculiar stars *(W. C. Seitter,
H. W. Duerbeck)*. Protostars, pre-main sequence objects
(I. Appenzeller). Planetary nebulae *(L. H. Aller)*. White dwarfs
(V. Weidemann). Compact objects *(M. Grewing, P. Kafka,
H. H. Fink)*. X-ray and γ-ray sources *(H. H. Fink, J. Trümper)*.
Double stars *(T. Herczeg)*. Star clusters and associations
(W. Seggewiss).

003.065 **Mysteries of the Universe.**   N. Henbest.
Van Nostrand Reinhold, New York − Marshall
Cavendish, London. 184 pp. Price $ 21.95, £ 8.95 (1981).
ISBN 0-442-23690-5. − Reviews in Sky Telesc., Vol. 63, 266
(1982); Spaceflight, Vol. 24, 96 (1982).

003.066 **Besiedelt die Menschheit das Weltall?**
D. B. Herrmann.
Urania-Verlag, Leipzig − Jena − Berlin, GDR. 128 pp. Price
M 4.50 (1981). − Reviews in Astron. Schule, 19. Jahrg., 45;
1982 *(H. Bernhard)*; Sky Telesc., Vol. 63, 481 (1982).

003.067 **Kosmische Weiten.** Geschichte der Entfernungsmessung im Weltall.   D. B. Herrmann.
2. Auflage. Johann Ambrosius Barth, Leipzig, 95 pp. Price
DM 14.00 (1981). − Review in Sterne Weltraum, 21. Jahrg.,
137; 1982 *(G. D. Roth)*.

003.068 **Von Sternen und Feuerrädern.**
D. B. Herrmann.
Verlag Junge Welt, Berlin, GDR. 63 pp. Price DM 10.50.
Review in Sterne Weltraum, 21. Jahrg., 133; 1982 *(A. Kunert)*.

003.069 **Starowolski's biographies of Copernicus. *(Studia
Copernicana, XXI)*.**   E. Hilfstein.
The Polish Academy of Sciences Press, Wrocław. 114 pp.
Price Zł 60.00 (1980). − Review in J. Hist. Astron., Vol. 13,
62 - 64; 1982 *(E. J. Aiton)*.

003.070 **Interplanetary dust.**   P. W. Hodge.
Gordon & Breach, New York − London. 8 + 280 pp.
Price $ 38.00 (1981). ISBN 0-677-03620-5. − Review in
Phys. Abstr., Vol. 85, Abstr. 38379 (1982).

003.071 **The quasar controversy resolved.**   F. Hoyle.
University College Cardiff Press, Cardiff. 80 pp.
Price £ 3.25 (1981). − Reviews in J. British Astron. Assoc.,
Vol. 92, 150; 1982 *(J. Mitton)*; Spaceflight, Vol. 24, 94 (1982).

003.072 **Jupiter.**   G. Hunt, P. Moore.
Rand McNally, 96 pp. Price $ 14.95 (1981). ISBN
528-81542-3. − Review in Sky Telesc., Vol. 64, 46 (1982).

003.073 **Quantum gravity 2.**
C. J. Isham, R. Penrose, D. W. Sciama (Editors).
Oxford University Press, Oxford. 5 + 669 pp. Price £ 28.00
(1981). − From J. British Astron. Assoc., Vol. 92, 154 (1982).

003.074 **The cosmic frontiers of general relativity.**
W. J. Kaufmann.
Translated from the English edition by N. V. Mitskevich.
Mir, Moskva. 349 pp. Price 1 Rbl. 20 Kop. (1981). In Russian.
Review in Priroda, No. 4, p. 120; 1982 *(O. I. Yakovlev)*.

003.075 **Planetary regularities in crater distribution on Mars,
the moon and Mercury (collected diagrams).**
D. A. Kazimirov, Zh. F. Rodionova, B. D. Sitnikov,
G. A. Poroshkova.
Prepr. Geol. inst. AN SSSR, Gos. Astron. inst. im.
P. K. Shternberga. Moskva, 1981, 67 pp. In Russian. − Reviews
in Ref. zh., 51. Astron., 4.51.44 (1982); 62. Issled. Kosm.
prostranstva, 5.62.414 (1982).

003.076 **Abu Nasr al-Farabi, 873 - 950.**
M. M. Khajrullaev.
Nauka, Moskva. 303 pp. (1982). In Russian. − Review in Ref.
zh., 51. Astron., 6.51.33 (1982).

003.077 **Mysterium cosmographicum − the secret of the
Universe.**   J. Kepler.
Translated by A. M. Duncan. Introduction and commentary
by E. J. Aiton. Abaris Books, New York. 267 pp. Price
$ 20.00 (1981). − Review in J. Hist. Astron., Vol. 13, 60 - 62;
1982 *(C. Wilson)*.

003.078 **Weather and climate on planets.**
K. Y. Kondratyev *(K. Ya. Kondrat'ev)*, G. E. Hunt.
Pergamon Press, Oxford − New York − Toronto − Sydney −
Paris − Frankfurt. 13 + 755 pp. Price $ 95.00, £ 50.00,
DM 265.00 (1982). ISBN 0-08-026493-X. − Reviews in
Nature, Vol. 298, 208; 1982 *(C. Leovy);* Observatory,
Vol. 102, 95; 1982 *(R. F. Griffin);* Sky Telesc., Vol. 63, 374
(1982).

003.079 **Cosmic rays and solar wind.**
G. F. Krymskij, A. I. Kuz'min, P. A. Krivoshapkin,
I. S. Samsonov, G. V. Skripin, I. A. Transkij, N. P. Chirkov.
Nauka, Novosibirsk. 224 pp. (1981). In Russian. − Review in
Ref. zh., 62. Issled. kosm. prostranstva, 5.62.77 (1982).

003.080 **Astronomy and national economy.**
A. A. Kulikov.
Nauka, Moskva. 164 pp. (1981). In Russian. − Review in Ref.
zh., 51. Astron., 3.51.34 (1982).

003.081 **Scientific space instrument making.**
T. I. Kurmanaliev (Editor).
Nauka, Moskva. 203 pp. (1981). In Russian. − Review in Ref.
zh., 62. Issled. kosm. prostranstva, 3.62.64 (1982).

003.082 **Venus.**   A. D. Kuz'min.
Nauka, Moskva. 93 pp. (1981). In Russian.
Review in Ref. zh., 51. Astron., 2.51.38 (1982).

003.083 **The cosmogonic chart on Phobos.**
D. G. Lahoz.
D. G. Lahoz, Toronto. 163 pp. Price $ 10.00 (1981). ISBN
0-9690985-3-7. − From Nature, Vol. 296, 691 (1982).

003.084 **Astronomy through the telescope.**   R. Learner.
Van Nostrand Reinhold, New York. 224 pp.
Price $ 29.95 (1981). ISBN 0-442-25839-9. Available from
Sky Publishing Corp. − Reviews in Sky Telesc., Vol. 63, 268
(1982); Sky Telesc., Vol. 63, 369 - 371; 1982 *(L. J. Robinson).*

003.085 **De geschiedenis van het astronomisch kunstuurwerk.**
A. Lehr.
Uitg. Martinus Nijhoff, 's-Gravenhage. 472 pp. Price ƒ 125.00
(1981). − Review in Zenit, 9. Jaarg., 90; 1982 *(M. J. Hagen).*

003.086 **La nouvelle révolution astronomique.**
J. Lequeux.
Edité par Hachette − Collection "La science en clair". 196 pp.
Price BF 524 (1981). − Review in Ciel Terre, Vol. 98, 116;
1982 *(J. Sauval).*

003.087 **How to buy and understand refracting telescopes.**
J. Levenson.
Levenson Press, P.O. Box 19606, Los Angeles, Calif. 90019.
96 pp. Price $ 18.95 (1981). ISBN 0-914442-09-0. − Review
in Sky Telesc., Vol. 63, 375 (1982).

003.088 **Discovering relativity four yourself.**   S. Lilley.
Cambridge University Press. 425 pp. Price £ 17.50
hbk, £ 7.95 pbk.(1981). − Review in Observatory, Vol. 102,
89; 1982 *(G. Pooley).*

003.089 **The friendly stars.**   M. E. Martin.
Van Nostrand Reinhold, New York. 140 pp. Price
$ 7.95 (1982). ISBN 0-442-21198-8. − Review in Sky Telesc.,
Vol. 64, 46 (1982).

003.090 **Regularities of the variations of the scattered light**
       and emission of the earth's twilight atmosphere.
T. G. Megrelishvili, with a preface by A. Kh. Khrgian.
Publishing House "Metsniereba", Tbilisi. 273 pp. Price 1Rbl.
90 Kop. (1981). In Russian.

003.091 **Il nostro sole.**   D. H. Menzel.
Faenza Ed., Faenza. 418 pp. Price L. 23 000 (1981).
Review in G. Astron., Vol. 7, 353 - 354; 1981 *(M. Rigutti).*

003.092 **The grand tour.**   R. Miller, W. K. Hartmann.
Workman Publishing, New York. 192 pp. Price
$ 19.95 cloth, $ 9.95 paper (1981). ISBN 0-89480-147-3 cloth,
ISBN 0-89480-146-5 paper. − Reviews in Sky Telesc., Vol. 63,
268 (1982); Sky Telesc., Vol. 63, 579 - 580; 1982 *(J. Lomberg).*

003.093 **The traveller's guide to the solar system.**
R. Miller, W. K. Hartmann.
Macmillan, London − New York. 192 pp. Price £ 8.95 (1981).
ISBN 0-333-32694-6. − Review in Spaceflight, Vol. 24, 142
(1982).

003.094 **Laser information systems of space apparatuses.**
I. V. Minaev, A. A. Mordovin, A. G. Sheremet'ev.
Mashinostroenie, Moskva. 269 pp. (1981). In Russian.
Review in Ref. zh., 62. Issled. kosm. prostranstva, 3.62.65
(1982).

003.095 **Solar activity and the earth.**
L. I. Miroshnichenko.
Nauka, Moskva. 145 pp. (1981). In Russian. − Review in Ref.
zh., 51. Astron., 3.51.41 (1982).

003.096 **The Moon.**   P. Moore.
Rand McNally. 96 pp. Price $ 14.95 (1981).
ISBN 528-81541-5. − Review in Sky Telesc., Vol. 64, 46
(1982).

003.097 **Voyages to Saturn.**   D. Morrison.
NASA SP-451. Scientific and Technical Information
Branch, National Aeronautics and Space Administration,
Washington, DC. 9 + 227 pp. (1982). Available from the
Superintendent of Documents, U. S. Government Printing
Office, Washington, DC. 20402. −
       Contents: The ringed planet. Pioneer to Saturn. The
Voyager mission. Encounter with Saturn. The last picture
show. The new Saturn system. Appendix A: Voyager science
teams. Appendix B: Voyager management teams. Appendix C:
Pictorial maps of the Saturnian satellites.

003.098 **The nature of matter.**   J. H. Mulvey (Editor).
Oxford University Press. 14 + 202 pp. Price $ 15.95
(1981). ISBN 0-19-851151-5. − Review in Sky Telesc., Vol. 63,
481 (1982).

003.099 **Catalogo dell'universo.**
P. Murdin, D. Allen, D. Malin.
Translated from the English edition. Editori Riuniti, Roma.
255 pp. Price L. 30 000 (1981). − Review in G. Astron., Vol. 7,
354 - 355; 1981 *(M. Rigutti).*

003.100 **Violent phenomena in the Universe.**   J. Narlikar.
Oxford University Press, Oxford. 11 + 218 pp.
Price £ 9.95 (1982). − From J. British Astron. Assoc., Vol. 92,
206 (1982).

003.101 **Quasi-periodic variations of the intensity and anisot-**
       ropy of cosmic radiation.
B. D. Naskidashvili, L. Kh. Shatashvili.
Metsniereba, Tbilisi. 130 pp. (1981). In Russian. − Review in
Ref. zh., 62. Issled. kosm. prostranstva, 3.62.68 (1982).

003.102 **Mars.**   I. M. Nicheva.
Narodn. astron. obs., Blagoev. r-n. Sofiya. 88 pp.
(1981). In Bulgarian. − Review in Ref. zh., 51. Astron.,
2.51.39 (1982).

003.103   **The fertile stars.**   B. O'Leary.
Everest House, 33 W. 60th St., New York, N. Y.
10023. 132 pp. Price $ 14.95 (1981). ISBN 0-89696-079-X.
Review in Sky Telesc., Vol. 63, 373; 1982 *(R. N. Watts, Jr.)*.

003.104   **Clefs pour l'Astronomie.**   J.-C. Pecker.
Editions Seghers. 318 pp. − Review in Astronomie,
Vol. 96, 268 - 269; 1982 *(S. Collin)*.

003.105   **The planet Jupiter: the observer's handbook.**
B. M. Peek, with a foreword by P. Moore.
Faber and Faber, London. 240 pp. Price £ 10.00 (1981).
Reviews in J. British Astron. Assoc., Vol. 92, 149 - 150; 1982
(W. E. Fox); Observatory, Vol. 102, 90; 1982 *(C. Anton)*;
Spaceflight, Vol. 24, 94 - 95 (1982).

003.106   **Patience dans l'azur: l'évolution cosmique.**
H. Reeves.
Editions du Seuil, 27 rue Jacob, Paris VIe, France. 268 pp.
Price FF 75.00, $ 12.00 (1981). ISBN 2-02-005924-X.
Reviews in Astronomie, Vol. 96, 321; 1982 *(G. Oudenot)*; Sky
Telesc., Vol. 64, 47 (1982).

003.107   **Cronache dell'universo.**   T. Regge.
Boringhieri, Torino. 180 pp. Price L. 5 000 (1981).
Review in G. Astron., Vol. 7, 349 - 351; 1981 *(M. Rigutti)*.

003.108   **Utilization of outer space and international law.**
G. C. M. Reijnen.
Elsevier Scientific Publishing Co., Amsterdam, Holland. 16 +
179 pp. Price US $ 60.00, Dfl. 150.00 (1981). − Review in
Astrophys. Space Sci., Vol. 84, 263; 1982 *(J. Kleczek)*.

003.109   **Motion of artificial and natural celestial bodies.**
G. S. Romashin (Editor).
Ural'sk. univ., Sverdlovsk. 158 pp. (1981). In Russian.
Review in Ref. zh., 62. Issled. kosm. prostranstva, 4.62.50
(1982).

003.110   **Cosmo.**   C. Sagan.
Mondadori Editore. 366 pp. Price L. 25 000
(1981). − Review in Orione, Vol. 3, 83; 1982 *(S. Baroni)*.

003.111   **Sonnenuhren.**   H. Schumacher, A. Peitz.
Verlag D. W. Callwey, München. 110 pp. Price
DM 52.00 (1981). − Review in Sterne Weltraum, 21. Jahrg.,
134, 1982 *(K. Schaifers)*.

003.112   **Modern cosmology.**   D. W. Sciama.
Cambridge University Press, Cambridge − New York.
14 + 214 pp. Price £ 5.95, $ 12.95 (1982).
ISBN 0-521-08069-X. − Reviews in J. British Astron. Assoc.,
Vol. 92, 203 (1982); Sky Telesc., Vol. 63, 374 (1982).

003.113   **Calcul astronomique pour amateurs.**   B. Serge.
Editions Masson, Paris. 154 pp (1981). − Review in
Orion 40. Jahrg., 111; 1982 *(R. Maeder)*.

003.114   **Extragalactic astronomy.**   J. L. Sérsic.
Geophysics and astrophysics monographs, Vol. 20.
D. Reidel Publishing Company, Dordrecht, Holland − Boston,
U.S.A. − London, England. 13 + 245 pp. Price Dfl. 120.00,
$ 49.50 (1982). ISBN 90-277-1321-9.
    Contents: Forms and structures. Normal galaxies. Active
galaxies. Galaxies and their environment. Measuring the Universe. Cosmology. Gravitational instability and galaxy formation. Notes and comments.

003.115   **The Paris Codex.** Decoding an Astronomical
Ephemeris.   G. M. Severin.
Transactions of the American Philosophical Society, Vol. 71,
Part 5. American Philosophical Society, Philadelphia. 102 pp.

Price $ 10.00 (1981). ISBN 0-87169-715-7. − From Science,
Vol. 216, 210 (1982).

003.116   **The Andromeda nebula.**   A. S. Sharov.
Nauka, Moskva. 447 pp. (1982). In Russian.
Review in Ref. zh., 51. Astron., 6.51.46 (1982).

003.117   **The unified model of the Universe.** The geometrically unified field solution.   S. Sheeter.
Process Press, Berkeley, Calif. 6 + 166 pp. Price $ 18.95 cloth,
$ 9.50 paper (1981). − From Science, Vol. 215, 660 (1982).

003.118   **Analysis of the 1980 - 81 apparition of Jupiter.**
P. C. Sherrod.
Mid-South Astronomical Research Society, Inc., P. O. Box 4145,
North Little Rock, Arkansas 72116. 59 pp. Price $ 5.00 (1981).
Review in Strolling Astron., Vol. 29, 126 - 127; 1982
(P. W. Budine).

003.119   **The physical universe: an introduction to astronomy.**
F. Shu.
University Science, Mill Valley, Calif. Price $ 28.00 (1982).
From Phys. Today, Vol. 35, No. 5, p. 90 (1982).

003.120   **The solar system and its strange objects.** Readings
from American Scientist.   B. J. Skinner (Editor).
Kaufmann, Los Altos, Calif. 4 + 194 pp. Price $ 9.95 (1981).
From Science, Vol. 216, 1258 (1982).

003.121   **The expanding Universe.**   R. Smith.
Cambridge University Press, Cambridge. 14 + 220 pp.
Price £ 19.00 (1982). − From J. British Astron. Assoc., Vol. 92,
206 (1982).

003.122   **Searching between the stars.**   L. Spitzer, Jr.
Yale University Press, New Haven − London.
15 + 179 pp. Price $ 25.00 (1982). ISBN 0-300-02709-5.
Review in Sky Telesc., Vol. 63, 584 (1982).
    Contents: The cosmic cycle of birth and death. The inter-
stellar medium as viewed in 1970. New windows on the Universe. Primordial hydrogen in the galactic disc. Clouds of molecular hydrogen. Heavy elements between the stars. A cloud
model of the interstellar gas.

003.123   **Physical processes in the interstellar medium.**
L. Spitzer, Jr.
Translated from the English edition. Mir, Moskva. 349 pp.
(1981). In Russian. − Review in Ref. zh., 51. Astron.,
3.51.45 (1982).

003.124   **Allgemeine Relativitätstheorie und relativistische
Astrophysik.**   N. Straumann.
Lecture Notes in Physics, Vol. 150. Springer-Verlag, Berlin −
Heidelberg − New York. 7 + 418 pp. (1981). ISBN
3-540-11182-4, ISBN 0-387-11182-4.
    Contents: Differentialgeometrische Hilfsmittel der allge-
meinen Relativitätstheorie. Allgemeine Relativitätstheorie: Das
Äquivalenzprinzip. Die Einsteinschen Feldgleichungen. Die
Schwarzschild-Lösung und die klassischen Tests der allgemei-
nen Relativitätstheorie. Schwache Gravitationsfelder. Die post-
Newtonsche Näherung. Relativistische Astrophysik: Neutronen-
sterne. Rotierende schwarze Löcher. Binäre Röntgenquellen.

003.125   **Classics in radio astronomy.**   W. T. Sullivan, III.
Studies in the history of modern science, Vol. 10.
D. Reidel Publishing Company, Dordrecht, Holland − Boston,
U.S.A. − London, England. 24 + 348 pp. Price Dfl. 140.00,
$ 59.50 (1982). ISBN 90-277-1356-1.
    Contents: Galactic background radiation. Techniques.
The solar system. Discrete sources. Spectroscopy.

003.126 **Hydrodynamic instabilities and the transition to turbulence.**   H. L. Swinney, J. P. Gollub (Editors). Springer Verlag, Berlin, Germany. 12 + 292 pp. (1981). ISBN 3-540-10390-2. – Review in Phys. Abstr., Vol. 85, Abstr. 23392 (1982).

003.127 **Theory of orbits. The restricted problem of three bodies.**   V. Szebehely. Translated from the English edition. Nauka, Moskva. 656 pp. (1982). In Russian. – Review in Ref. zh., 51. Astron., 6.51.36 (1982).

003.128 **Planetary science: a lunar perspective.** S. R. Taylor. Lunar and Planetary Institute, 3303 NASA Rd. 1, Houston, Tex. 77058. 481 pp. Price $ 39.95 (1982). ISBN 0-0942862-00-7. – Review in Sky Telesc., Vol. 64, 46 (1982).

003.129 **Wandel des Weltbildes – Astronomie, Physik und Meßtechnik in der Kulturgeschichte.** J. Teichmann. Kulturgeschichte der Naturwissenschaften und der Technik, Band 6. Deutsches Museum, München. 313 pp. Price DM 19.80 (1980). – Reviews in Phys. Bl., Vol. 38, 135; 1982 *(G. Hellbardt);* Phys. unserer Zeit, 13.Jahrg., 95 (1982).

003.130 **Essays in general relativity.** A Festschrift for Abraham Taub.   F. J. Tipler (Editor). Academic Press, New York. 18 + 236 pp. Price $ 30.00 (1980). Review in Science, Vol. 215, 53; 1982 *(R. Wald).*

003.131 **Living in space.**   J. Trefil. Scribner's New York. 132 pp. Price $ 10.95 (1981). From Phys. Today, Vol. 35, No. 4, p. 68 (1982).

003.132 **At the crossroads** ˙ **the earth and sky.** G. Urton. University of Texas Press, P. . Box 7819, Austin, Tex. 78712. 248 pp. Price $ 30.00 (1981) ˙ SBN 0-292-70349-X. Review in Sky Telesc., Vol. 6: 375 (1982).

003.133 **Light scattering by small particles.** H. C. van de Hulst. Dover Publications, Inc., New York. 10 + 470 pp. Price DM 35.40 (1981). ISBN 0-486-64228-3.

003.134 **Kijk op sterren en planeten.** Opzienbare ontdekkingen van astronomen en hun theorieën over het heelal.   R. van Helden. Zomer en Keuning, Ede. 100 pp. Price ƒ 19.50 (1981). Review in Zenit, 9. Jaarg., 43; 1982 *(A. Nagel).*

003.135 **Astronomisk Uppslagsbok.**   Å. Wallenquist. 3. edition. Bokförlaget Prisma, Stockholm. 286 pp. Price Sv. kr. 120.00 (1981). – Review in Astron. Tidsskr., Årg. 15, 48 (1982).

003.136 **Astronomie und Astrophysik – ein Grundkurs.** A. Weigert, H. J. Wendker. Physik-Verlag, Weinheim, FRG. 9 + 300 pp. Price DM 52.00 (1982). ISBN 3-87664-050-4. Contents: Sphärische Astronomie. Das Sonnensystem. Strahlung und Teleskope. Charakteristische Beobachtungsgrößen von Sternen. Die Außenschichten von Sonne und Sternen. Veränderliche Sterne. Innerer Aufbau und Entwicklung der Sterne. Interstellare Materie. Das Milchstraßensystem. Extragalaktische Systeme. Kosmologie.

003.137 **The first three minutes. Modern outlook on the origin of the Universe.**   S. Weinberg. Translated from the English edition (see 19.003.177).

Ehnergoizdat, Moskva. 209 pp. (1981). In Russian. – Review in Ref. zh., 51. Astron., 2.51.45 (1982).

003.138 **Mathematisches Hilfsbuch für Studierende und Freunde der Astronomie.**   W. Wepner. Treugesell-Verlag Dr. Vehrenberg KG, 4000 Düsseldorf 1. 279 pp. Price DM 26.80. – Review in Sterne Weltraum, 21. Jahrg., 224; 1982 *(M. Sarcander).*

003.139 **Het waarnemen van de zon.**   R. Wielinga. JWG (No. 64)/Stichting De Koepel, Price ƒ 5.00. Review in Zenit, 9. Jaarg., 215; 1982 *(A. Mak).*

003.140 **The total solar eclipse of 31 July 1981 in the Sakhalin region.**   V. K. Zakharov, A. A. Orlov. Prepr. Sakhalin. kompleks NII. Yuzhno-Sakhalinsk, 1981. 20 pp. In Russian. – Review in Ref. zh., 51. Astron., 3.51.37 (1982).

003.141 **Astronomy: the evolving universe.**   M. Zeilik. 3. edition. Harper and Row, New York. 623 pp. Price $ 26.95 (1982). ISBN 0-06-047376-2. – Review in Sky Telesc., Vol. 64, 47 (1982).

003.142 **Kalender und Chronologie.** Bekanntes und Unbekanntes aus der Kalenderwissenschaft. H. Zemanek. Verlag Oldenbourg, München – Wien. 156 pp. Price DM 29.80 (1981). – Review in Sternenbote, 25. Jahrg., 53 - 54 (1982).

003.143 **Geomagnetic activity and stability of the solar corpuscular field.**   I. D. Zosimovich. Nauka, Moskva. 191 pp. (1981). In Russian. – Review in Ref. zh., 51. Astron., 6.51.43 (1982).

003.144 **Astronomie pour garçons et filles.** Editions Les Deux Coqs d'Or, Paris. 93 pp. Price BF 322.00 (1981). – Review in Ciel Terre, Vol. 98, 115; *(N. Grevesse).*

003.145 **Astrophysical problems of propagation of radiation.** Tart. astrofiz. obs., Tallin. 31 pp. (1981). In Russian. Review in Ref. zh., 51. Astron., 3.51.38 (1982).

003.146 **Fire of life: the Smithsonian book of the sun.** W. W. Norton & Co., New York. 263 pp. Price $ 24.95 (1981). – Review in Sky Telesc., Vol. 64, 40; 1982 *(J. B. Zirker).*

003.147 **Planet guidebook,** Vol. 1 and 2. Japan Lunar and Planetary Observers Network (Editor). Seibundo Shinkosha Publishing Co., 5-1 Kandanishiki-cho, Chiyoda-ku, Tokyo 101, Japan. Vol. 1: 158 pp., Vol. 2: 286 pp. Price $ 35.00 (1981). – Review in Sky Telesc., Vol. 63, 481 (1982).

003.148 **Space images.** Van Nostrand Reinhold. 96 pp. Price $ 27.95 cloth, $ 15.95 paper (1982). ISBN 0-912810-37-8 cloth; ISBN 0-912810-36-X paper. – Review in Sky Telesc., Vol. 64, 46 (1982).

003.149 **The problem of solar-atmospheric relations. Sun – Atmosphere 1976 Experiment.** Sbornik statej. Tsentr. aehrol. obs. Gidrometeoizdat, Moskva. 92 pp. (1981). In Russian. – Review in Ref. zh., 51. Astron., 3.51.42 (1982).

004 History of Astronomy

004.001 The voice of astronomical history.
S. R. Weart, D. H. DeVorkin.
Sky Telesc., Vol. 63, 124 - 127 (1982).

004.002 Tycho country in Scandinavia.
D. Steel, M. Seb-Olsson.
Sky Telesc., Vol. 63, 233 - 235 (1982).

004.003 A Greek arithmetical method for finding oblique
ascensions.   O. Neugebauer.
J. Hist. Astron., Vol. 13, 19 - 22 (1982).

004.004 The Venus tablets: a fresh approach.
J. D. Weir.
J. Hist. Astron., Vol. 13, 23 - 49 (1982).

004.005 The first North American time ball.
I. R. Bartky, S. J. Dick.
J. Hist. Astron., Vol. 13, 50 - 54 (1982).

004.006 Nathaniel Bowditch's classical "black hole"
calculation of 1808.   J. W. Montgomery, Jr.
J. Hist. Astron., Vol. 13, 54 - 55 (1982).

004.007 Nuestro conocimiento de las estrellas a través del
tiempo.   J. Sahade.
Rev. Astron., Tomo 53, No. 219, p. 9 - 16 (1981).

004.008 Il y a 200 ans: une nouvelle planète.
G. Bodifée.
Astronomie, Vol. 96, 65 - 73 (1982).

004.009 Cent ans d'astronomie à travers "L'Astronomie".
J.-C. Pecker.
Astronomie, Vol. 96, 123 - 151 (1982).

004.010 La grande comète de 1882.
C. Bertaud.
Astronomie, Vol. 96, 162 (1982).

004.011 Saturne en 1882.   R. Verseau.
Astronomie, Vol. 96, 163 (1982).

004.012 Kalenderastronomie der Steinzeit/Odry.
H. Hindrichs.
Orion, 40. Jahrg., 4 - 10 (1982).

004.013 Was Galileo 2,000 years too late?
D. W. Hughes.
Nature, Vol. 296, 199 (1982).

004.014 The discovery of a new stone at Stonehenge.
M. W. Pitts.
Archaeoastronomy (U.S.A.), Vol. 4, No. 2, p. 16 - 21 (1981).

004.015 Navigation and astronomy — I: The first three
thousand years.   H. D. Howse.
J. British Astron. Assoc., Vol. 92, 53 - 60 (1982).

004.016 Cinquante années de radioastronomie: progrès,
découvertes et avenir.   J. P. Vallée.
J. R. Astron. Soc. Canada, Vol. 76, 1 - 18 (1982).
    The science of radio astronomy was born in January 1932.
A brief "tour guide" is presented, covering the achievements
of radio astronomy during the five decades which have elapsed
since then. These include a Nobel prize in 1974 to Ryle and
Hewish, a Nobel prize in 1978 to Penzias and Wilson.

004.017 Early days of astronomy at Toronto — Part II. The
Canadian Journal and the activities of the Toronto
Astronomical Club and Society 1868 - 1869.
H. Sawyer Hogg.
J. R. Astron. Soc. Canada, Vol. 76, 26 - 34 (1982).

004.018 Advocates and audience — aspects of popular
astronomy in England, 1750 - 1850.
I. Inkster.
J. British Astron. Assoc., Vol. 92, 119 - 123 (1982).

004.019 Atoms, astronomy and aeronomy.
M. J. Seaton.
Q. J. R. Astron. Soc., Vol. 23, 2 - 25 (1982).

004.020 An analysis of the solar observations of
Regiomontanus and Walther.   R. R. Newton.
Q. J. R. Astron. Soc., Vol. 23, 67 - 93 (1982).

004.021 L'accélération séculaire de la lune et le ralentissement
de la rotation de la terre.   F. Mignard.
Astronomie, Vol. 96, 175 - 186 (1982).

004.022 Who invented the astrolabe "Zarkala"?
A. Akhmedov, B. A. Rozenfel'd.
Obshchestv. nauk. v Uzbekistane, 1981, No. 8, p. 47 - 48. In
Russian. — Abstr. in Ref. zh., 51. Astron., 2.51.6 (1982).

004.023 Chinese studies in the history of astronomy, 1949—
1979.   Z. Xi.
ISIS, Vol. 72, 456 - 470 (1981).

004.024 Leonardo und die Astronomie.   G. Doebel.
Sterne Weltraum, 21. Jahrg., 72 - 75 (1982).

004.025 Maxwells Äthertheorien.
H.-G. Schöpf.
Astron. Nachr., Band 303, 29 - 37 (1982).

004.026 The Michelson-Morley-Miller experiments and the
Einsteinian synthesis.   L. S. Swenson, Jr.
Astron. Nachr., Band 303, 39 - 45 (1982).

004.027 Einstein and Michelson. The context of discovery
and the context of justification.
J. Stachel.
Astron. Nachr., Band 303, 47 - 53 (1982).

004.028 Die große Planetenbegegnung im Geburtsjahr Christi
nach der Bibel und astronomischen Keilschrifttexten.
K. Ferrari d'Occhieppo.
Mitt. Astron. Ges., Nr. 55, p. 135 - 137 (1982).

004.029 Arabic science. IV. Astrology and Astronomy. I.
N. Kokubu.
Rep. Univ. Electro-Commun., Vol. 31, 263 - 270 (1981). In
Japanese. — Abstr. in Phys. Abstr., Vol. 85, Abstr. 23400
(1982).

004.030 The origin of solar-terrestrial studies.
A. J. Meadows, J. E. Kennedy.
Vistas Astron., Vol. 25, 419 - 426 (1981).

004.031 Galileo and his medieval for-runners. On the
forming of the scientific principles of Galileo.
V. S. Shirokov.
Vopr. istor. estestvozn. i tekh., 1981, No. 3, p. 91 - 95. In
Russian. — Abstr. in Ref. zh., 51. Astron., 3.51.9 (1982).

004.032 On Kepler's role in the development of the conception of force. T. M. Kuk.
Ivano-Frank. inst. nefti i gaza. Ivano-Frankovsk., 1981. 6 pp. In Russian. – Abstr. in Ref. zh., 51. Astron., 3.51.11 (1982).

004.033 Where did the 1780 eclipse go? R. F. Rothschild.
Sky Telesc., Vol. 63, 558 - 560 (1982).

004.034 Possible astronomical alignments at Tsiping Pueblo, New Mexico. M. Zeilik, G. Sprinkle.
Proc. Southwest Reg. Conf., Vol. 7, (see 012.019), 97 - 99 (1982).

004.035 El aporte de Copérnico a la cosmología de su tiempo. C. J. Lavagnino.
Bol. Asoc. Argentina Astron., No. 18, p. 88 - 93 (1980).

004.036 Comparación entre el Almagesto de Ptolomeo y el De Revolutionibus de Copérnico.
C. J. Lavagnino.
Bol. Asoc. Argentina Astron., No. 18, p. 93 (1980). – Abstract.

004.037 The wondrous observatories of Jai Singh. D. Salwi.
New Scientist, Vol. 92, 726 - 730 (1981). – Abstr. in Phys. Abstr., Vol. 85, Abstr. 48145 (1982).

004.038 Materials on cosmogony of Middle Asia. V. A. Nikonov.
Onomastika Sredn. Azii, Frunze, Tom 2, 290 - 306 (1980). In Russian. – Abstr. in Ref. zh., 51. Astron., 4.51.11 (1982).

004.039 The revival of solar activity after Maunder minimum in reports and observations of E. Manfredi.
E. Baiada, R. Merighi.
Sol. Phys., Vol. 77, 357 - 362 (1982).
    The authors present information concerning observations of sunspots made at the beginning of the XVIIIth century. It concerns the years just after the end of the Maunder minimum. The data confirms the presence of a double maximum in 1704–1707, and shows a high asymmetry in sunspots latitude distribution, possibly related to the abnormal Sun activity in the second half of XVIIth century.

004.040 Five decades of radio astronomy. J. Grygar, M. Karlický.
Říše hvězd, Vol. 63, 45 - 48 (1982). In Czech.

004.041 Fourteenth century astronomy. G. Arrighi.
Atti Fond. Giorgio Ronchi, Vol. 36, 551 - 558 (1981). In Italian. – Abstr. in Phys. Abstr., Vol. 85, Abstr. 53119 (1982).

004.042 Astronomie in alter Zeit. J. Hamel.
Archenhold-Sternw. Berlin-Treptow, Vortr. Schr., Nr. 60, 52 pp. (1981).

004.043 Das Sirius-Rätsel oder ein hoffnungsloser Fall. D. B. Herrmann.
Archenhold-Sternw. Berlin-Treptow, Vortr. Schr., Nr. 61, 30 pp. (1982).

004.044 The ecliptical latitudes of Mars, Jupiter and Saturn according to Kushijar Dzhili as compared with modern calculations. A. F. Zausaev, Kh. F. Abdulla-Zade.
Astron. Tsirk., No. 1194, p. 4 - 6 (1981). In Russian.

004.045 The mandate of heaven. E. C. Krupp.
Bull. American Astron. Soc., Vol. 13, 793 (1981). Abstract.

004.046 Ancient astronomy in modern China. K. Brecher.
Bull. American Astron. Soc., Vol. 13, 793 (1981). – Abstract.

004.047 Lunar markings on Fajada Butte, Chaco Canyon, New Mexico.
A. Sofaer, R. M. Sinclair, L. E. Doggett.
Bull. American Astron. Soc., Vol. 13, 793 (1981). – Abstract.

004.048 Astronomical content of American Plains Indian winter counts. V. D. Chamberlain.
Bull. American Astron. Soc., Vol. 13, 793 - 794 (1981). Abstract.

004.049 The "Binding of the Years" and its cue for the nadir sun. E. C. Krupp.
Bull. American Astron. Soc., Vol. 13, 794 (1981). – Abstract.

004.050 Celestial alignments in Europe. L. B. Borst.
Bull. American Astron. Soc., Vol. 13, 794 (1981). Abstract.

004.051 Early spectrographic observations of the Crab Nebula. W. G. Hoyt.
Bull. American Astron. Soc., Vol. 13, 815 (1981). – Abstract.

004.052 Contributions of James E. Keeler to planetary research. D. E. Osterbrock.
Bull. American Astron. Soc., Vol. 13, 815 (1981). – Abstract.

004.053 "Spectrum very remarkable and inexpected:" P. J. C. Janssen and the solar eclipse of 1868.
S. L. Chapin.
Bull. American Astron. Soc., Vol. 13, 815 - 816 (1981). Abstract.

004.054 Quantum physics and the stars: Henry Norris Russell and the composition of the solar atmosphere.
D. H. Devorkin.
Bull. American Astron. Soc., Vol. 13, 816 (1981). – Abstract.

004.055 The Galileo affair in contemporary perspective. O. Gingerich.
Bull. American Astron. Soc., Vol. 13, 821 (1981). – Abstract.

004.056 Variable stars and cosmological speculation in the century after Descartes. W. Ashworth.
Bull. American Astron. Soc., Vol. 13, 830 (1981). – Abstract.

004.057 Laplace and invisible luminous stars. H. L. Poss.
Bull. American Astron. Soc., Vol. 13, 831 (1981). – Abstract.

004.058 The Metonic Cycle and the Saros: from Hyperborea to ancient Greece. J. H. Robinson.
Bull. American Astron. Soc., Vol. 13, 831 (1981). – Abstract.

004.059 Dating Ptolemy's lunar tables. S. J. Goldstein.
Bull. American Astron. Soc., Vol. 13, 831 (1981). Abstract.

004.060 An antebellum observatory at the University of Alabama. G. G. Byrd.
Bull. American Astron. Soc., Vol. 13, 831 (1981). – Abstract.

004.061 Prehistoric geometrical and astronomical knowledge of native Americans. J. A. Marshall.
Bull. American Astron. Soc., Vol. 13, 841 - 842 (1981). Abstract.

004.062 Pre-Copernican astronomy in Europe. M. C. Pande.

Indian J. Hist. Sci., Vol. 9, 123 - 137 (1974) = Uttar Pradesh State Obs. Repr. No. 113.

004.063 **Mayer, Herschel and Prévost on the solar motion.**
J. Hendry.
Ann. Sci., Vol. 39, 61 - 75 (1982).

004.064 **Seth Carlo Chandler and the observational origins of geodynamics.**   J. D. Mulholland, W. E. Carter.
High-precision Earth rotation and Earth-Moon dynamics.
Lunar distances and related observations, (see 012.035),
p. XV - XIX (1982).

004.065 **The Aristotelian theory of gravitation: a qualitative approach.**   V. P. Vizgin.
Priroda, 1982, No. 4, p. 97 - 104. In Russian.

004.066 **Aetas Hartwellianae.**   B. Warner.
Mon. Notes Astron. Soc. South. Africa, Vol. 41,
10 - 14 (1982).

004.067 **The motion of Venus, Mercury and the Sun in early Greek astronomy.**   B. L. van der Waerden.
Arch. Hist. Exact Sci., Vol. 26, 99 - 113 (1982).

004.068 **An investigation of the Ancient Star Catalog.**
D. Rawlins.
Publ. Astron. Soc. Pacific, Vol. 94, 359 - 373 (1982).
    Various tests are applied to Claudius Ptolemy's long-suspect claim that he himself observed the more than 1000 objects in the Ancient Star Catalog. An obvious consequence of observing under the influence of Ptolemy's large equinox-position error is found not to exist in the Catalog. The true observer's latitude, rough epoch, and probable identity are discovered. Hitherto-lost details of his fundamental work are reconstructed, including the earliest known accurate observation of the obliquity (135 B. C.).

004.069 **Astronomy and medicine.**
E. H. Strach.
J. British Astron. Assoc., Vol. 92, 164 - 169 (1982).

004.070 **La scoperta di Plutone.**   E. Badolati.
G. Astron., Vol. 7, 285 - 295 (1981).

004.071 **Aufbau und Entwicklung des Weltalls. I. Historische Wurzeln der Kosmologie.**   P. Ahnert.
Sterne, 58. Band, 131 - 138 (1982).

004.072 **Die Stellungnahme Keplers zur Gregorianischen Kalenderreform.**   H.-J. Felber.
Sterne, 58. Band, 139 - 146 (1982).

004.073 **Early days of astronomy at Toronto — Part III. The example of Ormsby MacKnight Mitchel and the Cincinnati Observatory.**   H. S. Hogg.
J. R. Astron. Soc. Canada, Vol. 76, 149 - 156 (1982).

004.074 **The Craig telescope of 1852.**   D. Steel.
Sky Telesc., Vol. 64, 12 - 13 (1982).

004.075 **William Herschel, Bath, and the Philosophical Society.**   R. Porter.
Uranus and the outer planets, (see 012.036), p. 23 - 34 (1982).

004.076 **Herschel's scientific apprenticeship and the discovery of Uranus.**   J. A. Bennett.
Uranus and the outer planets, (see 012.036), p. 35 - 53 (1982).

004.077 **Herschel and the construction of the heavens.**
M. Hoskin.
Uranus and the outer planets, (see 012.036), p. 55 - 66 (1982).

004.078 **The impact on astronomy of the discovery of Uranus.**   R. W. Smith.
Uranus and the outer planets, (see 012.036), p. 81 - 89 (1982).

004.079 **The structure of Aztec history.**
E. Umberger.
Archaeoastronomy *(U.S.A.)*, Vol. 4, No. 4, p. 10 - 18 (1981).

004.080 **Holes in the argument.**
A. Burl.
Archaeoastronomy *(U.S.A.)*, Vol. 4, No. 4, p. 19 - 25 (1981).

004.081 **Passage stellified: speculation upon archaeoastronomy in southeastern Zaire.**   A. F. Roberts.
Archaeoastronomy *(U.S.A.)*, Vol. 4, No. 4, p. 26 - 37 (1981).

004.082 **Venus dates revisited.**
M. P. Closs.
Archaeoastronomy *(U.S.A.)*, Vol. 4, No. 4, p. 38 - 41 (1981).

004.083 **Determination of the ecliptic coordinates of planets in the works of the scientists of the medieval Orient.**   A. F. Zausaev, Kh. F. Abdulla-zade.
Dokl. AN TadzhSSR, Tom 24, 417 - 421 (1981). In Russian.
Abstr. in Ref. zh., 51. Astron., 6.51.5 (1982).

004.084 **Sonnen- und Planetenbewegung nach Eudoxos.**
J. Teichmann.
Sterne Weltraum, 21. Jahrg., 247 - 250 (1982).

004.085 **The 'astronomical' chapters of the Ethiopic Book of Enoch (72 to 82).**
Translation and commentary by O. Neugebauer, with additional notes on the Aramaic fragments by M. Black.
Mat.-Fys. Medd., Vol. 40, No. 10, p. 1 - 42 (1981).
    Ethiopic literature has preserved the "Book of Enoch". Ten chapters of this work are concerned with astronomical concepts of a rather primitive character (variation in the length of daylight, illumination and rising amplitude of the moon, wind-directions, etc.), dominated by simple arithmetical patterns. The present paper gives a new translation of these chapters, followed by notes where the meaning of the text is not self-explanatory. An appendix, with additional notes, deals with related source material in the Qumran astronomical scrolls.

004.086 **Beiträge der Berliner Sternwarte zu bedeutenden Erkenntnissen auf dem Gebiet der klassischen Astronomie.**   W. Fricke.
Prepr. Astronomisches Rechen-Institut, Heidelberg, 19 pp. (1982). — Submitted to Astron. Nachr.
    The 150th Anniversary of the Berlin Observatory suggests a review of major contributions made by the observatory to the field of classical astronomy. The sensational discovery of the planet Neptune by Galle in 1846 on the basis of Leverrier's theory of perturbations of Uranus formed a prelude to research programmes of great importance and to new findings. Three research projects in the field of astrometry are described in more detail.

004.087 **Nicholas Copernicus and Giorgio Valla.**   E. Rosen.
Physis, Anno 23, 449 - 457 (1981).

004.088 **Astronomische Messungen bei den Griechen im 5. Jahrhundert v. Chr. und ihr Instrument.**
Á. Szabó.
Hist. Sci., No. 21, p. 1 - 26 (1981).

004.089 **A new, rational endeavour for understanding the Eratosthenes numerical result of the earth meridian measurement.**   M. Cimino.
Compendium in astronomy, (see 003.012), p. 11 - 21 (1982).

004.090 **Astronomical activity in the period 1600 to 1880.**
G. Teleki.
Compendium in astronomy, (see 003.012), p. 23 - 27 (1982).
Proceeding from the data presented in the Houzeau-Lancaster Bibliography (1882) the growth of the total sum of information in astronomy from 1600 to 1880 is analysed and estimated that the time interval, within which the number of data is doubled, were about 25 years. The growth of the total sum of the information in astronomy as a whole has been more rapid — by about a quarter — than that relating to the catalogues alone.

004.091 **Report of IAU Commission 41: History of astronomy (Histoire de l'astronomie).** M. A. Hoskin.
Trans. IAU, Vol. XVIIIA, (see 003.013), p. 551 - 552 (1982).

004.092 **Die Bilder und die Sachen. Eine Betrachtung über Herschel und seine Zeit.** K. Wälke.
Photorin, Heft 5, 16 - 27 (1982).

004.093 **Gotha 1798. Vorder- und Hintergründe des ersten Astronomen-Kongresses.** P. Brosche.
Photorin, Heft 5, 38 - 59 (1982).

Catalogue of the archives of the Royal Observatory, Edinburgh, 1764 - 1937. See Abstr. 002.055.

The view from planet earth.
See Abstr. 003.032.

Abu Nasr al-Farabi, 873 - 950.
See Abstr. 003.076.

Mysterium cosmographicum — the secret of the Universe. See Abstr. 003.077.

Classics in radio astronomy.
See Abstr. 003.125.

At the crossroads of the earth and sky.
See Abstr. 003.132.

Archaeoastronomers convene in Oxford.
See Abstr. 011.001.

Archaeoastronomy in the Americas. Papers from a conference, Sante Fé, N. M., 1979.   See Abstr. 012.056.

Interviews: the voice of astronomy history.
See Abstr. 013.037.

Computer aided instruction of ancient astronomy.
See Abstr. 014.013.

Solar rotation from 17th century records.
See Abstr. 072.057.

The pre-discovery observations of Uranus.
See Abstr. 101.014.

005 Biography

005.001 **An eclectic astronomer.**
K. Sugden.
Sky Telesc., Vol. 63, 27 - 29 (1982).

005.002 **Geoffrey Chaucer: amateur astronomer?**
T. Carter.
Sky Telesc., Vol. 63, 246 - 247 (1982).

005.003 **William Bradfield — elf kometen in tien jaar.**
R. J. Bouma.
Zenit, 9. Jaarg., 10 - 13 (1982).

005.004 **Karl Friedrich Zöllner. Zum 100. Todestag des deutschen Astrophysikers.** D. B. Herrmann.
Astron. Schule, 19. Jahrg., 12 - 13 (1982).

005.005 **En astronomisk studieresa i Europa på 1840-talet, I.** Utdrag ur J. M. Agardhs resedagbok 1843 - 1845.
C. Schalén, N. Hansson.
Astron. Tidsskr., Årg. 15, 1 - 13 (1982).

005.006 **Memorial dates of astronomy in 1982.**
A. I. Eremeeva.
Astronomical calendar. Yearbook. 1982, (see 047.005), p. 306 - 330 (1981). In Russian. — Abstr. in Ref. zh., 51. Astron., 2.51.12 (1982).

005.007 **Michelson in Potsdam.** R. S. Shankland.
Astron. Nachr., Band 303, 3 - 5 (1982).

005.008 **Albert A. Michelson an der Berliner Universität.**
J. Auth.
Astron. Nachr., Band 303, 7 - 14 (1982).

005.009 **Einstein and Michelson — artists in science.**
D. Michelson Livingston.
Astron. Nachr., Band 303, 15 - 16 (1982).

005.010 **Vier Michelson-Supplemente.**
H. Melcher.
Astron. Nachr., Band 303, 17 - 20 (1982).

005.011 **Manuel Johnson and the St. Helena Observatory.**
B. Warner.
Vistas Astron., Vol. 25, 383 - 409 (1981).

005.012 **Rudolf Minkowski.** A. Weigert.
Sterne Weltraum, 21. Jahrg., 188 - 189 (1982).

005.013 **Professor John Xanthakis — brief biographical note and list of publications.**
Compendium in astronomy, (see 003.012), p. VII - XII (1982).

005.014 **La obra de B. Gould anterior al Observatorio de Córdoba.** C. J. Lavagnino.
Bol. Asoc. Argentina Astron., No. 18, p. 99 - 101 (1980).

005.015 **F. A. Bredikhin,** on the occasion of the 150th anniversary of his birthday, 1981, December 8.
S. M. Poloskov.
Zemlya Vselennaya, 1982, No. 1, p. 45 - 48. In Russian.

005.016 **I. D. Zhongolovich,** on the occasion of the 90th anniversary of his birthday, 1982, February 20.
V. K. Abalakin.
Zemlya Vselennaya, 1982, No. 3, p. 43 - 47. In Russian.

005.017 **Ejnar Hertzsprung – Leben und Werk.**
D. B. Herrmann.
E. Hertzsprung: Zur Strahlung der Sterne. Ostwalds Klassiker der exakten Wissenschaften, Bd. 255, Akademische Verlagsgesellschaft Geest & Portig, Leipzig 1981, p. 7 - 25 = Mitt. Archenhold-Sternw. Berlin-Treptow, Nr. 119 (1981).

005.018 **K. E. Ciolkovskij im Spiegel westeuropäischer Raumfahrtliteratur. Ein Beitrag zur Wirkungsgeschichte der Ideen von Ciolkovskij.**   D. B. Herrmann.
NTM, Schriftenreihe für Geschichte der Naturwissenschaften, Technik und Medizin, Band 18, No. 2, p. 8 - 16 (1981) = Mitt. Archenhold-Sternw. Berlin-Treptow, Nr. 125 (1981).

005.019 **Die Selbstbildnisse des Nicolaus Copernicus.**
E. Sommerfeld.
Mitt. Archenhold-Sternw. Berlin-Treptow, Band 6, Nr. 134, 8 pp. (1981).

005.020 **A brief memoir of John A. Miller.**   J. S. Hall.
Bull. American Astron. Soc., Vol. 13, 814 - 815 (1981). – Abstract.

005.021 **Edward Singleton Holden and the organization of astronomical research.**   O. R. Butler.
Bull. American Astron. Soc., Vol. 13, 815 (1981). – Abstract.

005.022 **Antonia Maury: rainbows and pinwheels.**
B. L. Welther.
Bull. American Astron. Soc., Vol. 13, 816 (1981). – Abstract.

005.023 **Le 250ᵉ anniversaire de la naissance de Jérôme Lalande.**   P. de La Cotardière.
Astronomie, Vol. 96, 310 - 312 (1982).

Regiomontanus-Studien.   See Abstr. 003.057.

Starowolski's biographies of Copernicus.
See Abstr. 003.069.

Abu Nasr al-Farabi, 873 - 950. See Abstr. 003.076.

El aporte de Copérnico a la cosmología de su tiempo.
See Abstr. 004.035.

The wondrous observatories of Jai Singh.
See Abstr. 004.037.

006 Personal Notes

George E. D. Alcock received the Amateur Achievement Award.   H. C. Arp, A. Fraknoi.
Mercury, Vol. 10, 186, 189 (1981).

Robert N. Clayton received the V. M. Goldschmidt Medal 1981.   J. R. O'Neil.
Geochim. Cosmochim. Acta, Vol. 46, 1133 (1982).

G. Courtès received the Prix Janssen 1981.
Astronomie, Vol. 96, 84 (1982).

Éric Fossat received the Prix Petit d'Ormoy.
C. R. Acad. Sci. Paris, Tome 293 Suppl., 125 (1981). In French.

Riccardo Giacconi received the Bruce Medal.
H. C. Arp, A. Fraknoi.
Mercury, Vol. 10, 182 - 183 (1981).

Gérard Grec received the Prix Petit d'Ormoy.
C. R. Acad. Sci. Paris, Tome 293 Suppl., 125 (1981). In French.

Richard Kron received the Trumpler Award.
H. C. Arp, A. Fraknoi.
Mercury, Vol. 10, 183 - 184 (1981).

Hommage au Professeur Pierre Lacroute.
J. Delhaye.
Scientific aspects of the Hipparcos space astrometry mission, (see 012.041), p. VIII (1982).

Professor Sir Harrie Massey received the Gold Medal of the Royal Astronomical Society 1982.   A. W. Wolfendale.
Q. J. R. Astron. Soc., Vol. 23, 173 (1982).

Pierre Mein received the Prix Ancel.
C. R. Acad. Sci. Paris, Tome 293 Suppl., 120 (1981). In French.

Aden Baker Meinel, Frederic Ives medalist for 1980.
J. Opt. Soc. America, Vol. 72, 7 - 13 (1982).

Jean-François Minster received the F. W. Clarke Medal 1981.   S. R. Hart.
Geochim. Cosmochim. Acta, Vol. 46, 1135 - 1136 (1982).

Eh. R. Mustel', received the Belopol'skij prize in 1981.
Zemlya Vselennaya, 1982, No. 3, p. 47. In Russian.

P. Slavenas, on the occasion of the 80th anniversary of his birthday.
Vopr. istor. estestvozn. i tekh., 1981, No. 3, p. 172. In Russian. – Abstr. in Ref. zh., 51. Astron., 3.51.14 (1982).

Jean-Marie Souriau received the Prix Jaffé.
C. R. Acad. Sci. Paris, Tome 293 Suppl., 118 - 119 (1981). In French.

B. Šternberk, 85th birthday.
P. Andrle.
Říše hvězd, Vol. 63, 13 - 14 (1982). In Czech.

B. Šternberk, 85th birthday.   J. Bouška.
Vesmír, Vol. 61, 27 (1982). In Czech.

Professor Dr. Hans Straßl zum 75. Geburtstag.
W. Seitter.
Sterne, 58. Band, 3 - 5 (1982).

Dietrick E. Thomsen received the Klumpke-Roberts Award.   H. C. Arp, A. Fraknoi.
Mercury, Vol. 10, 184 - 186 (1981).

L. Webrová, 60th birthday.
Říše hvězd, Vol. 63, 58 - 59 (1982). In Czech.

007 Obituaries

Wendell C. DeMarcus died 1982 January 9.
F. Gabbard.
Phys. Today, Vol. 35, No. 4, p. 72 - 73 (1982).

V. A. Firsoff, 1910 - 1982.
P. Moore.
J. British Astron. Assoc., Vol. 92, 139 (1982).

Paul Herget, 1908 - 1981.   P. K. Seidelmann.
Phys. Today, Vol. 35, No. 1, p. 86 - 87 (1982).

Frank Malina, 1912 - 1981.   E. Kunovská.
Vesmír, Vol. 61, 49 - 51 (1982). In Czech.

Jeffrey Charles Percy Miller, 1906 August 31 -
1981 April 24.   D. H. Sadler.
Q. J. R. Astron. Soc., Vol. 23, 311 - 313 (1982).

Rolf Müller, 1898 Januar 26 - 1981 März 24.
F. Schmeidler.
Sterne, 58. Band, 109 - 111 (1982).

John Guy Porter, 1900 November 5 - 1981
September 13.   P. Moore.
J. British Astron. Assoc., Vol. 92, 137 (1982).

John Philip Manning Prentice, 1903 March 14 -
1981 October 6.   E. H. Collinson.
J. British Astron. Assoc., Vol. 92, 138 - 139 (1982).

Maria Luisa Righini Bonelli (1917 - 1981).
J. Hist. Astron., Vol. 13, 74 (1982).

Jan Schilt, 1894 February 3 - 1982 January 9.
P. van de Kamp.
Phys. Today, Vol. 35, No. 6, p. 67 - 68 (1982).

Carl Siegel, 1896 - 1981 April 4.
C. R. Acad. Sci. Paris, Tome 293 Suppl., 107 (1981). In
French.

Beatrice Muriel Hill Tinsley, 1941 January 27 -
1981 March 23.   R. B. Larson, L. L. Stryker.
Q. J. R. Astron. Soc., Vol. 23, 162 - 165 (1982).

Harold Clayton Urey: 1893 - 1981.
C. Sagan.
Icarus, Vol. 48, 348 - 352 (1981).

008 Observatories, Institutes

Reports, communications and publications of observatories and astronomical institutes are recorded in this section; included are numbered series of reprints. Whenever possible, the numbers of the abstracts referring to the publications are given. Observatories and institutes are listed in alphabetical order of their towns. In some cases observatory publications do not give the name of the town; the following list which gives names and towns of some institutions may serve as an aid in such cases.

Aarne Karjalainen Observatory	Oulu, Finland	Dudley Observatory	Schenectady, New York, USA
Algonquin Radio Observatory	Lake Traverse, Ontario, Canada	Dunsink Observatory	Dublin, Ireland
Allegheny Observatory	Pittsburgh, Pennsylvania, USA	Dyer Observatory, Vanderbilt University	Nashville, Tennessee, USA
Anglo-Australian Observatory	Epping, N. S. W., Australia	Ege University Observatory	Izmir, Turkey
Archenhold-Sternwarte	Berlin-Treptow, German Democratic Republic	Engelhardt Observatory	Kazan, USSR
		Erwin W. Fick Observatory, Iowa State University	Ames, Iowa, USA
Argentine Radioastronomy Institute	Pereyra, Iraola, Argentina	European Southern Observatory	La Silla, Chile
Arizona State University	Tempe, Arizona, USA	Fabra Observatory	Barcelona, Spain
Arthur J. Dyer Observatory	Nashville, Tennessee, USA	Felix Aguilar Observatory	San Juan, Argentina
Astronomical Latitude Station, Polish Academy of Sciences	Borowiec, Poland	Fernbank Observatory	Atlanta, Georgia, USA
Astronomisches Rechen-		Five College Observatories	Amherst, Massachusetts, USA
Institut	Heidelberg, F. R. Germany	Florida State University Radio Observatory	Tallahassee, Florida, USA
Bell Laboratories	Murray Hill, New Jersey, USA	Flower and Cook Observatories, University of Pennsylvania	Philadelphia, Pennsylvania, USA
Bell Telephone Laboratories	Holmdel, New Jersey, USA		
Bosscha Observatory	Lembang, Indonesia		
Boyden Observatory	Bloemfontein, South Africa		
Bureau International de l'Heure	Paris, France	George R. Wallace Jr. Astrophysical Observatory	Cambridge, Massachusetts, USA
Cajigal Observatory	Caracas, Venezuela		
California Institute of Technology	Pasadena, California, USA	Georgetown Observatory	Washington, D.C., USA
Carter Observatory	Wellington, New Zealand	Glavnaya Astronomicheskaya Observatoriya AN SSSR	Pulkovo, USSR
Cavendish Laboratory	Cambridge, England	Goddard Space Flight Center	Greenbelt, Maryland, USA
Centre de Données Stellaires	Strasbourg, France	Goethe Link Observatory, Indiana University	Bloomington, Indiana, USA
Centro Astronomico Hispano-Aleman (Max-Planck Institut für Astronomie Station)	Calar Alto, Almeria, Spain	H. M. Nautical Almanac Office, Royal Greenwich Observatory	Greenwich, England
Cerro Tololo Interamerican Observatory	La Serena, Chile	Hale Observatories	Pasadena, California, USA
Československá akademie věd, Astronomický ústav	Prague, Czechoslovakia	Harvard Radio Astronomy Station	Cambridge, Massachusetts, USA
Chamberlin Observatory, University of Denver	Denver, Colorado, USA	Haute Provence Observatory	Saint Michel, France
Columbia University, Department of Astronomy	New York, New York, USA	Haystack Observatory	Westford, Massachusetts, USA
Commonwealth Observatory	Canberra, Australia	Heinrich-Hertz Institut	Berlin-Adlershof, German Democratic Republic
Cornell University, Center for Radiophysics and Space Research	Ithaca, New York, USA	Herzberg Institute of Astrophysics	Victoria, B. C., Ottawa, Canada
Corralitos Observatory	Las Cruces, New Mexico, USA		
Crawford Hill Laboratory	Holmdel, New Jersey, USA	High Altitude Observatory, University of Colorado	Boulder, Colorado, USA
CSIRO, Radiophysics Division	Sydney, Australia	Hopkins Observatory	Williamstown, Massachusetts, USA
David Dunlap Observatory, University of Toronto	Richmond Hill, Ontario, Canada	Horn d'Arturo Observatory	Bologna, Italy
Dearborn Observatory	Evanston, Illinois, USA	Hvar Observatory	Zagreb, Yugoslavia
Department of Astronomy and Observatory, University of California	Los Angeles, California, USA	IBM Thomas J. Watson Research Center	Yorktown Heights, New York, USA
Department of Astronomy Swarthmore College	Swarthmore, Pennsylvania, USA	Indian Institute of Astrophysics	Bangalore, India
		Infrared Telescope Facility	Honolulu, Hawaii, USA
Department of Astronomy, University of Texas	Austin, Texas, USA	Institute for Astronomy, University of Hawaii	Honolulu, Hawaii, USA
Deutsches Hydrographisches Institut (DHI)	Hamburg, F. R. Germany	Institute for Theoretical Astronomy	Leningrad, USSR
Dominion Astrophysical Observatory	Victoria, B. C., Canada	Institute of Astronomy and Space Science, University of British Columbia	Vancouver, B. C., Canada
Dominion Observatory	Ottawa, Ontario, Canada	Institute of Theoretical Astrophysics, Blindern	Oslo, Norway
Dominion Radio Astrophysical Observatory	Penticton, B. C., Canada		

Instituto Argentino de
Radioastronomía   Villa Elisa, Provincia de
                                Buenos Aires, Argentina
Instituto de Astronomía y Física
del Espacio (IAFE)   Buenos Aires, Argentina
Instituto Venezolano de
Astronomia   Merida, Venezuela
Instituto y Observatorio de
Marina   San Fernando (Cádiz), Spain
International Latitude
Observatory   Mizusawa, Japan
IUE Observatory,
European Space Agency   Villafranca, Madrid, Spain
Jet Propulsion Laboratory,
California Institute of
Technology   Pasadena, California
Joint Institute for Laboratory
Astrophysics (JILA)   Boulder, Colorado, USA
Judson B. Coit Observatory   Boston, Massachusetts, USA
Kandilli Observatory   Istanbul, Turkey
Kansas University Observatory   Lawrence, Kansas, USA
Kapteyn Astronomical
Laboratory   Groningen, Netherlands
Karl-Schwarzschild-
Observatorium   Tautenburg, German
                                Democratic Republic
Kenneth Mees Observatory   Rochester, New York, USA
Kiepenheuer-Institut für Son-
nenphysik, formerly
Fraunhofer-Institut   Freiburg, F. R. Germany
Kitt Peak National Observatory Tucson, Arizona, USA
Kodaikanal Observatory   Bangalore, India
Korean National
Astronomical Observatory   Seoul, Korea
Kwasan and Hida Observatories   Kyoto, Japan
Lamont-Hussey Observatory   Bloemfontein, South Africa
Landessternwarte Heidelberg-
Königstuhl   Heidelberg, F. R. Germany
Las Campanas Observatory   Pasadena, California, USA
Lawrence Livermore Laboratory,
University of California   Livermore, California, USA
Leander McCormick Observatory,
University of Virginia   Charlottesville, Virginia, USA
Lee Observatory   Beirut, Lebanon
Leopold-Figl-Observatorium   Vienna, Austria
Leuschner Observatory   Berkeley, California, USA
Lick Observatory   Santa Cruz, (Mount Hamil-
                                ton), California, USA
Lindheimer Astronomical
Research Center   Evanston, Illinois, USA
Lockheed Palo Alto Research
Laboratory   Palo Alto, California, USA
Lockheed Solar Observatory   Saugus, California, USA
Lohrmann-Observatorium der
Technischen Universität
Dresden   Dresden, German Democratic
                                Republic
Louisiana State University
Observatory   Baton Rouge, Louisiana, USA
Lowell Observatory   Flagstaff, Arizona, USA
Lunar and Planetary Laboratory,
Catalina Station   Tucson, Arizona, USA
Max-Planck-Institut für
Astronomie   Heidelberg, F. R. Germany
Max-Planck-Institut für
Physik and Astrophysik   Munich, F. R. Germany
Max-Planck-Institut für
Radioastronomie   Bonn, F. R. Germany
McDonald Observatory   Fort Davis, Texas, USA
McMath Hulbert Observatory   Pontiac, Michigan, USA
Michigan State University
Observatory   East Lansing, Michigan, USA

Molonglo Radio Observatory,
University of Sydney   Sydney, Australia
Monterey Institut for Research
in Astronomy   Carmel Valley, California,
                                USA
Mount Cuba Observatory   Wilmington, Delaware, USA
Mount John Observatory   Lake Tekapo, New Zealand
Mount Stromlo Observatory   Canberra, Australia
Mount Wilson Observatory   Pasadena, California, USA
Mt. Laguna Observatory   San Diego, California, USA
Mullard Radio Astronomy
Observatory   Cambridge, England
Mullard Space Science
Laboratory   London, England
Multiple Mirror Telescope
Observatory   Tucson, Arizona, USA
Narrabri Observatory, University
of Sydney   Sydney, Australia
National Bureau of Standards   Washington, D.C., USA
National Radio Astronomy
Observatory   Charlottesville, Virginia, USA
                                Green Bank, West Virginia,
                                USA
                                Socorro, New Mexico, USA
                                Tucson, Arizona, USA
National Research Council
of Canada   Ottawa, Ontario, Canada
New Mexico State
University Observatory   Las Cruces, New Mexico, USA
Nicholas Copernicus
Observatory   Brno, Czechoslovakia
Nizamiah & Rangapur
Observatories   Hyderabad, India
Nuffield Radio Astronomy
Laboratories, Jodrell Bank
University of Manchester   Manchester, England
Oak Ridge Observatory
(formerly Harvard College
Observatory)   Cambridge, Massachusetts,
                                USA
Observatoire Royal de Belgique Uccle, Belgium
Observatories of the University
of Western Ontario   London, Canada
Observatório Astronômico
Antares   Feira de Santana, Brazil
Observatório Astronômico do
Instituto de Física da Universi-
dade Federal do Rio Grande
do Sul   Porto Alegre, Rio Grande
                                do Sul, Brazil
Observatorio de Cartuja   Granada, Spain
Observatorio del Ebro   Tortosa, Spain
Observatorio Nacional   Rio de Janeiro, Brazil
Observatorio Nacional de
Física Cósmica   San Miguel, Argentina
Observatory, University of
Michigan   Ann Arbor, Michigan, USA
Ohio State University
Radio Observatory   Columbus, Ohio, USA
Ole Roemer-Observatoriet   Aarhus, Denmark
Onsala Space Observatory   Göteborg, Sweden
Owens Valley Radio
Observatory   Big Pine, California, USA
Palomar Observatory   Pasadena, California, USA
Perkins Observatory, Ohio State
and Wesleyan Universities   Delaware, Ohio, USA
Physical Research Laboratory   Ahmedabad, India
Purple Mountain Observatory   Nanking, China
Radcliffe Observatory   Pretoria, South Africa
Raman Research Institute   Bangalore, India
Rattlesnake Mountain
Observatory   Richland, Washington, USA

Remeis-Sternwarte	Bamberg, F. R. Germany
Ritter Astrophysical Research Center of the University of Toledo	Toledo, Ohio, USA
Rosemary Hill Observatory	Gainesville, Florida, USA
Rothney Astrophysical Observatory	Calgary, Canada
Royal Aircraft Establishment, Geophysical Studies in Space Department	Farnborough, England
Royal Radar Establishment, Radio Astronomy Division	Malvern, England
Sacramento Peak Observatory	Sunspot, New Mexico, USA
Sagamore Hill Radio Observatory	Hamilton, Massachusetts, USA
San Fernando Observatory	Northridge, California, USA
Shaanxi Astronomical Observatory	Lintong, Sian, China
Siding Spring Observatory	Siding Spring, N. S. W., Australia
Smithsonian Astrophysical Observatory	Cambridge, Massachusetts, USA
Sonnenobservatorium Kanzelhöhe	Graz, Austria
South African Astronomical Observatory	Cape Town, South Africa
Specola Vaticana	Castel Gandolfo, Vatican
Specola di Padova	Asiago, Italy
Sproul Observatory	Swarthmore, Pennsylvania, USA
Sternberg Astronomical Institute	Moscow, USSR
Steward Observatory, University of Arizona	Tucson, Arizona, USA
W. Struve Tartu Astrophysical Observatory	Tartu, USSR
Tata Institute of Fundamental Research	Bombay, India
United States Naval Observatory	Washington, D. C., USA
University of Alabama	University, Alabama, USA
University of California	Berkeley, California, USA
University of Florida Observatories	Bronson, Florida, USA

University of Florida, Radio Observatory	Old Town, Florida, USA
University of Hawaii	Honolulu, Hawaii, USA
University of Illinois Observatory	Urbana, Illinois, USA
University of Kansas Observatory	Lawrence, Kansas, USA
University of Maryland	College Park, Maryland, USA
University of Michigan Observatories	Ann Arbor, Michigan, USA
University of Minnesota	Minneapolis, Minnesota, USA
University of South Florida Observatory	Tampa, Florida, USA
University of Texas, Department of Astronomy	Austin, Texas, USA
University of Washington, Astronomy Department	Seattle, Washington, USA
Uttar Pradesh State Observatory	Nainital, India
Van Vleck Observatory	Middletown, Connecticut, USA
Vatican Observatory	Castel Gandolfo, Vatican
Venezuelan Astronomical Institute	Merida, Venezuela
Wallace Astrophysical Observatory	Cambridge, Massachusetts, USA
Warner and Swasey Observatory	Cleveland, Ohio, USA
Washburn Observatory University of Wisconsin	Madison, Wisconsin, USA
West Melton Observatory	Christchurch, New Zealand
Wilhelm-Foerster Sternwarte	Berlin, F. R. Germany
Yale University Observatory	New Haven, Connecticut, USA
Yerkes Observatory	Williams Bay, Wisconsin, USA
Zentralinstitut für Astrophysik, Sternwarte Babelsberg	Potsdam-Babelsberg, German Democratic Republic
Zentralinstitut für Astrophysik, Sternwarte in Sonneberg	Sonneberg, German Democratic Republic
Zentralinstitut für solar-terrestrische Physik	Berlin-Adlershof, German Democratic Republic

008.001 Ames, Iowa

Iowa State University, Erwin W. Fick Observatory, Ames, Iowa 50011. − Report.
J. L. Russell, W. I. Beavers.
Bull. American Astron. Soc., Vol. 14, 229 - 231 (1982).

008.002 Amherst, Mass.

Five College Astronomy Department: Amherst College, Amherst, Massachusetts 01002; Hampshire College, Amherst, Massachusetts 01002; Mount Holyoke College, South Hadley, Massachusetts 01075; Smith College, Northampton, Massachusetts 01060; University of Massachusetts, Amherst, Massachusetts 01003. − Report.
Bull. American Astron. Soc., Vol. 14, 166 - 172 (1982).

008.003 Ann Arbor, Mich.

University of Michigan, Department of Astronomy, Ann Arbor, Michigan 48109. − Report.
Bull. American Astron. Soc., Vol. 14, 301 - 305 (1982).

008.004 Armagh

Armagh Observatory in 1979/80. Annual report covering the period 1 April, 1979 - 31 March, 1980.
M. de Groot.
Irish Astron. J., Vol. 14, 247 - 252 (1980).

Armagh Observatory Contribution, No. 105 (29.122.013).

008.005   Atlanta, Ga.

Georgia State University, Department of Physics and Astronomy, Atlanta, Georgia 30303. – Report for the period 1 October 1980 - 30 September 1981. H. A. McAlister. Bull. American Astron. Soc., Vol. 14, 173 - 176 (1982).

Fernbank Science Center, Atlanta, Georgia 30307. Report.   R. M. Williamon. Bull. American Astron. Soc., Vol. 14, 165 (1982).

008.006   Austin, Tex.

University of Texas at Austin, Department of Astronomy, Austin, Texas 78712; McDonald Observatory, Fort Davis, Texas 79734. – Report for the period 1 September 1980 - 31 August 1981. D. S. Evans, H. J. Smith, P. Vanden Bout. Bull. American Astron. Soc., Vol. 14, 494 - 526 (1982).

Department of Astronomy and McDonald Observatory of the University of Texas, Austin, Texas, Reprints Nos. 951 (29.094.024), 952 (29.131.076), 953 (29.032.012), 954 (29.064.104), 955 (27.008.008), 956 (29.131.083), 957 (29.131.084), 958 (29.122.048), 959 (29.131.070), 960 (30.158.034), 961 (30.125.301), 962 (29.131.107), 963 (29.122.061), 964 (29.141.187), 965 (29.154.018), 966 (29.008.008), 967 (29.131.086), 968 (29.131.228), 969 (29.131.229), 970 (29.065.086), 971 (29.131.068), 972 (30.160.031), 973 (29.122.170), 974, 975 (29.118.031), 976 (29.141.117), 977 (27.132.002), 978 (30.002.015), 979 (30.117.050), 980 (30.158.010), 981 (30.099.001), 982 (30.120.001), 983 (30.158.030), 984 (30.135.010), 985 (30.158.048), 986 (30.118.005), 987 (30.122.011), 988 (30.099.016), 989 (30.158.047), 990 (30.135.011), 991 (30.114.016), 992 (30.155.005), 993 (30.114.009), 994 (30.114.021), 995 (30.099.038), 996 (30.158.155), 997 (30.124.201), 998 (30.131.028), 999 (30.096.009), 1000 (30.131.147).

008.007   Basel

Preprint of the Astronomical Institute of the University of Basel, No. 3 (31.162.153).

008.008   Baton Rouge, La.

Louisiana State University Observatory, Baton Rouge, Louisiana 70803. – Report.   A. U. Landolt. Bull. American Astron. Soc., Vol. 14, 256 - 258 (1982).

008.009   Berkeley, Calif.

University of California, Berkeley, Los Angeles, San Diego, Lick Observatory, and Board of Studies. I. Berkeley Campus. – Report. Bull. American Astron. Soc., Vol. 14, 30 - 49 (1982).

008.010   Berlin

Heinrich-Hertz-Institut, Solare Beobachtungsergebnisse. Akademie der Wissenschaften der DDR, Zentralinstitut für Solar-Terrestrische Physik (Heinrich-Hertz-Institut), Berlin-Adlershof, HHI Solar Data, Vol. 32, 1981 September - December (31.072.061).

Mitteilungen der Archenhold-Sternwarte Berlin-Treptow, Nos. 119 (31.005.017), 120 (29.004.056), 125 (31.005.018), 134 (31.005.019).

Veröffentlichungen der Archenhold-Sternwarte, Berlin-Treptow, No. 10 (31.002.048).

Archenhold-Sternwarte, Berlin-Treptow, Vorträge und Schriften. Nos. 60 (31.004.042), 61 (31.004.043).

Blick in die Sternenwelt 1982 (31.047.020).

008.011   Bloemfontein

University of the Orange Free State: Boyden Observatory, Department of Astronomy.   A. H. Jarrett. Mon. Notes Astron. Soc. South. Africa, Vol. 41, 6 - 7 (1982).

008.012   Bloomington, Ind.

Goethe Link Observatory, Indiana University, Bloomington, Indiana 47405. – Report. Bull. American Astron. Soc., Vol. 14, 234 - 238 (1982).

008.013   Bonn

Veröffentlichungen der Astronomischen Institute Bonn, Nr. 95 (31.112.038).

008.014   Bordeaux

Publications de l'Observatoire de l'Université de Bordeaux (Floirac), Nouvelle Serie, Nos. 71 (20.061.006), 72 (21.100.501), 73 (22.098.006), 74 (25.133.003), 75 (21.106.001), 76 (21.106.010), 77 (25.031.601), 78 (27.031.612), 79, 80 (27.031.621), 81 (29.098.006), 82 (27.131.205), 83 (27.131.132), 84 (28.033.026), 85 (29.063.035), 86 (30.002.047), 87 (27.106.014), 88 (28.031.543), 89, 90 (30.106.056), 91 (29.033.044), 92 (30.106.028).

008.015   Borowiec

Circular, Time and Latitude Service, Nos. 158 - 159 (31.044.020).

008.016   Boulder, Colo.

High Altitude Observatory, National Center for Atmospheric Research, Boulder Colorado 80307. – Report for 1980. Bull. American Astron. Soc., Vol. 14, 195 - 205 (1982).

University of Colorado, Boulder, Colorado 80309.
Report for the period 1 October 1979 - 1 October 1981.
G. A. Dulk, J. M. Shull, J. Toomre.
Bull. American Astron. Soc., Vol. 14, 122 - 147 (1982).

008.017   Budapest

Communications from the Konkoly Observatory of the Hungarian Academy of Sciences, No. 75 (31.122.139).

008.018   Byurakan

Soobshcheniya Byurakanskoj Observatorii, Akademiya Nauk Armyanskoj SSR, Vyp. 53 (31.113.036; 31.116.021; 31.123.004; 31.114.111; 31.122.090; 31.121.023; 31.158.197; 31.160.045; 31.141.210; 31.082.043; 31.082.044; 31.098.041; 31.031.567; 31.034.042).

Byurakan Astrophysical Observatory, Armenia, USSR, Reprints, Nos. 271 (31.158.274), 272 (31.158.275), 273 (31.113.076), 274 (31.063.050), 275 (29.158.022), 276 (29.063.015), 277 (29.141.033), 278 (29.158.024; 29.134.006; 29.063.016), 279 (29.158.025), 280 (29.158.026), 281 (29.158.027), 282 (29.158.028), 283 (29.158.030), 284 (29.158.031), 285 (29.113.004), 286 (29.122.010), 287 (29.156.001), 288 (31.158.125), 289 (31.122.060), 291 (31.113.077), 292 (31.122.172), 293 (31.114.206), 294 (31.158.278), 296 (31.160.073), 297 (31.114.208), 298 (31.158.129), 299 (31.121.018), 300 (31.122.063; 31.114.065; 31.114.067), 301 (31.158.131), 302 (31.141.083), 303 (31.160.028), 304 (31.118.008).

008.019   Calar Alto

Max-Planck–Institut für Astronomy, D-6900 Heidelberg-Königstuhl, Federal Republic of Germany; Centro Astronomico Hispano-Aleman, Almeria, Spain. – Report for the period 1 January - 31 December 1980.
G. Münch.
Bull. American Astron. Soc., Vol. 14, 279 - 287 (1982).

008.020   Calgary

Rothney Astrophysical Observatory, The University of Calgary, Calgary, Alberta T2N 1N4, Canada. – Report.
E. F. Milone, T. A. Clark.
Bull. American Astron. Soc., Vol. 14, 467 - 472 (1982).

008.021   Cambridge, Mass.

Center for Astrophysics, Harvard College Observatory and Smithsonian Astrophysical Observatory, Cambridge, Massachusetts 02138. – Report.   G. B. Field.
Bull. American Astron. Soc., Vol. 14, 82 - 108 (1982).

Smithsonian Astrophysical Observatory, Special Report, No. 392 (31.012.045).

IAU Circulars, Nos. 3656 - 3707 (1982).

Minor Planet Circulars, (M. P. C.), Nos. 6573 - 6988 (1982).

008.022   Cape Town

South African Astronomical Observatory.
Report for year ending 1980 December 31.   M. W. Feast.
Q. J. R. Astron. Soc., Vol. 23, 134 - 150 (1982).

South African Astronomical Observatory. Report for the year ending 31 December 1981.   M. W. Feast.
Council for Scientific and Industrial Research (South Africa), Science and Engineering Research Council (United Kingdom), 54 pp. (1982). ISBN 0-7988-2404-2.

University of Cape Town: Department of Astronomy.   B. Warner.
Mon. Notes Astron. Soc. South. Africa, Vol. 41, 2 - 5 (1982).

Publications of the Department of Astronomy of the University of Cape Town, No. 4 (31.002.049).

University of Cape Town: Department of Applied Mathematics.   G. F. R. Ellis.
Mon. Notes Astron. Soc. South. Africa, Vol. 41, 5 (1982).

008.023   Carloforte-Cagliari

Rapporto annuale per il 1979 e 1980.
E. Proverbio.
Pubbl. Stn. Astron. Int. Latitudine, Carloforte-Cagliari, N. 88, 24 pp.

Circolari della Stazione Astronomica Internazionale di Latitudine, Carloforte-Cagliari, Nos. 15 (Ser. A (6)) (31.044.021), 16 (Ser. A (7)) (31.044.022), 17 (Ser. C (1)) (31.085.019).

Stazione Astronomica Internazionale di Latitudine, Carloforte-Cagliari, Note tecniche, N. 2 (31.021.035).

Pubblicazioni della Stazione Astronomica Internazionale di Latitudine, Carloforte-Cagliari, Nuova Serie, Nos. 57 (30.081.021), 59 (27.009.015), 68 (27.032.048), 71 (31.045.020), 75 (31.080.069), 76 (31.085.020), 79 (31.044.025), 80 (29.044.046), 81 (30.044.007), 82, 84 (31.045.021), 85 (30.002.054), 86 (31.015.029), 88 (31.008.023).

008.024   Carmel Valley, Calif.

Monterey Institute for Research in Astronomy, Carmel Valley, California 93924. – Report.
W. B. Weaver.
Bull. American Astron. Soc., Vol. 14, 320 - 321 (1982).

008.025   Castel Gandolfo

Annual report 1981.   G. V. Coyne.
Vatican Obs. Publ., Vol. 1, No. 20, p. 367 - 372 (1982).

Vatican Observatory Publications, Specola Vaticana, Città del Vaticano, Vol. 1, No. 20 (31.008.025).

008.026   Chapel Hill, N.C.

University of North Carolina, Chapel Hill, North Carolina 27514. – Report for the period 1 October 1979 - 30 September 1981.   B. W. Carney.
Bull. American Astron. Soc., Vol. 14, 427 - 429 (1982).

008.027   Charlottesville, Va.

Leander McCormick Observatory, University of Virginia, Charlottesville, Virginia 22903. – Report.
R. W. O'Connell.
Bull. American Astron. Soc., Vol. 14, 288 - 294 (1982).

National Radio Astronomy Observatory: Charlottesville, Virginia 22901; Green Bank, West Virginia 24944; Socorro, New Mexico 87801; Tucson, Arizona 85726. Report for the period July 1980 - June 1981.
M. S. Roberts, R. J. Havlen.
Bull. American Astron. Soc., Vol. 14, 370 - 410 (1982).

008.028   Chicago, Ill.

University of Chicago, Department of Astronomy and Astrophysics, Chicago, Illinois 60637. – Report.
D. N. Schramm, L. M. Hobbs.
Bull. American Astron. Soc., Vol. 14, 109 - 121 (1982).

008.029   Cleveland, Ohio

Warner and Swasey Observatory, Case Western Reserve University, Cleveland, Ohio 44106. – Report for the period 1 July 1980 - 30 June 1981.   P. Pesch.
Bull. American Astron. Soc., Vol. 14, 546 - 548 (1982).

008.030   College Park, Md.

University of Maryland, College Park, Maryland 20742. – Report for the period 1 September 1980 - 30 September 1981.
Bull. American Astron. Soc., Vol. 14, 264 - 278 (1982).

008.031   Columbus, Ohio

The Observatories of the Ohio State and Ohio Wesleyan Universities, Columbus, Ohio 43210 and Delaware, Ohio 43015. – Report.   E. R. Capriotti.
Bull. American Astron. Soc., Vol. 14, 430 - 438 (1982).

008.032   Copenhagen

Copenhagen University Observatory Reprint Nos. 409 (29.122.003), 410 (29.022.160), 411 (29.122.070), 412 (29.158.283), 413 (29.131.224), 414 (29.131.081), 415 (29.131.148), 416 (29.113.053), 417 (28.152.001), 418 (30.031.505), 419 (30.119.004), 420 (30.123.011), 421 (29.041.007), 422 (30.041.002), 423 (30.031.030), 424 (30.114.080), 425 (30.114.181), 426 (30.031.031), 427 (30.131.196), 428 (30.114.158), 429 (30.031.033),

430, 431, 432 (31.032.004), 433 (31.122.114), 434 (31.131.067), 435 (31.071.009), 436 (31.021.037).

008.033   Córdoba

Observatorio Astronómico de Córdoba. – Report.
Bol. Asoc. Argentina Astron., No. 18, p. 105 - 115 (1980).

008.034   Crimea

Chronicle.
Izv. Krymskoj Astrofiz. Obs., Tom 64, 201 - 202 (1981). In Russian. English translation in Bull. Crimean Astrophys. Obs., Vol. 64.

Izvestiya Ordena Trudovogo Krasnogo Znameni Krymskoj Astrofizicheskoj Observatorii, Akademiya Nauk SSSR, Tom 64 (31.114.075, 31.114.076, 31.114.077, 31.114.078, 31.113.021, 31.116.018, 31.124.601, 31.114.079, 31.114.080, 31.141.095, 31.158.157, 31.141.096, 31.072.024, 31.072.025, 31.074.018, 31.085.012, 31.077.013, 31.077.014, 31.034.024, 31.034.025, 31.034.026, 31.031.013, 31.031.014, 31.008.034).

008.035   Debrecen

Publications of Debrecen Heliophysical Observatory, Vol. 4, No. 2 (31.072.074).

008.036   Delaware, Ohio

The Observatories of the Ohio State and Ohio Wesleyan Universities, Columbus, Ohio 43210 and Delaware, Ohio 43015. Report.   E. R. Capriotti.
Bull. American Astron. Soc., Vol. 14, 430 - 438 (1982).

008.037   Dresden

Technische Universität Dresden, Lohrmann-Observatorium, Zirkular, Nos. 96 - 98 (31.045.022).

008.038   Dublin

Contributions from the Dunsink Observatory, Nos. 15 (26.113.034), 16 (21.122.042), 17 (27.098.134), 18 (30.103.801).

Dunsink Observatory Reprints Nos. 100 (25.009.014), 101 (21.098.065), 102 (26.031.604), 103 (25.098.062), 104 (28.004.001), 105 (25.102.025), 106 (29.013.028), 107 (30.013.022), 108 (30.032.024).

008.039   East Lansing, Mich.

Michigan State University, Department of Astronomy and Astrophysics and the Observatory, East Lansing, Michigan 48824. – Report.   T. R. Stoeckley.
Bull. American Astron. Soc., Vol. 14, 306 - 310 (1982).

008.040 Edinburgh

Occasional Reports of the Royal Observatory,
Edinburgh, Nos. 6 (31.114.194), 7 (31.114.195).

Catalogue of the archives of the Royal Observatory,
Edinburgh, 1764 - 1937.   See Abstr. 002.055.

008.041 El Segundo

The Aerospace Corporation, El Segundo,
California 90245. – Report.   C. J. Rice.
Bull. American Astron. Soc., Vol. 14, 1 - 5 (1982).

008.042 Epping

Anglo-Australian Observatory. – Report for the
year ending 1981 June 30.   D. C. Morton.
Q. J. R. Astron. Soc., Vol. 23, 260 - 271 (1982).

Anglo-Australian Observatory, Preprint Nos. 162
(31.132.070), 163 (31.132.071), 164 (31.158.262), 165
(31.141.271), 167 (31.158.263), 168 (31.131.222).

A. A. O. Newsletter, Nos. 20 - 21 (1982).

008.043 Evanston, Ill.

Lindheimer Astronomical Research Center and
Dearborn Observatory, Northwestern University, Evanston,
Illinois 60201. – Report.
Bull. American Astron. Soc., Vol. 14, 232 - 233 (1982).

008.044 Flagstaff, Ariz.

Lowell Observatory, Flagstaff, Arizona 86002.
Report for the period from 1 July 1980 - 30 June 1981.
A. A. Hoag, H. S. Horstman.
Bull. American Astron. Soc., Vol. 14, 259 - 263 (1982).

008.045 Fort Davis, Tex.

University of Texas at Austin, Department of
Astronomy, Austin, Texas 78712; McDonald Observatory,
Fort Davis, Texas 79734. – Report for the period 1September
1980 - 31 August 1981.
D. S. Evans, H. J. Smith, P. Vanden Bout.
Bull. American Astron. Soc., Vol. 14, 494 - 526 (1982).

008.046 Geneva

Publications de L'Observatoire de Genève, Série A,
Fasc. 85 (31.015.030).

008.047 Glasgow

Department of Astronomy, University of Glasgow.
Report for the period 1980 October to 1981 September.

P. A. Sweet.
Q. J. R. Astron. Soc., Vol. 23, 272 - 274 (1982).

008.048 Green Bank, W. Va.

National Radio Astronomy Observatory:
Charlottesville, Virginia 22901; Green Bank, West Virginia
24944; Socorro, New Mexico 87801; Tucson, Arizona 85726.
Report for the period July 1980 - June 1981.
M. S. Roberts, R. J. Havlen.
Bull. American Astron. Soc., Vol. 14, 370 - 410 (1982).

National Radio Astronomy Observatory, Green
Bank, Reprints, Series A, Nos. 1322 (30.158.125),
1323 (30.141.075), 1324 (30.141.076), 1325 (30.158.126),
1326 (30.131.083), 1327 (30.141.083), 1328 (30.141.084),
1329 (30.131.109), 1330 (29.112.017), 1331 (30.131.030),
1332 (30.141.102), 1333 (30.132.037), 1334 (30.131.120),
1335 (30.155.036), 1336 (30.141.105), 1337 (30.131.128),
1338 (30.002.064), 1339 (30.151.052), 1340 (30.156.009),
1341 (30.131.149), 1342 (30.132.045), 1343 (30.031.580),
1344 (30.131.150), 1345 (30.131.113), 1346 (30.131.114),
1347 (31.033.033), 1348 (30.135.049), 1349 (30.158.167),
1350 (30.141.160), 1351 (30.158.173), 1352 (30.158.203),
1353 (30.158.206), 1354 (30.064.079), 1355 (30.158.207),
1356 (30.141.137), 1357 (30.132.053), 1358 (30.160.055),
1359 (30.131.166), 1360 (30.041.014), 1361 (30.131.129),
1362 (30.131.179), 1363 (30.073.088), 1364 (30.141.153),
1365 (30.141.154), 1366 (30.131.183), 1367 (30.160.054),
1368 (30.158.238), 1369 (30.112.050), 1370 (31.033.034),
1371 (28.141.166), 1372 (31.033.035), 1373 (31.131.020).

National Radio Astronomy Observatory, Green
Bank, Reprints, Series B, Nos. 525 (30.158.011),
526 (31.131.103), 527 (31.141.127), 528 (31.141.199),
529 (31.141.144), 530 (31.131.118).

008.049 Greenwich

The Royal Greenwich Observatory. – Report for
the period 1979 October 1 to 1980 September 30.
F. G. Smith.
Q. J. R. Astron. Soc., Vol. 23, 151 - 161 (1982).

Royal Observatory Annals, Royal Greenwich
Observatory, Herstmonceux, No. 14 (31.002.054).

Royal Greenwich Observatory Bulletins, Nos. 186
(31.096.013), 188 (31.113.072).

Greenwich Time Report. Royal Greenwich Observa-
tory. Time and Latitude Service, 1981 April - September
(31.044.026).

008.050 Hamburg

Abhandlungen aus der Hamburger Sternwarte.
Band 10, Heft 3 (31.041.042; 31.041.043; 31.041.044;
31.041.045; 31.041.046; 31.002.070; 31.043.020; 31.013.
060; 31.041.047; 31.002.071; 31.141.307; 31.041.048;
31.158.103; 31.116.058; 31.116.059; 31.041.049; 31.141.
308; 31.041.050; 31.041.051; 31.041.052; 31.002.072;
31.002.073; 31.041.053; 31.041.054; 31.141.309; 31.002.
074; 31.041.055; 31.141.310; 31.141.311; 31.141.312).

Deutsches Hydrographisches Institut, Hamburg.
Zeit- und Breitendienst, 1981 July - December (31.044.027).

008.051 Hanscom, Mass.

Sagamore Hill Radio Observatory, Air Force
Geophysics Laboratory, Hanscom Air Force Base,
Massachusetts 01731. – Report.
E. W. Cliver, E. J. Eadon.
Bull. American Astron. Soc., Vol. 14, 486 - 487 (1982).

008.052 Heidelberg

Astronomisches Rechen-Institut, Heidelberg,
Mitteilungen Serie A, Nos. 135 (29.043.008), 136 (29.041.
077), 137 (29.041.075), 138 (31.041.004), 139 (31.098.074).

Astronomisches Rechen-Institut, Heidelberg,
Mitteilungen Serie B, Nos. 103 (30.041.017), 104 (30.141.
121), 105 (30.111.902), 106 (31.091.006), 107 (31.041.
002), 108 (31.111.004).

Astronomy and Astrophysics Abstracts. Vol. 30.
Literature 1981, Part 2 (31.002.099).

Max-Planck-Institut für Astronomie, D-6900 Heidel-
berg-Königstuhl, Federal Republic of Germany; Centro
Astronomico Hispano-Aleman, Almeria, Spain. – Report for
the period 1 January - 31 December 1980.
G. Münch.
Bull. American Astron. Soc., Vol. 14, 279 - 287 (1982).

008.053 Holmdel, N. J.

Bell Laboratories, Holmdel, New Jersey 07733.
Report.   R. W. Wilson.
Bull. American Astron. Soc., Vol. 14, 21 - 25 (1982).

008.054 Honolulu, Hawaii

University of Hawaii, Institute for Astronomy,
Honolulu, Hawaii 96822. – Report.
J. T. Jefferies.
Bull. American Astron. Soc., Vol. 14, 177 - 188 (1982).

The Infrared Telescope Facility, Honolulu, Hawaii
96822. – Report for the period 1 July 1980 - 30 June 1981.
Bull. American Astron. Soc., Vol. 14, 223 - 228 (1982).

008.055 Houston, Tex.

William Marsh Rice University, Department of Space
Physics and Astronomy, Houston, Texas 77001. – Report.
A. J. Dessler.
Bull. American Astron. Soc., Vol. 14, 458 - 460 (1982).

008.056 Istanbul

Publications of the Istanbul University Observatory,
Nos. 102 (31.022.159), 112 (31.114.196), 113 (31.114.197),
114 (31.072.075), 115 (31.113.073), 116 (31.113.074),
117 (31.114.198), 118 (31.153.038), 119 (31.072.076),
120 (31.072.077).

008.057 Ithaca, N.Y.

National Astronomy and Ionosphere Center, Cornell
University, Ithaca, Astronomy Publications, Nos. B81-1
(29.131.169), B81-2 (29.158.277), B81-3 (30.158.125),
B81-4 (30.158.206), B81-5 (30.122.122), B81-6 (30.131.121),
B81-7 (30.141.160), B81-8 (30.131.113), B81-9, B81-10
(30.031.592), B81-11 (30.158.238), B81-12 (30.132.042),
B81-13 (30.160.001), B81-14 (30.141.529), B81-15
(30.141.530), B81-16 (29.141.572), B81-17 (30.160.030),
B81-18 (30.160.044), B81-19 (30.125.048), B81-20
(30.141.004), B81-21 (26.158.048), B81-22 (30.159.017),
B81-23 (30.097.006), B81-24 (30.141.075), B81-25
(30.141.535), B81-26 (30.158.060), B81-27 (30.158.034),
B81-28 (30.141.515), B81-29 (30.141.527), B81-30
(30.158.062), B81-31 (30.141.041).

008.058 Jena

Mitteilungen der Universitäts-Sternwarte zu Jena,
Nos. 139 (26.160.058), 141 (28.034.025), 142 (29.021.037),
143 (27.123.014), 144 (31.036.009), 145 (28.160.067),
146 (28.022.149), 147 (29.158.237).

008.059 Kiel

Sonderdrucke der Sternwarte Kiel, Nos. 298,
299 (29.064.005), 300 (29.114.053), 301 (29.126.006),
302 (29.122.042), 303 (29.125.039), 304 (29.114.170),
305 (29.114.171), 306 (29.126.032), 307 (29.114.144),
308 (30.114.020), 309 (30.064.049), 310 (30.126.023),
311 (30.065.067), 312 (30.071.026), 313 (30.114.174),
314 (30.064.034), 315 (30.064.007), 316 (30.064.087),
317 (29.114.068), 318 (29.115.023), 319 (30.122.049),
320 (30.114.093), 321 (30.126.002).

008.060 Kiev

Astrometriya i Astrofizika, Akademiya Nauk
Ukrainskoj SSR, Glavnaya Astronomicheskaya Observatoriya,
Vyp. (Nos.) 43, 44, 45 (1981); 46 (1982).

008.061 La Plata

Observatorio Astronómico de La Plata. – Report.
J. Sahade.
Bol. Asoc. Argentina Astron., No. 18, p. 115 - 127 (1980).

008.062 **La Silla**

**European Southern Observatory. Annual report
1981.**   L. Woltjer.
Published by European South. Obs., Garching, F. R. Germany.
62 pp. (1982). ISSN 0531-4496.

ESO Scientific Preprint, Nos. 181 (31.160.067),
182 (31.132.072), 183 (31.160.068), 184 (31.142.094),
185 (31.141.272), 186 (31.142.095), 187 (31.158.264),
188 (31.002.056), 189 (31.122.165), 190 (31.158.265),
191 (31.160.069), 192 (31.098.076), 193 (31.114.199),
194 (31.101.010), 195 (31.135.039), 196 (31.158.266),
197 (31.125.046), 198 (31.141.273), 199 (31.122.166),
200 (31.122.167).

The ESO Scientific and Technical Committee.
P. Léna.
Messenger, No. 27, p. 1 - 2 (1982).

The ESO Observing Programmes Committee.
B. E. Westerlund.
Messenger, No. 28, p. 1 - 3 (1982).

The Messenger – El Mensajero, Nos. 27 - 28 (1982).

008.063 **Las Cruces, N.M.**

**New Mexico State University, Observatory and
Department of Astronomy, Las Cruces, New Mexico 88003-
4500.** – Report.   K. S. Anderson.
Bull. American Astron. Soc., Vol. 14, 416 - 418 (1982).

008.064 **Lembang**

**Contributions from the Bosscha Observatory
Lembang,** Nos. 68 (31.155.042), 69 (27.117.023),
70 (31.141.546), 71 (31.114.201), 72 (31.117.115),
73 (31.031.027), 74 (31.114.202), 75 (31.114.203).

008.065 **Leningrad**

Chronicle.
Byull. Inst. Teor. Astron., Leningrad, Tom 15, 132 - 134 (1981);
192 - 194 (1982). In Russian.

Byulleten' Instituta Teoreticheskoj Astronomii,
Akademiya Nauk SSSR, Tom 15, Nos. 1 (31.042.026;
31.042.027; 31.031.537; 31.031.538; 31.093.023; 31.042.028;
31.042.029; 31.052.010; 31.052.011; 31.031.539), 2
(31.047.006; 31.052.012; 31.042.030; 31.021.010;
31.099.046; 031.042.031; 31.052.013; 31.031.540;
31.008.065), 3 (31.042.032; 31.042.033; 31.091.031;
31.042.034; 31.097.084; 31.042.035; 31.042.036; 31.047.007;
31.008.065).

008.066 **Lintong**

Time and Frequency Services Bulletin, 1982,
Nos. 1 - 3 (31.044.033).

008.067 **Livermore, Calif.**

**Lawrence Livermore National Laboratory,
University of California, Livermore, California 94550.**
Report.   C. B. Tarter.
Bull. American Astron. Soc., Vol. 14, 239 - 242 (1982).

008.068 **London, Canada**

**The Observatories of the University of Western
Ontario, London, Ontario N6A 3K7, Canada.** – Report.
W. Wehlau.
Bull. American Astron. Soc., Vol. 14, 559 - 561 (1982).

008.069 **London, U.K.**

**Mullard Space Science Laboratory, University
College London.** – Report for the period 1980 October 1 to
1981 September 30.   R. L. F. Boyd.
Q. J. R. Astron. Soc., Vol. 23, 275 - 290 (1982).

008.070 **Los Alamos, N.M.**

**Los Alamos National Laboratory, University of
California, Los Alamos, New Mexico 87545.** – Report.
J. G. Hills.
Bull. American Astron. Soc., Vol. 14, 246 - 255 (1982).

008.071 **Los Angeles, Calif.**

**University of California, Berkeley, Los Angeles, San
Diego, Lick Observatory, and Board of Studies. II. Los
Angeles Campus.** – Report.
Bull. American Astron. Soc., Vol. 14, 49 - 56 (1982).

**University of Southern California, Department of
Astronomy, Los Angeles, California 90007.** – Report.
Bull. American Astron. Soc., Vol. 14, 490 - 491 (1982).

008.072 **Lund**

Reports from the Observatory of Lund, No. 18
(31.012.044).

008.073 **Lyon**

Observatoire de Lyon, Preprint No. 1 (31.123.021).

008.074 **Madison, Wis.**

**Washburn Observatory, University of Wisconsin at
Madison, Madison, Wisconsin 53706.** – Report.
L. R. Doherty.
Bull. American Astron. Soc., Vol. 14, 549 - 552 (1982).

008.075 Manchester, U.K.

Nuffield Radio Astronomy Laboratories, Jodrell Bank, University of Manchester. − Report for the year ending 1981 September 30.   B. Lovell.
Q. J. R. Astron. Soc., Vol. 23, 291 - 309 (1982).

008.076 Manila

Manila Observatory. Solar Division. Solar maps and activity. 1981 September - 1982 February (31.072.078).

008.077 Middletown, Conn.

Van Vleck Observatory, Wesleyan University, Middletown, Connecticut 06457. − Report.
A. R. Upgren.
Bull. American Astron. Soc., Vol. 14, 536 - 539 (1982).

008.078 Minneapolis, Minn.

University of Minnesota, Department of Astronomy, School of Physics and Astronomy, Minneapolis, Minnesota 55455. − Report.   T. W. Jones.
Bull. American Astron. Soc., Vol. 14, 311 - 319 (1982).

008.079 Mizusawa

Monthly Notes of the International Polar Motion Service, 1981 No. 12 - 1982 No. 4 (31.045.023).

Bulletins, Time Service of the Mizusawa Observatory, Vol. 25 (31.044.028).

Annual report of Geophysical Observations, Esashi Station, Japan 1979 (31.081.048).

008.080 Munich

Max-Planck-Institut für Physik und Astrophysik, Institut für Extraterrestrische Physik, Garching bei München, MPI-PAE/Extraterr. 173 (31.034.059).

008.081 Naini Tal

Uttar Pradesh State Observatory, Naini Tal. − Report for the period 1980 January 1 - 1980 December 31.
Bull. Astron. Soc. India, Vol. 9, 319 - 322 (1981).

Twenty-five years of Uttar Pradesh State Observatory, Naini Tal, India.   See Abstr. 009.017.

Uttar Pradesh State Observatory, Naini Tal, Reprints, Nos. 99 (14.121.094), 100 (22.071.017), 102 (22.064.059), 103 (22.114.044), 104 (14.121.077), 105 (14.121.076), 106 (17.122.042), 107 (17.153.017), 108 (17.121.028), 109 (17.121.029), 110 (18.071.005), 111 (18.122.125), 112 (18.122.137), 113 (31.004.062), 114 (18.071.101), 115 (18.071.102), 116 (18.073.026), 117 (18.114.137), 118 (18.072.022), 119 (18.122.093),

120 (18.122.109), 121 (19.122.145), 123 (19.072.084), 124 (20.113.020), 125 (20.072.019), 126 (20.071.009), 127 (20.115.014), 128 (20.122.117), 129 (20.153.027), 130 (20.124.353), 131 (31.085.024), 132 (21.153.031), 133 (21.072.006), 134 (21.103.304), 135 (21.121.027), 136 (21.114.540), 137 (22.153.008), 138 (22.072.034), 139 (22.123.029), 140 (22.153.001), 141 (22.114.047), 142 (22.072.028), 143 (22.072.042), 144 (22.122.108), 145 (25.120.033), 146 (25.123.048), 147 (26.122.052), 148 (26.071.012), 150 (25.072.056), 151 (26.124.019), 152 (26.071.019), 153 (26.114.067), 154 (26.117.002), 155 (26.122.093), 156 (26.103.101), 157 (26.072.011), 158 (26.022.046), 159 (26.123.035), 160 (25.072.055), 161 (25.072.057), 162 (25.119.129), 163 (26.151.003), 164 (26.153.022), 165 (26.119.048), 166 (27.117.015), 167 (27.119.017), 168 (27.122.068), 170 (26.072.053), 171 (27.120.034), 172 (27.119.091), 173 (27.122.195), 174 (27.122.029), 175 (27.122.030), 176 (27.122.033), 177 (27.122.220), 178 (28.071.002), 179, 180 (28.122.131), 181, 182 (28.122.027), 183 (29.079.108), 184 (28.122.122), 185 (31.032.041), 186 (30.074.061), 187 (30.073.065), 188 (28.114.137), 189 (28.117.072), 190 (29.117.081), 191 (29.122.057), 192 (29.119.035), 193 (30.119.013), 194 (29.119.027), 195 (29.122.033), 196 (29.153.008), 197 (30.072.051), 198 (30.122.164), 199 (30.114.055), 200 (30.022.151).

008.082 Nanking

Nanjing University Observatory, Papers Nos. 81-101 (29.141.522), 81-102 (29.066.033), 81-103 (30.141. 510), 81-104 (30.061.004), 81-105 (30.079.601), 81-106 (30.142.536), 81-107 (30.044.014), 81-108 (31.142.511), 81-109 (30.066.519), 81-110, 81-111 (30.031.539), 81-112, 81-113, 81-114 (29.158.105), 81-115 (29.062.051), 81-116 (29.065.035), 81-117 (30.063.045), 81-118 (30.062.093), 81-119, 81-120, 81-121, 81-122, 81.123, 81-124 (29.064. 035), 81-125, 81-126, 81-127, 81-128, 81-129, 81-130, 81-131, 81-132, 81-133, 81-134; 81-201 (30.052.003), 81-202 (30.042.018), 81-203 (31.042.044), 81-204 (31.042. 045), 81-205 (31.098.028), 81-206, 81-301 (29.043.003), 81-302 (31.041.024), 81-303 (31.045.012), 81-304 (31.081. 044), 81-305.

008.083 Nashville, Tenn.

Dyer Observatory, Vanderbilt University, Nashville, Tennessee 37235. − Report for the period 1 October 1980 - 30 September 1981.   A. M. Heiser.
Bull. American Astron. Soc., Vol. 13, 162 - 164 (1982).

008.084 Neuchâtel

Observatoire Cantonal de Neuchâtel. Rapport d'activité pour l'exercice 1981.
J. Bonanomi, G. Fischer.
Published by Observatoire Cantonal de Neuchâtel. 19 pp. (1982).

008.085 New Haven, Conn.

Yale University Observatory, New Haven, Connecticut 06511. − Report.   R. B. Larson.
Bull. American Astron. Soc., Vol. 14, 562 - 570 (1982).

008.086   New York, N.Y.

Columbia University in the City of New York, Department of Astronomy and Department of Physics, New York, New York 10027. – Report.   N. H. Baker.
Bull. American Astron. Soc., Vol. 14, 148 - 152 (1982).

008.087   Palo Alto, Calif.

Lockheed Palo Alto Research Laboratory, Palo Alto, California 94304. – Report.
Bull. American Astron. Soc., Vol. 14, 243 - 245 (1982).

008.088   Paris

Pré-publication Institut d'Astrophysique de Paris, Nos. 1 (31.125.047), 2 (31.155.043), 3 (31.063.049), 4 (31.073.118), 5 (31.153.039), 6 (31.077.048).

Bureau International de l'Heure. Annual report for 1981 (31.044.030).

Bureau International de l'Heure, (B.I.H.), Circular D182 - D187 (31.044.031), E11 (31.044.032).

008.089   Pasadena, Calif.

Annual report of the Director. The Mount Wilson and Las Campanas Observatories 1980 - 1981.
G. W. Preston.
Reprinted from Carnegie Institution of Washington Year Book 80, p. 583 - 650 (1981).

008.090   Peking

Publications of the Beijing Astronomical Observatory, No. 1/1981 (31.002.059; 31.002.060; 31.097.110; 31.044.036; 31.044.037; 31.044.038; 31.044.901).

008.091   Piscataway, N.J.

Rutgers University, Department of Physics and Astronomy, Piscataway, New Jersey 08854. – Report.
A. F. Cheng.
Bull. American Astron. Soc., Vol. 14, 473 - 474 (1982).

008.092   Potsdam

Beiträge der Berliner Sternwarte zu bedeutenden Erkenntnissen auf dem Gebiet der klassischen Astronomie.
See Abstr. 004.086.

Mitteilungen des Zentralinstituts für Physik der Erde, Potsdam, Nos. 821 (27.044.008), 899, 954, 955 (30.046.002), 958 (31.046.007).

008.093   Prague

Czechoslovak Academy of Sciences, Astronomical Institute, Circular of the Czechoslovak Observatories, Time and Latitude, 1981 April - September (31.044.029).

008.094   Princeton, N.J.

Princeton University Observatory, Princeton, New Jersey 08540. – Report.
Bull. American Astron. Soc., Vol. 14, 450 - 457 (1982).

008.095   Pulsnitz

Veröffentlichungen der Sternwarte Pulsnitz No. 19 (29.098.003).

008.096   Richmond Hill

David Dunlap Observatory, University of Toronto, Richmond Hill, Ontario L4C 4Y6, Canada. – Report for the period 1 July 1980 - 30 June 1981.   J. D. Fernie.
Bull. American Astron. Soc., Vol. 14, 153 - 161 (1982).

008.097   Rochester, N.Y.

C. E. Kenneth Mees Observatory, University of Rochester, Rochester, New York 14627. – Report.
J. L. Pipher.
Bull. American Astron. Soc., Vol. 14, 295 - 300 (1982).

008.098   Rohnert Park, Calif.

Sonoma State University, Department of Physics and Astronomy, Rohnert Park, California 94928. – Report for the period September 1980 - August 1981.
G. G. Spear, J. S. Tenn.
Bull. American Astron. Soc., Vol. 14, 488 - 489 (1982).

008.099   San Diego, Calif.

University of California, Berkeley, Los Angeles, San Diego, Lick Observatory, and Board of Studies. III. San Diego Campus. – Report.   E. M. Burbidge.
Bull. American Astron. Soc., Vol. 14, 57 - 65 (1982).

008.100   San Fernando

Memoria de las actividades en 1981.
A. Orte.
Inst. Obs. Marina, San Fernando (Cádiz). 23 pp. (1982).

008.101   Santa Cruz, Calif.

University of California, Berkeley, Los Angeles, San Diego, Lick Observatory, and Board of Studies. IV. Lick Observatory-Santa Cruz Campus. – Report.
D. E. Osterbrock.
Bull. American Astron. Soc., Vol. 14, 66 - 74 (1982).

Publications of the Lick Observatory, Vol. 23, Part 2 (31.122.168).

University of California, Berkeley, Los Angeles, San Diego, Lick Observatory, and Board of Studies. V. Board of Studies in Astronomy and Astrophysics, Santa Cruz Campus. Report.   W. G. Mathews.
Bull. American Astron. Soc., Vol. 14, 74 - 81 (1981).

008.102   Seattle, Wash.

University of Washington, Department of Astronomy, Seattle, Washington 98195. – Report.
B. Margon.
Bull. American Astron. Soc., Vol. 14, 553 - 558 (1982).

008.103   Sendai

Sendai Astronomiaj Raportoj, Nos. 225 (29.154. 007), 228 (29.064.031), 232 (29.162.091), 233 (30.065.100), 234 (30.131.211), 235 (30.103.241), 236 (30.008.070).

008.104   Shanghai

Time Service Annual Report 1980 (31.044.034).

008.105   Socorro, N.M.

National Radio Astronomy Observatory: Charlottesville, Virginia 22901; Green Bank, West Virginia 24944; Socorro, New Mexico 87801; Tucson, Arizona 85726. Report for the period July 1980 - June 1981.
M. S. Roberts, R. J. Havlen.
Bull. American Astron. Soc., Vol. 14, 370 - 410 (1982).

008.106   Sonneberg

Zentralinstitut für Astrophysik, Sonneberg. Mitteilungen über Veränderliche Sterne, Band 9, Heft 3 (31.123.022; 31.123.023; 31.123.024; 31.123.025; 31.123. 026; 31.123.027; 31.123.028; 31.133.018).

008.107   St. Andrews

Communications from the University Observatory, St. Andrews, Nos. 33 (27.113.049), 34 (29.113.002), 35 (29.114.008), 36 (29.117.054), 37 (29.117.136).

University Observatory, St. Andrews. Reprint Nos. 84 (29.119.003), 85 (29.131.027), 86 (29.114.140), 87 (29.114.068), 88 (29.122.028), 89 (30.122.049).

008.108   Stockholm

Stockholms Observatorium, Saltsjöbaden, Sweden, Report. No. 19 (31.021.037).

008.109   Stony Brook, N.Y.

State University of New York at Stony Brook, Stony Brook, New York 11794. – Report.   A. Yahil.
Bull. American Astron. Soc., Vol. 14, 419 - 426 (1982).

008.110   Strasbourg

Bulletin d'Information du Centre de Données Stellaires. No. 22 (1982).

008.111   Sunspot, N.M.

Sacramento Peak Observatory, Sunspot, New Mexico 88349. – Report.
Bull. American Astron. Soc., Vol. 14, 475 - 485 (1982).

008.112   Swarthmore, Penn.

Sproul Observatory and Swarthmore College, Swarthmore, Pennsylvania 19081. – Report for the period 1 July 1980 - 30 June 1981.   W. D. Heintz.
Bull. American Astron. Soc., Vol. 14, 492 - 493 (1982).

Sproul Observatory, Swarthmore, Pennsylvania, Reprints, Nos. 289, 290 (22.117.050), 291 (25.041.032), 292, 293 (25.111.014), 294 (25.008.109), 295 (26.118.017), 296, 297 (27.118.014), 298 (26.005.006), 299 (27.118.009), 300 (27.008.118), 301 (27.118.029), 302 (28.118.024), 303 (28.118.009), 304 (28.031.549), 305 (30.118.008), 306 (30.118.009), 307 (29.008.116), 308 (30.031.562), 309 (29.005.017).

008.113   Sydney

Sydney Observatory Papers, Nos. 90 (31.098.077), 91 (31.009.016).

008.114   Tempe, Ariz.

Arizona State University, Tempe, Arizona 85287. Report.   S. Starrfield.
Bull. American Astron. Soc., Vol. 14, 18 - 20 (1982).

008.115   Thessaloniki

Contributions from the Department of Geodetic Astronomy, University of Thessaloniki, Nos. 37 (30.122.172), 38 (25.151.069), 39 (30.122.024), 40 (31.122.113).

Publications of the Department of Geodetic Astronomy, University of Thessaloniki, Nos. 8 - 9.

008.116 Tokyo

Annals of the Tokyo Astronomical Observatory,
University of Tokyo, Second Series, Vol. 18, No. 4
(31.158.284; 31.032.039; 31.032.040; 31.073.120;
31.002.057; 31.002.058; 31.151.902).

Tokyo Astronomical Observatory, Reprints Nos.
607 (31.156.010), 608 (31.156.011), 609 (31.158.099),
610 (31.064.029), 611 (30.114.173), 612 (30.103.602),
613 (31.160.025), 614 (31.120.009), 615 (31.155.014),
616 (31.032.012), 617 (31.119.020), 618 (31.077.019),
619 (31.077.020), 620 (31.062.032), 621 (30.073.080),
622 (31.073.012), 623 (31.158.097), 624 (29.115.009).

Tokyo Astronomical Observatory, Time and
Latitude Bulletins, Vol. 55, No. 4 (31.044.035).

Tokyo Astronomical Observatory, Kiso Information
Bulletin, Vol. 1, No. 6 (1982).

Quarterly Bulletin on Solar Activity, Vol. 21 - 22,
Part I (31.072.079); Vol. 22, Part II (31.075.036); Vol. 21,
Part III (31.073.119); Vol. 22, Part IV (31.074.082).

Data Report of Hydrographic Observations. Series
of Astronomy and Geodesy, Maritime Safety Agency, Tokyo,
Japan, No. 16 (31.096.014).

008.117 Toledo, Ohio

Ritter Astrophysical Research Center, The University
of Toledo, Toledo, Ohio 43606. – Report.
N. D. Morrison.
Bull. American Astron. Soc., Vol. 14, 461 - 466 (1982).

008.118 Trieste

Publications of the Astronomical Observatory of
Trieste, Nos. 769, 770, 778 (30.119.117), 779 (31.112.030),
780 (31.114.153), 782 (31.064.060).

008.119 Tucson, Ariz.

National Radio Astronomy Observatory:
Charlottesville, Virginia 22901; Green Bank, West Virginia
24944; Socorro, New Mexico 87801; Tucson, Arizona 85726.
Report for the period July 1980 - June 1981.
M. S. Roberts, R. J. Havlen.
Bull. American Astron. Soc., Vol. 14, 370 - 410 (1982).

University of Arizona, Lunar and Planetary
Laboratory, Tucson, Arizona 85721. – Report for the period
1 October 1980 - 30 September 1981.
Bull. American Astron. Soc., Vol. 14, 9 - 17 (1982).

Multiple Mirror Telescope Observatory, Tucson,
Arizona 85721. – Report.
Bull. American Astron. Soc., Vol. 14, 322 - 325 (1982).

Preprints of the Steward Observatory, University of
Arizona, Tucson, Nos. 357 (31.158.267), 358 (31.116.054),
359 (31.031.601), 360 (31.158.268), 361 (31.034.060),
362 (31.031.602), 363 (31.113.075), 364 (31.141.274),
365 (31.155.044), 366 (31.126.037), 367 (31.131.223),
368 (31.124.121), 369 (31.158.269), 370 (31.126.038),

371 (31.032.035), 372 (31.158.270), 375 (31.032.036),
376 (31.141.275), 377 (31.131.224), 378 (31.133.019),
379 (31.114.204), 380 (31.122.169), 381 (31.031.603),
382 (31.032.037), 383 (31.158.271), 384 (31.141.276),
385 (31.032.530), 386 (31.133.020), 387 (31.032.038),
388 (31.121.033), 389 (31.158.272), 390 (31.133.021),
391 (31.160.071), 392 (31.133.022), 393 (31.141.277),
394 (31.118.028), 395 (31.132.073), 396 (31.158.273),
397 (31.131.225), 398 (31.121.034), 399 (31.141.278).

008.120 Turku

Turku University Observatory Informo, Nos.
59 (31.141.281), 60 (31.141.279), 61 (31.151.071),
62 (31.158.283), 63 (31.141.280), 64 (31.102.033),
65 (30.032.027).

Report Series, Department of Physical Sciences,
Institute of Astronomy, University of Turku, Nos. R 24
(31.141.279), R 25 (31.151.071), R 26 (31.158.283), R 27
(31.141.280), R 28 (31.102.033).

008.121 University, Ala.

The University of Alabama, Department of Physics
and Astronomy, University, Alabama 35486. – Report for the
period 15 August 1980 - 15 August 1981. G. G. Byrd.
Bull. American Astron. Soc., Vol. 14, 6 - 8 (1982).

008.122 University Park, Penn.

The Pennsylvania State University, Department of
Astronomy, University Park, Pennsylvania 16802. – Report
for the period 1 September 1980 - 31 August 1981.
S. Matsushima.
Bull. American Astron. Soc., Vol. 14, 439 - 449 (1982).

008.123 Uppsala

Annual report for 1981. Uppsala Astronomical
Observatory.
Uppsala Astron. Obs., Rep. No. 23, 26 pp. (1982).

Uppsala Astronomical Observatory, Report Nos.
22 (31.002.061), 23 (31.008.123).

008.124 Urbana, Ill.

University of Illinois at Urbana-Champaign,
Department of Astronomy, Urbana, Illinois 61801-3000.
Report for the period 1 September 1980 - 1 September 1981.
L. L. Smarr.
Bull. American Astron. Soc., Vol. 14, 209 - 222 (1982).

008.125 Vancouver

University of British Columbia, Department of
Geophysics and Astronomy, Vancouver, British Columbia
V6T 1W5, Canada. – Report.
Bull. American Astron. Soc., Vol. 14, 26 - 29 (1982).

008.126   Victoria

University of Victoria Observatory, Physics Department, University of Victoria, Victoria, British Columbia V8W 2Y2, Canada. – Report.
Bull. American Astron. Soc., Vol. 14, 540 - 542 (1982).

Dominion Astrophysical Observatory, Victoria, B. C.
J. B. Hutchings.
J. R. Astron. Soc. Canada, Vol. 76, 206 (1982).

Publications of the Dominion Astrophysical Observatory, Victoria, B. C., Vol. 16, No. 1 (31.031.604).

008.127   Villa Elisa

Actividades del Instituto Argentino de Radio-astronomía durante el año 1972. – Report.
K. Turner.
Bol. Asoc. Argentina Astron., No. 18, p. 136 - 138 (1980).

008.128   Villafranca

IUE Observatory, European Space Agency, Villafranca Satellite Tracking Station (VILSPA). Report for the year 1980 January 1 to December 31.
M. V. Penston, A. Heck.
Q. J. R. Astron. Soc., Vol. 23, 123 - 133 (1982).

008.129   Villanova, Penn.

Villanova University, Department of Astronomy, Villanova, Pennsylvania 19085. – Report.
Bull. American Astron. Soc., Vol. 14, 543 - 545 (1982).

008.130   Vilnius

Vilniaus Astronomijos Observatorijos Biuletenis (Bulletin of the Vilnius Astronomical Observatory), No. 58 (31.113.079; 31.113.080; 31.113.081; 31.113.082).

008.131   Warsaw

Warsaw University Observatory and Polish Academy of Science, N. Copernicus Astronomical Center, Reprint Nos. 445 (30.122.106), 446 (30.116.032), 447 (30.117.074), 448 (30.124.802), 449 (31.117.045), 450 (31.113.019), 451 (31.066.072), 452 (31.158.140).

008.132   Washington, D.C.

U. S. Naval Observatory, Washington, D. C. 20390. Report for the period 1 July 1980 - 30 June 1981.
G. Westerhout, R. A. Vohden.
Bull. American Astron. Soc., Vol. 14, 527 - 535 (1982).

United States Naval Observatory, Washington, D.C., Circular, No. 163 (31.043.003).

Publications of the United States Naval Observatory, Washington, Second Series, Vol. 24, Part V (31.118.029).

U. S. Naval Observatory, Washington, D.C. Time Service Publications. Series 4, Nos. 779 - 803; Series 6, Nos. 71 - 75; Series 7, Nos. 732 - 756; Series 14, No. 31 (31.044.039 - 31.044.042).

National Aeronautics and Space Administration: Ames Research Center; Goddard Space Flight Center; Jet Propulsion Laboratory; Johnson Space Center; Langley Research Center; Marshall Space Flight Center; Headquarters. Washington, D. C. 20546. – Report.
D. Gilman.
Bull. American Astron. Soc., Vol. 14, 326 - 364 (1982).

National Bureau of Standards, Washington, D. C. 20234. – Report for work undertaken since 1977.
J. L. Tech, F. J. Lovas, J. R. Fuhr.
Bull. American Astron. Soc., Vol. 14, 365 - 369 (1982).

Solar Physics Branch, E. O. Hulburt Center for Space Research, Naval Research Laboratory, Washington, D. C. 20375. – Report for the period 1 January 1980 - 30 September 1981.   G. E. Brueckner.
Bull. American Astron. Soc., Vol. 14, 411 - 415 (1982).

008.133   Westford, Mass.

Haystack Observatory, Northeast Radio Observatory Corporation (NEROC), Westford, Massachusetts 01886. Report for the period 1 July 1980 - 30 June 1981.
Bull. American Astron. Soc., Vol. 14, 189 - 194 (1982).

008.134   Williamstown, Mass.

Hopkins Observatory, Williams College, Williamstown, Massachusetts 01267. – Report for the 1980 - 1981 academic year.   J. M. Pasachoff.
Bull. American Astron. Soc., Vol. 14, 206 - 207 (1982).

008.135   Yorktown Heights, N.Y.

IBM Thomas J. Watson Research Center, Yorktown Heights, New York 10598. – Report.
P. Seiden.
Bull. American Astron. Soc., Vol. 14, 208 (1982).

008.136   Zagreb

Hvar Observatory Bulletin, Vol. 5, No. 1 (31.113. 078; 31.071.055; 31.073.121; 31.073.122).

008.137   Zelenchukskaya

Chronicle.
Astrofiz. Issled., Izv. Spets. Astrofiz. Obs., Tom 15, 164 - 166 (1982). In Russian.

Soobshcheniya Spetsial'noj Astrofizicheskoj Observatorii, Akademiya Nauk SSSR, Vyp. 31 (31.114.205; 31.160.072; 31.022.161), Vyp. 32 (31.012.029).

Astrofizicheskie Issledovaniya. Izvestiya Spetsial'noj Astrofizicheskoj Observatorii, Akademiya Nauk SSSR, Tom 15 (31.124.006; 31.114.200; 31.162.163; 31.151.067; 31.151.068; 31.160.070; 31.022.160; 31.033.025; 31.031.598; 31.033.026; 31.033.027; 31.033.028; 31.033.029; 31.033.030; 31.162.164; 31.008.137).

008.138   Zimmerwald

Mitteilungen der Satelliten-Beobachtungsstation Zimmerwald, Nos. 7 (31.046.008), 8 (31.021.038).

009 Notes on Observatories, Planetaria, Exhibitions

009.001   Our turn at Kitt Peak.
R. Genet, K. Kissell, G. Roberts.
Sky Telesc., Vol. 63, 240 - 242 (1982).

009.002   The Carlsberg Automatic Transit Circle on La Palma.
J. V. Clausen, L. Helmer, L. V. Morrison,
C. A. Murray, A. Orte, L. Quijano.
Observatory, Vol. 102, 9 - 10 (1982).

009.003   The work of the Royal Observatory Edinburgh.
R. J. Dodd.
South. Stars, Vol. 29, 57 - 62 (1981).

009.004   Double anniversary for the astronomers of the Moscow University.   Yu. P. Pskovskij.
Astronomical calendar. Yearbook. 1982, (see 047.005),
p. 283 - 299 (1981). In Russian. – Abstr. in Ref. zh., 51.
Astron., 2.51.14 (1982).

009.005   Lick Observatory's Chile Station.
R. P. S. Stone.
Sky Telesc., Vol. 63, 446 - 448 (1982).

009.006   Beobachten auf La Silla. Zum Arbeitsbesuch auf der ESO-Sternwarte in Chile.   R. Lukas.
Sterne Weltraum, 21. Jahrg., 14 - 16 (1982).

009.007   The 6-inch Cooke refractor in Toronto.
B. Beattie.
J. R. Astron. Soc. Canada, Vol. 76, 109 - 128 (1982).

009.008   L'Observatoire Européen de l'Hémisphère Sud (ESO).   H. Debehogne.
Ciel Terre, Vol. 98, 81 - 90 (1982).

009.009   La estación astrométrica austral.
S. Slaucitajs.
Bol. Asoc. Argentina Astron., No. 18, p. 96 - 99 (1980).

009.010   The activities of the Kiev Observatory in the years 1900 - 1940.
A. F. Bogorodskij, N. A. Chernega.
Vestn. Kiev. Univ., Vyp. 23, p. 3 - 15 (1981). In Russian.

009.011   CERGA: a modern French observatory.
M. Šidlichovský.
Kozmos, Vol. 13, 58 - 60 (1982). In Czech.

009.012   Observatory Lomnický Štít.
Říše hvězd, Vol. 63, 81 - 82 (1982). In Slovak.

009.013   The discoveries at the Kleť Observatory.
T. Fabini.
Kozmos, Vol. 13, 86 - 91 (1982). In Slovak.

009.014   The story of the Ondrejov Observatory.
P. Harmanec.
I. A. P. P. P. Commun., No. 7, p. 1 - 6 (1982).

009.015   Lake Afton Public Observatory.   D. R. Alexander.
Bull. American Astron. Soc., Vol. 13, 838 (1981).
Abstract.

009.016   Sydney Observatory 1958 to 1981.   H. Wood.
Sydney Obs. Pap., No. 91, 14 pp. (1981).

009.017   Twenty-five years of Uttar Pradesh State Observatory, Naini Tal, India.   M. C. Pande.
Published by Uttar Pradesh State Obs., Naini Tal, India.
46 pp. (1979).

009.018   Debrecen Heliophysical Observatory.   L. Dezsö.
Sol. Phys., Vol. 79, 195 - 199 (1982).
This article provides a brief summary of the progress in the solar physics research at the Debrecen Observatory in the last 12 years.

009.019   Rhodes University: Department of Physics and Electronics.   E. E. Baart.
Mon. Notes Astron. Soc. South. Africa, Vol. 41, 7 - 8 (1982).

009.020   University of South Africa: Department of Mathematics, Applied Mathematics and Astronomy.
P. D. Bennewith, J. Wolterbeek.
Mon. Notes Astron. Soc. South. Africa, Vol. 41, 8 - 9 (1982).

009.021   The Woodstock College Observatory.
P. Mozel.
J. R. Astron. Soc. Canada, Vol. 76, 168 - 180 (1982).

009.022   Iraq's new national observatory.   B. J. Bok.
Sky Telesc., Vol. 64, 33 - 35 (1982).

009.023   L'observatoire du Mont-Palomar.   D. Proust.
Bull. AFOEV, No. 20, p. 49 - 50 (1982).

009.024   De Bosscha-Sterrenwacht . Van thee tot sterrenkunde.   K. A. van der Hucht, C. L. M. Kerkhoven.
Zenit, 9. Jaarg., 292 - 300 (1982).

009.025   Ein Besuch bei der ESO-Zentrale in Garching bei München.   H. Hornung.
Orion, 40. Jahrg., 74 - 77 (1982).

009.026   75 Jahre Urania-Sternwarte Zürich.
E. Egli.
Orion, 40. Jahrg., 87 - 90 (1982).

009.027   L'Osservatorio Astrofisico di Catania. Dalla sua fondazione ad oggi.   S. Cristaldi.
Coelum, Vol. 50, 139 - 152 (1982).

009.028 The solar complex of SibIZMIR: observatories, instruments and the main results of the investigations. G. Ya. Smolkov (*Smol'kov*).
Sun and planetary system, (see 012.040), p. 123 - 124 (1982).

009.029 Astrometry at Skibotn Observatory, Norway: a brief status report.
B. R. Pettersen, K. Aksnes.
Rep. Obs. Lund, No. 18, (see 012.044), p. 118 - 119 (1982).

Planetarium: window to the universe.
See Abstr. 003.055.

Early days of astronomy at Toronto – Part III. The example of Ormsby MacKnight Mitchel and the Cincinnati Observatory. See Abstr. 004.073.

Les 21 ans de fonctionnement du plus grand télescope de Schmidt à Tautenburg.
See Abstr. 032.019.

El radiotelescopio del Instituto Argentino de Radioastronomía. See Abstr. 033.017.

010 Societies, Associations, Organizations

010.001 American Association of Variable Star Observers (AAVSO)

70th anniversary of the AAVSO.
J. A. Mattei, R. N. Mayall, D. Hoffleit, D. B. Pickering.
J. American Assoc. Variable Star Obs., Vol. 10, 98 - 105 (1981).

Meetings and activities of the Society, committee reports.
J. American Assoc. Variable Star Obs., Vol. 10, 106 - 115, 116 - 123 (1981).

The Journal of the American Association of Variable Star Observers, Vol. 10, No. 2 (1981).

010.002 American Astronomical Society (AAS)

The 159th meeting of the American Astronomical Society, held 10 - 13 January 1982 at Boulder, Colorado.
Abstracts of papers presented.
Bull. American Astron. Soc., Vol. 13, 755 - 923 (1981).

Late-paper abstracts from the 158th meeting of the American Astronomical Society, held 28 June - 1 July 1981 at Calgary, Alberta.
Bull. American Astron. Soc., Vol. 13, 924 - 926 (1981).

Annual reports of the AAS Divisions: Dynamical Astronomy Division *(W. H. Jefferys);* High Energy Astrophysics Division *(H. Bradt);* Historical Astronomy Division *(J. A. Eddy);* Planetary Sciences Division *(D. Morrison);* Solar Physics Division *(R. D. Chapman),*
Bull. American Astron. Soc., Vol. 13, 927 - 929 (1981).

Task Group on Education in Astronomy (TGEA): annual report for 1981. H. Shipman.
Bull. American Astron. Soc., Vol. 13, 929 - 930 (1981).

Late-paper abstracts from the 159th meeting of the American Astronomical Society, held 10 - 13 January 1982 at Boulder, Colorado.
Bull. American Astron. Soc., Vol. 14, 571 - 577 (1982).

13th regular meeting of the Division on Dynamical Astronomy, held 13 - 14 January 1982 at Boulder, Colorado.
Abstracts of papers presented.
Bull. American Astron. Soc., Vol. 14, 578 - 583 (1982).

AAS Photo-Bulletin, Issue 26, No. 1; Issue 27, No. 2 (1981).

Bulletin of the American Astronomical Society, Vol. 13, No. 4 (1981); Vol. 14, No. 1 (1982).

010.003 Asociación Argentina de Astronomía

Boletín de la Asociación Argentina de Astronomía. No. 18 (1980); Nos. 20 - 24, 26 (1981).

010.004 Association Française des Observateurs d'Etoiles Variables (A.F.O.E.V.)

La vie de l'Association. E. Schweitzer.
Bull. AFOEV, No. 19, p. 14; No. 20, p. 73 (1982).

Activité de l'A. F. O. E. V. en 1981.
E. Schweitzer.
Bull. AFOEV, No. 20, p. 55 - 66 (1982).

Bulletin de l'Association Française des Observateurs d'Etoiles Variables, Nos. 19, 20 (1982).

010.005 Association of Lunar and Planetary Observers (A.L.P.O.)

The Strolling Astronomer. The Journal of the Association of Lunar and Planetary Observers, Vol. 29, Nos. 5 - 6 (1982).

010.006 Astronomical Society of India

Bulletin of the Astronomical Society of India, Vol. 9, No. 4 (1981).

010.007 Astronomical Society of Japan

Publications of the Astronomical Society of Japan, Vol. 33, No. 4 (1981); Vol. 34, No. 1 (1982).

010.008   Astronomical Society of New York

Abstracts of papers given at the Fall meeting of the Astronomical Society of New York, held October 31, 1981, at Union College, Schenectady, New York.
News Lett. Astron. Soc. N. Y., Vol. 2, 1 (1982).

News Letter of the Astronomical Society of New York, Vol. 2, No. 1 (1982).

010.009   Astronomical Society of the Pacific (ASP)

Reminiscences on the occasion of *Mercury's* 10th anniversary.   J. G. Phillips.
Mercury, Vol. 11, 2 - 4 (1982).

Publications of the Astronomical Society of the Pacific, Vol. 93, No. 556 (1981/82); Vol. 94, Nos. 557, 558 (1982).

Mercury. The Journal of the Astronomical Society of the Pacific, Vol. 10, No. 6 (1981); Vol. 11, Nos. 1, 2 (1982).

010.010   Astronomical Society of Southern Africa (ASSA)

Notices.
Mon. Notes Astron. Soc. South. Africa, Vol. 41, 1 (1982).

Monthly Notes of the Astronomical Society of Southern Africa, Vol. 41, Nos. 1 - 4 (1982).

010.011   Astronomical Society of Western Australia (ASWA)

Journal of the Astronomical Society of Western Australia, Session 1981/82, September 1981 - March 1982.

010.012   Astronomische Gesellschaft (AG)

Mitteilungen des Vorstandes der Astronomischen Gesellschaft.
Mitt. Astron. Ges., Nr. 55, p. 8 - 11 (1982).

Internationale Astronomische Tagung Innsbruck 1981 mit 59. Ordentlicher Mitgliederversammlung der Astronomischen Gesellschaft vom 15. bis 18. September 1981.
Mitt. Astron. Ges., Nr. 55, p. 1 - 213 (1982).

Mitteilungen der Astronomischen Gesellschaft, Nr. 55 (1982).

010.013   British Astronomical Association (BAA)

Meetings and activities of the Association.
J. British Astron. Assoc., Vol. 92, 83 - 89, 140 - 143, 189 - 195 (1982).

Section reports.
J. British Astron. Assoc., Vol. 92, 74 - 76, 81, 90 - 92, 127 - 136, 170 - 189 (1982).

British Astronomical Association Circular, Nos. 617 - 621 (1981); 622 - 627 (1982).

Journal of the British Astronomical Association, Vol. 92, Nos. 2 - 4 (1982).

010.014   British Interplanetary Society (BIS)

JBIS. Journal of the British Interplanetary Society, Vol. 35, Nos.1 - 7 (1982).

Space Education. A publication of the British Interplanetary Society, Vol. 1, No. 3 (1982).

Spaceflight. A publication of the British Interplanetary Society, Vol. 24, Nos. 1 - 8 (1982).

010.015   European Space Agency (ESA)

ESA IUE Newsletter, No. 13 (1982).

ESA Bulletin, Nos. 29, 30 (1982).

010.016   International Amateur-Professional Photoelectric Photometry (I.A.P.P.P.)

The fall 1981 meeting of the Arizona Chapter (of I. A. P. P. P.).   J. L. Hopkins.
I. A. P. P. P. Commun., No. 7, p. 7 (1982).

The fall 1981 IAPPP Northwest Workshop.
E. Mannery.
I. A. P. P. P. Commun., No. 7, p. 19 - 22 (1982).

1981 IAPPP workshops on photoelectric photometry.
R. M. Genet.
I. A. P. P. P. Commun., No. 7, p. 23 - 24 (1982).

I. A. P. P. P. Communication No. 7 (1982).

010.017   International Astronomical Union (IAU)

International Astronomical Union, Information Bulletin. No. 47.   P. A. Wayman.
Published by D. Reidel Publishing Company, Dordrecht, Holland. 31 pp. (1982).
Contents: Introduction. The XVIIIth General Assembly. Executive Committee. Commissions. International Organizations. Symposia and colloquia. Meetings co-sponsored by the IAU. Other scientific meetings. IAU publications. Other publications. Membership.

International Astronomical Union, Information Bulletin. No. 48.   P. A. Wayman.
Published by D. Reidel Publishing Company, Dordrecht, Holland. 66 pp. (1982).
Contents: The Eighteenth General Assembly. Executive Committee. Commissions. International Organisations. Symposia and colloquia. Meetings co-sponsored by the IAU. Other scientific meetings. IAU publications. Other publications. Membership.

Reports on Astronomy. Transactions of the International Astronomical Union, Volume XVIIIA (1982).
See Abstr. 003.013.

XVIIIth General Assembly of the IAU.
J. Bouška.
Říše hvězd, Vol. 63, 104 - 105 (1982). In Czech.

Progress Report by IAU-Commission 24 Working Group on optical/radio astrometric sources for the establishment of an inertial reference frame.
A. N. Argue, C. de Vegt (Editors).
Abh. Hamburger Sternw., Band 10, 97 - 176 (1982).

Bulletin No. 27, IAU Commission 20, Working Group on Predictions of Occultations by Satellites and Minor Planets (31.098.078; 31.101.011; 31.097.109).

Circulaire d'Information, No. 86 (1982).

IAU Circulars, Nos. 3656 - 3707 (1982).

Commission 27 of the I. A. U. Information Bulletin on Variable Stars, Nos. 2062 - 2159 (1982).

Minor Planet Circulars, (M. P. C.), Nos. 6573 - 6988 (1982).

010.018 Irish Astronomical Association

The Irish Astronomical Association 1978/81.
D. E. Beesley.
Irish Astron. J., Vol. 14, 253 - 255 (1980).

010.019 Meteoritical Society

Abstracts of papers presented at the 44th annual meeting of the Meteoritical Society, August 17 - 21, 1981, Bern, Switzerland.
Meteoritics, Vol. 16, 287 - 409 (1981).

Meteoritics. The Journal of the Meteoritical Society, Vol. 16, No. 4 (1981); Vol. 17, No. 1 (1982).

010.020 Oriental Astronomical Association

The Heavens, Nos. 680 - 684, Vol. 63, Nos. 1 - 5 (1982).

010.021 Royal Astronomical Society (RAS)

Meetings of the Society.
Observatory, Vol. 102, 21 - 26, 61 - 77 (1982).

Meetings and activities of the Society.
Q. J. R. Astron. Soc., Vol. 23, 171 - 172, 323 - 324 (1982).

Geophysical Journal of the Royal Astronomical Society, Vol. 68, Nos. 1 - 3; Vol. 69, Nos. 1 - 3; Vol. 70, No. 1 (1982).

Monthly Notices of the Royal Astronomical Society, Vol. 198, Nos. 1 - 3; Vol. 199, Nos.1 - 3; Vol. 200, No. 1 (1982).

The Quarterly Journal of the Royal Astronomical Society, Vol. 23, Nos. 1, 2 (1982).

010.022 **Royal Astronomical Society of Canada (RAS Canada)**

Annual Report 1981. I. Halliday.
Suppl. J. R. Astron. Soc. Canada, 36 pp. (1982).
This Annual Report for 1981 summarizes the past year of activity for the Society and for its individual Centres.

The Journal of the Royal Astronomical Society of Canada, Vol. 76, Nos. 1 - 3 (1982).

National Newsletter. Supplement to the Journal of the Royal Astronomical Society of Canada, Vol. 76, Nos. 1 - 3 (1982).

010.023 **Royal Astronomical Society of New Zealand (RAS New Zealand)**

Southern Stars. Journal of the Royal Astronomical Society of New Zealand, Vol. 29, Nos. 3, 4 (1981).

010.024 **Schweizerische Astronomische Gesellschaft (SAG)**

Mitteilungen.
Orion, 40. Jahrg., 17 - 20, 51 - 56, 91 - 94 (1982).

BBSAG Bulletin, Nos. 58 - 60 (1982).

Orion. Zeitschrift der Schweizerischen Astronomischen Gesellschaft. Revue de la Société Astronomique de Suisse, 40. Jahrg., Nr. 188 - 190 (1982).

010.025 **Società Astronomica Italiana (S.A.It.)**

Giornale di Astronomia, Vol. 7, N. 4 (1981).

010.026 **Société Astronomique de France (SAF)**

Séances, commissions, activités de la Société.
Astronomie, Vol. 96, 3 - 6, 28 - 29, 34 - 42, 75 - 80, 87 - 94, 199 - 206, 253 - 263, 279 - 296 (1982).

Une commission des étoiles doubles à la S. A. F.
P. Muller.
Astronomie, Vol. 96, 3 - 6 (1982).

Cinquantenaire du Group d'Alsace de la Société Astronomique de France.   É. Schweitzer.
Astronomie, Vol. 96, 248 - 252 (1982).

L'Astronomie et Bulletin de la Société Astronomique de France, Vol. 96, janvier - juin (1982).

010.027 **Société Astronomique de Liège**

Le Ciel, Vol. 44, 1 - 144, janvier - juin (1982).

010.028 **Société Royale Belge d'Astronomie, de Météorologie et de Physique du Globe**

      Ciel et Terre. Bulletin de la Société Royale Belge d'Astronomie, de Météorologie et de Physique du Globe, Vol. 98, Nos. 1, 2 (1982).

010.029 **Vereinigung der Sternfreunde e.V.**

      Nachrichten der Vereinigung der Sternfreunde e.V. Sterne Weltraum, 21. Jahrg., 36 - 40, 87 - 88, 90, 129 - 131, 174 - 177, 218, 220 - 221, 266 - 267 (1982).

      Sonne. Mitteilungsblatt der Amateursonnenbeobachter, Jahrg. 6, Nos. 21, 22 (1982).

011 Reports on Colloquia, Congresses, Meetings, Symposia, Expeditions

011.001 **Archaeoastronomers convene in Oxford.**
      O. Gingerich.
Sky Telesc., Vol. 63, 7 - 10 (1982).

011.002 **Travellers' tales. A solar eclipse expedition to Siberia.** W. I. McLaughlin.
Spaceflight, Vol. 24, 114 - 118 (1982).

011.003 **Internationale Astronomische Tagung Innsbruck 1981.** W. Seggewiß.
Phys. Bl., Vol. 38, 20 - 21 (1982).

011.004 **After Pioneer – good science, bad news.**
      P. Campbell.
Nature, Vol. 296, 13 - 14 (1982).
      Report on the International Conference on the "Venus environment", held in Palo Alto, Calif., 1 - 6 November 1981.

011.005 **The big bang – "free lunch" at the Royal Society.**
      C. Hogan.
Nature, Vol. 296, 490 - 491 (1982).

011.006 **Ethnoastronomy and archaeoastronomy in the American tropics.** New York Academy of Sciences Conference, held in New York City from 30 March to 1 April, 1981. O. Gingerich.
Archaeoastronomy (U.S.A.), Vol. 4, No.2, p. 5 - 7 (1981).

011.007 **IAU Colloquium No. 68: a summary.**
      A. G. D. Philip.
News Lett. Astron. Soc. N. Y., Vol. 2, 25 - 27 (1982).

011.008 **Workshop on Interstellar Grains, 1981 April, Manchester.**
Q. J. R. Astron. Soc., Vol. 23, 94 - 96 (1982).

011.009 **SETI conference at Tallinn.**
      W. T. Sullivan, III.
Sky Telesc., Vol. 63, 350 - 353 (1982).

011.010 **Workshop on quasi-periodic climatic changes on Mars and earth.**
J. A. Cutts, A. D. Howard, J. B. Pollack, O. B. Toon.
Reports of Planetary Geology Program – 1981, (see 003.001), p. 345 - 346 (1981). – Abstract.

011.011 **Antarctic search for meteorites: preliminary report of the 1980/81 field season in South Victoria Land.**
L. Schultz, J. O. Annexstad, W. A. Cassidy, H. Y. McSween, L. A. Rancitelli, J. Schutt.
Meteoritics, Vol. 16, 384 (1981). – Abstract.

011.012 **Symposium "Origin and evolution of celestial bodies", Leningrad, 1981, May 20 - 21 on the occasion of the**

011.012 **100th anniversary of the Astronomical Observatory of the Leningrad University.** V. G. Gorbatskij.
Astron. Zh., Tom 59, 197 - 199 (1982). In Russian. English translation in Soviet Astron., Vol. 26, No. 1.

011.013 **All-Union conference "Star clusters and problems of stellar evolution". Sverdlovsk, 1981, April 13 - 17.**
A. S. Rastorguev, N. N. Samus'.
Astron. Zh., Tom 59, 199 - 201 (1982). In Russian. English translation in Soviet Astron., Vol. 26, No. 1.

011.014 **The great galactic centre mystery.** G. R. Riegler.
Nature, Vol. 297, 18 - 19 (1982).
      Report on a workshop on the galactic centre, held on 7 - 8 January 1982 at the California Institute of Technology.

011.015 **Be stars. IAU Symposium No. 98, Munich, April 6 - 10, 1981.** A. Slettebak.
Comments Astrophys., Vol. 9, 243 - 246 (1982).

011.016 **Binary and multiple stars as tracers of stellar evolution. I.A.U. Colloquium No. 69, Bamberg, August 31 - September 3, 1981.** F. B. Wood.
Comments Astrophys., Vol. 9, 247 - 249 (1982).

011.017 **International symposium on physics of the ionosphere and magnetosphere of the earth and solar wind. Tsakhkadzor, Armenian SSR, 1981, 27 - 30 May.**
Yu. M. Mikhajlov.
Geomagn. Aehron., Tom 22, 169 - 170 (1982). In Russian.

011.018 **Third All-Union school-seminar on mathematical models of the near cosmos. Divnogorsk, 1981, 22 - 28 June.** V. G. Pivovarov.
Geomagn. Aehron., Tom 22, 348 - 349 (1982). In Russian.

011.019 **1980 IAU Theory of Nutation: the final report of the IAU Working Group on Nutation.**
P. K. Seidelmann.
Celestial Mech., Vol. 27, 79 - 106 (1982).
      In 1979 the Seventeenth General Assembly of the International Astronomical Union (IAU) in Montreal, Canada, adopted the 1979 IAU Theory of Nutation upon the recommendation of this Working Group. Subsequently the International Union of Geodesy and Geophysics (IUGG) passed a resolution requesting that this action be reconsidered in favor of a theory based on a different Earth model. As a consequence of that reconsideration the 1980 IAU Theory of Nutation was adopted. The details of that theory and the history of its adoption are described.

011.020 **Tagung der Astronomischen Gesellschaft.**
      W. Seggewiß.
Phys. Bl., Vol. 38, 133 (1982).

011.021 **Hipparcos Kolloquium.**
S. Röser.
Phys. Bl., Vol. 38, 133 - 134 (1982).

011.022 **XXIInd Astrometric Conference of the USSR.**
Moscow, 1 - 5 June 1981.
D. N. Ponomarev.
Astron. Zh., Tom 59, 407 - 413 (1982). In Russian. English
translation in Soviet Astron., Vol. 26, No. 2.

011.023 **In, out and about meteorites.**
N. J. McNaughton, P. K. Swart.
Nature, Vol. 297, 453 - 454 (1982).
Report on the 13th Lunar and Planetary Science
conference, held in Houston, Texas on 15 - 19 March 1982.

011.024 **How does the theory of gravitation develop itself?**
5th All-Union conference on "Modern theoretical
and experimental problems of the theory of relativity and
gravitation", Moscow, July 1981.
Yu. S. Vladimirov.
Zemlya Vselennaya, 1982, No. 2, p. 56 - 58. In Russian.

011.025 **Twenty-fifth MIST meeting.**
P. A. Hadjiry, M. J. Laird.
Q. J. R. Astron. Soc., Vol. 23, 256 - 259 (1982).

011.026 **Seminar on extragalactic objects with active nuclei.**
1981, September 9 - 17.
R. D. Dagkesamanskij, Eh. A. Dibaj, L. I. Matveenko.
Astron. Zh., Tom 59, 617 - 619 (1982). In Russian. English
translation in Soviet Astron., Vol. 26, No. 3.

011.027 **14th Young European Radio Astronomers
Conference.** U. Klein.
Phys. Bl., Vol. 38, 160 (1982).

011.028 **Saturn briefing.** R. A. Kerr.
Science, Vol. 216, 1210 - 1211 (1982).
Report on the Saturn Conference held 11 to 15 May in
Tucson, Arizona.

011.029 **Astronomie- und Raumfahrtgeschichte auf dem
XVI. Weltkongreß der Wissenschaftshistoriker.**

D. B. Herrmann.
Sterne, 58. Band, 162 - 165 (1982).

011.030 **ESO workshop on "Ground-based observations of
Halley's comet".**
Messenger, No. 28, p. 21 - 22 (1982).

011.031 **IAU symposium no. 97 "Extragalactic radio sources".**
V. K. Kapahi.
Bull. Astron. Soc. India, Vol. 9, 323 - 326 (1981).

011.032 **The twentieth general assembly of URSI.**
N. V. G. Sarma.
Bull. Astron. Soc. India, Vol. 9, 326 - 327 (1981).

011.033 **The second Asian-Pacific regional meeting of
astronomy.** J. C. Bhattacharyya.
Bull. Astron. Soc. India, Vol. 9, 328 - 332(1981).

011.034 **Scientific sessions of the Department of General
Physics and Astronomy and the Department of
Nuclear Physics of the USSR Academy of Sciences, 27 - 28
May and 24 - 25 June 1981.**
Usp. fiz. nauk, Tom 136, 345 - 353 (1982). In Russian.
From Ref. zh., 51. Astron., 6.51.18 (1982).

011.035 **Internationale Astronomische Tagung Innsbruck
1981 mit 59. Ordentlichen Mitgliederversammlung
der Astronomischen Gesellschaft vom 15. bis 18. September
1981. Bericht über die Tagung.**
W. Seggewiß.
Mitt. Astron. Ges., Nr. 55, p. 7 - 8 (1982).

Erratum

011.901 **Relativistic astrophysics: the view from Texas in
Baltimore.** V. L. Trimble, S. P. Maran.
Astrophys. Lett., Vol. 22, 21 - 30 (1981). — See Abstr.
30.011.024.

012 Proceedings of Colloquia, Congresses, Meetings, Symposia

012.001 **Analytical methods and ephemerides: theory and
observations of the Moon and planets.** Proceedings
of a conference held in Namur, Belgium, 28 - 31 July 1980.
J. Chapront, J. Henrard, D. Schmidt (Editors).
Celestial Mech., Vol. 26, Nos. 1 - 3, p. 3 - 276 (1982). — The
individual contributions are included in their corresponding
subject categories — see abstracts 031.528, 031.529, 041.003,
042.010 - 042.016, 051.004, 091.028 - 091.030, 094.010 -
094.023, 097.080, 097.081, 099.039 - 099.042, 100.035.

012.002 **Regions of recent star formation.** Proceedings of the
symposium on "Neutral clouds near H II regions —
dynamics and photochemistry", held in Penticton, British
Columbia, June 24 - 26, 1981.
R. S. Roger, P. E. Dewdney (Editors).
Astrophysics and Space Science Library, Vol. 93, Proceedings.
D. Reidel Publishing Company, Dordrecht, Holland — Boston,
U.S.A. — London, England. 16 + 496 pp. Price Dfl. 140.00,

US $ 59.50 (1982). ISBN 90-277-1383-9. — The individual
contributions are included in their corresponding subject categories — see abstracts 034.018, 112.018, 112.019, 121.019,
131.087 - 131.124, 132.029 - 132.047, 156.012, 159.011.

012.003 **Proceedings of the meeting held by the Astronomical Science Group of Ireland at Armagh Observatory
on June 25th 1980.**
Irish Astron. J., Vol. 14, 216 - 240 (1980). — The individual
contributions are included in their corresponding subject
categories — see abstracts 051.003, 116.012, 118.014,
122.041.

012.004 **The sixth U. K. Geophysical Assembly, UKGA.**
Held at University College Cardiff, 5 - 7 April 1982.
With a preface by M. Brooks, J. Shaw, R. G. Pearce.
Geophys. J. R. Astron. Soc., Vol. 69, No. 1, p. 273 - 306
(1982). — The individual contributions within the subject

scope of Astronomy and Astrophysics Abstracts are included in their corresponding categories – see abstracts 081.012, 084.018, 084.019, 091.009, 106.007.

012.005 Radiative properties of hot dense matter. A conference held at the Naval Postgraduate School in Monterey, Calif., 17 - 20 November 1980.
B. Rozsnyai (Editor), with a preface by J. Davis, W. Huebner, C. Moser, B. Rozsnyai, A. Szoke.
J. Quant. Spectrosc. Radiat. Transfer, Vol. 27, No. 3, p. 209 - 385 (1982). – The individual contributions within the subject scope of Astronomy and Astrophysics Abstracts are included in their corresponding categories – see abstracts 022.026 - 022.033, 062.024, 062.025, 063.014 - 063.016.

012.006 XVIIIth All-Union meteorite conference. Chernogolovka, 15 - 17 September 1981. Abstracts.
Inst. geokhimii i analit. khimii, Moskva. 86 pp. (1981). In Russian. – From Ref. zh., 51. Astron., 2.51.40 (1982).

012.007 Radiation scattering in optical systems. Conference held at Huntsville, Ala., USA, 30 September – 1 October 1980.
Proc. Soc. Photo-Opt. Instrum. Eng., Vol. 257 (1980). Review in Phys. Abstr., Vol. 85, Abstr. 10546 (1982). See abstracts 031.005 - 031.011.

012.008 Proceedings of the Einstein centennial symposium on fundamental physics, held at Bogota, Colombia, 30 July – 5 August 1979.   S. M. Moore, A. M. Rodriguez-Vargas, A. Rueda, G. Violini (Editors). Published by Universidad de los Andes, Bogota, Colombia. 14 + 667 pp. (1981). – Review in Phys. Abstr., Vol. 85, Abstr. 10554 (1982). – See abstracts 066.037, 066.038, 162.051.

012.009 4th international conference on nuclei far from stability, held at Helsingor, Denmark, 7 - 13 June 1981.
CERN 81-09, Geneva, Switzerland, 2 Vol. 18 + 802 pp. (1981). Review in Phys. Abstr., Vol. 85, Abstr. 23372 (1982). See abstracts 061.021, 061.022.

012.010 Computer simulation systems for image processing, colloquium held at London, England, 2 October 1981.
IEE, London, England (1981). – See abstract 031.531.

012.011 A. M. Michelson - Colloquium, held at Potsdam, April 28 - 29, 1981.
Astron. Nachr., Band 303, 1 - 96 (1982). – The individual contributions are included in their corresponding subject categories – see abstracts 004.025 - 004.027, 005.007 - 005.010, 022.078 - 022.083.

012.012 Proceedings of the Johns Hopkins workshop on current problems in particle theory 5: Unified field theories and beyond. Conference held at Baltimore, Md., USA, 25 - 27 May 1981. Johns Hopkins Univ., Baltimore, Md., USA. 168 pp. (1981). – Review in Phys. Abstr., Vol. 85, Abstr. 28605 (1982). – See abstracts 162.062 - 162.065, 162.074.

012.013 Special topics in optical propagation. 28th meeting of the electromagnetic wave propagation panel, held at Monterey, Calif., USA, 6 - 10 April 1981.
P. Halley (Editor).
AGARD Conf. Proc., No. 300. AGARD, Neuilly-sur-Seine, France. 13 + 410 pp. (1981). – Review in Phys. Abstr., Vol. 85, Abstr. 28611 (1982). – See abstracts 082.020, 082.021, 084.027.

012.014 Recent progress in many-body theories. Proceedings of the second international conference, held at Oaxtepec, Mexico, 12 - 17 January 1981.
J. G. Zabolitzky, M. de Llano, M. Fortes, J. W. Clark (Editors).
Springer-Verlag, Berlin, Germany. 8 + 479 pp. (1981). ISBN 3-540-10710-X. – Review in Phys. Abstr., Vol. 85, Abstr. 28612 (1982). – See abstracts 022.088, 022.089.

012.015 Proceedings of the topical conference on low energy X-ray diagnostics, held at Monterey, Calif., USA, 8 - 10 June 1981.
AIP Conf. Proc., No. 75 (1981). – Review in Phys. Abstr., Vol. 85, Abstr. 33205 (1982). – See abstract 032.514.

012.016 Investigations of the upper atmosphere of the earth from results of observations of artificial earth satellites. Proceedings of a conference held in Dnepropetrovsk, 25 - 28 September 1979.
N. P. Erpylev, V. I. Kuryshev, M. A. Lur'e (Editors). Nablyud. Iskusstv. Nebesn. Tel, No. 79, 101 pp. (1981). In Russian. – The individual contributions are included in their corresponding subject categories – see abstracts 052.014 - 052.021, 054.004, 054.005, 082.022 - 082.025.

012.017 6th international conference on infrared and millimeter waves, held at Miami Beach, Fla., USA, 7 - 12 December 1981.
IEEE, New York, USA. 24 + 292 pp. (1981). – Review in Phys. Abstr., Vol. 85, Abstr. 38365 (1982). – See abstracts 033.009 - 033.013, 034.019 - 034.021, 077.011, 077.012.

012.018 Proceedings of the 27th international instrumentation symposium, held at Indianapolis, Ind., USA, 27 - 30 April 1981.
Instrumentation in the Aerospace Industry. Vol. 27. Advances in Test Measurement. Vol. 18. ISA, Research Triangle Park, NC, USA. 2. Vol., 784 pp. (1981). ISBN 0-87664-515-5. Review in Phys. Abstr., Vol. 85, Abstr. 38367 (1982). – See abstract 051.007.

012.019 Proceedings of the Southwest Regional Conference for Astronomy and Astrophysics. Vol. 7.
Albuquerque, New Mexico, May 23, 1981.
P. F. Gott, P. S. Riherd (Editors).
Southwest Regional Conference for Astronomy and Astrophysics, Dep. Phys., Texas Tech. Univ., Lubbock, Texas. 178 pp. (1982). ISSN 0147-2003. – The individual contributions are included in their corresponding subject categories – see abstracts 004.034, 014.010, 022.105, 032.518, 034.081, 072.029, 098.030, 099.050, 107.013, 117.051, 119.039 - 119.042, 131.138, 141.097, 141.098, 155.023.

012.020 Monde, Ringe und Asteroide im Planetensystem. Astronomisches Seminar, Universität Heidelberg, Sommersemester 1981.
H. Fechtig, C. Leinert, G. E. Morfill, H. Scholl (Editors). Published by Max-Planck-Institut für Kernphysik, Heidelberg, F. R. Germany. 2 + 191 pp. (1981). – The individual contributions are included in their corresponding subject categories– see abstracts 091.034 - 091.036, 098.032 - 098.034, 099.052.

012.021 Gauge theories, massive neutrinos and proton decay. Proceedings of Orbis Scientiae 1981. Conference held at Coral Gables, Fla., USA, 19 - 22 January 1981.
B. Kursunoglu; A. Perlmutter (Editors).
Plenum Press, New York. 9 + 392 pp. (1981). ISBN 0-306-40821-X. – Review in Phys. Abstr., Vol. 85, Abstr. 42794 (1982). – See abstracts 162.103, 162.105, 162.116.

012.022 Planetary exploration. A discussion held 4 and 5 November 1980 under the leadership of H. Massey, S. K. Runcorn, J. E. Guest, G. E. Hunt, M. M. Woolfson, with

concluding remarks by W. H. McCrea.
Philos. Trans. R. Soc. London A, Vol. 303, No. 1477, p. 213 - 381 (1981). — The individual contributions are included in their corresponding subject categories — see abstracts 091.040 - 091.042, 091.075, 093.029, 094.033, 098.035, 099.061, 100.055, 102.023, 105.175, 107.016, 107.017.

012.023   **Astrophysical parameters for globular clusters.**
International Astronomical Union. Colloquium No. 68, held at Union College, Schenectady, New York, 7 - 10 October, 1981.
A. G. D. Philip, D. S. Hayes (Editors), with a concluding summary by S. van den Bergh.
L. Davis Press, Inc.,Schenectady, New York, U.S.A. 16 + 614 pp. Price $ 25.00 (1981). ISBN 0-9607902-1-7. — The individual contributions are included in their corresponding subject categories — see abstracts 064.046, 065.030 - 065.033, 114.090 - 114.093, 122.073 - 122.076, 151.053, 154.021 - 154.076, 158.160, 159.016 - 159.025.

012.024   **Proceedings of the Twelfth Lunar and Planetary Science Conference**, Sect. 1: The moon. Sect. 2: Planets, asteroids and satellites. Houston, Texas, March 16 - 20, 1981.
R. B. Merrill, R. Ridings (Managing Editors), F. Hörz, W. Mendell, D. Phinney, W. C. Phinney (Editors).
Proceedings of Lunar and Planetary Science, Vol. 12, Part B. Geochim. Cosmochim. Acta, Suppl. 16. Pergamon Press, New York—Oxford—Toronto—Sydney—Frankfurt—Paris. 12 + 10 + 1825 + 20 + 2 pp. Price $ 175.00 (1982). ISBN 0-08-028074-9. — The individual contributions are included in their corresponding subject categories — see abstracts 022.116 - 022.121, 061.037, 091.043, 091.044, 093.030 - 093.032, 094.034 - 094.037, 094.533 - 094.595, 094.901, 097.089 - 097.095, 098.036, 099.062 - 099.067, 105.176 - 105.204, 105.901.

012.025   **Extragalactic radio sources.**
International Astronomical Union, Symposium No. 97, held in Albuquerque, U.S.A., August 3 - 7, 1981.
D. S. Heeschen, C. M. Wade (Editors), with an introductory lecture by J. H. Oort.
D. Reidel Publishing Company, Dordrecht, Holland — Boston, U.S.A. — London, England. 17 + 490 pp. Price Dfl. 125.00, $ 54.50 cloth, Dfl. 60.00, $ 26.00 paper (1982). ISBN 90-277-1384-7, ISBN 90-277-1385-5 (paper). — The individual contributions are included in their corresponding subject categories — see abstracts 062.066, 066.103, 125.301, 141.101 - 141.200, 142.043, 142.044, 158.161 - 158.174, 160.036 - 160.042, 161.005, 162.120.

012.026   **Applications of modern dynamics to celestial mechanics and astrodynamics.** Proceedings of the NATO Advanced Study Institute, held at Cortina d'Ampezzo, Italy, August 2 - 14, 1981. V. Szebehely (Editor).
NATO Advanced Study Institutes Series C, Vol. 82.
D. Reidel Publishing Company, Dordrecht, Holland — Boston, U.S.A. — London, England. 18 + 373 pp. Price Dfl. 100.00, $ 43.50 (1982). ISBN 90-277-1390-1. — The individual contributions are included in their corresponding subject categories — see abstracts 021.020 - 021.027, 022.123 - 022.131, 042.051 - 042.075, 043.001, 052.031 - 052.034, 094.038, 098.038, 098.039, 100.058, 100.059, 111.012, 151.054.

012.027   **Excitation and broadening in atomic spectra of astrophysical interest.** Troisième cycle inter-universitaire en astronomie et astrophysique organisé sous les auspices du F. N. R. S., Département d'Astrophysique. Université de l'Etat à Mons, 4 - 7 mai 1981.
S. Volonté, L. Houziaux (Editors).
Published by Institut d'Astrophysique, Université de Liège, 4200 Cointe-Ougrée, Belgium. 6 + 134 pp. (1981). The indi-

vidual contributions are included in their corresponding subject categories — see abstracts 022.132 - 022.135.

012.028   **Special issue on astronomical geodynamics.**
The second Annual Meeting on Astronomical Geodynamics, held 9 - 16 October 1979. Organized by the Institute of Geodesy and Geophysics of the Chinese Academy of Sciences at Wuhan. The Publishing House of Surveying and Mapping. 12 + 192 pp. (1981). In Chinese. — The individual contributions within the subject scope of Astronomy and Astrophysics Abstracts are included in their corresponding categories — see abstracts 002.038, 031.575, 031.576, 035.006, 041.023 - 041.025, 044.015 - 044.019, 045.009 - 045.018, 046.004, 081.043, 081.044, 094.040.

012.029   **Abstracts of the 4th conference of the Subcommission No. 4 "Magnetic Stars".** Committee of Multilateral Cooperation of the Academies of Sciences of the Socialist Countries "Physics and the evolution of stars". Special Astrophysical Observatory of the USSR Academy of Sciences, 1980, October 6 - 10.
V. E. Karachentseva, N. M. Chunakova, V. A. Lipovetskij, L. I. Snezhko (Editors), with concluding remarks by A. Z. Dolginov.
Soobshch. Spets. Astrofiz. Obs., Vyp. 32, 81 pp. (1981). The individual contributions are included in their corresponding subject categories — see abstracts 013.031, 021.032, 032.029, 034.050, 062.078, 063.043, 064.056, 064.057, 071.037, 111.014, 113.039 - 113.044, 114.114 - 114.127, 116.019, 116.023 - 116.038, 120.017, 121.024, 122.093 - 122.095, 126.030, 153.033.

012.030   **Be stars.**
International Astronomical Union. Symposium No. 98, held in Munich, Federal Republic of Germany, April 6 - 10, 1981.   M. Jaschek, H.-G. Groth (Editors).
D. Reidel Publishing Company, Dordrecht, Holland — Boston, U.S.A. — London, England. 15 + 523 pp. Price Dfl. 130.00 (1982). ISBN 90-277-1366-9 ISBN 90-277-1367-7 (pbk.). — The individual contributions are included in their corresponding subject categories — see abstracts 002.039 - 002.041, 013.032 - 013.034, 031.577, 064.058 - 064.062, 065.039, 112.027 - 112.035, 113.045 - 113.055, 114.128 - 114.157, 115.011 - 115.014, 116.039 - 116.045, 117.069, 117.070, 122.096, 122.097, 142.050 - 142.052.

012.031   **Scientific importance of high angular resolution at infrared and optical wavelengths.** ESO Conference, Garching, 24 - 27 March 1981.
M. H. Ulrich, K. Kjär (Editors).
Proceedings published and distributed by the European Southern Observatory, Karl-Schwarzschild-Str. 2, D-8046 Garching bei München. 7 + 443 pp. (1981). — The individual contributions are included in their corresponding subject categories — see abstracts 031.016 - 031.024, 031.578 - 031.590, 032.030, 033.021, 034.051, 091.045, 112.036, 112.037, 115.015 - 115.017, 131.169 - 131.171, 158.202 - 158.204.

012.032   **GIRL — German Infrared Laboratory.** Wissenschaftliche Instrumentierung, Beobachtungsprogramme und Mission. Ausarbeitung der Programm-Diskussion der Experimentatoren in der MPG-Tagungsstätte Schloß Ringberg, Tegernsee, Mai 1980.
D. Lemke, K. Proetel (Editors), with contributions by S. Drapatz, R. Hofmann, R. Katterloher, H. Trinks, D. Offermann, G. Lange, H. Rippel, D. Lemke, W. Martin, J. Riedinger, H. Elsässer, M. Grewing, P. Preussner, G. Klipping, H. D. Denner, U. Ruppert, K. Proetel, H. J. Bolle.
Available from D. Lemke, Max-Planck-Institut für Astronomie, Königstuhl, D-6900 Heidelberg 1 and from K. Proetel, DFVLR-BPT, D-5000 Köln 90. 243 pp. (1982).

012.033 **The comparative study of the planets.** Proceedings
of the NATO Advanced Study Institute held at
Vulcano (Aeolian Islands), Italy, September 14 - 25, 1981.
A. Coradini, M. Fulchignoni (Editors).
NATO Advanced Study Institutes Series C, Vol. 85. D. Reidel
Publishing Company, Dordrecht, Holland—Boston, U.S.A.—
London, England. 11 + 516 pp. Price Dfl. 140.00, $ 59.50
(1982). ISBN 90-277-1406-1. — The individual contributions
are included in their corresponding subject categories — see
abstracts 022.153 - 022.156, 046.005, 080.048, 081.045 -
081.047, 091.047 - 091.058, 094.043, 097.107, 098.074,
099.073 - 099.075, 100.064, 102.026, 103.203, 104.023,
105.208 - 105.210, 107.024 - 107.026.

012.034 **The most massive stars.** Proceedings of an ESO
workshop held at Garching, FR Germany, 23 - 25
November 1981.
S. D'Odorico, D. Baade, K. Kjär (Editors), with concluding
remarks by S. van den Bergh, R. M. Humphreys, C. Chiosi,
A. Maeder, R. P. Kudritzki, M. W. Feast.
Published and distributed by European Southern Observatory,
Karl-Schwarzschild-Str. 2, D-8046 Garching, FRG. 5 + 365 pp.
Price DM 50.00 (1982). — The individual contributions are
included in their corresponding subject categories — see
abstracts 065.044 - 065.048, 112.039, 112.040, 114.168 -
114.170, 115.018 - 115.023, 117.094, 117.095, 122.145 -
122.148, 131.179, 155.029, 155.030, 158.216, 158.217,
159.034, 162.160.

012.035 **High-precision Earth rotation and Earth-Moon
dynamics. Lunar distances and related observations.**
Proceedings of the 63rd Colloquium of the International
Astronomical Union, held at Grasse, France, May 22 - 27,
1981.   O. Calame (Editor).
Astrophysics and Space Science Library, Vol. 94.
D. Reidel Publishing Company, Dordrecht, Holland —
Boston, U.S.A. — London, England. 20 + 354 pp. Price
Dfl. 125.00, US $ 54.50 (1982). ISBN 90-277-1405-3. — The
individual contributions are included in their corresponding
subject categories — see abstracts 004.064, 031.608, 042.110,
043.006 - 043.010, 044.043 - 044.053, 045.027 - 045.031,
046.009, 081.053, 094.047 - 094.056, 141.287.

012.036 **Uranus and the outer planets.** Proceedings of the
IAU/RAS colloquium No. 60, held at the University
of Bath, U.K., 14 - 16 April 1981.
G. Hunt (Editor), with introductory remarks by
A. W. Wolfendale and closing remarks by P. A. Wayman.
Cambridge University Press, Cambridge—London—New York—
New Rochelle—Melbourne—Sydney. 9 + 307 pp. Price
DM 84.75 (1982). ISBN 0-521-24573-7. — The individual
contributions are included in their corresponding subject
categories — see abstracts 004.075 - 004.078, 032.540,
051.016, 101.014 - 101.025.

012.037 **Variations of the solar constant.** Proceedings of a
workshop held at Goddard Space Flight Center,
Greenbelt, Maryland, November 5 - 7, 1980.
S. Sofia (Editor).
NASA Conf. Publ., NASA CP-2191. 6 + 296 pp. (1981).
The individual contributions are included in their corresponding subject categories — see abstracts 032.541, 071.058,
072.088, 072.089, 080.075 - 080.093, 085.031 - 085.033,
113.091.

012.038 **Molecules in interstellar space.** Proceedings of a
Royal Society discussion meeting held on 20 and
21 May 1981.   A. Carrington, D. A. Ramsay (Editors), with
introductory remarks by G. Herzberg.
Philos. Trans. R. Soc. London, Ser. A, Vol. 303, No. 1480,
p. 463 - 631 (1981). Published as a special issue by the Royal
Society, London, Price £ 20.00 (1982). ISBN 0-85403-180-4.

The individual contributions are included in their corresponding subject categories — see abstracts 015.035, 015.036,
022.182, 022.183, 112.058, 131.243 - 131.250.

012.039 **Scientific and experimental aspects of the Giotto
mission.** Proceedings of an international meeting on
27 - 28 April 1981 at Noordwijkerhout, The Netherlands.
B. Battrick, J. Mort (Editors), with a foreword by R. Reinhard.
ESA SP-169. ESA Scientific and Technical Publications
Branch, ESTEC, Noordwijk, The Netherlands. 19 + 135 pp.
Price FF 80.00 (1981). ISSN 0379-6566. — The individual
contributions are included in their corresponding subject
categories — see abstracts 032.543 - 032.552, 051.017,
102.035 - 102.037, 103.510.

012.040 **Sun and planetary system.** Proceedings of the Sixth
European Regional Meeting in astronomy, held in
Dubrovnik, Yugoslavia, 19 - 23 October 1981.
W. Fricke, G. Teleki (Editors).
Astrophysics and space science library, Vol. 96. D. Reidel
Publishing Company, Dordrecht, Holland—Boston, U.S.A.—
London, England. 13 + 538 pp. Price cloth Dfl. 150.00,
US $ 65.00 (1982). ISBN 90-277-1429-0. — The individual
contributions are included in their corresponding subject
categories — see abstracts 009.028, 021.047, 031.625 -
031.628, 041.035, 041.036, 042.119 - 042.122, 043.012 -
043.014, 044.057 - 044.062, 045.033, 045.034, 046.012 -
046.014, 047.026, 052.041, 062.122, 071.060 - 071.066,
072.092 - 072.096, 073.140 - 073.145, 074.096, 074.097,
075.039, 077.060, 077.061, 080.097 - 080.108, 082.073 -
082.085, 084.104, 085.034, 091.063 - 091.070, 092.009,
094.061, 094.062, 097.114, 098.086 - 098.100, 099.087,
100.080, 102.039 - 102.049, 103.511, 103.741, 104.027 -
104.029, 106.033 - 106.038, 107.032, 107.033.

012.041 **Scientific aspects of the Hipparcos space astrometry
mission.** Proceedings of an international colloquium
jointly organised by the European Space Agency, Copenhagen
University Observatory and Observatoire de Strasbourg, held at
Strasbourg, 22 - 23 February 1982.
M. A. C. Perryman, T. D. Guyenne (Editors) with a preface by
E. Høg, C. Jaschek, M. A. C. Perryman and concluding remarks
by P. Lacroute.
ESA SP-177. ESA Scientific and Technical Publications Branch,
ESTEC, Noordwijk, The Netherlands. 8 + 242 pp. Price 80 FF.
(1982). The individual contributions are included in their
corresponding subject categories — see abstracts 032.553,
041.037 - 041.040, 043.015 - 043.019, 046.015, 051.018 -
051.026, 065.067, 066.144, 098.103, 111.019 - 111.028,
113.093 - 113.095, 115.035 - 115.037, 118.035 - 118.037,
121.040, 122.201, 131.256, 135.051, 153.049, 155.052,
155.053.

012.042 **Workshop on pulsating B stars.** Proceedings of a
workshop held at Nice Observatory, June 1 - 5,
1981. (Centenaire de l'Observatoire de Nice 1881 - 1981).
M. Auvergne, A. Baglin, D. Ducatel, J.-M. Le Contel,
P. J. Morel, J.-P. Sareyan, J.-C. Valtier, C. Sterken (Editors),
with concluding remarks by J.-P. Zahn.
Nice Observatory. 417 pp. Price $ 10.00 (1981). — The indivi-
dual contributions are included in their corresponding subject
categories — see abstracts 031.631, 031.632, 112.062,
112.063, 115.038, 117.037, 122.202 - 122.243.

012.043 **Gamma-ray astronomy.** A Royal Society discussion,
held in London, 27 - 28 November 1980.
H. Massey, A. W. Wolfendale, R. D. Wills (Editors), with
introductory remarks by A. W. Wolfendale and concluding
remarks by R. D. Wills.
Philos. Trans. R. Soc. London, Ser. A, Vol. 301, No. 1462,
p. 489 - 703 (1981). Published as a special issue by the
Royal Society, London, 9 + 211 pp. Price £ 24.00 (1981).

ISBN 0-85403-170-7. — The individual contributions are included in their corresponding subject categories — see abstracts 002.069, 032.554, 034.082, 034.083, 051.027, 066.145, 131.257, 141.551, 142,528 - 142.545, 157.012 - 157.014.

012.044   **The Nordic Astronomical Meeting in Lund.**
October 19 - 20, 1981.
G. Larsson-Leander (Editor), with an introduction by A. Elvius.
Rep. Obs. Lund, No. 18, 132 pp. (1982). — The individual contributions are included in their corresponding subject categories — see abstracts 009.029, 013.053 - 013.056, 031.634, 032.046, 032.047, 033.040, 034.085, 051.028, 064.089, 074.102, 075.041, 102.053, 113.096, 114.220, 114.221, 115.039, 118.038, 121.041, 122.244, 122.245, 125.054, 131.258 - 131.262, 133.027, 151.090, 153.050, 153.051, 154.020, 155.054, 155.055, 158.328.

012.045   **Second Cambridge workshop on cool stars, stellar systems, and the sun.** Vol. I. Proceedings of a work-shop held at the Harvard-Smithsonian Center for Astrophysics, Cambridge, MA 02138, October 21 - 23, 1981.
M. S. Giampapa, L. Golub (Editors).
Smithsonian Astrophys. Obs., Spec. Rep. No. 392, 9 + 262 pp. (1982). — The individual contributions are included in their corresponding subject categories — see abstracts 062.128, 064.090 - 064.099, 065.071, 072.099, 072.100, 073.148 - 073.150, 074.103 - 074.105, 075.042, 112.064, 114.222, 116.060, 119.130 - 119.133, 142.109.

012.046   **Cosmology and particles.**
Proceedings of the 16th Rencontre de Moriond Astrophysics Meeting, held at Les-Arcs-Savoie-France, 15 - 21 March 1981.
J. Audouze, P. Crane, T. Gaisser, D. Hegyi, J. Tran Thanh Van (Editors).
Editions Frontières, Dreux, France. 344 pp. Price $ 37.00 (1982). ISBN 2-86332-009-2. — The individual contributions are included in their corresponding subject categories — see abstracts 061.040, 142.110, 143.036 - 143.040, 158.329, 158.330, 162.200 - 162.211.

012.047   **Astrophysical cosmology.**
Proceedings of the study week on cosmology and fundamental physics, held at Città del Vaticano, September 28 - October 2, 1981.
H. A. Brück, G. V. Coyne, M. S. Longair (Editors), with an introductory survey by M. J. Rees.
Pontificiae Academiae Scientiarum Scripta Varia, No. 48, 35 + 600 pp. Price $ 58.00 (institutions), $ 43.00 (individuals), (1982). Available from Specola Vaticana, I–00120 Città del Vaticano. — The individual contributions are included in their corresponding subject categories — see abstracts 141.313 - 141.318, 142.111, 158.331 - 158.335, 160.081, 162.212 - 162.225.

012.048   **Extragalactic molecules.** Proceedings of a workshop held at the National Radio Astronomy Observatory Green Bank, West Virginia, November 2 - 4, 1981.
L. Blitz, M. L. Kutner (Editors).
Publications Division, NRAO, P. O. Box 2, Green Bank, West Virginia 24944. 12 + 214 pp. (1982). — The individual contributions are included in their corresponding subject categories — see abstracts 002.096, 062.130, 131.268 - 131.277, 132.080, 151.096, 156.021, 158.338 - 158.357, 159.040.

012.049   **Life in the Universe.** Proceedings of a conference, held at Moffett Field, Calif., June 1979.
J. Billingham (Editor).
MIT Press, Cambridge, Mass. 22 + 461 pp. Price $ 20.00, £ 14.00 cloth; $ 12.50, £ 8.75 paper (1982).

ISBN 0-262-02155-2 cloth; ISBN 0-262-52062-1 paper.
Review in Sky Telesc., Vol. 64, 47 (1982).

012.050   **Neutrino 81.** Proceedings of a conference, held at Maui, Hawaii, July 1981.
R. J. Cence, E. Ma, A. Roberts (Editors).
University of Hawaii, Department of Physics and Astronomy, Honolulu. Vol. 1: 18 + 510 pp, Vol. 2: 14 + 512 pp. Price $ 35.00 (1981). — Review in Science, Vol. 216; 617 - 618; 1982 *(D. Schramm, J. N. Fry)*.

012.051   **Compte rendu de l'Ecole d'été d'Astronomie,** Grasse, 27 August - 5 September 1980.
L. Gouguenheim (Editor).
Published by l'Observatoire de Paris. 450 pp. (1981).
Review in Astronomie, Vol. 96, 320; 1982 *(F. Chollet)*.

012.052   **To fulfill a vision.** Jerusalem Einstein Centennial Symposium on gauge theories and unification of physical forces. Conference held at Jerusalem, 20 - 23 March 1979.   Y. Ne'eman (Editor).
Addison-Wesley, Reading, MA, USA. 31 + 279 pp. (1981).
ISBN 0-201-05289-X. — Review in Phys. Abstr., Vol. 85, Abstr. 14983 (1982).

012.053   **Solar active regions.** A monograph from Skylab solar workshop III.   F. Q. Orrall (Editor).
Colorado Associated University Press, Boulder, Colo. 10 + 350 pp. Price $ 17.50 (1981). ISBN 0-87081-085-5.
Reviews in Nature, Vol. 296, 784; 1982 *(J. C. Brown)*; Science, Vol. 215, 1606 - 1607; 1982 *(R. Rosner)*.

012.054   **Tenth Texas Symposium on relativistic astrophysics.** Conference held at Baltimore, December 1980.
R. Ramaty, F. C. Jones (Editors).
Annals of the New York Academy of Sciences, Vol. 375. New York Academy of Sciences, New York. 10 + 468 pp. Price $ 100.00 (1981). ISBN 0-89766-139-7. — From Science, Vol. 216, 406 (1982).

012.055   **ULF pulsations in the magnetosphere.**
Reviews from the Special Sessions on geomagnetic pulsations at XVII General Assembly of the International Union for Geodesy and Geophysics, Canberra, December 1979.   D. J. Southwood (Editor).
Advances in Earth and Planetary Sciences, 11. Supplement Issue to Journal of Geomagnetism and Geoelectricity.
D. Reidel Publishing Company, Dordrecht, Holland — Boston, U.S.A. — London, England. 200 pp. Price Dfl. 60.00, $ 29.50 (1981). ISBN 90-277-1232-8. — From Nature, Vol. 296, 183 (1982).

012.056   **Archaeoastronomy in the Americas.** Papers from a conference, Santa Fé, N. M., 1979.
R. A. Williamson (Editor).
Ballena Press Anthropological Papers No. 22. Ballena Press, Los Altos, Calif., and Center for Archaeoastronomy, University of Maryland, College Park. 406 pp. Price $ 19.95 (1981).
From Science, Vol. 216, 1104 (1982).

012.057   **Einstein: 1879 - 1955.** — Colloque du centenaire, Collège de France, 6 - 9 juin 1979.
Editions du CNRS, Paris. 318 pp. Price FF 90.00 (1980).
Review in Ciel Terre, Vol. 98, 56; 1982 *(J. Demaret)*.

012.058   **Applications of space developments.**
31st International Astronautical Congress.
Conference held at Tokyo, Japan, 22 - 27 September 1980.
Acta Astronaut., Vol. 8, No. 5 - 6 (1981). — Review in Phys. Abstr., Vol. 85, Abstr. 6697 (1982).

012.059 **The physical basis of the ionosphere in the solar-terrestrial system.** Conference held at Pozzuoli, Italy. 28 - 31 October 1980.
AGARD Conference Proceedings No. 295. AGARD-CP-295.
AGARD, Neuilly-sur-Seine, France. 8 + 494 pp. (1981).
Review in Phys. Abstr., Vol. 85, Abstr. 18817 (1982).

012.060 **Physics of the Earth's interior.** Proceedings of the 78th course 'Enrico Fermi'.
A. M. Dziewonski, E. Boschi (Editors).
North-Holland, Amsterdam. 716 pp. Price $ 146.25 (1980).
ISBN 0-444-85461-4. — Review in Phys. Earth Planet. Inter., Vol. 28, 182 - 184; 1982 *(D. C. Tozer)*.

013 Reports on Astronomy in Various Countries and Particular Fields, International Cooperation

013.001 **L'astronomie des étoiles doubles à l'heure de l'astrométrie spatiale.** J. Dommanget.
Astronomie, Vol. 96, 15 - 27 (1982).

013.002 **Sterrenkunde in Finland.** G. W. E. Beekman.
Zenit, 9. Jaarg., 18 - 23 (1982).

013.003 **De toekomst van de Nederlandse sterrenkunde.**
G. Schilling.
Zenit, 9. Jaarg., 24 - 26 (1982).

013.004 **Hipparcos: Europe's astrometric satellite.**
N. Kidger.
Spaceflight, Vol. 24, 165 - 166 (1982).

013.005 **L'innovazione tecnica in campo amatoriale: traguardi e proposte.** S. Ghedini.
Astronomia, N. 3, p. 29 - 38 (1981).

013.006 **Nachbarn der Milchstraße.**
R. Wielebinski.
Umschau, 82. Jahrg., 190 - 192, 194 (1982).

013.007 **Astronomy in the next decade.**
R. A. Schorn.
Sky Telesc., Vol. 63, 339 - 342 (1982).

013.008 **Astronomy and federal $pending.**
L. J. Robinson.
Sky Telesc., Vol. 63, 343 (1982).

013.009 **Pieces of the sky.**
A. Chaikin.
Sky Telesc., Vol. 63, 344 - 34 (1982).

013.010 **Astronomy and astrophysics for the 1980s.**
G. B. Field.
Phys. Today, Vol. 35, No. 4, p. 46 - 52 (1982).

013.011 **Planetary data at the National Space Science Data Center.** R. W. Vostreys.
Reports of Planetary Geology Program — 1981, (see 003.001), p. 526 (1981). – Abstract.

013.012 **Planetary Geology Speakers Bureau.**
R. Greeley, R. D'Alli.
Reports of Planetary Geology Program — 1981, (see 003.001), p. 527 (1981). – Abstract.

013.013 **Towards an international space co-operation agency.**
S. K. Hurst.
Spaceflight, Vol. 24, 268 - 269 (1982).

013.014 **The need for inter-agency collaboration on missions to Halley's comet.** E. A. Trendelenburg.
ESA Bull., No. 29, p. 66 - 67 (1982).
Paper presented at the meeting "Space missions to Halley's comet and related activities", Padova, 13 - 15 September 1981.

013.015 **Vorbereitungen auf Komet Halleys Wiederkehr 1985/86.** J. Rahe, R. L. Newburn, Jr.
Mitt. Astron. Ges., Nr. 55, p. 19 (1982). – Abstract.

013.016 **Projekt Europäisches Astroplane (EAP).**
J. Schmid-Burgk.
Mitt. Astron. Ges., Nr. 55, p. 155 - 157 (1982).

013.017 **Operation Spacewatch.** A group of astronomers is preparing to scan the skies for asteroids.
M. M. Waldrop.
Science, Vol. 216, 42 (1982).

013.018 **Astronomy and astrophysics for the 1980's.**
M. M. Waldrop.
Science, Vol. 216, 282 (1982).

013.019 **De NVWS werkgroep Zon.** A. Mak.
Zenit, 9. Jaarg., 222 - 223 (1982).

013.020 **Development of rocket investigations of the circumterrestrial space in the USSR.**
L. A. Vedeshin.
Probl. kosm. issled. Moskva, 1981, p. 84 - 99. In Russian. Abstr. in Ref. zh., 62. Issled. kosm. prostranstva, 3.62.35 (1982).

013.021 **Astronomy in New Zealand.** G. L. Blow.
Sky Telesc., Vol. 63, 555 - 557 (1982).

013.022 **El desarrollo de la astronomía en la Argentina.**
C. J. Lavagnino.
Bol. Asoc. Argentina Astron., No. 18, p. 93 - 96 (1980).

013.023 **Sverdlovsk — the capital of USSR exact time in the years of the Great Patriotic War.**
P. G. Kulikovskij.
Dvizhenie iskusstv. i estestv. nebes. tel. Sverdlovsk, 1981, p. 3 - 7. In Russian. – Abstr. in Ref. zh., 51. Astron., 4.51.14 (1982).

013.024 **Proyecto infrarrojo del IAFE.**
E. Gandolfi, A. Quaglia, M. Pupareli, C. Alarcón, L. Cerrella, A. Ringuelet.
Bol. Asoc. Argentina Astron., No. 20 - 24, p. 188 (1981). Abstract.

013.025 **Rasgos estructurales para un plan moderno de estudios astronómicos.** C. J. Lavagnino.
Bol. Asoc. Argentina Astron., No. 20 - 24, p. 305 (1981). Abstract.

013.026 **Desarrollo de la investigación astronómica en el Río de La Plata.** C. J. Lavagnino.
Bol. Asoc. Argentina Astron., No. 20 - 24, p. 306 (1981). Abstract.

013.027 **Programa para el astrolabio OPL 01 en la futura estación de Río Grande.**
C. Mondinalli, R. Perdomo.
Bol. Asoc. Argentina Astron., No. 20 - 24, p. 319 (1981). Abstract.

013.028 **Observación en 21 cm de destelladores de rayos X.**
E. Bajaja, F. R. Colomb, W. G. L. Poppel, E. M. Arnal, R. Morras, C. A. Olano.
Bol. Asoc. Argentina Astron., No. 20 - 24, p. 433 (1981). Abstract.

013.029 **El programa espectroscópico sobre cúmulos abiertos y asociaciones en el Observatorio de La**

Plata.   H. Levato, S. Malaroda.
Bol. Asoc. Argentina Astron., No. 20 - 24, p. 479 (1981).
Abstract.

013.030   **Advances in astronomy in the year 1981.**
J. Grygar.
Říše hvězd, Vol. 63, 69 - 72, 99 - 103 (1982). In Czech.

013.031   **Suggestions on cooperative observations of Ap stars
(R. V. program).**   N. S. Polosukhina.
Soobshch. Spets. Astrofiz. Obs., Vyp. 32, (see 012.029),
p. 77 (1981).

013.032   **An observing campaign for systematic photoelectric
observations of bright Be stars.**
P. Harmanec, J. Horn, P. Koubský.
Be stars, (see 012.030), p. 269 - 274 (1982).

013.033   **Spectroscopic observing campaign.**
P. K. Barker.
Be stars, (see 012.030), p. 275 - 276 (1982).

013.034   **The continuing saga of the Be stars: a summary.**
T. P. Snow, Jr.
Be stars, (see 012.030), p. 509 - 519 (1982).
The paper, written in summary of IAU Symposium 98 on
the Be stars, contains an outline of the proceedings, organized
according to physical parameters of the stars, the circumstellar
environment, model explaining the data, and a summary of
the outstanding problems. Suggested future observations and
needed theoretical work are described.

013.035   **The World Space Foundation Asteroid Project.**
J. B. Child.
I. A. P. P. P. Commun., No. 7, p. 12 - 13 (1982).

013.036   **ILOC (*International Lunar Occultation Centre*)
news.**   D. W. Dunham.
Occultation Newsl., Vol. 2, 221 - 222 (1982).

013.037   **Interviews: the voice of astronomy history.**
S. R. Weart.
Bull. American Astron. Soc., Vol. 13, 831 - 832 (1981).
Abstract.

013.038   **Design considerations for a computer based
astronomical announcement service.**
J. B. Rafert, D. B. Caton.
Bull. American Astron. Soc., Vol. 13, 838 - 839 (1981).
Abstract.

013.039   **Interactive computer facilities for analyzing
IUE data.**   A. Boggess, S. R. Heap,
E. W. Brugel, J. M. Mead, F. H. Schiffer, T. P. Snow, Jr.
Bull. American Astron. Soc., Vol. 13, 839 - 840 (1981).
Abstract.

013.040   **Very long baseline interferometry: 1967 to 2000.**
M. Reid.
Bull. American Astron. Soc., Vol. 13, 897 (1981). – Abstract.

013.041   **A phase-coherent link between VLBI stations via
synchronous satellite.**
S. H. Knowles, W. B. Waltman, W. H. Cannon, D. A. Davidson,
W. T. Petrachenko, J. L. Yen, J. Popelar, D. N. Fort, J. Galt.
Bull. American Astron. Soc., Vol. 13, 899 (1981). – Abstract.

013.042   **Mark III VLBI: from "light bulb" to Ap. J. or
J. G. R.**
N. R. Vandenberg, D. B. Shaffer, T. A. Clark.
Bull. American Astron. Soc., Vol. 13, 899 - 900 (1981).
Abstract.

013.043   **Dynamics of the development of radio astronomy.
Part 1. Scientific researches of the years 1965 -
1980.**   Yu. P. Ilyasov.
Fiz. inst. AN SSSR. Prepr., 1981, No. 164, 35 pp. In Russian.
Abstr. in Ref. zh., 51. Astron., 5.51.44 (1982).

013.044   **The first three years of IUE.**
P. M. Gondhalekar, C. Jordan, A. J. Meadows,
K. Nandy, M. V. Penston, M. Pettini, M. C. W. Sandford,
J. A. J. Whelan, A. J. Willis, R. Wilson.
Rutherford Appleton Lab., Sci. Eng. Res. Council, RL-81-091,
55 pp. (1981).

013.045   **A long period eclipsing binary project – five years of
observations at ESO.**
P. Ahlin, A. Sundman.
Messenger, No. 28, p. 30 - 33 (1982).

013.046   **General survey of radio astronomy problems.**
R. Wielebinski.
Detrimental activities in space, (see 003.008), p. 23 - 24 (1982).

013.047   **Protection of radio astronomy against interference
from spacecraft after the World Administrative
Radio Conference, Geneva, 1979.**   B.-H. Grahl.
Detrimental activities in space, (see 003.008), p. 25 - 26 (1982).

013.048   **Decisions of WARC (*World Administrative Radio
Conference*), 1979.**   F. Horner.
Detrimental activities in space, (see 003.008), p. 27 (1982).

013.049   **Proposal for an international institute for space
sciences and electronics and for a giant equatorial
radio telescope as a collaborative effort of the developing
countries.**   G. Swarup.
Bull. Astron. Soc. India, Vol. 9, 269 - 277 (1981).

013.050   **Plans for upcoming asteroidal occultations.**
D. W. Dunham.
Occultation Newsl., Vol. 2, 202 - 204 (1982).

013.051   **Proposta all'Unione Astrofili Italiani.**
P. Bianucci.
Orione, Vol. 3, 70 - 72 (1982).

013.052   **The future of planetary exploration in the United
States.**   C. R. Chapman.
Strolling Astron., Vol. 29, 121 - 123 (1982).

013.053   **Research on galaxies in Uppsala – some current
projects.**   N. Bergvall, A. Ekman.
Rep. Obs. Lund, No. 18, (see 012.044), p. 20 - 22 (1982).

013.054   **Astrophysical global interferometry at Onsala
Space Observatory.**   L. B. Bååth.
Rep. Obs. Lund, No. 18, (see 012.044), p. 23 - 25 (1982).

013.055   **The rocket experiment PIRAT (*Pointed Infrared
Astronomical Telescope*).**
L. Nordh, G. Olofsson.
Rep. Obs. Lund, No. 18, (see 012.044), p. 38 - 40 (1982).

013.056   **Three polarimetric observing programmes at the
Metsähovi Observatory of the University of
Helsinki.**   T. Markkanen.
Rep. Obs. Lund, No. 18, (see 012.044), p. 72 - 73 (1982).

013.057   **Report of IAU Commission 6: Astronomical
telegrams (Télégrammes astronomiques).**
J. Hers.
Trans. IAU, Vol. XVIIIA, (see 003.013), p. 19 - 20 (1982).

013.058 **Report of IAU Commission 38: Exchange of astronomers** (Echange des astronomes).
J. Delhaye.
Trans. IAU, Vol. XVIIIA, (see 003.013), p. 527 - 528 (1982).

013.059 **Project MERIT.** Report on the Short Campaign and Grasse workshop with observations and results on earth-rotation during 1980 August - October.
G. A. Wilkins, M. Feissel (Editors), with contributions by
G. A. Wilkins, K. Yokoyama, F. Nouel, L. Aardoom,
J. D. Mulholland, W. J. Klepczynski, W. E. Carter, M. Feissel.
Published by the Royal Greenwich Observatory, on behalf of the IAU/IUGG Joint Working Group on the Rotation of the Earth. Royal Greenwich Observatory, Herstmonceux Castle, Hailsham, East Sussex, BN27 1RP, England. 10 + 117 pp. (1982).

013.060 **Present status of the Texas radio survey.**
P. Hemenway.
Abh. Hamburger Sternw., Band 10, 144 (1982).

Plans for a Large European Solar Telescope – LEST.
See Abstr. 032.047.

A note on the initial results and future plans of project MERIT. See Abstr. 044.058.

New impetus to the exploration of the solar system.
See Abstr. 091.063.

Space missions to Halley's comet and related activities. See Abstr. 103.503.

014 Teaching in Astronomy

014.001 **Divulgazione dell'astronomia nelle scuole.**
L. Prestinenza.
Astronomia, N. 3, p. 39 - 41 (1981).

014.002 **Teaching astronomy at university.**
D. McNally.
European J. Phys., Vol. 2, 117 - 126 (1981). – Abstr. in Phys. Abstr., Vol. 85, Abstr. 14792 (1982).

014.003 **The SC1 chart as a celestial cylinder.**
D. L. Manley.
Sky Telesc., Vol. 63, 463 - 465 (1982).

014.004 **The Helios planetarium. An astronomical teaching instrument.** J. Evans.
Space Educ., Vol. 1, 104 - 107 (1982).

014.005 **A demonstration model for near-polar orbits.**
J. R. Millburn.
Space Educ., Vol. 1, 108 - 111 (1982).

014.006 **Polaritäten im Fixsternbereich.** T. Schmidt.
Sterne Weltraum, 21. Jahrg., 28 - 30 (1982).

014.007 **Astronomie in der Schule – wie denn?**
T. Schmidt.
Sterne Weltraum, 21. Jahrg., 76 - 77 (1982).

014.008 **Die Entfernung der Planeten.** T. Schmidt.
Sterne Weltraum, 21. Jahrg., 118 - 119 (1982).

014.009 **Die Mobile Sternwarte Bochum.** T. H. Weyer.
Sterne Weltraum, 21. Jahrg., 162 - 163 (1982).

014.010 **Putting modern astrophysics into astronomy laboratory classes.** R. R. Robbins.
Proc. Southwest Reg. Conf., Vol. 7, (see 012.019), 85 - 95 (1982).

014.011 **Vědomosti žáků z astrofyziky.**
M. Široká, J. Široký.
Univerzita Palackého, Olomouc. With abstract in Russian and English. 96 pp. (1982). – Review in Říše hvězd, Vol. 63, 22 - 23 (1982). In Czech.

014.012 **Philosophy and the pseudosciences as useful content in introductory astronomy and physics courses.** S. P. Kanagy II.
Bull. American Astron. Soc., Vol. 13, 813 (1981). – Abstract.

014.013 **Computer aided instruction of ancient astronomy.**
N. A. Roughton.
Bull. American Astron. Soc., Vol. 13, 832 (1981). – Abstract.

014.014 **Tecniche di rilevamento di dati dalle immagini fotografiche: una esercitazione per la scuola media superiore.** E. De Cesaris, P. Fano, N. Lanciano, M. Poscolieri.
G. Astron., Vol. 7, 311 - 327 (1981).

014.015 **Progetto di insegnamento delle scienze sperimentali secondo i nuovi programmi per la scuola media statale.** E. Proverbio, S. Lai.
G. Astron., Vol. 7, 329 - 345 (1981).

014.016 **Report of IAU Commission 46: Teaching of astronomy** (Enseignement de l'astronomie).
D. G. Wentzel.
Trans. IAU, Vol. XVIIIA, (see 003.013), p. 633 - 634 (1982).

015 Miscellaneous Papers (Philosophical Aspects, Extraterrestrial Civilizations, etc.)

015.001 "L'Astronomie" ou le temps retrouvé.
B. Clouet.
Astronomie, Vol. 96, 153 - 161 (1982).

015.002 Can Venus be transformed into an earth-like planet?
S. J. Adelman.
J. British Interplanet. Soc., Vol. 35, 3 - 8 (1982).

015.003 Galactic extraterrestrial intelligence. I. – The constraint on search strategies imposed by the possibility of interstellar travel.   C. E. Singer.
J. British Interplanet. Soc., Vol. 35, 99 - 115 (1982).

015.004 The final question: paradigms for intelligent life in the universe.   M. A. G. Michaud.
J. British Interplanet. Soc., Vol. 35, 131 - 134 (1982).

015.005 Terraforming Venus.
S. J. Adelman.
Spaceflight, Vol. 24, 50 - 53 (1982).

015.006 How life began.   P. Cloud.
Nature, Vol. 296, 198 - 199 (1982).

015.007 On the possible mechanism of influence of solar activity on the human organism.   M. M. Kobrin.
Soln. Dannye 1981 Byull., No. 12, p. 86 - 90 (1982). In Russian.
A possibility is studied to explain the influence of powerful solar flares on the human organism as a consequence of the following processes: as a result of the rising flare X-ray emission the frequencies of the first and second mode of Schuman resonance grow up and become equal to the brain rhythms frequencies. So a resonance influence on the human "control system" takes place.

015.008 On the Einstein-Murphy interaction.
A. Held, P. Yodzis.
Gen. Relativ. Gravitation , Vol. 13, 873 - 882 (1981).
This paper is a first attempt to reconcile the two great concepts of twentieth century physics: Einstein's theory of general relativity, and Murphy's law, which states: "whatever can go wrong, will go wrong." A well-known folk lemma associated with this law maintains that "bread always falls butter side down." The editors of Astronomy and Astrophysics Abstracts propose that the authors should receive the Nobel prizes for physics, literature, and (a new one) for the promotion of food.

015.009 On the improbability of intelligent extraterrestrials.
A. Bond.
J. British Interplanet. Soc., Vol. 35, 195 - 207 (1982).

015.010 The most advanced civilization in the Galaxy is ours.
F. J. Tipler.
Mercury, Vol. 11, 5 - 11, 37 (1982).

015.011 The search for extraterrestrial civilizations – a new approach.   M. D. Papagiannis.
Mercury, Vol. 11, 12 - 16, 25 (1982).

015.012 Astrofili e astronomi.   E. Moltisanti.
Orione, Vol. 3, 14 - 21 (1982).

015.013 Call signs of extraterrestrial civilizations in form of a narrow spectral doublet.   P. V. Makovetskij.

Izv. vuzov. Radiofiz., Tom 23, 1378 - 1380 (1980). In Russian.
Abstr. in Ref. zh., 51. Astron., 2.51.2 (1982).

015.014 Infrared spectroscopy of micro-organisms near 3.4 μm in relation to geology and astronomy.
F. Hoyle, N. C. Wickramasinghe, S. Al-Mufti, A. H. Olavesen.
Astrophys. Space Sci., Vol. 81, 489 - 492 (1982).
Microorganisms sealed in KBr discs have an absorption spectrum over the $2.5 - 15$ μm waveband that shows thermal stability as they are heated in an inert atmosphere to temperatures of about 400°C. Microfossils tightly sealed within cavities in rocks could be endowed with similar properties of thermal stability. The observed absorption of interstellar material along the line of sight from the solar system to the galactic centre is remarkably similar to the spectrum of dry microorganisms over the $3.15 - 3.7$ μm waveband.

015.015 Ein Beitrag zu SETI.
G. M. Gruber, J. Pfleiderer.
Mitt. Astron. Ges., Nr. 55, p. 187 (1982). – Abstract.

015.016 Un atlas photographique avec une camera Schmidt de 14 cm.   D. Lemay.
J. R. Astron. Soc. Canada, Vol. 76, 129 - 137 (1982).

015.017 La Stella di Betlemme.   P. Custodi.
Coelum, Vol. 50, 77 - 82 (1982).

015.018 Wandel des Weltbildes.   J. Teichmann.
Sterne Weltraum, 21. Jahrg., 203 - 206 (1982).

015.019 A festival of planets.
Sky Telesc., Vol. 63, 564 - 565 (1982).

015.020 Search for intelligent life in the universe.
L. M. Gindilis.
Zemlya Vselennaya, 1982, No. 3, p. 48 - 53. In Russian.

015.021 Models of interstellar exploration.
D. G. Stephenson.
Q. J. R. Astron. Soc.,Vol. 23, 236 - 251 (1982).

015.022 The exceptional configuration of planets in the year 1982.   A. Hajduk.
Kozmos, Vol. 13, 4 - 5 (1982). In Slovak.

015.023 Grenzen der Naturwissenschaft.   G. M. Fasching.
Sternenbote, 25. Jahrg., 2 - 10 (1982).

015.024 Influence of solar activity on life duration of some categories of people.
N. I. Kozhevnikov, A. S. Sharov.
Astron. Tsirk., No. 1190, p. 4 - 6 (1981). In Russian.

015.025 Enjoyment in an astronomical occupation.
J. W. Evans.
Bull. American Astron. Soc., Vol. 13, 805 (1981). – Abstract.

015.026 On strong and weak strategies against astrology.
J. D. Mulholland.
Bull. American Astron. Soc., Vol. 13, 812 - 813 (1981). Abstract.

015.027 Careers for astronomers in the military.
M. S. Wilkerson.
Bull. American Astron. Soc., Vol. 13, 839 (1981). – Abstract.

015.028 A recent symbiotic approach to SETI observations: use of maps from the Westerbork Synthesis Radio Telescope. J. C. Tarter, F. P. Israel.
Bull. American Astron. Soc., Vol. 13, 840 (1981). – Abstract.

015.029 Civiltà extraterrestri e comunicazioni interstellari. E. Proverbio.
Atti Fond. Giorgio Ronchi, Anno XXXVI, N. 1 - 2, p. 196 - 237 (1981) = Pubbl. Stn. Astron. Int. Latitudine,Carloforte-Cagliari,N. 86.

015.030 Relation possible entre les alignements de planètes, l'activité solaire, le climat et l'épidémie de la peste noire au Moyen-Âge. B. Junod.
Arch. Sci. Genève, Vol. 34, 335 - 363 (1981) = Publ. Obs. Genève, Sér. A, Fasc. 85 (1982).

015.031 Astronomy, philosophy and common sense (Dialogue with readers on the gnoseological interpretation of scientific discoveries). A. Tursunov.
Vopr. filos., 1981, No. 11, p. 142 - 151. In Russian. – Abstr. in Ref. zh., 51. Astron., 5.51.1 (1982).

015.032 Will the real SETI please stand up? F. D. Drake.
Phys. Today, Vol. 35, No. 6, p. 9, 70 - 71 (1982).

015.033 Statistical publication histories of American astronomers. H. A. Abt.
Publ. Astron. Soc. Pacific, Vol. 94, 213 - 220 (1982).

015.034 Op zoek naar natuurlijke aardsatellieten. G. W. E. Beekman.
Zenit, 9. Jaarg., 302 - 306 (1982).

015.035 Possible impact of cosmochemistry on terrestrial biology: historical introduction. N. W. Pirie.
Philos. Trans. R. Soc. London, Ser. A, Vol. 303, (see 012.038), 589 - 594 (1981).

015.036 Organic matter in meteorites and Precambrian rocks: clues about the origin and development of living systems. J. Brooks.
Philos. Trans. R. Soc. London, Ser. A, Vol. 303, (see 012.038), 595 - 609 (1981).

015.037 Progressive development of the international cosmic law. G. P. Zhukov.
Mezhdunar. nauchn. sotrudnichestvo i pravov. vopr. osvoeniya kosmosa. Tr. 4-kh Nauchn. chtenij po kosmonavt., Moskva, 28 yanv. – 2 fevr., 1980. Moskva. 1980, p. 5 - 23. In Russian. Abstr. in Ref. zh., 62. Issled. kosm. prostranstva, 6.62.26 (1982).

015.038 Zur Frage der Reform des Osterdatums. M. Gossler.
Theologisch-praktische Quartalschrift, Jahrg. 130, Heft 2, p. 163 - 164 (1982). Oberösterreichischer Landesverlag, A-4020 Linz, Landstr. 41.

015.039 Episteme and gnosis: a tension in European thought. A. H. Batten.
Compendium in astronomy, (see 003.012), p. 3 - 9 (1982).

015.040 The colonization of the Galaxy – a key concept in the search for extraterrestrial intelligence. M. D. Papagiannis.
Compendium in astronomy, (see 003.012), p. 381 - 390 (1982).

The Drake equation, which precludes stellar colonization and assumes an independent origin for each stellar civilization, predicts a number $N = 10^5 - 10^6$ of advanced technological civilizations in the Galaxy, i.e., roughly one per million stars. The concept, on the other hand, of the colonization of the Galaxy predicts, either $N = 10^{10} - 10^{11}$ if the colonization has already occurred, with space colonies in orbit around every well behaved star of the Galaxy, or $N = 10^{-1} - 10^0$ if the colonization has not yet taken place. These conditions simplify considerably the search for extraterrestrial intelligence because it is sufficient to investigate only a small number of stars in our own vicinity, rather than millions of faraway stars as required by the Drake equation.

Contact with the stars. See Abstr. 003.023.

Plurality of worlds. See Abstr. 003.035.

Cosmology, physics, and philosophy. See Abstr. 003.047.

Extraterrestrials: where are they? See Abstr. 003.059.

Besiedelt die Menschheit das Weltall? See Abstr. 003.066.

The fertile stars. See Abstr. 003.103.

Utilization of outer space and international law. See Abstr. 003.108.

A new, rational endeavour for understanding the Eratosthenes numerical result of the earth meridian measurement. See Abstr. 004.089.

Life in the Universe. Proceedings of a conference held at Moffett Field, Calif., June 1979. See Abstr. 012.049.

Applied Mathematics, Physics

021 Mathematical Papers Related to Astronomy and Astrophysics, Computing, Data Processing

021.001 **Statistics under incomplete knowledge of data.**
J. Pfleiderer, P. Krommidas.
Mon. Not. R. Astron. Soc., Vol. 198, 281 - 288 (1982).
A different and generalized formulation and interpretation is proposed of the method used by Avni et al. to determine population distributions from preselected random samples with insufficient knowledge of the data, e. g. from intensity measurements with definite flux density results as well as upper limits. The information contained in inconclusive or inexact results is fully used.

021.002 **A new method for estimating the power spectrum of gapped data.** G. G. Fahlman, T. J. Ulrych.
Mon. Not. R. Astron. Soc., Vol. 199, 53 - 65 (1982).
The problem of estimating the power spectrum of a discrete time series which consists of individual segments separated by long gaps is considered. A maximum entropy approach is used to fill the gaps with a prediction based on the observed data segments. The method is described in detail and two numerical examples are presented to illustrate the application of the algorithm.

021.003 **In centrum te Reading.** CRAY I verricht 50 miljoen berekeningen per seconde.
B. Zwart.
Zenit, 9 Jaarg., 27 - 29 (1982).

021.004 **A direct analytic method of calculating the quadrupole parameters of a planetary magnetic field.** D. M. Willis.
Geophys. J. R. Astron. Soc., Vol. 68, 751 - 764 (1982).
Attention is drawn to a direct analytic method of calculating the quadrupole parameters of a planetary main magnetic field. Following a brief survey of the general theory of magnetic multipoles, an explicit algorithm is derived for calculating the quadrupole moment and the directions of the two quadrupole axes, given the five spherical harmonic coefficients of the second degree. It is shown that the direct analytic method of calculating the quadrupole parameters yield results for the geomagnetic quadrupole that are in exact agreement with those obtained by the more usual iterative procedure. It is also pointed out that a pseudo-quadrupole moment, which has been used to compare the quadrupole strengths of different planetary magnetic fields, is not strictly consistent with Maxwell's classical definition of a quadrupole moment. A precise physical definition of this pseudo-quadrupole moment is propounded.

021.005 **Astronomische Phänomenologie mit dem Taschenrechner.** H. Mucke.
Sterne, 58. Band, 30 - 47 (1982).

021.006 **Monte-Carlo simulation in astronomy.**
R. J. Dodd, H. T. MacGillivray.
South. Stars, Vol. 29, 78 - 86 (1981).

021.007 **Application of program calculators for express-accounting of the atmospheric extinction during photoelectric observations.** G. A. Terez, Eh. I. Terez.
Simferop. univ., Simferopol', 1981. 43 pp. In Russian.
Abstr. in Ref. zh., 51. Astron., 2.51.799 (1982).

021.008 **Blank noise tests for spectral analysis.**
L. M. A. Jeudy.
Astrophys. Space Sci., Vol. 81, 275 - 281 (1982).
Blank noise tests are derived for two types of spectral functions as defined by Vaníček. Their probability density has the beta-distribution which is easily related to Fisher distribution. Two examples are discussed.

021.009 **Statistik bei unvollkommener Kenntnis von Meßdaten.** P. Krommidas, J. Pfleiderer.
Mitt. Astron. Ges., Nr. 55, p. 188 (1982). – Abstract.

021.010 **On some problems of solving a system of conditional equations.** V. A. Izvekov.
Byull. Inst. Teor. Astron., Leningrad, Tom 15, 83 - 90 (1981). In Russian.
The system of conditional equations in the problem of improving planetary elements is analyzed. Significance of the right hand equations and that of the coefficients is evaluated. The influence of correlations between columns on the loss of significant orders of numbers in the process of solving the system is shown. Estimates of the required number of spare orders are given. Criteria for eliminating the correlating and unsignificant unknowns are substantiated. It is shown that the above technique permits safe insertion of a large number of unknowns of which after the analysis only a few are remaining determined in the best way.

021.011 **An algorithm for synthesizing the geomagnetic field.** S. R. C. Malin, D. R. Barraclough.
Comput. Geosci., Vol. 7, 401 - 405 (1981). – Abstr. in Phys. Abstr., Vol. 85, Abstr. 42257 (1982).

021.012 **Error propagation in the numerical solutions of the differential equations of orbital mechanics.**
V. R. Bond.
Celestial Mech., Vol. 27, 65 - 77 (1982).
The relationship between the eigenvalues of the linearized differential equations of orbital mechanics and the stability characteristics of numerical methods is presented. It is shown that the Cowell and Encke formulations with an independent variable related to the eccentric anomaly all have a real positive eigenvalue when linearized about the initial conditions. The real positive eigenvalue causes an amplification of the error of the solution when used in conjunction with a numerical integration method. In contrast an element formulation has zero eigenvalues and is numerically stable.

021.013 **The calculation of astronomical refraction in marine navigation.** G. G. Bennett.
J. Navig., Vol. 35, 255 - 259 (1982).

021.014 Examination of time series through randomly broken windows.
P. A. Sturrock, E. C. Shoub.
Astrophys. J., Vol. 256, 788 - 797 (1982).

In order to determine the Fourier transform of a quasi-periodic time series (linear problem) or the power spectrum of a stationary random time series (quadratic problem), it is desirable that data be recorded without interruption over a long time interval. In practice, this may not be possible. The effect of regular interruption such as the day-night cycle is well known. The authors investigate the effect of irregular interruption of data collection (the "breaking" of the window function) with the simplifying assumption that there is a uniform probability p that each interval of length τ, of the total interval of length $T = N\tau$, yields no data. For the linear case they find that the noise-to-signal ratio will have a (1σ) value less than ϵ if N exceeds $p^{-1}(1-p)\epsilon^{-2}$. For the quadratic case, the same requirement is met by the less restrictive requirement that N exceed $p^{-1}(1-p)\epsilon^{-1}$.

021.015 Technique of detection and analysis of stellar microvariability. I. Microvariability of XX Cam.
B. E. Zhilyaev, A. G. Totochava, L. M. Shul'man.
Astrometr. Astrofiz., Vyp. (No.) 43, p. 14 - 29 (1981). In Russian.

Methods of statistical analysis have been applied to series of photometric observations of stars. They enable the amplitude and temporal peculiarities of stellar variability to be determined when the signal-to-noise ratio is too small for the light curve to be derived. The paper deals with the method of detecting stellar variability by means of the dispersion and power spectral analysis of photometric series using rigorous statistical criteria. The analysis of XX Cam supergiant microvariability was based on observations in both U, B, V and narrow-band filters. This star has been shown to be a non-stationary microvariable with periods grouped in the following intervals: 14–17 min, 24–29 min. 50–200 min in the U, B, V filters.

021.016 Application of information criteria to evaluation of an astronomical observational system and its links.
B. O. Karapetyan, V. S. Oskanyan.
Astrometr. Astrofiz., Vyp. (No.) 44, p. 85 - 92 (1981). In Russian.

General principles of the information transfer in astronomical observational systems are discussed. The amount of transferred information is considered as a basic generalized criterion for the system evaluation. Some expressions are derived showing the dependence of the generalized information criteria on the working parameters of the system's separate links.

021.017 Automation of derivation of astronomical refraction formulae. Yu. V. Markov, V. I. Sergienko.
Astrometr. Astrofiz., Vyp. (No.) 45, p. 82 - 85 (1981). In Russian.

The results are given of using the analytical expression symbol treatment system designed to transform a system of non-linear differential equations for light beam movement in the atmosphere into a linear one. The described method allows to automatize the process of derivation of the astronomical refraction formulas for a three-dimensional model of the atmosphere.

021.018 Acceleration of numerical integration of the differential equations of motion of a solid body relative to the mass center. M. Yu. Belyaev, T. N. Tyan.
Kosm. Issled., Tom 20, 143 - 145 (1982). In Russian.

021.019 On the accuracy of solution of kinematic equations. I. Equation of errors. V. N. Branets.
Kosm. Issled., Tom 20, 184 - 190 (1982). In Russian.

021.020 Ergodic theory and area preserving mappings. R. W. Easton.
Application of modern dynamics to celestial mechanics and astrodynamics, (see 012.026), p. 267 - 276 (1982).

Much work has been done recently in numerical (computer) studies of area preserving mappings of the plane. Often the orbit of a single point seems to fill a region in the plane having positive area. Such a region is called an "ergodic zone". In this paper the author gives an exposition of some mathematical techniques which can be used to show that under suitable hypotheses, the closure of an orbit of a single point has positive Lebesque measure. He then applies these techniques to show that a family of mappings called linked twist maps are ergodic.

021.021 On some invariant manifold results and their applications. U. Kirchgraber.
Application of modern dynamics to celestial mechanics and astrodynamics, (see 012.026), p. 301 - 320 (1982).

A theory on the existence and the properties of invariant manifolds for a certain class of finite dimensional maps is described, with applications to averaging and to a problem in celestial mechanics.

021.022 On a family of continuous maps of the circle into itself related to the Van der Pol equation.
L. Alsedà, J. Llibre, R. Serra.
Application of modern dynamics to celestial mechanics and astrodynamics, (see 012.026), p. 325 (1982). — Abstract.

021.023 On the convergence of formal integrals in finite time. J. Martinez.
Application of modern dynamics to celestial mechanics and astrodynamics, (see 012.026), p. 353 (1982). — Abstract.

021.024 The equivalence of the generators of Deprit's and Giorgelli-Galgani's change of variables in differential systems. M. Rapaport.
Application of modern dynamics to celestial mechanics and astrodynamics, (see 012.026), p. 355 (1982). — Abstract.

021.025 Hopf-bifurcation in a nearly Hamiltonian system. F. Spirig.
Application of modern dynamics to celestial mechanics and astrodynamics, (see 012.026), p. 360 - 361 (1982). — Abstract.

021.026 A realization of dynamical systems in Boolean algebras. M. Ülküdas.
Application of modern dynamics to celestial mechanics and astrodynamics, (see 012.026), p. 362 (1982). — Abstract.

021.027 Completely integrable systems and singularity in the complex t–plane. H. Yoshida.
Application of modern dynamics to celestial mechanics and astrodynamics, (see 012.026), p. 365 (1982). — Abstract.

021.028 Estabilidad y precisión en la resolución numérica de algunos problemas gravitacionales
P. E. Zadunaisky.
Bol. Asoc. Argentina Astron., No. 20 - 24, p. 211 (1981). Abstract.

021.029 Pruebas del test estadístico de Warner y Robinson.
C. R. Fourcade, A. A. Puch, J. Colazo, J. R. Laborde.
Bol. Asoc. Argentina Astron., No. 20 - 24, p. 329 (1981). Abstract.

021.030 Programa atlas para el cálculo de modelos de atmósferas estelares. Z. López García.

Bol. Asoc. Argentina Astron., No. 20 - 24, p. 330 (1981).
Abstract.

021.031   **Adaptación del programa atlas a la computadora
IBM/360 de la Universidad Nacional de San Juan.**
Z. López García, A. Zaragoza.
Bol. Asoc. Argentina Astron., No. 20 - 24, p. 425 (1981).
Abstract.

021.032   **On a generation of the integro-exponential
function.**   E. A. Gussmann.
Soobshch. Spets. Astrofiz. Obs., Vyp. 32, (see 012.029),
p. 15 (1981).

021.033   **On-line computer system for photometric investigations.**   V. P. Kozhevnikov.
Astron. Tsirk., No. 1183, p. 3 - 5 (1981). In Russian.

021.034   **A Fortran subroutine for determining times of
minimum light.**   A. D. Mallama.
I. A. P. P. P. Commun., No. 7, p. 14 - 18 (1982).

021.035   **Sistema operativo per il microcomputer NASCOM.**
L. Mureddu.
Note tec. Stn. Astron. Int. Latitudine, Carlofote-Cagliari,
N. 2, 28 pp. (1981).

021.036   **Mathematical support of computers for photomet-
rical investigations.**
S. Yu. Gorda, A. A. Kalinin, N. V. Matkin.
Astron. Tsirk., No. 1192, p. 1 - 3 (1981). In Russian.

021.037   **DQPT: a computer program for solving non-LTE
problems for two-level atoms in one-dimensional
semi-infinite media with velocity fields.**
G. B. Scharmer, Å. Nordlund.
Stockholms Obs. Rep. No. 19, 37 pp. (1982).
   A FORTRAN program that solves non-LTE problems for
two-level atoms in one-dimensional semi-infinite media with
velocity fields is presented. The method of solution is based
on an iterative technique referred to as the depth-quadrature
perturbation technique (DQPT). This technique allows the
solution of radiative transfer problems with very small
amounts of computing time.

021.038   **Probleme der Parameterbestimmung in physikalischen Systemen.**   G. Beutler.
Mitt. Satell.-Beobachtungsstn. Zimmerwald, Nr. 8, 173 pp.
(1982).

021.039   **Expansion theory for the elliptic motion of
arbitrary eccentricity and semi-major axis. III.**
Analytical and computational developments of the functions
$H_1^{(m)}(\epsilon, \theta) = \cos m \{\cos^{-1}(1 - \epsilon \sin^2 \theta)\}; H_2(\epsilon, \xi, \theta) =$
$\log\{1 - \xi + \xi\epsilon \sin^2 \theta\}; H_3^{(q)}(\epsilon, \xi, \theta) = \{1 - \xi + \xi\epsilon \sin^2 \theta\}^q.$
M. A. Sharaf.
Astrophys. Space Sci., Vol. 84, 53 - 71 (1982).
   In this paper the expansions of the functions H_1, H_2, and
H_3 will be established analytically and computationally for
m positive integer, q any real number and ξ, ϵ are both
positive <1. Full recursive computational algorithms with
their numerical results will also be included.

021.040   **High-order multivalue integrators.**
D. L. Richardson.
Bull. American Astron. Soc., Vol. 14, 579 (1982). – Abstract.

021.041   **An algebraic manipulation system for the
APPLE II microcomputer.**   B. C. Brown.
Bull. American Astron. Soc., Vol. 14, 581 (1982). – Abstract.

021.042   **Space Telescope astrometry software.**
W. H. Jefferys, G. F. Benedict, P. D. Hemenway,
P. J. Shelus,
Bull. American Astron. Soc., Vol. 14, 581 (1982). – Abstract.

021.043   **On the accuracy of solution of kinematic equations.
2. Use of quasi-coordinates.**   V. N. Branets.
Kosm. Issled., Tom 20, 323 - 331 (1982). In Russian.

021.044   **The analytical calculation of the second spherical
exponential integral.**   T. Henning.
Astron. Nachr., Band 303, 125 - 126 (1982) = Mitt. Univ.-
Sternw. Jena, Nr. 157.

021.045   **The calculation of comet ephemerides.**
J. Tatum.
J. R. Astron. Soc. Canada, Vol. 76, 157 - 167 (1982).

021.046   **Computing analysis of light curves for eclipsing
binary stars. Part 2. Direct approach and its
programming for the Nairi K computer.**   M. I. Lavrov.
Kazan. univ. Kazan', 1981. 47 pp. In Russian. – Abstr. in Ref.
zh., 51. Astron., 6.51.770 (1982).

021.047   **Methods of computation of the perturbed motion
of small bodies in the solar system.**
Yu. V. Batrakov.
Sun and planetary system, (see 012.040), p. 415 - 419 (1982).
   In this review the following topics are considered: the
equations of motion and regularization, numerical and Taylor
integration methods, Encke's method and intermediate orbits.

021.048   **Construction of conditionally-periodic solutions of
canonical systems of differential equations with
multiple resonance.**   S. G. Zhuravlev.
Celestial Mech., Vol. 27, 179 - 189 (1982).
   A procedure of building of conditionally-periodic solutions of canonical systems of differential equations in a case
of multiple resonances is given. A modification of Delaunay–
Zeipel's method is used as a base of the procedure.

021.049   **Studies in the application of recurrence relations to
special perturbation methods. VI: Comparison with
classical single-step and multi-step methods of numerical
integration.**   H. E. Schwarz, I. W. Walker.
Celestial Mech., Vol. 27, 191 - 202 (1982).
   The numerical integration of the differential equations
describing dynamical systems has been shown in previous
papers of this series to be most effectively accomplished by an
explicit Taylor series method. In this paper the authors show
that one explicit Taylor series method, developed earlier in
this series and which appears to possess a high degree of
versatility, yields considerable gains in efficiency over classical
single-step and multi-step methods. For a given accuracy
criterion governing the local truncation error it is found that
the Taylor series method is generally twice as fast as the
classical multi-step method and up to twenty times faster than
the classical single-step method.

021.050   **Propagation of local errors in the solutions of the
differential equations for orbital elements.**
V. R. Bond.
Celestial Mech., Vol. 27, 203 - 210 (1982).
   The propagation of errors in the solutions of the differential equations for the orbital elements of perturbed two-
body motion is investigated. It is shown that the error in the
time-element grows linearly for differential equations for
orbital elements when only perturbations are present on the
right-hand side, cubically for formulations which have a two-
body term on the right-hand side, and linearly for formulations based upon extended phase space Hamiltonians.

021.051 Alternatives to least squares.
R. L. Branham, Jr.
Astron. J., Vol. 87, 928 - 937 (1982).

Three competing alternatives to the standard method of least squares are studied. The method of averages arranges the equations of condition into as many sets as there are unknowns and forms the sum of the equations in each set. The equations so obtained are solved for the unknowns. The method of minimum sum postulates that the best solution of an overdetermined system is the one that minimizes the sum of the absolute values of the residuals. And the Chebyshev method takes as the best solution the one that minimizes the largest residual in absolute value. Both the method of averages and the Chebyshev method prove inappropriate for the analysis of linear systems derived from observational data. The method of minimum sum, however, is very attractive. Use of simulated data that are concordant shows that the solution from minimum sum is only slightly inferior to that given by least squares. But analysis of a real data set with discordant data points demonstrates that the minimum-sum solution is superior to the least-squares solution.

021.052 Calculs astronomiques pour amateurs. J. Meeus.
Astronomie, Vol. 96, 43 - 45, 96 - 99, 211 - 213, 266 - 267 (1982).

Une banque de données pour l'étude des phénomènes de transfert radiatif dans les atmosphères planétaires: la banque "GEISA". See Abstr. 002.030.

Astrofysica voor calculators. See Abstr. 003.063.

Calcul astronomique pour amateurs. See Abstr. 003.113.

Mathematisches Hilfsbuch für Studierende und Freunde der Astronomie. See Abstr. 003.138.

Mark III VLBI: from "light bulb" to Ap. J. or J. G. R. See Abstr. 013.042.

Optimized computation of the Voigt and complex probability functions. See Abstr. 022.037.

LINPOS, an interactive program for stellar line position measurement. See Abstr. 031.604.

Kiso image detection system for photographic plates. See Abstr. 036.014.

Les géodésiques de l'ellipsoïde à trois axes inégaux, d'après Jacobi, Whittaker, Arnold. See Abstr. 042.057.

Exploding dynamical systems. See Abstr. 042.059.

Small divisors in the derivatives of Hansen's coefficients. See Abstr. 042.068.

Numerical modelling of the spectrophotometric investigation process of magnetic stars. See Abstr. 116.030.

022 Physical Papers Related to Astronomy and Astrophysics

022.001 Transition probabilities for forbidden lines in the
$2p^3$ configuration.   C. J. Zeippen.
Mon. Not. R. Astron. Soc., Vol. 198, 111 - 125 (1982).

Making use of the atomic structure computer program
SUPERSTRUCTURE, radiative transition probabilities are
calculated for the forbidden lines in the ground configuration
$1s^2 2s^2 2p^3$ of all the members of the isoelectronic sequence
from N I to Fe XX. Elaborate procedures, such as configuration interaction, semi-empirical term energy corrections,
special radial wave functions and relativistic corrections to the
magnetic dipole operator, are used in the calculation.

022.002 Transition probabilities for forbidden lines in the
$3p^3$ configuration.
C. Mendoza, C. J. Zeippen.
Mon. Not. R. Astron. Soc., Vol. 198, 127 - 139 (1982).

Making use of the atomic structure computer program
SUPERSTRUCTURE, radiative transition probabilities are
calculated for the forbidden lines in the ground configuration
$1s^2 2s^2 2p^6 3s^2 3p^3$ of all the members of the P I isoelectronic
sequence up to Ni XIV. Elaborate procedures such as configuration interaction, relativistic corrections to the magnetic dipole operator, special radial wave functions and semi-empirical
term energy corrections are used in the calculation.

022.003 Dielectronic satellite spectra for highly charged
helium-like ions – VI. Iron spectra with improved
inner-shell and helium-like excitation rates.
F. Bely-Dubau, J. Dubau, P. Faucher, A. H. Gabriel.
Mon. Not. R. Astron. Soc., Vol. 198, 239 - 254 (1982).

The atomic theory developed through earlier papers in
this series is extended in order to improve the understanding of
iron solar flare spectra in the region 1.85 - 1.88 Å. The new
work concerns impact excitation by distorted wave theory for
the inner-shell Fe XXIV transitions and the $1s^2$ - $1s$ $2l$ transitions in Fe XXV. In addition, rates are evaluated for contributions to the Fe XXV lines from cascade, radiative recombination, dielectronic recombination, and inner-shell ionization of
Fe XXIV.

022.004 Diffusion and viscosity coefficients for helium.
R. Roussel-Dupré.
Astrophys. J., Vol. 252, 393 - 401 (1982).

The first order Boltzmann-Fokker-Planck equation is
solved numerically to obtain diffusion and viscosity coefficients for a ternary gas mixture composed of electrons,
protons, and helium. The coefficients are tabulated for five
He/H abundances ranging from 0.01 to 10 and for both He II
and He III. For the astrophysically important gas mixtures, it
is concluded that the results of existing studies which employed Burgers's or Chapman and Cowling's coefficients will
remain substantially unaltered. The author also points to a
number of inconsistencies regarding diffusion, which have
appeared in the astrophysical literature.

022.005 Spectroscopic evidence for interstellar magnesium
oxide particles.   S. MacLean, W. W. Duley.
Astrophys. J., Lett., Vol. 252, L25 - L27 (1982).

Laboratory spectra of partially recrystallized MgO films
have been obtained in the wavelength range 300–135 nm. A
narrow spectral feature arising from an optical transition in
O^{2-} appears at 161 ± 1 nm. This feature is superposed on a
broad background due to interband transitions. The similarity
of the 161 nm feature to a band seen in interstellar extinction
at the same wavelength suggests that MgO may be a component of interstellar dust. A quantitative analysis shows that
10–30% of interstellar Mg is likely bound in MgO solids.

022.006 A reinvestigation of the rate of the $C^+ + H_2$ radiative
association reaction.   E. Herbst.
Astrophys. J., Vol. 252, 810 - 813 (1982).

New relevant experimental results and statistical theories
have prompted a reinvestigation of the rate coefficient
$k_{RA}(T)$ of the important interstellar reaction $C^+ + H_2 \rightarrow$
$CH_2^+ + h\nu$ in the 10–100 K temperature range. The new
results indicate that $10^{-16} \leqslant k_{RA} \leqslant 10^{-15}$ cm^3 s^{-1} at all
temperatures studied.

022.007 Rigorous and approximate scaling laws for the
photoionization cross-section of hydrogenic ions in
magnetic fields.
G. Wunner, H. Ruder, W. Schmitt, H. Herold,
M. R. C. McDowell.
Mon. Not. R. Astron. Soc., Vol. 198, 769 - 772 (1982).

A general form of the nuclear charge (Z) scaling law for the
photoionization cross-section in magnetic fields is derived. It
turns out that the cross-section for $Z > 1$ is scaled down to that
for $Z = 1$ taken at electron and photon energies which are no
longer connected by the conservation of energy. By specializ-
ing to the dipole approximation a simpler scaling law is obtained. The possibility of experimental tests is briefly discussed.
For a magnetic field of 4.7×10^{12} G and photon energies up to
50 keV the range of validity of the dipole approximation is
determined quantitatively.

022.008 MCDF calculation of wavelengths and intensities of
satellite lines in lithium-like ions.
J. Hata, I. P. Grant.
Mon. Not. R. Astron. Soc., Vol. 198, 1081 - 1088 (1982).

The MCDF-EAL relativistic self-consistent field method
has been used to compute wavelengths and radiative transition
rates for some 22 transitions connecting 19 levels of the configurations $1s^2 2s$, $1s^2 2p$, $1s\ 2s^2$, $1s\ 2s\ 2p$ and $1s\ 2p^2$ in
lithium-like ions. The results are in excellent agreement both
with experimental wavelength determinations in Fe XXIV and
V XXII spectra and with other theoretical predictions based on
the Breit-Pauli Hamiltonian.

022.009 Speed of light outside the solar system: a new test
using visual binary stars.
R. Gruber, D. Koo, J. Middleditch.
Publ. Astron. Soc. Pacific, Vol. 93, 777 - 782 (1981/82).

The authors have checked the constancy of the speed of
light outside our solar system with a new technique. The
semimajor axis of stellar binary systems was first derived from
parallax and visual binary data and then compared to the
semimajor axis derived from spectroscopic orbits. Although
only ten binary systems were found to have sufficiently
accurate data, the authors were still able to conclude that the
speed of light has been constant to within 10% for distances
up to 25 parsecs.

022.010 Precision measurement of relative oscillator
strengths Ti I – I. Transitions from levels
$a^3F_2(0.00eV)$ and a^3F_3 (0.02eV) and a^3F_4(0.05eV) mea-
sured with an accuracy of 0.5 per cent.
D. E. Blackwell, A. D. Petford, M. J. Shallis, S. Leggett.
Mon. Not. R. Astron. Soc., Vol. 199, 21 - 31 (1982).

The paper presents measures of the relative oscillator
strengths of 45 ground state lines of Ti I, for the range
$+0.215 > \log gf > -3.84$, made with an accuracy of
0.5 per cent using the Oxford furnace technique. The relative
measurements have been placed provisionally on an absolute
scale using the atomic beam measurements of Bell et al. and
the lifetimes measured by Roberts et al. This scale has an
accuracy of about 12 per cent. Comparisons are made with

the measured values of other experimenters and with the results of calculations.

022.011 Precision measurement of relative oscillator strengths – IX. Measures of Fe I transitions from levels a^5P_{1-3} (2.18–2.28 eV), a^3P_2 (2.28 eV), $a^3P_{0,1}$ (2.49–2.42 eV), $z^7D^0_{1-5}$ (2.48–2.40 eV) and a^3H_{4-6} (2.45–2.40 eV). D. E. Blackwell, A. D. Petford, M. J. Shallis, G. J. Simmons. Mon. Not. R. Astron. Soc., Vol. 199, 43 - 51 (1982).

Measurements are presented of the relative oscillator strengths of 52 lines of Fe I, with excitation energies of between 2.18 and 2.49 eV, and with $\log gf > -3.23$, made using the Oxford furnace technique. Comparisons are made with other experimental data and with the results of theoretical calculations.

022.012 Cometary NH: ultraviolet and submillimeter emission.   M. M. Litvak, E. N. R. Kuiper. Astrophys. J., Vol. 253, 622 - 633 (1982).

The purpose of this paper is to report the results of computer calculations on the molecule NH, an important constituent of cometary comae, and a probable, but an as yet undetected, interstellar species. Included are the emission rates in the many individual ultraviolet lines of the $A\ ^3\Pi_i - X\ ^3\Sigma^-$ 0–0 band for different heliocentric velocities of a typical comet, referenced to a solar distance of 1 AU. Semiquantitative results from the $X^3\Sigma^- - X^3\Sigma^-$ submillimeter wave lines are obtained for these cases.

022.013 Time-resolved spectroscopy of the C_2 Phillips system and revised interstellar C_2 abundances. P. Erman, D. L. Lambert, M. Larsson, B. Mannfors. Astrophys. J., Vol. 253, 983 - 988 (1982).

Radiative lifetimes of vibrational levels of the $C_2A^1\Pi_u$ state have been measured by the high frequency deflection technique. The mean lifetime for levels $v' = 3, 4, 6,$ and 7 is 11 μs. The results are used to deduce the absorption oscillator strengths of the Phillips $(A^1\Pi_u - X^1\Sigma_g^+)$ system. The value $f_{2,0} = (8.8 \pm 1.2) \times 10^{-4}$ has been used to revise published estimates of the interstellar C_2 column density.

022.014 The rotational spectra of $HOCO^+$, HOCN, HN_3, and HNCO from quantum mechanical calculations. D. J. DeFrees, G. H. Loew, A. D. McLean. Astrophys. J., Vol. 254, 405 - 411 (1982).

Ab initio molecular orbital theory has been used to determine the equilibrium geometries, rotational constants, and rotational spectra of four isoelectronic molecules. Two of these, $HOCO^+$ and HOCN, are candidate interstellar molecules. The other two, HNCO and HN_3, have known rotational constants. Theoretical rotational constants and spectra for the two unknown species were corrected with the mean experimental to theoretical ratios from the two known species. The results indicate that $HOCO^+$ is a better candidate for the source of a series of lines reported by Thaddeus, Guélin, and Linke than is HOCN.

022.015 Collisional excitation rates of complex atomic ions by electron impact.   R. E. H. Clark, N. H. Magee, Jr., J. B. Mann, A. L. Merts. Astrophys. J., Vol. 254, 412 - 418 (1982).

For many applications a large number of rate coefficients for electron impact excitation are needed. These rate coefficients may be obtained by integration of the collision strengths. The authors present here fits to collision strengths as a function of both the nuclear charge Z and X, the impact electron energy in threshold units. The collision strength fit form is readily integrated over a Maxwellian distribution and has fitted extensive distorted wave data to nearly 5% from singly ionized stages to molybdenum for 19 spin-allowed transitions of the isoelectric sequences of hydrogen, helium, lithium, beryllium, boron, sodium, and magnesium. The exci-

tation energies have been fitted as a function of Z to better than 1%.

022.016 Improved overlapping helium line profiles for stellar spectra studies.   A. Mazure, G. Nollez. Astrophys. J., Vol. 254, 823 - 825 (1982).

Stark profiles of the 4471 Å overlapping He I line are computed at densities of astrophysical interest using the Model Microfield Method (MMM) which provides an improved treatment of the "ion dynamics" effect and is in correct agreement with recent laboratory results. As a main result, MMM profiles show less structure than those calculated with the theory of Barnard, Cooper, and Smith usually used in stellar applications.

022.017 H_2 fluorescence spectrum from 1200 to 1700 Å by electron impact: laboratory study and application to Jovian aurora.   Y. L. Yung, G. R. Gladstone, K. M. Chang, J. M. Ajello, S. K. Srivastava. Astrophys. J., Lett., Vol. 254, L65 - L69 (1982) = Contrib. No. 3570 Div. Geol. Planet. Sci., Calif. Inst. Technol.

A combined experimental study of the fluorescence spectrum of H_2 at wavelengths of 1200 - 1700 Å by electron impact and its application to modeling the Jovian aurora have been carried out. On the basis of detailed atmospheric and radiative transfer modeling, the authors conclude that the recent IUE and Voyager observations are consistent with precipitation of electrons with energy in the range of 1 - 30 keV or other energetic particles that penetrate to number densities of $4 \times 10^{10} - 5 \times 10^{13}$ cm^{-3} in the atmosphere. The globally averaged energy flux and production of hydrogen atoms are 0.5 - 2 ergs cm^{-2} s^{-1} and 1 - 4 $\times 10^{10}$ atoms cm^{-2} s^{-1}, respectively.

022.018 Association of atomic oxygen and airglow excitation mechanisms.   P. C. Wraight. Planet. Space Sci., Vol. 30, 251 - 259 (1982).

An outline is presented of a method of calculating relative rates of association of oxygen atoms by energy transfer into the different bound states of molecular oxygen. The method takes account of the detailed form of the potential energy curves for the different states. Results are presented for seven bound states of O_2 at temperatures between 100 and 600 K and are compared with laboratory and airglow data.

022.019 The effect of high rotational quantum numbers on Franck-Condon factors.   R. Ramakrishna Reddy, P. Sambasiva Rao, T. V. Ramakrishna Rao. J. Quant. Spectrosc. Radiat. Transfer, Vol. 27, 103 - 105 (1982).

The effect of high rotational quantum numbers on Franck-Condon factors has been studied for some astrophysically important diatomic molecules by using the criterion of Murthy and Gowda.

022.020 The polarised frozen-core approximation: oscillator strengths for the boron isoelectronic sequence. R. P. McEachran, M. Cohen. J. Quant. Spectrosc. Radiat. Transfer, Vol. 27, 111 - 117 (1982).

Orbital wave functions of a large number of ns-, np- and nd-levels of the first five members of the boron isoelectronic sequence have been calculated using a frozen-core Hartree-Fock procedure augmented by an l-dependent core polarisation potential. The calculated ionisation energies are generally in very good agreement with observations. Electric dipole oscillator strengths have been derived from the calculated orbitals and energies and have been used to yield simple interpolation formulae for f-values of higher members of the sequence.

022.021 The polarized frozen core approximation: energies and oscillator strengths of C(I) and N(II).
R. P. McEachran, M. Cohen.
J. Quant. Spectrosc. Radiat. Transfer, Vol. 27, 119 - 126 (1982).

The frozen core and extended frozen core versions of the Hartree-Fock approximation have been improved significantly by means of l-dependent core polarization potentials. The resulting polarized frozen core procedure reproduces most of the observed $2pns$, $2pnp$ and $2pnd$ terms of C(I) and N(II) with great accuracy; over 60 unobserved terms in N(II) have been calculated. Length and velocity forms of electric dipole oscillator strengths, calculated with these polarized orbitals and the corresponding energies, are generally in excellent agreement, particularly for transitions between the more highly excited states.

022.022 Comments on the requirements for a general Stark broadening theory. R. L. Greene.
J. Quant. Spectrosc. Radiat. Transfer, Vol. 27, 185 - 190 (1982).

An analysis is made of requirements for a general theory of Stark broadening applicable to both strong and weak collisions. It is pointed out in particular that inclusion of the static and weak interaction limits is not sufficient to obtain a generally valid theory. At least one other condition is necessary — a dynamic correction for strong, primarily static, collisions. The kangaroo process model microfield method is examined from this point of view, and it is shown that the model conditional probability leads to an incorrect functional form for the lowest order dynamic correction.

022.023 Energy partitioning in the reaction $^{16}O(^1D) + H_2{}^{18}O \rightarrow {}^{16}OH + {}^{18}OH$ — VI. The u.v.-spectrum of ^{16}OH and ^{18}OH. M. Nuß, K.-H. Gericke, F. J. Comes.
J. Quant. Spectrosc. Radiat. Transfer, Vol. 27, 191 - 201 (1982).

Highly resolved observations of the u.v. spectrum of ^{16}OH and ^{18}OH have been made in the absorption mode by means of a narrow bandwidth dye laser. The spectrum of the transition from the electronic ground state to the first excited electronic state, $A^2\Sigma^+ \leftarrow X^2\Pi$, in the $0 \rightarrow 0$ vibrational band yielded molecular constants for these rotational systems in both isotopic modifications.

022.024 Semiclassical calculations of electron impact Stark widths of S(III), Cl(III) and S(IV) isolated lines.
M. S. Dimitrijević, N. Konjević.
J. Quant. Spectrosc. Radiat. Transfer, Vol. 27, 203 - 205 (1982).

The authors report results of semiclassical calculations of electron impact Stark widths of doubly-ionized sulfur and chlorine and of triply-ionized sulfur. Comparison with available experimental data yields an average ratio of measured to calculated line widths of 0.73 and 0.74 for S(III) and Cl(III), respectively, and of 0.50 for a single S(IV) line.

022.025 Velocity of light to be constant — official.
B. Marx.
Nature, Vol. 296, 110 (1982).

022.026 Calculation of gain at X-ray wavelengths resulting from optical pumping of helium-like ions.
W. E. Alley, G. Chapline, P. Kunasz, J. C. Weisheit.
J. Quant. Spectrosc. Radiat. Transfer, Vol. 27, (see 012.005), 257 - 266 (1982).

An equivalent two level scheme is described for calculating level populations of helium-like ions in dense optically thick plasmas. The scheme is used to show how population inversions due to optical pumping of $3P$ or $4P$ states are affected by line trapping. The importance of various line broadening mechanisms for determining gain is briefly discussed.

022.027 Density effects on the spectral emission of a high temperature argon plasma.
D. Duston, J. Davis.
J. Quant. Spectrosc. Radiat. Transfer, Vol. 27, (see 012.005), 267 - 279 (1982).

In order to explain the detailed features of radiation spectra obtained from dense argon plasmas, an ionization-radiation model for argon has been constructed which calculates time-integrated spectra as a function of plasma temperature, density and size.

022.028 Dielectronic satellite spectra from dense laser-produced plasmas.
J. F. Seely, R. H. Dixon, R. C. Elton, J. G. Lunney.
J. Quant. Spectrosc. Radiat. Transfer, Vol. 27, (see 012.005), 297 - 306 (1982).

The authors present the analysis of two sets of dielectronic satellite data. The carbon satellite radiation has been observed in a plasma produced from a laser-irradiated planar slab target. The corresponding silicon satellite radiation has been observed in the spectra from a series of imploded microballoon targets. In the analysis of both the carbon and the silicon satellite data, the electron density is determined from the Stark broadening of the Lyman resonance lines.

022.029 Level shifts and inelastic electron scattering in dense plasmas. J. Davis, M. Blaha.
J. Quant. Spectrosc. Radiat. Transfer, Vol. 27, (see 012.005), 307 - 313 (1982).

The authors obtained level shifts, transition probability coefficients and collision cross sections for Ne^{+9} in a plasma with electron density ranging from 10^{24} to 6×10^{24} cm^{-3} using a quantum-mechanical description of the plasma electrons.

022.030 Statistical mechanics of "atoms" in hot, partially ionized matter: the theory of the one electron Green function. M. W. C. Dharma-Wardana.
J. Quant. Spectrosc. Radiat. Transfer, Vol. 27, (see 012.005), 315 - 328 (1982).

The author has considered the problem of calculating the one electron Green function for hot, partially ionized condensed matter as a means of computing the statistical thermodynamics and transport properties of the system.

022.031 Problems in the use of statistical average atom potentials for estimating average degree of ionization. W. Zakowicz, I. J. Feng, R. H. Pratt.
J. Quant. Spectrosc. Radiat. Transfer, Vol. 27, (see 012.005), 329 - 333 (1982).

Consequences of a simple integral definition of electron charge bound to an ion are examined for Thomas–Fermi and Debye–Hückel–Thomas–Fermi (DHTF) average atom statistical potentials used to describe high temperature high density plasmas. A self-consistent scheme for calculating average degree of ionization within the DHTF approach is described.

022.032 INFERNO: a better model of atoms in dense plasmas. D. A. Liberman.
J. Quant. Spectrosc. Radiat. Transfer, Vol. 27, (see 012.005), 335 - 339 (1982).

A self-consistent field model of atoms in dense plasmas has been devised and incorporated in a computer program.

022.033 Electronic energy-levels in dense plasmas.
R. M. More.
J. Quant. Spectrosc. Radiat. Transfer, Vol. 27, (see 012.005), 345 - 357 (1982).

The author develops a simple parameterization of pressure ionization, discusses limitations of the Debye–Hückel model for plasma perturbations, and surveys an approximate description of X-ray spectra based on the WKB approximation.

022.034 Pressure-induced H_2 opacity in the 5-μm region.
D. Goorvitch, R. H. Tipping.
J. Quant. Spectrosc. Radiat. Transfer, Vol. 27, 397 - 404 (1982).

The authors have calculated the pressure-induced continuum opacity of H_2 in the 5-μm region to facilitate the analysis of the spectra of the outer planets in this spectral region.

022.035 Theoretical transition energies, lifetimes and fluorescence yields of multiply ionized silicon.
T. W. Tunnell, C. Can, C. P. Bhalla.
J. Quant. Spectrosc. Radiat. Transfer, Vol. 27, 405 - 416 (1982).

Theoretical X-ray transition energies, lifetimes and partial multiplet fluorescence yields are presented for all spectroscopic terms of electron configurations with a single K-shell vacancy and varying number of electrons in the L-shell and M_1-subshell for multiply-ionized silicon.

022.036 Water absorption lines, 931−961 nm: selected intensities, N_2-collision-broadening coefficients, self-broadening coefficients, and pressure shifts in air.
L. P. Giver, B. Gentry, G. Schwemmer, T. D. Wilkerson.
J. Quant. Spectrosc. Radiat. Transfer, Vol. 27, 423 - 436 (1982).

022.037 Optimized computation of the Voigt and complex probability functions. J. Humlíček.
J. Quant. Spectrosc. Radiat. Transfer, Vol. 27, 437 - 444 (1982).

Methods for computing the complex probability function $w(z) = e^{-z^2} \text{erfc}(-iz)$, which is related to the Voigt spectrum line profiles, are developed. The methods are simple, as demonstrated by a sample FORTRAN program.

022.038 Oscillator strengths of Cl(I) in the vacuum ultraviolet: the $^2D-^2P$ transitions. J. J. Schwab, J. G. Anderson.
J. Quant. Spectrosc. Radiat. Transfer, Vol. 27, 445 - 457 (1982).

022.039 Theoretical study of IR band intensities and electronic transition moments for the β and δ systems of NO. D. M. Cooper.
J. Quant. Spectrosc. Radiat. Transfer, Vol. 27, 459 - 465 (1982).

022.040 Rotational dependence of Franck-Condon factors for OH^+, NH^+, SiH, MgH^+, SiH^+, and NO^+.
P. D. Singh, A. A. de Almeida.
J. Quant. Spectrosc. Radiat. Transfer, Vol. 27, 471 - 479 (1982).

022.041 Fast neutron capture on ^{180}Hf and ^{184}W and the solar hafnium and tungsten abundance.
H. Beer, F. Käppeler, K. Wisshak.
Astron. Astrophys., Vol. 105, 270 - 277 (1982).

The capture cross sections of ^{180}Hf and ^{184}W were measured by the activation method and via direct detection of prompt gamma-rays, respectively. The Maxwellian averaged cross sections for kT = 30 keV were used to decompose the solar isotopic Hf- and W-abundances into s- and r-process contributions. Examination of the r-process contributions provided evidence that the abundances of Hf and W might be smaller than quoted in recent abundance compilations. Therefore it is proposed to reconsider the information from meteorite analyses and to perform new measurements if necessary.

022.042 Charge transfer ionization of Si^+ by H^+ at thermal energies.
M. Gargaud, R. McCarroll, P. Valiron.
Astron. Astrophys., Vol. 106, 197 - 200 (1982).

Reaction rates for ionization charge transfer of Si^+ in collision with H^+ and for recombination charge transfer of Si^{+2} in collision with atomic H are calculated over a wide range of thermal energies ($10 - 10^6$ K). The results, obtained using the molecular model, confirm that in coronal plasmas at temperatures where charge transfer ionization contributes to the ionization equilibrium, rotational mixing of the Σ and Π symmetry is of little practical importance. Only for collision energies in excess of 30 eV does rotational mixing begin to have an appreciable influence on the reaction rate.

022.043 Hyperfine structure measurement in Sc II.
A. Arnesen, R. Hallin, C. Nordling, Ö. Staaf, L. Ward, B. Jelénković, M. Kisielinski, L. Lundin, S. Mannervik.
Astron. Astrophys., Vol. 106, 327 - 331 (1982).

Hyperfine structure measurements have been performed on Sc II using the collinear laser-ion beam technique. The three metastable levels $^1D_2$, $^3P_1$, and $^3P_2$ in the $3d^2$ configuration and five levels $^1D_2^0$, $^3D_1^0$, $^3D_2^0$, $^3D_3^0$, and $^3P_1^0$ in the $3d4p$ configuration have been investigated. The hyperfine structure constants A and B for these eight levels have been determined. A comparison is made with hyperfine structure constants estimated with an earlier proposed approximate method.

022.044 Activity measurements by Knudsen cell mass spectrometry − the system Fe-Co-Ni and implications for condensation processes in the solar nebula.
D. G. Fraser, W. Rammensee.
Geochim. Cosmochim. Acta, Vol. 46, 549 - 556 (1982).

022.045 The ionization equilibrium of astrophysically abundant elements. J. M. Shull, M. Van Steenberg.
Astrophys. J., Suppl. Ser., Vol. 48, 95 - 107 (1982).

The authors present new calculations of ionization equilibrium fractions of 11 abundant elements (C, N, O, Ne, Mg, Si, S, Ar, Ca, Fe, Ni) as functions of temperature. They also tabulate convenient coefficients for fitting the rates of collisional ionization, radiative recombination, and dielectronic recombination. Many of the ionization rates are based on recent experimental measurements of cross sections for collisional ionization and autoionization following inner-shell excitation.

022.046 Io: cooling models for sulfur flows.
J. Fink, S. O. Park, R. Greeley.
Reports of Planetary Geology Program − 1981, (see 003.001), p. 36 - 37 (1981). − Abstract.

022.047 Laboratory modeling of sulfur flows on Io.
R. Greeley, J. Fink, S. O. Park.
Reports of Planetary Geology Program − 1981, (see 003.001), p. 38 - 40 (1981). − Abstract.

022.048 Experimental impact craters formed in viscous fluids. J. H. Fink, D. E. Gault, R. Greeley.
Reports of Planetary Geology Program − 1981, (see 003.001), p. 81 (1981). − Abstract.

022.049 Vertical structure of basaltic lava flows: implications for surface sampling and interpretation.
J. C. Aubele, L. S. Crumpler, W. E. Elston.
Reports of Planetary Geology Program − 1981, (see 003.001), p. 156 - 159 (1981). − Abstract.

022.050 Scale modeling of lava flow processes.
S. O. Park, J. H. Fink, R. Greeley.

Reports of Planetary Geology Program – 1981, (see 003.001), p. 160 (1981). – Abstract.

022.051 **Particles formed by fuel-coolant explosions.**
M. F. Sheridan.
Reports of Planetary Geology Program – 1981, (see 003.001), p. 167 - 168 (1981). – Abstract.

022.052 **Melt-water interactions: series II experimental design.** K. H. Wohletz, M. F. Sheridan.
Reports of Planetary Geology Program – 1981, (see 003.001), p. 169 - 171 (1981). – Abstract.

022.053 **Surface roughness effects on aeolian processes: wind tunnel experiments.**
L. M. Reding, S. Williams, R. Leach, B. R. White, R. Greeley.
Reports of Planetary Geology Program – 1981, (see 003.001), p. 195 - 196 (1981). – Abstract.

022.054 **Venusian surface wind tunnel.**
R. Greeley, J. Iversen, B. R. White, R. Leach, S. Williams.
Reports of Planetary Geology Program – 1981, (see 003.001), p. 200 (1981). – Abstract.

022.055 **Soil transport by winds on Venus.**
B. R. White, R. Greeley.
Reports of Planetary Geology Program – 1981, (see 003.001), p. 201 - 202 (1981). – Abstract.

022.056 **A method for modeling of small particle transport.**
J. Iversen, R. Greeley, J. B. Pollack.
Reports of Planetary Geology Program – 1981, (see 003.001), p. 203 - 204 (1981). – Abstract.

022.057 **An experimental study of the behaviour of electro-statically-charged fine particles in atmospheric suspension.** J. R. Marshall, D. Krinsley, R. Greeley.
Reports of Planetary Geology Program – 1981, (see 003.001), p. 208 - 210 (1981). – Abstract.

022.058 **An experimental investigation of Martian rock disintegration at the microlevel.**
J. R. Marshall, G. Stewart, D. Krinsley.
Reports of Planetary Geology Program – 1981, (see 003.001), p. 211 - 213 (1981). – Abstract.

022.059 **An experimental study of the erosion of basalt, obsidian and quartz by fine sand, silt and clay.**
G. Stewart, D. Krinsley, J. R. Marshall.
Reports of Planetary Geology Program – 1981, (see 003.001), p. 214 - 215 (1981). – Abstract.

022.060 **Vesiculation and lithification behavior of saline muds in near vacuum.** L. A. Johansen.
Reports of Planetary Geology Program – 1981, (see 003.001), p. 277 - 279 (1981). – Abstract.

022.061 **Groundwater sapping in sediments: theory and experiments.** A. D. Howard, C. McLane.
Reports of Planetary Geology Program – 1981, (see 003.001), p. 283 - 285 (1981). – Abstract.

022.062 **Some thermodynamic relationships governing the behavior of permafrost and frozen ground.**
D. M. Anderson.
Reports of Planetary Geology Program – 1981, (see 003.001), p. 292 - 296 (1981). – Abstract.

022.063 **Mars surface atmosphere exchange experiment: isothermal case.**
W. B. Banerdt, F. P. Fanale, R. S. Saunders.

Reports of Planetary Geology Program – 1981, (see 003.001), p. 355 - 357 (1981). – Abstract.

022.064 **Volatile release from Martian analog materials.**
R. K. Kotra, E. K. Gibson, Jr., M. A. Urbancic.
Reports of Planetary Geology Program – 1981, (see 003.001), p. 358 - 360 (1981). – Abstract.

022.065 **Alteration of rocks in hot CO_2 atmospheres: preliminary experimental results and application to Venus.** J. L. Gooding.
Reports of Planetary Geology Program – 1981, (see 003.001), p. 460 - 462 (1981). – Abstract.

022.066 **Reflectance spectroscopy of structural changes effected by the dehydration of goethite (α-FeOOH) and lepidocrocite (γ-FeOOH).**
R. V. Morris, H. V. Lauer, Jr.
Reports of Planetary Geology Program – 1981, (see 003.001), p. 472 - 474 (1981). – Abstract.

022.067 **Impact experiments and regolith buildup.**
W. K. Hartmann.
Reports of Planetary Geology Program – 1981, (see 003.001), p. 546 - 547 (1981). – Abstract.

022.068 **Classical rigid-ellipsoid model for collisions of H_2 with HC_7N and HC_9N.**
S. S. Bhattacharyya, A. S. Dickinson.
Astron. Astrophys., Vol. 107, 26 - 30 (1982).
The authors have used a rigid-shell potential and classical mechanics to calculate the cross sections for inelastic transitions in H_2 collisions with HC_7N and HC_9N. Corrections due to long-range forces are investigated and found to be significant primarily for the low Δj transitions, j being the rotational quantum number. Formulae are presented from which rate coefficients may be calculated for $10 \leqslant T \leqslant 50$ K, for $9 \leqslant j \leqslant 32$, $1 \leqslant \Delta j \leqslant 10$ for HC_7N and $12 \leqslant j \leqslant 33$, $1 \leqslant \Delta j \leqslant 14$ in HC_9N.

022.069 **Absolute transition probabilities in the spectra of Eu I and Eu II. II. Line intensity measurements.**
C. Karner, G. Meyer, F. Träger, G. zu Putlitz.
Astron. Astrophys., Vol. 107, 161 - 165 (1982).
The relative intensities of 27 optical transitions in the Eu II spectrum have been measured. All transitions originate from the decay of the excited states $z^9P_{3,4,5}$ and $z^7P_{4,3,2}$ respectively. Eu ions with a sufficient population of the excited states in question were produced in a continuous gas discharge. The emitted lines in the wavelength region from 350 nm - 750 nm were observed selectively by means of a grating monochromator.

022.070 **Absolute transition probabilities in the spectra of Eu II. III. Astrophysical applications.**
E. Biémont, C. Karner, G. Meyer, F. Träger, G. zu Putlitz.
Astron. Astrophys., Vol. 107, 166 - 171 (1982).
The authors discuss some astrophysical implications of a new scale of accurate oscillator strengths obtained for Eu II. A new value of the solar photospheric abundance of Europium is derived: $A_{Eu} = 0.51 \pm 0.08$, in the usual logarithmic scale. Some stellar implications of the new gf-values are also presented.

022.071 **Laser pyrolysis for light element and stable isotope studies.** S. J. Norris, P. W. Brown, C. T. Pillinger.
Meteoritics, Vol. 16, 369 (1981). – Abstract.

022.072 **Annealing experiments on experimentally shocked feldspar single crystals.** R. Ostertag.
Meteoritics, Vol. 16, 373 (1981). – Abstract.

022.073   On the homogeneous condensation of Fe, Si, FeO_x and SiO_x vapors, and implications for astronomical condensation.   J. R. Stephens, S. H. Bauer.
Meteoritics, Vol. 16, 388 - 389 (1981). – Abstract.

022.074   Simulation of lunar sputter erosion using heavy ion bombardment.
K. Thiel, U. Saßmannshausen, H. Külzer, W. Herr.
Meteoritics, Vol. 16, 392 (1981). – Abstract.

022.075   High sensitivity high precision stable carbon isotope analysis for extraterrestrial samples.
I. P. Wright, N. J. McNaughton, A. E. Fallick, C. T. Pillinger.
Meteoritics, Vol. 16, 405 - 406 (1981). – Abstract.

022.076   Laboratory measurements of the pure rotation $S(2)$ and $S(3)$ transitions in H_2.
D. E. Jennings, J. W. Brault.
Astrophys. J., Lett., Vol. 256, L29 - L31 (1982).

Frequencies and transition rates have been measured for $S(2)$ and $S(3)$ in the pure rotation ground-state spectrum of molecular hydrogen. For $S(2)$ the determined frequency, $814,4250 \pm 0.0005$ cm^{-1}, is 0.027 cm^{-1} lower than the value used by Beck, Lacy, and Geballe in their detection of this line in the Orion molecular cloud. The $S(3)$ frequency is 1034.67035 ± 0.00010 cm^{-1}. Transition probabilities measured for $S(2)$ and $S(3)$ are $(3.1 \pm 0.4) \times 10^{-9}$ s^{-1} and $(9.9 \pm 0.5) \times 10^{-9}$ s^{-1} respectively.

022.077 CO^+ comet-tail emission and chemiluminescence in collisions of N^+ ions with CO molecules in the energy range 10-300 eV.   H. Sasaki, H. Inoue, T. Ishikawa.
J. Phys. Soc. Japan, Vol. 50, 3491 - 3496 (1981). – Abstr. in Phys. Abstr., Vol. 85, Abstr. 15885 (1982).

022.078   Die Bedeutung des Experiments für den Fortschritt in der Physik.
R. Rompe, G. Albrecht.
Astron. Nachr., Band 303, 21 - 26 (1982).

022.079   Elektronische Rechentechnik im Experiment.
K. Lanius.
Astron. Nachr., Band 303, 27 - 28 (1982).

022.080   Non-locality, causality and aether in quantum mechanics.   J. P. Vigier.
Astron. Nachr., Band 303, 55 - 80 (1982).

022.081   Vacua and symmetries.
Z. Marić.
Astron. Nachr., Band 303, 81 - 84 (1982).

022.082   Interferenz- und Korrelationsphänomene in der Quantentheorie.   F. Kaschluhn.
Astron. Nachr., Band 303, 85 - 89 (1982).

022.083   Der Michelson-Versuch als experimentum crucis.
H.-J. Treder.
Astron. Nachr., Band 303, 91 - 96 (1982).

022.084   Radiative lifetimes for Pd I and the solar abundance of palladium.   E. Biémont, N. Grevesse,
M. Kwiatkowski, P. Zimmermann.
Astron. Astrophys., Vol. 108, 127 - 129 (1982).

New lifetime measurements obtained using laser excitation are reported for levels in the $4d^9 5p$ configuration of Pd I. These results are combined with branching ratios taken from Corliss and Bozman (1962) in order to provide a new set of oscillator strengths for eight transitions of solar interest. The photospheric abundance deduced from the study of these lines is $A_{Pd} = 1.69 \pm 0.04$.

022.085   Winkelverteilungen und Linienprofile von Zyklotronübergängen in superstarken Magnetfeldern.
H. Herold, H. Ruder, G. Wunner.
Mitt. Astron. Ges., Nr. 55, p. 33 - 35 (1982).

022.086   Franck-Condon factors and r-centroids for the $K^2 \Sigma$-$A^2\pi$ and $D^2\pi$-$A^2\pi$ band systems of the SiN molecule.   U. C. Joshi, M. Joshi, R. N. Singh.
Indian J. Phys. Part B, Vol. 55B, 335 - 338 (1981). – Abstr. in Phys. Abstr., Vol. 85, Abstr. 24679 (1982).

022.087   Lifetimes of excited levels of Nd I and Nd II. Oscillator strengths of spectral lines of Nd I.
V. N. Gorshkov, V. A. Komarovskij, A. L. Osherovich, N. P. Penkin.
Astrofizika, Tom 17, 799 - 806 (1981). In Russian. English translation in Astrophysics, Vol. 17, No. 4.

Lifetimes of 33 excited levels of Nd I and 11 levels of Nd II have been measured.

022.088   Pion condensation, equation of state of dense matter and neutron stars.   P. Haensel, M. Proszynski.
Recent progress in many-body theories, (see 012.014), p. 433 - 443 (1981). – Abstr. in Phys. Abstr., Vol. 85, Abstr. 29203 (1982).

022.089   Structure of baryonic system with pion condensation and its implication in neutron star problems.
R. Tamagaki.
Recent progress in many-body theories, (see 012.014), p. 444 - 452 (1981). – Abstr. in Phys. Abstr., Vol. 85, Abstr. 29204 (1982).

022.090   High-spin effects in superdense matter.
R. L. Bowers, A. M. Gleeson, R. D. Pedigo.
Nuovo Cimento B, Vol. 65B, Ser. 11, 427 - 441 (1981). Abstr. in Phys. Abstr., Vol. 85, Abstr. 29215 (1982).

022.091   The hydroxyl molecule: population inversion within the rotational levels of the $^2\pi$ state.
K. A. Hilsum, M. S. Shafik.
Indian J. Phys. Part B, Vol. 55B, 225 - 231 (1981). – Abstr. in Phys. Abstr., Vol. 85, Abstr. 29659 (1982).

022.092   Production of 6.13-MeV gamma rays from the $^{16}O(p, p'\gamma)^{16}O$ reaction at 23.7 and 44.6 MeV.
J. Narayanaswamy, P. Dyer, S. R. Faber, S. M. Austin.
Phys. Rev. C, Vol. 24, 2727 - 2730 (1981). – Abstr. in Phys. Abstr., Vol. 85, Abstr. 34039 (1982).

022.093   A calculation on the transition probability in the A'–X system of YO.   S. B. Rai, B. R. Yadav.
J. Sci. Res. Banaras Hindu Univ., Vol. 30, No. 1, p. 301 - 306 (1979 - 1980). – Abstr. in Phys. Abstr., Vol. 85, Abstr. 34498 (1982).

022.094   Spin-dependent polarizabilities of hydrogenic atoms in magnetic fields of arbitrary strength.
T. G. Castner, D. L. Dexter, S. D. Druger.
Phys. Rev. A, Vol. 24, 2897 - 2905 (1981). – Abstr. in Phys. Abstr., Vol. 85, Abstr. 34595 (1982).

022.095   Coming attractions in SUMs and cosmology.
F. Wilczek.
Comments Nucl. Part. Phys., Vol. 10, 175 - 185 (1981). Abstr. in Phys. Abstr., Vol. 85, Abstr. 38824 (1982).

022.096   Resonance strength measurements and thermonuclear reaction rates for $^{25}Mg(p, \gamma)^{26}$ Al. II.
M. R. Anderson, L. W. Mitchell, M. E. Sevior, S. R. Kennett, D. G. Sargood.

Nucl. Phys. A, Vol. A373, 326 - 340 (1982). — Abstr. in Phys. Abstr., Vol. 85, Abstr. 39028 (1982).

022.097   The $^{12}C + ^{12}C$ reaction at subcoulomb energies. II.
H. W. Becker, K. U. Kettner, C. Rolfs,
H. P. Trautvetter.
Z. Phys. A, Vol. 303, 305 - 312 (1981). — Abstr. in Phys. Abstr., Vol. 85, Abstr. 39070 (1982).

022.098   The redshift of hydrogen lines in a strong magnetic field.   S. H. Patil.
J. Phys. A, Vol. 14, 2251 - 2258 (1981). — Abstr. in Phys. Abstr., Vol. 85, Abstr. 39361 (1982).

022.099   Singularity of Störmer's problem.   D. Eschbach.
Celestial Mech., Vol. 27, 39 - 52 (1982). In French.

In previous papers (Eschbach, 1979, 1980, 1982) the author generalised the change of variables used by McGehee in the study of the triple collision. The object of this paper is to use the same technique for Störmer's problem, which studies the motion of a charged particle under the influence of a magnetic dipole.

022.100   Configuration mixing between $2p\,3d\,^1F^0$ and $2s\,4f\,^1F^0$ states in N IV.   R. Glass.
Mon. Not. R. Astron. Soc., Vol. 199, 435 - 439 (1982).

The author has used configuration interaction wavefunctions to look at the mixing between the $2p\,3d\,^1F^0$ and $2s\,4f\,^1F^0$ states in N IV. He concludes that the classification of these two states should be interchanged. He has also calculated the lifetimes of the $2s\,4f\,^1F^0$ and $2p\,3d\,^1F^0$ states and gives reasons why the lifetime of the $2p\,3d\,^1F^0$ state should be much shorter than the lifetime of the $2s\,4f\,^1F^0$ state.

022.101   Collisional excitation of OH by H_2 in the interstellar medium.   D. P. Dewangan, D. R. Flower.
Mon. Not. R. Astron. Soc., Vol. 199, 457 - 463 (1982).

The authors present the results of calculations of rate coefficients for rotational excitation of OH by para-H_2 over the temperature range believed to be relevant to the interstellar OH maser emission. The reliability and limitations of these results are discussed, as are their astrophysical implications. The calculations suggest that, if rotational excitation by H_2 is the primary population inversion mechanism, then relatively high kinetic temperatures ($T \gtrsim 60\,K$) are required in the maser regions.

022.102   A laboratory simulation of the interstellar 220 nanometer feature.
S. MacLean, W. W. Duley, T. J. Millar.
Astrophys. J., Lett., Vol. 256, L61 - L64 (1982).

Laboratory spectra of finely divided MgO powders show an absorption band due to an electronic transition of O^{2-} ions in low-coordination sites that simulates the interstellar feature at 220 nm. A similar feature is observed in SiO_2, and it is suggested that a possible source of the 220 nm interstellar feature is absorption by small grains of Mg-silicate composition. Some 30% of available Mg and Si would have to be in these small particles to account for the observed strength of the interstellar feature.

022.103   The hypothesis of neutron irradiation of protoplanetary matter and changes of the isotopic composition of some elements.   Yu. P. Bulashevich, R. L. Kharus.
Dokl. AN SSSR, Tom 261, 53 - 55 (1981). In Russian.
Abstr. in Ref. zh., 51. Astron., 3.51.159 (1982).

022.104   Bremsstrahlung of gravitons.
M. I. Krivoruchenko, B. V. Martem'yanov.
Zh. ehksp. teor. fiz., Tom 81, 1553 - 1555 (1981). In Russian.
Abstr. in Ref. zh., 51. Astron., 3.51.865 (1982).

022.105   Electron trajectories in the hydrogen minus ion, or if planets were repulsive.   K. B. Butterfield.
Proc. Southwest Reg. Conf., Vol. 7, (see 012.019), 73 - 83 (1982).

Electron trajectories have been calculated for the H^- ion using classical mechanics. Several categories of orbitals have been found, including non-elliptical orbits. Comparisons are made with Bohr-Sommerfield and quantum mechanical models.

022.106   Oscillator strengths for lines of neutral chromium.   R. I. Kostyk.
Astrometr. Astrofiz., Vyp. (No.) 45, p. 3 - 9 (1981). In Russian.

The solar microturbulence, macroturbulence and damping constant derived from Fraunhofer lines of neutral chromium are found to be: $v_{micro} = 0.8\,km\,s^{-1}$, $v_{macro} = 1.1\,km\,s^{-1}$, $\gamma = 2.0\,\gamma_6$. Oscillator strengths of 123 Cr I lines have been determined with an accuracy of 0.19 dex.

022.107   Oscillator strengths for lines of neutral nickel.   R. I. Kostyk.
Astrometr. Astrofiz., Vyp. (No.) 46, p. 58 - 61 (1982). In Russian.

Oscillator strengths of 175 Ni lines have been determined with the accuracy ±0.09 dex from central intensities of solar Fraunhofer lines.

022.108   Radiative lifetimes and quenching rate coefficients for directly excited rotational levels of
$OH(A^2\Sigma^+, v' = 0)$.   I. S. McDermid, J. B. Laudenslager.
J. Chem. Phys., Vol. 76, 1824 - 1831 (1982). — Abstr. in Phys. Abstr., Vol. 85, Abstr. 44004 (1982).

022.109   A schematic model of crater modification by gravity.   H. J. Melosh.
J. Geophys. Res., Vol. 87, 371 - 380 (1982).

Models are proposed to account for the formation of slump terraces, central peaks, peak rings, and concentric rings in large impact structures.

022.110   Photoabsorption into the $^3\Pi_u$ state of O_2.
A. C. Allison, S. L. Guberman, A. Dalgarno.
J. Geophys. Res., Vol. 87, 923 - 925 (1982).

Earlier calculations of the cross sections for absorption of ultraviolet radiation by molecular oxygen in the transition $1\,^3\Pi_u - X^3\Sigma_g^-$ are corrected and extended to shorter wavelengths. The calculations confirm the experimental identification of strong absorption in the region of 1350 Å, attributable to the $^3\Pi_u$ state.

022.111   Compressional wave velocities of a lunar regolith sample in a simulated lunar environment.
D. M. Johnson, A. L. Frisillo, J. Dorman, G. V. Latham,
D. Strangway.
J. Geophys. Res., Vol. 87, 1899 - 1902 (1982).

Ultrasonic compressional wave velocities have been measured in the laboratory for an Apollo 15 soil sample (15301, 38) under very low uniaxial stress and high vacuum conditions. The velocities measured range from 125 to 522 m/s. The velocities of the soil are stress dependent and are strongly affected by compaction history. Hertzian contact theory does not appear to fit the data adequately for the pressure range of the experiment. Moderate increases in temperature do not have a significant effect on the compressional wave velocities.

022.112   Radiative lifetimes and electronic quenching rate constants for single-photon-excited rotational levels of NO $(A^2\Sigma^+, v' = 0)$.   I. S. McDermid, J. B. Laudenslager.
J. Quant. Spectrosc. Radiat. Transfer, Vol. 27, 483 - 492 (1982).

022.113 **Spectroscopic line parameters of NH_3 and PH_3 in the far infrared.**
N. Husson, A. Goldman, G. Orton.
J. Quant. Spectrosc. Radiat. Transfer, Vol. 27, 505 - 515 (1982).

NH_3 and PH_3 rotation and rotation-inversion line parameters in the far to medium i.r. are calculated for remote sounding purposes of planetary atmospheres; 1607 lines of $^{14}NH_3$, 362 lines of $^{15}NH_3$ and 325 lines of PH_3 are compiled. The absolute intensity formulation has been reviewed in the case of rotation and rotation-inversion lines of molecules with C_{3v} symmetry. The justification of the general agreement between the authors, and comparisons with other published expressions are given.

022.114 **Electron impact ionization rate coefficients and cross sections for highly ionized iron.**
S. M. Younger.
J. Quant. Spectrosc. Radiat. Transfer, Vol. 27, 541 - 544 (1982).

022.115 **Self-broadening of the ammonia inversion lines.**
M. Cattani, Y. Yamamoto.
J. Quant. Spectrosc. Radiat. Transfer, Vol. 27, 563 - 567 (1982).

022.116 **Mare basin filling on the moon: laboratory simulations.**   R. Greeley, M. B. Womer.
Proc. Twelfth Lunar Planet. Sci. Conf., (see 012.024), p. 651 - 663 (1982).

022.117 **Noble gas trapping by laboratory carbon condensates.**   S. Niemeyer, K. Marti.
Proc. Twelfth Lunar Planet. Sci. Conf., (see 012.024), p. 1177 - 1188 (1982).

022.118 **Impact accretion experiments.**
V. Werle, H. Fechtig, E. Schneider.
Proc. Twelfth Lunar Planet. Sci. Conf., (see 012.024), p. 1641 - 1647 (1982).

022.119 **Impact cratering experiments in Bingham materials and the morphology of craters on Mars and Ganymede.**   J. H. Fink, R. Greeley, D. E. Gault.
Proc. Twelfth Lunar Planet. Sci. Conf., (see 012.024), p. 1649 - 1666 (1982).

022.120 **Xenon diffusion following ion implantation into feldspar: dependence on implantation dose.**
C. L. Melcher, D. S. Burnett, T. A. Tombrello.
Proc. Twelfth Lunar Planet. Sci. Conf., (see 012.024), p. 1725 - 1736 (1982).

022.121 **Effects of body shape on disk-integrated spectral reflectance.**   J. Gradie, J. Veverka.
Proc. Twelfth Lunar Planet. Sci. Conf., (see 012.024), p. 1769 - 1779 (1982).

Variations in scattering geometry have been shown to affect the shapes of spectral reflectance curves for a variety of powdered materials of planetary interest. Photometric data obtained over a range of incidence angle, emission angle, and wavelength have been used to determine which of several photometric functions most accurately describes these effects. The wavelength dependence of such functions implies that (1) there will be a difference in the shapes of the spectral reflectance curves of a flat sample and a spherical planet made of the same material, and (2) that there will be a difference between the shapes of the spectral reflectance of a spherical planet and an ellipsoidal one. The last fact can be applied to asteroids to demonstrate that elongated asteroids should appear redder at maximum light than at minimum light even if their surface material is laterally homogeneous.

022.122 **When are spectral reflectance curves comparable?**   J. Gradie, J. Veverka.
Icarus, Vol. 49, 109 - 119 (1982).

Spectral reflectance curves of flat laboratory samples of the carbonaceous chondrite Allende, a basalt, and the ordinary chondrite Bruderheim measured in a bidirectional geometry are shown to differ from those measured using an integrating sphere. In general, reflectance curves obtained by the bidirectional method are redder than those obtained with an integrating sphere. When spectral reflectance curves obtained by the two methods are compared to the reflectance curves expected for spherical and aspherical planets covered with the same materials, it is found that in general the integrating sphere measurements provide a better match to a planet at small phase angles. As the phase angle increases, bidirectional reflectance curves provide a closer match.

022.123 **Modern and old views in the dynamical foundations of classical statistical mechanics.**   L. Galgani.
Application of modern dynamics to celestial mechanics and astrodynamics, (see 012.026), p. 173 - 184 (1982).

A short review is offered giving dynamical foundations to some ideas in statistical mechanics which go back to L. Boltzmann and W. Nernst, concerning the distribution of energy in classical systems of weakly coupled oscillators.

022.124 **Quasi-random motions in a perturbed pendulum.**   A. Aubanell.
Application of modern dynamics to celestial mechanics and astrodynamics, (see 012.026), p. 328 (1982). – Abstract.

022.125 **On the applicability of the transition chains mechanism.**   A. Delshams.
Application of modern dynamics to celestial mechanics and astrodynamics, (see 012.026), p. 341 (1982). – Abstract.

022.126 **Applications of Hamilton's Law of Varying Action.**   D. L. Hitzl.
Application of modern dynamics to celestial mechanics and astrodynamics, (see 012.026), p. 346 (1982). – Abstract.

022.127 **Periodic orbits near homoclinic orbits.**   J. Llibre.
Application of modern dynamics to celestial mechanics and astrodynamics, (see 012.026), p. 350 (1982). – Abstract.

022.128 **Orbital behaviour in the vicinity of unstable periodic orbits in dynamical systems with three degrees of freedom.**   P. Magnenat.
Application of modern dynamics to celestial mechanics and astrodynamics, (see 012.026), p. 352 (1982). – Abstract.

022.129 **Stability of periodic orbits near a homoclinic orbit for analytical Hamiltonians with two degrees of freedom.**   C. Simó.
Application of modern dynamics to celestial mechanics and astrodynamics, (see 012.026), p. 357 (1982). – Abstract.

022.130 **A method for the investigation of the integrability of dynamical systems.**   A. Sivaramakrishnan.
Application of modern dynamics to celestial mechanics and astrodynamics, (see 012.026), p. 358 (1982). – Abstract.

022.131 **Reappearance of ordered motion in classical non-integrable Hamiltonian systems.**
R. L. Somorjai, M. K. Ali.
Application of modern dynamics to celestial mechanics and astrodynamics, (see 012.026), p. 359 (1982). – Abstract.

022.132 **Atomic transition probabilities and radiative lifetimes.**   E. Biemont.
Excitation and broadening in atomic spectra of astrophysical

interest, (see 012.027), p. 1 - 48 (1981).

The author presents a detailed review of experimental and theoretical advances realized during last years for a better knowledge of atomic transition probabilities and radiative lifetimes. Attention is mainly focussed upon results reported for neutral and singly ionized elements. Some solar implications are also briefly considered. The comprehensive bibliography was closed in april 1981.

022.133   **Elargissement des raies spectrales.**
N. Feautrier.
Excitation and broadening in atomic spectra of astrophysical interest, (see 012.027), p. 49 - 80 (1981).

022.134   **Atomic data for the interpretation of EUV astrophysical plasmas.**   H. E. Mason.
Excitation and broadening in atomic spectra of astrophysical interest, (see 012.027), p. 81 - 108 (1981).

The EUV spectral region covers approximately 200 - 800 Å. The UV and EUV wavelength regions contain spectral lines emitted by ions which exist in the electron temperature range $10^4 - 10^6$ K. This temperature range spans many interesting astrophysical plasmas. In the solar atmosphere, emission in this wavelength region comes from the transition zone which connects the low temperature chromosphere and the high temperature corona. Astrophysical plasmas which emit lines in this wavelength region include planetary nebulae, novae and quasars.

022.135   **Données atomiques et diagnostics necessaires à l'interprétation du spectre d'émission X du soleil actif** $(\lambda < 25 \text{ Å})$.   J. Dubau, M. Loulergue.
Excitation and broadening in atomic spectra of astrophysical interest, (see 012.027), p. 109 - 134 (1981).

022.136   **Excitación del He en condiciones solares.**
H. Molnar.
Bol. Asoc. Argentina Astron., No. 20 - 24, p. 143 (1981).
Abstract.

022.137   **Método de cálculo de poblaciones y líneas de hidrógeno fuera del equilibrio termodinámico local.**
J. M. Fontenla.
Bol. Asoc. Argentina Astron., No. 20 - 24, p. 243 (1981).
Abstract.

022.138   **Theoretical oscillator strengths for 21 spin-forbidden lines of C, N, O, Al, and Si.**
R. D. Cowan, L. M. Hobbs, D. G. York.
Astrophys. J., Vol. 257, 373 - 375 (1982).

Detailed calculations are reported of the oscillator strengths of 21 spin-forbidden lines of C II, C III, N I, N II, N III, O I, O III, Al II, and Si III, including 10 lines arising from seven ground levels. All 21 lines lie in the region $1160 \leqslant \lambda \leqslant 2670$ Å, and the oscillator strengths probably are accurate typically to ± 50%. These weak lines are particularly important in determining accurate elemental abundances in the interstellar gas.

022.139   **The rotational spectra of HCNH⁺ and COH⁺ from quantum mechanical calculations.**
D. J. DeFrees, G. H. Loew, A. D. McLean.
Astrophys. J., Vol. 257, 376 - 382 (1982).

Ab initio molecular orbital theory is used to determine the $J = 1 \rightarrow 0$ rotational frequency and the dipole moment of the candidate interstellar molecules HCNH⁺ and COH⁺. An empirically based relationship is used to correct for zero-point vibrational effects on the ab initio rotational constants. Systematic errors in the computed spectra are compensated for by using the mean ratio of experimental to theoretical rotational constants for C_2H_2, HCN, and HNC as a correction

factor. The predicted $J = 1 \rightarrow 0$ frequencies are 74.43 ± 0.3 GHz for HCNH⁺ and 88.64 ± 0.3 GHz for COH⁺.

022.140   **Radiative lifetimes for the A $^2\Pi$ and B $^2\Sigma^+$ electronic states of the CN molecule.**
D. C. Cartwright, P. J. Hay.
Astrophys. J., Vol. 257, 383 - 387 (1982).

Electronic transition moments, as a function of internuclear distance, have been calculated for the A $^2\Pi \leftrightarrow X\ ^2\Sigma^+$, B $^2\Sigma^+ \leftrightarrow X\ ^2\Sigma^+$, and B $^2\Sigma^+ \leftrightarrow A\ ^2\Pi$ transitions in the CN molecule by employing large configuration interaction wave functions. These ab initio electronic transition moments have been combined with vibrational wave functions determined from experimental (RKR) potential energy curves to produce spontaneous transition probabilities and absorption oscillator strengths for these three transitions.

022.141   **The infrared spectrum of a laboratory-synthesized residue: implications for the 3.4 micron interstellar absorption feature.**   M. H. Moore, B. Donn.
Astrophys. J., Lett., Vol. 257, L47 - L50 (1982).

Proton irradiation of low-temperature ice mictures of water, ammonia, and methane or propane results in the synthesis of an organic, room temperature residue whose dominant infrared signatures occur near 3.4 μm and 9.9 μm. The 3.4 μm laboratory-produced feature is compared with the interstellar 3.4 μm absorption feature in Sgr A – IRS 7 and OH 01 – 477.

022.142   **On the effective Landé factor of magnetic lines.**
E. Landi Degl'Innocenti.
Sol. Phys., Vol. 77, 285 - 289 (1982).

The effective Landé factor, $\bar{g}$, of a magnetic sensitive line can be calculated by means of the experimental Landé factors of the lower and upper levels of the atomic transition. Values of $\bar{g}$ for several iron lines of the solar spectrum are calculated and compared with the approximate values based on the L-S coupling scheme. Simple formulae are also derived to express the variance of the distributions of the σ and π components of a Zeeman multiplet around the respective centers of gravity.

022.143   **The empirical determination of damping constants in the solar photosphere. II: Results inferred from the wings of Fe I lines.**   E. A. (Eh. A.) Gurtovenko, G. L. Fedorchenko, N. N. Kondrashova.
Sol. Phys., Vol. 77, 291 - 297 (1982).

The authors determine empirical damping constants for 73 selected Fe I lines following the method of Gurtovenko and Kondrashova (1980), employing high-quality observations and the accurate list of Fe I oscillator strengths by Gurtovenko and Kostik (1980). The results show: (1) no increase of the enhancement factor to van der Waals broadening with excitation potential and with the frequency of the transition; (2) a substantial part of the commonly-used enhancement factor for weaker lines is not due to collisional damping, but to a misrepresentation of the inhomogeneous structure of the deep photosphere.

022.144   **Spin-orbit electric dipole transmissions in beryllium-like ions.**   R. Glass.
Sol. Phys., Vol. 78, 29 - 38 (1982).

The author has used configuration interaction wavefunctions to calculate energy levels, wavelengths, oscillator strengths and transition probabilities for spin-orbit electric dipole transitions between the $2s^2$, $2s2p$, and $2p^2$ states in beryllium-like ions ($Z = 6$–10). Some significant differences with previous calculations are obtained. A common set of radial functions is used.

022.145   **The millimeter wave spectrum and discharge chemistry of HC₅N.**

G. Winnewisser, M. Winnewisser, J. J. Christiansen.
Astron. Astrophys., Vol. 109, 141 - 144 (1982).

The laboratory millimeter wave spectrum of HC_5N has been measured up to 210 GHz, yielding refined ground state parameters $[B_0 = 1331.332714(47)$ MHz, $D_0 = 30.1017(58)$ Hz] and highly precise frequency predictions throughout the millimeter wave region. The discharge of any combination of hydrocarbons and a source of nitrogen such as HCN will generate HC_3N and HC_5N in observable quantities. Some of these reactions will probably be of astrophysical significance.

022.146 Intercambio de carga entre iones multicargados y helio neutro. L. Opradolce.
Bol. Asoc. Argentina Astron., No. 26, p. 191 - 192 (1981).

022.147 Equilibrio químico disociativo en envolturas circumestelares. C. A. Núñez, E. R. Iglesias.
Bol. Asoc. Argentina Astron., No. 26, p. 193 - 197 (1981).

022.148 Double ionization of atomic oxygen by electron impact.
D. L. Ziegler, J. H. Newman, K. A. Smith, R. F. Stebbings.
Planet. Space Sci., Vol. 30, 451 - 456 (1982).

Laboratory measurements of the cross-sections for double ionization of atomic oxygen by electrons are presented for energies from threshold to ~400 eV.

022.149 Measurement of cross sections of excitation of a simply charged iron ion by electron impact.
P. A. Kolosov, Yu. M. Smirnov.
Astron. Zh., Tom 59, 602 - 605 (1982). In Russian. English translation in Soviet Astron., Vol. 26, No. 3.

By the method of crossing bunches the values of excitation cross sections were measured for 21 lines of Fe II in the region 2170 - 2950 Å. Values of excitation cross sections are in the interval $(0.8 - 52.0) \times 10^{-18}$ cm^2.

022.150 Electron density diagnostic line ratios from the $n = 3$ lines of O V.
K. G. Widing, J. G. Doyle, P. L. Dufton, A. E. Kingston.
Astrophys. J., Vol. 257, 913 - 917 (1982).

New atomic physics calculations are presented for electron excitation rates for transitions between the $n = 2$ and $n = 3$ levels of O V. These are used to calculate theoretical line intensity ratios for the 192 Å, 215 Å, 220 Å, and 248 Å lines of O V. These line intensity ratios are electron density sensitive and provide valuable diagnostics at $T_e \sim 2 \times 10^5$ K for small impulsive flare events in which the transition zone ions are enhanced relative to the coronal ions. Two flares observed by the NRL spectroheliograph on Skylab, on 1973 December 22 and 1974 January 21, are studied, with electron densities of approximately 3×10^{11} cm^{-3} being deduced.

022.151 Probable abundance ratios for interstellar HCS_2^+, $HCOS^+$, and HCO_2^+.
W. Fock, T. McAllister.
Astrophys. J., Lett., Vol. 257, L99 - L101 (1982).

The proton affinities of CS_2, COS and CO_2 and the rate coefficients for the formation of their protonated derivatives, HCS_2^+, $HCOS^+$, and HCO_2^+ in mixtures with H_2 and CO have been used to dermine the ratios $n(HCX_2^+)/n(CX_2)$ for X = O or S in interstellar clouds. The results indicate that $n(HCOS^+)$ may be comparable to $n(HCO_2^+)$. Laboratory detection of the microwave spectra of HCS_2^+ and $HCOS^+$ in gas discharges may be hampered by the presence of H_2S as a by-product.

022.152 Laboratory measurements of amorphous silicate smokes and the infrared spectra of oxygen-rich stars.
J. A. Nuth, B. Donn.
Astrophys. J., Lett., Vol. 257, L103 - L105 (1982).

The infrared spectra of smokes condensed from $SiO-H_2$ and $Mg-SiO-H_2$ vapors and of smoke samples annealed for various times at 1000 K and 1250 K are presented. The spectra of these materials show features in the 8 - 25 μm interval in addition to the well-known bands near 10 and 20 μm. It is suggested that these features may correspond to weak structures that can be seen in the infrared spectra of oxygen-rich stars.

022.153 Experimental study of effects associated with macroscopic hypervelocity impacts.
G. Martelli, R. Bianchi, P. Cerroni, M. Coradini, R. Flavill, P. Hurren, P. N. Smith, F. Waldner.
The comparative study of the planets, (see 012.033), p. 333 - 357 (1982).

The authors present and discuss results from a series of hypervelocity impact experiments performed using explosive shaped charges. Primary and secondary cratering have been studied both at atmospheric pressure and "in vacuo" (p < 1 mm Hg) using basalt-like and clay targets. Magnetic phenomena associated with the impact produced plasma have also been recorded.

022.154 The "trumpet charge", a technique for producing macroscopic hypervelocity projectiles.
G. Martelli.
The comparative study of the planets, (see 012.033), p. 359 - 362 (1982).

A novel technique is proposed for the acceleration of macroscopic projectiles to velocities approaching 20 km/sec using explosive propulsion.

022.155 A possible mechanism for impact magnetisation of cratered surfaces. P. Cerroni, G. Martelli.
The comparative study of the planets, (see 012.033), p. 363 - 366 (1982).

A mechanism is proposed for the generation or amplification of magnetic fields during the expansion of impact-produced plasma clouds, the temperature gradient providing the thermal force required to excite azimuthal currents at the surface of the plasma clouds.

022.156 The effects of microparticle hypervelocity impacts on polished surfaces: tests for the choice of the Halley multicolour camera. M. Coradini, E. Flamini, P. Hurren, G. Martelli, P. N. Smith.
The comparative study of the planets, (see 012.033), p. 367 - 387 (1982).

In order to test the response to hypervelocity impacts of materials to be used in the construction of the mirror to be flown on the ESA GIOTTO probe as part of the multicolour camera which will image the Halley comet, the authors have bombarded reflecting surfaces with quartz particles of $10^{-6} - 10^{-7}$ gr travelling at velocities well in excess of 8 km/sec. Metallic and glassy mirrors were tested. Crater counts, morphological analysis of the impacted surfaces and theoretical considerations were used to evaluate the response of different materials to microparticle hypervelocity impacts.

022.157 Laboratory submillimeter spectroscopic studies of astrophysical interest. F. C. De Lucia, E. Herbst.
Bull. American Astron. Soc., Vol. 13, 827 (1981). — Abstract.

022.158 Ion beam measurements of electron excitation coefficients.
J. L. Kohl, G. P. Lafyatis, W. H. Parkinson.
Bull. American Astron. Soc., Vol. 13, 911 (1981). — Abstract.

022.159 A simple approximation on potential function and its application to the calculation of photoionization cross-section. Ç. Bolcal.
Istanbul Üniv. Fen Fak. Mec. Seri C, Vol. 45, 107 - 120 (1980) = Publ. Istanbul Univ. Obs. No. 102 (1981).

022.160 **On the mathematical properties of the Voigt profile of a spectral line.**   V. K. Khersonskij.
Astrofiz. Issled., Izv. Spets. Astrofiz. Obs., Tom 15, 75 - 87 (1982). In Russian.

The interrelation between the integral which describes the Voigt profile of a spectral line and the different special functions is considered. The expansion of this integral into infinite series under different values of the parameters characterising the spectral line is obtained. Its differential and integral properties and the methods of integration of the Voigt function as well as the functions which are determined by the Voigt function are discussed. For the illustration of the obtained results some integrals which play an important role in the radiation transfer theory are calculated.

022.161 **Emission of interstellar shocks. II. Density dependence of ionization and dielectronic recombination coefficients.**   I. S. Balinskaya, K. V. Bychkov.
Soobshch. Spets. Astrofiz. Obs., Vyp. 31, p. 49 - 95 (1981). In Russian.

The effect of metastable level excitation on the ionization coefficients by electron impact and dielectronic recombination coefficients is investigated.

022.162 **Interaction of charged particles with the field of a rotating magnetic dipole in the presence of electromagnetic radiation.**   A. K. Avetisyan.
Astrofizika, Tom 16, 285 - 303 (1980). In Russian. English translation in Astrophysics, Vol. 16, No. 2.

The interaction of charged particles with the field of a rotating magnetic dipole and an electromagnetic wave has been examined. The possibility of particle acceleration by the mentioned fields on the basis of relativistic equations of motion has been investigated. The classical effect of bunching as well as the quantum effect of modulation of the beam of charged particles at the frequencies of the rotating magnetic dipole and the electromagnetic wave are obtained. A calculation analysis of the obtained results is made, showing the possibility of applying these effects to pulsars.

022.163 **Metastable formations of nuclear matter.**   L. Sh. Grigoryan, G. S. Saakyan.
Astrofizika, Tom 16, 305 - 320 (1980). In Russian. English translation in Astrophysics, Vol. 16, No. 2.

022.164 **Spectra of Sn XX–Sn XXII in the extreme ultraviolet.**   M. A. Khan.
J. Opt. Soc. America, Vol. 72, 268 - 272 (1982).

022.165 **Infrared spectroscopy over the 2.9–3.9 μm waveband in biochemistry and astronomy.**
F. Hoyle, N. C. Wickramasinghe, S. Al-Mufti, A. H. Olavesen, D. T. Wickramasinghe.
Astrophys. Space Sci., Vol. 83, 405 - 409 (1982).

The infrared spectrum of the galactic centre source IRS 7 over the 2.9–3.9 μm waveband is interpreted as strong evidence for bacterial grains.

022.166 **Total ionization cross sections and rates for Na-like ions.**   D. H. Sampson.
Bull. American Astron. Soc., Vol. 14, 571 (1982). – Abstract.

022.167 **Laser fusion experiments relevant to problems in astrophysics.**   B. Yaakobi, T. Bristow.
Bull. American Astron. Soc., Vol. 14, 573 (1982). – Abstract.

022.168 **Laboratory simulation of 3.4 μm interstellar absorption feature.**   B. Donn, M. H. Moore.
Bull. American Astron. Soc., Vol. 14, 573 (1982). – Abstract.

022.169 **Computerized Lindstedt methods for systems with two degrees of freedom.**

R. Broucke, W. Presler.
Bull. American Astron. Soc., Vol. 14, 581 (1982). – Abstract.

022.170 **One-electron spectrum of Yb$^+$: relativistic energies, transition probabilities and dipole polarizability.**
J. Migdalek.
J. Quant. Spectrosc. Radiat. Transfer, Vol. 28, 61 - 69 (1982).

022.171 **Convergent calculations for electron impact broadening and shift of neutral helium lines.**
J. M. Bassalo, M. Cattani, V. S. Walder.
J. Quant. Spectrosc. Radiat. Transfer, Vol. 28, 75 - 80 (1982).

Using the convergent semiclassical method, the authors have calculated the electronic widths and shifts of 42 neutral He lines for an electronic density of $10^{16}/cm^3$ and $T = 5000$, 10,000, 20,000, and 40,000 K. To account for ions effects, the authors have calculated the Stark parameter A and parameter R.

022.172 **The N$_2$-broadened water vapor absorption line shape and infrared continuum absorption – I. Theoretical development.**   M. E. Thomas, R. J. Nordstrom.
J. Quant. Spectrosc. Radiat. Transfer, Vol. 28, 81 - 101 (1982).

Attenuation of infrared radiation in the troposphere is dominated by water vapor absorption. Most past work on the spectroscopy of water vapor has been on the analysis of the rotational and vibrational bands. The purpose of this study is to demonstrate the importance of far wing phenomena in characterizing H_2O continuum absorption. A total line shape for water vapor-nitrogen interactions valid under tropospheric conditions is derived.

022.173 **The N$_2$-broadened water vapor absorption line shape and infrared continuum absorption – II. Implementation of the line shape.**   M. E. Thomas, R. J. Nordstrom.
J. Quant. Spectrosc. Radiat. Transfer, Vol. 28, 103 - 112 (1982).

The total line shape model of the previous paper is tested using a set of experimental room temperature H_2O continuum measurements of high quality. Parameters of the far wing component of the total line shape are determined from near band experimental data. Grating spectrometer measurements from 300 to 650 cm^{-1} are used to determine unknown far wing parameters of the pure rotational band of H_2O. CO and HF laser measurements taken in the 5 and 3 μm regions are used to determine the far wing parameters of the ν_2 and ν_1, ν_3 fundamental bands, respectively. The total line shape model is applied to the 10 and 4 μm transmission windows with encouraging success.

022.174 **Transition probabilities of Mn(II) lines lying between 237 and 357 nm from arc emission measurements.**   T. Wujec, S. Weniger.
J. Quant. Spectrosc. Radiat. Transfer, Vol. 28, 113 - 129 (1982).

022.175 **Theoretical term energies and oscillator strengths for carbon.**   D. Hofsaess.
J. Quant. Spectrosc. Radiat. Transfer, Vol. 28, 131 - 160 (1982).

Thomas–Fermi–Hartree–Fock wave functions have been calculated for nl-terms of carbon C(I) ($n \leqslant 10$ and $l \leqslant 3$) and employed to calculate electric dipole oscillator strengths for all allowed transitions between these terms. The dipole length and velocity form of the calculated f-values agree well with the available measurements.

022.176 **The escape factor in atomic absorption spectroscopy.**   A. H. Bassyouni.
J. Quant. Spectrosc. Radiat. Transfer, Vol. 28, 161 - 167 (1982).

Values of the escape factor have been calculated from

atomic absorption measurements for a wide range of concentrations of Li, K, Cu, and Ag at the resonance lines 6707.84, 7664.01, 3247.54 and 3382.89 Å, respectively.

022.177   **The laboratory magnetosphere.**
P. J. Baum, A. Bratenahl.
Geophys. Res. Lett., Vol. 9, 435 - 438 (1982).
The UCR 10.9 m long T-1 terrella facility models collisionless hypermagnetosonic solar wind interaction with a planetary dipole field. Measurements demonstrate that T-1 has achieved all parameters necessary for limited simulation of the solar wind-earth interaction. The test time is significantly longer than existing experiments. The first magnetic field map is consistent with Dungey's model in the case of a southward IMF.

022.178   **Structures of molecular ions from microwave spectroscopy.**   R. C. Woods.
Ann. Israel Phys. Soc., Vol. 4, 221 - 227 (1980). – Abstr. in Phys. Abstr., Vol. 85, Abstr. 54835 (1982).

022.179   **Applications of neutron activation analysis of meteorite, lunar, and terrestrial samples.**
R. A. Schmitt.
Trans. American Nucl. Soc., Vol. 39, 68 - 69 (1981). – Abstr. in Phys. Abstr., Vol. 85, Abstr. 57667 (1982).

022.180   **Quantum assignments and intensity measures for methane between 1100 and 1800 cm^{-1}: a comparison between theory and experiment.**   B. L. Lutz,
C. Pierre, G. Pierre, J. P. Champion.
Astrophys. J., Suppl. Ser., Vol. 48, 507 - 530 (1982).
Theoretical line intensities are calculated and detailed rotational assignments are provided for the methane spectrum between 1100 and 1800 cm^{-1}, based on the Hamiltonian parameters obtained by Pierre et al. using the model of Champion and the laboratory spectra compiled in the atlas of Blatherwick et al. A line-by-line comparison of theoretical and experimental absolute intensities is carried out for more than 1200 lines listed in the atlas.

022.181   **The collision strength for the N III λ 1750 transition.**   H. Nussbaumer, P. J. Storey.
Astron. Astrophys., Vol. 109, 271 - 273 (1982).
The collision strength for electron excitation of the $N^{2+}\,2s^2\,2p\,^2P^0 - 2s2p^2\,^4P$ intersystem transition is calculated in LS coupling, in a six state close coupling approximation. Effective collision strengths are given for temperatures up to 60,000 K.

022.182   **Laboratory studies of isotope exchange in ion-neutral reactions: interstellar implications.**
D. Smith.
Philos. Trans. R. Soc. London, Ser. A, Vol. 303, (see 012.038), 535 - 542 (1981).
A laboratory study of isotope exchange in ion-molecule reactions has been carried out, the results of which indicate that fractionation of heavy isotopes can occur very efficiently at low temperatures. Consideration is given to reactions in which H–D, $^{12}C-^{13}C$, $^{14}N-^{15}N$ and $^{16}O-^{18}O$ exchange occurs, and it is shown how better estimates of the electron density and the temperature in interstellar clouds have been obtained from these laboratory data.

022.183   **Molecular spectroscopy prompted by astrophysical observations.**   A. H. Cook.
Philos. Trans. R. Soc. London, Ser. A, Vol. 303, (see 012.038), 551 - 563 (1981).

022.184   **A relaxation theory of Stark broadening by ions.**
R. L. Greene.
J. Quant. Spectrosc. Radiat. Transfer, Vol. 27, 639 - 651 (1982).
A theory of Stark broadening in plasmas is presented in which the effects of broadening by ion perturbers are treated in a manner similar to the relaxation theory or unified theory for electrons.

022.185   **The relativistic Doppler broadening of the line absorption profile.**   S. Kichenassamy,
R. Krikorian, A. Nikoghossian (*Nikogosyan*).
J. Quant. Spectrosc. Radiat. Transfer, Vol. 27, 653 - 655 (1982).
The classical results of Doppler broadening of the line absorption profile are generalized to a relativistic gas in thermal equilibrium by taking into account the relativistic variance of the volume absorption coefficients of the gas.

022.186   **Experimental Stark widths and shifts of two ultraviolet C(II) lines.**
A. Goly, S. Weniger.
J. Quant. Spectrosc. Radiat. Transfer, Vol. 27, 657 - 661 (1982).
Photoelectric measurements of Stark widths and shifts of the components of the C(II) multiplet UV13 at 2837 Å were made with a wall-stabilized arc in a gas mixture of argon, hydrogen and carbon dioxide. Plasma diagnostics have been carried out to determine the electron density and the temperature.

022.187   **Comment on the use of an exponential approximation for the Stark broadening of hydrogen lines.**
J. Cooper, E. W. Smith.
J. Quant. Spectrosc. Radiat. Transfer, Vol. 27, 665 - 666 (1982).

022.188   **Stable states of the electron-nuclear phase of cool dense magnetized matter.**   G. M. Nedyalkova,
V. S. Sekerzhitskij, S. S. Sekerzhitskij, G. A. Shul'man.
Researches on stellar physics, 1981, (see 003.010), p. 41 - 52 (1981). In Russian. – Abstr. in Ref. zh., 51. Astron., 6.51.119 (1982).

022.189   **The equation of state of cool superdense magnetized matter.**
V. S. Sekerzhitskij, S. S. Sekerzhitskij, G. A. Shul'man.
Researches on stellar physics, 1981, (see 003.010), p. 25 - 40 (1981). In Russian. – Abstr. in Ref. zh., 51. Astron., 6.51.121 (1982).

022.190   **Effect of gravitation on the proper energy of charged particles.**
A. I. Zel'nikov, V. P. Frolov.
Zh. ehksp. i teor. fiz., Tom 82, 321 - 335 (1982). In Russian. Abstr. in Ref. zh., 51. Astron., 6.51.142 (1982).

022.191   **Molecule formation by implantation in insulators.**
F. Rocard, J.-P. Bibring.
Phys. Rev. Lett., Vol. 48, 1763 - 1766 (1982).
Molecule synthesis upon implantation of 1 to 4 keV/amu ions in thin silica films and silicate grains is studied. It is shown that carbon implantation leads to the preferential synthesis of CO_2 at low fluences, CO becoming the dominant species at high fluences. The overall efficiency of the synthesis is $\sim 60\%$. CO_2 synthesized by implantation of solar-wind ions is observed in lunar dust grains, and this process may have important astrophysical implications.

022.192   **Transition probabilities for forbidden lines in the $3p^2$ configuration – II.**
C. Mendoza, C. J. Zeippen.
Mon. Not. R. Astron. Soc., Vol. 199, 1025 - 1032 (1982).
Making use of the atomic structure computer program SUPERSTRUCTURE, radiative transition probabilities are cal-

culated for the forbidden lines in the ground configuration $1s^2 2s^2 2p^6 3s^2 3p^2$ of all the members of the silicon isoelectronic sequence up to Ni XV. Elaborate procedures such as configuration interaction, semi-empirical term energy corrections, special radial wave functions and relativistic corrections to the magnetic dipole operator are used in the present calculation.

022.193   The O IV 1401 Å multiplet – calculation of the electron impact excitation rate coefficients.
M. A. Hayes.
Mon. Not. R. Astron. Soc., Vol. 199, 49P - 53P (1982).

The rate coefficients for the excitation of the metastable levels in O IV for electron temperatures $(1-4) \times 10^4$ K are presented. The total LS coupling rates and the rates for the fine structure transitions are given. The calculations were carried out in a close-coupling approximation including the effects of closed channel resonances.

022.194   Electron excitation rates for helium.
K. A. Berrington, W. C. Fon, A. E. Kingston.
Mon. Not. R. Astron. Soc., Vol. 200, 347 - 350 (1982).

Electron excitation cross-sections between $1^1 s$, $2^3 s$, $2^1 s$, $2^3 p$ and $2^1 p$ states of helium have been calculated recently using the R-matrix method. These cross-sections have been integrated over a Maxwellian distribution to give electron excitation rates.

022.195   On Franck-Condon factors and intensity distributions in some band systems of I_2, NS and PS molecules.
K. Raghuveer, N. A. Narasimham.
J. Astrophys. Astron., Vol. 3, 13 - 25 (1982).

Potential curves for the B and X states of I_2, NS and PS have been obtained by Rydberg-Klein-Rees (RKR) method. From these RKR potentials, Franck-Condon factors (FCFs) for the above band systems have been calculated using the best available molecular constants, tested for accuracy on the electronic transition moment (ETM)-r-centroid curve in the case of I_2 and used in the study of observed abnormal intensity distribution in some bands of NS. A brief outline of the method used in the calculations of the FCFs is given.

022.196   The atomic partition functions.   J.-h. Jin.
Acta Astron. Sinica, Vol. 23, 17 - 22 (1982). In Chinese.

022.197   [Ni II] emission under nebular conditions.
H. Nussbaumer, P. J. Storey.
Astron. Astrophys., Vol. 110, 295 - 299 (1982).

Transition probabilities and collision strengths are calculated for the energetically lowest 17 levels of Ni^+. The determination of electron density and temperature from intensity ratios of visual [Ni II] lines is discussed. Doubts about the assignment to [Ni II] of a strong spectral feature observed in the Crab nebula at 7378 Å are shown to be unfounded.

022.198   The behaviour of adiabatic invariants near resonances.   B. Barbanis.
Compendium in astronomy, (see 003.012), p. 31 - 42 (1982).

The author investigates numerically the changes of the adiabatic invariants in some resonant cases of time dependent one-dimensional dynamical systems. He finds that at each main resonance of the Mathieu equation there is a "resonant" adiabatic invariant which is much better conserved than other forms of adiabatic invariants. Then he studies the adiabatic invariant in the case of a simple pendulum with varying length.

022.199   The experimental curve of growth in function of different sets of oscillator strengths.
N. Gökdoğan, K. Avcioğlu, D. Koçer.
Compendium in astronomy, (see 003.012), p. 97 - 103 (1982).

In order to show the influence of the oscillator strengths

measurements on the experimental curve of growth, three curves are constructed for the solar iron, with three different sets of oscillator strengths. The excitation temperatures found in this way are different.

022.200   Classical phase space from a relativist's point of view.   S. Persides.
Compendium in astronomy, (see 003.012), p. 367 - 377 (1982).

A new formulation of phase space of classical mechanics is proposed based on a known four-dimensional formulation of the space-time. With space-time represented by a four-dimensional manifold S (with affine structure and stratified by Euclidean three dimensional spaces) the phase space $\bar{\Gamma}$ for a single particle is defined as the cotangent bundle of S. Thus $\bar{\Gamma}$ is an eight dimensional manifold and a vector bundle with symplectic structure and tensor fields transferred from S. Some remarks on the constants of motion are made. In the general case of a system with n degrees of freedom $\bar{\Gamma}$ is a manifold with $2n+2$ dimensions.

022.201   Report of IAU Commission 14: Atomic and molecular data (Données atomiques et moléculaires).
J. G. Phillips.
Trans. IAU, Vol. XVIIIA, (see 003.013), p. 115 - 151 (1982).

Geometric theory of diffraction.
See Abstr. 003.058.

Light scattering by small particles.
See Abstr. 003.133.

Hopf-bifurcation in a nearly Hamiltonian system.
See Abstr. 021.025.

Completely integrable systems and singularity in the complex t–plane.   See Abstr. 021.027.

Studies in the application of recurrence relations to special perturbation methods. VI: Comparison with classical single-step and multi-step methods of numerical integration.
See Abstr. 021.049.

The variational equations of Störmer's problem.
See Abstr. 042.017.

The adiabatic invariant: its use in celestial mechanics.
See Abstr. 042.054.

Quantisation in stable gravitational systems.
See Abstr. 042.104.

Intrinsic variational equations for conservative dynamical systems with n degrees of freedom.
See Abstr. 042.124.

Angle-averaged redistribution function in the laboratory frame.   See Abstr. 063.042.

On the establishment of internally consistent solar scales of oscillator strengths and abundances of chemical elements. III. Oscillator strengths obtained from equivalent widths of 360 FeI lines.   See Abstr. 071.005.

Infrared bands of C_2 in the solar photospheric spectrum.   See Abstr. 071.016.

Investigation of the physical characteristics of the photosphere from multiplet line profiles. I. LTE conditions.
See Abstr. 071.020.

Investigation of physical characteristics of the photosphere from multiplet line profiles. II. Deviation from local thermodynamical equilibrium.   See Abstr. 071.021.

S IV emission-line ratios in the sun. See Abstr. 071.030.

The solar O III spectrum. I. Photoexcitation of EUV lines by He II Lyman-α.   See Abstr. 076.013.

Tentative identification of CS$^+$ in comets. See Abstr. 102.015.

Detection of the $N = 3-2$ transition of CCH in Orion and determination of the molecular rotational constants.   See Abstr. 131.040.

Errata

022.901   Erratum: "Theoretical microwave spectral constants for C_3H^+ and C_4H^+" [Astrophys. J., Vol. 240, 968 - 970 (1980)].   S. Wilson, S. Green. Astrophys. J., Vol. 253, 989 (1982). — See Abstr. 28.022.051.

022.902   Erratum: 'Oscillator strengths for lines of the $\tilde{F}(0, 0, 0)-\tilde{X}(0, 0, 0)$ band of H_2O at 111.5 nanometers and the abundance of H_2O in diffuse interstellar clouds' [Astrophys. J., Vol. 250, 166 - 174 (1981)]. P. L. Smith, K. Yoshino, H. E. Griesinger, J. H. Black. Astrophys. J., Vol. 256, 798 - 799 (1982). — See Abstr. 30.022.104.

022.903   Erratum: 'Recombination coefficients for iron ions' [Astrophys. J., Vol. 249, 399 - 401 (1981)]. D. T. Woods, J. M. Shull, C. L. Sarazin. Astrophys. J., Vol. 257, 918 (1982). — See Abstr. 30.022.093.

Astronomical Instruments and Techniques

031 Astronomical Optics, Methods of Observation and Reduction

Astronomical Optics

031.001 Towards imaging with a speckle-interferometric optical synthesis telescope.
R. H. T. Bates, W. R. Fright.
Mon. Not. R. Astron. Soc., Vol. 198, 1017 - 1031 (1982).

Speckle images have been formed in the optical laboratory with a linear pupil (its length was 10 times its width) viewing through a fluctuating medium which introduced severe seeing. The images were recorded with a microprocessor-controlled CCD-camera system interfaced to a minicomputer. By slightly modifying several established processing techniques and combining them into a new synthesis procedure, faithful versions of diffraction-limited true images of simulated astronomical objects are reconstructed from sequences of speckle images.

031.002 The Barlow Lens.
S. J. Anderson.
J. British Astron. Assoc., Vol. 92, 135 - 136 (1982).

031.003 De Schmidt-camera.
H. G. J. Rutten, M. A. M. van Venrooij.
Zenit, 9. Jaarg., 171 - 175 (1982).

031.004 Short-exposition optical transmission function of a telescope with a receiving device of arbitrary shape.
A. D. Ryakhin.
Mater. 6-j Nauchn. konf. fak. fiz. i kvant. ehlektron. Moskva, 1981. 6 pp. In Russian. – Abstr. in Ref. zh., 51. Astron., 2.51.800 (1982).

031.005 General concepts and approach on making stray light calculations without the use of large computers.   F. F. Crandell.
Proc. Soc. Photo-Opt. Instrum. Eng., Vol. 257, (see 012.007), 29 - 38 (1980). – Abstr. in Phys. Abstr., Vol. 85, Abstr. 14771 (1982).

031.006 Stray radiation and the Infrared Astronomical Satellite (IRAS) telescope.
R. J. Noll, R. Harned, R. P. Breault, R. Malugin.
Proc. Soc. Photo-Opt. Instrum. Eng., Vol. 257, (see 012.007), 119 - 136 (1980). – Abstr. in Phys. Abstr., Vol. 85, Abstr. 14772 (1982).

031.007 Comparison of stray light mechanisms and performance in the Infrared Astronomy Satellite (IRAS) and Shuttle Infrared Telescope Facility (SIRTF) telescopes.   S. R. Lange, R. P. Breault, A. W. Greynolds.
Proc. Soc. Photo-Opt. Instrum. Eng., Vol. 257, (see 012.007), 137 - 147 (1980). – Abstr. in Phys. Abstr., Vol. 85, Abstr. 14773 (1982).

031.008 Straylight analysis of the German Infrared Laboratory (GIRL).   G. I. Geikas.
Proc. Soc. Photo-Opt. Instrum. Eng., Vol. 257, (see 012.007), 148 - 152 (1980). – Abstr. in Phys. Abstr., Vol. 85, Abstr. 14774 (1982).

031.009 Visible light scatter measurements of the Advanced X-ray Astronomical Facility (AXAF) mirror samples.   D. B. Griner.
Proc. Soc. Photo-Opt. Instrum. Eng., Vol. 257, (see 012.007), 217 - 222 (1980). – Abstr. in Phys. Abstr., Vol. 85, Abstr. 14775 (1982).

031.010 Measurements of X-ray scattering from Wolter type telescopes and various flat Zerodur mirrors.
B. Aschenbach, H. Bräuninger, G. Hasinger, J. Trümper.
Proc. Soc. Photo-Opt. Instrum. Eng., Vol. 257, (see 012.007), 223 - 229 (1980). – Abstr. in Phys. Abstr., Vol. 85, Abstr. 14776 (1982).

031.011 High resolution X-ray scattering measurements for Advanced X-ray Astrophysics Facility (AXAF).
M. V. Zombeck, C. C. Wyman, M. C. Weisskopf.
Proc. Soc. Photo-Opt. Instrum. Eng., Vol. 257, (see 012.007), 230 - 247 (1980). – Abstr. in Phys. Abstr., Vol. 85, Abstr. 14777 (1982).

031.012 Investigation of the influence of polution of the protective glass of a telescopic instrument on its resolving power.   V. P. Dubenskov, V. Ya. Vasil'ev, L. N. Popova-Dyumina, Yu. M. Marinchenko.
Opt.-mekh. prom-st', 1981, No. 10, p. 36 - 39. In Russian. Abstr. in Ref. zh., 51. Astron., 3.51.898 (1982).

031.013 On optical systems designed for large telescopes.
G. M. Popov, M. B. Popova.
Izv. Krymskoj Astrofiz. Obs., Tom 64, 188 - 197 (1981). In Russian. English translation in Bull. Crimean Astrophys. Obs., Vol. 64.

A numerical investigation computed for Cassegrain systems with 25-m primary mirror has been carried out. The shape of the secondary mirror is determined by exact formulae derived from conditions free from spherical aberration on the axis of the system. The diameter of the secondary mirror does not exceed 6 m.

031.014 Relative distribution of the efficiency over the surface of a concave grating with a blaze angle.
V. K. Prokof'ev.
Izv. Krymskoj Astrofiz. Obs., Tom 64, 198 - 200 (1981). In Russian. English translation in Bull. Crimean Astrophys. Obs., Vol. 64.

The relative distribution of the efficiency over the surface of concave grating has been calculated and compared with measurements of a grating with R = 250 mm and N = 600 mm^{-1}. The dependence of the efficiency distribution on the parameters R and N of the concave grating has been discussed.

031.015 Adaptive optics in astronomy.   Y. Fujimori.
Astron. Her., Vol. 74, 249 - 253 (1981). In Japanese. – From Phys. Abstr., Vol. 85, Abstr. 53094 (1982).

031.016 Atmospheric limitations to high angular resolution imaging. F. Roddier.
Scientific importance of high angular resolution at infrared and optical wavelengths, (see 012.031), p. 5 - 23 (1981).

The statistics of wavefront pertubations are reviewed and their implication on high angular resolution imaging is discussed. Expressions are given for the signal-to-noise ratios and the limiting magnitudes in several cases.

031.017 Active optics in astronomy. J. W. Hardy.
Scientific importance of high angular resolution at infrared and optical wavelengths, (see 012.031), p. 25 - 39 (1981).

The application of active optics technology to ground-based astronomy is reviewed with emphasis on the elimination of random wavefront disturbances that degrade angular resolution. The capabilities and limitations of current systems are discussed and examples are shown of real time compensation of stars and sunspots. An assessment is made of the future impact of this technology on the design of astronomical instruments.

031.018 Development of an active optical mirror for astronomical applications.
F. Merkle, K. Freischlad, J. Bille.
Scientific importance of high angular resolution at infrared and optical wavelengths, (see 012.031), p. 41 - 52 (1981).

It is reported about the development and testing of an active mirror and a microprocessor controlled feedback system. The mirror is made of a 0.5 micron aluminized polypropylen film and electrostatically elongated. In this first version, designed for a 75 cm RC-telescope, a Z80 microprocessor system is used for providing the feedback control. The image quality is controlled by a trial and error algorithm. For future systems a modal control is planned. Also if no complete phasefront correction is possible, telescope aberrations could be reduced, and the intensity for spectroscopic and speckle-interferometric measurements could be increased.

031.019 Active wavefront correction in large telescopes.
J. R. P. Angel.
Scientific importance of high angular resolution at infrared and optical wavelengths, (see 012.031), p. 53 - 60 (1981).

With the MMT in operation for nearly two years, one can already point to an impressive record of astronomical observations. It is only now though that one is more fully understanding and beginning to exploit its built-in potential for high resolution imaging by partial compensation for atmospheric distortion. In this paper the author wants to report on current progress being made in this direction. The techniques being developed are applicable to large telescopes in general, and could have a profound effect on the philosophy of designing such telescopes.

031.020 Honeycomb mirrors of borosilicate glass.
J. R. P. Angel, J. M. Hill.
Scientific importance of high angular resolution at infrared and optical wavelngths, (see 012.031), p. 61 - 65 (1981).

031.021 Image quality and high resolution in future telescopes. R. N. Wilson.
Scientific importance of high angular resolution at infrared and optical wavelengths, (see 012.031), p. 67 - 83 (1981).

The paper considers the possibilities of significant improvement in the optical quality of future big ground-based telescopes compared with what is being achieved today.

031.022 Coherent large telescopes. J. E. Nelson.
Scientific importance of high angular resolution at infrared and optical wavelengths, (see 012.031), p. 139 - 164 (1981).

A brief survey of existing telescopes and their evolution is made to see the trends that lead us to the present period of innovative telescope design. With this background, the author reviews the major telescope types and the critical design factors that must be considered in designing large telescopes for the future, if they are to be at all economical. After these generalities, the author describes as an example a project that he is involved with, the University of California Ten Meter Telescope Project. Finally a brief review of current work in progress on large telescopes is given.

031.023 Probability of diffraction-limited images in infrared through turbulence—experimental results.
J. P. Corteggiani, J. Gay, Y. Rabbia.
Scientific importance of high angular resolution at infrared and optical wavelengths, (see 012.031), p. 175 - 180 (1981).

031.024 Imaging by dilute apertures in the presence of atmospheric turbulence. D. S. Brown.
Scientific importance of high angular resolution at infrared and optical wavelengths, (see 012.031), p. 181 - 195 (1981).

031.025 A high resolution, grazing incidence, ultraviolet spectrograph. W. Cash, W. McClintock.
Bull. American Astron. Soc., Vol. 13, 884 (1981). – Abstract.

031.026 Estimate of the efficiency of methods for search for an extremum in adjustment of a composed telescopic mirror. I. A. Lapshina.
Sistemy upr. i ikh ehlementy. Leningrad, 1981, p. 121 - 125.
In Russian. – Abstr. in Ref. zh., 51. Astron., 5.51.914 (1982).

031.027 Plateholder-mirror alignment of Schmidt telescopes. S. Jutamulia, B. Hidayat.
Proc. Inst. Technol. Bandung, Lembang, Vol. 14, 1 - 5 (1981)= Contrib. Bosscha Obs. Lembang, No. 73 (1982).

For the plateholder mirror alignment, the authors produce an auxiliary point source which falls at the correcting lens. If the adjustment is correct, then the image of the first reflection, I_1, will coincide with the image of the third reflection I_3, as proposed by Dewhirst and Yates. However the results indicate that, due to off-axis position of the source, there exist aberrations at I_1 and I_3.

031.028 Reflectance and preparation of front-surface mirrors for use at various angles of incidence from the ultraviolet to the far infrared. G. Hass.
J. Opt. Soc. America, Vol. 72, 27 - 39 (1982).

031.029 Lie algebraic theory of geometrical optics and optical aberrations. A. J. Dragt.
J. Opt. Soc. America, Vol. 72, 372 - 379 (1982).

A new method, employing Lie algebraic tools, is presented for characterizing optical systems and computing aberrations. It represents the action of each separate element of a compound optical system, including all departures from Gaussian optics, by a certain operator. These operators can then be concatenated, following well-defined rules, to obtain a resultant operator that characterizes the entire system. New insight into the origin and possible correction of aberrations is provided. With some effort, it should be possible to produce, by manual calculations, explicit formulas for the third-, fourth and fifth-order aberrations of a general optical system including systems without axial symmetry. With the aid of symbolic manipulation computer programs, it should be possible to compute routinely explicit formulas for aberrations of seventh, eighth, and ninth order, and probably beyond.

031.030 Sky input horn for a far-infrared interferometer. M. S. Miller, W. L. Eichhorn, J. C. Mather.
Opt. Lett., Vol. 7, 210 - 211 (1982).

A unique optical design consisting of a compound parabolic concentrator and an elliptical concentrator joined at their throats is discussed. The optic has a number of advantages over conventional infrared telescopes for applications in interferometry and photometry. Ray-tracing results are presented.

031.031 Vacuum deposition of iridium on large astronomical mirrors for use in the far UV.
H. Herzig, R. S. Spencer.
Appl. Opt., Vol. 21, 15 - 16 (1982).

031.032 Actual blaze angle of the Bausch & Lomb R4 echelle grating.
R. A. Brown, R. L. Hilliard, A. L. Phillips.
Appl. Opt., Vol. 21, 167 - 168 (1982).

031.033 Segmented mirror polishing experiment.
R. A. Jones.
Appl. Opt., Vol. 21, 561 - 564 (1982).
An experiment was conducted to demonstrate the capability of polishing segmented mirrors. A mirror segment was figured with the computer controlled polisher using special software and tooling. This process rapidly reduced surface figure errors and produced a good figure to the segment's edges.

031.034 Optimum solution for spherical primary mirror with two and three aspheric corrector plates located near focus. Part 2.   A. B. Meinel, M. P. Meinel.
Appl. Opt., Vol. 21, 1323 - 1325 (1982).
The case for two aspheric plates is further investigated to include ray trace evaluation of various solutions, removing the possible two solution regions noted in Part 1. The study is extended to include one reflection plus one transmission aspheric and two reflection aspherics.

031.035 Encircled energy for systems with centrally obscured circular pupils.
J. J. Stamnes, H. Heier, S. Ljunggren.
Appl. Opt., Vol. 21, 1628 - 1633 (1982).
The encircled energy is computed for large f/No. aberration-free annular aperture systems with linear obscuration ratios between 0.0 and 0.9. The computations are based on an algorithm due to Hopkins. The results are displayed graphically as contour lines showing the fraction of the total energy

which falls within small circles centered on the optical axis in selected receiving planes parallel to the geometric focal plane.

031.036 Extrapolated least-squares optimization: a new approach to least-squares optimization in optical design.   E. D. Huber.
Appl. Opt., Vol. 21, 1705 - 1707 (1982).

031.037 Throughput of diffraction-limited field optics systems for infrared and millimetric telescopes.
R. H. Hildebrand, R. Winston.
Appl. Opt., Vol. 21, 1844 - 1846 (1982).
Telescopes for submillimeter wavelengths have point spread functions some millimeters or centimeters in diameter, but the detectors may be only fractions of a millimeter in size. Thus a field aperture and collecting optics are needed. The authors show how to optimize the aperture by a calculation of the effects of diffraction on signal and resolution as a function of size of the collecting aperture. Their calculations are compared to experimental results from observations of Mars at submillimeter wavelengths.

031.038 De optische kwaliteit van zes astrocamera's.
H. G. J. Rutten, M. A. M. van Venrooij.
Zenit, 9. Jaarg., 330 - 336 (1982).

031.039 Ray tracing and caustics from large reflectors used in spacecrafts.   P. S. Theocaris.
Compendium in astronomy, (see 003.012), p. 415 - 445 (1982).
The general equations of caustic surfaces formed by reflections of a point-light source on the reflecting surfaces of ellipsoid, paraboloid and hyperboloid form were derived for the case when the light source is placed at an arbitrary distance and angle from the axis of symmetry of the reflector. A study of the evolution of the shape and form of the caustics was undertaken for each type of reflector.

The telescope.   See Abstr. 003.021.

Astrophysical observations with high resolution X-ray telescopes.   See Abstr. 032.514.

Methods of Observation and Reduction

031.501 Bias of polarimetric estimators for binary star inclinations.
J. F. L. Simmons, C. Aspin, J. C. Brown.
Mon. Not. R. Astron. Soc., Vol. 198, 45 - 57 (1982).
It is shown that the polarimetric "modelling" used by previous authors to obtain the least squares fit to polarimetric binary data will tend to yield inferred values of inclination greater than the true value. This statistical bias is most pronounced at high noise levels and low inclinations when the inferred value and the formal linear error will have no bearing on the actual value. Errors for inclination which are established by formal techniques are seriously over optimistic except at extremely low noise levels.

031.502 A method for imaging radio meteor radiant distributions.   J. D. Morton, J. Jones.

Mon. Not. R. Astron. Soc., Vol. 198, 737 - 746 (1982).
The authors present some of the first maps of meteor radiant distributions obtained from a new microprocessor-based radar system. Although the present maps suffer from astigmatism, it is shown how this may be minimized in future implementations and also how the method can be extended to obtain velocities and hence meteor orbits.

031.503 A method for determination of the abundance of chemical elements in the atmospheres of cool giant stars.   N. S. Komarov, A. N. Shcherbak.
Astrometr. Astrofiz., Vyp. (No.) 43, p. 52 - 58 (1981). In Russian.
Synthetic spectra of models of stellar atmospheres were computed. The results obtained were used to test a modified curve-of-growth method which enables the multilayer structure of stellar atmospheres to be taken into account.

031.504 X- and γ-ray superfast photometry.
S. Bonazzola, M. Chevreton.
Astron. Astrophys., Vol. 105, 1 - 5 (1982).

The authors describe the importance of studying the fast time variations in the X- and γ-ray sources. They apply the theory of the photonic noise and show that a resolution time of order of few microseconds for the X-ray, and of some and few hours for γ sources can be achieved with the present, or future generation of satellites.

031.505 Excising terrestrial radio interference in low frequency radio astronomy.
B. L. Kasper, F. S. Chute, D. Routledge.
Mon. Not. R. Astron. Soc., Vol. 199, 345 - 354 (1982).

Interference from terrestrial transmitters is a major problem in low-frequency radio astronomy. Experimental work is described in which a 22.25-MHz interferometer provided baseband signals from which 128-channel spectra were calculated and processed by a microcomputer. Narrow-band interference was identified and excised from the in-phase and quadrature cross-spectra and the auto-spectra in a 52-kHz band in real time. An algorithm was developed in which robust estimation was used to give protection from statistical outliers. Low-level interference, which would not have been noticeable in records taken with a conventional receiving system, was detected and excised on most nights.

031.506 Sur quelques causes d'erreur dans les observations d'étoiles doubles: vraies et fausses étoiles doubles.
P. Baize.
Astronomie, Vol. 96, 7 - 14 (1982).

031.507 Die Sonnenflecken-Relativzahl.   R. Beck.
Orion, 40. Jahrg., 12 - 15 (1982).

031.508 Volgen.   S. Wadman.
Zenit, 9. Jaarg., 72 - 77 (1982).

031.509 Photogrammetric application of Viking orbital photography.
S. S. C. Wu, A. A. Elassal, R. Jordan, F. J. Schafer.
Planet. Space Sci., Vol. 30, 45 - 55 (1982).

Special techniques are described for the photogrammetric compilation of topographic maps and profiles from stereoscopic photographs taken by the two Viking Orbiter spacecraft. A series of contour maps of Mars is being compiled by these new methods using a wide variety of Viking Orbiter photographs, to provide the planetary research community with topographic information.

031.510 Restoration of astronomical images with the help of digital filtering methods.
A. A. Korovyakovskaya, Yu. P. Korovyakovskij.
Astron. Zh., Tom 59, 160 - 173 (1982). In Russian. English translation in Soviet Astron., Vol. 26, No. 1.

Digital methods for restoration of astronomical images made by the authors with the help of a measuring and computing technique available at the Special Astrophysical Observatory are described. With the processing of the image of the radio galaxy M 87 taken as an example an improvement of angular resolution of the given photograph by a factor of 2 - 3 is demonstrated, additional details in the jet structure are detected.

031.511 Recovering spectral information from unevenly sampled data: two machine-efficient solutions.
J. R. Kuhn.
Astron. J., Vol. 87, 196 - 202 (1982).

Astronomical data are often unevenly sampled. The problem of recovering a discrete Fourier transform is discussed, and two procedures for recovering an approximation to the transform are presented. Numerical examples suggest that the approach may be useful for many types of sample domains.

031.512 Analysis of the power spectra of galactic images and spiral structures.
W. Krakow, J. M. Huntley, P. E. Seiden.
Astron. J., Vol. 87, 203 - 217 (1982).

Several digitized images of galaxies M83, M51, and NGC 1365 have been analyzed using two-dimensional Fast Fourier Transform techniques. The second part of this study involves the Fourier analysis of computer-modeled ideal spirals with linear, logarithmic, and hyperbolic shapes as a function of radial distance. In order to approach the images of real galaxies, various modifications to these shapes have been evaluated.

031.513 Neutral hydrogen studies with a novel instrument.
J. M. Dickey, J. M. Benson.
Astron. J., Vol. 87, 278 - 305 (1982).

Observations of absorption and emission by the 21-cm line of neutral hydrogen have been made using the 300- and 140-ft telescopes as an interferometer. The authors describe the method of observation and present spectra at low and intermediate latitudes. The H I spin-temperature distribution obtained is similar to earlier studies at intermediate latitudes, but shows more cold gas at low latitudes. At low latitudes high optical depth clouds sometimes conceal large amounts of atomic gas. Correction factors for the column density observed by emission surveys can be as large as 1.8, i.e., almost half the gas has been missed.

031.514 How to measure the Sun like a star.   H. Tüg.
Astron. Astrophys., Vol. 105, 395 - 399 (1982).

The present paper describes a technique which allows to measure the global Sun like a bright star using a small Cassegrain telescope and its backend device. The method is based on the reduction of the solar brightness by high diffraction.

031.515 A direct UBV color measurement of the Sun.
H. Tüg, T. Schmidt-Kaler.
Astron. Astrophys., Vol. 105, 400 - 404 (1982).

A technique is described which allows the direct measurement of solar color indices. The Sun was compared with 25 bright stars including 3 primary and 4 secondary Johnson-Morgan UBV standards. The authors' final results are $+ 0^m_.686$ for $(B–V)_\odot$ and $+0^m_.183$ for $(U–B)_\odot$. A detailed error analysis leads to the conclusion that the accuracy is $\pm 0^m_.011$ for $(B–V)_\odot$ and $\pm 0^m_.020$ for $(U–B)_\odot$.

031.516 Possible measurement of the time delay between gravitational images of expanding double radiosources.   C. Vanderriest.
Astron. Astrophys., Vol. 106, L1 - L3 (1982).

If an expanding double radio-source is observed through a gravitational lens system, the apparent epoch of zero separation is different for the different images. This gives a direct measurement of the time delay between the images. For a constant velocity of expansion at the source, observations at only 2 epochs are theoretically needed. The feasibility of this method is discussed.

031.517 A model for constructing artificial integrated spectral lines and their Fourier transform properties relevant to the search for differential rotation of stars.
M. C. García-Alegre, M. Vázquez, H. Wöhl.
Astron. Astrophys., Vol. 106, 261 - 265 (1982) = Mitt. Kiepenheuer-Inst. Nr. 211.

A model is given for constructing artificial integrated spectral lines which allows the inclusion of all effects relevant to the synthesis of these lines by numerical methods. There are included noise, limb darkening, the centre-to-limb variation of the shapes and wavelength positions of the spectral line profiles, different laws of the differential rotation and a tilting of the rotating star. Fourier transformation of the artificial integrated spectral lines is applied to decide whether the different effects may influence the possibility of detecting stellar differential rotation by the method of Gray (1977).

031.518 An alternative procedure for extracting IUE low resolution spectra.
L. Crivellari, C. Morossi.
Astron. Astrophys., Vol. 106, 332 - 338 (1982).
    This paper presents an alternative procedure which extracts spectra from IUE low resolution images. It furnishes noticeably better results than the standard VILSPA extracting routine in the case of highly exposed images. As a by-product, some properties of the Point Spread Function perpendicular to the direction of dispersion are determined.

031.519 Video techniques applied to a telescope.
J. A. Gould.
J. British Astron. Assoc., Vol. 92, 69 - 72 (1982).

031.520 Analisi comparativa fra metodi visuali di timing delle occultazioni.   A. Filipponi.
Astronomia, N. 3, p. 20 - 28 (1981).

031.521 On the reduction of photographic observations of a star.   V. A. Kovalenko.
Geod., kartogr. i aehrofotosemka, L'vov, 1981, No. 34, p. 43 - 48. In Russian. — Abstr. in Ref. zh., 52. Geod. Aehrosemka, 2.52.95 (1982).

031.522 Remote sensing of fissure-fed basalt flows and their source areas: craters of the Moon Volcanic Field, Idaho.   L. Viglienzone, R. Greeley.
Reports of Planetary Geology Program — 1981, (see 003.001), p. 443 - 445 (1981). — Abstract.

031.523 Orthophoto mosaics and three dimensional transformations of Viking Orbiter pictures.
R. M. Batson, K. Edwards, B. A. Skiff.
Reports of Planetary Geology Program — 1981, (see 003.001), p. 493 - 495 (1981). — Abstract.

031.524 IUE data reduction. The parameterization of the motion of the IUE reseau grids and spectral formats as a function of time and temperature.
R. W. Thompson, B. E. Turnrose, R. C. Bohlin.
Astron. Astrophys., Vol. 107, 11 - 22 (1982).
    Variations of temperatures within the International Ultraviolet Explorer (IUE) cause shifts in the measured location of the fiducial reseau marks and in the location of the spectral format with respect to the reseau grid. Correlations of the systematic behavior of these motions as a function of time and of the camera head amplifier temperature have been found for the Long Wavelength Redundant and Short Wavelength Prime cameras.

031.525 New photographic method for the measurement of visual binaries.   M. Scardia, R. Pannunzio.
Astron. Astrophys., Vol. 107, 362 - 367 (1982).
    Preliminary results of a new photographic method for the measurement of visual double stars are given. This method, applied to four visual binaries, gives the parameters $\rho, \theta, \Delta m$ deduced from the measurements taken on two pairs of trails, obtained by photographic observations using the slow motions of the telescope.

031.526 The accuracy of very long baseline interferometry observables.   H. Nes.
Radio Sci., Vol. 16, 947 - 952 (1981). — Abstr. in Phys. Abstr., Vol. 85, Abstr. 14787 (1982).

031.527 Auswertung der Netz-Relativzahl.   E. Karkoschka.
Sonne, Jahrg. 6, 35 - 43 (1982).

031.528 The determination of the center of gravity of a planet from photographic plates.   J.-E. Arlot.
Celestial Mech., Vol. 26, (see 012.001), 199 - 205 (1982).
    The difficulty of the determination of the center of gravity of a planet depends on the fact that it has usually a large apparent diameter. On a photographic plate, the size of the image of a planet depends on the exposure time and on the focal length. Moreover, the phase angle (Sun—Planet—Earth) makes the image of the planet dissymetric and the surface features — when existing — make its brightness not uniform. For a planet the position of the satellites of which is well known (for example, Jupiter), tests for the determination of the center of gravity have been made using different methods.

031.529 A comparison of astrometric measurement techniques as applied to minor planets.
R. L. Duncombe, P. D. Hemenway.
Celestial Mech., Vol. 26, (see 012.001), 207 - 212 (1982).
    Differential and absolute minor planet positions previously applied to the study of a Fundamental Reference System have had accuracies of the order of ± ".2. Relative positions have been obtained to much higher accuracy, but that accuracy has not been applied directly to the formation of a celestial coordinate system. The regular motions of minor planets along long arcs in the sky are more accurately known than any single observed position. Thus, the dynamics of minor planets, coupled with new techniques of observation and reduction, will bring an independent component to bear on the problem of testing the Fundamental Reference System. Preliminary results based on measurements of Hypatia (238) are presented.

031.530 Temporal autocorrelation functions of solar speckle pattern.   C. Aime, S. Kadiri, F. Martin, G. Ricort.
Opt. Commun.,Vol. 39, 287 - 292 (1981). — Abstr. in Phys. Abstr., Vol. 85, Abstr. 23227 (1982).

031.531 The UCL interactive processing system for application to remote sensing studies.   G. Hunt.
Computer simulation systems for image processing, (see 012.010), p. 12/1 (1981). — Abstr. in Phys. Abstr., Vol. 85, Abstr. 23230 (1982).

031.532 Detection of solar and cosmic neutrinos by coherent scattering.   R. Opher.
Astron. Astrophys., Vol. 108, 1 - 6 (1982).
    The author examines the use of coherent scattering for the detection of low energy $p-p$ solar neutrinos and cosmic neutrinos using superconducting electrons.

031.533 Probleme der photographischen Positionsbestimmung mit dem 1.5 m Spiegelteleskop des L. Figl-Observatoriums.
E. Fotter, W. Nicholson, F. V. Prochazka.
Mitt. Astron. Ges., Nr. 55, p. 144 - 155 (1982).

031.534 Bildverarbeitung astronomischer Aufnahmen.
J. Rahe, D. A. Klinglesmith III.
Mitt. Astron. Ges., Nr. 55, p. 160 (1982). — Abstract.

031.535 Flächenphotometrie der Milchstraße im Visuellen.
G. Leuprecht, E. Mravlag, J. Pfleiderer.
Mitt. Astron. Ges., Nr. 55, p. 182 - 183 (1982).

031.536 Doppelkartierung von ausgedehnten Radioquellen als Mittel zur Verbesserung des Nullniveaus.
J. Pfleiderer.
Mitt. Astron. Ges., Nr. 55, p. 184 - 187 (1982).

031.537 VLBI solution of astrometric and geodynamical problems using a system of four radio telescopes.
I. D. Zhongolovich, V. I. Valyaev.

Byull. Inst. Teor. Astron., Leningrad, Tom 15, 10 - 17 (1981). In Russian.

This paper deals with a system of four radio telescopes forming a radio interferometric tetrahedron (RITET). The equations of relation resulting from geometric conditions as well as using autonomous atomic time for the whole RITET have been derived.

031.538 A VLBI method for simultaneous determination of arcs between superdistant radio sources as well as of several parameters of astrometric and geodynamical interest.
I. D. Zhongolovich, A. A. Malkov.
Byull. Inst. Teor. Astron., Leningrad, Tom 15, 18 - 26 (1981). In Russian.

The paper deals with a method of very accurate determination of the value of the arc between two superdistant radio sources. The method is based on the principle of radio interferometry by use of a set of radio telescopes located in four stations lying widely apart. In addition, some other very important quantities involved in astrometry, geodesy and geodynamics may be obtained. A numerical model example has been used to show in detail the aspects of application of the proposed method. All necessary formulae are given, and the expected accuracy of results has been estimated for two various observation methods, i. e. for differential and for alternating ones.

031.539 Evaluation of the accuracy of determination of the length of the arc between two quasars by VLBI technique using synchronous observations.
Eh. I. Yagudina.
Byull. Inst. Teor. Astron., Leningrad, Tom 15, 59 - 64 (1981). In Russian.

An investigation of the technique of determining the length of the arc between two quasars is conducted. Using different bases and real sources investigations of the conditions for obtaining the best results are conducted. It is shown that the accuracy of the determination of the arcs in given examples is equal to $0''.06$ when using observation as minimum of three sources in pairs (under the assumption that the errors of the measurement of τ is equal to 1ns, $f - 1$ mGc).

031.540 Use of some statistical methods for analyzing the uniformity of results of astronomical observations.
S. D. Shaporev.
Byull. Inst. Teor. Astron., Leningrad, Tom 15, 124 - 131 (1981). In Russian.

The paper presents some statistical criteria (parametric and nonparametric) as well as methods which can be utilized for analyzing the quality, uniformity and reliability of astronomical observations. Necessity and advisability of their use in practice is substantiated. Specification of computing program and examples of treating observations of seven short-period comets are given.

031.541 Comments on determination of division corrections.
R. L. Branham, Jr.
Astron. Astrophys., Vol. 108, L5 - L6 (1982).

Benevides-Soares and Boczko have proposed a method for the determination of division corrections of a divided circle that has great merit. The authors suggest some ways to deal with peaks in the power spectrum of the experimental errors. It is shown that elimination of such peaks should be accomplished by careful selection of the apertures rather than the suppression of reciprocal eigenvalues of the system of normal equations for the desired division corrections.

031.542 An outline of a computer program for two dimensional spectral classification.   H. Zekl.
Astron. Astrophys., Vol. 108, 380 - 386 (1982).

A method is presented to classify digitalized stellar spectra of normal stars of spectral classes B−K with respect to spectral class and absolute luminosity. The spectral class

can be calculated with a mean error of 0.7 subclasses, but only the luminosities of B- and A-stars of classes V-III can be calculated with fair accuracy, while for the later spectral types more work with better standard spectra is necessary.

031.543 A powerful method for star counting in the infrared.
A. B. Giles.
Mon. Not. R. Astron. Soc., Vol. 199, 483 - 491 (1982).

A familiar problem when mapping extended objects or star fields using a sky chopping infrared photometer system is the contamination of the signal beam by unknown objects passing through the background beam. A technique based on Fourier deconvolution methods is described here that is particularly suitable for the detection of point sources in crowded fields and provides a self-consistent best fit to the original data. The method is illustrated using infrared data obtained near the galactic centre.

031.544 The measurement of comet positions.
J. B. Tatum.
J. R. Astron. Soc. Canada, Vol. 76, 97 - 108 (1982).

031.545 On the reduction of observations of planetary satellites.   I. G. Chugunov.
Astron. vestn., Tom 15, 234 - 238 (1981). In Russian. – Abstr. in Ref. zh., 51. Astron., 3.51.992 (1982).

031.546 Application of preliminary digital filtration for processing of images of astronomical objects.
R. A. Vardanyan, M. S. Mirzoyan, G. A. Pogosyan.
Dokl. AN SSSR, Tom 72, 162 - 168 (1981). In Russian. Abstr. in Ref. zh., 51. Astron., 3.51.995 (1982).

031.547 Laser tracking data: signal detection, reduction of observations.   G. S. Kurbasova.
Astrometr. Astrofiz., Vyp. (No.) 43, p. 89 - 94 (1981). In Russian.

A method of laser tracking data processing is suggested. This method consists of detecting the useful signal against the intensive noise level using not highly accurate ephemerides, and of observational data reduction.

031.548 On problems of systematic errors existing in the astronomical determination of azimuth.   T. Han.
Acta Geod. Geophys., No. 3, p. 47 - 56 (1981). In Chinese.

031.549 UBV photographic photometry of Schmidt plates.
J. C. Muzzio, H. G. Marraco, A. Feinstein.
Bol. Asoc. Argentina Astron., No. 18, p. 17 - 20 (1980).

031.550 Observación de ocultaciones.
Rev. Astron., Tomo 54, No. 220, p. 3 - 20 (1982).

031.551 A comparison between estimations of Fried's parameter r_0 simultaneously obtained by measurements of solar granulation contrast and of the variance of angle-of-arrival fluctuations.
G. Ricort, J. Borgnino, C. Aime.
Sol. Phys., Vol. 75, 377 - 394 (1982).

Estimations of Fried's parameter r_0 are performed simultaneously using the same telescope, from the observed solar granulation contrast and from the variance of angle-of-arrival fluctuations. The results are well correlated and the mean ratio between the values for r_0 obtained from the 2 methods is found to be equal to 1.4. The sensitivity and accuracy of the methods are discussed in terms of their range of application (site testing, seeing measurements during solar observations...). Temporal power spectra of turbulent energy are computed and evidence of long period oscillations of about 10 min is found.

031.552 Quick matching technique to study the relationship between solar radius and luminosity variations.

S. Sofia, K. L. Chan.
Sol. Phys., Vol. 76, 145 - 153 (1982).

A simple matching technique is developed which allows to compute the response of the solar envelope to perturbations which occur within the solar convective region, and in timescales of importance to climate. This technique is applied to perturbation of the convective efficiency (α-mechanism), and of the non-gas component of the pressure in different regions of the convection zone (β-mechanism). The results indicate that while either perturbation affects the solar luminosity, the α-mechanism has almost no effect on the solar outer radius, regardless of the affected region, whereas the β-mechanism produces radius changes which may be large if the location of perturbation is deep enough.

031.553 **Gamma-ray-line astronomy.**
M. Leventhal, C. J. MacCallum.
Usp. fiz. nauk, Tom 135, 693 - 708 (1981). In Russian.
Abstr. in Ref. zh., 51. Astron., 4.51.595 (1982).

031.554 **Photography of settings of the moon from space as a method for studying the earth's atmosphere.**
A. S. Gurvich, V. Kan, L. I. Popov, V. V. Ryumin, S. A. Savchenko.
Issled. Zemli iz kosmosa, 1981, No. 6, p. 58 - 62. In Russian.
Abstr. in Ref. zh., 62. Issled. kosm. prostranstva, 4.62.572 (1982).

031.555 **Comments on reduction of photoelectric photometry.**   D. M. Popper.
Publ. Astron. Soc. Pacific, Vol. 94, 204 - 206 (1982).
Some critical details in the reduction of photometric observations to a standard system are discussed. Emphasis is on the avoidance of small systematic errors.

031.556 **Ampliación del primer Catálogo Astrolabio de San Juan, programa catálogo.**
E. Actis, A. Serafino.
Bol. Asoc. Argentina Astron., No. 20 - 24, p. 29 - 35 (1981).

031.557 **El ruido en densitogramas de espectros solares.**
T. Paneth.
Bol. Asoc. Argentina Astron., No. 20 - 24, p. 57 - 62 (1981).
Three of the most commonly used methods to avoid the effects of plate noise in densitograms of spectra photographs are compared.

031.558 **Nueva determinación de errores de trazos de ambos círculos del instrumento meridiano Repsold y su comparación con la realizada hace 60 años.**
R. A. Carestia.
Bol. Asoc. Argentina Astron., No. 20 - 24, p. 133 (1981).
Abstract.

031.559 **Calibración de las resistencias en un equipo para fotometría fotoeléctrica.**   L. A. Milone.
Bol. Asoc. Argentina Astron., No. 20 - 24, p. 183 (1981).

031.560 **Correlación en latitud entre el círculo meridiano y el astrolabio Danjon.**
R. A. Carestia, R. E. Orrego.
Bol. Asoc. Argentina Astron., No. 20 - 24, p. 205 (1981).
Abstract.

031.561 **Comportamiento del círculo meridiano en declinación. Error por ilusión optica.**
R. A. Carestia, R. E. Orrego.
Bol. Asoc. Argentina Astron., No. 20 - 24, p. 206 (1981).
Abstract.

031.562 **Extensión del primer catálogo Astrolabio de San Juan. Progràma de catálogo.**

A. Actis, A. Serafino.
Bol. Asoc. Argentina Astron., No. 20 - 24, p. 207 (1981).
Abstract.

031.563 **Un método para la determinación de períodos de estrellas variables.**   H. G. Marraco, J. C. Muzzio.
Bol. Asoc. Argentina Astron., No. 20 - 24, p. 228 (1981).
Abstract.

031.564 **Métodos fotométricos.**
C. De Franceschini, R. J. Terlevich.
Bol. Asoc. Argentina Astron., No. 20 - 24, p. 300 (1981).
Abstract.

031.565 **Análisis de los errores involucrados en la polarimetría.**   H. G. Marraco.
Bol. Asoc. Argentina Astron., No. 20 - 24, p. 412 (1981).
Abstract.

031.566 **Puntos de calibración para observaciones en la línea de 21 cm.**   E. Arnal, W. Poppel.
Bol. Asoc. Argentina Astron., No. 20 - 24, p. 492 (1981).
Abstract.

031.567 **On the calibration of objective prism spectrograms.**
A. T. Garibdzhanyan, S. M. Karapetyan.
Soobshch. Byurakan Obs., Vyp. 53, p. 131 - 134 (1982). In Russian.

031.568 **Ambiguity of magnetographic observations due to present-day methods of their calibration.**
V. G. Lozitskij.
Vestn. Kiev. Univ., Vyp. 23, p. 89 - 98 (1981). In Russian.

031.569 **Método de curvas de luz sintéticas aplicado a UZ Octantis.**   E. Lapasset, R. Sisteró.
Bol. Asoc. Argentina Astron., No. 26, p. 78 (1981). – Abstract.

031.570 **Photométrie photoélectrique dans le visible.**
F. Rufener.
Bull. Inf. Cent. Données Stellaires, No. 22, p. 2 - 11 (1982).

031.571 **Application of RGU-photometry to the study of galactic structure.**   A. Spaenhauer.
Bull. Inf. Cent. Données Stellaires, No. 22, p. 12 - 24 (1982).

031.572 **Photométrie photoélectrique différentielle d'étoiles variables.**   J.-C. Valtier.
Bull. Inf. Cent. Données Stellaires, No. 22, p. 37 - 40 (1982).
The differential photometry is a particular application of photometry which needs a great instantaneous precision (of the order of 1/1000 of magnitude) and the best possible time resolution. Attempts to satisfy these constraints are given, taking as an example the study of the short period variable Delta Scuti and Beta CMa stars. The author emphasizes the choice and knowledge of the instrumentation and data reduction.

031.573 **Compilation des données photométriques nécessaires pour la mission d'astrométrie spatiale HIPPARCOS.**   D. Egret.
Bull. Inf. Cent. Données Stellaires, No. 22, p. 63 - 66 (1982).

031.574 **An unidentified object in the constellation Cetus from 1911.**   G. M. Popovic.
Bull. Inf. Cent. Données Stellaires, No. 22, p. 96 - 97 (1982).

031.575 **Research on Doppler positioning techniques.**
C. H. Song.
Special issue on astronomical geodynamics, (see 012.028), p. 105 - 117 (1981). In Chinese.
The Doppler satellite positioning method has supplanted

all other satellite methods as a primary means of establishing a world geodetic system. Through placement of Doppler stations at locations where measurements by other techniques have been made, it is possible to relate local data not only to the Doppler coordinate system, but also to a desired geocentric coodinate system defined by the CIO pole and the BIH zero meridian.

031.576 **An analysis of the spectrum type systematic error with photoelectric astrolabe (Model I).**
T. G. Yang, J. Y. Xu.
Special issue on astronomical geodynamics, (see 012.028), p. 182 (1981). In Chinese and English. – Abstract.

031.577 **Far-ultraviolet colors of B stars with and without emission lines.**
J. Zorec, D. Briot, L. Divan.
Be stars, (see 012.030), p. 419 - 422 (1982).
A method free of interstellar reddening of comparing the energy distributions in the far-UV of Be/Shell stars to those of normal B stars is presented. The deviations of Be/Shell stars from the sequences of normal stars are correlated with other physical. parameters observed in these stars. The largest UV color differences to the sequence of normal stars are found for those Be/Shell stars having the largest IR color excesses.

031.578 **Speckle interferometry: review of the field and trends.**   A. Labeyrie.
Scientific importance of high angular resolution at infrared and optical wavelengths, (see 012.031), p. 87 - 93 (1981).

031.579 **Speckle interferometry, speckle holography, speckle spectroscopy, and reconstruction of high-resolution images from Space Telescope data.**   G. Weigelt.
Scientific importance of high angular resolution at infrared and optical wavelengths, (see 012.031), p. 95 - 114 (1981).
The author shows various applications of speckle interferometry (double stars up to 13^m, R136a in the 30 Doradus nebula, galactic nucleus of NGC 1140) and of speckle holography (a double star and a triple star). Image processing of Space Telescope data (2.4m-ST) with the speckle rotation method can yield images with a resolution of 0.01 arc second at $\lambda = 130nm$ (diffraction limit) despite of telescope aberrations. Speckle spectroscopy is a new method for reconstructing high resolution objective prism spectra from "spectrum speckle interferograms".

031.580 **On the deconvolution of brightness profiles of galaxies from seeing. Application to NGC 3379.**
O. Bendinelli, G. Parmeggiani, F. Zavatti.
Scientific importance of high angular resolution at infrared and optical wavelengths, (see 012.031), p. 115 - 122 (1981).

031.581 **Speckle interferometry in the infrared.**   P. Léna.
Scientific importance of high angular resolution at infrared and optical wavelengths, (see 012.031), p. 123 - 138 (1981).

031.582 **Rotation interferometry: a new technique for achieving high angular resolution.**
C. Roddier, F. Roddier, J. Vernin.
Scientific importance of high angular resolution at infrared and optical wavelengths, (see 012.031), p. 165 - 170 (1981).

031.583 **Multiple telescope infrared interferometry.**
C. H. Townes, E. C. Sutton.
Scientific importance of high angular resolution at infrared and optical wavelengths, (see 012.031), p. 199 - 223 (1981).
It seems clear that two or more telescopes can be effectively deployed as an interferometer or aperture synthesis system for the mid-infrared with baselines in the 5-1000 m

range. This would vastly extend the present possibilities for high spatial resolution and mapping in the infrared, and in addition provide precise astrometric measurements of broad utility.

031.584 **Multiple telescope interferometry.**   A. Labeyrie.
Scientific importance of high angular resolution at infrared and optical wavelengths, (see 012.031), p. 225 - 236 (1981).

031.585 **Interferometric connection of the Canada-France-Hawaii 3.6 metre telescope and the United Kingdom 3.8 metre telescope on Mauna Kea.**
W. A. Grundmann, G. J. Odgers, E. H. Richardson.
Scientific importance of high angular resolution at infrared and optical wavelengths, (see 012.031), p. 253 - 255 (1981).

031.586 **Interferometry with the Multiple Mirror Telescope and conventional telescopes.**   F. J. Low.
Scientific importance of high angular resolution at infrared and optical wavelengths, (see 012.031), p. 263 - 272 (1981).

031.587 **Coherent versus incoherent detection for interferometry at infrared wavelengths.**
T. de Graauw, H. van de Stadt.
Scientific importance of high angular resolution at infrared and optical wavelengths, (see 012.031), p. 273 - 283 (1981).

031.588 **Coherence and interferometry through optical fibers.**   C. Froehly.
Scientific importance of high angular resolution at infrared and optical wavelengths, (see 012.031), p. 285 - 293 (1981).

031.589 **Possible applications of long-baseline intensity interferometry.**   D. Dravins.
Scientific importance of high angular resolution at infrared and optical wavelengths, (see 012.031), p. 295 - 304 (1981).

031.590 **Angular momentum and star formation.**
P. A. Strittmatter.
Scientific importance of high angular resolution at infrared and optical wavelengths, (see 012.031), p. 351 - 365 (1981).
The author is mainly concerned with the importance of high angular resolution observations in studies of star formation and in particular with elucidating the role that angular momentum plays in the process. He also presents a brief report on recent high angular resolution observations made with the Steward Observatory speckle camera system.

031.591 **A new book on photoelectric photometry.**
R. M. Genet.
I. A. P. P. P. Commun., No. 7, p. 11 (1982).

031.592 **Reduction of results of latitude and longitude determination on computers.**   L. V. Neverov.
Geod. i kartogr., 1981, No. 11, p. 15 - 18. In Russian.
From Ref. zh., 52. Geod. Aehrosemka, 5.52.122 (1982).

031.593 **Power spectrum analysis of unevenly spaced data.**
J. D. Scargle.
Bull. American Astron. Soc., Vol. 13, 814 (1981). – Abstract.

031.594 **On the statistical uncertainties associated with line profile fitting.**
D. A. Landman, R. Roussel-Dupré, G. Tanigawa.
Bull. American Astron. Soc., Vol. 13, 841 (1981). – Abstract.

031.595 **On the calibration of luminosity criteria: numerical experiments.**   T. Lutz.
Bull. American Astron. Soc., Vol. 13, 844 - 845 (1981). Abstract.

031.596 **Extraction of the secondary's radial velocity curve from classically single-spectrum binaries.**
E. J. Devinney, Jr., C. S. Sutton.
Bull. American Astron. Soc., Vol. 13, 872 - 873 (1981).
Abstract.

031.597 **Post-facto dark current and gain determinations for solar data obtained with a diode array.**
G. W. Simon, L. J. November.
Bull. American Astron. Soc., Vol. 13, 878 (1981). – Abstract.

031.598 **Restoration of the convolution of the scattering pattern and the source on the record of the extended source.**   O. I. Krat.
Astrofiz. Issled., Izv. Spets. Astrofiz. Obs., Tom 15, 106 - 109 (1982). In Russian.
A method of cleaning records from the scattering pattern is suggested.

031.599 **Calibration and examination of a star imitator.**
D. I. Stepanov, E. S. Kupriyanov, L. B. Gusakovskaya, G. A. Ozeretskovskij.
Stavrop. gos. ped. inst. Stavropol', 1981, 28 pp. In Russian.
Abstr. in Ref. zh., 51. Astron., 5.51.940 (1982).

031.600 **Synthesis of an information-measuring system for astrophysical studies.**
V. I. Bogdanov, A. B. Bukach.
Mezhvuz. sb. nauchn. tr. Penz. politekh. inst., 1980, No. 6, p. 12 - 22. In Russian. – Abstr. in Ref. zh., 51. Astron., 5.51.947 (1982).

031.601 **High resolution imaging from the ground.**
N. J. Woolf.
Prepr. Steward Obs., No. 359, 60 pp. (1982).

031.602 **Multiple object fiber optic spectroscopy.**
J. M. Hill, J. R. P. Angel, J. S. Scott, D. Lindley, P. Hintzen.
Prepr. Steward Obs., No. 362, 10 pp. (1982).

031.603 **Differential imaging using charge-coupled device (CCD) imagers with on-chip charge storage.**
H. S. Stockman.
Prepr. Steward Obs., No. 381, 5 pp. (1982).

031.604 **LINPOS, an interactive program for stellar line position measurement.**   J. B. Rice.
Publ. Dominion Astrophys. Obs., Vol. 16, 1 - 10 (1981) = NRC No. 19956.
To promote the extension of wavelength-coincidence techniques to survey work on stars with thousands of lines, automated line identification and position measurement must be used. This paper discusses a program that recognizes data noise in calibrated and rectified PDS microdensitometer scans of high-dispersion spectra and then filters out noise and identifies and measures the positions of lines and line blends. Its greatest virtue perhaps is its interactive nature – allowing the operator to edit proceedings as the program runs.

031.605 **Suppression of atmospheric radio emission fluctuations in radio astronomical observations.**
M. N. Kaidanovski (*Kajdanovskij*), D. V. Korolkov (*Korol'kov*), A. A. Stotski (*Stotskij*).
Astrophys. Space Sci., Vol. 82, 317 - 341 (1982).
The fluctuations of radio emission of the atmosphere are a substantial limiting factor for observations of cosmic radio sources on centimetre and millimetre wavelengths. For suppressing these fluctuations the dual-beam method (differential receiving from two slightly separated direction) is widely used. Efforts for perfecting this method are going on as well as searches for other possibilities with the principal aim

of reducing limitations on sizes of sources under observation. These various methods of eliminating fluctuations in atmospheric radio emission are reviewed.

031.606 **The deconvolution of brightness profiles of galaxies from seeing: application to M32.**
O. Bendinelli, G. Parmeggiani, F. Zavatti.
Astrophys. Space Sci., Vol. 83, 239 - 246 (1982).
The integral equation relating to the brightness distribution in a galaxy and in its image formed by an optical system characterized by a Gaussian (or a sum of Gaussians) point-spread function (PSF) is derived. Since the solution of this equation, attainable by any classical method, is numerically unstable, an approximate and stable solution is obtainable by a first-order regularization in Tikhonov's sense. For the bright spike the application to M32 gives a radius of 2.1 arc sec a central surface brightness of 13.10 V mag arc sec^{-2} and a 12 V integrated magnitude.

031.607 **Superresolving image restoration using linear programming.**   R. Mammone, G. Eichmann.
Appl. Opt., Vol. 21, 496 - 501 (1982).

031.608 **An intercomparison of connected-element interferometer and lunar laser Earth rotation parameters.**
D. D. McCarthy.
High-precision Earth rotation and Earth-Moon dynamics.
Lunar distances and related observations, (see 012.035), p. 89 - 95 (1982).
It is now possible to begin to intercompare two of the sets of observations of the rotation of the Earth (UT1-UTC) obtained routinely and independently. These are the data obtained from laser ranging to the Moon and those derived from the connected-element interferometer information of the U.S. Naval Observatory. This has been done using data from July 1979 to March 1980 in comparison with Bureau International de l'Heure (BIH) data. Both sets of observations show that systematic errors do exist in the BIH data. The nature of the correlation of the two series is examined to establish possible models of the systematic errors.

031.609 **Über die Sektionsarbeit RR Lyrae Sterne oder wie finde ich Veränderliche dieses Typs bzw. auch andere am Himmel.**   W. Braune.
BAV Rundbrief, 31. Jahrg., 37 - 47 (1982).

031.610 **Radial velocity measurements using the image photon counting system.**   A. R. Walker.
Mon. Notes Astron. Soc. South. Africa, Vol. 41, 18 - 25 (1982).

031.611 **A note on atmospheric extinction corrections.**
A. P. Linnell.
Publ. Astron. Soc. Pacific, Vol. 94, 374 - 378 (1982).
Observations of variable stars typically must extend over large air mass ranges. On some occasions this requirement occurs in the presence of measurable variation in extinction. Commonly used extinction correction procedures do not handle this situation. This paper describes a procedure which assumes only that the extinction changes are smooth with time.

031.612 **"Least square fitting" and "CLEAN": a combination for analysis of one dimensional synthesis.**
F. Palagi.
Astron. Astrophys., Suppl. Ser., Vol. 49, 101 - 104 (1982).
Aperture synthesis maps of radio sources are generally processed by source modelling or CLEAN procedures. The two procedures are discussed, and a combination of them is proposed to give a more general tool for the source parameters' determination. Some results are shown in the case of one dimensional observations of a solar burst.

031.613 **Observing comets by radar.** P. Kamoun.
Int. Comet Q., Vol. 4, 3 - 4 (1982).

031.614 **Review of magnitude sources for visual cometary photometry. I.** D. W. E. Green, C. S. Morris.
Int. Comet Q., Vol. 4, 5 - 10 (1982).
The authors present recommendations for acceptable primary and secondary sources of comparison star magnitudes. The advantages and drawbacks of each reference are cited. Unacceptable sources are also discussed.

031.615 **The photometric reduction service on La Silla.** R. Barbier.
Messenger, No. 28, p. 33 - 34 (1982).

031.616 **Spectroscopie in de sterrenkunde.** R. Hoekstra.
Zenit, 9. Jaarg., 260 - 263 (1982).

031.617 **Automatic image classification.** S. A. Butchins.
Astron. Astrophys., Vol. 109, 360 - 365 (1982).
The paper is a description of an automatic image classification technique using parameters formed from the brightness profiles of the images. A cluster analysis program is used to combine the parameters and separate the images into stars and galaxies. The technique is tested on 12,823 images and compared to eye estimates. Further tests are then described and the results presented.

031.618 **Effect on optical astronomy of sunlight reflected from spacecraft.** F. G. Smith.
Detrimental activities in space, (see 003.008), p. 7 - 8 (1982).

031.619 **Effect of satellite trails on plates.** R. D. Eberst.
Detrimental activities in space, (see 003.008), p. 9 (1982).

031.620 **Effect of spacecraft debris on the observation of fast transient phenomena.** S. Hayakawa.
Detrimental activities in space, (see 003.008), p. 13 - 14 (1982).

031.621 **Effect of space activities on infrared astronomy.** M. S. Vardya.
Detrimental activities in space, (see 003.008), p. 17 - 19 (1982).

031.622 **Effect of space activities on radio astronomy.** L. H. Doherty.
Detrimental activities in space, (see 003.008), p. 28 - 30 (1982).

031.623 **Satellite interference with radio astronomy.** E. Argyle, C. H. Costain, P. E. Dewdney, J. A. Galt, T. Landecker, R. Roger.
Detrimental activities in space, (see 003.008), p. 31 - 32 (1982).

031.624 **Reduction techniques for échelle spectrograms.** J. B. Hearnshaw.
AAS Photo-Bull., No. 26, p. 9 - 13 (1981).

031.625 **CCD scanning for asteroids and comets.** T. Gehrels, R. S. McMillan.
Sun and planetary system, (see 012.040), p. 279 - 284 (1982).
The sky is to be electronically scanned for asteroids and comets that come close to the earth. This is to prevent their collision with the earth and to provide information on the origins of the solar system.

031.626 **Final refraction problems in time and latitude observations through classical techniques.**
C. Sugawa, I. Naito.
Sun and planetary system, (see 012.040), p. 471 - 474 (1982).
This paper discusses refraction effects remaining so far upon time and latitude observations through the classical

techniques such as the VZT, the PZT and the astrolabe from the meteorological points of view.

031.627 **The influence of ionospheric refraction on radio astronomy interferometry.** T. A. T. Spoelstra.
Sun and planetary system, (see 012.040), p. 493 - 496 (1982).
Observations with radio astronomy interferometers (VLBI and local interferometers) may suffer severe phase errors due to ionospheric refraction and its variations. Techniques to correct these phase errors are discussed. Results of the application of these correction procedures to observations done with the Westerbork Synthesis Radio Telescope are presented.

031.628 **Local geodynamics with two-colour instruments.** E. Tengström.
Sun and planetary system, (see 012.040), p. 511 - 518 (1982).

031.629 **Theory of radio occultation by Saturn's rings.** E. A. Marouf, G. L. Tyler, V. R. Eshleman.
Icarus, Vol. 49, 161 - 193 (1982).
The radio occultation technique is developed here as a new method for the study of the physical properties of planetary ring systems. Particular reference is made to geometrical and system characteristics of the Voyager dual-wavelength (13 and 3.6 cm) experiment at Saturn. The rings are studied based on the perturbations they introduce in the spectrum of coherent sinusoidal radio signals transmitted through the rings from a spacecraft in the vicinity of the planet to Earth.

031.630 **Implications of using broadband photometry for compositional remote sensing of icy objects.**
R. N. Clark.
Icarus, Vol. 49, 244 - 257 (1982).
Numerous recent investigations have been based on the premise that icy objects are "bright" and stony objects are "dark," and that the $J-H$ and $H-K$ colors of icy objects are different than those of stony objects. This study investigates the validity and limitations of assuming bright surfaces are icy and dark surfaces are stony, and the limitations of JHK colorimetry for distinguishing icy versus stony objects.

031.631 **Period determination techniques.** N. Pérez de la Blanca, R. Garrido.
Pulsating B stars, (see 012.042), p. 285 - 295 (1981).

031.632 **Some simple comments on methods used in the short period determination of variable stars.** Y. G. Biraud.
Pulsating B stars, (see 012.042), p. 297 - 298 (1981).

031.633 **Plotting errors in visual meteor observations.** J. Štohl, B. A. Lindblad.
Bull. Astron. Inst. Czechoslovakia, Vol. 33, 129 - 141 (1982).
Plotting errors in visual meteor observations are investigated. The data were collected during radar-visual recordings of the Perseid meteor shower at the Onsala Space Observatory in 1964 - 75. The study includes 401 meteors which were simultaneously recorded by two or more visual observers. In the error analysis the main emphasis is placed on the elevation angle of the plotted meteor paths.

031.634 **Can clouds be beaten photometrically?** A. Ardeberg, H. Lindgren.
Rep. Obs. Lund, No. 18, (see 012.044), p. 76 - 80 (1982).

031.635 **Interferometric measurements of stellar positions in the infrared.**
E. C. Sutton, S. Subramanian, C. H. Townes.
Astron. Astrophys., Vol. 110, 324 - 331 (1982).
Differential positional measurements have been made at wavelengths near 11 μm on stellar sources using an interfero-

metric technique. The nightly precision of these measurements is approximately 0."08, which is slightly better than the typical astrometric errors obtained at visible wavelengths using photographic zenith tubes. The present interferometric measurements are thought to be dominated by instrumental errors indicating that a significant improvement is possible in the precision of astrometric measurements. The limitation imposed by long term irregularities in the earth's atmosphere is estimated to be on the order of 0."01.

Very long baseline interferometry: 1967 to 2000. See Abstr. 013.040.

Towards imaging with a speckle-interferometric optical synthesis telescope.   See Abstr. 031.001.

Stellar interferometry: a widening frontier. See Abstr. 032.008.

The CCD/transit instrument (CTI) deep photometric and polarimetric survey — a progress report. See Abstr. 032.037.

Report of IAU Commission 9: Instruments and techniques (Instruments et techniques). See Abstr. 032.048.

Radio astronomy by very-long-baseline interferometry.   See Abstr. 033.037.

Low-light-level photometry at the Royal Greenwich Observatory.   See Abstr. 034.006.

A cautionary note on the use of Pickering-Racine prisms.   See Abstr. 034.032.

Photométrie stellaire par caméra électronique. See Abstr. 034.048.

Speckle spectroscopy using the Multi-Anode Microchannel Array detector system.   See Abstr. 034.053.

An image stacker for high-resolution spectroscopy on the Multiple Mirror Telescope.   See Abstr. 034.072.

NBS develops new clock synchronization technique. See Abstr. 035.003.

Report of IAU Commission 24: Photographic astrometry (Astrométrie photographique). See Abstr. 041.057.

Earth rotation information derived from MERIT and POLARIS VLBI observations.   See Abstr. 044.049.

Progress report on project MERIT. See Abstr. 044.050.

The determination of polar coordinates by the Doppler technique.   See Abstr. 045.013.

Cepstral analysis of interfering delay signals as applied to detection of gravitational lenses. See Abstr. 066.015.

Cepstral analysis of interfering delay signals as applied to detection of gravitational lenses. See Abstr. 066.093.

The flare of December 17, 1980 observed with high time resolution by a digital CCD camera. See Abstr. 073.127.

Indirect methods for measuring variations of the solar constant.   See Abstr. 080.080.

The effect of personal equation on measurements of the solar semi-diameter by transit circle observers. See Abstr. 080.098.

Topographic lineament analysis: possible stress indicators on planetary surfaces.   See Abstr. 091.021.

Color distribution fields of geomorphic features on Europa: initial results from a new technique. See Abstr. 099.022.

Report of IAU Commission 25: Stellar photometry and polarimetry (Photométrie et polarimétrie stellaires). See Abstr. 113.097.

On coherent properties of rotating star radiation. See Abstr. 116.016.

The eclipsing binary programme. See Abstr. 119.009.

Computation of the elements of eclipsing binary systems in the frequency-domain.   See Abstr. 119.023.

Computation of the elements of close eclipsing systems in the frequency-domain.   See Abstr. 119.027.

Contribution to Fourier analysis of the light curves of eclipsing variables.   See Abstr. 119.134.

Radial velocity variations of the δ Scuti variable β Cassiopeiae.   See Abstr. 122.181.

Methods of photometric abundance determination in globular clusters.   See Abstr. 154.025.

Influence of ellipticity on photometric profiles of elliptical galaxies.   See Abstr. 158.053.

032 Astronomical Instruments, Space Instrumentation

Astronomical Instruments

032.001 The Isaac Newton Telescope.
F. G. Smith, J. Dudley.
J. Hist. Astron., Vol. 13, 1 - 18 (1982).

032.002 Very large ground-based telescopes for optical and
IR astronomy. J. R. P. Angel.
Nature, Vol. 295, 651 - 657 (1982).
Optical and IR astronomers are taking a hard look at their
ground-based facilities and devising new ways of making more
economic, bigger and better telescopes. Features of instruments of the 15-m class are likely to include servo control to
compensate for atmospheric wavefront errors as well as structural deformation, large honeycomb mirror blanks and mirror
surfaces produced by economical techniques developed for
aspherics.

032.003 Photoguide complex of the horizontal solar telescope
of the Pulkovo Observatory. O. V. Nikonov,
A. P. Kulish, L. M. Zatsiorskij, Yu. S. Muzalevskij.
Soln. Dannye 1981 Byull., No. 9, p. 111 - 113 (1981). In
Russian.

032.004 The short term stability of the Brorfelde transit
circle. C. Fabricius.
Astron. Astrophys., Vol. 105, 413 - 416 (1982).
At the transit circle in Brorfelde a detailed study of the
stability of the circle reading system has been carried out. It is
proved that reference lines in the circle reading micrometers
can be evaded. The graduated circle has a translational displacement depending on the temperature. The nadir point of the
circle varies nearly $0\rlap{.}{''}15/^\circ C$ at higher temperatures, but less
and more irregularly at low temperatures. The sum of collimation free reading and inclination varies depending on
temperature near $0\rlap{.}{''}1/^\circ C$.

032.005 The challenge of deep sky photography.
J. Newton.
J. R. Astron. Soc. Canada, Vol. 76, 50 - 57 (1982).

032.006 A new spectrohelioscope.
B. G. W. Manning.
J. British Astron. Assoc., Vol. 92, 112 - 118 (1982).

032.007 Ein neues Fenster zum Universum. Größtes deutsches Teleskop kurz vor der Fertigstellung.
Umschau, 82. Jahrg., 193 (1982).

032.008 Stellar interferometry: a widening frontier.
A. Labeyrie.
Sky Telesc., Vol. 63, 334 - 338 (1982).

032.009 Measurement of the local refraction in the pavilion
of the meridian circle of the Engelhardt Astronomi-
cal Observatory by the autocollimation method.
S. S. Peruanskij, N. B. Kurochkina, V. N. Mukhametshina,
L. R. Valitova, O. G. Argat, E. S. Goncharova,
G. M. Narmatova, T. Yu. Flerova.
Kazan. univ. Kazan', 1981. 7 pp. In Russian. – Abstr. in Ref.
zh., 51. Astron., 2.51.94 (1982).

032.010 Vacuum aluminizing of telescope mirrors at U.P.
State Observatory, Naini Tal.
T. D. Padalia, B. B. Sanwal.
Vac News., Vol. 11, No. 3, p. 5 - 9 (1980). – Abstr. in Phys.
Abstr., Vol. 85, Abstr. 10322 (1982).

032.011 Tsubokawa's automatic astrolabe.
C. Kakuta, T. Tsubokawa.
Astron. Her., Vol. 74, 128 - 132 (1981). In Japanese. – Abstr.
in Phys. Abstr., Vol. 85, Abstr. 14769 (1982).

032.012 A scheme for determining division corrections of a
photoelectric meridian circle.
M. Miyamoto, H. Ishii.
Publ. Astron. Soc. Japan, Vol. 34, 117 - 140 (1982).
To determine the division corrections for the graduated
circle of a photoelectric meridian circle, a symmetric method
developed by Lévy and Høg is made suitable for a circle with
3600 divisions. It is found that division measurements, based
on some sets of three independent angles subtended by two
microscope pairs, permit a favorable least-squares solution for
the corrections.

032.013 The GEODSS difference. J. K. Beatty.
Sky Telesc., Vol. 63, 469 - 473 (1982).

032.014 3.5-m-Teleskop fertiggestellt. K. Bahner.
Sterne Weltraum, 21. Jahrg., 150 - 151 (1982).

032.015 Das 1 m-Teleskop der Manuel-Foster-Sternwarte in
Santiago: eine neue Beobachtungsmöglichkeit an
der Südhalbkugel. E. Heilmeier, N. Vogt.
Mitt. Astron. Ges., Nr. 55, p. 167 - 168 (1982).

032.016 The new technology telescopes.
M. M. Waldrop.
Science, Vol. 216, 280 - 281 (1982).
Astronomers are giving high priority to a new generation
of instruments built on a Brobdingnagian scale.

032.017 Le télescope Canada-France-Hawaii.
R. Cayrel.
Astronomie, Vol. 96, 227 - 236 (1982).

032.018 Richtfouten van de uuras. S. Wadman.
Zenit, 9. Jaarg., 212 - 214 (1982).

032.019 Les 21 ans de fonctionnement du plus grand
télescope de Schmidt à Tautenburg. J. Sauval.
Ciel Terre, Vol. 98, 91 - 100 (1982).

032.020 El nuevo telescopio solar del Departamento de
Física Solar. H. Molnar.
Bol. Asoc. Argentina Astron., No. 18, p. 79 - 81 (1980).

032.021 Preliminary investigations of the astronomical
flexure of the tube of a photographic vertical circle
by the autocollimation method.
B. K. Bagil'dinskij, V. D. Shkutov.
Astrometr. Astrofiz., Vyp. (No.) 46, p. 93 - 99 (1982). In
Russian.

032.022 The Pulkovo 30-inch refractor.
V. A. Gurikov.
Zemlya Vselennaya, 1982, No. 3, p. 62 - 64. In Russian.

032.023 Reformas efectuadas al astrolabio OPL 01 con
vistas a su utilización en la estación de Río Grande.
R. Platzeck, R. Pinciroli, C. Mondinalli, R. Perdomo.
Bol. Asoc. Argentina Astron., No. 20 - 24, p. 322 (1981).
Abstract.

032.024 Determinación del intervalo normal del círculo
meridiano Repsold 1907 del OAUNLP.

G. E. Crotti.
Bol. Asoc. Argentina Astron., No. 20 - 24, p. 410 (1981).
Abstract.

032.025 Resultados obtenidos de la comparación entre el
         cerro elegido como sitio del telescopio de 2.15 m
         y la Ciénaga.   F. López García, G. Sánchez.
Bol. Asoc. Argentina Astron., No. 20 - 24, p. 411 (1981).
Abstract.

032.026 Estabilidad del círculo meridiano del OAFA.
         W. L. Castro, R. A. Carestia.
Bol. Asoc. Argentina Astron., No. 20 - 24, p. 413 (1981).
Abstract.

032.027 Diseño, construcción y montaje del telescopio de
         600 mm del Observatorio Astronómico de
Mercedes.   L. Hordij, A. Di Palma,
M. A. De Laurenti, J. L. Marazzo, J. Gadagz.
Bol. Asoc. Argentina Astron., No. 20 - 24, p. 414 (1981).
Abstract.

032.028 The SibIZMIR Hα-cinematograph.
         V. G. Banin, Yu. A. Klevtsov, V. I. Skomorovskij,
V. D. Trifonov.
Soln. Dannye 1982 Byull., No. 1, p. 90 - 99 (1982). In Russian.
    A brief description of a chromospheric telescope
(Hα-cinematograph) developed and constructed at the
SibIZMIR is given. Factors that have determined the choice of
the functional diagram and parameters of the instrument are
considered.

032.029 Possibilities for spectroscopic observations in the
         Bulgarian National Astronomical Observatory.
D. Kolev.
Soobshch. Spets. Astrofiz. Obs., Vyp. 32, (see 012.029),
p. 37 - 38 (1981).

032.030 Two-telescope Michelson stellar interferometry at
         2.2 μ.
O. Citterio, G. Conti, G. P. Di Benedetto.
Scientific importance of high angular resolution at infrared
and optical wavelengths, (see 012.031), p. 237 - 245 (1981).

032.031 Use of a dynamic damping system to diminish the
         oscillations of the large vacuum solar telescope
tower.   A. K. Kitov.
Astron. Tsirk., No. 1192, p. 5 - 7 (1981). In Russian.

032.032 Some possibilities of applying coherent optics to
         processing of photographs obtained in the prime
focus of the 6-m telescope.
A. O. Bakrunov, O. S. Sladkov, M. F. Shabanov, I. V. Shchukin.
Astron. Tsirk., No. 1195, p. 1 - 3 (1981). In Russian.

032.033 On a kind of apodization for telescopes.
         A. V. Lenskij.
Astron. Zh., Tom 59, 591 - 593 (1982). In Russian. English
translation in Soviet Astron., Vol. 26, No. 3.
    A simple and sufficiently effective technique of apodiza-
tion is suggested for telescopes with central obscuration.

032.034 On the problem of optimum guiding of an
         astrograph.   V. I. Bojkov.
Sistemy upr. i ikh ehlementy. Leningrad, 1981, p. 125 - 128.
In Russian. − Abstr. in Ref. zh., 51. Astron., 5.51.919 (1982).

032.035 Scaling the MMT to 15 meters − similarities and
         differences.   N. J. Woolf, J. R. P. Angel,
J. Antebi, N. Carleton, L. Barr.
Prepr. Steward Obs., No. 371, 10 pp. (1982).

032.036 The performance of the Multiple Mirror Telescope.
         VII. Image shrinking in sub-arc second seeing at the
         MMT and 2.3 m telescopes.
N. J. Woolf, D. W. McCarthy, J. R. P. Angel.
Prepr. Steward Obs., No. 375, 7 pp. (1982).

032.037 The CCD/transit instrument (CTI) deep photometric
         and polarimetric survey − a progress report.
J. T. McGraw, H. S. Stockman, J. R. P. Angel, H. Epps,
J. T. Williams.
Prepr. Steward Obs., No. 382, 9 pp. (1982).

032.038 The performance of the Multiple Mirror Telescope.
         X. The first sub-millimeter phased array.
B. L. Ulich, C. L. Lada, N. R. Erickson, P. F. Goldsmith,
G. R. Huguenin.
Prepr. Steward Obs., No. 387, 7 pp. (1982).

032.039 PMC 190: a new photoelectric meridian circle of
         Tokyo Astronomical Observatory. I. Determination
of instrumental errors.   M. Yoshizawa, H. Yasuda.
Ann. Tokyo Astron. Obs., Second Ser., Vol. 18, 205 - 223
(1982).

032.040 PMC 190: a new photoelectric meridian circle of
         Tokyo Astronomical Observatory. II. Pinhole
system for solar and lunar observations.
M. Yoshizawa, H. Yasuda.
Ann. Tokyo Astron. Obs., Second Ser., Vol. 18, 224 - 235
(1982).

032.041 Vacuum aluminizing of telescope mirrors at
         U. P. State Observatory, Naini Tal.
T. D. Padalia, B. B. Sanwal.
Vac News, Vol. 11, No. 3, p. 5 - 10 = Uttar Pradesh State
Obs. Repr. No. 185.

032.042 Cost relationships for nonconventional telescope
         structural configurations.   A. B. Meinel.
J. Opt. Soc. America, Vol. 72, 14 - 20 (1982).
    Fast primary mirrors offer the possibility of using endo-
structure designs rather than the conventional exostructure
designs. A parametric analysis indicates that considerable cost
savings can be achieved in this manner for large optical telescopes. f/1 endostructures are compared with classical and
quasi-classical f/2.75 exostructure telescopes. The multiple
mirror telescope configuration is the lowest-cost design for
equal aperture area, but when equal light is considered, it is
comparable with the segmented monolith design.

032.043 Stand und Entwicklungstendenzen beim Bau
         großer optischer Teleskope.   R. Schielicke.
Sterne, 58. Band, 93 - 103 (1982).

032.044 Strumenti per l'astronomia ottica.
         H. Elsässer.
Orione, Vol. 3, 56 - 69 (1982).

032.045 LEST (Large European Solar Telescope). Report on
         a study for the Joint Organisation for Solar Observa-
tions.   O. Engvold, M. Hefter (Editors).
Sep. print Inst. Theor. Astrophys., Univ. Oslo, Blindern,
Norway. 7 + 253 pp. (1982).
    Contents: Seeing and telescope vibrations. Means for
reducing effects of seeing and telescope vibrations. Building
related measures. Telescope mounting. Triple mirror coelostat.
Alternative optical systems for LEST. A design for LEST.
Comments on the study and recommendations to JOSO board.

032.046 Comments on the 2.5 m telescope design study
         by Torben E. Andersen.   A. Reiz.
Rep. Obs. Lund, No. 18, (see 012.044), p. 11 - 12 (1982).

032.047 **Plans for a Large European Solar Telescope — LEST.**
O. Engvold.
Rep. Obs. Lund, No. 18, (see 012.044), p. 13 - 16 (1982).

032.048 **Report of IAU Commission 9: Instruments and techniques** (Instruments et techniques).
E. H. Richardson.
Trans. IAU, Vol. XVIIIA, (see 003.013), p. 43 - 54 (1982).

032.049 **Un método rápido y preciso para ajuste de monturas ecuatoriales.** T. Paneth.
Bol. Asoc. Argentina Astron., No. 18, p. 76 - 79 (1980).

Comments on determination of division corrections.
See Abstr. 031.541.

Interferometric connection of the Canada-France-Hawaii 3.6 metre telescope and the United Kingdom 3.8 metre telescope on Mauna Kea. See Abstr. 031.585.

Interferometry with the Multiple Mirror Telescope and conventional telescopes. See Abstr. 031.586.

Resultats d'observations du soleil à l'astrolabe du CERGA. See Abstr. 041.035.

The HAO solar diameter monitor.
See Abstr. 080.064.

Estudio de las condiciones de visibilidad para la observación solar en algunos puntos del país.
See Abstr. 082.032.

Space Instrumentation

032.501 **High resolution gamma-ray telescope using a coded aperture mask and drift chamber.**
J. N. Carter, D. Ramsden, G. M. Frye, Jr., T. L. Jenkins, R. Koga.
Mon. Not. R. Astron. Soc., Vol. 198, 33 - 43 (1982).

A method is described for improving the angular resolution of a high energy γ-ray telescope. It is shown that a coded aperture mask combined with a drift chamber pair detector can achieve a ~10 arcmin point spread function thereby resolving the structure of extended sources on a 10 arcmin scale and locating individual point sources to within a few arcmin, an improvement by more than an order of magnitude over existing instruments.

032.502 **An eye for tomorrow.**
L. J. Robinson.
Sky Telesc., Vol. 63, 128 (1982).

032.503 **SAR *(Synthetic Aperture Radar)* imaging: seeing the unseen.** M. Kobrick.
Sky Telesc., Vol. 63, 139 - 140 (1982).

032.504 **Unveiling Venus with VOIR.**
W. James.
Sky Telesc., Vol. 63, 141 - 144 (1982).

032.505 **Orbital astrophysical observatories.**
N. S. Yamburenko.
Astronomical calendar. Yearbook. 1982, (see 047.005), p. 229 - 245 (1981). In Russian. — Abstr. in Ref. zh., 51. Astron., 2.51.835 (1982).

032.506 **Possible construction principles of a gamma-telescope and of the logic of separating recorded particles.**
V. V. Akimov, S. A. Voronov, A. M. Gal'per, V. A. Grigor'ev, M. B. Dobriyan, L. F. Kalinkin, V. G. Kirillov-Ugryumov, T. I. Kurmanaliev, L. V. Kurnosova, B. I. Luchkov, A. S. Melioranskij, V. E. Nesterov, S. R. Tabaldyev, E. I. Chujkin.
Nauchn. kosm. priborostr. Moskva, 1981, p. 20 - 26. In Russian. — Abstr. in Ref. zh., 62. Issled. kosm. prostranstva, 2.62.62 (1982).

032.507 **Design, calibration and sensitivity of a passively anti-collimated (PAC) gamma-ray telescope.**
J. Benson, J. O. D. Jardim, I. M. Martin, U. B. Jayanthi, O. D. Aguiar.
Nucl. Instrum. Methods Phys. Res., Vol. 188, 613 - 617 (1981). — Abstr. in Phys. Abstr., Vol. 85, Abstr. 18703 (1982).

032.508 **Space from space *(Space Telescope).***
S. Greene.
Syst. Int., Vol. 9, No. 10, p. 74 - 75 (1981). — Abstr. in Phys. Abstr., Vol. 85, Abstr. 18707 (1982).

032.509 **Rocket-borne spectrometer to measure the near infrared absorption spectrum of the upper atmosphere.** A. Matsuzaki, Y. Nakamura, T. Itoh.
Rev. Sci. Instrum., Vol. 52, 1685 - 1689 (1981). — Abstr. in Phys. Abstr., Vol. 85, Abstr. 23142 (1982).

032.510 **Ein neues Zeitalter der Infrarot-Astronomie.**
D. Lemke.
Sterne Weltraum, 21. Jahrg., 62 - 65 (1982).

032.511 **Mirror telescope in X-ray astronomy.**
K. Makishima.
Astron. Her., Vol. 74, No. 6, p. 169 - 172 (1981). In Japanese. Abstr. in Phys. Abstr., Vol. 85, Abstr. 28341 (1982).

032.512 **Properties of the channel electron multiplier arrays (CEMAs) for the SOLEX solar X-ray spectrometer/spectroheliograph.** W. Eng, P. B. Landecker.
Nucl. Instrum. Methods Phys. Res., Vol. 190, 149 - 157 (1981). Abstr. in Phys. Abstr., Vol. 85, Abstr. 28348 (1982).

032.513 **Grating spectrometers for use with rockets and satellites.** T. Namioka, T. Harada, H. Noda.
Astron. Her., Vol. 74, No. 8, p. 220 - 226 (1981). In Japanese. — Abstr. in Phys. Abstr., Vol. 85, Abstr. 32840 (1982).

032.514 **Astrophysical observations with high resolution X-ray telescopes.** M. V. Zombeck.
AIP Conf. Proc., No. 75, (see 012.015), p. 200 - 209 (1981). Abstr. in Phys. Abstr., Vol. 85, Abstr. 32841 (1982).

032.515 **SAR: an instrument for planetary geodesy and navigation.**   S. N. Mohan, M. P. Ananda.
J. Astronaut. Sci., Vol. 29, 127 - 151 (1981). — Abstr. in Phys. Abstr., Vol. 85, Abstr. 42594 (1982).

032.516 **Astronomical two-channel spatial and spectral scanning spectrometer.**
V. D. Vdovichenko, S. M. Gajsin, K. S. Kuratov, S. S. Shumilin.
Astrofiz. inst. AN KazSSR. Alma-Ata, 1981, 13 pp. In Russian.
Abstr. in Ref. zh., 51. Astron., 3.51.942 (1982).

032.517 **Mass-spectrometer for investigation of the composition of the upper atmosphere: experience of measurements aboard the MR-12 rocket.**
Yu. P. Barzilovich, K. V. Grechnev, D. V. Latysh, B. I. Pilipenko, V. K. Semenov, Yu. A. Shul'chishin.
Ionos. issled. Moskva, 1981, No. 34, p. 112 - 120. In Russian.
Abstr. in Ref. zh., 62. Issled. kosm. prostranstva, 3.62.91 (1982).

032.518 **Neutrino astronomy using SOCRAS, a satellite observatory for cosmic ray air showers.**
J. Linsley.
Proc. Southwest Reg. Conf., Vol. 7, (see 012.019), 169 - 178 (1982).

032.519 **High-speed two-channel satellite electrophotometer operating in the regime of photon counting.**
M. V. Bratijchuk, V. P. Epishev, T. I. Laslo, Ya. M. Motrunich, I. F. Najbauer, V. R. Shumakov.
Astrometr. Astrofiz., Vyp. (No.) 46, p. 84 - 92 (1982). In Russian.
A high-speed two-channel electrophotometer is described and the results of its test are presented. The effective wavelength of the photometric system is 5590 Å.

032.520 **Space for the telescope. An account of NASA's projected Space Telescope.**   B. V. Barlow.
Contemp. Phys., Vol. 23, No. 1, p. 45 - 63 (1982). — Abstr. in Phys. Abstr., Vol. 85, Abstr. 48146 (1982).

032.521 **Wideband comet camera lens design.**   E. D. Huber.
Opt. Eng., Vol. 20, 941 - 946 (1981). — Abstr. in Phys. Abstr., Vol. 85, Abstr. 53091 (1982).

032.522 **Röntgensatellit. German X-ray satellite.**
G. Rausch, K. Frankenbach, W. Trogus, E. Bachor.
20th Goddard Memorial Symposium, held at GSFC, Greenbelt, Maryland, March 18 - 19, 1982, Paper AAS-82-123. 15 pp. (1982).
The German X-ray activities of the last ten years are now culminating in the development of a dedicated project — the German X-ray satellite "Röntgensatellit". The payload consists of the main instrument, a highly sensitive 80 cm aperture X-ray telescope and will be supplemented by a smaller UK instrument, the Wide Field Soft X-ray Camera. Very probably an US experiment, the High Resolution Imager will be included in the focal plane instrumentation of the main telescope. This paper gives a short description of the scientific and mission objectives and the satellite system aspects.

032.523 **A 1 meter class extreme ultraviolet telescope/spectrometer for stellar observations.**
R. Kimble, S. Bowyer, P. Jelinsky, M. Grewing, G. Kraemer, E. Schultz-Luepertz, C. Wulf-Mathies.
Bull. American Astron. Soc., Vol. 13, 813 - 814 (1981). Abstract.

032.524 **Status of the high resolution spectrograph for the Space Telescope.**
J. C. Brandt, M. Bottema, S. P. Maran.
Bull. American Astron. Soc., Vol. 13, 814 (1981). — Abstract.

032.525 **Faint Object Spectrograph update.**   R. J. Harms, J. R. Angel, F. Bartko, E. A. Beaver, R. C. Bohlin, E. M. Burbidge, A. F. Davidsen, H. C. Ford, B. Margon.
Bull. American Astron. Soc., Vol. 13, 840 (1981). — Abstract.

032.526 **The Shuttle Infrared Facility (SIRTF): technical progress and current status.**
M. W. Werner, F. C. Witteborn, L. A. Manning, L. S. Young.
Bull. American Astron. Soc., Vol. 13, 840 (1981). — Abstract.

032.527 **High resolution measurements of hard X-ray spectra of southern hemisphere sources.**
J. Tueller, T. Cline, W. Paciesas, B. J. Teegarden, D. Boclet, P. Durouchoux, J. M. Hameury.
Bull. American Astron. Soc., Vol. 13, 868 (1981). — Abstract.

032.528 **Low energy gamma-ray spectroscopy of the galactic center and other sources in the southern sky.**   W. S. Paciesas, T. L. Cline, B. J. Teegarden, J. Tueller.
Bull. American Astron. Soc., Vol. 13, 868 (1981). — Abstract.

032.529 **Multi-Anode Microchannel Array detector systems.**
J. G. Timothy, C. L. Joseph, R. L. Bybee.
Bull. American Astron. Soc., Vol. 13, 884 (1981). — Abstract.

032.530 **Performance of the spectropolarimeter for the Space Telescope Faint Object Spectrograph.**
R. G. Allen, J. R. P. Angel.
Prepr. Steward Obs., No. 385, 9 pp. (1982).

032.531 **Selection criteria for continuous channel electron multipliers in space applications.**   A. Urban.
Astrophys. Space Sci., Vol. 83, 3 - 14 (1982).
Criteria for selecting continuous channel electron multipliers were applied to fifteen Mullard units without significantly affecting their overall lifetime. The gain fatigue vs accumulated counts, the change of the pulse height distribution during lifetime, the gain degradation vs count rates as well as the gain vs operating voltage and the resolution vs operating voltage have been investigated using a tritium source. The results provide several criteria to determine good, marginal and poor multipliers.

032.532 **An upper limit on the aperture separation of ion drift meters.**   E. A. Bering, K. G. Weber, U. V. Fahleson.
Astrophys. Space Sci., Vol. 83, 37 - 49 (1982).
An upper limit has been calculated on the effective aperture separation or detector thickness of ion drift meters of two fundamental types. The limit applies to meters which compare currents collected by detectors with different view directions at the same retarding potential and to meters which measure the entire thermal ion distribution function. The limit was found to be important in two cases. First, in the F region on spacecraft with stringent electrostatic cleanliness requirements, the 10% error limit was found to be 40 cm. Second, in the E region, the limit was found to be $\lesssim 1$ cm.

032.533 **The HEAO-3 Cosmic Ray Isotope Spectrometer.**
M. Bouffard, J. J. Engelmann, L. Koch, A. Soutoul, N. Lund, B. Peters, I. L. Rasmussen.
Astrophys. Space Sci., Vol. 84, 3 - 33 (1982).
This paper describes the Cosmic Ray Isotope instrument launched aboard the HEAO-3 satellite on September 20, 1979. The primary purpose of the experiment is to measure the isotopic composition of cosmic ray nuclei from Be-7 to Fe-58 over the energy range 0.5 to 7 GeV/nucleon. In addition charge spectra will be measured between beryllium and tin over the energy range 0.5 to 25 GeV/nucleon. The instrument

consists of 5 Cerenkov counters, a 4 element neon flash tube hodoscope and a time-of-flight system. The determination of charge and energy for each particle is based on the multiple Cerenkov technique and the mass determination will be based upon a statistical analysis of particle trajectories in the geomagnetic field.

032.534 An image drift compensation system for a solar pointed space telescope.   J.-D. F. Bartoe.
Astrophys. Space Sci., Vol. 84, 115 - 132 (1982).

An image drift compensation system has been designed for a solar pointed space borne telescope and has performed successfully on two sounding rocket flights, yielding new scientific results. The system employs limb-sensing photodiodes at the telescope focal plane and provides drift compensation to better than 0.1 arc sec. A variation of the system will be used on the Shuttle/Spacelab 2 flight of the High Resolution Telescope and Spectrograph (HRTS) Instrument.

032.535 Space technology and the optical sciences.
H. W. Yates.
Appl. Opt., Vol. 21, 203 - 208 (1982).

The earth-orbiting satellites and the deep-space probes have provided for the optical sciences platforms from which to study the earth, the solar system, and the universe with truly revolutionary capability. For the terrestrial sciences the orbiting platforms for optical measurements in both low and geostationary orbits have given us a view of our planet and a global coverage never before possible. For the astronomical applications of optical instruments that "cataract of the telescopic eye," the atmosphere of the earth has been left behind and through proximity, including actual contact, we now have resolution and spectral coverage limited only by money and motive.

032.536 Calibration and efficiency of the Einstein objective grating spectrometer.   F. D. Seward,
T. Chlebowski, J. P. Delvaille, J. P. Henry, S. M. Kahn,
L. Van Speybroeck, J. Dijkstra, A. C. Brinkman, J. Heise,
R. Mewe, J. Schrijver.
Appl. Opt., Vol. 21, 2012 - 2021 (1982).

032.537 Balloon-borne scanning spectrometer system for atmospheric extinction studies in the 350 - 1100 nm spectral region.   D. A. Thompson. T. J. Pepin, R. W. Lane,
J. VanBaalen, H. R. Bauer, III.
Rev. Sci. Instrum., Vol. 53, 314 - 319 (1982). − Abstr. in Phys. Abstr., Vol. 85, Abstr. 53848 (1982).

032.538 High resolution X-ray scattering measurements for the Advanced X-ray Astrophysics Facility (AXAF).
M. V. Zombeck, C. C. Wyman, M. C. Weisskopf.
Opt. Eng., Vol. 21, No. 1, p. 63 - 72 (1982). − Abstr. in Phys. Abstr., Vol. 85, Abstr. 58254 (1982).

032.539 A gamma-ray telescope for the 1980's.
J. Carter, P. Charalambous, A. J. Dean, J. Stephen.
J. British Interplanet. Soc., Vol. 35, 291 - 296 (1982).

032.540 Uranus science with Space Telescope.
J. Caldwell.
Uranus and the outer planets, (see 012.036), p. 259 - 274 (1982).

The Space Telescope Observatory, scheduled for launch in 1985, is described. The advantages of the space environment and the consequent features of ST performance are given, with Uranus observations as examples. The first generation instruments, including two cameras, two spectrographs and a high speed photometer, are discussed. The Space Telescope Science Institute, which will manage the Observatory, is discussed briefly. The potential scientific interaction with the Voyager 2 encounter of Uranus is also considered.

032.541 An instrument to measure the solar spectrum from 170 to 3200 nm on board Spacelab.
G. Thuillier, P. C. Simon, R. Pastiels, D. Labs, H. Neckel.
Variations of the solar constant, (see 012.037), p. 165 - 173 (1981).

The instrument is composed of three double monochromators covering the range 170 to 3200 nm. The spectrometers have bandpasses of 1 nm up to 900 nm and 20 nm from 850 to 3200 nm with an accuracy 10^{-2} nm. Calibration lamps are included in the instrument to monitor any change of its sensitivity and wavelength scale.

032.542 Properties and performance of the MPI balloon borne Compton telescope.
V. Schönfelder, U. Graser, R. Diehl.
Astron. Astrophys., Vol. 110, 138 - 151 (1982).

A balloon-borne Compton telescope is described which allows one to perform astronomical γ-ray observations in the energy range 1–20 MeV. The properties of the telescope as determined from calibration measurements, Monte Carlo calculations and data from two balloon flights are described. A method is described by which the image of a celestial γ-ray source within the field of view can be constructed in the presence of the γ-ray background radiation at high balloon altitudes. The method is applied to the anticenter region of the galaxy which was observed during a balloon flight.

032.543 The Giotto magnetometer experiment.
F. M. Neubauer.
Scientific and experimental aspects of the Giotto mission, (see 012.039), p. 9 - 15 (1981).

032.544 A cometary plasma experiment for the mission to comet Halley.   A. Johnstone, D. Bryant,
T. Edwards, B. Hultquist, V. Formisano, L. Biermann,
R. Lüst, H. Schmidt, W. Feldman, P. Cerulli-Irelli,
M. Dobrowolny, A. Egidi, R. Terenzi, K. Jockers,
H. Rosenbauer, W. Studemann, B. Wilken, M. Wallis,
G. Haerendel, G. Paschmann, J. D. Winningham, H. Reme.
Scientific and experimental aspects of the Giotto mission, (see 012.039), p. 17 - 27 (1981).

032.545 The Copernic experiment to measure three-dimensional electron distribution and the composition of thermal positive ions including water clusters near comet Halley.   H. Rème, F. Cotin, C. d'Uston, J. A. Sauvaud,
A. Korth, A. K. Richter, K. A. Anderson, C. W. Carlson,
R. P. Lin, A. Wekhof, D. A. Mendis, A. Johnstone.
Scientific and experimental aspects of the Giotto mission, (see 012.039), p. 29 - 37 (1981).

032.546 An energetic particle detector for Giotto and its scientific objectives.   S. McKenna-Lawlor,
A. Thompson, D. O'Sullivan, E. Kirsch, D. B. Melrose,
K.-P. Wenzel.
Scientific and experimental aspects of the Giotto mission, (see 012.039), p. 39 - 44 (1981).

032.547 The particulate impact analyzer, an instrument to analyze small particles released by Halley's comet.
J. Kissel.
Scientific and experimental aspects of the Giotto mission, (see 012.039), p. 53 - 60 (1981).

032.548 A Dust Impact Detection System (DIDSY) for the Giotto Halley mission.
J. A. M. McDonnell, E. Grün, G. C. Evans, R. F. Turner,
J. G. Firth, W. C. Carey, H. Kuczera, W. M. Alexander,
D. H. Clark, R. J. L. Grard, M. S. Hanner, D. W. Hughes,
E. Igenbergs, B. A. Lindblad, J. C. Mandeville, G. Schwehm,
Z. Sekanina.

Scientific and experimental aspects of the Giotto mission, (see 012.039), p. 61 - 75 (1981).

032.549   **The Giotto ion mass spectrometer.**
H. Balsiger, J. Geiss, D. T. Young, H. Rosenbauer, R. Schwenn, W.-H. Ip, E. Ungstrup, M. Neugebauer, R. Goldstein, B. E. Goldstein, W. T. Huntress, E. G. Shelley, R. D. Sharp, R. G. Johnson, A. J. Lazarus, H. S. Bridge.
Scientific and experimental aspects of the Giotto mission, (see 012.039), p. 93 - 98 (1981).

032.550   **A Halley multicolour camera.**
H. U. Keller, C. Arpigny, C. Barbieri, P. Benvenuti, L. Biermann, R. M. Bonnet, S. Cazes, G. Colombo, C. B. Cosmovici, W. A. Delamere, W. F. Huebner, D. W. Hughes G. E. Hunt, C. Jamar, C. D. Mackay, D. Malaise, W. K. H. Schmidt, P. Seige, F. L. Whipple, K. Wilhelm.
Scientific and experimental aspects of the Giotto mission, (see 012.039), p. 105 - 117 (1981).

032.551   **The Halley optical probe experiment.**
A-C. Levasseur-Regourd, J. L. Weinberg, F. Giovane, D. W. Schuerman, P. Lamy, M. Festou, J. L. Bertaux, R. Dumont, A. Llebaria, R. H. Giese.
Scientific and experimental aspects of the Giotto mission, (see 012.039), p. 121 - 126 (1981).

032.552   **The Giotto Neutral Mass Spectrometer.**
D. Krankowsky, P. Lämmerzahl, P. Eberhardt, U. Herrmann, J. J. Berthelier, M. Sylvain, J. H. Hoffman, R. R. Hodges, U. von Zahn, H. U. Keller, M. Festou.
Scientific and experimental aspects of the Giotto mission, (see 012.039), p. 127 - 130 (1981).

032.553   **The satellite Hipparcos.**
J. Kovalevsky.
Scientific aspects of the Hipparcos space astrometry mission, (see 012.041), p. 15 - 20 (1982).
    The principle of Hipparcos astrometric satellite is recalled and its main features are presented. The expected precision of the measurements and the causes of errors are discussed. From them, resulting accuracies of positions, proper motions, parallaxes and magnitudes are derived.

032.554   **Prospects for γ-ray imaging telescopes.**
J. N. Carter, A. J. Dean, D. Ramsden.
Philos. Trans. R. Soc. London, Ser. A, Vol. 301, (see 012.043), 607 - 610 (1981).

Scientific space instrument making.
See Abstr. 003.081.

The effects of microparticle hypervelocity impacts on polished surfaces: tests for the choice of the Halley multicolour camera.   See Abstr. 022.156.

Stray radiation and the Infrared Astronomical Satellite (IRAS) telescope.   See Abstr. 031.006.

Comparison of stray light mechanisms and performance in the Infrared Astronomy Satellite (IRAS) and Shuttle Infrared Telescope Facility (SIRTF) telescopes.
See Abstr. 031.007.

Straylight analysis of the German Infrared Laboratory (GIRL).   See Abstr. 031.008.

Visible light scatter measurements of the Advanced X-ray Astronomical Facility (AXAF) mirror samples.
See Abstr. 031.009.

Measurements of X-ray scattering from Wolter type telescopes and various flat Zerodur mirrors.
See Abstr. 031.010.

High resolution X-ray scattering measurements for Advanced X-ray Astrophysics Facility (AXAF).
See Abstr. 031.011.

A high resolution, grazing incidence, ultraviolet spectrograph.   See Abstr. 031.025.

Report of IAU Commission 9: Instruments and techniques (Instruments et techniques).
See Abstr. 032.048.

Precision antenna of a space radio telescope.
See Abstr. 033.016.

Digest of celestial X-ray missions and experiments.
See Abstr. 051.014.

Proposed ultraviolet observations with Tycho.
See Abstr. 051.020.

CIRBS. Cosmic Infrared Background Satellite.
See Abstr. 051.029.

Prospective views in space astrometry.
See Abstr. 051.030.

Report of IAU Commission 44: Astronomy from space (L'astronomie à partir de l'espace).
See Abstr. 051.031.

Imaging properties of Hipparcos and the observation of multiple stars.   See Abstr. 118.035.

High-energy gamma-quanta from measurements aboard the Cosmos 856 and Cosmos 914 artificial earth satellites.   See Abstr. 142.512.

033 Radio Telescopes and Equipment

033.001 **30 m radiotelescoop voor millimeter-golf-lengten.**
J. W. M. Baars.
Zenit, 9. Jaarg., 4 - 8 (1982).

033.002 **Lunar radiointerferometer.**
V. S. Artyukh, V. I. Shishov.
Astron. Zh., Tom 59, 155 - 159 (1982). In Russian. English
translation in Soviet Astron., Vol. 26, No. 1.
It is suggested to use a radio source signal reflected from
the moon for constructing a lunar interferometer similar to a
sea interferometer. The baseline of the moon interferometer is
determined by the earth-moon distance.

033.003 **Brightness temperature calibration for 21-cm line
observations.**
P. M. W. Kalberla, U. Mebold, K. Reif.
Astron. Astrophys., Vol. 106, 190 - 196 (1982).
A telescope independent, absolute H I line brightness
temperature scale has been determined with an accuracy
better than 2%. From 21-cm line observations obtained with
the Effelsberg 100-m telescope and corrected for the telescope
stray radiation the authors derive reference data for telescopes
with $9' \le$ HPBW $\lesssim 35'$. The derived scale is consistent with
calibration data given by Penzias et al. (1970) and Wrixon and
Heiles (1973) for the 20-ft horn reflector antenna at Crawford
Hill.

033.004 **Millimetre wave astronomy.**
E.-J. Blum.
Phys. Technol., Vol. 12, No. 4, p. 162 - 169 (1981). – Abstr.
in Phys. Abstr., Vol. 85, Abstr. 18715 (1982).

033.005 **Das 25-m-Teleskop auf den Stockert.**
T. Schmidt-Kaler.
Sterne Weltraum, 21. Jahrg., 160 - 162 (1982).

033.006 **Das neue 3 m-Radioteleskop der Universität Köln.**
G. Winnewisser, B. Vowinkel, P. Wratil, T. Pauls.
Mitt. Astron. Ges., Nr. 55, p. 157 - 159 (1982).

033.007 **High-sensitive millimeter waveband radiometer with
a Josephson detector.**
A. G. Kislyakov, V. A. Kulikov, L. V. Matveets,
V. I. Chernyshev.
Pis'ma Astron. Zh., Tom 8, 253 - 256 (1982). In Russian.
English translation in Soviet Astron. Lett., Vol. 8.
A millimeter wavelength ($\lambda = 3$ - 4 mm) wideband radio-
meter with a Josephson superconducting point contact as
detector has been constructed for observations by the radio-
telescope RT-25 × 2. Test observations of the Sun, Moon and
Jupiter have been caried out.

033.008 **Exact gain measurement of large aperture antennas
using celestial radio sources.** T. Satoh, A. Ogawa.
IEEE Trans. Antennas Propag., Vol. AP-30, 157 - 161 (1982).
Abstr. in Phys. Abstr., Vol. 85, Abstr. 38057 (1982).

033.009 **A 0.4K bolometer receiver for millimeter astronomy.**
P. Ade, J. Davis, R. Howard, I. Nolt, J. Payne,
S. Predko, J. Radostitz.
Infrared and millimeter waves, (see 012.017), p. W-3-5/1-2
(1981). – Abstr. in Phys. Abstr., Vol. 85, Abstr. 42598
(1982).

033.010 **A portable, all-solid-state low noise receiver for
230 GHz.** J. W. Archer.
Infrared and millimeter waves, (see 012.017), p. W-3-6/1-2
(1981). – Abstr. in Phys. Abstr., Vol. 85, Abstr. 42599
(1982).

033.011 **A cryogenic receiver for 1 mm wavelength.**
N. R. Erickson.
Infrared and millimeter waves, (see 012.017), p. W-3-7/1-2
(1981). – Abstr. in Phys. Abstr., Vol. 85, Abstr. 42600
(1982).

033.012 **An ultra low-noise Schottky mixer receiver at
80-120 GHz.** A. V. Raisanen, N. R. Erickson,
J. L. R. Marrero, P. F. Goldsmith, C. R. Predmore.
Infrared and millimeter waves, (see 012.017), p. W-3-8/1-2
(1981). – Abstr. in Phys. Abstr., Vol. 85, Abstr. 42601
(1982).

033.013 **The transmission spectra of some potential radome
materials for near millimetre wavelength applications.** J. R. Birch, J. D. Dromey, R. L. T. Street.
Infrared and millimeter waves, (see 012.017), p. Th-3-3/1
(1981). – Abstr. in Phys. Abstr., Vol. 85, Abstr. 42603
(1982).

033.014 **Dynamics of development of radio astronomy.
Part 2. Making radio telescopes.** Yu. P. Ilyasov.
Fiz. inst. AN SSSR. Prepr., 1981, No. 165, 45 pp. In Russian.
Abstr. in Ref. zh., 51. Astron., 3.51.969 (1982).

033.015 **Multifrequency cryogenically cooled front-end
receivers for the Westerbork Synthesis Radio
Telescope.** J. L. Casse, E. E. M. Woestenburg, J. J. Visser.
IEEE Trans. Microwave Theory Tech., Vol. MTT-30, 201 - 209
(1982). – Abstr. in Phys. Abstr., Vol. 85, Abstr. 48147
(1982).

033.016 **Precision antenna of a space radio telescope.**
V. I. Kostenko, L. I. Matveenko.
Kosm. Issled., Tom 20, 149 - 151 (1982). In Russian.

033.017 **El radiotelescopio del Instituto Argentino de
Radioastronomía.** E. M. Filloy.
Bol. Asoc. Argentina Astron., No. 20 - 24, p. 115 - 125 (1981).

033.018 **El instrumento de síntesis de Westerbork y las
observaciones de galaxias en la línea de 21 cm del
hidrógeno neutro.** E. Bajaja.
Bol. Asoc. Argentina Astron., No. 20 - 24, p. 176 (1981).
Abstract.

033.019 **Polarímetro en 408 MHz del Observatorio de
La Plata.** R. J. Marabini.
Bol. Asoc. Argentina Astron., No. 20 - 24, p. 184 - 186 (1981).

033.020 **Condiciones del sitio y el desempeño de la antena
de 13.7 m de Itapetinga en f = 43 GHz (7 mm).**
P. Kaufmann, R. E. Schaal, J. C. Raffaelli.
Bol. Asoc. Argentina Astron., No. 20 - 24, p. 439 (1981).
Abstract.

033.021 **High dynamic range in radio interferometry.**
J. E. Noordam.
Scientific importance of high angular resolution at infrared
and optical wavelengths (see 012.031), p. 257 - 261 (1981).

033.022 **A 150 MHz phase-switching interferometer for
instructional use.** R. D. Dietz, A. Loomis.
Bull. American Astron. Soc., Vol. 13, 839 (1981). – Abstract.

033.023 Two-dimensional radioheliographic pictures of the sun's outer corona at 25.6 - 110.6 MHz.
M. R. Kundu, W. C. Erickson, P. J. Turner.
Bull. American Astron. Soc., Vol. 13, 891 (1981). – Abstract.

033.024 Frequency response of a synthesis array: performance limitations and design tolerances,
A. R. Thompson, L. R. D'Addario.
Bull. American Astron. Soc., Vol. 13, 900 (1981). – Abstract.

033.025 Automatic control system for the west sector of the circular mirror of the RATAN-600 radio telescope.   A. N. Angel'skij, G. S. Golubchin, Yu. K. Postoenko, V. D. Barmasov, A. M. Bechasnov, G. V. Zhekanis.
Astrofiz. Issled.,Izv. Spets. Astrofiz. Obs., Tom 15, 88 - 105 (1982). In Russian.

033.026 Investigation of the scattering background of RATAN-600 with the help of observations of the sun.   I. V. Ignat'eva, D. V. Korol'kov, O. I. Krat.
Astrofiz. Issled., Izv. Spets. Astrofiz. Obs., Tom 15, 110 - 116 (1982). In Russian.
   The scattering pattern of the north sector of RATAN-600, connected with systematic errors, has been measured with the help of observations of the sun. All measurements have been made at λ = 2.08, 3.9, 6.52 (8.2)cm.

033.027 Calculation of the power characteristic of the focusing system of the RATAN-600 radio telescope.
E. K. Majorova, A. A. Stotskij.
Astrofiz. Issled., Izv. Spets. Astrofiz. Obs., Tom 15, 117 - 125 (1982). In Russian.

033.028 Influence of the character of field distribution on the aperture on the parameters of a variable antenna periscope system.
E. K. Majorova, A. A. Stotskij.
Astrofiz. Issled., Izv. Spets. Astrofiz. Obs., Tom 15, 126 - 131 (1982). In Russian.
   Calculations of the characteristics of a variable profile antenna periscope system in dependence on different kind of the field distribution on the aperture of feed are given. The influence of an amplitude distribution shape with the constant level on the edge of the aperture and the influence of the square phase distribution on the aperture are considered.

033.029 Study of the accuracy of the reflecting surface of the RATAN-600 main mirror (the north sector).
S. Ya. Golosova, N. A. Esepkina, Yu. K. Zverev, G. N. Kalikhevich, D. V. Korol'kov, O. I. Krat, M. N. Naugol'naya, Yu. N. Parijskij, G. A. Pinchuk, N. S. Soboleva, A. A. Stotskij, O. N. Shivris.
Astrofiz. Issled., Izv. Spets. Astrofiz. Obs., Tom 15, 132 - 150 (1982). In Russian.

033.030 Polarization characteristics of the RATAN-600 radio telescope at an altitude of θ = 0° with aberration taken into account.   N. A. Esepkina, N. S. Bakhvalov, B. A. Vasil'ev, L. G. Vasil'eva, A. V. Temirova.
Astrofiz. Issled., Izv. Spets. Astrofiz. Obs., Tom 15, 151 - 160 (1982). In Russian.
   The polarization characteristics of the RATAN-600 radio telescope at an altitude θ_0 = 0°(south sector with a flat mirror) are determined when the primary feed is shifted from the focus. Formulas for antenna patterns are presented. It is shown that the south sector aberrations are considerably greater than the north sector aberrations at high altitudes.

033.031 Identification of the surface of the mirror antenna of a radio telescope.
T. N. Vasil'chenko, E. S. Kurisova.

Sistemy upr. i ikh ehlementy. Leningrad, 1981, p. 81 - 85. In Russian. – Abstr. in Ref. zh., 51. Astron., 5.51.959 (1982).

033.032 Estimate of the distortion of the geometry of the mirror antenna of a radio telescope.
V. V. Lavrent'ev.
Sistemy upr. i ikh ehlementy. Leningrad, 1981, p. 128 - 133. In Russian. – Abstr. in Ref. zh., 51. Astron., 5.51.960 (1982).

033.033 Internal twist and least-squares adjustment of four-cornered surface plates for reflector antennas.
S. von Hoerner.
IEEE Trans. Antennas Propag.,Vol. AP-29, 953 - 958 (1981) = Natl. Radio Astron. Obs., Green Bank, Repr. Ser. A, No. 1347.
   Surface plates with four adjustment screws, one at each corner, allow four degrees of freedom for their adjustment, including an internal nonplanar twist or warp. Thus the surface deformations of trapezoidal plates under enforced twists are investigated. A simple equation is derived, holding as a satisfactory approximation for various plate designs (solid plate , honeycomb, and skin-and-ribs). Measurements at four different experimental plates agree with it within 2 percent. This approximation is then used to develop a least-squares procedure for obtaining those adjustment amounts which minimize the root-mean-square (rms) deviation between the plate's surface and the desired telescope paraboloid. The resulting procedure is recommended for all telescopes with trapezoidal surface plates.

033.034 Low-noise 10.7 GHz cooled GaAs FET amplifier.   G. Tomassetti, S. Weinreb, K. Wellington.
Electron Lett., Vol. 17, 949 - 951 (1981) = Natl. Radio Astron. Obs., Green Bank, Repr. Ser. A 1370.

033.035 Frequency response of a synthesis array: performance limitations and design tolerances.
A. R. Thompson, L. R. D'Addario.
Radio Sci., Vol. 17, 357 - 369 (1982) = Natl. Radio Astron. Obs., Green Bank, Repr. Ser. A 1372.

033.036 The use of the large mm-wave antenna at Itapetinga in high-sensitivity solar research.
P. Kaufmann, F. M. Strauss, R. E. Schaal, C. Laporte.
Sol. Phys., Vol. 78, 389 - 399 (1982).
   Large dishes used in solar radio astronomy are becoming an essential tool for the analysis of low level activity and fine time structures in solar bursts. Some front-end and back-end arrangements have been added to the Itapetinga 13.7-m radome-enclosed antenna to allow for simultaneous 22 GHz and 44 GHz observations; 22 GHz right- and left-handed circular polarization (or two linear orthogonal), with sensitivities of the order of 0.03 s.f.u., and time resolution of 1 ms. This system is being used in a number of specific investigations, in SMM satellite related research, and in other internationally coordinated works.

033.037 Radio astronomy by very-long-baseline interferometry.   A. C. S. Readhead.
Sci. American, Vol. 246, No. 6, p. 38 - 47 (1982).
   Observations made simultaneously by radio telescopes thousands of miles apart can be combined with the aid of atomic clocks. The resolution of the observations is the highest ever achieved.

033.038 ESA's new standard 15 m S/X-band antenna – first installation at Villafranca.   P. Maldari.
ESA Bull., No. 30, p. 21 - 25 (1982).

033.039 Tidbinbilla two-element interferometer.
M. J. Batty, D. L. Jauncey, P. T. Rayner, S. Gulkis.
Astron. J., Vol. 87, 938 - 944 (1982).
   A phase-stable two-element interferometer has been

formed by linking the 64- and 34-m antennas of the Deep Space Network at Tidbinbilla, Australia. Utilizing the existing first-stage maser receivers at 2.3 GHz to yield a system temperature of ~20 K, the system has a 5σ detection sensitivity of 50 mJy in 1 s, with a rms confusion of ~6 mJy. The 195-m north-south baseline permits positional measurements to ~2 arcsec of sources stronger than ~100 mJy over the declination range $-80° \lesssim \delta \lesssim +30°$.

033.040   A proposed telescope for millimeter and submillimeter astronomy.   B. Rönnäng.
Rep. Obs. Lund, No. 18, (see 012.044), p. 17 - 19 (1982).

Classics in radio astronomy.
See Abstr. 003.125.

A phase-coherent link between VLBI stations via synchronous satellite.   See Abstr. 013.041.

Excising terrestrial radio interference in low frequency radio astronomy.   See Abstr. 031.505.

The accuracy of very long baseline interferometry observables.   See Abstr. 031.526.

Suppression of atmospheric radio emission fluctuations in radio astronomical observations.
See Abstr. 031.605.

Key development items in the realisation of a 300-500 GHz heterodyne spectrometer.   See Abstr. 034.020.

Die Arbeiten des Sonderforschungsbereiches 78 Satellitengeodäsie der Technischen Universität München im Jahre 1980.   See Abstr. 046.006.

Report of IAU Commission 40: Radio astronomy (Radio astronomie).   See Abstr. 141.322.

034 Auxiliary Instrumentation

034.001 **The Vienna Observatory echelle spectrograph.**
W. W. Weiss, M. Barylak, J. Hron, J. Schmiedmayer.
Publ. Astron. Soc. Pacific, Vol. 93, 787 - 794 (1981/82).
Design considerations for the Vienna Observatory echelle spectrograph are presented as well as a report on the optical and mechanical performance of this instrument.

034.002 **The 3-channel high-speed photometer of Beijing Observatory.**   D.-l. Wang, D.-s. Zhai, J.-m. Yu, J.-h. Sun, X.-y. Tang, F.-y. Zhao, J.-y. Hu.
Acta Astrophys. Sinica, Vol. 2, 56 - 62 (1982). In Chinese.

034.003 **An instrumental effect on spectral line profiles.**
R. C. Thompson.
J. Quant. Spectrosc. Radiat. Transfer, Vol. 27, 417 - 421 (1982).
The usual treatment of a scanning Fabry-Perot etalon set-up ignores any effect of the spectrograph on the spectral line shape. The author treats one such effect, namely an apparent baseline shift which arises because some light in the far wings of the line misses the spectrograph exit aperture. The author discusses the significance of this result and shows how conditions can be chosen such as to minimize this effect.

034.004 **An eyepiece off-set guiding device.**
R. C. Brooks.
J. British Astron. Assoc., Vol. 92, 73 (1982).

034.005 **Direct imaging and photometry with an intensified silicon vidicon detector.**   S. Jeffers.
J. R. Astron. Soc. Canada, Vol. 76, 19 - 25 (1982).
Direct imagery has been obtained by means of an intensified silicon vidicon at the Cassegrain focus of a 60-cm telescope. These data are presented and discussed together with an assessment of the photometric capabilities of the detector. The detective quantum efficiency of the detector is estimated to be 5 per cent.

034.006 **Low-light-level photometry at the Royal Greenwich Observatory.**   J. V. Jelley.
Observatory, Vol. 102, 30 - 36 (1982).

034.007 **Silicon "eyes" for Mt. John.**
D. A. Hall.
South. Stars, Vol. 29, 67 - 77 (1981).

034.008 **Field limiter for solar radiometer.**
C. Martin.
NASA Tech Briefs, Vol. 5, 429 (1981). – Abstr. in Phys. Abstr., Vol. 85, Abstr. 10320 (1982).

034.009 **A photometer for infrared astronomy.**
M. Roth, L. Carrasco, J. Franco, G. Resendiz.
Rev. Mexicana Fis., Vol. 27, No. 1, p. 39 - 54 (1980). In Spanish. – Abstr. in Phys. Abstr., Vol. 85, Abstr. 10324 (1982).

034.010 **Spectral, polarimetric and modulation instruments for the longwave infrared range.**
G. B. Sholomitskij, I. A. Maslov, S. A. Ignatenko, S. G. Namestnik, V. A. Soglasnova, V. D. Gromov.
Nauchn. kosm. priborostr. Moskva, 1981, p. 32 - 40. In Russian. – Abstr. in Ref. zh., 62. Issled. kosm. prostranstva, 2.62.63 (1982).

034.011 **Light source for absolute calibration of photometers of emission of the upper atmosphere.**
A. V. Rabinkov.

Nauchn. kosm. priborostr. Moskva, 1981, p. 40 - 45. In Russian. – Abstr. in Ref. zh., 62. Issled. kosm. prostranstva, 2.62.67 (1982).

034.012 **A two dimensional solar spectrometer.**
J. R. Brookes, G. R. Isaak, H. B. van der Raay.
J. Phys. E, Vol. 14, 1288 - 1290 (1981). – Abstr. in Phys. Abstr., Vol. 85, Abstr. 14770 (1982).

034.013 **Infrared instrumentation at ESO.**
A. F. M. Moorwood.
Messenger, No. 27, p. 11 - 14 (1982).

034.014 **Matching of transducers to resonant gravitational-wave antennas.**   G. V. Pallottino.
Nuovo Cimento C, Vol. 4C, Ser. 1, 237 - 283 (1981). – Abstr. in Phys. Abstr., Vol. 85, Abstr. 18709 (1982).

034.015 **Development of a linear diode array detector system for astronomical spectroscopy.**   A. W. Campbell, Thesis, Univ. Durham, England (1980). – Abstr. in Phys. Abstr., Vol. 85, Abstr. 18710 (1982).

034.016 **Das OPTRONICS S-3000 Microdensitometer in Freiburg (KIS).**
U. Großmann-Doerth, M. Knölker.
Mitt. Astron. Ges., Nr. 55, p. 168 - 169 (1982).

034.017 **Eine Apparatur zur Messung von Lichtblitzen im Nanosekundenbereich.**
G. Felkel, G. Schlemmer, D. Kuhn, J. Pfleiderer.
Mitt. Astron. Ges., Nr. 55, p. 179 (1982). – Abstract.

034.018 **A multi-purpose scanning Fabry-Pérot interferometer system.**
J. R. Roy, R. Arsenault, G. Joncas.
Regions of recent star formation, (see 012.002), p. 67 - 72 (1982).
The authors have built a self contained two-mode Fabry-Pérot interferometer system to obtain (1) radial velocity field maps and (2) photoelectric line profiles of extended emission line sources. The servo-controlled Fabry-Pérot is piezo-scanned by using capacitance micrometry to detect deviations from parallelism or changes in the absolute spacing. The imaging mode uses a 183mm focal length f/8 collimator which images the pupil on the Fabry-Pérot; the field with the characteristic fringes is reimaged by a 25 mm focal length f/0.95 refocussing lens. The photoelectric configuration uses two 200 mm f/3.5 lenses as collimator/camera optics.

034.019 **Submillimeter heterodyne astronomy from Mauna Kea.**   D. Buhl, G. Chin, H. R. Fetterman, D. D. Peck, B. J. Clifton, P. E. Tannenwald, G. A. Koepf.
Infrared and millimeter waves, (see 012.017), p. M-1-4/1-2 (1981). – Abstr. in Phys. Abstr., Vol. 85, Abstr. 42596 (1982).

034.020 **Key development items in the realisation of a 300-500 GHz heterodyne spectrometer.**
P. F. Clancy.
Infrared and millimeter waves, (see 012.017), p. W-3-2/1-2 (1981). – Abstr. in Phys. Abstr., Vol. 85, Abstr. 42597 (1982).

034.021 **A photopolarimeter for far infrared measurements.**
A. Blanco, F. D'Alessandro, S. Fonti, P. DeBernardis, S. Masi, F. Melchiorri, G. Dall'Oglio.
Infrared and millimeter waves, (see 012.017), p. W-3-11/1-2

(1981). – Abstr. in Phys. Abstr., Vol. 85, Abstr. 42602 (1982).

034.022 Optical devices for tuning of astronomical television apparatuses. G. M. Verzhbitskaya, T. V. Vik.
Opt.-mekh. prom-st', 1981, No. 10, p. 18 - 20. In Russian.
Abstr. in Ref. zh., 51. Astron., 3.51.965 (1982).

034.023 An automatic multi-channel device for measurement of radiocarbon concentration.
V. A. Vasil'ev.
Fiz.-tekh. inst. AN SSSR. Prepr., 1981, No. 726, 59 pp. In Russian. – Abstr. in Ref. zh., 51. Astron., 3.51.967 (1982).

034.024 Exact time registration system for observations of very high-energy γ-quanta.
V. G. Shitov, A. A. Stepanyan.
Izv. Krymskoj Astrofiz. Obs., Tom 64, 162 - 171 (1981). In Russian. English translation in Bull. Crimean Astrophys. Obs., Vol. 64.
A description of an exact time registration system is presented, which allows to register the moments of Čerenkov flashes occurrence with time resolution to 10^{-6} s and having high stability.

034.025 The choice of astatic systems to control the rotational velocity of a direct-current motor for altazimuth mountings of a γ-telescope. V. G. Shitov.
Izv. Krymskoj Astrofiz. Obs., Tom 64, 171 - 180 (1981). In Russian. English translation in Bull. Crimean Astrophys. Obs., Vol. 64.

034.026 Widening of the dynamical range of a data measuring system of a five-channel spectrophotometer.
L. V. Granitskij, A. B. Bukach, N. I. Bukach.
Izv. Krymskoj Astrofiz. Obs., Tom 64, 180 - 188 (1981). In Russian. English translation in Bull. Crimean Astrophys. Obs., Vol. 64.
A method to widen the dynamical range of a five-channel spectrograph with the help of registering devices located in the entrance has been described.

034.027 Automatic photometer for star observations in the UBVR system. N. I. Dorokhov, V. V. Egorov.
Astrometr. Astrofiz., Vyp. (No.) 43, p. 95 - 99 (1981). In Russian.
An automatic electrophotometer for photoelectric observations of stars in the UBVR system is described. According to the prescribed program the photometer makes measurements of light flux, records the results of measurements and data necessary for reduction with the help of a digitizer.

034.028 Use of an intensifier for laser range observations of artificial satellites.
M. K. Abele, V. I. Troyan, M. Ya. Tavadrus.
Astrometr. Astrofiz., Vyp. (No.) 43, p. 99 - 100 (1981). In Russian.
The INTERCOSMOS laser ranging system at the Helwan Astronomical Observatory was equipped with an image intensifier. This system enables the satellite STARLETTE of magnitude 11-12^m to be observed. During a month 72 laser range measurements were made.

034.029 Automatic digital two-coordinate microphotometer for input of photographic images into a computer.
V. G. Parusimov.
Astrometr. Astrofiz., Vyp. (No.) 45, p. 86 - 99 (1981). In Russian.
The paper deals with the automatic digital two-coordinate microphotometer developed at the Main Astronomical Observatory of the Ukrainian SSR Academy of Sciences. The device is used with a computer for astronomical investigations. Brief specifications of the device and some results of photometry and digital treatment of images are given.

034.030 Choice of the optical system for a microscope-micrometer. E. G. Zhilinskij, V. E. Pliss.
Astrometr. Astrofiz., Vyp. (No.) 46, p. 99 - 103 (1982). In Russian.

034.031 Laser mark.
B. A. Golovko, V. G. Tauber, V. G. Shamaev.
Astrometr. Astrofiz., Vyp. (No.) 46, p. 103 - 105 (1982). In Russian.
The paper is concerned with a design of a device to control the azimuth of a meridian circle's horizontal axis. The device consists of a laser with a focusing part and a corner reflector (mark). The tests proved a good efficiency and reliability of the device.

034.032 A cautionary note on the use of Pickering-Racine prisms. V. M. Blanco.
Publ. Astron. Soc. Pacific, Vol. 94, 201 - 203 (1982).
The value of Δm, the magnitude difference between primary and secondary images in photographic plates exposed with Pickering-Racine prisms, is found to depend, at least in some observations, on the magnitude of the primary image if the plates are measured with an iris-diaphragm astrophotometer. The effect may invalidate the extrapolation of photometric sequences when such measurements are used.

034.033 A cooled grating spectrometer using cylindrical optics. T. J. Jones, A. R. Hyland, M. A. Dopita, J. Hart, P. Conroy, J. Hillier.
Publ. Astron. Soc. Pacific, Vol. 94, 207 - 212 (1982).
A grating spectrometer for use from $1-5$ μm with all optics including two gratings cooled to cryogenic temperatures is described. In order to make the optical system as compact as possible, the novel design uses cylindrical teleconverter optics. This enables a resolving power of $R \sim 3 \times 10^3$ to be achieved at 2.3 μm with a 2.7 arc sec entrance aperture on the Mount Stromlo 1.9-meter telescope. Also described is a second-generation spectrometer, employing a cooled solid Fabry-Perot etalon for a higher resolution capability, that is currently under development.

034.034 Puesta en marcha de tubos de imágenes para uso astronómico. J. H. Calderón
Bol. Asoc. Argentina Astron., No. 20 - 24, p. 187 (1981). Abstract.

034.035 El polarímetro digital rotatorio del OALP.
R. Marabini, H. G. Marraco, J. C. Forte.
Bol. Asoc. Argentina Astron., No. 20 - 24, p. 189 (1981). Abstract.

034.036 Las primeras mediciones de polarización en el OALP.
H. G. Marraco.
Bol. Asoc. Argentina Astron., No. 20 - 24, p. 190 (1981). Abstract.

034.037 El fotómetro de alta resolución temporal del IAFE.
R. J. Terlevich, C. de Franceschini.
Bol. Asoc. Argentina Astron., No. 20 - 24, p. 191 (1981). Abstract.

034.038 Prisma "no objetivo".
R. Platzeck, J. C. Muzzio, H. G. Marraco.
Bol. Asoc. Argentina Astron., No. 20 - 24, p. 192 (1981). Abstract.

034.039 Sistema lógico del fotómetro de alta resolución temporal.

I. Czudnowski, A. M. Godel, J. C. Barberis.
Bol. Asoc. Argentina Astron., No. 20 - 24, p. 301 (1981).
Abstract.

034.040 El fotómetro digital multicanal de alta resolución temporal del IAFE.
R. J. Terlevich, C. De Franceschini, I. Czudnowski,
J. C. Barberis, A. M. Godel, C. A. Falcón, H. Russo.
Bol. Asoc. Argentina Astron., No. 20 - 24, p. 302 (1981).
Abstract.

034.041 Polarímetro digital rotatorio del Observatorio de La Plata.   R. J. Marabini.
Bol. Asoc. Argentina Astron., No. 20 - 24, p. 373 (1981).
Abstract.

034.042 On sensitivity improvement of modern astronomical TV systems.   Yu. V. Kuberskij, V. A. Malarev.
Soobshch. Byurakan Obs., Vyp. 53, p. 135 - 140 (1982). In Russian.
The problems connected with improvement of a television astronomical system on the basis of modern transmission tubes, image tubes and memory are considered.

034.043 A selective solar irradiance spectrometer.
B. J. Oranje.
Astron. Astrophys., Vol. 109, 32 - 36 (1982).
An optical device is described for the averaging of intensity over solid angle. This integrator serves to obtain irradiance spectra from the entire solar disk, or from smaller areas of arbitrary size and shape. The integrator consists of optical elements inserted between a conventional telescope and spectrometer. The loss in signal in comparison with intensity spectrometry is only about 50%, and there is no loss of spectral resolution. Illustrative results are shown from the Utrecht "the-sun-as-a-star" monitoring program of the disk-averaged Ca II H and K line cores.

034.044 Sistema de adquisición de datos del polarímetro del O. A. L. P.   R. J. Marabini, H. Marraco.
Bol. Asoc. Argentina Astron., No. 26, p. 137 - 138 (1981).

034.045 Detector para espectrógrafo con reticon del O. A. C.   R. J. Marabini.
Bol. Asoc. Argentina Astron., No. 26, p. 139 - 144 (1981).

034.046 Dispositivo foto-electrónico para registros de tránsitos del círculo meridiano.
R. A. Carestia, C. C. Mallamaci.
Bol. Asoc. Argentina Astron., No. 26, p. 145 - 148 (1981).

034.047 Fotómetro piloto para El Leoncito.
R. J. Marabini.
Bol. Asoc. Argentina Astron., No. 26, p. 149 (1981).
Abstract.

034.048 Photométrie stellaire par caméra électronique.
A. Blecha.
Bull. Inf. Cent. Données Stellaires, No. 22, p. 25 - 36 (1982).
A review of currently operating cameras is given. Comparison is made with recent CCD device. The yield of the data acquisition system camera-microdensitometer-image processing software in stellar photometry is more specifically examined.

034.049 An efficient faint object spectrograph.
C. G. Wynne.
Opt. Acta, Vol. 29, 137 - 141 (1982). – Abstr. in Phys. Abstr., Vol. 85, Abstr. 53090 (1982).

034.050 Hydrogen line magnetometer on spectrograph basis.   V. D. Bychkov, N. A. Vikul'ev,

O. Yu. Georgiev, Yu. V. Glagolevskij, I. D. Najdenov,
I. I. Romanyuk, V. G. Shtol'.
Soobshch. Spets. Astrofiz. Obs., Vyp. 32, (see 012.029),
p. 33 - 34 (1981).

034.051 A shearing, modulating interferometer.   E. Ribak.
Scientific importance of high angular resolution at infrared and optical wavelengths, (see 012.031), p. 171 -173 (1981).

034.052 Scanning photometer.
V. Yu. Terebizh.
Astron. Tsirk., No. 1188, p. 3 - 6 (1981). In Russian.

034.053 Speckle spectroscopy using the Multi-Anode Microchannel Array detector system.
H. Butcher, C. L. Joseph, J. G. Timothy.
Bull. American Astron. Soc., Vol. 13, 814 (1981). – Abstract.

034.054 The sensitivity of DUMAND (*Deep Undersea Muon and Neutrino Detection*) to extraterrestrial neutrino sources.   V. J. Stenger, J. G. Learned, V. Z. Peterson,
A. Roberts.
Bull. American Astron. Soc., Vol. 13, 814 (1981). – Abstract.

034.055 Image photon counting system.
L. Mertz, A. Title, T. D. Tarbell.
Bull. American Astron. Soc., Vol. 13, 840 - 841 (1981).
Abstract.

034.056 DUMAND (*Deep Undersea Muon and Neutrino Detection*) – an undersea neutrino telescope.
J. G. Learned, V. Z. Peterson, A. Roberts, V. J. Stenger.
Bull. American Astron. Soc., Vol. 13, 884 (1981). – Abstract.

034.057 The compact magnetograph: preliminary results.
B. A. Gillespie, A. M. Title.
Bull. American Astron. Soc., Vol. 13, 888 - 889 (1981).
Abstract.

034.058 100-ton scintillation counter for registration of antineutrino fluxes from collapsing stars in our Galaxy and for investigation of high-energy muon interactions.
V. I. Beresnev, R. I. Enikeev, G. T. Zatsepin, V. B. Korchagin,
P. V. Korchagin, A. S. Mal'gin, O. G. Ryazhskaya,
V. G. Ryasnyj, V. P. Talochkin, A. A. Chudin, V. F. Yakushev.
Pribory i tekh. ehksp., 1981, No. 6, p. 48 - 51. In Russian.
Abstr. in Ref. zh., 51. Astron., 5.51.977 (1982).

034.059 Ein durchstimmbarer Infrarot-Heterodyn-Empfänger für astronomische Beobachtungen.
D. Zasche.
Max-Planck-Inst. Phys. Astrophys., Inst. Extraterr. Phys., Garching bei München, MPI-PAE/Extraterr. 173, 117 pp. (1982). ISSN 0340-8922.

034.060 Design of a four channel simultaneous visual-infrared photometer.   P. R. M. Eisenhardt.
Prepr. Steward Obs., No. 361, 8 pp. (1982).

034.061 Interference filters with multiple peaks.
E. Pelletier, H. A. Macleod.
J. Opt. Soc. America, Vol. 72, 683 - 687 (1982).
A simple extension of techniques used in designing conventional single-peak narrow-band filters permits the design of filters having two or more peaks. The design procedure and monitoring techniques are discussed. An example of an actual filter produced by using this technique is presented.

034.062 Calgary's rapid alternate detection system.
R. M. Robb, E. F. Milone, F. M. Babott,

C. H. Hansen, D. Swadron.
Bull. American Astron. Soc., Vol. 14, 574 (1982). – Abstract.

034.063 **A second generation laser ranging system at**
McDonald Observatory.   P. J. Shelus.
Bull. American Astron. Soc., Vol. 14, 582 (1982). – Abstract.

034.064 **Low noise imaging photon counter for astronomy.**
L. Mertz, T. D. Tarbell, A. Title.
Appl. Opt., Vol. 21, 628 - 634 (1982).

The characteristics and performance of a Ranicon
photon-counting system combined with digital tape recording
are described. The most important features are a bialkali
photocathode response over 256 × 256 digital pixels, with
~100 × 100 resolvable pixels at 50% MTF, a dead time of
16 μsec/count, a maximum recordable count rate of
14,400/sec, and a background of <1 count/digital pixel/h.
A video cassette recorder serves for the digital recording
which retains the temporal sequence of the registered photons.
Astronomical applications will include low light level
quantitative imaging and speckle imaging.

034.065 **Echelle spectrographs at grazing incidence.**
W. Cash.
Appl. Opt., Vol. 21, 710 - 717 (1982).

It is shown that by using the conical diffraction mount
existing echelle gratings can be used at grazing incidence to
achieve high spectral resolution in the extreme UV and soft
X rays. Design considerations for grazing incidence echelle
spectrographs are examined, and two sample designs are
discussed. The first, for use in the extreme UV has a primary
mirror and an entrance slit to the spectrograph. The system
has resolution of 10^4, operates at any wavelength longward
of 100 Å, and covers 30% of the spectrum at a single setting.
The X-ray spectrograph uses objective gratings to obtain
spectral resolution of 2.8 × 10^4 over any factor of 2 in wavelength. It operates to wavelengths as short as 4 Å.

034.066 **Bolometer noise: nonequilibrium theory.**
J. C. Mather.
Appl. Opt., Vol. 21, 1125 - 1129 (1982).

New theoretical results for noise in cryogenic bolom-
eters are derived. Johnson noise is reduced by as much as 60%
by electrothermal feedback from the bias supply. Phonon
noise in the thermal link is reduced by as much as 30% relative
to the usual equilibrium formula. Photon noise in the
Rayleigh-Jeans limit is computed with attention to the
attenuation of the photon correlations in the light beam.
Basic results on bolometer responsivity, time constant, and
thermal properties are presented in a new and convenient
form. Excess 1/f and contact shot noise are also discussed.

034.067 **Evaluation of an InSb infrared detector at liquid**
N$_2$ and liquid He temperatures.
Y.-X. Zhang, F. O. Williamson.
Appl. Opt., Vol. 21, 2036 - 2040 (1982).

The performance of an InSb infrared detector has been
evaluated at liquid nitrogen and liquid helium cryogenic
temperatures. A significant improvement in sensitivity was
observed when the detection system was Johnson noise
limited. With 300-K background radiation and diffraction
limited throughout, the NEP of the detector is 1.5 × 10^{-16} W
Hz$^{-1/2}$ at 1.65 μm. Potential improvement is expected especially for astrophysical observations which are made in lower
photon background conditions.

034.068 **Mechanical filter for the suspension of gravitational**
wave antennas.   E. Coccia.
Rev. Sci. Instrum., Vol. 53, 148 - 153 (1982). – Abstr. in Phys.
Abstr., Vol. 85, Abstr. 53653 (1982).

034.069 **A study of the cosmic radiation underground using**
a large-volume anticoincidence detector.
R. J. Riley.
J. Phys. G, Vol. 8, 393 - 412 (1982). – Abstr. in Phys. Abstr.,
Vol. 85, Abstr. 58260 (1982).

034.070 **A reliable iron-arc light source.**
W. W. Weiss, A. Schalk.
Publ. Astron. Soc. Pacific, Vol. 94, 379 - 380 (1982).

The authors discuss design considerations and the
performance of an iron-arc comparison light source for spectroscopy.

034.071 **A grating spectrometer and Fabry-Perot interferom-**
eter for use in the 1 μm - 5 μm wavelength region.
S. E. Persson, T. R. Geballe, F. Baas.
Publ. Astron. Soc. Pacific, Vol. 94, 381 - 385 (1982).

An infrared spectrometer for the 1 μm < λ < 5 μm wavelength band is described. The instrument consists of a grating
spectrometer cooled to solid nitrogen temperature, which can
be used with a piezoelectrically scanned Fabry-Perot interferometer. Resolving powers between 10^2 and 10^5 can be achieved. This particular combination of spectrometer components
leads to high sensitivities longward of the 2 μm band, where
Fourier transform spectrometers are more severely limited by
noise from the thermal background.

034.072 **An image stacker for high-resolution spectroscopy**
on the Multiple Mirror Telescope.
F. H. Chaffee, Jr., D. W. Latham.
Publ. Astron. Soc. Pacific, Vol. 94, 386 - 389 (1982).

An image stacker has been in regular use with a conventional high-resolution Cassegrain echelle spectrograph on the
MMT since 1980. This device takes advantage of the unusual
optical configuration of the MMT and improves the slit efficiency by a factor of 2.5. Small lens-prisms are used at a
preslit to redirect the six individual telescope beams into a
single filled beam. A lens then reimages the preslit onto the input slit of the spectrograph and decreases the f/number of the
redirected beam by a factor of nearly 3.5.

034.073 **Spectrograph instrumental profiles: dependence on**
dispersion.   J. Andersen, D. Dravins.
Publ. Astron. Soc. Pacific, Vol. 94, 390 - 394 (1982).

Spectrograph instrumental profiles (including stray light
far away from the central peak) have been measured in blue
and red light for the three cameras in the coudé spectrograph
of the 1.52 m telescope at Observatoire de Haute-Provence.
The different dispersions 0.7, 1.2, and 2.0 nm mm^{-1} are obtained using the same ruled diffraction grating. On a linear
distance scale in the focal plane the profiles are rather similar
down to a 10^{-3} intensity level, but on a wavelength scale the
profiles improve with increasing dispersion, indicating the
presence of a stray light component other than that caused by
diffraction by grating irregularities. The effects of these
instrumental profiles on observed spectra are illustrated by
numerical convolutions with the solar spectrum.

034.074 **A note on the principle and nomenclature of**
heliostats, coelostats, siderostats.
L. M. Dougherty.
J. British Astron. Assoc., Vol. 92, 182 - 187 (1982).

034.075 **An introduction to transistors and diodes.**
H. R. Hatfield.
J. British Astron. Assoc., Vol. 92, 187 - 189 (1982).

034.076 **An electronic programmer for an integrator.**
S. K. Gupta.
J. R. Astron. Soc. Canada, Vol. 76, 181 - 184 (1982).

034.077 **TAURUS – the imaging Fabry-Perot at La Silla.**
P. D. Atherton, I. J. Danziger, R. A. E. Fosbury,
K. Taylor.
Messenger, No. 28, p. 9 - 11 (1982).

034.078 **Advances in photoelectric photometry.**
R. M. Genet.
J. American Assoc. Variable Star Obs., Vol. 10, 49 - 52 (1981).

034.079 **Ein Lichtleiter-gekoppelter Spektrograph für kleine
Teleskope.**   J. G. V. Schiffer.
Diplomarb., Landessternwarte Heidelberg, F. R. Germany.
78 pp. (1982).
    A fiber linked portable spectrograph for medium disper-
sion work is described. The optical system consists of two
single fibers each for star and background light, a f/6-colli-
mator, an image intensifier and a TV-guiding-system. The
resolution is 1 Å at a dispersion of 20 Å mm^{-1}.

034.080 **Ein Graphit-Hohlraumstrahler hoher Temperatur zur
Kalibration des Spacelab-Experimentes 1 ES 016.**
U. Reichert.
Diplomarb., Landessternwarte Heidelberg, F. R. Germany.
98 pp. (1982).
    A blackbody radiation source, which allows absolute
calibration of laboratory and satellite spectrometers with high
accuracy, is described. The cavity radiator has a large radiation outlet opening (50 mm^2) and can be operated at
temperatures of up to 3200 K.

034.081 **Astronomical applications of large, aberration-
limited reflectors on earth and in space.**
R. Benson, J. Linsley.
Proc. Southwest Reg. Conf., Vol. 7, (see 012.019), 161 - 168
(1982).

034.082 **The atmospheric Cherenkov technique in γ-ray
astronomy: the early days.**   J. V. Jelley.
Philos.Trans. R. Soc. London, Ser. A, Vol. 301, (see 012.043),
611 - 614 (1981).

034.083 **Gamma-rays above 100 GeV.**
K. E. Turver, T. C. Weekes.
Philos. Trans. R. Soc. London, Ser. A, Vol. 301, (see 012.043),
615 - 628 (1981).
    The Cherenkov light technique for the ground-based

detection of ultra-high energy γ-rays is described and some of
the most significant measurements are reported. Improvements in experiments leading to increases in sensitivity are
outlined and the aims of future work are discussed.

034.084 **Construction and primary observation of a
1 - 3 micron infrared photometer.**
P.-s. Chen, H. Gao, Y.-x. Hao, Q.-r. Chu, K.-p. Zhou.
Acta Astron. Sinica, Vol. 23, 89 - 94 (1982). In Chinese.

034.085 **QVANTOS – optical observations with nanosecond
resolution.**   D. Dravins.
Rep. Obs. Lund, No. 18, (see 012.044), p. 111 - 113 (1982).

    How to measure the Sun like a star.
See Abstr. 031.514.

    A direct UBV color measurement of the Sun.
See Abstr. 031.515.

    Bildverarbeitung astronomischer Aufnahmen.
See Abstr. 031.534.

    **Differential imaging using charge-coupled device
(CCD) imagers with on-chip charge storage.**
See Abstr. 031.603.

    **Report of IAU Commission 9: Instruments and
techniques (Instruments et techniques).**
See Abstr. 032.048.

    **Status of the Stanford gravitational wave experi-
ment.**   See Abstr. 066.067.

    **The atmospheric Cherenkov technique in searches
for exploding primordial black holes.**   See Abstr. 066.145.

    **Report of IAU Commission 25: Stellar photometry
and polarimetry (Photométrie et polarimétrie stellaires).**
See Abstr. 113.097.

    **Observations of the $J = 2 \rightarrow 1$ CO line in molecular
clouds near compact H^+ regions.**   See Abstr. 131.026.

    **Gamma-ray line investigations with the Durham
γ-ray spectrometer.**   See Abstr. 142.545.

035 Clocks and Frequency Standards

035.001 **A bit of porcelain.**   R. N. Mayall.
Sky Telesc., Vol. 63, 16 - 17 (1982).

035.002 **Atomic and gravitational clocks.**
V. M. Canuto, I. Goldman.
Nature, Vol. 296, 709 - 713 (1982).
Atomic and gravitational clocks are governed by the laws of electrodynamics and gravity respectively. While the strong equivalence principle (SEP) assumes that the two clocks have been synchronous at all times, recent planetary data seem to suggest a possible violation of the SEP. The authors' past analysis of the implications of an SEP violation on different physical phenomena revealed no disagreement. The concept of scale invariance, and the physical meaning of different systems of units, are now reviewed and the construction of two clocks that do not remain synchronous — whose rates are related by a non-constant function β_a — is demonstrated. The cosmological character of β_a is also discussed.

035.003 **NBS develops new clock synchronization technique.**
D. Allan.
Dimensions NBS, Vol. 65, No. 2, p. 21 - 22 (1981). — Abstr. in Phys. Abstr., Vol. 85, Abstr. 18679 (1982).

035.004 **Eine preisgünstige Quarz-Sternzeituhr für den Amateur.**   B. Egdorf, R. Streck.
Sterne Weltraum, 21. Jahrg., 84 - 85 (1982).

035.005 **Zeitskalenprobleme; jahreszeitliche Gangschwankungen von Atomuhren.**   G. Becker.
PTB Mitt., 92. Jahrg., 105 - 113 (1982).

035.006 **Analysis of the systematic errors of synthetic time signal corrections in China and corrections of BIH.**
F. M. Guo.
Special issue on astronomical geodynamics, (see 012.028), p. 56 - 61 (1981). In Chinese.
In this paper, the differences between the synthetic time signal corrections of China and those of BIH as well as the relations between them and the effect of solid earth tides are analysed. The author suggests that in the corrections of the synthetic time signal of China the effect of solid earth tides should be taken into account.

035.007 **Problems of time standards and time scales — PTB contribution to the CCDS 1980.**   G. Becker.
PTB Mitt., 92. Jahrg., 194 - 201 (1982).
This paper gives a review of the state of the art concerning the development of time standards and the establishment if time scales at the PTB. It then proposes and justifies the basing of the International Atomic Time scale (TAI) essentially on "primary clocks" as developed and operated by the NRC (Canada) and the PTB.

035.008 **Sincronizzazione mediante satelliti.**
L. Mureddu.
G. Astron., Vol. 7, 297 - 310 (1981).

De geschiedenis van het astronomisch kunstuurwerk.
See Abstr. 003.085.

Sonnenuhren.   See Abstr. 003.111.

Report of IAU Commission 31: Time (L'heure).
See Abstr. 044.064.

036 Photographic Materials and Techniques

036.001 **Neue Filme für die Astrofotografie.**
W. Maeder.
Orion, 40. Jahrg., 60 - 61 (1982).

036.002 **Beeldbewerking van astrofoto's.**
W. J. P. van Enckevort.
Zenit, 9. Jaarg., 178 - 180 (1982).

036.003 **Stereoskopische Himmelsaufnahmen.**
R. Mandler.
Sterne Weltraum, 21. Jahrg., 116 - 117 (1982).

036.004 **Maanfotografie voor beginners.** R. Wielinga.
Zenit, 9. Jaarg., 205 - 209 (1982).

036.005 **Use of nonlinear properties of photographic
materials for measurement of the contrast coefficient.** S. M. Gorskij, A. L. Matveev, A. A. Stromkov,
V. P. Tomarov, V. K. Fedorova.
Zh. nauchn. i prikl. fotogr. i kinematogr., Tom 26, 417 - 421
(1981). In Russian. – Abstr. in Ref. zh., 51. Astron., 3.51.960
(1982).

036.006 **Transformation of the characteristic curve of
photographic material taking into account background noise.** A. Eh. Rozenbush.
Astrometr. Astrofiz., Vyp. (No.) 45, p. 99 - 104 (1981). In
Russian.
It is proposed to take into account the photographic
plate background by means of transformation of the characteristic curve of photographic material. The transformation
needs a small correction for which the dependences on the
image density above the background, on the background
density as well as on the coefficient of emulsion contrast are
given. The question of increasing the sensibility by preliminary illumination of the emulsion is discussed.

036.007 **Detection of small features of photographic images
using a process of special development.**
O. M. Mikhajlova.
Astrometr. Astrofiz., Vyp. (No.) 45, p. 104 - 109 (1981). In
Russian.
The method of special development allowing detection
of a fine structure of images on the plates is applied for
improving the informativity of astrophotographs.

036.008 **Spectral sensitivity and resolving power of the new
infrared film I-1060V.**
I. I. Brejdo, B. I. Shapiro.
Astron. Tsirk., No. 1193, p. 6 - 8 (1981). In Russian.

036.009 **Bearbeitung von Astrofotografien mit dem Multi-
spektralprojektor MSP-4.**
R. Schielicke, V. Kroitzsch.
Bild und Ton, 34. Jahrg., Nr. 2, p. 48 - 49 (1981) = Mitt.
Univ.-Sternw. Jena, Nr. 145.

036.010 **Photographic image manipulation.** C. Madsen.
Messenger, No. 28, p. 19 - 21 (1982).

036.011 **Amateurfotografie van de zon.** L. Aerts.
Zenit, 9. Jaarg., 268 - 271 (1982).

036.012 **Off-axis volgen.** J. Gijsbers.
Zenit, 9. Jaarg., 327 - 329 (1982).

036.013 **Astronomical photography: its present status.**
J.-L. Heudier.
AAS Photo-Bull., No. 26, p. 3 - 8 (1981).

036.014 **Kiso image detection system for photographic plates.**
H. Maehara.
AAS Photo-Bull., No. 26, p. 14 - 16 (1981).

036.015 **Photographic enhancement of direct astronomical
images.** D. F. Malin.
AAS Photo-Bull., No. 27, p. 4 - 9 (1981).

036.016 **Photographic image enhancement of the λ Orionis
nebula.** W. V. Garner, P. G. Johnson.
AAS Photo-Bull., No. 27, p. 10 - 13 (1981).

036.017 **A reducing camera for a tube sensitometer.**
A. G. Smith, H. W. Schrader.
AAS Photo-Bull., No. 27, p. 15 - 16 (1981).

036.018 **Au-delà du rouge.** W. Maeder.
Orion, 40. Jahrg., 103 - 104 (1982).

036.019 **Photographie von Sonnenfinsternissen.**
F. Dorst.
Sterne Weltraum, 21. Jahrg., 261 - 262, 264 (1982).

Fotografia astronomica. See Abstr. 003.043.

The challenge of deep sky photography.
See Abstr. 032.005.

Positional Astronomy, Celestial Mechanics

041 Astrometry

041.001 **Analysis of lunar occultations – IV. Rotation of the FK4 reference frame.** L. V. Morrison.
Mon. Not. R. Astron. Soc., Vol. 198, 1119 - 1125 (1982).

Rotations of the FK4 reference frame in right ascension and about an axis through the equinoxes are determined from an analysis of timings of lunar occultations of FK4 stars made in the years 1800 - 1976. These results are found to be compatible with others deduced from the dynamics of the solar system and stellar kinematics. One implication of the results is that all values of Oort's constant B of galactic rotation, that are derived from the FK4 system of proper motions, should be increased algebraically by $-8 \, \text{km} \, \text{s}^{-1} \, \text{kpc}^{-1}$.

041.002 **Determination of the equinox and equator of the FK5.** W. Fricke.
Astron. Astrophys., Vol. 107, L13 - L16 (1982).

The determination of the FK5 equinox has been based on observations of the Sun and planets at mean epochs from 1900 to 1977 and on lunar occultations. Presented are the basic data and the solution for the equinox correction E(T). The values of E and its secular variation Ė indicate the correction + $0^s.035$ at 1950.0 to all right ascensions of the FK4 and the correction + $0^s.085$ per century to all proper motions in right ascension of the FK4. For the producers of the new ephemerides it is important to notice that these corrections will be applied in the construction of the system of the Fifth Fundamental Catalogue, the FK5. Recent observations of relevance for the improvement of the FK4 equator have not indicated the need for a significant correction. Therefore the equator of the FK4 will be maintained in the FK5.

041.003 **Dynamical equinox and analytical theory of the Sun.** G. A. Krasinsky (*Krasinskij*), M. L. Sveshnikov.
Celestial Mech., Vol. 26, (see 012.001), 171 - 177 (1982).

A concept of the dynamical equinox and its relation to the analytical form of the adopted theory of the Sun is discussed. Connection between the FK4 equinox and the dynamical equinox is determined by comparing two analytical theories of the Sun with solar meridian observations made at the U.S. Naval Observatory (1911–1971).

041.004 **A contribution to the FK 5: meridian observations 1964 to 1976 made at the Copenhagen Observatory in Brorfelde.** S. Röser, W. Fricke.
Mitt. Astron. Ges., Nr. 55, p. 139 - 143 (1982).

041.005 **Von der Sternhelligkeit abhängige Systemfehler des FK 4 am Südhimmel.**
R. Bien, H. Schwan.
Mitt. Astron. Ges., Nr. 55, p. 143 (1982). – Abstract.

041.006 **Entwicklung und Erprobung eines analytischen Verfahrens zur Herleitung eines Fundamentalsystems.** H. Schwan.
Mitt. Astron. Ges., Nr. 55, p. 144 (1982). – Abstract.

041.007 **Genauigkeiten radiointerferometrischer Positionen extragalaktischer Quellen.** H. G. Walter.
Mitt. Astron. Ges., Nr. 55, p. 163 (1982). – Abstract.

041.008 **Beobachtungsdaten von Sternen zur Herleitung des FK 5: Dokumentation und Aufbereitung.**
R. Hering, R. Jährling, H. Schwerdtfeger.
Mitt. Astron. Ges., Nr. 55, p. 170 - 174 (1982).

041.009 **Precise radio source positions from interferometric observations.** G. H. Kaplan, F. J. Josties,
P. E. Angerhofer, K. J. Johnston, J. H. Spencer.
Astron. J., Vol. 87, 570 - 576 (1982).

Observations made with the Green Bank radio interferometer during the period 1 October 1979 - 3 February 1980 were analyzed for the purpose of obtaining a set of very precise coordinates for 16 compact radio sources. The effects of the ionosphere have been removed from the data through the use of dual-frequency observations. Corrections for tropospheric effects were calculated from a standard model atmosphere. For sources with $\delta > 10°$, the positions derived from the observations have an internal precision of about $0''.01$. Agreement among this and similar catalogs of source positions is at the $0''.03$ level. A discussion of the effects of source structure on source position is presented.

041.010 **Position corrections to the FK4-stars derived from the results of observations with the Wuchang Time Observatory Danjon Astrolabe.**
Acta Geod. Geophys., No. 3, p. 27 - 46 (1981). In Chinese.

041.011 **Análisis estadístico de los resultados obtenidos en la observación del programa S.R.S. con el Círculo Meridiano Repsold de 190 mm.**
J. A. López, R. A. Carestía.
Bol. Asoc. Argentina Astron., No. 18, p. 2 - 4 (1980).

041.012 **Determination of star positions with similar weights in a unique system from observations with astrolabes.** E. I. Krejnin, S. A. Tolchel'nikova-Murri.
Astrometr. Astrofiz., Vyp. (No.) 44, p. 60 - 70 (1981). In Russian.

The method proposed 1978 for deriving star coordinates together with the mean longitudes and latitudes of astrolabes is discussed. It is equivalent to the least-squares solution of all the equations of condition obtained directly from observations with all the astrolabes participating in the coordinated programme of observations.

041.013 **Cosmic radioastrometry.** V. A. Alekseev.
Astrometr. Astrofiz., Vyp. (No.) 45, p. 74 - 77 (1981). In Russian.

A method of celestial coordinate system estimation is given. The astronomical approach to the problem of obtaining images of cosmic radio sources using a cosmic radiointerferometer is used.

041.014 **Correcciones a las declinaciones de 112 estrellas del FK4, deducidas de las observaciones con el**

astrolabio Danjon de San Juan, Argentina.
W. T. Manrique, E. Actis, A. Andreoni, J. Baldivieso.
Bol. Asoc. Argentina Astron., No. 20 - 24, p. 17 - 27 (1981).

041.015 Comparación de los resultados de las observaciones
para el programa S. R. S. entre círculo meridiano de
San Juan y círculo meridiano de El Leoncito.
R. A. Carestia.
Bol. Asoc. Argentina Astron., No. 20 - 24, p. 37 - 42 (1981).

041.016 Catálogo de 112 estrellas del FK4 y FK4 Supp.
observadas con astrolabio Danjon O. P. L. del
Observatorio Astronómico de San Juan.
W. Manrique, E. Actis, A. Andreoni, J. Baldivieso, A. Serafino.
Bol. Asoc. Argentina Astron., No. 20 - 24, p. 134 (1981).
Abstract.

041.017 Desvío promedio $\Delta\alpha_\delta$ del catálogo FK4.
R. A. Carestia, M. Gallego.
Bol. Asoc. Argentina Astron., No. 20 - 24, p. 320 (1981).
Abstract.

041.018 Resultados provisorios de las correcciones $\Delta\alpha$ y $\Delta\delta$
de 100 estrellas del programa de catálogo
observadas con astrolabio del O. A. F. A.
W. Manrique, A. Serafino, E. Actis.
Bol. Asoc. Argentina Astron., No. 20 - 24, p. 321 (1981).
Abstract.

041.019 · Procedimiento para la reducción en declinación de
series FK4 – resultados obtenidos.
R. A. Carestia, A. Zaragoza, M. Gallego.
Bol. Asoc. Argentina Astron., No. 20 - 24, p. 384 (1981).
Abstract.

041.020 Consideraciones sobre la reduccion de placas
astrométricas por el método de cuadrados
mínimos.   G. Iannini, J. J. Rodríguez, L. H. Gaitán.
Bol. Asoc. Argentina Astron., No. 20 - 24, p. 456 (1981).
Abstract.

041.021 Observación del catálogo F. K. Z.
R. Carestia, M. Gallego.
Bol. Asoc. Argentina Astron., No. 20 - 24, p. 459 (1981).
Abstract.

041.022 Correcciones preliminares a la ascención recta y
declinación de radiofuentes opticas.
W. Manrique, A. Serafino, E. Actis, J. Baldivieso, E. Alonso.
Bol. Asoc. Argentina Astron., No. 26, p. 59 - 60 (1981).

041.023 Danjon astrolabe mean residual as a function of
the spectral type.   Z. G. Li.
Special issue on astronomical geodynamics, (see 012.028), p.
82 - 88 (1981). In Chinese.
The author collected data of ten Danjon Astrolabe
Catalogues. The principle conclusions are: (1) The correction
of the individual positions in an astrolabe catalogue and the
mean residual as a function of the spectral type are mixed.
Therefore the accidental error of the mean residual as a function of the spectral type is very big. (2) Theoretical value of
the mean residual as a function of the spectral type in the
Danjon Astrolabe seems to be more certain than the value
derived by a statistical method.

041.024 On the new definition of the celestial pole.
B. X. Xu.
Special issue on astronomical geodynamics, (see 012.028),
p. 171 - 172 (1981). In Chinese and English. – Abstract.

041.025 The prospect of space astrometry and the significance of ground observations.   J. Y. Xu.

Special issue on astronomical geodynamics, (see 012.028),
p. 183 (1981). In Chinese and English. – Abstract.

041.026 On the Belgrade Catalogue of NPZT stars.
S. N. Sadzakova, V. A. Fomin.
Astron. Tsirk., No. 1194, p. 7 - 8 (1981). In Russian.

041.027 On new results of a study of the photographic irradiation effect in position observations of the
major planets.   B. S. Vozdvizhenskij.
Astron. Zh., Tom 59, 588 - 590 (1982). In Russian. English
translation in Soviet Astron., Vol. 26, No. 3.
The results of the study of the photographic irradiation
effect for DU-3 ORWO photoplates are presented and nomo-
grams of the image centre displacement of the "artificial
planet" are given.

041.028 Mark III VLBI: astrometry and epoch J2000.0.
C. Ma, T. A. Clark, D. B. Shaffer.
Bull. American Astron. Soc., Vol. 13, 899 (1981). – Abstract.

041.029 Observations of the sun at the CERGA astrolabe in
1980.
F. Laclare, M. Glentzlin, N. V. Leister, F. Chollet.
Astron. Astrophys., Suppl. Ser., Vol. 48, 371 - 373 (1982).
In French.
The results of a new campaign of solar position observations are reported. It is the first time that observations of the
sun have been carried out on the same day for 3 zenith
distances (30°, 45° and 60°), thanks to an instrumental
improvement related elsewhere.

041.030 Observations of Mars with the astrolabe of the
CERGA Observatory (February 1980 - May 1980).
J. Pham-Van, G. Dudognon, P. Granès, F. Mignard,
G. Vigouroux.
Astron. Astrophys., Suppl. Ser., Vol. 49, 105 - 106 (1982).
In French.

041.031 Observations of Jupiter with the astrolabe of the
CERGA Observatory (January 1978 - May 1979).
G. Vigouroux, C. Delmas, G. Guallino, F. Mignard,
J. Pham-Van.
Astron. Astrophys., Suppl. Ser., Vol. 49, 107 - 108 (1982).
In French.

041.032 On the discrepancy between the optical and radio
position of T Tauri.   C. de Vegt.
Astron. Astrophys., Vol. 109, L15 - L16 (1982).
A precise optical position and proper motion in the FK4
system of the nebular variable T Tauri has been derived from
astrograph plates covering more than 50 years of epoch
difference. The comparison with a recently derived VLA-radio
position by Cohen et al. yields a significant difference (optical-
radio) which confirms the offset of $\approx 0\rlap{.}''8$, quoted by these
authors. The new optical position is derived from short
exposure plates, showing no indication of disturbing
nebulosity. A systematic shift of the optical position due to
nebulosity is therefore questioned.

041.033 Comparison of precise optical and radio positions
for Cyg OB 2 members and P Cyg.   C. de Vegt.
Astron. Astrophys., Vol. 109, 282 - 284 (1982).
Precise optical positions and proper motions in the FK4-
system have been determined for four recently detected radio
stars in the Cyg OB2 association and the known radio star
P Cyg. A comparison with the VLA positions from Abott et al.
shows an excellent agreement of both declination systems in
that sky region, whereas a constant R.A. zero point difference
of −9 ms probably is present. The importance of determining
precise radio positions for the HIPPARCOS astrometry
satellite project is stressed.

041.034 Inertial frame determination using minor planets. A simulation of HIPPARCOS-observations.
S. Söderhjelm, L. Lindegren.
Astron. Astrophys., Vol. 110, 156 - 162 (1982).

The space astrometry project HIPPARCOS will measure positions, proper motions, and parallaxes for about 100,000 stars to an accuracy of a few milliarcsec. One way to define an inertial reference system is by observing minor planets, some 50 of which are bright enough to be observed by HIPPARCOS. Such observations, about 20 per year and planet, were simulated for the 4-year interval 1987–90. The accuracy of the determination of various parameters, including the rotation of the reference system are estimated.

041.035 Resultats d'observations du soleil à l'astrolabe du CERGA. F. Chollet.
Sun and planetary system, (see 012.040), p. 35 - 38 (1982).

The Solar Astrolabe is described, and the results of observations made in 1978 and 1979 are presented. Obtained were corrections to the orbital elements of the Sun and a correction to the equinox of the FK4.

041.036 Observations of the Sun and inner planets with the large meridian circle in Belgrade.
S. Sadžakov, M. Dačić, D. Šaletić, B. Ševarlić.
Sun and planetary system, (see 012.040), p. 445 - 446 (1982).

041.037 On the importance of the Hipparcos stellar net for photographic catalogue work. C. de Vegt.
Scientific aspects of the Hipparcos space astrometry mission, (see 012.041), p. 49 - 52 (1982).

Compared with the measuring accuracy of a single image on one plate, the Hipparcos positions provide for the first time practically "error free" reference points. The attainable model accuracy for large field catalogue work is discussed. Furtheron the application of Hipparcos and Tycho results to the reduction of important old epoch catalogues is investigated. Requirements for reference star observations by Hipparcos are quoted.

041.038 Radio stars as connecting link of the Hipparcos and VLBI reference frames. H. G. Walter.
Scientific aspects of the Hipparcos space astrometry mission, (see 012.041), p. 65 - 68 (1982).

The dual frequency emission of radio stars makes these objects potential candidates for relating the Hipparcos reference frame with the VLBI reference frame. Available radio and optical data are examined to establish the physical basis for this kind of frame connection.

041.039 US Naval Observatory parallaxes and the fundamental reference frame – their interaction with Hipparcos. G. Westerhout, J. A. Hughes.
Scientific aspects of the Hipparcos space astrometry mission, (see 012.041), p. 69 (1982).

New developments in the U.S. Naval Observatory parallax and fundamental astrometry programs are described. The interaction of these programs with the Hipparcos program is discussed.

041.040 Astrometric measurements required before Hipparcos launch. Y. Requième.
Scientific aspects of the Hipparcos space astrometry mission, (see 012.041), p. 207 - 209 (1982).

The r.m.s. error on the 1988.0 star positions to be collected in the Hipparcos Input Catalogue must be in each coordinate inferior to 1.0 arcsec for B < 10 and 1.5 arcsec for fainter stars. A conservative estimate (possibly biased later on by the real composition of the Hipparcos observing list) leads to remeasure about 4 000 Northern faint stars, 9 000 Southern faint stars and 20 or 25 000 Southern bright stars. Ground-based work is

also required on special objects: multiple stars, cluster stars, minor planets and objects for extragalactic reference link.

041.041 A method for compiling a general catalogue.
Y.-x. Zhu, L.-z. Lu, D.-j. Luo, Z.-f. Ling.
Acta Astron. Sinica, Vol. 23, 76 - 81 (1982). In Chinese.

041.042 A deep optical survey of fifteen benchmark sources.
A. N. Argue.
Abh. Hamburger Sternw., Band 10, 110 (1982).

041.043 Hipparcos – ST bootstrap stars for southern fields.
A. N. Argue, R. D. Baxter.
Abh. Hamburger Sternw., Band 10, 111 - 115 (1982).

041.044 Optical identifications for benchmark radio sources.
A. N. Argue, R. D. Baxter.
Abh. Hamburger Sternw., Band 10, 116 - 118 (1982).

041.045 Radio stars for precise astrometry.
C. de Vegt.
Abh. Hamburger Sternw., Band 10, 119 - 128 (1982).

041.046 Recent astrometric measurements using the Cambridge 5-km radio telescope. B. Elsmore.
Abh. Hamburger Sternw., Band 10, 129 - 130 (1982).

041.047 The precision of astrometric surveys of radio sources. H. G. Walter.
Abh. Hamburger Sternw., Band 10, 145 - 149 (1982).

Radio interferometric coordinates of extragalactic sources independently determined by different teams and instrumental equipment have been compared with a view to assess internal and external errors of the surveys. The average internal errors amount to a few milliseconds in right ascension and about $0\rlap{.}{''}03$ in declination. Mean differences with respect to a reference survey have been derived with mean errors as large as $0\rlap{.}{''}04$, especially for declinations.

041.048 On the connection of the radio and optical systems of position and proper motions. P. Brosche.
Abh. Hamburger Sternw., Band 10, 166 (1982). – Abstract.

041.049 Contribution des astrolabes au raccordement des systèmes de référence "optique" et "radio".
S. Débarbat.
Abh. Hamburger Sternw., Band 10, 168 (1982). – Abstract.

041.050 Investigation of systematic differences between the new Washington catalogs W5–50 and WL50, the Perth 70 and the AGK3R in their common zone of overlap, declinations −5° to +5°. T. Corbin, C. de Vegt.
Abh. Hamburger Sternw., Band 10, 169 (1982). – Abstract.

041.051 Comparison of precise optical and radio positions for Cyg OB2 members and P Cyg.
C. de Vegt.
Abh. Hamburger Sternw., Band 10, 170 (1982). – Abstract.

041.052 Precise optical positions of radio sources in the FK4-system. II. Results from 26 sources on the northern hemisphere. C. de Vegt, U. K. Gehlich.
Abh. Hamburger Sternw., Band 10, 170 (1982). – Abstract.

041.053 An attempt to compare the radio astronomical system of coordinates of quasars with FK4.
I. I. Kumkova.
Abh. Hamburger Sternw., Band 10, 172 (1982). – Abstract.

041.054 A comparison of the Smithsonian Astrophysical Observatory Catalogue (SAO) with AGK3R and

Perth 70.   C. Sullivan, A. N. Argue.
Abh. Hamburger Sternw., Band 10, 173 (1982). – Abstract.

041.055 Systematic differences between radio astrometric
         surveys.   H. G. Walter.
Abh. Hamburger Sternw., Band 10, 174 (1982). – Abstract.

041.056 Report of IAU Commission 8: Positional astronomy
         (Astronomie de position).   E. Høg.
Trans. IAU, Vol. XVIIIA, (see 003.013), p. 33 - 42 (1982).

041.057 Report of IAU Commission 24: Photographic
         astrometry (Astrométrie photographique).
H. Eichhorn.
Trans. IAU, Vol. XVIIIA, (see 003.013), p. 237 - 240 (1982).

KSS compiled catalogue of reference stars in the
declination zones from $-5°$ to $+90°$.   See Abstr. 002.006.

Catalogue of declinations of 100 equatorial stars of
A. A. Mikhajlov's list.   See Abstr. 002.014.

Choice of 315 stars in 87 areas with extragalactic
radio sources for meridian observations in the FK4 system.
See Abstr. 002.016.

Supplementary catalogue of the coordinates of 90
stars of the Moscow zenith zone and estimate of its accuracy.
See Abstr. 002.017.

Cuarto catálogo círculo meridiano San Juan
(FK4-SUR).   See Abstr. 002.019.

Segundo catálogo círculo meridiano San Juan
(FKSZ).   See Abstr. 002.020.

The Hyderabad (S) and Cordoba zones of the
astrographic catalogue: a preliminary investigation.
See Abstr. 002.023.

A preliminary catalogue of 366 stars observed with
the photoelectric astrolabe GD II No. 2.   See Abstr. 002.038.

The catalog of 1111 northern PZT stars compiled
from meridian circle catalogs in the 1950s, (NPZT$_{58}$).
See Abstr. 002.057.

Northern PZT stars catalog (NPZT$_{74}$).
See Abstr. 002.058.

General catalog of stars observed with astrolabes.
See Abstr. 002.059.

Preliminary catalogue of the photoelectric astrolabe
(PACP I).   See Abstr. 002.060.

Meridian observations made in Brorfelde
(Copenhagen University Observatory) 1969–1975. Positions
of 6427 stars brighter than 11.00 vis.mag.
See Abstr. 002.065.

The Texas data base.   See Abstr. 002.070.

Beiträge der Berliner Sternwarte zu bedeutenden
Erkenntnissen auf dem Gebiet der klassischen Astronomie.
See Abstr. 004.086.

Hipparcos: Europe's astrometric satellite.
See Abstr. 013.004.

Present status of the Texas radio survey.
See Abstr. 013.060.

A comparison of astrometric measurement tech-
niques as applied to minor planets.   See Abstr. 031.529.

Ampliación del primer Catálogo Astrolabio de San
Juan, programa catálogo.   See Abstr. 031.556.

Interferometric measurements of stellar positions
in the infrared.   See Abstr. 031.635.

The satellite Hipparcos.   See Abstr. 032.553.

Determination of positions and the reference sys-
tem.   See Abstr. 043.015.

Report by IAU Commission 24 Working Group
on Radio/Optical Astrometric Sources for the establishment
of an inertial reference frame.   See Abstr. 043.016.

Fiducial reference for the Hipparcos reference
system.   See Abstr. 043.017.

Catalogue Hipparcos et observations au sol.
See Abstr. 043.019.

HIPPARCOS and the dynamics of the solar system.
See Abstr. 051.004.

HIPPARCOS: space astrometry mission. A report
on the project. II.   See Abstr. 051.010.

Tycho, a planned astrometric and photometric
survey from space.   See Abstr. 051.019.

Proposed ultraviolet observations with Tycho.
See Abstr. 051.020.

Le programme Hipparcos.   See Abstr. 051.021.

The selection of stars for linking the Hipparcos
frame to extragalactic radio sources by Space Telescope
(progress report).   See Abstr. 051.023.

The astrometric satellite HIPPARCOS – a status
report.   See Abstr. 051.028.

Prospective views in space astrometry.
See Abstr. 051.030.

Problems of three-dimensional refraction in
astrometry.   See Abstr. 082.074.

Some results of the Mercury transit observations in
1970 and 1973 at Belgrade.   See Abstr. 092.009.

Motion of Mars: 1935–1976. See Abstr. 097.081.

Photographic position observations of Jupiter and
the Galilean satellites with a longfocal astrograph in 1975.
See Abstr. 099.051.

Hipparcos proper motions.   See Abstr. 111.019.

Arcsecond positions for milliarcsecond VLBI nuclei
of extragalactic radio sources. I. 546 sources.
See Abstr. 141.090.

Precise optical positions of 10 quasars.
See Abstr. 141.219.

VLBI measurements of radio source positions at
three U.S. stations.   See Abstr. 141.287.

042 Celestial Mechanics, Figures of Celestial Bodies

042.001   The calculation of characteristic exponents of triple collision of the three-body problem in the equilateral configuration.   F. Nahon, M. Irigoyen.
C. R. Acad. Sci. Paris, Tome 293, Sér. II, 1069 - 1072 (1981).
In French.

042.002   Some problems of motion of a body with changing dynamical structure.   V. R. Kolbut.
Pis'ma Astron. Zh., Tom 8, 57 - 59 (1982). In Russian. English translation in Soviet Astron. Lett., Vol. 8.
The problem of translatory-rotatory motion of a body in the gravitational field of a globe (Abul'naga and Barkin, 1979) is supplemented by the condition of changing dynamical structure. The model of the body with changing dynamical structure corresponds to a dynamically equivalent homogeneous deformed body.

042.003   A method to construct explicit solutions of a simplified version of the spatial circular restricted three-body problem.   V. Kh. Karaganchu, A. V. Karaganchu.
Pis'ma Astron. Zh., Tom 8, 125 - 128 (1982). In Russian. English translation in Soviet Astron. Lett., Vol. 8.
A Hamilton-Jacobi method is described to construct explicit solutions of a simplified version of the spatial circular restricted three-body problem. This version is obtained in terms of the true problem by use of Delaunay averaging for the perturbation function at a given precision of the Le Verrier expansion.

042.004   Plane resonance rotations of a dynamically symmetric satellite in the three-body problem.
P. S. Krasil'nikov.
Astron. Zh., Tom 59, 147 - 154 (1982). In Russian. English translation in Soviet Astron., Vol. 26, No. 1.
Some resonance effects of an attitude satellite motion in the elliptic restricted three-body problem are considered. A first-order approximation is constructed for the plane rotations of the satellite on the assumption that a) the ellipsoid of inertia is close to a sphere, b) the mass center of a satellite is moving along an arbitrary, almost-periodic orbit of a three-body problem. As an example some resonance rotations of a satellite in the vicinity of libration points as well as at large halo orbits in the earth-moon system are considered.

042.005   Orbital interactions: a new geometrical formalism.
R. Greenberg.
Astron. J., Vol. 87, 184 - 195 (1982).
A new analysis of the geometry of encounters of two bodies on independent Keplerian orbits about a third body avoids some of the approximations that had been made in earlier studies. Formulas for collision frequencies and for rates of change of orbital elements due to close approaches are derived. For most of the applications considered here, the results obtained using this new method agree with past results.

042.006   Density scaling of the angular momentum versus mass universal relationship.
L. Carrasco, M. Roth, A. Serrano.
Astron. Astrophys., Vol. 106, 89 - 93 (1982).
The relationship between the angular momentum and mass has been investigated for a large variety of astronomical objects, ranging from asteroids to clusters of galaxies using data taken from the literature. The authors find that the specific angular momentum follows an M^k power law $(2/3 \leqslant k \leqslant 3/4)$ for all objects under consideration. This power law can be explained in terms of mechanical equilibrium between gravitational and rotational energy. The existence of this relationship predicts a relationship,

$r \propto M^{0.33-0.5}$ for flat galaxies, in close agreement with observations.

042.007   Extremal tides: rigorous computation for the many-body case.   D. Rawlins.
Geophys. J. R. Astron. Soc., Vol. 69, 265 - 271 (1982).
A rigorous method is developed for the exact solution of the extrema of the total tidal field on a spherical celestial body, disturbed by a multiplicity of gravitating point masses distributed in three dimensions at distances large relative to the disturbed body's size. A short program is provided for convenient use of the method. As an illustration, maximal solar tides due to planetary attraction are calculated for the Solar System 1964–1991.

042.008   Studies of the stellar three-body problem.
S. Söderhjelm.
Astron. Astrophys., Vol. 107, 54 - 60 (1982).
The motions in point-mass triple star systems are studied by analytical and numerical methods. The existing analytical theories are found to be quite accurate as long as the ratio of orbital periods P_2/P_1 is above some 50:1. The theory for the secular (apse-node) motions is given in detail, with the elliptical functions expanded in more tractable Fourier series.

042.009   A generalization of Lagrange's formula for implicit function.   G. E. O. Giacaglia.
Publ. Astron. Soc. Japan, Vol. 33, 739 - 742 (1981).

042.010   Delaunay normalisations.   A. Deprit.
Celestial Mech., Vol. 26, (see 012.001), 9 - 21 (1982).
Too many terms are generated by a Delaunay normalisation when the perturbation is developed in powers of the eccentricity. Ways of bypassing the expansion are discussed. There are (1) Brouwer's method of implicit variables; (2) the preparation by canonical transformations; and (3) the application of representation theory for Lie algebras. Illustrations of the techniques are drawn from the main problem of satellite theory and from the $(1-1)$ resonance at the triangular equilibrium in the restricted problem of three bodies.

042.011   Computer implementation of an algorithm for the quadratic analytical solution of Hamiltonian systems.
R. A. Howland.
Celestial Mech., Vol. 26, (see 012.001), 23 (1982). – Abstract.

042.012   Asymptotic series for planetary motion in periodic terms in three dimensions.   P. J. Message.
Celestial Mech., Vol. 26, (see 012.001), 25 - 39 (1982).
For the 'planetary case' of the gravitational n-body problem in three dimensions, a sequence of Lie series contact transformations is used to construct asymptotic series representations for the canonical parameters of the instantaneous orbits in a Jacobi formulation. The series contain only periodic terms, the frequencies being linear combinations of those of the planetary orbits and those of the secular variations of the apses and nodes, and the series are in powers of the masses of the planets in terms of that of the primary, and of a quantity of the order of the excursions of the eccentricities and inclinations of the orbits. The treatment avoids singularities for circular and coplanar orbits. It follows that the major axes are given by series of periodic terms only, to all orders in the planetary masses.

042.013   Strongly perturbed quasi-periodic dynamical systems.
A. Sivaramakrishnan, W. H. Jefferys.
Celestial Mech., Vol. 26, (see 012.001), 41 - 49 (1982).
It is almost impossible to construct a general theory of

the motion of a strongly perturbed dynamical system using classical perturbation theory because this approach uses a reference orbit which is very different from the actual orbit. A general method, pioneered by Jefferys, is presented here. This method allows each quasi-periodic orbit to specify the coordinates to be used. The method is illustrated by a simple example.

042.014 A third-order intermediate orbit for planetary theory.   D. L. Richardson.
Celestial Mech., Vol. 26, (see 012.001), 187 - 195 (1982).

By use of a new canonical transformation procedure, a third-order intermediary for planetary motion is developed. The intermediary contains all contributions that arise from the assumption of circular, coplanar orbits for the disturbing masses. The results are expressible in terms of elliptic integrals of the first, second, and third kinds.

042.015 Application of a new algebraic manipulator theory.   T. C. Van Flandern.
Celestial Mech., Vol. 26, (see 012.001), 197 (1982). – Abstract.

042.016 General planetary theory extended to the case of resonance and application to the Galilean satellite system of Jupiter.   L. Duriez.
Celestial Mech., Vol. 26, (see 012.001), 231 - 255 (1982).
In French.

The author considers a system of planets defined by a given distribution of mean mean motions and masses: he represents the osculating elliptic elements of their heliocentric orbits by quasi-periodic functions of time, through a method adapted to the commensurability case; these functions are the sum of the general solution of a critical system, expressed in long-period terms, and of a particular solution. The author obtains a differential system with constant coefficients, whose linear part gives, as a first approximation, the great inequalities, the free oscillations and the libration; this solution agrees already with known results, but it should be improved by taking into account the non-linear parts and the solar terms in a new approximation.

042.017 The variational equations of Störmer's problem.   A. Hennawi.
Celestial Mech., Vol. 26, 277 - 283 (1982). In French.

A theory has already been established concerning all Lagrangians $L(q, \dot{q}, t)$ which possess the integrals or more generally invariant forms, originating for example from geometric invariances. This paper is a direct application to the variational equations of Störmer's problem that has captured the interest of many researchers in celestial mechanics.

042.018 Hill stability and distance curves for the general three-body problem.   C. Marchal, G. Bozis.
Celestial Mech., Vol. 26, 311 - 333 (1982).

The notion of Hill stability is extended from the circular restricted 3-body problem to the general three-body problem; it is even extended to systems of positive energy and the Hill's curves with their corresponding forbidden zones are generalized. The three limiting cases, restricted, planetary and lunar are analysed as well as some real stellar cases.

042.019 Integrals of motion in the elliptic three-dimensional restricted three-body problem.   E. Sarris.
Celestial Mech., Vol. 26, 353 - 360 (1982).

Three integrals of motion have been found in the three-dimensional elliptic restricted three-body problem for small eccentricity e' of the relative orbit of the primaries and small distance r and eccentricity e of the orbit of the third body around a primary. The integrals are given in the form of formal series in the mass-ratio μ, the eccentricities e', e and the coordinates and velocities. These integrals depend periodically on the time.

042.020 The gravitational field of a disk.
F. T. Krogh, E. W. Ng, W. V. Snyder.
Celestial Mech., Vol. 26, 395 - 405 (1982).

This note gives the gravitational potential of the disk $\{(x, y, z): x^2 + y^2 \leqslant \rho^2, z = 0\}$ and the gravitational field at a point (x, y, z). Formulas for a ring can be obtained as the difference of the authors' results for two different values of ρ. The results are obtained in terms of elliptic integrals and the authors indicate how these functions can be computed efficiently. Formulas necessary for the computation of partial derivatives are also given.

042.021 Numerical stabilization of Keplerian motion.   K. Zare.
Celestial Mech., Vol. 26, 407 - 412 (1982).

The time transformation $dt/ds = r^\alpha$ is studied in detail and numerically stabilized differential equations are obtained for $\alpha = 1, 2$, and $^3/_2$. The case $\alpha = 1$ corresponds to Baumgarte's results.

042.022 A comparison of the Bohlin–von Zeipel and Bohlin–Lie series methods in resonant systems.
A. H. Jupp.
Celestial Mech., Vol. 26, 413 - 422 (1982).

Whereas the Bohlin–von Zeipel procedure can be used successfully to construct formal solutions to some resonant dynamical systems, it is shown here that a direct Bohlin–Lie series approach seems not to be feasible. The fact that certain terms lose an order of magnitude on differentiation with respect to the momentum variable leads to a situation which precludes an accurate construction of the first-order term in the generating function. A simple remedy to this impasse is suggested, with particular reference to the Ideal Resonance Problem.

042.023 The rotation of a non-spherical satellite at libration points in the restricted circular three-body problem.
M. Šidlichovský.
Bull. Astron. Inst. Czechoslovakia, Vol. 33, 75 - 83 (1982).

The rotation of a satellite located at one of the libration points is investigated employing Andoyer's variables and the Lie-Hori theory. A first order solution is given for a tri-axial satellite in a non-resonant case. The resonant case is investigated under the assumption of axial symmetry of the satellite. In this case the problem is reduced to the Ideal Resonance Problem.

042.024 A direct method of computing small divisors in planetary theory.   R. Dvorak.
Astron. Astrophys., Vol. 108, 14 - 18 (1982).

A development of the inverse distance $1/\Delta$ between two planets is established, which is further used in combination with the Lagrange equations. The direct method developed is intermediate between a strict numerical approach (such as the Fourier analysis) and an analytical one (as in perturbation theories). Some examples of specific small divisors are given.

042.025 Integration constants and mean elements for all planets.   P. Bretagnon.
Astron. Astrophys., Vol. 108, 69 - 75 (1982). In French.

After having constructed a theory developed to the second order with respect to the masses for all planets (Bretagnon, 1980), and having determined terms of higher order for the outer planets through an iterative method (Bretagnon, 1981), the author has got better integration constants. With his solutions and new formulas of precession (Lieske et al., 1977), the author has determined the mean elements connected to the ecliptic of the date for all planets. Finally, he gives the integration constants related to J2000.

042.026 Intermediate orbits approximating the initial part of perturbed motion.   Yu. V. Batrakov.

Byull. Inst. Teor. Astron., Leningrad, Tom 15, 1 - 5 (1981). In Russian.

Intermediate non-perturbed orbits are constructed which ensure the second or third order contact to the real trajectory at initial time. Integration of the equations for deviations of the real motion from the intermediate one gives the perturbed motion.

042.027 **Statistically dependent normal places in orbital com-putations.**   B. B. Baghos.
Byull. Inst. Teor. Astron., Leningrad, Tom 15, 6 - 9 (1981). In Russian.

It is supposed that for the set of homogeneous data obtained during one satellite passage over a tracking station an interpolating polynomial has been constructed as a linear combination of discrete orthogonal Chebyshev polynomials. Also there is the supposition that the points of this polynomial have been found in which the curve of dispersion for the values of the polynomial has local minima. The values of the polynomial at these points, which are considered as normal places for this set of data, may serve as synthetic input data for orbital computations. The covariance matrix for the normal places of this kind has been obtained and it has been shown how to form normal equations for orbital parameters corrections from such synthetic data.

042.028 **On the regularization of Hill's problem.**
V. A. Kuz'minykh.
Byull. Inst. Teor. Astron., Leningrad, Tom 15, 33 - 36 (1981). In Russian.

A regularized system of equations of Hill's problem is obtained. The domain of expansion of the solution in Perron series has been estimated. A canonical regularized system of equations has been constructed.

042.029 **Les critériums des mouvements hyperboliques et hyperboliques-elliptiques dans la théorie de la capture.**   G. A. Merman.
Byull. Inst. Teor. Astron., Leningrad, Tom 15, 37 - 45 (1981). In Russian.

On propose des critériums nouveaux des mouvements hyperboliques et hyperboliques-elliptiques dans la théorie de la capture, applicables, quand le triangle formé par les trois corps n'est pas assez allongé.

042.030 **Construction of continuous Chebyshev approxima-tions for planetary coordinates.**
Yu. V. Batrakov, B. I. Kaminskij.
Byull. Inst. Teor. Astron., Leningrad, Tom 15, 76 - 82 (1981). In Russian.

Two methods for obtaining continuous approximations of functions by Chebyshev polynomials are considered. The first deals with an a priori fixed number of polynomials while the other allows to match any number of these one after another. Numerical results for one of the coordinates of Venus illustrate the methods.

042.031 **Quelques algorithmes nouveaux pour les solutions quasi-périodiques.**   G. A. Merman.
Byull. Inst. Teor. Astron., Leningrad, Tom 15, 96 - 113 (1981). In Russian.

On propose quelques méthodes nouvelles pour la construction des solutions quasi-périodiques des systèmes oscillants aux cas suivants: 1) pour les systèmes, qu'on rencontre dans la mécanique céleste (canonique ou non), 2) au voisinage de la position d'équilibre des systèmes canoniques dont les exposants peuvent être commensurables entre eux.

042.032 **On a representation of planetary coordinates by Chebyshev approximations which are continuous along with their first derivatives.**
Yu. V. Batrakov, M. A. Fursenko.

Byull. Inst. Teor. Astron., Leningrad, Tom 15, 137 - 141 (1982). In Russian.

A method has been developed and a computer program has been compiled for constructing Chebyshev approximations for planetary coordinates which are continuous along with their first derivatives at the ends of approximation intervals. The accuracy of approximations can be chosen a priori and it determines the order of the polynomials or the length of the intervals. A numerical example demonstrates the errors for continuous and discontinuous approximations.

042.033 **On the integration of functions of elliptic motion.**
N. N. Vasil'ev.
Byull. Inst. Teor. Astron., Leningrad, Tom 15, 142 - 144 (1982). In Russian.

For the integration of polynomial and trigonometric expressions involving the eccentric and true anomalies a new method is proposed. The method is based on recurrent relations resulting from the application of the method with indefinite coefficients.

042.034 **Method for determination of the parameter of an orbit from two positions of a celestial body using an auxiliary orbit.**   Yu. D. Medvedev.
Byull. Inst. Teor. Astron., Leningrad, Tom 15, 165 - 168 (1982). In Russian.

A method has been proposed for determining the parameter of the non-perturbed orbit from two known positions of a celestial body. The sector area of some auxiliary orbit is used instead of the conventional triangle area.

042.035 **Intermediate orbits with fourth order contact to trajectories of perturbed motion.**
V. G. Sokolov.
Byull. Inst. Teor. Astron., Leningrad, Tom 15, 176 - 181 (1982). In Russian.

A method has been given for constructing non-perturbed orbits in which the initial position vector and its successive time derivatives to the fourth order inclusive coincide with those of the real perturbed motion. Such orbits ensure good approximations to the initial parts of the real motion and these may be especially suitable in cases when the perturbations are not small.

042.036 **Using a uniform approximation of the coordinates of disturbing bodies in the Taylor-Steffensen numerical integration method.**   E. Z. Khotimskaya.
Byull. Inst. Teor. Astron., Leningrad, Tom 15, 182 - 189 (1982). In Russian.

It has been proposed to uniformly approximate by means of polynomials the coordinates of disturbing bodies for better efficiency of the numerical integration method developed by Myachin and Sizova (1970) when applied to the investigation of the motion of zero-mass bodies. The results are given of the uniform polynomial and rational approximation applied to the coordinates of the eight major planets of the solar system. The polynomials obtained are used for integration of the equations of motion of the minor planet 4 Vesta.

042.037 **Computation of gravitational forces and moments from external fields.**   C. Powell.
J. Astronaut. Sci., Vol. 29, 189 - 193 (1981). – Abstr. in Phys. Abstr., Vol. 85, Abstr. 42589 (1982).

042.038 **The origin of the Kirkwood gaps: a mapping for asteroidal motion near the 3/1 commensurability.**
J. Wisdom.
Astron. J., Vol. 87, 577 - 593 (1982).

A mapping of the phase space onto itself with the same low-order resonance structure as the 3/1 commensurability in the planar-elliptic restricted three-body problem is derived. This mapping is approximately 1000 times faster than the usual method of numerically integrating the averaged equations of motion. This mapping exhibits some surprising behavior that might provide a key to the origin of the Kirkwood gaps. A test asteroid placed in the gap may evolve for a million years with low eccentricity (< 0.05) and then suddenly jump to large eccentricity (> 0.3), becoming a Mars crosser. It is possible that the asteroid could then be removed by a close encounter with Mars. When the inclinations and the secular perturbations of Jupiter's orbit are included in the planar-elliptical mapping, evolutionary runs for 300 test asteroids near the 3/1 commensurability demonstrate the formation of a gap near the observed Kirkwood gap. Evidence is presented which shows that the proper width of the gap may develop when enough time is allowed for the evolutionary calculation.

042.039 Capture into resonance: an extension of the use of adiabatic invariants.   J. Henrard.
Celestial Mech., Vol. 27, 3 - 22 (1982).
The theory of the adiabatic invariant predicts the long term evolution of mechanical systems with slowly varying parameters. Unfortunately, it is not valid when the system goes across a critical trajectory. This case is important because it can lead to capture into resonance. Analysing the motion in the vicinity of the critical trajectory, the author is able to give general formulae for the probability of capture and to show that in general, the adiabatic invariant is conserved.

042.040 A note on a separation of the linearized equations of motion in the elliptic restricted problem.
V. R. Matas.
Celestial Mech., Vol. 27, 23 - 25 (1982).
Some additional remarks are made concerning an exact proof of the applicability of Hill's equations in the linear-stability question in the elliptic restricted problem of three bodies.

042.041 Periodic orbits of the elliptic restricted problem for the Sun–Jupiter–Saturn system.
J. H. Kwok, P. E. Nacozy.
Celestial Mech., Vol. 27, 27 - 38 (1982).
A systematic approach to generate periodic orbits in the elliptic restricted problem of three bodies is introduced. The approach is based on continuation from periodic orbits of the first and second kind in the circular restricted problem to periodic orbits in the elliptic restricted problem. Two families of periodic orbits of the elliptic restricted problem are found. The mass ratio of the primaries of these orbits is equal to that of the Sun–Jupiter system. The sidereal mean motions between the infinitesimal body and the smaller primary are in a 2 : 5 resonance. The linear stability of these periodic orbits is studied. The periodic orbit closest to the actual Sun–Jupiter–Saturn system is (linearly) stable.

042.042 Libration solutions in the averaged three-body problem.   Yu. V. Barkin, V. A. Kitova.
Izv. AN KazSSR. Ser. fiz.-mat., 1981, No. 5, p. 63 - 66. In Russian. – Abstr. in Ref. zh., 51. Astron., 3.51.61 (1982).

042.043 On the perturbation function in Andoyer variables.
V. G. Shkodrov.
Dokl. Bolg. AN, Vol. 34, 755 - 758 (1981). – Abstr. in Ref. zh., 51. Astron., 3.51.65 (1982).

042.044 Perturbation theory in extended phase space and its application.   L. Liu.
Acta Astron. Sinica, Vol. 22, 315 - 327 (1981). In Chinese.
The theory of perturbation in extended phase is discussed. It can be applied to establish the perturbation of an artificial satellite due to the geopotential.

042.045 Stabilization and time transformation.
T.-y. Huang, H. Ding.
Acta Astron. Sinica, Vol. 22, 328 - 335 (1981). In Chinese.

042.046 A modification of the Laplacian method for computing orbits.   P.-x. Xu.
Acta Astron. Sinica, Vol. 22, 346 - 349 (1981). In Chinese.

042.047 Aproximaciones al problema de los tres cuerpos (un caso particular de la dinámica estelar).
C. A. Altavista.
Bol. Asoc. Argentina Astron., No. 18, p. 23 - 29 (1980).

042.048 Motion of a material point in the gravitational field of a variable periodic star.   G. V. Kasatkin.
Astron. Zh., Tom 59, 389 - 396 (1982). In Russian. English translation in Soviet Astron., Vol. 26, No. 2.
The motion of a material point in the gravitational field of a variable periodic star of constant mass is considered. The force of gravitational attraction is characterized by a given force function U which is T-time-periodic, T being the pulsation period of the star. It is supposed that $U = U_0 + \epsilon U_1$, U_0 being the Newtonian potential, ϵU_1 a small perturbation.

042.049 On plane rotations of a satellite relative to the mass center in the vicinity of a collinear libration point.
A. P. Markeev, P. S. Krasil'nikov.
Kosm. Issled., Tom 20, 145 - 149 (1982). In Russian.

042.050 On the stability of the libration points in the restricted photogravitational circular three-body problem.   A. A. Perezhogin.
Kosm. Issled., Tom 20, 196 - 205 (1982). In Russian.

042.051 A qualitative study of stabilizing and destabilizing factors in planetary and asteroidal orbits.
J. D. Hadjidemetriou.
Application of modern dynamics to celestial mechanics and astrodynamics, (see 012.026), p. 25 - 44 (1982).
The stability of planetary systems with two planets is studied. The method is based on the mechanism by which instabilities develop in a planetary system with two massless planets when a Hamiltonian perturbation is applied. It is shown that instabilities develop at the resonances 1/3, 3/5, 5/7, ... and to a lesser extend at 1/2, 2/3, 3/4 ... At higher order resonances the instabilities are in most cases negligible. Particular Hamiltonian perturbations exist which stabilize an unstable resonant system.

042.052 Some aspects of motion in the general planar problem of three bodies; in particular in the vicinity of periodic solutions associated with near small-integer commensurabilities of orbital period.   P. J. Message.
Application of modern dynamics to celestial mechanics and astrodynamics, (see 012.026), p. 77 - 101 (1982).
Inquiry continues into the question as to which features of the motion of a system of mutually perturbing planets or satellites persist in the long term. In the case of three mutually perturbing bodies, one is led to a consideration of the properties of periodic solutions of the equations of motion, both because they respresent a class of motions of which the behaviour certainly is known for all time, once it is known for one period, and also because of the relation of periodic solutions to near-commensurabilities of orbital period.

042.053 The stability of n-body hierarchical dynamical systems.   A. E. Roy.
Application of modern dynamics to celestial mechanics and astrodynamics, (see 012.026), p. 103 - 130 (1982).
The equations of motion of n-body hierarchical dynamical system (HDS) in a generalized Jacobi coordinate system enable empirical stability parameters to be readily defined.

The magnitudes of these parameters, together with the ratios of successive radius vectors in the HDS may make it possible to compute the stability of the HDS, so providing a measure of the time interval in which there is an even chance of the status quo of the HDS being altered by mutual perturbations.

042.054 The adiabatic invariant: its use in celestial mechanics.   J. Henrard.
Application of modern dynamics to celestial mechanics and astrodynamics, (see 012.026), p. 153 - 171 (1982).

Many problems in the dynamical evolution of the Solar System can be modeled by some pendulum like Hamiltonian system with one degree of freedom and slowly varying parameters. The adiabatic invariant introduced in the context of quantum mechanics and of physics of nuclear particles is a very effective tool for the study of such problems. In this paper, the author describes the basic ideas of this theory and applies it to the problem of capture into resonance of Titan and Hyperion.

042.055 The KAM invariant and Poincaré's theorem.   T. Petrosky, I. Prigogine.
Application of modern dynamics to celestial mechanics and astrodynamics, (see 012.026), p. 185 - 199 (1982).

The relation between the KAM invariants and Poincaré's classical theorem on nonexistence of uniform invariants is discussed, and their compatibility is pointed out. It is also shown that the KAM invariants are regular invariants in the sense defined by Prigogine and his coworkers. Then, the relation between the ergodic problem and the KAM invariant is discussed.

042.056 Regularization of the singularities of the n-body problem.   C. Marchal.
Application of modern dynamics to celestial mechanics and astrodynamics, (see 012.026), p. 201 - 236 (1982).

The Easton-regularization of the singularities of the n-body problem has been discussed, and it was shown that, aside from the binary collisions, the singularities are generally not regularizable. The only positive new results concern multiple collisions of the rectilinear n-body problem when all colliding masses but perhaps one are infinitesimal.

042.057 Les géodésiques de l'ellipsoïde à trois axes inégaux, d'après Jacobi, Whittaker, Arnold.   F. Nahon.
Application of modern dynamics to celestial mechanics and astrodynamics, (see 012.026), p. 237 - 247 (1982).

The problem of determining the geodesics of an ellipsoid with three inequal axes presents three interesting features: (1) It can be solved by quadratures (Jacobi, 1858); (2) so it admits a second integral, the geometrical interpretation of which was shown by Whittaker; (3) for a critical value of this integral, all of the orbits are doubly asymptotic to the only one which is periodic. This description was given by Arnold. The author gives the proofs missing in Whittaker's and Arnold's papers and considers the limiting case of the billard-ball problem.

042.058 Coordonnées symétriques sur la variété de collision triple du problème plan des trois corps.   J. Waldvogel.
Application of modern dynamics to celestial mechanics and astrodynamics, (see 012.026), p. 249 - 266 (1982).

The planar problem of three bodies is described and a system of 7 differential equations (with 2 first integrals) for the 5-dimensional triple collision manifold T is obtained. The zero angular momentum solutions form a 4-dimensional invariant submanifold N ⊂ T represented by 6 differential equations with polynomial right-hand sides. The used variables are well suited for numerical studies of planar triple collision.

042.059 Exploding dynamical systems.   O. Gurel.
Application of modern dynamics to celestial mechanics and astrodynamics, (see 012.026), p. 277 - 299 (1982).

Various generic properties of "exploding" systems in terms of their multiple solutions are presented. Dynamical models with varying "explosion complexities" are discussed and the relationship between the stabilities and explosions are drawn. It is shown that a "stability index" in the sense of Poincaré may be introduced to identify the stability of the characteristics solutions and used as a tool of global analysis of exploding systems. Cardinality, stability and dimensionality issues of the systems are also considered, and illustrated with specific examples. Certain applications to celestial mechanics and astrodynamics are also conjectured and discussed.

042.060 Is celestial mechanics deterministic?   V. G. Szebehely.
Application of modern dynamics to celestial mechanics and astrodynamics, (see 012.026), p. 321 - 324 (1982).

It is shown that conditions for deterministic systems are neither satisfied by those dynamical systems which are encountered by actual physical situation, nor by problems of interest in celestial mechanics.

042.061 A review of semi-convergent series, periodic solutions and the vanishing Hessian in celestial mechanics.   C. A. Altavista.
Application of modern dynamics to celestial mechanics and astrodynamics, (see 012.026), p. 326 (1982). – Abstract.

042.062 Small bodies captured by a thin annular disk orbiting around a primary body.   V. Banfi.
Application of modern dynamics to celestial mechanics and astrodynamics, (see 012.026), p. 329 - 330 (1982) – Abstract.

042.063 Lie-algebraic methods in dynamics and celestial mechanics.   J. Baumgarte.
Application of modern dynamics to celestial mechanics and astrodynamics, (see 012.026), p. 331 - 332 (1982). – Abstract.

042.064 Parabolic escape and capture in the restricted three-body problem for larger values of the Jacobi constant.   A. Benseny.
Application of modern dynamics to celestial mechanics and astrodynamics, (see 012.026), p. 333 (1982). – Abstract.

042.065 Compatibility conditions for a non-quadratic integral of motion.   G. Bozis.
Application of modern dynamics to celestial mechanics and astrodynamics, (see 012.026), p. 334 - 335 (1982). – Abstract.

042.066 Contact systems and celestial mechanics.   J. Bryant.
Application of modern dynamics to celestial mechanics and astrodynamics, (see 012.026), p. 336 (1982). – Abstract.

042.067 Dimensions of the invariant manifolds associated with equilibrium points in the n-body problem.   J. Casasayas.
Application of modern dynamics to celestial mechanics and astrodynamics, (see 012.026), p. 337 (1982). – Abstract.

042.068 Small divisors in the derivatives of Hansen's coefficients.   S. Ferrer, R. Cid.
Application of modern dynamics to celestial mechanics and astrodynamics, (see 012.026), p. 342 (1982). – Abstract.

042.069 Periodic solutions of the restricted problem of three bodies with small values of the mass parameter.   G. Gómez.

Application of modern dynamics to celestial mechanics and astrodynamics, (see 012.026), p. 343 (1982). – Abstract.

042.070   **Non-universality for a class of bifurcations.**
D. C. Heggie.
Application of modern dynamics to celestial mechanics and astrodynamics, (see 012.026), p. 344 (1982). – Abstract.

042.071   **Application of Lie-series to numerical integration in celestial mechanics.**
A. Hanslmeier, R. Dvorak.
Application of modern dynamics to celestial mechanics and astrodynamics, (see 012.026), p. 345 (1982). – Abstract.

042.072   **About the triple collision manifold in the planar three-body problem.**   M. Irigoyen.
Application of modern dynamics to celestial mechanics and astrodynamics, (see 012.026), p. 348 - 349 (1982). – Abstract

042.073   **Some numerical results of a semi-analytic orbit theory using observed data.**   J. J. F. Liu.
Application of modern dynamics to celestial mechanics and astrodynamics, (see 012.026), p. 351 (1982). – Abstract.

042.074   **A numerical study of the asymptotic solutions to the family (C) of periodic orbits in the restricted problem of three bodies.**   R. Martinez, J. Llibre, C. Simo.
Application of modern dynamics to celestial mechanics and astrodynamics, (see 012.026), p. 354 (1982). – Abstract.

042.075   **The origin of the Kirkwood gaps: a mapping for asteroidal motion near the 3/1 commensurability.**
J. Wisdom.
Application of modern dynamics to celestial mechanics and astrodynamics, (see 012.026), p. 364 (1982). – Abstract.

042.076   **On the symmetry of the restricted three-body problem.**   L. G. Luk'yanov.
Vestn. MGU. Fiz., astron. Tom 22, No. 6, p. 55 - 59 (1981). In Russian. – Abstr. in Ref. zh., 51. Astron., 4.51.71 (1982).

042.077   **Investigation of forms of motion in Barrara's gravitational field.**
V. Ya. Konks, A. I. Prokof'ev.
MGU. Moskva, 1981. 12 pp. In Russian. – Abstr. in Ref. zh., 51. Astron., 4.51.82 (1982).

042.078   **On nearly circular orbits of the satellite of a spherical planet.**   V. Ya. Konks.
MGU. Moskva, 1981. 18 pp. In Russian. – Abstr. in Ref. zh., 51. Astron., 4.51.83 (1982).

042.079   **On a stochastic equivalence of the solutions of differential equations of orbital motion** depending on random parameters.   Yu. I. Trofimtsev.
Differents. ur-niya i ikh pril. Yakutsk, 1981, p. 190 - 194. In Russian. – Abstr. in Ref. zh., 51. Astron., 4.51.97 (1982).

042.080   **Necessary conditions for the existence of planar motions in the problem of translatory-rotational motion of two rigid bodies.**   V. V. Vidyakin.
Pis'ma Astron. Zh., Tom 8, 318 - 320 (1982). In Russian. English translation in Soviet Astron. Lett., Vol. 8.
Necessary conditions in compliance with which the centres of masses of two absolutely rigid bodies can perform translatory motion in the invariable plane are derived. It is proved that the necessary conditions are fulfilled in the case of regular motions of rigid bodies.

042.081   **Sobre un caso particular de un teorema de Lagrange para la solución por series del problema de los tres cuerpos.**   C. A. Altavista.

Bol. Asoc. Argentina Astron., No. 20 - 24, p. 212 (1981). Abstract.

042.082   **Sobre las cualidades de las soluciones de las ecuaciones diferenciales canónicas en un caso particular de la dinámica.**   C. A. Altavista.
Bol. Asoc. Argentina Astron., No. 20 - 24, p. 213 (1981). Abstract.

042.083   **Evolución dinámica de un sistema de tres cuerpos con masas finitas.**   F. López García.
Bol. Asoc. Argentina Astron., No. 20 - 24, p. 214 (1981). Abstract.

042.084   **Evolución dinámica en un sistema de tres cuerpos con masas finitas.**   F. López García.
Bol. Asoc. Argentina Astron., No. 20 - 24, p. 325 (1981). Abstract.

042.085   **El problema de las series formales de la mecánica celeste y las soluciones periódicas de Poincaré.**
C. A. Altavista.
Bol. Asoc. Argentina Astron., No. 20 - 24, p. 419 (1981). Abstract.

042.086   **Evolución dinámica de sistemas de tres cuerpos con masas finitas.**   F. López García, A. Zaragoza.
Bol. Asoc. Argentina Astron., No. 20 - 24, p. 420 (1981). Abstract.

042.087   **Sobre la representación de una función de varias variables en términos de una función de una unica variable.**   C. A. Altavista.
Bol. Asoc. Argentina Astron., No. 20 - 24, p. 421 (1981). Abstract.

042.088   **Sobre una forma particular de las ecuaciones de movimiento en el problema de los tres cuerpos.**
C. A. Altavista.
Bol. Asoc. Argentina Astron., No. 20 - 24, p. 422 (1981). Abstract.

042.089   **Conmensurabilidad triple en el sistema solar.**
S. Fernández.
Bol. Asoc. Argentina Astron., No. 20 - 24, p. 463 (1981). Abstract.

042.090   **On formal and periodic solutions of the three-body problem.**   C. A. Altavista.
Bol. Asoc. Argentina Astron., No. 26, p. 153 - 160 (1981).

042.091   **On formal integration of Lagrange's planetary equations of motion.**   C. A. Altavista.
Bol. Asoc. Argentina Astron., No. 26, p. 161 - 163 (1981).

042.092   **Resultados cualitativos en mecánica celeste.**
F. López García.
Bol. Asoc. Argentina Astron., No. 26, p. 164 (1981). Abstract.

042.093   **Approximación epicíclica de orbitas casi circulares en un potencial con simetría axial y variable con el tiempo.**   J. C. Muzzio.
Bol. Asoc. Argentina Astron., No. 26, p. 165 (1981). Abstract.

042.094   **Determination of the orbit from three observations.**   K. Sandler.
Říše hvězd, Vol. 63, 7 - 10, 29 - 32, 49 - 55 (1982). In Czech.

042.095   **Newtonian potentials and dynamics of a stratiform-heterogeneous ellipsoid.**   B. P. Kondrat'ev.

Astron. Zh., Tom 59, 458 - 470 (1982). In Russian. English translation in Soviet Astron., Vol. 26, No. 3.

The problem of finding the potential of a stratiform-heterogeneous ellipsoid in which constant density strata are presented by a family of coaxial in general nonsimilar ellipsoids is solved. The derived potentials have all the properties of Newtonian potentials. Some dynamical characteristics for the ellipsoid are found.

042.096 **A comparison of the pulsation and lateral modes of oscillations of a homogeneous sphere.**
A. S. Baranov.
Astron. Zh., Tom 59, 471 - 475 (1982). In Russian. English translation in Soviet Astron., Vol. 26, No. 3.

Periods of the non-linear pulsation and the lateral modes of oscillations of an ellipsoid with respect to a sphere are compared with some accuracy. It is found that in both cases the corrections to the frequency of the linearized oscillations appear to be equal. The applicability of the results to a number of astrophysical processes is discussed.

042.097 **Rotating post-Newtonian configurations of a homogeneous magnetized fluid similar to an ellipsoid.** A. N. Tsirulev, V. P. Tsvetkov.
Astron. Zh., Tom 59, 476 - 482 (1982). In Russian. English translation in Soviet Astron., Vol. 26, No. 3.

A method for determination of the equilibrium form of a rapidly rotating gravitating mass of a homogeneous fluid is developed supposing that the effects leading to deflections of the configuration from an ellipsoid are small. The gravitational potential of an arbitrary perturbed ellipsoid is calculated to a necessary accuracy and a system of equations for the determination of the coefficient involved in the equation of the surface is obtained.

042.098 **Mathematical model of the restricted three-body problem in a non-inertial reference system.**
N. P. Plakhtienko.
Prikl. mekh., Tom 17, No. 12, p. 103 - 107 (1981). In Russian. Abstr. in Ref. zh., 62. Issled. kosm. prostranstva, 5.62.138 (1982).

042.099 **Investigation of the motion of a body with variable mass in a Newtonian gravitational field with two fixed centres.** Ts. G. Chitaladze.
Tr. Tbilis. univ., Tom 218, 217 - 227 (1981). In Russian. Abstr. in Ref. zh., 62. Issled. kosm. prostranstva, 5.62.144 (1982).

042.100 **Motion in the plane of symmetry in the N-centres problem.** Z. F. Seidov.
Dokl. AN AzSSR, Tom 37, No. 12, p. 20 - 22 (1981). In Russian. – Abstr. in Ref. zh., 51. Astron., 5.51.78 (1982).

042.101 **Expansion theory for the elliptic motion of arbitrary eccentricity and semi-major axis. IV. Regularisation approach by using sectors independent variables.** M. A. Sharaf.
Astrophys. Space Sci., Vol. 84, 73 - 97 (1982).

In this paper a regularisation approach is proposed based on the idea of orbit segmentation into sectors. For this approach, general transformation equations to the sectors independent variables will be established. The author develops the formulation of general and arbitrary divisions of elliptic orbits through one function for all sectors so as to facilitate analytically and computationally the expansion theory in terms of the sectors independent variables. Numerical results for the parameters of some divisions are included, and finally a suggested function for the analyses is given with the corresponding elliptic equations.

042.102 **High precision planetary theories by Carpenter's method.** L. E. Doggett.
Bull. American Astron. Soc., Vol. 14, 580 (1982). – Abstract.

042.103 **Recent progress in the theory of the T. A.**
B. Garfinkel.
Bull. American Astron. Soc., Vol. 14, 581 (1982). – Abstract.

042.104 **Quantisation in stable gravitational systems.**
R. Wayte.
Moon Planets, Vol. 26, 11 - 32 (1982).

The similarities between gravitational and electric forces are used as a basis for explaining the stability of the various parts of the Solar System in terms of gravitational guiding waves, or quanta. This leads to commensurability and quantisation of all orbital and spin parameters. Ultimately, the theory proves that gravity is an electromagnetic quantum phenomenon.

042.105 **The construction of a general third-order planetary theory. Part I: Outline.** O. M. Kamel.
Moon Planets, Vol. 26, 47 - 60 (1982).

The author reviews in this part the outline of a third-order general planetary theory established through Von Zeipel's method and in terms of Poincaré's canonical variables. He considers his system to consist of the Sun as the primary body, one disturbed planet, and one disturbing planet.

042.106 **The rotation rates of accreting planetesimals.**
G. P. Horedt.
Moon Planets, Vol. 26, 89 - 92 (1982).

There are obtained upper limits for the relative velocity at infinity of accreting planetesimals for a nearly constant mass of the largest accreting planetesimal and also in the case of variable mass. The author concludes that while large planets cannot be brought to the stage of rotational instability by stochastic collisions, the asteroids could be brought if the relative velocities in the asteroid belt were larger than about 2 km s^{-1}.

042.107 **A postulate leading to the Titius–Bode law.**
R. Louise.
Moon Planets, Vol. 26, 93 - 96 (1982).

The revised Titius–Bode law (Balsano and Hughes, 1979) giving distances of planets from the Sun into integers that recall the Bohr law in the early quantum theory of hydrogen atom. The author found a formalism, similar to the Sommerfeld's one, accounting for the planetary distribution of the solar system.

042.108 **Solution of canonical equations which result from elimination of short-periodic terms in second-order planetary theory.** J. Meffroy.
Moon Planets, Vol. 26, 143 - 169 (1982).

The author solves the first order non-linear differential equation and calculates the two quadratures to which are reduced the canonical differential equations resulting from the elimination of the short period terms in a second order planetary theory carried out through Hori's method and slow Delaunay canonical variables with powers of eccentricities and of the sines of semi-inclinations greater than 3 are neglected and the eccentricity of the disturbing planet is identically equal to zero. The procedure can be extended to the case when the eccentricity of the disturbing planet is not identically equal to zero. In this latter general case, the author calculates the two quadratures expressing angular slow Delaunay canonical variables λ_1' and λ_2' of the disturbed planet and of the disturbing planet in terms of time t respectively.

042.109 **Saturn's rings and bimodality of Keplerian systems.** K. A. Hämeen-Anttila.
Moon Planets, Vol. 26, 171 - 196 (1982).

The correction terms which are introduced by non-zero size of the particles into the mechanics of Keplerian systems can be replaced by relatively simple approximations which agree with computer simulations. The theory of finite particles confirms the bimodality of collisional systems. In Saturn's rings the ringlets correspond to the 'degenerate' mode while the matter which fills the gaps is in the 'non-degenerate' state. The transition from one mode to the other which is needed to create a dense ring in a cloud of small particles follows from the growth of mass in the central body. This may be a recently-formed planet; but, more probably, the transition occurs in a loose pre-planetary disc.

042.110   **Expansion of the disturbing function by factorization.**   R. Broucke, W. Presler.
High-precision Earth rotation and Earth-Moon dynamics. Lunar distances and related observations, (see 012.035), p. 337 - 348 (1982).
    The authors present an expansion of the disturbing function for the third-body perturbations.

042.111   **Evolution of orbits in the planar restricted elliptical twice-averaged three-body problem.**
M. A. Vashkov'yak.
Kosm. Issled., Tom 20, 332 - 341 (1982). In Russian.

042.112   **Periodic solutions of the third kind in the problem of motion of a satellite of a "triaxial" rotating** solid body with account for perturbations from a sufficiently distant solid body.   A. S. Sarkisyan.
Kosm. Issled., Tom 20, 474 - 479 (1982). In Russian.

042.113   **Motion of the Jovian commensurability resonances and the character of the celestial mechanics in the** asteroid zone: implications for kinematics and structure.
M. Torbett, R. Smoluchowski.
Astron. Astrophys., Vol. 110, 43 - 49 (1982).
    The motion of the Jovian commensurability resonances during the early evolution of the solar system induced by the dissipation of the accretion disk results in fundamental differences in the celestial mechanics of objects over which a resonance passes from that observed for a stationary resonance. Objects experiencing resonance passage acquire irreversible increases of average eccentricity to large values accounting for the present-day random velocities of the asteroids. Semi-major axes are similarly irreversibly decreased by amounts capable of clearing the Kirkwood gaps. The gap widths are in agreement with observation.

042.114   **On the stability of the triangular points in the elliptic restricted problem.**   R. Meire.
Astron. Astrophys., Vol. 110, 152 - 155 (1982).
    New results are given for the linear stability of the triangular points in the elliptic restricted problem. For the first time, the author is able to show the existence of a stable region in the $\mu - e$ plane on a purely analytical basis. This is accomplished in a rather new kind of approach to the problem.

042.115   **Algorithm for constructing an approximate solution of a system of equations of celestial** mechanics with delaying argument.   G. D. Penev.
Vestn. LGU, 1982, No. 1, p. 93 - 100. In Russian. – Abstr. in Ref. zh., 51. Astron., 6.51.64 (1982).

042.116   **On the use of a semi-analytical method for calculation of perturbations in the motions of asteroids at** encounters.   J. L. Simovljević.
Glas. Srpska akad. nauka i umetn. Od. prirod.-mat. nauka, Tom 324, No. 47, p. 1 - 6 (1981). In Serbo-Croatian.
Abstr. in Ref. zh., 51. Astron. 6.51.70 (1982).

042.117   **On the use of vector elements for the calculation of particular perturbations of asteroidal orbits at** encounters.   J. L. Simovljević.
Glas. Srpska akad. nauka i umetn. Od. prirod.-mat. nauka, Tom 324, No. 47, p. 7 - 16 (1981). In Serbo-Croatian.
Abstr. in Ref. zh., 51. Astron., 6.51.71 (1982).

042.118   **Analytical construction of an optimum orbit in the presence of secular perturbations from nonsphericity** of a planet and outer bodies.   V. S. Novoselov.
Vestn. LGU, 1982, No. 1, p. 86 - 93. In Russian. – Abstr. in Ref. zh., 51. Astron., 6.51.81 (1982).

042.119   **A new method for expression of the perturbation function.**   V. G. Shkodrov.
Sun and planetary system, (see 012.040), p. 421 - 422 (1982).

042.120   **On the secular effects in the motion of a planetary satellite.**   V. G. Ivanova.
Sun and planetary system, (see 012.040), p. 429 - 430 (1982).
    A qualitative analysis has been made and secular disturbances of the first order in the motion of a trial body in a zonal gravitation field have been obtained.

042.121   **L'invariabilité séculaire des grands demiaxes des orbites planétaires.**   B. Popović.
Sun and planetary system, (see 012.040), p. 431 - 433 (1982).

042.122   **On linear stability of triangular libration points of the photogravitational restricted three-body problem when the more massive primary is an oblate spheroid.**
R. K. Sharma.
Sun and planetary system, (see 012.040), p. 435 - 436 (1982).

042.123   **Nonsemisimple 1:1 resonance at an equilibrium.**   J.-C. van der Meer.
Celestial Mech., Vol. 27, 131 - 149 (1982).
    Consider a Hamiltonian system of two degrees of freedom at an equilibrium. Suppose that the linearized vectorfield has eigenvalues $i\alpha, i\alpha, -i\alpha, -i\alpha (\alpha \in R, \alpha > 0)$ and is not semisimple. In this paper the author discusses the real normalization of the Hamiltonian function of such a system. He normalizes the Hamiltonian up to 4th order and shows how to compute its coefficients. For the planar restricted three body problem at L_4 the coefficient that plays an important role in the investigation of the qualitative behaviour of periodic solutions near the equilibrium is explicitly calculated.

042.124   **Intrinsic variational equations for conservative dynamical systems with n degrees of freedom.**
J. I. Palmore.
Celestial Mech., Vol. 27, 151 - 156 (1982).
    For a conservative dynamical system with n degrees of freedom the author shows that the equations of variation along an orbit may be written with respect to an orthonormal moving frame (a generalized Frenet frame) in which the tangential variation is given by a quadrature and the normal and $n - 2$ binormal variations are solutions of $n - 1$ coupled second order equations of the form of Hill's equation.

042.125   **Study of the n-tuple collision in the n-body problem with a homogeneous potential of degree $-k$ in the** case of $k > 0$ or $k = 2$.   D. Eschbach.
Celestial Mech., Vol. 27, 157 - 166 (1982). In French.
    In a preceding paper, the author used a change of variables which is simple and well adapted to the problem of triple collision and which is closely linked with the homogeneousness of the kinetic energy T (degree 2) and the potential U (degree -1). He now generalizes this change of variables to the case where U is homogeneous of degree $-k$ $(k > 0)$, in order to carry out the study of collision in these cases, bringing out the exceptional case $k = 2$.

042.126 Some new concepts in the n-body and 3-body
        problems.   A. Kyrala.
Celestial Mech., Vol. 27, 167 - 178 (1982).
    The n-body problem in 3-space for point masses of
arbitrary magnitude is approached by introduction of a
weighted harmonic mean separation and an rms velocity of the
particles. It is shown how these averages may be expressed in
terms of a single parameter for each of the cases of positive
and negative total energy. The general problems of escape and
collision are classified by the introduction of escape, rest and
collision polynomials. For systems with a non-null total
angular momentum it is shown how an rms angular momentum
may be constructed and used with a harmonic mean centroidal
moment of inertia. In the 3-body problem a graphical construction of the equipotentials equivalent to a numerical algorithm
is given. Finally the possibility of referencing 3-body motions
to the apex (point of least average separation) is discussed.

042.127 The study of planetary secular perturbations.
        J.- x. Zhang.
Acta Astron. Sinica, Vol. 23, 56 - 64 (1982). In Chinese.

042.128 Calculation of the first-order theory of Mars by
        Hansen's method.   D.- z. Xian, Z.- k. Yu.
Acta Astron. Sinica, Vol. 23, 65 - 68 (1982). In Chinese.

042.129 Hill stability for the general three-body problem.
        H. Ding, T.- y. Huang.
Acta Astron. Sinica, Vol. 23, 69 - 75 (1982). In Chinese.

042.130 Orbital patterns of interplanetary objects at close
        encounters with Jupiter.
A. Carusi, L. Kresák, G. B. Valsecchi.
Bull. Astron. Inst. Czechoslovakia, Vol. 33, 141 - 150 (1982).
    Orbital patterns of 180 objects experiencing low-velocity
encounters with Jupiter are examined. One group of these
objects consists of a chain of 80 fictitious comets supposed to
move along the orbit of P/Oterma before its approach to
Jupiter in 1934. The other group is a semi-random sample of
100 test objects of different eccentricity, with initial orbits
nearly tangent to that of Jupiter, either near perihelion or near
aphelion.

042.131 Report of IAU Commission 7: Celestial mechanics
        (Mécanique céleste).   Y. Kozai.
Trans. IAU, Vol. XVIIIA, (see 003.013), p. 21 - 32 (1982).

        Hemelmechanica: het tweelichamenprobleem.
See Abstr. 003.050.

        Motion of artificial and natural celestial bodies.
See Abstr. 003.109.

        Theory of orbits. The restricted problem of three
bodies.   See Abstr. 003.127.

        Kalender und Chronologie. Bekanntes und
Unbekanntes aus der Kalenderwissenschaft.
See Abstr. 003.142.

        Error propagation in the numerical solutions of the
differential equations of orbital mechanics.
See Abstr. 021.012.

        Ergodic theory and area preserving mappings.
See Abstr. 021.020.

        On some invariant manifold results and their
applications.   See Abstr. 021.021.

        Expansion theory for the elliptic motion of
arbitrary eccentricity and semi-major axis.III.
See Abstr. 021.039.

        High-order multivalue integrators.
See Abstr. 021.040.

        Construction of conditionally-periodic solutions of
canonical systems of differential equations with multiple
resonance.   See Abstr. 021.048.

        Studies in the application of recurrence relations to
special perturbation methods. VI: Comparison with classical
single-step and multi-step methods of numerical integration.
See Abstr. 021.049.

        Propagation of local errors in the solutions of the
differential equations for orbital elements.
See Abstr. 021.050.

        Singularity of Störmer's problem.
See Abstr. 022.099.

        Modern and old views in the dynamical foundations
of classical statistical mechanics.   See Abstr. 022.123.

        On the applicability of the transition chains
mechanism.   See Abstr. 022.125.

        Applications of Hamilton's Law of Varying Action.
See Abstr. 022.126.

        Periodic orbits near homoclinic orbits.
See Abstr. 022.127.

        Stability of periodic orbits near a homoclinic orbit
for analytical Hamiltonians with two degrees of freedom.
See Abstr. 022.129.

        A method for the investigation of the integrability
of dynamical systems.   See Abstr. 022.130.

        The behaviour of adiabatic invariants near reso-
nances.   See Abstr. 022.198.

        Dynamical equinox and analytical theory of the Sun.
See Abstr. 041.003.

        Reference systems for Earth dynamics.
See Abstr. 043.001.

        Dependence of the lunisolar perturbations in the
Earth rotation on the adopted Earth model.
See Abstr. 044.051.

        On the direct influence of the planets on the
precession and nutation of the Earth's axis of rotation.
See Abstr. 045.001.

        Three-dimensional subsatellite motion under air drag
and oblateness perturbations.   See Abstr. 052.004.

        On the motion of a satellite in resonance with its
rotating planet.   See Abstr. 052.005.

        On the symmetric difference quotient and its
application to the correction of orbits.   See Abstr. 052.007.

        On the development of an artificial satellite theory.
See Abstr. 052.032.

A comparison of Earth models by degree of their harmonic coefficients.   See Abstr. 081.002.

Relativistic perturbations for all the planets. See Abstr. 091.001.

On the invariable plane of the solar system. See Abstr. 091.006.

Orbital motion of the planets, theoretical and observational.   See Abstr. 091.028.

Theory of the inner planets.   See Abstr. 091.029.

The stability of the solar system. See Abstr. 091.050.

Progress in the analytical theories for the orbital motion of the Moon.   See Abstr. 094.010.

The ELP solution for the main problem of the Moon and some applications.   See Abstr. 094.011.

Comparison of SALE with numerical integration. See Abstr. 094.012.

The main problem of lunar theory solved by the method of Brown.   See Abstr. 094.013.

New approach to determining planetary perturbations in lunar theory.   See Abstr. 094.014.

Planetary perturbations of the Moon in ELP 2000. See Abstr. 094.015.

Perturbations by the oblateness of Earth and Moon. See Abstr. 094.016.

Perturbations by the oblateness of the Earth and by the planets in the motion of the Moon. See Abstr. 094.017.

Comments about the direct perturbations of Venus and Mars on the Moon's motion.   See Abstr. 094.018.

Physical libration of the Moon. See Abstr. 094.021.

The relativistic planetary perturbations and the orbital motion of the Moon.   See Abstr. 094.051.

Mars theory and its comparison with numerical integration.   See Abstr. 097.080.

Recent progress in the theory of the Trojan asteroids.   See Abstr. 098.038.

Motion at the second order resonances, 3 : 1 and 5 : 3.   See Abstr. 098.039.

On the stability of the asteroids. See Abstr. 098.106.

Secular and long period effects in the orbits of the Galilean satellites.   See Abstr. 099.040.

Determination of the semi-major axes of the Galilean satellites of Jupiter.   See Abstr. 099.042.

Astrometric observations of the satellites of the outer planets. V. The oppositions of 1978 - 1979, 1980, and 1981.   See Abstr. 099.089.

Literal theory of the ninth satellite of Saturn, Phoebe.   See Abstr. 100.035.

Perturbations in the motion of Saturn satellites. Part 4.   See Abstr. 100.068.

The evolution of the Earth-Moon system. See Abstr. 107.028.

The effect of orbital eccentricity on polarimetric binary diagnostics.   See Abstr. 117.009.

Spin nutation in binary systems due to general relativistic and quadrupole effects.   See Abstr. 117.011.

On the linear adiabatic oscillations of a uniformly and synchronously rotating component of a binary. See Abstr. 117.029.

Radiation pressure effect on dust surrounding the Pleiades bright stars.   See Abstr. 131.266.

Computer simulations of close encounters between binary and single stars: the effect of the impact velocity and the stellar masses.   See Abstr. 151.021.

Erratum

042.901   Erratum: "Non-linear axisymmetric oscillations of a homogeneous sphere" [Astron. Zh., Tom 57, 968 - 974 (1980). In Russian.]   A. S. Baranov. Astron. Zh., Tom 59, 204 (1982). In Russian. – See Abstr. 28.042.044.

043 Astronomical Constants, Reference Systems

043.001 **Reference systems for Earth dynamics.**
R. O. Vicente.
Application of modern dynamics to celestial mechanics and astrodynamics, (see 012.026), p. 131 - 144 (1982).
A comparison is presented of reference systems employed by classical and modern techniques of observing the Earth's rotation. Advantages and disadvantages of the systems so far employed are discussed. Any future system of reference for the Earth's dynamics should be well defined in order to avoid past ambiguities. The future of the Conventional International Origin and proposals about the Conventional Terrestrial System to be adopted by international agreement are presented.

043.002 **Transition year for astronomy.**
P. K. Seidelmann, G. H. Kaplan.
Bull. American Astron. Soc., Vol. 13, 874 (1981). – Abstract.

043.003 **The IAU resolutions on astronomical constants, time scales, and the fundamental reference frame.**
G. H. Kaplan (Editor).
United States Naval Obs. Circ., No. 163, 15 + A8 + B5 + C4 pp. (1981).
This Circular presents the resolutions recently adopted by the International Astronomical Union (IAU) regarding constants, time scales, and the new fundamental astronomical reference frame, the FK5. The resolutions are intended to apply to the reduction of observations taken on or after 1 January 1984, and to be used in the preparation of ephemerides for the years 1984 and beyond.

043.004 **Locating the equator and equinox of J2000.**
E. M. Standish.
Bull. American Astron. Soc., Vol. 14, 580 (1982). – Abstract.

043.005 **The use of minor planet dynamics toward improving the fundamental reference system.**
P. D. Hemenway, R. L. Duncombe.
Bull. American Astron. Soc., Vol. 14, 581 (1982). – Abstract.

043.006 **Is the gravitational constant changing?**
T. C. Van Flandern.
High-precision Earth rotation and Earth-Moon dynamics. Lunar distances and related observations, (see 012.035), p. 207 - 208 (1982). – Abstract.

043.007 **Relations between celestial and selenocentric reference frames.** J. Kovalevsky.
High-precision Earth rotation and Earth-Moon dynamics. Lunar distances and related observations, (see 012.035), p. 269 - 280 (1982).
The very great accuracy with which the motions of the Moon can now be monitored by laser ranging, differential VLBI and occultation observations, implies that the interpretation of the measurements is conditioned by the choice and the accurate knowledge of a selenocentric, a terrestrial and a celestial frame. Two different types of selenocentric reference frames can be envisioned. The present selenographic frames are discussed but the author proposes that one should introduce a system defined by a purely geometric means. Some consequences of such a choice are discussed.

043.008 **On the absolute orientation of the selenodetic reference frame.** V. S. Kislyuk.
High-precision Earth rotation and Earth-Moon dynamics. Lunar distances and related observations, (see 012.035), p. 281 - 286 (1982).
The selection of selenodetic reference coordinate system is an important problem in astronomy and selenodesy. For the

purposes of reduction of observations, planning and executing space missions to the Moon, it is necessary, in any case, to know the orientation of the adopted selenodetic reference system in respect to the inertial coordinate system.

043.009 **On the accuracy of the 1980 IAU Nutation Series.**
Ya. S. Yatskiv, S. M. Molodensky *(Molodenskij)*.
High-precision Earth rotation and Earth-Moon dynamics. Lunar distances and related observations, (see 012.035), p. 287 - 292 (1982).
(1) The 1980 IAU Nutation Series is based on the best Earth model, available presently. The accuracy of this series is of the order of $\pm 0\overset{''}{.}001$, which is better than those that have been obtained from long series of astronomical observations. (2) The effects of the oceans and core-mantle dissipative coupling appear to be remarkable and should be a subject of further study with using the new observational techniques.

043.010 **Comments on the effect of adopting new precession and equinox corrections.**
J. G. Williams, W. G. Melbourne.
High-precision Earth rotation and Earth-Moon dynamics. Lunar distances and related observations, (see 012.035), p. 293 - 303 (1982).
It is shown that the derived terrestrial longitude zero point and UT1 rate will be different when determined by classical optical and space (inertial) techniques so long as the precession constant and equinox offset and drift are imperfectly known. For the uncertainty of the new IAU precession constant, this inconsistency is expected to be nearly $0\overset{''}{.}04$ in longitude and 0.1 ms/yr in UT1 rate. The consistency of constants does not guarantee consistent results. The classical practice of defining Greenwich Mean Sidereal Time as invariant with respect to a moving equinox must be replaced by invariance with respect to a fixed equinox if the longitude and UT1 results from the inertial techniques are to be stable against changes in the precession constant.

043.011 **Redetermination of the Newtonian gravitational constant G.** G. G. Luther, W. R. Towler.
Phys. Rev. Lett., Vol. 48, 121 - 123 (1982).
The universal Newtonian gravitational constant is being redetermined at the National Bureau of Standards with use of the method of Boyes in which the period of a torsion pendulum is altered by the presence of two 10.5-kg tungsten balls. The difference in the squares of the frequencies with and without the balls is proportional to G. The resulting value of G is $(6.6726 \pm 0.0005) \times 10^{-11} \, \mathrm{m}^3 \, \mathrm{sec}^{-2} \mathrm{kg}^{-1}$.

043.012 **Effects of the non-rigidity of the Earth derived from astronomical observations.** N. Capitaine.
Sun and planetary system, (see 012.040), p. 185 - 188 (1982).
The non-rigidity of the Earth affects the astronomical observations in three ways: the tidal deflexion of the local vertical of the observer, the non-coincidence of the instantaneous rotation axis with the adopted one corresponding to a rigid Earth and periodic variations of the rotation rate. The principal tidal and nutational effects have been derived from latitude and time data of the Paris astrolabe from 1956.6 to 1979.0.

043.013 **Reference systems linkage from space.**
P. Farinella, A. Milani, A. M. Nobili, F. Sacerdote.
Sun and planetary system, (see 012.040), p. 197 - 198 (1982).

043.014 **Contribution of the Pulkovo Observatory to the improvement of orientation of the FK4 system using observations of selected minor planets.**

L. S. Koroleva, V. I. Orelskaya (*Orel'skaya*).
Sun and planetary system, (see 012.040), p. 449 - 451 (1982).

043.015 Determination of positions and the reference system.   W. Fricke.
Scientific aspects of the Hipparcos space astrometry mission, (see 012.041), p. 43 - 47 (1982).
 Presented is a review of the methods for the determination of positions and their purposes. Reference systems have been set up on the basis of absolute observations of stars; the inclusion of objects of the planetary system has yielded a "fundamental reference system" with the dynamical equinox and the equator as the zero points of the coordinates. Recent absolute and differential observations for the construction of the FK5 are discussed, and information is given on their accuracy. Reported are the main changes which will occur in the FK5. Finally, the relations are discussed between space astrometry and ground based observations including radio-interferometric observations of extragalactic sources.

043.016 Report by IAU Commission 24 Working Group on Radio/Optical Astrometric Sources for the establishment of an inertial reference frame.   A. N. Argue.
Scientific aspects of the Hipparcos space astrometry mission, (see 012.041), p. 57 - 59 (1982).

043.017 Fiducial reference for the Hipparcos reference system.   P. Hemenway, W. H. Jeffreys,
P. J. Shelus, R. L. Duncombe.
Scientific aspects of the Hipparcos space astrometry mission, (see 012.041), p. 61 - 63 (1982).
 Quasi-stellar objects (including BL Lacertae objects) are expected to form a cosmological reference system against which the Hipparcos instrumental system will be calibrated. One means of relating the bright Hipparcos stars (9 - 11 magnitude) directly to optically identified radio sources with accurate VLBI positions will be to observe a set of QSOs directly with respect to Hipparcos stars, using the astrometric capabilities of the NASA/ESA Space Telescope.

043.018 Inertial frame determination using minor planets – a proposal for Hipparcos observations.
S. Söderhjelm, L. Lindegren.
Scientific aspects of the Hipparcos space astrometry mission, (see 012.041), p. 191 - 192 (1982).
 The dynamical method defining an inertial system by minor planet observations is suggested for Hipparcos. A simulation experiment shows that interesting results can be expected if more than about ten asteroids are observed. The absolute rotation of the Hipparcos reference system should be measurable to within a few tenths of an arcsec per century, which is comparable to the estimated uncertainty of the FK 5 system of proper motions.

043.019 Catalogue Hipparcos et observations au sol.
F. Chollet, S. Débarbat.

Scientific aspects of the Hipparcos space astrometry mission, (see 012.041), p. 223 - 225 (1982).

043.020 Fiducial reference for the HIPPARCOS reference system using the Space Telescope.
P. Hemenway.
Abh. Hamburger Sternw., Band 10, 139 - 143 (1982).

043.021 Some problems on the detection of the nutation constant.   C.-l. Lu.
Acta Astron. Sinica, Vol. 23, 82 - 88 (1982). In Chinese.

 Candidate radio sources for a radio/optical reference catalog.   See Abstr. 002.071.

 Beiträge der Berliner Sternwarte zu bedeutenden Erkenntnissen auf dem Gebiet der klassischen Astronomie. See Abstr. 004.086.

 1980 IAU Theory of Nutation: the final report of the IAU Working Group on Nutation.   See Abstr. 011.019.

 Analysis of lunar occultations – IV. Rotation of the FK4 reference frame.   See Abstr. 041.001.

 Determination of the equinox and equator of the FK5.   See Abstr. 041.002.

 Dynamical equinox and analytical theory of the Sun. See Abstr. 041.003.

 Resultats d'observations du soleil à l'astrolabe du CERGA.   See Abstr. 041.035.

 Radio stars as connecting link of the Hipparcos and VLBI reference frames.   See Abstr. 041.038.

 US Naval Observatory parallaxes and the fundamental reference frame – their interaction with Hipparcos. See Abstr. 041.039.

 The new definition of universal time. See Abstr. 044.003.

 Astronomical and geophysical problems of the Earth's rotation.   See Abstr. 044.057.

 On the invariable plane of the solar system. See Abstr. 091.006.

 Recent results for the Moon's secular acceleration and their implication for the possible variation of G in Dirac's Large Number Hypothesis.   See Abstr. 094.063.

044 Time, Rotation of the Earth

044.001 The analemmas of the planets.
D. A. Harvey.
Sky Telesc., Vol. 63, 237 - 239 (1982).

044.002 Some terms of nutation derived from the BIH data.
N. Capitaine, N. Xiao.
Geophys. J. R. Astron. Soc., Vol. 68, 805 - 814 (1982).

The z-term of latitude and the w-term of UT, as computed by the BIH, are suitable for investigating imperfections in the representation of the real nutation in space. They have been previously used for deriving only the amplitude of the principal term of nutation. In this study they are used more generally for deriving the amplitudes of two terms of nutation and searching for a possible nutation in space due to the nearly-diurnal wobble.

044.003 The new definition of universal time.
S. Aoki, B. Guinot, G. H. Kaplan, H. Kinoshita,
D. D. McCarthy, P. K. Seidelmann.
Astron. Astrophys., Vol. 105, 359 - 361 (1982).

The planned improvement to the astronomical reference system, scheduled for 1 January 1984, will have profound effects on astrometry and timekeeping. In this paper, a new expression for the relationship between universal and sidereal time is developed for use with the new reference system.

044.004 Short period tidal variations of Earth rotation.
C. F. Yoder, J. G. Williams, M. E. Parke,
J. O. Dickey.
Ann. Géophys., Vol. 37, 213 - 217 (1981). − Abstr. in Phys. Abstr., Vol. 85, Abstr. 9905 (1982).

044.005 Love number _k_ determined from astronomical observations of the Earth's rotation.   J. Hefty.
Bull. Astron. Inst. Czechoslovakia, Vol. 33, 84 - 88 (1982).

Love number _k_ was estimated from measurements of universal time during the years 1967 - 1978 using two methods of spectral estimation: Fourier's transformation of the autocorrelation function and maximum entropy; the results of the two methods agree well. After dividing the time measurements into intervals of three years, the systematic differences in the constant _k_ were determined for the individual intervals. The values obtained from the analysis indicate that short-term changes in the Earth's rotation with periods of up to one month are probably due to other dynamic factors besides tidal deformations.

044.006 Why is the earth a poor timekeeper? Comments on "The earth's variable rotation: geophysical causes and consequences" by Kurt Lambeck.   W. Munk.
Geophys. J. R. Astron. Soc., Vol. 69, 831 - 835 (1982).

044.007 A comparison of recent theoretical results on the short-period terms in the length of day.
J. B. Merriam.
Geophys. J. R. Astron. Soc., Vol. 69, 837 - 840 (1982).

Recent results on the relation between the zonal lunisolar tide potentials and the changes in length of day they produce are examined and the differences explained.

044.008 Deceleration of the Earth's rotation from old solar tables.   S. J. Goldstein, Jr.
Celestial Mech., Vol. 27, 53 - 63 (1982).

The solar tables of ibn Yunis and of King Alfonso, those of Kepler, of G. D. and J.-J. Cassini, and of Lalande are compared with Newcomb's theory of the Sun to determine the deceleration of the Earth's rotation. Comparisons of mean motion and of longitude lead to separate determinations. A value for the deceleration, $\dot{P}/P = (5.0 \pm 2.7) \times 10^{-10}\,\mathrm{yr}^{-1}$, assumed to be independent of time is obtained for the period −146 through 1892.

044.009 Variations of the tidal torque in a "short" time scale.   P. Brosche, W. Hövel.
Naturwissenschaften, 69. Jahrg., 241 (1982).

A variation of the tidal torque of a factor 2 is produced by continental drift already within a few 10 million years.

044.010 On the solutions of the earth's rotation parameters from classical observations for time and latitude.
Acta Astron. Sinica, Vol. 22, 389 - 394 (1981). In Chinese.

044.011 Primer año de observación con el astrolabio de Danjon OPL 01 en Punta Indio.
C. A. Mondinalli, B. B. Jezieniecki.
Bol. Asoc. Argentina Astron., No. 20 - 24, p. 7 - 16 (1981).

044.012 Tablas para vincular relojes de tiempo solar y sideral.   T. Paneth.
Bol. Asoc. Argentina Astron., No. 20 - 24, p. 298 - 299 (1981).

044.013 Proyecto MERIT. Nuevas técnicas para la determinación de la rotación de la tierra.
R. A. Perdomo.
Bol. Asoc. Argentina Astron., No. 26, p. 52 - 58 (1981).

044.014 Resultados de tiempo y latitud en la Estación Astronómica Río Grande.
C. A. Mondinalli, R. A. Perdomo.
Bol. Asoc. Argentina Astron., No. 26, p. 61 - 66 (1981).

044.015 An explanation of the long-term retardation in the rotation of the Earth.   G. X. Song.
Special issue on astronomical geodynamics, (see 012.028), p. 40 - 44 (1981). In Chinese.

Long-term retardation in the rotation of the Earth is induced in terms of conservation of the angular momentum. It follows from the calculation that the angular acceleration caused in this way is about $1.56 \times 10^{-22}\,\mathrm{rad/sec^2}$.

044.016 The rate of rotation of the Earth and the atmospheric circulation.   Z. A. Li, W. Z. Ma, S. Y. Du.
Special issue on astronomical geodynamics, (see 012.028), p. 45 - 50 (1981). In Chinese.

In this paper, a relationship between the rate of the Earth's rotation and the atmospheric circulation for the 500 mb level in the northern hemisphere is discussed.

044.017 A preliminary analysis to the external stability of the Chinese Joint system.
S. J. Gong, S. F. Luo.
Special issue on astronomical geodynamics, (see 012.028), p. 51 - 55 (1981). In Chinese.

Differences between the Chinese Joint system and BIH system (UT1) in 1964−1978 are analysed. It is discovered that there are obvious annual and semi-annual terms in the Chinese system. The systems of 6 instruments in Asia and 3 instruments in North America are also analysed. An analogous result is obtained. This shows that the periodic terms in the Chinese Joint system are probably due to a regional effect.

044.018 Improvement of the Universal Time Service by means of AR series model.
D. W. Zhen, H. Y. Huang, D. C. Liao, S. F. Luo.
Special issue on astronomical geodynamics, (see 012.028), p. 168 (1981). In Chinese and English. − Abstract.

044.019 **The effect of environment on the time determination with transit.**   D. H. Li, J. Guo, H. Q. Liu.
Special issue on astronomical geodynamics, (see 012.028), p. 169 - 170 (1981). In Chinese and English. − Abstract.

044.020 **Time and Latitude Service, 1981 April - September.**
Circ. Time Latitude Serv., Nos. 158 - 159 (1981).

044.021 **Time Service for the year 1979.**   L. Mureddu.
Circ. Stn. Astron. Int. Latitudine, Carloforte-Cagliari, N. 15, (Ser. A(6)), 21 pp. (1980).

044.022 **Time Service for the year 1980.**   L. Mureddu.
Circ. Stn. Astron. Int. Latitudine, Carloforte-Cagliari, N. 16, (Ser A (7)), 22 pp. (1981).

044.023 **Analysis of systematic errors in earth rotation data.**
F. N. Withington, D. D. McCarthy, A. K. Babcock.
Bull. American Astron. Soc., Vol. 13, 876 (1981). − Abstract.

044.024 **Mark III VLBI: UT1, polar motion, and baselines.**
T. A. Clark, J. W. Ryan, D. B. Shaffer.
Bull. American Astron. Soc., Vol. 13, 899 (1981). − Abstract.

044.025 **Random and long periodic variations in the earth's motion.**   A. Poma, E. Proverbio.
J. Interdiscipl. Cycle Res., Vol. 12, 237 - 246 (1981) = Pubbl. Stn. Astron. Int. Latitudine, Carloforte-Cagliari, N. 79.

044.026 **Greenwich Time Report.** 1981 April - September.
A. Boksenberg.
R. Greenwich Obs., Time Latitude Serv., p. 257 - 286 (1981).

044.027 **Zeit- und Breitendienst.** Juli - Dezember 1981.
Deutsches Hydrogr. Inst., Hamburg, 13 pp. (1981/82).

044.028 **Bulletins, Time Service of the Mizusawa Observatory, Vol. 25, 1980.**   C. Kakuta.
Published by Int. Latitude Obs. Mizusawa, Mizusawa-Shi, Iwate-Ken, Japan. 34 pp. (1981). ISSN 0580-6585.

044.029 **Time and Latitude.** April - September 1981.
V. Ptáček, J. Vondrák, R. Weber.
Circ. Czechoslovak Obs., Time and Latitude, 40 pp. (1981).

044.030 **Bureau International de l'Heure. Annual report for 1981.**
Prepared with the participation of the Bureau International des Poids et Mesures. Bureau International de l'Heure, Paris, France. 5 + A14 + B53 + C11 + D102 pp. (1982).
    Contents: Part A: Explanation of tables and other information, Part B: Tables and figures, Part C: time signals, Part D: Earth rotation. Series of measurements by various techniques; comparisons.

044.031 **Bureau International de l'Heure (B.I.H.) Circular D.** January - June 1982.
B.I.H., Paris, Circ. D182 - D187 (1982).
    Coordinated Universal Time UTC. International Atomic Time. Universal Time and coordinates of the pole. Short term variation of UT1 and of the duration of the day. Data of individual networks. Informations on time signals.

044.032 **Bureau International de l'Heure (B.I.H.) Circular E.**
B.I.H., Paris, Circ E11 (1982).
    UTC time step on the 1st of July 1982.

044.033 **Time and Frequency Services Bulletin,** January - March 1982.
Published by Shaanxi Astronomical Observatory, Academia Sinica (1982).

044.034 **Time Service Annual Report 1980.**
Xu-Jia-Hui Sect., Shanghai Obs., Acad. Sinica, Shanghai, China. 144 pp. (1981).

044.035 **Time and Latitude Bulletins.** October - December 1981.
Tokyo Astron. Obs., Time Latitude Bull., Vol. 55, 57 - 78 (1982).

044.036 **Observational data of the photoelectric astrolabe Type II No. 2 in 1978 and 1979.**
Publ. Beijing Astron. Obs., No. 1, p. 53 - 90 (1981). In Chinese.

044.037 **Observational data of the impersonal astrolabe OPL No. 30 (1979).**
Publ. Beijing Astron. Obs., No. 1, p. 91 - 108 (1981). In Chinese.

044.038 **Time results of observations made at Beijing Observatory with the photoelectric transit instrument (PP II) during the year 1979−1980.**
Z.-g. Yao, S.-h. Jiang.
Publ. Beijing Astron. Obs., No. 1, p. 109 - 123 (1981). In Chinese.

044.039 **Daily time differences and relative phase values.** 1982 January - June.
U. S. Naval Obs., Washington, D. C. Time Serv. Publ., Ser. 4, Nos. 779 - 803 (1982).

044.040 **A.1 − UT1 data.** 1982 January - May.
U. S. Naval Obs. Washington, D. C. Time Serv. Publ., Ser. 6, Nos. 71 - 75 (1982).

044.041 **Preliminary times and coordinates of the pole.** 1982 January - June.
U. S. Naval Obs., Washington, D. C. Time Serv. Publ., Ser. 7, Nos. 732 - 756 (1982).

044.042 **Time service announcement UTC (USNO) time scale.**   G. M. R. Winkler.
U. S. Naval Obs., Washington, D. C. Time Serv. Publ., Ser. 14, No. 31 (1982).

044.043 **Combination of Earth rotation parameters obtained in 1980 by various techniques.**   M. Feissel.
High-precision Earth rotation and Earth-Moon dynamics. Lunar distances and related observations, (see 012.035), p. 3 - 10 (1982).
    In 1980, Earth rotation parameters have been measured by classical astrometry, Doppler and laser satellite techniques, Lunar Laser Ranging and radio-interferometry. The precision of the series and their systematic differences are investigated; a combination algorithm is applied to the series available throughout the year.

044.044 **Optical observations of time and latitude and the determination of the Earth's rotation parameters in 1980.**   S.-H. Ye.
High precision Earth rotation and Earth-Moon dynamics. Lunar distances and related observations, (see 012.035), p. 11 - 23 (1982).
    This paper is based on optical observations with 85 instruments from January to October 1980. The rotation parameters for every 5 days are solved. The results show that errors in time observations are greater, generally speaking, than those in latitude, and low frequency errors are greater than high frequency ones in time observations.

044.045 **Rotation of the Earth from lunar laser ranging.**
R. B. Langley, R. W. King, P. J. Morgan,

I. I. Shapiro.
High-precision Earth rotation and Earth-Moon dynamics.
Lunar distances and related observations, (see 012.035),
p. 25 - 29 (1982).

The authors have extended their estimates of variation
of latitude and UT0 from the McDonald Observatory lunar
laser ranging observations to span the period October 1970
to November 1980. They have compared their values of variation of latitude with those derived from the smoothed
Circular D pole positions published by the Bureau International de l'Heure. For the period covered by the MERIT
Short Campaign, they have also compared their smoothed
UT0 values with (unsmoothed) ones derived from daily UT1
and pole-position values obtained by the Goddard Space
Flight Center/Massachusetts Institute of Technology/
Haystack Observatory group from very-long-baseline inter-
ferometric observations spanning two one-weak periods.

044.046   **Earth rotation from a simultaneous reduction of**
LLR and LAGEOS laser ranging data.
P. J. Shelus, N. R. Zarate, R. J. Eanes.
High-precision Earth rotation and Earth-Moon dynamics.
Lunar distances and related observations, (see 012.035),
p. 31 - 40 (1982).

This investigation is seeking values for the Earth rotation
parameters averaged over 5 day intervals or less. In the case of
LLR these short-term effects are well-separated from any
unmodelled long term effects because it is believed that all
short term lunar orbital and librational effects down to the few
centimeter level are known. This is, of course, not yet the case
for LAGEOS. It is believed that this study is the first attempt
to obtain Earth rotation parameters by the simultaneous reduction of LLR and LAGEOS data at the observation level.

044.047   **Earth rotation in the EROLD framework.**
O. Calame.
High-precision Earth rotation and Earth-Moon dynamics.
Lunar distances and related observations, (see 012.035),
p. 41 - 51 (1982).

The project EROLD (Earth Rotation from Lunar
Distances) was conceived in 1974, in the COSPAR framework,
with the goal of demonstrating that the lunar distances technique might be an efficient candidate in a new-generation
service for the determination of Earth orientation.

044.048   **Intercomparison of lunar laser and traditional**
determinations of earth rotation.
H. F. Fliegel, J. O. Dickey, J. G. Williams.
High-precision Earth rotation and Earth-Moon dynamics.
Lunar distances and related observations, (see 012.035),
p. 53 - 88 (1982).

The rotational orientation of the earth (UT0 at
McDonald Observatory) has been determined from lunar
laser ranging (LLR) measurements for the interval 1971 to
1980. The results have been differenced from those obtained
by conventional means as published by the Bureau International de l'Heure (BIH), on its 1979 system. The difference
displays a quasi-seasonal signature, which the authors ascribe
to systematic errors in the conventional measurements. The
lunar data are well represented by a smooth curve, which
gives UT0 at McDonald with a precision of about 3/4 milliseconds or better, and UT1 to within 1 millisecond using BIH
polar coordinates. The amplitude spectrum of the UT1
differences (LLR-BIH) suggests that a moderately smoothed
version of UT1 will usually be a good representation of Earth
rotation.

044.049   **Earth rotation information derived from MERIT**
and POLARIS VLBI observations.
D. S. Robertson, W. E. Carter.
High-precision Earth rotation and Earth-Moon dynamics.
Lunar distances and related observations, (see 012.035),

p. 97 - 122 (1982).

In September and October 1980, the National Geodetic
Survey, jointly with the NASA and several other agencies and
institutions, conducted a series of astronomical radio interferometry (VLBI) observing sessions to support the
IAU/IUGG MERIT short campaign. This paper briefly traces
the planning, observing, and data processing activities, and
presents the Earth rotation information thus far derived from
the data.

044.050   **Progress report on project MERIT.**
G. A. Wilkins.
High-precision Earth rotation and Earth-Moon dynamics.
Lunar distances and related observations, (see 012.035),
p. 147 - 148 (1982).

Project MERIT is a special programme of international
collaboration to Monitor Earth-Rotation and Intercompare
the Techniques of observation and analysis. The MERIT
Short Campaign of observations was held during the period
1980 August 1 to 1980 October 31. The main objective of
the campaign was to provide a realistic test of the operational
arrangements that will be required during the MERIT Main
Campaign in 1983/4.

044.051   **Dependence of the lunisolar perturbations in the**
Earth rotation on the adopted Earth model.
N. Capitaine.
High-precision Earth rotation and Earth-Moon dynamics.
Lunar distances and related observations, (see 012.035),
p. 155 - 169 (1982).

The purpose of this paper is to review the theories and
results concerning the periodic lunar and solar fluctuations in
the rotational motion of the Earth for some adopted Earth
models. The motivation for this purpose is to get a better
understanding of the dynamical motion of a realistic Earth
model. Such an understanding is useful for using very precise
observations in order to determine some parameters of the
Earth's model and to obtain values which can improve the
knowledge of the Earth's structure and dynamical behaviour.

044.052   **Atmospheric angular momentum and the length of**
day.   R. B. Langley, R. W. King, I. I. Shapiro,
R. D. Rosen, D. A. Salstein.
High-precision Earth rotation and Earth-Moon dynamics.
Lunar distances and related observations, (see 012.035),
p. 171 - 172 (1982). – Abstract. – See Abstr. 30.044.015.

044.053   **Earth's rotation and polar motion based on global**
positioning system satellite data.
R. J. Anderle, L. K. Beuglass, J. T. Carr.
High-precision Earth rotation and Earth-Moon dynamics.
Lunar distances and related observations, (see 012.035),
p. 173 - 179 (1982).

Using current procedures, polar motion and Earth's
rotation can be computed from 7 days of observations from
four stations to four Global Positioning System Satellites to
an accuracy of 1.5 m and 0.3 msec/day, respectively. Improved computational techniques or instrument accuracy and/or
measurements from additional satellites or stations would
give significant improvements in accuracy.

044.054   **Non-uniformity of the rotation of the earth and**
motion of the poles.   N. S. Sidorenkov.
Priroda, 1982, No. 4, p. 82 - 91. In Russian.

044.055   **Danjon astrolabe observations at Rio de Janeiro:**
time and latitude.
A. H. Adrei, V. A. d'Ávila, J. L. Penna, M. Queiroz.
Astron. Astrophys., Suppl. Ser., Vol. 48, 491 - 501 (1982).

The Danjon Astrolabe OPL-33 has been used to obtain
time and latitude data at National Observatory, Rio de Janeiro.

Results from May 1977 to December 1979 are presented in FK 4 system.

044.056 Influence of the rotation of the earth on orbital clock lag relative to earth-ground clocks.
S. V. Rodichev.
Fiz. fak. MGU. Moskva, 1982. 15 pp. In Russian. − Abstr. in Ref. zh., 51. Astron., 6.51.94 (1982).

044.057 Astronomical and geophysical problems of the Earth's rotation.   J. Kovalevsky, Ya. S. Yatskiv.
Sun and planetary system, (see 012.040), p. 149 - 162 (1982).

Following a general description of the motion of the Earth about its center of mass, the main features of the new 1980 IAU nutation series are described, and the definition of the new Celestial Ephemeris Pole is justified. The use of excitation functions for polar motion and UT1 is explained. Then, the main characteristics of the short period components of the polar motion and UT1 are described, together with their geophysical implications.

044.058 A note on the initial results and future plans of project MERIT.   G. A. Wilkins.
Sun and planetary system, (see 012.040), p. 163 - 164 (1982).

044.059 Steps towards the determination of Earth rotation parameters not dependent on the observation technique.   M. Feissel.
Sun and planetary system, (see 012.040), p. 165 - 172 (1982).

The rotation of the Earth is currently measured by classical astrometry, Doppler and laser satellite tracking, laser ranging to the Moon, and radio interferometry. Several years long time series of pole coordinates and/or UT are available from most of these techniques. The various series are inter-compared and their stability in the time frame of years to days is estimated.

044.060 Secular and decade fluctuations in the Earth's rotation: 700 BC - AD 1978.
L. V. Morrison, F. R. Stephenson.
Sun and planetary system, (see 012.040), p. 173 - 178 (1982).

Secular and decade fluctuations in the Earth's rotation in the period 700 BC - AD 1978 are deduced from timings of lunar eclipses and occultations on the assumption that the Moon's tidal acceleration is $-26''/cy^2$. Besides the tidal deceleration of the Earth, it is found that there is an accelerative component which implies a fractional decrease in the moment of inertia of $8.4 \pm 1.0 \times 10^{-11}/yr$.

044.061 Oceanic tides and the rotation of the Earth.
P. Brosche.
Sun and planetary system, (see 012.040), p. 179 - 184 (1982).

Periodic effects of oceanic tides are near to the limits of modern observational techniques. The largest effects are due to the secular transfer of angular momentum and energy between the Earth's rotation and the lunar orbit. This interaction is mediated by oceanic tides. Because of the very complex resonance structure of the oceans, the changes due to continental drift give rise to considerable variations of the torque not only within the 100 million years time scale but even within 1-10 million years. These variations are about factors 2 or 3; if they appear in the sense of diminishing torques, they might settle the enigma of a narrow Earth-Moon-system at a time not far enough in the past.

044.062 Correlation between solar activity and Universal Time variations.   D. Dragutin.
Sun and planetary system, (see 012.040), p. 189 - 190 (1982).

044.063 Report of IAU Commission 19: Rotation of the earth (Rotation de la terre).   P. Pâquet.
Trans. IAU, Vol. XVIIIA, (see 003.013), p. 181 - 194 (1982).

044.064 Report of IAU Commission 31: Time (L'heure).
S. Iijima.
Trans. IAU, Vol. XVIIIA, (see 003.013), p. 369 - 382 (1982).

044.065 Paläogezeiten und Erdrotation.
J. Krohn, P. Brosche, J. Sündermann.
Geol. Rundsch., Band 70, 64 - 77 (1981).

Tidal friction of the hydrosphere causes a loss of rotational angular momentum of the Earth that is balanced by a gain of orbital angular momentum of the Moon. The rate of dissipation for a given ocean can be calculated by means of a hydrodynamical-numerical model. Tracing back the history of the Earth-Moon-system, due to the varying distribution of continents and oceans information is needed about the paleotides to determine the transfer of angular momentum. Results (corange-lines, cotidal-lines, residual currents) are presented for oceans of the Upper Cretaceous (70 m. y. b. p.) and the Middle Silurian (420 m. y. b. p.). The corresponding change of the Earth's rotational energy is indicated.

L'accélération séculaire de la lune et le ralentissement de la rotation de la terre.   See Abstr. 004.021.

Project MERIT.   See Abstr. 013.059.

An intercomparison of connected-element interferometer and lunar laser Earth rotation parameters.
See Abstr. 031.608.

Final refraction problems in time and latitude observations through classical techniques.
See Abstr. 031.626.

Analysis of the systematic errors of synthetic time signal corrections of China and corrections of BIH.
See Abstr. 035.006.

Problems of time standards and time scales − PTB contribution to the CCDS 1980.   See Abstr. 035.007.

Reference systems for Earth dynamics.
See Abstr. 043.001.

The IAU resolutions on astronomical constants, time scales, and the fundamental reference frame.
See Abstr. 043.003.

Comments on the effect of adopting new precession and equinox corrections.
See Abstr. 043.010.

Effects of the non-rigidity of the Earth derived from astronomical observations.
See Abstr. 043.012.

Annual variation of climate of far east continent and annual fluctuation of the latitude determination at time and latitude observatories in this area.
See Abstr. 045.009.

Astronomical evidence of relationships between polar motion, Earth rotation and continental drift.
See Abstr. 045.020.

Comparison of PZT observations between the ILOM (*Int. Latitude Obs. Mizusawa*) and the SAIL (*Stn. Astron. Int. Latitudine Carloforte-Cagliari*).
See Abstr. 045.021.

Polar motion and Earth rotation from LAGEOS laser ranging.   See Abstr. 045.027.

Activities of astro-geodynamics research in China.
See Abstr. 081.053.

Analytical description of the secular variations of
the geomagnetic field and the velocity of the rotation of the
earth.   See Abstr. 084.030.

Erratum

044.901   Erratum: 'The observation data of the photoelec-
          tric astrolabe type II No. 2 (1976 - 1977)' [Publ.
Beijing Astron. Obs., No. 3, p. 93 - 122 (1979)].
Publ. Beijing Astron. Obs., No. 1, p. 123 - 124 (1981).
See Abstr. 29.044.025.

045 Latitude Determination, Polar Motion

045.001   On the direct influence of the planets on the
          precession and nutation of the Earth's axis of
rotation.   J. Vondrák.
Bull. Astron. Inst. Czechoslovakia, Vol. 33, 26 - 32 (1982).
  The influence of perturbing forces of Venus, Mars and
Jupiter on the precessional and nutational motion of the
Earth's axis of rotation is derived. It is shown that this influence (in case of nutation) is as large as several parts in
10^{-4} seconds of arc. This value is larger than the smallest terms
retained in the nutation series, recently adopted by the IAU.

045.002   Influence of atmospheric models upon polar motion
          computed from satellite tracking data.
F. Nouel, A. Piuzzi.
Ann. Géophys., Vol. 37, 199 - 204 (1981). – Abstr. in Phys.
Abstr., Vol. 85, Abstr. 9904 (1982).

045.003   Geodetic and geophysical implications of the
          homogeneous reduction of pole coordinates
(1899–1979).   R. O. Vicente.
Bull. Géod., Vol. 56, 1 - 8 (1982).
  The new series ILS (H) of pole coordinates (1899.9–
1979.0) computed in a homogeneous system has been
employed for the determination of the period of Chandler's
component of polar motion. The comparison with a value
derived from a previous series ILS (VY) shows there is no
significant variation in the period in spite of the known
systematic errors affecting the ILS (VY) series. Any high
precision geodetic network adjustment has to take account of
the pole coordinates defined by the ILS (H) series.

045.004   Non-polar nature of some long periodic terms of
          astronomical latitude in Europe.   Z. Li.
Kexue Tongbao, Vol. 26, 330 - 333 (1981). – Abstr. in Phys.
Abstr., Vol. 85, Abstr. 14756 (1982).

045.005   Chandler wobble and earthquakes.
          M. Zhao, G. Song.
Kexue Tongbao, Vol. 26, 253 - 256 (1981). – Abstr. in Phys.
Abstr., Vol. 85, Abstr. 18356 (1982).

045.006   Statistical analysis of polar motion data. – I. Theory
          for the Chandler wobble analysis.
I. Okamoto, N. Kikuchi.
Geophys. J. R. Astron. Soc., Vol. 69, 669 - 719 (1982).

045.007   Group corrections of declinations of the interna-
          tional latitude program from observations with two
zenith telescopes in Kitab.   B. Makhmatgaziev.
Tsirk. Astron. inst. AN UzbSSR, 1981, No. 93, p. 14 - 20. In
Russian. – Abstr. in Ref. zh., 51. Astron., 3.51.92 (1982).

045.008   Analysis of the Chandler period from latitude
          observations in observatories of China.   Y. Qian.
Acta Geod. Geophys., No. 3, p. 65 - 70 (1981). In Chinese.

045.009   Annual variation of climate of far east continent
          and annual fluctuation of the latitude determina-
tion at time and latitude observatories in this area.
G. D. Zhang, S. H. Jiang, B. C. Zhang.
Special issue on astronomical geodynamics, (see 012.028), p.
62 - 67 (1981). In Chinese.
  The residual latitudes observed at Beijing Astronomical
Observatory, Tianjing International Latitude Station,
Shanghai Astronomical Observatory, Wuhan Time Station,
Shanxi Astronomical Observatory and Blagoveshchensk
Latitude Station are analysed. The analysis shows that the
annual fluctuations of the six stations are quite similar. The
annual variation of the atmosphere is the main cause of the
common fluctuation in latitude determinations of the area.

045.010   A discussion on the periodical fitting the instanta-
          neous polar coordinates and on the methods of
computing the polar coordinates with the data of only one
station.   Z. M. Wang, H. F. He, S. Q. Yang.
Special issue on astronomical geodynamics, (see 012.028), p.
68 - 74 (1981). In Chinese.
  The authors applied the harmonic analysis to the series
of the polar coordinates X_m and Y_m (1900.0–1969.9) which
are published by Proverbio (1973). Curves of the amplitude
and place change with the time were obtained. Then, by
means of periodogram, the authors analysed the periods of
these curves and calculated the amplitudes and phases with
respect to periods.

045.011   Some features of the local non-polar long periodic
          terms of latitude.   Z. X. Li.
Special issue on astronomical geodynamics, (see 012.028), p.
75 - 81 (1981). In Chinese.
  From an analysis of the spectral structures of the secular
variation of the 40 mean-latitude series and from a comparison of the amplitudes and the phases of the periodic term,
four kinds of local non-polar long periodic terms (6.62y.,
4.49y., 3.33y. and 2.53y.) are verified. Within the same range,
no polar periodic term is found.

045.012   Comments on detection of the amplitude of the
          nearly diurnal free wobble.   Z. X. Chen.
Special issue on astronomical geodynamics, (see 012.028), p.
89 - 91 (1981). In Chinese.

045.013   The determination of polar coordinates by the
          Doppler technique.   H. S. Wang, S. L. Zeng.
Special issue on astronomical geodynamics, (see 012.028),
p. 118 - 126 (1981). In Chinese.
  This paper describes the basic principle of polar motion
determination by the Doppler technique. Fundamental
formulae are derived. With the Doppler receiver of integral
type, a series of two least squares programs for the pole
determination is suggested.

045.014   Effect of the Moon tides on latitude observations at the Shanghai, Beijing and Wuhan Observatories.
T. Q. Han, J. Y. Xia.
Special issue on astronomical geodynamics, (see 012.028), p. 127 - 132 (1981). In Chinese.
   In this paper the authors have analysed the effect of the moon tides on latitude observations at the Shanghai Observatory, Beijing Observatory and Institute of Geodesy and Geophysics, Academia Sinica Wuhan, using the Danjon astrolabes No. 14,30, and 29, during 1968–1976, 1971–1976 and 1966–1976 respectively. The mean value of $1+k-1$ obtained from the tide wave M_2 is 0.71±0.19.

045.015   Variations of amplitude, period and phase of the Chandler's pole motion.   S. X. Wu, S. H. Wang.
Special issue on astronomical geodynamics, (see 012.028), p. 173 (1981). In Chinese and English. – Abstract.

045.016   Seasonal changes of the results of time latitude observations with photoelectric astrolabe (model I).
Y. Z. Liu, Z. M. Wang.
Special issue on astronomical geodynamics, (see 012.028), p. 174 (1981). In Chinese and English. – Abstract.

045.017   Possibility of 1.35 year period component existing in the motion of the Earth pole.   Y. Y. Dong.
Special issue on astronomical geodynamics, (see 012.028), p. 175 (1981). In Chinese and English. – Abstract.

045.018   The Chandler wobble and earthquakes.
M. Zhao, G. X. Song.
Special issue on astronomical geodynamics, (see 012.028), p. 185 (1981). In Chinese and English. – Abstract.

045.019   Periodogram of nearly diurnal latitude variations in Gorky.   L. D. Kovbasyuk.
Astron. Zh., Tom 59, 581 - 587 (1982). In Russian. English translation in Soviet Astron., Vol. 26, No. 3.
   The periodogram of nearly diurnal latitude variations (1961.5 - 1972.0) directly in the high-frequency region of the spectrum is given. The spectrum of the nearly diurnal latitude variations is very complicated. The main parameters of 18 harmonics with amplitudes exceeding 0″.01 are calculated.

045.020   Astronomical evidence of relationships between polar motion, Earth rotation and continental drift.
A. Poma, E. Proverbio.
Mechanism of Continental Drift and Plate Tectonics, P. A. Davies, S. K. Runcorn (Editors), Academic Press, London, p. 345 - 357 = Pubbl. Stn. Astron. Int. Latitudine, Carloforte-Cagliari, N. 71.

045.021   Comparison of PZT observations between the ILOM (Int. Latitude Obs. Mizusawa) and the SAIL (Stn. Astron. Int. Latitudine Carloforte-Cagliari).
S. Manabe, H. Kitago, G. Murakami, K. Iwadate, S. Uras, S. Pilloni, G. Alvito.
Report (56,3) of the International Latitude Observatory of Mizusawa, 1981, 17 pp. = Pubbl. Stn. Astron. Int. Latitudine, Carloforte-Cagliari, N. 84.
   Time and latitude results obtained with the PZT's at the ILOM and the SAIL are compared for the period of February, 1980. It is found that UT1 is consistent between the two observatories except for a constant offset of about 0.1 s. Latitude is also consistent. Therefore, both PZT's will provide useful data in the future ILS PZT chain.

045.022   Breitenbestimmungen 1981 Juli - 1982 März.
   Techn. Univ. Dresden, Lohrmann-Obs. Zirk. Nr. 96 - 98 (1981/82).

045.023   Monthly Notes of the International Polar Motion Service.
Mon. Notes Int. Polar Motion Serv., No. 12, p. 122 - 133 (1981), Nos. 1 - 4, p. 1 - 40 (1982).
   Announcement of latitude values observed at the collaborating stations during December 1981 - April 1982.

045.024   Meteorological excitation of the annual polar motion.   J. B. Merriam.
Geophys. J. R. Astron. Soc., Vol. 70, 41 - 56 (1982).

045.025   Polar wandering and the forced responses of a rotating, multilayered, viscoelastic planet.
R. Sabadini, D. A. Yuen, E. Boschi.
J. Geophys. Res., Vol. 87, 2885 - 2903 (1982).
   The purpose of the paper is to investigate the consequences of a layered earth model for the secular motion of the spin axis in the late Cenozoic glacial age.

045.026   Comments on 'Investigation of controversial polar motion features using homogeneous International Latitude Service data' by S. R. Dickman.
W. E. Carter, with a reply by S. R. Dickman.
J. Geophys. Res., Vol. 87, 2919 - 2922 (1982). – See Abstr. 30.045.001.

045.027   Polar motion and Earth rotation from LAGEOS laser ranging.   B. D. Tapley.
High-precision Earth rotation and Earth-Moon dynamics. Lunar distances and related observations, (see 012.035), p. 123 - 124 (1982). – Abstract.

045.028   Comparison of polar motion results using lunar laser ranging.   J. O. Dickey, H. F. Fliegel, J. G. Williams.
High-precision Earth rotation and Earth-Moon dynamics. Lunar distances and related observations, (see 012.035), p. 125 - 137 (1982).
   A comparison of polar motion results from three sources [Bureau International de l'Heure (BIH), Defense Mapping Agency Hydrographic/Topographic Center (DMAHTC-Doppler), the International Polar Motion Service (IPMS)] was performed using lunar laser ranging (LLR) data. The rms errors, both of the LLR data and of the determinations of polar motion by the three services, decreased in recent times.

045.029   The pole position in October 1980 as determined from LAGEOS laser data.
C. Reigber, H. Mueller, W. Wende.
High-precision Earth rotation and Earth-Moon dynamics. Lunar distances and related observations, (see 012.035), p. 139 (1982). – Abstract.

045.030   Comparison of polar motion data from the 1980 project MERIT short campaign.
I. I. Mueller, B. S. Rajal, Y. S. Zhu.
High-precision Earth rotation and Earth-Moon dynamics. Lunar distances and related observations, (see 012.035), p. 141 - 146 (1982).
   Polar motion coordinates obtained during the MERIT short campaign by various techniques, as well as their variations, have been compared directly and through a smoothing procedure.

045.031   Biases in pole position computed from data from different Navy Navigation Satellites.
R. J. Anderle, E. S. Colquitt, M. Tanenbaum, C. A. Malyevac.
High-precision Earth rotation and Earth-Moon dynamics. Lunar distances and related observations, (see 012.035), 313 - 327 (1982).
   Biases have been noted in pole positions computed from Doppler observations of different Navy Navigation Satellites.

Studies show that the differences in the orbits of the Navy Navigation Satellites, although small, are large enough so that uncertainties in knowledge of the earth's gravity field could produce the biases noted.

045.032 **Influence of improved declination values on PZT latitude determinations in Potsdam.**   M. Meinig.
Astron. Nachr., Band 303, 153 - 155 (1982) = Mitt. Zentralinst. Phys. Erde, Nr. 1027. In German.
The results of latitude determinations with the PZT at Potsdam Observatory in the years 1972–78 were used to improve the declinations of the observed stars. The latitude determinations were re-reduced with the calculated corrections of declinations in the system of the new catalog. The obtained results show a substantially better agreement and a higher stability with respect to the BIH system than the results in the system of the hitherto used star catalog.

045.033 **On investigations of mean latitude variations.**   B. Kołaczek, G. Teleki.
Sun and planetary system, (see 012.040), p. 191 - 194 (1982).
Nonpolar variations of mean latitude are discussed. Some examples of variations of mean latitude differences of stations, located in the vicinity or along a common meridian, are presented and discussed.

045.034 **Closing errors of the Belgrade latitude observations and temperature influences.**   M. Djokić.
Sun and planetary system, (see 012.040), p. 195 - 196 (1982).

045.035 **A modulation phenomenon on Chandler's polar motion.**   S. Wu, Y. Hua, S. Wang.
Sci. Sinica, Ser. A, Vol. 25, 413 - 421 (1982).

045.036 **Free wobble of the earth with a liquid core.**   C.-z. Zhang, T.-y. Huang.
Acta Geophys. Sinica, Vol. 25, 276 - 280 (1982). In Chinese.

1980 IAU Theory of Nutation: the final report of the IAU Working Group on Nutation.   See Abstr. 011.019.

An intercomparison of connected-element interferometer and lunar laser Earth rotation parameters.   See Abstr. 031.608.

Final refraction problems in time and latitude observations through classical techniques.   See Abstr. 031.626.

On the new definition of the celestial pole.   See Abstr. 041.024.

On the accuracy of the 1980 IAU Nutation Series.   See Abstr. 043.009.

Primer año de observación con el astrolabio de Danjon OPL 01 en Punta Indio.   See Abstr. 044.011.

Resultados de tiempo y latitud en la Estación Astronómica Río Grande.   See Abstr. 044.014.

Time and Latitude Service, 1981 April - September.   See Abstr. 044.020.

Mark III VLBI: UT1, polar motion, and baselines.   See Abstr. 044.024.

Random and long periodic variations in the earth's motion.   See Abstr. 044.025.

Zeit- und Breitendienst.   See Abstr. 044.027.

Time and Latitude.   See Abstr. 044.029.

Time and Latitude Bulletins.   See Abstr. 044.035.

Optical observations of time and latitude and the determination of the Earth's rotation parameters in 1980.   See Abstr. 044.044.

Rotation of the Earth from lunar laser ranging.   See Abstr. 044.045.

Earth's rotation and polar motion based on global positioning system satellite data.   See Abstr. 044.053.

Non-uniformity of the rotation of the earth and motion of the poles.   See Abstr. 044.054.

Danjon astrolabe observations at Rio de Janeiro: time and latitude.   See Abstr. 044.055.

Astronomical and geophysical problems of the Earth's rotation.   See Abstr. 044.057.

Report of IAU Commission 19: Rotation of the earth (Rotation de la terre).   See Abstr. 044.063.

Determination of coordinates for the Orroral lunar ranging station.   See Abstr. 046.009.

Secular trends in polar motions: a new tool for probing the viscosity of the lower mantle.   See Abstr. 081.045.

Activities of astro-geodynamics research in China.   See Abstr. 081.053.

046 Astronomical Geodesy, Satellite Geodesy, Navigation

046.001 **Stellar triangulation from an aircraft and measured directions and distances.** J. Kabelac.
Z. Vermessungswes., Vol. 106, 536 - 542 (1981). In German. Abstr. in Phys. Abstr., Vol. 85, Abstr. 37744 (1982).

046.002 **Final report on the observations and computations carried out in the Second European Doppler Observation Campaign (EDOC-2) for position determinations at 37 satellite tracking stations.**
C. Boucher, P. Paquet, P. Wilson.
Deutsche Geod. Komm. Bayerisch. Akad. Wiss., Reihe B: Angew. Geod., Heft Nr. 255, 75 + 37 pp. (1981) = Mitt. Nr. 161 Inst. Angew. Geod., Abt. II Deutsches Geod. Forschungsinst.
The motivation for EDOC−2, field operations, data pre-processing and computational methods are described. All solutions computed by the different computing centres are analysed and presented as differences with respect to the finally adopted solution. The results include computations using both broadcast and precise ephemerides. The comparison with ED−50 shows up obvious regional deficiencies in the terrestrial network.

046.003 **Simultaneous determination of azimuth and latitude from observations of a star.**
S. Pejchev, G. Vlev, D. Zhekov.
Izv. Glav. upr. geod., kartogr. i kadastr., 1981, No. 2, p. 11 - 17. In Bulgarian. − Abstr. in Ref. zh., 52. Geod. Aehrosemka, 4.52.99 (1982).

046.004 **Determination of the geocentric coordinates by means of the differential Doppler method on a single station.**
Special issue on astronomical geodynamics, (see 012.028), p. 184 (1981). In Chinese and English. − Abstract.

046.005 **High precision tracking of synchronous satellites for geophysical purposes.** L. Anselmo, B. Bertotti, P. Farinella, A. Milani, A. M. Nobili, F. Sacerdote.
The comparative study of the planets, (see 012.033), p. 195 - 202 (1982).
The possibility of tracking very accurately a geosynchronous satellite is very useful to improve the knowledge of the resonant geopotential coefficients (hence the knowledge of the geoid) and to determine with higher accuracy the radial departure of the sea surface from the geoid and eventually its seasonal or long period variations.

046.006 **Die Arbeiten des Sonderforschungsbereiches 78 Satellitengeodäsie der Technischen Universität München im Jahre 1980.** M. Schneider.
Veröff. Bayer. Komm. Int. Erdmessung, Bayer. Akad. Wiss., Astronomisch-Geodätische Arbeiten, Heft Nr. 41, 216 pp. (1981). ISSN 0340-7691. ISBN 3-7696-9784-7.
Contents: Richtungsmessungen, Entfernungsmessungen, Dopplermessungen, Aufbau einer Empfangsanlage für Radiointerferometrie, Mobiles Laserentfernungsmeßsystem, Figur- und Feldparameterbestimmung/Geopotential, Dynamik des Erde-Mond-Systems, Kinematik geodätischer Punktfelder.

046.007 **Ein Kriterium der inneren Genauigkeit bei der Richtungsbestimmung künstlicher Erdsatelliten mit dem SBG.** H. Pauscher.
Vermessungstechnik, Heft 3, 4 pp. (1981) = Mitt. Zentralinst. Phys. Erde, Nr. 958.

046.008 **NAVSTAR/Global Positioning System (GPS) (I).**
I. Bauersima.
Mitt. Satell.-Beobachtungsstn. Zimmerwald, Nr. 7, 30 pp. (1982).

046.009 **Determination of coordinates for the Orroral lunar ranging station.** P. Morgan, R. W. King.
High-precision Earth rotation and Earth-Moon dynamics. Lunar distances and related observations, (see 012.035), p. 305 - 311 (1982).
Using models of the Earth-Moon system developed from analysis of 10 years of laser ranging observations from the McDonald Observatory, the authors have analyzed observations obtained by the Orroral Lunar Ranging Station since April 1978. Observations performed during the MERIT Short Campaign (August - October 1980) using a 6 ns, single-mode pulse are apparently reliable. Using 27 single photoelectron events, obtained on 7 nights during this period, the authors have estimated the coordinates of Orroral, with respect to McDonald.

046.010 **Method for a combined determination of the latitude, longitude and azimuth.**
K. P. Bychkovskij.
Geod. i kartogr., 1982, No. 2, p. 15 - 19. In Russian. − Abstr. in Ref. zh., 52. Geod. Aehrosemka, 6.52.69 (1982).

046.011 **Simultaneous determination of the azimuth and latitude from observations of one star.**
O. Pejchev, G. V'lev, D. Zhakov.
Izv. Glav. upr. geod., kartogr. i kadastr., 1981, No. 2, p. 11 - 17. In Bulgarian. − Abstr. in Ref. zh., 52. Geod. Aehrosemka, 6.52.70 (1982).

046.012 **High precision tracking of geosynchronous satellites: oceanographic applications.**
P. Farinella, A. Milani, A. M. Nobili.
Sun and planetary system, (see 012.040), p. 199 - 200 (1982).

046.013 **Optical tracking of synchronous satellites for geophysical purposes.**
S. Catalano, P. Farinella, A. Milani, A. M. Nobili.
Sun and planetary system, (see 012.040), p. 201 - 202 (1982).

046.014 **Gravimetric geoid determination for the area of Greece.**
D. Arabelos, L. N. Mavridis, I. Tziavos.
Sun and planetary system, (see 012.040), p. 203 - 204 (1982).

046.015 **Geodetic applications of Hipparcos results.**
E. Groten.
Scientific aspects of the Hipparcos space astrometry mission, (see 012.041), p. 71 - 73 (1982).

Probleme der Parameterbestimmung in physikalischen Systemen. See Abstr. 021.038.

Research on Doppler positioning techniques. See Abstr. 031.575.

Reduction of results of latitude and longitude determination on computers. See Abstr. 031.592.

Taking into account the luni-solar perturbations in high-precision numerical integration of orbits of geodesic satellites. See Abstr. 052.022.

047 Ephemerides, Almanacs, Calendars, Chronology

047.001 A propos de la date de Pâques.   J. Sauval.
Ciel Terre, Vol. 98, 21 - 22 (1982).

047.002 400 Jahre Gregorianischer Kalender.
P. Gerber.
Orion, 40. Jahrg., 42 (1982).

047.003 International Geophysical Calendar for 1982.
J. Atmos. Terr. Phys., Vol. 44, 91 - 93 (1982).

047.004 Astronomical Calendar of the Sofia Observatory for the year 1982.   D. Rajkova, Z. Krajcheva,
Z. Ivanova, A. Antov; edited by A. Bonov.
Izdatelstvo na Blgarskata Akademiya na Naukite, Sofiya.
114 pp. Price 1.27 Lv. (1981). In Bulgarian.

047.005 Astronomical calendar. Yearbook. 1982. Variable part.   M. M. Dagaev (Editor).
Vses. astron.-geod. o-vo, vyp. 85. Nauka, Moskva. 336 pp.
(1981). In Russian. – Review in Ref. zh., 51. Astron., 2.51.34 (1982).

047.006 On a nautical almanac supplement for ephemerides of artificial celestial bodies and coordinates of natural radio sources.   V. M. Amelin, V. V. Terent'ev.
Byull. Inst. Teor. Astron., Leningrad, Tom 15, 65 - 70 (1981).
In Russian.
A brief survey of the attempts aimed at creating an almanac for artificial celestial bodies has been given. A simple graphic method to determine the position of a ship is proposed. This method is based on receiving the radio or laser signals, not requiring any cumbersome calculations on computers. The peculiarities of a form of the ephemeris for artificial celestial bodies as well as natural radio sources are also considered.

047.007 On the possibility of using photomethods for printing the Nautical Almanac.
V. P. Alekhin, V. M. Amelin, I. P. Kurbatov, V. V. Terent'ev.
Byull. Inst. Teor. Astron., Leningrad, Tom 15, 190 - 191 (1982).
In Russian.

047.008 Astronomical Yearbook of the USSR for the year 1985.   V. K. Abalakin (Editor).
Institut Teoreticheskoj Astronomii Akademii Nauk SSSR.
Izdatel'stvo Nauka, Leningradskoe Otdelenie, Leningrad.
736 pp. Price 10 Rbl. 60 Kop. (1982). In Russian.

047.009 The Gregorian calendar.   G. Moyer.
Sci. American, Vol. 246, No. 5, p. 104 - 111 (1982).

047.010 Die Gregorianische Kalenderreform.
P. Ahnert.
Astron. Schule, 19. Jahrg., 37 - 39 (1982).

047.011 The Indian Astronomical Ephemeris for the year 1982.
Prepared by Positional Astronomy Centre, India Meteorological Department, New Alipore, Calcutta-700053 under the supervision of S. A. Bandyopadhyay.
Published by the Controller of Publications, Civil Lines, Delhi.
18 + 502 pp. (1981).
Available from: The High Commission of India, India House, Aldwych, London, W. C. 2.

047.012 The Air Almanac 1982, July - December.
Air Publication 1602, issued by Her Majesty's Nautical Almanac Office, London; and Nautical Almanac Office, United States Naval Observatory, Washington.

United Kingdom Edition: Her Majesty's Stationery Office, London. United States Edition: Superintendent of Documents, US Government Printing Office, Washington, D.C. 20402.
p. 363 - 732, A1 - A104, F1 - F4. Price £17.50 (1982). ISBN 0-11-772267-7.

047.013 The Nautical Almanac 1983.
N.P. 314-83, issued by Her Majesty's Nautical Almanac Office, London. United Kingdom Edition: Her Majesty's Stationery Office, London. United States Edition: Superintendent of Documents, US Government Printing Office, Washington, D.C. 20402. A4 + 276 + 36 pp. Price £ 9.50 (1982). ISBN 0-11-772332-0.

047.014 Rocznik Astronomiczny na rok 1982.
Prepared under the supervision of J. Radecki.
Instytut Geodezji i Kartografii, Państwowe Przedsiębiorstwo Wydawnictw Kartograficznych, Warszawa, Vol. 37, 160 pp.
Price zł 81.00 (1981). ISBN 83-7000-034-7.

047.015 Anuário Astronômico 1982.
Published by Universidade de São Paulo, Instituto Astronômico e Geofísico, Caixa Postal 30 627, São Paulo, Brasil. 12 + 279 pp. (1981). ISSN 0080-6412.

047.016 Philippine Astronomical Handbook 1982.
Prepared by the Astronomical Observation Division of the National Geophysical and Astronomical Office under the supervision of R. L. Kintanar.
Published by Philippine Atmospheric, Geophysical and Astronomical Services Administration. 12 + 62 pp. (1981).
ISSN 0115-1207.

047.017 Tables of sunrise, sunset, twilight, moonrise and moonset 1982.
Prepared by the Astronomical Observation Division, National Geophysical and Astronomical Office under the supervision of R. L. Kintanar.
Published by Philippine Atmospheric, Geophysical and Astronomical Services Administration. 14 + 57 pp. ISSN 0115-3307.

047.018 Astronomiskais Kalendārs 1982.
J. Bikše, I. Daube, M. Dīriķis, J. Francmanis,
V. Freijs, J. Miezis (Editors).
Latvijas PSR Zinātņu Akadēmija, Radioastrofizikas Observatorija, Vissavienības Astronomijas un Ģeodēzijas Biedrības Latvijas Nodaļa. Zinātne, Riga. 208 pp. Price 55 Kop. (1981).

047.019 The triple conjunctions of Jupiter and Saturn.
J. Meeus.
Mercury, Vol. 11, 54 - 58 (1982).

047.020 Blick in die Sternenwelt 1982. Astronomischer Kalender der Archenhold-Sternwarte.
E. Rothenberg.
Archenhold-Sternw. Berlin-Treptow, 48 pp. (1981).

047.021 New planetary and lunar ephemerides.
E. M. Standish, P. K. Seidelmann.
Bull. American Astron. Soc., Vol. 13, 874 (1981). – Abstract.

047.022 Astronomical calendar. Constant part.
V. K. Abalakin (Editor).
7th revised edition. Nauka, Moskva. 704 pp. (1981). In Russian. – Abstr. in Ref. zh., 51. Astron., 5.51.50 (1982).

047.023   **The new fundamental ephemerides; comparisons
with the old ephemerides and with observations.**
K. F. Pulkkinen, E. J. Santoro, G. H. Kaplan,
T. C. Van Flandern, P. K. Seidelmann.
Bull. American Astron. Soc., Vol. 14, 579 - 580 (1982).
Abstract.

047.024   **The star Almanac for Land Surveyors for the year
1983.**
Prepared by H. M. Nautical Almanac Office. Published by
Her Majesty's Stationery Office, London, England. 16 + 80 pp.
Price £ 2.00 (1982). ISBN 0-11-886916-7.

047.025   **Nautisches Jahrbuch oder Ephemeriden und Tafeln
für das Jahr 1983**, zur Bestimmung der Zeit, Länge
und Breite auf See nach astronomischen Beobachtungen.
Edited by Deutsches Hydrographisches Institut, Hamburg.
132. Jahrg., 45 + 365 + 30 pp. (1982). ISSN 0077-6211.

047.026   **New planetary ephemerides back to 4000 B.C.**
R. Dvorak.
Sun and planetary system, (see 012.040), p. 441 - 442 (1982).
New planetary ephemerides for Jupiter and Saturn back
to 4000 B.C. are presented.

047.027   **Éphémérides Astronomiques pour l'An 1983.
Connaissance des Temps. Nouvelle série.**
B. Morando.
Bureau des Longitudes, 77 Avenue Denfert Rochereau,
75014 Paris. 41 + 126 pp. (1982). ISBN 2-11-080380-0, ISSN
0181-3048.

047.028   **Report of IAU Commission 4: Ephemerides**
(Ephémérides).   A. M. Sinzi, T. Lederle.
Trans. IAU, Vol. XVIIIA, (see 003.013), p. 1 - 13 (1982).

047.029   **Error in same-dating the Louisiana tricentennial,
its implications and a realistic calendar.**
A. A. Hirsch.
Proc. Louisiana Acad. Sci., Vol. 44, 132 - 142 (1981).
All tricentennials rooted in the 1600's arrive one day
early due to the three successive century decalations in the
Gregorian Reform Calendar, thereby causing an unsuspected
geometrical error distinct from the arithmetical error of 0.12
day at the end of a 400 year cycle. Other defects are cited.
Only a Natural Decalation Cycle of 128 days can date time
simply and properly.

047.030   **Astronomical Phenomena for the year 1984.**
Prepared by The Nautical Almanac Office, United
States Naval Observatory and Her Majesty's Nautical Almanac
Office, Royal Greenwich Observatory.
U. S. Government Printing Office, Washington; Her Majesty's
Stationery Office, London. 71 pp. (1981). ISBN 0-11-886911-6.

047.031   **Sterrengids 1982.**   W. Gielingh, J. Meeus.
Ed. Nederlandse Vereniging voor Weer- en Sterren-
kunde, Stichting "De Koepel", Utrecht. 159 pp. Price
BF 405.00, Dfl. 26.50 (1981). — Reviews in Ciel Terre,
Vol. 98, 54 - 55; 1982 *(P. Cugnon)*; Zenit, 9. Jaarg., 215; 1982
(C. Caljouw).

047.032   **Local planet visibility report.**   R. Mansfield.
Astronomical Data Service, 3922 Leisure Lane,
Colorado Springs, Colo. 80917, USA. 16 pp. Price $ 15.00
(1981). — Review in Sky Telesc., Vol. 63, 267 - 268 (1982).

047.033   **Hemelkalender 1982.**   J. Meeus.
Numéro spécial de Heelal (Bulletin de la V. V. S.).
80 pp. Price BF 150.00. — Review in Ciel Terre, Vol. 98, 54;
1982 *(R. Charles).*

047.034   **Astronomical Calendar 1982.**   G. Ottewell.
Furman University Physics Dept., Greenville, S. C.
29613, USA. 62 pp. Price $ 9.00 (1981). ISBN 0-934546-06-1.
Reviews in Sky Telesc., Vol. 63, 159 (1982); Strolling Astron.,
Vol. 29, 127; 1982 *(J. R. Smith).*

047.035   **The view from the earth 1982.**   G. Ottewell.
Astronomical Calendar, Furman University Physics
Dept., Greenville, S. C. 29613, USA. 34 pp. Price $ 4.00 (1981).
Reviews in Sky Telesc., Vol. 63, 374 (1982); Strolling Astron.,
Vol. 29, 127; 1982 *(J. R. Smith).*

047.036   **Astronomisk årsbok 1982.**
Bokförlaget INOVA, Stockholm. 96 pp. Price
Sv. kr. 20.00 (1981). — Review in Astron. Tidsskr., Årg. 15,
48 (1982).

**Astronomical content of American Plains Indian
winter counts.**   See Abstr. 004.048.

**The "Binding of the Years" and its cue for the
nadir sun.**   See Abstr. 004.049.

Transition year for astronomy.
See Abstr. 043.002.

**The IAU resolutions on astronomical constants,
time scales, and the fundamental reference frame.**
See Abstr. 043.003.

**The frequency of total and annular solar eclipses for
a given place.**   See Abstr. 079.001.

The JPL planetary ephemerides.
See Abstr. 091.030.

**Nouvelles théories des planètes et de la Lune dans
les éphémérides françaises.**   See Abstr. 091.070.

Numerical studies of the lunar orbit at CERGA.
See Abstr. 094.053.

**Comparison of lunar ephemerides (SALE and ELP)
with numerical integration.**   See Abstr. 094.054.

Ephemerides of minor planets for 1983.
See Abstr. 098.110.

Space Research

051 Extraterrestrial Research, Spaceflight Related to Astronomy and Astrophysics

051.001 **ESA gaat komeet Halley onderzoeken.**
F. W. Miedema.
Zenit, 9. Jaarg., 84 - 85 (1982).

051.002 **Science on the Space Shuttle.**
W. M. Neupert, P. M. Banks, G. E. Brueckner,
E. G. Chipman, J. Cowles, J. A. M. McDonnell, R. Novick,
S. Ollendorf, S. D. Shawhan, J. J. Triolo, J. L. Weinberg.
Nature, Vol. 296, 193 - 197 (1982).
The nine instrument packages on the third Shuttle flight
should yield important data for space plasma physicists, astro-
nomers, life scientists and engineers and also pave the way for
future scientific payloads.

051.003 **The rapid development of satellite UV astronomy.**
M. Burger.
Irish Astron. J., Vol. 14, (see 012.003), 238 - 240 (1980).

051.004 **HIPPARCOS and the dynamics of the solar system.**
J. Kovalevsky.
Celestial Mech., Vol. 26, (see 012.001), 213 - 220 (1982).
The HIPPARCOS program may contribute to dynamical
astronomy in two different ways: by determining the positions
of some bodies of the solar system or by improving the positions of the reference stars with respect to which observations
of members of the solar system are made. It is shown that only
minor planets may be observed validly by HIPPARCOS but
these observations alone cannot contribute significantly to the
determination of a reference frame. However, they can be useful for the improvement of orbits and as a complement to a
major observational effort in preparation of a new determination of dynamical system of reference.

051.005 **DISCO − Europe's solar astronomical satellite.**
A. Thomson.
Spaceflight, Vol. 24, 266 - 268 (1982).

051.006 **First three years of IUE.**
P. M. Gondhalekar.
Report RL-81-091, Rutherford Appleton Lab., Chilton, Oxon.,
England, 55 pp. (1981). − Abstr. in Phys. Abstr., Vol. 85,
Abstr. 38039 (1982).

051.007 **Comet nucleus impact probe.**
J. E. Chirivella.
Instrumentation in the Aerospace Industry, Vol. 27, (see
012.018), 17 - 43 (1981). − Abstr. in Phys. Abstr., Vol. 85,
Abstr. 42659 (1982).

051.008 **ESA's wetenschappelijke satellieten.**
C. Titulaer.
Zenit, 9. Jaarg., 202 - 204 (1982).

051.009 **Cosmic physics.**
V. M. Balebanov, A. V. Zakharov.
Zemlya Vselennaya, 1982, No. 2, p. 15 - 20. In Russian.

051.010 **HIPPARCOS: space astrometry mission. A report
on the project. II.** M. A. C. Perryman.
Bull. Inf. Cent. Données Stellaires, No. 22, p. 87 - 89 (1982).

051.011 **Astronautics in the year 1981.** A. Vítek.
Vesmír, Vol. 61, 137 - 142 (1982). In Czech.

051.012 **Die Venus-Missionen Venera 13 und Venera 14.**
N. Giesinger.
Sternenbote, 25. Jahrg., 74 - 80 (1982).

051.013 **La dernière née des sondes spatiales à la chasse d'un
astre errant . . .** P. Ringoet.
Ciel, Vol. 44, 32 - 40 (1982).

051.014 **Digest of celestial X-ray missions and experiments.**
M. C. Locke.
Bull. American Astron. Soc., Vol. 13, 883 (1981). − Abstract.

051.015 **The Solar Beacon, an experiment in solar
astrometry.** J. M. Beckers.
Bull. American Astron. Soc., Vol. 13, 890 (1981). − Abstract.

051.016 **The Voyager encounter with Uranus.** E. C. Stone.
Uranus and the outer planets, (see 012.036), p. 275 -
291 (1982).
The Voyager 2 spacecraft is targeted for an encounter
with Uranus in January, 1986. In addition to a brief description of the 11 scientific investigations and the Uranian
encounter geometry, the scientific capabilities of Voyager 2
are discussed for the general areas of the atmosphere, the
rings, the satellites, and the magnetosphere.

051.017 **Giotto − a mission to Halley's comet.**
R. Reinhard.
Scientific and experimental aspects of the Giotto mission,
(see 012.039), p. IX - XIX (1981).

051.018 **The space astrometry project: history, technical
evolution and future prospects.** P. Lacroute.
Scientific aspects of the Hipparcos space astrometry mission,
(see 012.041), p. 3 - 12 (1982). In French and English.

051.019 **Tycho, a planned astrometric and photometric
survey from space.** E. Høg, C. Jaschek,
L. Lindegren.
Scientific aspects of the Hipparcos space astrometry mission,
(see 012.041), p. 21 - 25 (1982).
In addition to the primary scientific result of extremely
accurate astrometric data to be obtained for 100 000 stars by
the main Hipparcos detection system it appears that exploitation of the satellite's "star mappers" (originally designed to be
used only for the attitude determination) can give further
results of great scientific value. The star mappers will observe
at least 400 000 stars brighter than B = 11 magnitude with an
accuracy at B = 10 of 0.03 arcsec for positions and 0.03 mag
for the B and V magnitudes.

051.020 Proposed ultraviolet observations with Tycho.
E. Høg.
Scientific aspects of the Hipparcos space astrometry mission,
(see 012.041), p. 27 - 28 (1982).

The Hipparcos star mappers were originally dedicated to
attitude determination of the satellite, but a modification has
recently been approved by ESA. This so-called Tycho experiment will observe at least 400 000 stars brighter than B = 11
with an accuracy at B = 10 of 0.03 arcsec for positions and
0.03 mag for the B and V magnitudes. It is proposed to introduce an ultraviolet slit in both star mappers and thus obtain
for the Tycho stars a U magnitude with an accuracy at B = 10
of 0.03 and 0.08 mag for the spectral types B0 and A0 to G0,
respectively.

051.021 Le programme Hipparcos.
P. Lacroute.
Scientific aspects of the Hipparcos space astrometry mission,
(see 012.041), p. 29 - 30 (1982).

051.022 Scientific organisation of the Hipparcos project.
M. A. C. Perryman.
Scientific aspects of the Hipparcos space astrometry mission,
(see 012.041), p. 31 - 38 (1982).

**051.023 The selection of stars for linking the Hipparcos
frame to extragalactic radio sources by Space
Telescope (progress report).**
A. N. Argue, R. D. Baxter, B. L. Morgan, H. Vine.
Scientific aspects of the Hipparcos space astrometry mission,
(see 012.041), p. 53 - 55 (1982).

A preliminary selection of 75 southern radio sources has
been made for tying the Hipparcos frame to extragalactic
objects, and suitable stars chosen for carrying out this tie-up
using Space Telescope. To test their astrometric quality, six
of these stars have been examined for structure using the
Imperial College, London, Speckle Interferometer. At the time
of observation one-third were resolved as double on the scale
0.1 - 1.0 arc sec.

**051.024 Advantages, implementation and cost estimate of a
long-life version of the Hipparcos payload.**
T. Schmidt-Kaler.
Scientific aspects of the Hipparcos space astrometry mission,
(see 012.041), p. 91 - 92 (1982).

051.025 Relativistic effects in Hipparcos data.
B. F. Schutz.
Scientific aspects of the Hipparcos space astrometry mission,
(see 012.041), p. 181 - 185 (1982).

Relativistic effects in the Hipparcos data affect both the
data analysis and the selection of programme stars. Light-
deflection by the Sun will be measureable over a considerable
fraction of the celestial sphere, which suggests that a very accurate test of general relativity is possible, and even that direct
measurement of the metric is feasible. Selection of distant
programme stars enhances the possibility of detecting a black
hole by the anomalous deflection it produces in a star's apparent position.

**051.026 The intended scientific uses of the Hipparcos satellite
by the Institute for Space Astrophysics, Frascati.**
V. Castellani.
Scientific aspects of the Hipparcos space astrometry mission,
(see 012.041), p. 201 - 202 (1982).

The foreseen use of the Hipparcos satellite by IAS is
briefly described. The main interests will be in stellar evolution, planetology and X-ray binaries.

051.027 Future prospects for γ-ray astronomy.
C. Fichtel.
Philos. Trans. R. Soc. London, Ser. A, Vol. 301, (see 012.043),

693 - 701 (1981). – Same as 30.051.018.

The promise of γ-ray astrophysics noted by theorists in
the late 1940s and 1950s is beginning to be realized. In the
future, satellites should carry instruments that will have over
an order of magnitude greater sensitivity than those flown
thus far, and, for at least some portions of the γ-ray energy
range, these detectors will also have substantially improved
energy and angular resolution. The information to be obtained from these experiments should greatly enhance our knowledge of several astrophysical phenomena including the very
energetic and nuclear processes associated with compact
objects, astrophysical nucleosynthesis, solar particle acceleration, the structure of the Galaxy, the origin and dynamic
pressure effects of the cosmic rays, high energy particles and
energetic processes in other galaxies, and the degree of matter-
antimatter symmetry of the Universe.

**051.028 The astrometric satellite HIPPARCOS – a status
report.**   L. Lindegren, S. Söderhjelm.
Rep. Obs. Lund, No. 18, (see 012.044), p. 74 - 75 (1982).

051.029 CIRBS. Cosmic Infrared Background Satellite.
Technical design study.
European Space Agency, SCI(82)1, 95 pp. (1982).

051.030 Prospective views in space astrometry.
P. Lacroute.
Compendium in astronomy, (see 003.012), p. 247 - 251
(1982).

The Hipparcos project of space astrometry was adopted
by the European Space Agency in March 1980. The corresponding satellite will be launched in 1986 and, after 2.5 years
of observation and about 2 years of reductions it will be
possible to obtain coordinates, proper motions and parallaxes
of 100.000 stars with uncertainty very much smaller than now.
It will be obtained a reference system much better than the
one now available and parallaxes more numerous and more
accurate.

**051.031 Report of IAU Commission 44: Astronomy from
space (L'astronomie à partir de l'espace).**
R. J. van Duinen.
Trans. IAU, Vol. XVIIIA, (see 003.013), p. 579 - 619 (1982).

051.032 Space Report.
Spaceflight, Vol. 24, 14 - 19, 62 - 66, 124 - 127,
154 - 160, 208 - 215, 259 - 265, 317 - 326 (1982).

**Between Sputnik and the Shuttle: new perspectives
on American astronautics.**   See Abstr. 003.040.

Hipparcos: Europe's astrometric satellite.
See Abstr. 013.004.

**Development of rocket investigations of the
circumterrestrial space in the USSR.**
See Abstr. 013.020.

**Compilation des données photométriques
nécessaires pour la mission d'astrométrie spatiale HIPPARCOS.**
See Abstr. 031.573.

Uranus science with Space Telescope.
See Abstr. 032.540.

**The prospect of space astrometry and the signifi-
cance of ground observations.**   See Abstr. 041.025.

**On the importance of the Hipparcos stellar net for
photographic catalogue work.**   See Abstr. 041.037.

Radio stars as connecting link of the Hipparcos and VLBI reference frames.   See Abstr. 041.038.

US Naval Observatory parallaxes and the fundamental reference frame − their interaction with Hipparcos. See Abstr. 041.039.

Astrometric measurements required before Hipparcos launch.   See Abstr. 041.040.

Reference systems linkage from space. See Abstr. 043.013.

Report by IAU Commission 24 Working Group on Radio/Optical Astrometric Sources for the establishment of an inertial reference frame.   See Abstr. 043.016.

Fiducial reference for the Hipparcos reference system.   See Abstr. 043.017.

Inertial frame determination using minor planets − a proposal for Hipparcos observations.   See Abstr. 043.018.

Catalogue Hipparcos et observations au sol. See Abstr. 043.019.

Cosmic investigations, achievements and perspectives.   See Abstr. 061.035.

A comparison of B and Be stars kinematics. See Abstr. 065.067.

Concerning relativistic astrometry. See Abstr. 066.144.

A summary of results from solar monitoring rocket flights.   See Abstr. 080.078.

Solar variability indications from Nimbus 7 satellite data.   See Abstr. 080.079.

Planets, asteroids and comets at high angular resolution.   See Abstr. 091.045.

Exploration of the asteroids. See Abstr. 098.003.

Search for binary asteroids with the ESA Space Astrometry Satellite Hipparcos.   See Abstr. 098.103.

Space missions to Halley's comet and related activities.   See Abstr. 103.503.

Hipparcos proper motions. See Abstr. 111.019.

Trigonometric parallaxes − the stellar content of the volume of space covered by Hipparcos and some of its astrophysical implications.   See Abstr. 111.020.

Astrophysical parameters for Ap and Am stars with Hipparcos.   See Abstr. 111.021.

Trigonometric parallaxes from the ground and from Hipparcos.   See Abstr. 111.022.

The luminosity and related parameters of variable and non-variable A, F stars.   See Abstr. 111.023.

Hipparcos and the determination of the helium content of some low-mass non-evolved halo and disk stars of the solar neighbourhood, inferred from the fine structure of the H-R diagram.   See Abstr. 111.024.

Absolute radii of single and multiple stars from the CADARS *(Catalogue of Apparent and Absolute Radii of Stars)*. See Abstr. 111.025.

Radial velocity measurements complementary to Hipparcos.   See Abstr. 111.027.

Proposition de mesure de Vitesse Radiales stellaires pour le programme Hipparcos.   See Abstr. 111.028.

Photometric aspects of Tycho. See Abstr. 113.093.

The new ground-based photometric measurements. See Abstr. 113.095.

Absolute magnitudes and other basic parameters of O and B stars.   See Abstr. 115.035.

Study of absolute magnitudes of B and Be stars. See Abstr. 115.036.

Observation of selected nearby cool carbon stars. See Abstr. 115.037.

Imaging properties of Hipparcos and the observation of multiple stars.   See Abstr. 118.035.

Caractéristiques des étoiles doubles observables par Hipparcos.   See Abstr. 118.036.

The interest of double-star observations by Hipparcos.   See Abstr. 118.037.

Mira-type variable stars and the impact of the Hipparcos mission.   See Abstr. 122.201.

Interstellar reddening and distribution of B stars in the solar neighbourhood: impact of the Hipparcos/Tycho mission.   See Abstr. 131.256.

Calibration of the distance scale of planetary nebulae.   See Abstr. 135.051.

Hipparcos could make the distance of the Hyades into a conventional constant.   See Abstr. 153.049.

The role of early-type stars in studies of galactic dynamics.   See Abstr. 155.053.

052 Astrodynamics, Navigation of Space Vehicles

052.001 **Change of the orbital inclination of satellite
           1974 - 70 A.**   L. Sehnal.
Bull. Astron. Inst. Czechoslovakia, Vol. 33, 16 - 26 (1982).
  The decrease of the orbital inclination of the satellite
1974 - 70 A shows some peculiarities which cannot be explained by the usual disturbing effects — odd zonal harmonics,
lunisolar perturbations and the rotation of the atmosphere. To
explain this, a lift force normal to the orbital plane is suggested since the shape of the satellite resembles a flat plate moving with a certain slope to the velocity vector. The theory of
the effect is developed and its use in case of diffuse reflection
with changing accomodation coefficient shows good agreement with the observed data.

052.002 **Spacecraft collisions.**   D. W. Hughes.
           Nature, Vol. 295, 100 (1982).

052.003 **Periodic motions of a satellite with a magnet in a
           polar elliptical orbit.**   M. Yu. Ovchinnikov.
Aehrofiz. i prikl. mat. Moskva, 1981, p. 102 - 104. In Russian.
Abstr. in Ref. zh., 62. Issled. kosm. prostranstva, 2.62.143
(1982).

052.004 **Three-dimensional subsatellite motion under air drag
           and oblateness perturbations.**   J. C. van der Ha.
Celestial Mech., Vol. 26, 285 - 309 (1982).
  The three-dimensional relative motion of a subsatellite
with respect to a reference station in an elliptical orbit is
studied. A general theory based on the variation of the relative
elements is formulated in order to incorporate arbitrary
perturbing forces acting on both satellites. The loss of precision inherent in the subtraction of almost identical quantities
is avoided by the consistent use of difference variables. In the
absence of perturbations exact analytical representations can
be obtained for the relative state parameters. The influences
of air drag and Earth's oblateness on the relative motion
trajectories are investigated and illustrated graphically for a
number of cases.

052.005 **On the motion of a satellite in resonance with its
           rotating planet.**   A. S. Sochilina.
Celestial Mech., Vol. 26, 337 - 352 (1982).
  The influence of resonance perturbations due to the
gravitational field of an oblate planet on its satellite whose
motion is commensurable with rotation of the planet has been
investigated. It has been shown that in special case of the
critical inclination or circular orbit the Lagrange equations can
be integrated for all resonance terms simultaneously. The
method is applied to the investigation of the motion of the
12-hour communication and navigation satellites of the
'Molniya' and 'Navstar' type. The computations have been
performed by the use of four models of the geopotential.

052.006 **On the secular decrease in the semimajor axis of
           Lageos's orbit.**   D. P. Rubincam.
Celestial Mech., Vol. 26, 361 - 382 (1982).
  The semimajor axis of the Lageos satellite's orbit is
decreasing secularly at the rate of 1.1 mm day^{-1}. Ten possible
mechanisms are investigated to discover which one (s), if any,
might be causing the orbit to decay. Charged particle drag with
the ions at Lageos's altitude is probably the principal cause of
the orbital decay.

052.007 **On the symmetric difference quotient and its
           application to the correction of orbits.**
S. A. Serafin.
Celestial Mech., Vol. 26, 383 - 393 (1982).
  The author considers the analytical foundations of

numerical applications of the symmetric difference quotient
for orbit corrections. It follows that the better results
obtained in numerical calculations obtained by replacing the
ordinary difference quotient with the symmetric difference
quotient, have not been obtained fortuitously.

052.008 **Differential correction for near-stationary satellites.**
           L. G. Taff, J. M. Sorvari.
Celestial Mech., Vol. 26, 423 - 431 (1982).
  This paper presents a new concept for the differential
correction of orbits. It is developed in detail for the case of
near-stationary artificial satellites.

052.009 **Trajectory optimization for lunar orbiter missions
           having low inclination.**   T. Nishimura, C. Suzuki.
Fujitsu Sci. Tech. J., Vol. 17, No. 3, p. 1 - 25 (1981). — Abstr.
in Phys. Abstr., Vol. 85, Abstr. 18699 (1982).

052.010 **The shadow equation without singularities with
           arbitrary orbit orientation.**   V. G. Sokolov.
Byull. Inst. Teor. Astron., Leningrad, Tom 15, 46 - 52 (1981).
In Russian.
  The equation for intersections of a satellite orbit with a
planet's shadow (the shadow equation) has been transformed
to an algebraic equation of the fourth degree with respect to
tg $((w-w^*)/2)$, w being the true longitude of the satellite,
w^* being the w value at the intersection of the orbit with the
terminator plane when a satellite is entering into the shadow
semi-space. It has been proved that the intersections of the
orbit with the shadow correspond to simple real positive roots
of the equation.

052.011 **Motion of an earth satellite. II. Nonlinear perturba-
           tions.**   A. M. Fominov.
Byull. Inst. Teor. Astron., Leningrad, Tom 15, 53 - 58 (1981).
In Russian.
  The formulae for nonlinear perturbations of the second
order for the orbital elements due to second harmonic of the
geopotential and air drag have been derived.

052.012 **Formulae for improvement of the orbits of close
           earth satellites with no singularities at zero inclina-
tion and eccentricity.**   Yu. V. Batrakov, T. K. Nikol'skaya.
Byull. Inst. Teor. Astron., Leningrad, Tom 15, 71 - 75 (1981).
In Russian.
  Linear conditional equations have been derived which
connect corrections of measured values (angular coordinates,
ranges, range rates) to those of orbital parameters. For the
latters the combinations of keplerian elements have been taken
which exclude singularities from the coefficients of conditional
equations at zero inclination and eccentricity. Coefficients of
these equations are given in closed form which makes it possible to use the equations for any elliptical orbits including
those with zero inclination and eccentricity. Secular perturbations due to non-sphericity of the earth have been taken into
account in these coefficients, so the equations can be used for
orbital improvement at long time intervals.

052.013 **On the influence of resonance perturbations due to
           the gravitational field of a planet on the motion of
its satellite.**   A. S. Sochilina.
Byull. Inst. Teor. Astron., Leningrad, Tom 15, 114 - 123 (1981).
In Russian.
  The influence of resonance perturbations due to the gravitational field of an oblate planet on the motion of its satellite
has been investigated. It has been shown that in the particular
case where an angular variable, the argument of perigee, is
constant, the Lagrange equations can simultaneously be inte-

grated for all resonance terms. In practice such cases are represented by the 12 hour communication and navigation satellites of the Molniya and Navstar type. The computations are performed for four models of the geopotential.

052.014 Some aspects of satellite aerodynamic drag.
M. G. Abramovskaya, V. P. Bass.
Nablyud. Iskusstv. Nebesn. Tel, No. 79, (see 012.016), p. 24 - 31 (1981). In Russian.

The paper gives results of numerical modelling of a hypersonic rarefied gas flow around various types of spacecraft. The free-molecule and near-free-molecule regimes are considered. Analyses have been made for different schemes of the flow-body surface interaction and some aspects of satellite drag during their orbital motion have been discussed.

052.015 Linear perturbations in the motion of an artificial satellite caused by air drag.   A. M. Fominov.
Nablyud. Iskusstv. Nebesn. Tel, No. 79, (see 012.016), p. 32 - 51 (1981). In Russian.

The problem of the construction of an artificial Earth satellite motion theory taking into account atmospheric drag is investigated. Formulae for the linear perturbations of the orbital elements due to air drag are derived.

052.016 An approach to determine changes in the draconistic period of rotation of a satellite under the influence of atmospheric drag.   N. M. Barabanov,
O. A. Karanov, G. M. Solov'ev, Eh. M. Khodkov.
Nablyud. Iskusstv. Nebesn. Tel, No. 79, (see 012.016), p. 52 - 61 (1981). In Russian.

Changes in the draconistic period of rotation of a satellite under the influence of atmospheric drag taking into account secular and short-period components as well as atmospheric perturbations and the earth's oblateness are investigated. Numerical values of the obtained correlations are presented.

052.017 Algorithm for determination of tangential acceleration of AES motion by the parallax method.
Yu. V. Surnin, Yu. V. Dement'ev.
Nablyud. Iskusstv. Nebesn. Tel, No. 79, (see 012.016), p. 62 - 64 (1981). In Russian.

A determination algorithm of the tangential acceleration of AES for polar and near-polar orbits was worked out. The accuracy of the tangential acceleration determination is expected to be similar to the accuracy of measurements.

052.018 The effect of rotation of bodies on their aerodynamic characteristics.   A. M. Yanshin.
Nablyud. Iskusstv. Nebesn. Tel, No. 79, (see 012.016), p. 67 - 74 (1981). In Russian.

The effect of rotations of bodies on their aerodynamic characteristics is investigated under the following assumptions. Axially symmetrical bodies arbitrarily rotating in a free molecular flow are studied. The scheme of the free-stream flow interaction with the surface of bodies is hyperthermal, gas-dynamical. General expressions are obtained for forces and momenta affecting the bodies of revolution. The results for disk, sphere and cylinder are obtained in finite form. The properties of matrices of additional forces and momenta depending on body rotation are investigated.

052.019 The effect of different atmospheric models on the accuracy of satellite positioning.
N. A. Sorokin, A. Horvath, L. B. Shmelev.
Nablyud. Iskusstv. Nebesn. Tel, No. 79, (see 012.016), p. 75 - 80 (1981). In Russian.

The paper estimates the effect of selection of atmospheric models CIRA 72, DTM and the simple stationary model BCA-60 on the accuracy of satellite positioning if its motion is predicted for up to 7 days. Besides the atmospheric drag the effect of the earth's compression is taken into account. The

accuracy of satellite positioning is analysed from the differences between radius-vectors of two orbits obtained with different atmospheric models.

052.020 On determining the rotational pole of an oblong mirror satellite from maximum brightness moments.
V. M. Grigorevskij, N. S. Zgonyajko, S. Ya. Kolesnik, M. Ya. Tavadrus.
Nablyud. Iskusstv. Nebesn. Tel, No. 79, (see 012.016), p. 81 - 87 (1981). In Russian.

A dependence has been found between the maximum mirror brightness moments of a cylindrical or conical satellite and its sidereal period and the rotational pole coordinates. A method has been developed for determining the above dynamical parameters of the satellite.

052.021 Results of an analysis of computation accuracy for positioning of low-perigee satellites and their orbital elements.   A. G. Kirichenko, T. V. Shabalova.
Nablyud. Iskusstv. Nebesn. Tel, No. 79, (see 012.016), p. 90 - 95 (1981). In Russian.

For 500 computed satellite positions the distribution of errors over the reference stars is shown. A satellite orbit is computed from observations carried out from May 1976 to October 1977. The evolution of the secular motion of the ascending mode is given.

052.022 Taking into account the luni-solar perturbations in high-precision numerical integration of orbits of geodesic satellites.   Yu. D. Stepi.
Redkol. zh. "Izv. vuzov. Geod. i aehrofotosemka". Moskva, 1981. 12 pp. In Russian. − Abstr. in Ref. zh., 62. Issled. kosm. prostranstva, 3.62.130 (1982).

052.023 Probability of collision and methods for collision avoidance of stationary satellites.   K. Takashi.
Rev. Radio Res. Lab., Vol. 26, 621 - 628 (1980). In Japanese. Abstr. in Phys. Abstr., Vol. 85, Abstr. 48132 (1982).

052.024 On quasi-stationary stable circular orbits of AES with a daily period.
M. L. Lidov, M. A. Vashkov'yak.
Kosm. Issled., Tom 20, 3 - 8 (1982). In Russian.

052.025 Dependence on time in the evolutionary motion of a rotating magnetized satellite.   Yu. A. Sadov.
Kosm. Issled., Tom 20, 19 - 29 (1982). In Russian.

052.026 Improvement of the parameters of motion of an orbital space vehicle with the help of navigation satellites.   V. I. Ogarkov, Yu. I. Bakulin.
Kosm. Issled., Tom 20, 41 - 47 (1982). In Russian.

052.027 Local method for determination of the orientation of stabilized AES.   I. G. Khatskevich.
Kosm. Issled., Tom 20, 140 - 143 (1982). In Russian.

052.028 Influence of a dissipative magnetic moment on the gravitational orientation of a rotating satellite.
V. A. Sarychev, V. V. Sazonov.
Kosm. Issled., Tom 20, 177 - 183 (1982). In Russian.

052.029 Qualitative investigation of the motion of a space vehicle with constant radial acceleration.
V. A. Ivanov.
Kosm. Issled., Tom 20, 191 - 195 (1982). In Russian.

052.030 Estimate of the influence of the dissipative magnetic moment from vortex currents on the fast rotation of a satellite.   V. A. Sarychev, V. V. Sazonov.
Kosm. Issled., Tom 20, 297 - 300 (1982). In Russian.

052.031   Perturbative effects of solar radiation pressure on
the orbital motion of high Earth satellites.
L. Anselmo, P. Farinella, A. Milani, A. M. Nobili.
Application of modern dynamics to celestial mechanics and
astrodynamics, (see 012.026), p. 327 (1982). — Abstract.

052.032   On the development of an artificial satellite theory.
S. Coffey.
Application of modern dynamics to celestial mechanics and
astrodynamics, (see 012.026), p. 338 (1982). — Abstract.

052.033   Some current astrodynamics developments at the
North American Aerospace Defense Command
(NORAD).   F. R. Hoots.
Application of modern dynamics to celestial mechanics and
astrodynamics, (see 012.026), p. 347 (1982). — Abstract.

052.034   Relativistic astrodynamics: problems in interstellar
flight.   G. Vulpetti.
Application of modern dynamics to celestial mechanics and
astrodynamics, (see 012.026), p. 363 (1982). — Abstract.

052.035   Canonical differential equations of the disturbed
motion of AES with account for the attraction of
the sun and moon.   D. Z. Koenov.
Dokl. AN TadzhSSR, Tom 24, 478 - 481 (1981). In Russian.
From Ref. zh., 62. Issled. kosm. prostranstva, 5.62.139
(1982).

052.036   On a statistical estimate of variations of the
revolution period of a satellite in a circular orbit.
T. D. Dzhuzumkulov, M. I. Kiselev.
Asimptotich. metod. teor. differents. i integro-differents.
uravnenij i ikh pril. Frunze, 1981, p. 63 - 67. In Russian.
Abstr. in Ref. zh., 62. Issled. kosm. prostranstva, 5.62.146
(1982).

052.037   Modelle des Strahlungsdrucks für die Theorie der
Satellitenbahnen.
H. Böhnhardt, H. Ruder, M. Schneider.
Deutsche Geod. Komm. Bayer. Akad. Wiss., Reihe A:
Theoretische Geodäsie, Heft Nr. 93, 73 pp. (1981). ISSN
0065-5309, ISBN 3-7696-8175-4.

052.038   On the secular evolution of rotational motion of a
satellite equipped with an electrical screen.
V. V. Beletskij, A. A. Khentov.
Kosm. Issled., Tom 20, 342 - 351 (1982). In Russian.

052.039   Simple method for improving the convergence of
Newton's method in the problem of determination
of the orbits of low-flying artificial earth satellites.
M. A. Degtyarev.
Kosm. Issled., Tom 20, 472 - 474 (1982). In Russian.

052.040   Determination of the orientation of space vehicles
from images of stars.
G. A. Avanesov, Ya. L. Ziman, V. A. Krasikov.
Kosm. Issled., Tom 20, 481 - 484 (1982). In Russian.

052.041   On the evolution of nearly circular orbits of
satellites with critical inclination.   A. S. Sochilina.
Sun and planetary system, (see 012.040), p. 217 - 218 (1982).
The present paper is dealing with an investigation of
"Navstar" type orbits including the influence of the critical
inclination, the principal terms of luni-solar perturbations and
resonance terms of the geopotential. The parameters of a
"Navstar" type orbit are the following: the eccentricity
$e = 0.005$, the inclination $i = 63°4$, the period of revolution is
commensurable with the Earth's rotation as 2:1.

052.042   Satellite attitude acquisition by momentum transfer–
the controlled wheel speed method.
F. R. Vigneron, D. A. Staley.
Celestial Mech., Vol. 27, 111 - 130 (1982).
An implementation of the momentum transfer method
for spacecraft attitude acquisition of momentum wheel
stabilized geostationary satellites is presented, in which the
wheel speed is varied in a predeterminable manner to reduce
the nutation usually associated with the method. The implementation is found to be capable of achieving the transfer
to the desired zero nutation end point with 5° to 20° of
residual nutation in practical situations without additional
nutation damping.

Motion of artificial and natural celestial bodies.
See Abstr. 003.109.

Acceleration of numerical integration of the differential equations of motion of a solid body relative to the
mass center.   See Abstr. 021.018.

Intermediate orbits approximating the initial part of
perturbed motion.   See Abstr. 042.026.

Statistically dependent normal places in orbital computations.   See Abstr. 042.027.

Perturbation theory in extended phase space and
its application.   See Abstr. 042.044.

Some numerical results of a semi-analytic orbit
theory using observed data.   See Abstr. 042.073.

Biases in pole position computed from data from
different Navy Navigation Satellites.   See Abstr. 045.031.

Evaluation of 15th-order harmonics in the geopotential from analysis of resonant orbits.   See Abstr. 081.051.

053 Lunar and Planetary Probes and Satellites

053.001   New starts to Venus.   S. A. Nikitin.
Priroda, 1982, No. 3, p. 104 - 105. In Russian.

053.002   On a possible scheme of a rocket vehicle in the
Venus atmosphere.
V. M. Linkin, G. M. Moskalenko, G. A. Skuridin.
Kosm. Issled., Tom 20, 128 - 139 (1982). In Russian.

054 Artificial Earth Satellites

054.001 **Bilan annuel: astronautique 1980.**
J. Vercheval.
Ciel Terre, Vol. 98, 39 - 50 (1982).

054.002 **Action of the neutral atmosphere and charged particles on the movement of a satellite — application to LAGEOS.** F. Mignard.
Ann. Géophys., Vol. 37, 247 - 252 (1981). In French. — Abstr. in Phys. Abstr., Vol. 85, Abstr. 10296 (1982).

054.003 **Geocentric coordinates of artificial earth satellites computed from visual observations. Satellite 63-53-6, March - October 1967.** V. V. Kondrashin, I. E. Kupriyanova, I. A. Kuznetsov, V. A. Smirnov.
Nablyud. Iskusstv. Nebesn. Tel, No. 75, 61 pp. (1976). In Russian.

054.004 **On the results of "Intercosmos" satellite tracking.** M. A. Lur'e.
Nablyud. Iskusstv. Nebesn. Tel, No. 79, (see 012.016), p. 88 - 89 (1981). In Russian.
Data are given on the number of Intercosmos satellite observations at optical satellite tracking stations of the Astronomical Council (USSR Academy of Sciences) in 1978 - 1979.

054.005 **Probability of visual discovery of satellites.** V. I. Kuryshev, M. P. Zamakhovskij.
Nablyud. Iskusstv. Nebesn. Tel, No. 79, (see 012.016), p. 96 - 101 (1981). In Russian.
A calculation is presented of visual discovery of satellites in different seasons, depending on the height of the object above the horizon.

054.006 **Observation of stationary artificial satellites with the SBG earth-satellite camera.**
S. I. Ignatovich, T. I. Laslo, V. E. Vash.
Astrometr. Astrofiz., Vyp. (No.) 45, p. 78 - 81 (1981). In Russian.
The method of geostationary artificial sky bodies observations by the SBG earth satellite camera is described. The results of astrometrical reduction and precision estimate are given.

054.007 **"Cosmos" satellites in space.** B. A. Pokrovskij.
Zemlya Vselennaya, 1982, No. 1, p. 22 - 26. In Russian.

054.008 **On high-speed electrophotometric observations of a flaring geostationary satellite.**
A. V. Mironov, O. N. Kovalenko.
Astron. Tsirk., No. 1194, p. 2 - 4 (1981). In Russian.

054.009 **The MAGSAT mission.** R. Langel, G. Ousley, J. Berbert, J. Murphy, M. Settle.
Geophys. Res. Lett., Vol. 9, 243 - 245 (1982).

054.010 **Satellite digest — 150 - 156.** Compiled by R. D. Christy.
Spaceflight, Vol. 24, 32 - 33, 80 - 81, 136 - 138, 185 - 186, 230 - 231, 282, 327 (1982).

The RAE table of earth satellites, 1957 - 1980. See Abstr. 002.107.

Change of the orbital inclination of satellite 1974 - 70 A. See Abstr. 052.001.

Theoretical Astrophysics

061 General Aspects (Nucleosynthesis, Neutrino Astronomy, etc.)

061.001 The thermal runaway r-process.
J. J. Cowan, A. G. W. Cameron, J. W. Truran.
Astrophys. J., Vol. 252, 348 - 355 (1982).
The authors have identified and discuss a variant of the
r-process which occurs in a degenerate helium gas containing
about 2% or more of ^{13}C, with the $^{13}C(\alpha, n)^{16}O$ reaction serving
as the energy and neutron production source. Thermal run-
away leads to a slowly rising neutron flux which causes a
buildup of heavier elements from lighter ones; in the final
stages of the thermal flash the nuclei are transformed into an
r-process distribution. It is of particular interest to know
whether the solar system r-process distribution can be produced starting with negligible amounts of heavier elements at the
beginning of the runaway. The authors have found that a series
of local helium flashes leads to this result.

061.002 Stellar weak interaction rates for intermediate-mass
nuclei. II. $A = 21$ to $A = 60$.
G. M. Fuller, W. A. Fowler, M. J. Newman.
Astrophys. J., Vol. 252, 715 - 740 (1982).
Astrophysical electron and positron emission, continuum
electron and positron capture rates, as well as the associated
neutrino energy loss rates are calculated for free nucleons and
226 nuclei with masses between $A = 21$ and 60. Measured
nuclear level information and matrix elements are used where
available. Unmeasured matrix elements for allowed transitions
are assigned as in Paper I. Simple shell model arguments are
used to estimate Gamow-Teller sum rules and collective state
resonance excitation energies. The important effect of neutron
shell closure blocking of electron capture on neutron-rich
nuclei, is included.

061.003 Neutron shell blocking of electron capture during
gravitational collapse.   G. M. Fuller.
Astrophys. J., Vol. 252, 741 - 764 (1982).
Current ideas on the nuclear equation of state predict
that early on in the collapse of the iron core of a massive star
the nuclei present will become so neutron rich that allowed
electron capture is blocked. This neutron shell blocking
phenomenon and several unblocking mechanisms operative at
high temperature and density, including forbidden electron
capture, are discussed in terms of the simple shell model. The
results of one-zone collapse calculations are presented which
suggest that the effect of neutron shell blocking is to produce
a larger core lepton fraction at neutrino trapping which may
lead to a larger inner-core mass and hence a stronger post-
bounce shock.

061.004 Transport properties of degenerate neutrinos in
dense matter.   B. T. Goodwin, C. J. Pethick.
Astrophys. J., Vol. 253, 816 - 838 (1982).
The authors calculate the transport coefficients of
degenerate neutrinos in dense matter, taking into account
both absorption and scattering of neutrinos by nucleons.
Exact solutions of the transport equation are used, and the
effect of neutron-proton interactions on the composition of
the matter is allowed for. The authors consider energy dissipation by neutrinos and show that under many circum-

stances diffusion of neutrinos is more important than neutrino
viscosity or heat conduction.

061.005 Mixed lattice phases in cold dense matter.
C. J. Jog, R. A. Smith.
Astrophys. J., Vol. 253, 839 - 841 (1982).
Over a wide density range, the ground state of cold
neutral matter in the absence of external magnetic fields is a
degenerate sea of electrons containing a lattice of nuclei. In
certain density regions, a phase composed of interpenetrating
cubic lattices of different nuclides is preferable to a body-
centered cubic lattice of any single nuclide. The arguments
supporting this result are first made assuming the electrons
to be a uniform background; the qualitative features remain
when screening and exchange effects are included.

061.006 Electrodisintegration and photodisintegration of
nuclei.   R. Schaeffer, H. Reeves, H. Orland.
Astrophys. J., Vol. 254, 688 - 698 (1982).
The physics of electron induced and photon induced
nuclear reactions leading to potentially observable effects
in astrophysical sites is reviewed. The authors consider in
particular the cases (e^- or γ^+ $^{12}C \rightarrow ^{12}C$, $^{11}B^*$, ^{11}B) for which
the excitation functions are given. A few astrophysical
implications are presented.

061.007 Lepton number violation, Majorana neutrinos, and
supernovae.   E. W. Kolb, D. L. Tubbs,
D. A. Dicus.
Astrophys. J., Lett., Vol. 255, L57 - L61 (1982).
Weak interaction theories with Majorana masses for neutrinos have lepton number violating reactions. The authors consider gravitational collapse of massive stars in such a model
and find that if the lepton nonconserving reactions occur, the
dynamics of collapse will be drastically different from the
low-entropy collapse that follows from the usual assumption
of lepton number conservation.

061.008 The mass of the neutrino from the dynamics of
groups of galaxies.   F. D. A. Hartwick.
Astrophys. J., Lett., Vol. 255, L91 - L92 (1982).
Assuming that neutrinos do possess a nonzero rest mass
as two recent experiments have suggested, the dynamics of
groups of galaxies have been used to set a lower limit on the
neutrino mass by allowing identification of the density at
which the neutrinos become phase-space limited. This mass is
$m_v \gtrsim 6.2 (H/50)^{1/2} g_v^{-1/4}$ eV, where H is the Hubble constant
and g_v the number of allowed helicity states of the neutrino.

061.009 Mass of a neutrino in physics of elementary particles
and in big-bang cosmology.
Ya. B. Zel'dovich, M. Yu. Khlopov.
Usp. fiz. nauk, Tom 135, 45 - 77 (1981). In Russian. – Abstr.
in Ref. zh., 51. Astron., 2.51.781 (1982).

061.010 The neutrino and its importance for astrophysics.
M. Guttormsen.
Fra Fys. Verden, Vol. 43, No. 2, p. 36 - 39 (1981). In

Norwegian. – Abstr. in Phys. Abstr., Vol. 85, Abstr. 10316 (1982).

061.011 Nucleosynthesis of lead isotopes.
C. J. Allègre, J. F. Minster, G. Manhes, J. Audouze.
Meteoritics, Vol. 16, 288 (1981). – Abstract.

061.012 Titanium s-process isotopic anomalies.
D. D. Clayton.
Meteoritics, Vol. 16, 303 (1981). – Abstract.

061.013 OB associations and the early solar system.
K. A. Olive, D. N. Schramm.
Meteoritics, Vol. 16, 371 (1981). – Abstract.

061.014 Massive neutrinos in strong gravitational fields.
F. De Felice, Y. Yunqiang.
Nuovo Cimento B, Vol. 65B, Ser. 11, 79 - 88 (1981). – Abstr. in Phys. Abstr., Vol. 85, Abstr. 14871 (1982).

061.015 Dirac neutrinos in early cosmology.
S. A. Bonometto, P. Cazzola, G. Sartori, F. Lucchin.
Nuovo Cimento B, Vol. 65 B, Ser. 11, 303 - 315 (1981). Abstr. in Phys. Abstr., Vol. 85, Abstr. 14960 (1982).

061.016 Right-handed and left-handed neutrinos and the two galactic populations of the Universe. Additional evidence for the neutrino mass. D. Fargion.
Nuovo Cimento B, Vol. 65B, Ser. 11, 316 - 326 (1981). Abstr. in Phys. Abstr., Vol. 85, Abstr. 14961 (1982).

061.017 On the question of neutrino emission from the Sun.
M. Sachs.
Lett. Nuovo Cimento, Vol. 32, Ser. 2, 307 - 310 (1981). Abstr. in Phys. Abstr., Vol. 85, Abstr. 15301 (1982).

061.018 Majorana masses, photon gas heating and cosmological constraints on neutrinos. Y. Hosotani.
Nucl. Phys. B, Vol. B191, 411 - 428 (1981). – Abstr. in Phys. Abstr., Vol. 85, Abstr. 15318 (1982).

061.019 Right-handed neutrino interactions in the early Universe. F. Antonelli, R. Konoplich, D. Fargion.
Lett. Nuovo Cimento, Vol. 32, Ser. 2, 289 - 294 (1981). Abstr. in Phys. Abstr., Vol. 85, Abstr. 15333 (1982).

061.020 Low energy consequences of intermediate mass neutrinos in π_{12} decays solar neutrino fluxes and neutrino oscillations. J. N. Ng.
Nucl. Phys. B, Vol. B191, 125 - 145 (1981). – Abstr. in Phys. Abstr., Vol. 85, Abstr. 19285 (1982).

061.021 Exotic nuclear beta transitions: astrophysical examples. K. Takahashi, K. Yokoi.
4th international conference on nuclei far from stability, (see 012.009), p. 351 - 358 (1981). – Abstr. in Phys. Abstr., Vol. 85, Abstr. 19508 (1982).

061.022 The astrophysical r-process and its dependence on properties of nuclei far from stability: beta strength functions and neutron capture rates.
H. V. Klapdor, J. Metzinger, T. Oda, F. K. Thielemann.
4th international conference on nuclei far from stability, (see 012.009), p. 341 - 350 (1981). – Abstr. in Phys. Abstr., Vol. 85, Abstr. 23218 (1982).

061.023 Solar-stellar astrophysics. S. D. Jordan.
Comments Astrophys., Vol. 9, 211 - 237 (1982).
The Sun is a relatively well studied astronomical object, offering to the observer the opportunity to study numerous physically significant processes on the small scale which governs much of the basic physics. Many of these processes occur, or are believed to occur, on other stars over a much broader range of physical conditions. There is a growing viewpoint among both solar and stellar astrophysicists that solar-stellar astrophysics is a single subject and should be treated as one. Some of the ideas related to this view are outlined here. In particular, the role of magnetic fields in controlling the detailed structure and dynamics of stellar atmospheres is discussed.

061.024 The heavy element yields of neutron capture nucleosynthesis. A. G. W. Cameron.
Astrophys. Space Sci., Vol. 82, 123 - 131 (1982).
An effort has been made to determine the contributions of the S- and R-processes of nucleosynthesis to the abundances of the heavy element isotopes. It has been concluded that the general previous assumption concerning the exclusive assignment of isobars to one or the other of these processes is probably in error. The R-process abundances are characterized by relatively small fluctuations in the abundances of odd and even mass numbers. If this is always true, and such is assumed here, then there are substantial S-process contributions to the abundances of 'R-process' isobars. This is consistent with transient flashing episodes in the S-process neutron production processes.

061.025 Galactic halos, globular clusters and massive neutrinos. R. Fabbri, R. Ruffini.
Astrophys. Space Sci., Vol. 82, 249 - 253 (1982).
Within the framework of a Gamow cosmology with massive neutrinos a scenario is proposed in which both galactic halos and globular clusters are formed due to the existence of a 'critical injection mass'. Galactic halos are formed at red shift $z \sim 10-100$ self-gravitating neutrinos, and globular clusters at $z \simeq 10^3$ by a critical injection mass of primordial plasma.

061.026 Neutrino mass and cosmological baryon excess in left-right symmetric grand unified theories.
M. Fukugita, T. Yanagita, M. Yoshimura.
Phys. Lett. B, Vol. 106B, 183 - 187 (1981). – Abstr. in Phys. Abstr., Vol. 85, Abstr. 23887 (1982).

061.027 Resonant reaction rates of $O^{16}(\alpha, \gamma)Ne^{20}$ affecting screening effect. A. E. M. Khairozzaman.
Proc. Indian Natl. Sci. Acad. Part A, Vol. 47, 264 - 268 (1981). Abstr. in Phys. Abstr., Vol. 85, Abstr. 24252 (1982).

061.028 Massive neutrinos and the stellar stopping power via the neutrino magnetic moment.
M. L. Rustgi, P. T. Leung, J. E. Turner, W. Brandt.
Phys. Rev. A, Vol. 24, 2425 - 2430 (1981). – Abstr. in Phys. Abstr., Vol. 85, Abstr. 28336 (1982).

061.029 Non-Doppler redshifts and energy decay of elementary particles. T. L. Chow.
Lett. Nuovo Cimento, Vol. 32, Ser. 2, 351 - 352 (1981). Abstr. in Phys. Abstr., Vol. 85, Abstr. 28564 (1982).

061.030 The rest mass of neutrinos and clustering in the early Universe. F. Lizhi, L. Yongzhen.
Lett. Nuovo Cimento, Vol. 32, Ser. 2, 129 - 132 (1981). Abstr. in Phys. Abstr., Vol. 85, Abstr. 28584 (1982).

061.031 Radiative decay lifetime of neutrinos and the evolution of the Universe after the recombination era.
Y. Rephaeli, A. S. Szalay.
Phys. Lett. B, Vol. 106B, 73 - 76 (1981). – Abstr. in Phys. Abstr., Vol. 85, Abstr. 28588 (1982).

061.032 A mean-field calculation of the equation of state of supernova matter. P. Bonche, D. Vautherin.

Nucl. Phys. A, Vol. A372, 496 - 526 (1981). – Abstr. in Phys. Abstr., Vol. 85, 38970 (1982).

061.033 Die Entstehung der chemischen Elemente.
W. Hillebrandt, W. Ober.
Naturwissenschaften, 69. Jahrg., 205 - 211 (1982).
The present understanding of nucleosynthesis will be reviewed with special attention given to nucleosynthetic processes in primordial stars and supernovae.

061.034 Five-reservoir model of the carbon cycle and some astrophysical phenomena.
G. E. Kocharov, V. M. Ostryakov, S. B. Chernov.
Fiz.-tekh. inst. AN SSSR. Prepr., 1981, No. 739, 25 pp. In Russian. – Abstr. in Ref. zh., 51. Astron., 3.51.155 (1982).

061.035 Cosmic investigations, achievements and perspectives.   R. Z. Sagdeev.
Probl. kosm. issled. Moskva, 1981, p. 3 - 12. In Russian. Abstr. in Ref. zh., 62. Issled. kosm. prostranstva, 3.62.1 (1982).

061.036 Spectrum of neutrino radiation of collapsing degenerate stellar cores. Monte-Carlo calculations.
Yu. L. Levitan, I. M. Sobol', M. Yu. Khlopov,
V. M. Chechetkin.
Astron. Zh., Tom 59, 334 - 345 (1982). In Russian. English translation in Soviet Astron., Vol. 26, No. 2.
The change of the hard part ($\epsilon > 10$ MeV) of the spectrum of neutrino radiation of collapsing degenerate stellar cores at small ($\tau \leqslant 10$) neutrino opacities of the matter is analyzed. Neutrino interaction in the degenerate matter is determined by the processes of neutrino scattering on nuclei in which neutrino energy is not changed, and of neutrino scattering on degenerate electrons in which neutrino energy may only decrease. Monte-Carlo calculations of the spectrum of neutrino radiation of a collapsing stellar core are presented.

061.037 Cosmic-ray-produced stable nuclides: various production rates and their implications.
R. C. Reedy.
Proc. Twelfth Lunar Planet. Sci. Conf., (see 012.024), p. 1809 - 1823 (1982).

061.038 Stellar weak interaction rates for intermediate mass nuclei. III. Rate tables for the free nucleons and nuclei with $A = 21$ to $A = 60$.
G. M. Fuller, W. A. Fowler, M. J. Newman.
Astrophys. J., Suppl. Ser., Vol. 48, 279 - 320 (1982).
Stellar electron and positron emission rates and continuum electron and positron capture rates, as well as the associated neutrino energy loss rates, are tabulated for the free nucleons and 226 nuclei with masses between $A = 21$ and 60. These rates were calculated in accordance with the procedure described in Papers I and II of this series and are presented here in tabular form on an abbreviated temperature and density grid. Results of these calculations on a detailed temperature and density grid are available in computer readable form on magnetic tape upon request to MJN. The stellar weak rate calculation procedure is reviewed, and the results are discussed. Comparison of the stellar weak rates to terrestrial decay rates are made where possible.

061.039 Cosmic processes and formation of minerals.
A. G. Zhabin.
Zemlya Vselennaya, 1982, No. 1, p. 59 - 61. In Russian.

061.040 Neutrinos of finite rest mass in astrophysics and cosmology.   R. Cowsik.
Cosmology and particles, (see 012.046), p. 157 - 187 (1982).
The author's early work on the role of neutrinos in astrophysics and cosmology indicating that 3 eV $< m_\nu < 35$ eV and

that $\tau_\nu/m_\nu > 10^{23}$ s/eV for radiative decays is presented and then extended to include the recently discovered τ-neutrino.

061.041 Kinetics of molecular hydrogen formation, thermochemical evolution of primordial matter of protogalaxies and characteristics of collapsing protostars of the first generation.   Yu. I. Izotov, I. G. Kolesnik.
Akad. nauk Ukrainskoj SSR, Inst. teor. fiz., Prepr., ITF-81-84R, 37 pp. Price 14 Kop. (1981). In Russian.
The molecular hydrogen formation in primeval gas of protogalaxies having $T \sim 10^4$ K and density $n \cong 0.1$ - 1 cm^{-3} is considered. Due to nonequilibrium conditions the relative number density of H_2 molecules up to $\sim 10^{-3}$ can be obtained. Cooling by molecular hydrogen leads to formation of cold fragments with masses $\sim 10^2$ - 10^4 $M_\odot$ and $T \sim 10^2$ K. The dynamical evolution of these fragments is considered and the conclusion is made that the first generation of stars may be massive with $M_* \sim 70$ $M_\odot$.

061.042 Effects of nucleon-nucleon interactions on scattering of neutrinos in neutron matter.
N. Iwamoto, C. J. Pethick.
Phys. Rev. D, Vol. 25, 313 - 329 (1982). – Abstr. in Phys. Abstr., Vol. 85, Abstr. 48975 (1982).

061.043 Elementary quantum – some consequences in physics and astrophysics of a minimal energy quantum.   H. Broberg.
Report ESA-STM-223, European Space Agency, Paris, France. 6 + 65 pp. (1981). – Abstr. in Phys. Abstr., Vol. 85, Abstr. 53078 (1982).

061.044 Cosmological bounds on the masses of stable, right-handed neutrinos.   K. A. Olive, M. S. Turner.
Phys. Rev. D, Vol. 25, 213 - 216 (1982). – Abstr. in Phys. Abstr., Vol. 85, Abstr. 53401 (1982).

061.045 s-process studies in the light of new experimental cross sections: distribution of neutron fluences and r-process residuals.   F. Käppeler, H. Beer, K. Wisshak, D. D. Clayton, R. L. Macklin, R. A. Ward.
Astrophys. J., Vol. 257, 821 - 846 (1982).
A best set of neutron-capture cross sections has been evaluated for the most important s-process isotopes. With this data base, s-process studies have been carried out using the traditional model which assumes a steady neutron flux and an exponential distribution of neutron irradiations. The calculated σN curve is in excellent agreement with the empirical σN-values of pure s-process nuclei. Simultaneously, good agreement is found between the difference of solar and s-process abundances and the abundances of pure r-process nuclei. The authors also discuss the abundance pattern of the iron group elements where the s-process results complement the abundances obtained from explosive nuclear burning. The results obtained from the traditional s-process model such as seed abundances, mean neutron irradiations, or neutron densities are compared to recent stellar model calculations which assume the He-burning shells of red giant stars as the site for the s-process.

061.046 On the origin of the e^+- e^- annihilation line from the galactic center.   R. Ramaty.
Bull. American Astron. Soc., Vol. 13, 908 - 909 (1981). Abstract.

061.047 Neutrino energy loss of nuclear matter.
L. Sh. Grigoryan.
Astrofizika, Tom 17, 398 - 402 (1981). In Russian. English translation in Astrophysics, Vol. 17, No. 2.

061.048 Neutrino oscillations in neutron star matter.
H. J. Haubold.

Astrophys. Space Sci., Vol. 82, 457 - 461 (1982).

The author discusses the hypothesis suggested by Mazurek (1979) that neutrino oscillations ($\nu_e \rightleftarrows \nu_\mu$) could transfer leptonic zero-point energy to baryons during the gravitational collapse of a massive star ($M \gtrsim 10 \, M_\odot$) and that subsequently the collapse ends in a stellar explosion. The estimate of the lengths of neutrino oscillations if occurring in vacuum or dense matter, respectively, shows, however, that vacuum oscillations can be suppressed in dense matter and should have no influence on the neutrino emission of neutron stars.

061.049 **Neutrino decay and spontaneous violation of lepton number.** J. Schechter, J. W. F. Valle.
Phys. Rev. D, Vol. 25, 774 - 783 (1982). − Abstr. in Phys. Abstr., Vol. 85, Abstr. 54154 (1982).

061.050 **The equation of state of hot dense matter and super-novae.** J. M. Lattimer.
Annu. Rev. Nucl. Part. Sci. Vol. 31, (see 003.007), 337 - 374 (1981). − Abstr. in Phys. Abstr., Vol. 85, Abstr. 54296 (1982).

061.051 **Average ^{186,187,188}Os(n, γ) cross sections and the age of the Galaxy via ^{187}Re decay to ^{187}Os.**
R. R. Winters, R. L. Macklin.
Phys. Rev. C, Vol. 25, 208 - 212 (1982). − Abstr. in Phys. Abstr., Vol. 85, Abstr. 54389 (1982).

061.052 **Total cross-section measurements for the production of nuclear gamma rays from light nuclei by low-energy deuterons.** F. E. Cecil, R. F. Fahlsing, R. A. Nelson.
Nucl. Phys. A, Vol. A376, 379 - 388 (1982). − Abstr. in Phys. Abstr., Vol. 85, Abstr. 54401 (1982).

061.053 **Production of ^{6}He, ^{6}Li, ^{7}Li and ^{7}Be in the $\alpha + a$ reaction between 60 - 160 MeV.** B. G. Glagola,
V. E. Viola, Jr., H. Breuer, N. S. Chant, A. Nadasen,
P. G. Roos, S. M. Austin, G. J. Mathews.
Phys. Rev. C, Vol. 25, 34 - 45 (1982). − Abstr. in Phys. Abstr., Vol. 85, Abstr. 54416 (1982).

061.054 **Photo-neutrino energy losses in strong magnetic fields.** B. G. Mazumder.
Indian J. Phys. Part A, Vol. 55A, 444 - 458 (1981). − Abstr. in Phys. Abstr., Vol. 85, Abstr. 58330 (1982).

061.055 **Does the standard hot-big-bang model explain the primordial abundances of helium and deuterium?**
N. C. Rana.
Phys. Rev. Lett., Vol. 48, 209 - 212 (1982).

It is shown that according to the standard hot-big-bang nucleosynthesis calculations, no single value of the ratio η of baryon to photon number density can explain the right abundances of both ^{4}He and ^{2}D. A slight neutrino degeneracy, however, provides a testable solution for η which is in conformity with the current estimate of the baryonic component of the density parameter $(\Omega_N)_0$ of the Universe.

061.056 **Comment on "Does the standard hot-big-bang model explain the primordial abundances of helium and deuterium?".** J. Bernstein.
Phys. Rev. Lett., Vol. 48, 774 (1982). − See Abstr. 31.061. 055.

061.057 **Large scale distribution of elements and gravity.**
W. Iwanowska.
Compendium in astronomy, (see 003.012), p. 315 - 322 (1982).

Present perspectives for the idea of the gravitational separation of elements as a primordial factor generating gradients of the chemical composition of matter in gravitating systems are discussed.

Neutrino 81. Proceedings of a conference held at Maui, Hawaii, July 1981. See Abstr. 012.050.

Production of 6.13-MeV gamma rays from the ^{16}O(p, p'γ)^{16}O reaction at 23.7 and 44.6 MeV.
See Abstr. 022.092.

The ^{12}C + ^{12}C reaction at subcoulomb energies. II.
See Abstr. 022.097.

Detection of solar and cosmic neutrinos by coherent scattering. See Abstr. 031.532.

The sensitivity of DUMAND (*Deep Undersea Muon and Neutrino Detection*) to extraterrestrial neutrino sources. See Abstr. 034.054.

Pion condensation in cold dense matter and neutron stars. See Abstr. 066.524.

Neutrinos from a standard solar model.
See Abstr. 080.002.

Solar neutrinos. See Abstr. 080.081.

Oxygen isotopes, rare-earth elements, supernovae and Wolf-Rayet stars. See Abstr. 105.043.

Nucleosynthesis in novae: a source of Ne-E and ^{26}Al? See Abstr. 124.002.

The supernova neutrino pulse shape in the scintillation detector. See Abstr. 125.017.

Ultraviolet background radiation and the search for decaying neutrinos. See Abstr. 142.110.

On massive-neutrino halos and galactic structures. See Abstr. 158.120.

Lifetime constraints on massive neutrinos from ultraviolet observations of clusters of galaxies. See Abstr. 160.057.

Massive neutrino decay and the photoionization of the intergalactic medium. See Abstr. 161.001.

Neutrinos of non-zero mass in Friedmann universes. See Abstr. 162.057.

Cosmological constraints on grand unified theories. See Abstr. 162.065.

Astrophysics and grand unification. See Abstr. 162.105.

Astrophysical tests for radiative decay of neutrinos and fundamental physics implications. See Abstr. 162.134.

The influence of a non-zero neutrino rest mass on the development of perturbations in an isotropic world. See Abstr. 162.155.

Scale-covariant gravitation and primordial nucleosynthesis. See Abstr. 162.158.

Baryon and massive neutrino normal modes in a massive neutrino dominated universe. See Abstr. 162.162.

Asymmetric lepton production in a universe with non-zero baryon number.   See Abstr. 162.168.

On the neutrino mass, the abundance of right-handed neutrinos and the closure of the Universe. See Abstr. 162.175.

Effects of anisotropy and dissipation on the primordial light-isotope abundances.   See Abstr. 162.194.

A GUT-ed tour through the early Universe. See Abstr. 162.200.

Cosmology and the neutron electric dipole moment. See Abstr. 162.201.

Elementary particle phase transitions in the very early Universe.   See Abstr. 162.202.

Constraints on neutrinos and axions from cosmology. See Abstr. 162.204.

Why and how to detect the cosmological neutrino background.   See Abstr. 162.205.

Elementary particle physics in the very early Universe.   See Abstr. 162.220.

Massive neutrinos in cosmology and galactic astronomy.   See Abstr. 162.221.

Some remarks on phase-density constraints on the masses of massive neutrinos.   See Abstr. 162.222.

062 Hydrodynamics, Magnetohydrodynamics, Plasma

062.001 Vortex funnels in accretion flows.
H. A. Scott, R. V. E. Lovelace.
Astrophys. J., Vol. 252, 765 - 774 (1982).
A study is made of the axisymmetric gravitational accretion of a perfect fluid with angular momentum. Flows are calculated using two coordinate expansions of the poloidal velocity potential. An example, the generalization of spherically symmetric inflow, is presented in detail. The flows possess empty vortex cores which become cylindrical far from the accreting object with a width proportional to the fluid's angular momentum.

062.002 Magnetohydrodynamic equilibrium. II. General integrals of the equations with one ignorable coordinate.   K. C. Tsinganos.
Astrophys. J., Vol. 252, 775 - 790 (1982).
The steady equations of hydromagnetics for the isentropic or nonisentropic flow of an inviscid magnetofluid of high electrical conductivity, with one ignorable coordinate in a general orthogonal system, are treated. Several integrals of the equations are established thereafter reducing them to a scalar, quasi-linear, second order, partial differential equation for the magnetic potential. Simple solutions of this final equation are presented.

062.003 Limitations on the upconversion of ion sound to Langmuir turbulence.
L. Vlahos, K. Papadopoulos.
Astrophys. J., Lett., Vol. 252, L75 - L80 (1982).
It is shown that the weak turbulence theory used by Tsytovich, Stenflo, and Wilhelmsson to evaluate the nonlinear transfer of ion acoustic waves to Langmuir waves is limited to the level of ion acoustic waves $W_s/nT < (m/M)^2$. The impossibility of accelerating electrons by such a process for any reasonable physical system is reaffirmed.

062.004 The effects of a simple shear layer on the growth of Kelvin-Helmholtz instabilities.   T. P. Ray.
Mon. Not. R. Astron. Soc., Vol. 198, 617 - 625 (1982).
The author studies Kelvin-Helmholtz instabilities for the case of a linear shear in an inviscid compressible homogeneous fluid. He finds that such a shear destabilizes zero growth vortex sheet modes when the Mach number (defined in terms of half the relative velocity across the shear) is sufficiently high. A comparison is drawn between the author's work for a linear shear, and earlier work on a hyperbolic tangent shear. Close parallels are seen in the results of these two velocity profiles.

062.005 The structure and variability of dynamo driven accretion discs.   R. E. Pudritz, G. G. Fahlman.
Mon. Not. R. Astron. Soc., Vol. 198, 689 - 706 (1982).
A turbulent dynamo operating in an accretion disc around a black hole can produce fields strong enough so that the Maxwell stress due to the fluctuation dominates. In this dynamo driven limit, enormous localized fluctuations can be expected because the Kepler flow energy density is efficiently tapped. The detailed radial structure of this model is calculated and applied to Cyg X-1.

062.006 Alfvénic fluctuations as asymptotic states of MHD turbulence.
R. Grappin, U. Frisch, J. Léorat, A. Pouquet.
Astron. Astrophys., Vol. 105, 6 - 14 (1982).
Satellite observations of the solar wind show a fully developed turbulent state in which the velocity and magnetic fluctuations are often very strongly correlated. To investigate a possible dynamical origin of this correlation, the authors apply a two-point closure method to study the time evolution of incompressible, homogeneous, mirror-symmetric MHD turbulence with non-zero velocity-magnetic fields correlation. Analytical and numerical evidence is presented that nonlinear interactions lead asymptotically to a state of maximal correlation. Implications of this result for the origin of strongly correlated turbulence in the solar wind are discussed.

062.007 The unsteady beam.   M. Nepveu.
Astron. Astrophys., Vol. 105, 15 - 20 (1982).
Variability of central objects in a radio galaxy may cause extra energy output statistically. This extra output does not destroy an associated beam, but can cause a non-linear Kelvin-Helmholtz instability. A constriction of the beam is brought about in a time given by the minimum nozzle circumference divided by the beam sound speed at the center. It travels along the beam at roughly the sound speed of the cool external medium. Subsequent constrictions may give rise to "patches" in radio maps.

062.008 Particle acceleration in modified shocks.
L. O'C. Drury, W. I. Axford, D. Summers.
Mon. Not. R. Astron. Soc., Vol. 198, 833 - 841 (1982).
Efficient particle acceleration in shocks must modify the shock structure with consequent changes in the particle acceleration. This effect is studied and analytic solutions are found describing the diffusive acceleration of particles with momentum independent diffusion coefficients in hyperbolic tangent type velocity transitions. If the input particle spectrum is a delta function, the shock smoothing replaces the truncated power-law downstream particle spectrum by a more complicated form, but one which has a power-law tail at high momenta. For a cold plasma this solution can be made completely self-consistent. Some problems associated with momentum dependent diffusion coefficients are discussed.

062.009 Magnetohydrodynamic Kelvin-Helmholtz instabilities in astrophysics — III. Hydrodynamic flows with shear layers.   A. Ferrari, S. Massaglia, E. Trussoni.
Mon. Not. R. Astron. Soc., Vol. 198, 1065 - 1079 (1982).
The authors discuss Kelvin-Helmholtz instabilities in pressure-confined two-dimensional flows (slabs) delimited by boundary layers with velocity and density gradients. It is found that the fastest growing modes in supersonic flows are produced by perturbations reflecting at the boundaries and have wavelengths of the order of the slab width. From a comparison of the results for the two-dimensional slab and three-dimensional cylinder the authors conclude that a two-dimensional treatment provides an adequate description of instabilities in fluid flows. In the final section the authors attempt a comparison of the results obtained with morphologies in collimated jets in extragalactic radio sources; general characteristics appear to be classifiable in terms of scale-lengths of the velocity and density gradients in the boundary layers.

062.010 Acceleration of a relativistic plasma by radiation pressure.   E. S. Phinney.
Mon. Not. R. Astron. Soc., Vol. 198, 1109 - 1118 (1982).
The greatly enhanced radiation pressure force felt by a relativistic plasma is accompanied by catastrophic Compton cooling and only in extreme conditions can it lead to acceleration to relativistic bulk velocities. The author solves the equations of motion in the optimal case and finds that the efficiency of acceleration is typically < 1 per cent. The complicating effects of expansion of the plasma, finite source size and scattering above the Klein-Nishina limit are described. The author ends with a short list of situations in which the phenomenon may nevertheless be of importance.

062.011 Extended adiabatic blast waves and a model of the soft X-ray background.
D. P. Cox, P. R. Anderson.
Astrophys. J., Vol. 253, 268 - 289 (1982).

The authors investigate the suggestion that much of the soft X-ray background may arise from the solar system being inside a large supernova blast wave propagating into the hot component of the interstellar medium. An analytical approximation is generated which models the development of an adiabatic spherical blast wave in a homogeneous ambient medium of finite pressure. An analytical approximation is also presented for the electron-temperature distribution resulting from Coulomb collisional heating. The dynamical, thermal, ionization, and spectral structures are calculated for blast waves of energy 5×10^{50} erg in a hot, low-density interstellar environment. It is shown that the B and C bands of the soft X-ray background are reproduced by such a model explosion if the ambient density is about 0.004 cm^{-3}, the blast radius is roughly 100 pc, and the solar system is located inside the shocked region. The age of such an explosion is roughly 10^5 a. The X-ray background in the M band, however, is considerably larger than the predicted flux and must be produced by an independent mechanism.

062.012 On an estimate of the dynamo-generated magnetic fields in late-type stars.
B. R. Durney, R. D. Robinson.
Astrophys. J., Vol. 253, 290 - 297 (1982).

The depth of the convection zone varies significantly with spectral type in late-type stars. The authors study the effect of this variation on the generation of magnetic fields by dynamo processes. The method is based upon the assumption that the fields are determined when the rise time of a magnetic flux tube due to buoyancy is equal to the e-folding amplification time for the stellar dynamo. The basic results indicate that for a given rotational period, the magnetic field increases with $(B - V)$ color, whereas the amplification time decreases, suggesting shorter cycle periods. For M stars and for rotational periods of the order of that of the Sun the magnetic field area coverage approaches 100% of the stellar surface, suggesting small dissipation scales for the magnetic field (flare stars).

062.013 The Parker instability in a self-gravitating gas layer.
B. G. Elmegreen.
Astrophys. J., Vol. 253, 634 - 654 (1982).

The dispersion relation for the Parker instability in a self-gravitating, exponential gas layer is derived and solved explicitly to give the growth time of a perturbation as a function of its dimensions and initial density. Self-gravity is important as an additional driving force for the Parker instability when a dimensionless density becomes comparable to the ratios of the magnetic and cosmic ray pressures to the thermal gas pressure. Self-gravity becomes more important than magnetic fields or cosmic rays in regions of higher gas density. For example, in a spiral density wave shock, where the gas density may be 5 cm^{-3}, the initial growth time of the combined instability is only 12 million years. The instability is dominated by self-gravitational forces at these densities. Cloud formation by this Parker-Jeans instability can be so fast that star formation may occur in a moderately compressed interstellar medium within only 20 million years.

062.014 The formation of giant cloud complexes by the Parker-Jeans instability.   B. G. Elmegreen.
Astrophys. J., Vol. 253, 655 - 665 (1982).

The Parker-Jeans instability is considered as a possible mechanism for forming the giant cloud complexes observed near OB associations. The author uses a previously derived dispersion relation to evaluate the masses and growth times of the dominant modes in this instability. The results show that massive clouds ($M \approx 10^6 M_\odot$) can form quickly (≈ 12 million yr) in the high density environments (5 cm^{-3})

associated with spiral density wave shocks. For densities larger than about 3 cm^{-3}, these clouds form primarily as a result of the self-gravitational forces in the interstellar medium. Lower mass clouds ($M \approx 10^5 M_\odot$) can form in lower density environments as a result of the pure Parker instability. The masses of the clouds that form when the density exceeds about 3 cm^{-3} are insensitive to the magnetic field strength, cosmic ray pressure, and ambient density.

062.015 Multi-ion resonances in finite temperature plasma.
D. D. Barbosa.
Astrophys. J., Vol. 254, 376 - 390 (1982).

This paper examines the dispersion properties of electrostatic waves in a multi-ion plasma. Beginning with a cold plasma formulation of the Buchsbaum, lower hybrid, and upper hybrid, resonances, the analysis incorporates the effects of finite temperature, finite parallel (to the magnetic field) wavenumber, and wave growth/damping in an ordered and systematic development of the equations. A number of numerical solutions and analytic formulae are given for an ion plasma consisting of H$^+$, O$^+$, and S$^+$ for application to Jupiter's magnetosphere. It is demonstrated how observations of plasma resonance lines may be used as a diagnostic of ion composition in magnetospheric plasma, and the first use of this technique is made to infer the hydrogen concentration in Jupiter's middle magnetosphere.

062.016 Oscillating dynamo in the presence of a fossil magnetic field. The solar cycle.
E. H. Levy, D. Boyer.
Astrophys. J., Lett., Vol. 254, L19 - L22 (1982).

Hydromagnetic dynamo generation of oscillating magnetic fields in the presence of an external, ambient magnetic field introduces a marked polarity asymmetry between the two halves of the magnetic cycle. The principle of oscillating dynamo interaction with external fields is developed, and a tentative application to the Sun is described. In the Sun a dipole moment associated with the stable fluid beneath the convection zone would produce an asymmetrical solar cycle.

062.017 An MHD instability in compact fluid objects.
D. Eichler.
Astrophys. J., Vol. 254, 683 - 687 (1982).

It is proved that spherical and nearly spherical magnetized fluids that contain purely poloidal fields having a nonzero vacuum component are in general unstable to MHD motions, as conjectured by Flowers and Ruderman. The exact growth rate is calculated for a simple geometry. It is suggested that the instability may occur in magnetic white dwarfs, with observable consequences.

062.018 Parametric excitation of Alfvén waves by magnetosonic waves with oblique propagation.
K. Yumoto, T. Saito.
Planet. Space Sci., Vol. 30, 199 - 207 (1982).

062.019 Nonlinear force-free magnetic fields.
B. C. Low.
Rev. Geophys. Space Phys., Vol. 20, 145 - 159 (1982).

Astrophysical magnetic fields in a low beta plasma can evolve in a quasi-steady state for extended periods of time which are much longer than the characteristic dynamical time scale. The evolution proceeds with the electric current flowing largely parallel to the magnetic field, so that as a first approximation the Lorentz force may be set to zero everywhere. This paper reviews in a unified picture the results of various theoretical investigations of this process. These investigations seek to uncover those circumstances under which extreme conditions develop to make the assumption of a quasi-steady state untenable, and the field must pass into a fast dynamical phase at this point. The astrophysical interest is in explaining how a solar magnetic field would occasionally break into a flare or other eruptions. Two types of mechanisms have been treated.

062.020 Magnetoid interaction with surrounding stars.
  L. M. Ozernoy (*Ozernoj*), V. V. Usov.
Nature, Vol. 296, 48 - 49 (1982).
  Lugg (1981) has recently analysed a magnetoid – a supermassive rotating highly magnetized star – situated at the centre of a compact galactic nucleus in the belief that previous investigations had taken no account of interaction with neighbouring stars. Such an interaction has been widely discussed, but there has been no confirmation of the main conclusion of Lugg that these stars can disrupt a magnetoid in as short a time as 10^4 yr. Here the authors demonstrate that this conclusion is erroneous.

062.021 Axisymmetric collapse of rotating, isothermal
  clouds.   A. P. Boss, J. G. Haber.
Astrophys. J., Vol. 255, 240 - 244 (1982).
  A numerical hydrodynamics code is used to calculate the axisymmetric collapse of isothermal clouds from uniform density and uniform rotation initial conditions. The authors report the results of 53 calculations covering the appropriate ranges of α and β, using either a constant pressure or constant volume boundary condition. Clouds with low initial values of α and β undergo significant collapse. Three end states are obtained: Bonnor-Ebert spheroids, rings and collapsing disks. The rings are formed with values of β typically less than the Maclaurin spheroid value for dynamic instability to ring formation.

062.022 Equation of motion and proper oscillations of a
  magnetic toroid.   A. A. Solov'ev.
Soln. Dannye 1981 Byull., No. 11, p. 93 - 98 (1982). In Russian.
  An equation of the motion of a magnetic toroid in a homogeneous outer medium is derived. This equation, together with the conditions of conservation of the magnetic flux in the rope-toroid, permits to describe the oscillations of the system near the equilibrium position $\kappa = 2$.

062.023 Particle drift, diffusion, and acceleration at shocks.
  J. R. Jokipii.
Astrophys. J., Vol. 255, 716 - 720 (1982).
  The gradient and curvature drifts implicit in the change of the ambient magnetic field at a hydromagnetic shock wave are incorporated into the "diffusive" theory of shock acceleration of charged particles. The conventional jump condition at the shock is modified by a term incorporating the large drift along the shock plane. This term vanishes identically for one-dimensional systems, but must be included in general for shocks which are finite in transverse extent or which have transverse structure.

062.024 An overview of the problems connected with
  theoretical calculations for hot plasmas.
B. F. Rozsnyai.
J. Quant. Spectrosc. Radiat. Transfer, Vol. 27, (see 012.005), 211 - 217 (1982).
  A brief description is given of the different quantum-mechanical models and approximations used in hot plasma calculations. The subject is elucidated by presentation of computed data for iron and astrophysical plasmas.

062.025 Iron plasma: sensitivity of photoelectric cross sec-
  tions to different models and general features of the
Fermi–Amaldi-modified model.
D. Shalitin, A. Ron, Y. Reiss, R. H. Pratt.
J. Quant. Spectrosc. Radiat. Transfer, Vol. 27, (see 012.005), 219 - 226 (1982).
  Photoelectric cross sections in several atomic models are presented as a function of temperature and density. The

models discussed are Thomas–Fermi (TF), Fermi–Amaldi-modified (FAM), and Debye–Hückel–Thomas–Fermi (DHTF). The authors also present some systematic results for the less known FAM potential model regarding predictions for: electrostatic potentials, bound electron level energies, pressures, and branching ratios of photoelectric cross sections. The pure iron plasma which was explored had a temperature in the range of 0.2-3 keV and a density in the range of 50-1000 g/cm^3.

062.026 The dynamics of fibril magnetic fields. I. Effect of
  flux tubes on convection.   E. N. Parker.
Astrophys. J., Vol. 256, 292 - 301 (1982).
  Observations have established the general fibril state of the magnetic field at the visible surface of the Sun. This extraordinary state of the field, in separate intense flux tubes, implies that the convection has a relatively permanent, closed topology, in spite of the large Reynolds number. This paper explores some of the effects of the separate flux tubes on the convective motions, generally pushing the convective cells toward aligning their downdrafts with the flux tubes so as to minimize the dissipation.

062.027 The dynamics of fibril magnetic fields. II. The mean
  field equations.   E. N. Parker.
Astrophys. J., Vol. 256, 302 - 315 (1982).
  A variety of scenarios for the origin and activity of magnetic fields in the convective zone of the Sun have been put forth in recent years. The present paper works out the mean field equations for intense thin fibrils of fixed cross section under the assumption of some significant local ordering. The aim is to establish to what degree new effects may arise as a result of the fibril state. It is shown that the fibril state of a mean field B_i enhances the tension in the field and the buoyancy of the field by the compression factor m of the field in the individual fibrils. New qualitative effects appear in the slip velocity u_i of the fibrils through the ambient fluid and in the neutral point reconnection of neighboring fibrils. One of the most important consequences is the simplification of the theory of turbulent diffusion by eliminating the looping and tangling that arises when a continuum field is carried in a chaotic turbulent flow.

062.028 Compressible convection in a rotating spherical shell.
  V. Induced differential rotation and meridional
circulation.   G. A. Glatzmaier, P. A. Gilman.
Astrophys. J., Vol. 256, 316 - 330 (1982).
  The authors describe solutions for the mean differential rotation and meridional circulation in a compressible, rotating, spherical fluid shell which are induced by the linear, anelastic solutions of their earlier papers for global convection. The mean solutions strongly depend on the density stratification, the rotation rate, the convective velocity distribution, and the amount of viscous diffusion relative to thermal diffusion. In order to obtain an equatorial acceleration which is large in amplitude relative to the meridional circulation, together with a small equator-pole temperature difference, when the density stratification is large as in the solar convection zone, at least one of two conditions must be met: either the effect of rotation must be large compared to the effects of viscous diffusion and buoyancy, or viscous diffusion must be small relative to thermal diffusion.

062.029 Nonlinear astrophysical dynamo: a marginally
  unstable case.   S. Hinata.
Astrophys. J., Lett., Vol. 256, L23 - L27 (1982).
  A simple set of nonlinear dynamo equations which incorporate the effect of Lorentz force to the differential rotation is derived and analyzed for a marginally unstable dynamo. The nonlinear equilibrium state is found within one mode approximation. The period of magnetic cycle found by

numerical calculations is the magnetic dissipation time, which is different from the linear result (period = generation time).

062.030   **High energy γ-rays from a relativistic plasma.**
    F. Giovannelli, S. Karakula, W. Tkaczyk.
Astron. Astrophys., Vol. 107, 376 - 377 (1982).
    Using the γ-ray production spectrum in the comoving plasma system for different temperatures by Giovannelli et al. (1981, 1982), the authors compare the results with those of Kolykhalov and Syunyaev (1979). The present results give wider spectra; so the authors briefly discuss the reasons for this discrepancy.

062.031   **Plasma heating in a sheared magnetic field.**
    V. Krishan.
J. Astrophys. Astron., Vol. 2, 379 - 385 (1981).
    The mechanism of spatial resonance of Alfvén waves for heating a collisionless plasma is studied in the presence of a twisted magnetic field. In addition to modifying the equilibrium condition for a cylindrical plasma, the azimuthal component of the magnetic field gives extra contribution to the energy deposition rate of the Alfvén waves. This new term clearly brings out the effects associated with the finite lifetime of the Alfvén waves. The theoretical system considered here conforms to the solar coronal regions.

062.032   **Equilibrium configuration of the magnetosphere of a star loaded with accreted magnetized mass.**
    Y. Uchida, B. C. Low.
J. Astrophys. Astron., Vol. 2, 405 - 419 (1981).
    Equilibrium configuration of the magnetosphere of a star loaded by the gravitationally accreted plasma having its own magnetic field is investigated. Axisymmetry around the star's magnetic axis is assumed for simplicity. It is shown that two distinct configurations appear for the cases of parallel and antiparallel magnetic field of the accreted plasma with respect to the star's magnetic moment.

062.033   **Structure and stability of rotating fluid disks around massive objects. I. Newtonian formulation.**
    D. K. Chakraborty, A. R. Prasanna.
J. Astrophys. Astron., Vol. 2, 421 - 437 (1981).
    In this paper the authors present a very general class of solutions for rotating fluid disks around massive objects (neglecting the self gravitation of the disk) with density as a function of the radial coordinate only and pressure being nonzero.

062.034   **Nonlinear interaction of magnetic field and convection in three-dimensional motion.**   N. Rudraiah.
Publ. Astron. Soc. Japan, Vol. 33, 721 - 737 (1981).
    Three-dimensional nonlinear convection in a horizontal Boussinesq layer of fluid heated from below is considered in the presence of an imposed vertical magnetic field. The study is based on local nonlinear analysis which is pivoted on the linear theory. The effect of magnetic field on convective heat transfer and the physically preferred cell patterns are determined.

062.035   **Numerical experiments on a model of penetrative convection.**   S. Yamaguchi.
Publ. Astron. Soc. Japan, Vol. 34, 99 - 116 (1982).
    In the present study, the Boussinesq equation of convection is assumed to be valid even though constant β_A is replaced by a function $\beta_A(\bar{T})$, and the horizontal extent of the medium, L, is fixed to a value 2 or 4. The aim of the study is to know the growth of the convective layer from the initial unstable layer, the vertical extension of the primary cell, and the character of the convection in the final state.

062.036   **On the onset of thermal convection in slowly rotating fluid shells.**   G. Geiger, F. Busse.

Geophys. Astrophys. Fluid Dyn., Vol. 18, 147 - 156 (1981). Abstr. in Phys. Abstr., Vol. 85, Abstr. 16311 (1982).

062.037   **Growth and decay of weak waves in radiative magnetogasdynamics.**   R. Shyam, J. Sharma, V. D. Sharma.
AIAA J., Vol. 19, 1246 - 1248 (1981). − Abstr. in Phys. Abstr., Vol. 85, Abstr. 20514 (1982).

062.038   **Generation of plasma vortices by fast magnetosonic waves.**   P. K. Shukla, M. Y. Yu, H. U. Rahman.
Plasma Phys., Vol. 23, 949 - 953 (1981). − Abstr. in Phys. Abstr., Vol. 85, Abstr. 20839 (1982).

062.039   **On the formation of protostellar disks.**
    P. Cassen, A. Moosman.
Icarus, Vol. 48, 353 - 376 (1981).
    An analysis is presented of the hydrodynamic aspects of the growth of protostellar disks from the accretion (or collapse) of a rotating gas cloud. The size, mass, and radiative properties of protostellar disks are determined by the distribution of mass and angular momentum in the clouds from which they are formed, as well as from the dissipative processes within the disks themselves. It is possible to construct models of the primitive solar nebula as an accretion disk, formed by the collapse of a slowly rotating protostellar cloud, and containing the minimum mass required to account for the planets.

062.040   **The gravitational instability of flow through porous medium for some systems of astrophysical interest.**
    R. C. Sharma, K. P. Thakur.
Astrophys. Space Sci., Vol. 81, 95 - 102 (1982).
    The gravitational instability of flow through a porous medium for some hydrodynamical and hydromagnetical systems of astrophysical interest is investigated. The effects of rotation, magnetic field, viscosity and finite electrical conductivity are studied for the gravitational instability through a porous medium. The effect of suspended particles on the instability is also considered. It is found that Jeans's criterion remains unchanged in the presence of porosity, viscosity, finite conductivity, rotation, magnetic field and suspended particles in the medium.

062.041   **Force-free fields in a three dimensional plasma torus.**   E. A. Evangelidis.
Astrophys. Space Sci., Vol. 81, 261 - 268 (1982).
    An analytic solution of the force-free field equation has been found for a three-dimensional torus.

062.042   **Approximate treatment of ideal blast waves with frozen-in magnetic field.**
    B. G. Verma, J. P. Vishwakarma, V. Sharan.
Astrophys. Space Sci., Vol. 81, 315 - 321 (1982).
    A blast wave, which is headed by a strong spherical shock wave, free or driven, moving into a medium permeated by a magnetic field and having a power law density distribution, is studied. The trajectory and associated quantities such as velocity of the shock are obtained analytically using a simple approximation technique.

062.043   **Hydromagnetic stability of rotating gas.**
    B. K. Shivamoggi.
Astrophys. Space Sci., Vol. 81, 329 - 331 (1982).
    Hydromagnetic stability of an inviscid perfectly conducting rotating gas is studied. The compressibility effects are shown not to promote stability of the flow under consideration.

062.044   **Strong-plane shock waves in optically-thin atmospheres in magnetogasdynamics.**
    J. B. Singh, S. K. Srivastava.
Astrophys. Space Sci., Vol. 81, 369 - 378 (1982).

In this paper similarity solutions for the propagation of strong-plane shock waves in optically-thin grey atmospheres are obtained in the presence of a magnetic field. Density and magnetic field are constant in the undisturbed gaseous medium in front of the shock. Planck's diffusion approximation has been taken into account in this problem and a comparative study has been made between the results of ordinary gasdynamics and magnetogasdynamics.

062.045 The radiation of energy by a current in a plasma.
P. C. W. Fung, K. Young.
Astrophys. Space Sci., Vol. 81, 453 - 470 (1982).

The derivation of the differential power emitted in any given direction by a current in a linear, homogeneous and non-absorbing plasma is reviewed in detail. The conventional derivation is shown to give the power emitted; a formalism for the power received is established by evaluating the Poynting vector in terms of the far field. It is pointed out that the two power expressions differ because the same energy dE is emitted in a time dt_e but received over a different time dt_r. The necessary steps for establishing a valid formalism for anisotropic media are briefly sketched.

062.046 Free convection and mass transfer effects on the unsteady laminar boundary-layer accelerated flow past a vertical porous limiting surface with uniform suction.
N. G. Kafousias, G. A. Georgantopoulos.
Astrophys. Space Sci., Vol. 81, 471 - 477 (1982).

An analysis of the mass transfer and free convection effects on the unsteady laminar accelerated flow of a viscous incompressible fluid past an infinite vertical porous limiting surface is presented when the free stream is accelerated and the limiting surface temperature and concentration changes with step-wise variations. Expressions for velocity and skin-friction are obtained by using Laplace transform, when the Prandtl number and the Schmidt number are equal to one.

062.047 A note on σ-stability in hydromagnetics.
J. A. Adam.
Astrophys. Space Sci., Vol. 82, 115 - 121 (1982).

The concept of σ-stability is reviewed and applied to the case of aligned magnetoatmospheric flow. A sufficient condition for σ-stability is derived, and the upper bound arising from this condition is investigated as a function of σ.

062.048 A comparative study of computational methods in cosmic gas dynamics.
G. D. van Albada, B. van Leer, W. W. Roberts, Jr.
Astron. Astrophys., Vol. 108, 76 - 84 (1982).

The objective of the article is to supply a comparison of a variety of numerical methods on a representative astrophysical flow problem. The methods considered include some that are widely used in astronomy, some that are widely used in other fields such as aerodynamics, and one that is rather new. All the methods are explicit and are especially suited for transonic and supersonic flows. This comparison aims to acquaint astronomers with the virtues and failings of typical numerical methods.

062.049 Modifikation des Bremsstrahlungsquerschnitts durch starke Magnetfelder.
G. Wunner, J. Lauer, H. Herold, H. Ruder.
Mitt. Astron. Ges., Nr. 55, p. 31 - 33 (1982).

062.050 Zweidimensionale Dynamomodelle auf der Basis magnetischer Flußröhren.   M. Schüssler.
Mitt. Astron. Ges., Nr. 55, p. 69 - 70 (1982).

062.051 Flute instability of a relativistic plasma.
T. D. Kaladze, A. B. Mikhajlovskij.
Astrofizika, Tom 17, 775 - 782 (1981). In Russian. English translation in Astrophysics, Vol. 17, No. 4.

A theoretical study of the flute instability of a relativistic plasma is given which is of interest in the theory of jump ejection of particles from pulsars. The boundaries of instability are established and the growth rate of perturbations for a plasma with arbitrary relativistic factor is calculated.

062.052 The magnetic field of a plasma stream flowing around a blunted body.
S. I. Alimarin, B. A. Tverskoj.
Geomagn. Aehron., Tom 22, 1 - 4 (1982). In Russian.

062.053 Pinned vorticity in rotating superfluids, with application to neutron stars.
D. Pines, J. Shaham, M. A. Alpar, P. W. Anderson.
Prog. Theor. Phys. Suppl., No. 69, p. 376 - 396 (1980).
Abstr. in Phys. Abstr., Vol. 85, Abstr. 23317 (1982).

062.054 Incompressible convection in a radiating atmosphere. I. General characteristics.
A. Legait.
Astron. Astrophys., Vol. 108, 287 - 295 (1982).

The differences between convection in a conductive and a radiating fluid are investigated. The Boussinesq approximation is used, a grey medium is assumed; the author solved the transfer integrals and introduced the thin layer approximation, which he compared with the Eddington approximation. The author points out the reliability of the thin layer approximation for the study of convection when a horizontal modal expansion is used. The most important differences with thermal convection are the importance of the boundary layers in the optically thin regions, and the existence of an upper limit for the ratio of the total flux to the radiative flux corresponding to radiative equilibrium.

062.055 On the generation of magnetic fields in late-type stars: a local time-dependent dynamo model.
R. D. Robinson, B. R. Durney.
Astron. Astrophys., Vol. 108, 322 - 325 (1982).

The authors assume that the magnetic field of late-type stars is generated in the lower part of the star's convection zone and study this generation mechanism with the help of local (in latitude) dynamo equations. For the spectral types G0, G5, K0, K5, M0, M2, and M5 the authors evaluate the magnetic field and period of the cycles as a function of rotation and compare them with the available observational data.

062.056 Broken symmetries and hydrodynamics of superfluid $^3P_2$-neutron-star matter.
H. Brand, H. Pleiner.
Phys. Rev. D, Vol. 24, 3048 - 3057 (1981). − Abstr. in Phys. Abstr., Vol. 85, Abstr. 38971 (1982).

062.057 A general formulation of the thin-shell approximation for axisymmetric, hypersonic, hydromagnetic flows.   J. L. Giuliani, Jr.
Astrophys. J., Vol. 256, 624 - 636 (1982).

Many gas dynamic problems in astrophysics involve the hypersonic expansion of a material shell supported by pressure forces. The thin-shell approximation basically consists of treating the shell as a two-dimensional surface. In this paper, a general system of time-dependent equations for the axisymmetric, hypersonic, hydromagnetic flow of a swept-up massive shell is formulated based upon the thin-shell approximation. The method of analysis incorporates previously considered physical aspects, such as the centrifugal correction, with the following new features: thermal pressure within the shell, finite thickness of the shell, and magnetic fields.

062.058 The dynamics of fibril magnetic fields. III. Fibril configurations in steady flows.   E. N. Parker.
Astrophys. J., Vol. 256, 736 - 745 (1982).

The goal of this paper is to provide an illustration of the equilibrium form of a hypothetical fibril field beneath the surface of the Sun. The known dynamical properties of fibrils and the paths of individual fibrils in a static atmosphere are reviewed briefly. There are given the equations for the path of individual fibrils in a horizontal flow, balancing the tension in the fibril against the buoyancy and aerodynamic drag. The equation is integrated for an isothermal atmosphere with a gentle vertical shear. The resulting fibril paths illustrate the forms that may be encountered a few thousand km beneath the visible surface of the Sun. It is pointed out that at depths of 10^4 km and more the fibrils are so strongly locked into the fluid motion as to be carried along bodily with the flow. At the surface of the Sun, on the other hand, the aerodynamic drag is so weak that the equilibrium path is essentially vertical, in agreement with observation.

062.059 The dynamics of fibril magnetic fields. IV. Trapping in closed convective rolls.
E. N. Parker.
Astrophys. J., Vol. 256, 746 - 760 (1982).

The equations of motion for a slender, buoyant flux tube extending horizontally along a closed convective roll are solved to illustrate the motion of the flux tube relative to the fluid. The principal effect of the buoyant rise u of the flux tube is to offset the path of the tube (which remains closed) from the closed circulation pattern of the fluid. Further investigation shows that an upward decrease of ρu causes the position of the tube to spiral inward to the dynamical equilibrium point where the convective downdraft is equal to the buoyant rise. The effect may be operative in the convective zone of the Sun and other stars where the meridional circulation is larger than the rate of rise of the standard fibril. Even if there is no long-term trapping of the field, the position of emergence of flux tubes at the surface of the Sun may be widely removed from their birthplace.

062.060 Pulse propagation in a magnetic flux tube.
I. C. Rae, B. Roberts.
Astrophys. J., Vol. 256, 761 - 767 (1982).

The linear development of a pulse as it propagates adiabatically along an isothermal magnetic flux tube embedded in a gravitationally stratified atmosphere is studied. It is shown that, for a quiescent environment, longitudinal disturbances in the tube are governed by an equation of the Klein-Gordon type. An impulsively generated disturbance results in a wave front propagating at the subsonic and subAlfvénic tube speed; the wave front trails a wake oscillating at the tube frequency. The results are illustrated for solar photospheric conditions.

062.061 Diffusion processes in Ap stars and white dwarfs.
V. A. Urpin, A. G. Muslimov.
Astron. Zh., Tom 59, 318 - 325 (1982). In Russian. English translation in Soviet Astron., Vol. 26, No. 2.

Diffusion processes in a completely ionized multi-component plasma are studied. The transport phenomena for ions as well as for electrons are considered. Detailed comparison of the obtained results with those of other authors is made. Some possible astrophysical consequences of the theory (surface layers of Ap stars and white dwarfs) are briefly discussed.

062.062 Damping and excitation of Langmuir waves in an inhomogeneous relativistic plasma; general theory and pulsar applications.   A. B. Mikhailovskii (*Mikhajlovskij*).
Plasma Phys., Vol. 24, No. 1, p. 1 - 18 (1982). – Abstr. in Phys. Abstr., Vol. 85, Abstr. 44901 (1982).

062.063 Adiabatic charged particle motion in rapidly rotating magnetospheres.
T. G. Northrop, T. J. Birmingham.
J. Geophys. Res., Vol. 87, 661 - 669 (1982).

062.064 Dependence of electron beam instability growth rates on the beam-plasma system parameters.
R. J. Strangeway.
J. Geophys. Res., Vol. 87, 833 - 841 (1982).

062.065 Magnification of pre-existing magnetic fields in impact-produced plasmas, with reference to impact craters.   P. Cerroni, G. Martelli.
Planet. Space Sci., Vol. 30, 395 - 398 (1982).

Magnetic fields generated by the electrical currents associated with thermal forces in an impact-produced plasma cloud are proposed as a possible explanation of the magnetic perturbation observed during hypervelocity impact events. Order of magnitude estimates for this effect show that this is compatible with experimental findings. The authors suggest that this effect may contribute to the magnetisation observed in the neighbourhood of lunar craters.

062.066 Compton rockets: radiative acceleration of a relativistic fluid.   S. L. O'Dell.
Extragalactic radio sources, (see 012.025), p. 365 - 366 (1982).

Unless a plasma moves at relativistic bulk speeds, the Compton radiative lifetime for relativistic electrons near a luminous object is less than the transit time from the source. Acceleration by adiabatic decompression is too slow to preserve much of the electrons' energy. However, the Compton-rocket thrust and the radiatively induced pressure gradient can accelerate a relativistic fluid to relativistic bulk speeds on a time scale comparable to that for radiative loss. Consequently, severe Compton losses are not only reduced by relativistic bulk motion, but can in fact effect such motion.

062.067 Propagation speeds and acoustic damping of waves in magnetic flux tubes.   H. C. Spruit.
Sol. Phys., Vol. 75, 3 - 17 (1982).

Propagation speeds are derived for the wave modes of a thin magnetic tube in an otherwise homogeneous magnetized or unmagnetized fluid. There are three types of wave, a (torsional) Alfvén wave and two waves which are specific for the thin tube. These are named the longitudinal and transversal tube waves, according to their polarization properties. They can be damped by radiating an MHD or acoustic wave into the surroundings of the tube. Conditions for occurrence of this acoustic damping, and the damping rates, are derived. The behavior of the waves in the solar convection zone and corona is discussed.

062.068 The effects on the MHD stability of field line tying to the end faces of a cylindrical magnetic loop.
C.-H. An.
Sol. Phys., Vol. 75, 19 - 34 (1982).

The author examines the magnetohydrodynamic (MHD) stability of a magnetic loop, taking into account field line tying at its foot points. He uses the ideal MHD energy equation to derive a stability equation for a specific class of perturbations. The author found that for a loop with large aspect ratio the field line tying effect is negligible to the $m = 1$ kink mode but important to the localized modes. The stability criterion for high m localized modes is derived and compared with the Suydam criterion. The result shows that for the perturbation of the class studied, there are two effects of field line tying; one is a field line bending effect which is always stabilizing and the other is a shear effect which is stabilizing or destabilizing depending on the sign of the gradient of potential magnetic field. The net effect of field line tying is determined by the sum of these two effects.

062.069 Coupling equations for a flow-wave field used to faculae heating.   X. Li, M. Song.
Sol. Phys., Vol. 75, 83 - 98 (1982).

In a binary system of a background fluid-wave field, the

wave effect may be important in some cases. From general properties of thermodynamics of the medium, the authors derive the coupling equations for the mean flow-wave field. For six wave modes the corresponding representation of the wave stress tensor is found. The representation for the Alfvén waves is applied to the faculae heating and a result consistent with observations is obtained.

062.070 **Thermal trigger for solar flares and coronal loops formation.**
B. V. Somov, S. I. Syrovatskii (*Syrovatskij*).
Sol. Phys., Vol. 75, 237 - 244 (1982).

A longitudinal stability is considered for a quasi-steady current sheet which is uniform along the current. In the MHD approximation, the stability problem is solved for the plane neutral sheet and small disturbances propagating along the current. The current sheet is shown to break-up into the system of cooler and more dense filaments due to radiative cooling. This process corresponds to the condensation mode of a thermal instability and can play a trigger role for a solar flare. Moreover, at the nonlinear stage of development, it can lead to the formation of very dense cold filaments surrounded by high-temperature low-density plasma inside the current sheet.

062.071 **Internal atmospheric hydromagnetic planetary-gravity waves in zonal wind-magnetic shears.**
O. M. El Mekki.
Sol. Phys., Vol. 75, 351 - 360 (1982).

Internal atmospheric hydromagnetic planetary-gravity waves propagating through a latitudinally sheared zonal flow and a zonal magnetic field sheared both latitudinally and vertically are studied. It is shown that the waves possess four critical latitudes.

062.072 **Wave propagation in a magnetically structured atmosphere. III: The slab in a magnetic environment.**
P. M. Edwin, B. Roberts.
Sol. Phys., Vol. 76, 239 - 259 (1982).

The propagation of waves in a magnetic slab embedded in a magnetic environment is investigated. Several different situations, representative of both photospheric and coronal conditions, are considered. In general, the structures are found to support both fast and slow, body and surface, waves. Under coronal conditions, for two dimensional propagation, disturbances propagate as fast and slow body waves.

062.073 **Magnetostatic atmospheres: a family of isothermal solutions.**   E. G. Zweibel, A. J. Hundhausen.
Sol. Phys., Vol. 76, 261 - 299 (1982).

Most models of large scale solar magnetic fields assume either that the fields are potential or that they are force free. The authors present a new, analytic, two parameter family of magnetic fields in equilibrium with isothermal plasma in a gravitational field. They discuss these models from the viewpoint of the insight into the balance of magnetic pressure gradient, and gravitational forces that they provide. They show that substantial deviations from the potential field configuration are obtained for plasma β of order unity, and they emphasize the variety of possible relationships between isobars and magnetic fieldlines.

062.074 **Compression of magnetic field in a viscous boundary layer.**   E. N. Parker.
Sol. Phys., Vol. 77, 3 - 11 (1982).

Galloway, Proctor, and Weiss have shown by numerical experiment that the magnetic field extending across a convective cell in a highly conducting viscous fluid may be concentrated into sheets with energy density $B^2/8\pi$ larger than the kinetic energy density $\frac{1}{2}\rho v^2$ of the convection by a factor $(\nu/\eta)^{1/2}$ or more. This paper employs conventional boundary layer theory for high Reynolds number to provide a simple

analytical example illustrating this remarkable effect of field concentration.

062.075 **Unstable poloidal magnetic fields in stars.**
W. Van Assche, R. J. Tayler, M. Goossens.
Astron. Astrophys., Vol. 109, 166 - 170 (1982).

The dynamic stability of an axisymmetric star with a purely poloidal magnetic field is studied by means of the energy method. It is shown that a poloidal magnetic field is always unstable at a simple neutral point, which is encircled by the magnetic field lines. The results obtained by Wright and by Markey and Tayler for particular field geometries are thus generalized. The σ-energy method is used to obtain the growth rates of instabilities in the neighbourhood of a simple neutral point. Numerical applications to weak poloidal fields in polytropes show that these fields are unstable with very short e-folding times.

062.076 **Rotating fluids in geophysics and planetary physics.**
R. Hide.
Q. J. R. Astron. Soc., Vol. 23, 220 - 235 (1982).

062.077 **Scattering and collapse of Langmuir waves driven by a weak electron beam.**
B. Hafizi, J. C. Weatherall, M. V. Goldman, D. R. Nicholson.
Phys. Fluids, Vol. 25, 392 - 401 (1982). – Abstr. in Phys. Abstr., Vol. 85, Abstr. 53203 (1982).

062.078 **Turbulence in regions where the magnetic field is perpendicular to the boundary.**
O. Lielausis.
Soobshch. Spets. Astrofiz. Obs., Vyp. 32, (see 012.029), p. 20 (1981).

062.079 **On calculation of the velocity of a strong shock wave in an inhomogeneous medium.**
B. I. Gnatyk.
Astron. Tsirk., No. 1195, p. 4 - 5 (1981). In Russian.

062.080 **Non-zero flow rate (accretional) semi-dynamo: two simple examples.**   Eh. M. Drobyshevskij.
Astron. Zh., Tom 59, 600 - 602 (1982). In Russian. English translation in Soviet Astron., Vol. 26, No. 3.

The communication presents two examples (nonstationary and stationary) of magnetic field generation in a fluid by processes having no self-exciting feedback, when the Joule dissipation of the magnetic field created by an impressed e.m.f. is compensated completely or partially by simple motion (V = const) of matter. The resultant field exceeds strongly the one generated by the e.m.f. action in a motionless medium.

062.081 **A stability criterion for many-parameter equilibrium families.**   R. D. Sorkin.
Astrophys. J., Vol. 257, 847 - 854 (1982).

Theorems are established which let one detect instabilities without recourse to the usual perturbation analysis. The method applies to any system whose stable equilibria maximize a functional S at fixed values of one or more parameters E^α. It generalizes the "turning point method" by inferring instability from the behavior in equilibrium of the E^α and of their conjugate parameters $\partial S/\partial E^\alpha$. The "cusp catastrophe" and the black hole equilibrium family illustrate the approach. In connection with the latter, an Appendix proves that the Gibbs free energy is an analytic function of its natural arguments, as would be expected if all the equilibria belonged to a single thermodynamic phase.

062.082 **A soliton gas model for astrophysical magnetized plasma turbulence.**
S. R. Spangler, J. P. Sheerin.
Astrophys. J., Vol. 257, 855 - 861 (1982).

The authors consider the astrophysical implications of a

turbulence model consisting of a "gas" of Alfvén solitons. Properties of such a model are: (1) a power law magnetic irregularity spectrum with index = −2, (2) identical spectral shapes for the density and magnetic field spectra, as is observed in the solar wind, and (3) a prediction that for high values of the plasma β, the density and magnetic field enhancements will be anticorrelated. The third of these properties may be of importance in the interpretation of radio astronomical Faraday rotation measurements.

062.083   On an estimate of the dynamo-generated magnetic fields in late-type stars.
B. R. Durney, R. D. Robinson.
Bull. American Astron. Soc., Vol. 13, 791 (1981). − Abstract.

062.084   The stability of radiative shock waves.
J. N. Imamura, R. A. Chevalier.
Bull. American Astron. Soc., Vol. 13, 794 (1981). − Abstract.

062.085   MHD of non-force-free constant pitch field.
H. M. Chang.
Bull. American Astron. Soc., Vol. 13, 837 (1981). − Abstract.

062.086   On the expulsion of magnetised plasmas through solar and stellar atmospheres.   B. C. Low.
Bull. American Astron. Soc., Vol. 13, 837 (1981). − Abstract.

062.087   Plasma in a temperature gradient. II. Analysis of the BGK equation.   E. C. Shoub.
Bull. American Astron. Soc., Vol. 13, 890 - 891 (1981). Abstract.

062.088   Stability of X-ray heated accretion flow.
R. F. Stellingwerf.
Bull. American Astron. Soc., Vol. 13, 902 (1981). − Abstract.

062.089   Helically symmetric hydromagnetic equilibria: exact solutions.   K. Tsinganos.
Bull. American Astron. Soc., Vol. 13, 907 (1981). − Abstract.

062.090   More on dynamos driven by global convection and differential rotation.   P. A. Gilman.
Bull. American Astron. Soc., Vol. 13, 907 (1981). − Abstract.

062.091   Global circulations driven by small scale convection: three-dimensional simulations of locally driven flows.   D. H. Hathaway, R. C. J. Somerville.
Bull. American Astron. Soc., Vol. 13, 907 - 908 (1981). Abstract.

062.092   A nonlinear dynamo.   S. Hinata.
Bull. American Astron. Soc., Vol. 13, 908 (1981). Abstract.

062.093   Non linear force free magnetic fields with chosen symmetry.   H. M. Chang, R. L. Carovillano.
Bull. American Astron. Soc., Vol. 13, 909 (1981). − Abstract.

062.094   Magnetic helicity: twisted fields in MHD.
G. Field.
Bull. American Astron. Soc., Vol. 13, 910 - 911 (1981). Abstract.

062.095   Nonlinear penetrative convection in a compressible medium.
N. Hurlburt, J. Toomre, J. M. Massaguer.
Bull. American Astron. Soc., Vol. 13, 912 (1981). − Abstract.

062.096   Evolution of MHD waves in a homogeneous iso-thermal medium.
G. S. Bisnovatyj-Kogan, S. B. Popov, Yu. P. Popov.
Astrofizika, Tom 17, 333 - 348 (1981). In Russian. English translation in Astrophysics, Vol. 17, No. 2.

The process of propagation and nonlinear transformation of magnetohydrodynamic (MHD) waves generated by an initial magnetic field perturbation is considered. One-dimensional equations of hydromagnetics in application to solar chromosphere physics have been applied. Making use of linear analysis, the evolution of a small perturbation in an isothermal medium in a homogeneous magnetic field and some energy characteristics depending on parameters are investigated. The evolution of the finite amplitude perturbation is numerically obtained in complete nonlinear formulation of the problem. The spectrum evolution and MHD-wave transformation are considered.

062.097   Effects of a helical magnetic field on the stability of a gravitating cylinder, II.   A. M. Karnik.
Astrophys. Space Sci., Vol. 82, 283 - 287 (1982).
The effect of an external magnetic field on an infinite gravitating cylinder is studied for axisymmetric perturbations. It is found that there is a destabilisation of the system due to the external field at very high values of the pitch of the field.

062.098   Collective effects in diffuse matter-antimatter plasma. I.   S. Rogers, W. B. Thompson.
Astrophys. Space Sci., Vol. 82, 409 - 413 (1982).
A separation mechanism is proposed which is effective in collisions between ionized but unmagnetized clouds of matter and anti-matter. This involves an electromagnetic instability which grows at the encounter layer, producing magnetic fields strong enough to separate the clouds after a very small interpenetration.

062.099   Collective effects in diffuse matter-antimatter plasma. II.   S. Rogers, W. B. Thompson.
Astrophys. Space Sci., Vol. 82, 415 - 421 (1982).
It is shown that even very slightly ionized clouds of matter and anti-matter can interpenetrate only a little on collision. Initial interpenetration produces fast electrons and positrons from annihilation. These, in turn, produce strong magneto-hydrodynamic shocks which give the small ionized component enough energy to ionize the neutral fraction and produce a Leidenfrost layer in about ten years after which interpenetration stops.

062.100   Determination of the electromagnetic field produced by a magnetic oblique-rotator. IV. Corotating-plasma solution (2).   O. Kaburaki.
Astrophys. Space Sci., Vol. 82, 441 - 456 (1982).
The inertial effect on the structure of the magnetosphere of a rotating star is investigated, in the corotation approximation for a surrounding quasi-neutral plasma. The equation of motion reduces to a usual static balance equation between the electromagnetic and the centrifugal forces, in the rotating frame. However the MHD condition, which can be regarded as a special form of the generalized Ohm's law, is modified by the inclusion of inertial effect, with a violation of the 'frozen-in' condition in case of a general (i.e., not restricted to corotation) plasma motion. The inertial effect on the electromagnetic field is summarized in a partial scalar potential named the non-Backus potential, which is proportional to the centrifugal potential in the corotation approximation.

062.101   Instability of the electromagnetic ordinary mode in counterstreaming plasmas.   B. K. Shivamoggi.
Astrophys. Space Sci., Vol. 82, 481 - 483 (1982).
The instability of a linearly-polarised electromagnetic ordinary mode in a plasma and propagating perpendicular to a uniform magnetic field caused by a counterstreaming of electrons along the latter is studied using a Vlasov plasma model. The results show that for weak magnetic fields, the thermal effects stabilise the ordinary mode; for strong magnetic fields, the thermal effects destabilise the ordinary mode.

062.102 Self-similar isothermal flows of increasing energy in magnetogasdynamics.
B. G. Verma, R. N. Lal Srivastava.
Astrophys. Space Sci., Vol. 83, 69 - 74 (1982).

Self-similar solution for isothermal flows driven by an expanding piston are investigated in the presence of a magnetic field. The total energy of the flow between the shock and the piston is taken to be dependent on time obeying a power law. The shock is assumed to be strong and propagating into a perfect gas at rest with non uniform density and magnetic field.

062.103 The instability of the electromagnetic ordinary modes in counterstreaming and counterrotating plasmas. B. K. Shivamoggi.
Astrophys. Space Sci., Vol. 83, 177 - 180 (1982).

The instability of a linearly-polarised electromagnetic ordinary mode in counterrotating plasmas and propagating perpendicular to a uniform magnetic field caused by a counterstreaming of electrons along the latter is studied using a cold-plasma model. It is found that: (1) In the presence of either a streaming or a rotation or both, the ordinary-wave propagation is possible even for frequencies less than the plasma frequency; (2) the Coriolis forces like the applied magnetic field stabilise the ordinary modes.

062.104 Effect of rotation on the stability of a gravitating cylinder. A. M. Karnik.
Astrophys. Space Sci., Vol. 83, 209 - 212 (1982).

The effect of rotation on the stability of an idealised gravitating cylinder is studied. It is seen that the range of unstable modes is increased with increasing angular velocities.

062.105 The effect of a magnetic field on an adiabatic explosion. A. R. Garlick.
Astrophys. Space Sci., Vol. 84, 205 - 223 (1982).

The author has investigated the problem of an adiabatic explosion into a uniform interstellar magnetic field both by solving the equations numerically and by making certain analytic approximations. The two approaches agree on the obvious qualitative features: the shock becomes oblate with respect to the magnetic field, while the hot internal material develops prolate density and temperature contours. These features can be explained by reasonably simple physical arguments. During the later stages of the evolution, which can only be followed numerically, the shock waves weaken and their eccentricity is determined by the relevant acoustic speeds. However, there is a concentration of accelerated material at the poles, headed by a somewhat stronger shock, which seems to lag behind the outer wave.

062.106 Note on the stability of parallel magnetic fields.
A. Satya Narayanan, K. Somasundaram.
Astrophys. Space Sci., Vol. 84, 247 - 250 (1982).

The stability characteristics of parallel magnetic fields when fluid motions are present along the lines of force is studied. The stability criterion for both symmetric ($m = 0$) and asymmetric ($m = 1$) modes are discussed and the results obtained by Trehan and Singh (1978) are amended in the present study.

062.107 Electromagnetic circularly-polarised modes in plasmas with electron-streaming along the applied magnetic field. B. K. Shivamoggi.
Astrophys. Space Sci., Vol. 84, 255 - 258 (1982).

The effect of the electron-streaming along the applied magnetic field on the electromagnetic circularly-polarised modes in a Vlasov plasma propagating along the applied magnetic field is studied. It is found that the growth or decay of the circularly-polarised modes is affected by the electron-streaming in the presence of the thermal effects, unlike the case with the thermal effects absent.

062.108 Connection of Alfvén velocity and phase delays in an one-dimensional irregular plasma.
M. B. Gokhberg, A. L. Krylov, A. E. Lifshits.
Geomagn. Aehron., Tom 22, 403 - 407 (1982). In Russian.

062.109 Comment on 'A transverse Kelvin-Helmholtz instability in magnetized plasma' by P. Kintner and N. D'Angelo.
J. D. Huba, with a reply by P. M. Kintner and N. D'Angelo.
J. Geophys. Res., Vol. 87, 2574 - 2576 (1982). – See Abstr. 19.062.050.

062.110 A note on the existence of Alfvén surface waves.
C. Uberoi.
Sol. Phys., Vol. 78, 351 - 354 (1982).

The Alfvén surface waves can arise due to the discontinuity in the Alfvén speed across the interface along which these waves propagate. This note studies the relationship between v_{A1} and v_{A2} which is required for the existence of Alfvén surface waves in low-β plasma.

062.111 Magneto-optical effects and the determination of vector magnetic fields from Stokes profiles.
M. Landolfi, E. Landi Degl'Innocenti.
Sol. Phys., Vol. 78, 355 - 364 (1982).

The analysis procedure proposed by Auer et al. (1977) for deducing magnetic field vectors from Stokes profiles has been tested to investigate the influence of magneto-optical effects on the deduced field parameters. The quality of the fit between synthetic profiles generated with the inclusion of magneto-optical effects and the profiles returned by the inversion routine is also investigated. The results show that magneto-optical effects should be included in the inversion routine especially to increase the accuracy of the deduced azimuth of the magnetic field.

062.112 Evolution of current sheets following the onset of enhanced resistivity.
T. G. Forbes, E. R. Priest, A. W. Hood.
J. Plasma Phys., Vol. 27, part 1, p. 157 - 176 (1982). – Abstr. in Phys. Abstr., Vol. 85, Abstr. 55708 (1982).

062.113 Comment on 'On exact equilibrium states in external gravitational fields of heated, self-gravitating gas clouds cooling by conduction and radiation' (by I. Lerche, B. C. Low, with their reply).
S. J. Wilson, F. S. Wan, I. Lerche, B. C. Low.
Physica D, Vol. 4D, 287 - 288 (1982). – Abstr. in Phys. Abstr., Vol. 85, Abstr. 58440 (1982). – See Abstr. 28.062.054.

062.114 Ion acoustic instability in the presence of plasma turbulence in the solar wind.
P. Revathy, S. R. Prabhakaran Nayar.
Sol. Phys., Vol. 79, 187 - 194 (1982).

The adiabatic theory of interaction between high and low frequency waves has been studied for the case of electron plasma oscillations and ion acoustic waves and the results are applied to the solar wind.

062.115 Collapse of accreting, rotating, isothermal, interstellar clouds. A. P. Boss, D. C. Black.
Astrophys. J., Vol. 258, 270 - 279 (1982).

The evolutions of the envelopes of collapsing, accreting, isothermal clouds have been numerically calculated for both spherically symmetric and rotating (axisymmetric) clouds. The results provide a cohesive picture of isothermal collapse, and their relationship to previous numerical calculations and similarity solutions is discussed. Even with a large initial rotation rate, the majority of the cloud envelope is accreted, in one case leaving behind a large-scale circulation current. The calculations are performed for both initially uniform density and centrally condensed clouds. Density and velocity profiles for a wide variety of observed systems are compared with those obtained in this study, providing a preliminary assessment of the stage of evolution and initial structure for the observed systems.

062.116 **The pair annihilation process in relativistic plasmas.**
R. Svensson.
Astrophys. J., Vol. 258, 321 - 334 (1982) = Lick Obs. Bull., No. 909.
Exact analytical expressions are derived in the Born approximation for the pair annihilation rate and spectral emissivity from isotropic, monoenergetic electrons and positrons of arbitrary energies. The annihilation rate and spectral emissivity for arbitrary particle distribution functions are given as double integrals. The spectral emissivity is determined exactly for the case where the positrons (electrons) have a Maxwell-Boltzmann distribution, while the electrons (positrons) are at rest. A discussion is given for the more complicated case in which both the electrons and the positrons have Maxwell-Boltzmann distributions at the same temperature. Asymptotic expressions are derived for the annihilation rate, the cooling rate, and the spectral emissivity at relativistic temperatures. Useful, approximate expressions, valid at temperatures larger than 10^8 K, are given for the annihilation rate and the cooling rate.

062.117 **Electron-positron pair equilibria in relativistic plasmas.**   R. Svensson.
Astrophys. J., Vol. 258, 335 - 348 (1982) = Lick Obs. Bull., No. 910.
The physical properties of a thermal relativistic plasma are studied assuming pair equilibrium and thermal balance. The radiation field is obtained by multiplying the spectral emissivities by the photon escape time. The effect of scatterings is approximately allowed for by treating them as coherent. All important pair and photon producing processes in a plasma with a Thomson scattering optical depth of order unity or less are included. The spectral emissivities and the cooling rates for bremsstrahlung and pair annihilation are discussed. The photon-photon, photon-particle, and particle-particle pair production rates, and the pair annihilation rate as well as the opacities are calculated. Simple approximate expressions are given for several rates.

062.118 **Diffusion of Keplerian motions by a stochastic force. I. A general formalism.**
P. Barge, R. Pellat, J. Millet.
Astron. Astrophys., Vol. 109, 228 - 232 (1982).
Many astrophysical problems involve an external field of force and a stochastic perturbative force. Generally they are studied in a statistical way with a Fokker-Planck equation describing the evolution of the distribution function of particles in phase space. This equation can be reduced to a diffusion equation through a formalism coming from the quasi-linear theory in plasma physics (Laval and Pellat, 1974). The method is a very general one and brings consequent simplifications. In this paper it is applied to two kind of astrophysical situations: (1) the diffusion by Alfvén waves of particles trapped in a magnetic field, (2) the diffusion of particles in Keplerian motion by any stochastic force.

062.119 **On local theories of time-dependent convection in the stellar pulsation problem. III. The effect of turbulent viscosity.**   G. Gonczi.
Astron. Astrophys., Vol. 110, 1 - 8 (1982).
The author solves the complete set of non adiabatic linear equations of pulsating stars with a turbulent viscosity term in the equation of momentum, representing the dynamical coupling between pulsation and convection. It is shown that this dynamical coupling can give the red edge of the Cepheid instability strip. It is also shown that the turbulent viscosity

has a very small influence on the pulsating velocity gradients because of the thermal dilatation; its influence on the pulsating eigenfunctions and on the instability coefficient is detailed.

062.120 **Effect of a strong magnetic field on the stability of equilibrium configurations of a hot relativistic plasma.**   M. A. Ivanov, S. V. Izmajlov.
Researches on stellar physics, 1981, (see 003.010), p. 3 - 14 (1981). In Russian. − Abstr. in Ref. zh., 51. Astron., 6.51.118 (1982).

062.121 **Statistical equilibrium of a hot nuclear plasma in a strong magnetic field.**   M. A. Ivanov.
Researches on stellar physics, 1981, (see 003.010), p. 15 - 25 (1981). In Russian. − Abstr. in Ref. zh., 51. Astron., 6.51.120 (1982).

062.122 **Gravito-magnetic explanation of the temperature of stellar coronae and of planetary Van Allen belts.**
E. Woyk (Chvojková).
Sun and planetary system, (see 012.040), p. 145 - 146 (1982).

062.123 **Hydromagnetic flows from accretion discs and the production of radio jets.**
R. D. Blandford, D. G. Payne.
Mon. Not. R. Astron. Soc., Vol. 199, 883 - 903 (1982).
The authors examine the possibility that energy and angular momentum are removed magnetically from accretion discs, by field lines that leave the disc surface and extend to large distances. They illustrate this mechanism by solving the equations of magnetohydrodynamics, assuming infinite conductivity, for axially symmetric, self-similar, cold magnetospheric flow from a Keplerian accretion disc. It is shown that a centrifugally driven outflow of matter from the disc is possible. At large distances from the disc, the toroidal component of the magnetic field becomes important and collimates the outflow into a pair of anti-parallel jets moving perpendicular to the disc. Close to the disc, the flow is probably driven by gas pressure in a hot magnetically dominated corona. The relevance of this mechanism for the evolution of accretion discs around massive black holes in galactic nuclei and the production of jets in extragalactic radio sources is described.

062.124 **On the nature of magnetic fields of stars and galaxies.**   Yu. G. Ignat'ev.
Gravitatsiya i teor. otnositel'nosti, Kazan', 1981, No. 18, p. 67 - 73. In Russian. − Abstr. in Ref. zh., 51. Astron., 4.51.155 (1982).

062.125 **The flux ejection dynamo effect.**
E. N. Parker.
Geophys. Astrophys. Fluid Dyn., Vol. 20, 165 - 189 (1982).
This paper points out a new dynamo effect arising in convective cells with strong asymmetry in the rotation of updrafts as against downdrafts. The creation of new magnetic flux arises from the ejection of reserve flux through the open boundary of the dynamo region. It is unlike the familiar α-effect in that individual components of the field may be amplified independently. The flux ejection dynamo may possibly contribute to the generation of field in the convective core of Earth and in the convective zone of the sun and other stars.

062.126 **On dynamo action in the high-conductivity limit.**
K.-H. Rädler.
Geophys. Astrophys. Fluid Dyn., Vol. 20, 191 - 211 (1982).
This paper builds on a speculation by Moffatt (1979) on an apparent conflict between two results of dynamo theory in the high conductivity limit. Some constraints on the mean electromotive force near the boundary of the conducting body are taken into account, which have not been recognized up to

now. In the framework of the second order correlation approximation it is shown that it is just these constraints that ensure the boundedness of the magnetic multipole moments in the high conductivity limit. The present theory also adds insight into dynamo process in cosmical objects, in a way that is briefly discussed.

062.127 **On the impossibility of mean-field dynamos with some spherical symmetry of the motions.**
J. Reichert.
Geophys. Astrophys. Fluid Dyn., Vol. 20, 213 - 226 (1982).
A spherical mean-field dynamo model is considered in which both the mean motion and the mean electro-motive force due to fluctuating motions show some spherical symmetry. It is shown that under some reasonable assumptions the magnetic field is bound to decay to zero.

062.128 **The ionization state in a gas with a non-Maxwellian electron distribution.** S. P. Owocki,
J. D. Scudder.
Smithsonian Astrophys. Obs., Spec. Rep. 392, (see 012.045), p. 107 - 112 (1982).

062.129 **Numerical treatment of the magnetohydrodynamic flow bounded by an infinite porous plate.**
C. L. Goudas, G. A. Katsiaris, M. A. Drymonitou,
A. J. Vernardis.
Compendium in astronomy, (see 003.012), p. 393 - 414 (1982).
The scheme and some numerical results obtained for the problem of one dimensional flow and heat transfer of an electrically conducting fluid, subject to an external transverse and constant magnetic field, when the fluid moves on the one side of an infinite non-conducting and porous plate which in general moves in the same direction with the fluid, is exposed. The results produced were found to be satisfactory by comparison with other results obtained from analytical treatment of cases corresponding to simplified body forces and conditions of motion.

062.130 **A two-fluid model for Population I.**
K. H. Prendergast.
Extragalactic molecules, (see 012.048), p. 193 - 196 (1982).
The author considers a strongly interacting system of gas and stars, under the assumptions that (1) stars form spontaneously wherever gas is present, (2) stars lose mass, (3) stars provide energy to the gas, and (4) the gas cools spontaneously.

Hydrodynamic instabilities and the transition to turbulence. See Abstr. 003.126.

Diffusion and viscosity coefficients for helium. See Abstr. 022.004.

Calculation of gain at X-ray wavelengths resulting from optical pumping of helium-like ions. See Abstr. 022.026.

Density effects on the spectral emission of a high temperature argon plasma. See Abstr. 022.027.

Dielectronic satellite spectra from dense laser-produced plasmas. See Abstr. 022.028.

Level shifts and inelastic electron scattering in dense plasmas. See Abstr. 022.029.

Statistical mechanics of "atoms" in hot, partially ionized matter: the theory of the one electron Green function. See Abstr. 022.030.

Problems in the use of statistical average atom potentials for estimating average degree of ionization. See Abstr. 022.031.

INFERNO: a better model of atoms in dense plasmas. See Abstr. 022.032.

Electronic energy-levels in dense plasmas. See Abstr. 022.033.

Charge transfer ionization of Si^+ by H^+ at thermal energies. See Abstr. 022.042.

Stark broadening in hot, dense, laser-produced plasmas: a review. See Abstr. 063.014.

Comments on the calculation of spectral line shifts induced by plasma perturbations. See Abstr. 063.015.

Spectrum diagnostics. The necessity for detailed non-LTE modeling of X-ray emission from dense plasmas. See Abstr. 063.016.

On the transport and propagation of relativistic electrons in galaxies. The effect of adiabatic deceleration in a galactic wind for the steady state case. See Abstr. 063.020.

The Fokker-Planck equation for the radiation transfer in a strongly magnetized plasma. See Abstr. 063.022.

Electron temperatures of astrophysical plasmas. See Abstr. 063.046.

Damping of electromagnetic wave resonances in a strongly magnetized plasma. See Abstr. 063.067.

Resonant electrodynamic heating of stellar coronal loops: an *LRC* circuit analog. See Abstr. 064.012.

The adiabatic stability of stars containing magnetic fields — V. Effect of Ohmic diffusion on stable fields. See Abstr. 065.003.

Polytropic stellar models with a core-dynamo magnetic field. See Abstr. 065.006.

Stability of toroidal flux tubes in stars. See Abstr. 065.013.

Stationary spherical symmetric accretion onto massive black holes: the radiation spectrum and luminosity. See Abstr. 066.003.

Relativistic thermal plasmas: pair processes and equilibria. See Abstr. 066.009.

Processes in relativistic plasmas. See Abstr. 066.014.

Local stability in general relativity. See Abstr. 066.026.

Kinetic theory in astrophysics and cosmology. See Abstr. 066.113.

Measurements of A-values for intersystem lines used in diagnosis of solar transition zone and other astrophysical plasmas. See Abstr. 073.112.

Coronal closed structures. IV. Hydrodynamical stability and response to heating perturbations. See Abstr. 074.001.

Electron heating by fast mode magnetohydrodynamic waves in the solar wind emanating from coronal holes.   See Abstr. 074.004.

The condensational instability in the solar transition region and corona.   See Abstr. 074.009.

Self-similar magnetohydrodynamics. I. The γ=4/3 polytrope and the coronal transient.   See Abstr. 074.012.

Heating of the corona and solar wind by switch-on shocks.   See Abstr. 074.013.

Resonant wave acceleration of minor ions in the solar wind.   See Abstr. 074.028.

Topological semi-dynamo. See Abstr. 075.007.

The evaporation of spherical clouds in a hot gas. III. Suprathermal evaporation.   See Abstr. 131.012.

Gravitationally driven instabilities in shock compressed gas layers.   See Abstr. 131.065.

Particle reacceleration and apparent radio source structure.   See Abstr. 141.030.

Helical and pinching instability of supersonic expanding jets in extragalactic radio sources. See Abstr. 141.222.

Stabilization of expanding supersonic jets. See Abstr. 141.245.

Fluid dynamics of relativistic beams flowing through channels and evolution of double radio sources. See Abstr. 141.282.

Hypersonic beam driven by high-energy particles in extragalactic radio sources.   See Abstr. 141.283.

An auroral precipitation model for the rapid X-ray burster.   See Abstr. 142.026.

The equilibrium and bifurcation of rotating stellar systems.   See Abstr. 151.023.

Stabilization of spiral densitiy waves in flat galaxies for a hydrodynamical model.   See Abstr. 151.054.

Gas dynamics of flow past galaxies. See Abstr. 151.094.

Two-fluid gravitational instabilities in the Galaxy. See Abstr. 151.096.

Relativistic hydromagnetic wave propagation and instability in an anisotropic universe. See Abstr. 162.135.

Erratum

062.901   Erratum: 'Higher order fluid equations for multicomponent nonequilibrium stellar (plasma) atmospheres and star clusters" [Astrophys. J., Vol. 239, 345 - 359 (1980)].   S. Cuperman, I. Weiss, M. Dryer. Astrophys. J., Vol. 258, 414 (1982). – See Abstr. 28.062. 017.

063 Radiative Transfer, Scattering

063.001 The polarization properties of magnetic accretion columns.
S. M. A. Meggitt, D. T. Wickramasinghe.
Mon. Not. R. Astron. Soc., Vol. 198, 71 - 82 (1982).

The authors present calculations of the intensity and polarization of cyclotron radiation from constant temperature magnetic accretion columns. The results are in agreement with the general observed properties of AM Herculis type systems. The linear polarization pulse is caused primarily by Faraday mixing and is characteristic of cyclotron radiation of high harmonic number. The rate of change of the position angle of linear polarization during the pulse gives a direct measure of the angle between the line of sight and the rotation axis of the magnetic white dwarf for a synchronous orbit.

063.002 Electron-ion coupling in rapidly varying sources.
P. W. Guilbert, A. C. Fabian, S. Stepney.
Mon. Not. R. Astron. Soc., Vol. 199, 19P - 21P (1982).

The Thomson depth of sources radiating at the efficiency limit must lie near unity. This allows to deduce a general condition on the maximum electron temperature attainable in such a source via two-body relaxation; $kT_e \lesssim 22$ keV. The application of this result to the rapid variability observed in NGC 6814 and Cygnus X-1 and to NGC 4151 is discussed.

063.003 Linear polarization of radio frequency lines in molecular clouds and circumstellar envelopes.
P. Goldreich, N. D. Kylafis.
Astrophys. J., Vol. 253, 606 - 621 (1982).

The authors predict that interstellar lines possess a few percent linear polarization provided that the optical depth in the source region is both anisotropic and of order unity and the radiative rates are at least comparable to the collision rates. Under circumstances in which the Zeeman splitting exceeds both the radiative and collisional rates the linear polarization is aligned either parallel or perpendicular to the projection of the magnetic field on the plane of the sky. This "strong magnetic field" limit is expected to apply to all frequency lines and to many of those far infrared lines which form between levels whose magnetic moments are comparable to the Bohr magneton. The "weak magnetic field" limit is relevant to most far-infrared lines formed between levels with magnetic moments of order the nuclear magneton. In this limit the polarization direction is determined by the orientation of the propagation direction with respect to the anisotropic optical depth.

063.004 The complete solution for radiative transfer problems with reflecting boundaries and internal sources. C. Devaux, C. E. Siewert, Y. L. Yuan.
Astrophys. J., Vol. 253, 773 - 784 (1982).

The F_N method is used to establish the complete solution for the radiation field basic to radiative transfer problems, in plane-parallel media, with reflecting boundaries and internal sources. An Lth order Legendre expansion of the phase function is used, azimuthal symmetry is not required, and numerical results are reported.

063.005 Synchro-Compton radiation from relativistic charges driven by a strong plane vacuum wave of elliptic polarization. C. Leubner.
Astrophys. J., Vol. 253, 859 - 872 (1982).

The classical astrophysical problem of the spectral and angular distribution of the incoherent radiation emitted by relativistic charges, when driven by a strong plane electromagnetic vacuum wave of elliptic polarization, is solved. The solution utilizes the asymptotic technique introduced by Bleistein and by Ursell to yield an accurate asymptotic approximation to the pertinent cross section that is uniformly valid for arbitrary initial momentum of the charged particles, for arbitrary intensity and polarization of the incident strong wave, for arbitrary direction of observation, and throughout the range of the emitted high harmonics of the incident frequency.

063.006 A note on Compton scattering.
D. D. Barbosa.
Astrophys. J., Vol. 254, 301 - 308 (1982).

Calculations on the energy exchange between free electrons and electromagnetic radiation are presented. The full Klein-Nishina cross section is used to evaluate average scattering coefficients for an electron moving with arbitrary velocity. A number of useful series expansions for the mean energy and mean square energy transfer rates are given. A Fokker-Planck equation which includes induced scattering is used to derive a generalized diffusion equation in frequency for multiple scattering of photons off nonrelativistic electrons.

063.007 A unified treatment of escape probabilities in static and moving media. I. Plane geometry.
D. G. Hummer, G. B. Rybicki.
Astrophys. J., Vol. 254, 767 - 779 (1982).

An expression giving the escape probability for photons in a spectral line formed in a planar atmosphere with an arbitrary monotonic velocity law is derived and evaluated. For a small velocity gradient, the usual static result is recovered; for large velocity gradients the Sobolev result is obtained, but only at optical depths sufficiently large that the "static" part of the escape probability is negligible. Extensive numerical results for the escape-probability function for a constant velocity gradient are given for Doppler, Voigt ($a = 10^{-3}, 10^{-2}$) and Lorentz profiles. The use of these results for flows with non-constant gradients is discussed.

063.008 Non-coherent scattering in subordinate lines: II. Collisional redistribution.
P. Heinzel, I. Hubený.
J. Quant. Spectrosc. Radiat. Transfer, Vol. 27, 1 - 14 (1982).

A suitable form of the laboratory-frame redistribution function for the resonance scattering in subordinate lines, allowing for the radiative as well as collisional broadening of both atomic levels involved, is derived.

063.009 Time-dependent emergent intensity from an anisotropically-scattering semi-infinite atmosphere.
B. D. Ganapol, W. L. Filippone.
J. Quant. Spectrosc. Radiat. Transfer, Vol. 27, 15 - 21 (1982).

The time-dependent emergent intensity for an arbitrarily anisotropically-scattering, semi-infinite medium is obtained in terms of the stationary collided intensities. A numerical example is presented for a Rayleigh scattering atmosphere.

063.010 Review of fundamental processes for matter-radiation interaction.
D. M. Heffernan, R. L. Liboff.
J. Quant. Spectrosc. Radiat. Transfer, Vol. 27, 55 - 77 (1982).

063.011 Radiative transfer in finite inhomogeneous plane-parallel atmospheres.
R. D. M. Garcia, C. E. Siewert.
J. Quant. Spectrosc. Radiat. Transfer, Vol. 27, 141 - 148 (1982).

The F_N method is used to deduce accurate numerical results for the exit distributions of radiation relevant to a finite, plane-parallel atmosphere with an exponentially varying albedo for single scattering.

063.012   Monte Carlo calculations of resonance radiative transfer through a semi-infinite atmosphere.
G. Slater, E. E. Salpeter, I. Wasserman.
Astrophys. J., Vol. 255, 293 - 302 (1982).

The results of Monte Carlo calculations of radiative transfer through a semi-infinite, plane-parallel atmosphere of resonant scatterers are presented. Accurate results for the mean number of scatters, the mean path length and the most probable frequency shift of escaping photons are derived. Approximate analytic calculations of these parameters are presented for a symmetric slab. Analogous calculations for an asymmetric slab are discussed.

063.013   Refined Monte Carlo method for simulating angle-dependent partial frequency redistributions.
J.-S. Lee.
Astrophys. J., Vol. 255, 303 - 306 (1982).

This paper describes a refined algorithm for generating emission frequencies from angle-dependent partial frequency redistribution functions R_{II} and R_{III}. The improved algorithm is based on a "rejection" technique which, for absorption frequencies $x < 5$, involves no approximations. The resulting procedure proves to be essential for effective studies of radiative transfer in optically thick or temperature varying media involving angle-dependent partial frequency redistributions.

063.014   Stark broadening in hot, dense, laser-produced plasmas: a review.
L. A. Woltz, C. A. Iglesias, C. F. Hooper, Jr.
J. Quant. Spectrosc. Radiat. Transfer, Vol. 27, (see 012.005), 233 - 242 (1982).

The paper analyzes a general expression for a line profile previously derived elsewhere and then discusses approximations that are frequently made for calculations appropriate for plasma conditions created in pellet implosion experiments. Two aspects of the authors' calculations will be stressed: (1) electric microfield distribution functions, (2) full-Coulomb electron-radiator interactions.

063.015   Comments on the calculation of spectral line shifts induced by plasma perturbations.   R. W. Lee.
J. Quant. Spectrosc. Radiat. Transfer, Vol. 27, (see 012.005), 249 - 251 (1982).

A discussion is presented of the method used to calculate line shifts. The method is analyzed to show that it depends on individual level shifts being combined to yield transition shift. The appropriateness of these calculations is commented upon.

063.016   Spectrum diagnostics. The necessity for detailed non-LTE modeling of X-ray emission from dense plasmas.   K. G. Whitney, P. C. Kepple.
J. Quant. Spectrosc. Radiat. Transfer, Vol. 27, (see 012.005), 281 - 296 (1982).

The authors discuss some of the radiative properties of steady-state argon at ion densities $> 10^{20}$ cm^{-3}, temperatures sufficient to produce K-shell emission (>300-400 eV), and for a spherical geometry. Under these conditions, argon radiates and self-absorbs helium-like and hydrogen-like K-series radiation from the Ar (XVII) and Ar (XVIII) ionization stages respectively.

063.017   A transparent atmosphere in the UV: results from darkening of Viking Lander UV chips.
A. Zent, E. A. Guinness, R. E. Arvidson, C. R. Spitzer.
Reports of Planetary Geology Program – 1981, (see 003.001), p. 453 - 454 (1981). – Abstract.

063.018   Bidirectional reflectance spectroscopy. III. Correction for macroscopic roughness.   B. W. Hapke.
Reports of Planetary Geology Program – 1981, (see 003.001), p. 475 (1981). – Abstract.

063.019   The strength of absorption bands in reflectance spectroscopy.   B. W. Hapke.
Reports of Planetary Geology Program – 1981, (see 003.001), p. 476 (1981). – Abstract.

063.020   On the transport and propagation of relativistic electrons in galaxies. The effect of adiabatic deceleration in a galactic wind for the steady state case.
I. Lerche, R. Schlickeiser.
Astron. Astrophys., Vol. 107, 148 - 160 (1982).

The solution of the steady-state transport equation describing the propagation of relativistic electrons perpendicular to the galactic plane in a galactic wind is discussed. The wind velocity is assumed to be zero in the plane and to increase with galactic height. The electrons undergo simultaneous diffusion, convection, adiabatic deceleration, radiative losses and injection. The authors contrast the behaviour of the resulting cooling in a convecting dynamical halo with that in (1) a purely convective halo and (2) also that in a static halo, by considering the total electron number density and the variation of the electron spectral index with galactic height. The solutions are applied to problems involving nonthermal radio spectra of external spiral galaxies.

063.021   On the theory of gamma-ray amplification through stimulated annihilation radiation.
R. Ramaty, J. M. McKinley, F. C. Jones.
Astrophys. J., Vol. 256, 238 - 246 (1982).

The theory of photon emission, absorption, and scattering in a relativistic plasma of positrons, electrons, and photons is studied. Expressions for the emissivities and absorption coefficients of pair annihilation, pair production, and Compton scattering are given and evaluated numerically. The conditions for negative absorption are investigated. In a system of photons and $e^{+}-e^{-}$ pairs, and emission line at ~ 0.43 MeV can be produced by grasar action (gamma-ray amplification through stimulated annihilation radiation) provided that the pair chemical potential exceeds ~ 1 MeV. At a temperature of $\sim 10^{9}$ K this requires a pair density $\gtrsim 10^{30}$ cm^{-3}, a value much larger than the thermodynamic equilibrium pair density at this temperature. This emission line could account for the observed lines at this energy from gamma-ray bursts, without a gravitational redshift.

063.022   The Fokker-Planck equation for the radiation transfer in a strongly magnetized plasma.
S. Bonazzola.
Astron. Astrophys., Vol. 108, 19 - 24 (1982).

The Fokker-Planck equation for the radiation transfer in a strongly magnetized plasma is obtained in the approximation $(KT/m_e c^2)$; $(h\nu/m_e c^2) \ll 1$.

063.023   Radiative transfer: analytic solution of difference equations.   W. Kalkofen, R. Wehrse.
Astron. Astrophys., Vol. 108, 42 - 48 (1982).

The calculus of finite differences is used to determine analytic solutions of the discretized equation of radiative transfer for coherent scattering in a medium with plane parallel geometry. The method is applied to the calculation of the radiation field in a one-dimensional medium with absorption ($\epsilon = 1$), conservative scattering ($\epsilon = 0$), or non-conservative scattering ($0 \leqslant \epsilon \leqslant 1$). The character of the solutions and the implications for numerical methods are discussed, and the extension of the method to noncoherent scattering with partial redistribution is indicated.

063.024   The scaling group of the radiative transfer equation.
B. H. J. McKellar, M. A. Box.
J. Atmos. Sci., Vol. 38, 1063 - 1068 (1981). – Abstr. in Phys. Abstr., Vol. 85, Abstr. 28201 (1982).

063.025 **N-stream approximations to radiative transfer.**
C. Acquista, F. House, J. Jafolla.
J. Atmos. Sci., Vol. 38, 1446 - 1451 (1981). − Abstr. in Phys.
Abstr., Vol. 85, Abstr. 28202 (1982).

063.026 **The shower model for compact synchrotron sources and an intrinsic red-shifting mechanism.**
P. F. Browne.
Phys. Lett. A, Vol. 86A, 391 - 396 (1981). − Abstr. in Phys.
Abstr., Vol. 85, Abstr. 28568 (1982).

063.027 **Polarization of radiation scattered by an inhomogeneous atmosphere.**
V. M. Loskutov, V. V. Sobolev.
Astrofizika, Tom 17, 97 - 108 (1981). In Russian. English
translation in Astrophysics, Vol. 17, No. 1.

The problem of determination of intensity and degree of polarization of light emerging from a semi-infinite medium with internal sources is considered. Nonconservative Rayleigh scattering is assumed. The results, which are presented in tables, are used to determine the degree of polarization of stellar radiation.

063.028 **Radiative pressure in spectral lines in a medium with axial-symmetric supersonic motions. III. Gas and dust systems with large-scale radiative coupling.**
V. P. Grinin.
Astrofizika, Tom 17, 109 - 123 (1981). In Russian. English
translation in Astrophysics, Vol. 17, No. 1.

The radiative pressure due to resonance-line radiation on dust of free electrons in a medium with axial-symmetric supersonic motions and large-scale radiative interaction is considered. The relations for the radial and tangential components of the pressure for two limiting cases are obtained, when the optical depth of the dust (or the electron scatter) is significantly more or less than 1. Criteria for the effective strengthening of the rotation by tangential pressure are presented.

063.029 **The thermalization length of resonance radiation with partial frequency redistribution.**
M. M. Basko.
Astrofizika, Tom 17, 125 - 139 (1981). In Russian. English
translation in Astrophysics, Vol. 17, No. 1.

The propagation of resonance-line radiation in an infinite homogeneous medium is studied when the frequency redistribution by resonance scattering is determined by the natural broadening of the line and by thermal Doppler shift. Monte-Carlo simulations are performed to calculate the mean displacement and the mean number of scatters experienced by resonance photons in a purely scattering medium within a time interval. The propagation of radiation in a coordinate space, as well as the line spectral evolution in Lorentz wings of the Voigt profile are shown to be of the diffusive type. The effect of recoil by resonance scattering is discussed.

063.030 **The radiation field of a plane atmosphere with anisotropic scattering. An invariance relation.**
Eh. G. Yanovitskij.
Astrofizika, Tom 17, 155 - 165 (1981). In Russian. English
translation in Astrophysics, Vol. 17, No. 1.

A plane atmosphere of optical thickness τ_0 illuminated by parallel rays is considered. A new invariance relation for the reduced source function $D(\tau, \mu, \tau_0)$ is obtained and its consequences are considered. A modification of van de Hulst's doubling method for direct calculation of the functions $X(\mu)$ and $Y(\mu)$ is given. It is shown that with the help of the modified doubling method one can easily calculate the internal radiation field.

063.031 **On a peculiarity of solution of conservative anisotropic scattering problems.**
M. A. Mnatsakanyan.
Astrofizika, Tom 17, 179 - 183 (1981). In Russian. English
translation in Astrophysics, Vol. 17, No. 1.

The possible independence of the solutions of anisotropic scattering problems on the Legendre-expansion coefficients of the indicatrix is analysed.

063.032 **Polarization of radiation multiply scattered in a plane layer.** V. M. Loskutov, V. V. Sobolev.
Astrofizika, Tom 17, 535 - 546 (1981). In Russian. English
translation in Astrophysics, Vol. 17, No. 3.

The problem of radiative transfer in a plane layer of finite optical thickness with nonconservative Rayleigh scattering is considered. To determine the emergent intensity use is made of the linear integral equations which generalize those found previously for a semi-infinite atmosphere. These equations are solved for the case of uniform distribution of the embedded sources. The degree of polarization is calculated and the results are tabulated.

063.033 **The Leningrad school of the theory of radiative transfer.** I. N. Minin.
Astrofizika, Tom 17, 585 - 618 (1981). In Russian. English
translation in Astrophysics, Vol. 17, No. 3.

A review of publications on the theory of radiative transfer as a section of theoretical astrophysics has been made. The attention has been concentrated on the results obtained by the scientists of the Leningrad University Astronomical Observatory. Monochromatic radiative transfer, radiative transfer with frequency redistribution and nonstationary radiative transfer are considered. Papers describing fundamental ideas and new methods have been specially pointed out. Results of primary importance for further development of the theory and its applications have been presented.

063.034 **Radiation field in multi-layered isotropically scattering atmospheres.** T. Viik.
Astrofizika, Tom 17, 735 - 748 (1981). In Russian. English
translation in Astrophysics, Vol. 17, No. 4.

A method of determining the radiation field in an isotropically scattering multi-layered atmosphere illuminated by parallel rays is described using some new invariance principles. It is shown that these principles can be used even in the case when the atmosphere considered is bounded by a reflecting bottom.

063.035 **Curvature radiation on longitudinal waves in the magnetosphere of a neutron star.**
V. E. Shaposhnikov.
Astrofizika, Tom 17, 749 - 763 (1981). In Russian. English
translation in Astrophysics, Vol. 17, No. 4.

The radiation of longitudinal waves with the refraction index $n \cong 1$, $|1-n| \ll 1$ from a relativistic charged particle moving along curved magnetic field lines is discussed. Reabsorption of these waves by curved stream particles is investigated. The optical thickness for curvature absorption in the case of a pulsar magnetosphere is estimated. The range of validity of the considered theory is given for a plasma with dispersion properties varying along the stream.

063.036 **An alternative derivation of the line transfer equation of an arbitrarily polarized radiation in the presence of a magnetic field, in non-LTE.** G. Mathys.
Astron. Astrophys., Vol. 108, 213 - 220 (1982).

The author presents an alternative method for the derivation of the transfer equation of the Stokes vector of arbitrarily polarized radiation in the presence of a magnetic field. He makes an application to the line transfer problem. The method is based upon quantum electrodynamics and allows to take into account departures from local thermodynamic equilibrium.

063.037 Time-dependent X-ray spectra of Compton-cooled plasmas.
P. W. Guilbert, A. C. Fabian, R. R. Ross.
Mon. Not. R. Astron. Soc., Vol. 199, 763 - 774 (1982).

Transmitted and reflected spectra of radiation Compton scattered by hot gas ($kT_e \cong 10$ and 511 keV) have been computed by two independent and complementary methods. A range of optical depths and of initial temperatures have been used and the gas temperature has been allowed to drop in response to Compton cooling. The authors show that quasi-power-law spectra always result from Comptonization whenever a significant fraction of the thermal energy of the gas is transferred to the radiation field. The results have applications to compact galactic X-ray sources, active galactic nuclei and X- and γ-ray bursts.

063.038 Emisiones no térmicas en radiofuentes compactas.
R. Terlevich, J. Frank, J. R. Albano.
Bol. Asoc. Argentina Astron., No. 18, p. 64 - 69 (1980).

063.039 Flux limiters and Eddington factors.
G. C. Pomraning.
J. Quant. Spectrosc. Radiat. Transfer, Vol. 27, 517 - 530 (1982).

The author presents a closure scheme for the first two angular moments of the time-dependent equation of transfer, either via an Eddington factor which leads to a telegrapher's description, or via a Fick's law which leads to a diffusion description. Points discussed include boundary conditions (both an extension of the classic Marshak–Milne condition and those arising from a boundary layer analysis), the flux limiting feature of the diffusion approximation, and the reduction of the theory to asymptotic diffusion theory in the steady state limit.

063.040 Radiative transfer with partial frequency redistribution in inhomogeneous atmospheres: application to the Jovian aurora.   G. R. Gladstone.
J. Quant. Spectrosc. Radiat. Transfer, Vol. 27, 545 - 556 (1982).

A direct finite difference numerical solution for the equation of radiative transfer by the Feautrier method is developed for use in planetary atmospheres. The procedure described here uses a plane-parallel atmosphere, and can treat partial frequency redistribution, inhomogeneity, external or internal sources, and various boundary conditions. Isotropic scattering is assumed, but in the case of no frequency redistribution, Rayleigh scattering can also be handled. A program utilizing this method is tested in a variety of situations against more powerful and elaborate methods. The case of the Lyman-α aurora on Jupiter is then considered, where the effects of partial frequency redistribution are shown to be of great importance. New results for the detailed line profiles for Lyman-α in the Jovian aurora are presented.

063.041 Radar backscattering from a rough rotating triaxial ellipsoid with applications to the geodesy of small asteroids.   R. F. Jurgens.
Icarus, Vol. 49, 97 - 108 (1982).

Radar geodesy of the asteroid 433 Eros was reported by R. F. Jurgens and R. M. Goldstein (1976). Their measurements were based on the spectral properties of a rough rotating triaxial ellipsoid. This paper presents the theory by which theoretical spectra based on a backscattering model of the form $\cos^n \theta$, or any reasonable backscattering model, can be computed.

063.042 Angle-averaged redistribution function in the laboratory frame.
M. Seitz, B. Baschek, R. Wehrse.
Astron. Astrophys., Vol. 109, 10 - 16 (1982).

Laboratory frame angle-averaged redistribution functions

are calculated for the case that upper and lower level are broadened by elastic and inelastic collisions as well as by radiative processes. The computations are based on the atomic-frame redistribution functions derived by Cooper and Ballagh (1978) in the framework of the density matrix formalism. The results are interpreted as a superposition of complete redistribution functions involving Gauss and Lorentz profiles and of δ-functions representing Rayleigh scattering.

063.043 Radiative transfer and model atmospheres of A and Ap stars.   I. Hubeny.
Soobshch. Spets. Astrofiz. Obs., Vyp. 32, (see 012.029), p. 23 - 24 (1981).

063.044 Asymptotic solution of the transport equation of linearly polarized X and gamma rays.
L. D. Pleshakov.
Astron. Zh., Tom 59, 503 - 511 (1982). In Russian. English translation in Soviet Astron., Vol. 26, No. 3.

The transport equation in the representation which allows to express all the parameters of the equation with spherical angles is considered. It is shown that this equation turns into the known equation of optics in the low-energy limit. An asymptotic solution of the transport equation of the plane-parallel problem with spatial azimuth symmetry is obtained.

063.045 Monte-Carlo method for multilevel problems of radiative transfer with account for continuum processes.   A. M. Sobolev.
Astron. Zh., Tom 59, 605 - 607 (1982). In Russian. English translation in Soviet Astron., Vol. 26, No. 3.

An algorithm is suggested for the solution of multilevel problems of radiative transfer for the cases when it is necessary to account for emission and absorption of photons in the continuum.

063.046 Electron temperatures of astrophysical plasmas.
A. Dalgarno, A. Sternberg.
Astrophys. J., Lett., Vol. 257, L87 - L90 (1982).

Electron temperatures of astrophysical plasmas are usually derived from the relative intensities of lines emitted by metastable levels of positive ions, particularly of O^{+2} and N^+, on the assumption that the levels are excited by electron impact. The populations of the metastable levels of O^{+2} are augmented by a radiative cascade following charge transfer of O^{+3} with hydrogen and are not always a reliable diagnostic of the temperature. An evaluation of the relative contributions of charge transfer to the ^1S and ^1D level populations shows that the temperatures derived from the O^{+2} line intensities tend to be overestimates. A quantitative measure of the effect of charge transfer can be obtained from the intensity of the O^{+2} line at 5592 Å. In planetary nebulae the effect on the derived electron temperature seems to be small. It will be larger in plasmas produced by nonthermal ionizing sources.

063.047 Radiative recombination coefficients at stellar densities.   D. G. Hummer, P. J. Storey.
Bull. American Astron. Soc., Vol. 13, 793 (1981). – Abstract.

063.048 Dissipation of charged particles from ionized atmospheres of stars and planets.
A. R. Bakhalbashyan, G. M. Nedyalkova, I. E. Turchinovich.
Issled. po fiz. zvezd. Leningrad, 1981, p. 81 - 89. In Russian. Abstr. in Ref. zh., 51. Astron., 5.51.142 (1982).

063.049 Line profile fluctuations in a turbulent atmosphere.
M. L. Loucif, C. Magnan.
Pré-publ. Inst. Astrophys. Paris, No. 3, 20 pp. (1982).
Submitted to Astron. Astrophys.

063.050 Green's function of an optically thick layer.
O. V. Pikichyan.

Astrofizika, Tom 16, 351 - 361 (1980). In Russian. English translation in Astrophysics, Vol. 16, No. 2.

The problem of the approximate determination of Green's function (GF) has been investigated for a slab of finite optical thickness. For the case of anisotropic scattering asymptotic formulas that quite simply express the GF of an optical thick slab in terms of GF for a semi-infinite medium and Milne's intensity have been obtained.

063.051 The radiation field in a plane atmosphere with anisotropic scattering. Separation of angular variables. Eh. G. Yanovitskij.
Astrofizika, Tom 16, 363 - 374 (1980). In Russian. English translation in Astrophysics, Vol. 16, No. 2.

A plane atmosphere of optical thickness τ_0 illuminated by parallel rays is considered. The intensity of radiation in the medium at optical depth τ is shown to be expressed in terms of the reduced source function $D^m(\tau, \mu_0; \tau_0)$ depending only on one angular variable. A singular equation for the function $D^m(\tau, \mu_0; \tau_0)$ is also obtained. A new scheme for the computation of the radiation field is given.

063.052 Comments on the evaluation of the inverse Compton flux, using $\epsilon_f \approx \gamma^2 \epsilon_i$.
S. Kichenassamy, R. A. Krikorian.
Astrofizika, Tom 17, 395 - 398 (1981).

063.053 Radiative transfer in a homogeneous magnetized plasma. N. A. Silant'ev.
Astrophys. Space Sci., Vol. 82, 363 - 395 (1982).

The general theory of radiative transfer in a homogeneous plasma with a strong magnetic field is developed. Linear and nonlinear equations generalizing the equations of scalar radiative transfer in isotropic media are derived. Explicit solutions are given for cases where the magnetic field is perpendicular to the surface. In particular, diffuse reflection of radiation from a semiinfinite medium is considered for several source configurations.

063.054 The surface cooling effect in radiative transfer.
C. K. B. Lee, G. C. Pomraning.
J. Quant. Spectrosc. Radiat. Transfer, Vol. 28, 21 - 27 (1982).

The halfspace radiative equilibrium problem with an incident beam of energy is analyzed by both the equation of transfer and the diffusion approximation.

063.055 On the computation of angular distributions of radiation in planetary atmospheres. K. Stamnes.
J. Quant. Spectrosc. Radiat. Transfer, Vol. 28, 47 - 51 (1982).

The discrete ordinate approximation to the radiative transfer equation is used to derive simple analytic expressions for the intensity in a vertically inhomogeneous plane parallel atmosphere. This approach allows one to compute the intensity at arbitrary depths and angles. The merit and soundness of these expressions, which are interpolatory in nature, are discussed and it is conjectured that they are superior to any other standard interpolation scheme as far as accuracy is concerned. The computational time also compares favorably with that of standard interpolation schemes such as cubic splines.

063.056 Scattering of radiation by a large particle with a random rough surface. S. Mukai, T. Mukai, R. H. Giese, K. Weiss, R. H. Zerull.
Moon Planets, Vol. 26, 197 - 208 (1982).

Intensity and polarization of scattered light by an absorbing spherical particle with a random rough surface and with a radius larger than the wavelength of radiation are investigated. Multiple reflections of incident light on the rough surface are treated based on the multiple scattering theory. Within the limits of our approximation the model gives a good agreement with typical scattering features by

irregular shaped particles derived by microwave analogue experiments and laser measurements, namely a backward enhancement of the intensity, and a reduction of magnitude of polarization.

063.057 Radiation transfer in stellar interiors.
R. Opher.
Astron. Astrophys., Vol. 109, 191 - 194 (1982).

It was recently shown that due to the plasmon-electron interaction the radiation transfer at the center of the sun (when free-free processes dominate the opacity) increases by a factor $Q(\equiv H_d/H) > 2$ (Opher, 1981), where $H_d(H)$ is the corrected (uncorrected) radiation transfer. The author analyses here the case when electron scattering dominates the opacity.

063.058 On the phase matrix basic to the scattering of polarized light. C. E. Siewert.
Astron. Astrophys., Vol. 109, 195 - 200 (1982).

A special representation of the scattering matrix, basic to studies of polarized light, that allows the components in a Fourier decomposition to be expressed analytically is reviewed. In order to provide a convenient method of using the representation for practical calculations, a basic set of orthogonality and recursive relations is reported and used to provide working formulas for the basic constants and basic matrices required in the established formalism. The final results are also used to provide an independent verification of known symmetry relations satisfied by the phase matrix.

063.059 Spectral line formation in YY Orionis envelopes: a multi-level hydrogen atom. U. Bastian.
Astron. Astrophys., Vol. 109, 245 - 257 (1982).

A powerful numerical method for the treatment of many-level Sobolev-type line transfer in the presence of non-local radiative coupling is presented. This method is applied to the formation of hydrogen lines in spherical infalling envelopes. The results are compared with observations of YY Orionis stars.

063.060 Radiative transfer: comparison of finite difference equations. W. Kalkofen, R. Wehrse.
Astron. Astrophys., Vol. 110, 18 - 22 (1982).

The authors compare the finite difference equations in several formulations of radiative transfer along a ray for prescribed source function in regard to stability and accuracy. They consider the transformations between the various forms of the differential equations and between the analogous finite difference equations. The authors investigate the stability of the solution of the finite difference equations, and judge their accuracy by the effect of the associated second difference operator in acting on a quadratic function of depth.

063.061 High order asymptotic expansions of the four kernel functions for line formation with the Voigt profile. D. G. Hummer.
J. Quant. Spectrosc. Radiat. Transfer, Vol. 27, 569 - 573 (1982).

Analytical expressions are given for the coefficients, as a function of the Voigt parameter a, in the asymptotic expansions of the kernel functions $K_1(\tau)$, $K_2(\tau)$, $M_1(\tau)$, and $M_2(\tau)$ that describe the transfer of radiation scattered with complete redistribution over a Voigt profile.

063.062 Non-coherent scattering in subordinate lines: III. Generalized redistribution functions. I. Hubený.
J. Quant. Spectrosc. Radiat. Transfer, Vol. 27, 593 - 609 (1982).

063.063 Non-linear transport problems and the renormalization group. M. A. Mnatsakanyan.
Dokl. AN SSSR, Tom 262, 856 - 860 (1982). In Russian. Abstr. in Ref. zh., 51. Astron., 6.51.112 (1982).

063.064   General invariant relations for investigation of
          radiative transfer problems in media with geometric
and physical characteristics of arbitrary complexity.
O. V. Pikichyan.
Dokl. AN SSSR, Tom 262, 860 - 863 (1982). In Russian.
Abstr. in Ref. zh., 51. Astron., 6.51.115 (1982).

063.065   General relativistic radiative transfer: the 14-moment
          approximation.   N. Udey, W. Israel.
Mon. Not. R. Astron. Soc., Vol. 199, 1137 - 1147 (1982).
     The equations of radiative hydrodynamics, with inclusion
of relaxation terms and couplings between the radiative flux,
viscous stresses and the acceleration and angular velocity of
the medium, are derived from the 14-moment approximation.
The physical mechanism responsible for the bulk viscosity is
discussed.

063.066   Light reflection from randomly oriented convex
          particles with rough surface.
R. Schiffer, K. O. Thielheim.
J. Appl. Phys., Vol. 53, 2825 - 2830 (1982).
     Light reflection from rough spheres being large compared
to the incident wavelength is considered – the resulting reflection pattern is identical to that one produced by randomly
oriented convex particles. Only the limiting case of short
wavelengths is treated. It turns out that the decisive quantity
is the probability distribution $P(\beta)$ of the directions of the
surface normal vectors. The resulting differential cross section shows the following behavior: with growing roughness
the cross section flattens for moderate scattering angles,
whereas a rise occurs in the region of backscattering.

063.067   Damping of electromagnetic wave resonances in a
          strongly magnetized plasma.
K. O. Thielheim, H. Wiese.
Lett. Nuovo Cimento, Ser. 2, Vol. 30, 327 - 330 (1981).

     Light scattering by small particles.
See Abstr. 003.133.

     Astrophysical problems of propagation of radiation.
See Abstr. 003.145.

     Elargissement des raies spectrales.
See Abstr. 022.133.

     On the mathematical properties of the Voigt profile
of a spectral line.   See Abstr. 022.160.

     The radiation of energy by a current in a plasma.
See Abstr. 062.045.

     Incompressible convection in a radiating atmo-
sphere. I. General characteristics.   See Abstr. 062.054.

     The pair annihilation process in relativistic plasmas.
See Abstr. 062.116.

     Electron-positron pair equilibria in relativistic
plasmas.   See Abstr. 062.117.

     An escape probability treatment of doublet
resonance lines in expanding stellar winds.
See Abstr. 064.013.

     The formation of resonance lines in locally non-
monotonic winds.   See Abstr. 064.014.

     Vertical structure of accretion disks.
See Abstr. 064.019.

     On hot star winds. I. Radiation-driven winds.
See Abstr. 064.025.

     On hot star winds. II. Energy transport – corona-
like temperature enhancements.   See Abstr. 064.026.

     Nonspherical stellar envelopes and winds: effects of
structure on radiative fluxes and apparent mass loss rates.
See Abstr. 064.031.

     Analytic treatment of polarization by arbitrary
scattering mechanisms in circumstellar envelopes – I. Single
stars.   See Abstr. 064.088.

     The effect of multiple Compton scattering on the
temperature and emission spectra of accreting black holes.
See Abstr. 066.004.

     Relativistic thermal plasmas: pair processes and
equilibria.   See Abstr. 066.009.

     Synchrotron radiation from spherically accreting
black holes.   See Abstr. 066.024.

     Two-dimensional nonlocal thermodynamic equilibri-
um transfer computations of resonance lines in quiescent
prominences.   See Abstr. 073.009.

     A study of the H-alpha line in late G and K super-
giants.   See Abstr. 114.218.

     Linear polarization of supernova light: a measure of
deviation from spherical symmetry.   See Abstr. 125.042.

     Gamma ray emission from interstellar clouds: a
plasma physical process capable of enhancing electron fluxes.
See Abstr. 131.015.

     Hydrogen line spectrum in quasars. II. A critical
discussion of model calculations for the broad line region.
See Abstr. 141.063.

     Radiative transfer in quasar emission line clouds.
See Abstr. 141.224.

     A probabilistic radiative transfer equation for finite
slab models of QSO emission line regions.
See Abstr. 141.225.

     Theoretical quasar emission line ratios. V. Balmer
continuum emission.   See Abstr. 141.226.

     Theoretical QSO emission line profiles.
See Abstr. 141.227.

     Pulsar optical emission as amplified synchrotron
emission.   See Abstr. 141.507.

     Radiation transport in accretion column of binary
X-ray sources.   See Abstr. 142.089.

     Compact gamma ray point sources: are gamma ray
sources optically thick at lower frequencies?
See Abstr. 142.506.

     Preinjection of cosmic rays and magnetic chemically
peculiar stars.   See Abstr. 143.013.

     Reverberation mapping of the emission line regions
of Seyfert galaxies and quasars.   See Abstr. 158.074.

Errata

063.901 **Erratum: "Non-coherent scattering in subordinate lines: a unified approach to redistribution functions".** [J. Quant. Spectrosc. Radiat. Transfer, Vol. 25, 483 - 499 (1981)]. P. Heinzel.
J. Quant. Spectrosc. Radiat. Transfer, Vol. 27, 109 (1982). See Abstr. 29.063.037.

063.902 **Erratum: 'Radiative transfer in hot plasma'** [Publ. Astron. Soc. Japan, Vol. 33, 77 - 90 (1981)]. I. Masaki.
Publ. Astron. Soc. Japan, Vol. 33, 749 (1981). – See Abstr. 29.063.025.

064 Stellar Atmospheres, Stellar Envelopes, Mass Loss, Accretion

064.001 The ionization structure in disklike circumstellar envelopes.   R. Poeckert, J. M. Marlborough.
Astrophys. J., Vol. 252, 196 - 200 (1982).

The ionization structure in circumstellar envelopes is shown to be dependent on the envelope density, the effective temperature of the underlying star, and the envelope geometry. Stars whose effective temperatures are below 25,000 K are not capable of completely ionizing dense envelopes. As a result, ionization structure exists within the envelope. The lowest degree of ionization occurs between the high density equatorial region and the lower density edge of the disk. The implications of such structure are discussed. In particular, it is shown that the usual assumption that ionization is primarily a radially dependent quantity is invalid, and that shell line profiles may be affected by the ionization structure.

064.002 The photochemistry of carbon-rich circumstellar shells.   P. J. Huggins, A. E. Glassgold.
Astrophys. J., Vol. 252, 201 - 207 (1982).

The effect of ambient ultraviolet photons on the chemical structure of carbon-rich, circumstellar envelopes is investigated with a simple formulation of the time-dependent, photochemical rate equations valid for optically thick shells. Molecules injected into the shielded inner envelope are broken down when they reach the outer regions where ambient ultraviolet photons can penetrate. A quantitative description of the abundance variations is obtained for the case of uniform expansion by detailed consideration of the shielding of the radiation by the dust and molecules of the envelope. Representative results are presented.

064.003 Theoretical models of homogeneous chromospheres for main sequence stars.   Z. Musielak.
Astron. Astrophys., Vol. 105, 23 - 36 (1982).

Theoretical chromosphere models based on the detailed balance between heating by mechanical flux and radiative loss are computed. Dissipation of acoustic, MHD and Alfvén waves is considered in the zone model of heating. Theoretical models are compared with semiempirical ones in order to define the free parameters for stars with weak and strong chromospheric activity. The results suggest the strength of the magnetic field might be responsible for the division of stars into stars with weak and strong chromospheric activity respectively. The increase of chromospheric activity with the decrease of effective temperature for main sequence stars is obtained.

064.004 Differential rotation, magnetic activity and X-ray emission of late type giants.
G. Belvedere, C. Chiuderi, L. Paternò.
Astron. Astrophys., Vol. 105, 133 - 139 (1982).

As suggested by Vaiana and Rosner, the conversion of magnetic into thermal energy may be responsible for the X-ray emission from late main sequence and giant stars. The proposed ingredients in this analysis are the differential rotation and dynamo action which are able to generate a magnetic activity at the star's surface which in turn gives rise to the observed X-ray emission. The authors compute a series of models of luminosity class III giant stars, and determine the surface X-ray flux. Comparing these results with those concerning the late type main sequence star models and with observations, it appears that the proposed mechanism is plausible.

064.005 Tides in differentially rotating convective envelopes. II. The tidal coupling.
E. T. Scharlemann.
Astrophys. J., Vol. 253, 298 - 308 (1982).

The tidal coupling between a star with an extended, differentially rotating convective envelope, and its companion

in a close binary system, is calculated from the tidal velocity field derived in an earlier paper. The derived coupling torque can be tested using observations of RS Canum Venaticorum systems, for which a photometric wave in the light curve provides an accurate stellar rotation rate, and for which observed orbital period changes require the stars in the systems to be coupled. The coupling torque is sufficient to explain the nearly synchronous rotation of the active star in RS CVn systems, despite the observed orbital period changes, but may not be able to explain the extreme tightness of the coupling implied by the very long periods for the migration of the photometric waves in the systems.

064.006 The evolution of viscous discs − II. Viscous variations.   G. T. Bath, J. E. Pringle.
Mon. Not. R. Astron. Soc., Vol. 199, 267 - 280 (1982).

The necessary conditions for mass flow through an accretion disc to be modulated by viscous variations are examined. The authors show that outburst behaviour is possible if the disc undergoes a global limit-cycle in which the viscosity is a two-valued function of disc structure. The authors attempt to model the eruptions of dwarf nova discs as a limit-cycle. The resulting light curves exhibit discontinuous changes in luminosity. If the viscosity is imposed, by some external agency, as a function of time then improved light curves are obtained. Nonetheless, it is not possible to obtain the whole range of dwarf nova light curves, and those which are fitted require a restricted range of conditions to be satisfied.

064.007 X-ray heating of the quiescent chromospheres of dMe stars.   L. E. Cram.
Astrophys. J., Vol. 253, 768 - 772 (1982).

The Einstein satellite has shown that dMe stars are surprisingly strong X-ray sources, with an X-ray luminosity that may be as large as 10% of the visual luminosity of the star. Atmospheric regions beneath the X-ray emitting coronae of these stars are thus bathed in an intense flux of X-rays and are consequently heated above the temperature that would exist in a radiative-equilibrium model. The authors show that such X-ray heating is an important factor in the energy balance of the quiescent chromospheres of dMe stars. This fact helps resolve the long-standing problem of the inadequacy of the acoustic-wave chromospheric heating hypothesis for such stars.

064.008 Thick accretion disks: self-similar, supercritical models.   M. C. Begelman, D. L. Meier.
Astrophys. J., Vol. 253, 873 - 896 (1982).

The authors generate self-similar models for geometrically thick, supercritical accretion disks, and study their structure and stability. By assumption, the models are characterized by near-equilibrium between gravity, centrifugal force, and radiation pressure. Slow nonazimuthal currents are driven by the viscous transport of angular momentum, which also dissipates binding energy. In contrast with thin disks, which are able to cool efficiently, the energy dissipated in a thick disk is partially trapped, resulting in large pressure gradients, sub-Keplerian angular velocities, and a shear stress which is no longer unidirectional. The assumption that hydrodynamic quantities scale as power laws in radius enables one to compute the disk structure as a function of angle from the rotation axis, given a model for the viscosity law.

064.009 A model of a thick disk with equatorial accretion.   B. Paczyński, M. A. Abramowicz.
Astrophys. J., Vol. 253, 897 - 907 (1982).

The authors construct a model of a geometrically thick

accretion disk orbiting a $10\,M_{\odot}$ black hole in which most of the interior is in convective equilibrium, and in which the accretion flow and heat generation are confined mostly to the layers close to the equatorial plane. Surfaces of constant angular momentum and surfaces of constant entropy coincide to form curved coaxial cylinders. Disks are marginally unstable dynamically along the cylinders (convection acts along them) and stable in all other directions. The authors consider two accretion rates: critical and twice critical.

064.010 Are Ap stars magnetic balloons?
E. N. Hubbard, D. S. P. Dearborn.
Astrophys. J., Vol. 254, 196 - 202 (1982).

The authors consider the effect of a large, global magnetic field on the structure of upper—main-sequence envelopes. They find that, for a simple stellar envelope model incorporating a 1000 gauss field, the additional $B^2/8\pi$ magnetic pressure component may be sufficient to expand the photosphere of a $2.0\,M_{\odot}$ star by about 20%. The implications of this effect for both Ap stars and for the position of the main sequence in cluster H-R diagrams are discussed.

064.011 Weber and Davis revisited: mass losing rotating magnetic winds.   P. K. Barker, J. M. Marlborough.
Astrophys. J., Vol. 254, 297 - 300 (1982).

The original Weber and Davis theory for the rotating magnetic solar wind is correct only in the case of zero photospheric mass loss. The authors have extended this theory to the case of nonzero photospheric mass loss, and they show how prior applications of the theory to the mass losing early-type stars are in conceptual and numerical error. One surprising result is that the azimuthal velocity structure may be affected in an interesting nonintuitive way.

064.012 Resonant electrodynamic heating of stellar coronal loops: an *LRC* circuit analog.   J. A. Ionson.
Astrophys. J., Vol. 254, 318 - 334 (1982).

This article addresses the important problem of electrodynamic coupling of $\beta < 1$ stellar coronal loops to underlying $\beta \gtrsim 1$ velocity fields. A rigorous analysis has revealed that the physics can be represented by a simple yet equivalent *LRC* circuit analog. This derived analog points to the existence of global structure oscillations which resonantly excite internal field line oscillations at a spatial resonance within the coronal loop. Although the width of this spatial resonance, as well as the induced currents and coronal velocity field, within the resonance region explicitly depend upon viscosity and resistivity, the resonant form of the generalized electrodynamic heating function is virtually independent of irreversibilities. This is a classic feature of high quality resonators that are externally driven by a broad-band source of spectral power.

064.013 An escape probability treatment of doublet resonance lines in expanding stellar winds.
G. L. Olson.
Astrophys. J., Vol. 255, 267 - 277 (1982).

When stellar winds have expansion velocities greater than the velocity separation between doublet line transitions, the line source function of the component with the longer wavelength depends on nonlocal values of the other component's source function. By using the Sobolev approximation and escape probability methods, the nonlocal radiative coupling for the case of spherically symmetric outflow with a monotonically increasing velocity can be solved explicitly and efficiently. This case is directly applicable to the doublet P Cygni type profiles observed in the ultraviolet spectra of O and B stars and leads to an improved determination of the line strengths. The nonlocal coupling also modifies the radiative forces that act on the mass outflow. Examples of how the nonlocal radiative coupling changes the line source functions, the line profiles, and the radiative forces are presented.

064.014 The formation of resonance lines in locally non-monotonic winds.   L. B. Lucy.
Astrophys. J., Vol. 255, 278 - 285 (1982).

In order to test the hypothesis that the X-ray emission by hot stars arises from shock distributed throughout their winds, the formation of lines by resonance scattering in winds with locally nonmonotonic velocity fields is investigated. The resulting P Cygni profiles for strong lines are shown to provide a useful diagnostic for nonmonotonic flow in the form of broad, deep absorption troughs that cannot be produced by a spherically symmetric wind with a monotonic velocity law. Such absorption troughs have in fact already been noted in the Copernicus spectra of several stars, and these troughs are therefore interpreted here as evidence for nonmonotonic flows. Moreover, the N V resonance line in the spectrum of ζ Puppis is shown quantitatively to be consistent with the predicted spatial frequency of X-ray emitting shocks.

064.015 X-ray emission from the winds of hot stars. II.
L. B. Lucy.
Astrophys. J., Vol. 255, 286 - 292 (1982).

The kinematic model adopted in Paper I for the unstable line-driven winds of early-type stars is modified and the resulting X-ray spectra compared with observation. In this revised phenomenological theory, radiatively driven shocks are conjectured to result from the amplification of unstable waves and to survive until shadowing by following shocks compels their decay. In consequence of this shock destruction mechanism, the revised theory predicts that X-ray emission continues far out into the nearly terminal flow; the effects of self-absorption are therefore not expected to be prominent, a result qualitatively consistent with recent analyses of Einstein data. Nevertheless the X-ray luminosities calculated for several O stars are too low and the spectra too soft. These failures are interpreted as implying a broad spectrum of shock strengths, with the bulk of the X-ray emission coming from the strongest shocks.

064.016 Intrinsic stellar mass flux and steady stellar winds.
R. L. T. Wolfson, T. E. Holzer.
Astrophys. J., Vol. 255, 610 - 616 (1982).

The conventional view that winds from stars like the Sun result from the existence of a hot corona and a low-pressure interstellar medium has recently been challenged by R. N. Thomas and colleagues. They suggest that the rates of mass loss from these and other stars are determined by conditions imposed on the flow at or below photospheric levels and that the warm chromosphere and hot corona of a solar-like star are simply consequences of this imposed photospheric flow and dissipation in the resultant stellar wind. The authors have examined this suggestion through the application of gas dynamic theories including dissipation. Extensive analytic and numerical calculations for both polytropic and thermally conductive flows, with viscosity included, indicate that the specification of an arbitrary intrinsic mass flux is not consistent with steady, radial, spherically symmetric flow in the absence of energy addition. The authors conclude that there is at present no theoretical support for the suggestion of Thomas and colleagues.

064.017 Heating of stellar chromospheres when magnetic fields are present.
P. Ulmschneider, R. F. Stein.
Astron. Astrophys., Vol. 106, 9 - 13 (1982).

From recent semi-empirical solar models of Vernazza, Avrett and Loeser (1981), from OSO-8 observations and from the stellar Mg II emission line fluxes of Basri and Linsky (1979) constraints on possible chromospheric heating mechanisms are derived. It is shown that a picture where non-magnetic regions are heated by acoustic shock waves and magnetic regions by slow mode shock waves appears to best satisfy the observational facts. For the high chromosphere this mechanism must be supplemented or replaced, possibly by Alfvén wave heating.

064.018 On the theory of thermally sustained stellar winds.
P. Souffrin.
Astron. Astrophys., Vol. 106, 14 - 15 (1982).

The influence on a stellar wind of a boundary layer structure at the stellar surface (Gonczi et al., 1977) is reconsidered. The discussion of the role of the conductive flux boundary layer by Couturier et al. (1979) is extended to include the role of the temperature boundary layer. As a result the constraint between the wind parameters exhibited by Couturier et al. is shown to be insensitive to the temperature at the lower level, so that the structure of the wind — including the transition zone — is ultimately dependent on only one parameter, namely the mechanical flux entering the transition zone. This conclusion is mechanism independent, although the structure of the wind is of course determined by the heating mechanism.

064.019 Vertical structure of accretion disks.
F. Meyer, E. Meyer-Hofmeister.
Astron. Astrophys., Vol. 106, 34 - 42 (1982).

The authors have computed the vertical structure of steady accretion disks with radiative and convective energy transfer. Accretion rates of 10^{-7}–$10^{-11} M_{\odot}$/yr were taken around a 1 $M_{\odot}$ central object, modeling disks in cataclysmic variables or around neutron stars. The geometrical shape of the disk produces regions shadowed from a central light source. The effect of irradiation is shown to be important for disks around neutron stars. Consequences are thicker disks which can explain the "missing eclipses" in low mass X-ray binaries and a new model for the 35^d cycle of Her X-1. The stability in low β-regions is investigated.

064.020 On the ionization and velocity structure of expanding circumstellar envelopes.
H. Drechsel, J. Rahe.
Astron. Astrophys., Vol. 106, 70 - 78 (1982).

The profiles of envelope lines formed in expanding atmospheres of luminous early-type and Wolf-Rayet stars by the scattering of the stellar continuum flux are sensitive to the variation of the optical depth in the line and of the velocity gradient of the stellar wind as a function of radial distance from the star. With appropriate model assumptions for the ionization and velocity structure of the circumstellar envelope a wide variety of line profiles can be obtained. A sample of theoretical profiles are presented. UV envelope line profiles observed in the spectra of stars with rapid mass loss via stellar wind, are compared with theoretically predicted profiles. Inferences from UV spectroscopic observations of the O7f supergiant UW CMa, the interacting early-type contact binary SV Cen, and the Wolf-Rayet star HD 192163 for models of the expanding envelopes are discussed.

064.021 Mass loss from α Cyg (A2Ia) derived from the profiles of low excitation Fe II lines.
H. Hensberge, H. J. G. L. M. Lamers, C. de Loore,
F. C. Bruhweiler.
Astron. Astrophys., Vol. 106, 137 - 150 (1982).

The low excitation Fe II lines in the spectral region 2000-3000 Å, arising from the ^6D and ^4F levels, have been studied in the spectrum of α Cyg. Calculations with the Sobolev approximation favour a stellar wind model. The mass loss rate corresponding to this model is 1 to $5 \times 10^{-9} M_{\odot} \, yr^{-1}$. It is shown that the contribution of blending photospheric absorption lines (e.g. Co II) to weaker P Cygni profiles has been largely underestimated previously. The authors' mass loss rate is a factor 500 smaller than the one derived from the infrared excess by Barlow and Cohen (1977).

064.022 NLTE model atmospheres for early-type stars of various chemical compositions and resulting

emission-line spectra for surrounding H II regions.
J. Borsenberger, G. Stasińska.
Astron. Astrophys., Vol. 106, 158 - 162 (1982).

The authors have computed several static NLTE plane-parallel model atmospheres with various chemical compositions, using the code of Mihalas (1972). The resulting emergent fluxes are taken to build model H II regions of similar composition. Variations of the elemental abundances in the atmospheres of the exciting stars are found to modify only very slightly the intensities of the lines emitted by the nebulae, contrary to what has been argued on the basis of LTE calculations.

064.023 Molecular abundances in IRC + 10216.
S. Lafont, R. Lucas, A. Omont.
Astron. Astrophys., Vol. 106, 201 - 213 (1982).

A new and general theoretical discussion of the abundances of molecules in the carbon circumstellar envelope IRC + 10216 is performed. The formation processes of the nearly 20 molecules observed in IRC + 10216 can probably be classified in four categories of approximately equal importance: "freezing" of LTE abundances, non-LTE gas reactions, grain processes, photodissociation.

064.024 Wind acceleration in early-type stars: the momentum problem and the terminal velocity.
N. Panagia, F. Macchetto.
Astron. Astrophys., Vol. 106, 266 - 273 (1982).

The processes of radiative acceleration of stellar winds in OB stars by single and multiple photon scattering are considered. Single scattering can be the dominant accelerating process for stars later than B2 and can account for terminal velocities up to 500 - 1000 km s^{-1}. Multiple scattering of photons in the approximate range 200 - 500 Å provides additional wind acceleration for stars earlier than B2 to reach terminal velocities of up to 2000 - 4000 km s^{-1}. A systematic increase of the terminal velocity as a function of the effective temperature is predicted. Observational data confirm this expectation quite well.

064.025 On hot star winds. I. Radiation-driven winds.
M. Leroy, J.-P. J. Lafon.
Astron. Astrophys., Vol. 106, 345 - 357 (1982).

Self consistent models of radiatively driven hot star winds are built up. The equations describing dynamics and radiative transfer are coupled and solved in a self consistent way using numerical iterations. The local radiative force depends on radiative transfer throughout the expanding gas. The authors have constructed 16 models of stars with T_{eff} = 50 000 K, M = 60 $M_{\odot}$, L = $9.7 \times 10^5 L_{\odot}$. The whole structure of the wind is controlled by the physics of the subsonic part of the wind, which is a thin zone close to the photosphere. The velocity profiles have a topology like that of profiles derived from observations.

064.026 On hot star winds. II. Energy transport — corona-like temperature enhancements.
M. Leroy, J.-P. J. Lafon.
Astron. Astrophys., Vol. 106, 358 - 361 (1982).

Energy transport in hot star winds is discussed. Radiation is so important that radiative equilibrium cannot be disrupted by temperature enhancements due to the dynamics or the properties of the expanding gas. However, if, as suggested in a previous paper (Leroy and Lafon, 1981), radiative equilibrium is disrupted by an additional energy input due to sources irrelevant to the expanding gas, this happens in a very thin "corona like" shell. A qualitative energy balance scheme is proposed in this case.

064.027 Meridional circulation in accretion disks.
W. Cabot.
News Lett. Astron. Soc. N. Y., Vol. 2, 36 (1982). − Abstract.

064.028 On the detection of abundance stratifications in peculiar stars through the curve of growth method.
G. Alecian.
Astron. Astrophys., Vol. 107, 61 - 65 (1982).

Theoretical curves of growth for manganese are computed for an atmosphere with T_{eff} = 12, 000 K and log g = 4. The author shows that the curves of growth may be strongly affected by the presence in the atmosphere of a plane parallel layer where manganese is overabundant. The theoretical results are compared with the observations of the Hg-Mn star υ Her. The existence of a radial inhomogeneity of manganese in the atmosphere of υ Her is discussed in the framework of the diffusion processes.

064.029 Thermally-driven stellar winds in late-type stars.
T. Watanabe.
Publ. Astron. Soc. Japan, Vol. 33, 679 - 699 (1981).

The thermodynamic structure of the outer atmospheres of late-type stars is investigated, the nonradiative heating term being schematically taken into account.

064.030 Structure around the inner edge of geometrically thin accretion disks.
S. Kato, J. Fukue, S. Inagaki, A. T. Okazaki.
Publ. Astron. Soc. Japan, Vol. 34, 51 - 63 (1982).

In geometrically thin, nearly Keplerian disks, there is an inner edge inside which the gas falls toward the black hole by dynamical instability due to a relativistic effect. The disk structure around this edge is examined in detail when the geometry is Schwarzschildian and the vertically averaged kinematic viscosity is constant. A steady disk model is found in which the disk structure changes continuously at the edge from a nearly Keplerian structure to a thin non-Keplerian structure of the infalling gas. The structure of this transition region is demonstrated.

064.031 Nonspherical stellar envelopes and winds: effects of structure on radiative fluxes and apparent mass loss rates.   J. Schmid-Burgk.
Astron. Astrophys., Vol. 108, 169 - 175 (1982).

Expressions are given for the radiative flux produced by extended stellar envelopes whose stratification deviates considerably from spherical symmetry. Among the structures considered are ellipsoidal configurations of arbitrary axial ratios, polar cones, equatorial belts and radial streamer systems. A relation between flux, total mass loss rate, source geometry and inclination angle is derived for the case where the envelope is a stellar wind zone; free-free emission serves as an example.

064.032 How disk accretion affects a rotating dipole.
U. Anzer, G. Börner, Y. Y. Zhou.
Mitt. Astron. Ges., Nr. 55, p. 28 - 31 (1982).

064.033 Zeitabhängige turbulente Akkretionsscheiben.
W. Duschl.
Mitt. Astron. Ges., Nr. 55, p. 161 - 163 (1982).

064.034 Modellrechnungen zur Erzeugung akustischer Energie bei späten Sternen.   H. U. Bohn.
Mitt. Astron. Ges., Nr. 55, p. 193 - 194 (1982).

064.035 The law of motion of strong shock waves in stellar envelopes.   I. A. Klimishin, B. I. Gnatyk.
Astrofizika, Tom 17, 547 - 555 (1981). In Russian. English translation in Astrophysics, Vol. 17, No. 3.

On the basis of calculations that were carried out by the authors approximate formulae for the determination of the velocity of a shock wave moving in nonuniform stellar envelopes are proposed. The use in cosmic gasdynamics of approximate analytical methods of Brinkley-Kirkwood (1947) and Whithem (1958) are discussed.

064.036 Model chromospheres of RS CVn stars: Balmer line profiles in λ Andromedae.
D. J. Mullan, L. E. Cram.
Astron. Astrophys., Vol. 108, 251 - 255 (1982).

The authors have constructed two models for the chromosphere of the RS CVn star λ And, one with a low-pressure transition zone, and one with a high pressure transition zone. The high pressure model predicts an Hα line profile which agrees fairly well with high resolution observations of λ And, without the need to include non-thermal line broadening. The low pressure model predicts an Hα profile which agrees very well with the observations, after application of macroturbulent broadening. The authors discuss methods which could distinguish between the two alternatives.

064.037 Onset of rapid mass loss in cool giant stars: magnetic field effects.   D. J. Mullan.
Astron. Astrophys., Vol. 108, 279 - 286 (1982).

The author asks the question: Where in the HR diagram can closed magnetic field loops exist in steady state in stellar atmospheres? To answer this, he applies a model derived from Pneuman (1968) for helmet streamers in the solar corona. Using a semi-empirical technique, the author finds that long-lived closed loops exist only below a certain boundary in the HR diagram. The region below this boundary is occupied by stars which are known to have hot coronae and slow mass loss. The author suggests that rapid mass loss sets in when closed field loops can no longer exist in steady state in the atmosphere.

064.038 Effect of spots on a star's radius and luminosity.
H. C. Spruit.
Astron. Astrophys., Vol. 108, 348 - 355 (1982).

The thermal response of a star following the sudden appearance of spots on its surface is calculated.

064.039 The flow of heat near a starspot.   H. C. Spruit.
Astron. Astrophys., Vol. 108, 356 - 360 (1982).

The thermal disturbance due to a single starspot of radius R and depth d in a polytropic convective envelope is calculated using a turbulent diffusion approximation. The effect of systematic fluid flows is neglected. Times large compared with the turbulent diffusion time scale and short compared with the Kelvin-Helmholtz time scale are considered so that the heat flow is quasi-static.

064.040 The influence of tidal effects on the structure of accretion discs in dwarf novae.   S. Kříž.
Mon. Not. R. Astron. Soc., Vol. 199, 725 - 734 (1982).

The author solves the equations for the structure of steady-state accretion discs taking into accout the angular momentum loss via tidal effects. He demonstrates that the surface density is substantially reduced in comparison with the standard model without tides. If the disc is optically thick the spectrum of radiation emerging from it due to the dissipation of potential energy is nearly the same as in the case of the standard model. The author also derives the theoretical radius of the outer edge of the disc and compares this tidal radius with observations.

064.041 Physical properties of thick supercritical accretion disks.   P. J. Wiita.
Astrophys. J., Vol. 256, 666 - 680 (1982).

Recent models of thick, super-Eddington accretion disks around black holes that were devised by Paczyński and Wiita are analyzed and expanded upon. The shape and ratio of disk luminosity to Eddington luminosity essentially depend upon the inner radius of the disk and the form of the (non-Keplerian) angular momentum distribution, while the size and actual luminosity grow linearly with the mass of the black hole. Two approaches are used to estimate the central temperatures and densities of such disks for different polytropic indices. Using either method, physical constraints involving mechanical equilibrium, the total mass of the disk, or the onset of nuclear fusion are found to stringently limit self-consistent models. In particular, it is difficult to build thick disks that are less massive than a nonrotating central black hole if its mass is greater than about $10^7 M_\odot$. However, somewhat more general models for the disk structure along with consideration of rotating black holes can probably provide enough luminosity to power the most active galactic nuclei.

064.042 **The structure and appearance of winds from super-critical accretion disks. II. Dynamical theory of supercritical winds.**   D. L. Meier.
Astrophys. J., Vol. 256, 681 - 692 (1982).
A general analytic theory of winds driven by super-Eddington luminosities is presented. The relevant parameters are the mass M of the central object, the radius r_i where luminosity and matter are injected, the ratio α'' of the free-fall time to the heating time at r_i, and the total luminosity L_T injected at r_i. Several different regimes of dynamical wind structure are identified, and the analytic expressions are shown to agree with the numerical results of Paper I in the appropriate case. In its general form the theory is the optically thick (to electron scattering) counterpart to optically thin radiation pressure-driven stellar winds.

064.043 **The stucture and appearance of winds from super-critical accretion disks. III. Thermal and spectral properties of supercritical winds.**   D. L. Meier.
Astrophys. J., Vol. 256, 693 - 705 (1982).
The thermal structure and emergent spectral properties of the supercritical winds outlined in Paper II are presented. Thermal emission due to free-free and Compton processes is considered, along with some nonthermal aspects, for the case when the photons are untrapped throughout the wind envelope. Useful figures and procedures for determining the injection region temperature, emergent spectral temperature, and rough spectral shape are outlined with some attention given to models with central objects of $1 M_\odot$ and $10^8 M_\odot$. A scenario is outlined where the central source of quasars is interpreted as a supercritical wind heated at its base on a nearly dynamical time scale.

064.044 **The structure and appearance of winds from super-critical accretion disks. IV. Analytic results with applications.**   D. L. Meier.
Astrophys. J., Vol. 256, 706 - 716 (1982).
The supercritical wind theory of Papers II and III is applied to winds driven by accretion phenomena. An expression for the total luminosity L_T is derived in terms of the accretion rate $\dot{M}_T$ and injection radius r_i. When applied specifically to accretion disks, the theory becomes the supercritical counterpart to standard subcritical accretion disk theory. Useful models for various values of the central object mass M, heating parameter α'', and $\dot{M}_T$ are presented graphically. Supercritical wind theory is also applied to several phenomena in which spherical outflow driven by super-Eddington luminosities has been proposed as an explanation. Specifically, applications are discussed for novae, symbiotic stars, P Cygni, X-ray binaries, SS 433 and the quasar PHL 5200.

064.045 **Results of a calculation of the parameters of ionization-dissociation shock waves moving in stellar envelopes.**   B. I. Gnatyk, I. A. Klimishin.
Astrometr. Astrofiz., Vyp. (No.) 44, p. 3 - 9 (1981). In Russian.
Some calculation results are given on the temperature and density jumps as well as the dissociation and ionization degree behind the front of a shock wave moving in an hydrogen stellar envelope, for a wide range of initial values of

particle concentrations, of temperature and velocity of the shock wave front.

064.046 **Model atmospheres for globular cluster stars.**   R. L. Kurucz.
Astrophysical parameters for globular clusters, (see 012.023), p. 289 - 292 (1981).

064.047 **Accretion and magnetism of stars.**   V. Yu. Shulikovskij.
Gravitatsiya i teor. otnositel'nosti, Kazan', 1981, No. 18, p. 110 - 117, 119. In Russian. — Abstr. in Ref. zh., 51. Astron., 4.51.156 (1982).

064.048 **Excitación de iones de carbono, nitrógeno y oxígeno en atmósferas estelares.**   L. A. M. Opradolce.
Bol. Asoc. Argentina Astron., No. 20 - 24, p. 151 (1981). Abstract.

064.049 **Formación de moléculas en las atmósferas estelares.**   L. A. Milone, C. López.
Bol. Asoc. Argentina Astron., No. 20 - 24, p. 345 (1981). Abstract.

064.050 **Time dependent models of grain-forming stellar atmospheres.**   J. E. J. Woodrow, J. R. Auman.
Astrophys. J., Vol. 257, 247 - 263 (1982).
Completely time-dependent models of the expanding atmospheres of cool, carbon-rich stars were calculated. The driving force for the expansion was radiation pressure on grains. The grains were assumed to have the structure of graphite. The stellar parameters adopted for the models were $M = 1.5 M_\odot$, $L = 1.94 \times 10^4 L_\odot$, $C/H = 1.22 \times 10^{-3}$, and $C/O = 1.76$. Two models were generated. Model 1 had $T^* = 2500$ K and model 2, $T^* = 2400$ K. The calculated mass loss rate for model 1 was $6.2 \times 10^{-9} M_\odot$ yr^{-1} and for model 2, $7.4 \times 10^{-8} M_\odot$ yr^{-1}. The mass flow approached a steady state in model 1, but in model 2 a small amplitude pulsation was superposed upon the outward flow. Initially this pulsation was very irregular, but after an elapsed time of 27×10^7 s the model had relaxed into a steady pulsation with a period of 6.4×10^7 s. The driving force for this pulsation appears to be an opacity-controlled feedback mechanism which operated between the grain-forming region of the model and the hydrogen dissociation zone.

064.051 **Wave-driven winds from cool stars. I. Some effects of magnetic field geometry.**
L. Hartmann, K. B. MacGregor.
Astrophys. J., Vol. 257, 264 - 268 (1982).
The wave-driven wind theory of Hartmann and MacGregor is extended to include effects due to non-radial divergence of the flow. Specifically, isothermal expansion within a flow tube whose cross-sectional area increases outward faster than r^2 near the stellar surface is considered. It is found that the qualitative conclusions of Hartmann and MacGregor concerning the physical properties of Alfvén wave-driven winds are largely unaffected. In particular, mass fluxes of similar magnitude are obtained and wave dissipation is still necessary to produce acceptably small terminal velocities.

064.052 **Models for stellar flares.**   L. E. Cram, D. T. Woods.
Astrophys. J., Vol. 257, 269 - 275 (1982).
The authors study the response of certain spectral signatures of stellar flares (such as Balmer line profiles and the broad-band continuum) to changes in atmospheric structure which might result from physical processes akin to those thought to occur in solar flares. While each physical process does not have a unique signature, some of the observed properties of stellar flares can be explained by a model which involves increased pressures and temperatures in the flaring

stellar chromosphere. The authors suggest that changes in stellar flare area, both with time and with depth in the atmosphere, may play an important role in producing the observed flare spectrum.

064.053 On the widths of the Ca II K emission in late-type stars. G. Severino.
Astron. Astrophys., Vol. 109, 90 - 94 (1982).

A simplified model for the formation of the Ca II K emission occurring in most late-type stars is used to determine the main parameters controlling the emission widths. The model accounts for partial redistribution effects. From the computed profiles the author infers that the K_2 peak separation, W_2, is, within a good degree of approximation, proportional to the amplitude of chromospheric turbulent velocities, as it was in old complete redistribution models. Moreover the claimed square-root dependence of the K_1 dip separation, W_1, on the mass column density at the temperature minimum, $m*$, is questioned.

064.054 Cálculo de modelos de "zonas de transición" en estrellas B.
J. M. Fontenla, M. Rovira, A. E. Ringuelet.
Bol. Asoc. Argentina Astron., No. 26, p. 169 (1981).
Abstract.

064.055 Modelos de vientos estelares en la región fotosférica.
J. M. Fontenla, A. D. Verga.
Bol. Asoc. Argentina Astron., No. 26, p. 173 - 175 (1981).

064.056 Stellar atmospheres with a magnetic field.
K. Stepień.
Soobshch. Spets. Astrofiz. Obs., Vyp. 32, (see 012.029), p. 16 - 17 (1981).

064.057 Synthetic variations of Balmer lines in an Ap star model atmosphere. J. Madej.
Soobshch. Spets. Astrofiz. Obs., Vyp. 32, (see 012.029), p. 24 - 25 (1981).

064.058 Model atmospheres of Be stars. R. Poeckert.
Be stars, (see 012.030), p. 453 - 483 (1982).

064.059 Hydrodynamical models of rotating magnetic winds.
P. K. Barker.
Be stars, (see 012.030), p. 485 - 488 (1982).
Numerical investigations of rotating stellar winds driven by a combination of magnetic and line radiation forces show a clear dichotomy: shallow winds have their azimuthal velocity enhanced by magnetic corotation, whereas steep winds may have a diminished circular velocity.

064.060 Gross structural pattern for the atmospheres of Be, and some closely related, stars.
V. Doazan, R. Stalio, R. N. Thomas.
Be stars, (see 012.030), p. 489 - 491 (1982).
The paper is an abstracted comparison of Struve's 1942 quasi-empirical model of the shell-atmosphere of Be and similar stars, based on visual data and quasi-radiative equilibrium, to that demanded by current visual + nonvisual data, requiring nonradiative + mass fluxes characterized by variability and individuality.

064.061 Theoretical surface brightness distributions and continuum polarization of rapidly rotating B stars.
G. Sonneborn.
Be stars, (see 012.030), p. 493 - 496 (1982).

064.062 On the Balmer progression phenomena in Be stars.
R. Hirata.
Be stars, (see 012.030), p. 497 (1982).

064.063 The effect of reflected and external radiation on stellar flux distributions. D. G. Hummer.
Astrophys. J., Vol. 257, 724 - 732 (1982).

The effect of radiation emitted or scattered by circumstellar material, such as a stellar wind, into the stellar photosphere is investigated on the basis of a gray model atmosphere generalized to include the effects of an external radiation field and a surface boundary condition describing the reflection of a specified fraction, depending on the frequency, of the outgoing radiation. Substantial modifications both to the temperature and flux distributions are found.

064.064 Synthetic Hα profiles and mass loss rates for Alpha Cygni. P. B. Kunasz.
Bull. American Astron. Soc., Vol. 13, 784 (1981). – Abstract.

064.065 Magnetically driven winds of Wolf-Rayet stars.
L. W. Hartmann, J. P. Cassinelli.
Bull. American Astron. Soc., Vol. 13, 785 (1981). – Abstract.

064.066 The magnetohydrodynamics of an expanding stellar envelope. B. C. Low.
Bull. American Astron. Soc., Vol. 13, 792 (1981). – Abstract.

064.067 Invalidity of the minimum flux corona concept.
R. Hammer.
Bull. American Astron. Soc., Vol. 13, 792 (1981). – Abstract.

064.068 Electrodynamic coupling in stellar atmospheres.
J. A. Ionson, J. Mosher.
Bull. American Astron. Soc., Vol. 13, 792 (1981). – Abstract.

064.069 Evidence for photospheric soft X-ray emission from Sirius B.
C. Martin, G. Basri, M. Lampton, S. Kahn.
Bull. American Astron. Soc., Vol. 13, 810 - 811 (1981). Abstract.

064.070 Line formation in accretion disks.
D. Carroll, D. G. Hummer, G. B. Rybicki.
Bull. American Astron. Soc., Vol. 13, 818 (1981). – Abstract.

064.071 The relation between coronal, chromospheric and magnetic activity: a case study.
G. S. Basri, F. M. Walter, G. Marcy.
Bull. American Astron. Soc., Vol. 13, 828 - 829 (1981). Abstract.

064.072 Constraints on the mass-loss rate of HR 1040 (A0Ia).
N. D. Morrison, P. B. Kunasz.
Bull. American Astron. Soc., Vol. 13, 829 (1981). – Abstract.

064.073 Accretion disk coronae. N. E. White, S. S. Holt.
Bull. American Astron. Soc., Vol. 13, 900 (1981). Abstract.

064.074 Heating mechanism for stellar winds of early-type stars. Eh. Ya. Vil'koviskij.
Astrofizika, Tom 17, 309 - 315 (1981). In Russian. English translation in Astrophysics, Vol. 17, No. 2.
It is shown that radiation pressure forces can move C III-type ions with high velocity (about 10^8 km/s) relative to a stellar wind plasma. This leads to the heating of the upper regions of expanding envelopes of early-type stars.

064.075 On radiation deficiency of shell stars in the Balmer continuum. Ya. N. Chkhikvadze.
Astrofizika, Tom 17, 317 - 326 (1981). In Russian. English translation in Astrophysics, Vol. 17, No. 2.
A statistical investigation of the distribution of some photometric parameters for shell and normal stars known at present is made. In terms of the results obtained the conclu-

sion is drawn on the presence of deficiency of ultraviolet radiation in shell stars. This is caused by an appreciable optical depth of gaseous envelopes in shell stars in the Balmer continuum. There is no difference in the radiation power in the gaseous envelopes of shell and normal Be stars. The belief that shell stars are of later spectral class than normal Be stars is confirmed.

064.076   **Strong spherical magnetogasdynamic shock waves in rotating stellar atmospheres.**
B. G. Verma, R. C. Srivastava, A. H. Khan.
Astrophys. Space Sci., Vol. 82, 307 - 309 (1982).

An analytic expression for the velocity of a magnetogasdynamic shock wave, propagating in a rotating stellar atmosphere has been obtained by using the method of characteristics and considering the effect of Coriolis force. It is shown that in the outer convective layer of the star both Coriolis force and magnetic field have a significant effect on the shock velocity.

064.077   **A model of two-stream non-radial accretion for binary X-ray pulsars.**   V. M. Lipunov.
Astrophys. Space Sci., Vol. 82, 343 - 361 (1982).

The general case of non-radial accretion is assumed to occur in real binary systems containing X-ray pulsars. The structure and the stability of the magnetosphere, the interaction between the magnetosphere and accreted matter, as well as evolution of neutron star in close binary system are examined within the framework of the two-stream model of nonradial accretion onto a magnetized neutron star. Observable parameters of X-ray pulsars are explained in terms of the model considered.

064.078   **A model of the atmospheric convection zone with a large increase of rotation velocity between its bottom and top.**   Yu. V. Vandakurov.
Astrophys. Space Sci., Vol. 83, 105 - 116 (1982).

A simple model of a slowly rotating stellar or planetary atmospheric convection zone is considered supposing that only growing convective perturbations contribute to the value of nonlinear azimuthal force maintaining the differential rotation of the zone. The angular velocity of rotation is assumed to be dependent only on the distance to the center. It turns out that in this model some resonance phenomenon is possible, due to which a structure can be formed with a large rotation velocity gradient. A hypothesis is put forward that fast rotation of the upper atmosphere of Venus is due to such resonance interaction.

064.079   **Plane-parallel vs spherical geometry vis-à-vis scale-heights.**   M. S. Vardya.
Astrophys. Space Sci., Vol. 84, 155 - 161 (1982).

Values of the surface gravity g_{min} have been computed above which plane-parallel approximation should provide satisfactory stellar model atmospheres.

064.080   **On iterative solutions of the LTE model atmosphere problem.**
D. Mihalas, R. P. Weaver, J. G. Sanderson.
J. Quant. Spectrosc. Radiat. Transfer, Vol. 28, 53 - 59 (1982).

The authors discuss iterative methods for solving the coupled radiative-transfer and energy-balance equations in the LTE model atmospheres problem including isotropic coherent scattering. They show that iterative solution of the grand matrix encountered in such problems is vastly more efficient than a direct solution, and is easily vectorized. The final computational effort is linear in the number of depths and frequencies considered, and thus this approach opens the door for the computation of both static and dynamic line-blanketed models using large numbers of depth-points and huge numbers of frequencies. The iterative methods discussed can be applied to line-formation problems with complete redistribution and to certain classes of problems with partial redistribution (e.g.

Compton scattering problems in the Fokker-Planck approximation).

064.081   **Hydrodynamics of X-ray induced stellar winds.**
R. A. London, B. P. Flannery.
Astrophys. J., Vol. 258, 260 - 269 (1982).

The authors present new theoretical models for X-ray induced stellar winds in binary systems. They numerically solve the hydrodynamic equations in one dimension, utilizing a simplified model of the physics of an X-ray heated plasma. The character of the solutions depends on three dimensionless parameters: (1) the ratio of the X-ray spectral temperature to the photospheric temperature; (2) the ratio of the X-ray temperature to a temperature characterizing the escape energy from the stellar surface; and (3) the ratio of the flow time to the heating time. The primary results of the calculations are the temperature and density profiles and the mass flux in the wind. The authors locate the known X-ray binary systems in the parameter space of their model and estimate the mass-loss rates from them.

064.082   **Time-dependent accretion onto magnetized white dwarfs.**   S. H. Langer, G. Chanmugam, G. Shaviv.
Astrophys. J., Vol. 258, 289 - 305 (1982).

The authors consider time-dependent accretion onto magnetized white dwarfs. A detailed description of a numerical method of solution to the hydrodynamical equations is given. The postshock flow is cooled by optically thin bremsstrahlung and is thermally unstable. As a result the shock height undergoes periodic oscillations. The authors consider the properties of this oscillation as a function of the accretion rate and of the mass and radius of the white dwarf. The structure of the accretion flow depends on a single scaling parameter. Below a critical accretion rate, which depends on the particular white dwarf, the nature of the accretion flow changes and the shock propagates up the accretion column indefinitely.

064.083   **A sufficient condition for the stability of atmospheres with magnetic fields.**
E. G. Zweibel.
Astrophys. J., Lett., Vol. 258, L53 - L56 (1982).

Using the MHD energy principle, the author shows that $P + B^2/8\pi$ = constant is a sufficient condition for the stability of magnetized fluid systems with the following properties: the gravitational acceleration g is uniform, the magnetic field lines lie in parallel planes aligned with g, and all quantities are uniform in the z-direction, perpendicular to the plane of the field lines. An example of a stable system is given, consisting of a vertical sheet of plasma supported against gravity by bowed field lines.

064.084   **Molecules in red-giant stars. I. Column densities in models for K and M stars.**
H. R. Johnson, A. J. Sauval.
Astron. Astrophys., Suppl. Ser., Vol. 49, 77 - 87 (1982).

Weighted column densities have been calculated for neutral atoms and ions of most elements and for many molecules in 6 selected model atmospheres of red-giant stars ($2500 \leqslant T_{eff} \leqslant 4000$ K) with solar composition. These comprise the most abundant molecules from a total of about 1600 compounds analyzed.

064.085   **On the theory of shock-heated atmospheres. III. Discussion of the formalism and application to stellar coronae.**   P. Souffrin.
Astron. Astrophys., Vol. 109, 205 - 212 (1982).

The two parameter model of Mangeney and Souffrin (1979) is extended to arbitrary shock strengths and to account for the coronal radiative losses.

064.086   Non-thermal emission from relativistic accretion disks: a simple model for axisymmetric inhomogeneous sources. S. Pineault.
Astron. Astrophys., Vol. 109, 294 - 300 (1982).

The thin relativistic disk model of Novikov and Thorne (1973) is used to investigate the properties of non-thermal spectra emitted by axisymmetric inhomogeneous sources under the assumption that fast particles are produced by the reconnection of magnetic fields. Spectra are computed for power law and monoenergetic electron distributions in the case where the acceleration and emission regions are coincident, the maximum electron energy being determined by a balance between energy gain and losses, and also in the case where the electrons are accelerated without significant losses before beginning to radiate. The model is applied to galactic X-ray sources and active galactic nuclei.

064.087   Extended static stellar atmospheres – VI. Search for a three-dimensional classification scheme for luminous M stars. M. Scholz, R. Wehrse.
Mon. Not. R. Astron. Soc., Vol. 200, 41 - 47 (1982).

Conventional classification criteria (continuous fluxes, metal-ion lines, TiO bands) are suited to define a three-dimensional classification of luminous middle to late M stars of normal surface compositon. This classification measures, besides T_{eff} and g_s, the geometric extension, d (roughly $\propto g_s^{-1} R^{-1}$), as a third fundamental parameter of the photosphere.

064.088   Analytic treatment of polarization by arbitrary scattering mechanisms in circumstellar envelopes – I. Single stars. J. F. L. Simmons.
Mon. Not. R. Astron. Soc., Vol. 200, 91 - 113 (1982).

An analytic treatment is given of the polarization of light produced by arbitrary spherically symmetric scattering mechanisms in optically thin, but otherwise arbitrary, circumstellar shells. Series expressions for the normalized Stokes parameters of the scattered light are obtained. In the first order approximation the author shows quite generally that the form of the polarization versus wavelength (λ) dependency is independent of the specific density distribution of scatterers in the envelope, and that the position angle is independent of the specific scattering mechanism. Scattering in envelopes of cool stars is discussed in detail for Mie scattering.

064.089   Convection in stellar atmospheres. D. Dravins, J. Lind.
Rep. Obs. Lund, No. 18, (see 012.044), p. 109 - 110 (1982).

064.090   Chromospheric structure in relation to radiation losses. R. G. Athay.
Smithsonian Astrophys. Obs., Spec. Rep. 392, (see 012.045), p. 3 - 14 (1982).

Under the assumptions that cool star chromospheres are heated by mechanical energy dissipation that depends quasilinearly on density and cooled by radiation loss, it is shown that the basic properties of chromospheres are determined by the ionization of hydrogen.

064.091   Dynamic phenomena in coronal flux tubes. J. T. Mariska, J. P. Boris.
Smithsonian Astrophys. Obs., Spec. Rep. 392, (see 012.045), p. 53 - 58 (1982).

064.092   The generation of acoustic energy from stellar convection zones. H. U. Bohn.
Smithsonian Astrophys. Obs., Spec. Rep. 392, (see 012.045), p. 67 - 72 (1982).

The heating of stellar chromospheres and coronae by the dissipation of acoustic waves remains an important heating mechanism in spite of many contradicting arguments. It is only in the lower solar chromosphere that short period acoustic wave heating seems to be undisputed. The arguments leading to the rejection of the so-called "acoustic heating theory" are derived mainly from comparisons of calculated acoustic energy fluxes with observational or theoretical requirements. However, before the acoustic heating theory is rejected, one should re-examine the boundary conditions of this theory, i.e. the amount of acoustic energy available for dissipation.

064.093   A heating mechanism for the chromospheres of M dwarf stars. M. S. Giampapa, L. Golub, R. Rosner, G. S. Vaiana, J. L. Linsky, S. P. Worden.
Smithsonian Astrophys. Obs., Spec. Rep. 392, (see 012.045), p. 73 - 79 (1982).

The systematic, detailed observational and theoretical investigation of the atmospheric structure of the dwarf M stars is especially important to the general field of stellar chromospheres and coronae. More specifically, the M dwarf stars constitute a class of objects for which the discrepancy between the predictions of the acoustic wave chromospheric/coronal heating hypothesis and the observations is most vivid. Conversely, they must therefore represent a class of stars where alternative atmospheric heating mechanisms, presumably magnetically related, are most clearly manifested. The authors propose to ascertain the validity of a recently advanced hypothesis to account for the origin of the chromospheric and transition region line emission in M dwarf stars.

064.094   Momentum and energy balance in late-type stellar winds. K. B. MacGregor.
Smithsonian Astrophys. Obs., Spec. Rep. 392, (see 012.045), p. 83 - 97 (1982).

The author briefly summarizes the observational evidence pertaining to the thermal and dynamical structure of the outer envelopes of cool stars. These results are then compared with the predictions of several theoretical models which have been proposed to account for mass loss (in the form of a stellar wind) from late-type stars.

064.095   Dependence of open stellar coronal regions on coronal heating. R. Hammer.
Smithsonian Astrophys. Obs., Spec. Rep. 392, (see 012.045), p. 121 - 126 (1982).

Models of open regions in hot stellar coronae are presented. For a given star these regions depend on the total amount of coronal heating and on the characteristic length over which this energy is dissipated.

064.096   Predictions of wave-driven wind models. L. Hartmann, E. Avrett.
Smithsonian Astrophys. Obs., Spec. Rep. 392, (see 012.045), p. 127 - 130 (1982).

064.097   Evidence for extended chromospheres surrounding red giant stars. R. E. Stencel.
Smithsonian Astrophys. Obs., Spec. Rep. 392, (see 012.045), p. 137 - 145 (1982).

There is now an increasing amount of both observational evidence and theoretical arguments that regions of partially ionized hydrogen extending several stellar radii are an important feature of red giant and supergiant stars. The purpose of this paper is to summarize this evidence and to examine the implications of the existence of extended chromospheres in terms of the nature of the outer atmospheres of, and mass loss from, cool stars.

064.098   Shock waves, atmospheric structure, and mass loss in Miras. L. A. Willson, J. N. Pierce.
Smithsonian Astrophys. Obs., Spec. Rep. 392, (see 012.045), p. 147 - 151 (1982).

064.099   Predicted magnitudes and colors from cool-star model atmospheres.

H. R. Johnson, T. Y. Steiman-Cameron.
Smithsonian Astrophys. Obs., Spec. Rep. 392, (see 012.045),
p. 239 - 244 (1982).

064.100   *uvbyβ* photometry of visual double stars: a compari-
son with stellar models and isochrones.
E. H. Olsen.
Astron. Astrophys., Vol. 110, 215 - 224 (1982).

For a sample of 14 physical binaries with late type giant
primaries *uvbyβ* photometry is combined with available calculated colours for stellar atmosphere models to derive T_e, g,
[Fe/H], B.C., and M_{bol} for both components. The accuracy and
reliability of the derived quantities are estimated. About one
third of the giant primaries seem to be more metal rich than
their secondaries. Comparisons with available isochrones
indicate that the mixing length parameter should be allowed to
decrease by more than 0.5 as the models evolve from the
ZAMS to the giant branch. Also, the position of the giant
primaries may indicate that mass loss occurs during the subgiant or early giant phase, but this indirect evidence is arguable.

064.101   **On mass loss from stars.**   N. Dallaporta.
Compendium in astronomy, (see 003.012), p. 219 -
231 (1982).
A general overview is drawn concerning the present day
status on the problem of mass loss from stars.

064.102   **Report of IAU Commission 36: Theory of stellar
atmospheres** (Thèorie des atmosphères stellaires).
G. Traving.
Trans. IAU, Vol. XVIIIA, (see 003.013), p. 479 - 497 (1982).

B stars with and without emission lines.
See Abstr. 003.003.

Landolt-Börnstein.   See Abstr. 003.064.

**Programa atlas para el cálculo de modelos de
atmósferas estelares.**   See Abstr. 021.030.

**Improved overlapping helium line profiles for
stellar spectra studies.**   See Abstr. 022.016.

**A model for constructing artificial integrated spec-
tral lines and their Fourier transform properties relevant to the
search for differential rotation of stars.**
See Abstr. 031.517.

Vortex funnels in accretion flows.
See Abstr. 062.001.

**The structure and variability of dynamo driven
accretion discs.**   See Abstr. 062.005.

**Equilibrium configuration of the magnetosphere of
a star loaded with accreted magnetized mass.**
See Abstr. 062.032.

On the formation of protostellar disks.
See Abstr. 062.039.

The stability of radiative shock waves.
See Abstr. 062.084.

**On the expulsion of magnetised plasmas through
solar and stellar atmospheres.**   See Abstr. 062.086.

**The polarization properties of magnetic accretion
columns.**   See Abstr. 063.001.

**Monte Carlo calculations of resonance radiative
transfer through a semi-infinite atmosphere.**
See Abstr. 063.012.

**An alternative derivation of the line transfer equa-
tion of an arbitrarily polarized radiation in the presence of
a magnetic field, in non-LTE.**   See Abstr. 063.036.

**Radiative transfer and model atmospheres of A and
Ap stars.**   See Abstr. 063.043.

**Radiative recombination coefficients at stellar
densities.**   See Abstr. 063.047.

**The helium to heavy element enrichment ratio,
$\Delta Y/\Delta Z$.**   See Abstr. 065.001.

**The evolution of massive stars losing mass and
angular momentum: origin of Wolf-Rayet stars.**
See Abstr. 065.010.

Spherical accretion with e^+-e^-- pair production.
See Abstr. 066.033.

**Thermonuclear processes on accreting neutron
stars: a systematic study.**   See Abstr. 066.513.

**Low-luminosity accretion onto magnetized neutron
stars.**   See Abstr. 066.516.

Chromospheric and coronal heating mechanisms.
See Abstr. 073.148.

**The thermal structure of solar coronal loops and
implications for physical models of coronae.**
See Abstr. 074.005.

Mass loss from massive stars.
See Abstr. 112.039.

**A model of the 10 micrometer silicate feature in the
spectra of BN-like IR point sources.**   See Abstr. 112.056.

**Dust and molecules in the shells of carbon stars —
the inverse greenhouse effect.**   See Abstr. 112.060.

**Radiative transfer in dust clouds — II. Circumstellar
dust shells around early M giants and supergiants.**
See Abstr. 112.061.

HD 115444 — a barium star of extreme Population II.
See Abstr. 114.008.

**Effective temperatures, and radii of luminous O and
B stars: a test for the accuracy of the model atmospheres.**
See Abstr. 114.009.

**On the search for transition zone lines in late A type
stars.**   See Abstr. 114.041.

The temperature of Arcturus.
See Abstr. 114.070.

On the spectrum of the Herbig Be star HD 200775.
See Abstr. 114.156.

NLTE analysis of massive O-stars.
See Abstr. 114.168.

Two lithium-rich supergiants.
See Abstr. 114.177.

Are discrepant asymmetry red giants necessarily hybrid stars?   See Abstr. 114.190.

Considerations regarding the C_2 bands in carbon stars.   See Abstr. 114.192.

Metallic-line stars in Praesepe open cluster: model atmosphere analysis and abundances.   See Abstr. 114.197.

A study of the H-alpha line in late G and K supergiants.   See Abstr. 114.218.

The theoretically expected X-ray luminosity and the binary nature of Wolf-Rayet runaway stars. See Abstr. 117.095.

Theoretical models for T Tauri mass loss. See Abstr. 121.032.

Nonradial pulsations in early-type B stars: g-modes or r-modes?   See Abstr. 122.022.

The hot subdwarfs revisited. See Abstr. 126.010.

Non-LTE analysis of subluminous O-stars. II. The hydrogen-deficient subdwarf O-star HD 127493. See Abstr. 126.019.

Spectral analysis of the OB subdwarf HD 149 382. See Abstr. 126.022.

X-ray and UV radiation from accreting nonmagnetic degenerate dwarfs. II.   See Abstr. 126.025.

Runaway expansion of giant shells driven by radiation pressure from field stars.   See Abstr. 131.036.

On the origin of planetary nebulae. See Abstr. 135.047.

Accretion disk coronae. See Abstr. 142.047.

Cosmic rays and gamma-rays from OB stars. See Abstr. 143.003.

065 Stellar Structure and Evolution

065.001 The helium to heavy element enrichment ratio,
$\Delta Y/\Delta Z$.   C. Chiosi, F. M. Matteucci.
Astron. Astrophys., Vol. 105, 140 - 148 (1982).
  The authors have studied models of chemical evolution of the solar vicinity which take into account occurrence of mass loss from stars, variation of the fraction ζ of mass corresponding to stars more massive than 1 $m_\odot$ with the metal content of gas, and inflow of gas with primordial chemical abundances. They have also studied $\Delta Y/\Delta Z$ over the galactic evolution and conclude that present day stellar evolution data and galactic evolution models can explain the large value of this ratio that has been determined observationally.

065.002 Evolutionary scenarios leading massive stars to WR stars: their mutual importance; the role of mixing.   A. Maeder.
Astron. Astrophys., Vol. 105, 149 - 158 (1982).
  Several scenarios leading to the formation of WR stars from O stars have been catalogued. As main physical processes the author recognizes in addition to binary mass transfer: mass loss at the red supergiant stage, internal mixing, mass loss during main sequence evolution.

065.003 The adiabatic stability of stars containing magnetic fields – V. Effect of Ohmic diffusion on stable fields.   R. J. Tayler.
Mon. Not. R. Astron. Soc., Vol. 198, 811 - 815 (1982).
  In previous papers it has been shown that dynamically stable strong magnetic fields in stellar interiors must have mixed poloidal/toroidal character. It is now shown that such stable fields probably diffuse into unstable configurations in a time which is less than a stellar main-sequence lifetime. This indicates that it is less likely than was previously thought that strong internal fields will survive the main-sequence lifetime of a star.

065.004 Core-overshooting at the He-flash and the colours of horizontal branch stars.
V. Castellani, A. Tornambè.
Mon. Not. R. Astron. Soc., Vol. 198, 861 - 864 (1982).
  Fresh carbon, mixed into H-rich layers by overshooting during the He-flash, can produce rather dramatic variations in the evolution of horizontal branch (HB) stars. It is shown that, under such a hypothesis, the range of surface temperatures covered during HB evolution can be greatly increased, and that a bimodal distribution can be attained.

065.005 More on carbon burning in electron-degenerate matter: within single stars of intermediate mass and within accreting white dwarfs.   I. Iben, Jr.
Astrophys. J., Vol. 253, 248 - 259 (1982).
  Carbon burning in highly electron-degenerate matter is followed in two astrophysically interesting cases: an intermediate-mass star that, on the asymptotic giant branch, has developed a large carbon-oxygen core; and an accreting white dwarf composed primarily of carbon and oxygen. Calculations are continued until heating at the edge of the region within which a dominant Urca pair is assumed to be mixed uniformly makes it impossible to achieve coincidence between this edge and the edge of an associated convectively unstable region within which convection is expected to promote mixing. Thereafter, global heating occurs at a rate that exceeds the rate of cooling by neutrino losses, and it does not appear that the convective Urca process can prevent a runaway toward a dynamic event.

065.006 Polytropic stellar models with a core-dynamo magnetic field.   D. Moss.
Mon. Not. R. Astron. Soc., Vol. 199, 321 - 330 (1982).
  Self-consistent $n = 3$ polytropic models with a steady α-effect dynamo operating in the inner 20 per cent by radius are constructed. These axisymmetric models are a first step towards developing self-consistent models of upper main sequence stars with a dynamo field generated in the convective core and pervading the radiative envelope.

065.007 FG Sagittae – missing link in de sterevolutie.   M. Drummen.
Zenit, 9. Jaarg., 30 - 33 (1982).

065.008 Accreting white dwarf models for Type I supernovae. I. Presupernova evolution and triggering mechanisms.   K. Nomoto.
Astrophys. J., Vol. 253, 798 - 810 (1982).
  The evolution of carbon-oxygen white dwarfs accreting helium in binary systems has been investigated from the onset of accretion up to the point at which a thermonuclear explosion occurs as a plausible explosion model for a Type I supernova. Although the accreted material has been assumed to be helium, the results should also be applicable to the more general case of accretion of hydrogen-rich material, since hydrogen shell burning leads to the development of a helium zone. It is found that the growth of a helium zone on the carbon-oxygen core leads to a supernova explosion which is triggered either by the off-center helium detonation for slow and intermediate accretion rates or by the carbon deflagration for slow and rapid accretion rates. Both helium detonation and carbon deflagration are possible for the case of slow accretion since, in this case, the initial mass of the white dwarf is an important parameter for determining the mode of ignition.

065.009 On the maximum extent of flash-driven convection. II. The core flash.   K. H. Despain.
Astrophys. J., Vol. 253, 811 - 815 (1982).
  It is demonstrated that the maximum extent of the flash-driven convective region during the core helium flash is determined by the condition $\tau_d \approx \tau_e$, where τ_d is the radiative diffusion time from the outer edge of the convective region to the outer envelope and τ_e is the growth time scale for the thermal runaway. This is the same condition that obtains in shell flashes. Numerical experiments with enhanced flashes have been performed in which the convective region approaches but does not penetrate the hydrogen shell though the approach is close enough that overshooting may lead to mixing. The possibility of such enhanced flashes actually occurring is discussed.

065.010 The evolution of massive stars losing mass and angular momentum: origin of Wolf-Rayet stars.
S. R. Sreenivasan, W. J. F. Wilson.
Astrophys. J., Vol. 254, 287 - 296 (1982).
  Evolutionary computations of massive Population I star models of mass ranging from 30 $M_\odot$ to 100 $M_\odot$ are reported. It is assumed that the models are rotating and subject to radiatively driven winds and the effects of nonthermal winds generated by shear turbulence. They are also assumed to spin down as a result of mass loss. When boundary conditions that reflect the changing chemical composition in the surface layers due to mass loss are incorporated into the interior integration of the structure equations, the models are shown to provide a consistent basis for comparison with the observations of Wolf-Rayet stars of the WN as well as WC sequence.

065.011 Structure of the deflagration front of carbon burning in degenerate stellar cores.

L. N. Ivanova, V. S. Imshennik, V. M. Chechetkin.
Pis'ma Astron. Zh., Tom 8, 17 - 25 (1982). In Russian. English
translation in Soviet Astron. Lett., Vol. 8.

The structure of the time-dependent deflagration front
of carbon burning is considered. This structure obtained in a
self-consistent hydrodynamic calculation is typical for thermonuclear burning of a degenerate carbon-oxygen stellar core.
The pertinent estimates lead to the conclusion that the heating
of matter due to weak shocks is important for the ignition of
the thermonuclear reaction in the deflagration burning regime.

065.012 Vibrational stability and critical mass of He stars.
A. Noels, C. Masereel.
Astron. Astrophys., Vol. 105, 293 - 295 (1982).

The critical mass of He stars is determined, using models
constructed with opacity coefficients computed from the
"Astrophysical Opacity Library". It is found to be $16\,M_\odot$ for
models with 2% of heavy elements and $11.5\,M_\odot$ for pure He
models, while the value obtained for pure He models with
electron scattering as the only opacity source still lies near
$9.4\,M_\odot$.

065.013 Stability of toroidal flux tubes in stars.
H. C. Spruit, A. A. van Ballegooijen.
Astron. Astrophys., Vol. 106, 58 - 66 (1982).

The stability of a magnetic flux tube in the equatorial
plane of a star is studied in a thin tube approximation. Only
adiabatic disturbances are considered. The tubes are unstable
to poleward motion. In addition, they are unstable to motions
within the equatorial plane if the superadiabaticity of the
stratification is large enough. The curvature of the tube in a
spherical geometry has a stabilizing effect but it is not strong
enough to stabilize flux tubes in the convective envelopes of
main sequence stars. The authors suggest that the toroidal
fields in a stellar dynamo occur at the interface between the
convection zone and the radiative interior rather than inside
the convection zone itself.

**065.014 Numerical function values for slowly rotating
partially degenerate semirelativistic isothermal
spheroids of arbitrary temperature.**
T. W. Edwards, M. P. Merilan.
Astrophys. J., Suppl. Ser., Vol. 47, 291 - 313 (1981).

Numerical function values are presented for the structural
properties of slowly rotating, partially degenerate, semirelativistic (PD-SR) isothermal spheroids of arbitrary but constant
temperature. Specifically, tabulations of the structural func-
tions $\theta(\xi)$, $\psi_0(\xi)$, $\psi_2(\xi)$, the scaled radius, mass, moment of
inertia, and density, the homology functions $u(\xi)$ and $v(\xi)$, the
expansion, ellipticity, and oblateness, and the rotational perturbations to the mass, density, and moment of inertia are
given for several combinations of the temperature parameter
$\zeta = kT/mc^2$, and central degeneracy parameter $\theta_0 = E_f/kT$.

**065.015 The evolution of massive stars. I. The influence of
mass loss on population I stars.**
W. M. Brunish, J. W. Truran.
Astrophys. J., Vol. 256, 247 - 258 (1982).

The authors have studied the evolution of 15, 30, and
$40\,M_\odot$ models with $Y = 0.28$ and $Z = 0.02$. All models have
been evolved through core helium burning up to the point of
carbon ignition. All models were evolved with and without the
inclusion of mass loss. The authors have used a moderate mass
loss rate with different efficiency factors for main-sequence
and blue supergiant stars. Their results indicate that all
massive stars ignite helium as blue supergiants. Stars with
$M_i \lesssim 30\,M_\odot$ evolve redward very slowly and spend less than
1% of their total lifetimes as red supergiants. Above $30\,M_\odot$,
the rate of redward evolution increases dramatically. The
$40\,M_\odot$ models spend up to 50% of their core helium-burning
lifetimes as red supergiants. The primary effect of mass loss is
to increase the rate of redward evolution during helium

burning for all masses. The effects of mass loss on the internal
structure, lifetimes, and surface abundances of the models are
also discussed.

**065.016 Overshooting from convective cores and the
occurrence of loops in the HRD.**
B. Matraka, C. Wassermann, A. Weigert.
Astron. Astrophys., Vol. 107, 283 - 291 (1982).

The evolution from the zero-age main sequence through
He-burning is calculated for stars of extreme Population I
and masses M between 4 and $8\,M_\odot$.

**065.017 On the possibility of stable quark and pion-condens-
ed stars.** B. Kampfer.
J. Phys. A, Vol. 14, L471 - L475 (1981). – Abstr. in Phys.
Abstr., Vol. 85, Abstr. 14867 (1982).

**065.018 Rotation and luminosity variations in post-main
sequence stars.** P. J. Wiita.
J. Astrophys. Astron., Vol. 2, 387 - 403 (1981).

Previous first-order analytic treatments of rotation
acting upon stellar equilibria are extended to include later,
post-helium burning, stages of stellar evolution. Strong differential rotation is capable of substantially increasing the photon
luminosities of post-main sequence stars, and thus accelerating
their evolution. On the other hand, uniform rotation reduces
the photon flux for a wide range of stellar interior types and
conditions. Similar conclusions are drawn regarding the effects
of rotation on the emission of neutrinos in pre-collapse phases
of evolution.

065.019 Not with a bang but a whimper.
S. Kwok.
Sky Telesc., Vol. 63, 449 - 451 (1982).

065.020 Stars which should not exist.
J. Lequeux.
Recherche, Vol. 12, 991 - 993 (1981). In French. – Abstr. in
Phys. Abstr., Vol. 85, Abstr. 18769 (1982).

**065.021 A comment on some oscillatory models of adiabatic
stars.** H. Knutsen.
Astrophys. Space Sci., Vol. 82, 209 - 212 (1982).

Three different oscillatory models of adiabatic stars are
reinvestigated. These are the homogeneous model, the inverse
square model and the Roche model. The ratio between the
amplitude of the oscillations and the distance from the center
is developed in a power series. For physical conclusions to be
drawn, it turns out to be crucial if the power series is divergent
or convergent. Mathematical arguments are given which show
that the power series are really divergent for all three models.

**065.022 Remarks on the evaluation of critical parameters in
self-gravitating systems.**
C. Bonoli, A. Curir.
Astrophys. Space Sci., Vol. 82, 241 - 246 (1982).

Using a simple model sketched by Thirring, thermodynamical quantities of a system of self-gravitating fermions
are investigated. The onset of a core-halo structure is considered as a phase transition; free energy degeneracy and related
critical parameters are found.

065.023 Active picture of rotation.
H. Ando.
Astron. Astrophys., Vol. 108, 7 - 13 (1982).

The author investigates the interaction between rotation
and wave from the view point of energy flow, and proposes an
active picture of rotation in the rotating stars. The energy
exchange rate between rotation and nonaxisymmetric wave is
estimated and its physical mechanism is discussed in the weakly
differential rotation. The author shows through the estimation
of the time scale that the rotation cycle suggested from the

active picture of rotation may be possible in the real stars, in which the solar "Maunder minimum" is discussed as a possible example.

065.024 **Vibrational instability of a 3000 $M_\odot$ star and the R136a problem.**
P. Ledoux, A. Noels, A. Boury.
Astron. Astrophys., Vol. 108, 49 - 50 (1982).

Recent published IUE observations of the central object of the Tarantula Nebula in LMC seem to support the hypothesis of the existence of a supermassive star. Linear calculations of radial oscillations of a 3000 $M_\odot$ star show an amplification time of 145 yr, very short compared to the characteristic evolutionary time. The escape velocity of the outer layers is reached after 2200 yr.

065.025 **Semiconvection in low-mass main sequence stars.**
R. A. Crowe, R. Mitalas.
Astron. Astrophys., Vol. 108, 55 - 60 (1982).

The paper discusses model calculations carried out by the authors in 1977 which reveal the presence of semiconvection (SC) in low-mass model stars of $M/M_\odot \geqslant 1.2$. The calculations show that the evolutionary tracks of these model stars, subsequent to the exhaustion of central hydrogen, are significantly modified by SC. Since low-mass stars of $1-2\ M_\odot$ are now leaving the hydrogen exhaustion phase in those open clusters older than the Hyades, it is important to take SC into account before comparing theory with observations of such clusters.

065.026 **The evolution of a 1 $M_\odot$ helium star.**
W.-Y. Law.
Astron. Astrophys., Vol. 108, 118 - 126 (1982).

Numerical calculations have been made for the evolution of a 1 $M_\odot$ helium star with initial chemical composition ($Y = 0.979$, $z = 0.021$) from the helium zero-age main sequence to the phase at which it was past its maximum radius.

065.027 **Eine bequeme Methode zur Berechnung stellarer Eigenfrequenzen − Anwendungen auf Sonnen-modelle.**   M. Knölker, M. Stix.
Mitt. Astron. Ges., Nr. 55, p. 138 - 139 (1982).

065.028 **Structure and oscillations of rotating polytropes.**
K. A. Sidorov.
Astrofizika, Tom 17, 783 - 797 (1981). In Russian. English translation in Astrophysics, Vol. 17, No. 4,

Chandrasekhar's first-order theory of rotationally distorted polytropes is reconsidered with a view to construct the sequence of rotating polytropes with fixed mass. It is found that the value of n at which dynamical instability sets in is reduced from $n = 3$ by rotation. A simple formula for the frequency of the pulsation mode of the rotating star is derived.

065.029 *R*-mode oscillations in uniformly rotating stars.
H. Saio.
Astrophys. J., Vol. 256, 717 - 735 (1982).

R-mode (or quasi-toroidal-mode) oscillations in uniformly and slowly rotating stars have been investigated. Perturbations of the gravitational potential have been included. Applying an adiabatic analysis to a polytrope of index $n = 3$, we confirm that gravitaitonal potential perturbations are small for r modes in most cases. A nonadiabatic analysis of oscillations with $l = |m|$ is applied to the models of massive zero-age main-sequence stars and ZZ Ceti type variable stars. All the modes examined for the massive zero-age main-sequence stars are stable whereas r modes in the ZZ Ceti model appear to be overstable. However, the neglect of coupling between oscillations and convection might be import for the latter stars because regions of excitation coincide with convection zones.

065.030 **The helium-core flash in globular-cluster stars.**
J. G. Mengel, A. V. Sweigart.
Astrophysical parameters for globular clusters, (see 012.023), p. 277 - 288 (1981).

The purpose of this paper is to study the characteristics of an off-center helium flash in a typical globular-cluster star. The authors adopt the standard assumptions of spherical symmetry and adiabatic convection and neglect any hydrodynamical effects.

065.031 **Hydrodynamic core helium flashes.**
P. W. Cole, R. G. Deupree.
Astrophysical parameters for globular clusters, (see 012.023), p. 293 - 299 (1981).

065.032 **Models for horizontal-branch stars with cores enriched in carbon and oxygen.**   P. Demarque.
Astrophysical parameters for globular clusters, (see 012.023), p. 301 - 308 (1981).

Evolutionary tracks have been constructed from the ZAHB for stars with helium cores enriched in carbon and oxygen. The C-O enriched models have the advantage over standard models of shortening the lifetime of HB evolution, thus bringing the helium abundance estimated by the R-method into better agreement with the helium determination from pulsation theory. However, the models seem to create new problems in the interpretation of observed HB morphologies due to the near disappearance of blueward loops in the evolutionary tracks. In particular, the observed period distributions of RR Lyrae variables in globular clusters become difficult to understand.

065.033 **A comparison of stellar models constructed with different stellar evolution programs.**
P. Demarque, J. B. Laird, D. A. VandenBerg.
Astrophysical parameters for globular clusters, (see 012.023), p. 319 - 324 (1981).

The purpose of this paper is to examine the sensitivity of the position of stellar models in the theoretical HR diagram as a function of uncertainties in (1) the physical assumptions and (2) the numerical approximations used. In particular, the authors intend to isolate the causes of differences in models constructed with different Henyey evolution programs and to establish some standard models.

065.034 **Helium shell flashes and evolution of accreting white dwarfs.**   M. Y. Fujimoto, D. Sugimoto.
Astrophys. J., Vol. 257, 291 - 302 (1982).

In a close binary system or in a dense cloud, gas may be accreted onto a carbon-oxygen white dwarf and will be processed into helium by hydrogen burning in an accreted envelope. As a result, a helium zone grows in mass, and a helium shell flash takes place just as in cores of red giant stars. Properties of such helium shell flashes are investigated both by a generalized theory of shell flash and by numerical computations. Conditions leading to such shell flashes are studied by computing the thermal evolution of the white dwarfs during the accretion. There exists a strong tendency for the strength to be determined mainly by the accretion rate: For fast accretion, the shell flashes are weak and triggered recurrently, while for the slow accretion, the helium shell flash, once triggered, develops into a detonation supernova.

065.035 **Diffusion and hydrogen shell burning on slowly accreting white dwarfs.**
M. Y. Fujimoto, J. W. Truran.
Astrophys. J., Vol. 257, 303 - 311 (1982).

The stability of hydrogen shell burning in accreted envelopes on white dwarfs is examined for the domain of low accretion rates. Account is taken of the effects of diffusion of CNO nuclei attributable to sedimentation associated with gravity and temperature gradients. The stable steady state for hydrogen shell burning is realized only when gas is accreted onto a hot white dwarf. For accretion onto cool white dwarfs, the stable regime is bypassed by recurrent shell flashes.

065.036 On the role of the accretion rate in nova outbursts.
D. Prialnik, M. Livio, G. Shaviv, A. Kovetz.
Astrophys. J., Vol. 257, 312 - 317 (1982).

The authors investigate the role of the accretion rate in determining the outcome of the accretion of hydrogen-rich material onto a 1.25 $M_\odot$ C/O white dwarf. It is found that there is an upper limit to the mass accretion rate that leads to a nova outburst. The upper limit is between 10^{-8} and $10^{-9} M_\odot$ yr^{-1} for a 1.25 $M_\odot$ white dwarf. Below this upper limit, the outburst characteristics are very similar for accretion rates down to $10^{-11} M_\odot$ yr^{-1}.

065.037 Models of stellar evolution and their use in calibrating distances and element abundances of stars.
T. Gehren.
Astron. Astrophys., Vol. 109, 187 - 190 (1982).

Recently published grids of stellar interior models fail to account for the temperature and luminosity of the Sun with reasonable values of mass, age and chemical composition. It is shown how this discrepancy may affect the determination of element abundances and distances of field stars and clusters.

065.038 Supernova theory.
G. E. Brown, H. A. Bethe, G. Baym.
Nucl. Phys. A, Vol. A375, 481 - 532 (1982). − Abstr. in Phys. Abstr., Vol. 85, Abstr. 53302 (1982).

065.039 The evolution of rapidly rotating B/Be stars.
A. S. Endal.
Be stars,(see 012.030), p. 299 - 302 (1982).

Rotation can significantly change the moment-of-inertia of a main sequence star. As a result, the ZAMS rotation rate need only be within ~30% of the critical value in order to reach critical rotation during the hydrogen burning stage. Calculations of the evolution of rotating stars show that the Be stars result from a normal (Maxwellian) distribution of B-star rotation velocities.

065.040 Multimode stellar pulsations. III. Resonances.
O. Regev, J. R. Buchler, M. Barranco.
Astrophys. J., Vol. 257, 715 - 723 (1982).

The authors include the treatment of resonances in the formalism of multitime asymptotic techniques previously devised to treat stellar pulsation (monomode and multimode) with a slow evolution in the absence of resonances. The evolution can be followed toward and through a resonance, and the effects of resonances can be investigated. The technique is illustrated by a simple example, and various aspects which may be relevant to stellar pulsation are discussed.

065.041 A theory of hydrogen shell flashes on accreting white dwarfs. I. Their progress and the expansion of the envelope.
M. Y. Fujimoto.
Astrophys. J., Vol. 257, 752 - 766 (1982).

By applying an analytical solution to the envelope, hydrogen shell flashes on accreting white dwarfs are computed semianalytically from their ignition to their final stage. When the mass of the white dwarf, M, and the mass of the hydrogen-rich envelope, ΔM_1, are specified, their progress is determined uniquely. Structural change and the resultant expansion of the envelope due to the shell flash depend mainly on the weight of overlying layer $P_1{}^*$ or the product of the column mass above the burning shell and the surface gravity of the white dwarf. When $P_1{}^*$ is low, the shell flash is weak, and the envelope settles in thermal equilibrium even before it expands to a solar radius. When $P_1{}^*$ is high enough, on the other hand, the structural change is so violent that the expansion is accelerated beyond the escape velocity and leads directly to a nova

explosion. These results indicate that nova explosions are phenomena which are related to the slow accretion onto massive white dwarfs. Even for the optimal case of the most massive white dwarfs, the accretion rate should be lower than $10^{-9} M_\odot$ yr^{-1}.

065.042 A theory of hydrogen shell flashes on accreting white dwarfs. II. The stable shell burning and the recurrence period of shell flashes.
M. Y. Fujimoto.
Astrophys. J., Vol. 257, 767 - 779 (1982).

By means of analytical solutions of the envelope, thermal properties of hydrogen shell burnings on accreting white dwarfs are studied and a general picture for their progress is presented which is described by two parameters, the accretion rate and the mass of the white dwarf. On a white dwarf, the thermal behavior of gas in the burning shell depends on the configuration of the envelope, which gives birth to two distinct types of stable configurations in thermal equilibrium, a high and a low state. In the high state, the nuclear shell burning makes up for the energy loss from the surface. In the low state, the nuclear burning is extinct, and the compressional heating by accreted gas balances the cooling through diffusion of heat. The shell flash is a phenomenon associated with the transition from the low to the high state driven by nuclear burning. This transition is a result of the increase in the envelope mass by the accretion. Conditions for the occurrence of recurrent hydrogen shell flashes are analysed and a recurrence period is estimated. The period may range from less than a year for a very massive white dwarf to several ten thousand years for low-mass white dwarfs.

065.043 Stellar core collapse: II. Inner core bounce and shock propagation.
K. A. Van Riper.
Astrophys. J., Vol. 257, 793 - 820 (1982).

The evolution of a collapsing core of a 15 $M_\odot$ star is followed, by numerical simulation, from a central density of 10^{14} g cm^{-3}, through homologous core bounce, until the outward motion of the shock front ceases a few milliseconds after bounce. The models use the equation of state of Lamb, Lattimer, Pethick, and Ravenhall below nuclear density and general relativistic hydrodynamics. The transport of neutrinos is treated by a leakage scheme. Homologous cores bounce at twice nuclear density. A shock wave forms at the surface of the bouncing core, and propagates outward in space until the postshock density is too low to trap the electron neutrinos, which are copiously produced by captures on the free protons in the dissociated matter behind the shock. These neutrinos carry away a large fraction of the internal energy gained by matter passing through the shock, leading to severe weakening of the shock. Several models are considered, showing that the final shock weakening occurs regardless of the neutrino trapping density and the equation of state above nuclear density.

065.044 Structure and evolution of massive stars.
C. Chiosi.
The most massive stars, (see 012.034), p. 27 - 48 (1981).

In this review the current knowledge of the observational properties and theoretical models of massive stars are outlined. Observational H−R diagrams are compared with theoretical models and the effects of mass loss on the evolution of these stars are discussed. The present picture of formation and evolution of Wolf-Rayet stars is briefly described.

065.045 Wolf-Rayet stars as a stage of the evolution of massive stars.
A. Maeder.
The most massive stars, (see 012.034), p. 173 - 189 (1981).

065.046 Effect of overshooting on theoretical yields.
F. Matteucci.
The most massive stars, (see 012.034), p. 293 - 296 (1981).

In current models of galactic evolution heavy element

enrichment takes place in massive stars (M $\geqslant$ 10 M$_\odot$). In this paper the effects of overshooting from convective cores and of mass loss by stellar winds on the production rate of He and heavy elements are discussed.

065.047   **Evolution and nucleosynthesis of primordial massive stars.**   M. F. El Eid, K. J. Fricke, W. W. Ober.
The most massive stars, (see 012.034), p. 303 - 314 (1981).

In recent years many arguments have been gathered in favour of the existence of a pregalactic generation of very massive stars. In this paper the authors report on evolution and nucleosynthesis calculations for massive primordial stars having no metals initially in the mass range 80 < M/M$_\odot$ < 500. The aim of this work is to determine the mass range in which the evolution terminates explosively and to assess the kind of enrichment caused by the stars in this mass range.

065.048   **Properties and cosmological consequences of very massive objects.**
B. J. Carr, J. R. Bond, W. D. Arnett.
The most massive stars, (see 012.034), p. 315 - 337 (1981).

A very massive object (VMO) is defined to be a star that goes pair unstable during its oxygen burning stage. This implies that it must have an initial mass in the range $10^2 - 10^4$ M$_\odot$. In the first part of this paper the evolution of these objects is outlined. In particular, it is explained why oxygen cores above a critical mass M$_c$ of order 100 M$_\odot$ must ultimately collapse into a black hole, whereas objects below this mass will explode. In the second part of the paper the cosmological consequences of pregalactic VMO's are discussed. It is demonstrated that they may play a crucial role in the problem of the "missing mass", in questions of element abundances originating from primordial nucleosynthesis and in explaining the 3K microwave background radiation.

065.049   **Neutrino damping of the reflected shock from stellar core collapse.**   K. A. Van Riper.
Bull. American Astron. Soc., Vol. 13, 791 (1981). − Abstract.

065.050   **Numerical study of the effects of ambipolar diffusion in collapsing magnetic protostars.**
D. C. Black, E. H. Scott.
Bull. American Astron. Soc., Vol. 13, 791 (1981). − Abstract.

065.051   **Gravitational energy release and mixing induced by the nuclear energy generation process.**
C. A. Rouse.
Bull. American Astron. Soc., Vol. 13, 791 (1981). − Abstract.

065.052   **Numerical studies of the non-spherical core carbon flash.**   E. Müller, W. D. Arnett.
Bull. American Astron. Soc., Vol. 13, 791 (1981). − Abstract.

065.053   **Finite propagation time in multi-dimensional thermonuclear runaways.**
B. A. Fryxell, S. E. Woosley.
Bull. American Astron. Soc., Vol. 13, 792 (1981). − Abstract.

065.054   **General solution to the Oppenheimer-Volkoff equation compatible with the second law of thermodynamics.**   D. W. Marks.
Bull. American Astron. Soc., Vol. 13, 792 (1981). − Abstract.

065.055   **Shell flashes as non-linear oscillations of a dissipative open system.**
S. Miyaji, M. Yasutomi, D. Sugimoto.
Bull. American Astron. Soc., Vol. 13, 792 (1981). − Abstract.

065.056   **R-mode oscillations in uniformly rotating stars.**
H. Saio.

Bull. American Astron. Soc., Vol. 13, 801 - 802 (1981). Abstract.

065.057   **The pulsational stability of massive stars and 30 Doradus.**   D. S. King.
Bull. American Astron. Soc., Vol. 13, 810 (1981). − Abstract.

065.058   **The collapse of low-mass stars.**
M. M. Basko, M. A. Rudzskij, Z. F. Seidov.
Astrofizika, Tom 16, 321 - 335 (1980). In Russian. English translation in Astrophysics, Vol. 16, No. 2.

The stability loss due to non-equilibrium beta-processes and the subsequent initial phase of the gravitational collapse has been calculated for two stellar pure-iron cores with masses 1.19 $M_\odot$ and 1.21 $M_\odot$.

065.059   **Rotating hot superdense stars.**   V. Balek.
Astrofizika, Tom 17, 349 - 358 (1981). In Russian. English translation in Astrophysics, Vol. 17, No. 2.

The deformation energy release due to the rotation of a superdense star is discussed. The energy of deformation of a neutron star is the main contribution in total energy during several hundred years. When the matter of neutron star contains π-condensate the corresponding time is approximately 10^4 years. In the latter case the thermal radiation of the star is practically unobservable during the most part of this time. The luminosity of white dwarfs, owing to the release of their deformation energy, is of the order of 0.08 $\tau_9^{-1}L_\odot$ without taking into account neutrino escape; τ_9 is the time measured in 10^9 years units during which the angular velocity of the white dwarf is close to its maximum value.

065.060   **The distribution law of stellar p-modes.**
J. Perdang.
Astrophys. Space Sci., Vol. 83, 311 - 333 (1982).

In this paper the author derives that the density of acoustic frequencies of a star obeys the asymptotic inequality g $(\omega) \geqslant K\omega^2$, where K is a model constant given in terms of an average sound velocity. If the local sound velocity does not vanish at the surface of the star, the inequality becomes an equality.

065.061   **Nonadiabatic stellar pulsation with slow, uniform rotation.**   B. W. Carroll, C. J. Hansen.
Bull. American Astron. Soc., Vol. 14, 577 (1982). − Abstract.

065.062   **Meridional circulation versus diffusion in stellar envelopes.**   G. Michaud.
Astrophys. J., Vol. 258, 349 - 353 (1982).

Tassoul and Tassoul have recently obtained self-consistent solutions for meridional circulation throughout chemically homogeneous stars. The meridional circulation velocities they determined are here compared to diffusion velocities of helium below the He II convection zone. It is shown that the He II convection zone can disappear only in stars with equatorial rotational velocities smaller than 90 km s^{-1}. This is in agreement with the cutoff velocity observed for the HgMn stars. This maximum velocity for the disappearance of the He II convection zone depends sensitively on gravity so that there is not one equatorial rotational velocity below which all stars should become HgMn. It depends on the evolutionary status and on how the star was slowed down. On the other hand, the rotational motion is turbulent, and the effect this turbulence has on chemical separation is poorly known.

065.063   **The effect of a magnetic field on the adiabatic oscillation of convective stellar models with radiation pressure.**   M. K. Das, J. Kar, J. N. Tandon.
Astrophys. J., Vol. 258, 354 - 366 (1982).

Using an unrestricted second order virial tensor analysis, the characteristic frequencies of small adiabatic oscillations of a convective stellar model in the presence of a weak poloidal magnetic field, with significant radiation pressure, have been calculated. In the absence of the magnetic field, the effect of increase in radiation pressure is to decrease the frequencies of radial modes, in spite of the increase in central condensation of the system. The characteristic frequencies of the transverse shear mode, the toroidal mode, and the pulsation modes show a general increase with the increase in the field strength. The marginal stability of the configuration in the presence of the magnetic field has also been discussed.

065.064 Physical processes in stars on late stages of their evolution.   G. S. Bisnovatyj-Kogan.
Astron. Nachr., Band 303, 131 - 137 (1982).
    A review of the physical processes on late stages of stellar evolution and in neutron stars is given. Different types of accretion into black holes are considered. The models of γ-ray bursts and Cyg X-1 are discussed briefly.

065.065 An exploding 10 $M_\odot$ star: a model for the Crab supernova.   W. Hillebrandt.
Astron. Astrophys., Vol. 110, L3 - L6 (1982).
    The gravitational collapse of the inner core of a 10 $M_\odot$ star at the end of its thermonuclear evolution is computed. Although the model presented may not explain the average type II supernova, its properties are consistent with those of Crab-like events.

065.066 Expected broadband linear polarization from cool stars with magnetic structures.
E. Landi Degl'Innocenti.
Astron. Astrophys., Vol. 110, 25 - 29 (1982).
    A theoretical investigation is presented of the broadband linear polarization that is expected to arise from cool stars with magnetic structures through the mechanism of magnetic intensification. Some polarization diagrams are presented and the concept of stochastic mean is introduced for the linear polarization to be expected from a random distribution of magnetic regions in two activity belts equidistant from the stellar equator. The relevance of linear polarization observations as a diagnostic tool to deduce the magnetic configuration of the star is discussed in some detail.

065.067 A comparison of B and Be stars kinematics.
D. Briot.
Scientific aspects of the Hipparcos space astrometry mission, (see 012.041), p. 89 - 90 (1982).
    The evolutionary stage of "classical" Be stars as compared with B stars without emission is not yet explained in a satisfactory way. Some of the hypotheses which exist about this subject are reviewed. A kinematical compared study of B and Be stars from data given by Hipparcos is expected to provide some information about this subject.

065.068 Contracting members of double stars.
P. Lindroos.
Messenger, No. 27, p. 4 - 7 (1982).

065.069 Turbulent convection and pulsational stability of variable stars. Nonpulsating stars in the cepheid strip.
D. Xiong.
Sci. Sinica, Ser. A, Vol. 25, 295 - 301 (1982).
    The time-dependent convection theory is applied to study the cepheids. The growth rate for small amplitude pulsations has been investigated in nonadiabatic linear approximation. The results show that a star within the cepheid strip is pulsationally stable when its luminosity is lower than a critical value depending on stellar mass.

065.070 The evolution of the physical characteristics of theoretical stellar models with variable G (Brans-
Dicke cosmological theory) in isochrones of one, three and five billion years.   A. D. Pinotsis, P. G. Laskarides.
Bull. Astron. Inst. Czechoslovakia, Vol. 33, 180 - 187 (1982).
    Theoretical isochrones with variable G for one, three and five billion years are calculated under the assumption of the Brans-Dicke cosmological theory. The evolution of the central and surface characteristics of the individual mass models is studied for each isochrone and compared to the evolution of the characteristics of the models in the standard (G = constant) case. This comparison helps in understanding stellar evolution with variable G under any cosmological theory with constant mass of the model.

065.071 Convective dynamos for rotating stars.
P. A. Gilman.
Smithsonian Astrophys. Obs., Spec. Rep. 392, (see 012.045), p. 165 - 179 (1982).

065.072 Report of IAU Commission 35: Stellar constitution (Constitution des étoiles).   R. J. Tayler.
Trans. IAU, Vol. XVIIIA, (see 003.013), p. 457 - 478 (1982).

    Landolt-Börnstein.   See Abstr. 003.064.

    Symposium "Origin and evolution of celestial bodies", Leningrad, 1981, May 20 - 21 on the occasion of the 100th anniversary of the Astronomical Observatory of the Leningrad University.   See Abstr. 011.012.

    All-Union conference "Star clusters and problems of stellar evolution". Sverdlovsk, 1981, April 13 - 17.
See Abstr. 011.013,

    Stellar weak interaction rates for intermediate-mass nuclei. II. A = 21 to A = 60.   See Abstr. 061.002.

    Neutron shell blocking of electron capture during gravitational collapse.   See Abstr. 061.003.

    Transport properties of degenerate neutrinos in dense matter.   See Abstr. 061.004.

    Mixed lattice phases in cold dense matter.
See Abstr. 061.005.

    Lepton number violation, Majorana neutrinos, and supernovae.   See Abstr. 061.007.

    Massive neutrinos in strong gravitational fields.
See Abstr. 061.014.

    Resonant reaction rates of $O^{16}(\alpha, \gamma)Ne^{20}$ affecting screening effect.   See Abstr. 061.027.

    Massive neutrinos and the stellar stopping power via the neutrino magnetic moment.   See Abstr. 061.028.

    Spectrum of neutrino radiation of collapsing degenerate stellar cores. Monte-Carlo calculations.
See Abstr. 061.036.

    Stellar weak interaction rates for intermediate mass nuclei. III. Rate tables for the free nucleons and nuclei with A = 21 to A = 60.   See Abstr. 061.038.

    s-process studies in the light of new experimental cross sections: distribution of neutron fluences and r-process residuals.   See Abstr. 061.045.

    Unstable poloidal magnetic fields in stars.
See Abstr. 062.075.

Radiation transfer in stellar interiors.
See Abstr. 063.057.

Differential rotation, magnetic activity and X-ray emission of late type giants.   See Abstr. 064.004.

uvbyβ photometry of visual double stars: a comparison with stellar models and isochrones.
See Abstr. 064.100.

Critical-mass limits for collapsed stars in a generalized theory of gravitation.   See Abstr. 066.050.

The solar and stellar convection zones.
See Abstr. 080.019.

The inner structure of the sun and of solar-type stars.   See Abstr. 080.029.

Are there more than nine planets in the universe? Is the theory of stellar evolution wrong?
See Abstr. 107.023.

CNO in halo stars.   See Abstr. 114.091.

The chemical compositions and evolutionary state of the subgiant CH stars.   See Abstr. 114.174.

Physical properties of B stars.
See Abstr. 115.038.

The evolution of highly compact binary stellar systems.   See Abstr. 117.020.

The velocity-mass correlation of the O-type stars: model results.   See Abstr. 117.024.

Detached → contact scenario for the origin of W UMa stars.   See Abstr. 117.061.

On the stability and evolution of contact binaries. I.
See Abstr. 117.062.

On the stability of age-zero contact binaries. II.
See Abstr. 117.064.

Evolution of massive close binary systems.
See Abstr. 117.094.

The theoretically expected X-ray luminosity and the binary nature of Wolf-Rayet runaway stars.
See Abstr. 117.095.

The precession of gaseous stars.
See Abstr. 117.138.

Opacity and nonlinear effects on theoretical BL Herculis models.   See Abstr. 122.012.

Nonlinear models of classical Cepheids endowed with tangled magnetic fields.   See Abstr. 122.028.

On a possible connection between mass loss and instability.   See Abstr. 122.235.

Instability mechanisms in Beta Cephei stars.
See Abstr. 122.236.

Nonradial pulsations of Upsilon Orionis and A supergiants.   See Abstr. 122.238.

Linear and nonlinear theory study of Alpha Virginis.
See Abstr. 122.239.

Type I supernovae. I. Analytic solutions for the early part of the light curve.   See Abstr. 125.004.

Accreting white dwarf models for type I supernovae. II. Off-center detonation supernovae.   See Abstr. 125.035.

Steady nuclear burning on white dwarfs.
See Abstr. 126.001.

Hydrogen-driving and the blue edge of compositionally stratified ZZ Ceti star models.
See Abstr. 126.004.

On the surface compositions of magnetic white dwarfs.   See Abstr. 126.008.

The hot subdwarfs revisited.
See Abstr. 126.010.

Das Massenspektrum bei der Sternentstehung.
See Abstr. 131.086.

From red giants to planetary nebulae.
See Abstr. 135.043.

Two component model of initial mass function.
See Abstr. 151.049.

Comments on the origin of the carbon and nitrogen variations within NGC 6752 and 47 Tucanae.
See Abstr. 154.006.

An evolutionary upper limit for the effective temperatures of horizontal-branch stars and the occurrence of gaps in their observed distribution.   See Abstr. 154.048.

Ages of the oldest star clusters from new synthetic color-magnitude diagrams.   See Abstr. 154.051.

CNO isotopes and galactic chemical evolution.
See Abstr. 155.005.

The distribution of WR and supergiant stars on the galactic plane.   See Abstr. 155.029.

Metallicity and the N_{RSG}/N_G ratio.
See Abstr. 155.030.

Pregalactic stars: precursors to galaxy formation.
See Abstr. 162.160.

Erratum

065.901   Erratum: 'A study of simple polytropes. I. Fundamentals and classification of solutions' [Publ. Astron. Soc. Japan, Vol. 33, 273 - 298 (1981)].   H. Kimura. Publ. Astron. Soc. Japan, Vol. 33, 749 (1981). — See Abstr. 30.065.019.

066 Relativistic Astrophysics, Gravitation Theory, Background Radiation, Black Holes, Neutron Stars

066.001 Electrodynamics in curved spacetime: 3 + 1 formulation. K. S. Thorne, D. Macdonald.
Mon. Not. R. Astron. Soc., Vol. 198, 339 - 343, Microfiche MN 198/1 (1982).

This paper re-expresses the equations of curved-spacetime electrodynamics in terms of a 3 + 1 (space + time) split, in which the key quantities are three-dimensional vectors (electric field E, magnetic field B, etc.) that lie in hypersurfaces of constant time t. After developing the 3 + 1 formalism for general spacetimes, the paper specializes to the spacetime outside a stationary but rotating black hole. The Znajek-Damour boundary conditions at the hole's horizon are re-expressed in 3 + 1 language.

066.002 Black-hole electrodynamics: an absolute-space/ universal-time formulation.
D. Macdonald, K. S. Thorne.
Mon. Not. R. Astron. Soc., Vol. 198, 345 - 382 (1982).

This paper reformulates and extends the Blandford-Znajek theory of a stationary, axisymmetric magnetosphere anchored in a black hole and in its accretion disc. Such a magnetosphere should transfer much of the rotational energy of the hole and orbital energy of the disc into an intense flux of electromagnetic energy — which in turn might be the energizer for quasars and active galactic nuclei. The reformulation is done by replacing the relativist's 'unified spacetime' viewpoint with an equivalent Galilean-type 'absolute-space-plus-universal-time' viewpoint, and by replacing the electromagnetic field tensor $F_{\mu\nu}$ with electric and magnetic fields E and B that reside in the absolute space outside the black hole.

066.003 Stationary spherical symmetric accretion onto massive black holes: the radiation spectrum and luminosity. R. Z. Yahel.
Astrophys. J., Vol. 252, 356 - 374 (1982).

The author derives the hydrodynamic equations of a nonmagnetic fluid in a static, Schwarzschild field taking into account electron-positron creation and annihilation. He solves these equations coupled to the radiation intensity equation and obtains consistent solutions that describe the infalling plasma. He finds that the efficiency of conversion of gravitational energy into radiation energy can be a few percent provided the mass of the central object is greater than $10^7 M_\odot$. The spectrum of the emergent radiation is roughly described by a power law with index $\gtrsim 1$ in the hard X-ray range ($E \lesssim 0.1$ MeV) and is found to decrease exponentially for $E > E_0$, where E_0 depends on the photon-photon optical depth of the cloud. The relevance of the models to the recent observations of the quasars 3C 273 and QSO 0241 + 622 and the Seyfert galaxy NGC 4151 is discussed.

066.004 The effect of multiple Compton scattering on the temperature and emission spectra of accreting black holes. L. Maraschi, R. Roasio, A. Treves.
Astrophys. J., Vol. 253, 312 - 317 (1982).

Spherical accretion onto a black hole with magnetic field dissipation is considered. For an optical depth close to 1, multiple Compton scattering is the dominant cooling mechanism near the hole. The effect is introduced in the energy balance equation using at each shell the spectral index, α, of the Comptonized emission for homogeneous spheres with optical depths equivalent to that of the shell. In the inner region of the flow, the temperature obtained with this procedure differs significantly from that computed without taking into accout Comptonization losses. In the cases examined the resulting spectra are nearly power laws over a wide frequency range with $1 \lesssim \alpha \lesssim 0.5$. The spectral indexes have a much weaker dependence on optical depth than in the case of sources of fixed temperature.

066.005 Does the galactic synchrotron radio background originate in old supernova remnants? S. Sarkar.
Mon. Not. R. Astron. Soc., Vol. 199, 97 - 108 (1982).

Observations of the galactic synchrotron radio background indicate that the emission arises in localized regions of high emissivity. Various lines of evidence suggest that these are the radiative shells of old supernova remnants in which the synchrotron emissivity is enhanced due to compression of the interstellar magnetic field along with the correlated increase in the energy density of cosmic ray electrons. The intensity of the background can then be understood without needing to evoke higher magnetic fields or cosmic ray electron fluxes. Other explanations for the origin of the background are briefly discussed. This result casts doubt on the existence of large-scale cosmic ray gradients inferred from galactic gamma ray observations.

066.006 Los agujeros negros y sus implicancias astrofísicas. Parte II. D. L. Block.
Rev. Astron., Tomo 53, No. 219, p. 2 - 5 (1981).

066.007 Gravitational lens formulae in the background of Robertson-Walker metric. C.-m. Xu, Y.-g. Sheng.
Acta Astrophys. Sinica, Vol. 2, 8 - 18 (1982). In Chinese.

066.008 Dalle nane bianche ai buchi neri.
R. Balbinot, R. Bergamini, B. Giorgini.
Coelum, Vol. 50, 11 - 30, 71 - 76 (1982).

066.009 Relativistic thermal plasmas: pair processes and equilibria. A. P. Lightman.
Astrophys. J., Vol. 253, 842 - 858 (1982).

The author investigates the equilibria of relativistic, thermal plasmas, taking into account electron-positron creation and annihilation and photons produced within the plasma. By including pair-producing photon processes and effects due to the finite size of the plasma, he extends and generalizes the earlier work of Bisnovatyi-Kogan, Zel'dovich and Sunyaev.

066.010 A new test of general relativity: gravitational radiation and the binary pulsar PSR 1913+16.
J. H. Taylor, J. M. Weisberg.
Astrophys. J., Vol. 253, 908 - 920 (1982).

Observations of pulse arrival times from the binary pulsar PSR 1913+16 between 1974 September and 1981 March are now sufficient to yield a solution for the component masses and the absolute size of the orbit. The authors find the total mass to be almost equally distributed between the pulsar and its unseen companion, with $m_p = 1.42 \pm 0.06\, M_\odot$ and $m_c = 1.41 \pm 0.06\, M_\odot$. These values are used, together with the well determined orbital period and eccentricity, to calculate the rate at which the orbital period should decay as energy is lost from the system via gravitational radiation. According to the general relativistic quadrupole formula, one should expect for the PSR 1913+16 system an orbital period derivative $\dot{P}_b = (-2.403 \pm 0.005) \times 10^{-12}$. The observations yield the measured value $\dot{P}_b = (-2.30 \pm 0.22) \times 10^{-12}$. The excellent agreement provides compelling evidence for the existence of gravitational radiation, as well as a new and profound confirmation of the general theory of relativity.

066.011 Parity nonconservation and the origin of cosmic magnetic fields.   A. Vilenkin, D. A. Leahy.
Astrophys. J., Vol. 254, 77 - 81 (1982).

Three mechanisms of cosmic magnetic field generation are discussed: (1) asymmetric decay of particles emitted by rotating black holes; (2) asymmetric proton emission by black holes due to weak radiative corrections, and (3) equilibrium parity-violating currents. It is shown that all three mechanisms can produce a seed field sufficiently strong to account for the present galactic fields.

066.012 Orbital perturbations of a gravitationally bound two-body system with the passage of gravitational waves.   L. A. Nelson, W. Y. Chau.
Astrophys. J., Vol. 254, 735 - 747 (1982).

The perturbed orbital equations governing the behavior of a nonrelativistic, elliptical, self-gravitating binary system with the impinging of weak, monochromatic gravitational plane waves are studied with solutions for the radial deviation expressed in terms of the observable, "coordinate" angle ϕ. Explicit expressions for the components of the perturbing force in terms of the Eulerian angles are also presented in the case of oblique incidence. Complete analytical solutions, however, are obtained only for (a) normal incidence and small eccentricity and (b) the z-component of a circular orbit at oblique incidence, and are compared with those of other authors.

066.013 Vorticity-free rings orbiting black holes. I. The metric.   M. A. Abramowicz.
Astrophys. J., Vol. 254, 748 - 754 (1982).

The author explicitly integrates one of the Einstein field equations describing the stationary and axially symmetric spacetime, whose source is a perfect-fluid, vorticity-free ring of matter orbiting a black hole. This reduces the number of unknown metric functions to three.

066.014 Processes in relativistic plasmas.   R. J. Gould.
Astrophys. J., Vol. 254, 755 - 766 (1982).

The problem of the establishment and maintenance of a Boltzmann distribution in particle kinetic energies is discussed for a highly relativistic plasma. It is shown that thermalization of the electron gas by binary collisions (Møller scattering) is not sufficiently effective to maintain the equilibrium distribution when other processes are considered which perturb the equilibrium. These processes include bremsstrahlung losses and electron-positron pair production in electron-electron and electron-ion collisions. Perturbations of the Boltzmann distribution by synchrotron radiation are also evaluated. Thermalization by means of processes like interaction with plasma waves is discussed briefly. The opacity of the relativistic plasma is computed for Compton scattering, pair production in the fields of electrons and ions, bremsstrahlung, and synchrotron self-absorption.

066.015 Cepstral analysis of interfering delay signals as applied to detection of gravitational lenses.
Eh. L. Afrajmovich.
Pis'ma Astron. Zh., Tom 8, 136 - 138 (1982). In Russian. English translation in Soviet Astron. Lett., Vol. 8.

It is shown that gravitational lenses can be detected from measurement of the difference in the time of emission propagation from source to observer along different paths even if these paths cannot be angularly resolved. With this purpose it is suggested to carry out cepstral analysis of time variations of intensity of single sources of variable emission simultaneously in optical and radio waves.

066.016 Quantum conformal fluctuations in a singular spacetime.   T. Padmanabhan, J. V. Narlikar.
Nature, Vol. 295, 677 - 678 (1982).

066.017 Pancake detonation of stars by black holes in galactic nuclei.   B. Carter, J. P. Luminet.
Nature, Vol. 296, 211 - 214 (1982).

For black holes in the mass range 10^4 - $10^7 M_\odot$, individual stars penetrating well inside the Roche radius will undergo compression to a short-lived pancake configuration very similar to that produced by a high velocity symmetric collision of the kind likely to occur in the neighbourhood of black holes in the higher mass range $\gtrsim 10^9 M_\odot$. Thermonuclear energy release ensuing in the more extreme events may be sufficient to modify substantially the working of the entire accretion process.

066.018 Gravitational lenses.   C. Alcock.
Nature, Vol. 295, 284 (1982).

066.019 Propagation of high-frequency gravitational waves in vacuum: nonlinear effects.   T. Elster.
Gen. Relativ. Gravitation , Vol. 13, 731 - 745 (1981).

Linear and nonlinear perturbations of vacuum spacetimes are examined in the approximation of high frequencies. Propagation equations determining the amplitudes of these perturbations are derived. Finally, the relative motion in a system of test particles under the influence of both main wave and second harmonic as seen by a free falling observer is investigated. If inertial forces are absent, the second harmonic gives rise to oscillations in two transverse-traceless modes and in a transverse-scalar mode. The amplitude of the latter mode is proportional to the shear of the ray congruence along which the perturbations propagate.

066.020 Radiation reaction and energy loss in the post-Newtonian approximation of general relativity.
R. A. Breuer, E. Rudolph.
Gen. Relativ. Gravitation , Vol. 13, 777 - 793 (1981).

The authors calculate the radiation reaction force within the Anderson-Decanio scheme of a post-Newtonian expansion of general relativity modified such that no infinite expressions appear up to the order considered. They show that for quasi-periodic motions of bodies in bound systems, i.e., motions which differ little from Keplerian motion, the dominant contribution to the change of the Newtonian energy of the system equals the power loss given by Einstein's "quadrupole formula."

066.021 Necessary and sufficient conditions for trivial solutions in supergravity.
P. C. Aichelburg, H. K. Urbantke.
Gen. Relativ. Gravitation , Vol. 13, 817 - 828 (1981).

066.022 Gravitation in flat space-time.   W. Petry.
Gen. Relativ. Gravitation , Vol. 13, 865 - 872 (1981).

A previously studied Lorentz-covariant theory of gravitation is given in generally covariant form, i.e., the theory holds for arbitrary reference frames. Flat space-time is a natural condition for the conservation of energy and momentum. The energy-momentum tensor of matter and gravitation is the source of the gravitational field.

066.023 Thermalization of starlight by elongated grains: could the microwave background have been produced by stars?   E. L. Wright.
Astrophys. J., Vol. 255, 401 - 407 (1982).

The author considers the possibility of the microwave background being produced by stars after the big bang. The critical problem for this hypothesis is the source of the long-wavelength opacity for observed wavelengths greater than 10 cm. The author shows that free-free opacity cannot thermalize the background. Spherical dust grains also fail, but needle-shaped conducting grains can provide sufficient opacity to produce the observed spectrum with a metal abundance $Z \sim 10^{-7}$.

066.024 Synchrotron radiation from spherically accreting black holes. J. R. Ipser, R. H. Price.
Astrophys. J., Vol. 255, 654 - 673 (1982).

Spherical accretion onto a Schwarzschild black hole, of gas with frozen-in magnetic field, is studied numerically and analytically for a range of hole masses and accretion rates in which synchrotron emission is the dominant radiative mechanism. At small radii the equipartition of magnetic, kinetic, and gravitational energy is assumed to apply, and the gas is heated by dissipation of infalling magnetic energy, turbulent energy, etc. The models can be classified into three types: (a) synchrotron cooling negligible, (b) synchrotron cooling important but synchrotron self-absorption negligible, (c) synchrotron cooling and self-absorption important.

066.025 On a new integral of motion in relativistic galactic dynamics. N. Spyrou, H. Varvoglis.
Astrophys. J., Vol. 255, 674 - 690 (1982).

The authors establish the general theoretical framework necessary for the construction in the post-Newtonian approximation of general relativity of a formal third integral of motion valid for a test particle moving in the four-dimensional spacetime of a gravitating source, composed of a bounded perfect-fluid mass with a plane of symmetry and with axisymmetric and time-independent density distributions of mass and velocities. Relativistic contributions are negligible for normal galaxies but nonnegligible for elliptical galaxies of large mass and small linear dimensions.

066.026 Local stability in general relativity. H. E. Kandrup.
Astrophys. J., Vol. 255, 691 - 704 (1982).

This paper presents a unified treatment of the problem of local adiabatic stability for stationary, axisymmetric, perfect fluid configurations in general relativity. Criteria for both dynamical and secular stability are obtained via energy principles, without reference to the existence of normal modes. In general, the criteria that obtain are straightforward generalizations of Newtonian results. However, one new purely relativistic effect does arise: if Eulerian variations in pressure cannot be neglected, there exists the possibility of an additional secular instability toward viscosity unless one imposes a "causality" assumption, namely that the velocity of sound be less than unity.

066.027 Thermodynamic transport properties on dense stars. T. W. Edwards.
News Lett. Astron. Soc. N. Y., Vol. 2, 35 (1982). – Abstract.

066.028 A classical model for gravitation. P. C. Wagener.
South African J. Phys., Vol. 4, No. 2, p. 31 - 33 (1981). Abstr. in Phys. Abstr., Vol. 85, Abstr. 6780 (1982).

066.029 Mass scale of grand unification in N = 8 extended supergravity. M. Gluck, E. Reya.
Phys. Lett. B, Vol. 105B, 30 - 32 (1981). – Abstr. in Phys. Abstr., Vol. 85, Abstr. 6786 (1982).

066.030 The equations of conformal supergravity. B. M. Zupnik.
Phys. Lett. B, Vol. 105B, 153 - 154 (1981). – Abstr. in Phys. Abstr., Vol. 85, Abstr. 6787 (1982).

066.031 The trace anomaly in coloured black holes. E. Sanchez-Velasco.
Phys. Lett. B, Vol. 105B, 41 - 42 (1981). – Abstr. in Phys. Abstr., Vol. 85, Abstr. 10474 (1982).

066.032 Double Compton process and the spectrum of the microwave background.
L. Danese, G. De Zotti.
Astron. Astrophys., Vol. 107, 39 - 42 (1982).

The authors discuss the role of the double Compton process in smoothing out early distortions of the cosmic microwave background spectrum.

066.033 Spherical accretion with $e^+ - e^-$-pair production. W. P. Brinkmann.
Astron. Astrophys., Vol. 107, 48 - 50 (1982).

Stationary, spherically symmetric accretion of a completely ionized optically thin hydrogen plasma onto a Schwarzschild black hole is considered taking into account electron-positron pair production by binary collisions of the plasma particles. It is shown that there exists an upper limit for the plasma temperature of about 2×10^{11} K, thus putting strong constraints on γ-ray production via π°-mesons.

066.034 Upper limits of a cosmic infrared background flux as determined by X- and gamma-ray observations of M 87. R. Schlickeiser, M. Harwit.
Astron. Astrophys., Vol. 107, 186 - 189 (1982).

Upper limits on the energy density of infrared photons in the radio lobe regions of M 87 are derived using measurements of the X-ray and gamma-ray emission. The calculations are based on an inverse Compton scattering model initiated by radio-flux producing electrons. It is shown that the energy density of infrared photons in the radio lobe regions is smaller than $2\,eV\,cm^{-3}$.

066.035 Slowly rotating fluid spheres in general relativity with and without radiation. S. S. Bayin.
Phys. Rev. D, Vol. 24, 2056 - 2065 (1981). – Abstr. in Phys. Abstr., Vol. 85, Abstr. 10672 (1982).

066.036 Generalized plane gravitational waves. A. Kellner.
Int. J. Theor. Phys., Vol. 20, 433 - 441 (1981). – Abstr. in Phys. Abstr., Vol. 85, Abstr. 10677 (1982).

066.037 Geodesic coordinates in the Schwarzschild field. R. Gutreau.
Fundamental physics, (see 012.008), p. 357 - 436 (1981). Abstr. in Phys. Abstr., Vol. 85, Abstr. 10680 (1982).

066.038 What does the work in a gravitational field? R. A. Vera.
Fundamental physics, (see 012.008), p. 597 - 626 (1981). Abstr. in Phys. Abstr., Vol. 85, Abstr. 10690 (1982).

066.039 A gravitational model for a matter-free torsion ball. T. Dereli, R. W. Tucker.
J. Phys. A, Vol. 14, 2957 - 2967 (1981). – Abstr. in Phys. Abstr., Vol. 85, Abstr. 10692 (1982).

066.040 Proposed optical test of metric gravitation theories. M. O. Scully, M. S. Zubairy, M. P. Haugan.
Phys. Rev. A, Vol. 24, 2009 - 2016 (1981). – Abstr. in Phys. Abstr., Vol. 85, Abstr. 10697 (1982).

066.041 The space-time metric inside a black hole. P. F. Gonzalez-Diaz.
Lett. Nuovo Cimento, Vol. 32, Ser. 2, 161 - 163 (1981). Abstr. in Phys. Abstr., Vol. 85, Abstr. 14872 (1982).

066.042 Can charged black holes have a 'superhair'? . P. C. Aichelburg, R. Guven.
Phys. Rev. D, Vol. 24, 2066 - 2076 (1981). – Abstr. in Phys. Abstr., Vol. 85, Abstr. 14873 (1982).

066.043 The component gauges in supergravity.
U. Lindstrom, A. Karlhede, M. Rocek.
Nucl. Phys. B, Vol. B191, 549 - 573 (1981). — Abstr. in Phys.
Abstr., Vol. 85, Abstr. 15100 (1982).

066.044 Background of gravitational-wave antennas of
possible terrestrial origin. I.   E. Amaldi, E. Coccia,
S. Frasca, I. Modena, P. Rapagnani, F. Ricci, G. V. Pallottino,
G. Pizzella, P. Bonifazi, C. Cosmelli, U. Giovanardi, V. Iafolla,
S. Ugazio, G. Vannaroni.
Nuovo Cimento C, Vol. 4C, Ser. 1, 295 - 308 (1981). — Abstr.
in Phys. Abstr., Vol. 85, Abstr. 15102 (1982).

066.045 Background of gravitational-wave antennas of
possible terrestrial origin. II.   E. Amaldi, S. Frasca,
G. V. Pallottino, G. Pizzella, P. Bonifazi.
Nuovo Cimento C, Vol. 4C, Ser. 1, 309 - 323 (1981). — Abstr.
in Phys. Abstr., Vol. 85, Abstr. 15103 (1982).

066.046 Relativistic quadrupole moment.
B. M. Barker, G. M. O'Brien, R. F. O'Connell.
Phys. Rev. D, Vol. 24, 2332 - 2335 (1981). — Abstr. in Phys.
Abstr., Vol. 85, Abstr. 18987 (1982).

066.047 Gravitation radiation from cosmic strings.
A. Vilenkin.
Phys. Lett. B, Vol. 107B, 47 - 50 (1981). — Abstr. in Phys.
Abstr., Vol. 85, Abstr. 18988 (1982).

066.048 Spherically symmetric collapse in quantum gravity.
V. P. Frolov, G. A. Vilkovisky:
Phys. Lett. B, Vol. 106B, 307 - 313 (1981). — Abstr. in Phys.
Abstr., Vol. 85, Abstr. 19000 (1982).

066.049 Geometrical first order supergravity in five space
time dimensions.
R. D'Auria, E. Maina, T. Regge, P. Fre.
Ann. Physics, Vol. 135, 237 - 269 (1981). — Abstr. in Phys.
Abstr., Vol. 85, Abstr. 19001 (1982).

066.050 Critical-mass limits for collapsed stars in a generaliz-
ed theory of gravitation.   J. W. Moffat.
Lett. Nuovo Cimento, Vol. 32, Ser. 2, 277 - 280 (1981).
Abstr. in Phys. Abstr., Vol. 85, Abstr. 23217 (1982).

066.051 Local thermodynamics and stress tensor of the
Hawking radiation.   T. Zannias, W. Israel.
Phys. Lett. A, Vol. 86A, 82 - 84 (1981). — Abstr. in Phys.
Abstr., Vol. 85, Abstr. 23320 (1982).

066.052 Problems connected with the tangential metric
coefficient in the Schwarzschild metric: a possible
solution.   P. Voráček.
Astrophys. Space Sci., Vol. 81, 85 - 94 (1982).
   Four situations are shown where the Schwarzschild
metric cannot be used or is subject to insurmountable
problems. The first is the question of a metric useful for PPN-
formalism checking different gravitational theories. The second
problem occurs in connection with Mach's principle, when the
flatness of the spacetime inside a massive hollow sphere is a
generally accepted solution. The metrical discontinuity on the
same spherical shell is a third problem. The fourth one is the
anisotropy of the mass-energy of a test particle in the gravitational field. Principles and methods for solution of these
problems are outlined.

066.053 Massive oscillators as cosmic energy sources.
K. M. V. Apparao, J. V. Narlikar.
Astrophys. Space Sci., Vol. 81, 397 - 410 (1982).
   The usual picture in which a massive object undergoes a
gravitational collapse to become a black hole and ultimately
end up in space-time singularity, is modified with the introduc-

tion of a negative energy force of repulsion effective only at a
short range. It is shown that the object executes oscillations
between states of high and low densities. From the view point
of high energy astrophysics, such a massive oscillator combines
some of the attractive features of black holes and white holes.
It is suggested that the energy production and spectral features
of quasars, BL-Lacs and the active galactic nuclei might be
accounted for by postulating the existence of massive oscillators.

066.054 Equations of motion of n massive-charged particles
in general relativity theory.
C. G. Kostakis, M. Antonacopoulos.
Astrophys. Space Sci., Vol. 82, 149 - 159 (1982).
   Starting with the Einstein-Maxwell field equations, the
authors obtain the post-Newtonian equations of motion of
n massive-charged particles in general relativity.

066.055 Is collapse of a deformed star always effectual for
gravitational radiation?   T. Nakamura, M. Sasaki.
Phys. Lett. B, Vol. 106B, 69 - 72 (1981). — Abstr. in Phys.
Abstr., Vol. 85, Abstr. 23514 (1982).

066.056 Quantum theory of wormholes.
P. Hajicek.
Phys. Lett. B, Vol. 106B, 77 - 80 (1981). — Abstr. in Phys.
Abstr., Vol. 85, Abstr. 23530 (1982).

066.057 A review of the group manifold approach and its
application to conformal supergravity.
L. Castellani, P. Fre, P. van Nieuwenhuizen.
Ann. Physics, Vol. 136, 398 - 434 (1981). — Abstr. in Phys.
Abstr., Vol. 85, Abstr. 23537 (1982).

066.058 Classical solutions of the equations of supergravity.
R. J. Finkelstein, J. Kim.
J. Math. Phys., Vol. 22, 2228 - 2234 (1981). — Abstr. in Phys.
Abstr., Vol. 85, Abstr. 23538 (1982).

066.059 Quantization of two-dimensional supergravity and
critical dimensions for string models.
E. S. Fradkin, A. A. Tseytlin (*Tsejtlin*).
Phys. Lett. B, Vol. 106B, 63 - 68 (1981). — Abstr. in Phys.
Abstr., Vol. 85, Abstr. 23539 (1982).

066.060 General relativistic collapse of rotating stars.
T. Nakamura, H. Sato.
Phys. Lett. A, Vol. 86A, 318 - 320 (1981). — Abstr. in Phys.
Abstr., Vol. 85, Abstr. 28493 (1982).

066.061 The radiation damping on binary stars due to emis-
sion of gravitational waves.   N. Hu, D.-H. Zhang,
H.-G. Ding.
Acta Phys. Sinica, Vol. 30, 1003 - 1010 (1981). In Chinese.
Abstr. in Phys. Abstr., Vol. 85, Abstr. 28508 (1982).

066.062 Shear-free collapse with heat flow.
E. N. Glass.
Phys. Lett. A, Vol. 86A, 351 - 352 (1981). — Abstr. in Phys.
Abstr., Vol. 85, Abstr. 28587 (1982).

066.063 All nontwisting N's with cosmological constant.
A. Garcia Diaz, J. F. Plebanski.
J. Math. Phys., Vol. 22, 2655 - 2658 (1981). — Abstr. in Phys.
Abstr., Vol. 85, Abstr. 28727 (1982).

066.064 Quantum field theory in curved space-time. Massive
and massless vector fields.
H. Ceccatto, A. Foussats, H. Giacomini, O. Zandron.
Phys. Rev. D, Vol. 24, 2576 - 2585 (1981). — Abstr. in Phys.
Abstr., Vol. 85, Abstr. 28739 (1982).

066.065 **Dynamical calculation of bound-state supermultiplets in N = 8 supergravity.**
M. T. Grisaru, H. J. Schnitzer.
Phys. Lett. B, Vol. 107B, 196 - 200 (1981). − Abstr. in Phys. Abstr., Vol. 85, Abstr. 28741 (1982).

066.066 **Off-shell central charges and linearised N = 8 supergravity.** J. G. Taylor.
Phys. Lett. B, Vol. 107B, 217 - 222 (1981). − Abstr. in Phys. Abstr., Vol. 85, Abstr. 28742 (1982).

066.067 **Status of the Stanford gravitational wave experiment.** M. S. McAshan, W. M. Fairbank, P. F. Michelson, R. C. Taber.
Physica B, C, Vol. 107 B + C, No. 1 - 3, p. 23 - 25 (1981). Abstr. in Phys. Abstr., Vol. 85, Abstr. 28744 (1982).

066.068 **The variation of G: a modern look.**
V. M. Canuto.
An. Acad. Brasileira Cienc., Vol. 53, 269 - 278 (1981). Abstr. in Phys. Abstr., Vol. 85, Abstr. 32825 (1982).

066.069 **Radiation reaction and angular momentum loss in small angle gravitational scattering.**
T. Damour, N. Deruelle.
Phys. Lett. A, Vol. 87A, 81 - 84 (1981). − Abstr. in Phys. Abstr., Vol. 85, Abstr. 33392 (1982).

066.070 **Linearised N = 2 superfield supergravity.**
V. O. Rivelles, J. G. Taylor.
J. Phys. A, Vol. 15, 163 - 175 (1982). − Abstr. in Phys. Abstr., Vol. 85, Abstr. 33408 (1982).

066.071 **Hierarchical scales and 'family' of black holes.**
K. Namsrai.
Int. J. Theor. Phys., Vol. 20, 749 - 754 (1981). − Abstr. in Phys. Abstr., Vol. 85, Abstr. 33769 (1982).

066.072 **A model of a thin accretion disk around a black hole.** B. Paczyński, G. Bisnovatyi-Kogan (*Bisnovatyj-Kogan).*
Acta Astron., Vol. 31, 283 - 291 (1981).
An "alpha disk" model is presented with a radial heat flow due to accretion flow, and a radial pressure gradient taken into account. Even though the model ignores some important physical effects, it removes infinities present in a standard thin disk model at the inner disk boundary at $r_{in} = 3 r_g$. The model has its inner boundary somewhat closer to the black hole, and the radial flow is transonic at that place.

066.073 **Radiation from relativistic particles in nongeodesic motion in a strong gravitational field.**
A. N. Aliev, D. V. Galtsov (*Gal'tsov*).
Gen. Relativ. Gravitation, Vol. 13, 899 - 912 (1981).
The scalar and electromagnetic radiation emitted by relativistic particles moving along the stable nongeodesic trajectories in the Kerr gravitational field are described. Two particular models of the nongeodesic motion are developed involving a slightly charged rotating black hole and a rotating black hole immersed in an external magnetic field.

066.074 **Gravitational theories with de Sitter constant vacuum solution.** G. Cognola, R. Soldati.
Gen. Relativ. Gravitation, Vol. 13, 923 - 937 (1981).

066.075 **Gravity as an internal Yang-Mills gauge field theory of the Poincaré group.** J. Hennig, J. Nitsch.
Gen. Relativ. Gravitation, Vol. 13, 947 - 962 (1981).
In the framework of affine bundles the authors present gravity as an "internal" gauge field theory of the Poincaré group. The resulting geometry is a Riemann-Cartan space-time carrying torsion and curvature. In order to admit a nontrivial action of the translation group they formally extend the matter Lagrangian to affine field variables.

066.076 **Poincaré-invariant gravitational field and equations of motion of two pointlike objects: the postlinear approximation of general relativity.** L. Bel, T. Damour, N. Deruelle, J. Ibanez, J. Martin.
Gen. Relativ. Gravitation, Vol. 13, 963 - 1004 (1981).
Using a fast-motion approximation method the authors obtain the second-order gravitational field and equations of motion for two pointlike objects in algebraically closed form. A regularization procedure is used which is shown to guarantee the consistency of the approximation scheme. The equations of motion are then transformed within the framework of relativistic predictive mechanics into a system of ordinary differential equations.

066.077 **Black hole in a gravitational field.**
R. M. Kerns, W. J. Wild.
Gen. Relativ. Gravitation, Vol. 14, 1 - 3 (1982).
An exact solution of the vacuum Einstein field equations representing a Schwarzschild black hole in an external gravitational field is derived using a formalism developed by Ernst.

066.078 **An estimate of the probability of observing a gravitational lens effect.** S. Hacyan.
Astrophys. Lett., Vol. 22, 97 - 99 (1982).
An estimate is made of the absolute probability that the image of a source at a given redshift be observed as multiple due to a gravitational lens effect.

066.079 **Evidence for cosmic censorship.**
E. N. Glass, A. Harpaz.
Phys. Rev. D, Vol. 24, 3038 - 3043 (1981). − Abstr. in Phys. Abstr., Vol. 85, Abstr. 38525 (1982).

066.080 **Gravitational vacuum hypothesis and cosmology with variable particle number.** K. P. Staniukovich (*Stanyukovich*), V. N. Melnikov (*Mel'nikov*), K. A. Bronnikov.
Int. J. Theor. Phys., Vol. 20, 831 - 841 (1981). − Abstr. in Phys. Abstr., Vol. 85, Abstr. 38537 (1982).

066.081 **Uniqueness of the propagator in spacetime with cosmological singularity.** C. Charach, L. Parker.
Phys. Rev. D, Vol. 24, 3023 - 3037 (1981). − Abstr. in Phys. Abstr., Vol. 85, Abstr. 38550 (1982).

066.082 **Comments on different derivations of N = 2 supergravity using the group manifold.**
P. van Nieuwenhuizen.
Phys. Rev. D, Vol. 24, 3058 - 3064 (1981). − Abstr. in Phys. Abstr., Vol. 85, Abstr. 38552 (1982).

066.083 **Interacting supergravity in ten dimensions: the role of six-index gauge field.** A. H. Chamseddine.
Phys. Rev. D, Vol. 24, 3065 - 3072 (1981). − Abstr. in Phys. Abstr., Vol. 85, Abstr. 38553 (1982).

066.084 **Mechanical-transfer function and Brownian-noise measurements at T = 4.2K of a small (M = 20.3 kg) gravitational-wave antenna using double 'four-point' mechanical suspensions.** P. Bonifazi, E. Coccia, P. Rapagnani.
Nuovo Cimento C, Vol. 4C, Ser. 1, 408 - 416 (1981). − Abstr. in Phys. Abstr., Vol. 85, Abstr. 38554 (1982).

066.085 **Background of gravitational-wave antennas of possible terrestrial origin. III.** E. Amaldi, E. Coccia, S. Frasca, F. Ricci, P. Bonifazi, V. Iafolla, G. Natali, G. V. Pallottino, G. Pizzella.
Nuovo Cimento C, Vol. 4C, Ser. 1, 441 - 457 (1981). − Abstr. in Phys. Abstr., Vol. 85, Abstr. 38555 (1982).

066.086 Rigidly rotating disk as a source of the Kerr geometry.   C. A. Lopez.
Nuovo Cimento B, Vol. 66B, Ser. 11, 17 - 33 (1981). – Abstr. in Phys. Abstr., Vol. 85, Abstr. 42707 (1982).

066.087 Bound states in quantum evaporation of black holes.   L. A. Kofman.
Phys. Lett. A, Vol. 87A, 281 - 284 (1982). – Abstr. in Phys. Abstr., Vol. 85, Abstr. 42708 (1982).

066.088 On the equivalence of the relativistic theories of gravitation.   M. Ferraris, J. Kijowski.
Gen. Relativ. Gravitation, Vol. 14, 165 - 180 (1982).

066.089 The force law for the dynamic two-body problem in the second post-Newtonian approximation of general relativity.   R. A. Breuer, E. Rudolph.
Gen. Relativ. Gravitation, Vol. 14, 181 - 211 (1982).

066.090 On the contribution of a stochastic background of gravitational radiation to the timing noise of pulsars.
B. Mashhoon.
Mon. Not. R. Astron. Soc., Vol. 199, 659 - 666 (1982).
The influence of a stochastic and isotropic background of gravitational radiation on the timing measurements of pulsars is investigated. Using pulsar timing data, significant upper limits may be set on the energy density of a stochastic background with a simple spectrum provided it is dominated by waves of period longer than ~ 1 yr but shorter than the period of pulsar observations.

066.091 Linsen im Weltraum. Teil 1: Der Gravitationslinseneffekt und Beobachtungen an einem Zwillingsquasar.
U. Borgeest, S. Refsdal.
Sterne Weltraum, 21. Jahrg., 199 - 202 (1982).

066.092 Self-gravitating accretion disk models for active galactic nuclei: self-consistent α-models for the broad emission-line region.   S. N. Shore, R. L. White.
Astrophys. J., Vol. 256, 390 - 396 (1982).
The authors present a formulation of an α-model accretion disk which includes the effects of self-gravitation, radiation pressure, and variable opacity source. The heating is assumed due to turbulent stresses in a differentially rotating, massive disk. The results are compared with previous work, and it is shown that earlier studies did not properly include the effects of the disk mass on vertical structure. These models should more properly represent the structure of the accretion regions around massive black holes in galactic nuclei. A discussion of consequences for Seyfert I/II galaxies and for the Milky Way nuclear region is included.

066.093 Cepstral analysis of interfering delay signals as applied to detection of gravitational lenses.
E. L. Afraimovich (*Eh. L. Afrajmovich*).
Astron. Astrophys., Vol. 105, L5 - L6 (1982).
It is shown that gravitational lenses can be detected by measuring the difference time of emission propagation from source to observer along different paths even if these paths cannot be angularly resolved. The method offered makes it possible to expand substantially the class of sources which can be used to detect gravitational lenses.

066.094 The energy-momentum tensor and the Reissner-Nordström metric.
I. Gottlieb, C. Mociutchi, N. Ionescu-Pallas.
Rev. Roumaine Phys., Vol. 26, 1057 - 1066 (1981). – Abstr. in Phys. Abstr., Vol. 85, Abstr. 42915 (1982).

066.095 A null tetrad analysis of the Ernst metric.
S. K. Bose, E. Esteban.

J. Math. Phys., Vol. 22, 3006 - 3009 (1981). – Abstr. in Phys. Abstr., Vol. 85, Abstr. 42917 (1982).

066.096 Incomplete black-hole evaporation.
P. F. Gonzalez-Diaz.
Lett. Nuovo Cimento, Vol. 33, Ser. 2, 127 - 130 (1982). Abstr. in Phys. Abstr., Vol. 85, Abstr. 42918 (1982).

066.097 Linearized N = 2 superfield supergravity.
S. J. Gates, Jr., W. Siegel.
Nucl. Phys. B, Vol. B195, 39 - 60 (1982). – Abstr. in Phys. Abstr., Vol. 85, Abstr. 42926 (1982).

066.098 Ten-dimensional Maxwell-Einstein supergravity, its currents and the issue of its auxiliary fields.
E. Bergshoeff, M. De Roo, B. De Wit, P. Van Nieuwenhuizen.
Nucl. Phys. B, Vol. B195, 97 - 136 (1982). – Abstr. in Phys. Abstr., Vol. 85, Abstr. 42927 (1982).

066.099 The NUT metric in synchronous coordinates.
S. Gupta.
Phys. Lett. A, Vol. 87A, 220 - 223 (1982). – Abstr. in Phys. Abstr., Vol. 85, Abstr. 48143 (1982).

066.100 Primordial black holes and super-massive stars (in galactic nuclei).   F. Hagio.
Prog. Theor. Phys., Vol. 66, 1504 - 1507 (1981). – Abstr. in Phys. Abstr., Vol. 85, Abstr. 48273 (1982).

066.101 Stability of gravity with a cosmological constant.
L. F. Abbott, S. Deser.
Nucl. Phys. B, Vol. B195, 76 - 96 (1982). – Abstr. in Phys. Abstr., Vol. 85, Abstr. 48370 (1982).

066.102 Gibt es Oberflächeneffekte der Gravitation?
P. A. Thießen, H.-J. Treder.
Gerlands Beitr. Geophys., Band 91, 97 - 107 (1982).

066.103 Black holes and the origin of radio sources.
K. S. Thorne, R. D. Blandford.
Extragalactic radio sources, (see 012.025), p. 255 - 262 (1982).
Powerful extragalactic radio sources might be fuelled by energy release near a massive black hole. Some relativistic effects which may be relevant to this process are described in this review. The "3 + 1" formulation of relativistic gravity is used. Specifically described are the gravitational field near a black hole, Lense-Thirring and geodetic precession, electromagnetic energy extraction of the spin energy of a black hole and the structure of accretion tori around a black hole.

066.104 Trajectory of a light ray in second order according to the gravitational constant.   A. V. Ul'din.
Univ. druzhby narodov im. P. Lumumby. Moskva, 1981, 7 pp. In Russian. – Abstr. in Ref. zh., 51. Astron., 4.51.163 (1982).

066.105 On some properties of an ideal isentropic fluid in general relativity.   G. G. Matveev.
Gravitatsiya i teor. otnositel'nosti, Kazan, 1981, No. 18, p. 82 - 86. In Russian. – Abstr. in Ref. zh., 51. Astron., 4.51.165 (1982).

066.106 Sobre la existencia de horizontes 4d en relatividad general.   H. A. Dottori.
Bol. Asoc. Argentina Astron., No. 20 - 24, p. 196 (1981). Abstract.

066.107 The suppression of gravitational radiation from finite-size stars falling into black holes.
M. P. Haugan, S. L. Shapiro, I. Wasserman.
Astrophys. J., Vol. 257, 283 - 290 (1982).
The authors show that the gravitational radiation

emitted in a head-on collision between a finite-size star and a black hole can be substantially less than might be expected on the basis of results for point mass-blackhole collisions. First the suppression for gravitational radiation from a dust cloud of rest mass m freely falling into a Schwarzschild black hole of mass M $\gg$ m is calculated. The calculation is then applied to a realistic star tidally disrupted by a massive black hole. Head-on, free-fall collisions of main-sequence stars, white dwarfs and neutron stars with black holes of mass 10 M$_\odot$, 10^3 M$_\odot$, and 10^6 M$_\odot$ are considered.

066.108 Non-zero electromagnetic radiation in gravitational fields.   F. E. Khlystun.
Vestn. Kiev. Univ., Vyp. 23, p. 15 - 22 (1981). In Russian.

066.109 Instability of flat space at finite temperature.   D. J. Gross, M. J. Perry, L. G. Yaffe.
Phys. Rev. D, Vol. 25, 330 - 355 (1982). − Abstr. in Phys. Abstr., Vol. 85, Abstr. 48587 (1982).

066.110 Relativistic effects in the solar system. II. The deflexion of starlight by the Sun.   E. R. Bagge.
Atomkernenerg. Kerntech., Vol. 40, No. 1, p. 47 - 50 (1982). Abstr. in Phys. Abstr., Vol. 85, Abstr. 53072 (1982).

066.111 General relativistic collapse of rotating supermassive stars.   T. Nakamura, H. Sato.
Prog. Theor. Phys., Vol. 66, 2038 - 2051 (1981). − Abstr. in Phys. Abstr., Vol. 85, Abstr. 53076 (1982).

066.112 Stimulated emission processes near a black hole.   L. H. Ford.
J. Phys. A, Vol. 15, 825 - 830 (1982). − Abstr. in Phys. Abstr., Vol. 85, Abstr. 53313 (1982).

066.113 Kinetic theory in astrophysics and cosmology.   J. R. Ray.
Astrophys. J., Vol. 257, 578 - 586 (1982).
The author presents results associated with exact solution of the Einstein-Boltzmann and Einstein-Maxwell-Boltzmann equations. An important aspect of the paper is the generalization of Ehler's Killing vector approach for the distribution function to charged particles.

066.114 Self-absorbed and comptonized synchrotron radiation from spherically accreting black holes.
J. R. Ipser, R. H. Price.
Bull. American Astron. Soc., Vol. 13, 821 - 822 (1981). Abstract.

066.115 Electron-positron jet models.   M. L. Burns.
Bull. American Astron. Soc., Vol. 13, 823 (1981). Abstract.

066.116 Optically thick relativistic accretion onto a black hole.   P. A. Vitello.
Bull. American Astron. Soc., Vol. 13, 845 (1981). − Abstract.

066.117 Propagation of light in a Maxwell-like gravitational field.   V. Majerník.
Astrophys. Space Sci., Vol. 82, 473 - 476 (1982).
The author proposes a coupling between the gravitational field and electromagnetic fields by adding the four-vector of the gravitational potential to the differential operators of Maxwell's basic field equations. It is shown that by this coupling all tests of Einstein's theory of gravitation connected with the propagation of light in a gravitational field can be correctly calculated.

066.118 General covariance, accelerated frames and the particle concept.   T. Padmanabhan.
Astrophys. Space Sci., Vol. 83, 247 - 268 (1982).

The definition of particle states in various accelerated frames is considered. It is shown that in any realistically accelerated system, quantum field theory can be formulated without any ambiguity. The author further shows that the definition of a particle based on Green's function techniques does not always agree with the definition based on explicit quantization. He analyses the standard accelerated detector results from this point of view and shows that the uncertainty principle imposes a rigorous bound on these detection processes.

066.119 Some astrophysical consequences of the extended Maxwell-like gravitational field equations.
V. Majerník.
Astrophys. Space Sci., Vol. 84, 191 - 204 (1982).
The Maxwell-like gravitational field equations have been generalized and coupled through the gravitational four-potential to the electromagnetic Maxwell equations. It is shown that this has several astrophysical consequences, among which are the following: (1) the gravitational instability of a system of mass bodies manifesting itself by a Hubble-like motion on cosmological scales, (2) the possible change of light intensity propagating through a large distance, (3) instability of a planetary system on cosmological time scales, due to the momentum increase of moving bodies in a generalized gravitational field.

066.120 Light deflection during solar eclipses.   J. Bouet.
Sol. Phys., Vol. 78, 385 - 387 (1982).
A simultaneity is observed between fluctuations in ellipticity of the solar corona and variation of the light deflection by the sun, during eclipses.

066.121 Moving platform experiments.   S. Marinov.
Indian J. Phys. Part B, Vol. 55B, 403 - 418 (1981). − Abstr. in Phys. Abstr., Vol. 85, Abstr. 53607 (1982).

066.122 H-space with a cosmological constant.   C. R. LeBrun.
Proc. R. Soc. London Ser. A, Vol. 380, 171 - 185 (1982). Abstr. in Phys. Abstr., Vol. 85, Abstr. 53617 (1982).

066.123 Gravitational radiation of a particle falling towards a black hole. I. The case of a non-rotating black hole.
Y. Tashiro, H. Ezawa.
Prog. Theor. Phys., Vol. 66, 1612 - 1626 (1981). − Abstr. in Phys. Abstr., Vol. 85, Abstr. 53621 (1982).

066.124 On a semi-relativistic treatment of the gravitational radiation from a mass thrusted into a black hole.
R. Ruffini, M. Sasaki.
Prog. Theor. Phys., Vol. 66, 1627 - 1638 (1981). − Abstr. in Phys. Abstr., Vol. 85, Abstr. 53622 (1982).

066.125 Cylindrically symmetric Zel'dovich fluid distributions in general theory of relativity.
G. Mohanty, R. N. Tiwari, J. R. Rao.
Int. J. Theor. Phys., Vol. 21, 105 - 119 (1982). − Abstr. in Phys. Abstr., Vol. 85, Abstr. 53625 (1982).

066.126 Rotating charged dust in general relativity.   A. K. Raychaudhuri.
J. Phys. A, Vol. 15, 831 - 840 (1982). − Abstr. in Phys. Abstr., Vol. 85, Abstr. 53626 (1982).

066.127 Spontaneously generated gravity and the Second Law of Thermodynamics.   P. C. W. Davies.
Phys. Lett. B, Vol. 110B, 111 - 113 (1982). − Abstr. in Phys. Abstr., Vol. 85, Abstr. 53629 (1982).

066.128 Exact solution of a rotating dyon black hole.
M. Kasuya.
Phys. Rev. D, Vol. 25, 995 - 1001 (1982). — Abstr. in Phys. Abstr., Vol. 85, Abstr. 53632 (1982).

066.129 Gauge formulation of gravitation theories. I. The Poincaré, de Sitter, and conformal cases.
E. A. Ivanov, J. Niederle.
Phys. Rev. D, Vol. 25, 976 - 987 (1982). — Abstr. in Phys. Abstr., Vol. 85, Abstr. 53637 (1982).

066.130 Gauge formulation of gravitation theories. II. The special conformal case.   E. A. Ivanov, J. Niederle.
Phys. Rev. D, Vol. 25, 988 - 994 (1982). — Abstr. in Phys. Abstr., Vol. 85, Abstr. 53638 (1982).

066.131 Twisted symmetry breaking on the projective hypersphere: a model of the small cosmological constant.
S. C. Unwin.
J. Phys. A, Vol. 15, 841 - 848 (1982). — Abstr. in Phys. Abstr., Vol. 85, Abstr. 53641 (1982).

066.132 A new torsion balance for studies in gravitation and cosmology.   R. Cowsik.
Indian J. Phys. Part B, Vol. 55B, 497 - 512 (1981). — Abstr. in Phys. Abstr., Vol. 85, Abstr. 53655 (1982).

066.133 On the validity of the geodesic motion near a black hole: a clarification.   N. Dadhich, T. Padmanabhan.
Lett. Nuovo Cimento, Vol. 33, Ser. 2, 317 - 318 (1982). Abstr. in Phys. Abstr., Vol. 85, Abstr. 58248 (1982).

066.134 Acceleration radiation and the generalized second law of thermodynamics.   W. G. Unruh,
R. M. Wald.
Phys. Rev. D, Vol. 25, 942 - 958 (1982). — Abstr. in Phys. Abstr., Vol. 85, Abstr. 58380 (1982).

066.135 Superposition of Planckian spectra and the distortions of the cosmic microwave background radiation.   M. Alexanian.
Astrophys. J., Vol. 258, 43 - 45 (1982).
    A fit of the spectrum of the cosmic microwave background radiation (CMB) by means of a positive linear superposition of Planckian spectra implies an upper bound to the photon spectrum. The observed spectrum of the CMB gives a weighting function with a normalization greater than unity.

066.136 Composition independence of the possible finite-range gravitational force.   Y. Fujii.
Gen. Relativ. Gravitation, Vol. 13, 1147 - 1155 (1981).

066.137 Hydrodynamics in the O_4 gravity.
T. Obata, H. Oshima, J. Chiba.
Gen. Relativ. Gravitation, Vol. 13, 1161 - 1176 (1981).
    The authors develop hydrodynamics in a new geometrical gravitational theory, called O_4 gravity, which they recently proposed. According to this formulation, matter is not necessarily conserved. The nonconservation of matter might have been considerable in an early era of cosmological evolution.

066.138 Black hole emissions and phase transitions.
A. Curir.
Gen. Relativ. Gravitation, Vol. 13, 1177 - 1184 (1981).
    The nonthermal emission of a rotating black hole is related to the inner horizon of the hole. The transition from the Kerr black hole state to the naked singularity state is considered as a phase transition. An arrangement of the energetics of Kerr black holes into a two-phase thermodynamics is suggested.

066.139 Solitonic gravitational waves in Bianchi II cosmologies. 1. The general framework.
V. Belinski (*Belinskij*), M. Francaviglia.
Gen. Relativ. Gravitation, Vol. 14, 213 - 229 (1982).
    The inverse scattering method is applied to a class of space-times belonging to the Bianchi types I–VII. Solitonic perturbations corresponding to one or two poles on an arbitrary cosmological background are described in detail. The fundamental matrix ψ_0 is explicitly calculated for a Bianchi II background, thus providing the first known example of a nondiagonal case.

066.140 Stationary, spherically symmetric solutions of Jordan's unified theory of gravity and electromagnetism.   P. Dobiasch, D. Maison.
Gen. Relativ. Gravitation, Vol. 14, 231 - 242 (1982).
    All stationary, spherically symmetric solutions of Jordan's unified theory of gravity and electromagnetism are constructed. Conditions for the solutions are given to represent black holes with nonvanishing mass, electric and magnetic charge.

066.141 Is quantum gravity deterministic and/or time symmetric?   D. N. Page.
Gen. Relativ. Gravitation, Vol. 14, 299 - 302 (1982).
    S. W. Hawking suggests that quantum gravity introduces a new level of uncertainty into physics by turning pure states into mixed states. Although the evidence for this information loss is based upon a semiclassical approximation and hence is not conclusive, it is interesting to examine the implications. As originally formulated, Hawking's proposal violates CPT invariance by singling out one direction of time in which pure states turn into mixed states. An alternate hypothesis is suggested whereby the theory could be time symmetric and yet allow a loss of information. In this model the universe as a whole would be an open system, and even its density matrix would not have a deterministic evolution. The question remains of how much uncertainty there actually is in quantum gravity.

066.142 Calculation of Newton's gravitational constant in infrared-stable Yang-Mills theories.   A. Zee.
Phys. Rev. Lett., Vol. 48, 295 - 298 (1982).
    Newton's gravitational constant G is calculated in a class of scale-invariant gauge theories with an infrared fixed point. The sign of G depends on the coefficients in the renormalization-group β function.

066.143 Stationary spherical accretion into black holes – II. Theory of optically thick accretion.
R. A. Flammang.
Mon. Not. R. Astron. Soc., Vol. 199, 833 - 867 (1982).
    In this paper, the problem of spherical, steady-state, optically thick accretion into black holes is solved. The author analyses the integral curves of the differential equations describing the problem. He finds a one-parameter family of critical points, where the inflow velocity equals the isothermal sound speed. Physical solutions must pass through one of these critical points. The author obtains a complete set of boundary conditions which the solution must satisfy at the horizon of the black hole, and shows that these, plus the requirement that the solution passes through a critical point, determine a unique solution to the problem.

066.144 Concerning relativistic astrometry.
S. V. M. Clube.
Scientific aspects of the Hipparcos space astrometry mission, (see 012.041), p. 187 - 189 (1982).
    The possible achievement of all sky milliarcsecond precision in stellar positions with Hipparcos will provide opportunities for testing fundamental aspects of relativity theory. Deviations from positions predicted by general relativity are ambiguous in the case of gravitational light deflection since modifi-

cation of the field equation may be involved, but in the case of aberration, should it be a function of the earth's absolute motion, no such ambiguity arises.

066.145 The atmospheric Cherenkov technique in searches for exploding primordial black holes.
S. Danaher, D. J. Fegan, N. A. Porter, T. C. Weekes.
Philos. Trans. R. Soc. London, Ser. A, Vol. 301, (see 012.043), 665 - 667 (1981). − Same as 30.066.132.
The Cherenkov technique has been used with a number of detectors, ranging from 1.5 m^2 mirrors to the Central Receiver Test Facility of 8400 m^2. Limits have been set to the flux of primordial black holes for various models of the evaporation process.

066.146 An investigation of some properties of the metric field outside a rotating and charged object under VGM.   Q.-h. Peng.
Acta Astron. Sinica, Vol. 23, 1 - 9 (1982). In Chinese.

066.147 Linsen im Weltraum. Teil 2: Astrophysikalische Anwendungen des Gravitationslinseneffekts.
U. Borgeest, S. Refsdal.
Sterne Weltraum, 21. Jahrg., 244 - 246 (1982).

066.148 On the time scales of the pair production processes in astrophysics.   A. A. Zdziarski.
Astron. Astrophys., Vol. 110, L7 - L10 (1982).
The problem of the e^+e^- pair production by photon-photon, photon-particle, and particle-particle collisions in a black hole spherically symmetric accretion is considered. It is shown that these processes are much too slow to lead to any appreciable pair density in an optically thin accretion flow. The rates of pair production in the binary particle collisions are calculated. The maximum temperature of an optically thin stationary plasma is lower than that previously calculated by Bisnovatyi-Kogan, Zel'dovich and Sunyaev (1971).

066.149 Non linear Einstein−Maxwell differential equations.   G. Antonacopoulos, C. G. Kostakis.
Compendium in astronomy, (see 003.012), p. 349 - 359 (1982).
Starting with the Einstein-Maxwell field equations in general relativity the authors construct the general differential equations governing the components of the metric tensor. These equations allow to find h_{ij} in various orders.

Cosmology, physics, and philosophy.
See Abstr. 003.047.

The cosmic frontiers of general relativity.
See Abstr. 003.074.

Allgemeine Relativitätstheorie und relativistische Astrophysik.   See Abstr. 003.124.

How does the theory of gravitation develop itself?
See Abstr. 011.024.

The big bang − "free lunch" at the Royal Society.
See Abstr. 021.005.

Possible measurement of the time delay between gravitational images of expanding double radio-sources.
See Abstr. 031.516.

Matching of transducers to resonant gravitational-wave antennas.   See Abstr. 034.014.

Mechanical filter for the suspension of gravitational wave antennas.   See Abstr. 034.068.

Atomic and gravitational clocks.
See Abstr. 035.002.

Is the gravitational constant changing?
See Abstr. 043.006.

Relativistic effects in Hipparcos data.
See Abstr. 051.025.

The structure and variability of dynamo driven accretion discs.   See Abstr. 062.005.

Numerical experiments on a model of penetrative convection.   See Abstr. 062.035.

Compton rockets: radiative acceleration of a relativistic fluid.   See Abstr. 062.066.

A stability criterion for many-parameter equilibrium families.   See Abstr. 062.081.

Synchro-Compton radiation from relativistic charges driven by a strong plane vacuum wave of elliptic polarization.   See Abstr. 063.005.

On the theory of gamma-ray amplification through stimulated annihilation radiation.   See Abstr. 063.021.

General relativistic radiative transfer: the 14-moment approximation.   See Abstr. 063.065.

A model of a thick disk with equatorial accretion.
See Abstr. 064.009.

Structure around the inner edge of geometrically thin accretion disks.   See Abstr. 064.030.

Physical properties of thick supercritical accretion disks.   See Abstr. 064.041.

Properties and cosmological consequences of very massive objects.   See Abstr. 065.048.

Relativistic perturbations for all the planets.
See Abstr. 091.001.

Relativistic effects in the Solar System. I. Relativistic pathway modifications of the planets.
See Abstr. 092.006.

The relativistic planetary perturbations and the orbital motion of the Moon.   See Abstr. 094.051.

Recent results for the Moon's secular acceleration and their implication for the possible variation of G in Dirac's Large Number Hypothesis.   See Abstr. 094.063.

Dalle nane bianche ai buchi neri: storia di alcuni concetti di astrofisica relativistica.   See Abstr. 126.042.

The nature of the light variations in the double QSO Q0957+561.   See Abstr. 141.043.

Discovery of a third gravitational lens.
See Abstr. 141.047.

Mechanisms for jets.   See Abstr. 141.148.

Supercritical accretion and its possible relation to quasars and radio sources.   See Abstr. 141.157.

Modelling the gravitational lens of the double quasar.   See Abstr. 141.197.

Superluminal velocities of compact radio sources: a gravitational lens effect.   See Abstr. 141.198.

Gravitational lenses and the apparent association of QSOs and bright galaxies.   See Abstr. 141.235.

High energy electrons in pulsar magnetospheres. See Abstr. 141.506.

Gravitational radiation and the binary pulsar. See Abstr. 141.527.

Could primordial black holes be the source of the cosmic ray antiprotons?   See Abstr. 143.024.

Star clusters containing massive, central black holes. IV. Galactic tidal fields.   See Abstr. 151.016.

The effect of gravitational radiation on the secular stability of a rotating, axisymmetric galaxy. See Abstr. 151.043.

Interaction of stars with the accretion disc around a massive black hole in the nuclei of active galaxies and quasars.   See Abstr. 158.090.

Active galactic nuclei and particle acceleration in accretion disks around massive black holes. See Abstr. 158.337.

X-ray observations of Abell 2218 and implications for the Sunyaev-Zel'dovich effect.   See Abstr. 160.050.

Background radiation fields as a probe of the large-scale matter distribution in the Universe. See Abstr. 162.006.

Inhomogeneous cosmology: gravitational radiation in Bianchi backgrounds.   See Abstr. 162.007.

Cosmic strings.   See Abstr. 162.048.

The phase structure of the early Universe in the minimal SU(5) grand unified theory.   See Abstr. 162.053.

The behavior of null geodesics in a class of rotating space-time homogeneous cosmologies.   See Abstr. 162.084.

Primordial black holes and the cosmic baryon number − II.   See Abstr. 162.093.

Anisotropy in nonprimordial cosmic background radiation.   See Abstr. 162.096.

A cosmological version of flatness conditions in general relativity.   See Abstr. 162.102.

Gravitational lenses and cosmological evolution. See Abstr. 162.120.

Multipole anisotropy of the cosmic background radiation in density wave models.   See Abstr. 162.136.

Gravitational wave backgrounds and the early Universe.   See Abstr. 162.137.

Two cosmological solutions of Regge calculus. See Abstr. 162.140.

Fate of wormholes created by first-order phase transition in the early universe.   See Abstr. 162.147.

Production of magnetized black holes and wormholes by first-order phase transition in the early universe. See Abstr. 162.149.

Horizon-free universe.   See Abstr. 162.161.

Black hole formation in the early universe. See Abstr. 162.167.

Cosmology of Nordström's first theory of gravitation.   See Abstr. 162.172.

Geodesic instability and internal time in relativistic cosmology.   See Abstr. 162.178.

Colliding gravitational waves in expanding cosmologies.   See Abstr. 162.179.

Cosmology for grand unified theories with radiatively induced symmetry breaking.   See Abstr. 162.192.

Gravitational lenses and cosmological evolution. See Abstr. 162.196.

Kerr-Newman metric in cosmological background. See Abstr. 162.198.

Anisotropy of the cosmic microwave background radiation.   See Abstr. 162.207.

Population III objects and the shape of the cosmological background radiation.   See Abstr. 162.209.

Distortion of the microwave background by dust. See Abstr. 162.210.

The nature and origin of large-scale density fluctuations.   See Abstr. 162.215.

Neutron Stars

066.501 **Model calculation of neutron stars with pion condensations.**
Yu. A. Berezin, O. E. Dmitrieva, N. N. Yanenko.
Pis'ma Astron. Zh., Tom 8, 86 - 89 (1982). In Russian. English translation in Soviet Astron. Lett., Vol. 8.

It is shown numerically that in neutron stars with pion condensation in a time of the order of 1 ms occurs a separation into a core and an envelope with a sharp boundary between them. Yet there is no any blowing off of the envelope.

066.502 **On supercritical disk accretion onto magnetized neutron stars.** V. M. Lipunov.
Astron. Zh., Tom 59, 87 - 91 (1982). In Russian. English translation in Soviet Astron., Vol. 26, No. 1.

The interaction between a supercritical accretion disk and a magnetized neutron star (NS) is considered. The maximum accretion rate onto a magnetized NS is found. The evolution of an NS is considered.

066.503 **Neutron star evolutionary sequences.**
M. B. Richardson, H. M. Van Horn, K. F. Ratcliff, R. C. Malone.
Astrophys. J., Vol. 255, 624 - 653 (1982).

The authors present detailed numerical calculations of the evolution of neutron stars that are cooling through the range of central temperatures from about 10^{10} to 10^{7} K. The calculations are solutions of the full set of general relativistic equations that describe the evolution of a spherical star. The best current expressions for transport processes and neutrino emission rates are employed. In the treatment of thermal properties of neutron star matter the effects of nucleon superfluidity in the inner crust and core have been included and models with and without a pion condensate at high densities have been constructed. The consequences of crystallization of the crust are also investigated. It is found that localized neutrino cooling is so rapid that heat transport within the star cannot keep pace, and the temperature distribution is not even approximately isothermal. The pion condensate greatly accentuates this effect. It is concluded that present observations are not sufficient to distinguish between the various models of neutron star cooling.

066.504 **Some remarks on the spectra of X-ray bursts.**
J. van Paradijs.
Astron. Astrophys., Vol. 107, 51 - 53 (1982).

The method of determining the apparent radius of a neutron star using black-body and modified black-body spectra of X-ray burst emission is discussed. It is argued that a classical stellar atmosphere in hydrostatic and radiative equilibrium represents a better approximation of the emitting regions observed during burst decay. Some results of preliminary calculations of hot neutron-star atmospheres are presented. A possible way to determine both the mass and radius of neutron stars from X-ray burst observations is indicated.

066.505 **Changing orientation of dipole and spin axes in binary X-ray pulsars.** Y.-M. Wang, M. Robnik.
Astron. Astrophys., Vol. 107, 222 - 228 (1982).

It is shown that the inclination angle between the dipole and spin axes of a neutron star in an X-ray binary system should increase during spinup episodes and decrease during spindown, on a timescale comparable with the change in spin period. This will lead to secular variations in the pulse shapes, and, under the assumption that a typical X-ray pulsar has spun down on net during its history of interaction with its mass-losing companion, may account for an apparent preponderance of "single-pulse" profiles.

066.506 **Neutrino cyclotron radiation from superfluid vortexes in neutron stars: a new mechanism for pulsar spin down.** Q.-H. Peng, K.-L. Huang, J.-H. Huang.
Astron. Astrophys., Vol. 107, 258 - 266 (1982).

The authors discuss a new type of neutrino emission mechanism of neutron stars — the neutrino (cyclotron) radiation emitted by neutron superfluid vortexes in the interior of neutron stars — calculate its power, and derive the increased rate of the pulsar spin down.

066.507 **General relativistic effects on the cooling of neutron stars.** C. Kindl, N. Straumann.
Helvetica Phys. Acta, Vol. 54, 214 - 218 (1981). – Abstr. in Phys. Abstr., Vol. 85, Abstr. 14870 (1982).

066.508 **Helium shell flash on accreting neutron stars: effects of hydrogen-rich envelope and recurrence of X-ray bursts.** T. Hanawa, D. Sugimoto.
Publ. Astron. Soc. Japan, Vol. 34, 1 - 20 (1982).

Complete cycles of shell flashes have been computed numerically through accretion, pre-flash, flash, and the succeeding accretion phases. The case of an intermediate accretion rate has been investigated where a steady hydrogen-burning shell is formed. A helium zone grows below the hydrogen-burning shell and the helium shell flash is triggered when a critical amount of helium is accumulated. After the flash, the steady hydrogen-burning shell is formed again.

066.509 **Remarks on the beta stability in neutron stars.**
J. Boguta.
Phys. Lett. B, Vol. 106B, 255 - 258 (1981). – Abstr. in Phys. Abstr., Vol. 85, Abstr. 23319 (1982).

066.510 **Slow rotation of neutron stars according to the Jordan - Brans - Dicke theory of gravitation.**
V. I. Rejzlin.
Astrofizika, Tom 17, 187 - 192 (1981). In Russian. English translation in Astrophysics, Vol. 17, No. 1.

The problem of slow rotation in the frame work of Jordan - Brans - Dicke theory of gravitation is considered. The equation describing the rotation of equilibrium configurations is obtained in first-order approximation of the angular velocity. The vacuum solution of this equation as well as the form of the moment of inertia are found. The numerical results are compared with the observational data on the pulsar PSR 0532 in the Crab nebula.

066.511 **p-wave superfluidity in neutron stars and ^{3}He.**
J. A. Sauls, D. L. Stein.
Physica B, C, Vol. 107 B + C, No. 1 - 3, p. 55 - 56 (1981). Abstr. in Phys. Abstr., Vol. 85, Abstr. 33001 (1982).

066.512 **Broken symmetries and hydrodynamics of superfluid $^3P_2$-neutron star matter.**
H. Brand, H. Pleiner.
Physica B, C, Vol. 107 B + C, No. 1 - 3, p. 53 - 54 (1981). Abstr. in Phys. Abstr., Vol. 85, Abstr. 33061 (1982).

066.513 **Thermonuclear processes on accreting neutron stars: a systematic study.**
A. Ayasli, P. C. Joss.
Astrophys. J., Vol. 256, 637 - 665 (1982).

The authors have carried out a series of model calculations for the evolution of the surface layers of an accreting neutron star. They systematically varied the neutron star mass, radius, core temperature, and surface magnetic field strength, as well as the accretion rate onto the neutron star surface and the metallicity of the accreting matter, in order to determine the effects of these parameters on the properties of thermonuclear flashes in the surface layers and the emitted X-ray bursts that result from such flashes. The models include the following features: (1) All significant general relativistic corrections to the equations of stellar structure and evolution are taken into account. (2) A simplified but adequate nuclear reaction network is introduced that takes into account proton-capture, alpha-capture and beta decays, involving nuclei with atomic masses up to A = 56. (3) The heat flow into and out of the neutron star core prior to and during a thermonuclear flash is followed in detail, in order to determine accurately the conditions required for the thermal equilibrium of the core. Calculated X-ray burst have properties that in general fall within the range of observed properties of type I X-ray bursts. Possible causes of the most serious remaining discrepancies are discussed.

066.514 On the possibility of observing iron line emission from the surface of magnetized neutron stars.
R. Z. Yahel.
Astron. Astrophys., Vol. 109, 1 - 3 (1982).

The author solves the radiation problem of the Fe XXVI Ly α photons in the emission zone of strongly magnetized neutron stars. It is shown that for the cosmic abundance of Fe and for plausible thermal conditions in this zone, the possibility of observing atomic line transitions in the hard X-ray spectrum is very unlikely. Absorption features may be observed, however, provided the electron temperature close to the stellar surface is relatively low (i.e. less than about 3 keV), or the Fe abundance is abnormally high compared to the cosmic abundance (i.e. bigger by a factor of ≈ 10).

066.515 Gamma ray bursts and neutron stars.
Z. Mikulášek.
Říše hvězd, Vol. 63, 92 - 94 (1982). In Czech.

066.516 Low-luminosity accretion onto magnetized neutron stars.   S. H. Langer, S. Rappaport.
Astrophys. J., Vol. 257, 733 - 751 (1982).

The authors have studied the behavior of matter accreting at low rates ($\dot{M} < 10^{16}$ g s^{-1}) onto the polar caps of a highly magnetized ($B \sim 10^{12}$ gauss) neutron star. They have found flow solutions for the case in which the matter undergoes a stationary, collisionless shock. The electron and ion fluids are treated separately, and the ion temperature is found to be much higher than the electron temperature throughout the flow. At these low accretion rates, the emitted radiation is assumed to exert no significant pressure on the infalling matter and is further assumed to escape from the column without significant degradation in energy. The authors find that cyclotron emission is the dominant energy loss mechanism and can yield continuum spectra resembling those observed from X-ray pulsars. They compute a number of relations among the accretion rate, the surface magnetic field, the shock height, and the characteristic electron and ion temperatures.

066.517 Two dimensional time dependent accretion onto magnetic neutron stars. I. Dynamics with radiation pressure.   R. I. Klein, J. Arons, S. M. Lea.
Bull. American Astron. Soc., Vol. 13, 902 (1981). – Abstract.

066.518 On hard X-ray spectra of accreting neutron stars.
V. V. Zheleznyakov.
Astrophys. Space Sci., Vol. 83, 81 - 103 (1982).

The formation of the spectra of X-ray pulsars and gamma bursters is investigated. Interpretation of a hard X-ray spectrum of pulsars containing cyclotron lines is feasible on the basis of an isothermal model of a polar spot heated due to accretion to a neutron star. The part played by the accreting column in the case of strong accretion ($\cong 10^{19}$ el cm^{-3}) needed for sustaining the high level of X-ray emission from a neutron star-pulsar is studied. The spectra of gamma-bursters recorded by 'Venus 11' and 'Venus 12' are interpreted in terms of a two-layer model of a polar hot spot. Estimates are given of the distance to some of the bursters, of the emission

measure from a high-temperature layer responsible for continuum radiation and of the dispersion measure of a colder layer forming cyclotron lines in absorption. The problem of measuring the magnetic fields of neutron stars taking account of the gravitational redshift and the quantum recoil effect in emission and in absorption is discussed.

066.519 Relativistic ejection from compact stars with a strong magnetic field.
I. G. Mitrofanov, A. I. Tsygan.
Astrophys. Space Sci., Vol. 84, 35 - 51 (1982).

The acceleration of charged particles by radiation in the strong magnetic field of a compact star is considered. Different regimes of ejection, the dependence on intensity, spectrum, angular distribution and polarization of accelerating radiation, and the influence of the opacity of ejecting plasma are analyzed. The energy of ejected plasma is shown to increase up to relativistic values. A possible connection of relativistic ejection with the origin of gamma-ray bursts and other astrophysical consequences are discussed.

066.520 Modelling of rotating neutron stars using exact solutions of Einstein's equations.
V. G. Pisarenko, A. N. Kryshtal (*Krishtal*), Yu. A. Selivanov.
Acta Astronaut., Vol. 8, 831 - 838 (1981). – Abstr. in Phys. Abstr., Vol. 85, Abstr. 58375 (1982).

066.521 Neutrino cooling of neutron stars by percolating quarks.   M. Kiguchi, K. Sato.
Prog. Theor. Phys., Vol. 66, 725 - 728 (1981). – Abstr. in Phys. Abstr., Vol. 85, Abstr. 58376 (1982).

066.522 Magnetic vortices in a rotating $^3P_2$ neutron superfluid.   J. A. Sauls, D. L. Stein, J. W. Serene.
Phys. Rev. D, Vol. 25, 967 - 975 (1982). – Abstr. in Phys. Abstr., Vol. 85, Abstr. 58377 (1982).

066.523 The rheology of neutron stars. Vortex-line pinning in the crust superfluid.
P. W. Anderson, M. A. Alpar, D. Pines, J. Shaham.
Philos. Mag. A, Vol. 45, 227 - 238 (1982). – Abstr. in Phys. Abstr., Vol. 85, Abstr. 58378 (1982).

066.524 Pion condensation in cold dense matter and neutron stars.   P. Haensel, M. Prószyński.
Astrophys. J., Vol. 258, 306 - 320 (1982).

The authors study possible influence, on the neutron star structure, of a pion condensation occurring in cold dense matter. Several equations of state with pion-condensed phase are considered. The models of neutron stars are calculated and confronted with existing observational data on pulsars. Such a confrontation appears to rule out the models of dense matter with an abnormal self-bound state, and therefore it seems to exclude the possibility of the existence of abnormal superheavy neutron nuclei and abnormal neutron stars with a liquid pion-condensed surface. The authors also consider the collapse of a normal neutron star to a configuration with pion-condensed core. This could happen when pion condensation implies the first order phase transition in dense matter.

066.525 Matter accreting neutron stars.   P. Mészáros.
NASA Tech. Memo., NASA TM-83835, 20 pp. (1981).

Some of the fundamental neutron star parameters, such as the mass and the magnetic field strength, have been experimentally determined in accreting neutron star systems. The author reviews some of the relevant data and the models used to derive useful information from them, concentrating mainly on X-ray pulsars. He discusses the latest advances in our understanding of the radiation mechanisms and the transfer in the strongly magnetized polar cap regions.

Rigorous and approximate scaling laws for the photoionization cross-section of hydrogenic ions in magnetic fields.   See Abstr. 022.007.

Pion condensation, equation of state of dense matter and neutron stars.   See Abstr. 022.088.

Structure of baryonic system with pion condensation and its implication in neutron star problems. See Abstr. 022.089.

High-spin effects in superdense matter. See Abstr. 022.090.

Spin-dependent polarizabilities of hydrogenic atoms in magnetic fields of arbitrary strength. See Abstr. 022.094.

Neutrino oscillations in neutron star matter. See Abstr. 061.048.

Modifikation des Bremsstrahlungsquerschnitts durch starke Magnetfelder.   See Abstr. 062.049.

Pinned vorticity in rotating superfluids, with application to neutron stars.   See Abstr. 062.053.

Broken symmetries and hydrodynamics of super-fluid $^3P_2$-neutron-star matter.   See Abstr. 062.056.

A model of two-stream non-radial accretion for binary X-ray pulsars.   See Abstr. 064.077.

Physical processes in stars on late stages of their evolution.   See Abstr. 065.064.

Electromagnetic cascades in pulsars. See Abstr. 141.501.

Pair production and pulsar cutoff in magnetized neutron stars with nondipolar magnetic geometry. See Abstr. 141.510.

X-ray bursts and shell flashes on accreting neutron stars.   See Abstr. 142.032.

X-ray synchrotron nebulae and the origin of neutron stars.   See Abstr. 142.091.

The gamma-ray burster puzzle. See Abstr. 142.510.

Errata

066.901 Erratum: "On gravitational lenses and the cosmo-logical evolution of quasars" [Astrophys. J., Lett., Vol. 248, L95 - L99 (1981)].   Y. Avni. Astrophys. J., Lett., Vol. 253, L95 (1982). – See Abstr. 30.066.051.

066.902 Erratum: "The Einstein-Maxwell field equations. II." [Astrophys. Space Sci., Vol. 77, 383 - 389 (1981)].   D. D. Dionysiou. Astrophys. Space Sci., Vol. 82, 255 (1982). – See Abstr. 29.066.053.

066.903 Erratum: "Spectral shifts near compact objects" [Astrophys. Space Sci., Vol. 79, 515 - 519 (1981)]. K. Lake, E. Myra. Astrophys. Space Sci., Vol. 82, 255 (1982). – See Abstr. 30.066.112.

Sun

071 Photosphere, Spectrum

071.001 Solar luminosity variation. IV. The photospheric lines, 1976–1980. W. Livingston, H. Holweger.
Astrophys. J., Vol. 252, 375 - 385 (1982).

Kitt Peak full disk spectrophotometric records covering the period 1976 - 1980 have been analyzed to study the behavior of seven spectrum lines sensitive to photospheric parameters. A secular decrease of equivalent widths ranging from 0 to 2.3% is observed. The authors conclude that the line weakenings are due to global variations of surface properties. Model atmosphere analysis suggests that the observed response pattern reflects a slight flattening of the lower photospheric temperature gradient, corresponding to a 15% increase in mixing length, at constant, effective temperature. The associated increase in the efficiency of convection can be reconciled with a constant luminosity if the change is assumed to occur only in the outer ~100 km of the convection zone.

071.002 Note on the interpretation of Fe I lines (2.18–2.49 eV) in the solar spectrum.
D. E. Blackwell, M. J. Shallis, G. J. Simmons.
Mon. Not. R. Astron. Soc., Vol. 199, 33 - 36 (1982).

Fourteen Fe I lines ($2.18 \, eV < \chi < 2.4 \, eV$) in the centre of disc solar spectrum are analysed using Oxford oscillator strengths of 1 per cent accuracy. The iron abundances and microturbulence velocities given by these lines are discussed. The five lines with excitation energies 2.40–2.48 eV apparently give a higher abundance than the nine lines of excitation energy 2.18–2.22 eV.

071.003 Interpretation of Ti I lines of excitation energy 0.0–0.05 eV in the solar spectrum; use of new oscillator strengths of accuracy 0.5 per cent.
D. E. Blackwell, M. J. Shallis, G. J. Simmons.
Mon. Not. R. Astron. Soc., Vol. 199, 37 - 42 (1982).

An analysis is made of the seven most suitable 0 eV Ti I lines in the spectrum of the centre of the solar disc, using oscillator strengths of 0.5 per cent relative accuracy, and the solar model atmospheres of Holweger & Müller and Vernazza et al. The solar abundance of titanium is found to be $\log A = 5.08$ and the microturbulence is 1.00 km s^{-1}, using the preferred atmosphere of Holweger & Müller.

071.004 Table of solar diatomic molecular lines. IV. Spectral range: 7600-8100 Å.
R. Boyer, P. Sotirovski, J. W. Harvey.
Astron. Astrophys., Suppl. Ser., Vol. 47, 145 - 157 (1982).

The present publication is an extension of results already published covering 6100 - 6600, 6600 - 7100 and 7100 - 7600 Å. According to this investigation, equivalent widths of molecular lines become increasingly difficult to measure beyond 8030 Å. As no valuable quantitative information is to be expected from the study of the near infrared part of the spectrograms, the analysis stops at 8100 Å.

071.005 On the establishment of internally consistent solar scales of oscillator strengths and abundances of chemical elements. III. Oscillator strengths obtained from equivalent widths of 360 FeI lines.
E. A. (*Eh. A.*) Gurtovenko, R. I. Kostik (*Kostyk*).

Astron. Astrophys., Suppl. Ser., Vol. 47, 193 - 197 (1982).

Oscillator strengths for 360 selected Fraunhofer lines were determined using the observed equivalent widths of those lines. The R.M.S. error of the *gf*-values obtained amounts to 0.06–0.07 dex. The comparison of oscillator strengths calculated from equivalent widths (w) and central intensity (d) shows the reliability of the methods used and of the underlying data. The systematic increase of the difference $\log gf_w - \log gf_d$ for strong lines of high excitation potential is noticeable.

071.006 On the phenomenon of chains in the photospheric granulation. L. D. Parfinenko.
Soln. Dannye 1981 Byull., No. 10, p. 101 - 106 (1981). In Russian.

The problem of bright chains of granulation is studied. The contrast of the chains exceeds the mean contrast of granulation by several per cent (to 10%). The lifetime of bright chains is twice as long as the mean lifetime of granulation. A weaker effect of the chain structure of the photospheric granulation is noted.

071.007 Interpretation of the CH molecular line alteration in the center and on the limb of the solar disk. I. Mean optical depths of formation and profiles of CH absorption lines in the solar spectrum. D. V. Erofeev, Yu. A. Solonskij.
Soln. Dannye 1981 Byull., No. 12, p. 93 - 99 (1982). In Russian.

Absorption lines of (0, 0), (1, 1), (2, 2) bands of transitions $A^2 \Delta - X^2 \Pi$ and of (0, 0), (1, 1), (1, 0) bands of transition $B^2 \Sigma^- - X^2 \Pi$ for the CH molecule in the solar spectrum are investigated. Recommendations for use of spectral molecular data and physical parameters of the photosphere are given.

071.008 Center to limb observations of sodium lines in the solar spectrum. A. K. Pierce, C. Slaughter.
Astrophys. J., Suppl. Ser., Vol. 48, 73 - 93 (1982).

Center to limb ($\cos \theta = 1.0$ to 0.1) intensities are given for the profiles of 25 sodium lines distributed from λ3302 to λ22083 in the solar spectrum. Central intensities and half-widths are tabulated for five additional sodium lines.

071.009 Numerical simulations of the solar granulation. I. Basic equations and methods. Å. Nordlund.
Astron. Astrophys., Vol. 107, 1 - 10 (1982).

The hydrodynamical and radiative transfer equations that govern the evolution of granular convection patterns are discussed. The anelastic approximation of the continuity equation is used as a convenient way of excluding pressure waves from the problem. Suitable numerical methods are developed and used to numerically simulate the solar granulation.

071.010 Die Farbenindizes der Sonne im UBV-System. H. Tüg, T. Schmidt-Kaler.
Mitt. Astron. Ges., Nr. 55, p. 18 - 19 (1982). – Abstract.

071.011 Modelle photosphärischer Magnetfeld-Konzentrationen.

W. Deinzer, G. Hensler, D. Schmitt, M. Schüssler, E. Weißhaar.
Mitt. Astron. Ges., Nr. 55, p. 65 - 68 (1982).

071.012 Limb-Effekt und Asymmetrien solarer Spektral-
linien zwischen 4800 und 6500 Angström in
Fouriertransformspektren   H. Balthasar, H. Wöhl.
Mitt. Astron. Ges., Nr. 55, p. 71 (1982). – Abstract.

071.013 Einfluß von Temperaturinhomogenitäten in der
Sonnenatmosphäre auf die Bestimmung von
Elementhäufigkeiten.
W. Hermsen, H. Holweger, W. Mattig.
Mitt. Astron. Ges., Nr. 55, p. 91 (1982). – Abstract.

071.014 Vertikale Struktur der solaren Photosphäre II. Das
Geschwindigkeitsfeld kleiner räumlicher Strukturen.
A. Nesis, C. J. Durrant.
Mitt. Astron. Ges., Nr. 55, p. 92 (1982). – Abstract.

071.015 Messung des Geschwindigkeitsfeldes der solaren
Supergranulation mit einem 100 × 100 Photo-
diodenarray.   G. Küveler, H. Wöhl.
Mitt. Astron. Ges., Nr. 55, p. 92 - 93 (1982). – Abstract.

071.016 Infrared bands of C_2 in the solar photospheric
spectrum.   J. W. Brault, L. Delbouille, N. Grevesse,
G. Roland, A. J. Sauval, L. Testerman.
Astron. Astrophys., Vol. 108, 201 - 205 (1982).
        Lines of the C_2 Phillips system have been successfully
searched for on new tracings of high resolution solar spectra.
From a rather large number of lines of the (0,0), (1,0), and
(0,1) bands, the authors derive empirical values for the band
oscillator strengths. These solar f-values are discussed and
compared with recent laboratory data.

071.017 On the radial rotation law in the solar supergranula-
tion layer.   A. Gailitis (*Gajlitis*), G. Rüdiger.
Astrophys. Lett., Vol. 22, 89 - 96 (1982).
        Solar observations pose the non-trivial problem of under-
standing the maintenance of subrotation by supergranulation.
Subrotation means that such horizontal rotation fields possess
inward directed non-diffusive zonal momentum flux, $V < 0$.
By means of a non-Boussinesq, slowly rotating turbulence
model the authors find that density stratification produces
negative (even negative-definite) V, independent of the assumed
intensity anisotropy. An order of magnitude estimate provides
the observationally required values for those (inviscid) turbu-
lences whose correlation lengths approach the density scale
height. This well-known mixing-length condition is, in fact,
realized in the supergranulation layer of the Sun.

071.018 Determination of the physical conditions in
continuum emission grains.
Eh. A. Baranovskij, A. N. Koval'.
Izv. Krymskoj Astrofiz. Obs., Tom 64, 127 - 132 (1981).
In Russian. English translation in Bull. Crimean Astrophys.
Obs., Vol. 64.
        An observed dependence of the emission of continuum
emission grains (c. e. g.) on the wavelength in the continuum
and in the wings of the K Ca II and D_2 Na I lines is obtained.
This dependence is compared with that calculated for different
depths of c. e. g. locations. The conclusion is made that the observed features of c. e. g. are well explained by a model in
which the temperature in the photospheric layers (τ 10^{-2} to
1.7) is enhanced by 200 - 300 K.

071.019 Titanium abundance in the solar photosphere.
        Eh. A. Gurtovenko, G. L. Fedorchenko,
V. A. Sheminova.
Astrometr. Astrofiz., Vyp. (No.) 43, p. 59 - 62 (1981). In
Russian.
        The titanium abundance in the solar photosphere has

been derived from equivalent widths of 38 weak Fraunhofer
lines of Ti I.

071.020 Investigation of the physical characteristics of the
photosphere from multiplet line profiles. I. LTE
conditions.   V. I. Troyan.
Astrometr. Astrofiz., Vyp. (No.) 44, p. 19 - 23 (1981). In
Russian.
        Oscillator-strengths ratios for 48 pairs of 20 multiplets of
Fe I, Fe II, Ti I, Ni I, V I, Ca I, Na I were defined assuming
LTE.

071.021 Investigation of physical characteristics of the
photosphere from multiplet line profiles. II. Devia-
tion from local thermodynamical equilibrium.   V. I. Troyan.
Astrometr. Astrofiz., Vyp. (No.) 45, p. 9 - 13 (1981). In
Russian.
        The dependence of the source function in the line and
the non-LTE factor on the depth τ_s is determined. Deviation
from LTE must be taken into account for the depths $\tau_s < 0.20$.
The values of oscillator strength ratios obtained in previous
papers are confirmed.

071.022 Non-LTE analysis of potassium K I in the solar
spectrum. I. Initial estimates of level populations.
N. G. Shchukina.
Astrometr. Astrofiz., Vyp. (No.) 45, p. 13 - 20 (1981). In
Russian.
        Tentative estimates of the non-LTE level populations
and source function (λ = 7699 Å) for potassium K I in the
solar atmosphere are discussed. The four-level model atom and
the HSRA model solar atmosphere are used. The conclusion is
made that LTE-assumption is not valid for the resonance
line of K I.

071.023 An additional identification of cyanogen lines in the
solar spectrum in the wavelength interval from
4145 Å to 4190 Å. Weak lines not indicated in Rowland's
tables of 1966.   G. A. Porfir'eva.
Astron. Zh., Tom 59, 372 - 375 (1982). In Russian. English
translation in Soviet Astron., Vol. 26, No. 2.
        The Liège solar atlas has been used. On the high-resolution
spectra many weak lines not indicated in Rowland's table of
solar lines, 1966, were revealed. The wavelengths of these solar
lines were approximately evaluated, about 30 of them have
been identified with the lines of the CN molecule.

071.024 On the size and structure of bright solar Ca⁺-network
cells depending on the heliographic position.
R. Brune, H. Wöhl.
Sol. Phys., Vol. 75, 75 - 78 (1982).
        From photographic recordings of some hundred bright
Ca⁺-network cells on the solar disk the authors find evidence
for a smaller size of polar cells as compared to equatorial cells
by a factor of about 0.9. They do not find an indication of a
dependence of the structure of the cells on the heliographic
position.

071.025 Gamma radiation and photospheric white-light flare
continuum.   H. S. Hudson, B. N. Dwivedi.
Sol. Phys., Vol. 76, 45 - 61 (1982).
        Recent gamma-ray observations of solar flares have
provided a better means for estimating the heating of the solar
atmosphere by energetic protons. Such heating has been
suggested as the explanation of the continuum emission of
the white-light flare. The authors have analyzed the effects on
the photosphere of high-energy particles capable of producing
the intense gamma-ray emission observed in the 1978 July 11
flare. Using a simple energy-balance argument and taking into
account hydrogen ionization conclusions are obtained. It
remains energetically possible, within observational limits,
that high-energy protons could cause sufficient heating of the

upper photosphere to produce detectable excess continuum, but emission from the vicinity of $\tau = 1$ is not significant.

071.026   Quiet Sun observations of the Al I autoionization lines λ1932 and λ1936.
J. W. Cook, O. Kjeldseth Moe.
Sol. Phys., Vol. 76, 109 - 116 (1982).
The authors present quiet Sun observations obtained during a rocket flight of the Al I autoionization lines λ1932 and λ1936 at solar pointings ranging from $\mu = 0.73$ out to the visible limb. Absolute intensities are estimated to be accurate to approximately ±20%. These lines progressively weaken with decreasing μ but never go into emission before finally disappearing with the continuum just beyond the visible solar limb. The observations are compared with LTE line profiles computed through the quiet Sun atmosphere of Vernazza et al. (1976). The authors discuss several areas of disagreement between the synthetic and observed profiles.

071.027   Measurements of the granule-intergranular lane contrast at 5200 Å and 6300 Å.
C. E. Alissandrakis, C. J. Macris, T. G. Zachariadis.
Sol. Phys., Vol. 76, 129 - 136 (1982).
The authors present measurements of the granule-intergranular lane intensity ratio at 5200 Å and 6300 Å, at the center of the disk. The observations were obtained at Pic du Midi and Sacramento Peak observatories between 1967 and 1978. The measurements were corrected for smearing using a two-dimensional model of the brightness distribution and of the instrumental profile. The authors attempt to investigate the change of contrast with time.

071.028   The equatorial photospheric rotation rate.
T. L. Duvall, Jr.
Sol. Phys., Vol. 76, 137 - 143 (1982).
The equatorial photospheric rotation rate has been observed on 14 days in 1978–1980. The resulting rotation rate, $\omega = 14.14 \pm 0.04°$/day, is 2% slower than the rate as observed for long-lived sunspots.

071.029   Verificación de la posición del eje de rotación de la fotósfera solar basada en la estadística de manchas.
T. Paneth.
Bol. Asoc. Argentina Astron., No. 20 - 24, p. 339 - 341 (1981).

071.030   S IV emission-line ratios in the sun.
P. L. Dufton, A. Hibbert, A. E. Kingston, G. A. Doschek.
Astrophys. J., Vol. 257, 338 - 344 (1982).
New atomic data are presented for transitions between the two $3s^2\,3p\ ^2P^0$ levels and three $3s3p^2\,^4P^e$ levels of S IV. Electric dipole transition probabilities have been calculated, using configuration interaction wavefunctions and the Breit-Pauli approximation for five intercombination transitions and are found to differ by typically a factor of 2 from previous values. Electron impact collision strengths are presented for transitions between and among the $^2P^0$ and $^4P^e$ levels, calculated using the R-matrix method. These atomic data are used to predict S IV level populations and emission-line intensity ratios for electron densities and temperatures appropriate to the solar transition region. Excellent agreement is found with observations obtained using the NRL slit spectrograph aboard Skylab.

071.031   The wavelength variation of the granule/intergranule contrast.   R. J. Bray.
Sol. Phys., Vol. 77, 299 - 302 (1982).
Modern measurements of the granule contrast are reviewed and compared with a curve showing the wavelength variation to be expected on the assumption of black-body emission as well as with the predictions of recent inhomogene-

ous models. The difference in effective temperature between granules and intergranular lanes is $\approx 270–280$ K.

071.032   Nonlinear simulations of solar rotation effects in supergranules.   D. H. Hathaway.
Sol. Phys., Vol. 77, 341 - 356 (1982).
Nonlinear calculations for the three-dimensional and time dependent convective flow in a plane parallel layer of fluid are carried out with parameter values appropriate for supergranules on the Sun. A rotation vector is used which is tilted from the the vertical to represent various latitudes. For the incompressible fluid used in this model the solar rotation produces turning motions sufficient to completely twist a fluid column in about one day. The tilted rotation vector produces anisotropies and systematic Reynolds stresses which drive mean flows. The resulting flows produce a rotation rate which increases inward and a meridional circulation with poleward flow along the outer surface.

071.033   Fine structure of the radial velocity field in the solar photosphere.   L. M. Pravdyuk.
Soln. Dannye 1982 Byull., No. 2, p. 103 - 112 (1982). In Russian.
The structure of the radial velocity field and its variations with height in the solar photosphere is studied using a spectrogram taken with the Pamir solar telescope on the 1 August 1978.

071.034   Solar limb brightening at the extreme limb from photoelectric eclipse observations.
W. A. Rosen, H. L. Poss.
Sol. Phys., Vol. 78, 17 - 27 (1982).
Photoelectric observations of the light intensity from the solar crescent before and after totality were made during the eclipses of 7 March, 1970 and 26 February, 1979. Effective wavelengths were determined by interference filters of 20 nm bandwidth. To obtain the limb darkening function, the resulting intensity curves were analyzed by an extension of the method of Julius. The limb darkening function at 433 nm was obtained for the region $0.937 < \sin\theta < 0.9999$. Additional data were obtained at 600 nm for $0.994 < \sin\theta < 0.9999$. The curves at both wavelengths show a distinctive limb brightening effect at $\sin\theta = 0.999$.

071.035   Damping constant and velocity field in the solar photosphere.   R. I. Kostyk.
Akad. nauk Ukrainskoj SSR, Inst. teor. fiz., Prepr., ITF-81-20R 46 pp. Price 16 Kop. (1981). In Russian.
The empirically determined damping constants for Fe, Ti, Cr, Ni lines are equal to $1.3\ \gamma_6$, $1.7\ \gamma_6$, $1.6\ \gamma_6$, $1.6\ \gamma_6$, respectively. In the region of creation of weak and medium strong lines the microturbulence increases with height, the macroturbulence decreases, the total field (vertical component) is depth-independent.

071.036   Curvas de crecimiento empíricas para el Sol.
M. M. Villada, L. A. Milone.
Bol. Asoc. Argentina Astron., No. 26, p. 89 - 93 (1981).

071.037   On the radial rotation law in the solar supergranulation layer.   A. Gajlitis, G. Rüdiger.
Soobshch. Spets. Astrofiz. Obs., Vyp. 32, (see 012.029), p. 76 (1981).

071.038 ^{3}He in the sun.
L. M. Kozlova, G. F. Sitnik.
Astron. Tsirk., No. 1170, p. 1 - 2 (1981). In Russian.

071.039   On the thermal shift of the image of the solar spectrum.   Kh. I. Abdusamatov.
Astron. Tsirk., No. 1194, p. 1 - 2 (1981). In Russian.

071.040 On the intensity of the photospheric continuum and the faculae at the limb from eclipse observations.
L. A. Akimov, I. L. Belkina, N. P. Dyatel.
Astron. Zh., Tom 59, 552 - 562 (1982). In Russian. English translation in Soviet Astron., Vol. 26, No. 3.
The absolute integrated and surface brightness distributions of the photospheric continuum ($\lambda \sim$ 5870 Å) and faculae at the extreme limb are obtained from July 10, 1972 solar eclipse slitless spectrograms. Some possible reasons of the limb brightening in the surface brightness distributions of the photosphere are discussed. The active regions are approximately 300 km higher than the photosphere. A schematic model of a photospheric facula is given.

071.041 Relationships between Ca II H line fine structure and the integrated solar H line.
L. Damé, L. Cram.
Bull. American Astron. Soc., Vol. 13, 829 - 830 (1981). Abstract.

071.042 Excitation of the chlorine I line at 1351Å.
R. A. Shine, B. E. Woodgate, T. R. Ayres.
Bull. American Astron. Soc., Vol. 13, 830 (1981). – Abstract.

071.043 Observation of photospheric gravity waves.
R. T. Stebbins, P. R. Goode, H. A. Hill.
Bull. American Astron. Soc., Vol. 13, 858 (1981). – Abstract.

071.044 Solar Ca II H and K line variations: the sun as a star.
J. W. Cook.
Bull. American Astron. Soc., Vol. 13, 877 (1981). – Abstract.

071.045 The evolution of an average solar granule.
R. C. Altrock, S. Musman.
Bull. American Astron. Soc., Vol. 13, 879 (1981). – Abstract.

071.046 The size of facular points in the quiet photosphere.
R. Muller.
Bull. American Astron. Soc., Vol. 13, 879 (1981). – Abstract.

071.047 A photometric study of heat flow inhomogeneities at the solar photosphere.
P. Foukal, L. A. Fowler, T. Duvall, Jr.
Bull. American Astron. Soc., Vol. 13, 879 (1981). – Abstract.

071.048 The temperature contrast in mesogranulation.
L. J. November.
Bull. American Astron. Soc., Vol. 13, 879 (1981). – Abstract.

071.049 The solar spectrum 3069 Å - 2095 Å an extention of Rowland's preliminary table of solar spectrum wavelengths from the echelle spectrograph flown in 1961 and 1964. C. E. Moore, R. Tousey, C. M. Brown.
Bull. American Astron. Soc., Vol. 13, 879 - 880 (1981). Abstract.

071.050 An improved theoretical solar photospheric model.
R. L. Kurucz.
Bull. American Astron. Soc., Vol. 13, 888 (1981). – Abstract.

071.051 ATM observations of H I Lyman α above the quiet solar limb. D. Roussel-Dupré.
Bull. American Astron. Soc., Vol. 13, 909 - 910 (1981). Abstract.

071.052 A search for granular induced wave modes.
S. L. Keil.
Bull. American Astron. Soc., Vol. 13, 911 (1981). – Abstract.

071.053 Observations of photospheric line profiles in plages.
S. R. Walton.
Bull. American Astron. Soc., Vol. 13, 913 (1981). – Abstract.

071.054 Height dependence of steady flows determined from coordinated SMM and SPO observations.
K. B. Gebbie, F. Hill, J. Toomre, L. J. November, G. W. Simon, J. B. Gurman, R. A. Shine, B. E. Woodgate.
Bull. American Astron. Soc., Vol. 13, 914 (1981). – Abstract.

071.055 A comment on the character of the photospheric granular net. L. Hejna.
Hvar Obs. Bull., Vol. 5, No. 1, p. 25 - 30 (1981).

071.056 On the centre-to-limb variation and latitude dependence of the asymmetry and wavelength shift of the solar line λ 5576. P. N. Brandt, E. H. Schröter.
Sol. Phys., Vol. 79, 3 - 18 (1982) = Mitt. Kiepenheuer-Inst. Nr. 207.
Low noise photoelectric measurements of the line profile of the $g = 0$ Fe line λ 5576.097 combined with determinations of the wavelength shift of its centre calibrated by use of an I_2 absorption tube are reported. Measurements taken at various limb distances ($1.0 \leqslant \cos \vartheta \leqslant 0.2$) and along 4 different diameters of the Sun are used to investigate the behaviour of the line asymmetry (C-shape) and wavelength shift of the line centre as functions of $\cos \vartheta$ and of latitude and to search for possible pole-equator differences. An accuracy of approx. 0.8 mÅ r.m.s. is achieved for the determination of the centre of the solar line relative to the iodine lines and of 0.3 mÅ to 1 mÅ r.m.s. for the relative variations of the C-shape. The analysis shows a significant difference between the limb-effect curves along polar and equatorial diameters for $\cos \vartheta \leqslant 0.4$ and changes of the C-shape for $0.9 \geqslant \cos \vartheta \geqslant 0.6$ with a rather strong indication of a latitude dependence of the C-shape.

071.057 Morphological and evolutionary features of Ellerman bombs. H. Kurokawa, I. Kawaguchi, Y. Funakoshi, Y. Nakai.
Sol. Phys., Vol. 79, 77 - 84 (1982).
Morphological and evolutionary features of Ellerman bombs were studied with Hα filtergrams of two active regions very close to the solar limb. The authors quantitatively determined the elongated or spike-like shape of the bomb. The first maximum brightness of a typical bomb is attained, on average in about 2 min. Bombs grow longer in the first brightening phase and their mean upward velocity explains the blue asymmetry of Hα emission profiles of moustaches.

071.058 Observed variability in the Fraunhofer line spectrum of solar flux, 1975 - 1980.
W. Livingston, H. Holweger, O. R. White.
Variations of the solar constant, (see 012.037), p. 95 - 109 (1981).
Over the past five years double-pass spectrometer observations of the "sun-as-a-star" have revealed significant changes in line intensities. The photospheric component has weakened linearly with time 0 to 2.3%. From a lack of correlation between these line weakenings and solar activity indicators like sunspots and plage the authors infer a global variation of surface properties. The behavior of photospheric and chromospheric lines is markedly different, with the possibility of secular change for the former.

071.059 Absolute measurement of the bisector of the 6301.5091 Fe I line in the solar spectrum.
F. Cavallini, G. Ceppatelli, A. Righini.
Astron. Astrophys., Vol. 109, 233 - 237 (1982).
A measurement of the absolute bisector of the 6301.5091 Fe I solar line is presented. The bisector agrees with that previously obtained by Adam et al. (1976). The wavelength of the line coincides with previous measurements, suggesting that, at the disk center, the line is stable within 5 m/s for an integration area of $2'$.

071.060   Granulation, supergranulation and atmospheric
          waves.   V. A. Krat.
Sun and planetary system, (see 012.040), p. 81 - 87 (1982).
   A short review of the last years investigations of the fine
and large scale structure of the solar atmosphere and
connected with this structure atmospheric waves is given.

071.061   Some photospheric characteristics of the sun as a
          star.   E. A. (Eh. A.) Gurtovenko.
Sun and planetary system, (see 012.040), p. 99 - 100 (1982).

071.062   Stark broadening of heavy ion solar lines.
          M. S. Dimitrijević.
Sun and planetary system, (see 012.040), p. 101 - 102 (1982).

071.063   The analysis and observation of the neutral spectrum
          of potassium in the solar atmosphere.
N. G. Stchukina (Shchukina).
Sun and planetary system, (see 012.040), p. 103 - 104 (1982).

071.064   Mesoturbulence in the solar atmosphere.
          R. I. Kostik (Kostyk).
Sun and planetary system, (see 012.040), p. 105 - 106 (1982).

071.065   Voigt profile for the CO emission lines in the solar
          sunspot spectrum.
C. J. Macris, B. C. Petropoulos.
Sun and planetary system, (see 012.040), p. 107 - 108 (1982).

071.066   Solar radio granulation at microwaves and its
          optical identification.
G. B. Gelfreikh (Gel'frejkh), V. M. Bogod, A. N. Korzhavin,
E. V. (Eh. V.) Kononovich, O. B. Smirnova, S. V. Startsev,
V. V. Piotrovich.
Sun and planetary system, (see 012.040), p. 109 - 112 (1982).

   Fast neutron capture on ^{180}Hf and ^{184}W and the
solar hafnium and tungsten abundance.
See Abstr. 022.041.

   Absolute transition probabilities in the spectra of
Eu II. III. Astrophysical applications.   See Abstr. 022.070.

   Radiative lifetimes for Pd I and the solar abundance
of palladium.   See Abstr. 022.084.

   On the effective Landé factor of magnetic lines.
See Abstr. 022.142.

   The empirical determination of damping constants
in the solar photosphere. II. Results inferred from the wings
of Fe I lines.   See Abstr. 022.143.

   The experimental curve of growth in function of
different sets of oscillator strengths.   See Abstr. 022.199.

   How to measure the Sun like a star.
See Abstr. 031.514.

   A direct UBV color measurement of the Sun.
See Abstr. 031.515.

   Temporal autocorrelation functions of solar speckle
pattern.   See Abstr. 031.530.

   A comparison between estimations of Fried's
parameter r_0 simultaneously obtained by measurements of
solar granulation contrast and of the variance of angle-of-
arrival fluctuations.   See Abstr. 031.551.

   An instrument to measure the solar spectrum from
170 to 3200 nm on board Spacelab.   See Abstr. 032.541.

   A selective solar irradiance spectrometer.
See Abstr. 034.043.

   Growth and decay of weak waves in radiative magnetogasdynamics.   See Abstr. 062.037.

   Wave propagation in a magnetically structured
atmosphere. III: The slab in a magnetic environment.
See Abstr. 062.072.

   Appearance and evolution of facular points in the
quiet photosphere.   See Abstr. 072.073.

   Skylab observations of H I Lyman-alpha.
See Abstr. 073.030.

   Modelos de flares fotosféricos y cromosféricos.
See Abstr. 073.054.

   Magnetic fields observed to affect granular
convection.   See Abstr. 075.026.

   Variations in photospheric limb darkening as a
diagnostic of changes in solar luminosity.
See Abstr. 080.004.

   Relationship between photospheric and chromo-
spheric $\Omega(\vartheta, t)$ deduced by temporarily and spatially correlated tracers.   See Abstr. 080.104.

   The sun among the stars. V. A second search for
solar spectral analogs. The Hyades' distance.
See Abstr. 114.010.

Erratum

071.901   Erratum: 'Empirical limb effect curves for the Fe I
          lines λ5250 and λ5576' [Sol. Phys., Vol. 71, 233 -
236 (1981)].   D. H. Bruning.
Sol. Phys., Vol. 76, 199 (1982). – See Abstr. 29.071.037.

072 Sunspots, Faculae, Activity Cycles, Solar Patrol

072.001 Photoelectric observations of propagating sunspot oscillations.
B. W. Lites, O. R. White, D. Packman,
Astrophys. J., Vol. 253, 386 - 392, plate 8 (1982).

The Sacramento Peak Observatory Vacuum Tower Telescope and diode array were used to make repeated intensity and velocity images of a large, isolated sunspot in both a chromospheric (λ8542 Ca II) and a photospheric (λ5576 Fe I) line. The movie of the digital data for the chromospheric line shows clearly a relationship between the propagating umbral disturbances and the running penumbral waves. The oscillations at any given point in the sunspot are very regular, and the phase relationship between the velocity and intensity of the chromospheric oscillations is radically different than that for the quiet Sun. Preliminary interpretation of the phase relationship involves acoustic waves with wave vector directed downwards along the magnetic field lines.

072.002 Some characteristics of the space distribution of sunspot active regions for cycle 20.
B.-s. Tang.
Acta Astrophys. Sinica, Vol. 2, 74 - 76 (1982). In Chinese.

072.003 Transition region oscillations in sunspots.
J. B. Gurman, J. W. Leibacher, R. A. Shine,
B. E. Woodgate, W. Henze.
Astrophys. J., Vol. 253, 939 - 948 (1982).

Time series observations of the profile of the C IV resonance line λ1548.19 have been obtained in eight sunspots with the Ultraviolet Spectrometer and Polarimeter on the Solar Maximum Mission. All of the sunspots display significant oscillations in line-of-sight velocity with frequencies in the range from 5.8 mHz to 7.8 mHz. Significant intensity oscillations were observed at the same periods in four of the time series; the maximum intensity is in phase with maximum blueshift. Estimates of the intensity variation (~10%) expected from adiabatic acoustic waves with the observed velocity amplitudes (0.8−3.5 km s⁻¹) are in reasonable agreement with the measured intensity amplitudes (6−9%).

072.004 The 22-year solar cycle: a heliospheric oscillation?
D. A. Gurnett, N. D'Angelo.
Planet. Space Sci., Vol. 30, 307 - 312 (1982).

A new mechanism is proposed for the origin of the 22-year solar cycle in which the solar cycle is caused by a large-scale oscillation of the heliosphere. In its simplest terms the oscillation is directly analogous to an LC oscillator, with the heliospheric current system providing the inductance, and accumulated charge near the heliosphere boundary providing the capacitance. Estimates of the oscillation period using reasonable parameters are close to 22 years.

072.005 Radiodiagnostics of preflare current sheets in active regions of the sun. V. D. Kuznetsov.
Astron. Zh., Tom 59, 108 - 118 (1982). In Russian. English translation in Soviet Astron., Vol. 26, No. 1.

The possibility of discovery and diagnostics of preflare current sheets in active regions by means of radio observations is considered. Measuring of the radio emission spectra in the mm- and cm-wave range with the help of a narrow-beam antenna allows to discover and to carry out the diagnostics of the current sheet.

072.006 On motions of small sunspots and pores in a sunspot group. L. M. Selezneva.
Soln. Dannye 1981 Byull., No. 9, p. 94 - 97 (1981). In Russian.

Results are given of a 40-min series of frames of sunspot group N 222 obtained on 11 August 1978 with the aim of determining motions of its sunspots and pores in relation to the group center. Quasiperiodic oscillations of relative distances between the features with periods from 13.3 to 6.7 min have been found. Their probable origin is discussed.

072.007 Depolarization of the π-components of Fe I 6213 Å and 6337 Å lines in sunspot spectra.
G. F. Vyal'shin, E. S. Kulagin.
Soln. Dannye 1981 Byull., No. 9, p. 97 - 102 (1981). In Russian.

Depolarization of the center-shifted π-components of the magnetic split of the Fe I 6213 Å and 6337 Å lines in the spectrum of a sunspot umbra in the central zone is studied. On all the unblended lines parts of π-components show an intensive circular polarization. All the lines are suspected to be subject more or less to depolarization of π-components. The effect may be explained in terms of fine structure of the magnetic field of a sunspot. The angular resolution in the element is averaged and a line of magnetic split is formed from the magnetic fields of different azimuth inclinations.

072.008 A comparison of variations in the indices of sunspots and solar radio emission in the 19th and 20th solar cycles. Yu. I. Vitinskij, N. N. Petrova.
Soln. Dannye 1981 Byull., No. 9, p. 102 - 107 (1981). In Russian.

The variation in the correlation coefficients between indices of relative numbers and total area of sunspots and density of the radio emission flux at 2800 and 3750 MHz for 1952 - 1976 is considered. It is shown that the character of this variation is essentially different in the 19th and 20th solar cycle. This is believed to be connected with the secular solar cycle.

072.009 On double maxima of the secular cycles and a long-term forecast of solar activity. V. F. Chistyakov.
Soln. Dannye 1981 Byull., No. 9, p. 107 - 111 (1981). In Russian.

An analysis of the sunspot activity in the XVI - XX centuries shows that the secular minima recur every 90 - 130 years, and its maxima have two peaks. A forecast of the next secular cycles was made with the help of some time regularities. The next secular minimum will take place during cycle N 24, and the secular maximum will be in the second half of the 21st century.

072.010 Peculiarities of distribution of event intensity in different phases of the solar cycle.
M. N. Belovskij, Yu. P. Ochelkov, A. V. Ustinov.
Soln. Dannye 1981 Byull., No. 10, p. 115 - 118 (1981). In Russian.

The intensity distribution of events related to solar flares has been constructed using observations in the epoch of growth, maximum and decrease of the solar activity. It has been found that the intensity distribution of microwave and X-ray bursts does not depend on the cycle phase and solar cycle. In the intensity distribution of the solar cosmic ray events the parameter determining the average intensity of events changes. The maximum average intensity of the solar cosmic ray events is observed during decrease of the solar activity.

072.011 On two-dimensional radiative transfer in a sunspot.
Yu. A. Nagovitsyn.
Soln. Dannye 1981 Byull., No. 11, p. 89 - 93 (1982). In Russian.

An approximate analytical solution is given for the problem of the two-dimensional radiative transfer in a sunspot.

072.012 **Structure of the boundary layer of a sunspot. I. Equilibrium state.**   A. A. Solov'ev.
Soln. Dannye 1981 Byull., No. 12, p. 71 - 75 (1982). In Russian.

It is shown that the boundary layer model of a sunspot disagrees with the laws of hydrodynamics if the magnetic field in the layer is not taken into account. The equilibrium hydrostatic state is considered, when the gas density in the layer equals that of the surroundings, and the temperature difference is determined by a formula.

072.013 **On light bridges of sunspots.**
L. D. Parfinenko.
Soln. Dannye 1981 Byull., No. 12, p. 79 - 86 (1982). In Russian.

An investigation of the fine structure of light bridges has been made. The classification of light bridges is made more strict. Only light bridges belong physically to sunspots. They consist of filaments of an anomalous penumbra with separate bright grains. These bridges are located along the neutral line of the magnetic field.

072.014 **Polarized horseshoes around sunspots at 6 centimeter wavelength.**   K. R. Lang, R. F. Willson.
Astrophys. J., Lett., Vol. 255, L111 - L117 (1982).

Horseshoe-shaped structures associated with sunspot penumbrae have been detected in 6 cm synthesis maps of circular polarization taken with the Westerbork synthesis radio telescope. The high degree of circular polarization (ρ_c = 95%) of the horseshoes requires gyroresonant emission. The absence of polarized emission above the sunspot umbrae is also explained by gyroresonance theory. In sharp contrast to the polarized emission, the total intensity of the 6 cm radiation is enhanced above the sunspot umbrae and exhibits a remarkable correlation with the longitudinal magnetic field of the underlying photosphere. Brightness temperatures of $T_B \approx 10^6$ K are found above sunspot umbrae in coronal regions where the longitudinal magnetic field strength H_l = 600 - 900 gauss.

072.015 **Diode laser heterodyne observations of silicon monoxide in sunspots.**   D. A. Glenar,
D. Deming, D. E. Jennings, T. Kostiuk, M. J. Mumma.
News Lett. Astron. Soc. N. Y., Vol. 2, 37 (1982). – Abstract.

072.016 **Solar activity in 1978.**
R. S. Gnevysheva.
Astronomical calendar. Yearbook. 1982, (see 047.005), p. 157 - 169 (1981). In Russian. – From Ref. zh., 51. Astron., 2.51.374 (1982).

072.017 **The solar cycle.**
G. Newkirk, Jr., K. Frazier.
Phys. Today, Vol. 35, No. 4, p. 25 - 34 (1982).

The mysteries of the 11-year period of sunspot activity are yielding to new approaches, such as magnetic-dynamo modeling and a seismology that can detect photosphere pulsations as small as a few meters in amplitude.

072.018 **A morphological study of some umbral fine structures.**   D. Soltau.
Astron. Astrophys., Vol. 107, 211 - 213 (1982) = Mitt. Kiepenheuer-Inst. No. 208.

Photographic analysis of a high resoluted white-light photograph of a big sunspot (Mount Wilson No. 20551) shows that there is a morphological relation between bright penumbral filaments and umbral dots. Within the umbra evidence is found for the existence of features which are considerably bigger than umbral dots and which may be identified with earlier found "umbral granulation".

072.019 **The two components in the distribution of sunspot groups with respect to their maximum areas.**
M. H. Gokhale, K. R. Sivaraman.
J. Astrophys. Astron., Vol. 2, 365 - 377 (1981).

072.020 **Some indices of sunspot groups in the 11-year cycle No. 20.**   M. Kopecký.
Bull. Astron. Inst. Czechoslovakia, Vol. 33, 65 - 72 (1982).

A continuation of the "Greenwich Catalogues of Large Sunspot Groups" for the years 1965 - 1976 has been compiled and it has been used to discuss the question of sequences of large sunspot groups in the 11-year cycle No. 20. The course of the number of sunspot groups of various types has been determined and discussed, and it has been proved that the decrease of the average importance of the sunspot groups has continued, compared to the 11-year cycles Nos 18 and 19, in accordance with the 80-year cycle.

072.021 **The evolution of the proton flare active region McMath 11128.**
Q.-f. Yin, D.-y. Yiao, Y. Ma, H.-g. Jia.
Acta Sci. Nat. Univ. Pekinensis, No. 4, p. 67 - 70 (1980). In Chinese. – Abstr. in Phys. Abstr., Vol. 85, Abstr. 23270 (1982).

072.022 **Magnetfeld, Intensität und Strömung in Penumbra-Feinstrukturen.**
G. Stellmacher, E. Wiehr, M. Knölker.
Mitt. Astron. Ges., Nr. 55, p. 68 - 69 (1982). – Abstract.

072.023 **Some studies on sunspots in relation to Hα flares and radiobursts.**   T. K. Das, M. K. Das Gupta,
S. K. Sarkar.
Indian J. Radio Space Phys., Vol. 10, 203 - 205 (1981). Abstr. in Phys. Abstr., Vol. 85, Abstr. 42675 (1982).

072.024 **Some properties of the rotation of sunspots.**
S. I. Gopasyuk.
Izv. Krymskoj Astrofiz. Obs., Tom 64, 108 - 118 (1981). In Russian. English translation in Bull. Crimean Astrophys. Obs., Vol. 64.

The results of an investigation of the sunspots rotation within the group crossing the central meridian on October 4, 1974 are presented.

072.025 **On a connection between solar active regions.**
M. B. Ogir'.
Izv. Krymskoj Astrofiz. Obs., Tom 64, 118 - 127 (1981). In Russian. English translation in Bull. Crimean Astrophys. Obs., Vol. 64.

A photometric investigation of 19 solar active regions observed on the Crimean observatory coronograph in September 20, 22, 27 and October 3, 1978 was carried out. For each day of observations the flare activity in all these regions was compared. The obtained results speak in favour of the reality of sympathetic flares and, consequently, of the existence of the physical connection between very distant active regions. These connections are characterized by fine and rather rapid variations in time. The possible cause of the sympathetic flares appearance is either the simultaneous change of the magnetic fields in different active regions or the propagation of hydromagnetic waves in the upper chromosphere or in the corona.

072.026 **Force-free magnetic fields of four-polar sunspots and magnetic energy of the big active region of August 1972.**   Y. Lin, Z. Wang.
Sci. Sinica, Ser. A, Vol. 25, 89 - 98 (1982).

The magnetic field structure of the big active region of August 1972 is discussed using the approximation of the four-polar force-free magnetic field. The analytic formulae of three dimensional components of the magnetic field, the magnetic energy and the extractable energy for a four-polar constant symmetric field are derived.

072.027 Molecular formation in sunspots.
H. M. Lee, D. W. Kim, H. S. Yun, R. Beebe, R. Davis.
J. Korean Astron. Soc., Vol. 14, 19 - 35 (1981).

Calculations of molecular number densities as a function of optical depth in selected umbral, penumbral and photospheric models predict penumbral enhancement of diatomic molecules containing carbon atoms, strong umbral enhancement of oxides, and moderate umbral enhancement of hydrides. The role of CO formation in an oxygen rich atmosphere is discussed.

072.028 An analysis of selected molecular lines in sunspots.
H. M. Lee, H. S. Yun, Y. B. Lee.
J. Korean Astron. Soc., Vol. 14, 79 - 87 (1981).

Theoretical profiles of selected rotational lines of C_2, CH, CN, TiO and MgH are computed by using the current models of sunspot umbrae and penumbrae. It is found that the lines of the diatomic carbides are enhanced in penumbrae relative to umbrae, while MgH lines are more strongly enhanced in umbrae than in penumbrae and the quiet photosphere. The results are discussed with respect to selecting lines suitable for studying the structure of sunspots.

072.029 Simultaneous observations of high resolution spectra over a sunspot umbra.   H. A. Beebe, W. E. Baggett, J. Christou, R. Olson, R. Suggs, H. S. Yun.
Proc. Southwest Reg. Conf., Vol. 7, (see 012.019), 13 (1982). Abstract.

072.030 Comments on filament disintegration and its relation to other aspects of solar activity.
H. W. Dodson, R. Hedeman, M. Rovira de Miceli.
Bol. Asoc. Argentina Astron., No. 18, p. 51 - 55 (1980).

072.031 Eu, La and Sm in sunspot spectra.
H. Molnar.
Bol. Asoc. Argentina Astron., No. 18, p. 55 - 59 (1980).

072.032 Características de una región inusual activa del Sol entre los días 29 de julio y 10 de agosto de 1972.
R. J. Marabini.
Bol. Asoc. Argentina Astron., No. 18, p. 60 - 63 (1980).

072.033 Sunspot numbers.
Sky Telesc., Vol. 63, 97, 215, 317, 430, 531, 638; Vol. 64, 101 (1982).

072.034 Non-linear force-free magnetic field and field distribution over a sunspot.   A. A. Solov'ev.
Astron. Zh., Tom 59, 380 - 388 (1982). In Russian. English translation in Soviet Astron., Vol. 26, No. 2.

A force-free magnetic flux rope with uniform twisting has been considered. The cross-section radius of any magnetic surface varies along the symmetry axis of the system. The field distribution derived as a solution of a non-linear differential equation for the force-free field satisfies the well known similarity condition. The basic properties of the model are derived.

072.035 Developing forecasting charts for sunspot numbers.
S. G. Kapoor, S. M. Wu.
J. Geophys. Res., Vol. 87, 9 - 16 (1982).

A new statistical modeling approach is used to fit models to 200 years of sunspot data and the adequate models for yearly, monthly, and daily sunspot numbers are employed to obtain the minimum mean squared error forecasts. The analysis of new models revealed a long-term 76-year period, a short-term 27-year period, and a 2.5-year period due to stratospheric winds in addition to a commonly known 11-year

periodicity of the sunspots. The forecasting charts are developed in a format that can be easily read to obtain the long-term (yearly average) and short-term (daily average) predictions of sunspot numbers.

072.036 Five-minute oscillations as a subsurface probe of sunspot structure.
J. H. Thomas, L. E. Cram, A. H. Nye.
Nature, Vol. 297, 485 - 487 (1982).

Observations are reported which show that the 5-min. oscillations in a sunspot umbra actually split into several individual modes of different period. The authors interpret these modes of oscillation as the response of the sunspot to forcing by the 5-min p-modes in the surrounding quiet atmosphere. Also, they show how detailed observations of the multiple 5-min modes in a sunspot may be used as a probe of the structure of a sunspot beneath the solar surface.

072.037 Motions and lifetimes of the penumbral bright grains in sunspots.   K. Tönjes, H. Wöhl.
Sol. Phys., Vol. 75, 63 - 69 (1982).

It is confirmed that the penumbral bright grains are moving towards the sunspots umbra. The authors find different proper motions of 0.08 to 0.33 km s^{-1} for different penumbrae and different reduction methods. The lifetimes of these bright grains are about 1.5 to 3 hr depending on the position in the penumbra.

072.038 About the foreshortening effect on sunspot umbral dots.   A. Adjabshirzadeh, S. Koutchmy.
Sol. Phys., Vol. 75, 71 - 74 (1982).

Using high-resolution pictures of the core of a unipolar sunspot observed with several cos θ values, the authors studied the center limb effect on the form of the bright umbral dots. The ratio of the apparent sizes in radial and tangential direction do not show the foreshortening effect typically observed in granular structures.

072.039 On changes of the rotation velocities of stable, recurrent sunspots and their interpretation with a flux tube model.
H. Balthasar, M. Schüssler, H. Wöhl.
Sol. Phys., Vol. 76, 21 - 28 (1982) = Mitt. Kiepenheuer-Inst. Nr. 201.

The angular rotation velocities of stable, recurrent sunspots were investigated using data from the Greenwich Photoheliographic Results 1940 until 1968. The authors found constant rotation velocities during the passages on the solar disk with errors of about ±4 m s^{-1}. During their lifetime these spots show a decreasing braking of their rotation velocities from 0.8 to 0.3 m s^{-1} per day. A plausible interpretation is found by assuming the spots to be coupled to a slowly rising subsurface flux tube and a rotation velocity which increases with depth.

072.040 Recurrence of solar activity: evidence for active longitudes.   R. S. Bogart.
Sol. Phys., Vol. 76, 155 - 165 (1982).

The autocorrelation coefficients of the daily Wolf sunspot numbers over a period of 128 years reveal a number of interesting features of the variability of solar activity. In addition to establishing periodicities for the solar rotation, the solar activity cycle, and perhaps the 'Gleissberg Cycle', the authors suggest that active longitudes do exist, but with much greater strength and persistence in some solar cycles than in others. There is evidence for a variation in the solar rotation period, as measured by sunspot number, of as much as two days between different solar cycles.

072.041 On the dissolution of sunspot groups.
S. G. Wallenhorst, R. Howard.
Sol. Phys., Vol. 76, 203 - 209 (1982).

The behavior of magnetic fluxes from active regions is investigated for times near sunspot disappearance. It is found that the magnetic fluxes decrease on or near the date the spot vanishes. The authors investigate this effect, and conclude that it is actually due to changes in the field, rather than through dissipation of the active region fields. This is important in considerations of the large-scale behavior of solar magnetic fields.

072.042   **Cambios de la polarización de radiobursts en la región McMath 13403.**   R. Marabini.
Bol. Asoc. Argentina Astron., No. 20 - 24, p. 144 (1981). Abstract.

072.043   **Regiones activas MacMath 12136 y 12139. Acoplamiento magnético de larga escala. Radioburst del 9 de Diciembre de 1972.**
R. Marabini, A. González Thomas.
Bol. Asoc. Argentina Astron., No. 20 - 24, p. 236 (1981). Abstract.

072.044   **Sonnenfleckenrelativzahlen des S. I. D. C. (Sunspot Index Data Center).**
Sterne Weltraum, 21. Jahrg., 50, 94, 138, 182, 226, 274 (1982).

072.045   **A new magneto-hydrostatic theory of sunspots.**
V. A. Osherovich.
Sol. Phys., Vol. 77, 63 - 68 (1982).
The structure of a stationary sunspot of circular shape is considered. Schluter-Temesvary theory, based on the 'similarity assumption' is criticized. It is shown that this theory does not describe the observed inclination of magnetic field lines in a sunspot. A new assumption is proposed taking into account field lines which return to the photosphere. On the basis of this assumption, the main equation of the new theory is obtained and the results compare well with observations.

072.046   **On the width distribution of penumbral filaments in sunspots.**   J. A. Bonet, J. D. Ponz,
M. Vazquez.
Sol. Phys., Vol. 77, 69 - 75 (1982).
The mean width and distribution of penumbral filaments of a sunspot have been estimated, using white light photographs. Three areas corresponding to the penumbra of a sunspot have been analysed. Data were collected during the solar eclipse of June 1973. The photometric profiles of the Moon limb over the photosphere have been analysed to obtain useful information on both atmospheric and instrumental perturbation on each exposure.

072.047   **Analysis of the high resolution Mg XI X-ray spectra. II. Physical parameters of the plasma in active region McMath 14352.**   M. Siarkowski, J. Sylwester,
G. Bromboszcz, V. V. Korneev, S. L. Mandelshtam,
(*Mandel'shtam*), S. N. Oparin, A. M. Urnov, I. A. Zhitnik,
S. Vasha.
Sol. Phys., Vol. 77, 183 - 203 (1982).
In this paper, the second in a series dealing with high-resolution spectra (9.14–9.33 Å) measured on board the INTERCOSMOS-16 satellite, the analysis of the physical conditions in the coronal part of the McMath 14352 active region is performed. The temperature structure of the emitting plasma is investigated on the basis of the photon fluxes measured in six selected wavelength bands involving the resonance, intercombination, and forbidden lines of the Mg XI ion and a number of satellite lines. Relative line intensities are discussed in terms of the active region plasma density.

072.048   **Structure of the sunspot boundary layer. II. Convective instability.**   A. A. Solov'ev.

Soln. Dannye 1982 Byull., No. 1, p. 86 - 90 (1982). In Russian.
Hydrostatic equilibrium of the sunspot boundary layer with the temperature difference $\Delta T = TH^2/(8\pi\rho)$ between the layer and surroundings is shown to be convectively unstable.

072.049   **On active longitudes of sunspot groups.**
Yu. I. Vitinskij.
Soln. Dannye 1982 Byull., No. 2, p. 113 - 119 (1982). In Russian.
Active longitudes of the total sunspot area and the number of sunspot groups with the average area 500 m.v.h. have been detected. The main indices (characteristics) of the longitudinal distribution of the number of large sunspot groups have been determined. A relationship between the longitudinal distribution of sunspot groups with the 80-90 year solar cycle has been found.

072.050   **Investigation of the distribution of spot groups and flares near sector boundaries (1964 - 1973).**
V. V. Tel'nyuk-Adamchuk.
Vestn. Kiev. Univ., Vyp. 23, p. 58 - 65 (1981). In Russian.

072.051   **Distribution of the indices of solar activity along the sectors of one polarity (1964 - 1973).**
V. V. Tel'nyuk-Adamchuk.
Vestn. Kiev. Univ., Vyp. 23, p. 66 - 72 (1981). In Russian.

072.052   **The angle of inclination of the sunspot symmetry axis to the solar surface.**
S. O. Obashev, R. Kh. Gainullina (*Gajnullina*),
T. M. Minasyants, G. S. Minasyants.
Sol. Phys., Vol. 78, 59 - 66 (1982).
On the basis of photoheliograms the Wilson effect for the whole solar disk is investigated, the east and west parts of the disk being studied separately. 111 sunspots of regular shape at different heliocentric angles were measured. To study the dependence of the Wilson effect on the heliocentric angle, all observations within an angular interval of $10°$ were averaged. The dependence thus derived is described by two sinusoids having the zero point shifted along both axes. The shift of the zero Wilson effect to the west, i.e., a shift along the heliocentric angle axis, can be caused by the deviation of the sunspot axis to the east from the normal to the solar surface. On the 'line of sight-normal' plane the angle corresponding to this deviation is $\psi = 34° \pm 14°$.

072.053   **Distribution of sunspots according to their magnetic fluxes.**   T. K. Das, M. K. Das Gupta.
Sol. Phys., Vol. 78, 67 - 70 (1982).
The occurrence frequency distribution of sunspots in different magnetic flux values has been examined. The number of sunspots decreases as $\phi^{-1.9}$ for sunspots with magnetic flux greater than 3×10^{21} Maxwell, where ϕ is the said flux of a sunspot.

072.054   **Structure and physics of solar faculae. I. Principles and observational procedures from ground-based instruments and OSO-8 satellite.**
S. Dumont, Z. Mouradian, J.-C. Pecker.
Sol. Phys., Vol. 78, 71 - 81 (1982).
This paper reviews the motivation and the principles of the measurements which aim at a systematic exploration, from photosphere to corona, of the facular regions (plages). The sequence of observation actually achieved is described. Preliminary results are given, in a purely indicative way.

072.055   **Structure and physics of solar faculae. II. The non-thermal velocity field above faculae.**
Z. Mouradian, S. Dumont, J.-C. Pecker, E. Chipman,
G. E. Artzner, J. C. Vial.
Sol. Phys., Vol. 78, 83 - 100 (1982).
The OSO-8 satellite enabled to study various charac-

teristics of the profiles of Si II, Si IV, C IV, and O VI lines above active areas of the Sun, as well as above quiet areas, and to derive some physical properties of the transition region between chromosphere and corona (CCT): (1) The study of the lines shows a general tendency for the microvelocity fields on the average to be nearly constant for the heights corresponding to $T > 10^5$ K. (2) A multicomponent model of the CCT is necessary, and its geometry is far from being a set of plane-parallel columns. It is similar to an association of moving knots within the non-moving principal component of the matter. (3) The proportion of mass, in the knots relative to that in the non-moving component, is several times larger in active regions than in quiet regions. (4) In the knots, the non-thermal microvelocity fields are smaller in active regions and seem to decrease for T increasing above 10^5 K, contrary to what happens in the steady principal component.

072.056 **Progressive brightenings observed in the wing of Hα line.**
I. Kawaguchi, H. Kurokawa, Y. Funakoshi, Y. Nakai.
Sol. Phys., Vol. 78, 101 - 105 (1982) = Contrib. Kwasan and Hida Obs., Univ. Kyoto, No. 249.

In an active region, several points were observed to brighten progressively on the monochromatic image of Hα−1.2Å. The phenomena were interpreted as small flares or subflares. The propagation velocity was measured in two cases and discussed in terms of the multiple loop activation observed in EUV radiation.

072.057 **Solar rotation from 17th century records.**
B. D. Yallop, C. Hohenkerk, L. Murdin, D. H. Clark.
Q. J. R. Astron. Soc., Vol. 23, 213 - 219 (1982).

072.058 **When do sunspot groups occur at the greatest distances from the solar equator?**
M. Kopecký.
Říše hvězd, Vol. 63, 73 - 74 (1982). In Czech.

072.059 **A new look on the solar activity. (Detection of the torsional oscillations).** D. L. Dimitrov.
Říše hvězd, Vol. 63, 89 - 92 (1982). In Czech.

072.060 **Provisional sunspot-numbers for December 1981 − May 1982.**
Yamamoto Circ., Nos. 1968, 1970, 1973 - 1976 (1982).

072.061 **Solare Beobachtungsergebnisse. Solar data. Solar radio emission.** 1981 September - December.
HHI Sol. Data, Vol. 32, 132 - 198 (1981).

072.062 **The magnetohydrostatic model of the facula.**
T. Fla.
Bull. American Astron. Soc., Vol. 13, 835 (1981). − Abstract.

072.063 **Evershed effect in return-flux sunspot model.**
V. A. Osherovich.
Bull. American Astron. Soc., Vol. 13, 838 (1981). − Abstract.

072.064 **Five-minute oscillations in sunspots.**
J. H. Thomas, L. E. Cram, A. H. Nye.
Bull. American Astron. Soc., Vol. 13, 858 (1981). − Abstract.

072.065 **SMM/UVSP observations of oscillations and other properties in a sunspot.**
W. Henze, E. Tandberg-Hanssen, E. J. Reichmann, R. A. Shine, B. E. Woodgate, J. B. Gurman, R. G. Athay.
Bull. American Astron. Soc., Vol. 13, 858 (1981). − Abstract.

072.066 **A search for microwave counterparts of umbral flashes.**
D. McConnell, M. R. Kundu, E. J. Schmahl, B. Lites.
Bull. American Astron. Soc., Vol. 13, 858 (1981). − Abstract.

072.067 **The sunspot atmosphere at 10^5 K.**
J. B. Gurman, B. E. Woodgate.
Bull. American Astron. Soc., Vol. 13, 880 (1981). − Abstract.

072.068 **Observations of small magnetic features in sunspots and active regions.** J. K. Lawrence.
Bull. American Astron. Soc., Vol. 13, 882 (1981). − Abstract.

072.069 **On long-term periodicities in the sunspot cycle.**
S. G. Wallenhorst.
Bull. American Astron. Soc., Vol. 13, 905 (1981). − Abstract.

072.070 **Large-scale patterns in solar activity during the ascending phase of cycle 21.**
V. Gaizauskas, K. Harvey, J. Harvey, C. Zwaan.
Bull. American Astron. Soc., Vol. 13, 906 (1981). − Abstract.

072.071 **Temperature-field relation and current distribution in the return-flux model.**
A. Skumanich, V. A. Osherovich, T. Flaa.
Bull. American Astron. Soc., Vol. 13, 910 (1981). − Abstract.

072.072 **Active region flows in the transition region.**
R. G. Athay, J. B. Gurman, W. Henze.
Bull. American Astron. Soc., Vol. 13, 914 - 915 (1981). Abstract.

072.073 **Appearance and evolution of facular points in the quiet photosphere.** R. Muller.
Bull. American Astron. Soc., Vol. 13, 925 (1981). − Abstract.

072.074 **A note on sunspot group development, spot motions and solar rotation.** L. Dezsö, Á. Kovács.
Publ. Debrecen Heliophys. Obs., Vol. 4, 35 - 59 (1981).

On the basis of long daily series of accurate photographic sunspot positions over some consecutive days it is shown through a few examples that there is the possibility of distinguishing between two kinds of apparent spot motions, i.e. motions relative to the surrounding photosphere and those due to solar rotation. It could also be seen from these limited instances that the development of a sunspot group and flare occurrences are closely associated with spot motions and the rate of solar rotation can already be determined from a few days observations.

072.075 **On sunspot umbra models.** M. Hotinli.
Istanbul Üniv. Fen Fak. Mec. Seri C, Vol. 45, 39 - 54 (1980) = Publ. Istanbul Univ. Obs. No. 114 (1981).

072.076 **The sunspot observations made in 1974.**
H. Menteşe.
Istanbul Üniv. Fen Fak. Mec. Seri C, Vol. 45, 177 - 186 (1980) = Publ. Istanbul Univ. Obs. No. 119 (1981).

This paper gives the heliographic coordinates for the sunspot groups observed in 1974 and some other results.

072.077 **The sunspot observations made in 1975.**
D. Koçer.
Istanbul Üniv. Fen Fak, Mec. Seri C, Vol. 45, 187 - 203 (1980) = Publ. Istanbul Univ. Obs. No. 120 (1981).

This paper gives the heliographic coordinates for the sunspot groups observed in 1975 and some other results.

072.078 **Solar maps and activity.** 1981 September - 1982 February. F. J. Heyden, V. L. Badillo.
Manila Obs., Sol. Div.

072.079 **Sunspots,** July 1979 - December 1980.
Q. Bull. Sol. Act., Vol. 21, Part I, p. 5 - 7 (1979); Vol. 22, Part I, p. 1 - 7 (1980).

072.080 **Evidence of interdependence within 22-year solar cycles.**   B. P. Tritakis.
Astrophys. Space Sci., Vol. 82, 463 - 471 (1982).

In this paper the author demonstrates some strong indications that the evolution and the final modulation of a solar cycle has a close dependence on some basic parameters of the previous solar cycle. The evidence of such an interdependence supports the notion that 11-year solar cycles are not independent.

072.081 **Comparison of estimated and observed active region intensity balance.**   G. A. Chapman, A. D. Meyer.
Bull. American Astron. Soc., Vol. 14, 573 (1982). – Abstract.

072.082 **Intensity ratios of spectral cores Ca II K to H across a sunspot.**   H. S. Yun, H. A. Beebe.
Sol. Phys., Vol. 78, 347 - 349 (1982).

Ratios of the intensity in the core of the Ca II K line to the intensity of the H line core across a sunspot (SPO 5007) were determined from measurements of spectra made simultaneously with the echelle spectrograph at the Vacuum Tower Telescope, Sacramento Peak Observatory. The measured values averaged over the neighboring nonspot region, the penumbra and the umbra are found to be 1.13 ± 0.04, 1.19 ± 0.05, and 1.25 ± 0.03, respectively.

072.083 **Umbral oscillations in a detailed model umbra.**   J. H. Thomas, M. A. Scheuer.
Sol. Phys., Vol. 79, 19 - 29 (1982).

The authors' theory of umbral oscillations as resonant modes of magneto-atmospheric waves (Scheuer and Thomas, 1981) is extended and confirmed by calculating the resonant modes in a much more detailed model of the umbral atmosphere. The depths of forcing required to produce observed oscillation periods (roughly 140 to 185 s) are in good agreement with the depths of overstable convection found in other studies (Moore, 1973; Mullan and Yun, 1973).

072.084 **The chromosphere above a sunspot umbra.**   H. A. Beebe, W. E. Baggett, H. S. Yun.
Sol. Phys., Vol. 79, 31 - 39 (1982).

The authors examined a specific part of a sunspot umbra which is thought to be coolest over the spot. An optimum model representing such a region is presented and its physical properties are discussed.

072.085 **On the latitude drift of sunspot groups and solar rotation.**   J. Tuominen, J. Kyröläinen.
Sol. Phys., Vol. 79, 161 - 172 (1982).

The N–S drift of sunspot groups has been studied in a different way than previously, using positions of recurrent groups of the years 1874–1976. The existence of the meridional motions, the general shape of the drift curves, and the dissimilarity between these curves around sunspot maxima and minima, are all confirmed. In addition, also for the angular velocity of the Sun the same material gives differences around the times of sunspot maxima and minima.

072.086 **Daily maps of the sun and of magnetic fields of sunspots.**
Soln. Dannye 1981 Byull., No. 9, p. 1 - 93; No. 10, p. 1 - 90; No. 11, p. 1 - 88; No. 12, p. 1 - 70; 1982 Byull., No. 1, p. 1 - 79; No. 2, p. 1 - 102 (1981/1982). In Russian.

072.087 **Der Wilson-Effekt am Beispiel einer H-Gruppe.**   J. Jahn.
Sonne, Jahrg. 6, 70 - 73 (1982).

072.088 **On the seat of the solar cycle.**   D. Gough.
Variations of the solar constant, (see 012.037), p. 185 - 206 (1981).

This paper is a discussion of some of the issues that have been raised in connection with the seat of the solar cycle. Is the cycle controlled by a strictly periodic oscillator that operates in the core, or is it a turbulent dynamo confined to the convection zone and possibly a thin boundary layer beneath it? It appears that the ratio $W = \delta \ln R / \delta \ln L$ increases with the depth of the disturbance that produces the variations, so that imminent observations might determine whether or not the principal dynamical processes are confined to only the outer layers of the sun.

072.089 **Variation of the solar He I 10830 Å line: 1977 - 1980.**   J. W. Harvey.
Variations of the solar constant, (see 012.037), p. 265 - 272 (1981).

072.090 **Brackett-gamma line and facula models.**   B. M. Tripathi, M. C. Pande.
Bull. Astron. Soc. India, Vol. 9, 287 - 290 (1981).

An attempt has been made to show that the profile of the Brackett-gamma line can distinguish between different facula models. It is suggested that low noise high resolution scans of Brackett lines can help improve the existing facula models.

072.091 **Le tre "leggi" del ciclo delle macchie solari.**   G. Bonacina.
Orione, Vol. 3, 73 - 76 (1982).

072.092 **Subphotospheric velocity fields inferred by sunspots motions.**   M. Ternullo.
Sun and planetary system, (see 012.040), p. 79 - 80 (1982).

072.093 **Motions in sunspots like torsional oscillations.**   S. I. Gopasyuk.
Sun and planetary system, (see 012.040), p. 125 - 126 (1982).

072.094 **Sunspots proper motions and positions of H-alpha filaments in the active region of August 1979.**   Z. B. Korobova.
Sun and planetary system, (see 012.040), p. 127 - 128 (1982).

072.095 **On the umbral dots.**   I. Sattarov.
Sun and planetary system, (see 012.040), p. 129 - 130 (1982).

072.096 **Temperature and steady flows in slender magnetic tubes.**   E. Ribes.
Sun and planetary system, (see 012.040), p. 143 - 144 (1982).

072.097 **Some dynamical effects of the sunspot magnetic flux ring.**   D.-q. Zhou, J.-k. Qian.
Acta Astron. Sinica, Vol. 23, 23 - 28 (1982). In Chinese.

072.098 **Solar radio SVC radiation in an active region of bipolar sunspots.**   R.-y. Zhao, S.-j. Qian.
Acta Astron. Sinica, Vol. 23, 34 - 38 (1982). In Chinese.

072.099 **Solar activity – the sun as an X-ray star.**   L. Golub.
Smithsonian Astrophys. Obs., Spec. Rep. 392, (see 012.045), p. 39 - 52 (1982).

072.100 **High-sensitivity circular polarimetry of the sun, $0.5 - 1.7 \mu$.**   J. C. Kemp.
Smithsonian Astrophys. Obs., Spec. Rep. 392, (see 012.045), p. 191 - 197 (1982).

Using an isometric (symmetrical) photoelastic-modulator polarimeter and a Cassegrain telescope, the author has obtained broadband polarization measures of the sun, with unprecedented sensitivity.

072.101   Relations of solar activity indices.
          J. Kleczek, J. Olmr.
Compendium in astronomy, (see 003.012), p. 105 - 110
(1982).
          Currently used and some newly introduced activity in-
dices have been used for individual active regions from the
period January 1967 to June 1972. It is shown to what degree
the indices are interrelated.

072.102   Why the total solar radio flux at a wavelength of
          10 cm cannot fully replace the Wolf relative sunspot
number.   M. Kopecký.
Compendium in astronomy, (see 003.012), p. 111 - 115
(1982).
          The radio flux at a wavelength of 10 cm is an independent,
objectively determined solar activity index which, however,
cannot have a priori the same time course as the Wolf sunspot
number R or as the total sunspot area, and cannot replace
them in an equivalent way.

072.103   A correlation between various indices of solar
          activity and solar cycles.
Yu. I. Vitinsky (*Vitinskij*).
Compendium in astronomy, (see 003.012), p. 139 - 148
(1982).
          A variation with time of various parameters of relation-
ship between relative numbers and total areas of sunspots is
considered on the basis of Zürich and Greenwich data for
1879-1964. A significant difference of these indices on the
descending and ascending branch of the 11-year cycle is found.
Their variation is essentially different in the even and odd
11-year cycles. Secular variations in parameters of relationship
W-S were found for the ascending branch of 11-year cycles.
An interpretation is given of the revealed time variations in the
relationship between W and S.

072.104   Report of IAU Commission 10: Solar activity
          (Activité solaire).   V. Bumba.
Trans. IAU, Vol. XVIIIA, (see 003.013), p. 55 - 92 (1982).

072.105   L'activité solaire.   M.-J. Martres, G. Zlicaric.
          Astronomie, Vol. 96, 45 - 47, 94 - 95, 166 - 167,
219, 271, 322 - 323 (1982).

          The revival of solar activity after Maunder minimum
in reports and observations of E. Manfredi.
See Abstr. 004.039.

          Solar active regions. A monograph from Skylab solar
workshop III.   See Abstr. 012.053.

          On the possible mechanism of influence of solar
activity on the human organism.   See Abstr. 015.007.

          A calculation on the transition probability in the
A'−X system of YO.   See Abstr. 022.093.

          Die Sonnenflecken-Relativzahl.
See Abstr. 031.507.

          Auswertung der Netz-Relativzahl.
See Abstr. 031.527.

          Correlation between solar activity and Universal
Time variations.   See Abstr. 044.062.

          Oscillating dynamo in the presence of a fossil
magnetic field. The solar cycle.   See Abstr. 062.016.

          Coupling equations for a flow-wave field used to
faculae heating.   See Abstr. 062.069.

          MHD of non-force-free constant pitch field.
See Abstr. 062.085.

          Table of solar diatomic molecular lines. IV. Spectral
range: 7600-8100 Å.   See Abstr. 071.004.

          The size of facular points in the quiet photosphere.
See Abstr. 071.046.

          Zur Struktur der Chromosphäre über Sonnenflecken.
See Abstr. 073.033.

          Rotational modulation of Ca K flux ratio and sun-
spot number.   See Abstr. 073.047.

          A theoretical model of the solar chromosphere over
a sunspot.   See Abstr. 073.140.

          Increasing trend of flare activity before the proton
flare April 30, 1976.   See Abstr. 073.151.

          Solar active region SD 55/1975 in the frame of the
background magnetic field development.
See Abstr. 075.001.

          Magnetic field configurations associated with
polarity intrusion in a solar active region. I. The force-free
fields.   See Abstr. 075.017.

          On the gradient of the magnetic field above sunspots
determined from optical and radio astronomical measurements.
See Abstr. 075.019.

          Activity and structure of vector magnetic fields of
complex sunspots containing δ-classifications
See Abstr. 075.024.

          Active region magnetic fields.
See Abstr. 075.025.

          Open magnetic fields and the solar cycle.
See Abstr. 075.030.

          The measurement of magnetic fields in the solar
atmosphere above sunspots using gyroresonance emission.
See Abstr. 075.037.

          Vector magnetic field evolution, energy storage,
and associated photospheric velocity shear within a flare-
productive active region.   See Abstr. 075.038.

          The XUV structure of solar active regions.
See Abstr. 076.008.

          Soft X-ray emission from active regions shortly
before solar flares.   See Abstr. 076.010.

          Extreme ultraviolet spectra of solar active regions
and their analysis.   See Abstr. 076.011.

          EUV observations of high-speed downflows over
sunspots.   See Abstr. 076.021.

          Observations of ring structure in a sunspot associated
source at 6 centimeter wavelength.
See Abstr. 077.001.

          Centimeter wavelength observations of active regions
and flares with a few arc second resolution.
See Abstr. 077.065.

Solar neutrinos, solar oscillation and the eleven year solar cycle.   See Abstr. 080.016.

Solar irradiance variations due to active regions. See Abstr. 080.027.

Torsional waves on the Sun and the activity cycle. See Abstr. 080.036.

The effects of sunspots on solar irradiance. See Abstr. 080.038.

Relation between solar activity and the moment of beginning of the seasonal maximum of the cesium-137 concentration in the ground layer of the atmosphere. See Abstr. 080.040.

Solar motion and solar activity. See Abstr. 080.059.

Calculated solar constant variations from 1874 through 1981 caused by active regions. See Abstr. 080.071.

Short and long term variations in the "solar constant".   See Abstr. 080.090.

A prospectus for a theory of variable variability. See Abstr. 080.093.

Solar rotation and meridional motions derived from sunspot groups.   See Abstr. 080.106.

Predominant periods in the time series of Drought Area Index for the Western High Plains AD 1700 to 1962. See Abstr. 085.033.

Does the 80-year sunspot period affect the duration of sunshine in Central Europe?   See Abstr. 085.036.

Jupiter's gravitational effects on the solar wind and geomagnetic activity in different epochs of solar activity. See Abstr. 099.012.

Dependence of the radiation dose aboard the Salyut 6 station on the indices of solar and geomagnetic activity. See Abstr. 143.022.

073 Chromosphere, Flares, Prominences

073.001 Interaction of a collisionless conduction front with the chromosphere and solar hard X-ray bursts.
D. F. Smith, D. W. Harmony.
Astrophys. J., Vol. 252, 800 - 809 (1982).

The interaction of a collisionless conduction front with the transition region and chromosphere is investigated in a one-dimensional fluid approach as an explanation of elementary flare bursts in hard X-rays. It is shown that, for finite energy injection times, material boiled off of the chromosphere rises into the corona and eventually quenches the X-ray emission. Softer (10−20 keV) X-rays should come primarily from near the chromosphere, while harder (90−100 keV) X-rays should come primarily from higher in the corona. Ion heating to 10^8 K and upward mass motions to 1000 km s^{-1} should be observed.

073.002 Optically thick lines in a quiescent prominence: profiles of Lyα, Lyβ (H I), k and h (Mg II), and K and H (Ca II) lines with the OSO 8 LPSP instrument.
J. C. Vial.
Astrophys. J., Vol. 253, 330 - 352 (1982).

A full set of observations of the resonance lines of Mg II, Ca II, H I (and O VI) and the Lyβ (H I) line has been obtained with the LPSP instrument on OSO 8. The observing modes (images, profiles) and the procedures of analysis are described, and special attention is paid to the intensity calibrations. Typical line profile parameters (widths, intensities) are summarized. The Mg II lines are measured in the whole prominence only. The full profiles of Lyα and Lyβ (H I) lines are observed for the first time in a quiescent prominence. Both are reversed, with a separation of 0.35−0.40 Å for Lyα and 0.33 Å for Lyβ. The Lyα/Lyβ ratio is lower than the chromospheric one. The ratio of Lyα to Ca K intensities does not vary very much (less than a 1.6 factor) and indicates a ionization degree of ∼3. Both features indicate that the ionizing Lyman continuum penetrates fully in the prominence.

073.003 Observations of solar flare transition zone plasmas from the Solar Maximum Mission. C.-C. Cheng,
E. C. Bruner, E. Tandberg-Hanssen, B. E. Woodgate,
R. A. Shine, P. J. Kenny, W. Henze, G. Poletto.
Astrophys. J., Vol. 253, 353 - 366 (1982).

The authors present observations of the 1980 April 8 flare in the Si VI 1402 Å and O IV 1401 Å lines, obtained with the Ultraviolet Spectrometer and Polarimeter on board the Solar Maximum Mission satellite. They derive from the Si IV and O IV lines the spatial and temporal evolution of density and mass motion in the transition zone plasmas (∼10^5 K) in the preflare stages as well as during the impulsive phase of the flare. Although there are intensity enhancements in various bright points in the 10^5 K plasma prior to the flare, the locations of the flare are unpredictable. The impulsive Si IV/O IV flare burst reaches its peak in a time scale of less than one minute. The intensity enhancement is accompanied by an equally drastic increase in density. Large mass motions are associated with the density increases.

073.004 The vertical propagation of waves in the solar atmosphere. II. Phase delays in the quiet chromosphere and cell-network distinctions.
B. W. Lites, E. G. Chipman, O. R. White.
Astrophys. J., Vol. 253, 367 - 385 (1982).

The differences in the phase of the velocity oscillations between a pair of chromospheric Ca II lines was measured using the Vacuum Tower Telescope at the Sacramento Peak Observatory. The observed phase differences indicate that the acoustic modes are trapped or envanescent, rather than propagating, in the chromosphere. The authors find systematic distinctions in the phase delays between quiet network and cell interior regions for both intensity and velocity oscillations in photospheric and chromospheric lines. The theory of linear perturbations in an isothermal atmosphere is invoked to interpret these differences.

073.005 The bright mass ejection on the solar disk on October 14, 1980. Z.-x. Shi.
Acta Astrophys. Sinica, Vol. 2, 63 - 70 (1982). In Chinese.

073.006 The energy spectrum of 20 keV−20 MeV electrons accelerated in large solar flares.
R. P. Lin, R. A. Mewaldt, M. A. I. Van Hollebeke.
Astrophys. J., Vol. 253, 949 - 962 (1982).

The authors present IMP 6, 7, and 8 measurements of the energy spectrum of 20 keV to 20 MeV electrons observed from large solar flares. To minimize propagation effects, only events from flares located at W30° to W90° solar longitude were chosen for study. The authors find that every event shows the same spectral shape: a double power law with a smooth transition around 100−200 keV and power law exponents of 0.6−2.0 below and 2.4−4.3 above. The spectra are generally similar to those inferred from hard X-ray and microwave measurements, and the peak >20 keV electron flux is well correlated with the peak microwave emission at 10 cm. These findings indicate that the observed electron spectra are representative of the accelerated electron spectra at the Sun. The shock waves observed in these large flare events are likely to be the accelerating agent.

073.007 Magnetic structure of a flaring region producing impulsive microwave and hard X-ray bursts.
M. R. Kundu, E. J. Schmahl, T. Velusamy.
Astrophys. J., Vol. 253, 963 - 974, plates 13, 14 (1982).

The authors have synthesized 6 cm "snapshot" maps from VLA observations of the 1B/M1 flare of 1980 June 25. The spatial and temporal resolutions during the 9 minutes of the impulsive phase were 1″× 2″ and 10 s, respectively. Concurrent hard X-ray observations were made of the burst, but these had no spatial resolution. The 6 cm burst was fully resolved into at least eight components, many of which were bipolar in nature. The four strongest components form a magnetic arcade which bridged the photospheric magnetic neutral line, and the Hα flare ribbons appear as "footprints" to these magnetic structures. The 6 cm emitting region occupies a large fraction of the flaring loops including their tops.

073.008 Relationships between the energetics of impulsive and gradual emissions from solar flares.
C. J. Crannell, J. T. Karpen, R. J. Thomas.
Astrophys. J., Vol. 253, 975 - 982 (1982).

The gradual soft X-ray emissions associated with a homogeneous set of solar flares have been investigated in the context of a thermal model proposed to explain the impulsive components. The results of this investigation are consistent with the hypothesis that the hard X-ray and microwave emissions are produced by bulk heating of a common thermal source. The quantitative relationships require an additional source to explain the soft X-ray observations, consistent with previous results.

073.009 Two-dimensional nonlocal thermodynamic equilibrium transfer computations of resonance lines in quiescent prominences. J. C. Vial.
Astrophys. J., Vol. 254, 780 - 795 (1982).

The author uses the two-dimensional transfer code of Mihalas, Auer, and Mihalas to compute emergent profiles of resonance lines of H I, Mg II, and Ca II. The model (a uniform slab limited in two directions) and the thermodynamic and radiative quantities are described. Two-dimensional and one-dimensional profiles are compared. A good agreement between computed and observed profiles is found for Lyα and Ca II, but not for Mg II lines. Radial velocities improve the agreement. The author discusses necessary improvements in the computations and the observations of resonance lines in prominences.

073.010   **K_{2V}/K_{2R} asymmetries in the sun and stars.**
K. R. Sivaraman.
Astrophys. J., Vol. 254, 814 - 815 (1982).

The K_{2V}/K_{2R} asymmetry in the self-reversed emission peaks of the Ca II K line in the sun is the result of the redward displacement of K_3. This redward displacement is explained as caused by the dark condensations in K_3 and not due to the supergranulation flow pattern. Arguments are presented to show that such asymmetries in the spectra of stars provide evidence for the presence of these structures in their chromospheres.

073.011   **Spatial structure of $\gtrsim 100$ keV X-ray sources in solar flares.**
S. R. Kane, E. E. Fenimore, R. W. Klebesadel, J. G. Laros.
Astrophys. J., Lett., Vol. 254, L53 - L57 (1982).

Stereoscopic X-ray observations of the 1979 November 5 solar flares permitted the first measurements of the altitude structure of the impulsive and gradual sources of $\gtrsim 100$ keV X-rays. It is found that the brightness of the impulsive as well as the gradual hard X-ray sources decreases rapidly with increase in altitude above the photosphere. Most of the X-ray emission seems to originate at altitudes $\lesssim 2500$ km, indicating that models of hard X-ray sources with low ($\lesssim 10^{10}$ cm^{-3}) ion density are not consistent with the observations.

073.012   **High-resolution solar flare X-ray spectra obtained with rotating spectrometers on the Hinotori satellite.**
K. Tanaka, T. Watanabe, K. Nishi, K. Akita.
Astrophys. J., Lett., Vol. 254, L59 - L63 (1982).

High-resolution solar X-ray line spectra obtained with rotating crystal spectrometers on the Hinotori are presented. The wavelength ranges recorded by the two spectrometers are 1.72 - 1.95 Å and 1.83 - 1.89 Å. Resonances and satellites of Fe XXVI and Fe XXV, Kα and Kβ, are observed together with the spectral features from Fe XIX to Fe XXV. For an impulsive X2.2 flare that occurred on 1981 April 2, time profiles of various line intensities, line widths, and plasma parameters are derived.

073.013   **Large scale solar magnetic fields at the site of flares, the greatness of flares, and solar-terrestrial disturbances.**   H. W. Dodson, E. R. Hedeman, E. C. Roelof.
Geophys. Res. Lett., Vol. 9, 199 - 202 (1982).

Major solar flares during 1967 - 1970 are significantly more likely to occur in active regions whose inferred overlying large-scale (~100,000 km) magnetic flux is oriented preferentially north-to-south than for south-to-north or indeterminant orientations. This purely solar effect may be the dominant cause of previously reported correlations between southward solar active region magnetic fields and enhancements in geomagnetic disturbances, solar wind velocities and solar flare proton fluxes.

073.014   **Oscillatory motions in two quiescent prominences.**
V. S. Bashkirtsev, N. I. Kobanov, G. P. Mashnich.
Pis'ma Astron. Zh., Tom 8, 103 - 105 (1982). In Russian.
English translation in Soviet Astron. Lett., Vol. 8.

Oscillations of the line-of-sight velocities with periods of 82.2 min and 76.7 min were detected in two quiescent promi-

nences with coordinates $\varphi = -75°$W and $\varphi = -18°$W respectively.

073.015   **Analysis of the optical spectra of solar flares.
I. – The flare of April 30, 1976.**
E. Acampa, R. Falciani, A. M. Sambuco, L. A. Smaldone.
Astron. Astrophys., Suppl. Ser., Vol. 47, 485 - 503 (1982).

A new interactive procedure to reduce solar flare spectra developed in the framework of the SMY activities is described. This method allows one to obtain the values of the continuum emission of the flare and several parameters of the line emission such as wavelength and intensity of the maximum emission of the line, its halfwidth, the total emissivity within the line and the line identification. The procedure has been tested on existing observational material of the Sacramento Peak Observatory and the results are given for the April 30, 1976 flare spectra.

073.016   **Observational constraints for a theoretical model describing the soft X-ray flare.**
U. Feldman, C.-C. Cheng, G. A. Doschek.
Astrophys. J., Vol. 255, 320 - 324 (1982).

High-resolution solar flare X-ray spectra have recently been obtained from X-ray spectrometer experiments on the Air Force spacecraft P78-1. Interpretation of the spectra has produced new results concerning the physical conditions and time behavior of the thermal soft X-ray emitting plasma at temperatures near 20×10^6 K. The authors argue that soft and hard X-ray events are not causally related to each other. They probably occur in different plasma volumes. The source of the preflare plasma appears to be in the cooler parts of the solar atmosphere, perhaps transition region loops with initial temperatures of 10^5 K and densities of 10^{11} cm^{-3}. Continuous energy input, rather than sequential activation of loops, is required to explain the observations. Compression coupled with chromospheric ablation may produce the high densities in coronal flare loops.

073.017   **Doppler wavelength shifts of ultraviolet spectral lines in solar active regions.**
U. Feldman, L. Cohen, G. A. Doschek.
Astrophys. J., Vol. 255, 325 - 328 (1982).

Doppler shifts are measured for solar UV emission lines formed in the lower transition region of active regions. The wavelength shifts are measured relative to lines of neutral and singly ionized species formed in the chromosphere. Measurements are made for active regions on the disk and near the limb. Measurements are made over time intervals of about 20 minutes at the same solar location and at different locations in active regions, both on the disk and near the limb.

073.018   **A high-resolution measurement of the 2.223 MeV neutron capture line in a solar flare.**
T. A. Prince, J. C. Ling, W. A. Mahoney, G. R. Riegler, A. S. Jacobson.
Astrophys. J., Lett., Vol. 255, L81 - L84 (1982).

An intense solar flare lasting 40 s was observed by the HEAO 3 γ-ray spectrometer on 1979 November 9 at 3:05 UT. The flare was observed in four high-resolution germanium detectors as well as in five CsI shield detectors over an energy range of 100 keV to above 5 MeV. Of particular interest is a line feature at 2.2248 ± 0.0010 MeV. The precise energy measurement provides unambiguous evidence that this is the $^1H(n, \gamma)^2H$ line resulting from neutron capture on hydrogen.

073.019   **Structure of the impulsive phase of solar flares from microwave observations.**   V. Petrosian.
Astrophys. J., Lett., Vol. 255, L85 - L89 (1982).

Variation of the microwave intensity and spectrum due to synchrotron radiation from semirelativistic particles injected at the top of a closed magnetic loop has been described. Using the recent high spatial resolution X-ray observations from the HXIS experiment of the *SMM* and observations by the VLA, it is shown that the high microwave brightness observed at the top of the flare loop can come about if (1) the magnetic field from top to footpoints of the loop does not increase very rapidly, and (2) the accelerated particles injected in the loop have a nearly isotropic pitch angle distribution.

073.020 On "seasonal" variations of solar flare activity.
     I. E. Pogodin.
Soln. Dannye 1981 Byull., No. 10, p. 91 - 96 (1981). In Russian.
     On the basis of the consideration of the 1955 - 1974 data on solar flares it is shown that so-called "seasonal" variations of the flare activity are caused by its enhancements which don't cover more than 15 - 20% of the period under study in the same months.

073.021 Spectral studies of the chromospheric Hα spicules.
     T. G. Gadzhiev.
Soln. Dannye 1981 Byull., No. 10, p. 96 - 101 (1981). In Russian.
     Fifty Hα spicules were studied in time sequences. Total intensities, half-widths, radial velocities, lifetimes of the spicules were determined. The periods and amplitudes in the time variations of these characteristics were also found.

073.022 Determination of the optical thickness of promi-
     nences and flares on the basis of Balmer line profiles
by the least-squares method.
N. A. Yakovkin, M. N. Pasechnik, S. V. Pasechnik.
Soln. Dannye 1981 Byull., No. 10, p. 106 - 110 (1981). In Russian.
     A new method is proposed for a determination of τ_0 and $\Delta\lambda_D$. The method enables to use any number of profile widths of all the observed Balmer lines. The theoretical profiles constructed with τ_0 and $\Delta\lambda_D$ determined by least squares have minimum deviations from observations.

073.023 On the development of calcium plage areas.
     N. Yu. Bocharova, V. A. Burov, G. S. Ivanov-
Kholodnyj, I. V. Mironova, A. A. Nusinov.
Soln. Dannye 1981 Byull., No. 10, p. 111 - 115 (1981). In Russian.
     An analysis of the main features of the evolution in calcium plage areas was developed on the basis of the 1972 - 1978 observations. It is shown that variations of the plage area S as function of time t can be described by a general law. A semi-empirical formula for the description of the plage area in the process of its evolution was obtained.

073.024 Populations of hydrogen levels in prominences.
     E. G. Rudnikova.
Soln. Dannye 1981 Byull., No. 11, p. 105 - 114 (1982). In Russian.
     The statistical equilibrium equations for a 10-level hydrogen atom in the approximation of volume-average parameters are solved. Populations of the 2 - 10 levels, the Menzel coefficients for the second level and the optical thickness in the Hα line calculated with $n_e = 10^{10} - 10^{12}$ cm, $T_e = 5000 - 12000°$K, $\ell = 10^6 - 10^{10}$ cm and $v_t = 10$ km/s are tabulated.

073.025 Rotation of chromospheric filaments in January -
     September 1958.   A. S. Ugol'nikov.
Soln. Dannye 1981 Byull., No. 12, p. 76 - 79 (1982). In Russian.
     The rotation of chromospheric filaments was studied on Hα-spectroheliograms. The law of the filament rotation for the period January - September 1958 was determined as $\xi = 14.67 - 2.60 \sin^2 \varphi$. The results indicate the filament rotation rate to be higher at solar activity maximum than the mean in the cycle. A higher rotation rate of filaments than that of sunspots is confirmed. Time variations of the rotation rate and south-north asymmetry were not discovered.

073.026 Limb flare observations.   Eh. I. Tetruashvili.
     Soln. Dannye 1981 Byull., No. 12, p. 90 - 92 (1982). In Russian.
     The limb flare of November 6, 1979 has been observed on the east solar limb within 6^h - 9^h UT. The observed spectral lines are given in a table as well as equivalent widths and relative intensities. The electron concentration is estimated according to continuous emission and measure of the latter.

073.027 Some lines of the chromosphere and corona in the
     magnetic field.
L. B. Demkina, G. M. Nikol'skij.
Soln. Dannye 1981 Byull., No. 12, p. 105 - 111 (1982). In Russian.
     Properties of some bright spectral lines of the chromosphere and the corona in the magnetic field are discussed. Effective Landé factors are calculated taking into account different contributions of individual line components to the magnetic split. The distribution of the circular polarization along the whole profile with different widths of the contour is given. The effect of the optical thickness of the line on the distribution of the circular polarization along the profile is considered.

073.028 Solar transition region response to variations in the
     heating rate.   J. T. Mariska, J. P. Boris,
E. S. Oran, T. R. Young, Jr., G. A. Doschek.
Astrophys. J., Vol. 255, 783 - 796 (1982).
     The authors have examined the response of a numerical model for the upper chromosphere, transition region, and corona to variations in the energy input. The numerical model solves the set of one-dimensional two-fluid hydrodynamic equations in a simple vertical magnetic flux tube. The atmosphere responds to both the increase and decrease in energy deposition by smoothly readjusting the temperature gradient and the amount of material in the region of peak radiating efficiency to radiate away energy being deposited. At no time during this readjustment a departure from a thin laminar transition region structure is seen. In addition, a time-dependent description of the nonequilibrium ionization of all of the ionization stages of oxygen has been included.

073.029 Direct measurements of impulsive extreme ultra-
     violet and hard X-ray solar flare emission.
D. M. Horan, R. W. Kreplin, G. G. Fritz.
Astrophys. J., Vol. 255, 797 - 805 (1982).
     Direct, simultaneous measurements of the 100−1030 Å EUV and 15−150 keV X-ray emission during solar flares having impulsive phases are reported. Time histories of the energy fluxes and flux ratios for three flares were obtained from satellite data. Separation of the impulsive phase from the gradual phase of the flares provides a test of the partial precipitation model for the EUV and hard X-ray emission, with which the data are generally found to be consistent. A possible conflict with the model could be resolved by assuming that the impulsive EUV source moves deeper into the chromosphere during the early phases of the burst.

073.030 Skylab observations of H I Lyman-alpha.
     D. Roussel-Dupré.
Astrophys. J., Vol. 256, 284 - 291 (1982).
     The author presents the H I Lyman-alpha profiles obtained with the Skylab SO82-B spectrograph at spectral and spatial resolutions of 60 mÅ and 2″ X 60″. Profile variations from center to limb, at different heights above the limb, and over a cell and network boundary are illustrated. The Lyα core exhibits weak or no limb darkening, while the line wings show weak limb darkening. Above the limb, the Lyα core integrated intensity and line width remain essentially constant up to +12″ and may result from resonance scattering of chromospheric radiation by coronal neutral hydrogen or from spicules. On the disk, intensities in the line core appear to

correlate well with the Hα network; however, the Lyα wing intensities appear to correlate better with the transition region lines.

073.031 Chromospheric effects of XUV radiation emitted during solar flares.
M. E. Machado, J.-C. Hénoux.
Astron. Astrophys., Vol. 108, 61 - 68 (1982).

The authors show that X and UV radiation emitted from the flare corona and transition zone affects both the energy balance and ionization balance of chromosphere layers. The effect of X-rays (1-300 Å) is most important in producing temperature increases through the chromosphere. UV lines radiation (1000–1600 Å), on the other hand, produces strong variations in the ionization balance of some elements. The authors discuss their results in terms of the implications of recent flare model atmosphere calculations. They also show that UV radiation does not produce substantial heating through the chromosphere and its effect is negligible in the energy balance equation compared to that of soft X-rays and accelerated particles.

073.032 Radio imaging of solar flares using the Very Large Array: new insights into flare process.
M. R. Kundu, E. J. Schmahl, T. Velusamy, L. Vlahos.
Astron. Astrophys., Vol. 108, 188 - 194 (1982).

The authors present an interpretation of VLA observations of microwave bursts in an attempt to distinguish between certain models of flares. The VLA observations provide information about the pre-flare magnetic field topology and the existence of mildly relativistic electrons accelerated during flares. Examples are shown of changes in magnetic field topology in the hour before flares. Because of the observed diversity of magnetic field topologies in microwave bursts, the authors believe that the magnetic energy must be dissipated in more than one way. The VLA observations are clearly providing means for sorting out the diverse flare models.

073.033 Zur Struktur der Chromosphäre über Sonnenflecken.
F. Kneer, W. Mattig, M. von Uexküll.
Mitt. Astron. Ges., Nr. 55, p. 65 (1982). – Abstract.

073.034 Chromosphärische Helligkeitsoszillationen.
F. Kneer, G. Newkirk, M. von Uexküll.
Mitt. Astron. Ges., Nr. 55, p. 70 (1982). – Abstract.

073.035 Triggering of solar flare by magnetosonic waves in a neutral sheet plasma. J.-I. Sakai, H. Washimi.
Report IPPJ-540, Nagoya Univ., Japan, 30 pp. (1981).
Abstr. in Phys. Abstr., Vol. 85, Abstr. 28438 (1982).

073.036 Impulsive and gradual hard X-ray sources in a solar flare. N. Vilmer, S. R. Kane, G. Trottet.
Astron. Astrophys., Vol. 108, 306 - 313 (1982).

Recent observations of a relatively large hard X-ray burst with well developed impulsive and gradual components are reported. The hard X-ray burst was associated with a relatively modest optical flare on 14 August 1979. During the impulsive phase, the X-ray maxima at different energies were essentially simultaneous. However, the gradual hard X-ray component exhibited energy-dependent time delays in temporal features observed before only in the largest flares such as the one on 4 August 1972. The present observations are consistent with a relatively high density (thick-target) source for the impulsive component and a medium density ($n \approx 6 \times 10^{10}\,cm^{-3}$) source with efficient electron trapping for the gradual hard X-ray component.

073.037 Solar flare. X-ray spectra from the Solar Maximum Mission flat crystal spectrometer.
K. J. H. Phillips, J. W. Leibacher, C. J. Wolfson,
J. H. Parkinson, B. C. Fawcett, B. J. Kent, H. E. Mason,
L. W. Acton, J. L. Culhane, A. H. Gabriel.
Astrophys. J., Vol. 256, 774 - 787 (1982).

High-resolution solar X-ray spectra obtained with the flat crystal spectrometer aboard the Solar Maximum Mission from two solar flares and a nonflaring active region are analyzed. The 1–22 Å region was observed during the flare on 1980 August 25, while smaller spectral regions were repeatedly covered during the 1980 November 5 flare. Voigt profiles were fitted to spectral lines to derive accurate wavelengths and to resolve blends. During the August 25 flare, 205 lines were found in the range 5.68–18.97 Å, identifications being provided for all but 40 (mostly weak) lines. Upper limits to flare densities are derived from various line ratios, the hotter ($T \approx 10^7\,K$) ions giving $N_e < \sim 10^{12}\,cm^{-3}$ for the August 25 flare.

073.038 Hydrodynamic simulation of surge prominences.
V. N. Dermendjiev.
Dokl. Bolg. AN, Vol. 34, 915 - 918 (1981). – Abstr. in Ref. zh., 51. Astron., 3.51.369 (1982).

073.039 Espectros de espículas cromosféricas.
H. Molnar.
Bol. Asoc. Argentina Astron., No. 18, p. 85 - 87 (1980).

073.040 Anisotropy of solar flare activity.
Eh. I. Nesmyanovich.
Astrometr. Astrofiz., Vyp. (No.) 46, p. 61 - 67 (1982). In Russian.

Total energy indices distributed within 10° longitude zones are considered for 53000 flares on the sun over the period 1957 - 1973. The distribution appears to be irregular. for subflares this may be explained by the "visibility function", but for bright flares the distribution will be of a more complicated character with a statistically significant maximum at $\lambda_E = 50$ - 60° and with a minimum at $\lambda_W = 20$ - 30°. Other hypotheses are suggested to explain the asymmetry of the flare distribution.

073.041 Numerical hydrodynamics of the jet phenomena in the solar atmosphere. I. Spicules.
Y. Suematsu, K. Shibata, T. Nishikawa, R. Kitai.
Sol. Phys., Vol. 75, 99 - 118 (1982).

The authors present a spicule model whose eruption occurs as a result of a sudden pressure enhancement at a bright point located at the root of the spicule.

073.042 The vertical filamentary structures of quiescent prominences. B. C. Low.
Sol. Phys., Vol. 75, 119 - 131 (1982).

The author presents a simple magnetostatic theory of the thin vertical filaments that make up the quiescent prominence plasma as revealed by fine spatial resolution Hα photographs. A class of exact equilibrium solutions is obtained describing a horizontal row of long vertical filaments whose weights are supported by bowed magnetic field lines. The role of the magnetic field in supporting and thermally shielding the filament plasma is illustrated. It is found that the filament can have a sharp transition perpendicular to the local field, whereas the transition in the direction of the local field is necessarily diffuse.

073.043 Intensity of lines from low-lying levels in C II, N III, O IV, Ne VI, Mg VIII, Si X, and Si II.
S. Chandra.
Sol. Phys., Vol. 75, 133 - 137 (1982).

The intensities of the lines for the transition $^2P^o_{3/2} - ^2P^o_{1/2}$ in C II, N III, O IV, Ne VI, Mg VIII, Si X, and Si II in the chromosphere-corona transition region are investigated. It is found that in the transition region the intensity (which may be expressed as a function of temperature only) increases with the charge on the ion for a sequence.

073.044 **Second-stage acceleration in a limb-occulted flare.**
H. S. Hudson, R. P. Lin, R. T. Stewart.
Sol. Phys., Vol. 75, 245 - 261 (1982).

The authors present a study of the limb-occulted flare event of 22 July 1972, for which especially comprehensive observations exist. This paper concentrates on the second stage particle acceleration; thus the authors emphasize the hard X-ray, energetic particle and radio observations. The analysis shows that second stage acceleration is physically distinct from the impulsive phase, and is characterized by continuous and widespread electron acceleration to high energies, most likely by the type II burst shock wave.

073.045 **A qualitative interpretation of 7 August 1972 impulsive phase flare Hα line profiles.**
R. C. Canfield.
Sol. Phys., Vol. 75, 263 - 275 (1982).

The purpose of this paper is to show that existing models of the formation of the Hα line during flares appear to provide clear qualitative evidence that heating of the Hα-forming regions of the flare chromosphere in the bright Hα kernels observed during the impulsive phase of solar flares is not due primarily to heating by Coulomb collisions of a power-law distribution of 10−100 keV electrons with chromospheric material; instead some shorter-range process, perhaps conduction or optically thick radiative transfer, seems to be favored.

073.046 **Prompt injection of relativistic protons from the September 1, 1971 solar flare.** E. W. Cliver.
Sol. Phys., Vol. 75, 341 - 345 (1982).

The September 1, 1971 flare in McMath region 11482 was projected to have occurred ~30° behind the west limb. An anisotropic ground level effect began < 30 min after the inferred explosive phase of the flare. The rapid injection of relativistic protons onto the earth spiral field line to a shock wave associated with an observed type II burst.

073.047 **Rotational modulation of Ca K flux ratio and sunspot number.** R. W. Stimets, C. Londono.
Sol. Phys., Vol. 76, 167 - 180 (1982).

The work reported in this paper consists of the daily monitoring of the Ca K flux of integrated sunlight during the period 1 July 1979−18 February 1980, analysis of the data by both power spectrum and autocorrelation methods, and comparison of the Ca K flux variations with those of sunspot number.

073.048 **Observed mass motions in limb prominences.**
A. H. Lategan, A. H. Jarrett.
Sol. Phys., Vol. 76, 323 - 330 (1982).

Line shifts occurring on Fabry-Perot interferograms of four limb prominences obtained at Boyden Observatory are reported and analyzed in terms of mass motions of the emitting parts of the prominences. The contribution of these motions to the energy balance of prominences is discussed.

073.049 **Slow-shock heating and the Kopp-Pneuman model for 'post'-flare loops.**
P. J. Cargill, E. R. Priest.
Sol. Phys., Vol. 76, 357 - 375 (1982).

The heating of 'post'-flare loops in the Kopp-Pneuman (1976) model is reconsidered. In that kinematic model the loops are heated by gas-dynamic shocks. However, in a full dynamic model they would be replaced by slow magneto-hydrodynamic shocks, which may provide more heating due to the additional release of magnetic energy. It is shown from a local compressible analysis that such shock waves can account for the observed temperatures and also for the observed upward loop speeds. A full dynamic model would require a sophisticated numerical computation, and so a simple global analytic model is developed instead. It is incompressible and includes a strong solar-wind inflow along the reconnecting field lines. As the upflow increases, the loops become more compressed and the Alfvén waves approach one another.

073.050 **The Hα/Hβ ratio in solar flares.**
H. Zirin, M. Liggett, A. Patterson.
Sol. Phys., Vol. 76, 387 - 392 (1982).

The authors have measured the ratio of Hα to Hβ central intensities in the peak kernels of 14 flares, using simultaneous filtergrams. The ratio is typically 1.0 with some scatter. By contrast, in stellar flares the ratio is about 0.8.

073.051 **Estudio de espectros de protuberancias.**
J. M. Fontenla, M. Rovira de Miceli.
Bol. Asoc. Argentina Astron., No. 20 - 24, p. 51 - 56 (1981).

073.052 **Cálculo de poblaciones y líneas del He I en la cromósfera solar.** H. Molnar, J. M. Fontenla.
Bol. Asoc. Argentina Astron., No. 20 - 24, p. 63 - 69 (1981).

073.053 **Estudio espectral de espículas cromosféricas.**
H. Molnar, M. Rovira de Miceli.
Bol. Asoc. Argentina Astron., No. 20 - 24, p. 71 - 79 (1981).

073.054 **Modelos de flares fotosféricos y cromosféricos.**
M. E. Machado, J. L. Linsky.
Bol. Asoc. Argentina Astron., No. 20 - 24, p. 145 (1981).
Abstract.

073.055 **Análisis del flare en luz blanca del 7 de agosto de 1972. Espectro y estructura atmosférica.**
M. E. Machado, D. M. Rust.
Bol. Asoc. Argentina Astron., No. 20 - 24, p. 147 (1981).
Abstract.

073.056 **Procesos de emisión y su relación con la estructura y origen de los flares solares.** M. E. Machado.
Bol. Asoc. Argentina Astron., No. 20 - 24, p. 148 (1981).
Abstract.

073.057 **Espectro y estructura de las fulguraciones solares.**
M. E. Machado.
Bol. Asoc. Argentina Astron., No. 20 - 24, p. 235 (1981).
Abstract.

073.058 **Estudio de la radiación de He I en una protuberancia.**
H. Molnar, H. Grossi Gallegos, J. R. Seibold.
Bol. Asoc. Argentina Astron., No. 20 - 24, p. 239 (1981).
Abstract.

073.059 **On the origin of solar spicules.**
J. V. Hollweg.
Astrophys. J., Vol. 257, 345 - 353 (1982).

The nonlinear evolution of vertical motions on intense solar magnetic flux tubes, is considered. It is shown that a quasi-impulsive source in the photosphere can excite a train of upward-propagating rebound shocks in the chromosphere. The rebound shock train is the nonlinear development of oscillations of the atmosphere at its natural frequency. The rebound shocks impinge on the transition region and thrust the underlying chromosphere upward. It is found that the rebound shock train leads naturally to structures which can be identified with the solar spicules.

073.060 **Electron pitch angle scattering and the impulsive phase microwave and hard X-ray emission from solar flares.** G. D. Holman, M. R. Kundu, K. Papadopoulos.
Astrophys. J., Vol. 257, 354 - 360 (1982).

Observations and theoretical considerations have led to a model for impulsive phase flare emission involving the heating and acceleration of thermal electrons in the coronal part of a magnetic loop. The bulk of the heated gas is confined between conduction fronts, but particles with velocities a few times greater than the thermal velocity can escape into the lower part of the loop. The authors show that, when the electron gyrofrequency exceeds the plasma frequency, the escaping electrons are unstable to the generation of electrostatic plasma waves which scatter the particles in pitch angle to a nearly isotropic distribution. This scattering can enhance the microwave emission from the upper part of the loop and can lead to one or two breaks in the impulsive phase hard X-ray spectrum.

073.061   Fine structure of motions in a quiescent prominence.
M. Sh. Gigolashvili, Yu. D. Zhugzhda.
Sol. Phys., Vol. 77, 95 - 108 (1982).

The Ca^+ K line of the quiescent prominence of 15 October 1969 is studied in detail. The behaviour of various fragments of the prominence with the line-of-sight Doppler velocities is investigated for different places of the prominence along the spectrum height. Due to superposition of the fragments and the radial velocities a complex non-gaussian profile of the spectral line is formed. Decomposition of the non-gaussian profile into suitable gaussian ones is made with a computer yielding objectivity and high accuracy. It is found that the components into which the Ca^+ K line decomposes at each level, are traced while proceeding from one photometric section to another throughout the spectral height.

073.062   Velocity fields in quiescent prominences.
E. Jensen.
Sol. Phys., Vol. 77, 109 - 119 (1982).

Three quiescent prominences were observed in the Ca II K-line and a fourth one also in the H-line. These data are used to study the distribution of the line-of-sight velocity component. The range in observed velocities varies considerably between the prominences. For the best observed prominence more than 70% of the kinetic energy is in the supersonic range. In the other cases none or only an insignificant part of the observations exceed the velocity of sound. Considerable deviations from gaussian distributions are apparent for the smallest velocities. If it is assumed that we have to do with MHD-turbulence, a characteristic relationship should exist between velocity and eddy size. When supersonic velocities are present, compressibility effects may severely alter this relationship.

073.063   Numerical hydrodynamics of the jet phenomena in the solar atmosphere. II. Surges.
K. Shibata, T. Nishikawa, R. Kitai, Y. Suematsu.
Sol. Phys., Vol. 77, 121 - 151 (1982).

One-dimensional hydrodynamic simulations of surges are performed in order to make clear their origin and structure. Surges are regarded as the jets resulting from a sudden pressure increase at the base of the model atmosphere. The height of the explosion (h_0), and the strength of the sudden pressure increase (p/p_0) at h_0 are regarded as free parameters. Simulations are performed for values in the ranges of 540 km $\leqslant h_0 \leqslant$ 1920 km and 3 $\leqslant p/p_0 \leqslant$ 30. It was found that for a fixed p/p_0 there exists a critical height (h_c) in h_0, which separates the jet (surge) models into two types. For $h_0 > h_c$, jets are produced directly by the pressure gradient force near h_0, and made of the matter ejected from the explosion itself. The essential hydrodynamic structure of this type is the same as that in a shock tube. For $h_0 < h_c$, jets are produced by the shock wave which is generated by the pressure enhancement and which has propagated through the chromosphere. General properties of both types are investigated in detail.

073.064   A possible explanation of non-steady-state appearances in X-ray spectra of solar flares.
K. N. Koshelev, E. Ya. Kononov.
Sol. Phys., Vol. 77, 177 - 181 (1982).

A model of essentially transient ionization of plasma is suggested to explain some features in observed spectra of solar flares, which cannot be understood if stationary conditions are assumed.

073.065   Remote flare brightenings and type III reverse slope bursts.   F. Tang, R. L. Moore.
Sol. Phys., Vol. 77, 263 - 276 (1982).

The authors present two large flares which were exceptional in that each produced an extensive chain of Hα emission patches in remote quiet regions more than 10^5 km away from the main flare site. They were also unusual in that a large group of the rare type III reverse slope bursts accompanied each flare. The observations suggest that this is no coincidence, but that the two phenomena are directly connected. The onset of about half of the remote Hα emission patches were found to be nearly simultaneous with RS bursts. The authors propose: (1) that the remote Hα brightenings were initiated by direct heating of the chromosphere by RS burst electrons traveling in closed magnetic loops connecting the flare site to the remote patches; and (2) that after onset, the brightenings were heated by thermal conduction by slower thermal electrons ($kT \sim 1$ keV) which immediately follow the RS burst electrons along the same loops.

073.066   The relation between the surges and solar radio emission.   I. N. Garczyńska, B. Rompolt, A. O. Benz, C. Slottje, A. Tlamicha, C. Zanelli.
Sol. Phys., Vol. 77, 277 - 283 (1982).

120 limb surges which have been observed from September 1966 to November 1977 are investigated. The evolution of surges was compared with the radio data during the surges. A correlation between radio bursts and the surges was found, particularly with chains of type I radio bursts, which is the first reliable correlation found of these bursts with non-radio events. The authors compared the maximum height reached by a surge with the frequencies of the radio bursts emitted at the same time and the maximum velocity of the rising surge with the frequency drift of type I chains. No such a correlation was however found. They discuss the possibility that surges are the result of a sudden energy input into the chromosphere related to the type I source in the corona.

073.067   Evidence of redshifts in the average solar line profiles of C IV and Si IV from OSO-8 observations.
D. Roussel-Dupré, R. A. Shine.
Sol. Phys., Vol. 77, 329 - 340 (1982).

Measurements of the C IV 1548 Å and Si IV 1393 Å lines show that the mean profiles are redshifted at disk center. The authors measure an apparent average downflow of material in the 50000 to 100000 K temperature range which is weighted by the emission measure in these lines. The magnitude of the redshift varies from $6-17$ km s^{-1} with a mean of 12 km s^{-1} and is persistent at least on the order of months. It is possible that material observed to be downflowing in C IV and Si IV has its origins in the upward moving spicule material.

073.068   Flares and filament activation.
Yu. M. Slonim.
Soln. Dannye 1982 Byull., No. 1, p. 80 - 86 (1982). In Russian.

Filament activation is connected not only with large two-ribbon flares but also with small ones, among which there are sometimes single emission features. These small flares have a relatively long lifetime and are located mostly at one end of the filament. They are followed by long microwave bursts.

073.069   Spectrophotometry of a mean-intensity flare-ejection on the solar limb on June 16, 1967.
P. N. Polupan.
Vestn. Kiev. Univ., Vyp. 23, p. 26 - 31 (1981). In Russian.

073.070   On the nature of self-reversal of hydrogen spectral line profiles in chromospheric flares.

V. A. Ostapenko, M. Yu. Zel'dina.
Vestn. Kiev. Univ., Vyp. 23, p. 32 - 40 (1981). In Russian.

073.071   Investigation of some physical characteristics of
          chromospheric flares of different intensity.
V. A. Ostapenko, P. N. Polupan.
Vestn. Kiev. Univ., Vyp. 23, p. 40 - 48 (1981). In Russian.

073.072   On a connection between flare activity of sunspot
          groups and the sectorial boundaries of the inter-
planetary magnetic field.   V. M. Efimenko.
Vestn. Kiev. Univ., Vyp. 23, p. 84 - 89 (1981). In Russian.

073.073   Impulsive phase of flares in soft X-ray emission.
          E. Antonucci, A. H. Gabriel, L. W. Acton,
J. L. Culhane, J. G. Doyle, J. W. Leibacher, M. E. Machado,
L. E. Orwig, C. G. Rapley.
Sol. Phys., Vol. 78, 107 - 123 (1982).
    Observations using the Bent Crystal Spectrometer
instrument on the Solar Maximum Mission show that
turbulence and blue-shifted motions are characteristic of the
soft X-ray plasma during the impulsive phase of flares, and are
coincident with the hard X-ray bursts observed by the Hard
X-ray Burst Spectrometer. A method for analysing the Ca XIX
and Fe XXV spectra characteristic of the impulsive phase is
presented. Non-thermal widths and blue-shifted components in
the spectral lines of Ca XIX and Fe XXV indicate the presence
of turbulent velocities exceeding 100 km s^{-1} and upward
motions of 300–400 km s^{-1}. The April 10, May 9, and June
29, 1980 flares are studied.

073.074   Comparison of observed Ca XIX and Ca XVIII
          relative line intensities with current theory.
C. Jordan, N. J. Veck.
Sol. Phys., Vol. 78, 125 - 135 (1982).
    A comparison is made between Ca XIX and Ca XVIII
line ratios observed in solar flares with the Bent Crystal
Spectrometer on the Solar Maximum Mission satellite and
currently available atomic data. Close agreement is found with
the excitation rates recently published by Pradhan et al.(1981).
The observations show little dependence of line ratios on
electron temperature, supporting a further conclusion that
cascade contributions to the 2^3P and 2^3S levels are not
significant.

073.075   Rising motion of a behind-the-limb flare at 35 GHz.
          T. Kato, T. Omodaka, M. Fujishita, K.-A. Kawabata,
H. Ogawa.
Sol. Phys., Vol. 78, 137 - 140 (1982).
    Interferometer observation of a behind-the-limb flare on
7 September, 1977, at 35 GHz (λ = 8.6 mm) shows that the
microwave non-thermal radio source of the burst is located in
the coronal region at the height higher than 7000 km above
the photosphere and rises gradually with the velocity of about
30 km s^{-1}.

073.076   Spectrum of the white-light solar flare of Septem-
          ber 5, 1981.   Z.-y. Wang, X.-z. Chen.
Astron. Circ., No. 11, 4pp. (1981). In Chinese and English.

073.077   Direct evidence for chromospheric evaporation in a
          well-observed compact flare.   R. C. Canfield,
L. W. Acton, T. A. Gunkler, H. S. Hudson, A. L. Kiplinger,
J. W. Leibacher.
Bull. American Astron. Soc., Vol. 13, 819 (1981). – Abstract.

073.078   Models of electron-heated solar flare chromo-
          spheres.   P. J. Ricchiazzi, R. C. Canfield.
Bull. American Astron. Soc., Vol. 13, 819 - 820 (1981).
Abstract.

073.079   Radiative processes in white-light flares.
          L. Dame, L. Cram.
Bull. American Astron. Soc., Vol. 13, 820 (1981). – Abstract.

073.080   The continuous opacity function in a white-light
          flare.   D. F. Neidig.
Bull. American Astron. Soc., Vol. 13, 820 (1981). – Abstract.

073.081   The significance of multiple flares.
          K. T. Strong, B. R. Dennis, J. W. Leibacher,
C. J. Wolfson.
Bull. American Astron. Soc., Vol. 13, 820 (1981). – Abstract.

073.082   A statistical study of solar flares and permanent
          filament changes.   D. M. Rust.
Bull. American Astron. Soc., Vol. 13, 820 (1981). – Abstract.

073.083   The radiative instability as a mechanism for bright
          point flares.   J. M. Davis, D. F. Webb.
Bull. American Astron. Soc., Vol. 13, 821 (1981). – Abstract.

073.084   The structure of the lower transition region.
          S. K. Antiochos.
Bull. American Astron. Soc., Vol. 13, 835 (1981). – Abstract.

073.085   Inhibition of heat conduction into the transition
          region by magnetic construction.
J. Dowdy, R. Moore, S. T. Wu.
Bull. American Astron. Soc., Vol. 13, 835 (1981). – Abstract.

073.086   Acoustic energy flux in the solar transition zone.
          M. E. Bruner, G. Poletto.
Bull. American Astron. Soc., Vol. 13, 835 (1981). – Abstract.

073.087   Solar transition region response to heating rate
          variations.
E. S. Oran, J. T. Mariska, J. P. Boris, T. R. Young,
G. A. Doschek.
Bull. American Astron. Soc., Vol. 13, 836 (1981). – Abstract.

073.088   HRTS (*High Resolution Telescope and Spectro-
          graph*) observations of the solar chromosphere and
transition zone.
K. P. Dere, J.-D. F. Bartoe, G. E. Brueckner.
Bull. American Astron. Soc., Vol. 13, 845 (1981). – Abstract.

073.089   Temporal features in a solar flare observed in
          mm-waves, cm-waves and X-rays with fine time
resolution.   P. Kaufmann, F. M. Strauss, J. E. R. Costa,
B. R. Dennis, A. Kiplinger, K. J. Frost, L. E. Orwig.
Bull. American Astron. Soc., Vol. 13, 846 (1981). – Abstract.

073.090   A formula for forecasting the probability of erup-
          tion of quiescent filaments.
S. F. Martin, V. W. Lawrence.
Bull. American Astron. Soc., Vol. 13, 847 (1981). – Abstract.

073.091   Gyrosynchrotron masering in solar flares.
          P. J. Morrison, G. D. Holman, M. R. Kundu.
Bull. American Astron. Soc., Vol. 13, 860 (1981). – Abstract.

073.092   Velocities of shock waves generated by solar flares.
          A. Maxwell.
Bull. American Astron. Soc., Vol. 13, 861 (1981). – Abstract.

073.093   Solar proton flares with weak impulsive phases.
          E. W. Cliver, S. W. Kahler, P. S. McIntosh.
Bull. American Astron. Soc., Vol. 13, 861 (1981). – Abstract.

073.094   Evolution of electron and proton temperatures in a
          flaring loop: I. A case of thermal heating of

electrons. F. Nagai, E. Tandberg-Hanssen, S. T. Wu.
Bull. American Astron. Soc., Vol. 13, 861 (1981). – Abstract.

073.095 Impulsive events in quiescent prominences.
G. D. Toot, J. M. Malville.
Bull. American Astron. Soc., Vol. 13, 862 (1981). – Abstract.

073.096 MHD simulation of 1980 June 29 (1821 UT) flare
and coronal transient. S. T. Wu, S. Wang,
M. Dryer, A. J. Poland, D. Sime, C. J. Wolfson, L. E. Orwig.
Bull. American Astron. Soc., Vol. 13, 862 (1981). – Abstract.

073.097 Submillimeter continuum observations of solar
plages. J. T. Jefferies, E. E. Becklin,
C. Lindsey, F. Q. Orrall, I. Gatley, M. Werner.
Bull. American Astron. Soc., Vol. 13, 881(1981). – Abstract.

073.098 The relationship between the microwave and hard
X-ray sources in a solar flare.
K. A. Marsh, H. Zirin, P. Hoyng, B. R. Dennis.
Bull. American Astron. Soc., Vol. 13, 889 (1981). – Abstract.

073.099 The pre-eruption phase of filaments observed in
He I 10830. K. L. Harvey.
Bull. American Astron. Soc., Vol. 13, 890 (1981). – Abstract.

073.100 Absolute light curves of solar flares in Hα.
K. Topka, G. J. Hurford.
Bull. American Astron. Soc., Vol. 13, 890 (1981). – Abstract.

073.101 Gamma ray observations of the white light flare of
1 July 1980. J. M. Ryan, E. L. Chupp,
S. M. Matz, E. Rieger, C. Reppin, G. Kanbach, G. Share.
Bull. American Astron. Soc., Vol. 13, 902 - 903 (1981).
Abstract.

073.102 The comparison of the theoretical and observational
ratios of gamma ray fluences in the 4 - 7 MeV
channel to the 2.22 MeV line in 8 solar flares. R. Ramaty.
Bull. American Astron. Soc., Vol. 13, 903 (1981). – Abstract.

073.103 The spectrum of prompt nuclear deexcitation gam-
ma ray lines from solar flares.
D. J. Forrest, B. M. Gardner, S. M. Matz, E. L. Chupp,
C. Reppin, E. Rieger, G. Kanbach, G. Share.
Bull. American Astron. Soc., Vol. 13, 903 (1981). – Abstract.

073.104 A comparison of gamma ray line and X-ray brems-
strahlung time profiles in several flares.
B. M. Gardner, D. J. Forrest, M. C. Zolcinski, E. L. Chupp,
E. Rieger, C. Reppin, G. Kanbach, G. Share.
Bull. American Astron. Soc., Vol. 13, 903 (1981). – Abstract.

073.105 A search for rapid oscillations in hard X-ray solar
flares.
A. L. Kiplinger, B. R. Dennis, K. J. Frost, L. E. Orwig.
Bull. American Astron. Soc., Vol. 13, 903 (1981). – Abstract.

073.106 Hard X-ray structure of flaring loops.
J. Leach, V. Petrosian.
Bull. American Astron. Soc., Vol. 13, 903 - 904 (1981).
Abstract.

073.107 Association between gradual hard X-ray emission
and metric continua during large flares.
L. Klein, K. Anderson, M. Pick, G. Trottet, N. Vilmer,
S. Kane.
Bull. American Astron. Soc., Vol. 13, 904 (1981). – Abstract.

073.108 Time-series of radio, UV, soft, and hard X-ray data
from flare of November 1, 1980.
E. Tandberg-Hanssen, P. Kaufmann, E. J. Reichmann,

D. L. Teuber.
Bull. American Astron. Soc., Vol. 13, 904 (1981). – Abstract.

073.109 Evidence for compressive heating in a solar flare
from coincident hard X-ray and microwave observa-
tions. D. A. Batchelor, C. J. Crannell, B. R. Dennis,
A. Magun, H. Wiehl.
Bull. American Astron. Soc., Vol. 13, 904 - 905 (1981).
Abstract.

073.110 Observation of vertical phase delays of chromo-
spheric oscillations above sunspot umbrae.
B. W. Lites.
Bull. American Astron. Soc., Vol. 13, 909 (1981). – Abstract.

073.111 Heliocentric angular dependence for gamma ray
flares observed with the SMM satellite.
M. C. Zolcinski, D. J. Forrest, B. M. Gardner, E. L. Chupp,
C. Reppin, E. Rieger, G. Kanbach, G. Share.
Bull. American Astron. Soc., Vol. 13, 910 (1981). – Abstract.

073.112 Measurements of A-values for intersystem lines
used in diagnosis of solar transition zone and other
astrophysical plasmas.
B. C. Johnson, R. D. Knight, P. L. Smith, H. S. Kwong,
W. H. Parkinson.
Bull. American Astron. Soc., Vol. 13, 910 (1981). – Abstract.

073.113 Interpretation of the "second-step" acceleration in
the impulsive phase of a solar flare.
T. Bai, H. S. Hudson.
Bull. American Astron. Soc., Vol. 13, 912 (1981). – Abstract.

073.114 On the mechanism of solar spicules.
M. L. Blake, P. A. Sturrock.
Bull. American Astron. Soc., Vol. 13, 914 (1981). – Abstract.

073.115 On the origin of solar spicules.
J. V. Hollweg.
Bull. American Astron. Soc., Vol. 13, 914 (1981). – Abstract.

073.116 Thermal bifurcation in solar calcium plages.
T. R. Ayres, J. L. Linsky, L. Testerman, J. Brault.
Bull. American Astron. Soc., Vol. 13, 915 (1981). – Abstract.

073.117 Strong plasma turbulence and its application to
solar flares.
V. M. Tomozov, V. N. Tsytovich.
Issled. po geomagn., aehron. i fiz. Solntsa, Moskva, 1981,
No. 57, p. 28 - 37. In Russian. – Abstr. in Ref. zh., 51.
Astron., 5.51.378 (1982).

073.118 Measurements of the magnetic field in solar promi-
nences with a spectrally scanning magnetograph.
G. M. Nikolsky (Nikol'skij), I. S. Kim, S. Koutchmy.
Pré-publ. Inst. Astrophys. Paris, No. 4, 20 pp. (1982).
Submitted to Sol. Phys.

073.119 Eruption chromosphérique, July - December 1979.
Q. Bull. Sol. Act., Vol. 21, Part III, p. 81 - 163
(1979).

073.120 Solar flare soft X-ray spectra from the HINOTORI.
I. Iron line spectra and their time variations of
seven X-class flares. K. Tanaka, K. Akita, T. Watanabe,
H. Miyazaki, K. Kumagai, M. Miyashita, K. Nishi, F. Moriyama,
Ann. Tokyo Astron. Obs., Second Ser., Vol. 18, 237 - 338
(1982).

073.121 Observations carried out at Hvar Observatory
during SMY (Solar Maximum Year) August 1979 -

February 1981.   V. Ruždjak, B. Vršnak, N. Novak.
Hvar Obs. Bull., Vol. 5, No. 1, p. 31 - 40 (1981).

073.122   Post SMY (*Solar Maximum Year*) observations at Hvar Observatory, May - June 1981.
V. Ruždjak, B. Vršnak, N. Novak.
Hvar Obs. Bull., Vol. 5, No. 1, p. 41 - 43 (1981).

073.123   Observations of sudden changes of magnetic structure in a flare.
R. L. Moore, G. J. Hurford, H. P. Jones, S. R. Kane.
Bull. American Astron. Soc., Vol. 14, 572 (1982). − Abstract.

073.124   Magnesium prominences observed at 5172 Å at the San Fernando Observatory.   A. D. Herzog.
Bull. American Astron. Soc., Vol. 14, 572 (1982). − Abstract.

073.125   The role of the big flare syndrome in correlations of solar energetic proton fluxes and associated microwave burst parameters.   S. W. Kahler.
J. Geophys. Res., Vol. 87, 3439 - 3448 (1982).
In the previous studies correlating $E > 10$ MeV proton fluxes and spectra with various associated microwave burst parameters, the resulting high correlations were assumed to reflect a common acceleration process for the protons and the microwave-emitting electrons. The author suggests and tests an alternative explanation for these correlations, which he terms the big flare syndrome, that states that, statistically, energetic flare phenomena are more intense in larger flares, regardless of the detailed physics.

073.126   The impulsive ana gradual phases of a solar limb flare as observed from the Solar Maximum Mission satellite.   A. I. Poland, M. E. Machado, C. J. Wolfson, K. J. Frost, B. E. Woodgate, R. A. Shine, P. J. Kenny, C. C. Cheng, E. A. Tandberg-Hanssen, E. C. Bruner, W. Henze.
Sol. Phys., Vol. 78, 201 - 213 (1982).
Simultaneous observations of a solar limb flare in the X-ray and ultraviolet regions of the spectrum are presented. Temporal and spectral X-ray observations were obtained for the 25−300 keV range while temporal, spectral, and spatial X-ray observations were obtained for the 30−0.3 keV range. The ultraviolet observations were images with a $10''$ spatial resolution in the lines of O V and Fe XXI. The observations provide limitations for current flare models and will provide the data needed for initial conditions in modeling the concurrent coronal transient.

073.127   The flare of December 17, 1980 observed with high time resolution by a digital CCD camera.
N. Kämpfer, W. Schöchlin.
Sol. Phys., Vol. 78, 215 - 224 (1982).
An electro-optical system for the digital registration of Hα-images with high time resolution (1.6s) is described. Data are digitized with a resolution of 8 bit and written on magnetic tape by means of a minicomputer. Image analysis on a large computer eliminates instrumental effects, calibrates the data and reproduces them in graphical form. The analysis of a first flare (December 17, 1980; 12:09 UT) shows that the different flare kernels brighten at different times and pulsate in diameter and intensity. The decay of the flare is slower if fainter regions are included. This supports the idea that an impulsive flare confined in a magnetic loop spreads out over a larger area during the gradual phase. Comparison with microwave observations shows the same risetime for both spectral ranges which indicates that they are excited by the same agent.

073.128   On the color of the 26 February 1981 white light flare.   D. F. Neidig, R. O. Beck.
Sol. Phys., Vol. 78, 225 - 228 (1982).
The color observed in the 26 February 1981 white light flare could be attributed to Hα if this line were more than three times stronger than in any flare previously observed, implying either unusually high flare densities or anomalous Stark broadening at lower density.

073.129   Energetics of two-ribbon solar flares.
G. W. Pneuman.
Sol. Phys., Vol. 78, 229 - 241 (1982).
A theory of two-ribbon solar flares is presented which identifies the primary energy release site with the tops of the flare loops. Based upon the supposition that the energy release at the loop tops is in the form of Joulean dissipation of magnetic energy at the rising reconnection site, a quantitative model of the energy release process is developed based upon an analytic reconnecting magnetic field geometry believed to represent the basic process. Predicted curves of energy density vs time are compared with X-ray observations taken aboard Skylab for the events of 29 July, 13 August, and 21 August in 1973.

073.130   Study of the post-flare loops on 29 July 1973. IV. Revision of T and n_e values and comparison with the flare of 21 May 1980.   Z. Švestka, H. W. Dodson-Prince, S. F. Martin, O. C. Mohler, R. L. Moore, J. T. Nolte, R. D. Petrasso.
Sol. Phys., Vol. 78, 271 - 285 (1982).
The authors present revised values of temperature and density for the flare loops of 29 July 1973 and compare the revised parameters with those obtained aboard the SMM for the two-ribbon flare of 21 May 1980.

073.131   Why are spicules absent over plages and long under coronal holes?   K. Shibata, Y. Suematsu.
Sol. Phys., Vol. 78, 333 - 345 (1982).
One-dimensional hydrodynamic simulations are performed in order to examine the influence of initial atmospheric structures on the dynamics of spicules. Spicules are produced by the shock wave (MHD slow mode shock) which originates from a bright point appearance (sudden pressure increase) at the network in the photosphere or in the low chromosphere. Simulation results well reproduce the observational facts that spicules are absent over plages and long under coronal holes.

073.132   The spectrum of solar flares.
S. V. Krishna Rao, V. V. Sekhara Prasad.
Indian J. Phys. Part B, Vol. 55B, 510 - 511 (1981). − Abstr. in Phys. Abstr., Vol. 85, Abstr. 58317 (1982).

073.133   Spatial and temporal evolution of soft and hard X-ray emission in a solar flare.
M. E. Machado, A. Duijveman, B. R. Dennis.
Sol.Phys., Vol. 79, 85 - 106 (1982).
The authors study the spatial and temporal characteristics of the 3.5 to 30.0 keV emission in a solar flare on April 10, 1980. Key results of the investigation are: (a) continuous energy release is needed to substain the increase of the emission through the rising phase of the flare, before and after the impulsive phase in hard X-rays. (b) The observational parameters characterizing the impulsive burst show that it is most likely associated with non-thermal processes (particle acceleration). (c) The continuous energy release is associated with strong chromospheric evaporation.

073.134   On the directivity of X rays of solar flares.
M. N. Belovskij, Yu. P. Ochelkov, N. K. Pereyaslova, A. V. Ustinov.
Kosm. Issled., Tom 20, 417 - 421 (1982). In Russian.

073.135   Solar flares, proton showers, and the Space Shuttle.
D. M. Rust.
Science, Vol. 216, 939 - 946 (1982).

073.136 Comparison of theoretically predicted and observed Solar Maximum Mission X-ray spectra for the 1980 April 13 and May 9 flares.
D. F. Smith, L. E. Orwig.
Astrophys. J., Vol. 258, 367 - 372 (1982).

A method for predicting the hard X-ray spectrum in the 10−100 keV range for compact flares during their initial rise is developed on the basis of a thermal model. Observations of the flares of 1980 April 13, 4:05 U.T., and 1980 May 9, 7:12 U.T. are given and their combined spectra from the Hard X-ray Burst Spectrometer and Hard X-ray Imaging Spectrometer on the Solar Maximum Mission are deduced. Constraints on the cross sectional area of the supposed emitting arch are obtained from data from the Hard X-ray Imaging Spectrometer. A power-law spectrum is predicted for the rise of the flare of April 13 for initial arch densities less than 10^{10} cm^{-3} and also for the flare of May 9 for initial arch densities less than 5.4×10^{10} cm^{-3}. In both cases power-law spectra are observed.

073.137 Loop coalescence in flares and coronal X-ray brightening.   T. Tajima, F. Brunel, J. Sakai.
Astrophys. J., Lett., Vol. 258, L45 - L48 (1982).

Characteristics of solar flares, such as their impulsive nature, time scale, heating, high-energy particle spectrum, and γ-ray oscillations, as well as recent X-ray photographs of coronal brightening, are explained by the nonlinear coalescence instability of current loops.

073.138 An explanation for the systematic flow of plasma in the solar transition region.
J. P. Boris, J. T. Mariska.
Astrophys. J., Lett., Vol. 258, L49 - L52, plate L2 (1982).

Using numerical simulations, the authors show that the systematic flow of plasma along a coronal magnetic flux tube is easily produced by a change in the spatial dependence of the heating rate from a symmetric deposition which supports a stationary equilibrium to a time-independent asymmetric deposition. The velocity of the flow is roughly proportional to the heating asymmetry and is directed to the side of the loop away from the bulk of the energy deposition.

073.139 Die Beobachtung des Flashspektrums während der Sonnenfinsternis vom 31. Juli 1981.
M. Rätz, K. Kirsch, K. Reichenbächer, W. Weise, D. Böhme.
Sterne, 58. Band, 82 - 85 (1982).

073.140 A theoretical model of the solar chromosphere over a sunspot.   M. Marik.
Sun and planetary system, (see 012.040), p. 113 - 114 (1982).

073.141 Space-borne and ground-based observations of a solar active region and a flare.   F. Chiuderi Drago.
Sun and planetary system, (see 012.040), p. 115 - 118 (1982).

073.142 Turbulent velocity fields in quiescent prominences.
E. Jensen, O. Engvold.
Sun and planetary system, (see 012.040), p. 131 - 132 (1982).

073.143 Determination of the electron density in a quiescent prominence.   M. Sh. Gigolashvili.
Sun and planetary system, (see 012.040), p. 133 - 134 (1982).

073.144 A polarization investigation of prominences.
Ts. S. Khetsuriani.
Sun and planetary system, (see 012.040), p. 135 - 136 (1982).

073.145 The formation of solar prominences.
C. Chiuderi.
Sun and planetary system, (see 012.040), p. 137 - 138 (1982).

073.146 Ein Weißlichtflare in neuem Licht.
H. Hornung.
Sterne Weltraum, 21. Jahrg., 256 - 258 (1982).

073.147 June - July 1974 proton-flare region. II. Regularities in the magnetic field area and related sunspot flare-activity development.   V. Bumba, L. Hejna, Le Bach Yen.
Bull. Astron. Inst. Czechoslovakia, Vol. 33, 160 - 173 (1982).

The studies of changes of the background magnetic field areas with time demonstrate the physical unity of this whole complex evolutionary process and its relation to the cyclic magnetic field generation in the Sun.

073.148 Chromospheric and coronal heating mechanisms.
J. Leibacher, R. F. Stein.
Smithsonian Astrophys. Obs., Spec. Rep. 392, (see 012.045), p. 23 - 37 (1982).

073.149 Is energy conserved at the foot of the solar chromosphere?   W. Kalkofen.
Smithsonian Astrophys. Obs., Spec. Rep. 392, (see 012.045), p. 59 - 66 (1982).

Current empirical models of the solar atmosphere have kinetic temperatures that are too low at the temperature minimum to balance radiative heating and cooling. If there is additional energy input from the dissipation of hydrodynamic waves the apparent imbalance is aggravated. It is suggested that the problem lies in the assumption of a static upper photosphere. It is proposed that the mechanical waves, which further out cause the chromospheric temperature rise, traverse the temperature minimum region with large amplitude and produce the apparent non-conservation of energy as well as other difficulties of the empirical models through non-linear, time-dependent effects.

073.150 Solar plages and the interpretation of stellar Ca II H and K line variations in late type dwarfs.
J. W. Cook.
Smithsonian Astrophys. Obs., Spec. Rep. 392, (see 012.045), p. 181 - 189 (1982).

073.151 Increasing trend of flare activity before the proton flare April 30, 1976.   L. Křivský.
Compendium in astronomy, (see 003.012), p. 117 - 118 (1982).

A number of examples have been given earlier to document that the identification of the increasing trend on the summation curves of the flare index from an active region, where a flare with an emission of cosmic or sub-cosmic radiation is later generated, can be used for forecasting these energetic flares. Using the active region McMath 14149, S 09°, W 47°, CMP Apr. 27, 1976 the author is able to demonstrate another similar case, in which the onset of a steep flare trend was associated with the generation of a small satellite group of C and D type sunspots adjacent to an old H-type group which outlived a live active region from the preceding solar rotations.

Electron density diagnostic line ratios from the $n = 3$ lines of O V.   See Abstr. 022.150.

Limitations on the upconversion of ion sound to Langmuir turbulence.   See Abstr. 062.003.

Propagation speeds and acoustic damping of waves in magnetic flux tubes.   See Abstr. 062.067.

Thermal trigger for solar flares and coronal loops formation.   See Abstr. 062.070.

Evolution of current sheets following the onset of enhanced resistivity.   See Abstr. 062.112.

Heating of stellar chromospheres when magnetic fields are present. See Abstr. 064.017.

A sufficient condition for the stability of atmospheres with magnetic fields. See Abstr. 064.083.

Observations of photospheric line profiles in plages. See Abstr. 071.053.

The evolution of the proton flare active region McMath 11128. See Abstr. 072.021.

Some studies on sunspots in relation to H_α-flares and radiobursts. See Abstr. 072.023.

Investigation of the distribution of spot groups and flares near sector boundaries (1964 - 1973). See Abstr. 072.050.

Structure and physics of solar faculae. II. The non-thermal velocity field above faculae. See Abstr. 072.055.

The chromosphere above a sunspot umbra. See Abstr. 072.084.

Sunspots proper motions and positions of H-alpha filaments in the active region of August 1979. See Abstr. 072.094.

Report of IAU Commission 10: Solar activity (Activité solaire). See Abstr. 072.104.

The cooling and condensation of flare coronal plasma. See Abstr. 074.008.

The role of spicules in heating the solar atmosphere. See Abstr. 074.016.

On the constancy of $^{20}Ne/^{22}Ne$ in the solar wind and secular changes in the flare/wind ratio. See Abstr. 074.022.

On the thermal stability of hot coronal loops: the coupling between chromosphere and corona. See Abstr. 074.031.

Enthalpy flux cooling of the solar corona. See Abstr. 074.046.

The dynamics of coronal flare loops: I. Gasdynamics. See Abstr. 074.055.

Non-linear development of the radiative hydro-dynamic instability in empirical solar loop models. See Abstr. 074.062.

Association of coronal transient phenomena with disk flare activity from SMM Coronagraph/Polarimeter data. See Abstr. 074.067.

Magnetic reconnection and coronal transients. See Abstr. 074.091.

A numerical simulation of cooling coronal flare plasma. See Abstr. 074.093.

Magnetic energy storage and conversion in the solar atmosphere. See Abstr. 075.009.

Reversed-polarity regions. See Abstr. 075.010.

Flare build-up in preflare magnetic loops and non-linear force-free magnetic fields. See Abstr. 075.012.

Vector magnetic field evolution, energy storage, and associated photospheric velocity shear within a flare-productive active region. See Abstr. 075.038.

Observations of a post-flare radio burst in X-rays. See Abstr. 076.009.

The relation between the fluxes of soft and hard X-rays, and its relevance to flare energetics. See Abstr. 076.018.

2.2 MeV gamma-ray line observed during two SN solar flares. See Abstr. 076.023.

Connection of the radio brightness variations of the sun with hills of the magnetic field and flocculi during the minimum of solar activity. See Abstr. 077.014.

On the microwave radiation of solar flares. See Abstr. 077.016.

The flare-related depression of the noise storm on May 5, 1978. See Abstr. 077.021.

Study of the character and parameters of wave motions in the transition layer and low solar corona by radio observations. See Abstr. 077.023.

Observations of low-frequency radio bursts connected with the proton flare of April 24, 1981. See Abstr. 077.037.

Flare buildup at 6 cm wavelength, in UV and $H\alpha$. See Abstr. 077.041.

VLA observations of the evolution of a solar burst source structure at 6 centimeter wavelength. See Abstr. 077.054.

Centimeter wavelength observations of active regions and flares with a few arc second resolution. See Abstr. 077.065.

Influence of the dynamics of structural formations in the interplanetary medium on propagation of charged particles generated in solar flares at longitudes 46 - 85° W in September - November 1973. See Abstr. 078.008.

Airborne total eclipse observation of the extreme solar limb at 400 μm. See Abstr. 079.301.

Relationship between photospheric and chromospheric $\Omega\,(\vartheta, t)$ deduced by temporarily and spatially correlated tracers. See Abstr. 080.104.

Report of IAU Commission 12: Radiation and structure of the solar atmosphere (Radiation et structure de l'atmosphère solaire). See Abstr. 080.109.

Use of frequency measurement on satellite signals for computing differential Doppler and solar flare detection. See Abstr. 083.002.

Severe geomagnetic storms and their sources on the sun. See Abstr. 084.001.

Solar flare isotopic pattern — a new component to consider the isotopic composition of the solar system. See Abstr. 094.526.

Solar flare neon: clues from implanted noble gases in lunar soils and rocks. See Abstr. 094.571.

Evidence for annealing of solar flare tracks in certain gas-rich meteorites. See Abstr. 105.098.

Solar flare shocks in interplanetary space and solar flare particle events. See Abstr. 106.019.

Erratum

073.901 An interacting loop model for solar flare bursts.
A. G. Emslie.
Astrophys. Lett., Vol. 22, 41 - 47 (1981). — See Abstr. 30.073.100.

074 Corona, Solar Wind

074.001 Coronal closed structures. IV. Hydrodynamical stability and response to heating perturbations.
G. Peres, R. Rosner, S. Serio, G. S. Vaiana.
Astrophys. J., Vol. 252, 791 - 799 (1982).

The authors have studied the response of magnetically confined atmospheres (loops) to perturbations in (1) the temperature and density distribution, and (2) the local heating rate by means of a one-dimensional time-dependent hydro-dynamical code which incorporates the full energy, momentum and mass conservation equations. These studies extend the linear instability analysis of Habbal and Rosner into the finite-amplitude regime, and generalize the confined atmosphere models of Serio et al. to the time-dependent domain. The relevance of the results to the current understanding of the physical mechanisms responsible for energy deposition in the solar chromosphere and corona is discussed.

074.002 Comparison of the 530.3 nm coronal brightness estimates of the Wendelstein Solar Observatory and of the 530.3 nm coronal measurements of Alma-Ata, Kislovodsk, Lomnický Štít, Norikura, with the Wendelstein 530.3 nm – coronal photometer.   C. Spannagl.
Bull. Astron. Inst. Czechoslovakia, Vol. 33, 60 - 62 (1982).

The scale of the Wendelstein visual coronal estimates at 530.3 nm is compared with the measurements of the Wendel-stein 530.3 nm – coronal photometer and the quality of the estimates is discussed. Furthermore, the 530.3 nm coronal measurements of Alma-Ata, Kislovodsk, Lomnický Štít and Norikura are compared with the photoelectric measurements at 530.3 nm of Wendelstein.

074.003 Diagnostic of coronal heating processes based on the emission measure of UV lines.
G. Torricelli-Ciamponi, G. Einaudi, C. Chiuderi.
Astron. Astrophys., Vol. 105, L1 - L4 (1982).

The authors use the emission measure of solar UV lines to deduce information on coronal loops heating mechanisms. The method turns out to be quite effective. The results obtained are discussed and their implications pointed out.

074.004 Electron heating by fast mode magnetohydro-dynamic waves in the solar wind emanating from coronal holes.   S. R. Habbal, E. Leer.
Astrophys. J., Vol. 253, 318 - 322 (1982).

Fast mode magnetohydrodynamic waves, propagating outwards from the Sun in coronal hole regions, will dissipate primarily through collisionless interaction with electrons rather than with protons. This dissipation can lead to higher electron than proton temperatures in the accelerating region of the solar wind, if the waves carry a sufficiently large energy flux.

074.005 The thermal structure of solar coronal loops and implications for physical models of coronae.
J. C. Raymond, P. Foukal.
Astrophys. J., Vol. 253, 323 - 329, plate 7 (1982).

The authors analyze EUV spectra of three active region loops observed above the solar limb with the S055 spectrom-eter on Skylab. The lengths, peak temperatures, and pressures of the loops are typical of the X-ray coronal loops to which static models have been applied. The authors find that the physical parameters of the coronal loop plasma derived from EUV spectra and raster pictures are not well represented by the static models. Several line ratios in the loop spectrum indicate departures from ionization equilibrium caused by rapid cooling. The authors discuss the source of this cooling material with reference to several models of loop dynamics.

074.006 The solar coronal X-ray spectrum from 15.4 to 23.0 Å: lines from highly ionized calcium and chromium and their usefulness as plasma diagnostics.
D. L. McKenzie, P. B. Landecker.
Astrophys. J., Vol. 254, 309 - 317 (1982).

Observations by the SOLEX experiment from the USAF *P78-1* satellite are used to examine the 15.4−23.0 Å X-ray spectrum from both flaring and nonflaring solar active region plasmas. High sensitivity enabled the detec-tion of more than 60 lines in this wavelength region. Most of the lines expected to be strongest from Cr XV−XVI and Ca XV−XVIII were detected. Line fluxes are used to estimate electron densities for flaring and nonflaring solar active regions. From the Ca XV line flux the flare electron density is estimat-ed at $\sim 4 \times 10^6$ K. The coronal abundances of Cr and Ca with respect to O are estimated and are found to be
$A(Cr)/A(O) = 0.0036 \pm 0.0018$ and
$A(Ca)/A(O) = 0.015 \pm 0.005$.

074.007 The initiation of a coronal transient.
B. C. Low, R. H. Munro, R. R. Fisher.
Astrophys. J., Vol. 254, 335 - 342 (1982).

This paper analyzes the coronal transient/eruptive promi-nence event of 1980 August 5 observed by the Mauna Loa ex-periment system. This event yielded data on the early develop-ment of the transient in the low corona between $1.2\,R_\odot$ and $2.2\,R_\odot$. The transient's initial appearance in the form of a ris-ing density-depleted structure, prior to the eruption of the associated prominence, can be explained as an effect of mag-netic buoyancy. The height versus base length relationship of the evolving transient resembles, remarkably well, the theoret-ical predictions obtained from a quasi-static model of a mag-netically buoyant loop system.

074.008 The cooling and condensation of flare coronal plasma.   S. K. Antiochos, P. A. Sturrock.
Astrophys. J., Vol. 254, 343 - 348 (1982).

The authors investigate a model for the decay of flare heated coronal loops in which rapid radiative cooling at the loop base creates strong pressure gradients which, in turn, generate large (supersonic) downward flows. Hence, the coronal material cools and "condenses" onto the flare chromo-sphere. The important features of this model which distinguish it from previous models of flare cooling are: (1) Most of the thermal energy of the coronal plasma may be lost by mass motion rather than by conduction or coronal radiation. (2) Flare loops are not isobaric during their decay phase, and large downward velocities are present near the footpoints. (3) The differential emission measure q has a strong temperature dependence, $q \propto T^{3.5}$.

074.009 The condensational instability in the solar transition region and corona.
E. S. Oran, J. T. Mariska, J. P. Boris.
Astrophys. J., Vol. 254, 349 - 360 (1982).

The authors investigate the stability of plasmas at tem-peratures and densities typical of the solar transition region and corona using both a linear analysis and nonlinear time-dependent numerical simulations. The nonlinear regime is characterized by a bifurcation of the plasma into a cool dense condensation surrounded by a hot tenuous corona. The con-densation may then be accelerated by forces in the plasma such as those arising from gravity or differential heating. The results of the detailed simulation show that the transition region is a dynamically stable structure which is the result of the nonlinear evolution of the condensational instability.

074.010 **Analysis of coronal H I Lyman alpha measurements from a rocket flight on 1979 April 13.**
G. L. Withbroe, J. L. Kohl, H. Weiser, G. Noci, R. H. Munro.
Astrophys. J., Vol. 254, 361 - 370 (1982).

Measurements of the profiles of resonantly scattered hydrogen Lyman-α coronal radiation have been used to determine hydrogen kinetic temperatures from 1.5 to 4 $R_\odot$ from Sun center in a quiet region of the corona. Proton temperatures derived from the line widths decrease with height from 2.6×10^6 K at $r = 1.5\ R_\odot$ to 1.2×10^6 K at $r = 4\ R_\odot$. Comparison of measured Lyman-α intensities with those calculated using a representative model for the radial variation of the coronal electron density provides information on the magnitude of the electron temperature gradient and suggests that the solar wind flow was subsonic for $r < 4\ R_\odot$ in the observed region.

074.011 **Ultraviolet continuum absorption ($\lesssim 1000$ Å) above the quiet sun transition region.**
G. A. Doschek, U. Feldman.
Astrophys. J., Vol. 254, 371 - 375 (1982).

The authors investigate Lyman continuum absorption shortward of 912 Å in the quiet Sun solar transition region by combining spectra obtained from the Apollo Telescope Mount experiments on Skylab. The most recent atomic data are used to compute line intensities for lines that fall on both sides of the Lyman limit. Lines of O III, O IV, O V, and S IV are considered. The computed intensity ratios of most lines from O IV, O V, and S IV agree with the experimental ratios to within a factor of 2. The discrepancies show no apparent wavelength dependence.

074.012 **Self-similar magnetohydrodynamics. I. The $\gamma=4/3$ polytrope and the coronal transient.**
B. C. Low.
Astrophys. J., Vol. 254, 796 - 805 (1982).

The full set of ideal magnetohydrodynamic (MHD) equations for a $\gamma=4/3$ polytrope admits self-similar solutions which can be derived by analytic methods. An axisymmetric magnetic field in a stratified stellar atmosphere is assumed. The general properties of these self-similar solutions are discussed in connection with the coronal transient as an illustration. The solutions admit a large variety of magnetic structures, including those in the form of loops, moving through the corona with large scale coherence and sharp small scale features. The Lagrangian velocity of an individual transient feature is found to accelerate or decelerate according to a positive or negative gain in momentum by the plasma in a self-similar distribution of the Lorentz force, the pressure gradient, and the gravitational force. The physical implications are discussed, arguing in favor of the transient beginning as a fully nonlinear MHD motion that ejects both magnetic field and plasma out of the gravitational bond of the Sun.

074.013 **Heating of the corona and solar wind by switch-on shocks.** J. V. Hollweg.
Astrophys. J., Vol. 254, 806 - 813 (1982).

The author examines the possibility that the corona is heated by a train of weak switch-on shocks which are formed in the chromosphere from a train of Alfven waves, and which subsequently enter the corona from below. It is shown that most of the energy in the shock train can be dissipated within one or two solar radii above the coronal base. A train of switch-on shocks therefore represents a viable coronal heating mechanism. The results are generalized to switch-on shocks in the solar wind. It is shown that such shocks can dissipate rapidly, but it is concluded that they are not the dominant factor governing the evolution of the solar wind turbulence.

074.014 **Effect of the neutral component of the solar wind on the interaction of the solar system with the interstellar gas flux.** M. A. Gruntman.

Pis'ma Astron. Zh., Tom 8, 48 - 51 (1982). In Russian. English translation in Soviet Astron. Lett., Vol. 8.

An interaction of the solar wind's neutral component and the ionised component of the interstellar gas is shown likely to affect essentially the interstellar gas characteristics. Taking this effect into consideration may change both conceptions on processes in the solar wind termination region and interpretation of the measurements of Lα radiation from the interstellar gas.

074.015 **The steady global corona..**
R. S. Steinolfson, S. T. Suess, S. T. Wu.
Astrophys. J., Vol. 255, 730 - 742 (1982).

A model is developed for the formation of the steady coronal structure consisting of coronal streamers and coronal holes. A coronal streamer consists of closed magnetic field lines near the solar surface with overlying and adjacent open field lines. The open field region has the characteristics of a coronal hole. The atmosphere is stationary in the closed region and flowing outward in the open region (the coronal hole). The steady coronal structure is obtained by starting the numerical calculation with a state consisting of a polytropic, hydrodynamic solution to the steady state radial equation of motion coupled with a dipole magnetic field. The numerical solution of the complete time-dependent equations then asymptotically approaches a steady state. Global coronal configurations are calculated for values of the plasma beta (ratio of thermal pressure to magnetic pressure) varying from 0.1 to 100.

074.016 **The role of spicules in heating the solar atmosphere.**
R. G. Athay, T. E. Holzer.
Astrophys. J., Vol. 255, 743 - 752 (1982).

From observations of downflowing material at transition region temperature, together with reasonable assumptions about the fate of spicules after they disappear from view in the visual spectrum, it is shown that the rise and fall of spicular material can supply the thermal energy required by radiative losses from the transition region and upper chromosphere in the network. If sufficient heat is added to spicules, in conjunction with their acceleration, the spicule phenomenon may also play a major role in the production and maintenance of much of the solar corona. Thus, the processes whereby spicules are generated and heated may be of central importance to the energy balance of the outer solar atmosphere.

074.017 **Coronal emission-line polarization from the statistical equilibrium of magnetic sublevels. II. Fe XIV 5303 Å.**   L. L. House, C. W. Querfeld, D. E. Rees.
Astrophys. J., Vol. 255, 753 - 763 (1982).

Coronal magnetic fields influence the intensity and linear polarization of light scattered by coronal Fe XIV ions. To interpret polarization measurements of Fe XIV 5303 Å coronal emission requires a detailed understanding of the dependence of the emitted Stokes vector on coronal magnetic field direction, electron density, and temperature and on height of origin. The required dependence is included in the solutions of statistical equilibrium for the ion which are solved explicitly for 34 magnetic sublevels in both the ground and four excited terms. The full solutions are reduced to equivalent simple analytic forms which clearly show the required dependence on coronal conditions.

074.018 **The formation and interpretation of the Fe XIII 10747 Å coronal emission line.**
C. W. Querfeld.
Astrophys. J., Vol. 255, 764 - 773 (1982).

The intensity and linear polarization of the Fe XIII 10747 Å coronal emission line depends on the magnetic field direction, electron density, temperature, and height of emitting coronal volumes. The interpretation of coronal emission-line observations requires that the effects of these variables on the coronal emission-line Stokes vector be simply described. The Sahal-Brechot and House solutions of statistical equilibrium equations for Fe XIII include fully the effects of electron and proton collisions in the ground term and of electron collisional excitation of excited terms which decay radiatively to the ground term. The solutions are recast in a simple analytic form which displays the explicit dependence of coronal line emission on the relevant coronal conditions.

074.019 Equilibria and stability of coronal magnetic arches.
R. L. T. Wolfson.
Astrophys. J., Vol. 255, 774 - 782 (1982).

This paper explores the magnetostatic equilibria of coronal magnetic arches using a current sheet model and a nonlinear minimization technique. Two classes of solutions are found, representing arches which are either stable against spreading of the arch footpoints. Physically, stability properties of the arches are related to the form of magnetic flux redistribution occurring as the footpoints spread. The results suggest that although coronal transients may erupt spontaneously, they are more likely when an impulsive event occurs under a magnetic arch.

074.020 Non-adiabatic expansion of low-temperature solar wind radial temperature gradients.
A. Geranios.
J. Geophys., Vol. 49, 192 - 197 (1981). — Abstr. in Phys. Abstr., Vol. 85, Abstr. 10294 (1982).

074.021 Effects of thermal conductivity on large-scale spiral waves in the solar wind.   S. Wang, L. Fang.
Kexue Tongbao, Vol. 26, 618 - 622 (1981). — Abstr. in Phys. Abstr., Vol. 85, Abstr. 10295 (1982).

074.022 On the constancy of ^{20}Ne/^{22}Ne in the solar wind and secular changes in the flare/wind ratio.
R. Wieler, P. Signer.
Meteoritics, Vol. 16, 401 - 402 (1981). — Abstract.

074.023 Global properties of the solar wind. III. Density and temperature fluctuations.
M. Eyni, R. Steinitz.
Astrophys. J., Vol. 256, 259 - 262 (1982).

The authors demonstrate that large-scale interactions in the solar wind can be studied through the use of quantities which permit the coverage of a continuous distance range and all velocities probed. The method is applied to data from Helios 1. It is confirmed that the solar wind is more uniform at higher flow speeds. It appears that a positive correlation between proton temperature and density develops with heliocentric distance for the intermediate flow speeds.

074.024 EUV spectroscopic plasma diagnostics for the solar wind acceleration region.
J. L. Kohl, G. L. Withbroe.
Astrophys. J., Vol. 256, 263 - 270 (1982).

Existing empirical information about the physical conditions in the solar wind acceleration region beyond a few tenths $R_\odot$ above the solar surface is extremely limited. The authors discuss plasma diagnostics that can be implemented through coronagraphic measurements of ultraviolet coronal lines from ions such as N V, O VI, Ne VIII, Mg X, and Si XII. They illustrate how profiles and intensities of the collisionally excited and resonantly scattered components of these spectral lines can be used to probe the physical conditions (temperatures, densities, mass flow velocities, and chemical abundances) in the solar wind acceleration region out to $4-8\,R_\odot$ from Sun center and beyond.

074.025 On the solar type III radio burst emission process.
D. G. Wentzel.
Astrophys. J., Vol. 256, 271 - 283 (1982).

This paper investigates the problem whether solar type III radio emission can be caused by plasma solitons evolving in one dimension. The high energy threshold for initiating nonlinear plasma phenomena in the corona is discussed. Conditions for stabilizing coronal electron beams by nonlinear plasma interactions are analyzed. It is concluded that the large variety of observed type III radio bursts implies a range of important plasma phenomena much larger than presently considered by theoretical models.

074.026 Simulation of three-dimensional solar wind disturbances and resulting geomagnetic storms.
K. Hakamada, S.-I. Akasofu.
Space Sci. Rev., Vol. 31, 3 - 70 (1982).

The authors have succeeded in simulating, for the first time, the three-dimensional structure of the two-stream condition of the solar wind, its disturbances generated by solar flares, as well as the resulting geomagnetic storms. The results demonstrate the usefulness of the kinematic method developed in this paper. The uniqueness of this method is that one can simulate closely a self-consistent data set, either theoretical or observational, so that their results can be reasonably self-consistent.

074.027 Magnetically closed regions in the solar wind.
A. Geranios.
Astrophys. Space Sci., Vol. 81, 103 - 122 (1982).

Interplanetary plasma and magnetic field data collected by Helios-1, Helios-2 and IMP-8 satellites over the periods December 1974 - December 1976, January 1976 - December 1976 and December 1974 - December 1976, respectively, are analysed. From this analysis, the author identified 85 cases in which the proton temperature was very low. In 50 of these cases, the interplanetary magnetic field showed characteristic variations favorable for closed structures in the solar wind. By using the calculated radial temperature gradients as a function of the solar wind speed and the heliocentric distance 'cold' protons in the neighborhood of the Sun (0.3AU) could be identified.

074.028 Resonant wave acceleration of minor ions in the solar wind.   J. F. McKenzie, E. Marsch.
Astrophys. Space Sci., Vol. 81, 295 - 314 (1982).

This paper extends previous work on the acceleration of minor ions in the solar wind to include the effects of wave acceleration and heating arising from minor ions interacting via the gyroresonance with ion cyclotron waves. Resonant wave acceleration is made up of two contributions, the first, and generally the more important, is a 'local' acceleration which is proportional to the wave power and the number of resonant particles; while the other contribution is basically 'fluid dynamic' in character, arises from the inhomogeneity of the medium and is proportional to the radial gradient of the resonant wave power. Under suitable circumstances both contributions exhibit the feature that heavier ions receive greater acceleration than lighter ones.

074.029 Long-period observations of the solar wind plasma between 0.3 and 1.0 AU — solar activity.
A. Geranios.
Astrophys. Space Sci., Vol. 81, 333 - 343 (1982).

Hourly interplanetary plasma data measured by Helios-1 satellite over the period 10 December 1974 - 31 December 1977 are analysed. This analysis showed that the slow solar wind first increases its speed with heliocentric distance and then becomes more or less constant; the mean speed in the range 0.3 to 1.0 AU is 350 km s^{-1} for the slow solar plasma, while for the fast the mean value is between 650 and 700 km s^{-1}. It seems, particularly in the neighbourhood of the earth, that an additional (intermediate) component appears at 450 km s^{-1}. During the phase of enhanced solar activity (11—yr solar cycle) only the slow solar wind is present, while at solar minimum all three types of the solar wind are equally represented.

074.030 **On a quasi-stationary electromagnetic field in the solar wind.**
I. I. Alekseev, I. S. Veselovskij, A. P. Kropotkin.
Geomagn. Aehron., Tom 22, 5 - 9 (1982). In Russian.

074.031 **On the thermal stability of hot coronal loops: the coupling between chromosphere and corona.**
N. P. M. Kuin, P. C. H. Martens.
Astron. Astrophys., Vol. 108, L1 - L4 (1982).

The authors consider the interaction of the hot plasma in coronal loops with the underlying chromospheric plasma, and find stable static equilibria if the coupling between corona and chromosphere is sufficiently strong. However, for typical coronal loop conditions the interaction is not strong enough for perfect stabilisation and an oscillatory solution is found with a period of about a day. The latter solution is very similar to the static solution during most of the time and is relatively cool only during a short while. The authors tentatively identify this cyclic behaviour with the observed upflows and downflows in the solar corona.

074.032 **Possible evidence for coronal Alfvén waves.**
J. V. Hollweg, M. K. Bird, H. Volland, P. Edenhofer, C. T. Stelzried, B. L. Seidel.
J. Geophys. Res., Vol. 87, 1 - 8 (1982).

The 2.29 GHz S band carrier signals of the two Helios spacecraft are used to probe the magnetic and density structures of the solar corona inside 0.05 AU. In this paper the authors analyze the observed fluctuations of the electron content and Faraday rotation. A simple statistical ray analysis is employed. The authors conclude that (1) the observed Faraday rotation fluctuations cannot be solely due to electron density fluctuations in the corona unless the coronal magnetic field is some 5 times stronger than suggested by current estimates, and (2) the observed Faraday rotation fluctuations are consistent with the hypothesis that the sun radiates Alfvén waves with sufficient energies to heat and accelerate high-speed solar wind streams.

074.033 **Solar wind helium ions: observations of the Helios solar probes between 0.3 and 1 AU.**
E. Marsch, K.-H. Mühlhäuser, H. Rosenbauer, R. Schwenn, F. M. Neubauer.
J. Geophys. Res., Vol. 87, 35 - 51 (1982).

A survey of solar wind helium ion velocity distributions and derived parameters as measured by the Helios solar probes between 0.3 and 1 AU is presented.

074.034 **Solar wind protons: three-dimensional velocity distributions and derived plasma parameters measured between 0.3 and 1 AU.**
E. Marsch, K.-H. Mühlhäuser, R. Schwenn, H. Rosenbauer, W. Pilipp, F. M. Neubauer.
J. Geophys. Res., Vol. 87, 52 - 72 (1982).

A survey of solar wind three-dimensional proton velocity distributions as measured by the Helios solar probes between 0.3 and 1 AU is presented. A variety of nonthermal features like temperature anisotropies, heat fluxes, or proton double streams has been observed. The observations indicate that local heating or considerable proton heat conduction occurs in the solar wind. Some consequences of nonthermal features of proton distributions for plasma instabilities are discussed as well as kinetic processes that may shape the observed distributions.

074.035 **A sub-Alfvénic solar wind: interplanetary and magnetosheath observations.**
J. T. Gosling, J. R. Asbridge, S. J. Bame, W. C. Feldman, R. D. Zwickl, G. Paschmann, N. Sckopke, C. T. Russell.

J. Geophys. Res., Vol. 87, 239 - 245 (1982).

Many years of observation have established that the solar wind flow at 1 AU normally is both supersonic and super-Alfvénic. However, for portions of an ~5-hour period on November 22, 1979, the solar wind flow speed (~320 km s^{-1}) observed at ISEE 3 was considerably less than the Alfvén speed (~540 km s^{-1}). At the same time ISEE 1 and 2 made plasma and field measurements both within the magnetosphere and within the disturbed solar wind flow adjacent to the magnetosphere. The authors present and discuss the ISEE 3 evidence for sub-Alfvénic flow and examine the nature of the disturbed flow adjacent to the magnetosphere. This examination suggests that the earth's bow wave retained its shock-like character when the solar wind flow was sub-Alfvénic.

074.036 **Enhanced scintillations associated with high speed streams in the solar wind.**
A. Basu, A. C. Das.
J. Geophys. Res., Vol. 87, 1688 - 1690 (1982).

074.037 **Density and temperature determination of neutral hydrogen in coronal structures.**
R. M. Bonnet, G. Tsiropoula.
Sol. Phys., Vol. 75, 139 - 143 (1982).

High-resolution filtergrams in Lα have been obtained with a rocket borne instrument and evidence several loop shaped structures which can be seen as absorption features over the solar surface. The optical thickness of these coronal structures is measured with respect to nearby unabsorbed disk features. Their shape and dimension being known from the images, the determination of the neutral hydrogen temperature and density is possible. It is found that temperatures below 10^5 K and densities of a few 10^4 hydrogen atoms cm^{-3} are compatible with the opacities observed in the structures.

074.038 **Polar coronal plumes.**
S. T. Suess.
Sol. Phys., Vol. 75, 145 - 159 (1982).

Polar coronal plumes are modeled using concentrations of magnetic flux at $1.01 R_\odot$, and assuming the field is current-free, or a potential field. Identifying the density enhancement of plumes with magnetic flux concentration produces good agreement between $1.01 R_\odot$ and $1.10 R_\odot$, for model conditions of a large background magnetic field and a plume separation of 50 000 to 70 000 km at the base. Beyond $1.10 R_\odot$, both plumes and the potential field diverge very nearly as r^2.

074.039 **Coronal holes: mass loss driven by magnetic reconnection.**   D. J. Mullan, I. A. Ahmad.
Sol. Phys., Vol. 75, 347 - 350 (1982).

The authors propose that bubbles of matter ejected from magnetic reconnection sites in polar plumes drive the solar wind in coronal holes.

074.040 **Resistive tearing mode in coronal neutral sheets.**
L. Janicke.
Sol. Phys., Vol. 76, 29 - 43 (1982).

The apparent stability of coronal neutral sheets with respect to the resistive tearing mode has been attributed by previous authors to the influence of a weak normal component of the confining magnetic field. To check this hypothesis a normal mode analysis is performed applying rigorously singular perturbation technique. If parameters for a typical neutral sheet in the middle corona (0.5 solar radii) are inserted, the result is that no stabilization by a normal component occurs, if the value of the growth time predicted by the one-dimensional theory is far shorter than ten minutes – independent of the values assumed for the width of the neutral sheet or the resistivity.

074.041 **The evolution and the secondary maximum of the green line intensity.**

J. Xanthakis, B. Petropoulos, H. Mavromichalaki.
Sol. Phys., Vol. 76, 181 - 190 (1982).

A new relation has been given in order to calculate the intensity of the green line of the solar corona at 5303 Å as a function of the number of proton events N_p and the $I_\alpha(R)$ index of solar activity. This relation is available for the 19th and 20th solar cycles. Moreover there is given a theoretical justification of this relationship taking into account as a new parameter the evolution of the coronal magnetic field during the solar cycle.

074.042 The stability and uniqueness of coronal loops.
I. J. D. Craig, T. D. Robb, M. D. Rollo.
Sol. Phys., Vol. 76, 331 - 355 (1982).

A hydrodynamic model of high resolution is used to examine the stability of coronal loops to finite amplitude perturbations. The loop is heated by means of a low-amplitude energy input and its subsequent dynamic relaxation is followed. The initial atmosphere is generated by solving the time independent form of the hydrodynamic equations. It is shown that the loop structure depends critically on the balance between the radiative losses and the quiescent heating at the base of the transition zone, i.e. on the concavity of the temperature profile in this region. The dynamic evolution of the loop is then investigated for two classes of lower boundary conditions. The observational consequences of the analysis are discussed.

074.043 Results of polarization radiation of the solar corona during maximal activity of the sun.
S. I. Avdyushin, A. F. Bogomolov, E. I. Zajtsev, V. A. Krylov, S. P. Leonenko, B. A. Poperechenko.
Dokl. AN SSSR, Tom 261, 833 - 835 (1981). In Russian.
Abstr. in Ref. zh., 51. Astron., 4.51.379 (1982).

074.044 Prediction of the structure of the solar corona.
G. M. Nikol'skij.
Zemlya Vselennaya, 1982, No. 3, p. 29 - 30. In Russian.

074.045 Non-equilibrium ionization in coronal loops.
G. Borrini, G. Noci.
Sol. Phys., Vol. 77, 153 - 166 (1982).

The ionization conditions in coronal loops are investigated in the temperature range $2 \times 10^5 - 2 \times 10^6$ K, assuming velocity, density and temperature distributions computed for a siphon model of a pure hydrogen plasma. It is found that the deviation from equilibrium ionization is large for subsonic-supersonic flow if the density is less than 5×10^9 cm^{-3}, with the exception of the lower part of the first leg of very cool loops ($T \approx 2 \times 10^5$ K).

074.046 Enthalpy flux cooling of the solar corona.
S. G. Wallenhorst.
Sol. Phys., Vol. 77, 167 - 175 (1982).

The differential emission measure profile for quiet and flaring solar regions is considered, using a model in which the principal downflow of heat is due to the enthalpy of downward-flowing material, rather than conduction. It is found that the emission measure profile for quiet solar regions is matched well by a downward particle number flux which decreases with temperature. This would be expected if this particle flux is due to heated spicular material falling back onto the chromosphere. In flaring regions, however, a particle flux which increases with temperature is required to explain the steep emission measure profile. This could be a result of mass motions downward out of the flaring loops.

074.047 Coronal holes as observed with the Pulkovo large radio telescope.   N. G. Peterova.
Soln. Dannye 1982 Byull., No. 1, p. 100 - 103 (1982). In Russian.

An identification of several coronal holes has been made using solar observations with the Pulkovo Large Radio

Telescope in the wavelength range 4.4 - 9.0 cm during the Skylab mission (May - December 1973).

074.048 Morphology and dynamics of coronal formations in red and green lines.
Yu. B. Kolesnik, O. S. Popov.
Vestn. Kiev. Univ., Vyp. 23, p. 48 - 58 (1981). In Russian.

074.049 The Hanle effect of the coronal Lα line of hydrogen: theoretical investigation.
V. Bommier, S. Sahal-Bréchot.
Sol. Phys., Vol. 78, 157 - 178 (1982).

The use of the Lα line of hydrogen is proposed for Hanle effect studies in the solar corona. The processes of formation of the coronal Lα of hydrogen are reviewed and the domain of sensitivity of the Hanle effect is given. The quantum theory is extended to an atom having a hyperfine structure (case of hydrogen). The polarization degree of Lα as a function of the height above the limb is computed in zero magnetic field, and after that the Hanle effect is calculated. Analytical formulae are provided which can then be entered in a solar model for integration over the emission volume in order to derive the magnetic field diagnostic.

074.050 The solar wind plasma.
H. Rosenbauer.
Solar system plasmas and fields, (see 003.005), p. 47 - 50 (1982).

074.051 MHD resonators in the solar corona: the radio emission modulation effect.
V. V. Zajtsev, A. V. Stepanov.
Astron. Zh., Tom 59, 563 - 570 (1982). In Russian. English translation in Soviet Astron., Vol. 26, No. 3.

Type II solar radio bursts data are shown to suggest the existence of Alfvén velocity minimum at a height of $\sim 1 R_\odot$ in the corona. The domain of a low Alfvén velocity is a resonator for the fast magnetosonic waves. The eigenmodes of the resonator are determined. The main mode period is about a few minutes.

074.052 On the calculation of the solar wind parameters from ground-based geomagnetic measurements.
A. D. Bazarzhapov.
Issled. po geomagn., aehron. i fiz. Solntsa, Moskva, 1982, No. 58, p. 126 - 134. In Russian. − Abstr. in Ref. zh., 62. Issled. kosm. prostranstva, 5.62.494 (1982).

074.053 On the two-fluid polytropic solar wind model.
D. Summers.
Astrophys. J., Vol. 257, 881 - 886 (1982).

The steady, spherically symmetric, polytropic, two-fluid solar wind model in which the electron and proton gases possess the distinct, constant polytropic indices α_e and α_p is investigated. The model equations reduce to a single differential equation for the fluid velocity which involves α_e, α_p, and a parameter β which depends on the mass flux and constants related to the entropy of the electron and proton gases. All solutions to the model are sought for the range $1 \leqslant \alpha_e < \alpha_p \leqslant 5/3$. The model permits 14 distinct solution types, and the solution topologies for the velocity are sketched to illustrate these cases.

074.054 Time-dependent solar wind ionization.
S. P. Owocki, A. J. Hundhausen.
Bull. American Astron. Soc., Vol. 13, 812 (1981). − Abstract.

074.055 The dynamics of coronal flare loops: I. Gasdynamics.
C.-C. Cheng, G. A. Doschek, J. P. Boris, J. T. Mariska, E. S. Oran.
Bull. American Astron. Soc., Vol. 13, 819 (1981). − Abstract.

074.056 **The dynamics of coronal flare loops: II. Comparison to observations.**   G. A. Doschek, C.-C. Cheng, J. P. Boris, J. T. Mariska, E. S. Oran.
Bull. American Astron. Soc., Vol. 13, 819 (1981). – Abstract.

074.057 **Collisionless conduction front propagation in large loops and solar hard X-ray bursts.**
D. F. Smith, D. W. Harmony.
Bull. American Astron. Soc., Vol. 13, 819 (1981). – Abstract.

074.058 **Mass flows in the solar corona as a diagnostic of the coronal heating function.**
J. T. Mariska, J. P. Boris.
Bull. American Astron. Soc., Vol. 13, 836 (1981). – Abstract.

074.059 **Thermoelectric effects on coronal heat balance.**
D. L. Book.
Bull. American Astron. Soc., Vol. 13, 836 (1981). – Abstract.

074.060 **Enthalpy flux cooling of the solar corona.**
S. G. Wallenhorst.
Bull. American Astron. Soc., Vol. 13, 836 (1981). – Abstract.

074.061 **Inhibition of coronal transient loop formation in strong magnetic field.**
C. Sawyer, W. J. Wagner, R. M. E. Illing, L. L. House.
Bull. American Astron. Soc., Vol. 13, 836 (1981). – Abstract.

074.062 **Non-linear development of the radiative hydrodynamic instability in empirical solar loop models.**
C.-H. An, R. C. Canfield, A. N. McClymont, G. H. Fisher.
Bull. American Astron. Soc., Vol. 13, 837 (1981). – Abstract.

074.063 **Cross-B electric field in coronal loops.**
S. Hinata.
Bull. American Astron. Soc., Vol. 13, 837 (1981). – Abstract.

074.064 **Radio wave scattering in the outer solar corona.**
W. C. Erickson, M. J. Mahoney, W. M. Cronyn.
Bull. American Astron. Soc., Vol. 13, 841 (1981). – Abstract.

074.065 **Frequency and location of coronal transients observed with the Coronagraph/Polarimeter aboard the Solar Maximum Mission satellite, preliminary results.**
E. Hildner, R. M. E. Illing, W. J. Wagner, L. L. House, C. B. Sawyer, C. L. Hyder.
Bull. American Astron. Soc., Vol. 13, 861 - 862 (1981). Abstract.

074.066 **Hα ejecta in the outer corona.**
L. L. House, R. M. E. Illing, C. Sawyer, W. J. Wagner.
Bull. American Astron. Soc., Vol. 13, 862 (1981). – Abstract.

074.067 **Association of coronal transient phenomena with disk flare activity from SMM Coronagraph/Polarimeter data.**
R. M. E. Illing, L. L. House, W. J. Wagner, C. Sawyer.
Bull. American Astron. Soc., Vol. 13, 862 (1981). – Abstract.

074.068 **Simultaneous and overlapping white-light and soft X-ray observations of the solar corona.**
D. G. Sime, D. Webb.
Bull. American Astron. Soc., Vol. 13, 878 (1981). – Abstract.

074.069 **Solar flares and coronal heating: a new magnetic energy release mechanism of current loops.**
J. Chen, M. L. Xue.
Bull. American Astron. Soc., Vol. 13, 889 - 890 (1981). Abstract.

074.070 **A flare-associated coronal transient.**
D. Friend, R. H. Munro, R. R. Fisher, M. McCabe.
Bull. American Astron. Soc., Vol. 13, 890 (1981). – Abstract.

074.071 **MHD stability of coronal loops with radiative energy loss.**   C.-H. An.
Bull. American Astron. Soc., Vol. 13, 891 (1981). – Abstract.

074.072 **Striations of plasma in the solar corona: the gXB instability.**   S. Migliuolo.
Bull. American Astron. Soc., Vol. 13, 891 (1981). – Abstract.

074.073 **The solar corona near the maximum of cycle 21.**
R. R. Fisher, P. Seagraves.
Bull. American Astron. Soc., Vol. 13, 905 (1981). – Abstract.

074.074 **The total solar eclipse of 31 July 1981.**
L. Lacey, R. R. Fisher.
Bull. American Astron. Soc., Vol. 13, 905 (1981). – Abstract.

074.075 **Solar coronal observations of plasma diagnostic X-ray lines of O VII and Ne IX.**
D. L. McKenzie, P. B. Landecker.
Bull. American Astron. Soc., Vol. 13, 911 (1981). – Abstract.

074.076 **The active corona II: north sector. Images from the SMM Coronagraph/Polarimeter.**
R. M. E. Illing, W. J. Wagner, L. L. House, C. Sawyer.
Bull. American Astron. Soc., Vol. 13, 911 - 912 (1981). Abstract.

074.077 **Solar coronal X-ray spectra of calcium and chromium from 15.4 to 23.0 Ångstroms.**
P. B. Landecker, D. L. McKenzie.
Bull. American Astron. Soc., Vol. 13, 912 (1981). – Abstract.

074.078 **Coronal outflow velocities: 1980 rocket measurements.**
R. H. Munro, J. L. Kohl, H. Weiser, G. L. Withbroe.
Bull. American Astron. Soc., Vol. 13, 912 (1981). – Abstract.

074.079 **Results of Ly-α coronagraphic observations following the 1980 eclipse.**
H. Weiser, J. L. Kohl, R. H. Munro, G. L. Withbroe.
Bull. American Astron. Soc., Vol. 13, 913 (1981). – Abstract.

074.080 **Coronal jets, strong shock wave heating of the corona and a cloud model of the solar wind.**
G. E. Brueckner, J. -D. F. Bartoe.
Bull. American Astron. Soc., Vol. 13, 913 (1981). – Abstract.

074.081 **The dynamics of accelerating coronal bullets.**
J. T. Karpen, E. S. Oran, J. P. Boris, J. T. Mariska, G. E. Brueckner.
Bull. American Astron. Soc., Vol. 13, 913 - 914 (1981). Abstract.

074.082 **Corona, January - December 1980.**
Q. Bull. Sol. Act., Vol. 22, Part IV, p. 1 - 30 (1980).

074.083 **A mechanism for a class of solar coronal disturbances.**   S. T. Wu, Y. Q. Hu, S. Wang, M. Dryer, E. Tandberg-Hanssen.
Astrophys. Space Sci., Vol. 83, 189 - 194 (1982).
In this paper, a new ideal magnetohydrodynamic (MHD) model is used to examine the dynamical response of the upper solar atmosphere to injection of cold mass from the photosphere akin to a surge. A significant new physical phenomenon is revealed: the formation of an almost stationary loop prominence in the atmosphere as a consequence of the ejected material. Simultaneously with the formation of this new loop, the simulation exhibits MHD waves that propagate

outward (i.e., away from the loop) to excite coronal material. It is conjectured that these waves may trigger a class of coronal disturbances.

074.084 The observation of a coronal transient directed at earth.
D. J. Michels, R. A. Howard, M. J. Koomen, N. R. Sheeley, Jr.
Bull. American Astron. Soc., Vol. 14, 572 (1982). — Abstract.

074.085 Radial variation of the solar wind speed between 1 and 15 AU.
H. R. Collard, J. D. Mihalov, J. H. Wolfe.
J. Geophys. Res., Vol. 87, 2203 - 2214 (1982).

Pioneer 10 and 11 solar wind speeds measured between 1.4 and 15.2 AU are compared with IMP 6, 7, and 8 solar wind speeds measured at 1 AU for six radial alignment periods between 1973 and 1978. The authors show how the solar wind speed-time profile changes in character and how the average properties of the solar wind speed vary as it is observed at increasingly greater distances from the sun.

074.086 Statistical properties of low-frequency magnetic field fluctuations in the solar wind from 0.29 to 1.0 AU during solar minimum conditions: HELIOS 1 and HELIOS 2.   K. U. Denskat, F. M. Neubauer.
J. Geophys. Res., Vol. 87, 2215 - 2223 (1982).

To obtain information on the temporal and spatial evolution of MHD waves and discontinuities in the solar wind, the authors studied by means of statistical methods magnetic field fluctuations measured by the two HELIOS spacecraft in the frequency range between 2.4×10^{-5} Hz and 1.3×10^{-2} Hz at distances from the sun between 0.29 AU and 1.0 AU.

074.087 Voyager observations of solar wind proton temperature: 1-10 AU.   P. R. Gazis, A. J. Lazarus.
Geophys. Res. Lett., Vol. 9, 431 - 434 (1982).

Solar wind proton temperatures are measured simultaneously by the Voyager 1 and 2 spacecraft, far from Earth, and the IMP 8 spacecraft in Earth orbit. The average value of the proton temperature from 1 to 9 AU is observed to decrease as $r^{-0.7 \pm 0.2}$ which is slower than would be expected for adiabatic expansion. A detailed look at the solar wind stream structure shows that considerable heating occurs at the interface between high and low speed streams.

074.088 A magnetohydrodynamic theory of coronal loop transients.   T. Yeh.
Sol. Phys., Vol. 78, 287 - 316 (1982).

The author presents a magnetohydrodynamic theory of coronal loop transients. This comprehensive theory deals with both physical and geometrical attributes of transient loops for their dynamical evolution. The new theory is based on a generalized Archimedes' principle of magnetohydrodynamic buoyancy force (Yeh, 1982). It explains how a magnetic rope is ejected from the solar surface as a result of magnetic unwinding. It also explains how an ejected coronal loop is accelerated by the buoyancy force and expanded by the self-induced force.

074.089 Radar studies of the non-spherically symmetric solar corona.
S. P. Owocki, G. A. Newkirk, D. G. Sime.
Sol. Phys., Vol. 78, 317 - 331 (1982).

The authors review the results of radar studies of the sun made at El Campo, Texas 1961 - 69 with particular emphasis on the record of observed solar radar cross sections. Using ray traces which include the effects of refraction, absorption, and scattering in non-spherically symmetric models of the corona, the authors investigate the role of focusing by large-scale coronal geometries in enhancing the radar cross section. They conclude that the present dataset does not support the

hypothesis that radar observations of the sun will be useful in determining the properties of large-scale coronal features.

074.090 Correlation of high latitude coronal holes with solar wind streams far above or below the ecliptic.
K. B. Baker, M. D. Papagiannis.
Sol. Phys., Vol. 78, 365 - 372 (1982).

For the 2.5 year period from January 1, 1977 to June 30, 1979, the authors have correlated the positions of high latitude coronal holes, obtained from the He 10830 Å synoptic maps, with the velocities of solar wind streams, determined from interplanetary scintillation, that would have originated from these coronal holes. From 24 cases analyzed the authors find that these high latitude coronal holes are often, but not always, correlated with high speed solar wind streams.

074.091 Magnetic reconnection and coronal transients.
U. Anzer, G. W. Pneuman.
Sol. Phys., Vol. 79, 129 - 147 (1982).

The authors begin with the premise that coronal transients are indeed produced through magnetic reconnection occurring in the lower corona. A self consistent and sufficiently detailed model is developed to illustrate this process and the results of the model are shown to agree   quite well with observation. Flare loops accompanying two-ribbon flares as well as similar soft X-ray enhancements associated with eruptive prominence events indicate that mangetic reconnection is the fundamental process in these events. Since transients also occur with prominence eruptions and two-ribbon flares (Munro et al., 1979), the authors propose that these structures are propelled outward into interplanetary space by magnetic reconnection.

074.092 Magnetic measurements of coronal holes during 1975—1980.
K. L. Harvey, N. R. Sheeley, Jr., J. W. Harvey.
Sol. Phys., Vol. 79, 149 - 160 (1982).

Photospheric magnetic fluxes and average field strengths have been measured beneath 33 coronal holes observed on 63 occasions during 1975—1980. The principal result is that low-latitude holes contained 3 times more flux near sunspot maximum than near minimum despite the fact that their sizes were essentially the same. Average magnetic field strengths ranged from 3—36 G near sunspot maximum compared to 1—7 G near minimum. Evidently the low-latitude coronal holes received a proportion of the extra flux that was available at low latitudes near sunspot maximum.

074.093 A numerical simulation of cooling coronal flare plasma.   G. A. Doschek, J. P. Boris, C.-C. Cheng, J. T. Mariska, E. S. Oran.
Astrophys. J., Vol. 258, 373 - 383 (1982).

The authors have simulated the cooling of coronal flare plasma ($T_e > 10^7$ K) using a numerical model of a vertical magnetic flux tube containing an idealized flare chromosphere, transition region, and corona. The model solves the set of one-dimensional, two-fluid hydrodynamic equations. The cooling of the flux tube is calculated for a specific case beginning with an initial atmosphere in hydrostatic equilibrium and a maximum temperature of about 18×10^6 K. The behavior of temperature, density, and velocity is calculated as a function of height as the system cools. The expected spectral line emission from the system in X-ray lines of Fe XXV, Fe XXIV, Fe XXII, O VIII, and O VII is also calculated and compared to recent observational results. Some observational results can be explained as a consequence of simple cooling of flare flux tubes. The expected spectral line emission from certain transition region lines is also briefly considered.

074.094 Very Large Array observations of coronal loops at 20 centimeter wavelength.

K. R. Lang, R. F. Willson, J. Rayrole.
Astrophys. J., Vol. 258, 384 - 387 (1982).

Looplike coronal structures have been observed in
VLA synthesis maps at 20 cm wavelength with an angular
resolution of $3.''7 \times 5''$. The 20 cm loops extend across regions
of opposite magnetic polarity in the underlying photosphere.
The absence of detectable circular polarization suggests that
the 20 cm loops are optically thick with brightness temperatures equal to coronal electron temperatures of millions of
degrees. The authors interpret the 20 cm emission in terms
of the optically thick bremsstrahlung of thermal electrons
trapped within magnetic loops which are in hydrostatic
equilibrium. The semilength $L \approx 5 \times 10^9$ cm, maximum
electron temperature $T_e(\max) \approx 3 \times 10^6$ K, pressure
$p \approx 2$ dyn cm^{-2}, and electron density $N_e \approx 2.5 \times 10^9$ cm^{-3}
of the 20 cm loops are all characteristic of the coronal loops
detected at X-ray wavelengths.

074.095　**An estimation of the amount of heating in solar
　　　　　coronal loops II. Cooling through conduction-driven
evaporation.　　U. Narain.**
Bull. Astron. Soc. India, Vol. 9, 278 - 286 (1981).

A theory of cooling of solar coronal loops through conduction-driven evaporation in presence of a source of heating
is presented.

074.096　**Magnetohydrodynamics and thermodynamics of
　　　　　coronal loops.　　G. Einaudi.**
Sun and planetary system, (see 012.040), p. 141 - 142 (1982).

074.097　**Near infrared emission from the solar corona.**
　　　　　A. Mampaso, C. Sánchez Magro, J. Buitrago.
Sun and planetary system, (see 012.040), p. 257 - 258 (1982).

074.098　**Doppler shift measurements on the green coronal
　　　　　line — evidence for largescale macroscopic mass
motion.**
J. N. Desai, T. Chandrasekhar, P. D. Angreji.
J. Astrophys. Astron., Vol. 3, 69 - 77 (1982).

Fabry-Perot interferometric observations on the green
coronal line (λ 5303 Å) carried out during the total solar
eclipse of 1980 February 16 have yielded relative Doppler
shift velocities with an accuracy of $\pm$ 7 km s^{-1}. The values
show a peak in the 30 - 50 km s^{-1} range indicating largescale
macroscopic mass motion in the solar maximum corona.

074.099　**High-speed streams from coronal holes and the
　　　　　accelerating mechanism of the solar wind.**
W.-R. Hu.
Geophys. Astrophys. Fluid Dyn., Vol. 19, 311 - 330 (1982).

The general Mach number equation is derived, and the
influence of typical energy forms in the solar wind is analysed
in detail. It shows that the accelerating process of the solar
wind is influenced critically by the form of heating in the corona, and that the transonic mechanism is mainly the result of
the adjustment of the variation of the cross section of flowing
tubes and the heat source term. The accelerating mechanism
for both the high-speed stream from the coronal hole and the
normal solar wind is similar.

074.100　**The energetics of steady-state flows in the solar
　　　　　corona.　　P. J. Cargill, E. R. Priest.**
Geophys. Astrophys. Fluid Dyn., Vol. 20, 227 - 245 (1982).

The energetics of steady-state flows in coronal loops are
examined as an example of flows in flux tubes. The equations
of continuity, momentum, energy and state are solved numerically. A steady flow is found to remove the symmetry present
in static loops and to lower the maximum loop temperature.
Also, the possibility of a catastrophe, which can exist in static
loops as non-equilibrium, is found to be enhanced by the presence of a flow.

074.101　**Effect of pressure gradients and line-tying on the
　　　　　magnetic stability of coronal loops.**
A. W. Hood, E. R. Priest, G. Einaudi.
Geophys. Astrophys. Fluid Dyn., Vol. 20, 247 - 263 (1982).

The authors study the stability of an idealised magnetostatic coronal loop, incorporating both the effect of line-tying,
due to the dense photosphere, and of pressure gradients. From
the marginally stable case, the critical conditions separating
instability from stability are derived. It is found that stretching
or twisting a loop eventually makes it kink unstable.

074.102　**Correlation between polar coronal holes and solar
　　　　　wind high-velocity streams.　　B. A. Lindblad.**
Rep. Obs. Lund, No. 18, (see 012.044), p. 124 - 128 (1982).

074.103　**Models of transition region and coronal plasma in
　　　　　solar "loop" structures.　　J. C. Raymond,**
R. Rosner.
Smithsonian Astrophys. Obs., Spec. Rep. 392, (see 012.045),
p. 15 - 22 (1982).

The theory of coronal loops has been extensively developed in the last few years. The authors discuss comparisons between the simple version of the theory and observations before
turning to more recent theoretical work.

074.104　**1980 rocket coronagraph measurements of the solar
　　　　　wind acceleration region.　　G. L. Withbroe,**
J. L. Kohl, R. H. Munro, H. Weiser.
Smithsonian Astrophys. Obs., Spec. Rep. 392, (see 012.045),
p. 99 - 105 (1982).

Spectroscopic measurements of temperatures, densities
and flow velocities in the solar wind acceleration region provide critical empirical constraints on solar/stellar wind theory.
Preliminary results of an analysis of H I Lyman-alpha and
white light measurements made on 16 February 1980 in a
polar coronal region are reported.

074.105　**Modification of average coronal properties in the
　　　　　presence of periodic temperature and density variations near the base.　　S. T. Suess.**
Smithsonian Astrophys. Obs., Spec. Rep. 392, (see 012.045),
p. 113 - 120 (1982).

Time-dependent flow in the solar corona has been modeled using an implicit time-differencing solution to the equations
of motion for single-fluid, spherically symmetric, radial flow.

Cosmic rays and solar wind.
See Abstr. 003.079.

**Two-dimensional radioheliographic pictures of the
sun's outer corona at 25.6 - 110.6 MHz.**
See Abstr. 033.023.

**Alfvénic fluctuations as asymptotic states of
MHD turbulence.　　See Abstr. 062.006.**

Plasma heating in a sheared magnetic field.
See Abstr. 062.031.

**Generation of plasma vortices by fast magnetosonic
waves.　　See Abstr. 062.038.**

**Propagation speeds and acoustic damping of waves
in magnetic flux tubes.　　See Abstr. 062.067.**

**Thermal trigger for solar flares and coronal loops
formation.　　See Abstr. 062.070.**

**Wave propagation in a magnetically structured
atmosphere. III: The slab in a magnetic environment.**
See Abstr. 062.072.

Ion acoustic instability in the presence of plasma turbulence in the solar wind.   See Abstr. 062.114.

Resonant electrodynamic heating of stellar coronal loops: an *LRC* circuit analog.   See Abstr. 064.012.

Dynamic phenoma in coronal flux tubes. See Abstr. 064.091.

Light deflection during solar eclipses. See Abstr. 066.120.

Structure and physics of solar faculae. II. The non-thermal velocity field above faculae.   See Abstr. 072.055.

Report of IAU Commission 10: Solar activity (Activité solaire).   See Abstr. 072.104.

Some lines of the chromosphere and corona in the magnetic field.   See Abstr. 073.027.

Solar transition region response to variations in the heating rate.   See Abstr. 073.028.

Intensity of lines from low-lying levels in C II, N III, O IV, Ne VI, Mg VIII, Si X, and Si II. See Abstr. 073.043.

The structure of the lower transition region. See Abstr. 073.084.

Inhibition of heat conduction into the transition region by magnetic construction.   See Abstr. 073.085.

Acoustic energy flux in the solar transition zone. See Abstr. 073.086.

Solar transition region response to heating rate variations.   See Abstr. 073.087.

HRTS (*High Resolution Telescope and Spectrograph*) observations of the solar chromosphere and transition zone.   See Abstr. 073.088.

Velocities of shock waves generated by solar flares. See Abstr. 073.092.

MHD simulation of 1980 June 29 (1821 UT) flare and coronal transient.   See Abstr. 073.096.

Measurements of A-values for intersystem lines used in diagnosis of solar transition zone and other astrophysical plasmas.   See Abstr. 073.112.

An explanation for the systematic flow of plasma in the solar transition region.   See Abstr. 073.138.

Chromospheric and coronal heating mechanisms. See Abstr. 073.148.

The coronal field lines of an evolving bipolar magnetic region.   See Abstr. 075.005.

Computation of inner coronal magnetic fields from longitudinal field components on a spherical photosphere. See Abstr. 075.011.

Model for flare loops, fast motions, and opening of magnetic field in the corona.   See Abstr. 075.013.

Effect of flare-site magnetic field on solar wind speed and geomagnetic activity.   See Abstr. 075.041.

Fast solar hard X-ray bursts and large-scale coronal structures.   See Abstr. 076.003.

The XUV structure of solar active regions. See Abstr. 076.008.

Radio observation of sudden shock wave deceleration above solar flare.   See Abstr. 077.009.

Study of the character and parameters of wave motions in the transition layer and low solar corona by radio observations.   See Abstr. 077.023.

Radio echo and sporadic radiation scattering in the solar corona.   See Abstr. 077.031.

The change of the radio radius of the sun connected with the emergence of a coronal hole at the solar limb. See Abstr. 080.032.

Tunneling and interference of Alfvén waves. See Abstr. 080.037.

Decrease of the radio radius when a coronal hole emerges at the limb of the quiet sun. See Abstr. 080.047.

Report of IAU Commission 12: Radiation and structure of the solar atmosphere (Radiation et structure de l'atmosphère solaire).   See Abstr. 080.109.

Soft X-rays from the sunlit earth's atmosphere. See Abstr. 082.057.

Observations of penetrated solar wind plasma elements in the plasma mantle.   See Abstr. 084.003.

Dynamo process governing solar wind-magnetosphere energy coupling.   See Abstr. 084.055.

The electrical field in the magnetosphere's tail depending on the level of geomagnetic activity and intensity E_y in the solar wind.   See Abstr. 084.077.

Geomagnetic variation and field-aligned currents at northern high-latitudes, and their relations to the solar wind parameters.   See Abstr. 084.105.

The solar wind interaction. See Abstr. 093.010.

Comet-like interaction of Venus with the solar wind. III. The atomic oxygen corona.   See Abstr. 093.046.

Fractionation in solar system krypton and xenon, and their isotopic compositions in the solar wind. See Abstr. 094.517.

Correlation between solar wind ^{4}He distribution and noble gas fractionation in lunar ilmenites. See Abstr. 094.518.

Jupiter's gravitational effects on the solar wind and geomagnetic activity in different epochs of solar activity. See Abstr. 099.012.

Role of high frequency turbulence in cometary plasma tails.   See Abstr. 102.002.

Coronagraph observations of a sun-directed comet, Aug. 30 - 31, 1979: images, analysis and photometry. See Abstr. 102.032.

Correlation relations between the components of the interplanetary magnetic field and solar wind velocity. See Abstr. 106.012.

The effect of solar-wind convection on charged particle transport in interplanetary space. See Abstr. 106.024.

Radial evolution of power spectra of interplanetary Alfvénic turbulence.   See Abstr. 106.027.

Statistical properties of MHD fluctuations associated with high-speed streams from Helios-2 observations. See Abstr. 106.029.

Plasma-dust interactions in the solar and cometary environment.   See Abstr. 106.035.

Report of IAU Commission 49: The interplanetary plasma and the heliosphere (Plasma interplanétaire et l'héliosphère).   See Abstr. 106.041.

Erratum

074.901 Erratum: 'Observations of solar-wind helium' [Fundam. Cosmic Phys., Vol. 7, 131 - 199 (1981)]. M. Neugebauer. Fundam. Cosmic Phys., Vol. 7, 312 (1982). – See Abstr. 29.074.086.

075 Magnetic Fields

075.001 **Solar active region SD 55/1975 in the frame of the background magnetic field development.** V. Bumba, M. Klvaňa, K. Pflug. Bull. Astron. Inst. Czechoslovakia, Vol. 33, 36 - 46 (1982).

The relation of active region SD 55/1975 to the background magnetic field development is demonstrated. The studied active region represents only a small part of a more general evolutionary process of the magnetic field: it appears during the early stages of a new "impulse of activity" in the main active magnetic longitude as a substantial component of a complex of activity. All its morphological changes in the photosphere, as well as in the chromosphere are clearly related to the subsequent formations of new secondary magnetic fluxes and to the strong renewal of activity at the end of the observational interval.

075.002 **Solitons in solar magnetic flux tubes.** B. Roberts, A. Mangeney. Mon. Not. R. Astron. Soc., Vol. 198, 7P - 11P (1982).

It is suggested on theoretical grounds that solar intense magnetic flux tubes can support the propagation of solitons. Under photospheric conditions, a "tube soliton" may propagate with a speed of about 7 km s^{-1} and an "external soliton" with a speed of some 11 km s^{-1}. The authors speculate that the tube soliton may be manifest in the chromosphere as a spicule.

075.003 **Evidence for solar magnetic loops beyond 1 AU.** E. T. Sarris, S. M. Krimigis. Geophys. Res. Lett., Vol. 9, 167 - 170 (1982).

The authors present observations of energetic particles injected by solar flares into extended solar magnetic loop-like structures. From the development of the angular distributions of the intensities of energetic protons ($E_p \geqslant 300$ keV, $E_p \geqslant 25$ MeV) and electrons ($E_e \geqslant 220$ keV), the authors have inferred that energetic particles are bouncing between two magnetic mirrors and have obtained for the first time estimates of the extent of magnetic loops to distances ~3.5 AU from the sun.

075.004 **An atlas of theoretical Stokes profiles for solar disk observations.** P. Arena, E. Landi Degl'Innocenti. Astron. Astrophys., Suppl. Ser., Vol. 48, 81 - 83, 2 microfiches (1982).

An atlas of theoretical Stokes profiles for solar disk observations of magnetic regions has been prepared. The atlas is especially meant to familiarize observers with the wide variety of polarimetric profiles which can result in observations according to the intensity and direction of the magnetic field vector and to the strength of the line.

075.005 **The coronal field lines of an evolving bipolar magnetic region.**   N. R. Sheeley, Jr. Astrophys. J., Vol. 255, 316 - 319 (1982).

A simple potential field model is presented to illustrate that loops of magnetic flux rise upward through the corona during the relatively short growth phase of a bipolar magnetic region but contract back to the Sun's surface during the much longer decay phase of the photospheric region. To reconcile this behavior with the unidirectional, solar-wind–driven convection of flux outward from the Sun, one must postulate the existence of an X-type neutral line in the middle corona where open field lines can be converted back to closed ones.

075.006 **The structure of the solar magnetic field below the photosphere. I. Adiabatic flux tube models.** A. A. van Ballegooijen. Astron. Astrophys., Vol. 106, 43 - 52 (1982).

The author investigates the structure and evolution of adiabatic flux tubes, rooted in the stable layer below the convective zone. The hydrostatic equilibrium field strength $B(z)$ of a thin adiabatic flux tube is derived as function of depth z. The transverse forces on flux loops in the Sun are considered. The drift velocity of adiabatic flux tubes is estimated. A timescale of a few days is obtained for the development of new active regions. Radiative diffusion may change the thermal structure of flux tubes on a timescale of one month. Finally a model for the disappearance of flux tubes from the solar surface is described in which loops are pulled down through the convective zone.

075.007 **Topological semi-dynamo.**   Eh. M. Drobyshevskij, Eh. N. Kolesnikova, V. S. Yuferev. Fiz.-tekh. inst. AN SSSR. Prepr., 1981, No. 724, 33 pp. In Russian. – Abstr. in Ref. zh., 51. Astron., 2.51.304 (1982).

075.008 **On the moving configuration of magnetic plasma.** W. Hu.

Kexue Tongbao, Vol. 26, 572 - 573 (1981). — Abstr. in Phys. Abstr., Vol. 85, Abstr. 10381 (1982).

075.009 Magnetic energy storage and conversion in the solar atmosphere. D. S. Spicer.
Space Sci. Rev., Vol. 31, 351 - 435 (1982).
A review of the theoretical problems associated with preflare magnetic energy storage and conversion is presented. The review consists of three parts; preflare magnetic energy storage, magnetic energy conversion mechanisms, and preflare triggers.

075.010 Reversed-polarity regions.
F. Tang.
Sol. Phys., Vol. 75, 179 - 188 (1982).
The author presents results of a statistical study of reversed-polarity regions. RPRs, collected over the past 11 years, 1969 - 1979. The 58 RPRs studied have a lifespan comparable to normal active regions and have no tendency to rotate toward a more normal alignment. They seem to have stable configurations with no apparent evidence suggesting stress due to their anomalous magnetic alignment. Magnetic complexity in RPRs is the key to flare productivity just as it is in normal regions. The RPRs differ from normal regions in the frequency of having complex spots, particularly the long-lived complex spots, in them.

075.011 Computation of inner coronal magnetic fields from longitudinal field components on a spherical photosphere. G. Elwert, K. Müller, L. Thür, P. Balz.
Sol. Phys., Vol. 75, 205 - 227 (1982).
A method is developed which for a certain day permits the approximate calculation of closed small and large scale magnetic field lines. The method is applied to observations of September 5 and September 7, 1973. The projected magnetic field lines are compared with the loop structures which are visible in XUV pictures taken on these days.

075.012 Flare build-up in preflare magnetic loops and nonlinear force-free magnetic fields. Q.-R. Su.
Sol. Phys., Vol. 75, 229 - 236 (1982).
In this paper the author extends B. C. Low's study on nonlinear force-free magnetic fields. Based on Low's mathematical method, a revised boundary-value problem of the two-dimensional nonlinear force-free magnetic field is solved analytically. The solution shows that higher magnetic loops evolve towards preflare loops when the gradient of longitudinal magnetic field at the photospheric level and the angle ('shear') included between the magnetic field line and magnetic neutral line increase with time. The density, temperature and the current density are higher in the preflare loops than in the highlying magnetic loops.

075.013 Model for flare loops, fast motions, and opening of magnetic field in the corona.
S. I. Syrovatskii (*Syrovatskij*).
Sol. Phys., Vol. 76, 3 - 20 (1982).
A model is presented for the penetration into the corona of a new magnetic field of a developing bipolar region and for its interaction with an old large-scale coronal field. An important feature of the model is a reconnection of the old and new field inside the current sheet arising along the zero line of the total magnetic field calculated in the potential approximation. The magnetic reconnection and accumulation of plasma inside the current sheet can explain the appearance of dense coronal loops and the energy source at their tops. The plasma together with the magnetic lines is flowed into the sheet from both its sides. This fact explains the appearance of coronal cavities above the loops.

075.014 Green's function methods for potential magnetic fields. T. Sakurai.

Sol. Phys., Vol. 76, 301 - 321 (1982).
The Green's function method to calculate potential magnetic field on the Sun is extended to the following three cases: (a) The field component along the line of sight, which is not generally normal to the flat boundary plane, is specified; (b) the line of sight component on a spherical boundary surface is specified; (c) the normal component on a spherical surface is specified, together with the condition that the field becomes approximately radial on an outer spherical surface (the so-called source surface). Properties of these Green's functions are examined, and the applicability of these methods to solar magnetic data is discussed.

075.015 Thermodynamical properties of unresolved magnetic flux tubes. I: A diagnostic method based on circular polarization ratios in line pairs.
E. Landi Degl'Innocenti, M. Landolfi.
Sol. Phys., Vol. 77, 13 - 26 (1982).
The authors propose a diagnostic method, based on the observation of circular polarization signals in line pairs, to derive the thermodynamical properties of unresolved magnetic elements in the solar atmosphere. The concept of response function for the ratio of circular polarization signals in two lines is introduced and its main properties are analyzed. Some detailed calculations for suitably selected line pairs are presented.

075.016 On the relative roles of unipolar and mixed-polarity fields. R. G. Giovanelli.
Sol. Phys., Vol. 77, 27 - 42 (1982).
Away from plages, solar magnetic fields may be classified as unipolar or as of mixed polarity, though the distinction is strictly arbitrary. The dividing line used is $0.4 \le |\bar{B}_{minor}/\bar{B}_{major}| \le 1$, where average fields of major and minor polarities are measured over large areas. Some of their statistical properties and cyclical variations are detailed. In unipolar regions, $3 \lesssim \bar{B}_{major} \lesssim 50$ G, $\bar{B}_{minor} \approx 0.1 \bar{B}_{major}$, and $|\bar{B}| \approx 1.1 \bar{B}_{major}$. In regions of mixed polarity, $3.5 \lesssim |\bar{B}| \lesssim 10$ G. Below latitudes of $\pm 60\%$, mixed polarities predominate for about 5 yr around sunspot minimum. For several years around sunspot maximum, unipolar fields fill the $20°-40°$ zone completely, and occupy about 75% of the $0°-20°$ and $40°-60°$ zones. The polar unipolar fields are weak on the whole with small regions having stronger fields.

075.017 Magnetic field configurations associated with polarity intrusion in a solar active region. I. The force-free fields. B. C. Low.
Sol. Phys., Vol. 77, 43 - 61 (1982).
This paper presents a new class of exact solutions describing the non-linear force-free field above a spatially localized photospheric bipolar magnetic region. An essential feature is the variation in all three Cartesian directions and this could not be modelled adequately with previously known symmetric force-free fields. Sequences of force-free fields are constructed and analyzed to simulate the slow growth of a pair of spots on the photosphere.

075.018 Simulation of the magnetic structure of the inner heliosphere by means of a non-spherical source surface. R. H. Levine, M. Schulz, E. N. Frazier.
Sol. Phys., Vol. 77, 363 - 392 (1982).
The authors develop and implement a significant new tool for constructing models of the coronal and interplanetary magnetic field from observations of the photospheric magnetic field. This tool, the non-spherical source surface, represents a significant improvement over previous models of solar magnetic field structure and is suitable for the study of the interplanetary magnetic field and its solar origin.

075.019 On the gradient of the magnetic field above sunspots determined from optical and radio astronomical

measurements.　　V. M. Bogod, G. F. Vyal'shin,
G. B. Gel'frejkh, N. S. Petrova.
Soln. Dannye 1982 Byull., No. 1, p. 104 - 109 (1982). In Russian.
The results of measurements of the magnetic field on 9, 11
and 13 August 1979 in sunspot group N357 by the radio astronomical method at the bottom of the corona and by the
optical method at the photospheric (Fe line) and chromospheric levels (Hα line) are given.

075.020　On the influence of incompensated Doppler
　　　　　shifts of spectral lines on magnetographic measurements of magnetic fields.　　V. G. Lozitskij, T. T. Tsap.
Astron. Tsirk., No. 1192, p. 3 - 5 (1981). In Russian.

075.021　Surface magnetic fields and the solar luminosity.
　　　　　D. S. P. Dearborn, J. B. Blake.
Astrophys. J., Vol. 257, 896 - 900 (1982).
Changes in the solar surface magnetic field alter the
structure of the superadiabatic zone (outer $10^{-7} M_\odot$) and
produce luminosity fluctuations. The processes modeled in
this paper include both the effect of magnetic pressure and the
redistribution of the radiative flux. It is found that increases
in the solar magnetic field (or sunspot area) cause a global
decrease in the solar luminosity. The "lost" energy is stored as
gravitational potential energy in the convection zone.

075.022　Can magnetic transients be explained by line
　　　　　emission?　　A. Patterson, H. Zirin.
Bull. American Astron. Soc., Vol. 13, 821 (1981). – Abstract.

075.023　Modeling of energy buildup for a flare-productive
　　　　　region.
S. T. Wu, Y. Q. Hu, K. Krall, M. J. Hagyard, J. B. Smith, Jr.
Bull. American Astron. Soc., Vol. 13, 821 (1981). – Abstract.

075.024　Activity and structure of vector magnetic fields of
　　　　　complex sunspots containing δ-classifications.
S. R. Patty, M. J. Hagyard.
Bull. American Astron. Soc., Vol. 13, 881 (1981). – Abstract.

075.025　Active region magnetic fields.　　J. B. Smith, Jr.,
K. T. Strong, E. J. Schmahl, M. R. Kundu,
K. R. Krall, R. D. Bentley.
Bull. American Astron. Soc., Vol. 13, 881 (1981). – Abstract.

075.026　Magnetic fields observed to affect granular convec-
　　　　　tion.　　W. Livingston, D. Carbon.
Bull. American Astron. Soc., Vol. 13, 881 (1981). – Abstract.

075.027　Magnetic canopies in unipolar regions.
　　　　　H. P. Jones, R. G. Giovanelli.
Bull. American Astron. Soc., Vol. 13, 881 - 882 (1981).
Abstract.

075.028　Flux changes in small magnetic regions.
　　　　　P. R. Wilson, G. W. Simon.
Bull. American Astron. Soc., Vol. 13, 882 (1981). – Abstract.

075.029　HRTS II observations of a solar bi-polar region.
　　　　　R. Roussel-Dupré, J. Wrathall, K. Nicolas,
J. D. F. Bartoe, G. E. Brueckner.
Bull. American Astron. Soc., Vol. 13, 889 (1981). – Abstract.

075.030　Open magnetic fields and the solar cycle.
　　　　　R. H. Levine.
Bull. American Astron. Soc., Vol. 13, 905 (1981). – Abstract.

075.031　Rebirth of the polar holes at solar maximum.
　　　　　D. F. Webb, J. M. Davis.
Bull. American Astron. Soc., Vol. 13, 906 (1981). – Abstract.

075.032　Rotation of solar magnetic fields.
　　　　　H. B. Snodgrass, D. H. Bruning.
Bull. American Astron. Soc., Vol. 13, 906 (1981). – Abstract.

075.033　Buoyancy instabilities at the base of the solar
　　　　　convection zone.　　J. H. M. M. Schmitt, R. Rosner.
Bull. American Astron. Soc., Vol. 13, 907 (1981). – Abstract.

075.034　The structure of a force free magnetic flux tube.
　　　　　K. A. Wear, S. K. Antiochos, P. A. Sturrock.
Bull. American Astron. Soc., Vol. 13, 915 (1981). – Abstract.

075.035　A model for a fine-structural magnetic element in
　　　　　the quiet solar atmosphere.
V. E. Merkulenko.
Issled. po geomagn., aehron. i fiz. Solntsa, Moskva, 1981,
No. 57, p. 53 - 74. In Russian. – Abstr. in Ref. zh., 51.
Astron., 5.51.355 (1982).

075.036　Solar magnetic fields, January - December 1980.
　　　　　Q. Bull. Sol. Act., Vol. 22, Part II, p. 1 - 8 (1980).

075.037　The measurement of magnetic fields in the solar
　　　　　atmosphere above sunspots using gyroresonance
emission.　　Sh. B. Akhmedov, G. B. Gelfreikh (Gel'frejkh),
V. M. Bogod, A. N. Korzhavin.
Sol. Phys., Vol. 79, 41 - 58 (1982).
An analysis of the local sources (LS) structure of the
S-component of solar radio emission confirms the presence
of a core component which is characterized by strong circular
polarization and a steep growing spectrum at shorter centimeter wavelengths. The spectral and polarization observations
of LS made with RATAN-600 using high resolution in the
wavelength range 2.0–4.0 cm, allow the authors to measure
the maximum magnetic fields of the corresponding sunspots
at the height of the chromosphere-corona transition region
(CCTR). This method is based on determining the short wavelength limit of gyroresonance emission of the LS and relating
it to the third harmonic of gyrofrequency. An analysis of a
large number of sunspots and their LS (core component) has
shown a good correlation between radio magnetic fields near
the CCTR and optical photospheric ones.

075.038　Vector magnetic field evolution, energy storage,
　　　　　and associated photospheric velocity shear within
a flare-productive active region.　　K. R. Krall, J. B. Smith, Jr.
M. J. Hagyard, E. A. West, N. P. Cummings.
Sol. Phys., Vol. 79, 59 - 75 (1982).
The evolution of vector photospheric magnetic fields has
been studied in concert with photospheric spot motions for a
flare-productive active region. Over a three-day period, sheared
photospheric velocity fields inferred from spot motions are
compared both with changes in the orientation of transverse
magnetic fields and with the flare history of the region. Maps
of vertical current density suggest that parallel (as contrasted
with antiparallel) currents flow along the stressed magnetic
loops. For the active region, a constant-α, force-free magnetic
field at the photosphere is ruled out by the observations.

075.039　The beginning of observations of large-scale solar
　　　　　magnetic fields at the Sayan Observatory: instrument, plans, preliminary results.
V. M. Grigoryev (Grigor'ev), V. S. Peshcherov, M. L. Demidov.
Sun and planetary system, (see 012.040), p. 119 - 122 (1982).

075.040　Fe I λ 5324.19 Å line forms in the solar magnetic
　　　　　field and the theoretical calibration of the solar magnetic field telescope.　　G.-x. Ai, W. Li, H.-q. Zhang.
Acta Astron. Sinica, Vol. 23, 39 - 48 (1982). In Chinese.

075.041　Effect of flare-site magnetic field on solar wind
　　　　　speed and geomagnetic activity.

H. Lundstedt, P. B. Duffy, J. M. Wilcox, P. H. Scherrer.
Rep. Obs. Lund, No. 18, (see 012.044), p. 129 - 132 (1982).

075.042 **Magnetic fields on the sun.**
R. Howard.
Smithsonian Astrophys. Obs., Spec. Rep. 392, (see 012.045), p. 155 - 164 (1982).
Synoptic observations of solar magnetic fields are discussed.

075.043 **Solar magnetic fields and activity.** V. Bumba.
Compendium in astronomy, (see 003.012), p. 81 - 96 (1982).
In the present paper the author is trying to summarize the experience of a solar observer dealing with the dynamics of solar background magnetic fields formation, including the development of new active regions magnetic fluxes. Single active regions occurrences, complexes of activity and proton-flare region related large-scale regular magnetic patterns growths are investigated. The meaning and importance of magnetic active longitudes are indicated. The physical background of magnetic fields generation in connection with convection, differential rotation and with the photospheric surface kinematics of the background fields is discussed.

Variability of the general magnetic field of the sun as a source of cosmic ray modulation.
See Abstr. 003.027.

Oscillating dynamo in the presence of a fossil magnetic field. The solar cycle. See Abstr. 062.016.

Nonlinear force-free magnetic fields.
See Abstr. 062.019.

The dynamics of fibril magnetic fields. I. Effect of flux tubes on convection. See Abstr. 062.026.

The dynamics of fibril magnetic fields. II. The mean field equations. See Abstr. 062.027.

Zweidimensionale Dynamomodelle auf der Basis magnetischer Flußröhren. See Abstr. 062.050.

The dynamics of fibril magnetic fields. III. Fibril configurations in steady flows. See Abstr. 062.058.

The dynamics of fibril magnetic fields. IV. Trapping in closed convective rolls. See Abstr. 062.059.

Pulse propagation in a magnetic flux tube.
See Abstr. 062.060.

The effects on the MHD stability of field line tying to the end faces of a cylindrical magnetic loop.
See Abstr. 062.068.

Magnetostatic atmospheres: a family of isothermal solutions. See Abstr. 062.073.

Compression of magnetic field in a viscous boundary layer. See Abstr. 062.074.

More on dynamos driven by global convection and differential rotation. See Abstr. 062.090.

Magneto-optical effects and the determination of vector magnetic fields from Stokes profiles.
See Abstr. 062.111.

Modelle photosphärischer Magnetfeld-Konzentrationen. See Abstr. 071.011.

On the dissolution of sunspot groups.
See Abstr. 072.041.

Observations of small magnetic features in sunspots and active regions. See Abstr. 072.068.

Report of IAU Commission 10: Solar activity (Activité solaire). See Abstr. 072.104.

Large scale solar magnetic fields at the site of flares, the greatness of flares, and solar-terrestrial disturbances. See Abstr. 073.013.

Equilibria and stability of coronal magnetic arches. See Abstr. 074.019.

Magnetic reconnection and coronal transients. See Abstr. 074.091.

Magnetic measurements of coronal holes during 1975−1980. See Abstr. 074.092.

Connection of the radio brightness variations of the sun with hills of the magnetic field and flocculi during the minimum of solar activity. See Abstr. 077.014.

Is the sun an oblique magnetic rotator?
See Abstr. 080.007.

The rotation-magnetism-convection coupling in the sun. See Abstr. 080.103.

Report of IAU Commission 12: Radiation and structure of the solar atmosphere (Radiation et structure de l'atmosphère solaire). See Abstr. 080.109.

076 UV, X, Gamma Radiation

076.001 **Balloon observations of solar ultraviolet irradiance at solar minimum.**
P. C. Simon, R. Pastiels, D. Nevejans.
Planet. Space Sci., Vol. 30, 67 - 71 (1982).
Balloon observations of solar irradiance between 200 and 240 nm have been performed in 1976 and 1977 corresponding to minimum conditions of solar activity. Ultraviolet spectra have been recorded for different zenith angles at an altitude of 41 km by means of a spectrometer with a spectral bandpass of 0.4 nm. Solar irradiances at 1 a.u. confirm previous values obtained by balloon. They are compared with other measurements and discussed in term of possible long-term variability.

076.002 **Inferring solar UV variability from the atmospheric tide.**   N. S. Cooper.
Nature, Vol. 296, 131 - 132 (1982).
A new method of determining variations in the solar UV output is presented. The method is based on the fact that the semi-diurnal (12 h) atmospheric tide is dominated by the component forced by ozone absorption of UV. This tide shows very little interannual variability. The failure to find any solar cycle-related variation in the semi-diurnal tide suggests that the UV varies by $<2\%$ over the sunspot cycle.

076.003 **Fast solar hard X-ray bursts and large-scale coronal structures.**   G. M. Simnett.
Astrophys. J., Vol. 255, 721 - 729 (1982).
The conditions at the Sun at the times corresponding to a selected set of 22 fast impulsive hard X-ray bursts reported by Crannell et al. are examined. It is suggested that one of the bursts must arise from a precipitating beam of subrelativistic electrons; the source of the electrons is postulated to be in a region very remote from the X-ray site on the basis of type III and other radio data. The connection is via a coronal magnetic loop extending to ~3 $R_\odot$ above the photosphere.

076.004 **On the origin of solar hard X-ray pulsations.**
V. V. Zajtsev, A. V. Stepanov.
Pis'ma Astron. Zh., Tom 8, 248 - 252 (1982). In Russian.
English translation in Soviet Astron. Lett., Vol. 8.
Quasi-periodic pulsations of solar flare hard X-ray with periods ~1 s are suggested to be connected with the MHD oscillations of the energy release region near the top of a flaring loop.

076.005 **The solar spectrum of O IV, including photoexcitation by Fe IX 171.07 Å.**   S. O. Kastner.
Astron. Astrophys., Vol. 108, 361 - 368 (1982).
The extreme and far ultraviolet doublet spectrum of O IV, emitted from the solar transition region, is calculated taking account of expected photo-excitation by Fe IX at the wavelength 171.07 Å. Four multiplets are shown to be sensitive to such photoexcitation, of which two in particular are potentially observable and could provide an estimate of the local Fe IX radiation field.

076.006 **Gamma bursts: experimental data in favour of heliospheric origin.**   A. V. Kuznetsov.
Kosm. Issled., Tom 20, 89 - 96 (1982). In Russian.

076.007 **Measurement of the profile of solar He I resonance lines.**   E. Phillips, D. L. Judge, R. W. Carlson.
J. Geophys. Res., Vol. 87, 1433 - 1438 (1982).
The profile of the helium resonance line at λ 584 Å from the full solar disk has been investigated through analysis of data obtained by sounding rockets flown in 1977 and 1980.

076.008 **The XUV structure of solar active regions.**
K. P. Dere.
Sol. Phys., Vol. 75, 189 - 203 (1982).
XUV spectroheliograms of 2 active regions are studied. The images are due to lines emitted at temperatures between 8×10^4 K and 2×10^6 K and thus are indicative of transition region and coronal structures.

076.009 **Observations of a post-flare radio burst in X-rays.**
Z. Švestka, R. T. Stewart, P. Hoyng, W. Van Tend, L. W. Acton, A. H. Gabriel, C. G. Rapley, A. Boelee, E. C. Bruner, C. De Jager, H. Lafleur, G. Nelson, G. M. Simnett, H. F. Van Beek, W. J. Wagner.
Sol. Phys., Vol. 75, 305 - 329 (1982).
More than six hours after the two-ribbon flare of 21 May 1980, the hard X-ray spectrometer aboard the SMM imaged an extensive arch above the flare region which proved to be the lowest part of a stationary post-flare noise storm recorded at the same time at Culgoora.

076.010 **Soft X-ray emission from active regions shortly before solar flares.**   C. J. Wolfson.
Sol. Phys., Vol. 76, 377 - 386 (1982).
Detailed examinations of the variations in the intensity of soft X-ray emission prior to many solar flares are presented. In addition, these preflare intensity variations are contrasted with the variations typically observed for the same active regions in the absence of a flare. It is shown that a 5–20 min preflare brightening phase is not typically observed. These observations are discussed in context with other complimentary investigations and theoretical models.

076.011 **Extreme ultraviolet spectra of solar active regions and their analysis.**   K. P. Dere.
Sol. Phys., Vol. 77, 77 - 93 (1982).
In this paper, intensities of extreme ultraviolet lines emitted from solar active regions are presented and analyzed to derive pressures, velocities and differential emission measures as a function of temperature and position. The data has been obtained by three experiments: the NRL Skylab XUV spectroheliograph (170–630 Å), the HCO OSO-VI XUV spectroheliometer (280–1370 Å), and the NRL High Resolution Telescope and Spectrograph (1170–1710 Å).

076.012 **The role of betatron acceleration in complex solar bursts.**   J. T. Karpen.
Sol. Phys., Vol. 77, 205 - 230 (1982).
A thorough discussion of the betatron model is presented by Brown and Hoyng (1975). The present work is concentrated on the application of the original betatron model to the selected set of OSO-5 hard X-ray events. The basic characteristics of the betatron model are described. The analysis technique and criteria for testing the model are outlined. The results of the application of the betatron model to the multiple-spike bursts and the two-stage bursts are given. The interpretation of these results is presented. The conclusions on the role of betatron acceleration in solar flares are summarized.

076.013 **The solar O III spectrum. I. Photoexcitation of EUV lines by He II Lyman-α.**
A. K. Bhatia, S. O. Kastner, W. E. Behring.
Astrophys. J., Vol. 257, 887 - 895 (1982).
A detailed calculation of solar O III EUV line intensities is described, including the process of photoexcitation by He II Lyman-α. Excellent agreement with observed intensities is found for a photoexciting radiation field of intensity 9.8×10^{-8} ergs cm^{-2} s^{-1} sr^{-1} per Hz. The presence of some O III lines in flares is discussed. The process of charge exchange

does not appear to be a significant factor in formation of the O III spectrum.

076.014   EUV observations of solar mass loss from the lower solar atmosphere.
G. J. Rottman, F. Q. Orrall, J. A. Klimchuk.
Bull. American Astron. Soc., Vol. 13, 812 (1981). − Abstract.

076.015   Modeling of solar irradiance variations in the UV.
L. Oster, S. Sofia, K. H. Schatten.
Bull. American Astron. Soc., Vol. 13, 877 (1981). − Abstract.

076.016   Variability of UV solar spectral irradiance from 160 - 400 nm about solar maximum.
D. F. Heath, H. Park.
Bull. American Astron. Soc., Vol. 13, 877 (1981). − Abstract.

076.017   The non-thermal interpretation of short time-scale solar hard X-ray bursts.   A. G. Emslie.
Bull. American Astron. Soc., Vol. 13, 904 (1981). − Abstract.

076.018   The relation between the fluxes of soft and hard X-rays, and its relevance to flare energetics.
B. R. Dennis, K. J. H. Phillips.
Bull. American Astron. Soc., Vol. 13, 904 (1981). − Abstract.

076.019   On the correlation between hard X-rays and type III radio bursts during flares.
V. Petrosian, J. Leach.
Bull. American Astron. Soc., Vol. 13, 905 (1981). − Abstract.

076.020   High resolution UV filtergrams of the sun.
B. Foing, R. M. Bonnet, E. C. Bruner, L. W. Acton.
Bull. American Astron. Soc., Vol. 13, 911 (1981). − Abstract.

076.021   EUV observations of high-speed downflows over sunspots.   J. A. Klimchuk, G. J. Rottman.
Bull. American Astron. Soc., Vol. 13, 914 (1981). − Abstract.

076.022   Solar spectral irradiance, 120 to 190 nm, October 13, 1981 - January 3, 1982.
G. J. Rottman, C. A. Barth, R. J. Thomas, G. H. Mount,
G. M. Lawrence, D. W. Rusch, R. W. Sanders, G. E. Thomas,
J. London.
Geophys. Res. Lett., Vol. 9, 587 - 590 (1982).
   Beginning on October 13, 1981 a two channel spectrometer aboard the Solar Mesosphere Explorer has been obtaining daily measurements of full disc solar irradiance. These observations cover the spectral interval 120 to 305 nm with ~ .75 nm spectral resolution. The authors present analyses of Lyman-alpha, the integrated Schumann-Runge continuum (130 - 175 nm), and the integrated Schumann-Runge bands (175 - 190 nm). All three show a clear variability related primarily to the 27-day solar rotation period. Correlations of these three values of solar irradiance to ground-based indices of solar activity, 10.7 cm flux and sunspot number, are presented.

076.023   2.2 MeV gamma-ray line observed during two SN solar flares.   K. R. Rao, I. M. Martin,
J. O. D. Jardim, U. B. Jayanthi.
Sol. Phys., Vol. 79, 121 - 127 (1982).
   The aim of the present work is to show the presence of the 2.22 MeV line during a SN flare and calculate the neutron production responsible for this line.

076.024   Identification of Fe I lines in the ultraviolet solar spectrum.   M. K. McCabe, H. C. McAllister.
Astrophys. J., Suppl. Ser., Vol. 48, 437 - 454 (1982).
   High-resolution ultraviolet echelle spectrograms of the Sun obtained from a rocket-borne spectrograph have been used for the identification of Fe I absorption lines in the wavelength range 1770−2020 Å. Wavelength measurements,

with a precision of ~ 15 mÅ, are tabulated for 425 Fe I features, along with available laboratory wavelengths or predicted wavelengths calculated from atomic energy level data. The classification system of Moore has been extended to include possible additional multiplets with 126 newly classified lines; 234 unclassified lines were observed and identified with laboratory Fe I spectra.

   Atomic data for the interpretation of EUV astrophysical plasmas.   See Abstr. 022.134.

   Données atomiques et diagnostics necessaires à l'interpretation du spectre d'émission X du soleil actif (λ < 25 Å).   See Abstr. 022.135.

   Gamma radiation and photospheric white-light flare continuum.   See Abstr. 071.025.

   S IV emission-line ratios in the sun.
See Abstr. 071.030.

   Peculiarities of distribution of event intensity in different phases of the solar cycle.   See Abstr. 072.010.

   Analysis of the high resolution Mg XI X-ray spectra. II. Physical parameters of the plasma in active region McMath 14352.   See Abstr. 072.047.

   Solar activity − the sun as an X-ray star.
See Abstr. 072.099.

   Interaction of a collisionless conduction front with the chromosphere and solar hard X-ray bursts.
See Abstr. 073.001.

   Magnetic structure of a flaring region producing impulsive microwave and hard X-ray bursts.
See Abstr. 073.007.

   Spatial structure of $\gtrsim$ 100 keV X-ray sources in solar flares.   See Abstr. 073.011.

   High-resolution solar flare X-ray spectra obtained with rotating spectrometers on the Hinotori satellite.
See Abstr. 073.012.

   Observational constraints for a theoretical model describing the soft X-ray flare.   See Abstr. 073.016.

   A high-resolution measurement of the 2.223 MeV neutron capture line in a solar flare.   See Abstr. 073.018.

   Direct measurements of impulsive extreme ultraviolet and hard X-ray solar flare emission.
See Abstr. 073.029.

   Chromospheric effects of XUV radiation emitted during solar flares.   See Abstr. 073.031.

   Impulsive and gradual hard X-ray sources in a solar flare.   See Abstr. 073.036.

   Solar flare X-ray spectra from the Solar Maximum Mission flat crystal spectrometer.   See Abstr. 073.037.

   Electronic pitch angle scattering and the impulsive phase microwave and hard X-ray emission from solar flares.
See Abstr. 073.060.

   A possible explanation of non-steady-state appearances in X-ray spectra of solar flares.   See Abstr. 073.064.

Evidence of redshifts in the average solar line profiles of C IV and Si IV from OSO-8 observations. See Abstr. 073.067.

Impulsive phase of flares in soft X-ray emission. See Abstr. 073.073.

Comparison of observed Ca XIX and Ca XVIII relative line intensities with current theory. See Abstr. 073.074.

Direct evidence for chromospheric evaporation in a well-observed compact flare.   See Abstr. 073.077.

Temporal features in a solar flare observed in mm-waves, cm-waves and X-rays with fine time resolution. See Abstr. 073.089.

The relationship between the microwave and hard X-ray sources in a solar flare.   See Abstr. 073.098.

Gamma ray observations of the white light flare of 1 July 1980.   See Abstr. 073.101.

The comparison of the theoretical and observational ratios of gamma ray fluences in the 4 - 7 MeV channel to the 2.22 MeV line in 8 solar flares.   See Abstr. 073.102.

The spectrum of prompt nuclear deexcitation gamma ray lines from solar flares.   See Abstr. 073.103.

A comparison of gamma ray line and X-ray bremsstrahlung time profiles in several flares. See Abstr. 073.104.

A search for rapid oscillations in hard X-ray solar flares.   See Abstr. 073.105.

Hard X-ray structure of flaring loops. See Abstr. 073.106.

Association between gradual hard X-ray emission and metric continua during large flares. See Abstr. 073.107.

Time-series of radio, UV, soft, and hard X-ray data from flare of November 1, 1980.   See Abstr. 073.108.

Evidence for compressive heating in a solar flare from coincident hard X-ray and microwave observations. See Abstr. 073.109.

Heliocentric angular dependence for gamma ray flares observed with the SMM satellite. See Abstr. 073.111.

Solar flare soft X-ray spectra from the HINOTORI. I. Iron line spectra and their time variations of seven X-class flares.   See Abstr. 073.120.

The impulsive and gradual phases of a solar limb flare as observed from the Solar Maximum Mission satellite. See Abstr. 073.126.

Spatial and temporal evolution of soft and hard X-ray emission in a solar flare.   See Abstr. 073.133.

On the directivity of X rays of solar flares. See Abstr. 073.134.

Comparison of theoretically predicted and observed Solar Maximum Mission X-ray spectra for the 1980 April 13 and May 9 flares.   See Abstr. 073.136.

Loop coalescence in flares and coronal X-ray brightening.   See Abstr. 073.137.

The solar coronal X-ray spectrum from 15.4 to 23.0 Å: lines from highly ionized calcium and chromium and their usefulness as plasma diagnostics. See Abstr. 074.006.

The dynamics of coronal flare loops: II. Comparison to observations.   See Abstr. 074.056.

Collisionless conduction front propagation in large loops and solar hard X-ray bursts.   See Abstr. 074.057.

Solar coronal observations of plasma diagnostic X-ray lines of O VII and Ne IX.   See Abstr. 074.075.

Solar coronal X-ray spectra of calcium and chromium from 15.4 to 23.0 Ångstroms. See Abstr. 074.077.

About the relation between radio and soft X-ray emission in case of very weak solar activity. See Abstr. 077.007.

Direct inference of solar burst primary magnetic transient leading fast time structures at mm-waves and hard X-rays.   See Abstr. 077.039.

VLA observations of large scale microwave brightening following a flare.   See Abstr. 077.040.

Positional characteristics of meter-decameter wavelength bursts associated with hard X-ray bursts. See Abstr. 077.052.

UV radiation from the young sun and oxygen and ozone levels in the prebiological palaeoatmosphere. See Abstr. 082.016.

Interpretations and implications of γ-ray lines from solar flares, the galactic centre and γ-ray transients. See Abstr. 142.544.

A search for 2.22 MeV gamma ray line transients with the SMM Gamma Ray Spectrometer. See Abstr. 157.007.

077 Radio, Infrared Radiation

077.001 **Observations of ring structure in a sunspot associated source at 6 centimeter wavelength.**
C. E. Alissandrakis, M. R. Kundu.
Astrophys. J., Lett., Vol. 253, L49 - L52, plates L1 - L4 (1982).

The authors report the detection of a new kind of sunspot associated source in which the emission comes predominantly from a ring structure with size between that of the umbra and the penumbra. The absence of emission from the center of the spot is interpreted in terms of the orientation of the magnetic field and the presence of low temperature material above the umbra.

077.002 **Observations of the spectrum and location of the sources of two solar spike radio bursts.**
G. P. Chernov.
Pis'ma Astron. Zh., Tom 8, 106 - 108 (1982). In Russian. English translation in Soviet Astron. Lett., Vol. 8.

A comparison is made of simultaneous observations of two spike bursts in the noise storm on July 19, 1971. Estimation of the spike source sizes from the extension of the noise storm source gives a value of about 2'. The origin of the spike sources seems to be connected with fragmentation of type I sources.

077.003 **Solar type I noise storms and newly emerging magnetic flux.**
D. S. Spicer, A. O. Benz, J. D. Huba.
Astron. Astrophys., Vol. 105, 221 - 228 (1982).

A new model of solar type I radio bursts is presented based on the assumption that newly emerging magnetic flux can become capable of driving weak collisionless shocks in the front of the flux as it emerges. Using the assumption that the collisionless shock is maintained near marginal stability with respect to various collisionless flute like instabilities, the authors compute the micro-turbulence level of lower hybrid waves excited by the shock. The resulting mode coupling between lower hybrid and upper hybrid waves is then suggested as a radiation mechanism for the type I bursts.

077.004 **Detection of a 192s oscillatory component of the sun at 8.6 mm wavelength.** R. Bocchia.
Astron. Astrophys., Vol. 106, 79 - 88 (1982). In French.

Using the two-element interferometer at 8.6 mm wavelength, power spectra of the complex visibility function and power spectra of fringe amplitude of the solar radio radiation have been obtained.

077.005 **A geometrical model for solar local radio sources.**
R. Zhao.
Kexue Tongbao, Vol. 26, 531 - 535 (1981). − Abstr. in Phys. Abstr., Vol. 85, Abstr. 10380 (1982).

077.006 **Search for harmonic emission in solar type I radio bursts.** M. Jaeggi, A. O. Benz.
Astron. Astrophys., Vol. 107, 88 - 92 (1982).

The authors have made a statistical analysis of the harmonic emission of type I bursts, based upon the latest plasma wave theories for the emission mechanism. No systematic harmonic emission is found within the detection limit. This is also the case for a superposed epoch analysis of many bursts.

077.007 **About the relation between radio and soft X-ray emission in case of very weak solar activity.**
E. Fürst, A. O. Benz, W. Hirth.
Astron. Astrophys., Vol. 107, 178 - 185 (1982).

The authors report on centimeter and decimeter observations of very small solar soft X-ray events detected by the GOES-2 satellite on March 2, 1979. Four coinciding events in microwaves, decimetric radio and soft X-rays are presented. The analysis shows that the radio emission (both microwave and decimetric) is non-thermal and preceeds the thermal X-ray maximum by a few seconds. In a summarizing discussion the authors show that the events can be interpreted by the concept of efficient acceleration. Numerical estimates are based on the assumption that the observed soft X-rays originated in sources of the size of typical kernels.

077.008 **A model of cyclotron-resonance radiation for the SVC of solar radio emission.** S. Qian, R. Zhao.
Kexue Tongbao, Vol. 26, 384 (1981). − Abstr. in Phys. Abstr., Vol. 85, Abstr. 14838 (1982).

077.009 **Radio observation of sudden shock wave deceleration above solar flare.** M. Karlický, K. Jiřička, O. Kepka, L. Křivský, A. Tlamicha.
Bull. Astron. Inst. Czechoslovakia, Vol. 33, 72 - 75, plates 1 - 3 (1982).

In three different cases of type II burst observations sudden changes in frequency drift were observed in the initial phase which indicate a quick deceleration of the corresponding shock wave. The described phenomenon was interpreted as a sudden shock wave deceleration and supposed to be accompanied by the formation of a turbulent structure in the shock wave. The discussion indicates that the sudden shock wave deceleration process has probably a connection with the second phase of acceleration in the flare. The magnitudes of magnetic fields in the corona were determined on the basis of proposed interpretation.

077.010 **Evidence of primary and secondary bursts in solar type III emission.** A. O. Benz, R. Treumann, N. Vilmer, A. Mangeney, M. Pick, A. Raoult.
Astron. Astrophys., Vol. 108, 161 - 168 (1982).

Simultaneous observations of groups of metric type III radio bursts at high spectral, spatial, and temporal resolution are presented. They have occurred on different days and above the photospheric limb. It is found that they tend to be clustered in pairs. The two bursts in each pair have considerably different properties. The primary bursts have generally higher frequency drifts, stronger polarization, higher starting frequency, smaller size, and sometimes larger distance from the center of the Sun. The hypothesis of fundamental and harmonic emission encounters serious difficulties. The observations fit better with electron beams propagating along two different (primary and secondary) paths. The authors are led to a model where the secondary beam originates from electrons with a different acceleration region being triggered by the first energy release and having a smaller energy input and thus smaller beam velocity.

077.011 **Far-IR observations of active regions at sunspots with a balloon-borne 60 cm telescope.**
F. Cartier, F. Kneubühl, D. Huguenin, E. A. Müller.
Infrared and millimeter waves, (see 012.017), p. W-3-1/1-2 (1981). − Abstr. in Phys. Abstr., Vol. 85, Abstr. 42679 (1982).

077.012 **Sub-mm heterodyne observations of the Sun, Moon, Jupiter and Orion at 691 GHz.**
H. P. Roser, G. V. Schultz, R. Wattenbach.
Infrared and millimeter waves, (see 012.017), p. W-3-3/1-2 (1981). − Abstr. in Phys. Abstr., Vol. 85, Abstr. 42680 (1982).

077.013 **Radio brightness distribution over the solar disk.**
Eh. E. Dubov, L. S. Lyubimkov.

Izv. Krymskoj Astrofiz. Obs., Tom 64, 140 - 151 (1981). In
Russian. English translation in Bull. Crimean Astrophys. Obs.,
Vol. 64.

The brightness temperature distribution over the solar
disk has been computed at wavelengths 0.5, 1, 3, 5, 10, 30 and
50 cm using a non-homogeneous model of the chromosphere
and of the transition zone. The following features of the radio
brightness distribution have been determined: contrast and
relative dimensions of supergranula boundaries, limb brightening, parameters of the bright ring near the limb and radio
diameter of the sun. Computed values and observational data
have been compared.

077.014 **Connection of the radio brightness variations of the**
         sun with hills of the magnetic field and flocculi
during the minimum of solar activity.
A. F. Bachurin, A. S. Dvoryashin, N. N. Eryushev,
L. I. Tsvetkov.
Izv. Krymskoj Astrofiz. Obs., Tom 64, 152 - 162 (1981). In
Russian. English translation in Bull. Crimean Astrophys. Obs.,
Vol. 64.

The radio emission of the sun during the minimum of
solar activity (November 2 - 12, 1976) is considered. A connection of the radio emission regions with flocculae and magnetic hills of the weak field is established. The brightness
temperature of regions with increased radio emission increases
with the magnetic flux density and with the decrease of the
distance between the hills with opposite polarity.

077.015 **Possible origin of solar bursts not correlated on time**
         at different frequencies.   R. J. Marabini.
Bol. Asoc. Argentina Astron., No. 18, p. 47 - 49 (1980).

077.016 **On the microwave radiation of solar flares.**
         A. S. Grebinskij, A. P. Sedov.
Astron. Zh., Tom 59, 365 - 371 (1982). In Russian. English
translation in Soviet Astron., Vol. 26, No. 2.

The theory of gyrosynchrotron radiation of nonthermal
relativistic particles in a magneto-active plasma is considered.
This allowed to construct a model of a microwave source including the self-absorption and the influence of the thermal
plasma of the solar atmosphere. A comparison of the model
with experimental data has been made demonstrating the efficiency of the approximation used. An estimate is obtained for
the magnetic field in the microwave source region.

077.017 **Some peculiarities of statistical distributions of**
         solar radio bursts.   I. E. Pogodin.
Astron. Zh., Tom 59, 376 - 379 (1982). In Russian. English
translation in Soviet Astron., Vol. 26, No. 2.

An empirical statistical method for estimating the optical
depth of radio sources is suggested. On the basis of an analysis
of the statistical distributions of the parameters of corpuscular
and electromagnetic radiation from solar flares it is shown that
solar microwave radio bursts are generated by the electrons
that reach the earth's orbit with energies greater than 40 keV.

077.018 **Fundamental wave of type III solar radio bursts and**
         whistler waves.   T. Takakura.
Sol. Phys., Vol. 75, 277 - 292 (1982).

It is demonstrated by a numerical simulation that both
the whistler waves and plasma waves are excited by a common
solar electron beam. The excitation of the whistler waves is
ascribed to the loss-cone distribution which arises at a later
phase of the passage of the beam at a given height due to a
velocity dispersion in the electron beam with a finite length.
It is highly probable that the fundamental of type III bursts
are caused by the coalescence of the whistler waves and the
plasma waves excited by a common electron beam, although
the plasma waves must suffer induced scatterings by thermal
ions to have small wave numbers before the coalescence occur.

077.019 **A long-enduring multi-source burst at 17 GHz and**
         its relation to a type IV_{m-dm} burst with spectral
fine features.   T. Kosugi.
Sol. Phys., Vol. 75, 293 - 304 (1982).

The author presents a multi-source microwave burst as a
typical example of well-developed long-enduring bursts. He
describes the microwave observations in some detail and compares them with the optical observations and the dynamic spectrum observations at metric-to-decimetric wavelengths. Discussions made with emphasis on the facts that two distinct types
of sources exist concurrently in the long-enduring microwave
burst and that one of them correlate well to the associated
type IV burst with spectral fine features at metric-to-decimetric
wavelengths.

077.020 **Differences of observed characteristics between**
         impulsive bursts and post-burst increases.
K. Kai, T. Kosugi, H. Nakajima.
Sol. Phys., Vol. 75, 331 - 339 (1982).

The change of source characteristics during the transition
from the impulsive phase to the post-burst phase is investigated for cm bursts on a statistical basis.

077.021 **The flare-related depression of the noise storm on**
         May 5, 1978.   A. Böhme, A. Krüger.
Sol. Phys., Vol. 76, 63 - 75 (1982).

Though a number of flares is capable to trigger the
emission of a noise storm, in some other rare cases flares may
also lead to a depression of the radio flux of a pre-existing
noise storm. Details of this phenomenon are demonstrated at
the flare of May 5, 1978 which can be regarded as an in-
structive example. Using extensive observations during the
solar cycle No. 20 a number of further noise storm depressions
could be detected whereas chance coincidences with flares
could be ruled out by a statistical treatment. Possible mechanisms related to the noise storm depression effect are briefly
discussed.

077.022 **Time variability and structure of quiet Sun sources**
         at 6 cm wavelength.
F. T. Erskine, M. R. Kundu.
Sol. Phys., Vol. 76, 221 - 237 (1982).

The authors present the results of a detailed study of the
quiet Sun emitting regions at 6 cm, and identify them with
specific regions observed at optical wavelengths. Their study is
based upon both two-dimensional and one-dimensional
synthesis observations, using the WSRT with a spatial resolution as good as 6 arc sec.

077.023 **Study of the character and parameters of wave**
         motions in the transition layer and low solar corona
by radio observations.   M. M. Kobrin, S. D. Snegirev.
Pis'ma Astron. Zh., Tom 8, 308 - 311 (1982). In Russian.
English translation in Soviet Astron. Lett., Vol. 8.

Simultaneous observations were made of integral flux
fluctuations of the "quiet" sun radiation generated in the
transition layer ($\lambda = 3$ cm) and in the low corona ($\lambda = 30$ cm).
From measurements of phase relations between the train-type
intensity oscillations of the radio emission it is found that for
periods 45 - 180 s the propagation of disturbances was always
observed from the chromosphere into the corona with a delay
time of 40 - 200 s. Estimates made when using a number of
models for the "quiet" sun give velocities of disturbance
propagation of 7 - 40 km/s which correspond to the velocity of
sound and slow MHD waves.

077.024 **A partial eclipse of sun on 4 January 1973 at**
         73.5 cm.   R. J. Marabini.
Bol. Asoc. Argentina Astron., No. 20 - 24, p. 45 - 50 (1981).

077.025 **Polarización de los radiobursts solares**
         correspondientes a la región activa MacMath 13225.

R. Marabini.
Bol. Asoc. Argentina Astron., No. 20 - 24, p. 237 (1981).
Abstract.

077.026 Morfología de bursts en 400 MHz y sus relaciones físicas. R. Marabini, J. Ladaga.
Bol. Asoc. Argentina Astron., No. 20 - 24, p. 238 (1981).
Abstract.

077.027 A study of the parameters of individual type-I bursts. A. Kattenberg, G. van der Burg.
Sol. Phys., Vol. 77, 231 - 248 (1982).

Hundreds of type-I bursts in the digitally recorded solar noise storms were studied and the results are presented in this paper. The authors try to find patterns in the dynamic spectra of individual bursts or statistical relations between some of the parameters that can be defined to describe the spectra. The bursts are reduced and represented in a standard way. Graphical representations are inspected by eye and compared qualitatively. Numerical burst parameters are studied statistically. The authors describe the properties of instantaneous burst spectra. The polarization properties of bursts and the relation between burst- and continuum-polarization are studied to some extent.

077.028 Positions of type II fundamental and harmonic sources in the 30–100 MHz range.
H. S. Sawant, T. E. Gergely, M. R. Kundu.
Sol. Phys., Vol. 77, 249 - 254 (1982).

Observations of a type II burst with fundamental and harmonic structure were made, with an interferometer operating in the range 120–20 MHz with time and frequency resolution of 1 s and 100 kHz, respectively. The type II burst was preceded by a type III-type V, and the associated flare of importance SN was located at S 20 W 73. The interferometric data show that the fundamental and harmonic were coincident in position. Further, the type III positions as a function of frequency were practically the same as those of the type II burst. The implications of these results are discussed.

077.029 A note on the mechanism of broad band absorptions and emissions in the type IV decimeter continuum. A. D. Fokker.
Sol. Phys., Vol. 77, 255 - 261 (1982).

The broad band absorptions and emissions in the type IV decimeter continuum are remarkable for the simultaneity of their occurrence at widely different frequencies and for their very short durations. During a given event there is a continuous succession of different patterns of broad band features. It is suggested that the different aspects of the broad band features are variations on a common underlying process. The process would seem to be the screening of a large part of the outgoing beam of radiation, which supposedly is ducted along a tube or channel of relatively low density. As an agency which causes the screening, one may think of shocks or solitons that impinge transversely on the channel along which the radiation propagates. The various patterns of broad band absorptions and emissions would then be related to structural features of series or trains of passing shocks and/or solitons.

077.030 Numerical simulation of type III solar radio bursts on meter- and hectometer-waves. T. Takakura.
Sol. Phys., Vol. 78, 141 - 156 (1982).

A numerical simulation of type III bursts is made by the use of a fully numerical scheme. Although the electron distribution function is one-dimensional in velocity space, the plasma waves are cylindrically symmetric two-dimensional in k-space. It is confirmed that the previous simulation made by the use of a semi-analytical method assuming the plateau distribution of the electron distribution is qualitatively correct, but the number density of the electron beam to have a typical

type III burst was overestimated by a factor of about 3. It is demonstrated that a tentative neglection of a term for the induced scattering of plasma waves into nonresonant k-range gives no remarkable effect on the energy loss of the electron beam, though the scattering is strong.

077.031 Radio echo and sporadic radiation scattering in the solar corona. E. P. (*Eh. P.*) Abranin,
L. L. Bazelyan, V. V. Zaitsev (*Zajtsev*), V. O. Rapoport,
Ya. G. Tsybko.
Sol. Phys., Vol. 78, 179 - 186 (1982).

Some properties of solar radio bursts observed at the Earth are mainly due to propagation effects in the corona. A radio echo of short-time narrow-band bursts is observed by a decameter radioheliograph on the basis of UTR-2 antenna. Propagation effects are manifested in the marked regular change of the burst intensity-time profile at 25 MHz during a half-rotation of the Sun. A displacement of limb diffuse bursts deep into the solar atmosphere of $1.5 - 2 R_\odot$ has been also found during the burst lifetime.

077.032 Type II solar radio events observed in the interplanetary medium. I: General characteristics.
H. V. Cane, R. G. Stone, J. Fainberg, J. L. Steinberg, S. Hoang.
Sol. Phys., Vol. 78, 187 - 198 (1982).

Fifteen type II solar radio events have been identified in the 2 MHz to 30 kHz frequency range by the radio astronomy experiment on the ISEE-3 satellite over the period from September 1978 to December 1979. Dynamic spectra of a number of events are presented. Where possible, the 15 events have been associated with an initiating flare, ground-based radio data, the passage of a shock at the spacecraft and the sudden commencement of a geomagnetic storm. The general characteristics of kilometric type II bursts are discussed.

077.033 The intense solar radio outburst observed at Purple Mountain Observatory on October 12, 1981.
Y. Liu, H.-a. Wu.
Astron. Circ., No. 12, p. 1 - 2, 13 - 14, 16 (1981). In Chinese and English.

077.034 The intense microwave burst observed at Beijing Observatory on October 12, 1981.
C.-m. Hu, Z.-x. Shi, S.-z. Jin.
Astron. Circ., No. 12, p. 2, 14 - 15, 17 (1981). In Chinese and English.

077.035 Application of the method of distribution of simultaneous ejections to researches of the relation between solar radio emission fluctuations at the frequencies 755 MHz and 612 MHz. M. E. Paupere.
Astron. Tsirk., No. 1184, p. 6 - 8 (1981). In Russian.

077.036 On quasi-periodic components of solar radio emission fluctuations at the frequency f = 755 MHz during June - September 1979. E. A. Aver'yanikhina,
G. A. Ozolins, M. E. Paupere, M. K. Eliass.
Astron. Tsirk., No. 1185, p. 2 - 4 (1981). In Russian.

077.037 Observations of low-frequency radio bursts connected with the proton flare of April 24, 1981.
V. P. Grigor'eva, A. K. Yangalov.
Astron. Tsirk., No. 1196, p. 6 - 8 (1981). In Russian.

077.038 Time delays in solar bursts measured in the mm-cm range of wavelengths.
P. Kaufmann, J. E. R. Costa, F. M. Strauss.
Bull. American Astron. Soc., Vol. 13, 845 (1981). – Abstract.

077.039 Direct inference of solar burst primary magnetic transient leading fast time structures at mm-waves and hard X-rays.

P. Kaufmann, F. M. Strauss, J. E. R. Costa, B. R. Dennis.
Bull. American Astron. Soc., Vol. 13, 846 (1981). – Abstract.

077.040   **VLA observations of large scale microwave brightening following a flare.**
M. Bobrowsky, M. R. Kundu, D. Rust.
Bull. American Astron. Soc., Vol. 13, 846 (1981). – Abstract.

077.041   **Flare buildup at 6 cm wavelength, in UV and Hα.**
E. J. Schmahl, M. R. Kundu, B. Woodgate, R. Shine.
Bull. American Astron. Soc., Vol. 13, 846 (1981). – Abstract.

077.042   **Occultation of a compact solar microwave source by a type II burst associated shock front.**
G. J. Hurford, F. Tang.
Bull. American Astron. Soc., Vol. 13, 860 (1981). – Abstract.

077.043   **An active region-type III burst association.**
B. V. Jackson.
Bull. American Astron. Soc., Vol. 13, 861 (1981). – Abstract.

077.044   **The sun at 1.4 GHz as observed with the VLA.**
G. A. Dulk, D. E. Gary.
Bull. American Astron. Soc., Vol. 13, 878 (1981). – Abstract.

077.045   **A comprehensive study of the sun in the submillimeter continuum.**   E. E. Becklin,
J. T. Jefferies, C. Lindsey, F. Q. Orrall, I. Gatley, M. Werner.
Bull. American Astron. Soc., Vol. 13, 880 (1981). – Abstract.

077.046   **Submillimeter observations of the extreme solar limb obtained in the total eclipse of 1981 July 31.**
C. Lindsey, E. E. Becklin, J. T. Jefferies, F. Q. Orrall, I. Gatley, M. Werner.
Bull. American Astron. Soc., Vol. 13, 880 (1981). – Abstract.

077.047   **Observations of the center-to-limb intensity of the quiet sun at 30 - 200 μm.**   F. Q. Orrall,
E. E. Becklin, J. T. Jefferies, C. Lindsey, I. Gatley, M. Werner.
Bull. American Astron. Soc., Vol. 13, 880 - 881 (1981). Abstract.

077.048   **Observation of global 160-min infrared (differential) intensity variation of the sun.**
V. A. Kotov, S. Koutchmy, O. Koutchmy.
Pré-publ. Inst. Astrophys. Paris, No. 6, 30 pp. (1982).
Submitted to Sol. Phys.

077.049   **Simultaneous mapping of microwave burst sources with the VLA at 2, 6 and 20 cm wavelengths.**
M. R. Kundu, T. Velusamy, D. McConnell.
Bull. American Astron. Soc., Vol. 14, 572 (1982). – Abstract.

077.050   **Source structure of gradual rise and fall bursts at 17 GHz.**   K. Kai, T. Kosugi, H. Nakajima.
Sol. Phys., Vol. 78, 243 - 252 (1982).
The authors investigate GRF (gradual rise and fall) bursts on a statistical basis, particularly the spatial structure in order to know in which part of flares high temperature plasmas exist.

077.051   **Spectral characteristics of solar S bursts.**
D. McConnell.
Sol. Phys., Vol. 78, 253 - 269 (1982).
The observed spectral properties of solar radio S bursts have been presented. The bursts were observed in the 30–82 MHz range and their main characteristics were: (1) Narrow instantaneous bandwidth (123 ± 56 kHz); (2) short duration measured at a single frequency (49 ± 34 ms); (3) drift towards lower frequencies at about $\frac{1}{3}$ the type III burst drift rate; (4) and the tendency of some S bursts to appear as a series of fringes separated by about 100 kHz.

077.052   **Positional characteristics of meter-decameter wavelength bursts associated with hard X-ray bursts.**
M. R. Kundu, T. E. Gergely, S. R. Kane.
Sol. Phys., Vol. 79, 107 - 119 (1982).
Several type III, type II, and type IV bursts were observed on April 25 and 26, 1979 with the Clark Lake Radio Observatory's E–W and N–S swept frequency interferometers in the range 20–110 MHz. The radio bursts were associated with hard X-ray bursts in the energy range 26–154 keV, as observed by ISEE-3.

077.053   **Fundamental emission for type III bursts in the interplanetary medium: the role of ion-sound turbulence.**   D. B. Melrose.
Sol. Phys., Vol. 79, 173 - 185 (1982).
It is argued (a) that the onset times of type III radio emission and of the streaming electrons implies that type III bursts in the interplanetary medium are generated predominantly at the fundamental, (b) that in view of recent observations of ion-sound waves in the interplanetary medium the theory of the generation of the bursts should be revised to take account of these waves, and (c) the revised theory favours fundamental emission. A detailed discussion of the effect of ion-sound waves on type III bursts is given.

077.054   **VLA observations of the evolution of a solar burst source structure at 6 centimeter wavelength.**
T. Velusamy, M. R. Kundu.
Astrophys. J., Vol. 258, 388 - 392 (1982).
Evolutionary changes in the total intensity and polarization structure of a 6 cm radio burst source have been observed with the VLA, over time scales ranging from 10 s to several minutes. This burst was associated with a 2B/M1 flare observed on 1980 May 14. The 6 cm burst consisted of a gradual phase of 30 minutes duration and a strong impulsive phase of duration less than 2 minutes. Synthesized maps of total intensity and polarization were obtained with spatial resolution of $2'' \times 3''$ and with temporal resolutions of 5 minutes during the gradual phase and 10 s during the impulsive phase. The sequence of polarization maps suggests a complex magnetic field structure undergoing rapid changes. Most importantly, they show the development of two bipolar regions or quadrupole structure just prior to the impulsive energy release.

077.055   **'Plasma emission' without Langmuir waves.**
D. B. Melrose.
Australian J. Phys., Vol. 35, 67 - 86 (1982).
Recent observations have confirmed that the level of Langmuir waves associated with type III streams of electrons in the interplanetary medium is usually too low to account for the observed radio emission by the accepted 'plasma emission' processes, and it has been suggested that emission mechanisms which do not require Langmuir waves should be explored. Four such mechanisms are discussed.

077.056   **Metre wavelength solar radio bursts with periodic modulation.**   G. R. A. Ellis.
Australian J. Phys., Vol. 35, 87 - 90 (1982).
Observations of solar radio bursts in the frequency range 125–150 MHz are described. It is found that their properties are consistent with those observed for similar bursts in the 30–55 MHz range, except that all, rather than just a small proportion, showed periodic intensity maxima or fringes in the frequency-time plane.

077.057   **Probing the radio Sun.**   M. R. Kundu.
Sky Telesc., Vol. 64, 6 - 9 (1982).

077.058   **Fine structure near the starting frequency of solar type III radio bursts.**
A. O. Benz, P. Zlobec, M. Jaeggi.

Astron. Astrophys., Vol. 109, 305 - 313 (1982).

The authors have systematically analyzed the period in time and frequency adjacent to the beginning of type III bursts digitally recorded at Bleien during the second half of 1980. A surprisingly high percentage (10%, possibly more than 20%) of the type III bursts show fine structure in the form of narrow-banded spikes of 0.05 s and less duration, which form "clusters" of relatively large bandwidth. These spikes are classified and discussed in detail.

077.059 Magneto-ionic conditions and energetics of simultaneous solar microwave bursts at 2.8, 10 and 19.3 GHz on 1980 June 4. S. S. Degaonkar, S. K. Alurkar, R. V. Bhonsle, O. P. N. Calla, B. Lokanadham.
Bull. Astron. Soc. India, Vol. 9, 291 - 296 (1981).

077.060 On the detection of the periodic oscillations of the inclination of the frequency spectrum of the S-component of solar radio emission.
A. R. Abbasov, L. F. Golubeva, L. B. Tzirulnik.
Sun and planetary system, (see 012.040), p. 97 - 98 (1982).

077.061 Radiation from surface wave packet and type II solar radio emission. V. M. Čadež.
Sun and planetary system, (see 012.040), p. 139 - 140 (1982).

The author discusses a possibility of generating the electromagnetic radiation by a potential surface wave packet in an inhomogeneous plasma under physical conditions similar to those existing in the solar corona regions where the type II radio emission originates.

077.062 Ultra-fast fine structures of a microwave outburst. R. Zhao, S. Jin.
Sci. Sinica, Ser. A, Vol. 25, 422 - 429 (1982).

077.063 Spectral correlations between type IV solar radio bursts and associated proton energies.
C.-s. Li, S.-y. Jiang, X.-w. Zheng.
Acta Astron. Sinica, Vol. 23, 29 - 33 (1982). In Chinese.

077.064 Coronal radio emission activity and its relation to planetary geomagnetic activity in the course of cycle No. 20. L. Křivský, A. Prigancová.
Bull. Astron. Inst. Czechoslovakia, Vol. 33, 174 - 179 (1982).

Since the pattern of some geomagnetic indices, characterizing the degree of disturbance of the geomagnetic field during 11-year solar cycles, does not agree well with the usual indices of solar activity, an attempt was made to correlate the number of days with the geomagnetic index $A_p \geqslant 40$ and the occurrence of decametric radio bursts on 29.5 MHz, originating in the solar corona. Considerable agreement was observed between the two values in semi-annual steps within the whole last solar cycle No. 20 (1965 - 1976).

077.065 Centimeter wavelength observations of active regions and flares with a few arc second resolution.
M. R. Kundu.
Compendium in astronomy, (see 003.012), p. 119 - 137 (1982).

In recent years large aperture synthesis instruments such as the Westerbork Synthesis Radio Telescope (WSRT) and the Very Large Array (VLA) with resolution of a few arc seconds at cm wavelengths have been used for solar observations. These observations have provided valuable information on the structures of radio emissive regions in active regions, and flares on scales of a few arc seconds. The synthesis observations give the total intensity (I) and circular polarization (V) maps for active regions. The maps can be used for comparison of the magnetic field structure in the transition zone and the low corona with the photospheric magnetic field.

Two-dimensional radioheliographic pictures of the sun's outer corona at 25.6 - 110.6 MHz.
See Abstr. 033.023.

The use of the large mm-wave antenna at Itapetinga in high-sensitivity solar research. See Abstr. 033.036.

Scattering and collapse of Langmuir waves driven by a weak electron beam. See Abstr. 062.077.

Solar radio granulation at microwaves and its optical identification. See Abstr. 071.066.

Radiodiagnostics of preflare current sheets in active regions of the sun. See Abstr. 072.005.

A comparison of variations in the indices of sunspots and solar radio emission in the 19th and 20th solar cycles. See Abstr. 072.008.

Peculiarities of distribution of event intensity in different phases of the solar cycle. See Abstr. 072.010.

Polarized horseshoes around sunspots at 6 centimeter wavelength. See Abstr. 072.014.

Some studies on sunspots in relation to H_α-flares and radiobursts. See Abstr. 072.023.

Regiones activas MacMath 12136 y 12139. Acoplamiento magnetico de larga escala. Radioburst del 9 de Diciembre de 1972. See Abstr. 072.043.

Solare Beobachtungsergebnisse. Solar data. Solar radio emission. See Abstr. 072.061.

A search for microwave counterparts of umbral flashes. See Abstr. 072.066.

Solar radio SVC radiation in an active region of bipolar sunspots. See Abstr. 072.098.

Why the total solar radio flux at a wavelength of 10 cm cannot fully replace the Wolf relative sunspot number. See Abstr. 072.102.

Magnetic structure of a flaring region producing impulsive microwave and hard X-ray bursts. See Abstr. 073.007.

Structure of the impulsive phase of solar flares from microwave observations. See Abstr. 073.019.

Radio imaging of solar flares using the Very Large Array: new insights into flare process. See Abstr. 073.032.

Second-stage acceleration in a limb-occulted flare. See Abstr. 073.044.

Electron pitch angle scattering and the impulsive phase microwave and hard X-ray emission from solar flares. See Abstr. 073.060.

Remote flare brightenings and type III reverse slope bursts. See Abstr. 073.065.

The relation between the surges and solar radio emission. See Abstr. 073.066.

Rising motion of a behind-the-limb flare at 35 GHz. See Abstr. 073.075.

Temporal features in a solar flare observed in mm-waves, cm-waves and X-rays with fine time resolution. See Abstr. 073.089.

Submillimeter continuum observations of solar plages.   See Abstr. 073.097.

The relationship between the microwave and hard X-ray sources in a solar flare.   See Abstr. 073.098.

Association between gradual hard X-ray emission and metric continua during large flares. See Abstr. 073.107.

Evidence for compressive heating in a solar flare from coincident hard X-ray and microwave observations. See Abstr. 073.109.

The role of the big flare syndrome in correlations of solar energetic proton fluxes and associated microwave burst parameters.   See Abstr. 073.125.

On the solar type III radio burst emission process. See Abstr. 074.025.

Coronal holes as observed with the Pulkovo large radio telescope.   See Abstr. 074.047.

Very Large Array observations of coronal loops at 20 centimeter wavelength.   See Abstr. 074.094.

The measurement of magnetic fields in the solar atmosphere above sunspots using gyroresonance emission. See Abstr. 075.037.

On the correlation between hard X-rays and type III radio bursts during flares.   See Abstr. 076.019.

Estimates of the exponent of the energy spectrum of protons from data on solar microwave radio bursts. See Abstr. 078.006.

Fast speed recording of the July 31, 1981 solar eclipse at the ten centimeter band.   See Abstr. 079.106.

The change of the radio radius of the sun connected with the emergence of a coronal hole at the solar limb. See Abstr. 080.032.

Decrease of the radio radius when a coronal hole emerges at the limb of the quiet sun. See Abstr. 080.047.

078 Cosmic Radiation

078.001 On the formation of the diffuse component of solar electron streams.   B. N. Levin. Astron. Zh., Tom 59, 99 - 107 (1982). In Russian. English translation in Soviet Astron., Vol. 26, No. 1.
The expansion of a fast electron stream limited in space is considered. The distribution function of hot electrons during the expansion becomes unstable relative to the generating excitation. The mechanism considered may be significant in the case of formation of the diffuse component of the solar electron stream.

078.002 Cosmic-ray variations related to solar, geomagnetic and interplanetary disturbances (23 March - 7 April, 1976).   A. Geranios, H. Mavromichalaki. Astrophys. Space Sci., Vol. 82, 133 - 148 (1982).
During the second interval of the Study of Travelling Interplanetary Phenomena a series of solar, interplanetary, geomagnetic and cosmic-ray events have occurred. These are surprising events, since this period falls into the minimum of the solar activity of the past solar cycle. The present analysis is concentrated on Forbush decreases, cosmic-ray increases, geomagnetic variations and the related solar wind disturbances recorded by the heliocentric satellites Helios-1, 2 and the geocentric IMP-8, in the period 23 March - 7 April, 1976.

078.003 Low energy solar particles observed during the pre-phase of large particle events. E. Kirsch, N. Martinic, E. Keppler. J. Geophys., Vol. 50, 45 - 50 (1981). – Abstr. in Phys. Abstr., Vol. 85, Abstr. 28436 (1982).

078.004 Integral multiplicity of generation for the neutron component and precision of calculation of the solar cosmic ray spectrum.

V. M. Bednazhevskij, L. I. Miroshnichenko. Geomagn. Aehron., Tom 22, 125 - 126 (1982). In Russian.

078.005 Rise-time, equivalent diffusion coefficient, and mean free-path of solar protons.   G. Zhang. Kexue Tongbao, Vol. 26, 715 - 718 (1981). – Abstr. in Phys. Abstr., Vol. 85, Abstr. 42559 (1982).

078.006 Estimates of the exponent of the energy spectrum of protons from data on solar microwave radio bursts.   I. M. Chertok. Geomagn. Aehron., Tom 22, 182 - 186 (1982). In Russian.

078.007 Characteristics of proton streams of solar flares in the region of an auroral oval depending on the magnetosphere's disturbance. N. A. Mikirova, N. K. Pereyaslova. Geomagn. Aehron., Tom 22, 187 - 191 (1982). In Russian.

078.008 Influence of the dynamics of structural formations in the interplanetary medium on propagation of charged particles generated in solar flares at longitudes 46 - 85° W in September - November 1973. P. V. Vakulov, N. I. Vologdin, B. M. Kuzhevskij, Yu. V. Mineev, E. S. Spir'kova, I. P. Shestopalov. Kosm. Issled., Tom 20, 73 - 81 (1982). In Russian.

078.009 Isotopic anomalies in cosmic rays: effects of pre-acceleration in collapsing magnetic neutral sheets. D. Mullan. Bull. American Astron. Soc., Vol. 14, 577 (1982). – Abstract.

078.010 Ionic charge state distribution of helium, carbon, oxygen, and iron in an energetic storm particle enhancement.   D. Hovestadt, B. Klecker, H. Höfner,

M. Scholer, G. Gloeckler, F. M. Ipavich.
Astrophys. J., Lett., Vol. 258, L57 - L62 (1982).

The authors have analyzed the ionic charge state distribution of He, C, O, and Fe for an energetic storm particle (ESP) event on 1978 September 28 - 29. The data are obtained with the electrostatic analyzer-proportional counter sensor ULEZEQ of the Max-Planck-Institut/University of Maryland experiment aboard the ISEE 3 spacecraft. The He^+/He^{++} ratio is lowered from 0.057 ± 0.007 during the period before the ESP event to a value of 0.016 ± 0.004 during the ESP event. In parallel, the mean charge states of O and Fe are increased by $\sim 3\%$ and $\sim 16\%$ respectively. The temporal variations are discussed in terms of a first-order Fermi acceleration of the preexisting solar flare particles by a propagating interplanetary shock wave.

Comets and corpuscular radiation of the sun.
See Abstr. 003.018.

Geomagnetic activity and stability of the solar corpuscular field. See Abstr. 003.143.

Peculiarities of distribution of event intensity in different phases of the solar cycle. See Abstr. 072.010.

The energy spectrum of 20 keV−20 MeV electrons accelerated in large solar flares. See Abstr. 073.006.

The role of the big flare syndrome in correlations of solar energetic proton fluxes and associated microwave burst parameters. See Abstr. 073.125.

Solar cosmic-ray-produced radionuclides in meteorites. See Abstr. 105.229.

Origins of the low-energy relativistic interplanetary electrons. See Abstr. 106.023.

Report of IAU Commission 49: The interplanetary plasma and the heliosphere (Plasma interplanétaire et l'héliosphère). See Abstr. 106.041.

The charge and isotopic composition of $Z = 6-14$ cosmic ray nuclei at their source. See Abstr. 143.001.

The heliospheric intensity gradients of the anomalous He^4 and the galactic cosmic-ray components. See Abstr. 143.026.

079 Solar Eclipses

079.001 The frequency of total and annular solar eclipses for a given place. J. Meeus.
J. British Astron. Assoc., Vol. 92, 124 - 126 (1982).

079.002 Der Verlauf der zentralen Sonnenfinsternisse im Alpenraum für die Zeit von 1400 bis 2400 n. Chr.
R. A. Gubser.
Orion, 40. Jahrg., 78 - 86 (1982).

079.003 The many eclipses of 1982. F. Espenak.
Strolling Astron., Vol. 29, 89 - 96 (1982).

Where did the 1780 eclipse go?
See Abstr. 004.033.

Photographie von Sonnenfinsternissen.
See Abstr. 036.019.

Analysis of geomagnetic effect of previous 5 solar eclipses occurring in China during past 50 years.
See Abstr. 084.083.

Catalogue of observations of occultations of stars by the Moon for the years 1623 to 1942 and solar eclipses for the years 1621 to 1806. See Abstr. 096.013.

Solar eclipse 1981 July 31

079.101 A sunspot maximum corona.
R. R. Fisher.
Sky Telesc., Vol. 63, 18 - 19 (1982).

079.102 Total solar eclipse of July 31, 1981.
V. G. Surdin.
Priroda, 1982, No. 1, p. 108 - 109. In Russian.

079.103 Soviet-French observations of the total solar eclipse.
S. Koutchmy, G. M. Nikol'skij.
Zemlya Vselennaya, 1982, No. 1, p. 65 - 67. In Russian.

079.104 The University College London 1981 solar eclipse expedition to Siberia. J. H. Parkinson.
Q. J. R. Astron. Soc., Vol. 23, 252 - 255 (1982).

079.105 Observation of the total solar eclipse of 31 July 1981. V. Rušin.
Říše hvězd, Vol. 63, 5 - 7, 28 - 29, 33 - 36 (1982). In Slovak.

079.106 Fast speed recording of the July 31, 1981 solar eclipse at the ten centimeter band. Q.-j. Fu,
S.-z. Jin, R.-y. Zhao, L.-p. Zheng, Y.-y. Liu, X.-c. Li, Z.-j. Chen
Astron. Circ., No. 13, 5 pp. (1982). In Chinese and English.

079.107 Studies of the July 1981 total solar eclipse from 37,000 feet. M. E. Hults, R. L. Scott.
Bull. American Astron. Soc., Vol. 13, 891 (1981). − Abstract.

079.108 Die totale Sonnenfinsternis vom 31. Juli 1981 − Beobachtungsort Mariinskoje.
T. Brosowski, T. Marold, J. Rose, H. Sanke.
Sterne, 58. Band, 67 - 73 (1982).

079.109 Die totale Sonnenfinsternis vom 31. Juli 1981 − Beobachtungsort Bratsk.
J. Rendtel, I. Rendtel, A. Knöfel, F. Andreas, R. Kollar.
Sterne, 58. Band, 74 - 81 (1982).

079.110 Luminance of the sky in the zenith during the total solar eclipse of July 31, 1981.
M. Minarovjech, M. Rybanský.
Bull. Astron. Inst. Czechoslovakia, Vol. 33, 189 - 190 (1982).

The total solar eclipse of 31 July 1981 in the Sakhalin region. See Abstr. 003.140.

Travellers' tales. A solar eclipse expedition to Siberia.   See Abstr. 011.002.

Die Beobachtung des Flashspektrums während der Sonnenfinsternis vom 31. Juli 1981.
See Abstr. 073.139.

The total solar eclipse of 31 July 1981.
See Abstr. 074.074.

Solar eclipse 1983 June 11

079.201   **Total solar eclipse of 1983 June 11.**
Astron. Her., Vol. 74, No. 3, p. 77 - 81 (1981). In Japanese. – Abstr. in Phys. Abstr., Vol. 85, Abstr. 23206 (1982).

079.202   **Total eclipse of June 1983.**   A. D. Fiala.
Nature, Vol. 297, 536 (1982).

079.203   **Darkness at noon.**
Sky Telesc., Vol. 64, 30 - 32 (1982).

Solar eclipse 1979 February 26

079.301   **Airborne total eclipse observation of the extreme solar limb at 400 μm.**   T. A. Clark, R. T. Boreiko.
Sol. Phys., Vol. 76, 117 - 128 (1982).

The total solar eclipse of February 26, 1979 was monitored at far infrared wavelengths. The resultant eclipse curve for radiation within a bandwidth of 20 cm^{-1} centered upon 25 cm^{-1} (400 μm) was measured and analysed at an equivalent angular resolution of 1 arc sec over a 100 arc sec region adjacent to the limb to provide information on the intensity distribution of continuum radiation close to this limb. The curve has been compared to predictions derived from models of the solar atmosphere for the specific geometry of this eclipse. The absence of the significant limb brightening predicted by the HSRA model (in which homogeneity within the source region is assumed) is in substantial agreement with lower resolution results from mountain altitudes. This result is interpreted as further evidence for the presence in the Sun's lower chromosphere of significant inhomogeneity with a scale size of at least 1000 km at this depth.

080 Atmosphere, Figure, Internal Constitution, Neutrinos, Rotation, etc.

080.001 On the interpretation of five-minute oscillations in solar spectrum line shifts.
J. Christensen-Dalsgaard, D. O. Gough.
Mon. Not. R. Astron. Soc., Vol. 198, 141 - 171 (1982).

It is argued that p modes in the Sun are excited on average to a surface amplitude which, except for modes of highest degree, is a function of frequency alone. This hypothesis appears to be consistent with almost all the observational data in the frequency range 2 - 4 mHz. The sharp-line component in the power spectra of whole-disc measurements first reported by Claverie et al. is due to a near coincidence of the eigenfrequencies of modes of low degree. The amplitudes of modes of higher degree, which are responsible for the five-minute oscillations are roughly in accord with the hypothesis stated above.

080.002 Neutrinos from a standard solar model.
B. W. Filippone, D. N. Schramm.
Astrophys. J., Vol. 253, 393 - 398 (1982).

A detailed estimate is presented of the possible uncertainty range for the neutrino flux from a standard solar model. Using present estimated errors in the key input parameters, detailed solar models are calculated to give an uncertainty in the theoretical ν_e capture rate in both the ongoing ^{37}Cl experiment and the proposed experiment using ^{71}Ga. It is found that with the most recent experimental value of Davis − 2.2 ± 0.4 SNU − and the best-estimate mean model of the present work − 7.0 ± 3.0 SNU − the theoretical to experimental ratio is 3.2 ± 1.5, or 1.5 standard deviations from agreement. The prediction for ^{71}Ga is 111 ± 13 SNU. The results are discussed, in light of recent experimental and theoretical work on neutrino oscillations, to determine if the present situation provides any evidence for such oscillations.

080.003 Modeling solar variability. R. L. Gilliland.
Astrophys. J., Vol. 253, 399 - 405 (1982).

Results of detailed numerical models yielding the amplitudes of the atmospheric variations of solar radius, luminosity, and temperature that arise from rapid perturbations to properties of the convective envelope are presented. The numerical models predict significantly different surface responses for perturbations of the temperature gradient at the convective envelope base, variations of convective efficiency, and the effects of magnetic flux tubes, than was previously found based on simple hydromagnetic arguments.

080.004 Variations in photospheric limb darkening as a diagnostic of changes in solar luminosity.
W. A. Rosen, P. V. Foukal, R. L. Kurucz, A. K. Pierce.
Astrophys. J., Lett., Vol. 253, L89 - L93, plate L6 (1982).

The authors report first results of a program for investigating possible correlations between changes of photospheric limb darkening and variations in the solar constant. They explain the method and observing procedures, and note that a significant change observed in the limb-darkening function in 1980 September corresponds to an increase in the solar constant reported at the same time from satellite radiometric data. This change in limb darkening can be interpreted as a net decrease in photospheric temperature gradient, accompanied by a small increase in effective temperature.

080.005 Solar irradiance modulation by active regions from 1969 through 1980. K. H. Schatten, N. Miller, S. Sofia, L. Oster.
Geophys. Res. Lett., Vol. 9, 49 - 51 (1982).

The solar irradiance variations resulting from sunspot deficits and facular excesses in emission have been calculated from 1969 through 1980. Agreement appears to exist between the calculations and the major features seen with the Nimbus 7 cavity pyrheliometer and with both the major and minor features detected by The Solar Maximum Mission ACRIM experiment. The 12-year irradiance variations calculated by the authors suggest a larger variance with increased solar activity, and little change in the average irradiance with solar activity. The largest excursions over these 12 years show a 0.4% variation.

080.006 On the roughness of the solar limb seen in the continuum Eh. E. Dubov, V. I. Makarov.
Astron. Zh., Tom 59, 186 - 189 (1982). In Russian. English translation in Soviet Astron., Vol. 26, No. 1.

Near the very solar limb, the continuum splits into separate strips considered by the authors to be a manifestation of rough structure of the limb (the "hilly" surface of the sun). The distances between the strips correspond to the sizes of supergranules. The height of the hills is about 400 km.

080.007 Is the sun an oblique magnetic rotator?
G. R. Isaak.
Nature, Vol. 296, 130 - 131 (1982).

The recent observation of rotational splitting of the nonradial p modes of $l = 1$ and $l = 2$ (with $n \sim 20$) of the 5-min global oscillation of the sun was interpreted in terms of rotational splitting associated with a rapid rotation of the solar interior. Here the author attempts to correlate the size of the observed rotational splitting with the enigmatic 12.2-day variation in the measurement of solar oblateness discovered by Dicke previously and to interpret the observed near equality of the amplitudes of the m-components of the $l = 1$ and $l = 2$ rotationally split modes. This leads to the first empirical evidence that an asymmetric magnetic rotator with megagauss magnetic fields exists in the interior of the sun.

080.008 Determination of solar radial velocities using the photometric automatic apparatus Zenith 2.
V. S. Kirichuk, M. V. Kushnir, N. S. Yakovenko.
Soln. Dannye 1981 Byull., No. 11, p. 99 - 105 (1982). In Russian.

Spectrograms of high spatial resolution of the center of the solar disc are studied. The distribution of radial velocities along the spectral slit with a spacing of $0\overset{''}{.}18$ has been derived for 16 Fraunhofer lines of various intensities. The accuracy of the determination of the radial velocity is ± 30 m/sec.

080.009 On some spatial fluctuations in the solar atmosphere.
N. S. Petrova, M. N. Stoyanova, K. S. Tavastsherna.
Soln. Dannye 1981 Byull., No. 12, p. 99 - 105 (1982). In Russian.

A statistical reduction of a spectrogram of high resolution has been done. For several Fraunhofer lines and Hα power spectra of brightness fluctuations and radial velocities have been obtained. The following periods of fluctuations have been detected: $3\overset{''}{.}3$, $4\overset{''}{.}7$ and $6\overset{''}{.}9$. The obtained results are discussed.

080.010 On the modal structure of the solar oscillations.
R. F. Stein.
Astron. Astrophys., Vol. 105, 417 - 418 (1982).

The very different frequency structure of the high and low order modes of the observed solar oscillations is shown to be due to the difference in their cavities. The low order mode cavities extend deep into the solar interior, while the high order mode cavities are confined near the surface of the Sun. As a result the vertical component of the wave vector has a very different dependence on depth in the two cases.

080.011   **The sun's rotation derived from sunspots 1970–1979.**   G. Lustig.
Astron. Astrophys., Vol. 106, 151 - 152 (1982).
Results are presented for the solar rotation from 1970 to 1979. Observations of the sun were made at the Solar-Observatory Kanzelhöhe (1526 m altitude). The sidereal rotation rate was calculated to be: $\omega(B) = (14.27 \pm 0.02) - (1.84 \pm 0.12) \times \sin^2 B$.

080.012   **The disquieting sun: how big, how steady?**   L. J. Robinson.
Sky Telesc., Vol. 63, 354 - 357 (1982).

080.013   **On rotation of the Sun before the Maunder minimum *(auroral data)*.**   Y. Lin, N. Dai, M. Chen, J. Zhang.
Kexue Tongbao, Vol. 26, 526 - 530 (1981). – Abstr. in Phys. Abstr., Vol. 85, Abstr. 10379 (1982).

080.014   **The solar neutrino problem.**   J. B. Taylor, J. W. Connor.
Astron. Astrophys., Vol. 107, L1 - L2 (1982).
A recent suggestion (Opher 1981) that distortion of the electron distribution by electron-plasmon interactions might resolve the solar neutrino problem is shown to be based on an incomplete analysis of the interactions. The full analysis is consistent with an undistorted Maxwellian distribution – leaving the solar neutrino problem unresolved.

080.015   **Observation of nonacoustic, 5 minute period, vertical traveling waves in the photosphere of the sun.**   H. A. Hill, P. R. Goode, R. T. Stebbins.
Astrophys. J., Lett., Vol. 256, L17 - L21 (1982).
Nonacoustic, radially propagating traveling waves have been observed in the solar photosphere. These traveling waves have a period of 278 ± 41 s. The vertical wavelength (~ 500 km) and phase velocity (~ 2 km s^{-1}) of the waves are among their properties deduced from the data. It is also observed that the waves have outgoing phase part of the time and ingoing phase the remainder of the time. The traveling waves are interpreted to be gravity waves. Their role in the heating of the chromosphere is discussed.

080.016   **Solar neutrinos, solar oscillation and the eleven year solar cycle.**   P. Raychaudhuri.
Speculations Sci. Technol., Vol. 4, 271 - 274 (1981). – Abstr. in Phys. Abstr., Vol. 85, Abstr. 14839 (1982).

080.017   **Theoretical eigenfrequencies of solar oscillations of low harmonic degree *l* in five-minute range.**
H. Shibahashi, Y. Osaki.
Publ. Astron. Soc. Japan, Vol. 33, 713 - 719 (1981).
Theoretical eigenfrequencies of the standard solar model with the normal abundance of elements are calculated for radial and nonradial p-mode oscillations with low harmonic index l ($l = 1, 2, 3,$ and 4). It is found that theoretical eigenfrequencies for the standard model agree fairly well with observed peaks in the power spectra for the full-disk five-minute oscillation of the sun. Eigenfrequencies of nonradial p-modes with high degree l for the same model are also presented for comparison with observations of the conventional five-minute oscillation with shorter horizontal wavelengths.

080.018   **Jet phenomena in the solar atmosphere.**   K. Shibata.
Astron. Her., Vol. 74, No. 1, p. 10 - 15 (1981). In Japanese. Abstr. in Phys. Abstr., Vol. 85, Abstr. 18744 (1982).

080.019   **The solar and stellar convection zones.**   H.-s. Yang.
Acta Sci. Nat. Univ. Pekinensis, No. 4, p. 47 - 54 (1980). In Chinese. – Abstr. in Phys. Abstr., Vol. 85, Abstr. 23269 (1982).

080.020   **Turbulent diffusion and the solar neutrino problem.**   E. Schatzman.
Report CERN-81-11, CERN, Geneva, Switzerland. 5 + 23 pp. (1981). – Abstr. in Phys. Abstr., Vol. 85, Abstr. 23294 (1982).

080.021   **Sonnenrotation des photosphärischen und umbralen Plasmas.**   A. Koch, H. Wöhl.
Mitt. Astron. Ges., Nr. 55, p. 91 - 92 (1982). – Abstract.

080.022   **^{81}Br and ^{79}Br as detectors of solar neutrinos.**   J. N. Bahcall.
Phys. Rev. C, Vol. 24, 2216 - 2221 (1981). – Abstr. in Phys. Abstr., Vol. 85, Abstr. 29258 (1982).

080.023   **When the Sun vibrates.**   E. Fossat.
Recherche, Vol. 12, 1280 - 1282 (1981). In French. – Abstr. in Phys. Abstr., Vol. 85, Abstr. 42678 (1982).

080.024   **Magnetic fields, convection and solar luminosity variability.**   W. C. Livingston.
Nature, Vol. 297, 208 - 209 (1982).
The author demonstrates from Fourier transform spectrometer observations of Fraunhofer line asymmetry that granular convection is retarded in the presence of surface magnetism. He then presents full disk observations which suggest a lessening of convection over the past 5 yr. Finally he notes the continued temporal decrease in the spectroscopic temperature of the low photosphere and considers how a change of convection and the lowering temperature may be related.

080.025   **On solar models and their periods of oscillation.**   J. Christensen-Dalsgaard.
Mon. Not. R. Astron. Soc., Vol. 199, 735 - 761 (1982).
Accurate calculations of solar models and their oscillation periods are needed if the observed oscillation periods of the Sun are to be used to infer properties of the solar interior. The paper describes a new programme for calculation of solar evolution sequences; an analysis of the numerical accuracy of the computed models is given, and the effects of changing the opacity interpolation or opacity tables are investigated. In addition selected periods of adiabatic oscillations for the models are calculated and the errors in these periods estimated.

080.026   **Solar neutrino production of technetium-97 and technetium-98.**   G. A. Cowan, W. C. Haxton.
Science, Vol. 216, 51 - 54 (1982).
It may be possible to determine the boron-8 solar neutrino flux, averaged over the past several million years, from the concentration of technetium-98 in molybdenite. The mass spectrometry of this system is greatly simplified by the absence of stable technetium isotopes, and the presence of the fission product technetium-99 provides a monitor of uranium-induced backgrounds. This geochemical experiment could provide the first test of nonstandard solar models that suggest a relation between the chlorine-37 solar neutrino puzzle and the recent ice age.

080.027   **Solar irradiance variations due to active regions.**   L. Oster, K. H. Schatten, S. Sofia.
Astrophys. J., Vol. 256, 768 - 773 (1982).
The authors have been able to reproduce the variations of the solar irradiance observed by ACRIM to an accuracy of better than ± 0.4 W m^{-2}, assuming that during the 6 month observation period in 1980 the solar luminosity was constant. The improvement over previous attempt is primarily due to

the inclusion of faculae. The reproduction scheme uses simple geometrical data on spot and facula areas, and conventional parameters for the respective fluxes and angular dependencies. It is interesting that the time average of the integrated excess emission (over directions) of the faculae cancels out the integrated deficit produced by the spots, within an accuracy of about 10%.

080.028 Global vibrations and the inner structure of the sun. Helioseismology. A. B. Severnyj.
Vestn. AN SSSR, 1981, No. 11, p. 62 - 68. In Russian. – Abstr. in Ref. zh., 51. Astron., 3.51.446 (1982).

080.029 The inner structure of the sun and of solar-type stars. A. I. Laptukhov.
Inst. zem. magn., ionos. i rasprostr. radiovoln. AN SSSR. Prepr., 1981, No. 34, 52 pp. In Russian. – Abstr. in Ref. zh., 51. Astron., 3.51.447 (1982).

080.030 Campo de Velocidades en el limbo solar. T. Paneth.
Bol. Asoc. Argentina Astron., No. 18, p. 49 - 51 (1980).

080.031 Departure from local thermodynamical equilibrium in the solar atmosphere: results of investigations. N. G. Shchukina.
Astrometr. Astrofiz., Vyp. (No.) 44, p. 24 - 34 (1981). In Russian.
Investigations of the effects of departure from LTE on the solar Fraunhofer spectrum are summarized. Departure from LTE is emphasized to occur in lines of all strengths formed at all depths above $\tau_c = 1$ (τ_c = continuum optical depth).

080.032 The change of the radio radius of the sun connected with the emergence of a coronal hole at the solar limb. V. N. Borovik, M. A. Livshits.
Astron. Zh., Tom 59, 355 - 364 (1982). In Russian. English translation in Soviet Astron., Vol. 26, No. 2.
A careful selection of radio scans of the sun obtained daily at the Pulkovo Large Radio Telescope in the centimetre wavelength range in 1973, as well as in 1974 - 76, allowed to study the effects associated with coronal holes against the quiet sun's background. The holes are formations of weak contrast against this background. It is shown by two independent methods that by the emergence of a coronal hole at the limb a 3 - 6 percent decrease of the radio radius at the wavelength 9 cm is observed. The effect is confirmed by the statistical investigation of widths of radio scans obtained in 1974 - 76.

080.033 Observation of additional low-degree 5-min modes of solar oscillation. P. H. Scherrer, J. M. Wilcox, J. Christensen-Dalsgaard, D. Gough.
Nature, Vol. 297, 312 - 313 (1982).
By measuring the difference between the shifts in the Fe 5,124 spectrum line from light integrated from a central circular portion of the solar disk and from an annular portion exterior to it, the authors have detected high-order solar oscillations with degrees $l = 3, 4$ and 5.

080.034 Alfvén waves in the solar atmosphere. III. Nonlinear waves on open flux tubes. J. V. Hollweg, S. Jackson, D. Galloway.
Sol. Phys., Vol. 75, 35 - 61 (1982).
The purpose of this paper is to consider some nonlinear aspects of the propagation of Alfvén waves on solar magnetic flux tubes. First, the steepening of Alfvén waves to form shocks in the chromosphere is investigated. Second, the interaction of a chromospheric switch-on shock with the transition region is examined. Finally, the authors investigate the extent to which the nonlinear forces exerted on the plasma by the Alfvén waves can drive mass motions in the chromosphere and

corona. The results of these computations are examined in terms of their applicability to chromospheric heating, coronal heating, and spicule formation.

080.035 Comment on the paper 'A new resonance in the solar atmosphere' by Joseph V. Hollweg. P. Venkatakrishnan, S. S. Hasan.
Sol. Phys., Vol. 75, 79 - 81 (1982). – See Abstr. 26.080.007.
In the absence of genuine forcing terms, there is no resonance between linear fast MHD and gravito-acoustic waves.

080.036 Torsional waves on the Sun and the activity cycle. B. J. LaBonte, R. Howard.
Sol. Phys., Vol. 75, 161 - 178 (1982).
Some properties of the recently-discovered torsional oscillations of the Sun are presented. The detailed relation of this velocity feature to magnetic activity gives evidence that these motions represent a fundamental oscillation within the Sun that is responsible for the solar activity cycle and that they are not a natural consequence of an α-ω dynamo. A new torsional oscillation with wave number 1 hemisphere^{-1} is demonstrated to exist on the Sun.

080.037 Tunneling and interference of Alfvén waves. Y. D. Žugžda (*Yu. D. Zhugzhda*), V. Locāns.
Sol. Phys., Vol. 76, 77 - 108 (1982).
The propagation and interference of Alfvén waves in magnetic regions is studied. A multilayer approximation of the standard models of the solar atmosphere is used. The low-frequency Alfvén waves ($P > 1$ s) are able to transfer the energy from sunspots into the corona by tunneling only. The interference and resonance of Alfvén waves are found to be important to wave propagation through the magnetic coronal arches. The transmission coefficient of Alfvén waves into the corona increases sharply on the resonance frequences. To take into account the wave absorption in the corona, a method of equivalent schemes is developed. The heating of a coronal arch by Alfvén waves is discussed.

080.038 The effects of sunspots on solar irradiance. H. S. Hudson, S. Silva, M. Woodard, R. C. Willson.
Sol. Phys., Vol. 76, 211 - 219 (1982).
Sunspots have an obvious direct effect upon the visible radiant energy falling upon the Earth. The authors show how to estimate this effect and compare it quantitatively with recent observations of the solar total irradiance. Since the sunspot effect on irradiance produces an asymmetry of the solar radiation, rather than (necessarily) a variation of the total luminosity, the authors have estimated the sunspot population on the invisible hemisphere. This extrapolation allows to estimate the true luminosity deficit produced by sunspots.

080.039 Proper oscillations in models of the present-day sun with mixed internal layers. E. A. Gavryuseva, G. T. Zatsepin, Yu. S. Kopysov.
Kratk. soobshch. po fiz., 1981, No. 10, p. 46 - 51. In Russian. Abstr. in Ref. zh., 51. Astron., 4.51.370 (1982).

080.040 Relation between solar activity and the moment of beginning of the seasonal maximum of cesium-137 concentration in the ground layer of the atmosphere. A. S. Avramenko, K. P. Makhon'ko.
Fiz. atmos. Vil'nyus, 1981, No. 7. In Russian. – Abstr. in Ref. zh., 51. Astron., 4.51.464 (1982).

080.041 Verificación de la posición del eje de rotación del sol. T. Paneth.
Bol. Asoc. Argentina Astron., No. 20 - 24, p. 87 - 88 (1981).

080.042 Formación de líneas espectrales fuera de equilibrio termodinámico en la atmósfera solar. J. M. Fontenla, H. Molnar.

Bol. Asoc. Argentina Astron., No. 20 - 24, p. 146 (1981). Abstract.

080.043 Overstability of acoustic modes and the solar five-minute oscillations.
H. M. Antia, S. M. Chitre, D. Narasimha.
Sol. Phys., Vol. 77, 303 - 327 (1982).

The stability of linear convective and acoustic modes in solar envelope models is investigated by incorporating the thermal and mechanical effects of turbulence through the eddy transport coefficients. With a reasonable value of the turbulent Prandtl number it is possible to obtain the scales of motion corresponding to granulation, supergranulation and the five-minute oscillations. Several of the acoustic modes trapped in the solar convection zone are found to be overstable and the most unstable modes, spread over a region centred predominantly around a period of 300 s with a wide range of horizontal length scales, are in reasonable accord with the observed power-spectrum of the five-minute oscillations.

080.044 Modulation of the velocity of emergence of convective elements on the solar surface by tidal forces of planets.   P. R. Romanchuk.
Vestn. Kiev. Univ., Vyp. 23, p. 22 - 25 (1981). In Russian.

080.045 A magnetic core in the sun? The solar rotator.
R. H. Dicke.
Sol. Phys., Vol. 78, 3 - 16 (1982).

The previously found solar distortion rotating rigidly and wave-like on the surface with a $\sim$12 day period is interpreted as the shape of the gravitational potential induced by the solar core distorted by an internal magnetic field and rotating rigidly with this period. The distortion does not have a symmetry axis and the necessary magnetic field is not compatible with the axial symmetry required of a quasi-static field locked in the rotating core. It is concluded that if the solar distortion is due to such a process the core is oscillating with a very long period, a toroidal oscillation with a period of the order of years.

080.046 Damping constant and turbulence in the solar atmosphere.   R. I. Kostik (Kostyk).
Sol. Phys., Vol. 78, 39 - 57 (1982).

In the region of the formation of weak and medium-strong lines, the microturbulence increases with height, the macroturbulence decreases, and the total velocity field (vertical component) is depth-independent. The empirical damping constants for Fe, Ti, Cr, Ni lines are equal $1.3\gamma_6$, $1.7\gamma_6$, $1.6\gamma_6$, $1.6\gamma_6$, respectively. The correlation length (the Kubo-Anderson process has been used) in the solar photosphere is $520-550$ km.

080.047 Decrease of the radio radius when a coronal hole emerges at the limb of the quiet sun.
V. N. Borovik, M. A. Livshits.
Astron. Tsirk., No. 1187, p. 1 - 5 (1981). In Russian.

080.048 The internal structure of the sun, its pulsations and the neutrino problem.   L. Paternò.
The comparative study of the planets, (see 012.033), p. 151 - 160 (1982).

The theory of the internal structure of the sun is reviewed. The various attempts to reconcile the solar neutrino flux as measured by the Davis experiment with the predictions of the standard solar model are critically examined. The importance of the 5 min solar oscillations in this context is outlined.

080.049 Linear analysis of single- and double-tearing modes in the solar atmosphere.
R. S. Steinolfson, G. Van Hoven, T. Tachi.
Bull. American Astron. Soc., Vol. 13, 837 (1981). – Abstract.

080.050 Non-local effects of radiative transfer on radiative hydrodynamic stability.
G. H. Fisher, R. C. Canfield, A. N. McClymont.
Bull. American Astron. Soc., Vol. 13, 846 - 847 (1981). Abstract.

080.051 Observations of p-mode oscillations in the total solar irradiance.
M. Woodard, H. Hudson, R. C. Willson.
Bull. American Astron. Soc., Vol. 13, 858 - 859 (1981). Abstract.

080.052 Observations at SCLERA of symmetry properties of nonradial solar oscillations.
R. J. Bos, H. A. Hill.
Bull. American Astron. Soc., Vol. 13, 859 (1981). – Abstract.

080.053 Global solar oscillations: high order modes with degree 3 to 5.   P. H. Scherrer, J. M. Wilcox.
Bull. American Astron. Soc., Vol. 13, 859 (1981). – Abstract.

080.054 The effect of subsurface inhomogeneities on the solar k-w diagram.   A. H. Nye, L. E. Cram.
Bull. American Astron. Soc., Vol. 13, 859 (1981). – Abstract.

080.055 Further comparisons between observed and theoretical eigenfrequencies for low-degree solar p -mode oscillations.   E. J. Rhodes, Jr., R. K. Ulrich.
Bull. American Astron. Soc., Vol. 13, 859 (1981). – Abstract.

080.056 On the frequencies of the solar 5 min oscillations of high degree.
J. Christensen-Dalsgaard, D. O. Gough.
Bull. American Astron. Soc., Vol. 13, 859 - 860 (1981). Abstract.

080.057 Solar five-minute oscillations as probes of velocity and temperature fields.
F. Hill, J. Toomre, L. J. November.
Bull. American Astron. Soc., Vol. 13, 860 (1981). – Abstract.

080.058 Magneto-acoustic-gravity waves on the sun.
R. F. Stein.
Bull. American Astron. Soc., Vol. 13, 860 (1981). – Abstract.

080.059 Solar motion and solar activity.   J. B. Blizard.
Bull. American Astron. Soc., Vol. 13, 876 (1981). Abstract.

080.060 Solar brightness variations as seen in λ5250.
D. H. Bruning, B. J. LaBonte.
Bull. American Astron. Soc., Vol. 13, 876 - 877 (1981). Abstract.

080.061 Solar irradiance modulation by active regions from 1969 through 1980.
K. H. Schatten, N. Miller, L. Oster, S. Sofia.
Bull. American Astron. Soc., Vol. 13, 877 (1981). – Abstract.

080.062 The effects of solar activity on the total solar irradiance.
H. S. Hudson, M. Woodard, R. C. Willson.
Bull. American Astron. Soc., Vol. 13, 877 - 878 (1981). Abstract.

080.063 Large scale order in the sun.   C. L. Wolff.
Bull. American Astron. Soc., Vol. 13, 878 (1981). Abstract.

080.064 The HAO solar diameter monitor.   T. M. Brown.
Bull. American Astron. Soc., Vol. 13, 878 (1981). Abstract.

080.065   Solar calibration of stellar rotation tracers.
          B. LaBonte.
Bull. American Astron. Soc., Vol. 13, 889 (1981). – Abstract.

080.066   Absolute spectroscopic measurements of the solar
          equatorial rotation rate.   J. L. Snider.
Bull. American Astron. Soc., Vol. 13, 906 (1981). – Abstract.

080.067   Observations of velocity fields at the solar poles.
          D. Guenther, L. Cram, B. Durney.
Bull. American Astron. Soc., Vol. 13, 906 (1981). – Abstract.

080.068   Nonlinear anelastic models of solar convection.
          J. Toomre, J. Latour, J.-P. Zahn.
Bull. American Astron. Soc., Vol. 13, 907 (1981). – Abstract.

080.069   Analogical and numerical models for the calcula-
          tion of global solar radiation.
E. Proverbio, T. Zanzu.
Energia Solare e Nuove Prospettive, Vol. 3, Atti della 18.
Conferenza Internazionale, Milano 23 - 27 sett. 1979, 13 pp.=
Pubbl. Stn. Astron. Int. Latitudine, Carloforte-Cagliari, N. 75.

080.070   Models of the present state of the sun with mixed
          central zone.   E. A. Gavryuseva.
Kratk. soobshch. po fiz., 1981, No. 10, p. 52 - 57. In Russian.
Abstr. in Ref. zh., 51. Astron., 5.51.452 (1982).

080.071   Calculated solar constant variations from 1874
          through 1981 caused by active regions.
D. V. Hoyt, J. A. Eddy.
Bull. American Astron. Soc., Vol. 14, 575 (1982). – Abstract.

080.072   Solar atmospheric dynamics. II. Nonlinear models
          of the photospheric and chromospheric oscillations.
J. Leibacher, P. Gouttebroze, R. F. Stein.
Astrophys. J., Vol. 258, 393 - 403 (1982).
The authors investigate the one-dimensional, nonlinear
dynamics of the solar atmosphere and describe models of the
observed photospheric (300 s) and chromospheric (200 s)
oscillations. These are resonances of acoustic wave cavities
formed by the variation of the temperature and ionization
between the subphotospheric, hydrogen convection zone
and the chromosphere-corona transition region. The dependence of the oscillations upon the excitation and boundary
conditions leads the authors to conclude that for the observed
amplitudes, the modes are independently excited and, as
trapped modes, transport little if any mechanical flux. In the
upper photosphere and lower chromosphere, where the two
modes have comparable energy density, interference between
them leads to apparent vertical phase delays which might be
interpreted as evidence of an energy flux.

080.073   The influence of partial ionization and scattering
          states on the solar interior structure.
R. K. Ulrich.
Astrophys. J., Vol. 258, 404 - 413 (1982).
The equation of state for the solar interior is normally
assumed to be a fully ionized gas corrected by the Debye-
Hückel coulomb interaction, partial degeneracy, and radiation
pressure. The assumption of full ionization is dropped in this
paper, and the influence of scattering states is included. The
theory of scattering states appears to be new to astrophysics.
The effect of scattering states eliminates the need to invoke a
process of "pressure ionization" for which no satisfactory
theory exists. Six solar models which include varying forms of
the equation of state are discussed. The Saha equation without scattering states gives a neutrino counting rate of
7.41 SNU for the ^{37}Cl experiment, while assumed ionization
for $T > 3 \times 10^5$ K gives 8.87 SNU, and the Saha equation with
the lowest order effect of scattering states (Planck-Larkin
equation) gives 8.83 SNU. Inclusion of the second virial

coefficient due to scattering states brings the result to
9.02 SNU.

080.074   Nachweismöglichkeiten solarer Neutrinos.
          C. Friedemann.
Sterne, 58. Band, 158 - 161 (1982).

080.075   Review of ground-based measurements.
          R. J. Angione.
Variations of the solar constant, (see 012.037), p. 11 - 29
(1981).
Early measurements of the solar constant made from the
ground are described and discussed with particular emphasis
on the Smithsonian program. A brief description is given of
the monitoring program now operating at San Diego State.

080.076   Change in the solar constant between 1968 and
          1978.   J. J. Kosters, D. G. Murcray.
Variations of the solar constant, (see 012.037), p. 31 - 36
(1981).
Solar irradiance measurements made from a balloon on
January 27, 1978 and February 10, 1980 show a change of
0.4% over similar measurements made in 1968. This change
is greater than the uncertainty of the measurement and is felt
to be the result of a change in the solar constant.

080.077   The variability of the solar output.
          C. Fröhlich.
Variations of the solar constant, (see 012.037), p. 37 - 44
(1981).
A review of recent solar constant determinations and
measurements of its spectral distribution is presented. For
the period from 1966 to 1980 a mean value of 1367 Wm^{-2} is
determined. Short term solar variations and their spectral
dependance have been deduced from measurements during
four hours on June 20, 1980 from 34 km altitude with amplitudes of ± 500 ppm at 368 nm, of ± 200 ppm at 500 nm and
± 150 ppm at 778 nm. The power spectrum shows a weak
peak at about 3.2 mHz, which corresponds to the frequency
of the 5-minutes solar oscillation.

080.078   A summary of results from solar monitoring rocket
          flights.   C. H. Duncan.
Variations of the solar constant, (see 012.037), p. 45 - 58
(1981).
Three rocket flights to measure the solar constant and
provide calibration data for sensors aboard Nimbus 6, 7, and
Solar Maximum Mission (SMM) spacecraft have been accomplished. The values obtained by the rocket instruments for the
solar constant in SI units are: 1367 Wm^{-2} on 29 June 1976;
1372 Wm^{-2} on 16 November 1978; and 1374 Wm^{-2} on 22 May
1980. The uncertainty of the rocket measurements is ± 0.5%.

080.079   Solar variability indications from Nimbus 7 satellite
          data.   J. R. Hickey, B. M. Alton, F. J. Griffin,
H. Jacobowitz, P. Pellegrino, E. A. Smith, T. H. Vonder Haar,
R. H. Maschhoff.
Variations of the solar constant, (see 012.037), p. 59 - 72
(1981).
The cavity pyrheliometer sensor of the Nimbus 7 Earth
Radiation Experiment (ERB) has indicated low-level variability of the total solar irradiance. The variability appears to be
inversely correlated with common solar activity indicators in
an "event" sense. The limitations of the measuring system and
available data sets are described.

080.080   Indirect methods for measuring variations of the
          solar constant.   S. Sofia.
Variations of the solar constant, (see 012.037), p. 73 - 79
(1981).
The various techniques thus far used to measure or infer
variations of the solar constant, S, are reviewed. The difference

between the methods that measure δS, and those that measure variations in the solar luminosity, δL, is discussed. It is shown that the past practice of simply relating δS to δL by geometrical arguments is not valid because of anisotropy of the solar radiation. The observations of changes in the solar diameter support the existence of structurally induced variations of the solar luminosity on timescales of tens of years, which are clearly significant in our understanding of climatic variations.

080.081   Solar neutrinos.
J. N. Bahcall.
Variations of the solar constant, (see 012.037), p. 81 - 93 (1981).

The topics which are covered are, in order: an overview of the subject of solar neutrinos, a brief summary of the theory of stellar evolution, a description of the main sources of solar neutrinos, a brief summary of the results of the Brookhaven [37]Cl experiment, an analysis of the principal new solar neutrino experiments that have been proposed, a discussion of how solar neutrino experiments can be used to detect the collapse of stars in the Galaxy, and finally, a description of how the proposed [71]Ga experiment can be used to decide whether the origin of the present discrepancy between theory and observation lies in our conventional solar models or our conventional physics.

080.082   The horizontal and vertical semi-diameters of the sun observed at the Cape of Good Hope (1834 - 1887) and Paris (1837 - 1906): a report on work in progress.
C. Smith, D. Messina.
Variations of the solar constant, (see 012.037), p. 111 - 116 (1981).

Cape and Paris meridian observations of the solar limbs which permit an estimate to be made of the solar semi-diameter are being surveyed, sampled, and compared with Greenwich and U.S. Naval Observatory observations. Significant systematic errors have been found in the Paris work and have been correlated with changes of instruments and observers. Preliminary results from the more stable Cape series indicate that work should continue on the compilation of data from Cape observations of the sun.

080.083   Eclipse radius measurements.   D. W. Dunham,
J. B. Dunham, A. D. Fiala, S. Sofia.
Variations of the solar constant, (see 012.037), p. 117 - 121 (1981).

The authors have improved methods for predicting the path edges and reducing observations of total solar eclipses for determining variations of the solar radius. Recently-analyzed observations of the 1925 January eclipse show a 0."7 (arc second) decrease in the solar radius during the past fifty years.

080.084   SCLERA *(Santa Catalina Laboratory for Experimental Relativity by Astrometry)* solar diameter
observations.   H. A. Hill, T. P. Caudell, R. J. Bos.
Variations of the solar constant, (see 012.037), p. 123 - 128 (1981).

The recent use of computer-controlled photoelectric devices has led to the systematic removal of observer bias and atmospheric seeing from measurements of the solar diameter, but a new set of problems has surfaced due to variations in the sun itself. Careful attention must be given to the definition of an edge on the solar disk in order to distinguish between observed diameter variations due to physical shape changes and those due to variability in other solar properties which affect the edge definition. Programs which propose to accurately measure intrinsic solar diameters must be prepared to detect changes in other properties of the sun at the extreme limb.

080.085   Solar radius measurements.
T. L. Duvall, Jr., H. P. Jones.
Variations of the solar constant, (see 012.037), p. 129 - 130 (1981).

Preliminary results of solar radius measurements made during 1979 - 1980 are discussed. Variability in the radius measurements of 0.4 π is found, of unknown origin.

080.086   Estimating short-term solar variations by a simple envelope matching technique.
K. L. Chan, S. Sofia.
Variations of the solar constant, (see 012.037), p. 131 - 136 (1981).

A simple matching technique is explained which allows the computation of the response of the solar surface to perturbations which occur at any depth within the convective envelope of the Sun. This technique was applied to a perturbation of the convective efficiency (α-mechanism), and of the non-gas component of the pressure (β-mechanism) in different regions of the convection zone. The results indicate that either perturbation affects the solar luminosity.

080.087   Thermal perturbation of the sun.
L. W. Twigg, A. S. Endal.
Variations of the solar constant, (see 012.037), p. 137 - 142 (1981).

An investigation of thermal perturbations of the solar convective zone via changes in the mixing length parameter have been carried out, with a view toward understanding the possible solar radius and luminosity changes that have been cited in the literature.

080.088   Effects of changes in convective efficiency on the solar radius and luminosity.   A. V. Sweigart.
Variations of the solar constant, (see 012.037), p. 143 - 164 (1981).

A sequence of solar models has been constructed in order to investigate the sensitivity of the solar radius and luminosity to small changes in the ratio α of the mixing length l to the pressure-scale height H_p throughout the solar convective envelope. These calculations gave the following results. (1) A perturbation in α produces immediate changes in the solar radius and luminosity. Initially ΔL and $\Delta\alpha$ are related by $\Delta L/L = 0.30\Delta\alpha/\alpha$, (2) the value of the ratio $W = \Delta\log R/\Delta\log L$ is strongly time dependent, (3) the ratio $H = (\Delta\log L)^{-1} \times d \Delta\log R/dt$ is much less time dependent and is a more suitable means for relating the changes in the solar radius and luminosity, (4) both of these ratios imply that for any reasonable change in the solar luminosity the corresponding change in the solar radius is negligible.

080.089   Evolutionary variations of solar luminosity.
A. S. Endal.
Variations of the solar constant, (see 012.037), p. 175 - 183 (1981).

Theoretical arguments for a 30% increase in the solar luminosity over the past 4.7 billion years are reviewed. The effect of the solar luminosity increase on the terrestrial climate is briefly considered. It appears unlikely that an enhanced greenhouse effect, due to reduced gases (NH_3, CH_4), can account for the long-term paleoclimatic trends.

080.090   Short and long term variations in the "solar constant".   K. H. Schatten.
Variations of the solar constant, (see 012.037), p. 207 - 217 (1981).

Short and long term variations in the solar constant are examined theoretically. The variations observed by the Solar Maximum Mission, lasting several days and associated with the passage of sunspot groups, strikingly demonstrates the well known lack of a "bright ring" effect around sunspots. This suggests that sunspot magnetic fields do not simply block the heat flowing upward into the photosphere. Rather, it is suggested that gravitational draining occurs; this cools sunspots and transports downward the heat that would otherwise flow into the photosphere.

080.091 Solar pulsations and long-term solar variability.
P. R. Goode, J. D. Logan, H. A. Hill.
Variations of the solar constant, (see 012.037), p. 229 - 233 (1981).

The seismology of the solar atmosphere is important in relating changes in luminosity to variations in other observables. This approach has already led to the identification of properties which were not previously observed or recognized. Equally important results from solar seismology are expected in the future.

080.092 Observations of large-scale motions on the Sun.
B. LaBonte.
Variations of the solar constant, (see 012.037), p. 235 - 240 (1981).

Recent observations of large-scale mass motions on the Sun are discussed. The principal large-scale velocity flows are convection, rotation, meridional flow, and torsional and radial oscillations.

080.093 A prospectus for a theory of variable variability.
S. Childress, E. A. Spiegel.
Variations of the solar constant, (see 012.037), p. 273 - 291 (1981).

The authors propose that the kind of stellar variability exhibited by the sun in its magnetic activity cycle should be considered as a prototype of a class of stellar variability. The signature includes long 'periods' (compared to that of the radial fundamental mode), erratic behavior, and intermittency. As other phenomena in the same variability class they nominate the luminosity fluctuations of ZZ Ceti stars and the solar 160^m oscillation. The authors discuss the possibility that analogous physical mechanisms are at work in all these cases, namely instabilities driven in a thin layer.

080.094 The solar structure and the low l five-minute oscillation. I.
M. Gabriel, R. Scuflaire, A. Noels.
Astron. Astrophys., Vol. 50 - 53 (1982).

The parameters X, Z, and l/H_p have been obtained by fitting the theoretical frequencies with the values observed by Claverie et al. (1981). A normal chemical composition is found. To get accurate values for X and Z, the ratio Z/X must be imposed. They are then in good agreement with observations. The mixing length obtained is larger than predicted by standard solar models. The observed rotational splitting favours slow or moderately fast internal rotation laws compatible with Hill and Stebbins oblateness measurements.

080.095 Solar emission lines produced in the wake of a shock wave. II. Line profiles.
D. R. Flower, G. Pineau des Forêts.
Astron. Astrophys., Vol. 110, 163 - 168 (1982).

The authors consider the influence of the passage of a shock wave on the profiles of lines in its wake. The time dependent ionisation equations are solved following a fluid particle which has undergone the shock compression and the profiles of the C IV $\lambda1548.20$ and O V $\lambda1218.35$ lines are evaluated. Comparison of the computed line profiles with existing observations proves inconclusive.

080.096 On a possibility of applying the approximation of single scattering to the study of the solar aureole.
V. M. Prokhorov.
Red. zh. "Izv. AN SSSR. Fiz. atmos. i okeana". Moskva, 1982, 10 pp. In Russian. – Abstr. in Ref. zh., 51. Astron., 6.51.728 (1982).

080.097 The star "Sun". J.-C. Pecker.
Sun and planetary system, (see 012.040), p. 25 - 34 (1982). – Review paper.

080.098 The effect of personal equation on measurements of the solar semi-diameter by transit circle observers.
C. Smith, D. Messina.
Sun and planetary system, (see 012.040), p. 39 - 43 (1982).

Systematic trends in the semi-diameter of the Sun from measurements made with the Cape Observatory transit circle in 1834 and from 1861 to 1871 and 1884 to 1892 (the only years for which raw observations of the solar limbs were available to the authors) should probably be attributed to large variations in the personal equation of the individual observers rather than to any real variation of the solar diameter.

080.099 A direct UBV color measurement of the Sun.
H. Tüg, T. Schmidt-Kaler.
Sun and planetary system, (see 012.040), p. 45 - 46 (1982).

080.100 Nuclear astrophysics of the Sun.
G. E. Kocharov.
Sun and planetary system, (see 012.040), p. 47 - 52 (1982).

An analysis is given of the state of the art and prospects of investigations of high energy processes in solar matter.

080.101 The solar neutrino puzzle. G. Marx.
Sun and planetary system, (see 012.040), p. 53 - 58 (1982).

Experiments indicate a significant solar neutrino flux, but the detection rate is smaller than predicted by the standard solar model. The paper discusses the possible reasons of this discrepancy and the line of future actions.

080.102 Solar models and neutrino problem.
N. Kiziloğlu, D. Ezer.
Sun and planetary system, (see 012.040), p. 59 - 62 (1982).

080.103 The rotation-magnetism-convection coupling in the sun. M. Stix.
Sun and planetary system, (see 012.040), p. 63 - 70 (1982).

080.104 Relationship between photospheric and chromospheric Ω (ϑ, t) deduced by temporarily and spatially correlated tracers. R. A. Zappalà, F. Zuccarello.
Sun and planetary system, (see 012.040), p. 71 - 72 (1982).

In previous works the authors pointed out that solar mean angular velocity measured by tracers (sunspot-groups and K faculae) depends not only on latitude but also on the age of the tracer. On the basis of those results it seemed interesting to analyze the motion of "spatially" and "temporarily" correlated photospheric and chromospheric features, whose ages ranged from 2 - 9 days; thus one could obtain information on the kinematic behaviour of the same young active region in two different layers of the solar atmosphere.

080.105 Line-of sight velocity field of synodic solar rotation.
A. Kubičela, M. Karabin.
Sun and planetary system, (see 012.040), p. 73 - 76 (1982).

The angular velocity of synodic solar rotation has been treated as a vector. A colinear and a perpendicular component with respect to the solar rotation axis has been found. Consequently, a factor cos $7.^{\circ}25$ has been introduced into the apparent effect of the Earth's orbital motion. An alternating term has been also added in the expression of the synodic line-of-sight velocity of an arbitrary photospheric point.

080.106 Solar rotation and meridional motions derived from sunspot groups.
J. Tuominen, I. Tuominen, J. Kyröläinen.
Sun and planetary system, (see 012.040), p. 77 - 78 (1982).

Latitudinal and longitudinal motions of sunspot groups have been studied using the positions of recurrent sunspot groups of 103 years published by Greenwich Observatory. In order to avoid any limb effects, only positions close to the central meridian have been used.

080.107 **A noisy sun?**   J. Perdang.
Sun and planetary system, (see 012.040), p. 89 - 92 (1982).
The author presents theoretical and observational arguments suggesting that the solar oscillations contain a 'chaotic' component. On the basis of experiments on nonlinear mode coupling he conjectures that the structure of the low frequency end of the observed power spectra is indicative of such chaotic motions.

080.108 **Spectral analysis of wave motions in the region of temperature minimum of the sun's atmosphere.**
V. E. Merkulenko, V. I. Polyakov, V. S. Loskutnikov.
Sun and planetary system, (see 012.040), p. 93 - 96 (1982).
The authors present some results of an analysis of the spatial spectrum and two-dimensional distribution of horizontally running waves of five-minute oscillations.

080.109 **Report of IAU Commission 12: Radiation and structure of the solar atmosphere** (Radiation et structure de l'atmosphère solaire).   Y. Uchida.
Trans. IAU, Vol. XVIIIA, (see 003.013), p. 93 - 114 (1982).

080.110 **On the variability of the solar diameter.**
A. D. Wittmann, J. A. Bonet Navarro, H. Wöhl.
The physics of sunspots. Proceedings of the 5th SPO Summer Workshop, held at Sacramento Peak National Observatory, Sunspot, New Mex., December 1981, L. E. Cram, J. H. Thomas (Editors), p. 424 - 433.
The authors have made visual and photoelectric measurements of the solar semidiameter R. The best value from all their observations is R = (960.0 ± 0.1)'', i.e. the canonical value of 959.63'' (Auwers, 1891) is slightly too small. The visual semidiameter R_{vis} = (960.2 ± 0.1)'' is only slightly larger than the photoelectric semidiameter, i.e. the value of 961.18'' adopted by the Astronomical Almanac (Ephemeris) is much too large, and "irradiation" amounts to only a few tenths of a second of arc. There is no evidence of a significant secular variation of the solar semidiameter: dR/dt = (0.000 ± 0.001)''/ yr. The amplitude of any hypothetical variation in phase with the activity cycle (Secchi-Rosa law) is less than 0.2''.

Quantum physics and the stars: Henry Norris Russell and the composition of the solar atmosphere. See Abstr. 004.054.

Detection of solar and cosmic neutrinos by coherent scattering.   See Abstr. 031.532.

Quick matching technique to study the relationship between solar radius and luminosity variations. See Abstr. 031.552.

Post-facto dark current and gain determinations for solar data obtained with a diode array. See Abstr. 031.597.

Observations of the sun at the CERGA astrolabe in 1980.   See Abstr. 041.029.

Extremal tides: rigorous computation for the many-body case.   See Abstr. 042.007.

The Solar Beacon, an experiment in solar astrometry.   See Abstr. 051.015.

On the question of neutrino emission from the Sun. See Abstr. 061.017.

Low energy consequences of intermediate mass neutrinos in π_{12} decays solar neutrino fluxes and neutrino oscillations.   See Abstr. 061.020.

Solar-stellar astrophysics.   See Abstr. 061.023.

Massive neutrinos and the stellar stopping power via the neutrino magnetic moment.   See Abstr. 061.028.

Compressible convection in a rotating spherical shell. V. Induced differential rotation and meridional circulation. See Abstr. 062.028.

On the expulsion of magnetised plasmas through solar and stellar atmospheres.   See Abstr. 062.086.

Nonlinear penetrative convection in a compressible medium.   See Abstr. 062.095.

Electrodynamic coupling in stellar atmospheres. See Abstr. 064.068.

Eine bequeme Methode zur Berechnung stellarer Eigenfrequenzen — Anwendungen auf Sonnenmodelle. See Abstr. 065.027.

Models of stellar evolution and their use in calibrating distances and element abundances of stars. See Abstr. 065.037.

Gravitational energy release and mixing induced by the nuclear energy generation process. See Abstr. 065.051.

Solar luminosity variation. IV. The photospheric lines, 1976 - 1980.   See Abstr. 071.001.

Einfluß von Temperaturinhomogenitäten in der Sonnenatmosphäre auf die Bestimmung von Elementhäufigkeiten.   See Abstr. 071.013.

Nonlinear simulations of solar rotation effects in supergranules.   See Abstr. 071.032.

Observed variability in the Fraunhofer line spectrum of solar flux, 1975 - 1980.   See Abstr. 071.058.

Granulation, supergranulation and atmospheric waves.   See Abstr. 071.060.

Five-minute oscillations as a subsurface probe of sunspot structure.   See Abstr. 072.036.

On the latitude drift of sunspot groups and solar rotation.   See Abstr. 072.085.

On the seat of the solar cycle. See Abstr. 072.088.

Subphotospheric velocity fields inferred by sunspots motions.   See Abstr. 072.092.

Thermodynamic properties of unresolved magnetic flux tubes. I: A diagnostic method based on circular polarization ratios in line pairs.   See Abstr. 075.015.

Surface magnetic fields and the solar luminosity. See Abstr. 075.021.

On the detection of the periodic oscillations of the inclination of the frequency spectrum of the S-component of solar radio emission. See Abstr. 077.060.

Indirect methods for measuring variations of the solar constant. See Abstr. 080.080.

Solar spectral irradiance and atmospheric transmission at Mauna Loa Observatory. See Abstr. 082.070.

Sensitivity of the earth's climate to changes in the solar constant. See Abstr. 085.031.

Monitoring solar-type stars. See Abstr. 113.091.

Earth

081 Structure, Figure, Gravity, Orbit, etc.

081.001 **The object at the centre of the Earth.**
J. M. Herndon.
Naturwissenschaften, 69. Jahrg., 34 - 37 (1982).
The foundation is described for understanding the chemical composition of the inner core, the core and the lower mantle of the Earth.

081.002 **A comparison of Earth models by degree of their harmonic coefficients.**
L. Pospíšilová, J. Klokočník, J. Kostelecký.
Bull. Astron. Inst. Czechoslovakia, Vol. 33, 33 - 36, plates 1 - 4 (1982).
Comprehensive solutions for the Earth's gravitational potential (the Earth models) are compared by means of the power spectra of the harmonic coefficients. A total of 13 Earth models were tested. This method of comparison by degree of the harmonic coefficients complements the comparison by order of them (via the lumped coefficients), which has been performed by the authors in their previous study.

081.003 **Spherical harmonic representation of the gravity field from dynamic satellite data.**
S. M. Klosko, C. A. Wagner.
Planet. Space Sci., Vol. 30, 5 - 28 (1982).

081.004 **Lumped geopotential harmonics of order 29, from analysis of the orbit of Cosmos 837 rocket.**
H. Hiller.
Planet. Space Sci., Vol. 30, 73 - 80 (1982).

081.005 **Comments on "Intercomparisons of earth models by means of lumped coefficients" by Klokočník and Pospíšilová.** K.-R. Koch, B. H. Chovitz.
Planet. Space Sci., Vol. 30, 215 (1982). – See Abstr. 30.081.007.

081.006 **The geoid: effect of compensated topography and uncompensated oceanic trenches.**
C. G. Chase, M. K. McNutt.
Geophys. Res. Lett., Vol. 9, 29 - 32 (1982).

081.007 **Chemical inhomogeneity of the mantle: geochemical considerations.** L.-g. Liu.
Geophys. Res. Lett., Vol. 9, 124 - 126 (1982).

081.008 **The structure of the lowermost mantle determined by short-period P-wave amplitudes.**
L. J. Ruff, D. V. Helmberger.
Geophys. J. R. Astron. Soc., Vol. 68, 95 - 119 (1982).

081.009 **Stable regions in the earth's liquid core.**
D. Gubbins, C. J. Thomson, K. A. Whaler.
Geophys. J. R. Astron. Soc., Vol. 68, 241 - 251 (1982).

081.010 **Excitation of normal modes on non-rotating and rotating earth models.** B. F. Chao.
Geophys. J. R. Astron. Soc., Vol. 68, 295 - 315 (1982).
The method of spectral decomposition for linear operators, formulated in Dirac's bra-ket notation, gives the excitation formulae for the normal modes of infinitesimal oscillations of a non-rotating earth. The formalism is then extended, in parallel with Lancaster's λ-matrix treatment, to obtain the corresponding formulae for a rotating earth. The algebraic structure of these formulae is carefully examined based in Chao's group-theoretical results, and some particular cases of earth models and sources are discussed.

081.011 **Surface waves and free oscillations in a regionalized earth model.** J. H. Woodhouse, T. P. Girnius.
Geophys. J. R. Astron. Soc., Vol. 68, 653 - 673 (1982).
The linearized equation is derived which relates observed long-period seismic waveforms to the aspherical perturbations of a spherically symmetric earth model. This is accomplished by formulating the theory of spectral splitting in the time domain. It is shown to be possible greatly to simplify the resulting equations in a way which makes it apparent that for each modal multiplet the 'scattered' field depends only upon three local functionals of earth structure.

081.012 **Geoid anomalies and mantle convection.**
D. McKenzie.
Geophys. J. R. Astron. Soc., Vol. 69, (see 012.004), 276 (1982). – Abstract.

081.013 **Variational solution of long-period oscillations of the earth.** W. Moon.
Geophys. J. R. Astron. Soc., Vol. 69, 431 - 458 (1982).

081.014 **Normal modes of the viscoelastic earth.**
D. A. Yuen, W. R. Peltier.
Geophys. J. R. Astron. Soc., Vol. 69, 495 - 526 (1982).

081.015 **The interactions of Earth tides and oceanic tides.**
P. Melchior.
Ann Géophys., Vol. 37, 189 - 197 (1981). In French. – Abstr. in Phys. Abstr., Vol. 85, Abstr. 9903 (1982).

081.016 **The geology of Dyngjufjöll Ytri crater, north central Iceland.** D. B. Eppler, M. C. Malin.
Reports of Planetary Geology Program – 1981, (see 003.001), p. 186 - 187 (1981). – Abstract.

081.017 **Formation and evolution of playa ventifacts, Amboy, California.** S. Williams, R. Greeley.
Reports of Planetary Geology Program – 1981, (see 003.001), p. 197 - 199 (1981). – Abstract.

081.018 **Field modeling of the response of various desert surfaces to the long- and short-term effects of the wind – Mars applications.** J. F. McCauley, M. J. Grolier, C. S. Breed, D. J. MacKinnon, G. H. Billingsley.
Reports of Planetary Geology Program – 1981, (see 003.001), p. 238 - 240 (1981). – Abstract.

081.019 **Serrated eolian deposits in China's northwestern deserts and their comparisons to dark splotches on Mars.** F. El-Baz, L. S. Manent.

Reports of Planetary Geology Program – 1981, (see 003.001), p. 241 - 243 (1981). – Abstract.

081.020 **Dune forms in the Great Sand Sea and application to Mars.**   F. El-Baz, M. Mainguet.
Reports of Planetary Geology Program – 1981, (see 003.001), p. 244 - 246 (1981). – Abstract.

081.021 **Eolian processes in Iceland's cold deserts.**
M. C. Malin, D. B. Eppler.
Reports of Planetary Geology Program – 1981, (see 003.001), p. 247 - 250 (1981). – Abstract.

081.022 **Production of fine silt and clay during natural eolian abrasion.**   D. Krinsley, J. R. Marshall,
J. F. McCauley, C. S. Breed, M. J. Grolier.
Reports of Planetary Geology Program – 1981, (see 003.001), p. 251 - 254 (1981). – Abstract.

081.023 **Sapping processes and the development of theatre-headed valleys.**
J. E. Laity, R. S. Saunders.
Reports of Planetary Geology Program – 1981, (see 003.001), p. 280 - 282 (1981). – Abstract.

081.024 **Analysis towards a dynamic origin for the formation of subglacial longitudinal grooving in sediment or bedrock.**   D. E. Thompson.
Reports of Planetary Geology Program – 1981, (see 003.001), p. 297 - 298 (1981). – Abstract.

081.025 **Exhumed topography – a review of some principles.**
D. D. Rhodes.
Reports of Planetary Geology Program – 1981, (see 003.001), p. 316 - 318 (1981). – Abstract.

081.026 **Australian analogs to geomorphic features on Mars.**
V. R. Baker.
Reports of Planetary Geology Program – 1981, (see 003.001), p. 329 - 330 (1981). – Abstract.

081.027 **Oceanic ridges, transforms, trenches would be seen in PV altimetry data – even under Venusian ambient conditions.**   R. E. Arvidson.
Reports of Planetary Geology Program – 1981, (see 003.001), p. 371 (1981). – Abstract.

081.028 **New York-Pennsylvania rock cities: a Martian comparison.**   C. A. Baskerville.
Reports of Planetary Geology Program – 1981, (see 003.001), p. 394 - 398 (1981). – Abstract.

081.029 **Rock and soil mapping and change detection from LANDSAT multispectral scanner data – clues to limits of interpretability from Viking Orbiter color data.**
R. E. Arvidson, P. A. Jacobberger, D. Rashka.
Reports of Planetary Geology Program – 1981, (see 003.001), p. 455 - 456 (1981). – Abstract.

081.030 **Regolith development in Mars-like environments.**
E. K. Gibson, Jr., R. Bustin, S. J. Wentworth.
Reports of Planetary Geology Program – 1981, (see 003.001), p. 463 - 465 (1981). – Abstract.

081.031 **Chemical weathering and diagenesis in a soil profile in Antarctica: implications for the Martian regolith.**
D. S. McKay, S. J. Wentworth, R. V. Morris.
Reports of Planetary Geology Program – 1981, (see 003.001), p. 466 - 468 (1981). – Abstract.

081.032 **Weathering of silicate minerals in Antarctic Dry Valleys: implications for volatile-regolith interac-**

tions on Mars   S. J. Wentworth, D. S. McKay.
Reports of Planetary Geology Program – 1981, (see 003.001), p. 469 - 471 (1981). – Abstract.

081.033 **Seafloor instabilities on continental slopes and Mars analogs.**   D. Nummedal.
Reports of Planetary Geology Program – 1981, (see 003.001), p. 544 - 545 (1981). – Abstract.

081.034 **Crustal evolution of the earth.**
D. M. Shaw.
Meteoritics, Vol. 16, 386 (1981). – Abstract.

081.035 **A review of methods of comparison of 'resonant' lumped geopotential coefficients.**
J. Klokočník.
Bull. Astron. Inst. Czechoslovakia, Vol. 33, 89 - 104, plates 4 - 8 (1982).
The comparisons between the gravitational field characteristics derived solely from resonant phenomena in orbits of close Earth artificial satellites and coefficients reconstituted from comprehensive Earth gravitational field models as well as mutual comparisons of the resonant results involve: the application of the lumped coefficients for variable orbital inclination, of a rough symmetry of the lumped values, and a test of one-day vs two-day resonances. Emphasis is put on the numerical examples, using recent data.

081.036 **A model of Earth's structure inferred from eigen-periods of torsional oscillation. III.**   H. Oda.
Tôhoku Geophys. J. Sci. Rep. Tôhoku Univ. Fifth Ser., Vol. 27, No. 2, p. 57 - 70 (1980). – Abstr. in Phys. Abstr., Vol. 85, Abstr. 18394 (1982).

081.037 **Inviscid, frozen-flux velocity components at the top of Earth's core from magnetic observations. I. A new methodology.**   E. R. Benton.
Geophys. Astrophys. Fluid Dyn., Vol. 18, 157 - 174 (1981). Abstr. in Phys. Abstr., Vol. 85, Abstr. 18399 (1982).

081.038 **On the application of the Pontryagin's maximum principle for identification of the radiogenic heat distribution in the Earth's interior.**   R. N. Singh.
Int. J. Eng. Sci., Vol. 19, 1601 - 1604 (1981). – Abstr. in Phys. Abstr., Vol. 85, Abstr. 22936 (1982).

081.039 **Application of Stokes' formula for temporal gravity variations.**   P. Biro.
Z. Vermessungswes., Vol. 106, 523 - 531 (1981). In German. Abstr. in Phys. Abstr., Vol. 85, Abstr. 37743 (1982).

081.040 **The inverse problem of constructing a gravimetric geoid.**
V. Zlotnicki, B. Parsons, C. Wunsch.
J. Geophys. Res., Vol. 87, 1835 - 1848 (1982).
The authors present a computational procedure based on linear inverse theory for estimating geoidal heights from incomplete sets of data.

081.041 **Geopotential harmonics of order 29, 30 and 31 from analysis of resonant orbits.**
D. G. King-Hele, D. M. C. Walker.
Planet. Space Sci., Vol. 30, 411 - 425 (1982).

081.042 **New earth's gravitational field models (GEM 10B and 10C).**   J. Klokočník.
Říše hvězd, Vol. 63, 94 - 99 (1982). In Czech.

081.043 **On the formulation of the statical equations in the liquid core of the Earth.**
G. X. Song, M. Zhao.
Special issue on astronomical geodynamics, (see 012.028), p.

92 - 99 (1981). In Chinese.

In the SNREI Earth model having a liquid core, the statical equations for deformations of the Earth under external load cannot be solved without assuming the Adams-Williamson condition. In the existing literature it is supposed that the hydrostatic pressure after deformation is composed of the two terms. In this paper a third term in the hydrostatic pressure after deformation is proposed.

081.044 **The fluid core of the Earth and the free nutation of the core.**   C. Z. Zhang.
Special issue on astronomical geodynamics, (see 012.028), p. 100 - 104 (1981). In Chinese.

081.045 **Secular trends in polar motions: a new tool for probing the viscosity of the lower mantle.**
R. Sabadini, E. Boschi, D. A. Yuen.
The comparative study of the planets, (see 012.033), p. 181 - 194 (1982).

Transient flow in the mantle, induced by glacial cycles, produce small but discernible variations in the earth's rotational wobble and in the length of the day. These geophysical observables, which arise solely as a consequence of strain fields endowed with low order spherical harmonics ($l = 0$ and $l = 2$), can be useful in distinguishing the viscosity structure of the lower mantle, potentially more so than inferences drawn from relative sea level and gravity anomaly data, whose spectral contents are dominated by much higher angular orders ($l \gtrsim 6$).

081.046 **Lava flows on Etna, a morphometric study.**
R. Lopes, J. E. Guest.
The comparative study of the planets, (see 012.033), p. 441 - 458 (1982).

081.047 **A petrological model on magma evolution of vulcano eruptive complex (Aeolian Islands – Italy).**
G. Castellet y Ballarà, R. Crescenzi, A. Pompili, R. Trigila.
The comparative study of the planets, (see 012.033), p. 459 - 476 (1982).

081.048 **Tidal observations at the Esashi Earth Tides Station for the year 1979. Annual report of geophysical observations.**
Published by Int. Latitude Obs. Mizusawa, Japan. 95 pp. (1981). ISSN 0579-5958.

081.049 **Isostatic geoid anomalies on a sphere.**
F. A. Dahlen.
J. Geophys. Res., Vol. 87, 3943 - 3947 (1982).

081.050 **On the outer layers of the core and geomagnetic secular variation.**
J.-L. Le Mouël, V. Courtillot.
J. Geophys. Res., Vol. 87, 4103 - 4108 (1982).

The observed correlation between the decade variations in the earth's rotation and geomagnetic secular variation probably can be accounted for by electromagnetic core-mantle coupling as suggested by Bullard long ago. The new observations, combined with a simplified earth model, provide some indications on the thickness of the outer core layers which are responsible for geomagnetic secular variation.

081.051 **Evaluation of 15th-order harmonics in the geopotential from analysis of resonant orbits.**
D. G. King-Hele, D. M. C. Walker.
Proc. R. Soc. London Ser. A, Vol. 379, 247 - 288 (1982).
Abstr. in Phys. Abstr., Vol. 85, Abstr. 57953 (1982).

081.052 **The large-number hypothesis and the Earth's expansion.**   S. Yabushita.
Moon Planets, Vol. 26, 135 - 141 (1982).

The equation of state of the terrestrial material obtained from seismic data is adopted to construct three zone earth models under hypothesis of variable constant of gravity G as proposed by Dirac. Three hypotheses are investigated: variable G without creation, creation such that m (mass) $\propto G^{-1}$, and multiplicative creation, $m \propto G^{-2}$. It is shown that, with the currently accepted value of the Hubble constant, $\dot{R} = 7 \times 10^{-3}$ cm yr^{-1}, $\dot{R} = 1.9 \times 10^{-2}$ cm yr^{-1} and $\dot{R} = 2.5 \times 10^{-2}$ cm yr^{-1} for each hypothesis.

081.053 **Activities of astro-geodynamics research in China.**   S.-H. Ye.
High-precision Earth rotation and Earth-Moon dynamics. Lunar distances and related observations, (see 012.035), p. 181 - 188 (1982).

In recent years, efforts of astro-geodynamics research in China have mainly been concerned with the exploration of the regularity and mechanism of the rotational and crustal motion of the Earth as well as with the possible relationships between astronomical factors and earthquakes. In the meantime, observational devices are under modification and some new techniques have been established.

081.054 **Degree variances of the earth's potential, topography and its isostatic compensation.**   R. H. Rapp.
Bull. Géod., Vol. 56, 84 - 94 (1982).

A spherical harmonic expansion of the earth's gravitational potential and equivalent rock topography to degree and order 180 is described.

081.055 **Marées terrestres.**   P. Melchior (Editor).
Bull. Inf., (Obs. R. Belgique, Bruxelles), No. 87, p. 5566 - 5624 (1982).

081.056 **Terra e luna.**   M. Poscolieri.
Coelum, Vol. 50, 113 - 123 (1982).

081.057 **Ascending droplets in the Earth's core.**   S. Franck.
Phys. Earth Planet. Inter., Vol. 27, 249 - 254 (1982) = Mitt. Zentralinst. Phys. Erde, Nr. 1025.

This paper is concerned with some new problems of the dynamics and energetics of the Earth's core. The model of the so-called gravitationally-powered dynamo is investigated under the assumption of liquid immiscibility in the Fe—S system as a possible core material. In this way the growing inner core causes nucleation of small FeS-droplets that ascend under the release of gravitational potential energy. This energy is enough to drive a dynamo with a toroidal magnetic field of mean size.

The evolving Earth.   See Abstr. 003.028.

Evolution of the earth.
See Abstr. 003.039.

Physics of the Earth's interior. Proceedings of the 78th course 'Enrico Fermi'.   See Abstr. 012.060.

Love number k determined from astronomical observations of the Earth's rotation.   See Abstr. 044.005.

Dependence of the lunisolar perturbations in the Earth rotation on the adopted Earth model.
See Abstr. 044.051.

On the direct influence of the planets on the precession and nutation of the Earth's axis of rotation.
See Abstr. 045.001.

Biases in pole position computed from data from different Navy Navigation Satellites.   See Abstr. 045.031.

High precision tracking of geosynchronous satellites: oceanographic applications.   See Abstr. 046.012.

Optical tracking of synchronous satellites for geophysical purposes.   See Abstr. 046.013.

Gravimetric geoid determination for the area of Greece.   See Abstr. 046.014.

Pole-strength of the earth from MAGSAT and magnetic determination of the core radius. See Abstr. 084.092.

Coincidence of some magnetic and gravity field characteristics.   See Abstr. 084.104.

Geology and geophysics. See Abstr. 093.005.

Results from lunar laser ranging data analysis. See Abstr. 094.050.

Deep gravitational CREEP deformation: earth analogue of a Mars chaos area.   See Abstr. 097.053.

082 Atmosphere (Refraction, Scintillation, Extinction, Airglow, Site Testing)

082.001 **A new interpretation of the 1304 Å triplet airglow intensity ratio with the fine structure levels $O(^3P_J)$ in local thermodynamic equilibrium.** T. Ogawa.
Planet. Space Sci., Vol. 30, 39 - 44 (1982).

The intensity ratios of the 1304 Å triplet airglow of atomic oxygen observed by Fastie and Crosswhite (1964) are interpreted on the basis of the radiative transfer formulation for a model with complete frequency redistribution in a Voigt line profile. A model for the fine structure levels $O(^3P_J)$ in local thermodynamic equilibrium is favorable to the observed intensity ratios, as far as a Voigt profile is applicable. In view of large cross sections as calculated theoretically by Allison and Burke (1969), the mutual relaxation among the ^3P_J levels should occur rapidly enough to allow the population in the ^3P_J levels to be in thermodynamic equilibrium with the ambient neutral gases.

082.002 **Excessive absorption by atmospheric water vapor in the infrared and $5-17$ cm^{-1} regions.**
H. R. Carlon.
J. Atmos. Terr. Phys., Vol. 44, 19 - 23 (1982).

Measurements of 'excessive' absorption by atmospheric water vapor, i. e. absorption in excess of water monomer absorption and droplet scattering predictions, are analyzed for the infrared and $5-17$ cm^{-1} regions. Both fair weather and fog data are considered.

082.003 **Magnetic activity (K_p) and quasi-periodic variations of intensity of the oxygen green and red lines.**
K. Misawa, I. Takeuchi.
J. Atmos. Terr. Phys., Vol. 44, 179 - 182 (1982).

082.004 **Anomalous atmospheric spectral features between 300 and 310 nm interpreted in light of new ozone absorption coefficient measurements.**
R. D. McPeters, A. M. Bass.
Geophys. Res. Lett., Vol. 9, 227 - 230, with a correction p. 498 (1982).

Continuous scan data from the solar backscattered ultraviolet instrument on Nimbus 7, after known scattering and ozone absorption effects have been subtracted, reveals real structure in the atmospheric albedo between 300 and 310 nm, a region in which spectral anomalies have been reported in ground based observations. The authors find that these spectral anomalies are largely explained as structure at the one to five percent level in the ozone absorption coefficient as measured by Bass and Paur.

082.005 **Photometric characteristics of the night sky on the Crimea.** V. M. Lyutyj, A. S. Sharov.
Astron. Zh., Tom 59, 174 - 181 (1982). In Russian. English translation in Soviet Astron., Vol. 26, No. 1.

The results of investigation of the night sky brightness and colors are given. The observations were carried out at the Crimean Astrophysical Observatory and the Southern Station of the Sternberg Astronomical Institute.

082.006 **Weather at Boyden Observatory from 1952 to 1980.**
A. H. Jarrett.
Irish Astron. J., Vol. 14, 212 - 215 (1980).

082.007 **Separation of the atmospheric extinction into components.**
T. V. Smolyaninova, G. A. Terez, Eh. I. Terez.
Simferop. univ., Simferopol', 1981. 26 pp. In Russian. − Abstr. in Ref. zh., 51. Astron., 2.51.797 (1982).

082.008 **Spectral atmospheric transparency at the Astrophysical Observatory of the Kishinev University.** I. M. Naku, V. A. Chernobaj.
Kishinev. univ., Kishinev, 1981. 19 pp. In Russian. − Abstr. in Ref. zh., 51. Astron., 2.51.798 (1982).

082.009 **Absorption of infrared radiation by atmospheric water vapor in the region 4.3 to 5.5 microns: preliminary measurements.** A. Ben-Shalom, A. D. Devir, S. G. Lipson, U. P. Oppenheim.
Opt. Eng., Vol. 20, 746 - 748 (1981). − Abstr. in Phys. Abstr., Vol. 85, Abstr. 10165 (1982).

082.010 **The evolution of primordial and radiogenic noble gases from the Earth.** G. Turner.
Meteoritics, Vol. 16, 394 - 395 (1981). − Abstract.

082.011 **Lower atmosphere and solar seeing: an experiment of simultaneous measurements of nearby turbulence by thermal radiosondes, by angle of arrival statistics and image motion observation.**
J. Borgnino, G. Ceppatelli, G. Ricort, A. Righini.
Astron. Astrophys., Vol. 107, 333 - 337 (1982).

The site "Roque de Los Muchachos" on the island of San Miguel de la Palma has been tested using optical and in situ measurements. Estimations of Fried's parameter r_0 have been obtained from the variance of the angle-of-arrival fluctuations observed with a 10-cm telescope. These optical values have been found in good agreement with the results obtained from in situ soundings performed up to an altitude of about 700 m. r_0 estimations have been also deduced from image motion measurements and a comparison of the results has shown that the two optical monitors in use give coherent measurements.

082.012 **Evaluation of aerosol-share on the absorption of direct solar radiation.** J. Lukac.
Contrib. Geophys. Inst. Slovak Acad. Sci. Ser. Meteorol., Vol. 3, 7 - 13 (1981). In Russian. − Abstr. in Phys. Abstr., Vol. 85, Abstr. 14600 (1982).

082.013 **An efficient method for computing the absorption of solar radiation by water vapor.**
M.-D. Chou, A. Arking.
J. Atmos. Sci., Vol. 38, 798 - 807 (1981). − Abstr. in Phys. Abstr., Vol. 85, Abstr. 14601 (1982).

082.014 **Water vapour absorption in the $3.5-4.2$ μm atmospheric window.** I. J. Barton.
Q. J. R. Meteorol. Soc., Vol. 107, 967 - 972 (1981). − Abstr. in Phys. Abstr., Vol. 85, Abstr. 23112 (1982).

082.015 **Short-period oscillations of intensity and rotational temperature of the OH (6−2) band.**
I. Takeuchi, K. Misawa.
Ann. Géophys., Vol. 37, 315 - 319 (1981). − Abstr. in Phys. Abstr., Vol. 85, Abstr. 23153 (1982).

082.016 **UV radiation from the young sun and oxygen and ozone levels in the prebiological palaeoatmosphere.**
V. M. Canuto, J. S. Levine, T. R. Augustsson, C. L. Imhoff.
Nature, Vol. 296, 816 - 820 (1982).

UV measurements of young T-Tauri stars, resembling the sun at an age of a few million years, have recently been made with the International Ultraviolet Explorer. They indicate that young stars emit up to 10^4 times more UV than the present sun. The implications for the origin and evolution of O_2 and O_3 in the prebiological palaeoatmosphere are presented here. The results of photochemical calculations indicate that the O_2 surface mixing ratio was a factor 10^4-10^6 times greater than the standard value of 10^{-15}. This new value reconciles the simultaneous existence of oxidized iron and reduced uranium.

082.017 Untersuchung von Referenzlinien zur solaren
Dopplermessung.   U. Thiele, H. Wöhl.
Mitt. Astron. Ges., Nr. 55, p. 93 (1982). – Abstract.

082.018 An approximation to multiple scattering in the
Earth's atmosphere: almucantar radiance formulation.   M. A. Box, A. Deepak.
J. Atmos. Sci., Vol. 38, 1037 - 1048 (1981). – Abstr. in Phys. Abstr., Vol. 85, Abstr. 28200 (1982).

082.019 A fast line-by line method for atmospheric absorption computations: the Automatized Atmospheric
Absorption Atlas.   N. A. Scott, A. Chedin.
J. Appl. Meteorol., Vol. 20, 802 - 812 (1981). – Abstr. in Phys. Abstr., Vol. 85, Abstr. 28204 (1982).

082.020 Optical C_N^2 remote sensing in the upper atmosphere by multidimensional analysis of stellar scintillation.
J. Vernin, M. Azouit.
Special topics in optical propagation, (see 012.013), p. 21/1 - 9 (1981). – Abstr. in Phys. Abstr., Vol. 85, Abstr. 32647 (1982).

082.021 Infrared radiance model of the upper atmosphere.
T. C. Degges, H. J. P. Smith.
Special topics in optical propagation, (see 012.013), p. 23/1 - 14 (1981). – Abstr. in Phys. Abstr., Vol. 85, Abstr. 32699 (1982).

082.022 Theoretical model construction of the earth's
atmosphere.   M. Ya. Marov.
Nablyud. Iskusstv. Nebesn. Tel, No. 79, (see 012.016), p. 5 - 8 (1981). In Russian.

082.023 Simulation of atmospheric nonstationary physicochemical processes.   G. I. Zmievskaya.
Nablyud. Iskusstv. Nebesn. Tel, No. 79, (see 012.016), p. 9 - 12 (1981). In Russian.

082.024 Correlation between densities of the upper atmosphere determined from satellite drag data and red
oxygen emission intensity.
Yu. L. Truttse, V. D. Belyavskaya.
Nablyud. Iskusstv. Nebesn. Tel, No. 79, (see 012.016), p. 13 - 23 (1981). In Russian.
The authors have investigated the correlation between night values of the red oxygen emission intensity of λ 6300 Å and the density of the neutral atmosphere obtained from satellite tracking data. The coefficient of correlation between them is equal to 0.7 under quiet geomagnetic conditions.

082.025 On the calculation of air density values in the upper
layers of the atmosphere.
A. M. Fominov, L. L. Filenko.
Nablyud. Iskusstv. Nebesn. Tel, No. 79, (see 012.016), p. 65 - 66 (1981). In Russian.

082.026 Temporal and latitudinal 5577 Å airglow variations.
L. L. Cogger, R. D. Elphinstone, J. S. Murphree.
Canadian J. Phys., Vol. 59, 1296 - 1307 (1981). – Abstr. in Phys. Abstr., Vol. 85, Abstr. 38014 (1982).

082.027 The effect of the ionospheric plasma on general
circulation in the upper atmosphere. II.
A. G. Khantadze, A. I. Gvelesiani.
Geomagn. Aehron., Tom 22, 66 - 69 (1982). In Russian.

082.028 The escape of energetic albedo particles from the
atmosphere into cosmic space.   N. L. Grigorov.
Geomagn. Aehron., Tom 22, 197 - 204 (1982). In Russian.

082.029 On the nature of global atmospheric aerosols.
S. P. Golenetskij, S. G. Malakhov, V. V. Stepanok.
Astron. vestn., Tom 15, 226 - 233 (1981). In Russian. – Abstr. in Ref. zh., 51. Astron., 3.51.203 (1982).

082.030 An analytical model for air density in the region
from 160 to 2000 km of CIRA 1972.
B.-k. Lu, C.-a. Sun.
Acta Astron. Sinica, Vol. 22, 336 - 345 (1981). In Chinese.

082.031 Wind effect on the observational results with a
Danjon astrolabe at Wuchang Time Observatory.
Y. Sun.
Acta Geod. Geophys., No. 3, p. 57 - 64 (1981). In Chinese.

082.032 Estudio de las condiciones de visibilidad para la
observación solar en algunos puntos del país.
A. L. Peretti Hollemaert, E. A. Marquevich, J. M. Fontenla,
R. C. Estol.
Bol. Asoc. Argentina Astron., No. 18, p. 82 - 85 (1980).

082.033 Self-absorption of the N_2 Lyman-Birge-Hopfield
bands in the far ultraviolet dayglow.
R. R. Conway.
J. Geophys. Res., Vol. 87, 859 - 866 (1982).
Absorption by N_2 of the Lyman-Birge-Hopfield (LBH) bands in the far ultraviolet dayglow has been calculated using a line-by-line synthesis of individual bands. A band transmission function which can be used in the radiative transfer theory of any band in the system is presented for rotational and kinetic temperatures from 200 to 1000 K. Analysis of rocket observations of the LBH emissions in the 1250–1400 Å region of the dayglow is performed.

082.034 The direct and scattered solar flux within the
stratosphere.   J. R. Herman, J. E. Mentall.
J. Geophys. Res., Vol. 87, 1319 - 1330 (1982).
Solar ultraviolet flux data were obtained within the atmosphere by using Fastie-Ebert double monochromators carried on a balloon-borne gondola and a rocket payload. Both the direct and scattered components of the solar ultraviolet flux at wavelengths from 190 to 320 nm were measured at the balloon float altitude of 40 km. The nearly identical spectrometer carried on the rocket flight measured the direct solar flux from 60 to 38 km during a parachute descent. The ozone column content above 40 km and the temperature profile and ozone density below 40 km are deduced from the scattered and direct solar flux components.

082.035 A global model of the neutral thermosphere in
magnetic coordinates based on OGO 6 data.
C. G. Stehle, J. S. Nisbet, E. Bleuler.
J. Geophys. Res., Vol. 87, 1615 - 1622 (1982).

082.036 Airglow measurement looking downward from
orbit at selected darker fields of view.   T. A. Croft.
J. Geophys. Res., Vol. 87, 1669 - 1675 (1982).
The practicability of performing useful observations of airglow near nadir is examined by study of digital records from the DMSP visual scanner and comparable records from the VAE instrument on AE-C.

082.037 The association of visible airglow features with a
gravity wave.   E. B. Armstrong.
J. Atmos. Terr. Phys., Vol. 44, 325 - 336 (1982).
A faintly visible pattern was observed by eye in the northern region (towards the equator) of the clear, moonless sky on one night. Photographic and photometric observations

indicated that the pattern was not due to auroral activity, but was probably caused by the effect of a gravity wave, launched from the troposphere, on the OI 557.7 nm airglow emission. A period of 57 min, a wavelength of 244 km and a speed of 72 m s^{-1} were obtained from the photographs. Possible ways in which the gravity wave produced the pattern are considered.

082.038 New emissions of the upper atmosphere as a consequence of the anthropogenic action on the ionosphere.   V. I. Krasovskij, Z. Ts. Rapoport, A. I. Semenov.
Kosm. Issled., Tom 20, 237 - 243 (1982). In Russian.

082.039 Resultados preliminares en el estudio de la visibilidad solar.
A. L. Peretti Hollemaert, R. C. Estol, E. A. Marquevich.
Bol. Asoc. Argentina Astron., No. 20 - 24, p. 81 - 86 (1981).

082.040 Comparación entre los valores de la refracción dados por las tablas de Pulkovo y por la serie de Laplace.   G. E. Crotti.
Bol. Asoc. Argentina Astron., No. 20 - 24, p. 385 (1981). Abstract.

082.041 Modificaciones introducidas en el baño de mercurio del círculo meridiano Repsold 1907 del OAUNLP y en el calculo de la refracción para la reducción de las observaciones.   G. E. Crotti.
Bol. Asoc. Argentina Astron., No. 20 - 24, p. 409 (1981). Abstract.

082.042 Estudio de las condiciones observacionales en El Leoncito.   J. Sanguin.
Bol. Asoc. Argentina Astron., No. 20 - 24, p. 415 (1981). Abstract.

082.043 Research of the atmosphere's transparency with an infrared nonlinear up-converter.
Yu. K. Melik-Alaverdyan, R. A. Muradyan, A. N. Fradkin.
Soobshch. Byurakan Obs., Vyp. 53, p. 112 - 117 (1982). In Russian.
Measurements of atmospheric transparency in the infrared region (1.6 - 2.1 μ) in Byurakan and Selim Station were made. The content of water vapor in the atmosphere in terms of condensed water is determined.

082.044 On variations of the atmospheric extinction in Byurakan.   R. S. Asatryan, G. Kh. Khachatryan, Eh. M. Fajnberg, Zh. V. Khachatryan, G. A. Pogosyan.
Soobshch. Byurakan Obs., Vyp. 53, p. 118 - 123 (1982). In Russian.
This paper presents the results of atmospheric extinction variation measurements during one observational night in the V–system.

082.045 Solar site-testing campaign of JOSO on the Canary Islands in 1979.   P. N. Brandt, H. Wöhl.
Astron. Astrophys., Vol. 109, 77 - 89 (1982) = Mitt. Kiepenheuer-Inst. Nr. 210.
The JOSO effort since 1968 to find an excellent site for solar observations culminating in a 8 month testing campaign in 1979 on the Canary Islands of Tenerife and La Palma is described.

082.046 The effects of atmospheric turbulence in optical astronomy.   F. Roddier.
Progress in optics, Vol. 19, (see 003.004), 283 - 376 (1981).

082.047 An analysis of the OI 1304 A dayglow using a Monte Carlo resonant scattering model with partial frequency redistribution.   R. R. Meier, J.-S. Lee.
Planet. Space Sci., Vol. 30, 439 - 450 (1982).
A Monte Carlo model of the atomic oxygen 1304 A airglow triplet has been developed which accurately describes the transport of resonance radiation under very optically thick conditions. Partial frequency redistribution, temperature gradients, pure absorption, and multilevel scattering are accounted for properly. Analysis of a recent rocket experiment which observed the 1304 A dayglow shows that all features of the data can be explained by photoelectron impact excitation and resonant scattering of sunlight.

082.048 Line shape of the non-thermal 6300 Å O(^{1}D) emission.
G. A. Schmitt, V. J. Abreu, P. B. Hays.
Planet. Space Sci., Vol. 30, 457 - 461 (1982).
The line shape of the non-thermal O(^{1}D) 6300 Å emission is calculated using the two population model of Schmitt, Abreu and Hays (1981). The calculated line shapes simulate observations made from a space platform at different zenith angles and altitudes. The non-thermal line shapes observed at zenith angles other than the local vertical have been obtained by using the Addition theorem for spherical harmonics of a Legendre polynomial expansion of the non-thermal population distribution function.

082.049 On the interaction of auroral protons with the earth's atmosphere.   M. H. Rees.
Planet. Space Sci., Vol. 30, 463 - 472 (1982).

082.050 Changes in the concentration of mesospheric ozone during the total solar eclipse.
V. V. Agashe, S. M. Rathi.
Planet. Space Sci., Vol. 30, 507 - 513 (1982).
During the total solar eclipse of 16 February 1980, ground based measurements of dayglow radiance of $O_2(^1\Delta_g)$ band at 1270 nm were obtained. These data have been analysed to estimate the changes in mesospheric ozone concentration during the eclipse.

082.051 Submillimeter atmospheric transmission in Shorbulak at the Eastern Pamirs.
G. B. Sholomitskij, I. A. Maslov, V. M. Grozdilov.
Astron. Zh., Tom 59, 594 - 599 (1982). In Russian. English translation in Soviet Astron., Vol. 26, No. 3.

082.052 Results of a pilot program to measure light pollution.
A. R. Upgren, T. E. Armandroff, D. W. Dawson.
Bull. American Astron. Soc., Vol. 13, 813 (1981). – Abstract.

082.053 On the temperature inversion of the thermosphere.
A. G. Kolesnik, S. S. Korolev, S. G. Pasynkov.
Geomagn. Aehron., Tom 22, 435 - 439 (1982). In Russian.

082.054 On the nature of wave variations of night luminescence of hydroxyl in the upper atmosphere.
N. M. Gavrilov, V. A. Yudin.
Geomagn. Aehron., Tom 22, 444 - 449 (1982). In Russian.

082.055 A global study of $O_2(^1\Delta_g)$ airglow: day and twilight.   J. F. Noxon.
Planet. Space Sci., Vol. 30, 545 - 557 (1982).
Aircraft measurements of $O_2(^1\Delta)$ emission made over a 10-yr period provide information on the variation of ozone with latitude and season in the altitude region 50-90 km.

082.056 Dayglow emissions of the O_2 Herzberg bands and the Rayleigh backscattered spectrum of the earth.
J. E. Frederick, R. B. Abrams.
Planet. Space Sci., Vol. 30, 575 - 580 (1982).
The authors examine the contribution of the Herzberg bands to the ultraviolet spectrum of the earth, with the goal of determining their potential significance as a contaminant in backscatter ultraviolet measurements.

082.057   Soft X-rays from the sunlit earth's atmosphere.
D. L. McKenzie, H. R. Rugge, P. A. Charles.
J. Atmos. Terr. Phys., Vol. 44, 499 - 508 (1982).
The HEAO-1 A-2 experiment low energy proportional counters have been used to measure the X-ray spectrum of the sunlit earth in the energy range 0.2–0.8 keV. The X-rays arise by coherent scattering of, or fluorescence of atmospheric constituents by, solar coronal X-rays incident on the atmosphere. The observed spectra were compared with calculations in order to derive the coronal temperature and emission measure, parameters that characterize the incident solar spectrum.

082.058   Field aligned airglow observations of transequatorial bubbles in the tropical F-region.   B. A. Tinsley.
J. Atmos. Terr. Phys., Vol. 44, 547 - 557 (1982).

082.059   Simplified computation of synthetic spectra in the far-infrared.   P. Rabache, B. Rebours.
Infrared Phys., Vol. 22, 1 - 7 (1982).
A simplified computation method is suggested for determining the spectral luminance from the stratosphere, observed in the far-infrared. The spectral range concerned lies between 10 and 200 cm^{-1}. Observations are conducted from balloon and satellite altitudes and are aimed at the terrestrial limb. The calculations are carried out for the same height and sight angle conditions as in recent experiments. The synthetic spectra and measured spectra are compared. The computation parameters are optimized and the computation times are discussed.

082.060   Determination of atmospheric precipitable water vapour and turbidity parameters from diurnal infrared hygrometer and turbidimeter data.
A. Borghesi, E. Bussoletti, G. Falcicchia, A. Minafra.
Infrared Phys., Vol. 22, 149 - 155 (1982).
The authors present the preliminary results of a site analysis performed in Italy to identify places suitable for the installation of i.r. telescopes. The experimental arrangement consists of a near-i.r. photometer and a Volz turbidimeter mounted on an equatorial base controlled by a sun follower. Day-time atmospheric water vapour content and turbidity parameters have been measured. An empirical method which uses the dew-point temperature has been adopted to evaluate the night-time atmospheric vapour content.

082.061   The contribution of singly scattered photons to the optically thick resonance radiation field of the helium geocorona.   H. J. Fahr, T. Smid.
J. Geophys. Res., Vol. 87, 2487 - 2499 (1982).
Analysis of the geocoronal He-I 58.4-nm resonance radiation field with rocket borne gas absorption cell photometers has revealed singly scattered photons to attain steadily increasing importance in the radiation signals if photons close to the line center are effectively suppressed. In order to confirm this quantitatively, the authors have reinvestigated the radiation transport problem and have given the spectral intensity in terms of contributions from different scattering orders.

082.062   Stability of ammonia in the primitive terrestrial atmosphere.   J. F. Kasting.
J. Geophys. Res., Vol. 87, 3091 - 3098 (1982).

082.063   Stratospheric measurements of collision-induced absorption by molecular oxygen.
C. P. Rinsland, M. A. H. Smith, R. K. Seals, Jr., A. Goldman, F. J. Murcray, D. G. Murcray, J. C. Larsen, P. L. Rarig.
J. Geophys. Res., Vol. 87, 3119 - 3122 (1982).
Collision-induced absorption by the fundamental vibration-rotation band of O_2 has been studied in high-resolution (0.02 cm^{-1}) stratospheric solar absorption spectra. The data were recorded during sunset with the University of Denver

balloon-borne interferometer from a float altitude of 33 km. The O_2 continuum has been identified in the 1400-1700 cm^{-1} region in spectra obtained at tangent altitudes below 22 km. Measurements of transmittances in narrow intervals nearly free of atmospheric line absorption are in good agreement with values calculated with the O_2 absorption coefficients as compiled by Timofeyev and Tonkov (1978). Absorption by other atmospheric species contributes only a small fraction of the total absorption at the measured frequencies.

082.064   Solar irradiance in the stratosphere: implications for the Herzberg continuum absorption of O_2.
J. E. Frederick, J. E. Mentall.
Geophys. Res. Lett., Vol. 9, 461 - 464 (1982).
The third Solar Absorption Balloon Experiment performed measurements of the attenuated solar irradiance between 200 and 210 nm as the payload ascended from 32 to 39 km. Comparison of these data with calculations based on ozone and pressure measurements made from the payload shows more solar radiation reaching the middle stratosphere than is predicted by O_2 and O_3 cross sections which are widely used in photochemical modeling. Consideration of the uncertainties in the balloon data and laboratory experiments indicates that the discrepancy can best be removed by adopting Herzberg continuum cross sections for O_2 which are equal to or slightly less than the smallest values which have appeared in the literature.

082.065   Rocket observation of the NII 2143Å dayglow.
C. A. Barth, R. E. Steele.
Geophys. Res. Lett., Vol. 9, 559 - 561 (1982).
The NII 2143Å doublet has been observed in the earth's dayglow with a rocket spectrometer at an altitude of 200 km. The slant column emission rate of this airglow is 333 ± 67 Rayleighs at a zenith angle of 87.6°. The likely excitation mechanisms are photoionization of molecular nitrogen by solar extreme ultraviolet radiation and photoelectron impact excitation of molecular nitrogen.

082.066   The evolution of the earth's atmosphere and oceans.
C. E. Melton, A. A. Giardini.
Geophys. Res. Lett., Vol. 9, 579 - 582 (1982).
A model describing the evolution of the present atmosphere and hydrosphere has been developed.

082.067   Atlas of stratospheric IR absorption spectra.
A. Goldman, R. D. Blatherwick, F. J. Murcray, J. W. VanAllen, F. H. Murcray, D. G. Murcray.
Appl. Opt., Vol. 21, 1163 - 1164 (1982).

082.068   Measurements of the wavelength dependence and other properties of stellar scintillation at Mauna Kea, Hawaii.   J. C. Dainty, B. M. Levine, B. J. Brames, K. A. O'Donnell.
Appl. Opt., Vol. 21, 1196 - 1200 (1982).
The variance of intensity of stellar scintillation has been measured as a function of wavelength using photon counting and on-line digital analysis techniques. The experimental data are consistent with that predicted by the theory of Tatarski. Measurements of the temporal correlation function of intensity and the higher moments of the probability density function of scintillation are also described. Time scales in the 1.7–10-msec range were observed, and the observed higher moments were consistently lower than those predicted by a log normal distribution. All the measurements were made at Mauna Kea Observatory, Hawaii.

082.069   Variable-wavelength solar-blind Raman lidar for remote measurement of atmospheric water-vapor concentration and temperature.
K. Petri, A. Salik, J. Cooney.
Appl. Opt., Vol. 21, 1212 - 1218 (1982).

082.070 **Solar spectral irradiance and atmospheric transmission at Mauna Loa Observatory.**
G. E. Shaw.
Appl. Opt., Vol. 21, 2006 - 2011 (1982).

082.071 **Negative ions in the disturbed stratosphere and their influence on ozone.** S. I. Kozlov.
Kosm. Issled., Tom 20, 399 - 406 (1982). In Russian.

082.072 **On internal gravitational waves in the atmosphere from a jet stream-type source.**
A. I. Grachev, S. V. Zagorujko, A. K. Matveev,
M. I. Mordukhovich, E. P. Chunchuzov.
Polyar. siyaniya i svechenie noch. neba. Moskva, 1981, No. 29, p. 80 - 83. In Russian. – Abstr. in Ref. zh., 62. Issled. kosm. prostranstva, 6.62.448 (1982).

082.073 **Aerosol stratification of the stratosphere and mesosphere.** G. G. Mateshvili.
Sun and planetary system, (see 012.040), p. 213 - 214 (1982).

082.074 **Problems of three-dimensional refraction in astrometry.** G. Teleki, J. Saastamoinen.
Sun and planetary system, (see 012.040), p. 455 - 462 (1982).
  After discussion of the basic questions, it is demonstrated that it is possible – and indeed necessary – to determine spatial refractional changes appearing in the classical astrometric observations at larger zenith distances. Making use of the local corrections of this kind to high precision astrometric observations is indispensable. Suggestions concerning further researches and application are given. A new global atmospheric model is presented.

082.075 **The use of Lidar to obtain three-dimensional refraction data.** J. A. Hughes, S. DeLateur.
Sun and planetary system, (see 012.040), p. 463 - 470 (1982).
  The need for a real time (or near real time) detailed knowledge of the atmospheric structure necessary for determining astronomical refraction in general is exhibited. Examples involving water vapor and isopycnic tilts are given. The possibility of determining the necessary parameters by means of active atmospheric probing using Light Detection and Ranging (Lidar) methods is reviewed.

082.076 **Astronomical refraction calculated from aerological data in Japan.** R. Fukaya, H. Yasuda.
Sun and planetary system, (see 012.040), p. 475 - 476 (1982).

082.077 **Experimental model for diurnal astronomical refraction.** A. Poma, E. Proverbio, S. Mancuso.
Sun and planetary system, (see 012.040), p. 477 - 482 (1982).
  A technique for the determination of an empirical model of diurnal refraction based on the measure of the effects of differential refraction of the solar disk is presented. Preliminary results are compared with those calculated from the model of standard atmosphere.

082.078 **Astrometric site selection.** G. Teleki.
  Sun and planetary system, (see 012.040), p. 483 - 491 (1982).
  The requirements for the most convenient astrometric observing sites are discussed. Among properties of a good astrometric observing site, besides those demands set forth for the astrophysical sites, there must be also: the maximum stability of the ground, of the plumb line and of the atmospheric layers of constant refractivity; as homogeneous as possible temperature field around the instruments and pavilions; clear days during the year distributed as evenly as possible. The author comes forward with the suggestion bearing on the most favourable locations for the best astrometric observations.

082.079 **Disregard of some ray tracing principles in optical and radio spectral bands.** E. Woyk (*Chvojková*).
Sun and planetary system, (see 012.040), p. 497 - 498 (1982).

082.080 **Refraction effects on geodetic measurements in three-dimensional terrestrial nets.** L. Hradilek.
Sun and planetary system, (see 012.040), p. 499 - 502 (1982).
  The biasing influence of terrestrial refraction on geodetic three-dimensional nets can be substantially reduced by an appropriate design of the observational procedures complemented by relevant mathematical models expressing refraction and statistical tests eliminating major bias.

082.081 **Experimental investigation of refraction above water crossings.** V. S. Milovanović.
Sun and planetary system, (see 012.040), p. 503 - 504 (1982).

082.082 **Atmospheric turbulence and its effects on direction measurements.** F. K. Brunner.
Sun and planetary system, (see 012.040), p. 505 - 510 (1982).
  Optical direction measurements to terrestrial and extraterrestrial targets are affected by atmospheric turbulence. The known formulae for the variance and the spectrum of the angle-of-arrival fluctuations and experimental results are summarised.

082.083 **Terrestrial refraction and vertical temperature gradient.** L. N. Mavridis.
Sun and planetary system, (see 012.040), p. 519 - 522 (1982).
  Current work on the determination of the coefficient of terrestrial refraction and its diurnal and seasonal variation is reviewed. For the computation of the refractive effects on the basis of meteorological measurements the vertical temperature gradient is needed. Recent empirical determinations of this gradient and corresponding theoretical atmospheric models are discussed.

082.084 **Terrestrial refraction and vertical temperature gradient in the area of Thessaloniki.**
A. I. Gounaris, L. N. Mavridis, A. L. Papadimitriou.
Sun and planetary system, (see 012.040), p. 523 - 528 (1982).

082.085 **A study of the refractivity N of the air in the area of Athens.** P. Savaidis.
Sun and planetary system, (see 012.040), p. 529 - 530 (1982).

082.086 **Report of IAU Commission 21: Light of the night sky (Lumière du ciel nocturne).** H. Tanabe.
Trans. IAU, Vol. XVIIIA, (see 003.013), p. 211 - 218 (1982).

082.087 **Report of IAU Commission 50: Identification and protection of existing and potential observatory sites (Protection des sites d'observatoires existants et potentiels).** F. G. Smith.
Trans. IAU, Vol. XVIIIA, (see 003.013), p. 667 - 668 (1982).

082.088 **Water vapour content over two Indian sites tested for a millimeter-wave radio observatory.**
U. N. Maiya, P. Dierich.
Proc. Indian Acad. Sci. Earth Planet. Sci., Vol. 90, 281 - 290 (1981).

082.089 **Marcha del proyecto sobre visibilidad solar.**
A. L. Peretti Hollemaert, R. C. Estol.
Bol. Asoc. Argentina Astron., No. 20 - 24, p. 240 (1981). Abstract.

  Regularities of the variations of the scattered light and emission of the earth's twilight atmosphere.
See Abstr. 003.090.

Application of program calculators for express-accounting of the atmospheric extinction during photoelectric observations.   See Abstr. 021.007.

The calculation of astronomical refraction in marine navigation.   See Abstr. 021.013.

Association of atomic oxygen and airglow excitation mechanisms.   See Abstr. 022.018.

Water absorption lines, 931–961 nm: selected intensities, N_2-collision-broadening coefficients, self-broadening coefficients, and pressure shifts in air. See Abstr. 022.036.

Radiative lifetimes and quenching rate coefficients for directly excited rotational levels of $OH(A^2\Sigma^+, v'=0)$. See Abstr. 022.108.

The N_2-broadened water vapor absorption line shape and infrared continuum absorption – I. Theoretical development.   See Abstr. 022.172.

The N_2-broadened water vapor absorption line shape and infrared continuum absorption – II. Implementation of the line shape.   See Abstr. 022.173.

A comparison between estimations of Fried's parameter r_0 simultaneously obtained by measurements of solar granulation contrast and of the variance of angle-of-arrival fluctuations.   See Abstr. 031.551.

High resolution imaging from the ground. See Abstr. 031.601.

Suppression of atmospheric radio emission fluctuations in radio astronomical observations. See Abstr. 031.605.

A note on atmospheric extinction corrections. See Abstr. 031.611.

Final refraction problems in time and latitude observations through classical techniques. See Abstr. 031.626.

Local geodynamics with two-colour instruments. See Abstr. 031.628.

The performance of the Multiple Mirror Telescope. X. The first sub-millimeter phased array. See Abstr. 032.038.

The rate of rotation of the Earth and the atmospheric circulation.   See Abstr. 044.016.

Annual variation of climate of far east continent and annual fluctuation of the latitude determination at time and latitude observatories in this area. See Abstr. 045.009.

The scaling group of the radiative transfer equation. See Abstr. 063.024.

N-stream approximations to radiative transfer. See Abstr. 063.025.

Ascending droplets in the Earth's core. See Abstr. 081.057.

Investigation of a connection between disturbances of the height of the E_S layer and intensity of night luminescence [O I] 5577 Å.   See Abstr. 083.029.

Recent auroral and airglow measurements in the infrared.   See Abstr. 084.027.

Dissipation of the primordial terrestrial atmosphere due to irradiation of the solar far-UV during T Tauri stage. See Abstr. 107.014.

083 Ionosphere

083.001 Effects of IMF polarity on the *F*2-region.
L. Třísková.
J. Atmos. Terr. Phys., Vol. 44, 37 - 41 (1982).
On the basis of data from nineteen ionospheric stations in both hemispheres it is shown that the effect of the sectorial structure of the IMF appears in the *F*2-region as a response to the changes in geomagnetic activity caused by the changes of the IMF polarity. It is most pronounced in the nighttime *foF*2 in equinoxial periods at solar maximum.

083.002 Use of frequency measurement on satellite signals for computing differential Doppler and solar flare detection.   P. F. Checcacci, P. Spalla, P. P. Tiezzi.
J. Atmos. Terr. Phys., Vol. 44, 195 - 199 (1982).
During 1975 - 76 ionospheric measurements were made at the I.R.O.E. station located in Tuscany. Using the ATS-6 radio beacon, total electron content measurements and solar flares detection were carried out, by means of a new kind of Doppler differential method. Some results are presented.

083.003 Ionospheric troughs in Antarctica.
J. R. Dudeney, M. J. Jarvis, R. I. Kressman, M. Pinnock, A. S. Rodger, K. H. Wright.
Nature, Vol. 295, 307 - 308 (1982).
The authors report the first observations of the dynamics of both the mid-latitude and high-latitude troughs made by the Advanced Ionospheric Sounder (AIS) at Halley Station, Antarctica.

083.004 Gravity waves seeding ionospheric irregularities.
J. Röttger.
Nature, Vol. 296, 111 - 112 (1982).

083.005 Ionospheric scintillation modelling.
P. K. Pasricha, B. M. Reddy.
Indian J. Radio Space Phys., Vol. 10, 153 - 159 (1981).
Abstr. in Phys. Abstr., Vol. 85, Abstr. 23165 (1982).

083.006 Variation of electron collision frequency in the ionosphere.   S. M. Datta, K. M. Aggarwal, C. S. G. K. Setty.
Indian J. Radio Space Phys., Vol. 10, 160 - 163 (1981).
Abstr. in Phys. Abstr., Vol. 85, Abstr. 23166 (1982).

083.007 On the geophysical invariance of the lower part of the ionospheric F region.
G. S. Ivanov-Kholodnyj, Yu. K. Kalinin.
Geomagn. Aehron., Tom 22, 23 - 27 (1982). In Russian.

083.008 Sturcture of the day-time equatorial ionosphere.
G. F. Deminova.
Geomagn. Aehron., Tom 22, 28 - 32 (1982). In Russian.

083.009 Using data on non-coherent scattering for defining the aeronomic parameters at altitudes of the F2 region. Annual variations of atomic oxygen.
A. V. Mikhajlov, G. I. Ostrovskij.
Geomagn. Aehron., Tom 22, 60 - 65 (1982). In Russian.

083.010 On pulsating cosmic (radio) noise absorption.
N. D'Angelo.
Ann. Géophys., Vol. 37, 417 - 421 (1981). – Abstr. in Phys. Abstr., Vol. 85, Abstr. 42514 (1982).

083.011 Plasma physics in the ionosphere. II. The ionosphere in the plasma laboratory.   J. Trulsen.
Fra Fys. Verden, Vol. 43, No. 3, p. 67 - 71 (1981). In Norwegian. – Abstr. in Phys. Abstr., Vol. 85, Abstr. 42515 (1982).

083.012 Seasonal and solar control of ionospheric absorption at a low latitude station.
K. M. Kotadia, G. Datta, G. M. Chhipa.
Indian J. Radio Space Phys., Vol. 10, 171 - 175 (1981).
Abstr. in Phys. Abstr., Vol. 85, Abstr. 42516 (1982).

083.013 Effect of interplanetary magnetic field on apparent drift speed at low latitude stations.
M. Agrawal, S. K. Vijayvergia, R. K. Rai.
Indian J. Radio Space Phys., Vol. 10, 201 - 202 (1981).
Abstr. in Phys. Abstr., Vol. 85, Abstr. 42518 (1982).

083.014 Interplanetary magnetic field effects in the E-region drifts at low latitudes.   G. D. Vyas, H. Chandra.
Indian J. Radio Space Phys., Vol. 10, 206 - 208 (1981).
Abstr. in Phys. Abstr., Vol. 85, Abstr. 42519 (1982).

083.015 On dynamics of the lower ionosphere during auroral disturbances from very long frequency data.
G. F. Remenets, M. I. Beloglazov.
Geomagn. Aehron., Tom 22, 205 - 210 (1982). In Russian.

083.016 On an isolated internal gravitational wave in the F region of the ionosphere.
G. F. Deminova, M. G. Deminov, L. M. Erukhimov, O. N. Savina, L. A. Yudovich.
Geomagn. Aehron., Tom 22, 211 - 215 (1982). In Russian.

083.017 A model of distribution of the electric field in the ionosphere caused by the azimuthal component of the interplanetary magnetic field.   V. M. Uvarov.
Geomagn. Aehron., Tom 22, 216 - 219 (1982). In Russian.

083.018 Dependence of the electron concentration between the E and F layers on the zenith angle of the sun.
V. N. Ivanov, V. I. Kubov, I. A. Nasyrov.
Geomagn. Aehron., Tom 22, 310 - 311 (1982). In Russian.

083.019 Ionospheric currents in the southern polar cap depending on the sign of the azimuthal component of the interplanetary magnetic field.
V. O. Papitashvili, V. A. Popov.
Geomagn. Aehron., Tom 22, 314 - 315 (1982). In Russian.

083.020 On an effective model of the quiet auroral ionosphere for very long frequencies.
M. I. Beloglazov, I. N. Zabavina.
Geomagn. Aehron., Tom 22, 319 - 321 (1982). In Russian.

083.021 Some contributions of the ISIS program towards advances in knowledge of low latitude ionospheric phenomena.   L. L. Cogger.
Space Sci. Rev., Vol. 31, 437 - 452 (1982).
The ISIS (International Satellites for Ionospheric Studies) program yielded four scientific satellites, Alouette 1 and 2, and ISIS 1 and 2. This review is limited to the scientific contribution of the program to advances in knowledge relating to the portion of our atmosphere and ionosphere that lie within the plasmasphere. Topside ionograms form the principal data base from which global distributions of electron density and other geophysical parameters can be obtained. Ionospheric ducts and bubbles have also been studied with the aid of ionograms. The utilization of other instruments has facilitated investigations of ion composition, electron temperature and airglow variations.

083.022 Model calculations of the velocity of convection in the polar ionosphere and of magnetic effects of longitudinal currents and their comparison with the results of direct experiments.   N. I. Masevich.
Kosm. Issled., Tom 20, 264 - 276 (1982). In Russian.

083.023 Variations of the $[O^+]/[N^+]$ relation in the polar cusp depending on geomagnetic activity.
M. N. Vlasov, V. A. Telegin, A. P. Yaichnikov.
Kosm. Issled., Tom 20, 308 - 310 (1982). In Russian.

083.024 Tidal dynamo in the upper atmosphere. Validity and limit of the conventional theory.
S. Kato, T. Tsuda.
J. Geomagn. Geoelectr., Vol. 33, 383 - 397 (1981). – Abstr. in Phys. Abstr., Vol. 85, Abstr. 52998 (1982).

083.025 On the nature of the geophysical invariance of the lower part of the ionospheric F-region.
G. S. Ivanov-Kholodnyj, Yu. K. Kalinin.
Geomagn. Aehron., Tom 22, 378 - 382 (1982). In Russian.

083.026 Calculation of space-time distribution of the maximum ionization of the polar F2-layer.
A. S. Besprozvannaya, T. M. Krupitskaya, L. N. Makarova, V. M. Uvarov, K. E. Chernin, A. V. Shirochkov.
Geomagn. Aehron., Tom 22, 383 - 391 (1982). In Russian.

083.027 On a possibility of generating mid-latitude inner gravitational waves in the upper ionosphere.
O. N. Savina.
Geomagn. Aehron., Tom 22, 392 - 395 (1982). In Russian.

083.028 Mathematical model of ionosphere-plasmasphere interactions.   V. M. Polyakov, M. A. Koen, L. D. Ryazanova, I. M. Sidorov, G. V. Khazanov.
Geomagn. Aehron., Tom 22, 396 - 402 (1982). In Russian.

083.029 Investigation of a connection between disturbances of the height of the E_S layer and intensity of night luminescence [O I] 5577 Å.
T. A. Gorbunova, N. Yu. Kuznetsova, G. M. Shved.
Geomagn. Aehron., Tom 22, 430 - 434 (1982). In Russian.

083.030 Influence of an electromagnetic drift on the distribution of ionization at night time in the E- and F-regions.   M. A. Koen, G. V. Khazanov, D. V. Khazanov.
Geomagn. Aehron., Tom 22, 494 - 495 (1982). In Russian.

083.031 Seasonal differences in the display of the negative phase of disturbances of the mid-latitude F2-region of the ionosphere.   B. E. Serebryakov.
Geomagn. Aehron., Tom 22, 495 - 497 (1982). In Russian.

083.032 Zonal winds in the ionospheric F_2-region.
A. I. Semenov.
Geomagn. Aehron., Tom 22, 497 - 498 (1982). In Russian.

083.033 Influence of convection on the temperature regime of the polar ionosphere.
G. I. Mingaleva, T. V. Syrnikova, V. S. Mingalev, V. A. Vlaskov, Yu. G. Mizun.
Geomagn. Aehron., Tom 22, 512 - 515 (1982). In Russian.

083.034 Radio wave scintillations in the ionosphere.
K. C. Yeh, C.-H. Liu.
Proc. IEEE, Vol. 70, 324 - 360 (1982).

083.035 Global morphology of ionospheric scintillations.
J. Aarons.
Proc. IEEE, Vol. 70, 360 - 378 (1982).

083.036 On a feature of the sporadic E-layer.
E. P. Datsko, O. I. Maksimenko, V. I. Moskalyuk.
Kosm. issled. na Ukraine, Kiev, Tom 15, 62 - 67 (1981). In Russian. – Abstr. in Ref. zh., 62. Issled. kosm. prostranstva, 6.62.402 (1982).

The influence of ionospheric refraction on radio astronomy interferometry.   See Abstr. 031.627.

The effect of the ionospheric plasma on general circulation in the upper atmosphere. II.
See Abstr. 082.027.

New emissions of the upper atmosphere as a consequence of the anthropogenic action on the ionosphere. See Abstr. 082.038.

Europe probes the auroral atmosphere.
See Abstr. 084.013.

The relationship between field-aligned currents and the auroral electrojets: a review.   See Abstr. 084.023.

084 Aurorae, Geomagnetic Field, Magnetosphere

084.001   **Severe geomagnetic storms and their sources on the sun.**   S. Krajčovič, L. Křivský.
Bull. Astron. Inst. Czechoslovakia, Vol. 33, 47 - 59 (1982).
The purpose of this project was to compile a catalogue of severe geomagnetic storms and to co-ordinate them with source flares in cycles Nos 19 and 20 (1954 - 1976).

084.002   **Solitary auroral arc generation.**   S. V. Leontyev (*Leont'ev*), W. B. Lyatsky (*V. B. Lyatskij*).
Planet. Space Sci., Vol. 30, 1 - 4 (1982).
A theory of the solitary auroral arc generation is suggested. It is based on the existence of inclined currents over the arc and on the generation by those currents of a field-aligned electric field at some distance from the ionosphere.

084.003   **Observations of penetrated solar wind plasma elements in the plasma mantle.**
R. Lundin, B. Aparicio.
Planet. Space Sci., Vol. 30, 81 - 91 (1982).

084.004   **A study of the dynamics of a discrete auroral arc.**   G. Marklund, I. Sandahl, H. Opgenoorth.
Planet. Space Sci., Vol. 30, 179 - 197 (1982).
High resolution electric field and particle data, obtained by the S23L1 rocket crossing over a discrete prebreakup arc in January 1979, are studied in coordination with ground observations (Scandinavian Magnetometer Array–SMA, TV and all-sky cameras) in order to clarify the electrodynamics of the arc and its surroundings.

084.005   **Rapid fluctuations of random nature in auroral optical emissions.**   C. I. Haldoupis, K. Måseide.
J. Atmos. Terr. Phys., Vol. 44, 61 - 69 (1982).
Rapid intensity fluctuations in the optical emissions, $N_2^+(4278\ Å)$, $Hβ(4861\ Å)$, and $OI(5577\ Å)$ have been observed with rocket borne photometers during a passage through a homogeneous auroral arc. The measurements indicate that relatively weak (2–5%) intensity fluctuations of random nature, with dominant frequencies in the 2–20 Hz range, do exist in the optical emissions of quiet-form auroras. These variations have well defined averaged spectral characteristics and apparently differ from the strong quasi-periodic type of variations seen during pulsing auroras. It is argued that the observed fluctuations, which are usually composed of weak short-lived microbursts, reflect the temporal and spatial micro-structure which apparently exists always within auroral arcs.

084.006   **Optical measurements of a nightside poleward expanding aurora.**   P.-E. Sandholt, A. Egeland, K. Henriksen, R. Smith, P. Sweeney.
J. Atmos. Terr. Phys., Vol. 44, 71 - 79 (1982).
The authors present coordinated optical observations from two sites. They concentrate on the poleward expanding aurora during a substorm of 27 January 1979. Special emphasis is placed on the triangulated height measurements of different auroral emissions together with the neutral wind and temperature at the *F*-region height based on observations of the 6300 Å oxygen line. The findings are related to simultaneous ground- and satellite-based observations near the equatorward edge of the instantaneous auroral oval.

084.007   **Photometric and interferometric observations of the SAR arc event of March 5/6, 1981.**
T. Watanabe, J. S. Kim.
Geophys. Res. Lett., Vol. 9, 64 - 67 (1982).

084.008   **The palaeomagnetic field as inferred from magnetic studies on magmatic arc zones.**
J. Urrutia-Fucugauchi.
Geophys. J. R. Astron. Soc., Vol. 68, 49 - 56 (1982).

084.009   **International geomagnetic reference field 1980: a report by IAGA Division I Working Group 1.**
N. W. Peddie.
Geophys. J. R. Astron. Soc., Vol. 68, 265 - 268 (1982).

084.010   **Geomagnetic variation anomalies and deflection of telluric currents.**   J. L. Le Mouel, M. Menvielle.
Geophys. J. R. Astron. Soc., Vol. 68, 575 - 587 (1982).
The authors show that most of the abnormal variations of the transient magnetic field have deflection and canalization of telluric currents as main sources.

084.011   **Magnetic messages from the earth's core.**
D. Gubbins.
Nature, Vol. 295, 15 - 16 (1982).

084.012   **Chorus-related electrostatic bursts in the earth's outer magnetosphere.**
L. A. Reinleitner, D. A. Gurnett, D. L. Gallagher.
Nature, Vol. 295, 46 - 48 (1982).
The authors report recent studies of wideband plasma wave data from the ISEE 1 and ISEE 2 spacecraft which reveal that whistler-mode chorus emissions in the earth's outer magnetosphere are often accompanied by high-frequency bursts of electrostatic noise.

084.013   **Europe probes the auroral atmosphere.**
H. Rishbeth.
Nature, Vol. 295, 93 - 94 (1982).

084.014   **Substorms and the growth phase problem.**
S. W. H. Cowley.
Nature, Vol. 295, 365 - 366 (1982).

084.015   **Overshoots in planetary bow shocks.**
C. T. Russell, M. M. Hoppe, W. A. Livesey.
Nature, Vol. 296, 45 - 48 (1982).
An overshoot in the magnitude of the magnetic field is shown to be a consistent feature of supercritical collisionless shocks, Mach number $\gtrsim 3$, throughout the solar system. Data from Jupiter and Saturn allow to study the bow shock at high Mach numbers rarely, if ever, observed at earth. These combined planetary data show that the planetary bow shocks, at least as characterized by their overshoots, form part of a continuum, differences being dependent mainly on the varying solar wind conditions at each of the planets.

084.016   **Geomagnetic cores and secular effects.**
D. H. Tarling.
Nature, Vol. 296, 394 - 395 (1982).

084.017   **Intensity of the geomagnetic field near Loyang, China between 500 BC and AD 1900.**
Q. Y. Wei, D. J. Li, G. Y. Cao, W. S. Zhang, S. P. Wang.
Nature, Vol. 296, 728 - 729 (1982).
There are remarkably few data on the intensity of the geomagnetic field in early China. New results reported here indicate that the total intensity of the magnetic field of the earth in the region near Loyang has changed considerably during the past 2,400 yr (500 BC - AD 1900). The maximum value reached about 1,700 yr ago is as much as 54% higher than the present one.

084.018 The long period electromagnetic response of the
earth.   R. G. Roberts.
Geophys. J. R. Astron. Soc., Vol. 69, (see 012.004), 287
(1982). — Abstract.

084.019 Fully defined palaeomagnetic data and its expression
of geomagnetic field behaviour.
N. Roberts, J. Shaw.
Geophys. J. R. Astron. Soc., Vol. 69, (see 012.004), 288
(1982). — Abstract.

084.020 Correlation of geomagnetic activity indices ap with
the solar wind speed and the southward interplanetary magnetic field.   H. Schreiber.
J. Geophys., Vol. 49, 169 - 175 (1981). — Abstr. in Phys.
Abstr., Vol. 85, Abstr. 10276 (1982).

084.021 Spherical harmonic analysis of geomagnetic tides,
1964 - 1965.   D. E. Winch.
Philos. Trans. R. Soc. London A, Vol. 303, 1 - 104 (1981).
Abstr. in Phys. Abstr., Vol. 85, Abstr. 14391 (1982).

084.022 Towards a unified theory of discrete auroras.
J. R. Kan.
Space Sci. Rev., Vol. 31, 71 - 117 (1982).
The purpose of this paper is to review recent theoretical
developments in auroral research, leading toward a unified
theory of discrete auroras.

084.023 The relationship between field-aligned currents and
the auroral electrojets: a review.   Y. Kamide.
Space Sci. Rev., Vol. 31, 127 - 243 (1982).
This paper attempts to review the main results of the
last decade of research in such diverse fields as electric fields
and currents in the high-latitude ionosphere and field-aligned
currents and their relationship to the large-scale distribution
of auroras and auroral precipitation. It also contains discussions on some efforts in synthesizing the vast amount of the
observations to construct an empirical model which connects
the ionospheric currents with field-aligned currents.

084.024 The plasma mantle: composition and other characteristics observed by means of the Prognoz-7
satellite.   R. Lundin, B. Hultqvist, N. Pissarenko
(Pisarenko), A. Zackarov (Zakharov).
Space Sci. Rev., Vol. 31, 247 - 345 (1982).
The main content of this paper is a presentation and
discussion of the first direct observations by a spacecraft —
Prognoz-7 — of the ion composition in the high-latitude
boundary layer of the magnetosphere, the plasma mantle.

084.025 Dynamics of the outer radiation belts in relation to
polar substorms and injections of hot plasma at
geostationary orbit.   J. A. Sauvaud, J. R. Winckler.
Ann. Géophys., Vol. 37, 261 - 269 (1981). In French. — Abstr.
in Phys. Abstr., Vol. 85, Abstr. 23185 (1982).

084.026 The aurora.   S.-I. Akasofu.
American Sci., Vol. 69, 492 - 499 (1981). — Abstr.
in Phys. Abstr., Vol. 85, Abstr. 28261 (1982).

084.027 Recent auroral and airglow measurements in the
infrared.   A. T. Stair, Jr., R. Nadile, J. C. Ulwick,
K. D. Baker, D. J. Baker.
Special topics in optical propagation, (see 012.013), p. 19/1 - 6
(1981). — Abstr. in Phys. Abstr., Vol. 85, Abstr. 32703
(1982).

084.028 On the generation of acousto-gravitational waves by
auroral electrojets.   V. V. Kulikov.
Geomagn. Aehron., Tom 22, 45 - 50 (1982). In Russian.

084.029 Interconnection of ring current precipitating elec-
trons and auroral luminescences in the morning
sector of the magnetosphere from data of ASE Cosmos 900.
A. V. Dronov, M. I. Panasyuk, Eh. N. Sosnovets,
L. V. Tverskaya, V. I. Tulupov, O. V. Khorosheva.
Geomagn. Aehron., Tom 22, 85 - 89 (1982). In Russian.

084.030 Analytical description of the secular variations of
the geomagnetic field and the velocity of the
rotation of the earth.   S. I. Braginskij.
Geomagn. Aehron., Tom 22, 115 - 122 (1982). In Russian.

084.031 On the distribution of polar aurorae.
V. B. Lyatskij.
Geomagn. Aehron., Tom 22, 149 - 151 (1982). In Russian.

084.032 On a connection of geomagnetic variations in the
polar cap with the interplanetary magnetic field in
a solar activity cycle.   V. O. Papitashvili.
Geomagn. Aehron., Tom 22, 151 - 154 (1982). In Russian.

084.033 On a possible mechanism of increasing the longitudinal electric field and formation of V-shaped
structures of the potential in the magnetosphere.
V. Yu. Trakhtengerts, A. Ya. Fel'dshtejn.
Geomagn. Aehron., Tom 22, 157 - 159 (1982). In Russian.

084.034 Analytical models of the secular variation of the
geomagnetic field 1975, 1977 and 1980.
G. I. Kolomijtseva, T. N. Bondar'.
Geomagn. Aehron., Tom 22, 164 - 166 (1982). In Russian.

084.035 On a method of representation of data on the
secular geomagnetic variations and some errors
connected with it.   Yu. D. Kalinin.
Geomagn. Aehron., Tom 22, 168 (1982). In Russian.

084.036 The aurora: an electrical discharge phenomenon
surrounding the Earth.   S.-I. Akasofu.
Rep. Prog. Phys., Vol. 44, 1123 - 1149 (1981). — Abstr. in
Phys. Abstr., Vol. 85, Abstr. 42512 (1982).

084.037 Numerical models of quiet and disturbed geomag-
netic field in the cislunar part of the magnetosphere.
N. A. Tsyganenko.
Ann. Géophys., Vol. 37, 381 - 391 (1981). — Abstr. in Phys.
Abstr., Vol. 85, Abstr. 42543 (1982).

084.038 Generation of a quiet auroral arc.
S. V. Leont'ev, V. B. Lyatskij.
Geomagn. Aehron., Tom 22, 233 - 238 (1982). In Russian.

084.039 Measurements of temporal variations of the bright-
ness of aurorae and emission layers with the
scientific orbital station Salyut 6.
S. V. Avakyan, G. A. Dolgopolov, L. S. Evlashin,
A. S. Ivanchenkov, V. V. Kovalenko, G. M. Kolesnikov,
T. A. Kornilova, G. M. Kravchenko, A. I. Lazarev,
A. N. Svirskij, G. V. Starkov, S. A. Chernous.
Geomagn. Aehron., Tom 22, 239 - 242 (1982). In Russian.

084.040 Investigation of the altitude distribution of hydrogen
luminescence in aurorae with the help of rockets on
the Franz-Josef Land.   T. M. Tarasova, O. I. Yagodkina,
L. S. Evlashin, A. I. Lazarev, M. V. Orlova.
Geomagn. Aehron., Tom 22, 243 - 247 (1982). In Russian.

084.041 Altitudes of radio aurorae.
E. E. Timofeev, Yu. G. Miroshnikov, T. N. Khor'kova,
T. V. Miroshnikova.
Geomagn. Aehron., Tom 22, 248 - 251 (1982). In Russian.

084.042 The vertical component of the geoelectrical field at high latitudes and processes in the magnetosphere.
O. I. Bandilet, S. P. Chernysheva, V. M. Sheftel'.
Geomagn. Aehron., Tom 22, 252 - 256 (1982). In Russian.

084.043 Wave processes in the period of the magnetospheric disturbance of February 15, 1978.
T. V. Gajvoronskaya, L. A. Yudovich, L. Lois.
Geomagn. Aehron., Tom 22, 312 - 313 (1982). In Russian.

084.044 Angular distribution and polarization of auroral X-ray emission. E. G. Berezhko.
Geomagn. Aehron., Tom 22, 321 - 323 (1982). In Russian.

084.045 The influence of anisotropy of the electron temperature on the distribution of charged particle concentration in the plasmasphere.
Yu. V. Konikov, G. V. Khazanov.
Geomagn. Aehron., Tom 22, 323 - 325 (1982). In Russian.

084.046 On two types of variations of the magnetic field in the near-polar region with the northern direction of the interplanetary magnetic field.
P. V. Sumaruk, Ya. I. Fel'dshtejn.
Geomagn. Aehron., Tom 22, 332 - 334 (1982). In Russian.

084.047 Spectral structure of the secular geomagnetic variations for the last century.
Yu. D. Kalinin, T. S. Rozanova.
Geomagn. Aehron., Tom 22, 342 - 344 (1982). In Russian.

084.048 Variations of the boundary of capture of relativistic electrons in the earth's magnetosphere.
I. N. Senchuro, P. I. Shavrin.
Kosm. Issled., Tom 20, 156 - 158 (1982). – In Russian.

084.049 On the relationships between interplanetary quantities and the global auroral electrojet index.
A. Meloni, A. Wolfe, L. J. Lanzerotti.
J. Geophys. Res., Vol. 87, 119 - 127 (1982).

084.050 Chatanika radar observations of the electrostatic potential distribution of an auroral arc.
O. de la Beaujardière, R. Vondrak.
J. Geophys. Res., Vol. 87, 797 - 809 (1982).

084.051 Is there an increase of geomagnetic activity preceding total lunar eclipses?
A. C. Fraser-Smith.
J. Geophys. Res., Vol. 87, 895 - 898 (1982).
    The possibility that there is a lunar modulation of geomagnetic activity around the time of total lunar eclipses is investigated.

084.052 Solar wind energy transfer through the magnetopause of an open magnetosphere.
L. C. Lee, J. G. Roederer.
J. Geophys. Res., Vol. 87, 1439 - 1444 (1982).
    The authors examine the process of energy and particle transfer through the magnetopause and integrate the results to achieve some quantitative understanding of the total power transfer from the solar wind to the magnetosphere.

084.053 A classification of polar cap auroral arcs.
S. Ismail, C.-I. Meng.
Planet. Space Sci., Vol. 30, 319 - 330 (1982).
    Global auroral imagery obtained by DMSP satellites during the years 1972–1979 over both the northern and southern high latitude polar regions were examined to study the morphology of the discrete arcs known as polar cap arcs.

084.054 Scaling relations governing magnetospheric energy transfer.
V. M. Vasyliunas, J. R. Kan, G. L. Siscoe, S.-I. Akasofu.
Planet. Space Sci., Vol. 30, 359 - 365 (1982).

084.055 Dynamo process governing solar wind-magnetosphere energy coupling. J. R. Kan, S.-I. Akasofu.
Planet. Space Sci., Vol. 30, 367 - 370 (1982).

084.056 A two-dimensional model of the magnetosphere including a current system.
N. Kömle, H. Biernat, H. Rucker.
Gerlands Beitr. Geophys., Band 91, 108 - 120 (1982).
    The authors demonstrate the influence of the current system on the shape of the magnetopause and on the topology of the magnetospheric magnetic field lines.

084.057 Geomagnetic secular variation as a precursor of climatic change.
V. Courtillot, J. L. Le Mouel, J. Ducruix, A. Cazenave.
Nature, Vol. 297, 386 - 387 (1982).
    The authors point out a correlation between variations in the earth's magnetic field, the earth's rotation rate and some climatic indicators, thus suggesting a possible long term influence of core motions on climate. The authors suggest that geomagnetic secular variation can be used to forecast a climatic change in the 1990s.

084.058 Diffuse auroral zone. VI. Penetration of electrons and protons into the daytime sector.
T. M. Mulyarchik, Yu. I. Gal'perin, V. A. Gladyshev, L. M. Nikolaenko, J.-A. Sauvaud, J. Crasnier, Ya. I. Fel'dshtejn.
Kosm. Issled., Tom 20, 244 - 263 (1982). In Russian.

084.059 Formation of streams of energetic ions in a geostationary orbit. M. I. Panasyuk.
Kosm. Issled., Tom 20, 277 - 288 (1982). In Russian.

084.060 Investigation of the plasma mantle of the earth's magnetosphere. I. Double structure.
Eh. M. Dubinin, A. V. Zakharov, N. F. Pisarenko, R. Lundin, B. Hultqvist.
Kosm. Issled., Tom 20, 289 - 296 (1982). In Russian.

084.061 Peculiarities of geomagnetic variations in connection with the structure of interplanetary magnetic fields.
Eh. I. Nesmyanovich.
Vestn. Kiev. Univ., Vyp. 23, p. 98 - 105 (1981). In Russian.

084.062 A brief panorama. J. Lemaire.
Solar system plasmas and fields, (see 003.005), p. 3 - 10 (1982).
    Contents: The magnetosphere of the earth. The interplanetary medium. Magnetospheres of other planets. Perspectives and future projects.

084.063 Discovering the earth's magnetosphere.
W. I. Axford.
Solar system plasmas and fields, (see 003.005), p. 11 - 12 (1982).

084.064 The geomagnetic field and its extension into space.
W. P. Olsen.
Solar system plasmas and fields, (see 003.005), p. 13 - 17 (1982).

084.065 Electric fields in the magnetosphere.
C.-G. Fälthammar.
Solar system plasmas and fields, (see 003.005), p. 19 - 24 (1982).

084.066   Plasma waves in the earth's magnetosphere.
D. Jones.
Solar system plasmas and fields, (see 003.005), p. 25 - 31
(1982).

084.067   Low energy particles in the magnetosphere.
C. R. Chappell.
Solar system plasmas and fields, (see 003.005), p. 33 - 38
(1982).

084.068   Hot plasmas in the magnetosphere.
T. E. Eastman, L. A. Frank.
Solar system plasmas and fields, (see 003.005), p. 39 - 42
(1982).

084.069   High energy particles in the magnetosphere.
G. A. Paulikas, J. B. Blake.
Solar system plasmas and fields, (see 003.005), p. 43 - 46
(1982).

084.070   Geomagnetic reversals and tidal friction.
P. Brosche.
Naturwissenschaften, 68. Jahrg., 139 - 140 (1981).
The resonance structure of the oceans leads to a very
complex time behaviour of the tidal torque; this could be the
cause for magnetic reversals.

084.071   La magnétosphère terrestre.   M. Roth.
Ciel, Vol. 44, 8 - 18 (1982).

084.072   Proton streams in the upper atmosphere at various
geomagnetic latitudes.
A. P. Babaev, V. A. D'yachenko, Yu. M. Zhuchenko,
V. A. Lipovetskij, A. N. Makeev, L. S. Novikov, M. A. Savel'ev,
V. F. Tulinov, V. V. Tulyakov, V. M. Fejgin.
Probl. soln.-atmos. svyazej, Ehksp. Solntse - atmos. 1976.
Moskva, 1981, p. 75 - 80. In Russian. − Abstr. in Ref. zh., 62.
Issled. kosm. prostranstva, 5.62.607 (1982).

084.073   Results of measurements of geoactive corpuscular
radiation aboard the artificial satellite Meteor 25 in
July - August 1976.
A. P. Babaev, V. A. D'yachenko, T. A. Zhuchenko,
Yu. M. Zhuchenko, V. I. Lazarev, V. A. Lipovetskij,
B. V. Mar'in, M. A. Savel'ev, M. V. Tel'tsov, V. F. Tulinov,
V. M. Fejgin.
Probl. soln.-atmos. svyazej, Ehksp. Solntse - atmos. 1976.
Moskva, 1981, p. 81 - 86. In Russian. − Abstr. in Ref. zh., 62.
Issled. kosm. prostranstva, 5.62.608 (1982).

084.074   Pulsations of the geomagnetic field in the
0.1 - 0.6 Hz region observed in the polar auroral
zone.   B. V. Dovbnya, O. D. Zotov, V. F. Ruban,
A. R. Polyakov, A. S. Potapov.
Issled. po geomagn., aehron. i fiz. Solntsa, Moskva, 1982,
No. 58, p. 33 - 40. In Russian. − Abstr. in Ref. zh., 62.
Issled. kosm. prostranstva, 5.62.654 (1982).

084.075   Some peculiarities of motion of high-energy charged
particles in the geomagnetic field.
E. V. Gorchakov, V. I. Severinov, M. V. Ternovskaya.
Geomagn. Aehron., Tom 22, 357 - 361 (1982). In Russian.

084.076   Processes of heating of cold geomagnetospheric
plasma.   I. V. Kovalevskij.
Geomagn. Aehron., Tom 22, 455 - 459 (1982). In Russian.

084.077   The electrical field in the magnetosphere's tail
depending on the level of geomagnetic activity and
intensity E_y in the solar wind.
M. I. Pudovkin, V. V. Osipov, M. A. Shukhtina, S. A. Zajtseva.
Geomagn. Aehron., Tom 22, 460 - 464 (1982). In Russian.

084.078   On properties of a convective stream of a magnetospheric plasma.   L. M. Alekseeva.
Geomagn. Aehron., Tom 22, 476 - 479 (1982). In Russian.

084.079   Inner and outer parts of the field of the 60-year
variations of the geomagnetic field.
Yu. D. Kalinin, T. S. Rozanova.
Geomagn. Aehron., Tom 22, 480 - 482 (1982). In Russian.

084.080   Space-time distribution of longitudinal currents in
the daytime sector of high latitudes depending on
conditions in the interplanetary magnetic field.
R. G. Afonina, B. A. Belov, V. Yu. Gajdukov, N. I. Gershenzon,
A. E. Levitin, D. S. Faermark, Ya. I. Fel'dshtejn.
Geomagn. Aehron., Tom 22, 516 - 519 (1982). In Russian.

084.081   Intensity of the geomagnetic field in the Caucasus
for the last 2000 years.   K. S. Burakov,
S. P. Burlatskaya, I. E. Nachasova, Z. A. Chelidze.
Geomagn. Aehron., Tom 22, 523 - 524 (1982). In Russian.

084.082   Monopoly.   B. M. Hodder.
Geophys. J. R. Astron. Soc., Vol. 70, 217 - 228
(1982).
A model for the geomagnetic secular variation field is
given, consisting of a series of magnetic monopoles at the
surface of the Earth's core. A monopole model is calculated
from observatory secular change data for the epoch 1957.5 −
1962.5 and its usefulness assessed.

084.083   Analysis of geomagnetic effect of previous 5 solar
eclipses occurring in China during past 50 years.
K.-K. Tschu, J.-X. Zhang, C.-F. Liu.
Planet. Space Sci., Vol. 30, 587 - 594 (1982).
Results and analysis of geomagnetic observations during
previous 5 solar eclipses occurred in China are summarized.
The methods of evaluation for eclipse effects on the geomagnetic field are briefly described both for the quiet and
disturbed days. The discussion of the data is made with
reference to Chapman's theoretical consideration on optical
eclipse effect, together with the quiet-day overhead current
systems in the upper atmosphere.

084.084   Self-consistent theory of time-dependent convection
in the earth's magnetotail.
K. Schindler, J. Birn.
J. Geophys. Res., Vol. 87, 2263 - 2275 (1982).

084.085   Extremely high latitude auroras.
M. S. Gussenhoven.
J. Geophys. Res., Vol. 87, 2401 - 2412 (1982).

084.086   Analysis of nitrogen and oxygen far ultraviolet
auroral emissions.   R. R. Meier, R. R. Conway,
P. D. Feldman, D. J. Strickland, E. P. Gentieu.
J. Geophys. Res., Vol. 87, 2444 - 2452 (1982).
A far ultraviolet rocket observation of an auroral arc has
been analyzed by using laboratory measured emission cross
sections and self-consistent models of the atmospheric composition and energetic electron flux.

084.087   The ultraviolet spectrum of an aurora 530-1520 Å.
P. D. Feldman, E. P. Gentieu.
J. Geophys. Res., Vol. 87, 2453 - 2458 (1982).
Ultraviolet spectra between 530 and 1520 Å of an active
auroral arc were obtained at 6.5-Å instrumental resolution.
Several new emission features are identified, including several
bands of the N_2 Birge-Hopfield $(1, v'')$ progression and numer-
ous N I multiplets.

084.088   Mapping of auroral X-rays from rocket overflights.
R. A. Goldberg, J. R. Barcus, L. A. Treinish,

R. R. Vondrak.
J. Geophys. Res., Vol. 87, 2509 - 2524 (1982).

In March 1978, two Nike Tomahawk payloads were launched from Poker Flat, Alaska, to observe the structure of bremsstrahlung X rays and precipitating particles during both nighttime and daytime aurorally disturbed conditions. The data have been used to produce maps of relative X-ray count rates, e-folding energies, and X-ray energy flux by building images from the detector scan induced by rocket precession and translation. It was found that for the geophysical conditions and energy range studied, the emitting radiation can be characterized by a two component spectrum with bright regions enhanced mainly from the low-energy component.

084.089 **Auroral nitric oxide concentration and infrared emission.**   W. P. Reidy, T. C. Degges, A. G. Hurd,
A. T. Stair, Jr., J. C. Ulwick.
J. Geophys. Res., Vol. 87, 3591 - 3598 (1982).

Since 1972, the Air Force Geophysics Laboratory has launched a series of rockets to study emission in the 5.3-μm and 2.7-μm infrared bands of nitric oxide at high altitudes. The instrumentation varied from one probe to another; infrared fixed band radiometers and CVF spectrometers, narrow band photometers and instrumentation to measure the incident energy spectrum were each included on various occasions. Optical photometric data were also obtained with ground-based instruments. The authors have used the available data to model the time history and chemistry of the auroral deposition.

084.090 **The magnetic earth as seen from MAGSAT, initial results.**   R. A. Langel.
Geophys. Res. Lett., Vol. 9, 239 - 242 (1982).

084.091 **A geomagnetic field spectrum.**
R. A. Langel, R. H. Estes.
Geophys. Res. Lett., Vol. 9, 250 - 253 (1982).

Selected MAGSAT data are used to derive a spherical harmonic model of the earth's internal magnetic field of degree and order 23.

084.092 **Pole-strength of the earth from MAGSAT and magnetic determination of the core radius.**
C. V. Voorhies, E. R. Benton.
Geophys. Res. Lett., Vol. 9, 258 - 261 (1982).

084.093 **Frozen-flux upper limits to the magnitudes of geomagnetic Gauss coefficients, based on MAGSAT observations.**   E. R. Benton, M. C. Coulter.
Geophys. Res. Lett., Vol. 9, 262 - 264 (1982).

084.094 **A problem in representing the core magnetic field of the earth using spherical harmonics.**
H. M. Carle, C. G. A. Harrison.
Geophys. Res. Lett., Vol. 9, 265 - 268 (1982).

084.095 **Initial scalar magnetic anomaly map from MAGSAT.**   R. A. Langel, J. D. Phillips, R. J. Horner.
Geophys. Res. Lett., Vol. 9, 269 - 272 (1982).

084.096 **Initial vector magnetic anomaly map from MAGSAT.**   R. A. Langel, C. C. Schnetzler, J. D. Phillips,
R. J. Horner.
Geophys. Res. Lett., Vol. 9, 273 - 276 (1982).

084.097 **MAGSAT scalar anomaly distribution: the global perspective.**   H. Frey.
Geophys. Res. Lett., Vol. 9, 277 - 280 (1982).

084.098 **Magnetic anomaly maps from 40°N to 83°N derived from MAGSAT satellite data.**
R. L. Coles, G. V. Haines, G. Jansen van Beek, A. Nandi,

J. K. Walker.
Geophys. Res. Lett., Vol. 9, 281 - 284 (1982).

084.099 **Application of dipole modeling to magnetic anomalies.**
D. Schmitz, J. B. Frayser, J. C. Cain.
Geophys. Res. Lett., Vol. 9, 307 - 310 (1982).

084.100 **Evaluation of high latitude disturbances with MAGSAT (the importance of the MAGSAT geomagnetic field model).**
L. J. Zanetti, T. A. Potemra, M. Sugiura.
Geophys. Res. Lett., Vol. 9, 365 - 368 (1982).

084.101 **Investigation of the plasma mantle of the earth's magnetosphere. 2. Ion composition.**
A. V. Zakharov, Eh. M. Dubinin, N. F. Pisarenko, R. Lundin,
B. Hultqvist.
Kosm. Issled., Tom 20, 387 - 398 (1982). In Russian.

084.102 **"Where and when can I see the Aurora Borealis?"**
R. J. Livesey.
J. British Astron. Assoc., Vol. 92, 180 - 181 (1982).

084.103 **Die Kluft im Magnetschirm. Polarlichter am Übergang zwischen Atmosphäre und Weltraum.**
D. Möhlmann.
Umschau, 82. Jahrg., 432 - 434 (1982).

084.104 **Coincidence of some magnetic and gravity field characteristics.**   G. Barta.
Sun and planetary system, (see 012.040), p. 205 - 208 (1982).

084.105 **Geomagnetic variation and field-aligned currents at northern high-latitudes, and their relations to the solar wind parameters.**   A. E. Levitin, R. G. Afonina,
B. A. Belov, Ya. I. Feldstein (*Fel'dshtejn*).
Philos. Trans. R. Soc. London, Ser. A, Vol. 304, 253 - 301 (1982).

It has been shown that the geomagnetic field at high latitudes can be represented by a sum of fields controlled by components of the interplanetary magnetic field (IMF) and by the velocity and density of the solar wind plasma. A correlation model is proposed by means of which hourly-mean values of the three components (X, Y, Z) of the geomagnetic field vector at high latitudes may be determined from hourly-mean values of the components B_x, B_y and B_z of the IMF. The space-time distribution of field-aligned currents has been reconstructed from the horizontal component of the ground level geomagnetic field variation. The correlation model of field-aligned currents has been obtained for high latitudes.

084.106 **Geomagnetic and solar data.**   H. E. Coffey (Editor).
J. Geophys. Res., Vol. 87, 284, 926, 1733, 2580, 3628 (1982).

Geomagnetic activity and stability of the solar corpuscular field.   See Abstr. 003.143.

ULF pulsations in the magnetosphere.
See Abstr. 012.055.

A direct analytic method of calculating the quadrupole parameters of a planetary magnetic field.
See Abstr. 021.004.

An algorithm for synthesizing the geomagnetic field.   See Abstr. 021.011.

The laboratory magnetosphere.
See Abstr. 022.177.

The magnetic field of a plasma stream flowing around a blunted body.   See Abstr. 062.052.

Simulation of three-dimensional solar wind disturbances and resulting geomagnetic storms. See Abstr. 074.026.

A sub-Alfvénic solar wind: interplanetary and magnetosheath observations.   See Abstr. 074.035.

Coronal radio emission activity and its relation to planetary geomagnetic activity in the course of cycle No. 20. See Abstr. 077.064.

Characteristics of proton streams of solar flares in the region of an auroral oval depending on the magnetosphere's disturbance.   See Abstr. 078. 007.

On dynamics of the lower ionosphere during auroral disturbances from very long frequency data. See Abstr. 083.015.

Mathematical model of ionosphere-plasmasphere interactions.   See Abstr. 083.028.

On a possible relation of semi-annual temperature and wind fluctuations in the high-latitude stratosphere with the indices of solar and geomagnetic activity. See Abstr. 085.022.

Amplitude variations of the annual temperature run in connection with solar and geophysical agents. See Abstr. 085.023.

Measurements of the lunar induced magnetic moment in the geomagnetic tail: evidence for a lunar core? See Abstr. 094.034.

Jupiter's gravitational effects on the solar wind and geomagnetic activity in different epochs of solar activity. See Abstr. 099.012.

Investigation of statistical properties of a connection of the interplanetary and the geomagnetic fields with the method of multiple coherence functions. See Abstr. 106.011.

Notes on the origin of the B_z-component of the interplanetary magnetic field and the geoefficiency index in Akasofu's model.   See Abstr. 106.013.

On using a phase diagram for investigation of the connection of the interplanetary magnetic field with ground variations of the geomagnetic field.   See Abstr. 106.014.

Dependence of the radiation dose aboard the Salyut 6 station on the indices of solar and geomagnetic activity. See Abstr. 143.022.

Trajectory parameterization − a new approach to the study of the cosmic ray penumbra.   See Abstr. 143.032.

085 Solar-terrestrial Relations

085.001 **The influence of solar ultraviolet variability on climate.**   J. W. Chamberlain.
Planet. Space Sci., Vol. 30, 147 - 150 (1982).

085.002 **The evolution of ozone with changing solar activity.**
D. K. Chakrabarty, P. Chakrabarty.
Geophys. Res. Lett., Vol. 9, 76 - 78 (1982).

085.003 **Solar activity and extremal temperatures in Moscow.**
I. F. Nikulin.
Soln. Dannye 1981 Byull., No. 11, p. 115 - 119 (1982). In Russian.
Daily thermal records for Moscow during the last 100 years are considered. The distribution of extremal temperatures with the phase of the solar cycle is studied.

085.004 **The Sun and the terrestrial environment.**
R. Gendrin.
Recherche, Vol. 12, 942 - 951 (1981). In French. – Abstr. in Phys. Abstr., Vol. 85, Abstr. 18749 (1982).

085.005 **Stratospheric condensation nuclei variations may relate to solar activity.**
D. J. Hofmann, J. M. Rosen.
Nature, Vol. 297, 120 - 124 (1982).
Observations of increases of stratospheric condensation nuclei suggest a photo-initiated sulphuric acid vapour formation process in spring in polar regions. It is proposed that the sulphuric acid rapidly forms condensation nuclei through attachment to negatively charged multi-ion complexes and that the process may be modulated through variations in solar activity.

085.006 **The effect of solar wind velocity on location and structure of aurorae in the midnight sector.**
V. G. Vorob'ev, V. L. Zverev.
Geomagn. Aehron., Tom 22, 81 - 84 (1982). In Russian.

085.007 **On the influence of preflare pulsations of solar radiation on the earth's magnetosphere.**
M. M. Kobrin, V. I. Malygin, S. D. Snegirev.
Geomagn. Aehron., Tom 22, 156 - 157 (1982). In Russian.

085.008 **Influence of solar wind parameters on geomagnetic activity.**   V. G. Vorob'ev, V. L. Zverev.
Geomagn. Aehron., Tom 22, 257 - 262 (1982). In Russian.

085.009 **Solar wind velocity and geomagnetic moment variations.**   Yu. D. Kalinin, T. S. Rozanova.
Geomagn. Aehron., Tom 22, 278 - 280 (1982). In Russian.

085.010 **Geomagnetic negative sudden impulses after the powerful isolated flares of 1957 - 1978.**
K. G. Ivanov, N. V. Mikerina.
Geomagn. Aehron., Tom 22, 325 - 327 (1982). In Russian.

085.011 **Geomagnetic activity in even and odd eleven-year cycles.**   M. A. Livshits.
Geomagn. Aehron., Tom 22, 335 - 337 (1982). In Russian.

085.012 **Oscillations in the magnetosphere of the earth with a period 160 min caused by the pulsations of the sun.**
B. M. Vladimirskij, V. P. Bobova, V. N. Repin, V. K. Veretennikova.
Izv. Krymskoj Astrofiz. Obs., Tom 64, 132 - 140 (1981). In Russian. English translation in Bull. Crimean Astrophys. Obs., Vol. 64.
Optical observations of solar pulsations with period 160 min together with measurements of the amplitudes envelope of the micropulsating geomagnetic field Pc 3.4 were reduced by cosinor analysis. It has been found that the amplitudes of Pc 3.4 are modulated by a period 160 min with stable phase.

085.013 **Total ozone – solar activity relationship.**
C. A. Reber, F. T. Huang.
J. Geophys. Res., Vol. 87, 1313 - 1318 (1982).
Nearly 6 years of global ozone monthly mean data from the Nimbus 4 BUV instrument are compared with monthly values of solar activity using 10.7-cm flux ($F(10.7)$) as a parameter.

085.014 **Solar cycle and seasonal variations in the efficiency of solar flares in producing sudden ionospheric disturbances.**   V. Letfus, E. M. Apostolov.
J. Atmos. Terr. Phys., Vol. 44, 359 - 362 (1982).

085.015 **Some peculiarities of cyclone activity in connection with the 11-year solar activity cycle.**
M. A. Nuzhdina.
Vestn. Kiev. Univ., Vyp. 23, p. 72 - 83 (1981). In Russian.

085.016 **What will the mean weather be in the first half of the next century?**   M. Kopecký.
Říše hvězd, Vol. 63, 1 - 2 (1982). In Czech.

085.017 **Consequences of solar-related processes on the earth's environment and man's devices.**
R. Gendrin, V. Domingo.
Solar system plasmas and fields, (see 003.005), p. 71 - 77 (1982).

085.018 **Perspectives and future projects.**
D. J. Williams.
Solar system plasmas and fields, (see 003.005), p. 79 - 82 (1982).
The paper deals with solar-terrestrial relations.

085.019 **Misura della radiazione solare globale e diffusa rilevata nell'anno 1980.**   T. Zanzu.
Circ. Stn. Astron. Int. Latitudine, Carloforte-Cagliari, N. 17 (Ser. C (1)), 17 pp. (1981).

085.020 **Sunshine and direct solar radiation upon a tilted collector.**   E. Proverbio, T. Zanzu.
Energia Solare e Nuove Prospettive, Vol. 3, Atti della 18. Conferenza Internazionale, Milano 23 - 27 sett. 1979, 15 pp. = Pubbl. Stn. Astron. Int. Latitudine, Carloforte-Cagliari, N. 76.

085.021 **Annual variation of magnetic and ionospheric disturbances, asymmetry of the activity of the solar hemispheres and the interplanetary magnetic field.**
Z. Ts. Rapoport.
Inst. zem. magn., ionos. i rasprostr. radiovoln AN SSSR. Prepr., 1981, No. 46, 22 pp. In Russian. – Abstr. in Ref. zh., 51. Astron., 5.51.411 (1982).

085.022 **On a possible relation of semi-annual temperature and wind fluctuations in the high-latitude stratosphere with the indices of solar and geomagnetic activity.**
V. G. Kadiyarova, I. A. Shcherba.
Tr. Glav. geofiz. obs., 1981, No. 458, p. 24 - 29. In Russian. Abstr. in Ref. zh., 51. Astron., 5.51.412 (1982).

085.023 **Amplitude variations of the annual temperature run in connection with solar and geophysical agents.**
V. F. Loginov, V. A. Molodykh.

Tr. Glav. geofiz. obs., 1981, No. 458, p. 62 - 71. In Russian.
Abstr. in Ref. zh., 51. Astron., 5.51.415 (1982).

085.024   vlf sudden phase anomalies during flares of
August 1972.
R. C. Dubey, S. K. Gupta, M. C. Pande.
Indian J. Radio Space Phys., Vol. 6, 67 - 68 (1977) = Uttar
Pradesh State Obs. Repr. No. 131.

085.025   The heliomagnetic cycle in geomagnetic activity.
N. S. Zaretskij.
Geomagn. Aehron., Tom 22, 450 - 454 (1982). In Russian.

085.026   A high time resolution study of the solar wind-
magnetosphere energy coupling function.
S.-I. Akasofu, J. F. Carbary, C.-I. Meng, J. P. Sullivan,
R. P. Lepping.
Planet. Space Sci., Vol. 30, 537 - 543 (1982).

085.027   Further evidence for the dependence of fast atmospheric light pulsations on solar activity.
O. T. Tümer.
J. Geophys. Res., Vol. 87, 2569 - 2570 (1982).
The millisecond time scale diffuse atmospheric light
emissions showing 10-kHz damped oscillations were studied for
a period of 4 years from June 12, 1972, to April 1, 1976. The
experimental setup consisted of a wide-angle photomultiplier
system. New evidence was found relating these light pulses to
solar activity. They were concentrated in the autumn and
winter months.

085.028   Effect of solar flare and sunspot numbers on the
intensity of 5577 Å line in the night airglow.
N. Kundu, S. N. Ghosh.
Indian J. Phys. Part B, Vol. 55B, 343 - 352 (1981). – Abstr. in
Phys. Abstr., Vol. 85, Abstr. 58194 (1982).

085.029   The type dependence of sudden field anomaly at low
frequencies.   R. K. Kaul.
Indian J. Phys. Part B, Vol. 55B, 435 - 442 (1981). – Abstr. in
Phys. Abstr., Vol. 85, Abstr. 58195 (1982).

085.030   The sun's influence on the earth's atmosphere and
interplanetary space.   J. V. Evans.
Science, Vol. 216, 467 - 474 (1982).

085.031   Sensitivity of the earth's climate to changes in the
solar constant.   G. R. North.
Variations of the solar constant, (see 012.037), p. 1 - 9
(1981).
A brief review of climate sensitivity to solar variations is
presented with special attention to simplified models. A number of uncertainties remain in the understanding of climate
and these are elaborated upon. Especially vexing are possible
feedbacks which might operate on long time scales and are
therefore not testable directly.

085.032   The combined solar and tidal influence in climate.
P. R. Bell.
Variations of the solar constant, (see 012.037), p. 241 - 255
(1981).

085.033   Predominant periods in the time series of Drought
Area Index for the Western High Plains
AD 1700 to 1962.   P. R. Bell.
Variations of the solar constant, (see 012.037), p. 257 - 264
(1981).

085.034   Interplanetary magnetic field and solar cycle
modulation of the tropospheric circulation over
West Mediterranean.   B. P. Tritakis.
Sun and planetary system, (see 012.040), p. 215- 216 (1982).

085.035   A possible astronomical factor for major seismic
events – the photospheric magnetic field. F.-m. Hu.
Acta Geophys. Sinica, Vol. 25, 270 - 275 (1982). In Chinese.

085.036   Does the 80-year sunspot period affect the duration
of sunshine in Central Europe?
M. Kopecký, J. Reichrt.
Bull. Astron. Inst. Czechoslovakia, Vol. 33, 190 - 191 (1982).

085.037   Caractères météorologiques du climat, et activité
solaire: quelques remarques.   J.-C. Pecker.
Compendium in astronomy, (see 003.012), p. 151 - 160
(1982).
The variation of pluviosity and air temperature in 8
stations of the equatorial zone and in Paris was studied using
various techniques including Fourier analysis. The correlation
with solar activity is rather weak. This correlation might however become stronger if the solar activity at a given time with
the pluviosity and the air temperature at a later time is compared.

085.038   A suggested approach to research on the sun-weather
problem.   W. O. Roberts, R. H. Olson.
Compendium in astronomy, (see 003.012), p. 161 - 167
(1982).
The authors have tried to show that there are both
theoretical and empirical reasons to suppose that some sort of
mechanism involving atmospheric electricity is likely to be a
key to understanding sun-weather relationships.

085.039   Solar activity and Indian weather/climate.
H. N. Bhalme, R. S. Reddy, D. A. Mooley,
B. V. Ramana Murty.
Proc. Indian Acad. Sci. Earth Planet. Sci., Vol. 90, 245 - 262
(1981).

Solar activity and the earth.   See Abstr. 003.095.

The problem of solar-atmospheric relations. Sun –
Atmosphere 1976 Experiment.   See Abstr. 003.149.

The origin of solar-terrestrial studies.
See Abstr. 004.030.

The physical basis of the ionosphere in the solar-
terrestrial system. Conference held at Pozzuoli, Italy. 28 - 31
October 1980.   See Abstr. 012.059.

Report of IAU Commission 10: Solar activity
(Activité solaire).   See Abstr. 072.104.

Large scale solar magnetic fields at the site of flares,
the greatness of flares, and solar-terrestrial disturbances.
See Abstr. 073.013.

Simulation of three-dimensional solar wind
disturbances and resulting geomagnetic storms.
See Abstr. 074.026.

Review of ground-based measurements.
See Abstr. 080.075.

Evolutionary variations of solar luminosity.
See Abstr. 080.089.

Erratum

085.901   Erratum: "A relation of disastrous weather events
with solar activity" [(Soln. Dannye 1981 Byull.,
No. 5, p. 97 - 100 (1981)].   N. A. Lebedeva.
Soln. Dannye 1982 Byull., No. 1, p. 110 (1982). In Russian.
See Abstr. 30.085.010.

Planetary System

091 Physics of the Planetary System (Dynamics, Figure, Rotation, Interiors, Atmospheres, Magnetic Fields, etc.)

091.001 **Relativistic perturbations for all the planets.**
J.-F. Lestrade, P. Bretagnon.
Astron. Astrophys., Vol. 105, 42 - 52 (1982). In French.
The purpose of this paper is to present the relativistic perturbations in the osculating elements of all planets due to the theory of general relativity. Only the gravitational field of the sun is taken into account for these relativistic effects calculated in the post-Newtonian approximation.

091.002 **Planetogonic implications of angular momenta in satellite systems.** D. Möhlmann.
Gerlands Beitr. Geophys., Band 91, 90 - 91 (1982).

091.003 **Geodetic singularities in the gravity field of a non-homogeneous planet.** F. Bocchio.
Geophys. J. R. Astron. Soc., Vol. 68, 643 - 652 (1982).
An inverse geodetic singularity problem is considered for a non-homogeneous spherical planet. The singularity condition is expressed in terms of the density distribution and of the geometrical parameters of the configuration; the condition for the density distribution is deduced which gives rise to singularities of parabolic type in the external gravity field of the planet. The structure of the gravity field in the neighbourhood of the singularities is investigated in detail together with the behaviour of the gradients of the disturbances in the geodetic coordinates.

091.004 **On conditions for the existence of intersections of a satellite orbit with a conical shadow.**
V. G. Sokolov.
Astron. Zh., Tom 59, 142 - 146 (1982). In Russian. English translation in Soviet Astron., Vol. 26, No. 1.
The equation which determines the points of intersection of a satellite orbit with the conical shadow of a planet (shadow equation) has been transformed to an algebraic equation of the fourth power with respect to tg $(\overline{w}/2)$, $\overline{w}$ being the satellite longitude counted from the point of the terminator plane at which the satellite enters into the shadow semi-space. With such choice of the longitude the roots corresponding to intersections of the orbit with the shadow are always positive and restricted.

091.005 **Particle acceleration at planetary bow shock waves.**
M. M. Hoppe, C. T. Russell.
Nature, Vol. 295, 41 - 42 (1982).
We can extend our understanding of collisionless shocks by comparing their behaviour in a variety of plasma conditions at several different planets. One property of such shocks is the occurrence of upstream magnetohydrodynamic waves associated with particle beams accelerated at these shocks, and flowing back towards the sun. The authors report observations of one of these classes of wave at Mercury, Venus, Earth and Jupiter.

091.006 **On the invariable plane of the solar system.**
G. Burkhardt.
Astron. Astrophys., Vol. 106, 133 - 136 (1982).
The position of the invariable plane of the solar system in space was determined using the Development Ephemeris number 102. The inclination and ascending node of the invariable plane with respect to the coordinate systems ecliptic-equinox and equator-equinox of the two epochs B 1950.0 and J2000.0 are given.

091.007 **Innerer Aufbau und Dynamik der erdartigen Planeten.** U. Walzer.
Sterne, 58. Band, 6 - 21 (1982) = Mitt. Zentralinst. Phys. Erde Nr. 927.

091.008 **New investigations of planets and their satellites.**
L. V. Ksanfomaliti.
Astronomical calendar. Yearbook. 1982, (see 047.005), p. 170 - 187 (1981). In Russian. – Abstr. in Ref. zh., 51. Astron., 2.51.28 (1982).

091.009 **Uses of volcanic eruption processes as probes of the structure of planetary crusts.**
L. Wilson, J. W. Head III.
Geophys. J. R. Astron. Soc., Vol. 69, (see 012.004), 302 (1982). – Abstract.

091.010 **Nitrogen isotopes in the solar system.**
J. Geiss, P. Bochsler.
Geochim. Cosmochim. Acta, Vol. 46, 529 - 548 (1982).
Measurements of the isotopic composition of nitrogen in the solar system are summarized. The authors show that the 30% change, during the last 3 to 4 billion years, of $^{15}N/^{14}N$ in solar-wind-bearing lunar soils and breccias probably does not reflect changes in this ratio at the solar surface. Such changes, whether by spallation or thermonuclear reactions are ruled out by comparing the yields of ^{15}N with those of other rare isotopes such as ^{9}Be, ^{11}B, ^{3}He or ^{13}C, even if an arbitrary degree of solar mixing is introduced. Moreover, the authors calculate that the solar activity required for producing significant amounts of ^{15}N by spallation at the solar surface should have resulted in a particle bombardment of the Moon of an intensity that would have produced amounts of spallation isotopes (e. g. ^{15}N, ^{21}Ne, ^{38}Ar, ^{131}Xe) several orders of magnitude in excess of what is actually found in the whole regolith.

091.011 **Solid state convection in icy satellites: effects of phase transitions upon stability.**
R. Reynolds, C. Alexander, A. Summers, P. Cassen.
Reports of Planetary Geology Program – 1981, (see 003.001), p. 59 - 61 (1981). – Abstract.

091.012 **Lobate and multilobate ejecta deposits: a mechanism for their emplacement and its implications for the water content of the Martian subsurface.** A. Woronow.
Reports of Planetary Geology Program – 1981, (see 003.001), p. 85 - 86 (1981). – Abstract.

091.013 **Impact basins: stages in basin formation and evolution.** J. W. Head, S. C. Solomon.

Reports of Planetary Geology Program – 1981, (see 003.001), p. 111 - 113 (1981). – Abstract.

091.014 **Viscous relaxation of impact basin topography: implications for the moon and Venus.**
S. C. Solomon, R. P. Comer, S. K. Stephens, J. W. Head.
Reports of Planetary Geology Program – 1981, (see 003.001), p. 114 - 116 (1981). – Abstract.

091.015 **A size: rank model for basin rings.**
R. J. Pike.
Reports of Planetary Geology Program – 1981, (see 003.001), p. 123 - 125 (1981). – Abstract.

091.016 **Planetary impact basin peak-ring spacing: a comparison.** J. M. Boyce.
Reports of Planetary Geology Program – 1981, (see 003.001), p. 126 - 128 (1981). – Abstract.

091.017 **Planetary megaregoliths.** R. A. De Hon.
Reports of Planetary Geology Program – 1981, (see 003.001), p. 129 - 131 (1981). – Abstract.

091.018 **Numerical taxonomy of central volcanoes on the planets.** R. J. Pike, G. D. Clow.
Reports of Planetary Geology Program – 1981, (see 003.001), p. 138 - 140 (1981). – Abstract.

091.019 **Streamlined islands: an analysis of their minimum-drag shape.** P. D. Komar.
Reports of Planetary Geology Program – 1981, (see 003.001), p. 266 - 268 (1981). – Abstract.

091.020 **Comparative rates of geologic processes on various terrestrial type bodies.** J. A. Cutts.
Reports of Planetary Geology Program – 1981, (see 003.001), p. 331 - 333 (1981). – Abstract.

091.021 **Topographic lineament analysis: possible stress indicators on planetary surfaces.**
D. U. Wise, M. L. Allison.
Reports of Planetary Geology Program – 1981, (see 003.001), p. 377 - 379 (1981). – Abstract.

091.022 **Clay minerals on planetary surfaces: a cautionary note regarding their identification by VIS/NIR spectral remote sensing.** J. L. Gooding.
Reports of Planetary Geology Program – 1981, (see 003.001), p. 457 - 459 (1981). – Abstract.

091.023 **Globes of the planets.**
R. M. Batson, J. L. Inge.
Reports of Planetary Geology Program – 1981, (see 003.001), p. 479 - 480 (1981). – Abstract.

091.024 **Outer solar system nomenclature.**
T. Owen.
Reports of Planetary Geology Program – 1981, (see 003.001), p. 530 - 531 (1981). – Abstract.

091.025 **Geology of small bodies: prospectus for the Planetary Geology Program Workshop.**
C. R. Chapman.
Reports of Planetary Geology Program – 1981, (see 003.001), p. 532 - 534 (1981). – Abstract.

091.026 **A comparison of some explosive volcanic eruption processes on the earth, moon, Mars, Venus and Io.**
L. Wilson, J. W. Head.
Reports of Planetary Geology Program – 1981, (see 003.001), p. 537 - 538 (1981). – Abstract.

091.027 **Agglutinates, agglutinate recycling and planetary regolith.** A. Basu, D. S. McKay.
Meteoritics, Vol. 16, 292 - 293 (1981). – Abstract.

091.028 **Orbital motion of the planets, theoretical and observational.** P. K. Seidelmann.
Celestial Mech., Vol. 26, (see 012.001), 149 - 160 (1982).
A review is presented of the progress that has taken place since 1976 in the knowledge of the orbital motion of the planets. Analytical theories, numerical integrations, inter-comparisons of theories and ephemerides, observational data, the fundamental constants for the theories and the calculation of ephemerides are all discussed separately. In addition, the prospects for future developments in each area are also discussed.

091.029 **Theory of the inner planets.** P. Bretagnon.
Celestial Mech., Vol. 26, (see 012.001), 161 - 167 (1982). In French.
In the construction of planetary theories for the whole of the solar system undertaken at the Bureau des Longitudes, the aim is to obtain the precision of $0.''001$ over several centuries for the inner planets; $0.''01$ over one century, $0.''1$ over 1000 years for the outer planets. To get these precisions one must compute the perturbations at least to the 3rd order of the masses for the inner planets and to the 6th order of the masses for the outer planets. The author has used an iterative method which has given the perturbations up to the 6th order of the masses for the outer planets and has built the perturbations up to the 3rd order with respect to the masses for all the planets.

091.030 **The JPL planetary ephemerides.**
E. M. Standish, Jr.
Celestial Mech., Vol. 26, (see 012.001), 181 - 186 (1982).
This paper discusses the JPL planetary ephemerides, concentrating on what the ephemerides represent and which facets are well-determined. It also presents numerical estimates of the elements, mentions the expected progress in the future and discusses possible uses for the ephemerides themselves.

091.031 **Improvement of the ephemerides of the inner planets and the moon using radar measurements, lunar laser ranging and meridian observations 1961 - 1980.**
G. A. Krasinskij, E. V. Pit'eva, M. L. Sveshnikov, E. S. Sveshnikova.
Byull. Inst. Teor. Astron., Leningrad, Tom 15, 145 - 164 (1982). In Russian.
A description of a numerical theory of the motions of the inner planets and the moon is given. The theory is compared with high accuracy radar measurements of the inner planets made during 1961 - 1980 and lunar laser ranging of the corner reflector Apollon-15. The disagreement with radar measurements makes for Mercury 3, for Venus 1.3 and for Mars 0.6 km. The mean square error of representation of lunar laser ranging equals 9 m. The connection between the coordinate systems of the constructed theory and FK 4 is determined by comparing it with meridian observations of the sun. For three dates the coordinates and velocities of the major planets and the moon are given. These data may be used as initial values for numerical integration.

091.032 **New frontiers of planetary explorations.**
L. V. Ksanfomaliti.
Probl. kosm. issled. Moskva, 1981, p. 35 - 47. In Russian.
Abstr. in Ref. zh., 51. Astron., 3.51.168 (1982).

091.033 **Weather and climate on planets.**
K. Ya. Kondrat'ev.
Probl. kosm. issled. Moskva, 1981, p. 48 - 59. In Russian.
Abstr. in Ref. zh., 51. Astron., 3.51.180 (1982).

091.034 Die Problematik der Entstehung der Monde und ihrer Bahnentwicklung am Beispiel des Erdmondes und der Marsmonde. W. Sanns.
Monde, Ringe und Asteroide im Planetensystem, (see 012.020), p. 1 - 28 (1981).

091.035 Oberflächen und Aufbau der Monde. B. Moßbacher.
Monde, Ringe und Asteroide im Planetensystem, (see 012.020), p. 29 - 74 (1981).

091.036 Beobachtung von Planetenringen. H. Mandel.
Monde, Ringe und Asteroide im Planetensystem, (see 012.020), p. 139 - 170 (1981).

091.037 A climatic theory of temperature distribution of certain planetary atmospheres. Y.-h. Lu, J.-p. Chao.
Sci. Atmos. Sinica, Vol. 5, 145 - 156 (1981). – Abstr. in Phys. Abstr., Vol. 85, Abstr. 48166 (1982).

091.038 On the possibility of existing of global electromagnetic resonances on the planets of the solar system. A. P. Nikolaenko, L. M. Rabinovich.
Kosm. Issled., Tom 20, 82 - 88 (1982). In Russian.

091.039 The range and unity of planetary circulations. G. P. Williams, J. L. Holloway, Jr.
Nature, Vol. 297, 295 - 299 (1982).
Altering the rotation rate, obliquity and diurnal period of an Earth-like model atmosphere produces a wide range of circulation forms, some of which resemble those observed on Venus, Mars, Jupiter, Saturn and (perhaps) on Uranus and Neptune. These unified solutions suggest: that Jupiter and Saturn resemble a larger, faster-spinning Earth and possess a stress-bearing or momentum-exchanging sublayer; that easterly winds prevail in Uranus' summer hemisphere; and that Venus resembles a slowly rotating Earth if diurnal heating variations are included.

091.040 Magnetic fields and charged particles around major planets and their satellites. M. G. Kivelson.
Philos. Trans. R. Soc. London A, Vol. 303, (see 012.022), 247 - 260 (1981).
The theme of this paper is planetary magnetic fields and magnetospheres. The properties of Jupiter's and Saturn's internal fields are described and the uncertainty inherent in published field models is stressed. The recent progress in measuring and interpreting synchrotron radiation from Jupiter is described, following which some features of Io's torus are reviewed. The possibility that Io has an intrinsic magnetic field is considered and the properties of Io's putative magnetosphere, embedded within the Jovian magnetosphere, are discussed.

091.041 The origin of planetary rings. S. F. Dermott.
Philos. Trans. R. Soc. London A, Vol. 303, (see 012.022), 261 - 279 (1981).
Recent spacecraft and ground-based observations have revealed the presence of narrow rings encircling the planets Jupiter, Saturn and Uranus. The structure of the Saturnian F ring has been resolved by Voyager 1 and appears to be determined by the action of two small neighbouring satellites which were also imaged by the spacecraft. All nine Uranian rings are extremely narrow and some are appreciably eccentric. The resolution of the numerous, but well defined dynamical problems posed by these narrow rings must precede any discussion of the origin of rings.

091.042 Constraints on the origin and interior structure of the major planets. W. B. Hubbard.
Philos. Trans. R. Soc. London A, Vol. 303, (see 012.022), 315 - 326 (1981).
From fitting models to the external gravity field of the major planets, Uranus, Neptune, Jupiter and Saturn, one finds that certain interior characteristics may be common to all four. For Uranus and Neptune, a model with a central iron-silicate core of about $4M_E$ (M_E= mass of Earth), an 'ice' layer of H_2O, CH_4 and NH_3 in solar proportions of ca. $10M_E$, and an H_2–He atmosphere of ca. $M_E - 2M_E$ gives a good fit to available constraints, including heat flow measurements. Models of Jupiter and Saturn have cores very similar to those of Uranus and Neptune. Modes of origin consistent with these features are discussed. Such model predict a considerable enrichment of deuterium relative to primordial solar abundances in Uranus and Neptune. Such enrichment is not observed in Uranus; implications for interior structure and origin are discussed.

091.043 Natural radioactivity of the moon and planets. Yu. A. Surkov.
Proc. Twelfth Lunar Planet. Sci. Conf., (see 012.024), p. 1377 - 1386 (1982).

091.044 A possible common origin for the rare gases on Venus, Earth, and Mars. C. J. Hostetler.
Proc. Twelfth Lunar Planet. Sci. Conf., (see 012.024), p. 1387 - 1393 (1982).
The rare gas concentrations in the terrestrial planets and the carbonaceous chondrites could have been derived from a single source – an early solar wind. The fractionation processes responsible for the varying concentration patterns in this model are differential ionization and the relative separation of neutrals from ions due to interaction with the solar magnetic field.

091.045 Planets, asteroids and comets at high angular resolution. T. Encrenaz.
Scientific importance of high angular resolution at infrared and optical wavelengths, (see 012.031), p. 307 - 317 (1981).

091.046 The magnetospheres of Saturn, Mercury, Venus and Mars. Y. C. Whang, K. I. Gringauz.
Solar system plasmas and fields, (see 003.005), p. 61 - 69 (1982).

091.047 The shape of small solar system bodies: gravitational equilibrium vs. solid-state interactions. P. Farinella, F. Ferrini, A. Milani, A. M. Nobili, P. Paolicchi, V. Zappalà.
The comparative study of the planets, (see 012.033), p. 71 - 77 (1982).
The authors discuss the new data on the shape of small planetary satellites and asteroids, comparing them with the theory of equilibrium figures for self-gravitating masses. This comparison gives interesting informations on the structure, density and strength of these bodies.

091.048 Ages of the solar system: isotopic dating. G. Turner.
The comparative study of the planets, (see 012.033), p. 85 - 94 (1982).
This paper gives a survey of the methods whereby measurements of isotopic abundances in radioactive decay chains have been used to determine the chronology of the origin and evolution of the solar system. Results from radioactive dating are compared with data from other age determination methods for solar system bodies.

091.049 Solar system cratering chronology and dating of the surface structures of the terrestrial-type planets. G. Neukum.
The comparative study of the planets, (see 012.033), p. 101 - 116 (1982).

The ancient impact record of the terrestrial-type planets Mercury, Mars, earth's moon, and of the satellites of Jupiter and Saturn is discussed on the basis of data from spacecraft imagery. The mass-velocity distribution of the impactors seems to have been the same or very similar in the inner part of the solar system and probably also at Jupiter and Saturn. Ancient impact rates appear to have been comparable. The time dependence of the impact rate in the earth-moon system and by analogy at the other terrestrial-type planets is in accordance with a smooth rapid decay during the first 1000 million years of solar system history rather than with a peak in impact rate (cataclysm) at 4000 million years ago.

091.050 The stability of the solar system.   A. E. Roy.
The comparative study of the planets, (see 012.033), p. 117 - 124 (1982).
The empirical stability criteria approach to the stability of hierarchical dynamical n-body systems is described. Application of this approach is made to the problem of the long-term stability of the solar system's hierarchical systems.

091.051 Heat transfer and the development of internal structure in the terrestrial planets.   D. C. Tozer.
The comparative study of the planets, (see 012.033), p. 161 - 180 (1982).
A theory of planetary differentiation is developed which incorporates the tendency of convective heat transfer to maintain very high average viscosities throughout planetary interiors under a wide range of conditions.

091.052 Atmospheric evolution.   H.-J. Bolle.
The comparative study of the planets, (see 012.033), p. 203 - 220 (1982).
The abundances of atmospheric gases in planetary atmospheres impose certain boundary conditions upon planetary formation theories. One of the most important is that the present volatiles did probably not undergo a hot phase during early accretion and must have been added at a later stage if there was such a hot phase. A short overview is given of the present knowledge about the structure of the planetary atmospheres.

091.053 Circulation systems in planetary atmospheres.   H.-J. Bolle.
The comparative study of the planets, (see 012.033), p. 221 - 252 (1982).
The physical processes which generate motions in planetary atmospheres are shortly discussed. The application of the basic principles is illustrated by (1) the relation between the thermal structure of a planetary atmosphere and its wind system and by (2) the need for CO_2 transports in the Martian atmosphere. The importance of radiative processes for the observed phenomena can be estimated by means of a scale analysis. An overview is given on the circulation systems observed in different planetary atmospheres and their possible mechanisms.

091.054 Elements of comparative magnetoplanetology.   F. Mariani.
The comparative study of the planets, (see 012.033), p. 261 - 290 (1982).
In this review the present state of our knowledge of the magnetic fields and plasma environments of the planets is summarized. Mechanisms capable of generating magnetic fields outside and inside a planet are discussed and the interaction of planetary magnetospheres with the solar wind is described. An overall view of the results of observations onboard space probes and satellites is presented.

091.055 Geologic overview of the terrestrial planets.   R. E. Arvidson.
The comparative study of the planets, (see 012.033), p. 391 -

407 (1982).
The terrestrial moons and planets of our solar system are composed largely of rocky and metallic materials, and preserve geologic records. The moons of the giant planets, especially the larger moons, can be thought of as terrestrial bodies, albeit exotic ones. The terrestrial bodies are close enough in bulk physical and chemical parameters to make meaningful comparisons between them. On the other hand, they are different enough so that each one has evolved a unique geological record. In this paper, the Moon is discussed first, followed by Mars, then Venus, and the paper ends with a discussion of the evolution of the Galilean Satellites.

091.056 Faulting and fracturing of planetary surfaces.   D. U. Wise.
The comparative study of the planets, (see 012.033), p. 409 - 418 (1982).
Investigation of faulting and fracturing of planetary surfaces involves the determination of past stress systems and their effect on the materials of those surfaces and to the upper layers of the planet. The stress tensor and its relationship to fault types using theories of E. M. Anderson are described. Fault systems, stress systems, and possible ways of dating and generating these stresses on regional and planetary scales are discussed. These principles are applied to the Alba Volcano of Mars.

091.057 Aeolian modification of planetary surfaces.   R. Greeley.
The comparative study of the planets, (see 012.033), p. 419 - 434 (1982).
Any planet or satellite having a dynamic atmosphere and a solid surface has the potential for aeolian processes. A survey of the Solar System shows that wind plays an important, and in some regions, the key role in surface modifications. Most deserts and many coastal areas on Earth are subject to aeolian processes. From the extremely dense, hot atmosphere of Venus to the low atmospheric density of Mars and the extremely cold environment of Titan, there is the opportunity to study a single geological process under a wide range of environments to derive fundamental knowledge of how aeolian processes operate.

091.058 Some thermodynamic relationships governing the behavior of permafrost and frozen ground.   D. M. Anderson.
The comparative study of the planets, (see 012.033), p. 435 - 440 (1982).
It is well established that water-ice occurs in the surface materials of Mars and that temperature regimes are such that ice-rich permafrost may be present in many localities. Water-ice is also thought to be a major constituent of the surface materials on Europa, Ganymede, Callisto, and several of the moons of Saturn.

091.059 Some results from a reduction of radar, laser and optical observations of the inner planets and of the moon.   G. A. Krasinskij, E. V. Pit'eva, M. L. Sveshnikov, E. S. Sveshnikova.
Dokl. AN SSSR, Tom 261, 1320 - 1324 (1981). In Russian. Abstr. in Ref. zh., 51. Astron., 5.51.121 (1982).

091.060 The gravito-electrodynamics of charged dust in planetary magnetospheres.   D. A. Mendis, H. L. F. Houpis, J. R. Hill.
J. Geophys. Res., Vol. 87, 3449 - 3455 (1982).
The dynamics of small electrically charged dust grains within the rigidly corotating regions of planetary magnetospheres such as those of Jupiter and Saturn is considered.

091.061 Interiors of the giant planets.   D. J. Stevenson.
Annu. Rev. Earth Planet. Sci., Vol. 10, (see 003.009).

257 - 295 (1982) = Contrib. No. 3659 Div. Geol. Planet. Sci., Calif. Inst. Technol., Pasadena, CA 91125.

091.062 Regoliths on small bodies in the solar system.
K. R. Housen, L. L. Wilkening.
Annu. Rev. Earth Planet. Sci., Vol. 10, (see 003.009), 355 - 376 (1982).
Contents: The lunar regolith. Regoliths on the meteorite parent bodies. Asteroidal regoliths. Regoliths on other small bodies.

091.063 New impetus to the exploration of the solar system.
W. Fricke.
Sun and planetary system, (see 012.040), p. 9 - 11 (1982).

091.064 The solar system — known facts and unsolved problems. Z. Kopal.
Sun and planetary system, (see 012.040), p. 13 - 21 (1982).

091.065 Precession and nutation influence on the harmonic coefficients of a planetary potential.
V. Oumlensky, V. Shkodrov.
Sun and planetary system, (see 012.040), p. 209 - 212 (1982).
In this paper an analysis of the precession and nutation influence on the coefficients of spherical harmonics is made. The greatest influence on the accuracy of the zonal coefficients is exerted by the deviation of the z-axis of the inertial system with respect to the instantaneous one. In this case the relative error is proportional to the order of the harmonic coefficient. The x-axis deviation does not exert any influence.

091.066 A physical detail relevant to the Savić-Kašanin theory of behaviour of materials under high pressure.
V. Čelebonović.
Sun and planetary system, (see 012.040), p. 249 - 250 (1982).
P. Savić and R. Kašanin have proposed a theory of behaviour of materials under high pressure. Their theory can be applied to the explanation of the internal structures of planets and stars. In the present paper, a simple method for the calculation of the internal temperatures of the terrestrial planets is proposed.

091.067 Associative ionization and sodium in the atmospheres of planetary system bodies.
V. Vujnović.
Sun and planetary system, (see 012.040), p. 259 - 260 (1982).

091.068 On physical interpretation of hypsometric characteristics of the Moon and planets. I. V. Gavrilov.
Sun and planetary system, (see 012.040), p. 261 - 262 (1982).
Hypsometric characteristics of the planets contain important information on the origin and evolution of these bodies. With the use of earth based and space observations, during the past two decades many hypsometric data have been obtained for the Moon, Mars, Mercury and some other planets and their satellites. For the purposes of comparative planetology it is very important that hypsometric characteristics of these bodies were homogeneous from point of view of its physical and geometric interpretation.

091.069 Review of the dynamics of satellites and planetary rings. P. J. Message.
Sun and planetary system, (see 012.040), p. 353 - 359 (1982).
Some of the new discoveries and theoretical work of the most recent few years are reviewed. Many new satellites have been discovered in the systems of Jupiter and Saturn, and the planets Jupiter, Saturn and Uranus have been found to possess a number of very narrow rings.

091.070 Nouvelles théories des planètes et de la Lune dans les éphémérides françaises.
P. Bretagnon, J. Chapront, M. Chapront-Touzé.

Sun and planetary system, (see 012.040), p. 437 - 440 (1982).
New theories for the motions of the Moon and the planets, which were developed at the Bureau des Longitudes, are compared with observations and with numerical methods. These new theories will be introduced in 'Connaissance des Temps' from 1984 on.

091.071 Inclination of rotation axis of planets.
M.- c. Zhang.
Acta Astron. Sinica, Vol. 23, 49 - 55 (1982). In Chinese.

091.072 Significant achievements in the Planetary Geology Program, 1981.
P. J. Mouginis-Mark (Editor), M. J. Cintala, W. E. Elston, F. P. Fanale, E. K. Gibson, J. S. King, P. D. Komar, B. K. Lucchitta, P. Thomas, J. Veverka, D. U. Wise (Contributing Authors).
NASA Contract. Rep., NASA CR-3554, 5 + 59 pp. (1982).
Recent developments in planetology research as reported at the 1982 NASA Planetary Geology Program Principal Investigators meeting are summarized. Important developments are summarized in topics ranging from solar system evolution, comparative planetology, and geologic processes, to techniques and instrument development for future exploration.

091.073 Report of IAU Commission 16: Physical study of planets and satellites (Étude physique de planètes et satellites). B. A. Smith.
Trans. IAU, Vol. XVIIIA, (see 003.013), p. 175 - 179 (1982).

091.074 Report of the IAU working group for planetary system nomenclature (Nomenclature du système planétaire). P. M. Millman.
Trans. IAU, Vol. XVIIIA, (see 003.013), p. 669 (1982).

091.075 Constitution of terrestrial planets.
H. Wänke.
Philos. Trans. R. Soc. London A, Vol. 303, (see 012.022), 287 - 302 (1981).

Une banque de données pour l'étude des phénomènes de transfert radiatif dans les atmosphères planétaires: la banque "GEISA". See Abstr. 002.030.

Introduction to planetary physics. See Abstr. 003.016.

Weather and climate on planets. See Abstr. 003.078.

The traveller's guide to the solar system. See Abstr. 003.093.

Planet guidebook, Vol. 1 and 2. See Abstr. 003.147.

The ecliptical latitudes of Mars, Jupiter and Saturn according to Kushijar Dzhili as compared with modern calculations. See Abstr. 004.044.

Determination of the ecliptic coordinates of planets in the works of the scientists of the medieval Orient. See Abstr. 004.083.

Planetary data at the National Space Science Data Center. See Abstr. 013.011.

A direct analytic method of calculating the quadrupole parameters of a planetary magnetic field. See Abstr. 021.004.

Pressure-induced H_2 opacity in the 5-μm region. See Abstr. 022.034.

Experimental impact craters formed in viscous fluids.   See Abstr. 022.048.

Vertical structure of basaltic lava flows: implications for surface sampling and interpretation. See Abstr. 022.049.

Scale modeling of lava flow processes. See Abstr. 022.050.

Particles formed by fuel-coolant explosions. See Abstr. 022.051.

Melt-water interactions: series II experimental design.   See Abstr. 022.052.

Some thermodynamic relationships governing the behavior of permafrost and frozen ground. See Abstr. 022.062.

Impact experiments and regolith buildup. See Abstr. 022.067.

A schematic model of crater modification by gravity.   See Abstr. 022.109.

Spectroscopic line parameters of NH_3 and PH_3 in the far infrared.   See Abstr. 022.113.

Quantum assignments and intensity measures for methane between 1100 and 1800 cm^{-1}: a comparison between theory and experiment.   See Abstr. 022.180.

Remote sensing of fissure-fed basalt flows and their source areas: craters of the Moon Volcanic Field, Idaho. See Abstr. 031.522.

The determination of the center of gravity of a planet from photographic plates.   See Abstr. 031.528.

The UCL interactive processing system for application to remote sensing studies.   See Abstr. 031.531.

Dynamical equinox and analytical theory of the Sun. See Abstr. 041.003.

On new results of a study of the photographic irradiation effect in position observations of the major planets. See Abstr. 041.027.

Asymptotic series for planetary motion in periodic terms in three dimensions.   See Abstr. 042.012.

A third-order intermediate orbit for planetary theory.   See Abstr. 042.014.

Using a uniform approximation of the coordinates of disturbing bodies in the Taylor-Steffensen numerical integration method.   See Abstr. 042.036.

Periodic orbits of the elliptic restricted problem for the Sun—Jupiter—Saturn system.   See Abstr. 042.041.

Quantisation in stable gravitational systems. See Abstr. 042.104.

The rotation rates of accreting planetesimals. See Abstr. 042.106.

A postulate leading to the Titius—Bode law. See Abstr. 042.107.

Adiabatic charged particle motion in rapidly rotating magnetospheres.   See Abstr. 062.063.

Rotating fluids in geophysics and planetary physics. See Abstr. 062.076.

Solar motion and solar activity. See Abstr. 080.059.

A brief panorama.   See Abstr. 084.062.

New approach to determining planetary perturbations in lunar theory.   See Abstr. 094.014.

Support of topographic and other loads on the moon and on the terrestrial planets.   See Abstr. 094.035.

Report of IAU Commission 20: Positions and motions of minor planets, comets and satellites (Positions et mouvements des petites planètes, des comètes et des satellites). See Abstr. 098.107.

Impact cratering mechanics. See Abstr. 105.210.

The origin of impacting populations in the inner and outer solar system.   See Abstr. 107.005.

Volatile substances in terrestrial planets and their atmospheres and formation of planetary atmospheres. See Abstr. 107.015.

Equation of state experiments and theory relevant to planetary modelling.   See Abstr. 107.016.

Some remarks on the formation of terrestrial planets. See Abstr. 107.024.

Origin and evolution of the giant planets. See Abstr. 107.025.

092 Mercury

092.001 **Mercurio.**   R. Casacchia.
Coelum, Vol. 50, 1 - 10 (1982).

092.002 **Een lange, hete dag.** Op Mercurius en Venus.
K. Velt.
Zenit, 9. Jaarg., 115 - 117 (1982).

092.003 **Mercury's history revisited.**
M. A. Leake, C. R. Chapman, S. J. Weidenschilling,
D. R. Davis, R. Greenberg.
Reports of Planetary Geology Program — 1981, (see 003.001),
p. 408 - 410 (1981). — Abstract.

092.004 **Early fluctuations in the radius of Mercury.**
B. M. Cordell.
Mod. Geol., Vol. 7, 209 - 216 (1981). — Abstr. in Phys. Abstr.,
Vol. 85, Abstr. 18724 (1982).

092.005 **Principal features of structure and evolution of**
Mercury's surface.   G. N. Katterfel'd.
Geomorfologiya, 1982, No. 1, p. 13 - 21. In Russian. — Abstr.
in Ref. zh., 51. Astron., 5.51.177 (1982).

092.006 **Relativistic effects in the Solar System. I. Relativis-**
tic pathway modifications of the planets.
E. R. Bagge.
Atomkernenerg. Kerntech., Vol. 39, 223 - 228 (1981).
Abstr. in Phys. Abstr., Vol. 85, Abstr. 58227 (1982).

092.007 **Global volcanism and tectonism on Mercury:**
comparison with the Moon.
P. G. Thomas, P. Masson, L. Fleitout.

Earth Planet. Sci. Lett., Vol. 58, 95 - 103 (1982).
Both morphologic and tectonic studies indicate that
Mercury and the Moon have quite different internal histories,
despite their apparently similar morphologies. The evaluation
of the volcanic surfaces indicates a decreasing volcanism on
Mercury at the largest impacting time, despite short and
local reactivations. On the Moon, the basaltic volcanism was
increasing at the same time and continued for 1 billion years.
That indicates a strongly different thermal evolution for these
two planetary bodies.

092.008 **Far -field tectonics associated with a large impact**
basin: applications to Caloris on Mercury and
Imbrium on the Moon.   L. Fleitout, P. G. Thomas.
Earth Planet. Sci. Lett., Vol. 58, 104 - 105 (1982).
Lithospheric readjustments after the formation of a
large impact crater have been computed. The models predict
tectonic perturbations over a major portion of the surface of
the planet. The weight of the ejecta and the mechanical
perturbations in the crater area give rise to membrane stresses.

092.009 **Some results of the Mercury transit observations in**
1970 and 1973 at Belgrade.
V. Benishek-Protitch (*Protic-Benisek*).
Sun and planetary system, (see 012.040), p. 447 - 448 (1982).

Planetary regularities in crater distribution on Mars,
the moon and Mercury (Collected diagrams).
See Abstr. 003.075.

On some invariant manifold results and their
applications.   See Abstr. 021.021.

093 Venus

093.001 **Venus: the mystery continues.**
J. K. Beatty.
Sky Telesc., Vol. 63, 134 - 138 (1982).

093.002 **Venus − Zwilling der Erde oder Außenseiterin im Sonnensystem?** C. Bloess.
Umschau, 82. Jahrg., 48 - 49 (1982).

093.003 **Plasma clouds above the ionopause of Venus and their implications.**
L. H. Brace, R. F. Theis, W. R. Hoegy.
Planet. Space Sci., Vol. 30, 29 - 37 (1982).

Early Pioneer Venus orbiter measurements by the Electron Temperature Probe have revealed wavelike structures at the ionopause and clouds of plasma above the ionopause, features which may represent ionospheric plasma at different stages in its removal by solar wind−ionosphere interaction processes. Continuing operation of the orbiter through three Venus years has now provided enough additional examples of these features to permit their morphologies to be examined in some detail.

093.004 **Magnetic field and plasma wave observations in a plasma cloud at Venus.** C. T. Russell,
J. G. Luhmann, R. C. Elphic, F. L. Scarf, L. H. Brace.
Geophys. Res. Lett., Vol. 9, 45 - 48 (1982).

Pioneer Venus magnetic field and plasma wave data are examined in a particularly clear example of a plasma cloud above the Venus ionosphere.

093.005 **Geology and geophysics.** G. E. McGill.
Nature, Vol. 296, 14 - 15 (1982). − Presented at the International Conference on the "Venus environment", held in Palo Alto, Calif.,1 - 6 November 1981.

093.006 **Structure and energetics of the atmosphere.**
F. W. Taylor.
Nature, Vol. 296, 16 - 17 (1982). − Presented at the International Conference on the "Venus environment", held in Palo Alto, Calif.,1 - 6 November 1981.

093.007 **Dynamics of the atmosphere.** L. S. Elson.
Nature, Vol. 296, 17 (1982). − Presented at the International Conference on the "Venus environment", held in Palo Alto, Calif.,1 - 6 November 1981.

093.008 **Clouds and hazes.** R. G. Knollenberg.
Nature, Vol. 296, 18 (1982). − Presented at the International Conference on the "Venus environment", held in Palo Alto, Calif.,1 - 6 November 1981.

093.009 **Structure and dynamics of the ionosphere.**
A. F. Nagy, L. H. Brace.
Nature, Vol. 296, 19 (1982). − Presented at the International Conference on the "Venus environment", held in Palo Alto, Calif.,1 - 6 November 1981.

093.010 **The solar wind interaction.**
P. A. Cloutier, C. T. Russell.
Nature, Vol. 296, 20 (1982). − Presented at the International Conference on the "Venus environment", held in Palo Alto, Calif.,1 - 6 November 1981.

093.011 **A Soviet view of the venusian surface.** L. Wilson.
Nature, Vol. 296, 607 - 608 (1982).

093.012 **Variations in wind speeds in the upper atmosphere of Venus.**

C. Boyer, G. Coupinot, J. Hecquet, J.-L. Pieplu.
C. R. Acad. Sci. Paris, Tome 294, Sér. II, 257 - 260 (1982).
In French.

093.013 **Venus observation programme of the terrestrial planets section.** C. J. R. Lord.
J. British Astron. Assoc., Vol. 92, 74 - 76 (1982).

093.014 **Escape of hydrogen from Venus.**
M. B. McElroy, M. J. Prather, J. M. Rodriguez.
Science, Vol. 215, 1614 - 1615 (1982).

Recombination of O_2^+ represents a source of fast oxygen atoms in Venus' exosphere, and subsequent collisions of oxygen atoms with hydrogen atoms lead to escape of about 10^7 hydrogen atoms per square centimeter per second. Escape of deuterium atoms is negligible, and the ratio of deuterium to hydrogen should increase with time. It is suggested that the mass-2 ion observed by Pioneer Venus is D^+, which implies a ratio of deuterium to hydrogen in the contemporary atmosphere of about 10^{-2}, an initial ratio of 5×10^{-5}, and an original H_2O abundance not less than 800 grams per square centimeter.

093.015 **Venus gravity fields.** W. L. Sjogren, M. Ananda,
B. G. Williams, P. W. Birkeland, P. S. Esposito,
R. N. Wimberly, S. J. Ritke.
Ann. Géophys., Vol. 37, 179 - 184 (1981). − Abstr. in Phys. Abstr., Vol. 85, Abstr. 10346 (1982).

093.016 **What do hypsograms tell about planetary tectonics?**
M. C. Malin.
Reports of Planetary Geology Program − 1981, (see 003.001), p. 369 - 370 (1981). − Abstract.

093.017 **The roughness of the Venusian surface: a progress report.** G. G. Schaber.
Reports of Planetary Geology Program − 1981, (see 003.001), p. 429 - 431 (1981). − Abstract.

093.018 **Report from a torrid planet.**
J. K. Beatty.
Sky Telesc., Vol. 63, 452 - 453 (1982).

093.019 **Venus: chemical weathering of igneous rocks and buffering of atmospheric composition.**
S. Nozette, J. S. Lewis.
Science, Vol. 216, 181 - 183 (1982).

Data from the Pioneer Venus radar mapper, combined with measurements of wind velocity and atmospheric composition, suggest that surface erosion on Venus varies with altitude. Calcium- and magnesium-rich weathering products are produced at high altitudes by gas-solid reactions with igneous minerals, then removed into the hotter lowlands by surface winds. These fine-grained weathering products may then rereact with the lower atmosphere and buffer the composition of the observed gases carbon dioxide, water vapor, sulfur dioxide, and hydrogen fluoride in some regions of the surface. This process is a plausible mechanism for the establishment in the lowlands of a calcium-rich mineral assemblage, which had previously been found necessary for the buffering of these species.

093.020 **Cloud motions on Venus: global structure and organization.** S. S. Limaye, V. E. Suomi.
J. Atmos. Sci., Vol. 38, 1220 - 1235 (1981). − Abstr. in Phys. Abstr., Vol. 85, Abstr. 18725 (1982).

093.021 Origin of bright ring-shaped craters in radar images of Venus.
J. A. Cutts, T. W. Thompson, B. H. Lewis.
Icarus, Vol. 48, 428 - 452 (1981).
    The surface of Venus viewed in Arecibo radar images has a small population of bright ring-shaped features. These features are interpreted as the rough or blocky deposits surrounding craters of impact or volcanic origin. Population densities of these bright ring features are small compared with visually identified impact craters on the surface of the Moon and volcanic craters on Io. However, they are comparable to the short-lived radar-bright haloes associated with ejecta deposits of young craters on the Moon. This suggests that bright radar signatures of the deposits around Venusian craters are obliterated by an erosional or sedimentary process.

093.022 Infrared properties of haze particles of Venus.
S. Mukai, T. Mukai.
Icarus, Vol. 48, 482 - 487 (1981).
    The authors examine the infrared properties of haze particles of Venus by comparing the computed infrared radiation, based on the haze particles' properties in the visible defined by Kawabata et al. (1980), with the Earth-based infrared observations. Consequently, the optical constant in the infrared which is the most probable candidate as haze particle material is predicted.

093.023 Construction of a numerical theory of the motion of Venus over the time span 1961 - 1972. III. Reduction computations.   V. A. Izvekov.
Byull. Inst. Teor. Astron., Leningrad, Tom 15, 27 - 32 (1981). In Russian.
    The paper presents a method for reducing the computed coordinates to the form accepted in current publications. Corrections for precession, aberration, parallax and the lunar inequality are taken into account in rectangular coordinates. Small effects usually neglected in computations of O−C are dealt with in the present study.

093.024 Venere.   R. Bianchi.
Coelum, Vol. 50, 63 - 69 (1982).

093.025 Optical properties of the Venus atmosphere and radiative heat exchange.
K. Ya. Kondrat'ev, N. I. Moskalenko.
Astron. vestn., Tom 15, 196 - 210 (1981). In Russian. − Abstr. in Ref. zh., 51. Astron., 3.51.194 (1982).

093.026 On the magnetic barrier at Venus.
    V. G. Pivovarov, N. V. Erkaev, A. S. Volokitin,
T. K. Breus, S. V. Ivanova.
Kosm. Issled., Tom 20, 97 - 103 (1982). In Russian.

093.027 Subcloud aerosol of the Venus atmosphere.
    Yu. M. Golovin, E. A. Ustinov.
Kosm. Issled., Tom 20, 104 - 110 (1982). In Russian.

093.028 Holes in the nightside ionosphere of Venus.
    L. H. Brace, R. F. Theis, H. G. Mayr, S. A. Curtis,
J. G. Luhmann.
J. Geophys. Res., Vol. 87, 199 - 211 (1982).
    Measurements of electron density and temperature by the Pioneer-Venus orbiter electron temperature probe have been employed to examine the characteristics and morphology of ionospheric holes in the antisolar ionosphere of Venus.

093.029 Pioneer Venus atmospheric observations.
    F. W. Taylor, J. T. Schofield, S. P. Bradley.
Philos. Trans. R. Soc. London, Ser. A, Vol. 303, (see 012.022), 215 - 223 (1981).
    Some selected results from the scientific experiments on the recent Pioneer missions to Venus are reviewed, with partic-ular emphasis on data from the infrared remote sensing experiment on the orbiter. Various aspects of the structure, dynamics and energy budget of the atmosphere, as revealed by the measurements, are presented for discussion.

093.030 Landing induced dust clouds on Venus and Mars.
    J. B. Garvin.
Proc. Twelfth Lunar Planet. Sci. Conf., (see 012.024), p. 1493 - 1505 (1982).

093.031 Density constraints on the composition of Venus.
    K. A. Goettel, J. A. Shields, D. A. Decker.
Proc. Twelfth Lunar Planet. Sci. Conf., (see 012.024), p. 1507 - 1516 (1982).
    The composition of Venus is constrained most directly by the mean density of the planet, because the moment of inertia factor is unknown and seismic data do not yet exist for the interior. The density of Venus is estimated to be $1.0\pm0.4\%$ less dense than the density of an Earthlike Venus (i.e., a planet with the mass of Venus, but identical to Earth in composition and structure, with internal temperatures equal to temperatures in Earth at corresponding depths, except near the hot surface). Five models which have attempted to explain the density difference between Venus and Earth are reviewed: iron fractionation, basalt/eclogite ratio, oxidation state, multicomponent mixing, and equilibrium condensation. All of these models are capable of explaining the observed density difference, either with the model assumptions as published or with modest adjustments in model parameters.

093.032 Metal chloride and elemental sulfur condensates in the Venusian troposphere: are they possible?
V. L. Barsukov, I. L. Khodakovsky (*Khodakovskij*),
V. P. Volkov, Yu. I. Sidorov, V. A. Dorofeeva, N. E. Andreeva.
Proc. Twelfth Lunar Planet. Sci. Conf., (see 012.024), p. 1517 - 1532 (1982).

093.033 Atomic oxygen in the Venus upper atmosphere.
    V. S. Sholokhov, M. S. Burgin.
Kosm. Issled., Tom 20, 312 - 314 (1982). In Russian.

093.034 Evidence for high-altitude haze thickening on the dark side of Venus from 10-micron heterodyne spectroscopy of CO_2.   D. Deming, F. Espenak,
D. Jennings, T. Kostiuk, M. Mumma.
Icarus, Vol. 49, 35 - 48 (1982).
    The 10.86-μm $P(44)$ and 10.33-μm $R(8)$ lines of $^{12}C^{16}O_2$ were observed on Venus with an infrared heterodyne spectrometer. Modeling of the line profile indicates that a discrete, optically thick, cloud deck occurs at 45 mbar pressure, in essential agreement with current understanding of the Venusian cloud structure.

093.035 Structure of the Venus mesosphere and lower thermosphere from measurements during entry of the Pioneer Venus probes.   A. Seiff, D. B. Kirk.
Icarus, Vol. 49, 49 - 70 (1982).
    Data on the thermal structure of the nightside middle atmosphere of Venus, from 84 to 137 km altitude, have been obtained from analysis of deceleration measurements from the third Pioneer Venus small probe which entered the atmosphere near the midnight meridian at 27°S latitude. The temperature structure is essentially diurnally invariant up to 100 km, above which the nightside structure diverges sharply from the dayside toward lower temperatures. Very large diurnal pressure differences develop above 100 km. The data are compared with the measurements of G. M. Keating, J. Y. Nicholson, and L. R. Lake (1980), with theoretical thermal structure models of Dickinson, and with data obtained by Russian Venera spacecraft.

093.036 **Deducing the age of the dense Venus atmosphere.**
R. Kahn.
Icarus, Vol. 49, 71 - 85 (1982).

The author shows how crater size–density counts may
be used to help constrain the history of the Venus atmosphere, based on the predictions of simple but reasonable
models for crater production, surface erosion, and the effects
of atmospheric drag and breakup on incident meteors in the
Venus atmosphere. Once a large fraction of Venus surface has
been imaged at kilometer resolution it could be possible to
make an early determination of the age of the Venus atmosphere.

093.037 **On the possibility of plate tectonics on Venus.**
G. W. Brass, C. G. A. Harrison.
Icarus, Vol. 49, 86 - 96 (1982).

Several arguments have been put forward suggesting that
Venus has no plate tectonics. The authors examine some of
these arguments and suggest that because conditions on the
surface of Venus are very different from those on Earth, the
arguments should be reconsidered.

093.038 **Basic features of the carbon cycle on Venus.**
I. L. Khodakovskij, V. P. Volkov, Yu. I. Sidorov,
N. E. Andreeva.
Vses. soveshch. po geokhimii ugleroda, Moskva, 14 - 16 dek.,
1981. Tez. dokl. Moskva, 1981, p. 304 - 308. In Russian.
From Ref. zh., 51. Astron., 4.51.192 (1982).

093.039 **On the structure of Venus atmospheric circulation
and possible irregularities of the planetary rotation
rate.** G. S. Golitsyn.
Pis'ma Astron. Zh., Tom 8, 312 - 317 (1982). In Russian.
English translation in Soviet Astron. Lett., Vol. 8.

It is shown that in the atmosphere of Venus a mechanism
of inertial instability can act causing a decay of the axisymmetric circulation into a number of cells in the meridional
plane. This structure is able to provide a zero mean torque of
the atmospheric friction at the surface. A possibility is pointed
out of noticeable variations of the rotation angular velocity due
to temporal changes of the atmospheric circulation intensity.

093.040 **Radar investigations of Venus.**
G. M. Petrov.
Zemlya Vselennaya, 1982, No. 1, p. 8 - 16. In Russian.

093.041 **The surface of Venus.**
L. V. Ksanfomaliti.
Zemlya Vselennaya, 1982, No. 1, p. 16 - 19. In Russian.

093.042 **Venus was wet: a measurement of the ratio of
deuterium to hydrogen.** T. M. Donahue,
J. H. Hoffman, R. R. Hodges, Jr., A. J. Watson.
Science, Vol. 216, 630 - 633 (1982).

The deuterium-hydrogen abundance ratio in the Venus
atmosphere was measured while the inlets to the Pioneer
Venus large probe mass spectrometer were coated with
sulfuric acid from Venus' clouds. The ratio is $(1.6\pm0.2)\times10^{-2}$.
The hundredfold enrichment of deuterium means that at
least 0.3 percent of a terrestrial ocean was outgassed on
Venus, but is consistent with a much greater production.

093.043 **The internal evolution of Venus and the Galilean
satellites.** S. C. Solomon.
Nature, Vol. 298, 15 - 16 (1982).

093.044 **On possible condensates in the Venus main cloud
layer.**
V. P. Volkov, Yu. I. Sidorov, I. L. Khodakovskij,
V. L. Barsukov.
Geokhimiya, 1982, No. 1, p. 3 - 22. In Russian. – Abstr. in
Ref. zh., 51. Astron., 5.51.181 (1982).

093.045 **Improved Venus ionopause altitude calculation and
comparison with measurement.**
W. C. Knudsen, K. L. Miller.
J. Geophys. Res., Vol. 87, 2246 - 2254 (1982).

The authors calculate the altitude of the Venus ionopause
following the inviscid fluid approach initially introduced
by Spreiter et al. (1970) and later modified by Spreiter and
Stahara (1980) but incorporate several improved approximations. The calculated altitude is compared with median
altitudes measured by the Pioneer Venus retarding potential
analyzer.

093.046 **Comet-like interaction of Venus with the solar
wind. III. The atomic oxygen corona.**
M. K. Wallis.
Geophys. Res. Lett., Vol. 9, 427 - 430 (1982).

Suprathermal atomic oxygen constituting an extensive
exospheric corona give rise to new heavy ions within the
solar plasma flowing around Venus. Because their gyro-radii
are large compared with transverse scales, the O^+ ions do not
simply effect a drag due to 'mass loading' of the flow. Some
precipitate into the ionosphere and provide a novel ion source.
Also, the back-reaction on the flow not only decelerates it,
but also deviates it laterally and essentially asymmetrically
about the planet.

093.047 **Venus: limited extension and volcanism along zones
of lithospheric weakness.** G. G. Schaber.
Geophys. Res. Lett., Vol. 9, 499 - 502 (1982).

Three global-scale zones of possible tectonic origin are described as occurring along broad, low rises within the Equatorial Highlands on Venus. The two longest of these tectonic
zones, the Aphrodite-Beta and Themis-Atla zones, extend for
21,000 and 14,000 km, respectively. Several lines of evidence
indicate that Beta and Atla Regiones, located at the only two
intersections of the three major tectonic zones, are dynamically supported volcanic terranes associated with currently active
volcanism. Rift valleys south of Aphrodite Terra and between
Beta and Phoebe Regiones are characterized by 75- to 100 km
widths, raised rims, and extensions of only a few tens of kilometers, about the same magnitudes as continental rifts on
the Earth. Horizontal extension on Venus was probably
restricted by an early choking-off of plate motion by high
crustal and upper-mantle temperatures, and the subsequent
loss of water and an asthenosphere.

093.048 **Venus nightside ionosphere: a model with keV
electron impact ionization.** S. Kumar.
Geophys. Res. Lett., Vol. 9, 595 - 598 (1982).

A model of the "full up" nightside ionosphere of Venus
is presented for the equatorial midnight region. The author
proposes keV electron impact as the strongest source of
ionization. The observed altitude profiles of CO_2^+, O^+, O_2^+, H^+,
and H_2^+ (or D^+) can be reproduced with this model.

093.049 **Outstanding achievement in investigation of Venus.**
S. A. Nikitin.
Priroda, 1982, No. 6, p. 2 - 3. In Russian.

093.050 **Investigation of the aerosol of the Venus cloud
layer aboard the Venera 12 automatic interplane-
tary station (Preliminary data).** Yu. A. Surkov,
F. F. Kirnozov, V. I. Gur'yanov, V. N. Glazov,
A. G. Dunchenko, S. S. Kurochkin, V. N. Rasputnyj,
Eh. G. Kharitonova, L. P. Tatsij, V. L. Gimadov.
Kosm. Issled., Tom 20, 435 - 441 (1982). In Russian.

093.051 **Peculiarities of the method of analyzing the
aerosol of Venus clouds.** I. V. Petryanov,
B. M. Andrejchikov, B. N. Korchuganov, E. I. Ovsyankin,
B. I. Ogorodnikov, V. I. Skitovich, V. K. Khristianov.
Kosm. Issled., Tom 20, 442 - 448 (1982). In Russian.

093.052 The elongation of Venus: 1980 August.
C. J. R. Lord.
J. British Astron. Assoc., Vol. 92, 177 - 179 (1982).

**093.053 Venus: halide cloud condensation and volatile
element inventories.**   J. S. Lewis, B. Fegley, Jr.
Science, Vol. 216, 1223 - 1225 (1982).
Several recently suggested Venus cloud condensates,
including aluminum chloride and halides, oxides, and sulfides
of arsenic and antimony, are assessed for their thermodynamic
and geochemical plausibility.

**093.054 Airborne spectroscopy and spacecraft radiometry
of Venus in the far infrared.**
H. H. Aumann, J. V. Martonchik, G. S. Orton.
Icarus, Vol. 49, 227 - 243 (1982).
In March 1979, the spectrum of Venus was recorded in
the far infrared from the G. P. Kuiper Airborne Observatory.
The brightness temperature was observed to be $275°K$ near
110 cm^{-1}, dropping to $230°K$ near 270 cm^{-1}. Radiance
calculations yield results consistently brighter than the observations. Supplementing the spectral data, Pioneer Venus OIR
data provide the constraint that any additional infrared opacity
must be contained in the upper cloud. The most likely possibility for supplementing the far-infrared opacity is a population of large particles ($\bar{r} \geqslant 1$ μm) in the upper cloud with
number densities less than 1 particle cm^{-3}.

Weather and climate on planets.
See Abstr. 003.078.

Venus.   See Abstr. 003.082.

After Pioneer – good science, bad news.
See Abstr. 011.004.

Venusian surface wind tunnel.
See Abstr. 022.054.

Soil transport by winds on Venus.
See Abstr. 022.055.

**Alteration of rocks in hot CO_2 atmospheres:
preliminary experimental results and application to Venus.**
See Abstr. 022.065.

Unveiling Venus with VOIR.
See Abstr. 032.504.

**SAR: an instrument for planetary geodesy and
navigation.**   See Abstr. 032.515.

**Construction of continuous Chebyshev approxima-
tions for planetary coordinates.**   See Abstr. 042.030.

**On a possible scheme of a rocket vehicle in the
Venus atmosphere.**   See Abstr. 053.002.

**A model of the atmospheric convection zone with
a large increase of rotation velocity between its bottom and
top.**   See Abstr. 064.078.

**Formation and evolution of playa ventifacts,
Amboy, California.**   See Abstr. 081.017.

**Oceanic ridges, transforms, trenches would be seen
in PV altimetry data – even under Venusian ambient
conditions.**   See Abstr. 081.027.

**Viscous relaxation of impact basin topography: im-
plications for the moon and Venus.**
See Abstr. 091.014.

Een lange, hete dag. Op Mercurius en Venus.
See Abstr. 092.002.

**Impact crater ejecta morphologies on the moon and
Venus.**   See Abstr. 094.506.

The lunar occultation of Venus, 1980 October 5.
See Abstr. 096.003.

**Characterization of rock populations on the sur-
faces of Mars, Venus, and earth: a summary.**
See Abstr. 097.018.

**Topographic reduction of Mars and Venus radar
observations.**   See Abstr. 097.084.

094 Moon (Dynamics, General Aspects, Local Properties)

Moon (Dynamics, General Aspects)

094.001 The moon: its figure and orbital evolution.
A. B. Binder.
Geophys. Res. Lett., Vol. 9, 33 - 36 (1982).

A first order analysis of current selenodetic and selenochronological data indicates that the present lunar figure was formed $3.0 \pm 0.5 \times 10^9$ years ago at an earth-moon distance of 20.4 ± 2.3 earth radii. This datum and data from a variety of sources are used to define the time evolution of the earth-moon distance. It is shown that there was a sudden increase in this distance about 2×10^9 years ago when the Q of the earth must have decreased by a factor of 5. This change in Q correlates well with the onset of rapid continental growth and hence the expansion of shallow sea- and continental shelf waters as shown in the geological record for that time.

094.002 Limits on the lunar temperature profile.
L. L. Hood, C. P. Sonett.
Geophys. Res. Lett., Vol. 9, 37 - 40 (1982).

Limits on the selenotherm are estimated using (1) a preferred set of bounds on the lunar electrical conductivity profile; (2) published laboratory conductivity vs. temperature data for an olivine and several aluminous orthopyroxenes; and (3) estimates for the Al_2O_3 content of the deep interior suggested by bulk composition models. The inferred limits are narrowest in the depth range 450 to 1350 km and are in accord with independent geophysical constraints. Thermal history models which yield present-day selenotherms that are in best agreement with these limits are those which permit subsolidus convection at depths greater than 800 km in the moon.

094.003 Correlation of the albedo with the polarization properties of the moon (heterogeneity of the relative porosity of the surface of the western part of the moon's near side).
V. V. Novikov, Yu. G. Shkuratov, A. P. Popov, M. V. Goryachev.
Astron. Zh., Tom 59, 129 - 136 (1982). In Russian. English translation in Soviet Astron., Vol. 26, No. 1.

Two new parameters of the lunar surface are considered, characterizing the global heterogeneity of the surface porosity and the chemico-mineralogical composition of the soil. The new parameters are studied on the basis of the correlation between the albedo and the maximum degree of polarization.

094.004 On an estimate of exponential dispersions of the gravitational field of the moon.
P. M. Zazulyak, V. E. Zinger, V. V. Kirichuk.
Geod., kartogr. i aehrofotosemka, L'vov, 1981, No. 34, p. 25 - 29. In Russian. – Abstr. in Ref. zh., 52. Geod. Aehrosemka, 2.52.201 (1982).

094.005 Kinematics of basin subsidence, grabens, and lunar expansion.
G. E. McGill, M. P. Golombek.
Reports of Planetary Geology Program – 1981, (see 003.001), p. 402 - 404 (1981). – Abstract.

094.006 Photogrammetric compilation of the global map of the Moon.
S. S. C. Wu.
Reports of Planetary Geology Program – 1981, (see 003.001), p. 497 - 499 (1981). – Abstract.

094.007 Geochemical anomalies on the lunar surface: implications for early volcanism and the origin of light plains.
B. R. Hawke, P. D. Spudis, P. E. Clark.
Reports of Planetary Geology Program – 1981, (see 003.001), p. 548 - 550 (1981). – Abstract.

094.008 Constraints on the origin of the moon and the metal content of the eucrite parent body deduced from the partitioning behavior of tungsten.
H. E. Newsom, M. J. Drake.
Meteoritics, Vol. 16, 366 (1981). – Abstract.

094.009 The intensity of the ancient lunar magnetic field, revisited.
D. W. Strangway, N. Sugiura.
Meteoritics, Vol. 16, 389 - 390 (1981). – Abstract.

094.010 Progress in the analytical theories for the orbital motion of the Moon.
M. Chapront-Touze.
Celestial Mech., Vol. 26, (see 012.001), 53 - 62 (1982).

During the last 15 years, an important effort has been done by several authors to produce new and precise analytical solutions for the Moon's orbital motion. This construction is still in a very growing phase. Achieved works are described and compared. Some indications are given about what remains to be done.

094.011 The ELP solution for the main problem of the Moon and some applications.
M. Chapront-Touze.
Celestial Mech., Vol. 26, (see 012.001), 63 - 69 (1982).

First derivatives of the ELP 2000 solution of the main problem have been obtained. They are used for computation of the lunar motion perturbations due to the Earth's oblateness and to secular terms in the solar eccentricity and perigee.

094.012 Comparison of SALE with numerical integration.
H. Kinoshita.
Celestial Mech., Vol. 26, (see 012.001), 71 - 73 (1982).

Recently analytical solutions of the main problem of the Moon have been obtained by Henrard (1979), for SALE, Chapront (1979), for ELP, and Schmidt (1980). Chapront and Henrard (1980) compared in detail amplitudes of periodic terms and secular motions of angular variables in their theories. The authors compared SALE over one year with numerical integration. Initial position and velocity for numerical integration are calculated from the theory.

094.013 The main problem of lunar theory solved by the method of Brown.
D. S. Schmidt.
Celestial Mech., Vol. 26, (see 012.001), 75 (1982). – Abstract.

094.014 New approach to determining planetary perturbations in lunar theory.
V. A. Brumberg, T. V. Ivanova.
Celestial Mech., Vol. 26, (see 012.001), 77 - 81 (1982).

The technique of the general planetary theory has been proposed for constructing a theory of motion of the Moon. This method enables one to elaborate the consistent theory of motion of the principal planets and the Moon, which is of particular importance for determining planetary perturbations in lunar motion. As an initial approximation for lunar motion, an intermediate orbit generalizing the Hill's variational curve has been built.

094.015 Planetary perturbations of the Moon in ELP 2000.
J. Chapront, M. Chapront-Touze.
Celestial Mech., Vol. 26, (see 012.001), 83 - 94 (1982).

A new solution for the planetary perturbations of the Moon has been built in the frame of ELP 2000, using Bretagnon's planetary theories, and achieved at the first order. The internal precision of computation is 2×10^{-6} arcsec. First-order planetary perturbations, in the direct case (Venus &

Mars), have been compared to Standaert's solution. The major discrepancy reaches 70 cm in the longitude of Venus. Perturbations of the second order with respect to planetary masses, have been undertaken and illustrations are given. Finally, new values for the perigee and node motions are proposed.

094.016 **Perturbations by the oblateness of Earth and Moon.**
J. Henrard.
Celestial Mech., Vol. 26, (see 012.001), 95 (1982). – Abstract.

094.017 **Perturbations by the oblateness of the Earth and by the planets in the motion of the Moon.**   Y. Kubo.
Celestial Mech., Vol. 26, (see 012.001), 97 - 112 (1982).
Perturbations in the motion of the Moon are computed for the effect by the oblateness of the Earth and for the indirect effect of planets. Based on Delaunay's analytical solution of the main problem, the computations are performed by a method of Fourier series operation.

094.018 **Comments about the direct perturbations of Venus and Mars on the Moon's motion.**   D. Standaert.
Celestial Mech., Vol. 26, (see 012.001), 113 - 119 (1982).
In a previous paper (Standaert, 1980) the author described an algorithm to compute the direct perturbation of the planets on the Moon's motion. The first results permit to present some complements and comments about these computations. The algorithm is based upon the Lie transform method and is implemented using Chapront's ELP as solution of the main problem with the partial derivatives of Henrard's Semi-Analytical Lunar Ephemeris, and Bretagnon's mean Keplerian orbit. An analysis of truncation errors in intermediate results is presented. The effects of second-order terms in the masses are investigated.

094.019 **A comparison of numeric and semi-analytic lunar libration models.**   R. J. Cappallo, D. H. Eckhardt.
Celestial Mech., Vol. 26, (see 012.001), 125 - 127 (1982).
The numerical lunar libration model of Cappallo (1980) has been compared with the semi-analytic model of Eckhardt (1982). Both models have also been compared with independently-derived, but similar, lunar libration models.

094.020 **Semi-analytic model of the Moon's rotation.**
D. H. Eckhardt.
Celestial Mech., Vol. 26, (see 012.001), 129 (1982). – Abstract

094.021 **Physical libration of the Moon.**   M. Moons.
Celestial Mech., Vol. 26, (see 012.001), 131 - 142 (1982).
The aim of this paper is to present a semi-analytical theory of the libration of the Moon obtained by the application of the Lie transform method to the Hamiltonian of the problem: free rotation of a rigid Moon around its center of mass plus perturbations due to Earth and Sun.

094.022 **Orbit and rotation of the Moon by numerical integration.**   C. Oesterwinter.
Celestial Mech., Vol. 26, (see 012.001), 143 (1982). – Abstract.

094.023 **Evidence for lunar librations near resonance.**
R. J. Cappallo, R. W. King, C. C. Counselman III, I. I. Shapiro.
Celestial Mech., Vol. 26, (see 012.001), 145 (1982). – Abstract

094.024 **When were the tectonically-magmatic processes on the moon finished?**
Priroda, 1982, No. 3, p. 119. In Russian.

094.025 **Impact cratering and regolith dynamics.**
F. Horz.
Phys. Chem. Earth, Vol. 10, 3 - 15 (1980). – Abstr. in Phys. Abstr., Vol. 85, Abstr. 42614 (1982).

094.026 **Orbital chemistry – lunar surface analysis from the X-ray and gamma ray remote sensing experiments.**
I. Adler, J. I. Trombka.
Phys. Chem. Earth, Vol. 10, 17 - 43 (1980). – Abstr. in Phys. Abstr., Vol. 85, Abstr. 42615 (1982).

094.027 **The irradiation history of the lunar soil.**
G. Crozaz.
Phys. Chem. Earth, Vol. 10, 197 - 214 (1980). – Abstr. in Phys. Abstr., Vol. 85, Abstr. 42618 (1982).

094.028 **Constraints on the Moon's origin from the partitioning behaviour of tungsten.**
H. E. Newsom, M. J. Drake.
Nature, Vol. 297, 210 - 212 (1982).
The authors conclude that a metal content consistent with present geophysical constraints on the size of a metallic lunar core could account for the depletion of W observed in lunar rocks. The low W/La ratio of the Moon cannot be used as evidence for the formation of the Moon from the Earth's mantle by fission following terrestrial core formation.

094.029 **The megarelief of the near side of the moon.**
I. V. Gavrilov, V. S. Kislyuk, L. A. Karasev.
Astron. vestn., Tom 15, 211 - 215 (1981). In Russian. – Abstr. in Ref. zh., 51. Astron., 3.51.272 (1982).

094.030 **On the selenodetic basis for lunar cartography.**
I. V. Gavrilov, L. A. Karasev.
Astrometr. Astrofiz., Vyp. (No.) 44, p. 71 - 80 (1981). In Russian.
Nets of the selenodetic reference points and hypsometric characteristics of the lunar surface are the selenodetic basis for mapping of the moon. The paper deals with the classification and distribution on the lunar surface of the existing selenodetic data.

094.031 **On the determination of the mean inclination of the lunar equator to the ecliptic by the position angles method.**   V. S. Kislyuk.
Astrometr. Astrofiz., Vyp. (No.) 44, p. 80 - 84 (1981). In Russian.

094.032 **Die Bestimmung geodätisch-geodynamischer Parameter mit Hilfe von Laserdistanzmessungen zum Mond.**   L. Ballani.
Gerlands Beitr. Geophys., Band 91, 121 - 140 (1982) = Mitt. Zentralinst. Phys. Erde, Nr. 995.
After some general remarks on basic facts and the recent state of lunar laser range measurements, modelling of laser signal time delay is concisely treated in the frame of Earth–Moon dynamics.

094.033 **The free librations of a dissipative Moon.**
C. F. Yoder.
Philos. Trans. R. Soc. London A, Vol. 303, (see 012.022), 327 - 338 (1981).
Dissipation in the Moon produces a small offset, ca. 0.23″, of the Moon's rotation axis from the plane defined by the ecliptic and lunar orbit normals. Both solid body tidal friction and viscous fluid friction at a core-mantle interface are plausible mechanisms. In this paper the author discusses the merits of each and find that solid friction requires a low lunar tidal Q, ca. 28, while turbulent fluid friction requires a core of radius ca. 330 km to cause the signature observed by lunar laser ranging.

094.034 **Measurements of the lunar induced magnetic moment in the geomagnetic tail: evidence for a lunar core?**   C. T. Russell, P. J. Coleman, Jr., B. E. Goldstein.
Proc. Twelfth Lunar Planet. Sci. Conf., (see 012.024), p. 831 - 836 (1982).

094.035 **Support of topographic and other loads on the moon and on the terrestrial planets.**
R. J. Willemann, D. L. Turcotte.
Proc. Twelfth Lunar Planet. Sci. Conf., (see 012.024), p. 837 - 851 (1982).

094.036 **Mascons and loading of the lunar lithosphere.**
S. Pullan, K. Lambeck.
Proc. Twelfth Lunar Planet. Sci. Conf., (see 012.024), p. 853 - 865 (1982).

094.037 **Correction of lunar seismograms for instrumental and near-surface effects and constraints on the velocity structure of the lunar interior.** P. Horvath.
Proc. Twelfth Lunar Planet. Sci. Conf., (see 012.024), p. 867 - 889 (1982).

094.038 **Modern lunar theory.** J. Kovalevsky.
Application of modern dynamics to celestial mechanics and astrodynamics, (see 012.026), p. 59 - 76 (1982).
Formal solutions of the main problem of the lunar theory are presented and the effects of small divisors are described. Semi-numerical solutions are also described.

094.039 **Globes of the moon.**
Zh. F. Rodionova, V. I. Chikmachev.
Zemlya Vselennaya, 1982, No. 1, p. 20 - 21. In Russian.

094.040 **The possible contribution to the research on the secular acceleration of the Earth-Moon system with the Chinese ancient eclipse records.** S. X. Wu.
Special issue on astronomical geodynamics, (see 012.028), p. 33 - 39 (1981). In Chinese.
The secular tidal friction of the Earth—Moon system goes through the whole age of the evolution of the Earth. During the long geological age, this tidal friction must be changed. It would be difficult to obtain the timely information if only the data of two-three centuries were used. Moreover, the long periodic and irregular variations would also cause some false phenomena. Therefore the investigation in the application of the ancient observation records to this problem is necessary. It is to expect that the rich materials of the Chinese ancient astronomical observations (the solar and lunar eclipses, the occultation of planets and stars by the Moon) would make a better contribution to the investigation on the secular acceleration of the Earth—Moon system.

094.041 **Distribution of the phase brightness gradient for two spectral intervals over the lunar surface. Preliminary survey.**
L. A. Akimov, Yu. G. Shkuratov.
Astron. Tsirk., No. 1167, p. 3 - 6 (1981). In Russian.

094.042 **Relationship between the albedo and polarization properties of the moon. Application of digital image processing.** Yu. V. Kornienko, Yu. G. Shkuratov, V. I. Bychinskij, D. G. Stankevich.
Astron. Zh., Tom 59, 571 - 577 (1982). In Russian. English translation in Soviet Astron., Vol. 26, No. 3.
The albedo and the maximum polarization coefficient are used for preparing bi-parametric maps of the region of the lunar surface. It is based on analysing the diagram of the albedo — maximum polarization coefficient by digital image processing.

094.043 **Lunar palaeomagnetism.** S. K. Runcorn.
The comparative study of the planets, (see 012.033), p. 291 - 294 (1982).
Although the Moon possesses no general magnetic field today, it is inferred from lunar palaeomagnetism, studied both in returned samples and by magnetometers and satellites around the Moon, implies the existence of an early lunar field

generated by a core dynamo. This field decayed from 1 Gauss 4 Ga ago to 0.02 Gauss 3.2 Ga ago: raising profound questions about the early heat sources in the solar system.

094.044 **Modelling the effect of tidal friction in the lunar orbital motion.** J. D. Mulholland.
Bull. American Astron. Soc., Vol. 14, 582 (1982). – Abstract.

094.045 **The evolution of impact basins: viscous relaxation of topographic relief.**
S. C. Solomon, R. P. Comer, J. W. Head.
J. Geophys. Res., Vol. 87, 3975 - 3992 (1982).
The authors evaluate the hypothesis that viscous relaxation has been an important process for modifying the topographic profiles of ancient large impact basins on the moon. The authors adopt a representative topographic profile of the Orientale basin, the youngest large impact basin on the moon, as an estimate of the initial topography of older basins of similar horizontal dimensions, and they predict the topographic profiles that would result from viscous relaxation according to a number of simple analytical representations of the rheological response of the moon to surface topography.

094.046 **Post-Imbrian global lunar tectonism: evidence for an initially totally molten Moon.** A. B. Binder.
Moon Planets, Vol. 26, 117 - 133 (1982).
The compressional thermoelastic stresses generated in a Moon which was initially totally molten, as is the case of a Moon formed by fission, are up to 3.5 kbar in the outer few km of the crust at present. These stresses are well within the range needed to cause thrust faulting in the outer 4 km of the crust. According to this model there should be modest scale (10 km), young (< 0.5 to 1×10^9 yr old) thrust fault scarps in the highlands. Photoselenological investigations confirm that scarps with the expected age and geometric characteristics are found in the highlands. Thus the currently available photoselenological data support the stress model derived for an initially totally molten Moon.

094.047 **Dissipation in the moon: a review of the experimental evidence and physical implications.**
R. W. King.
High-precision Earth rotation and Earth-Moon dynamics. Lunar distances and related observations, (see 012.035), p. 191 - 192 (1982). – Abstract.

094.048 **Planetary and Earth figure perturbations in the librations of the Moon.** D. H. Eckhardt.
High-precision Earth rotation and Earth-Moon dynamics. Lunar distances and related observations, (see 012.035), p. 193 - 198 (1982).
Comparisons of numeric, semi-analytic and analytic lunar libration models indicate that the planetary and Earth figure perturbations used to supplement the semi-analytic and analytic models are inadequate. Using the ELP-2000 of Chapront and Chapront-Touzé, an improved solution of these perturbations is developed. For the libration in longitude there are numerous terms with periods near that of the free libration; their amplitudes remain ill-defined.

094.049 **Modelling the effect of Earth tides in the lunar orbital motion.** J. D. Mulholland, O. Calame.
High-precision Earth rotation and Earth-Moon dynamics. Lunar distances and related observations, (see 012.035), p. 199 - 206 (1982).
The effect of tidal friction in the lunar orbit is one of the classic examples of ad hoc modelling of an unknown physical phenomenon. The two basic calculational approaches in current use are developed in some detail, and numerical tests and comparisons are presented. Although attention is normally concentrated on the acceleration in orbital longitude produced as a result of terrestrial dissipation, it is shown that

the variation in Earth-Moon distance due to Earth tides is far from negligible.

094.050 Results from lunar laser ranging data analysis.
J. O. Dickey, J. G. Williams, C. F. Yoder.
High-precision Earth rotation and Earth-Moon dynamics.
Lunar distances and related observations, (see 012.035), p. 209 - 216 (1982).

The lunar laser range data taken at McDonald Observatory between August 1969 and May 1980 has been analyzed. The simple rms residual for the 2954 ranges is 31 cm. Results of the analysis include $GM_{earth} = 398600.45 \pm 0.02$ km^3/sec^2 and a secular acceleration of the lunar orbital mean longitude of $\dot{n} = -23.8 \pm 1.5''$/century^2 which yields a Q of 12.3 at semidiurnal frequencies. The lunar harmonic C_{30} is $(-8.7 \pm 1.1) \times 10^{-6}$ and the lunar rotational dissipation $k_{2m} T = (4.7 \pm 0.5) \times 10^{-3}$ day. Also resulting from the solution are geocentric coordinates of McDonald accurate to 30 cm, including the first value for the longitude with the new IAU constants and a dynamical equinox.

094.051 The relativistic planetary perturbations and the orbital motion of the Moon.
J.-F. Lestrade, J. Chapront, M. Chapront-Touzé.
High-precision Earth rotation and Earth-Moon dynamics.
Lunar distances and related observations, (see 012.035), p. 217 - 225 (1982).

In this contribution, a review of the calculation of the lunar orbit in the relativistic framework is drawn up. Then, the particular dependence of the lunar orbit motion upon the relativistic perturbations of the motions of the planets themselves is put forth.

094.052 Analytical theories of the motion of the Moon.
J. Henrard.
High-precision Earth rotation and Earth-Moon dynamics.
Lunar distances and related observations, (see 012.035), p. 227 - 232 (1982).

Almost every aspect of the analytical theory of the motion of the Moon has been reinvestigated. This paper is a review of these investigations. The improvement of the I.L.E. (the best known earlier theory based upon the work of Brown) is spectacular, but it is still too early to assess the exact value of these theories with respect to numerical integration.

094.053 Numerical studies of the lunar orbit at CERGA.
O. Calame.
High-precision Earth rotation and Earth-Moon dynamics.
Lunar distances and related observations, (see 012.035), p. 233 - 243 (1982).

A few years ago, it was decided that it would be desirable to support the Lunar Laser data reduction with the construction of a new ephemeris by numerical integration. This was considered valuable not only from the scientific point of view, but also as an operational tool for the EROLD and MERIT programs. Early in 1980 this ephemeris, called ECT 18, was adopted in the EROLD computations for the Annual Reports of BIH and for the needs of MERIT Operating Center and Analysis Center at CERGA.

094.054 Comparison of lunar ephemerides (SALE and ELP) with numerical integration.   H. Kinoshita.
High-precision Earth rotation and Earth-Moon dynamics.
Lunar distances and related observations, (see 012.035), p. 245 - 255 (1982).

SALE and ELP, analytical theories of the main problem of the Moon, are developed by Henrard (1979) and Chapront-Touzé (1980), respectively. Both theories are compared with numerical integration over one year, which covers about 13 revolutions of the Moon's orbit. The root-mean-square residuals in the distance of SALE is about 10 cm for

series truncated at 10^{-5} arcsecond and 1.2 cm for series truncated at 10^{-6} arcsecond. ELP is also compared with 20 years of numerical integration and the root-mean-square residuals in the distance is about 1.5 cm.

094.055 Comparison of ELP-2000 to a JPL numerical integration.   J. Chapront, M. Chapront-Touzé.
High-precision Earth rotation and Earth-Moon dynamics.
Lunar distances and related observations, (see 012.035), p. 257 - 264 (1982).

This contribution is an outline of the main results obtained by the authors in comparing their solution ELP-2000, to a JPL numerical integration, LE-51.

094.056 Note about a new evaluation of the direct perturbations of the planets on the Moon's motion.
D. Standaert.
High precision Earth rotation and Earth-Moon dynamics.
Lunar distances and related observations, (see 012.035), p. 265 - 266 (1982).

The aim of this paper is to present the principal features of a new evaluation of the direct perturbations of the Moon's motion. Using the method already published in Celestial Mechanics (Standaert, 1980), the author computes "a first-order theory" aiming at an accuracy of the order of the meter for all periodic terms of period less than 3 500 years.

094.057 A comparison between the density of distribution of craters on the moon and its topography.
T. P. Skobeleva, B. D. Sitnikov.
Astron. vestn., Tom 16, 27 - 30 (1982). In Russian. – Abstr. in Ref. zh., 51. Astron., 6.51.218 (1982).

094.058 Mapping the relative density parameter of the surface layer of the eastern sector of the near side of the moon.   V. V. Novikov, A. P. Popov.
Astron. vestn., Tom 16, 31 - 33 (1982). In Russian. – Abstr. in Ref. zh., 51. Astron., 6.51.219 (1982).

094.059 Observations of some transient phenomena on the moon (Collection 5).
P. V. Florenskij, V. M. Chernov.
Astron. vestn., Tom 16, 60 - 62 (1982). In Russian. – Abstr. in Ref. zh., 51. Astron., 6.51.220 (1982).

094.060 The megarelief of the near side of the moon.
I. V. Gavrilov, V. S. Kislyuk, L. A. Karasev.
Astron. vestn., Tom 15, 211 - 215 (1981). In Russian. Abstr. in Ref. zh., 52. Geod. Aehrosemka, 6.52.124 (1982).

094.061 The lunar photometric constant in the system of true full moon.   V. V. Shevchenko.
Sun and planetary system, (see 012.040), p. 263 - 264 (1982).

094.062 Relativistic dynamics of the Earth-Moon system.
V. A. Brumberg, T. V. Ivanova.
Sun and planetary system, (see 012.040), p. 423 - 428 (1982).

Equations of motion of the Earth and Moon convenient for numerical integration have been derived within the framework of the mass-point PPN formalism taking into account parameters β, γ and coordinate parameter α. The first degree terms in lunar motion have been found by analytical iterations with the aid of the Poissonian processor up to order m^7. Variational terms and mean motions of the perigee and node have been obtained up to order m^8.

094.063 Recent results for the Moon's secular acceleration and their implication for the possible variation of G in Dirac's Large Number Hypothesis.   L. V. Morrison.
Compendium in astronomy, (see 003.012), p. 361 - 366 (1982).

Recent data on the secular acceleration of the Moon

imply a rate of variation of G between zero and −6 parts in 10^{11} per year with Dirac's additive creation hypothesis.

Planetary regularities in crater distribution on Mars, the moon and Mercury (Collected diagrams). See Abstr. 003.075.

The Moon.   See Abstr. 003.096.

Planetary science: a lunar perspective. See Abstr. 003.128.

L'accélération séculaire de la lune et le ralentissement de la rotation de la terre.   See Abstr. 004.021.

Plane resonance rotations of a dynamically symmetric satellite in the three-body problem. See Abstr. 042.004.

On the absolute orientation of the selenodetic reference frame.   See Abstr. 043.008.

Oceanic tides and the rotation of the Earth. See Abstr. 044.061.

Die Arbeiten des Sonderforschungsbereiches 78 Satellitengeodäsie der Technischen Universität München im Jahre 1980.   See Abstr. 046.006.

Sub-mm heterodyne observations of the Sun, Moon, Jupiter and Orion at 691 GHz.   See Abstr. 077.012.

Terra e luna.   See Abstr. 081.056.

Viscous relaxation of impact basin topography: implications for the moon and Venus. See Abstr. 091.014.

Improvement of the ephemerides of the inner planets and the moon using radar measurements, lunar laser ranging and meridian observations 1961 - 1980. See Abstr. 091.031.

Die Problematik der Entstehung der Monde und ihrer Bahnentwicklung am Beispiel des Erdmondes und der Marsmonde.   See Abstr. 091.034.

Some results from a reduction of radar, laser and optical observations of the inner planets and of the moon. See Abstr. 091.059.

Nouvelles théories des planètes et de la Lune dans les éphémérides françaises.   See Abstr. 091.070.

Global volcanism and tectonism on Mercury: comparison with the Moon.   See Abstr. 092.007.

Far-field tectonics associated with a large impact basin: applications to Caloris on Mercury and Imbrium on the Moon.   See Abstr. 092.008.

Lunar eclipses, lunar luminescence and transient lunar phenomena.   See Abstr. 095.002.

The evolution of the Earth-Moon system. See Abstr. 107.028.

Moon (Local Properties)

094.501 Copernicus crater central peak: lunar mountain of unique composition.   C. M. Pieters.
Science, Vol. 215, 59 - 61 (1982).
Olivine is identified as the major mafic mineral in a central peak of Copernicus crater. Information on the mineral assemblages of such unsampled lunar surface material is provided by near infrared reflectance spectra (0.7 to 2.5 micrometers) obtained with Earth-based telescopes. The composition of the deep-seated material comprising the Copernicus central peak is unique among measured areas. Other lunar terra areas and the wall of Copernicus exhibit spectral characteristics of mineral assemblages comparable to the feldspathic breccias returned by the Apollo missions, with low-calcium orthopyroxene being the major mafic mineral.

094.502 Untersuchungen zur Klassifizierung von Plagioklas-, Pyroxen- und Olivinkristalliten und zur elektrischen Leitfähigkeit des Regoliths der Proben Luna 24.101.1, 24.158.1 und 24.191.1.
R. Wäsch, A. Kraft, J. v. Faber, I. Bauer.
Gerlands Beitr. Geophys., Band 91, 1 - 12 (1982).
In the present paper results of X-ray investigations aiming at the determination of lunar minerals of the Luna 24 regolith are presented. Furthermore, measurements of the electrical conductivity of regolith under high pressures and temperatures are described, and the results are discussed.

094.503 Limneus: un mutamento sulla luna?
W. Ferreri.
Orione, Vol. 3, 3 - 13 (1982).

094.504 Graphite in Luna 24 regolith.
N. A. Ashikhmina, O. A. Bogatikov, T. L. Evstigneeva, I. D. Samotoin, V. A. Stepanchikov, D. I. Frikh-Khar.
Dokl. AN SSSR, Tom 260, 989 - 992 (1981). In Russian.
Abstr. in Ref. zh., 51. Astron., 2.51.233 (1982).

094.505 The isotopic composition of zirconium in terrestrial and extraterrestrial samples: implications for extinct ^{92}Nb.   J.-F. Minster, C. J. Allègre.
Geochim. Cosmochim. Acta, Vol. 46, 565 - 573 (1982).

094.506 Impact crater ejecta morphologies on the moon and Venus.   T. W. Thompson, J. A. Cutts.
Reports of Planetary Geology Program − 1981, (see 003.001), p. 108 - 110 (1981). − Abstract.

094.507 The Nectaris basin.   J. L. Whitford-Stark.
Reports of Planetary Geology Program − 1981, (see 003.001), p. 117 - 119 (1981). − Abstract.

094.508 The nature of lunar basin ejecta deposits inferred from Apollo highland landing site geology.
P. D. Spudis.
Reports of Planetary Geology Program − 1981, (see 003.001), p. 120 - 122 (1981). − Abstract.

094.509 The origins of lunar dark-halo craters: implications for volcanic and impact processes.

B. R. Hawke, J. F. Bell.
Reports of Planetary Geology Program – 1981, (see 003.001),
p. 135 - 137 (1981). – Abstract.

094.510   **Lunar sinuous rille formation by thermal erosion: eruption conditions, rates and durations.**
J. W. Head, L. Wilson.
Reports of Planetary Geology Program – 1981, (see 003.001),
p. 161 - 163 (1981). – Abstract.

094.511   **Relative ages of lunar basins (II); Serenitatis.**
D. E. Wilhelms.
Reports of Planetary Geology Program – 1981, (see 003.001),
p. 405 - 407 (1981). – Abstract.

094.512   **^{39}Ar-^{40}Ar-age of norite 78236.**
U. Aeschlimann, P. Eberhardt, J. Geiss, N. Grögler,
K. Marti.
Meteoritics, Vol. 16, 287 (1981). – Abstract.

094.513   **Partitioning of trace elements among coexisting opaque oxides in lunar basalts using a proton probe microanalyzer.**   H. Blank, R. Nobiling, K. Traxel,
A. El Goresy.
Meteoritics, Vol. 16, 294 - 295 (1981). – Abstract.

094.514   **Provenance of the Apollo 16 feldspathic fragmental breccias at North Ray crater.**
R. Borchardt, D. Stöffler.
Meteoritics, Vol. 16, 296 - 297 (1981). – Abstract.

094.515   **Nitrogen concentrations and isotopic ratios from separated lunar soils.**   P. W. Brown, C. T. Pillinger.
Meteoritics, Vol. 16, 298 (1981). – Abstract.

094.516   **Solar flare implanted noble gases detected in plagioclase separates from lunar soils.**
P. Etique, P. Signer.
Meteoritics, Vol. 16, 313 - 314 (1981). – Abstract.

094.517   **Fractionation in solar system krypton and xenon, and their isotopic compositions in the solar wind.**
U. Frick, R. O. Pepin.
Meteoritics, Vol. 16, 316 - 317 (1981). – Abstract.

094.518   **Correlation between solar wind ^{4}He distribution and noble gas fractionation in lunar ilmenites.**
J. Kiko, N. Mahninger, W. Rittershausen, T. Kirsten.
Meteoritics, Vol. 16, 339 - 340 (1981). – Abstract.

094.519   **Comparative chemistry of size fractions from the Apollo and Luna sites.**   J. C. Laul, J. J. Papike.
Meteoritics, Vol. 16, 347 (1981). – Abstract.

094.520   **A lunar troctolite with unique evidence of remelting and vesiculation.**   U. B. Marvin, D. Walker.
Meteoritics, Vol. 16, 355 - 356 (1981). – Abstract.

094.521   **The origin of Apollo 16 dimict breccias.**
J. P. McKinley, G. J. Taylor, K. Keil.
Meteoritics, Vol. 16, 356 - 357 (1981). – Abstract.

094.522   **Noble gases and nitrogen in lunar breccia 79035.**
R. O. Pepin, U. Frick.
Meteoritics, Vol. 16, 376 - 377 (1981). – Abstract.

094.523   **Stratigraphy and evolution of the upper highland crust near North Ray, Apollo 16.**
D. Stöffler, R. Ostertag, W.-U. Reimold, R. Borchardt.
Meteoritics, Vol. 16, 389 (1981). – Abstract.

094.524   **The significance of rust in lunar rocks and meteorites.**   L. A. Taylor, R. H. Hunter, H. Y. McSween.
Meteoritics, Vol. 16, 391 - 392 (1981). – Abstract.

094.525   **Eucrites and the moon: regularities among europium "anomalies".**   P. H. Warren.
Meteoritics, Vol. 16, 399 (1981). – Abstract.

094.526   **Solar flare isotopic pattern – a new component to consider the isotopic composition of the solar system.**   A. Yaniv, T. Kirsten.
Meteoritics, Vol. 16, 407 - 408 (1981). – Abstract.

094.527   **The evolution of the lunar Nectaris multiring basin.**
J. L. Whitford-Stark.
Icarus, Vol. 48, 393 - 427 (1981).
    Nectaris is an 820-km-diameter, multiring impact basin located on the near side of the Moon. The transient cavity is estimated to have been less than 90 km in depth and materials were excavated from a depth of less than 30 km. About 2 km thickness of impact melt is believed to line the cavity center. The impact event probably took place at about $3.98 \pm 0.03 \times 10^9$ years ago. Inter-ring plains deposits were deposited after the formation of the Nectaris basin. The most persuasive origin for the smooth plains is one of extrusives overlain by a thin veneer of ejecta.

094.528   **Lunar soil study by EPR and microwaves.**
R. Baican, B. Baican.
Stud. Cercet. Fiz., Vol. 33, 613 - 626 (1981). In Rumanian.
Abstr. in Phys. Abstr., Vol. 85, Abstr. 28362 (1982).

094.529   **Exposure ages and erosion rates for lunar rocks.**
D. S. Burnett. D. S. Woolum.
Phys. Chem. Earth, Vol. 10, 63 - 101 (1980). – Abstr. in Phys. Abstr., Vol. 85, Abstr. 42616 (1982).

094.530   **Lunar Rb-Sr chronology.**
L. E. Nyquist.
Phys. Chem. Earth, Vol. 10, 103 - 142 (1980). – Abstr. in Phys. Abstr., Vol. 85, Abstr. 42617 (1982).

094.531   **Mineralogy, petrology and chemistry of ANT (*anorthositic-noritic-troctolitic*) suite rocks from the lunar highlands.**   M. Prinz, K. Keil.
Phys. Chem. Earth, Vol. 10, 215 - 237 (1980). – Abstr. in Phys. Abstr., Vol. 85, Abstr. 42619 (1982).

094.532   **Petrology, mineralogy and chemistry of KREEP basalt.**   C. Meyer, Jr.
Phys. Chem. Earth, Vol. 10, 239 - 260 (1980). – Abstr. in Phys. Abstr., Vol. 85, Abstr. 42620 (1982).

094.533   **Magnesian feldspathic basalts and KREEP from Luna 24 core sample 24114.**
G. Kurat, A. Kracher.
Proc. Twelfth Lunar Planet. Sci. Conf., (see 012.024), p. 1 - 19 (1982).

094.534   **Foraging westward for pristine nonmare rocks: complications for petrogenetic models.**
P. H. Warren, G. J. Taylor, K. Keil, C. Marshall, J. T. Wasson.
Proc. Twelfth Lunar Planet. Sci. Conf., (see 012.024),
p. 21 - 40 (1982).

094.535   **On compositional variations among lunar anorthosites.**
L. A. Haskin, M. M. Lindstrom, P. A. Salpas, D. J. Lindstrom.
Proc. Twelfth Lunar Planet. Sci. Conf., (see 012.024),
p. 41 - 66 (1982).

094.536 A comparative Rb−Sr, Sm−Nd, and K−Ar study of shocked norite 78236: evidence of slow cooling in the lunar crust?   L. E. Nyquist, W. U. Reimold, D. D. Bogard, J. L. Wooden, B. M. Bansal, H. Wiesmann, C.-Y. Shih.
Proc. Twelfth Lunar Planet. Sci. Conf., (see 012.024), p. 67 - 97 (1982).

094.537 Troctolite 76535: a study in the preservation of early isotopic records.
M. Caffee, C. M. Hohenberg, B. Hudson.
Proc. Twelfth Lunar Planet. Sci. Conf., (see 012.024), p. 99 - 115 (1982).

094.538 Petrology and petrogenesis of lunar breccia 12013.
J. E. Quick, O. B. James, A. L. Albee.
Proc. Twelfth Lunar Planet. Sci. Conf., (see 012.024), p. 117 - 172 (1982) = Contrib. No. 3617, Div. Geol. Planet. Sci., Calif. Inst. Technol.

094.539 A reexamination of the Rb-Sr isotopic systematics of lunar breccia 12013.
J. E. Quick, O. B. James, A. L. Albee.
Proc. Twelfth Lunar Planet. Sci. Conf., (see 012.024), p. 173 - 184 (1982).

094.540 Distribution and provenance of lunar highland rock types at North Ray Crater, Apollo 16.
D. Stöffler, R. Ostertag, W. U. Reimold, R. Borchardt, J. Malley, A. Rehfeldt.
Proc. Twelfth Lunar Planet. Sci. Conf., (see 012.024), p. 185 - 207 (1982).

094.541 Petrologic and age relations of the Apollo 16 rocks: implications for subsurface geology and the age of the Nectaris Basin.   O. B. James.
Proc. Twelfth Lunar Planet. Sci. Conf., (see 012.024), p. 209 - 233 (1982).

094.542 Petrology of suevitic lunar breccia 67016.
M. D. Norman.
Proc. Twelfth Lunar Planet. Sci. Conf., (see 012.024), p. 235 - 252 (1982).

094.543 Rust and schreibersite in Apollo 16 highland rocks: manifestations of volatile-element mobility.
R. H. Hunter, L. A. Taylor.
Proc. Twelfth Lunar Planet. Sci. Conf., (see 012.024), p. 253 - 259 (1982).

094.544 Rusty rock 66095: a paradigm for volatile-element mobility in highland rocks.
R. H. Hunter, L. A. Taylor.
Proc. Twelfth Lunar Planet. Sci. Conf., (see 012.024), p. 261 - 280 (1982).

094.545 Simplified model evaluation of cooling rates for glass-containing lunar compositions.
D. R. Uhlmann, H. Yinnon, C.-Y. Fang.
Proc. Twelfth Lunar Planet. Sci. Conf., (see 012.024), p. 281 - 288 (1982).

094.546 Carbon isotopic ratios in some low-δ^{15}N lunar breccias.   R. H. Becker, S. Epstein.
Proc. Twelfth Lunar Planet. Sci. Conf., (see 012.024), p. 289 - 293 (1982) = Contrib. No. 3619, Div. Geol. Planet. Sci., Calif. Inst. Technol.

094.547 Aspects of the history of 66095 based on trace elements in clasts and whole rock.
S. Jovanovic, G. W. Reed, Jr.
Proc. Twelfth Lunar Planet. Sci. Conf., (see 012.024), p. 295 - 304 (1982).

094.548 Geochemical studies of rocks from North Ray Crater, Apollo 16.
M. M. Lindstrom, P. A. Salpas.
Proc. Twelfth Lunar Planet. Sci. Conf., (see 012.024), p. 305 - 322 (1982).

094.549 The significance of Cl/P_2O_5 ratios from lunar samples.   L. A. Taylor, R. H. Hunter.
Proc. Twelfth Lunar Planet. Sci. Conf., (see 012.024), p. 323 - 331 (1982).

094.550 The significance of Cl/P_2O_5 ratios from lunar samples − a response.
G. W. Reed, Jr., S. Jovanovic.
Proc. Twelfth Lunar Planet. Sci. Conf., (see 012.024), p. 333 - 336 (1982). − See also 094.549.

094.551 The significance of Cl/P_2O_5 ratios from lunar samples: a reply.   L. A. Taylor, R. H. Hunter.
Proc. Twelfth Lunar Planet. Sci. Conf., (see 012.024), p. 337 (1982). − See abstracts 094.549 and 094.550.

094.552 Glasses of impact origin from Apollo 11, 12, 15, and 16: evidence for fractional vaporization and mare/highland mixing.   J. W. Delano, D. H. Lindsley, R. Rudowski.
Proc. Twelfth Lunar Planet. Sci. Conf., (see 012.024), p. 339 - 370 (1982).

094.553 The lunar regolith: comparative studies of the Apollo and Luna sites. Petrology of soils from Apollo 17, Luna 16, 20, and 24.
S. B. Simon, J. J. Papike, J. C. Laul.
Proc. Twelfth Lunar Planet. Sci. Conf., (see 012.024), p. 371 - 388 (1982).

094.554 The lunar regolith: comparative studies of the Apollo and Luna sites. Chemistry of soils from Apollo 17, Luna 16, 20, and 24.
J. C. Laul, J. J. Papike, S. B. Simon.
Proc. Twelfth Lunar Planet. Sci. Conf., (see 012.024), p. 389 - 407 (1982).

094.555 The relationship of the lunar regolith $< 10 \ \mu m$ fraction and agglutinates. Part I: A model for agglutinate formation and some indirect supportive evidence.
J. J. Papike, S. B. Simon, C. White, J. C. Laul.
Proc. Twelfth Lunar Planet. Sci. Conf., (see 012.024), p. 409 - 420 (1982).

094.556 The relationship of the lunar regolith $< 10\mu m$ fraction and agglutinates. Part II: Chemical composition of agglutinate glass as a test of the "fusion of the finest fraction" (F^3) model.   R. J. Walker, J. J. Papike.
Proc. Twelfth Lunar Planet. Sci. Conf., (see 012.024), p. 421 - 432 (1982).

094.557 Regolith maturation on the earth and the moon with an example from Apollo 15.
A. Basu, D. S. McKay, S. A. Griffiths, G. Nace.
Proc. Twelfth Lunar Planet. Sci. Conf., (see 012.024), p. 433 - 449 (1982).

094.558 Core segment 15008: regolith stratigraphy at Apennine Front station 2 using multispectral imaging.
C. M. Pieters, A. Meloy, J. S. Nagle, B. R. Hawke.
Proc. Twelfth Lunar Planet. Sci. Conf., (see 012.024), p. 451 - 461 (1982).

094.559 Depositional history of core 15008/7: some implications regarding slope processes.   J. S. Nagle.

Proc. Twelfth Lunar Planet. Sci. Conf., (see 012.024),
p. 463 - 473 (1982).

094.560 **Petrology of Apollo 15 station 9A surface and drive tube soils.**
S. A. Griffiths, A. Basu, D. S. McKay, G. Nace.
Proc. Twelfth Lunar Planet. Sci. Conf., (see 012.024),
p. 475 - 484 (1982).

094.561 **The Apollo 15 regolith: comparative petrology of drive tube 15010/15011 and drill core section 15003.**   R. J. Walker, J. J. Papike.
Proc. Twelfth Lunar Planet. Sci. Conf., (see 012.024),
p. 485 - 508 (1982).

094.562 **The Apollo 15 regolith: chemical modeling and mare/highland mixing.**
R. J. Walker, J. J. Papike.
Proc. Twelfth Lunar Planet. Sci. Conf., (see 012.024),
p. 509 - 517 (1982).

094.563 **An ancient avalanche deposit in the Taurus-Littrow Valley is part of the long depositional history of the Apollo 17 drill core.**   J. S. Nagle.
Proc. Twelfth Lunar Planet. Sci. Conf., (see 012.024),
p. 519 - 528 (1982).

094.564 **Distribution and evolution of Zn, Cd, and Pb in Apollo 16 regolith samples and the average U–Pb ages of the parent rocks.**   E. H. Cirlin, R. M. Housley.
Proc. Twelfth Lunar Planet. Sci. Conf., (see 012.024),
p. 529 - 540 (1982).

094.565 **Double drive tube 74001/2: a two-stage exposure model based on noble gases, chemical abundances and predicted production rates.**   O. Eugster, N. Grögler, P. Eberhardt, J. Geiss, W. Kiesl.
Proc. Twelfth Lunar Planet. Sci. Conf., (see 012.024),
p. 541 - 558 (1982).

094.566 **Carbon and carbon-14 in lunar soil 14163.**
E. L. Fireman, R. W. Stoenner.
Proc. Twelfth Lunar Planet. Sci. Conf., (see 012.024),
p. 559 - 565 (1982).

094.567 **Studies of lunar regolith dynamics using measurements of cosmogenic radionuclides in lunar rocks, soils and cores.**
J. S. Fruchter, J. H. Reeves, J. C. Evans, R. W. Perkins.
Proc. Twelfth Lunar Planet. Sci. Conf., (see 012.024),
p. 567 - 575 (1982).

094.568 **Compositional trends in Apollo 16 soils.**
R. L. Korotev.
Proc. Twelfth Lunar Planet. Sci. Conf., (see 012.024),
p. 577 - 605 (1982).

094.569 **Apollo 11 breccias and soils: aluminous mare basalts or multi-component mixtures?**
J. M. Rhodes, D. P. Blanchard.
Proc. Twelfth Lunar Planet. Sci. Conf., (see 012.024),
p. 607 - 620 (1982).

094.570 **A model for synthesis of methane in lunar soil.**
A. S. Tamhane.
Proc. Twelfth Lunar Planet. Sci. Conf., (see 012.024),
p. 621 - 625 (1982).

094.571 **Solar flare neon: clues from implanted noble gases in lunar soils and rocks.**
C. M. Nautiyal, J. T. Padia, M. N. Rao, T. R. Venkatesan.

Proc. Twelfth Lunar Planet. Sci. Conf., (see 012.024),
p. 627 - 637 (1982).

094.572 **Selenographic distribution of apparent crater depth.**
R. A. De Hon.
Proc. Twelfth Lunar Planet. Sci. Conf., (see 012.024),
p. 639 - 650 (1982).

094.573 **Remote sensing studies of lunar dark-halo impact craters: preliminary results and implications for early volcanism.**   B. R. Hawke, J. F. Bell.
Proc. Twelfth Lunar Planet. Sci. Conf., (see 012.024),
p. 665 - 678 (1982).

094.574 **The Reiner Gamma Formation: composition and origin as derived from remote sensing observations.**
J. F. Bell, B. R. Hawke.
Proc. Twelfth Lunar Planet. Sci. Conf., (see 012.024),
p. 679 - 694 (1982).

094.575 **Coordination chemistry of iron in glasses contributing to remote-sensed spectra of the moon.**
M. D. Dyar, R. G. Burns.
Proc. Twelfth Lunar Planet. Sci. Conf., (see 012.024),
p. 695 - 702 (1982).

094.576 **Observations of silicate reststrahlen bands in lunar infrared spectra.**   A. E. Potter, Jr., T. H. Morgan.
Proc. Twelfth Lunar Planet. Sci. Conf., (see 012.024),
p. 703 - 713 (1982).

094.577 **The Balmer Basin: regional geology and geochemistry of an ancient lunar impact basin.**
T. A. Maxwell, C. G. Andre.
Proc. Twelfth Lunar Planet. Sci. Conf., (see 012.024),
p. 715 - 725 (1982).

094.578 **Compositional variation in the Hadley Apennine region.**   P. E. Clark, B. R. Hawke.
Proc. Twelfth Lunar Planet. Sci. Conf., (see 012.024),
p. 727 - 749 (1982).

094.579 **Thorium concentrations in the lunar surface: V. Deconvolution of the central highlands region.**
A. E. Metzger, M. I. Etchegaray-Ramirez, E. L. Haines.
Proc. Twelfth Lunar Planet. Sci. Conf., (see 012.024),
p. 751 - 766 (1982).

094.580 **Regional chemical setting of the Apollo 16 landing site and the importance of the Kant Plateau.**
C. G. Andre, F. El-Baz.
Proc. Twelfth Lunar Planet. Sci. Conf., (see 012.024),
p. 767 - 779 (1982).

094.581 **Chemical mixing model studies of lunar orbital geochemical data: Apollo 16 and 17 highlands compositions.**   P. D. Spudis, B. R. Hawke.
Proc. Twelfth Lunar Planet. Sci. Conf., (see 012.024),
p. 781 - 789 (1982).

094.582 **On compositional modeling of lunar highlands soils, including application to the orbiting gamma-ray experimental data.**   L. A. Haskin, R. L. Korotev.
Proc. Twelfth Lunar Planet. Sci. Conf., (see 012.024),
p. 791 - 808 (1982).

094.583 **High spatial resolution Mg/Al maps of the western Crisium and Sulpicius Gallus regions.**
E. Schonfeld.
Proc. Twelfth Lunar Planet. Sci. Conf., (see 012.024),
p. 809 - 816 (1982).

094.584 **Sources of lunar magnetic anomalies and their bulk directions of magnetization: additional evidence from Apollo orbital data.**   L. L. Hood.
Proc. Twelfth Lunar Planet. Sci. Conf., (see 012.024), p. 817 - 830 (1982).

094.585 **Adsorption and excess fission xenon.**
F. A. Podosek, T. J. Bernatowicz, F. E. Kramer.
Proc. Twelfth Lunar Planet. Sci. Conf., (see 012.024), p. 891 - 901 (1982).

094.586 **An origin for the REE characteristics of KREEP.**
L. P. Gromet, P. C. Hess, M. J. Rutherford.
Proc. Twelfth Lunar Planet. Sci. Conf., (see 012.024), p. 903 - 913 (1982).

094.587 **A chemical study of individual green glasses and brown glasses from 15426: implications for their petrogenesis.**   M.-S. Ma, Y.-G. Liu, R. A. Schmitt.
Proc. Twelfth Lunar Planet. Sci. Conf., (see 012.024), p. 915 - 933 (1982).

094.588 **Compositional variations among Apollo 15 green glass spheres.**   T. L. Grove.
Proc. Twelfth Lunar Planet. Sci. Conf., (see 012.024), p. 935 - 948 (1982).

094.589 **Mechanism for the extrusion of KREEP.**
D. N. Shirley, J. T. Wasson.
Proc. Twelfth Lunar Planet. Sci. Conf., (see 012.024), p. 965 - 978 (1982).

094.590 **Progressive differentiation of a growing moon.**
J. V. Smith.
Proc. Twelfth Lunar Planet. Sci. Conf., (see 012.024), p. 979 - 990 (1982).

094.591 **Soret separation of lunar liquid.**
D. Walker, C. E. Lesher, J. F. Hays.
Proc. Twelfth Lunar Planet. Sci. Conf., (see 012.024), p. 991 - 999 (1982).

094.592 **Preliminary modeling of high pressure partial melting: implications for early lunar differentiation.**
J. Longhi.
Proc. Twelfth Lunar Planet. Sci. Conf., (see 012.024), p. 1001 - 1018 (1982).

094.593 **Thermal and impact histories of pyroxenes in lunar eucrite-like gabbros and eucrites.**
H. Takeda, H. Mori, T. Ishii, M. Miyamoto.
Proc. Twelfth Lunar Planet. Sci. Conf., (see 012.024), p. 1297 - 1313 (1982).

094.594 **Secondary cratering effects on lunar microterrain: implications for the micrometeoroid flux.**
R. J. Allison, J. A. M. McDonnell.
Proc. Twelfth Lunar Planet. Sci. Conf., (see 012.024), p. 1703 - 1716 (1982).

094.595 **On the estimation of lunar paleointensities – studies of synthetic analogues of stably magnetized samples.**
J. R. Dunn, M. Fuller, D. A. Clauter.
Proc. Twelfth Lunar Planet. Sci. Conf., (see 012.024), p. 1747 - 1758 (1982).

094.596 **Remote X-ray spectral element analysis of lunar surface rocks based on artificial excitation of the characteristic radiation by an electron beam.**
E. K. Kolesnikov, A. P. Kuryshev.
Vestn. LGU. Ser. mat., mekh., astron., Leningrad, 1982. 33 pp.

In Russian. – Review in Ref. zh., 62. Issled. kosm. prostranstva, 4.62.278 (1982).

094.597 **Akaganeit in lunar soil.**
N. A. Ashikhmina, O. A. Bogatikov, D. I. Frikh-Khar, T. L. Evstigneeva, G. N. Muravitskaya, V. A. Stepanchikov.
Dokl. AN SSSR, Tom 261, 735 - 737 (1981). In Russian.
Abstr. in Ref. zh., 62. Issled. kosm. prostranstva, 4.62.293 (1982).

094.598 **Crystalline structure of the spinel from regolith of the Luna 24 automatic station.**
M. T. Dmitrieva, G. B. Bokij, N. A. Ashikhmina.
Dokl. AN SSSR, Tom 261, 738 - 740 (1981). In Russian.
Abstr. in Ref. zh., 62. Issled. kosm. prostranstva, 4.62.294 (1982).

094.599 **Mare Orientale.**   V. V. Shevchenko.
Zemlya Vselennaya, 1982, No. 3, p. 70 - 72. In Russian.

094.600 **Lava flooding of ancient planetary crusts: geometry, thickness, and volumes of flooded lunar impact basins.**   J. W. Head.
Moon Planets, Vol. 26, 61 - 88 (1982).
A technique is described and developed which provides volume estimates by artificially flooding unflooded lunar topography characteristic of certain geological environments, and tracking the area covered, lava thicknesses, and lava volumes. Comparisons of map patterns of incompletely buried topography in these artificially flooded areas are then made to lava-flooded topography on the Moon in order to estimate the actual lava volumes. This technique is applied to two areas related to lunar impact basins; the relatively unflooded Orientale basin, and the Archimedes–Apennine Bench region of the Imbrium basin. It is concluded that early flooding of the basin interior places a major load on the lithosphere in the same geographic region where mascon gravity anomalies are concentrated.

094.601 **Zircon saturation in lunar basalts and granites.**
J. E. Dickinson, Jr., P. C. Hess.
Earth Planet. Sci. Lett., Vol. 57, 336 - 344 (1982).

094.602 **On the location of acid-hydrolysable carbon in lunar soil fines.**   A. E. Fallick, I. P. Wright, C. T. Pillinger, A. Stephenson, R. V. Morris.
Earth Planet. Sci. Lett., Vol. 59, 28 - 32 (1982).

094.603 **Observing Luna Incognita in 1982.**
J. E. Westfall.
Strolling Astron., Vol. 29, 117 - 118, 120 - 121 (1982).

Laser pyrolysis for light element and stable isotope studies.   See Abstr. 022.071.

Simulation of lunar sputter erosion using heavy ion bombardment.   See Abstr. 022.074.

Compressional wave velocities of a lunar regolith sample in a simulated lunar environment.
See Abstr. 022.111.

Mare basin filling on the moon: laboratory simulations.   See Abstr. 022.116.

Xenon diffusion following ion implantation into feldspar: dependence on implantation dose.
See Abstr. 022.120.

Applications of neutron activation analysis of meteorite, lunar, and terrestrial samples. See Abstr. 022.179.

Magnification of pre-existing magnetic fields in impact-produced plasmas, with reference to impact craters. See Abstr. 062.065.

Post-Imbrian global lunar tectonism: evidence for an initially totally molten Moon.   See Abstr. 094.046.

Pu-Nd-Xe dating: systematics in ^{244}Pu fission and REE spallation components.   See Abstr. 105.075.

The significance of W in planetary differentiation processes: evidence from new data on eucrites. See Abstr. 105.176.

Microchrons: the ^{87}Rb–^{87}Sr dating of microscopic samples.   See Abstr. 105.178.

A review of data on paleomagnetic determination from magnetization of meteorites and lunar rocks. See Abstr. 105.207.

Terrestrial, meteoritic, and lunar titanium isotopic ratios revaluated: evidence for correlated variations. See Abstr. 105.227.

Erratum

094.901 Erratum: 'Record of the solar corpuscular radiation in minerals from lunar soils: a comparative study of noble gases and tracks' [Proc. Eleventh Lunar Planet. Sci. Conf., p. 1369 - 1393 (1980)]. R. Wieler, P. Etique, P. Signer, G. Poupeau. Proc. Twelfth Lunar Planet. Sci. Conf., (see 012.024) (1982). See Abstr. 30.094.560.

095 Lunar Eclipses

095.001 **Photometrie der Mondfinsternis vom 9. Januar 1982.** D. Böhme. Orion, 40. Jahrg., 57 - 59 (1982).

095.002 **Lunar eclipses, lunar luminescence and transient lunar phenomena.**   J. S. Shepherd. J. British Astron. Assoc., Vol. 92, 66 - 68 (1982). Lunar eclipses offer a unique opportunity to study lunar emissions other than reflected sunlight. A possible connection is drawn between this so-called luminescence and transient lunar phenomena.

095.003 **El eclipse de Luna del 24/25 de Mayo de 1975.** A. Feinstein, J. C. Forte, L. Cabrera. Bol. Asoc. Argentina Astron., No. 20 - 24, p. 275 (1981). Abstract.

095.004 **Partial eclipse of the moon 1981 July 17.** T. P. Cooper. Mon. Notes Astron. Soc. South. Africa, Vol. 41, 14 - 15 (1982).

095.005 **Total lunar eclipse: July 5 - 6, 1982.** J. E. Westfall. Strolling Astron., Vol. 29, 106, 108 - 110 (1982).

095.006 **Two reports from England on the January 9, 1982 total lunar eclipse.**   W. H. Haas. Strolling Astron., Vol. 29, 128 - 129 (1982).

The many eclipses of 1982. See Abstr. 079.003.

Is there an increase of geomagnetic activity preceding total lunar eclipses?   See Abstr. 084.051.

Extended USNO total occultation predictions for early 1982, including the total lunar eclipse of 1982 January 9. See Abstr. 096.015.

096 Lunar and Planetary Occultations

096.001 **Aldebarans bedekking nader bekeken.**
D. Schmidt.
Zenit, 9. Jaarg., 82 - 83 (1982).

096.002 **Illinois occultation summary. II. 1979.**
R. Radick, D. Lien.
Astron. J., Vol. 87, 170 - 174 (1982).
The authors present results from the third year of a program undertaken to record lunar occultations at the University of Illinois Prairie Observatory. The 32 events summarized include 17 observations of stars brighter than 7th mag. 13 reappearances, four angular diameter measurements (three for Aldebaran), and seven observations of binary stars or stars which may be double.

096.003 **The lunar occultation of Venus, 1980 October 5.**
G. W. Amery.
J. British Astron. Assoc., Vol. 92, 132 - 134 (1982).
Report of a rare occultation of Venus.

096.004 **Fresnel zone structure in planetary occultations.**
T. A. Croft.
Radio Sci., Vol. 16, 1463 - 1472 (1981). − Abstr. in Phys. Abstr., Vol. 85, Abstr. 42624 (1982).

096.005 **Observations of star occultations by the moon at the Astronomical Observatory of the Kiev University in 1979.** A. K. Osipov, V. I. Mazur, A. A. Zhitetskij.
Vestn. Kiev. Univ., Vyp. 23, p. 109 - 113 (1981). In Russian.

096.006 **Possible occultation by Pluto on 1982 April 15.**
IAU Circ., Nos. 3674, 3688 (1982).

096.007 **Occultation by Uranus and its rings.**
IAU Circ., Nos. 3690, 3695 (1982).

096.008 **Occultation of Uranus and his rings by BD −20° 51615 (9^{m}9).**
Yamamoto Circ., No. 1974 (1982).

096.009 **Occultations during the total lunar eclipse of 1982 July 6.** D. W. Dunham.
Occultation Newsl., Vol. 2, 214 - 219, 230 (1982).

096.010 **Grazing occultations.** D. W. Dunham.
Occultation Newsl., Vol. 2, 220 - 221 (1982).

096.011 **Radio observations of the forthcoming lunar occultations of the Crab Nebula.**
F. P. Maloney, S. T. Gottesman.
Bull. American Astron. Soc., Vol. 13, 843 (1981). − Abstract.

096.012 **Observation of lunar occultations from orbit.**
T. J. Sherrill.
Bull. American Astron. Soc., Vol. 13, 874 (1981). − Abstract.

096.013 **Catalogue of observations of occultations of stars by the Moon for the years 1623 to 1942 and solar eclipses for the years 1621 to 1806.**
L. V. Morrison, M. R. Lukac, F. R. Stephenson.
R. Greenwich Obs. Bull. No. 186, 54 pp. + 2 microfiches (1981).

096.014 **Occultation observation in 1980.**
Data Rep. Hydrogr. Obs., Ser. Astron. Geod. Tokyo, No. 16, p. 1 - 45 (1982).
In 1980, timing data of 821 lunar occultations, including 334 photoelectric observations, of reliable quality were obtained at four astronomical stations of JHD. 172 visual timing data reported by cooperators in Japan were also adopted. Reduction and analysis give the following provisional result: ET $(j=2)$ − TAI = 30^{s}50 ± 0^{s}04 (m.e.), $\Delta B = -0''37 ± 0''03$ (m.e.) for 1980.5 on the FK4 system.

096.015 **Extended USNO total occultation predictions for early 1982, including the total lunar eclipse of 1982 January 9.** D. W. Dunham.
Occultation Newsl., Vol. 2, 188 - 189 (1982).

096.016 **Planetary occultation predictions for 1982.**
D. W. Dunham.
Occultation Newsl., Vol. 2, 191 - 196, 204 - 212 (1982).

096.017 **More planetary occultations during 1982.**
A. Lowe, D. W. Dunham.
Occultation Newsl., Vol. 2, 198 - 200 (1982).

096.018 **Observations of asteroidal occultations and appulses.**
D. W. Dunham.
Occultation Newsl., Vol. 2, 200 - 202 (1982).

096.019 **17/11/1981: Venere occulta Nunki. Osservazioni visuali e fotografiche all'oss. "G. Horn d'Arturo".**
C. Frisoni, R. Di Luca.
G. A. A. B., N. 65, p. 4 - 10 (1982).

096.020 **Cloudcroft occultation summary. I. December 1978 - March 1980.** R. R. Radick, J. L. Africano, W. F. Rogers, T. J. Schneeberger, E. T. Tyson.
Astron. J., Vol. 87, 885 - 892 (1982).
The authors present results from the first 16 months of a program undertaken to record lunar occultations at Cloudcroft Observatory. The 85 events summarized include 38 observations of stars brighter than seventh magnitude, 26 reappearances, nine angular diameter measurements, and 11 observations of binary stars or stars which may be double.

Southern Astrographic Catalog data added to USNO C-catalog for extended-coverage total occultation predictions.
See Abstr. 002.047.

Analisi comparativa fra metodi visuali di timing delle occultazioni. See Abstr. 031.520.

On the reduction of observations of planetary satellites. See Abstr. 031.545.

Observación de ocultaciones.
See Abstr. 031.550.

Analysis of lunar occultations − IV. Rotation of the FK4 reference frame. See Abstr. 041.001.

Structure of the Martian atmosphere from ϵ Gem occultation observations. See Abstr. 097.105.

Occultation of AGK3 −1°1676 by Mars on 1983 December 1. See Abstr. 097.109.

Occultation of SAO 042418 by (344) Desiderata.
See Abstr. 098.053.

Occultation by (146) Lucina.
See Abstr. 098.065.

Video record of secondary occultation obtained at Meudon Observatory during (146) Lucina appulse on 1982 April 18.   See Abstr. 098.071.

Planetary occultation predictions for 1982. See Abstr. 098.072.

Predictions of occultations by minor planets 1983. See Abstr. 098.078.

Report of IAU Commission 20: Positions and motions of minor planets, comets and satellites (Positions et mouvements des petites planètes, des comètes et des satellites). See Abstr. 098.107.

The 15 August 1980 occultation by the Uranian system: structure of the rings and temperature of the upper atmosphere.   See Abstr. 101.010.

Occultation of a 15th magnitude star by Pluto/ Charon on 1983 April 24.   See Abstr. 101.011.

An occultation angular diameter in H-alpha light. See Abstr. 115.001.

Occultation angular diameters of α Tauri by least squares and deconvolution.   See Abstr. 115.026.

097 Mars, Mars Satellites

097.001 **Astrolabe observations of Mars and phase effect.**
F. Chollet.
C. R. Acad. Sci. Paris, Tome 294, Sér. II, 191 - 193 (1982).
In French.
Recently, an analysis of the results of observations of Mars with astrolabes shows that the classical phase effect correction is not sufficient. The author gives here a new formulation of this correction, based on a more realistic model of planetary images, which is in very good accordance with the observations.

097.002 **Organic and inorganic interpretations of the martian UV−IR reflectance spectrum.**
K. D. Pang, S. F. S. Chun, J. M. Ajello, Z. Nansheng, L. Minji.
Nature, Vol. 295, 43 - 46 (1982).
The authors consider why no organic molecules were detected at the Viking landing sites, whether the sterility of the two sites is representative of the entire planet and if there are locations on Mars more conducive to the formation and preservation of organics. There is no reason to believe that organic molecules are preferentially preserved anywhere on the planet.

097.003 **Martian ejecta flow craters.**
V. M. Horner, R. Greeley.
Reports of Planetary Geology Program − 1981, (see 003.001)
p. 75 - 77 (1981). − Abstract.

097.004 **The energy line: a heuristic model for Martian rampart ejecta sheets.** M. F. Sheridan.
Reports of Planetary Geology Program − 1981, (see 003.001),
p. 87 - 88 (1981). − Abstract.

097.005 **Martian crater morphology and evolution: radar results.** R. S. Saunders, L. E. Roth.
Reports of Planetary Geology Program − 1981, (see 003.001),
p. 89 - 90 (1981). − Abstract.

097.006 **The Martian cratering record.**
M. Gurnis.
Reports of Planetary Geology Program − 1981, (see 003.001),
p. 91 - 92 (1981). − Abstract.

097.007 **Three simple classes of Martian crater ejecta − I. Global relationships of class abundance to latitude and terrain.**
K. R. Blasius, J. A. Cutts, B. H. Lewis, A. V. Vetrone.
Reports of Planetary Geology Program − 1981, (see 003.001),
p. 93 - 95 (1981). − Abstract.

097.008 **Three simple classes of Martian crater ejecta − II. Global relationships of crater radius to latitude and terrain.** K. R. Blasius, J. A. Cutts, B. H. Lewis, A. V. Vetrone.
Reports of Planetary Geology Program − 1981, (see 003.001),
p. 96 - 98 (1981). − Abstract.

097.009 **Martian crater-form topography on Tempe Terra.**
P. A. Davis, D. J. Roddy, N. E. Witbeck.
Reports of Planetary Geology Program − 1981, (see 003.001),
p. 99 - 101 (1981). − Abstract.

097.010 **The distribution of crater ejecta and central peaks on Mars.** L. A. Johansen.
Reports of Planetary Geology Program − 1981, (see 003.001),
p. 102 - 104 (1981). − Abstract.

097.011 **Phobos, Deimos, and the moon: comparison of crater ejecta patterns.**
S. Lee, P. Thomas, J. Veverka.
Reports of Planetary Geology Program − 1981, (see 003.001),
p. 105 - 107 (1981). − Abstract.

097.012 **Quantitative analysis of Olympus Mons.**
S. S. C. Wu, P. A. Garcia, R. Jordan, F. J. Schafer.
Reports of Planetary Geology Program − 1981, (see 003.001),
p. 141 - 143 (1981). − Abstract.

097.013 **Thickness distribution of Tharsis volcanic materials.**
R. A. De Hon.
Reports of Planetary Geology Program − 1981, (see 003.001),
p. 144 - 146 (1981). − Abstract.

097.014 **Eruptive styles of Martian volcanoes.**
P. J. Mouginis-Mark.
Reports of Planetary Geology Program − 1981, (see 003.001),
p. 147 - 149 (1981). − Abstract.

097.015 **Tharsis volcano burial.**
J. L. Whitford-Stark.
Reports of Planetary Geology Program − 1981, (see 003.001),
p. 150 - 152 (1981). − Abstract.

097.016 **Albedo changes in lava.**
J. L. Whitford-Stark.
Reports of Planetary Geology Program − 1981, (see 003.001),
p. 153 - 155 (1981). − Abstract.

097.017 **Theoretical analyses of Martian explosive eruption mechanisms.** L. Wilson, J. W. Head.
Reports of Planetary Geology Program − 1981, (see 003.001),
p. 164 - 166 (1981). − Abstract.

097.018 **Characterization of rock populations on the surfaces of Mars, Venus, and earth: a summary.**
J. B. Garvin, P. J. Mouginis-Mark, J. W. Head.
Reports of Planetary Geology Program − 1981, (see 003.001),
p. 177 - 179 (1981). − Abstract.

097.019 **Martian sediments: evidence for sand on Mars.**
A. R. Peterfreund, R. Greeley, D. Krinsley.
Reports of Planetary Geology Program − 1981, (see 003.001),
p. 205 - 207 (1981). − Abstract.

097.020 **Clay aggregates on earth, Mars and Io.**
D. Nummedal.
Reports of Planetary Geology Program − 1981, (see 003.001),
p. 216 - 218 (1981). − Abstract.

097.021 **Dust and sand movement on Mars: present activity and its relation to sediment deposits.**
P. Thomas.
Reports of Planetary Geology Program − 1981, (see 003.001),
p. 219 - 221 (1981). − Abstract.

097.022 **Comparison of aeolian activity in Elysium and Tharsis.** S. Lee, P. Thomas, J. Veverka.
Reports of Planetary Geology Program − 1981, (see 003.001),
p. 222 - 224 (1981). − Abstract.

097.023 **Eolian stratigraphy of the west central equatorial region of Mars: Viking Lander 1 and Orbiter color observations.** E. L. Strickland, III.
Reports of Planetary Geology Program − 1981, (see 003.001),
p. 225 - 227 (1981). − Abstract.

097.024 Sketch map of the eolian units of the west central
    equatorial region of Mars.   E. L. Strickland,III.
Reports of Planetary Geology Program − 1981, (see 003.001),
p. 228 (1981). − Abstract.

097.025 Wind-modifications of the Chryse channels.
    D. Nummedal.
Reports of Planetary Geology Program − 1981, (see 003.001),
p. 229 - 231 (1981). − Abstract.

097.026 Crescent-shaped pits on Mars.
    D. D. Rhodes, T. Neal.
Reports of Planetary Geology Program − 1981, (see 003.001),
p. 232 - 234 (1981). − Abstract.

097.027 Eolian erosion of poorly consolidated sedimentary
    blankets on Mars: a small-scale terrestrial analog.
G. A. Brook.
Reports of Planetary Geology Program − 1981, (see 003.001),
p. 235 - 237 (1981). − Abstract.

097.028 Chronology of channels in Chryse Planitia Mars.
    J. A. Cutts, B. H. Lewis, K. R. Blasius.
Reports of Planetary Geology Program − 1981, (see 003.001),
p. 257 - 259 (1981). − Abstract.

097.029 Chaotic terrain and Chryse basin outflow channels:
    structural control and evolution.
J. A. Cutts, J. Helu, K. R. Blasius.
Reports of Planetary Geology Program − 1981, (see 003.001),
p. 260 - 261 (1981). − Abstract.

097.030 Chryse hydrographic basin, Mars − a progress
    report.   H. Masursky, A. L. Dial, Jr.,
M. E. Strobell, G. S. Downs, T. Thompson.
Reports of Planetary Geology Program − 1981, (see 003.001),
p. 262 - 265 (1981). − Abstract.

097.031 Etched plains and braided ridges of the south polar
    region of Mars: features produced by basal melting
of ground ice?   A. D. Howard.
Reports of Planetary Geology Program − 1981, (see 003.001),
p. 286 - 288 (1981). − Abstract.

097.032 Geomorphic implications from Martian ground ice.
    L. A. Rossbacher, S. Judson.
Reports of Planetary Geology Program − 1981, (see 003.001),
p. 289 - 291 (1981). − Abstract.

097.033 A composite origin for Martian outflow channels.
    B. K. Lucchitta.
Reports of Planetary Geology Program − 1981, (see 003.001),
p. 299 - 301 (1981). − Abstract.

097.034 Geomorphic mapping of Capri Chasma.
    J. C. Boothroyd, B. S. Timson.
Reports of Planetary Geology Program − 1981, (see 003.001),
p. 302 - 304 (1981). − Abstract.

097.035 Preliminary comparison of inselbergs in the
    Cerberus region of Mars to terrestrial isolated hills
in arid, humid, and glacial terrains.
L. S. Manent, F. El-Baz.
Reports of Planetary Geology Program − 1981, (see 003.001),
p. 305 - 307 (1981). − Abstract.

097.036 Subsidence depressions on Martian plateau terrains.
    J. J. Fagan, D. Weiss, J. Steiner, O. L. Franke.
Reports of Planetary Geology Program − 1981, (see 003.001),
p. 308 - 311 (1981). − Abstract.

097.037 A morphological comparison of depressional
    features in plateau materials of the Deuteronilus
Mensae region based on ellipsoidal characteristics.
J. Steiner, C. Sodden, D. Weiss, J. J. Fagan, O. L. Franke.
Reports of Planetary Geology Program − 1981, (see 003.001),
p. 312 - 315 (1981). − Abstract.

097.038 A preliminary survey of slope and related features
    at and near the boundary between the plateau-
fretted terrain and northern plains of Mars.
O. L. Franke, J. Steiner, D. Weiss, J. J. Fagan.
Reports of Planetary Geology Program − 1981, (see 003.001),
p. 319 - 320 (1981). − Abstract.

097.039 Modifications of escarpments along channels and
    plateaus on Mars.   R. C. Kochel, V. R. Baker.
Reports of Planetary Geology Program − 1981, (see 003.001),
p. 321 - 323 (1981). − Abstract.

097.040 Evolution of the spur and gully topography on the
    Valles Marineris wall scarps.   P. C. Patton.
Reports of Planetary Geology Program − 1981, (see 003.001),
p. 324 - 325 (1981). − Abstract.

097.041 More on landslides − Valles Marineris.
    B. K. Lucchitta, K. L. Kaufman, D. J. Tosline.
Reports of Planetary Geology Program − 1981, (see 003.001),
p. 326 - 328 (1981). − Abstract.

097.042 Topography and stratigraphy of Martian polar
    layered deposits.
J. A. Cutts, K. R. Blasius, A. D. Howard.
Reports of Planetary Geology Program − 1981, (see 003. 001),
p. 337 - 339 (1981). − Abstract.

097.043 Uniform deposition rate and climate modulated
    deposition rate models of Martian polar layered
deposits.   J. A. Cutts, K. R. Blasius.
Reports of Planetary Geology Program − 1981, (see 003.001),
p. 340 - 341 (1981). − Abstract.

097.044 Diagnostic stratigraphic relationships in areas of
    complex topography on polar layered deposits.
A. D. Howard, J. A. Cutts, K. R. Blasius.
Reports of Planetary Geology Program − 1981, (see 003.001),
p. 342 - 344 (1981). − Abstract.

097.045 Mars: theoretical and experimental studies of
    regolith-atmospheric-cap CO_2 exchange and climate
change.
F. P. Fanale, W. B. Banerdt, R. S. Saunders, J. R. Salvail.
Reports of Planetary Geology Program − 1981, (see 003.001),
p. 347 - 350 (1981). − Abstract.

097.046 Viking MAWD (*Mars Atmospheric Water Detector*)
    observations and regolith water vapor sources on
Mars.   R. L. Huguenin, S. M. Clifford, B. W. Hapke.
Reports of Planetary Geology Program − 1981, (see 003.001),
p. 351 - 354 (1981). − Abstract.

097.047 Martian center of mass − center of figure offset.
    L. E. Roth, M. Kobrick, G. S. Downs,
R. S. Saunders, G. Schubert.
Reports of Planetary Geology Program − 1981, (see 003.001),
p. 372 - 374 (1981). − Abstract.

097.048 Mars structural studies.
    R. S. Saunders, W. B. Banerdt, R. J. Phillips.
Reports of Planetary Geology Program − 1981, (see 003.001),
p. 375 - 376 (1981). − Abstract.

097.049 **Ridge orientations in the Tharsis province of Mars: deviations from Tharsis-related trends.**
T. A. Maxwell, T. R. Watters.
Reports of Planetary Geology Program – 1981, (see 003.001), p. 380 - 382 (1981). – Abstract.

097.050 **Ridge-rille intersections in the Tharsis province of Mars.**   T. R. Watters, T. A. Maxwell.
Reports of Planetary Geology Program – 1981, (see 003.001), p. 383 - 385 (1981). – Abstract.

097.051 **Statistical approach to the fracture pattern of the Tharsis region of Mars.**
R. Salvatori, R. Bianchi, M. Coradini, M. Fulchignoni.
Reports of Planetary Geology Program – 1981, (see 003.001), p. 386 - 388 (1981). – Abstract.

097.052 **The basal scarp of Olympus Mons.**   E. C. Morris.
Reports of Planetary Geology Program – 1981, (see 003.001), p. 389 - 390 (1981). – Abstract.

097.053 **Deep gravitational CREEP deformation: earth analogue of a Mars chaos area.**   C. A. Baskerville.
Reports of Planetary Geology Program – 1981, (see 003.001), p. 391 - 393 (1981). – Abstract.

097.054 **Deformed impact craters on Mars.**
C. G. Andre, F. El-Baz.
Reports of Planetary Geology Program – 1981, (see 003.001), p. 399 - 401 (1981). – Abstract.

097.055 **A Viking solution to a Mariner stratigraphic problem.**
D. H. Scott.
Reports of Planetary Geology Program – 1981, (see 003.001), p. 411 - 413 (1981). – Abstract.

097.056 **Progress report: Tharsis lava flow map series.**
D. H. Scott.
Reports of Planetary Geology Program – 1981, (see 003.001), p. 414 - 415 (1981). – Abstract.

097.057 **Crater counts on Olympus Mons.**
C. Neal.
Reports of Planetary Geology Program – 1981, (see 003.001), p. 416 - 418 (1981). – Abstract.

097.058 **Valles Marineris – faults, volcanic rocks, channels, basin beds.**   B. K. Lucchitta.
Reports of Planetary Geology Program – 1981, (see 003.001), p. 419 - 421 (1981). – Abstract.

097.059 **Preliminary observations of the detailed stratigraphy across the highland-lowlands boundary.**
D. Weiss, J. J. Fagan, J. Steiner, O. L. Franke.
Reports of Planetary Geology Program – 1981, (see 003.001), p. 422 - 425 (1981). – Abstract.

097.060 **Characteristics of Mars north polar region from bistatic radar.**
R. A. Simpson, G. L. Tyler, H. T. Howard.
Reports of Planetary Geology Program – 1981, (see 003.001), p. 432 - 434 (1981). – Abstract.

097.061 **Radar studies of the cratered hemisphere of Mars.**
P. J. Mouginis-Mark, S. H. Zisk.
Reports of Planetary Geology Program – 1981, (see 003.001), p. 435 - 437 (1981). – Abstract.

097.062 **Surface properties of Mars determined from high resolution infrared and visual data.**
J. Zimbelman, R. Greeley.

Reports of Planetary Geology Program – 1981, (see 003.001), p. 446 - 448 (1981). – Abstract.

097.063 **Multivariate classification of surficial units on Mars from Viking Orbiter color and infrared data.**
E. A. Guinness, R. E. Arvidson, A. Zent.
Reports of Planetary Geology Program – 1981, (see 003.001), p. 449 (1981). – Abstract.

097.064 **Recent weathering of rocks at the Viking landing sites: evidence from enhanced images and spectral estimate ratios.**   E. L. Strickland, III.
Reports of Planetary Geology Program – 1981, (see 003.001), p. 450 - 452 (1981). – Abstract.

097.065 **The control network of Mars: September 1981.**
M. E. Davies.
Reports of Planetary Geology Program – 1981, (see 003.001), p. 483 (1981). – Abstract.

097.066 **Mars atlases.**   R. M. Batson, R. L. Tyner.
Reports of Planetary Geology Program – 1981, (see 003.001), p. 486 (1981). – Abstract.

097.067 **1 : 2,000,000 scale controlled photomosaics of Mars.**   R. M. Batson, R. L. Tyner.
Reports of Planetary Geology Program – 1981, (see 003.001), p. 487 - 488 (1981). – Abstract.

097.068 **Mars 1 : 2 million contour mapping problems with Viking Orbiter photographs.**
S. S. C. Wu, R. Jordan, F. J. Schafer.
Reports of Planetary Geology Program – 1981, (see 003.001), p. 489 - 490 (1981). – Abstract.

097.069 **Special purpose Mars mapping.**   R. M. Batson.
Reports of Planetary Geology Program – 1981, (see 003.001), p. 491 - 492 (1981). – Abstract.

097.070 **Revisions of 1 : 5,000,000 scale Mars maps.**
R. M. Batson, P. M. Bridges.
Reports of Planetary Geology Program – 1981, (see 003.001), p. 496 (1981). – Abstract.

097.071 **Review and highlights of Mars geologic mapping – western hemisphere.**   D. H. Scott.
Reports of Planetary Geology Program – 1981, (see 003.001), p. 500 (1981). – Abstract.

097.072 **Geologic mapping of Mangala Vallis from Viking Orbiter survey mission data.**
E. Stofan, D. C. Pieri, R. S. Saunders.
Reports of Planetary Geology Program – 1981, (see 003.001), p. 501 - 502 (1981). – Abstract.

097.073 **Geologic mapping of Martian valley systems I: Nirgal Vallis and vicinity.**   D. C. Pieri, T. Parker.
Reports of Planetary Geology Program – 1981, (see 003.001), p. 503 (1981). – Abstract.

097.074 **Geologic mapping of plains material in Mare Acidalium quadrangle (MC-4), Mars.**
N. E. Witbeck, J. R. Underwood, Jr.
Reports of Planetary Geology Program – 1981, (see 003.001), p. 504 - 506 (1981). – Abstract.

097.075 **The third Mars year of imaging at the Mutch Memorial Station.**   S. D. Wall, D. C. Pieri.
Reports of Planetary Geology Program – 1981, (see 003.001), p. 515 - 517 (1981). – Abstract.

097.076 **Viking Lander imaging experiment – update and new observations, II.**   K. L. Jones, S. K. LaVoie.
Reports of Planetary Geology Program – 1981, (see 003.001), p. 518 - 519 (1981). – Abstract.

097.077 **Some observations of changes – Viking Landers 1 and 2.**   H. J. Moore, R. E. Hutton, C. R. Spitzer.
Reports of Planetary Geology Program – 1981, (see 003.001), p. 520 - 522 (1981). – Abstract.

097.078 **Global inventory of Martian impact craters: status report.**   D. J. Roddy, A. S. McEwen, A. Wasserman, C. L. Mardock, H. O. Webb, P. A. Davis, L. A. Soderblom.
Reports of Planetary Geology Program – 1981, (see 003.001), p. 551 - 552 (1981). – Abstract.

097.079 **Martian impact crater forming processes and crater physical dimensions.**
D. J. Roddy, D. A. Arthur, P. A. Davis, L. A. Soderblom.
Reports of Planetary Geology Program – 1981, (see 003.001), p. 553 - 555 (1981). – Abstract.

097.080 **Mars theory and its comparison with numerical integration.**   H. Kinoshita, H. Nakai.
Celestial Mech., Vol. 26, (see 012.001), 169 (1982). – Abstract.

097.081 **Motion of Mars: 1935–1976.**   Y. Niimi.
Celestial Mech., Vol. 26, (see 012.001), 179 (1982). – Abstract.

097.082 **Martian polar caps.**
K. Iwasaki.
Astron. Her., Vol. 74, No. 3, p. 64 - 69 (1981). In Japanese. Abstr. in Phys. Abstr., Vol. 85, Abstr. 23246 (1982).

097.083 **Mars 1980. Oppositionsbericht des Arbeitskreises Planetenbeobachter.**
M. Bendzulla, H. Haug.
Sterne Weltraum, 21. Jahrg., 124 - 126 (1982).

097.084 **Topographic reduction of Mars and Venus radar observations.**   E. V. Pit'eva.
Byull. Inst. Teor. Astron., Leningrad, Tom 15, 169 - 175 (1982). In Russian.
A method for the topographic reduction of Mars and Venus time delay measurements is described. The method makes use of the topographic maps which have become available recently. Compressed time delay normal points for Mars measurements 1964 - 1980 are presented.

097.085 **Mars residual north polar cap: earth-based spectroscopic confirmation of water ice as a major constituent and evidence for hydrated minerals.**
R. N. Clark, T. B. McCord. .
J. Geophys. Res., Vol. 87, 367 - 370 (1982).
A new reflectance spectrum of the Martian north polar cap is analyzed, and it shows water ice absorption features. The spectra show that other materials are probably present as well. These other materials may be hydrated minerals, as indicated by weak absorptions in the spectrum at 1.9, 2.3, and possibly 1.4 μm, but the materials seem to be otherwise spectrally bland.

097.086 **Long cloud observations on Mars and implications for boundary layer characteristics over slopes.**
R. Kahn, P. Gierasch.
J. Geophys. Res., Vol. 87, 867 - 880 (1982).
The authors present a self-consistent interpretation of an atmospheric cloud phenomenon that provides some insight into the qualitative behavior of the martian atmospheric boundary layer in steeply sloping terrain. The cloud phe-

nomenon has been called a bore wave and occurs in the saddle region between pairs of high volcanoes (e.g., Arsia and Pavonis Mons) during the morning hours in late spring and early summer. The authors argue that if the cloud is formed by the interaction of downslope winds generated on the slopes of the large volcanoes, then the shape of the cloud, its location and apparent velocity, the wavelength of several long clouds that form parallel to the leading cloud, and the size and spatial distribution of associated groups of smaller ripplelike clouds can be explained. The model sets some constraints on the depth and velocity of the boundary layer flows, and places a condition on the static stability of the atmosphere directly above the boundary layer.

097.087 **Aureole deposits of the Martian volcano Olympus Mons.**   E. C. Morris.
J. Geophys. Res., Vol. 87, 1164 - 1178 (1982).
A pyroclastic origin is proposed for the aureole deposits that surround Olympus Mons.

097.088 **Ignimbrites of Amazonis Planitia region of Mars.**   D. H. Scott, K. L. Tanaka.
J. Geophys. Res., Vol. 87, 1179 - 1190 (1982).
A series of postulated ignimbrite units is mapped in the Amazonis, Memnonia, and Aeolis quadrangles of Mars.

097.089 **Mars' global properties: maps and applications.**   H. H. Kieffer, P. A. Davis, L. A. Soderblom.
Proc. Twelfth Lunar Planet. Sci. Conf., (see 012.024), p. 1395 - 1417, 18 plates (1982).
Mars data sets from many sources have been put into a common digital format and global maps at constant scale have been produced. These formatted data sets form the basis of the Mars Consortium. They currently include maps of geology, volcanic units, channels, wind features, topography, gravity, Viking approach and apoapsis color, predawn temperature residuals, thermal inertia, radiometric and solar albedo, water vapor abundance, 1.4 μm albedo, crater abundance, Earth-based radar and photographic telescope observations, and terrestrial spectral observations. The generation of these data sets is briefly described.

097.090 **High resolution visual, thermal, and radar observations in the northern Syrtis Major region of Mars.**
J. R. Zimbelman, R. Greeley.
Proc. Twelfth Lunar Planet. Sci. Conf., (see 012.024), p. 1419 - 1429 (1982).

097.091 **Late-stage summit activity of martian shield volcanoes.**   P. J. Mouginis-Mark.
Proc. Twelfth Lunar Planet. Sci. Conf., (see 012.024), p. 1431 - 1447 (1982).

097.092 **Mars: a large highland volcanic province revealed by Viking images.**   D. H. Scott, K. L. Tanaka.
Proc. Twelfth Lunar Planet. Sci. Conf., (see 012.024), p. 1449 - 1458 (1982).

097.093 **A secondary origin for the central plateau of Hebes Chasma.**   C. Peterson.
Proc. Twelfth Lunar Planet. Sci. Conf., (see 012.024), p. 1459 - 1471 (1982).

097.094 **Spectral reflectance of weathered terrestrial and martian surfaces.**
D. L. Evans, T. G. Farr, J. B. Adams.
Proc. Twelfth Lunar Planet. Sci. Conf., (see 012.024), p. 1473 - 1479 (1982).
Laboratory reflectance spectra of Hawaiian samples were compared to martian surface color units seen in Viking Orbiter images. Spectral characteristics of the samples are consistent with some of the martian units and with several

telescopic spectra of Mars. Variations in the spectral characteristics of the Hawaiian rocks can be directly related to the
age and history of the samples. It appears that externally
derived ferric oxide-rich coatings, and in some cases primary
oxidation, can explain some of the variations seen in color
characteristics of martian global units observed by Viking
Orbiter.

**097.095   Mars weathering analogs: secondary mineralization
in Antarctic basalts.   J. L. Berkley.**
Proc. Twelfth Lunar Planet. Sci. Conf., (see 012.024),
p. 1481 - 1492 (1982).

097.096   Relief and tectonic development of Mars.
Ya. G. Kats, N. V. Makarova.
Vestn. MGU. Geol., 1981, No. 6, p. 17 - 29. In Russian.
Abstr. in Ref. zh., 51. Astron., 4.51.200 (1982).

**097.097   Post-Viking models for the structure of the summer
atmosphere of Mars.   A. Seiff.**
The Mars reference atmosphere, (see 003.006), p. 3 - 17
(1982).
   It is the purpose of the paper to propose a basic model
for the structure of the Martian atmosphere up to 100 km altitude.

097.098   Martian meteorological variability.
C. Leovy.
The Mars reference atmosphere, (see 003.006), p. 19 - 44
(1982).
   The paper deals with Martian "weather": the variability
of wind, temperature, and surface pressure. Because the atmospheric opacity is intimately linked to the temperature distribution, properties and distribution of the atmospheric aerosol
and cloud are also briefly discussed.

**097.099   Properties of dust in the Martian atmosphere and
its effect on temperature structure.**
J. B. Pollack.
The Mars reference atmosphere, (see 003.006), p. 45 - 56
(1982).
   Recent spacecraft missions to Mars have afforded an
opportunity to define a number of properties of the dust
particles suspended in the Martian atmosphere and to determine their influence on the vertical temperature structure. In
this report, the author draws primarily upon analyses of observations taken with the Viking lander experiment of the
Mariner 9 Orbiter to obtain the desired properties. These
results are used in a 1D radiative-convective model to obtain a
first order estimate of the atmospheric temperature profile,
and these predictions are compared with relevant Viking observations.

**097.100   Mars' atmospheric behavior from Viking infra-red
thermal mapper measurements.**
T. Z. Martin, M. M. Kieffer, E. D. Miner.
The Mars reference atmosphere, (see 003.006), p. 57 - 65
(1982).

097.101   The clouds of Mars as seen by Viking.
P. B. James.
The Mars reference atmosphere, (see 003.006), p. 67 - 74
(1982).

097.102   The composition of the Martian atmosphere.
T. Owen.
The Mars reference atmosphere, (see 003.006), p. 75 - 80
(1982).

**097.103   Water vapor in the Martian atmosphere: a discussion
of the Viking data.   P. E. Doms.**
The Mars reference atmosphere, (see 003.006), p. 81 - 85
(1982).

097.104   Mars' upper atmosphere: mean and variations.
A. J. Stewart, W. B. Hanson.
The Mars reference atmosphere, (see 003.006), p. 87 - 101
(1982).
   The authors briefly review the published experimental
material concerning the upper atmosphere and ionosphere of
Mars. They propose a mean model for the upper atmosphere
and ionosphere and discuss the extent of known variations
from this mean.

**097.105   Structure of the Martian atmosphere from ∊ Gem
occultation observations.   W. B. Hubbard.**
The Mars reference atmosphere, (see 003.006), p. 103 - 106
(1982).
   The author has collected information about Martian
atmospheric scale heights derived from observations of the
occultation of ϵ Gem by Mars on April 8, 1976. The observations give data in the altitude range ~50 to 80 km. A rough,
unweighted average of results so far available yields a temperature of ~165°. Excursions of ~±40°K about this mean may be
present as a function of both altitude and areographic coordinates.

097.106   Mars.
IAU Circ., No. 3659 (1982).

**097.107   Morphology and network patterns of Martian
valleys.   D. C. Pieri.**
The comparative study of the planets, (see 012.033), p. 477 -
483 (1982).
   Martian valleys exhibit morphologies and network
patterns which do not provide compelling evidence of formation by rainfall. Rather, cuspate amphitheater terminations
and sparse, open networks, often strongly structurally
controlled, argue for erosion by subsurface water.

**097.108   Ancient and modern slopes in the Tharsis region of
Mars.**
P. J. Mouginis-Mark , S. H. Zisk, G. S. Downs.
Nature, Vol. 297, 546 - 550 (1982).
   The directions of lava flows in the Tharsis region of Mars
are used to identify regional palaeo-slopes, vent areas and
local topography. A comparison is made between these flow
directions and the present day radar-measured topography;
good agreement between these data sets indicates that little
intra-regional tectonic deformation has occurred following
the emplacement of the preserved lavas.

**097.109   Occultation of AGK3 −1°1676 by Mars on 1983
December 1.   G. E. Taylor.**
Bull. No. 27, IAU Comm. 20, Working Group on Predictions
of Occultations by Satellites and Minor Planets, 1 pp. (1982).

**097.110   Observations de Mars et de Jupiter à l'astrolabe
OPL No. 30 de l'Observatoire de Beijing.**
C.-y. Chiao.
Publ. Beijing Astron. Obs., No. 1, p. 44 - 52 (1981). In
Chinese.

**097.111   The seasonal and global behavior of water vapor in
the Mars atmosphere: complete global results of the
Viking atmospheric water detector experiment.**
B. M. Jakosky, C. B. Farmer.
J. Geophys. Res., Vol. 87, 2999 - 3019 (1982).
   The water vapor content of the Mars atmosphere was
measured from the Viking Orbiter Mars Atmospheric Water
Detectors for a period of more than 1 Martian year, from
June 1976 through April 1979. Results are presented in the
form of global maps of column abundance for 24 periods

throughout each Mars year. The data reduction incorporates spatial and seasonal variations in surface pressure and supplements earlier published versions of less complete data.

097.112 Mars: near-infrared spectral reflectance of surface regions and compositional implications.
T. B. McCord, R. N. Clark, R. B. Singer.
J. Geophys. Res., Vol. 87, 3021 - 3032 (1982).
Reflectance spectra (0.65–2.50 μm) are presented for 11 Martian areas. The spectral resolution is $\sim 1^1/_2$% and the spatial resolution is 1000–2000 km. The high photometric quality of these data combined with increased near-infrared spectral coverage compared to previous regional observations provide new information about the spectral behavior and therefore the composition and physical nature of Martian surface materials.

097.113 Observing Mars in the 1980s: the Mars programme of the Terrestrial Planets Section.   R. McKim.
J. British Astron. Assoc., Vol. 92, 170 - 176 (1982).

097.114 Ephémérides et observations de Mars à l'astrolabe.
S. Débarbat, M. Sanchez, M. Standish.
Sun and planetary system, (see 012.040), p. 443 - 444 (1982).
Observations of Mars carried out with astrolabes around the oppositions of 1975 and 1978 have been analysed. Compared were the observations with the ephemerides developed by JPL and with those based on various analytical theories.

097.115 Radar measurement of small-scale surface texture: Syrtis Major.   R. A. Simpson,
G. L. Tyler, J. K. Harmon, A. R. Peterfreund.
Icarus, Vol. 49, 258 - 283 (1982).
The purpose of this paper is to examine and compare the available data on Syrtis Major, especially in light of lunar experience, with the goal of inferring the small-scale (centimeter to decameter) structure, in order that this information can be applied to the more general problem of origin and present state. The authors review the radar data which have been acquired, introducing three previously unpublished sets of observations obtained during the 1978 Mars opposition and arrive at a self-consistent model for the surface in Syrtis Major which includes structure on all scales. It is shown that the Moon is not, in fact, a good radar analog for Syrtis Major.

097.116 Mars–Rima Tenuis.
British Astron. Assoc. Circ., No. 625 (1982).

097.117 Observing Mars IX – the 1981 - 1982 aphelic apparition.   C. F. Capen, D. C. Parker, J. Beish.
Strolling Astron., Vol. 29, 110 - 117 (1982).

097.118 Ridge systems of Mars.   A. W. Gifford.
Advances in planetary geology, (see 003.011),
p. 221 - 363 (1981).
The purpose of the present study is to systematically map and describe the distribution of ridges for the planet Mars, including morphology and occurrence on various terrain types, and to measure planet wide ridge orientations. Based on this information, a comparison with lunar ridges, and a review of criteria used to support proposed modes of formation, the various mechanisms hypothesized for ridge origins can be evaluated.

097.119 The weather on Mars on the basis of the measurements carried out by the Viking mission.
C. J. Macris, B. C. Petropoulos.
Compendium in astronomy, (see 003.012), p. 201 - 208 (1982).
This paper summarizes some new results concerning the Mars atmosphere obtained after the Viking mission. On the basis of the results of the measurements made by the Viking 2 lander and Viking orbiter, the values of pressure and density corresponding to the altitudes from 28 to 100 km and different molecular weights have been computed. The computed values have been compared with the ones measured by Viking 2.

Viking Orbiter Stereo Imaging Catalog: Second Edition.   See Abstr. 002.007.

Archival storage of digital data on videotapes and videodisks.   See Abstr. 002.008.

Viking Orbiter Stereo Imaging Catalog.
See Abstr. 002.079.

The Viking Mosaic Catalog.   See Abstr. 002.080.

The channels of Mars.   See Abstr. 003.020.

The surface of Mars.   See Abstr. 003.026.

Planetary regularities in crater distribution on Mars, the moon and Mercury (Collected diagrams).
See Abstr. 003.075

Weather and climate on planets.   See Abstr. 003.078.

The cosmogonic chart on Phobos.
See Abstr. 003.083.

Mars.   See Abstr. 003.102.

Workshop on quasi-periodic climatic changes on Mars and earth.   See Abstr. 011.010.

Surface roughness effects on aeolian processes: wind tunnel experiments.   See Abstr. 022.053.

A method for modeling of small particle transport.
See Abstr. 022.056.

An experimental study of the behaviour of electrostatically-charged fine particles in atmospheric suspension.
See Abstr. 022.057.

An experimental investigation of Martian rock disintegration at the microlevel.   See Abstr. 022.058.

An experimental study of the erosion of basalt, obsidian and quartz by fine sand, silt and clay.
See Abstr. 022.059.

Vesiculation and lithification behavior of saline muds in near vacuum.   See Abstr. 022.060.

Groundwater sapping in sediments: theory and experiments.   See Abstr. 022.061.

Mars surface atmosphere exchange experiment: isothermal case.   See Abstr. 022.063.

Volatile release from Martian analog materials.
See Abstr. 022.064.

Reflectance spectroscopy of structural changes effected by the dehydration of goethite (α-FeOOH) and lepidocrocite (γ-FeOOH).   See Abstr. 022.066.

Impact cratering experiments in Bingham materials and the morphology of craters on Mars and Ganymede. See Abstr. 022.119.

Photogrammetric application of Viking orbital photography.   See Abstr. 031.509.

Orthophoto mosaics and three dimensional transformations of Viking Orbiter pictures. See Abstr. 031.523.

Observations of Mars with the astrolabe of the CERGA Observatory (February 1980 - May 1980). See Abstr. 041.030.

Calculation of the first-order theory of Mars by Hansen's method.   See Abstr. 042.128.

A transparent atmosphere in the UV: results from darkening of Viking Lander UV chips. See Abstr. 063.017.

The geology of Dyngjufjöll Ytri crater, north central Iceland.   See Abstr. 081.016.

Formation and evolution of playa ventifacts, Amboy, California.   See Abstr. 081.017.

Field modeling of the response of various desert surfaces to the long- and short-term effects of the wind — Mars applications.   See Abstr. 081.018.

Serrated eolian deposits in China's northwestern deserts and their comparisons to dark splotches on Mars. See Abstr. 081.019.

Dune forms in the Great Sand Sea and application to Mars.   See Abstr. 081.020.

Eolian processes in Iceland's cold deserts. See Abstr. 081.021.

Production of fine silt and clay during natural eolian abrasion.   See Abstr. 081.022.

Sapping processes and the development of theatre-headed valleys.   See Abstr. 081.023.

Analysis towards a dynamic origin for the formation of subglacial longitudinal grooving in sediment or bedrock. See Abstr. 081.024.

Exhumed topography — a review of some principles. See Abstr. 081.025.

Australian analogs to geomorphic features on Mars. See Abstr. 081.026.

New York-Pennsylvania rock cities: a Martian comparison.   See Abstr. 081.028.

Rock and soil mapping and change detection from LANDSAT multispectral scanner data — clues to limits of interpretability from Viking Orbiter color data. See Abstr. 081.029.

Regolith development in Mars-like environments. See Abstr. 081.030.

Chemical weathering and diagenesis in a soil profile in Antarctica: implications for the Martian regolith. See Abstr. 081.031.

Weathering of silicate minerals in Antarctic Dry Valleys: implications for volatile-regolith interactions on Mars. See Abstr. 081.032.

Seafloor instabilities on continental slopes and Mars analogs.   See Abstr. 081.033.

Die Problematik der Entstehung der Monde und ihrer Bahnentwicklung am Beispiel des Erdmondes und der Marsmonde.   See Abstr. 091.034.

Landing induced dust clouds on Venus and Mars. See Abstr. 093.030.

A search for terrestrial analogs to Martian lobed impact craters.   See Abstr. 105.021.

SNC meteorites: igneous rocks from Mars? See Abstr. 105.201.

098 Minor Planets

098.001 **The secular motion of Pallas.**
D. B. Taylor.
Mon. Not. R. Astron. Soc., Vol. 199, 255 - 265 (1982).

The secular motion of Pallas has been studied by numerically integrating the Lagrange planetary equations for 700 000 yr. Two terms with periods of approximately 15 000 yr and 190 000 yr dominate the secular elements. The former corresponds to a perturbation with argument 2ω and the latter to a perturbation with argument $\omega - \omega_J$, where ω and ω_J are the arguments of perihelion of Pallas and Jupiter respectively, defined with Jupiter's plane as the reference plane. Pallas is found to have a slow retrograde motion for the longitude of perihelion and this is unusual for asteroids. It circulates with a period of 544 600 yr.

098.002 **Los Troyanos.**   J. C. Muzzio.
Rev. Astron., Tomo 53, No. 219, p. 6 - 8 (1981).

098.003 **Exploration of the asteroids.**
J. R. French, Jr., N. D. Hulkower.
J. British Interplanet. Soc., Vol. 35, 167 - 171 (1982).

098.004 **Statistics of a complete population of bright and faint minor planets.**
S. G. Zhuravlev, V. N. Kiryushenkov.
Astron. Zh., Tom 59, 137 - 141 (1982). In Russian. English translation in Soviet Astron., Vol. 26, No. 1.

A statistical analysis of distributions of the complete population of the 3886 known bright and faint minor planets is presented. 2095 minor planets from the Ephemeris volume for 1980 are considered as bright minor planets. 1791 minor planets from the Palomar-Leiden Survey are considered as faint minor planets. Some common features and peculiarities of the distributions of bright and faint minor planets are revealed.

098.005 **Positions of asteroids obtained during 1978.**
V. Zappalà, C.-I. Lagerkvist, G. de Sanctis.
Astron. Astrophys., Suppl. Ser., Vol. 47, 447 - 450 (1982).

Precise positions are presented for 78 asteroids observed during 1978 at the Uppsala Southern Station.

098.006 **Positions of the minor planets 102 Miriam, 1024 Hale and 1687 Glarona obtained in May and June 1980 with the GPO, ESO, La Silla.**
H. Debehogne, L. E. Machado, J. F. Caldeira, G. G. Vieira, E. R. Netto.
Astron. Astrophys., Suppl. Ser., Vol. 47, 463 - 465 (1982).

098.007 **Studies of small asteroids. II. Positions of asteroids obtained during 1980 with the ESO Schmidt telescope.**   C.-I. Lagerkvist.
Astron. Astrophys., Suppl. Ser., Vol. 47, 513 - 521 (1982).

098.008 **Positions of asteroids obtained during 1976−1979 with the Uppsala astrograph and with the Kvistaberg Schmidt telescope.**
B. Pettersson, G. Hahn, C.-I. Lagerkvist.
Astron. Astrophys., Suppl. Ser., Vol. 47, 533 - 534 (1982).

Precise positions are presented for 36 asteroids observed with the 20 cm astrograph and with the 1 m Schmidt telescope.

098.009 **Positions of minor planets.**   G. Soulié.
Astron. Astrophys., Suppl. Ser., Vol. 47, 611 - 613 (1982). In French.

The observed asteroids belong to the Selected Minor Planets list. The negatives were obtained on Kodak IIaO plates with the Bordeaux 13 inch-photographic refractor.

098.010 **A revised rotation period for the asteroid 164 Eva.**
H. J. Schober.
Astron. Astrophys., Suppl. Ser., Vol. 48, 57 - 62 (1982).

A new rotation period of $P = 13^h 66 \pm 0^h 01$ $(= 0^d 569 \pm 0^d 001)$ was derived with a completely observed double wave lightcurve and a maximum light variation of $\Delta m = 0^m 36$. Using a phase coefficient of 0.039 mag/deg absolute magnitudes $\bar{V}(1, 0) = 9.06$ and $V_0(1, 0) = 8.88$ were computed; the colors $B - V = 0.69$ and $U - B = 0.35$ are in agreement with those listed in Bowell et al. (1979).

098.011 **Asteroid rotation rates depend on diameter and type.**   S. F. Dermott, C. D. Murray.
Nature, Vol. 296, 418 - 421 (1982).

The rotational frequency of main-belt asteroids is shown to depend on both asteroidal type and diameter. If asteroids of any one diameter are considered, then, on average, M asteroids rotate faster than S asteroids which in turn rotate faster than C asteroids. For all three types, although the dispersions of the frequencies are large, the authors prove that the mean frequency increases linearly with the mean diameter.

098.012 **Quadruple extrema in the complex lightcurve of the asteroid 37 Fides?**   H. J. Schober.
Astron. Astrophys., Vol. 105, 419 - 421 (1982).

The asteroid 37 Fides was reobserved at ESO in Aug. 1979 and a period of rotation $P = 14^h 66 \pm 0^h 03$ $(0^d 611 \pm 0^d 001)$ was deduced, which rules out the published value of $P = 4^h 0$? by Scaltriti and Zapalà (1978). The lightcurve amplitude is now $0^m 16$. 37 Fides shows a complicated lightcurve with four distinct and different maxima and four minima. Absolute magnitudes $\bar{V}(1, 0) = 7.42$ and $V_0(1, 0) = 7.36$ were derived, the colors are $B - V = 0.84$ and $U - B = 0.40$ with no variation during all rotational phases exceeding the scatter of 0.02.

098.013 **The structure of the asteroid belt.**
S. F. Dermott, C. D. Murray.
News Lett. Astron. Soc. N. Y., Vol. 2, 15 - 24 (1982).

098.014 **Nouvelles des earth-grazers − 4.**
M.-A. Combes, J. Meeus.
Astronomie, Vol. 96, 187 - 198 (1982).

098.015 **Minor planets: discovery, number, name.**
V. A. Shor.
Astronomical calendar. Yearbook. 1982, (see 047.005), p. 202 - 224 (1981). In Russian. − Abstr. in Ref. zh., 51. Astron., 2.51.181 (1982).

098.016 **The stochastic evolution of asteroidal regoliths and the origin of brecciated and gas-rich meteorites.**
K. R. Housen.
Advances in planetary geology, (see 003.011), p. 3 - 217 (1981).

Contents: The cratering, collisional, and charged-particle environment of asteroids. Previous modeling. The stochastic evolution of asteroidal regoliths. Results. Implications for recent models and the origin of meteorites. Summary and future work.

098.017 **Survey for bright Mars-crossing asteroids.**
E. M. Shoemaker, C. S. Shoemaker, E. F. Helin, S. J. Bus, R. F. Wolfe.

Reports of Planetary Geology Program – 1981, (see 003.001), p. 17 - 19 (1981). – Abstract.

098.018 **Thermal models and observational rotational studies of asteroids: implications for the asteroid-meteorite connection.** M. J. Gaffey.
Meteoritics, Vol. 16, 317 (1981). – Abstract.

098.019 **The asteroids 36 Atalante and 48 Doris: rotation, UBV-photometry, and lightcurves.**
H. J. Schober, A. Schroll.
Astron. Astrophys., Vol. 107, 402 - 405 (1982).
The asteroids 36 Atalante and 48 Doris were observed during their opposition in 1978. For 36 Atalante a period of rotation was derived with $P = 9^h 93 \pm 0^h 01$. The maximum amplitude of the double mode lightcurve was found to be $0^m 15$. Mean colors are $B-V = 0.70 \pm 0.01$ and $U-B = 0.37 \pm 0.01$. 48 Doris shows a rotation rate of $P = 11^h 900 \pm 0^h 005$. The authors state a double mode light-curve with a maximum amplitude of at least $0^m 30$. Mean colors are $B-V = 0.71 \pm 0.01$ and $U-B = 0.45 \pm 0.02$. Absolute magnitudes were computed for both asteroids.

098.020 **The 1.7- to 4.2-μm spectrum of asteroid 1 Ceres: evidence for structural water in clay minerals.**
L. A. Lebofsky, M. A. Feierberg, A. T. Tokunaga, H. P. Larson, J. R. Johnson.
Icarus, Vol. 48, 453 - 459 (1981).
A high-resolution Fourier spectrum (1.7 - 3.5 μm) and medium-resolution spectrophotometry (2.7 - 4.2 μm) were obtained for asteroid 1 Ceres. The presence of the 3-μm absorption feature due to water of hydration was confirmed. The 3-μm feature is compared with the 3-μm bands due to water of hydration in clays and salts. It is concluded that the spectrum of Ceres shows a strong absorption at 2.7 - 2.8 μm due to structural OH groups in clay minerals. The dominant minerals on the surface of Ceres are therefore hydrated clay minerals structurally similar to terrestrial montmorillonites. There is also a narrow absorption feature at 3.1 μm which is attributable to a very small amount of water ice on Ceres. This is the first evidence for ice on the surface of an asteroid.

098.021 **Die Identität A899 OF = (1) Ceres.**
L. D. Schmadel.
Sterne Weltraum, 21. Jahrg., 114 - 115 (1982).

098.022 **Photoelectric photometry of three dark asteroids.**
H. Debehogne, G. De Sanctis, V. Zappalà.
Astron. Astrophys., Vol. 108, 197 - 200 (1982).
Lightcurves of the dark and average-size asteroids 53 Kalypso, 156 Xanthippe, and 313 Chaldaea were obtained from March 11 to 17, 1981. These observations were planned to enlarge the present knowledge of average-size and small objects, for which the statistics are poor. Rotation periods and the $B-V$ and $U-B$ color indices of the three asteroids are obtained.

098.023 **Eigenelemente für Asteroiden vom Hilda-Typ.**
J. Schubart.
Mitt. Astron. Ges., Nr. 55, p. 17 (1982). – Abstract.

098.024 **The diameter of (51) Nemausa.**
L. K. Kristensen.
Mitt. Astron. Ges., Nr. 55, p. 17 (1982). – Abstract.

098.025 **37 Fides – ein neues Puzzle bezüglich der Rotationsperiode?** H. J. Schober.
Mitt. Astron. Ges., Nr. 55, p. 17 (1982). – Abstract.

098.026 **Farbindexmessungen von 3 Juno, 36 Atalante und 48 Doris während ihrer Rotationen.**
H. J. Schober, A. Schroll.
Mitt. Astron. Ges., Nr. 55, p. 18 (1982). – Abstract.

098.027 **Photographic positions of minor planets and comets.**
J. Bem, B. Szczodrowska-Kozar.
Acta Astron., Vol. 31, 369 - 371 (1981).
The topocentric positions of minor planets Ceres, Pallas, Hebe, Iris and comets Encke (1980), Bradfield (1979 c), Meier (1978 f), and Kohler (1977 m) are given.

098.028 **Theoretical calculation of the light curve of a three-axis asteroid.** J.-q. Zheng, H.-n. Zhou.
Acta Astron. Sinica, Vol. 22, 370 - 377 (1981). In Chinese.

098.029 **The rotation of (16) Psyche.**
X.-h. Zhou, X.-y. Yang.
Acta Astron. Sinica, Vol. 22, 378 - 382 (1981). In Chinese.

098.030 **Single body models for asteroids having suspected binary lightcurves.** J. V. Lambert.
Proc. Southwest Reg. Conf., Vol. 7, (see 012.019), 127 - 138 (1982).

098.031 **Observaciones de asteroides en La Plata.**
F. Muñoz.
Bol. Asoc. Argentina Astron., No. 18, p. 4 - 5 (1980).

098.032 **Asteroide: Entstehung und Bahnentwicklung.**
H. Gropp.
Monde, Ringe und Asteroide im Planetensystem, (see 012.020), p. 75 - 96 (1981).

098.033 **Physik der Asteroiden.** J. Adam.
Monde, Ringe und Asteroide im Planetensystem, (see 012.020), p. 97 - 116 (1981).

098.034 **Asteroiden – Quelle für Meteorite und Staub.**
L. Smolik.
Monde, Ringe und Asteroide im Planetensystem, (see 012.020), p. 117 - 137 (1981).

098.035 **The smaller bodies of the solar system.**
L. Biermann.
Philos. Trans. R. Soc. London A, Vol. 303, (see 012.022), 351 - 352 (1981).

098.036 **The stochastic variability of asteroidal regolith depths.** K. R. Housen.
Proc. Twelfth Lunar Planet. Sci. Conf., (see 012.024), p. 1717 - 1724 (1982).
Modeling the depth of regolith on asteroids is approached from a statistical point of view. It is demonstrated that average values are not good descriptors of regolith depth on asteroids. Large deviations from the average can be expected to occur due to both large variations in depth over the surface of a body and to the fact that each asteroid has a unique regolith. The large variability associated with regolith depth severely limits the power of regolith models in predicting parent body size for the brecciated meteorites.

098.037 **Theoretical meteor radiants of Apollo, Amor, and Aten asteroids.** J. D. Drummond.
Icarus, Vol. 49, 143 - 153 (1982).
A compilation of theoretical meteor radiants is presented for all numbered (through 2525) asteroids which approach the Earth's orbit to within 0.20 AU. On the basis of orbital similarity, asteroids associated with current meteor streams and Prairie Network fireballs are listed; plausible associations with medieval fireball radiants are also given. The best defunct comet candidates in terms of meteoric evidence appear to be 2101 Adonis and 2201 1947XC. Asteroids which may be either extinct comets or perturbed main belt asteroids

accompanied by collisional debris (represented by fireballs) are 1917 Cuyo, 2202 Pele, 2061 Anza, and 2340 Hathor.

098.038 **Recent progress in the theory of the Trojan asteroids.** B. Garfinkel.
Application of modern dynamics to celestial mechanics and astrodynamics, (see 012.026), p. 145 - 152 (1982).
This paper summarizes the author's previous publications on the subject in the light of the numerical integrations by Deprit and Henrard. The period of libration is expressed as a function of the mass parameter and the normalized Jacobian constant. Brown's conjecture regarding the termination of the tadpole branch of the family at L_3 is refined, and a heuristic proof of its validity is offered.

098.039 **Motion at the second order resonances, 3 : 1 and 5 : 3.** G. Colombo, F. A. Franklin.
Application of modern dynamics to celestial mechanics and astrodynamics, (see 012.026), p. 339 - 340 (1982). – Abstract.

098.040 **Spectroscopic evidence for undifferentiated S-type asteroids.**
M. A. Feierberg, H. P. Larson, C. R. Chapman.
Astrophys. J., Vol. 257, 361 - 372 (1982).
High-resolution Fourier spectra (0.9—2.5 μm) have been obtained for S-type asteroids 3 Juno, 5 Astraea, 7 Iris, 8 Flora, 9 Metis, 12 Victoria, 18 Melpomene, 27 Euterpe, 29 Amphitrite, 39 Laetitia, and 433 Eros. These data are combined with 0.3—1.1 μm spectrophotometry for compositional analysis. All 11 spectra show olivine and pyroxene absorption features, with olivine/pyroxene abundance ratios between 1 and 10. The range of silicate compositions observed in these asteroids overlaps with the compositions of ordinary and carbonaceous chondrites but does not approach the compositions of the most common differential meteorites. S-type asteroids have spectral reflectances consistent with undifferentiated compositions, and some of them may be ordinary chondrite parent bodies.

098.041 **Determination of the diameters of minor planets by a statistical method.**
R. A. Vardanyan, O. Kh. Torosyan, L. G. Akhverdyan.
Soobshch. Byurakan Obs., Vyp. 53, p. 124 - 130 (1982). In Russian.
The type (C or S) of 756 minor planets and their diameters, determined by a statistical method, are presented.

098.042 **Asteroids.** Ľ. Kresák.
Kozmos, Vol. 13, 78 - 82 (1982). In Slovak.

098.043 **Observations of minor planets.**
Minor Planet Circ., (M. P. C.), Nos. 6582 - 6628, 6658 - 6696, 6753 - 6814, 6902 - 6938 (1982).
Observations made at the following stations are published: Bergedorf, Brorfelde, Bucharest, Burgsolms, Calar Alto, Catalina Stat. Lunar Planet. Lab., Caussols, Centro Astron. Yebes, Cerro el Roble, Crimea (46th - 48th report), European Southern Obs., Floirac, Geisei, Goethe Link Obs., Göttingen, Haute Provence, Hemingford Abbots, JCPM Hamatonbetsu Stat., JCPM Oi Stat., Kleť, La Plata, La Seyne-sur-Mer, Leiden, Lincoln Lab., Lowell Obs., Lowell Obs. Anderson Mesa Stat., Madonna di Dossobuono (Verona), Mt. John Obs., Mt. Lemmon, Mt. Palomar, Nice, Oak Ridge Obs., Piszkestetö (Hungary), Purple Mountain Obs., Quonochontaug, Reintal, S. Vittore (Bologna), San Fernando, Siding Spring, Stakenbridge, Steward Obs., Stony Ridge Obs., Tautenburg, Tokai, Tokyo Obs. Kiso Stat., Turku, U. S. Naval Obs., Wallace Obs., Wrocław, Zimmerwald, Zo-Se.

098.044 **Identifications and identification changes of minor planets.**

Minor Planet Circ., (M. P. C.), Nos. 6574 - 6580, 6655, 6752, 6840, 6899 - 6900 (1982).

098.045 **Ephemerides of minor planets and comets.**
Minor Planet Circ., (M. P. C.), Nos. 6651 - 6652, 6712 - 6750, 6836 - 6838, 6896 - 6898, 6957 - 6988 (1982).

098.046 **New names of minor planets.**
Minor Planet Circ., (M. P. C.), Nos. 6647 - 6650, 6831 - 6835, 6954 - 6957 (1982).

098.047 **Orbital elements of one-opposition minor planets.**
Minor Planet Circ., (M. P. C.), Nos. 6628 - 6630, 6696 - 6697, 6815 - 6817, 6878 - 6880, 6938 - 6939 (1982).

098.048 **Critical list of minor planets.**
Minor Planet Circ., (M. P. C.), Nos. 6653 - 6654 (1982).

098.049 **Double designations.**
Minor Planet Circ., (M. P. C.), No. 6840 (1982).

098.050 **Orbital elements of numbered minor planets.**
Minor Planet Circ., (M. P. C.), Nos. 6573 - 6988 (1982).
The minor planets are listed according to their definitive number. Newly numbered objects are indicated by an asterisk. The names of the computers are given behind the respective M. P. C. numbers. (10) 6630 L. D. Schmadel; (330)* 6939 - 6940 C. M. Bardwell; (699) 6896 D. W. E. Green; (829) 6630 L. D. Schmadel; (2149) 6697 B. G. Marsden; (2526)* - (2527)* 6631 L. D. Schmadel; (2528)* - (2532)* 6631 - 6633 S. Nakano; (2533)* - (2544)* 6633 - 6637 B. G. Marsden; (2545)* - (2559)* 6640 - 6644 C. M. Bardwell; (2560)* - (2571)* 6697 - 6701 B. G. Marsden; (2572)* - (2582)* 6703 - 6706 C. M. Bardwell; (2583)* - (2590)* 6709 - 6710, 6818 S. Nakano, T. Urata; (2591)* - (2603)* 6819 - 6824 C. M. Bardwell; (2604)* - (2614)* 6825 - 6829 B. G. Marsden; (2615)* - (2622)* 6880 - 6882 S. Nakano, T. Urata; (2623)* - (2636)* 6883 - 6888 C. M. Bardwell; (2637)* - (2649)* 6890 - 6894 B. G. Marsden; (2650)* - (2663)* 6940 - 6944 C. M. Bardwell; (2664)* - (2674)* 6946 - 6949 B. G. Marsden; (2675)* 6952 - 6953 S. Nakano, T. Urata; (2676)* 6954 L. K. Kristensen.

098.051 **Orbital elements of unnumbered minor planets.**
Minor Planet Circ., (M. P. C.), Nos. 6573 - 6988 (1982).
The unnumbered minor planets are listed according to their preliminary designation. Objects from the Palomar-Leiden Survey are sorted by number. The names of the computers are given behind the respective M. P. C. numbers. [1932 BG] 6706 - 6707, [1932 CK] 6888, [1934 CY] 6944, [1935 OK] 6888 C. M. Bardwell; [1937 AC] 6701, [1938 DH$_2$] 6949 B. G. Marsden; [1938 WA] 6882 - 6883 S. Nakano, T. Urata; [1941 UL] 6894 B. G. Marsden; [1948 PK] 6944 C. M. Bardwell; [1950 DH] 6883 S. Nakano, T. Urata; [1951 AJ] 6707 C. M. Bardwell; [1953 EE] 6894 B. G. Marsden; [1965 AK$_1$] 6707 C. M. Bardwell; [1965 UU$_1$] 6637 B. G. Marsden; [1966 DH] 6945 C. M. Bardwell; [1969 TC$_3$] 6953 S. Nakano, T. Urata; [1971 TJ$_2$] 6638 B. G. Marsden; [1972 AR] 6707, [1972 UA] 6824 C. M. Bardwell; [1974 WB] 6949, [1976 UU] 6949 B. G. Marsden; [1977 JP] 6707 C. M. Bardwell; [1977 VM$_1$] 6829, [1977 YA] 6949, [1978 LB] 6638, [1978 PA] 6638 B. G. Marsden; [1978 SZ$_2$] 6953, [1978 WN$_{14}$] 6953 S. Nakano, T. Urata; [1979 FR$_2$] 6638 B. G. Marsden; [1979 HD$_5$] 6824 C. M. Bardwell; [1979 MU$_2$] 6639, [1979 MB$_6$] 6639, [1979 OA] 6950, [1979 OB] 6950, [1979 QK$_2$] 6829 - 6830 B. G. Marsden; [1979 QP$_8$] 6883 S. Nakano, T. Urata; [1979 SF] 6830 B. G. Marsden; [1980 GA] 6645 C. M. Bardwell; [1980 LB] 6639, [1980 OD]

6894 B. G. Marsden; [1980 OF] 6645 C. M. Bardwell;
[1980 PM] 6702, [1980 WA] 6702, [1980 YH] 6894 - 6895,
[1981 AD] 6895, [1981 CW] 6950, [1981 EE] 6895
B. G. Marsden; [1981 EV] 6883 S. Nakano, T. Urata;
[1981 OF] 6945 C. M. Bardwell; [1981 OG] 6896
L. K. Kristensen; [1981 PB] 6950 B. G. Marsden; [1981 PG]
6945 C. M. Bardwell; [1981 QA] 6702, 6950, [1981 QB]
6702, 6895, [1981 QC] 6830, 6950 B. G. Marsden; [1981 QF$_2$]
6711 S. Nakano, T. Urata; [1981 RG$_1$] 6951 B. G. Marsden;
[1981 SA] 6645, [1981 SH] 6645, [1981 SJ$_1$] 6646
C. M. Bardwell; [1981 TK] 6951 B. G. Marsden; [1981 UL]
6711 S. Nakano, T. Urata; [1981 VA] 6702, [1981 VB]
6951 B. G. Marsden; [1981 WA] 6711, [1981 WT] 6711
S. Nakano, T. Urata; [1981 WW] 6646, [1981 WY] 6646,
[1981 WP$_1$] 6646, [1981 WR$_1$] 6646 - 6647 C. M. Bardwell;
[1981 WV$_1$] 6818 - 6819 S. Nakano, T. Urata; [1981 YD]
6647 C. M. Bardwell; [1981 YE] 6819 S. Nakano, T. Urata;
[1981 YH$_1$] 6951, [1982 BB] 6703, 6895, 6951 B. G. Marsden:
[1982 BC$_1$] 6708, [1982 BH$_1$] 6708 C. M. Bardwell;
[1982 BJ$_1$] 6830, [1982 BK$_1$] 6830 - 6831 B. G. Marsden;
[1982 BM$_1$] 6708, [1982 BT$_1$] 6888, [1982 BX$_1$] 6825,
[1982 BY$_1$] 6825, [1982 BZ$_2$] 6708 C. M. Bardwell;
[1982 BD$_3$] 6951, [1982 DA] 6952, [1982 DB] 6831, 6952
B. G. Marsden; [1982 DJ] 6825 C. M. Bardwell; [1982 DV]
6831, 6952, [1982 HR] 6895, 6952, [4017 P-L] 6639
B. G. Marsden; [6558 P-L] 6889 C. M. Bardwell.

098.052 **1978 DA.**
IAU Circ., Nos. 3658, 3666 (1982).

098.053 **Occultation of SAO 042418 by (344) Desiderata.**
IAU Circ., No. 3659 (1982).

098.054 **1982 BB**
IAU Circ., Nos. 3660, 3665, 3669, 3686 (1982).

098.055 **(433) Eros.**
IAU Circ., No. 3661 (1982).

098.056 **1982 CA.**
IAU Circ., No. 3666 (1982).

098.057 **1982 DA.**
IAU Circ., Nos. 3669, 3673, 3675, 3680, 3702
(1982).

098.058 **(330) Adalberta.**
IAU Circ., No. 3672 (1982).

098.059 **1982 DB.**
IAU Circ., Nos. 3675, 3677, 3686 (1982).

098.060 **1982 DV.**
IAU Circ., Nos. 3675, 3678, 3679, 3683, 3686,
3694, 3706 (1982).

098.061 **1982 EA.**
IAU Circ., No. 3678 (1982).

098.062 **1982 FT.**
IAU Circ., Nos. 3685, 3694 (1982).

098.063 **(1009) Sirene.**
IAU Circ., No. 3691 (1982).

098.064 **1982 HR.**
IAU Circ., Nos. 3692, 3693, 3702 (1982).

098.065 **Occultation by (146) Lucina.**
IAU Circ., Nos. 3692, 3699 (1982).

098.066 **Object Helin 1982 CA.**
Yamamoto Circ., No. 1971 (1982).

098.067 **1982 DV.**
Yamamoto Circ., No. 1973 (1982).

098.068 **1982 DB.**
Yamamoto Circ., No. 1973 (1982).

098.069 **Orbital elements of 20 unnumbered asteroids.**
J.-x. Zhang, J.-x. Yang, S.-l. Wei, Q. Wang,
Y.-l. Ge, C.-l. Yuan.
Astron. Circ., No. 12, p. 1, 3 - 12, 16 (1981). In Chinese and
English.

098.070 **Bemerkenswerte Daten aus der Geschichte himmels-**
kundlicher Forschung: die Kleinplaneten.
W. Valentin.
Sternenbote, 25. Jahrg., 38 - 44 (1982).

098.071 **Video record of secondary occultation obtained at**
Meudon Observatory during (146) Lucina appulse
on 1982 April 18.   D. W. Dunham, J. Lecacheux.
Occultation Newsl., Vol. 2, 219 - 220 (1982).

098.072 **Planetary occultation predictions for 1982.**
Occultation Newsl., Vol. 2, 226 - 234 (1982).

098.073 **On lost minor planets.**
S. G. Zhuravlev, V. N. Kiryushenkov.
Astron. Zh., Tom 59, 578 - 580 (1982). In Russian. English
translation in Soviet Astron., Vol. 26, No. 3.
The basic parameters of twenty lost minor planets were
compared with those of almost 4000 known minor planets to
discover the twins, i. e. the minor planets which are in a certain
sense close to the lost minor planets. It was found that at
least two lost minor planets, No. 878 and 1026, have twins,
i. e. the minor planets 6085 PL and No. 2130 respectively.

098.074 **Dynamics of the asteroids.**   H. Scholl.
The comparative study of the planets, (see 012.033),
p. 125 - 130 (1982).
The large majority of the known asteroids is situated in a
belt between the orbits of Mars and Jupiter. Obviously, the
boundaries of the belt are due to these two planets. The
dynamics of the belt asteroids is at present mainly determined
by Jupiter. In the past, collisions among asteroids played an
important role which is indicated by the Hirayama families.
Those asteroids which cross the orbits of a planet might have
suffered or will suffer drastic changes in their orbits or will
even collide with that planet unless particular protection
mechanisms prevent such close approaches.

098.075 **Radio observations of 1 Ceres.**
K. J. Johnston, P. K. Seidelmann, C. M. Wade.
Bull. American Astron. Soc., Vol. 13, 875 (1981). − Abstract.

098.076 **On the reality of minor planet (330) Adalberta.**
R. M. West, C. Madsen, L. D. Schmadel.
ESO Sci. Prepr. No. 192, 12 pp. (1982). − Submitted to
Astron. Astrophys.

098.077 **Precise observations of minor planets at Sydney**
Observatory during 1980.   N. R. Lomb.
J. Proc. R. Soc. New South Wales, Vol. 114, 1 - 5 (1981) =
Sydney Obs. Pap. No. 90.
Positions of 2 Pallas, 3 Juno, 6 Hebe, 7 Iris,
18 Melpomene, 40 Harmonia, 51 Hemausa and 532 Herculina
obtained with the 23-cm camera are given.

098.078 **Predictions of occultations by minor planets 1983.**
G. E. Taylor.

Bull. No. 27, IAU Comm. 20, Working Group on Predictions of Occultations by Satellites and Minor Planets, 3 pp. (1982).

098.079   **Identifications of minor planets.**
B. G. Marsden.
Bull. American Astron. Soc., Vol. 14, 580 - 581 (1982).
Abstract.

098.080   **Report on occultation of SAO 77636 by 15-Eunomia on 1982 March 30.**   M. D. Overbeek.
Mon. Notes Astron. Soc. South. Africa, Vol. 41, 25 - 26 (1982).

098.081   **Positions of asteroids (1981).**
H. Debehogne, G. De Sanctis, V. Zappalà.
Astron. Astrophys., Suppl. Ser., Vol. 48, 449 - 451 (1982).

244 positions of 11 asteroids were obtained from plates taken in 1981 by means of the GPO (f = 4 m, D = 40 cm) of the European Southern Observatory at La Silla (Chile). 5 new asteroids were discovered. The reductions were made by dependences method using 5 or 6 reference stars.

098.082   **Recovery of the long lost minor planet (843) Nicolaia after 65 years.**   L. D. Schmadel, L. Kohoutek.
Astron. Nachr., Band 303, 139 - 141 (1982) = Mitt. Astron. Rechen-Inst. Heidelberg, Ser. B.

The long lost minor planet (843) Nicolaia was recovered with the 1 m ESO Schmidt telescope after 65 years. This was due to the careful examination of the old 1916 plates and the wellbalanced computation of search ephemerides. An improved orbit based on 19 right ascensions and 20 declinations from the two apparitions 1916 and 1981 is given.

098.083   **Observations of minor planets.**
F. Börngen, K. Kirsch.
Astron. Nachr., Band 303, 143 - 147 (1982).

The authors present 266 positions of 77 minor planets determined from plates taken in 1972 and in 1979—1980 with the Schmidt telescope 1340/2000/4000 of Karl Schwarzschild Observatory Tautenburg.

098.084   **Orbit improvement of minor planet (1019) Strackea.**
W. Landgraf.
Astron. Nachr., Band 303, 149 - 152 (1982). In German.

098.085   **Collisioni fra asteroidi e loro evoluzione.**
V. Zappalà.
Orione, Vol. 3, 47 - 55 (1982).

098.086   **Surface properties of asteroids.**   H. J. Schober.
Sun and planetary system, (see 012.040), p. 265 - 268 (1982).

From spectrophotometry, polarimetry, IR-radiometry and UBV-photometry surface properties of asteroids are derived, such as: diameters, spectral reflectivity, albedo; based on observable parameters a classification into taxonomic types (Bowell et al. 1978) is made and an interpretation in terms of the mineralogy of the surface and meteoritic analogs can be done. From accurate photometry during rotational cycles irregularities on the surfaces are found. A short summary of our present knowledge of asteroid surfaces is given.

098.087   **Earth-crossing asteroids: new discoveries.**
E. F. Helin.
Sun and planetary system, (see 012.040), p. 269 - 276 (1982).

A total of 43 Earth-crossing asteroids are now known. Twenty-five were discovered or recovered in the last decade. There were only six numbered Earth-crossing asteroids prior to 1970. Since then, twenty-one have been numbered. The Aten asteroids, a new group of Earth-crossing asteroids, have orbits smaller than that of the Earth. The largest Earth crosser, Hephaistos, has C-type UBV colors and is probably about

10 km in diameter. The smallest, Hathor, has unusual UBV colors and a probable diameter of about 200 m.

098.088   **Synthetic lightcurves of asteroidal binary systems.**
F. Scaltriti, V. Zappalà, E. Anderlucci.
Sun and planetary system, (see 012.040), p. 277 - 278 (1982).

The purpose of this research is to investigate, in a statistical sense, the effect of the presence of a satellite on a lightcurve. The model consists of two spheres with the same albedo revolving in circular orbit in a plane containing the sun and the observer.

098.089   **Color variations of asteroids during rotation.**
H. J. Schober, A. Schroll.
Sun and planetary system, (see 012.040), p. 285 - 286 (1982).

A list of asteroids is presented, observed frequently for color variations during rotation as an evidence for spots on the surface. Out of a sample of 49 asteroids there is evidence now for ten objects to exhibit spotted surfaces.

098.090   **Results from occultations by minor planets.**
G. E. Taylor.
Sun and planetary system, (see 012.040), p. 287 - 288 (1982).

098.091   **On the rotations of M asteroids.**
C.-I. Lagerkvist, H. Rickman.
Sun and planetary system, (see 012.040), p. 289 - 290 (1982).

The authors' conclusion is that both M asteroids and those designated CMEU are characterized by high spin rates and large lightcurve amplitudes.

098.092   **Asteroid rotation rates: comparison between theory and observations.**
P. Paolicchi, P. Farinella, V. Zappalà.
Sun and planetary system, (see 012.040), p. 291 - 294 (1982).

Harris' (1979) theory on the collisional evolution of the asteroid spin rates is compared with the observational evidence provided by the statistics of photometric data. The resulting discrepancies can be qualitatively explained as due to different physical processes, occurred in the frame of the asteroid collisional history and connected in particular with the outcomes of catastrophic impact events.

098.093   **Asteroid collisional evolution: outcomes of catastrophic impacts.**
P. Paolicchi, P. Farinella, V. Zappalà.
Sun and planetary system, (see 012.040), p. 295 - 298 (1982).

The authors discuss how catastrophic collisional events have influenced some outstanding properties of asteroids, like their structure, rotation, shape and family membership. The ultimate outcomes of the collisional evolution are found to be strongly dependent on the asteroid size, because of the different importance of self-gravitational forces and of the different probability of impacts with high projectile-to-target mass ratio.

098.094   **An improved representation of the average opposition magnitudes of asteroids.**
Z. Knežević, V. Zappalà.
Sun and planetary system, (see 012.040), p. 299 - 302 (1982).

It is well-known that the asteroid mean opposition magnitude $-$ B(a, 0) $-$ represents a crude approximation of asteroid average opposition brightness, especially for objects of high eccentricity and/or inclination. The authors propose a new simple method of calculating the "quasi-median opposition magnitude", which very closely matches the median apparent brightness of an asteroid in opposition.

098.095   **A coordinate program for pole determination of asteroids.**   V. Zappalà, F. Scaltriti.
Sun and planetary system, (see 012.040), p. 303 - 304 (1982).

098.096 **The worldwide photoelectric campaign for the asteroid 51 Nemausa.**
H. J. Schober, L. K. Kristensen.
Sun and planetary system, (see 012.040), p. 305 - 306 (1982).

098.097 **Some characteristics of the asteroid belt structure.**
M. Kuzmanoski.
Sun and planetary system, (see 012.040), p. 403 - 404 (1982).

098.098 **Resonances in the motions of minor planets and their use for the determination of masses.**
H. Scholl.
Sun and planetary system, (see 012.040), p. 405 - 410 (1982).
The secular behavior of resonant orbits is described on the basis of Greenberg's and Schubart's theories. The author reviews the results of numerical investigations obtained by Froeschlé and Scholl and the results obtained recently by Wisdom, who found a mapping for resonant motion. In addition he reviews the possibilities for mass determinations of planets and of minor planets using resonant motion.

098.099 **Stability of real Hecuba and Hilda asteroids.**
T. Kiang.
Sun and planetary system, (see 012.040), p. 411 - 412 (1982).

098.100 **First-order theory of Ceres, Pallas, Juno, and Vesta.**
W. Höppner.
Sun and planetary system, (see 012.040), p. 413 - 414 (1982).

098.101 **Asteroidal size distribution.** D. W. Hughes.
Mon. Not. R. Astron. Soc., Vol. 199, 1149 - 1157 (1982).
Asteroids with diameters $D > 260$ km have a mass distribution index of 1.58 ± 0.25 whereas asteroids with $248 > D > 130$ km have a mass distribution index of 2.02 ± 0.14. It is thought that the transition region $260 > D > 248$ km separates the 'original' asteroids from the collisional fragments of this initial population.

098.102 **Infrared (JHK) photometry of asteroids.**
G. J. Veeder, D. L. Matson, C. Kowal.
Astron. J., Vol. 87, 834 - 839 (1982).
The authors report JHK (1.2, 1.6 and 2.2 μ) photometry for 23 asteroids of various spectral types. All of the TRIAD asteroid classes were sampled. The C and S classes have distinct, infrared color domains. Most R-class asteroids fall within the S-color domain and the observed M asteroids fall within the C-color domain. The E asteroids have neutral colors. Asteroid 446 Aeternitas was discovered to have an unusual J–H color (0.88 mag).

098.103 **Search for binary asteroids with the ESA Space Astrometry Satellite Hipparcos.** H. J. Schober.
Scientific aspects of the Hipparcos space astrometry mission, (see 012.041), p. 169 - 171 (1982).
There is high evidence for a number of asteroids to show binary nature or to have satellites. A proposal is presented and discussed to use the ESA Space Astrometry Satellite Hipparcos to confirm or find possibly the binary nature of selected asteroid-targets or their satellite(s) by direct imaging with the high astrometric accuracy provided by the Hipparcos-system.

098.104 **Minor planet 2100 (Ra-Shalom).**
British Astron. Assoc. Circ., No. 619 (1981).

098.105 **On the reality of minor planet (330) Adalberta.**
R. M. West, C. Madsen, L. D. Schmadel.
Astron. Astrophys., Vol. 110, 198 - 202 (1982).
The lost minor planet (330) Adalberta was discovered photographically by Wolf in Heidelberg in March 1892, and has not been seen since. A reexamination of the original plates indicates that Wolf most probably measured two stellar images and that therefore this planet is non-existent.

098.106 **On the stability of the asteroids.**
J. D. Hadjidemetriou.
Compendium in astronomy, (see 003.012), p. 67 - 78 (1982).
It is proved that all the nearly circular orbits of asteroids at the resonances with Jupiter of the form $(2n+1)/(2n-1)$, i.e. $3/1$, $5/3$, . . . are unstable and gaps in the distribution of the asteroids must be expected there. Since these resonant unstable orbits are very close to each other and have an accumulation point at the orbit of Jupiter, the space near Jupiter must be empty of asteroids. The resonant asteroid orbits of the form $(n+1)/n$, i.e. $2/1$, $3/2$, $4/3$, . . . are not necessarily all unstable, and concentrations of asteroids could appear, as is the case with the $3/2$ resonance.

098.107 **Report of IAU Commission 20: Positions and motions of minor planets, comets and satellites** (Positions et mouvements des petites planètes, des comètes et des satellites). G. Sitarski, E. Roemer.
Trans. IAU, Vol. XVIIIA, (see 003.013), p. 195 - 210 (1982).

098.108 **Catalogue of orbits of unnumbered minor planets.**
B. G. Marsden, C. M. Bardwell.
Published by Minor Planet Center, Smithsonian Astrophysical Observatory, 60 Garden Street, Cambridge, Mass. 02138, U. S. A. 106 pp. Price $ 8.00 (1982).
This catalogue is a considerably updated edition of the earlier publication 'Elements of unnumbered minor planets' (Minor Planet Center 1961). It contains elliptical orbital elements for 2471 unnumbered minor planets observed between 1900 and 1982 and is complete through the batch of Minor Planet Circulars issued on 1982 April 8.

098.109 **Catalogue of discoveries and identifications of minor planets.** B. G. Marsden, C. M. Bardwell.
Published by Minor Planet Center, Smithsonian Astrophysical Observatory, 60 Garden Street, Cambridge, Mass. 02138, U. S. A. 111 pp. Price $ 8.00 (1982).
This catalogue gives brief references to the discovery announcements of the unnumbered minor planets and identifications with numbered minor planets where they are known to exist. The catalogue represents an updated version of the work by Strobel (1963).

098.110 **Ephemerides of minor planets for 1983.**
Yu. V. Batrakov (Editor).
Published by Institut Teoreticheskoj Astronomii Akademii Nauk SSSR. Izdatel'stvo Nauka, Leningradskoe Adelenie, Leningrad. 270 pp. Price 3 Rbl. 10 Kop. (1982). In Russian and English.
Contents: Introduction, elements, opposition dates, ephemerides, ephemerides of bright planets, ephemerides of some unusual planets, critical list.

Operation Spacewatch. A group of astronomers is preparing to scan the skies for asteroids.
See Abstr. 013.017.

The World Space Foundation Asteroid Project.
See Abstr. 013.035.

Plans for upcoming asteroidal occultations.
See Abstr. 013.050.

Methods of computation of the perturbed motion of small bodies in the solar system. See Abstr. 021.047.

Effects of body shape on disk-integrated spectral reflectance. See Abstr. 022.121.

A comparison of astrometric measurement techniques as applied to minor planets.   See Abstr. 031.529.

CCD scanning for asteroids and comets.
See Abstr. 031.625.

Inertial frame determination using minor planets. A simulation of HIPPARCOS-observations.
See Abstr. 041.034.

Using a uniform approximation of the coordinates of disturbing bodies in the Taylor-Steffensen numerical integration method.   See Abstr. 042.036.

The origin of the Kirkwood gaps: a mapping for asteroidal motion near the 3/1 commensurability.
See Abstr. 042.038.

A qualitative study of stabilizing and destabilizing factors in planetary and asteroidal orbits.
See Abstr. 042.051.

The origin of the Kirkwood gaps: a mapping for asteroidal motion near the 3/1 commensurability.
See Abstr. 042.075.

Motion of the Jovian commensurability resonances and the character of the celestial mechanics in the asteroid zone: implications for kinematics and structure.
See Abstr. 042.113.

The use of minor planet dynamics toward improving the fundamental reference system.
See Abstr. 043.005.

Contribution of the Pulkovo Observatory to the improvement of orientation of the FK4 system using observations of selected minor planets.   See Abstr. 043.014.

Inertial frame determination using minor planets — a proposal for Hipparcos observations.   See Abstr. 043.018.

Radar backscattering from a rough rotating triaxial ellipsoid with applications to the geodesy of small asteroids. See Abstr. 063.041.

On the computation of angular distributions of radiation in planetary atmospheres.   See Abstr. 063.055.

Planetary occultation predictions for 1982.
See Abstr. 096.016.

Observations of asteroidal occultations and appulses. See Abstr. 096.018.

On the similarity of orbits of associated comets, asteroids and meteoroids.   See Abstr. 102.012.

Phase functions of polarization and brightness and the nature of cometary atmosphere particles.
See Abstr. 102.013.

The influx of comets and asteroids to the earth.
See Abstr. 102.023.

Report of IAU Commission 15: Physical study of comets, minor planets and meteorites (L'étude physique des comètes, des petites planètes et des météorites).
See Abstr. 102.055.

Position observations of comet Stephan-Oterma (1980g) and of asteroid 751.   See Abstr. 103.721.

Cosmic spherules and asteroid collisions.
See Abstr. 106.007.

Formation and collisional evolution of small bodies: effects of two-material systems on large-scale geologic structure.   See Abstr. 107.004.

099 Jupiter, Jupiter Satellites

099.001 **The early evolution of Jupiter in the absence of solar tidal forces.** N. Schofield, M. M. Woolfson.
Mon. Not. R. Astron. Soc., Vol. 198, 947 - 959 (1982).

The early evolution of a Jupiter-like protoplanet is simulated by constructing a physically detailed computer-based model which solves the equations of hydrodynamics and radiative energy transfer for the spherically symmetric case. The model is specifically developed to study the initial and boundary conditions relevant to the capture theory for the origin of the solar system. It is found that the absence of an external medium promotes the rapid expansion of surface material which is enhanced by solar irradiation. Only when the Jeans Criterion, ratio $2\Pi/\Omega$, is less than 0.8 does a "spontaneous" hydrodynamic collapse of the interior allow a substantial proportion of the protoplanet to condense to planetary densities.

099.002 **The early evolution of Jupiter in the tidal field of the sun.** N. Schofield, M. M. Woolfson.
Mon. Not. R. Astron. Soc., Vol. 198, 961 - 973 (1982).

An intricate point-mass simulation of early protoplanetary evolution in the tidal field of the sun, and under physical constraints appropriate to the capture theory, shows that a Jupiter-like protoplanet which is moving on a highly eccentric orbit passes through a series of prolate spheroids until self-gravitation restores the protoplanet towards sphericity. The sun-facing tidal protuberance initially lags the radius vector resulting in a prograde rotation being induced. The fast internal hydrodynamic collapse coupled with a slowly moving surface permits a significant proportion of the inner protoplanet to contract to Jupiter dimensions while the surface material gains specific angular momentum comparable to that of the Galilean satellites.

099.003 **Plasma characteristics of the Io torus.** J. S. Morgan, C. B. Pilcher.
Astrophys. J., Vol. 253, 406 - 421, with a correction Vol. 254, 420 (1982).

The authors have obtained simultaneous low spectral resolution measurements of the following seven forbidden emission lines of the Jovian plasma torus: [S II] $\lambda\lambda$4069, 4076; [S II] $\lambda\lambda$6716, 6731; [S III] λ3722; [O II] $\lambda\lambda$3726, 3729. Significant changes in the characteristics of these emissions occur on a time scale of days. The [S II] and [O II] lines yield independent and consistent values for the average electron density. Variations about this average are larger than the measurement uncertainties. A few unusual spectra in which the blue but not the red [S II] lines are present indicate the existence of transient conditions of high plasma density and low electron temperature.

099.004 **On the dynamics of Jupiter's atmosphere.** G. E. Hunt.
Vistas Astron., Vol. 25, 235 - 243 (1982).

Observations of the atmosphere of Jupiter by the imaging and infrared instruments on the Voyager spacecraft have been analysed to provide new insight into the meteorology of Jupiter.

099.005 **The inner satellites of Jupiter.** J. Veverka, P. Thomas, S. Synott.
Vistas Astron., Vol. 25, 245 - 262 (1982).

In 1979 images of Jupiter's surroundings returned by Voyagers 1 and 2 led to the discovery of a ring around the planet and of three satellites within the orbit of Io. Two of these, 1979 J1 and 1979 J2, are closer to Jupiter than Amalthea. The authors concentrate on Amalthea and the recently discovered small satellites. Collectively, these will be called the inner satellites, to distinguish them from Jupiter's Galilean satellites and from the planet's outer satellites: J6 through J13.

099.006 **Jupiter and Io: a binary magnetosphere.** F. L. Scarf, F. V. Coroniti, C. F. Kennel, D. A. Gurnett.
Vistas Astron., Vol. 25, 263 - 314 (1982).

The magnetosphere of Jupiter and Io has recently been traversed by two Voyager spacecraft. This review outlines qualitatively the physical picture emerging from ongoing data analysis and theoretical interpretation. The authors discuss observations of the Io torus EUV emissions and Jupiter's aurora. They next turn to Jupiter's middle magnetosphere, concentrating on the observations of corotating ions, their ambiguities, and their implications. They then return to the classical question of Jupiter's interaction with the solar wind, as manifested by its magnetic tail. Finally, they attempt to construct a unifying conceptual picture.

099.007 **The atmosphere and ionosphere of Jupiter.** S. K. Atreya, T. M. Donahue.
Vistas Astron., Vol. 25, 315 - 335 (1982).

The properties of the atmosphere of Jupiter are reviewed in the light of observations carried out by the Voyager mission. Solar occultation measurements in the ultraviolet show that the temperature of the upper atmosphere is 1100 ± 200K. A very large solar cycle variation in the atomic hydrogen Lyman alpha airglow emission rate is discussed. The equatorial electron density profile determined by Voyager radio occultation can be explained if ion molecule reactions between H^+ and vibrationally excited H_2 at high temperatures are fast enough. Temperature dependence of these reactions also accounts for an observed ionospheric diurnal variation.

099.008 **A new look at the ionosphere of Jupiter in light of the UVS occultation results.** J. C. McConnell, J. B. Holberg, G. R. Smith, B. R. Sandel, D. E. Shemansky, A. L. Broadfoot.
Planet. Space Sci., Vol. 30, 151 - 167 (1982).

Simple photochemical models cannot reconcile Jupiter's ionospheric electron density profiles with the observed neutral atmosphere. The location of the peak electron density predicted when the neutral atmosphere determined by the Voyager Ultraviolet Spectrometer is combined with simple models falls about 1000 km lower than the peak determined by radio occultation. The locations and magnitudes of the peaks in electron density can be accounted for by including the effects of vertical transport of ions in the ionospheric models. In view of the complex relationship between the ionosphere and neutral atmosphere, an attempt to infer one from the other cannot succeed. However, combining independent information on the two leads to new insights into the coupling of the neutral atmosphere, the ionosphere and the magnetosphere.

099.009 **Quadrupole and octupole parameters of Jupiter's main magnetic field.** D. M. Willis, A. D. Osborne.
Geophys. J. R. Astron. Soc., Vol. 68, 765 - 776 (1982).

The classical theory of multipoles is used to calculate the true quadrupole and octupole parameters for six different models of Jupiter's main magnetic field. These six magnetic-field models, which are based on measurements made by the Pioneer 10 and Pioneer 11 spacecraft, are specified in terms of the fifteen spherical harmonic coefficients required to define the Jovian dipole (3), quadrupole (5) and octupole (7). The set of five equations for the quadrupole parameters and the set of seven equations for the octupole parameters are each solved iteratively to give the corresponding true multipole moment and the directions of the associated multipole axes.

099.010 The tectonics of Ganymede.
E. M. Parmentier, S. W. Squyres, J. W. Head, M. L. Allison.
Nature, Vol. 295, 290 - 293 (1982).

The surface of Ganymede is composed of two distinct terrain types showing abundant evidence of global-scale tectonics. There is geological evidence consistent with the formation of younger bright terrain by the emplacement of relatively clean ice into rift zones developed in the more silicate-rich dark terrain.

099.011 Convective growth rates of equatorial features in the jovian atmosphere.
G. E. Hunt, J.-P. Müller, P. Gee.
Nature, Vol. 295, 491 - 494 (1982).

The first results of measurements of divergence and vertical velocity of cloud features in the atmosphere of Jupiter are presented from analyses of Voyager images, performed using the interactive planetary imaging processing system at University College London. Active equatorial plumes are found to have typical divergences of $\sim(1-2) \times 10^{-5} \text{ s}^{-1}$, and vertical velocities of 20–40 cm s^{-1}, during the observational periods examined. These plumes are considerably more active than the bright albedo features in the turbulent regions of the jovian north and south equatorial belts.

099.012 Jupiter's gravitational effects on the solar wind and geomagnetic activity in different epochs of solar activity.   V. D. Reshetov.
Soln. Dannye 1981 Byull., No. 9, p. 114 - 118 (1981). In Russian.

It is known that the geomagnetic disturbance strengthens with the lag of 4 - 9 months after Jupiter's opposition. This lag decreases with the solar activity level. This phenomenon is considered as the development of convection sectors of waves in the solar wind under Jupiter's gravitational forces. The scheme of the phenomenon development gives the opportunity to estimate the solar wind velocity in the different epochs of the solar activity between the earth's orbit and Jupiter's orbit in the interval 430 - 870 km sec^{-1}.

099.013 The far-ultraviolet spectra and geometric albedos of Jupiter and Saturn.   J. T. Clarke, H. W. Moos, P. D. Feldman.
Astrophys. J., Vol. 255, 806 - 818 (1982).

Photometrically calibrated spectra (1200–1940 Å) of Jupiter and Saturn, compiled from IUE observations over the period 1978 December through 1980 July, are presented along with the resulting wavelength variation of the geometric albedos of these planets. Airglow emission features from both planets at H I Lyα (1216 Å), C I (1657 Å), and the H$_2$ Lyman and Werner bands (1230–1608 Å) are identified; probable excitation mechanisms for these emissions are discussed. Relative to Jupiter, Saturn shows the C$_2$H$_2$ absorption bands more strongly and has a lower value of the albedo for $\lambda < 1750$ Å. A model of the atmospheric absorption was constructed using experimental photoabsorption cross sections and assuming homogeneous mixing in order to investigate the abundances of absorbing constituents in the upper part of the lower atmosphere.

099.014 Arc structures in the Jovian decametric emission observed from the Earth and from Voyager.
C. H. Barrow, A. Lecacheux, Y. Leblanc.
Astron. Astrophys., Vol. 106, 94 - 97 (1982).

Jovian decametric arc structures have been compared, for similar configurations of central meridian longitude

CML (1965.0) and Io-phase, observed from the Earth close to opposition and from the Voyager-1 and Voyager-2 spacecraft before each encounter. Linear relationships have been found between the positions in CML of certain structural features and the corresponding values of jovigraphic latitude D_E for each observation. The slopes of the lines are opposite for the A- and the B-sources. The events showing these effects are Io-correlated, in contrast to the well-known relationship, from occurrence probability considerations, between D_E and the position in CML of the Non-Io A-source.

099.015 De magnetosfeer van Jupiter.   I. de Pater.
Zenit, 9. Jaarg., 152 - 156 (1982).

099.016 Constraints on Galilean satellite geophysics from photometric and geomorphic observations.
M. C. Malin.
Reports of Planetary Geology Program – 1981, (see 003.001), p. 27 (1981). – Abstract.

099.017 The ejection of material from Io.
C. B. Pilcher.
Reports of Planetary Geology Program – 1981, (see 003.001), p. 28 (1981). – Abstract.

099.018 Variable features on Io.
R. J. Terrile, T. V. Johnson, L. A. Soderblom, R. G. Strom.
Reports of Planetary Geology Program – 1981, (see 003.001), p. 29 - 31 (1981). – Abstract.

099.019 Volcanologic constraints on models of Io volcanism.
L. S. Crumpler, R. G. Strom.
Reports of Planetary Geology Program – 1981, (see 003.001), p. 32 - 35 (1981). – Abstract.

099.020 Geomorphology of Ra Patera Io: a quantitative approach to sulfur volcanism.
D. C. Pieri, S. M. Baloga, R. M. Nelson, C. Sagan.
Reports of Planetary Geology Program – 1981, (see 003.001), p. 41 - 43 (1981). – Abstract.

099.021 Voyager photometry of Europa.
B. Buratti, J. Veverka.
Reports of Planetary Geology Program – 1981, (see 003.001), p. 44 - 46 (1981). – Abstract.

099.022 Color distribution fields of geomorphic features on Europa: initial results from a new technique.
T. A. Meier.
Reports of Planetary Geology Program – 1981, (see 003.001), p. 47 - 49 (1981). – Abstract.

099.023 Morphotectonic maps of the grooved terrain on Ganymede.   P. L. Masson.
Reports of Planetary Geology Program – 1981, (see 003.001), p. 50 (1981). – Abstract.

099.024 A mechanical analysis of extensional instability on Ganymede.   J. H. Fink, R. C. Fletcher.
Reports of Planetary Geology Program – 1981, (see 003.001), p. 51 - 53 (1981). – Abstract.

099.025 Sequential development of grooved terrain and polygons on Ganymede.
M. P. Golombek, M. L. Allison.
Reports of Planetary Geology Program – 1981, (see 003.001), p. 54 - 56 (1981). – Abstract.

099.026 Possible geologic implications of Ganymede crater densities.   J. B. Plescia.

Reports of Planetary Geology Program – 1981, (see 003.001), p. 57 - 58 (1981). – Abstract.

099.027 **Geophysical evolution of Ganymede and Callisto. I.**
W. B. McKinnon.
Reports of Planetary Geology Program – 1981, (see 003.001), p. 62 - 64 (1981). – Abstract.

099.028 **Geophysical evolution of Ganymede and Callisto. II.**
W. B. McKinnon.
Reports of Planetary Geology Program – 1981, (see 003.001), p. 65 - 67 (1981). – Abstract.

099.029 **Geophysical evolution of Ganymede and Callisto. III.**
W. B. McKinnon.
Reports of Planetary Geology Program – 1981, (see 003.001), p. 68 - 70 (1981). – Abstract.

099.030 **Crater obliteration by relaxation is not an important process on Callisto.**   A. Woronow, R. G. Strom.
Reports of Planetary Geology Program – 1981, (see 003.001), p. 71 - 72 (1981). – Abstract.

099.031 **Ganymede rampart craters.**
V. M. Horner, R. Greeley.
Reports of Planetary Geology Program – 1981, (see 003.001), p. 82 - 84 (1981). – Abstract.

099.032 **The control networks of the satellites of Jupiter and Saturn.**   M. E. Davies.
Reports of Planetary Geology Program – 1981, (see 003.001), p. 481 - 482 (1981). – Abstract.

099.033 **Voyager cartography.**
R. M. Batson, P. M. Bridges, K. F. Mullins.
Reports of Planetary Geology Program – 1981, (see 003.001), p. 484 - 485 (1981). – Abstract.

099.034 **The Galilean Satellite Geological Mapping Program.**
B. K. Lucchitta.
Reports of Planetary Geology Program – 1981, (see 003.001), p. 507 (1981). – Abstract.

099.035 **A geologic map of Europa.**
B. K. Lucchitta, L. A. Soderblom.
Reports of Planetary Geology Program – 1981, (see 003.001), p. 508 - 510 (1981). – Abstract.

099.036 **Geologic mapping of Europa.**
D. C. Pieri, K. Hiller.
Reports of Planetary Geology Program – 1981, (see 003.001), p. 511 (1981). – Abstract.

099.037 **Crater frequency distributions on Ganymede.**
S. S. Mims, D. Nummedal.
Reports of Planetary Geology Program – 1981, (see 003.001), p. 539 - 540 (1981). – Abstract.

099.038 **New constants for the Sampson-Lieske theory of the Galilean satellites of Jupiter.**   J.-E. Arlot.
Astron. Astrophys., Vol. 107, 305 - 310 (1982).
New constants for the Sampson-Lieske theory of the motion of the Galilean satellites of Jupiter are calculated using a large set of photographic observations: 8856 individual observations. Ephemerides are built and a comparison is made with recent observations of some mutual phenomena of the Galilean satellites in 1979.

099.039 **Comparison of Sampson–Lieske theory of the Galilean satellites of Jupiter with observations.**
R. Biancale, S. Ferraz-Mello, M. Tsuchida.
Celestial Mech., Vol. 26, (see 012.001), 225 - 228 (1982).

The aim of the present work is to compare photographic observations of the Galilean satellites of Jupiter with the theory developed by Sampson at the beginning of the century and corrected and implemented recently by Lieske. The results lead to the suggestion that important long period defects still exist in the theory of Sampson–Lieske. This is not surprising, due to the difficulties of the computation of the long-period inequalities in mean longitudes, even in a first-order theory.

099.040 **Secular and long period effects in the orbits of the Galilean satellites.**   B. C. Brown.
Celestial Mech., Vol. 26, (see 012.001), 229 (1982). – Abstract.

099.041 **Early eclipses of the Galilean satellites.**
J. H. Lieske.
Celestial Mech., Vol. 26, (see 012.001), 257 - 263 (1982).
A brief summary of the development of the theory of motion of the Galilean satellites is presented. Over 7700 eclipse observations have been collected and reduced using the Ephemeris E–2. They are of great potential in improving the ephemerides of the satellites and can yield important information on the evolution of the Galilean system.

099.042 **Determination of the semi-major axes of the Galilean satellites of Jupiter.**   D. T. Vu.
Celestial Mech., Vol. 26, (see 012.001), 265 - 270 (1982). In French.
In non-resonant cases, a constant part coming from the perturbations can be easily separated from the observed mean motion. Another part created by resonance must be separated from the observed mean motion in the case of the first three Galilean satellites. The determination of it gives better value of the semi-major axis. In this investigation, an analytical process is chosen to avoid a mixture of orders in successive expansions and integrations. The main terms entering in the computation of the libration are the great inequalities of the first three satellites.

099.043 **Sodium remote from Io.**
R. A. Brown, N. M. Schneider.
Icarus, Vol. 48, 519 - 535 (1981).
The authors find that faint sodium emission originating in the middle Jupiter magnetosphere has two distinct kinematical components. The "normal" signature of atoms on bound orbits with large apojoves seems always to be present, and the authors suggest these atoms are an extension of the bright, near-Io sodium cloud. The "fast" signature, with speeds up to at least 100 km sec^{-1}, is seen only occasionally, and the authors suggest it is due to an interaction of the near-Io sodium cloud with the corotating, heavy-ion plasma. Both elastic and charge-exchange collisions seem consistent with the observed kinematical and temporal signatures. Elastic collisions seem marginally more capable of producing the high observed sodium atom speeds.

099.044 **Ground-based observations of the Jovian ring and inner satellites.**
D. C. Jewitt, G. E. Danielson, R. J. Terrile.
Icarus, Vol. 48, 536 - 539 (1981).

099.045 **Examination of Le Verrier and Gaillot's theories of Jupiter and Saturn.**   R. Dvorak.
Sitzungsber. Österreich.Akad. Wiss. Math.-Naturwiss. Kl. Abt. II, Vol. 189, No. 1- 3, p. 129 - 138 (1980). In German.
Abstr. in Phys. Abstr., Vol. 85, Abstr. 28405 (1982).

099.046 **Influence of the short-period perturbations on the motion of the Galilean satellites of Jupiter.**
E. N. Lemekhova, T. F. Sokolova.
Byull. Inst. Teor. Astron., Leningrad, Tom 15, 91 - 95 (1981). In Russian.

An analysis of the short-period perturbations of the Galilean satellites of Jupiter on the base of the analytic expressions obtained in Marsden's theory is carried out. An estimation of the influence of these perturbations on the true jovicentric longitude, latitude and radius vector is given for the time span of 1000 hours. The change of these perturbations depending on the values of the masses of the satellites is investigated.

099.047 The volcanism of Io.
G. Israel, V. Le Buisson.
Europhys. News, Vol. 12, No. 10, p. 9 - 11 (1981). – Abstr. in Phys. Abstr., Vol. 85, Abstr. 32899 (1982).

099.048 The Great Red Spot and the magnetic field of Jupiter.   L. I. Miroshnichenko.
Priroda, 1982, No. 3, p. 87. In Russian.

099.049 Positions of Jupiter and satellites in 1978.
H. Debehogne, N. Havaut.
Acta Astron., Vol. 31, 387 - 390 (1981).

77 astronomical positions of Jupiter and Galilean satellites are given as obtained from photographic observations made at Uccle, in December 1978.

099.050 A comparison of the atmospheres of Jupiter and Saturn.   R. F. Beebe.
Proc. Southwest Reg. Conf., Vol. 7, (see 012.019), 109 (1982). Abstract.

099.051 Photographic position observations of Jupiter and the Galilean satellites with a longfocal astrograph in 1975.   T. I. Levitskaya.
Astrometr. Astrofiz., Vyp. (No.) 46, p. 80 - 83 (1982). In Russian.

099.052 Entstehung des Jupiterrings.   P. Kaiser.
Monde, Ringe und Asteroide im Planetensystem, (see 012.020), p. 171 - 191 (1981).

099.053 Jovimagnetic secular variation.
J. E. P. Connerney, M. H. Acuña.
Nature, Vol. 297, 313 - 315 (1982).

The authors report new limits on jovimagnetic secular variations found by comparison of a jovian internal field model obtained from the Voyager 1 magnetic field observations at epoch 1979.2 with the epoch 1974.9 Pioneer 11 O_4 model. No significant secular variation of either the magnitude or position of the jovidipole was found for the years 1974.9 to 1979.2, although a small Earth-like variation cannot be ruled out.

099.054 Time-dependent calculations of Jupiter's ionosphere.
R. H. Chen.
J. Geophys. Res., Vol. 87, 167 - 170 (1982).

Time-dependent calculations of the vertical distribution of protons in Jupiter's ionosphere show that the accumulation of protons in the topside ionosphere produced from solar ionizing radiation overwhelms the loss to vibrationally excited molecular hydrogen at vibrational temperatures as high as $1600°K$. At $2500°K$ the ionization is decreased over the entire planet with little diurnal variation. For Voyager 1 then, unless the H_2 vibrational temperature is as high as thousands of degrees and the topside density of H_2 is asymmetric and larger by orders of magnitude, dynamical processes are more likely causes of the low electron densities seen in the nightside upper ionosphere. A calculation of the H_3^+ density profile showed that the distribution above the turbopause is controlled by diffusion.

099.055 Io's hot plasma torus – a synoptic view from Voyager.   B. R. Sandel, A. L. Broadfoot.

J. Geophys. Res., Vol. 87, 212 - 218 (1982).

A study of the morphology of Io's hot plasma torus has encompassed hundreds of Voyager UVS measurements of torus intensity. The long-term average state of the torus can be characterized by an axial asymmetry in the brightness of the prominent S III 685-Å feature manifested as an enhancement in brightness whose peak is fixed near 1900 local time. No long-term correlation of brightness with magnetic longitude is present. On time scales of a few Jupiter rotations, the torus can differ markedly from its average axial asymmetry, and its brightness is correlated with magnetic longitude for short times.

099.056 The injection of energy into the Io plasma torus.
D. E. Shemansky, B. R. Sandel.
J. Geophys. Res., Vol. 87, 219 - 229 (1982).

Voyager EUV observations of the Io plasma torus showing a Jupiter local time asymmetry have been interpreted as a 10-hour periodicity in electron temperature in the corotating reference frame. The plasma shows two remarkable morphological characteristics. First, the intensity periodicity showed no tendency to change over the $\cong 0.5$-year period of reduced data, indicating it to be a permanent feature of the torus. Second, deviations from the mean behavior, such as short-term magnetic longitude effects, all appear to be caused by electron temperature effects. Thus no local plasma mass variations have been detected during the observational period. These characteristics and the variability of the energy source required to drive the observed asymmetry are discussed.

099.057 The probability distribution of S^+ gyrospeeds in the Io torus.   R. A. Brown.
J. Geophys. Res., Vol. 87, 230 - 234 (1982).

The author reports a measured probability distribution for S^+ gyrospeeds in the Io torus derived from [S II] 6716 Å, 6731 Å emission line profiles recorded at axial distance $5.9 R_J$ from Jupiter.

099.058 Photochemistry of SO_2 in the atmosphere of Io and implications on atmospheric escape.   S. Kumar.
J. Geophys. Res., Vol. 87, 1677 - 1684 (1982).

Photochemical models of Io's atmosphere are presented with the assumption that SO_2 is the major gas and that the SO_2 surface pressure is controlled by vapor-pressure equilibrium at the surface. Photolysis of SO_2 leads to efficient production of SO, O_2, S, and O. Of these products, O and S are likely to be the dominant constituents in the upper atmosphere, and the atmospheric escape is expected to be in atomic form.

099.059 Modified disc model of Jupiter's magnetosphere.
Z. X. Liu.
J. Geophys. Res., Vol. 87, 1691 - 1695 (1982).

The paper establishes a set of static magnetohydrodynamic equations in which a differential rotation of the magnetosphere and a wavy magnetodisc structure are considered. Solutions are obtained that include magnetic field, pressure, density, and temperature. By using the theoretical formulas of the magnetic field and pressure, the author calculates the thickness of the plasma sheet.

099.060 A simple mechanical model of Valhalla basin, Callisto.   H. J. Melosh.
J. Geophys. Res., Vol. 87, 1880 - 1890 (1982).

The Valhalla basin on Callisto is a multiringed structure that extends over much of the satellite's surface. Although its apperance differs in detail from lunar multiringed basins, its origin may also be attributed to lithospheric fragmentation accompanying collapse of the transient crater formed by an impact event. The paper explores the mechanics of the collapse process by treating the lithosphere as a thin elastic-

Von Mises plastic sheet (plane geometry) or shell (spherical geometry).

099.061 Atmospheres of Jupiter and Saturn.
G. E. Hunt.
Philos. Trans. R. Soc. London A, Vol. 303, (see 012.022), 225 - 245 (1981).

In this paper the author reviews current knowledge of the atmospheres of Jupiter and Saturn, making use of the extensive telescopic studies, International Ultraviolet Explorer Satellite observations and the measurements made during the recent Pioneer and Voyager flybys which have been supported by detailed theoretical studies. A detailed discussion is given of the composition of these atmospheres and the abundance ratios which provide insight into their original state and their evolution.

099.062 An Io thermal model with intermittent volcanism.
G. J. Consolmagno.
Proc. Twelfth Lunar Planet. Sci. Conf., (see 012.024), p. 1533 - 1542 (1982).

A two-dimensional time-dependent thermal model of Io was constructed to examine the relationship between enhanced tidal heating, heat flow within the body, and surface volcanism. From the models one can conclude that (1) a simple homogeneous body cannot maintain enhanced tidal heating in a thin crust, but such heating is possible if the crust contains an insulating layer, as of sulfur; (2) this sulfur layer will be subject to melting by heat from the interior, resulting in hotspots which can expel heat much more rapidly than tidal heating inputs it; and (3) the location of these hotspots will vary in time and space; they can appear anywhere on the surface, even if tidal heating is concentrated at the poles, and they seem to occur roughly 10% of the time.

099.063 Microstructure and particulate properties of the surfaces of Io and Ganymede: comparison with other solar system bodies.
K. D. Pang, K. Lumme, E. Bowell.
Proc. Twelfth Lunar Planet. Sci. Conf., (see 012.024), p. 1543 - 1553 (1982).

The phase curves of Io and Ganymede between $0°$ and $40°$ solar phase angle have been compiled from Voyager photo-polarimeter and groundbased observations. Modeling the data with the Lumme-Bowell theory allowed the authors to determine improved estimates of the surface texture and particulate properties. Accurate V-band zero-phase geometric albedos, phase integrals, and Bond albedos are also obtained.

099.064 Structures on Europa.
B. K. Lucchitta, L. A. Soderblom, H. M. Ferguson.
Proc. Twelfth Lunar Planet. Sci. Conf., (see 012.024), p. 1555 - 1567 (1982).

Lineation patterns on Europa in the area covered by high resolution Voyager 2 images were divided into four classes and analyzed for their systematic trends. The classes consist of dark, wedge-shaped bands in the west-central part of the area; triple bands associated with diagonally trending, global lineations; gray, concentric, curvilinear bands in the southeastern part; and ridges near the terminator. The gray bands are probably old, the wedge-shaped bands and ridges are young, and the triple bands have developed during most of the time of formation of Europa's present crust. Among the patterns most persistently reactivated is a global grid system of diagonally trending conjugate shear fractures, and possibly an old dichotomy consisting of one region dominated by straight fractures and another dominated by curvilinear fractures.

099.065 The sputter-generation of planetary coronae: Galilean satellites of Jupiter.   C. C. Watson.
Proc. Twelfth Lunar Planet. Sci. Conf., (see 012.024), p. 1569 - 1583 (1982).

Energetic particle fluxes which impinge upon and sputter planetary surfaces or atmospheres are effective agents for the generation or modification of atmospheric coronae. The structure of such coronae is determined by the characteristics of the sputtered particle flux, and hence of the sputtering mechanisms involved. A collisional cascade type interaction tends to produce a more extended exosphere with a greater fraction of gravitationally unbound molecules than does a thermal spike sputtering mechanism, although the total mass of the generated atmosphere might exceed an exospheric level in the latter case.

099.066 Tectonic deformation of Galileo Regio and limits to the planetary expansion of Ganymede.
W. B. McKinnon.
Proc. Twelfth Lunar Planet. Sci. Conf., (see 012.024), p. 1585 - 1597 (1982).

099.067 Dark-ray craters on Ganymede.   J. Conca.
Proc. Twelfth Lunar Planet. Sci. Conf., (see 012.024), p. 1599 - 1606 (1982).

The distribution and ejecta ray pattern of dark-ray craters on Ganymede are found to show no systematic variation with terrain type. However, the distribution is correlated with latitude and longitude, with dark-ray craters strongly concentrated on Ganymede's trailing hemisphere increasing toward the antapex of orbital motion.

099.068 The Jovian ammonia abundance from interferometric observations of limb darkening at 3.4 mm.
F. Valdes, W. J. Welch, D. Haber.
Icarus, Vol. 49, 17 - 26 (1982).

The interferometer visibility of Jupiter, observed at a wavelength of 3.4 mm, is used to determine the global limb darkening of the planet's brightness. From a single-parameter fit to the visibility curve, the authors find an ammonia-to-molecular hydrogen mixing ratio of $6.4[+5.1, -1.9] \times 10^{-5}$, which corresponds to $35[+28, -10]\%$ of the solar nitrogen abundance if all of the nitrogen is in the form of ammonia. The fitting procedure uses a simple model atmosphere for the Jovian atmosphere which is based on other observations of the planet. The dependence of the result on the various model parameters is studied.

099.069 About the distances of Jupiter's satellites.
J. Bouška.
Acta Univ. Carolinae Math. Phys., Vol. 23, No. 1, p. 67 - 68 (1982).

The known satellites of Jupiter do not move around the planet in arbitrary distances but they make evidently five groups. It cannot be excluded that in the distance $r \approx 52 \, R_J$ one or more not yet discovered small satellites may exist.

099.070 A novel view on the system of Jupiter.
V. Pohánka.
Kozmos, Vol. 13, 11 - 21 (1982). In Slovak.

099.071 The magnetosphere of Jupiter.
A. J. Dessler.
Solar system plasmas and fields, (see 003.005), p. 55 - 59 (1982).

099.072 The C/H ratio in Jupiter from the Voyager infrared investigation.   D. Gautier, B. Bezard, A. Marten,
J. P. Baluteau, N. Scott, A. Chedin, V. Kunde, R. Hanel.
Astrophys. J., Vol. 257, 901 - 912 (1982).

From a selection of Voyager IRIS spectra corresponding to cloud-free areas of Jupiter, the authors have determined the CH_4/H_2 volume ratio in the atmosphere of this planet as equal to $(1.95 \pm 0.22) 10^{-3}$ which corresponds to 2.07 ± 0.24 times the solar value of Lambert (C/H = 4.7×10^{-4}). Estimate of errors includes both instrument noise and systematic uncertainties. Implications of this result on the formation and evolution of Jupiter are discussed.

099.073 Ganymede and Callisto: accumulation heat content.
A. Coradini, C. Federico, P. Lanciano.
The comparative study of the planets, (see 012.033), p. 61 - 70 (1982).
Using the accumulation theory for the formation of the two outermost regular satellites of Jupiter, the accumulation heat content has been obtained. The early thermal profiles show the onset of convection just at the end of the formation process. The fraction of the gravitational energy converted into thermal energy is found to be larger for Callisto than for Ganymede.

099.074 The present picture of Io's electrodynamic coupling with the magnetosphere of Jupiter.
M. Dobrowolny.
The comparative study of the planets, (see 012.033), p. 295 - 309 (1982).
The recent Voyager 1 measurements have greatly added to the knowledge of Io's electrodynamic interactions with the magnetosphere of Jupiter. The author discusses the present physical picture of these interactions and its relation to the decametric radiation from Jupiter.

099.075 Stratigraphic relationships among the upper layers of the outer Galilean satellites, inferred from the investigation of their ray systems.   M. Poscolieri.
The comparative study of the planets, (see 012.033), p. 485 - 494 (1982).

099.076 On the possibility of origin of microorganisms on the Jupiter satellites.
L. O. Kolokolova, A. F. Steklov.
Mikrobiol. zh., Tom 43, 791 - 796 (1981). In Russian.
Abstr. in Ref. zh., 51. Astron., 5.51.207 (1982).

099.077 Galilean satellites: the value of early eclipses.
J. H. Lieske.
Bull. American Astron. Soc., Vol. 14, 580 (1982). − Abstract.

099.078 Discovery of an Io-correlated energy source for Io's hot plasma torus.   B. R. Sandel, A. L. Broadfoot.
J. Geophys. Res., Vol. 87, 2231 - 2240 (1982).
The authors report the discovery of a correlation of the brightness of the ansae of the torus with the apparent orbital phase of Io. By using their understanding of the radiative cooling of the torus, the authors show that this observation is consistent with and most naturally explained by a flow of energy into the plasma electrons that excite the EUV emissions. This energy flow is localized in azimuth to the region of the torus near Io, leading to a correlation of torus brightness with the orbital phase of Io.

099.079 Light ion concentrations in Jupiter's inner magnetosphere.
R. L. Tokar, D. A. Gurnett, F. Bagenal, R. R. Shaw.
J. Geophys. Res., Vol. 87, 2241 - 2245 (1982).
Voyager 1 plasma wave instrument observations of lightning-generated whistlers are combined with Voyager 1 plasma instrument heavy ion ($8 \leqslant A/Z \leqslant 64$) charge concentrations to investigate the concentration of light ions ($A/Z < 8$) along the whistler propagation path. Two models for light ion concentration over dipole L shells for L between 5.2 and 6.2 are obtained, one giving a constant concentration along the field line and the other corresponding to an exponential density distribution. Near the equator the light ion concentration ranges from only about 1 to 10% of the heavy ion concentration, while outside of the torus, at distances of from 1 to 1.5 R_J above the equator, the light ions are the dominant species.

099.080 Voyager 1 assessment of Jupiter's planetary magnetic field.
J. E. P. Connerney, M. H. Acuña, N. F. Ness.
J. Geophys. Res., Vol. 87, 3623 - 3627 (1982).
A new estimate of Jupiter's planetary magnetic field is obtained from the Voyager 1 observations of the Jovian magnetosphere. An explicit model for the magnetodisc current system is combined with a spherical harmonic model of the planetary field with both sets of parameters determined simultaneously by using a nonlinear generalized inverse methodology.

099.081 Voyager spacecraft images of Jupiter and Saturn.
M. M. Birnbaum.
Appl. Opt., Vol. 21, 214 - 227 (1982).

099.082 An explanation of the persistence of the Great Red Spot of Jupiter.   A. Kyrala.
Moon Planets, Vol. 26, 105 - 107 (1982).
An argument is given basing the persistence of the Great Red Spot of Jupiter on compensation of the natural decay of vorticity by collision with a portion of the vortices shed by the South boundary of the South Tropical Zone. The latter are deviated northward by Coriolis acceleration. The GRS itself is regarded as a Rankine vortex with a central depression revealing the coloration of a layer below.

099.083 Very low frequency (VLF) hiss in Jupiter's magnetosphere.   P. N. Khosa, Lalmani, M. M. Ahmad.
Moon Planets, Vol. 26, 227 - 233 (1982).
The particle energy required to generate the observed VLF hiss in the Jovian magnetosphere has been computed under longitudinal and transverse resonance condition. It is shown that the minimum energy required by electrons to generate VLF hiss under the longitudinal resonance condition lies in the range of 100 eV−1 keV for the wave frequencies of 2−10 kHz, while the corresponding energy range for the transverse resonance condition for the same frequency range comes out to be 8 keV−40 keV. Further, the average radiated power by the Čerenkov process in Jupiter's magnetosphere at $L = 5.6$ Rj by electrons of energy 10 eV, 100 eV, and 1 keV for the wave frequency of 5 kHz has also been computed.

099.084 Peculiarities of formation of thermal radiation from Jupiter.   S. V. Kazmenko, Yu. M. Timofeev.
Kosm. Issled., Tom 20, 449 - 459 (1982). In Russian.

099.085 La planète Jupiter en 1978 - 1979.
R. Néel.
Astronomie, Vol. 96, 279 - 296 (1982).

099.086 Lexique des abréviations utilisées pour Jupiter par la Commission des Surfaces Planétaires.
R. Néel.
Astronomie, Vol. 96, 297 (1982).

099.087 Magnetospheres of Jupiter and Saturn.
N. F. Ness.
Sun and planetary system, (see 012.040), p. 243 - 248 (1982).
This paper is concerned with the study of the magnetospheres of these planets: the immediate region of their environment in which the planetary magnetic field is the dominant physical force, and also with the boundary of these magnetospheres, which is formed by the interaction with the solar wind. The attention will be restricted to the recent results obtained by the Voyagers.

099.088 Conversion of para and ortho hydrogen in the Jovian planets.   S. T. Massie, D. M. Hunten.
Icarus, Vol. 49, 213 - 226 (1982).
A mechanism is proposed which partially equilibrates the para and ortho rotational levels of molecular hydrogen in

the atmospheres of Jupiter, Saturn, and Uranus. Catalytic reactions between the free-radical surface sites of aerosol particles and hydrogen molecules yield significant equilibration near 1 bar pressure, if the efficiency of conversion per collision is between 10^{-8} and 10^{-10} and the effective eddy mixing coefficient is 10^4 cm^2/sec. At lower pressures the ortho-para ratio retains the value at the top of the cloud layer, except for a very small effect from conversion in the thermosphere. The influence of conversion on the specific heat and adiabatic lapse rate is also investigated.

099.089   **Astrometric observations of the satellites of the outer planets. V. The oppositions of 1978 - 1979, 1980, and 1981.**   J. R. Rohde, P. A. Ianna, L. C. Stayton, F. H. Levinson.
Astron. J., Vol. 87, 698 - 717 (1982).
   The authors present accurate photographic positions of the Galilean satellites and satellites of Saturn obtained during the oppositions of 1978 - 1979, 1980, and 1981, as part of a continuing program at the Leander McCormick Observatory. A total of 1032 pairs of spherical-equatorial coordinates for the equator and equinox of 1950.0 and 1316 intersatellite positions are reported, all obtained with the 67-cm refractor.

099.090   **Discussion of the photographic observations of the Galilean satellites in the period 1913 - 1928.**
M. Tsuchida, S. Ferraz-Mello, R. Biancale.
Astron. J., Vol. 87, 924 - 927 (1982).
   The series of photographic observations of the four great satellites of Jupiter taken in the period 1913 - 1928 are discussed. It is shown that the residuals are in general small (0".06 - 0".08 in mutual distances). The series are compared to modern ephemerides (Sampson-Lieske theory). These observations are of good quality and may help in improving the current ephemerides.

099.091   **Synoptic report of the 1977 - 78 apparition of the planet Jupiter.**   P. K. Mackal.
Strolling Astron., Vol. 29, 94 - 103 (1982).

099.092   **The morphology and evolution of Ganymede and Callisto.**   S. W. Squyres.
Advances in planetary geology, (see 003.011), p. 367 - 718 (1981).
   In this thesis, Voyager images are used to investigate the surface characteristics, geologic processes, and internal evolution of Ganymede and Callisto.

099.093   **Voyager encounters with Jupiter's magnetosphere: results of the low energy charged particle (LECP) experiment.**   S. M. Krimigis.
Compendium in astronomy, (see 003.012), p. 191 - 200 (1982).
   The two Voyager spacecraft encountered the planet Jupiter and its magnetosphere during March and July 1979. The measurements have revealed that the Jovian magnetosphere is filled with a hot (20-40 keV) multicomponent plasma consisting primarily of hydrogen, helium, oxygen, and sulfur ions which is moving in the corotation direction out to the magnetopause on the dayside (50-80 R$_J$) and to a distance of 100-150 R$_J$ on the nightside. Beyond $\sim$150 R$_J$ the plasma flow changes abruptly to a generally anti-solar direction and continues in that direction to $\gtrsim$ 360 R$_J$ at the 3 AM meridian. Hot plasma velocities in this region range from $\sim$300 to $\gtrsim$ 1000 km/sec, and the composition is similar to that in the inner part of the magnetosphere. An overall phenomenological model is presented and discussed.

Jupiter.   See Abstr. 003.072.

Weather and climate on planets.
See Abstr. 003.078.

The planet Jupiter: the observer's handbook.
See Abstr. 003.105.

Analysis of the 1980 - 81 apparition of Jupiter.
See Abstr. 003.118.

H$_2$ fluorescence spectrum from 1200 to 1700 Å by electron impact: laboratory study and application to Jovian aurora.   See Abstr. 022.017.

Io : cooling models for sulfur flows.
See Abstr. 022.046.

Laboratory modeling of sulfur flows on Io.
See Abstr. 022.047.

Impact cratering experiments in Bingham materials and the morphology of craters on Mars and Ganymede.
See Abstr. 022.119.

Observations of Jupiter with the astrolabe of the CERGA Observatory (January 1978 - May 1979).
See Abstr. 041.031.

General planetary theory extended to the case of resonance and application to the Galilean satellite system of Jupiter.   See Abstr. 042.016.

Multi-ion resonances in finite temperature plasma.
See Abstr. 062.015.

Radiative transfer with partial frequency redistribution in inhomogeneous atmospheres: application to the Jovian aurora.   See Abstr. 063.040.

Sub-mm heterodyne observations of the Sun, Moon, Jupiter and Orion at 691 GHz.   See Abstr. 077.012.

Overshoots in planetary bow shocks.
See Abstr. 084.015.

Review of the dynamics of satellites and planetary rings.   See Abstr. 091.069.

The internal evolution of Venus and the galilean satellites.   See Abstr. 093.043.

Albedo changes in lava.   See Abstr. 097.016.

Clay aggregates on earth, Mars and Io.
See Abstr. 097.020.

Observations de Mars et de Jupiter à l'astrolabe OPL No. 30 de l'Observatoire de Beijing.
See Abstr. 097.110.

The dynamics of close planetary satellites and rings.
See Abstr. 100.058.

Nodal regression of the Quadrantid meteor stream: an analytic approach.   See Abstr. 104.016.

Origins of the low-energy relativistic interplanetary electrons.   See Abstr. 106.023.

Origin of regular satellites.   See Abstr. 107.026.

On the origin and initial temperature of Jupiter and Saturn.   See Abstr. 107.034.

Cosmic rays from Jupiter.
See Abstr. 143.021.

100 Saturn, Saturn Satellites

100.001   **Charting the moons of Saturn – II.**
Sky Telesc., Vol. 63, 35 - 37 (1982).

100.002   **Voyager 2 encounter with the Saturnian system.**
E. C. Stone, E. D. Miner.
Science, Vol. 215, 499 - 504 (1982).

100.003   **A new look at the Saturn system: the Voyager 2
images.**   B. A. Smith, L. Soderblom, R. Batson,
P. Bridges, J. Inge, H. Masursky, E. Shoemaker, R. Beebe,
J. Boyce, G. Briggs, A. Bunker, S. A. Collins, C. J. Hansen,
T. V. Johnson, J. L. Mitchell, R. J. Terrile, A. F. Cook II,
J. Cuzzi, J. B. Pollack, G. E. Danielson, A. P. Ingersoll,
M. E. Davies, G. E. Hunt, D. Morrison, T. Owen, C. Sagan,
J. Veverka, R. Strom, V. E. Suomi.
Science, Vol. 215, 504 - 537 (1982).
Voyager 2 photography has complemented that of
Voyager 1 in revealing many additional characteristics of
Saturn and its satellites and rings.

100.004   **Photopolarimetry from Voyager 2: preliminary
results on Saturn, Titan, and the rings.**
A. L. Lane, C. W. Hord, R. A. West, L. W. Esposito,
D. L. Coffeen, M. Sato, K. E. Simmons, R. B. Pomphrey,
R. B. Morris.
Science, Vol. 215, 537 - 543 (1982).
The Voyager 2 photopolarimeter was reprogrammed
prior to the August 1981 Saturn encounter to perform
orthogonal-polarization, two-color measurements on Saturn,
Titan, and the rings. Saturn's atmosphere has ultraviolet limb
brightening in the mid-latitudes and pronounced polar darkening north of 65° N. Titan's opaque atmosphere shows strong
positive polarization at all phase angles (2.7° to 154°), and no
single-size spherical particle model appears to fit the data. A
single radial stellar occultation of the darkened, shadowed
rings indicated a ring thickness of less than 200 meters at
several locations and clear evidence for density waves caused
by satellite resonances.

100.005   **Infrared observations of the Saturnian system from
Voyager 2.**   R. Hanel, B. Conrath, F. M. Flasar,
V. Kunde, W. Maguire, J. Pearl, J. Pirraglia, R. Samuelson,
D. Cruikshank, D. Gautier, P. Gierasch, L. Horn,
C. Ponnamperuma.
Science, Vol. 215, 544 - 548 (1982).
During the passage of Voyager 2 through the Saturn
system, infrared spectral and radiometric data were obtained
for Saturn, Titan, Enceladus, Tethys, Iapetus, and the rings.
Combined Voyager 1 and Voyager 2 observations of temperatures in the upper troposphere of Saturn indicate a seasonal
asymmetry between the northern and southern hemispheres,
with superposed small-scale meridional gradients. Comparison
of high spatial resolution data from the two hemispheres poleward of 60° latitude suggests an approximate symmetry in the
small-scale structure, consistent with the extension of a
symmetric system of zonal jets into the polar regions.

100.006   **Extreme ultraviolet observations from the Voyager 2
encounter with Saturn.**   B. R. Sandel,
D. E. Shemansky, A. L. Broadfoot, J. B. Holberg, G. R. Smith,
J. C. McConnell, D. F. Strobel, S. K. Atreya, T. M. Donahue,
H. W. Moos, D. M. Hunten, R. B. Pomphrey, S. Linick.
Science, Vol. 215, 548 - 553 (1982).
Combined analysis of helium (584 angstroms) airglow
and the atmospheric occultations of the star δ Scorpii imply a
vertical mixing parameter in Saturn's upper atmosphere of K
(eddy diffusion coefficient) ~ 8×10^7 square centimeters per
second, an order of magnitude more vigorous than mixing in

Jupiter's upper atmosphere. Atmospheric H_2 band absorption
of starlight yields a preliminary temperature of 400 K in the
exosphere and a temperature near the homopause of ~ 200 K.

100.007   **Radio science with Voyager 2 at Saturn: atmosphere
and ionosphere and the masses of Mimas, Tethys,
and Iapetus.**   G. L. Tyler, V. R. Eshleman, J. D. Anderson,
G. S. Levy, G. F. Lindal, G. E. Wood, T. A. Croft.
Science, Vol. 215, 553 - 558 (1982).
Voyager 2 radio occultation measurements of Saturn's
atmosphere probed to the 1.2-bar pressure level, where the
temperature was 143 ± 6 K and the lapse rate apparently
equaled the dry adiabatic value of 0.85 K per kilometer. The
tropopause at both mid-latitude occultation locations
(36.5°N and 31°S) was at a pressure level of about 70 milli-
bars and a temperature of approximately 82 K. The stratospheric structures were very similar with the temperature rising
to about 140 K at the 1-millibar pressure level. Radio measurements of the masses of Tethys and Iapetus yield (7.55 ± 0.90)
$\times 10^{20}$ and (18.8 ± 1.2) $\times 10^{20}$ kilograms respectively; the
Tethys-Mimas resonance theory then provides a derived mass
for Mimas of (0.455 ± 0.054) $\times 10^{20}$ kilograms. These values
for Tethys and Mimas represent major increases from previously accepted ground-based values, and appear to reverse a
suggested trend of increasing satellite density with orbital
radius in the Saturnian system.

100.008   **Magnetic field studies by Voyager 2: preliminary
results at Saturn.**   N. F. Ness, M. H. Acuña,
K. W. Behannon, L. F. Burlaga, J. E. P. Connerney,
R. P. Lepping.
Science, Vol. 215, 558 - 563 (1982).
Further studies of the Saturnian magnetosphere and
planetary magnetic field by Voyager 2 have substantiated the
earlier results derived from Voyager 1 observations in 1980.
The magnetic field is primarily that of a centered dipole. No
evidence was found in the data or the analysis for any large-
scale magnetic anomaly in the northern hemisphere which
could be associated with the periodic modulation of
Saturnian kilometric radiation radio emissions.

100.009   **Plasma observations near Saturn: initial results from
Voyager 2.**   H. S. Bridge, F. Bagenal,
J. W. Belcher, A. J. Lazarus, R. L. McNutt, J. D. Sullivan,
P. R. Gazis, R. E. Hartle, K. W. Ogilvie, J. D. Scudder,
E. C. Sittler, A. Eviatar, G. L. Siscoe, C. K. Goertz,
V. M. Vasyliunas.
Science, Vol. 215, 563 - 570 (1982).
Results of measurements of plasma electrons and
positive ions made during the Voyager 2 encounter with
Saturn have been combined with measurements from
Voyager 1 and Pioneer 11 to define more clearly the configura-
tion of plasma in the Saturnian magnetosphere.

100.010   **Low-energy hot plasma and particles in Saturn's
magnetosphere.**   S. M. Krimigis, T. P. Armstrong,
W. I. Axford, C. O. Bostrom, G. Gloeckler, E. P. Keath,
L. J. Lanzerotti, J. F. Carbary, D. C. Hamilton, E. C. Roelof.
Science, Vol. 215, 571 - 577 (1982).
The low-energy charged particle instrument on
Voyager 2 measured low-energy electrons and ions
(energies ≳ 22 and ≳ 28 kiloelectron volts, respectively) in
Saturn's magnetosphere. The magnetosphere structure and
particle population were modified from those observed during
the Voyager 1 encounter in November 1980 but in a manner
consistent with the same global morphology.

100.011 Energetic charged particles in Saturn's magnetosphere: Voyager 2 results.   R. E. Vogt,
D. L. Chenette, A. C. Cummings, T. L. Garrard, E. C. Stone,
A. W. Schardt, J. H. Trainor, N. Lal, F. B. McDonald.
Science, Vol. 215, 577 - 582 (1982).
Results from the cosmic-ray system on Voyager 2 in
Saturn's magnetosphere are presented. During the inbound
pass through the outer magnetosphere, the $\geqslant$ 0.43-million-
electron-volt proton flux was more intense, and both the
proton and electron fluxes were more variable, than previously
observed. These changes are attributed to the influence on the
magnetosphere of variations in the solar wind conditions.

100.012 Planetary radio astronomy observations from
Voyager 2 near Saturn.   J. W. Warwick,
D. R. Evans, J. H. Romig, J. K. Alexander, M. D. Desch,
M. L. Kaiser, M. Aubier, Y. Leblanc, A. Lecacheux,
B. M. Pedersen.
Science, Vol. 215, 582 - 587 (1982).
Planetary radio astronomy measurements obtained by
Voyager 2 near Saturn have added further evidence that
Saturnian kilometric radiation is emitted by a strong dayside
source at auroral latitudes in the northern hemisphere and by
a weaker source at complementary latitudes in the southern
hemisphere. These emissions are variable because of Saturn's
rotation and, on longer time scales, probably because of
influences of the solar wind and Dione.

100.013 Voyager 2 plasma wave observations at Saturn.
F. L. Scarf, D. A. Gurnett, W. S. Kurth,
R. L. Poynter.
Science, Vol. 215, 587 - 594 (1982).
The first inbound Voyager 2 crossing of Saturn's bow
shock and the last outbound crossing had similar plasma wave
signatures. However, many other aspects of the plasma wave
measurements differed considerably during the inbound and
outbound passes, suggesting the presence of effects associated
with significant north-south or noon-dawn asymmetries, or
temporal variations.

100.014 Titan.   T. Owen.
Sci. American, Vol. 246, No. 2, p. 76 - 85 (1982).
The largest moon of Saturn is the only moon in the solar
system with a substantial atmosphere. The chemistry of the
atmosphere may resemble that of the earth's atmosphere
before life arose.

100.015 The moons of Saturn.
L. A. Soderblom, T. V. Johnson.
Sci. American, Vol. 246, No. 1, p. 72 - 75, 78 - 86 (1982).

100.016 Quelques photographies de Saturne, de ses
anneaux et de ses satellites.   J. Sauval.
Ciel Terre, Vol. 98, 13 - 20 (1982).

100.017 Le passage de Voyager 2 près de Saturne.
S. Dumont.
Astronomie, Vol. 96, 55 - 64 (1982).

100.018 The new Saturn system.   D. Morrison.
Mercury, Vol. 10, 162 - 181 (1981).

100.019 Saturnusmanen in kaart gebracht.
O. Namba.
Zenit, 9. Jaarg., 52 - 56 (1982).

100.020 Kleine maantjes van Saturnus.
O. Namba.
Zenit, 9. Jaarg., 100 - 105 (1982).

100.021 Letzte Live-Übertragung vom Saturn – ein Bildbe-
richt.   H. W. Köhler.
Phys. Bl., Vol. 38, 42 - 45 (1982).

100.022 A possible link between the rotation of Saturn and
its ring structure.
F. A. Franklin, G. Colombo, A. F. Cook.
Nature, Vol. 295, 128 - 130 (1982).
The authors present evidence that two previously unidentified, yet conspicuous gaps in Saturn's rings lie at
distances corresponding to 2/3 and 4/3 of the planet's rotation period. Gaps such as these can be produced in a ring of
large bodies or small uncharged particles only by a non-
axisymmetric gravitational field, a fact that is relevant to
models of planetary interiors.

100.023 Crater densities and geological histories of Rhea,
Dione, Mimas and Tethys.
J. B. Plescia, J. M. Boyce.
Nature, Vol. 295, 285 - 290 (1982).
Crater density determinations for parts of the surfaces of
several of Saturn's icy satellites including Rhea, Dione, Mimas,
Tethys and 1980S3 reveal significant variations. These data,
combined with observations of surface morphology, indicate
that each satellite has undergone extensive post heavy bom-
bardment geological evolution.

100.024 Variazioni ottiche delle dimensioni degli anelli di
Saturno.   E. Sassone Corsi, P. Sassone Corsi.
Astronomia, N. 3, p. 3 - 9 (1981).

100.025 Measurements of stratosphere aerosol on Saturn
using an eclipse of Titan.
D. W. Smith, R. W. Shorthill, P. E. Johnson, E. Budding,
A. S. Asaad.
News Lett. Astron. Soc. N. Y., Vol. 2, 29 (1982). – Abstract.

100.026 Saturns nye ringer og måner.   K. Aksnes.
Astron. Tidsskr., Årg. 15, 14 - 21 (1982).

100.027 Saturnmånene vi kjente fra för.   Ö. Hauge.
Astron. Tidsskr., Årg. 15, 22 - 26 (1982).

100.028 Implications of Voyager results for the history of
the Saturn system.   J. B. Pollack.
Reports of Planetary Geology Program – 1981, (see 003.001),
p. 3 - 4 (1981). – Abstract.

100.029 Crater densities of the Saturnian satellites: Rhea,
Dione, and Mimas.   J. B. Plescia, J. M. Boyce.
Reports of Planetary Geology Program – 1981, (see 003.001),
p. 5 - 6 (1981). – Abstract.

100.030 Crater densities of the Saturn satellites: Enceladus,
Iapetus, and Tethys.   J. B. Plescia, J. M. Boyce.
Reports of Planetary Geology Program – 1981, (see 003.001),
p. 7 - 9 (1981). – Abstract.

100.031 Thermal evolution of the icy Saturnian satellites.
G. Schubert, K. Ellsworth.
Reports of Planetary Geology Program – 1981, (see 003.001),
p. 10 - 12 (1981). – Abstract.

100.032 Voyager photometry of Saturn's satellites.
J. Veverka, P. Thomas, J. Gradie, T. V. Johnson,
D. Morrison.
Reports of Planetary Geology Program – 1981, (see 003.001),
p. 13 - 14 (1981). – Abstract.

100.033 Photometric analysis of the outer ring E of Saturn.
A. Dollfus.
Meteoritics, Vol. 16, 308 - 309 (1981). – Abstract.

100.034 Faint satellites of outer planets.   C. Veillet.
Messenger, No. 27, p. 25 - 29 (1982).

**100.035 Literal theory of the ninth satellite of Saturn,
Phoebe.   A. Bec-Borsenberger.**
Celestial Mech., Vol. 26, (see 012.001), 271 - 276 (1982). In
French.
The author studies a theory for the ninth satellite of
Saturn, Phoebe, based on the literal solution obtained in the
main problem of the lunar theory. These series were computed by solving, by successive approximations, the Lagrange's
equations. The series are computed to the complete seventh
order and a great part of the perturbations to the eighth and
ninth order.

**100.036 Longitudinal variability of methane and ammonia
bands on Saturn.**
A. L. Cochran, W. S. Cochran.
Icarus, Vol. 48, 488 - 495 (1981).
A total of 129 spectra of the center of the disk of Saturn
were obtained in March 1980 in order to search for possible
longitudinal variations in the CH_4 6190 and NH_3 6450 bands.
The spectra were reduced to reflectivities, and the band equivalent widths were measured using a blind, automated continuum fitting technique. The methane and ammonia equivalent
widths are well correlated with each other. There are some latitude regions in which both bands show enhanced absorption
and some latitudes in which both bands are weaker than the
average. The continuum reflectivity is quite constant with latitude and shows no apparent correlation with molecular band
equivalent width.

**100.037 Optical reflectance polarimetry of Saturn's globe
and rings. IV. Aerosols in the upper atmosphere of
Saturn.   R. Santer, A. Dollfus.**
Icarus, Vol. 48, 496 - 518 (1981).
From 1958 to 1976 the degree and direction of polarization of the light at Saturn's disk center were measured.
Measurements were also recorded at limb, terminator, and pole.
In addition, extensive regional polarization measurements were
collected over Saturn's disk and several polarization maps were
produced. These data were analyzed on the basis of Mie scattering theory and of transfer theory in planetary atmospheres.
A model of the Saturn upper atmosphere aerosol structure is
derived.

**100.038 Ringplanet Saturn. Unser Kenntnisstand 1982.
Teil 1, 2. H. W. Köhler.**
Sterne Weltraum, 21. Jahrg., 110 - 114, 154 - 156 (1982).

100.039 Zur Chemie der Titanatmosphäre.   R. Fröböse.
Sterne Weltraum, 21. Jahrg., 157 - 159 (1982).
The Voyager 2 UV spectrometer observed a stellar
occultation by the rings of Saturn, which located ring features
with an accuracy of 12 km. A high-resolution (3 km) optical
depth atlas of the rings shows at least nine features, including
four density wave patterns, identified with satellite resonances.
Analysis of these density wave patterns yields the first surface
mass densities for the A ring and, together with the authors'
optical depth atlas, a total ring mass of 6.4×10^{-8} Saturn
masses.

**100.040 Identification of resonance features within the rings
of Saturn.**
J. B. Holberg, W. T. Forrester, J. J. Lissauer.
Nature, Vol. 297, 115 - 120 (1982).

**100.041 Dynamical features in the northern hemisphere of
Saturn from Voyager 1 images.**
G. E. Hunt, D. Godfrey, J.-P. Müller, R. F. T. Barrey.
Nature, Vol. 297, 132 - 134 (1982).
Results of the analysis of Voyager measurements of

divergence and inferred vertical velocity of convective cloud
features in the northern hemisphere of Saturn are presented.
The winds in this region have been measured by tracking the
cloud elements observed in sequences of images. The derived
zonal winds imply that these convective features reside at the
maximum of strong easterly jets centred at 41°N planeto-
centric latitude and that the jet is barotropically unstable.

**100.042 Apparent thickness and scattering properties of
Saturn's rings from March 1980 observations.**
B. Sicardy, J. Lecacheux, P. Laques, R. Despiau, A. Auge.
Astron. Astrophys., Vol. 108, 296 - 305 (1982).
Ground-based observations of Saturn at the times of
the transits of the Sun and the Earth through the plane of
Saturn's rings are presented. The reduction of the electronographic plates indicate that, when observed and lit at low tilt
angle ($\lesssim 0°.5$), the rings are not well described as an homogeneous scattering layer. The analysis of the plates shows that a
residual flux remains in the edge-on configuration. This residual flux gives an equivalent thickness of the ring system:
$z_0 = 1.1 (+0.9/-0.5)$ km. The non-homogeneous behaviour of
the rings, the residual flux, as well as a transitory asymmetry
of brightness between the rings' ansae may be interpreted as
a result of a wide size-distribution of particles in the rings, up
to diameters of a few kilometers.

**100.043 Viscous origin of Saturn's ring structure.
F. C. Michel.**
Astrophys. Lett., Vol. 22, 101 - 102 (1982).
The author proposes that Saturn's disk is subdivided into
numerous fine ringlets as a direct result of internal disk viscosity. The viscosity automatically leads to separation of ringlets
at the edge of an initially uniform disk, leading to the decomposition of the disk into ringlets and the consequent removal
of viscosity as an important evolutionary factor.

100.044 De ringen van Saturnus.   O. Namba.
Zenit, 9. Jaarg., 196 - 201 (1982).

**100.045 New interpretation of the Saturn rings.
V. D. Davydov.**
Inst. kosm. issled. AN SSSR. Prepr., 1981, No. 645, 52 pp.
In Russian. – Abstr. in Ref. zh., 62. Issled. kosm.
prostranstva, 3.62.350 (1982).

**100.046 Voyager 2 at Saturn.
R. Berry, R. Burnham.**
Astronomy, Vol. 9, No. 11, p. 6 - 30 (1981). – Abstr. in Phys.
Abstr., Vol. 85, Abstr. 48179 (1982).

**100.047 Titan's highly variable plasma environment.
D. A. Wolf, F. M. Neubauer.**
J. Geophys. Res., Vol. 87, 881 - 885 (1982).
The authors have calculated the location of the stagnation point of Saturn's magnetosphere $r_{st} = 16 - 26\, R_S$
assuming a terrestrial type magnetosphere. Typical plasma
parameters along the orbit of Titan are shown for high solar
wind pressure. The magnetoplasma incident on Titan will be
generally be subsonic, subalfvénic in the tail, transsonic,
transalfvénic in the outer magnetosphere, and subsonic, super-
alfvénic in the magnetosheath. During crossings of the
Saturnian magnetopause or bow shock by Titan the authors
expect abrupt changes of the flow direction and stagnation
pressure and rapid associated changes in Titan's uppermost
atmosphere.

**100.048 Titan's upper atmosphere: composition and tem-
perature from the EUV solar occultation results.**
G. R. Smith, D. F. Strobel, A. L. Broadfoot, B. R. Sandel,
D. E. Shemansky, J. B. Holberg.
J. Geophys. Res., Vol. 87, 1351 - 1359 (1982).
The temperature and composition of the upper atmo-

sphere of Titan have been inferred by observing an occultation of the sun by Titan, using the Voyager 1 ultraviolet
spectrometer. The temperature is 176 ± 20 K near the evening
terminator and 196 ± 20 K near the morning terminator. The
major constituent is N_2 with a density of $2.7 \pm 0.2 \times 10^8$ cm^{-3}
at 3840 km. The mixing ratio of CH_4 is $8 \pm 3\%$ at a radial
distance of 3700 km near the evening terminator. The
acetylene mixing ratio above 3400 km is measured at the 1 to
2% level. A layer of absorbing molecules, possibly polymers,
is present at both morning and evening terminators. A simple
photochemical model is discussed.

100.049 EUV emission from Titan's upper atmosphere: Voyager 1 encounter.
D. F. Strobel, D. E. Shemansky.
J. Geophys. Res., Vol. 87, 1361 - 1368 (1982).
The authors describe observations and analysis of the
Voyager 1 EUV emission spectra of Titan. The observations in
the 500-Å-1700-Å region are dominated by the emission spectrum of electron-excited N_2. The spectra contain other
emitters, hydrogen in particular, and features that have not
been positively identified. The authors examine the N_2 emission in detail and discuss the other features briefly.

100.050 The induced magnetosphere of Titan.
N. F. Ness, M. H. Acuna, K. W. Behannon,
F. M. Neubauer.
J. Geophys. Res., Vol. 87, 1369 - 1381 (1982).
The Voyager 1 spacecraft had a close encounter (miss
distance = 6970 km) with Titan (diameter = 5140 km) on
November 12, 1980, while this large satellite was located within the Saturnian magnetosphere at a local solar time of 1330.
No clear evidence was found for any intrinsic magnetic field
nor for the development of a bow shock wave as the corotating
Saturnian magnetoplasma convected past Titan. However, a
strong electrodynamic interaction was evidenced with the
observation of a very well developed, induced bipolar magnetic
tail. The paper presents the results of the analysis of the
magnetic field data, which suggest an interpretation qualitatively described by the draping of the Saturnian magnetic
field around the ionosphere of Titan.

100.051 Titan's ion exosphere observed from Voyager 1.
R. E. Hartle, E. C. Sittler, Jr., K. W. Ogilvie,
J. D. Scudder, A. J. Lazarus, S. K. Atreya.
J. Geophys. Res., Vol. 87, 1383 - 1394 (1982).
The authors present an interpretation of the ion and
electron observations made close to Titan by the plasma
science instrument on Voyager 1. Based upon the idea that
the interaction between Titan and the rotating magnetospheric plasma of Saturn closely resembles that between
Venus and the solar wind, simple models of ion pickup in the
ion exosphere outside Titan's magnetic tail and ion flow
within the tail have been used. The ion observations have been
compared with synthetic spectra constructed from both
cycloidal and Maxwellian distribution functions. Present
knowledge of the atmosphere is used to estimate the minimum
ionopause radius, and the neutral exosphere is represented by
a spherically symmetric model composed of H and N_2.

100.052 The structure of Titan's wake from plasma wave observations.
D. A. Gurnett, F. L. Scarf, W. S. Kurth.
J. Geophys. Res., Vol. 87, 1395 - 1403 (1982).
The authors present a detailed analysis of the plasma
wave emissions detected by the Voyager 1 plasma wave
instrument in the vicinity of Titan.

100.053 Planetary radio astronomy observations during the Voyager 1 Titan flyby.
G. Daigne, B. M. Pedersen, M. L. Kaiser, M. D. Desch.
J. Geophys. Res., Vol. 87, 1405 - 1409 (1982).
During the Voyager 1 Titan flyby, unusual radio emissions were observed by the planetary radio astronomy
experiment in the 20- to 97-kHz frequency range. The authors
show that Titan itself is not the source of the observed radio
emission. Rather, they attribute the emission features to
modification of the normal Saturn kilometric radiation by
propagation effects in enhanced density structures within the
Titan wake. Furthermore, spiky emission observed in the
magnetic wake of Titan are interpreted in terms of local electrostatic instabilities at the electron plasma frequency. From
these measurements the authors derive a range of electron
densities in the wake region and they discuss the consistency
of the results.

100.054 Effects of Titan on trapped particles in Saturn's magnetosphere. C. G. Maclennan, L. J. Lanzerotti,
S. M. Krimigis, R. P. Lepping, N. F. Ness.
J. Geophys. Res., Vol. 87, 1411 - 1418 (1982).
The authors present the results of an analysis of the low-
energy charged particles measured by Voyager 1 as it passed
close by Titan on November 12, 1980.

100.055 An outer ring to Saturn and an orbit to Janus.
A. Dollfus.
Philos. Trans. R. Soc. London A, Vol. 303, (see 012.022),
281 - 283 (1981).
New results emerging from observations of Saturn in
1979 - 1980, when the rings were seen edge on, are reported.

100.056 415 μm brightness temperature of Titan.
R. F. Loewenstein, R. H. Hildebrand.
Astron. Astrophys., Vol. 110, L18 - L19 (1982).
Broadband submillimeter measurements indicate that the
415 μm brightness temperature of Titan is 68 ± 8.5 K if the
observed radiation originates from a layer 2600 km in radius.
This temperature lies within 1 standard deviation below the
minimum tropopause temperature determined by Voyager.

100.057 Motions in Saturn's atmosphere: observations before Voyager encounters. A. Sanchez Lavega.
Icarus, Vol. 49, 1 - 16 (1982).
A summary is presented of all the telescopic observations
of the clouds in Saturn's atmosphere made prior to the
Voyager encounters with the planet. Saturn displays a pattern
of belts and zones like Jupiter, although they are practically
constant in latitude and without significant changes over the
last century. The most obvious features of Saturn are the great
white spots (GWS) with diameters of 20,000 km or more. A
possible subatmospheric origin for the GWS is proposed.

100.058 The dynamics of close planetary satellites and rings.
K. Aksnes.
Application of modern dynamics to celestial mechanics and
astrodynamics, (see 012.026), p. 1 - 20 (1982).
The observational history and the evolving dynamical
theories are reviewed for the rings of Saturn, Uranus, and
Jupiter with particular emphasis on very recent results from
the Voyager space missions. An account is then given of three
Jovian satellites and eight Saturnian ones discovered from
ground or space in 1979 and 1980. The coupled motions of
some of these satellites and their possible interactions with
the rings are explained in some detail.

100.059 The motion of Saturn's co-orbiting satellites 1980S1 and 1980S3. G. Colombo.
Application of modern dynamics to celestial mechanics and
astrodynamics, (see 012.026), p. 21 - 23 (1982).
The coorbiting pair 1980S1 and 1980S3 are the most
unusual Trojan-like objects, with mean orbital radius of
151×10^3 km and angular motion of $518°.3$/day. They have,
unlike other 1 : 1 librators, comparable masses. Both masses
are, however, very small compared with Saturn's mass M. The

most unusual feature of this motion is the large amplitude of
the librations of both satellites with respect to the mean orbit.
An available code for studying the dynamics of three-body
systems, Saturn, S1 and S3, was used.

100.060 **News about Saturn.**   D. Yu. Gol'dovskij.
Zemlya Vselennaya, 1982, No. 3, p. 34 - 36. In
Russian.

100.061 **Recent data on Saturn.**   P. Lála.
Říše hvězd, Vol. 63, 74 - 76, 81 (1982). In Czech.

100.062 **Voyager 2 and Saturn.**   P. Lála.
Vesmír, Vol. 61, 18 - 22 (1982). In Czech.

100.063 **Satellites of Saturn.**
IAU Circ., Nos. 3656, 3660, 3685, 3695, 3707
(1982).

100.064 **The case for a bimodal size distribution in Titan's
upper haze layer.**   P. H. Smith.
The comparative study of the planets, (see 012.033), p. 253 -
260 (1982).
   Recent flybys of Titan have provided strong evidence
that the upper haze layer, or layers, on Titan are both highly
polarizing and forward scattering. Spherical particles tend to
have either one quality or the other but not both. A search of
prolate spheroids has been carried out and a class of particles
defined in which large particles (x = 3.4) can be highly
polarizing. It is found that even these particles cannot explain
the spacecraft measurements as large phase angles. A second
nucleation mode of tiny aerosols is proposed which can
provide the back scattering necessary to match the geometric
albedo. Without this second size mode the particles must be
so bright that the polarization is lost through multiple scattering.

100.065 **New geometric albedos of Titan and Neptune.**
G. W. Lockwood, B. L. Lutz, D. T. Thompson.
Bull. American Astron. Soc., Vol. 13, 839 (1981). — Abstract.

100.066 **Spectrophotometry of Saturn from 60 to 180
microns.**
G. Melnick, R. W. Russell, T. R. Gosnell, M. Harwit.
Bull. American Astron. Soc., Vol. 13, 874 (1981). — Abstract.

100.067 **Zonal harmonic model of Saturn's magnetic field
from Voyager 1 and 2 observations.**
J. E. P. Connerney, N. F. Ness, M. H. Acuña.
Nature, Vol. 298, 44 - 46 (1982).
   An axisymmetric octupole model of Saturn's planetary
magnetic field is proposed. This three parameter model,
characterized by the Schmidt-normalized spherical harmonic
coefficients $g_1^0 = 21,535$ nT, $g_2^0 = 1,642$ nT and $g_3^0 = 2,743$ nT,
is extremely efficient in representing the main magnetic field
of Saturn and reconciling the in situ magnetic field observations obtained by Pioneer 11 with those obtained by the
Voyager 1 and 2 spacecraft.

100.068 **Perturbations in the motion of Saturn satellites.
Part 4.**   I. G. Chugunov.
Mordov. univ. Saransk. 1981, 14 pp. In Russian. — Abstr. in
Ref. zh., 51. Astron., 5.51.89 (1982).

100.069 **Narrow-band absolute spectrophotometry of
Saturn at λ 0.6 - 1.1 μm from 1980 observations.**
K. S. Kuratov.
Astrofiz. inst. AN KazSSR. Alma-Ata, 1982. 55 pp. In
Russian. — Abstr. in Ref. zh., 51. Astron., 5.51.213 (1982).

100.070 **Detection of a cloud surrounding Saturn's A and
B rings.**   W. A. Baum, T. J. Kreidl.
Bull. American Astron. Soc., Vol. 14, 575 (1982). — Abstract.

100.071 **The spokes in Saturn's rings: a new approach.**
J. F. Carbary, P. F. Bythrow, D. G. Mitchell.
Geophys. Res. Lett., Vol. 9, 420 - 422 (1982).
   The authors propose that zonal winds in Saturn's atmosphere cause superrotation of the ionosphere and by virtue of
field-aligned currents generate a potential of ~16 kV across the
B-ring. Such a potential can momentarily polarize ~1-10 μm
size ice particles in a radial sense and thus create the spectacular
spokes. A resulting current flowing radially through the B-ring
would give rise to a power of ~10^7 W, which is close to that
dissipated in an SED burst.

100.072 **On the nature of particles in Saturn's spokes.**
M. F. Thomsen, C. K. Goertz, T. G. Northrop,
J. R. Hill.
Geophys. Res. Lett., Vol. 9, 423 - 426 (1982).
   The authors adopt the hypothesis of Hill and Mendis
(1981) that the spoke particles are electrostatically repelled
from larger parent bodies, and within that context the authors
pursue several lines of reasoning that enable one to place
constraints on important properties of the spokes, specifically
the size of the spoke particles and the surface potential of the
parent bodies.

100.073 **Volcanism and igneous processes in small icy
satellites.**   D. J. Stevenson.
Nature, Vol. 298, 142 - 144 (1982).
   The small saturnian satellites exhibit a remarkable
diversity of surficial features. The author argues that the most
likely cause of endogenic processes in these bodies is the
igneous activity associated with a low melting point NH_3-H_2O
magma. Radiogenic heating probably suffices to provide the
necessary melting in Tethys, Dione, Rhea and Iapetus whereas
tidal heating is needed for Enceladus and possibly Mimas.

100.074 **Voyager radio studies of the Saturn system, and
selected interpretations.**   R. Eshleman.
Trans. American Nucl. Soc., Vol. 39, 66 (1981). — Abstr. in
Phys. Abstr., Vol. 85, Abstr. 58303 (1982).

100.075 **On the excess thermal fluxes of Titan and Saturn.**
E. M. Drobyshevski (*Eh. M. Drobyshevskij*).
Moon Planets, Vol. 26, 33 - 46 (1982).
   The assumption of a recent (3.5—10 thousand years ago)
explosion of the electrolysis products of Titan's ices suggests a
common explanation for many peculiar features of Saturn's
system. In particular, the high excess luminosity of Saturn is
due, possibly, to an accretion of a part of the material lost by
Titan. The accretion of this material from Saturn's rings is
continuing at present. The crucial experiment to support the
possibility of a recent Titan's ice explosion would be detection of excess heat flux.

100.076 **Saturn's ringlets.**   J. A. Bastin, D. H. Smith.
Moon Planets, Vol. 26, 97 - 100 (1982).
   This paper suggests that Saturn's magnetic field is, in
part, responsible for the very fine-scale radial features, or
ringlets, seen in the ring-system. The planet's dipole field
interacts with slight radial variations in plasma density, and
the operation of an instability segregates the magnetic flux
and plasma in the ring-plane into narrow alternating zones.
The authors suggest that this mechanism may act by itself to
give rise to the inner ringlets. At greater radial distances they
believe it amplifies gravitational resonances.

100.077 **The isolated non-circular ringlets of Saturn.**
J. R. Hill, D. A. Mendis.
Moon Planets, Vol. 26, 217 - 226 (1982).

100.078   Evidence for the ordered orientation of small bodies in the Saturn rings and for the identity of "spokes" from the appearance of magnetic anomalies.
V. D. Davydov.
Kosm. Issled., Tom 20, 460 - 471 (1982). In Russian.

The authors show that the combined effect of electrodynamic and gravitational forces can account for a number of features observed by Voyagers 1 and 2 in the isolated fine dust rings of Saturn. These include (a) the appearance and disappearance of the braids in the F-ring, (b) the eccentricities of the F-ring and the ringlets within the Encke and Cassini divisions and a gap in the C-ring, and (c) the kinks in the eccentric Encke ring. They may also account for the very existence of these rings.

100.079   Saturnus' hoofdringen. Symfonie uit een nieuwe wereld.   O. Namba.
Zenit, 9. Jaarg., 318 - 326 (1982).

100.080   On the lifetime of E-ring grains and their nature.
A. Cheng, L. J. Lanzerotti, V. Pirronello.
Sun and planetary system, (see 012.040), p. 251 - 252 (1982).

100.081   Braided rings of Saturn – their topology.
R. Pratap.
J. Astrophys. Astron., Vol. 3, 5 - 11 (1982).

The general topology of the braiding in Saturn's F ring is explained by invoking the theory of invariant surfaces to which a plasma would confine itself. This surface, in the framework of self-consistent fields, is indeed generated by two helicoids turning in opposite directions and braided.

100.082   Observation and photometry of an outer ring of Saturn.   A. Dollfus, S. Brunier.
Icarus, Vol. 49, 194 - 204 (1982).

A faint outer ring (E ring), which lies outside the classical rings A, B, C, and F, has been detected out to eight Saturn radii. The authors first observed it on November 1, 1979, and thereby confirmed the 1966 observation by Feibelman. Photometry of the edge-on ring E lineament shows a strong brightness increase at small phase angles, which is compatible with scattering by particles of several microns in radius. The excess reflectivity in blue compared to the B ring implies a significant contribution of small particles in the scattering process. The E ring is probably a flat structure with a condensation centered at a distance of $4\,R_s$, but without a simple axial symmetry. It is probably shaped by segments or lumps and may have streamerlike structures.

100.083   Far-infrared brightness temperature of Saturn's disk and rings.   R. R. Daniel, S. K. Ghosh,
K. V. K. Iyengar, T. N. Rengarajan, S. N. Tandon, R. P. Verma.
Icarus, Vol. 49, 205 - 212 (1982).

The authors present far-infrared observations of Saturn in the wavelength band 76–116 μm, using a balloon-borne 75-cm telescope launched on 10 December 1980 from Hyderabad, India. Normalizing with respect to Jupiter, they find the average brightness temperature of the disk-ring system to be 90 ± 3°K. Correcting for the contribution from rings using experimental information on the brightness temperature of rings at 20 μm, they find the brightness temperature of the disk, to be 96.9 ± 3.5° K.

100.084   Spatially resolved reflectivities of Saturn: 3000 - 6000 Å.   W. D. Cochran.
Astron. J., Vol. 87, 718 - 723 (1982).

Spectrophotometric observations of seven spatially resolved regions on the disk of Saturn from 3000 to 6000 Å are reported. The data show an asymmetry between the northern and southern hemispheres in the equivalent width of the CH_4 5430-Å band and in the continuum reflectivity. These spectra do not contain the peak in reflectivity near 4700 Å

present in the results of Bergstralh et al. (1981). The data exhibit general agreement with the model by Podolak and Danielson (1977).

100.085   The periodic variation of spokes in Saturn's rings.
C. C. Porco, G. E. Danielson.
Astron. J., Vol. 87, 826 - 833 (1982).

The discovery of a periodic variation in spoke activity in Saturn's rings from the analysis of Voyager images is reported. Maximum spoke activity is most likely to be observed on the morning half of the rings when a particular magnetic field sector coincides with this area. The results suggest that the fundamental period of spoke variation is that of Saturn's magnetic field, and that spoke activity is associated with the region of the field which gives rise to the Saturn kilometric radiation. Passage of this region through Saturn's shadow may play a significant role in the creation and/or rejuvenation of spokes.

100.086   Saturn.
British Astron. Assoc. Circ., No. 619 (1981).

100.087   Saturn central meridian ephemeris: 1982.
J. E. Westfall.
Strolling Astron., Vol. 29, 102 - 105, 107, 109 (1982).

Voyages to Saturn.   See Abstr. 003.097.

Saturn briefing.   See Abstr. 011.028.

Theory of radio occultation by Saturn's rings.
See Abstr. 031.629.

The adiabatic invariant: its use in celestial mechanics.
See Abstr. 042.054.

Saturn's rings and bimodality of Keplerian systems.
See Abstr. 042.109.

Overshoots in planetary bow shocks.
See Abstr. 084.015.

The gravito-electrodynamics of charged dust in planetary magnetospheres.   See Abstr. 091.060.

Review of the dynamics of satellites and planetary rings.   See Abstr. 091.069.

The far-ultraviolet spectra and geometric albedos of Jupiter and Saturn.   See Abstr. 099.013.

The control networks of the satellites of Jupiter and Saturn.   See Abstr. 099.032.

Voyager cartography.   See Abstr. 099.033.

Examination of Le Verrier and Gaillot's theories of Jupiter and Saturn.   See Abstr. 099.045.

A comparison of the atmospheres of Jupiter and Saturn.   See Abstr. 099.050.

Atmospheres of Jupiter and Saturn.
See Abstr. 099.061.

Voyager spacecraft images of Jupiter and Saturn.
See Abstr. 099.081.

Magnetospheres of Jupiter and Saturn.
See Abstr. 099.087.

Conversion of para and ortho hydrogen in the
Jovian planets.   See Abstr. 099.088.

Astrometric observations of the satellites of the
outer planets. V. The oppositions of 1978 - 1979, 1980, and
1981.   See Abstr. 099.089.

Formation and collisional evolution of small bodies:
effects of two-material systems on large-scale geologic
structure.   See Abstr. 107.004.

Origin of regular satellites.
See Abstr. 107.026.

On the origin and initial temperature of Jupiter and
Saturn.   See Abstr. 107.034.

Errata

100.901 Erratum: "Measurement of stratospheric aerosol on
         Saturn using an eclipse of Titan" [Icarus, Vol. 46,
424 - 428 (1981)].   D. W. Smith, R. W. Shorthill,
P. E. Johnson, E. Budding, A. S. Asaad.
Icarus, Vol. 48, 540 (1981). – See Abstr. 30.100.042.

100.902 Erratum: "The phosphine abundance on Saturn
         derived from new 10-micrometer spectra" [Icarus,
Vol. 42, 79 - 85 (1980)].
A. T. Tokunaga, H. L. Dinerstein, D. F. Lester, D. M. Rank.
Icarus, Vol. 48, 540 (1981). – See Abstr. 27.100.051.

100.903 Erratum: "Saturn's E ring. I. CCD observations of
         March 1980" [Icarus, Vol. 47, 84 - 96 (1981)].
W. A. Baum, T. Kreidl, J. A. Westphal, G. E. Danielson,
P. K. Seidelmann, D. Pascu, D. G. Currie.
Icarus, Vol. 48, 540 (1981). – See Abstr. 30.100.058.

101 Uranus, Neptune, Pluto, Transplutonian Planets

101.001 Occultation by a possible third satellite of Neptune.
         H. J. Reitsema, W. B. Hubbard, L. A. Lebofsky,
D. J. Tholen.
Science, Vol. 215, 289 - 291 (1982).
    The 24 May 1981 close approach of Neptune to an
uncataloged star was photoelectrically monitored from two
observatories separated by 6 kilometers parallel to the
occultation track. An 8.1-second drop in signal, recorded
simultaneously at both sites, is interpreted as resulting from
the passage of a third satellite of Neptune in front of the star.
From the duration of the event, the derived minimum diameter for an object sharing Neptune's motion is 180 kilometers.
If the object was in Neptune's equatorial plane and there are
no significant errors in the prediction ephemeris, the object
was located at a distance of 3 Neptune radii from Neptune's
center.

101.002 Photometric study of Pluto near the perihelion. I.
         UBV photometry.
V. M. Lyutyj, V. P. Tarashchuk.
Pis'ma Astron. Zh., Tom 8, 109 - 114 (1982). In Russian.
English translation in Soviet Astron. Lett., Vol. 8.
    Results of photoelectric UBV observations of Pluto are
given. Significant variability of the color index $U-B$ is found.
There is an UV excess near the minimum light. Eclipsing
phenomena in the binary Pluto-Charon system seem to have
been observed. The changes of the shape of the light curve and
the $U-B$ variations are discussed.

101.003 Observations of Uranus made with the Danjon
         Astrolabe of Santiago, Chile, during 1979.
F. Noël, S. Barros.
Astron. Astrophys.,Suppl. Ser., Vol. 47, 481 - 482 (1982).
    This paper gives the results of the observations of Uranus
made with the Danjon Astrolabe of Santiago,Chile, during
1979. The residuals in zenith distance of the planet, the
number of the fundamental stars group to which the residuals
are referred as well as its weight are given among other
pertinent data.

101.004 Radial widths, optical depths, and eccentricities of
         the Uranian rings.   P. D. Nicholson,
K. Matthews, P. Goldreich.

Astron. J., Vol. 87, 433 - 447 (1982).
    Observations of the stellar occultation by the Uranian
rings of 15/16 August 1980 are used to estimate radial widths
and normal optical depths for segments of rings 6, 5, 4, α, β,
η, γ, and δ. A review of published data confirms the existence
of width-radius relations for rings α and β, and indicates that
the optical depths of these two rings vary inversely with their
radial widths. Masses are obtained for rings α and β. Elliptical
models for rings 6, 5, 4, α, β, and ϵ are presented.

101.005 New ring signals beyond the ϵ Uranian ring.
         D. Chen, Z. Wu, X. Yang, Z. Xie, Q. Cheng,
S. Jiang.
Kexue Tongbao, Vol. 26, 627 - 628 (1981). – Abstr. in Phys.
Abstr., Vol. 85, Abstr. 10354 (1982).

101.006 Wie kalt ist Pluto?
         W. J. Altenhoff, H. J. Wendker.
Mitt. Astron. Ges., Nr. 55, p. 18 (1982). – Abstract.

101.007 Positions of Uranus and 4 satellites in 1980 obtained
         at ESO La Silla.
H. Debehogne, R. R. Freitas Mourao, G. Vieira.
Acta Astron., Vol. 31, 391 - 392 (1981).
    34 positions of Uranus and satellites Ariel, Umbriel,
Titania, Oberon are given as obtained from photographic obser-
vations made at ESO La Silla, in May 1980, with the Astro-
graph GPO.

101.008 On the D/C ratio in the atmosphere of Uranus.
         T. Encrenaz, M. Combes.
Icarus, Vol. 49, 27 - 34 (1982).
    A method for deriving mixing ratios in the outer planets,
mostly independent of scattering processes, is applied to
Uranus. It is shown that scattering processes play a major role
in the line formation in the atmospheres of Uranus and
Neptune; consequently, abundance ratios derived from a
reflecting-layer model can be questionable. The authors derive
for Uranus D/C $< 6 \times 10^{-3}$, which is significantly smaller than
their result on Jupiter. The simplest explanation implies a
C/H enrichment by at least a factor of 6 relative to the solar
value.

101.009   Polarimetry of Pluto.
        M. Breger, W. D. Cochran.
Icarus, Vol. 49, 120 - 124 (1982).
        The polarization of Pluto has been measured for a range of solar phase angles from 0.8 to 1.8°. A mean linear polarization of 0.29 ± 0.01 % (error of the mean) was found. No dependence of both the amount of polarization and position angles with rotational phase or solar phase angle could be detected. The positional angles of polarization agree with calculated position angles of the defect of illumination and are therefore parallel to the scattering plane. The observed polarization cannot be explained as resulting purely from a surface material which is similar to asteroidal surfaces. A hypothesis of polarization from a thin atmosphere, in addition to the surface polarization, is advanced.

101.010   The 15 August 1980 occultation by the Uranian system: structure of the rings and temperature of the upper atmosphere.   B. Sicardy, M. Combes, A. Brahic, P. Bouchet, C. Perrier, R. Courtin.
ESO Sci. Prepr. No. 194, 36 pp. (1982). – Submitted to Icarus.

101.011   Occultation of a 15th magnitude star by Pluto/ Charon on 1983 April 24.   G. E. Taylor.
Bull. No. 27, IAU Comm. 20, Working Group on Predictions of Occultations by Satellites and Minor Planets, 1 pp. (1982).

101.012   The effect of Triton's mass on the determination of the J_2 of Neptune.   A. W. Harris.
Bull. American Astron. Soc., Vol. 14, 580 (1982). – Abstract.

101.013   Gravitational resonances and the rings of Uranus.
        G. A. Steigmann.
J. British Astron. Assoc., Vol. 92, 157 - 163 (1982).
        The close association between the orbital radii of seven of the nine known narrow rings encircling Uranus, and the gravitational resonances generated by its satellites, suggests that resonance theory might be applied to this system. If the radii of the rings are determined by resonances then the β and δ rings indicate the probable existence of an as yet undiscovered sixth satellite of Uranus. This satellite is predicted to move in a near-circular orbit of radius 105 221 km with a period of 1 024 days and to possess a diameter of approximately 220 km.

101.014   The pre-discovery observations of Uranus.
        E. G. Forbes.
Uranus and the outer planets, (see 012.036), p. 67 - 80 (1982).

101.015   The origin of Uranus: compositional considerations.
        M. Podolak.
Uranus and the outer planets, (see 012.036), p. 93 - 109 (1982).
        Several cosmogonic theories are examined for their ability to explain the details of Uranus' composition as inferred from observations and interior models. Suggestions are made as to how future work may enable us to decide among competing scenarios.

101.016   Internal structure of Uranus.
        J. J. MacFarlane, W. B. Hubbard.
Uranus and the outer planets, (see 012.036), p. 111 - 124 (1982).
        The authors present an updated study of Uranus interior models using current information about the planet's gravity field and rotation rate. The most plausible model, both from the point of view of recent data and cosmogony, has a central core of iron and magnesium silicates, an outer envelope of liquid water, methane, and ammonia, and a deep "atmosphere" of almost four earth masses of hydrogen, helium, and methane. The "atmosphere" contains a gravitationally nonnegligible amount of methane – about 40% by mass. All plausible

models are most consistent with a rotation period of ~15 to 16 hours.

101.017   The magnetosphere of Uranus.   W. I. Axford.
        Uranus and the outer planets, (see 012.036), p. 125 - 142 (1982).

101.018   The rotation of Uranus.   R. M. Goody.
        Uranus and the outer planets, (see 012.036), p. 143 - 153 (1982).
        The weighted mean of the author's data and that obtained from the planetary figure is $T = 16.31 \pm 0.27\,h$.

101.019   An introductory review of our present understanding of the structure and composition of Uranus' atmosphere.   M. J. S. Belton.
Uranus and the outer planets, (see 012.036), p. 155 - 171 (1982).

101.020   Imaging of Uranus and Neptune.
        B. A. Smith, H. J. Reitsema.
Uranus and the outer planets, (see 012.036), p. 173 - 179 (1982).

101.021   Meteorology of the outer planets.   G. E. Hunt.
        Uranus and the outer planets, (see 012.036), p. 181 - 191 (1982).

101.022   The satellites of Uranus.   D. P. Cruikshank.
        Uranus and the outer planets, (see 012.036), p. 193 - 210 (1982).
        The Uranian satellite system contains five known members, all of which are difficult to study owing to their faintness and proximity to Uranus. The photometry of these objects is not in a satisfactory state, nor is the photovisual spectrophotometry. Near infrared work reveals water ice or frost on the satellite surfaces, perhaps in a very pure state. The satellites are most similar to Ganymede in terms of the strength of the ice bands. Most of the data and theories are consistent with bodies with radii in the range 160 - 520 km (similar to the larger asteroids) and albedos on the order of 0.5, consistent with ice and snow. The mean densities are probably similar to those of the icy Saturn satellites, about 1.3 g/cm³.

101.023   The rings of Uranus.   A. Brahic.
        Uranus and the outer planets, (see 012.036), p. 211 - 236 (1982).

101.024   Rings of Uranus: a review of occultation results.
        J. L. Elliot.
Uranus and the outer planets, (see 012.036), p. 237 - 256 (1982).

101.025   Uranus among the outer planets.   T. Owen.
        Uranus and the outer planets, (see 012.036), p. 295 - 302 (1982).

101.026   Au-delà de Saturne.   R. Dejaiffe.
        Rev. Quest. Sci., Tome 153, No. 1, p. 11 - 30 (1982).

        Il y a 200 ans: une nouvelle planète.
See Abstr. 004.008.

        La scoperta di Plutone.   See Abstr. 004.070.

        Herschel's scientific apprenticeship and the discovery of Uranus.   See Abstr. 004.076.

        Uranus science with Space Telescope.
See Abstr. 032.540.

The Voyager encounter with Uranus.
See Abstr. 051.016.

Review of the dynamics of satellites and planetary rings. See Abstr. 091.069.

Possible occultation by Pluto on 1982 April 15.
See Abstr. 096.006.

Occultation by Uranus and its rings.
See Abstr. 096.007.

Occultation of Uranus and his rings by BD −20° 51615 (9ᵐ9). See Abstr. 096.008.

Conversion of para and ortho hydrogen in the Jovian planets. See Abstr. 099.088.

Faint satellites of outer planets.
See Abstr. 100.034.

The dynamics of close planetary satellites and rings.
See Abstr. 100.058.

New geometric albedos of Titan and Neptune.
See Abstr. 100.065.

Erratum

101.901   Erratum: 'Analytical two-dimensional model for a pole-on magnetosphere' [Planet. Space Sci., Vol. 29, 1101 - 1107 (1981)].
H. Biernat, N. Kömle, H. Rucker.
Planet. Space Sci., Vol. 30, 313 (1982). − See Abstr. 30.101. 012.

102 Comets (Origin, Structure, Atmospheres, Dynamics)

102.001   **On the folding phenomenon of comet tail rays.**
A. I. Ershkovich.
Mon. Not. R. Astron. Soc., Vol. 198, 297 - 302 (1982).
    The folding phenomenon of the comet tail rays is shown to be compatible with the Ferraro isorotation law if the comet tail magnetic field has no azimuthal component: $B_\phi = 0$. The magnetic field B of about 20 - 40 γ in the coma and $B \lesssim 10\,\gamma$ in the distant tail is estimated under typical solar wind conditions at 1 AU.

102.002   **Role of high frequency turbulence in cometary plasma tails.**   B. Buti.
Astrophys. J., Lett., Vol. 252, L43 - L47 (1982).
    The problem of solar wind/cometary tail interaction is reinvestigated by appropriately accounting for the presence of the high frequency electrostatic turbulence observed in interplanetary space. The turbulence has a drastic stabilizing effect on the hydromagnetic waves and permits the wavy structure in the tail only beyond some critical distance from the head of the comet; this is completely in agreement with the observations.

102.003   **Temporal variations in the cometary mass distribution.**   D. W. Hughes, P. A. Daniels.
Mon. Not. R. Astron. Soc., Vol. 198, 573 - 582 (1982).
    The large majority of cometary mass is contained in the nucleus. This paper investigates the variation of the mass distribution of comets as they evolve. Theoretical predictions are compared with the observations of two specific cometary groups − firstly, comets which have orbital periods greater than 200 yr and which are assumed to have been recently perturbed into the inner solar system and secondly, comets which have periods less than 200 yr. These are members of the Jovian and Saturnian family and have been in the inner solar system for a considerable time.

102.004   **Visibility of comet nuclei.**   E. P. Ney.
Science, Vol. 215, 397 - 398 (1982).
    Photography of the nucleus of comet Halley is the goal of several planned space missions. The nucleus of a comet is surrounded by a cloud of dust particles. If this cloud is optically thick, it will prevent observation of the nuclear surface. Broadband photometry of nine comets has been analyzed to determine the visibility of their nuclei. Only in the case of comet West near perihelion was the dust dense enough to interfere with imaging. Comparison of the visual brightness of the well-observed comets with that of Halley in 1910 leads to the conclusion that the nucleus of Halley can be imaged without significant obscuration by the dust.

102.005   **Kelvin-Helmholtz instabilities in cometary ion tails.**
T. Ray.
Planet. Space Sci., Vol. 30, 245 - 250 (1982).
    The author interprets observations of wave-like phenomena in cometary ion tails in terms of the interaction of the tails with the solar wind through Kelvin-Helmholtz helical instabilities. The calculations are applied to three comets, Comet Kohoutek (1973f), Comet Arend Roland (1957 III) and Comet Morehouse (1908c). Whilst some disagreement is found with earlier work, it is nevertheless shown that, assuming typical parameters, the growth rate of the K−H helical mode should be significant for wavelengths approximately equal to the tail radius.

102.006   **The capture of interstellar comets.**
M. J. Valtonen, K. A. Innanen.
Astrophys. J., Vol. 255, 307 - 315 (1982).
    A large number of numerical orbit integrations of particles interacting with the Sun-Jupiter system have been performed in order to derive empirical, velocity-dependent expressions for the capture cross section of the system. By integrating these with appropriate velocity distribution functions, bounds can be set on whether comets can be of interstellar origin. These indicate that a significant accumulation of interstellar comets into an "Oort cloud" could occur only with a prolonged transit of the solar system through a relatively large, dense cloud of interstellar debris (comets) having a mean internal velocity dispersion of $\sim$1.0 km s^{-1} and at a relative velocity not exceeding 0.5 km s^{-1}. At these levels, a significant antisymmetry of cometary aphelia in the direction of the relative solar motion is also to be expected. Finally, it is pointed out that suitable binary stars must be exceedingly efficient traps for both protostellar and interstellar debris.

102.007 Finding comets.
D. W. Hughes.
J. British Astron. Assoc., Vol. 92, 61 - 65 (1982).
This paper investigates the statistics of cometary discovery and concludes that the discovery rate is increasing, not because more comets are coming close to the Earth, or because more people are actively seeking comets as time passes, but simply because more and more faint comets are being found.

102.008 Apparition of comets in 1980.
K. I. Churyumov.
Astronomical calendar. Yearbook. 1982, (see 047.005), p. 188 - 201 (1981). In Russian. − Abstr. in Ref. zh., 51. Astron., 2.51.262 (1982).

102.009 The consequences of wasting by sublimation on comet nuclei.   F. L. Whipple.
Reports of Planetary Geology Program − 1981, (see 003.001), p. 26 (1981). − Abstract.

102.010 What shall we look for in comets?
E. Anders.
Meteoritics, Vol. 16, 289 - 290 (1981). − Abstract.

102.011 The ultraviolet bands of the CO_2^+ ion in comets.
M. C. Festou, P. D. Feldman, H. A. Weaver.
Astrophys. J., Vol. 256, 331 - 338 (1982).
Eight comets have been studied with the IUE spectrographs. The existence of the CO_2^+ ion in a comet is confirmed by the presence of the 2890 Å doublet in at least three of these objects. Spatial and spectral resolution obtained in comets Bradfield (1979 X) and Seargent (1978 XV) allow to discuss the production mechanisms of this ion. The spectra show new ionic features in the 3100−3400 Å range, which are attributed to resonance fluorescence of the Fox-Duffendack-Barker system of the CO_2^+ ion and, near 3350 Å, to the OH^+ ion.

102.012 On the similarity of orbits of associated comets, asteroids and meteoroids.   L'. Kresák.
Bull. Astron. Inst. Czechoslovakia, Vol. 33, 104 - 110 (1982).
Distribution characteristics of the D-parameter, measuring the degree of geometrical similarity between heliocentric orbits, are computed for different types of evolutionary associations of interplanetary objects. These include: (1) changes of orbits by planetary perturbations, (2) dispersion of the products of disintegration of common parent bodies, and (3) the structure of subsystems of objects held together by stability conditions. The results are interfaced with those obtained for the distribution in the Tisserand invariant and in the heliocentric energy. Some conclusions are drawn about the evolutionary processes which may be responsible for the observed distributions.

102.013 Phase functions of polarization and brightness and the nature of cometary atmosphere particles.
N. N. Kiselev, G. P. Chernova.
Icarus, Vol. 48, 473 - 481 (1981).
The dependence of brightness and polarization of cometary atmospheres on the phase angle is studied. The similarity between the phase curves of comets, minor planets, and the zodiacal cloud is pointed out. The dependences found correspond to dielectric particles with dimensions greater than 1 μm.

102.014 A model of a comet coma with interstellar molecules in the nucleus.
L. Biermann, P. T. Giguere, W. F. Huebner.
Astron. Astrophys., Vol. 108, 221 - 226 (1982).
The coma of a comet is modeled assuming the icy nucleus contains interstellar molecules. This composition overcomes discrepancies between observation and earlier model predictions for CN, C_2, C_3, and NH_2 abundances. It is found

that the abundances of CN, C_2, and C_3-bearing compounds in the nucleus must be constrained to trace amounts in order to account for the observed column densities. NH_3 also cannot be abundant by more than about 1%. The model gives good agreement with observed relative ranges of neutral species in the coma as well as with their observed relative intensity dependence on heliocentric distance. The size of the nucleus and the heliocentric ranges considered are relevant to comet Halley.

102.015 Tentative identification of CS^+ in comets.
P. D. Singh.
Astron. Astrophys., Vol. 108, 369 - 372 (1982).
Weak discrete emission features in $\lambda\lambda 2400-2800$ wavelength range observed in several comets are tentatively assigned as due to electronic bands of CS^+ $(B^2\Sigma^+ - X^2\Sigma^+)$ system. Using the recently determined lifetime of the CS^+ molecule in the $B^2\Sigma^+(v' = 0)$ state, absolute transition probabilities and oscillator strengths for the electronic bands of the CS^+ $(B^2\Sigma^+ - X^2\Sigma^+)$ and the CS^+ $(B^2\Sigma^+ - A^2\Pi_i)$ systems are obtained.

102.016 Gibt es die Oortsche Kometenwolke wirklich?
V. Kasten.
Sterne Weltraum, 21. Jahrg., 196 - 198 (1982).

102.017 Scattering of Lα radiation in the atmospheres of comets.   L. O. Kolokolova.
Astrometr. Astrofiz., Vyp. (No.) 44, p. 35 - 39 (1981). In Russian.
The scattering of Lα solar radiation in a cometary hydrogen atmosphere is considered. The formulas obtained may be used for determining the profile of the Lα cometary line and for constructing isophotes of Lα radiation of the comet.

102.018 Intensity and linear polarization degree of radiation resonantly scattered in the D_2 Na I line of a cometary atmosphere.   G. F. Chernyj.
Astrometr. Astrofiz., Vyp. (No.) 44, p. 39 - 42 (1981). In Russian.
D_2 Na I line intensities and the values of linear polarization degrees in cometary atmospheres are given. Analytical expressions are presented for these quantities.

102.019 Is an alternative model of the cometary nucleus possible?   L. M. Shul'man.
Astrometr. Astrofiz., Vyp. (No.) 44, p. 43 - 51 (1981). In Russian.
Two alternative models of a cometary nucleus are analysed: the dust cloud model which is continuously proposed by Lyttleton and the core-mantle model. The models are shown to be less realistic than the icy one.

102.020 A two-layer model of a cometary nucleus.
L. M. Shul'man.
Astrometr. Astrofiz., Vyp. (No.) 45, p. 21 - 34 (1981). In Russian.
Physical properties of the two-layer (ice core and dust mantle) model of a cometary nucleus are considered. The contact thermoconductivity of the dust mantle is shown to exceed both the radiative and the gaseous ones. The hypothesis is proposed that the dust mantle will not be thrown out at small heliocentric distances, as it is often supposed, but the mantle will be fluidized by the gas blown through it.

102.021 Interpretation of integral photometry data of comets.   L. M. Shul'man.
Astrometr. Astrofiz., Vyp. (No.) 45, p. 35 - 45 (1981).
The connection between physical and photometric properties of comets is revised. A more precise relationship is obtained for the gas component of the total visual brightness of the comet. The total brightness of the dust component

decreases with the heliocentric distance faster than that of the gas component does. Therefore, the total visual brightness of a comet at large heliocentric distances is due to the parent molecules gas, but not to the dust, as it is often supposed.

102.022 **Contribution of dust grains originated by condensation to the integral brightness of a comet.**
L. M. Shul'man.
Astrometr. Astrofiz., Vyp. (No.) 46, p. 68 - 72 (1982). In Russian.

The kinetic equation is solved for the grains distribution function. The grains are supposed to grow due to condensation or polymerization. An expression is obtained describing the connection between the total brightness of a comet, the radius of its nucleus and heliocentric distance. It is shown that the total brightness of the dust coma, originated by condensation, decreases with heliocentric distance fastly and depends strongly on the cometary nucleus size. Therefore, only the brightest comets at smallest heliocentric distances are expected to have dust formed by condensation or polymerization.

102.023 **The influx of comets and asteroids to the earth.**
D. W. Hughes.
Philos. Trans. R. Soc. London A, Vol. 303, (see 012.022), 353 - 368 (1981).

This paper reviews the influx rates of the minor objects in the Solar System to the Earth's surface and contrasts the populations of the Earth impactions with the populations in for example the comet cloud and the asteroid belt.

102.024 **The role of the critical ionization velocity phenomena in the production of inner coma cometary plasma.**   V. Formisano, A. A. Galeev, R. Z. Sagdeev.
Planet. Space Sci., Vol. 30, 491 - 497 (1982).

The critical velocity triggering anomalous ionization of the neutral gas by plasma flow is calculated for the model based on the lower hybride instability. It depends strongly on the plasma and gas parameters, defining the instability development of the ionized atoms beam in the counter-streaming plasma. In particular, the possible role of the critical ionization mechanism for Halley's comet is examined.

102.025 **Le nuage de cometes de Oort.**
Ciel, Vol. 44, 58 - 60 (1982).

102.026 **On the orbital evolution of short-period comets having low-velocity encounters with Jupiter.**
A. Carusi, G. B. Valsecchi.
The comparative study of the planets, (see 012.033), p. 131 - 148 (1982).

The origin of short-period comets in the framework of the Oort cloud hypothesis is discussed. This hypothesis can explain the appearance of "new" long-period comets, via the gravitational effects of passing stars. It seems improbable that a single close encounter with a giant planet would be sufficient to transform a long-period comet into a short-period one. More probably these comets are those members of a vast population of bodies revolving in the outer regions of the Solar System in "chaotic orbits", the aphelia of which have accumulated around the orbit of Jupiter by the maximum possible reduction of energy.

102.027 **A comet on collision course with the sun: observations on August 30 - 31, 1979.**
R. A. Howard, M. J. Koomen, D. J. Michels, N. R. Sheeley, Jr.
Bull. American Astron. Soc., Vol. 13, 875 (1981). — Abstract.

102.028 **A comet on collision course with the sun: orbital data deduced from the observations of August 30 - 31, 1979.**

N. R. Sheeley, Jr., R. A. Howard, M. J. Koomen, D. J. Michels.
Bull. American Astron. Soc., Vol. 13, 875 (1981). — Abstract.

102.029 **A comet on collision course with the sun: dynamical interpretation of the observations of August 30 - 31, 1979.**
D. J. Michels, N. R. Sheeley, Jr., R. A. Howard, M. J. Koomen.
Bull. American Astron. Soc., Vol. 13, 875 (1981). — Abstract.

102.030 **CO (CO_2) and H_2O abundances in comets.**
S. Wyckoff, P. A. Wehinger, E. Leibowitz, T. Pitman.
Bull. American Astron. Soc., Vol. 13, 875 (1981). — Abstract.

102.031 **Do comets have satellites?**   T. C. Van Flandern.
Bull. American Astron. Soc., Vol. 13, 875 - 876 (1981). — Abstract.

102.032 **Coronagraph observations of a sun-directed comet, Aug. 30 - 31, 1979: images, analysis and photometry.**
M. J. Koomen, D. J. Michels, N. R. Sheeley, Jr., R. A. Howard.
Bull. American Astron. Soc., Vol. 13, 891 (1981). — Abstract.

102.033 **On capture of comets into the solar system.**
M. J. Valtonen.
Rep. Ser., Dep. Phys. Sci., Univ. Turku, Turku-FTL-R 28, 8 pp. (1982) = Turku Univ. Obs. Informo No. 64. ISBN 951-642-153-9, ISSN 0356-9896. — Submitted to Observatory.

102.034 **Evaluation of infrared line emission from constituent molecules of cometary nuclei.**
T. Yamamoto.
Astron. Astrophys., Vol. 109, 326 - 330 (1982).

Infrared fluxes of the line emission from candidate constituent molecules of a cometary nucleus are estimated for future observations with evaluations of other infrared sources such as zodiacal light, thermal emission of the nucleus, and infrared continuum from dust in a coma. It is shown that there are many detectable emission lines in the near infrared. Of these, the emissions from H_2O, CO, CO_2, CH_4, and OH are particularly remarkable. Some observational items interesting in cometary physics are presented.

102.035 **Plasma flow and magnetic fields in comets.**
H. U. Schmidt, R. Wegmann.
Scientific and experimental aspects of the Giotto mission, (see 012.039), p. 3 - 7 (1981).

Plasma flow and field amplification resulting from the interaction of the solar wind with a comet are studied by model calculations for the rates $G = 7 \times 10^{26}$ to 2×10^{30} of total cometary gas production. The influence of the magnetic field is investigated first in an axisymmetric model. A three-dimensional MHD-calculation shows how the magnetic field induces an asymmetry. A sudden change of direction in the interplanetary field produces streamer-like structures.

102.036 **Size distribution, structure, and chemical composition of cometary dust particles.**   H. Fechtig.
Scientific and experimental aspects of the Giotto mission, (see 012.039), p. 47 - 52 (1981).

A short review of observations of cometary dust is presented both for Earth-bound measurements and for space measurements.

102.037 **Dynamics and chemistry of cometary comae: a pre-Giotto view.**   W.-H. Ip.
Scientific and experimental aspects of the Giotto mission, (see 012.039), p. 79 - 91 (1981).

102.038 **Electromagnetic instabilities in cometary tails.**
A. M. Tupchienko, A. K. Yukhimuk.
Geofiz. zh., Tom 4, 86 - 88 (1982). In Russian. — Abstr. in Ref. zh., 51. Astron., 6.51.231 (1982).

102.039 Perihelion asymmetry in the photometric parameters of long-period comets at large heliocentric
distances.   J. Svoreň.
Sun and planetary system, (see 012.040), p. 321 - 322 (1982).

102.040 Ultraviolet spectroscopy of comets.
J. Rahe.
Sun and planetary system, (see 012.040), p. 323 - 330 (1982).

Ultraviolet observations are for cometary studies
especially important since the four basic atomic elements,
H, O, C, and N have their strongest resonance transitions between 1200 and 1700 Å. The most prominent features in all
cometary spectra are the Lyman-alpha emission of atomic
hydrogen at 1216 Å, and the (0, 0) band of OH near 3090 Å.
Recent spectroscopic ultraviolet observations of comets are
presented, and the results derived from them, are discussed.

102.041 Outbursts of comets.
S. Gąska, P. Gronkowski.
Sun and planetary system, (see 012.040), p. 337 - 338 (1982).

The aim of this report is an attempt to explain the sudden
brightness variations of the following comets: Schwassmann–
Wachmann (1925 II), Kritzinger (1914a), and Ikeya-Seki
(1965f).

102.042 Dynamical evolution and disintegration of comets.
Ľ. Kresák.
Sun and planetary system, (see 012.040), p. 361 - 370 (1982).

Current concepts of the origin and evolution of comets
are reviewed. The present structure of the system of comets is
controlled by the dynamical evolution of its individual
members limited by their physical aging by disintegration.
Where the lifetimes are short, as in the Jupiter family of short-
period comets, an equilibrium between elimination and
replenishment is established. The role of different destructive
processes and the resulting survival times are discussed.

102.043 A dynamical study of possible birthplaces of comets.
J. A. Fernández.
Sun and planetary system, (see 012.040), p. 371 - 374 (1982).

The orbital evolution of comets with different assumed
birthplaces, namely the Uranus-Neptune region, the Jupiter
region and 'in-situ' from where 'new' comets seem to come, is
followed numerically throughout the solar system lifetime.
Perturbations of the four giant planets and random passing
stars are considered. Several evolutionary paths are thus
defined for the comet orbits. The efficiency of each one of
them in bringing comets into the planetary region is discussed
in connection with the origin of comets.

102.044 An interstellar origin for comets.
W. M. Napier.
Sun and planetary system, (see 012.040), p. 375 - 378 (1982).

102.045 Strong perturbations at close encounters with
Jupiter.   A. Carusi, G. B. Valsecchi.
Sun and planetary system, (see 012.040), p. 379 - 384 (1982).

The phenomena occurring at a close encounter of a
minor body with Jupiter are reviewed and the relations between the characteristics of the pre-encounter orbit and the
possible outcomes of the interaction are discussed.

102.046 Statistics of close encounters of minor bodies with
the outer planets.   A. Carusi, G. B. Valsecchi.
Sun and planetary system, (see 012.040), p. 385 - 388 (1982).

For each of the four outer planets a sample of 1000
fictitious minor bodies in heliocentric orbits has been
generated, and a single close encounter of each minor body
with the corresponding planet has been computed. In this
paper the authors present the distributions of the initial and
final orbital elements, together with the statistics of various
types of orbital evolutions.

102.047 Keplerian estimates of pre-discovery encounters
with Jupiter for short-period comets.
H. Rickman, J. Karm.
Sun and planetary system, (see 012.040), p. 389 - 390 (1982).

Orbital perturbations at close encounters with Jupiter
are known to have a predominant influence on the population
of short-period comets. For the purpose of identifying such
encounters and roughly estimating the approach dates and
distances one can often use unperturbed, Keplerian orbits.
The authors have made such estimates for all the 118 short-
period comets known at the end of 1980. Such encounters
occurred for nearly half the comets.

102.048 The motions of comets near the 2/1 resonance with
Jupiter.   S. Vaghi, H. Rickman.
Sun and planetary system, (see 012.040), p. 391 - 394 (1982).

The authors present the results of orbital integrations in a
9-body model for comets P/Lexell, P/Tempel-Swift,
P/Wirtanen, P/Kohoutek, P/West-Kohoutek-Ikemura,
P/Haneda-Campos and P/Wild 2. All these comets have spent
some time near the 2/1 resonance with Jupiter. The authors
study in particular the time evolution of the orbital energy in
order to evaluate the role of a repeated encounter mechanism.

102.049 Temporary satellite captures by Jupiter for orbits
resembling the one of comet P/Gehrels 3.
H. Rickman, A. M. Malmort.
Sun and planetary system, (see 012.040), p. 395 - 396 (1982).

102.050 The formation of comets by radiation pressure in
the outer protosun.   J. G. Hills.
Astron. J., Vol. 87, 906 - 910 (1982).

The pressure due to the radiation from the Sun and
neighboring protostars may have forced the coagulation into
comets of the dust grains in the collapsing layers of the protosun at $r = (1-5) \times 10^3$ AU. The grains were forced together by
their self-shielding, which results in the radiation pressure due
to photons coming from the direction of strong concentrations
of dust being less than the pressure due to photons coming
from a direction having a low concentration of dust. This
causes the dust to drift toward regions of already strong dust
concentration. The formation of comets under these conditions is consistent with the low rotation period of new
comets and their extremely volatile chemical constituents.

102.051 On the distribution of orbits among long-period
comets.   R. S. Bogart, P. D. Noerdlinger.
Astron. J., Vol. 87, 911 - 917 (1982).

The distribution of the axes defining the planes and
orientations of the orbits of 542 long-period comets are analyzed. The directions of the perihelia and of the oriented plane
normals show significant nonuniformity in their distributions.
The preferred direction of perihelia near the apex of solar
motion is refined to an (1σ) error circle of $20°6$. Planes of
preference (and avoidance) are found from the distribution
ellipsoids of the three orbital axes, with the perihelion
directions lying preferentially along the galactic plane. The
distribution ellipsoids of all three sets of orbital axes exhibit
anisotropies roughly twice those expected for random distributions.

102.052 On the reality and genetic association of comet
groups and pairs.   L. Kresák.
Bull. Astron. Inst. Czechoslovakia, Vol. 33, 150 - 160 (1982).

102.053 Origin of comets.   M. Valtonen, K. Innanen.
Rep. Obs. Lund, No. 18, (see 012.044), p. 120 -
123 (1982).

The theory of interstellar origin of comets has been
studied. The rate of captures of interstellar comets by the
Sun–Jupiter system has been calculated under various assump-

tions of the internal velocity dispersion α of the cometary medium and its mean speed u_0 relative to the Sun.

102.054 On the global size of Oort's cloud of cometary nuclei and their total number.   L. Biermann. Compendium in astronomy, (see 003.012), p. 183 - 190 (1982).

The size of Oort's cloud and the total number of the cometary nuclei it contains are rediscussed on the basis of the new data presented in Marsden, Sekanina and Everhart's paper of 1978 and of Oort's working model of the "cloud".

102.055 Report of IAU Commission 15: Physical study of comets, minor planets and meteorites (L'étude physique des comètes, des petites planètes et des météorites). B. D. Donn. Trans. IAU, Vol. XVIIIA, (see 003.013), p. 153 - 174 (1982).

Comets and corpuscular radiation of the sun. See Abstr. 003.018.

The calculation of comet ephemerides. See Abstr. 021.045.

Methods of computation of the perturbed motion of small bodies in the solar system.   See Abstr. 021.047.

Cometary NH: ultraviolet and submillimeter emission.   See Abstr. 022.012.

CO⁺ comet-tail emission and chemiluminescence in collisions of N^+ ions with CO molecules in the energy range 10-300 eV.   See Abstr. 022.077.

The measurement of comet positions. See Abstr. 031.544.

Review of magnitude sources for visual cometary photometry. I.   See Abstr. 031.614.

Wideband comet camera lens design. See Abstr. 032.521.

Orbital patterns of interplanetary objects at close encounters with Jupiter.   See Abstr. 042.130.

The smaller bodies of the solar system. See Abstr. 098.035.

Theoretical meteor radiants of Apollo, Amor, and Aten asteroids.   See Abstr. 098.037.

Report of IAU Commission 20: Positions and motions of minor planets, comets and satellites (Positions et mouvements des petites planètes, des comètes et des satellites). See Abstr. 098.107.

Peculiarities in the ionic tail of Comet Ikeya-Seki (1965f).   See Abstr. 103.541.

Interplanetary gas. XXVIII. A study of the three-dimensional properties of interplanetary sector boundaries using disconnection events in cometary plasma tails. See Abstr. 106.008.

Plasma-dust interactions in the solar and cometary environment.   See Abstr. 106.035.

Interstellar planetesimals – I. Dissipation of a primordial cloud of comets by tidal encounters with massive nebulae.   See Abstr. 107.001.

Spiral arms, comets and terrestrial catastrophism. See Abstr. 107.003.

103 Comets (Individual Objects)

103.001 Observations of comets.
British Astron. Assoc. Circ., No. 617 (1981).
Concerning comets: 1980 XI Encke, 1980h Tuttle, 1980g Meier, 1980g Stephan-Oterma, 1980u Panther.

103.002 Die Sichtbarkeiten der Kometen des Jahreswechsels 1980/81.   A. Kammerer.
Sterne Weltraum, 21. Jahrg., 25 - 28 (1982).

103.003 Observations of comets.
Minor Planet Circ., (M. P. C.), Nos. 6580 - 6582, 6656 - 6658, 6752 - 6753, 6841 - 6842, 6900 - 6902 (1982).
Observations made at the following stations are published: Caussols, Centro Astron. Yebes, Cerro el Roble, Chamberlin Obs., Geisei, Hemingford Abbots, Hoher List, JCPM Hamatonbetsu, Stat., Kleť, Kvistaberg, La Seyne-sur-Mer, Lowell Obs., Lowell Obs. Anderson Mesa Stat., Mt. John Obs., Mt. Palomar, Oak Ridge Obs., S. Vittore (Bologna), Siding Spring, Stakenbridge, Wrocław. Concerning observations of the following comets: 1910 II Halley, 1972 VIII Heck-Sause, 1972 IX Sandage, 1973 XII Kohoutek, 1974 II Schwassmann-Wachmann 1, 1974 XII van den Bergh, 1975 VII Smirnova-Chernykh, 1975 VIII Lovas, 1976 II

Wolf, 1976 III Gunn, 1976 IV Bradfield, 1976 VIII Harrington-Abell, 1976 IX Lovas, 1977 VII Gehrels 3, 1977 IX West, 1977 XIV Kohler, 1978 II Tempel 1, 1978 VIII Whipple, 1978 XIV Ashbrook-Jackson, 1978 XXI Meier, 1978 XXIII Clark, 1978 XXIV van Biesbroeck, 1979 III Giacobini-Zinner, 1980 X Stephan-Oterma, 1980 XI Encke, 1980 XII Meier, 1980b Bowell, 1980i Borrelly, 1980n Reinmuth 2, 1980u Panther, 1981c Elias, 1981d Bus, 1981e Finlay, 1981f Gehrels 2, 1981g González, 1981h Kearns-Kwee, 1981i Slaughter-Burnham, 1981j Swift-Gehrels, 1981k Howell, 1981l Väisälä 1, 1982a Grigg-Skjellerup, 1982b du Toit-Hartley, 1982d Tempel 2, 1982e d'Arrest, 1982f Churyumov-Gerasimenko.

103.004 Ephemerides of comets.
Minor Planet Circ., (M. P. C.), Nos. 6837 - 6838 (1982).
Concerning ephemerides of the following comets: 1958 IV Oterma, 1975 VII Smirnova-Chernykh, 1980 XI Encke.

103.005 Orbital elements of comets.
Minor Planet Circ., (M. P. C.), Nos. 6573 - 6988

(1982).

The comets are listed according to their Roman numeral designation or preliminary designation. The names of the authors are given behind the respective M. P. C. numbers.
1980b Bowell 6889, 1981c Elias 6889, 1981d Bus 6945, 1981g González 6889, 1981k Howell 6889, 1982c du Toit-Hartley 6890 B. G. Marsden.

103.006   Possible comet Shcherbanovskij.
IAU Circ., No. 3656 (1982).

103.007   Probable comet.
IAU Circ., Nos. 3657, 3662 (1982).

103.008   Probable comet.
Yamamoto Circ., Nos. 1968, 1971 (1982).

103.009   Possible comet Shcherbanovskij.
Komet. Tsirk., Kiev, No. 284 (1982). In Russian.

103.010   Observations of comets at Agassiz Station during 1981.   R. E. McCrosky, C.-Y. Shao.
Int. Comet Q., Vol. 4, 10 - 13 (1982).

103.011   Tabulation of comet observations.
Int. Comet Q., Vol. 4, 19 - 35 (1982).
Concerning comets: 1948 XI Eclipse comet, 1951 I Minkowski, 1951 II Pajdušáková, 1951 IV Tuttle-Giacobini-Kresák, 1951 VII Kopff, 1952 I Wilson-Harrington, 1952 II Wolf-Harrington, 1952 III Schaumasse, 1952 V Mrkos, 1952 VI Peltier, 1953 I Harrington, 1953 II Mrkos, 1953 III Mrkos-Honda, 1954 II Pajdušáková, 1954 VII Pons-Brooks, 1954 VIII Vozarova, 1954 IX Encke, 1954 X Abell, 1954 XII Kresák-Peltier, 1955 VII Perrine-Mrkos, 1972 V Tempel 1, 1972 VI Giacobini-Zinner, 1977 XIV Kohler, 1979 X Bradfield, 1980 X Stephan-Oterma, 1980 XIII Tuttle, 1980 XV Bradfield, 1980 XX Encke, 1980 i Borrelly, 1981 h Kearns-Kwee, 1981 j Swift-Gehrels, P/ Schwassmann-Wachmann 1.

103.012   Observations de comètes faibles avec IUE.
M. Festou.
Astronomie, Vol. 96, 299 - 302 (1982).

103.013   Observations des membres de la commission des comètes (fin 1980 et 1981).
C. Bertaud.
Astronomie, Vol. 96, 313 - 318 (1982).

103.014   Short-period comets in 1982: Grigg-Skjellerup, Perrine-Mrkos, Väisälä 1, d'Arrest, Tempel-Swift, Churyumov-Gerasimenko, Gunn, Neujmin 3.
N. A. Belyaev, K. I. Churyumov.
Komet. Tsirk., Kiev, No. 284 (1982). In Russian.

103.015   Recoveries of periodic comets.
British Astron. Assoc. Circ., No. 620 (1981).
Concerning comets: 1981e Finlay, 1981f Gehrels, 1981h Kearns-Kwee, 1981i Slaughter-Burnham, 1981j Swift-Gehrels.

103.016   Recent observations of comets.
British Astron. Assoc. Circ., No. 622 (1982).
Concerning comets: 1981j Swift-Gehrels, 1981h Kearns-Kwee.

103.017   Recoveries of periodic comets.
British Astron. Assoc. Circ., No. 627 (1982).
Concerning comets: 1982e d'Arrest, 1982f Churyumov-Gerasimenko.

103.018   Comet digest.   J. E. Bortle.
Sky Telesc., Vol. 63, 98, 215, 315, 427, 532, 634; Vol. 64, 101 (1982).

Photographic positions of minor planets and comets.
See Abstr. 098.027.

Comet 1979 XI Howard-Koomen-Michels

103.101   Observations of a comet on collision course with the sun.
D. J. Michels, N. R. Sheeley, Jr., R. A. Howard, M. J. Koomen.
Science, Vol. 215, 1097 - 1102 (1982).
The authors report the observation of a new comet, apparently the first to be discovered from a spacecraft and the first observed to collide with the sun. Preliminary analysis has provided a set of orbital elements of modest accuracy and suggests that the comet may be one of the Kreutz sungrazers. Disintegration products from the encounter caused a major change in the coronal brightness distribution, which persisted for more than one full day. Indications of the encounter on the solar surface have not yet been identified but are a distinct possibility.

103.102   Comet Howard-Koomen-Michels 1979 XI.
J. Bouška.
Říše hvězd, Vol. 63, 25 - 28 (1982). In Czech.

Comet 1973 XII Kohoutek

103.121   Observaciones del cometa Kohoutek (1973 f) en La Plata.   C. Rogati, F. Muñoz.
Bol. Asoc. Argentina Astron., No. 18, p. 6 (1980).

103.122   Observaciones del cometa Kohoutek en 1420 MHz.
F. R. Colomb, R. Morras, G. L. Poppel.
Bol. Asoc. Argentina Astron., No. 20 - 24, p. 179 (1981). Abstract.

Comet 1981c Elias

103.141   New comet Elias 1981c.
British Astron. Assoc. Circ., No. 617 (1981).

Orbital elements of comets.
See Abstr. 103.005.

Comet 1981d Bus

103.161   New comet Bus 1981d.
British Astron. Assoc. Circ., No. 617 (1981).

Orbital elements of comets.
See Abstr. 103.005.

Comet 1980 XV Bradfield

103.201   Waarnemingsresultaten van komeet Bradfield.
H. Feijth.
Zenit, 9. Jaarg., 13 - 16 (1982).

103.202   Bradfields helderheidsverloop.
R. J. Bouma.
Zenit, 9. Jaarg., 16 - 17 (1982).

103.203   Near infrared spectroscopy of comet Bradfield
1980t.   C. Barbieri, C. Bonoli, F. Bortoletto.
The comparative study of the planets, (see 012.033), p. 317 - 322 (1982).

Three spectra of comet Bradfield 1980t taken with the RETICON system at Asiago Observatory are discussed; an estimate of the absolute fluxes and a discussion of the principal features in the spectra is presented.

103.204   Spectrophotometry of Comet Bradfield (1980t)
during post-perihelion period.
P. S. Goraya, B. K. Sinha, U. S. Chaubey, B. B. Sanwal.
Moon Planets, Vol. 26, 3 - 9 (1982).

Spectrum scans of the head of Comet Bradfield (1980t) covering the wavelength range $\lambda\lambda$365–640 nm were made on two nights when the heliocentric distance of the comet varied from 0.55 to 0.58 AU. The emission features of the CN band at λ388 nm and Swan band sequence of C_2 at λ474 nm, λ516 nm, and λ563 nm are identified and absolute fluxes in these bands as well as in the continuum are derived. The continuum energy distribution curves of the comet have been compared with those of the Sun and the star β Crv (G5 III). An estimate of the number of C_2 and CN molecules in the head of the comet has been made.

Comet 1976 VI West

103.211   The spectrum of comet West 1976 VI.
V. K. Rozenbush.
Astrometr. Astrofiz., Vyp. (No.) 43, p. 63 - 66 (1981). In Russian.

Spectral observations of the comet West, 1976 VI, were carried out in the wavelength range 3800–6800 Å on March 15 - April 4, 1976. 37 cometary emissions were identified. The relative energy distribution in the cometary spectrum was constructed. The total energy fluxes in separate emission bands were determined.

103.212   Structure of the tails of comet West 1976 VI.
Yu. V. Sizonenko.
Astrometr. Astrofiz., Vyp. (No.) 45, p. 46 - 52 (1981). In Russian.

The structure of the comet West 1976 VI tail has been studied using plates obtained on March 9 - 12, 1976. The location of structural details in the plane of the cometary orbit as well as the motion of bands of the II type tail have been determined. The velocity of the solar wind was calculated by the main gas flux orientation. The Alfvén wave lengths in the I type tail were measured. The mean wave length decreases gradually from March 9 till March 12.

103.213   Probable detection of Cameron band emission in
Comet West spectrum.   A. M. Smith.
Bull. American Astron. Soc., Vol. 13, 839 (1981). – Abstract.

Comet 1981g González

103.221   New comet González 1981g.
British Astron. Assoc. Circ., No. 620 (1981).

Orbital elements of comets.
See Abstr. 103.005.

Periodic comet Howell

103.241   New periodic comet Howell 1981k.
British Astron. Assoc. Circ., No. 621 (1981).

Orbital elements of comets.
See Abstr. 103.005.

Periodic comet du Toit-Hartley

103.261   Periodic comet du Toit-Hartley (1945 II =
1982b, 1982c).
IAU Circ., Nos. 3663 - 3665, 3668, 3672, 3674, 3678, 3682, 3694 (1982).

103.262   Comet du Toit-Hartley (1945 II = 1982b, 1982c).
Yamamoto Circ., Nos. 1970, 1971, 1972 (1982).

103.263   Comet P/du-Toit-Hartley 1945 II = 1982c, 1982b.
British Astron. Assoc. Circ., Nos. 624, 625 (1982).

Orbital elements of comets.
See Abstr. 103.005.

Periodic comet Kearns-Kwee

103.281   Periodic comet Kearns-Kwee (1981h).
IAU Circ., No. 3656 (1982).

103.282   P/Kearns-Kwee (1981h).
Yamamoto Circ., No. 1968 (1982).

Periodic comet Schwassmann-Wachmann 1

103.301   Spectrophotometry of Comet Schwassmann-
Wachmann 1. II. Its color and CO^+ emission.
A. L. Cochran, W. D. Cochran, E. S. Barker.
Astrophys. J., Vol. 254, 816 - 822, plate 11 (1982).

The authors obtained spectra of Comet Schwassmann-Wachmann 1 during the winter 1980 - 1981. On 1981 February 7 UT the comet was in outburst and had a one-armed spiral shape. On that night, spectra of the "nucleus" and of several locations along the arm were obtained. All of the spectra were reduced to relative reflectances. The quiescent state of the comet may be modeled, using Mie scattering theory, by a distribution of relatively large particles with a small variance in particle size. Most of the positions observed during outburst required the addition of a distribution of smaller size particles. As the outburst progressed, more and more of the smaller particles were required. The data may be fitted equally well by either H_2O or CO_2 ice particles. The data show that the CO^+ emission turns on and off on a rapid time scale.

103.302   Some remarks about the outbursts of comet
Schwassmann-Wachmann 1 1925 II.
S. Grudzińska.
Mitt. Astron. Ges., Nr. 55, p. 189 - 191 (1982).

103.303   Periodic comet Schwassmann-Wachmann 1.
IAU Circ., Nos. 3660, 3663, 3687, 3689 (1982).

103.304   P/Schwassmann-Wachmann 1.
Yamamoto Circ., Nos. 1968 - 1971, 1974, 1975 (1982).

Periodic comet Grigg-Skjellerup

103.321   **Periodic comet Grigg-Skjellerup (1982a).**
IAU Circ., Nos. 3659, 3695, 3698, 3700, 3705
(1982).

103.322   **P/Grigg-Skjellerup (1982a).**
Yamamoto Circ., Nos. 1969, 1975, 1976 (1982).

103.323   **Periodic comet Grigg-Skjellerup 1982a.**
British Astron. Assoc. Circ., Nos. 624, 626 (1982).

103.324   **Short-period comet Grigg-Skjellerup.**
Komet. Tsirk., Kiev, No. 284 (1982). In Russian.

Periodic comet Swift-Gehrels

103.361   **Periodic comet Swift-Gehrels (1981j).**
IAU Circ., No. 3667 (1982).

Periodic comet d'Arrest

103.381   **Periodic comet d'Arrest (1982e).**
IAU Circ., No. 3697 (1982).

103.382   **Periodic comet d'Arrest (1982e).**
Yamamoto Circ., No. 1975 (1982).

Comet 1980b Bowell

103.401   **Comet Bowell (1980b): an active-looking dormant
object?**   Z. Sekanina.
Astron. J., Vol. 87, 161 - 169 (1982).
It is argued that solid particles in the coma and tail of
the distant Comet Bowell are not recent ejecta and that the
comet has probably been dormant. As a by-product, the position of the spin axis of Comet Bowell — the first among "new"
comets — is determined.

103.402   **A comet in a million.**
D. W. E. Green, B. G. Marsden.
Sky Telesc., Vol. 63, 366 - 367 (1982).

103.403   **Comet Bowell (1980b).**
IAU Circ., Nos. 3662, 3670, 3683, 3697, 3699
(1982).

103.404   **Comet Bowell (1980b).**
Yamamoto Circ., Nos. 1969, 1972, 1974, 1976
(1982).

103.405   **Comet Bowell 1980b.**
British Astron. Assoc. Circ., Nos. 617 (1981), 624
(1982).

Orbital elements of comets.   See Abstr. 103.005.

Periodic comet Churyumov-Gerasimenko

103.421   **Periodic comet Churyumov-Gerasimenko (1982f).**
IAU Circ., No. 3700 (1982).

103.422   **Periodic comet Churyumov-Gerasimenko (1982f).**
Yamamoto Circ., No. 1976 (1982).

Periodic comet Gunn

103.441   **Periodic comet Gunn.**
IAU Circ., No. 3700 (1982).

103.442   **Short-period comet Gunn.**
Komet. Tsirk., Kiev, No. 284 (1982). In Russian.

Periodic comet Neujmin 3

103.461   **Periodic comet Neujmin 3.**
IAU Circ., No. 3704 (1982).

Comet 1982g Austin

103.481   **Comet Austin (1982g).**
IAU Circ., Nos. 3705, 3706 (1982).

103.482   **Comet Austin (1982g).**
Yamamoto Circ., Nos. 1976, 1977 (1982).

103.483   **New comet Austin 1982g.**
British Astron. Assoc. Circ., No. 627 (1982).

Periodic comet Halley

103.501   **On the brightness of Halley's comet.**
I. Ferrin.
Astron. Astrophys., Vol. 107, L7 - L9 (1982).
Periodic comet Halley was not a "well behaved" comet
photometrically in its apparition of 1910. Its light curve
shows fluctuations of long and short duration that suggest
that this comet has a surface with uneven distribution of
volatiles. Moreover the comet shows a brightness decrease
after perihelion and a sharp increase at 1.0 AU from the sun.
The author's reduction of the available data shows that this
anomaly is significant, since four independent data sets show
it. He gives an explanation for it, in terms of a break up of
the nucleus. As a by product, the influence of twilight on
visual observations of this comet is derived.

103.502   **The return of Halley's comet.**   M. J. Hendrie.
Spaceflight, Vol. 24, 242 - 248 (1982).

103.503   **Space missions to Halley's comet and related
activities.**   R. Reinhard.
ESA Bull., No. 29, p. 68 - 83 (1982).
Paper presented at the meeting "Space missions to
Halley's comet and related activities", Padova, 13 - 15
September 1981.

103.504   **Komet Halley 1985/86.**   J. Rahe.
Sterne Weltraum, 21. Jahrg., 21 - 24 (1982).

103.505   **Halley's comet: energy and perturbations.**
L. Buffoni, A. Manara, M. Scardia.
Astron. Astrophys., Vol. 108, 141 - 142 (1982).
Mechanical energy and the planetary perturbations of
Halley's comet are provided. In the next passage there will be a
slight decrease in the mechanical energy of the comet with a
corresponding decrease of the period. The main perturbation is
that of Jupiter, except for two periods in which the greatest
action will be exerted by Venus and by Earth.

103.506   **A very simple method for calculating the ephemeris
of comet Halley.**

V. V. Radzievskij, L. A. Alekseeva.
Astron. vestn., Tom 15, 239 - 243 (1981). In Russian. – Abstr.
in Ref. zh., 51. Astron., 3.51.89 (1982).

103.507 **Periodic comet Halley.**
IAU Circ., No. 3688 (1982).

103.508 **Comet Halley ephemeris development.**
D. K. Yeomans.
Bull. American Astron. Soc., Vol. 14, 580 (1982). – Abstract.

103.509 **Halley's comet.**
R. L. Newburn, Jr., D. K. Yeomans.
Annu. Rev. Earth Planet. Sci., Vol. 10, (see 003.009), 297 -
326 (1982).
In this review, the authors begin by covering the one
truly quantitative aspect possible for P/Halley, its motion, in a
step-by-step historical development. Then, following a brief
look at the meteoroid streams associated with Halley, the
authors discuss the observations of 1910 in the dual context
of the state of cometary astronomy then and now. They
derive what facts one can obtain from those observations and
extend them by analogy with comets observed recently.

103.510 **Notes on the P/Halley nucleus.**   F. L. Whipple.
Scientific and experimental aspects of the Giotto
mission, (see 012.039), p. 101 - 103 (1981).

103.511 **Halley's comet and plans for its observation during
its return in 1986.**
J. Rahe, J. C. Brandt, L. D. Friedman, R. L. Newburn.
Sun and planetary system, (see 012.040), p. 307 - 320 (1982).
The authors briefly discuss the general characteristics
and properties of Halley's comet, as they can be derived from
earlier apparitions. The major part of this paper is devoted to
the 1986 apparition.

103.512 **The light curve of periodic comet Halley 1910 II.**
C. S. Morris, D. W. E. Green.
Astron. J., Vol. 87, 918 - 923 (1982).
Photometric parameters for periodic comet Halley 1910 II
have been derived from 144 total visual magnitude estimates.
The pre-perihelion data are best represented by an absolute
magnitude of 5.47 and a power-law exponent of 4.44; post-
perihelion results show that the absolute magnitude brightened
to 4.94, and the exponent decreased to 3.07. Only small fluctuations in brightness about the power-law solutions are noted.
Based on these results, a forecast of the visual brightness of
periodic comet Halley's 1985 - 1987 apparition is presented.

Kometen kommer!   See Abstr. 003.025.

**ESO workshop on "Ground-based observations of
Halley's comet".**   See Abstr. 011.030.

**Scientific and experimental aspects of the Giotto
mission.**   See Abstr. 012.039.

**The need for inter-agency collaboration on missions
to Halley's comet.**   See Abstr. 013.014.

**Vorbereitungen auf Komet Halleys Wiederkehr
1985/86.**   See Abstr. 013.015.

**The role of the critical ionization velocity phenom-
ena in the production of inner coma cometary plasma.**
See Abstr. 102.024.

**The total mass and structure of the meteor stream
associated with comet Halley.**   See Abstr. 104.027.

Periodic comet Väisälä 1

103.521 **P/Väisälä 1 (1981 *l*)**
Yamamoto Circ., No. 1968 (1982).

103.522 **Rediscovery of short-period comet Väisälä 1 (1981*l*).**
Komet. Tsirk., Kiev, No. 284 (1982). In Russian.

103.523 **Periodic comet Väisälä (1) 1981*l*.**
British Astron. Assoc. Circ., No. 622 (1982).

Comet 1965 VIII Ikeya-Seki

103.541 **Peculiarities in the ionic tail of Comet Ikeya-Seki
(1965f).**   V. Krishan, K. R. Sivaraman.
Moon Planets, Vol. 26, 209 - 215 (1982).
Direct photographs of Comet Ikeya-Seki obtained on
four consecutive days from October 29 to November 1, 1965,
are used for an analysis of the multiple helical structures in
the ionised tail. The formation of these structures is explained
on the basis of plasma instabilities excited in the tail containing twisted magnetic fields. The growth rate of the modes
excited at the mode rational surface agrees well with the
observed results. This model also accounts for the presence of
harmonic structures seen in the tail of the comet.

Comet 1980 XII Meier

103.561 **Kometenduo Meier en Panther.**   H. Feijth.
Zenit, 9. Jaarg., 314 - 316 (1982).

Comet 1980u Panther

103.581 **Comet Panther 1980u.**
British Astron. Assoc. Circ., No. 617 (1981).

Kometenduo Meier en Panther.
See Abstr. 103.561.

Comet 1979 X Bradfield

103.601 **Production of CS and S in Comet Bradfield
(1979 X).**   W. M. Jackson, J. B. Halpern,
P. D. Feldman, J. Rahe.
Astron. Astrophys., Vol. 107, 385 - 389 (1982).
High and low resolution ultraviolet spectra of carbon
monosulfide (CS) in Comet Bradfield (1979 X) were
obtained with the IUE satellite. The high resolution
rotational profile of the (0,0) band at 257.5 nm can be fitted
with a theoretical profile derived assuming a Boltzmann
temperature of 70 K. Spatial plots of the low resolution data
for both S and CS show that these emissions are concentrated
toward the cometary nucleus. The results that have been
obtained are consistent with a Haser model for CS and S
where the parent molecule is CS_2. A very rapid variation of
CS brightness with heliocentric distance is found.

103.602 **CO in the IUE spectrum of Comet Bradfield (1979*l*).**
P. S. Murty.
Moon Planets, Vol. 26, 101 - 104 (1982).
New identifications are reported of the fourth positive
bands and Cameron bands of CO in the IUE satellite spectrum
of Comet Bradfield (1979*l*). Although the predicted band
intensities as well as the band identifications in Comet West

(1976 VI) support the proposed assignments, VUV cometary spectra of higher resolution are necessary for conformation.

Periodic comet Swift-Tuttle

103.641 **Steelt komeet Swift-Tuttle de show?**
R. J. Bouma, E. P. Bus.
Zenit, 9. Jaarg., 110 - 114 (1982).

Periodic comet Gehrels 3

Temporary satellite captures by Jupiter for orbits resembling the one of comet P/Gehrels 3.
See Abstr. 102.049.

Periodic comet Tempel 2

103.701 **Periodic comet Tempel 2 (1982d).**
IAU Circ., Nos. 3676, 3698 (1982).

103.702 **Periodic comet Tempel 2 (1982d).**
Yamamoto Circ., No. 1973 (1982).

103.703 **Comet P/Temple 2 1982d.**
British Astron. Assoc. Circ., No. 625 (1982).

Comet nucleus impact probe.
See Abstr. 051.007.

Periodic comet Stephan-Oterma

103.721 **Position observations of comet Stephan-Oterma (1980g) and of asteroid 751.**
S. S. Smirnov.
Komet. Tsirk., Kiev, No. 284 (1982). In Russian.

Periodic comet Boethin

103.741 **On the long-term orbital evolution of comet P/Boethin.** D. Benest, R. Bien, H. Rickman.
Sun and planetary system, (see 012.040), p. 397 - 400 (1982).
The possible effect of nongravitational forces on the motion of comet P/Boethin is investigated. Besides, the orbital period is varied. The authors find in all cases that the comet librates temporarily around the 1/1 resonance with Jupiter as a remote Jovian satellite. A time interval of 2000 years backward and forward is treated. It is found that the long-term orbital evolution of comet P/Boethin is likely to be governed by major perturbations at close encounters with Jupiter.

Comet 1981b Bus

103.761 **New comet Bus 1981b.**
British Astron. Assoc. Circ., No. 617 (1981).

Periodic comet Encke

103.801 **Comet Encke: radar detection of nucleus.**
P. G. Kamoun, D. B. Campbell, S. J. Ostro, G. H. Pettengill, I. I. Shapiro.
Science, Vol. 216, 293 - 295 (1982).
The nucleus of the periodic comet Encke was detected in November 1980 with the Arecibo Observatory's radar system (wavelength, 12.6 centimeters). The echoes in the one sense of circular polarization received imply a radar cross section of 1.1 ± 0.7 square kilometers. The estimated bandwidth of these echoes combined with an estimate of the rotation vector of Encke yields a radius for the nucleus of $1.5^{+2.3}_{-1.0}$ kilometers. The uncertainties given are dependent primarily on the range of model considered for the comet and for the manner in which its nucleus backscatters radio waves. Should this range prove inadequate, the true value of the radius of the nucleus might lie outside the limits given.

Periodic comet Perrine-Mrkos

103.901 **Short-period comet Perrine-Mrkos 1968 VIII.**
Komet. Tsirk., Kiev, No. 284 (1982). In Russian.

104 Meteors, Meteor Streams

104.001 The annual variation in the activity of the Geminid shower and the theory of the dispersal of meteoroid clusters.   J. Jones.
Mon. Not. R. Astron. Soc., Vol. 198, 23 - 32 (1982).

The author finds an intrinsic fluctuating component in the activity of the Geminid shower which is explained in terms of a number of distinct streamlets produced from the fragments of a disintegrated comet. Comparison with theory indicates that the particles in these streamlets were ejected uniformly round one orbit of the parent cometary fragment rather than close to perihelion and supports the notion of an insulating layer on the cometary surface.

104.002 Meteor shower flux determination from radar observations. I. The case when the applied physical theory does not consider deceleration.   P. Pecina.
Bull. Astron. Inst. Czechoslovakia, Vol. 33, 1 - 10 (1982).

Kaiser's and Belkovich's theories of flux determination are briefly discussed. A new variant of the shower-flux determination theory, which uses meteor-echo overdense durations as measurable quantities, is then developed. The principal formula, enabling the desired relation to be derived, contains the flux as a norming factor and may be applied to showers as well as to the sporadics. The developed theory is applied to the Geminid shower observations 1959 - 1978.

104.003 Meteor shower flux determination from radar observations. II. The case when the applied physical theory considers deceleration.   P. Pecina.
Bull. Astron. Inst. Czechoslovakia, Vol. 33, 11 - 16 (1982).

The paper deals with determining the meteor shower flux when the deceleration of meteor bodies is taken into account. The flux formula is derived and the procedure described. The flux values for the Geminid shower show that if deceleration is neglected, the computed flux is systematically underestimated, but since the difference is only a few percent, in comparison with other possible effects the deceleration can actually be neglected.

104.004 Meteor trajectory data from visual observations.
S. S. Mims.
J. Meteor Res., Vol. 1, 5 - 8 (1982).

104.005 The Upsilon Pegasid meteor shower.
H. R. Povenmire.
J. Meteor Res., Vol. 1, 9 - 11 (1982).

104.006 Fireball observations from the Institute for Astronomical Research, 1975 - 1980.
R. W. James, S. S. Mims.
J. Meteor Res., Vol. 1, 12 - 15 (1982).

104.007 Osservazione delle Perseidi 1980.
M. Eltri, E. Stomeo.
Astronomia, N. 3, p. 10 - 19 (1981).

104.008 The stability of the node of the Perseid meteor stream.   D. W. Hughes, B. Emerson.
Observatory, Vol. 102, 39 - 42 (1982).

104.009 Vector model of a diffraction radio echo from a meteor train.   K. V. Kostylev.
Meteorn. rasprostr. radiovoln, Kazan', 1981, No. 17, p. 3 - 8.
In Russian. − Abstr. in Ref. zh., 51. Astron., 2.51.264 (1982).

104.010 Using diffraction pictures for measurement of underdense meteor trains.   N. S. Andrianov.

Meteorn. rasprostr. radiovoln, Kazan', 1981, No. 17, p. 9 - 13.
In Russian. − Abstr. in Ref. zh., 51. Astron., 2.51.265 (1982).

104.011 Method of fluctuations in meteor astronomy problems.   G. V. Andreev, R. G. Lazarev, L. N. Rubtsov.
Bull. Astron. Inst. Czechoslovakia, Vol. 33, 111 - 116 (1982).

The article proposes a probability model describing the behaviour of a number of fixed meteors as a combination of sporadic and stream meteors. The application of the method of fluctuations (Agekyan, 1957) to this model enabled a general differential equation of fluctuations of the number of observed meteors to be obtained in recurrent form with respect to the initial moments. This equation can be applied to some statistical problems of meteor astronomy. In particular, the present article shows the application of this equation to the problem of separating the stream meteors from the observed number of fixed meteors in the period of Leonids activity.

104.012 Distribution of sporadic meteor bodies according to kinetic energies depending on the position of the apex.   Z. M. Ioffe, T. P. Antropova, L. N. Rubtsov.
Astron. vestn., Tom 15, 248 - 249 (1981). In Russian. − Abstr. in Ref. zh., 51. Astron., 3.51.308 (1982).

104.013 Community of origin and evolution of meteor bodies causing radio and photographic meteors.
V. N. Lebedinets, A. V. Manokhina.
Astron. vestn., Tom 15, 216 - 219 (1981). In Russian. − Abstr. in Ref. zh., 51. Astron., 3.51.312 (1982).

104.014 Cosmic matter influx to the earth: estimate based on spectrophotometric observations of meteors.
V. A. Smirnov.
Astron. vestn., Tom 15, 220 - 225 (1981). In Russian. − Abstr. in Ref. zh., 51. Astron., 3.51.313 (1982).

104.015 Meteor influx rate to the earth.
I. N. Kovshun.
Astron. vestn., Tom 15, 244 - 247 (1981). In Russian. − Abstr. in Ref. zh., 51. Astron., 3.51.314 (1982).

104.016 Nodal regression of the Quadrantid meteor stream: an analytic approach.   C. D. Murray.
Icarus, Vol. 49, 125 - 134 (1982).

By using a high-order expansion of the disturbing function the author shows how the behavior of the longitude of ascending node of the Quadrantid stream is a result of both secular and resonant effects. The analysis illustrates how the proximity of the stream's orbit to the 2:1 commensurability with Jupiter dominates the short-term variations in orbital elements.

104.017 Theoretical twin meteor showers. J. D. Drummond.
Icarus, Vol. 49, 135 - 142 (1982).

Attention is drawn to, and theoretical radiants calculated for, 27 hypothetical twin showers to Cook's (1973) working list of meteor streams. Of these twin showers, 7 are previously known, 1 is a night-time twin to the ϵ Geminids, 9 are found to be contained among Sekanina's (1976) radar systems, and the remaining 10 are undetected daytime showers. Minimum radii are computed for all of Cook's streams and are used to assess the possibility of detecting the twin showers.

104.018 Formation of the Geminid meteor stream and primary stage of its development. Model calculation. I.
A. M. Kazantsev, L. M. Sherbaum.
Vestn. Kiev. Univ., Vyp. 23, p. 105 - 109 (1981). In Russian.

104.019 **Meteor showers.**   Ľ. Kresák.
Kozmos, Vol. 13, 61 - 62 (1982). In Slovak.

104.020 **Shall we see a Perseid meteor stream?**
M. Kresáková.
Kozmos, Vol. 13, 39 - 41 (1982). In Slovak.

104.021 **April Lyrid meteors.**
IAU Circ., No. 3691 (1982).

104.022 **1982 Quadrantids.**
Yamamoto Circ., No. 1968 (1982).

104.023 **Structure of small meteoroids deduced from two
station low light level TV observations of meteor
trails.**   M. A. Hapgood, P. Rothwell.
The comparative study of the planets, (see 012.033), p. 311 -
315 (1982).
The authors report on a study of meteor trails in the
earth's upper atmosphere, which were imaged with low light
level TV cameras simultaneously from two stations. They
determined the characteristics of 98 meteor trails, 39
recorded in the South of England during the Perseid meteor
showers of 1977 and 1978, and 59 recorded in Northern
Norway, in October/November 1977.

104.024 **Did the Perseids peak in 1980?**   J. A. Russell.
Sky Telesc., Vol. 64, 10 - 11 (1982).

104.025 **On the effect of the astronomical factor on orbital
distribution of meteor bodies.**
V. N. Lebedinets, A. V. Manokhina.
Astron. vestn., Tom 16, 37 - 42 (1982). In Russian. – Abstr.
in Ref. zh., 51. Astron., 6.51.247 (1982).

104.026 **On the process of formation of Trowbridge tubes in
meteor trains.**
O. V. Dobrovol'skij, A. V. Blokhin, G. G. Novikov.
Astron. vestn., Tom 16, 43 - 45 (1982). In Russian. – Abstr.
in Ref. zh., 51. Astron., 6.51.248 (1982).

104.027 **The total mass and structure of the meteor stream
associated with comet Halley.**   A. Hajduk.
Sun and planetary system, (see 012.040), p. 335 - 336 (1982).
The meteor stream associated with comet Halley has
been studied on the basis of a long series of visual and radar
observations during the Eta Aquarid and Orionid meteor
shower periods. It appeared that the stream exhibited
inhomogeneities in both directions, across it and along the
orbit. A stable zone of higher density and variable filaments
have been detected. The total mass of the stream was determined at 5×10^{14} kg.

104.028 **Sarajevo Observatory fireball patrol.**
V. Miličević, M. Muminović.
Sun and planetary system, (see 012.040), p. 349 - 350 (1982).

104.029 **On the displacement of meteor showers activity.**
P. B. Babadzhanov, Yu. V. Obrubov.
Sun and planetary system, (see 012.040), p. 401 - 402 (1982).
The influence of planetary perturbations on the orbits of
meteor streams causes the appearance of new meteor showers
and disappearance of the known ones. The authors consider
another consequence of planetary perturbations: the regular
displacement of dates of the maximum activity of meteor
showers.

104.030 **High-resolution radar studies of the Geminid meteor
shower.**   J. Jones, J. D. Morton.

Mon. Not. R. Astron. Soc., Vol. 200, 281 - 291 (1982).
Observations of the 1980 Geminid shower made with a
new high-resolution imaging meteor-radar system are described.
1013 Geminid echoes were observed and at its peak the rate of
shower meteors was about 13 times that of the sporadic
meteor background. The authors find a corrected radiant of
RA = 112°3, dec = 33°05, and no evidence of a second centre
of activity near RA = 90°0, dec = 32°0, as has been previously
postulated. A mean value of 1.69 ± 0.07 is found for the mass
distribution index, corresponding to meteors of mean magnitude +8.0 ± 0.5.

104.031 **The evolution of the orbit of the Geminid meteor
stream.**   K. Fox, I. P. Williams, D. W. Hughes.
Mon. Not. R. Astron. Soc., Vol. 200, 313 - 324 (1982).
The Geminids were first reliably recorded in 1862.
Observations of this stream over the last 118 years indicate
that the ascending node is stationary at an Ω value of
261.38 ± 0.11. A theoretical analysis of the effect of planetary
induced gravitational perturbation on the stream results in a
predicted decrease in the ascending node by 1.57 deg
century^{-1}, The general rule that streams with inclinations
greater than 90° have positive $d\Omega/dt$ values whereas streams
with inclinations less than 90° have negative values is
disobeyed by the Geminids. Various possible explanations for
this discrepancy are discussed and one of these turns out to
be very satisfactory.

104.032 **Bright fireball: 1981 May 25.**
British Astron. Assoc. Circ., No. 618 (1981).

104.033 **1981 Delta Aquarids and Perseids.**
British Astron. Assoc. Circ., No. 618 (1981).

104.034 **Fireball – 1981 July 30.**
British Astron. Assoc. Circ., No. 620 (1981).

104.035 **Fireball – 1981 August 2.**
British Astron. Assoc. Circ., No. 620 (1981).

104.036 **1981 Orionids.**
British Astron. Assoc. Circ., No. 621 (1981).

104.037 **Bright fireballs.**
British Astron. Assoc. Circ., No. 625 (1982).

104.038 **The Lyrids – unexpectedly intense shower.**
British Astron. Assoc. Circ., No. 626 (1982).

104.039 **Report of IAU Commission 22: Meteors and interplanetary dust** (Météores et la poussière interplanétaire).   W. G. Elford.
Trans. IAU, Vol. XVIIIA, (see 003.013), p. 219 - 236 (1982).

**A method for imaging radio meteor radiant distribu-
tions.**   See Abstr. 031.502.

Plotting errors in visual meteor observations.
See Abstr. 031.633.

**Theoretical meteor radiants of Apollo, Amor, and
Aten asteroids.**   See Abstr. 098.037.

**On the similarity of orbits of associated comets,
asteroids and meteoroids.**   See Abstr. 102.012.

105 Meteorites, Meteorite Craters

105.001 A corundum-rich inclusion in the Murchison carbonaceous chondrite.   M. Bar-Matthews, I. D. Hutcheon, G. J. MacPherson, L. Grossman.
Geochim. Cosmochim. Acta, Vol. 46, 31 - 41 (1982).

A corundum-hibonite inclusion, BB-5, has been found in the Murchison carbonaceous chondrite. This is the first reported occurrence of corundum as a major phase in any refractory inclusion, even though this mineral is predicted by thermodynamic calculations to be the first condensate from a cooling gas of solar composition.

105.002 The concentration and isotopic composition of hydrogen, carbon and nitrogen in carbonaceous meteorites.   F. Robert, S. Epstein.
Geochim. Cosmochim. Acta, Vol. 46, 81 - 95 (1982) = Contrib. No. 3609 Div. Geol. Planet. Sci., Calif. Inst. Technol.

Concentrations and isotopic compositions were determined for H_2, N_2 and C extracted by stepwise pyrolysis from powdered meteorites, from residues of meteorites partially dissolved with aqueous HF, and from residues of meteorites reacted with HF-HCl solutions. The meteorites treated were the carbonaceous chondrites, Orgueil, Murray, Murchison, Renazzo and Cold Bokkeveld.

105.003 Carbon, hydrogen and nitrogen isotopes in solvent-extractable organic matter from carbonaceous chondrites.   R. H. Becker, S. Epstein.
Geochim. Cosmochim. Acta, Vol. 46, 97 - 103 (1982) = Contrib. No. 3520 Div. Geol. Planet. Sci., Calif. Inst. Technol.

Solvent extractions were done on the carbonaceous chondrites Murray, Murchison, Orgueil and Renazzo, using CCl_4 and CH_3OH. The combined data for C, H and N isotopes makes it highly unlikely that the CH_3OH-soluble components are derivable from, or simply related to, the insoluble organic polymer found in the same meteorites. A relationship is suggested between the event that formed hydrous minerals in CI 1 and CM 2 meteorites and the introduction of water-soluble (methanol-soluble) organic compounds.

105.004 A dendritic solidification model to explain Ge-Ni variations in iron meteorite chemical groups.
C. Narayan, J. I. Goldstein.
Geochim. Cosmochim. Acta, Vol. 46, 259 - 268 (1982).

Segregation during solidification is responsible for the secondary fractionation of trace elements in iron meteorite chemical groups. This study examines the consequences of dendritic segregation on the Ge-Ni fractionation in iron meteorite chemical groups.

105.005 Basic nitrogen-heterocyclic compounds in the Murchison meteorite.
P. G. Stoks, A. W. Schwartz.
Geochim. Cosmochim. Acta, Vol. 46, 309 - 315 (1982).

A fragment of the Murchison (C2) carbonaceous meteorite was analyzed for basic, N-heterocyclic compounds, by dual detector capillary gas chromatography as well as capillary gas chromatography/mass spectrometry, using two columns of different polarity. In the formic acid extract 2,4,6-trimethylpyridine, quinoline, isoquinoline, 2-methylquinoline and 4-methylquinoline were positively identified.

105.006 Evidence for the three-stage cooling history of olivine-porphyritic fluid droplet chondrules.
H. N. Planner, K. Keil.
Geochim. Cosmochim. Acta, Vol. 46, 317 - 330 (1982).

Three haplochondritic melts were experimentally examined to investigate thermal parameters in the pre-accretionary chondrule environment. A fractional crystallization model, designed to incorporate the effects of lattice diffusion in olivine but not volume diffusion of magnesium or iron in the melt, was found to yield results in excellent agreement with olivine formed in experimental charges having a melt composition most similar to chondrules. Application of this numerical model to three olivine-porphyritic chondrules from the Tieschitz (H3) chondrite indicates that their mean olivine phenocryst compositions are too fayalitic to have been derived from a continuous cooling event. Partial equilibration of the phenocrysts with the melt in a high-temperature isothermal environment is postulated to explain the measured mean values.

105.007 Metamorphic reactions in mesosiderites: origin of abundant phosphate and silica.
G. E. Harlow, J. S. Delaney, C. E. Nehru, M. Prinz.
Geochim. Cosmochim. Acta, Vol. 46, 339 - 348 (1982).

The high modal abundances of merrillite $[Ca_3(PO_4)_2]$ and tridymite in most mesosiderites are not the result of igneous fractionation but are attributed to redox reactions between silicates and P-bearing Fe-Ni metal within a limited T-fO_2 range at low pressure. The Emery mesosiderite is the most tridymite- and merrillite-rich mesosiderite so it is used as the model for this study.

105.008 Der Arizona-Meteoritenkrater.   W. Lüthi.
Orion, 40. Jahrg., 40 - 41 (1982).

105.009 Micrometeorieten: nog steeds vol raadsels.   J. van Diggelen.
Zenit, 9. Jaarg., 106 - 109 (1982).

105.010 De meteorietkraters van Faugères en Steinheim.   W. Buijze.
Zenit, 9. Jaarg., 132 - 133 (1982).

105.011 On the probable composition of the El'gygytgyn meteorite.   A. A. Val'ter, I. F. Barchuk, V. S. Bulkin, A. F. Ogorodnik, Eh. Yu. Kotishevskaya.
Pis'ma Astron. Zh., Tom 8, 115 - 120 (1982). In Russian. English translation in Soviet Astron. Lett., Vol. 8.

By means of chemical, emission spectral and neutron activation analyses, the excesses of average contents in impactites in comparison with target rocks are determined. The measurements correspond to the achondrite (ureilite) type of the meteorite that formed the crater.

105.012 Meteorite erosion and cosmic ray variations.   D. W. Hughes.
Nature, Vol. 295, 279 - 280 (1982).

105.013 Lower mass limits for impacted meteoroids.   S. S. Mims.
J. Meteor Res., Vol. 1, 1 - 4 (1982).

Crater-forming collisions are an important aspect of geological, meteorical, and cosmogonical studies of the solar system. By considering the largest impact structures on the terrestrial planets and satellites, lower mass limits for meteoroids at the upper end of the size distribution can be found.

105.014 Classification and elucidation of meteor acoustics.   C. H. Annett.
Irish Astron. J., Vol. 14, 199 - 211 (1980).

105.015 On preatmospheric radii of the Tagalyn Bulan meteorite.
G. I. Dolivo-Dobrovol'skaya, V. D. Kolomenskij, L. K. Levskij, D. Lkhagvasurehn, O. Otgonsurehn, V. P. Perelygin,

S. G. Stetsenko, D. S. Yadav, A. P. Sharma.
Obedin. inst. yader. issled. Dubna, Prepr., 1981, No. P14-81-550,
p. 1 - 5. In Russian. – Abstr. in Ref. zh., 51. Astron., 2.51.275
(1982).

105.016 **Ordered mixed-layer structures in the Mighei
carbonaceous chondrite matrix.**
I. D. R. Mackinnon.
Geochim. Cosmochim. Acta, Vol. 46, 479 - 489 (1982).

105.017 **Evidence for aqueous alteration in a carbonaceous
xenolith from the Plainview (H5) chondrite.**
S. Nozette, L. L. Wilkening.
Geochim. Cosmochim. Acta, Vol. 46, 557 - 563 (1982).

105.018 **The Blue Angel: I. The mineralogy and petrogenesis
of a hibonite inclusion from the Murchison
meteorite.**
J. T. Armstrong, G. P. Meeker, J. C. Huneke, G. J. Wasserburg.
Geochim. Cosmochim. Acta, Vol. 46, 575 - 595 (1982) = Div.
Geol. Planet. Sci., Calif. Inst. Technol., Pasadena, Calif.,
Contrib. No. 3522 (370).
A detailed mineralogic, chemical, and petrologic study of
the Blue Angel, a relatively large (~1.5 mm) hibonite-con-
taining inclusion from the Murchison meteorite, was performed in an attempt to understand the mechanisms of formation
and modification of hibonite-rich inclusions.

105.019 **The compositional classification of chondrites: II.
The enstatite chondrite groups.**
D. W. Sears, G. W. Kallemeyn, J. T. Wasson.
Geochim. Cosmochim. Acta, Vol. 46, 597 - 608 (1982).
The authors present new data from a neutron activation
analysis of four enstatite chondrites including the taxonomi-
cally important St. Sauveur, and discuss the classification of
enstatite chondrites.

105.020 **L'olivine dans les howardites: origine, et implications
pour le corps parent de ces météorites achondriti-
ques.** C. Desnoyers.
Geochim. Cosmochim. Acta, Vol. 46, 667 - 680 (1982).

105.021 **A search for terrestrial analogs to Martian lobed
impact craters.**
J. F. McHone, R. Greeley.
Reports of Planetary Geology Program – 1981, (see 003.001),
p. 78 - 80 (1981). – Abstract.

105.022 **Tetrataenite and the origin of magmatic meteorites.**
J. F. Albertsen, N. O. Roy-Poulsen, L. Vistisen.
Meteoritics, Vol. 16, 288 (1981). – Abstract.

105.023 **Hydrothermal alteration of melt rocks in impact
craters.** C. C. Allen, K. Keil.
Meteoritics, Vol. 16, 288 - 289 (1981). – Abstract.

105.024 **Some observations on the mechanism of meteorite
concentration in Victoria Land, Antarctica.**
J. O. Annexstad.
Meteoritics, Vol. 16, 290 (1981). – Abstract.

105.025 **Liquid-vapor inclusions in achondritic meteorites.**
L. D. Ashwal, D. W. Mogk, S. C. Bergman,
E. K. Gibson, Jr., D. J. Henry, J. L. Warner, R. Lee-Berman.
Meteoritics, Vol. 16, 290 - 291 (1981). – Abstract.

105.026 **Electron microscopy and metallographic studies of
artificially shocked Kernouvé chondrite.**
J. R. Ashworth, A. W. R. Bevan.
Meteoritics, Vol. 16, 291 (1981). – Abstract.

105.027 **Structure, mineralogy and mineral chemistry of
silicate inclusions in the Elga octahedrite.**
G. V. Baryshnikova, E. G. Osadchij.
Meteoritics, Vol. 16, 292 (1981). – Abstract.

105.028 **Radiogenic noble gases and the thermal history of
Kirin.**
F. Begemann, Z. Li, S. Schmitt-Strecker, H. Weber, Z. Xu.
Meteoritics, Vol. 16, 293 (1981). – Abstract.

105.029 **Multiple exposure history of chondrites.**
N. Bhandari, M. B. Potdar, K. M. Suthar.
Meteoritics, Vol. 16, 293 (1981). – Abstract.

105.030 **Trace element contents of Antarctic meteorites.**
S. Biswas, T. M. Walsh, M. E. Lipschutz.
Meteoritics, Vol. 16, 294 (1981). – Abstract.

105.031 **Impact melting and brecciation of the Cachari
eucrite 3.0 Gy ago.** D. D. Bogard, G. J. Taylor,
K. Keil, M. R. Smith, R. A. Schmitt, J. Danon.
Meteoritics, Vol. 16, 296 (1981). – Abstract.

105.032 **Cosmic ray track studies of Jilin (Kirin) chondrite.**
M. Bourot-Denise, P. Pellas.
Meteoritics, Vol. 16, 297 (1981). – Abstract.

105.033 **On the oriented sheen in ataxites.**
V. F. Buchwald.
Meteoritics, Vol. 16, 298 - 299 (1981). – Abstract.

105.034 **Usti nad Orlici (Kerhartice), a L6 chondrite from
Czechoslovakia.**
M. Bukovanská, P. Jakeš, A. El Goresy.
Meteoritics, Vol. 16, 299 - 300 (1981). – Abstract.

105.035 **Laboratory annealing studies of thermoluminescence
in the Kirin (H5) chondrite.**
R. K. Bull, S. A. Durrani.
Meteoritics, Vol. 16, 300 - 301 (1981). – Abstract.

105.036 **Cm/U, Th/U, and $^{235}U/^{238}U$ in meteorites.**
J. H. Chen, G. J. Wasserburg.
Meteoritics, Vol. 16, 301 (1981). – Abstract.

105.037 **Microanalysis by Raman spectroscopy of carbon in
chondrites.**
M. Christophe-Michel-Lévy, A. Lautié.
Meteoritics, Vol. 16, 301 - 302 (1981). – Abstract.

105.038 **Composition and metallography of two recently
recovered Wabar meteorite specimens.**
R. S. Clarke, Jr., E. Jarosewich, A. A. Almohandis.
Meteoritics, Vol. 16, 303 (1981). – Abstract.

105.039 **Refractory inclusions in the Mokoia C3(V) carbonaceous chondrites.** R. E. Cohen.
Meteoritics, Vol. 16, 304 (1981). – Abstract.

105.040 **Noble gases in mineral fractions of E-chondrites.**
J. Crabb.
Meteoritics, Vol. 16, 304 - 305 (1981). – Abstract.

105.041 **Awaruite (Ni_3Fe) in the genomict LL chondrite
Parambu: formation under high FO_2.**
J. Danon, M. Christophe-Michel-Lévy, C. Jehanno, K. Keil,
C. B. Gomes, R. B. Scorzelli, I. Souza Azevedo.
Meteoritics, Vol. 16, 305 (1981). – Abstract.

105.042 **The Guangrao olivine-hypersthene chondrite.**
D. Wang, W. Hou, X. Zhou.
Meteoritics, Vol. 16, 306 (1981). – Abstract.

105.043 **Oxygen isotopes, rare-earth elements, supernovae and Wolf-Rayet stars.**   D. S. Dearborn, W. V. Boynton.
Meteoritics, Vol. 16, 306 - 307 (1981). − Abstract.

105.044 **Umbarger and Summerfield: two new chondrites from Texas.**
B. D. Dod, E. Olsen, D. Eatough, P. Sipiera.
Meteoritics, Vol. 16, 307 (1981). − Abstract.

105.045 **Metal-silicate fractionation in chondrites.**
R. T. Dodd.
Meteoritics, Vol. 16, 307 - 308 (1981). − Abstract.

105.046 **Experimental trace element partitioning applied to iron meteorites.**   M. J. Drake, J. Jones.
Meteoritics, Vol. 16, 309 - 310 (1981). − Abstract.

105.047 **Chemistry of the Shergotty parent body.**
G. Dreibus, H. Palme, W. Rammensee, B. Spettel, H. Wänke.
Meteoritics, Vol. 16, 310 (1981). − Abstract.

105.048 **Ries crater, Germany: petrography of the suevite and conclusions on crater formation.**
W. v. Engelhardt, G. Graup.
Meteoritics, Vol. 16, 311 (1981). − Abstract.

105.049 **Preatmospheric size of Dhurmsala, ALHA 77003, and other meteorites: Mn-53 and Al-26 studies on core samples.**   P. Englert, U. Herpers, W. Herr.
Meteoritics, Vol. 16, 311 - 312 (1981). − Abstract.

105.050 **Late crystallisation of the natural Fe-Ni-S-P system − evidence from Cape York troilite inclusions.**
K. H. Esbensen, V. F. Buchwald.
Meteoritics, Vol. 16, 313 (1981). − Abstract.

105.051 **Nitrogen thermochemistry in solar composition material.**   M. B. Fegley, Jr.
Meteoritics, Vol. 16, 314 (1981). − Abstract.

105.052 **Thermal experiments on study of metal-troilite relationships in ordinary chondrites.**
A. V. Fisenko, Z. A. Lavrentjeva, A. K. Lavrukhina.
Meteoritics, Vol. 16, 315 (1981). − Abstract.

105.053 **Trace element concentration in Abee density separates and minerals.**   R. M. Frazier, W. V. Boynton.
Meteoritics, Vol. 16, 315 (1981). − Abstract.

105.054 **Carbonates and magnetites in the Renazzo chondrite.**   K. Fredriksson, B. Mason, R. Beauchamp, G. Kurat.
Meteoritics, Vol. 16, 316 (1981). − Abstract.

105.055 **Correlations between physical and petrological properties of chondrules.**
J. L. Gooding, K. Keil, T. Fukuoka, R. A. Schmitt.
Meteoritics, Vol. 16, 318 (1981). − Abstract.

105.056 **Stable carbon isotope measurements of ordinary chondrites.**   M. M. Grady, P. K. Swart, C. T. Pillinger.
Meteoritics, Vol. 16, 319 (1981). − Abstract.

105.057 **An unequilibrated inclusion in the Romero (H4) chondrite.**   A. L. Graham.
Meteoritics, Vol. 16, 319 (1981). − Abstract.

105.058 **Origin and derivation of stony-iron meteorites.**
R. Greenberg, C. R. Chapman.
Meteoritics, Vol. 16, 320 (1981). − Abstract.

105.059 **Manicouagan impact structure: its original dimensions and form.**   R. A. F. Grieve, J. W. Head, III.
Meteoritics, Vol. 16, 320 - 321 (1981). − Abstract.

105.060 **The refractory component in Semarkona chondrules and the fractionation of refractory elements during the formation of ordinary chondrites.**
J. N. Grossman, J. T. Wasson.
Meteoritics, Vol. 16, 321 - 322 (1981). − Abstract.

105.061 **Lead isotopic studies of Antarctic chondrites.**
B. B. Hanan, G. R. Tilton.
Meteoritics, Vol. 16, 322 (1981). − Abstract.

105.062 **Phosphorus in mesosiderite metal: a IIIAB correlation.**   G. E. Harlow, J. S. Delaney, M. Prinz, C. E. Nehru.
Meteoritics, Vol. 16, 322 - 323 (1981). − Abstract.

105.063 **Ion microscopy of chondritic components: a search for prechondritic materials.**
A. Havette, J. C. Lorin, G. Slodzian.
Meteoritics, Vol. 16, 324 (1981). − Abstract.

105.064 **^{53}Mn and ^{26}Al in observed chondrite falls with high exposure ages.**   W. Herr, U. Herpers, P. Englert.
Meteoritics, Vol. 16, 324 - 325 (1981). − Abstract.

105.065 **^{40}Ar-^{39}Ar dating of Chainpur chondrules: search for excessive ages.**   I. Herrwerth, A. El Goresy, E. K. Jessberger, T. Kirsten, N. Müller, E. Pernicka.
Meteoritics, Vol. 16, 325 (1981). − Abstract.

105.066 **Chemical composition of differentiated Antarctic meteorites.**   J. Hertogen.
Meteoritics, Vol. 16, 326 (1981). − Abstract.

105.067 **Kirin, its irradiation ages and the reconstruction of its preatmospheric size.**
G. Heusser, Z. Ouyang.
Meteoritics, Vol. 16, 326 - 327 (1981). − Abstract.

105.068 **Basaltic clasts in the Garland diogenite.**
R. H. Hewins.
Meteoritics, Vol. 16, 328 (1981). − Abstract.

105.069 **Titanium isotopic anomalies in meteorites: endemic, ubiquitous, ephemeral, or incompletely characterized?** H. R. Heydegger, J. J. Foster, W. Compston.
Meteoritics, Vol. 16, 328 - 329 (1981). − Abstract.

105.070 **^{39}Ar loss during neutron irradiation and the aging of Allende inclusions.**   J. C. Huneke, I. M. Villa.
Meteoritics, Vol. 16, 329 - 330 (1981). − Abstract.

105.071 **H-group chondrites: they accreted hot and petrologic type is inversely related to cooling rate.**
R. Hutchison, A. W. R. Bevan, S. O. Agrell, J. R. Ashworth.
Meteoritics, Vol. 16, 330 (1981). − Abstract.

105.072 **Terrestrial-type xenon in meteoritic troilite.**
G. Hwaung, L. L. Oliver, O. K. Manuel.
Meteoritics, Vol. 16, 331 (1981). − Abstract.

105.073 **Unmasking "extra" ^{40}Ar in ALHA 77005 by the laser extraction technique.**   E. K. Jessberger, O. A. Schaeffer, R. Warasila, R. Walker, T. Labotka.
Meteoritics, Vol. 16, 331 - 332 (1981). − Abstract.

105.074 Search for extinct ^{36}Cl in Allende.
J. Jordan, E. Pernicka.
Meteoritics, Vol. 16, 332 - 333 (1981). – Abstract.

105.075 Pu-Nd-Xe dating: systematics in ^{244}Pu fission and REE spallation components.
D. T. Jost, K. Marti, E. Sutter.
Meteoritics, Vol. 16, 334 (1981). – Abstract.

105.076 Noble metal sites in Allende – chemical evidence.
S. Jovanovic, G. W. Reed, Jr.
Meteoritics, Vol. 16, 334 - 335 (1981). – Abstract.

105.077 Mineralogy, chemistry and noble gases in silicate inclusions of the Campo del Cielo iron meteorite.
J. Jung, L. Schultz.
Meteoritics, Vol. 16, 335 - 336 (1981). – Abstract.

105.078 Apatite in Orgueil. Carrier phase for Neon-E?
M. H. A. Jungck, F. O. Meier, P. Eberhardt.
Meteoritics, Vol. 16, 336 - 337 (1981). – Abstract.

105.079 The composition and formation of the carbonaceous chondrites. G. W. Kallemeyn.
Meteoritics, Vol. 16, 337 (1981). – Abstract.

105.080 Tracks and cosmogenic isotopes in the Tsarev chondrite. L. L. Kashkarov, V. D. Gorin,
A. V. Fisenko, G. V. Kalinina, A. K. Lavrukhina.
Meteoritics, Vol. 16, 337 - 338 (1981). – Abstract.

105.081 Trace elements in refractory inclusions from the Murchison C2 chondrite.
I. Kawabe, L. Grossman, T. Tanaka, A. M. Davis.
Meteoritics, Vol. 16, 338 - 339 (1981). – Abstract.

105.082 The Ybbsitz – meteorite.
W. Kiesl, F. Kluger.
Meteoritics, Vol. 16, 339 (1981). – Abstract.

105.083 Are CAI's condensates or distillation residues? Evidence from a comprehensive survey of fine- to medium-grained inclusions in the Allende meteorite.
A. S. Kornacki.
Meteoritics, Vol. 16, 341 - 342 (1981). – Abstract.

105.084 Spinel chondrules: further clues to ordinary chondrite precursor rocks.
A. Kracher, F. Brandstätter, G. Kurat.
Meteoritics, Vol. 16, 342 - 343 (1981). – Abstract.

105.085 Rusty ornans.
G. Kurat, F. Brandstätter, H. Palme,
M. Christophe-Michel-Lévy.
Meteoritics, Vol. 16, 343 - 344 (1981). – Abstract.

105.086 Extraterrestrial siderophiles in abyssal sediments.
F. T. Kyte.
Meteoritics, Vol. 16, 345 (1981). – Abstract.

105.087 Differential thermal /DTA/ and thermogravimetric /TG/ study of selected meteoritic material.
B. Lang, A. Grodziński, L. Stoch.
Meteoritics, Vol. 16, 345 (1981). – Abstract.

105.088 Unusual textures and minerals in enstatite chondrites. J. W. Larimer, B. M. P. Trivedi,
J. A. Connolly.
Meteoritics, Vol. 16, 346 - 347 (1981). – Abstract.

105.089 On distribution of siderophile elements in metal phase of ordinary and enstatite chondrites.
A. K. Lavrukhina, A. Yu. Ljulj (Lyul'), G. V. Baryshnikova.
Meteoritics, Vol. 16, 347 - 348 (1981). – Abstract.

105.090 X-ray non-destructive examination of stony meteorites. F. A. Levi.
Meteoritics, Vol. 16, 348 - 349 (1981). – Abstract.

105.091 Noble-gas carriers in Allende (C3V), Hamlet (LL3), and Murchison (C2). R. S. Lewis, M. Ebihara,
E. Anders, J. Matsuda.
Meteoritics, Vol. 16, 349 - 350 (1981). – Abstract.

105.092 Mineralogy of the Kirin chondrite consortium samples. S. Liu, Z. Ouyang, A. El Goresy.
Meteoritics, Vol. 16, 350 - 351 (1981). – Abstract.

105.093 Allende and Leoville inclusions: anomalous magnesium along with normal potassium.
J. C. Lorin, A. Havette, G. Slodzian, P. Lablanquie.
Meteoritics, Vol. 16, 351 - 352 (1981). – Abstract.

105.094 A comparative study of the Murray CM2 carbonaceous chondrite.
I. D. R. Mackinnon, M. B. Baker, T. V. V. King.
Meteoritics, Vol. 16, 352 - 353 (1981). – Abstract.

105.095 Compositional variations among members of the IIIAB Cape York and IVA Gibeon showers.
D. J. Malvin, J. T. Wasson, V. F. Buchwald, K. H. Esbensen.
Meteoritics, Vol. 16, 353 (1981). – Abstract.

105.096 More on the (U-Th-Pb) systematics of the Juvinas achondrite: evidence of primordial lead component.
G. Manhes, C. J. Allègre, A. Provost.
Meteoritics, Vol. 16, 353 - 354 (1981). – Abstract.

105.097 The Abee consortium: probing the parent body of enstatite meteorites. K. Marti.
Meteoritics, Vol. 16, 354 (1981). – Abstract.

105.098 Evidence for annealing of solar flare tracks in certain gas-rich meteorites. B. J. Martinek.
Meteoritics, Vol. 16, 354 - 355 (1981). – Abstract.

105.099 A unique type 3 ordinary chondrite containing graphite-magnetite aggregates – Allan Hills A77011.
S. G. McKinley, E. R. D. Scott, G. J. Taylor, K. Keil.
Meteoritics, Vol. 16, 357 - 358 (1981). – Abstract.

105.100 Deuterium/hydrogen ratios from unequilibrated ordinary chondrites. N. J. McNaughton,
C. T. Pillinger, J. Borthwick, A. E. Fallick.
Meteoritics, Vol. 16, 358 - 359 (1981). – Abstract.

105.101 Igneous layering in an achondritic meteorite.
H. Y. McSween, Jr., A. M. Reid.
Meteoritics, Vol. 16, 359 (1981). – Abstract.

105.102 Chemical composition of Ne-E rich phases in Orgueil. F. O. Meier, M. H. A. Jungck,
P. Eberhardt.
Meteoritics, Vol. 16, 359 - 360 (1981). – Abstract.

105.103 On the production of cosmogenic nuclides in meteorites by solar protons.
R. Michel, G. Brinkmann, R. Stück, W. Herr.
Meteoritics, Vol. 16, 360 - 361 (1981). – Abstract.

105.104 Allan Hills 78084: ^{10}Be and noble gas contents.
R. K. Moniot, T. H. Kruse, W. Savin, T. Milazzo,
G. Herzog, R. Warasila.
Meteoritics, Vol. 16, 361 (1981). – Abstract.

105.105 The Penwell stony-iron meteorite.
C. B. Moore, C. F. Lewis, J. C. Clark.
Meteoritics, Vol. 16, 361 - 362 (1981). — Abstract.

105.106 Evolution of the Moore County pyroxenes as
viewed by an analytical transmission electron
microprobe (ATEM).   H. Mori, H. Takeda.
Meteoritics, Vol. 16, 362 - 363 (1981). — Abstract.

105.107 Th/U microdistribution in enstatite meteorites.
M. T. Murrell, D. S. Burnett.
Meteoritics, Vol. 16, 363 (1981). — Abstract.

105.108 Orthopyroxenites in howardites and mesosiderites
contrasted with diogenites: minor minerals and their
implications.   C. E. Nehru, J. S. Delaney, S. Frishman,
G. E. Harlow, M. Prinz.
Meteoritics, Vol. 16, 364 - 365 (1981). — Abstract.

105.109 Titanium isotopic anomalies in Cl and C2 chondrites.   S. Niemeyer, G. W. Lugmair.
Meteoritics, Vol. 16, 366 - 367 (1981). — Abstract.

105.110 Cosmogenic nuclides in Yamato meteorites.
K. Nishiizumi, M. Imamura, M. Honda, O. Nitoh,
K. Nagao, K. Takaoka.
Meteoritics, Vol. 16, 367 - 368 (1981). — Abstract.

105.111 Sims measurement of magnesium isotopic ratios in
chondrites.   H. Nishimura, J. Okano.
Meteoritics, Vol. 16, 368 - 369 (1981). — Abstract.

105.112 Tetrataenite in the Estherville mesosiderite.
P. Novotny, J. I. Goldstein, D. B. Williams.
Meteoritics, Vol. 16, 370 (1981). — Abstract.

105.113 Caswellsilverite, $NaCrS_2$, a new mineral in the
Norton County enstatite achondrite.
A. Okada, K. Keil.
Meteoritics, Vol. 16, 370 - 371 (1981). — Abstract.

105.114 Spallogenic ^{53}Mn in the Kirin (Jilin) chondrite.
G. Osadnik, U. Herpers, W. Herr, P. Englert.
Meteoritics, Vol. 16, 371 - 372 (1981). — Abstract.

105.115 Cosmogenic noble gases in diamond-bearing
ALHA77283 iron meteorite.
U. Ott, F. Begemann, H. P. Löhr.
Meteoritics, Vol. 16, 373 (1981). — Abstract.

105.116 Iridium-rich phases in Ornans.
H. Palme, F. Wlotzka.
Meteoritics, Vol. 16, 373 - 374 (1981). — Abstract.

105.117 Correlated isotope anomalies on the wing of the
iron abundance peak.
D. A. Papanastassiou, G. J. Wasserburg, F. R. Niederer.
Meteoritics, Vol. 16, 375 (1981). — Abstract.

105.118 Fission tracks in the Marjalahti pallasite.
P. Pellas, C. Perron, G. Crozaz, V. P. Perelygin,
S. G. Stetsenko.
Meteoritics, Vol. 16, 375 - 376 (1981). — Abstract.

105.119 A study of refractory siderophile elements in iron
meteorites.   E. Pernicka, J. T. Wasson.
Meteoritics, Vol. 16, 377 - 378 (1981). — Abstract.

105.120 Incorporation of olivines in the matrix of carbonaceous chondrites: when did it all happen?
R. S. Rajan, G. Poupeau, A. S. Tamhane, J. Gooding,

T. R. Watters.
Meteoritics, Vol. 16, 378 (1981). — Abstract.

105.121 Relict grains in chondrules.
E. Rambaldi.
Meteoritics, Vol. 16, 378 - 379 (1981). — Abstract.

105.122 Reduction and metal forming processes in carbonaceous chondrites.   W. Rammensee, H. Wänke.
Meteoritics, Vol. 16, 379 - 380 (1981). — Abstract.

105.123 Evolution of carbonaceous chondrite parent bodies.
M. N. Rao, C. M. Nautiyal, J. T. Padia,
T. R. Venkatesan.
Meteoritics, Vol. 16, 380 (1981). — Abstract.

105.124 Local bulk Ni-concentration variation in iron
meteorites.   K. L. Rasmussen.
Meteoritics, Vol. 16, 380 - 381 (1981). — Abstract.

105.125 Cosmic-ray-produced deuterium, carbon-13, and
nitrogen-15 in meteorites.   R. C. Reedy.
Meteoritics, Vol. 16, 381 (1981). — Abstract.

105.126 The Dimmitt H chondrite regolith breccia and
implications for the structure of the H chondrite
parent body.   A. E. Rubin, E. R. D. Scott, G. J. Taylor.
K. Keil.
Meteoritics, Vol. 16, 382 - 383 (1981). — Abstract.

105.127 Petrogenetic considerations for ureilites.
G. Ryder.
Meteoritics, Vol. 16, 383 - 384 (1981). — Abstract.

105.128 Three CO3 chondrites from Antarctica — comparison of carbonaceous and ordinary type 3 chondrites.   E. R. D. Scott, G. J. Taylor, P. Maggiore, K. Keil,
S. G. McKinley, H. Y. McSween, Jr.
Meteoritics, Vol. 16, 385 (1981). — Abstract.

105.129 Chemical composition, petrography, mineralogy
and rare gases in the chondrite Ogi.
M. Shima, S. Murayama, H. Yabuki, M. Shima.
Meteoritics, Vol. 16, 386 (1981). — Abstract.

105.130 Isotopic composition of Xe and Kr in the white
inclusions of carbonaceous chondrite Efremovka
(C3V).   J. A. Shukolyukov, Dang Vu Minh,
V. I. Simonovsky, M. M. Fugzan, M. A. Nazarov, M. I. Korina.
Meteoritics, Vol. 16, 387 (1981). — Abstract.

105.131 Neutron-capture reactions in meteorites.
M. S. Spergel, R. C. Reedy, O. W. Lazareth,
P. W. Levy.
Meteoritics, Vol. 16, 387 - 388 (1981). — Abstract.

105.132 The impact melt rocks at the Ries crater.
V. Stähle.
Meteoritics, Vol. 16, 388 (1981). — Abstract.

105.133 Magnetic properties of the Abee meteorite.
N. Sugiura, D. W. Strangway.
Meteoritics, Vol. 16, 390 (1981). — Abstract.

105.134 Yamato-75032, a missing link between diogenites
and eucrites.   H. Takeda, H. Mori.
Meteoritics, Vol. 16, 390 - 391 (1981). — Abstract.

105.135 Al-, K-bearing phyllosilicates in a Ca-Al-rich inclu-
sion from the Murchison C2 meteorite.
K. Tomeoka, P. R. Buseck.
Meteoritics, Vol. 16, 392 - 393 (1981). — Abstract.

105.136 The isotopic composition of cosmic-ray produced lithium in iron meteorites.   H. Voshage.
Meteoritics, Vol. 16, 395 (1981). − Abstract.

105.137 Rare gas components in Abee clast separates.
J. F. Wacker.
Meteoritics, Vol. 16, 396 (1981). − Abstract.

105.138 Shock-induced mobilization of trace elements.
T. M. Walsh, M. E. Lipschutz.
Meteoritics, Vol. 16, 396 - 397 (1981). − Abstract.

105.139 Possible origins of type A and type B Ca-Al-rich inclusions in Allende.   D. A. Wark.
Meteoritics, Vol. 16, 398 - 399 (1981). − Abstract.

105.140 Noble gases in the Kirin H5-chondrite.
H. Weber, Z. Li, Z. Xu, F. Begemann.
Meteoritics, Vol. 16, 399 - 400 (1981). − Abstract.

105.141 The Cape York iron meteorite: trace element content in metal and sulphide.
H. H. Weinke, W. Kiesl, R. Gijbels, V. F. Buchwald.
Meteoritics, Vol. 16, 400 (1981). − Abstract.

105.142 Magnetic properties and paleointensity of seven H chondrites.   M. Westphal, H. Whitechurch.
Meteoritics, Vol. 16, 400 (1981). − Abstract.

105.143 Orbits and sources of unrecovered meteorites.
G. W. Wetherill, R. O. ReVelle.
Meteoritics, Vol. 16, 400 - 401 (1981). − Abstract.

105.144 Aqueous alteration of a CAI and ameboid inclusion in a Plainview carbonaceous xenolith.
L. L. Wilkening, S. Nozette.
Meteoritics, Vol. 16, 402 (1981). − Abstract.

105.145 Metallographic cooling rates for equilibrated, ordinary chondrites.   J. Willis, J. I. Goldstein.
Meteoritics, Vol. 16, 402 - 403 (1981). − Abstract.

105.146 Classification of unequilibrated ordinary chondrites.
F. Wlotzka.
Meteoritics, Vol. 16, 403 - 404 (1981). − Abstract.

105.147 Major element compositions and Rb-Sr, Sm-Nd isotopic systematics of Allan Hills polymict eucrites.   J. L. Wooden, R. Brown, B. Bansal, C.-Y. Shih, H. Wiesmann, L. E. Nyquist, A. M. Reid.
Meteoritics, Vol. 16, 404 - 405 (1981). − Abstract.

105.148 A comparison of the composition and morphology of spherules from meteorite craters with those of deep sea spherules.   F. W. Wright, P. W. Hodge.
Meteoritics, Vol. 16, 405 (1981). − Abstract.

105.149 ^{26}Al in several Japanese chondrites.
S. Yabuki, M. Shima, N. Takaoka.
Meteoritics, Vol. 16, 406 (1981). − Abstract.

105.150 Large lithic materials of LL-group chondrites in the Yamato-79 meteorites.
K. Yanai, H. Takeda, G. Sato, H. Kojima.
Meteoritics, Vol. 16, 407 (1981). − Abstract.

105.151 Determination of major elements in some Chinese chondrites by 14 MeV neutron activation analysis.
W. Yi, D. Wang.
Meteoritics, Vol. 16, 408 - 409 (1981). − Abstract.

105.152 Electron microscopy of Kirin meteorite.
P. Zhang, K. Tao, D. Yang, X. Chen.
Meteoritics, Vol. 16, 409 (1981). − Abstract.

105.153 Microtektite discovered in Beijing.
D. Li, A. Wang, Z. Xie, H. Wang, Q. Liu, J. Pei, Y. Huang.
Kexue Tongbao, Vol. 26, 382 - 383 (1981). − Abstr. in Phys. Abstr., Vol. 85, Abstr. 14814 (1982).

105.154 Not quite full circle? − non-racemic amino acids in the Murchison meteorite.   C. T. Pillinger.
Nature, Vol. 296, 802 (1982).

105.155 Discontinuous precipitation reaction in the metal of Richardton chondrite.
H. J. Axon, V. J. Grokhovsky.
Nature, Vol. 296, 835 - 837 (1982).
A metallographic and microprobe study was conducted on the H5 chondritic meteorite Richardton. The observations indicate that initially well annealed and slowly cooled metal experienced an incomplete episode of discontinuous precipitation in the temperature range 350–400°C, perhaps after mild deformation at some of the metal-silicate interfaces. There were no indications of externally imposed deformation after the 350°C episode.

105.156 Distribution and enantiomeric composition of amino acids in the Murchison meteorite.
M. H. Engel, B. Nagy.
Nature, Vol. 296, 837 - 840 (1982).
Using improved chromatographic and mass spectrometric procedures the authors have been able to amplify and resolve the partially racemized amino acids, Glu, aspartic acid (Asp), proline (Pro), leucine (Leu) and Ala, in an interior sample of a Murchison meteorite stone. Water-extractable amino acids were more racemized than those recovered by digesting the water-extracted meteorite in 6 M HCl. This is the first report of amino acids in a carbonaceous meteorite which, based on currently accepted criteria, appear to be indigenous but for unknown reasons are not racemic.

105.157 Meteoritics: evidence for chemical fractionation in the early solar system.   R. Hutchison.
Nature, Vol. 297, 20 - 21 (1982).

105.158 Accretionary dark rims in unequilibrated chondrites.   T. V. V. King, E. A. King.
Icarus, Vol. 48, 460 - 472 (1981).
Textural and qualitative EDX investigations of dark-rimmed particles in six low petrologic type chondrites indicate that the rims accreted on host particles over a wide range of temperatures prior to initial accumulation and lithification of the meteorites in which the rimmed particles are now contained. Many dark rims are enriched in moderately volatile trace elements such as Na, Cl, P, and K, relative to the host particles and matrix.

105.159 Meteorite und Astrophysik. Häufigkeitsanomalien und die Entstehung des Sonnensystems.
F. Begemann.
Sterne Weltraum, 21. Jahrg., 106 - 109 (1982).

105.160 The cosmic-ray-exposure age of meteorites.
M. Shima.
Astron. Her., Vol. 74, No. 6, p. 160 - 168 (1981). In Japanese. − Abstr. in Phys. Abstr., Vol. 85, Abstr. 28416 (1982).

105.161 A calculation of the energy released from a meteorite explosion based on the infrasound data.
X.-k. Hu.

Acta Acust., No. 3, p. 142 - 147 (1981). – Abstr. in Phys. Abstr., Vol. 85, Abstr. 32960 (1982).

105.162   Olivine orientation in the ALHA 77005 achondrite.
J. L. Berkley, K. Keil.
American Mineral., Vol. 66, 1233 - 1236 (1981). – Abstr. in Phys. Abstr., Vol. 85, Abstr. 42663 (1982).

105.163   Comparing investigations on the surface structures of irghizites and pyroclastics by SEM.
K. Heide, G. Völksch, P. W. Florenski (*P. V. Florenskij*).
Meteoritics, Vol. 17, 1 - 7 (1982).

The surface structures of irghizites from the Zhamanshin crater in Kazakhstan, USSR, play an important role in the discussion of their genesis. These surface structures were compared with those of typical tektites (australites) and pyroclastics (obsidians, lapilli) based on investigation by electron microscope. The results of these investigations indicate that there are no unambiguous genetical relatioships between the morphology of irghizites and the surface features of tektites and pyroclastics.

105.164   Tibooburra, a new Australian meteorite find, and other carbonaceous chondrites of high petrologic grade.   M. J. Fitzgerald, A. L. Jaques.
Meteoritics, Vol. 17, 9 - 26 (1982).

Tibooburra, a new meteorite find from western New South Wales, belongs to the Vigarano subgroup of the carbonaceous chondrites and, on the basis of its opaque mineralogy, appears to be oxidised. Petrological evidence suggests that, like the Allende meteorite, Tibooburra is a CV3 chondrite which has experienced greater metamorphic effects than other CV3 meteorites. Tibooburra has a bulk composition intermediate between the CO and less altered chondrites. This transitional nature is exhibited by several elements and is convincingly displayed by the multivariate techniques of cluster analysis and principal component analysis.

105.165   The Madiun, Indonesia, chondrite.
C. G. R. Reid, K. Fredriksson.
Meteoritics, Vol. 17, 27 - 30 (1982).

105.166   A study of the relative rates of meteorite falls on the Earth's surface.   I. Halliday, A. A. Griffin.
Meteoritics, Vol. 17, 31 - 46 (1982).

Meteorite camera networks have provided reliable data on typical orbits for meteorites. Using an adopted distribution of 20 orbits the authors determine the relative rates of meteorite falls over the surface of the earth taking account of the important effects due to the earth's gravity. The data are then used to study the expected variation in rates as a function of daylight, twilight or night conditions; time of day; season of the year; and geographic latitude.

105.167   Uranium in the silicate inclusions of stony-iron and iron meteorites.
G. Crozaz, S. F. Sibley, D. R. Tasker.
Geochim. Cosmochim. Acta, Vol. 46, 749 - 754 (1982).

The microdistribution of U has been studied, using fission track techniques, in eleven mesosiderites, seven pallasites and four iron meteorites with silicate inclusions. When concentrated, U is usually found in phosphates: merrillite and/or chlorapatite. As in stony meteorites, the U concentrations in a given phosphate phase are highly variable from meteorite to meteorite and sometimes also exhibit variations in the same meteorite.

105.168   Origin of rapidly solidified metal-troilite grains in chondrites and iron meteorites.   E. R. D. Scott.
Geochim. Cosmochim. Acta, Vol. 46, 813 - 823 (1982).

In this paper cooling rates during solidification are derived for occurrences of rapidly solidified metal inclusions in chondrites with the aim of elucidating their thermal history and origin. In addition, these occurrences in chondrites are compared with examples of dendritic structures in iron meteorites, and with rapidly solidified shock-melted sulfides, which occur in nearly all types of meteorites.

105.169   Antarktische Meteorite.   L. Schultz.
Naturwissenschaften, 69. Jahrg., 220 - 225 (1982).

105.170   Impaktverdächtige Strukturen in Venezuela.
P. Brosche.
Sterne Weltraum, 21. Jahrg., 194 - 195 (1982).

105.171   Eyewitnesses evidence of the Tunguska fall.
N. V. Vasil'ev, A. F. Kovalevskij, S. A. Razin, L. E. Ehpiktetova.
Tomsk. univ. Tomsk, 1981. 304 pp. In Russian. – Abstr. in Ref. zh., 51. Astron., 3.51.330 (1982).

105.172   Isotopically anomalous tellurium in Allende: another relic of local element synthesis.
L. L. Oliver, R. V. Ballad, J. F. Richardson, O. K. Manuel.
J. Inorg. Nucl. Chem., Vol. 43, 2207 - 2216 (1981). – Abstr. in Phys. Abstr., Vol. 85, Abstr. 48189 (1982).

105.173   Neutron activation analysis of some meteorites from Romania.   V. Znamirovschi, C. Cosma, M. Salagean, V. Cojocaru.
Rev. Roumaine Phys., Vol. 26, 745 - 750 (1981). – Abstr. in Phys. Abstr., Vol. 85, Abstr. 48190 (1982).

105.174   Applications of nuclear tracks in extraterrestrial materials.   S. Rajan.
Trans. American Nucl. Soc., Vol. 39, 63 - 64 (1981). – Abstr. in Phys. Abstr., Vol. 85, Abstr. 48191 (1982).

105.175   Isotopic variations in primitive meteorites.
R. N. Clayton.
Philos. Trans. R. Soc. London A, Vol. 303, (see 012.022), 339 - 349 (1981).

New observations of isotopic variations in meteorites continue to be made, and attempts to put them into appropriate astrophysical contexts naturally follow. The presence of large internal ^{16}O variability in ordinary chondrites greatly extends the range of meteorite types in which this phenomenon has been observed. These results may lead to identification of major gas and dust reservoirs in the cloud from which the Solar System formed.

105.176   The significance of W in planetary differentiation processes: evidence from new data on eucrites.
H. Palme, W. Rammensee.
Proc. Twelfth Lunar Planet. Sci. Conf., (see 012.024), p. 949 - 964 (1982).

105.177   Carbon-14 ages of Allan Hills meteorites and ice.
E. L. Fireman, T. Norris.
Proc. Twelfth Lunar Planet. Sci. Conf., (see 012.024), p. 1019 - 1025 (1982).

105.178   Microchrons: the $^{87}Rb-^{87}Sr$ dating of microscopic samples.   D. A. Papanastassiou, G. J. Wasserburg.
Proc. Twelfth Lunar Planet. Sci. Conf., (see 012.024), p. 1027 - 1038 (1982).

105.179   A unique type 3 ordinary chondrite containing graphite-magnetite aggregates – Allan Hills A77011.
S. G. McKinley, E. R. D. Scott, G. J. Taylor, K. Keil.
Proc. Twelfth Lunar Planet. Sci. Conf., (see 012.024), p. 1039 - 1048 (1982).

105.180   The Elga meteorite: silicate inclusions and shock metamorphism.   E. G. Osadchii (*Osadchij*), G. V. Baryshnikova, G. V. Novikov.
Proc. Twelfth Lunar Planet. Sci. Conf., (see 012.024), p. 1049 - 1068 (1982).

105.181   SEM, optical, and Mössbauer studies of submicrometer chromite in Allende.   R. M. Housley.
Proc. Twelfth Lunar Planet. Sci. Conf., (see 012.024), p. 1069 - 1077 (1982).

105.182   Origin of rims on coarse-grained inclusions in the Allende meteorite.
G. J. MacPherson, L. Grossman, J. M. Allen, J. R. Beckett.
Proc. Twelfth Lunar Planet. Sci. Conf., (see 012.024), p. 1079 - 1091 (1982).

105.183   Petrogenesis of light and dark portions of the Leighton gas-rich chrondritic breccia.
H. Y. McSween, Jr., S. Biswas, M. E. Lipschutz.
Proc. Twelfth Lunar Planet. Sci. Conf., (see 012.024), p. 1093 - 1103 (1982).

105.184   Mineralogical aspects of terrestrial weathering effects in chondrites from Allan Hills, Antarctica.
J. L. Gooding.
Proc. Twelfth Lunar Planet. Sci. Conf., (see 012.024), p. 1105 - 1122 (1982).

105.185   Conditions of formation of pyroxene excentroradial chondrules.
R. H. Hewins, L. C. Klein, B. V. Fasano.
Proc. Twelfth Lunar Planet. Sci. Conf., (see 012.024), p. 1123 - 1133 (1982).

105.186   A revision of metallographic cooling rate curves for chondrites.   J. Willis, J. I. Goldstein.
Proc. Twelfth Lunar Planet. Sci. Conf., (see 012.024), p. 1135 - 1143 (1982).

105.187   Ordinary chondrite parent body: an internal heating model.
M. Miyamoto, N. Fujii, H. Takeda.
Proc. Twelfth Lunar Planet. Sci. Conf., (see 012.024), p. 1145 - 1152 (1982).

105.188   Electron microscopy of carbonaceous matter in Allende acid residues.   G. R. Lumpkin.
Proc. Twelfth Lunar Planet. Sci. Conf., (see 012.024), p. 1153 - 1166 (1982).

105.189   Carbon in the Allende meteorite: evidence for poorly graphitized carbon rather than carbyne.
P. P. K. Smith, P. R. Buseck.
Proc. Twelfth Lunar Planet. Sci. Conf., (see 012.024), p. 1167 - 1175 (1982).

105.190   Silicon and oxygen isotopes in selected Allende inclusions.   R. H. Becker, S. Epstein.
Proc. Twelfth Lunar Planet. Sci. Conf., (see 012.024), p. 1189 - 1198 (1982) = Contrib. No. 3623, Div. Geol. Planet. Sci., Calif. Inst. Technol.

105.191   An estimate of atmospheric contamination of Allende coarse-grained inclusion 3529Z.
R. Warasila, O. A. Schaeffer, K. Frank.
Proc. Twelfth Lunar Planet. Sci. Conf., (see 012.024), p. 1199 - 1207 (1982).

105.192   Irradiation records of Acapulco and other small meteorites derived from ^{53}Mn and rare gas measurements.   P. Englert, U. Herpers, W. Herr.
Proc. Twelfth Lunar Planet. Sci. Conf., (see 012.024), p. 1209 - 1215 (1982).

105.193   Stable NRM and mineralogy in Allende: chondrules.   P. J. Wasilewski, C. Saralker.
Proc. Twelfth Lunar Planet. Sci. Conf., (see 012.024), p. 1217 - 1227 (1982).

105.194   The composition of natural remanent magnetization of an Antarctic chondrite, ALHA 76009 (L_6).
T. Nagata, M. Funaki.
Proc. Twelfth Lunar Planet. Sci. Conf., (see 012.024), p. 1229 - 1241 (1982).

105.195   The magnetic properties of the Abee meteorite: evidence for a strong magnetic field in the early solar system.   N. Sugiura, D. W. Strangway.
Proc. Twelfth Lunar Planet. Sci. Conf., (see 012.024), p. 1243 - 1256 (1982).

105.196   Howardites and polymict eucrites: regolith samples from the eucrite parent body. Petrology of Bholgati, Bununu, Kapoeta, and ALHA 76005.
M. Fuhrman, J. J. Papike.
Proc. Twelfth Lunar Planet. Sci. Conf., (see 012.024), p. 1257 - 1279 (1982).

105.197   Ion probe analysis of plagioclase in three howardites and three eucrites.   I. M. Stelle, J. V. Smith.
Proc. Twelfth Lunar Planet. Sci. Conf., (see 012.024), p. 1281 - 1296 (1982).

105.198   Metamorphism in mesosiderites.   J. S. Delaney, C. E. Nehru, M. Prinz, G. E. Harlow.
Proc. Twelfth Lunar Planet. Sci. Conf., (see 012.024), p. 1315 - 1342 (1982).

105.199   Roaldite, a new nitride in iron meteorites.   H. P. Nielsen, V. F. Buchwald.
Proc. Twelfth Lunar Planet. Sci. Conf., (see 012.024), p. 1343 - 1348 (1982).

105.200   Complementary rare earth element patterns in unique achondrites, such as ALHA 77005 and shergottites, and in the earth.
M.-S. Ma, J. C. Laul, R. A. Schmitt.
Proc. Twelfth Lunar Planet. Sci. Conf., (see 012.024), p. 1349 - 1358 (1982).

105.201   SNC meteorites: igneous rocks from Mars?   C. A. Wood, L. D. Ashwal.
Proc. Twelfth Lunar Planet. Sci. Conf., (see 012.024), p. 1359 - 1375 (1982).

105.202   A method for estimating the initial impact conditions of terrestrial cratering events, exemplified by its application to Brent crater, Ontario.
R. A. F. Grieve, M. J. Cintala.
Proc. Twelfth Lunar Planet. Sci. Conf., (see 012.024), p. 1607 - 1621 (1982) = Contrib. No. 934, Earth Phys. Branch, Dep. Energy, Mines, Resources, Ottawa.

105.203   Some key issues on isotopic anomalies: astrophysical history and aggregation.   D. D. Clayton.
Proc. Twelfth Lunar Planet. Sci. Conf., (see 012.024), p. 1781 - 1802 (1982).
   The author describes ways in which astrophysical history is utilized to understand isotopic anomalies in meteorites, concentrating on those elements, oxygen, titanium, and noble gases, whose anomalies are ubiquitous in carbonaceous meteorites. Attention is given to ideas of nucleosynthesis and

chemical condensation that seem capable of generating understanding.

105.204   **Comments on: "Xe¹²⁹ and the origin of CCF xenon in meteorites" by R. S. Lewis and E. Anders.**
D. Heymann.
Proc. Twelfth Lunar Planet. Sci. Conf., (see 012.024), p. 1803 - 1807 (1982).

105.205   **Isotopic anomalies in meteorites.**   E. R. D. Scott.
Nature, Vol. 297, 361 - 362 (1982).

105.206   **Isotopically distinguishable carbon phases in the Allende meteorite.**
P. K. Swart, M. M. Grady, C. T. Pillinger.
Nature, Vol. 297, 381 - 383 (1982).
The stepwise oxidation of the Allende meteorite reported here has shown that there are at least three isotopically distinct carbon phases. The authors attempt to correlate the isotopic signatures recognized with phases whose existence has been postulated to explain noble gas systematics and to provide some insight into the nature and location of these phases.

105.207   **A review of data on paleomagnetic determination from magnetization of meteorites and lunar rocks.**
E. G. Gus'kova.
Postoyan. geomagn. pole, magn. gorn. porod i paleomagn. 2-j Vses. sezd po geomagn. Tbilisi, 16 - 21 noyab. 1981. Tez. dokl. Ch. 2. Tbilisi, 1981, p. 15. In Russian. – Abstr. in Ref. zh., 51. Astron., 4.51.180 (1982).

105.208   **Meteorites and cosmochemistry.**   G. Turner.
The comparative study of the planets, (see 012.033), p. 79 - 84 (1982).
An outline of the major features of meteorite chemistry, mineralogy and petrology is given. The importance of meteorite observations for some of the current ideas of the origin of the solar system is discussed. Furthermore, recent discoveries of isotopic anomalies in meteorites are briefly analysed.

105.209   **Do the age differences given by relative or absolute chronologies of the most ancient meteorites correspond to real age differences?**   P. Pellas.
The comparative study of the planets, (see 012.033), p. 95 - 100 (1982).
Recent results from absolute and relative chronologies of the most ancient meteorites are reviewed in order to analyze if they are significant or not. Use of the various chronometers to analyze the same meteoritic sample is shown to be an interesting approach to retrace the prehistory of meteorites and their environments.

105.210   **Impact cratering mechanics.**   J. Pohl.
The comparative study of the planets, (see 012.033), p. 323 - 331 (1982).
Impact cratering by high-velocity projectiles is an important process in the evolution of planets. Observations on planets and satellites with solid surfaces show that small impact craters are bowl-shaped and deep. Larger impact structures are shallow, they have overall flat interiors with several characteristic morphological elements, such as central peaks or rings. In addition to observations the cratering process is investigated by impact and explosion experiments and by theoretical continuum mechanics calculations.

105.211   **A large Cretaceous meteorite crater in Kansas.**
N. Sperling.
Bull. American Astron. Soc., Vol. 13, 839 (1981). – Abstract.

105.212   **On the conditions of formation of carbides in iron and enstatite meteorites.**
M. I. Petaev, I. L. Khodakovskij.
Vses. soveshch. po geokhimii ugleroda, Moskva, 14 - 16 dek., 1981. Tez. dokl. Moskva, 1981, p. 308 - 311. In Russian. Abstr. in Ref. zh., 51. Astron., 5.51.324 (1982).

105.213   **The Logajskij meteoritic crater and its formations (structure, lithologically-stratigraphic differentiation, palinologic characteristic, age and genesis).**
N. V. Veretennikov, A. S. Makhnach, A. F. Burlak, V. I. Avkhimovich, G. I. Il'kevich.
Mater. po geol. kristal. fundam. i osad. chekhla Belorussii. Minsk: Nauk. i tekh., 1981, p. 201 - 224. In Russian. – Abstr. in Ref. zh., 51. Astron., 5.51.330 (1982).

105.214   **Crystallization of a S-saturated Fe, Ni-melt, and the origin of iron meteorite groups IAB and IIICD.**
A. Kracher.
Geophys. Res. Lett., Vol. 9, 412 - 415 (1982).
New data on trace element partitioning in the Fe-Ni-S-P system suggest that iron meteorites of groups IAB and IIICD come from one or more S-rich parent bodies. Heating of the IAB parent body melted all troilite, but caused only very minor partial melting of silicates. Cocrystallization of metal and troilite produced the high-Ni "tail" of group IAB. The trace element patterns are consistent with a parent body with CI sulfur and siderophile abundances. Compositionally the parent body was intermediate between ordinary and E4 chondrites, its thermal history was in between ordinary chondrite and fully differentiated parent bodies.

105.215   **Meteorite falls.**   D. W. Hughes.
Nature, Vol. 298, 116 - 117 (1982).

105.216   **Evidence for a major meteorite impact on the earth 34 million years ago: implication for eocene extinctions.**   R. Ganapathy.
Science, Vol. 216, 885 - 886 (1982).

105.217   **Sorption of noble gases by solids, with reference to meteorites. I. Magnetite and carbon.**
J. Yang, R. S. Lewis, E. Anders.
Geochim. Cosmochim. Acta, Vol. 46, 841 - 860 (1982).
To simulate trapping of meteoritic noble gases by solids, 18 samples of Fe_3O_4 were synthesized in a noble gas atmosphere.

105.218   **Sorption of noble gases by solids, with reference to meteorites. II. Chromite and carbon.**
J. Yang, E. Anders.
Geochim. Cosmochim. Acta, Vol. 46, 861 - 875 (1982).
The authors studied trapping of noble-gases by chromite and carbon: two putative carriers of primordial noble gases in meteorites. Nineteen samples were synthesized in a Ne-Ar-Kr-Xe atmosphere.

105.219   **Sorption of noble gases by solids, with reference to meteorites. III. Sulfides, spinels, and other substances; on the origin of planetary gases.**
J. Yang, E. Anders.
Geochim. Cosmochim. Acta, Vol. 46, 877 - 892 (1982).
To simulate trapping of noble gases by meteorites, the authors reacted 15 FeCr or FeCrNi alloy samples with CO, H_2O or H_2S at 350-720 K, in the presence of noble gases.

105.220   **The Nilpena ureilite, an unusual polymict breccia: implications for origin.**
A. L. Jaques, M. J. Fitzgerald.
Geochim. Cosmochim. Acta, Vol. 46, 893 - 900 (1982).

105.221 Fine, nickel-poor Fe-Ni grains in the olivine of unequilibrated ordinary chondrites.
E. R. Rambaldi, J. T. Wasson.
Geochim. Cosmochim. Acta, Vol. 46, 929 - 939 (1982).

105.222 Evidence for primitive nebular components in chondrules from the Chainpur chondrite.
J. N. Grossman, J. T. Wasson.
Geochim. Cosmochim. Acta, Vol. 46, 1081 - 1099 (1982).

105.223 Cosmogenic nuclides in the Kirin chondrite.
M. Honda, K. Nishiizumi, M. Imamura, N. Takaoka, O. Nitoh, K. Horie, K. Komura.
Earth Planet. Sci. Lett., Vol. 57, 101 - 109 (1982).

105.224 The U-Th-Pb age of equilibrated L chondrites and a solution to the excess radiogenic Pb problem in chondrites. D. M. Unruh.
Earth Planet. Sci. Lett., Vol. 58, 75 - 94 (1982).
Two approaches have been used: (1) the chondrite-troilite apparent initial Pb isotopic compositions were used to approximate the mixture of indigenous initial Pb and terrestrial Pb in the whole-rock sample, and (2) a single-stage (concordant) model was applied using the assumption that the excess radiogenic Pb in these samples was terrestrial. Data for L5 and L6 chondrites yield a 4551 ± 7 My age using the former correction and a 4550 ± 5 My age using the latter one. Corrected data for one L4 chondrite, Tennasilm, yield a 4552 ± 13 My age which is indistinguishable from that of the L5 - L6 chondrites. However, the other L4 chondrite, Bjurböle, yields a 4590 ± 6 My age.

105.225 Cosmogenic ^{21}Ne and ^{22}Ne depth profiles in chondrites. N. Bhandari, M. B. Potdar.
Earth Planet. Sci. Lett., Vol. 58, 116 - 128 (1982).
The production rate profiles of ^{21}Ne and ^{22}Ne as a function of depth in meteoroids due to spallation by solar flare cosmic rays and galactic cosmic rays are calculated and their dependence on size and composition of meteoroids has been evaluated.

105.226 Evidence for ^{244}Pu fission tracks in hibonites from Murchison carbonaceous chondrite.
R. S. Rajan, A. S. Tamhane.
Earth Planet. Sci. Lett., Vol. 58, 129 - 135 (1982).

105.227 Terrestrial, meteoritic, and lunar titanium isotopic ratios revaluated: evidence for correlated variations.
H. R. Heydegger, J. J. Foster, W. Compston.
Earth Planet. Sci. Lett., Vol. 58, 406 - 418 (1982).

105.228 Dating of meteorite falls using thermoluminescence: application to Antarctic meteorites.
S. W. S. McKeever.
Earth Planet. Sci. Lett., Vol. 58, 419 - 429 (1982).
The date of fall of a meteorite may be estimated from its thermoluminescence (TL) and in this paper the principle of a method of utilising TL to determine the terrestrial ages of eight Antarctic meteorites (Allan Hills-77) is described. The TL in a meteorite is primarily induced by cosmic ray irradiation in space and once the meteorite is on the Earth's surface, it is shielded from further cosmic ray irradiation. Under these conditions the TL will decay at a rate governed by the thermal stability of the TL and by the environmental temperature. Only upper limits to the terrestrial ages can be calculated because the TL at the time of the fall to Earth is highly variable from sample to sample.

105.229 Solar cosmic-ray-produced radionuclides in meteorites. R. Michel, G. Brinkmann, R. Stück.
Earth Planet. Sci. Lett., Vol. 59, 33 - 48 (1982).

Meteorites. A petrologic-chemical synthesis.
See Abstr. 003.037.

Antarctic search for meteorites: preliminary report of the 1980/81 field season in South Victoria Land.
See Abstr. 011.011.

In, out and about meteorites. See Abstr. 011.023.

Noble gas trapping by laboratory carbon condensates. See Abstr. 022.117.

When are spectral reflectance curves comparable?
See Abstr. 022.122.

Experimental study of effects associated with macroscopic hypervelocity impacts. See Abstr. 022.153.

The "trumpet charge", a technique for producing macroscopic hypervelocity projectiles. See Abstr. 022.154.

A possible mechanism for impact magnetisation of cratered surfaces. See Abstr. 022.155.

Applications of neutron activation analysis of meteorite, lunar, and terrestrial samples. See Abstr. 022.179.

Impact cratering and regolith dynamics.
See Abstr. 094.025.

The isotopic composition of zirconium in terrestrial and extraterrestrial samples: implications for extinct ^{92}Nb.
See Abstr. 094.505.

The significance of rust in lunar rocks and meteorites. See Abstr. 094.524.

Thermal and impact histories of pyroxenes in lunar eucrite-like gabbros and eucrites. See Abstr. 094.593.

The stochastic evolution of asteroidal regoliths and the origin of brecciated and gas-rich meteorites.
See Abstr. 098.016.

Asteroiden — Quelle für Meteorite und Staub.
See Abstr. 098.034.

The stochastic variability of asteroidal regolith depths. See Abstr. 098.036.

Spectroscopic evidence for undifferentiated S-type asteroids. See Abstr. 098.040.

Report of IAU Commission 15: Physical study of comets, minor planets and meteorites (L'étude physique des comètes, des petites planètes et des météorites).
See Abstr. 102.055.

Report of IAU Commission 22: Meteors and interplanetary dust (Météores et la poussière interplanétaire).
See Abstr. 104.039.

Erratum

105.901 Erratum: 'Allen Hills A77219, the first Antarctic mesosiderite' [Proc. Eleventh Lunar Planet. Sci. Conf., p. 1027 - 1045 (1980)].
W. N. Agosto, R. H. Hewins, R. S. Clarke, Jr.
Proc. Twelfth Lunar Planet. Sci. Conf., (see 012.024) (1982).
See Abstr. 30.105.043.

106 Interplanetary Matter, Interplanetary Magnetic Field, Zodiacal Light

106.001 **"Flip-flop" of electric potential of dust grains in space.**   N. Meyer-Vernet.
Astron. Astrophys., Vol. 105, 98 - 106 (1982).

It is shown that, contrary to what is generally believed, the charge on a cosmic grain is not always unique. This phenomenon has some important consequences: (1) Identical grains in the same environment (but with different histories) may have opposite charges; (2) a small change in the environment may induce large and rapid charge's variation; (3) one may conjecture that in suitable conditions, the charge could oscillate. Several consequences on the grain accretion, dynamics, and radio-emission are briefly outlined. The results may be relevent for the discharge-like radio-emission recently discovered by Voyager I near Saturn.

106.002 **Spectral properties of hydromagnetic fluctuations near 4 and 5 a.u.**   G. D. Parker.
Planet. Space Sci., Vol. 30, 57 - 66 (1982).

The purpose of the paper is to investigate the spectral properties of magnetic fluctuations in the quiet solar wind near 4–5 a.u.

106.003 **The inner zodiacal light from pictures obtained on board the Salyut 6 spaceship.**
G. M. Grechko, N. B. Divari, G. M. Nikol'skij, Yu. V. Romanenko.
Pis'ma Astron. Zh., Tom 8, 52 - 56 (1982). In Russian. English translation in Soviet Astron. Lett., Vol. 8.

Pictures of the inner zodiacal light were obtained on board the Salyut 6 spaceship. Photometry of one set with 30-s exposure time was carried out. The brightness distribution along the ecliptic at the elongations $\epsilon \geqslant 22°\!.5$ is given by the law $I \propto \epsilon^{-k}$, where $k = 3.2$. The north part of the zodiacal light is by 10 - 20 per cent brighter than the south one.

106.004 **Observing interplanetary disturbances from the ground.**
G. R. Gapper, A. Hewish, A. Purvis, P. J. Duffett-Smith.
Nature, Vol. 296, 633 - 636 (1982).

The scintillation of celestial radio sources due to small-scale turbulence along lines of sight through the interplanetary medium provides a convenient, ground-based method of monitoring disturbances in interplanetary space. With the sensitive 3.6-hectare Array at Cambridge the authors have carried out a new programme in which ~900 sources were observed each day for more than 1 yr. When the long-term average scintillation behaviour of each source had been accurately determined they found that transient disturbances could be clearly distinguished. The authors detected large clouds of enhanced turbulence moving out from the sun and were able to track them to distances beyond the earth's orbit.

106.005 **Stability and symmetry of zodiacal light polarization in the antisolar hemisphere.**
C. Leinert, B. Planck.
Astron. Astrophys., Vol. 105, 364 - 368 (1982).

Observations of zodiacal light polarization in the antisolar hemisphere performed by the Helios 1 and 2 spaceprobes were searched for variations and anomalies reported earlier from ground-based photometry. No change in polarized intensity or direction of polarization was found. The polarization pattern was found symmetric to the antisolar point. The angle of polarization is perpendicular to the scattering plane except for the viewing directions closest to the antisolar point ($\beta = \pm 16°$, $\epsilon = 161°$) where is a tendency for negative polarization.

106.006 **Interplanetary scintillations of maser sources of the water vapour line.**

D. F. Blums, N. A. Lotova, R. L. Sorochenko.
Dokl. AN SSSR, Tom 260, 570 - 573 (1981). In Russian. Abstr. in Ref. zh., 51. Astron., 2.51.356 (1982).

106.007 **Cosmic spherules and asteroid collisions.**
D. W. Parkin.
Geophys. J. R. Astron. Soc., Vol. 69, (see 012.004), 301 (1982). – Abstract.

106.008 **Interplanetary gas. XXVIII. A study of the three-dimensional properties of interplanetary sector boundaries using disconnection events in cometary plasma tails.**   M. B. Niedner, Jr.
Astrophys. J., Suppl. Ser., Vol. 48, 1 - 50 (1982).

On the assumption that they are caused by magnetic reconnection at interplanetary sector boundary crossings, disconnection events (DEs) of the cometary plasma tail have been used to probe the three-dimensional structure of sector boundaries to the high latitude regions presently inaccessible to spacecraft. The study utilizes 72 DEs in 29 comets extending back to 1892 and provides information about the latitudinal extent and tilt properties of sector boundaries over a wide range of solar cycle phase and over eight solar cycles. The results are compared with the predictions of the warped sheet model of interplanetary sector structure. The behavior of interplanetary sector patterns at Earth during 1965 - 1968 is also examined. It is concluded that the concept of one sinusoidal, nearly equatorial neutral sheet may be an oversimplification, and that the geometry of the sectored fields at 1 AU may depend as much on the distribution of equatorial to mid-latitude solar sources to the interplanetary field as on the strength of the polar fields alone.

106.009 **Solar wind pressure on interplanetary dust.**
T. Mukai, T. Yamamoto.
Astron. Astrophys., Vol. 107, 97 - 100 (1982).

The pseudo-Poynting-Robertson effect on an interplanetary grain due to solar wind bombardments is examined with careful consideration for sputtering of a grain and for the velocity dispersion of the solar wind particles. It is found that for water-ice and obsidian grains with radii in the range 0.01 - 100 μm, the retarding force due to solar wind is stronger than, and of the same order as, respectively, that due to solar radiation. On the other hand, for magnetite this drag force is always less than that due to solar radiation. In addition, since the wind flow generally comes from the east of the sun, a grain in a prograde orbit suffers a large retarding force compared with a grain in a retrograde orbit.

106.010 **Zodiakallichtstruktur im Gegenscheingebiet.**
C. Winkler, T. Schmidt-Kaler, W. Schlosser.
Mitt. Astron. Ges., Nr. 55, p. 165 - 166 (1982).

106.011 **Investigation of statistical properties of a connection of the interplanetary and the geomagnetic fields with the method of multiple coherence functions.**
L. B. Volkomirskaya, A. N. Zajtsev, S. V. Panfilov.
Geomagn. Aehron., Tom 22, 90 - 94 (1982). In Russian.

106.012 **Correlation relations between the components of the interplanetary magnetic field and solar wind velocity.**   V. O. Papitashvili, L. I. Gromova.
Geomagn. Aehron., Tom 22, 126 - 128 (1982). In Russian.

106.013 **Notes on the origin of the B_z-component of the interplanetary magnetic field and the geoefficiency index in Akasofu's model.**   M. I. Pudovkin, D. I. Ponyavin.
Geomagn. Aehron., Tom 22, 154 - 155 (1982). In Russian.

106.014   On using a phase diagram for investigation of the connection of the interplanetary magnetic field with ground variations of the geomagnetic field.
L. B. Volkomirskaya, A. N. Zajtsev.
Geomagn. Aehron., Tom 22, 327 - 330 (1982). In Russian.

106.015   Nonlocal plasma turbulence associated with interplanetary shocks.   C. F. Kennel, F. L. Scarf, F. V. Coroniti, E. J. Smith, D. A. Gurnett.
J. Geophys. Res., Vol. 87, 17 - 34 (1982).

106.016   The IMF sector pattern through the solar minimum: two spacecraft observations during 1974–1978.
U. Villante, F. Mariani, P. Francia.
J. Geophys. Res., Vol. 87, 249 - 253 (1982).

The hourly average values of the IMF as observed by Helios 1 and 2 in the period December 1974 through March 1978 have been examined to deduce how the sector pattern seems to evolve with time and heliographic latitude ($\pm 7.2°$ from the solar equator). The dominant two-sector pattern observed after 1972 extends to the whole declining phase of solar cycle 20, and the corresponding source regions seem to corotate with an angular velocity smaller than the equatorial one. The latitudinal dependence of the IMF polarity (sunward field lines below the solar equator) is observed. Gross scale implications for the shape and location of the sector boundary surface are discussed.

106.017   Structure of current sheets in the sector boundaries: Helios 2 observations during early 1976.
U. Villante, R. Bruno.
J. Geophys. Res., Vol. 87, 607 - 612 (1982).

The authors examined the orientation and structure of 14 current sheets that were observed in sector boundary regions during the primary mission of Helios 2. They found that the field rotation through the current sheets tends to occur at large angles with respect to the ecliptic plane. Most of the observed structures can be interpreted as tangential discontinuities oriented at large angles with respect to the ecliptic plane. The authors also discuss the implications of a comparison between present results and the global configuration of the sector boundary surface.

106.018   Interplanetary magnetic clouds at 1 AU.
L. W. Klein, L. F. Burlaga.
J. Geophys. Res., Vol. 87, 613 - 624 (1982).

Magnetic clouds are defined as regions with a radial dimension ≈ 0.25 AU (at 1 AU) in which the magnetic field strength is high and the magnetic field direction changes appreciably by means of rotation of one component of $\underline{B}$ nearly parallel to a plane. The magnetic field geometry in such a magnetic cloud is consistent with that of a magnetic loop, but it cannot be determined uniquely. Forty-five clouds were identified in interplanetary data obtained near earth between 1967 and 1978; at least one cloud passed the earth every 3 months. Three classes of clouds were identified, corresponding to the association of a cloud with a shock, a stream interface, or a CME.

106.019   Solar flare shocks in interplanetary space and solar flare particle events.
P. Evenson, P. Meyer, S. Yanagita.
J. Geophys. Res., Vol. 87, 625 - 631 (1982).

The authors examine data on particle events resulting from three solar flares, two of which generated interplanetary shocks, to determine the effect of the shocks on the population of protons at energies $\gtrsim 30$ MeV.

106.020   The bidirectional particle event of October 12, 1977, possibly associated with a magnetic loop.
F. J. Kutchko, P. R. Briggs, T. P. Armstrong.
J. Geophys. Res., Vol. 87, 1419 - 1431 (1982).

A fortuitous set of interplanetary and solar events which occurred on October 12, 1977, allowed the observations by several interplanetary spacecraft, IMP 7 and 8, of features in charged particle angular distributions which suggest the presence of a closed magnetic loop. This loop must have extended beyond the orbit of earth and have been connected to particle sources and/or mirrors at both ends. The salient feature of the anisotropies which calls attention to the possible presence of a loop is the occurrence of a trapping-type electron distribution peaked at 90° to the magnetic field, while the ions are strongly field aligned and bidirectional.

106.021   Latitudinal and solar-cycle dependence of the interplanetary magnetic field predominant polarity.
X. Moussas, B. Tritakis.
Sol. Phys., Vol. 75, 361 - 375 (1982).

A study of the predominant interplanetary magnetic field (IMF) polarity is made, for the time period 1957 - 1977. The examination of the mean positive and negative sector width for time periods (semesters) for which the Earth was in northern and southern heliolatitudes shows that the predominant polarity for every semester follows, up to a certain extent, the Rosenberg-Coleman effect. However, the statistical support is not satisfactory. The same conclusion was pointed out by a similar study of data grouped over various phases of the solar cycle. Additionally the relative frequency of positive (negative) IMF polarity days, appeared over a mean solar rotation, shows that the general pattern of the mean IMF has a tendency to reoccur in the homologous (corresponding) phases of different solar cycles.

106.022   Magnetic fields in the interplanetary medium.
L. F. Burlaga.
Solar system plasmas and fields, (see 003.005), p. 51 - 54 (1982).

106.023   Origins of the low-energy relativistic interplanetary electrons.   J. H. Eraker.
Astrophys. J., Vol. 257, 862 - 880 (1982).

Observations of electrons in the energy range from 1.75 to 25 MeV by the University of Chicago instrument on the Pioneer 10 spacecraft over the heliocentric radial distance $R = 1.0-21.5$ AU are presented. It is confirmed that, at $R < 11$ AU, the dominant component of the interplanetary relativistic flux less than 25 MeV, except during occasional solar flare events, is the Jovian electron intensity modulated every $25-27$ days. Beyond 11 AU this recurrent modulation of the Jovian electrons was not observed, and it became possible to conduct a sensitive search for the galactic low-energy electron flux. It is concluded that the minimum intensity $1-25$ MeV electrons in the equatorial regions of the heliosphere to at least 21 AU are (primarily) of Jovian origin. An upper limit for the galactic electron flux at 12 MeV is estimated by assuming that the 12 MeV flux at 21.5 AU is of galactic origin and by applying standard assumptions regarding solar modulation. It is found that the upper limit for the flux of 12 MeV galactic electrons is a factor of 50 or more below the minimum intensity 12 MeV electron flux at 1 AU.

106.024   The effect of solar-wind convection on charged particle transport in interplanetary space.
J. A. Earl.
Bull. American Astron. Soc., Vol. 13, 812 (1981). – Abstract.

106.025   Characteristics of interplanetary type III radio storms.   J. L. Bougeret, J. Fainberg, R. G. Stone.
Bull. American Astron. Soc., Vol. 13, 860 - 861 (1981). Abstract.

106.026   Factors controlling degree of correlation between ISEE 1 and ISEE 3 interplanetary magnetic field measurements.

N. U. Crooker, G. L. Siscoe, C. T. Russell, E. J. Smith.
J. Geophys. Res., Vol. 87, 2224 - 2230 (1982).

The degree of correlation between ISEE 1 and ISEE 3 IMF measurements is highly variable. Approximately 200 two-hour periods when the correlation was good and 200 more when the correlation was poor are used to determine the relative control of several factors over the degree of correlation.

106.027 Radial evolution of power spectra of interplanetary Alfvénic turbulence.
B. Bavassano, M. Dobrowolny, F. Mariani, N. F. Ness.
J. Geophys. Res., Vol. 87, 3617 - 3622 (1982).

The radial evolution of the power spectra of the MHD turbulence within the trailing edge of high-speed streams in the solar wind has been investigated with the magnetic field data of HELIOS 1 and 2 for heliocentric distances between 0.3 and 0.9 AU.

106.028 Galactic and zodiacal light surface brightness measurements with the Atmosphere Explorer satellites.   V. J. Abreu, J. H. Yee, P. B. Hays.
Appl. Opt., Vol. 21, 2287 - 2290 (1982).

Galactic and zodiacal light surface maps are presented at 7320, 6300, 5577, 5200, and 4278 Å. These were prepared from measurements made with the Visible Airglow Experiment on board the Atmosphere Explorer-C, -D, and -E satellites.

106.029 Statistical properties of MHD fluctuations associated with high-speed streams from Helios-2 observations.
B. Bavassano, M. Dobrowolny, G. Fanfoni, F. Mariani, N. F. Ness.
Sol. Phys., Vol. 78, 373 - 384 (1982).

A variance analysis of Helios-2 magnetic data has been used to derive several statistical properties of MHD fluctuations associated with the trailing edge of a given stream observed in different solar rotations. Such properties are derived both as a function of distance from the sun and as a function of the frequency range included in the sample. The most noticeable result is that the radial gradients of various parameters, such as anisotropy and normalized power of the fluctuations, depend from frequency range.

106.030 Results of a measurement of electron streams with energies $\geqslant$ 40 keV not connected with solar flares aboard the automatic interplanetary station Mars 7.
N. V. Alekseev, P. V. Vakulov, N. I. Vologdin, Yu. I. Logachev.
Kosm. Issled., Tom 20, 422 - 428 (1982). In Russian.

106.031 Stability of the zodiacal light from minimum to maximum of the solar cycle (1974–1981).
C. Leinert, I. Richter, B. Planck.
Astron. Astrophys., Vol. 110, 111 - 114 (1982).

No variation of zodiacal light brightness or polarization was found from observations with the Helios space probes between December 1974 and July 1981. This interval includes minimum and maximum of solar cycle 21. The data give an upper limit of 2% for possible variations of zodiacal light intensity and 0.01 for variations of polarization with solar cycle. However, the average density of interplanetary plasma at high heliographic latitudes is changing and markedly enhanced during solar maximum.

106.032 Search for short term variations of zodiacal light and optical detection of interplanetary plasma clouds.   I. Richter, C. Leinert, B. Planck.
Astron. Astrophys., Vol. 110, 115 - 120 (1982).

The zodiacal light data recorded from the Helios space probes were searched for short term variations. No variations caused by meteor streams or by dust particles released in the

orbits of comets, no enhancements or depletions near planets and no variations correlated with solar activity were found. The only detected short term variations are caused by plasma clouds, with number and size increasing with increasing solar activity. The number density of electrons in interplanetary space appears to increase from sunspot minimum to sunspot maximum. These results confirm the smoothness of the distribution of interplanetary dust and show that the particles responsible for zodiacal light are not appreciably influenced by flares, interplanetary plasma or magnetic fields.

106.033 H$_2$ enrichment of interplanetary medium.
V. Pirronello, G. Strazzulla, G. Foti.
Sun and planetary system, (see 012.040), p. 253 - 254 (1982).

106.034 Study of the interplanetary dust at high ecliptic latitudes: Doppler-Fizeau shifts.
R. Robley, A. Bücher, S. Koutchmy.
Sun and planetary system, (see 012.040), p. 255 - 256 (1982).

106.035 Plasma-dust interactions in the solar and cometary environment.
H. J. Fahr, H. W. Ripken, G. Lay.
Sun and planetary system, (see 012.040), p. 331 - 333 (1982).

By zodiacal light observations, direct dust particle detections, white light F-corona observations, and infrared corona observations it has become evident that interplanetary dust particles are present even very close to the sun, i.e. up to solar distances of about 4 solar radii. These particles, being either of a siliceous or a carbonaceous type, are known to have an amorphous rather than a highly organized cristalline structure. Within the solar wind domain they give rise to a specific form of a plasma-solid body interaction that is described in detail.

106.036 On the electric charge of interplanetary grains.
J.-P. J. Lafon.
Sun and planetary system, (see 012.040), p. 339 - 341 (1982).

In this communication the author describes a work which emphasizes the importance of surface effects in what concerns the mechanism of grain charging in the interplanetary plasma. Numerical computations performed under various conditions of interplanetary plasma show that the photoemission and the thermoemission of electrons can govern the charge and even reverse it at radial distances of a few solar radii.

106.037 Observations of zodiacal light at Abu-Simbel, Egypt, during the period 1975-1979.   A. S. Asaad.
Sun and planetary system, (see 012.040), p. 343 - 344 (1982).

106.038 The brightness integral equation. A different approach to the study of the zodiacal light.
J. Buitrago, R. Gómez, F. Sánchez.
Sun and planetary system, (see 012.040), p. 345 - 347 (1982).

The authors present preliminary results of a new method developed for obtaining the three-dimensional distribution of interplanetary dust from observations performed in the ecliptic. This new approach is essentially based on the analytical transformation of the zodiacal light brightness integral into a Volterra integral equation of the second kind and its subsequent numerical solution.

106.039 Helios zodiacal light measurements – a tabulated summary.   C. Leinert, I. Richter, E. Pitz, M. Hanner.
Astron. Astrophys., Vol. 110, 355 - 357 (1982).

Intensity, colour and polarization of the average zodiacal light as observed by the Helios zodiacal light experiment between December 1974 and January 1981 are given in tabulated form.

106.040 Observations of zodiacal light in Egypt and its variations.   A. S. Asaad.

Compendium in astronomy, (see 003.012), p. 171 - 182 (1982).

Observations of brightness and colour of zodiacal light in blue, yellow and red regions at Abu Simbel during the period 1975 - 1979 are given. Dependence of observations on geographical latitude and elevation above sea level, besides lunar and solar variations are discussed. Possible explanations for short and long period variations are mentioned.

106.041 **Report of IAU Commission 49: The interplanetary plasma and the heliosphere** (Plasma interplanétaire et l'héliosphère). H. J. Fahr.
Trans. IAU, Vol. XVIIIA, (see 003.013), p. 651 - 665 (1982).

Interplanetary dust. See Abstr. 003.070.

Light scattering by small particles. See Abstr. 003.133.

On a connection between flare activity of sunspot groups and the sectorial boundaries of the interplanetary magnetic field. See Abstr. 073.072.

Velocities of shock waves generated by solar flares. See Abstr. 073.092.

Effects of thermal conductivity on large-scale spiral waves in the solar wind. See Abstr. 074.021.

Magnetically closed regions in the solar wind. See Abstr. 074.027.

Long-period observations of the solar wind plasma between 0.3 and 1.0 AU − solar activity. See Abstr. 074.029.

Radial variation of the solar wind speed between 1 and 15 AU. See Abstr. 074.085.

Voyager observations of solar wind proton temperature: 1 - 10 AU. See Abstr. 074.087.

Correlation of high latitude coronal holes with solar wind streams far above or below the ecliptic. See Abstr. 074.090.

Evidence for solar magnetic loops beyond 1 AU. See Abstr. 075.003.

Simulation of the magnetic structure of the inner heliosphere by means of a non-spherical source surface. See Abstr. 075.018.

Fundamental emission for type III bursts in the interplanetary medium: the role of ion-sound turbulence. See Abstr. 077.053.

'Plasma emission' without Langmuir waves. See Abstr. 077.055.

Cosmic-ray variations related to solar, geomagnetic and interplanetary disturbances (23 March - 7 April, 1976). See Abstr. 078.002.

Influence of the dynamics of structural formations in the interplanetary medium on propagation of charged particles generated in solar flares at longitudes 46 - 85° W in September - November 1973. See Abstr. 078.008.

Ionic charge state distribution of helium, carbon, oxygen, and iron in an energetic storm particle enhancement. See Abstr. 078.010.

Report of IAU Commission 21: Light of the night sky (Lumière du ciel nocturne). See Abstr. 082.086.

Effects of IMF polarity on the $F2$-region. See Abstr. 083.001.

Effect of interplanetary magnetic field on apparent drift speed at low latitude stations. See Abstr. 083.013.

Interplanetary magnetic field effects in the E-region drifts at low latitudes. See Abstr. 083.014.

A model of distribution of the electric field in the ionosphere caused by the azimuthal component of the interplanetary magnetic field. See Abstr. 083.017.

Ionospheric currents in the southern polar cap depending on the sign of the azimuthal component of the interplanetary magnetic field. See Abstr. 083.019.

Correlation of geomagnetic activity indices ap with the solar wind speed and the southward interplanetary magnetic field. See Abstr. 084.020.

On a connection of geomagnetic variations in the polar cap with the interplanetary magnetic field in a solar activity cycle. See Abstr. 084.032.

On two types of variations of the magnetic field in the near-polar region with the northern direction of the interplanetary magnetic field. See Abstr. 084.046.

On the relationships between interplanetary quantities and the global auroral electrojet index. See Abstr. 084.049.

Peculiarities of geomagnetic variations in connection with the structure of interplanetary magnetic fields. See Abstr. 084.061.

A brief panorama. See Abstr. 084.062.

The geomagnetic field and its extension into space. See Abstr. 084.064.

Space-time distribution of longitudinal currents in the daytime sector of high latitudes depending on conditions in the interplanetary magnetic field. See Abstr. 084.080.

Geomagnetic variation and field-aligned currents at northern high-latitudes, and their relations to the solar wind parameters. See Abstr. 084.105.

Annual variation of magnetic and ionospheric disturbances, asymmetry of the activity of the solar hemispheres and the interplanetary magnetic field. See Abstr. 085.021.

The sun's influence on the earth's atmosphere and interplanetary space. See Abstr. 085.030.

Voyager encounters with Jupiter's magnetosphere: results of the low energy charged particle (LECP) experiment. See Abstr. 099.093.

Phase functions of polarization and brightness and the nature of cometary atmosphere particles. See Abstr. 102.013.

Report of IAU Commission 22: Meteors and interplanetary dust (Météores et la poussière interplanétaire). See Abstr. 104.039.

Microchrons: the $^{87}Rb-^{87}Sr$ dating of microscopic samples.   See Abstr. 105.178.

The galactic cosmic-ray radial intensity gradient and large-scale modulation in the heliosphere. See Abstr. 143.006.

Separation of the steady-state cosmic ray equation of transport: a generalization.   See Abstr. 143.011.

On the coefficient of diffusion of galactic cosmic rays in the interplanetary space.   See Abstr. 143.017.

Reversal of the cosmic ray density gradient perpendicular to the ecliptic plane.   See Abstr. 143.023.

The heliospheric intensity gradients of the anomalous He^4 and the galactic cosmic-ray components. See Abstr. 143.026.

Magnetic field configuration of the heliosphere and spiral galaxies.   See Abstr. 151.015.

107 Cosmogony

107.001   Interstellar planetesimals – I. Dissipation of a primordial cloud of comets by tidal encounters with massive nebulae.   W. M. Napier, M. Staniucha. Mon. Not. R. Astron. Soc., Vol. 198, 723 - 735 (1982).
The Oort cometary cloud is subject to tidal disruption during close encounters with massive nebulae. The process is studied by numerical integration of ~33000 comet orbits, using realistic models of the Oort cloud and the interstellar medium. It is found that the cloud is rapidly cleared of long period comets, and it is concluded that such comets are probably interstellar in origin. A capture event or tidal disturbance may have taken place $\lesssim$ a few 10^7 yr ago.

107.002   Der Bauplan für Sonne und Planeten. Moderne Ideen zur Entstehung des Sonnensystems. H. J. Fahr. Umschau, 82. Jahrg., 30, 32 - 34 (1982).

107.003   Spiral arms, comets and terrestrial catastrophism. S. V. M. Clube, W. M. Napier. Q. J. R. Astron. Soc., Vol. 23, 45 - 66 (1982).
The authors review a hypothesis of terrestrial catastrophism in which comets grow in molecular clouds and are captured by the Sun as it passes through the spiral arms of the Galaxy. Assuming that comets are a major supplier of the Earth-crossing (Apollo) asteroid population, the latter fluctuates correspondingly and leads to episodes of terrestrial bombardement. Changes in the rotational momentum of core and mantle, generated by impacts, lead to episodes of magnetic field reversal and tectonic activity, while surface phenomena lead to ice-ages and mass extinctions. An episodic geophysical history with an interstellar connection is thus implied. If comets in spiral arms are necessary intermediaries in the process of star formation, the theory also has implications relating to early solar system history and galactic chemistry.

107.004   Formation and collisional evolution of small bodies: effects of two-material systems on large-scale geologic structure.   D. R. Davis, C. R. Chapman, R. Greenberg, S. J. Weidenschilling. Reports of Planetary Geology Program – 1981, (see 003.001), p. 20 - 22 (1981).–Abstract.

107.005   The origin of impacting populations in the inner and outer solar system. R. Strom, A. Woronow. Reports of Planetary Geology Program – 1981, (see 003.001), p. 23 - 25 (1981). – Abstract.

107.006   Aluminium-26 and proton exposure in early solar system.   R. I. Kuznetsova, A. K. Lavrukhina. Meteoritics, Vol. 16, 344 - 345 (1981). – Abstract.

107.007   Spallation of Sr, Y and Zr targets and cosmogenic krypton. S. Regnier, B. Lavielle, M. Simonoff, G. N. Simonoff. Meteoritics, Vol. 16, 382 (1981). – Abstract.

107.008   Chemical heating during protostellar collapse: the thermal evolution of the solar nebula. B. M. P. Trivedi, J. W. Larimer. Meteoritics, Vol. 16, 393 - 394 (1981). – Abstract.

107.009   Primary and secondary objects: a new concept of the early days of the solar nebula. H. Wänke, W. Rammensee. Meteoritics, Vol. 16, 397 - 398 (1981). – Abstract.

107.010   The birth of the planets. N. Henbest. New Scientist, Vol. 92, 173 - 176 (1981). – Abstr. in Phys. Abstr., Vol. 85, Abstr. 23234 (1982).

107.011   Numerical experiments on the stability of preplanetary disks. P. M. Cassen, B. F. Smith, R. H. Miller, R. T. Reynolds. Icarus, Vol. 48, 377 - 392 (1981).
Gravitational stability of gaseous protostellar disks is relevant to theories of planetary formation. Stable gas disks favor formation of planetesimals by the accumulation of solid material; unstable disks allow the possibility of direct condensation of gaseous protoplanets. The authors present the results of numerical experiments designed to test the stability of thin disks against large-scale, self-gravitational disruption.

107.012   Some aspects of spin formation of planets. S. I. Ipatov. Inst. prikl. mat. AN SSSR. Prepr., 1981, No. 102, 28 pp. In Russian. – Abstr. in Ref. zh., 51. Astron., 3.51.165 (1982).

107.013   The effect of a planet's size on the evolution of its atmosphere.   M. H. Hart. Proc. Southwest Reg. Conf., Vol. 7, (see 012.019), 111 - 126 (1982).

107.014   Dissipation of the primordial terrestrial atmosphere due to irradiation of the solar far-UV during T Tauri stage.   M. Sekiya, C. Hayashi.

Prog. Theor. Phys., Vol. 66, 1301 - 1316 (1981). — Abstr. in Phys. Abstr., Vol. 85, Abstr. 48009 (1982).

107.015   Volatile substances in terrestrial planets and their atmospheres and formation of planetary atmospheres.
M. N. Izakov.
Kosm. Issled., Tom 20, 111 - 127 (1982). In Russian.

107.016   Equation of state experiments and theory relevant to planetary modelling.
M. Ross, H. C. Graboske, Jr., W. J. Nellis.
Philos. Trans. R. Soc. London A, Vol. 303, (see 012.022), 303 - 313 (1981).

In recent years there have been a number of static and shockwave experiments on the properties of planetary materials. The highest pressure measurements, and the ones most relevant to planetary modelling, have been obtained by shock compression. Theoretical models have been developed for computing the equations of state of materials used in planetary studies. A compelling feature that has followed from the use of improved material properties is a simplification in the planetary models.

107.017   Recent developments on the problem of the origin of the solar system.   H. Reeves.
Philos. Trans. R. Soc. London A, Vol. 303, (see 012.022), 369 - 375 (1981).

Relics of the molecular cloud origins of the solar system are found in the deuterated molecules of meteorites. The situation is summarized and discussed in conjunction with the isotopic anomalies of heavier elements, to obtain an overall view of the whole event.

107.018   On kinetic limits in physico-chemical models of the condensation process of the protoplanetary cloud (Example of carbon-containing combinations).
I. L. Khodakovskij, R. A. Mendybaev, A. K. Lavrukhina.
Vses. soveshch. po geokhimii ugleroda, Moskva, 14 - 16 dek., 1981. Tez. dokl. Moskva, 1981, p. 300 - 304. In Russian.
From Ref. zh., 51. Astron., 4.51.175 (1982).

107.019 ^{26}Al in the dynamical and biological history of the solar system.   L. S. Marochnik.
Inst. kosm. issled. AN SSSR. Prepr., 1981, No. 664, 12 pp. In Russian. — Abstr. in Ref. zh., 51. Astron., 4.51.176 (1982).

107.020 ^{26}Al and the formation of the planetary system.
L. S. Marochnik.
Inst. kosm. issled. AN SSSR. Prepr., 1981, No. 665, 8 pp. In Russian. — Abstr. in Ref. zh., 62. Issled. kosm. prostranstva, 4.62.314 (1982).

107.021   Origin of the earth and planets.
V. S. Safronov, E. L. Ruskol.
Zemlya Vselennaya, 1982, No. 3, p. 6 - 11. In Russian.

107.022   OB associations and the nonuniversality of the cosmic abundances: implications for cosmic rays and meteorites.   K. A. Olive, D. N. Schramm.
Astrophys. J., Vol. 257, 276 - 282 (1982).

The formation of the solar system inside an OB association is examined with particular attention to the elemental abundances which would have been ejected by the association's first few supernovae. It is found that the solar system material may have been significantly contaminated by these supernovae and thus the average interstellar composition may differ from the solar system composition. The authors find that many of the so-called isotopic and elemental abundance anomalies (e.g., Ne, C, O, s-process/r-process, etc.) found in meteoritic inclusions and in cosmic rays may be more representative of the average interstellar abundance. In other words, it may be that the average solar system abundances are what is "anomalous".

107.023   Are there more than nine planets in the universe? Is the theory of stellar evolution wrong?
G. Field.
Mercury, Vol. 11, 42 - 47, 71 (1982).

107.024   Some remarks on the formation of terrestrial planets.
A. Coradini, C. Federico, G. Magni.
The comparative study of the planets, (see 012.033), p. 3 - 24 (1982).

The process of terrestrial planet formation in a low mass protoplanetary disk is examined. Emphasis is placed on the role played by the gas during coagulation and settling of grains to the central plane of the system. The effects of turbulence are also taken into account. Gravitational instability in a gas-grain thin disk is shown to be a possible mechanism of planetesimal formation. The theory by which terrestrial planets formed by accumulation of planetesimals in a gas free protoplanetary swarm is briefly reviewed.

107.025   Origin and evolution of the giant planets.
P. Bodenheimer.
The comparative study of the planets, (see 012.033), p. 25 - 48 (1982).

Two major hypotheses concerning the origin of the giant planets are discussed: (A) a protoplanet forms in the solar nebula as a gravitationally unstable gaseous subcondensation and evolves as a chemically homogeneous object until a later stage when a solid core may form; (B) a solid core forms first by accumulation of planetesimals, after which solar-composition gas accretes onto the core and eventually becomes unstable to collapse. In general, under either of these scenarios, the evolution falls into three phases: (1) an early cool phase in hydrostatic equilibrium, (2) a hydrodynamic collapse, and (3) a final phase of hydrostatic contraction and cooling to the present state.

107.026   Origin of regular satellites.
S. J. Weidenschilling.
The comparative study of the planets, (see 012.033), p. 49 - 59 (1982).

The regular satellites of Jupiter and Saturn are generally believed to have accreted within cooling circumplanetary nebulae. Small silicate bodies are lost into the planet by gas drag before ice can condense. Larger silicate protosatellites survive by exerting tidal torques on the gas, clearing low-density "tunnels" around their orbits. The nebula is thus divided into series of gas rings depleted in silicates. Cooling eventually allows ice condensation, yielding another generation of icy bodies. Collisional accretion of these objects accounts for stochastic density variations of Saturn's inner satellites.

107.027   Tides and the evolution of the earth—moon system.
D. J. Webb.
Geophys. J. R. Astron. Soc., Vol. 70, 261 - 271 (1982).

A model of the tides in a hemispherical ocean is used to investigate the effect of changes in the earth's rotation rate on the power dissipated by the ocean tides. The results obtained are then used in an idealized astronomical model to investigate how they affect the history of the earth—moon system.

107.028   The evolution of the Earth-Moon system.
D. G. Finch.
Moon Planets, Vol. 26, 109 - 114 (1982).

The tidally-induced couple acting on the Moon, due to friction between the oceans and their beds, is calculated as a function of the Earth-Moon separation. The function is found to be proportional to $1 + d/R^3$, and not the previously used $1/R^6$. By use of this new function it is found that the present rate of lunar recession gives an acceptable history for the system if it is assumed the Moon was initially in a close geostationary orbit 4 billion years ago, when perturbed by the condensation of the Earth's core.

107.029 Is the position of the solar system in the Galaxy exceptional?   L. S. Marochnik.
Priroda, 1982, No. 6, p. 24 - 30. In Russian.

107.030 Ouderdomsbepalingen.   J. van Diggelen.
Zenit, 9. Jaarg., 244 - 252 (1982).

107.031 Dynamical constraints on the formation and evolution of planetary bodies.
A. W. Harris, W. R. Ward.
Annu. Rev. Earth Planet. Sci., Vol. 10, (see 003.009), 61 - 108 (1982).
    The authors review a number of inferences as to the origin of planetary bodies based on the present dynamical state of the solar system, and point out some of the limitations that apply to these conclusions.

107.032 Excitation in the early solar nebula – new experimental findings.   G. Arrhenius, M. J. Corrigan, R. W. Fitzgerald, C. Schimmel.
Sun and planetary system, (see 012.040), p. 221 - 232 (1982).
    The isotope distribution in meteorites suggests that molecular excitation processes similar to those observed today in circumstellar regions and dark interstellar clouds were operating in the early solar nebula. Laboratory model experiments together with these observations give evidence on the thermal state of the source medium from which refractory meteoritic dust formed.

107.033 Physical processes of relevance to the formation of the planetary system.   H. J. Völk.
Sun and planetary system, (see 012.040), p. 233 - 242 (1982).
    The processes mainly reviewed are grain evaporation and condensation, grain-grain coagulation, grain diffusion, sedimentation and radial drifts, as well as self-gravitational fragmentation of dust disks. All this is done within the framework of a turbulent protostellar accretion disk. It is shown that such a model can in principle be consistent with a number of cosmochemical observations. However, steady turbulent disks probably do not allow effective grain sedimentation due to losses by radial drifts. An intermittent turbulence is suggested to allow the formation of large seed bodies onto which the disk material can accrete to form planetesimals.

107.034 On the origin and initial temperature of Jupiter and Saturn.   V. S. Safronov, E. L. Ruskol.
Icarus, Vol. 49, 284 - 296 (1982).
    A two-stage growth of the giant planets, Jupiter and Saturn, is considered, which is different from the model of contraction of large gaseous protoplanets. The authors' model is based on the application of the mechanism of accumulation of the terrestrial planets from solid bodies as an initial stage of the formation process of Jupiter and Saturn. To this end the solid nuclei of these planets reach a certain critical mass at which an effective accretion of gas begins. The temperatures of the growing planets are discussed.

107.035 On the remarkable position of the solar system in the Galaxy.   L. S. Marochnik.
Dokl. AN SSSR, Tom 261, 571 - 574 (1981). In Russian.
From Ref. zh., 51. Astron., 4.51.171 (1982).

107.036 Mass loss from the protoplanetary nebula.   G. P. Horedt.
Astron. Astrophys., Vol. 110, 209 - 214 (1982).
    The mass loss from the protoplanetary nebula is discussed in terms of energy balance considerations by taking into account also loss from the innermost boundary of the nebula.

Within the limits of the adopted simple blow-off approximation the intervals during which the nebula disappears are of order 10^9 yr for heating of the nebula by extreme ultraviolet radiation from the Sun, and about 10^6 yr when the nebula interacts with a T-Tauri-like solar wind. An analytical model is developed for the evolution of the inner and outer boundary of the nebula until its complete disappearance. Temperatures occurring in the outer layers of the nebula due to EUV-heating are determined in the appendix.

Structure and evolutionary history of the solar system.   See Abstr. 003.017.

Activity measurements by Knudsen cell mass spectrometry – the system Fe-Co-Ni and implications for condensation processes in the solar nebula.   See Abstr. 022.044.

The hypothesis of neutron irradiation of protoplanetary matter and changes of the isotopic composition of some elements.   See Abstr. 022.103.

The rotation rates of accreting planetesimals.   See Abstr. 042.106.

OB associations and the early solar system.   See Abstr. 061.013.

On the formation of protostellar disks.   See Abstr. 062.039.

Planetogonic implications of angular momenta in satellite systems.   See Abstr. 091.002.

Constraints on the origin and interior structure of the major planets.   See Abstr. 091.042.

Ages of the solar system: isotopic dating.   See Abstr. 091.048.

Solar system cratering chronology and dating of the surface structures of the terrestrial-type planets.   See Abstr. 091.049.

Atmospheric evolution.   See Abstr. 091.052.

Constraints on the Moon's origin from the partitioning behaviour of tungsten.   See Abstr. 094.028.

The stochastic evolution of asteroidal regoliths and the origin of brecciated and gas-rich meteorites.   See Abstr. 098.016.

The early evolution of Jupiter in the absence of solar tidal forces.   See Abstr. 099.001.

The early evolution of Jupiter in the tidal field of the sun.   See Abstr. 099.002.

Ganymede and Callisto: accumulation heat content.   See Abstr. 099.073.

The origin of Uranus: compositional considerations.   See Abstr. 101.015.

The capture of interstellar comets.   See Abstr. 102.006.

The formation of comets by radiation pressure in the outer protosun.   See Abstr. 102.050.

On the global size of Oort's cloud of cometary nuclei and their total number.   See Abstr. 102.054.

Meteoritics: evidence for chemical fractionation in the early solar system.   See Abstr. 105.157.

Meteorite und Astrophysik. Häufigkeitsanomalien und die Entstehung des Sonnensystems.
See Abstr. 105.159.

The magnetic properties of the Abee meteorite: evidence for a strong magnetic field in the early solar system. See Abstr. 105.195.

Meteorites and cosmochemistry.
See Abstr. 105.208.

Do the age differences given by relative or absolute chronologies of the most ancient meteorites correspond to real age differences?   See Abstr. 105.209.

Una comparación entre la estrella de Barnard y el sistema solar.   See Abstr. 118.015.

Stars

111 Parallaxes, Proper Motions, Radial Velocities, Space Motions, Distances

111.001 Studies of late-type dwarfs. III. Radial velocities and spectral characteristics for 74 stars.
A. P. Cowley, F. D. A. Hartwick.
Astrophys. J., Vol. 253, 237 - 241 (1982).

Seventy-four radial velocities and spectral types are given for a sample of late type dwarfs taken mainly from the U.S. Naval Observatory parallax catalogs. Fifty-two of the stars had no previously published velocities. Spectral types have been assigned, and space motions are computed using the newly determined radial velocities and previously published parallaxes and proper motions. Several high velocity stars are found, but one particularly interesting star is subluminous by approximately 2 mag while having a space motion of 384 km s^{-1}.

111.002 Radial velocities of southern HR stars, II.
W. I. Beavers, J. J. Eitter.
Publ. Astron. Soc. Pacific, Vol. 93, 765 - 768 (1981/82).

Photoelectric measurements of the radial velocities of an additional 106 southern HR stars are reported. This study detects 15 new possible velocity variables and one new double-line spectroscopic binary. Follow-up observations of suspected velocity variables in the first group of southern HR stars are reported.

111.003 On the correction of stellar proper motions for accidental errors.
D. K. Karimova, E. D. Pavlovskaya.
Pis'ma Astron. Zh., Tom 8, 95 - 102 (1982). In Russian. English translation in Soviet Astron. Lett., Vol. 8.

By means of numerical experiments it is shown that the proper motions of population I stars more distant than 1 kpc (with the exception of those determined with high precision $\epsilon_\mu < 0''.0010$) cannot be used without correction for accidental errors. The proper motions of stars with distances $0.5 < r < 1.0$ kpc must be corrected if $\epsilon_\mu > 0''.0040$. The proper motions of the nearest stars ($r < 0.5$ kpc) as well as those of halo population stars may be used without correction.

111.004 Photometric parallaxes of nearby main-sequence stars with annual proper motion of $0''.7$ or more derived from Eggen's B, V and R, I data. W. Gliese.
Astron. Astrophys.,Suppl. Ser., Vol. 47, 471 - 480 (1982).

In two papers Eggen (1979, 1980) has given U, B, V and/or R, I photometry for more than 800 stars with annual proper motion of $0''.7$ or more. Mean colour-luminosity relations M_V-$(B-V)$, M_V-$(R-I)$, and M_R-$(R-I)$ for main-sequence stars in the range $+4 < M_V < +14$ have been derived using for calibration only stars with good percent accuracy of their trigonometric parallaxes ($\sigma_\pi/\pi < 0.14$, where σ_π is the standard error). From these relations photometric parallaxes $\geqslant 0''.040$ have been determined for nearly 90 objects with still unknown distances, the vast majority of them being red dwarfs.

111.005 Parallaxes and proper motions. XIV.
L. A. Breakiron, A. R. Upgren, E. W. Weis.
Astron. J., Vol. 87, 141 - 144 (1982).

Parallaxes and proper motions are presented for 19 stars.

Results are given to the nearest ten-thousandth of an arcsecond and separately in x and y coordinates along with a combined solution.

111.006 U.S. Naval Observatory parallaxes of faint stars. List VI. C. C. Dahn, R. S. Harrington, B. Y. Riepe, J. W. Christy, H. H. Guetter, V. V. Kallarakal, M. Miranian, R. L. Walker, F. J. Vrba, A. V. Hewitt, W. S. Durham, H. D. Ables.
Astron. J., Vol. 87, 419 - 427 (1982).

Trigonometric parallaxes, relative proper motions, and photoelectric photometry are presented for 97 stars in 93 systems.

111.007 Parallax studies of four selected fields.
J. L. Russell, G. D. Gatewood, T. F. Worek.
Astron. J., Vol. 87, 428 - 432 (1982).

Astrometric studies have been completed for four central stars—Wolf 294, 39 Com, Case 2, and Vega— and selected reference stars in each of those regions. The parallax of Wolf 294 was measured as 0.1806 ± 0.0033 arcsec. 39 Com is a visual/astrometric binary with an estimated period of 125 yr and a parallax of 0.0297 ± 0.0044 arcsec. C 2 was measured as a white dwarf from a Warner and Swasey Observatory spectral survey; its white dwarf status is confirmed by its parallax of 0.0186 ± 0.0109 arcsec. Vega's parallax was measured as 0.1243 ± 0.0049 arcsec.

111.008 Radial velocities of HD 187399.
N. L. Ivanova, A. N. Khotnyanskij.
Astrofizika, Tom 17, 819 - 823 (1981). In Russian. English translation in Astrophysics, Vol. 17, No. 4.

Five spectra of HD 187399 in 1978 were obtained. The radial velocities of H, He I, Mg II, Si II, Fe II and K Ca II are determined.

111.009 Investigation of catalogues of absolute proper motions of stars compiled in accordance with the KSZ plan. S. P. Rybka.
Astrometr. Astrofiz., Vyp. (No.) 43, p. 75 - 83 (1981). In Russian.

The differences of the systems of four catalogues of proper motions with respect to galaxies are given (Pulkovo, Tashkent, Moscow and Goloseevo). The systematic catalogue differences are shown to depend on stellar magnitude. The mean square errors of absolute proper motions in each of the above catalogues are estimated. Some problems of compiling a general absolute proper motion catalogue are discussed.

111.010 Investigation of stellar motions in the field of the open cluster NGC 6913. Eh. A. Gerts.
Astrometr. Astrofiz., Vyp. (No.) 45, p. 56 - 62 (1981). In Russian.

On the basis of relative proper motions of the stars in the field of NGC 6913 229 possible cluster members were determined by the maximum likelihood method. Their characteristics and the absolute proper motion of the cluster are given.

111.011 **On the determination of stellar proper motions from a sequence of plates.** A. I. Yatsenko.
Astrometr. Astrofiz., Vyp. (No.) 45, p. 63 - 68 (1981). In Russian.

The application of Schlesinger's method of determination of the stellar proper motions from a sequence of plates taken over 19 years is described. A catalogue is given of relative proper motions of 54 stars in the region of NGC 581.

111.012 **Perturbations in stellar paths.** P. van de Kamp.
Application of modern dynamics to celestial mechanics and astrodynamics, (see 012.026), p. 45 - 57 (1982).

An overview is given concerning problems and results connected with searches for perturbations of stellar motions.

111.013 **The high-velocity stars: Galactic senior citizens.** H. J. Augensen.
Mercury, Vol. 11, 48 - 53 (1982).

111.014 **Variation of radial velocity of the Ap star ϵ UMa.** A. Woszczyk, M. Jasinski.
Soobshch. Spets. Astrofiz. Obs., Vyp. 32, (see 012.029), p. 39 (1981).

111.015 **On the effect of the magnitude equation in determination of proper motions of stars.** A. N. Deutsch.
Astron. Tsirk., No. 1191, p. 4 - 7 (1981). In Russian.

111.016 **Systematic effects in trigonometric parallaxes: comparisons with spectroscopic and cluster parallaxes.** R. B. Hanson, T. E. Lutz.
Bull. American Astron. Soc., Vol. 13, 876 (1981). – Abstract.

111.017 **First results of investigations of proper motions of stars on the Main Astronomical Observatory of the Academy of Sciences of the Ukrainian SSR.** I. G. Kolchinskij.
Visn. AN URSR, 1981, No. 11, p. 14 - 22. In Ukrainian. Abstr. in Ref. zh., 51. Astron., 5.51.116 (1982).

111.018 **Radial velocity observations with the 36″ telescope at Cerro San Cristobal, Santiago, Chile.** C. Sterken, N. Vogt.
Messenger, No. 28, p. 12 - 13 (1982).

111.019 **Hipparcos proper motions.** R. Wielen.
Scientific aspects of the Hipparcos space astrometry mission, (see 012.041), p. 77 - 82 (1982).

The author compares the expected accuracy and number of Hipparcos proper motions with ground-based measurements. He discusses the use of the Hipparcos proper motions for improving the knowledge of galactic kinematics and evolution. Objects with accurate distances and good age determinations (e.g. open star clusters, classical Cepheids, RR Lyrae stars) should be well represented in the observing list.

111.020 **Trigonometric parallaxes – the stellar content of the volume of space covered by Hipparcos and some of its astrophysical implications.** A. Blaauw.
Scientific aspects of the Hipparcos space astrometry mission, (see 012.041), p. 97 - 100 (1982).

The author shows how the Hipparcos parallax program will provide the new base for the calibration of luminosity criteria, for studies of the fine structure in the colour-luminosity diagram, and for the use of highly accurate luminosities of individual stars. He also refers to the open clusters within reach of the parallax program, and to the determination of accurate stellar masses from visual double stars.

111.021 **Astrophysical parameters for Ap and Am stars with Hipparcos.** M. Gerbaldi, D. Briot, C. Burkhart, D. Egret, M. Floquet, A. Gomez, S. Grenier, B. Hauck, C. Jaschek, M. Jaschek, N. Morguleff, P. North, P. Renson, C. van t'Veer.
Scientific aspects of the Hipparcos space astrometry mission, (see 012.041), p. 111 - 113 (1982).

Hipparcos provides a very good opportunity to obtain distances and proper motions for Ap and Am stars. Distances based upon a geometrical method and not upon an astrophysical one are very important since the colours and spectra of these stars are affected by their peculiarities. In addition, one can derive the radius, which is an appropriate parameter for determining the age of Ap and Am stars and which will enable one to discuss their evolutionary status.

111.022 **Trigonometric parallaxes from the ground and from Hipparcos.** C. A. Murray.
Scientific aspects of the Hipparcos space astrometry mission, (see 012.041), p. 115 - 119 (1982).

The present status of trigonometric parallax measurements from the ground, and their statistical interpretation, are reviewed. The expected high accuracy of absolute parallaxes to be obtained from Hipparcos, for a total of some 10^5 stars, provides the opportunity for forming a complete census of all stars brighter than $M_V \approx 5.5$ within 100 parsecs of the Sun. Trigonometric parallaxes of OB stars and cepheids should give a direct check on cosmic distances, to within about ten percent.

111.023 **The luminosity and related parameters of variable and non-variable A, F stars.** E. Antonello, M. Fracassini, L. E. Pasinetti, L. Pastori.
Scientific aspects of the Hipparcos space astrometry mission, (see 012.041), p. 123 - 124 (1982).

A better knowledge of the trigonometric parallaxes allows one to define with more accuracy the various groups of stars, which occupy the same small zone of the HR diagram at the bottom of the instability strip: Am, Am:, A normal, δ Del variables, δ Scuti variables and dwarf Cepheids variables. In particular, it shall give new indications about the possible differences of physical parameters between variable and non-variable stars.

111.024 **Hipparcos and the determination of the helium content of some low-mass non-evolved halo and disk stars of the solar neighbourhood, inferred from the fine structure of the H-R diagram.** M. N. Perrin.
Scientific aspects of the Hipparcos space astrometry mission, (see 012.041), p. 125 - 128 (1982).

Hipparcos is expected to furnish accurate parallaxes for about a hundred low mass non evolved halo and disk stars of the solar neighbourhood, thus providing an accurate location of these stars in the H-R diagram. This essential contribution linked to internal stellar structure and stellar atmosphere computations is the basis of an approach to the helium content determination of these stars.

111.025 **Absolute radii of single and multiple stars from the CADARS (Catalogue of Apparent and Absolute Radii of Stars).** M. Fracassini, L. E. Pasinetti.
Scientific aspects of the Hipparcos space astrometry mission, (see 012.041), p. 165 - 167 (1982).

A better knowledge of the trigonometrical parallaxes will improve the values of the absolute radii of stars. The authors propose a list of single and multiple stars of the "Catalogue of Apparent and Absolute Radii of Stars (CADARS)" by Fracassini et al. for the ESA Space Astrometry Mission Hipparcos.

111.026 **Detectable perturbations in the proper motions of the nearest stars caused by Jupiter like companions?**

W. Gliese.
Scientific aspects of the Hipparcos space astrometry mission, (see 012.041), p. 193 - 194 (1982).

Ranges and periods of perturbations in the proper motions of the nearest stars caused by Jupiter like companions in various orbits are given.

111.027   Radial velocity measurements complementary to Hipparcos.   M. Mayor.
Scientific aspects of the Hipparcos space astrometry mission, (see 012.041), p. 217 - 220 (1982).

To make full use of the astrometrical data obtained from Hipparcos some related astrophysical data should be gathered from ground-based observations. Jointly with proper motions and parallaxes the radial velocities are of evident need for all the studies on kinematics and dynamics of the Galaxy.

111.028   Proposition de mesure de Vitesse Radiales stellaires pour le programme Hipparcos.
C. Fehrenbach, M. Duflot.
Scientific aspects of the Hipparcos space astrometry mission, (see 012.041), p. 221 - 222 (1982).

111.029   Proper motion survey with the forty-eight inch Schmidt telescope. LIV. White dwarfs statistics of the NLTT catalogue.   W. J. Luyten.
Sep. print University of Minnesota, Minneapolis, Minn., USA. 4 pp. (1980).

111.030   Proper motion survey with the forty-eight inch Schmidt telescope. LV. First supplement to the NLTT catalogue.   W. J. Luyten, H. S. Hughes.
Sep. print University of Minnesota, Minneapolis, Minn., USA. 16 pp. (1980).

The present publication gives data for 398 stars with motions larger than $0.''179$ annually.

111.031   Proper motion survey with the 48-inch Schmidt telescope. LVII. The stars of large proper motion.
W. J. Luyten.
Sep. print University of Minnesota, Minneapolis, Minn., USA. 6 pp. (1981).

111.032   Proper motion survey with the 48-inch Schmidt telescope. LVIII. The proper motion of BD +12° 584. Notes on the Lowell Proper Motion Survey. Plagiarism in astronomy.   W. J. Luyten.
Sep. print University of Minnesota, Minneapolis, Minn., USA. 10 pp. (1981).

111.033   Proper motion survey with the 48-inch Schmidt telescope. LIX. A catalogue of 929 possible candidates for Hyades membership.
W. J. Luyten, G. Hill. S. Morris.
Sep. print University of Minnesota, Minneapolis, Minn., USA. 17 pp. (1981).

111.034   Report of IAU Commission 30: Radial velocities (Vitesses radiales).   M. Duflot.
Trans. IAU, Vol. XVIIIA, (see 003.013), p. 361 - 368 (1982).

Errors or omissions in star-identifications in the General Catalogue of Trigonometric Stellar Parallaxes. See Abstr. 002.028.

Catalogue of radial velocities from Herstmonceux and Kottamia 1964 - 1971.   See Abstr. 002.054.

Radial velocities from objective-prism plates in the direction of the Large Magellanic Cloud. See Abstr. 002.063.

Suggestions on cooperative observations of Ap stars (R. V. program).   See Abstr. 013.031.

Report of IAU Commission 24: Photographic astrometry (Astrométrie photographique). See Abstr. 041.057.

Models of stellar evolution and their use in calibrating distances and element abundances of stars. See Abstr. 065.037.

Photometry, space density, and kinematics of a sample of Sanduleak's north galactic pole M stars. See Abstr. 113.013.

Fundamental data for southern stars. (Seventh list). See Abstr. 113.072.

***uvby* photometry in McCormick proper motion fields.**   See Abstr. 113.089.

The absolute magnitudes of G5—M3 stars near the giant branch.   See Abstr. 115.006.

Absolute magnitudes and other basic parameters of O and B stars.   See Abstr. 115.035.

Radial velocity and magnetic field measurements of the A-type supergiant ν Cep (HD 207260). See Abstr. 116.035.

Proper motions of Herbig-Haro objects. See Abstr. 121.029.

Parallaxes for pre-main sequence Herbig Ae/Be stars. See Abstr. 121.040.

Mira-type variable stars and the impact of the Hipparcos mission.   See Abstr. 122.201.

The role of early-type stars in studies of galactic dynamics.   See Abstr. 155.053.

112 Circumstellar Matter (Shells, Dust, Masers, Stellar Winds, etc.)

112.001 **Mass loss from evolved stars. I. Observations of 17 stars in the CO(2–1) line.**
G. R. Knapp, T. G. Phillips, R. B. Leighton, K. Y. Lo, P. G. Wannier, H. A. Wootten, P. J. Huggins.
Astrophys. J., Vol. 252, 616 - 634 (1982).
Emission in the $J = 2-1$ line of CO has been observed from the molecular envelopes produced by mass loss from 17 evolved stars, four of which are detected as CO sources for the first time. The sizes of several molecular envelopes have been measured. Approximate mass loss rates for all of the stars detected in the CO lines are derived from simple model calculations, and vary between $\sim 10^{-7} M_\odot$ yr^{-1} and $\sim 10^{-4} M_\odot$ yr^{-1}. Values of mass loss rates found from these models are used to examine the hypothesis that mass loss is driven by radiation pressure from the central star. This mechanism appears to work for the Mira variables (with [O]>[C]), but the mass loss rates from many carbon stars appear to be too high to be driven by radiation pressure.

112.002 **Interpretation of the line profiles of the 8 micron band of silicon monoxide from VY Canis Majoris.**
D. Van Blerkom, X. Mao.
Astrophys. J., Lett., Vol. 252, L73 - L74 (1982).
Observations of the 8 μm transitions of silicon monoxide from the expanding circumstellar envelope of VY Canis Majoris have previously been interpreted in terms of line formation in a thin shell of gas surrounding an opaque core of dust which emits a continuous spectrum. The authors show that the use of this model is the cause of the apparent displacement of the stellar velocity derived from the 8 μm lines from the value based on the microwave SiO data and on the midpoint of the OH maser emission. No such displacement occurs if a model of line formation is employed in which emission from the entire envelope contributes to the emergent profile.

112.003 **Infrared excess and line emission in Be stars.**
A. D. Neto, J. A. de Freitas Pacheco.
Mon. Not. R. Astron. Soc., Vol. 198, 659 - 668 (1982).
Observations of a selected sample of Be stars in the JKL bands and in Hα are presented. The data indicate that both infrared excess and Hα originate in a common region, which the authors identify as being a disc-shaped envelope around the equatorial region, responsible also for the observed optical continuum polarization. The authors present a method to estimate the electron density of such an envelope using the continuum and the line emission data.

112.004 **VLBI observations of the main line OH masers in VY Canis Majoris.** J. M. Benson, R. L. Mutel.
Astrophys. J., Vol. 253, 199 - 207 (1982).
The authors have conducted a series of VLBI observations of the 1665 MHz and 1667 MHz OH masers in the VY Canis Majoris circumstellar shell. The resulting spatial maps show maser clouds nearly 5×10^{15} cm in diameter and individual maser spots as small as 2×10^{14} cm across. The strongest masers have brightness temperatures of 10^{12} K and are highly circularly polarized. The kinematic model for the masing region is an expanding ($V_{exp} \sim 20$ km s^{-1}), rotating ($V_{rot} \lesssim 3$ km s^{-1}) material lying in a disklike region 8×10^{15} cm from the central star. A Zeeman doublet feature has flared up in the 1665 MHz spectra and permits a direct measurement of the circumstellar magnetic induction, $B \sim 1$ milligauss.

112.005 **Evidence for mass loss at polar latitudes in the Be stars ω Orionis and 66 Ophiuchi.** G. J. Peters.
Astrophys. J., Lett., Vol. 253, L33 - L37 (1982).

IUE observations of the "pole-on" Be stars ω Ori and 66 Oph have revealed the unexpected presence of high velocity (-250 to -850 km s^{-1}), relatively narrow absorption components to the resonance lines of C IV, Si IV, and Si III. The C IV features show structure indicative of multiple shells or clouds. If ω Ori and 66 Oph are indeed viewed pole-on, then these observations suggest that substantial matter is being ejected from the polar regions of these stars. The nature of these unusual high velocity features and the column densities and mass loss rates implied by them are discussed.

112.006 **The extension of OB star winds to lower luminosities.**
T. P. Snow, Jr.
Astrophys. J., Lett., Vol. 253, L39 - L42 (1982).
Mass loss rates derived for a number of early-to-mid-B main-sequence stars have been compared with extrapolations of recently derived empirical correlations between $\dot{M}$ and stellar parameters for luminous OB stars. The order of magnitude of the rates for these stars is consistent with a simple dependence on luminosity over the range from Of to mid-B, and is not consistent with previously derived correlations involving a dependence on mass or radius. It appears that the winds in Be stars may represent a straightforward extension of the OB star wind phenomenon to lower luminosities.

112.007 **A different type of maser star?**
D. F. Dickinson, A. St. Clair Dinger.
Astrophys. J., Vol. 254, 136 - 142 (1982).
A systematic survey of short-period, semiregular variable stars has been made resulting in the detection of six new water masers. Of the 14 short-period maser stars now known, nine are classified as SRb variables. All are very late spectral type SRb's, typically M7. It is demonstrated that these stars have characteristic properties distinguishing them from the Mira and long-period semiregular variables. The short-period semiregular variables may therefore represent a very different type of maser star.

112.008 **Circumstellar molecular emission of evolved stars and mass loss: IRC + 10216.**
J. Kwan, R. A. Linke.
Astrophys. J., Vol. 254, 587 - 593 (1982).
Using CO $J=2\rightarrow1$ and $1\rightarrow0$ data, the authors revise an earlier model of the circumstellar envelope of IRC +10216. They determine that the mass loss rate is $\dot{M} \approx 4 \times 10^{-5} (d/200 \text{ pc})^2 M_\odot$ yr^{-1}, and [CO]/[H$_2$] $\approx 6 \times 10^{-4}$.

112.009 **The remarkable spectrum of some material ejected by Eta Carinae.**
K. Davidson, N. R. Walborn, T. R. Gull.
Astrophys. J., Lett., Vol. 254, L47 - L51, plates L5 - L7 (1982).
Outlying condensations of ejecta from η Carinae have nebular emission-line spectra, which are easier to analyze than the spectrum of the central object. Here the authors report some ground-based and IUE observations of the brightest outer condensation. It appears to be nitrogen rich; five ionization states of nitrogen but none of carbon or oxygen are observed. This is perhaps the most definite known clue to the evolutionary status of this very massive star.

112.010 **Detection of HC$_{11}$N in IRC+10°216.** M. B. Bell, P. A. Feldman, S. Kwok, H. E. Matthews.
Nature, Vol. 295, 389 - 391 (1982).
The observations of three rotational transitions ($J = 70 \rightarrow 69$, $71 \rightarrow 70$, and $72 \rightarrow 71$) of HC$_{11}$N in the microwave emission spectrum of the circumstellar envelope of the cool carbon star IRC + 10°216 are reported. With these observ-

ations, taken during March and May 1981 at the Haystack Observatory, $HC_{11}N$ becomes the largest and heaviest molecule yet detected outside the earth's atmosphere.

112.011 New circumstellar cyanoacetylene sources.
P. R. Jewell, L. E. Snyder.
Astrophys. J., Lett., Vol. 255, L69 - L73 (1982).

A search for the $J = 10-9$ transition of cyanoacetylene (HC_3N) in circumstellar molecular envelopes has resulted in new detections of CIT 6 (definite) and GL 3068 (probable). The 17 sources surveyed included a mixture of carbon-rich, oxygen-rich, and S-type stars, and the previously known source IRC +10216. Four circumstellar shells have now been identified as sources of HC_3N emission. IRC +10216, CIT 6, and GL 3068 are cool, carbon-rich giants; the remaining object, GL 2688, which is not included in this study, consists of a much hotter supergiant surrounded by a carbon-rich nebula. The authors were unable to detect HC_3N from any oxygen-rich shell, including W Hydrae, once reported as an HC_3N source.

112.012 Further observations of the microwave inversion lines of ammonia in IRC+10°216.   M. B. Bell,
S. Kwok, H. E. Matthews, P. A. Feldman.
Astron. J., Vol. 87, 404 - 407 (1982).

Further observations are reported of the microwave inversion transitions of NH_3 arising in the circumstellar envelope of IRC+10°216. The results confirm an earlier probable detection of the (1,1) line and a feature which may be the (2,2) line. Interpretation of the data is complicated by an apparent blueshift of the (2,2) line.

112.013 Mass loss rates in the open cluster IC 1805.
F. Llorente de Andrés, G. Burki,
J. A. Ruiz del Arbol.
Astron. Astrophys., Vol. 107, 43 - 47 (1982).

The profiles of the ultraviolet C IV resonance lines of 6 O-type stars in IC 1805 observed by the IUE satellite are examined. The mass loss rates are determined by means of the fitting of theoretical line profiles on the observed lines.

112.014 High sensitivity molecular line observations of IRC + 10216.   H. Olofsson,
L. E. B. Johansson, Å. Hjalmarson, Nguyen-Quang-Rieu.
Astron. Astrophys., Vol. 107, 128 - 144 (1982).

The authors present high-sensitivity observations of a number of molecular millimeter wave transitions in the envelope surrounding the carbon star IRC + 10216. The goal is to study the elemental abundances and the chemistry in the stellar atmosphere, the mass-loss rate and its possible evolution with time, the kinetics, the thermal structure, and the chemistry of the gas/dust envelope, and finally to understand the detailed radiative transfer and excitation processes in the molecular shell.

112.015 Der Massenverlust von SK-67/111.
O. Stahl, B. Wolf, A. Cassatella, M. Friedjung,
G. Muratorio, O. Ricciardi.
Mitt. Astron. Ges., Nr. 55, p. 52 - 53 (1982). – Abstract.

112.016 Ionisationsstruktur und Geschwindigkeitsfelder expandierender zirkumstellarer Hüllen.
H. Drechsel.
Mitt. Astron. Ges., Nr. 55, p. 55 (1982). – Abstract.

112.017 Stellar winds in the association Cyg OB2.
C. Leitherer, B. Wolf.
Mitt. Astron. Ges., Nr. 55, p. 55 - 56 (1982). – Abstract.

112.018 Dynamical effects of stellar winds and shocked gas on interstellar clouds.   J. M. Shull.

Regions of recent star formation, (see 012.002), p. 91 - 105 (1982).

Stellar winds are a ubiquitous feature of nearly all stars. Their energy input may affect the dynamics of interstellar gas on scales from 10^{15} to 10^{21} cm. New observations suggest that powerful winds are also present during the pre-main sequence phase of massive stars. This review summarizes recent observations and theories of winds and high velocity gas, with special emphasis on inhomogeneities in O-star winds, shocked molecular clouds, H_2O masers, and Herbig-Haro objects.

112.019 Infrared atomic hydrogen line formation in luminous stars.   J. H. Krolik, H. A. Smith.
Regions of recent star formation, (see 012.002), p. 161 - 166 (1982).

Infrared atomic hydrogen lines observed in luminous stars, generally attributed to compact circumstellar H II regions, can also be formed in the winds likely to emanate from these stars. The authors show how the mass-loss rate may be estimated from line fluxes, and also how constraints on the velocity profile may be deduced. Winds capable of producing lines of the observed flux should also be weak, but detectable, sources of radio emission.

112.020 Stellar wind in the nucleus of IC 2149.
M. Perinotto, P. Benvenuti, M. Cerruti-Sola.
Astron. Astrophys., Vol. 108, 314 - 321 (1982).

Low and high resolution spectra of the planetary nebula IC 2149 obtained with the IUE satellite, reveal the existence of a stellar wind in the central star of the nebula. The terminal velocity is found from the C IV λ 1549 profile to be 1440 ± 100 km s^{-1}. An effort to evaluate the associated mass loss rate gives values around $10^{-8}M_\odot$ yr^{-1}.

112.021 Conditions for novalike optically thick winds.
M. Friedjung.
Acta Astron., Vol. 31, 373 - 381 (1981).

The nature of optically thick winds accelerated by the radiation pressure of an object above the Eddington limit is reconsidered, using the theory of Ruggles and Bath (1979). Except perhaps when the object is very close to the limit at the base of the wind, a condition relating the photospheric flux of radiation and the flux of kinetic energy is found. This condition is approximately obeyed by novae, supporting the idea that they have such winds. The condition is not satisfied by the symbiotic star Z Andromedae, but may be satisfied by η Carinae probably related to Hubble-Sandage variables.

112.022 Comparison of winds in the Small Magallanic Cloud and galactic early-type stars.
F. C. Bruhweiler, S. B. Parsons, J. D. Wray.
Astrophys. J., Lett., Vol. 256, L49 - L54 (1982).

IUE low-resolution spectra of 11 OB stars in the Small Magellanic Cloud reveal considerably lower wind velocities than corresponding galactic stars. The reduced effectiveness of radiative acceleration due to lower abundances of heavy elements in the SMC is believed to be responsible for this effect. The absorption strengths of the SMC P Cygni profiles are somewhat weaker in Si IV and much weaker in C IV. Crude estimates of relative mass-loss rates suggest that the observed differences may be dominated by the differing relative abundances of Si and C, although a lower degree of wind ionization or lower mass-loss rates for the SMC stars cannot be ruled out.

112.023 Observations of o Andromedae. III. A model.
R. Poeckert, A. F. Gulliver, J. M. Marlborough.
Publ. Astron. Soc. Pacific, Vol. 94, 87 - 96 (1982).

The authors present a model for the Be star o And which attempts to account for the observed spectroscopic, photometric, and polarimetric changes during the recent shell episode. The model suggests that the waning of the shell was a two-phased process. The initial phase was one in which the density in the inner regions of the circumstellar envelope decreased resulting in a decrease in polarization. Subsequently, the density in the outer regions decreased resulting in the fading of the shell absorption lines. A simple model in which the expansion velocity is independent of time and the mass-loss rate decreases, resulting in a density decrease which propagates outward, fails to satisfy the observed changes in o And. A more complex scenario which does fit the data is proposed.

112.024 **High-spectral-resolution observations of the 7.7 μm feature in HD 44179.**
R. W. Russell, G. Gull, S. Beckwith, N. J. Evans II.
Publ. Astron. Soc. Pacific, Vol. 94, 97 - 101 (1982).

Observations of the moon and HD 44179 were obtained in the wavelength range of $7.5-8.5$ μm at a resolving power of ~800. The spectrum of the moon shows absorptions caused by telluric methane. Use of the moon as a calibrator is effective in removing these atmospheric lines. The spectrum of HD 44179 shows that the 7.7 μm emission feature does not break up into discrete, resolved emission features. Instead, it must be a broad, apparently continuous, emission feature.

112.025 **CN abundance variations in the shell of IRC + 10216.** A. Wootten, S. M. Lichten, R. Sahai, P. G. Wannier.
Astrophys. J., Vol. 257, 151 - 160 (1982).

First observations of the $N = 2-1$ line of CN at 226.9 GHz are reported. High-resolution maps of IRC + 10216 and OMC-1 show that the peak in both sources is compact but resolved in the 30″ beam. In IRC + 10216 the emission extends ~70″ at the line center velocity. A model for CN excitation and the absence of infrared vibration-rotation absorption lines suggest the presence of a source of CN molecules in the outer circumstellar shell. Examination of chemical, expansion, and photodissociation time scales in the outer shell leads to the identification of HCN photodissociation by the interstellar radiation field as a plausible source. Maps and spectra were also made in OMC-1 and OMC-2. In OMC-1, CN emission lies along a narrow ridge, peaking in the BN/KL region.

112.026 **Masing and nonmasing silicon monoxide emission from evolved stars.** R. S. Wolff, E. R. Carlson.
Astrophys. J., Vol. 257, 161 - 170 (1982).

Observations of the $J = 3-2$ and $J = 2-1$ rotational transitions in the ground ($v = 0$) and first vibrational ($v = 1$) states of SiO toward six evolved stars are reported. Comparisons of the line widths of the two ground state rotational transitions indicate that the thermal emission regions are not coextensive in all stars, with the $J = 3-2$ line exhibiting a greater velocity dispersion. The masing ($v = 1$) $J = 3-2$ and $J = 2-1$ transitions show marked variability in both line shape and intensity. Repeated observations of the $J = 2-1$ ($v = 1$) transition indicate that the integrated line intensities vary in phase with the IR fluxes of the periodically variable stars. A search for emission in the $J = 2-1$ line of the ($v = 2$) excited state yielded negative results in all cases.

112.027 **On the radiation deficiency of shell stars in the Balmer continuum.**
J. N. (Ya. N.) Chkhikvadze.
Be stars, (see 012.030), p. 141 - 146 (1982).

The paper gives the results of a statistical comparison of shell stars with normal Be stars to establish the differences in the parameters of the gaseous envelopes, such as the radiation power in the Paschen continuum and the optical depth of the envelopes in the Balmer continuum.

112.028 **The variable shell phase of HD 184279 between 1976 and 1980.**

D. Ballereau, A. M. Hubert-Delplace.
Be stars, (see 012.030), p. 171 - 175 (1982).

The variable shell B0.5 star HD 184279 presents variations of radial velocity and line profiles with a cycle of about 4 years. Some correlations between photometric and spectroscopic data are found. A comparison with 48 Lib and ζ Tau is given.

112.029 **A spectrographic study of the shell star EW Lac.**
G. Scholz.
Be stars, (see 012.030), p. 177 - 179 (1982).

38 Zeeman spectrograms of the Be star EW Lac (HD 217050) were obtained in three successive years, 1978, 1979 and 1980. The investigated lines show long-time variations of the radial velocities, line widths, and line intensities. No hints at the occurrence of a global magnetic field larger than 150 Gauss were found.

112.030 **Rotational velocity versus mass loss in Be stars.**
V. Doazan, M. L. Franco, R. Stalio, R. N. Thomas.
Be stars, (see 012.030), p. 319 - 324 (1982).

C IV and Si IV resonance line profiles of 21 Be, B-shell and normal stars are studied with the aim of detecting evidences for mass loss. The authors found that almost all sampled stars are loosing mass. A relation between an estimated lower limit for the rate of mass loss and the observed rotational velocity was searched but not found.

112.031 **Stellar winds and mass-loss rates from Be stars.**
T. P. Snow, Jr.
Be stars, (see 012.030), p. 377 - 385 (1982).

Resonance-line profiles of SiIII and SiIV lines in 22 B and Be stars have been analyzed in the derivation of mass-loss rates. Of the 19 known Be or shell stars in the sample group, all but one show evidence of winds. It is argued that for stars of spectral type B1.5 and later, SiIII and SiIV are the dominant stages of ionization, and this conclusion, together with theoretical fits to the line profiles, leads to mass-loss rates between 10^{-11} and 3×10^{-9} for the stars. The mass-loss rates from Be stars are apparently insufficient to affect stellar evolution.

112.032 **Mass loss from π Aquarii.** J. A. de Freitas Pacheco.
Be stars, (see 012.030), p. 391 - 392 (1982).

Fe III lines from metastable states have been detected in the UV spectrum of π Aqr (HD 212571). The line profiles are asymmetric and they are probably formed in a expanding envelope. From the observed line widths and using the Sobolev approximation the author has derived a mass loss rate of about $2.5 \times 10^{-9} M_\odot$ y^{-1}.

112.033 **The expanding atmosphere of HD 218393.**
F. Paterson-Beeckmans, A. M. Hubert-Delplace, H. Hubert, D. Ballereau.
Be stars, (see 012.030), p. 393 - 397 (1982).

Visible and far ultraviolet high resolution spectra of the B2 binary star HD 218393 have been obtained at one day interval, shortly before the phase of maximum outwards velocity, adopting the period of 38.908 days. Both spectra show the existence of an extended atmosphere, accelerated outwards. The SiIV resonance lines show some indication of mass loss.

112.034 **The peculiar Be star HD 87643.**
J. A. de Freitas Pacheco, D. Gilra.
Be stars, (see 012.030), p. 399 - 400 (1982).

IUE observations as well as optical spectra of the peculiar star HD 87643 were obtained. The observed optical Fe II lines are probably formed in a expanding envelope and the rate of mass loss derived from the analysis of these lines is about $7 \times 10^{-7} M_\odot$ y^{-1}.

112.035 Evidence for mass loss at polar latitudes in ω Ori and 66 Oph.   G. J. Peters.
Be stars, (see 012.030), p. 401 - 404 (1982).
IUE observations of the "pole-on" Be stars ω Ori and 66 Oph have revealed the unexpected presence of high velocity relatively narrow absorption components in the resonance lines of C IV, Si III, and Si IV. The C IV features show structure indicative of multiple shells or clouds. The nature of these unusual features and the column densities and mass loss rates implied by them are discussed.

112.036 Observation of stellar winds and chromospheres in the visible and infrared.   E. Schatzman.
Scientific importance of high angular resolution at infrared and optical wavelengths, (see 012.031), p. 367 - 377 (1981).
The importance of the observation of stellar winds and chromospheres is stressed both from a physical and an astrophysical point of view. Indications are given concerning the effect of mass loss for stellar evolution and space. Estimates are then given of what can possibly be observed: size of expanding layers, variability, clumpiness. The importance of narrow band observations in various spectral lines is noticed.

112.037 Circumstellar envelopes.   B. Zuckerman.
Scientific importance of high angular resolution at infrared and optical wavelengths, (see 012.031), p. 379 - 389 (1981).

112.038 Zur Natur von OH/IR-Objekten, Infrarotsternen mit OH-Maser-Emission.   D. Engels.
Veröff. Astron. Inst. Bonn, Nr. 95, 91 pp. (1982).
1612 MHz OH masers showing a characteristic double-peaked velocity profile (OH/IR sources) were discovered by radio surveys and could not be identified optically. They are associated with infrared objects (OH/IR stars) of which the evolutionary stage was unknown. This work analyses the nature of these OH/IR stars. They have been observed several years in the infrared spectral region between 1 and 30 microns to prove their variability. By comparison with known variable stars, showing the same characteristic OH maser emission, they must be considered as extension of the optically known Mira stars toward greater masses and longer periods.

112.039 Mass loss from massive stars.   C. de Jager.
The most massive stars, (see 012.034), p. 67 - 80 (1981).
A comparison is made betweeen the various methods for determining stellar rates of mass loss, and the difficulties involved in their application are discussed. The results from these methods are compared. A general review of stellar mass loss rates is given and a few remarkable objects are described. The terminal stellar wind velocity, the upper limit of the rate of mass loss, the stellar wind equations, and the acceleration mechanisms are discussed. The domains of the Hertzsprung-Russell diagram, where these various mechanisms are important, are described.

112.040 Free-free emission, anomalous extinction and mass loss rates of O-type stars.
P. S. Thé, M. Arens, M. Groot, K. P. Simon.
The most massive stars, (see 012.034), p. 81 - 84 (1981).
A sample of about 100 O-type stars, of which ANS ultraviolet measurements are available, were observed on the Walraven WULBV system, the Cousins VRI system and the ESO JHKLM system. For all of these stars UBV measurements (Johnson system) are available. These measurements will be used for the study of the mass loss rate of O-type stars. In the present paper the problems concerning the mass loss rate determination of O-type stars located in dusty gaseous nebulosities are discussed.

112.041 Winds from hot stars.   P. S. Conti.
Bull. American Astron. Soc., Vol. 13, 783 (1981).
Abstract.

112.042 The winds of O and Of stars.   G. L. Olson.
Bull. American Astron. Soc., Vol. 13, 783 (1981).
Abstract.

112.043 Mass loss from OB stars and the formation of interstellar shells.   D. C. Abbott, J. H. Bieging, E. Churchwell.
Bull. American Astron. Soc., Vol. 13, 783 - 784 (1981).
Abstract.

112.044 Mass loss rates of Wolf-Rayet stars from radio continuum observations.
J. H. Bieging, D. C. Abbott, E. B. Churchwell.
Bull. American Astron. Soc., Vol. 13, 785 (1981). – Abstract.

112.045 Circumstellar ammonia in o Ceti.   A. L. Betz, J. J. Ottusch.
Bull. American Astron. Soc., Vol. 13, 808 (1981). – Abstract.

112.046 Resolution of the radio emission from P Cygni's wind.   R. L. White, R. H. Becker.
Bull. American Astron. Soc., Vol. 13, 835 (1981). – Abstract.

112.047 A search for circumstellar material around rapidly rotating non-emission line B stars.
W. R. Oegerle, R. S. Polidan.
Bull. American Astron. Soc., Vol. 13, 856 (1981). – Abstract.

112.048 Analysis of the Balmer lines in Z CMa: evidence for a decelerating stellar wind.
S. A. Drake, R. K. Ulrich.
Bull. American Astron. Soc., Vol. 13, 885 (1981). – Abstract.

112.049 Observations of a low velocity stellar wind in the Be star η Cen.   R. S. Polidan.
Bull. American Astron. Soc., Vol. 13, 885 (1981). – Abstract.

112.050 Line blocking in stellar winds.   J. Castor, D. Friend, D. C. Abbott.
Bull. American Astron. Soc., Vol. 13, 886 (1981). – Abstract.

112.051 OH shells in circumstellar envelopes.   P. J. Huggins, A. E. Glassgold.
Bull. American Astron. Soc., Vol. 13, 895 (1981). – Abstract.

112.052 Long-term radial velocity variations in some Be stars.   E. Antonello, M. Fracassini, L. E. Pasinetti, L. Pastori.
Astrophys. Space Sci., Vol. 83, 381 - 390 (1982).
All the radial velocities (RV) available in the literature since the beginning of the century for the Be stars EW Lac, 28 Tau, ζ Tau, KX And, KY And, CX Dra and 88 Her, are assembled to search for eventual periodic phenomena. The following conclusions have been drawn: EW Lac could be a spectroscopic binary with a period of about 40 years; 28 Tau shows some indications of regular long-term RV variations on the time scale of decades; a probable recurrent shell activity of ζ Tau could be hypothesized.

112.053 The structure and variability of Hα emission in early-type supergiants.   D. Ebbets.
Astrophys. J., Suppl. Ser., Vol. 48, 399 - 414 (1982).
Line profiles and equivalent widths are presented for Hα and Hβ in a sample of O and B supergiants. By making reference to model atmospheres and stellar wind theory, emission line luminosities are derived and shown to correlate with the rate of mass loss. Variability is observed to occur on a dynamical time scale of $1^d - 10^d$. The nature of the varia-

tions suggests that the velocity of outflowing material fluctuates with an amplitude of ~100 km s⁻¹ and that the fluctuations are not globally symmetric. No evidence for rapid variations with time scales shorter than 6 hr was found. Previously published Hα profiles indicate that for the most luminous stars changes in emission line luminosity by as much as a factor of 3 have occurred on time scales of years or longer. Variations of the mass loss rates of 10% to 30% were inferred for all stars which were observed 3 or more times. In most stars the photospheric absorption lines remained unchanged, but in several cases the He I λ6678 profile was asymmetric and varied with the same time scale as the Hα emission.

112.054 The shell episode of 59 Cygni (1974–1975)
P. K. Barker.
Astrophys. J., Suppl Ser., Vol. 49, 89 - 110, plates 1 - 3 (1982).

High quality spectroscopic observations have been obtained of the very early rapidly rotating Be star 59 Cyg during an active phase characterized by spectacular emission-line variations and a 160 day shell episode. A detailed spectroscopic history of the star is given for the period prior to the current outburst: the 1974–75 (second) shell. For the current outburst, line profiles and equivalent width and velocity data provide a comprehensive description of the nature and time scale of the spectral variations in the optical region throughout pre-shell, shell, and post-shell phases. Line profile components due to the circumstellar envelope alone are extracted from the data. Shell absorption lines are shown to display variations in strength and velocity which have a time scale and form radically different from the variations seen in Pleione and o And, the only other stars for which episodes of this type have been well observed.

112.055 Narrow components in the profiles of ultraviolet resonance lines: evidence for a two-component stellar wind for O and B stars?
H. J. G. L. M. Lamers, R. Gathier, T. P. Snow, Jr.
Astrophys. J., Vol. 258, 186 - 200 (1982).

The spectra of 26 early-type stars were studied for the presence of narrow shifted absorption features, superposed on the wide P Cygni profiles of the UV resonance lines. Seventeen out of the 26 stars show such narrow absorptions. The central velocity, v_s, of these features is almost the same for different ions within one star, with no evidence for a relation between v_s and the ionization potential. The velocities of the narrow absorptions for different stars are correlated with the terminal velocity, v_∞, of the stellar wind and have a typical value of $0.75 v_\infty$. The narrow components have a hydrogen column density of about 10^{21} to 10^{22} cm⁻², and contribute about 10% to 60% of the total column density of the wind. The degree of ionization derived from the narrow components is higher than the mean ionization of the wind. Several possible explanations are considered. The observations may be explained by assuming that the wind consists of a two-component gas: a "normal" component which is accelerated in the wind from $v \approx 0$ to v_∞, and which produces the general P Cygni profile, and a less dense component which occurs only at a distance $r \gtrsim 2 R_*$ and has a smaller outward velocity than the normal component. The difference in velocity between the two components may generate shocks which are responsible for the production of the observed X-rays and for the superionization of the wind.

112.056 A model of the 10 micrometer silicate feature in the spectra of BN-like IR point sources.
T. Henning.
Astron. Nachr., Band 303, 117 - 124 (1982) = Mitt. Univ.-Sternw. Jena, Nr. 150.

A model representing the profile of the silicate absorption band at 9.8 μm in the spectra of Becklin-Neugebauer objects has been developed, solving the equation of radiative transfer. The model based on the condition of radiative equilibrium, on an analytical temperature distribution and on the assumption of constant density in the circumstellar shell allows the evaluation of the optical depth in the band's centre and the investigation of the effect of the temperature variation at the inner boundary of the envelope. The optical depth in the band's centre is calculated for the BN point source.

112.057 Magnetic structure in cool stars. V. Chromospheric and transition-region emission from giants.
B. J. Oranje, C. Zwaan, F. Middelkoop.
Astron. Astrophys., Vol. 110, 30 - 36 (1982).

The authors present IUE spectra (1200–1900 Å) of five G- and K-type giants, selected on the basis of Ca II H and K line-core emission strength. They derive emission-line fluxes per unit area at the stellar surface, and extend their data with data published elsewhere. The authors plot the transition-region flux F_{tr} against the chromospheric flux F_{chrom} and find a surprisingly tight relation F_{tr} (:) $F_{chrom}^{1.44 \pm 0.02}$, which extends over two orders of magnitude in F_{chrom} and three orders in F_{tr}. The fluxes F_{tr} and F_{chrom} are found to correlate with rotation rate, for both single and binary stars. The authors discuss these results in terms of discrete solar-like magnetic structure.

112.058 The IRC + 10216 circumstellar shell.
S. T. Ridgway, J. J. Keady.
Philos. Trans. R. Soc. London, Ser. A, Vol. 303, (see 012.038), 497 - 502 (1981).

Spectroscopic studies of IRC + 10216 reveal numerous molecular species and at least two grain species. To define the composition of the ejected material and confirm the activity of chemical processes in the expanding shell, it is necessary to integrate infrared and microwave spectroscopic results with multi-spectral spatial mapping.

112.059 A survey of 3 μm emission features in stellar spectra.
D. A. Allen, D. W. T. Baines, J. C. Blades, D. C. B. Whittet.
Mon. Not. R. Astron. Soc., Vol. 199, 1017 - 1024 (1982).

Spectra between 2.9 and 4.0 μm were taken of 56 emission-line stars and stellar planetary nebulae, and of the bipolar nebulae Roberts 22, M 2-9 and Mz 3. In seven (possibly eight) cases emission features were found between 3.2 and 3.6 μm. The emission features cover a broad range of wavelengths, and have disparate band shapes, indicating that no single material is responsible in any star. The features are observed both in carbon-rich WC 10 stars and in the OH source Roberts 22. Spectra of the WC 10 stars between 1.4 and 2.4 μm also show ionic lines which are attributed to H I and C II.

112.060 Dust and molecules in the shells of carbon stars – the inverse greenhouse effect.
E. M. McCabe.
Mon. Not. R. Astron. Soc., Vol. 200, 71 - 82 (1982).

The expanding, cooling atmosphere of the carbon star IRC + 10216 provides the first evidence for the inverse greenhouse effect: calculations using the Planck mean absorption coefficient show that silicon carbide grains close to the star would be cooler, and graphite grains would be hotter, than if they were blackbodies, thus inhibiting graphite grain formation and enhancing silicon carbide grain formation. As a result SiC is expected to condense close to the star, but graphite to condense much further out. The abundance of 18 molecules observed in the extended envelope suggest that, at about $T = 1250$ K, C has not condensed but SiC has. This is based on a molecular freeze-out model, in which reactions suddenly cease as the expanding gas cools. It fits in with recent infrared models which suggest that graphite is not present within $5 R_*$ of the stellar surface.

112.061 Radiative transfer in dust clouds – II. Circumstellar dust shells around early M giants and supergiants.
M. Rowan-Robinson, S. Harris.

Mon. Not. R. Astron. Soc., Vol. 200, 197 - 215 (1982).

The solution of the equation of radiative transfer through dust shells, developed by Rowan-Robinson (1980), is applied to a sample of 27 stars of spectral types M0 to M4 and having appreciable infrared excess. The models, consisting of a star in a spherically symmetric circumstellar dust shell (CDS), are compared with the available observational data. All the CDS can be successfully modelled with an $n(r) \propto r^{-2}$ density distribution, consistent with steady outflow at a constant velocity. The dust grains used are dirty silicate grains.

112.062   Observational results on mass loss in B stars.
V. Doazan, R. N. Thomas.
Pulsating B stars, (see 012.042), p. 85 - 88 (1981).

112.063   Spectral type dependence of the characteristics of Be stars.   D. Briot, J. Zorec.
Pulsating B stars, (see 012.042), p. 109 - 118 (1981).

112.064   A comparison of circumstellar gas and dust in M giants and supergiants.
W. Hagen, D. F. Dickinson, R. M. Humphreys, R. E. Stencel.
Smithsonian Astrophys. Obs., Spec. Rep. 392, (see 012.045), p. 231 - 238 (1982).

112.065   Circumstellar shells in M17.
R. Chini.
Astron. Astrophys., Vol. 110, 332 - 335 (1982).

JHKL photometry of heavily reddened early type stars within M17 indicates the existence of protostellar dust shells. The dust temperatures range from 1800 K to 3600 K. Various optical depths of the shells suggest different evolutionary stages of the embedded objects.

Landolt-Börnstein.   See Abstr. 003.064.

The continuing saga of the Be stars: a summary.
See Abstr. 013.034.

Equilibrio químico disociativo en envolturas circumestelares.   See Abstr. 022.147.

Laboratory measurements of amorphous silicate smokes and the infrared spectra of oxygen-rich stars.
See Abstr. 022.152.

Linear polarization of radio frequency lines in molecular clouds and circumstellar envelopes.
See Abstr. 063.003.

Intrinsic stellar mass flux and steady stellar winds.
See Abstr. 064.016.

On the ionization and velocity structure of expanding circumstellar envelopes.   See Abstr. 064.020.

Modelos de vientos estelares en la región fotosférica.
See Abstr. 064.055.

The effect of reflected and external radiation on stellar flux distributions.   See Abstr. 064.063.

Synthetic Hα profiles and mass loss rates for Alpha Cygni.   See Abstr. 064.064.

Constraints on the mass-loss rate of HR 1040 (A0Ia).
See Abstr. 064.072.

Report of IAU Commission 36: Theory of stellar atmospheres (Thèorie des atmosphères stellaires).
See Abstr. 064.102.

Wolf-Rayet stars as a stage of the evolution of massive stars.   See Abstr. 065.045.

Infrared photometry of southern Be stars.
See Abstr. 113.015.

A study of Be star variability.
See Abstr. 113.046.

Far-infrared photometry of optical emission-line stars. II.   See Abstr. 113.085.

International Ultraviolet Explorer observations of Alpha Scorpii.   See Abstr. 114.007.

P-Cygni profiles observed in the ultraviolet and visible spectra of Wolf-Rayet stars.   See Abstr. 114.012.

IUE observations of R Aquarii.
See Abstr. 114.014.

International Ultraviolet Explorer spectroscopy of hot stars in the LMC and SMC: the SMC extinction law, stellar flux distributions, and details of the stellar winds.
See Abstr. 114.032.

Detection of further red giants with "hybrid" atmospheres and a possible correlation with double circumstellar Mg II and Ca II lines.   See Abstr. 114.046.

Spectroscopy and infrared photometry of Cyg OB 2 stars: velocity law and mass-loss rates.
See Abstr. 114.049.

Analysis of the IUE and optical spectra of the peculiar Be star HD 87643.   See Abstr. 114.050.

IUE low-dispersion spectra of six luminous stars in symmetric nebulae.   See Abstr. 114.073.

Ultraviolet, visual, and infrared observations of the WC7 variable HD 193793.   See Abstr. 114.074.

Spectroscopic study of Pleione in 1977–1979.
See Abstr. 114.137.

Ultraviolet observations, stellar winds, and mass loss for Be stars.   See Abstr. 114.146.

UV observations of γ Cas: intermittent mass-loss enhancement.   See Abstr. 114.152.

Spectral energy distribution (119–685 nm) in 16 shell stars and a tentative model for accreting Be stars.
See Abstr. 114.155.

Early-type emission-line stars.
See Abstr. 114.223.

The Eta Carinae event.   See Abstr. 115.021.

The status of the corona/wind boundary line in the cool star region of the HR diagram.
See Abstr. 115.024.

Radio observations of early-type emission-line stars and related objects.   See Abstr. 116.001.

Polarimetry and physics of Be star envelopes.
See Abstr. 116.039.

High resolution maps of IRC + 10216 in the
89 GHz (J = 1 - 0) line of HCN.   See Abstr. 116.051.

Hot wind from γ² Velorum observed in the ultraviolet carbon lines.   See Abstr. 117.002.

Ultraviolet and optical spectrum studies of Lambda
Andromedae: evidence for atmospheric inhomogeneities.
See Abstr. 117.006.

Outer atmospheres of cool stars. X. HR 1099 at
quadrature.   See Abstr. 117.018.

A new investigation of the T Tauri star RU Lupi – II.
The physical conditions of the line emitting region.
See Abstr. 121.002.

The chromosphere and wind of the Herbig Ae star,
AB Aurigae.   See Abstr. 121.005.

Time series infrared spectroscopy of the Mira
variable χ Cygni.   See Abstr. 122.007.

Variations du profil de la raie Hα dans Kappa Dra
et 25 Ori.   See Abstr. 122.226.

Asymmetric absorption line profiles in Be stars.
See Abstr. 122.228.

Development of the λ2200 extinction feature in
posteruptive novae.   See Abstr. 124.001.

Rotational temperatures of cyanodiacetylene in
Sagittarius B2, TMC-1, and IRC +10216.
See Abstr. 131.043.

A search for interstellar and stellar iron monoxide.
See Abstr. 131.077.

Interstellar grain size spectrum and circumstellar
grain-grain collisions.   See Abstr. 131.082.

CO J = 3 → 2 and far infrared continuum observations of L1551, Orion KL and IRC +10216.
See Abstr. 131.105.

The physical state of the gas towards HD 93206.
See Abstr. 131.127.

Nearly simultaneous observations of vibrationally
excited $J = 1 \to 0, J = 2 \to 1, J = 3 \to 2$, and $J = 4 \to 3$ SiO
masers.   See Abstr. 131.136.

Observations of stellar wind blown bubbles.
See Abstr. 131.180.

High velocity CO toward Ae and Be stars associated
with nebulosity.   See Abstr. 131.181.

Detection of rotationally excited OH emission from
Vy 2 - 2, and studies of carbon chain molecules in cool
circumstellar shells.   See Abstr. 131.218.

Vibrationally excited molecular hydrogen in
circumstellar clouds and the interstellar CH⁺ abundance.
See Abstr. 131.227.

Wolf-Rayet stars in extragalactic H II regions.
II. NGC 604 – a giant H II region dominated by many Wolf-
Rayet stars.   See Abstr. 132.050.

Wolf-Rayet stars in extragalactic H II regions:
II. NGC 604 – a giant H II region dominated by many Wolf-
Rayet stars.   See Abstr. 132.072.

A search for the infrared counterpart of type II OH
masers. II. Statistical analysis.   See Abstr. 133.004.

New infrared counterparts of southern Type II OH
maser sources.   See Abstr. 133.009.

Galactic ring nebulae associated with Wolf-Rayet
stars. IV. The ring nebulae S308 and its interstellar environment.   See Abstr. 134.005.

Galactic ring nebulae associated with Wolf-Rayet
stars. V. The stellar wind blown bubbles.
See Abstr. 134.006.

Galactic ring nebulae associated with Wolf-Rayet
stars. VI. NGC 3199, anon (MR 26), RCW 58, and RCW 104.
See Abstr. 134.007.

Ring nebulae associated with Wolf-Rayet stars in
the Large Magellanic Cloud. II. Kinematics of DEM 45, 137,
165, 174, and 208.   See Abstr. 134.008.

A new search for nebulae surrounding Wolf-Rayet
stars.   See Abstr. 135.001.

From red giants to planetary nebulae.
See Abstr. 135.043.

Einstein detection of X rays from the Alpha Centauri
system.   See Abstr. 142.007.

X-ray emission from the Carina nebula and the
associated early stars.   See Abstr. 142.039.

X-ray luminosities of B supergiants estimated from
ultraviolet resonance lines.   See Abstr. 142.040.

Stellar winds in X-ray binaries.
See Abstr. 142.065.

Ofpe/WN9 circumstellar shells in the Large Magellanic Cloud.   See Abstr. 159.014.

Erratum

112.901   An expanding motion in the ionized envelope of
        HBV 475.   S. Tamura.
Astrophys. Lett., Vol. 22, 35 - 40 (1981). – See Abstr.
30.112.048.

113 Photometric Properties

113.001 *UBV* sequences for South Polar Cap galaxies.
 D. A. Hanes, G. R. Grieve.
Mon. Not. R. Astron. Soc., Vol. 198, 193 - 197 (1982).
 Photoelectric sequences in the *UBV* system are presented
for the South Polar Cap galaxies NGC 55, 247, 253 and 7793,
as well as for a central field around NGC 1399 in the Fornax
cluster of galaxies. Finder charts and photometry are presented
for 74 stars in the range $7 < V < 14.2$ mag.

113.002 Transformation revisions for the four-color *uvby*
 system.   E. C. Olson.
Publ. Astron. Soc. Pacific, Vol. 93, 783 - 786 (1981/82).
 The author discusses the original transformation equations to the four-color system, given by Crawford and Barnes
(1970), and proposes a simple revision for the metal index m_1.
He also discusses transformations in $(v-y)$ and $(u-b)$, and
demonstrates the effect of the proposed revisions on the
primary eclipse light curve of a totally-eclipsing Algol binary.

113.003 Infrared observations of HM Sagittae.
 O. G. Taranova, B. F. Yudin.
Pis'ma Astron. Zh., Tom 8, 90 - 94 (1982). In Russian. English
translation in Soviet Astron. Lett., Vol. 8.
 Photometric *UVRJHKLMN* observations of the peculiar
star HM Sge were performed in 1978 - 1981. The observations
revealed the IR-flux variations with an amplitude $\Delta K > 1^m$ and
characteristic time ~ 500 days, as well as optical brightness
fluctuations with an amplitude $\leqslant 0^m\!.5$. Any correlation between
optical and IR variations of the brightness is not discovered.

113.004 UBV photometry of the X-ray binary HD 77581 =
 4U 0900 −40.   A. M. Cherepashchuk.
Pis'ma Astron. Zh., Tom 8, 158 - 162 (1982). In Russian.
English translation in Soviet Astron. Lett., Vol. 8.
 Results of new photoelectric UBV observations of
HD 77581 obtained in February - June 1980 are presented.
The long-term variability of the shape and amplitude of the
optical light curves with period $93^d\!.3$ is confirmed. This variability seems to be connected with the forced precession of the
rotational axis of the optical star B0.5Ib leading to a change
of orientation of the eclipsing gaseous streams and accretion
structure surrounding the relativistic object.

113.005 Kuwano's peculiar object is a novalike (symbiotic?)
 binary with a red giant. Photometric and polarimet-
ric observations.
T. S. Belyakina, Yu. S. Efimov, E. P. Pavlenko, V. I. Shenavrin.
Astron. Zh., Tom 59, 1 - 5 (1982). In Russian. English translation in Soviet Astron., Vol. 26, No. 1.
 The results of photometric and polarimetric observations
of Kuwano's peculiar object (N Vul 1979) from August 1979
to December 1980 are presented.

113.006 Detailed comparison of stellar spectrophotometric
 data obtained at the Sternberg Astronomical
Institute with photometry in the band V of the UBV system
and 13-colour photometry of Johnson and Mitchell.
I. N. Glushneva, S. L. Ovchinnikov.
Astron. Zh., Tom 59, 65 - 72 (1982). In Russian. English translation in Soviet Astron., Vol. 26, No. 1.
 Spectrophotometric data obtained at the Sternberg Astronomical Institute are compared in detail with the photometry
in the band V of the UBV system and the 13-colour photometry of Johnson and Mitchell for the spectrophotometric
catalogue as a whole and separately for groups of stars observed with different amplifiers.

113.007 Photometric observations of symbiotic stars in the
 UBVRJHKLMN system. 3. AX Per, AG Dra, BF Cyg,
V443 Her and YY Her.   O. G. Taranova, B. F. Yudin.
Astron. Zh., Tom 59, 92 - 98 (1982). In Russian. English translation in Soviet Astron., Vol. 26, No. 1.
 Photometry of AX Per, AG Dra, BF Cyg, V443 Her and
YY Her is presented. The observations did not reveal any
variations of IR brightness greater than $0^m\!.3$ for these stars. At
the same time considerable $(\Delta U > 1^m)$ changes in the ultraviolet of AX Per and AG Dra were noticed. The physical parameters of the individual components (hot star, cold star and
ionized gas) of the systems AX Per, AG Dra and BF Cyg are
estimated.

113.008 Geneva photometric boxes. II. The reddening
 towards the galactic poles.   B. Nicolet.
Astron. Astrophys., Suppl. Ser., Vol. 47, 199 - 210 (1982).
 Photometric boxes allow a very accurate estimation of
individual reddenings for B- and early A-type stars. A
catalogue containing 129 stars with galactic latitudes higher
than 30° is given. A small but significant reddening is found
in the direction of both the north and south galactic poles:
$E_{B-V} \approx 0.04$.

113.009 Picture gallery: a structured presentation of OAO-2
 photometric data supported by OAO-2 spectro-
photometric data and *UBV*, ANS and TD1 observations.
J. Koornneef, M. R. Meade, P. R. Wesselius, A. D. Code,
R. J. van Duinen.
Astron. Astrophys., Suppl. Ser., Vol. 47, 341 - 418 (1982).
 Stellar fluxes for 531 stars in the wavelength range
$\lambda\lambda\ 5500-1330$ Å are presented in the form of graphs. OAO-2
medium band interference filter photometry for all the stars
is supplemented by OAO-2 spectrophotometry of 213 stars,
UBV photometry from various sources for most of the stars,
ANS photometry (3300−1550 Å) for 363 objects, and the
TD1 fluxes at 2740, 2365, 1965, and 1565 Å for 488 stars.
The stellar magnitudes are on an absolute energy basis. The
merging of the four photometric systems, as well as the way
the objects are ordered, should be helpful for such studies as
interstellar reddening, luminosity effects, bandwidth effects,
and comparisons with model stellar atmospheres. The agreement between the various ultraviolet photometric systems
for early type stars is generally better than 0.10 mag. A list
of stars with observed photometric properties which are
indicative of stellar or interstellar anomalies is also provided.

113.010 *VBLUW* photometry of Magellanic Cloud super-
 and hypergiants, made in 1977 up to 1979.
A. M. van Genderen, F. van Leeuwen, J. Brand.
Astron. Astrophys., Suppl. Ser., Vol. 47, 591 - 594 (1982).
 VBLUW photometry (Walraven system) of SMC and
LMC super- and hypergiants (= super-supergiants) is presented.
The values for V and $B-V$ of the *UBV* system (with subscript
J) are also given. A short discussion is given on the stability
of the photometric parameters within the last 10 to 25 y by
comparing stars in common with other authors.

113.011 UV photometric data on standard A, F and Am
 stars observed by S2/68.   C. Van't Veer-Menneret,
R. Faraggiana, C. Burkhart, Y. Oberto.
Astron. Astrophys., Suppl. Ser., Vol. 47, 595 - 599 (1982).
 The present paper lists the data derived from the stellar
UV fluxes of the S2/68 experiment and used for analysis and
interpretation in a previous paper (Van't Veer et al., 1980)
published in the main journal.

113.012 **Three-colour photometry of a field in the galactic anticentre section near NGC 2360.**
C. Morales Durán.
Astron. Astrophys., Suppl. Ser., Vol. 48, 139 - 152 (1982).

A galactic field of 0.89 square degrees containing 2420 stars near to the open clusters NGC 2360, NGC 2353 and NGC 2345 has been studied photometrically in the *UBV* system down to a limiting magnitude of $V = 15.95$. The interstellar reddening function, the density functions for different groups of stars and luminosity functions have been determined up to the limiting distance. The interstellar reddening in this direction is caused by one absorbing cloud lying at a distance of about 1400 pc. The density functions for the main sequence stars of given intervals in absolute magnitude as well as for the late-type giants are all monotonously decreasing, except perhaps for the nearer distance intervals. The luminosity functions for the distance intervals from 0.4–0.8 kpc and from 0.8–1.2 kpc have the same slope as that for the solar neighbourhood.

113.013 **Photometry, space density, and kinematics of a sample of Sanduleak's north galactic pole M stars.**
P. Pesch, C. C. Dahn.
Astron. J., Vol. 87, 122 - 130 (1982).

Photoelectric observations (V and $B - V$ on the Johnson-Morgan system and $V - I$ on the Kron-Mayall system) are presented for a majority of the probable dwarf M stars found by Sanduleak. Visual absolute magnitudes are determined. Distances are then derived and used to determine both space densities and, in combination with published proper motions, tangential motions.

113.014 **Stellar luminosity functions in the *R, I, J,* and *K* bands obtained by transformation from the visual band.** G. A. Mamon, R. M. Soneira.
Astrophys. J., Vol. 255, 181 - 190 (1982).

The authors transform the stellar luminosity function that has been measured in the visual band into the $R, I, J,$ and K bands, where it has not yet been obtained directly. The transformation is accomplished by subdividing the total visual function, which includes all stars, into sub–luminosity functions for each luminosity class (supergiants through white dwarfs), applying the known $(V–D)$ color, $D = R, I, J, K,$ for each spectral type, and then summing the resultant transformed sub–luminosity functions into a total luminosity function for the band D. The possibility of a systematic error resulting from the existence of a very red stellar population not accounted for in the visual band luminosity function is considered. Application of these luminosity functions and sub-luminosity functions to the study of galactic structure, including star count models in the $R, I, J,$ and K bands is briefly discussed.

113.015 **Infrared photometry of southern Be stars.**
J. Dachs, W. Wamsteker.
Astron. Astrophys., Vol. 107, 240 - 246 (1982).

Infrared broadband photometry in the JHKLM bands (from $\lambda = 1.2$ to 5 microns) is presented for 46 bright southern early-type emission-line stars and supplemented by N-band photometry (at $\lambda = 10$ microns) for 8 of them. For Be stars earlier than about B6e infrared excesses as measured by the (J–M) colour index are found to be strongly correlated to the equivalent widths of Hα emission-lines measured three months earlier. This confirms interpretation of infrared excesses in early-type Be stars as hydrogen free-free emission from circumstellar envelopes. Time variations of infrared magnitudes are derived and are discussed for several stars including μ Cen and χ Oph. By studying correlations of infrared variations to visual photometric variations it is concluded that infrared as well as visual light variations of early-type Be stars are due to variations of density and extent of the circumstellar envelopes around these stars.

113.016 **Photometry of southern stars in the uvby system.**
J. Manfroid.
Bull. Soc. R. Sci. Liège, Vol. 49, 425 - 432 (1980). – Abstr. in Phys. Abstr., Vol. 85, Abstr. 14845 (1982).

113.017 **UBVR photometry of the eclipsing binary system CV Ser and characteristics of the Wolf-Rayet star.**
N. A. Lipunova.
Pis'ma Astron. Zh., Tom 8, 242 - 247 (1982). In Russian. English translation in Soviet Astron. Lett., Vol. 8.

UBVR light curves for the system CV Ser (WC8 + O8) are given. Supposing pure Thomson scattering and the companion star being a point object moving inside the extended spherical envelope of the WC8 star the orbit inclination angle as well as physical characteristics of the WC8 star have been determined. In the Hertzsprung-Russell diagram the core of the WC8 star is located to the left of the main sequence just near the sequence of uniform helium stars.

113.018 **Classification properties of the Vilnius-Geneva photometric system. II. Stars with peculiar chemical composition.** P. North, B. Hauck, V. Straižys.
Astron. Astrophys., Vol. 108, 373 - 379 (1982).

Photometric properties of subdwarfs, metal-deficient giants, metallic-line and peculiar A-type stars in the joint Vilnius-Geneva photometric system are investigated. It is shown that this system preserves the properties of the original Vilnius and Geneva systems in isolation of peculiar types of stars and in determination of their metallicities.

113.019 ***UBVRI* photometry of late-type stars.**
S. M. Rucinski.
Acta Astron., Vol. 31, 363 - 367 (1981).

The note presents photometric observations of 6 brighter late type stars (mostly fast-rotating late-type giants similar to FK Com) and of 6 faint late-type dwarfs (sp. types later than dM3.5e) obtained at Kitt Peak in March 1979.

113.020 ***uvby*β photometry of visual double stars: absolute magnitudes of intrinsically bright stars.**
E. H. Olsen.
Astron. Astrophys., Suppl. Ser., Vol. 48, 165 - 218 (1982).

The main purpose of the present study is to derive absolute visual magnitudes, M_v, for intrinsically bright stars and evolved stars with as few assumptions as possible. The justification for this is to provide a large body of independent and accurate M_v values, which may be used in calibrations of photometric systems, especially for late type giant stars. *uvby*β photometry has been obtained for 398 members of visual multiple stellar systems with at least one main sequence component. A secondary purpose of the present study is to investigate binary systems with main sequence primaries and secondary components off the main sequence. Originally, it was hoped that binaries contradicting the current theory of stellar evolution might be found, but certain systems of this kind have not been uncovered. However, several systems, in which at least one component may be in the pre-main sequence contraction stage, are pointed out.

113.021 **Light variations of the star HD 29697.**
P. F. Chugajnov.
Izv. Krymskoj Astrofiz. Obs., Tom 64, 51 - 54 (1981). In Russian. English translation in Bull. Crimean Astrophys. Obs., Vol. 64.

Variability is discovered from photoelectric observations of the star HD 29697 of the spectral type K3 V with Ca II emission lines H and K. The magnitudes varied in the ranges $7.95 - 8\overset{m}{.}07$ in V, $9.03 - 9\overset{m}{.}18$ in B, $9.96 - 10\overset{m}{.}14$ in U with a possible period of $1\overset{d}{.}3290$. The star probably belongs to the BY Dra type.

113.022 **Comparison of spectrophotometric and photometric data of stars.**   V. F. Gopka, N. S. Komarov.
Astrometr. Astrofiz., Vyp. (No.) 43, p. 40 - 43 (1981). In Russian.

The spectrophotometric catalogue compiled at the Odessa Astronomical Observatory is compared with both Johnson's and Mitchell's catalogues and that of the Vilnius Observatory.

113.023 **On the homogeneity of systems of spectrophotometric standards.**
E. A. Depenchuk, N. S. Komarov.
Astrometr. Astrofiz., Vyp. (No.) 46, p. 17 - 22 (1982). In Russian.

The homogeneity of systems of spectrophotometric standards is discussed on the basis of comparison with photoelectric data of the Arizona Observatory and Johnson's UBV system. The necessity is shown of a new absolute calibration of α Lyr and other primary standards in the ultraviolet and blue spectral regions.

113.024 **Infrared observations of V1016 Cygni.**
B. F. Yudin.
Astron. Zh., Tom 59, 307 - 311 (1982). In Russian. English translation in Soviet Astron., Vol. 26, No. 2.

During 1978 - 1980 the photometric observations of V1016 Cyg in the *UBVRJHKLMN* system were performed. The observations show that the optical brightness of the star (*UBVR*) was constant with an accuracy $\pm 0^m2$, but its IR radiation varied with an amplitude $\Delta K \approx 1^m$ at a timescale $\sim 500^d$. The form of the IR spectrum of V1016 Cyg, when the star is at the maximum of its IR brightness, evidently indicates the presence of a cold component in this system. The cold component is surrounded by a dust shell. The luminosity of the hot star is approximately four times greater than that of the cold component at maximum.

113.025 *VI* **photometry of selected SAO stars.**
E. R. Craine, W. W. G. Scharlach.
Publ. Astron. Soc. Pacific, Vol. 94, 67 - 69 (1982).

Johnson *V*- and *I*-band photoelectric photometry has been obtained for 158 SAO stars in a continuing program in support of the Near Infrared Photographic Sky Survey. These data are utilized in the calibration of the survey photographs and are presented here to assist other programs reliant upon access to photometry in the photographic infrared.

113.026 **Algunas propiedades del diagrama (B–V) vs. (R–I).**
L. A. Milone.
Bol. Asoc. Argentina Astron., No. 20 - 24, p. 171 - 172 (1981).

113.027 **La estrella central del objeto Fourcade-Figueroa.**
C. R. Fourcade.
Bol. Asoc. Argentina Astron., No. 20 - 24, p. 224 (1981). Abstract.

113.028 **Fotometría UBV y H beta de estrellas débiles de tipo temprano en Norma.**
J. C. Muzzio, J. C. Forte.
Bol. Asoc. Argentina Astron., No. 20 - 24, p. 258 (1981). Abstract.

113.029 **Fotometría UBV de H beta de estrellas OB débiles en Crux y Centaurus.**
J. C. Muzzio, A. Feinstein, A. M. Orsatti.
Bol. Asoc. Argentina Astron., No. 20 - 24, p. 259 (1981). Abstract.

113.030 **Búsqueda de objetos con emisión H-alfa en Ara.**
J. C. Muzzio, M. Rabolli, I. Vega.
Bol. Asoc. Argentina Astron., No. 20 - 24, p. 389 (1981). Abstract.

113.031 **Estudio de objetos con emisión H-alfa en Crux.**
R. E. Martínez, J. C. Muzzio, S. Waldhausen.
Bol. Asoc. Argentina Astron., No. 20 - 24, p. 392 (1981). Abstract.

113.032 **Fotometría infrarroja de estrellas tempranas y peculiares.**   A. Feinstein, H. G. Marraco.
Bol. Asoc. Argentina Astron., No. 20 - 24, p. 397 (1981). Abstract.

113.033 **Colores intrínsecos de las estrellas de tipo temprano.**   A. G. de Moreno.
Bol. Asoc. Argentina Astron., No. 20 - 24, p. 469 (1981). Abstract.

113.034 **Fotometría multicolor de estrellas Be.**
H. G. Marraco, A. Feinstein.
Bol. Asoc. Argentina Astron., No. 20 - 24, p. 473 (1981). Abstract.

113.035 **Yellow supergiant reddenings from *BVRI* data.**
J. D. Fernie.
Astrophys. J., Vol. 257, 193 - 197 (1982).

Using the Johnson *BVRI* system of photometry, it is shown that the expression
$E_{B-V} = -0.255 + 1.727 (R-I) - 0.475 (B-V)$ will reproduce the color excesses given by Parsons and Bell for stable F and G supergiants with an rms deviation of less than 0.03 mag. The same expression is applicable to classical Cepheids at mean light. For observations of Cepheids at a particular phase ϕ a correction term, depending on ϕ and the amplitude A of the visual lightcurve, has been derived. Color excesses determined in this manner are listed for 41 Cepheids and 17 stable supergiants.

113.036 **Results of UBV electrophotometric observations of red supergiants.**   G. V. Abramyan.
Soobshch. Byurakan Obs., Vyp. 53, p. 3 - 39 (1982). In Russian.

The results of electrophotometric observations of red supergiants carried out 1973 - 1977 are presented. The homogeneity of the obtained observational data allowed to determine some statistical relations for brightness and colour variations. It is shown that in the case of many supergiants there exists a correlation between brightness and colour variations. The position of red supergiants on the diagram of the brightness gradient is determined. The mean and the maximum rates of brightness variations are determined.

113.037 **Comentarios adicionales sobre la estrella Sanduleak-Seggewiss (NGC 6231-92).**
L. A. Milone.
Bol. Asoc. Argentina Astron., No. 26, p. 101 (1981). Abstract.

113.038 **The intrinsic colours of stars in the ultraviolet.**
D. J. Carnochan.
Bull. Inf. Cent. Données Stellaires, No. 22, p. 75 - 78 (1982).

113.039 **Behaviour of Ap stars on the two-colour diagrams of the Vilnius system.**   I. Iliev.
Soobshch. Spets. Astrofiz. Obs., Vyp. 32, (see 012.029), p. 48 - 49 (1981).

113.040 **On the low-amplitude variability of B- and A-stars.**
E. Želwanowa.
Soobshch. Spets. Astrofiz. Obs., Vyp. 32, (see 012.029), p. 57 - 61 (1981).

113.041 **Rapid variability of Ap stars (Review of observational data).**   N. S. Polosukhina.

Soobshch. Spets. Astrofiz. Obs., Vyp. 32, (see 012.029), p. 67 - 68 (1981).

113.042 **Rapid light variations of the star HD 184905.** K. Panov.
Soobshch. Spets. Astrofiz. Obs., Vyp. 32, (see 012.029), p. 73 (1981).

113.043 **Short-time light variation of HD 219749.** G. Hildebrandt.
Soobshch. Spets. Astrofiz. Obs., Vyp. 32, (see 012.029), p. 74 - 75 (1981).

113.044 **The light variability of 17 and 21 Com.** I. F. Alaniya, A. G. Totochava, B. E. Zhilaev, O. P. Abuladze.
Soobshch. Spets. Astrofiz. Obs., Vyp. 32, (see 012.029), p. 75 - 76 (1981).

113.045 **Some photometric characteristics of Be stars.** E. E. Mendoza V.
Be stars, (see 012.030), p. 3 - 17 (1982).
Homogeneous observational data on four photometric systems which cover the spectral range from 0.14 to 4 microns through narrow, intermediate and broad passband filters are used to review and derive some photometric characteristics of Be stars. The author finds that 60% of the stars under discussion have ultraviolet and infrared excesses.

113.046 **A study of Be star variability.** J. Dachs.
Be stars, (see 012.030), p. 19 - 22 (1982).
Variations of Balmer emission line strength for Be stars are compared to simultaneously measured photometric light-curves in the visual and infrared regions of the spectrum. Three different types of correlations between photometric and spectrometric variations are described. Variable intensity of continuous recombination radiation from ionized hydrogen in the circumstellar envelope is shown to be a major source of light variations for Be stars.

113.047 **Statistical analysis of the data available for Be stars.** D. Egret.
Be stars, (see 012.030), p. 27 - 31 (1982).
A comprehensive catalogue of more than 1100 Be stars with known MK classification has been prepared at the Strasbourg Stellar Data Center. The compilation of photometric and spectroscopic data available for these stars has made possible a general statistical approach of the Be star spectral group.

113.048 **The Vatican emission star survey: review and comments.** P. Cardon de Lichtbuer.
Be stars, (see 012.030), p. 45 - 48 (1982).

113.049 **Light variations in several broad-lined B stars.** M. Jerzykiewicz, C. Sterken.
Be stars, (see 012.030), p. 49 - 52 (1982).

113.050 **Correlation between spectrum characteristics and photometric behaviour of Be stars.** G. A. Ponomareva.
Be stars, (see 012.030), p. 65 - 67 (1982).
Based upon a sample of 140 B0−A0 emission-line stars contained in "A Photographic Atlas of Be stars" statistical relationships between variable stars and Be stars are examined. It is found that 44% of all the stars of the sample show some kind of photometric variations. The stars with narrow metallic absorption lines display relatively more intense photometric activity. Proceeding from this conclusion, the photometric variability of two stars HD 50138 and HD 193182 can be predicted.

113.051 **Optical variations of the Be star HDE 245770/ A 0535+26.** A. Guarnieri, C. Bartolini, A. Piccioni, A. Giangrande, F. Giovannelli.
Be stars, (see 012.030), p. 69 - 74 (1982).
Results of photoelectric observations of HDE 245770 spread over six years are presented. No variations greater than some hundredths of magnitudes are evident in contrast with the ancient light history, which is more complex. A sensible variation of the colour indices is present on a time scale of tens of years. The temporal analysis of these data shows some evidence of orbital periodicity.

113.052 **Infrared photometry of Be stars.** A. Feinstein.
Be stars, (see 012.030), p. 235 - 240 (1982).

113.053 **Search for variability in near infrared fluxes of peculiar emission-line objects.** P. Bouchet, J. P. Swings.
Be stars, (see 012.030), p. 241 - 245 (1982).

113.054 **Infrared emission from four Be stars optical counterparts of galactic X-ray sources.** P. Persi, M. Ferrari-Toniolo, G. L. Grasdalen.
Be stars, (see 012.030), p. 247 - 251 (1982).
Preliminary results of infrared observations from 2.3 up to 10 and 20 microns of the Be-X-ray stars X Per, γ Cas and HDE 245770, indicate the presence of an ionized circumstellar disk. X Per and γ Cas show a variable infrared excess at 10 μ suggesting variability in the stellar wind. LS I+65° 010 presents an anomalous infrared energy distribution for a Be star.

113.055 **Radial-velocity and photometric variations of o And: critical evaluation of possible periods.** J. Horn, P. Koubský, J. Arsenijević, J. Grygar, P. Harmanec, J. Krpata, S. Kříž, K. Pavlovski.
Be stars, (see 012.030), p. 315 - 318 (1982).

113.056 **SVS 2132 − a distant star of constant brightness.** M. M. Zakirov.
Astron. Tsirk., No. 1168, p. 7 - 8 (1981). In Russian.

113.057 **Brightness variation of He2-442.** B. F. Yudin.
Astron. Tsirk., No. 1169, p. 8 (1981). In Russian.

113.058 **Photoelectric photometry of Ap stars in the galactic cluster NGC 2516: preliminary results.** P. North, F. Rufener, P. Bartholdi.
Inf. Bull. Variable Stars, No. 2103, 3 pp. (1982).

113.059 **BV photometry of Betelgeuse Oct. 1979 to Apr. 1981.** K. Krisciunas.
Inf. Bull. Variable Stars, No. 2104, 5 pp. (1982).

113.060 **No rapid variability observed for the Be stars HD 58050 and β CMi.** H. Božić, M. Muminović, K. Pavlovski, M. Stupar, P. Harmanec, J. Horn, P. Koubský.
Inf. Bull. Variable Stars, No. 2123, 4 pp. (1982).

113.061 **The light variability of the Be star HD 58050 (OT Gem).** E. Poretti.
Inf. Bull. Variable Stars, No. 2129, 3 pp. (1982).

113.062 **The variability of BD +60°562.** C. Bartolini, P. Custodi, F. Dell'Atti, A. Guarnieri, A. Piccioni.
Inf. Bull. Variable Stars, No. 2139, 2 pp. (1982).

113.063 **Johnson BVR magnitudes for selected comparison stars.** R. H. van Gent.
Inf. Bull. Variable Stars, No. 2140, 4 pp. (1982).

113.064 **Photometry of the shell star BU Tau (Pleione) 1980 - 1982.**
U. Hopp, S. Witzigmann, E. H. Geyer.
Inf. Bull. Variable Stars, No. 2148, 5 pp. (1982).

113.065 **Relative photometry of bright stars of NGC 6913 in 1978 and 1979.**   R. M. Raznik.
Astron. Tsirk., No. 1195, p. 6 - 8 (1981). In Russian.

113.066 **Voyager absolute far-ultraviolet spectrophotometry of hot stars.**   J. B. Holberg, W. T. Forrester,
D. E. Shemansky, D. C. Barry.
Astrophys. J., Vol. 257, 656 - 671 (1982).
Voyager observations in the 912–1200 Å spectral region are used to indirectly intercompare absolute stellar spectrophotometry from previous experiments. Measurements of hot stars obtained by the Voyager 1 and 2 ultraviolet spectrometers show considerably higher 912–1200 Å continuum fluxes than the recent observations of Brune et al. and Carruthers et al. By intercomparing the various observations and adjusting the calibration below 1050 Å to the pure hydrogen line-blanketed model atmospheres of Wesemael et al. a revised Voyager calibration is constructed. The revised calibration is in good agreement with current model atmosphere fluxes for both early-type stars and DA white dwarfs and will be used for Voyager astronomical observations.

113.067 **Faint blue objects at high galactic latitude: summary of four Palomar Schmidt fields.**
P. D. Usher, A. Warnock III, R. F. Green.
Bull. American Astron. Soc., Vol. 13, 800 (1981). – Abstract.

113.068 **Intrinsic dispersion and binary frequency of M dwarfs.**   R. G. Probst.
Bull. American Astron. Soc., Vol. 13, 803 (1981). – Abstract.

113.069 **A search for light variations of the brightest Pleiades stars.**
J. E. Brolley, A. N. Cox, E. K. Hodson, S. W. Hodson.
Bull. American Astron. Soc., Vol. 13, 832 (1981). – Abstract.

113.070 **The photometric variability of main sequence stars.**   R. R. Radick, L. W. Hartmann, D. M. Mihalas,
S. P. Worden.
Bull. American Astron. Soc., Vol. 13, 832 - 833 (1981). Abstract.

113.071 **Comparison of predicted and observed magnitudes and colors of cool stars.**
T. Y. Steiman-Cameron, H. R. Johnson.
Bull. American Astron. Soc., Vol. 13, 886 (1981). – Abstract.

113.072 **Fundamental data for southern stars. (Seventh list).**   R. M. Catchpole, D. S. Evans, D. H. P. Jones,
D. L. King, R. E. Wallis.
R. Greenwich Obs. Bull. No. 188, 38 pp. (1982).
Newly determined radial velocities, magnitudes and colours on the B, V system, are given for 339 stars in the Southern hemisphere, including several stars of large parallax.

113.073 **A combined RGU-photometry of six fields in the direction to the Scutum Cloud.**   S. Karaali.
Istanbul Üniv. Fen Fak. Mec. Seri C, Vol. 45, 65 - 92 (1980) = Publ. Istanbul Univ. Obs. No. 115 (1981).

113.074 **Revised RGU photometry of an anticentre field around NGC 2129.**   L. Topaktaş.
Istanbul Üniv. Fen Fak. Mec. Seri C, Vol. 45, 93 - 106 (1980) = Publ. Istanbul Univ. Obs. No. 116 (1981).
A galactic anticentre field of 1.09 square degrees containing 1959 stars around galactic cluster NGC 2129 discussed by Wick (1965) has been reduced to the present RGU system

and rediscussed. The interstellar reddening function, the density functions for different groups of stars and the luminosity functions have been redetermined. The reddening is caused by two interstellar clouds with distances of 360 and 670 pc.

113.075 **A catalog of red stars near L1454.**   R. Duerr, E. R. Craine.
Prepr. Steward Obs., No. 363, 16 pp. (1982).

113.076 **Slow flares in stellar aggregates. II.**   Eh. S. Parsamyan.
Astrofizika, Tom 16, 231 - 241 (1980). In Russian. English translation in Astrophysics, Vol. 16, No. 2.
Photometric data on nine slow flares in the Pleiades, Praesepe and Hyades are given. By using known data on slow flares in aggregates the slow-flare curves of brightness are divided into three types. It is shown that there may be a combination of slow and rapid flares.

113.077 **Photographic colorimetry of stellar flares in the Pleiades and Orion aggregates.**
L. V. Mirzoyan, O. S. Chavushyan, N. D. Melikyan,
R. Sh. Natsvlishvili, G. B. Oganyan, V. V. Ambaryan,
A. T. Garibdzhyan.
Astrofizika, Tom 17, 197 - 213 (1981). In Russian. English translation in Astrophysics, Vol. 17, No. 2.
The results of synchronous colorimetric observations of stellar flares in the UBV system are presented. During observations of the Pleiades and Orion regions for 53 hours of effective time in all flares were detected: 13 in Pleiades and 6 in Orion. 4 new flare stars were discovered, all in Orion. 5 flares in aggregates were measured in three bands U, B, V for the first time: on the stars No. 2, 105, 156 and 540 in Pleiades and T129 in Orion.

113.078 **Simulated numerical transformation between similar wide band photometric systems.**
K. Pavlovski.
Hvar Obs. Bull., Vol. 5, No. 1, p. 1 - 23 (1981).
Transformation of an instrumental UBV photometric system to standard one is simulated by numerical computations. The observed transformation coefficients can be satisfactorily reproduced by synthesized transformation. Effects which can affect response functions of the photometric systems and such computations are discussed.

113.079 **Photoelectric photometry of the halo population and other peculiar stars in the Vilnius photometric system.**   J. Sperauskas, A. Bartkevičius, K. Zdanavičius.
Bull. Vilnius Astron. Obs., No. 58, p. 3 - 31 (1981). In Russian.
A catalogue of 377 halo population and other peculiar stars observed in the Vilnius photometric system in 1967 – 1977 is presented.

113.080 **Photoelectric photometry and classification of 19 halo population stars in the Vilnius system.**
A. Bartkevičius, S. Bartašiūtė, K. Zdanavičius.
Bull. Vilnius Astron. Obs., No. 58, p. 32 - 40 (1981). In Russian.
The results of photoelectric photometry of 19 stars belonging mainly to the halo population are presented. Color indices of the Vilnius system and V magnitudes transformed to the UBV system are given in a table.

113.081 **Photoelectric photometry of stars at high galactic latitudes in the Vilnius system.**   S. Bartašiūtė.
Bull. Vilnius Astron. Obs., No. 58, p. 41 - 53 (1981). In Russian.
A catalog with the results of photoelectric photometry in the Vilnius seven-color system is presented. It contains 195

stars down to V = 12.2 with large proper motions, high velocity stars, and stars with peculiarities in their spectra, located in an area of 900 □° near the South Galactic Pole, with galactic latitudes between −44° and −82°.

113.082 Colour excess ratios for metal-deficient, barium and CH stars.   A. Bartkevičius, Z. Sviderskienė.

Bull. Vilnius Astron. Obs., No. 58, p. 54 - 64 (1981). In Russian.

Colour excess ratios in the UPXYZVS, uvby, and UBV systems were computed for 10 extreme subdwarfs, 5 metal-deficient giants, 2 CH and 3 barium stars.

113.083 TiO band strengths in metal-rich globular clusters IV. The solar neighborhood sequence.

J. R. Mould, M. J. Siegel.

Publ. Astron. Soc. Pacific, Vol. 94, 223 - 225 (1982).

TiO band strengths and near infrared color indices have been measured in a sample of nearby late-type giants for the purpose of comparison with observations in metal-rich globular clusters. Transformations to the Wing narrow-band system are determined.

113.084 *UBVRI* standard stars in the E-regions.   J. A. Graham.

Publ. Astron. Soc. Pacific, Vol. 94, 244 - 265 (1982).

Photometry on the *UBVRI* Kron-Cousins system is presented for 102 stars covering the magnitude range 7 to 16 in the nine Harvard E-regions. These stars, at declination close to −45°, are suitable for use as Southern Hemisphere standards in the photometry of faint stars and galaxies. Positions accurate to a few arc seconds as well as identification charts are given for each star.

113.085 Far-infrared photometry of optical emission-line stars. II.   P. M. Harvey, B. A. Wilking.

Publ. Astron. Soc. Pacific, Vol. 94, 285 - 288 (1982).

The authors report far-infrared photometry of dust clouds around a sample of eight optically bright stars, five of which are believed to be pre-main-sequence and three of which are probably post-main-sequence. They find energy distributions consistent with the results of an earlier study of eight other stars where a good correlation was found between the existence of cool, far-infrared-emitting dust clouds and the presumed youth of the central star. The post-main-sequence objects, as in the previous study, were found to have energy distributions falling steeply from 10 to 100 μm implying very small amounts of cool dust relative to the amount of dust at a few hundred K.

113.086 *BV* photometry of a sample of faint red stars in the direction of the south galactic pole.

G. Alcaino, P. S. Thé.

Publ. Astron. Soc. Pacific, Vol. 94, 335 - 340 (1982).

Using existing direct photographic plates, *BV* photometry of a sample of 99 red objects (13.74 < *V* < 19.48) from Staller el al.'s (1981) blink survey shows that most of them are indeed M-type stars: $(B-V)_0 \geqslant 1.4$.

113.087 Photoelectric photometry of faint M-type stars in the direction of the south galactic pole.

P. Pesch.

Publ. Astron. Soc. Pacific, Vol. 94, 345 - 346 (1982).

Photoelectric *V* (Johnson-Morgan *UBV* system) and (*V–I*) (Kron-Mayall *PVI* system) photometry is presented for 54 faint M-type stars from Pesch and Sanduleak's catalog of probable dwarf stars of type M3 and later in the direction of the south galactic pole.

113.088 DDO photometry of G and K stars.   M. G. Pastoriza, T. B. Storchi, S. H. B. Livi.

Publ. Astron. Soc. Pacific, Vol. 94, 347 - 349 (1982).

Spectral types, stellar population, and an estimation of metal abundances [Fe/H] were obtained for a sample of G and K peculiar stars from a list of Stock and Wroblewski and of metal-deficient stars from a list of Bond using DDO photometry.

113.089 *uvby* photometry in McCormick proper motion fields.   J. Degewij.

Astron. Astrophys., Suppl. Ser., Vol. 48, 481 - 484 (1982).

High precision *uvby* photometry is reported for 50 F2 to G2 stars, having 9.4 < *V* < 12.3 mag, selected in southern galactic polar regions of the McCormick proper motion fields. Each star was measured on six different nights. Compared with earlier data (Blaauw et al., 1976), the brighter stars systematically show slightly smaller (~0.02 mag) m_1 indices. Single measurements are given for 98 stars in 8 McCormick fields at intermediate southern galactic latitudes.

113.090 High angular resolution *uvbyβ* observations of stars earlier than G0 in the intermediate and low latitude areas SA 128 and SA 156.   J. Knude.

Astron. Astrophys., Suppl. Ser., Vol. 49, 69 - 72 (1982).

The *V*, *b-y*, m_1, c_1, and *β* values on the standard system, for the magnitude limited sample are presented. In the two areas, SA 128 and SA 156, *uvbyβ* data have been obtained for 64 and 131 stars respectively. Both areas size 15 square degrees, thus allowing investigation of the $E(b-y)$ variation on scales smaller than 30'.

113.091 Monitoring solar-type stars.   G. W. Lockwood.

Variations of the solar constant, (see 012.037), p. 219 - 227 (1981).

Old UBV and recent uvby photometry of solar-type dwarfs and other standard stars yield an upper limit of variability (determined by observational errors) of about 0.004 mag rms. A factor two improvement in this upper limit is achievable.

113.092 A search for Hα-emission objects in a region in Vela.   E. I. Vega.

Astron. J., Vol. 87, 794 - 802 (1982).

A search for Hα-emission objects covering an area of 42 deg² in the Vela region of the Milky Way resulted in the discovery of 108 objects. UBV photographic photometry and approximate spectral classification on objective-prism plates of 1360 Å mm⁻¹ at Hγ was also obtained for the objects discovered; two of them are early-type supergiants and 25 are recognized as OB stars. The limiting magnitude of the survey is about *R* = 14.5 mag, and more than half of the objects were not previously known.

113.093 Photometric aspects of Tycho.   C. Jaschek.

Scientific aspects of the Hipparcos space astrometry mission, (see 012.041), p. 133 - 135 (1982).

The existing photometric surveys and their completeness are examined. The discovery completeness of bright variable stars is examined. It is found that Tycho can be expected to discover a large number of new variables. A more detailed discussion is given for eclipsing binaries, and again it is shown that Tycho will contribute with a large number of discoveries. Finally arguments are given in favour of the use of broad band B and V filters on Tycho.

113.094 Ground-based photometric data.   J.-C. Mermilliod, M. Mermilliod.

Scientific aspects of the Hipparcos space astrometry mission, (see 012.041), p. 139 - 141 (1982).

The present state of the photometric data base prepared at the Institut d'Astronomie de l'Université de Lausanne is described in connection with the formation of the Input Catalogue for the Hipparcos mission. The so-called General

Catalogue of Photometric Data is presented. A test sample of 43667 HD stars brighter than ptg=8.5 is analysed. About 54% of the stars in this sample have neither UBV nor uvby data. The proportion of available data per system is also given for six photometries.

113.095 The new ground-based photometric measurements.
M. Grenon.
Scientific aspects of the Hipparcos space astrometry mission, (see 012.041), p. 211 - 215 (1982).

The main objectives of the new ground-based observations related to Hipparcos mission are reviewed. For the prelaunch phase, the new observations which are needed for the compilation of the Input Catalogue are described and possible solutions are discussed. The contribution of multicolour photometry to a full exploitation of Hipparcos astrometric results is described for various stellar samples in the cases of stars with and without significant parallaxes from the space.

113.096 Cool stars and the *uvby* system.
A. Ardeberg, H. Lindgren.
Rep. Obs. Lund, No. 18, (see 012.044), p. 81 - 87 (1982).

113.097 Report of IAU Commission 25: Stellar photometry and polarimetry (Photométrie et polarimétrie
stellaires).   J. A. Graham.
Trans. IAU, Vol. XVIIIA, (see 003.013), p. 241 - 256 (1982).

Catalogue of measurements in the DDO photo-electric photometric system (magnetic tape).
See Abstr. 002.003.

Geneva photometric boxes. O. Announcement of the catalogue (microfiches and magnetic tape).
See Abstr. 002.064.

Homogeneous catalogue of red and infrared magnitudes in the photoelectric photometric system of Kron (magnetic tape).   See Abstr. 002.066.

Landolt-Börnstein.   See Abstr. 003.064.

Technique of detection and analysis of stellar microvariability. I. Microvariability of XX Cam.
See Abstr. 021.015.

Far-ultraviolet colors of B stars with and without emission lines.   See Abstr. 031.577.

Automatic image classification. See Abstr. 031.617.

Infrared excess and line emission in Be stars.
See Abstr. 112.003.

Free-free emission, anomalous extinction and mass loss rates of O-type stars.   See Abstr. 112.040.

Circumstellar shells in M17.   See Abstr. 112.065.

Spectral classification by objective prism and microdensitometer.   See Abstr. 114.018.

Strongly reddened Hg-Mn star HD 29647.
See Abstr. 114.027.

Variability and mass loss in the extreme supergiant ζ^1 Sco.   See Abstr. 114.042.

The temperature of Arcturus. See Abstr. 114.070.

Ultraviolet, visual, and infrared observations of the WC7 variable HD 193793.   See Abstr. 114.074.

Spectroscopic and photometric study of HR 5023. See Abstr. 114.121.

The relationship between spectrum and light variations in the Ap stars HD 119213, HD 170000 and HD 173650. See Abstr. 114.125.

Report of IAU Commission 45: Stellar classification (Classification stellaire).   See Abstr. 114.225.

The properties of R 136, the central object in 30 Doradus.   See Abstr. 115.020.

Galactic and extragalactic Wolf-Rayet stars of Pop. I.   See Abstr. 115.022.

Five-color band ultraviolet photometry of fourteen close binaries.   See Abstr. 117.031.

Infrared photometry of Mira variables in the Baade windows and the distance to the Galactic Centre. See Abstr. 122.001.

Red supergiant variables and their possible use as distance indicators.   See Abstr. 122.147.

Photometric campaign on Be stars.
See Abstr. 122.243.

UBVRI photometry of the optical counterparts of X-ray sources in Einstein deep survey fields.
See Abstr. 142.048.

UV and visible photometry of the brightest Pleiades stars.   See Abstr. 153.005.

DDO photometry of giants in the open cluster IC 4651.   See Abstr. 153.007.

Evidence of helium abundance differences between the Hyades stars and field stars, and between Hyades stars and Coma cluster stars.   See Abstr. 153.016.

Stellar content of young open clusters. II. Be stars. See Abstr. 153.029.

Abundances, temperatures and reddenings of field and cluster population II giants.   See Abstr. 154.055.

Field horizontal-branch and low metallicity stars in the Strömgren and Geneva systems.   See Abstr. 154.075.

Photometry of faint stars in globular clusters. V. The turn-off point of the cluster M10 = NGC 6254. See Abstr. 154.089.

Three-colour photometry of the Milky-Way field HD 95540.   See Abstr. 155.007.

RGU three colour photometry of a field in Centaurus.   See Abstr. 155.008.

Three-colour photometry of a field near the galactic centre (SA 133 F).   See Abstr. 155.048.

Carbon stars in the Carina dwarf spheroidal galaxy. See Abstr. 158.042.

OB stars in the Small Magellanic Cloud.
See Abstr. 159.026.

114 Spectra, Temperatures, Chemical Composition, etc.

114.001 A differential curve-of-growth analysis of the candidate barium star 93 Herculis.
G. R. Smith, D. L. Harmer.
Mon. Not. R. Astron. Soc., Vol. 198, 273 - 280 (1982).

A differential curve-of-growth analysis of 93 Herculis relative to ε Virginis shows that the star is generally metal deficient, [Fe/H] = −0.3, but barium and other heavy elements are slightly enhanced, [Ba/Fe] = +0.5.

114.002 The spectrum of Canopus.
J. B. Hearnshaw, K. Desikachary.
Mon. Not. R. Astron. Soc., Vol. 198, 311 - 320, Microfiche MN 198/1 (1982).

High dispersion échelle spectra have been obtained for the bright southern supergiant Canopus. This paper presents the measurements of the equivalent widths of 1506 absorption lines covering the spectral range 3060 - 6850 Å.

114.003 Ultraviolet observations of stellar chromospheric activity. L. Hartmann, A. K. Dupree, J. C. Raymond.
Astrophys. J., Vol. 252, 214 - 229 (1982).

The authors present a survey of stellar emission line fluxes detected with IUE in the wavelength region 1175−1950 Å, obtained in the low dispersion mode, as well as Mg II fluxes (λ2800) measured at high dispersion. They have estimated the uncertainty in the flux measurements for detected lines and have carefully constructed upper limits to the fluxes of undetected high temperature lines. From these data, they are able to infer the temperature structure of the outer atmospheres of cool stars and to investigate the variation of emission line strength and the general presence of hot outer atmospheric gas throughout the H−R diagram.

114.004 High resolution observations of magnesium II 2800 Å in Alpha Centauri A: the density of interstellar magnesium II and the stellar chromospheric profiles.
W. R. Oegerle, Y. Kondo, R. E. Stencel, E. J. Weiler.
Astrophys. J., Vol. 252, 302 - 304 (1982).

High resolution scans (< 0.10 Å) of the Mg II h and k lines in α Cen A have been obtained with the Copernicus satellite. The profiles are virtually identical with the solar profiles except for the presence of an absorption feature near line center in the h and k lines of α Cen A. The authors find that this absorption feature can be explained by interstellar absorption of Mg II along the line of sight. The average density of Mg II has been found to be 2.75 (± 0.7) ×10^{-7} cm^{-3}, which is in good agreement with the previously determined values in the solar vicinity in the direction of α CMa and α Lyr.

114.005 Elimination of C_3 as the Bond-Neff depression opacity source in HR 774. S. R. Baird.
Astrophys. J., Vol. 252, 305 - 307 (1982).

The pseudocontinuum of C_3 has been proposed by Fix as the source of the Bond-Neff depression in HR 774. In such a case the strongest rotational lines of the 4050 Å vibrational band of C_3 should be clearly evident at high dispersion. A 2.2 Å mm^{-1} plate of HR 774 is examined in the 4050−4060 Å region where the strongest lines should form. The absence of C_3 rotational lines indicates that C_3 cannot be the major contributor to the Bond-Neff depression in HR 774.

114.006 International Ultraviolet Explorer observations of the central stars of the planetary nebulae NGC 6853 and NGC 7293.
R. C. Bohlin, J. P. Harrington, T. P. Stecher.
Astrophys. J., Vol. 252, 635 - 643 (1982).

Absolute flux measurements have been obtained by the IUE in the wavelength range from λ1200 to λ3100 for the central stars of the planetary nebulae NGC 6853 and NGC 7293. The effective temperatures of 144,000 K for NGC 6853 and 123,000 K for NGC 7293 are determined by the Zanstra method and model atmosphere calculations with standard helium abundances. Additional stellar parameters derived for NGC 6853 and NGC 7293, respectively, are surface gravities for log g = 7.7 and 8.0 for a 1 $M_\odot$ star, stellar radii of 0.023 and 0.017 $R_\odot$, and luminosities of 200 and 60 $L_\odot$.

114.007 International Ultraviolet Explorer observations of Alpha Scorpii. A. P. Bernat.
Astrophys. J., Vol. 252, 644 - 652 (1982).

IUE observations of the absorption spectrum of α^2 Sco B2.5 V are presented. These lines arise from matter being lost by the red supergiant primary, α^1 Sco M1.5 Iab, which is passing in front of the B star. There is a factor of 30 spread in the hydrogen column densities derived from different elements on the assumption of solar abundance ratios. Possible causes are discussed.

114.008 HD 115444 − a barium star of extreme Population II.
R. & R. Griffin, B. Gustafsson, T. Vieira.
Mon. Not. R. Astron. Soc., Vol. 198, 637 - 658 (1982).

The star HD 115444 is shown to be a very metal-poor red giant. A model atmosphere analysis based on colours, energy distributions and moderate-resolution spectrograms has been performed differentially with respect to HD 122563. The following parameters result for HD 115444: T_{eff} = 4800 K, log g = 1.6, M/H = −2.95, ξ_t = 2.6 km s^{-1}. The star is therefore one of the most metal-poor stars known to exist. However, its s-element abundances are high relative to iron in comparison with the sun, and even more so in comparison with other extreme Population II giants. Europium, which is considered to be mainly an r-process element, is also significantly enhanced. Some possible implications of the pattern of abundances in HD 115444 are discussed.

114.009 Effective temperatures, and radii of luminous O and B stars: a test for the accuracy of the model atmospheres. H. Remie, H. J. G. L. M. Lamers.
Astron. Astrophys., Vol. 105, 85 - 97 (1982).

The effective temperatures and angular diameters of 37 early type stars ranging in spectral type from O8 to B2 and with luminosity class Ia to V, and for ζ Pup (O4 If) are derived from the integrated total flux and the infrared flux, using the method of Blackwell and Shallis (1977). The observed flux distribution for supergiants does not agree with the predictions for the line-blanketed model atmospheres. The models can be modified to agree with the observations. The derived values of T_{eff} have an accuracy of 1000 to 1500 K, and the accuracy in θ is 7 to 10%. The basic parameters: T_{eff}, L, M, R and log g, are given for each star. The effective temperatures for supergiants and for ζ Pup are significantly lower than the temperatures derived from the observed ratio of He II/He I lines. This discrepancy is possibly due to the effect of a stellar wind.

114.010 The sun among the stars. V. A second search for solar spectral analogs. The Hyades' distance.
J. Hardorp.
Astron. Astrophys., Vol. 105, 120 - 132 (1982).

Photoelectric violet spectra of G-type stars are compared, at 20 Å resolution, with those of Jupiter satellites and the blue sky. The search for a match of the solar spectrum is thereby completed in the whole sky down to the limit of the Bright Star Catalogue. It shows once again that solar analogs are rare. A distance modulus for the Hyades of m-M = 3.30 is derived

from matching observed masses of visual binaries to the primary of the double-lined spectroscopic eclipsing binary van
Bueren 22, whose mass was recently determined by McClure
(1981). This mass also indicates that the Hyades must have
nearly solar helium abundance. The sun fits the mass-luminosity relation of the Hyades within 0^m16 if evolutionary and
metal abundance effects are taken into account.

114.011 **A search for C_2^- features in the hydrogen-poor
carbon star HD 182040.**   G. Wallerstein.
Astron. Astrophys., Vol. 105, 219 - 220 (1982).

The author has searched for rotational lines of the 0–0
vibrational band of the $B^2 \Sigma_u^+ - X^2 \Sigma_g^+$ electronic transition of
the C_2^- ion in the spectrum of the hydrogen-deficient carbon
star HD 182040. Significant wavelength coincidences were not
found. The possibility that photo-detachment of C_2^- is a significant opacity source in the blue-violet region of cool carbon
stars has not been eliminated by the lack of detectable C_2 lines
in a relatively hot carbon star.

114.012 **P-Cygni profiles observed in the ultraviolet and
visible spectra of Wolf-Rayet stars.**   A. J. Willis.
Mon. Not. R. Astron. Soc., Vol. 198, 897 - 920 (1982).

High resolution ($\Delta\lambda \sim 0.1 - 0.2$ Å) IUE ultraviolet spectra
of ten WR stars have been analysed to investigate P-Cygni
profiles in both WN and WC spectra. Measured velocity parameters of these profiles for each star, combined with measurements from visible spectra, are examined for trends with
IP and EP of the relevant ionic species and transitions. For
each star a well-defined correlation is found between the
central-displaced absorption velocity (V_0) and EP, demonstrating lower degrees of spectral excitation at higher stellar wind
velocities.

114.013 **WC stars in 30 Doradus.**   M. M. Phillips.
Mon. Not. R. Astron. Soc., Vol. 198, 1053 - 1057
(1982).

Two Wolf-Rayet stars belonging to the carbon sequence
have been observed in the core of 30 Doradus. These are the
first WC stars to be identified in 30 Dor, where previously
more than 10 WN stars had been found. The absolute luminosities of the two WC stars may be significantly greater than
those of the early-type WC stars found in other parts of the
Large Magellanic Cloud.

114.014 **IUE observations of R Aquarii.**   H. M. Johnson.
Astrophys. J., Vol. 253, 224 - 229 (1982).

High- and low-dispersion observations made in 1980 are
compared with earlier observations. Certain ultraviolet lines
are apparently constant in the face of Mira pulsation and rotation and orbital motion of the hypothetical hot companion.
Other line features and the ultraviolet continuum vary
significantly but not in phase with the Mira. The Mg II doublet
profile continued to show an asymmetry that is attributable
to 200 km s^{-1} mass ejection. The suggestion of a hot companion photosphere under circumstellar olivine-smoke extinction is discussed.

114.015 **The Wolf-Rayet star HD 193077: evidence for a
low-mass companion and the possibility of a third
body.**   R. Lamontagne, A. F. J. Moffat, G. Koenigsberger,
W. Seggewiss.
Astrophys. J., Vol. 253, 230 - 236 (1982).

The Wolf-Rayet component of HD 193077 is shown to
consist of a WN 6 star orbited by an unseen, low-mass,
probably neutron star with a period of 2.3238±0.0001 days.
This system in turn may be in mutual orbit around a rapidly
rotating, but modest, OB star in a period of 1763±15 days.

114.016 **Metal abundances in the SC star UY Cen.**
R. M. Catchpole.
Mon. Not. R. Astron. Soc., Vol. 199, 1 - 20 (1982).

The composition of the SC star UY Cen is investigated by
the curve of growth method by using Johnson's model atmospheres. The two methods are in good agreement. The
principal results are, that with respect to the Sun, Li is
0.8 ± 1.0 dex underabundant, the elements Na to Fe, with the
possible exception of V, are normal while Co and all heavier
elements, with the exception of Rb, Eu and Dy, are overabundant by more than 0.6 dex. The heavy element abundances are compared with Cowley & Downs' recent neutron
exposure calculations. A revised $^{12}C/^{13}C$ is given.

114.017 **The strong 3.3 micron emission line in Wolf-Rayet
stars.**   P. M. Williams.
Mon. Not. R. Astron. Soc., Vol. 199, 93 - 96 (1982).

A number of Wolf-Rayet stars have been found to show
in their spectra a strong emission feature at 3.28 μm, the wavelength of the 'unidentified' feature observed in some nebular
spectra. From comparison of the strength of this line from
stars of different spectral type and excitation, it is identified
with the C IV (11−10) transition group and shown not to be
connected with the circumstellar dust associated with some
Wolf-Rayet stars.

114.018 **Spectral classification by objective prism and
microdensitometer.**
B. D. Kelly, J. A. Cooke, D. Emerson.
Mon. Not. R. Astron. Soc., Vol. 199, 239 - 243 (1982).

Photoelectric photometry has been obtained for 39 stars
within a one square degree area covering the range
$13.5 < V < 17.1$, $-0.22 < B - V < 1.50$. Microdensitometer
scans of spectra of the same objects from a widened
UK Schmidt low dispersion objective prism plate have been
compared with the photometry. The results imply that micro-
densitometry can enable spectral classification with an accuracy of one or two spectral sub-types to be carried out on these
plates.

114.019 **Magnesium emission variability among late-type
giant stars.**   D. J. Mullan, R. E. Stencel.
Astrophys. J., Vol. 253, 716 - 726 (1982).

The authors have investigated the variability of the emission cores in the h and k lines of Mg II in a sample of 21
late-type giants. They have used high resolution line profiles
obtained by IUE to determine not only the total emission
intensity, but also the ratio of the intensities of the shortward
and longward components of the doubly-reversed emission.
They argue that variations in the total emission intensity are
indicative of variations in chromospheric heating, while variations in the ratio can be interpreted in terms of variations in
structures in the corona and outer atmosphere.

114.020 **Precise radial velocities. I. A preliminary search for
oscillations in Arcturus.**   M. A. Smith.
Astrophys. J., Vol. 253, 727 - 734 (1982).

High speed, serial Reticon spectra of Arcturus have been
obtained over 10.5 hr on three nights to search for an analog
of the solar 5-minute oscillations. The observational technique
is to reference small, apparent shifts of stellar lines against
those of telluric O_2 lines ($\lambda 6300$) interspersed through the
spectrum. By subtracting out the spurious telluric line shifts
introduced by guiding, etc., the stellar shift errors can be
reduced to ±6 to ±8 m s^{-1} (rms) for a few hours and to even
smaller values over short intervals. A possible signal is occasionally seen at a period of about 90 minutes. Otherwise,
our negative results suggest that any nonradial oscillations on
Arcturus must have a semiamplitude less than or equal to
5 m s^{-1}, if its dominant period is between 15 minutes and
half a day.

114.021 **Determination of the ratios of $^{12}C/^{13}C$ and C/O in
the carbon stars V460 Cygni and TX Piscium.**
H. R. Johnson, G. T. O'Brien, J. L. Climenhaga.

Astrophys. J., Vol. 254, 175 - 190 (1982).

Values of the ratio of carbon to oxygen (C/O) and the $^{12}C/^{13}C$ isotopic ratio have been determined for two N-type carbon stars: V460 Cygni and TX Piscium. The values have been determined by comparing synthetic spectra based on recently constructed model atmospheres for luminous carbon stars with high-resolution photographic spectra of V460 Cyg and TX Psc in the wavelength range from 7950 to 8065 Å. The observed and synthetic spectra match most closely for values of C/O = 1.02 and $^{12}C/^{13}C$ = 24(V460 Cyg) and 27 (TX Psc), while outer limits appear to be $1.00 \leqslant C/O \leqslant 1.05$ and $15 \leqslant {}^{12}C/^{13}C \leqslant 40$ for both stars.

114.022 **Possible iron abundance variations among super-**
ficially normal A stars. C. R. Cowley, R. L. Sears,
G. C. L. Aikman, K. Sadakane.
Astrophys. J., Vol. 254, 191 - 195 (1982).

High-dispersion spectra of a sample of superficially normal late B and early A stars have been examined by the method of wavelength coincidence statistics (WCS). The sample was selected primarily on the basis of low projected rotational velocities. Two lists of approximately 70 strong and weak Fe II lines were used. The line intensities were predicted for 10,000 K. The WCS are influenced by $\nu \sin i$, effective temperature, and iron abundance in a predictable way. On the basis of WCS for stars having published fine analyses, the authors suggest that several stars in the sample may be deficient in iron. The kinematic properties of the sample indicate a young population.

114.023 **The O3 stars.** N. R. Walborn.
Astrophys. J., Lett., Vol. 254, L15 - L17, plate L4 (1982).

A brief review of the 10 known objects in this earliest spectral class is presented. Two new members are included: HD 64568 in NGC 2467 (Puppis OB2), which provides the first example of an O3 V ((f*)) spectrum; and Sk −67°22 in the Large Magellanic Cloud, which is intermediate between types O3 If * and WN6-A. In addition, the spectrum of HDE 269810 in the LMC is reclassified as the first of type O3 III (f *). The absolute visual magnitudes of these stars are rediscussed.

114.024 **Carbon, nitrogen, and oxygen abundances in Sirius**
and Vega. D. L. Lambert, S. W. Roby, R. A. Bell.
Astrophys. J., Vol. 254, 663 - 669 (1982).

Carbon, nitrogen, and oxygen abundances are obtained from C I, N I, and O I high excitation permitted lines in the spectra of the standard A star Vega and the "hot" Am star Sirius. Vega has normal abundances. Relative to Vega, Sirius is C deficient by 0.60 dex, N enhanced by 0.22 dex, and O deficient by 0.27 dex.

114.025 **Mapping of Fe and Cr on the surface of the Ap star**
ϵ UMa.
W. Wehlau, J. Rice, N. E. Piskunov, V. L. Khokhlova.
Pis'ma Astron. Zh., Tom 8, 30 - 37 (1982). In Russian. English translation in Soviet Astron. Lett., Vol. 8.

The distribution of Fe and Cr over the surface of ϵ UMa (HD 112185) was determined by the method of an inverse problem solution (Goncharskij et al., 1977) using high precision profiles of Fe I, Fe II, Cr I and Cr II lines. Both elements are concentrated in two large regions near the equator of rotation of the star.

114.026 **The Hα emission line in the spectrum of the star**
HDE 245770 = A 0535 + 26.
O. Eh. Aab, L. V. Bychkova, I. M. Kopylov,
R. N. Kumajgorodskaya.
Pis'ma Astron. Zh., Tom 8, 179 - 182 (1982). In Russian. English translation in Soviet Astron. Lett., Vol. 8.

A search has been made for rapid variability of the parameters of the Hα emission line in the spectrum of HDE 245770

with a period of 103.84 s. The analysis of observations made during 8 periods has shown that the parameters of the Hα line do not change within the measurement errors.

114.027 **Strongly reddened Hg-Mn star HD 29647.**
V. Straižys, Yu. V. Glagolevskij, I. I. Romanyuk,
V. D. Bychkov.
Pis'ma Astron. Zh., Tom 8, 183 - 186 (1982). In Russian. English translation in Soviet Astron. Lett., Vol. 8.

The heavily reddened star HD 29647 is investigated using 6-meter telescope spectrograms and photometric observations in the Vilnius seven-color system. The parameters $T_e = 15600$ K and log g = 3.70 from hydrogen lines and Balmer jump were obtained. HD 29647 is a peculiar star of the Hg-Mn type.

114.028 **Kuwano's peculiar object is a novalike (symbiotic?)**
binary with a red giant. Spectral observations.
R. E. Gershberg, V. I. Krasnobabtsev, P. P. Petrov,
K. K. Chuvaev.
Astron. Zh., Tom 59, 6 - 14 (1982). In Russian. English translation in Soviet Astron., Vol. 26, No. 1.

Results of spectral observations of Kuwano's peculiar object (PU Vul 1979) from August 1979 till October 1980 are given. The list of identified lines, estimates of radial velocities and measured equivalent widths of spectral lines and intensity jumps near molecular band limits are presented.

114.029 **IUE ultraviolet spectrophotometry of 15 galactic**
Wolf-Rayet stars. H. Nussbaumer, W. Schmutz,
L. J. Smith, A. J. Willis.
Astron. Astrophys., Suppl. Ser., Vol. 47, 257 - 294 (1982).

Data are presented both as spectral plots and tabulations of the observed absolute flux distributions (in erg cm^{-2} s^{-1} Å^{-1}) covering the wavelength range $\lambda\lambda$1150-3090. The characteristic of these UV WR spectra are discussed, and equivalent widths of the most prominent emission lines have been measured. It is shown that the WR stars follow the normal galactic interstellar extinction law. Colour excesses, E_{B-V}, are derived for each star. The intrinsic WR visible-UV continua yield black body colour temperatures in the range 20000–40000 K.

114.030 **The spectra of late-type dwarfs and sub-dwarfs in**
the near ultraviolet. I. Line identifications.
J. E. Beckman, L. Crivellari, P. L. Selvelli.
Astron. Astrophys., Suppl. Ser., Vol. 47, 295 - 317 (1982).

Spectra from IUE of 6 main sequence and near main sequence stars from classes A5 to G8, between 2700 Å and 2900 Å wavelength are described. In toto, some 250 absorption lines are identified. Observed differences in those atomic and ionic multiplets present in each spectrum are shown to be due either to the regular progression in spectral class, or to the smearing effect of stellar rotation.

114.031 **Study of H_d profile in 72 Be stars.**
Y. Andrillat, C. Fehrenbach.
Astron. Astrophys., Suppl. Ser., Vol. 48, 93 - 136 (1982). In French.

The authors present a catalogue of H_d profiles of 72 Be stars, observed with the "Multiphot" system of Haute Provence Observatory. The structure of the emission of H_d line appears to be both very complex and variable and, in most cases, a central absorption is present. The measurements of widths at mid-height of emission intensity show and confirm a correlation with the rotational velocity. Line-profiles have been also calculated by means of Gauss functions. The wing extension of H_d line is also determined. It shows that there exist correlations with the mid-height width, with $\nu \sin i$ and with the spectral type. The Be stars of group I are then well separated from those of the other groups. This result agrees with the hypothesis of the broadening of H_d line from electron scattering in the envelope.

114.032 International Ultraviolet Explorer spectroscopy of hot stars in the LMC and SMC: the SMC extinction law, stellar flux distributions, and details of the stellar winds.
J. B. Hutchings.
Astrophys. J., Vol. 255, 70 - 78 (1982).

Data are presented from high-dispersion observations of 7 stars in each of the Magellanic Clouds, and from low-dispersion observations of 14 more in each cloud. The SMC ultraviolet extinction curve is found to be much steeper than in the Galaxy or the LMC. Stellar effective temperatures and luminosities are derived for all stars and found to be similar to those in the Galaxy. Stellar wind phenomena are not always present and are generally weak when present in the SMC, and stronger in the LMC though still weak compared with galactic star phenomena. The implied consequences of these findings are discussed briefly.

114.033 Effective temperatures of A and F stars.
E. Böhm-Vitense.
Astrophys. J., Vol. 255, 191 - 199 (1982).

Effective temperatures of late A and early F stars have been determined from the observed fluxes in the visual at 1900 Å and at 1420 Å. Observed ratios were compared with those calculated by R. Kurucz. A correction of the theoretical fluxes at 1900 Å by $-0.10 \leqslant \Delta \log F_\lambda(1900) \leqslant 0$ brings the T_{eff} obtained from the different ratios into reasonable agreement. For late A stars, the T_{eff} determined in this way agree well with those obtained from the optical region. For F stars, however, the T_{eff} obtained from the UV are higher than those obtained from the optical region, if radiative equilibrium models are used for the comparison. The author suggests that this discrepancy may be due to the effects of temperature, pressure, and absorption coefficient inhomogeneities caused by convection.

114.034 Observations of spectral line asymmetries and convective velocities in F, G, and K stars.   D. F. Gray.
Astrophys. J., Vol. 255, 200 - 209 (1982).

Line asymmetries are found to exist generally in F, G, and K stars. Line bisectors were measured for 11 relatively unblended lines in each of 27 stars, and an average line bisector was determined for each star. The average line bisectors show systematic changes in velocity span and shape with spectral type. The bisector characteristics are interpreted in terms of granulation-type motion. Convective velocities are deduced as a function of spectral type. The height of penetration of convection is found to be lowest in solar-type stars with convection essentially dying out before reaching the top of their photospheres. In nonsolar-type stars, larger convective velocities produce greater height penetration.

114.035 A K giant with an unusually high abundance of lithium: HD 112127.
G. Wallerstein, C. Sneden.
Astrophys. J., Vol. 255, 577 - 584 (1982).

The strong-lined K giant star HD 112127 exhibits an extremely strong lithium resonance line in its spectrum. In most other respects the atmosphere of the star resembles the well-studied, strong-lined star μ Leo. A model atmosphere analysis has been carried out for HD 112127 with particular emphasis on the light element abundances. Iron peak elements show essentially solar composition, while the CNO group is enhanced by about 0.2 dex. The carbon isotope ratio is near 20, while the lithium content of this star is approximately the maximum seen in unevolved stars. Schemes for the production of lithium in a relatively low luminosity giant are discussed.

114.036 On the spectrum of the Herbig Be star HD 200775.
B. Baschek, M. Beltrametti, J. Köppen, G. Traving.
Astron. Astrophys., Vol. 105, 300 - 305 (1982).

On the basis of high-dispersion spectra and other observational material, HD 200775 is reanalyzed. By comparing the depths in the wings of Hγ and Hδ and the equivalent widths of He I lines with predictions from model atmospheres the authors find T_{eff} = 17,000 ± 1000 K and $\log g$ [cm s^{-2}] = = 3.6 ± 0.3 corresponding to a spectral type of B 3 V. This is consistent with the continuum energy distribution if one allows for some circumstellar emission. The interstellar absorption features at 4430 and 2200 Å are of unusual weakness. The strength of the Hα emission requires a substantial rate of photoionization from the first excited level ($n = 2$) of hydrogen which probably arises from a compact envelope.

114.037 Wolf-Rayet stars in extragalactic H II regions: discovery of a peculiar WR in IC 1613/# 3.
S. D'Odorico, M. Rosa.
Astron. Astrophys., Vol. 105, 410 - 412 (1982).

Wolf-Rayet bands centered at λ4650 and λ5810 Å have been detected in the spectrum of the core of an H II region in the dwarf, metal poor irregular galaxy IC 1613. The emission features show both WC and WN characteristics and unusually large equivalent widths (>300 Å). A WR star, be it a single, peculiar object or a multiple system, can alone account for the ionization of the H II region.

114.038 A carbon star in the globular cluster Lindsay 102.
A. C. Danks.
Astron. Astrophys., Vol. 106, 4 - 6 (1982).

A new carbon star is identified in the SMC globular cluster Lindsay 102. From a low-dispersion IDS spectrum the star is tentatively classified as C5,4. Photometry of Lindsay 102 through a 60″ diaphragm gives $V = 14.82$, $B-V = 0.52$ and these colours in combination with the cluster's large distance from the bar of the SMC are used to infer that the age of the cluster is intermediate 2×10^8 to 1×10^9 yr or possibly older than 1×10^9 yr.

114.039 Absorption line symmetries for two HgMn stars.
J. B. Rice, W. H. Wehlau.
Astron. Astrophys., Vol. 106, 7 - 8 (1982).

Observations of the profiles of several absorption lines in the spectra of ι CrB and ϕ Her have been made to search for possible asymmetries that have been suggested by the diffusion theory for Ap stars. No asymmetries attributable to such effects were found. The profiles for ι CrB are not inconsistent with the binary nature of that star.

114.040 The variable shell star HR 5999. VI. Strong chromospheric and transition region emission lines in the ultraviolet spectrum of a Herbig Ae star.
H. R. E. Tjin A Djie, P. S. Thé, M. Hack, P. L. Selvelli.
Astron. Astrophys., Vol. 106, 98 - 104 (1982).

High and low resolution IUE spectra of the Herbig Ae-type star HR 5999 (sp. $A7$ IIIe) show strong emission lines of H, O I, C II, C IV, Mg II and possibly C III, N I and O V, and numerous shell absorption lines, mainly of Fe II, Cr II, Mn II and Ni II. The presence of strong emission lines due to hot gas suggests that HR 5999 has a chromosphere and a corona, probably induced and maintained by a convection zone below the photosphere of this fast rotating (~180 km s^{-1}) $3 M_\odot$ star. It is for the first time that such emission line spectra are observed in an A-type star, resembling in strength those of T Tauri stars and late type stars with chromospheres and coronae. These facts confirm the pre-main sequence character of HR 5999.

114.041 On the search for transition zone lines in late A type stars.   L. Crivellari, F. Praderie.
Astron. Astrophys., Vol. 107, 75 - 87 (1982).

The authors attempted to obtain information on the transition region in late A type stars from IUE observations. Low resolution spectra have been converted in absolute fluxes by means of a careful correction of parasite effects, one of them being specially important when the energy distribution presents a large jump, as is the case for stars later than A5, due to Si I discontinuities. Upper bounds on the integrated emission at $\lambda = 1400$ Å (Si IV), $\lambda = 1550$ Å (C IV), and $\lambda = 1640$ Å (He II) have been derived from these low resolution spectra; the two high resolution spectra of the sample (15 Vul, γ Boo) allowed to decrease these upper bounds by a factor 30. No emission being detected at high resolution, and a fortiori at low resolution, the authors can only place upper limits on relevant quantities for the transition region.

114.042   Variability and mass loss in the extreme supergiant ζ^1 Sco.   G. Burki, A. Heck, L. Bianchi, A. Cassatella.
Astron. Astrophys., Vol. 107, 205 - 210 (1982).

ζ^1 Sco, a B1/2 Ia supergiant, was surveyed during two years by means of the IUE satellite (UV spectroscopy) and of the Geneva 7-colour system (optical photometry). The photometric variability is different from that characterizing the best known pulsating stars. The shape of the UV resonance lines strongly varies with time. In addition to the global variation of the P Cygni profiles, narrow shifted absorption features are observable, the intensity of which is also variable. The authors suggest that the wind of ζ^1 Sco is constituted by two components: a P Cygni material and a puff material. In this model, the production of puff material is stimulated by the vibrational instability which is also the cause of the photometric variability.

114.043   The chemical composition of late-type supergiants. IV. Homogeneous abundances and galactic metallicity trends.   R. E. Luck.
Astrophys. J., Vol. 256, 177 - 188 (1982).

In a series of papers investigating the chemical composition of G and K Ib supergiants Luck and Bond have concluded that supergiants in the solar neighborhood have about twice the iron content of the sun ([Fe/H] = +0.3) and that supergiants between 7.7 and 10.2 kpc from the galactic center show a steep radial metallicity gradient, $d[\text{Fe/H}]/dR = -0.24$ kpc^{-1}. In the present paper the author re-analyzes the data for 19 northern supergiant stars using detailed integrations through model stellar atmospheres and improved atomic data. An increase in microturbulent velocities of about 50% with respect to the older values leads to a decrease in [Fe/H] values by an average of 0.2 dex, bringing the solar neighborhood supergiants to within 0.1 dex of the sun and local F-type dwarfs. These changes lead to a major revision of the galactic metallicity gradient as derived from supergiants. The new value is $d[\text{Fe/H}]/dR = -0.13$ (± 0.03) kpc^{-1}. This value is compared to other recent results for young disk objects with the result that all young disk objects imply a metallicity gradient of the order of $d[\text{M/H}]/dR \sim -0.1$ kpc^{-1}.

114.044   Lithium in late-type giants. II. 31 M giants and supergiants.   R. E. Luck, D. L. Lambert.
Astrophys. J., Vol. 256, 189 - 205 (1982).

High resolution, high signal-to-noise spectra have provided Li abundances for 31 M giants and supergiants. The spectrum around the Li I 6707 Å doublet is depressed by unresolved TiO lines. A spectrum synthesis technique was developed to account for the TiO line blanketing and to extract the Li abundance. The Li abundances in the sample of 25 giants show a large scatter about a mean $\log \epsilon(\text{Li}) \sim -0.2$ with several stars showing no detectable Li or $\log \epsilon(\text{Li}) \lesssim -1.0$. A parallel analysis of the Al I 6696 and 6698 Å lines gives a mean abundance [Al/H] = -0.3 consistent with the idea that the M giants are somewhat older than the Sun. Four of the six M supergiants have the Li/Al ratio expected for rather massive evolved stars. Two (α Ori and 119 Tau) stars show no detectable Li line. The Li deficiency may result from mass loss at the main-sequence phase or internal mixing and nuclear processing leading to Li destruction.

114.045   Nitrogen anomalies in O-type stars: a new spectroscopic criterion.
G. F. Bisiacchi, J. A. López, C. Firmani.
Astron. Astrophys., Vol. 107, 252 - 257 (1982).

A spectroscopic analysis of 95 O-type stars has been carried out in search of nitrogen anomalies. In addition to previous spectroscopic criteria for the ON classification, it has been found that the N III λ 4514 line is a good indicator of the abundance variations of N in this kind of stars. For the first time anomalies have been investigated for several early O stars. Some evidence is found supporting the idea that the ON phenomenon is a continuous one rather than a marked discontinuity between the "normal" O stars and those showing nitrogen overabundance.

114.046   Detection of further red giants with "hybrid" atmospheres and a possible correlation with double circumstellar Mg II and Ca II lines.   D. Reimers.
Astron. Astrophys., Vol. 107, 292 - 299 (1982).

In addition to the three known hybrid atmosphere stars (α, β Aqr, α TrA) the authors have detected three further red giants (θ Her, ι Aur, δ TrA) which have both transition layer lines (N V, C IV, Si IV, C II) and cool winds. It is shown that all six known "hybrid" stars have double circumstellar Ca II and Mg II absorption lines, one high velocity component at -70 to -140 km s^{-1}, one low velocity component at -5 to -20 km s^{-1}. Together with six further stars with double circumstellar lines, these stars occupy the dividing region in the HR-diagram which separates stars with transition layer lines from stars with circumstellar lines.

114.047   The sun among the stars. VI. The solar analog HD 44594.
J. Hardorp, H. Tüg, T. Schmidt-Kaler.
Astron. Astrophys., Vol. 107, 311 - 312 (1982).

Scanner observations at 20 Å resolution from 3250 to 8750 Å are reported for HD 44594 and HD 20630. The authors see no difference from the solar flux distribution of Neckel and Labs (1981). The violet spectrum of HD 44594 is indistinguishable from that of the sun, thus this star is confirmed as a solar analog.

114.048   Activity of HBV 475 from its spectral variations.   S. Tamura.
Publ. Astron. Soc. Japan, Vol. 33, 701 - 711 (1981).

The author analyzes the yearly intensity variations of emission lines and the activity of HBV 475.

114.049   Spectroscopy and infrared photometry of Cyg OB 2 stars: velocity law and mass-loss rates.
C. Leitherer, H. Hefele, O. Stahl, B. Wolf.
Astron. Astrophys., Vol. 108, 102 - 110 (1982).

Spectroscopic observations of the most luminous stars of the Cyg OB 2 association were carried out. In addition infrared observations of 24 member stars of Cyg OB 2 in the HKLMN wavelength bands were obtained. For most program stars the authors derived infrared excesses due to free-free radiation of their expanding envelopes. By combining the infrared data with recent radio observations of five Cyg OB 2 stars an empirical velocity law of the stellar wind in these O stars was derived. This law agrees remarkably well with the prediction of the radiation-pressure driven stellar wind theory. Using this empirical velocity law, the authors derived mass-loss rates for the program stars with measured infrared excess.

114.050   Analysis of the IUE and optical spectra of the peculiar Be star HD 87643.
J. A. de Freitas Pacheco, D. P. Gilra, S. R. Pottasch.
Astron. Astrophys., Vol. 108, 111 - 117 (1982).

The authors present IUE observations of the peculiar Be star HD 87643. They interpret the broad absorptions seen at 2400 Å, 2600 Å, and 2750 Å as being produced by Fe II transitions from the ground-state and the low lying metastable states. The optical lines of Fe II seen in emission are most probably excited by continuum fluorescence. This interpretation allows to estimate the rate of mass-loss through the analysis of the iron lines. This mass-loss rate is two to three orders of magnitude higher than the values deduced from the UV lines for other Be stars, leading to the conclusion that a very active short-lived phase in the star's evolution is observed.

114.051 The spectrum of the WC-O VI star ST 3 in the yellow range. F. Thévenin, A. Pitault.
Astron. Astrophys., Vol. 108, 195 - 196 (1982).

A spectrum of the WC-O VI star ST 3 from 4500 Å to 6000 Å has been taken. It displays a rather strong O VI line and typical features of the WC 4 subtype. This spectrum is compared to those of two other WC-O VI stars, Sand 3 and Sand 4.

114.052 Line profiles of Herbig Ae, Be stars.
I. Appenzeller, U. Finkenzeller.
Mitt. Astron. Ges., Nr. 55, p. 20 (1982). – Abstract.

114.053 Absorption lines in the visual spectrum of the "continuous" central star of the PN NGC 3242.
R. P. Kudritzki, R. H. Méndez, K. P. Simon.
Mitt. Astron. Ges., Nr. 55, p. 36 (1982). – Abstract.

114.054 Spektrale Feinanalyse des extremen Heliumsterns BD + 10°2179. U. Heber.
Mitt. Astron. Ges., Nr. 55, p. 41 - 42 (1982). – Abstract.

114.055 The helium-rich star HD 58260.
K. P. Simon, D. Groote, J. P. Kaufmann.
Mitt. Astron. Ges., Nr. 55, p. 43 (1982). – Abstract.

114.056 New observations and improved model for σ Ori E.
D. Groote, K. Hunger.
Mitt. Astron. Ges., Nr. 55, p. 43 - 50 (1982).

114.057 Non-LTE analysis of OBN-stars.
D. Schönberner, R. P. Kudritzki, K. P. Simon.
Mitt. Astron. Ges., Nr. 55, p. 50 (1982). – Abstract.

114.058 LB 3459 – an sdO-type eclipsing binary system: non-LTE analysis of the primary.
R. P. Kudritzki, K. P. Simon, A. E. Lynas-Gray, D. Kilkenny, P. W. Hill.
Mitt. Astron. Ges., Nr. 55, p. 50 (1982). – Abstract.

114.059 Non-LTE analyses of the O3 stars HD 93128, HD 93129A and HD 303308.
K. P. Simon, G. Jonas, R. P. Kudritzki, J. Rahe.
Mitt. Astron. Ges., Nr. 55, p. 51 (1982). – Abstract.

114.060 R 136a – ein supermassiver Stern.
T. Schmidt-Kaler, J. V. Feitzinger.
Mitt. Astron. Ges., Nr. 55, p. 51 - 52 (1982).

114.061 Der Emissionslinienstern R 50 in der Kleinen Magellanschen Wolke.
F. J. Zickgraf, B. Wolf, M. Friedjung, G. Muratorio, A. Cassatella, O. Ricciardi, B. Viotti.
Mitt. Astron. Ges., Nr. 55, p. 54 (1982). – Abstract.

114.062 Statistik der Mn-Hg-Sterne.
H. Schneider.
Mitt. Astron. Ges., Nr. 55, p. 169 (1982). – Abstract.

114.063 Space and ground-based spectrophotometric observations of a group of F, G and K type stars.
R. A. Epremyan.
Astrofizika, Tom 17, 495 - 508 (1981). In Russian. English translation in Astrophysics, Vol. 17, No. 3.

The results of spectrophotometric measurements of 22 F, G and K type stars in the wavelength region 2400 - 4800 Å are presented. The observed energy distributions in the continuous spectra of these stars are in good accordance with the latest theoretical star photosphere models by Kurucz. A combination of space observations with the ground-based observations gives a possibility for the determination of the brightness classes for the components of double systems.

114.064 On rapid and slow variations in the spectrum of SU Aur. L. V. Timoshenko.
Astrofizika, Tom 17, 519 - 534 (1981). In Russian. English translation in Astrophysics, Vol. 17, No. 3.

Using 51 spectrograms of SU Aur rapid oscillations of the spectral parameters taking place irregularly during some minutes and with variable amplitude, oscillations from night to night and seasonal oscillations of these parameters have been found. An analysis of the photometric data on SU Aur has also been carried out.

114.065 Apparent distribution of carbon stars in the region $115° \leqslant l \leqslant 126°, -5° \leqslant b \leqslant +5°$.
O. M. Kurtanidze, M. G. Nikolashvili.
Astrofizika, Tom 17, 576 - 579 (1981). In Russian. English translation in Astrophysics, Vol. 17, No. 3.

Forty new carbon stars are revealed by a low-dispersion spectral survey of six regions situated in Cas. The distribution of all known carbon stars in this region is discussed.

114.066 On the number of Hα emission stars in the region of the Orion nebula. Eh. S. Parsamyan.
Astrofizika, Tom 17, 579 - 583 (1981). In Russian. English translation in Astrophysics, Vol. 17, No. 3.

The total number of Hα emission stars in the region of the Orion nebula on the basis of new observational data is estimated. It is shown that the total number of Hα stars is of the order of 450.

114.067 Criteria for the $^{12}C/^{13}C$ ratio of carbon stars by line profiles of ^{13}CN. Y. Fujita.
Proc. Japan Acad. Ser. B, Vol. 57, 169 - 173 (1981). – Abstr. in Phys. Abstr., Vol. 85, Abstr. 38220 (1982).

114.068 Spectral investigation of the Ap star Iota Cassiopeiae.
M. Jasiński, M. Muciek, A. Woszczyk.
Acta Astron., Vol. 31, 321 - 338 (1981).

The variations of radial velocity and equivalent widths with phase are found. The curves for iron and chromium are double waves. For Ca II the variation of radial velocity and equivalent width is sinusoidal. A map of abundance distribution of iron is computed, on which iron lies in a band along a great circle. The curve of growth analysis leads to the determination of the variation of Fe II abundance from 8.1 to 8.6 (log $N_H = 12$) and indicates that the microturbulence varies on the surface of the star.

114.069 On the spectral variation of the Ap star Epsilon Ursae Majoris.
A. Woszczyk, T. Michałowski.
Acta Astron., Vol. 31, 339 - 349 (1981).

Spectral changes in the period of light variation and rapid variation of different spectral features have been discussed on the basis of 88 spectra of ϵ UMa obtained at the Toruń Observatory with a dispersion of 28 Å/mm. Variations of hydrogen Balmer lines from H_8 to H_γ as well as of Ca II K line have been found. The present spectrophotometry is unable to confirm or exclude the reality of the rapid variation of Balmer and Ca II K lines in the spectrum of this star.

114.070 The temperature of Arcturus. U. Frisk, R. A. Bell, B. Gustafsson, H. L. Nordh, S. G. Olofsson.

Mon. Not. R. Astron. Soc., Vol. 199, 471 - 481 (1982).

New photometry is presented for Arcturus (K2 III) in the wavelength region 0.4–2.2 μm. A comparison with detailed model atmospheres leads to an effective temperature of 4375 K with an estimated maximum uncertainty of about 50 K. The model atmosphere adopted is found to reproduce the red and infrared flux from Arcturus very well, but the blue-violet quasi-continuous flux is too high, which suggests an extra opacity source, not considered in the ordinary model calculations.

114.071 Mass loss from Be stars derived from UV spectra.
J. A. de Freitas Pacheco.
Mon. Not. R. Astron. Soc., Vol. 199, 591 - 600 (1982).

High resolution IUE spectra of αEri, δCen and πAqr are presented. A detailed analysis is made for πAqr where strong asymmetries are present not only in the Si IV and C IV resonance lines but also in the lines of Fe III originated from low-lying metastable stars. The author has derived a mass-loss rate for πAqr and made a rough estimate for the rate of δCen.

114.072 Outer atmospheres of cool stars. XI. High-dispersion IUE spectra of five late-type dwarfs and giants.
T. R. Ayres, J. L. Linsky, G. S. Basri, W. Landsman, R. C. Henry, H. W. Moos, R. E. Stencel.
Astrophys. J., Vol. 256, 550 - 558 (1982).

The authors present high-dispersion, far-ultraviolet (1150–2000 Å) spectra of five late-type dwarfs and giants obtained with the International Ultraviolet Explorer. The chromospheric ($T \lesssim 10^4$ K) emission lines in the giants tend to be about twice as broad as the corresponding features of the dwarf star spectra, suggesting a width-luminosity relation similar to the Wilson-Bappu effect for Ca II H and K. No evidence has been found of asymmetric or shifted emission line profiles that may indicate the presence of warm ($T \leqslant 10^5$ K) stellar winds. In particular, broad C IV profiles seem to be typical of active chromosphere giant stars and are unlikely to indicate an expanding stellar wind. It is concluded that opacity must be an important broadening enhancement mechanism in active chromosphere giant stars. The line strength of the He II $\lambda1640$ emission correlates well with soft X-ray fluxes of the sample stars, as predicted by photoionization-recombination models for He II.

114.073 IUE low-dispersion spectra of six luminous stars in symmetric nebulae. H. M. Johnson.
Astrophys. J., Vol. 256, 559 - 567 (1982).

The stars and nebulae are HD 156738 and HDE 319703 A, respectively centered in a pair of symmetric nebulae among the NGC 6334 group, AG Car in its nebula, HDE 250550 in nebula 8 of a catalog by Herbig, 209 BAC in M1-67, and HD 89358 in NGC 3199. These include two O stars, two WN stars, an unstable B supergiant, and a ZAMS B star. Spectral line identifications and types, and several parameters of mass loss, are tabulated. When the present mass loss rates are compared with previous results from other methods, there is an outstanding difference only for WN stars, since $1-3 \times 10^{-7} M_\odot$ yr^{-1} is derived here.

114.074 Ultraviolet, visual, and infrared observations of the WC7 variable HD 193793.
E. L. Fitzpatrick, B. D. Savage, M. L. Sitko.
Astrophys. J., Vol. 256, 578 - 593 (1982).

HD 193793, a WC 7 star that has undergone a recent episode of pronounced infrared brightening, was observed in the ultraviolet at low- and high-resolution with the IUE satellite. These data are combined with nearly simultaneous visual and infrared spectrophotometry to produce an absolute energy distribution extending from 0.12 μm to 12.5 μm. The ultraviolet data indicate a "normal" WC7-type star. The P Cygni lines of C IV, Si IV, and C III imply a wind terminal speed of about 3000 km s^{-1}. The interstellar line spectrum

of HD 193793 resembles that found for heavily reddened stars, although the C IV and Si IV lines are probably formed in the nebular surroundings. There is no evidence for peculiar molecular and/or atomic abundances as might be expected for an object that has recently formed circumstellar dust. The ultraviolet extinction as inferred from the normal appearing extinction bump near 2200 Å implies $E(B-V) = 0.85$. With this value of $E(B-V)$, the dereddened ultraviolet to visual energy distribution is consistent with a star having $T_{eff} \approx 43{,}000$ K. The infrared evidence for dust formation is convincing.

114.075 Model analysis of F-type supergiants. I. Microturbulence distribution and element abundances in the atmospheres of the stars γ Cyg and α UMi.
A. A. Boyarchuk, L. S. Lyubimkov.
Izv. Krymskoj Astrofiz. Obs., Tom 64, 3 - 12 (1981). In Russian. English translation in Bull. Crimean Astrophys. Obs., Vol. 64.

Using model atmospheres the authors have analysed spectra of the stars γ Cyg and α UMi. The analysis of about 100 Fe I lines has shown that the microturbulent velocity ξ_t in the atmospheres of the stars cannot be constant. The dependence of ξ_t on the optical depth τ_{5000} has been determined. The following values of effective temperature and surface gravity were obtained for the deduced microturbulence distribution: $T_{eff} = 5950$ K and lg $g = 1.2$ for γ Cyg; $T_{eff} = 6050$ K and lg $g = 1.95$ for α UMi. Abundances of 25 elements were determined.

114.076 Spectrophotometric investigation of the lithium star ξ Boo A. A. A. Boyarchuk, I. Ehglitis.
Izv. Krymskoj Astrofiz. Obs., Tom 64, 13 - 25 (1981). In Russian. English translation in Bull. Crimean Astrophys. Obs., Vol. 64.

A differential analysis of the chemical composition of the lithium star ξ Boo A and the normal G star β Aql was carried out. The rare earth elements probably are overabundant in the atmosphere of the lithium star.

114.077 Photoelectric studies on time variations in the λ 4254.35 Å (Cr I) line in the spectrum of the Ap star α^2 CVn. V. M. Kuvshinov, S. I. Plachinda.
Izv. Krymskoj Astrofiz. Obs., Tom 64, 25 - 36 (1981). In Russian. English translation in Bull. Crimean Astrophys. Obs., Vol. 64.

Profiles of the λ4254.35 Å (Cr I) line of α^2 CVn were obtained. The analysis of the observational data shows the effectiveness of the described method. Some evidence for possible spectral variation of the star with characteristic time of an order of several hours, being not correlated with star rotation, have been obtained.

114.078 On rapid variations in the spectrum of the magnetic star 53 Cam.
N. S. Polosukhina, K. K. Chuvaev, V. P. Malanushenko.
Izv. Krymskoj Astrofiz. Obs., Tom 64, 37 - 50 (1981). In Russian. English translation in Bull. Crimean Astrophys. Obs., Vol. 64.

The paper is dedicated to preliminary results in the study of rapid variations in the spectrum of the magnetic star 53 Cam (A2, 6^m3). Spectrograms obtained on the 6-m telescope in April 1977 show strong peculiar changes in K Ca II, Sr II, Eu II and Si II lines. The range of rapid variation or activity is obviously changing not only with the phase, but from cycle to cycle as well. The period of variations $P = 21^m0$ has been established, using the data analysis of the equivalent widths in K Ca II lines observed in January 24 - 25, 1978.

114.079 On the behaviour of emission in the spectrum of γ Cassiopeia observed in September - November 1977 and in September - October 1979.

T. S. Galkina.
Izv. Krymskoj Astrofiz. Obs., Tom 64, 72 - 80 (1981). In
Russian. English translation in Bull. Crimean Astrophys. Obs.,
Vol. 64.

The profiles of the Hα, Hβ, Hγ and λ5876 Å He I lines
were analysed in the spectrum of the peculiar Be star
γ Cassiopeia associated with the variable X-ray source
MX 0053 + 60. It is found that compared with preceding years
the emission intensity has been increasing in September 1977.
It is noticed that the emission intensity has been sharply increasing in the period from September 18 to 28 1979, which
is evidently connected with powerful ejection from the star. It
is suggested that the X-ray emission occurs together with
powerful ejection of hot plasma from the star.

114.080   Spectral investigation of the eclipsing binary
V822 Aql.   T. M. Rachkovskaya.

Izv. Krymskoj Astrofiz. Obs., Tom 64, 81 - 91 (1981). In
Russian. English translation in Bull. Crimean Astrophys. Obs.,
Vol. 64.

The spectrum of the eclipsing binary V822 Aql has been
studied using 30 spectrograms obtained in 1968 - 1979. The
behaviour of profile variations in the hydrogen, λ4471 and
λ4026 He I and λ4481 Mg II lines within the orbital periods
indicates the possible presence of unstable processes in the
system. The spectral class of each component of V822 Aql has
been estimated. The following parameters have been obtained
from the velocity curves: the semi-amplitude of the radial velocity and the velocity of the system.

114.081   Technique of detection and analysis of stellar
microvariability. II. Variations of absorption
features in the spectrum of XX Cam.   B. E. Zhilyaev,
V. A. Oshchepkov, A. G. Totochava, L. M. Shul'man.

Astrometr. Astrofiz., Vyp. (No.) 43, p. 30 - 40 (1981). In
Russian.

A statistical analysis is made for the equivalent width
variations of some absorption lines and blends in XX Cam
spectrograms. Some significant variations are found in the
absorption spectra of XX Cam. The statistical power spectrum
displays harmonics with 25.6 and 136 min periods, which
cannot be considered as a coherent process and, apparently,
involve strongly damping harmonic fluctuations.

114.082   Investigation of spectral peculiarities in the variable
star RR Lyrae.

Yu. S. Romanov, Z. N. Fenina, S. V. Vasil'eva.
Astrometr. Astrofiz., Vyp. (No.) 43, p. 43 - 52 (1981). In
Russian.

The physical features of RR Lyrae displaying the
Blazhko effect were studied by a quantitative spectral classification method. It was found on the basis of the analysis of
93 spectrograms that the parameters ΔS and p are of a complicated variable character.

114.083   Note on the spectrum of the Wolf-Rayet star
HD 62910.   V. S. Niemela.

Bol. Asoc. Argentina Astron., No. 18, p. 7 - 9 (1980).

Line identifications on the blue-violet spectral region
of HD 62910 confirm its classification as an intermediate
object between the WN and WC sequences of the Wolf-Rayet
stars.

114.084   Southern B-type stars with emission at H-alpha.
B. Kucewicz.

Bol. Asoc. Argentina Astron., No. 18, p. 10 (1980).

114.085   Spectrophotometric investigation of selected
variable stars of R Coronae Borealis type.

A. Eh. Rozenbush.
Astrometr. Astrofiz., Vyp. (No.) 44, p. 13 - 18 (1981). In
Russian.

The energy distribution in the spectra of selected
R Coronae Borealis stars (XX Cam, UV Cas, V482 Cyg,
CL Sge, SV Sge) is obtained from objective prism spectra. For
some of these stars spectral types and absolute magnitudes
have been determined. UBV magnitudes are given for comparison stars.

114.086   Spectral classification of stars up to 15ᵐ0 around
the cluster NGC 6823.   V. I. Kuznetsov.

Astrometr. Astrofiz., Vyp. (No.) 45, p. 53 - 56 (1981). In
Russian.

A spectral classification of 289 stellar spectra up to
15ᵐ0 around the cluster NGC 6823 was carried out.

114.087   Energy distribution in a continuous spectrum of
five spectrophotometric standards.

E. A. Depenchuk, N. S. Komarov.
Astrometr. Astrofiz., Vyp. (No.) 46, p. 15 - 17 (1982). In
Russian.

The energy distribution is obtained in the continuum
spectrum of β CMi, θ Leo, α Oph, α Peg, β Tau for the wavelength range from 3200 to 7700 Å using α Lyr as the main
standard.

114.088   Methods of spectral classification of stars.
V. I. Kuznetsov.

Astrometr. Astrofiz., Vyp. (No.) 46, p. 43 - 57 (1982). In
Russian.

Different visual, quantitative and photometric methods
of spectral classification of stars are considered from the
standpoint of their application to studying the galactic structure. The method of stellar spectral classification based on
unwidened low-dispersion spectrograms is recommended for
wide application.

114.089   An analysis of the accuracy of modern optical
stellar spectrophotometry.

E. I. Hagen-Thorn, E. V. Ruban.
Astron. Zh., Tom 59, 346 - 354 (1982). In Russian. English
translation in Soviet Astron., Vol. 26, No. 2.

A comparison of the data of various optical spectrophotometric catalogues is made. The values of the reduction functions are given which permit to reduce the data of different
authors to a single system. The problem of the internal agreement of catalogues is considered.

114.090   Lithium abundance in Population II dwarfs.
M. Spite, F. Spite.

Astrophysical parameters for globular clusters, (see 012.023),
p. 59 - 64 (1981).

The determination of the abundance of ^{7}Li in the very
old Population II stars may bring much useful information;
for example the variation of this abundance with the effective
temperature of the star can provide many clues about stellar
interior structure of Population II dwarfs which are probably
similar to the dwarfs of globular clusters. With the coudé
spectrograph of the C.F.H. Telescope in Hawaii, using a
RETICON as receptor the authors obtained spectra centered
at about λ 6700 Å of seven among the brightest Population II
dwarfs. The temperature of these stars are spread from
$\theta_{eff} = 0.86$ to $\theta_{eff} = 1.00$ and the mean metal abundance
relative to the sun from [Fe/H] = -1.1 to [Fe/H] = -2.4. In
all these stars, but HD 103095 (the coolest star) the authors
observe a rather strong lithium line.

114.091   CNO in halo stars.   B. Barbuy.

Astrophysical parameters for globular clusters,
(see 012.023), p. 85 - 88 (1981).

The author has selected some population II stars for
which the three elements, C, N and O were available, in order
to study the behavior of CNO as a whole. The sum (C + N + O)
is more representative of the interstellar matter from which the

star formed, i.e., even if the CNO tri-cycle is operative and mixing mechanisms bring the processed material from inner layers to the surface, the sum (C + N + O) must remain constant.

114.092   **The extreme metal-weak star CD −38°245.**
M. S. Bessell, J. Norris.
Astrophysical parameters for globular clusters, (see 012.023), p. 137 - 143 (1981).

114.093   **Carbon abundances in giant stars of the Draco dwarf galaxy.**
T. D. Kinman, D. F. Carbon, N. B. Suntzeff, R. P. Kraft.
Astrophysical parameters for globular clusters, (see 012.023), p. 451 (1981). − Abstract.

114.094   **Spectroscopic detection of chemical inhomogeneity of surfaces of Ap stars in the case of absence of their spectral variability.**   V. L. Khokhlova.
Pis'ma Astron. Zh., Tom 8, 302 - 305 (1982). In Russian.
English translation in Soviet Astron. Lett., Vol. 8.
It is shown that the determined "average" abundances in Ap-stars with a spotty surface depend on the intensity of the line used for analysis, due to nonlinear character of the curve of growth. A method to reveal chemical inhomogeneity is proposed for the cases when no spectral variability is seen.

114.095   **Copernicus observations of the N V resonance doublet in 53 early-type stars.**
D. C. Abbott, R. C. Bohlin, B. D. Savage.
Astrophys. J., Suppl. Ser., Vol. 48, 369 - 393 (1982).
UV spectra in the wavelength interval 1170−1270 Å are presented for 53 early-type stars ranging in spectral type from O6.5 to B2.5 IV. The sample includes four Wolf-Rayet stars, seven known Oe−Be stars, and six galactic halo OB stars. A qualitative analysis of the stellar N V doublet reveals that: (1) N V is present in all stars hotter and more luminous than type B0 for the main sequence, B1 for giants, and B2 for supergiants; (2) shell components of N V and an unidentified absorption feature at 1230 Å are present in about half of the stars; (3) the column density of N V is well correlated with bolometric luminosity over the spectral range O6 to B2; and (4) the ratio of emission to absorption equivalent width is a factor of 2 smaller in the main sequence stars than in supergiants, which suggests that the wind structure changes as a star evolves.

114.096   **Analysis of the spectra of Sirius and Vega in the 3600 to 4200 Angstrom region.**
R. A. Bell, L. A. Dreiling.
Publ. Astron. Soc. Pacific, Vol. 94, 50 - 54 (1982).
Flux constant line-blanketed model atmospheres have been used to compute theoretical spectra in the region of the Balmer limit. These spectra have been compared with observations of Sirius and Vega. The agreement is satisfactory, the mean difference between theory and observation being about 8%. The effect of the differences on theoretical color calculations is examined.

114.097   **Carbon and nitrogen abundances in extremely metal-deficient red giants.**
R. P. Kraft, N. B. Suntzeff, G. E. Langer, D. F. Carbon, C. F. Trefzger, E. Friel, R. P. S. Stone.
Publ. Astron. Soc. Pacific, Vol. 94, 55 - 66 (1982) = Lick Obs. Bull., No. 907.
Low-resolution image-dissector scanner spectra have been obtained for 64 metal-poor ([Fe/H] $\lesssim$ −1.4) halo giants selected from Bond's objective prism survey. Metallicities estimated from H and K line strengths, following Suntzeff, are shown to agree well with those derived by Bond on the basis of the Strömgren photometric index m_1. Estimates of carbon and nitrogen abundances were obtained from a comparison of observed and synthetic spectra in the regions of the NH($\sim \lambda 3360$) and CH($\sim \lambda 4300$) bands, respectively. Abundance estimates [C/Fe] and [N/Fe] for halo field stars have been obtained using the same abundance methodology and using the same spectrographic equipment as had been employed in the study of 89 red-giant members of metal-poor globular clusters M3, M13, and M92. Derived abundances for the halo giants are compared with abundances in the globular clusters. Anomalies in carbon depletion and nitrogen enhancement are noted and possible reasons for these anomalies are discussed.

114.098   **Observaciones de campos de velocidades en estrellas tempranas.**   V. S. Niemela.
Bol. Asoc. Argentina Astron,, No. 20 - 24, p. 221 (1981). Abstract.

114.099   **Estrellas tempranas con emisión en Hα.**
B. Kucewicz.
Bol. Asoc. Argentina Astron., No. 20 - 24, p. 225 (1981). Abstract.

114.100   **La medición de las líneas de Balmer en estrellas Be.**
A. Feinstein.
Bol. Asoc. Argentina Astron., No. 20 - 24, p. 226 (1981). Abstract.

114.101   **Determinación de abundancias en estrellas de tipo K III.**   Z. Lopez Garcia.
Bol. Asoc. Argentina Astron., No. 20 - 24, p. 230 (1981). Abstract.

114.102   **Búsqueda de estrellas ráfagas en el hemisferio sur.**
Z. López García, F. López García.
Bol. Asoc. Argentina Astron., No. 20 - 24, p. 232 (1981). Abstract.

114.103   **Investigación espectroscópica de la estrella RY Sgr.**
L. A. Milone, M. Villada de Arnedo.
Bol. Asoc. Argentina Astron., No. 20 - 24, p. 333 (1981). Abstract.

114.104   **Búsqueda de estrellas ráfagas en el hemisferio sur.**
Z. López García, F. López García, G. Sánchez.
Bol. Asoc. Argentina Astron., No. 20 - 24, p. 336 (1981). Abstract.

114.105   **La estrella de Mn-Hg en IC 4665.**
S. Malaroda, H. Levato.
Bol. Asoc. Argentina Astron., No. 20 - 24, p. 391 (1981). Abstract.

114.106   **Estrellas tempranas con H-Alfa en emisión.**
B. Kucewicz.
Bol. Asoc. Argentina Astron., No. 20 - 24, p. 394 (1981). Abstract.

114.107   **Parámetros espectrofotométricos y modelos incompletos de atmósferas estelares.**
C. E. Lopez.
Bol. Asoc. Argentina Astron., No. 20 - 24, p. 470 (1981). Abstract.

114.108   **Espectrofotometría de baja dispersión de estrellas de tipo G.**   H. Moreno, A. G. de Moreno.
Bol. Asoc. Argentina Astron., No. 20 - 24, p. 472 (1981). Abstract.

114.109   **La estrella Ap HD 34 452.**
Z. López García.
Bol. Asoc. Argentina Astron., No. 20 - 24, p. 476 (1981). Abstract.

114.110 **Photoelectric measures of chromospheric H and K and He in giant stars.**   O. C. Wilson.
Astrophys. J., Vol. 257, 179 - 192 (1982).

The coudé scanner of the 100 inch telescope was used to measure the fluxes at the centers of H and K of Ca II and at the position of He, and its antiposition, in about 200 late-type giant stars. For the large majority of class III giants, it was found that the total chromospheric radiation of H and K and of He is constant in amount from G8 to early M-type stars with respect to the energy in the V band, and that the two Ca II lines together emit about 3 times as much energy as does He. A comparison is made between the chromospheric emissions in the Hyades main sequence and the normal giants, and some quite large differences are pointed out. A sample of main-sequence stars was also observed, and evidence is given that main-sequence stars with H and K emission also have emission in He.

114.111 **The spectrum of P Cyg.**   N. L. Ivanova, M. B. Babaev, A. A. Gusejnzade, E. B. Zvereva.
Soobshch. Byurakan Obs., Vyp. 53, p. 79 - 87 (1982). In Russian.

The results of an investigation of the energy distribution in the continuum, the measurements of radial velocities and line profiles in the spectrum of P Cyg are given.

114.112 **Identificación de elementos en HD 213918.**   S. Malaroda.
Bol. Asoc. Argentina Astron., No. 26, p. 97 (1981). – Abstract

114.113 **Errors in the HD identification in "Spectral Survey of the Southern Milky Way III" by Loden, L. O., Loden, K., Nordström, B., Sundman, A.**
K. Loden, A. Sundman.
Bull. Inf. Cent. Données Stellaires, No. 22, p. 119 - 120 (1982). – See Abstr. 17.114.008.

114.114 **On a possible explanation of observed underabundance of He in the atmospheres of Bp stars.**
V. L. Khokhlova.
Soobshch. Spets. Astrofiz. Obs., Vyp. 32, (see 012.029), p. 14 (1981).

114.115 **Determination of T_{eff} and lg g from Kurucz models.**   V. D. Bychkov, V. S. Lebedev.
Soobshch. Spets. Astrofiz. Obs., Vyp. 32, (see 012.029), p. 22 (1981).

114.116 **On the influence of abundance of easily ionized atoms on the spectral characteristics of Ap stars.**
V. V. Leushin, V. V. Sokolov.
Soobshch. Spets. Astrofiz. Obs., Vyp. 32, (see 012.029), p. 35 - 36 (1981).

114.117 **On the influence of helium diffusion on the spectral lines of light elements in CU Vir.**
J. Madej, L. Rossi.
Soobshch. Spets. Astrofiz. Obs., Vyp. 32, (see 012.029), p. 36 (1981).

114.118 **Spectrophotometry of α^2 Canum Venaticorum-type magnetic variables in the Torun Observatory. The case of ι Cas, 53 Cam, ϵ UMa, α^2 CVn, β CrB and 73 Dra.**
A. Burnicki, L. Gawenda, J. Gertner, T. Michalovski, A. Woszczyk, D. Zaremba.
Soobshch. Spets. Astrofiz. Obs., Vyp. 32, (see 012.029), p. 40 (1981).

114.119 **A spectroscopic study of the magnetic variable star Iota Cassiopeiae.**
A. Burnicki, M. Jasinski, M. Muciek, A. Woszczyk.

114.110 Soobshch. Spets. Astrofiz. Obs., Vyp. 32, (see 012.029), p. 41 - 42 (1981).

114.120 **New results of an investigation of the star CQ UMa.**   Z. Mikulàšek.
Soobshch. Spets. Astrofiz. Obs., Vyp. 32, (see 012.029), p. 42 (1981).

114.121 **Spectroscopic and photometric study of HR 5023.**   J. Zverko.
Soobshch. Spets. Astrofiz. Obs., Vyp. 32, (see 012.029), p. 43 - 44 (1981).

114.122 **Spectrophotometry of the Ap stars 53 Cam and 52 Her.**   A. N. Khotnyanskij.
Soobshch. Spets. Astrofiz. Obs., Vyp. 32, (see 012.029), p. 47 (1981).

114.123 **On line identification for ϵ UMa.**   S. Hubrig.
Soobshch. Spets. Astrofiz. Obs., Vyp. 32, (see 012.029), p. 48 (1981).

114.124 **Relations between the He-peculiar stars and the Ap stars.**   W. Schöneich.
Soobshch. Spets. Astrofiz. Obs., Vyp. 32, (see 012.029), p. 50 - 56 (1981).

114.125 **The relationship between spectrum and light variations in the Ap stars HD 119213, HD 170000 and HD 173650.**   B. Musielok.
Soobshch. Spets. Astrofiz. Obs., Vyp. 32, (see 012.029), p. 66 - 67 (1981).

114.126 **Rapid variability of the Hγ line in the Ap star β CrB.**   G. N. Alekseev, V. M. Kuvshinov.
Soobshch. Spets. Astrofiz. Obs., Vyp. 32, (see 012.029), p. 72 - 73 (1981).

114.127 **Search for hydrogen line variability of three Ap stars in the region 0.01 - 1 Hz (ϵ UMa, α And, 53 Cam).**
G. N. Alekseev, V. D. Bychkov, V. S. Lebedev, V. G. Shtol'.
Soobshch. Spets. Astrofiz. Obs., Vyp. 32, (see 012.029), p. 73 (1981).

114.128 **Correlations between BCD parameters of the continuous spectrum and the Balmer decrement of Be stars.**   L. Divan, J. Zorec, D. Briot.
Be stars, (see 012.030),p. 53 - 56 (1982).

114.129 **Intrinsic reddening of Be stars and its relation with Hα emission intensities.**
L. Divan, V. Doazan, J. Zorec.
Be stars, (see 012.030), p. 57 - 59 (1982).

The intrinsic reddening and its relation with Hα emission has been determined for 32 Be/shell stars observed simultaneously at high dispersion at Hα and between $\lambda\lambda 3000-$ and 6200 Å with the Chalonge spectrograph.

114.130 **BCD spectrophotometry of the Be-shell star 88 Her.**   L. Divan, J. Zorec.
Be stars, (see 012.030), p. 61 - 63 (1982).

114.131 **Spectroscopic observations of Be stars in the photographic and visual regions.**   A. Slettebak.
Be stars, (see 012.030), p. 109 - 124 (1982).

The author limits this review paper to a discussion of the "classical" Be stars. These are defined as stars of luminosity classes III to V, usually rapid rotators, which show normal B-type spectra with superposed Balmer (and sometimes Fe II) emission. Included also, however, will be the Oe stars and the A-type shell stars, which seem to represent extensions of the classical Be phenomenon to higher and lower temperatures, respectively.

114.132 **Statistical properties of Be stars.**
A. M. Hubert-Delplace, M. Jaschek, H. Hubert, M. T. Chambon.
Be stars, (see 012.030), p. 125 - 130 (1982).
The emission features of 140 Be stars described in "An Atlas of Be stars" are briefly reviewed. The time scale of the emission variations are determined for about 35 stars. For several stars the authors estimated, on the Hβ line, the V/R ratio of the two emission components.

114.133 **Results of a new survey for early-type emission stars.** D. J. MacConnell.
Be stars, (see 012.030), p. 131 - 133 (1982).

114.134 **Observation de la raie Hα dans les étoiles Be.**
Y. Andrillat, C. Fehrenbach.
Be stars, (see 012.030), p. 135 - 139 (1982).
The author discusses the discovery, on objective-prism plates, of 846 stars showing Hα in emission on non-banded spectra. These stars have not previously been reported to have emission and are located primarily along the southern galactic plane. Twenty-six percent of the stars are known to be of type A0 or earlier.

114.135 **Intensifier-dissector-scanner observations of the bright northern Be stars.** P. K. Barker.
Be stars, (see 012.030), p. 147 - 150 (1982).
Early results are presented from an extensive program of regular Balmer-line spectroscopy; during eight months in 1980 Hα profiles were obtained of over 100 northern Be stars. Transient emission events have been observed in the very early rapid rotator 59 Cygni; sequential variations in Hα are quasi-periodic.

114.136 **Search for long-period radial velocity variations in some Be stars.** L. Pastori, E. Antonello, M. Fracassini, L. E. Pasinetti.
Be stars, (see 012.030), p. 155 - 159 (1982).

114.137 **Spectroscopic study of Pleione in 1977 - 1979.**
R. Hirata, J. Katahira, J. Jugaku.
Be stars, (see 012.030), p. 161 - 165 (1982).
Shell spectra of Pleione in 1977–1979 are characterized by further increase in their strengths and by the development of the blue-winged profiles without noticeable variation of their radial velocities. The MgII resonance lines at $\lambda 2800$Å have the blue-shifted components with a velocity of -35 km/s relative to the other shell lines. The development of the shell structure is derived. The mass loss rate was $7 \times 10^{-11} M_\odot$/yr.

114.138 **Radial velocity variations in 69 Orionis.**
M. Bossi, G. Guerrero, L. Mantegazza.
Be stars, (see 012.030), p. 185 - 188 (1982).

114.139 **On periodic variations in the spectrum of the B0e star X Persei associated with the X-ray source 3U 0352+30.** T. S. Galkina.
Be stars, (see 012.030), p. 189 - 194 (1982).

114.140 **Recent changes of the Be star HD 58050.**
A. M. Hubert-Delplace, H. Hubert, D. Ballereau, M. T. Chambon.
Be stars, (see 012.030), p. 195 - 199 (1982).
Emission line variations in 1961–1981 of the B2 star HD58050 are reported. Brightness variations are recalled. An estimation of the veiling effect given by continuous emission of the envelope, observed in November 1980 in the Balmer lines, is given.

114.141 **R 81: P Cygni of the LMC.**
O. Stahl, B. Wolf, M. J. H. De Groot, C. Sterken.
Be stars, (see 012.030), p. 201 - 203 (1982). — Abstract.

114.142 **On the problem of the chemical composition of β Lyrae.** V. Bahýl'.
Be stars, (see 012.030), p. 205 - 208 (1982).

114.143 **Spectroscopic observations of Be stars especially in the infrared.** L. Houziaux, Y. Andrillat.
Be stars, (see 012.030), p. 211 - 228 (1982).

114.144 **Le spectre des étoiles Oe dans le rouge et le proche infrarouge.**
Y. Andrillat, J. M. Vreux, M. Dennefeld.
Be stars, (see 012.030), p. 229 - 233 (1982).

114.145 **A preliminary digital analysis of the spectrum of β Lyrae.** E. E. Mendoza V.
Be stars, (see 012.030), p. 253 - 255 (1982).

114.146 **Ultraviolet observations, stellar winds, and mass loss for Be stars.** J. M. Marlborough.
Be stars, (see 012.030), p. 361 - 376 (1982).

114.147 **Variation of anomalous stages of ionization with spectral type for Be stars.**
J. M. Marlborough, G. J. Peters.
Be stars, (see 012.030), p. 387 - 390 (1982).

114.148 **Recent changes in the ultraviolet spectrum of the Be star HR 2855.** G. J. Peters.
Be stars, (see 012.030), p. 411 - 414 (1982).
High resolution IUE observations of the Be star HR 2855 (HD 58978) spaced five months apart have revealed striking variations in the resonance lines of C IV and N V.

114.149 **The active UV phase of 59 Cyg.**
V. Doazan, C. Grady, L. V. Kuhi, J. M. Marlborough, T. P. Snow, R. N. Thomas.
Be stars, (see 012.030), p. 415 - 418 (1982).
Coordinated UV and visual observations of 59 Cyg in 1978–81 show strong mass ejection activity and strong variability in displacements and profiles of superionized lines during the new Be phase, starting from a "quasi normal B" phase in 1977, and increasing irregularly through 1981 to a low and then moderate Hα emission. These data show that visual data alone cannot describe the activity of the star.

114.150 **The problem of X Persei.** R. Viotti, M. Ferrari-Toniolo, A. Giangrande, P. Persi, G. B. Baratta.
Be stars, (see 012.030), p. 423 - 426 (1982).

114.151 **The spectrum of HD 51585 in the blue and in the ultraviolet.** L. Houziaux, Y. Andrillat, A. Heck, K. Nandy.
Be stars, (see 012.030), p. 427 - 430 (1982).
New visible and ultraviolet spectra of the peculiar emission line star HD 51585 are described. Interstellar lines and the λ 2200 feature are rather weak. A colour excess E(B-V) = 0.33 is derived. The extinction curve resembles the one obtained from LMC stars.

114.152 **UV observations of γ Cas: intermittent mass-loss enhancement.** H. F. Henrichs.
Be stars, (see 012.030), p. 431 - 435 (1982).

114.153 **IUE observations of 17 Lep.**
P. Molaro, P. L. Selvelli, R. Stalio.
Be stars, (see 012.030), p. 437 - 442 (1982).
The spectrum of 17 Lep is dominated by numerous and strong shell lines of once-ionized metals. The shell shows a

composite radial velocity structure which indicates the presence of multiple components with a velocity range between −50 and −250 km s⁻¹. Unlike in the visible, where these multiple components are present only during outburst, in the ultraviolet such components seem to be a permanent feature.

114.154 **Simultaneous IUE and ground-based spectroscopic observations of the variable LMC star R 71.**
B. Wolf, I. Appenzeller, O. Stahl.
Be stars, (see 012.030), p. 443 (1982). − Abstract.

114.155 **Spectral energy distribution (119−685 nm) in 16 shell stars and a tentative model for accreting Be stars.** M. J. Plavec, J. J. Dobias, J. L. Weiland, R. P. S. Stone.
Be stars, (see 012.030), p. 445 - 450 (1982).
I.U.E. low-dispersion spectra and spectral scans made with the Lick Observatory IDS scanners have been combined for 16 shell stars. Eleven objects can be represented by Kurucz model atmospheres, although some of them display strong shell-type line spectra. Five among them are known binaries. The six remaining objects (all interacting binaries) display complex spectra. A model involving continuum and line radiation from a hydrogen cloud surrounding the accreting component is proposed.

114.156 **On the spectrum of the Herbig Be star HD 200775.**
B. Baschek, M. Beltrametti, J. Köppen, G. Traving.
Be stars, (see 012.030), p. 499 (1982). − Abstract.

114.157 **Spectroscopic investigations of Herbig-Ae-Be-stars.**
U. Finkenzeller.
Be stars, (see 012.030), p. 501 - 507 (1982).
Radial velocities and remarks on peculiarities are given for the star HD 200775, which seems to represent a typical Herbig-Ae-Be star fairly well. A catalogue of about 60 supposed Herbig-Ae-Be-stars is also presented.

114.158 **Observations spectroscopiques et photographiques d'une étoile variable à très courte période.**
C. C. Huang.
Inf. Bull. Variable Stars, No. 2114, 5 pp. (1982).

114.159 **Spectrum variability of HD 147010.**
N. Kameswara Rao, R. Rajamohan.
Inf. Bull. Variable Stars, No. 2121, 2 pp. (1982).

114.160 **A search for short-term radial-velocity variations of o And.**
D. Baade, H. Pollok, J. D. Schumann, H. W. Duerbeck.
Inf. Bull. Variable Stars, No. 2125, 6 pp. (1982).

114.161 **New Hα-emission stars in the region of γ Cygni.**
K. P. Tsvetkova, M. K. Tsvetkov.
Inf. Bull. Variable Stars, No. 2134, 2 pp. (1982).

114.162 **Spectrophotometric observations of the star MWC 17 in January 1982.** R. Gravina.
Inf. Bull. Variable Stars, No. 2135, 4 pp. (1982).

114.163 **o Andromedae.**
IAU Circ., No. 3658 (1982).

114.164 **Energy distribution and equivalent widths of Balmer lines in the spectra of rapidly rotating B and A stars.**
I. N. Glushneva.
Astron. Zh., Tom 59, 523 - 535 (1982). In Russian. English translation in Soviet Astron., Vol. 26, No. 3.
The characteristics of the continuum (spectrophotometric gradients, Balmer jumps) and equivalent widths of Balmer lines Hα, Hβ, Hγ in the spectra of rapidly rotating B and A stars are investigated. A comparison of these characteristics for slowly rotating stars and with theoretical models for rapidly rotating stars is presented.

114.165 **Calcium in the atmospheres of peculiar stars.**
V. V. Leushin.
Astron. Zh., Tom 59, 543 - 551 (1982). In Russian. English translation in Soviet Astron., Vol. 26, No. 3.
The calcium abundance in the atmospheres of Ap stars has been analyzed. The abundances were obtained by model atmosphere analysis and from scientific publications. The Ca I and Ca II spectra of the Ap stars have been investigated. The equivalent widths of the K Ca II line in Ap star spectra are in agreement with $B−V$ in the average. The calcium abundance is normal in the average in the atmospheres of Ap stars. The decrease and increase of the K Ca II line intensity in Ap star spectra may account for the difference of structures of the atmospheres of normal and peculiar stars.

114.166 **Investigation of energy distribution in the spectra of ten B−A stars.** A. P. Ipatov.
Astron. Zh., Tom 59, 607 - 610 (1982). In Russian. English translation in Soviet Astron., Vol. 26, No. 3.
The energy distribution in the spectra of ten B−A stars having $V ≈ 8^m$ is obtained. The photometric accuracy of calibration of the absolute energy distribution in the spectra with respect to Vega is in the mean better than 3 - 4 per cent, thus allowing to use these stars as secondary photometric standards.

114.167 **A note on the temperature of HD 101065.**
A. Przybylski.
Astrophys. J., Lett., Vol. 257, L83 - L86 (1982).
The claim of Kurtz and Wegner that the lines of the Paschen series have an equivalent width of up to 4 Å in HD 101065 cannot be confirmed. Their strength is only about 1% of that amount. Consequently, the temperature of HD 101065 cannot be as high as 7400 K as Kurtz and Wegner claim. Several criteria show that HD 101065 is a late F star, most likely an F8 star.

114.168 **NLTE analysis of massive O-stars.**
R. P. Kudritzki.
The most massive stars, (see 012.034), p. 49 - 66 (1981).
The NLTE-effects in photospheres of massive O-stars are summarized and the "spectroscopic" temperature scale for the earliest O-spectral types is discussed. The results of detailed NLTE analyses of four O3-stars and the O4f-star ζ Pup are presented.

114.169 **On the nature of the very luminous Wolf-Rayet stars in M33.** S. D'Odorico, M. Rosa.
The most massive stars, (see 012.034), p. 191 - 197 (1981).
Spectroscopic and photometric data for the Wolf-Rayet objects discovered by Conti and Massey in giant HII regions of M33 are compared with the properties of the cluster at the center of 30 Doradus in the LMC. It is concluded that the M33 objects are likely to be clusters of young stars with many WR components unresolved at the distance of M33. The existence of objects like R 136 in these clusters is considered unlikely.

114.170 **Wolf-Rayet stars in emission-line galaxies.**
D. Kunth.
The most massive stars, (see 012.034), p. 207 - 209 (1981).

114.171 **H/He ratios for WN stars in the Galaxy and LMC.**
D. N. Perry, P. S. Conti.
Bull. American Astron. Soc., Vol. 13, 784 - 785 (1981). Abstract.

114.172 **The optical spectrum of R136A.**
D. Ebbets, P. Conti.
Bull. American Astron. Soc., Vol. 13, 785 (1981). – Abstract.

114.173 **Near infrared spectra of Algol B.** H. Zirin.
Bull. American Astron. Soc., Vol. 13, 804 (1981). Abstract.

114.174 **The chemical compositions and evolutionary state of the subgiant CH stars.** H. E. Bond, R. E. Luck.
Bull. American Astron. Soc., Vol. 13, 810 (1981). – Abstract.

114.175 **High-resolution studies of the archetype K giant Arcturus with IUE and Einstein: a sensitive search** for high-temperature emission.
T. R. Ayres, T. Simon, J. L. Linsky.
Bull. American Astron. Soc., Vol. 13, 811 (1981). – Abstract.

114.176 **Radial velocities of hot secondary stars from IUE spectra.** S. B. Parsons.
Bull. American Astron. Soc., Vol. 13, 817 (1981). – Abstract.

114.177 **Two lithium-rich supergiants.** R. E. Luck.
Bull. American Astron. Soc., Vol. 13, 827 (1981). Abstract.

114.178 **A statistical search for medium Z elements in the ultraviolet spectrum of κ Cancri.**
J. P. Davidson, D. J. Bord.
Bull. American Astron. Soc., Vol. 13, 827 - 828 (1981). Abstract.

114.179 **The ultraviolet spectra of three N-type carbon stars.**
H. R. Johnson, G. T. O'Brien.
Bull. American Astron. Soc., Vol. 13, 828 (1981). – Abstract.

114.180 **Magnesium enrichment in carbon stars.**
J. H. Goebel, J. D. Bregman, M. Cohen.
Bull. American Astron. Soc., Vol. 13, 828 (1981). – Abstract.

114.181 **Observations of line profile variability in λ Cep.**
C. A. Grady, T. P. Snow, J. G. Timothy.
Bull. American Astron. Soc., Vol. 13, 829 (1981). – Abstract.

114.182 **Absolute spectrophotometry of Wolf-Rayet stars in the Local Group.** P. Massey.
Bull. American Astron. Soc., Vol. 13, 829 (1981). – Abstract.

114.183 **Determination of the surface gravity of Arcturus from MgH lines.** R. A. Bell, B. Gustafsson.
Bull. American Astron. Soc., Vol. 13, 829 (1981). – Abstract.

114.184 **The L(400) magnitude.** R. F. Wing, C. P. Rinsland.
Bull. American Astron. Soc., Vol. 13, 840 (1981). Abstract.

114.185 **Abnormalities in the spectrum of the Herbig emission star HK Ori.**
R. Davis, S. Strom, K. Strom.
Bull. American Astron. Soc., Vol. 13, 855 (1981). – Abstract.

114.186 **Six meter spectra of M 92 blue horizontal-branch stars.** A. G. D. Philip, N. Samus.
Bull. American Astron. Soc., Vol. 13, 884 (1981). – Abstract.

114.187 **Beryllium abundances in Hg-Mn stars.**
A. M. Boesgaard, W. D. Heacox, S. C. Wolff,
J. Borsenberger, F. Praderie.
Bull. American Astron. Soc., Vol. 13, 884 (1981). – Abstract.

114.188 **IUE échelle mode observations contrasting coronal and non-coronal late type giant and supergiant stars.**

A. Brown, R. E. Stencel, J. L. Linsky, C. Jordan, O. Engvold.
Bull. American Astron. Soc., Vol. 13, 885 (1981). – Abstract.

114.189 **A metallicity calibration of absorption-line strengths in K giant stars.**
S. M. Faber, E. D. Friel, D. Burstein.
Bull. American Astron. Soc., Vol. 13, 885 - 886 (1981). Abstract.

114.190 **Are discrepant asymmetry red giants necessarily hybrid stars?** D. J. Mullan, R. E. Stencel.
Bull. American Astron. Soc., Vol. 13, 886 (1981). – Abstract.

114.191 **UV and visual spectra of field blue horizontal branch stars.**
D. P. Huenemoerder, K. S. de Boer, A. D. Code.
Bull. American Astron. Soc., Vol. 13, 925 - 926 (1981). Abstract.

114.192 **Considerations regarding the C$_2$ bands in carbon stars.** J. H. Goebel, D. Goorvitch, D. Cooper,
J. D. Bregman, F. C. Witteborn.
Bull. American Astron. Soc., Vol. 13, 926 (1981). – Abstract.

114.193 **Steps towards the abundance scale. I. The abundance of iron in the Hyades.**
R. Cayrel, G. Cayrel de Strobel, B. Campbell.
Bull. American Astron. Soc., Vol. 13, 926 (1981). – Abstract.

114.194 **Spectrophotometry of cool stars in the near infrared. V. Results for a region in Monoceros.**
F. Smriglio, K. Nandy.
Occas. Rep. R. Obs. Edinburgh, No. 6, 14 pp. (1981).
Spectral types and infrared magnitudes of M stars brighter than V = 15^{m}0 have been obtained for a galactic region of 21.5 square degrees in the direction of Monoceros centred on R. A. (1950) = 6^{h}40^m and Dec. (1950) = 10°50′ ($l^{II} \sim 100°$). The classification is based on the relative strengths of molecular bands visible in the near infrared in low dispersion objective prism spectra. The space distribution of M stars in this direction shows a correlation with spiral arm structure as outlined by other spiral arm tracers.

114.195 **Spectrophotometry of cool stars in the near infrared. VI. Results for a region in the direction of Cepheus.** F. Smriglio, K. Nandy.
Occas. Rep. R. Obs. Edinburgh, No. 7, 16 pp. (1981).
Infrared magnitudes and spectral types of M and C stars brighter than I = 13^m have been obtained for a region of about 18 square degrees in the direction of Cepheus, centred at l^{II} = 103° and b^{II} = 0°. The spectral classification is based on the relative strengths of the TiO bands visible in the infrared low dispersion objective prism spectra. The observed space distribution of early M-type stars shows a good correlation with the galactic spiral features as revealed by the traditional spiral tracers, but the stars later than M4 are evenly distributed in arm and interarm regions. The space densities of M2-M4 stars reveal an asymmetry with respect to the galactic equator, which appears also in the distribution of OB stars, and supergiants.

114.196 **Spectroscopic study of the star Beta Comae Berenices (G0).** H. H. Menteşe.
Istanbul Üniv. Fen Fak. Mec. Seri C, Vol. 45, 29 - 38 (1980) = Publ. Istanbul Univ. Obs. No. 112 (1981).
In this paper the blue spectrum of the star β Com was studied and its effective temperature and surface gravity were determined by means of the theoretical profiles calculated by de Jager and Neven (1967 - 1968). The best agreement was obtained for T$_{eff}$ = 5940°K and log g = 4.

114.197 **Metallic-line stars in Praesepe open cluster: model atmosphere analysis and abundances.** D. Koçer.
Istanbul Üniv. Fen Fak. Mec. Seri C, Vol. 45, 9 - 27 (1980) = Publ. Istanbul Univ. Obs. No. 113 (1981).

From medium-high resolution (12.4 Å mm^{-1}) spectra in the wavelength region 3600–4950 Å, metallic-line stars HD 73730 and HD 73618 are analyzed by means of model atmospheres in convective equilibrium without line-blanketing. Comparison of observed and computed quantities (Balmer lines, some metal lines) leads to atmospheric models and parameters. The abundance analysis have confirmed that the Ca and Sc are underabundant, the group of Fe are normal, Y is overabundant and Eu is drastically underabundant.

114.198 **The symbiotic star Cl Cygni.**
F. Şenel Yildizdoğdu.
Istanbul Üniv. Fen Fak. Mec. Seri C, Vol. 45, 121 - 136 (1980) = Publ. Istanbul Univ. Obs. No. 117 (1981).

Eleven spectra of the symbiotic star Cl Cygni, taken at Asiago Observatory with the 122 cm telescope, from 1969 to 1974, have been reduced and are discussed. During this period the star was passing through a stage of intense activity. Line identification, variation of their relative intensities and spectral changes, connected with the characteristics of the light curve, are reported. The spectra show the presence of a strong gM4e component which overlaps the emission spectrum of a hot B-type star.

114.199 **Spectrophotometry of Wolf-Rayet star candidates in M33.** E. J. Wampler.
ESO Sci. Prepr. No. 193, 12 pp. (1982). – Submitted to Astron. Astrophys.

114.200 **Some parameters of the Ap star ε UMa.**
Yu. V. Glagolevskij, V. D. Bychkov, I. Kh. Iliev, I. D. Najdenov, I. I. Romanyuk, V. G. Shtol', G. A. Chuntonov.
Astrofiz. Issled., Izv. Spets. Astrofiz. Obs., Tom 15, 14 - 20 (1982). In Russian.

The parameters of the continuous spectrum in the ultraviolet and visible spectral regions correspond to the effective temperature T_e = 9800 K and the hydrogen profiles correspond to T_e = 9500 K and lg g = 3.5. The analysis of the intensities of metallic lines gives T_e = 9650 K. The star belongs to "weak peculiar" stars having intensified lines of Cr, EU, Mn. On the basis of $v \sin i$ = 35 km/s and period of rotation P = 5^{d}09 and lg g = 3.5 the radius of the star R = 5$R_\odot$, the inclination angle i = 44° of the star to the line of sight and the absolute magnitude M_v = −0^{m}3 are found. The magnetic field was measured from H_γ and Fe II 4520.2 lines.

114.201 **A search for galactic emission-line objects.**
H. Maehara.
Contrib. Bosscha Obs. Lembang, No. 71, 10 pp. (1982).

A search for emission-line objects is carried out in a low galactic field (centered at α_{1950} = 18^{h}20^m, δ_{1950} = −5°00′) using both Kiso and Bosscha Schmidt telescopes. A detection is made on the objective-prism plates, and positions and magnitudes are determined on the direct plate. Of 19 detected objects, five are newly discovered which are classified as faint early-type stars and planetary nebulae.

114.202 **A galactic window at l = 355°, b = −1° and its implications for further observations.**
K. Hamajima, K. Ishida, T. Ichikawa, B. Hidayat, M. Raharto.
Contrib. Bosscha Obs. Lembang, No. 74, 9 pp. (1982).

114.203 **Survey of red giants near the galactic center direction.** T. Ichikawa, K. Hamajima, K. Ishida, B. Hidayat, M. Raharto.
Contrib. Bosscha Obs. Lembang, No. 75, 10 pp. (1982).

114.204 **On the Balmer emission lines of the Herbig Be star HD 200775.**
J. Köppen, U. Finkenzeller, R. Mundt, M. Beltrametti.
Prepr. Steward Obs., No. 379, 13 pp. (1982).

114.205 **Some results of an investigation of the Mn-Hg-star 33 Gem. I.**
N. M. Chunakova, V. D. Bychkov, Yu. V. Glagolevskij.
Soobshch. Spets. Astrofiz. Obs., Vyp. 31, p. 5 - 22 (1981). In Russian.

21 Zeeman spectrograms of 33 Gem = HD 49606 have been obtained with the purpose of detection of a magnetic field. The preliminary period of the field variation (3^{d}099) has been estimated. The magnetic curve amplitude is 3 kGs and the shape of the curve is sinusoidal. The effective temperature and the gravitation force on the surface of 33 Gem have been detected by comparison of experimental and theoretical Hγ profiles; they are T_e = 15000°, lg g = 3.5. Using the evolutionary tracks on the diagram (T_e, lg g) the mass of the star has been found. The radius of the star is 5.8 $R_\odot$ and the rotation velocity $v \sin i$ ∼30 km/s. 33 Gem is an Mn-Hg-P-star and it has some features inherent to Si λ4200 stars.

114.206 **Wide-band spectrophotometry of carbon stars.**
A. Kh. Avetisyan, V. P. Zalinyan, Yu. K. Melik-Alaverdyan, R. Kh. Oganesyan, G. M. Tovmasyan.
Astrofizika, Tom 17, 225 - 230 (1981). In Russian. English translation in Astrophysics, Vol. 17, No. 2.

The results of spectrophotometry of 18 carbon stars are presented. It is shown that intensities of ultraviolet radiation are very different even for stars of the same spectral subclass.

114.207 **Origin of the optical continuum of flares on red dwarfs.**
M. M. Katsova, A. G. Kosovichev, M. A. Livshits.
Astrofizika, Tom 17, 285 - 300 (1981). In Russian. English translation in Astrophysics, Vol. 17, No. 2.

The dynamical response of a red dwarf chromosphere intensively heated during 10 seconds has been studied. The ionization and radiative losses of hydrogen have been determined for optically thin and thick Lα-line layers separately. The temperature jump propagates downwards and the shock is formed in front of this thermal wave. The radiative losses being substantial, the density jump in this shock wave is great.

114.208 **Connection between ultraviolet excess and lithium content in late giants.** Yu. K. Melik-Alaverdyan.
Astrofizika, Tom 17, 327 - 332 (1981). In Russian. English translation in Astrophysics, Vol. 17, No. 2.

A correlation between ultraviolet excess and lithium content in M giants is discovered. A model in which excessible ultraviolet radiation and lithium are of the same origin is suggested.

114.209 **The resonance lines of Hg II in Hg-rich stars.**
D. S. Leckrone.
Bull. American Astron. Soc., Vol. 14, 571 (1982). – Abstract.

114.210 **Atomic and molecular abundances towards HD 24534 from ultraviolet and optical data.**
D. J. Lien, L. E. Snyder, R. M. Crutcher.
Bull. American Astron. Soc., Vol. 14, 576 (1982). – Abstract.

114.211 **IUE observations of Si and C lines and comparison with non-LTE models.** L. W. Kamp.
Astrophys. J., Suppl. Ser., Vol. 48, 415 - 436 (1982).

Measurements of the equivalent widths and other properties of the line profiles of 24 photospheric lines of Si II, Si III, Si IV, C II, C III, and C IV in the spectral range between 1175 and 1725 Å are presented for seven B stars and two O stars of luminosity classes IV and V. The observed properties of the photospheric absorption lines were determined by fitting smoothed profiles to the observed spectra, making use of second-order polynomials in up to six segments. The observed line profiles are compared with theoretical profiles for lines of Si II, Si III, and Si IV computed by Kamp using non-LTE theory and models, and with similar profiles for some C II and C III lines computed by York using line-blanketed model atmospheres. It is concluded that, on the whole, the present theory of line formation in the photosphere, when used with solar abundances, represents fairly well the observed ultraviolet photospheric lines of the silicon and carbon ions in the atmospheres of main sequence stars of types B5 to O9.

114.212 Problems in the temperature classification of the red giant stars. P. C. Keenan.
Publ. Astron. Soc. Pacific, Vol. 94, 299 - 303 (1982).

Effective temperatures determined by Tsuji by the infrared-flux method were shown by him to remain nearly constant in the range M0 to M2. It is easily shown that this is not due to any luminosity effect in the spectral types. Rather, revised MK types based primarily on TiO band strengths in this range are shown to be more sensitive to temperature changes than are the commonly used color indices.

114.213 Periodic spectrum variations in helium-rich stars. N. R. Walborn.
Publ. Astron. Soc. Pacific, Vol. 94, 322 - 327 (1982).

Spectroscopic observations of four helium-rich stars are presented. In HD 37776, antiphase variations of Si III and He I have been found, which represent another point of similarity to the Ap phenomenon. The remarkable Hα emission variations in σ Ori E are illustrated with uniform phase coverage, and strict periodicity over a five-year interval is shown. A radial-velocity study of HD 64740 shows constancy to within the accuracy of the observations. Finally, δ Ori B is confirmed as a helium-rich star.

114.214 Discoveries on southern, red-sensitive objective-prism plates IV: extension to higher latitudes.
D. J. MacConnell.
Astron. Astrophys., Suppl. Ser., Vol. 48, 355 - 361 (1982).

Lists are presented of 107 non-banded Hα-emission stars, 12 M-type Hα-emission stars, 5 suspected planetary nebulae, 10 carbon stars, 121 S/MS stars, 3 M dwarfs, and a peculiar star not previously reported.

114.215 Equivalent width measurements in galactic supergiant and in Small Magellanic Cloud star spectra.
P. Dubois.
Astron. Astrophys., Suppl. Ser., Vol. 48, 375 - 381 (1982).

Measurements of equivalent width are made in spectra of 40 galactic supergiants and 21 Small Magellanic Cloud stars. These measurements confirm the results of spectral classification in the SMC (Dubois et al., 1977) and show a general weakness of the metallic lines in the SMC star spectra. This weakness is not the same for all the metals and some cases may be attributable to physical phenomena which occur in the atmospheres of these luminous stars.

114.216 Near infrared observations of O stars.
Y. Andrillat.
Messenger, No. 28, p. 22 - 25 (1982).

114.217 Dependence of the Wilson−Bappu effect on stellar atmospheric parameters.
T. E. Lutz, B. E. J. Pagel.
Mon. Not. R. Astron. Soc., Vol. 199, 1101 - 1111 (1982).

The relation discovered by Wilson & Bappu (1957) between the width W_0 of the $Ca^+ K_2$ emission component and the absolute visual magnitudes of F5 to M-type stars seems to represent a fundamental relationship between some chromospheric property and the global parameters of the underlying star. In this paper the authors make a fresh attempt to determine the relationship between W_0 and the fundamental stellar atmospheric parameters: the effective temperature T_e, the surface gravity g and the metal abundance [Fe/H].

114.218 A study of the H-alpha line in late G and K supergiants. S. V. Mallik.
J. Astrophys. Astron., Vol. 3, 39 - 61 (1982).

A spectroscopic study of Hα has been carried out to investigate the properties of expanding chromospheres of late G and K supergiants. Spectra of 23 stars brighter than $V = 6.0$ have been obtained. The Hα profiles are all asymmetric in the sense that the absorption core is shifted to the blue by amounts ranging between −4 and −24 km s^{-1}. Hα profiles were theoretically computed using radiative transfer in spherically symmetric expanding atmospheres. Their analysis shows that the Hα line is formed in a region with velocity increasing outward. The computed equivalent widths and line core displacements were matched with those observed to obtain hydrogen column densities and expansion velocities. From these, the rates of mass loss in these stars were determined to be in the range of 10^{-6} - $10^{-7} M_\odot$ yr^{-1}.

114.219 Aldebaran. Eine Beschreibung.
F. Gondolatsch.
Sterne Weltraum, 21. Jahrg., 239 - 243 (1982).

114.220 Transition zones in F stars. M. Saxner.
Rep. Obs. Lund, No. 18, (see 012.044), p. 97 - 99 (1982).

114.221 CNO abundances in red giant stars.
P. Kjaergaard, B. Gustafsson, G. A. H. Walker, L. Hultquist.
Rep. Obs. Lund, No. 18, (see 012.044), p. 100 (1982).

114.222 Speckle interferometry of α Ori: preliminary results.
L. Goldberg, E. K. Hege, E. N. Hubbard, P. A. Strittmatter, W. J. Cocke.
Smithsonian Astrophys. Obs., Spec. Rep. 392, (see 012.045), p. 131 - 135 (1982).

114.223 Early-type emission-line stars.
S. N. Svolopoulos.
Compendium in astronomy, (see 003.012), p. 285 - 293 (1982).

114.224 Report of IAU Commission 29: Stellar spectra (Spectres stellaires). W. K. Bonsack.
Trans. IAU, Vol. XVIIIA, (see 003.013), p. 343 - 360 (1982).

114.225 Report of IAU Commission 45: Stellar classification (Classification stellaire). A. Slettebak.
Trans. IAU, Vol. XVIIIA, (see 003.013), p. 621 - 632 (1982).

Some errata in the Fifth General Catalogue of MK Spectral Classifications. See Abstr. 002.029.

MK spectral classifications. Fifth General Catalogue. See Abstr. 002.078.

B stars with and without emission lines. See Abstr. 003.003.

Landolt-Börnstein. See Abstr. 003.064.

Franck-Condon factors and r-centroids for the $K^2 \Sigma$-$A^2 \pi$ and $D^2 \pi$-$A^2 \pi$ band systems of the SiN molecule. See Abstr. 022.086.

A calculation on the transition probability in the A'—X system of YO.   See Abstr. 022.093.

A method for determination of the abundance of chemical elements in the atmospheres of cool giant stars. See Abstr. 031.503.

IUE data reduction. The parameterization of the motion of the IUE reseau grids and spectral formats as a function of time and temperature. See Abstr. 031.524.

An outline of a computer program for two dimensional spectral classification.   See Abstr. 031.542.

LINPOS, an interactive program for stellar line position measurement.   See Abstr. 031.604.

Possibilities for spectroscopic observations in the Bulgarian National Astronomical Observatory. See Abstr. 032.029.

Mass loss from α Cyg (A2Ia) derived from the profiles of low excitation Fe II lines.   See Abstr. 064.021.

On the detection of abundance stratifications in peculiar stars through the curve of growth method. See Abstr. 064.028.

On the widths of the Ca II K emission in late-type stars.   See Abstr. 064.053.

Evidence for extended chromospheres surrounding red giant stars.   See Abstr. 064.097.

Wolf-Rayet stars as a stage of the evolution of massive stars.   See Abstr. 065.045.

K_{2V}/K_{2R} asymmetries in the sun and stars. See Abstr. 073.010.

Solar plages and the interpretation of stellar Ca II H and K line variations in late type dwarfs. See Abstr. 073.150.

Infrared excess and line emission in Be stars. See Abstr. 112.003.

Comparison of winds in the Small Magellanic Cloud and galactic early-type stars.   See Abstr. 112.022.

The structure and variability of Hα emission in early-type supergiants.   See Abstr. 112.053.

The shell episode of 59 Cygni (1974—1975). See Abstr. 112.054.

Narrow components in the profiles of ultraviolet resonance lines: evidence for a two-component stellar wind for O and B stars?   See Abstr. 112.055.

A study of Be star variability. See Abstr. 113.046.

Statistical analysis of the data available for Be stars. See Abstr. 113.047.

Voyager absolute far-ultraviolet spectrophotometry of hot stars.   See Abstr. 113.066.

DDO photometry of G and K stars. See Abstr. 113.088.

A search for Hα-emission objects in a region in Vela. See Abstr. 113.092.

Spectrophotometric studies of Ap stars. See Abstr. 115.007.

On the reality of a boundary in the H-R diagram between late-type stars with and without high temperature outer atmospheres.   See Abstr. 115.010.

Galactic and extragalactic Wolf-Rayet stars of Pop. I.   See Abstr. 115.022.

On the reality of a boundary in the H-R diagram between late-type stars with and without high temperature outer atmospheres.   See Abstr. 115.029.

Spectrophotometric studies of Am stars. See Abstr. 115.030.

Physical properties of B stars. See Abstr. 115.038.

HD 15144, a magnetic Ap star with a possible intrinsic periodic variation of the magnetic field. See Abstr. 116.005.

Magnetic structure in cool stars. IV. Rotation and Ca II H and K emission of main-sequence stars. See Abstr. 116.013.

Simultaneous spectroscopic and polarimetric observations of π Aqr.   See Abstr. 116.040.

The strongly polarized P Cygni star with infrared excess CPD—52°9243.   See Abstr. 116.041.

SS 433: how the moving lines move. See Abstr. 117.052.

A brightening of the symbiotic variable SY Muscae. See Abstr. 117.065.

Spectra of OB eclipsing binaries. See Abstr. 119.016.

Spectroscopic observations of β Lyrae. II. The Hα profile.   See Abstr. 119.118.

Mass loss, linear polarization variability, and duplicity of the luminous B2 supergiant HD 80077. See Abstr. 120.024.

A new investigation of the T Tauri star RU Lupi—I. Observation and immediate analysis. See Abstr. 121.001.

YY Orionis line profiles in the spectrum of RW Aurigae.   See Abstr. 121.010.

An usually short stable period of absorption line asymmetries and V/R variations in the spectrum of the Be star 28 CMa.   See Abstr. 122.010.

Spectral classification of select δ Scuti stars. II. See Abstr. 122.015.

On the variability of the Ap stars 53 Cam, 41 Tau, β CrB and α^2 CVn in the K(Ca II) and Hγ lines. See Abstr. 122.095.

Carbon and oxygen abundances of field RR Lyrae stars. I. Carbon abundances.   See Abstr. 122.198.

The sharp interstellar-like features in the spectrum of the nearby white dwarf G191 - B2B.
See Abstr. 126.036.

A new interstellar component in the spectrum of HD 72127 A.   See Abstr. 131.006.

The interstellar carbon abundance toward Delta Scorpii.   See Abstr. 131.007.

The physical state of the gas towards HD 93206.
See Abstr. 131.127.

High-velocity interstellar resonance lines in the UV spectra of Wolf-Rayet stars.   See Abstr. 131.182.

The outlying condensations around Eta Carinae.
See Abstr. 134.033.

The spectrum of FG Sge in 1979-1980. I.
$\lambda\lambda$ 3700-5000 Å.   See Abstr. 135.045.

X-ray luminosities of B supergiants estimated from ultraviolet resonance lines.   See Abstr. 142.040.

Spectral morphology in the open cluster Collinder 228.   See Abstr. 153.002.

Bp and Ap stars in the moving cluster Scorpius-Centaurus.   See Abstr. 153.033.

Supergiants and variable stars in the χ and h Persei cluster region.   See Abstr. 153.034.

The anticorrelation of carbon and nitrogen on the horizontal branch of 47 Tucanae.   See Abstr. 154.003.

The cyanogen distributions in NGC 3201, M55, and M71.   See Abstr. 154.004.

A comment on the metal abundance of the globular cluster M71.   See Abstr. 154.007.

Carbon, nitrogen and heavy-element abundances on the red giant branch of Omega Centauri.
See Abstr. 154.032.

The cyanogen distributions of the giants in ten globular clusters and the role of mixing and primordial abundance variations.   See Abstr. 154.034.

The metallicity of M 71.   See Abstr. 154.035.

A comparison of blue spectra of M5 and M71 giant stars.   See Abstr. 154.036.

Carbon and nitrogen abundances in giant stars of the globular clusters M3 and M13.   See Abstr. 154.037.

High-dispersion spectra of BHB stars in M 4 and NGC 6397.   See Abstr. 154.038.

C and N abundances in giant stars of the metal-poor globular cluster M 15.   See Abstr. 154.093.

Carbon stars in the Carina dwarf spheroidal galaxy.
See Abstr. 158.042.

Discovery of carbon stars in the Draco dwarf spheroidal galaxy.   See Abstr. 158.043.

OB stars in the Small Magellanic Cloud.
See Abstr. 159.026.

Erratum

114.901   Erratum: 'Infrared lines of O I and Ca II in Be stars with Paschen emission lines' [Astron. Astrophys., Vol. 103, 1 - 4 (1981)]. 'Paschen lines in Be stars. II. Study of Paschen emission lines' [Astron. Astrophys., Vol. 103, 5 - 18 (1981)].   D. Briot.
Astron. Astrophys., Vol. 105, 422 (1982). – See abstracts 30.114.134, 30.114.135.

115 Luminosities, Masses, Diameters, HR and other Diagrams

115.001   An occultation angular diameter in H-alpha light.
N. M. White, T. J. Kreidl, L. Goldberg.
Astrophys. J., Vol. 254, 670 - 675 (1982).
    The lunar occultation of 119 Tauri, spectral type M2 Ib, was observed in continuum light and in the light of the Hα absorption line. The restored strip-brightness distributions indicate that the Hα light comes from a region having at least twice the diameter of that producing the continuum light.

115.002   Relationship between colorimetric characteristics of flare stars in aggregates and the aggregate's age.
G. A. Gurzadyan.
Pis'ma Astron. Zh., Tom 8, 38 -42 (1982). In Russian. English translation in Soviet Astron. Lett., Vol. 8.
    An empirical relationship is found between the mean remoteness from the main sequence (along the $U-B$ axis) of flare star systems and Hα emission star systems on the two-

color diagram $(U-B)$ - $(B-V)$ and the aggregate's age or the age of nonstable stars included in the aggregate system. The ages of four T Tauri-type stars are estimated.

115.003   A list of stars with large expected angular diameters.
F. Ochsenbein, J. L. Halbwachs.
Astron. Astrophys.,Suppl. Ser., Vol. 47, 523 - 531 (1982).
    A calibration of the stellar angular diameter from the spectral classification was used to provide a list of 627 stars with the largest expected angular diameters within each spectral class.

115.004   Linear radii of 51 red giants in galactic clusters.
L. A. Coleman.
Astron. J., Vol. 87, 369 - 377 (1982).
    The linear radii of 51 red giants in the galactic clusters Hyades, Praesepe, M37, M67, NGC 7789, and Stock 2 have

been determined using the Barnes-Evans surface brightness relation and cluster-fitting distances. Masses were obtained for the 28 stars for which surface gravities were obtainable from published DDO photometry. The results indicate that revisions of the bolometric correction and DDO surface gravity calibrations are desirable. The mass determination appears to be a sensitive indicator of cluster membership requiring only photometry.

115.005   On the radius determination of the variable F-type supergiant BL Tel(F).   A. M. van Genderen.
Astron. Astrophys., Vol. 105, 250 - 253 (1982).

The radius of the variable F type supergiant in the eclipsing system BL Tel is derived with the aid of the Baade-Wesselink method, the visual surface brightness-colour index relation technique and with a third method based on a relation between pulsational amplitude and light amplitude. The three methods give similar results viz. $R = 200 \pm 40 R_\odot$ in accordance with the value derived from the orbital light curve. An ambiguity arose since the value for the pulsational constant $Q (0.10 \pm 0.03)$ is twice as high as those for radial pulsators. This suggests a non-radial pulsation (cf. Maeder, 1980), while the techniques mentioned above, resulting in an acceptable result, are assumed to be only applicable to radial pulsators.

115.006   The absolute magnitudes of G5–M3 stars near the giant branch.
D. Egret, P. C. Keenan, A. Heck.
Astron. Astrophys., Vol. 106, 115 - 120 (1982).

The absolute magnitudes of stars on the red giant branch (G-K-M) have been determined using both trigonometric and statistical parallaxes, from a sample of 212 stars classified in the Revised MK System (Keenan and Pitts, 1980). The results of both methods are summarized. A good agreement is found and the difference between trigonometric and statistical parallaxes is found not to be greater than $\pm 0\rlap{.}{''}002$. The computed absolute magnitudes and space motions are listed.

115.007   Spectrophotometric studies of Ap stars.
G. S. D. Babu, B. S. Shylaja.
Astrophys. Space Sci., Vol. 81, 269 - 274 (1982).

Based on the observed energy distribution curves of about a hundred Ap stars, the various relationships among their physical parameters, namely, the temperature, colour index, bolometric correction and bolometric magnitude have been studied. The bolometric corrections are independent of parallax measurements: the Ap stars as well as the normal stars follow the same sequence of bolometric corrections when related to temperature. The Ap stars appear to be slightly evolved and their position in the HR diagram indicates the hydrogen shell burning phase. The mass range of Ap stars is similar to that of normal A stars.

115.008   A study on masses of O stars in H II regions.
J. O. Woo.
J. Korean Astron. Soc., Vol. 14, 43 - 49 (1981). In Korean.

Making use of log T_{eff}-M_{bol} diagram along with Stothers' stellar evolutionary tracks the author estimates the masses of 107 O stars in H II regions. It is found that the author's estimated masses of O stars range from $20 M_\odot$ to $120 M_\odot$ and about 20% of them falls in the mass range above $60 M_\odot$ in agreement with earlier findings.

115.009   La masa estelar en asociaciones jóvenes.
J. C. Forte.
Bol. Asoc. Argentina Astron., No. 20 - 24, p. 467 (1981). Abstract.

115.010   On the reality of a boundary in the H-R diagram between late-type stars with and without high temperature outer atmospheres.
T. Simon, J. L. Linsky, R. E. Stencel.

Astrophys. J., Vol. 257, 225 - 246 (1982).

The authors test the hypothesis originally proposed by Linsky and Haisch that a boundary exists in the H-R diagram separating yellow giants ($V-R \lesssim 0.80$), which typically show evidence of $10^5 K$ plasma in their outer atmospheres, from red giants and supergiants ($V-R > 0.80$), which typically show little or no evidence of any plasma hotter than $10^4 K$. They present and discuss IUE 1150–2000 Å low-resolution spectra of 39 late-type stars of luminosity classes I-IV. The presence of the C IV $\lambda 1549$ Å emission line is taken as indicator for the existence of high-temperature ($\sim 10^5 K$) plasma in the outer atmospheres of the stars. The authors conclude that the Linsky-Haisch transition region boundary is a real phenomenon in the sense that single stars to the right of the boundary, with the exception of one hybrid K star, contain significantly less $10^5 K$ plasma compared with single stars to the left of the boundary. They speculate that a radiative instability may play an important role in the existence of such a boundary in the H-R diagram. Problems arising from the appearance of He I $\lambda 10830$ emission and absorption in stars to the right of the Linsky-Haisch boundary are discussed.

115.011   Composite colour-magnitude and colour-colour diagrams for Be stars in open clusters.
J.-C. Mermilliod.
Be stars, (see 012.030), p. 23 - 26 (1982).

115.012   Absolute magnitudes and intrinsic colours of non-supergiant Be stars.   J. R. Kozok.
Be stars, (see 012.030), p. 33 - 36 (1982).

101 normal Be stars, probable members of 56 galactic clusters and OB-associations, and more than 20 extreme Be stars in the Large Magellanic Cloud were used to derive intrinsic colours of O9–B9 (III–V)e stars. Furthermore, the correlation between the intrinsic colour $(U-B)_0$ and the absolute magnitude of non-supergiant Be stars was studied.

115.013   Luminosity classification of Be stars by Balmer line narrow band photometry.   W. Zeuge.
Be stars, (see 012.030), p. 37 - 40 (1982).

The absolute luminosity of most Be stars can be determined by using Balmer line narrow band photometry with an accuracy of about 0.4 mag. The few cases in which this method fails can be detected.

115.014   Long-term variation of Be stars on the color-magnitude diagram.   R. Hirata.
Be stars, (see 012.030), p. 41 - 44 (1982).

The ratios $\alpha = \Delta V / \Delta (B-V)$ and $\beta = \Delta (U-B)/\Delta (B-V)$ for the long-term variation of Be stars have been derived. The inclination effect on α is found. The dependence of α on the spectral type is also inferred. The signs of α and β are essentially the same.

115.015   Giant stars angular diameter measurements with the stellar interferometer at CERGA.
M. Faucherre, D. Bonneau.
Scientific importance of high angular resolution at infrared and optical wavelengths, (see 012.031), p. 247 - 251 (1981).

The interferometer at CERGA has now proved able to yield astrophysical data. The authors analyse the performance of this device and give diameters of α Ari and β Gem.

115.016   Binary stars; remarks on the determination of stellar masses.   A. Blaauw.
Scientific importance of high angular resolution at infrared and optical wavelengths, (see 012.031), p. 391 - 395 (1981).

115.017   High angular resolution observation of single and close binary stars with the CERGA interferometer.
F. Vakili, D. Bonneau, L. Koechlin.
Scientific importance of high angular resolution at infrared

and optical wavelengths, (see 012.031), p. 399 - 404 (1981).

The 35 meter baseline of the two-telescope interferometer at CERGA is operational since March 1980. Interference fringes have been observed on single and double stars with the maximum angular resolution of the instrument (1 milliarcsecs). Some preliminary astrophysical results relative to these observations are presented. Future projects for an extended base of 67 meters are discussed.

115.018 The observed upper limit to stellar luminosities in galaxies. R. M. Humphreys.
The most massive stars, (see 012.034), p. 5 - 26 (1981).

In this paper the author discusses the observed H−R diagrams for the most luminous stars and what they may tell us about the evolution of massive stars in our Milky Way and other galaxies of the Local Group. First the observational data are reviewed and then two questions are considered: (1) what is the observed upper bound of stellar luminosities and what does it mean regarding the evolution of massive stars, and (2) how the luminosities of the brightest stars may depend on the parent galaxy.

115.019 R 136a in 30 Dor: an overluminous, supermassive object. T. Schmidt-Kaler, J. V. Feitzinger.
The most massive stars, (see 012.034), p. 105 - 121 (1981).

The optical and UV observations of the peculiar central object R 136a are summarized: V = 10.77, B−V = − 0.03, $E_{B-V} = 0.46$, $A_V = 2.16$, $M_V = − 10.0$, $M_b = − 15.2$ or $L = 9 \times 10^7 L_\odot$. Speckle photometry shows that R 136a has a 2^m fainter companion, and that the diameter of R 136a1 is less than $0''.08 = 0.02$ pc. The temperature is 60000 K. The authors conclude that the object is most probably a supermassive star of $3.2 \times 10^3 M_\odot$ and 90 $R_\odot$ with an optically thick, strong stellar wind of 3600 km/s velocity, possibly rotating with v sini = 2200 km/s, a mass loss of the order of $3 \times 10^{-3} M_\odot$/yr, and an age of the order of 10^6 yrs. The alternative that R 136a1 is a supergiant accretion disk cannot be completely eliminated by the observations presently known, but seems very improbable in the light of the low X-ray intensity.

115.020 The properties of R 136, the central object in 30 Doradus. N. Panagia, E. G. Tanzi, M. Tarenghi.
The most massive stars, (see 012.034), p. 123 - 129 (1981).

The authors report preliminary infrared photometry (1.25 to 3.6 μm) of R 136, the central object of the 30 Doradus Nebula, obtained as part of a current study of the infrared emission of hot, luminous stars in the Galaxy and in the Magellanic Clouds.

115.021 The Eta Carinae event. R. Viotti, C. D. Andriesse.
The most massive stars, (see 012.034), p. 141 - 146 (1981).

η Car is a very luminous (~5 × $10^6 L_\odot$), very massive star which belongs to a unique O3 association. Since more than 100 years it is loosing mass at a rate of ~ 0.075 $M_\odot$ per year. The authors suggest that it is a massive star in the post main sequence evolutionary stage and that similar objects should be observed in massive H II regions of external galaxies.

115.022 Galactic and extragalactic Wolf-Rayet stars of Pop. I. K. A. van der Hucht.
The most massive stars, (see 012.034), p. 157 - 172 (1981).

The author reviews the occurrence of Pop. I Wolf-Rayet stars, their intrinsic and physical parameters, their masses and mass loss, the ring nebulae around some of them, their galactic distribution, and subsequent consequences for their evolution. Trends of M_V, H/He ratio, binary mass ratio, ring nebula class, and galactic distribution, with subtype are noted. It appears that WR stars evolve from 'late' to 'early' subtypes.

115.023 M supergiants and the distance to M101. R. M. Humphreys, S. E. Strom.
The most massive stars, (see 012.034), p. 245 - 249 (1981).

Results of a search for bright M supergiants in the spiral galaxy M101 are reported. These stars are used as standard candles in establishing an accurate distance to the galaxy. A distance modulus $(V_0 - M_V) = 28.6^{+0.2}_{-0.3}$ mag, corresponding to a distance of 5.25 Mpc, is derived.

115.024 The status of the corona/wind boundary line in the cool star region of the HR diagram.
B. M. Haisch, T. Simon.
Bull. American Astron. Soc., Vol. 13, 784 (1981). − Abstract.

115.025 Speckle interferometry of the WC9 star Roberts 80 at 3.8 and 4.8 microns.
R. D. Wolstencroft, H. M. Dyck, T. Simon.
Bull. American Astron. Soc., Vol. 13, 808 (1981). − Abstract.

115.026 Occultation angular diameters of α Tauri by least squares and deconvolution.
N. M. White, T. J. Kreidl.
Bull. American Astron. Soc., Vol. 13, 830 (1981). − Abstract.

115.027 The initial mass function for O-type stars.
C. D. Garmany.
Bull. American Astron. Soc., Vol. 13, 841 (1981). − Abstract.

115.028 Can the relative ages of Orion population objects be derived from their emission line strengths?
W. B. Weaver.
Bull. American Astron. Soc., Vol. 13, 857 (1981). − Abstract.

115.029 On the reality of a boundary in the H-R diagram between late-type stars with and without high temperature outer atmospheres.
T. Simon, J. L. Linsky, R. E. Stencel.
Bull. American Astron. Soc., Vol. 13, 885 (1981). − Abstract.

115.030 Spectrophotometric studies of Am stars.
G. S. D. Babu, B. S. Shylaja.
Astrophys. Space Sci., Vol. 83, 367 - 371 (1982).

The relationships among the various physical parameters of 25 Am stars − namely, the effective temperatures, radii and bolometric magnitudes − have been studied. Their effective temperatures are in the range of 7200 K to 9700 K; the radii, 1.5 $R_\odot$ to 2.5 $R_\odot$; the bolometric magnitudes, 0.75 mag to 2.25 mag; and the masses, 1.5 $M_\odot$ to 2.25 $M_\odot$. The Am stars appear redder than their normal counterparts, the blanketing in the blue and UV regions being the major cause. They are located in the neighbourhood of the upper edge of the zero-age Main Sequence band.

115.031 Speckle observations of R136a.
J. Meaburn, J. C. Hebden, B. L. Morgan, H. Vine.
Mon. Not. R. Astron. Soc., Vol. 200, 1P - 5P (1982).

It has recently been suggested that R136a is a single, supermassive star (2500 $M_\odot$) with a surface temperature of 60000 K. Speckle observations have now been made with the 3.9-m Anglo-Australian telescope which definitely puts an upper limit of $\cong 0.02$ arcsec to its diameter and which possibly just resolve it as a disc of $\cong 0.02$ arcsec ($\cong 5 \times 10^{-3}$ pc).

115.032 Angular diameters by the lunar occultation technique. IV. α Leo and the Cepheid ζ Gem.
S. T. Ridgway, G. H. Jacoby, R. R. Joyce, M. J. Siegel, D. C. Wells.
Astron. J., Vol. 87, 680 - 684 (1982).

The 18 June 1980 occultation of α Leo was recorded by two telescopes. The uniform disk diameters obtained from measurements by the 1.3- and 4-m telescopes were 1.46 ± 0.37 and 1.42 ± 0.10 milliarcsec, respectively. A measurement of the Cepheid ζ Gem on 10 April 1981 at phase 0.5961 yielded a uniform disk diameter of 1.81 ± 0.31 milliarcsec and T_{eff} = 5140 ± 400 K.

115.033 Angular diameters by the lunar occultation technique. V. 26 late-type stars.
S. T. Ridgway, G. H. Jacoby, R. R. Joyce, M. J. Siegel, D. C. Wells.
Astron. J., Vol. 87, 808 - 817 (1982).
   Near-infrared occultation observations have yielded 37 new angular diameter measurements for 26 late-type stars. The list includes three carbon stars and five giants in the range M4 - M8. A multichannel photometer has been tested successfully on five events.

115.034 Lunar occultation stellar angular diameter measurements. III.
W. I. Beavers, R. R. Cadmus, Jr., J. J. Eitter.
Astron. J., Vol. 87, 818 - 825 (1982).
   Additional stellar angular diameter measurements obtained by two-color photoelectric lunar occultation observations are reported. Within this set of 21 candidate events, nine yield measurable angular diameters, five are interpreted to be of unresolved stars, and seven proved to be too noisy for reliable analysis.

115.035 Absolute magnitudes and other basic parameters of O and B stars.   L. Divan, J. Zorec.
Scientific aspects of the Hipparcos space astrometry mission, (see 012.041), p. 101 - 104 (1982).
   The two parameters λ_1 and D of the BCD system, already calibrated in absolute magnitudes by Chalonge and Divan have been calibrated in effective temperature and bolometric magnitudes for O and B stars. Verification and/or improvements awaited from Hipparcos parallaxes are discussed.

115.036 Study of absolute magnitudes of B and Be stars.   D. Briot, A. M. Hubert-Delplace.
Scientific aspects of the Hipparcos space astrometry mission, (see 012.041), p. 107 - 110 (1982).
   The authors review some of the difficulties encountered in the determination of absolute magnitudes of Be stars. Some relevant physical hypotheses are briefly examined. Finally they describe what is expected from Hipparcos and the related programs which are desirable.

115.037 Observation of selected nearby cool carbon stars.   J. Bergeat.
Scientific aspects of the Hipparcos space astrometry mission, (see 012.041), p. 121 - 122 (1982).
   The essential aim of this program is to determine fundamental data such as luminosities and radii for cool carbon stars (CCS). Their loci in the theoretical HR diagram could then be confronted to predictions for evolutionary tracks, mixing or "dredge-up" lines, and instability zones.

115.038 Physical properties of B stars.
E. Schatzman, A. Baglin.
Pulsating B stars, (see 012.042), p. 21 - 44 (1981).

115.039 Fundamental data for Wolf-Rayet stars.
I. Lundström, B. Stenholm.
Rep. Obs. Lund, No. 18, (see 012.044), p. 101 - 104 (1982).

Landolt-Börnstein.   See Abstr. 003.064.

On the calibration of luminosity criteria: numerical experiments.   See Abstr. 031.595.

X-ray heating of the quiescent chromospheres of dMe stars.   See Abstr. 064.007.

Effect of spots on a star's radius and luminosity. See Abstr. 064.038.

Structure and evolution of massive stars. See Abstr. 065.044.

Wolf-Rayet stars as a stage of the evolution of massive stars.   See Abstr. 065.045.

Astrophysical parameters for Ap and Am stars with Hipparcos.   See Abstr. 111.021.

Hipparcos and the determination of the helium content of some low-mass non-evolved halo and disk stars of the solar neighbourhood, inferred from the fine structure of the H-R diagram.   See Abstr. 111.024.

Absolute radii of single and multiple stars from the CADARS *(Catalogue of Apparent and Absolute Radii of Stars)*. See Abstr. 111.025.

Stellar luminosity functions in the *R, I, J,* and *K* bands obtained by transformation from the visual band. See Abstr. 113.014.

Effective temperatures, and radii of luminous O and B stars: a test for the accuracy of the model atmospheres. See Abstr. 114.009.

The O3 stars.   See Abstr. 114.023.

International Ultraviolet Explorer spectroscopy of hot stars in the LMC and SMC: the SMC extinction law, stellar flux distributions, and details of the stellar winds. See Abstr. 114.032.

NLTE analysis of massive O-stars. See Abstr. 114.168.

Aldebaran. Eine Beschreibung. See Abstr. 114.219.

On the masses of Wolf-Rayet components in close binary systems.   See Abstr. 117.115.

Rediscussion of eclipsing binaries. XIII. DI Herculis, a B-type system with an eccentric orbit. See Abstr. 119.003.

Mass loss, linear polarization variability, and duplicity of the luminous B2 supergiant HD 80077. See Abstr. 120.024.

Spectral classification of select δ Scuti stars. II. See Abstr. 122.015.

Radii and masses of classical cepheids. See Abstr. 122.162.

The luminosity function of very low mass stars. See Abstr. 126.005.

An unbiased survey of field star X-ray emission. See Abstr. 142.014.

Comparisons of the HR diagrams of the youngest clusters in the Galaxy, the LMC and SMC. Evidence for a large MS widening.   See Abstr. 153.008.

Structure and distance modulus of the Small Magellanic Cloud from blue supergiants. See Abstr. 159.034.

116 Magnetic Fields, Polarization, Figure, Rotation, Radio Radiation

116.001 Radio observations of early-type emission-line stars and related objects. C. R. Purton, P. A. Feldman, K. A. Marsh, D. A. Allen, A. E. Wright.
Mon. Not. R. Astron. Soc., Vol. 198, 321 - 338, Microfiche MN 198/1 (1982).

Radio observations are presented of 325 early-type emission-line stars in both the northern and southern hemispheres. Forty-four of these have been detected as radio sources. The radio emission appears to be thermal in origin, and 13 of the detected stars have radio spectra of the approximate form $S \propto \nu$, indicating the presence of a circumstellar envelope formed by prolonged mass outflow.

116.002 The origin of stellar angular momentum.
S. C. Wolff, S. Edwards, G. W. Preston.
Astrophys. J., Vol. 252, 322 - 336 (1982).

Rotational velocities have been measured for 306 early B-type stars. The distribution of apparent rotational velocities peaks at low values of $\nu \sin i$ and decreases monotonically with increasing $\nu \sin i$. Physical mechanisms that produce slow rotation though a substantial loss of angular momentum after the star-formation process is complete are discussed. A possibility is that stars form from material with initially low angular momentum and that the observed distribution of rotational velocities is the consequence of gravitational encounters within proto-cluster clouds.

116.003 The bimodal distribution of rotational velocities of late B-type stars in galactic clusters.
B. N. G. Guthrie.
Mon. Not. R. Astron. Soc., Vol. 198, 795 - 810 (1982).

The overall distribution of projected rotational velocities $\nu \sin i$ for 195 late B-type stars in 13 clusters at low galactic latitudes is clearly bimodal with a scarcity of values between 80 and 160 km s^{-1}. The sharply peaked distribution for the rich cluster NGC 2516 implies some degree of alignment of the axes of its rapidly rotating members, and there is probably a general tendency for the axes of rotation of cluster stars to be roughly aligned perpendicular to the galactic plane. If so, the initial values of ν were distributed around 230 km s^{-1}. The distribution of $\nu \sin i$ for field stars is not bimodal and does not depend on galactic latitude; the data are consistent with their axes of rotation being randomly orientated and their initial distribution of ν being Maxwellian. For both cluster and field stars the proportion of slow rotators increases with age; this is partly due to tidal braking in binaries and rapid braking of magnetic stars through interaction with interstellar clouds.

116.004 Lower atmosphere changes inferred from variations in Betelgeuse's linear polarization.
D. P. Hayes.
Publ. Astron. Soc. Pacific, Vol. 93, 752 - 755 (1981/82).

Variations in Betelgeuse's linear polarization were intensively monitored over the nearly eight-month interval which comprised the 1980–81 observing season. As was the case in the preceding observing season, significant ordered polarization changes were detected. These variations are explained as being due to the waxing and waning, and motion of large-scale elements of the star's lower atmosphere, and are likely to be a manifestation of nonradial pulsations.

116.005 HD 15144, a magnetic Ap star with a possible intrinsic periodic variation of the magnetic field.
W. K. Bonsack.
Publ. Astron. Soc. Pacific, Vol. 93, 756 - 764 (1981/82).

HD 15144 (HR 710) is a single-line spectroscopic binary Ap star with a period of 3^{d}00, which has a variable magnetic field but no obvious variations in line strengths or light. New spectrograms are combined with previous observations by Babcock to derive a probable magnetic period of 15^{d}9. An apparent rotational velocity of 7.7 km s^{-1} is obtained and combined with elements of the orbit to estimate the mass of the secondary. The most plausible model for the system appears to be one in which the rotation period of the Ap star is 3^{d}00, the system is seen nearly pole-on, the secondary is a subluminous star of approximately 1.2 solar masses, and the magnetic variation is due to intrinsic changes of the field strength or geometry. An alternative model is also discussed.

116.006 On the coronae of rapidly rotating stars. III. An improved coronal rotation-activity relation in late type dwarfs. F. M. Walter.
Astrophys. J., Vol. 253, 745 - 751 (1982).

The author examines the rotation-activity relation in G dwarfs. Previous authors have claimed that X-ray luminosity varies as some power of the rotational velocity. However, a single power law does not provide an acceptable fit to all the data. The data are well described by either an exponential dependence of L_x/L_{bol} upon Ω, or by a power law with a break in slope near a 12 day period. This break occurs at the same rotational period as the bifurcation in Vaughan and Preston's chromospheric study of G dwarfs. The author speculates that the two breaks are related, and may be due to a change in the mode of the stellar dynamo at a rotation period of about 12 days corresponding to an age of 10^9 years for single G dwarfs.

116.007 Detection of microwave emission from both components of the red dwarf binary EQ Pegasi.
K. Topka, K. A. Marsh.
Astrophys. J., Vol. 254, 641 - 645 (1982).

The authors report on the detection at 4.9 GHz of the late main sequence binary EQ Pegasi (dM3.5e + dM4.5e) with the VLA. Both components were detected, at flux levels of 0.69 mJy and 0.4 mJy, respectively. Thermal gyroresonance emission from the quiescent coronae of these stars appears to explain the observations, provided coronal magnetic fields in excess of 300 gauss exist over a region that has a length scale of at least twice the radii of these stars. Support for this model is provided by the unlikelihood of both stars flaring simultaneously, and by the fact that the emission was confined to each star within the observational uncertainty of a few AU.

116.008 Search for strong magnetic field stars.
Yu. V. Glagolevskij, V. D. Bychkov, I. Kh. Iliev, I. I. Romanyuk, N. M. Chunakova.
Pis'ma Astron. Zh., Tom 8, 26 - 29 (1982). In Russian. English translation in Soviet Astron. Lett., Vol. 8.

The longitudinal magnetic field of 9 peculiar stars possessing an extreme supposed surface field has been measured from Zeeman spectrograms. A strong magnetic field has been detected in 4 investigated stars.

116.009 Is the magnetic field also responsible for the phenomenon SS 433?
E. Woyk (Chvojkova).
Astron. Zh., Tom 59, 182 - 183 (1982). In Russian. English translation in Soviet Astron., Vol. 26, No. 1.

The role of a magnetic field in the model of the object SS 433 is outlined. A high-energy plasma, strongly compressed round a stellar core due to thermonuclear reactions on one side and by strong magnetic field lines on the other side, can at best slightly grow in dimensions, but an outburst from the core can occur merely in the direction of the magnetic axis. The greater the field, the narrower are the two polar jets. The shock brought about by the first polar burst brings the whole frozen-in plasma to oscillate with radius, and the corresponding pressure variation contributes to the observed velocity variation of the two polar plasma jets.

116.010 Periodic radio emission from LS I +61°303.
A. R. Taylor, P. C. Gregory.
Astrophys. J., Vol. 255, 210 - 216 (1982).

The authors report the discovery of periodic radio emission from the highly variable radio star LS I +61°303. Based on an analysis of 144 flux density measurements, from 1977 August to 1981 March, they derive a period of 26.52 ± 0.04 days. Comparison of the radio light curves at 5.0 and 10.5 GHz rules out occultation as explaining the periodic variation. The implications of the radio emission are discussed in the context of a binary model, which also accounts for the X-ray and possible γ-ray emission associated with the source.

116.011 Radio emission from AM Herculis-type binaries.
G. Chanmugam, G. A. Dulk.
Astrophys. J., Lett., Vol. 255, L107 - L110 (1982).

The authors report the discovery of radio emission from the magnetic cataclysmic variable AM Her. The VLA was used to search for 4.9 GHz radiation from AM Her and the similar binary EF Eri. AM Her was detected with a flux density of 0.67 ± 0.052 mJy; no circular polarization was detected. EF Eri was not detected, the upper limit (3 σ) to its flux density being about 0.2 mJy. The data for AM Her are consistent with a model in which the radiation is due to gyrosynchrotron emission from electrons, of a few hundred keV of energy, which are trapped in the magnetosphere of the white dwarf.

116.012 The surface magnetic field structure of Ap stars.
M. J. Stift.
Irish Astron. J., Vol. 14, (see 012.003), 216 - 219 (1980).

116.013 Magnetic structure in cool stars. IV. Rotation and
Ca II H and K emission of main-sequence stars.
F. Middelkoop.
Astron. Astrophys., Vol. 107, 31 - 35 (1982).

The measured Ca II H and K emission indices of stars are converted to surface fluxes $(F_H + F_K)$. This conversion largely eliminates the dependence on spectral type in the relation between the uncorrected Ca II H and K emission and the rotational velocity of a star which has been found by Vaughan et al. This relation holds for single main-sequence stars as well as for short-period binaries indicating that the enhanced emission in short-period binaries is a result of rapid rotation enforced by tidal coupling. A plot of surface fluxes against v sin i values suggests a color-dependent discontinuity in the relation. This discontinuity may explain the two branches in $(F_H + F_K)$ for $(B-V) < 1.00$ among main-sequence stars. From the relation between $(F_H + F_K)$ and rotational velocity it is deduced that the average rotational velocity of late-type main-sequence Hyades decreases with decreasing effective temperature.

116.014 Radio emission from young stars.
M. Felli, G. F. Gahm, R. H. Harten, R. Liseau,
N. Panagia.
Astron. Astrophys., Vol. 107, 354 - 361 (1982).

Observations at 6 cm of 8 young stars with the Westerbork Synthesis Radio Telescope are presented. Four stars were detected at 3–4σ signal-to-noise levels. It is demonstrated that the radio emission cannot be explained as due to nonthermal emission from solar-like flare events. The radio emission can be produced by thermal processes in an ionized circumstellar envelope in a state of expansion or accretion. The implications of mass loss or mass accretion are investigated and several arguments in favour of mass loss are found.

116.015 Radio observations of pre-main-sequence stars:
results and interpretation.
C. Bertout, C. Thum.
Astron. Astrophys., Vol. 107, 368 - 375 (1982).

The authors report detection of nine pre-main-sequence stars at either 15 or 23 GHz. They show that the radio spectrum not only allows derivation of the ionized envelope properties and the mass-flow rate, but is also useful for understanding the stars' ultraviolet, optical and near-infrared continua. A detailed interpretation is presented for T Tauri and DG Tauri. In particular, the authors conclude that T Tauri's stellar wind probably originates in a very hot $(T_e \gtrsim 10^6 \, \text{K})$ region.

116.016 On coherent properties of rotating star radiation.
A. V. Mandzhos, S. V. Khmil'.
Astrofizika, Tom 17, 141 - 153 (1981). In Russian. English translation in Astrophysics, Vol. 17, No. 1.

The complex coherence degree at a spectral line of rotating star radiation is calculated. The possibility of determining the star's rotational velocities and axis orientations by modern stellar interferometers is established.

116.017 β Lyrae — a magnetic binary star.
M. Yu. Skul'skij.
Pis'ma Astron. Zh., Tom 8, 238 - 241 (1982). In Russian. English translation in Soviet Astron. Lett., Vol. 8.

A magnetic field of mean effective intensity $H_e = -1.5$ kGs has been discovered in the atmosphere of the bright component of β Lyrae. The magnetic field has complex spatial structure. During the orbital period the magnetic field intensity varies by a factor of two.

116.018 Polarimetric observations of polar AN UMa.
Yu. S. Efimov, N. M. Shakhovskoj.
Izv. Krymskoj Astrofiz. Obs., Tom 64, 55 - 66 (1981). In Russian. English translation in Bull. Crimean Astrophys. Obs., Vol. 64.

Polarimetric observations were made of the short-period variable AN UMa and phased using the revised ephemeris of the authors for 1977 - 1978. The inclination of the rotation axis measured with respect to the line of sight and the angle between the magnetic axis and the axis of rotation are derived. The observed variations of the peak linear polarization appear to be caused by variations of non-polarized light.

116.019 Surface structure of Ap stars.   V. L. Khokhlova.
Soobshch. Spets. Astrofiz. Obs., Vyp. 32,
(see 012.029), p. 63 (1981).

116.020 Polarización en sistemas binarios.
H. Luna, H. Levato.
Bol. Asoc. Argentina Astron., No. 20 - 24, p. 474 (1981).
Abstract.

116.021 The interstellar component of linear polarization of
the radiation of 12 red supergiants.
G. V. Abramyan.
Soobshch. Byurakan Obs., Vyp. 53, p. 40 - 76 (1982). In Russian.

The results of the author's electropolarimetric and UBV electrophotometric observations as well as the published data of other authors are used to determine the interstellar polarization component of the radiation of 12 red supergiants.

116.022 Observaciones polarimétricas de HD 162679.
H. Luna.
Bol. Asoc. Argentina Astron., No. 26, p. 76 (1981). – Abstract.

116.023 Some problems in the theory of magnetic stars.
L. A. Pustil'nik.
Soobshch. Spets. Astrofiz. Obs., Vyp. 32, (see 012.029),
p. 9 - 10 (1981).

116.024 Processes in stellar magnetospheres.
A. Z. Dolginov.
Soobshch. Spets. Astrofiz. Obs., Vyp. 32, (see 012.029),
p. 11 - 12 (1981).

116.025 On the influence of differential rotation of a star on the symmetry of its magnetic field.
K.- H. Rädler.
Soobshch. Spets. Astrofiz. Obs., Vyp. 32, (see 012.029),
p. 12 (1981).

116.026 A new method of determination of magnetic fields of hot stars. Yu. N. Gnedin, N. A. Silant'ev.
Soobshch. Spets. Astrofiz. Obs., Vyp. 32, (see 012.029),
p. 13 (1981).

116.027 On the magnetic field in the variable star RR Lyrae.
Yu. S. Romanov, S. N. Udovichenko.
Soobshch. Spets. Astrofiz. Obs., Vyp. 32, (see 012.029),
p. 15 - 16 (1981).

116.028 Remarks concerning the equator-symmetric rotator model. L. Oetken.
Soobshch. Spets. Astrofiz. Obs., Vyp. 32, (see 012.029),
p. 18 (1981).

116.029 Does γ Equ have a magnetic cycle like the sun?
F. Krause, G. Scholz.
Soobshch. Spets. Astrofiz. Obs., Vyp. 32, (see 012.029),
p. 19 (1981).

116.030 Numerical modelling of the spectrophotometric investigation process of magnetic stars.
V. S. Lebedev.
Soobshch. Spets. Astrofiz. Obs., Vyp. 32, (see 012.029),
p. 21 (1981).

116.031 The magnetic field of the peculiar star HD 119213.
Yu. V. Glagolevskij, I. I. Romanyuk, K. I. Kozlova,
V. D. Bychkov, V. S. Lebedev.
Soobshch. Spets. Astrofiz. Obs., Vyp. 32, (see 012.029),
p. 26 (1981).

116.032 Magnetic field of ε UMa.
Yu. V. Glagolevskij, V. D. Bychkov, I. D. Najdenov,
I. I. Romanyuk, G. A. Chuntonov, V. G. Shtol', I. Kh. Iliev.
Soobshch. Spets. Astrofiz. Obs., Vyp. 32, (see 012.029),
p. 27 - 28 (1981).

116.033 Some results of the investigation of the Mn-Hg-star 33 Gem.I.
N. M. Chunakova, V. D. Bychkov, Yu. V. Glagolevskij.
Soobshch. Spets. Astrofiz. Obs., Vyp. 32, (see 012.029),
p. 28 - 29 (1981).

116.034 Magnetic field measurement of the peculiar star ω Her (HD 148112).
I. A. Aslanov, N. A. Salomatina.
Soobshch. Spets. Astrofiz. Obs., Vyp. 32, (see 012.029),
p. 29 (1981).

116.035 Radial velocity and magnetic field measurements of the A-type supergiant ν Cep (HD 207260).
G. Scholz, E. Gerth.
Soobshch. Spets. Astrofiz. Obs., Vyp. 32, (see 012.029),
p. 30 (1981).

116.036 New results of an investigation of the magnetic star 52 Her. E. Gerth.
Soobshch. Spets. Astrofiz. Obs., Vyp. 32, (see 012.029),
p. 31 (1981).

116.037 β Lyrae is a magnetic star.
M. Yu. Skul'skij.
Soobshch. Spets. Astrofiz. Obs., Vyp. 32, (see 012.029),
p. 31 - 32 (1981).

116.038 On a possible interpretation of short- and long-period variations of the effective magnetic field strength of magnetic stars by gyroscopic motions. E. Gerth.
Soobshch. Spets. Astrofiz. Obs., Vyp. 32, (see 012.029),
p. 32 (1981).

116.039 Polarimetry and physics of Be star envelopes.
G. V. Coyne, I. S. McLean.
Be stars, (see 012.030), p. 77 - 93 (1982).
A review of the most recent developments in polarization studies of Be stars is presented. New polarization techniques for high-resolution spectropolarimetry and for near infrared polarimetry are described and a wide range of new observations are discussed. The physical significance of the observational material is discussed in the light of recent theoretical models. Emphasis is placed on the physical and geometrical parameters of Be star envelopes.

116.040 Simultaneous spectroscopic and polarimetric observations of π Aqr. K. Metz.
Be stars, (see 012.030), p. 95 - 99 (1982).

116.041 The strongly polarized P Cygni star with infrared excess CPD−52°9243. J. P. Swings.
Be stars, (see 012.030), p. 101 - 102 (1982). − Abstract.

116.042 Polarization in peculiar emission-line objects.
R. Barbier, J. P. Swings.
Be stars, (see 012.030), p. 103 - 106 (1982).
The authors report on visual polarimetric measurements of B[e] and related stars, and on an unsuccessful search for correlating the observational data with characteristics of the objects or of their surrounding dust shells.

116.043 Optical spectroscopy of HD102567 (4U1145-61).
E. J. Pacheco, C. Chevalier, S. A. Ilovaisky.
Be stars, (see 012.030), p. 151 - 154 (1982).
Analysis of optical spectra of HD102567 (HEN 715), the optical counterpart of the X-ray source 4U1145-61, is presented. Estimates of the star's rotation velocity, inclination angle, distance and envelope characteristics are given. Some consequences of the possible existence of a 190^d orbital period for this Be/X-ray source are discussed.

116.044 Rotation, expansion and duplicity of Be stars.
P. Harmanec.
Be stars, (see 012.030), p. 279 - 297 (1982).

116.045 Determination of the inclination of rotational axes and rotational velocity from the line profiles of rotating stars. M. Ruusalepp.
Be stars, (see 012.030), p. 303 - 310 (1982).
For a grid of rotating hot stars the computed line widths at the half maximum depth for the lines HeI λ 4471 and MgII λ 4481 are given. Based on the correspondent observational data the inclination of the rotational axes i and the reduced rotational velocity w have been found for 19 B and Be stars.

116.046 Eclipsing AM Herculis-type magnetic binary.
IAU Circ., No. 3680 (1982).

116.047 AM Herculis and AN Ursae Majoris.
IAU Circ., No. 3698 (1982).

116.048   **On standard polarized stars.**
J.-C. Hsu, M. Breger.
Bull. American Astron. Soc., Vol. 13, 813 (1981). − Abstract.

116.049   **Detection of magnetic fields in late-type stars with the KPNO McMath telescope and MAMA detector system.**   J. G. Timothy, C. L. Joseph, J. L. Linsky.
Bull. American Astron. Soc., Vol. 13, 828 (1981). − Abstract.

116.050   **Rotational velocities of Herbig Ae/Be stars.**
R. Davis, K. Strom, S. Strom.
Bull. American Astron. Soc., Vol. 13, 856 (1981). − Abstract.

116.051   **High resolution maps of IRC + 10216 in the 89 GHz (J = 1 - 0) line of HCN.**
W. J. Welch, B. Chapman, J. H. Bieging.
Bull. American Astron. Soc., Vol. 13, 865 (1981). − Abstract.

116.052   **On an estimate of the dynamo-generated magnetic fields in late-type stars.**
B. R. Durney, R. D. Robinson.
Bull. American Astron. Soc., Vol. 13, 906 (1981). − Abstract.

116.053   **Spectropolarimetry of OB stars.**
O. L. Lupie, K. H. Nordsieck.
Bull. American Astron. Soc., Vol. 13, 925 (1981). − Abstract.

116.054   **Magnetic balloons II. A very massive star.**
E. N. Hubbard, D. S. P. Dearborn.
Prepr. Steward Obs., No. 358, 19 pp. (1982).

116.055   **The magnetic field surface structure of HD 215441.**
M. Goossens, W. Van Assche, R. Demoitié,
L. Gadeyne.
Astrophys. Space Sci., Vol. 83, 213 - 224 (1982).
The oblique rotator model with axisymmetric magnetic fields containing a toroidal component is adopted for the description of the magnetic variations observed in Ap stars. The toroidal component is taken as a non-linear function of the stream function of the poloidal component. A scheme is presented for the determination from H_e and H_s observations of the parameters that describe the surface distribution of the magnetic field components. This scheme is succesfully applied to the magnetic observations of HD 215441.

116.056   **Observational evidence against differential rotation in F stars.**   D. F. Gray.
Astrophys. J., Vol. 258, 201 - 208 (1982).
Seven F stars on or near the main sequence are investigated for differential rotation by looking at the Fourier transforms of their line profiles. It is found that the amplitudes and widths of the sidelobes are completely consistent with rigid body rotation and macroturbulence alone. This evidence indicates a differential effect substantially smaller than the solar case, in contradiction to some theoretical values which are several times larger.

116.057   **A statistical study of rotational velocities of the stars.**
I. Fukuda.
Publ. Astron. Soc. Pacific, Vol. 94, 271 - 284 (1982).
Rotational velocities of the stars are studied statistically on the basis of the data listed in the new catalog of stellar rotational velocities. First, an overall feature of mean rotational velocities of various groups of stars is shown. The results are essentially not different from those in earlier works. Next, the characteristic range of equatorial rotational velocities for each group is discussed using a simple model distribution function. The analysis shows that the rotational velocities of the normal stars are distributed over a relatively narrow range, so that the intrinsically slow rotators whose rotational velocities are roughly less than 100 km s^{-1} are seldom found among normal

main-sequence stars and that early Be stars share a fairly wide range of rotational velocities with normal B stars.

116.058   **Observations optiques de la radiosource β Persei.**
S. Débarbat.
Abh. Hamburger Sternw., Band 10, 167 (1982). − Abstract.

116.059   **Cinq premieres campagnes d'observations systema- tiques de β Persei a l'astrolabe de l'Observatoire de Paris.**   S. Débarbat, F. Chollet, L. B. F. Clauzet, M. Feissel, S. K. Lam, P. Texier, M. Tomas, J. Vanhollebeke.
Abh. Hamburger Sternw., Band 10, 168 (1982). − Abstract.

116.060   **Polarimetry and photometry of two RS CVn stars.**
M. S. Barbour, J. C. Kemp.
Smithsonian Astrophys. Obs., Spec. Rep. 392, (see 012.045), p. 255 - 259 (1982).

116.061   **Polarimetric observations of HD 199178 − an FK Comae type star.**   V. Piirola, O. Vilhu.
Astron. Astrophys., Vol. 110, 351 - 354 (1982).
Linear polarization observations *(UBVRI)* of HD 199178, an FK Com type star, a rapidly rotating single late type giant, are reported. The polarization shows roughly a λ^{-4} dependence and is variable with a periodicity of 4d. Magnetic active regions (spots or faculae), asymmetrically distributed along the longitude, are the probable cause of the polarization (Rayleigh scattering and/or saturation in the transverse Zeeman effect) varying with the stellar rotation period.

On an estimate of the dynamo-generated magnetic fields in late-type stars.   See Abstr. 062.012.

On the generation of magnetic fields in late-type stars: a local time-dependent dynamo model.
See Abstr. 062.055.

Turbulence in regions where the magnetic field is perpendicular to the boundary.   See Abstr. 062.078.

On an estimate of the dynamo-generated magnetic fields in late-type stars.   See Abstr. 062.083.

Are Ap stars magnetic balloons?
See Abstr. 064.010.

Stellar atmospheres with a magnetic field.
See Abstr. 064.056.

The relation between coronal, chromospheric and magnetic activity: a case study.   See Abstr. 064.071.

Strong spherical magnetogasdynamic shock waves in rotating stellar atmospheres.   See Abstr. 064.076.

The evolution of rapidly rotating B/Be stars.
See Abstr. 065.039.

Expected broadband linear polarization from cool stars with magnetic structures.   See Abstr. 065.066.

Solar calibration of stellar rotation tracers.
See Abstr. 080.065.

A spectrographic study of the shell star EW Lac.
See Abstr. 112.029.

Rotational velocity versus mass loss in Be stars.
See Abstr. 112.030.

Mass loss rates of Wolf-Rayet stars from radio continuum observations.   See Abstr. 112.044.

Resolution of the radio emission from P Cygni's wind.   See Abstr. 112.046.

Magnetic structure in cool stars. V. Chromospheric and transition-region emission from giants. See Abstr. 112.057.

Relations between the He-peculiar stars and the Ap stars.   See Abstr. 114.124.

Some results of an investigation of the Mn-Hg-star 33 Gem. I.   See Abstr. 114.205.

Reappraisal of the polarimetric data for seven binaries.   See Abstr. 117.048.

Radio and optical observations of the R Aquarii jet. See Abstr. 117.123.

Duplicity in the solar neighborhood. II. Spectroscopic orbits for four bright stars HD 21018, HD 30021, HD 158837, and HD 190658.   See Abstr. 120.005.

Evidence for starspots on DK Draconis (HR 4665). See Abstr. 120.022.

Mass loss, linear polarization variability, and duplicity of the luminous B2 supergiant HD 80077. See Abstr. 120.024.

T Tau-type stars and magnetic stars. See Abstr. 121.024.

The line profile variations of Spica. See Abstr. 122.081.

On the rapid spectrum variability of the magnetic star 53 Cam.   See Abstr. 122.094.

A rotation study of BY Dra stars. See Abstr. 122.164.

VLA observations of quiescent and flare emission from dMe stars.   See Abstr. 122.175.

Interstellar polarization in the immediate solar neighbourhood.   See Abstr. 131.016.

Einstein observations of BY Draconis variables and RS CVn binaries.   See Abstr. 142.038.

Radio observations of coronal X-ray sources. See Abstr. 142.097.

The angular momentum history of the Hyades K giants.   See Abstr. 153.003.

117 Close Binaries (Observations, Theory)

117.001   Gravitational spin precession in binary systems.
A. J. S. Hamilton, C. L. Sarazin.
Mon. Not. R. Astron. Soc., Vol. 198, 59 - 70 (1982).

The paper examines the precessional behaviour of spins in a binary, modelled as a system of three non-dissipative coupled angular momenta, namely the orbital angular momentum and the two spin angular momenta. It is found that the precessional behaviour consists of a periodic nutation super-imposed on an overall precession of the system about the axis of total angular momentum. Attention is focused on precession in close binaries containing compact objects, including the binary pulsar PSR 1913 + 16 and a binary model of SS 433; it is shown that nutation is such systems will generally be too small to be observable.

117.002   Hot wind from γ^2 Velorum observed in the ultra-violet carbon lines.
Y. Kondo, W. A. Feibelman, D. K. West.
Astrophys. J., Vol. 252, 208 - 213 (1982).

The close binary system γ^2 Vel, consisting of a WC8 and an O9 I star, has been observed in the ultraviolet at various phases. Preliminary analysis shows that the temperature of the stellar wind from this binary increases outward. This suggests that an additional source of energy other than radiation pressure may exist for the wind. The mass flow occurs in all directions from the binary. However, there is evidence of an increase in the mass flow through the third Lagrangian point, although from the extant data the authors are unable to decide whether or not a similar increase occurs through the second Lagrangian point. There is also evidence of a mass surge, which probably is independent of orbital phase.

117.003   2A 0311 − 227 (EF Eridani): radial velocities of two emission line components.
P. Young, D. P. Schneider, W. L. W. Sargent, A. Boksenberg.
Astrophys. J., Vol. 252, 269 - 276, plate 9 (1982).

The magnetic binary 2A 0311 − 227 has been observed for over three orbits on each of two nights with 1 Å spectral resolution and 120 s temporal resolution. The radial velocity variations of the strong Balmer, He I, and He II emission lines in the region $\lambda\lambda 4110$-5050 were studied. The emission lines are distinctly from cycle to cycle. For about 50% of the time the lines consist of a single broad component (sometimes with hints of substructure) phased with the 81 minute orbital period. At other times a second "sharp" emission component appears in all lines. Other components were seen transiently. The authors have not been able to explain the multiple emission components. They suspect that they arise from magneti-cally funneled gas streams.

117.004   Further photometric observations of 2A 0311−227.
G. Williams, W. A. Hiltner.
Astrophys. J., Vol. 252, 277 - 284 (1982).

Simultaneous V, B, and U photometry is presented for this AM Herculis type system. A Fourier analysis has failed to detect the 6 minute quasi-periodic oscillation reported earlier. It has, however, revealed the presence of oscillations at 20.2 and 13.7 minutes. These are interpreted in terms of a non-sinusoidal light curve. U, B, and V light curves are given and discussed in terms of current models for AM Her systems, $B-V$ and $U-B$ color curves are also presented, as is the behavior of the system in the color-color plane.

117.005   The long-term starspot activity on V711 Tauri.
J. D. Dorren, E. F. Guinan.
Astrophys. J., Vol. 252, 296 - 301 (1982).

The $2\overset{d}{.}838$ RS Canum Venaticorum binary V711 Tauri (HR 1099) exhibits striking changes in its light curve from season to season. An analysis of all the available V-band and $\lambda 6585$ light curves is made on the assumption that light variations are due to the presence of starspots on one member of the system. Radio flares occur during an interval in which the spots appear to be changing rapidly in size and location.

117.006   Ultraviolet and optical spectrum studies of Lambda Andromedae: evidence for atmospheric inhomoge-neities.   S. L. Baliunas, A. K. Dupree.
Astrophys. J., Vol. 252, 668 - 680 (1982).

To pursue the study of solar phenomena in cool stars, the authors have investigated chromospheric activity in λ Andromedae (HD 222107). This binary, whose primary star is G7–G8 IV–III, shows strong chromospheric emissions and is related to the RS CVn-type systems. The authors present the first quantiative measurements of chromospheric and solar-type transition-region emissions as a function of the variable starspot and active-region phenomena in an RS CVn star. The presence of optically darker starpots in λ And coincides with the brightening of both Ca II K emission and the ultraviolet transition-region lines. The ultraviolet and optical spectra show attributes of starspots, active regions, and mass flow.

117.007   A first look at the eclipsing cataclysmic variable Lanning 10.
K. Horne, H. H. Lanning, R. H. Gomer.
Astrophys. J., Vol. 252, 681 - 689 (1982).

The authors present photometry and spectroscopy of a new cataclysmic variable star, Lanning 10. The star is shown to be an eclipsing binary with a $7^h 42^m$ orbital period. Blue flickering is observed to occur on a time scale of 10 minutes. The light curve is analyzed with a numerical simulation of the eclipse of the accretion disk by the lobe-filling star. The disk nearly spans the Roche lobe of the white dwarf. The spectra show broad Hα and He I $\lambda 6678$ emission lines from the accre-tion disk and a weak absorption feature from the red dwarf. Radial velocity measurements give the mass function $f(M) = 0.46 \, M_\odot$.

117.008   Ultraviolet spectroscopy of ϕ Persei.
C. R. Kitchin.
Mon. Not. R. Astron. Soc., Vol. 198, 457 - 472 (1982).

Ultraviolet observations at high dispersion of ϕ Persei obtained on the IUE satellite are described. The interstellar lines are separated from the stellar and shell lines, and all the lines are identified. The velocity of the interstellar medium in the direction of ϕ Per is estimated to be about 34 km s^{-1}. A model of the system based on the ultraviolet stellar and shell lines together with some visual and infrared spectra is developed.

117.009   The effect of orbital eccentricity on polarimetric binary diagnostics.
J. C. Brown, C. Aspin, J. F. L. Simmons, I. S. McLean.
Mon. Not. R. Astron. Soc., Vol. 198, 787 - 794 (1982).

The assumption of corotation implicit in all previous modelling of phase-locked polarization variations from close binaries, is relaxed and the simple case of a localized scattering region in an eccentric orbit about a point light source is developed. The authors find that where previously only second harmonic variations of polarization were present first and third harmonics are added when the eccentricity $e \neq 0$. The authors show that erroneous model parameter values (such as the orbital inclination) can occur if a circular orbit model is assumed when analysing the data from an eccentric orbit bina-ry. Extension of the equations to fitting of noisy data is briefly discussed and for illustration is applied to the polarimetric data for Cygnus X-1.

117.010 **The flicker spectrum of AE Aquarii.**
Y. P. Elsworth, J. F. James.
Mon. Not. R. Astron. Soc., Vol. 198, 889 - 896 (1982).

The technique of High-Speed Fourier Transform photometry has been used to observe the flickering activity on AE Aqr up to the frequency limit imposed by photon shot noise. The flicker spectrum exhibits an inverse relation between amplitude and frequency to the observable limit of 0.5 Hz. Some conclusions are drawn from this about the nature of the mechanism responsible for the flickering. The presence of appreciable amounts of flickering above 0.1 Hz places a limit on the opacity of the emitting region.

117.011 **Spin nutation in binary systems due to general relativistic and quadrupole effects.**
B. M. Barker, G. G. Byrd, R. F. O'Connell.
Astrophys. J., Vol. 253, 309 - 311 (1982).

The authors consider two spinning compact objects in a binary system where the orbital angular momentum is much greater than the spin angular momenta. The amplitude and frequency of the spin nutation of body 2 (and body 1 by interchanging indices) due to general relativistic and quadrupole effects are given. The spin nutation frequency of either body is found to be equal to the difference in the two spin precession frequencies.

117.012 **Spectroscopy of the X-ray cataclysmic binary IE 0643.0−1648.** J. B. Hutchings, A. P. Cowley, D. Crampton, G. Williams.
Publ. Astron. Soc. Pacific, Vol. 93, 741 - 746 (1981/82).

Moderate (2Å) resolution spectroscopic data of this recently discovered cataclysmic variable reveal that its binary period is either 0.22 or 0.18 day. The component stars are a white dwarf of $0.9-1.0\,M_\odot$ and a mass-losing companion of $0.4-0.5\,M_\odot$, viewed at $i \sim 45°$. Active dwarf nova behavior obscures the study of phase-dependent quantities. The authors find no evidence for high magnetic fields.

117.013 **AU Mon: a semidetached binary system.**
G. Giuricin, F. Mardirossian, M. Mezzetti.
Mon. Not. R. Astron. Soc., Vol. 199, 131 - 134 (1982).

Using Wood's model the authors have analysed Lorenzi's photoelectric light-curve of the single-lined spectrum eclipsing binary AU Mon. Photometric elements have been obtained. The B5 (hotter and brighter) primary turns out to be considerably smaller than its cooler F-type companion, which fills its Roche lobe for plausible values of the mass ratio ($q \sim 0.2$). Estimates of the absolute elements of AU Mon reveal that the primary is a main sequence star, whilst the secondary is distinctly oversized and overluminous for its mass and spectral type.

117.014 **Relativistic jets in SS 433.** B. Margon.
Science, Vol. 215, 247 - 252 (1982).

A variety of recent optical, radio, and X-ray observations have confirmed the hypothesis that the peculiar star SS 433 is ejecting two narrow, opposed, highly collimated jets of matter at one-quarter the speed of light. This unique behavior is probably driven by mass exchange between a relatively normal star and a compact companion, either a neutron star or a black hole. However, numerous details regarding the energetics, radiation, acceleration, and collimation of the jets remain to be understood. This phenomenon may well be a miniature example of similar collimated ejection of gas by active extragalactic objects such as quasars and radio galaxies.

117.015 **Ultraviolet observations of four symbiotic stars.**
A. G. Michalitsianos, M. Kafatos, W. A. Feibelman, R. W. Hobbs.
Astrophys. J., Vol. 253, 735 - 744 (1982).

Observations were obtained with the IUE of four symbiotic stars. The UV spectra of YY Her, SY Mus, CL Sco, and BX Mon are characterized by varying degrees of thermal excitation. The authors have analyzed these low resolution spectra in terms of line-blanketed model atmospheres of early A, B, and F type stars in order to identify the nature of the hot companion in these systems. The expected emission from early main sequence stars does not fully explain the observed distribution of UV continuum energy over the entire IUE spectral range (1200–3200 Å). More likely the observed continuum may be originating from an accretion disk and/or hot subdwarf that photoionizes circumstellar material.

117.016 **A radial velocity study of 4U 2129+47: a low mass X-ray binary system.**
J. R. Thorstensen, P. A. Charles.
Astrophys. J., Vol. 253, 756 - 759 (1982).

The authors have detected radial velocity variations of the He II λ4686 line in the low luminosity X-ray binary 4U 2129+47, in phase with the 5.24 hour binary period. The observed K velocity of 216 ± 70 km s^{-1} implies a mass for the compact object of 0.53 (+1.0, −0.4) solar masses. This confirms that the system mass is low; the compact object may be a degenerate dwarf. The λ4686 line is quite broad and may originate in an accretion disk.

117.017 **Ultraviolet observations of the 1980 eclipse of the symbiotic star CI Cygni.** R. E. Stencel, A. G. Michalitsianos, M. Kafatos, A. A. Boyarchuk.
Astrophys. J., Lett., Vol. 253, L77 - L82 (1982).

Secular and eclipse variations of UV lines and continua during the course of nearly a full orbit of the symbiotic binary CI Cygni are presented. High-excitation resonance lines have brightened on an orbital time scale and show minimal effects of eclipse, while intercombination lines have faded and show pronounced but nontotal eclipse effects. The data are discussed in terms of mass transfer from the extended cool envelope of the red giant to a compact secondary.

117.018 **Outer atmospheres of cool stars. X. HR 1099 at quadrature.** T. R. Ayres, J. L. Linsky.
Astrophys. J., Vol. 254, 168 - 174 (1982).

The authors report high dispersion, far-ultraviolet (1150–2000 Å) spectra of the active chromosphere, RS CVn binary HR 1099 = V711 Tauri (K0 IV + G5 V) obtained with the IUE. Observations were taken near the opposite orbital quadratures (maximum radial velocity separation). Emission features produced by high temperature species, such as C II and C IV, are very bright, exhibit structure, change significantly in the one week interval separating the two exposures, and generally follow the radial velocity motion of the K subgiant primary. The authors conclude, in agreement with previous studies, that chromospheric and transition region emission in RS CVn binaries is genuinely a stellar, rather than a system, phenomenon.

117.019 **The 1979 minimum state of AN Ursae Majoris.**
J. Liebert, S. Tapia, H. E. Bond, A. D. Grauer.
Astrophys. J., Vol. 254, 232 - 241 (1982).

The authors report spectrophotometry, photometry, and circular polarimetry of the magnetic binary AN UMa during 1979 February/March when it reached a very low state of $B \sim 19-20$, 2 mag fainter than ever observed previously. The observations are compared with earlier data when the system was in a high state and are interpreted within the model of a cyclotron funnel source with variable mass accretion onto the magnetic degenerate primary star.

117.020 **The evolution of highly compact binary stellar systems.**
S. Rappaport, P. C. Joss, R. F. Webbink.
Astrophys. J., Vol. 254, 616 - 640 (1982).

The authors present a new theoretical treatment of the evolution of highly compact binary systems consisting of a collapsed star and a low-mass companion. The low-mass star is assumed to transfer mass onto the collapsed primary star due to the decay of the orbit resulting from gravitational radiation. It is demonstrated that in these circumstances the low-mass secondary is well represented by an $n = 3/2$ polytrope. Using this approximation, the effects of varying several physical parameters, including the stellar masses, the composition of the secondary, mass and angular momentum losses from the system are investigated. The evolution of the system is followed until the secondary has been almost completely consumed. The results yield an explanation for the existence of a cutoff in the orbital period distribution of ~ 80 min among cataclysmic variables and clarify possible generic relationships among cataclysmic variables, low-mass X-ray binaries and "soft" X-ray transient sources.

117.021 Orbital period and radial velocity curve for V 436 Centauri.   R. L. Gilliland.
Astrophys. J., Vol. 254, 653 - 657 (1982).

High time resolution spectrophotometry has been used to determine the orbital period of V 436 Cen. The derived radial velocity curve demonstrates that the 90.0 ± 0.3 minute orbital period is $\sim 2\%$ shorter than the previously determined superhump period. The orbital and superhump periods thus follow the usual relation found in other SU UMa type dwarf novae. At 90.0 minutes V 436 Cen has the shortest orbital period of the classical (i.e., excluding WZ Sge) dwarf novae.

117.022 Hα emission in HR 1099.
D. A. Fraquelli.
Astrophys. J., Lett., Vol. 254, L41 - L45 (1982).

Hα spectrophotometry of HR 1099, a system that exhibits the RS CVn phenomena, is presented. A correlation is found between the Hα flux and the log of the radio flux. The Hα emission-line profiles are double, with a component present from each star. From the Hα emission integrals, the thickness of the chromosphere is estimated. A comparison is made with the Sun.

117.023 Determination of parameters of W UMa systems. II: TW Cet, S Ant, U Peg, ER Ori.
G. Russo, C. Sollazzo, C. Maceroni, L. Milano.
Astron. Astrophys., Suppl. Ser., Vol. 47, 211 - 216 (1982).

Photoelectric lightcurves of four W UMa systems have been solved using the Wilson-Devinney computer code, modified in order to run on a medium-size minicomputer. The solutions confirm the contact hypothesis for these four systems, three of which (TW Cet, U Peg, ER Ori) turned out to be of Binnendijk's W-type (deeper minimum produced by the eclipse of the less massive star) and another one (S Ant) of A-type. A comparison with previous solutions is made, and new absolute elements are derived for all four systems.

117.024 The velocity-mass correlation of the O-type stars: model results.   R. C. Stone.
Astron. J., Vol. 87, 90 - 97 (1982).

This paper presents new model results describing the evolution of massive close binaries from their initial ZAMS to post-supernova stages. These new results allow explicitly for mass loss from the binary system occurring during the core hydrogen- and helium-burning stages of the primary binary star as well as during the Roche lobe overflow. Model results are given for several reasonable choices for these rates. All of the models consistently predict an increasing relation between the peculiar space velocities and masses for runaway OB stars which agrees well with the observed correlations.

117.025 On the evolutionary scenario of massive close binaries with primary masses between $20 M_\odot$ and $160 M_\odot$.   D. Vanbeveren.
Astron. Astrophys., Vol. 105, 260 - 269 (1982).

Massive close binary evolution (case B) is reconsidered.

The masses of the primary stars range between $20 M_\odot$ and $160 M_\odot$. A general comparison is made between theory and observations of O-type stars, X-ray binaries and WR stars.

117.026 On the evolutionary state of the W Ursae Majoris contact binaries.   W. Van Hamme.
Astron. Astrophys., Vol. 105, 389 - 394 (1982).

The author compares the observed spectral type of 17 A-type and 14 W-type W UMa binaries with the expected type when their primary components are normal main sequence objects. For both groups the observed spectral type appears on average to be shifted towards the later spectral types. This shift corresponds to an excess of the mass and radius of the primary components with respect to the main sequence mass and radius in conformity with their observed spectral type. It is argued that neither for the W- nor all the A-type systems this excess should be explained as a result of a normal nuclear evolution of the system. Not all W UMa binaries of type A can be considered as evolved.

117.027 The 6-day photometric and spectroscopic periods in SS 433.   J. J. Matese, D. P. Whitmire.
Astron. Astrophys., Vol. 106, L9 - L11 (1982).

The effects of nutation and pseudoregular precession in misaligned binary systems are investigated and applied to SS 433. In the context of slaved disc type models, geometrical considerations alone predict beam angles should oscillate with a period of 6.06 d, as observed by Newsom and Collins (1981) and Wagner et al. (1981). These angular oscillations, combined with the 164 d precession period, predict a photometric oscillation of 6.30 d, as reported by Mazeh et al. (1981). If periodic mass flow enhancement is important 6.06 d oscillations in beam speed and disc intensity are expected as reported by Newsom and Collins (1981) and Kemp et al. (1981).

117.028 Evidence of variable migration rate and a past direction reversal of the RS CVn wave-like distortion.   C. Blanco, S. Catalano, E. Marilli, M. Rodonò.
Astron. Astrophys., Vol. 106, 311 - 313 (1982).

By comparing recent and older photometric data on RS CVn evidence of a systematically variable velocity and a past migration reversal of the wave-like distortion in the light curve is presented. The importance of collecting uninterrupted data on the migration and amplitude of the photometric wave is emphasized. A possible scenario, based on the spot-model proposed by Eaton and Hall (1979), is presented. It can offer a qualitative interpretation of the presently puzzling behaviour of several RS CVn systems.

117.029 On the linear adiabatic oscillations of a uniformly and synchronously rotating component of a binary.
L. Martens, P. Smeyers.
Astron. Astrophys., Vol. 106, 317 - 326 (1982).

A mathematical analysis is made of the solutions of the system of equations that govern linear adiabatic oscillations of a uniformly and synchronously rotating component of a binary. The tidal axis is taken perpendicular to the axis of rotation. The solutions of the equations are examined up to the second order with respect to the angular velocity. Special attention is devoted to the degeneracy of the eigenvalue problem of the linear adiabatic oscillations of a spherically symmetric star (the zeroth order approximation). In particular, it is shown that the tidal perturbation already affects the eigenfunctions at the first order approximation.

117.030 Researches of massive close binary systems of early spectral type. 3. Orbital elements of the eclipsing binary SX Aurigae.   I. I. Bondarenko, L. F. Istomin.
Perem. Zvezdy, Tom 21, 445 - 449 (1980). In Russian.

The light curve by Klavitter (1973) in the V system of the close binary SX Aurigae is settled. The orbital elements of the system are determined by differential correction. The absolute elements of the system are obtained in combination with Popper's (1943) spectroscopic elements.

117.031 Five-color band ultraviolet photometry of fourteen close binaries.
Y. Kondo, G. E. McCluskey, C.-C. Wu.
Astrophys. J., Suppl. Ser., Vol. 47, 333 - 338 (1981).

The authors present photometric observations obtained with the ANS in five ultraviolet wavelength regions for 14 close binaries. The binaries TT Hya, RX Cas, and SX Cas exhibit a substantial excess of far-ultraviolet flux. This is probably the result of mass transfer phenomena in these systems.

117.032 The ultraviolet spectrum of CH Cygni during the outburst started in 1977.
M. Hack, P. L. Selvelli.
Astron. Astrophys., Vol. 107, 200 - 204 (1982).

The ultraviolet spectrum of the symbiotic star CH Cygni during the outburst phase started in 1977 and still going on is described, and the spectroscopic mechanisms able to explain the emission lines are discussed. A mechanism able to explain both the presence of the intersystem line λ 1641.3 O I and the anomalous intensity ratio of the λ 1303 O I triplet is proposed.

117.033 The monoenergetic beams of SS 433.
M. Milgrom, S. F. Anderson, B. Margon.
Astrophys. J., Vol. 256, 222 - 226 (1982).

The widths of the Doppler-shifted emission features in SS 433 contain information on the monoenergicity of the ejected beams, as well as the finite angular size of, and possible microscopic random motions within, the ejected jets. The authors present a theoretical framework to extract this information from the observed line widths. The model is applied to analyze 150 spectra of SS 433 obtained over 3 yr, covering more than four cycles of the 164 day period. The expected periodic line width variations are detected in the data, not only via a satisfactory least-squares fit of the model to the observations, but also in Fourier power spectrum analysis of the line widths, independent of model assumptions. The relative spread of velocities in the region of the jets responsible for the optical emission proves to be very small, on the order of a few percent.

117.034 Detection of hydrogen α periodicity in X Persei.
T. Mazeh, R. R. Treffers, S. S. Vogt.
Astrophys. J., Lett., Vol. 256, L13 - L16 (1982).

High spectral resolution observations of the Hα profile in the Be star X Persei are presented. Fluctuations with a period consistent with the 13.9 min period of the associated X-ray pulsar were detected. The amplitude of the fluctuations is about 2% of the continuum. Evidence that the amplitude of the fluctuations varies across the Hα profile was also found.

117.035 Period changes in detached close binary systems due to anisotropic ejection of mass.
W. Van Hamme.
Astron. Astrophys., Vol. 107, 397 - 401 (1982).

Close binary systems with at least one component of moderate or late spectral type can show chromospheric activity and consequently anisotropic mass loss. The author presents a mathematical elaboration of a method to relate quantitatively such a mass loss with the corresponding intrinsic period change. Applied to the detached system TX Her, the results point to real, irregular period changes due to mass ejection from the surface of the secondary component, rather than periodic apparent changes due to the light time effect in a triple system.

117.036 The period behaviour of the detached close binary system TX Herculis. W. Van Hamme.
Astron. Astrophys., Vol. 107, 409 - 411 (1982).

An updated period study of the detached close binary system TX Her is presented. It was the author's hope to decide whether the pattern in the $(O-C)$ diagram is purely periodic and due to a light time effect in a triple system, or cyclic but non-periodic so that we are dealing with irregular real period variations. Statistically, a slight preference for the latter case appears.

117.037 Binaries among B stars.
P. Harmanec.
Pulsating B stars, (see 012.042), p. 99 - 107 (1981).

117.038 HZ Her – possible line profiles.
Z. Šíma.
Bull. Astron. Inst. Czechoslovakia, Vol. 33, 122 - 127 (1982).

On the basis of a semi-analytical method the profiles of spectral lines are calculated for the system HZ Her. Mass transfer in the form of a gas stream flowing from L_1 or a ring around the compact component is taken into account. The interpretation of such lines is discussed.

117.039 Secular change in the '164-day' period of SS 433.
G. W. Collins, II, G. H. Newsom.
Astrophys. Space Sci., Vol. 81, 199 - 208 (1982).

In this paper the authors present evidence that the '164-day' period in the wavelengths of the 'moving' lines in SS 433 has been decreasing at the surprisingly rapid rate of $\dot{P} = -0.010 \pm 0.002$. An ephemeris of the moving lines for the 1981 observing season is provided for the case $\dot{P} = -0.010$.

117.040 HR 4975: a possible early-type contact system with unequal components.
C. Waelkens, P. Bartholdi.
Astron. Astrophys., Vol. 108, 51 - 54 (1982).

From 295 measurements in the Geneva seven-colour photometric system, it is found that the suspected variable star HR 4975 is in fact a close binary system. The colour variations and the very short period (0.64254d) indicate a possible contact configuration. It is also argued that the two components are probably unequal in size and that the system is near to the ZAMS. Theoretical implications are briefly discussed.

117.041 Numerische Berechnung von Akkretionsscheiben in engen Doppelsternsystemen. G. Hensler.
Mitt. Astron. Ges., Nr. 55, p. 27 (1982). – Abstract.

117.042 AE Phe 1978. Lichtkurve eines W UMa-Veränderlichen im ruhigen Stadium. K. Walter.
Mitt. Astron. Ges., Nr. 55, p. 71 - 72 (1982). – Abstract.

117.043 UBV-Photometrie von Kontakt-Doppelsternen.
H. Mauder, N. Kappelmann.
Mitt. Astron. Ges., Nr. 55, p. 72 - 77 (1982).

117.044 The variable, single-line WN8 star HD 86161: another Wolf-Rayet star with a low-mass companion.
A. F. J. Moffat, V. S. Niemela.
Astron. Astrophys., Vol. 108, 326 - 333 (1982).

Extensive intermediate-band photo-electric photometry suggests that HD 86161 is an ellipsoidal variable with a period of 10.73 d. Together with high-dispersion spectroscopic data, which indicate a very low-amplitude radial velocity orbit with this period, the unseen companion to the WN8 star has a most probable mass in the range $0.5-1.2\,M_\odot$. This may be a white dwarf, or more likely, a neutron star. The Wolf-Rayet envelope shows stronger variability the further one proceeds outwards to lower wind densities.

117.045 Radiation from the discs in cataclysmic variables: the chromosphere. A. Schwarzenberg-Czerny.
Acta Astron., Vol. 31, 241 - 265 (1981).

A simple model is developed of the illumination of the concave surface of an accretion disc by a central source of UV-radiation which produces an optically thin ionized emission layer (the chromosphere of the disc). The emission predicted by the model is sufficient to explain the observed spectra of CV with stationary discs. Colours and continuous spectra of the disc are calculated using a simple model of the chromosphere and models of stellar atmospheres. The models are in agreement with observations of nova-like stars, Z Cam stars at standstill and dwarf novae during outburst.

117.046   Photoelectric observations and light curve variations of VW Cephei.
J. M. Kreiner, M. Winiarski.
Acta Astron., Vol. 31, 351 - 361 (1981).
452 photoelectric observations of VW Cep obtained in three successive nights are reported. The light curve of VW Cep changes from night to night.

117.047   System of Roche coordinates for tidally distorted stellar models.   V. P. Singh.
Proc. Indian Natl. Sci. Acad. Part A, Vol. 47, 417 - 427 (1981).
Abstr. in Phys. Abstr., Vol. 85, Abstr. 42725 (1982).

117.048   Reappraisal of the polarimetric data for seven binaries.   C. Aspin, J. F. L. Simmons.
Mon. Not. R. Astron. Soc., Vol. 199, 601 - 609 (1982).
The authors use the canonical model of polarimetric variations from a binary star system to find the optimum inclinations and associated confidence intervals for seven binaries: Algol, AO Cas, HD 47129, σOri E (B and U filter data), u Her, U Sge and V444 Cygni. They also consider the estimation of the amount of circumstellar material in a binary from polarimetric observations.

117.049   HD 81410: a new RS CVn binary.
A. V. Raveendran, M. V. Mekkaden, S. Mohin.
Mon. Not. R. Astron. Soc., Vol. 199, 707 - 714 (1982).
B and V photometry of HD 81410 obtained on 34 nights during 1981 January 3 - March 14 is presented. The amplitudes are found to be ~0.15 mag both in B and V. It is found that a 12.86833 day period satisfies all the available photometric data. Orbital elements of the system have been derived from the published radial velocity measurements. The orbital period, strong Ca II H and K emission and the spectral types of the components indicate that HD 81410 is a member of the RS CVn group.

117.050   PG 1550 + 191: a new AM Herculis type binary system.   J. Liebert, H. S. Stockman,
R. E. Williams, S. Tapia, R. F. Green, D. Rautenkranz, D. H. Ferguson, P. Szkody.
Astrophys. J., Vol. 256, 594 - 604 (1982).
The authors report the discovery of a new AM Herculis type magnetic variable, found from a sample of cataclysmic variable candidates in the Palomar Green Survey. The orbital period, as defined by linear and circular polarimetry, is 113.56 minutes. The maximum of linear polarization was generally weak (~2%-5%), while the circular polarization varies between 0% and -12%. Simultaneous $UBVR$ photometry shows mild flickering and smooth variations which are roughly synchronous with the polarization. Infrared (JHK) photometry indicates an energy distribution similar to AM Her. The emission-line spectrum shows the usual strong He II, He I, H, and high-excitation lines, split into broad and narrow components. The radial velocities of the narrow emission components have semiamplitude K = 170 km s^{-1} and a positive-going zero crossing at magnetic phase 0.57. The broad emission component shows a semiamplitude K = 230–330 km s^{-1} with a zero crossing near phase 0.35.

117.051   Searching for companions of Be stars.
J. S. Price, C. A. Bordner.
Proc. Southwest Reg. Conf., Vol. 7, (see 012.019), 45 - 49 (1982).

117.052   SS 433: how the moving lines move.
S. A. Grandi, R. P. S. Stone.
Publ. Astron. Soc. Pacific, Vol. 94, 80 - 86 (1982).
Several sets of spectra taken on consecutive nights of the unique emission-line object SS 433 are discussed in an attempt to answer the question: How do the moving lines move? The moving lines definitely appear to be composed of a set of independent components (or "bullets") that probably represent separate, but correlated, ejection events in the twin beams. Spectra are presented showing changes in these components on a time scale of hours. There is no evidence for light-travel time effects in the motions of the moving lines. Finally, examples of both an apparent ~6-day variation and stochastic jitter are shown in the moving line data.

117.053   The UV spectrum of the symbiotic binary AR Pavonis during eclipse egress.
J. B. Hutchings, A. P. Cowley.
Publ. Astron. Soc. Pacific, Vol. 94, 107 - 112 (1982).
The UV spectrum of AR Pav ($\lambda\lambda 1200-3200$) was observed during egress from eclipse, with the IUE satellite. The far UV continuum shows little or no eclipse, while at $\lambda 3200$ the light curve is like the V band. The continuum appears to be the sum of an M giant, a hot (power law?) continuum, and an ~4000 K thermal continuum. Only the latter changed during the observations. Emission lines are not eclipsed outside of phase 0.041 (at which phase the V continuum is eclipsed by 1 magnitude). These findings are discussed in terms of models for the system.

117.054   ANS spectrophotometry: the bright X-ray binaries Hercules X-1 (HZ Herculis) and Cygnux X-1 (HDE 226868).   C.-C. Wu, J. A. Eaton, A. V. Holm, M. Milgrom, G. Hammerschlag-Hensberge.
Publ. Astron. Soc. Pacific, Vol. 94, 149 - 156 (1982).
The authors have obtained complete ultraviolet light curves for HZ Her immediately after a well-defined X-ray turn on in 1975. The features of this light curve correspond well with features seen in the optical. The absolute level of emission at maximum light at 1550 Å agrees well with the calculations of the X-ray reflection effect of Milgrom and Salpeter. The authors find that at 1550 Å roughly 20% of the light at maximum comes from the accretion disk. Similar observations of Cyg X-1 allow to determine the degree of reddening, equivalent to $E(B-V) = 0.95 \pm 0.07$, from the strength of the 2200 Å interstellar absorption feature. The distance inferred from this amount of reddening requires the optical star to be luminous and, therefore, massive. Observations obtained during the May 1975 X-ray transition show ultraviolet excess when compared to the data obtained six months later.

117.055   The orbital period of the symbiotic star AX Persei.
S. J. Kenyon.
Publ. Astron. Soc. Pacific, Vol. 94, 165 - 168 (1982).
The author has used the photometric data of Mjalkovskij (1977) to analyze the quiescent light curve of the symbiotic star AX Per. The minima in the photographic and photovisual light curves are given by Min = JD 2436679.4 + 681.6 E. The mean photographic magnitude is 13.05 with an amplitude of 0.^{m}7, while the mean photovisual magnitude is 11.38 with an amplitude of 0.^{m}36. The variation may be qualitatively understood in terms of the reflection of light from a hot star off of a red-giant companion. The 681.6-day period thus represents the binary period of AX Per.

117.056   AC Cancri: a new cataclysmic variable.
A. Okazaki, M. Kitamura, A. Yamasaki.

Publ. Astron. Soc. Pacific, Vol. 94, 162 - 164 (1982).

From spectroscopic observations the short-period eclipsing binary AC Cnc is found to show broad emission lines of H and He II superimposed on a continuum spectrum. The measured halfwidths of emission lines give the velocity dispersion of 550 km s^{-1}. The magnitude and color indices of $V = 13.80$, $(B-V) = +0.31$, and $(U-B) = -0.62$ are obtained for this star. These results strongly suggest that AC Cnc is a new cataclysmic variable.

117.057 Evolution of massive close binary systems.
A. G. Masevich, A. V. Tutukov.
Zemlya Vselennaya, 1982, No. 1, p. 27 - 31. In Russian.

117.058 Transferencia de materia en sistemas binarios cerrados. Parte II. J. Zorec.
Bol. Asoc. Argentina Astron., No. 20 - 24, p. 91 - 112 (1981).

117.059 Modelo de hot-spot en estrellas W UMa.
E. Lapasset, R. F. Sisteró.
Bol. Asoc. Argentina Astron., No. 20 - 24, p. 334 (1981).
Abstract.

117.060 IUE observations of the peculiar star RX Puppis.
M. Kafatos, A. G. Michalitsianos, W. A. Feibelman.
Astrophys. J., Vol. 257, 204 - 213 (1982).

The authors have obtained the first high-dispersion observations of RX Pup in the wavelength region 1200–2000 Å with the IUE. RX Pup has been classified a symbiotic star and has been compared to slow novae as to η Carinae. The anomalies observed in high-excitation lines in RX Pup such as He II, O III], C III], C IV, and Si III] that show split line profiles, Doppler displaced multiple components, and possible inverse P Cygni profiles in N III] and N IV] suggest dynamic activity in circumstellar material that probably has the form of rings and/or gas streamers between the cool giant and the hot companion. The authors find electron densities in the line-emitting region in the range 10^9-10^{11} cm^{-3}, temperatures in the range 10,000–20,000 K, and linear sizes $\lesssim$ a few $\times 10^{13}$ cm. The photoionizing radiation may be due to the presence of an unseen, hot subdwarf with most probable effective temperature in the range 75,000–90,000 K. Alternatively it may be due to an accretion disk around a secondary with boundary layer temperature $\sim 10^5$ K.

117.061 Detached → contact scenario for the origin of W UMa stars. O. Vilhu.
Astron. Astrophys., Vol. 109, 17 - 22 (1982).

The scenario where contact binaries of W UMa type are produced from detached ones by gradual angular momentum loss is studied. The classical rotation-age dependence $(V_{rot} \sim t^{-1/2})$ gives $dJ/dt \sim P^{-3}$ for single solar type main sequence stars rotating slower than with a 3 d period. A similar trend probably continues to even shorter periods but perhaps with a different exponent: $dJ/dt \sim P^{-\alpha}(\alpha < 3)$. Assuming (1) spin-orbit coupling, (2) a contact life-time 5×10^8 yr, (3) the initial period distribution of new binaries has a smooth cut-off around 2 d, (4) $\alpha = 1.5$ for $P < 3$ d, then this process has good chances of being realistic and of representing the correct production mechanism for W UMa stars.

117.062 On the stability and evolution of contact binaries. I.
T. Rahunen.
Astron. Astrophys., Vol. 109, 66 - 76 (1982).

The stability and evolution of zero-age and evolved contact binaries is investigated. A zero-age system of $1.0\,M_\odot$ and $0.6\,M_\odot$ is found to be unstable and the subsequent evolution occurs in thermal cycles with a period of about 6×10^6 yr. During the cycles the system is permanently in contact and the entropy difference is large. Nuclear evolution seems to remove the instability. After about 10^9 yr the system is found

to be stable and it evolves towards more extreme mass ratios and equal entropies. The adopted form of the equation describing the energy transport between the two stars and the relevance of the results for the theory of W UMa stars are discussed.

117.063 Observations and analysis of the light curve of AE Phoenicis in 1978. K. Walter.
Astron. Astrophys., Vol. 109, 107 - 116 (1982).

UBV observations of the W UMa variable AE Phe, obtained with the Bochum-Telescope at La Silla in 1978, are presented. Two basic periods, one near 55^d, the other near 95^d, each clearly belonging to one component, were found. The two periods are interpreted to be the precessional periods of the rotational axes of the components. Obviously AE Phe in 1978 was found in a state of low activity.

117.064 On the stability of age-zero contact binaries. II.
J. Hazlehurst, W. Höppner, S. Refsdal.
Astron. Astrophys., Vol. 109, 117 - 122 (1982).

The secular stability of a contact binary system of $0.85\,M_\odot + 0.5\,M_\odot$ is investigated. The system, which is chemically unevolved ("age-zero") is found to be secularly stable. Using the results of this and a previous paper, the authors suggest that evolution of contact binaries can occur in either "cyclic" (secularly unstable) or "non-cyclic" (secularly stable) modes, depending on the system parameters. According to either picture, a problem arises in connection with the ages of the four W UMa systems in NGC 188.

117.065 A brightening of the symbiotic variable SY Muscae.
A. G. Michalitsianos, M. Kafatos, W. A. Feibelman, G. Wallerstein.
Astron. Astrophys., Vol. 109, 136 - 140 (1982).

The symbiotic variable SY Muscae has been observed with IUE in September 1980 and June 1981 and in the photographic region in May 1981. The entire ultraviolet spectrum brightened between September and June by about a factor of 5. The optical spectrum is dominated by permitted lines. The increase in ultraviolet continuum and line emission may be due to enhanced mass transfer from the cool star whose period is 623 d and whose maximum was predicted to occur very close to the time of the June 1981 observations. Alternatively the hot star and much of the emitting gas could have been in eclipse in September 1980.

117.066 Variaciones en la intensidad del doblete ultravioleta del Mg II en estrellas simbióticas.
J. Sahade, E. Brandi.
Bol. Asoc. Argentina Astron., No. 26, p. 72 (1981). – Abstract.

117.067 Observaciones IUE de estrellas simbióticas.
J. Sahade, E. Brandi.
Bol. Asoc. Argentina Astron., No. 26, p. 73 (1981). – Abstract.

117.068 Symbiotic stars. H. Maehara.
Astron. Her., Vol. 74, 256 - 260 (1981). In Japanese.
From Phys. Abstr., Vol. 85, Abstr. 53271 (1982).

117.069 Be stars as interacting binaries. G. J. Peters.
Be stars, (see 012.030), p. 311 - 314 (1982).

117.070 Ultraviolet observations of interacting binary Be stars. G. J. Peters, R. S. Polidan.
Be stars, (see 012.030), p. 405 - 409 (1982).

Initial results from the analysis of a series of timed, high resolution IUE observations of HR 2142, ϕ Per, CX Dra, KX And, AU Mon, and TT Hya are presented.

117.071 MT Cassiopeiae, another contact binary with complete eclipses. M. Hoffmann.
Inf. Bull. Variable Stars, No. 2063, 3 pp. (1982).

117.072 **HD 8152 – a new W Ursae Majoris type eclipsing binary.** M. B. K. Sarma, K. R. Radhakrishnan.
Inf. Bull. Variable Stars, No. 2073, 3 pp. (1982).

117.073 **Photoelectric light curve of 44 Bootis.**
R. M. Genet, K. E. Kissell, G. C. Roberts.
Inf. Bull. Variable Stars, No. 2074, 2 pp. (1982).

117.074 **Two (O-C) residuals of W Ursae Majoris system.**
E. Hamzaoğlu, V. Keskin, T. Eker.
Inf. Bull. Variable Stars, No. 2083, 2 pp. (1982).

117.075 **1981 UBVR photometry of UV Psc.** M. Zeilik, R. Elston, G. Henson, P. Schmolke, P. Smith.
Inf. Bull. Variable Stars, No. 2089, 4 pp. (1982).

117.076 **1981 UBVR photometry of RT And.** M. Zeilik, R. Elston, G. Henson, P. Schmolke, P. Smith.
Inf. Bull. Variable Stars, No. 2090, 4 pp. (1982).

117.077 **High speed photometry of H0139-68.**
M. S. Cropper.
Inf. Bull. Variable Stars, No. 2096, 2 pp. (1982).

117.078 **Photoelectric observations of VW Cephei.**
T. Abe.
Inf. Bull. Variable Stars, No. 2108, 3 pp. (1982).

117.079 **The spectral type of RW Doradus.**
W. P. Bidelman, N. Sanduleak.
Inf. Bull. Variable Stars, No. 2122 (1982).

117.080 **H alpha variations and the near infrared spectrum of CI Cygni.**
G. B. Baratta, P. A. Mengoli, R. Viotti.
Inf. Bull. Variable Stars, No. 2126, 3 pp. (1982).

117.081 **New minimum times of the W UMa star SW Lacertae.**
U. Hopp, M. Hoffmann, S. Witzigmann.
Inf. Bull. Variable Stars, No. 2142, 3 pp. (1982).

117.082 **Photoelectric observation of W UMa (BD +56° 1400).**
M. Hamzaoğlu, E. Hamzaoğlu, T. Eker.
Inf. Bull. Variable Stars, No. 2151, 2 pp. (1982).

117.083 **New photoelectric light curves of VW Cephei.**
H. A. Mahdy, M. A. Soliman.
Inf. Bull. Variable Stars, No. 2153, 4 pp. (1982).

117.084 **Photoelectric minima and new light curve of V566 Ophiuchi.** H. A. Mahdy, M. A. Soliman.
Inf. Bull. Variable Stars, No. 2154, 4 pp. (1982).

117.085 **New photoelectric times of minima of W Ursae Majoris.**
U. Hopp, S. Witzigmann, M. Kiehl.
Inf. Bull. Variable Stars, No. 2156, 2 pp. (1982).

117.086 **SY Muscae.**
IAU Circ., Nos. 3657, 3687 (1982).

117.087 **V1343 Aquilae.**
IAU Circ., No. 3670 (1982).

117.088 **Shallow partial eclipses of a star by a disk or sphere-like component and the parameters of Cyg X-1 = V1357 Cyg.**
E. A. Karitskaya, N. G. Bochkarev.
Astron. Tsirk., No. 1184, p. 1 - 6 (1981). In Russian.

117.089 **He I line emission and the helium abundance in cataclysmic variables.**
R. E. Williams, D. H. Ferguson.
Astrophys. J., Vol. 257, 672 - 685 (1982).

Time-resolved spectroscopy of a sample of eclipsing cataclysmic variables indicates that the He I emission lines are formed with the Balmer lines in the outer regions of the accretion disk. Additional observations of lower inclination non-eclipsing systems demonstrate that the He I emission is frequently characterized by relatively high singlet to triplet intensity ratios and an inverted line decrement, an unusual situation which can be explained in terms of emission from an optically thick (in the lines) region in local thermodynamic equilibrium. Calculations of steady state accretion disk structure show that the observed characteristics of the He I and H I lines can be reproduced in normal disk models, but only if the helium abundance is assumed to be very high, i.e., $He/H \gtrsim 100$ by number. Enhanced helium abundances are thus implied for some cataclysmic variable secondaries, suggesting that they are either highly evolved cores or that they have an outer layer of processed material which was accreted from the white dwarf during earlier evolution of the system.

117.090 **The low state of AM Herculis: observations from 0.12 to 10 microns.**
P. Szkody, J. C. Raymond, R. W. Capps.
Astrophys. J., Vol. 257, 686 - 694 (1982).

Observations of AM Her during a low state over a wavelength range from 0.12 to 10 μm are reported. These include *IUE* ultraviolet spectra, light curves at $U, B, V, R, J, H, K,$ and magnitudes at $L, M,$ and N. The UV observations reveal a nearly Rayleigh-Jeans continuum spectral distribution and broad Lyα-absorption from a hot ($T_{eff} \approx 50,000$ K) white dwarf. Of the strong emission lines present in the high state, only weak C IV (1550 Å) and Mg II (2800 Å) features remain. The optical light curves are markedly different from the high state, while the infrared light curves are similar in appearance to the high state. The infrared variations cannot be explained solely by the ellipsoidal variations of a secondary star which is heated by an accretion column. The large available wavelength range is used to constrain the relative contributions of the white dwarf, the red dwarf, and the accretion columns.

117.091 **Generalized Roche potential for misaligned binary systems: properties of the critical lobe.**
Y. Avni, N. Schiller.
Astrophys. J., Vol. 257, 703 - 714 (1982).

The Roche potential for binary systems where the stellar rotation axis is not aligned with the orbital revolution axis is studied. The properties of the critical equipotential lobe and of the internal Lagrangian point that defines this lobe are derived. It is found that, as the degree of misalignment varies, internal Lagrangian points and external Lagrangian points may switch their roles. A systematic, practical method to identify the internal Lagrangian point and to calculate the volume of the critical lobe is developed, and numerical results are presented for a wide range of parameters for binary systems with circular orbits.

117.092 **Two-second variability in AM Herculis binaries.**
J. Middleditch.
Astrophys. J., Lett., Vol. 257, L71 - L75 (1982).

AN UMa and the newly identified AM Her star E1405-451 both show excesses of optical variability in the broad frequency range from 0.4 to 0.8 Hz (2.5 - 1.25 s). The optical fluxes from AN UMa and E1405-451 must vary by 2.4% and 1.2% rms, respectively, in order to produce the level of noise observed in the power spectra. Time series observations of AM Her made during both its "bright" and "faint" phases show that no such noise feature is present to less than 0.25% and 1.3% rms of the stellar light.

117.093   **Discovery of optical variability in the hard X-ray source HD 8357.**
D. S. Hall, G. W. Henry, H. Louth.
Astrophys. J., Lett., Vol. 257, L91 - L92 (1982).

Photometry obtained at three different observatories on 27 different nights in the 1980 - 1981 season shows this RS CVn binary to be variable. The light curve is sinusoidal in shape, the total range in V is 0.08 mag, the photometric period is $12^d3 \pm 0^d1$, and an epoch of minimum light is JD $2,444,504.5 \pm 0^d1$.

117.094   **Evolution of massive close binary systems.**
C. de Loore, J. P. De Grève.
The most massive stars, (see 012.034), p. 85 - 102 (1981).

Starting from the two basic aspects of massive close binary systems (MCBS), their massiveness and their interacting binary nature, the classical evolutionary scheme is reviewed with respect to recent theoretical and observational developments. Accent is given to the structure and nature of the secondary, i.e. the originally less massive star and its influence on the history of the system. The appearance and the importance of different groups of MCBS such as Wolf-Rayet systems and X-ray binaries is discussed.

117.095   **The theoretically expected X-ray luminosity and the binary nature of Wolf-Rayet runaway stars.**
D. Vanbeveren, W. Van Rensbergen, C. de Loore.
The most massive stars, (see 012.034), p. 199 - 205 (1981).

In this paper two problems concerning Wolf-Rayet + compact star binaries are discussed:(1) WR stars are losing mass by stellar wind; can one expect to see hard X-ray radiation? (2) the number of WR + compact star systems compared to the number of WR + OB binaries is estimated including a critical discussion on the binary survival probability during the SN explosion.

117.096   **Photoelectric observation of W UMa (BD +56°1400).**
E. Hamzaoğlu, V. Keskin, T. Eker.
Inf. Bull. Variable Stars, No. 2102, 3 pp. (1982).

117.097   **Nodding motions of accretion discs in SS 433 and Her X-1.**
J. I. Katz, S. F. Anderson, B. Margon, S. A. Grandi.
Bull. American Astron. Soc., Vol. 13, 801 (1981). – Abstract.

117.098   **Departures from the kinematic model of SS 433.**
G. W. Collins II, G. H. Newsom.
Bull. American Astron. Soc., Vol. 13, 801 (1981). – Abstract.

117.099   **Periods and period changes in SS 433.**
S. F. Anderson, B. Margon, S. A. Grandi, R. A. Downes.
Bull. American Astron. Soc., Vol. 13, 801 (1981). – Abstract.

117.100   **A photometric and spectroscopic study of a newly discovered cataclysmic variable, H2215-086.**
A. W. Shafter, D. M. Targan.
Bull. American Astron. Soc., Vol. 13, 802 (1981). – Abstract.

117.101   **UBVR photometry of short-period RS CVn systems.**
M. Zeilik, D. Carriaga, R. Elston, D. Gonzales, G. Henson, P. Smith, P. Smolke.
Bull. American Astron. Soc., Vol. 13, 802 (1981). – Abstract.

117.102   **Near-infrared TiO band observations of RS CVn stars.**   H. Nations, L. W. Ramsey.
Bull. American Astron. Soc., Vol. 13, 803 (1981). – Abstract.

117.103   **The hot components of symbiotic stars.**
S. J. Kenyon.
Bull. American Astron. Soc., Vol. 13, 804 (1981). – Abstract.

117.104   **NOISARS in AM Her systems.**
J. Middleditch.
Bull. American Astron. Soc., Vol. 13, 816 (1981). – Abstract.

117.105   **An ultraviolet study of the white dwarf and K2V components of the close eclipsing binary V471 Tauri.**   E. F. Guinan, E. M. Sion.
Bull. American Astron. Soc., Vol. 13, 817 (1981). – Abstract.

117.106   **Near infrared fluxes in RS CVn type binary stars.**
S. A. Naftilan.
Bull. American Astron. Soc., Vol. 13, 833 (1981). – Abstract.

117.107   **Starspot development and color variation on VW Cephei.**   A. P. Linnell.
Bull. American Astron. Soc., Vol. 13, 833 (1981). – Abstract.

117.108   **On the turn-ons of Hercules X-1 and the periodicities in the radial velocity variations of SS 433.**
A. M. Levine, J. G. Jernigan.
Bull. American Astron. Soc., Vol. 13, 833 (1981). – Abstract.

117.109   **Radiative acceleration of astrophysical jets: line-locking in SS 433.**
P. R. Shapiro, M. Milgrom, M. J. Rees.
Bull. American Astron. Soc., Vol. 13, 833 - 834 (1981). Abstract.

117.110   **A model for the beams of SS 433.**
D. Kazanas, A. K. Harding.
Bull. American Astron. Soc., Vol. 13, 834 (1981). – Abstract.

117.111   **The X-ray lobes of SS 433.**
M. G. Watson, J. E. Grindlay, F. D. Seward.
Bull. American Astron. Soc., Vol. 13, 834 (1981). – Abstract.

117.112   **X-ray variability of SS 433.**   J. Grindlay, D. Band, F. Seward, D. Leahy, M. Weisskopf.
Bull. American Astron. Soc., Vol. 13, 834 (1981). – Abstract.

117.113   **Results of an IUE program of monitoring the ultraviolet emission line fluxes of four binary systems: HR 1099, II Peg, AR Lac and BY Dra.**
J. L. Linsky, T. Simon, M. Narstad, M. Rodono, C. Blanco, S. Catalano, E. Marilli, A. D. Andrews, C. J. Butler, P. B. Byrne.
Bull. American Astron. Soc., Vol. 13, 872 (1981). – Abstract.

117.114   **The chromosphere and corona of the contact binary: Epsilon Coronae Austrinae.**
A. K. Dupree, M. Dussault.
Bull. American Astron. Soc., Vol. 13, 873 (1981). – Abstract.

117.115   **On the masses of Wolf-Rayet components in close binary systems.**   W. Sutantyo, L. Dermawan.
Contrib. Bosscha Obs. Lembang, No. 72, 7 pp. (1982).

Massey (1981) indicates that the average mass of the WR components in double line WR binaries is about 20 solar masses. The authors argue that if such large mass WR stars eventually explode as supernova, the systems will generally be accelerated to velocities considerably in excess of 50 - 65 km/s. Such large velocities are incompatible with the moderate galactic z values of massive X-ray binaries and single line WR stars/binaries which are believed to be massive binaries in the post supernova stage. They conclude that if the estimate of Massey for the masses of WR stars is correct, a considerable amount of mass (about 10 solar masses) should have been lost from the star before the explosion.

117.116   **CW 1103+254.**
IAU Circ., No. 3696 (1982).

117.117 Evolutionary computations of the early-type contact binary SV Centauri.
M. Nakamura, Y. Nakamura.
Astrophys. Space Sci., Vol. 83, 163 - 170 (1982).

Evolutionary models of the early-type contact binary SV Centauri are recalculated with contact condition taken into account. With the initial masses of 13.4 and 7.0 $M_\odot$ for the component stars, the observed features such as the rate of mass transfer, the degree of contact, and the positions of both components in the H-R diagram can be reproduced. In agreement with a previous paper this indicates that the binary system SV Cen is actually in the rapid phase of mass transfer preceding the reversal of the mass ratio. In contrast to the steadily increasing rate of mass transfer shown in the previous paper, however, the rate of mass transfer suddenly turns to decrease as soon as the system evolves into the contact phase.

117.118 HR 7275: a new variable star.   R. E. Fried,
    J. A. Eaton, D. S. Hall, G. W. Henry, L. P. Lovell,
K. Krisciunas, C. R. Chambliss, P. K. Detterline,
H. J. Landis, H. Louth, D. R. Skillman.
Astrophys. Space Sci., Vol. 83, 181 - 188 (1982).

Differential photometry of the K1 IV-III RS CVn-type binary HR 7275 in 1978, 1979, and 1980 at nine different observatories shows it definitely to be variable, thus confirming the suspicion of Herbst. The photometric period determined two ways was $27^d_.91$ or $27^d_.65$, thus about 3% shorter than the spectroscopically determined orbital period of $28^d_.59$. The total variation observed during the three years was $0^m_.22$ in the V. The light curve was always asymmetrical, with a stillstand on the rising branch in 1978 but on the falling branch in 1980.

117.119 The effect of the stellar wind velocity on the eccentric-orbit binary model.
T. Okuda. S. Sakashita.
Astrophys. Space Sci., Vol. 83, 441 - 444 (1982).

The previous eccentric-orbit binary model for the recurrent X-ray sources is modified by taking account of the velocity field of the stellar wind. The resultant formula of the light curves for the X-ray sources may have an useful application to the observed X-ray sources.

117.120 RW Sextantis, a disk with a hot, high-velocity wind.
    J. L. Greenstein, J. B. Oke.
Astrophys. J., Vol. 258, 209 - 216 (1982).

The flickering blue variable, RW Sex (−7° 3007), is a cataclysmic variable currently stabilized at maximum. The continuous spectrum observed spectrophotometrically, from 10,000 to 1150 Å, is dominated by an accretion disk, with flat spectrum in the ultraviolet, except at $\lambda > 5000$ Å, where a blackbody near 7000 K is seen. A distance of 400 parsecs is derived, if the latter arises from an F type main-sequence star. The accretion rate required is near $10^{-8} M_\odot \mathrm{yr}^{-1}$. Only weak emission is seen, except for Lyα; strong, broad UV absorption lines are seen with centers displaced up to −3000 km s^{-1}, with terminal velocities up to −4500 km s^{-1}, the velocity of escape from a white dwarf. The star is a weak X-ray source. The low X-ray flux may arise from absorption within an unusually dense, hot wind from the innermost portions of the disk. The suggested geometry is conical, with the disk seen from its pole. The estimated mass-loss rate is near $10^{-12} M_\odot \mathrm{yr}^{-1}$. There is considerable resemblance to the ejection mechanism from quasars but on a smaller scale.

117.121 Time-resolved spectroscopy of the accretion disk in RW Tauri.   R. H. Kaitchuck, R. K. Honeycutt.
Astrophys. J., Vol. 258, 224 - 235 (1982).

Time-resolved spectroscopy of the Algol RW Tauri during 12 eclipses has determined the following characteristics of the disk. The disk is smaller than previously believed

(rarely exceeding 1.5 times the radius of the central star) and shows significant variations in size in one orbit. The emission-line widths are always at least twice that expected from broadening by rotational motion in a disk. A detailed study of one particularly favorable eclipse shows $\bar{n}_e$ to be approximately 10^{11} cm^{-3}. The trailing side of the disk appears to be nearly isothermal, of approximately constant density, and with an apparent radius that is likely set by the ionization limit. The rotational velocity field is highly non-Keplerian; on the leading side of the disk, the highest rotational velocity is at the outer edge.

117.122 Far-ultraviolet observations of MV Lyrae.
    L. Chiappetti, L. Maraschi, E. G. Tanzi, A. Treves.
Astrophys. J., Vol. 258, 236 - 239 (1982).

MV Lyrae was observed in low state ($m_B \approx 18$) with the IUE from 1200 to 3100 Å. Weak emission and absorption features are apparent over a well defined continuum, the shape of which is close to, but appreciably flatter than, that given by a Rayleigh-Jeans law. The energy distribution obtained combining the ultraviolet observations with published optical and near infrared data, can be interpreted as a high-temperature blackbody($T = 6 - 7 \times 10^4$ K), with emitting area $A \approx 10^{18}$ cm^2, consistent with the whole surface of the white dwarf, plus the emission of a red dwarf companion. This model predicts negligible soft X-ray flux. Alternatively, a large portion of the ultraviolet emission could be ascribed to a hot spot with a corresponding soft X-ray flux of the order of that observed in 1977 and in 1979.

117.123 Radio and optical observations of the R Aquarii jet.
    R. J. Sopka, G. Herbig, M. Kafatos,
A. G. Michalitsianos.
Astrophys. J., Lett., Vol. 258, L35 - L39, plate L1 (1982).

VLA observations at 6 cm and Lick Observatory optical plates of R Aquarii indicate the existence of a jetlike feature extending $7'' - 10''$ from the central star. A wide field map at 6 cm shows an unresolved compact radio source which lies close to the axis defined by the jet at a distance of ∼3′ from R Aqr. Episodic mass transfer in this symbiotic variable could explain the erratic outbursts that R Aqr is known to undergo. Formation of an accretion disk and the accompanying radio-optical jet may characterize the observed outbursts in this system.

117.124 *IUE* and optical observations of MV Lyrae at intermediate and low states.
P. Szkody, R. A. Downes.
Publ. Astron. Soc. Pacific, Vol. 94, 328 - 334 (1982).

The authors have obtained *IUE* and optical spectra of MV Lyr during its normal low state (V ∼ 14th mag) and its very low state (V ∼ 17th mag). The normal low state shows UV and optical emission lines similar to dwarf novae, and a flux distribution that can be accounted for by a steady-state disk with a temperature range between 3700 K - 34,000 K ($\dot{M} \sim 10^{-9} M_\odot \mathrm{yr}^{-1}$). In the very low state, the prominent UV emission lineshave disappeared and the optical lines have narrowed appreciably. The flux distribution is compatible with a hot ($T \geqslant 50,000$ K) white dwarf, a cooler disk with $\dot{M} \sim 5 \times 10^{-11} M_\odot \mathrm{yr}^{-1}$, and a late-type secondary.

117.125 The eclipsing binary UV Lyncis.
    N. L. Markworth, E. J. Michaels, Jr.
Publ. Astron. Soc. Pacific, Vol. 94, 350 - 355 (1982).

Photoelectric observations of UV Lyn have yielded complete light curves in the *UBV* system on four closely spaced, superior quality nights. These new light curves are used to improve the ephemeris and to provide new solutions by the computer technique of Wilson and Devinney. It is found that UV Lyn is an over-contact binary. Spectroscopic observations will be needed to choose between the two candidate solutions, but arguments are presented to show that the solution with a mass ratio 0.52 is preferred.

117.126 Determination of parameters of W UMa systems. III: CC Com, YY Eri, V 502 Oph and TY Pup.
C. Maceroni, L. Milano, G. Russo.
Astron. Astrophys., Suppl. Ser., Vol. 49, 123 - 128 (1982).
The lightcurves of four W UMa binaries have been solved using a minicomputer version of the Wilson and Devinney (1971) code, and the absolute elements have been computed. Three systems (CC Com, YY Eri, V 502 Oph) are found to be of W-type, and another one (TY Pup) of A-type.

117.127 Le stelle simbiotiche. A. Vittone.
G. Astron., Vol. 7, 277 - 284 (1981).

117.128 Die RS-Canum-Venaticorum-Doppelsterne.
S. Rössiger.
Sterne, 58. Band, 147 - 157 (1982).

117.129 Photometric observations of RS Canum Venaticorum stars. J. R. Percy, D. L. Welch.
J. R. Astron. Soc. Canada, Vol. 76, 185 - 204 (1982).

117.130 Satellite observations of cataclysmic variables.
K. O. Mason, F. A. Cordova.
Sky Telesc., Vol. 64, 25 - 29 (1982).

117.131 Une étoile particulière: AM Herculis.
M. Verdenet.
Bull. AFOEV, No. 20, p. 51 - 54 (1982).

117.132 The symbiotic star CI Cygni.
J. A. Mattei.
J. American Assoc. Variable Star Obs., Vol..10, 92 - 95 (1981).
The spectroscopic and optical characteristics of symbiotic stars are summarized. Long-term visual and photographic behavior of the symbiotic star, CI Cygni, is discussed.

117.133 Planetary nebulae with close binary nuclei-corrections to angular momentum loss.
J. Salzman, M. Livio, G. Shaviv.
Astron. Astrophys., Vol. 109, 201 - 204 (1982).
Corrections to the angular momentum per unit mass loss through L_2 in a close binary system are calculated. Near L_2 an expansion of the hydrodynamic equations following the procedure developed by Lubow and Shu is solved numerically. Then the single particle ballistic orbits are followed. The calculation is found to introduce only small corrections to the restricted three body problem results. The effects of radiation pressure on the angular momentum per unit mass loss is evaluated in a very approximate way.

117.134 The ellipsoidal binary V470 Cygni.
G. Russo, L. Milano, C. Maceroni.
Astron. Astrophys., Vol. 109, 368 - 370 (1982).
The yellow and blue photoelectric lightcurves of the double lined spectrum binary V 470 Cyg (Ebbighausen et al. 1975) have been analyzed by means of the Wilson and Devinney (1971) method. The system turns out to be a detached one, with two almost equal B2−B3 components, which do not eclipse each other because of the very low inclination. At primary minimum, the bigger and more massive star is in front.

117.135 Tidal evolution in close binary systems for high eccentricity. P. Hut.
Astron. Astrophys., Vol. 110, 37 - 42 (1982).
The weak friction model for tidal interaction in a close binary system was investigated in a previous paper, where a classification was given of the different types of tidal evolution behavior. Here the analysis of highly eccentric orbits is refined

by introducing a coordinate transformation, regular across the transition between elliptic and hyperbolic orbits.

117.136 Synchronization in binary stars.
R. Rajamohan, P. Venkatakrishnan.
Bull. Astron. Soc. India, Vol. 9, 309 - 318 (1981).
The authors find that the tidal effects during the main sequence lifetime of early type stars cannot explain the observed rotational behaviour of close binary stars. They have shown the existence of two sequences of stars, a synchronous or a peculiar sequence and a non-synchronous or a normal sequence. An initial distribution of magnetic fields of the stars might explain the existence of the two sequences.

117.137 A multiwavelength study of the AM Herculis type binary 2A 0311-227. J. Bailey, J. H. Hough, D. J. Axon, I. Gatley, T. J. Lee, P. Szkody, G. Stokes, G. Berriman.
Mon. Not. R. Astron. Soc., Vol. 199, 801 - 815 (1982).
The authors present the results of extensive photometric and polarimetric observations of the AM Herculis type binary 2A 0311-227 made during the 1979 observing season. Light curves at wavelengths from 0.36 to 2.2 μm and polarization data from 0.44 to 1.25 μm have been obtained. The observations show substantial wavelength dependence in the form of the light curves and polarization curves. The authors find that many features of the observations are consistent with recent calculations of the polarized cyclotron radiation from magnetic accretion columns. They discuss a model for the system in terms of a single accretion column which is optically thick to cyclotron radiation in the infrared becoming optically thin at shorter wavelengths.

117.138 The precession of gaseous stars.
J. C. B. Papaloizou, J. E. Pringle.
Mon. Not. R. Astron. Soc., Vol. 200, 49 - 69 (1982).
The authors discuss the precession of a gaseous star in a close binary system. They consider the case when the precessional amplitude is small and much less than the tidal distortion of the star justifying the treatment of the problem as one in linear oscillation theory. The authors find a close parallel with the problem of apsidal motion. The problem is complicated by the fact that the stellar oscillation spectrum is dense at frequencies of interest. Nevertheless it is shown that precession can still occur for time-scales of physical interest. The analogy with apsidal motion allows one to show that precession is damped faster than, or on a comparable time-scale to, the damping of eccentricity by tidal effects. Thus one expects to find no evidence of precession in the close binary systems, such as Her X-1, in which tidal circularization has already occurred.

117.139 Spectrophotometry of PG 1550+191 at red wavelengths. J. Echevarría, D. H. P. Jones, R. Costero.
Mon. Not. R. Astron. Soc., Vol. 200, 23P - 26P 1982).
The spectrum of PG 1550+191, the new AM Herculis variable discovered by Stockman et al. (1981), has been observed from λλ 4400 to 8300 Å and shows titanium oxide bands. These bands can be associated with an M dwarf whose spectrum is superimposed on a continuum, together with an emission spectrum. Lower and upper limits for the distance are 61 and 583 pc respectively.

117.140 Spot activity in the RS CVn binary HR 1099.
M. V. Mekkaden, A. V. Raveendran, S. Mohin.
J. Astrophys. Astron., Vol. 3, 27 - 38 (1982).
UBV photometry of HR 1099 obtained during the 1979 - 80 and 1980 - 81 observing seasons is presented. An analysis of the available data shows that the brightness at the light curve maximum increases as the wave amplitude increases, while the brightness at the light minimum remains almost the same. In terms of the starspot model it implies that there is always a

hemisphere of the active component that is nearly "saturated" with spots and that spots occupy a larger fraction of the stellar surface when the wave amplitude is smaller. It is found that the two-spot model proposed by Dorren and Guinan (1982) is inadequate to describe all the observed photometric peculiarities of HR 1099.

117.141 Photometric and spectroscopic observations of the optical counterpart of H2215−086.

A. W. Shafter, D. M. Targan.
Astron. J., Vol. 87, 655 - 664 (1982).

The authors present photometric and spectroscopic observations of the optical counterpart of the X-ray source H2215−086 which reveal it to be a short-period binary system. In addition to the orbital period there is a 21-min periodicity. The authors suspect that this 21-min optical pulsation is due to hard radiation from the magnetic poles of a compact star which has been reprocessed by material which is at rest in a frame corotating with the binary. This interpretation predicts that hard X-ray pulsations should be seen at the beat period between the orbital and 21-min periods, corresponding to the rotation period of the compact object.

117.142 Spectrophotometry of the RS CVn stars. I. The F, G, and K standards.

S. J. Adelman, S. N. Shore.
Astron. J., Vol. 87, 665 - 669 (1982).

The authors present energy distributions $\lambda\lambda 4032-10800$ and effective temperatures for 14 MK standards between F3 and K4. The stars were chosen to cover the spectral range of the components of RS CVn stars and the evolved members of Algol systems.

117.143 The interacting early-type contact binary SV Centauri.

H. Drechsel, J. Rahe, W. Wargau, B. Wolf.
Astron. Astrophys., Vol. 110, 246 - 262 (1982).

Extensive optical and UV observations of the interacting early-type contact binary SV Cen (B1V + B6.5III) were obtained between 1978 and 1981. Orbital elements and absolute dimensions were derived from the optical photoelectric and spectroscopic data. The unsteady period changes can be explained by mass loss through an outer Lagrangean point. The UV spectroscopy yielded the detection of an expanding circumbinary envelope. Finally, the authors consider the evolutionary stage of the system.

117.144 New evidence of strong UV radiation in TT Ari.

W. Wargau, H. Drechsel, J. Rahe, N. Vogt.
Astron. Astrophys., Vol. 110, 281 - 286 (1982).

Three SWP and one LWR spectra of the cataclysmic variable TT Ari have been obtained with the IUE satellite during three different brightness phases, including the brightness minimum of November 1980. The absolute continuum intensity distribution in the UV and optical spectral region is compared with several steady state optically thick accretion disk models. It is shown that the flux distribution cannot be represented by any of these approaches. The strong excess UV radiation is in good agreement with a $F_\nu \propto \nu^2$ power law.

117.145 Visual and near infrared photometry of 2A 0311-227.

C. Motch, J. van Paradijs, H. Pedersen, S. A. Ilovaisky, C. Chevalier.
Astron. Astrophys., Vol. 110, 316 - 323 (1982).

The authors present the results of extensive $V, R,$ and I photometric observations of the AM Her type system 2A 0311-227. They find an orbital period of $0.0562659 \pm 4 \times 10^{-7}$d and suggest, within the framework of the "bare bones" model of White (1980) and Schneider and Young (1980), emission from an optically thick accretion column.

117.146 Report of IAU Commission 42: Close binary stars (Étoiles doubles serrées).   B. Warner.

Trans. IAU, Vol. XVIIIA, (see 003.013), p. 553 - 577 (1982).

An atlas of southern and equatorial dwarf novae. See Abstr. 002.062.

Landolt-Börnstein.   See Abstr. 003.064.

Bias of polarimetric estimators for binary star inclinations.   See Abstr. 031.501.

The polarization properties of magnetic accretion columns.   See Abstr. 063.001.

Tides in differentially rotating convective envelopes. II. The tidal coupling.   See Abstr. 064.005.

Vertical structure of accretion disks. See Abstr. 064.019.

Model chromospheres of RS CVn stars: Balmer line profiles in λ Andromedae.   See Abstr. 064.036.

A model of two-stream non-radial accretion for binary X-ray pulsars.   See Abstr. 064.077.

Hydrodynamics of X-ray induced stellar winds. See Abstr. 064.081.

Orbital perturbations of a gravitationally bound two-body system with the passage of gravitational waves. See Abstr. 066.012.

Infrared observations of V1016 Cygni. See Abstr. 113.024.

The Wolf-Rayet star HD 193077: evidence for a low-mass companion and the possibility of a third body. See Abstr. 114.015.

Absorption line symmetries for two HgMn stars. See Abstr. 114.039.

IUE observations of 17 Lep. See Abstr. 114.153.

Report of IAU Commission 29: Stellar spectra (Spectres stellaires).   See Abstr. 114.224.

High angular resolution observation of single and close binary stars with the CERGA interferometer. See Abstr. 115.017.

Periodic radio emission from LS I +61°303. See Abstr. 116.010.

Radio emission from AM Herculis-type binaries. See Abstr. 116.011.

β Lyrae − a magnetic binary star. See Abstr. 116.017.

Rotation, expansion and duplicity of Be stars. See Abstr. 116.044.

Revised photometric elements of the eclipsing binary EE Aquarii.   See Abstr. 119.018.

Algol.   See Abstr. 119.022.

Computation of the elements of close eclipsing systems in the frequency-domain.   See Abstr. 119.027.

Revised photometric data for six eclipsing binaries. See Abstr. 119.120.

Z Chamaeleontis: evidence for an eccentric disk during supermaximum?   See Abstr. 122.006.

On the nature of Hα outbursts in the RS Canum Venaticorum binary SZ Piscium.   See Abstr. 122.014.

Kuwano's peculiar object is a novalike (symbiotic?) binary with a red giant. Discussion of observational results. See Abstr. 122.072.

H2215 - 086, king of the DQ Herculis stars. See Abstr. 122.154.

The old-nova GK Per. II. Optical outbursts. See Abstr. 124.201.

A photometric and polarimetric investigation of the old nova RR Pictoris.   See Abstr. 124.701.

The complex emission-line structure in the magnetic white dwarf binary 2A 0311−227 (EF Eridani). See Abstr. 126.003.

An analysis of the cyclotron spectrum and polarization properties of VV Puppis.   See Abstr. 126.007.

A new model for white dwarf supernovae. See Abstr. 126.013.

Planetary nebulae with close binary central stars. See Abstr. 135.002.

SS433 − observing evolution in a precessing, relativistic jet.   See Abstr. 141.144.

The compact radio structure of SS433. See Abstr. 141.145.

SS433: periodic changes in the radio structure of scale size 10^{16} cm.   See Abstr. 141.146.

A study of Sk 160/SMC X-1. See Abstr. 142.001.

The discovery of 50 minute periodic absorption events from 4U 1915−05.   See Abstr. 142.015.

Discovery of a 50 minute binary period and a likely 22 magnitude optical counterpart for the X-ray burster 4U 1915−05.   See Abstr. 142.016.

The low mass X-ray binary 2A 1822−371. See Abstr. 142.024.

Ultraviolet spectrophotometry of 2A 1822−371: a bulge on the accretion disk.   See Abstr. 142.025.

HEAO 1 observations of the long-term variability of Hercules X-1.   See Abstr. 142.029.

Mass determination of massive X-ray binaries. See Abstr. 142.031.

Einstein observations of BY Draconis variables and RS CVn binaries.   See Abstr. 142.038.

Accretion disk coronae. See Abstr. 142.047.

Optical identification of X-ray source H0139−68 with an AM Herculis-type system.   See Abstr. 142.093.

Visual and near infrared photometry of 2A 0311-227. See Abstr. 142.094.

The noncompact binary X-ray source 4U 2129+47. See Abstr. 142.100.

Photoelectric photometry of 4U 2129 + 47. See Abstr. 142.102.

The orbital period of 2S 1223−624 (GX301−2). See Abstr. 142.103.

An interpretation of the X-ray spectrum of 4U 1822−37.   See Abstr. 142.104.

Errata

117.901   Erratum: "Instability in detached binary systems of the main sequence with solar-type components. II. BH Virginis" [Perem. Zvezdy, Tom 20, 577 - 587 (1978)]. R. A. Botsula. Perem. Zvezdy, Tom 21, 463 (1980). In Russian. − See Abstr. 27.117.032.

117.902   On the photometric behavior of SS 433 in 1979 and 1980.   T. Mazeh, E. M. Leibowitz, O. Lahav. Astrophys. Lett., Vol. 22, 55 - 61 (1981). − See Abstr. 30.117.115.

118 Visual Binaries, Multiple Stars, Astrometric Binaries

118.001   **Observations of binary stars by speckle inter-**
          ferometry – III.
B. L. Morgan, G. K. Beckmann, R. J. Scaddan, H. A. Vine.
Mon. Not. R. Astron. Soc., Vol. 198, 817 - 824 (1982).
   The results of 25 measurements of 21 objects are presented. The objects include long-period spectroscopic binaries from
the 6th Catalogue of Batten (1968) and close visual binary
systems from the 3rd Catalogue of Finsen and Worley (1970).
Eight of the objects have not been previously resolved by
speckle interferometry, and six of the objects are members of
the Hyades cluster. These results are sufficiently accurate to
demonstrate that past errors by visual observers cannot explain
the apparent deviation of Hyades stars from the normal mass-
luminosity relationship; however, if the distance modulus
calculated by Hanson is assumed, it is not necessary to assume
a separate relationship. γ Tau is resolved into two components
for the first time.

118.002   **Orbits of three southern visual binaries.**
          C. E. Worley.
Publ. Astron. Soc. Pacific, Vol. 93, 772 - 773 (1981/82).
   Orbits of the visual binaries RST 321, HU 225, and
OL 22 are presented. The latter is a first determination.

118.003   **Wide binaries in the solar neighborhood.**
          J. M. Retterer, I. R. King.
Astrophys. J., Vol. 254, 214 - 220 (1982).
   The data on wide binaries provided by Bahcall and
Soneira are used to augment the data from other surveys of
binaries. The authors then interpret all these data using a
dynamical theory for the behavior of wide binaries that experience encounters with field stars. The sharp drop in binary
frequency beyond 2×10^4 AU is a consequence of encounters,
and the shape of the drop-off follows from the theory.

118.004   **Interferometric observations of binary stars. I.**
          A. A. Tokovinin.
Pis'ma Astron. Zh., Tom 8, 43 - 47 (1982). In Russian. English
translation in Soviet Astron. Lett., Vol. 8.
   For 68 visual binaries position angles and separations are
measured at the 60-cm reflector with a photoelectric phase-
grating interferometer. The method of data processing is described, random and systematic errors are estimated. From the
analysis of residuals from ephemerides it is shown that for
close pairs ($\rho \leqslant 0\rlap{.}''2$) the existing orbits predict systematically
greater separations.

118.005   **Interferometric observations of binary stars. II.**
          A. A. Tokovinin.
Pis'ma Astron. Zh., Tom 8, 187 - 192 (1982). In Russian.
English translation in Soviet Astron. Lett., Vol. 8.
   For 83 visual binaries ($0\rlap{.}''065 \leqslant \rho \leqslant 1\rlap{.}''42$) position angles
and separations are measured. 20 pairs show significant devia-
tions from the ephemerides, 9 stars remained unresolved.

118.006   **Photographic measures of visual double stars.**
          R. Pannunzio, F. Siciliano.
Astron. Astrophys., Suppl. Ser., Vol. 47, 159 - 165 (1982).
   157 photographic measures of selected visual double
stars are given; for this programme the 105 cm astrometric
reflector of the Astronomical Observatory of Torino was
used. Further improvements to the reduction method, with
respect to that used in previous papers, have been made.

118.007   **Orbital elements of visual binary stars ADS 221 and**
          ADS 1762.   M. Scardia.
Astron. Astrophys., Suppl. Ser., Vol. 47, 167 - 170 (1982). In
French.

The orbital elements of the visual binary stars ADS 221
and ADS 1762 are given. At last, the dynamical parallaxes and
total masses of the systems have been calculated.

118.008   **The orbits of the visual double stars ADS 10621**
          and ADS 15650.   P.-J. Morel.
Astron. Astrophys., Suppl. Ser., Vol. 47, 217 - 219 (1982).
   The orbital elements, the ephemeris and the residuals
from observations for the two visual double stars ADS 10621
and ADS 15650 are given.

118.009   **Photometric and astrometric observations of close**
          visual binaries.   K. D. Rakos, R. Albrecht,
H. Jenkner, T. Kreidl, R. Michalke, D. Oberlerchner,
E. Santos, A. Schermann, A. Schnell, W. Weiss.
Astron. Astrophys., Suppl. Ser., Vol. 47, 221 - 235 (1982).
   The first photoelectric sequence for the magnitude
differences of 215 close visual binaries in the Johnson UBV
and Strömgren $uvby$ systems have been established. In addition, the position angle and the separation of 140 stars were
measured. Finally, for 134 stars, new photoelectric measurements of the combined integral brightness of both components in the UBV and Strömgren systems were made. The
measurements were carried out using the area scanning
technique.

118.010   **Orbits of 16 visual binaries.**   W. D. Heintz.
          Astron. Astrophys., Suppl. Ser., Vol. 47,
569 - 573 (1982).
   Elements and masses of the pairs ADS 2980, 3114,
4876, 6276, 7555, 8486, 9532, 9654, 9836, 10007, 10196,
10542, 14360, h 3494, Fin 326, and Cou 14 have been
determined with recent observations.

118.011   **Parallaxes, mass ratios, masses, and planetary detec-**
          tion capability from 60-yr Sproul plate series on
          five visual binaries.   J. L. Hershey.
Astron. J., Vol. 87, 145 - 151 (1982).
   Parallaxes, mass ratios, and masses have been determined
for five visual binaries from 60-yr time interval Sproul plate
series totaling over 700 nights of observation. Ho 296 serves
as a test of the ability of the Sproul refractor to detect a long-
period low-amplitude orbit comparable to the perturbation of
a nearby star by a planetary companion. Upper limits for dark
companions to the components of Σ 2398 are at the level of
one to several Jupiter masses for periods ranging from a few
years to several decades.

118.012   **Wolf 630: a peculiar astrometric binary.**
          E. W. Weis.
Astron. J., Vol. 87, 152 - 154 (1982).
   Analysis of the measures of 55 plates taken at the Van
Vleck Observatory indicates a semimajor axis of the photo-
centric orbit of $0\rlap{.}''052 \pm 0\rlap{.}''007$, in good agreement with
previous results from the Sproul Observatory. A perturbation
of comparable amplitude is present in most of the older
parallax series of other observatories. This implies that
component B is twice as massive as component A.

118.013   **Is 21 Ari = COU 79 a multiple system?**
          P. Couteau, P.-J. Morel.
Astron. Astrophys., Vol. 105, 323 - 325 (1982).
   With the main sequence mass luminosity relation, the
dynamical parallax computed from the elements of a preliminary orbit, leads to masses of $1.8\,M_\odot$ for the components of
the visual pair COU 79 = 21 Ari. This result is not consistent with
the F6 V spectral type. This discrepancy can be explained
either by an error in the luminosity class, or, more likely, by

the presence of a third body of one solar mass closer than 0''.03 to one of the components.

118.014   Data inferred from visual binaries.
E. L. van Dessel, J. Dommanget.
Irish Astron. J., Vol. 14, (see 012.003), 226 - 237 (1980).

118.015   Una comparación entre la estrella de Barnard y el sistema solar.   C. J. Lavagnino.
Bol. Asoc. Argentina Astron., No. 18, p. 70 - 71 (1980).

118.016   Orbits of 25 visual binaries.
G. A. Starikova.
Pis'ma Astron. Zh., Tom 8, 306 - 307 (1982). In Russian.
English translation in Soviet Astron. Lett., Vol. 8.

From new observations and from available published data revised orbits of 25 visual binary stars are obtained. The new orbits are in better agreement with observations.

118.017   Speckle interferometric measurements of binary stars. VI.   H. A. McAlister, E. M. Hendry.
Astrophys. J., Suppl. Ser., Vol. 48, 273 - 278 (1982) = Astron. Contrib. Georgia State Univ., No. 56.

Three hundred and fifty-four measurements of 169 binary stars observed by means of speckle interferometry with the 2.1-m telescope at KPNO are presented. Measured separations range from 0''.066 to 4''.09. The previously unknown binary HD 46100 is resolved for the first time.

118.018   Planetary systems close to the nearest stars − perspectives for search.   L. V. Ksanfomaliti.
Zemlya Vselennaya, 1982, No. 3, p. 65 - 69. In Russian.

118.019   Las nuevas binarias astrométricas y el sistema solar.   C. J. Lavagnino.
Bol. Asoc. Argentina Astron., No. 20 - 24, p. 217 (1981). Abstract.

118.020   Estadística de bin as.   O. Ferrer.
Bol. Asoc. Argent ⎯ ᵢ Astron., No. 20 - 24, p. 398 (1981). − Abstract.

118.021   Determinación de diferencias de magnitud en dobles visuales.   O. Ferrer.
Bol. Asoc. Argentina Astron., No. 20 - 24, p. 399 (1981). Abstract.

118.022   Photométrie d'étoiles doubles.
E. Oblak.
Bull. Inf. Cent. Données Stellaires, No. 22, p. 41 - 50 (1982).

118.023   Double stars.   Z. Komárek.
Říše hvězd, Vol. 63, 10 - 13 (1982).   In Slovak.

118.024   Orbites nouvelles.
Circ. Inf., No. 86 (1982).

118.025   Etoiles doubles nouvelles.
W. D. Heintz, P. Couteau, G. M. Popovic, P. Muller.
Circ. Inf., No. 86 (1982).

118.026   X-ray and UV observations of the rapidly rotating triple system HD 165590.
R. A. Stern, A. Skumanich.
Bull. American Astron. Soc., Vol. 13, 812 (1981). − Abstract.

118.027   Reduction of astrometric binaries for low mass companions.   J. L. Russell.
Bull. American Astron. Soc., Vol. 13, 873 (1981). − Abstract.

118.028   Infrared detection of the low mass companion to Zeta Aquarii B.   D. W. McCarthy, F. J. Low,
S. G. Kleinmann, D. V. Arganbright.
Prepr. Steward Obs., No. 394, 15 pp. (1982).

118.029   Photographic measures of double stars.
F. J. Josties, V. V. Kallarakal, G. G. Douglass, J. W. Christy.
Publ. United States Naval Obs., Washington, Second Ser., Vol. 24, Part V, 76 pp. (1978).

The series of photographic measures of double stars reported here covers approximately a four year period and is complete through the winter observing season 1976-77. This is a continuation of three earlier series. The paper lists 1931 measures of $\Delta\alpha \cos \delta$ and $\Delta\delta$ on 1796 plates of 253 double stars.

118.030   New double stars (17th series) discovered at Nice.
P. Couteau.
Astron. Astrophys., Suppl. Ser., Vol. 48, 443 - 447 (1982). In French.

The author gives a list of 150 double stars discovered at the 50 cm refractor.

118.031   Orbit of the visual binary Burnham 1164 = ADS 1156 = B 765 (95 Piscium).
R. R. de Freitas Mourão, H. Debehogne.
Astron. Nachr., Band 303, 157 - 159 (1982). In French.

The physical and orbital elements of the binary star ADS 1156 = B 1164 ($\alpha_{1900} = 01^h 22^m 0$, $\delta_{1900} = +04°50'$) are determined. The absolute visual magnitudes of the two components are $3^m 72$ resp. $3^m 97$, the masses $1.43\,M_\odot$ resp. $1.33\,M_\odot$.

118.032   A catalogue of variable and visual double stars.
D. Proust, F. Ochsenbein, B. R. Pettersen.
J. American Assoc. Variable Star Obs., Vol. 10, 56 (1981). Abstract.

118.033   Spectrophotometric analysis of the atmosphere of the bright component of the binary system v Sgr.
V. V. Kravtsov, V. V. Leushin.
Rostov. n/D. univ. Rostov n/D., 1981, 21 pp. In Russian.
Abstr. in Ref. zh., 51. Astron., 6.51.432 (1982).

118.034   A study of the intensive 40-yr Sproul plate series on Lalande 21185 and BD + 5°1668.
J. L. Hershey, S. L. Lippincott.
Astron. J., Vol. 87, 840 - 844 (1982).

New measurements of the entire intensive plate series of the stars Lal 21185 and BD + 5°1668 yield upper limits of very small mass for dark companions. Simulations of planetary effects illustrate the high detectability of planetary masses and suggest that the complexities of a multibody planetary system need not be a serious obstacle to detection.

118.035   Imaging properties of Hipparcos and the observation of multiple stars.   L. Lindegren.
Scientific aspects of the Hipparcos space astrometry mission, (see 012.041), p. 147 - 152 (1982).

The image of a compound object such as a double or multiple star can to a certain extent be recovered from observations with the astrometry satellite Hipparcos. Examples are given of the effective "beam pattern" for Hipparcos and of the resulting resolution and signal-to-noise ratio. The preliminary results indicate that many new binaries with separations in the range 0.05 - 0.5 arcsec can be discovered with the satellite.

118.036   Caractéristiques des étoiles doubles observables par Hipparcos.   P. Bacchus.
Scientific aspects of the Hipparcos space astrometry mission, (see 012.041), p. 153 - 159 (1982).

On détermine les conditions à remplir par une étoile double pour qu'elle puisse être traitée par Hipparcos comme une étoile simple, sans erreur notable. Ces conditions sont satisfaites par environ 40% des doubles d'orbite connue. L'observation peut aussi être traitée en tenant compte du caractère double, ce qui permet dans les cas favorables d'obtenir les paramètres astrométriques et spectrophotométriques du couple. Des tests sur le signal fourni par Hipparcos permettent de détecter les étoiles doubles, jusque vers une séparation de 0,"1 ou 0,"2 selon le Δm.

118.037   The interest of double-star observations by
          Hipparcos.   J. Dommanget.
Scientific aspects of the Hipparcos space astrometry mission, (see 012.041), p. 161 - 164 (1982).

118.038   Orbital evolution of triple stars.
          S. Söderhjelm.
Rep. Obs. Lund, No. 18 (see 012.044), p. 88 (1982).

118.039   Report of IAU Commission 26: Double stars
          (Etoiles doubles).   O. G. Franz.
Trans. IAU, Vol. XVIIIA, (see 003.013), p. 257 - 262 (1982).

118.040   On the orbital and radial motions of α Centauri.
          W. D. Heintz.
Observatory, Vol. 102, 42 - 43 (1982).

A catalogue of photoelectric magnitudes and colours of visual double and multiple systems.
See Abstr. 002.061.

Landolt-Börnstein.   See Abstr. 003.064.

L'astronomie des étoiles doubles à l'heure de l'astrométrie spatiale.   See Abstr. 013.001.

Speed of light outside the solar system: a new test using visual binary stars.   See Abstr. 022.009.

Sur quelques causes d'erreur dans les observations d'étoiles doubles: vraies et fausses étoiles doubles.
See Abstr. 031.506.

New photographic method for the measurement of visual binaries.   See Abstr. 031.525.

Studies of the stellar three-body problem.
See Abstr. 042.008.

Molecular abundances in IRC + 10216.
See Abstr. 064.023.

uvbyβ photometry of visual double stars: a comparison with stellar models and isochrones.   See Abstr. 064.100.

Contracting members of double stars.
See Abstr. 065.068.

uvbyβ photometry of visual double stars: absolute magnitudes of intrinsically bright stars.   See Abstr. 113.020.

The visual double W UMa binary BV and BW Draconis.   See Abstr. 119.006.

Discovery of an infrared companion to T Tauri.
See Abstr. 121.009.

A preflare diminution in the quiescent flux of EQ Pegasi.   See Abstr. 122.004.

The diameter of Mira.   See Abstr. 122.032.

Photoelectric observations of 20 Leo (HR 3889).
See Abstr. 122.127.

Mira B.   See Abstr. 122.193.

Two unusual cepheids in multiple systems — photometry, radial velocities and IUE spectra.
See Abstr. 122.244.

More variable stars among bright visual binaries.
See Abstr. 123.032.

A study of ultraviolet spectroscopic and light variations in the X-ray binaries LMC X-4 and SMC X-1.
See Abstr. 142.028.

Circumstellar matter in young clusters. IV. Photometric measurements of the 2200 hump.
See Abstr. 153.011.

119 Eclipsing Binaries

119.001 Further spectrographic observations of GG Carinae.
C. A. Hernández, L. López, J. Sahade,
A. D. Thackeray.
Publ. Astron. Soc. Pacific, Vol. 93, 747 - 751 (1981/82).

Spectrograms of GG Car appear to confirm that we are dealing with a 31-day period binary system. The component that is behind at eclipse is surrounded by an expanding, thick, extended envelope that does not allow us to "see" the lines arising in the stellar photosphere. The whole system appears to be surrounded by an outer envelope.

119.002 Minima and photometry for southern hemisphere eclipsing binaries. A. D. Mallama.
Publ. Astron. Soc. Pacific, Vol. 93, 774 - 776 (1981/82).

Photoelectric data for ten prominent systems are reported in this paper. The heliocentric times of minimum determined from V and B observations, respectively, are presented.

**119.003 Rediscussion of eclipsing binaries. XIII.
DI Herculis, a B-type system with an eccentric orbit.** D. M. Popper.
Astrophys. J., Vol. 254, 203 - 213 (1982).

The properties of the components of the detached middle B-type binary, DI Her, are derived from a new series of Lick spectrograms and the Martynov-Khalliulin photometry. The components lie along the lower envelope of the distributions of stars with well determined masses and radii in the mass-radius and color-absolute magnitude planes. They are also close to the empirical zero-age main sequence based on open clusters.

119.004 Light curve of the eclipsing system CX Cep and characteristics of the Wolf-Rayet star.
N. A. Lipunova, A. M. Cherepashchuk.
Astron. Zh., Tom 59, 73 - 86 (1982). In Russian. English translation in Soviet Astron., Vol. 26, No. 1.

Photoelectric B V R observations of the eclipsing Wolf-Rayet binary CX Cep have been obtained. The physical characteristics of the WN 5 star, the core radius and the brightness temperature of the core are determined. The characteristics are close to those of the WN 5 star in the eclipsing Wolf-Rayet binary V 444 Cyg. The results of the interpretation of the light curves of the two eclipsing Wolf-Rayet binaries V 444 Cyg and CX Cep confirm the conclusions of the modern theory of evolution of massive close binary systems.

119.005 The variable light curve of BH Virginis.
M. Hoffmann.
Astron. Astrophys.,Suppl. Ser., Vol. 47, 561 - 568 (1982).

Photoelectric observations of the short period eclipsing binary BH Vir in B and two selected narrow bands are presented. Night-to-night variations, as well as a larger change of the light curve compared with earlier published observations, indicate activity of the system. Relations to the RS CVn group of stars are discussed briefly.

119.006 The visual double W UMa binary BV and BW Draconis.
E. H. Geyer, M. Hoffmann, M. T. Karimie.
Astron. Astrophys., Suppl. Ser., Vol. 48, 85 - 91, 1 microfiche (1982).

The eclipsing binaries BV and BW Dra have been observed photoelectrically from 1974 to 1980. BV Dra has a fairly regular and constant light curve, whereas the shorter-period system BW Dra shows frequent activity in its light curve. Orbital elements of BV Dra have been determined which show that its primary minimum is a transit. Also a new astrometric observation of the visual pair BV and BW Dra is reported.

119.007 The symbiotic star CI Cygni: s-process episode or accretion event?
S. J. Kenyon, R. F. Webbink, J. S. Gallagher, J. W. Truran.
Astron. Astrophys., Vol. 106, 109 - 111 (1982).

Evidence that the symbiotic star CI Cygni is an eclipsing binary is reviewed. It is shown that the "s-process episode" described by Audouze et al. (1981) during its 1975 outburst arises from superposition of normal gM 4 absorption features on the continuum of the hot component during eclipse ingress. The peculiar velocity displacements of absorption lines with different excitation potentials during this episode are identified as signatures of an optically thick accretion disk. The data presented by Audouze et al., and the shape of the light curve thus provide evidence that the outburst is accretion-powered.

119.008 LB 3459 — an O-type subdwarf eclipsing binary system. Non-LTE analysis of the primary.
R. P. Kudritzki, K. P. Simon, A. E. Lynas-Gray, D. Kilkenny, P. W. Hill.
Astron. Astrophys., Vol. 106, 254 - 260 (1982).

A non-LTE analysis is presented for the primary component of the sdO eclipsing binary LB 3459. The results of the analysis give $T_{eff} = 40,000^{+3000}_{-2000}$ K, $\log g = 5.3 \pm 0.2$ (cgs) and a surprisingly low helium abundance of $y = 0.3^{+0.2}_{-0.1}$ % by numbers. LB 3459 appears to be the result of mass exchange during the first giant phase. The primary most likely has a mass of $0.3\,M_\odot$ and consists of a hydrogen shell and degenerate helium core. Diffusion is suggested as the cause of helium depletion in the photosphere. The secondary has a mass of $0.04\,M_\odot$ and a radius of $0.09\,R_\odot$, which is close to a degenerate configuration of solar composition.

119.009 The eclipsing binary programme.
J. E. Isles.
J. British Astron. Assoc., Vol. 92, 76 - 80 (1982).

119.010 Photoelectric observations of BO Cephei.
V. I. Kardopolov, N. A. Shutemova.
Perem. Zvezdy, Tom 21, 305 - 309 (1980). In Russian.

Results of UBV photoelectric photometry of BO Cep obtained in June–September 1978 are presented. The amplitude of BO Cep is found to be about 0^m4 in V. Irregular variations of the star brightness are still questionable. BO Cep is expected to be an eclipsing variable star.

**119.011 Derivation and application of simple criteria to a wide classification of eclipsing variable stars.
I. Derivation of the criteria for a wide classification of eclipsing variable stars.** M. A. Svechnikov, L. F. Istomin, O. A. Grekhova.
Perem. Zvezdy, Tom 21, 399 - 412 (1980). In Russian.

New simple criteria for classification of eclipsing binaries according to their physical properties were developed on the basis of the card catalogue by Svechnikov (1969) and observational data related to depths of minima, periods and spectra of the main components. The new criteria would permit to classify close binaries with the confidence level more then 90%.

119.012 Derivation and application of simple criteria to a wide classification of eclipsing stars. II. Classification of eclipsing variables included in the GCVS (3rd edition) and three supplements. M. A. Svechnikov, L. F. Istomin, O. A. Grekhova.
Perem. Zvezdy, Tom 21, 413 - 443 (1980). In Russian.

On the basis of the criteria derived in part I of the authors' work a classification of 4704 eclipsing binaries included in GCVS (3rd edition) and three supplements was carried out. The percentage of different types of eclipsing variables is shown in a table.

119.013   RX Cassiopeiae.   D. Ya. Martynov,
G. V. Zajtseva, M. I. Kumsiashvili.
Perem. Zvezdy, Tom 21, 451 - 454 (1980). In Russian.

119.014   Photoelectric two-colour observations of MR Cygni in 1974.   V. P. Murnikova, O. P. Paramonova.
Perem. Zvezdy, Tom 21, 455 - 462 (1980). In Russian.
Photoelectric observations of the eclipsing binary MR Cygni in the BV system are given. 789 values of light in the B filter and 795 in the V filter have been obtained. The moment of primary minimum has been determined.

119.015   RS Ursae Minoris.   H. Grzelczyk.
BAV Rundbrief, 31. Jahrg., 18 (1982).

119.016   Spectra of OB eclipsing binaries.
D. M. Popper.
Astrophys. J., Suppl. Ser., Vol. 47, 339 - 355 (1981).
Microphotometer tracings are reproduced of the spectra of 26 OB-type close binaries in the wavelength range $\lambda\lambda 4300 - 4500$ Å. Most of the systems are included for which masses and radii, based directly on spectrographic and photometric orbits, have been published. The tracings demonstrate the difficulty of finding systems for which lines of the two components are clearly present but are not blended with each other, and, hence also demonstrate the probable presence of systematic errors in most published masses and radii of OB eclipsing binaries.

119.017   On excitation through radiative pumping of the Fe II UV-mult. 191 $\lambda\lambda 1785-88$ Å observed with IUE during the eclipse of 32 Cyg.   K. Hempe, D. Reimers.
Astron. Astrophys., Vol. 107, 36 - 38 (1982).
The authors show by means of high-resolution IUE spectra of 32 Cyg during and shortly after eclipse that the Fe II UV-multiplet 191 is pumped via the Fe II resonance UV-multiplet 9 $\lambda\lambda 1260-76$ Å. The authors show directly from observations (1) that all absorbed multiplet 9 photons are emitted as Fe II UV mult. 191 lines, (2) that the observed line intensity ratio 1785/1787/1788 is consistent with the pumping mechanism.

119.018   Revised photometric elements of the eclipsing binary EE Aquarii.   G. Russo, C. Sollazzo.
Astron. Astrophys., Vol. 107, 197 - 199 (1982).
The V lightcurves observed by Williamon (1974) and by Padalia (1979) of the eclipsing binary EE Aquarii, classified as a β Lyrae-type star, have been solved using a minicomputer version of the Wilson and Devinney (1971) method for light curve synthesis. An estimate of the absolute dimensions of the system has been performed. The system results to be composed of an A8V primary and a K3-K4 secondary which fills its lobe.

119.019   Energy distribution in the strongly interacting binary system SX Cassiopeiae.
M. J. Plavec, J. L. Weiland, R. H. Koch.
Astrophys. J., Vol. 256, 206 - 221 (1982).
One of several anomalies of the long-period eclipsing binary system SX Cassiopeiae is the presence of strong Balmer emission lines, inconsistent with the usual spectral classification of the components as A6 III + G 6 III. In a search for a hotter source, the authors obtained IUE spectra and optical scans covering the interval 110−680 nm. The IUE spectra showed two unexpected features: (1) strong emission lines of C IV, N V, Si IV etc., requiring even much higher temperatures than the Balmer lines; (2) a continuum several magnitudes brighter than the extrapolated continuum of an A6 star. A chromospheric origin of the emission is unlikely. The far-

ultraviolet flux excess is removed if the spectral types are revised to B7 + K3 III. A considerable amount of continuous hydrogen radiation is found, probably associated with a disk seen edge-on; this disk is probably responsible for strong line and continuous absorption of the light of the B7 star.

119.020   Discovery of an eclipse in the unique binary system BE Ursae Majoris.
H. Ando, A. Okazaki, S. Nishimura.
Publ. Astron. Soc. Japan, Vol. 34, 141 - 146 (1982).

119.021   The mystery of Epsilon Aurigae.
F. J. Reddy.
Sky Telesc., Vol. 63, 460 - 462 (1982).

119.022   Algol.   Y. Hosokawa.
Astron. Her., Vol. 74, No. 3, p. 70 - 75 (1981). In Japanese. − Abstr. in Phys. Abstr., Vol. 85, Abstr. 23321 (1982).

119.023   Computation of the elements of eclipsing binary systems in the frequency-domain.   Z. Kopal.
Astrophys. Space Sci., Vol. 81, 123 - 177 (1982).
The aim of the present paper is to translate the essential parts of the theory of Fourier analysis of the light changes of eclipsing variables into more practical terms and to describe numerical procedures with examples for this analysis. The scope of the present paper is restricted to an exposition of the analysis of light changes caused by eclipses of spherical stars; while between minima due to this cause the light of the system should remain sensibly constant.

119.024   The almost contact system RU Eridani.
G. Russo.
Astrophys. Space Sci., Vol. 81, 209 - 213 (1982).
The V and B light curves of the eclipsing binary RU Eridani have been analysed in order to obtain a solution based on the Roche model. The system turns out to consist of an evolved K1 secondary which appears to fill the lobe, capable of containing its mass, and of an F0 V primary, which seems almost to fill its own. An estimate of the absolute elements of RU Eri has been made, on the assumption that the primary has a mass corresponding to its spectral type.

119.025   Photometric elements and evolutionary status of the eclipsing binary DI Pegasi.   U. S. Chaubey.
Astrophys. Space Sci., Vol. 81, 283 - 293 (1982).
The light curves of eclipsing binary DI Pegasi in $U, B,$ and V filters have been presented and discussed. Photometric elements of the system have been determined. Using the colour indices, the author estimates absolute dimensions and discusses the evolutionary status of the system.

119.026   DE Draconis: spectroscopic orbit, rotation effect and physical model.
D. P. Hube, J. S. Couch.
Astrophys. Space Sci., Vol. 81, 357 - 368 (1982).
New spectroscopic observations are used for the determination of orbital elements and for a study of the rotation effect of DE Draconis. The primary component is found to rotate much faster than the synchronous rate and with its axis of rotation possibly inclined with respect to the perpendicular to the orbital plane. The system is fully detached and the stars are relatively young and unevolved.

119.027   Computation of the elements of close eclipsing systems in the frequency-domain.   Z. Kopal.
Astrophys. Space Sci., Vol. 81, 411 - 451 (1982).
The aim of the present paper is to generalize the methods for computation of the elements of eclipsing binary systems in the frequency-domain, summarized in a recent paper by the author, to the case of close systems, in which photometric

proximity effects become conspicuous and must be taken into account before the methods previously outlined become directly applicable.

119.028 Revised photometric elements of three semi-detached binary systems.
L. Milano, G. Russo, S. Marcozzi, A. D'Orsi.
Astrophys. Space Sci., Vol. 82, 189 - 198 (1982).
The V and B light curves of three Algol-type eclipsing binaries (UZ Cyg, VW Cyg, AQ Peg) have been analysed using the Wilson and Devinney model. The authors find that all of them are semi-detached systems, with A-type primary components and K-type evolved secondary which fill the corresponding lobe, although VW Cyg should perhaps be considered as an sd-d system.

119.029 Photoelectric observations of BV Draconis.
P. Rovithis, H. Rovithis-Livaniou.
Astrophys. Space Sci., Vol. 82, 229 - 240 (1982).
In this paper the two B and V light curves of the close eclipsing binary BV Draconis, obtained with the 48-in reflector of the National Observatory of Athens, are presented and discussed. Also, a new period of the system is given.

119.030 SV Centauri − Struktur und Wechselwirkungsprozesse.
H. Drechsel, J. Rahe, W. Wargau, B. Wolf.
Mitt. Astron. Ges., Nr. 55, p. 79 - 80 (1982). − Abstract.

119.031 Flare-Aktivität beim Fleckendoppelstern SV Cam.
L. Patkós.
Mitt. Astron. Ges., Nr. 55, p. 82 - 85 (1982).

119.032 Die spektroskopische Bahn des Bedeckungsveränderlichen AR Cas. M. Gaida, W. Seggewiß.
Mitt. Astron. Ges., Nr. 55, p. 163 - 164 (1982). − Abstract.

119.033 Der Bedeckungsveränderliche ER Vulpeculae.
H. W. Duerbeck, M. T. Karimie, J. D. Schumann.
Mitt. Astron. Ges., Nr. 55, p. 164 (1982). − Abstract.

119.034 The abundance of He in the atmosphere of V 380 Cyg (the evolutionary status of the system).
V. V. Leushin, L. I. Snezhko.
Astrofizika, Tom 17, 563 - 572 (1981). In Russian. English translation in Astrophysics, Vol. 17, No. 3.
The equivalent widths of hydrogen and helium lines of the bright component of the system V 380 Cyg have been analysed. Strengthening of all the absorption lines in the phase of coming from eclipse of the second component is found and interpreted by the presence of envelopes. The helium abundance is determined. A conclusion is made on the necessity of the evolutionary status revision for the system V 380 Cyg.

119.035 A recent time of minimum for V356 Sgr.
D. S. Hall, G. W. Henry, W. H. Murray.
Acta Astron., Vol. 31, 383 - 386 (1981).
A fragmentary light curve based on photoelectric observations obtained in 1979 yields a mid point for primary eclipse of JD(hel.) = 2444158.027. This indicates that the orbital period is not undergoing any change large enough to be measurable at this time.

119.036 AO Camelopardalis.
E. F. Milone, D. H. Piggott, S. L. Morris.
J. R. Astron. Soc. Canada, Vol. 76, 90 - 96 (1982) = Publ. Rothney Astrophys. Obs. No. 17.
The first UBV light curves of the system AO Cam are presented and reveal a W UMa light curve with a period of $0^{d}329917$. Fourier coefficients for the full B and V light curves were used to place approximate empirical limits on the inclination ($75° \lesssim i \lesssim 80°$), fill-out parameter ($f \sim 0.8$) and

mass-ratio ($0.7 \lesssim q \lesssim 0.8$) following the methods of Rucinski. The spectral type of the system is about G5.

119.037 RZ Ophiuchi − Minimum 1981.
M. Grünanger, W. Herzner.
Sterne Weltraum, 21. Jahrg., 214 - 215 (1982).

119.038 A model for 0921−63: a second halo X-ray source.
A. P. Cowley, D. Crampton, J. B. Hutchings.
Astrophys. J., Vol. 256, 605 - 611 (1982).
Spectroscopic and photometric observations reveal the orbital period of 0921−63 is $8^{d}99$. The He II and H emission are formed near the degenerate neutron star. The secondary, weakly seen in the spectrum, appears to be an F−G giant of $\sim 1 M_{\odot}$. The system eclipses so that ultimately precise radii and orbital inclination can be determined. The galactic location (~ 1.5 kpc below the plane) and its low mass (total mass $\sim 2 M_{\odot}$) suggests 0921−63 is a halo object.

119.039 UBVR photometry of short period RS CVn binaries.
M. Zeilik, R. Elston, G. Henson, P. Smith.
Proc. Southwest Reg. Conf., Vol. 7, (see 012.019), 15 - 19 (1982).

119.040 X-ray eclipses in AR Lac: preliminary results.
D. M. Gibson, F. M. Walter.
Proc. Southwest Reg. Conf., Vol. 7, (see 012.019), 21 - 27 (1982).

119.041 Preliminary light curve analysis of the eclipsing binary AG Virginis. R. W. Olson.
Proc. Southwest Reg. Conf., Vol. 7, (see 012.019), 29 - 34 (1982).

119.042 Infrared photometry of Beta Lyrae and Algol.
M. Zeilik, G. Henson, P. Smith, E. Budding.
Proc. Southwest Reg. Conf., Vol. 7, (see 012.019), 35 - 43 (1982).

119.043 Análise das curvas de luz B e V da binária eclipsante BV 590. G. R. Quast, J. Barroso, Jr.
Bol. Asoc. Argentina Astron., No. 18, p. 11 - 13 (1980).

119.044 Approximation of light loss curves for eclipsing systems by special polynomials.
A. A. Rubashevskij.
Astrometr. Astrofiz., Vyp. (No.) 46, p. 29 - 35 (1982). In Russian.
A method is suggested for approximation of symmetrical light loss curves for eclipsing systems by special polynomials smoothed by derivatives. The method is illustrated by the example of the narrow-band curve of V444 Cyg in the continuum.

119.045 HD 191765 − a possible binary Wolf-Rayet star with a low-mass companion.
I. I. Antokhin, A. A. Aslanov, A. M. Cherepashchuk.
Pis'ma Astron. Zh., Tom 8, 290 - 296 (1982). In Russian. English translation in Soviet Astron. Lett., Vol. 8.
A periodic ($P = 7^{d}44 \pm 0^{d}10$) photometric and spectroscopic variability of the star HD 191765 (WN6), connected with the ring-like nebula S109 is discovered. It is shown that this WR star is an eclipsing binary system with a low-mass companion.

119.046 Radius variations in the classical Algol binary U Sagittae. E. C. Olson.
Publ. Astron. Soc. Pacific, Vol. 94, 70 - 75 (1982).
Five-color standardized photometry of the totally-eclipsing Algol binary U Sge in primary eclipse has been obtained from 1978 to 1981. Light curves show evidence of

small brightness variations in the subgiant, and of some transient mass transfer. Among 16 Algol-like binaries monitored for about two years, U Sge is unique in showing radius changes of ±2% to 3%, probably in the cool subgiant.

119.047 Ground-based and IUE spectral observations of AU Monocerotis.   J. Sahade, O. E. Ferrer.
Publ. Astron. Soc. Pacific, Vol. 94, 113 - 121 (1982).

This paper reports on the results of ground-based and IUE observations of the eclipsing binary AU Mon made in January 1979. The behavior of the object has changed, in regard to the main outflow characteristics, relative to the earlier observations by Sahade and Cesco in 1944. The ultraviolet spectrum displays only absorption features; and the presence of the resonance lines of N V suggests that there is an increase in excitation temperature at a certain distance from the star and, consequently, the existence of nonthermal sources of energy, probably arising from shock waves generated by the mass outflow.

119.048 The close binary V757 Centauri.
M. A. Cerruti, R. F. Sisteró.
Publ. Astron. Soc. Pacific, Vol. 94, 189 - 194 (1982).

The W UMa system V757 Cen has been observed photoelectrically in the UBV system. Full light curves were constructed from 1135 V, 1104 B, and 1104 U individual observations. An improved ephemeris based on about 12,000 cycles is given. A first classical solution for partial eclipses, determined with complementary spectroscopic data, suggests the system to be a contact W UMa star consisting of a larger component of spectral type G1–G2 and an F9–G0 companion.

119.049 A spectrographic orbit for the eclipsing binary V757 Centauri.
M. A. Cerruti, V. S. Niemela.
Publ. Astron. Soc. Pacific, Vol. 94, 195 - 197 (1982).

Twenty-two spectrograms obtained at the Cerro Tololo Inter-American Observatory, Chile, have been used to determine, for the first time, the spectrographic orbital elements of the eclipsing binary V757 Cen. Both components of the binary system are of the spectral type about G0 V, and the physical parameters are found to be similar to the other W UMa binaries.

119.050 HD 227586: a variable comparison star of V453 Cygni?   H. L. Cohen.
Publ. Astron. Soc. Pacific, Vol. 94, 198 - 200 (1982) = Contrib. Dep. Astron. Univ. Florida No. 41.

Published UBV Photoelectric observations, made in 1966, of the massive eclipsing binary V453 Cyg suggest intrinsic variability with the system $0\overset{m}{.}026$ brighter in June-July compared to August–October. Analysis of additional photoelectric photometry, also made in 1966, between the comparison star, HD 227586, and a check star, HD 227611, indicate the variability is apparently associated with HD 227586 rather than the eclipsing binary. These results show HD 227586 was $0\overset{m}{.}018$, $0\overset{m}{.}022$, and $0\overset{m}{.}024$ fainter during the first two months. Since these values are small and published elements based on the 1966 data are already adjusted by $0\overset{m}{.}026$, no changes in the V453 Cyg elements are needed. Additional UBV photographic photometry of HD 227586 obtained in 1979 also suggests that the variability of HD 227586 may not be intrinsic but due to the sporadic inclusion into the photometer diaphragm of a faint companion 17"5 to the east. A second, fainter star 9"8 to the southwest adds a consistent $0\overset{m}{.}004$, $0\overset{m}{.}008$, $0\overset{m}{.}009$, in U, B, and V, respectively, to the measured brightness of HD 227586.

119.051 Las binarias W UMa: UZ Oct, HD 123732 y MW Pav.
M. E. Castore de Sisteró, R. F. Sisteró, E. Lapasset.
Bol. Asoc. Argentina Astron., No. 20 - 24, p. 137 (1981). Abstract.

119.052 Análisis del sistema MW Pav.
E. Lapasset.
Bol. Asoc. Argentina Astron., No. 20 - 24, p. 227 (1981). Abstract.

119.053 Curvas de luz de UZ Octantis.
M. E. Castore de Sisteró, R. F. Sisteró, B. Candellero.
Bol. Asoc. Argentina Astron., No. 20 - 24, p. 231 (1981). Abstract.

119.054 El espectro de h 4866 B.
J. Sahade, O. Ferrer.
Bol. Asoc. Argentina Astron., No. 20 - 24, p. 390 (1981). Abstract.

119.055 Optical fluxes of hot and cool components of Algol-like binaries.   E. C. Olson.
Astrophys. J., Vol. 257, 198 - 203 (1982).

A search for persistent nonstellar anomalies has been made in 16 Algol-like binaries with orbital periods $\sim1^d$ to 12^d. Five-color standardized photometry outside eclipse and at mid-primary eclipse shows that in 14 systems, the continuum optical flux distribution of each star is normal. Two dissimilar systems, RS Cephei ($P = 12\overset{d}{.}4$) and AI Draconis ($P = 1\overset{d}{.}2$), show large ultraviolet excesses at mid-eclipse suggestive of hot continuum sources.

119.056 Reinvestigación de la variable de eclipse μ^1 Scorpii.
J. Sahade, L. García de Ferrer.
Bol. Asoc. Argentina Astron., No. 26, p. 69 (1981). – Abstract.

119.057 El sistema eclipsante AU Monocerotis.
O. Ferrer.
Bol. Asoc. Argentina Astron., No. 26, p. 70 (1981). – Abstract.

119.058 Análisis de la fotometriá en el lejano ultravioleta de Beta Lyrae.   E. Lapasset.
Bol. Asoc. Argentina Astron., No. 26, p. 74 (1981). – Abstract.

119.059 Fotometriá fotoeléctrica UBV de V 758 Centauri.
S. L. Lípari.
Bol. Asoc. Argentina Astron., No. 26, p. 77 (1981). – Abstract.

119.060 Intérét de la campagne d'observation TYCHO pour l'étude des binaires à éclipse.   J. L. Halbwachs.
Bull. Inf. Cent. Données Stellaires, No. 22, p. 53 - 62 (1982).

This paper analyses the impact of the TYCHO project upon eclipsing binary studies.

119.061 Eclipsing binaries observations on the Czechoslovak public observatories.   J. Šilhán.
Říše hvězd, Vol. 63, 55 - 58 (1982). In Czech.

119.062 Revision of the photometric elements of DI Her.
M. I. Lavrov.
Astron. Tsirk., No. 1183, p. 5 - 6 (1981). In Russian.

119.063 Photometric elements of 14 eclipsing binaries in Delphini.
V. G. Karetnikov, O. G. Lakinskaya.
Astron. Tsirk., No. 1183, p. 7 - 8 (1981). In Russian.

119.064 The period of DN UMa (HR 4560): a bright eclipsing binary.   A. Giménez, J. A. Quesada.
Inf. Bull. Variable Stars, No. 2068, 3 pp. (1982).

119.065 Observations of the 1981 eclipse of RZ Oph.
J. Papoušek, M. Vetešník.
Inf. Bull. Variable Stars, No. 2070, 2 pp. (1982).

119.066 **V 1068 Cygni – a long-period RS CVn star?**
S. Rössiger.
Inf. Bull. Variable Stars, No. 2075, 3 pp. (1982).

119.067 **Photoelectric photometry of θ^1 Orionis A =
V1016 Ori.**   J. R. Sowell, D. S. Hall.
Inf. Bull. Variable Stars, No. 2076, 2 pp. (1982).

119.068 **Preliminary photometric orbital elements of
1 Persei.**   A. Gaspani.
Inf. Bull. Variable Stars, No. 2077 (1982).

119.069 **Photometric observations of 1 Persei.**   J. R. Percy.
Inf. Bull. Variable Stars, No. 2085, 3 pp. (1982).

119.070 **Photoelectric times of minima of eclipsing binaries.**
T. E. Margrave.
Inf. Bull. Variable Stars, No. 2086, 3 pp. (1982).

119.071 **Spectroscopic observations of V 471 Tauri
(BD+16°516).**   E. Hamzaoğlu, F. Sabbadin.
Inf. Bull. Variable Stars, No. 2092, 2 pp. (1982).

119.072 **Photoelectric minima times of DO Cassiopeiae and
BX Andromedae.**
P. Rovithis, H. Rovithis-Livaniou.
Inf. Bull. Variable Stars, No. 2094, 2 pp. (1982).

119.073 **Photoelectric observations of the eclipsing binary
IM Aurigae.**
N. Güdür, Ö. Gülmen, C. Sezer.
Inf. Bull. Variable Stars, No. 2098, 3 pp. (1982).

119.074 **Photoelectric photometry of NN Cephei.**
Ö. Gülmen, N. Güdür, C. Sezer.
Inf. Bull. Variable Stars, No. 2099, 2 pp. (1982).

119.075 **Photoelectric minima and light curves of V 478
Cygni.**   C. Sezer, N. Güdür, Ö. Gülmen.
Inf. Bull. Variable Stars, No. 2100, 3 pp. (1982).

119.076 **Photoelectric photometry of the eclipsing binary
DM Persei.**   C. Sezer.
Inf. Bull. Variable Stars, No. 2101, 3 pp. (1982).

119.077 **1981 UBVR photometric observations of ER Vul.**
M. Zeilik, R. Elston, G. Henson, P. Schmolke,
P. Smith.
Inf. Bull. Variable Stars, No. 2107, 4 pp. (1982).

119.078 **HD 26337: a new RS CVn variable star.**
F. C. Fekel, D. S. Hall, G. W. Henry, H. J. Landis,
T. R. Renner.
Inf. Bull. Variable Stars, No. 2110, 3 pp. (1982).

119.079 **HD 136905: a new RS CVn variable star.**
E. W. Burke, J. E. Baker, F. C. Fekel, D. S. Hall,
G. W. Henry.
Inf. Bull. Variable Stars, No. 2111, 4 pp. (1982).

119.080 **Central star of planetary nebula NGC 2346: new
eclipsing binary.**   L. Kohoutek.
Inf. Bull. Variable Stars, No. 2113, 4 pp. (1982).

119.081 **Photometric observations of the eclipsing variable
ZZ Cyg.**
M. S. Frolov, E. N. Pastukhova, A. V. Mironov.
Inf. Bull. Variable Stars, No. 2117, 3 pp. (1982).

119.082 **Photoelectric times of minima of eclipsing binaries.**
H. D. Kennedy.
Inf. Bull. Variable Stars, No. 2118, 2 pp. (1982).

119.083 **Photoelectric timings of primary minimum of
TV Cas and the method.**   A. C. de Landtsheer.
Inf. Bull. Variable Stars, No. 2119, 2 pp. (1982).

119.084 **Photoelectric observation of V 836 Cygni.**
S. Bozkurt.
Inf. Bull. Variable Stars, No. 2124, 3 pp. (1982).

119.085 **Photoelectric minimum timings of Algol.**
G. A. Bower.
Inf. Bull. Variable Stars, No. 2127, 4 pp. (1982).

119.086 **Photoelectric minima of BW Draconis.**
P. Rovithis, H. Rovithis-Livaniou.
Inf. Bull. Variable Stars, No. 2137, 2 pp. (1982).

119.087 **1981 UBVR photometry of CG Cyg.**   M. Zeilik,
R. Elston, G. Henson, P. Schmolke, P. Smith.
Inf. Bull. Variable Stars, No. 2138, 4 pp. (1982).

119.088 **HD 219634 – a massive new eclipsing binary.**
A. F. Gulliver, D. P. Hube, A. Lowe.
Inf. Bull. Variable Stars, No. 2146, 3 pp. (1982).

119.089 **H-alpha photometry of Lambda Andromedae.**
R. Elston. M. Zeilik, G. Henson, P. Schmolke,
P. Smith.
Inf. Bull. Variable Stars, No. 2150, 3 pp. (1982).

119.090 **Times of minima for eight eclipsing binaries.**
E. Derman, N. Yilmaz, S. Engin, Z. Aslan,
C. Aydin, Z. Tüfekçioğlu.
Inf. Bull. Variable Stars, No. 2159, 2 pp. (1982).

119.091 **A campaign to observe eclipsing binaries with long
periods.**   R. Diethelm.
I. A. P. P. P. Commun., No. 7, p. 8 - 10 (1982).

119.092 **91st - 93rd list of minima of eclipsing binaries.**
Compiled by S. Amsler, M. Andrakakou,
R. Boninsegna, R. Cadalbert, D. Delhaye, P. Dokic,
D. P. Elias, R. Germann, T. Grüebler, M. Häfliger,
L. Horowitz, J. Kukan, J.-F. Le Borgne, D. Leyman,
R. Leyman, K. Locher, S. Mammoliti, G. Mavrofridis,
C. Mouillard, G. Mouillard, D. Mourikis, I. Nikolaou,
H. Peter, B. Staub, A. Stucky, G. Stefanopoulos, N. Stoikidis,
W. Zwing, L. Capol, R. Diethelm, R. Häring, D. Hunn, J. Kägi,
K. Kocian, M. Kohl, C. Maranta, B. Biedermann, S. Ferrand,
N. Hasler, A. Kaiser, J.-P. Liégeois, P. Louis, T. Schildknecht.
BBSAG Bull., No. 58, p. 1 - 5, No. 59, p. 1 - 4, No. 60, p. 1 - 5
(1982).

119.093 **DD Geminorum: amplitude much larger than
catalogued.**   K. Locher.
BBSAG Bull., No. 58, p. 6 (1982).

119.094 **ES Tauri: new, totally different period.**
K. Locher.
BBSAG Bull., No. 58, p. 6 - 7 (1982).

119.095 **DQ Monocerotis: detection of the period.**
K. Locher.
BBSAG Bull., No. 59, p. 4 - 5 (1982).

119.096 **Tsvetkova's star in Cygnus: lightcurve parameters.**
K. Locher.
BBSAG Bull., No. 60, p. 6 (1982).

119.097 **List of minima and accurate determination of mean
minimum for VW Cep.**   E. Poretti.
GEOS Circ., EB 6, 4 pp. (1981).

119.098 Characteristics of the Wolf-Rayet star in the eclipsing binary system CX Cep.
N. A. Lipunova, A. M. Cherepashchuk.
Astron. Tsirk., No. 1190, p. 1 - 4 (1981). In Russian.

119.099 Period change and mass loss in the eclipsing binary system CQ Cep by a Wolf-Rayet star.
Eh. A. Antokhina, N. A. Lipunova, A. M. Cherepashchuk.
Astron. Tsirk., No. 1191, p. 1 - 4 (1981). In Russian.

119.100 An investigation of the light curve anomalies of the short period eclipsing binary VW Cephei.
G. P. McCook, E. F. Guinan.
Bull. American Astron. Soc., Vol. 13, 802 (1981). — Abstract.

119.101 Spectral energy distribution and FUV emission lines in Algols.
M. J. Plavec, J. J. Dobias, J. L. Weiland.
Bull. American Astron. Soc., Vol. 13, 802 (1981). — Abstract.

119.102 EX Hydrae: spectroscopic and photometric variations.   R. L. Gilliland.
Bull. American Astron. Soc., Vol. 13, 802 - 803 (1981). Abstract.

119.103 Spectral energy distribution in β Lyrae.
M. J. Plavec, J. L. Weiland, J. J. Dobias.
Bull. American Astron. Soc., Vol. 13, 803 (1981). — Abstract.

119.104 A coordinated ultraviolet-optical-infrared observing campaign for the 1982 - 84 eclipse of Epsilon Aurigae.   R. M. Genet, R. E. Stencel.
Bull. American Astron. Soc., Vol. 13, 804 (1981). — Abstract.

119.105 Ultraviolet spectroscopy of the 1981 eclipse of 32 Cygni.
R. E. Stencel, R. D. Chapman, Y. Kondo, R. F. Wing.
Bull. American Astron. Soc., Vol. 13, 830 (1981). — Abstract.

119.106 Starspot observability: observations and models.
D. B. Caton.
Bull. American Astron. Soc., Vol. 13, 833 (1981). — Abstract.

119.107 The structure of the coronae of AR Lacertae.
F. Walter, D. M. Gibson, G. S. Basri.
Bull. American Astron. Soc., Vol. 13, 833 (1981). — Abstract.

119.108 Hα observations of FK Comae.   B. W. Bopp.
Bull. American Astron. Soc., Vol. 13, 872 (1981).
Abstract.

119.109 Light curve variability in ER Vul.   B. J. Hrivnak.
Bull. American Astron. Soc., Vol. 13, 896 (1981).
Abstract.

119.110 V 356 Sgr revisited.   J. Ziolkowski.
Bull. American Astron. Soc., Vol. 13, 924 - 925 (1981). — Abstract.

119.111 WY Sagittae.
IAU Circ., No. 3707 (1982).

119.112 Light curves of SZ Piscium for 1977 and 1978.
J. A. Eaton, F. Scaltriti, M. Cerruti-Sola,
M. B. K. Sarma, B. D. Ausekar, S. Catalano, M. Rodono.
Astrophys. Space Sci., Vol. 82, 289 - 306 (1982).
The authors present a relatively complete V-band light curve of SZ Psc for 1978 and a partial light curve for 1977. From the 1978 light curve they derive a new time of primary minimum, JD$_\odot$2443823.674±0.001. The hotter component of this system is a F5—8 main-sequence star, the cooler component a K3—4 star well above the main sequence. The system

is detached with the larger component filling only 82% of its Roche lobe. The distortion wave in this RS CVn-type binary seems not to migrate regularly as do those in many other such systems, but rather seems to change phase and amplitude more erratically.

119.113 *BV* photometry of the W UMa-star 44i Bootis.
U. Hopp, S. Witzigmann.
Astrophys. Space Sci., Vol. 83, 171 - 176 (1982).
Two new B and V light curves and four minima of 44i Boo are given. The actual period value is discussed. Light-curve disturbances are observed and it is shown that these light-curve activities are a more or less continuous phenomena not connected to the sudden period changes.

119.114 GW Cephei, a W UMa type system.   M. Hoffmann.
Astrophys. Space Sci., Vol. 83, 195 - 202 (1982).
Photoelectric light curves in B and V are presented for GW Cep. The light elements (P = 0^d319) have been refined. Orbital elements have been determined according to the Russell—Merrill method. They show great similarity with those of at least four other well known W UMa systems, e.g. W UMa itself. The similarities of that group are discussed.

119.115 The very short period W UMa system V1022 Ophiuchi.   M. Hoffmann.
Astrophys. Space Sci., Vol. 83, 203 - 207 (1982).
Photoelectric light curves of V1022 Oph in B and V are presented. It has been confirmed that it is a W UMa type eclipsing binary with one of the shortest periods known ($P < 0^d$25). Some general properties of the system have been derived.

119.116 Photoelectric light and colour variations in V471 Tauri.   O. Tümer, C. Ibanoğlu, M. Kurutaç, Z. Tunca.
Astrophys. Space Sci., Vol. 83, 269 - 278 (1982).
A new investigation of the variations in the light curves and in the period of the eclipsing binary V471 Tau is presented. The collected observational data have been re-examined and it was found that (1) the decrease in the period of the system slows down and (2) that the mean brightness of the system has been increasing and this is greater at the longer wavelength. For the last seven years the increase in the brightness is estimated to be 0.15 mag in B and 0.18 mag in V bands respectively.

119.117 BV and narrow-band observations of the eclipsing binary AM Leonis.   M. Hoffmann, U. Hopp.
Astrophys. Space Sci., Vol. 83, 391 - 403 (1982).
Photometric observations obtained between 1977 and 1981 are reported. The period of AM Leo did not change during the last 20 years. Light curve variability was found to affect prominently the depths and shapes of the minima. The effects of the complications on broad and narrow band colour indices are discussed.

119.118 Spectroscopic observations of β Lyrae. II. The Hα profile.   A. Sanyal.
Publ. Astron. Soc. Pacific, Vol. 94, 341 - 344 (1982).
Image Isocon observations of the Hα profile of β Lyr indicate that: (a) the emission strength undergoes phase-correlated variations; (b) the nature of the correlation between emission strength and emission width in the hemisphere preceding the B8 star is different from that in the hemisphere following it; (c) the orbital V/R variations in 1974 - 75 are different from those in 1971. A model of the system containing an accretion disk, a hot region between the two stars, a gas stream flowing from the B8 star to the unseen companion in the hemisphere following the unseen companion, and a rotating shell-velocity-ellipse qualitatively explains the observations.

119.119 Evidence for a third component in the U CrB system. R. H. Van Gent.
Astron. Astrophys., Suppl. Ser., Vol. 48, 457 - 480 (1982).

Differences between observed and predicted times of minima of the eclipsing binary U CrB, covering the period 1870-1975, have been analyzed and evidence ·is given for the presence of a third component, orbiting the eclipsing binary in approximately 75 years in a highly eccentric orbit. An explanation of the O-C curve in terms of apsidal rotation can be disproved. Also, quasi-periodic impulsive mass-transfer, which spectroscopic data seem to support, and which up to now was considered to be the most likely mechanism, leads to contradicting results. Possibilities of photometric, astrometric or spectroscopic confirmation of the postulated third component are also discussed.

119.120 Revised photometric data for six eclipsing binaries. G. Giuricin, F. Mardirossian, M. Mezzetti.
Astron. Astrophys., Suppl. Ser., Vol. 49, 89 - 98 (1982).

Using Wood's (1972) model the authors have reanalyzed the lightcurves of six eclipsing binaries (V523 Cas, GT Cep, RV Crv, RU Eri, UV Psc, and V499 Sco), whose general properties have not yet been discussed adequately. For all binaries (except UV Psc) they have derived photometric elements appreciably different from those previously deduced from the same observational material.

119.121 Three colour photoelectric observations of the eclipsing binary TT Her.   R. Burchi, A. Dipaolantonio, S. Mancuso, L. Milano, A. Vittone.
Astron. Astrophys., Suppl. Ser., Vol. 49, 129 - 135 (1982).

Three colour photoelectric observations of the eclipsing binary TT Her are presented. A preliminar analysis of the period variability is presented.

119.122 Zeta-Aurigae-Sterne: Hinweis auf bevorstehende Minima.   D. Böhme.
Sterne, 58. Band, 104 - 108 (1982).

119.123 A study of the light curve of R Aquarii, 1933 to 1981.   P. M. Garnavich.
J. American Assoc. Variable Star Obs., Vol. 10, 60 - 63 (1981).

The light curve of R Aquarii over a period containing two eclipses was analyzed. An attempt was made to reduce light variations caused by the Mira component, in order to better determine a period for the eclipses.

119.124 A contribution to the controversy regarding the period of EG Sagittarii.
E. P. Belserene, G. G. Smith.
J. American Assoc. Variable Star Obs., Vol. 10, 71 - 74 (1981).

119.125 UU Ophiuchi revisited.
M. Taylor.
J. American Assoc. Variable Star Obs., Vol. 10, 83 - 84 (1981).

119.126 Evidence for apsidal motion in BM Monocerotis.
B. E. Schaefer.
J. American Assoc. Variable Star Obs., Vol. 10, 85 - 89 (1981).

119.127 Some background information on R Aquarii as an eclipsing system.   E. B. Weston.
J. American Assoc. Variable Star Obs., Vol. 10, 96 (1981).

119.128 HD 134518: a main sequence detached eclipsing binary.
G. Giuricin, F. Mardirossian, M. Mezzetti.
Astron. Astrophys., Vol. 109, 366 - 367 (1982).

Using Wood's (1972) model the authors have analyzed Przybylski's (1981) *UBV* lightcurves of the eclipsing binary HD 134518. The photometric solutions, coupled with plausible assumptions about its masses, indicate that this binary is a detached main sequence system; it consists of a late A star, accompanied by a considerably fainter and smaller K star.

119.129 V 463 Cyg: revised photometric elements.
G. Giuricin, F. Mardirossian, S. Ferluga.
Bull. Astron. Inst. Czechoslovakia, Vol. 33, 187 - 189 (1982).

119.130 Observations of FK Comae stars.
B. W. Bopp.
Smithsonian Astrophys. Obs., Spec. Rep. 392, (see 012.045), p. 207 - 217 (1982).

119.131 On the enigma of FK Comae.
F. Walter, G. Basri.
Smithsonian Astrophys. Obs., Spec. Rep. 392, (see 012.045), p. 219 - 224 (1982).

119.132 A flare event in the peculiar giant FK Comae.
L. W. Ramsey, H. L. Nations.
Smithsonian Astrophys. Obs., Spec. Rep. 392, (see 012.045), p. 225 - 230 (1982).

119.133 High-resolution IUE observations of the 1981 eclipse of 32 Cyg.
D. Reimers, A. Che, K. Hempe.
Smithsonian Astrophys. Obs., Spec. Rep. 392, (see 012.045), p. 245 - 254 (1982).

119.134 Contribution to Fourier analysis of the light curves of eclipsing variables.   Z. Kopal.
Compendium in astronomy, (see 003.012), p. 233 - 245 (1982).

The aim of the present paper is to evaluate analytically the weights needed to express the moments A_{2m} of the light curves – which constitute the basic quantities from which all elements of the respective eclipsing system can subsequently be evaluated as weighted means of the coefficients a_n of an empirical Fourier cosine series representing the observed light curve. It will, in particular, be shown that the weight coefficients ψ_i defining even moments ($A_0, A_2, A_4, A_6, \ldots$) are expressible in a closed form in terms of simple trigonometric functions.

Landolt-Börnstein.   See Abstr. 003.064.

A long period eclipsing binary project – five years of observations at ESO.   See Abstr. 013.045.

A Fortran subroutine for determining times of minimum light.   See Abstr. 021.034.

Computing analysis of light curves for eclipsing binary stars. Part 2. Direct approach and its programming for the Nairi K computer.   See Abstr. 021.046.

Método de curvas de luz sintéticas aplicado a UZ Octantis.   See Abstr. 031.569.

Perturbations in stellar paths.
See Abstr. 111.012.

UBVR photometry of the eclipsing binary system CV Ser and characteristics of the Wolf-Rayet star.
See Abstr. 113.017.

LB 3459 – an sdO-type eclipsing binary system: **non-LTE analysis of the primary.**   See Abstr. 114.058.

Spectral investigation of the eclipsing binary V822 Aql.   See Abstr. 114.080.

Near infrared spectra of Algol B.
See Abstr. 114.173.

On the radius determination of the variable F-type supergiant BL Tel(F).   See Abstr. 115.005.

On the magnetic field in the variable star RR Lyrae.
See Abstr. 116.027.

β Lyrae is a magnetic star.
See Abstr. 116.037.

Polarimetry and photometry of two RS CVn stars.
See Abstr. 116.060.

AU Mon: a semidetached binary system.
See Abstr. 117.013.

Ultraviolet observations of the 1980 eclipse of the symbiotic star CI Cygni.   See Abstr. 117.017.

Observations and analysis of the light curve of AE Phoenicis in 1978.   See Abstr. 117.063.

Time-resolved spectroscopy of the accretion disk in RW Tauri.   See Abstr. 117.121.

Spectroscopic evidence for an atmospheric eclipse of δ Sagittae.   See Abstr. 120.001.

Variation of the Hβ emission lines of YY Geminorum. II. Change of sectorial structures of active regions. See Abstr. 120.009.

Cepheid binaries – II. New southern examples.
See Abstr. 122.192.

BT Monocerotis: an eclipsing nova.
See Abstr. 124.101.

Wolf-Rayet stars in the Magellanic Clouds. II. The peculiar eclipsing binary HD 5980 in the SMC.
See Abstr. 159.030.

Erratum

119.901   Flare activity of the RS CVn star SV Cam.
L. Patkős.
Astrophys. Lett., Vol. 22, 1 - 3 (1981). – See Abstr. 30.119.040.

120 Spectroscopic Binaries

120.001   Spectroscopic evidence for an atmospheric eclipse of δ Sagittae.   A. H. Batten, W. A. Fisher.
Publ. Astron. Soc. Pacific, Vol. 93, 769 - 771 (1981/82).
High-dispersion spectrograms obtained in 1980 and 1981 confirm earlier reports that δ Sge, a system of the ζ Aur type, undergoes at least atmospheric eclipses approximately every 3725 days. The new observations also reduce the uncertainty surrounding the orbital elements.

120.002   Orbit of the spectroscopic binary HR 1883.
M. M. Dworetsky.
Mon. Not. R. Astron. Soc., Vol. 199, 303 - 307 (1982).
HR 1883 (HD 36881; B9, m_v = 5.60) is a single-lined spectroscopic binary with $P \simeq 1857$ days and an unusually large mass function $f(m)$ = 0.72 $M_\odot$.

120.003   Spectroscopic binary orbits from photoelectric radial velocities. Paper 42: HD 181602.
R. F. Griffin.
Observatory, Vol. 102, 1 - 4 (1982).

120.004   The Hyades binary HD 27130 and the mass-luminosity relation and distance of the Hyades cluster.
R. D. McClure.
Astrophys. J., Vol. 254, 606 - 615 (1982).
Masses have been determined for the components of the Hyades spectroscopic binary HD 27130, which is found to be both double-line and eclipsing. They are significantly more massive than stars of the same luminosity that lie on the field star mass-luminosity relation. Since the solar neighborhood stars should be evolved off the zero-age (Hyades) main sequence, and since the Hyades may be metal rich, this result is to be expected without the necessity of postulating a helium abundance difference. By combining the mass obtained for HD 27130 with the visual binary data for Hyades stars, a dynamical parallax can be obtained. This results in a distance

modulus of 3.47 ± 0.05, a value that is considerably larger than most astrometric determinations.

120.005   Duplicity in the solar neighborhood. II. Spectroscopic orbits for four bright stars HD 21018, HD 30021, HD 158837, and HD 190658.
P. B. Lucke, M. Mayor.
Astron. Astrophys., Vol. 105, 318 - 322 (1982).
Orbital elements have been determined for four stars from the bright star catalog; HD 21018, HD 30021, HD 158837, and HD 190658. The rotation of these stars has been determined, and HD 21018 is shown to be a giant in rapid non-synchronous rotation. HD 190658 is an M2III giant which could fill its Roche lobe and eventually show eclipses. HD 190658 is also a high velocity star ($V_0 = -112.1$ km s^{-1}) and has the peculiarity of having the shortest period (198 days) of any binary of type M III. It should also be noted that these stars are all members of triple systems.

120.006   Spectroscopic orbits for three double-lined binaries in the Hyades field, 22°669, vA 771, and vB 166.
R. F. Griffin, M. Mayor, J. E. Gunn.
Astron. Astrophys., Vol. 106, 221 - 228 (1982).
22°669, vA 771, and vB 166 have been suspected of being members of the Hyades. All were discovered to be double-lined spectroscopic binaries with large amplitudes and short periods. Their orbits were established during a single intensive observing run with the CORAVEL radial-velocity spectrometer. 22°669 proves to be a Hyades member; vA 771 is not a member of the Hyades cluster but belongs to the Hyades moving group; vB 166 is unrelated to the Hyades.

120.007   Spectroscopic binary orbits from photoelectric radial velocities. Paper 43: HR 7024.
R. F. Griffin.
Observatory, Vol. 102, 27 - 30 (1982).

**120.008   Contribution to the study of composite spectra.
II. A, Am, Ap spectroscopic binaries.**
N. Ginestet, M. Jaschek, J. M. Carquillat, A. Pédoussaut.
Astron. Astrophys., Vol. 107, 215 - 221 (1982). In French.

Using the file of spectroscopic binaries of the Toulouse
Observatory the authors have re-examined some results concerning spectroscopic binaries (SB) among A and Am stars.
The authors present the distributions in magnitude, spectral
types and orbital periods and some UBV diagrams. They found
very few SB's later than A4; as a consequence 70% of the
A-type SB's known are early A-type stars (A0–A3). The
authors repeated the work of Abt and Bidelman (1969) concerning the distribution over spectral types and orbital periods.
Using their same ranges they found that 85% of the SB's
are Am's, 15% are A's.

**120.009   Variation of the Hβ emission lines of YY Geminorum.
II. Change of sectorial structures of active regions.**
K. Kodaira, K. Ichimura.
Publ. Astron. Soc. Japan, Vol. 34, 21 - 37 (1982).

Sixty-three image-tube spectrograms of YY Gem are
analyzed to yield the radial-velocity curves and the variations
in the intensities and the widths of Hβ emission lines during
the quiescent phase at epochs 1980 February 11 - 16, 1981
January 14 - 15, and 1981 March 11. The emission-line
intensity of component A varied in a single-wave mode over an
orbital period. The pattern of the intensity variation of component B changed within a few years. The ratio of the
amplitudes of radial-velocity curves (K_A/K_B) of Hβ emission
was found to be 0.91 in February 1980 but 1.01 in January
1981. This modulation in the ratio is interpreted as the results
of the varying inhomogeneous distributions of emission
intensities over the stellar surfaces which are inferred from the
observed intensity variations under the assumption of
synchronous rotation. A ratio $K_A/K_B = 1.00 \pm 0.01$ is proposed
as the actual value which would be observed if the effects of
inhomogeneities were negligible.

**120.010   Masses and luminosities for the giant spectroscopic/
speckle interferometric binaries Gamma Persei and
Phi Cygni.**   H. A. McAlister.
Astron. J., Vol. 87, 563 - 569 (1982) = Astron. Contrib.
Georgia State Univ., No. 60.

Speckle interferometric measurements of the spectroscopic binaries γ Per and φ Cyg are combined with the spectroscopic orbits to determine the elements of the apparent
orbits and the masses, distances, and luminosities of the four
components. φ Cyg is a rare example of a binary star whose
components are similarly evolved off the main sequence. The
newly determined masses of $(2.50 \pm 0.09)\,M_\odot$ and
$(2.39 \pm 0.08)\,M_\odot$ and distance of 71.9 ± 5.5 pc place the
G8III - IV components of φ Cyg at the point of beginning the
ascent of the red giant branch. The orbit of γ Per is confirmed
to be highly inclined, and the masses of $4.7\,M_\odot$ and $2.8\,M_\odot$
and distance of 73.8 pc suggest that a classification of
G8II-III + B9V is more appropriate than the existing classification of G8III + A3V.

**120.011   Spectroscopic binary orbits from photoelectric
radial velocities. Paper 44: ε Aquilae.**
R. F. Griffin.
Observatory, Vol. 102, 82 - 85 (1982).

**120.012   Distribution of masses of primary components and
major semi-axes of orbits of spectroscopic binary
stars.**   E. I. Popova, A. V. Tutukov, L. R. Yungel'son.
Pis'ma Astron. Zh., Tom 8, 297 - 301 (1982). In Russian.
English translation in Soviet Astron. Lett., Vol. 8.

The distribution of 333 double-line spectroscopic binaries in the $a - M_1$ diagram has been studied. In most of these
stars mass exchange did not occur. It has been found that the
number of systems with $a/R_\odot \leqslant 6(M_1/M_\odot)^{1/3}$ and $M_1 \leqslant 1.5\,M_\odot$

per unit of lg a (a is the major semi-axis) is ~60 times lower
than that of wider systems. Stars with $M_1 \geqslant 1.5\,M_\odot$ and
$a/R_\odot \leqslant 6(M_1/M_\odot)^{1/3}$ are practically absent.

**120.013   HD 208095: disappearance of the last of the over-
massive detached binaries?   D. M. Popper.**
Publ. Astron. Soc. Pacific, Vol. 94, 76 - 79 (1982).

The late B-type spectroscopic binary, HD 208095, is
removed from the diminishing ranks of spectroscopic binaries
with apparently anomalous masses by means of analysis of a
new set of spectrograms. The minimum masses (3.5 and 3.2
$M_\odot$) are close to the masses expected for the spectral types.
The more massive star has the slightly earlier type. While the
line widths in the spectrum of the cooler component of this
detached system are consistent with equality of orbital and
rotational periods, the lines of the hotter component are
considerably broader.

**120.014   A preliminary orbit for HR 2577: a K + Be binary
system.   E. M. Hendry.**
Publ. Astron. Soc. Pacific, Vol. 94, 169 - 171 (1982).

A spectroscopic period of 58 years has been found for
the K2 II + B3 IVe composite system HR 2577 (HD 50820).
Preliminary orbital elements were calculated for this double-
lined pair, giving equal masses for the two stars. The B3 star is
seen to undergo secular changes in the appearance of its
Balmer line emission which seem to be unrelated to the binary
motion. It is predicted that the system may be capable of
being resolved by the technique of speckle interferometry.

**120.015   La binaria espectroscópica del tipo Wolf-Rayet,
HD 90657.   V. S. Niemela.**
Bol. Asoc. Argentina Astron., No. 20 - 24, p. 222 (1981).
Abstract.

**120.016   La estrella central de NGC 1360: una binaria
espectroscópica en una nebulosa planetaria.**
R. H. Méndez, V. S. Niemela.
Bol. Asoc. Argentina Astron., No. 20 - 24, p. 335 (1981).
Abstract.

**120.017   Investigation of the spectroscopic binary Ap star
ET And.   M. Ouhrabka.**
Soobshch. Spets. Astrofiz. Obs., Vyp. 32, (see 012.029),
p. 45 - 46 (1981).

**120.018   The spectroscopic orbit of the barium star Zeta
Capricorni.   R. B. Culver.**
Bull. American Astron. Soc., Vol. 13, 873 (1981). – Abstract.

**120.019   Ultraviolet and optical studies of binaries with
luminous cool primaries and hot companions.
II. BVRI observations.   S. B. Parsons, T. J. Montemayor.**
Astrophys. J., Suppl. Ser., Vol. 49, 175 - 181 (1982).

Johnson system $BVRI$ measurements are presented for
117 stars, most of them with no previous R and I photometry
and many with no previous B and V. The prime objects for
measurement are unresolved or nearly unresolved binaries
containing a late-type giant or supergiant and an early-type
companion. Other objects on the program include suspected
binaries and other F–G giants and supergiants lacking at least
R and I magnitudes. The variable F and G supergiants 1 Mon,
89 Her, HR 7308, HR 8157, HR 8752, and ρ Cas and the
eclipsing systems W Ser and ST Aqr were observed; HR 8752
showed significant dimming and cooling over 125 days.
Several of Halliwell's candidates for nearby stars were measured.

120.020   HR 3626: double-line spectroscopic binary.
W. I. Beavers, J. J. Salzer.
Publ. Astron. Soc. Pacific, Vol. 94, 356 - 358 (1982).
Orbital elements are determined for the double-lined

spectroscopic binary HR 3626, which had previously been recognized as being only a single-lined system. A mass ratio of 1.26 ± 0.02 is determined, and a magnitude difference $\Delta m_B = 1.0 \pm 0.2$ is inferred from the photoelectric radial velocity dips. This leads to the conclusion that the system consists of a pair of G dwarf or subgiant stars.

120.021 **Spectroscopic binaries near the North Galactic Pole. Paper 3: HD 105341.**   R. F. Griffin.
J. Astrophys. Astron., Vol. 3, 1 - 4 (1982).

Photoelectric radial-velocity measurements show that HD 105341 is a spectroscopic binary with a near-circular orbit and a period of 194 days.

120.022 **Evidence for starspots on DK Draconis (HR 4665).**   E. F. Guinan, G. P. McCook, J. L. Fragola, W. C. O'Donnell, S. Tomczyk, A. G. Weisenberger.
Astron. J., Vol. 87, 893 - 898 (1982).

The long-period RS CVn-type binary DK Dra was observed on 24 nights from February through September of 1978 with pairs of intermediate- and narrowband interference filters centered near the wavelengths of the Balmer Hα line and of the O I λ 7774 triplet. The light variation appears to be quasi-sinusoidal with a period of $64^d.6 \pm 0^d.4$. No evidence of either Hα or O I emission was found. Theoretical light curves were generated with a starspot program and fit to the observed mean light curves. Satisfactory fits to the data were obtained with a minimum of two spots, each spot having a temperature ~1100 ± 150 K cooler than the photosphere. The two spots covered about 8% of the total stellar surface during the interval investigated.

120.023 **Spectroscopic orbits for two very high velocity halo stars: HD 111980 and HD 149414.**
M. Mayor, C. Turon.
Astron. Astrophys., Vol. 110, 241 - 245 (1982).

Spectroscopic orbits for two subdwarf stars have been computed from measurements performed with the CORAVEL spectrometer. Both their metallicities relative to the Sun (Fe/H = −1.2 and −1.4) and their mean radial velocities show that they are halo stars. These two F and G halo stars are the first spectroscopic binaries with such extreme values for velocities and metallicities for which orbits have ever been determined.

120.024 **Mass loss, linear polarization variability, and duplicity of the luminous B2 supergiant HD 80077.**
G. Knoechel, A. F. J. Moffat.
Astron. Astrophys., Vol. 110, 263 - 271 (1982).

Photographic coudé spectroscopy, linear polarimetry and wide band photoelectric photometry reveal that HD 80077 may be a single-line spectroscopic binary with a low mass, possibly compact companion. A period of 21.2 ± 0.2 d is found from the polarimetry. Several independent observational constraints lead to an absolute bolometric magnitude of −10.4 ± 0.5 for HD 80077, putting it among the brightest known B-type supergiants in the Galaxy.

Landolt-Börnstein.   See Abstr. 003.064.

Radial velocities of southern HR stars, II.
See Abstr. 111.002.

Long-term radial velocity variations in some Be stars.   See Abstr. 112.052.

The Wolf-Rayet star HD 193077: evidence for a low-mass companion and the possibility of a third body.
See Abstr. 114.015.

Radial velocities of hot secondary stars from IUE spectra.   See Abstr. 114.176.

HD 15144, a magnetic Ap star with a possible intrinsic periodic variation of the magnetic field.
See Abstr. 116.005.

Observations of binary stars by speckle interferometry — III.   See Abstr. 118.001.

Orbital motion of the pulsating star V644 Her.
See Abstr. 122.025.

A simultaneous photometric and radial velocity study of short-period southern Cepheids. III. An analysis for binaries.   See Abstr. 122.178.

Cepheid binaries— II. New southern examples.
See Abstr. 122.192.

The line profile variations of Spica.
See Abstr. 122.208.

C1: a white-dwarf—red-dwarf spectroscopic binary.
See Abstr. 126.009.

Wolf-Rayet stars in the Magellanic Clouds. I. The WN3 binary AB 6 in the SMC.   See Abstr. 159.029.

121 Early-stage Stars (T Tauri Stars, Herbig-Haro Objects, etc.)

121.001 **A new investigation of the T Tauri star RU Lupi – I.**
Observation and immediate analysis.
M. T. V. T. Lago, M. V. Penston.
Mon. Not. R. Astron. Soc., Vol. 198, 429 - 443 (1982).

New spectra of the southern T Tauri star, RU Lupi, are presented and a list of emission lines compiled. Attempts to distinguish different emitting regions on the basis of radial velocity are unsuccessful but the line widths define several different regions in the stellar wind.

121.002 **A new investigation of the T Tauri star RU Lupi – II.**
The physical conditions of the line emitting region.
M. T. V. T. Lago.
Mon. Not. R. Astron. Soc., Vol. 198, 445 - 456 (1982).

High dispersion spectra of the T Tauri star RU Lupi are analysed to obtain information on the physical conditions in the line emitting region. A non-uniform temperature is indicated through the region. The profiles of the Balmer lines are analysed to yield the distance and velocity of the region where these lines are produced, while a new analysis of the fluorescence-enhanced lines gives the range of values of electron, proton and neutral hydrogen densities. From the forbidden lines inferences are made on the values of the electron density and temperature in the lower density region. Finally the existence is suggested of several distinct line emitting regions with well determined physical conditions.

121.003 **A Herbig–Haro object in the core of M16**
(NGC 6611). J. Meaburn, N. J. White.
Mon. Not. R. Astron. Soc., Vol. 199, 121 - 129 (1982).

A bright knot emitting strong [S II] and [N II] lines has been positively identified as an HH object within a neutral lane in M16. This has been called M16 HH 1. Two dominant components of collisionally ionized plasma (shock velocity $\simeq 85$ km s^{-1} and pre-shock density $\simeq 34$ cm^{-3}) originate within this object with values of heliocentric radial velocity of $\simeq +10$ km s^{-1} and $\simeq +90$ km s^{-1}. A cavity with flowing walls produced by the wind from a cool star is proposed to explain these observations.

121.004 **VLA observations of mass loss from T Tauri stars.**
M. Cohen, J. H. Bieging, P. R. Schwartz.
Astrophys. J., Vol. 253, 707 - 715 (1982).

The authors have detected radio emission at 4.885 GHz from 6 of 24 pre-main-sequence stars surveyed with the Very Large Array (VLA). The detected stars include V410 Tau, T Tau, DG Tau, LkHα 101, L1551 IRS5, and Z CMa. In most cases, the radio maps show unresolved cores ($<0.5''$) and faint extended structures with sizes of $1''-2''$. Mass loss rates derived under the assumption of uniform spherical winds span the range $\sim 3 \times 10^{-7}$ to $\sim 4 \times 10^{-5}\, M_\odot$ yr^{-1}. However, the outflows are probably highly anisotropic, so these estimates are at best upper limits.

121.005 **The chromosphere and wind of the Herbig Ae star,**
AB Aurigae. F. Praderie, A. Talavera,
P. Felenbok, J. Czarny, A. M. Boesgaard.
Astrophys. J., Vol. 254, 658 - 662 (1982).

The Herbig Ae variable, AB Aur, has been observed with the CFHT coudé spectrograph at 95 mÅ resolution for five consecutive nights in the blue spectral region and with the IUE satellite in the high resolution, long wavelength mode. The Ca II K line shows an asymmetric and variable profile and shows weak emission in the absorption core at least some of the time. An exceptional event, characterized by the appearance of blue components in Ca II K and Balmer lines, occurred on 1980 October 26. The characteristic line emission and line positions, breadth, and variability show the presence of an active and variable chromosphere with mass outflow.

121.006 **Some results of simultaneous spectroscopic and**
photometric observations of RW Aurigae.
A. B. Bukach, V. P. Grinin, P. P. Petrov, N. I. Shakhovskaya.
Pis'ma Astron. Zh., Tom 8, 172 - 178 (1982). In Russian.
English translation in Soviet Astron. Lett., Vol. 8.

Three most typical spectra of the T Tauri-type star RW Aur corresponding to essentially different levels of its luminosity are described. The strengthening of the absorption spectrum of the envelope and the appearance of a G-type photospheric spectrum is observed at light maximum. The profiles of Balmer lines show systematic changes while passing from Hα to higher lines of the series.

121.007 **S CrA and CoD-35°10525, two bright young stars.**
C. Bertout, L. Carrasco, R. Mundt, B. Wolf.
Astron. Astrophys., Suppl. Ser., Vol. 47, 419 -439 (1982).

The two YY Orionis stars S CrA and CoD-35°10525 were observed during twelve consecutive nights at the coudé focus of the 1.52 m ESO telescope at La Silla, Chile, and simultaneously at the 1.5 m photometric telescope of the Observatorio Astronomico Nacional in San Pedro Martir, Baja California, Mexico, using a medium-narrow band 13-color photometric system. This paper describes the extensive data gained during these observing runs. Rapid photometric variations occur in the ultraviolet part of the spectrum and correlate with changes in the appearance of the line spectrum. Extensive line identifications are given for both stars. Radial velocities of emission and absorption lines are tabulated, and the line profile variations are studied in detail.

121.008 **A linear polarization survey of T Tauri stars.**
P. Bastien.
Astron. Astrophys., Suppl. Ser., Vol. 48, 153 - 164 (1982), with a correction p. 513 - 518.

A two-band linear polarization survey of the T Tauri stars brighter than about thirteenth magnitude and north of $-30°$ declination has been carried out between 1976 and 1979. A few fainter T Tauri stars, three FU Orionis stars, and one Herbig emission star have also been included. Polarization variability has been detected in at least 35% of the stars for which two or more observations are available. Correlations have been found between polarization and the color excess $E(B-V)$, between polarization and average infrared color indices, especially $\langle V \rangle$-$\langle L \rangle$, and between polarization and position of the stars in the HR diagram. The correlations with the infrared color indices indicate that the infrared excess at $3.5\,\mu$m and longward can be attributed to absorption of stellar radiation by dust and reemission in the infrared for at least 25% of the stars.

121.009 **Discovery of an infrared companion to T Tauri.**
H. M. Dyck, T. Simon, B. Zuckerman.
Astrophys. J., Lett., Vol. 255, L103 - L106 (1982).

The authors have obtained one-dimensional speckle interferometry of T Tauri at 2.2 μm, 3.8 μm, and 4.8 μm. The visibility curves indicate that T Tau is a double star with a north-south separation of $0.''61 \pm 0.''04$. Both components are unresolved and have very different colors. In the authors' model, the cooler northern component accounts for the infrared excess between 2.2 μm and 10 μm. Indirect arguments suggest that it is the visible southern component which drives the outflow evident in H$_2$ and CO spectra and which accounts for the $\sim 10^{-7} M_\odot$ yr^{-1} mass-loss rate.

121.010 YY Orionis line profiles in the spectrum of RW Aurigae. I. Appenzeller, B. Wolf.
Astron. Astrophys., Vol. 105, 313 - 317 (1982).

The authors discuss three high resolution coudé spectrograms of the bright T Tauri star RW Aur, which were obtained in January 1981 on three consecutive nights. One spectrogram shows well defined absorption features longward of various strong emission lines. This seems to confirm earlier reports by other authors that RW Aur is a member of the YY Orionis subclass of the T Tauri stars. Also present on one spectrogram is a slightly redshifted G-type absorption spectrum which may be produced either by absorption in the outer layers of the circumstellar envelope or by a superposition of the envelope emission lines on a late type photospheric spectrum.

121.011 On the absence of coronal line emission from Orion population stars. G. F. Gahm, J. Krautter.
Astron. Astrophys., Vol. 106, 25 - 28 (1982).

14 young stars of the Orion population were searched for forbidden coronal line emission of [Fe XIV] λ 5303 Å and [Fe X] λ 6374.5 Å. No star shows these lines in emission. For several stars it is demonstrated that the intensity of any 10^4 K corona is less than 10^4 times that of the Sun, while emission lines of e.g. C IV and Si IV forming in regions with temperatures of 10^5 K are enhanced by 10^4-10^6 times the solar values. Earlier suggestions that strong-line T Tauri stars have intensive coronal emission absorbed in extensive circumstellar gas are not supported by the present investigation.

121.012 Circumstellar dust shells associated with T Tauri stars: another progress report.
A. E. Rydgren, J. T. Schmelz, F. J. Vrba.
News Lett. Astron. Soc. N. Y., Vol. 2, 13 (1982). – Abstract.

121.013 Results of the joint program observations of the T Tauri-type star DI Cephei. V. P. Grinin,
Yu. S. Efimov, V. I. Krasnobabtsev, N. I. Shakhovskaya,
N. M. Shakhovskoj, A. G. Shcherbakov, G. V. Zajtseva,
E. A. Kolotilov, G. I. Shanin, N. N. Kiselev, Ch. G. Gyulaliev,
I. R. Salmanov.
Perem. Zvezdy, Tom 21, 247 - 271 (1980). In Russian.

The results of observations of the T Tauri-type star DI Cep under the 1974–75 joint program are presented. The photometric activity of the star in the UBV bands, the polarimetric variability in the UBVOR bands and the spectroscopic variations in the region from H and K Ca II lines to near infrared has been studied. The correlation between B–V, U–B colours and visual star brightness as well as the wavelength dependence of the polarization degrees has been obtained. Calculations of the equivalent widths and relative intensities of the hydrogen, calcium and oxygen lines have been performed. A comparison of the joint observational results was made. Preliminary estimates of some parameters of the stars and circumstellar envelopes have been carried out.

121.014 Polarimetric observations of RY Tauri and T Tauri.
Yu. S. Efimov.
Perem. Zvezdy, Tom 21, 273 - 284 (1980). In Russian.

Results of five-colour observations of the optical linear polarization of RY Tau and T Tau are presented. The observations were carried out in 1976–1978. The light variations in V and variations of percentage of polarization for RY Tau are in opposite sense with the maximum range of variations in the ultraviolet region. RY Tau seems to have a flat, disk-like gaseous dust envelope. The dust in the envelope of T Tau consists of larger particles than that of RY Tau.

121.015 Line profiles of T Tauri stars: clues to the nature of the mass flow. L. Hartmann.
Astrophys. J., Suppl. Ser., Vol. 48, 109 - 126 (1982).

A set of high resolution observations of the Hα, Na D, He I λ5876, and Hγ lines in several T Tauri stars is presented.

Repeated observations of some of these objects yield some indication of the importance of time-dependent effects. The line profiles should be useful for detailed radiative transfer modeling. A cursory survey of the profiles suggests that "turbulent" velocities in some cases may not be negligible in comparison with the bulk flow speeds and that many emission lines, such as Na D, may be formed in a narrow layer near the stellar photosphere.

121.016 Observations of rapid line profile variability in the spectra of T Tauri stars.
R. Mundt, M. S. Giampapa.
Astrophys. J., Vol. 256, 156 - 167 (1982).

The authors present high spectral and temporal resolution observations of six T Tauri stars, obtained with the Multiple Mirror Telescope echelle spectrograph and Reticon detector. The spectral resolution of the observations is $\lambda/\Delta\lambda = 2.4 \times 10^4$ and the temporal resolutions range from 300 s to 900 s. Short-term variability is exhibited only by the Hγ line profile in the line spectrum of the T Tauri star RW Aurigae at time scales as short as 10 minutes. The rapid line profile variability manifested by RW Aur is discussed within the context of two physical processes, namely (1) variable mass infall into an accretion shock, and (2) flare activity.

121.017 Evidence for a characteristic maximum temperature in the circumstellar dust associated with T Tauri stars. A. E. Rydgren, J. T. Schmelz, F. J. Vrba.
Astrophys. J., Vol. 256, 168 - 176 (1982).

Nearly simultaneous *BVRI* and *JHKL* photometry is presented for 26 T Tauri and related young stars in the Taurus and NGC 2264 regions. Additional *UBVRI* photometry of 56 T Tauri and related stars in these two regions is also given. The authors find evidence for a correlation between the color excess $E(V-I)$ and the IR color $H-K$ for T Tauri stars in both regions. This is most easily understood as a circumstellar reddening effect and suggests that some of the observed $V-I$ reddening in typical T Tauri stars is not interstellar in origin. After correcting for likely interstellar reddening, it is found that the intrinsic loci of the Taurus region T Tauri stars in the $(J-H, H-K)$ and $(H-K, K-L)$ diagrams are remarkably narrow. These observed loci are consistent with circumstellar dust shell models having a maximum dust temperature of approximately 1300 K.

121.018 On the evolutionary connection between FUORs and T Tau-type stars.
Yu. K. Melik-Alaverdyan.
Astrofizika, Tom 17, 557 - 562 (1981). In Russian. English translation in Astrophysics, Vol. 17, No. 3.

A suggestion that the T Tau-state of a star is caused by increased opacity of external layers of the star is advanced. It is suggested that the increase of opacity is caused by the presence of the product of decay of protostellar matter in external layers of T Tau-type stars.

121.019 High velocity CO emission around T Tauri stars.
S. Edwards, R. L. Snell.
Regions of recent star formation, (see 012.002), p. 141 - 146 (1982).

T Tauri winds are detected via their interactions with ambient cloud material. The magnitude of the mass flows are calculated and dynamical effects on molecular clouds are discussed.

121.020 New objects resembling Herbig-Haro ones.
A. L. Gyul'budagyan.
Pis'ma Astron. Zh., Tom 8, 232 - 233 (1982). In Russian. English translation in Soviet Astron. Lett., Vol. 8.

A list of 21 objects appearing as Herbig-Haro objects which were found in the Palomar Sky Survey prints is presented.

121.021 The variability of RY Lupi.
A. Evans, M. F. Bode, D. C. B. Whittet, J. K. Davies, D. Kilkenny, D. W. T. Baines.
Mon. Not. R. Astron. Soc., Vol. 199, 37P - 43P (1982).

Simultaneous optical and infrared broadband photometry $(0.35 \leqslant \lambda(\mu m) \leqslant 3.5)$ and $H\beta$ index photometry have been obtained for the T Tauri star RY Lupi. The results favour a model in which the variability of the star is caused by variable circumstellar extinction.

121.022 V 1057 Cygni et les variables du type FU Orionis.
C. Bertaud, E. Stram.
Astronomie, Vol. 96, 237 - 247 (1982).

121.023 On the dust shells of T Tau-type stars.
Yu. K. Melik-Alaverdyan, G. G. Tovmasyan.
Soobshch. Byurakan Obs., Vyp. 53, p. 93 - 98 (1982). In Russian.

The parameters of the dust shells of seven T Tau-type stars are determined. It is shown that there are definite correlations between masses, sizes and temperatures of the dust shells.

121.024 T Tau-type stars and magnetic stars.
R. E. Gershberg.
Soobshch. Spets. Astrofiz. Obs., Vyp. 32, (see 012.029), p. 14 (1981).

121.025 Energy distribution in the spectrum of FU Ori.
N. I. Bondar, A. B. Bukach, N. I. Shakhovskaya.
Inf. Bull. Variable Stars, No. 2136, 3 pp. (1982).

121.026 Recent infrared photometry of V1057 Cygni.
T. Simon, R. D. Wolstencroft, H. M. Dyck, R. R. Joyce.
Inf. Bull. Variable Stars, No. 2155, 4 pp. (1982).

121.027 Hydrodynamic models of Herbig-Haro objects.
M. T. Sandford II, R. W. Whitaker.
Bull. American Astron. Soc., Vol. 13, 805 (1981). – Abstract.

121.028 High resolution, absolute flux profiles of the Mg II h and k lines for T Tauri stars.
M. S. Giampapa, C. Morossi, M. Ramella, C. L. Imhoff.
Bull. American Astron. Soc., Vol. 13, 811 (1981). – Abstract.

121.029 Proper motions of Herbig-Haro objects.
B. F. Jones.
Bull. American Astron. Soc., Vol. 13, 855 - 856 (1981). Abstract.

121.030 Shock wave simulations of the spectra of low excitation Herbig-Haro objects.
R. D. Schwartz, M. A. Dopita, L. Binette.
Bull. American Astron. Soc., Vol. 13, 856 (1981). – Abstract.

121.031 Spectroscopic indications for FU Orionis-like events in T Tauri stars. R. Mundt.
Bull. American Astron. Soc., Vol. 13, 856 (1981). – Abstract.

121.032 Theoretical models for T Tauri mass loss.
L. Hartmann, S. Edwards, E. H. Avrett.
Bull. American Astron. Soc., Vol. 13, 856 (1981). – Abstract.

121.033 Far-infrared observations of FU Orionis.
H. A. Smith, H. A. Thronson, Jr., C. J. Lada, D. A. Harper, R. F. Loewenstein, J. Smith.
Prepr. Steward Obs., No. 388, 23 pp. (1982).

121.034 H-H1 and H-H2: the results of an eruptive event in the Cohen-Schwartz star?
R. Mundt, L. Hartmann.
Prepr. Steward Obs., No. 398, 32 pp. (1982).

121.035 Physical conditions in emission regions and mechanisms of activity of T Tauri-type stars.
V. P. Grinin.
Astrofizika, Tom 16, 243 - 256 (1980). In Russian. English translation in Astrophysics, Vol. 16, No. 2.

A comparison between theoretical and observed Balmer decrements of T Tau-type stars has been carried out. It has been shown that the emission of purely chromospheric origin is not typical for the majority of stars of this type. However, it is presented in the spectra of about 1/3 of the stars as a component. But the main part of the radiation in the hydrogen lines is formed in circumstellar envelopes. The question of the origin of ultraviolet emission of T Tau-type stars has been considered, using the star DF Tau as an example. The comparison between the theoretical and observed UBV-color radiation of DF Tau flares shows that the source of flare emission is a hot spot on the surface of the star which is likely to be formed during mass accretion from the circumstellar envelope. Possible mechanisms of the activity of T Tau-type stars are discussed with respect to the results of spectral and photometrical observations.

121.036 The Balmer decrements of T Tau stars.
N. A. Katysheva.
Astrofizika, Tom 17, 301 - 307 (1981). In Russian. English translation in Astrophysics, Vol. 17, No. 2.

An analysis of the Balmer decrements (b.d.) of T Tau stars from the new catalogue by Cohen and Kuhi (1979) has been carried out. A tendency to the displacement of average b.d. from stars with spectral classes G5 - K5 to K5 - M5 stars is obtained. The comparison of observed b.d.'s with theoretical ones calculated on the basis of Sobolev's method (1947) has been made.

121.037 Far-infrared observations of FU Orionis.
H. A. Smith, H. A. Thronson, Jr., C. J. Lada, D. A. Harper, R. F. Loewenstein, J. Smith.
Astrophys. J., Vol. 258, 170 - 176 (1982).

The far-infrared flux from FU Orionis was measured using the NASA Kuiper Airborne Observatory, with a beam size of 49". The far-infrared dust temperature was 15 ± 5 K, and the total far-infrared luminosity was $1.1 \pm 0.2 L_\odot$. The dust is extended in the east-west direction, in the same direction as the gas cloud found by earlier ^{12}CO mapping. The gas temperature (^{12}CO) was measured with a 66" beam and found to be $T_g \approx 13.3 \pm 2.0$ K. The lack of enhancement in the gas and dust temperatures around FU Ori suggests that the star interacts only weakly with the surrounding B35 cloud. The results also confirm the presence of a hotter circumstellar dust component whose spectrum fits a ~ 225 K gray body. Most of the radiation from the star does not interact with this hotter dust, a fact which supports arguments that it may comprise a disk of material.

121.038 The case for anisotropic mass loss from T Tauri stars.
M. Cohen.
Publ. Astron. Soc. Pacific, Vol. 94, 266 - 270 (1982).

Observational evidence and theoretical arguments are assembled in favor of directed, rather than isotropic, mass flows from T Tauri stars. These arguments include the structure of T Tauri nebulosities, radio continuum maps of the stars, the motions of Herbig-Haro objects away from their exciting T Tauri stars, bipolar CO flows, and particularly the spatial coincidence of an optical jet and radio emission associated with an embedded, probably T Tauri star in the dark cloud L1551.

121.039 Mass spectra of young stars. R. B. Larson.
Mon. Not. R. Astron. Soc., Vol. 200, 159 - 174 (1982).

In an effort to understand the factors that determine the initial mass function of stars, the available data on the mass spectra of young stars in different regions have been studied for possible correlations with the properties of the associated molecular clouds and the spatial distributions of the stars. The most massive stars appear to form in the dense cores of forming clusters or associations, and the mass of the most massive young star increases systematically with the mass of the associated molecular cloud. The data are consistent with a picture in which molecular cloud cores grow by accretion and become progressively more massive and condensed, while forming stars in larger and more condensed clusters and with a mass spectrum that increasingly favours massive stars. In this picture the more massive stars form by accumulation processes, rather than by fragmentation, in the dense core regions of protoclusters.

121.040 Parallaxes for pre-main sequence Herbig Ae/Be stars.
    P. S. Thé, H. R. E. Tjin A Djie, F. Praderie, C. Catala.
Scientific aspects of the Hipparcos space astrometry mission, (see 012.041), p. 105 - 106 (1982).
    Herbig Ae/Be stars are young irregularly variable objects associated with nebulosities located near the galactic plane. Their spectrum exhibit emission lines and a strong IR excess, usually explained as due to thermal radiation by a dust shell. They are most likely high mass ($2-5$ M$_\odot$) counterparts of the T Tauri stars. No accurate groundbased distance determinations exist of these stars.

121.041 Molecular fluorescence from youngsters.
    R. Liseau.
Rep. Obs. Lund, No. 18, (see 012.044), p. 93 - 96 (1982).

    Landolt-Börnstein.   See Abstr. 003.064.

    On the discrepancy between the optical and radio position of T Tauri.   See Abstr. 041.032.

    Spectral line formation in YY Orionis envelopes: a multi-level hydrogen atom.   See Abstr. 063.059.

    Dynamical effects of stellar winds and shocked gas on interstellar clouds.   See Abstr. 112.018.

    A model of the 10 micrometer silicate feature in the spectra of BN-like IR point sources.   See Abstr. 112.056.

    Relationship between colorimetric characteristics of flare stars in aggregates and the aggregate's age.   See Abstr. 115.002.

    Radio observations of early-type emission-line stars and related objects.   See Abstr. 116.001.

    Radio emission from young stars.   See Abstr. 116.014.

    Radio observations of pre-main-sequence stars: results and interpretation.   See Abstr. 116.015.

    Search for short-time light variations in DF Tauri.   See Abstr. 122.062.

    Brightness fading of the FU Orionis star V1515 Cygni.   See Abstr. 122.100.

    Report of IAU Commission 27: Variable stars (Étoiles variables).   See Abstr. 122.250.

    Excess line emission in protostellar objects.   See Abstr. 131.158.

    Low-mass star formation in the dense interior of Barnard 18.   See Abstr. 131.177.

    The H I cloud surrounding the emission-line star LkHα 101 in the region of NGC 1579.   See Abstr. 132.014.

    The variable infrared source near HH100.   See Abstr. 133.026.

    Quasars and T Tau-type stars: a complete analogy.   See Abstr. 141.217.

122 Intrinsic Variables (Pulsating Variables, Spectrum Variables, etc.)

122.001 **Infrared photometry of Mira variables in the Baade windows and the distance to the Galactic Centre.**
I. S. Glass, M. W. Feast.
Mon. Not. R. Astron. Soc., Vol. 198, 199 - 214 (1982).

JHKL infrared observations are presented for 70 Mira variables in the Galactic Centre windows, Sgr I and NGC 6522. Distances were derived using the period-luminosity relation of Glass and Lloyd Evans for LMC Miras. An analysis of the Mira data yields A_v = 2.0 mag, R_0 = 8.8 kpc. The data can be combined with the analysis by Oort and Plaut of RR Lyrae variables in the NGC 6522 field to yield A_v =1.5 mag, R_0 = 9.2 ($\pm$ 0.6) kpc which is adopted as best estimate.

122.002 **Coordinated X-ray, optical, and radio observations of flaring activity on YZ Canis Minoris.**
S. Kahler, L. Golub, F. R. Harnden, Jr., W. Liller, F. Seward,
G. Vaiana, B. Lovell, R. J. Davis, R. E. Spencer,
D. R. Whitehouse, P. A. Feldman, M. R. Viner, B. Leslie,
S. M. Kahn, K. O. Mason, M. M. Davis, C. J. Crannell,
R. W. Hobbs, T. J. Schneeberger, S. P. Worden,
R. A. Schommer, S. S. Vogt, B. R. Pettersen, G. D. Coleman,
J. T. Karpen, M. S. Giampapa, E. K. Hege, V. Pazzani,
M. Rodono, G. Romeo, P. F. Chugainov *(Chugajnov)*.
Astrophys. J., Vol. 252, 239 - 249 (1982).

A coordinated search for flares from the dMe star YZ Canis Minoris was performed in 1979 October using the Einstein Observatory and ground-based optical and radio telescopes. An event was detected in the optical, radio, and X-ray wavebands on October 25, and a second optical event on October 27 was seen as a marginal (2 σ) X-ray enhancement. The properties of the first event are discussed in detail, and it is shown that the similarities to solar flares are considerable.

122.003 **Ultraviolet spectrum variability of UX Ursae Majoris.**
A. V. Holm, R. J. Panek, F. H. Schiffer III.
Astrophys. J., Lett., Vol. 252, L35 - L37 (1982).

IUE spectra show UX Ursae Majoris to be a spectacular ultraviolet line spectrum variable on time scales of months. Time-resolved spectra around the orbit show that the continuum is strongly eclipsed, whereas the line emission is not. Features in the light curve which are identified with a hot spot are not prominent in the far-UV. The UV line emission probably originates in a wind.

122.004 **A preflare diminution in the quiescent flux of EQ Pegasi.** M. S. Giampapa, J. L. Africano,
A. Klimke, J. Parks, R. J. Quigley, R. D. Robinson,
S. P. Worden.
Astrophys. J., Lett., Vol. 252, L39 - L42 (1982).

The authors report the occurrence of a remarkable flare event on EQ Peg as recorded by high speed photometry in the Johnson *U* band: a stellar flare event is immediately preceded by a well-defined decline in the quiescent flux of the star. The *U* band flux decays to a minimum level that is 75% of the stellar quiescent flux, and the duration of the so-called negative flare event is 2.7 minutes. The authors present a description of the observation and qualitatively discuss hypotheses that may eventually account for this phenomenon.

122.005 **The Oosterhoff period groups and the age of globular clusters. IV. Field RR Lyrae stars: age of the galactic disk.** A. Sandage.
Astrophys. J., Vol. 252, 574 - 581 (1982).

In this paper, the author continues his analysis of the Oosterhoff period shifts in RR Lyrae variables by applying it to field stars in the galactic disk. It is shown that the correla-
tion of [Fe/H] with these period shifts for field variables is the same as that found for variables in globular clusters. It is concluded that the oldest disk stars began to form long before the cluster NGC 188 was born and are as old as the halo globular clusters. The implications for the [Fe/H] distribution in the early history of galactic disk formation by halo collapse are discussed.

122.006 **Z Chamaeleontis: evidence for an eccentric disk during supermaximum?** N. Vogt.
Astrophys. J., Vol. 252, 653 - 667 (1982).

Spectroscopic and photometric observations of the eclipsing dwarf nova Z Cha are presented. From the spectroscopic data, combined with published eclipse light curve characteristics, the principal system parameters of Z Cha are derived, in particular an orbital inclination i = 79$^\circ \pm 2^\circ$, stellar masses M_1 = 0.85 $\pm$ 0.15 $M_\odot$ and M_2 = 0.17 $\pm$ 0.05 $M_\odot$, binary separation, stellar radii, as well as size and location of the hot spot. Photoelectric high speed photometry, obtained during the 1978 March supermaximum, revealed partial eclipses, as well as superhumps whose period P_s = 0$.^d$07725 exceeds the orbital period P_0 by 3.7%. None of the previously suggested models for supermaxima and superhumps fits all observational constraints available now. Therefore, a new, consistent solution is suggested, implying an eccentric ring which surrounds the inner accretion disk during superoutburst.

122.007 **Time series infrared spectroscopy of the Mira variable χ Cygni.**
K. H. Hinkle, D. N. B. Hall, S. T. Ridgway.
Astrophys. J., Vol. 252, 697 - 714 (1982).

A time series of 32 high resolution infrared spectra spanning more than three cycles of the S-type Mira variable χ Cygni is analyzed. Most spectra are of the 1.6–2.5 μm region; five are of a narrow region at 4.6 μm. Using information derived from the vibration-rotation bands of CO, the authors show that χ Cygni is characterized by a regularly pulsating photosphere and complex, circumstellar envelope. The structure of the circumstellar envelope appears to be dominated by a stationary layer which is inferred from its 800 K temperature to be at a distance of the order of 10 stellar radii from the photosphere. This layer in χ Cygni built up rapidly late in 1975 and steadily diminished over the next three cycles.

122.008 **The light and spectrum variations of VX Sagittarii, an extremely cool supergiant.**
G. W. Lockwood, R. F. Wing.
Mon. Not. R. Astron. Soc., Vol. 198, 385 - 404 (1982).

Narrow-band photometry covering more than two cycles of its 732-day variation strongly supports the interpretation that VX Sgr is an extremely cool supergiant star. The spectral type from near-infrared bands of TiO and VO ranges from M5.5 at maximum light to M9.8 at minimum, while the brightness varies by 1.2 mag in I(104) and by more than 6 mag in V; in these respects VX Sgr resembles a cool Mira variable. Unlike a Mira, however, VX Sgr shows strong bands of CN throughout the cycle, corresponding to a Ia luminosity classification. The TiO band strength is normal for the colour if the interstellar extinction is A_V = 1.5 mag; the VO bands, however, are abnormally strong for the colour. Spectral scans at 8 Å resolution support the interpretation of VX Sgr as an M-type supergiant. The authors find that the radius varies from 6.3 AU at maximum light to approximately 9 AU at minimum. VX Sgr appears to be a Population I member of the Sagittarius arm at a distance of about 1500 pc.

122.009 Atmospheric kinematics of high velocity long period variables.
L. A. Willson, G. Wallerstein, C. A. Pilachowski.
Mon. Not. R. Astron. Soc., Vol. 198, 483 - 516 (1982).

The authors have analysed radial velocities of atomic absorption lines of three long period variables, RT Cyg, Z Oph and S Car, in order to understand velocity gradients and discontinuities in their atmospheres. Phase coverage is from five days before maximum to 73 days after maximum for RT Cyg, from 17 days before to 44 days after maximum for Z Oph, and at 9 days before maximum for S Car. On a few spectrograms double lines were seen. All spectrograms were analysed by a four-parameter regression programme to yield the dependence of the radial velocity on the excitation potential, first ionization potential, wavelength and line strength, as indicators of the depth of line formation. The data were analysed to yield the velocity discontinuity across shock waves and velocity gradients between shock waves. Consistent parameters are obtained if these stars are fundamental mode pulsators with total masses in the range of $0.5-1.0\,M_\odot$ and effective radii in the range of $0.85-1.5 \times 10^{13}$ cm.

122.010 An unusually short stable period of absorption line asymmetries and V/R variations in the spectrum of the Be star 28 CMa.   D. Baade.
Astron. Astrophys., Vol. 105, 65 - 75 (1982).

In 28 CMa (B2 - 3, IV - Ve, $m_V \cong 3\overset{m}{.}8$) the by far shortest stable period known of Be stars has been detected. The 1.365-d periodicity affects the profiles of absorption lines (changing asymmetry), the radial velocities of emission lines and their V/R ratios. Because of numerous similarities to β Cephei stars, an attempt has been made to develop a model of nonradial pulsations for 28 CMa. It is shown that P_2^2-pulsations can explain not only all of the observed spectroscopic variations of 28 CMa, but also the much more frequently observed long "period" V/R variations of other Be stars.

122.011 On the period of the pulsating hydrogen-deficient star BD + 13°3224.
D. Kilkenny, A. E. Lynas-Gray.
Mon. Not. R. Astron. Soc., Vol. 198, 873 - 879 (1982).

Recent observations of maxima of the pulsating early-type hydrogen-deficient star BD + 13°3224 indicate a decrease in period since the discovery observations. The data are best fitted by a decrease rate of 46×10^{-10} day cycle^{-1}, a value which is comparable with the rate derived from a theoretical study of the fast evolutionary phase of a post-giant helium star. Four-colour data are averaged to give mean colours at 0.1 phase intervals around the pulsation cycle.

122.012 Opacity and nonlinear effects on theoretical BL Herculis models.
S. W. Hodson, A. N. Cox, D. S. King.
Astrophys. J., Vol. 253, 260 - 267 (1982).

Linear and nonlinear pulsation models for BL Herculis variables have been constructed to investigate the resonance which seems to occur when the ratio of the second overtone (Π_2) to fundamental (Π_0) radial periods is near 0.5. This resonance is shown to affect the shapes of the light and velocity curves and produce bumps on either ascending or descending light just as for classical Cepheids. Linear theory predicts the resonance to occur at periods between 1.7 and 3.0 days for $0.55\,M_\odot$ and between 2.1 and 4.0 days for $0.75\,M_\odot$ stars at the red and blue edges, respectively, of the instability strip. These periods are rather independent of the composition and opacity tables. However, observations show the resonance to be about 1.7 days for all BL Her variables. Nonlinear calculations indicate that the linear theory predictions are not reliable just at $\Pi_2/\Pi_0 = 0.5$, and the predicted resonance occurs always at the proper period as observed.

122.013 Period distributions of irregularly variable RR Lyrae stars.   H. A. Smith.
Publ. Astron. Soc. Pacific, Vol. 93, 721 - 727 (1981/82).

Period distributions have been determined for irregularly variable RR Lyrae stars in M3, M5, M15, ω Cen, and the Draco dwarf spheroidal galaxy. Irregular variability is more frequent among the shorter-period ab-type RR Lyraes ($P < 0\overset{d}{.}65$) and, at least in M15 and ω Cen, among the longer period c-type variables. These period distributions are consistent with some of the mode-mixing explanations for apparently irregular variability. Borkowski's suggestion that the Blazhko effect in ab-type RR Lyraes is a nonlinear mixture of the fundamental and third overtone radial pulsation modes accounts for the occurrence of irregular ab variables only at the shorter periods, but yields RR Lyrae star masses larger than those predicted by stellar evolution theory.

122.014 On the nature of $H\alpha$ outbursts in the RS Canum Venaticorum binary SZ Piscium.
L. W. Ramsey, H. L. Nations.
Publ. Astron. Soc. Pacific, Vol. 93, 732 - 734 (1981/82).

The authors describe spectroscopic observations of an $H\alpha$ emission episode on the RS CVn binary SZ Psc in 1979. These data are compared and contrasted with a recently reported 1978 outburst. The authors discuss these and previous data in the context of the surface phenomena or starspot model for RS CVn stars.

122.015 Spectral classification of select δ Scuti stars. II.
R. Peniche, J. H. Peña, M. A. Hobart.
Publ. Astron. Soc. Pacific, Vol. 93, 735 - 740 (1981/82).

Through spectroscopic observations a reclassification of several δ Scuti stars has been carried out. With this new classification new values of M_v have been obtained that fix the position of these stars in both the H−R and the PLCR diagrams.

122.016 The rates of change of the fundamental and overtone periods of SX Phe.
D. W. Coates, L. Halprin, K. Thompson.
Mon. Not. R. Astron. Soc., Vol. 199, 135 - 139 (1982).

Using existing data for SX Phe, the authors calculate the rates of change in the fundamental period and the overtone period assuming that both are changing at a constant rate. The method used is applicable to any pulsating variable star whose times of maximum light exhibit measurable multiperiodicity.

122.017 Infrared photometry of Mira variables in the LMC and the pulsational properties of Miras.
I. S. Glass, M. W. Feast.
Mon. Not. R. Astron. Soc., Vol. 199, 245 - 253 (1982).

Infrared J, H, K observations are presented for eleven Mira variables in the Large Magellanic Cloud. Except for a known carbon variable and an object with poor colours, the variables lie in the region of the $(J-H)-(H-K)$ diagram populated by galactic Me Miras. They also fit galactic $(J-K)-\log P$ and $\log T_{BB}-\log P$ relations. Period−luminosity relations in M_J, M_H, M_K and M_{bol} are all well defined. A $\log T_{BB}-\log P$ relation is derived for galactic Me Miras. Combining this with the period−luminosity relation leads to the conclusion that the stars are pulsating in the first overtone with masses ranging from $\sim 0.8\,M_\odot$ at a period of 150 day to $\sim 1.5\,M_\odot$ at 500 day.

122.018 On the changes of the period of DY Pegasi.
S.-y. Jiang, N.-s. Zhao.
Acta Astrophys. Sinica, Vol. 2, 44 - 48 (1982). In Chinese.

122.019 Further discussion concerning the mass loss rates from Mira variables.   J. Sun, S.-m. Wu, Y. Fan.
Acta Astrophys. Sinica, Vol. 2, 49 - 55 (1982). In Chinese.

122.020   The period-luminosity relation. IV. Intrinsic relations and reddenings for the Large Magellanic Cloud Cepheids.   B. F. Madore.
Astrophys. J., Vol. 253, 575 - 579 (1982).

Reddening-independent parameters, specifically period, amplitude, and the Wesenheit function, are used to probe the intrinsic calibration of the period-luminosity relation for Cepheids with photoelectric B, V photometry in the Large Magellanic Cloud. These relations can be inverted to determine reddenings to the individual Cepheids confirming earlier indications that detectable absorption exists within the LMC itself and that the longer-period Cepheids are systematically more heavily obscured. Independent of the detailed reddening values it is also shown that amplitude varies systematically across the instability strip such that at all periods amplitude increases toward the cool (faint) edge.

122.021   Pulsational mode-typing in line profile variables. IV. Selected δ Scuti stars.   M. A. Smith.
Astrophys. J., Vol. 254, 242 - 262 (1982).

Variations of 129 new high-quality line profiles have been compared with models to distinguish between radial and nonradial pulsation in nine δ Scuti stars. Even for secondary and tertiary amplitude modes, the data show a widespread consistency with recent multifrequency analyses of photometric data. A significant aspect of this study is the finding that the oscillations in δ Scuti stars are sufficiently adiabatic that the spectroscopic and photometric amplitudes can be compared and matched to nonradial pulsation models.

122.022   Nonradial pulsations in early-type B stars: g-modes or r-modes?   M. A. Smith.
Astrophys. J., Vol. 254, 708 - 712 (1982).

A variety of considerations suggest that the nonradial oscillations present in 53 Per stars and several β Cep stars do not arise from Rossby (r-) waves, but rather from spheroidal waves, such as g-modes.

122.023   Peculiarities of the cepheid distribution in the Large Magellanic Cloud.
Yu. N. Efremov, E. D. Pavlovskaya.
Pis'ma Astron. Zh., Tom 8, 9 - 16 (1982). In Russian. English translation in Soviet Astron. Lett., Vol. 8.

The distribution of cepheids in the LMC is shown to be very nonuniform. Only the cepheids with lg $P < 0.9$ are concentrated in the main body (bar) of the LMC. Therefore about 3×10^7 years ago the bar had ceased to be the locus of star formation and now star formation is going on in large regions randomly distributed in the LMC. The difference of periods (ages) of cepheids decreases with decreasing distances between them.

122.024   Superflare of EV Lacertae.
G. Sh. Rojzman, V. S. Shevchenko.
Pis'ma Astron. Zh., Tom 8, 163 - 164 (1982). In Russian. English translation in Soviet Astron. Lett., Vol. 8.

A great flare of EV Lacertae with amplitude of $\Delta U = 6^{m}4$ and duration more than 4.5 hours is described. The occurrence of "slow" flares of field red dwarfs is argued.

122.025   Orbital motion of the pulsating star V644 Her.
C. Bardin, M. Imbert.
Astron. Astrophys., Suppl. Ser., Vol. 47, 319 - 322 (1982). In French.

The δ Scuti type star V644 Her, first known as a spectroscopic binary, was observed from 2 March to 11 November 1980 with the CORAVEL photoelectric spectrometer at the Observatoire de Haute-Provence. The orbital elements and the period have been determined. Assuming for this young disk variable an absolute magnitude of $M_v = 2.0$, which is consistent with an F2IV spectral type, and a mass of about 1.5 $M_{\odot}$, the authors find the secondary component to be cooler than

an F6−8V star. The system is separated, with a distance of 30 $R_{\odot}$ between the two components, and an inclination greater than 30°. Its distance from us is about 66-76 or 72-83 pc, depending whether or not we take the brightness of the secondary component into account.

122.026   Photoelectric photometry of Cepheid variables with periods between one and three days.
R. Diethelm, G. A. Tammann.
Astron. Astrophys., Suppl. Ser., Vol. 47, 335 - 339 (1982).

122.027   The period and photometry of BC Draconis.
L. Szabados, R. S. Stobie.
Astron. Astrophys., Suppl. Ser., Vol. 47, 541 - 545 (1982).

New BV photoelectric photometry of BC Draconis is presented and frequency analysed. BC Draconis appears to be a normal RR Lyrae-type variable of period 0.719576 days.

122.028   Nonlinear models of classical Cepheids endowed with tangled magnetic fields.   R. Stothers.
Astrophys. J., Vol. 255, 227 - 231 (1982).

The effect of tangled magnetic fields has been included in a new study of full-amplitude models of classical Cepheids. As compared with nonmagnetic models, the magnetic models have longer periods, larger amplitudes, and earlier phases of the small secondary bump that appears on the velocity and light curves. The induced changes of period and of bump phase yield better agreement with observations if Cepheids have normal evolutionary masses. But the predicted amplitudes are larger than those observed; moreover, the inferred value of the mean ratio of magnetic pressure to thermodynamic pressure falls significantly below the value needed to explain the period ratios of the double-mode Cepheids.

122.029   An RR Lyrae survey with the Lick astrograph. V. A survey of three fields at intermediate latitudes towards the galactic anticenter.
T. D. Kinman, C. T. Mahaffey, C. A. Wirtanen.
Astron. J., Vol. 87, 314 - 352 (1982).

Photometric data, positions, and finding charts are given for 62 variables in three astrograph fields covering 84 square degrees at galactic latitudes −18°, +26°, and +36° towards the anticenter. These variables have mean photographic magnitudes in the range $11.2 < m_{pg} < 17.9$. Photoelectric observations are given for all the RR Lyrae stars and their $B-V$ colors at minimum light are used in a discussion of the extinctions in these fields. It is found that the RR Lyrae stars in these anticenter fields and at the north galactic pole show a significant deficiency of very metal-poor members (with [Fe/H] < -1.6) if compared with samples of globular clusters or a sample of nearby subdwarfs, and the reasons are discussed.

122.030   Metal abundances of RR Lyrae variables in selected galactic star fields. III. The Lick astrographic fields near the galactic anticenter.   D. Butler, E. Kemper, R. P. Kraft, N. B. Suntzeff.
Astron. J., Vol. 87, 353 - 359 (1982) = Lick Obs. Bull. No. 906.

Values of the Preston metal-abundance index Δs and corresponding metallicities [Fe/H] are derived for 26 of the 28 stars classified as RR Lyraes that have been discovered in three anticenter intermediate-latitude fields of the Lick astrograph survey. The mean metallicity and metallicity spread of these stars are essentially identical with those derived earlier from a somewhat larger sample of RR Lyraes in the north galactic pole. In both the anticenter and polar regions, the stars extend to galactocentric distances of 25 kpc or more. The entire sample of 61 RR Lyraes shows no direct evidence for an abundance gradient and the extremes of metallicity at any given [Fe/H] range over about 1 dex.

Problems in deducing the true halo abundance gradient from RR Lyraes are discussed.

122.031   The colors of the pulsations and flickering of SY Cancri during outburst.
J. Middleditch, F. A. Córdova.
Astrophys. J., Vol. 255, 585 - 595 (1982).

The spectra of the short period (~30 s) oscillations and flickering of the dwarf nova SY Cnc have been determined by simultaneously measuring the optical flux in three broad, contiguous spectral bands during an optical outburst of the star. The spectrum of the oscillations rises too rapidly toward short wavelengths to be consistent with any simple thermal model. The colors of the flickering are even more extreme than those of the pulsations. No timing differences were detected to a fraction of a second between the ultraviolet cyan and red bands of the pulsations; thus, the optical oscillations appear to arise from a single physical location in the binary system.

122.032   The diameter of Mira.
D. Bonneau, R. Foy, A. Blazit, A. Labeyrie.
Astron. Astrophys., Vol. 106, 235 - 239 (1982).

The authors have measured the angular diameter of Mira using the digital speckle interferometer at the 3.60 m ESO telescope, at time of maximum light in 1978 and 1979. They confirm that the diameter varies abruptly with wavelength, in relation with the TiO spectrum. They estimate the true radius to be $\theta_c = 28 \pm 6$ milliarcsec, or $R_c = 230\,R_\odot$ assuming a distance of 77 pc, and discuss how dependent this value is upon the model representing the atmosphere. The diameter measured in strong TiO features was larger during the low-luminosity maximum in 1978 than during the normal maximum in 1979.

122.033   U Geminorum, 1970 - 79.
D. R. B. Saw.
J. British Astron. Assoc., Vol. 92, 127 - 131 (1982).

An analysis is given of visual observations of U Geminorum for the period 1970 - 79. Rates of rise and fall are discussed, together with relationships between various outburst parameters. Variation at minimum is compared with photoelectric measurements. From 1904 to 1979, the mean interval between outbursts for 61 alternating short and long maxima is 100.4 days.

122.034   Brightness of BH Cephei in June-September 1978.
A. I. Zheleznyakova, V. I. Kardopolov.
Perem. Zvezdy, Tom 21, 301 - 304 (1980). In Russian.

BH Cep has been monitored photoelectrically in June–September 1978. The results of observations in the UBV system are given. The data show no constancy in stellar brightness in the observational period. On JD 2443692 and JD 2443732 the starlight of BH Cep was observed in minimum.

122.035   On the character of brightness variations of RZ Pisces.   V. I. Kardopolov, V. V. Sakhanenok, N. A. Shutemova.
Perem. Zvezdy, Tom 21, 310 - 313 (1980). In Russian.

Results of UBV photoelectric observations of the irregular variable star RZ Psc made in August–September 1978 are given. The data show no constancy in stellar brightness and stellar color indices in the observational period. All peculiarities of RZ Psc light variations can be probably explained by superposition of different types of variability.

122.036   Optical and infrared photometry of four R Coronae Borealis-type stars.   V. I. Shenavrin.
Perem. Zvezdy, Tom 21, 315 - 319 (1980). In Russian.

The results of UBVRJHKLMN photometry of the stars XX Cam, UV Cas, SU Tau, SV Sge are presented. No dust shell was found in XX Cam. The infrared excess was detected in UV Cas, where it was interpreted as the radiation of a dust

shell with T = 900°K. The spectral energy distribution of SU Tau corresponds to Planck radiation with T = 750°K. Interstellar absorption in the direction of SV Sge was found to be $A_V = 3\overset{m}{.}1$.

122.037   Spatial distribution of flare stars in the Pleiades derived by the orthogonal expansion method.
E. L. Kosarev.
Perem. Zvezdy, Tom 21, 321 - 363 (1980). In Russian.

A new orthogonal expansion method (OEM) for recovery of spatial star distribution in globular clusters is presented. The first part of the paper consists of the detailed description of this method and the results of its testing. In the second part of the paper the OEM is applied to recovery of spatial density of 441 flare stars in the Pleiades cluster. It is shown that their surface density has elliptic symmetry with small eccentricity. The spatial density has a maximum in the centre of the cluster with magnitude about $(1.6-2.5)\ pc^{-3}$. The shape of the spatial density looks like a Gauss curve with scale parameter about 3.5 pc.

122.038   A study of the period change of RR Lyrae-type variables in the globular cluster M92.
B. V. Kukarkin, N. P. Kukarkina.
Perem. Zvezdy, Tom 21, 365 - 389 (1980). In Russian.

A study of period changes of 10 RR Lyr-type variables in the globular cluster M92 (NGC 6341) is given. Maxima of the variables were summarized on the interval JD 2424000–42000.

122.039   Determination of the period of the Blažhko effect of the variable star V5 in the globular cluster M3.
K. Panov.
Perem. Zvezdy, Tom 21, 391 - 398 (1980). In Russian.

The method of Goranskij (1976) was applied for searching for the period of the Blažhko effect of the variable star V5 in the globular cluster M3. For the period of the Blažhko effect a value of $194\overset{d}{.}551$ was obtained. The Blažhko effect is pronounced mainly in the variation of the level of maximum light.

122.040   Warum beobachten wir Mirasterne?
M. Fernandes.
BAV Rundbrief, 31. Jahrg., 6 - 15 (1982).

122.041   Ultra-violet spectroscopy of flare stars.
P. B. Byrne, C. J. Butler, A. D. Andrews.
Irish Astron. J., Vol. 14, (see 012.003), 219 - 226 (1980).

122.042   A simultaneous photometric and radial velocity study of short-period southern Cepheids. II. The photometry.   W. Gieren.
Astrophys. J., Suppl. Ser., Vol. 47, 315 - 332 (1981).

About 600 photoelectric UBVRI observations on the Kron-Cousins system and the resulting light and color curves are presented for 15 southern short-period Cepheids. These observations were obtained simultaneously with the photoelectric radial velocities which were presented in Paper I. The internal accuracy of the individual Cepheid observations is ~0.005 mag in V and all of the colors.

122.043   The chromospheres of classical Cepheids. I. Low resolution IUE spectra.
E. G. Schmidt, S. B. Parsons.
Astrophys. J., Suppl. Ser., Vol. 48, 185 - 198 (1982).

The authors have undertaken a program to study the behavior of chromospheres in classical Cepheids using both ground-based and satellite-based observations. Five stars, δ Cep, η Aql, β Dor, ζ Gem, and l Car, have been included. The first part of this investigation concerns low dispersion spectra from the IUE satellite. The low dispersion spectra have been used to determine continuum magnitudes at various ultraviolet wavelengths, and light curves are shown. The amplitude increases with decreasing wavelength except around 1550 Å. Line emission in the short wavelength region (1100 to 1900 Å) is observed in three of the Cepheids and is found to be phase dependent.

122.044 Lithium and barium in R CrB and XX Cam.
K. Hunger, D. Schönberner, W. Steenbock.
Astron. Astrophys., Vol. 107, 93 - 96 (1982).

9 Å/mm Coudé-spectrograms have been fine analyzed for Li and Ba. Lithium is (slightly) overabundant with respect to the cosmic value in R CrB (+0.5 ± 0.4 dex) and deficient in XX Cam (< −1.1 dex), while barium is overabundant in both stars: +1.2 ± 0.3 dex in R CrB, and +0.7 ± 0.2 dex in XX Cam.

122.045 The pulsation of the outer layers of the Beta Cephei-type variable BW Vul.
M. Burger, C. de Jager, G. H. J. van den Oord, N. Sato.
Astron. Astrophys., Vol. 107, 320 - 325 (1982).

Eleven high-resolution spectrograms of BW Vul were obtained with the IUE, covering 1.6 period. The shape of the line profiles varies over the pulsational cycle, the lines are broadest shortly before zero radial velocity, in agreement with the picture that at the end of the contraction phase both a downfalling component and a stationary component are seen. The radial velocity curve of the C IV lines is different from the photospheric one and shows that both the photosphere and the C IV layers are accelerated upward impulsively. In about one hour the acceleration decreases to zero, whereafter the whole atmosphere falls down with a nearly constant downward acceleration of 2100 cm s^{-2}. The varying asymmetry of the C IV line profiles indicates the occurrence of mass loss due to the pulsation of the atmosphere, a mass loss superimposed on the continuous mass loss of this star.

122.046 Mg II h and k line observations of Delta Scuti variables.
M. Fracassini, L. E. Pasinetti.
Astron. Astrophys., Vol. 107, 326 - 332 (1982).

The authors report the results of IUE observations of seven Delta Scuti variables and a comparison star. The Mg II h and k lines are analysed and compared with those of normal stars observed by other authors. τ Cyg(F0IV), β Cas(F2IV) and ρ Pup(F6IIp) show evident emissions; τ Peg(A5IV), k^2 Boo(A7IV), o^1 Eri(F2III) and ι Cyg(A5V, comparison star) show doubtful emissions. The visual binary and suspected Delta Scuti variable τ Cyg, and the peculiar Delta Scuti variable ρ Pup show a peculiar behaviour. A relation between the full width emissions and the period of pulsation seems probable.

122.047 Profile variations of the Si III (4452 and 4568) lines and Mg II (4481) doublet in γ Peg.
J.-M. Le Contel, P.-J. Morel.
Astron. Astrophys., Vol. 107, 406 - 408 (1982).

The temporal evolution of spectral features from high resolution spectra of the β CMa variable γ Peg is described. Though these observations are almost consistent with the bouncing shell model of Smith and McCall (1978), the authors emphasize on a shock wave model as suggested by the observation of simultaneous supersonic blue and red emission components.

122.048 Helium opacity bump and excitation mechanisms of Beta Cephei pulsations.
U. Lee, Y. Osaki.
Publ. Astron. Soc. Japan, Vol. 34, 39 - 50 (1982).

The Stellingwerf mechanism of helium opacity bump for the pulsational driving of the Beta Cephei stars is re-examined. The continuous opacity in the relevant temperature range (around $T = 1.5 \times 10^5$ K) is calculated with a sufficiently fine mesh interval in $\log_{10} T$, and the opacity derivatives, which are needed for the pulsational stability analysis, are obtained from

it. It is found that the opacity bump due to the He$^+$-ionization edge is not large enough to destabilize stars as a whole, although it contributes locally to the excitation of pulsations. However, the effect of line opacity, which may contribute as much as fifty percent to the total opacity, is not taken into account in this analysis. Other possibilities of envelope excitation mechanisms for β Cephei pulsations are also examined.

122.049 Observations of the small amplitude β Cephei stars.
M. Kubiak.
Messenger, No. 27, p. 17 - 19 (1982).

122.050 La variable WW Vul est-elle du type R CrB?
D. Böhme.
Bull. AFOEV, No. 19, p. 2 - 5 (1982).

122.051 A new cepheid variable, HD 200925.
T. D. Padalia, S. K. Gupta.
Astrophys. Space Sci., Vol. 81, 251 - 260 (1982).

Photoelectric observations of the star HD 200925 in the standard UBV system have been secured and analysed. The period is determined to be 0$.^d$267394. From the period and shape of the light and colour curves, the star HD 200925 appears to be a dwarf cepheid. The physical parameters have been derived. The mass derived for this star is found to agree well with the value inferred from the evolutionary tracks. The star appears to be a post-Main Sequence star in the hydrogen shell burning stage of evolution. The spectral class is assigned to be F2 III.

122.052 The dwarf cepheid NJL 79 in Omega Centauri.
H. E. Jørgensen.
Astron. Astrophys., Vol. 108, 99 - 101 (1982).

A faint dwarf cepheid NJL 79 with $\langle V \rangle = 16.8$ has been found at a distance of only 6' from the centre of the globular cluster Omega Centauri. The period is approximately 91 min. The magnitude is within 0.1 the same as for the other known dwarf cepheid NJL 220 in the field of Omega Centauri placing the two variables at approximately the same distance. Adopting NJL 79 to be a member of Omega Centauri the author obtains $M_V = 3.0$, which is typical for population II dwarf cepheids. Using the pulsation equation the author obtains a mass in the range $0.6 - 1.0 M_\odot$.

122.053 Spectroscopic observations of the S Dor variable R71.
B. Wolf, I. Appenzeller, O. Stahl.
Mitt. Astron. Ges., Nr. 55, p. 53 (1982). − Abstract.

122.054 Systemparameter für südliche Zwergnovae aufgrund von Radialgeschwindigkeitsmessungen.
W. Wargau, N. Vogt.
Mitt. Astron. Ges., Nr. 55, p. 77 - 78 (1982). − Abstract.

122.055 UV-Spektroskopie des kataklysmischen Veränderlichen TT Ari.
W. Wargau, H. Drechsel, J. Rahe, G. Klare, B. Wolf, J. Krautter, N. Vogt.
Mitt. Astron. Ges., Nr. 55, p. 78 - 79 (1982). − Abstract.

122.056 IR measurements of U Gem and RR Pic.
K. Beuermann, D. Groote, J. P. Kaufmann.
Mitt. Astron. Ges., Nr. 55, p. 80 - 82 (1982).

122.057 T$_{eff}$, g-Eichung der UBV-Photometrie des RR Lyrae-Sternes SU Draconis.
S. Barcza.
Mitt. Astron. Ges., Nr. 55, p. 85 - 91 (1982).

122.058 Die Perioden des Beta-Cephei-Sterns IL Vel (HD 80383).
U. Haug.
Mitt. Astron. Ges., Nr. 55, p. 167 (1982). − Abstract.

122.059 **Timeshift-relation für klassische Cepheiden.**
E. W. Elst.
Mitt. Astron. Ges., Nr. 55, p. 195 - 199 (1982).

122.060 **Flare stars in the Pleiades. VI.**
L. V. Mirzoyan, O. S. Chavushyan, G. B. Oganyan,
V. V. Ambaryan, A. T. Garibdzhanyan, N. D. Melikyan,
R. Sh. Natsvlishvili.
Astrofizika, Tom 17, 71 - 85 (1981). In Russian. English
translation in Astrophysics, Vol. 17, No. 1.
The results of photographic observations of stellar flares
in the Pleiades region carried out mainly during 1976 - 1979
are given. On the basis of these observations 17 new flare
stars have been found.

122.061 **A phenomenological model of the antiflare star
RZ Psc.**   A. F. Pugach.
Astrofizika, Tom 17, 87 - 96 (1981). In Russian. English translation in Astrophysics, Vol. 17, No. 1.
A model of the phenomenon which leads to the light
fading of the brightness of I_S (A) stars is discussed. Two
hypotheses have been used: 1. The appearance of some obscur-
ing matter over the star disk causes the light fading. 2. Short-
wave excess of radiation emerges in the zone lying over the
region of light absorption.

122.062 **Search for short-time light variations in DF Tauri.**
V. S. Shevchenko, N. A. Shutemova.
Astrofizika, Tom 17, 509 - 517 (1981). In Russian. English
translation in Astrophysics, Vol. 17, No. 3.
10-hour patrol observations has not shown any light
variations of DF Tau more than 0^m05 min^{-1} with time resolution of 10 s and 20 s. There are light fluctuations up to 0^m1 in
U during 1 - 2 hours and up to 1^m7 in U from night to night.
The 17 nights $UBVRI$ observations have shown that the star is
active.

122.063 **New variable stellar objects with UV continuum.**
V. A. Lipovetskij, D. A. Stepanyan.
Astrofizika, Tom 17, 573 - 576 (1981). In Russian. English
translation in Astrophysics, Vol. 17, No. 3.
The authors have discovered five new variable stars in the
course of a search for galaxies with ultraviolet continuum.

122.064 **An analysis of the light curve of SU Aurigae in
1900–1979.**   L. V. Timoshenko.
Astrofizika, Tom 17, 727 - 733 (1981). In Russian. English
translation in Astrophysics, Vol. 17, No. 4.
An analysis of the photographic observations of the
light of SU Aur made from 1900 to 1979 has been carried
out. The following results have been found: a) a great number
of non-periodic weakenings of light by 1^m ; b) sharp short-
time increase of the light of flare type; c) the possibility of a
pseudo-cycle with an interval of 12–16 years.

122.065 **Viscous boundary layer and hard X-rays from
dwarf novae.**   R. Tylenda.
Acta Astron., Vol. 31, 267 - 281 (1981).
An approach to the problem of optically thin, viscous
boundary layer formed between an accretion disc and a degenerate dwarf has been proposed. It has been assumed that both
dissipation of kinetic energy and transport of angular momentum are due to turbulent viscosity. The kinematic viscosity has
been parametrized by means of the critical Reynolds number,
R. It has been found that the viscous boundary layer can radiate hard X-rays provided that the accretion rate is sufficiently
low and viscosity is sufficiently effective. For a 1 $M_\odot$ degenerate dwarf and $R \cong 10^3$ hard X rays are expected if
$\dot{M} \lesssim 10^{16}$ g s^{-1}. The results have been compared with observations of SS Cyg at minumum.

122.066 **Photometry of red variables in 47 Tucanae.**
M. W. Fox.
Mon. Not. R. Astron. Soc., Vol. 199, 715 - 723 (1982).
$BVRI$ observations of known and suspected variables in
the globular cluster 47 Tuc are described. Twelve period
determinations were made, seven of which are new. The non-
Mira stars are found to vary in semi-regular manner, regardless
of period. Values of T_{eff}, M_{bol} and Q, the pulsation constant,
are derived by combining the $BVRI$ with $JHKL$ observations
made near the same epoch. The results strongly favour pulsation in an overtone mode for both the Mira and semi-regular
variables in 47 Tuc.

122.067 **On twelve RR Lyrae-type stars.**
V. P. Tsesevich.
Astrometr. Astrofiz., Vyp. (No.) 43, p. 3 - 13 (1981). In
Russian.
Elements of light variation of 12 RR-Lyrae stars were
obtained. The periods of IM, V672, V597 Aquilae, BH and
DI Lyrae proved to be variable. V672 Aquilae displays the
Blazhko phenomenon.

122.068 **On the red edge of the RR Lyrae instability strip.**
D.-r. Xiong.
Acta Astron. Sinica, Vol. 22, 350 - 356 (1981). In Chinese.
The thermodynamic coupling between radial pulsations
and convection is studied. The growth rate for small-amplitude
pulsation has been calculated in linear nonadiabatic approximation for 11 series of RR Lyrae models. The dependence of
the red edge of their instability strip on the mass, luminosity,
helium abundance and convective parameter has been
investigated.

122.069 **Photographic investigation of the variability of
XY Per.**   S. I. Avramenko, A. F. Pugach.
Astrometr. Astrofiz., Vyp. (No.) 44, p. 9 - 13 (1981). In
Russian.
From an analysis of photographic observations it is
concluded that XY Per belongs to the type of antiflare stars.
For the last twenty years its brightness was characterized by
irregular variations superposed by random light fadings of
different depth. The duration of the light fadings ranges from
several days to several months. When the brightness of the
variable decreases, its colour index increases essentially.

122.070 **On the brightness variation of the variable star
RV Ari from observations with the SBG earth-
satellite camera.**   V. E. Vash, S. I. Ignatovich.
Astrometr. Astrofiz., Vyp. (No.) 45, p. 69 - 73 (1981). In
Russian.
Methods and results of the variable star RV Ari observa-
tions carried out with the SBG earth satellite camera are given.
The periods of brightness variation of the star are determined
by a maximum entropy spectral analysis.

122.071 **Application of the integral method to plotting of
the light curve and for determining the mean radius
of XZ Cyg.**   A. S. Gadun, L. P. Zajkova, Yu. S. Romanov.
Astrometr. Astrofiz., Vyp. (No.) 46, p. 23 - 29 (1982). In
Russian.
The light curves of XZ Cyg have been computed for two
phases of the Blazhko effect from radius and temperature
variations. Comparison of the calculated and observed light
curves made it possible to obtain the mean radii. A specified
integral method was used to calculate light curves of pulsating
stars.

122.072 **Kuwano's peculiar object is a novalike (symbiotic?)
binary with a red giant. Discussion of observational
results.**   T. S. Belyakina, R. E. Gershberg, Yu. S. Efimov,
V. I. Krasnobabtsev, E. P. Pavlenko, P. P. Petrov, K. K. Chuvaev,
V. I. Shenavrin.

Astron. Zh., Tom 59, 302 - 306 (1982). In Russian. English translation in Soviet Astron., Vol. 26, No. 2.

Photometric, polarimetric and spectral observations permit to conclude that Kuwano's object is a binary system that consists of an M giant and of a low-luminosity star. During the 1979 flare the absolute magnitude of the weak component has increased up to about -6^m, the M giant had apparently small variations as well. The distance to the object is estimated to be 5 - 7 kpc, and it is located certainly out of the galactic plane. Similarities between Kuwano's object and slow novae and symbiotic stars are noted.

122.073 RR Lyrae variable pulsations and the Oosterhoff groups.
A. N. Cox, S. W. Hodson, S. P. Clancy.
Astrophysical parameters for globular clusters, (see 012.023), p. 337 - 347 (1981).

The authors conclude that Oosterhoff group I clusters have 0.55 $M_\odot$ stars and group II clusters have 0.65 $M_\odot$ stars. The Y value is always about 0.29. Mean log $L/L_\odot$ values are 1.66 and 1.78 giving $M_{bol} = 0.60$ and 0.30 for the RR Lyrae variables in these two groups of clusters.

122.074 The masses and pulsations of BL Herculis variables.
S. W. Hodson, A. N. Cox, D. S. King.
Astrophysical parameters for globular clusters, (see 012.023), p. 363 - 368 (1981).

The BL Herculis variables are primarily Population II stars found in galactic halo globular clusters, dwarf spheroidal galaxies, and the Small Magellanic Cloud. This class of variables pulsates in the radial fundamental mode at periods between 1 and 3 days with a few as long as 8 days. Their evolution places them between the RR Lyrae and the W Virginis variables. There central He is exhausted and energy is supplied by both He and H burning shells. The masses of BL Her variables must be nearer to 0.55 $M_\odot$ than 0.75 $M_\odot$ if the bump phase transition (resonance) is to be located anywhere near the observed period range of $1^d5 - 1^d7$.

122.075 Period changes in BL Her stars in globular clusters.
A. Wehlau, D. Bohlender.
Astrophysical parameters for globular clusters, (see 012.023), p. 547 - 550 (1981).

122.076 Dwarf cepheids in Omega Centauri?
H. E. Jørgensen.
Astrophysical parameters for globular clusters, (see 012.023), p. 581 - 584 (1981).

122.077 Flare star in M42 region.
British Astron. Assoc. Circ., No. 617 (1981).

122.078 Spectroscopy of the dwarf nova RX Andromedae at minimum.   J. B. Hutchings, B. Thomas.
Publ. Astron. Soc. Pacific, Vol. 94, 102 - 106 (1982).

Spectrographic data on RX And have been obtained with 1.5 Å resolution and a time resolution of ~0.03 in phase, covering 50% of the orbit cycle, during three nights in 1979. The system appears to have been at or near minimum at all times. Results are presented of detailed velocity measures, line-profile studies, and phase-related spectroscopic changes. A somewhat shorter period is derived than that indicated by Kraft's (1962) study, and the authors find no definite evidence for the previously derived orbital eccentricity. Line profiles show fine structure at some phases, and a broad wing extending shortward. The secondary spectrum is detected. Parameters for the system are discussed, using a revised mass function value.

122.079 A $uvby\beta$ photometric study of RR Lyrae stars I.
M. J. Siegel.
Publ. Astron. Soc. Pacific, Vol. 94, 122 - 136 (1982).

The first results of a high-phase resolution $uvby\beta$ photometric program for RR Lyrae stars are presented. Data are given for the three metal-poor RRab stars SU Dra, RX Eri, and RR Lyr and for four comparison stars. Intrinsic colors were derived using Crawford's dereddening procedure. The strong phase variation of $E(b-y)$ suggests that RR Lyrae stars present too great an extrapolation from the mostly dwarf-star data used in Crawford's calibrations. The atmospheric grids of Bell and Manduca were used to find the variations of effective temperatures and surface gravities with phase. The Baade-Wesselink method was applied in order to find absolute magnitudes.

122.080 A polarization study of dwarf novae and nova-like objects.   P. Szkody, J. J. Michalsky, G. M. Stokes.
Publ. Astron. Soc. Pacific, Vol. 94, 137 - 142 (1982).

Linear polarization measurements for four dwarf novae (SS Cyg, RX And, U Gem, and AH Her) and six nova-like variables (AE Aqr, V426 Oph, UX UMa, CI Cyg, EZ Peg, and TT Ari) were obtained to study variability associated with the outburst cycle. No polarization changes are apparent for SS Cyg or RX And throughout their outburst cycles but a significant difference is noted for AE Aqr. RX And shows evidence for variability on orbital time scales.

122.081 The line profile variations of Spica.
G. A. H. Walker, K. Moyles, S. Yang, G. G. Fahlman.
Publ. Astron. Soc. Pacific, Vol. 94, 143 - 148 (1982).

Spectral series of high signal-to-noise ratio (~800) of Spica for 1979 March 13 and 1981 April 14 (UT) are presented. Reticon arrays were used as detectors and the spectra cover Hα and He I λ6678. Distinct features which could be associated with nonradial β Cep pulsation modes combined with stellar rotation are seen moving linearly through the He I and O II λ6721 lines of the primary with accelerations ~0.007 km s^{-2}. Some of the features appear to change during the series. The FWHM of the primary line is 255 km s^{-1} on both nights which is much greater than anyone has measured before and not compatible with the acceleration of the features and the known radius and inclination of the star.

122.082 R Coronae Borealis near maximum light.
J. D. Fernie.
Publ. Astron. Soc. Pacific, Vol. 94, 172 - 176 (1982).

$UBVRI$ photometry of R CrB near maximum light over several seasons is presented. Light fluctuations of about 0^m1 or 0^m2 are confirmed, but the light curve is very irregular in shape, amplitude, and length. A quasi- or characteristic period of 46 ± 5 days is found. Combined with a theoretical mass of $0.8 \pm 0.2 M_\odot$ and a variety of period-radius-mass relations, this period yields a radius of $73 \pm 9 R_\odot$ and log $g = 0.61 \pm 0.15$. The $BVRI$ color data suggest $T_{eff} = 6500 \pm 400$ K and $E_{B-V} = 0.00$, which, with the above radius, indicate $M_{bol} = -5.1 \pm 0.4$.

122.083 HR 1225: new observations and period search.
D. L. DuPuy, G. Collins, D. N. Swingler.
Publ. Astron. Soc. Pacific, Vol. 94, 177 - 181 (1982).

Four nights of photoelectric observations of the δ Scuti star HR 1225 have been obtained. The light curve has a variable amplitude envelope, indicating that more than one period is present. A search for periodicities using the Jurkevich method and Fourier analysis suggests periods of 0^d156 and 0^d097, in agreement with an earlier study.

122.084 High-speed photometry of the cepheid TT Aquilae.
L. P. Connolly, J. L. Africano, A. Klimke, S. P. Worden.
Publ. Astron. Soc. Pacific, Vol. 94, 182 - 188 (1982).

High-speed differential photometry is presented for the classical cepheid TT Aql. There is good coverage over most phases of the light curve. An irregularity in the light curve during the ascending branch stand-still is possibly identified. The period for this cepheid is briefly discussed.

122.085   **Breve comentario sobre las estrellas variables V Coronae Austrinae y W Mensae.**
L. A. Milone.
Bol. Asoc. Argentina Astron., No. 20 - 24, p. 139 - 140 (1981).

122.086   **Búsqueda de cefeidas de largo período en Norma.**
A. L. Cabrera, J. C. Muzzio.
Bol. Asoc. Argentina Astron., No. 20 - 24, p. 229 (1981). Abstract.

122.087   **La variable de helio HD 184927.**
H. Levato, S. Malaroda.
Bol. Asoc. Argentina Astron., No. 20 - 24, p. 393 (1981). Abstract.

122.088   **Determinación de movimientos propios de algunas estrellas R CrB.**
L. A. Milone, M. V. de Arnedo.
Bol. Asoc. Argentina Astron., No. 20 - 24, p. 455 (1981). Abstract.

122.089   **The Cepheid distance scale: a new application for infrared photometry.**
R. McGonegal, R. A. McLaren, C. W. McAlary, B. F. Madore.
Astrophys. J., Lett., Vol. 257, L33 - L36 (1982).
It is shown that near-infrared photometry of Cepheid variables provides a powerful and practical means of calibrating the distance scale to nearby galaxies. Compared with similar work in the blue, random-phase observations in the near-infrared produce a factor of 2.5 decrease in the apparent width of the period-luminosity relation. This is attributed to a substantially decreased effect of differential reddening at long wavelengths, to the low sensitivity of the infrared flux to metallicity variations, and furthermore to the fact that the cyclical luminosity variations are also greatly reduced in the infrared.

122.090   **On the total number of irregular variables in the T1 Mon association.**   L. K. Erastova.
Soobshch. Byurakan Obs., Vyp. 53, p. 88 - 92 (1982). In Russian.
A method of estimation of the total number of irregular variable stars in a stellar aggregate is presented. It is used for determination of the total number of these variables in the Monoceros stellar aggregate. The total number of this type variables brighter than $m_{pg} = 18^m.5 - 19^m$ with an amplitude $\geqslant 1^m$ exceeds 120.

122.091   **Antiflare stars.**   A. F. Pugach.
Akad. nauk Ukrainskoj SSR, Inst. teor. fiz., Prepr., ITF-81-128R, 57 pp. Price 21 Kop. (1981). In Russian.
The results of photometric observations of antiflare stars are examined. The stars have phenomenological and morphological resemblance but differ from that of T Tau and R CrB in many observational features. The analysis of the data available favours the suggestion that antiflare stars form an independent subgroup within the group of irregular variable stars.

122.092   **Observaciones de AU Mic durante la campaña internacional del IUE.**
H. Marraco, H. Luna, F. López García.
Bol. Asoc. Argentina Astron., No. 26, p. 75 (1981). – Abstract.

122.093   **The determination of spot patterns on Ap stars from light curves.**   A. Hempelmann.
Soobshch. Spets. Astrofiz. Obs., Vyp. 32, (see 012.029), p. 64 - 65 (1981).

122.094   **On the rapid spectrum variability of the magnetic star 53 Cam.**
N. S. Polosukhina, K. K. Chuvaev, V. P. Malanushenko, I. Tuominen.
Soobshch. Spets. Astrofiz. Obs., Vyp. 32, (see 012.029), p. 68 - 69 (1981).

122.095   **On the variability of the Ap stars 53 Cam, 41 Tau, β CrB and α^2 CVn in the K(Ca II) and Hγ lines.**
V. M. Kuvshinov, S. I. Plachinda.
Soobshch. Spets. Astrofiz. Obs., Vyp. 32, (see 012.029), p. 70 - 71 (1981).

122.096   **An unusually stable and short spectroscopic period of the Be star 28 CMa.**   D. Baade.
Be stars, (see 012.030), p. 167 - 170 (1982).

122.097   **A preliminary report on simultaneous ultraviolet and optical observations of Lambda Eridani.**
C. T. Bolton.
Be stars, (see 012.030), p. 181 - 184 (1982).

122.098   **On the ageing (*period gradient*) of cepheids across the spiral arm Carina-Sagittarius.**
Yu. N. Efremov, G. R. Ivanov.
Astron. Tsirk., No. 1166, p. 1 - 3 (1981). In Russian.

122.099   **On a possible anomaly of pulsating masses of low-amplitude cepheids (s-type).**   G. R. Ivanov.
Astron. Tsirk., No. 1166, p. 3 - 4 (1981). In Russian.

122.100   **Brightness fading of the FU Orionis star V1515 Cygni.**   E. A. Kolotilov, P. P. Petrov.
Astron. Tsirk., No. 1167, p. 1 - 3 (1981). In Russian.

122.101   **The interesting variable star UU Her.**
D. D. Saselov.
Astron. Tsirk., No. 1167, p. 6 - 8 (1981). In Russian.

122.102   **On cyclic variations of the brightness of V1504 Cyg.**
N. E. Kurochkin.
Astron. Tsirk., No. 1169, p. 3 - 5 (1981). In Russian.

122.103   **FH Scuti – an R CrB-type star.**
V. P. Tsesevich.
Astron. Tsirk., No. 1169, p. 5 - 6 (1981). In Russian.

122.104   **Flare stars in Orion.**   R. Sh. Natsvlishvili.
Inf. Bull. Variable Stars, No. 2062, 3 pp. (1982).

122.105   **Linear polarization of the late-type variable stars μ Cep and Mira in year 1981.**   D. P. Hayes.
Inf. Bull. Variable Stars, No. 2064, 4 pp. (1982).

122.106   **Flare stars in the Pleiades.**   M. K. Tsvetkov, H. S. Chavushian (*O. S. Chavushyan*).
Inf. Bull. Variable Stars, No. 2067, 2 pp. (1982).

122.107   **Photoelectric UBV observations of PU Vul in 1981.**   A. Purgathofer, A. Schnell.
Inf. Bull. Variable Stars, No. 2071, 4 pp. (1982).

122.108   **Period and light curve of UW Gruis from photoelectric observations.**
A. Bernard, M. Burnet.
Inf. Bull. Variable Stars, No. 2072, 4 pp. (1982).

122.109   **HD 13831 a new Beta Cephei star.**
R. Garrido, A. J. Delgado.
Inf. Bull. Variable Stars, No. 2080, 2 pp. (1982).

122.110 *o* And: a new active episode?   M. Bossi,
G. Guerrero, L. Mantegazza, M. Scardia.
Inf. Bull. Variable Stars, No. 2082, 4 pp. (1982).

122.111 On the radial pulsations of the Delta Scuti stars
Sigma Octantis and B Octantis.
Ts. G. Tsvetkov.
Inf. Bull. Variable Stars, No. 2084, 2 pp. (1982).

122.112 The Delta Scuti variable HR 1287.
T. E. Margrave.
Inf. Bull. Variable Stars, No. 2087, 3 pp. (1982).

122.113 Photoelectric observations of the flare star
DO Cep in 1975.   M. E. Contadakis,
F. M. Mahmoud, L. N. Mavridis, D. Stavridis.
Inf. Bull. Variable Stars, No. 2088, 3 pp. (1982).

122.114 PS4452-1347 (star No. 1347 measured on Palomar
Schmidt plate No. 4452) − a new RR Lyr variable
star of Bailey type c.   T. B. Andersen.
Inf. Bull. Variable Stars, No. 2091, 4 pp. (1982).

122.115 S Eridani − a Delta Scuti variable.
D. W. Coates, L. Halprin, T. T. Moon, K. Thompson
Inf. Bull. Variable Stars, No. 2093, 3 pp. (1982).

122.116 The optical variability of PU Vulpeculae (Kuwano's
object) in 1979 - 1981.
E. A. Kolotilov, T. S. Belyakina.
Inf. Bull. Variable Stars, No. 2097, 4 pp. (1982).

122.117 On the nova-like objects in the central region of
M31.   A. S. Sharov.
Inf. Bull. Variable Stars, No. 2105 (1982).

122.118 HD 65227: a new short period Cepheid of very
small amplitude.   O. J. Eggen.
Inf. Bull. Variable Stars, No. 2106, 3 pp. (1982).

122.119 Periodic light variations of the dwarf nova
CN Orionis.   R. Schoembs.
Inf. Bull. Variable Stars, No. 2116, 2 pp. (1982).

122.120 Observations of early-type ultra-short period
variables.   L. A. Balona.
Inf. Bull. Variable Stars, No. 2120, 3 pp., with a correction
in No. 2124 (1982).

122.121 Photoelectric observations of the flare star
EV Lac in 1981.   K. P. Panov, I. Pamukchiev,
P. Christov, G. Asteriadis, L. N. Mavridis.
Inf. Bull. Variable Stars, No. 2128, 4 pp. (1982).

122.122 Photoelectric observations of the flare star EV Lac
in 1981.
L. N. Mavridis, G. A. Asteriadis, M. K. Tsvetkov.
Inf. Bull. Variable Stars, No. 2133, 4 pp. (1982).

122.123 Confirmation of flare activity on G9-8 by photoelectric photometry.   B. R. Pettersen.
Inf. Bull. Variable Stars, No. 2141, 3 pp. (1982).

122.124 Photoelectric flare observations of Gliese 867 B.
B. B. Sanwal.
Inf. Bull. Variable Stars, No. 2143, 3 pp. (1982).

122.125 Photoelectric observations of ϑ^2 Tau (HR 1412).
E. Antonello, L. Mantegazza.
Inf. Bull. Variable Stars, No. 2144, 2 pp. (1982).

122.126 Refinement of the fundamental frequency of
pulsation of Delta Scuti.   T. T. Moon, D. M. Keay.
Inf. Bull. Variable Stars, No. 2145, 4 pp. (1982).

122.127 Photoelectric observations of 20 Leo (HR 3889).
E. Antonello, L. Mantegazza.
Inf. Bull. Variable Stars, No. 2152, 2 pp. (1982).

122.128 A new β Cephei star in Harvard Standard Region E4.
A. W. J. Cousins.
Inf. Bull. Variable Stars, No. 2158, 3 pp. (1982).

122.129 RY Sagittarii.
IAU Circ., Nos. 3662, 3683, 3696 (1982).

122.130 Eruptive variable in the Large Magellanic Cloud.
IAU Circ., No. 3662 (1982).

122.131 KR Aurigae.
IAU Circ., Nos. 3674, 3689 (1982).

122.132 TT Arietis.
IAU Circ., No. 3683 (1982).

122.133 AM Herculis.
IAU Circ., Nos. 3689, 3693, 3703 (1982).

122.134 AM Herculis.
Yamamoto Circ., No. 1977 (1982).

122.135 GR 304 Ursae Maioris: rather an intrinsic variable.
K. Locher.
BBSAG Bull., No. 60, p. 7 (1982).

122.136 131 new maxima of the light variation of
DY Herculis.   J. - F. Le Borgne.
GEOS Circ., RR 2, 3 pp. (1980).

122.137 133 times of maximum and first ephemeris for the
RRc star VZ Draconis.   A. Figer.
GEOS Circ., RR 3, 7 pp. (1982).

122.138 WY Gem: a new semi-regular variable with a period
of 169 days.   A. Buzzoni.
GEOS Circ., SR 2, 8 pp. (1981).

122.139 Period changes in dwarf cepheids, II. YZ Bootis,
XX Cygni, and DY Herculis.
B. Szeidl, H. A. Mahdy.
Commun. Konkoly Obs. Hungarian Acad. Sci., No. 75, 35 pp.
(1981).

The period changes in the dwarf cepheids YZ Boo,
XX Cyg and DY Her are discussed and O-C diagrams of these
stars are constructed. The period of XX Cyg changed abruptly in
1942 ($\Delta P = +87 \times 10^{-9}$ day = +0.0075 sec). Besides the sudden
increase of the period, it has also shown small fluctuations.
With regard to YZ Boo there may be a slight continuous
increase (at a rate $\beta = +10.6 \times 10^{-13}$ day cycle^{-1} = +3.2 × 10^{-2}
sec century^{-1}) in its period. The period of DY Her has shown a
definite continuous decrease at a rate $\beta = -37.2 \times 10^{-13}$ day
cycle $^{-1}$ = -7.9×10^{-2} sec century^{-1} during the time interval
covered by photoelectric observations (1951−1979).

122.140 Synchronous observations of CH Cygni on
λ 0.36 μm and λ 2.2 μm.
O. G. Taranova, V. I. Shenavrin.
Astron. Tsirk., No. 1185, p. 4 - 7 (1981). In Russian.

122.141 On mean light oscillations of SU Aurigae.
L. V. Timoshenko.
Astron. Tsirk., No. 1185, p. 7 - 8 (1981). In Russian.

122.142 **RR Lyrae variables in globular clusters.**
    Z. I. Kadla, A. N. Gerashchenko,
Yu. K. Vinogradova, N. V. Yablokova.
Astron. Tsirk., No. 1187, p. 5 - 8 (1981). In Russian.

122.143 **Ultraviolet light variations of Z And in 1978 - 1981.**
    O. G. Taranova, B. F. Yudin.
Astron. Tsirk., No. 1188, p. 7 - 8 (1981). In Russian.

122.144 **Dust in the envelope of CI Cyg.**
    B. F. Yudin.
Astron. Tsirk., No. 1192, p. 7 - 8 (1981). In Russian.

122.145 **New observational results on S Dor variables.**
    I. Appenzeller, B. Wolf.
The most massive stars, (see 012.034), p. 131 - 139 (1981).

    The class of bright blue stars in extragalactic systems
which are called "S Dor" or "Hubble-Sandage" variables
contains some of the most luminous stars known. Two such
objects are members of the LMC. One is the prototype object
S Dor, the other one is HDE 269006 = R 71. The authors
obtained a series of high resolution coudé spectrograms of
these two stars and various related LMC objects. In addition
they observed the UV spectra of the two LMC S Dor variables
using the IUE satellite.

122.146 **On the variability of the A2-hypergiant HD 160529.**
    C. Sterken.
The most massive stars, (see 012.034), p. 147 - 153 (1981).

    HD 160529 is an extreme supergiant in the direction of
the galactic center. Fast-photometry observations have been
performed in the photometric B range during four 45 min
intervals on separate nights. Evidence is presented which
clearly shows that in HD 160529 considerable irregular variations are superimposed on a long-term semi-periodical mean
variation.

122.147 **Red supergiant variables and their possible use as
    distance indicators.**   M. W. Feast.
The most massive stars, (see 012.034), p. 217 - 226 (1981).

    Recent progress in the use of Mira variables as distance
indicators is first briefly reviewed. Infrared studies of red
supergiant variables in the Magellanic Clouds are then
summarized and the possibility is considered of using them
as distance indicators and to establish the upper limit to the
masses of evolved stars. The present situation regarding the
use of the brightest red supergiants as distance indicators is
discussed.

122.148 **Infrared Mira variables as distance indicators.**
    D. Engels.
The most massive stars, (see 012.034), p. 243 - 244 (1981).

122.149 **Properties of Mira variables.**   L. A. Willson.
    Bull. American Astron. Soc., Vol. 13, 803 (1981).
Abstract.

122.150 **The behavior of Hα in the cepheid variables
    Zeta Gem and X Cygni.**
G. Wallerstein, T. Jacobsen.
Bull. American Astron. Soc., Vol. 13, 804 (1981). – Abstract.

122.151 **HR 1225: a Delta Scuti star with three periods?**
    D. L. Dupuy, G. Collins, D. N. Swingler.
Bull. American Astron. Soc., Vol. 13, 804 (1981). – Abstract.

122.152 **HZ-22: the mystery of the missing secondary star.**
    A. Young, S. Wentworth.
Bull. American Astron. Soc., Vol. 13, 816 - 817 (1981).
Abstract.

122.153 **IUE observations of the pulsating white dwarf
    G29 - 38.**
F. H. Schiffer III, A. V. Holm, R. J. Panek.
Bull. American Astron. Soc., Vol. 13, 817 (1981). – Abstract.

122.154 **H2215 - 086, king of the DQ Herculis stars.**
    J. Patterson, J. Steiner.
Bull. American Astron. Soc., Vol. 13, 817 (1981). – Abstract.

122.155 **Simultaneous X-ray and optical photometry of the
    cataclysmic variable TT Arietis.**
K. Jensen, F. A. Cordova, J. Middleditch, K. Mason,
A. Grauer, K. Horne, R. Gomer.
Bull. American Astron. Soc., Vol. 13, 818 (1981). – Abstract.

122.156 **X-ray and optical measurements of the cataclysmic
    variable CH UMa.**
R. H. Becker, A. S. Wilson, S. H. Pravdo, G. A. Chanan.
Bull. American Astron. Soc., Vol. 13, 818 (1981). – Abstract.

122.157 **Dwarf nova eruptions.**   B. L. Everson.
    Bull. American Astron. Soc., Vol. 13, 818 (1981).
Abstract.

122.158 **BV photometry of R CrB.**   K. Krisciunas.
    Bull. American Astron. Soc., Vol. 13, 832 (1981).
Abstract.

122.159 **Nova-like Hα sources in the center of M31.**
    R. Ciardullo, H. Ford, G. Jacoby.
Bull. American Astron. Soc., Vol. 13, 842 (1981). – Abstract.

122.160 **New double-mode RR Lyrae variables in the
    globular cluster M15.**
S. W. Hodson, A. N. Cox, S. P. Clancy.
Bull. American Astron. Soc., Vol. 13, 870 (1981). – Abstract.

122.161 **Masses and evolution stages of globular cluster
    population II cepheids.**   A. N. Cox, S. P. Clancy.
Bull. American Astron. Soc., Vol. 13, 871 (1981). – Abstract.

122.162 **Radii and masses of classical cepheids.**
    W. Gieren.
Bull. American Astron. Soc., Vol. 13, 871 (1981). – Abstract.

122.163 **Experimental envelope models for cepheids.**
    N. R. Simon.
Bull. American Astron. Soc., Vol. 13, 871 (1981). – Abstract.

122.164 **A rotation study of BY Dra stars.**
    D. R. Soderblom, S. S. Vogt, G. D. Penrod.
Bull. American Astron. Soc., Vol. 13, 873 (1981). – Abstract.

122.165 **UV, optical and IR observations of the Cepheid
    R Muscae.**   W. Eichendorf, A. Heck, B. Caccin,
G. Russo, C. Sollazzo.
ESO Sci. Prepr. No. 189, 17 pp. (1982). – Submitted to
Astron. Astrophys.

122.166 **IUE observations of dwarf novae during active
    phases.**   G. Klare, J. Krautter, B. Wolf, O. Stahl,
N. Vogt, W. Wargau, J. Rahe.
ESO Sci. Prepr. No. 199, 21 pp. (1982). – Submitted to
Astron. Astrophys.

122.167 **HR 2724 – a new bright variable in the δ Scuti
    instability strip.**   D. Baade, O. Stahl.
ESO Sci. Prepr. No. 200, 12 pp. (1982). – Submitted to
Astron. Astrophys.

122.168 Spectroscopic and photometric data for AE Aquarii.
G. Chincarini, M. F. Walker.
Publ. Lick Obs., Vol. 23, pt. 2, 34 pp. (1981).

122.169 Cataclysmic variable candidates from the Palomar Green Survey.
R. F. Green, D. Ferguson, J. Liebert, M. Schmidt.
Prepr. Steward Obs., No. 380, 13 pp. (1982).

122.170 Critical comments on G. A. Gurzadyan's papers on flare stars.   R. E. Gershberg.
Astrofizika, Tom 16, 375 - 381 (1980). In Russian. English translation in Astrophysics, Vol. 16, No. 2.

122.171 Reply to Gershberg's "Critical comments".
G. A. Gurzadyan.
Astrofizika, Tom 16, 383 - 391 (1980). In Russian. English translation in Astrophysics, Vol. 16, No. 2.

122.172 Slow brightness variations of BY Dra.
A. S. Melkonyan, K. Olah, A. V. Oskanyan, Jr., V. S. Oskanyan.
Astrofizika, Tom 17, 215 - 224 (1981). In Russian. English translation in Astrophysics, Vol. 17, No. 2.
The results of investigations of slow brightness changes of BY Dra are presented. The periodic character of light variation − period about 3.8 days and amplitude 0.02 mag. is proved. A break down of light changes lasting a few days is noticed. After that interval the light changes renew, but with another phase. This phenomenon reminds the light variations of δ Scu type stars.

122.173 Stellar atmospheric velocity fields: the Beta Cephei variables γPeg and βCep.   C. De Jager, N. Sato, M. Burger, L. Neven.
Astrophys. Space Sci., Vol. 83, 411 - 415 (1982).
The shape parameters of a number of selected ultraviolet lines in spectra of the Beta Cephei stars γ Peg and β Cep have been analyzed to determine the principal parameters of the atmospheric velocity field. The authors find for both stars a fairly high value ($\sim$5 km s^{-1}) for the microturbulent line-of-sight velocity component, which confirms an earlier result based on lower resolution UV spectra. Macroturbulent and rotational velocities are virtually zero in the atmosphere of γ Peg; for βCep one finds $v_{rot} \sin i$ = 40 km s^{-1}.

122.174 Visible and/or near infrared spectral variations of V1016 Cygni, HM Sagittae and MWC 349.
Y. Andrillat, F. Ciatti, J. P. Swings.
Astrophys. Space Sci., Vol. 83, 423 - 435 (1982).
The recent spectral evolution of the three peculiar emission-line objects with IR excess V1016 Cyg, HM Sge, and MWC 349 is described, on the basis of data obtained at the Asiago (blue and red spectral regions) and Haute Provence (near infrared) Observatories.

122.175 VLA observations of quiescent and flare emission from dMe stars.   P. L. Fisher, D. M. Gibson.
Bull. American Astron. Soc., Vol. 14, 574 (1982). − Abstract.

122.176 A detailed analysis of three R CrB stars.
P. L. Cottrell, D. L. Lambert, D. Schönberner.
Bull. American Astron. Soc., Vol. 14, 576 (1982). − Abstract.

122.177 Photometry of the dwarf nova CN Orionis.
B. Warner.
Mon. Notes Astron. Soc. South. Africa, Vol. 41, 15 - 18 (1982).

122.178 A simultaneous photometric and radial velocity study of short-period southern Cepheids. III. An analysis for binaries.   W. Gieren.
Astrophys. J., Suppl. Ser., Vol. 49, 1 - 26 (1982).

New radial velocity curves and five-color photometry are used to analyze for the possible binary nature of 15 classical Cepheids. Comparison with earlier radial velocity data reveals one new spectroscopic binary, AZ Cen, and confirms the spectroscopic binary nature of V350 Sgr, previously claimed by Lloyd Evans. Three more stars, R TrA, AP Sgr, and V496 Aql, are probably spectroscopic binaries. Various photometric tests confirm the binary nature of AZ Cen and strongly suggest AG Cru and V496 Aql to have companions as well. While the companions of AZ Cen and AG Cru are early-type stars, photometry suggests a G-type giant companion to V496 Aql. For three more Cepheids, AP Sgr, V350 Sgr, and BB Sgr, the photometric properties give moderate evidence for the existence of companions (blue companions to AP Sgr and V350 Sgr, a red companion in the case of BB Sgr).

122.179 Ultraviolet spectroscopy of the nova-like variable V3885 Sagittarii (= CD −42° 14462).
E. F. Guinan, E. M. Sion.
Astrophys. J., Vol. 258, 217 - 223 (1982).
Low-dispersion long and short wavelength IUE spectra of the nova-like system V3885 Sgr were obtained on 1979 August 24 UT. The UV spectrum exhibits strong, high-temperature absorption lines of C III, C IV, He II, N IV, N V, and Si IV which all have relatively large negative velocities of 600−1000 km s^{-1} at their line centers. The C IV (λ1550) features shows a P Cygni-type profile with a blueshifted absorption line; the presence of the P Cygni-type feature strongly suggests a significant mass outflow from the system in the form of a hot stellar wind. The observed continuum fluxes are fitted with theoretical accretion disk model fluxes, and a tentative physical interpretation of the observed spectral properties is presented.

122.180 The nature of the dwarf cepheid XX Cygni.
M. D. Joner.
Publ. Astron. Soc. Pacific, Vol. 94, 289 - 298 (1982).
Photometry of XX Cyg in the $uvby$ and β systems has been secured and analyzed. A reddening value, $E(b-y)$ = +0^{m}057, is derived from the photometry. Intrinsic $(b-y)$ and c_1 values are used to determine a mean effective temperature, $\langle T_{eff} \rangle$ = 7530 K, and a mean surface gravity, $\langle \log g \rangle$ = 3.66. XX Cyg is found to be a metal-poor dwarf cepheid with [Fe/H] = −0.49. A radius of 2.4 $R_\odot$ has been found from the Wesselink method. The radius and effective temperature indicate that $\langle M_{bol} \rangle$ = +1.7. A mass of 1.0 $M_\odot$ is found from the radius and effective gravity. The radial-velocity data yield a mean radial velocity of −108 km s^{-1} and a total velocity range of 37 km s^{-1}.

122.181 Radial velocity variations of the δ Scuti variable β Cassiopeiae.
S. Yang, G. A. H. Walker, G. G. Fahlman, B. Campbell.
Publ. Astron. Soc. Pacific, Vol. 94, 317 - 321 (1982).
Precision radial velocities with a time resolution of 15 minutes $(P/10)$ have been measured for β Cas $(P=2^h30^m)$ over almost two cycles using the HF absorption-cell technique. The precision of ± 0.2 km s^{-1} is limited by line-profile variations rather than by the technique. The period of 0.0976 day derived from the Ca II velocity curve agrees with the photometric period of 0.104 day. Some details are given of the HF reference line technique. It promises to be a powerful probe of atmospheric motions in δ Scuti and δ Delphini stars.

122.182 The 1981 light curve of the unique cepheid HR 7308.
J. R. Percy, R. P. Ford.
J. American Assoc. Variable Star Obs., Vol. 10, 53 - 55 (1981).

122.183 Period and amplitude of V1670 Sagittarii.
D. Gabuzda.
J. American Assoc. Variable Star Obs., Vol. 10, 64 - 65 (1981).

The period and light curve of V1670 Sagittarii are consistent with its classification as a type II Cepheid. The amplitude is variable. Short- and long-term behavior of the period are discussed.

122.184 A recent light curve of TW Capricorni.
J. A. DeYoung.
J. American Assoc. Variable Star Obs., Vol. 10, 66 - 68 (1981).

122.185 Period determination for MY Scuti.
S. Davis.
J. American Assoc. Variable Star Obs., Vol. 10, 69 - 70 (1981).

**122.186 Photometry of Y Canum Venaticorum and
R Coronae Borealis.**   K. Krisciunas.
J. American Assoc. Variable Star Obs., Vol. 10, 75 - 80 (1981).

122.187 Period changes in the RR Lyrae star IM Aquilae.
E. A. Lada, E. P. Belserene.
J. American Assoc. Variable Star Obs., Vol. 10, 81 - 82 (1981).

**122.188 Revised period and amplitude change for
V802 Cygni.**   L. N. Ventura.
J. American Assoc. Variable Star Obs., Vol. 10, 90 - 91 (1981).

122.189 The peculiar classical cepheid HR 7308.
G. Burki, M. Mayor, W. Benz.
Astron. Astrophys., Vol. 109, 258 - 270 (1982).
The authors analyse 246 radial velocity measurements of the cepheid HR 7308 obtained during 3 yr of monitoring with the spectrophotometer CORAVEL. In addition, photometry in the B and V bands by Breger and Greenberg (1980) allows to apply a Baade-Wesselink method. The surface of the CORAVEL dip confirms that HR 7308 is a population I (classical) cepheid. However, the star exhibits several characteristics which are unique with respect to this class of stars.

**122.190 UV, optical and IR observations of the Cepheid
R Muscae.**   W. Eichendorf, A. Heck, B. Caccin,
G. Russo, C. Sollazzo.
Astron. Astrophys., Vol. 109, 274 - 278 (1982).
Ultraviolet IUE spectra, optical 12 Å mm $^{-1}$ red and blue spectra, optical BV photometry and infrared $JHKL$ photometry of the 7.5-d classical Cepheid R Muscae = HD 110311 is presented and discussed together with some earlier photometric and radial velocity data.

**122.191 The pulsation of the outer layers of the Beta Cephei
star σ Sco.**
M. Burger, C. de Jager, G. H. J. van den Oord.
Astron. Astrophys., Vol. 109, 289 - 293 (1982).
The pulsation of the outer layers of the β Cephei-type variable σ Sco is investigated on the basis of 17 ultraviolet spectrograms (1200–2000 Å). The authors confirm a pulsational model established earlier for BW Vul.

122.192 Cepheid binaries – II. New southern examples.
T. L. Evans.
Mon. Not. R. Astron. Soc., Vol. 199, 925 - 941 (1982).
A homogeneous set of 769 radial velocity observations of 50 classical Cepheids in the period 1966–70 is used to detect possible binaries by comparison with earlier velocity curves. Seven stars are considered probable binaries: U Aql, AX Cir, R Mus, BF Oph, R TrA, V 350 Sgr and AH Vel. The binary frequency is found to be 18 per cent. Orbits were determined for the known binaries S Mus and V 636 Sco and the period of FF Aql was confirmed. S Mus has the shortest orbital period so far reported for a Cepheid, 506 day. Published photometric data and some new infrared photometry permit detection of some of the companions. Ultraviolet spectroscopy of a blue companion from a satellite appears to

offer the best method of determining the dynamical mass of a Cepheid.

122.193 Mira B.   D. J. Stickland, A. Cassatella, D. Ponz.
Mon. Not. R. Astron. Soc., Vol. 199, 1113 - 1118 (1982).
A high resolution IUE spectrum of Mira taken at the time of the 1980 maximum shows sharp emission features from the M star and broad ones attributable to a large volume of gas around the hot companion. Some information on the emission from the vicinity of Mira B is available from line widths and ratios.

**122.194 Beat Cepheid studies – II. Atmospheres of beat
Cepheids.**   S. L. Barrell.
Mon. Not. R. Astron. Soc., Vol. 200, 127 - 137 (1982).
The author describes a differential curve of growth analysis of the atmospheres of nine beat Cepheids, yielding microturbulence and iron abundance results which are typical of Population I Cepheids. Surface gravities, are combined with previously published Wesselink radii to predict spectroscopic masses for five beat Cepheids. These are close to predicted evolutionary masses, which are in turn similar to Population I Cepheid evolutionary masses. It is concluded that beat Cepheids are spectroscopically indistinguishable from Population I Cepheids.

**122.195 Beat Cepheid studies – III. Photometric search for
additional beat Cepheids.**   S. L. Barrell.
Mon. Not. R. Astron. Soc., Vol. 200, 139 - 151 (1982).
A search for additional beat Cepheids has been carried out using existing and newly-obtained photometric data. Frequency analysis of nine selected suspects reveals that they are all single-mode pulsators and so the number of known beat Cepheids remains at eleven. One of the suspects is identified as a short period Population II bump Cepheid. A proposed cause of the double-mode pulsations, mode switching, is shown to be improbable.

**122.196 Ultraviolet observations of the hydrogen-deficient
variable star MV Sagittarii.**
N. K. Rao, K. Nandy.
J. Astrophys. Astron., Vol. 3, 79 - 92 (1982).
IUE observations of the hydrogen-deficient irregular variable star MV Sgr obtained in 1980 June - October and also in 1979 November are discussed. The dereddened continuum can be fitted to a theoretical energy distribution of a helium star model with T_{eff} = 18000 K and log g = 2.5, similar to that of BD + 10° 2179. A model is proposed in which a cool companion star, surrounded by dust, occasionally blows gas towards the hotter hydrogen-poor B star. This model explains the irregular light variations and the spectroscopic phenomena.

**122.197 The variable stars in the globular cluster NGC 6864
(M75).**   G. Pinto, L. Rosino, C. M. Clement.
Astron. J., Vol. 87, 635 - 639 (1982).
The variable stars of the globular cluster NGC 6864 (M75) have been studied on plates taken at the Asiago Astrophysical Observatory, the Radcliffe Observatory, and the Cerro Tololo and Las Campanas Observatories between 1954 and 1980. Ten variables (three of which were discovered in the course of the present study) have been found to be of the RR Lyrae type; maximum and minimum magnitudes, elements, and light curves have been obtained for six of them. The remaining variables, very likely of the irregular or semiregular type, show only small brightness fluctuations.

**122.198 Carbon and oxygen abundances of field RR Lyrae
stars. I. Carbon abundances.**   D. Butler,
A. Manduca, D. Deming, R. A. Bell.
Astron. J., Vol. 87, 640 - 649 (1982).

From an analysis of KPNO 4-m echelle plates and simultaneous $uvby\beta$ photometry, the authors have determined carbon abundances and carbon-to-iron ratios for a large number of field RR Lyrae stars having [Fe/H] $\gtrsim -1.2$. It is found that these field RR Lyrae stars — stars which are known to be in an advanced evolutionary state — have carbon-to-iron ratios which are similar to those of unevolved stars.

122.199 A spectrophotometric survey of cataclysmic variable stars.   J. B. Oke, R. A. Wade.
Astron. J., Vol. 87, 670 - 679 (1982).

The authors present spectrophotometry from 0.32 to 1.1 μm for 31 cataclysmic variable stars. Old novae, nova-like variables, and dwarf novae in outburst are characterized by blue continua and Balmer jumps in absorption. Quiescent dwarf novae have Balmer jumps in emission; some of these objects show clear signs of a late-type stellar component, while others do not. Measurements of the integrated fluxes for the Balmer emission lines in quiescent dwarf novae are presented.

122.200 An investigation of period changes in cluster BL Herculis stars.
A. Wehlau, D. Bohlender.
Astron. J., Vol. 87, 780 - 791 (1982).

A considerable amount of unpublished material made available by a number of observers has been combined with already published measures in order to determine rates of period change for BL Herculis stars in galactic globular clusters. There are 20 known stars with periods ranging from 1.13 to 7.90 days and it was possible to investigate period changes for 12 of these. In all cases the observations span at least 65 yr. The period changes observed may represent evolutionary changes for stars evolving away from the horizontal branch toward the asymptotic giant branch.

122.201 Mira-type variable stars and the impact of the Hipparcos mission.
M. O. Mennessier, J. Guibert, Nguyen-Q-Rieu.
Scientific aspects of the Hipparcos space astrometry mission, (see 012.041), p. 137 - 138 (1982).

The accuracy of presently available parallaxes and proper motions is extremely poor and concerns quite a limited number of objects. The authors indicate the improvement of the present situation which can be expected from the astrometric satellite, and discuss the questions raised by the preparation of observing lists and estimate of accuracy in the case of large amplitude variables.

122.202 Intrinsic variability in B stars. Observational facts.
J.-M. Le Contel, J.-P. Sareyan, J.-C. Valtier.
Pulsating B stars, (see 012.042), p. 45 - 63 (1981).

122.203 Notes on the variability and evolutionary status of OB stars and B supergiants.
A. Maeder, F. Rufener.
Pulsating B stars, (see 012.042), p. 65 - 72 (1981).

122.204 uvby β photometry of young star clusters containing β CMa stars.   R. R. Shobbrook.
Pulsating B stars, (see 012.042), p. 75 - 83 (1981).

122.205 Line broadening in short period B variables.
J.-P. Sareyan, M. Giudicelli, D. Egret.
Pulsating B stars, (see 012.042), p. 89 - 98 (1981).

122.206 The evolutionary state of the Beta Cephei stars.
J. R. Percy.
Pulsating B stars, (see 012.042), p. 119 - 125 (1981).

122.207 Fast time-resolution observations of Hα and ionized carbon line profiles in BW Vulpeculae.

A. Young, I. Furenlid, H. R. Rymer.
Pulsating B stars, (see 012.042), p. 129 - 139 (1981).

122.208 The line profile variations of Spica.
G. A. H. Walker, K. Moyles, S. Yang, G. G. Fahlman.
Pulsating B stars, (see 012.042), p. 141 - 149 (1981).

122.209 Emission and absorption components in high resolution spectra of γ Peg.
J.-M. Le Contel, P. J. Morel.
Pulsating B stars, (see 012.042), p. 151 - 156 (1981).

122.210 IUE spectra of Beta Cephei stars.
J. Rountree-Lesh.
Pulsating B stars, (see 012.042), p. 157 - 159 (1981).

122.211 Radial-velocity and light curves of Gamma Pegasi.
P. Koubský, P. Harmanec, J. Krpata, F. Ždárský.
Pulsating B stars, (see 012.042), p. 161 - 166 (1981).

122.212 Simultaneous spectroscopic and photoelectric observations of V2052 Oph and σ Sco.
M. Kubiak, W. Seggewiss.
Pulsating B stars, (see 012.042), p. 167 - 171 (1981).

122.213 Ultraviolet photometry of 12 and 16 Lacertae.
J. Rountree-Lesh.
Pulsating B stars, (see 012.042), p. 173 - 180 (1981).

122.214 The outer layers of the β Cep type variables BW Vul and σ Sco.
M. Burger, C. de Jager, G. H. J. van den Oord.
Pulsating B stars, (see 012.042), p. 181 - 186 (1981).

Ultraviolet high-resolution observations were obtained with the IUE in order to study the pulsation properties of the outer layers of two β Cep type stars. The star BW Vul was selected because of its relatively large amplitude in light and the large amplitude in radial velocity variation. The star σ Sco was chosen because asymmetric lines with varying asymmetry had already been observed with Copernicus, suggesting the presence of a stellar wind.

122.215 Photometric variations of HD 43818.
S. F. González-Bedolla.
Pulsating B stars, (see 012.042), p. 187 - 190 (1981).

Differential photoelectric photometry in the U, B, V filters of Johnson's system have been carried out in the star HD 43818; the variability of this star discovered by Hill is confirmed.

122.216 Photometry of the β Cephei suspects HD 43078 and HD 43818.
M. Jerzykiewicz, J.-M. Le Contel.
Pulsating B stars, (see 012.042), p. 191 - 194 (1981).

122.217 Beta Cephei stars in the open cluster NGC 3293.
L. A. Balona, C. Engelbrecht.
Pulsating B stars, (see 012.042), p. 195 - 202 (1981).

The galactic cluster NGC 3293 is extraordinarily rich in β Cep variables: twenty eight stars were monitored for variability of which nine are of the β Cep type. They lie in a distinctive region of the cluster sequence where no constant stars are found. The implications of these results on various problems connected with β Cep stars are discussed.

122.218 Short period variability in Iota Herculis.
S. F. González-Bedolla.
Pulsating B stars, (see 012.042), p. 203 - 208 (1981).

122.219 Observations of B variables.
R. Garrido, A. J. Delgado.
Pulsating B stars, (see 012.042), p. 209 - 214 (1981).

122.220 **Light variability of KP Persei.**
T. Jarzebowski. M. Jerzykiewicz.
Pulsating B stars, (see 012.042), p. 215 - 216 (1981).

122.221 **Variability of Be stars.**
R. Hirata, A.-M. Hubert-Delplace.
Pulsating B stars, (see 012.042), p. 217 - 225 (1981).
Long-term photometric and spectroscopic variations of
Be stars are summarized briefly. Time scales of variations in
magnitude and emission-line intensity are about ten years or
more, while that of V/R variation is around seven years. A
typical pattern observed in HD 28497, γ Cas and Pleione is
shown.

122.222 **Short-period photometric variability of Be stars.**
J. R. Percy.
Pulsating B stars, (see 012.042), p. 227 - 232 (1981).
The photometric variability of the following Be stars has
been studied on time scales of hours to days, and is described
and discussed: HR 1679 (λ Eri), HR 2142, HR 7647 (25 Cyg),
HR 7708 (28 Cyg), HR 8731 (EW Lac) and HR 9070.

122.223 **Light variations in broad-lined B stars.**
M. Jerzykiewicz, C. Sterken.
Pulsating B stars, (see 012.042), p. 233 - 236 (1981).

122.224 **Observations of HD 58050.**
A. Figer.
Pulsating B stars, (see 012.042), p. 237 - 246 (1981).

122.225 **Thirteen color photometry of variable Be stars.**
M. Alvarez.
Pulsating B stars, (see 012.042), p. 247 - 253 (1981).
The 13-color photometric study of 16 variable Be stars
shows some of the difficulties associated with the spectral
classification of these objects. A list of short time and/or
small amplitude variable stars ('suspected' β Cephei type) is
given.

122.226 **Variations du profil de la raie Hα dans Kappa Dra
et 25 Ori.**   Y. Andrillat, C. Fehrenbach.
Pulsating B stars, (see 012.042), p. 255 - 260 (1981).

122.227 **Spectral variations in ζ Ophiuchi related to a disc,
rotation, or non-radial pulsations?**
G. A. H. Walker, S. Yang, G. G. Fahlman.
Pulsating B stars, (see 012.042), p. 261 - 266 (1981).
More extensive spectroscopic observations in 1979 confirm earlier results that there are features moving through the
He I and H I lines with an acceleration ~0.045 km s^{-2}. These
features are best explained by irregular shadowing of one
hemisphere by an equatorial disc of material orbiting the star.
It is not clear however that an interpretation in terms of non-
radial pulsations coupled with rotation can be ruled out.

122.228 **Asymmetric absorption line profiles in Be stars.**
D. Baade.
Pulsating B stars, (see 012.042), p. 267 - 276 (1981).

122.229 **Photometric variability of B-type supergiants: a
preliminary report.**   J. R. Percy.
Pulsating B stars, (see 012.042), p. 277 - 282 (1981).

122.230 **Mode identification in pulsating B stars.**
J. Rountree-Lesh.
Pulsating B stars, (see 012.042), p. 301 - 307 (1981).

122.231 **The van Hoof effect and nonradial pulsations of
β Cephei stars.**   D. Baade.
Pulsating B stars, (see 012.042), p. 313 - 315 (1981).

122.232 **Modal characteristics in β Cephei and 53 Persei stars.**
M. A. Smith.
Pulsating B stars, (see 012.042), p. 317 - 327 (1981).

122.233 **Polarization variation in the Beta Canis Majoris
stars.**   A. P. Odell, S. Tapia.
Pulsating B stars, (see 012.042), p. 329 - 336 (1981).

122.234 **Quasi toroidal modes in B stars.**
G. Berthomieu, J. Provost.
Pulsating B stars, (see 012.042), p. 337 - 343 (1981).

122.235 **On a possible connection between mass loss and
instability.**   E. Schatzman.
Pulsating B stars, (see 012.042), p. 347 - 349 (1981).

122.236 **Instability mechanisms in Beta Cephei stars.**
J. R. Percy.
Pulsating B stars, (see 012.042), p. 351 - 356 (1981).

122.237 **On the excitation mechanism in Beta Cephei
variables.**   W. Dziembowski, M. Kubiak.
Pulsating B stars, (see 012.042), p. 357 - 358 (1981).
Abstract.

122.238 **Nonradial pulsations of Upsilon Orionis and A
supergiants.**   A. N. Cox.
Pulsating B stars, (see 012.042), p. 359 - 363 (1981).

122.239 **Linear and nonlinear theory study of Alpha Virginis.**
A. N. Cox, S. W. Hodson, S. P. Clancy.
Pulsating B stars, (see 012.042), p. 365 - 375 (1981).
Nonlinear radiation hydrodynamic calculations using a
model for α Virginis, a β Cephei star, have been made to see
if the cause of the recurrent radial pulsation epochs can be
discovered. A review of the various proposals to make these
stars pulsate concludes that the excitation mechanism must
be in the central convective core or variable composition regions.

122.240 **A possible excitation mechanism of B-type variables.**
H. Ando.
Pulsating B stars, (see 012.042), p. 377 - 379 (1981).

122.241 **Envelope ionization mechanisms and
BW Vulpeculae.**
A. N. Cox, S. W. Hodson, S. P. Clancy.
Pulsating B stars, (see 012.042), p. 381 - 388 (1981).
Envelope ionization variations during the pulsations of
β Cephei variables are known to be insufficient to drive pulsations in these stars in the presence of strong deeper radiative
damping. It is suggested that increased envelope helium,
which might occur by accretion of matter from an involved
companion, or just as self homogenization, gives some helium
ionization driving and a smaller decay rate. Perhaps
BW Vulpeculae has such helium enriched surface layers which
explain its unique large amplitude, even though the helium
lines do appear normal.

122.242 **The resonance theory of pulsating stars.**
M. Takeuti.
Pulsating B stars, (see 012.042), p. 389 - 394 (1981).

122.243 **Photometric campaign on Be stars.**
P. Harmanec, J. Horn, P. Koubský.
Pulsating B stars, (see 012.042), p. 397 - 404 (1981).

122.244 **Two unusual cepheids in multiple systems −
photometry, radial velocities and IUE spectra.**
G. Henriksson.
Rep. Obs. Lund, No. 18, (see 012.044), p. 89 - 92 (1982).

122.245 **Stellar flare activity: a comparison between the time scales of variations in flare continua and in the Hα and Hβ emission lines.**   B. R. Pettersen.
Rep. Obs. Lund, No. 18, (see 012.044), p. 114 - 117 (1982).

122.246 **Does 28 CMa have a photometric period differing from its spectroscopic period?**   D. Baade.
Astron. Astrophys., Vol. 110, L15 - L17 (1982).

Results of recent photometric and spectroscopic observations of the variable Be star 28 CMa are presented. The emission strength has decreased noticeably since 1979. This may be accompanied by a decrease of the radial velocity amplitudes derived from the absorption lines. Qualitatively all previously described variations (Baade, 1982) remained the same. The spectroscopic period, P = 1.37 day, is stable. Unexpectedly, the same period could not be detected during 5 1/2 consecutive nights of *uvby* photometry. Instead, a 0.435 day periodicity of very small amplitude (0^{m}007) was found in all filters which in turn could not be verified in the radial velocity data. The meaning of this photometric near-constancy is briefly discussed.

122.247 **High dispersion spectroscopy of the LMC star S Doradus during maximum light.**
O. Stahl, B. Wolf.
Astron. Astrophys., Vol. 110, 272 - 280 (1982).

Extensive high dispersion spectroscopic observations of the luminous Hubble-Sandage variable S Dor of the LMC have been carried out during its maximum phase from 1973 to 1981 at ESO, La Silla. The spectrum is characterized by strong P Cygni type profiles of the Balmer lines H$_\alpha$–H$_9$ and of singly ionized metallic lines. The continuum energy distribution can be explained by a superposition of an early A-type supergiant photospheric energy distribution and the radiation of a dense ($N_e \approx 10^{10}$ cm^{-3}) rather cool ($T_e \approx 9000$ K) envelope. The mass loss of S Dor during maximum phase was derived to be $\dot{M} \approx 5 \times 10^{-5} M_\odot$ yr^{-1}. The ejection of discrete shells may play an important role for the mass loss.

122.248 **Long-term changes of the flare stars EV Lac and BY Dra.**
L. N. Mavridis, G. Asteriadis, F. M. Mahmoud.
Compendium in astronomy, (see 003.012), p. 253 - 276 (1982).

The long term changes of the quiet-state luminosity and the flare activity of the flare stars EV Lac and BY Dra during the decade 1971-80 have been studied. The quiet-state luminosity of the star EV Lac shows long-term fluctuations with an amplitude of 0^{m}3 and a period of about 5 years. The flare activity of this star shows similar variations, which in most of the cases run parallel to the fluctuations of the quiet-state luminosity. The quiet-state luminosity of the star BY Dra shows also long-term fluctuations. The current cycle has a duration of at least 14 years and an amplitude of more than 0^{m}3. Long-term changes are also indicated for the flare activity of this star.

122.249 **On the problem of Cepheid amplitude dependence on the star position in the instability strip.**
N. S. Nikolov, G. R. Ivanov.
Compendium in astronomy, (see 003.012), p. 277 - 284 (1982).

The important problem of the Cepheid amplitude behaviour within the instability strip is reviewed. Particularly, an attempt is made to answer whether the noted differences between the photometrically and spectroscopically derived Cepheid intrinsic colors B−V are responsible for the discrepancy between the results of the investigations of the problem, and the answer is negative. The data lead to the conclusion, that the amplitudes of the galactic Cepheids at a fixed period increase toward the cool edge of the instability strip.

122.250 **Report of IAU Commission 27: Variable stars (Étoiles variables).**   J. D. Fernie.
Trans. IAU, Vol. XVIIIA, (see 003.013), p. 263 - 302 (1982).

Catalogue of the positions of red variables. See Abstr. 002.015.

Supplement to the Catalogue of RR Lyrae-type stars.   See Abstr. 002.042.

Landolt-Börnstein.   See Abstr. 003.064.

Technique of detection and analysis of stellar microvariability. I. Microvariability of XX Cam. See Abstr. 021.015.

Photométrie photoélectrique différentielle d'étoiles variables.   See Abstr. 031.572.

Period determination techniques. See Abstr. 031.631.

Some simple comments on methods used in the short period determination of variable stars. See Abstr. 031.632.

On local theories of time-dependent convection in the stellar pulsation problem. III. The effect of turbulent viscosity.   See Abstr. 062.119.

The evolution of viscous discs – II. Viscous variations.   See Abstr. 064.006.

Effect of spots on a star's radius and luminosity. See Abstr. 064.038.

The flow of heat near a starspot. See Abstr. 064.039.

The influence of tidal effects on the structure of accretion discs in dwarf novae.   See Abstr. 064.040.

Models for stellar flares. See Abstr. 064. 052.

Shock waves, atmospheric structure, and mass loss in Miras.   See Abstr. 064.098.

Overshooting from convective cores and the occurrence of loops in the HRD.   See Abstr. 065.016.

A comment on some oscillatory models of adiabatic stars.   See Abstr. 065.021.

R-mode oscillations in uniformly rotating stars. See Abstr. 065.029.

Multimode stellar pulsations. III. Resonances. See Abstr. 065.040.

R-mode oscillations in uniformly rotating stars. See Abstr. 065.056.

The distribution law of stellar p-modes. See Abstr. 065.060.

Nonadiabatic stellar pulsation with slow, uniform rotation.   See Abstr. 065.061.

Turbulent convection and pulsational stability of variable stars. Nonpulsating stars in the cepheid strip. See Abstr. 065.069.

Report of IAU Commission 35: Stellar constitution (Constitution des étoiles).   See Abstr. 065.072.

The luminosity and related parameters of variable and non-variable A, F stars.   See Abstr. 111.023.

A different type of maser star? See Abstr. 112.007.

Observations of o Andromedae. III. A model. See Abstr. 112.023.

Infrared observations of HM Sagittae. See Abstr. 113.003.

Light variations of the star HD 29697. See Abstr. 113.021.

Yellow supergiant reddenings from BVRI data. See Abstr. 113.035.

Slow flares in stellar aggregates. II. See Abstr. 113.076.

Photographic colorimetry of stellar flares in the Pleiades and Orion aggregates.   See Abstr. 113.077.

IUE observations of R Aquarii. See Abstr. 114.014.

The variable shell star HR 5999. VI. Strong chromospheric and transition region emission lines in the ultraviolet spectrum of a Herbig Ae star.   See Abstr. 114.040.

Variability and mass loss in the extreme supergiant ζ^1 Sco.   See Abstr. 114.042.

Activity of HBV 475 from its spectral variations. See Abstr. 114.048.

On rapid and slow variations in the spectrum of SU Aur.   See Abstr. 114.064.

Investigation of spectral peculiarities in the variable star RR Lyrae.   See Abstr. 114.082.

Spectrophotometric investigation of selected variable stars of R Coronae Borealis type. See Abstr. 114.085.

Rapid variability of the Hγ line in the Ap star β CrB.   See Abstr. 114.126.

Relationship between colorimetric characteristics of flare stars in aggregates and the aggregate's age. See Abstr. 115.002.

On the radius determination of the variable F-type supergiant BL Tel(F).   See Abstr. 115.005.

Detection of microwave emission from both components of the red dwarf binary EQ Pegasi. See Abstr. 116.007.

AM Herculis and AN Ursae Majoris. See Abstr. 116.047.

Spectroscopy of the X-ray cataclysmic binary IE 0643.0−1648.   See Abstr. 117.012.

Orbital period and radial velocity curve for V 436 Centauri.   See Abstr. 117.021.

Binaries among B stars.   See Abstr. 117.037.

Radiation from the discs in cataclysmic variables: the chromosphere.   See Abstr. 117.045.

A photometric and spectroscopic study of a newly discovered cataclysmic variable, H2215-086. See Abstr. 117.100.

Results of an IUE program of monitoring the ultraviolet emission line fluxes of four binary systems: HR 1099, II Peg, AR Lac and BY Dra. See Abstr. 117.113.

RW Sextantis, a disk with a hot, high-velocity wind. See Abstr. 117.120.

Radio and optical observations of the R Aquarii jet. See Abstr. 117.123.

IUE and optical observations of MV Lyrae at intermediate and low states. See Abstr. 117.124.

The symbiotic star CI Cygni: s-process episode or accretion event?   See Abstr. 119.007.

Flare-Aktivität beim Fleckendoppelstern SV Cam. See Abstr. 119.031.

EX Hydrae: spectroscopic and photometric variations.   See Abstr. 119.102.

A flare event in the peculiar giant FK Comae. See Abstr. 119.132.

Variation of the Hβ emission lines of YY Geminorum. II. Change of sectorial structures of active regions. See Abstr. 120.009.

Evidence for starspots on DK Draconis (HR 4665). See Abstr. 120.022.

New flare stars in the γ Cygni region. See Abstr. 123.013.

Flare stars in Orion.   See Abstr. 123.014.

On the statistics of ZZ Ceti stars. See Abstr. 126.037.

Einstein observations of BY Draconis variables and RS CVn binaries.   See Abstr. 142.038.

The Oosterhoff period groups and the age of globular clusters. III. The age of the globular cluster system. See Abstr. 154.001.

Asymptotic giant branch variable stars. See Abstr. 154.039.

Are there variable stars in the LMC star clusters NGC 2210 and NGC 1786?   See Abstr. 154.043.

He abundance in RR-Lyrae rich clusters. See Abstr. 154.047.

Metal abundances and ages for some Magellanic Cloud variable stars. See Abstr. 159.033.

Errata

122.901 Erratum: "Variable stars in the globular cluster
M13" [Perem. Zvezdy, Tom 21, 169 - 174 (1979)].
R. M. Rusev, T. S. Ruseva.
Perem. Zvezdy, Tom 21, 463 (1980). In Russian. – See
Abstr. 30.122.138.

122.902 Erratum: "Polarization models for hot nonradial
pulsators" [Acta Astron., Vol. 30, 193 - 214 (1980)].

P. A. Stamford, R. D. Watson.
Acta Astron., Vol. 31, 394 (1981). – See Abstr. 28.122.016.

122.903 Erratum: "A survey of variable yellow supergiants
in the southern Milky Way" [Publ. Astron. Soc.
Pacific, Vol. 93, 351 - 360 (1981)]. A. Arellano Ferro.
Publ. Astron. Soc. Pacific, Vol. 93, 817 (1981/82). – See
Abstr. 30.122.028.

123 Variable Stars (Surveys, Lists of Observations, Charts, etc.)

123.001 Das neue BAV-Programm bedeckungsveränderlicher
Sterne.
BAV Rundbrief, 31. Jahrg., 1 - 5 (1982).

123.002 Une étoile mysterieuse dans Andromède.
E. Schweitzer, E. Cifuentes-Torres.
Bull. AFOEV, No. 19, p. 5 - 7 (1982).

123.003 Tableaux des observations faites par les sociétaires
de l'AFOEV en octobre, novembre et décembre
1981.
Bull. AFOEV, No. 19, p. 15 - 47 (1982).

123.004 Variable stars in the vicinity of M13. I.
Eh. Ya. Oganesyan.
Soobshch. Byurakan Obs., Vyp. 53, p. 77 - 78 (1982). In
Russian.
A list of 35 variable sta found during UBV photometry
of blue objects in the vicinit f M13 is presented. The bright-
nesses of these stars in B col and the intervals of their
changes are determined.

123.005 List of variable stars discovered in the USSR and
preliminary SVS designations.
Astron. Tsirk., No. 1169, p. 6 - 7 (1981). In Russian.

123.006 List of variable stars discovered in the USSR and
preliminary SVS designations.
Astron. Tsirk., No. 1171, p. 8 (1981). In Russian.

123.007 Variable stars in the General Catalogue of
Trigonometric Parallaxes. D. Hoffleit.
Inf. Bull. Variable Stars, No. 2066, 4 pp. (1982).

123.008 CX Aurigae: a K-dwarf variable.
E. P. Belserene, W. P. Bidelman.
Inf. Bull. Variable Stars, No. 2081 (1982).

123.009 29 Draconis: a new variable star.
D. S. Hall, G. W. Henry, H. Louth, T. R. Renner,
S. N. Shore.
Inf. Bull. Variable Stars, No. 2109, 3 pp. (1982).

123.010 Variable stars in the northern luminous stars
catalogues. W. P. Bidelman.
Inf. Bull. Variable Stars, No. 2112, 5 pp. (1982).

123.011 Variable stars in the Pleiades cluster II.
J. J. M. Meys, P. Alphenaar, F. van Leeuwen.
Inf. Bull. Variable Stars, No. 2115, 8 pp. (1982).

123.012 New variable star in the γ Cygni region.
K. P. Tsvetkova.
Inf. Bull. Variable Stars, No. 2130, 2 pp. (1982).

123.013 New flare stars in the γ Cygni region.
K. P. Tsvetkova.
Inf. Bull. Variable Stars, No. 2131, 2 pp. (1982).

123.014 Flare stars in Orion.
M. K. Tsvetkov, A. G. Tsvetkova.
Inf. Bull. Variable Stars, No. 2132 (1982).

123.015 Flare stars in the Praesepe region.
G. Oganjan (Oganyan), I. Jankovics, J. Kelemen.
Inf. Bull. Variable Stars, No. 2149 (1982).

123.016 New variable stars in Cygnus. L. Dahlmark.
Inf. Bull. Variable Stars, No. 2157, 7 pp. (1982).

123.017 The American Association of Variable Star Observers,
Bulletin 45 (1982). Maxima and minima of long
period variables. 1982 annual predictions. J. A. Mattei.
AAVSO Bull. 45, 8 pp. (1982).

123.018 New variable star SVS 2490 in the Pleiades.
L. K. Erastova.
Astron. Tsirk., No. 1190, p. 6 - 7 (1981). In Russian.

123.019 Information on photoelectric observations of
variable stars deposited at the Odessa Astronomical
Observatory. V. P. Tsesevich, E. N. Makarenko.
Astron. Tsirk., No. 1191, p. 7 - 8 (1981). In Russian.

123.020 List of variable stars discovered in the USSR given
preliminary SVS designations.
Astron. Tsirk., No. 1193, p. 8 (1981). In Russian.

123.021 New variable stars in the direction of the bright
cloud B in Sagittarius.
A. Terzan, A. Bijaoui, K. H. Ju, C. Ounnas.
Obs. Lyon Prepr. No, 1, 37 pp. (1982).
621 new variable stars have been detected on plates
obtained with the Schmidt telescopes of the Mount Palomar
Observatory (1968) and of the European Southern Observatory (1976 - 1980).

123.022 Beobachtung von 12 wenig bekannten veränderlichen Sternen. L. Meinunger.
Mitt. Veränderl. Sterne (MVS), Band 9, 59 - 63 (1981).

123.023 Mehrfarben-Beobachtungen von V 5 im Kugel-
haufen M 3.   I. Meinunger.
Mitt. Veränderl. Sterne (MVS), Band 9, 63 - 66 (1981).

123.024 Photoelektrische Messungen von 11 langsam und
unregelmäßig veränderlichen, nicht roten Sternen.
L. Meinunger.
Mitt. Veränderl. Sterne (MVS), Band 9, 67 - 76 (1981).

123.025 Untersuchungen zum Problem der Identität von
CV Aqr auf Sonneberger Platten.   B. Fuhrmann.
Mitt. Veränderl. Sterne (MVS), Band 9, 77 - 78 (1981).

123.026 Photographische Beobachtungen von vier Veränderlichen in Lyra.   H. Geßner.
Mitt. Veränderl. Sterne (MVS), Band 9, 78 - 79 (1981).

123.027 Beobachtungsergebnisse des Arbeitskreises
"Veränderliche Sterne" im Kulturbund der DDR
(Teil VIII).
Mitt. Veränderl. Sterne (MVS), Band 9, 80 - 85 (1981).

123.028 Neuer Veränderlicher S 10848 Ophiuchi.
H. Geßner.
Mitt. Veränderl. Sterne (MVS), Band 9, 85 (1981).

123.029 Sektion Photographische Veränderlichenbeob-
achtung.   E. Goercke.
BAV Rundbrief, 31. Jahrg., 48 - 50 (1982).

123.030 List of 333 variable, microvariable or suspected
variable stars detected in the Geneva photometry.
F. Rufener, P. Bartholdi.
Astron. Astrophys., Suppl. Ser., Vol. 48, 503 - 511 (1982).

123.031 Tableaux des observations reçues à l'AFOEV en
janvier, février et mars 1982.
Bull. AFOEV, No. 20, p. 75 - 103 (1982).

123.032 More variable stars among bright visual binaries.
D. Hoffleit.
J. American Assoc. Variable Star Obs., Vol. 10, 57 - 59 (1981).

123.033 Southern hemisphere objective-prism discoveries.
W. P. Bidelman, D. J. MacConnell.
Astron. J., Vol. 87, 792 - 793 (1982).

This paper reports 27 southern objects of astrophysical
interest discovered on moderate-dispersion Curtis Schmidt
objective-prism plates obtained at Cerro Tololo. About half of
the objects are previously unclassified red variables, but the list
also includes two new bright stars showing composite spectra,
several interesting peculiar A stars, and a possible new
RR Lyrae star.

Étoiles variables et documentation automatique.
See Abstr. 002.010.

An atlas of southern and equatorial dwarf novae.
See Abstr. 002.062.

Charts for southern variables. Series No. 14.
See Abstr. 002.076.

A catalogue of variable and visual double stars.
See Abstr. 118.032.

Photoelectric two-colour observations of MR Cygni
in 1974.   See Abstr. 119.014.

Supergiants and variable stars in the χ and h Persei
cluster region.   See Abstr. 153.034.

124 Novae

124.001 Development of the λ2200 extinction feature in
posteruptive novae.   M. F. Bode, A. Evans.
Astrophys. J., Vol. 254, 263 - 270 (1982).

The authors describe the changes in the λ2200 graphite
feature that might be expected to occur in classical novae
during the posteruptive period. Four models are considered,
including the rapid grain formation model of Clayton and
Wickramasinghe, and three variants of the preexisting grain
model of Bode and Evans. The results suggest that these
models could easily be distinguished using the IUE satellite or
similar facility, and that the detailed form of the λ2200 feature
could be an accurate guide to grain size in circumstellar shells.

124.002 Nucleosynthesis in novae: a source of Ne-E and
^{26}Al?   W. Hillebrandt, F.-K. Thielemann.
Astrophys. J., Vol. 255, 617 - 623 (1982).

In the framework of recent nova models, nucleosynthetic
products of explosive hydrogen burning are computed. It is
shown that nova condensates are likely to contain isotopic
anomalies in both ^{22}Ne from ^{22}Na decay and ^{26}Mg from ^{26}Al
decay. Results are compared with the so-called Ne-E anomaly
in carbonaceous chondrites, a component highly enriched
in ^{22}Ne.

124.003 Role of the thermonuclear instability in recent
models of stellar novae.   L. Secco.
Nuovo Cimento B, Vol. 65B, Ser. 11, 345 - 403 (1981). In
Italian. – Abstr. in Phys. Abstr., Vol. 85, Abstr. 33043
(1982).

124.004 The ultraviolet spectrum of the old novae HR Del,
GK Per, RR Pic, and RS Oph.
L. Rosino, A. Bianchini, P. Rafanelli.
Astron. Astrophys., Vol. 108, 243 - 248 (1982).

Spectroscopic observations of three old novae (HR Del,
GK Per, RR Pic) and a recurrent nova (RS Oph), all of them
at minimum, have been carried out with the IUE satellite.

124.005 Periodic eruptions in T CrB?
L. H. Palmer, J. L. Africano.
Inf. Bull. Variable Stars, No. 2069 (1982).

124.006 Processes in nova envelopes in the phase of
diffuse-enhanced and Orion spectra.
V. S. Bychkova.
Astrofiz. Issled., Izv. Spets. Astrofiz. Obs., Tom 15, 3 - 13
(1982). In Russian.

The interaction of a patchy stellar wind with the

principal envelope explains simultaneously the observed systems of rather narrow lines with different Doppler shifts as well as the strong variations of these shifts and the temporal disappearance of some line systems.

124.007   Nova speed class and infrared development.
M. F. Bode, A. Evans.
Mon. Not. R. Astron. Soc., Vol. 200, 175 - 181 (1982).

By considering the effect of changes in the temperature of the underlying pseudophotosphere of classical novae on grain absorption efficiencies the authors derive the time at which grain growth can commence. As the bolometric luminosity, ejection velocity and rate of change of the pseudo-photospheric temperature are all dependent on the speed class of a given nova they find that the condensation time is also dependent on this factor. The form of the dependence leads to a straightforward explanation of why fast novae do not seem to be able to form grains as readily as slower novae.

On the role of the accretion rate in nova outbursts.
See Abstr. 065.036.

A theory of hydrogen shell flashes on accreting white dwarfs. I. Their progress and the expansion of the envelope.   See Abstr. 065.041.

Conditions for novalike optically thick winds.
See Abstr. 112.021.

Report of IAU Commission 42: Close binary stars (Étoiles doubles serrées).   See Abstr. 117.146.

A polarization study of dwarf novae and nova-like objects.   See Abstr. 122.080.

A spectrophotometric survey of cataclysmic variable stars.   See Abstr. 122.199.

Novae as sources of nitrogen in galaxies.
See Abstr. 158.273.

Nova Monocerotis 1939 = BT Monocerotis

124.101   BT Monocerotis: an eclipsing nova.
E. L. Robinson, R. E. Nather, S. O. Kepler.
Astrophys. J., Vol. 254, 646 - 652 (1982).

Photometric observations of BT Mon (= Nova Mon 1939) demonstrate that it is an eclipsing binary system with an orbital period of 0.3338141 days. The authors show that the accretion disk in BT Mon is exceptionally large and luminous. Its radius is within 60% of the radius of its Roche lobe and is at least 3 times larger than the radius of a zero-viscosity disk; and its absolute visual magnitude is 4.0 ± 1.0, which requires a mass flux greater than $2 \times 10^{-8} M_\odot$ yr^{-1}. The late-type star in BT Mon cannot simultaneously fit a main-sequence mass-radius relation and a main-sequence mass-luminosity relation, in the sense that it is underluminous for a normal main-sequence star.

Nova CP Puppis

124.121   Spectroscopic analysis of the extended shells around the novae CP Puppis and T Pyxidis.
R. E. Williams.
Prepr. Steward Obs., No. 368, 36 pp. (1982).

Nova T Pyxidis

Spectroscopic analysis of the extended shells around the novae CP Puppis and T Pyxidis.
See Abstr. 124.121.

Nova Vulpeculae 1979 = PU Vulpeculae

Kuwano's peculiar object is a novalike (symbiotic?) binary with a red giant. Photometric and polarimetric observations.   See Abstr. 113.005.

Kuwano's peculiar object is a novalike (symbiotic?) binary with a red giant. Spectral observations.
See Abstr. 114.028.

Nova Persei 1901 = GK Persei

124.201   The old-nova GK Per. II. Optical outbursts.
A. Bianchini, F. Sabbadin, E. Hamzaoglu.
Astron. Astrophys., Vol. 106, 176 - 178 (1982).

The photographic and spectroscopic behaviour of the old-nova GK Per (1901) during optical outbursts can be explained by an increase of the mass transfer rate from the late-type secondary onto the primary. In particular, the observed luminosity variation of the primary during the Feb-Apr 1981 brightening needs an accretion rate changing from 10^{18} to 10^{20} g s^{-1}. A temperature enhancement of the primary during outbursts is indicated by the observed B−V and 4686/ Hβ variations.

Nova Vulpeculae 1670 = CK Vulpeculae

124.221   The recovery of CK Vulpeculae (nova 1670) − the oldest "old nova".
M. M. Shara, A. F. J. Moffat.
Astrophys. J., Lett., Vol. 258, L41 - L44 (1982).

Comparison of narrow-band Hα with continuum plates has yielded the location of CK Vulpeculae, a slow nova which reached maximum brightness $m_B \approx 3$ in 1670 and again in 1671. CK Vul affords us the opportunity to study a nova remnant almost two centuries older than any other known. Associated with the central object are two nebulous blobs radiating mostly Hα and [N II]. These nebulae are probably ejecta of the nova sweeping up interstellar matter.

Nova Coronae Austrinae 1981

124.301   Spectrophotometry of Nova Coronae Austrinae 1981.   N. Brosch.
Astron. Astrophys., Vol. 107, 300 - 304 (1982).

The spectrum of Nova Coronae Austrinae 1981 was found to be similar to that of V1500 Cygni. The outburst ejected material in two polar caps and an equatorial torus roughly aligned in the line of sight. The ejection velocity was about 2.2×10^3 km s^{-1}. This fast nova was apparently located in, or beyond the galactic bulge. The interstellar matter of the galactic disk attenuates the nova light by about 0.3 mag while at most 1.4 mag of extinction are probably contributed by bulge material.

124.302   Rediscussion of Nova CrA 1981 distance.
J. A. R. Caldwell.
Inf. Bull. Variable Stars, No. 2147, 4 pp. (1982).

Nova Herculis 1934 = DQ Herculis

124.401   **Physical conditions in the shell around the old nova DQ Herculis 1934.**   H. Itoh.
Publ. Astron. Soc. Japan, Vol. 33, 743 - 747 (1981).

The shell around DQ Herculis may have been ionized up to the present by the ultraviolet radiation from the central white dwarf. Under this assumption, the following observational facts are explained: (1) At present, the gas with C^{++}, N^{++}, and O^{++} ions has a temperature lower than 500 K, while the gas with N^+ and O^+ ions has a much higher temperature. (2) In the 1940's, the gas with O^{++} ion was hot enough to emit [O III] λ 5007 Å line.

Nova Serpentis 1970 = FH Serpentis

124.501   **Spectroscopy of the nova FH Serpentis (1970) during the early decline stage.**
S. Štefl, J. Grygar.
Bull. Astron. Inst. Czechoslovakia, Vol. 33, 116 - 122 (1982).

Eleven spectrograms obtained between February 21 and March 30, 1970 in the spectral region from 480 to 660 nm with an original dispersion of 1.0 to 1.5 nm mm^{-1} were used for line identification and spectral development study of the nova FH Ser in the early decline stage. Finally, the variations of the strengths of the emission peaks of the lines of Hβ and Fe II were found and the distance of the nova, r = 900 pc, was derived.

Nova Cygni 1975 = V1500 Cygni

124.601   **Photometric researches on short-period brightness variability of nova Cygni 1975 (V1500 Cyg) in the continuum.**   E. P. Pavlenko, V. V. Prokof'eva.
Izv. Krymskoj Astrofiz. Obs., Tom 64, 67– 72 (1981). In Russian. English translation in Bull. Crimean Astrophys. Obs., Vol. 64.

About 3800 TV pictures of nova Cyg 1975 with patrol time of about 36.5 hours were obtained in September 1977. The mean light curve was constructed with $P = 0^{d}2760$. The difference of rising branches is confident with probability 0.95. Individual light curves show strong brightness fluctuations ranging from minutes to days. A phase dependence of the brightness fluctuations has been found with reliability 0.99. A model of a close binary system with inclination of 60° satisfies this dependence.

Nova Pictoris 1925 = RR Pictoris

124.701   **A photometric and polarimetric investigation of the old nova RR Pictoris.**   R. Haefner, K. Metz.
Astron. Astrophys., Vol. 109, 171 - 178 (1982) = Veröff. Sternw. München, Band 7, Nr. 32.

Extensive photometric and polarimetric observations of the old nova RR Pic are reported. Using published radial velocity measurements it is possible to interpret the light and colour curves within the basic model for cataclysmic variables.

Nova Aquilae 1918 = V603 Aquilae

124.801   **Photometric observations of V603 Aquilae.**   T. Herczeg.
Inf. Bull. Variable Stars, No. 2078, 6 pp. (1982).

Nova Aquilae 1982

124.901   **Nova Aquilae 1982.**
IAU Circ., Nos. 3661, 3663, 3673, 3676, 3682, 3689, 3696 (1982).

124.902   **Nova Aquilae 1982.**
Yamamoto Circ., Nos. 1969, 1971, 1973, 1975 (1982).

124.903   **Nova Aquilae 1982.**
British Astron. Assoc. Circ., No. 623 (1982).

125 Supernovae, Supernova Remnants

125.001 **Detection of radio emission from optically identified supernova remnants in M31.**
J. R. Dickel, S. D'Odorico, M. Felli, M. Dopita.
Astrophys. J., Vol. 252, 582 - 588 (1982).

The Very Large Array was used to conduct a radio search at a wavelength of 20 cm for ten optically identified supernova remnants (SNRs) in M31. Five SNRs were detected, and for the other objects, upper limits to the emission were determined. On the average, the surface brightness, Σ, of an SNR in M31 appears to be fainter than that of an SNR in the Galaxy. It is suggested that the median Σ at a given diameter is higher in late-type spirals than in Sb systems.

125.002 **Velocity dispersions of knots in the Cygnus Loop and IC 443.** P. Shull, Jr., R. A. R. Parker, T. R. Gull, R. J. Dufour.
Astrophys. J., Vol. 253, 682 - 695 (1982).

Very high resolution spectroscopy of optical emission lines indicates that the velocity dispersions of knots in Cygnus Loop and IC 443 filaments result primarily from turbulence within the emitting regions. Line-of-sight velocity dispersions (half-widths at half-maximum) for the observed knots are on the order of $10-30$ km s^{-1}, including associated thermal velocity dispersions of roughly 15 km s^{-1} for hydrogen, 6 km s^{-1} for oxygen, and 4 km s^{-1} for nitrogen. The knots themselves move randomly relative to each other with speeds of $10-30$ km s^{-1}.

125.003 **The ultraviolet spectrum of the Crab Nebula.**
K. Davidson, T. R. Gull, S. P. Maran, T. P. Stecher, R. A. Fesen, R. A. Parise, C. A. Harvel, M. Kafatos, V. L. Trimble.
Astrophys. J., Vol. 253, 696 - 706, plates 9 - 12 (1982).

Data from 65 hours of observation of the Crab Nebula with the IUE are reported, together with new ground-based spectrophotometry. The authors have measured the important C IV λ1549, He II λ1640, and C III] λ1908 emission line intensities and placed upper limits on other ultraviolet features for the brightest filamentary region in the nebula. They have also measured some ultraviolet continuum surface brightnesses at two places in the Crab. The emission lines imply an average ionic abundance ratio $n(C^{+2})/n(O^{+2})$ in the range from 0.4 to 1.5 in the observed gaseous condensations. The elemental abundance ratio of carbon to oxygen is probably in the same range. The large helium abundance, small carbon and oxygen abundances, and presence of a neutron star in the Crab Nebula suggest that the presupernova star had a mass close to 8 $M_\odot$ when it was on the main sequence.

125.004 **Type I supernovae. I. Analytic solutions for the early part of the light curve.** W. D. Arnett.
Astrophys. J., Vol. 253, 785 - 797 (1982).

Analytic solutions for light curves, effective temperatures, and broad-band colors of Type I supernovae are presented. The method is generalized to include effects of finite (large) initial radius and increasing transparency to γ-rays and to thermal photons. A theoretical construct, the "blackbody supernova," is introduced. Many observed features of Type I supernovae are shown to be reproduced by the theory. For a given composition it is shown that the homogeneity of spectral evolution is a necessary consequence of the thermonuclear model but only a possible consequence of the gravitational collapse model.

125.005 **Variation of the radio flux from young supernova remnants.**
V. P. Ivanov, I. T. Bubukin, K. S. Stankevich.
Pis'ma Astron. Zh., Tom 8, 83 - 85 (1982). In Russian. English

translation in Soviet Astron. Lett., Vol. 8.

Measurements of radio fluxes from Cassiopeia A, the Crab nebula and the supernova remnant SN 1572 relative to those from the radio galaxies Cygnus A and Virgo A were made in 1981 at 31.5 cm wavelength. The decrease of the Cassiopeia A flux has slowed 2.2 times down to (0.413 ± 0.08) per cent per year during the last decade. In 1981 the Crab nebula flux has diminished by (3.5 ± 1.0) per cent as compared with observations before 1972. The 17-year mean of the annual rate of the SN 1572 flux decrease amounts (0.5 ± 0.15) per cent.

125.006 **Neutral hydrogen in the vicinity of galactic radio sources. Supernova remnant W 28.**
A. P. Venger, I. V. Gosachinskij, V. G. Grachev, T. M. Egorova, N. F. Ryzhkov, V. K. Khersonskij.
Astron. Zh., Tom 59, 20 - 26 (1982). In Russian. English translation in Soviet Astron., Vol. 26, No. 1.

The results of emission and absorption observations of H I in the vicinity of the radio source W 28 are presented. The distance to the supernova remnant is estimated using the H I absorption line and it is shown that its compact H II regions are at the same distance. The parameters of the supernova remnant are derived and some conclusions on a possible genetic connection between H II regions and the supernova remnant are made.

125.007 **Is the remnant of SN 1006 composed of iron?**
A. C. Fabian, G. C. Stewart, W. Brinkmann.
Nature, Vol. 295, 508 - 509 (1982).

The authors conclude by considering the cup-shape appearance of the remnant of SN 1006, as it appears in the X-ray. This could partly be due to the exploding star having been a member of a close binary system. The flow and subsequent shocks surrounding the companion could well have slowed the ejecta down in one direction, provided that the ejecta was roughly at its final velocity by the time it had crossed the binary orbit (say ~10^{11} cm).

125.008 **X-ray spectral classification of supernova remnants in the Large Magellanic Cloud.** D. H. Clark, I. R. Tuohy, K. S. Long, A. E. Szymkowiak, M. A. Dopita, D. S. Mathewson, J. L. Culhane.
Astrophys. J., Vol. 255, 440 - 446 (1982).

The solid state spectrometer on the Einstein Observatory was used to measure the 0.6−4.5 keV X-ray spectra of six prominent supernova remnants in the Large Magellanic Cloud, namely, N132D, N63A, N49, 0525−66.0, N157B, and 0540−69.3. Thermal emission is detected from the first four remnants and is similar in nature to that observed from young galactic SNRs. In contrast, N157B and 0540−69.3 have featureless X-ray spectra which are well described by power-law models. The data support a synchrotron origin for the X-rays from N157B and 0540−69.3.

125.009 **A search for hot dust in the fast moving knots in Cassiopeia A.** H. L. Dinerstein, M. W. Werner, R. W. Capps, E. Dwek.
Astrophys. J., Vol. 255, 552 - 556 (1982).

The fast moving knots in the Cassiopeia A supernova remnant have unusual abundances, implying that they are composed of material from the interior of the star and are possible locations for grains formed by thermal condensation during the expansion of the remnant. The authors report here negative results from a search at 3.8 μm for emission from dust associated with these knots. These results set upper limits on the mass of hot dust and the dust-to-gas ratio in the knots.

125.010   An X-ray study of two Crablike supernova remnants: 3C 58 and CTB 80.
R. H. Becker, D. J. Helfand, A. E. Szymkowiak.
Astrophys. J., Vol. 255, 557 - 563 (1982).

The instruments on the Einstein Observatory have been used to study the X-ray properties of two Crablike SNRs: 3C 58 and CTB 80. Images of the objects are similar, showing an extended plateau of emission surrounding a pronounced central maximum consistent with the presence of a point source. The X-ray spectrum of 3C 58 is consistent with a power law of spectral index 0.5 modified by absorption in a hydrogen column density N_H of $(2 \pm 0.5) \times 10^{21} \, cm^{-2}$. Quantitative comparisons of the X-ray and radio properties for six crablike remnants indicate significant differences which may reflect an evolutionary sequence.

125.011   The radio morphology of supernova remnants.
P. A. Shaver.
Astron. Astrophys., Vol. 105, 306 - 312 (1982).

The radio structure of 89 supernova remnants is examined. The data suggest that shell-type remnants are initially highly spherical and relatively complete; they become more incomplete and distorted through interaction with the interstellar medium. In general the emission is strongest on the side nearest the galactic plane. Any toroidal component, which may arise due to rotation of the original massive star (Bodenheimer and Woosley, 1980), probably accounts for less than 10% of the radio emission.

125.012   G33.2 − 0.6, an old supernova remnant with a spectral break.   W. Reich.
Astron. Astrophys., Vol. 106, 314 - 316 (1982).

Sensitive 2.7 GHz and 4.75 GHz observations with the Effelsberg 100-m telescope have resulted in the detection of the weak shell-type source G33.2 − 0.6. Its nonthermal spectral index of $\alpha = 0.7 \pm 0.2$ $(S \sim \nu^{-\alpha})$ classifies it as a supernova remnant (SNR). The source is at the limits of published low frequency surveys, suggesting a spectral break as seen in a few other old supernova remnants. This type of source is of special interest as an aid to understanding the evolution of SNR and their interaction with the interstellar medium.

125.013   Some thoughts on SN 1006: not the Schweizer-Middleditch star.
M. P. Savedoff, H. M. Van Horn.
News Lett. Astron. Soc. N. Y., Vol. 2, 28 (1982). − Abstract.

125.014   The Schweizer-Middleditch star: not a stellar remnant of SN 1006.
M. P. Savedoff, H. M. Van Horn.
Astron. Astrophys., Vol. 107, L3 - L4 (1982).

The blue star found by Schweizer and Middleditch (1980) near the center of SN 1006 cannot be a stellar remnant of that supernova, unless it is considerably more exotic than is justified by existing observational data. Conventional stellar evolution theory shows that the time to cool to the observed effective temperature is $\sim 10^6$ years, far in excess of the 975 year age of the supernova remnant.

125.015   Type I supernova − thermonuclear versus collapse.
S. A. Colgate, A. G. Petschek.
Nature, Vol. 296, 804 - 805 (1982).

125.016   Observations of Cassiopeia A at 20 and 25 MHz with the URAN-1 interferometer.
V. P. Bovkoon (*Bovkun*), S. Ya. Braude, A. V. Megn (*Men'*).
Astrophys. Space Sci., Vol. 81, 221 - 230 (1982).

Results of observations of Cassiopeia A at decametric wavelengths on radiointerferometers with baselines of 2840 and 3550 wavelengths are presented. As has been found, the model of the large-scale structure, deduced from the measurements in the meter and decimeter wavebands, is inconsistent with the results obtained at decameter wavelengths. The data can be reconciled by assuming a compact component with angular dimensions about 4″. The contribution of that component to the total flux is 6.8 ± 2.4% and 3.7 ± 1.4% at 20 and 25 MHz, respectively.

125.017   The supernova neutrino pulse shape in the scintillation detector.   Z. F. Seidov.
Astrophys. Space Sci., Vol. 81, 483 - 488 (1982).

The temporal structure of the neutrino scintillation detector response to a supernova explosion signal is calculated, taking into account the duration and the spectrum of the supernova neutrino radiation and also the neutrino rest-mass.

125.018   Thermodynamic properties of supernova matter.
D. Hartmann, M. El Eid.
Mitt. Astron. Ges., Nr. 55, p. 58 - 60 (1982).

125.019   Radiokontinuumsbeobachtungen der alten Supernovaüberreste G127.1 + 0.5 und G126.2 + 1.6.
R. Steube, W. Reich.
Mitt. Astron. Ges., Nr. 55, p. 60 - 62 (1982).

125.020   Hochauflösende Radiokarten von Supernova-Überresten.   U. R. Buczilowski.
Mitt. Astron. Ges., Nr. 55, p. 62 - 65 (1982).

125.021   The effects of non-equilibrium ionization on the X-ray emission of supernova remnants.
E. H. B. M. Gronenschild, R. Mewe.
Astron. Astrophys., Suppl. Ser., Vol. 48, 305 - 331 (1982).

The authors investigated how non-equilibrium ionization affects the emergent X-ray spectrum from supernova remnants which are assumed to obey the evolution as described by Sedov (1959). They found that the continuum spectrum shows enhanced two-photon decay radiation mainly due to oxygen and suppressed recombination radiation of bare oxygen nuclei. The line emission is strongly intensified especially at wavelengths longer than about 10 Å which causes an apparent softening of the spectrum compared to equilibrium conditions. It is shown that the standard analysis of X-ray spectra observed from supernova remnants by fitting the spectrum with an isothermal equilibrium spectrum will lead to erroneous values for the basic parameters of the remnants. In general the derived temperature is too low, as is the resulting explosion energy, whereas the calculated ambient density will be too large.

125.022   Soft X-ray observation of supernova remnant SN1006.
C. M. F. Galas, D. Venkatesan, G. Garmire.
Astrophys. Lett., Vol. 22, 103 - 108 (1982).

The observations of the supernova remnant SN 1006 by the Low Energy Detectors on HEAO−1 reveal that the position of the centroid of the X-ray emission is located SE of the center of the remnant. The spectrum of SN 1006 is obtained after subtracting the contributions of both the background and the Lupus Loop. The procedure is discussed in detail. To fit the spectrum analytically, a soft thermal component with oxygen lines is required. This indicates the existence of a plasma at $2 \times 10^6 \, °K$ within the remnant.

125.023   On two types of supernova remnants.
F. Kh. Sakhibov, M. A. Smirnov.
Pis'ma Astron. Zh., Tom 8, 281 - 285 (1982). In Russian. English translation in Soviet Astron. Lett., Vol. 8.

On the basis of 57 calibration sources the space distribution of supernova remnants (SNR), the evolution of their radio emission and possible selection effects are considered. While plerions and shell SNRs with central source are distributed in the Galaxy with the height scale $z \sim 150$ pc, the distribution of pure shell SNRs is more narrow with $z \sim 50$ pc. Moreover these two types of SNRs differ in the evolution of radioemission.

125.024 Supernova having created the Crab nebula.
Yu. P. Pskovskij.
Zemlya Vselennaya, 1982, No. 3, p. 36 - 40. In Russian.

125.025 Estudio del remanente de supernova G 261.9 + 5.5 (PKS 0902 − 38) en la linea de 21 cm.
F. R. Colomb, G. M. Dubner.
Bol. Asoc. Argentina Astron., No. 20 - 24, p. 369 (1981).
Abstract.

125.026 Observaciones en la línea de 21 cm del remanente de supernova G 261.9 + 5.5.
F. R. Colomb, G. M. Dubner.
Bol. Asoc. Argentina Astron., No. 20 - 24, p. 434 (1981).
Abstract.

125.027 A multiwavelength comparison of Cassiopeia A and Tycho's supernova remnant.
J. R. Dickel, S. S. Murray, J. Morris, D. C. Wells.
Astrophys. J., Vol. 257, 145 - 150, plate 5 (1982).
    A comparison of high resolution radio, optical, and X-ray images of two young supernova remnants (SNR), Cas A and Tycho, shows significant differences between them. Cas A probably broke into many small knots at the time of the initial explosion, whereas Tycho's SNR appears to be a more uniformly expanding blast wave.

125.028 Dynamical determination of the mass of ejected shells of Cassiopeia A-like supernovae.
F. Kh. Sakhibov.
Astron. Tsirk., No. 1171, p. 1 - 2 (1981). In Russian.

125.029 On the distance to the galactic supernova remnants Vela XYZ and HB 3.
F. Kh. Sakhibov, M. A. Smirnov.
Astron. Tsirk., No. 1171, p. 2 - 4 (1981). In Russian.

125.030 Supernovae in anonymous galaxies.
IAU Circ., No. 3683 (1982).

125.031 Possible supernova near NGC 1332.
IAU Circ., No. 3684 (1982).

125.032 Supernova in anonymous galaxy.
IAU Circ., Nos. 3693, 3696 (1982).

125.033 Supernova in anonymous galaxy.
IAU Circ., No. 3702 (1982).

125.034 On continuous spectra of supernova stars.
A. K. Kolesov, V. V. Sobolev.
Astron. Zh., Tom 59, 417 - 423 (1982). In Russian. English translation in Soviet Astron., Vol. 26, No. 3.
    The influence of electron scattering on the energy distribution in supernova continuous spectra is investigated. The problem of transfer of radiation through the supernova envelope is solved for two cases: 1) when the envelope is a homogeneous sphere and 2) when it consists of plane parallel layers. As a result the spectrophotometric gradient is found as a function of characteristics of the envelope. It is shown that the spectrophotometric temperature can strongly differ from the surface temperature.

125.035 Accreting white dwarf models for type I supernovae. II. Off-center detonation supernovae.
K. Nomoto.
Astrophys. J., Vol. 257, 780 - 792 (1982).
    Supernova models based on accreting carbon-oxygen white dwarfs are presented. In these models, the explosion is triggered by the strong helium shell flash which occurs when the accretion forms a helium layer of substantial mass. It is found that the helium flashes give rise to detonation supernovae of the following types, depending on the accretion rate: For intermediate accretion rates the helium flash is so strong that it produces double detonation waves, namely, a helium detonation wave that propagates outward and a carbon detonation wave that propagates inward. For the case with a slower accretion rate the helium flash is strong enough to form a helium detonation but too weak to initiate a carbon detonation. Then only a single helium detonation wave propagates outward. The double detonation waves incinerate most of the white dwarf material into ^{56}Ni. The released nuclear energy is large enough to disrupt the star completely; i.e., no compact star is left. On the other hand, the single detonation wave either leaves a white dwarf remnant behind or disrupts the star completely, depending on the conditions in the white dwarf.

125.036 Evolution of pulsar-driven supernova remnants.
S. P. Reynolds, R. A. Chevalier.
Bull. American Astron. Soc., Vol. 13, 794 (1981). − Abstract.

125.037 The effects of ejecta on the X-ray luminosities of supernova remnants.
K. S. Long, M. A. Dopita, I. R. Tuohy.
Bull. American Astron. Soc., Vol. 13, 795 (1981). − Abstract.

125.038 X-ray, optical and UV observations of the supernova remnant in NGC 4449.
W. P. Blair, J. C. Raymond, R. P. Kirshner, P. F. Winkler.
Bull. American Astron. Soc., Vol. 13, 795 (1981). − Abstract.

125.039 MSH 15-5(2) − a SNR containing 2 compact X-ray sources. F. Seward, P. Murdin.
Bull. American Astron. Soc., Vol. 13, 795 (1981). − Abstract.

125.040 Radio observations of the supernova remnant, W28.
M. D. Andrews, J. P. Basart, R. C. Lamb.
Bull. American Astron. Soc., Vol. 13, 795 - 796 (1981).
Abstract.

125.041 High resolution X-ray spectroscopy of the supernova remnant N132D in the Large Magellanic Cloud.
C. J. Berg, C. R. Canizares, P. F. Winkler, T. H. Markert, C. W. Waite, Jr., F. K. Perkins.
Bull. American Astron. Soc., Vol. 13, 796 (1981). − Abstract.

125.042 Linear polarization of supernova light: a measure of deviation from spherical symmetry.
P. G. Sutherland, P. R. Shapiro.
Bull. American Astron. Soc., Vol. 13, 796 (1981). − Abstract.

125.043 Astrometric motions in Cas A.
P. E. Angerhofer, R. A. Perley.
Bull. American Astron. Soc., Vol. 13, 841 (1981). − Abstract.

125.044 Radial distribution of relative emission-line intensities in the Cygnus Loop.
R. A. Fesen, T. R. Gull, R. P. Kirshner.
Bull. American Astron. Soc., Vol. 13, 887 (1981). − Abstract.

125.045 Detection of infrared emission from Cas A.
H. L. Dinerstein, M. W. Werner, E. Dwek, R. W. Capps.
Bull. American Astron. Soc., Vol. 13, 895 - 896 (1981).
Abstract.

125.046 Spectrophotometry in the galactic supernova remnants Kepler, RCW 86 and RCW 103.
E. M. Leibowitz, I. J. Danziger.
ESO Sci. Prepr. No. 197, 23 pp. (1982). − Submitted to Mon. Not. R. Astron. Soc.

125.047 **A spectrophotometric study of Kepler supernova-remnant.** M. Dennefeld.
Pré-publ. Inst. Astrophys. Paris, No. 1, 24 pp. (1982).
Submitted to Astron. Astrophys.

125.048 **Supernova in anonymous galaxy.**
IAU Circ., No. 3707 (1982).

125.049 **Multicoloured X-ray image of the Cygnus Loop.**
W. H.-M. Ku, M. H. Vartanian, C. J. Hailey.
Bull. American Astron. Soc., Vol. 14, 576 - 577 (1982).
Abstract.

125.050 **A high-resolution X-ray image of Puppis A: in-homogeneities in the interstellar medium.**
R. Petre, C. R. Canizares, G. A. Kriss, P. F. Winkler.
Astrophys. J., Vol. 258, 22 - 30, plates 1 - 3 (1982).
The authors present a 0.1–4 keV image of the Puppis A supernova remnant constructed from 11 Einstein Observatory HRI exposures. The image displays a complex morphology which presumably reflects the structure of the shocked interstellar medium. There is a density gradient of a factor greater than 4 across the $\sim$30 pc diameter of the remnant, perpendicular to the galactic plane. A shell of X-ray emission surrounds the northern half of Puppis A, coincident with the radio shell. Brightness profiles across the edge of the shell show a sharp rise, which indicates that the hot plasma has been heated directly by the blast wave rather than evaporated from clouds. A wealth of interior structure suggests inhomogeneities over a wide range of sizes 0.1–5 pc, but with a typical density contrast of no more than a factor of 2 about an average of $\sim$1 cm^{-3}. Isolated clouds of density 10–30 cm^{-3} are responsible for the two brightest X-ray features, but represent only a small fraction of the mass in Puppis A.

125.051 **A search for radio emission from six historical supernovae in the galaxies NGC 5236 and NGC 5253.** J. J. Cowan, D. Branch.
Astrophys. J., Vol. 258, 31 - 34, plates 4 - 6 (1982).
A search for radio emission at 20 cm from the six historical supernovae in NGC 5253 and NGC 5236 (M83) has been carried out with the VLA. None of the supernovae were detected. The 3 σ upper limits are $\sim$0.9 mJy. In addition to the extended emission from the central regions of the two galaxies, two point sources were detected in the NGC 5236 field. Extrapolation of the 21 cm $\log N - \log S$ relation indicates that they are unlikely to be background sources. Both sources lie along the inner edge of the northern spiral arm, but neither coincides with H II regions in NGC 5236. The authors suggest that they may be supernova remnants emitting at 20 cm at about the Cas A level.

125.052 **The Hubble diagram for type I supernovae.**
D. Branch.
Astrophys. J., Vol. 258, 35 - 40 (1982).
Apparent magnitudes of type I supernovae (SN I) and distances to their parent galaxies as determined by de Vaucouleurs are used to investigate the relation between absolute magnitude and the light curve decay rate as defined by Pskovskii. The range in absolute magnitude among SN I is estimated to be $\sim$2.0 mag, the mean internal extinction of SN I in spiral galaxies is 0.5 mag, and the adopted distance scale is found to be nonlinear, in the sense of being too compressed, in the interval $24.1 \leqslant \mu_0 \leqslant 31.5$.

125.053 **The fate of dust grains in a shock wave originated by a SN explosion.**
S. D'Odorico, A. F. M. Moorwood.
Messenger, No. 28, p. 29 - 30 (1982).

125.054 **What can we learn from UV observations of supernovae?** C. Fransson.
Rep. Obs. Lund, No. 18, (see 012.044), p. 105 - 108 (1982).

Creation of a supernova database.
See Abstr. 002.044.

A mean-field calculation of the equation of state of supernova matter. See Abstr. 061.032.

The equation of state of hot dense matter and supernovae. See Abstr. 061.050.

Accreting white dwarf models for Type I supernovae. I. Presupernova evolution and triggering mechanisms.
See Abstr. 065.008.

Helium shell flashes and evolution of accreting white dwarfs. See Abstr. 065.034.

Supernova theory. See Abstr. 065.038.

Stellar core collapse: II. Inner core bounce and shock propagation. See Abstr. 065.043.

An exploding 10 $M_\odot$ star: a model for the Crab supernova. See Abstr. 065.065.

Report of IAU Commission 35: Stellar constitution (Constitution des étoiles). See Abstr. 065.072.

Does the galactic synchrotron radio background originate in old supernova remnants? See Abstr. 066.005.

Oxygen isotopes, rare-earth elements, supernovae and Wolf-Rayet stars. See Abstr. 105.043.

A new model for white dwarf supernovae.
See Abstr. 126.013.

Neutral hydrogen towards Tycho's supernova remnant. See Abstr. 131.096.

Report of IAU Commission 34: Interstellar matter and planetary nebulae (Matière interstellaire et nébuleuses planétaires). See Abstr. 131.278.

Galactic ring nebulae associated with Wolf-Rayet stars. VII. The nebula G2.4 + 1.4. See Abstr. 134.004.

The nearest stellar wind bubble or fossil supernova remnant. See Abstr. 134.014.

Star counts in the Lagoon, Trifid and W28 complex.
See Abstr. 134.019.

The Crab Nebula. I. Spectrophotometry of the filaments. See Abstr. 134.040.

The Crab Nebula. II. A photoionization model analysis for the filaments. See Abstr. 134.041.

H I absorption distances to four point sources near supernova remnants. See Abstr. 141.015.

Further observations of radio sources from the BG survey. I. The non-thermal sources near l = 94°.
See Abstr. 141.017.

Radio and X-ray observations of compact sources in or near supernova remnants. See Abstr. 141.051.

A continuum study of galactic radio sources in the constellation of Monoceros. See Abstr. 141.213.

G127.11 + 0.54 – the compact radio source at the center of the SNR G127.1 + 0.5. See Abstr. 141.230.

A new, fast X-ray pulsar in the supernova remnant MSH 15 - 52. See Abstr. 142.041.

Could the cosmic-ray knee at 10^{15} eV be due to the finite time for supernova shock acceleration? See Abstr. 143.028.

Spiral arms and a supernova-dominated interstellar medium. See Abstr. 151.005.

Radio emission from supernova remnants in the galaxy M 33. See Abstr. 158.018.

Abundance gradients in M31: comparison of results from supernova remnants and H II regions. See Abstr. 158.037.

Redshifts of parent galaxies of supernovae. See Abstr. 158.264.

Redshifts of parent galaxies of supernovae. See Abstr. 158.299.

The cosmic distance scale: methods for determining the distance to supernovae. See Abstr. 162.017.

Supernova in NGC 4536

125.101 Interpretation of the maximum light spectrum of a type I supernova.
D. Branch, R. Buta, S. W. Falk, M. L. McCall,
P. G. Sutherland, A. Uomoto, J. C. Wheeler, B. J. Wills.
Astrophys. J., Lett., Vol. 252, L61 - L64 (1982).

A high quality optical spectrum of the 1981 type I supernova in NGC 4536 at maximum light is well represented by a synthetic spectrum consisting of resonance scattering lines of Ca II, Si II, S II, Mg II, and O I, superposed on a continuum. The assumption of LTE at the photosphere leads to relative abundances of Ca, Si, S, Mg, and O that are consistent with solar abundances for a range of continuum temperatures.

125.102 Photometric observations of the supernova in NGC 4536. D. Yu. Tsvetkov.
Pis'ma Astron. Zh., Tom 8, 219 - 221 (1982). In Russian. English translation in Soviet Astron. Lett., Vol. 8.

Results of photographic and photoelectric observations of the supernova 1981 in NGC 4536 are presented. Parameters of light curves, color excess and absolute magnitude at maximum light are determined.

Supernova in NGC 7704

125.121 Supernova in NGC 7704.
L. K. Erastova.
Astron. Tsirk., No. 1188, p. 6 - 7 (1981). In Russian.

Supernova in NGC 6946 (1980k)

125.201 Detection of X-rays during the outburst of supernova 1980k.

C. R. Canizares, G. A. Kriss, E. D. Feigelson.
Astrophys. J., Lett., Vol. 253, L17 - L21 (1982).

The authors have detected X-ray emission from SN 1980k in NGC 6946 ~ 35 days after maximum light using the IPC on the Einstein Observatory. The absorption corrected X-ray flux of SN 1980k was ~ 0.03 μJy at 0.24×10^{18} Hz (1 keV), corresponding to a luminosity of ~ 2×10^{39} ergs s^{-1} (0.2 - 4 keV) at 10 Mpc. This is 3500 times weaker than the optical luminosity at that time. The data are compatible with a thermal X-ray spectrum with $kT > 0.5$ keV or a power-law energy spectrum with index > -3.

125.202 Upper mass limit for the stellar progenitor to the 1980k supernova in NGC 6946.
L. A. Thompson.
Astrophys. J., Lett., Vol. 257, L63 - L66 (1982).

A deep photograph of the nearby spiral galaxy NGC 6946 was taken 49 days before maximum light of the galaxy's recent Type II supernova explosion. Since there is no detectable image at the position of the supernova, an upper limit can be placed on the mass of the stellar progenitor. Although the mass estimate is somewhat uncertain due to the lack of a faint stellar sequence, the progenitor was probably $\lesssim 18 \, M_\odot$.

125.203 The evolution of the infrared spectrum of the 1980k supernova in NGC 6946.
E. Dwek, M. F. A'Hearn, E. E. Becklin, R. W. Capps,
C. M. Telesco, A. T. Tokunaga, C. G. Wynn-Williams,
H. L. Dinerstein, M. W. Werner, I. Gatley.
Bull. American Astron. Soc., Vol. 13, 795 (1981). – Abstract.

Extragalactic radio supernovae in NGC 4321 and NGC 6946. See Abstr. 125.301.

Radio supernovae in NGC 4321 and NGC 6946. See Abstr. 125.302.

The interstellar spectrum of the supernova 1980k in NGC 6946. See Abstr. 131.128.

Supernova in NGC 4321 (1979c)

125.301 Extragalactic radio supernovae in NGC 4321 and NGC 6946. R. A. Sramek, K. W. Weiler,
J. M. van der Hulst.
Extragalactic radio sources, (see 012.025), p. 391 - 392 (1982).

The supernovae SN1979c in NGC 4321 and SN1980k in NGC 6946 have both been detected at centimeter wavelengths at the VLA. The radio emission turns on very rapidly, but may be delayed by as much as a year with respect to the optical outburst. In both supernovae, the 20 cm radiation peaks after the 6 cm, and the radio emission has a very slow post-maximum decay.

125.302 Radio supernovae in NGC 4321 and NGC 6946.
R. A. Sramek, J. M. van der Hulst, K. W. Weiler.
Bull. American Astron. Soc., Vol. 13, 795 (1981). – Abstract.

Supernova in NGC 1316

125.401 BV photometry of supernova in NGC 1316.
E. W. Olszewski.
Inf. Bull. Variable Stars, No. 2065 (1982).

Supernova in MCG−5−28−17

125.501 **Possible supernova in MCG −5−28−17.**
IAU Circ., No. 3661 (1982).

Supernova in NGC 2268

125.601 **Supernova in NGC 2268.**
IAU Circ., Nos. 3667, 3671, 3678 (1982).

Supernova in NGC 5679

125.701 **Supernova in NGC 5679A, B.**
IAU Circ., No. 3681 (1982).

Supernova in NGC 4185

125.801 **Supernova in NGC 4185.**
IAU Circ., No. 3683 (1982).

Supernova in NGC 4490

125.901 **Supernova in NGC 4490.**
IAU Circ., Nos. 3689, 3690 (1982).

126 Low-luminosity Stars, Subdwarfs, White Dwarfs, Degenerate Stars

126.001 **Steady nuclear burning on white dwarfs.**
J. C. B. Papaloizou, J. E. Pringle, J. MacDonald.
Mon. Not. R. Astron. Soc., Vol. 198, 215 - 220 (1982).
The authors investigate the steady nuclear burning of accreted hydrogen on a white dwarf. For a given CNO-abun-
dance there is a maximum luminosity for which steady burning is possible. For normal solar abundances the maximum luminosity is $\sim 5 \times 10^{-2} L_{\odot}$ corresponding to an accretion rate of
$\sim 8 \times 10^{-13} M_{\odot} \text{yr}^{-1}$.

126.002 **The interpretation of the spectra of two magnetic
degenerates.**   J. L. Greenstein, J. B. Oke.
Astrophys. J., Vol. 252, 285 - 295 (1982).
Ultraviolet and optical fluxes have been measured for
two magnetic degenerates, Feige 7 (Gr 267) and
Grw +70°8247 (EG 129), from 10,000 to 1150 Å. A method
is developed for fitting blackbodies and models to this wide
range of fluxes. Three hypotheses are advanced to explain the
flux deviations from smoothness: (1) cyclotron absorptions
at 4 to 8×10^8 gauss; (2) stationary Zeeman components of
hydrogen in a larger field; (3) C I absorption lines and a continuum.

126.003 **The complex emission-line structure in the magnetic
white dwarf binary 2A 0311−227 (EF Eridani).**
J. B. Hutchings, A. P. Cowley, D. Crampton, W. A. Fisher,
M. H. Liller.
Astrophys. J., Vol. 252, 690 - 696 (1982).
All phases of the 81 minute orbit of the magnetic X-ray
binary 2A 0311−227 are covered in a series of single-trailed
spectrograms of duration 25 minutes each, and resolution
2 Å. Superposition of the data shows that an extremely complex line-structure and variation exists, which is phase-locked
with the orbit. An improved value for the orbital period is
derived by comparing observations separated by 9 months.
Models for explaining the complex gas streaming are briefly
discussed.

126.004 **Hydrogen-driving and the blue edge of composition-
ally stratified ZZ Ceti star models.**
D. E. Winget, H. M. Van Horn, M. Tassoul, C. J. Hansen,
G. Fontaine, B. W. Carroll.
Astrophys. J., Lett., Vol. 252, L65 - L68 (1982).
The authors report preliminary results of fully nonadia-

batic g-mode pulsation calculations for evolutionary white
dwarf models incorporating stratified H/He/C envelopes.

126.005 **The luminosity function of very low mass stars.**
R. G. Probst, R. W. O'Connell.
Astrophys. J., Lett., Vol. 252, L69 - L72 (1982).
The authors report results from an infrared search for
very low mass stars in binary systems with white dwarf primaries. The search was designed to be complete to $M_v \sim 20.7$. It
indicates that the initial luminosity function declines steeply
for $M_v > 14$ and that the total mass in stars with
$16 \lesssim M_v \lesssim 21$ is $\rho \lesssim 0.005 M_{\odot} \text{pc}^{-3}$ in the solar neighborhood.
It appears unlikely that very low mass stars make a substantial
contribution to the mass of either the disk or massive halo
components of the Galaxy.

126.006 **Infrared photometry of cool white dwarfs.**
D. T. Wickramasinghe, D. A. Allen, M. S. Bessell.
Mon. Not. R. Astron. Soc., Vol. 198, 473 - 482 (1982).
The authors present the results of a search for the effects
of pressure induced H_2 dipole opacity on the infrared JHK
magnitudes of cool white dwarfs. LHS 1126 is found to be a
very cool ($T_e \sim 4250$ K) DC white dwarf with a H rich atmospheric composition dominated by H_2 dipole opacity in the
infrared. The authors' JHK photometry also favours a H rich
atmospheric composition for the DK white dwarfs LP 658-2
and W489. The surprisingly high proportion of hydrogen rich
white dwarfs in the sample appears to suggest that the
mechanism which inhibits the accretion of hydrogen in the
hotter helium stars becomes less effective at low ($T_e \lesssim 5500$ K)
temperatures. The importance of the H_3^+ ion in cool hydrogen
rich white dwarf atmospheres is pointed out and it is
suggested that the opacity due to this ion may be responsible
for the blanketing observed in the U and B magnitudes of
some cool white dwarfs.

126.007 **An analysis of the cyclotron spectrum and polarization properties of VV Puppis.**
D. T. Wickramasinghe, S. M. A. Meggitt.
Mon. Not. R. Astron. Soc., Vol. 198, 975 - 983 (1982).
The authors present an analysis of the cyclotron spectrum
of VV Puppis using a model which allows for polarization and
radiative transfer effects. It is shown that the cyclotron features arise in emission from a region with $T \cong 10 \text{keV}$,

$B = 3.2 \times 10^7$ G and $\Lambda = 10^5$, which emits strongly polarized radiation in the optical. The emission region is physically thin (height $\ll$ cross-sectional radius) and is dominated by cyclotron cooling. There is also evidence for a second component of radiation which is essentially unpolarized and dominates in the blue region of the spectrum.

126.008 On the surface compositions of magnetic white dwarfs.   G. Michaud, G. Fontaine.
Astrophys. J., Lett., Vol. 253, L29 - L32 (1982).

Magnetic white dwarfs have mixed spectra (of hydrogen and helium) much more frequently than nonmagnetic white dwarfs. Many authors have suggested that this is caused by the slowing down of gravitational settling by the magnetic field. This is shown not to be the case. The magnetic field can effectively stop diffusion only at very small optical depths. Instead it is suggested that the magnetic field stops the selective wind of hydrogen postulated by Michaud and Fontaine to explain the near complete absence of hydrogen from nonmagnetic, non-DA white dwarfs.

126.009 C1: a white-dwarf–red-dwarf spectroscopic binary.   H. H. Lanning.
Astrophys. J., Vol. 253, 752 - 755 (1982).

The star C1 is shown to be a spectroscopic binary with a period of 16 hours and a semiamplitude $K = 116$ km s^{-1}. The primary is a DA white dwarf, the secondary is a late main-sequence M star ($\sim$dM2e). A lower limit for the mass of the white dwarf of $\sim 0.4\,M_\odot$ at an inclination of $i \sim 90°$ is derived. Though coincident with the Ursa Major Cluster, C1 is confirmed as a nonmember.

126.010 The hot subdwarfs revisited.   F. Wesemael, D. E. Winget, W. Cabot, H. M. Van Horn, G. Fontaine.
Astrophys. J., Vol. 254, 221 - 231 (1982).

The properties and evolutionary status of the hot B and O subdwarfs are reinvestigated using recent homogeneous grids of hot, high-gravity model atmospheres and detailed convective envelope calculations appropriate to the hot subdwarf region of the H-R diagram discussed by Greenstein and Sargent. An alternative method of analysis of the hot subdwarfs is outlined, where comparisons between observations and theoretical evolutionary tracks are made as close to the fundamental observational results as possible. The properties of stars evolving along the extended horizontal branch are discussed. The consequences of interpreting the sdB/sdO transition as a result of convective mixing along that branch are analyzed.

126.011 The pulsation periods of the pulsating white dwarf G117–B15A.   S. O. Kepler, E. L. Robinson, R. A. Nather, J. T. McGraw.
Astrophys. J., Vol. 254, 676 - 682 (1982).

G117–B15A is a pulsating DA white dwarf, or ZZ Ceti-type star. Using high speed photometry accumulated over the 5 year interval from 1975 to 1980, the authors have disentangled the unusually complex variations of its light curve. G117–B15A has six pulsation modes simultaneously excited with periods of 107.6 s, 119.8 s, 126.2 s, 215.2 s, 271.0 s, and 304.4 s. The 215.2 s pulsation has the largest semiamplitude, 0.022 mag, of the six pulsations and dominates the light curve. The upper limit to the rate of change of the period of this large amplitude pulsation is $|\dot{P}| < 7.8 \times 10^{-14}$ s s^{-1} at the 68% confidence level.

126.012 New spectral types for 33 white dwarfs previously thought to be of classes DC and C_2.
G. Wegner, F. H. Yackovich.
Astron. J., Vol. 87, 155 - 160 (1982).

The visible spectra of 33 white dwarfs have been re-examined with the following results: two are cool narrow-lined normal DA, five have weak He I lines and thus appear to be cool DB stars, 12 show the Swan bands of the C_2 molecule, while five have others such as CH or broad unidentified shallow absorptions in their spectra. One of these latter objects, EG 250, is a known magnetic white dwarf. Out of the original sample of 28 DC stars, only nine now remain with continuous spectra and it is suggested that many DC will have weak features in their spectra if observed using a high enough signal-to-noise ratio.

126.013 A new model for white dwarf supernovae.   R. Canal, J. Isern, J. Labay.
Nature, Vol. 296, 225 - 226 (1982).

The authors show that carbon–oxygen white dwarfs in close binary systems are likely progenitors of neutron stars and type I supernovae (SN I). This conclusion is supported by different effects associated with crystallization of the carbon–oxygen mixture at relatively low temperatures ($T \lesssim 5 \times 10^7$ K) and high densities ($\rho \gtrsim 5 \times 10^9$ g cm^{-3}).

126.014 A spectrophotometric analysis of the hot helium-rich white dwarf HD 149499 B.
E. M. Sion, E. F. Guinan, F. Wesemael.
Astrophys. J., Vol. 255, 232 - 239 (1982).

A comprehensive analysis is presented of the hot helium-rich white dwarf HD 149499 B based upon IUE ultraviolet spectra along with available optical spectra and photometry. Line strengths, line profiles, and continuum fluxes are analyzed in terms of a new grid of hot, high-gravity, mixed-composition stellar atmospheres. The authors derive the following parameters for HD 149499 B: $T_e = 55{,}000$ K$\pm$ 5000 K; log g near 8; log (N_{He}/N_H) $\geqslant 0$, with a value as large as 4 possible; $M_v = +8.95$; $L = 1.35\,L_\odot$; log ($R/R_\odot$) $= -1.89$. They thus confirm Wegner's classification of HD 149499 B as a DO star.

126.015 The influence of the C_2 absorption bands on the U, B, V magnitudes of carbon white dwarfs.
F. Durret, G. Vauclair.
Astron. Astrophys., Vol. 106, 67 - 69 (1982).

The blanketing due to C_2 absorption bands in two carbon white dwarfs (G 47−18 and G 99−37) is shown to affect both B and V colors almost identically. The position of these two stars in the M_V–$(B-V)$ plane is therefore not affected by color blanketing. The color-absolute magnitude relation is critically examined for carbon white dwarfs with trigonometric parallaxes. Two DC white dwarfs for which ultraviolet observations have revealed strong absorption due to carbon have been added to the calibration. The validity of such a calibration is discussed.

126.016 Sirius B: une naine blanche encore méconnue.   F. Wesemael, G. Fontaine.
J. R. Astron. Soc. Canada, Vol. 76, 35 - 49 (1982).

The authors review our knowledge of Sirius B. They discuss in detail recent observations of the white dwarf in the ultraviolet, as well as the theoretical problems posed by the interpretation of X-ray observations of the system.

126.017 The O type subdwarf ROB 162 in the globular cluster NGC 6397.
V. Caloi, V. Castellani, N. Panagia.
Astron. Astrophys., Vol. 107, 145 - 147 (1982).

The far UV spectrum of the O type subdwarf in NGC 6397 is discussed together with available information in the visible range. A substantial discrepancy is found between the determinations of the effective temperature from the continuum and the H-spectrum on the one side, and the He-lines on the other.

126.018 The peculiar structure of very cool non-DA white dwarf atmospheres.   S. Kapranidis, K. H. Böhm.

Astrophys. J., Vol. 256, 227 - 233 (1982).

Model atmospheres of very cool ($T_{eff} < 4000$ K) non-DA degenerate stars have a rather peculiar structure provided the influence of trace elements on the equation of state and the opacity can be neglected. Below an almost completely transparent outermost layer the actual photosphere always appears just in those layers in which pressure ionization sets in and the resulting free electrons permit a buildup of opacity. The total density variation in such a photosphere is much smaller than in other stellar atmospheres. As a prototype atmosphere of this type the authors have calculated a flux constant model for T_{eff} = 2750 K, $g = 10^8$ cm s^{-2} in which the radiative, convective, and conductive flux are taken into account simultaneously, and all three turn out to be important. Pressure effects in the equation of state and the opacities are included. The resulting frequency distribution of the surface flux and the infrared colors are discussed.

126.019 Non-LTE analysis of subluminous O-stars. II. The hydrogen-deficient subdwarf O-star HD 127493.
K. P. Simon.
Astron. Astrophys., Vol. 107, 313 - 319 (1982).

A detailed analysis of the sdO HD 127493 (= CD 22° 10510 = BD −22° 3804) is performed by means of non-LTE model atmospheres and non-LTE line formation computations. The fit of the equivalent widths of 14 lines and blends and ten profiles of H, He I and He II determines the parameters of the stellar atmosphere. Suggestions are made for the evolutionary state of field subdwarf O-stars.

126.020 Extremely metal-poor subdwarfs.   T. Gehren.
Messenger, No. 27, p. 22 - 25 (1982).

126.021 Häufigkeitsanomalien in Weißen Zwergen mit Kohlenstoffbanden.   I. Bues.
Mitt. Astron. Ges., Nr. 55, p. 56 - 58 (1982).

126.022 Spectral analysis of the OB subdwarf HD 149 382.
B. Baschek, R. P. Kudritzki, M. Scholz, K. P. Simon.
Astron. Astrophys., Vol. 108, 387 - 405 (1982).

The ultraviolet (1160–3230 Å) and the blue spectrum of the OB subdwarf HD 149 382 is analyzed. Fitting the observed continuum, hydrogen lines and helium lines to fluxes of unblanketed non-LTE models yields T_{eff} = (35,000 ± 2000) K, $g_s = 10^{5.5 \pm 0.3}$ cm s^{-2}, $\log\epsilon_{He}$ = 10.6±0.2. The microturbulence, estimated from carbon and sulfur lines, is ξ = (5 ± 3) km s^{-1}. LTE analyses of heavy-element lines yield, compared to the sun, deficiencies of carbon and oxygen and normal nitrogen abundance, indicating processing of the material by the CNO cycle. Calcium is probably overabundant, phosphorus and sulfur are normal, aluminum and silicon are deficient. A tentative explanation of the abundance pattern in terms of the diffusion hypothesis is suggested.

126.023 Spectroscopic and photometric observations of white dwarfs.   D. Koester, V. Weidemann.
Astron. Astrophys., Vol. 108, 406 - 411 (1982).

Image dissector scanner observations and ubvy colours are reported for a variety of white dwarfs of spectral types DB, DC, C_2, and DA as well as some white dwarf candidates. G 268−40 is shown to be the third C_2 star with lines of neutral carbon in the visible spectrum.

126.024 Spectrophotometry of the very low luminosity dwarf star RG 0050–2722.
J. Liebert, D. H. Ferguson.
Mon. Not. R. Astron. Soc., Vol. 199, 29P - 30P (1982).

A near-infrared spectrophotometric scan confirms the discovery by Reid and Gilmore of a very low luminosity dwarf star. The spectrum is compared with that of VB 10.

126.025 X-ray and UV radiation from accreting nonmagnetic degenerate dwarfs. II.
N. D. Kylafis, D. Q. Lamb.
Astrophys. J., Suppl. Ser., Vol. 48, 239 - 272 (1982).

The authors report the results of detailed numerical calculations of X-ray and UV emission from accreting nonmagnetic degenerate dwarfs. The results are also valid for magnetic degenerate dwarfs so long as bremsstrahlung cooling is more important than cyclotron cooling. The calculations span the entire range of accretion rates and stellar masses and include the important, but previously unexplored, regime at moderate and high accretion rates. The authors find that the maximum hard X-ray luminosity for degenerate dwarfs undergoing spherical accretion is 2.2 × 10^{36} ergs s^{-1}. The temperature characterizing the X-ray spectra produced by degenerate dwarfs strongly depends not only on the stellar mass but also on the accretion rate. The resulting correlation between spectral temperature and luminosity may be an important signature of degenerate dwarf X-ray sources. The authors apply the calculations to cataclysmic variables such as AM Her and SS Cyg and discuss the implications of the results for Cyg X-2 and the Sco X-1–like sources.

126.026 Búsqueda de estrellas OB débiles en Vela, Crux, Circinus y Norma.
J. C. Muzzio, A. M. Orsatti.
Bol. Asoc. Argentina Astron., No. 20 - 24, p. 261 (1981). Abstract.

126.027 Búsqueda de estrellas OB débiles en la Vía Láctea austral.   J. C. Muzzio, A. M. Orsatti.
Bol. Asoc. Argentina Astron., No. 20 - 24, p. 395 (1981). Abstract.

126.028 Fotometría UBV y H beta de estrellas OB débiles en la Vía Láctea austral.   J. C. Muzzio.
Bol. Asoc. Argentina Astron., No. 20 - 24, p. 401 (1981). Abstract.

126.029 IUE observation of UV absorption in the spectrum of the C_2 white dwarf L1363-3.
G. Vauclair, V. Weidemann, D. Koester.
Astron. Astrophys., Vol. 109, 7 - 9 (1982).

Strong ultraviolet absorption features have been observed in a short wavelength IUE spectrum of the weak carbon white dwarf L1363−3. The spectrum is quite similar to those obtained for the weak C_2 white dwarf L145−141 and the DC white dwarf G33−49 and shows strong absorption at λ1920Å and between λ1520Å and λ1680Å. Identification with carbon absorption features is discussed.

126.030 Thermomagnetic instability in degenerate cores of white dwarfs.   A. Z. Dolginov, V. A. Urpin.
Soobshch. Spets. Astrofiz. Obs., Vyp. 32, (see 012.029), p. 17 - 18 (1981).

126.031 Twin white dwarfs.
IAU Circ., No. 3703 (1982).

126.032 Einstein observations of hot DB white dwarfs.
G. Fontaine, T. Montmerle, G. Michaud.
Astrophys. J., Vol. 257, 695 - 702 (1982).

The authors report the observations of 10 hot DB white dwarfs obtained with the IPC instrument aboard the Einstein Observatory. No flux was detected in nine objects yielding upper limits for the soft X-ray luminosity that range from 2 × 10^{28} to 8 × 10^{28} ergs s^{-1} in the ~0.1–5 keV bandwidth. A possible detection, at the 2.2 σ level, was obtained for GD 205 (WD 1709+23) which corresponds to a luminosity $L_x \sim 5.2-11.6 \times 10^{28}$ erg s^{-1}. For these helium-rich white dwarfs, the present results remain compatible with the models which require the presence of a corona causing hydrogen-rich stellar winds that may explain the strong hydrogen deficiency of these objects.

126.033 **Diffusion of carbon from the cores to the atmospheres of white dwarf stars.** D. O. Muchmore.
Bull. American Astron. Soc., Vol. 13, 810 (1981). − Abstract.

126.034 **High resolution IUE spectra of the hot helium rich white dwarf HD 149499B.**
E. M. Sion, E. F. Guinan.
Bull. American Astron. Soc., Vol. 13, 811 (1981). − Abstract.

126.035 **He II in the spectrum of the hot white dwarf HZ43: photospheric or interstellar?**
R. F. Malina, G. Basri, S. Bowyer.
Bull. American Astron. Soc., Vol. 13, 873 (1981). − Abstract.

126.036 **The sharp interstellar-like features in the spectrum of the nearby white dwarf G191 - B2B.**
F. C. Bruhweiler, Y. Kondo.
Bull. American Astron. Soc., Vol. 13, 924 (1981). − Abstract.

126.037 **On the statistics of ZZ Ceti stars.**
G. Fontaine, J. T. McGraw, D. S. P. Dearborn, J. Gustafson, P. Lacombe.
Prepr. Steward Obs., No. 366, 37 pp. (1982).

126.038 **Spectrophotometry of the very low luminosity dwarf star RG0050-2722.**
J. Liebert, D. H. Ferguson.
Prepr. Steward Obs., No. 370, 6 pp. (1982). − Submitted to Mon. Not. R. Astron. Soc.

126.039 **Search for variability in hydrogen-poor stars. I. Preliminary results of photoelectric observations** for six stars. C. Bartolini, A. Bonifazi, F. D'Antona, F. Fusi Pecci, L. Oculi, A. Piccioni, R. Serra.
Astrophys. Space Sci., Vol. 83, 287 - 310 (1982).
Photoelectric observations of six objects are reported as a first step of a long-term project devoted to search for variability of a large sample of hydrogen-poor stars. The observed stars show phenomena of microvariability with an amplitude of the order of $0.^m1$ or less. Two extreme helium stars have been examined: a period in the range of $0.^d162 - 0.^d164$ has been found for BD + 10°2179, and P = $0.^d1079962$ for BD + 13°3324. The mass-losing O subdwarf (sdO) BD + 37°443 presents short-term fluctuations with a time-scale of several minutes and long-term variations on a scale of months. The sdO star BD + 75°325 is probably non-variable. The high gravity sdO BD + 25°4655, which is very close to the white dwarf stage, also presents variability on a time-scale of about 13 minutes. The variability of the intermediate helium star HD 37776 is confirmed.

126.040 **Time-dependent accretion onto magnetized white dwarfs.**
S. H. Langer, G. Chanmugam, G. Shaviv.
Bull. American Astron. Soc., Vol. 14, 574 (1982). − Abstract.

126.041 **White dwarfs − the dying stars.** D. Koester.
Messenger, No. 28, p. 25 - 28 (1982).

126.042 **Dalle nane bianche ai buchi neri: storia di alcuni concetti di astrofisica relativistica.**
R. Balbinot, R. Bergamini, B. Giorgini.
Coelum, Vol. 50, 125 - 138 (1982);

Catalogue of white dwarfs.
See Abstr. 002.012.

Landolt-Börnstein. See Abstr. 003.064.

Spin-dependent polarizabilities of hydrogenic atoms in magnetic fields of arbitrary strength.
See Abstr. 022.094.

An MHD instability in compact fluid objects.
See Abstr. 062.017.

Evidence for photospheric soft X-ray emission from Sirius B. See Abstr. 064.069.

Time-dependent accretion onto magnetized white dwarfs. See Abstr. 064.082.

More on carbon burning in electron-degenerate matter: within single stars of intermediate mass and within accreting white dwarfs. See Abstr. 065.005.

Helium shell flashes and evolution of accreting white dwarfs. See Abstr. 065.034.

Diffusion and hydrogen shell burning on slowly accreting white dwarfs. See Abstr. 065.035.

On the role of the accretion rate in nova outbursts.
See Abstr. 065.036.

A theory of hydrogen shell flashes on accreting white dwarfs. I. Their progress and the expansion of the envelope. See Abstr. 065.041.

A theory of hydrogen shell flashes on accreting white dwarfs. II. The stable shell burning and the recurrence period of shell flashes. See Abstr. 065.042.

Dalle nane bianche ai buchi neri.
See Abstr. 066.008.

Relativistic ejection from compact stars with a strong magnetic field. See Abstr. 066.519.

Proper motion survey with the forty-eight inch Schmidt telescope. LIV. White dwarfs statistics of the NLTT catalogue. See Abstr. 111.029.

Radio emission from AM Herculis-type binaries.
See Abstr. 116.011.

A first look at the eclipsing cataclysmic variable Lanning 10. See Abstr. 117.007.

The flicker spectrum of AE Aquarii.
See Abstr. 117.010.

Two-second variability in AM Herculis binaries.
See Abstr. 117.092.

An ultraviolet study of the white dwarf and K2V components of the close eclipsing binary V471 Tauri.
See Abstr. 117.105.

IUE observations of the pulsating white dwarf G29 - 28. See Abstr. 122.153.

The Schweizer-Middleditch star: not a stellar remnant of SN 1006. See Abstr. 125.014.

Accreting white dwarf models for type I supernovae. II. Off-center detonation supernovae. See Abstr. 125.035.

Zentralsterne Planetarischer Nebel und Subdwarf-O-Sterne. See Abstr. 135.016.

Faint blue objects at high galactic latitude. II. Palomar Schmidt field centered on selected area 29. See Abstr. 141.064.

Faint blue objects at high galactic latitude. III. Palomar Schmidt field centered on Selected Area 28. See Abstr. 141.290.

Masses of white dwarf progenitors from open cluster studies.   See Abstr. 153.006.

Discussion of a deep blink survey of faint red objects toward the south galactic pole.   See Abstr. 155.003.

Interstellar Matter, Nebulae

131 Interstellar Matter, Star Formation

131.001 Galactic H I and dust in the region $l^{II} = 301°$, $-18° > b > -31°$.
D. J. King, K. N. R. Taylor, M. I. Darby.
Mon. Not. R. Astron. Soc., Vol. 198, 255 - 258 (1982).
    The results of an optical survey of nebulosity at the South Celestial Pole have been combined with the data from a high resolution H I survey to show the presence of an interesting correlation between the gas and dust in this region. A dust filament is shown to run parallel to, but displaced towards the galactic plane from, a ridge of enhanced H I emission. It is suggested that this results from the passage of a shock wave through the system, resulting in the spatial displacement of H I to higher velocities but leaving the dust grains behind presumably through coupling to the local magnetic field in this area.

131.002 OH observations of IRS 1 in NGC 7538.
    H. R. Dickel, A. H. Rots, W. M. Goss, J. R. Forster.
Mon. Not. R. Astron. Soc., Vol. 198, 265 - 272 (1982).
    The OH masers associated with IRS 1 in NGC 7538 have been observed with the Very Large Array at a spatial resolution of 1.2 arcsec and a velocity resolution of 0.65 km s⁻¹. At 1720 MHz, two components are found to lie on the southern edge of the ultracompact H II region. Three spatial components are measured at 1665 MHz: one is on the northern edge of IRS 1, one is on the western edge and the third is midway between the centre and the 1720-MHz OH masers to the south. The 1720-MHz and 1665-MHz OH masers, together with the two H_2CO masers, lie roughly on a ring which is centred on IRS 1 and whose outer radius is ~ 0.4 arcsec. The positions and velocities of the maser spots are interpreted in terms of a simple shell model.

131.003 An ultraviolet study of high velocity interstellar lines in the Carina Nebula.
N. R. Walborn, J. E. Hesser.
Astrophys. J., Vol. 252, 156 - 171 (1982).
    With high resolution observations from the International Ultraviolet Explorer, an analysis has been made of interstellar absorption lines in the spectra of stars within the Carina Nebula (NGC 3372). The high velocity structure is most pronounced in the low ionization ultraviolet lines (single ionized metals and O I). The extremely strong Mg II and C II, C II* lines reveal a number of new components with velocities even higher than those seen optically, and the total velocity range observed is 550 km s⁻¹. The velocities correspond to temperatures of a few million degrees, suggesting a relationship between the interstellar line motions and the surprising Einstein Observatory X-ray results in this region.

131.004 The Zeeman effect in 21 centimeter line radiation: methods and initial results.
T. H. Troland, C. Heiles.
Astrophys. J., Vol. 252, 179 - 192 (1982).
    The authors report upon a search for the Zeeman effect in the 21 cm line of galactic neutral hydrogen (H I) using the 26 m radio telescope at Hat Creek Observatory. They have carefully investigated the causes and remedies for instrumental circular polarization. The field strengths derived for H I regions

are low, and magnetic energy densities are smaller than energy densities associated with turbulent or other macroscopic motions but larger than those of thermal motions.

131.005 Is the degree of ionization always relevant for ambipolar diffusion in interstellar clouds?
T. C. Mouschovias.
Astrophys. J., Vol. 252, 193 - 195 (1982).
    The author reexamines the well-known result that ambipolar diffusion proceeds faster at smaller degrees of ionization. He concludes that, in interstellar clouds of moderate density, the degree of ionization is not relevant to ambipolar diffusion. In much denser clouds in which gravitational forces can dominate, the degree of ionization does not uniquely determine the rate at which ambipolar diffusion progresses; the specification of two additional dimensionless parameters is necessary. The underlying physics is discussed.

131.006 A new interstellar component in the spectrum of HD 72127 A.
L. M. Hobbs, G. Wallerstein, E. M. Hu.
Astrophys. J., Lett., Vol. 252, L17 - L20, plate L1 (1982).
    New high-dispersion observations are reported of the very strong, broad interstellar K line of Ca II in the spectrum of HD 72127 A, a star located near a filament of the Vela supernova remnant. When compared with similar observations made in 1977, the new data reveal temporal variability of the interstellar absorption. All five Ca II components which were seen in both years show very large column-density ratios $N(Ca\ II)/N(Na\ I) \geqslant 9$. Marked differences in the structure of the K line between the two early-type components of this binary star are confirmed.

131.007 The interstellar carbon abundance toward Delta Scorpii.
L. M. Hobbs, D. G. York, W. Oegerle.
Astrophys. J., Lett., Vol. 252, L21 - L23 (1982).
    The spin-forbidden interstellar absorption line of C II at $\lambda_{air} = 2324.69$ Å is detected in the spectrum of δ Scorpii, with an equivalent width $W_\lambda = 0.6 \pm 0.3$ (2 σ) mÅ. The resulting interstellar abundance of gaseous carbon, derived from a newly calculated oscillator strength $f = 6.7 \times 10^{-8}$, is smaller than the solar carbon abundance by a factor of 3. This deficiency can provide most of the mass in the grains required to explain the corresponding interstellar extinction.

131.008 Thermal SiO as a probe of high velocity motions in regions of star formation.   D. Downes, R. Genzel, Å. Hjalmarson, L. Å. Nyman, B. Rönnäng.
Astrophys. J., Lett., Vol. 252, L29 - L33 (1982).
    New observations of the $v = 0, J = 2 \to 1$ line of SiO at 86.8 GHz show a close association of the thermal SiO emission and infrared and maser sources in regions of star formation. In addition to SiO emission with low velocity dispersion ($\Delta v \leqslant 7$ km s⁻¹), the authors report the first detection of high velocity ("plateau") emission toward W49 and W51. The low velocity SiO component may come from the core of the molecular cloud which contains the infrared and maser

sources. The "plateau" may indicate mass outflow from stars within the infrared clusters.

131.009   Tidal stability of gas clouds in the Large Magellanic Cloud and M101.   L. Blitz, A. E. Glassgold.
Astrophys. J., Vol. 252, 481 - 486 (1982).

The authors examine the tidal stability of the giant atomic gas clouds in the Large Magellanic Cloud (LMC) and in M101. The giant atomic clouds in the Large Magellanic Cloud are found to be unstable against tidal disruption by the gravity of the LMC, but the clouds in M101 are in approximate tidal balance. It is unlikely that there is sufficient unobserved molecular gas to stabilize the LMC clouds, although they may have substantial molecular cores which are tidally stable. The time scale for tidal disruption of the LMC clouds is of the order of $2-3 \times 10^7$ yr. Because most of the clouds have associated H II regions, the onset of star formation in these clouds must occur rapidly.

131.010   Motions of the cloud medium behind large scale galactic shocks.   C. Yuan, C. Y. Wang.
Astrophys. J., Vol. 252, 508 - 523 (1982).

Mechanisms of decelerating the cloud medium in a large-scale galactic shock are studied. It is shown that the process of cloud-cloud collisions, which results in diffusive momentum transport and hence gives rise to turbulent viscosity, is very effective in slowing down the cloud medium to the postshock velocity of the intercloud medium. The drag force exerted by the slow-moving intercloud medium alone cannot effectively decelerate the cloud medium in the shock. The internal structure of the shock of the cloud medium is analyzed and its thickness is found to be of the order of 100 pc.

131.011   Collapse models for dark interstellar clouds.
K. R. Villere, D. C. Black.
Astrophys. J., Vol. 252, 524 - 528 (1982).

Properties of self-consistent numerical hydrodynamic models are compared with observed properties of several dark clouds. The results are consistent with the view that these clouds are undergoing gravitational collapse. The clouds appear to have ages comparable to their free-fall times. Derived cloud masses range between 10 and $10^3 M_\odot$.

131.012   The evaporation of spherical clouds in a hot gas. III. Suprathermal evaporation.
S. A. Balbus, C. F. McKee.
Astrophys. J., Vol. 252, 529 - 552 (1982).

The authors present a model for the evaporation of spherical clouds embedded in a very hot, tenuous ambient medium in the regime where the range of the hot electrons is larger than the outflow column density. The hot gas is treated as a freely permeating suprathermal component of the evaporating gas, interacting thermally via Coulomb collisions. The authors have included the effects of diffusive and saturated thermal conduction. Magnetic effects have been ignored.

131.013   The spectral dependence of dust emissivity at millimeter wavelengths.   P. R. Schwartz.
Astrophys. J., Vol. 252, 589 - 593 (1982).

Millimeter wavelength ($\nu \leqslant 300$ GHz) observations of the three objects NGC 2264, S 140, and NGC 6334N are analyzed and the power-law spectral dependence, β, of dust emissivity derived. β is found to be steeper than quadratic, and it is concluded that silicate core grains with thin ice mantles are a likely grain material.

131.014   The diffuse interstellar bands. V. High-resolution observations.   G. H. Herbig, D. R. Soderblom.
Astrophys. J., Vol. 252, 610 - 615 (1982) = Lick Obs. Bull., No. 899.

The well-known diffuse interstellar bands $\lambda\lambda 6195, 6613$

and two new features at $\lambda\lambda 6993, 7223$ have been observed with an echelle scanner in several reddened B-type and early A-type supergiants. At spectral resolutions of 2.6 to 4.6 km s^{-1} there is no indication of breakdown of any of these four diffuse bands into fine structure. These high-resolution data do not permit a clear choice between the free molecule and the very small particle views of the carrier of the diffuse band spectrum. None of these features show the shortward absorption wings predicted by the Purcell-Shapiro theory of an impurity embedded in grains of a single size and concentration.

131.015   Gamma ray emission from interstellar clouds: a plasma physical process capable of enhancing electron fluxes.   G. E. Morfill.
Mon. Not. R. Astron. Soc., Vol. 198, 583 - 588 (1982).

Cos B measurements suggest that electrons play a dominant role in γ-ray production in our Galaxy. A process is discussed which can significantly enhance relativistic electron intensities inside molecular clouds. If this process is as common in our Galaxy as the theory suggests, bremsstrahlung emission will be significantly enhanced too.

131.016   Interstellar polarization in the immediate solar neighbourhood.   J. Tinbergen.
Astron. Astrophys., Vol. 105, 53 - 64 (1982).

About 180 stars within 35 pc have been observed for interstellar linear polarization with a precision of about 7×10^{-5} (degree of polarization). The results, combined with those of Piirola (1977), establish the following points: (1) Within 35 pc of the Sun the dust content is very low indeed: visual extinction over 35 pc is $A_v = 0.002$ mag or less. (2) The polarizations are inconsistent with a uniform magnetic field throughout the 35-pc sphere observed. (3) There is evidence for a region with a very regular field, near $l = 0°$, $b = -20°$. Its distance is less than 20 pc, its angular extent $30° - 60°$. (4) Six stars in the sample are not suitable for use as zero-polarization standards. (5) Stars of spectral type F0 and later are suspected of intrinsic linear polarization at the 10^{-4} level. This polarization seems to be variable in time.

131.017   The Lick galaxy counts, the local interstellar absorption and molecular hydrogen.
A. W. Strong, F. Lebrun.
Astron. Astrophys., Vol. 105, 159 - 163 (1982).

An interpretation of the relationships between the Lick galaxy counts and atmospheric and galactic extinction leads to a simple picture in which apparent variations in the gas-to-dust ratio are mainly due to molecular hydrogen. Parameters entering into these relationships are re-evaluated. Finally this allows to map the large-scale distribution of molecular hydrogen in the local interstellar medium.

131.018   Interstellar extinction in the Perseus arm.
D. H. Morgan, A. McLachlan, K. Nandy.
Mon. Not. R. Astron. Soc., Vol. 198, 779 - 785 (1982).

Individual UV extinction curves for nine stars located in the direction of h and χ Persei in the distance range 0.6 - 4.4 kpc are presented. The sample has been enlarged by the use of available UV and visible data for 32 reddened stars with distances $r \geqslant 2$ kpc located in the same direction. It is found that the mean interstellar extinction curve for the stars located in the Perseus arm shows significantly higher extinction shortward of 2000 Å than the galactic mean.

131.019   Interstellar masers: some theoretical considerations.
D. Field.
Mon. Not. R. Astron. Soc., Vol. 198, 991 - 1006 (1982).

By applying the methods of quantum electronics, expressions are obtained which allow the calculation of interstellar maser brightness temperatures, and the results are compared with recent observations of W3(OH) at 1665 and 1667 MHz.

The predicted brightness temperatures are in good agreement with the high values ($> 10^{12}$ K) observed. H_2O masers are also briefly considered.

131.020 The energetics of molecular clouds. V. The S37 molecular cloud.
N. J. Evans II, G. N. Blair, D. Nadeau, P. Vanden Bout.
Astrophys. J., Vol. 253, 115 - 130 (1982).

The S37 molecular cloud has been observed at the frequencies of several molecular transitions and at infrared wavelengths from 2 to 125 μm. The molecular cloud coincides with a region of visual obscuration near S37; the temperature and column density peak near the reflection nebulae vdB 118 and 119. Line splitting and possible self-absorption are observed near vdB 119. A plethora of 2 μm sources are observed in the direction of the cloud, but no bright 10 μm sources are seen. Far-infrared emission is observed over an extensive region in the cloud. The molecular cloud has a size exceeding 9 pc and a mass of at least $9 \times 10^3 M_\odot$.

131.021 Carbon and oxygen X-ray line emission from the interstellar medium.
H. W. Schnopper, J. P. Delvaille, R. Rocchia, C. Blondel, C. Cheron, J. C. Christy, R. Ducros, L. Koch, R. Rothenflug.
Astrophys. J., Vol. 253, 131 - 135 (1982).

A rocket-borne system consisting of three lithium-drifted silicon detectors was used to obtain a soft X-ray spectrum (0.3–1.0 keV) from a 1 sr region which includes a portion of the North Polar Spur. Emission lines from carbon (C V, C VI) and oxygen (O VII, O VIII) are clearly present. The spectrum is well fitted by a two-component, modified Kato Model with $T = 1.1 \times 10^6$ K for the local interstellar medium and $T = 3.8 \times 10^6$ K for the North Polar Spur.

131.022 Velocity, reddening, and temperature structure of the H_2 emission in Orion.
N. Z. Scoville, D. N. B. Hall, S. G. Kleinmann, S. T. Ridgway.
Astrophys. J., Vol. 253, 136 - 148 (1982).

The authors present FTS spectra of the H_2 emission in the core of the Orion molecular cloud. The data cover a wavelength region (2.02–2.44 μm) sufficient to encompass six transitions from the H_2 ($v = 1$) state, at high spectral resolution [19 km/s at S(1)] and high spatial resolution (3.''75) at 12 points which sample the entire H_2 emitting region. The spatial distribution of the foreground extinction is derived and $A_{2.1\,\mu m}$ is shown to be much smaller than previously estimated. The total luminosity and mass of emitting H_2 are estimated to be $\sim 150 L_\odot$ and $\sim 2 \times 10^{-2} M_\odot$. The kinematics inferred from the H_2 line profiles are consistent with most of the emission arising from a shock-excited region at the interface between the near side of the plateau source and the surrounding molecular cloud.

131.023 Detection of the torsionally excited state of methanol in Orion A.
F. J. Lovas, R. D. Suenram, L. E. Snyder, J. M. Hollis, R. M. Lees.
Astrophys. J., Vol. 253, 149 - 153 (1982).

The authors report the detection of torsionally excited methanol (CH_3OH, $v_t = 1$) in Orion A. Three emission lines have been observed in the region of 93 GHz to 100 GHz. These coincide with laboratory measurements for the $1_0-2_1 E$, $6_1-5_0 E$ and blended $2_1-1_1 E$ and $2_0-1_0 E$ transitions of CH_3OH in its torsionally excited state which lies near 200 cm^{-1} (~ 290 K) above the ground state.

131.024 The motion and distribution of the vibrationally excited H_2 in the Orion molecular cloud.
D. Nadeau, T. R. Geballe, G. Neugebauer.
Astrophys. J., Vol. 253, 154 - 166 (1982).

Observations of line profiles of vibrationally excited H_2 gas in the Orion molecular cloud are presented. The $v = 1 \rightarrow 0 S(1)$, $v = 1 \rightarrow 0 S(0)$ and $v = 2 \rightarrow 1 S(1)$ lines,

emitted at wavelengths near 2 μm, have been observed with a spectral resolution of 20 km s^{-1}. The region has been mapped extensively in the $v = 1 \rightarrow 0 S(1)$ line with spatial resolutions of 10'' and 5'', and the line has been monitored at a few positions over a period of 15 months. The results are interpreted by a model of a radially expanding flow of gas colliding with the surrounding molecular cloud. The H_2 line emission is compared to the CO and H_2O maser emission and to the infrared continuum sources.

131.025 The reliability of finite difference and particle methods for fragmentation problems.
R. A. Gingold, J. J. Monaghan.
Mon. Not. R. Astron. Soc., Vol. 199, 115 - 119 (1982).

The authors reply to criticisms made by Boss & Bodenheimer of the use of the SPH particle method for fragmentation problems.

131.026 Observations of the $J = 2 \rightarrow 1$ CO line in molecular clouds near compact H^+ regions.
P. W. Riley, L. T. Little, A. T. Brown, R. E. Hills, R. Padman, D. Vizard, J. C. G. Lesurf, N. J. Cronin.
Mon. Not. R. Astron. Soc., Vol. 199, 197 - 209 (1982).

Observations of the $J = 2 \rightarrow 1$ transitions of ^{12}CO and ^{13}CO at 230 and 220 GHz in 13 molecular clouds near compact H^+ regions have been made at UKIRT using an uncooled Schottky diode mixer and a digital autocorrelation spectrometer. A comparison between ^{12}CO and ^{13}CO spectra reveals a variety of self-absorption effects. The asymmetry observed in six sources out of ten is most easily explained if the clouds are collapsing; there is no clear evidence for expansion. The ^{13}CO linewidths are systematically wider than those from the NH_3 cores, suggesting that the velocity dispersion in the sources increases with distance from the centre.

131.027 Dense cloud chemistry – I. Direct and indirect effects of grain surface reactions.
T. J. Millar.
Mon. Not. R. Astron. Soc., Vol. 199, 309 - 319 (1982).

A detailed chemical network involving nearly 200 gas-phase reactions and more than 20 grain surface reactions has been used to describe the chemistry of 95 species in dense interstellar clouds. The network is chosen so as to study the chemistry of sulphur- and silicon-bearing species in some detail. The author finds that direct formation on grains is an important production mechanism for many species while for others grain reactions have indirect effects which can dominate the formation processes.

131.028 Fragmentation in rotating interstellar gas clouds.
D. Wood.
Mon. Not. R. Astron. Soc., Vol. 199, 331 - 343 (1982).

A large number of numerical calculations on the collapse and fragmentation of rotating interstellar gas clouds are presented. The initial conditions chosen can be parameterized by their rotational and thermal energies; the calculations covered the entire parameter space available. Collapses from higher thermal energies fragmented through a bar-like mode, whereas collapses from lower thermal energies led to a strong ring-like wave from the axis; the latter collapses fragmented through non-axisymmetric instabilities of a higher wavenumber.

131.029 The galactic fountain, observations of extragalactic radio sources, and the cosmic ray halo.
A. N. Hall.
Mon. Not. R. Astron. Soc., Vol. 199, 355 - 374 (1982).

The author shows that the occurrence of MHD instabilities in the galactic fountain can lead to the suppression of interstellar scintillations of compact extragalactic radio sources, and can also account for the size of the cosmic ray halo, and for the diffusion coefficient of cosmic rays within the halo. He places limits on the parameters describing the conditions in the hot component of the interstellar medium and the galactic fountain, and he finds the rates at which energy in turbulent motions of the gas must be dissipated in order that the various observations may be accounted for.

131.030 $J = 4 \to 3$ HCN observations of M17 SW, DR21(OH), W 51 and NGC 6334.
G. J. White, J. P. Phillips, J. E. Beckman, N. J. Cronin.
Mon. Not. R. Astron. Soc., Vol. 199, 375 - 383 (1982).

The authors report observations of NGC 6334N, M17 SW, W 51 and DR 21(OH) in the $J = 4 \to 3$ transition of HCN at 354 GHz. These results are combined with previously obtained $J = 1 \to 0$ HCN data to determine molecular abundances $X(HCN) \sim 4 \times 10^{-11} - 4 \times 10^{-12}$, and characteristic line excitation zone densities $n(H_2) \sim 10^6$ cm^{-3}. These results are at variance with previous analyses based on $J = 1 \to 0$ results alone, and the predictions of certain models of the cloud chemistry. The reasons for these discrepancies are discussed.

131.031   The isotope abundance ratio [H_2 ^{12}CO]/[H_2 ^{13}CO] towards the Galactic Centre.
F. F. Gardner, J. B. Whiteoak.
Mon. Not. R. Astron. Soc., Vol. 199, 23P - 28P (1982).

Observations of the 5-GHz transitions of H_2 ^{13}CO and H_2 ^{12}CO with a 4.6 arcmin beam at 12 positions near the Galactic Centre indicate that the low isotope abundance ratios [H_2 ^{12}CO]/[H_2 ^{13}CO] found for the massive molecular clouds near Sgr A and Sgr B2 are not atypical; low values are found for all clouds believed to be located near the nucleus. Large-scale H_2 ^{12}CO features show well defined ranges of ratio. These vary from $11-19$ for the feature nearest Sgr A to $23-55$ for the 'molecular ring'.

131.032   Dust clouds of Sagittarius.
D. F. Malin.
Sky Telesc., Vol. 63, 254 - 259 (1982).

131.033   The hunter and the starcloud.   M. M. Waldrop.
Science, Vol. 215, 647 - 650 (1982).

The great nebula in Orion is a tiny bright patch on an immense dark cloud; that cloud is the best place to study star formation.

131.034   Carbon monoxide broad wings and self-reversals in NGC 2071.   S. M. Lichten.
Astrophys. J., Vol. 253, 593 - 600 (1982).

The author reports and analyzes CO spectra of NGC 2071 which exhibit broad wings and prominent self-reversed line profiles. These wings provide evidence for high-velocity bipolar gas flow powered by a strong stellar wind along the axis of a disk. The self-reversed line profiles are due in part to radiative transfer effects in the main body of the cloud, which may be contracting; however, a significant amount of self-absorption also originates in cooler foreground gas at the cloud periphery.

131.035   A study of the diffuse interstellar gas near the Pleiades.   S. R. Federman.
Astrophys. J., Vol. 253, 601 - 605 (1982).

The interstellar gas toward the Pleiades was studied by observing lines of CH, CH$^+$, and K I. New detections of CH and K I, and of CH$^+$ and K I in the directions of 20 Tau and η Tau, respectively, are reported. Evidence for a moderately strong shock of velocity $10-15$ km^{-1} was found for the line of sight toward 20 Tau, where the CH line is blueshifted by $3-4$ km s^{-1} with respect to the CH$^+$ line. A reexamination of the observed distribution of H_2 among its rotational levels indicates that collisions occurring in the shock are largely responsible for populating levels with $J > 2$.

131.036   Runaway expansion of giant shells driven by radiation pressure from field stars.
B. G. Elmegreen, W.-H. Chiang.
Astrophys. J., Vol. 253, 666 - 678 (1982).

Radiation pressure from field stars can exert an outward force on a large shell of gas and dust in the interstellar medium. This radiative force increases with increasing shell size, so a sufficiently large shell can expand at an ever-increasing speed up to a kiloparsec or more in size. The supershells that are observed in our Galaxy and in other galaxies could have originated as smaller shells around OB associations or star complexes and then have grown to their kiloparsec sizes by radiation pressure from background starlight. The formation times for such supershells are between 50 and 100 million years, so the OB associations that initially triggered their growth may not be visible anymore. Background starlight alone can give a giant shell a kinetic energy of 10^{51} ergs or more.

131.037   High spectral and spatial resolution observations of the 12.28 micron emission from H_2 in the Orion Molecular Cloud.   S. C. Beck, E. E. Bloemhof, E. Serabyn, C. H. Townes, A. T. Tokunaga, J. H. Lacy, H. A. Smith.
Astrophys. J., Lett., Vol. 253, L83 - L87 (1982).

The pure rotational $S(2)$ line of molecular hydrogen at 12.28 μm has been looked for in 44 positions in the Orion Molecular Cloud with 6″ beams and 35 km s^{-1} spectral resolution; it was detected in 27 positions. Emission has been observed over a velocity range of $\pm$ 100 km s^{-1}. The lines are approximately symmetric and have full widths at half-maximum ranging from 100 km s^{-1} down to the resolution limit. The distribution of intensities and line shapes is largely consistent with that seen in the 2 μm hydrogen transitions. However, unexpectedly complex line profiles and point-to-point variations in line shapes appear, particularly in the region near IRc9.

131.038   The local interstellar medium.
R. M. Crutcher.
Astrophys. J., Vol. 254, 82 - 87 (1982).

Analysis of the velocities of optical interstellar lines shows that the Sun is immersed in a coherently moving local interstellar medium whose velocity vector agrees with that of the interstellar wind observed through backscatter of solar H Lyα and He λ584 photons. The local interstellar medium consists of both cool clouds and warm intercloud medium gas, has a mass of perhaps $\sim$ 30 $M_\odot$, does not have severe depletion of trace elements from the gas phase, and appears to be material which has been shocked and accelerated by stellar winds and supernovae associated with the Sco-Oph OB association.

131.039   Observations of interstellar zinc.
D. G. York, M. Jura.
Astrophys. J., Vol. 254, 88 - 93 (1982).

The authors have performed IUE observations of interstellar zinc toward 10 stars. They find that zinc is, at most, only slightly depleted in the interstellar medium; its abundance may serve as a tracer of the true metallicity in the gas. The local interstellar medium has abundances that apparently are homogeneous to within a factor of 2 when integrated over paths of about 500 pc, and this result is important for understanding the history of nucleosynthesis in the solar neighborhood.

131.040   Detection of the $N = 3-2$ transition of CCH in Orion and determination of the molecular rotational constants.   L. M. Ziurys, R. J. Saykally, R. L. Plambeck, N. R. Erickson.
Astrophys. J., Vol. 254, 94 - 99 (1982).

The authors report the detection of five hyperfine components of the $N = 3-2$ transition of CCH in Orion A. From an analysis of this data combined with $N = 1-0$ data from previous observations, the rotational constants (B_0 and D_0) were determined, and the values of the fine structure (γ) and hyperfine constants (b, c) were improved. The frequencies of additional rotational transitions of CCH up to $N = 5-4$ are predicted. A rotational temperature and column density are estimated on the basis of the 3−2 data.

131.041 The $^{12}CO/^{13}CO$ abundance ratio toward ζ Ophiuchi.
P. G. Wannier, A. A. Penzias, E. B. Jenkins.
Astrophys. J., Vol. 254, 100 - 107 (1982).

A system of ultraviolet absorption lines has been used to determine the $^{12}CO/^{13}CO$ abundance ratio toward ζ Ophiuchi as well as certain physical properties of the interstellar material. The Copernicus and IUE observations imply a rather low CO isotope ratio of 55 ± 11 and a ^{12}CO column density of $(2.4 ± 0.3) \times 10^{15}$ cm^{-2}. Further analysis implies a rather small velocity dispersion of 0.46 ± 0.15 km s^{-1} and a kinetic temperature in excess of 70 K.

131.042 Interstellar C_2 molecules toward Zeta Ophiuchi.
L. M. Hobbs, B. Campbell.
Astrophys. J., Vol. 254, 108 - 110 (1982).

Ten weak interstellar absorption lines of the (2−0) Phillips band of C_2 near λ8760 are detected in the spectrum of ζ Ophiuchi. All of the lines have equivalent widths $W_\lambda < 2$ mÅ and originate from the six lowest rotational levels of C_2. The resulting total column density is $N(C_2) \approx 1.7 \times 10^{13}$ cm^{-2}, and the excitation temperature is $T = 130 ± 10$ K.

131.043 Rotational temperatures of cyanodiacetylene in Sagittarius B2, TMC-1, and IRC + 10216.
D. E. Jennings, K. Fox.
Astrophys. J., Vol. 254, 111 - 115 (1982).

Four consecutive HC_5N transitions $J = 7−6, 8−7, 9−8$, and 10−9 have been recorded for each of the three sources Sgr B2, TMC 1, and IRC + 10216. The observed line brightness temperatures have been used to derive HC_5N rotational temperatures: 28 K for Sgr B2, 5.5 K for TMC-1, and 13 K for IRC + 10216. These are consistent with temperatures previously estimated for these molecular clouds. The $J = 55−54$ transition of $HC_{11}N$ was searched in TMC-1 but was not found.

131.044 A model of Taurus Molecular Cloud 1 based on HC_3N observations.
L. W. Avery, J. M. MacLeod, N. W. Broten.
Astrophys. J., Vol. 254, 116 - 125 (1982).

The authors have observed five rotational transitions of cyanoacetylene in the dark cloud TMC 1 and, using a Monte Carlo radiative transfer code, have produced a model of this source which is consistent with these observations. The interaction between radiative trapping and population equilibrium was solved for the 15 lowest rotational levels of HC_3N, assuming a cylindrical geometry for the source. The model, representing a compact condensation embedded well within Heiles' Cloud 2, is characterized by a relatively high excitation core and a lower excitation halo. The column density of HC_3N in TMC 1 is estimated to be 1.1×10^{14} cm^{-2}.

131.045 Isotope ratios in interstellar formaldehyde from 6 centimeter observations.
M. L. Kutner, D. E. Machnik, K. D. Tucker, W. Massano.
Astrophys. J., Vol. 254, 538 - 542 (1982).

The authors report observations of 6 cm absorption by $H_2^{13}CO$ and $H_2C^{18}O$ in Sgr A, Sgr B2, W 51, W 33, and Ori B. These provide more accurate isotope ratios than have been previously reported because both isotopes have been observed in the same manner, and the continuum sources were monitored on a regular basis. The authors find a double ratio, $H_2^{13}CO/H_2C^{18}O$, close to the terrestrial value of 5.6 in the galactic plane sources, and enhanced to 7.6 in the galactic center sources.

131.046 Near-infrared spectroscopy of moderate luminosity sources: OMC-2 IRS 3 and IRS 4.
H. A. Thronson, Jr., R. I. Thompson.

Astrophys. J., Vol. 254, 543 - 549 (1982).

The two brighter near-infrared sources in Orion Molecular Cloud 2, IRS 3 and 4, were observed at moderate spectral resolution in the near-infrared. At least eight members of the $V=1\rightarrow0$ and $V=2\rightarrow1$ rotation-vibration series of H_2 were found in IRS 4. The line ratios and intensities argue that the source is shock-excited with an excitation temperature of $T_{ex} = 3000 ± 800$ K. The values determined for extinction and excitation of IRS 4 are consistent with previous determinations of a middle B ZAMS object, extinguished by 10−15 visual magnitudes. The region of H_2 emission, however, is more probably extinguished by 20−30 magnitudes.

131.047 Interpretation of the H_2O maser outburst in Orion.
V. S. Strel'nitskij.
Pis'ma Astron. Zh., Tom 8, 165 - 171 (1982). In Russian.
English translation in Soviet Astron. Lett., Vol. 8.

It is shown that the H_2O maser that flared up in Orion was partly unsaturated. The anti-correlation between the line width and intensity, the asymmetry of the profile and the changes of the visibility function within it are explained by blending of two components, one of which has experienced a flare. From the observed polarization properties the upper limit to the electron density, the strength of the magnetic field and its direction within the source are deduced.

131.048 A search for hydrogen recombination lines in the meter wavelength range and restrictions on the parameters of ionized interstellar gas.
V. I. Ariskin, S. A. Kolotovkina, E. E. Lekht, G. M. Rudnitskij, R. L. Sorochenko.
Astron. Zh., Tom 59, 38 - 43 (1982). In Russian. English translation in Soviet Astron., Vol. 26, No. 1.

The recombination lines H 392α, 393α, and 394α have been searched for towards the radio sources Sgr A, M 17, and W 51. The upper limit on the line-to-continuum ratio is 4×10^{-4} for Sgr A and M 17, and 1.4×10^{-3} for W 51. The parameters of the ionized interstellar gas in the direction of the sources studied are estimated.

131.049 Star formation and fragmentation processes in the dark clouds complex in Taurus.
U. A. Nurmanova.
Astron. Zh., Tom 59, 61 - 64 (1982). In Russian. English translation in Soviet Astron., Vol. 26, No. 1.

An examination of the published data on the physical conditions in the dark clouds complex in Taurus shows that star formation efficiency is very low, about 1%. The gravitational collapse of the complex seems to be halted by a magnetic field and rotation resulting in such a low efficiency of star formation.

131.050 Formation of dense solid particles in a protoplanetary cloud.   V. S. Kessel'man.
Astron. Zh., Tom 59, 119 - 128 (1982). In Russian. English translation in Soviet Astron., Vol. 26, No. 1.

The process of growth of solid particles in a protoplanetary cloud is considered in terms of the theory of sintering. The effects of the composition, shape, and temperature of coagulating dust grains on the limiting dimensions of growing solid particles are discussed.

131.051 H_2O masers in W49N. I. Maps.
R. C. Walker, D. N. Matsakis, J. A. Garcia-Barreto.
Astrophys. J., Vol. 255, 128 - 142 (1982).

A multiple point fringe rate map of H_2O masers in W49N is presented that shows the locations of 386 separate features. The principal results, readily apparent in the map, are the clear separation of positive and negative high-velocity features and the presence of isolated features outside the centers of activity noted in earlier observations. The distribution of features suggests that the masers are seen in regions where material, which has been accelerated near a central star to velocities of up to a few hundred km s^{-1}, is interacting with a surrounding stationary or slowly moving medium.

131.052 Isotope-selective photodestruction of carbon monoxide.   J. Bally, W. D. Langer.
Astrophys. J., Vol. 255, 143 - 148 (1982).

Observations of the molecular cloud boundary layer near the H II region S68 reveal an overabundance of ^{12}CO and ^{13}CO relative to $C^{18}O$ consistent with a simple model of isotope-selective photodestruction of the rarer CO species. Self-shielding and the isotopic shift of the UV dissociative transitions increase the lifetime of the more abundant isotopes of carbon monoxide in a UV irradiated environment. As a result, large variations occur in the abundance ratios of CO isotopes in the surface layer of clouds and near internal UV sources.

131.053 Determination of density structure in dark clouds from CS observations.
R. L. Snell, W. D. Langer, M. A. Frerking.
Astrophys. J., Vol. 255, 149 - 159 (1982).

The authors present CS $J = 1-0$ and $J = 2-1$ observations of the three dark clouds TMC-1, L134 N, and B335. The $J = 2-1$ CS lines toward L134 N and B335 are self-absorbed; there is no evidence for self-absorption in the $J = 1-0$ lines of L134 N and B335, nor evidence for self-absorption in the CS lines in TMC-1. Maps of these three clouds have been obtained in the $J = 1-0$ CS transition. These observations have shown that B335 is centrally condensed and TMC-1 and L134 N are fragmented into two or more condensations. These condensations have densities greater than 10^4 cm^{-3} and masses between 2 and 10 $M_\odot$.

131.054 A study of DCO$^+$ emission regions in interstellar clouds.
A. Wootten, R. B. Loren, R. L. Snell.
Astrophys. J., Vol. 255, 160 - 175 (1982).

Observations of the $J = 1-0$ lines of HCO$^+$, H^{13}CO$^+$, and DCO$^+$ and of the $J = 2-1$ line of DCO$^+$ taken with similar beam sizes are used to construct models of the DCO$^+$ emission regions in eleven molecular clouds. The densities and extents of the regions as determined from these DCO$^+$ observations are similar to the densities and extents of the same regions modeled on formaldehyde observations. The DCO$^+$/H^{13}CO$^+$ abundance ratio ranges from 0.2 to 6 in the sample. These observations, interpreted in the light of newly measured reaction rates, constrain electron abundances in some regions to be as low as $X(e) < 1 \times 10^{-7}$. Independent limits placed on $X(e)$ using ^{13}CO and H^{13}CO$^+$ observations yield more sensitive limits, $X(e) < 6 \times 10^{-8}$. The abundance ratio has been used to test the predicted temperature dependence of Watson's theory for DCO$^+$ formation. The DCO$^+$/HCO$^+$ ratio declines with increasing temperature as predicted.

131.055 Collision-induced dissociation of H$_2$ and CO molecules.   W. Roberge, A. Dalgarno.
Astrophys. J., Vol. 255, 176 - 180 (1982).

A discussion is presented of the collision-induced dissociation of molecular hydrogen and carbon monoxide in shock-heated regions of the interstellar medium. The modifications to the dissociation rate caused by radiative stabilization are considered, and estimates are presented of the effective rates of collision-induced dissociation as functions of density and temperature in gases of atomic and molecular hydrogen. Below a certain density radiative stabilization greatly reduces the rate of collision-induced dissociation with substantial consequences for the thermal and chemical evolution of the postshock gas.

131.056 Detection of the $N = 1 \to 0$ transition of C$_4$H.
M. B. Bell, T. J. Sears, H. E. Matthews.

Astrophys. J., Lett., Vol. 255, L75 - L79 (1982).

Five hyperfine components of the $N = 1 \to 0$ rotational transition of C$_4$H have been detected toward the dust cloud TMC-1. Using these and previously published data, rotation, spin-rotation, and hyperfine constants have been determined for the molecule. An upper limit is presented for the intensity of the strongest component in IRC +10°216.

131.057 Automated star counts in the dark cloud L 1454.
R. Duerr, E. R. Craine.
Astron. J., Vol. 87, 408 - 418 (1982).

The Near Infrared Photographic Sky Survey (NIPSS) is a data base with broad potential not only for optical identification of point infrared sources, but also for a variety of studies related to the distribution of red stars. It has been suggested that these applications could be greatly aided by digitization of the data base, for which a feasibility study is under way. As a part of this study a NIPSS visual and near-infrared photographic pair, encompassing the dark cloud L 1454, was digitized. Star counts in this region suggest the existence of two clouds along the line of sight and allowed the distances, extinctions, and masses of the clouds to be estimated. In addition, maps of the region as a function of the redness of the constituent stars were generated. These, when compared to the extinction maps, allow a discussion of the star-formation properties of the region.

131.058 The gaseous galactic halo as inferred from the line spectra of the galaxies Markarian 509 and Fairall 9.
D. G. York, J. C. Blades, L. L. Cowie, D. C. Morton, A. Songaila, C.-C. Wu.
Astrophys. J., Vol. 255, 467 - 474 (1982).

Narrow interstellar absorption lines of S II λ1259.52, Si II λ1260.42, and Fe II λ1608.46 due to gas in the disk and the halo of our Galaxy have been detected in the spectrum of the Seyfert galaxy Mrk 509 with the IUE. This gas is also seen at higher resolution in the Ca II and Na I absorption lines in two components at LSR velocities of +6 and +62 km s^{-1}. Si II λ1260.42 absorption from the galactic disk and from the Magellanic Stream or the halo of the SMC have been detected with the IUE in the spectrum of Fairall 9. The observations of these two objects when combined with existing results are shown to be consistent with a corotating galactic halo having a height of less than 10 kpc at the Sun.

131.059 CO observations around galactic longitude $l = 45°$.
F. P. Israel.
Astrophys. J., Vol. 255, 475 - 488 (1982).

An area of about 1°.5 × 2°.0, centered on $l = 45°.5$ and $b = 0°.0$, was mapped in the ^{12}CO line at intervals of one beamwidth. A total of 22 individual CO cloud complexes was identified; the brightest of these is associated with a massive H I cloud and the H II region complex G45.5+0.1. This object most likely represents a small OB star cluster in its early stages of development. All other H II regions in the area mapped are likewise associated with CO maxima. The survey results indicate a structure of the Sagittarius arm generally in agreement with that derived in earlier studies; they also indicate that the Galaxy as a whole contains of the order of 3500 molecular clouds larger than 10 pc in diameter.

131.060 The formation and destruction of HeH$^+$ in astrophysical plasmas.
W. Roberge, A. Dalgarno.
Astrophys. J., Vol. 255, 489 - 496 (1982).

A discussion is presented of the formation and destruction mechanisms of the molecular ion HeH$^+$ in astrophysical plasmas. Calculations are made of the steady state abundance of HeH$^+$ in planetary nebulae and in dense clouds subjected to X-ray and XUV ionization. The excitation processes are investigated, and estimates are made of the emission-line intensities at 149.13, 3.364, and 3.607 μm. Emission at

149.13 μm from H II blisters and at 3.364 μm from high density planetary nebulae may be detectable.

131.061 Upper limits for interstellar boron and beryllium abundances toward Zeta Ophiuchi.
D. G. York, M. Meneguzzi, T. P. Snow.
Astrophys. J., Vol. 255, 524 - 526 (1982).

Observations of interstellar boron and beryllium in the ζ Oph spectrum were made using Copernicus. While neither element was detected, the boron limit can be used to derive $[B/H] < 3 \times 10^{-11}$ in the main interstellar cloud seen toward ζ Oph. Since $[B/H] = 1.5 \pm 0.7 \times 10^{-10}$ in the interstellar gas between the Sun and κ Ori, and since in stars of Type A and B, $[B/H] \sim 2 \times 10^{-10}$, the authors conclude that boron is depleted in the main ζ Oph cloud and that the abundance of boron in the interstellar medium is variable, probably by at least a factor of 5.

131.062 Molecular hydrogen emission from W51.
S. Beckwith, B. Zuckerman.
Astrophys. J., Vol. 255, 536 - 540 (1982).

The authors have detected emission from the $v = 1 \rightarrow 0\, S(1)$ quadrupole transition of H_2 toward the cluster of intense infrared and H_2O maser sources in W51 NORTH. The apparent luminosity of this line in W51 NORTH is only about 4% of the luminosity of the same line toward the Kleinmann-Low infrared cluster in Orion; however, additional line-of-sight extinction and spatial extent of the source may account for the lower apparent power in W51.

131.063 High-velocity molecular gas in the dark cloud L1529. S. M. Lichten.
Astrophys. J., Lett., Vol. 255, L119 - L122 (1982).

High-velocity CO wings have been detected near a dense condensation in the dark cloud L1529. Embedded stars or young stellar objects nearby have been identified as likely energy sources responsible for the unusual velocity dispersion (30 km s^{-1}) in the wings; these objects may also be responsible for the more moderate, yet still supersonic velocity dispersion (2 - 3 km s^{-1}) typical of the CO line cores. The observations of CO and high-excitation molecules in L1529 are consistent with cloud models in which stellar winds play a major role in the energetics and internal kinematics of dark clouds.

131.064 Very high Rydberg states ($n \approx 600$) of carbon in the interstellar gas. C. M. Walmsley, W. D. Watson.
Astrophys. J., Lett., Vol. 255, L123 - L127 (1982).

Calculations are presented for populations of very high electronic states with principal quantum numbers 200 - 900 associated with radio recombination lines of neutral carbon atoms at temperatures below 100 K. The results demonstrate that the carbon $n\alpha$ transitions with $n \approx 630$ - 640, which are coincident with the absorption features towards Cas A near 26 MHz will be in absorption and will have a line strength and line width compatible with observations. Major deviations from hydrogenic populations, which would yield emission recombination lines, occur for carbon under the physical conditions of the observed gas due to a dielectronic-like recombination process. The strength of the very high $n\alpha$ absorption lines of carbon is amplified by a substantial factor (10 - 100) due to the dielectronic-like process.

131.065 Gravitationally driven instabilities in shock compressed gas layers. G. L. Welter.
Astron. Astrophys., Vol. 105, 237 - 241 (1982).

In previous studies of the stability properties of shock compressed gas layers Elmegreen and Elmegreen (1978) and Welter and Schmid-Burgk (1981) made the computational simplification of treating the layer as stationary and pressure bounded rather than moving and shock bounded. In the present paper the author reconsiders the problem, taking proper account of the layer's motion. The results verify that the simpler procedure yields fairly accurate results.

131.066 On the angular momentum of colliding interstellar clouds. G. P. Horedt.
Astron. Astrophys., Vol. 106, 29 - 33 (1982).

There are determined limits for the expected angular velocity and the angular momentum change due to cumulative and binary collisions of spherical clouds. The obtained angular momenta are generally uppermost limits. The expected changes of rotation rates are several orders of magnitude larger for interstellar clouds and protostellar fragments of low mass, in comparison to large-mass clouds. Because of the expected large angular momentum changes cloud collisions could be responsible for some observed random orientations of the inclinations of the rotation axes of interstellar clouds and of stars.

131.067 The correlation between diffuse far ultraviolet background and line of sight hydrogen column: dust scattering and H_2 fluorescence. P. Jakobsen.
Astron. Astrophys., Vol. 106, 375 - 377 (1982).

It is shown that the correlation found by several groups between diffuse far-ultraviolet background intensity at intermediate latitudes and line of sight H I column density, can be understood in terms of the combined contributions from scattering of galactic plane starlight off high latitude interstellar dust and fluorescence of molecular hydrogen forming in this material. Enhanced background intensity in some directions may be due to rapid H_2 photodissociation in overpressured interstellar material or to the existence of a dust component having a very high ultraviolet albedo and isotropic phase-scattering function.

131.068 Infrared observations of dust in space.
W. J. Forrest.
News Lett. Astron. Soc. N. Y., Vol. 2, 5 - 11 (1982).

131.069 Observation af interstellare grundstoffer.
M. Winther.
Astron. Tidsskr., Årg. 15, 27 - 33 (1982).

131.070 The interstellar extinction law in dark nebulae connected with two T-associations in Taurus.
U. A. Nurmanova.
Perem. Zvezdy, Tom 21, 285 - 300 (1980). In Russian.

The author has analysed the extinction law in Taurus on the basis of multi-colour photometry and spectrophotometry of the star 72 Tau and multi-colour polarimetry of four background stars. The extinction law in the dust cloud complex in Taurus and in other complexes is normal with $R = A_V/E_{B-V}$ about 3. The spectrophotometry of two stars is used to investigate the reddening curve in the range $\lambda\lambda 3425 - 6810$ Å; the broadband structure is indicated.

131.071 Model calculations of the chemical abundance of interstellar clouds. A. A. Rejtblat.
Inst. kosm. issled. AN SSSR, Prepr., 1981, No. 624, 72 pp. In Russian. – Abstr. in Ref. zh., 51. Astron., 2.51.620 (1982).

131.072 Giant molecular-cloud complexes in the galaxy.
L. Blitz.
Sci. American, Vol. 246, No. 4, p. 72 - 80 (1982).

Consisting almost entirely of hydrogen molecules, they are the most massive objects in the galaxy. They also give rise to most of the galaxy's stars. A decade ago their presence was unknown.

131.073 Neutral hydrogen observations in the direction of extended background radio sources.
H. E. Payne, E. E. Salpeter, Y. Terzian.
Astrophys. J., Suppl. Ser., Vol. 48, 199 - 218 (1982).

Neutral hydrogen emission-absorption observations are reported for a wide galactic latitude range. Background continuum sources were primarily chosen to cover a broad range of angular sizes. These data are compared with those of Dickey, Salpeter, and Terzian which consist primarily of small background sources. The authors find no obvious correlation between the size of the background source and any of the observed H I quantities. These results put restrictions on the small scale structure of the interstellar neutral hydrogen.

131.074 Formaldehyde emission from DR21 (OH).
T. L. Wilson, J. Martin-Pintado, F. F. Gardner, C. Henkel.
Astron. Astrophys., Vol. 107, L10 - L12 (1982).

The discovery of H_2CO emission from the 2_{11}-2_{12} K-doublet line at 14.5 GHz is reported. The emission is centered at $\alpha = 20^h 37^m 13.8^s$, $\delta = +42°11'55''$ (1950.0), which is within $15''$ of the position of the OH and H_2O emission source DR21 (OH). The extent of the line emitting source is $\lesssim 30''$. From excitation considerations, the authors believe that this emission has a quasi-thermal origin.

131.075 The state of ionization in dense molecular clouds.
M. Guélin, W. D. Langer, R. W. Wilson.
Astron. Astrophys., Vol. 107, 107 (1982).

Observations of five isotopes of HCO^+, including the first detection in an astronomical source of the very rare isotope $D^{13}CO^+$, and extensive mappings of three of them, have allowed a careful determination of the DCO^+/HCO^+ abundance ratio throughout regions of the cool molecular clouds TMC1, TMC2, L183 (L134N), L1450 (NGC 1333), and NGC 2264. The ratio is found to be large and almost constant in the interior of the clouds, implying that the fractional electron abundance is smaller than $1 - 2 \times 10^{-7}$ throughout the cloud centers. Lower limits to the electron abundance, only a factor of 15 smaller than the upper limits, are derived from the observed HCO^+ column density. The low fractional ionization of the gas inside the clouds suggests that metals, such as Fe and Mg, may be severely depleted.

131.076 The molecular cloud complex in the vicinity of IC 5146.
W. H. McCutcheon, R. S. Roger, R. L. Dickman.
Astrophys. J., Vol. 256, 139 - 150 (1982).

CO observations of the molecular cloud complex in the vicinity of IC 5146 reveal three regions of enhanced emission, all lying around the periphery of the Sharpless region S125. The most intense region has a peak CO line temperature of 41 K. The continued existence of this hot spot can be accounted for by an embedded protostar of luminosity class between B0.5 and B1. A comparison with H I observations of about the same spatial and velocity resolution shows that the surrounding H I cloud emission is generally strong where the CO emission is weak. The velocity of the H II region, S125, is blueshifted with respect to the gas behind it by 5.6 km s^{-1}. This has yielded an estimate for the expansion age of the H II region of about 10^5 yr, a factor of 30 smaller than the age given for the stellar cluster.

131.077 A search for interstellar and stellar iron monoxide.
A. J. Merer, C. M. Walmsley, E. Churchwell.
Astrophys. J., Vol. 256, 151 - 155 (1982).

A search for the lowest energy rotational transition ($J = 5-4$) of iron monoxide (FeO) near 153 GHz in the lowest and first excited vibrational states has been conducted toward a variety of objects. FeO was not detected. Our upper limit on the FeO column density in Orion and other warm molecular clouds is $N(FeO) < 2 \times 10^{12}$ cm^{-2}. In Orion, $< 2 \times 10^{-7}$ of the cosmic iron abundance can be in the form of FeO and $[FeO]/[H_2] < 10^{-11}$. It is likely that most Fe is incorporated into interstellar dust grains. The remaining fraction in the gas phase probably favors atomic iron rather than FeO.

131.078 Observations of neutral carbon in the NGC 1977 bright rim.
A. Wootten, T. G. Phillips, C. A. Beichman, M. Frerking.
Astrophys. J., Lett., Vol. 256, L5 - L8, plate L1 (1982).

Strong neutral carbon emission at 610 μm (492 GHz) has been detected from a bright-rimmed cloud abutting the H II region NGC 1977. The similarity of velocity and width between ^{13}CO and C I lines suggests that both lines originate in the same region. A model for the density and temperature structure of the cloud, based on ^{13}CO and ^{12}CO observations, has been used to estimate the carbon abundance. The variation in the relative abundance distributions of CO and C I confirms the importance of photodissociation in the chemistry of molecular clouds, and of the C I line to studies of the interaction of hot stars with clouds.

131.079 Interstellar titanium abundances toward 19 high-latitude stars.   C. E. Albert.
Astrophys. J., Lett., Vol. 256, L9 - L12 (1982).

The $\lambda 3384$ absorption line of interstellar Ti II has been observed at high resolution along lines of sight toward nine nearly aligned pairs of foreground disk and background halo stars, the latter having z-distance greater than 500 pc. The primary result is the great strength of Ti II absorption perpendicular to the plane of the Galaxy. Compared with the well-studied disk abundances, there is apparently a vertical gradient of titanium depletion above the galactic plane. The gaseous titanium abundance relative to hydrogen increases with z-distance and with LSR velocity by factors exceeding 20 and can reach nearly solar proportions.

131.080 H_2O masers — survey of the galactic plane. II.
M. A. Braz, E. Scalise, Jr.
Astron. Astrophys., Vol. 107, 272 - 275 (1982).

The authors report the result of a search for H_2O maser emission sources performed towards radio continuum peaks of southern galactic H II regions, type I OH emission sources and OH emission sources for which the type is still unknown. 16 new sources displaying H_2O maser emission were found. Comments on individual sources are presented.

131.081 Fluxes of energetic particles and the ionization rate in very dense interstellar clouds.
T. Umebayashi, T. Nakano.
Publ. Astron. Soc. Japan, Vol. 33, 617 - 635 (1981).

The authors investigate the propagation of the primary and secondary energetic protons, neutrons, electrons, and photons in a very dense cloud (protostar). Solving the one-dimensional transport equations they obtain the energy spectra and intensities of these particles as functions of the depth x (in g cm^{-2}) from the surface of the protostar. Using the calculated intensities the authors investigate the ionization rates of the hydrogen molecule by these particles.

131.082 Interstellar grain size spectrum and circumstellar grain-grain collisions.   J. Dorschner.
Astrophys. Space Sci., Vol. 81, 323 - 328 (1982).

In this paper, grain-grain collisions, which were recently suggested by Biermann and Harwit to occur in cool circumstellar envelopes and to be responsible for the interstellar grain size spectrum, are investigated. On the basis of the author's fragmentation theory, it is shown that as a result of such collisions size distributions of the type $n(a) \propto a^{-p}$ arise. In the steady-state case the exponent p ranges from 3.4 to 3.7. This result matches well with grain size spectra derived from the interstellar extinction curve.

131.083 Dependence on reddening of interstellar column densities in the direction of O stars.
H. H. Menteşe.
Astrophys. Space Sci., Vol. 82, 173 - 187 (1982).

The author has investigated 14 O-type stars with IUE high resolution SWP spectra in order to improve the relations of element depletions with $E(B - V)$, and to look for other possible relations with two stellar parameters: namely, the rate of mass loss and rotational velocity. The stars were chosen so as to cover several directions in the Galaxy, as well as a wide range in interstellar reddening. The author found a clear inverse trend relating the abundance of elements to interstellar reddening, and a crude relation between the increasing depletion and condensation temperature.

131.084 **A relaxation oscillation model for bursts of star formation in nuclei of galaxies.**
H.-H. Loose, K. J. Fricke.
Mitt. Astron. Ges., Nr. 55, p. 100 - 101 (1982).

131.085 **Interstellar masers: the influence of the geometrical shape on the radiation properties.**   E. Bettwieser.
Mitt. Astron. Ges., Nr. 55, p. 120 (1982). – Abstract.

131.086 **Das Massenspektrum bei der Sternentstehung.**
H. Zinnecker.
Mitt. Astron. Ges., Nr. 55, p. 160 - 161 (1982).

131.087 **Excited OH in absorption towards W3 (OH).**
C. M. Walmsley, A. Winnberg, T. L. Wilson, A. Baudry.
Regions of recent star formation, (see 012.002), p. 81 - 82 (1982).

131.088 **A calculation of molecule abundances behind slow shocks.**   G. F. Mitchell, T. J. Deveau.
Regions of recent star formation, (see 012.002), p. 107 - 116 (1982).

Post-shock abundances of 105 chemical species are followed after the passage of a 10 km s⁻¹ shock through an interstellar cloud of initial density $10^4 cm^{-3}$. The authors find significant enhancement in the column densities of H, H_2O, NH_3, and HS. The column densities of most ions decrease in abundance by rather large factors.

131.089 **Gravitational instabilities in shock compressed gas layers.**   G. L. Welter.
Regions of recent star formation, (see 012.002), p. 117 - 122 (1982).

In previous studies of the stability properties of shock compressed gas layers, Elmegreen and Elmegreen and Welter and Schmid-Burgk made the computational simplification of treating the layer as stationary and pressure bounded rather than moving and shock bounded. In the present paper the problem is reconsidered taking proper account of the layer's motion. The results verify that the simpler procedure yields fairly accurate results.

131.090 **Magnetic fields and the evolution of shocked gas clouds.**   J. Nittmann.
Regions of recent star formation, (see 012.002), p. 123 - 128 (1982).

The author is studying the influence of large scale interstellar magnetic fields on the early evolution of a high density gas cloud which is hit by a strong shock wave. The incident shock is assumed, a priori, to be driven by a spiral density wave. Results are presented for the flows which develop in the interstellar gas with magnetic field strength of $1\mu G$ and $3\mu G$, respectively.

131.091 **High velocity gas in molecular clouds.**
R. L. Snell, S. Edwards.
Regions of recent star formation, (see 012.002), p. 133 - 139 (1982).

A new source of high velocity molecular gas to the south of the H II region NGC 2068 (M78) near the Herbig-Haro

objects HH 25 - 26 is reported. The redshifted and blueshifted CO wings seen in this region are spatially separated indicating that this is another region of bi-polar mass outflow from a young stellar object. The authors compare this region to the other known regions of bi-polar mass outflow and discuss the implications of these energetic outflows.

131.092 **Observations and interpretation of the line profiles of excited H_2 in Orion.**   T. R. Geballe, D. Nadeau.
Regions of recent star formation, (see 012.002), p. 147 - 153 (1982).

The profiles of the v = 1–0 S(1) line of H_2 in the Orion molecular cloud show a variation from the center of the emission region where the profiles are broad and have extended blue wings, to the periphery where the profiles are narrow and have a peak at the velocity of the molecular cloud. These observations can be understood by a model of a radially expanding supersonic flow of gas colliding with the surrounding molecular cloud. It is concluded that the high velocity emission originates in the flow and the low velocity emission originates in the molecular cloud.

131.093 **High velocity H_2 line emission in the NGC 2071 region.**   S. E. Persson, T. R. Geballe, T. Simon, C. J. Lonsdale, F. Baas.
Regions of recent star formation, (see 012.002), p. 155 - 159 (1982).

The line profile of the v = 1 → 0 S(1) line of H_2 in NGC 2071 is ~ 100 km s⁻¹ wide, and asymmetric; it resembles S(1) profiles seen toward Orion KL. The NGC 2071 region represents the second detection of high velocity H_2 emission in a region showing signs of ongoing star formation.

131.094 **The NGC 7538 region: the distribution and dynamics of molecules compared with those of H I and H⁺.**
H. R. Dickel, J. R. Dickel, W. J. Wilson.
Regions of recent star formation, (see 012.002), p. 175 - 180 (1982).

CO maps and preliminary H_2S and H_2CO data for the molecular cloud associated with the H II region NGC 7538 are compared with the distributions of ionized and neutral hydrogen. South of the optical H II region is a ridge of high ^{13}CO column density with cold, self-absorbed H I gas just beyond it. The percentage of the hydrogen in atomic form varies from ~ 0.1% in the dense region to ~ 0.8% in the outskirts. The lower-density region of expanding gas seen next to the H II region in the southwest is attributed to the passage of a molecular dissociation wave.

131.095 **Neutral hydrogen observations of the Puppis Window.**   J. G. Stacy, P. D. Jackson.
Regions of recent star formation, (see 012.002), p. 185 - 191 (1982).

A 21-cm neutral hydrogen survey has been carried out in a region of the galactic disk known as the "Puppis Window" at $l = 245°$. The authors have attempted to establish correlations between distinct H I features and optical spiral tracers. The presence of many H I shells and detailed filamentary structure provides strong evidence for a highly turbulent interstellar medium in this region.

131.096 **Neutral hydrogen towards Tycho's supernova remnant.**   J. S. Albinson, S. F. Gull.
Regions of recent star formation, (see 012.002), p. 193 - 199 (1982).

The radio remnant 3C10 of Tycho's supernova (AD 1572) has been observed with the Cambridge Half-Mile-Telescope. Absorption, spin temperature and column density profiles for the H I along the line of sight to 3C10 are derived. Velocity diagrams reveal H I emission from the Local, the Perseus and the Outer spiral arms. The distance to 3C10 is estimated to be in the range 2.0 - 2.5 kpc.

131.097 **Observations of CO J = 3 → 2 emission from molecular clouds.**
G. J. White, J. P. Phillips, G. D. Watt.
Regions of recent star formation, (see 012.002), p. 237 - 244 (1982).

Spectral observations in the CO J = 3 → 2 line are reported for a sample of molecular clouds in star formation regions. Data were obtained at the 3.8 UKIRT with the QMC heterodyne receivers. The hot-centered clouds show evidence of self-absorption or high-velocity gas. The self-absorbed line centers are slightly red-shifted with respect to optically thin emission lines. Clear evidence for rotation is seen for NGC 2071. Detailed contour maps in the J = 3 → 2 transition are presented for NGC 2068 and NGC 2023.

131.098 **Properties of giant molecular clouds in the galactic molecular ring.** W. L. H. Shuter, A. Szabo.
Regions of recent star formation, (see 012.002), p. 245 - 247 (1982).

The authors present results on the physical properties of giant molecular clouds based on a fully sampled J = 1 → 0 CO survey at $b = 0°$ from $l = 29° - 46°$ using the 4.6-m millimetre wave telescope at the University of British Columbia. The telescope beam-width was 2.6' and the velocity resolution (FWHM) was 2.6 km s^{-1}.

131.099 **Infrared and maser sources in regions of star formation.** R. Genzel, D. Downes.
Regions of recent star formation, (see 012.002), p. 251 - 286 (1982).

This review discusses the mass loss phenomena associated with compact infrared and maser sources in regions of star formation. Most of the compact $2 - 20$ μm sources associated with H_2O, and OH (and SiO) masers and ultra-compact H II regions in regions of formation of stars more massive than a few $M_\odot$ seem to have mass outflow at velocities between 10 and 250 km s^{-1}. The authors discuss the evidence for these motions in several sources (in particular, Orion–KL), and their dependence on luminosity and evolutionary stage. The mass loss phase in the early evolutionary stages of stars appears to cover the whole spectrum of stellar masses and to be of long duration.

131.100 **High velocity CO line wings and the dynamics of star forming molecular cloud cores.** J. Bally.
Regions of recent star formation, (see 012.002), p. 287 - 293 (1982).

Preliminary results of a search for high velocity CO in relatively nearby (d < 3 kpc) star forming cloud cores are presented. Detailed CO, CS, and VLA observations of the molecular outflow region associated with NGC 2071 suggest that a disk constrained stellar wind is responsible for the bipolar high velocity CO flow.

131.101 **Kinematics of molecular gas in Orion from observations of the ^{13}CO J = 2 → 1 line.**
P. F. Goldsmith, R. Arquilla, F. P. Schloerb, N. Z. Scoville.
Regions of recent star formation, (see 012.002), p. 295 - 299 (1982).

The authors have obtained spectra of the J = 2 → 1 transition of ^{13}CO covering a 16 arcminute by 33 arcminute region centered on the KL object in the Orion molecular cloud. These spectra reveal a high degree of complexity in the emission at many positions, which suggests that the molecular gas has been significantly perturbed by the H II region and Trapezium stars located in front of it.

131.102 **Molecular hydrogen emission from broad wing cloud cores.** A. P. Lane, J. Bally.
Regions of recent star formation, (see 012.002), p. 301 - 306 (1982).

The authors report observations of 2 μ line emission from vibrationally excited H_2 in the vicinity of the high velocity molecular flows associated with Cepheus A, NGC 2071, and GL 961. The luminosity and spatial extent of the H_2 emission in these regions are compared with those of other known H_2 sources associated with high velocity molecular cloud cores.

131.103 **Asymmetric broad HCO$^+$ line wings in cores of molecular clouds.**
Å. Sandqvist, A. Wootten, R. B. Loren, P. Friberg, Å. Hjalmarson.
Regions of recent star formation, (see 012.002), p. 307 - 314 (1982).

High resolution observations of HCO$^+$ in the cores of the W3, NGC 2071 and Cep MC–1 molecular clouds reveal asymmetric line wings extending over velocity ranges of 20, 35 and 55 km s^{-1}, respectively. Red and blue wings of the profiles are enhanced on opposite sides of infrared objects embedded in the cores of NGC 2071 and Cep MC–1. The lines attain their maximum breadths at the positions of these infrared objects. W3, on the other hand, is more complex and evidence is presented which shows the presence of two molecular clouds in the core, one centered near IRS 5 and the other near IRS 4, the latter exhibiting a blue-shifted wing.

131.104 **Some of the problems raised by CO and HCO$^+$ observations in the Rho Ophiuchi cloud.**
M. Pérault, E. Falgarone.
Regions of recent star formation, (see 012.002), p. 315 - 321 (1982).

The central parts of the Rho Ophiuchi dark cloud have been observed in the J = 1–0 transition of HCO$^+$ and H^{13}CO$^+$ with the new cooled receiver which equips a 2.5 m antenna at the Bordeaux Observatory.

131.105 **CO J = 3 → 2 and far infrared continuum observations of L1551, Orion KL and IRC +10216.**
J. P. Phillips, G. J. White, P. A. R. Ade, C. T. Cunningham, E. I. Robson, G. D. Watt.
Regions of recent star formation, (see 012.002), p. 323 - 328 (1982).

Observations of the sources L1551, Orion KL and IRC +10216 in the J = 3 → 2 transitions of CO are reported, together with photometry at wavelengths λ 377, 811 and 1136 μm. All results were acquired with the United Kingdom Infrared Telescope at Mauna Kea, Hawaii, with instrumental beam-sizes of 86 arcsecs for the photometry, and 60 arcsecs for the CO spectra.

131.106 **CO in the Horsehead Nebula.**
A. A. Stark, J. Bally.
Regions of recent star formation, (see 012.002), p. 329 - 334 (1982).

Carbon monoxide observations of molecular gas at the ionization front associated with IC 434 show a corrugated structure with a periodicity of 1.4 pc along the entire 9 pc length of the front. The CO distribution may be explained by the Rayleigh-Taylor instability resulting from the rocket acceleration of the Orion B molecular cloud by Lyman continuum radiation from the Ori OB I association.

131.107 **Small scale clumping in the Orion Molecular Cloud.**
P. Bastien, J. Bieging, C. Henkel, R. N. Martin, T. Pauls, C. M. Walmsley, T. L. Wilson, L. M. Ziurys.
Regions of recent star formation, (see 012.002), p. 335 - 336 (1982). – Abstract.

131.108 **The ON–1 CO cloud complex – onset of star formation.** F. P. Israel, H. A. Wootten.
Regions of recent star formation, (see 012.002), p. 337 - 342 (1982).

Molecular line observations of the ON–1 region show the presence of a large CO cloud complex with overall dimensions of 25 × 60 pc at a distance of 1.4 kpc to the Sun. The complex consists of two major parts, CON−1 East and CON−1 West with sizes of respectively 20 × 40 pc and 15 × 35 pc. ON−1 itself is associated with a compact CO cloud of mean size 1.2 pc and a mass of about 750 $M_\odot$.

131.109 CO observations of the molecular cloud encompassing Sharpless 222.
R. A. Christie, W. H. McCutcheon, C. P. Chan.
Regions of recent star formation, (see 012.002), p. 343 - 348 (1982).

^{12}CO observations around Sharpless 222 reveal a very wide region of CO emission at a temperature of about 10 K. There is a ridge of enhanced emission along the southern portion of the molecular cloud. H I emission associated with S222 lies above this ridge.

131.110 Chemistry relevant to molecular clouds near H II regions.   W. D. Watson, C. M. Walmsley.
Regions of recent star formation, (see 012.002), p. 357 - 377 (1982).

The successes and failures of models of interstellar chemistry are discussed with particular reference to molecular clouds, such as Orion, which are in the immediate neighborhood of H II regions. Some recent laboratory measurements at temperatures below 100 K are reviewed and their astrophysical consequences discussed. From a survey of abundance measurements towards Orion, it is concluded that composition differences between hot H II region clouds and cold dark clouds are not great. The influence upon the chemistry of shocks and mass outflow from young protostars is discussed. It is concluded that although some anomalies do exist, gas phase ion-molecule type schemes can account for most of the qualitative features of the observations.

131.111 Selective photodestruction of CO isotopic species.   W. D. Langer, J. Bally.
Regions of recent star formation, (see 012.002), p. 379 - 384 (1982).

Observations of the relative abundances of the CO isotopes near the H II region S68 show enhanced $^{12}CO/C^{18}O$ and $^{13}CO/C^{18}O$ ratios. These large ratios can be explained by selective photodestruction of the isotopes of CO in which line absorption and self-shielding play a crucial role.

131.112 X-ray ionization and the chemistry of the Orion molecular cloud.   J. H. Krolik, T. R. Kallman.
Regions of recent star formation, (see 012.002), p. 385 - 389 (1982).

The collection of unusually strong stellar X-ray sources in the vicinity of the Orion molecular cloud together bathe the gas with such an intensity of X-rays that they, rather than cosmic rays, dominate the ionization and heating of the gas. The authors present estimates of the ionization rate and the elevation in temperature, and discuss the consequences for the gas chemistry.

131.113 Methyl cyanide as a probe of the temperature and density in Sgr B2; quasi-equilibrium in molecular rotational levels.   R. A. Linke, S. E. Cummins, S. Green, P. Thaddeus.
Regions of recent star formation, (see 012.002), p. 391 - 397 (1982).

Observations of 14 rotational transitions of methyl cyanide have been used in conjunction with new estimates of the collisional excitation cross sections und H_2 impact to determine the kinetic temperature and density in the central 2 arcminutes of Sgr B2. An H_2 temperature near 90 K and an H_2 density range of $6 - 16 \times 10^4$ cm^{-3} were obtained.

131.114 High spatial resolution observations of HCN in S 235B.   G. Sandell, B. Höglund, A. G. Kislyakov.

Regions of recent star formation, (see 012.002), p. 399 - 406 (1982).

HCN has been mapped with 42″ spacing (=HPBW) around the BN type IR source S 235B (=IRS 4). The HCN hyperfine structure is well resolved and departs from LTE. H ^{13}CN has also been detected in two positions. The HCN column density toward S 235B is $> 8.0 \times 10^{13}$ cm^{-2}.

131.115 A model for the formaldehyde maser near NGC 7538 − IRS 1.   W. Boland, T. de Jong.
Regions of recent star formation, (see 012.002), p. 407 - 408 (1982).

The population of the 6 cm H_2CO transition can be inverted by the radio continuum radiation of a nearby compact H II region. The H II region must be very compact with emission measures of $10^8 - 10^{10}$ cm^{-6} pc. The model does explain the observed maser emission near NGC 7538 − IRS 1 if a large formaldehyde abundance $\sim 8 \times 10^{-7}$ is assumed.

131.116 H_2CO near compact H II regions − new WSRT results.   J. R. Forster, W. M. Goss, H. R. Dickel.
Regions of recent star formation, (see 012.002), p. 409 - 418 (1982).

Aperture synthesis observations in the 6 cm λ H_2CO line are presented for five fields containing ultra-compact H II regions associated with OH maser sources. The H_2CO optical depths measured towards these sources are large compared to other sources in the fields. Neutral gas densities greater than 10^6 cm^{-3} are suggested.

131.117 Is a BN-type object the energy input to the NH$_3$ cloud in NGC 2071?   G. Calamai, M. Felli.
Regions of recent star formation, (see 012.002), p. 419 - 423 (1982).

An extended molecular cloud has been found close to the reflection nebula NGC 2071 in the northern part of the Orion complex. The J, K = (1, 1) and (2, 2) inversion lines of NH$_3$ at 23.69 GHz and 23.72 GHz respectively were mapped with the 25-m radiotelescope of the S. R. C. Appleton Laboratory at Chilbolton Observatory.

131.118 VLA observations of OH masers and associated ultracompact continuum sources.
B. E. Turner.
Regions of recent star formation, (see 012.002), p. 425 - 432 (1982).

The author has observed 22 OH masers and associated continuum sources with the VLA at λ18 cm, and deduces that type I masers are always associated with ultracompact continuum sources, while type II(a) (1720 MHz) masers probably never are. Implications for the type of pumping for each category of maser are discussed.

131.119 Molecular line mapping of OMC−1.
F. P. Schloerb, P. F. Goldsmith, N. Z. Scoville.
Regions of recent star formation, (see 012.002), p. 439 - 444 (1982).

The authors present new, fully sampled maps of the molecular emission from CO, ^{13}CO, and HCN in the Orion Molecular Cloud. The high resolution of the maps (< 1 arcmin) reveals considerable structure in the molecular gas that is related to the H II region surrounding the Trapezium.

131.120 An upper limit to the atomic carbon abundance in the Orion plateau.
C. A. Beichman, T. G. Phillips, H. A. Wootten, M. Frerking.
Regions of recent star formation, (see 012.002), p. 445 - 452 (1982).

Observations made of the atomic carbon line at 492 GHz toward OMC−1 show no evidence for the high velocity dispersion wings observed for many molecular rotational lines. The 3σ upper limit to the CI column density, N_{CI}, is

$6.9 \times 10^{17} \mathrm{cm}^{-2}$ for velocities $\geqslant 4 \mathrm{~km~s}^{-1}$ from the line center. Atomic carbon is apparently depleted by a factor as large as five in the hot plateau gas, relative to its abundance in other molecular clouds.

131.121   Observations of neutral carbon in the NGC 1977 bright rim.
A. Wootten, T. G. Phillips, C. A. Beichman, M. Frerking.
Regions of recent star formation, (see 012.002), p. 453 - 461 (1982).

Neutral carbon emission has been observed at 610 μm (492 GHz) from the region of an interface between the H II region NGC 1977 and the northern extension of the dense OMC-2 cloud in Orion.

131.122   Observations of CO in TMC 1.
R. Braun, W. H. McCutcheon, W. L. H. Shuter.
Regions of recent star formation, (see 012.002), p. 463 - 467 (1982).

Observations of CO in TMC1, the molecular ridge in the large dust cloud Heiles' 2, give evidence for a local contraction or expansion centered in position and velocity on the HC_5N maximum. The estimated velocity gradient of this motion is $\sim 1.5 \mathrm{~km~s}^{-1} \mathrm{pc}^{-1}$ if the cloud is assumed to be at a distance of 115 pc.

131.123   CO emission associated with cold neutral hydrogen.
W. L. Peters III, F. N. Bash.
Regions of recent star formation, (see 012.002), p. 469 - 477 (1982).

CO observations are compared with neutral hydrogen observations. It is found that a third of the CO clouds had H I self-absorption counterparts. A statistical analysis indicates that the correlation is significant and consistent with the view that molecular clouds contain enough atomic hydrogen to noticeably affect the atomic hydrogen emission line profiles.

131.124   On the correlation of CH abundance and extinction in dark nebulae.   G. Sandell, L. E. B. Johansson.
Regions of recent star formation, (see 012.002), p. 479 - 484 (1982).

The correlation of CH abundance and extinction has been investigated in individual dark nebulae as well as in a sample of clouds. The authors find a good correlation in individual clouds; however, the relation between CH column density, $N_{CH}(\mathrm{cm}^{-2})$, and blue extinction, A_B (mag), varies from cloud to cloud: $N_{CH} = (2-7) \times 10^{13} A_B$. The reasons for the cloud to cloud variations are discussed.

131.125   Origin of amorphous interstellar grains.
J. Seki, H. Hasegawa.
Prog. Theor. Phys., Vol. 66, 903 - 912 (1981). − Abstr. in Phys. Abstr., Vol. 85, Abstr. 38279 (1982).

131.126   Star formation in galactic spiral arms.
G. Gilmore.
Nature, Vol. 297, 179 - 180 (1982).

131.127   The physical state of the gas towards HD 93206.
B. Y. Welsh, C. K. Thomas.
Mon. Not. R. Astron. Soc., Vol. 199, 385 - 397 (1982).

The Carina star HD 93206 has been observed at high resolution using the IUE satellite and high velocity interstellar absorption components have been detected at −260 and −185 km s^{-1}. A line profile fitting analysis has been carried out on the non-neutral atomic I-S lines to determine cloud component column densities and subsequently to determine the physical condition and element abundance of the associated I-S gas. It is proposed that stellar winds from early-type stars or supernovae outbursts are responsible for the very high velocity gas motions.

131.128   The interstellar spectrum of the supernova 1980k in NGC 6946.   M. Pettini, P. Benvenuti, J. C. Blades, A. Boggess, A. Boksenberg, M. Grewing, A. Holm, D. L. King, N. Panagia, M. V. Penston, B. D. Savage, W. Wamsteker, C.-C. Wu.
Mon. Not. R. Astron. Soc., Vol. 199, 409 - 423 (1982).

The authors report high resolution ultraviolet observations, augmented by medium and high resolution optical spectra, of supernova 1980k in the Scd galaxy NGC 6946, aimed at probing the intervening interstellar media along this very extended line of sight. They detect interstellar absorption lines of Mg I, Mg II, Fe II, Mn II, Ca II and Na I. Significant absorption by supernova material is not responsible; absorption by the interstellar medium in NGC 6946 is intrinsically blended with local foreground gas. The blueward extent of the ultraviolet absorption lines can be produced by differential Galactic rotation and may imply that interstellar gas exists up to 40−50 kpc along the line of sight, or 8−10 kpc above the plane.

131.129   The dust distribution in Bok globules.
I. P. Williams, H. C. Bhatt.
Mon. Not. R. Astron. Soc., Vol. 199, 465 - 470 (1982).

The authors discuss the notion that grains settle to the centre of Bok globules. In some globules these are replaced by grains driven into the globules by radiation pressure. In this case, the authors derive the density variation and calculate the expected variation in A_V. The observed variation in A_V is in excellent agreement with the calculated one. It is also shown that examples may be found where grain replacement was not efficient and where it was too efficient, producing a totally opaque body.

131.130   Chemical composition of gas and grains in dense interstellar clouds.   J. Lequeux.
Geochim. Cosmochim. Acta, Vol. 46, 777 - 782 (1982).

This paper is a short review of the present knowledge on the composition of dense clouds where stars form, with extensions to the less dense interstellar medium and to the most likely sites of grain formation. The problem is complex and intricate, and it is still difficult to answer such fundamental questions as the lifetime of the dense clouds or the lifetime of the interstellar grains.

131.131   Search for interstellar superheavy hydrogen.
M. Jura, D. G. York.
Science, Vol. 216, 54 - 55 (1982).

Models for fundamental physical interactions allow for the existence of stable or nearly stable elementary particles much heavier than the proton. Stellar spectra were searched for a positively charged superheavy particle, X^+, which, with a bound electron, should appear as apparently superheavy neutral hydrogen in the interstellar medium. An upper limit for the abundance of X relative to normal hydrogen in the line of sight toward the bright star γ Cassiopeiae is 2×10^{-8}.

131.132   The formation and properties of grains in the interstellar medium.
S. J. Czyzak, J. P. Hirth, R. G. Tabak.
Vistas Astron., Vol. 25, 337 - 382 (1981).

The present state of the observational data and theories on interstellar grains is reviewed. The models which are based to some degree on the observational data are examined in detail. The nucleation and growth of grains, the grain temperature, and the electric charge on the grains are critically discussed. A discussion of the observational and experimental data on the grains obtained to date is reviewed. Particular reference is made to graphite and to the likelihood that in the interstellar medium disordered carbon grains are more likely than graphite, based on present theoretical and experimental data of crystal formation and growth.

131.133 Star formation: the influence of velocity fields and turbulence. J. H. Hunter, Jr., R. C. Fleck, Jr.
Astrophys. J., Vol. 256, 505 - 513 = Contrib. No. 40 Dep. Astron. Univ. Florida.

It is shown that the Jeans mass for gravitational collapse can be very much reduced by the influence of velocity fields, even when allowance is made for non-isothermal gas behavior. The authors examine the role of turbulence in establishing the initial stellar mass function and show that the flattening and/or turnover at the low mass end may be a signature of interstellar turbulence. They consider also the implications of primordial turbulence for the formation of stars in the early universe.

131.134 Does fragmentation occur on protostellar mass scales during the dynamic collapse phase? J. Silk.
Astrophys. J., Vol. 256, 514 - 522 (1982).

The fragmentation of a uniform, nearly pressure free, collapsing spheroidal cloud is studied, both from rest and from an initial state of uniform expansion. It is found that, relative to the case of a spherical collapse, the fragmentation efficiency is considerably increased. Fluctuations that are oblate relative to the main flow grow more rapidly than relatively prolate fluctuations. These conclusions are not significantly modified by the effects of a finite uniform pressure, provided that the initial configuration is well above the Jeans limit. While inclusion of more realistic effects associated with density gradients, magnetic fields, cooling, and opacity are likely to modify the cloud evolution, it seems difficult to avoid the conclusion that fragmentation down to protostellar mass scales should occur during the dynamical collapse either of interstellar or primordial clouds.

131.135 Detection of pedestal features in dark clouds: evidence for formation of low mass stars. M. A. Frerking, W. D. Langer.
Astrophys. J., Vol. 256, 523 - 529 (1982).

A survey of 180 dark, cold molecular clouds in ^{12}CO has been made to search for unusual pedestal features resembling those observed in giant molecular clouds associated with embedded stellar sources. Low-intensity line wings with characteristic dispersions of 3 to 10 km s^{-1} (considerably smaller than pedestal features of giant molecular clouds) were detected in four of these sources. The authors suggest that these features probably arise from stellar winds associated with embedded low-mass stars.

131.136 Nearly simultaneous observations of vibrationally excited $J = 1 \to 0$, $J = 2 \to 1$, $J = 3 \to 2$, and $J = 4 \to 3$ SiO masers. P. R. Schwartz, B. Zuckerman, J. M. Bologna.
Astrophys. J., Lett., Vol. 256, L55 - L59 (1982).

The authors have detected the SiO $J = 3 \to 2$ v = 2 and $J = 4 \to 3$ v = 1 and 2 masers and made nearly simultaneous observations of the strongest SiO maser transitions in a selection of late-type stars and Orion.

131.137 Physical conditions in dark interstellar clouds: magnetic field strength and density. S. S. Hong.
J. Korean Astron. Soc., Vol. 14, 37 - 42 (1981).

In order to know how the magnetic field increases with density in interstellar clouds, the author has analyzed observations of extinction and polarization for stars in the ρ Oph molecular cloud complex.

131.138 The ionization structure of externally illuminated clouds. K. S. Anderson.
Proc. Southwest Reg. Conf., Vol. 7, (see 012.019), 61 - 71 (1982).

131.139 Estudio en 21 cm de tres nubes oscuras. M. B. Gordon, S. L. Garzoli.
Bol. Asoc. Argentina Astron., No. 18, p. 30 - 36 (1980).

131.140 Evolution of isothermal clouds under varying external pressure. Yu. I. Izotov, I. G. Kolesnik.
Astrometr. Astrofiz., Vyp. (No.) 46, p. 3 - 15 (1982). In Russian.

The self-consistent problem is solved on the evolution of optically thin clouds in quasi-equilibrium under varying external pressure. It is assumed that the function of volume cooling and variation of external pressure are arbitary and the temperature of the cloud is constant in volume, but varies in time. A system of equations describing the behaviour of the cloud parameters is obtained. The regimes of the clouds evolution are studied as functions of rate of the pressure variation and cooling. The conditions are formulated for the cloud stability loss and self-gravitation contraction. Possible applications of this research are shown.

131.141 3.4-mm HCN and continuum survey of dark clouds. A. B. Burov, V. N. Voronov, I. I. Zinchenko, A. A. Krasil'nikov, Eh. P. Kukina.
Astron. Zh., Tom 59, 267 - 275 (1982). In Russian. English translation in Soviet Astron., Vol. 26, No. 2.

19 dark clouds have been observed in the HCN line ($J = 1 - 0$) and in the continuum at 3.4 mm wavelength. HCN emission has been detected in the direction of nine of them. Upper limits on the continuum brightness temperature have been found. Dust and HCN column densities are estimated.

131.142 Observations of the variability of H_2O sources associated with star formation regions. E. E. Lekht, M. I. Pashchenko, G. M. Rudnitskij, R. L. Sorochenko.
Astron. Zh., Tom 59, 276 - 285 (1982). In Russian. English translation in Soviet Astron., Vol. 26, No. 2.

The results of observations of sources of maser emission in the 1.35 cm line of the water vapour molecule in star formation regions for the period from November 1979 to June 1981 are presented. The data obtained for four sources, NGC 2071 (G 205.1 - 14.1), W 44 C, W 75 N, and Cep A, are considered in detail. For all these objects considerable variations of the H_2O line profile as a whole, as well as of flux density and radial velocity of separate features were observed. The evolutionary stage and the dynamics of the sources studied are discussed.

131.143 Disk accretion due to dynamical friction (a model of the dynamical evolution of giant molecular clouds). V. M. Lipunov.
Astron. Zh., Tom 59, 286 - 289 (1982). In Russian. English translation in Soviet Astron., Vol. 26, No. 2.

The model of nonstationary disk accretion due to dynamical friction is suggested. The model is applied to an analysis of the influence of dynamical friction on the motion of the giant molecular clouds in the Galaxy. A comparison between theoretical results and the observed distribution of gas in the galactic disk imposes hard restrictions on the physical parameters of a mean molecular cloud.

131.144 Survey of dark nebulae in the HCN line and in the continuum at 3.4 mm. A. B. Burov, V. N. Voronov, I. I. Zinchenko, A. A. Krasil'nikov, Eh. P. Kukina.
Inst. prikl. fiz. AN SSSR. Prepr., 1981, No. 32, 22 pp. In Russian. − Abstr. in Ref. zh., 51. Astron., 4.51.653 (1982).

131.145 The kinetic chemistry of dense interstellar clouds. T. E. Graedel, W. D. Langer, M. A. Frerking.
Astrophys. J., Suppl. Ser., Vol. 48, 321 - 368 (1982).

A detailed model of the time-dependent chemistry of dense interstellar clouds has been developed to study the dominant chemical processes in carbon and oxygen isotope fractionation, formation of nitrogen-containing molecules, evolution of product molecules as a function of cloud density and temperature, and other topics of interest. The full computation involves 328 individual reactions (expanded to 1067 to study carbon and oxygen isotope chemistry); photo-degradation processes are unimportant in these dense clouds and are excluded. The authors describe the formulation of the chemical model and the calculation of the abundances of the dominant isotopes of the carbon- and oxygen-bearing molecules. Special attention is paid to comparison of the results with observations and to the implications of the results for future observational work. These goals are pursued with calculations that cover a wide range of densities and times; the extensive results are presented compactly for species of particular observational interest by three-dimensional graphical displays.

131.146 **Búsqueda de H neutro en nubes oscuras.**
M. B. Gordon, W. G. L. Poppel.
Bol. Asoc. Argentina Astron., No. 20 - 24, p. 289 (1981).
Abstract.

131.147 **Estudio del H interestelar a bajas latitudes en la zona $-12° \leqslant l \leqslant +12°$.**
M. L. Franco, W. G. L. Poppel.
Bol. Asoc. Argentina Astron., No. 20 - 24, p. 293 (1981).
Abstract.

131.148 **Estudio de una extensa nube de H neutro en la zona de Scorpius-Ophiucus.**
C. A. Olano, W. G. L. Poppel.
Bol. Asoc. Argentina Astron., No. 20 - 24, p. 294 (1981).
Abstract.

131.149 **Evolución química de nubes moleculares.**
E. Iglesias.
Bol. Asoc. Argentina Astron., No. 20 - 24, p. 346 (1981).
Abstract.

131.150 **Distribución del hidrógeno neutro en el hemisferio sur.**
F. R. Colomb, W. Poppel, C. Heiles.
Bol. Asoc. Argentina Astron., No. 20 - 24, p. 361 (1981).
Abstract.

131.151 **Búsqueda de H I en glóbulos de Bok.**
E. M. Arnal, T. Gergely.
Bol. Asoc. Argentina Astron., No. 20 - 24, p. 363 (1981).
Abstract.

131.152 **Ondas de choque en nubes moleculares. I.**
Evolución de las abundancias químicas.
E. R. Iglesias.
Bol. Asoc. Argentina Astron., No. 20 - 24, p. 429 (1981).
Abstract.

131.153 **Estudio de la región $-12° \leqslant l \leqslant 12°, +3° \leqslant b \leqslant +17°$ en la línea de 21 cm.**
M. L. Franco, W. G. L. Poppel.
Bol. Asoc. Argentina Astron. No. 20 - 24, p. 440 (1981).
Abstract.

131.154 **Atlas de hidrógeno galáctico en la región $320° \leqslant l \leqslant 345°, +18° \leqslant b \leqslant +26°$.**
C. A. Olano, W. G. L. Poppel, E. R. Vieira.
Bol. Asoc. Argentina Astron., No. 20 - 24, p. 441 (1981).
Abstract.

131.155 **Distribución del hidrógeno neutro a $|b| \geqslant 10°$.**
F. R. Colomb, W. G. L. Poppel, C. Heiles.

Bol. Asoc. Argentina Astron., No. 20 - 24, p. 494 (1981).
Abstract.

131.156 **Measurements of CH and CH⁺ in diffuse interstellar clouds.**   S. R. Federman.
Astrophys. J., Vol. 257, 125 - 134 (1982).
     A survey of CH and CH⁺ absorption was made for directions with known amounts of atomic and molecular hydrogen. Both radicals are detected only toward directions with substantial amounts of H_2 [$N(H_2) \gtrsim 10^{18}$ cm⁻²]. The column density of CH varies linearly with $N(H_2)$; $N(CH^+)$ is independent of $N(H_2)$. The CH line is shifted in velocity from the CH⁺ line in ~50% of the lines of sight. A model which incorporates the presence of a shock best describes the chemical and the dynamical information available from the data.

131.157 **Components in the interstellar medium toward ϵ Persei and δ Persei.**   E. R. Martin, D. G. York.
Astrophys. J., Vol. 257, 135 - 144 (1982).
     The authors analyze the lines of sight toward ϵ Persei and δ Persei with a procedure that gives velocity components for various interstellar ions. The column densities found for ions expected to be relatively undepleted are used to estimate the column density of neutral hydrogen in each component. The velocities found correspond well with those determined from previous optical studies when the optical components can be resolved. Whenever possible the authors calculate electron density, calcium and titanium depletion, molecular hydrogen excitation temperature, and hydrogen volume density for each component. Toward each star there is one dominant component with high column density, low LSR velocity, a large depletion in Ca and Ti, and a low H_2 excitation temperature. The H_2 results also indicate that the dominant component has a high hydrogen volume density. The components at higher velocities are characterized by lower column densities and less Ca and Ti depletion, relative to the dominant component.

131.158 **Excess line emission in protostellar objects.**   R. I. Thompson.
Astrophys. J., Vol. 257, 171 - 178 (1982).
     Observations of the infrared hydrogen Brackett lines have revealed that many protostellar objects show a significant excess of line flux over the flux expected from an H II region around a ZAMS star of the same luminosity. The line excess appears to be confined to objects with luminosities less than $10^5 L_\odot$. The observations which lead to these conclusions are presented and evaluated. Three possible mechanisms for the excess line flux are discussed: excess UV radiation due to the accretion of natal material onto the star, extended atmospheres due to stellar winds, and deviations from the Menzel case B recombination line strengths due to high density and optical depth in the H II region.

131.159 **The mass of hot, shocked CO in Orion: first observations of the J = 17 → J = 16 transition at 153 microns.**
G. J. Stacey, N. T. Kurtz, S. D. Smyers, M. Harwit, R. W. Russell, G. Melnick.
Astrophys. J., Lett., Vol. 257, L37 - L40 (1982).
     The authors have observed the Kleinmann-Low Nebula in Orion and detected the J = 17→J = 16 transition of CO at a flux level of 7×10^{-17} W cm⁻². Their best estimate for the total mass of hot, $T \gtrsim 750$ K, carbon monoxide in the nebula is 8×10^{30} g. From this value they assess the total hydrogen mass at this temperature to be ~ 1.5 $M_\odot$. A puzzling apparent deficit of oxygen in the nebula is discussed.

131.160 **Astronomical study of the C_3N and C_4H radicals: hyperfine interactions and rho-type doubling.**
M. Guélin, P. Friberg, A. Mezaoui.
Astron. Astrophys., Vol. 109, 23 - 31 (1982).
     The authors report the detection in TMC 1 of low order

rotational transitions arising from the molecular species identified with C_3N and C_4H by Guélin and Thaddeus (1977) and Guélin, Green, and Thaddeus (1978). The hyperfine structure of the transitions is completely resolved for both species. An analysis of this structure confirms the identifications of C_3N and C_4H and yields a precise determination of the fine and hyperfine constants. The C_4H $N = 2-1$ line is surprisingly strong in TMC 1 and is detected in all seven dark clouds which were surveyed. The C_4H/HC_3N abundance ratio is found to be about the same ($\cong 3$) in these clouds. This ratio, in contrast, is smaller in hotter clouds such as SgrB 2.

131.161 **Star formation in the NH₃ cloud of the NGC 2071 region.**
G. Calamai, M. Felli, S. Giardinelli.
Astron. Astrophys., Vol. 109, 123 - 130 (1982).
The authors present a detailed ammonia map of an extended molecular cloud in the Orion complex, close to the reflection nebula NGC 2071. The kinetic temperature ($T_K \approx 20$ K), the inferred density ($n(H_2) \gtrsim 10^4$ cm^{-3}) and the total derived mass ($M \approx 10^3$ $M_\odot$) classify it as a typical molecular cloud in which star formation may occur. The peak of the ammonia cloud is located in a region where H_2O maser emission and a small hot infrared source, having the typical characteristics of a BN-type object, have been found. The authors have detected an unresolved radio source at 6 cm at the level of 6 mJy at the position of the IR peak. The existence of an H II region points out that star formation has already occurred in the center of the molecular cloud.

131.162 **Contribution to the warm intercloud medium to the diffuse ultraviolet background.**
J. M. Deharveng, M. Joubert, P. Barge.
Astron. Astrophys., Vol. 109, 179 - 181 (1982).
The far UV emission from the warm intercloud medium (mainly two photon continuum and emission lines) is evaluated and compared to observations of the diffuse UV radiation. Whereas this emission may account for the UV surface brightness of faint extended H II regions, it is found unable to contribute significantly to the general diffuse UV background.

131.163 **Observación de H I en filamentos obscuros.**
J. C. Cersósimo, W. G. L. Poppel.
Bol. Asoc. Argentina Astron., No. 26, p. 105 (1981).
Abstract.

131.164 **Estudio de la distribución y cinemática del H I en la región** $270° < l < 320°$ **y** $3° < b < 17°$.
C. A. Olano.
Bol. Asoc. Argentina Astron., No. 26, p. 106 (1981).
Abstract.

131.165 **Relevamiento de nubes de alta velocidad.**
E. Bajaja, C. Cappa, J. Cersósimo, N. Loiseau, C. Martín, R. Morras, C. Olano, W. G. L. Poppel.
Bol. Asoc. Argentina Astron., No. 26, p. 107 (1981).
Abstract.

131.166 **Observaciones de H I conectadas con un estallido en rayos X detectado en la zona de Lupus.**
F. R. Colomb, G. M. Dubner.
Bol. Asoc. Argentina Astron., No. 26, p. 112 - 116 (1981).

131.167 **Estructura fina en nubes de alta velocidad.**
E. Bajaja, R. Morras.
Bol. Asoc. Argentina Astron., No. 26, p. 187 (1981).
Abstract.

131.168 **Photodissociation rates of interstellar molecules in an intercloud region.**
Acta Univ. Carolinae Math. Phys., Vol. 23, No. 1, p. 69 - 72 (1982).

Photodissociation rates of interstellar molecules CO, HCN, NH_3, H_2CO and HC_3N are derived for a region with low interstellar extinction.

131.169 **Protostars and their envelopes.** H. W. Yorke.
Scientific importance of high angular resolution at infrared and optical wavelengths, (see 012.031), p. 319 - 340 (1981).

131.170 **The impact of high angular resolution observations on the study of pre-main-sequence objects.**
H. J. Habing.
Scientific importance of high angular resolution at infrared and optical wavelengths, (see 012.031), p. 341 - 349 (1981).
The usefulness of high angular resolution observations is demonstrated mainly by discussing recent VLBI results on H_2O masers in Orion.

131.171 **Comparison of IR maps with a 2″ map of NH₃ for the Orion Molecular Cloud.**
T. L. Wilson, T. A. Pauls, R. N. Martin, J. Bieging.
Scientific importance of high angular resolution at infrared and optical wavelengths, (see 012.031), p. 397 - 398 (1981).
Abstract.

131.172 **Interstellar absorption and distance scale in the Galaxy.** K. A. Barkhatova, O. P. Pyl'skaya.
Astron. Tsirk., No. 1168, p. 5 - 7 (1981). In Russian.

131.173 **OH 205.1-14.1.**
IAU Circ., No. 3700 (1982).

131.174 **18-cm emission from W49N at negative radial velocities.** M. I. Pashchenko.
Astron. Tsirk., No. 1185, p. 1 - 2 (1981). In Russian.

131.175 **An upper limit on the electron concentration in Heiles' dust cloud 2 determined from decameter-wave observations.**
E. A. Abramenkov, V. V. Krymkin, A. A. Rejtblat, V. I. Slysh.
Astron. Zh., Tom 59, 500 - 502 (1982). In Russian. English translation in Soviet Astron., Vol. 26, No. 3.
The attempts to observe the effect of absorption of the decametric background emission (at the frequencies 14.7, 16.7 20 MHz) by Heiles dust cloud 2 put an upper limit on the emission measure of the cloud $\approx 10^{-2}$ cm^{-6} pc and the electron density $\approx 10^{-1}$ cm^{-3}.

131.176 **Observations of 2 μm molecular hydrogen emission from NGC 2071, Cepheus A, and GL 961.**
J. Bally, A. P. Lane.
Astrophys. J., Vol. 257, 612 - 619 (1982).
The infrared emission line at $\lambda 2.12 \mu$m from the $v=1-0$ $S(1)$ transition of molecular hydrogen has been detected and partially mapped in the vicinity of the high-velocity molecular flows associated with NGC 2071, Cepheus A, and GL 961. The H_2 emission regions are extended on a scale at least 0.2 to 0.3 pc, comparable to or somewhat greater than the size of the Orion H_2 source. The H_2 surface brightness, uncorrected for possible extinction, is, however, at least an order of magnitude lower in NGC 2071, Cep A, and GL 961 than in Orion. The source GL 490 was mapped over a region subtending several arcminutes and does not exhibit $S(1)$ line emission above a flux level of 1×10^{-20} W cm^{-2}.

131.177 **Low-mass star formation in the dense interior of Barnard 18.** P. C. Myers.
Astrophys. J., Vol. 257, 620 - 632 (1982).
CO observations, visible obscuration, and near-infrared estimates of extinction in the star-forming cloud Barnard 18 indicate that the relatively dense [$n(H_2) \lesssim 3000$ cm^{-3}], visibly opaque region is associated with a less dense

$[n(H_2) \lesssim 700 \text{ cm}^{-3}]$, nearly transparent region of greater spatial extent (~ 2 pc $\times$ 5 pc). A two-component model of the cloud and measurements of stellar extinction give estimates of the line-of-sight stellar positions with respect to the gas. These indicate that the dense inner gas has nine known stars, star formation efficiency $\sim 2\%$, and star space density ~ 8 pc^{-3}; the less dense outer gas has eight known stars, star formation efficiency $\sim 1\%$, and star space density ~ 0.9 pc^{-3}. Therefore, these young ($\sim 3 \times 10^5$ yr), low-mass stars are preferentially located in the dense inner gas.

131.178 **Star formation in the M17 SW giant molecular cloud.**
D. T. Jaffe, G. G. Fazio.
Astrophys. J., Lett., Vol. 257, L77 - L81 (1982).
The authors have used a far-infrared survey of the M17 SW giant molecular cloud complex, excerpted from the survey of Jaffe, Stier, and Fazio, to investigate OB star formation mechanisms. There are 13 far-infrared sources within the 20 km s^{-1} M17 SW giant molecular cloud complex. The authors summarize the far-infrared, radio continuum, CO, and H$_2$O observations of these 13 sources. They discuss and evaluate three mechanisms for the formation of OB stars within the giant molecular cloud and conclude that sequential formation of the OB clusters probably cannot explain the sources within the complex. Both external triggering by a spiral shock and stochastic formation are reasonable explanations of the observations.

131.179 **Collective formation of massive stars in galaxies.**
H.-H. Loose, K. J. Fricke.
The most massive stars, (see 012.034), p. 269 - 292 (1981).
Numerical simulations of collective star formation in the gravitational field of a background star distribution are presented. Two types of solutions describing bursts of star formation and very localized continuous star formation leading to compact cores are found. A versatile analytical treatment of these processes is described and applied to star formation bursts in external galaxies like the compact dwarf galaxies and to the formation of compact nuclei as observed in the Seyfert phenomenon. Massive stars are the driving agent in this evolutionary scenario. The effects of contagious star formation and of the thresholds for the onset of star formation are discussed.

131.180 **Observations of stellar wind blown bubbles.**
Y. H. Chu, R. R. Treffers.
Bull. American Astron. Soc., Vol. 13, 783 (1981). – Abstract.

131.181 **High velocity CO toward Ae and Be stars associated with nebulosity.**
N. Calvet, J. Canto, L. F. Rodriguez.
Bull. American Astron. Soc., Vol. 13, 784 (1981). – Abstract.

131.182 **High-velocity interstellar resonance lines in the UV spectra of Wolf-Rayet stars.** J. N. Heckathorn.
Bull. American Astron. Soc., Vol. 13, 784 (1981). – Abstract.

131.183 **A CO survey of the southern galactic plane.**
W. H. McCutcheon, B. J. Robinson, J. B. Whiteoak.
Bull. American Astron. Soc., Vol. 13, 787 (1981). – Abstract.

131.184 **Massive star formation in NGC 6946.**
K. Degioia-Eastwood, G. L. Grasdalen, S. E. Strom, K. M. Strom.
Bull. American Astron. Soc., Vol. 13, 797 (1981). – Abstract.

131.185 **The inhibition of star formation in barred spiral galaxies.** A. D. Tubbs.
Bull. American Astron. Soc., Vol. 13, 797 (1981). – Abstract.

131.186 **Multifrequency photometry of the Orion Ridge at far infrared wavelengths.**

J. Smith, D. A. Harper, R. Hildebrand.
Bull. American Astron. Soc., Vol. 13, 808 (1981). – Abstract.

131.187 **H I Zeeman measurements at five high-latitude positions.**
C. Heiles, T. Troland, M. Stevens.
Bull. American Astron. Soc., Vol. 13, 824 (1981). – Abstract.

131.188 **The profile of the λ4430 diffuse interstellar band, observed with the MAMA detector.**
T. P. Snow, J. G. Timothy, S. Saar.
Bull. American Astron. Soc., Vol. 13, 825 (1981). – Abstract.

131.189 **Effects of a weak shock on depletion in the ζ Ophiuchi cloud.**
K. A. Meyers, T. P. Snow, S. R. Federman, M. Breger.
Bull. American Astron. Soc., Vol. 13, 825 (1981). – Abstract.

131.190 **The production of HCO⁺ in a shocked interstellar cloud.** G. F. Mitchell, T. J. Deveau.
Bull. American Astron. Soc., Vol. 13, 825 (1981). – Abstract.

131.191 **MHD shocks in the Orion Molecular Cloud.**
D. Chernoff, C. McKee, D. Hollenbach.
Bull. American Astron. Soc., Vol. 13, 826 (1981). – Abstract.

131.192 **Evidence for a moderate velocity interstellar shock propagating through intercloud gas.**
R. J. Reynolds.
Bull. American Astron. Soc., Vol. 13, 826 (1981). – Abstract.

131.193 **Shock processing of interstellar grains.**
C. G. Seab, J. M. Shull, T. P. Snow.
Bull. American Astron. Soc., Vol. 13, 826 (1981). – Abstract.

131.194 **Turbulence and the stellar mass function.**
R. C. Fleck.
Bull. American Astron. Soc., Vol. 13, 826 (1981). – Abstract.

131.195 **Diffuse clouds, Alfvén waves, and the pressure of the intercloud medium.** J. C. Higdon.
Bull. American Astron. Soc., Vol. 13, 826 (1981). – Abstract.

131.196 **The evaporation of spherical clouds – revisited.**
J. L. Giuliani.
Bull. American Astron. Soc., Vol. 13, 827 (1981). – Abstract.

131.197 **Ambipolar diffusion in shock-compressed interstellar clouds.** M. J. Ridgway, D. C. Black.
Bull. American Astron. Soc., Vol. 13, 827 (1981). – Abstract.

131.198 **A compact radio source associated with the Becklin-Neugebauer object.**
J. M. Moran, G. Garay, M. J. Reid, R. Genzel.
Bull. American Astron. Soc., Vol. 13, 852 - 853 (1981).
Abstract.

131.199 **Observations of HNCO in the Sgr A molecular clouds.** J. T. Armstrong, A. H. Barrett.
Bull. American Astron. Soc., Vol. 13, 853 (1981). – Abstract.

131.200 **Observations of NH₃ toward the Sgr A molecular clouds.**
P. T. P. Ho, S. Vogel, J. T. Armstrong, A. H. Barrett.
Bull. American Astron. Soc., Vol. 13, 853 (1981). – Abstract.

131.201 **Observation of interstellar ammonia ice.**
R. F. Knacke, S. McCorkle, R. C. Puetter,
E. F. Erickson, W. Krätschmer.
Bull. American Astron. Soc., Vol. 13, 854 (1981). – Abstract.

131.202 **Ammonia observations of regions with bipolar mass outflow.**
J. M. Torrelles, L. F. Rodriguez, J. Canto, J. M. Moran, P. T. P. Ho, J. Marcaide.
Bull. American Astron. Soc., Vol. 13, 854 (1981). — Abstract.

131.203 **Low frequency recombination line observations toward 3C123.**
H. E. Payne, E. E. Salpeter, Y. Terzian.
Bull. American Astron. Soc., Vol. 13, 854 (1981). — Abstract.

131.204 **An IUE abundance survey of interstellar heavy elements.** J. M. Shull, M. Van Steenberg.
Bull. American Astron. Soc., Vol. 13, 855 (1981). — Abstract.

131.205 **Protostellar mass and angular momentum loss.**
K. B. MacGregor, L. Hartmann.
Bull. American Astron. Soc., Vol. 13, 855 (1981). — Abstract.

131.206 **Millimeter molecular astronomy.** M. L. Kutner.
Bull. American Astron. Soc., Vol. 13, 862 - 863 (1981). — Abstract.

131.207 **HCO synthesis maps of velocity structure in K3 - 50.**
S. N. Vogel, W. J. Welch.
Bull. American Astron. Soc., Vol. 13, 864 (1981). — Abstract.

131.208 **Millimeter and optical dark cloud observations.**
R. M. Crutcher.
Bull. American Astron. Soc., Vol. 13, 864 (1981). — Abstract.

131.209 **Low mass star formation in the dense interior of Barnard 18.** P. C. Myers.
Bull. American Astron. Soc., Vol. 13, 864 (1981). — Abstract.

131.210 **Interstellar turbulence and the origin of isolated dark globules.**
C. M. Leung, M. L. Kutner, K. N. Mead.
Bull. American Astron. Soc., Vol. 13, 864 (1981). — Abstract.

131.211 **Mesoturbulence in molecular clouds: a numerical approach.** R. Dickman, S. Kleiner.
Bull. American Astron. Soc., Vol. 13, 864 - 865 (1981). Abstract.

131.212 **VLA observations of H₂CO clumps in DR 21.**
H. R. Dickel, A. F. Lubenow, J. R. Forster, A. H. Rots, W. M. Goss.
Bull. American Astron. Soc., Vol. 13, 886 - 887 (1981). Abstract.

131.213 **X-ray ionization effects in the Orion molecular cloud.** T. Kallman, J. Krolik.
Bull. American Astron. Soc., Vol. 13, 887 (1981). — Abstract.

131.214 **Ultraviolet observations of a non-radiative shock wave.** J. C. Raymond, R. Fesen, T. R. Gull.
Bull. American Astron. Soc., Vol. 13, 887 (1981). — Abstract.

131.215 **The peculiar extinction of Herschel 36.**
J. Hecht, H. L. Helfer, J. Wolf, B. Donn, J. L. Pipher.
Bull. American Astron. Soc., Vol. 13, 887 - 888 (1981). Abstract.

131.216 **A search for high density gas in broad-wing molecular sources.** H. A. Thronson, C. J. Lada.
Bull. American Astron. Soc., Vol. 13, 888 (1981). — Abstract.

131.217 **Evolution of inhomogeneities in molecular clouds.**
D. Gilden, J. Scalo.
Bull. American Astron. Soc., Vol. 13, 888 (1981). — Abstract.

131.218 **Detection of rotationally excited OH emission from Vy 2 - 2, and studies of carbon chain molecules in cool circumstellar shells.** P. R. Jewell, L. E. Snyder.
Bull. American Astron. Soc., Vol. 13, 895 (1981). — Abstract.

131.219 **Infrared photometry of twin-line OH masers.**
R. D. Gehrz, G. L. Grasdalen, J. A. Hackwell, S. G. Kleinmann.
Bull. American Astron. Soc., Vol. 13, 895 (1981). — Abstract.

131.220 **The detection of N V absorption by Milky Way halo gas toward the SMC.**
B. D. Savage, E. L. Fitzpatrick.
Bull. American Astron. Soc., Vol. 13, 895 (1981). — Abstract.

131.221 **W50 as a probe of the interstellar medium.**
A. Königl.
Bull. American Astron. Soc., Vol. 13, 896 (1981). — Abstract.

131.222 **Interstellar absorption lines in the direction of extragalactic objects.** J. C. Blades, D. C. Morton.
Anglo-Australian Obs. Prepr. No. 168, 22 pp. (1982). — Submitted to Mon. Not. R. Astron. Soc.

131.223 **Excess line emission in protostellar objects.**
R. I. Thompson.
Prepr. Steward Obs., No. 367, 29 pp. (1982).

131.224 **Airborne observations of the Orion molecular hydrogen emission spectrum.**
D. S. Davis, H. P. Larson, H. A. Smith.
Prepr. Steward Obs., No. 377, 55 pp. (1982).

131.225 **M supergiants and star formation at the Galactic Center.**
M. J. Lebofsky, G. H. Rieke, A. T. Tokunaga.
Prepr. Steward Obs., No. 397, 19 pp. (1982).

131.226 **Structure and characteristics of turbulent interstellar clouds.** L. N. Arshutkin, I. G. Kolesnik.
Astrofizika, Tom 17, 359 - 370 (1981). In Russian. English translation in Astrophysics, Vol. 17, No. 2.
The influence of turbulence on the structure and parameters of interstellar clouds is considered. The results of model calculations for spheric-symmetrical clouds with turbulent pressure are given.

131.227 **Vibrationally excited molecular hydrogen in circumstellar clouds and the interstellar CH⁺ abundance.** A. Freeman, D. A. Williams.
Astrophys. Space Sci., Vol. 83, 417 - 422 (1982).
The production of CH^+ in dense interstellar clouds under intense UV irradiation is discussed. A model applicable to the cloud towards the star 20 Tau is described.

131.228 **Segregation of dust grains in dark clouds.**
H. C. Bhatt, J. N. Desai.
Astrophys. Space Sci., Vol. 84, 163 - 172 (1982).
Gravitational settling of dust grains in dark clouds has been considered. It has been shown that such a process gives rise to a modification of the grain size distribution. Starting with a simple model of uniform spherical cloud and normal interstellar grain size distribution for the dust the authors derive expressions for the modified grain size distribution function, average grain size and extinction as functions of distance from the cloud's center and the age of the cloud. The mean grain size increases toward the center of the cloud as does the extinction. Results of the numerical evaluation of these quantities have been discussed with their implications for the observations of anomalous reddening and polarization within dark clouds and Bok globules.

131.229 'Interstellar absorptions at $\lambda = 3.2\,\mu$m and 3.3 μm.
S. Al-Mufti, A. H. Olavesen, F. Hoyle,
N. C. Wickramasinghe.
Astrophys. Space Sci., Vol. 84, 259 - 261 (1982).
   Unidentified interstellar absorption bands at $\lambda = 3.2\,\mu$m
and 3.3 μm might be due to amino-acids in bacterial grains.

131.230 **Ambipolar diffusion in collapsing interstellar clouds.**
T. C. Mouschovias, E. V. Paleologou.
Bull. American Astron. Soc., Vol. 14, 571 (1982). – Abstract.

131.231 **The nearby interstellar medium.**   P. C. Frisch.
Bull. American Astron. Soc., Vol. 14, 574 (1982).
Abstract.

131.232 **Cyanoacetylene as a density probe of molecular**
clouds.
P. A. Vanden Bout, R. B. Loren, R. L. Snell, A. Wootten.
Bull. American Astron. Soc., Vol. 14, 575 (1982). – Abstract.

131.233 **Hyperfine anomalies of HCN in cold dark clouds.**
C. M. Walmsley, E. Churchwell, A. Nash,
E. Fitzpatrick.
Bull. American Astron. Soc., Vol. 14, 576 (1982). – Abstract.

131.234 **UV radiation field inside dense clouds: possibilities**
and implications.
S. S. Prasad, S. P. Tarafdar.
Bull. American Astron. Soc., Vol. 14, 576 (1982). – Abstract.

131.235 **Molecules in space.**   M. Elitzur.
Ann. Israel Phys. Soc., Vol. 4, 271 - 286 (1980).
Abstr. in Phys. Abstr., Vol. 85, Abstr. 58435 (1982).

131.236 **Sharp fronts in interstellar medium.**
K. K. Ghosh, S. N. Ghosh.
Indian J. Phys. Part B, Vol. 55B, 443 - 446 (1981). – Abstr.
in Phys. Abstr., Vol. 85, Abstr. 58436 (1982).

131.237 **Dissipation of supersonic turbulence in interstellar**
clouds.   J. M. Scalo, W. A. Pumphrey.
Astrophys. J., Lett., Vol. 258, L29 - L33 (1982).
   The dissipation of supersonic turbulent motions in interstellar clouds is reconsidered using N-body simulations of systems of interacting gas fragments. The adopted collision rules
are based on recent hydrodynamical studies of cloud collisions.
It is shown that the preponderance of oblique collisions which
dissipate relatively little kinetic energy results in a dissipation
time scale which exceeds the geometrical collision time scale
by a factor of about 20 or more. The authors also discuss the
process of turbulent virialization, whereby cloud gravitational
potential energy is converted into random internal motions,
leading to a slow cloud contraction controlled by the turbulent
dissipation rate. The cloud lifetime can be extended significantly beyond the free-fall time by this process, without the need
for any additional source of turbulent energy.

131.238 **Emission globules in Musca.**   N. J. Irvine.
Publ. Astron. Soc. Pacific, Vol. 94, 239 - 243 (1982).
   A group of Bok globules in Musca exhibit weak emission
by virtue of their proximity to α Mus. The association of star
and nebulae permits an unusually good distance determination
for the globules. A faint Hα emission star is coincident with
one of the globules.

131.239 **The interstellar 2200 Å band: a catalogue of**
equivalent widths.
J. Gürtler, R. Schielicke, J. Dorschner, C. Friedemann.
Astron. Nachr., Band 303, 105 - 116 (1982) = Mitt. Univ.-
Sternw. Jena, Nr. 151.
   Using the data of the Ultraviolet Bright-Star Spectro-
photometric Catalogue, the equivalent width of the interstellar

absorption band at 2200 Å is derived for 422 stars. A preliminary statistical analysis confirms the good correlation between the equivalent width of the 2200 Å band and the colour
excess. The wavelength position of the band centre seems to
vary to a small extent.

131.240 **High-resolution synthesis maps of 86 GHz SO emis-**
sion toward Orion-KL.
R. L. Plambeck, M. C. H. Wright, B. Baud, J. Bieging,
P. T. P. Ho, S. N. Vogel, W. J. Welch.
Bull. American Astron. Soc., Vol. 13, 865 (1981). – Abstract.

131.241 **Interstellar grain explosions: molecule cycling**
between gas and dust.   L. B. d'Hendecourt,
L. J. Allamandola, F. Baas, J. M. Greenberg.
Astron. Astrophys., Vol. 109, L12 - L14 (1982).
   In dense molecular clouds all small condensible molecules
should have been frozen onto the dust forming mantles.
Although these accretion mantles are observed, a significant
number of such molecules (CO, H_2O etc.) have also been
detected in the gas suggesting that an equilibrium between
accretion on and ejection from grains exists. Laboratory
evidence for molecule ejection from vacuum ultraviolet
photolyzed grain mantle analogs is shown to provide the basic
mechanism for this equilibrium.

131.242 **Further ($^{12}C/^{13}C$) ratios from formaldehyde: a**
variation with distance from the galactic center.
C. Henkel, T. L. Wilson, J. Bieging.
Astron. Astrophys., Vol. 109, 344 - 351 (1982).
   The authors present new observational material, and also
apply corrections for photon trapping and for hyperfine
splitting in narrow lines to the uncorrected $H_2^{12}CO/H_2^{13}CO$
ratios of Gardner and Whiteoak (1979). The authors combine
these results with data obtained at Effelsberg to check for the
presence of a gradient in the $^{12}C/^{13}C$ ratio with distance from
the galactic center and for source to source variations.

131.243 **Radio observations of molecules in the interstellar**
gas.   P. Thaddeus.
Philos. Trans. R. Soc. London, Ser. A, Vol. 303, (see 012.038),
469 - 486 (1981).

131.244 **Near infrared spectroscopy of protostars.**
N. Z. Scoville.
Philos. Trans. R. Soc. London, Ser. A, Vol. 303, (see 012.038),
487 - 496 (1981).
   Observational study of protostars and their immediate
environs has recently become possible as a result of advances
in infrared spectroscopy, especially in the near infrared
($\lambda = 2-5\,\mu$m). Data on the BN object covering the CO, ^{13}CO,
and H_2 vibrational bands and the H II lines are presented.

131.245 **Chemical processes in the interstellar medium: on**
the nature of the carrier of the diffuse interstellar
bands.   G. H. Exarhos, J. Mayer, W. Klemperer.
Philos. Trans. R. Soc. London, Ser. A, Vol. 303, (see 012.038),
503 - 511 (1981).
   The great progress of the past decade in molecular astronomy, primarily by millimetre radio observations, has focused
attention upon the synthesis of observed gas-phase species.
The authors briefly discuss chemical modelling of the gas
phase. The major portion of the article is devoted to the
discussion of the nature of the carrier of the diffuse interstellar
bands. The paper suggests that the carrier is transition metal
ions interacting with the polysulphide ions S_2^- and S_3^- in oxidic
lattices.

131.246 **Chemical processes in the shocked interstellar gas.**
A. Dalgarno.
Philos. Trans. R. Soc. London, Ser. A, Vol. 303, (see 012.038),
513 - 522 (1981).

A review is presented of molecular formation and destruction processes in shocked interstellar gas, and the chemistry of some particular regions of the interstellar medium is briefly discussed.

131.247 Interstellar deuterium chemistry.
R. D. Brown, E. Rice.
Philos. Trans. R. Soc. London, Ser. A, Vol. 303, (see 012.038), 523 - 533 (1981).

An interstellar reaction scheme of the type described by E. Iglesias (1977) has been extended to include deuterium chemistry and also isomeric forms of some molecules.

131.248 A search for interstellar H_3^+.
T. Oka.
Philos. Trans. R. Soc. London, Ser. A, Vol. 303, (see 012.038), 543 - 549 (1981).

Based on the results of recent laboratory observation of the infrared ν_2 fundamental band of the H_3^+ molecular ion, the possibility of observing this important ion in interstellar space is discussed. An observation of this spectrum has been attempted.

131.249 High-resolution interstellar spectroscopy and star formation.   G. Winnewisser.
Philos. Trans. R. Soc. London, Ser. A, Vol. 303, (see 012.038), 565 - 579 (1981).

During the past several years, high spatial and spectral resolution molecular spectroscopy has greatly contributed to our knowledge of the physics, dynamics and chemistry of interstellar molecular clouds and thus has led to a better understanding of the conditions that lead to star formation. According to their physical properties, molecular clouds can be grouped into four different types: (1) the dark clouds, (2) the molecular clouds associated with H II regions, (3) the 'protostellar' (or maser) environment, and (4) the molecular envelopes of late-type stars. Typical examples of the four types are discussed.

131.250 Magnetic fields in dense interstellar clouds.
R. D. Davies.
Philos. Trans. R. Soc. London, Ser. A, Vol. 303, (see 012.038), 581 - 587 (1981).

The author explores the evidence for the strength and arrangement of magnetic fields in interstellar clouds ranging in density from ca. 1 to ca. $10^8 \, cm^{-3}$. It is concluded that molecular clouds with densities in this range contain magnetic fields whose strength is determined by the compression of the intrinsic interstellar field consequent upon cloud collapse.

131.251 CO $J = 3 \rightarrow 2$ and $J = 2 \rightarrow 1$ observations of the Kleinmann–Low nebula.
J. P. Phillips, G. J. White, G. D. Watt.
Mon. Not. R. Astron. Soc., Vol. 199, 1033 - 1043 (1982).

An area of sky centred on the Kleinmann–Low nebula has been mapped in the $J = 3 \rightarrow 2$ and $J = 2 \rightarrow 1$ lines of CO. There appears evidence for asymmetry of the 'pedestal' emission to the west of the source centre, probably indicative of an irregular mass distribution. This is discussed in the context of two models of the high-velocity gas, namely high-velocity non-isotropic outflow, and rotation within a compact cloud. Although the present observations do not enable to discriminate between these, the presence of optically thin CO emission in this source is shown to lead to severe structural constraints on any rotational model.

131.252 Observations of the $J = 4 \rightarrow 3$ transition of HCO^+ in OMC 1, GL 961, Mon R 2 and NGC 2071.
R. Padman, P. F. Scott, A. S. Webster.
Mon. Not. R. Astron. Soc., Vol. 200, 183 - 189 (1982).

The authors have observed the HCO^+ $J = 4 \rightarrow 3$ transition in OMC 1 and three other galactic sources which have self-reversed CO profiles. None of the sources show self-absorption in this line and, in NGC 2071 and Mon R2, the line wings, seen in CO and HCO^+ $1 \rightarrow 0$, have disappeared. Minimum masses between 200 and 1200 $M_\odot$ are derived for the high-density cores; these lower limits will only be reduced significantly if HCO^+ is much more abundant than has previously been thought. A new compact, optically thin source is postulated in OMC 1.

131.253 Observations of H_2O maser emission in southern galaxies.   R. A. Batchelor, D. L. Jauncey, J. B. Whiteoak.
Mon. Not. R. Astron. Soc., Vol. 200, 19P - 21P (1982).

The first observations in a search for H_2O masers in southern galaxies involved objects containing previously reported maser emission. The maser in NGC 4945 was confirmed, but no emission exceeding 0.5 Jy was detected in NGC 253 or in N159 of the Large Magellanic Cloud. The maser in NGC 4945 is associated with the nucleus of the galaxy and has a luminosity ($165 L_\odot$) at least an order of magnitude greater than any maser of our Galaxy.

131.254 VLBI observations of the maser emission in the four ground-state OH transitions from W3(OH) and Sgr B2.   J. E. Fouquet, M. J. Reid.
Astron. J., Vol. 87, 691 - 694 (1982).

The authors have obtained accurate, relative positions of the maser emission in all four ground-state OH transitions from W3(OH) with VLBI observations. All ground-state OH maser emission comes from the vicinity of the compact H II region. A comparison of total-power spectra taken over nearly a 15-yr time span shows some changes, but suggests that most OH maser features have lifetimes greater than several decades. The four ground-state OH masers of Sgr B2 were totally resolved and have apparent sizes greater than 0.05 arcsec. Such large sizes are probably a result of interstellar plasma scattering.

131.255 The wavelength dependence of interstellar linear polarization: stars with extreme values of λ_{max}.
B. A. Wilking, M. J. Lebofsky, G. H. Rieke.
Astron. J., Vol. 87, 695 - 697 (1982).

Infrared polarimetry of 13 stars with predominantly extreme values of λ_{max} has been obtained. Combining these new data with the 20 stars observed by Wilking et al. (1980) permits a more accurate description of the variation of the full width normalized linear polarization curve with λ_{max}. Infrared polarimetry of κ Cas is combined with existing ultraviolet and optical polarimetry to suggest that the broadening of the normalized linear polarization curve in the near ultraviolet is significant for stars with short λ_{max}.

131.256 Interstellar reddening and distribution of B stars in the solar neighbourhood: impact of the Hipparcos/Tycho mission.   J. Guibert, D. Egret.
Scientific aspects of the Hipparcos space astrometry mission, (see 012.041), p. 195 - 198 (1982).

131.257 Gamma-rays from cosmic-ray irradiated molecular clouds.   M. R. Issa, T.-P. Li.
Philos. Trans. R. Soc. London, Ser. A, Vol. 301, (see 012.043), 533 - 535 (1981).

An examination is made of the contribution to the number of apparently discrete sources of γ-rays from cosmic rays interacting with molecular clouds in the Galaxy. Attention is directed to specific nearby clouds and to clouds in general, the latter by a Monte-Carlo analysis.

131.258 Molecular observations of and in dark clouds.
P. Friberg.
Rep. Obs. Lund, No. 18, (see 012.044), p. 41 - 42 (1982).

131.259 Spectral scan of Orion A and IRC+10216 — astrophysics and chemistry in a star formation region and in the envelope of a carbon rich star.
Å. Hjalmarson.
Rep. Obs. Lund, No. 18, (see 012.044), p. 43 (1982).

131.260 HCO⁺ in molecular clouds with star formation.
Aa. Sandqvist, A. Wootten, R. B. Loren,
P. Friberg, Å. Hjalmarson.
Rep. Obs. Lund, No. 18, (see 012.044), p. 44 - 45 (1982).

131.261 Observations of H_2O masers with the Metsähovi 13.7-m radio telescope.   K. Mattila, L. Malkamäki,
N. Holsti, G. Sandell, R. Anttila, M. Toriseva.
Rep. Obs. Lund, No. 18, (see 012.044), p. 46 - 47 (1982).

131.262 Millimeter wave studies of circumstellar clouds.
H. Olofsson.
Rep. Obs. Lund, No. 18, (see 012.044), p. 48 - 50 (1982).

131.263 Ortho-to-para ratios in interstellar ammonia.
T. L. Wilson, W. Batrla, T. A. Pauls.
Astron. Astrophys., Vol. 110, L20 - L22 (1982).
Spectra toward the continuum source W33 for the four lowest metastable rotation-inversion doublets of ammonia are presented. From the presence of narrow absorption features in the $(J,K) = (1,1)$, $(2,2)$, and $(4,4)$ lines of para-NH_3 and the absence of any absorption in the $(3,3)$ line of ortho-NH_3, the authors conclude that either the excitation temperature of the $(3,3)$ line is considerably larger than that of the para species or the population in the $(3,3)$ doublet levels is slightly inverted. This behavior is not evident in all sources studied and this suggests that specific physical conditions have an influence on this phenomenon.

131.264 Dynamics of the supergiant shell LMC 2 in the Large Magellanic Cloud.   A. Caulet,
L. Deharveng, Y. M. Georgelin, Y. P. Georgelin.
Astron. Astrophys., Vol. 110, 185 - 197 (1982).
The velocity field of the supergiant shell LMC 2 was derived from sixteen Hα photographic Fabry-Perot interferograms. 967 radial velocities have been obtained over the ionized filaments and the nearby giant H II regions. The photon flux necessary for the ionization of LMC 2 has been estimated from its integrated flux in the radio continuum; it has been compared to the flux of ionizing photons which is likely to be emitted by the inner stellar associations. The supergiant shell might result from the combined action of stellar winds and supernova explosions. Its age is thus of about 10^7 yr.

131.265 Can giant molecular clouds form in spiral arms?
F. Casoli, F. Combes.
Astron. Astrophys., Vol. 110, 287 - 294 (1982).
The authors study quantitatively the efficiency of molecular cloud collisions and coalescence in forming Giant Molecular Clouds (GMC). Taking as an hypothesis that these GMC have a short lifetime $(4 \times 10^7 \text{yr})$ it is shown that they still have time to be formed substantially in spiral arms. The ensemble of molecular clouds never reaches a steady state in its rotation around the galactic center and cannot be considered as a fluid. The mass and velocity distributions derived are compatible with available molecular observations.

131.266 Radiation pressure effect on dust surrounding the Pleiades bright stars.   P. Bouvier.
Compendium in astronomy, (see 003.012), p. 43 - 54 (1982).
The action of the radiation pressure from a B-star on a dust grain is reexamined in the celestial mechanics approximation of a two-body problem. When the author turns to the reflection nebulae associated with the bright stars of the Pleiades cluster he is faced with several observational difficulties; nevertheless, in the case of the Merope nebula the existence of dust in the close neighbourhood of star 23 Tau is indirectly confirmed.

131.267 On the origin of nebulae.
V. A. Ambartsumian (*Ambartsumyan*).
Compendium in astronomy, (see 003.012), p. 211 - 218 (1982).
The author gives a short review of ideas on the origin of nebulae and shows that the unified picture of the origin of all nebulae in the Galaxy from masses ejected by dense bodies seems now more attractive than ever.

131.268 Review of current state of observations of extragalactic molecules.   L. J. Rickard.
Extragalactic molecules, (see 012.048), p. 1 - 11 (1982).

131.269 GMCs in M 31.   L. Blitz.
Extragalactic molecules, (see 012.048), p. 93 - 100 (1982).
In an attempt to compare properties of Giant Molecular Clouds (GMCs) in M 31 with those in the Milky Way, the author has made CO observations of 54 H II regions from the Baade and Arp catalogue of H II regions with the 11 m telescope at NRAO.

131.270 One perspective on the workshop "Extragalactic molecules".   F. J. Kerr.
Extragalactic molecules, (see 012.048), p. 153 - 154 (1982).

131.271 The large scale distribution of molecular clouds in the outer Galaxy.   K. N. Mead.
Extragalactic molecules, (see 012.048), p. 155 - 156 (1982).
The author used the NRAO 36-ft telescope to survey the first quadrant of the Galaxy for CO emission from molecular clouds outside the solar circle. Three strips, at b = 1.3, 1.5 and 1.7° were observed in the longitude range 55° to 95° in 0.1° steps. The latitude range corresponds to the centroid latitude of negative velocity H I. A crude estimate of the H_2 mass in the outer Galaxy can be made from the data.

131.272 Physical properties of molecular clouds in the outer Galaxy.   M. L. Kutner.
Extragalactic molecules, (see 012.048), p. 157 - 158 (1982).

131.273 Can spiral arms form giant molecular clouds?
F. Combes.
Extragalactic molecules, (see 012.048), p. 159 - 163 (1982).
The author builds a collisional model to show that, if the gas density is enhanced in spiral arms, giant molecular clouds have time to be formed substantially in the arms, even if their lifetime is as short as 4×10^7 years. A collisional steady-state is reached neither in arms nor in interarms for molecular clouds, but giant clouds are much more concentrated in spiral arms than the whole interstellar matter. The author's model takes into account coalescence of molecular clouds, fragmentation, and disruption of giant clouds after star formation.

131.274 Comparison of J = 2−1 vs. J = 1−0 CO emission in the galactic plane — preliminary results.
W. L. Peters.
Extragalactic molecules, (see 012.048), p. 175 (1982).
Frank Bash and the author have observed the 230 GHz line of ^{12}CO in the galactic plane in the longitude range 20 to 35° at a spacing of 0.2 in longitude. The spectra were compared with the Burton and Gordon (1976) J = 1−0 CO survey. A comparison of the J = 2−1 line profiles with the corresponding J = 1−0 line profiles has been made.

131.275 Relating CO line flux to molecular column density.
J. Young, N. Scoville.
Extragalactic molecules, (see 012.048), p. 177 - 181 (1982).
The authors examine observational data for various

clouds observed in the Milky Way to derive an empirical relationship between the observed CO intensity integrals and the H_2 column density obtained from either extinction estimates, ^{13}CO measurements, or Virial theorem analysis.

131.276 Correcting H I surveys for self-absorption.
J. Dickey.
Extragalactic molecules, (see 012.048), p. 183 - 188 (1982).

In an attempt to measure directly the effect of self-absorption on estimates of the H I column density from emission surveys an interferometer system has been developed (Dickey and Benson 1982). A typical low latitude spectrum (toward W3) is shown. The author can estimate how much H I is concealed inside the optically thick gas assuming a single temperature for the H I at each velocity. For the 45 directions observed in the Green Bank experiment in most cases at high and intermediate latitudes not much correction needs to be made; in directions below $|b| = 3°$, however, there are cases where the situation is different.

131.277 Implications of collisionally-supported giant molecular clouds for spiral galactic structure and massive star formation. D. Leisawitz, F. Bash.
Extragalactic molecules, (see 012.048), p. 197 (1982).

A study has been made of the environment experienced by ballistically-moving giant molecular clouds (GMCs) as they encounter a hydrodynamically-flowing interstellar medium. If the galactic distributions of small colliding clouds are like those observed for H I, the model predictions are in agreement with several observational characteristics of the galaxy and M 81. Radial rates of star formation are reproduced for both galaxies.

131.278 Report of IAU Commission 34: Interstellar matter and planetary nebulae (Matière interstellaire et nébuleuses planétaires). V. Radhakrishnan.
Trans. IAU, Vol. XVIIIA, (see 003.013), p. 413 - 456 (1982).

A bibliography of observations of molecular clouds in galaxies. See Abstr. 002.096.

The Orion complex: a case study of interstellar matter. See Abstr. 003.051.

Searching between the stars. See Abstr. 003.122.

Physical processes in the interstellar medium. See Abstr. 003.123.

Light scattering by small particles. See Abstr. 003.133.

Spectroscopic evidence for interstellar magnesium oxide particles. See Abstr. 022.005.

Cometary NH: ultraviolet and submillimeter emission. See Abstr. 022.012.

Time-resolved spectroscopy of the C_2 Phillips system and revised interstellar C_2 abundances. See Abstr. 022.013.

The rotational spectra of $HOCO^+$, HOCN, HN_3, and HNCO from quantum mechanical calculations. See Abstr. 022.014.

Classical rigid-ellipsoid model for collisions of H_2 with HC_7N and HC_9N. See Abstr. 022.068.

Laboratory measurements of the pure rotation $S(2)$ and $S(3)$ transitions in H_2. See Abstr. 022.076.

The hydroxyl molecule: population inversion within the rotational levels of the $^2\pi$ state. See Abstr. 022.091.

Collisional excitation of OH by H_2 in the interstellar medium. See Abstr. 022.101.

A laboratory simulation of the interstellar 220 nanometer feature. See Abstr. 022.102.

Theoretical oscillator strengths for 21 spin-forbidden lines of C, N, O, Al, and Si. See Abstr. 022.138.

The rotational spectra of $HCNH^+$ and COH^+ from quantum mechanical calculations. See Abstr. 022.139.

The infrared spectrum of a laboratory-synthesized residue: implications for the 3.4 micron interstellar absorption feature. See Abstr. 022.141.

The millimeter wave spectrum and discharge chemistry of HC_5N. See Abstr. 022.145.

Probable abundance ratios for interstellar HCS_2^+, $HCOS^+$, and HCO_2^+. See Abstr. 022.151.

Emission of interstellar shocks. II. Density dependence of ionization and dielectronic recombination coefficients. See Abstr. 022.161.

Infrared spectroscopy over the 2.9–3.9 μm waveband in biochemistry and astronomy. See Abstr. 022.165.

Laboratory simulation of 3.4 μm interstellar absorption feature. See Abstr. 022.168.

Structures of molecular ions from microwave spectroscopy. See Abstr. 022.178.

Laboratory studies of isotope exchange in ion-neutral reactions: interstellar implications. See Abstr. 022.182.

Molecular spectroscopy prompted by astrophysical observations. See Abstr. 022.183.

Angular momentum and star formation. See Abstr. 031.590.

Extended adiabatic blast waves and a model of the soft X-ray background. See Abstr. 062.011.

The Parker instability in a self-gravitating gas layer. See Abstr. 062.013.

The formation of giant cloud complexes by the Parker-Jeans instability. See Abstr. 062.014.

Axisymmetric collapse of rotating, isothermal clouds. See Abstr. 062.021.

Collapse of accreting, rotating, isothermal, interstellar clouds. See Abstr. 062.115.

A two-fluid model for Population I. See Abstr. 062.130.

Linear polarization of radio frequency lines in molecular clouds and circumstellar envelopes. See Abstr. 063.003.

Parity nonconservation and the origin of cosmic magnetic fields.   See Abstr. 066.011.

Effect of the neutral component of the solar wind on the interaction of the solar system with the interstellar gas flux.   See Abstr. 074.014.

OB associations and the nonuniversality of the cosmic abundances: implications for cosmic rays and meteorites.   See Abstr. 107.022.

Dynamical effects of stellar winds and shocked gas on interstellar clouds.   See Abstr. 112.018.

High-spectral-resolution observations of the 7.7 μm feature in HD 44179.   See Abstr. 112.024.

CN abundance variations in the shell of IRC + 10216.   See Abstr. 112.025.

Zur Natur von OH/IR-Objekten, Infrarotsternen mit OH-Maser-Emission.   See Abstr. 112.038.

Mass loss from OB stars and the formation of interstellar shells.   See Abstr. 112.043.

Geneva photometric boxes. II. The reddening towards the galactic poles.   See Abstr. 113.008.

Revised RGU photometry of an anticentre field around NGC 2129.   See Abstr. 113.074.

A catalog of red stars near L1454. See Abstr. 113.075.

High angular resolution $uvby\beta$ observations of stars earlier than G0 in the intermediate and low latitude areas SA 128 and SA 156.   See Abstr. 113.090.

High resolution observations of magnesium II 2800 Å in Alpha Centauri A: the density of interstellar magnesium II and the stellar chromospheric profiles.   See Abstr. 114.004.

The spectrum of HD 51585 in the blue and in the ultraviolet.   See Abstr. 114.151.

Ultraviolet spectroscopy of φ Persei. See Abstr. 117.008.

Far-infrared observations of FU Orionis. See Abstr. 121.037.

Mass spectra of young stars. See Abstr. 121.039.

New variable stars in the direction of the bright cloud B in Sagittarius.   See Abstr. 123.021.

A high-resolution X-ray image of Puppis A: inhomogeneities in the interstellar medium. See Abstr. 125.050.

The sharp interstellar-like features in the spectrum of the nearby white dwarf G191 - B2B. See Abstr. 126.036.

Anatomy of a region of star formation: infrared images of S106 (AFGL 2584).   See Abstr. 132.004.

Radio sources in NGC 6334. See Abstr. 132.009.

Structure and evolution of molecular clouds near H II regions. II. The disk constrained H II region, S106. See Abstr. 132.011.

Infrared and radio observations of W51: another Orion-KL at a distance of 7 kiloparsecs? See Abstr. 132.013.

The H I cloud surrounding the emission-line star LkHα 101 in the region of NGC 1579. See Abstr. 132.014.

Anomalous motions of H I clouds. See Abstr. 132.018.

The radio H II regions associated with Cep A. See Abstr. 132.020.

The gas dynamics of H II regions. VI. H II regions in collapsing massive molecular clouds. See Abstr. 132.024.

H II regions in collapsing massive molecular clouds. See Abstr. 132.030.

Optical and millimeter wavelength study of the complex Sh2−147/Sh2−153.   See Abstr. 132.034.

The H II region W40 and its molecular cloud. See Abstr. 132.035.

H II bubbles and shocks in molecular clouds. See Abstr. 132.036.

Radiation-hydrodynamics of H II regions and molecular clouds.   See Abstr. 132.039.

Atomic and ionized hydrogen in Cepheus OB3. See Abstr. 132.041.

A CO survey of 372 optical H II regions. See Abstr. 132.043.

H I and CO observations of distant H II regions in the galactic anticenter.   See Abstr. 132.044.

Dynamics of CO clouds around H II regions in the outer Galaxy.   See Abstr. 132.045.

CO J = 2 → 1 observations of southern galactic plane H II regions.   See Abstr. 132.046.

Formation of a B0.5 star due to the interaction of a shock wave with a molecular cloud in IC 1805. See Abstr. 132.047.

H II regions and star formation in M83 and M33. See Abstr. 132.061.

Far infrared maps of M 43.   See Abstr. 132.062.

VLA observations of the H II regions and OH masers in Sgr B2 and G 34.3 + 0.1. See Abstr. 132.067.

A high-resolution study of Sgr B2 at cm-wavelengths.   See Abstr. 132.075.

The structure of Orion B (NGC 2024): a recombination line and continuum map.   See Abstr. 132.076.

Star formation at a front: G 134.2+0.8.
See Abstr. 132.078.

The distribution of the $^2\pi_{3/2} J = 3/2$ OH maser emission associated with W49N.   See Abstr. 132.079.

Formation of OB clusters.
See Abstr. 132.080.

10 and 20 micron images of regions of star formation.   See Abstr. 133.001.

A high resolution submillimeter map of OMC-1.
See Abstr. 133.002.

Infrared spectra of protostars: composition of the dust shells.   See Abstr. 133.003.

Speckle interferometry of molecular cloud sources at 4.8 μm.   See Abstr. 133.007.

A high resolution far-infrared survey of a section of the galactic plane. II. Far-infrared, CO, and radio continuum results.   See Abstr. 133.008.

Triple structure of infrared source 3 in the Monoceros R2 molecular cloud.   See Abstr. 133.013.

Near-infrared observations of the Far-Infrared Source V region in NGC 6334.   See Abstr. 133.024.

The variable infrared source near HH100.
See Abstr. 133.026.

An atlas of interstellar Ca II and Na I profiles in the Carina Nebula.   See Abstr. 134.013.

Ultraviolet surface brightness measurements of the Barnard Loop nebula and the Orion reflection nebulosity.
See Abstr. 134.022.

Radio continuum observations of cometary nebulae.
See Abstr. 134.045.

COS-B gamma-ray measurements, cosmic rays and the local interstellar medium.   See Abstr. 142.507.

Using gaseous disks to probe the geometric structure of elliptical galaxies.   See Abstr. 151.002.

Spiral arms and a supernova-dominated interstellar medium.   See Abstr. 151.005.

Coherent galactic oscillations.
See Abstr. 151.010.

Waves of sequential star formation.
See Abstr. 151.064.

Three dimensional stochastic self-propagating star formation (SSPSF) in spiral galaxies.   See Abstr. 151.075.

The structure and evolution of nuclear discs.
See Abstr. 151.083.

Propagating star formation and the structure and evolution of galaxies.   See Abstr. 151.084.

R associations. VI. The reddening law in dust clouds and the nature of early-type emission stars in nebulosity from a study of five associations.   See Abstr. 152.001.

UV and visible photometry of the brightest Pleiades stars.   See Abstr. 153.005.

Infrared dust emission from globular clusters.
See Abstr. 154.013.

Bursts of star formation in the galactic centre.
See Abstr. 155.012.

Strukturen im galaktischen Halogas in Richtung zur Großen Magellanschen Wolke.   See Abstr. 155.017.

The distribution of free electrons in the inner Galaxy from pulsar dispersion measures.
See Abstr. 155.028.

Metallicity and the N_{RSG}/N_G ratio.
See Abstr. 155.030.

[O III] imagery of the galactic plane.
See Abstr. 155.040.

The post-explosion shock propagation in the central region of our Galaxy.   See Abstr. 155.045.

Three-colour photometry of a field near the galactic centre (SA 133 F).   See Abstr. 155.048.

Westerbork observations of H I absorption in the direction of Sgr A.   See Abstr. 155.051.

A high resolution far-infrared survey of a section of the galactic plane. I. The nature of the sources.
See Abstr. 156.001.

Far IR emission of the galactic plane at high longitudes.   See Abstr. 156.002.

Widespread galactic OH emission at 1720 MHz: a new tracer of spiral arms?   See Abstr. 156.005.

Where are stars formed in the Galaxy?
See Abstr. 156.008.

Changes of the star formation rate and the initial mass function with galactic radius.   See Abstr. 156.012.

Spectrum of the galactic magnetic fields.
See Abstr. 156.017.

Far-infrared observations of Sagittarius A: the luminosity and dust density in the central parsec of the Galaxy.   See Abstr. 156.019.

Widespread galactic OH emission at 1720 MHz: a new tracer of spiral arms.   See Abstr. 156.021.

The γ-ray emissivity of the local interstellar medium from correlations with gas at intermediate latitudes.
See Abstr. 157.012.

Low-latitude galactic γ-ray emission: a probe, not a proof.   See Abstr. 157.013.

Hydroxyl absorption toward galactic nuclei.
See Abstr. 158.005.

Star formation and chemical abundances in clumpy irregular galaxies.   See Abstr. 158.008.

Models for far-infrared emission from normal galaxies.   See Abstr. 158.038.

A new thermometer for external galaxies.
See Abstr. 158.063.

The inhibition of star formation in barred spiral galaxies.   See Abstr. 158.075.

Hot stars in the bulge of M31: upper limit to the star formation rate.   See Abstr. 158.082.

Neutral hydrogen in M31. II. Kinematical properties and density waves.   See Abstr. 158.101.

Search for redshifted CH_2O, H_2O, O_2, and NH_3 in Seyfert galaxies and quasars.   See Abstr. 158.178.

Galaxy morphology, starclusters and the mass spectrum of star formation.   See Abstr. 158.216.

The initial mass function of massive stars and the evolution of galaxies.   See Abstr. 158.217.

VLA observations of massive star formation in spiral galaxy cores.   See Abstr. 158.288.

The amount of dust and the distance of NGC 253. See Abstr. 158.302.

CO in Seyferts.   See Abstr. 158.338.

A search for carbon monoxide in clumpy irregular galaxies.   See Abstr. 158.339.

CO radial distributions in IC 342 and NGC 6946. See Abstr. 158.340.

The molecular cloud distribution in NGC 891 between 3 and 15 kpc.   See Abstr. 158.342.

The molecular distribution in M51.
See Abstr. 158.343.

CO observations of NGC 3628.
See Abstr. 158.344.

CO emission and the optical disk in the giant Sc galaxy M 101.   See Abstr. 158.345.

CO in the arms of Andromeda.
See Abstr. 158.348.

CO survey: comparison with morphological type. See Abstr. 158.349.

Ammonia in the spiral galaxy IC 342.
See Abstr. 158.350.

The far infrared disk of M 51.
See Abstr. 158.353.

Ultraviolet absorption by interstellar gas near the LMC star HD 36402 in the interstellar bubble N51D. See Abstr. 159.009.

The gas to dust ratio and the near-infrared extinction law in the Large Magellanic Cloud.
See Abstr. 159.010.

Carbon monoxide in the Magellanic Clouds.
See Abstr. 159.011.

H_2O maser survey in the Magellanic Clouds.
See Abstr. 159.013.

Infrared interstellar extinction in the LMC.
See Abstr. 159.038.

Errata

131.901   Erratum: "Observations of $^{14}N/^{15}N$ in the galactic
        disk" [Astrophys. J., Vol. 247, 522 - 529 (1981)].
P. G. Wannier, R. A. Linke, A. A. Penzias.
Astrophys. J., Vol. 254, 419 (1982). − See Abstr. 30.131.001.

131.902   Erratum: 'Grain formation behind shock fronts and
        the origin of isotopically anomalous meteoritic
inclusions' [Astrophys. J., Vol. 251, 820 - 831 (1981)].
B. G. Elmegreen.
Astrophys. J., Vol. 258, 414 (1982). − See Abstr. 30.131.180.

132 H I, H II Regions

132.001 An H II region in NGC 6744: spectrophotometry and chemical abundances. D. L. Talent.
Astrophys. J., Vol. 252, 594 - 600 (1982).

Spectrophotometry of emission lines in the λλ3700−6800 spectral range is presented for an H II region in an outer arm of NGC 6744, a southern hemisphere galaxy of type SAB(r)bc II. The electron temperature, ionic abundances and total relative number abundances are derived by comparison to standard models. The NGC 6744 H II region abundances are compared to similar data for H II regions in the SMC, LMC, and the Perseus arm of the Galaxy.

132.002 High-resolution X-ray observations of the Orion Nebula. W. H.-M. Ku, G. Righini-Cohen, M. Simon.
Science, Vol. 215, 61 - 64 (1982).

Observations of the Trapezium region in the Orion Nebula obtained with the high-resolution X-ray imaging instrument on board the Einstein Observatory reveal at least 58 sources of X-ray emissions. All but two of the sources can be identified with visible stars. The strongest X-ray source is the star θ^1 C, which excites the emission nebula. Its X-ray luminosity is 6×10^{32} ergs per second. The rest of the X-ray sources may be identified with stars of all spectral types. Strong X-ray emission is not observed from members of the infrared cluster embedded within the Orion molecular cloud.

132.003 High resolution far-infrared observations of the evolved H II region M16.
B. McBreen, G. G. Fazio, D. T. Jaffe.
Astrophys. J., Vol. 254, 126 - 131 (1982).

M16 is an evolved, extremely density bounded H II region, which now consists only of a series of ionization fronts at molecular cloud boundaries. The source of ionization is the OB star cluster (NGC 6611) which is about 5×10^6 years old. The authors used the CFA/UA 102 cm balloon-borne telescope to map this region and detected three far-IR sources embedded in an extended ridge of emission. Source I is an unresolved far-IR source embedded in a molecular cloud near a sharp ionization front. An H_2O maser is associated with this source, but no radio continuum emission has been observed. The other two far-IR sources (II and III) are associated with ionized gas-molecular cloud interfaces, with the far-IR radiation arising from dust at the boundary heated by the OB cluster.

132.004 Anatomy of a region of star formation: infrared images of S106 (AFGL 2584).
R. D. Gehrz, G. L. Grasdalen, M. Castelaz, C. Gullixson, D. Mozurkewich, J. A. Hackwell.
Astrophys. J., Vol. 254, 550 - 561, plate 10 (1982).

Infrared images of the young object S106 (AFGL 2584) have been produced at 3.6 μm, 10 μm (N band), and 19.5 μm (Q band) with 5″ spatial resolution using the Wyoming infrared telescope. Photometry from 2.3 μm to 23 μm is presented for eight compact sources which were identified in the infrared nebula. There is remarkable spatial coincidence among compact optical, infrared, and radio knots suggesting that the gas and dust in the nebula are well mixed and that Lyman photons may be important in heating the nebular dust. The authors conclude that a single source (IRS 4) is responsible for exciting the entire optical, infrared, and radio structure of S106. The exciting object is probably a single luminous star of spectral type O7−O9.

132.005 High-velocity H II regions delineating a central bar in our Galaxy?
J. L. Caswell, R. F. Haynes.
Astrophys. J., Lett., Vol. 254, L31 - L33 (1982).

The authors report the discovery near the galactic center of three discrete H II regions with very high velocities. They discuss these in the context of other new evidence favoring a central elliptical disk or bar structure at the center of our Galaxy. Late-type stars detected as OH masers may also serve to delineate the bar structure.

132.006 A new model for giant H II shells.
J. C. Raymond.
Nature, Vol. 295, 282 (1982).

132.007 Association of PSR0740 − 28 with an H I shell in Puppis. J. G. Stacy, P. D. Jackson.
Nature, Vol. 296, 42 - 44 (1982).

During a survey of the interstellar atomic hydrogen distribution in the Puppis region of the Milky Way, the authors detected numerous, sometimes expanding, H I shells and filaments, indicative of past energetic explosive events in this area of the Galaxy. One of these shells, GS241 −01 +15, appears to be physically associated with the young pulsar PSR0740 − 28.

132.008 The bipolar nebula S106: photometric, polarimetric, and spectropolarimetric observations.
H. J. Staude, R. Lenzen, H. M. Dyck, G. D. Schmidt.
Astrophys. J., Vol. 255, 95 - 102, plate 4 (1982).

Photometric, polarimetric, and spectropolarimetric observations of the bipolar nebula S106 and its surroundings are presented. The distance of the molecular complex surrounding S106 is found to be $d = (600 \pm 100)$ pc. The IR polarization of the central star appears to be due to aligned grains in the equatorial dust lane of the bipolar nebula. The dust embedded in source 1, the thick bright knot in the southern lobe, reflects a stellar continuum from the exciting star as well as line emission from a compact central nebula. The extinction between the central star and source 1 is derived, and some constraints to the topology and nature of the scattering grains are determined. From the emission line spectrum observed in source 1, a spectral type O9 V of the central star is deduced. No evidence for shock excitation of the emission lines is seen.

132.009 Radio sources in NGC 6334.
L. F. Rodríguez, J. Cantó, J. M. Moran.
Astrophys. J., Vol. 255, 103 - 110 (1982).

Using the VLA at 6 cm the authors mapped the ridge of star formation in NGC 6334. Six radio sources were detected, ranging in angular size from very compact (∼0.″3) to extended (>40″). Five of these sources are probably H II regions. The shapes observed in some of the H II regions can be explained qualitatively in terms of hydrodynamical models that describe the evolution of young H II regions in diverse gaseous environments. One of the sources, NGC 6334 (B), has an angular diameter of about 0.″3 and brightness temperature of 3×10^5 K at 6 cm. Observations of this source at 20, 6, 2, and 1.3 cm showed that its angular size is proportional to λ^2 and its flux density spectrum is flat. These properties may be explained by a model of a nonhomogeneous synchrotron source in a thermal plasma or possibly by plasma scattering. The source is similar to the compact source in the galactic center.

132.010 H II regions surrounding high galactic latitude O stars. R. J. Reynolds, P. M. Ogden.
Astron. J., Vol. 87, 306 - 313 (1982).

Scans of Hα and [N II] λ 6584 were obtained in directions toward 13 O stars that have galactic latitudes greater than $10°$, and z distances between 60 pc and 1000 pc. The associated H II regions have emission measures that range from 500 cm^{-6} pc to less than 2 cm^{-6} pc and radii that extend from about 20 pc to more than 200 pc from the star. The derived rms electron densities for these regions range from about 3 cm^{-3} for $|z| < 100$ pc to less than 0.1 cm^{-3} for $|z| > 600$ pc.

132.011 Structure and evolution of molecular clouds near H II regions. II. The disk constrained H II region, S106.   J. Bally, N. Z. Scoville.
Astrophys. J., Vol. 255, 497 - 509 (1982).

Observations of the molecular gas surrounding the biconical H II region S106 reveal a ~100 $M_\odot$ disk constraining the development of the ionized region. Two CO peaks are seen in the equatorial plane of the biconical nebula on opposite sides of a near infrared source thought to be the exciting star for the H II region. The authors develop a model for S106 with a B0 exciting star situated in a dense neutral disk which dominates the dynamics of the nebula. Isothermal expansion of freshly ionized material from the disk drives a supersonic flow of plasma which supplies gas to the two lobes of the H II region. The observed biconical morphology, electron density distribution, and high-velocity outflow of ionized gas can be reproduced by the model.

132.012 Infrared emission line studies of the structure and excitation of H II regions.
J. H. Lacy, S. C. Beck, T. R. Geballe.
Astrophys. J., Vol. 255, 510 - 523, plates 5, 6 (1982).

Maps of five H II regions in one or more of the infrared fine-structure lines of Ne II (12.8 μm), Ar III (9.0 μm), and S IV (10.5 μm) have been obtained with angular resolutions ranging from 4″ to 7″. The observations are used to discuss the morphology and excitation of these nebulae. Considerable diversity is found in the structures of the nebulae, probably resulting from differences in their ages and the circumstances of their formation. In all cases, more ionizing luminosity than would be provided by a single dominant ionizing star appears to be required.

132.013 Infrared and radio observations of W51: another Orion-KL at a distance of 7 kiloparsecs?
R. Genzel, E. E. Becklin, C. G. Wynn-Williams, J. M. Moran, M. J. Reid, D. T. Jaffe, D. Downes.
Astrophys. J., Vol. 255, 527 - 535 (1982).

The authors report high-resolution infrared and radio continuum observations of the W51 region. The bright infrared source W51 IRS 2 has at least three components with different physical and evolutionary properties. The spatial distribution and the near-infrared spectra of the components in IRS 2 are remarkably similar to, but more luminous than, those found in Orion, where an H II region of comparable linear size is also located close to a cluster of compact infrared sources. The characteristics of the nearby W51 NORTH H_2O maser source, and the detection of 2 μm H_2 quadrupole emission in IRS 2 indicate that the mass loss phenomena found in Orion-KL also exist in W51. In contrast to W51 NORTH, there is no compact 20 μm source associated with W51 MAIN, the second strong H_2O maser in this region. The spatial distribution of 20 μm continuum radiation from the neighboring infrared source W51 IRS 1 follows closely that of the radio continuum emission and may be interpreted as arising from dust heated by Lyα within the extended H II region.

132.014 The H I cloud surrounding the emission-line star LkHα 101 in the region of NGC 1579.
P. E. Dewdney, R. S. Roger.
Astrophys. J., Vol. 255, 564 - 576 (1982).

The reflection nebula NGC 1579 is situated in a dark cloud and is illuminated by the highly obscured emission-line star LkHα 101. New aperture synthesis observations at 21 cm (resolution 2′× 3.5 by 0.67 km s^{-1} in radial velocity) show a continuum source of 176 mJy centered on the star and ~85 $M_\odot$ of atomic hydrogen in a surrounding cloud ~3.5 pc

across at the assumed distance of 800 pc. Models of the H I emission cloud based on two-step dissociation of H_2 by ultraviolet radiation from the central star indicate that the density in the medium is much higher on the near side of the star than on the other. Unlike H I zones associated with more diffuse H II regions, no expansion of the atomic gas is detected. This reinforces other evidence that the star and its H II and H I regions are newly formed.

132.015 Observations of NGC 604 over six decades in frequency.
F. P. Israel, I. Gatley, K. Matthews, G. Neugebauer.
Astron. Astrophys., Vol. 105, 229 - 235 (1982).

Maps in the near-infrared, H-alpha and radio continuum have been obtained of NGC 604, the brightest H II region complex in M 33, as well as a measurement of the UV spectrum of the ionizing star cluster. The data show that NGC 604 consists of about ten components of typical size 10 pc, embedded in a much larger envelope with dimensions of 225 pc. At least one component is optically almost completely obscured; it is located close to another object that is only detected in the near-infrared and whose nature is uncertain.

132.016 Optical study of the W51 complex.
C. Goudis, H. Hippelein.
Astron. Astrophys., Vol. 105, 329 - 334 (1982).

The results of an interferometric study of the W51 complex in the light of the [S III] emission line at 9531 Å are presented. With angular and spectral resolutions of 1′.4 and 8.5 km s^{-1} respectively, the distributions of surface brightness and radial velocity over the radio emitting areas have been derived. In the direction of W51 A two emission maxima are found which coincide with the components g and h of the radio source G 49.5 − 0.4. A third [S III] emission region coincides in position and radial velocity, $V_{LSR} \cong +66$ km s^{-1}, with the radio source W51 B G 48.9 −0.3 lying in front of W51 A. From comparison of the [S III] line intensities with radio continuum data the visual extinction A_V has been mapped. An empirical model of the W51 complex is finally proposed.

132.017 Photoelectric heating of H II regions.
W. J. Maciel, S. R. Pottasch.
Astron. Astrophys., Vol. 106, 1 - 3 (1982).

The heating of H II regions by electrons ejected from grains after absorption of stellar photons as well as by recombination Lα photons is considered. The nebula is heated further by photoionization of H and He, and cooled essentially by collisional excitation of low lying levels of abundant ions. It is found that the present mechanism can influence the thermal structure of the nebula between ~20% and 60% of the Strömgren radius, depending on the assumed electron density.

132.018 Anomalous motions of H I clouds.   P. A. Shaver, V. Radhakrishnan, K. R. Anantharamaiah, D. S. Retallack, W. Wamsteker, A. C. Danks.
Astron. Astrophys., Vol. 106, 105 - 108 (1982).

In a comparison of H I absorption line velocities and radio recombination line velocities for 38 H II regions of known distance, it is found that the H I absorption extends beyond the recombination line velocity in the large majority of cases. This is shown to be the result of the overlapping of velocities due to the chaotic motions of two populations of H I clouds. The absorption generally extends all the way to the source and can be used to estimate the velocity corresponding to the distance of the source itself. But it can be misleading if used to resolve the kinematic distance ambiguity when the recombination line velocity is within ~10−20 km s^{-1} of the tangential velocity.

132.019 An H I absorption determination of the distance of W31.
P. M. W. Kalberla, W. M. Goss, T. L. Wilson.
Astron. Astrophys., Vol. 106, 167 - 170 (1982).

Using the Westerbork Synthesis Radio Telescope, H I absorption observations of G 10.2–0.3 and G 10.3–0.1 in W31 have been made. No detectable absorption at velocities >45 km s^{-1} is present and the lower limit for the distance to these sources is 5.5 kpc. The authors suggest that the complex is located in the near side of the 4-kpc arm and has a peculiar velocity of ~30 km s^{-1}. There are indications for small scale structure in the optical depth distribution with a scale size of ~1 pc.

132.020 The radio H II regions associated with Cep A.
V. A. Hughes, J. G. A. Wouterloot.
Astron. Astrophys., Vol. 106, 171 - 173 (1982).

Recent observations of Cep A at λ 21 cm, using the Westerbork Synthesis Radio Telescope, have shown the presence of two main H II regions. One contains compact H II regions and H$_2$O and OH masers, and the other contains the supposed Herbig-Haro Object GGD37. The results should clear up some confusion regarding Cep A.

132.021 On the distance to the giant galactic H II region NGC 3603. J. Melnick, P. Grosbøl.
Astron. Astrophys., Vol. 107, 23 - 25 (1982).

Electronographic *UBV* photometry of stars in the ionizing cluster of NGC 3603 is used to determine a distance of 5.3 (+1.6/−1.2) kpc to the nebula. This estimate is significantly different from previous photometric determinations. It also differs from the kinematic distance of 7 - 8 kpc obtained from the observed radial velocity confirming the existence of large-scale regional motions near the nebula.

132.022 Infalling clouds with very high velocities: a collision with the Milky Way in the anticenter.
I. F. Mirabel.
Astrophys. J., Vol. 256, 112 - 119 (1982).

Neutral hydrogen clouds in the anticenter and inner direction of the Galaxy that are infalling with very high velocities were observed with the Arecibo telescope. In the anticenter, the authors find direct evidence for the collision of a very high velocity cloud (V_{GSR} = −210 km s^{-1}) with Milky Way material. The observations indicate that a complex of very high velocity clouds has reached the outer Galaxy and, at its closer angular distance from the galactic plane, is interacting with matter associated with the disk. The infall of a single cloud deposits a typical energy of 5 × 10^{52} ergs on a small region of the galactic disk. The complete complex would transfer an energy of 10^{53}–10^{54} ergs. The complex in the anticenter is a fragment of the Magellanic Stream that has fallen as far as a few kiloparsecs from the Sun, whereas the complex in the direction of the inner Galaxy is situated beyond the galactic center and may be part of a distinct Magellanic Stream in the southern hemisphere.

132.023 Atomic and ionized hydrogen near IC 5146 (S125).
R. S. Roger, J. A. Irwin.
Astrophys. J., Vol. 256, 127 - 138, plates 11, 12 (1982).

The authors report synthesis observations in continuum and H I line emission at 21 cm of a 2° field containing the emission nebula IC 5146. The resolution is 2.̍0 × 2.̍7 by 0.66 km s^{-1} in radial velocity. The H II region is approximately 8′ in diameter, and its continuum emission is remarkably symmetric about the exciting star BD +46° 3474. At the presumed distance of 960 pc, the emission indicates a total H II mass of 9.8 ± 1 $M_{\odot}$ within an extreme radius of 1.4 pc. The H I line emission is also confined to a region around the star but is less symmetrically distributed. On average, it extends to 3.5 pc from the star, but to the south at certain velocities it reaches to 6 pc. There is no extension along the

dust lane leading to the northwest. The total mass of H I associated with IC 5146 is 445 ± 105 $M_{\odot}$. A model of the region is outlined where the exciting star (B0 V) and its surrounding H II region are located 3 pc on the near side from the center of the parent molecular cloud.

132.024 The gas dynamics of H II regions. VI. H II regions in collapsing massive molecular clouds.
H. W. Yorke, P. Bodenheimer, G. Tenorio-Tagle.
Astron. Astrophys., Vol. 108, 25 - 41 (1982).

Two-dimensional numerical calculations of the evolution of H II regions associated with self-gravitating, massive molecular clouds are presented. Parameters include the birthplace of the exciting star (inside or outside the cloud), the stellar motion, and the mass and rotation of the molecular cloud. These results together with observational data concerning molecular clouds allow to conclude that massive molecular clouds with OB stars can be evaporated in 10^7 yr.

132.025 Line splitting in H II regions.
H. W. Yorke, G. Tenorio-Tagle, P. Bodenheimer.
Mitt. Astron. Ges., Nr. 55, p. 24 - 25 (1982). − Abstract.

132.026 11 cm-Beobachtungen der H II Region um λ Orionis. C. Crezelius, W. Reich.
Mitt. Astron. Ges., Nr. 55, p. 25 - 27 (1982).

132.027 Eine H I Region mit optischer Emission.
W. Goerigk, U. Mebold.
Mitt. Astron. Ges., Nr. 55, p. 130 - 133 (1982).

132.028 Cold H I clouds in the Galaxy.
Y. Fukui, F. Sato, T. Hasegawa.
Astron. Her., Vol. 74, No. 8, p. 216 - 219 (1981). In Japanese. Abstr. in Phys. Abstr., Vol. 85, Abstr. 33104 (1982).

132.029 The dynamical evolution of H II regions in non-uniform environments. G. Tenorio-Tagle.
Regions of recent star formation, (see 012.002), p. 1 - 14 (1982).

Recent developments and results in modeling the dynamical evolution of H II regions in star formation regions are reviewed. It is demonstrated that when star formation occurs near the edges of giant molecular cloud complexes the expansion of a newly formed H II region will lead to a "champagne" flow of ionized material in the boundary region of the parent molecular cloud. Consequences of this flow for the evolution of the H II/molecular cloud complexes are outlined. Observational evidence related to the "champagne" flow of ionized gas is briefly discussed.

132.030 H II regions in collapsing massive molecular clouds.
H. W. Yorke, P. Bodenheimer, G. Tenorio-Tagle.
Regions of recent star formation, (see 012.002), p. 15 - 23 (1982).

Results of two-dimensional numerical calculations of the evolution of H II regions associated with self-gravitating, massive molecular clouds are presented. Depending on the location of the exciting star, a champagne flow can occur concurrently with the central collapse of a nonrotating cloud. Partial evaporation of the cloud at a rate of about 0.005 M$_{\odot}$/yr results. When 100 O-stars are placed at the center of a freely falling cloud of 3 × 10^5 M$_{\odot}$ no evaporation takes place. Rotating clouds collapse to disks and the champagne flow can evaporate the cloud at a higher rate (0.01 M$_{\odot}$/yr).

132.031 The dynamical evolution of an H II region.
R. H. Harten, M. Felli.
Regions of recent star formation, (see 012.002), p. 25 - 30 (1982).

Radio aperture synthesis observations combined with single dish measurements of a large sample (77) of H II regions allow to test the applicability of models of star formation and H II region evolution. Strong evidence is found in support of the 'champagne flow' model of H II region evolution. There is no strong statistical evidence in support of the theory of successive star formation proposed by Elmegreen and Lada although there are several individual cases which do support it.

132.032   Radio continuum observations of W1.
H. E. Matthews.
Regions of recent star formation, (see 012.002), p. 31 - 38 (1982).

Observations are presented of the continuum emission from the thermal radio source W1 (S 171, NGC 7822). The complete region has been accurately mapped at $\lambda 11$cm using the 100–m telescope in both total power and polarized radiation. The central part of the object also has been observed with the Westerbork Synthesis Radio Telescope at $\lambda 49$, $\lambda 21$, and $\lambda 6$cm. In the light of these observations, the question of the existence of a non-thermal component of W1 is discussed, and indications for secondary star formation processes are examined.

132.033   Interaction of the H II region S 236 with the surrounding medium.
A. Falchi, G. Tofani, R. H. Harten.
Regions of recent star formation, (see 012.002), p. 39 - 42 (1982).

The H II region S 236 has been observed in the continuum at 50 cm wavelength and in the recombination line H166α at 10 selected positions. A mean electron temperature T_e = 7600 K is derived for the nebula and the measured velocity pattern is explained qualitatively in terms of a gas flow from the molecular cloud.

132.034   Optical and millimeter wavelength study of the complex Sh2−147/Sh2−153.
M. Heydari-Malayeri, C. Kahane, R. Lucas, G. Testor.
Regions of recent star formation, (see 012.002), p. 43 - 51 (1982).

Sh2−147/Sh2−153 is a vast H II region and molecular cloud complex located in the Perseus arm at $l \approx 109°$. The authors present millimeter observations in the molecular lines ^{13}CO, HCO^+, HCN and $H^{13}CO^+$ of the complex as well as detailed monochromatic (Hα) photography and spectrography of some of the H II regions.

132.035   The H II region W40 and its molecular cloud.
R. M. Crutcher, Y.-H. Chu.
Regions of recent star formation, (see 012.002), p. 53 - 60 (1982).

Extensive new and published older observations of the H II region W40 and its associated molecular cloud have been analyzed. The H II region is near the center of an extensive dark cloud which is contracting at 3 km/s along the line of sight. The H II region is adjacent to the high-density, warm core of the molecular cloud and is responsible for heating it. The ionized hydrogen is expanding away from the high-density molecular gas at velocities up to 8 km/s and is just beginning to break through holes in the near side of the dark cloud.

132.036   H II bubbles and shocks in molecular clouds.
T. J. Mazurek.
Regions of recent star formation, (see 012.002), p. 61 - 66 (1982).

Supersonic gas velocities and large density gradients are observed in compact H II regions. Possible causes of such flows are examined analytically. It is shown that density gradients in either the H II region or the exterior molecular cloud can naturally give supersonic gas speeds. Cases discussed include the swelling of uniform H II bubbles into clouds with decreasing densities, isothermal shock propagation within the ionized region, and homologous expansion of non-uniform H II bubbles.

132.037   Fine structure lines in H II regions interacting with molecular clouds.   D. A. Naylor, R. Emery,
B. Fitton, I. Furniss, R. E. Jennings, K. J. King.
Regions of recent star formation, (see 012.002), p. 73 - 80 (1982).

This paper presents observations of far infrared fine structure emission lines of OIII at 52 and 88 μm, OI at 63 μm and NIII at 57 μm, obtained using a balloon-borne telescope with a Michelson interferometer system. The sources M42 and M17 have been mapped with a beam of 1.6' FWHM. Physical quantities derived from these measurements are presented and some of the results are discussed.

132.038   Ultra-compact H II regions embedded in infrared sources.   S. Kwok.
Regions of recent star formation, (see 012.002), p. 83 - 89 (1982).

The nature and evolutionary state of weak ultracompact H II regions associated with infrared sources is discussed. It is demonstrated that these H II regions can occur either in a site of star formation where a young star starts to ionize the material of a surrounding molecular cloud or when a star is in transition from a red-giant to a planetary nebula and an ionized nebular shell interacts with the remnant circumstellar matter of the red-giant stage. Observational evidence for both cases is presented.

132.039   Radiation-hydrodynamics of H II regions and molecular clouds.   M. T. Sandford II, R. W. Whitaker,
R. I. Klein.
Regions of recent star formation, (see 012.002), p. 129 - 132 (1982).

Two-dimensional calculations of ionization-shock fronts surrounding neutral cloud clumps reveal that a radiation-driven implosion of the clump can occur. The implosion of a cloud clump results in the formation of density enhancements that may eventually form low mass stars. The smaller globules produced may become Herbig-Haro objects, or maser sources.

132.040   Atomic hydrogen zones associated with H II regions.
R. S. Roger.
Regions of recent star formation, (see 012.002), p. 167 - 174 (1982).

The broad H I zones detected in 21-cm emission near three H II regions, NGC 281, IC 5146 and NGC 1579, are described and compared. The formation of such zones in low and medium density gas by dissociation of H_2 with UV radiation from the exciting star, is discussed. The H I spin temperatures are probably in the range 100K - 300K and for the two most evolved sources the zones are expanding.

132.041   Atomic and ionized hydrogen in Cepheus OB3.
P. E. Dewdney.
Regions of recent star formation, (see 012.002), p. 181 - 184 (1982).

Aperture synthesis maps of 21-cm continuum radiation and H I emission in a field centered on the H II region S 155 and the Cepheus OB3 association are presented. The data were obtained with the Penticton synthesis telescope. Possible explanations for an apparent depression in H I emission towards S 155 are discussed.

132.042   The velocities of the neutral and ionized components of H II regions.
M. Fich, R. R. Treffers, L. Blitz.
Regions of recent star formation, (see 012.002), p. 201 - 208 (1982).

The authors investigate statistically the relative velocities of the ionized gas in H II regions and the neutral gas in the molecular clouds associated with them for 151 optical H II regions in the Galaxy. The velocity of the molecular cloud is determined from CO observations at 2.6 mm and the ionized component from Hα spectra. The mean velocity difference between the molecular cloud and the ionized gas is 1.4 ± 0.4 km s^{-1} with a dispersion of 4.6 km s^{-1}. Both values are important for determining the mean flow velocity and they indicate that the bulk of the ionized gas in most H II regions flow away from the associated clouds at velocities of $5 - 10$ km s^{-1}.

132.043   A CO survey of 372 optical H II regions.
L. Blitz, M. Fich, A. A. Stark.
Regions of recent star formation, (see 012.002), 209 - 212 (1982).

Optical H II regions have been surveyed in the $J = 1 - 0$ transition of CO. The results indicate that at least 70% of all H II regions have associated molecular clouds. The survey provides an order-of-magnitude improvement in measured radial velocities toward H II region/molecular cloud complexes. With the CO rotation curve kinematic distances for H II regions as large as 20 kpc from the galactic center are now available.

132.044   H I and CO observations of distant H II regions in the galactic anticenter.
E. J. Grayzeck, P. D. Jackson, J. R. Sewall.
Regions of recent star formation, (see 012.002), p. 213 - 220 (1982).

The authors present CO and H I observations of the distant H II regions S241, S242, S259 and S271 located near the galactic anticenter. Line profiles are discussed and velocity fields are derived. Estimates of the amount of neutral material surrounding the ionized regions are given.

132.045   Dynamics of CO clouds around H II regions in the outer Galaxy.   P. D. Jackson, J. R. Sewall.
Regions of recent star formation, (see 012.002), p. 221 - 230 (1982).

This paper reports radio observations in the $J = 1 \rightarrow 0$ line of CO of molecular clouds surrounding distant H II regions in the outer Galaxy. Contour maps were obtained for seven H II regions from the Sharpless list. Column densities of ^{13}CO are derived and total cloud masses are estimated. Detailed radial velocity data are presented.

132.046   CO $J = 2 \rightarrow 1$ observations of southern galactic plane H II regions.   G. J. White, J. P. Phillips.
Regions of recent star formation, (see 012.002), p. 231 - 236 (1982).

Results of a survey of southern galactic H II regions in the $J = 2 \rightarrow 1$ sub-mm line of CO are reported. The observations were performed at the 2.5 m telescope at Las Campanas with the QMC heterodyne line receiver. Regions were selected where the presence of water masers indicates recent star-formation activity. Line profiles and radial velocities for the individual sources are presented.

132.047   Formation of a B0.5 star due to the interaction of a shock wave with a molecular cloud in IC 1805.
V. A. Hughes.
Regions of recent star formation, (see 012.002), p. 349 - 355 (1982).

An isolated B0.5 star has been observed, using radio techniques, in the optically obscured region ahead of a shock front which is propagating into a molecular cloud. The shock is driven chiefly by stellar winds from an OB-association.

132.048   On the nature of galactic "supershells" of neutral hydrogen.   I. V. Gosachinskij.
Pis'ma Astron. Zh., Tom 8, 214 - 218 (1982). In Russian. English translation in Soviet Astron. Lett., Vol. 8.

It is shown that most of shells and "supershells" typical of the galactic H I distribution are unlikely to be connected with supernova remnants or stellar winds but seem to be part of the normal structure of spiral arms.

132.049   Results of a radio survey for new compact H II regions.   J. E. Wink, W. J. Altenhoff, P. G. Mezger.
Astron. Astrophys., Vol. 108, 227 - 242 (1982).

The paper contains an investigation of 91 sources which appear to be point sources when observed with the angular resolution of the 100-m telescope at 4.9 GHz and which are expected to be H II regions. These sources (or subsamples of these sources, respectively) were observed in the continuum at 5, 15, and 86 GHz with single dish telescopes, and at 2.7 and 8.1 GHz with a three-element interferometer. Another subsample was searched for its H 90 α and H 76 α recombination line emission. For various reasons a number of previously known compact H II regions was included in some of the single-dish and interferometer observations. Five out of the 91 sources of the basic search list were found to be non-thermal sources and another was identified with a known planetary nebula. Observations related to be the remaining 85 sources from the basic search list are consistent with the hypothesis that these sources are H II regions. Distance estimates were obtained for 50 of the 85 H II regions and their physical parameters were determined.

132.050   Wolf-Rayet stars in extragalactic H II regions. II. NGC 604 – a giant H II region dominated by many Wolf-Rayet stars.   M. Rosa, S. D'Odorico.
Astron. Astrophys., Vol. 108, 339 - 343 (1982).

The unusually large numbers of WR stars, the numerous shells of ionized material and the supersonic motions observed in NGC 604 are discussed within the theory of interstellar wind-driven bubbles. It is shown, that the kinetic energies involved can be easily supplied by the strong mass loss phenomena of the WR stars.

132.051   An H II region near NML Cygnus.
H. J. Habing, W. M. Goss, A. Winnberg.
Astron. Astrophys., Vol. 108, 412 - 414 (1982).

An extended continuum radio source near NML Cygnus has been detected at 21 cm. The source has a thermal spectrum between 2.8 and 21 cm and is associated with a faint H II region 35″ to the west of NML Cygnus. The authors argue that the H II region probably is physically associated with NML Cygnus. The H II region may have resulted from mass lost previously by the M supergiant and now ionized by a nearby B1–B2 star.

132.052   A catalogue of model H II regions.
G. Stasińska.
Astron. Astrophys., Suppl. Ser., Vol. 48, 299 - 304 (1982).

The author presents an extensive grid of model H II regions, covering a wide range of stellar effective temperatures (30 000 K to 55 000 K) and abundances (from 5 $Z \odot$ to $Z \odot/50$, with $Z \odot$ denoting the cosmic abundances). The gas densities vary from 10 to 10^3 cm^{-3}, including several two-component models.

132.053   Radio recombination lines from high emission measure nebulae.   F. J. Lockman.
Astrophys. J., Vol. 256, 543 - 549 (1982).

High frequency radio recombination lines from high emission measure nebulae, like the compact components of H II regions and perhaps some planetary nebulae, may have optical depths that approach -1. Under these circumstances, the hydrogen line peak will be enhanced, and its line width will narrow. Recombination lines from other species, like helium, have smaller optical depth and are not amplified as strongly. Thus, at high frequencies, the apparent helium abundance from a bright, compact nebula can appear anomalously low. This phenomenon might contribute to the low apparent helium abundances in Sgr B2, W3(OH), and DR 21.

132.054   Distribution of neutral hydrogen in the region of the Rosette nebula and the Monoceros OB 2 stellar association.   I. V. Gosachinskij, V. K. Khersonskij.
Astron. Zh., Tom 59, 237 - 245 (1982). In Russian. English translation in Soviet Astron., Vol. 26, No. 2.

The results of the observation of the H I radio line are presented. A thin H I envelope with diameter about 130 pc surrounding the Monoceros supernova remnant, the association Mon OB 2, and the Rosette nebula, is detected. The envelope mass is $2 \times 10^5 M_\odot$. The envelope is expanding with the velocity of about 20 km/s. Under the assumption that the envelope is the result of a single supernova explosion the expansion energy and the supernova remnant age are determined. Some conclusions on the possible connection of the envelope with the objects which are observed in this region are made.

132.055   Estudio de una pequeña estructura de hidrógeno neutro con alta velocidad expulsada del núcleo de la galaxia.   I. F. Mirabel, M. Franco.
Bol. Asoc. Argentina Astron., No. 20 - 24, p. 175 (1981). Abstract.

132.056   Regiones H II en NGC 1566.
J. L. Sérsic, J. H. Calderón.
Bol. Asoc. Argentina Astron., No. 20 - 24, p. 355 (1981). Abstract.

132.057   Complejo de H I a latitudes intermedias.
E. Bajaja, F. R. Colomb, R. Morras.
Bol. Asoc. Argentina Astron., No. 20 - 24, p. 367 (1981). Abstract.

132.058   La estructura del campo magnético en NGC 3372.
H. G. Marraco, J. C. Forte.
Bol. Asoc. Argentina Astron., No. 20 - 24, p. 481 (1981). Abstract.

132.059   H I vinculado con regiones H II.
C. Cappa de Nicolau, W. G. L. Poppel.
Bol. Asoc. Argentina Astron., No. 26, p. 205 (1981). Abstract.

132.060   Infalling clouds with very high velocities: a collision with the Milky Way in the anticenter.
I. F. Mirabel.
Bull. American Astron. Soc., Vol. 13, 787 (1981). – Abstract.

132.061   H II regions and star formation in M 83 and M 33.
K. S. Rumstay, M. Kaufman.
Bull. American Astron. Soc., Vol. 13, 797 - 798 (1981). Abstract.

132.062   Far infrared maps of M 43.   J. Smith,
D. A. Harper, R. F. Loewenstein, W. Glaccum.
Bull. American Astron. Soc., Vol. 13, 808 (1981). – Abstract.

132.063   An infrared line and continuum study of W43.
D. Lester, H. Dinerstein. M. Werner, D. Watson,
R. Genzel, C. Townes, J. Storey, P. Harvey.
Bull. American Astron. Soc., Vol. 13, 808 (1981). – Abstract.

132.064   The ratio of extinction to reddening in four compact H II regions.   P. Pişmiş.
Bull. American Astron. Soc., Vol. 13, 825 (1981). – Abstract.

132.065   Recombination line observation of compact H II regions with the VLA.
G. Garay, M. J. Reid, J. M. Moran.

Bull. American Astron. Soc., Vol. 13, 853 - 854 (1981). Abstract.

132.066   S201 far ultraviolet imagery of Messier 8 and the Sagittarius star field.
G. R. Carruthers, T. L. Page.
Bull. American Astron. Soc., Vol. 13, 857 (1981). – Abstract.

132.067   VLA observations of the H II regions and OH masers in Sgr B2 and G 34.3 + 0.1.
J. M. Benson, K. J. Johnston.
Bull. American Astron. Soc., Vol. 13, 887 (1981). – Abstract.

132.068   Spherically symmetric models of the H II region in G 333.6 – 0.2.
R. H. Rubin, D. J. Hollenbach, E. F. Erickson.
Bull. American Astron. Soc., Vol. 13, 888 (1981). – Abstract.

132.069   OH and CO molecules in H I clouds.
I. Kazes, J. Crovisier, J. M. Dickey, J. Brillet.
Bull. American Astron. Soc., Vol. 13, 924 (1981). – Abstract.

132.070   A high excitation optically obscured H II region in the nucleus of NGC 5253.
D. K. Aitken, P. F. Roche, M. C. Allen, M. M. Phillips.
Anglo-Australian Obs. Prepr. No. 162, 6 pp. (1982). – Submitted to Mon. Not. R. Astron. Soc.

132.071   Detection of molecular hydrogen emission from G333.6–0.2.   J. W. V. Storey.
Anglo-Australian Obs. Prepr. No. 163, 9 pp. (1982). – Submitted to Mon. Not. R. Astron. Soc.

132.072   Wolf-Rayet stars in extragalactic H II regions: II. NGC 604 – a giant H II region dominated by many Wolf-Rayet stars.   M. Rosa, S. D'Odorico.
ESO Sci. Prepr. No. 182, 20 pp. (1982). – Submitted to Astron. Astrophys.

132.073   The Bubble Nebula: far-infrared and radio molecular observations of NGC 7635.
H. A. Thronson, Jr., C. J. Lada, P. M. Harvey, M. W. Werner.
Prepr. Steward Obs., No. 395, 27 pp. (1982).

132.074   Are high-velocity H I clouds galactic objects?
Yu. A. Shchekinov.
Astrofizika, Tom 16, 265 - 272 (1980). In Russian. English translation in Astrophysics, Vol. 16, No. 2.

It has been shown that available observational data is not sufficient to say whether high-velocity H I clouds are galactic or extragalactic. Namely, observational sharp edges and fine structure of high-velocity clouds can be explained both by galactic and extragalactic models.

132.075   A high-resolution study of Sgr B2 at cm-wavelengths.   T. B. Pyatunina.
Astrophys. Space Sci., Vol. 84, 143 - 153 (1982).

The radio source Sgr B2 was studied with the radio telescope of the USSR Academy of Sciences, RATAN-600, at wavelengths 13, 8.2, 3.9, 2.08, and 1.38 cm. It is shown that the source contains no less than four ultracompact knots with electron densities exceeding 2×10^4 cm^{-3}. An extended source with a flux density at 1.38 cm equal to 12.2 f.u. and an angular size 58 arc sec was found in a nuclear region. Excitation parameters of all compact knots are shown to correspond to stars of a spectral type earlier than O6.

132.076   The structure of Orion B (NGC 2024): a recombination line and continuum map.
E. Krügel, C. Thum, J. Martin-Pintado, V. Pankonin.
Astron. Astrophys., Suppl. Ser., Vol. 48, 345 - 353 (1982).

Radio recombination line and continuum observations

of the compact H II region Orion B are presented. The angular resolution is 1 arcmin. The distribution of H76α over the whole continuum source and, to a lesser extent, that of He76α and C76α are given. A geometrical model is suggested. Ionized carbon is emitted primarily from a region just outside the ionization front in the south.

132.077 Radio, infrared, and optical observations of compact H II regions. IV. The nebula S235 B.
J. Krassner, J. L. Pipher, S. Sharpless, T. Herter.
Astron. Astrophys., Vol. 109, 223 - 227 (1982).

The compact H II region S235 B has been observed at radio, infrared, and optical wavelengths. The radio continuum observations have yielded an upper limit to the 2.7 GHz flux. The infrared spectrum shows the recombination lines of hydrogen and the unidentified 3.3 μm, 8.7 μm, and 11.3 μm emission features. Multicomponent fits to the 8−13 μm spectrum, and optical photographic photometry, indicate a moderate amount of extinction. Optical photographs reveal a stellar core surrounded by a small, uniform nebula. The data are consistent with an ultracompact core surrounded by a diffuse halo.

132.078 Star formation at a front: G 134.2 + 0.8.
V. A. Hughes, M. R. Viner.
Astron. J., Vol. 87, 685 - 690 (1982).

The Westerbork Synthesis Radio Telescope has been used at λ 6 cm to obtain higher-resolution maps (7.″9 × 9.″0) of the H II region G 134.2+0.8. The H II region is produced by a young B0.5 star which has formed in the dense region ahead of a shock front which is being driven by stellar winds, and which is moving into a molecular cloud. The front has passed over the star which is now situated in the ionized shell behind the front. The resulting drop in ambient density probably accounts for the anomalously low electron density of the H II region. The parameters of the region are determined.

132.079 The distribution of the $^2\pi_{3/2}J = 3/2$ OH maser emission associated with W49N.
K. J. Johnston, S. S. Hansen.
Astron. J., Vol. 87, 803 - 807 (1982).

The OH emission in the $^2\pi_{3/2}J = 3/2$ $F = 1\rightarrow 1$ and $F = 2 \rightarrow 2$ transitions associated with W49N was found to lie in an area 10″× 8″. The radiation appears to surround the intense H_2O masers and the compact continuum source. The majority of the spectral features which lie between 10 and 20 km s⁻¹ are clumped in an area of diameter less than 1″. The features between −2 and +7 km s⁻¹ may be caused by an outflow of material from the continuum position, while the features between 10 and 20 km s⁻¹ are probably associated with mass outflow from a compact object located at their position.

132.080 Formation of OB clusters.
P. T. P. Ho, A. D. Haschick.
Extragalactic molecules, (see 012.048), p. 189 - 191 (1982).

Continuum aperture synthesis observations of four compact H II regions W33, G10.6−0.4, G19.6−0.2, and G20.1−0.1 at 20, 6 and 2 cm were made using the VLA. In all cases, the authors find that the morphology consists of multiple structures on the scale of $\leqslant 0.1$ pc which can be interpreted as the Strömgren spheres of a cluster of OB stars. If this cluster interpretation is correct, the authors find that the similar size scales of the H II regions and their close spatial distribution imply star formation; the entire cluster of stars have formed essentially at the same time.

The Orion complex: a case study of interstellar matter.   See Abstr. 003.051.

NLTE model atmospheres for early-type stars of various chemical compositions and resulting emission-line spectra for surrounding H II regions.   See Abstr. 064.022.

The pulsational stability of massive stars and 30 Doradus.   See Abstr. 065.057.

Circumstellar shells in M17.   See Abstr. 112.065.

Wolf-Rayet stars in extragalactic H II regions: discovery of a peculiar WR in IC 1613/# 3.   See Abstr. 114.037.

IUE low-dispersion spectra of six luminous stars in symmetric nebulae.   See Abstr. 114.073.

A study on masses of O stars in H II regions.   See Abstr. 115.008.

Neutral hydrogen in the vicinity of galactic radio sources. Supernova remnant W 28.   See Abstr. 125.006.

OH observations of IRS 1 in NGC 7538.   See Abstr. 131.002.

The Zeeman effect in 21 centimeter line radiation: methods and initial results.   See Abstr. 131.004.

The spectral dependence of dust emissivity at millimeter wavelengths.   See Abstr. 131.013.

The energetics of molecular clouds. V. The S37 molecular cloud.   See Abstr. 131.020.

$J = 4 \rightarrow 3$ HCN observations of M17 SW, DR21(OH), W 51 and NGC 6334.   See Abstr. 131.030.

H_2O masers in W49N. I. Maps.   See Abstr. 131.051.

Isotope-selective photodestruction of carbon monoxide.   See Abstr. 131.052.

CO observations around galactic longitude $l = 45°$.   See Abstr. 131.059.

The formation and destruction of HeH⁺ in astrophysical plasmas.   See Abstr. 131.060.

Molecular hydrogen emission from W51.   See Abstr. 131.062.

Neutral hydrogen observations in the direction of extended background radio sources.   See Abstr. 131.073.

The molecular cloud complex in the vicinity of IC 5146.   See Abstr. 131.076.

Observations of neutral carbon in the NGC 1977 bright rim.   See Abstr. 131.078.

Excited OH in absorption towards W3 (OH).   See Abstr. 131.087.

The NGC 7538 region: the distribution and dynamics of molecules compared with those of H I and H⁺.   See Abstr. 131.094.

Neutral hydrogen observations of the Puppis Window.   See Abstr. 131.095.

Observations of CO $J = 3 \rightarrow 2$ emission from molecular clouds.   See Abstr. 131.097.

Infrared and maser sources in regions of star formation.   See Abstr. 131.099.

CO in the Horsehead Nebula.
See Abstr. 131.106.

CO observations of the molecular cloud encompassing Sharpless 222.   See Abstr. 131.109.

High spatial resolution observations of HCN in S 235B.   See Abstr. 131.114.

H_2CO near compact H II regions – new WSRT results.   See Abstr. 131.116.

Is a BN-type object the energy input to the NH_3 cloud in NGC 2071?   See Abstr. 131.117.

VLA observations of OH masers and associated ultracompact continuum sources.   See Abstr. 131.118.

Observations of neutral carbon in the NGC 1977 bright rim.   See Abstr. 131.121.

The mass of hot, shocked CO in Orion: first observations of the $J = 17 \rightarrow J = 16$ transition at 153 microns.
See Abstr. 131.159.

Star formation in the NH_3 cloud of the NGC 2071 region.   See Abstr. 131.161.

The impact of high angular resolution observations on the study of pre-main-sequence objects.
See Abstr. 131.170.

HCO synthesis maps of velocity structure in K3 - 50.
See Abstr. 131.207.

VLA observations of H_2CO clumps in DR 21.
See Abstr. 131.212.

VLBI observations of the maser emission in the four ground-state OH transitions from W3(OH) and Sgr B2.
See Abstr. 131.254.

Report of IAU Commission 34: Interstellar matter and planetary nebulae (Matière interstellaire et nébuleuses planétaires).   See Abstr. 131.278.

10 and 20 micron images of regions of star formation.   See Abstr. 133.001.

Infrared spectra of protostars:composition of the dust shells.   See Abstr. 133.003.

A cluster of near-infrared sources in the neutral intrusions within M16 (NGC 6611).   See Abstr. 133.005.

A high resolution far-infrared survey of a section of the galactic plane. II. Far-infrared, CO, and radio continuum results.   See Abstr. 133.008.

On the infrared sources 1 and 2 in NGC 7538.
See Abstr. 133.012.

OI (63 μ) and C II (157 μ) line emission from photodissociation regions.   See Abstr. 133.015.

Near-infrared observations of the Far-Infrared Source V region in NGC 6334.   See Abstr. 133.024.

Ring nebulae associated with Wolf-Rayet stars in the Large Magellanic Cloud. II. Kinematics of DEM 45, 137, 165, 174, and 208.   See Abstr. 134.008.

Ultraviolet absorption by highly ionized atoms in the Orion Nebula.   See Abstr. 134.010.

Radio continuum observations of cometary nebulae.
See Abstr. 134.045.

8–13 μm spectrophotometry of compact planetary nebulae and emission line objects.   See Abstr. 135.049.

A continuum study of galactic radio sources in the constellation of Monoceros.   See Abstr. 141.213.

Extended γ-ray sources and active regions in the Galaxy: the Carina and Orion complexes.
See Abstr. 142.529.

The galactic abundance gradient.
See Abstr. 155.026.

Infrared mapping of the galactic plane. II. Medium-resolution maps of the Cygnus X region.
See Abstr. 156.003.

Star formation and chemical abundances in clumpy irregular galaxies.   See Abstr. 158.008.

The optical structure of II Zw 40.
See Abstr. 158.012.

Abundance gradients in M31: comparison of results from supernova remnants and H II regions.
See Abstr. 158.037.

H I line studies of galaxies: I-General catalogue of 21-cm line data.   See Abstr. 158.049.

Gradients in the physical conditions of M101 and the pregalactic helium abundance.   See Abstr. 158.065.

Giant ringlike H II regions and the distance to M101.
See Abstr. 158.068.

Giant ring-like H II regions and the distance to M 101.
See Abstr. 158.257.

The carbon abundance in the Magellanic Clouds from IUE observations of H II regions.   See Abstr. 159.001.

Erratum

132.901   On the far ultraviolet flux distribution of the Orion Nebula.   G. R. Carruthers,
H. M. Heckathorn.
Astrophys. Lett., Vol. 22, 5 - 11 (1981). – See Abstr. 30.132.036.

133 Infrared Sources

**133.001 10 and 20 micron images of regions of star forma-
tion.**   J. A. Hackwell, G. L. Grasdalen,
R. D. Gehrz.
Astrophys. J., Vol. 252, 250 - 268 (1982).

A fast-mapping technique, developed for use on the
Wyoming Infrared Telescope, has been used to make infrared
images of regions of star formation. The authors present
10 μm and 20 μm images of Mon R2, W 51 IRS 1 and IRS 2,
S 140, and NGC 7538; and 5 μm images of W 51 IRS 2 and
NGC 7538. The most luminous infrared sources are associated
with radio continuum sources. The observational technique
also provides highly accurate positional information for the
objects observed, and the authors make positional comparisons
between near-infrared objects, radio continuum sources, OH
and H_2O masers, and far-infrared sources.

133.002 A high resolution submillimeter map of OMC-1.
J. Keene, R. H. Hildebrand, S. E. Whitcomb.
Astrophys. J., Lett., Vol. 252, L11 - L15 (1982).

The authors have mapped the 400 μm emission from the
central region of OMC-1 ($\sim 3' \times 5'$) with $35''$ resolution. This
region contains two emission peaks with sizes of ~ 0.5,
separated by ~ 1.5, which probably represent distinct density
condensations in the molecular cloud. Comparison of the
observations with earlier far-infrared observations shows the
two condensations to have similar optical depths and dust
masses. A bar of 400 μm emission is found $\sim 15''$ SE of the
ionization front near θ^2 A Ori, indicating a sharp increase in
dust density in the neutral matter outside the ionized region.

**133.003 Infrared spectra of protostars: composition of the
dust shells.**   S. P. Willner, F. C. Gillett,
T. L. Herter, B. Jones, J. Krassner, K. M. Merrill, J. L. Pipher,
R. C. Puetter, R. J. Rudy, R. W. Russell, B. T. Soifer.
Astrophys. J., Vol. 253, 174 - 187 (1982).

Nearly complete 2 to 13 μm spectra are presented for
13 compact infrared sources associated with molecular clouds,
as well as partial spectra of six additional objects. The spectra
resemble blackbodies with superposed absorption features
from 2.8 to 3.5 μm, at 6.0 and 6.8 μm, and in the silicate band
centered near 9.7 μm. Correlations among the features are
studied in an attempt to confirm possible identifications.

**133.004 A search for the infrared counterpart of type II OH
masers. II. Statistical analysis.**
T. J. Jones, A. R. Hyland, J. L. Caswell, I. Gatley.
Astrophys. J., Vol. 253, 208 - 223 (1982).

The results of a search for the infrared counterpart to
type II OH/IR masers found in sensitivity limited radio surveys
are presented. The IR sources are, on the whole, very red,
many having $K-L$ colors greater than +7. The OH peak
velocity separation ΔV is found to correlate with intrinsic
bolometric luminosity, increasing luminosity corresponding
to increasing ΔV. Many of the low ΔV sources have intrinsic
luminosities well below $10^4 L_\odot$, lower than expected from an
extrapolation of the classical Mira phenomena to higher
masses and longer periods. The sources showing the low
H_2O, high CO absorption band strengths typical of M supergiants usually have high ΔV's and high luminosities, consistent with their identification as extreme Population I M
supergiants. The low ΔV, low luminosity sources have absorption band strengths typical of Mira variables. Several high ΔV,
high luminosity stars also show Mira characteristics in their
band strengths, implying the existence of massive M stars with
cooler and more extended atmospheres than are presently
known to exist.

**133.005 A cluster of near-infrared sources in the neutral
intrusions within M16 (NGC 6611).**
J. R. Walsh, N. J. White.
Mon. Not. R. Astron. Soc., Vol. 199, 9P - 13P (1982).

Eight near-infrared sources have been discovered in the
vicinity of the neutral intrusions in the core of the H II, H I
and molecular complex M16 (NGC 6611). *JHKL* photometry
of four of these sources is presented and indicates emission
from hot (~ 1000K) circumstellar dust. The brightest infrared
source, M16 IRS1, is coincident in position with a radio continuum peak.

**133.006 The double structure of W3−IRS 5 as determined
from high-resolution spatial scans.**
G. Neugebauer, E. E. Becklin, K. Matthews.
Astron. J., Vol. 87, 395 - 399 (1982).

Slit scans over the infrared source W3−IRS 5 from
4.8 μm to 12.7 μm are interpreted in terms of two sources of
luminosity separated by $0.9''$ in declination. The data as a
function of wavelength show that the two sources must have
remarkably similar properties.

**133.007 Speckle interferometry of molecular cloud sources
at 4.8 μm.**   H. M. Dyck, R. R. Howell.
Astron. J., Vol. 87, 400 - 403 (1982).

The authors have observed spatial structure at 4.8 μm
in W3−IRS 5, Mon R2−IRS 3, and S 140−IRS 1, the first
two of which are clearly double. In addition, for four compact
sources they can set low upper limits to the angular diameters
which are near those for blackbodies having the same fluxes
and color temperatures observed for those sources. With a
factor-of-2 higher resolution one may be able to distinguish
between thermal emission or reddening as the source of the
observed spectral shape.

**133.008 A high resolution far-infrared survey of a section of
the galactic plane. II. Far-infrared, CO, and radio
continuum results.**   M. T. Stier, D. T. Jaffe, G. G. Fazio,
W. G. Roberge, C. Thum, T. L. Wilson.
Astrophys. J., Suppl. Ser., Vol. 48, 127 - 143 (1982).

The authors have surveyed 7.5 deg^2 of the galactic plane
at 70 μm with a $\sim 1'$ beam. The region lies between $l^{II} = 10°$
and $l^{II} = 16°$ and includes the M17 and W33 complexes. The
weakest of the 42 sources detected had a flux density of
350 Jy at 70 μm. Detailed far-infrared, ^{12}CO, ^{13}CO, and radio
continuum observations of these sources are presented. The
derivation of the important physical parameters of the sources
and their surrounding molecular clouds is discussed. Furthermore, properties of the individual regions are investigated and
maps of selected sources are presented.

**133.009 New infrared counterparts of southern Type II OH
maser sources.**   N. Epchtein, Nguyen-Quang-Rieu.
Astron. Astrophys., Vol. 107, 229 - 234 (1982).

A search for infrared counterparts of 66 Type II OH
maser sources in the southern galactic plane has resulted in
the detection of infrared objects in the error box of 61 OH
positions. JHKL photometry of these sources has been performed. Discussion based on the infrared colour indices and
variability of some sources allowed the authors to identify
23 objects as the probable counterparts of the OH masers;
most of the remaining sources are likely to be reddened field
stars. Mid-infrared observations of the 7 brightest OH−IR
sources have shown that the energy distribution can be fitted
by 480 to 1250 K blackbody curves; in one case a strong
silicate absorption band has been found. The intrinsic luminosities, estimated using the kinematic distances derived from the

OH maser line velocities, are consistent with the values corresponding to M giant or supergiant stars.

133.010 On balloon-borne observations in infrared astronomy.   H.-c. Zou.
Wuli, Vol. 10, 204 - 209 (1981). In Chinese. – Abstr. in Phys. Abstr., Vol. 85, Abstr. 23353 (1982).

133.011 Model for the Becklin-Neugebauer infrared point source.   I. G. Kolesnik, S. G. Kravchuk.
Pis'ma Astron. Zh., Tom 8, 234 - 237 (1982). In Russian. English translation in Soviet Astron. Lett., Vol. 8.

A model for the Becklin-Neugebauer source is proposed assuming this object to be a forming massive star at the time when the radiation pressure of the star-like core with mass $\sim 14 M_\odot$ and luminosity $\sim 10^4 L_\odot$ leads to the formation of a dense shell near the dust sublimation radius in the extended envelope.

133.012 On the infrared sources 1 and 2 in NGC 7538.   H. Elsässer, K. Birkle, C. Eiroa, R. Lenzen.
Astron. Astrophys., Vol. 108, 274 - 278 (1982).

The authors present recent observations in the visible and near infrared of two point-like sources in NGC 7538 near the positions of IRS 1 and 2. Arguments are raised toward identifying these stars with the excitation sources of the associated H II regions. Photometric measurements are used to derive obscuration and reddening of the two stars. The authors find that the star associated with IRS 1, an H II region with a deep 10 μm absorption feature and containing maser sources, has visual extinction comparable to that of IRS 2, which lacks 10 μm absorption and maser sources, in disagreement with the often postulated universal relation between extinction and 10 μm absorption optical depth.

133.013 Triple structure of infrared source 3 in the Monoceros R2 molecular cloud.   D. W. McCarthy.
Astrophys. J., Lett., Vol. 257, L93 - L97 (1982).

The infrared object Mon R2 IRS 3 exhibits a triple structure containing two sources, approximately 0.'10 in diameter, centered on a fainter source approximately 2″ in diameter. The separation, position angle, and brightness ratio of the double components are 0.'87 (830 AU), 13°5, and 0.41 respectively; these parameters are wavelength independent from 2.2 to 12.5 μm. The double source is interpreted as two early B-type stars. The larger source is tentatively identified as an envelope of scattered light. Over a 7 month period, photometric variations were detected in the primary component of the double and in the larger envelope.

133.014 Extended far-infrared sources near DR 21.   M. F. Campbell, W. F. Hoffmann, H. A. Thronson, D. Niles, R. Nawfel.
Bull. American Astron. Soc., Vol. 13, 809 (1981). – Abstract.

133.015 OI (63 μ) and C II (157 μ) line emission from photodissociation regions.   D. J. Hollenbach.
Bull. American Astron. Soc., Vol. 13, 809 (1981). – Abstract.

133.016 High resolution infrared spectra of S106.   R. I. Thompson.
Bull. American Astron. Soc., Vol. 13, 809 (1981). – Abstract.

133.017 Far-IR, compact sources in the galactic plane.   T. Kelsall, M. G. Hauser, R. F. Silverberg, E. Dwek, M. T. Stier, D. Y. Gezari.
Bull. American Astron. Soc., Vol. 13, 809 - 810 (1981). Abstract.

133.018 Zum Helligkeitsanstieg von IRC + 10420 = V 1302 Aquilae.   H. Geßner.
Mitt. Veränderl. Sterne (MVS), Band 9, 86 (1981).

133.019 Triple structure of source three in the Mon R2 molecular cloud.   D. W. McCarthy.
Prepr. Steward Obs., No. 378, 18 pp. (1982).

133.020 The energetic molecular outflow near AFGL 961: millimeter-wave and infrared observations.   C. J. Lada, T. N. Gautier III.
Prepr. Steward Obs., No. 386, 30 pp. (1982).

133.021 Far-infrared sources in Cygnus X: an extended emission complex at DR 21, and unresolved sources at S 106 and ON 2.   M. F. Campbell, W. F. Hoffmann, H. A. Thronson, Jr., D. Niles, R. Nawfel, M. Hawrylycz.
Prepr. Steward Obs., No. 390, 30 pp. (1982).

133.022 Polarization of compact sources in the galactic center.   M. J. Lebofsky, G. H. Rieke, M. R. Deshpande, J. C. Kemp.
Prepr. Steward Obs., No. 392, 16 pp. (1982).

133.023 Gas condensation as a cause of the infrared excess in the spectra of cold nebulae.   M. E. Perel'man.
Astrofizika, Tom 17, 383 - 393 (1981). In Russian. English translation in Astrophysics, Vol. 17, No. 2.

It is shown that the occurrence of the IR radiation excess near 10 μm can be attributed to processes of neutral gas cloud condensation formed in these objects. It is shown that in such clouds the latent energy of the phase transitions, especially at the early stages of sublimation, must be released in non-equilibrium IR radiation with frequencies determined by the types of the interatomic (intermolecular) bindings appearing under condensation. These estimations lead to frequencies, band-widths and intensities of the IR radiation close to the observed ones. The significance of the transition energy consideration in stellar evolution is also demonstrated.

133.024 Near-infrared observations of the Far-Infrared Source V region in NGC 6334.   J. Fischer, R. R. Joyce, M. Simon, T. Simon.
Astrophys. J., Vol. 258, 165 - 169 (1982).

The authors have observed a very red near-infrared source at the center of NGC 6334 FIRS V, a far-infrared source suspected of variability by McBreen et al. The near-infrared source has deep ice and silicate absorption bands, and its half-power size at 20 μm is $\sim 15'' \times 10''$. Over the past 2 years the authors have observed no variability in the near-infrared flux. They have also detected an extended source of H_2 line emission in this region. The total luminosity in the $H_2 v = 1 - 0\ S(1)$ line, uncorrected for extinction along the line of sight, is 0.3 $L_\odot$. Detection of emission in high-velocity wings of the $J = 1 - 0$ ^{12}CO line suggests that the H_2 emission is associated with a supersonic gas flow.

133.025 Near-infrared slit scans of molecular cloud sources. II.   H. M. Dyck, H. J. Staude.
Astron. Astrophys., Vol. 109, 320 - 325 (1982).

The authors present 2.2 and 3.8 μm slit scans of five molecular cloud sources showing sub-arc sec structure. W3/IRS9 is a double with a separation $\Delta = 2''.4$ and position angle $PA = 29°$. The brighter component itself is resolved and is more extended north-south than east-west. S235/IRS4 is extended east-west. Three sources in NGC 7538 are extended with complex geometry exhibited by the IRS1-IRS2 association.

133.026 The variable infrared source near HH100.   D. J. Axon, D. A. Allen, J. Bailey, J. H. Hough, M. J. Ward, R. F. Jameson.
Mon. Not. R. Astron. Soc., Vol. 200, 239 - 245 (1982).

The authors report photometric observations which clearly show that the infrared source associated with the Herbig-Haro object HH100 is highly variable (up to 2.5 mag at K). They argue that such a large variation can only be a consequence of either dramatic variations in the extinction of the source due to the motions of a patchy circumstellar dust shell or fluctuations in the temperature of the underlying protostar. To account for the observations solely on the basis of extinction changes a two component model of the source comprising of a hot star and a $\sim$800 K dust shell is required. In addition an abnormally flat reddening law is necessary to match the $(J-H)$ and $(H-K)$ colour changes.

133.027 **Results from balloon observations at far-infrared wavelengths.**   L. Nordh, M. Fridlund.
Rep. Obs. Lund, No. 18, (see 012.044), p. 35 - 37 (1982).

A powerful method for star counting in the infrared. See Abstr. 031.543.

A model of the 10 micrometer silicate feature in the spectra of BN-like IR point sources.   See Abstr. 112.056.

Circumstellar shells in M17. See Abstr. 112.065.

Discovery of an infrared companion to T Tauri. See Abstr. 121.009.

Infrared Mira variables as distance indicators. See Abstr. 122.148.

OH observations of IRS 1 in NGC 7538. See Abstr. 131.002.

$J = 4 \rightarrow 3$ **HCN observations of M17 SW, DR21(OH), W51 and NGC 6334.**   See Abstr. 131.030.

Near-infrared spectroscopy of moderate luminosity sources: OMC-2 IRS 3 and IRS 4.   See Abstr. 131.046.

Molecular hydrogen emission from W51. See Abstr. 131.062.

Infrared and maser sources in regions of star formation.   See Abstr. 131.099.

Excess line emission in protostellar objects. See Abstr. 131.158.

High-resolution synthesis maps of 86 GHz SO emission toward Orion-KL.   See Abstr. 131.240.

Infrared and radio observations of W51: another Orion-KL at a distance of 7 kiloparsecs? See Abstr. 132.013.

Ultra-compact H II regions embedded in infrared sources.   See Abstr. 132.038.

cm-wavelength observations of the linear polarization and total flux density of extragalactic red and infrared objects.   See Abstr. 141.257.

A high resolution far-infrared survey of a section of the galactic plane. I. The nature of the sources. See Abstr. 156.001.

Infrared mapping of the galactic plane. II. Medium-resolution maps of the Cygnus X region. See Abstr. 156.003.

Far infrared survey of extended molecular cloud H II region complexes along the galactic plane. See Abstr. 156.007.

A high-excitation optically obscured H II region in the nucleus of NGC 5253.   See Abstr. 158.146.

Observations of far-infrared emission from late-type galaxies.   See Abstr. 158.352.

134 Emission Nebulae, Reflection Nebulae

134.001   Optical polarization of the cometary nebula NGC 2261.
M. R. Gething, R. F. Warren-Smith, S. M. Scarrott, R. G. Bingham.
Mon. Not. R. Astron. Soc., Vol. 198, 881 - 888 (1982).

Optical linear polarization maps in two wavebands are presented for Hubble's variable nebula, NGC 2261. The data confirm that NGC 2261 is a reflection nebula illuminated throughout by the star R Mon. Across the head of the cometary nebula and including R Mon is a band of polarization that is produced by extinction by aligned grains rather than simple scattering. It is suggested that there is an equatorial disc around R Mon and that dust grains within are aligned by a toroidal magnetic field which is a trapped remnant of the field within the cloud out of which R Mon formed.

134.002   The γ Cassiopeiae nebulae.
R. Poeckert, S. van den Bergh.
Publ. Astron. Soc. Pacific, Vol. 93, 703 - 706 (1981/82).

Intercomparison of red and blue plates shows that γ Cas is surrounded by both emission and reflection nebulosity. The authors suggest that the reflection nebulosity lies in the equatorial plane of γ Cas where it is shielded from ionizing Lyman continuum radiation by the disk-like envelope that is believed to surround this Be star. Intercomparison of direct plates taken 27 years apart shows no evidence for motion or significant structural changes in either the emission or reflection nebulosity near γ Cas.

134.003   A large shell nebula in NGC 55.
J. A. Graham, D. G. Lawrie.
Astrophys. J., Lett., Vol. 253, L73 - L75 (1982).

The authors report the optical detection of a large shell nebula, at least several hundred parsecs across, in the nearby galaxy NGC 55. The observation was made in the light of the [O III]λ5007 emission line using an image-tube camera in conjunction with a narrow-band interference filter. Spectroscopic observations confirm its gaseous nature and its association with the galaxy. The observed nebulous arc is roughly centered on a bright stellar knot which appears to be an unusually active region of star formation. It is suggested that the shell is similar in form to the neutral hydrogen supershells observed by Heiles in our own Galaxy.

134.004   Galactic ring nebulae associated with Wolf-Rayet stars. VII. The nebula G2.4 + 1.4.
R. R. Treffers, Y.-H. Chu.
Astrophys. J., Vol. 254, 132 - 135 (1982).

Narrow band interference filter photographs and Fabry-Perot spectra are presented for the nebula G2.4 + 1.4. In Hα and [O III]λ5007, the nebula shows a complex double shell structure, which appears quite different in [N II]λ6583 and [S II]λ6732. The Fabry-Perot spectra show components with velocities V_{LSR} between −20 and +25 km s^{-1}. It is concluded that G2.4 + 1.4 is a supernova remnant near the galactic center that is illuminated by the peculiar WC star LSS 4368.

134.005   Galactic ring nebulae associated with Wolf-Rayet stars. IV. The ring nebula S308 and its interstellar environment.   Y.-H. Chu, T. R. Gull, R. R. Treffers, K. B. Kwitter, T. H. Troland.
Astrophys. J., Vol. 254, 562 - 568 (1982).

The authors have obtained narrow-band filter photographs, [O III]λ5007 Fabry-Perot line profiles and 21 cm maps of the vicinity of the Sharpless H II region S308 and adjacent nebulae. S308 is a ring nebula surrounding the WN5 star HD 50896. An expansion velocity of 60 km s^{-1} and an age of 7×10^4 years are derived for S308. The total kinetic energy

and thermal energy in the shell of S308 can account for only 5% of the total stellar wind energy input. The H II regions S303 and S304 exterior to S308 are also excited by the star HD 50896, showing that S308 is optically thin to UV photons.

134.006   Galactic ring nebulae associated with Wolf-Rayet stars. V. The stellar wind blown bubbles.
R. R. Treffers, Y.-H. Chu.
Astrophys. J., Vol. 254, 569 - 577 (1982).

Fabry-Perot observations at Hα and [O III]λ5007 are presented for the Wolf-Rayet ring nebulae NGC 2359, NGC 6888, and the nebulae around MR 100. The measured expansion velocity of these nebulae are 18, 75, and 50 km s^{-1}, respectively. The kinetic energy of the shells is 1% or less of the kinetic energy from the wind over the lifetime of the shell. This energy is more than a factor of 20 less than expected from the theory of stellar wind blown bubbles. The observed dynamics agrees better with the momentum conserving expansion than the adiabatic stage of wind driven expansion.

134.007   Galactic ring nebulae associated with Wolf-Rayet stars. VI. NGC 3199, anon (MR 26), RCW 58, and RCW 104.   Y.-H. Chu.
Astrophys. J., Vol. 254, 578 - 586 (1982).

The author has obtained narrow-band interference filter photographs and high resolution Fabry-Perot spectra for four galactic ring nebulae associated with Wolf-Rayet stars— NGC 3199, anon (MR 26), RCW 58, and RCW 104. All of these four nebulae show interaction between the stellar wind and the ambient interstellar medium. NGC 3199, anon (MR 26), and RCW 104 are classified as W-type nebulae. RCW 58, having a prominent ring of stellar ejecta, is classified as an E-type nebula. For most W-type nebulae, the kinetic energy in the shell is only about 1% of the total mechanical energy input from the stellar wind.

134.008   Ring nebulae associated with Wolf-Rayet stars in the Large Magellanic Cloud. II. Kinematics of DEM 45, 137, 165, 174, and 208.   Y.-H. Chu.
Astrophys. J., Vol. 255, 79 - 86 (1982).

The author has obtained high resolution Fabry-Perot data for five ring nebulae around Wolf-Rayet (W-R) stars in the Large Magellanic Cloud (LMC)—DEM 45, 137, 165, 174, and 208. Here he reports on their kinematics. DEM 45 and DEM 165 have internal motion less than 10 km s^{-1}. DEM 137 and DEM 208 are shells expanding at velocities of about 15 to 20 km s^{-1}, and their dynamical ages are several million years. DEM 174 is confirmed to be an ordinary H II region with a hole in the center and a large amount of internal motion. Four of these LMC W-R ring nebulae are 10 to 20 times larger than those in the Galaxy.

134.009   High-resolution photographs in the rocket ultraviolet of the Orion Nebula.
R. C. Bohlin, J. K. Hill, T. P. Stecher, A. N. Witt.
Astrophys. J., Vol. 255, 87 - 94, plates 2, 3 (1982).

The Orion Nebula has been imaged through four UV filters centered near 1400, 1820, 2240, and 2620 Å by a rocket borne telescope with 10″ resolution. The nebular images are presented in absolute units of specific intensity using contours to display the dynamic range of about 300. Ultraviolet color gradients for the nebula are given for each color relative to the 1820 Å image. The total flux in the nebula in each bandpass is about 2.5 times the stellar flux from the Trapezium as determined by IUE, indicating that the scattering efficiency of the dust is high throughout the UV. The UV morphology is determined by the distribution of the dust with respect to the illuminating stars since the contribution from two-photon and Balmer continuum emission is small. The observed color gradients indicate that reddening occurs toward the outer parts of the nebula and that the scattering by the dust is fairly symmetric about the θ^1 and θ^2 Ori region.

134.010   Ultraviolet absorption by highly ionized atoms in the Orion Nebula.
J. Franco, B. D. Savage.
Astrophys. J., Vol. 255, 541 - 551 (1982).

The IUE was used to obtain high-resolution, far-UV spectra of θ^1 A, θ^1 C, θ^1 D, and θ^2 A Orionis. The interstellar absorption lines in these spectra are discussed with an emphasis on the high-ionization lines of C IV and Si IV. θ^2 A Ori has interstellar C IV and Si IV absorption of moderate strength at the velocity found for normal H II region ions. θ^1 C Ori has very strong interstellar C IV and Si IV absorption at velocities blueshifted by ~25 km s^{-1} from that found for the normal H II region ions. The authors consider the possible origin of the high-ionization lines by three processes: X-ray ionization, collisional ionization, and UV photoionization.

134.011   The kinematical structure of the bipolar nebula AFGL 618.   U. Carsenty, J. Solf.
Astron. Astrophys., Vol. 106, 307 - 310 (1982).

The authors report on the peculiar radial velocity structure discovered in the visible lobes of the bipolar nebula AFGL 618. The velocity data are used to construct a geometrical and kinematical model of the bipolar nebula. A kinematical age of less than 600 yr is derived for the visible nebula. The results support the view that AFGL 618 represents an early phase in the rapid evolutionary transition from the red giant stage to that of a planetary nebula.

134.012   Polarimetric observations of S106.
M. G. Lacasse.
News Lett. Astron. Soc. N. Y., Vol. 2, 30 - 34 (1982).

134.013   An atlas of interstellar Ca II and Na I profiles in the Carina Nebula.   N. R. Walborn.
Astrophys. J., Suppl. Ser., Vol. 48, 145 - 160 (1982).

Optical interstellar absorption line profiles in the spectra of 22 stars within the Carina Nebula, observed with the 4 m echelle spectrograph at resolving powers of $3 - 6 \times 10^4$, are presented. Radial velocities, equivalent widths, and representative column densities are given for resolved velocity components. As many as 12 distinct components spanning several hundred km s^{-1} are seen in a single line of sight. At least two kinds of motions are implied by the interstellar lines: a systematic expansion of the entire nebula at 15 - 20 km s^{-1}, and smaller scale "bubbles" or "bullets" with much higher velocities. A census of the energy available from the stellar winds of the known O and WN-A stars over the lifetime of the association shows that these stellar winds are in principle capable of generating the high-energy phenomena observed in the nebula.

134.014   The nearest stellar wind bubble or fossil supernova remnant.   P. G. Johnson.
Astrophys. Space Sci., Vol. 82, 213 - 221 (1982).

A new series of ultra-deep, wide-angle Hα filter photographs at higher angular resolution than hitherto achieved, have been obtained covering the Orion-Taurus-Eridanus region. These reveal new, both diffuse and filamentary nebulosities near the Barnard Arc Nebula, which are correlated with H I features in the region. Mechanisms capable of producing the observed features are discussed. Both a stellar wind bubble and a fossil supernova remnant are considered possible explanations.

134.015   JHK-Photopolarimetrie früher Sterne in M17.
H. J. Staude, H. M. Dyck.
Mitt. Astron. Ges., Nr. 55, p. 20 (1982). − Abstract.

134.016   Der Bipolare Nebel S106: Photometrische, Polarimetrische und Spektropolarimetrische Beobachtungen.
H. J. Staude, R. Lenzen, H. M. Dyck, G. D. Schmidt.
Mitt. Astron. Ges., Nr. 55, p. 20 (1982). − Abstract.

134.017   Das Geschwindigkeitsfeld des bipolaren Nebels S106.
J. Solf.
Mitt. Astron. Ges., Nr. 55, p. 21 - 22 (1982).

134.018   Spektroskopische Untersuchungen von expandierenden Gasnebeln.   U. Carsenty, J. Solf.
Mitt. Astron. Ges., Nr. 55, p. 22 - 24 (1982).

134.019   Star counts in the Lagoon, Trifid and W28 complex.
H. Hartl, H. M. MacGillivray, W. J. Zealey.
Mitt. Astron. Ges., Nr. 55, p. 175 - 179 (1982).

134.020   New cometary nebulae.
A. L. Gyul'budagyan.
Pis'ma Astron. Zh., Tom 8, 222 - 223 (1982). In Russian.
English translation in Soviet Astron. Lett., Vol. 8.

A list of 13 cometary nebulae found in the Palomar Sky Survey prints is presented.

134.021   Search for ring nebulae around Of stars.
T. A. Lozinskaya, A. I. Lomovskij.
Pis'ma Astron. Zh., Tom 8, 224 - 231 (1982). In Russian.
English translation in Soviet Astron. Lett., Vol. 8.

A survey of surroundings of 72 Of stars ($\delta > -43°$) on the Palomar Sky atlas has been carried out in order to find ring nebulae produced by stellar wind. For ~ 50 per cent of the stars ring nebulae around them, if existing at all, cannot be observed because of high interstellar absorption and/or bright Hα background emission. 14 Of stars reveal ring-like structure of surrounding H II regions.

134.022   Ultraviolet surface brightness measurements of the Barnard Loop nebula and the Orion reflection nebulosity.   D. H. Morgan, K. Nandy, G. I. Thompson.
Mon. Not. R. Astron. Soc., Vol. 199, 399 - 408 (1982).

Ultraviolet surface brightness maps of the Orion region at 2740, 2350, 1950 and 1550 Å have been constructed from data obtained with the S2/68 ultraviolet sky-survey telescope. Barnard's Loop is seen in these maps superimposed upon an extended region of high ultraviolet brightness. The latter is interpreted as diffuse reflection from dust with a scattering phase function that is more isotropic at 1550 Å than at longer wavelengths.

134.023   Radio emission of the Crab nebula and anisotropic relativity of Boltyanskij.   L. K. Nikolaev.
Differents. uravneniya, Tom 17, 2097 - 2100 (1981). In Russian. − Abstr. in Ref. zh., 51. Astron., 3.51.721 (1982).

134.024   Low-frequency observations of the emission nebula NGC 2264.
E. A. Abramenkov, V. V. Krymkin.
Astron. Zh., Tom 59, 263 - 266 (1982). In Russian. English translation in Soviet Astron., Vol. 26, No. 2.

Observations of the H II region NGC 2264 have been carried out at frequencies 12.6, 14.7, 16.7, 20, 25 MHz. Maps of the brightness distribution over the source are presented for each observation frequency. The H II region is seen reliably in absorption at all frequencies. The electron temperature of the H II nebula, the emission measure and the electron density have been determined.

134.025   Monochromatic photographs of Cygnus Loop with a fiber glass image tube.
T. G. Sitnik, T. A. Birulya, G. M. Lyapunov.
Pis'ma Astron. Zh., Tom 8, 286 - 289 (1982). In Russian.

English translation in Soviet Astron. Lett., Vol. 8.

The nebulae NGC 6992 - 6995 and NGC 6960 (the east and west parts of the supernova remnant Cygnus Loop) were photographed in the [O III] 5007 Å line by an image tube with fiber glass input and output. The nebula NGC 6960 was also observed in the [S II] 6717, 6732 Å doublet. The images of NGC 6960 in various wavelengths differ by structure details.

134.026 Detección de la línea H166 en la nebulosa de Carina.   I. F. Mirabel, S. A. Acero.
Bol. Asoc. Argentina Astron., No. 20 - 24, p. 279 - 287 (1981).

134.027 Observaciones de prisma objetivo, UBVRIK, H beta y polarimétricas de NGC 5367.
H. G. Marraco, J. C. Forte.
Bol. Asoc. Argentia Astron., No. 20 - 24, p. 400 (1981). Abstract.

134.028 H 166 α en Carina.
I. N. Azcárate, J. C. Cersósimo, F. R. Colomb.
Bol. Asoc. Argentina Astron., No. 26, p. 179 - 184 (1981).

134.029 Radial velocities in the nebula S 119.
T. G. Sitnik.
Astron. Tsirk., No. 1193, p. 5 - 6 (1981). In Russian.

134.030 Ring nebulae around Of stars – a new type of emission nebulae.
T. A. Lozinskaya, V. M. Lyutyj.
Astron. Tsirk., No. 1196, p. 1 - 4 (1981). In Russian.

134.031 Six wavelength 8 - 13 micron images of Eta Carinae.
J. A. Hackwell, R. D. Gehrz, G. L. Grasdalen.
Bull. American Astron. Soc., Vol. 13, 809 (1981). – Abstract.

134.032 Variations in the [S II] and [O II] doublet across the Hourglass region of M8.
B. T. Lynds, E. J. O'Neil, Jr.
Bull. American Astron. Soc., Vol. 13, 825 (1981). – Abstract.

134.033 The outlying condensations around Eta Carinae.
K. Davidson, N. R. Walborn, T. R. Gull.
Bull. American Astron. Soc., Vol. 13, 825 (1981). – Abstract.

134.034 VLA observations of bipolar structure in Hb12.
R. T. Newell, R. M. Hjellming.
Bull. American Astron. Soc., Vol. 13, 853 (1981). – Abstract.

134.035 Search for extended X-ray emission surrounding the Crab Nebula.
P. Gorenstein, F. R. Harnden, Jr., M. Mitchell, F. D. Seward.
Bull. American Astron. Soc., Vol. 13, 865 (1981). – Abstract.

134.036 The Einstein objective grating X-ray spectrum of the Crab Nebula.
S. M. Kahn, F. D. Seward, T. Chlebowski, J. Dijkstra.
Bull. American Astron. Soc., Vol. 13, 865 (1981). – Abstract.

134.037 An interfilament medium in the Crab Nebula?
S. A. Balbus.
Bull. American Astron. Soc., Vol. 13, 866 (1981). – Abstract.

134.038 An interpretation of polarimetric observations of two bipolar reflection nebulae.
N. V. Voshchinnikov.
Astrofizika, Tom 16, 257 - 263 (1980). In Russian. English translation in Astrophysics, Vol. 16, No. 2.

The polarization maps of the peculiar objects CRL 2688 and M1-92 obtained by Schmidt et al. (1978) are interpreted with the use of the first-order theory of scattering by dust grains in a homogeneous nebula. It is shown that graphite grains and silicate core-ice mantle grains in an optically thin nebula model fairly represent the observations of CRL 2688 and M1-92 respectively.

134.039 Recent observations of H_2 in the Orion Nebula.
T. N. Gautier III, E. T. Young.
Bull. American Astron. Soc., Vol. 14, 576 (1982). – Abstract.

134.040 The Crab Nebula. I. Spectrophotometry of the filaments.   R. A. Fesen, R. P. Kirshner.
Astrophys. J., Vol. 258, 1 - 10 (1982).

New spectrophotometry for ten positions in the Crab Nebula is presented which provides relative line intensities for the wavelength region 3700 to 7400 Å. Electron temperatures determined from [O III] line intensities range from 11,000 to 18,300 K, with generally lower temperatures indicated from the lines of [O II], [N II], and [S II]. Electron densities estimated from the ratio of the [S II] λλ6717, 6731 lines range from 550 to 3500 cm⁻³ with a typical value of 1300 cm⁻³ for the filaments measured. The Balmer decrement for many filaments is consistent with pure recombination, but a few filaments exhibit slightly steeper decrements. Helium line intensities vary considerably among filaments with a few northern filaments showing He I λ5876 line emission that is weaker by a factor of 3. A wide range of ionization states is present within some filaments as demonstrated by lines of [Fe II], [Fe III], [Fe V], and [Fe VII].

134.041 The Crab Nebula. II. A photoionization model analysis for the filaments.
R. B. C. Henry, G. M. MacAlpine.
Astrophys. J., Vol. 258, 11 - 21 (1982).

Detailed photoionization models are employed to interpret recent spectrophotometric observations of Crab Nebula filaments. Large measured filament-to-filament variations in the He I λ5876/Hβ ratio are probably due to differences in relative helium abundance. Preliminary evidence suggests that helium may be distributed anisotropically among the filaments, with its abundance being several times lower in the northeastern sector than in the remaining regions. In terms of mass fractions relative to solar values, carbon may be normal, while nitrogen, oxygen, neon, and sulfur appear to be underabundant by factors of 2 to 3 in some filaments. The models indicate that [S II] λλ6716, 6731 emission is produced in that portion of a filament dominated by H⁰ and He⁰, and the observed strengths of these lines are due to the large volume of such gas in the nebula. Upon combining this result with observations of the total flux of Hβ and [S II] from the Crab, a lower limit to the mass of filamentary gas of 1.2 M⊙ is derived.

134.042 Eta Carinae's numbered days.
N. R. Walborn, T. R. Gull, K. Davidson.
Sky Telesc., Vol. 64, 16 - 17 (1982).

134.043 Optical colours and polarization of a model reflection nebula IV. Mixture of grains in the case of the star within the nebula.   G. A. Shah.
Bull. Astron. Soc. India, Vol. 9, 297 - 308 (1981).

The UBV colour differences and polarization that can be caused by mixtures of interstellar grains in a model reflection nebula have been presented. The geometrical configuration of the nebula is adapted in the form of a homogeneous plane-parallel slab with the star within the nebula. The materials of the grain species include ice, graphite, enstatite silicate and silicon carbide. The author has considered separately one-, two-, and three-component mixtures of the grains in various proportions of their number densities.

134.044 The physical conditions within the poly-polar nebula NGC 6302–III.   J. F. Barral, J. Cantó, J. Meaburn, J. R. Walsh.
Mon. Not. R. Astron. Soc., Vol. 199, 817 - 832 (1982).

IUE observations of the ultraviolet emission lines and continuum from the wind-driven poly-polar nebula NGC 6302 have been combined with those obtained of the visible emission lines to investigate the physical parameters of the ionized gas. Several relationships consolidate the view that this nebula is predominantly ionized radiatively by a very hot central star. However, the 'poly-polar' appearance and complex, high velocity flows of ionized material from the nebular core strongly suggest the presence of an energetic stellar wind from the central, but obscured star.

134.045　Radio continuum observations of cometary nebulae.
　　　　　K. C. Turner, Y. Terzian.
Astron. J., Vol. 87, 881 - 884 (1982).
The class of objects known as "cometary nebulae" is investigated through their radio emissions. Observations made with the Arecibo radio telescope at λ 12 and λ 21 cm are reported. Several objects show thermal radio emission characteristic of ionized gases.

The Orion complex: a case study of interstellar matter.　　　See Abstr. 003.051.

Early spectrographic observations of the Crab Nebula.　　　See Abstr. 004.051.

[Ni II] emission under nebular conditions.
See Abstr. 022.197.

Radio observations of the forthcoming lunar occultations of the Crab Nebula.　　　See Abstr. 096.011.

On the number of Hα emission stars in the region of the Orion nebula.　　　See Abstr. 114.066.

The ultraviolet spectrum of the Crab Nebula.
See Abstr. 125.003.

Supernova having created the Crab nebula.
See Abstr. 125.024.

Velocity, reddening, and temperature structure of the H$_2$ emission in Orion.　　　See Abstr. 131.022.

Detection of the torsionally excited state of methanol in Orion A.　　　See Abstr. 131.023.

The motion and distribution of the vibrationally excited H$_2$ in the Orion molecular cloud.
See Abstr. 131.024.

The hunter and the starcloud.
See Abstr. 131.033.

Carbon monoxide broad wings and self-reversals in NGC 2071.　　　See Abstr. 131.034.

Runaway expansion of giant shells driven by radiation pressure from field stars.　　　See Abstr. 131.036.

Observations and interpretation of the line profiles of excited H$_2$ in Orion.　　　See Abstr. 131.092.

Kinematics of molecular gas in Orion from observations of the ^{13}CO J = 2 → 1 line.　　　See Abstr. 131.101.

Small scale clumping in the Orion Molecular Cloud. See Abstr. 131.107.

X-ray ionization and the chemistry of the Orion Molecular Cloud.　　　See Abstr. 131.112.

Molecular line mapping of OMC−1.
See Abstr. 131.119.

An upper limit to the atomic carbon abundance in the Orion plateau.　　　See Abstr. 131.120.

The mass of hot, shocked CO in Orion: first observations of the J = 17 → J = 16 transition at 153 microns. See Abstr. 131.159.

Radiation pressure effect on dust surrounding the Pleiades bright stars.　　　See Abstr. 131.266.

High resolution infrared spectra of S 106.
See Abstr. 133.016.

8−13 μm spectrophotometry of compact planetary nebulae and emission line objects.　　　See Abstr. 135.049.

Line feature around 73 keV from the Crab Nebula. See Abstr. 142.002.

On the optical counterparts of the X-ray sources in the Orion nebula.　　　See Abstr. 142.035.

X-ray emission from the Carina nebula and the associated early stars.　　　See Abstr. 142.039.

On the evidence for high energy γ-ray emission from the Orion nebula stemming from COS-B observations. See Abstr. 142.532.

R associations. VI. The reddening law in dust clouds and the nature of early-type emission stars in nebulosity from a study of five associations.　　　See Abstr. 152.001.

135 Planetary Nebulae

135.001 **A new search for nebulae surrounding Wolf-Rayet stars.**   J. N. Heckathorn, F. C. Bruhweiler, T. R. Gull.
Astrophys. J., Vol. 252, 230 - 238, plates 3 - 8 (1982).
The authors have used a comprehensive narrow band emission line survey of the Milky Way to search for nebulosities surrounding Wolf-Rayet stars. Fifteen ring nebulae have been definitely identified, including five previously unreported shell structures. An additional 30 nebulosities have been classified as "probable" or "possible" ring nebulae. Angular diameter, sharpness or diffuseness, and level of brightness in three emission line bandpasses (Hα +[N II], [O III], and [S II]) have been determined.

135.002 **Planetary nebulae with close binary central stars.**
M. Livio.
Astron. Astrophys., Vol. 105, 37 - 41 (1982).
A model recently proposed for the formation of planetary nebulae with close binary nuclei predicts a characteristic nonspherical morphology. The observational situation concerning planetary nebulae with binary nuclei is surveyed. The shapes of those planetary nebulae that have been observed or suggested to contain a double nucleus are examined and are generally found to agree with the predicted one, although some difficulties are indicated. A number of objects, possessing exactly the predicted morphology are presented as possible candidates for containing a close binary nucleus.

135.003 **The ionization structure of the Ring Nebula. II. Ultraviolet observations.**   T. Barker.
Astrophys. J., Vol. 253, 167 - 173 (1982).
Ultraviolet observations of emission line intensities have been made in four positions in the Ring Nebula previously studied optically. There is good agreement for O and N abundances, and the importance of charge-transfer reactions to the ionization of Ne is confirmed. The Ne/O abundance of 0.25 ±0.03 is close to the average value for planetary nebulae found by Kaler; the C/H abundance is $(3.9 \pm 0.4) \times 10^{-4}$, also in good agreement with other measurements in planetaries. The most surprising result is that the C^{2+} abundance inferred from the optical λ4267 recombination line is as much as 10 times higher than that measured from the UV lines.

135.004 **Ultraviolet spectroscopy of planetary nebulae in the Magellanic Clouds.**
S. P. Maran, L. H. Aller, T. R. Gull, T. P. Stecher.
Astrophys. J., Lett., Vol. 253, L43 - L47 (1982).
Ultraviolet spectra of three high excitation planetary nebulae in the Magellanic Clouds (LMC P40, SMC N2, SMC N5) were obtained with the International Ultraviolet Explorer. The results are analyzed together with new wavelength spectrophotometry of LMC P40 and published data on SMC N2 and SMC N5 to investigate chemical composition and, in particular, to make the first reliable estimates of the carbon abundance in extragalactic planetary nebulae.

135.005 **Ionization structure and partial obscuration of the planetary nebulae NGC 3132 and NGC 3242.**
A. Condal, C. Pritchet, G. G. Fahlman, G. A. H. Walker.
Publ. Astron. Soc. Pacific, Vol. 93, 695 - 702 (1981/82).
The authors discuss their observations of the ionization structure and reddening variations in the two planetaries NGC 3132 and NGC 3242.

135.006 **The planetary nebula NGC 6826.**
W. A. Feibelman.
Publ. Astron. Soc. Pacific, Vol. 93, 719 - 720 (1981/82).
NGC 6826 is established as one of only three known triple-shell, "giant halo" type planetary nebulae. Some confusing errors concerning its size and structure in the recent literature are discussed.

135.007 **Observations of 196 southern planetary nebulae at 408 MHz.**   M. R. Calabretta.
Mon. Not. R. Astron. Soc., Vol. 199, 141 - 150 (1982).
The Molonglo Radio Telescope, operating at 408 MHz with halfpower beamwidth 3 arcmin, has been used to observe 196 planetary nebulae south of declination +18°. Flux densities are presented for the 43 nebulae which were detected, and flux density limits are given for the remainder. There is no evidence of non-thermal radio emission from any of the planetary nebulae in this survey.

135.008 **Ultraviolet spectra of planetary nebulae − VIII. The C/O abundance ratio in the Ring nebula.**
D. R. Flower.
Mon. Not. R. Astron. Soc., Vol. 199, 15P - 18P (1982).
The Ring nebula (NGC 6720) has been observed with the International Ultraviolet Explorer satellite. Measurements of line intensities at a position on the ring, offset from the central star, lead to a value of the abundance ratio C/O ≃ 1.2, a result which is insensitive to the adopted value of the electron temperature in the nebula.

135.009 **A look at some unstable stars.**
D. F. Malin.
Sky Telesc., Vol. 63, 22 - 26 (1982).

135.010 **Bubbles from dying stars.**
J. B. Kaler.
Sky Telesc., Vol. 63, 129 - 133 (1982).

135.011 **Physical variations in the planetary nebula IC 4997.**
J. Kiser, C. T. Daub.
Astrophys. J., Vol. 253, 679 - 681 (1982).
The planetary nebula IC 4997 has a spectrum which shows a variation in the [O III] λ4363/Hγ line intensity ratio that ranges from 1.6 in 1938 to 0.68 in 1962. Approximate values for the electron temperature and electron density of IC 4997 in 1962 are calculated from line intensity ratios of O III and N II ions. The equation of thermal equilibrium is used to calculate the electron temperature and electron density of the nebula in 1938. The changes in electron temperature and electron density in IC 4997 between 1938 and 1962 are then shown to be possible consequences of the physical expansion of the nebula and not necessarily due, as has previously been suggested, to a change in the temperature of the central star.

135.012 **Observations and morphological study of ring planetary nebulae in [O III].**   R. Louise.
Astron. Astrophys.,Suppl. Ser., Vol. 47, 575 - 589 (1982).
In French.
Ten ring planetary nebulae are observed with a narrowband interference filter (Δλ ≈ 10 Å) centered on the [O III] λ = 5007 Å emission line: NGC 40, NGC 1514, NGC 2392, NGC 6543, NGC 6781, NGC 6826, NGC 7354, NGC 7048, NGC 7009, NGC 7662. Photometric and geometric parameters, as defined by Louise (1974), are derived for each nebula. These parameters fit well with a shell model for most of the observed nebulae. Isophotic contours are derived in arbitrary scale.

135.013 **Spectral variations and evidence for edge and/or line locking mechanism(s) in the low-excitation planetary nebula HD 138403.**

A. Surdej, J. Surdej, J. P. Swings.
Astron. Astrophys., Vol. 105, 242 - 249 (1982).

Within the last decades, striking variations occured in the visual spectrum of the low-excitation planetary nebula HD 138403 ≡ 315−13°1: the [O III], He II λ 4686 and other high excitation lines of e.g. He I which were reported absent (or very faint) in 1950 are now present in emission. In addition, an [O III] nebula has been discovered around the central star. Furthermore, faint red and blue emission satellites have been found around the central Balmer ($H_\beta - H_9$) and [O II] λλ 3726, 3729 lines, with a mean velocity separation of +122 and −126 km s^{-1}, respectively. These spectral features are best interpreted as evidence for mass-outflow from the central object. It is also argued that the presence of these satellites directly reflects the formation of a bipolar structure around the central nucleus.

135.014   The distance to the planetary nebula NGC 7027.
S. R. Pottasch, W. M. Goss, E. M. Arnal, R. Gathier.
Astron. Astrophys., Vol. 106, 229 - 234 (1982).

Various direct methods for determining the distance to NGC 7027 are discussed. The result is a distance of between 1 and 1.5 kpc. In an appendix the authors present an attempt to measure 21-cm emission from NGC 7027. A low upper limit of 0.04 $M_\odot$ for neutral hydrogen is found.

135.015   Planetary nebulae in local group galaxies. IX. Velocity modulated photographs of the center of
M31.   D. G. Lawrie, H. C. Ford.
Astrophys. J., Vol. 256, 120 - 126, plates 7 - 10 (1982).

The authors used a sequence of velocity-modulated photographs to find and measure the radial velocities of faint planetary nebulae in the center of M31. The photographs were made with a velocity modulating camera (VMC) which consists of a temperature-tuned 2.1 Å (FWHM) [O III] λ5007 interference filter, a cooled, two-stage image intensifier, and a calibrating photomultiplier. The authors found 42 planetary nebulae within 250 pc of the nucleus of M31. From these 19 are new identifications. Radial velocities and relative brightnesses are measured for 32 of the nebulae. The results are used to derive a luminosity function for planetary nebulae and to estimate their total number in M31.

135.016   Zentralsterne Planetarischer Nebel und Subdwarf-O-Sterne.   R. P. Kudritzki.
Sterne Weltraum, 21. Jahrg., 17 - 20 (1982).

135.017   Die Emissionslinienspektren der Planetarischen Nebel NGC 6572 und VV 68.
G. Krämer, M. Grewing, E. Schulz-Lüpertz.
Mitt. Astron. Ges., Nr. 55, p. 36 - 41 (1982).

135.018   On the origin of planetary nebulae.
L. S. Pilyugin, G. S. Khromov.
Astrofizika, Tom 17, 167 - 178 (1981). In Russian. English translation in Astrophysics, Vol. 17, No. 1.

The basic well-established data on planetary nebulae and their nuclei together with data on their evolution are used to revise the problem of their origin. It is shown that the empirics does contradict the popular theoretical schemes of the formation of planetary nebulae by means of long-term mass loss from the central stars, or by the series of minor ejections from them. More realistic is the hypothesis of one-time single ejection of the massive nebula.

135.019   E (B−V) extinction values for 24 planetary nebulae derived from IUE data.
W. A. Feibelman.
Astron. J., Vol. 87, 555 - 557 (1982).

Excellent agreement was found for E(B−V) values for nine planetary nebulae observed in common by IUE and ANS, derived from a strong λ 2200 absorption feature. E(B−V) values for seven additional objects with moderately strong λ 2200 absorption were determined. A third group of eight planetary nebulae for which the λ 2200 absorption feature is marginal or absent had E(B−V) extinctions determined from the observed He II I(λ 1640)/I(λ 4686) ratio. A comparison with optical and radio data is included.

135.020   Ultraviolet spectra of planetary nebulae − VI. NGC 7662.
J. P. Harrington, M. J. Seaton, S. Adams, J. H. Lutz.
Mon. Not. R. Astron. Soc., Vol. 199, 517 - 564 (1982).

A detailed study of NGC 7662 is based on UV results obtained from 15 IUE spectra and on observations of other workers at optical, IR and radio wavelengths. Improved techniques are used to extract IUE data for an extended source. The central star is fainter than has been previously supposed (by more than two magnitudes). The blackbody He II Zanstra temperature of 113 000 K is consistent with the UV colour temperature. Two models are discussed.

135.021   Electron temperature mapping of planetary nebulae.
N. K. Reay, S. P. Worswick.
Mon. Not. R. Astron. Soc., Vol. 199, 581 - 589 (1982).

The linear response of the electronographic process provides a simple method for producing relative intensity ratios of emission lines across a complete object. Using narrow-band interference filters to isolate the temperature sensitive lines of O^{++}, variations in electron temperature have been mapped across the planetary nebulae NGC 3242, 6543, 6826, 7662 and IC 3568.

135.022   Measurements of expansion velocities in planetary nebulae.
G. J. Robinson, N. K. Reay, P. D. Atherton.
Mon. Not. R. Astron. Soc., Vol. 199, 649 - 657 (1982).

The authors have used a scanning Fabry-Perot interferometer on the 1.5-m flux collector, Tenerife, to obtain [O III] λ 5007 Å emission line profiles at the centres of 33 planetary nebulae. Small young nebulae were chosen to form the majority of the sample, and the shell expansion velocities V_{exp} were deduced from the observations. These new data, along with published expansion velocity data for other nebulae, are used to investigate the relationship between V_{exp} and the nebular radii R_n. Two possible relationships are shown to exist, and comparison with theoretical radiation pressure-driven nebular shell models implies two distinct evolutionary sequences.

135.023   Can planetary nebulae rotate?
V. P. Grinin.
Astron. Zh., Tom 59, 326 - 333 (1982). In Russian. English translation in Soviet Astron., Vol. 26, No. 2.

It is shown that the inclination of spectral lines observed in some planetary nebulae in the case when the spectrograph slit is aligned along the large axis, what now is ascribed to the inhomogeneous expansion of envelopes, in reality might be consistent, as was originally proposed by Campbell and Moore, with the rotation of nebulae around their small axes. It is supposed that the source of the initial rotation of a protoplanetary nebula is the rotation of a central star.

135.024   Red/blue intensity ratios in expanding planetary nebulae.   J. R. Doughty, J. B. Kaler.
Publ. Astron. Soc. Pacific, Vol. 94, 43 - 49 (1982).

The authors have measured the red-to-blue intensity ratios of the Doppler-split emission-line components of six bright planetary nebulae from high-dispersion coudé plates taken by O. C. Wilson. Three, NGC 6210, NGC 6818, and NGC 7662, exhibit noticeable changes in this ratio as a function of ionization potential, and presumably in mass distribution as a function of distance from the central star. Of these, NGC 6210 consistently shows r/b ratios significantly greater than unity. The authors attempt to interpret wavelength variations in the ratio in terms of internal dust: the results are marginal, and serve to place upper limits on internal optical depth. They give back-to-front electron density ratios for three of the nebulae; two, NGC 6210 and NGC 7009 show significant differences from unity.

135.025   La estrella central de NGC 3132.
R. H. Méndez.
Bol. Asoc. Argentina Astron., No. 20 - 24, p. 138 (1981). Abstract.

135.026   Las estrellas centrales de NGC 246 y NGC 1360.
R. H. Méndez, V. S. Niemela.
Bol. Asoc. Argentina Astron., No. 20 - 24, p. 223 (1981). Abstract.

135.027   Nebulosas planetarias con estrellas centrales de tipo espectral A.   R. Méndez.
Bol. Asoc. Argentina Astron., No. 20 - 24, p. 468 (1981). Abstract.

135.028   Observaciones de tres estrellas centrales de nebulosas planetarias.
R. H. Méndez, V. S. Niemela.
Bol. Asoc. Argentina Astron., No. 20 - 24, p. 475 (1981). Abstract.

135.029   The expansion velocity field within the planetary nebulae NGC 40 and NGC 7026.
F. Sabbadin, E. Hamzaoglu.
Astron. Astrophys., Vol. 109, 131 - 135 (1982).

The Hβ, [O III], Hα and [N II] expansion velocity fields within the planetary nebulae NGC 40 and NGC 7026 have been obtained from Echelle spectrograms. The expansion of the main part of NGC 40 resembles that of a shell with mean radius 15$''$ and thickness 5$''$. External wisps and condensations (north and south of the nebula) present peculiar velocities. No simple spatial model can be obtained for NGC 7026; the complex gas motion indicates that emissions come from localized material almost symmetrically placed with respect to the central star.

135.030   Abundances in the planetary nebula NGC 6853,
S. R. Pottasch, D. P. Gilra, P. R. Wesselius.
Astron. Astrophys., Vol. 109, 182 - 186 (1982).

IUE ultraviolet spectra of several portions of planetary nebula NGC 6853 have been analysed. Combining these measurements with those made from the ground, the authors are able to determine the relative abundances of C, N, O, and Ne, without making any important assumption concerning the ionization of any of these elements. C, O, and Ne have solar abundances whereas N is overabundant by a factor of three. The UV nebular continuum is also discussed.

135.031   Chemical composition of high-excitation planetaries.
L. H. Aller, S. J. Czyzak.
Proc. Natl. Acad. Sci. USA, Vol. 78, 5266 - 5270 (1981). Abstr. in Phys. Abstr., Vol. 85, Abstr. 53353 (1982).

135.032   Photoelectric UBV observations of the variability of planetary nebulae during 1968 - 1980.
E. B. Kostyakova, V. P. Arkhipova.
Astron. Tsirk., No. 1166, p. 4 - 7 (1981). In Russian.

135.033   NGC 2346.
IAU Circ., No. 3667 (1982).

135.034   10 and 20 micron images of IC 418.
A. F. Bentley, J. A. Hackwell, G. L. Grasdalen, R. D. Gehrz.

Bull. American Astron. Soc., Vol. 13, 808 - 809 (1981). Abstract.

135.035   4 - 8 μm spectroscopy of NGC 7027.
R. W. Russell, S. V. Beckwith, J. Wyant, N. J. Evans II, A. Natta.
Bull. American Astron. Soc., Vol. 13, 809 (1981). – Abstract.

135.036   The spectrum of NGC 7027 from 5 to 8 microns.
J. D. Bregman, H. L. Dinerstein, J. H. Goebel, D. F. Lester, F. C. Witteborn, D. M. Rank.
Bull. American Astron. Soc., Vol. 13, 852 (1981). – Abstract.

135.037   The hydrogen depleted planetary nebulae – Abell 30 and Abell 78.   G. H. Jacoby, H. C. Ford.
Bull. American Astron. Soc., Vol. 13, 854 - 855 (1981). Abstract.

135.038   Ultraviolet spectroscopy of extragalactic planetary nebulae with the International Ultraviolet
Explorer.   S. P. Maran, L. H. Aller, T. R. Gull, T. P. Stecher.
Bull. American Astron. Soc., Vol. 13, 855 (1981). – Abstract.

135.039   Study of planetary nebula K 1 - 2 and its variable nucleus.   L. Kohoutek, G. F. O. Schnur.
ESO Sci. Prepr. No. 195, 12 pp. (1982). – Submitted to Mon. Not. R. Astron. Soc.

135.040   Expanding shells within the Helix nebula (NGC 7293)?   J. Meaburn, N. J. White.
Astrophys. Space Sci., Vol. 82, 423 - 439 (1982).

Filter photographs and Fabry–Pérot interferograms of the Helix nebula reveal evidence of approximately spherical expansion. Heliocentric radial velocities from +44 to -51 km s^{-1} have been detected over the centre of the nebula. A 'dumb-bell' configuration, viewed at a small angle to the longest axis, is proposed to explain many of these observations.

135.041   Theoretical models of planetary nebulae. V. Objects of intermediate-to-high excitation.   L. H. Aller.
Astrophys. Space Sci., Vol. 83, 225 - 238 (1982).

Theoretical models are calculated for 15 planetary nebulae of medium-to-high excitation, following procedures previously described. Initial stellar energy distributions are adopted from Cassinelli (1971), but are subsequently modified to obtain the best representation of optical spectra for the selected objects. Other adjustable parameters include the stellar radius,R_*, the nebular density, N_H, the truncation radius, r_c, for the nebular shell, and the chemical composition. Excitation-sensitive ratios are usually well-represented as are the actual observed intensities of spectral lines. Forbidden lines arising from $3p^3$ configurations offer difficulties. For this sample of nebulae, the mean abundances seem to agree well with those found in an earlier study where the models were used as interpolation devices.

135.042   CCD spectra of planetary nebulae in the 6000 Å - 10000 Å region.
P. N. Kupferman, G. E. Danielson, D. C. Jewitt.
Bull. American Astron. Soc., Vol. 14, 573 (1982). – Abstract.

135.043   From red giants to planetary nebulae.
S. Kwok.
Astrophys. J., Vol. 258, 280 - 288 (1982).

The transition from red giants to planetary nebulae is studied by comparing the spectral characteristics of red giant envelopes and planetary nebulae. Observational and theoretical evidence both suggest that remnants of red giant envelopes may still be present in planetary nebula systems and should have significant effects on their formation. The dynamical effects of the interaction of stellar winds from central stars of planetary nebulae with the remnant red giant envelopes are evaluated and the mechanism found to be capable of producing the observed masses and momenta of planetary nebulae. The observed mass-radii relation of planetary nebulae may also be best explained by the interacting winds model. The possibility that red giant mass loss, and therefore the production of planetary nebulae, is different between Population I and II systems is also discussed.

135.044   A search for SiO emission from planetary nebulae.
H. A. Thronson, Jr., C. J. Lada.
Publ. Astron. Soc. Pacific, Vol. 94, 226 - 228 (1982).

The authors have searched unsuccessfully for thermal, 86.8 GHz, J = 2 → 1 SiO emission from seven planetary nebulae. The observed 2 σ limit was $T_A^* \leqslant 0.02$ K.

135.045   The spectrum of FG Sge in 1979-1980. I.
           $\lambda\lambda$ 3700-5000 Å.
A. Acker, M. Jaschek, F. Gleizes.
Astron. Astrophys., Suppl. Ser., Vol. 48, 363 - 369 (1982).

An analysis of the spectrum of FG Sge in 1979 and 1980 is given. About 600 absorption lines and 500 emission features were measured. In general the spectrum is characterized by the absence or extreme weakness of the lines belonging to iron peak elements. On the other hand lines belonging to rare earths are numerous and strong. The spectral classification which can be attributed to the star in 1979 and 1980 is G8-K0 Ia. The spectrum does not seem to evolve toward that of a Ba-type star, as predicted.

135.046   High spectral resolution observations of [S II] lines
           in the planetary nebula IC 418 at the CES spectro-
graph.   R. Louise, E. Maurice.
Messenger, No. 28, p. 28 - 29 (1982).

135.047   On the origin of planetary nebulae.
H. Nussbaumer.
Astron. Astrophys., Vol. 110, L1 - L2 (1982).

It is suggested, that magnetic flux, randomly emerging from stellar surfaces, may be partially or fully responsible for certain types of stellar winds.

135.048   Internal motions in planetary nebulae.
F. Sabbadin, E. Hamzaoglu.
Astron. Astrophys., Vol. 110, 105 - 110 (1982).

The [O III] expansion velocity has been derived for twenty six northern planetary nebulae. The status of the existing observational data on the expansion velocity in planetary nebulae is presented.

135.049   8−13 μm spectrophotometry of compact planetary
           nebulae and emission line objects.
D. K. Aitken, P. F. Roche.
Mon. Not. R. Astron. Soc., Vol. 200, 217 - 237 (1982).

The authors present 8−13 μm spectra of 24 planetary nebulae and other emission line objects, 19 of which are published here for the first time. This brings the total of planetary nebulae for which such spectra are available to 23 including four classed as very low excitation (VLE) objects and three as WC-11 objects. On the basis of these spectra, roughly one-third of the planetaries are identified as having oxygen-rich (silicate) and rather less than one-third carbon-rich (silicon carbide) dust grains. Of the oxygen-rich nebulae, three are VLEs while the rest are very compact and, on various grounds, considered to be young.

135.050   Ultraviolet spectra of planetary nebulae − VII. The
           abundance of carbon in the very low excitation
nebula He 2−131.   S. Adams, M. J. Seaton.
Mon. Not. R. Astron. Soc., Vol. 200, 7P - 12P (1982).

Fluxes in the multiplets [O II] λ 2470 and C II] λ 2326 have been measured for the VLE nebula He 2−131 = HD 138403 using IUE high-dispersion spectra. An analysis

similar to that of Harrington et al. (1980) for IC 418 gives C/O = 0.3 for He 2−131, compared with C/O = 1.3 for IC 418 and 0.6 for the Sun.

135.051   Calibration of the distance scale of planetary
           nebulae.   A. Acker.
Scientific aspects of the Hipparcos space astrometry mission, (see 012.041), p. 199 - 200 (1982).

The distances of planetary nebulae, faint and distant stars, forming an unhomogeneous group, are practically unknown. The trigonometric parallax determined by Hipparcos for some of these objects may allow the calibration of different distance scales, and thus lead to a better knowledge of the physical parameters and the evolution of planetary nebulae.

Landolt-Börnstein.   See Abstr. 003.064.

FG Sagittae − missing link in de sterevolutie.
See Abstr. 065.007.

Not with a bang but a whimper.
See Abstr. 065.019.

Mass loss from evolved stars. I. Observations of
17 stars in the CO(2−1) line.   See Abstr. 112.001.

Stellar wind in the nucleus of IC 2149.
See Abstr. 112.020.

A survey of 3 μm emission features in stellar spectra.
See Abstr. 112.059.

International Ultraviolet Explorer observations of
the central stars of the planetary nebulae NGC 6853 and
NGC 7293.   See Abstr. 114.006.

Activity of HBV 475 from its spectral variations.
See Abstr. 114.048.

Absorption lines in the visual spectrum of the
"continuous" central star of the PN NGC 3242.
See Abstr. 114.053.

Discoveries on southern, red-sensitive objective-
prism plates IV: extension to higher latitudes.
See Abstr. 114.214.

Planetary nebulae with close binary nuclei-correc-
tions to angular momentum loss.   See Abstr. 117.133.

Central star of planetary nebula NGC 2346: new
eclipsing binary.   See Abstr. 119.080.

La estrella central de NGC 1360: una binaria
espectroscopica en una nebulosa planetaria.
See Abstr. 120.016.

The formation and destruction of HeH⁺ in astro-
physical plasmas.   See Abstr. 131.060.

Detection of rotationally excited OH emission from
Vy 2 - 2, and studies of carbon chain molecules in cool
circumstellar shells.   See Abstr. 131.218.

Report of IAU Commission 34: Interstellar matter
and planetary nebulae (Matière interstellaire et nébuleuses
planétaires).   See Abstr. 131.278.

Radio recombination lines from high emission
measure nebulae.   See Abstr. 132.053.

The kinematical structure of the bipolar nebula AFGL 618.   See Abstr. 134.011.

VLA observations of bipolar structure in Hb12. See Abstr. 134.034.

Erratum

135.901   **Erratum: "Dust in planetary nebulae"** [Astrophys. J., Vol. 248, 189 - 194 (1981)].   A. Natta, N. Panagia.
Astrophys. J., Vol. 254, 419 (1982). — See Abstr. 30.135.008.

Radio Sources, X-ray Sources, Cosmic Radiation

141 Radio Sources, Quasars, Pulsars

Radio Sources, Quasars

141.001 A polarization burst in the BL Lac object AO 0235 + 164.
C. D. Impey, P. W. J. L. Brand, S. Tapia.
Mon. Not. R. Astron. Soc., Vol. 198, 1 - 9 (1982).

Simultaneous infrared and optical polarimetry and photometry have been obtained for AO 0235 + 164 covering a five
night period. The object underwent a polarization burst during
which the 2.2 μm polarization rose from 17.5 to 28.7 per cent
and fell again to 14.9 per cent. At its peak the degree of optical polarization was 43.9 per cent. The data show the degree
of polarization to increase towards shorter wavelengths. The
large changes in polarization are not accompanied by large
changes in flux. Implications of the luminosity, polarization
and variability are discussed.

141.002 Observations of the spectra of Q0122 −380 and Q1101 −264. R. F. Carswell, J. A. J. Whelan,
M. G. Smith, A. Boksenberg, D. Tytler.
Mon. Not. R. Astron. Soc., Vol. 198, 91 - 110 (1982).

The spectra of the QSOs Q0122 −380 (z_{em} = 2.181) and
Q1101 −264 (z_{em} = 2.143) are described and discussed. Wavelengths and equivalent widths for the absorption lines in the
range 3200 - 5200 Å are given for each object, and a number of
absorption systems are suggested. Evidence is presented for
velocity structure in some of these systems. The Lyα absorption line densities in these and other QSOs are compared. For
the small data sample available the density function is consistent with that expected from a uniform space distribution of
comoving absorbers.

141.003 A revised statistical estimate of the counts of faint radio sources at 5 GHz.
J. V. Wall, P. A. G. Scheuer, I. I. K. Pauliny-Toth, A. Witzel.
Mon. Not. R. Astron. Soc., Vol. 198, 221 - 237 (1982).

A $P(D)$ experiment with the 100-m reflector of MPIfR at
5 GHz has produced an estimate of the counts of faint extragalactic radio sources at flux densities ~1 mJy. The surface
densities of sources are found to be substantially lower than
those derived from a previous experiment of this type carried
out with the Parkes Radio Telescope, and an instrumental
error in the earlier work is identified and discussed. The new
results are in agreement with simple extrapolations of direct
source counts, and the cosmological implications are briefly
considered.

141.004 Radio emission from the 2006 −56 region.
W. M. Goss, R. D. Ekers, D. J. Skellern, R. M. Smith.
Mon. Not. R. Astron. Soc., Vol. 198, 259 - 264 (1982).

The steep-spectrum radio source 2006 − 56 has been mapped with the Fleurs Synthesis Telescope at 1415 MHz with a
resolution of 50 arcsec. The radio source is associated with a
rich irregular cluster and consists of two possible head-tail
objects and diffuse emission with a size of ~2 Mpc
(H = 50 km s^{-1} Mpc^{-1}). Detection of the new head-tail sources
adds to the unbiased sample of tailed sources needed to
examine the importance of galaxy environment.

141.005 A high-resolution study of the absorption spectra of three QSOs: evidence for cosmological evolution in the Lyman-alpha lines. P. Young, W. L. W. Sargent,
A. Boksenberg.
Astrophys. J., Vol. 252, 10 - 31 (1982).

High-resolution (0.8 Å FWHM) spectroscopic observations over the wavelength range from 3260 Å to 4900 Å are
presented for the QSOs Q0002+051 (z_{em} =1.899),
Q0421+019 (z_{em}=2.051), and the gravitationally lensed QSO,
Q1115+080(z_{em}= 1.725). Line profile fits to the absorption
spectra are used to determine column densities and velocity
dispersions within the absorbing clouds. A modest amount of
cosmological evolution is suggested in the sense that there are
more Lyα absorption lines at higher redshifts.

141.006 Do quasars have cosmologically long lifetimes?
G. A. Chanan.
Astrophys. J., Vol. 252, 32 - 36 (1982).

Turner and Tyson have independently suggested that the
apparent evolution of quasars may be an artifact caused by
(unseen) gravitational lenses; some of the problems inherent in
the usual picture of space density evolution are thereby avoided. The author discusses how these problems may be similarly
avoided without invoking any such gravitational effects: apparent (and unreal) density evolution follows as an immediate
consequence if quasar lifetimes are of the order of 3 × 10^9
years. If the lifetimes are indeed this long, quasars may occur
much less frequently than previously thought but, at the same
time, the local density of quasars may have been grossly underestimated.

141.007 Quasar pancakes. W. G. Mathews.
Astrophys. J., Vol. 252, 39 - 53 (1982) = Lick Obs.
Bull., No. 896.

The theory of radiatively accelerated pancake and simple
clouds for the broad emission line region in quasars and
Seyfert galaxies is extended to include momentum absorbed
from X-ray photons. Several simple assumptions are made
concerning net line emission rates from the clouds. The model
predicts that the flat-topped nonlogarithmic line profiles might
be particularly sensitive to far-UV or X-ray continuum flux
variations, conforming to the observed situation (3C 390.3
and NGC 7603). The model provides dynamical explanations
for the restricted range of cloud plasma densities
$10^8 \lesssim n \lesssim 10^{10}$ cm^{-3} and ionization parameters $\Gamma \lesssim 10^{-2\pm1}$.

141.008 A high-resolution spectroscopic study of Q0119−046 and the nature of absorption complexes with $z_{abs} > z_{em}$.
W. L. W. Sargent, P. Young, A. Boksenberg.
Astrophys. J., Vol. 252, 54 - 68 (1982).

High-resolution (0.8 Å FWHM) spectroscopic observations over the wavelength range from 3260 Å to 4920 Å are
presented for the QSO Q0119−046(z_{em} =1.937). The authors
find 61 absorption lines in this object and identify seven main
absorption line redshifts ranging from z_{abs} =0.6577 to 1.9751.
There is a complex of three main systems with $z_{abs} > z_{em}$, with
"infall" velocities relative to the QSO ranging from −2780 to
−3870 km s^{-1}. The authors discuss the nature of the

$z_{abs} > z_{em}$ complex in Q0119−046 and a similar complex in Q1115+080. The most likely of several possible explanations for such complexes is that they arise from a (possibly collapsing) cluster of galaxies of which the QSO is a member.

141.009   Discovery of a narrow line quasar.
J. Stocke, J. Liebert, T. Maccacaro, R. E. Griffiths, J. E. Steiner.
Astrophys. J., Vol. 252, 69 - 74 (1982).
    The authors report the discovery of a stellar object with $z = 0.338$ and X-ray and optical luminosities typical of quasars but which has only narrow permitted and forbidden emission lines over the observed spectral range. This object could be a prototype for a new class of quasar analogous to high luminosity Seyfert type 2 galaxies. But if such a class of objects exists, it cannot comprise more than $\sim 10\%$ of all quasars.

141.010   Strong radio sources in bright spiral galaxies. II.
          Rapid star formation and galaxy-galaxy interactions.
J. J. Condon, M. A. Condon, G. Gisler, J. J. Puschell.
Astrophys. J., Vol. 252, 102 - 124 (1982).
    Statistically complete samples comprising 33 bright spiral galaxies that are strong radio sources were selected. Most of the sources appear to be coextensive with regions of intense star formation. The typical radio luminosity, $\approx 10^{21}$ W Hz^{-1} sr^{-1} at 1413 MHz, of the resolved sources can be explained by synchrotron radiation from supernova remnants (SNRs) produced at the rate of $\lesssim 1$ yr^{-1}. Five of the radio sources are $\leqslant 1$ pc in size and probably are not related to star formation of SNRs. Nearly all of the strong, extended radio sources are found in galaxies with nearby companions, so most of the episodic bursts of star formation are apparently triggered by galaxy-galaxy interactions.

141.011   A high redshift BL Lacertae object: PKS 0215+015.
C. M. Gaskell.
Astrophys. J., Vol. 252, 447 - 454 (1982)   = Lick Obs. Bull., No. 897.
    Spectrophotometry and polarimetry of PKS 0215+015 show it to have all the commonly accepted defining properties of a BL Lac object. Over the past 30 years PKS 0215+015 has varied by over 4 mag. It has been observed to show variable polarization of up to 37%. Although PKS 0215+015 has so far shown no emission lines, it has three definite absorption line systems with redshifts of 1.3449, 1.5494, and 1.6494. PKS 0215+015 is shown to be similar to previously known lower redshift BL Lac objects.

141.012   Four QSOs found in a survey of faint objects.
D. C. Morton, K. P. Tritton.
Mon. Not. R. Astron. Soc., Vol. 198, 669 - 672 (1982).
    Four QSOs with redshifts $z = 0.6185, 0.6267, 0.850$ and 1.665 have been found in the central 33.4 arcmin square of the field at 22^h05, $-18°91$ surveyed by Savage and Bolton, and Krug, Morton and Tritton. The first two QSOs are separated by 6.8 arcmin and may be associated with groups of faint galaxies.

141.013   Jodrell Bank MTRLI observations of nine core-
          dominated sources at 408 MHz.
I. W. A. Browne, M. J. L. Orr, R. J. Davis, A. Foley, T. W. B. Muxlow, P. Thomasson.
Mon. Not. R. Astron. Soc., Vol. 198, 673 - 688 (1982).
    The Jodrell Bank MTRLI has been used to map nine flat-spectrum core-dominated radio sources at 408 MHz. Most of the sources consist of a compact core with a flat spectrum and an extended 'jet' component with a steeper spectrum. Evidence is presented that in several of the sources there is, in addition to the core and the jet which are visible on the maps, a halo component of relatively low brightness and very steep spectrum. The observations are interpreted within the framework of relativistic beam models and it is concluded that

core-dominated quasars are most likely to be classical double radio quasars seen end-on. The diffuse halo is identified with the superposed outer lobes, and the jet with one of the relativistic beams transporting energy to the lobes.

141.014   High-accuracy linear and circular polarization
          measurements at 21 cm.
I. de Pater, K. W. Weiler.
Mon. Not. R. Astron. Soc., Vol. 198, 747 - 755 (1982).
    New high-accuracy linear and circular polarization measurements have been obtained for 27 small-diameter radio sources at λ 21 cm (1415 MHz). From these and other observed properties of the sources, estimates of the average internal magnetic field strengths in the sources are made by applying the uniform synchrotron emission model to the measured circular polarization and by using equipartion arguments. These two values are compared and found to be in agreement to within an order of magnitude. Also, the magnetic fields estimated from circular polarization measurements at two different wavelengths (λ 49 and λ 21 cm) are compared and found to be in rough agreement but with indications of differences between variable and non-variable sources. A comparison of the magnitudes of linear and circular polarization in sources shows no correlations.

141.015   H I absorption distances to four point sources near
          supernova remnants.
J. H. van Gorkom, W. M. Goss, E. R. Seaquist, W. S. Gilmore.
Mon. Not. R. Astron. Soc., Vol. 198, 757 - 765 (1982).
    H I absorption observations have been made of four point sources near supernova remnants (SNR) from the list of Ryle et. al. The sources are 0503+466 (near HB9), 1849+005 (near G33.6+0.1), 1910+052 (near SS 433 and W50), and 2013+370 (near G74.9+1.2). The first three are probably not associated with the SNR while the latter is ambiguous. The observation of 1910+052 indicates that an upper limit to the distance of SS 433 is 4.7 kpc. The distance to SS 433 determined from H I absorption is, thus, 3.7−4.7 kpc.

141.016   QSO counts: a complete survey of stellar objects to
          $B = 23$.   D. C. Koo, R. G. Kron.
Astron. Astrophys., Vol. 105, 107 - 119 (1982).
    The authors discuss photometry from Mayall prime focus plates in the four bands U, J, F, N. They concentrate here on those images classified by size as stellar and discuss the statistics of those objects which do not lie in the domain of galactic stars in color-color diagrams. Since the volume of $U−J$, $J−F, F−N$ space that is occupied by common stars is small, this technique should yield a complete sample of all extragalactic objects of stellar appearance. The limit $J \lesssim 23$ is set by increasing errors in both classification and colors. The QSO counts flatten for $J \gtrsim 21$, as anticipated by constraints from radio and X-ray observations.

141.017   Further observations of radio sources from the
          BG survey. I. The non-thermal sources near
$l = 94°$.
F. Mantovani, M. Nanni, C. J. Salter, P. Tomasi.
Astron. Astrophys., Vol. 105, 176 - 183 (1982).
    This paper is the first of a series which will study Galactic radio sources from the BG catalogue. New observations will be presented. These were made with the Effelsberg 100-m telescope at 1720 MHz and 10.7 GHz and with the full Bologna Cross telescope at 408 MHz. Here, four extended radio sources near $l = 94°$ are examined. CTB104A and 3C434.1 seem to be confirmed as being SNR, while a similar classification for NRAO 655 remains a possibility. CTB104A is found to possess a highly unusual morphology and parallels are drawn with the structure of the peculiar remnant W50. The source 4C50.55 is shown to be an extragalactic object of large angular diameter.

141.018 Bright extragalactic radio sources at 2.7 GHz – II. Observations with the Cambridge 5-km telescope.
J. A. Peacock, J. V. Wall.
Mon. Not. R. Astron. Soc., Vol. 198, 843 - 860 (1982).

The paper presents observations with the Cambridge 5-km telescope of 51 radio sources taken from a complete sample of 168 extragalactic sources having $S(2.7 \text{ GHz}) > 1.5$ Jy. The measurements were made at a frequency of either 5 or 2.7 GHz. Of these 51 sources, only nine were resolved; the three most extended sources were also observed at 1400 and 408 MHz with the Cambridge One-Mile telescope. Since the other 117 sources have been studied previously with the 5-km telescope, structural information is now available on the whole 2.7-GHz sample and allows a comparison to be made with analogous samples selected at low frequencies.

141.019 A search for interplanetary scintillation of Cygnus A at 81.5 MHz.
S. C. Tsien, P. J. Duffett-Smith.
Mon. Not. R. Astron. Soc., Vol. 198, 941 - 945 (1982).

IPS observations of Cygnus A at 81.5 MHz with the Cambridge 3.6 hectare array set an upper limit on the scintillation of 0.07 per cent of the total flux density. This suggests that the hotspots seen at high frequencies are much less prominent at low frequencies.

141.020 Evidence for a decrease in the space density of quasars at $z \gtrsim 3.5$.   P. S. Osmer.
Astrophys. J., Vol. 253, 28 - 37, plate 1 (1982).

The author reports the results of a survey of emission line quasars with redshift $3.5 \leqslant z \leqslant 4.7$ made with a red-blazed grating at the CTIO 4 m telescope. No quasars with $z > 3.5$ have been found. Enough quasars at lower redshift are found, however, for establishing the sensitivity limits of the survey. It is shown that if known quasars from the Hoag-Smith survey in the redshift range $2.5 \leqslant z \leqslant 3.5$ were moved out to $z > 3.5$, they would have easily been detected. It is concluded that the apparent space density of quasars must significantly decrease at $3.7 \leqslant z \leqslant 4.7$.

141.021 Variability of the compact radio source at the galactic center.   R. L. Brown, K. Y. Lo.
Astrophys. J., Vol. 253, 108 - 114 (1982).

The authors report dual frequency radio observations of the compact radio source at the galactic center that were made on 25 epochs over a period of more than 3 years. They find the object to be variable by 20–40% on all time scales from years to days. The spectral index on each epoch is found to be positive, $S \propto \nu^{a}$ with $0.08 < a < 0.55$. The spectrum appears to be flattening with time owing largely to a steady secular increase in the mean 2695 MHz flux density. For a source of constant brightness, the increase in 2695 MHz flux density suggests that the source diameter increased by about 12% in 2 years.

141.022 4C 18.68: a QSO with precessing radio jets?
A. C. Gower, J. B. Hutchings.
Astrophys. J., Lett., Vol. 253, L1 - L5 (1982).

High resolution VLA radio maps at 20 cm and 6 cm wavelengths of the quasar 4C 18.68 reveal an extended halo ($\sim 20''$) containing complex curved structures extending east and west from the central source. The central source has a flat spectrum, while the spectrum generally steepens with distance from the center of the structure. The details of the structure and polarization of the emission suggest relativistic ejection in opposing directions by a precessing or rotating double jet with a period of $\sim 5 \times 10^{4}$ years, consistent with the presence of two interacting massive bodies in the central source.

141.023 Spectroscopy of the QSO pair PKS 0254–334.
A. E. Wright, D. C. Morton, B. A. Peterson, D. L. Jauncey.

Mon. Not. R. Astron. Soc., Vol. 199, 81 - 91 (1982).

The radio QSO PKS 0254–334/R with z_{em} = 1.9130 is located 60 arcsec from the radio-quiet QSO 0254–334/2 with z_{em} = 1.8640. Spectroscopy with a resolution of ~ 2Å has identified nine absorption-line systems in 0254–334/2 – four definite, four probable and one possible. The Lα absorption line is extraordinarily weak relative to the heavy-element lines in three of these systems, implying that hydrogen is almost absent or nearly totally ionized. Absorption by Mg II at a common redshift of about 0.213 is possibly present in both members of the QSO pair. This absorption may be caused by two separate galaxies in a cluster or by a single cloud of gas covering both QSOs.

141.024 The derivation of hotspot parameters from the integrated spectra of double radio sources.
J. A. Peacock.
Mon. Not. R. Astron. Soc., Vol. 199, 295 - 301 (1982).

Although various studies of the hotspots at the outer edges of powerful 3CR sources have suggested positive correlations between source luminosity, hotspot size, and the fraction of emission in hotspots, these results have been inconclusive, mainly due to the limited angular resolution of the observations. Another way of studying hotspots is by their effect on the integrated radio spectra of the sources concerned: on the assumption that low-frequency spectral curvature is due to synchrotron self-absorption in the hotspots, parameters for the hotspots may be derived. The results for a sample of 3CR sources are consistent with the correlations found previously.

141.025 Laboratory exercises in astronomy – Quasars.
D. B. Hoff.
Sky Telesc., Vol. 63, 20 - 21 (1982).

141.026 Radio maps of the sky.
G. Haslam, R. Wielebinski, W. Priester.
Sky Telesc., Vol. 63, 230 - 232 (1982).

141.027 High resolution maps of the southern radio galaxies 1331–09 and 1417–20.
J.-s. Chen, B.-l. Liang, Z.-x. Cui, Z.-l. Zou, H.-j. Su.
Acta Astrophys. Sinica, Vol. 2, 19 - 29 (1982). In Chinese.

141.028 Sneller dan een foton?   G. Schilling.
Zenit, 9. Jaarg., 57 - 59 (1982).

141.029 Milli-arcsecond jets in radio galaxies: interpretation.
R. Linfield.
Astrophys. J., Vol. 254, 465 - 471 (1982).

Some implications of observed parsec scale jets in double radio galaxies are considered. It is concluded that in all cases observed thus far, the asymmetry reflects an intrinsic asymmetry on a parsec scale. If the jet in Cyg A is pressure confined, it must be surrounded by gas hotter than 5×10^{7} K.

141.030 Particle reacceleration and apparent radio source structure.   J. A. Eilek.
Astrophys. J., Vol. 254, 472 - 482 (1982).

The radio galaxy model which uses magnetohydrodynamic turbulence generated by surface instabilities to reaccelerate the radiating electrons has striking consequences for apparent source structure. Strong wave damping in the plasma results in a narrow turbulent edge. Particles accelerated in this edge must diffuse across field lines into the radio source; this predicts strong limb brightening in some cases. The structure of this edge and diffusion into the source are described. The relevance of this model to jets, radio tails, and standard double sources is discussed.

141.031 The large-scale radio structure of 3C 120.
B. Balick, T. M. Heckman, P. C. Crane.
Astrophys. J., Vol. 254, 483 - 488 (1982).

VLA maps and video camera images of 3C 120 are reported. The radio structure at 6 and 20 cm extends ~25–30″(~18 kpc) west of the nucleus. The extended structure, which has a much steeper spectral index ($\alpha \sim -0.6$) than the nucleus ($\alpha \sim 0$), appears to be a large-scale extension of the milli-arcsecond structure observed with VLBI techniques. Thus, in many respects, 3C 120 is a low luminosity member of the "D2" class of radio sources. The observations are discussed in the context of the relativistic beam models used to describe the structure of radio sources which exhibit superluminal expansion and the 100 kpc structures observed in most radio galaxies.

141.032 **The broad and narrow lines in the spectrum of the quasar 3C 351.**   H. Netzer, B. J. Wills, D. Wills.
Astrophys. J., Vol. 254, 489 - 493 (1982).

3C 351 is the only quasar so far in which the broad and narrow components of Lyα and Hβ can be easily separated. The combination of Lyα from previous IUE data and the new optical observations suggests that Lyα/H$\beta \approx 12$ in both components. By comparing this line ratio with the results of a photoionization calculation for the narrow-line region, the authors find a line-of-sight reddening $E_{B-V} \approx 0.25$, most of which is probably intrinsic. Applying this reddening to the observed broad emission lines gives an intrinsic intensity ratio Lyα/H$\beta \sim 52$, indicating that collisional enhancement of Hβ may not be important.

141.033 **Far-infrared photometry of compact extragalactic objects: detection of 3C 345.**
P. M. Harvey, B. A. Wilking, M. Joy.
Astrophys. J., Lett., Vol. 254, L29 - L30 (1982).

The authors report the first detection of a quasar, 3C 345, in the far-infrared spectral region. The observed flux density of 2.2 ± 0.5 Jy at 100 μm and an upper limit of 0.5 ± 0.6 Jy at 50 μm clearly define the overall energy distribution and show the quasar to be a powerful far-infrared source. The observations place some limits on possible emission mechanisms for the far-infrared and millimeter wavelength emission and allow a determination of the total luminosity.

141.034 **Fine structure investigation of 3C84 and 3C345 at 18 cm wavelength.**
L. I. Matveenko, V. I. Kostenko, I. G. Moiseev, J. D. Romney, N. Bartel, L. Padrielli, A. Fikarra, F. Mantovani.
Pis'ma Astron. Zh., Tom 8, 148 - 152 (1982). In Russian.
English translation in Soviet Astron. Lett., Vol. 8.

In February 1980 the angular structures of 3C 84 and 3C 345 were investigated by means of the global radiointerferometer network.

141.035 **On the nature of emission of the star-gas-dust complex in the radio source W 1.**
V. A. Udal'tsov, A. V. Kovalenko.
Astron. Zh., Tom 59, 27 - 37 (1982). In Russian. English translation in Soviet Astron., Vol. 26, No. 1.

The brightness distribution of the radio source W 1 at 102 MHz has been investigated. It is shown that W 1 is genetically connected with the stellar association Cep IV as well as with the extended emission nebula GS 285 which consists of numerous nebulae, including two bright ones, Sharpless 171 and NGC 7822. The radio emission of the nebula S 171 is shown to be thermal. By two independent methods, the distance to S 171 has been evaluated to be 840 pc. The emission of NGC 7822 is mainly thermal. The extended nebula GS 285 is a thermal source. For the nebulae S 171, NGC 7822, and GS 285 the emission measures, electron concentrations, and masses have been estimated in the case $T_e = 6000$ K.

141.036 **A 408 MHz all-sky continuum survey. II. The atlas of contour maps.**   C. G. T. Haslam, C. J. Salter, H. Stoffel, W. E. Wilson.

Astron. Astrophys., Suppl. Ser., Vol. 47, 1 - 143 (1982).

The authors present an atlas of the radio continuum brightness of the whole sky at a frequency of 408 MHz. A set of both α, δ as well as l, b maps is shown. The angular resolution is 0°.85. The atlas combines data from four different surveys using the Jodrell Bank MkI, Bonn 100 meter, Parkes 64 meter and Jodrell Bank MkIA telescopes.

141.037 **Integrated linear polarization of extragalactic radio sources at 10.5 GHz (λ 2.86 cm). – II.**
M. Simard-Normandin, P. P. Kronberg, S. Button.
Astron. Astrophys., Suppl. Ser., Vol. 48, 137 - 138 (1982).

Measurements of integrated linear polarization and flux density at 10.5 GHz (λ 2.86 cm) are presented for 68 extragalactic radio sources. These results, made with the ARO 46 metre telescope, complement and extend a previous set of measurements made at this and nearby wavelengths in a continuing program to establish further Faraday rotation measures and to define the polarization-wavelength relationship for radiogalaxies and quasars.

141.038 **Ion-supported tori and the origin of radio jets.**
M. J. Rees, M. C. Begelman, R. D. Blandford, E. S. Phinney.
Nature, Vol. 295, 17 - 21 (1982).

While apparently supplying tremendous power to their extended radio-emitting regions, the nuclei of most radio galaxies emit little detectable radiation. It is proposed that at the centre of each is a spinning black hole surrounded by a torus of gas too hot and tenuous to radiate efficiently. The torus anchors magnetic fields which extract rotational energy from the hole in the form of two collimated beams of relativistic particles and fields. These in turn drive the observed radio jets and hot spots. A large supply of accreting gas is thus unnecessary and radio galaxies may be interpreted as starved quasars.

141.039 **A galaxy for quasar 3C48.**   R. F. Carswell.
Nature, Vol. 296, 395 - 396 (1982).

141.040 **Detection of the underlying galaxy in the QSO 3C48.**   T. A. Boroson, J. B. Oke.
Nature, Vol. 296, 397 - 399 (1982).

Spectra have been obtained of the faint nebulosity north and south of the centre of the QSO 3C48. In addition to the emission lines previously known, a continuum dominated by hot stars is seen at both positions. This suggests that the host galaxy is a spiral and may explain why searches for features indicative of an old stellar population in QSOs have been unsuccessful. The redshift of the underlying galaxy is the same as that of the permitted lines in the QSO but differs from the redshift of the forbidden lines in the QSO by 500 km s^{-1}.

141.041 **IUE observations of variability and differences in the UV spectra of double quasar 0957 + 561 A, B.**
P. M. Gondhalekar, R. Wilson.
Nature, Vol. 296, 415 - 418 (1982).

The first UV observations of the double quasar were made with IUE in December 1979, and have been supplemented by data obtained during September and December 1980. Here the authors present these new data together with an analysis based on the images being a gravitational lens effect.

141.042 **Absolute spectrophotometry of very large redshift quasars.**   J. B. Oke, D. G. Korycansky.
Astrophys. J., Vol. 255, 11 - 19 (1982).

Absolute spectral energy distributions have been obtained with the multichannel spectrometer of 19 very large redshift quasars. There is a drop in the flux below Lyα and a further drop below Lyβ. A drop or change of slope occurs near the Lyman limit. All of these changes, including the

Lyman jump, can be attributed to absorbing clouds at different redshifts. The covering factor produced by the broad emitting line region is small. If the continuum can be extrapolated to 100 Å, then the emission-line intensities of Ly α and λ1640 of He II are both consistent with each other and with recombination; the covering factor is then about 0.08. Ly α and all other measured emission lines, except λ1550 of C IV, have intensities which are proportional to the continuum flux.

141.043 The nature of the light variations in the double QSO Q0957+561.   W. C. Keel.
Astrophys. J., Vol. 255, 20 - 24, plate 1 (1982) = Lick Obs. Bull., No. 904.

Photographic photometry of Q0957+561 shows that during the 1980 and 1981 observing seasons, the northern image remained at nearly constant brightness, while the southern one showed a rapid brightening followed by a slow, nearly exponential fading. The speed of variation and shape of the light curve suggest that the observed variations are due to intrinsic variability of the imaged QSO, rather than to effects of stars in the lens galaxy. The differential time delay must be greater than 2.7 years. Archival plates taken early in this century show both components brighter by about 0.5 mag, demonstrating variability on long time scales.

141.044 The 3000 Å bump in quasars.   S. A. Grandi.
Astrophys. J., Vol. 255, 25 - 38 (1982).

The λ3000 bump — a broad humplike excess of emission superposed on a quasar's power-law continuum — is modeled via synthetic spectra. The authors find that a likely combination of expected emission components (Balmer lines, optically thin Balmer continuum emission, two-photon emission, and Fe II emission) combined with a small amount of dust reddening can adequately account for the λ3000 bump seen in many quasars. However, some quasars show excess emission in their λ3000 bumps; their continua keep rising shortward of the Balmer edge. The authors rule out emission from a single-temperature blackbody and enhanced two-photon emission as explanations for this excess emission. However, partially optically thick Balmer continuum emission from very deep emission-line clouds can apparently supply this excess emission.

141.045 Compact radio sources: the dependence of variability and polarization on spectral shape.
L. Rudnick, T. W. Jones.
Astrophys. J., Vol. 255, 39 - 47 (1982).

VLA observations have been made at 20, 6, and 2 cm of an unbiased sample of 40 flat-spectrum radio sources selected from the S4 (6 cm strong source) survey. The authors have explored the polarization properties of these sources and examined their variability over a 6–8 year baseline at 6 cm by comparing their flux density values with the original S4 measurements. Most of the flux density in these sources is from regions unresolved ($\lesssim$ 0''2) on the VLA. Properties of the sources can be divided fairly clearly according to three simple spectral shape categories: straight, simple-convex ('humped'), and complex sources. Physical implications and possible explanations of this classification scheme are discussed.

141.046 Time-dependent narrow emission-line profiles of quasars and active galactic nuclei.
E. R. Capriotti, C. B. Foltz.
Astrophys. J., Vol. 255, 48 - 56 (1982).

The narrow-line emitting regions of quasars and active nuclei of galaxies are assumed to consist of material undergoing gravitational infall due to acceleration by centrally located mass concentrations. Two cases are considered. In one, the material is assumed to be in the form of optically thick, similar clouds which emit line radiation monochromatically, isotropically and in inverse proportion to the square of the distance from the center of the system. In the other case, the material is assumed to be homogeneous, isothermal and to have the same ionization structure everywhere. The material is assumed to be excited by an ionizing continuum created by a supernova-like outburst of radiation. Line profiles are computed for various combinations of epoch after outburst, continuum decay-times, and spectral resolution. The computed profiles compare well with many observed [O III] profiles.

141.047 Discovery of a third gravitational lens.
D. W. Weedman, R. J. Weymann, R. F. Green, T. M. Heckman.
Astrophys. J., Lett., Vol. 255, L5 - L9, plate L1 (1982).

The two components of a quasar pair of separation 7''3, discovered in a low-dispersion spectroscopic survey with the Canada-France-Hawaii telescope, were observed with the Multiple Mirror telescope and found to have redshifts of 2.152 and 2.147, with uncertainties of ±0.005. This pair, with approximate magnitudes of 19.5 and 21, is interpreted as the third known example of gravitational lensing. It is emphasized that survey techniques such as that used to discover this pair have found over 1000 quasars, and the authors are puzzled as to why pairs with smaller separations have not been found in this way.

141.048 Apparent superluminal motion in the quasar NRAO 140.   A. P. Marscher, J. J. Broderick.
Astrophys. J., Lett., Vol. 255, L11 - L15 (1982).

Very long baseline interferometer (VLBI) measurements of the compact radio structure in the quasar NRAO 140 ($z = 1.258$) have been obtained at three epochs at a wavelength of 2.8 cm. These observations indicate that the two most compact radio components are separating at an angular rate of 0.10–0.14 milli-arcsec per year. For cosmological distances and $H_0 = 50$ and $q_0 = 0$, this corresponds to a velocity of separation (in the quasar's rest frame) of 10 ± 2 times the speed of light, c; for $H_0 = 100$ and $q_0 = 1$, the value is $(3.1 \pm 0.6) c$. Other interpretations of the temporal changes in correlated flux density and closure phase are discussed and are considered unlikely.

141.049 Infrared photometry of the ultracompact radio source in NGC 6334.   P. M. Harvey.
Astrophys. J., Lett., Vol. 255, L55 - L56 (1982).

Photometry of the ultracompact radio source in NGC 6334 has been obtained at 1.6, 2.2, and 10 μm. The infrared-to-radio energy distribution is qualitatively similar to that of many quasars. The ratio of observed infrared-to-radio flux is a factor of 300 less than for the compact radio source in the galactic center with which the NGC 6334 object has been compared.

141.050 A confusion-limited extragalactic source survey at 4.755 GHz. III. Accurate positions and optical identifications.   J. J. Condon, J. E. Ledden.
Astron. J., Vol. 87, 219 - 241 (1982).

Accurate positions of sources stronger than 15 mJy at 4.755 GHz from a survey covering almost 0.01 sr of extragalactic sky were measured with the VLA. Optical identifications were made. The spectral index α(1.415,4.755) distributions of the identified QSO's and galaxies are significantly different from the corresponding distributions of strong-source identifications. The relative scarcity of faint flat-spectrum ($\alpha < +0.5$) QSO's is probably caused by source evolution which cuts off at a high ($z > 3$) redshift, while the lack of faint galaxies with $\alpha \geqslant +0.8$ results from the spectral index-luminosity correlation of radio galaxies and the limited sensitivity of the PSS prints.

141.051 Radio and X-ray observations of compact sources in or near supernova remnants.
E. R. Seaquist, W. S. Gilmore.
Astron. J., Vol. 87, 378 - 386 (1982).

The authors present VLA multifrequency radio observations of six compact radio sources from the list of nine objects proposed by Ryle et al. (1978) as a new class of radio star, possibly the stellar remnants of supernovae. They also present the results of a search for X-ray emission from four of these objects with the Einstein observatory. The radio observations provide information on spectra, polarization, time variability, angular structure, and positions for these sources. No X-ray emission was detected from any of the four objects observed. The measurements provide no compelling arguments to consider any of the six objects studied as radio stars.

141.052 Redshifts, first and second order clustering properties, and refined radio parameters of 4C radio galaxies in poor clusters. S. A. Gregory, J. O. Burns.
Astrophys. J., Vol. 255, 373 - 381 (1982).

By determining 16 new redshifts and confirming two others, the authors have completed distance determinations for a complete sample of 22 4C radio galaxies in poor Zwicky clusters. Using the new redshifts, they have refined the absolute radio parameters such as the radio powers at 2.7 GHz ($P_{2.7}$) and source sizes. They have also determined richness, compactness, and Bautz-Morgan types for the clusters by carefully examining the cluster fields on the Palomar Sky Survey. We find that $P_{2.7}$ has a similar distribution to that of rich clusters and that there is a weak correlation of $P_{2.7}$ with the Bautz-Morgan classification which measures properties of the dominant cluster galaxy. This sample of poor clusters indicates that the radio galaxy phenomenon is related more to the properties of the central, dominant galaxies than to the extended cluster environment.

141.053 VLA observations of an unbiased sample of extragalactic X-ray sources.
E. D. Feigelson, T. Maccacaro, G. Zamorani.
Astrophys. J., Vol. 255, 392 - 400 (1982).

The authors present results of a Very Large Array survey at 6 cm of an unbiased sample of extragalactic X-ray sources serendipitously discovered with the Einstein X-ray Observatory. Fourteen of 42 X-ray sources are detected in the radio down to a limiting sensitivity of 1−2 mJy. Detections include eight of 31 examined active galactic nuclei (AGN-quasars or Seyferts), five of eight clusters of galaxies and one BL Lac object. The radio detection rate of X-ray selected AGNs is higher than that of optically selected AGNs, suggesting that a physical link between radio and X-ray emission is present. The radio properties of the X-ray selected clusters are similar to those of optically selected Abell clusters.

141.054 A new double jet model for 3C 449.
R. H. Lupton, J. R. Gott III.
Astrophys. J., Vol. 255, 408 - 412 (1982).

The radio source 3C 449 contains a point source with two radio jets extending from it and is associated with a cD galaxy at redshift z = 0.0181. Both jets undergo periodic oscillations in position angle as a function of radius from the point source. The whole configuration shows a reflection symmetry about an axis through the point source and therefore could not be explained by a precessing beam model. The model proposed here is that of a massive object orbiting ~1 kpc from the center of the cD galaxy and emitting twin beams, with $V_{beam} \approx 6.25 \, V_{orb}$. It is this periodic orbital velocity component added to the beam velocity which produces the observed beam oscillations and explains the observed mirror symmetry of the beams. The predicted beam velocity is in agreement with independent radio determinations.

141.055 The spectral flux distributions of sources in an optically selected sample of QSOs: $10^{13} - 10^{15}$ Hz.
R. W. Capps, M. L. Sitko, W. A. Stein.
Astrophys. J., Vol. 255, 413 - 418 (1982).

Data on a number of QSOs in an optically selected sample that was observed at $\lambda \approx 6$ cm with the VLA have been extended to 3.5 μm and to 10 μm and 20 μm in a few cases. The authors combine these results with data published earlier in an examination of potential differences between spectral flux distributions of radio-loud and radio-quiet QSOs.

141.056 The extended radio structure of compact extragalactic sources.
R. A. Perley, E. B. Fomalont, K. J. Johnston.
Astrophys. J., Lett., Vol. 255, L93 - L97 (1982).

High dynamic range maps made with the VLA at 1.5 and 4.9 GHz of 21 luminous, core-dominated radio sources have clearly established the basic asymmetric dispostion of the associated arcsecond-extended radio emission. This is in contrast with the generally symmetric brightness distributions of the less luminous, larger angular size sources. The asymmetry in this radio emission, many kiloparsecs from the radio core, may be caused by a Doppler enhancement produced by a relativistic bulk flow, or the asymmetry may be real.

141.057 Radio observations of the giant quasar 4C 34.47.
W. J. Jägers, W. J. M. van Breugel, G. K. Miley, R. T. Schilizzi, R. G. Conway.
Astron. Astrophys., Vol. 105, 278 - 283 (1982).

Multifrequency measurements of the giant radio source associated with the quasar 4C 34.47 are presented. The source, which is ~7′ in total extent with a prominent central core, has been mapped at 4995, 1415, and 610 MHz with the WSRT, and is shown to be edge-brightened. Significant linear polarization has been detected from the outer hot spots with directions which indicate that the magnetic field bends around the hot spots. The central core of the source is variable. *VLBI* observations on European baselines of the core show it to be extended along the large scale axis of the source.

141.058 On the quasar surface density.
P. Véron, M. P. Véron.
Astron. Astrophys., Vol. 105, 405 - 409 (1982).

The authors discuss all available data on the change of the surface density of quasars with limiting magnitude. A plate taken under excellent seeing conditions with the CFH 3.6 m telescope shows that all the faint UV excess objects found by Formiggini et al. (1980) are starlike. This contradicts the finding of Bonoli et al. (1980) that 2/3 of these objects are extended. The low surface densities obtained by Savage and Bolton (1979) and by Arp and Surdej (1981) are probably due to errors in the magnitude scales used. These surface densities show the expected flattening of the log $(N(<B))/B$ curve starting at $B \sim 20$.

141.059 On symmetric structure in compact radio sources.
R. B. Phillips, R. L. Mutel.
Astron. Astrophys., Vol. 106, 21 - 24 (1982).

The authors call attention to a group of compact double radio sources ($\lesssim 1$ kpc) discovered by *VLBI*. The relativistic beam model for compact radio sources is discussed with respect to compact doubles. Morphology and spectral characteristics of these sources appear to be consistent with those expected if the angle between the beams and observer's line of sight is large ($\theta \gtrsim 30°$). Radio emission from compact doubles appears to come from slowly advancing ($v \lesssim 0.1 \, c$), isotropically radiating regions which have formed at the beam ends.

141.060 An assessment of the detectability of X-ray emission from winds in active galactic nuclei and quasars.
M. Beltrametti, J. Drew.

Astron. Astrophys., Vol. 106, 153 - 157 (1982).

The authors show that hot ($T \gtrsim 10^8$ K) winds, which may be produced by quasars and active galactic nuclei, may be detectable in the X-ray part of the spectrum by means of their bremsstrahlung emission. It is deduced that only winds that are moderately or highly opaque to electron scattering are likely to affect the observed X-ray continuum in these objects and that the chances of detection are much better for strong radio or infra-red emitters. The model is compatible with the observed X-ray continuum of the quasar, 0241+622.

141.061 Profiles of [O III] lines in QSOs.
G. K. Miley, T. M. Heckman.
Astron. Astrophys., Vol. 106, 163 - 166 (1982).

High spectral resolution measurements have been made of the [O III] λ5007 emission line profiles for fifteen low-redshift QSOs brighter than 17th magnitude. The resolution was ~3 Å FWHM, corresponding to velocity resolutions from 115 to 150 km s^{-1}. The observed line-widths vary from ~300 to 800 km s^{-1} and their distribution is similar to that seen in Seyfert 1 galaxies. Possible explanations for the observed difference between the [O III] profiles of Seyfert galaxies and QSOs are briefly considered.

141.062 The periodicity in the distribution of quasar redshifts and the density perturbations in the early Universe.
L.-Z. Fang, Y.-Q. Chu, Y. Liu, C. Cao.
Astron. Astrophys., Vol. 106, 287 - 292 (1982).

The distribution of emission line redshifts for quasars has been studied by means of the statistical analysis of powerspectrum. It has been demonstrated that the distribution with respect to an argument is periodic. The existence of the periodicity is not an undoubted evidence for the non-cosmological origin of quasar redshifts. It is because the periodicity might be explained as remains of density wave perturbations in the early Universe. The main consequences of this model have been preliminarily verified.

141.063 Hydrogen line spectrum in quasars. II. A critical discussion of model calculations for the broad line region.
S. Collin-Souffrin, S. Dumont, J. Tully.
Astron. Astrophys., Vol. 106, 362 - 374 (1982).

The authors review the most important, recent observations of hydrogen spectra in quasars and Seyfert galaxies. They show that the standard photoionized model produces Hα/Hβ line ratios which are too large and propose that broad lines can be produced by a composite model comprising two physically distinct regions: (1) A standard photoionized region. (2) A hot ($T_{eff} \gtrsim 10^4$ K), dense ($n_e \sim 10^{11}$ cm^{-3}) ionized medium which is collisionally heated. This region emits almost all the observed flux in the 3000 Å bump as well as a large fraction of the Balmer and Paschen lines and of the optical Fe II lines.

141.064 Faint blue objects at high galactic latitude. II. Palomar Schmidt field centered on selected area 29.
P. D. Usher, D. Mattson, A. Warnock III.
Astrophys. J., Suppl. Ser., Vol. 48, 51 - 71 (1982).

The second part of a list of ultraviolet excess objects is given. The objects have been selected from a Palomar Schmidt field centered on Kapteyn selected area 29. The population of objects belonging to color classes 1A, 1, and 1B should be comprised primarily of quasars and white dwarfs and should be statistically complete to about $B = 18.5$ mag. One of the listed objects is a 5 mag optical variable; a journal of observations is given, derived from Harvard archival and contemporary Palomar Schmidt plates.

141.065 Optical identification/flux density relationship for radio galaxies.
G. Swarup, C. R. Subrahmanya, K. L. Venkatakrishna.
Astron. Astrophys., Vol. 107, 190 - 196 (1982).

Optical identification statistics have been summarized over a flux-density range of about 1000 : 1 at 408 MHz for identifications of galaxies made on the Palomar Sky Survey prints from the 3CR, Ooty-occultation, Bologna and 5C surveys. The percentage identification of galaxies seen on PSS, PI (S), decreases from about 60% for a flux density of about 15 Jy at 408 MHz to about 15% at 1 Jy and ~ 10% at 25 mJy. The observed PI (S) relation is compared with the predictions of theoretical models of radio luminosity function derived by Wall et al. (1977) and Robertson (1980) based on radio source counts and luminosity distribution of strong radio sources.

141.066 Theoretical studies of compact radio sources. I. Synchrotron radiation from relativistic flows.
S. P. Reynolds.
Astrophys. J., Vol. 256, 13 - 37 (1982).

Static models of compact radio sources often suffer from confinement difficulties, and it can be argued that outflow of the radiating material is most likely relativistic. Dynamics of such outflows are considered in spherical and jet geometries. Gravity is shown to be usually unimportant; for steady solar-wind type acceleration the central object must be a black hole and the sonic point is close to the Schwarzschild radius. Force-free steady streaming is therefore considered; simple power laws result for behavior of various physical quantities. Analytic expressions are obtained for the spectra of synchrotron radiation from various types of flows. These results are applied to compact nuclear radio sources in M81, M87, and M104. The only models which can reproduce the observations are opaque jets and winds dominated by the inertia of nonrelativistic protons. Observations of Faraday depolarization can test the idea of relativistic motion by putting limits on the amount of internal thermal gas present. VLBI measurements of source size as a function of frequency can discriminate between jets and winds.

141.067 Theoretical studies of compact radio sources. II. Inverse-Compton radiation from anisotropic photon and electron distributions: general results and spectra from relativistic flows.
S. P. Reynolds.
Astrophys. J., Vol. 256, 38 - 53 (1982).

A general expression is derived for the inverse Compton emissivity due to arbitrary distributions of photons and electrons. This expression is used to treat the anisotropies characteristic of synchrotron self-Compton radiation in relativistic flows. Detailed numerical calculations have been performed, producing self-consistent inverse Compton spectra from relativistic winds, including the effects of cutoffs in the electron distribution. Analytic expressions are derived for the inverse Compton spectra of relativistic winds and jets. These results are applied to compact radio and X-ray sources in the nuclei of M81, M87, and M104; jet models are fairly well constrained while wind models produce far too little X-radiation, and can probably be ruled out.

141.068 Multifrequency VLBI observations of the nucleus of NGC 1275.
S. C. Unwin, R. L. Mutel, R. B. Phillips, R. P. Linfield.
Astrophys. J., Vol. 256, 83 - 91 (1982).

VLBI observations of the compact nucleus of NGC 1275 (3C 84) have been made at 1.7, 5.0, and 10.7 GHz, using arrays of four and five telescopes. Hybrid maps derived from these observations reveal complex structure on scales between ~1 and ~14 milli-arcsec, with strong frequency dependence. Two regions with strongly inverted spectra show up from comparison of the 5.0 and 10.7 GHz maps. Assuming that these represent synchrotron self-absorption, the authors conclude that either there are at least two sources of relativistic electrons in this object, or one of the features is a knot in a jet ejected from the central core. These components are able to produce the observed hard X-ray flux from the Perseus cluster by inverse Compton scattering of synchrotron-emitting photons. Significant structure changes occurred at 10.7 GHz between 1980 April and 1980 July.

141.069 Observations of Paschen α in the broad-line radio galaxy 3C 445.   R. J. Rudy, A. T. Tokunaga.
Astrophys. J., Lett., Vol. 256, L1 - L3 (1982).

Near-infrared spectrophotometric observations of Paschen α are presented for the broad-line radio galaxy 3C 445. The measured line flux is 3.4×10^{-13} ergs cm^{-2} s^{-1}, approximately 60% of the value of Hα and much larger than can be explained by recombination theory or current radiative transfer models. However, the Pα/Hα/Hβ ratios are consistent with Case B values and those predicted by the models of Canfield and Puetter when reddened by $E_{B-V} \sim 1.0$ mag. This indicates that dust extinction plays a major role in determining the line ratios. The derived value of reddening implies an Hα luminosity of 4×10^{43} ergs s^{-1}, comparable to the brightest Seyfert galaxies.

141.070 Determination of physical parameters in the radio source 5C 4.81.   J. Roland.
Astron. Astrophys., Vol. 107, 267 - 271 (1982).

The author shows how it is possible to determine physical parameters in the tail of the head-tail radio galaxy 5C 4.81. He deduces that the energies of the proton and electron components of the cosmic rays inside the radio tail are the same.

141.071 Formation of double radio sources.   K. Morita.
Publ. Astron. Soc. Japan, Vol. 34, 65 - 87 (1982).

The author studies the gas motion induced by the continuous or impulsive mass ejection associated with energy release at the galactic nucleus, especially in the early phase ($t \sim 10^5$ yr). The calculations are performed by the use of a time-dependent two-dimensional hydrodynamic method. The model considered is that the central region of the galactic nuclei consists of a spherical stellar system with a highly concentrated density distribution and a differentially rotating isothermal gas. In addition, the author considers the time evolution of the radio and X-ray luminosities and how such collimated flows will be observed at these frequency bands.

141.072 A statistical analysis of QSOs' redshifts: the luminosity and evolution of QSOs.
Y. Bian, X. Tang, S. Cao, X. Xiao, Y. Liu, S. Jiang.
Kexue Tongbao, Vol. 26, 429 - 433 (1981). − Abstr. in Phys. Abstr., Vol. 85, Abstr. 23352 (1982).

141.073 High-red-shift molecular clouds and absorption-line spectra of quasars.
D. A. Varshalovich, S. A. Levshakov.
Comments Astrophys., Vol. 9, 199 - 209 (1982).

Absorption-line systems with the redshifts $z_a = 1.5$–3.1, which contain a great deal of the H$_2$ and CO molecular lines along with the atom and ion lines, are found by reanalyzing the published optical spectra of the far quasars ($z_e > 2$). It is shown that molecular clouds existed in an early epoch when the cosmological age of the universe was only 10%−30% of the present age. It is most probable that the molecular clouds belonged to some distant invisible intervening galaxies because the chemical composition of these clouds and their ionization degree are similar to those of interstellar diffuse clouds (with $N(H) \gtrsim 10^{19}$ cm^{-2}) in our galaxy.

141.074 Is there a β/L relation for any type of QSO absorption line system?   A. S. Trew.
Astrophys. Space Sci., Vol. 82, 223 - 228 (1982).

The possible correlation noted between the intrinsic quasar luminosity and the absorption line expulsion velocity is re-examined using homogeneous data sets for metal and Lα only line systems. The method of analysis is chosen to enable

any reasonable form of correlation to be found. No correlation is detected at a confidence level > 10%.

141.075 The distribution of quasar absorption line systems − selection effects.   C. M. Gaskell.
Astrophys. Space Sci., Vol. 82, 247 - 248 (1982).

The reported correlation between the absorption-line and emission-line redshifts of quasars is shown to be due to selection effects and thus to favour neither the 'intrinsic' nor 'intervening' hypotheses for the origin of quasar absorption lines.

141.076 New study on quasars and isotropy of H_0.   H. J. Reboul.
Astron. Astrophys., Vol. 108, 85 - 88 (1982).

A first sample of 132 quasars selected by means of their radio spectral index had lead (Reboul, 1980) to an upper limit to the dipole component of a generalized Hubble modulus HM* and had − a posteriori − revealed that HM* appeared to be at a minimum in the general direction of the Virgo cluster. A new and independent sample of 334 quasars has been selected by means of the same criteria. Its study gives two main results: (1) A better limit to any dipole component of HM* which, translated in H_0, is: $H_0(-20\%; +25\%)$. (2) A confirmation of the − now a priori − tested hypothesis: HM* minimum (~ -0.1) in the general direction of the Virgo cluster at the 0.95 level of confidence. While (1) seems, for the moment, without cosmological significance, the author discusses the possible explanations of (2).

141.077 VLBI observations of 12 compact radio sources north of declination 70°.   A. Eckart, P. Hill, K. J. Johnston, I. I. K. Pauliny-Toth, J. H. Spencer, A. Witzel.
Astron. Astrophys., Vol. 108, 157 - 160 (1982).

The authors present models of the structure of 12 compact radio sources with declinations > 70°, based on VLBI measurements at 5 GHz. At least 7 of these sources are BL Lac objects and 6 of them contain more than 75% of their total flux density at 5 GHz in a region smaller than about 5 mas. These sources may also be used to establish an almost absolute reference frame.

141.078 Mehr-Frequenz-Spektralindexuntersuchungen an drei geschweiften Radiogalaxien.
H. Andernach.
Mitt. Astron. Ges., Nr. 55, p. 97 - 98 (1982). − Abstract.

141.079 A spectral index − flux density relation for extragalactic radio sources.
Gopal-Krishna, H. Steppe.
Mitt. Astron. Ges., Nr. 55, p. 101 - 105 (1982).

141.080 11 cm Beobachtungen variabler Radioquellen.   P. Steffen. W. Reich, K. Reif.
Mitt. Astron. Ges., Nr. 55, p. 105 - 107 (1982).

141.081 Der Stockert-21 cm-Radiokontinuumssurvey.   W. Reich.
Mitt. Astron. Ges., Nr. 55, p. 133 - 135 (1982).

141.082 Further observations of variable radio sources with the RATAN-600 radio telescope.
V. A. Majzel', M. G. Mingaliev, S. A. Pustil'nik, S. A. Trushkin.
Astrofizika, Tom 17, 445 - 454 (1981). In Russian. English translation in Astrophysics, Vol. 17, No. 3.

Radio spectra at epoch 1977.23 of the variable radio sources Simeiz 0528 + 13, PKS 0735 + 17, OJ 287, 3C 273, 3C 279 and PKS 1510 −08 are presented as a result of observations with RATAN-600 at wavelengths 2.08, 3.9, 6.52 and 8.2 cm. Comparison of these spectra is made with the spectra obtained at epochs 1976.25 and 1976.75. Simeiz 0528 +13 has shown in late March 1977 a fast increase of flux density at λ 2.08 cm with characteristic time $\tau \lesssim 10$ days. For this source the light curve for centimeter range was constructed using all known measurements during the last 10 years. This light curve gives evidence that Simeiz 0528 +13 is one of the most active radio sources.

141.083 Spectrophotometric studies of a QSO.
M. A. Kazaryan, Eh. E. Khachikyan.
Astrofizika, Tom 17, 661 - 666 (1981). In Russian. English translation in Astrophysics, Vol. 17, No. 4.

The results of spectrophotometric studies of the object N 102 from the list of galaxies with UV excess (Kazaryan, 1979) are presented. The profiles, equivalent widths and the relative intensities of the emission lines have been derived.

141.084 Search for extended structures in the vicinity of the radio sources 3C 120 and 3C 273.
N. S. Soboleva, A. B. Berlin, N. A. Nizhel'skij, E. E. Spangenberg.
Pis'ma Astron. Zh., Tom 8, 205 - 209 (1982). In Russian. English translation in Soviet Astron. Lett., Vol. 8.

Results of observations of the vicinity of the radio sources 3C 273 and 3C 120 at 3.9, 7.6 and 8.2 cm wavelengths are presented. The existence of one of the two details (Reich et al., 1980), at the distance of ~ 8' from the quasar 3C 273 is confirmed. New extended details (~ 1') near the source 3C 120 were found on both sides of the source at the distance of about 1.'3.

141.085 A radio continuum survey of the northern sky at 1420 MHz – Part I. W. Reich.
Astron. Astrophys., Suppl. Ser., Vol. 48, 219 - 297 (1982).

The first part of the Stockert 21 cm radio continuum survey of the whole northern sky is presented. These observations cover the area north of declination +20°. The results are given in the form of contour maps. The effective resolution of the survey is about 35', while the effective sensitivity is about 50 mK T_B (full beam).

141.086 The physical nature of the blue objects in the field of 88 Leonis. L. Erculiani Abati.
Astron. Astrophys., Suppl. Ser., Vol. 48, 333 - 343 (1982).

128 blue objects of a field (25 square degrees) in the magnitude range $15.5 < m \leqslant 17.5$ and colour range $U-B \leqslant 0.0$ have been classified from objective prism spectra. Most of them are stars, 9 % are quasars or suspected quasars and 13% are compact galaxies or suspected compact galaxies.

141.087 Rotating compact radio sources and angular momentum transfer from the nucleus to outlying gas in active galaxies. A. P. Marscher, E. M. Burbidge.
Astrophys. Lett., Vol. 22, 83 - 87 (1982).

The authors demonstrate that, if the recently reported polarization variations in compact extragalactic radio sources actually represent physical rotations of the radio components, the angular momentum carried out of the central regions is $10^{60 \pm 3}$ erg s yr^{-1}. In rotating magnetoid ("spinar") models, the central object loses angular momentum to its surroundings at a rate consistent with the lower end of this range. Accretion disks, however, are much less "efficient" losers of angular momentum. In either case, it is not clear how the radio-emitting plasma could obtain the lost angular momentum. The authors propose that once the radio components have reached the outlying interstellar gas, the interaction causes angular momentum transfer from the radio plasma to the gas. The rotational motions of the outlying gas in 3C 120, and possibly in NGC 1052 and NGC 4278 might be explained in this way.

141.088 Quenchings and outbursts of extragalactic radio sources: nine years of 3.3 mm measurements and comparisons with centimeter-wave variations.
E. E. Epstein, W. G. Fogarty, J. Mottmann, E. Schneider.
Astron. J., Vol. 87, 449 - 461 (1982).

The extragalactic variable radio sources 3C 84, 3C 120, OJ 287, 3C 273, and BL Lac have been observed frequently at 3.3 mm for up to nine years; the results are compared with 19-, 28-, 38-, and 45-mm observations. The 3-mm variations are often rapid; significant information is lost if observations are not made every 1 - 3 days. The correlation between the 3-mm and centimeter-wave variations of these five sources ranges from high to almost nonexistent; in general, the flatter the spectrum, the better the correlation.

141.089 The highly polarized galaxy 3C 76.1.
J. P. Vallée.
Astron. J., Vol. 87, 486 - 493 (1982).

The high degree of linear polarization integrated over the galaxy 3C 76.1 ranks this source among the most polarized sources in the sky, and its slow decrease with increasing wavelength ranks this source among the least depolarizing sources in the sky. Observations of the galaxy 3C 76.1 at high resolution (from 0.5 to 6.2 arcsec) have been obtained with the Very Large Array in New Mexico, at a wavelength of λ6.1 cm. The VLA data reveal several components, of sizes of a few arcseconds, located roughly along a central line at a position angle of about 135° and passing through the center of the optical galaxy, much like two blobby, discontinuous but oppositely directed jets. Here the magnetic field structure is running transverse to this central line, prior to the bending of this line into a Z-shaped curve. Past the bending points, the magnetic field structure is inferred to run parallel to the radio structure there.

141.090 Arcsecond positions for milliarcsecond VLBI nuclei of extragalactic radio sources. I. 546 sources.
D. D. Morabito, R. A. Preston, M. A. Slade, D. L. Jauncey.
Astron. J., Vol. 87, 517 - 527 (1982).

VLBI measurements of time delay and fringe frequency at 2290 MHz on baselines of 10^4 km have been used to determine the positions of the milliarcsecond nuclei in 546 extragalactic radio sources. Estimated accuracies generally range from ~0.''5 to 1.''0 in both right ascension and declination, with ~ 6% of the sources having estimated uncertainties $\geqslant 3.''5$ in at least one position component. The observed sources are part of an all-sky VLBI catalog of milliarcsecond radio sources.

141.091 Search for short-term variability in nonthermal radio sources.
Z. Abraham, P. Kaufmann, L. C. L. Botti.
Astron. J., Vol. 87, 532 - 536 (1982).

A long series of 22-GHz daily observations (24 days) of the radio sources 3C 273, Cen A, Sgr A, and OV −236 was carried out in July 1980. None of these sources presented any short-term variability larger than the accuracy of the measurements (about 10%). Cen A presented the highest flux level in relation to the past seven years, and it decreased consistently during the period of observation at a rate of 0.2 ± 0.02 Jy/day. The flux density of OV −236 increased in the same period at a rate of 0.1 ± 0.02 Jy/day.

141.092 A complete sample of extragalactic radio sources at 1 Jy at 408 MHz – I. The radio observations.
J. R. Allington-Smith.
Mon. Not. R. Astron. Soc., Vol. 199, 611 - 631 (1982).

A complete sample of 59 B2 radio sources with flux densities between 1 and 2 Jy at 408 MHz has been studied. The radio observations are described; these cover a wide range of angular scales from 1 arcsec to several arcmin, and a range of frequencies 0.4–5 GHz. The sample contains a wide variety of source types.

141.093 The clustering of quasars from an objective-prism survey. A. Webster.
Mon. Not. R. Astron. Soc., Vol. 199, 683 - 705 (1982).

The positions and redshifts of 108 quasars from the Cerro Tololo objective-prism survey are subjected to Fourier Power Spectrum Analysis in a search for clustering in their spatial distribution. It is found that, on the whole, these quasars are not clustered but are scattered in space independently at random. The sole exception is a group of four quasars at $z = 0.37$ which, with a size of about 100 Mpc, may therefore be the largest known structure in the Universe.

141.094 A method for estimating the masses of some quasars.   C. C. Dyer, R. C. Roeder.
Astrophys. J., Vol. 256, 386 - 389 (1982).

The gravitatiońal distortion effect of a nearby quasar on the image of a distant quasar in nearly the same direction is discussed as a method for estimating the mass of the nearer quasar. Account is taken of the effect of a cluster of galaxies in which the nearer quasar may be situated. Such "optically double" quasars yield upper mass limits when the image of the distant quasar is not completely resolved. The pair 1038+528 is considered as an example of the method and yields a limiting mass of $7 \times 10^{13} M_\odot$ for simple assumptions, consistent with the estimate of Bahcall and Tremaine.

141.095 On the variability of 3C 273, OJ 287 and PKS 0735 + 17 in radio and optical emission.
V. A. Efanov, I. G. Moiseev, N. S. Nesterov, N. M. Shakhovskoj. Izv. Krymskoj Astrofiz. Obs., Tom 64, 91 - 98 (1981). In Russian. English translation in Bull. Crimean Astrophys. Obs., Vol. 64.

It is shown that variations of the radio flux at wavelength 1.35 cm of the sources 3C 273, OJ 287 and PKS 0735 + 17 are related to variations of linearly polarized optical emission both in daily and in slow (during several years) variations. The maxima of polarized emission correspond to the minima of the radio flux as a rule.

141.096 Observations of radio sources with RT-22 of the Crimean Astrophysical Observatory and RT-14 of RHUT in the millimeter wave range.
V. A. Efanov, I. G. Moiseev, N. S. Nesterov, M. Tiuri, S. Urpo. Izv. Krymskoj Astrofiz. Obs., Tom 64, 103 - 108 (1981). In Russian. English translation in Bull. Crimean Astrophys. Obs., Vol. 64.

The results of observations of several radio sources are presented. The measurements were made simultaneously with the 22-m radio telescope of the Crimean Astrophysical Observatory at $\lambda = 1.35$ cm and the 14-m radio telescope of the Radio Laboratory of Helsinki University of Technology at $\lambda = 8.15$ mm. The existence of rapid radio emission fluctuations at both wavelengths of 3C 84, OJ 287 and BL Lac in May - June and of 3C 273 and BL Lac in April 1978 is remarked. The radio spectra of quasars OH 471 and OX 057 are shown and attention is drawn to millimeter wave emission excess in the spectrum of OH 471.

141.097 Dual frequency VLA observations of 3C338: an unusually steep-spectrum galaxy at the center of Abell 2199.   E. Schwendeman, J. O. Burns, R. A. White.
Proc. Southwest Reg. Conf., Vol. 7, (see 012.019), 139 - 149 (1982).

141.098 High energy quasar spectra and the γ ray background.   J. A. Eilek, M. Kafatos.
Proc. Southwest Reg. Conf., Vol. 7, (see 012.019), 153 - 159 (1982).

141.099 Interpretation of the results of observations of flux variations of extragalactic sources in the decimeter wavelength range.   N. Ya. Shapirovskaya.
Astron. Zh., Tom 59, 246 - 252 (1982). In Russian. English translation in Soviet Astron., Vol. 26, No. 2.

New data on the variability of extragalactic sources in the decimetre range are interpreted in terms of a hypothesis of scintillations on inhomogeneities of interstellar structures of the Galaxy, i. e. loops, spurs, ridges. The behaviour of the scintillation index in a broad frequency band is explained by the theory of scintillations on large irregularities in the weak-focusing regime at the strong-focusing boundary.

141.100 2.2-micron mapping of the nuclear region of NGC 5128 (Centaurus A).
J. R. Walsh, N. J. White. Observatory, Vol. 102, 78 - 81 (1982).

The nuclear region of the radio galaxy NGC 5128 (Centaurus A) has been mapped at K (2.2 μm) with a 12″ beam. The distribution of emission at 2.2 μm is compared with that at 10 μm along the dust lane and it is found that, whilst the 10-μm flux increases again at distances $\gtrsim 1$ kpc from the nucleus, the 2.2-μm flux continues to decrease.

141.101 Large scale emission from radio galaxies.
H. Andernach, R. Wielebinski. Extragalactic radio sources, (see 012.025), p. 13 - 20 (1982).

The present state of investigating large radio galaxies with aperture synthesis telescopes is reviewed. The morphology of sources is outlined in some detail and the most important problems in interpreting the physical nature of the various types of sources are described.

141.102 Evolutionary tracks of extended radio sources.   J. E. Baldwin.
Extragalactic radio sources, (see 012.025), p. 21 - 24 (1982).

A diagram of radio luminosity versus overall physical size has been prepared for the radio sources of the 3 CR 166 sample. Effects of evolutionary sequences for radio galaxies are indicated in this diagram.

141.103 Rotating gas and the shapes of radio sources.   L. S. Sparke.
Extragalactic radio sources, (see 012.025), p. 25 - 26 (1982).

This paper discusses how the rotation of a radio galaxy affects the distribution of gas within it, and consequently the radio structure in elliptical and Seyfert galaxies.

141.104 Aperture synthesis observations of Cygnus A at 86.2 GHz.   M. Birkinshaw, M. C. H. Wright.
Extragalactic radio sources, (see 012.025), p. 27 - 28 (1982).

Recent observations of Cygnus A with the Hat Creek interferometer at 86.2 GHz limit the spectral curvature of the hotspots and show that the diffuse lobe emission has a spectral index of about 1.5. The central component is, at most, weakly variable and its spectrum shows a distinct break to a spectral index near 0.65 at about 20 GHz.

141.105 The spectral index distribution of Cygnus A.   P. F. Scott.
Extragalactic radio sources, (see 012.025), p. 29 - 31 (1982).

High dynamic range maps of Cygnus A at 2.7 and 5 GHz have been used to investigate the variation of spectral index over the extended parts of the source. Although both components show a steepening of spectral index away from the hotspots there is a marked asymmetry between the two components.

141.106 Observations of radio galaxies and QSR with RATAN-600.   N. S. Soboleva, Y. N. (*Yu. N.*) Parijskij.
Extragalactic radio sources, (see 012.025), p. 33 - 34 (1982). Abstract.

141.107 A possible evolutionary feature in the spectra of radio galaxies.   O. B. Slee.
Extragalactic radio sources, (see 012.025), p. 35 - 38 (1982).

The presence of minima in the spectral slopes of many of the closer radio galaxies is discussed. The correlation between spectral index and redshift is probably caused by these minima. Their implications for evolutionary models of radio galaxies are described.

141.108 High resolution VLA observations of quasars with distorted radio structure.
J. Stocke, W. Christiansen, J. Burns.
Extragalactic radio sources, (see 012.025), p. 39 - 40 (1982). Abstract.

141.109 The radio spectrum across three tailed radio galaxies. H. Andernach.
Extragalactic radio sources, (see 012.025), p. 41 - 42 (1982).
The complex, low luminosity radio galaxies 3C 40, HB13, 2247+11 with angular extent $20' < \Theta < 30'$ have been mapped with the Effelsberg 100m-telescope at frequencies 2.7, 4.9 and 10.7 GHz.

141.110 VLA and optical mapping of the quasar PKS 0812+020. F. D. Ghigo, L. Rudnick, K. J. Johnston, P. A. Wehinger, S. Wyckoff.
Extragalactic radio sources, (see 012.025), p. 43 - 44 (1982).
Observations are reported of the remarkable object PKS 0812+020, the first quasar found to have optical emission in one of its radio lobes. A distinct radio jet is also seen, and there is radio and optical evidence that the quasar is near the center of a galaxy cluster.

141.111 What bends wide-angle tailed radio sources?
J. O. Burns, J. A. Eilek, F. N. Owen.
Extragalactic radio sources, (see 012.025), p. 45 - 46 (1982).

141.112 Nuclear ejection – one side at a time.
L. Rudnick.
Extragalactic radio sources, (see 012.025), p. 47 - 49 (1982).
Examination of the structures of extragalactic radio sources shows a distinct asymmetry in addition to the more general symmetries which are well known. The most likely explanation for the observed asymmetries is that ejection from the active nucleus occurs in only one direction at a time. This direction then switches back and forth to form the large scale double structures. Implications of this picture for the nuclear engine and the radio source environment are discussed.

141.113 Time dependent energy supply in radio sources and morphology of radio lobes.
W. A. Christiansen, A. G. Pacholczyk, J. S. Scott.
Extragalactic radio sources, (see 012.025), p. 51 - 52 (1982).

141.114 Multi-frequency polarization studies of radio galaxies. G. G. Pooley.
Extragalactic radio sources, (see 012.025), p. 53 - 54 (1982). Abstract.

141.115 A complete sample of radio galaxies.
P. A. Shaver, I. J. Danziger, R. D. Ekers, R. A. E. Fosbury, W. M. Goss, D. Malin, A. F. M. Moorwood, J. V. Wall.
Extragalactic radio sources, (see 012.025), p. 55 - 57 (1982).
The authors report preliminary results of a multi-wavelength study of a complete sample of radio galaxies. The sample is comprised of 93 radio sources from the Parkes 11 cm catalog which are identified with galaxies of 17th magnitude or brighter in the declination zone $-17°$ to $-40°$. The objective is to cross-correlate the radio, infrared, optical, and other properties of a properly defined sample of radio galaxies.

141.116 Extended structure in high-redshift radio sources.
P. J. Duffett-Smith, A. Purvis.
Extragalactic radio sources, (see 012.025), p. 59 - 60 (1982).

The authors have compared measurements of several hundred 3C and 4C radio sources at large redshifts to investigate how radio-source structure changes over a factor of $5-10$ in luminosity. The results show that for $z \gtrsim 0.6$: (1) most sources (both 3C and 4C) have hotspots about 3.5 kpc in size ($H_0 = 50$ km s^{-1} Mpc^{-1}); (2) lower-luminosity sources (bottom of 4C) have less-extended outer lobes.

141.117 Extended optical line emission associated with radio galaxies. W. van Breugel, T. Heckman.
Extragalactic radio sources, (see 012.025), p. 61 - 64 (1982).
Using the Video Camera and the High Gain Video Spectrometer the authors are carrying out a program at Kitt Peak to search for optical line emission associated with the jets and lobes of radio galaxies. Several sources have been found in which extended optical line emission is clearly related to the non-thermal radio emission.

141.118 Extended emission lines in radio galaxies.
R. A. E. Fosbury.
Extragalactic radio sources, (see 012.025), p. 65 - 67 (1982).
The author summarizes spectroscopic observations of three southern radio galaxies which show optical emission lines from regions tens of kiloparsecs in extent.

141.119 Emission lines: sign of a new energy source?
D. S. De Young.
Extragalactic radio sources, (see 012.025), p. 69 - 70 (1982).
The occurrence of optical emission line regions coexistent with extended radio sources is discussed. A model is briefly described where the emission line gas arises from the entrainment of interstellar matter into the material ejected from the nuclear source.

141.120 Optical inverse Compton emission in extragalactic radio sources. S. E. Okoye, O. Obinabo.
Extragalactic radio sources, (see 012.025), p. 71 - 73 (1982).
In this paper the reported detection of weak optical emission in the lobes of three 3C sources – 3C 265, 3C 285 and 3C 390.3 – is reappraised in the framework of three optical emission mechanisms – synchrotron, synchrotron inverse Compton and blackbody inverse Compton.

141.121 Proton-proton collisions in extragalactic radio sources. S. E. Okoye, P. N. Okeke.
Extragalactic radio sources, (see 012.025), p. 75 - 76 (1982).
The mutual interactions of fast particles in relativistic plasma beams or jets of radio galaxies are investigated. In particular the effects of proton-proton collisions with their production of pions and muons are considered.

141.122 A preliminary examination of the effect of cluster gas on tailed radio galaxies. D. E. Harris.
Extragalactic radio sources, (see 012.025), p. 77 - 83 (1982).
From a comparison of X-ray and radio data for 20 clusters which contain tailed radio galaxies, the author finds evidence for the effects of buoyant forces on the low brightness parts of radio tails. In several cases, the width of the tail increases markedly in low gas density regions, strengthening the case for thermal gas confinement of radio tails. Three examples of enhanced X-ray emission around radio galaxies which are further than 2 Mpc from their cluster centers are found.

141.123 The radio emission of interacting galaxies.
E. Hummel, J. M. van der Hulst, J. H. van Gorkom, C. G. Kotanyi.
Extragalactic radio sources, (see 012.025), p. 93 - 94 (1982).
The authors present VLA maps of the radio structure in three galaxies with a peculiar radio morphology (NGC 4038/39, NGC 4410/IC 790, NGC 4438). In at least one of these (NGC 4438) the radio structure is better inter-

preted in terms of interaction with a diffuse intergalactic gas rather than in terms of a galaxy-galaxy interaction.

141.124   Radio and X-ray structure of Centaurus A.
E. D. Feigelson.
Extragalactic radio sources, (see 012.025), p. 107 - 114 (1982).

Recent studies of the nearby radio galaxy Centaurus A with the Very Large Array and the Einstein X-Ray Observatory reveal complex radio and X-ray structures. A prominent one-sided jet comprised of resolved knots located 0.2–6 kpc from the nucleus is seen in both radio and X-rays. The X-ray emission is probably synchrotron. Inverse Compton emission is not a likely explanation though a thermal model in which the nucleus ejects dense $10^5 M_\odot$ clouds cannot be excluded. An elongated X-ray region is also found near the "middle" radio lobe and optical H II regions ~30 kpc NE of the nucleus. Conditions around the active nucleus, the absence of X-rays from the inner radio lobes, and X-ray evidence for a hot interstellar medium are briefly discussed.

141.125   Emission regions in Centaurus A.
R. M. Price, J. A. Graham.
Extragalactic radio sources, (see 012.025), p. 115 - 116 (1982).

141.126   VLBI observations of the nucleus of Centaurus A.
R. A. Preston, A. E. Wehrle, D. D. Morabito, D. L. Jauncey, M. Batty, R. F. Haynes, A. E. Wright, G. D. Nicolson.
Extragalactic radio sources, (see 012.025), p. 119 - 120 (1982).

VLBI observations at 2.3 GHz of the nucleus of Centaurus A have been made at a network of three southern hemisphere observatories.

141.127   Systematics of large-scale radio jets.
A. H. Bridle.
Extragalactic radio sources, (see 012.025), p. 121 - 128 (1982).

Radio jets occur in sources with a wide range of radio luminosities, and in 70% to 80% of nearby radio galaxies. There may be two basic types of large-scale (>1 kpc) jet: longitudinal field-dominated one-sided jets in sources with luminous radio cores, and transverse field-dominated two-sided jets in sources with weak radio cores. The large-scale jets that have been observed at high linear resolution are well collimated within a few kpc of their cores, then flare and recollimate further out. Their brightness-radius evolution is often "subadiabatic".

141.128   Radio and X-ray observations of large scale jets in quasars.   J. F. C. Wardle, R. I. Potash.
Extragalactic radio sources, (see 012.025), p. 129 - 131 (1982).

The authors report observations of large scale radio jets in six quasars by using the VLA at 5 GHz. Since the six jets were discovered among a sample of only thirteen quasars, large scale jets are fairly common in quasars. Three of the six quasars with jets have been detected in soft X-rays with the IPC on board the Einstein Observatory.

141.129   4C 18.68: a QSO with precessing radio jets?
A. C. Gower, J. B. Hutchings.
Extragalactic radio sources, (see 012.025), p. 133 - 134 (1982).

The radio source 4C 18.68, identified with the QSO 2305+187 (z = 0.313), was observed at 20 cm and 6 cm with the VLA in its highest-resolution configuration.

141.130   The quasar jet 4C 32.69 at 1.4 GHz.
J. W. Dreher.
Extragalactic radio sources, (see 012.025), p. 135 - 136 (1982).

One of the most luminous radio jet sources known is 4C 32.69, identified with a z = 0.67 QSO. A 1.4 GHz VLA map (beam size 1.4"×2") is presented and compared with the 5 GHz map by Potash and Wardle (1980).

141.131   Bent jets in radio quasars.   S. G. Neff.
Extragalactic radio sources, (see 012.025), p. 137 - 138 (1982).

The author presents four examples of quasars with bent radio jets. The observations were made with the VLA at 1.635 MHz and 4.885 MHz.

141.132   The jets in 3C 449 revisited.
T. J. Cornwell, R. A. Perley.
Extragalactic radio sources, (see 012.025), p. 139 - 140 (1982).

141.133   Recent WSRT and VLA observations of the jet radio galaxy NGC 6251.   A. G. Willis, R. G. Strom, R. A. Perley, A. H. Bridle.
Extragalactic radio sources, (see 012.025), p. 141 - 144 (1982).

NGC 6251, a 14th mag elliptical galaxy, has large-scale radio emission with a total angular extent of ~1.1°. The galaxy has been observed with the Westerbork telescope at 49 cm wavelength and a beam size of 55". The central 4' regime of the bright main jet has been extensively mapped with the VLA.

141.134   Kiloparsec scale structure in high luminosity radio sources observed with MTRLI.
P. N. Wilkinson.
Extragalactic radio sources, (see 012.025), p. 149 - 156 (1982).

Results of observations with the Jodrell Bank Multi-Telescope-Radio-Link Interferometer at 408 MHz and 1666 MHz are presented for high-luminosity core dominated radio sources. In a majority of these sources the extended emission is dominated by a one-sided jet.

141.135   A suggested classification and explanation for hot-spots in some powerful radio sources.
P. P. Kronberg, T. W. Jones.
Extragalactic radio sources, (see 012.025), p. 157 - 160 (1982).

141.136   Hot-spots in luminous extragalactic radio sources.
R. A. Laing.
Extragalactic radio sources, (see 012.025), p. 161 - 162 (1982).

141.137   Morphology and power of radio sources.
P. A. G. Scheuer.
Extragalactic radio sources, (see 012.025), p. 163 - 165 (1982).

141.138   The radio jet of 3C273.   R. G. Conway.
Extragalactic radio sources, (see 012.025), p. 167 - 168 (1982).

A map of the radio jet in the quasar 3C273 at 408 MHz is presented, which has been obtained with the Jodrell Bank MTRLI at a resolution of 0.9".

141.139   Relativistic beaming and quasar statistics.
I. W. A. Browne, M. J. L. Orr.
Extragalactic radio sources, (see 012.025), p. 169 - 172 (1982).

The predictions of a scheme which attributes the observed differences between flat and steep spectrum quasars to projection and the effects of relativistic beaming are explored. It is concluded that the statistical properties of quasars are entirely consistent with such a scheme provided the mean Lorentz factor in the central components of quasars is ~5.

141.140 The radio core in 3C 236.
E. B. Fomalont, A. H. Bridle, G. K. Miley.
Extragalactic radio sources, (see 012.025), p. 173 - 174 (1982).

The 2 kpc steep-spectrum radio core in the giant radio galaxy 3C 236 has been mapped with 0".1 resolution using the VLA. The core morphology is substantially different from other radio cores and suggests that the flow of energy from the galactic nucleus may be continuous on one side and "blobby" on the other side.

141.141 The arcsecond morphology of compact radio sources. R. A. Perley.
Extragalactic radio sources, (see 012.025), p. 175 - 176 (1982).

As part of the VLA calibration program, 404 small angular size sources have been observed at both 6 and 20 cm with resulting resolutions of 0.4" and 1.2" respectively. Use of self-calibration techniques has allowed a search for associated extended structure to a level of ~0.3% of the peak. Preliminary analysis of the results is reported.

141.142 Highly polarized emission from the E-hotspot in DA240. S. C. Tsien, R. Saunders.
Extragalactic radio sources, (see 012.025), p. 177 - 178 (1982).

The hotspot in the eastern lobe of the nearby radio galaxy DA 240 (z = 0.0356) has been observed with the Cambridge 5-km telescope at 2.7 and 5.0 GHz. A detailed polarization map of the hotspot at 5.0 GHz is presented.

141.143 Westerbork observations of low luminosity radio sources. P. Parma.
Extragalactic radio sources, (see 012.025), p. 193 - 194 (1982).

Radio maps obtained with the WSRT at 5 GHz of the radio galaxies 0326+39 and 1321+31 are presented.

141.144 SS433 – observing evolution in a precessing, relativistic jet. R. M. Hjellming, K. J. Johnston.
Extragalactic radio sources, (see 012.025), p. 197 - 204 (1982).

VLA maps of intensity and linear polarization at 6 cm have been obtained for SS433 at various epochs. Data are compared with the model of a precessing twin-jet ejection mechanism, where radio emitting material is ejected with velocity 0.26 c from the central object in a 'corkscrew' manner. From the observed proper motion of 3" per year in the radio structure a distance of 5.5 kpc is derived for SS433. The central point source in VLA maps is unpolarized, whereas the extended structure is linearly polarized up to 20%.

141.145 The compact radio structure of SS433.
R. T. Schilizzi, I. Fejes, J. D. Romney, G. K. Miley, R. E. Spencer, K. J. Johnston.
Extragalactic radio sources, (see 012.025), p. 205 - 206 (1982).

Preliminary results of a campaign of VLBI observations of SS433 extending from January 1980 to June 1981 are presented.

141.146 SS433: periodic changes in the radio structure of scale size 10^{16} cm. A. E. Niell, T. G. Lockhart, R. A. Preston, D. C. Backer.
Extragalactic radio sources, (see 012.025), p. 207 - 208 (1982).

From observations at three epochs using the antennas at Hat Creek, Big Pine and Goldstone at 2.3 GHz the authors have made hybrid VLBI maps of SS433. The angular resolution is about two days of travel time along the jets. The maps show that outside the core the jets are dominated by knots with brightness temperatures greater than 10^6 K.

141.147 Radiative acceleration of astrophysical jets: line-locking in SS433. P. R. Shapiro, M. Milgrom, M. J. Rees.
Extragalactic radio sources, (see 012.025), p. 209 - 210 (1982).

Observations of SS433 are consistent with the view that the Doppler-shifted line emission originates in a pair of oppositely-directed, precessing jets in which a gas outflow is maintained at the remarkably time- and space-invariant speed of 0.26c. A radiative acceleration mechanism is described for the jets and a detailed, numerical, relativistic flow calculation presented which explain this terminal velocity as the result of "line-locking".

141.148 Mechanisms for jets. M. J. Rees.
Extragalactic radio sources, (see 012.025), p. 211 - 222 (1982).

The evidence is now compelling that "jets" delineate the channels along which power is supplied from galactic nuclei into extended radio sources. Jets (often apparently one-sided) have been discovered inside many symmetrical double sources. And M87, familiar optically as a "one-sided jet" is now found to have weak double radio lobes. The VLA has resolved ~70 jets in extended sources; there are now many instances where small jet-like structures are found on the VLBI scale; indirect arguments indicate that there is directed outflow on still smaller scales, and that the primary collimation may occur right down in the central "powerhouse" (scales $\lesssim 10^{15}$ cm). Some possible mechanisms in galactic nuclei that could set up a collimated outflow are discussed.

141.149 Viscous dissipation in jets. M. C. Begelman.
Extragalactic radio sources, (see 012.025), p. 223 - 225 (1982).

The slow decline of surface brightness along many large-scale jets indicates that their internal pressures are determined largely by dissipation. If dissipation arises from a viscous interaction between a jet and its environment, then the observed degree of collimation enables one to constrain the nature of the viscous stress. Simple phenomenological models of the stress account for the frequently observed "gaps" and provide a means of slowing down jets without their becoming decollimated.

141.150 Simple formula for radio jet surface brightness.
R. N. Henriksen.
Extragalactic radio sources, (see 012.025), p. 227 - 228 (1982).

141.151 Instabilities in pressure confined beams and morphology of extended radio sources.
A. Ferrari, S. Massaglia, E. Trussoni, L. Zaninetti.
Extragalactic radio sources, (see 012.025), p. 229 - 230 (1982).

141.152 Connections between turbulence and jet morphology. G. Benford.
Extragalactic radio sources, (see 012.025), p. 231 - 232 (1982).

141.153 Particle acceleration in radio sources with internal turbulence. J. A. Eilek, R. N. Henriksen.
Extragalactic radio sources, (see 012.025), p. 233 - 234 (1982).

141.154   **Jets from discs and doughnuts.**   P. M. Allan.
Extragalactic radio sources, (see 012.025), p.
235 - 236 (1982).

Various authors have suggested that there is a close connection between jets in radio galaxies and the precessing
beams of SS433, and that the underlying mechanism for
forming the jets is the same on both scales. The author
examines a possible model for generating gas jets close to the
central object in either of these two cases.

141.155   **Vortex accretion funnel/relativistic beam models of
double radio sources.**   H. A. Scott,
R. V. E. Lovelace.
Extragalactic radio sources, (see 012.025), p. 237 - 238
(1982).

Some aspects of a self-consistent model of extragalactic
double radio sources are discussed. The model includes gravitational accretion of a fluid with angular momentum as power
source, the production of hydrodynamic or relativistic beam
jets, and the formation of expanding radio components.

141.156   **The nature of the energy source in radio galaxies
and active galactic nuclei.**
F. Pacini, M. Salvati.
Extragalactic radio sources, (see 012.025), p. 247 - 253
(1982).

Recent observational evidence regarding the central
energy sources and the processes which extract energy from
these objects in extragalactic radio sources is reviewed. It is
concluded that there is now reasonable evidence that the
energy source is associated with a supermassive configuration
which has undergone gravitational collapse down to a black
hole stage or its proximity. A choice between accretion and
electrodynamic models of energy extraction cannot yet be
made, but the electrodynamic models seem to be able to
account for the general energy budget and the gross spectral
distribution of radiation.

141.157   **Supercritical accretion and its possible relation to
quasars and radio sources.**   D. L. Meier.
Extragalactic radio sources, (see 012.025), p. 263 - 264
(1982).

This paper reviews some of the properties of supercritical
accretion which make it a possible model for the quasar
optical central source and for producing the jet in radio
sources. Some of the problems with this scenario can be
remedied with alternative, but related, models.

141.158   **X-ray and optical observations of quasars.**
H. Tananbaum, H. L. Marshall.
Extragalactic radio sources, (see 012.025), p. 269 - 277
(1982).

A preliminary report on Einstein X-ray observations of
a sample of quasars in a 1.7 square degree field is given. The
intent of this program of X-ray and optical observations is to
improve the estimate of the contribution of quasars to the
2 keV extragalactic X-ray background and to reconcile
number counts of discrete X-ray sources with reported optical
number counts of quasars.

141.159   **The milliarcsecond structure of radio galaxies and
quasars.**   A. C. S. Readhead, T. J. Pearson.
Extragalactic radio sources, (see 012.025), p. 279 - 288
(1982).

Hybrid maps of the nuclei of radio galaxies and quasars
show a variety of morphologies. Among compact sources, two
structures are common: an asymmetric, "core-jet" morphology (eg, 3C 273), and an "equal double" morphology
with two separated, similar components (eg, CTD 93). The
nuclei of extended, double radio galaxies generally have a
core-jet morphology with the jet directed toward one of the
outer components.

141.160   **High resolution observations of the quasar 3C147.**
E. Preuss, W. Alef, I. Pauliny-Toth,
K. I. Kellermann.
Extragalactic radio sources, (see 012.025), p. 289 - 290
(1982).

The authors report preliminary results of observations of
the quasar 3C147 which were made at 6 cm with a resolution
of about 1 milliarcsecond using a VLB interferometer system
with four antennas in the USA and one in Europe.

141.161   **Structural evolution in the nucleus of NGC 1275.**
J. D. Romney, W. Alef, I. I. K. Pauliny-Toth,
E. Preuss, K. I. Kellermann.
Extragalactic radio sources, (see 012.025), p. 291 - 292
(1982).

Recent observations are reported which demonstrate
structural evolution in the compact radio nucleus of NGC
1275 on a milliarcsec scale. The measurements were
performed at 2.8 cm wavelength with VLBI arrays of seven
stations in 1979 and five stations in 1981.

141.162   **VLBI observations of M87.**
M. J. Reid, J. H. M. M. Schmitt, F. N. Owen,
R. S. Booth, P. N. Wilkinson, D. B. Shaffer, K. J. Johnston,
P. E. Hardee.
Extragalactic radio sources, (see 012.025), p. 293 - 294
(1982).

The authors conducted an 8-station intercontinental
VLBI experiment in order to study the nucleus and jet of M87
at 1666.6 MHz in right circular polarization.

141.163   **Compact radio sources: their use and size.**
K. J. Johnston.
Extragalactic radio sources, (see 012.025), p. 295 - 296
(1982).

141.164   **Spectral shapes of compact extragalactic radio
sources.**   S. R. Spangler.
Extragalactic radio sources, (see 012.025), p. 297 - 299
(1982).

Possible mechanisms for producing the observed broad
radio spectra of compact extragalactic radio sources are
discussed. The explanations considered are: (1) superposition
of the spectra of sub-components, (2) inhomogeneous
synchrotron sources, and (3) synchrotron radiation from
"non-standard" energetic electron spectra. These three models
are compared with results of spectral and VLBI observations.

141.165   **Polarization of the compact radio structure of
3C 454.3.**   W. D. Cotton, B. J. Geldzahler,
I. I. Shapiro.
Extragalactic radio sources, (see 012.025), p. 301 - 303
(1982).

The authors present the results of the first successful
VLBI measurements of the polarized emission from the extragalactic compact radio source 3C 454.3.

141.166   **A millimetre/submillimetre study of optically
selected quasars.**   W. A. Sherwood, G. V. Schultz,
E. Kreysa, H.-P. Gemünd.
Extragalactic radio sources, (see 012.025), p. 305 - 306
(1982).

141.167   **Changes in the H I absorption line spectrum of
AO 0235+164.**   M. M. Davis, A. M. Wolfe.
Extragalactic radio sources, (see 012.025), p 311 - 312
(1982).

141.168   **Theoretical models to explain the variable 21 cm
absorption spectrum in AO 0235+164.**
A. M. Wolfe.

Extragalactic radio sources, (see 012.025), p. 313 - 315 (1982).

141.169 **Variable radio sources.** R. Fanti, L. Padrielli, M. Salvati.
Extragalactic radio sources, (see 012.025), p. 317 - 324 (1982).

Flux variations of flat spectrum compact extragalactic radio sources are analysed and a detailed comparison with predictions from theoretical models is presented. Models are based on an expanding synchrotron radiating plasma of relativistic electrons and magnetic fields partially opaque to its own radiation. In particular, an account is given of theoretical implications and possible explanations of low frequency variability in compact sources, where the inverse Compton limit may be exceeded for the synchrotron source.

141.170 **Interstellar scintillations as a tool for investigations of hyperfine structure in extragalactic radio sources.**
L. M. Ozernoy (*Ozernoj*), V. I. Shishov.
Extragalactic radio sources, (see 012.025), p. 325 - 326 (1982).

Attempts to use interstellar scintillations (ISS) to study the fine angular structure of quasars and galactic nuclei have not yet met with any success. The main reason for this lies in the comparatively large angular size of the nuclear structure which has been investigated. If dependencies on wavelength for the scintillation index, flux and angular size of a source are taken into account, it appears that centimeter wave band rather than decimeter wave range used in previous work is most appropriate for observing ISS.

141.171 **Radio flux flicker of extragalactic sources.**
D. S. Heeschen.
Extragalactic radio sources, (see 012.025), p. 327 - 328 (1982).

Compact sources display a "flicker" in their intrinsic centimeter wavelength radiation, with an amplitude of about 2% and a characteristic time scale of a few days.

141.172 **Broadband studies of compact sources.**
T. W. Jones, L. Rudnick.
Extragalactic radio sources, (see 012.025), p. 329 - 330 (1982).

Progress is reported on a program to study the polarization, spectral and variability properties of compact sources at centimeter and millimeter wavelengths. Source characteristics divide according to the broadband shapes of their spectra.

141.173 **Polarization variability of some compact radio sources.** M. M. Komesaroff, D. K. Milne, P. T. Rayner, J. A. Roberts, D. J. Cooke.
Extragalactic radio sources, (see 012.025), p. 331 - 333 (1982).

141.174 **cm-wavelength fluxes and polarizations of compact extra-galactic X-ray sources.** M. F. Aller, H. D. Aller, P. E. Hodge.
Extragalactic radio sources, (see 012.025), p. 335 - 336 (1982).

cm-wavelength observations of 15 BL Lac objects are presented. The degree of radio-wavelength variability is compared with the strength of the emission at optical and X-ray wavelengths.

141.175 **Rotating structures in extragalactic variable radio sources.**
H. D. Aller, P. E. Hodge, M. F. Aller.
Extragalactic radio sources, (see 012.025), p. 337 - 338 (1982).

Four sources have now been found by the Michigan variability program which exhibit large amplitude rotations in polarization position angle with time. The most straightforward explanation for the phenomenon is a physical rotation in the radio emitting region.

141.176 **Depolarization of extragalactic radio sources.**
M. Inoue, H. Tabara.
Extragalactic radio sources, (see 012.025), p. 339 - 340 (1982).

141.177 **The optical polarization of QSOs.**
R. L. Moore.
Extragalactic radio sources, (see 012.025), p. 341 - 344 (1982).

Recent optical polarization studies indicate that there are two distinct types of QSOs – "normal" QSOs with $P \leqslant 1\%$, and highly polarized QSOs (HPQs) with $P > 3\%$. The HPQs are very similar to BL Lac objects, yet still share some properties of normal QSO's (e.g. strong emission lines). The results reported here generally support the relativistic beaming model for QSOs.

141.178 **Superluminal radio sources.**
M. H. Cohen, S. C. Unwin.
Extragalactic radio sources, (see 012.025), p. 345 - 354 (1982).

Several compact radio sources studied with VLBI show apparent transverse velocities much greater than the speed of light. This review is a summary of VLBI observations of these "superluminal" sources as of August 1981. The physical models proposed to explain this phenomenon are also discussed.

141.179 **Superluminal expansion of 3C 273.**
T. J. Pearson, S. C. Unwin, M. H. Cohen, R. P. Linfield, A. C. S. Readhead, G. A. Seielstad, R. S. Simon, R. C. Walker.
Extragalactic radio sources, (see 012.025), p. 355 - 356 (1982).

Hybrid maps at 10.65 GHz of the core of 3C 273 at five epochs are presented, which were obtained with arrays of 4 or 5 VLBI antennas. A linear expansion rate between emission knots of $v/c \cong 5.3/h$ is derived by assuming $H_0 = 100\ h$ km s^{-1} Mpc^{-1} and $q_0 = 0.05$.

141.180 **Superluminal expansion of the quasar 3C 345.**
S. C. Unwin.
Extragalactic radio sources, (see 012.025), p. 357 - 358 (1982).

VLBI hybrid maps of the compact radio structure in 3C 345 at 5.0 and 10.7 GHz show a core-jet morphology, with two components in the jet separating from the core with apparent transverse velocities $v/c \cong 8/h$ (for $H_0 = 100\ h$ km/s/ Mpc and $q_0 = 0.05$). These "knots" decay as they move down the jet, with lifetimes $\cong 2-3$ years.

141.181 **Superluminal motion in NRAO 140 and a possible future method for constraining H_0 and q_0.**
A. P. Marscher, J. J. Broderick.
Extragalactic radio sources, (see 012.025), p. 359 - 360 (1982).

VLBI observations of the quasar NRAO 140 at 2.8 cm wavelength are reported. An evolution of angular separation between compact components has been observed which corresponds to apparent superluminal velocities. A prescription is outlined for the determination of upper limits on the cosmological parameters H_0 and q_0.

141.182 **Superluminal expansion in 3C 179.**
R. W. Porcas.
Extragalactic radio sources, (see 012.025), p. 361 - 362 (1982).

The detection of apparent faster-than-light motion
(v = 7.6c) in the core of 3C 179 poses some problems for the
simple relativistic jet explanation.

141.183   Relativistic jets as radio and X-ray sources.
A. Königl.
Extragalactic radio sources, (see 012.025), p. 363 - 364
(1982).
   Various predictions and implications of the relativistic-jet
model for compact extragalactic sources are reviewed in the
light of recent radio and X-ray observations.

141.184   VLA observations of the Palomar bright quasar
survey.   D. B. Shaffer, R. F. Green, M. Schmidt.
Extragalactic radio sources, (see 012.025), p. 367 - 368
(1982).
   The quasar survey of Green and Schmidt of some
10000 square degrees of the northern sky contains about 100
quasars in the B magnitude range 13.1−16.5. The authors
observed 94 of these quasars with the VLA and detected radio
emission from 27 of them. They conclude that bright quasars
are definitely more likely to be detectable radio sources than
intrinsically faint ones.

141.185   Optical spectra and radio properties of quasars.
B. J. Wills.
Extragalactic radio sources, (see 012.025), p. 373 - 374
(1982).

141.186   Characteristics of nebulosity associated with Parkes
quasars.   P. A. Wehinger, S. Wyckoff, T. Gehren.
Extragalactic radio sources, (see 012.025), p. 375 - 376
(1982).
   A sample of 13 out of 15 low redshift (0.1 ⩽ z ⩽ 0.6)
radio-loud quasars have been resolved on large-scale sky-
limited ($\mu_R \sim 26.5$ mag sec^{-2}) Kodak IIIa-F photographs
obtained with the ESO 3.6-m telescope.

141.187   Evidence for relativistic motion in the millisecond
structure of BL Lac.
R. L. Mutel, R. B. Phillips.
Extragalactic radio sources, (see 012.025), p. 385 - 386
(1982).
   The authors have begun a series of VLBI observations to
monitor the milliarcsecond structure of BL Lac at λ6 and
λ2.8 cm wavelengths, using a five element VLBI array.

141.188   Radio observations of the galactic center.
D. C. Backer.
Extragalactic radio sources, (see 012.025), p. 389 - 390
(1982).

141.189   The angular size distribution of radio sources at low
flux densities.   A. Downes.
Extragalactic radio sources, (see 012.025), p. 393 - 400
(1982).

141.190   The evolution of linear sizes.
V. K. Kapahi, C. R. Subrahmanya.
Extragalactic radio sources, (see 012.025), p. 401 - 410
(1982).
   The evidence for possible evolutionary effects in the
linear sizes of extragalactic radio sources is discussed by
analyzing the angular size (θ) − redshift (z) relation. Because
of the strong correlation between redshift z and radio luminosity P in flux limited radio samples it is difficult to decide
whether the observed decrease in sizes with z is caused by an
epoch dependence of linear sizes l or by an inverse correlation
between P and l. It is concluded that the available data at
408 MHz down to a flux limit of 55mJy appear to require
evolution in linear sizes according to $l \sim (1 + z)^{-n}$ with

$1 \leqslant n \leqslant 1.5$. Size evolution is needed in addition to any P−l
correlation implicit in the 3CR data.

141.191   Hot-spots and radio lobes of quasars.
G. Swarup, R. P. Sinha, C. J. Salter.
Extragalactic radio sources, (see 012.025), p. 411 - 412
(1982).
   A number of investigations have been made of the
dependence of largest angular size (LAS) of quasars on their
redshifts z. The authors report here preliminary results for 28
quasars out of a sample of 35 observed in the full 35-km
configuration of the VLA at a wavelength of 6 cm.

141.192   The optical and infrared properties of 3CR radio
galaxies.   S. J. Lilly, M. S. Longair.
Extragalactic radio sources, (see 012.025), p. 413 - 422
(1982).
   For the last 20 years, a huge observational effort has been
devoted to the identification of all 3CR radio sources lying in
directions away from the galactic plane. The aim of the present
paper is to assess the reliability of the identifications of distant
radio galaxies and to explore their properties using recent
infrared observations.

141.193   A study of small angular size Ooty sources.
T. K. Menon.
Extragalactic radio sources, (see 012.025), p. 433 - 434
(1982).
   The author has selected an unbiased sample of 73 sources
with angular size ⩽ 4″ from the Ooty survey and measured
their flux densities at 5 GHz and 2.7 GHz using the 300 ft
telescope of NRAO at Green Bank.

141.194   A comparison of the structures of 3CR quasars and
blank field radio sources.
F. N. Owen, J. J. Puschell, R. A. Laing.
Extragalactic radio sources, (see 012.025), p. 435 -
436 (1982).

141.195   Space distribution of quasars based on optically
selected samples.
M. Schmidt, R. F. Green.
Extragalactic radio sources, (see 012.025), p. 437 - 440
(1982).

141.196   Cosmological evolution of QSOs and radio galaxies
from radio-selected samples.
J. V. Wall, C. R. Benn.
Extragalactic radio sources, (see 012.025), p. 441 - 449
(1982).
   The authors describe recent advances in observation and
analysis which lead to improved understanding of the spatial
distribution of QSOs and radio galaxies.

141.197   Modelling the gravitational lens of the double
quasar.   P. K. Moore, S. M. Harding.
Extragalactic radio sources, (see 012.025), p. 461 - 462
(1982).
   Recent VLBI observations of the double quasar
0957+561 A, B are used for producing a refined model of the
gravitational lens hypothesis.

141.198   Superluminal velocities of compact radio sources:
a gravitational lens effect.   J. M. Barnothy.
Extragalactic radio sources, (see 012.025), p. 463 - 464
(1982).

141.199   Symmetry in radio galaxies.   R. D. Ekers.
Extragalactic radio sources, (see 012.025), p.
465 - 474 (1982).
   One of the most striking features of radio galaxies is the
predominance of a symmetrical double lobed structure. The

author reviews large-scale and small-scale symmetries in these sources, discusses mirror symmetry and inversion symmetry and gives a statistical comparison between the radio ejection axis and the optical rotation axis. He argues that from simple morphological considerations important constraints can be derived for theories of radio galaxies.

141.200   A consequence of the asymmetry of jets in quasars and active nuclei of galaxies.
I. S. Shklovsky (*Shklovskij*).
Extragalactic radio sources, (see 012.025), p. 475 - 481 (1982).

It is concluded that many, if not most, jets are truly one-sided. The hypothesis that the powerful radio emission of quasars and radio galaxies is caused by ejections of "plasmoids" originating in supercritical accretion on massive black holes is discussed. Because of asymmetry in the ejection of plasmoids from the thick accretion disks which form around massive black holes, the latter acquire considerable recoil momentum and should escape from the nuclei of the galaxies with large velocities. This provides a possibility for explaining a number of evolutionary effects and an approach to solving the problem of "dead" quasars.

141.201   Ionized gas clouds and the nature of apparent variability of a "point" radio source at the galactic center.   L. M. Ozernoj, V. I. Shishov.
Pis'ma Astron. Zh., Tom 8, 275 - 280 (1982). In Russian. English translation in Soviet Astron. Lett., Vol. 8.

Clouds of ionized gas recently discovered at the center of our Galaxy are proposed to be responsible for the observed "halo-core" structure of the "point" radio source. This hypothesis implies that the core flux is probably variable, which resolves the contradiction between observations made by Kellermann et al. (1977) and Lo et al. (1981).

141.202   Relevamiento del cielo austral en la línea de 21 cm y a bajas velocidades.
F. R. Colomb, W. G. L. Poppel, C. Heiles.
Bol. Asoc. Argentina Astron., No. 20 - 24, p. 178 (1981). Abstract.

141.203   Compact companions to QSOs.
A. Stockton.
Astrophys. J., Vol. 257, 33 - 39, plate 2 (1982).

Three cases of nearly stellar objects found at small angular separations from low-redshift QSOs are discussed. New spectra for two of these, together with previous results for the third, show that all three have redshifts within 200 km s^{-1} of those of their respective QSOs and that the one near PKS 2135−147 almost merits being called a QSO in its own right. It is argued (1) that the actual separations of these compact objects from the QSOs are of the same order as the projected separations, i.e., a few kpc, (2) that similar compact objects may well exist in the neighborhoods of a significant fraction (~20%) of all low-redshift QSOs, and (3) that the presence of these close companions is related to the QSO activity. A plausible, self-consistent case can be made for regarding such objects as remnant cores of galaxies that have interacted with the galaxies in which the QSOs reside and for supposing that the QSO activity results from fresh material being brought into the nuclei of these galaxies during the interaction.

141.204   Detection of neutral hydrogen emission and optical nebulosity in the low redshift QSO 0351 + 026.
G. D. Bothun, W. Romanishin, B. Margon, R. A. Schommer, G. A. Chanan.
Astrophys. J., Vol. 257, 40 - 46 (1982).

Radio and optical observations of the extragalactic, serendipitously discovered X-ray source 0351 + 026 have revealed a remarkably complex system. The optical spectros-copy, photometry, and imagery are consistent with an active nucleus ($z = 0.036$) with the spectrum of a QSO or type 1 Seyfert, embedded in a low luminosity ($M_V \sim -18$) host galaxy of colors similar to M31. The H I observations reveal a tremendous amount of neutral gas associated with the system, $M_{HI}/L_B = 15$ in solar units, and the velocity width of the feature is 1500 km s^{-1}. Both parameters substantially exceed those seen in other galaxies and interacting pairs. The unique H I properties of the system combined with the presence of an active nucleus with QSO-like features is a tantalizing, but poorly understood, combination. At slightly larger redshift and thus inferior angular scale, this object would be indistinguishable from a QSO in its optical and X-ray characteristics.

141.205   Rapid structural variations in 3C 120.
R. C. Walker, G. A. Seielstad, R. S. Simon, S. C. Unwin, M. H. Cohen, T. J. Pearson, R. P. Linfield.
Astrophys. J., Vol. 257, 56 - 62 (1982).

VLBI observations of the radio galaxy 3C 120 at 2.8 and 6.0 cm between 1978.24 and 1980.27 are presented. The source consists of a highly variable, linear emission region with a relatively stable component at one end, suggestive of a "core-jet" morphology. The observations were not frequent enough and did not have sufficient dynamic range to clearly show the evolution of individual components, with the probable exception of one bright component which appeared to separate from the "core" at an apparent rate of 3.7 ± 0.8 times the speed of light ($H_0 = 55$ km s^{-1} Mpc^{-1}) during 1979.

141.206   Synchrotron brightness distribution of turbulent radio jets.
R. N. Henriksen, A. H. Bridle, K. L. Chan.
Astrophys. J., Vol. 257, 63 - 74 (1982).

The authors introduce the notion of radio jets as turbulent mixing regions. They further propose that the essential small-scale viscous dissipation in these jets is by Lighthill emission of MHD waves and by their subsequent strong damping due, at least partly, to gyroresonant acceleration of suprathermal particles. A formula relating the synchrotron surface brightness of a radio jet to the turbulent power input is deduced on very general and simple ground from our physical postulates, and it is tested against the data for NGC 315 and 3C 31 (NGC 383). The predicted brightness depends essentially on the collimation behavior of the jet, and, to a lesser extent, on the Chan and Henriksen picture of a "high" nozzle with accelerating flow. The conditions for forming a large-scale jet at a high nozzle from a much smaller scale jet are discussed. In particular, the optimum condition for retaining the memory of the initial jet direction is given as that of a turbulent jet.

141.207   Deep CCD images of 3C 273.
J. A. Tyson, W. A. Baum, T. Kreidl.
Astrophys. J., Lett., Vol. 257, L1 - L5 (1982).

The authors have obtained wide dynamic range images of the field around the quasar 3C 273 by occulting the stellar part of the quasar image in a coronagraphic CCD camera. These observations reveal an elliptical nebulosity of the same shape as detected photographically by Wyckoff et al. After scaled comparison-star subtraction, most of the flux in the nebulosity falls in the red, appears morphologically similar to NGC 4889 (a first-ranked E galaxy in the Coma cluster), and is found to have an absolute magnitude $M_v \sim -22.5$ (based on $H_0 \sim 100$, $q_0 \sim 0$) if it is at the redshift distance of 3C 273.

141.208   Rapid expansion of BL Lacertae.
R. B. Phillips, R. L. Mutel.
Astrophys. J., Lett., Vol. 257, L19 - L21 (1982).

BL Lacertae has expanded rapidly along its "historic" compact structure axis between 1980.93 and 1981.44. A straightforward interpretation of the expansion indicates that

BL Lacertae has ejected at least one component with an apparent transverse velocity of more than $5c$. An additional component has either been ejected with an even larger velocity or is possibly a temporarily quiescent region which has again become visible. The structure changes seem to be taking place with a characteristic time scale of a few months.

141.209 On acceleration of jets by radiation pressure.
T. Piran.
Astrophys. J., Lett., Vol. 257, L23 - L25 (1982).

Acceleration by radiation pressure could, in principle, explain the observed relativistic velocity of jets in extragalactic radio sources. The radiative acceleration mechanism is based on radiation beams that are produced and collimated in funnels (vortices) along the rotation axis of a thick accretion disk. These radiation beams accelerate particles to form the relativistic jets. In order to accelerate particles to $\gamma > 1.5$, the funnels must be short and very steep. However, such funnels cannot satisfy Rayleigh's stability criterion. Therefore, the radiative acceleration mechanism cannot accelerate particles to high relativistic velocities.

141.210 On a possibility of fine structure study of scintillating radio sources.   V. G. Panadzhyan.
Soobshch. Byurakan Obs., Vyp. 53, p. 108 - 111 (1982). In Russian.

The frequency dependence of the maximum value of the interplanetary scintillation index of the radio source PKS 1148 - 00 is studied. The obtained results are used for analysing the fine structure of scintillating radio sources.

141.211 Historical light variations in quasars measured in Turku.   L. O. Takalo.
Astron. Astrophys., Vol. 109, 4 - 6 (1982).

Historical light variations in four quasars were measured using plates taken in the Turku University Observatory since 1938. Measurements are compared to those made from Harvard plate collection.

141.212 Quasars in a control field far from bright galaxies.
H. Arp, J. Surdej.
Astron. Astrophys., Vol. 109, 101 - 106 (1982).

In order to further investigate the background density of quasars, a field far from bright galaxies has been selected. A total of 137 ultraviolet objects were found. Spectra of 25 of these candidates were obtained with the SIT spectrograph on the 5 m Palomar telescope. Twelve turned out to be quasars. This yields a value of 6.7 ± 0.9 deg^{-2} quasars down to $B = 20$ mag, over the 19.0 deg^2 area of the field. Large fluctuations of the quasar density (up to 100%) are present in this field over small areas approximately 1.5 deg^2 in size and may be correlated with groups of intermediate brightness galaxies which are in this field.

141.213 A continuum study of galactic radio sources in the constellation of Monoceros.
D. A. Graham, C. G. T. Haslam, C. J. Salter, W. E. Wilson.
Astron. Astrophys., Vol. 109, 145 - 154 (1982).

Radio continuum maps have been made of a large region in the constellation of Monoceros at 2700 MHz. The authors present studies of the extended Galactic radio features G 206.9+2.3, the Monoceros Nebulosity, and the Rosette Nebula. The radio properties of the sources are studied and the spectra of PKS 0646+06 and the Monoceros Nebulosity found to be in excellent agreement with those expected for supernova remnants. The present observations of the Rosette Nebula are combined with available low-frequency measurements to deduce an electron temperature of 4100 K for the Nebula. An average emissivity of 3.3 K/pc at 38 MHz is derived for the foreground Galactic continuum radiation between the Sun and the Rosette Nebula.

141.214 Radio continuum in nearby galaxies.
R. Wielebinski.
Bol. Asoc. Argentina Astron., No. 26, p. 129 - 134 (1981).

141.215 Faster than light?   O. Obůrka.
Říše hvězd, Vol. 63, 3 - 5 (1982). In Czech.

141.216 Spatial distribution of QSOs and periodicity in their redshifts.
O. A. Pushkarev, N. G. Makarenko.
Astron. Tsirk., No. 1183, p. 1 - 3 (1981). In Russian.

141.217 Quasars and T Tau-type stars: a complete analogy.
N. E. Kurochkin.
Astron. Tsirk., No. 1193, p. 1 - 4 (1981). In Russian.

141.218 The spectra of 6 unknown radio sources.
V. K. Konnikova.
Astron. Tsirk., No. 1194, p. 6 - 7 (1981). In Russian.

141.219 Precise optical positions of 10 quasars.
K. V. Kuimov, O. D. Solov'eva.
Astron. Tsirk., No. 1195, p. 3 - 4 (1981). In Russian.

141.220 The continuum of QSOs and the nature of the broad 3600 Å emission feature.
R. C. Puetter, E. M. Burbidge, H. E. Smith, W. A. Stein.
Astrophys. J., Vol. 257, 487 - 498 (1982).

Models to explain the shape of the 3600 Å blue emission feature in QSO spectra are derived that depend only on a nonthermal, featureless power-law and Balmer continuum from emission-line clouds. These models require a rather flat power-law component with $F(\nu) \propto \nu^{-0.5}$ to explain data observed on 3C 273 at visual and ultraviolet wavelengths. The required synchrotron–self-Compton radiation is consistent with X-ray and γ-ray emission. The models are an alternative to those requiring an entirely new component of visual and ultraviolet emission from a thermal accretion region.

141.221 X-ray, optical, and radio properties of quasars.
G. R. Blumenthal, W. C. Keel, J. S. Miller.
Astrophys. J., Vol. 257, 499 - 508 (1982) = Lick Obs. Bull., No. 908.

The authors have examined a sample of 26 low-redshift quasars for relationships between X-ray luminosity and optical spectroscopic features. All quasars were observed with the Einstein Observatory and with the IDS on the Lick 3 meter telescope. The authors find evidence for correlations between quasar X-ray luminosity and both optical continuum luminosity and Hβ luminosity. In the latter case, there is a smooth relationship connecting quasars, Seyfert 1, and Seyfert 2 galaxies. For the quasars in this sample, there is also a strong correlation between optical continuum luminosity and both the Hβ luminosity and equivalent width. Overall, the authors found few strong correlations between optical spectroscopic quantities and X-ray properties of quasars. Some of the implications of these results for models of quasars and quasar emission line regions are discussed.

141.222 Helical and pinching instability of supersonic expanding jets in extragalactic radio sources.
P. E. Hardee.
Astrophys. J., Vol. 257, 509 - 526 (1982).

The jets that are observed in extragalactic radio sources expand as distance from the nucleus increases. The author considers the stability properties of an expanding jet. He finds that an expanding jet, like a jet of constant radius, is unstable to Kelvin-Helmholtz perturbations. Wave amplitudes increase linearly rather than exponentially along the jet. The long-wavelength perturbations which can affect the structure in observable fashion can be stabilized by convection along the expanding jet. If a jet appears distorted by a pinching or a helical wave, then jet Mach number and density relative to the external medium can be estimated from observation of the wavelength.

141.223 Multifrequency VLA observations of 3C 388: evidence for an intermittent jet?
J. O. Burns, W. A. Christiansen, D. H. Hough.
Astrophys. J., Vol. 257, 538 - 558 (1982).
VLA observations of the double-lobed radio galaxy 3C 388 were performed at 20, 6, and 2 cm. These data were used to make maps of total and polarized intensity with a resolution of ~1″ at 6 cm and ~3″ at 20 and 2 cm. In addition, 3″ resolution distributions of rotation measure, fractional polarization, depolarization, and spectral index were constructed. The radio maps show that 3C 388 is very asymmetrical on the small scale (~5″), although the overall lobe structure and luminosities are comparable on the large scale (~20″). As part of this asymmetrical fine-scale structure, the authors find that 3C 388 contains a curious jetlike feature embedded within the western lobe. No comparable jet is found in the eastern lobe. Unlike previously discovered jets, the 3C 388 jet is totally within the radio lobe, not between the radio galaxy nucleus and the extended lobe. Relativistic Doppler projection effects are probably not the cause of the one-sidedness of the jet. The orientation of the lobes/jet with respect to the nucleus and the steeper spectral index in the eastern lobe support a picture in which the jet mechanism is inherently intermittent.

141.224 Radiative transfer in quasar emission line clouds.
E. H. Avrett, R. Loeser, P. A. Pinto.
Bull. American Astron. Soc., Vol. 13, 788 (1981). – Abstract.

141.225 A probabilistic radiative transfer equation for finite slab models of QSO emission line regions.
P. J. Ricchiazzi, R. C. Puetter, E. N. Hubbard, R. C. Canfield.
Bull. American Astron. Soc., Vol. 13, 788 (1981). – Abstract.

141.226 Theoretical quasar emission line ratios. V. Balmer continuum emission.   R. C. Puetter, P. D. LeVan.
Bull. American Astron. Soc., Vol. 13, 788 (1981). – Abstract.

141.227 Theoretical QSO emission line profiles.
E. N. Hubbard, R. C. Puetter.
Bull. American Astron. Soc., Vol. 13, 788 - 789 (1981).
Abstract.

141.228 The spectrum of the broad absorption-line QSO 0932 + 501.   V. T. Junkkarinen.
Bull. American Astron. Soc., Vol. 13, 789 (1981). – Abstract.

141.229 The λ 3000 feature in QSOs.
G. A. Shields, J. B. Oke.
Bull. American Astron. Soc., Vol. 13, 790 (1981). – Abstract.

141.230 G127.11 + 0.54 – the compact radio source at the center of the SNR G127.1 + 0.5.
B. J. Geldzahler, D. B. Shaffer.
Bull. American Astron. Soc., Vol. 13, 796 (1981). – Abstract.

141.231 Deep imaging of quasars.
M. Malkan, B. Margon, G. Chanan.
Bull. American Astron. Soc., Vol. 13, 800 (1981). – Abstract.

141.232 A study of high-redshift quasars.
G. M. MacAlpine, F. R. Feldman.
Bull. American Astron. Soc., Vol. 13, 800 (1981). – Abstract.

141.233 Another close quasar pair.   D. W. Weedman.
Bull. American Astron. Soc., Vol. 13, 805 (1981).
Abstract.

141.234 The association of QSOs with groups of galaxies.
H. B. French, J. E. Gunn.
Bull. American Astron. Soc., Vol. 13, 805 (1981). – Abstract.

141.235 Gravitational lenses and the apparent association of QSOs and bright galaxies.   W. C. Keel.
Bull. American Astron. Soc., Vol. 13, 805 - 806 (1981).
Abstract.

141.236 Imaging of the resolved component of quasars.
H. K. C. Yee, H. S. Stockman, R. F. Green.
Bull. American Astron. Soc., Vol. 13, 806 (1981). – Abstract.

141.237 Properties of quasar galaxies.
P. A. Wehinger, S. Wyckoff, T. Gehren.
Bull. American Astron. Soc., Vol. 13, 806 (1981). – Abstract.

141.238 The 6-cm surface density of radio sources below one milli-Jansky.
E. B. Fomalont, K. I. Kellermann, J. V. Wall.
Bull. American Astron. Soc., Vol. 13, 807 (1981). – Abstract.

141.239 Observations at 6 cm of the 611 MHz Arecibo source sample.
C. R. Lawrence, C. L. Bennett, J. Hewitt, B. F. Burke.
Bull. American Astron. Soc., Vol. 13, 807 (1981). – Abstract.

141.240 VLA source counts at 6 cm and 2 cm.
C. L. Bennett, C. R. Lawrence, J. Hewitt, B. F. Burke.
Bull. American Astron. Soc., Vol. 13, 807 (1981). – Abstract.

141.241 Accurate positions and optical identifications of Parkes radio sources.
D. L. Jauncey, M. J. Batty, S. Gulkis, A. Savage.
Bull. American Astron. Soc., Vol. 13, 807 (1981). – Abstract.

141.242 Compton-heated winds from accretion disks in QSOs and binary X-ray sources.
C. F. McKee, G. A. Shields, M. Begelman.
Bull. American Astron. Soc., Vol. 13, 822 (1981). – Abstract.

141.243 Spherical accretion onto quasars.
R. A. London, J. H. Krolik.
Bull. American Astron. Soc., Vol. 13, 822 (1981). – Abstract.

141.244 Can relativistic beaming reconcile the Einstein Observatory X-ray data with the synchrotron self-Compton predictions in compact extragalactic sources?
G. Madejski, D. A. Schwartz, W. H.- M. Ku.
Bull. American Astron. Soc., Vol. 13, 822 (1981). – Abstract.

141.245 Stabilization of expanding supersonic jets.
P. E. Hardee.
Bull. American Astron. Soc., Vol. 13, 823 (1981). – Abstract.

141.246 Intrinsically asymmetric nuclear jets in Cygnus A.
P. J. Wiita, D. J. Saikia.
Bull. American Astron. Soc., Vol. 13, 823 (1981). – Abstract.

141.247 The magnetic jet theory of the double sources in radio galaxies.   W.- R. Hu.
Bull. American Astron. Soc., Vol. 13, 823 - 824 (1981).
Abstract.

141.248 VLA scaled-array observations of the radio galaxy 3C166.   S. R. Spangler, A. H. Bridle.
Bull. American Astron. Soc., Vol. 13, 842 (1981). – Abstract.

141.249 The kiloparsec structures of six superluminal radio sources.
R. A. Perley, E. B. Fomalont, K. J. Johnston.
Bull. American Astron. Soc., Vol. 13, 842 (1981). – Abstract.

141.250 **Superluminal behavior in BL Lacertae.**
R. B. Phillips, R. L. Mutel.
Bull. American Astron. Soc., Vol. 13, 842 (1981). – Abstract.

141.251 **The effects of galaxy dynamics and mergers on the creation of extended radio emission associated with cD galaxies.**   E. Schwendeman, J. O. Burns, R. A. White.
Bull. American Astron. Soc., Vol. 13, 843 (1981). – Abstract.

141.252 **Circular polarization observations at 4.8 and 8.0 GHz.**   P. E. Hodge.
Bull. American Astron. Soc., Vol. 13, 843 (1981). – Abstract.

141.253 **A VLA 1411 MHz radio survey: radio detections in a sample of optically selected ultraviolet-excess stellar objects.**   K. J. Mitchell, J. J. Condon.
Bull. American Astron. Soc., Vol. 13, 843 (1981). – Abstract.

141.254 **Why is the compact radio source in M87 so stable?**   D. L . Jones.
Bull. American Astron. Soc., Vol. 13, 843 - 844 (1981). Abstract.

141.255 **The polarization variability of 3C279 at cm-wavelengths: a source with two preferred polarization position angles.**   H. D. Aller, M. F. Aller, P. E. Hodge.
Bull. American Astron. Soc., Vol. 13, 844 (1981). – Abstract.

141.256 **An extended radio source in the nuclear region of M31.**
R. M. Hjellming, L. Smarr, L. Van Speybroeck.
Bull. American Astron. Soc., Vol. 13, 844 (1981). – Abstract.

141.257 **cm-wavelength observations of the linear polarization and total flux density of extragalactic red and infrared objects.**   M. F. Aller, H. D. Aller, P. E. Hodge.
Bull. American Astron. Soc., Vol. 13, 844 (1981). – Abstract.

141.258 **The QSO number density and contribution to the 2 keV X-ray background.**
H. L. Marshall, H. Tananbaum, Y. Avni, G. Zamorani, A. Braccesi, J. Zitelli, J. P. Huchra.
Bull. American Astron. Soc., Vol. 13, 848 (1981). – Abstract.

141.259 **Detection of neutral hydrogen emission and optical nebulosity in the low-redshift QSO 0351 + 026.**
G. D. Bothun, W. Romanishin, B. Margon, R. A. Schommer, G. A. Chanan.
Bull. American Astron. Soc., Vol. 13, 848 (1981). – Abstract.

141.260 **X-ray, radio, and optical properties of quasars.**   G. Blumenthal, W. C. Keel, J. S. Miller.
Bull. American Astron. Soc., Vol. 13, 849 (1981). – Abstract.

141.261 **X-ray observation of a complete sample of 3CR radio galaxies.**   G. Fabbiano, G. Trinchieri, L. Miller, M. Longair, M. Elvis, E. J. Schreier.
Bull. American Astron. Soc., Vol. 13, 849 (1981). – Abstract.

141.262 **X-ray emission associated with a tailed radio galaxy.**   D. E. Harris, P. E. Dewdney, C. H. Costain.
Bull. American Astron. Soc., Vol. 13, 870 (1981). – Abstract.

141.263 **H I absorption against the nucleus and jet in Centaurus A (NGC 5128).**
J. M. van der Hulst, W. F. Golisch, A. D. Haschick.
Bull. American Astron. Soc., Vol. 13, 893 - 894 (1981). Abstract.

141.264 **Kinematics of the superluminal quasar 3C 345.**   S. C. Unwin.
Bull. American Astron. Soc., Vol. 13, 897 (1981). – Abstract.

141.265 **Structural variations at 329 MHz in 3C 147.**   R. S. Simon.
Bull. American Astron. Soc., Vol. 13, 898 (1981). – Abstract.

141.266 **Hybrid maps of 3C 84 and 3C 345 at 22.2 GHz.**   A. C. S. Readhead, D. H. Hough, M. S. Ewing, R. C. Walker, J. D. Romney.
Bull. American Astron. Soc., Vol. 13, 898 (1981). – Abstract.

141.267 **Detection of milliarcsecond radio cores in all members of a complete sample of radio quasars.**   R. A. Preston, D. D. Morabito, D. L. Jauncey.
Bull. American Astron. Soc., Vol. 13, 898 (1981). – Abstract.

141.268 **A search at the milliJansky level for milliarcsecond cores in a complete sample of radio galaxies.**
A. E. Wehrle, R. A. Preston, D. L. Meier, M. V. Gorenstein, A. E. E. Rogers, I. I. Shapiro, A. Rius.
Bull. American Astron. Soc., Vol. 13, 898 (1981). – Abstract.

141.269 **Mark III VLBI: high sensitivity mapping of weak radio sources.**   M. V. Gorenstein, I. I. Shapiro, N. L. Cohen, E. E. Falco, J. M. Marcaide, A. E. E. Rogers, A. R. Whitney, H. F. Hinteregger, R. J. Cappallo, R. W. Porcas, R. A. Preston.
Bull. American Astron. Soc., Vol. 13, 898 (1981). – Abstract.

141.270 **Magnetic field estimates for QSO absorption line clouds.**   P. P. Kronberg, J. J. Perry.
Bull. American Astron. Soc., Vol. 13, 925 (1981). – Abstract.

141.271 **1107+036: an unusual QSO-galaxy pair.**   H. S. Murdoch, R. W. Hunstead, H. C. Arp, J. C. Blades, J. J. Condon, E. M. Burbidge.
Anglo-Australian Obs. Prepr. No. 165, 41 pp. (1982). – Submitted to Astrophys. J.

141.272 **Quasars in the universe.**   L. Woltjer, G. Setti.
ESO Sci. Prepr. No. 185, 23 pp. (1982). – Submitted to Proc. Vatican Study Week on "Cosmology and Fundamental Physics", Rome 1981.

141.273 **Optical spectroscopy of 28 southern radio galaxies.**   I. J. Danziger, W. M. Goss.
ESO Sci. Prepr. No. 198, 23 pp. (1982). – Submitted to Mon. Not. R. Astron. Soc.

141.274 **A thermal wind model for the broad emission line region of quasars.**   R. J. Weymann, J. S. Scott, A. V. R. Schiano, W. A. Christiansen.
Prepr. Steward Obs., No. 364, 51 pp. (1982).

141.275 **Energy transport in radio sources: the relationship between radio jets and extended structures.**
W. A. Christiansen, J. S. Scott.
Prepr. Steward Obs., No. 376, 70 pp. (1982).

141.276 **Extended radio galaxies. II. Physics of the trail region in double lobed sources.**
W. A. Christiansen, A. G. Pacholczyk, J. S. Scott.
Prepr. Steward Obs., No. 384, 47 pp. (1982).

141.277 **Abrupt cutoffs in the optical-infrared spectra of nonthermal sources.**
G. H. Rieke, M. J. Lebofsky, W. Z. Wisniewski.
Prepr. Steward Obs., No. 393, 23 pp. (1982).

141.278 **Collimation of radio sources: jet-like structures in the plasmon theory.**
W. A. Christiansen, A. G. Pacholczyk, J. S. Scott.
Prepr. Steward Obs., No. 399, 24 pp. (1982).

141.279 **Diffusion of electrons in radio galaxies.**
E. Valtaoja.
Rep. Ser., Dep. Phys. Sci., Univ. Turku, Turku-FTL-R-24,
30 pp. (1982) = Turku Univ. Obs. Informo No. 60 . ISBN
951-642-111-3, ISSN 0356-9896.— Submitted to Astron.
Astrophys.

141.280 **Flux density measurements of bright extragalactic**
        sources at 36.8 GHz.   E. Salonen, H. Lehto,
S. Urpo, P. Teerikorpi, H. Teräsranta, S. Haarala, E. Valtaoja,
L. Tähtinen, A. Sillanpää, M. Tiuri, M. Valtonen.
Rep. Ser., Dep. Phys. Sci., Univ. Turku, Turku-FTL-R 27, 43pp.
(1982) = Turku Univ. Obs. Informo No. 63. ISBN 951-642-
152-0, ISSN 0356-9896.

141.281 **Historical light variations in quasars measured in**
        Turku.   L. O. Takalo.
Turku Univ. Obs. Informo No. 59, 5 pp. (1981). — Submitted
to Astron. Astrophys.

141.282 **Fluid dynamics of relativistic beams flowing**
        through channels and evolution of double radio
sources.   M. Yokosawa.
Astrophys. Space Sci., Vol. 83, 335 - 366 (1982).
    The dynamical evolution of a relativistic beam ejected
from a galactic centre is studied using the similarity method
for the relativistic winds flowing through channels. The expansion phase is divided into two stages: a relativistic expansion and a non-relativistic expansion stage. The flow structures
of the relativistic wind are given. The evolution of extragalactic
double radio sources is considered. The relative position of the
hot spot in the radio map is presented at each stage of the
expansion and is compared with observational radio maps.
The time variation of the radio emission is predicted.

141.283 **Hypersonic beam driven by high-energy particles in**
        extragalactic radio sources.   M. Yokosawa.
Astrophys. Space Sci., Vol. 84, 225 - 242 (1982).
    A mechanism is proposed for the formation of collimat-
ed beams in radio galaxies. The collimated flows which are
non-thermally driven by high energy particles and magnetohydrodynamic (MHD) waves are presented. Cool gas accretes
onto the galactic nucleus and the accretion matter can confine the wave zone around the nucleus in which the high
energy particles are completely locked to the MHD waves. A
complete set of hydrodynamic equations which contain the
energy transfer of high energy particles and MHD waves is
presented. One-dimensional flows which are in pressure
equilibrium with the surrounding accretion matter are calculated. On the basis of the collimated beams driven by high
energy particles, the radio morphology of the double radio
sources is discussed.

141.284 **Sub-millimeter measurements of the anisotropy of**
        the cosmic microwave background.
E. L. Wright, M. Halpern, R. Weiss.
Bull. American Astron. Soc., Vol. 14, 576 (1982). — Abstract.

141.285 **The redshift of CL 4.**
S. Tapia, D. A. Turnshek.
Bull. American Astron. Soc., Vol. 14, 577 (1982). — Abstract.

141.286 **Extended radio structure of BL Lac type objects.**
D. Stannard, B. K. McIlwrath.
Nature, Vol. 298, 140 - 142 (1982).
    The authors report observations which show that the
BL Lacs do indeed possess the general core-halo morphology
predicted by the Blazar models, although there is an important difference to the quasars in that compact jet-like
features are much less conspicuous.

141.287 **VLBI measurements of radio source positions at**
        three U.S. stations.   S.-H. Ye.
High-precision Earth rotation and Earth-Moon dynamics.
Lunar distances and related observations, (see 012.035),
p. 329 - 336 (1982).
    Results of VLBI measurements of 14 radio source
positions at three U.S. stations during the MERIT short
campaign are presented. Comparisons with other solutions
are given, together with comparisons between several radio
source catalogues.

141.288 **Nebulae around quasars.**   B. V. Komberg.
Priroda, 1982, No. 5, p. 37 - 39. In Russian.

141.289 **C IV absorption in an unbiased sample of 33 QSOs:**
        evidence for the intervening galaxy hypothesis.
P. Young, W. L. W. Sargent, A. Boksenberg.
Astrophys. J., Suppl. Ser., Vol. 48, 455 - 506 (1982).
    The authors present new observations of 27 QSOs at
redshifts $z_{em} \approx 2$ with 2.5 Å resolution and with uniform
signal-to-noise ratio. Using some previously observed objects
they define an unbiased, homogeneous sample of 33 QSOs
with which they investigate the properties of the absorption
line systems. Particular attention is paid to C IV $\lambda\lambda1548$,
1550 doublets. They find that: 1. Three QSOs out of 33 have
broad absorption troughs due to material ejected by the QSO
at velocities up to $0.1c$. 2. In the 30 QSOs without troughs
the C IV lines are randomly distributed in redshift in a manner
consistent with absorption from intervening galaxies. 3. Contrary to previous studies, there is no statistically significant
excess of absorption systems near the emission redshifts of the
nontroughed QSOs. 4. The two-point correlation function for
C IV lines shows a positive signal for velocity splittings
$v < 2000$ km s^{-1}, in a manner consistent with galaxy-galaxy
correlations extrapolated back to $z \approx 2$. It is suggested that
the fine splittings frequently observed in C IV doublets at
high spectral resolution are produced by galaxy clustering
rather than by multiple clouds in single galaxies. 5. The distribution of C IV absorption in the three QSOs with troughs is
markedly nonuniform. There is an excess of systems with
velocities relative to the QSO of $v < 0.1c$. 6. In Q1309−056
one observes "sharp" absorption systems which are abnormally diffuse and have high ionization states. A statistical study
shows that these systems also arise from material ejected by
the QSO.

141.290 **Faint blue objects at high galactic latitude.**
        III. Palomar Schmidt field centered on Selected
Area 28.   P. D. Usher, K. J. Mitchell.
Astrophys. J., Suppl. Ser., Vol. 49, 27 - 52 (1982).
    The third part of the list of ultraviolet excess objects is
comprised of 1179 objects selected from a Palomar Schmidt
three-color plate centered on Kapteyn Selected Area 28. The
population belonging to color classes 1A, 1, and 1B should
consist primarily of quasars and white dwarfs and should be
statistically complete to at least $B = 18.5$ mag.

141.291 **Theoretical quasar emission-line ratios. VI. A**
        probabilistic radiative transfer equation for finite
slab atmospheres.   R. C. Puetter, E. N. Hubbard,
P. J. Ricchiazzi, R. C. Canfield.
Astrophys. J., Vol. 258, 46 - 52 (1982).
    Previous papers in this series have been based on an
approximation in which the line ratios were inferred from
those computed for a semi-infinite cloud model. In this
paper the authors present a superior method, which permits
the treatment of the emission-line clouds as slab atmospheres
of finite thickness. In common with the previous semi-infinite
approach, it is based on photon escape probabilities, yet it
recognizes the important distinction between the photon
escape probability and the flux divergence. This distinction
is neglected in all existing models of energy balance in QSO

emission-line clouds. The present method reduces to the previous one in the semi-infinite case. It is shown that the method provides a solution that departs from the exact solution by at most a few tens of percent in cases of physical interest, while retaining all the advantages of the previous method.

141.292 Evidence for 200 second variability in the X-ray flux of the quasar 1525 + 227.
T. Matilsky, C. Shrader, H. Tananbaum.
Astrophys. J., Lett., Vol. 258, L1 - L5 (1982).

Sixteen hundred seconds of Einstein Observatory IPC data for 1525 + 227, a bright, nearby ($m_v = 16.4$, $z = 0.253$) quasar show significant intensity fluctuations on a time scale of 200 seconds. Similar fluctuations were not detected 13 months later during a follow-up 12,500 second observation. Interpretation of the time variations requires $H_0 \geqslant 100\,\mathrm{km\,s^{-1}\,Mpc^{-1}}$ to obtain even marginal consistency with standard isotropic emission models. An alternative interpretation invokes anisotropic emission probably due to the presence of relativistic motions with associated beaming effects.

141.293 Multifrequency high resolution observations of the large radio galaxy B2 1321 + 31.
R. Fanti, C. Lari, P. Parma, A. H. Bridle, R. D. Ekers, E. B. Fomalont.
Astron. Astrophys., Vol. 110, 169 - 178 (1982).

Observations at 49.21 and 6 cm of the total and polarized intensity of the jets and extended lobes of the large radio galaxy B2 1321+31 are presented.

141.294 Cosmological implications of the redshift distribution of QSO absorption systems.
P. Khare-Joshi, J. J. Perry.
Mon. Not. R. Astron. Soc., Vol. 199, 785 - 800 (1982).

In order to investigate the intervening galaxies hypothesis the authors have used the observational data on QSO absorption redshifts, as compiled by Perry, Burbidge and Burbidge (1978), Drew (1978) and Weyman et al. (1979), to study various selection effects likely to affect the distribution of absorption redshifts and, then, to determine the probable number distribution of absorbers per redshift interval of 0.1, as a function of z. The distribution obtained, assuming all the observed absorption to be intervening, is found to be statistically incompatible with the redshift distribution of galaxies with constant cross-section for any Friedman cosmology with zero cosmological constant and $q_0 \geqslant 0$. Various criteria for eliminating absorption systems of intrinsic origin are discussed.

141.295 An infrared study of quasars.
A. R. Hyland, D. A. Allen.
Mon. Not. R. Astron. Soc., Vol. 199, 943 - 952 (1982).

New high quality near infrared data are presented for a significant number of quasars with redshifts out to 3.12. The colours ($J-H$) and ($H-K$) are highly redshift dependent, becoming larger to smaller redshifts. When the effects of emission lines in the bandpasses have been taken into account, these colours demand the presence of excess continuum emission in the $1-3$ μm region (in the rest frame), over and above the continuum components present at shorter and longer wavelengths. This excess emission in the $1-3$ μm region is interpreted in terms of thermal emission from hot dust ($T \sim 1200$ K), which is shown to exist predominantly in regions where the broad emission lines are formed.

141.296 An investigation of the properties of double radio sources using the Spearman partial rank correlation coefficient.   J. T. Macklin.
Mon. Not. R. Astron. Soc., Vol. 199, 1119 - 1136 (1982).

The methods available for analysing correlations between two variables in the presence of a third are reviewed, and it is concluded that the Spearman partial rank correlation coefficient makes the most efficient use of the available data. This is therefore used to study the correlations of the properties of double radio sources with size and redshift z, and with z and luminosity P. Results for a sample of 3CR sources show that larger sources are more diffuse, but their spectral indices are not steeper. If all sources have similar expansion speeds, these results provide information about how sources evolve, since then larger sources are older ones. The possibility of determining the density parameter Ω of the Universe from the relation between z, P and physical size is considered, and the extension of the Spearman partial rank correlation coefficient to correlations involving more than three parameters is discussed.

141.297 Nuclear jets in Cygnus A.   D. J. Saikia, P. J. Wiita.
Mon. Not. R. Astron. Soc., Vol. 200, 83 - 89 (1982).

The authors analyse published VLBI data on Cygnus A at 10.65 and 15 GHz in an attempt to explain the intensity and spectra of the jets in its nucleus. Because one knows that the angle between the plane of the source and the plane of the sky is only $\sim 6°$, the straightforward beaming picture with concomitant Doppler shifts does not seem to allow for the apparently observed differences between the flux densities and the spectra of the jet and the counter-jet. External thermal absorption can explain the asymmetries within the assumed errors, although such an absorbing cloud cannot simultaneously explain the collimation in the twin-beam model of Blandford & Rees. The asymmetry could also be due to the possibility that the jet is no longer being fed and that the counter-jet is just being inflated, a variant of the flip-flop picture, but the authors feel that this too is improbable. One is left with the somewhat unpleasant conclusion that intrinsic asymmetries between the jets are likely to exist.

141.298 Masses of quasars.   A. Sołtan.
Mon. Not. R. Astron. Soc., Vol. 200, 115 - 122 (1982).

Quasar masses are investigated assuming that accretion on to massive black holes is the ultimate source of energy produced by quasars. Lower limit for the total energy emitted and the mass accumulated in black holes in 1 Gpc^3 is calculated using various data on quasar counts and bolometric luminosities. The energy produced is at least $8.5 \times 10^{66}\,\mathrm{erg\,Gpc^{-3}}$. This result is independent of the cosmological model. Assuming that quasars reside in nuclei of giant galaxies it is shown that minimum masses of dead quasars are of the order of $10^8 M_\odot$, close to the observational threshold for ground-based telescopes.

141.299 Spectroscopy of 33 QSO candidates from the Jodrell Bank 966-MHz survey.
D. Walsh, R. F. Carswell.
Mon. Not. R. Astron. Soc., Vol. 200, 191 - 195 (1982).

Thirty-three QSO candidates from the Jodrell Bank 966-MHz survey have been observed spectroscopically. Twenty-five are confirmed as QSOs and redshifts are given for 23 of them. The remaining eight objects are galactic stars.

141.300 The Lyα/Hα ratio in high-redshift quasars.
D. A. Allen, J. R. Barton, P. R. Gillingham, R. F. Carswell.
Mon. Not. R. Astron. Soc., Vol. 200, 271 - 279 (1982).

The authors present Hα measurements in eight quasars with redshifts between 2.1 and 2.8. These bring the total number of high-redshift quasars with measured Lyα/Hα ratios to 12. The authors find a mean value for this ratio of 1.5, with a spread of values of about a factor of 2 either side of the mean. A correlation exists between the Lyα/Hα ratio and the continuum spectral index; this exactly matches a known correlation between the Lyα equivalent width and the spectral index. Various model explanations for the Lyα/Hα ratio are discussed.

141.301 Polarization observations of DA240: structure of a hotspot. S. C. Tsien.
Mon. Not. R. Astron. Soc., Vol. 200, 377 - 384 (1982).

The author presents observations of the large-scale structure of the giant radio galaxy DA 240 at 0.15 and 1.4 GHz, and high-resolution observations of the hotspot in its eastern lobe at 2.7 and 5.0 GHz. It is shown that the large-scale structure of DA 240 at 0.15 GHz is closely similar to that at 1.4 GHz. An age of about 4×10^7 yr is estimated from the spectral index. The hotspot contains a compact component of $\lesssim 1 \times 2.5$ kpc^2 in size, and is strongly polarized, up to 50−60 per cent in its south-east region.

141.302 Observations of X-ray quasars at 90 GHz.
F. N. Owen, J. J. Puschell.
Astron. J., Vol. 87, 595 - 601 (1982).

Observations at 90 GHz are reported of 51 quasars previously detected by the Einstein X-ray observatory. Unlike strong millimeter quasars, these generally selected quasars do not show a strong correlation between their millimeter and X-ray flux densities.

141.303 2.3 GHz accurate positions and optical identifications for selected Parkes radio sources.
D. L. Jauncey, M. J. Batty, S. Gulkis, A. Savage.
Astron. J., Vol. 87, 763 - 773 (1982).

Accurate radio positions of 74 extragalactic radio sources contained in the Parkes 2.7 GHz survey were measured at 2.3 GHz with the Tidbinbilla interferometer, located near Canberra, Australia. Optical identifications have been made on the basis of positional coincidence alone, without regard to color or morphology, using the UK Schmidt telescope deep IIIa-J sky survey plates. Identifications are suggested and accurate optical positions have been measured for 62 objects brighter than magnitude 22.5 on the IIIa-J plates.

141.304 Radio two-color spectral distributions.
E. Pacht.
Astron. J., Vol. 87, 774 - 779 (1982).

Radio two-color diagrams have been constructed for quasars and radio galaxies as a function of redshift, and for three strong complete radio samples as a function of survey frequency. Isometric ellipses are used to define the spectral distributions on the two-color diagrams.

141.305 Einstein X-ray observations of QSO's with absorption-line systems.
V. T. Junkkarinen, A. P. Marscher, E. M. Burbidge.
Astron. J., Vol. 87, 845 - 848 (1982).

The authors report the detection of X-ray emission from eight QSO's, plus an upper limit to the X-ray flux from one QSO, using the Einstein X-ray Observatory (HEAO-2). Each object in the sample contains at least one absorption-line system that has been identified in its optical spectrum. The results are combined with those of other investigators to form a sample of 44 absorption-line QSO's (with $z_e > 1.2$) which have been observed in the X-ray. This sample cannot be distinguished, in terms of X-ray properties, from one which consists of QSO's in which no absorption systems have been identified.

141.306 The positions, structures, and polarizations of 404 compact radio sources. R. A. Perley.
Astron. J., Vol. 87, 859 - 880 (1982).

Accurate positions of 404 compact radio sources used as calibrators by the VLA are presented. In addition, the structure and polarization of each source at both 4885 and 1465 MHz are given. Eighty-five percent of the sources have spectral indices flatter than 0.5; all of these are dominated by an unresolved core. Half of these flat-spectrum sources contain nearby, associated diffuse structure at a level exceeding ∼0.4% of the core brightness at 20 cm.

141.307 Optical identifications of reference frame benchmark radio sources. A. N. Argue, C. Sullivan.
Abh. Hamburger Sternw., Band 10, 166 (1982). − Abstract.

141.308 An improved optical position of 3C 273B in the FK4-system. C. de Vegt, U. K. Gehlich.
Abh. Hamburger Sternw., Band 10, 169 (1982). − Abstract.

141.309 A VLA survey of strong radio sources.
J. Ulvestad, K. Johnston, R. Perley, E. Fomalont.
Abh. Hamburger Sternw., Band 10, 173 (1982). − Abstract.

141.310 Precise optical positions of radio sources in the southern hemisphere. H. G. Walter, R. M. West.
Abh. Hamburger Sternw., Band 10, 175 (1982). − Abstract.

141.311 Compact radio sources at declinations $>67°$.
E. Waltman, K. J. Johnston, J. H. Spencer, I. Pauliny-Toth, J. Schraml, A. Witzel.
Abh. Hamburger Sternw., Band 10, 175 (1982). − Abstract.

141.312 Optical positions of benchmark radio sources south of +5° declination. R. M. West, H. G. Walter.
Abh. Hamburger Sternw., Band 10, 176 (1982). − Abstract.

141.313 The space distribution of quasars.
M. Schmidt, R. F. Green.
Astrophysical cosmology, (see 012.047), p. 281 - 292 (1982).

The space density of quasars increases steeply with distance out to a redshift of three at least. This phenomenon is a consequence of the variation of the quasar luminosity function with time, which reflects the collective result of the births and subsequent evolution of quasars.

141.314 Quasars in the universe. L. Woltjer, G. Setti.
Astrophysical cosmology, (see 012.047), p. 293 - 314 (1982).

141.315 The distribution of quasar redshifts.
L. Z. Fang.
Astrophysical cosmology, (see 012.047), p. 345 - 348 (1982). Abstract.

141.316 Evidence from deep radio surveys for cosmological evolution. H. van der Laan, R. A. Windhorst.
Astrophysical cosmology, (see 012.047), p. 349 - 371 (1982).

141.317 Some aspects of the cosmological evolution of extragalactic radio sources. M. S. Longair.
Astrophysical cosmology, (see 012.047), p. 373 - 382 (1982).

141.318 On evolutionary models of radio sources.
G. Swarup, C. R. Subrahmanya, V. K. Kapahi.
Astrophysical cosmology, (see 012.047), 383 - 391 (1982).

It is well known that evolutionary effects in the properties of extragalactic radio sources dominate the geometrical effects of reasonable world models. The authors comment on two aspects of the evolution of radio sources, viz. the use of optical identification data to constrain possible changes in the radio luminosity function with epoch and the evidence for evolution in the linear sizes of double radio sources.

141.319 Photometry of 0957 + 561; detection of short period variations. C. Vanderriest, A. Bijaoui, P. Félenbok, G. Lelièvre, J. Schneider, G. Wlérick.
Astron. Astrophys., Vol. 110, L11 - L14 (1982). In French.

Absolute photometry is necessary in order to find the time delay Δt between 0957 + 561 A and B by cross-correlation of their light curves. On several electronographic plates obtained with the Canada-France Hawaii Telescope (3.6 m) and the 1.93 m telescope in Haute Provence Observatory images A and B were individually measured with an accuracy of a few hundredths of magnitude. The authors found an amplitude of variation (in V color) about 0.20 magnitude for both images over a 1.5 year period. Moreover, the flux ratio A/B can be determined with a much better accuracy ($<$ 1%); it varies by about 7% in 3 days (26 - 29 April 1981) and 3% in 24 hours (29 - 30 April), these variations being identical in B and V.

141.320 **Multifrequency comparison of the total intensity and polarization distributions for 3C 31, 3C 66B, and 3C 129.**   W. van Breugel.
Astron. Astrophys., Vol. 110, 225 - 237 (1982).

Westerbork observations at 6 cm, 21 cm, and 49 cm of the three well known extended radio galaxies 3C 31, 3C 66B, and 3C 129 are discussed. The magnetic field distributions of these sources are presented. In all three sources the magnetic field seems to be structured by the combined effects of expansion of jets and shearing or compression along their boundaries. Intercomparison of the 6 cm and 21 cm polarization data in 3C 66B and 3C 129 allowed to estimate the relatively large Galactic foreground rotation measures in the directions of these sources which were taken into account for the derivation of the magnetic field directions. The spectral index distributions (6 cm/21 cm and 21 cm/49 cm) of 3C 66B and 3C 129 are also presented.

141.321 **Spectral index behaviour of low frequency variable radio sources.**   F. Mantovani.
Astron. Astrophys., Vol. 110, 345 - 347 (1982).

Results are given of observations made at 6 cm and 2.8 cm using the Effelsberg 100 m radiotelescope, of 20 sources showing flux density variability at low frequency. The computed spectral indices of these variable sources show that 13 have flat spectra and 7 steep spectra. A comparison with a previous publication is then given.

141.322 **Report of IAU Commission 40: Radio astronomy (Radio astronomie).**   G. Swarup.
Trans. IAU, Vol. XVIIIA, (see 003.013), p. 529 - 549 (1982).

Choice of 315 stars in 87 areas with extragalactic radio sources for meridian observations in the FK4 system.
See Abstr. 002.016.

Dumbbell galaxies and precessing radio jets.
See Abstr. 002.068.

The quasar controversy resolved.
See Abstr. 003.071.

Classics in radio astronomy.
See Abstr. 003.125.

Five decades of radio astronomy.
See Abstr. 004.040.

Neutral hydrogen studies with a novel instrument.
See Abstr. 031.513.

Possible measurement of the time delay between gravitational images of expanding double radio-sources.
See Abstr. 031.516.

Evaluation of the accuracy of determination of the length of the arc between two quasars by VLBI technique using synchronous observations.   See Abstr. 031.539.

Exact gain measurement of large aperture antennas using celestial radio sources.   See Abstr. 033.008.

Precise radio source positions from interferometric observations.   See Abstr. 041.009.

Cosmic radioastrometry.   See Abstr. 041.013.

Optical identifications for benchmark radio sources.   See Abstr. 041.044.

The selection of stars for linking the Hipparcos frame to extragalactic radio sources by Space Telescope (progress report).   See Abstr. 051.023.

The unsteady beam.   See Abstr. 062.007.

Magnetohydrodynamic Kelvin-Helmholtz instabilities in astrophysics − III. Hydrodynamic flows with shear layers.   See Abstr. 062.009.

Hydromagnetic flows from accretion discs and the production of radio jets.   See Abstr. 062.123.

Emisiones no térmicas en radiofuentes compactas.   See Abstr. 063.038.

Stationary spherical symmetric accretion onto massive black holes: the radiation spectrum and luminosity.   See Abstr. 066.003.

Gravitational lenses.   See Abstr. 066.018.

Upper limits of a cosmic infrared background flux as determined by X- and gamma-ray observations of M 87.   See Abstr. 066.034.

Linsen im Weltraum. Teil 1: Der Gravitationslinseneffekt und Beobachtungen an einem Zwillingsquasar.   See Abstr. 066.091.

Self-gravitating accretion disk models for active galactic nuclei: self-consistent α-models for the broad emission-line region.   See Abstr. 066.092.

Black holes and the origin of radio sources.   See Abstr. 066.103.

Electron-positron jet models.   See Abstr. 066.115.

Faint blue objects at high galactic latitude: summary of four Palomar Schmidt fields.   See Abstr. 113.067.

The galactic fountain, observations of extragalactic radio sources, and the cosmic ray halo.   See Abstr. 131.029.

Neutral hydrogen observations in the direction of extended background radio sources.   See Abstr. 131.073.

Comparison of IR maps with a 2″ map of NH_3 for the Orion Molecular Cloud.   See Abstr. 131.171.

A compact radio source associated with the Becklin-Neugebauer object.   See Abstr. 131.198.

W50 as a probe of the interstellar medium.   See Abstr. 131.221.

Radio sources in NGC 6334.   See Abstr. 132.009.

Optical study of the W51 complex.   See Abstr. 132.016.

Radio emission of the Crab nebula and anisotropic relativity of Boltyanskij.   See Abstr. 134.023.

Large scale X-ray and radio structures associated with compact extragalactic sources.   See Abstr. 142.003.

X-ray quasars and the X-ray background. See Abstr. 142.004.

PKS 2155 – 304: relativistically beamed synchrotron radiation from a BL Lacertae object.   See Abstr. 142.005.

The luminosity of serendipitous X-ray QSOs. See Abstr. 142.008.

X-ray emission from Centaurus A. See Abstr. 142.043.

Luminosity indicators in X-ray selected QSOs. See Abstr. 142.068.

A model for the quiescent radio emission from Cyg X-3.   See Abstr. 142.070.

The luminosity of serendipitous X-ray QSOs: implications for the diffuse X-ray background radiation. See Abstr. 142.074.

Deep X-ray surveys of two quasar grism fields. See Abstr. 142.075.

Radio observations of coronal X-ray sources. See Abstr. 142.097.

A model for elliptical radio galaxies with dust lanes. See Abstr. 151.001.

Cosmic density waves and its observable vestige. See Abstr. 151.046.

Radio sources in globular clusters? See Abstr. 154.099.

Neutral hydrogen emission and absorption in three active Irr II galaxies.   See Abstr. 158.003.

The spectrum of 3C 390.3 and the hydrogen line problem.   See Abstr. 158.013.

Centimetre wavelengths radio studies of clumpy irregular galaxies.   See Abstr. 158.015.

Winds from dwarf galaxies and $L\alpha$ absorption features in quasar spectra.   See Abstr. 158.020.

The galaxy components of BL Lacertae objects, N systems, and quasi-stellar objects.   See Abstr. 158.027.

High-resolution 1666-MHz observations of the nucleus of NGC 4151.   See Abstr. 158.028.

The ultraviolet excess of Seyfert 1 galaxies and quasars.   See Abstr. 158.035.

Galactic gas and the shapes of radio sources. See Abstr. 158.040.

Kinematics of gas clouds in the nuclei of Seyfert galaxies and quasars.   See Abstr. 158.045.

The active radio galaxy 1413+135. See Abstr. 158.061.

Radio continuum observations of the spiral galaxies NGC 2841, NGC 5055, and NGC 7331. See Abstr. 158.069.

The short-term radio variability of BL Lacertae objects.   See Abstr. 158.073.

Reverberation mapping of the emission line regions of Seyfert galaxies and quasars.   See Abstr. 158.074.

The properties of AP Librae from UBV photoelectric photometry.   See Abstr. 158.077.

The radio continuum properties of S0 galaxies. See Abstr. 158.086.

Interaction of stars with the accretion disc around a massive black hole in the nuclei of active galaxies and quasars.   See Abstr. 158.090.

The optical spectrum of the radio galaxy PKS 2152–69.   See Abstr. 158.108.

A survey of the distribution of $\lambda2.8$ cm radio continuum in nearby galaxies. II. NGC 6946. See Abstr. 158.111.

Zur Beziehung zwischen optischen und Radiohelligkeiten von Galaxien.   See Abstr. 158.119.

Cosmic jets.   See. Abstr. 158.147.

NGC 4258: a bent jet in a spiral galaxy. See Abstr. 158. 161.

Seyfert galaxies.   See Abstr. 158.162.

Radio continuum observations of the nuclei of nearby galaxies.   See Abstr. 158.163.

Radio emission from the Seyfert galaxy NGC 5548. See Abstr. 158.164.

Radio observations of Markarian 8. See Abstr. 158.165.

Infrared observations of radio galaxies. See Abstr. 158.166.

Galactic centers and twin-jets. See Abstr. 158.167.

Detection of a broad H I absorption feature at 5300 km sec^{-1} associated with NGC 1275 (3C84). See Abstr. 158.168.

A search for H I in elliptical galaxies with nuclear radio sources.   See Abstr. 158.169.

Optical spectra of radio-loud and radio-quiet active galactic nuclei.   See Abstr. 158.170.

BL Lac objects and their associated galaxies. See Abstr. 158.171.

Mark III VLBI observations of the nucleus of M81 at 2.3 and 8.3 GHz.   See Abstr. 158.172.

Redshift estimates for distant radio galaxies based on broadband photometry.   See Abstr. 158.173.

Colors of radio galaxies at high redshifts.
See Abstr. 158.174.

Search for redshifted CH_2O, H_2O, O_2, and NH_3 in Seyfert galaxies and quasars.   See Abstr. 158.178.

The extended radio source in the center of M31.
See Abstr. 158.196.

Optical observations of active galaxies and quasars at high angular resolution.   See Abstr. 158.203.

Highly compact structures in galactic nuclei and quasars.   See Abstr. 158.204.

Near-infrared photometry of distant radio galaxies: spectral flux distributions and redshift estimates.
See Abstr. 158.215.

The heating of dust in the broad line regions of active galaxies and quasars.   See Abstr. 158.218.

VLA detection of the H 110α line in M82.
See Abstr. 158.234.

Mark III VLBI observations of the nucleus of M 81 at 2.3 and 8.3 GHz.   See Abstr. 158.261.

Observations of Paschen α in the broad-line radio galaxy 3C 445.   See Abstr. 158.268.

Hβ luminosity of extragalactic objects and evolution of galaxies.   See Abstr. 158.286.

Extended H I-envelopes around spiral galaxies: NGC 2655 and NGC 2715.   See Abstr. 158.305.

Infrared studies of a sample of 3C radio galaxies.
See Abstr. 158.314.

CO observations of the SAB galaxies NGC 157, 2903, 4321, and 5248, and the Seyfert galaxy NGC 1068.
See Abstr. 158.322.

Complete samples of active extragalactic objects. I. A 1411 MHz VLA survey centered on $\alpha = 12^h 04^m$, $\delta = +11°30'$.
See Abstr. 158.325.

The colours of faint radio galaxies.
See Abstr. 158.334.

Evidence for the cosmological evolution of the stellar content of radio galaxies.   See Abstr. 158.335.

The radio structure of the nuclear region of NGC 1365.   See Abstr. 158.336.

VLA detection of H 110α in M 82.
See Abstr. 158.354.

Report of IAU Commission 28: Galaxies (Galaxies).
See Abstr. 158.358.

Radio and optical observations of 9 nearby Abell clusters: A262, A347, A569, A576, A779, A1213, A1228, A2162, A2666.   See Abstr. 160.006.

A Westerbork survey of clusters of galaxies.
XIV. Abell 779 and Abell 1314.   See Abstr. 160.014.

Quasar-generating superclusters: an explanation for a clumpy quasar sky?   See Abstr. 160.020.

Radio continuum observations at 5 GHz of Shakhbazyan's compact groups of galaxies.
See Abstr. 160.025.

Radio-optical studies of a complete sample of Abell clusters.   See Abstr. 160.036.

Studies of a complete sample of Abell clusters at 1400 MHz.   See Abstr. 160.037.

Two peculiar radio galaxies in A1367.
See Abstr. 160.038.

Radio observations at 1.4 GHz of Abell clusters.
See Abstr. 160.039.

Radio evolution in high redshift clusters.
See Abstr. 160.042.

Massive neutrino decay and the photoionization of the intergalactic medium.   See Abstr. 161.001.

Intergalactic Lyman-alpha absorption lines in a close pair of high-redshift QSOs.   See Abstr. 161.003.

The intergalactic medium.   See Abstr. 161.005.

On the ionization balance of helium in the intergalactic medium.   See Abstr. 161.008.

Quasars as probes of the distant and early universe.
See Abstr. 162.008.

The Hubble diagram of quasars and the deceleration parameter q_0.   See Abstr. 162.009.

The age of galaxies.   See Abstr. 162.117.

Problems of metagalactic astronomy.
See Abstr. 162.130.

Gravitational lenses and cosmological evolution.
See Abstr. 162.196.

Report of IAU Commission 47: Cosmology (Cosmologie).   See Abstr. 162.226.

Pulsars

141.501 **Electromagnetic cascades in pulsars.**
J. K. Daugherty, A. K. Harding.
Astrophys. J., Vol. 252, 337 - 347 (1982).

The development of pair-photon cascades initiated by high energy electrons above a pulsar polar cap is simulated numerically. The calculation uses the energy of the primary electron, the magnetic field strength, and the period of rotation as parameters and follows the curvature radiation emitted by the primary, the conversion of this radiation to e^+e^- pairs in the intense fields and the quantized synchrotron radiation by the secondary pairs. Gamma-ray and pair spectra are calculated for cascades in different parts of the polar cap and with different acceleration models. The synchrotron radiation from secondary pairs makes an important contribution to the γ-ray spectrum above 25 MeV, and the final γ-ray and pair spectra are insensitive to the height of the accelerating region, as long as the acceleration of the primary electrons is not limited by radiation reaction.

141.502 **The non-aligned pulsar magnetosphere: an illustrative model for small obliquity.**
L. Mestel, Y.-M. Wang.
Mon. Not. R. Astron. Soc., Vol. 198, 405 - 427 (1982).

The electromagnetic field outside a pulsar of small obliquity is approximated by Goldreich–Julian (GJ) conditions out to the light-cylinder and by an outgoing vacuum wave beyond, matched by the appropriate surface charge-current distribution. The energy supply for the wave requires current flow between the pulsar and the light-cylinder. As in the earlier proposal for the aligned rotator, the cold electrons carrying the current achieve relativistic energies near the light-cylinder; the consequent inertial and radiation damping forces enable the electrons to drift across the field-lines and so complete their circuits back to the pulsar. It is hypothesized that low-obliquity pulsars are essentially emitters of a plasma-modified low-frequency wave and of gamma-radiation near the light-cylinder. Illustrative models are constructed as perturbations about an analogous approximate model for the aligned case.

141.503 **Relativistic beaming of the pulsar emission.**
H. Ardavan.
Mon. Not. R. Astron. Soc., Vol. 198, 627 - 635 (1982).

It is shown that relativistic beaming is a feature of the radiation from all quasi-steady pulsar magnetospheres, irrespective of whether or not such magnetospheres contain a corotating zone. This stems from the fact that the source distribution in a quasi-steady magnetosphere has a rotating pattern and as such predominantly radiates into a beam whose width in the longitude reflects the azimuthal distribution of the emissivity of the charges and currents in the vicinity of the light cylinder.

141.504 **Crab pulsar infrared fluxes and pulse shapes.**
A. J. Penny.
Mon. Not. R. Astron. Soc., Vol. 198, 773 - 778 (1982).

New *JHKL*, 1 ms time resolution, observations of the Crab pulsar strengthen the evidence for a 1/3 power law fall-off in the spectrum from the visible to the infrared. The shape of the infrared light curve is the same as the visible light curve, except that there is a depression of the peaks of the pulses, relative to the wings, of about 20 ± 10 per cent at H and K. If this depression is due to self-absorption, then the amount of self-absorption is 9 ± 5 per cent. If the sharp fall-off in flux towards shorter wavelengths in the visible continues into the ultraviolet, then the sharpness of the peak in the spectrum implies an excess of emitting particles of the corresponding energy.

141.505 **A transient 77 keV emission feature from the Crab pulsar.**
M. S. Strickman, J. D. Kurfess, W. N. Johnson.
Astrophys. J., Lett., Vol. 253, L23 - L27 (1982).

During a balloon-borne observation of the Crab pulsar, a transient emission line at 77 keV, pulsed at the pulsar frequency, was detected for the first 25 minutes of a 3 hour observation. This line then disappeared, although the continuum remained unchanged. The authors discuss the implications of a cyclotron origin of this feature.

141.506 **High energy electrons in pulsar magnetospheres.**
A. A. da Costa, F. D. Kahn.
Mon. Not. R. Astron. Soc., Vol. 199, 211 - 217 (1982).

Magnetospheres of pulsars contain charged particles. The motion of electrons and positrons in a simplified cylindrical model is discussed for a pulsar with the period of the Crab. It will be shown that this motion near the light cylinder provides a mechanism for the production of incoherent radiation at gamma ray frequencies. Such radiation has been observed in the Crab and the Vela pulsars.

141.507 **Pulsar optical emission as amplified synchrotron emission.** R. J. Stoneham.
Mon. Not. R. Astron. Soc., Vol. 199, 219 - 228 (1982).

Particles streaming relativistically along magnetic field lines through a slower-moving electron-positron plasma can gain finite pitch angles by the two-stream instability. Synchrotron radiation from these particles can be amplified for appropriate particle distributions. A synchrotron laser operating along the open magnetic field lines at a distance of 3.5×10^7 cm from the surface is suggested as the optical emission mechanism for the Crab pulsar.

141.508 **A possible evolutionary way of pulsars.**
G.-j. Qiao, X.-j. Wu, X.-y. Xia.
Acta Astrophys. Sinica, Vol. 2, 35 - 43 (1982). In Chinese.

141.509 **Northern hemisphere pulsar survey: a third radio pulsar in a binary system.** M. Damashek,
P. R. Backus, J. H. Taylor, R. K. Burkhardt.
Astrophys. J., Lett., Vol. 253, L57 - L60 (1982).

The pulsar survey begun by Damashek, Taylor, and Hulse has been completed. A total of 50 pulsars north of declination $+20°$ were detected in the survey, 23 of which were not previously known. Timing observations have revealed that one of the new pulsars, PSR 0655+64, is a member of a binary system with an orbital period of 24^h41^m and very small orbital eccentricity. In this Letter the authors report parameters of the six new pulsars not published previously, as well as details on the orbit and other characteristics of PSR 0655+64.

141.510 **Pair production and pulsar cutoff in magnetized neutron stars with nondipolar magnetic geometry.**
J. J. Barnard, J. Arons.
Astrophys. J., Vol. 254, 713 - 734 (1982).

The authors construct models of the open field zone in pulsars in which they assume space-charge limited, time independent, electron flow from the surface of a conducting magnetized neutron star with dipolar and quadrupolar components to the magnetic field. They find that the radius of curvature of the field can be much less than in a pure dipole field and can be sufficient to account quantitatively for the cutoff line in the P-$\dot{P}$ diagram if systematic departures from the basic assumptions about the outer magnetosphere occur.

141.511 **100 nanosecond time resolution observations of PSR 1133 + 16.** N. Bartel, T. H. Hankins.
Astrophys. J., Lett., Vol. 254, L35 - L39 (1982).

Ultrashort intensity variations with time scales occasionally as short as 2.5 μs and flux densities of 300 - 500 Jy were detected from pulsar PSR 1133 + 16 at 1.72 GHz. The significance of the pulses was revealed in a statistical analysis which the authors consider to be applicable to any exponentially distributed process about a time varying mean. The 100 ns time resolution was achieved by applying a dispersion removal filter before detection, but no significant unresolved pulses have been discovered.

141.512 Mechanism of optical radiation of the pulsar NP 0531−21.
G. Z. Machabeli, D. M. Sakhokiya.
Pis'ma Astron. Zh., Tom 8, 78 - 82 (1982). In Russian. English translation in Soviet Astron. Lett., Vol. 8.

A plasma mechanism of optical radiation of the pulsar NP 0531−21 is considered. It is shown that plasma particles diffusion and effective pitch-angle appear as a result of the interaction of relativistic electron-positron plasma particles with Langmuir waves.

141.513 The role of relativistic electron-positron plasma turbulence in the prepulse formation of the radio spectrum of the pulsar NP 0531 + 21.
M. Eh. Gedalin, G. Z. Machabeli.
Pis'ma Astron. Zh., Tom 8, 153 - 157 (1982). In Russian. English translation in Soviet Astron. Lett., Vol. 8.

The decay of a low-frequency electromagnetic linearly polarized wave into two in a turbulent magnetized electron-positron plasma near a pulsar is investigated. The radiation spectrum is obtained, formed due to the above interaction. The results are used to interpret basic properties of the radio emission prepulse of the NP 0531 + 21 pulsar.

141.514 Galactic distribution, birthrate and evolution of pulsar luminosity.
O. Kh. Gusejnov, F. K. Kasumov, I. M. Yusifov.
Astron. Zh., Tom 59, 51 - 60 (1982). In Russian. English translation in Soviet Astron., Vol. 26, No. 1.

The radial distribution of pulsars in the Galaxy is investigated. It is shown that this distribution has a circular structure with the average width 7 - 8 kpc. The surface density distribution of pulsars inside the circle reaches maximum at $R = 5.2 \pm 0.4$ kpc and then decreases rapidly, showing a small gradient in the direction to the galactic centre. The total number of pulsars with the luminosity $3 \times 10^{26} \leqslant L \leqslant 3 \times 10^{29}$ erg/s is estimated and their birthrate in the Galaxy is obtained. The evolution of pulsar luminosity with the age is investigated.

141.515 γ-quanta capture by magnetic field and pair creation suppression in pulsars.
A. E. Shabad, V. V. Usov.
Nature, Vol. 295, 215 - 217 (1982).

141.516 High-energy γ-ray light curve of PSR0531 + 21.
R. D. Wills, K. Bennett, G. F. Bignami, R. Buccheri, P. A. Caraveo, W. Hermsen, G. Kanbach, J. L. Masnou, H. A. Mayer-Hasselwander, J. A. Paul, B. Sacco.
Nature, Vol. 296, 723 - 726 (1982).

In the first 6 yr of operation COS B has made five observations (each of 30 - 40 days duration) in which PSR0531 + 21 was within 15° of the centre of the field of view. The total data from the five observations contain the first evidence of interpulse emission between the two peaks in the energy range 50 - 3,000 MeV.

141.517 Improved parameters for 67 pulsars from timing observations.
P. R. Backus, J. H. Taylor, M. Damashek.
Astrophys. J., Lett., Vol. 255, L63 - L67 (1982).

The authors report the results of a series of pulse arrival time observations for 67 pulsars, made over the interval from 1978 October to 1981 February. Accurate periods, period derivatives, and positions have been obtained for the majority of the sources. Together with other recent measurements, these results bring the number of known pulsar period derivatives to 292; the statistical implications of this sample of data are briefly discussed.

141.518 Pulsar restlessness as induced by an external source.
M. Cerdonio, S. Vitale.
Lett. Nuovo Cimento, Vol. 32, Ser. 2, 302 - 306 (1981).
Abstr. in Phys. Abstr., Vol. 85, Abstr. 18775 (1982).

141.519 Some constraints on the evolutionary history of the binary pulsar PSR 1913 + 16.
G. Srinivasan, E. P. J. van den Heuvel.
Astron. Astrophys., Vol. 108, 143 - 147 (1982).

The aim of the paper is to examine in the framework of a spiral-in scenario the possible ways in which the binary pulsar PSR 1913 + 16 could have obtained its present characteristics - notably its low surface field strength − and to discuss possible implications of this history for the origin of radio pulsars in general.

141.520 Das Mode-Switching Phänomen in den Pulsaren PSR 0329 + 54 und 1237 + 25.
N. Bartel, D. Morris, W. Sieber, T. H. Hankins.
Mitt. Astron. Ges., Nr. 55, p. 35 (1982). − Abstract.

141.521 Equation of state of the A-e phase of a pulsar crust with account for superstrong magnetic field action.
G. V. Gadiyak, M. S. Obrekht, N. N. Yanenko.
Astrofizika, Tom 17, 765 - 774 (1981). In Russian. English translation in Astrophysics, Vol. 17, No. 4.

The results of an equation of state of matter (iron) calculation in a magnetic field of the order of 10^{12} G is presented for the pressure range $10^3 - 10^8$ a.u. There are a number of anomalous features of the equation of state.

141.522 Spectrum of the coherent radio emission from pulsars.
H. Ardavan.
Mon. Not. R. Astron. Soc., Vol. 199, 667 - 682 (1982).

The power-law form of pulsar spectra is here derived solely from the observational data on micropulses and the quasi-steady time dependence of the magnetosphere. The values that can be assumed by the spectral index, -1, $-5/3$, -3 or $-11/3$, together with the high-frequency break and the low-frequency turnover of the spectrum are all features of the radiation from a randomly fluctuating electric current whose distribution on the large scale has a rotating pattern. The correlation between these and certain other features of a quasi-steady magnetosphere are also pointed out.

141.523 The flux of the Crab pulsar at 74 MHz from 1971 to 1981.
B. J. Rickett, J. H. Seiradakis.
Astrophys. J., Vol. 256, 612 - 616 (1982).

Interplanetary scintillation observations of the Crab pulsar at 74 MHz reveal that the steep decline of the pulsar flux reported for 1971−1975 has ceased and that the flux has remained steady from 1977 to 1981 at about 4.3% of the Crab Nebula flux (85 Jy). The 6 month average level of the "microturbulence" in the solar wind (rms density deviations on a scale of 200 km) has not varied by more than ±20% over 10 years of sunspot cycle).

141.524 Pulsar magnetogyro ratios and pulsar evolution. II. Implications of an expanded sample.
J. F. Woodward.
Astrophys. J., Vol. 256, 617 - 623 (1982).

The evolutionary implications of the (P, γ) diagram for pulsars are briefly discussed. The (P, γ) diagram for 288 pulsars is presented, and it is noted that the chief features present in the (P, γ) diagram for 84 pulsars are still present. A statistical test employing the $(P, \dot{P})$ diagram that enables one to discriminate between the magnetic dipole decay-alignment evolutionary model and the γ enhancement model suggested by the (P, γ) diagram is described. The results of this test favor the γ enhancement evolutionary model.

141.525 **Formation of electron-positron plasma in radio pulsars.** A. I. Tsygan.
Fiz.-tekh. inst. AN SSSR. Prepr., 1981, No. 737, 18 pp. In Russian. – Abstr. in Ref. zh., 51. Astron., 3.51.585 (1982).

141.526 **Star formation function and genetics of pulsars origin.**
O. Kh. Gusejnov, F. K. Kasumov, I. M. Yusifov.
Astron. Zh., Tom 59, 312 - 317 (1982). In Russian. English translation in Soviet Astron., Vol. 26, No. 2.
The function of star formation and the genetics of pulsars origin is discussed. It is shown that the pulsar progenitors are stars of the main sequence with masses $> 5\,M_\odot$ almost for all types of the function of masses discussed in the literature. Pulsars are genetically connected with supernova explosions (mainly of type II). The probability of a "quiet collapse" is extremely small. Thus, the hypothesis of pulsars formation from objects of the extreme flat population of the Galaxy is confirmed on more complete and statistically homogeneous material.

141.527 **Gravitational radiation and the binary pulsar.** V. Trimble.
Nature, Vol. 297, 357 - 358 (1982).

141.528 **Glitches and pinned vorticity in the Crab pulsar.** M. A. Alpar, P. W. Anderson, D. Pines, J. Shaham.
Proc. Natl. Acad. Sci. USA, Vol. 78, 5299 - 5301 (1981). Abstr. in Phys. Abstr., Vol. 85, Abstr. 53307 (1982).

141.529 **Theory of pulsar magnetospheres.** F. C. Michel.
Rev. Mod. Phys., Vol. 54, 1 - 66 (1982). – Abstr. in Phys. Abstr., Vol. 85, Abstr. 53308 (1982).

141.530 **Radio pulsar in the Large Magellanic Cloud.**
IAU Circ., No. 3703 (1982).

141.531 **Radio pulsar in MSH 15-32.**
IAU Circ., No. 3704 (1982).

141.532 **Crab pulsar.**
IAU Circ., No. 3705 (1982).

141.533 **Detection of the regular, "abnormally" directed drift of subpulses of the pulsar PSR 0320 + 39.**
V. A. Izvekova, A. D. Kuz'min, Yu. P. Shitov.
Astron. Zh., Tom 59, 536 - 542 (1982). In Russian. English translation in Soviet Astron., Vol. 26, No. 3.
A regular, highly organized drift of subpulses of the long-period pulsar PSR 0320 + 39 has been detected at the Pushchino Radio Astronomical Observatory of the Lebedev Physical Institute at 102.5 MHz. The drift period is $P_d = 8.5\,P$ and the band spacing is $S_d = 31.7$ ms.

141.534 **JPL pulsar timing observations: spinups in PSR 0525 + 21.** G. S. Downs.
Astrophys. J., Lett., Vol. 257, L67– L70 (1982).
Twelve years of observation at 2.4 GHz reveal two small jumps ΔP in period P for PSR 0525 + 21, where $\Delta P/P \sim -10^{-9}$ and $P \sim 3.745$ s. The short-term behavior following the larger jump suggests an exponential decay. Long-term changes in the period derivative of about 0.03% clearly persist after each jump.

141.535 **Multi-frequency pulsar polarimetry: Arecibo 1400 MHz observations.** D. R. Stinebring, J. M. Cordes, J. M. Rankin, J. M. Weisberg, V. Boriakoff.
Bull. American Astron. Soc., Vol. 13, 850 (1981). – Abstract.

141.536 **Polarization of radio emission from the Vela pulsar.** S. Krishnamohan, G. S. Downs.
Bull. American Astron. Soc., Vol. 13, 850 (1981). – Abstract.

141.537 **Frequency independence of the pulsar main pulse – interpulse separation.**
T. H. Hankins, L. A. Fowler.
Bull. American Astron. Soc., Vol. 13, 850 - 851 (1981). Abstract.

141.538 **On the spectral behavior of component width of pulsar average profiles.** J. M. Rankin.
Bull. American Astron. Soc., Vol. 13, 851 (1981). – Abstract.

141.539 **Pulsar microstructure correlations over 2 GHz.** D. C. Ferguson, V. Boriakoff.
Bull. American Astron. Soc., Vol. 13, 851 (1981). – Abstract.

141.540 **Pulsar disk electrodynamics.** F. C. Michel.
Bull. American Astron. Soc., Vol. 13, 851 (1981). Abstract.

141.541 **Current flow patterns in nonneutral beam models of pulsar magnetospheres.**
D. F. Smith, L. A. Muth, J. Arons.
Bull. American Astron. Soc., Vol. 13, 851 (1981). – Abstract.

141.542 **Wave propagation in the open flux zone of pulsar magnetospheres.** J. J. Barnard, J. Arons.
Bull. American Astron. Soc., Vol. 13, 851 (1981). – Abstract.

141.543 **HEAO C-1 observations of the Crab Nebula pulsar.** W. A. Mahoney, J. C. Ling, G. R. Riegler, A. S. Jacobson.
Bull. American Astron. Soc., Vol. 13, 866 (1981). – Abstract.

141.544 **A search for long term variability in the Crab pulsar.**
M. R. Pelling, G. Jung, R. Schwartz, R. P. Lin, K. Hurley.
Bull. American Astron. Soc., Vol. 13, 882 (1981). – Abstract.

141.545 **Interpretations of the pulsed emission from the Crab pulsar.**
F. K. Knight, J. L. Matteson, L. E. Peterson, R. E. Rothschild, D. E. Gruber.
Bull. American Astron. Soc., Vol. 13, 883 (1981). – Abstract.

141.546 **The origin of high velocity pulsars.** W. Sutantyo.
Contrib. Bosscha Obs. Lembang, No. 70, 15 pp. (1982).
Theories of the origin of high velocity pulsars are critically reviewed. The author argues that if observations confirm the alignment between the velocity vectors and the spin axis, mostly, only neutron stars which originate from single stars can be detected as runaway radio pulsars. Those only form some 1/4 of all neutron stars. On the other hand, if the velocity vectors are proven to be perpendicular to the spin axis, high velocity pulsars may have originated from both, single stars and binary systems, or only from binary systems, depending on the nature of the explosion.

141.547 **On the origin of pulsar 'nulling'.** P. N. Okeke, C. E. Akujor.
Astrophys. Space Sci., Vol. 84, 243 - 246 (1982).
The authors show that, depending on the type of pulsar evolution model adopted, the phenomenon of nulling is found to be either intrinsic to all pulsars, occurring during a particular stage of its life, or may be peculiar to only a class of pulsars.

141.548 **The diffuse gamma-ray background and the pulsar magnetic window.** C. S. Shukre,

V. Radhakrishnan.
Astrophys. J., Vol. 258, 121 - 130 (1982).

The authors have investigated in detail the hypothesis advanced earlier that the spark discharges in the Ruderman and Sutherland model are triggered by diffuse background gamma rays. They find that such triggering can be effective, but only within a narrow range around 2.5×10^{12} gauss of the surface magnetic field. The position of this magnetic window is quite insensitive both to the radius of the neutron star and to the scaling with radius of the polar gap size. A long period cutoff at 4s for pulsars also follows as a result. Assuming that gamma-ray triggering plays a role and that the fields are dipolar, those neutron stars with surface magnetic fields within the window will have a higher probability of functioning as pulsars. This offers a natural explanation for the peaking (around this value) of derived pulsar magnetic fields which has been a puzzle for many years.

141.549   **Pulse-interpulse interaction in pulsar PSR 1822—09.**
L. A. Fowler, G. A. E. Wright.
Astron. Astrophys., Vol. 109, 279 - 281 (1982).

In an analysis of single-pulse data from both the main pulse and the interpulse of the pulsar PSR 1822-09 the authors have exploited the subpulse properties to test ideas concerning the site of the radio emission in the pulsar magnetosphere. The authors find that the interpulse appears to participate in the mode-change, but not in a manner which conforms simply with any current model for pulsar emission.

141.550   **PSR 1133+16: determination of the dispersion measure and the locations of the emitting regions.**
N. S. Kardashev, N. Ya. Nikolaev, A. Yu. Novikov, M. V. Popov, V. A. Soglasnov, A. D. Kuzmin (*Kuz'min*), T. V. Smirnova, N. Bartel, W. Sieber, R. Wielebinski.
Astron. Astrophys., Vol. 109, 340 - 343 (1982).

Clear correlation has been found for the microstructure of PSR 1133+16 observed simultaneously at 67.5 MHz, 78.9 MHz, and 102.55 MHz. The time delay follows the dispersion relation indicating a dispersion measure of $DM = 4.8413 \pm 0.0003$ cm^{-3} pc. This precise determination of dispersion measure makes it possible to measure the retardation and aberration time of pulses emitted simultaneously at 102.55 and 1700 MHz, from which the radii of the emission regions at these frequencies may be computed. The limits of the Lorentz factor and the electron density of the emitting plasma can also be estimated.

141.551   **On the secular variation of the γ-ray emission from PSR 0531 + 21.**   R. D. Wills.
Philos. Trans. R. Soc. London, Ser. A, Vol. 301, (see 012.043), 537 - 539 (1981).

141.552   **The ancestry of pulsars.**   M. N. McMorris.
      Physis, Anno 23, 473 - 484 (1981).

An examination is made of the background to, and the discovery of, pulsars with a view to establishing the continuity of 'pulsar' research over more than three decades: the story is taken up to about 1972.

141.553   **H I en los alrededores de pulsares.**
F. R. Colomb, J. C. Testori.
Bol. Asoc. Argentina Astron., No. 26, p. 201 (1981).
Abstract.

Interaction of charged particles with the field of a rotating magnetic dipole in the presence of electromagnetic radiation.   See Abstr. 022.162.

Pinned vorticity in rotating superfluids, with application to neutron stars.   See Abstr. 062.053.

Damping and excitation of Langmuir waves in an inhomogeneous relativistic plasma; general theory and pulsar applications.   See Abstr. 062.062.

Curvature radiation on longitudinal waves in the magnetosphere of a neutron star.   See Abstr. 063.035.

A new test of general relativity: gravitational radiation and the binary pulsar PSR 1913+16. See Abstr. 066.010.

On the contribution of a stochastic background of gravitational radiation to the timing noise of pulsars. See Abstr. 066.090.

Slow rotation of neutron stars according to the Jordan - Brans - Dicke theory of gravitation. See Abstr. 066.510.

Association of PSR0740 – 28 with an H I shell in Puppis.   See Abstr. 132.007.

Report of IAU Commission 40: Radio astronomy (Radio astronomie).   See Abstr. 141.322.

Transient emission of ultra-high energy pulsed γ rays from Crab pulsar PSR0531.   See Abstr. 142.508.

On the nature of the galactic 2 CG γ-ray sources. See Abstr. 142.528.

Detection of pulsed γ-rays at energies above 300 GeV from pulsars – T.I.F.R. experiments. See Abstr. 142.535.

High energy γ-rays from the direction of the Crab pulsar.   See Abstr. 142.539.

The distribution of free electrons in the inner Galaxy from pulsar dispersion measures. See Abstr. 155.028.

Erratum

141.901   **One millimeter continuum observations of high redshift quasars.**
D. J. Ennis, B. T. Soifer, G. Neugebauer, M. Werner.
Astrophys. Lett., Vol. 22, 13 - 20 (1981). – See Abstr. 30.141.098.

141.902   **Erratum: 'Theoretical quasar emission-line ratios. III. Flux divergence and photon escape'** [Astrophys. J., Vol. 248, 82 - 86 (1981)].   R. C. Canfield, R. C. Puetter, P. J. Ricchiazzi.
Astrophys. J., Vol. 256, 798 (1982). – See Abstr. 30.141.034.

142 UV Sources, X-ray Sources, X-ray Background, Gamma-ray Sources, Gamma-ray Background

UV Sources, X-ray Sources, X-ray Background

142.001 **A study of Sk 160/SMC X-1.** I. D. Howarth.
Mon. Not. R. Astron. Soc., Vol. 198, 289 - 296 (1982).

The extragalactic X-ray binary system Sk 160/SMC X-1 is analysed within the context of constraints provided by spectroscopic and X-ray data. From a light-curve analysis evidence is found for a large, optically thick accretion disc which contributes ~5 per cent of the total luminosity of the system at V. The most likely source for this luminosity appears to be reprocessed X-rays, which also act on the stellar photosphere. The system lies at a distance of 45 ± 5 kpc, outside the main body of the SMC.

142.002 **Line feature around 73 keV from the Crab Nebula.** R. K. Manchanda, A. Bazzano, C. D. La Padula, V. F. Polcaro, P. Ubertini.
Astrophys. J., Vol. 252, 172 - 178 (1982).

An observation of the Crab Nebula was made during a transmediterranean balloon flight launched on 1979 August 26. The data in the 20–150 keV energy region were obtained from a multiwire, high-pressure xenon, proportional counter. The results indicate a line feature around 73 keV in the X-ray spectrum of the Crab Nebula. The continuum emission from the source fits a power law of $16.5E^{-2.3}$, while the line intensity is measured to be $(5 \pm 1.5) \times 10^{-3}$ photons cm^{-2}s^{-1} for an equivalent width of 3 keV. The authors briefly discuss the result in the context of cyclotron emission process as the possible mechanism for the observed line.

142.003 **Large scale X-ray and radio structures associated with compact extragalactic sources.**
P. Biermann, K. Fricke, K. J. Johnston, H. Kühr, I. I. K. Pauliny-Toth, P. A. Strittmatter, M. Urbanik, A. Witzel.
Astrophys. J., Lett., Vol. 252, L1 - L5 (1982).

Knots of X-ray emission have been detected within 20′ of five compact sources initially selected from the MPIfR north polar (S5) 5 GHz survey. Two of the knots have also been detected at centimeter wavelengths and probably have nonthermal spectra. They appear to be associated with the compact sources. While the apparent association may be due to colocation of the sources in a distant supercluster, it is suggested that the association may be similar to that found in extended radio sources.

142.004 **X-ray quasars and the X-ray background.**
A. K. Kembhavi, A. C. Fabian.
Mon. Not. R. Astron. Soc., Vol. 198, 921 - 934 (1982).

The Einstein X-ray observations of a sample of 202 radio- and optically-selected quasars are analysed. Correlations between X-ray, optical and radio luminosities are examined. The contribution of radio-loud quasars to the 2-keV X-ray background is estimated using high-frequency radio-source counts, and the contribution due to radio-quiet, optically bright quasars using optical counts. It is shown that radio-loud quasars and radio-quiet optically bright quasars together contribute ~15 per cent of the observed 2-keV X-ray background. The contribution of optically faint radio-quiet quasars is uncertain.

142.005 **PKS 2155−304: relativistically beamed synchrotron radiation from a BL Lacertae object.**

C. M. Urry, R. F. Mushotzky.
Astrophys. J., Vol. 253, 38 - 46 (1982).

The newly discovered BL Lac object, PKS 2155−304, has been observed with the medium and high energy detectors of the HEAO 1 A-2 experiment. Variability by a factor of 2 in less than a day is confirmed. Two spectra, obtained a year apart, while the satellite was in scanning mode, are well fitted by simple power laws with energy spectral index $\alpha_1 \sim 1.4$. A third spectrum, of higher statistical quality, obtained while the satellite was pointed at this source, has two components, and one gets an acceptable fit using a two power-law model, with indices $\alpha_1 = 2.0$ and $\alpha_2 = -1.5$. An interpretation of the overall spectrum from radio through X rays in terms of a synchrotron self-Compton model gives a good description of the data if one allows for relativistic beaming.

142.006 **On syntheses of the X-ray background with power-law sources.** G. De Zotti, E. A. Boldt, A. Cavaliere, L. Danese, A. Franceschini, F. E. Marshall, J. H. Swank, A. E. Szymkowiak.
Astrophys. J., Vol. 253, 47 - 52 (1982).

The authors discuss if and under what conditions the combined emission from power-law sources can mimic the XRB spectrum in the range 3–50 keV measured with the A-2 experiment on HEAO 1. They confirm that a good fit can be obtained, but the required spectral properties of component sources differ from those observed for local active galactic nuclei. Strong constraints are deduced for the low luminosity extension and for the evolution of such loud objects. It is shown that any other class of sources significantly contributing to the X-ray background must be characterized by an energy spectral index $\gamma \lesssim 0.4$, the mean index of the XRB (3–15 keV), and must exhibit steeper spectra at somewhat higher energies.

142.007 **Einstein detection of X rays from the Alpha Centauri system.** L. Golub, F. R. Harnden, Jr., R. Pallavicini, R. Rosner, G. S. Vaiana.
Astrophys. J., Vol. 253, 242 - 247, plate 6 (1982).

The authors report detection of quiescent X-ray emission from the stellar components of the α Cen system: α Cen A (G2 V) and α Cen B (K1 V). Contrary to previous theoretical expectations (e.g., Mewe), both stars are X-ray emitters and at about the same level: $L_x = 1.2 \times 10^{27}$ and 2.8×10^{27} ergs s^{-1} for A and B, respectively. Comparison with previous chromospheric and transition region measurements indicates that α Cen A and B may have changed in relative strength in recent years. The coronal temperature of the combined Cen AB source, which is dominated ($\sim {}^2/_3$ of the total) by the K star, is $(2.1 \pm 0.4) \times 10^6$ K, similar to that of the average solar corona. The applicability of coronal loop models and the conclusions which can be drawn on the basis of such models are discussed, using the Sun and the α Cen stars as examples.

142.008 **The luminosity of serendipitous X-ray QSOs.**
B. Margon, G. A. Chanan, R. A. Downes.
Astrophys. J., Lett., Vol. 253, L7 - L11 (1982).

The authors have identified the optical counterparts of 47 serendipitously discovered *Einstein Observatory* X-ray sources with previously unreported quasi-stellar objects. The mean ratio of X-ray to optical luminosity of this sample agrees reasonably well with that derived from X-ray observations of previously known QSOs. However, despite the fact that the authors' limiting magnitude $V = 18.5$ should permit detection of typical QSOs to $z = 0.9$, the mean redshift of the sample is only $z = 0.42$. Thus the mean luminosity of these objects, $M_V = -24$, differs significantly from that of previous QSO surveys with similar optical thresholds. Possible explanations of this result and its implications for the luminosity function and space density of QSOs are discussed.

142.009   Spectrophotometry of an X-ray source near M 33.
C. A. Christian, R. A. Schommer.
Astrophys. J., Lett., Vol. 253, L13 - L15 (1982).

The authors have obtained IIDS spectrophotometry of an object identified with the X-ray source near M 33 observed by Long et al. The source has an apparent radial velocity of $10,800 \pm 800 \, \text{km s}^{-1}$, an integrated spectral type similar to a late G or early K giant, and an optical image with noticeably elongated isophotes, suggesting that the source is an elliptical galaxy at $z = 0.03$ rather than a globular cluster.

142.010   Should there be any XUV lines of Her X-1?
Z.-q. Li.
Acta Astrophys. Sinica, Vol. 2, 71 - 73 (1982). In Chinese.

142.011   On the nature of Circinus X-1.
A. N. Argue, C. Sullivan.
Observatory, Vol. 102, 4 - 6 (1982).

142.012   A complete X-ray sample of the high-latitude
($|b| > 20°$) **sky from HEAO 1 A-2: log N–log S**
and luminosity functions.   G. Piccinotti, R. F. Mushotzky,
E. A. Boldt, S. S. Holt, F. E. Marshall, P. J. Serlemitsos,
R. A. Shafer.
Astrophys. J., Vol. 253, 485 - 503 (1982).

The HEAO 1 experiment A-2 has performed a complete X-ray survey of the 8.2 sr of the sky at $|b| > 20°$ down to a limiting sensitivity of $\sim 3.1 \times 10^{-11} \text{ergs cm}^{-2} \text{s}^{-1}$ in the 2–10 keV band. Of the 85 detected sources, 17 have been identified with galactic objects, 61 have been identified with extragalactic objects, and 7 remain unidentified. The log N–log S relation for the nongalactic objects is well fitted by the Euclidean relationship. The authors have used the X-ray spectra of these objects to construct the log N–log S relation in physical units. The complete sample of identified sources has been used to construct X-ray luminosity functions for clusters of galaxies and active galactic nuclei.

142.013   A medium sensitivity X-ray survey using the
Einstein Observatory: the log N–log S relation for
extragalactic X-ray sources.   T. Maccaro, E. D. Feigelson,
M. Fener, R. Giacconi, I. M. Gioia, R. E. Griffiths,
S. S. Murray, G. Zamorani, J. Stocke, J. Liebert.
Astrophys. J. Vol. 253, 504 - 511 (1982).

Results are presented from an X-ray survey of $\sim 50 \, \text{deg}^2$ of the high galactic latitude sky at sensitivities in the range 7×10^{-14} to $5 \times 10^{-12} \text{ergs cm}^{-2} \text{s}^{-1}$ (0.3–3.5 keV) carried out with the IPC aboard the Einstein Observatory. The complete sample consists of 63 sources detected at or above the 5 σ level, 48 of which are certainly or most likely of extragalactic origin. The number-flux relation is derived for the extragalactic population. The content of the sample is analyzed in terms of types of sources and is found to be significantly different from the content of similar samples selected at higher fluxes. In particular, the Einstein Observatory medium sensitivity sample of extragalactic sources is dominated by active galactic nuclei (AGNs–quasars or Seyfert galaxies), while samples selected at higher fluxes and higher energies are dominated by clusters of galaxies. Thus the number-flux relation for extragalactic sources may be interpreted, to a first approximation, as the sum of two different distributions with flatter and steeper slopes describing clusters and AGNs, respectively.

142.014   An unbiased survey of field star X-ray emission.
D. J. Helfand, J.-P. Caillault.
Astrophys. J., Vol. 253, 760 - 767 (1982).

To establish the prevalence of X-ray emission among field stars, the authors have conducted a systematic search of over 1700 stars brighter than 10th magnitude which fell within fields observed by the Einstein Observatory imaging proportional counter. Fewer than 4% of the stars were detected at a limiting flux value of $\sim 10^{-13} \text{ergs cm}^{-2} \text{s}^{-1}$. The authors use the observed distribution of X-ray to optical flux ratios for both the detected and undetected stars to limit the X-ray luminosity function for field stars of all spectral types. They find that, while stars are the dominant population of soft X-ray sources observed by Einstein, they probably do not contribute significantly to the diffuse soft X-ray background.

142.015   The discovery of 50 minute periodic absorption
events from 4U 1915–05.
N. E. White, J. H. Swank.
Astrophys. J., Lett., Vol. 253, L61 - L66 (1982).

The authors demonstrate that the steady flux from 4U 1915–05 undergoes periodic absorption dips with a period of 50 minutes. This period most likely represents the underlying orbital period of the system. Variations in the depth and duration of these events suggest that they are caused by a bulge on the edge of the accretion disk, at the point where the gas stream impacts the disk. The mass-losing star in this system is probably a low-mass white dwarf. The spectrum of the dips indicates that the metallicity of the absorbing material is at least a factor of 17 below solar values.

142.016   Discovery of a 50 minute binary period and a likely
22 magnitude optical counterpart for the X-ray
burster 4U 1915–05.   F. M. Walter, S. Bowyer,
K. O. Mason, J. T. Clarke, J. P. Henry, J. Halpern,
J. E. Grindlay.
Astrophys. J., Lett., Vol. 253, L67 - L71, plate L5 (1982).

The authors have observed absorption dips which recur with a period of 2985 s from the X-ray burst source 4U 1915–05 (= MXB 1916–05). They suggest that the dips are caused by obscuration of the X-ray source by material at the point where the gas stream from the companion meets the accretion disk, and that the 2985 s periodicity is the orbital period of the binary system. Assuming this model is appropriate, these observations represent the first direct evidence of the binary nature of X-ray burst sources. The authors suggest a new optical identification of 4U 1915–05: a 22d magnitude candidate observed in a CCD image of the optical field at the arcsecond-accurate HRI X-ray position.

142.017   Balloon observations of galactic high-energy X-ray
sources.   G. S. Maurer, W. N. Johnson,
J. D. Kurfess, M. S. Strickman.
Astrophys. J., Vol. 254, 271 - 278 (1982).

A number of X-ray sources were observed with the Naval Research Laboratory hard X-ray observatory during balloon flights on 1976 May 11 and 1977 November 24. The primary objective of the observations was to characterize the temporal and spectral behaviour of the sources in the energy range from 20 to 250 keV. Results of the observations include: evidence for possible phase-dependent spectral variations in the emission from GX 1 + 4; flaring in the emission from GX 301–2; detection of unmodulated X-ray emission up to ~ 35 keV from GX 304–1; and an unexpectedly hard spectrum above 20 keV from Cyg X-2.

142.018   IUE observations of the X-ray burst source
4U/MXB 1735 - 44.
G. Hammerschlag-Hensberge, J. E. McClintock, J. van Paradijs.
Astrophys. J., Lett., Vol. 254, L1 - L5 (1982).

The authors have observed the ultraviolet (1250 - 1950 Å) energy distribution of the X-ray burst source 4U/MXB 1735 - 44 with the IUE. The source was detected with an average flux $1.1 \times 10^{-15} \text{ergs cm}^{-2} \text{s}^{-1} \text{Å}^{-1}$. Assuming that the optical and ultraviolet emission of MXB 1735 - 44 is due to reprocessing of X-rays in an accretion disk, the ratio of X-ray to bolometric (optical + ultraviolet) flux has been used to make an estimate of the thickness of the disk. For an assumed X-ray albedo of 0.5, the authors find that the disk extends to less than $\sim$ 6° from the orbital plane.

142.019 Optical polarization observations of the X-ray transient A0538 - 66.
G. C. Clayton, I. Thompson.
Astrophys. J., Lett., Vol. 254, L7 - L9 (1982).

The authors present unfiltered optical linear polarization observations of the X-ray transient A0538 - 66 during the period 1981 March 5 - 14. An outburst was observed, with $\Delta m \sim 1.5$ mag, peaking on March 10 (UT). The preoutburst polarization was only marginally detected at $0.28 \pm 0.11\%$, but increased sharply to $1.8 \pm 0.1\%$ coincident with the optical outburst.

142.020 X-ray observations of the 1980 Cygnus X-1 'high state'.   Y. Ogawara, K. Mitsuda, K. Masai,
J. V. Vallerga, L. R. Cominsky, J. M. Grunsfeld, J. S. Kruper, G. R. Ricker.
Nature, Vol. 295, 675 - 676 (1982).

Observations of intensity transitions in the X-ray emission from Cygnus X-1 were made by the Hakucho satellite during the summer of 1980 and are reported here. In addition, the authors report the results of hard (20–120 keV) X-ray observations of Cyg X-1. These data were obtained during the 'high' intensity state by balloon-borne mercuric iodide detectors, and are compared with data obtained at similar energies approximately 1 month earlier during the 'low' intensity state.

142.021 A Compton-cooled feedback mechanism for Cygnus X-1.
P. W. Guilbert, A. C. Fabian.
Nature, Vol. 296, 226 - 228 (1982).

The authors consider the effect of relaxing the assumption of constant temperature because a Compton-cooled gas, in which the gas temperature drops in response to the comptonization of soft photons, can readily produce a two-state time-averaged spectrum from variations in the soft flux. They speculate on a feedback mechanism that may effectively lock the source in either of the two states.

142.022 Cygnus X-3 observed at photon energies above 500 GeV.
R. C. Lamb, C. P. Godfrey, W. A. Wheaton, T. Tümer.
Nature, Vol. 296, 543 - 544 (1982).

The authors have used the twin 11-m diameter mirrors of the NASA Jet Propulsion Laboratory's solar energy facility to observe high energy γ rays from Cyg X-3 using the atmospheric Cerenkov technique. They report data from $\sim 10^5$ air shower events obtained 29 August - 6 September 1981 with an approximate threshold energy of 500 GeV.

142.023 Hakucho observations of X-ray bursts from 4U 1702–42.   K. Makishima, H. Inoue,
K. Koyama, M. Matsuoka, T. Murakami, M. Oda, Y. Ogawara, T. Ohashi, N. Shibazaki, Y. Tanaka, S. Hayakawa, H. Kunieda, F. Makino, K. Masai, F. Nagase, Y. Tawara, S. Miyamoto, H. Tsunemi, K. Yamashita, I. Kondo.
Astrophys. J., Lett., Vol. 255, L49 - L53 (1982).

X-ray bursts were observed with the Hakucho satellite from a source designated XB 1702–429, which coincides in position with a medium intensity Uhuru source 4U/2S 1702–429 within an accuracy of about 0°.2. In 1979 this burst source was burst active. Its activity was much lower in 1980. Brief descriptions are given of the properties of bursts and persistent X-ray emissions from this source. Its relation

to previous X-ray burst observations in the same sky region is discussed.

142.024 The low mass X-ray binary 2A 1822–371.
A. P. Cowley, D. Crampton, J. B. Hutchings.
Astrophys. J., Vol. 255, 596 - 602 (1982).

The optical counterpart of the X-ray source 2A 1822–371 has been followed spectroscopically through its 5.6 hr period. Velocity variations of a weak, rather broad He II λ4686 emission show that it must be formed primarily in a disk surrounding the degenerate star. The line profile variation additionally suggests the presence of an emission hot spot or stream, reminiscent of those observed in cataclysmic variables. The orbital period is further refined to be 0.232108 days, with the X-ray and optical eclipses probably being coincident, contrary to earlier claims. The primary appears to be a neutron star, probably of about 1 solar mass. The secondary has a mass of about 0.3 $M_\odot$, but its radius appears to be about twice as large as that of a main-sequence star of that mass.

142.025 Ultraviolet spectrophotometry of 2A 1822–371: a bulge on the accretion disk.
K. O. Mason, F. A. Córdova.
Astrophys. J., Vol. 255, 603 - 609 (1982).

The X-ray source 2A 1822–371 has been observed with the IUE satellite over an 8 hour period. The data provide evidence that the shape of the 5.57 hr modulation evolves smoothly with energy between extremes defined by the optical and X-ray curves. The far-UV light curve is more deeply modulated than the X-ray light curve. The combined ultraviolet and the UBV band optical data can be fitted with a single blackbody of temperature 2.7×10^4 K, or an optically thick disk model with parameters $T_* = 1.2 \times 10^5$ K and $R_{out}/R_{in} \sim 30$. Emission lines of C IV 1550 Å and N V 1240 Å are detected in the UV spectrum. In the present paper, the 5.57 hr optical, X-ray and ultraviolet modulation of 2A 1822–371 is interpreted as the result of partial occultation of the emitting region by a companion star and a bulge on the outer accretion disk. X-ray heating of the bulge will probably also contribute to the modulation at optical and ultraviolet wavelengths.

142.026 An auroral precipitation model for the rapid X-ray burster.   G. T. Davidson.
Astrophys. J., Vol. 255, 705 - 715 (1982).

It is proposed that the rapid X-ray burster may be a compact object with a slowly rotating magnetosphere. Because the outer regions of slowly rotating magnetospheres are subject mainly to electric and magnetic forces, magnetospheric substorms and other types of unstable behavior are expected to occur frequently and the X-ray burster should exhibit phenomena analogous to auroral activity in the earth's magnetosphere. Repeated bursts of particle precipitation from a magnetospheric trapping region are expected when particle injection in a substorm results in saturation of the trapping region with more matter than it can contain in equilibrium. Saturation leads to a rapid growth of plasma waves which are turned off when particles are removed by precipitation onto the neutron star. A semiregular pattern of radiation burst may result when the precipitating particles are stopped at the surface of the star.

142.027 Hard X-ray emission (15–150 keV) from the region of 4U0515+38.
P. Ubertini, A. Bazzano, C. D. La Padula, V. F. Polcaro.
Astron. Astrophys., Vol. 106, 174 - 175 (1982).

The authors report the observation of a new hard X-ray source located at R. A. $5^h 16^m$ and declination 38 deg, emitting in the range 15–150 keV.

142.028 A study of ultraviolet spectroscopic and light variations in the X-ray binaries LMC X-4 and

SMC X-1.   M. van der Klis, G. Hammerschlag-Hensberge,
J. M. Bonnet-Bidaud, S. A. Ilovaisky, M. Mouchet,
W. M. Glencross, A. J. Willis, J. A. van Paradijs,
E. J. Zuiderwijk, C. Chevalier.
Astron. Astrophys., Vol. 106, 339 - 344 (1982).

Low-resolution IUE spectra of the massive X-ray binaries
LMC X-4 and SMC X-1 have been obtained between 1200 and
3200 Å at several epochs in August 1979 and April 1980.
Comparison of these observations with theoretical light curves
for an X-ray heated tidally distorted star indicate the existence of a disk around the compact star. The resonance-line
doublets of N V, C IV and Si IV show marked changes with
orbital phase, being particularly weak when the X-ray source
is in front of the primary. These changes can be understood in
terms of an anisotropic ionization structure in the expanding
atmosphere of the primary caused by the presence of the
X-ray source companion.

**142.029   HEAO 1 observations of the long-term variability of
Hercules X-1.**
A. Gorecki, A. Levine, M. Bautz, F. Lang, F. A. Primini,
W. H. G. Lewin, W. A. Baity, D. E. Gruber, R. E. Rothschild.
Astrophys. J., Vol. 256, 234 - 237 (1982).

High energy X-ray (13–80 keV) observations of Her X-1
covering two complete 35^d cycles in 1978 are reported. Three
high ON states and two low ON states of the X-ray intensity
were observed. During the time interval of an expected third
low ON state the source was not detected at the strength of
the other two low ON states. The results are interpreted in the
context of precessing tilted accretion disk–periodic mass
transfer models.

142.030   The hard X-ray spectrum of Cygnus X-1.
H. Steinle, W. Voges, W. Pietsch, C. Reppin,
J. Trümper, E. Kendziorra, R. Staubert.
Astron. Astrophys., Vol. 107, 350 - 353 (1982).

The hard X-ray source and black-hole candidate
Cygnus X-1 was observed in the high-energy X-ray range
(15−160 keV) by balloon borne scintillation detectors in
1975, 1976, and 1977. During all three observations the
source was in the "low-state" in the low energy range. The
spectra coincide within the error limits above 30 keV. The
existence of a spectral break near 80 keV is firmly
established. The data are well fitted by a spectrum expected
from comptonization of soft photons in a hot cloud having
an electron temperature of ~30 keV and a Thomson scattering optical length of τ ~5.

142.031   Mass determination of massive X-ray binaries.
C. de Loore, M. Mouchet, E. L. van Dessel,
M. Burger.
Messenger, No. 27, p. 14 - 17 (1982).

**142.032   X-ray bursts and shell flashes on accreting neutron
stars.**   S. Miyaji.
Astron. Her., Vol. 74, No. 2, p. 36 - 41 (1981). In Japanese.
Abstr. in Phys. Abstr., Vol. 85, Abstr. 18776 (1982).

142.033   The rapid burster.   Y. Tawara.
Astron. Her., Vol. 74, No. 1, p. 4 - 9 (1981). In
Japanese. – Abstr. in Phys. Abstr., Vol. 85, Abstr. 18802
(1982).

142.034   The pulse-period distribution of binary X-ray pulsars.
H. C. Bhatt.
Astrophys. Space Sci., Vol. 81, 379 - 385 (1982).

The pulse-period distribution of binary X-ray pulsars has
been considered. A gap in this distribution, in the period range
P ~ 10 s to P ~ 100 s has been explained in terms of the
character of mass transfer in the X-ray binary systems. It is
shown that this gap arises because the rotating magnetised
neutron stars in these systems are slowed down by accretion

torques, either to $P \lesssim 10$ s when the mass transfer is by means
of Roche-lobe overflow in low mass binaries, or to
$P \gtrsim 100$ s by stellar winds in massive binaries.

**142.035   On the optical counterparts of the X-ray sources in
the Orion nebula.**   L. Bianchi.
Astrophys. Space Sci., Vol. 82, 161 - 166 (1982).

A great step forward in the knowledge of the X-ray emission from the Orion nebula has recently been made possible
by HEAO-2 observations. On the basis of these new data, and
with the purpose of suggesting further analysis and observations, this note aims to contribute to on-going discussions
about plausible optical counterparts, by considerations on
those observational traits which have not yet been taken into
account.

142.036   Coherent oscillations from X-ray burst sources.
M. Livio.
Astrophys. Space Sci., Vol. 82, 167 - 172 (1982).

The possible origin of coherent oscillations in X-ray
bursters is discussed. Such oscillations with a period of the
order of 10 ms have recently been observed. Nevertheless, it
is shown that nonradial pulsations of a neutron star or
torsional oscillations of a neutron star's crust seem to be an
attractive possibility.

**142.037   Observation of hard X-ray line emission from
Her X-1.**   V. F. Polcaro, A. Bazzano,
C. La Padula, P. Ubertini, G. Vialetto, R. K. Manchanda,
S. V. Damle.
Astron. Astrophys., Vol. 108, 249 - 250 (1982).

The authors present the results of a hard X-ray measurement of the binary source Her X-1, carried out with a balloon
borne X-ray telescope. The source was observed during the
"Mid-on" state. The data confirm the previously reported
high energy emission line overimposed on the low energy
thermal spectrum.

**142.038   Einstein observations of BY Draconis variables and
RS CVn binaries.**   J.-P. Caillault.
Astron. J., Vol. 87, 558 - 562 (1982).

Observations of five BY Draconis variables and four
RS Canum Venaticorum systems with the Einstein observatory were presented. All nine stars were detected. Although the
BY Draconis stars have an average X-ray luminosity one to
two orders of magnitude lower than the RS CVn binaries, at
least three of these systems fit the L_x/L_{bol} -vs-period relation
first described by Walter and Bowyer for RS CVn's. Since one
of the BY Draconis stars is almost certainly single, this result
lends strong support to the suggestion that rapid rotation, and
not duplicity, is the determining factor in defining the level of
stellar activity in such stars. In addition, the X-ray data are
consistent with the suggestion that the BY Draconis stars are
in the final stages of pre-main-sequence evolution.

**142.039   X-ray emission from the Carina nebula and the
associated early stars.**
F. D. Seward, T. Chlebowski.
Astrophys. J., Vol. 256, 530 - 542 (1982).

New Einstein observations of the Carina nebula show the
extent of the diffuse X-ray nebula to be 1°.1 E-W by 0°.7 N-S.
The diffuse X-ray emission is generally associated with
optically bright regions, but the correlation is not exact. Emission is strongest from regions bordering the apparently overlying dust lane indicating an interaction between the energetic
material of the optical nebula and the region containing the
dust. The diffuse X-rays are from hot gas with T ~ 10^7 K and
n_e 0.1–1 cm^{-3}. The energy to heat this gas probably comes
from strong stellar winds of the early stars embedded in the
nebula. X-rays from 15 O and W-R stars in the Carina OB1
association have been detected. Data are consistent with the
hypothesis that $L_x \approx 2.0 \times 10^{-7} L_{bol}$ for all O stars in this

region with remarkably little scatter. X-ray emission from the three W-R stars observed is significantly different.

142.040 X-ray luminosities of B supergiants estimated from ultraviolet resonance lines.
N. Odegard, J. P. Cassinelli.
Astrophys. J., Vol. 256, 568 - 577 (1982).

Superionization in the winds of O and B stars has been explained as a result of Auger ionization by X-rays produced in coronal zones in their outer atmospheres. A recent survey of supergiants by the Einstein Observatory detected X-ray emission from stars as late as B2 but obtained only upper limits for later B supergiants. These stars show strong lines of high stages of ionization such as C IV and Si IV in the ultra-violet. The authors use the observed strengths of these lines to estimate X-ray luminosities for a group of 20 B supergiants. The C IV $\lambda 1548$ line is a useful X-ray diagnostic for spectral classes from B 1.5 to B6, and Si IV $\lambda 1394$ and Si IV $\lambda 1403$ are found to be useful from B3 to B9. It is found that L_x/L_{bol} decreases from $10^{-7.2}$ observed for B0 and B1 stars to roughly $10^{-8.5}$ for late B supergiants.

142.041 A new, fast X-ray pulsar in the supernova remnant MSH 15 - 52.   F. D. Seward, F. R. Harnden, Jr.
Astrophys. J., Lett., Vol. 256, L45 - L47 (1982).

A pulsing X-ray source has been discovered within the shell of the supernova remnant MSH 15 - 52. The period is 0.150 s, and the rate of increase of period with time is the highest measured for any pulsar. These characteristics and the fact that the pulsar is surrounded by a small, bright nebula indicate that this object is very similar to the Crab pulsar.

142.042 Spectroscopic elements of the orbits of X-ray binaries.   A. A. Aslanov, A. M. Cherepashchuk.
Astron. Zh., Tom 59, 290 - 301 (1982). In Russian. English translation in Soviet Astron., Vol. 26, No. 2.

The analysis of all published spectroscopic data for five X-ray binaries with O−B supergiants has been carried out. The spectroscopic elements of orbits for the systems Cyg X-1, 3U 1700 - 37, 3U 0900 - 40, SMC X-1, Cen X-3 are obtained and an estimation of the vicinity effects (ellipticity, reflection etc.) is made.

142.043 X-ray emission from Centaurus A.
J. Terrell.
Extragalactic radio sources, (see 012.025), p. 117 - 118 (1982).

Observations of 3−12 keV X-ray emission from NGC 5128 (Cen A) were made by Vela spacecraft over the period 1969−1979. These data are in good agreement with previously reported data, but are much more complete. Numerous peaks of X-ray intensity occurred during the period 1973−1975, characterized by rapid increases and equally rapid decreases (in less than 10 days). Thus it seems probable that most of the X-ray flux from the nucleus of Cen A came from a single source of small size.

142.044 X-ray emission from BL Lac objects: comparison to the synchrotron self-Compton models.
D. A. Schwartz, G. Madejski, W. H.-M. Ku.
Extragalactic radio sources, (see 012.025), p. 383 - 384 (1982).

142.045 The hard X-ray flux from γ Cassiopeiae during 1970−73.   G. J. Peters.
Publ. Astron. Soc. Pacific, Vol. 94, 157 - 161 (1982).

X-ray observations of the Be star γ Cas obtained with the *Uhuru* satellite from 1970 December 27 to 1973 February 15 are presented. Although variable, the X-ray source (MX0053+60) persisted throughout the lifetime of the *Uhuru* satellite with an average daily flux (2−6 keV) of 4.1 cts s^{-1} or 4.0 X 10^{32} ergs s^{-1} if the star is located at a distance of 220 pc.

Variations on time scales of months, days, and hours were observed but no pulsation period was found.

142.046 Evidence for coherent emission with a 12 milli-second period during a burst from MXB 1728−34.
D. Sadeh, E. T. Byram, T. A. Chubb, H. Friedman, R. L. Hedler, J. F. Meekins, K. S. Wood, D. J. Yentis.
Astrophys. J., Vol. 257, 214 - 224 (1982).

Four bursts of MXB 1728−34 have been observed with the HEAO A-1 instrument using 5 ms timing resolution. The rapid initial rise has been resolved in all four bursts. Fourier analysis indicates the presence of a periodic component with a 12.2 ms period during one of the bursts. Epoch-folding analysis confirms this period and further indicates that the period decreases with time, with $dP/dt = -10^{-6}$. Attempts to explain the 12 ms period as a neutron star spin period meet with severe difficulties. The observed periodic emission is most likely associated with the Kepler orbital period in a viscous accretion disk surrounding a black hole of $\sim 10\ M_\odot$.

142.047 Accretion disk coronae.
N. E. White. S. S. Holt.
Astrophys. J., Vol. 257, 318 - 337 (1982).

Recent observations of partial X-ray eclipses from 4U 1822−37 have shown that the central X-ray source in this system is diffused by a large Compton-thick accretion disk corona (ADC). The authors show that another binary, 4U 2129 + 47, also displays a partial eclipse and contains an ADC. The possible origin of an ADC is discussed and a simple hydrostatic evaporated ADC model is developed which, when applied to 4U 1822−37, 4U 2129+47, and Cyg X-3, can explain their temporal and spectral properties. The quasi-sinusoidal modulation of all three sources can be reconciled with the partial occultation of the ADC by a bulge at the edge of the accretion disk which is caused by the inflowing material A consequence of the model is that any accreting neutron star X-ray source in a semidetached binary system which is close to its Eddington limit is likely to contain an optically thick ADC.

142.048 *UBVRI* photometry of the optical counterparts of X-ray sources in Einstein deep survey fields.
W. Liller, G. Alcaino.
Astrophys. J., Lett., Vol. 257, L27 - L31 (1982), plates L1 - L3.

The authors have carried out five-color photometry of the brighter optical counterparts of X-ray sources in the Einstein Deep Survey Fields in Cetus, Eridanus, and Pavo. Most of the observed objects appear to be stars showing no signs of variability; exceptions include one star that varies in brightness over a small range and at least one variable quasar with a redshift $z = 1.96$.

142.049 Discovery of fast optical activity in the X-ray source GX 339-4.
C. Motch, S. A. Ilovaisky, C. Chevalier.
Astron. Astrophys., Vol. 109, L1 - L4 (1982).

The authors present here the first results of fast optical photometry of the GX 339-4 counterpart obtained in late May 1981 when the object was found in an unusually bright state (V = 15.4). The optical flux displays 20 second quasi-oscillations of 30−40% full amplitude together with very short time scale activity. The authors observe flares as short as 10−20 milliseconds during which the flux can be increased by a factor up to 5. The overall optical time behaviour is shown to be very similar to the X-ray time behaviour of Cyg X-1. Optical flares are likely to come from the hot inner part of the accretion disc. Possible interpretations of the 20 second quasi-oscillations are discussed.

142.050 X-ray observations of Be stars.
S. Rappaport, E. P. J. van den Heuvel.
Be stars, (see 012.030), p. 327 - 346 (1982).

Be star binaries with neutron star companions are shown to constitute a major class of X-ray sources. Some general observational and interpretive techniques of X-ray astronomy are reviewed. Data for 12 Be/X-ray binary systems are summarized.

142.051 Be components in X-ray binaries.
C. de Loore, M. Burger, E. L. van Dessel, M. Mouchet.
Be stars, (see 012.030), p. 347 - 351 (1982).

142.052 Are classical Be stars sources of hard X-rays?
G. J. Peters.
Be stars, (see 012.030), p. 353 - 357 (1982).
A point summation technique was used to search the UHURU data base for X-ray emission from classical Be stars. Of the thirty-two stars considered, only three (γ Cas, HR 4009, and HD 187399) were detected at the 3.3σ level or higher.

142.053 New minimum of X-ray source KR Aurigae.
V. Popov.
Inf. Bull. Variable Stars, No. 2095 (1982).

142.054 Optical counterpart of H0139-68.
IAU Circ., No. 3658 (1982).

142.055 2S 1553-542.
IAU Circ., No. 3667 (1982).

142.056 A0538-66.
IAU Circ., No. 3671 (1982).

142.057 E1405-451 and E1013-477.
IAU Circ., Nos. 3684, 3685 (1982).

142.058 3A 0729+103.
IAU Circ., No. 3687 (1982).

142.059 GX 301-2 = 4U 1223-62.
IAU Circ., Nos. 3688, 3703 (1982).

142.060 Soft X-ray transient.
IAU Circ., No. 3695 (1982).

142.061 1E 2259+586.
IAU Circ., No. 3701 (1982).

142.062 Forced precession of the optical components in X-ray binary systems.   E. A. Karitskaya.
Astron. Tsirk., No. 1186, p. 1 - 5 (1981). In Russian.

142.063 Optical eclipses and the precession effects in the X-ray binary system HD 77581 = 4U 0900-40.
T. S. Khruzina, A. M. Cherepashchuk.
Astron. Zh., Tom 59, 512 - 522 (1982). In Russian. English translation in Soviet Astron., Vol. 26, No. 3.
Long-period variability of the amplitude and shape of the optical light curves of the binary HD 77581 has been discovered from the analysis of all published photometric data. The 93.3-day period is presumably the period of the forced precession of the rotational axis of the optical star. It is shown that the system HD 77581 appears to be an eclipsing binary in the optical range with the amplitude of the ellipticity effect $\sim 0^m.04$ and the depth of the eclipse reaching $\sim 0^m.04$.

142.064 The nuclear X-ray source in M81.
M. Elvis, L. Van Speybroeck.
Astrophys. J., Lett., Vol. 257, L51 - L55 (1982).
The Einstein Observatory High Resolution Imager has detected a pointlike X-ray source coincident with the radio and optical nucleus of the nearby galaxy M81 (NGC 3031). It has a luminosity of $\sim 1.7 \times 10^{40}$ ergs s^{-1} (0.5 - 4.5 keV). The au-

thors discuss the evidence that this is a low-luminosity Seyfert 1 galaxy from comparisons of X-ray, optical, and radio data and conclude that this is very likely the case. If this is so, it implies that the active galaxy X-ray luminosity function extends two orders of magnitude below its previous observed limit.

142.065 Stellar winds in X-ray binaries.   D. B. Friend.
Bull. American Astron. Soc., Vol. 13, 785 (1981).
Abstract.

142.066 A flux limited X-ray survey of the galactic plane.
P. Hertz, J. E. Grindlay.
Bull. American Astron. Soc., Vol. 13, 787 (1981). – Abstract.

142.067 Map of the soft X-ray sky from SAS-3 observations.
F. J. Marshall.
Bull. American Astron. Soc., Vol. 13, 788 (1981). – Abstract.

142.068 Luminosity indicators in X-ray selected QSOs.
R. A. Downes, G. A. Chanan, S. F. Anderson, B. Margon.
Bull. American Astron. Soc., Vol. 13, 799 (1981). – Abstract.

142.069 Active galaxy contribution to the diffuse X-ray background.
R. E. Rothschild, D. E. Gruber, J. L. Matteson, W. A. Baity, R. F. Mushotzky, F. Primini.
Bull. American Astron. Soc., Vol. 13, 800 (1981). – Abstract.

142.070 A model for the quiescent radio emission from Cyg X-3.   W. T. Vestrand.
Bull. American Astron. Soc., Vol. 13, 818 (1981). – Abstract.

142.071 The pulsation spectrum of H2252-035 from 2.2 to 0.36 μ.   F. A. Cordova, K. O. Mason, K. Horne.
Bull. American Astron. Soc., Vol. 13, 818 (1981). – Abstract.

142.072 Discovery of two X-ray emitting Be stars with the HEAO-1 scanning modulation collimator.
D. A. Schwartz, A. Ferrara, M. Garcia, J. Patterson, J. Steiner, R. E. Doxsey, M. D. Johnston, J. McClintock, R. Remillard, I. McHardy, J. P. Pye, R. S. Warwick, M. Watson.
Bull. American Astron. Soc., Vol. 13, 834 (1981). – Abstract.

142.073 Optical spectroscopy of the X-ray source H2252-035.
J. T. Clarke, K. O. Mason, S. Bowyer.
Bull. American Astron. Soc., Vol. 13, 834 (1981). – Abstract.

142.074 The luminosity of serendipitous X-ray QSOs: implications for the diffuse X-ray background radiation.   B. Margon.
Bull. American Astron. Soc., Vol. 13, 847 (1981). – Abstract.

142.075 Deep X-ray surveys of two quasar grism fields.
G. A. Kriss, C. R. Canizares.
Bull. American Astron. Soc., Vol. 13, 848 (1981). – Abstract.

142.076 Discovery of 13.5 s X-ray pulsations from LMC X-4.
R. L. Kelley, J. G. Jernigan, A. Levine, L. D. Petro, S. Rappaport.
Bull. American Astron. Soc., Vol. 13, 866 (1981). – Abstract.

142.077 A study of pulsing X-ray sources.
S. Naranan, W. Darbro, R. F. Elsner, D. Leahy, M. C. Weisskopf, P. G. Sutherland, J. E. Grindlay, S. M. Kahn.
Bull. American Astron. Soc., Vol. 13, 866 (1981). – Abstract.

142.078 Search for periodic pulsation in four globular cluster X-ray sources.
D. Leahy, W. Darbro, R. F. Elsner, S. Naranan, M. C. Weiss-

kopf, P. G. Sutherland, P. Ghosh, J. E. Grindlay, S. M. Kahn.
Bull. American Astron. Soc., Vol. 13, 866 (1981). – Abstract.

142.079 **Microsecond time resolution observations of
Cygnus X-1.**
M. C. Weisskopf, W. A. Darbro, R. F. Elsner, D. Leahy,
S. Naranan, P. G. Sutherland, J. E. Grindlay, S. M. Kahn.
Bull. American Astron. Soc., Vol. 13, 866 - 867 (1981).
Abstract.

142.080 **Observation of high energy gamma rays from
Cygnus X-3.**
C. P. Godfrey, R. C. Lamb, W. A. Wheaton, T. Tümer.
Bull. American Astron. Soc., Vol. 13, 867 (1981). – Abstract.

142.081 **A spatial model for the soft X-ray background.**
D. N. Burrows, D. McCammon, W. T. Sanders,
W. L. Kraushaar.
Bull. American Astron. Soc., Vol. 13, 882 (1981). – Abstract.

142.082 **A Compton-cooled model for Cygnus X-1.**
P. W. Guilbert, A. C. Fabian.
Bull. American Astron. Soc., Vol. 13, 883 (1981). – Abstract.

142.083 **High energy gamma ray observations of Cygnus X-3.**
T. C. Weekes, S. Danaher, D. Fegan, N. A. Porter.
Bull. American Astron. Soc., Vol. 13, 896 - 897 (1981).
Abstract.

142.084 **The 1980 optical outburst of 4U0115 + 63.**
L. Cominsky, G. Kriss, S. Rappaport, R. Remillard,
G. Williams, J. Thorstensen.
Bull. American Astron. Soc., Vol. 13, 900 (1981). – Abstract.

142.085 **Observations of the transient Cen X-4 during
quiescence.**
L. Petro, C. Canizares, G. Kriss, J. McClintock, R. Remillard.
Bull. American Astron. Soc., Vol. 13, 900 (1981). – Abstract.

142.086 **X-ray burst source with degenerate companion.**
J. H. Swank, N. E. White.
Bull. American Astron. Soc., Vol. 13, 901 (1981). – Abstract.

142.087 **Discovery of a 50 minute binary period and a
likely 22^m optical counterpart for the X-ray burster**
4U1915 – 05.   S. Bowyer, F. M. Walter, K. O. Mason,
J. T. Clarke, J. P. Henry, J. Halpern, J. Grindlay.
Bull. American Astron. Soc., Vol. 13, 901 (1981). – Abstract.

142.088 **A fast, bright, high galactic latitude X-ray transient.**
C. Ambruster, K. Wood, J. Meekins, D. Yentis,
H. Smathers, E. Byram, T. Chubb, H. Friedman.
Bull. American Astron. Soc., Vol. 13, 901 (1981). – Abstract.

142.089 **Radiation transport in accretion column of binary
X-ray sources.**   R. Lieu.
Bull. American Astron. Soc., Vol. 13, 901 - 902 (1981).
Abstract.

142.090 **Population II X-ray sources.**   A. Finzi.
Bull. American Astron. Soc., Vol. 13, 902 (1981).
Abstract.

142.091 **X-ray synchrotron nebulae and the origin of
neutron stars.**   D. J. Helfand.
Bull. American Astron. Soc., Vol. 13, 908 (1981). – Abstract.

142.092 **Discovery of 69 ms periodic X-ray pulsations in
A0538–66.**   G. K. Skinner, D. K. Bedford,
R. F. Elsner, D. Leahy, M. C. Weisskopf, J. Grindlay.
Nature, Vol. 297, 568 - 570 (1982).
The authors report the detection, during observations

with the Einstein Observatory at the time of an outburst, of
X-ray pulsations with a period of 69.2126 ms from
A0538–66. The observed rate of change of period shows the
source to be an eccentric binary system in which the two
components approach close to each other during the outbursts
and may even become immersed in a common envelope.

142.093 **Optical identification of X-ray source H0139–68
with an AM Herculis-type system.**
N. Visvanathan, A. Pickles.
Nature, Vol. 298, 41 - 44 (1982).
The authors report the identification of the soft X-ray
source H0139–68 with a star at RA = 01 h 39 min 37.5 s,
dec = −68°08'32" (1950). Time-resolved spectrophotometry
of the star over 3 h shows a V magnitude variation between
14.9 and 16.4 with a period of ~110 min. The optical
characteristics of the candidate star are similar to those of the
magnetic cataclysmic variable AM Herculis.

142.094 **Visual and near infrared photometry of 2A 0311-227.**
C. Motch, J. van Paradijs, H. Pedersen,
S. A. Ilovaisky, C. Chevalier.
ESO Sci. Prepr. No. 184, 21 pp. (1982). – Submitted to
Astron. Astrophys.

142.095 **The origin of the X- and γ-ray backgrounds.**
G. Setti, L. Woltjer.
ESO Sci. Prepr. No. 186, 33 pp. (1982). – Submitted to
Vatican Study Week on "Cosmology and Fundamental
Physics", Rome 1981.

142.096 **X-ray observation of two gamma-ray source fields
with the Einstein observatory.**
K. P. Singh, K. M. V. Apparao, R. K. Manchanda.
Astrophys. Space Sci., Vol. 82, 477 - 479 (1982).
Two gamma ray source fields CG 189+1 and CG 075+00,
centered on the corresponding UHURU Point X-ray source
positions, were observed with the IPC onboard Einstein
Observatory. An unidentified weak X-ray source is detected
from the direction of CG 189+1 only.

142.097 **Radio observations of coronal X-ray sources.**
D. M. Gibson, P. L. Fisher, D. J. Helfand.
Bull. American Astron. Soc., Vol. 14, 575 (1982). – Abstract.

142.098 **The temperature of thermal X-ray and γ-ray
sources.**   R. J. Gould.
Astrophys. J., 258, 131 - 134 (1982).
A framework is developed for the accurate determination of the temperature of a source, assuming that it is optically thin and emits a bremsstrahlung spectrum. The temperature
can be measured from the shape of the spectral distribution,
and an explicit relation is derived for the temperature in terms
of this slope and a sum of correction terms to the limiting
form of the bremsstrahlung formula. For the case where there
is a distribution of temperatures in the plasma, an additional
correction is derived in terms of the rms variation in the
plasma temperature. Some results for highly relativistic
plasmas are given.

142.099 **Ultraviolet spectra of the X-ray transient
A0538–66.**   J. C. Raymond.
Astrophys. J., Vol. 258, 240 - 244 (1982).
Ultraviolet spectra were obtained before and after the
optical spectrum changed from absorption to emission lines.
The continuum brightness dropped by a factor of 2. Spectra
with bright continua show N V, Si IV, and C IV in absorption, while the spectrum with fainter continuum has these
lines in emission. The emission lines of N V and Si IV are
anomalously bright compared with C IV. The brighter spectra
indicate an effective temperature of 25,000 K, while the
fainter continuum indicates a somewhat lower temperature.

Probable interstellar absorption lines due to galactic halo and the LMC interstellar medium are observed.

142.100   The noncompact binary X-ray source 4U 2129+47.
J. E. McClintock, R. A. London, H. E. Bond, A. D. Grauer.
Astrophys. J., Vol. 258, 245 - 253 (1982).

The authors observed the 5.2 hr X-ray binary 4U 2129+47 for a full orbital cycle using the Einstein IPC and MPC detectors. They also made simultaneous photometric observations. The two most important findings are : (1) the shapes of the 5.2 hr X-ray light curves are independent of energy ($1 < E < 7$ keV); (2) a partial X-ray eclipse occurred which was centered on the time of optical minimum and which lasted 20% of the orbital period. During this interval the X-ray intensity varied smoothly by a factor of 3, and the light curve was symmetric relative to the time of minimum. These findings argue that the X-ray emitting region is extended ($\sim 0.5\,R_\odot$) and highly ionized. The authors present a model in which an accretion disk corona scatters radiation from a central accreting neutron star.

142.101   Properties of X-ray bursts from MXB 1636−53.
T. Ohashi, H. Inoue, K. Koyama, K. Makishima, M. Matsuoka, T. Murakami, M. Oda, Y. Ogawara, N. Shibazaki, Y. Tanaka, I. Kondo, S. Hayakawa, H. Kunieda, F. Makino, K. Masai, F. Nagase, Y. Tawara, S. Miyamoto, H. Tsunemi, K. Yamashita.
Astrophys. J., Vol. 258, 254 - 259 (1982).

MXB 1636−53 was observed from Hakucho in four separate periods between 1979 April and July. Its persistent component was at $\sim 1 \times 10^{-9}$ ergs cm^{-2} s^{-1} in June, about half the intensity as observed by SAS 3 in 1977 January. The X-ray bursts from MXB 1636−53 are found to exhibit a wide variety of profiles from burst to burst, and the peak flux of burst fluctuated by a factor of 6. The blackbody radius in the decay portion of the bursts is constant for all the bursts with largely different profiles.

142.102   Photoelectric photometry of 4 U 2129 + 47.
R. Calafat, R. Canal, J. Núñez, J. Torra.
Astron. Astrophys., Vol. 110, 23 - 24 (1982).

Results from *UBV* photoelectric photometry of the optical counterpart of the weak X-ray source 4 U 2129 + 47 are presented. They confirm the gross features shown by the spectrophotometric observations, the 5.26 h period and the effects of X-ray heating on the companion. A comparison with previous observations, pointing towards a long-term variability of the source, is made.

142.103   The orbital period of 2S 1223−624 (GX301−2).
M. G. Watson, R. S. Warwick, R. H. D. Corbet.
Mon. Not. R. Astron. Soc., Vol. 199, 915 - 924 (1982).

The authors use observations by the Ariel 5 Sky Survey Instrument (SSI) over a four-year period to show that the binary X-ray pulsar GX 301−2 flares in X-rays every 41.4 day Published pulse timing measurements confirm this as the orbital period of the system and also indicate considerable orbital eccentricity ($e \approx 0.47$). Comparison of the SSI, and other observations, with an ephemeris derived from the pulse timing data indicates that the flares occur a few days prior to periastron passage. The observations are discussed in terms of a model of this system in which accretion of the stellar wind from the primary star on to the neutron star secondary varies around the orbit.

142.104   An interpretation of the X-ray spectrum of 4U 1822−37.
A. C. Fabian, P. W. Guilbert, R. R. Ross.
Mon. Not. R. Astron. Soc., Vol. 199, 1045 - 1051 (1982).

White et al. (1981) have suggested that the partial X-ray eclipse observed by them in the binary X-ray source 4U 1822−37 is due to an extended scattering region of size $\sim 10^{10}$ cm surrounding a compact source. The authors have simulated the observed X-ray spectrum by radiative transfer through a shell of gas and determined the distance of that gas from the source. This is shown to be consistent with simple estimates of an evaporated atmosphere above an accretion disc. The partial eclipse can then be explained in an ad hoc manner by appending an extensive low density envelope to the outer disc.

142.105   Einstein observations of the confused 2A 2315−428 region.
P. A. Charles, M. M. Phillips.
Mon. Not. R. Astron. Soc., Vol. 200, 263 - 270 (1982).

The authors report an observation of the 2A 2315−428 field with the Einstein Observatory imaging proportional counter which revealed four distinct X-ray sources. By far the strongest in this energy band (0.2−3.8 keV) is associated with the rich cluster Sérsic 159-03 which contains a cD galaxy. This source may account for as much as 40 per cent of the observed 2A flux, the remainder of which is almost certainly produced by the much weaker, but more heavily cut-off galaxy NGC 7582. The authors also detect a weak X-ray flux from NGC 7552 and reanalyse previous observations to show that this source is not variable.

142.106   Evidence for periodicity in GX349+2 and GX17+2.
T. Ponman.
Mon. Not. R. Astron. Soc., Vol. 200, 351 - 360 (1982).

Evidence is presented suggesting periodicities of 6.44 day and 8.71 day in the X-ray intensities of GX17+2 and GX349+2. In the former case the periodicity is apparently independent of the large flares which occur in the source, whilst in the latter the periodicity resides entirely in the flares and seems to require a physical rather than a geometrical explanation.

142.107   Discovery of a new BL Lacertae object (1E 1402.3 + 0416) with the Einstein Observatory.
J. Stocke, J. Liebert, H. Stockman, J. Danziger, J. Lub, T. Maccacaro, R. Griffiths, P. Giommi.
Mon. Not. R. Astron. Soc., Vol. 200, 27P - 32P (1982).

One (and only one) BL Lac object, designated 1E 1402.3 + 0416, has been so far discovered as part of an optical identification program for a complete sample of faint X-ray sources detected with the Einstein Observatory. Consistent with earlier X-ray discovered BL Lacs, this object is blue and radio weak. The percentage of BL Lacs in the faint extragalactic X-ray sample is smaller than the percentage at higher X-ray fluxes. This suggests that BL Lacs do not evolve in a manner similar to quasars and are, therefore, not substantial contributors to the X-ray background.

142.108   On the problem of cosmic transition radiation to dust grains.
F. A. Agaronyan, A. S. Ambartsumyan.
Astrofizika, Tom 17, 807 - 818 (1981). In Russian. English translation in Astrophysics, Vol. 17, No. 4.

The possible contribution of X-ray transition radiation (XTR), from dust grains to cosmic X rays is considered. It is shown that in the energy range $\lesssim 5$ keV the intensity of XTR may essentially exceed that of the bremsstrahlung. The contribution of XTR in the diffuse (isotropic) X-ray background as well as in the radiation from compact X-ray sources appears negligible.

142.109   Stellar contributions to the diffuse soft X-ray background.
J. Bookbinder, Y. Avni, L. Golub, R. Rosner, G. Vaiana.
Smithsonian Astrophys. Obs., Spec. Rep. 392, (see 012.045), p. 201 - 205 (1982).

142.110 Ultraviolet background radiation and the search for decaying neutrinos.   R. C. Henry.
Cosmology and particles, (see 012.046), p. 211 - 229 (1982).

The spectrum of the observed far-ultraviolet background at high galactic latitudes provides superficial evidence of radiation from neutrino decay, but the spectrum is so uncertain that conclusions are not possible. A limit of ~300 photons $(cm^2 sec\, ster\, Å)^{-1}$ is set on any non-stellar ultraviolet flux above latitude 20°. The disagreement between the Berkeley and the Johns Hopkins ultraviolet background radiation data is analysed.

142.111 The origin of the X-ray and γ-ray backgrounds.
G. Setti, L. Woltjer.
Astrophysical cosmology, (see 012.047), p. 315 - 343 (1982).

142.112 Report of IAU Commission 48: High energy astrophysics (Astrophysique de grande énergie).
F. Pacini.
Trans. IAU, Vol. XVIIIA, (see 003.013), p. 649 (1982).

Second catalogue of X-ray sources.
See Abstr. 002.013.

X- and γ-ray superfast photometry.
See Abstr. 031.504.

High resolution measurements of hard X-ray spectra of southern hemisphere sources.   See Abstr. 032.527.

Electron-ion coupling in rapidly varying sources.
See Abstr. 063.002.

X-ray heating of the quiescent chromospheres of dMe stars.   See Abstr. 064.007.

Accretion disk coronae.
See Abstr. 064.073.

A model of two-stream non-radial accretion for binary X-ray pulsars.   See Abstr. 064.077.

Hydrodynamics of X-ray induced stellar winds.
See Abstr. 064.081.

Report of IAU Commission 36: Theory of stellar atmospheres (Thèorie des atmosphères stellaires).
See Abstr. 064.102.

Physical processes in stars on late stages of their evolution.   See Abstr. 065.064.

Some remarks on the spectra of X-ray bursts.
See Abstr. 066.504.

Changing orientation of dipole and spin axes in binary X-ray pulsars.   See Abstr. 066.505.

On hard X-ray spectra of accreting neutron stars.
See Abstr. 066.518.

Matter accreting neutron stars.
See Abstr. 066.525.

UBV photometry of the X-ray binary HD 77581 = 4U 0900 − 40.   See Abstr. 113.004.

Infrared emission from four Be stars optical counterparts of galactic X-ray sources.   See Abstr. 113.054.

On periodic variations in the spectrum of the B0e star X Persei associated with the X-ray source 3U 0352+30. See Abstr. 114.139.

High-resolution studies of the archetype K giant Arcturus with IUE and Einstein: a sensitive search for high-temperature emission.   See Abstr. 114.175.

Optical spectroscopy of HD102567 (4U1145-61).
See Abstr. 116.043.

Eclipsing AM Herculis-type magnetic binary.
See Abstr. 116.046.

Spectroscopy of the X-ray cataclysmic binary IE 0643.0−1648.   See Abstr. 117.012.

A radial velocity study of 4U 2129+47: a low mass X-ray binary system.   See Abstr. 117.016.

The evolution of highly compact binary stellar systems.   See Abstr. 117.020.

Detection of hydrogen α periodicity in X Persei.
See Abstr. 117.034.

HZ Her − possible line profiles.
See Abstr. 117.038.

ANS spectrophotometry: the bright X-ray binaries Hercules X-1 (HZ Herculis) and Cygnus X-1 (HDE 226868). See Abstr. 117.054.

Shallow partial eclipses of a star by a disk or sphere-like component and the parameters of Cyg X-1 = V1357 Cyg.   See Abstr. 117.088.

Discovery of optical variability in the hard X-ray source HD 8357.   See Abstr. 117.093.

Nodding motions of accretion discs in SS 433 and Her X-1.   See Abstr. 117.097.

A photometric and spectroscopic study of a newly discovered cataclysmic variable, H2215-086.
See Abstr. 117.100.

On the turn-ons of Hercules X-1 and the periodicities in the radial velocity variations of SS 433.
See Abstr. 117.108.

The X-ray lobes of SS 433.
See Abstr. 117.111.

X-ray variability of SS 433.
See Abstr. 117.112.

Photometric and spectroscopic observations of the optical counterpart of H2215−086.   See Abstr. 117.141.

Visual and near infrared photometry of 2A 0311-227.   See Abstr. 117.145.

Report of IAU Commission 42: Close binary stars (Étoiles doubles serrées).   See Abstr. 117.146.

X-ray and UV observations of the rapidly rotating triple system HD 165590.   See Abstr. 118.026.

A model for 0921−63: a second halo X-ray source.
See Abstr. 119.038.

X-ray eclipses in AR Lac: preliminary results.
See Abstr. 119.040.

Viscous boundary layer and hard X-rays from
dwarf novae.   See Abstr. 122.065.

Simultaneous X-ray and optical photometry of the
cataclysmic variable TT Arietis.   See Abstr. 122.155.

X-ray and optical measurements of the cataclysmic
variable CH UMa.   See Abstr. 122.156.

X-ray spectral classification of supernova remnants
in the Large Magellanic Cloud.   See Abstr. 125.008.

An X-ray study of two Crablike supernova remnants:
3C 58 and CTB 80.   See Abstr. 125.010.

The effects of non-equilibrium ionization on the
X-ray emission of supernova remnants.
See Abstr. 125.021.

Soft X-ray observation of supernova remnant
SN 1006.   See Abstr. 125.022.

The effects of ejecta on the X-ray luminosities of
supernova remnants.   See Abstr. 125.037.

X-ray, optical and UV observations of the supernova remnant in NGC 4449.   See Abstr. 125.038.

MSH 15-5(2) − a SNR containing 2 compact X-ray
sources.   See Abstr. 125.039.

High resolution X-ray spectroscopy of the supernova remnant N132D in the Large Magellanic Cloud.
See Abstr. 125.041.

Multicoloured X-ray image of the Cygnus Loop.
See Abstr. 125.049.

A high-resolution X-ray image of Puppis A:
inhomogeneities in the interstellar medium.
See Abstr. 125.050.

The complex emission-line structure in the magnetic
white dwarf binary 2A 0311−227 (EF Eridani).
See Abstr. 126.003.

Einstein observations of hot DB white dwarfs.
See Abstr. 126.032.

Carbon and oxygen X-ray line emission from the
interstellar medium.   See Abstr. 131.021.

Observaciones de H I conectadas con un estallido
en rayos X detectado en la zona de Lupus.
See Abstr. 131.166.

High-resolution X-ray observations of the Orion
Nebula.   See Abstr. 132.002.

Search for extended X-ray emission surrounding the
Crab Nebula.   See Abstr. 134.035.

The Einstein objective grating X-ray spectrum of
the Crab Nebula.   See Abstr. 134.036.

VLA observations of an unbiased sample of extragalactic X-ray sources.   See Abstr. 141.053.

An assessment of the detectability of X-ray emission
from winds in active galactic nuclei and quasars.
See Abstr. 141.060.

Radio and X-ray structure of Centaurus A.
See Abstr. 141.124.

X-ray and optical observations of quasars.
See Abstr. 141.158.

cm-wavelength fluxes and polarizations of compact
extra-galactic X-ray sources.   See Abstr. 141.174.

Relativistic jets as radio and X-ray sources.
See Abstr. 141.183.

Detection of neutral hydrogen emission and
optical nebulosity in the low redshift QSO 0351 + 026.
See Abstr. 141.204.

X-ray, optical, and radio properties of quasars.
See Abstr. 141.221.

Compton-heated winds from accretion disks in
QSOs and binary X-ray sources.   See Abstr. 141.242.

Can relativistic beaming reconcile the Einstein
Observatory X-ray data with the synchrotron self-Compton
predictions in compact extragalactic sources?
See Abstr. 141.244.

The QSO number density and contribution to the
2 keV X-ray background.   See Abstr. 141.258.

X-ray, radio, and optical properties of quasars.
See Abstr. 141.260.

X-ray observation of a complete sample of 3CR
radio galaxies.   See Abstr. 141.261.

X-ray emission associated with a tailed radio galaxy.
See Abstr. 141.262.

Evidence for 200 second variability in the X-ray flux
of the quasar 1525 + 227.   See Abstr. 141.292.

Observations of X-ray quasars at 90 GHz.
See Abstr. 141.302.

Einstein X-ray observations of QSO's with
absorption-line systems.   See Abstr. 141.305.

A transient 77 keV emission feature from the
Crab pulsar.   See Abstr. 141.505.

γ-ray sources as comptonized X-ray sources.
See Abstr. 142.513.

Hard X-rays from the gamma-ray source CG 195 + 4.
See Abstr. 142.515.

Gamma-ray bursts from X-ray bursts and a hot
corona.   See Abstr. 142.516.

Very high energy γ-rays from Cygnus X-3.
See Abstr. 142.536.

An experiment to detect γ-rays of energy above
1000 GeV from Cygnus X-3.   See Abstr. 142.537.

Observations of Cygnus X-3 at energies above
1000 GeV.   See Abstr. 142.538.

Cosmic rays and gamma-rays from OB stars.
See Abstr. 143.003.

An X-ray survey of the Pleiades.
See Abstr. 153.036.

Evidence for extended X-ray emission from globular clusters.   See Abstr. 154.005.

Possible diffuse X-ray emission from globular clusters.   See Abstr. 154.022.

High resolution X-ray and optical studies of globular clusters.   See Abstr. 154.023.

On the origin of the 1 keV diffuse X-ray background.   See Abstr. 157.010.

The extreme Seyfert galaxy associated with the X-ray source 3A 0557 - 383.   See Abstr. 158.021.

X-ray emission from elliptical galaxies.
See Abstr. 158.089.

The X-ray spectrum and time variability of narrow emission line galaxies.   See Abstr. 158.096.

X-ray observations of peculiar galaxies with the Einstein Observatory.   See Abstr. 158.151.

X-ray observations with the Einstein Observatory of emission-line galaxies.   See Abstr. 158.192.

X-ray maps of peculiar galaxies.
See Abstr. 158.229.

Interpretation of the redshift distribution of X-ray selected active galactic nuclei.   See Abstr. 158.239.

Two X-ray-selected BL Lacertae-type objects.
See Abstr. 158.240.

X-ray spectra of active galactic nuclei with the Einstein Observatory.   See Abstr. 158.241.

X-ray observations of emission-line galaxies with the Einstein Observatory.   See Abstr. 158.242.

Observations of NGC 4151 at 2 keV - 2 MeV from HEAO-1.   See Abstr. 158.243.

Extended soft X-ray emission from NGC 4151.
See Abstr. 158.244.

X-ray emission from Centaurus A.
See Abstr. 158.245.

X-ray coronae around galaxies.
See Abstr. 158.259.

Discovery of a new BL Lacertae object (1E1402.3+0416) with the Einstein Observatory.
See Abstr. 158.267.

X-ray emission from clusters of galaxies containing classical double radio sources.   See Abstr. 160.010.

Abell 2069: an X-ray cluster of galaxies with multiple subcondensations.   See Abstr. 160.017.

A morphological classification of clusters of galaxies from Einstein images.   See Abstr. 160.041.

X-ray observations of Abell 2218 and implications for the Sunyaev-Zel'dovich effect.   See Abstr. 160.050.

Extended X-ray emission from poor clusters with central dominant galaxies.   See Abstr. 160.054.

Low X-ray luminosity cD cluster A 2670.
See Abstr. 160.062.

Limits to gas temperature variations in the central regions of the Centaurus cluster of galaxies.
See Abstr. 160.064.

HEAO A-1 X-ray survey of southern clusters.
See Abstr. 160.066.

Optical investigations of two X-ray clusters of galaxies: 0430.6–6133 and 0626.7–5426.
See Abstr. 160.067.

Radio and X-ray observations of the Abell 2241 galaxy clusters.   See Abstr. 160.068.

The detection of hot intergalactic gas in the NGC 3607 group of galaxies with the Einstein satellite.
See Abstr. 161.004.

The intergalactic medium.   See Abstr. 161.005.

A detailed X-ray study of the cooling intracluster gas in A 496.   See Abstr. 161.007.

Gamma-ray Sources, Gamma-ray Background

142.501   Observation of two gamma-ray bursts by Vela X-ray detectors.   J. Terrell, E. E. Fenimore, R. W. Klebesadel, U. D. Desai.
Astrophys. J., Vol. 254, 279 - 286 (1982).

Bursts of X-rays coincident in time with two gamma-ray burst events were observed by the 3–12 keV collimated X-ray detectors on the Vela spacecraft. Both of these observations show recurrence on a time scale of hundreds of seconds. For one of these events (GB 720514) the X-ray detection gives an improved position as well as information on the spectrum late in the outburst. The other event (GB 740723) is of special interest because the source, not previously located, is consistent in direction with the binary pulsar SMC X-1 in the Small Magellanic Cloud.

142.502   Precise source location of the anomalous 1979 March 5 gamma-ray transient.   T. L. Cline, U. D. Desai, B. J. Teegarden, W. D. Evans, R. W. Klebesadel, J. G. Laros, C. Barat, K. Hurley, M. Niel, G. Vedrenne, I. V. Estulin (*Ehstulin*), V. G. Kurt, G. A. Mersov, V. M. Zenchenko, M. C. Weisskopf, J. Grindlay.
Astrophys. J., Lett., Vol. 255, L45 - L48 (1982).

Refinements in the source direction analysis of the observations of the unusual 1979 March 5 gamma-ray transient are presented. The final results from the interplanetary gamma-ray burst network produce a 0.1 arcmin^2 error box. It is nested inside the initially determined 2 arcmin^2 source region of Evans et al. that identified the supernova remnant N49 in the Large Magellanic Cloud as a possible source. This smaller source location is within both the optical and X-ray contours of N49 although not positioned at either contour center.

142.503   High energy gamma rays from cosmic ray nucleons.   R. Schlickeiser.
Astron. Astrophys., Vol. 106, L5 - L8 (1982).

The production rate of high-energy (> 10 MeV) gamma rays due to interactions of cosmic ray nucleons with the interstellar gas is calculated. Additionally to π°-decay the author considers the contribution of bremsstrahlung gamma rays from secondary electrons which simultaneously are produced, and shows that this contribution is significant because the confinement time of secondary electrons in the galactic matter disk is longer than $\cong (10^7/n_0)$ yr (n_0: interstellar gas density). Spectral data on the galactic gamma-ray emission as well as the absolute value of the integral (> 100 MeV) gamma-ray production rate thus are consistent with a cosmic-ray nucleon origin.

142.504   Gamma ray astronomy – a new window on the universe.   A. W. Wolfendale.
J. British Astron. Assoc., Vol. 92, 105 - 111 (1982).

142.505   Infrared scans of gamma ray burst source regions.   K. M. V. Apparao, D. Allen.
Astron. Astrophys., Vol. 107, L5 - L6 (1982).

The authors have made infrared scans covering the recently determined error region of the gamma-ray burst event of 6 April, 1979 and the radio sources in the error region of the 19 November, 1978 event, using the 3.9 m Anglo-Australian telescope. No sources brighter than J = 17.5 mag were detected. This result has implications for the models of gamma-ray bursts.

142.506   Compact gamma ray point sources: are gamma ray sources optically thick at lower frequencies?   R. Schlickeiser.

Astron. Astrophys., Vol. 107, 378 - 384 (1982).

The author discusses the hypothesis that some of the unidentified gamma ray point sources are objects which emit only at the highest frequencies but are optically thick against stimulated Compton scattering or nonthermal electron bremsstrahlung reabsorption at lower frequencies. From the calculated absorption coefficients it follows that such a scenario is only possible for inverse Compton scattering under extreme conditions which may hold in the core of powerful sources. This would mean these gamma ray point sources are very energetic and compact objects.

142.507   COS-B gamma-ray measurements, cosmic rays and the local interstellar medium.   F. Lebrun, G. F. Bignami, R. Buccheri, P. A. Caraveo, W. Hermsen, G. Kanbach, H. A. Mayer-Hasselwander, J. A. Paul, A. W. Strong, R. D. Wills.
Astron. Astrophys., Vol. 107, 390 - 396 (1982).

A study of the relation between galaxy counts and gamma rays measured by the COS-B satellite clearly shows that there exists a bidimensional correlation between these quantities. Though the 21-cm measurements are more accurate than galaxy counts they show a weaker correlation with gamma rays. A significant fraction of the local gas in molecular form, linked to the dust, can account for this situation: gamma rays and galaxy counts being total gas tracers. The gamma-ray emissivity per hydrogen atom is derived leading to an estimate of the spectrum of the cosmic-ray electrons contributing to the gamma-ray emission through bremsstrahlung interactions.

142.508   Transient emission of ultra-high energy pulsed γ rays from Crab pulsar PSR0531.   A. I. Gibson, A. B. Harrison, I. W. Kirkman, A. P. Lotts, J. H. Macrae, K. J. Orford, K. E. Turver, M. Walmsley.
Nature, Vol. 296, 833 - 835 (1982).

A new experiment has been established to measure the energy spectrum of ultra-high energy γ rays (E>2,000 GeV) from a number of celestial objects. Observations of PSR0531 are based on 34h of exposure between 25 September and 2 November 1981, and show overall an integral pulsed flux; most important, they also show strong evidence for a burst of pulsed emission of ~15 min duration. The present work is the first to demonstrate short duration (15 min) emission and to offer an explanation of previous apparently discordant results.

142.509   Seyfert galaxies and the cosmic γ-ray diffuse background.   L. Bassani, R. C. Butler, A. J. Dean, G. Di Cocco, F. Perotti, G. Villa.
Astrophys. Space Sci., Vol. 82, 199 - 207 (1982).

Three active galaxies, generally classified as Seyferts, have been discovered recently to be powerful, low energy γ-ray sources. The similarity between their spectral characteristics and those of the cosmic background at γ-ray energies suggests that these objects could make a significant contribution to this diffuse flux. This contribution has been assessed using two different number densities of γ-ray-emitting Seyfert galaxies based on optical and X-ray data.

142.510   The gamma-ray burster puzzle.   R. A. Schorn.
Sky Telesc., Vol. 63, 560 - 562 (1982).

142.511   A model of cosmic γ-ray burst event on 5 March 1979 (II).   D.-y. Wang, A.-a. Xu, Q.-y. Qu, Z.-q. Li.
Acta Astron. Sinica, Vol. 22, 364 - 369 (1981).   In Chinese.

142.512   High-energy gamma-quanta from measurements aboard the Cosmos 856 and Cosmos 914 artificial earth satellites.   I. D. Blokhintsev, V. A. Volzhenskaya,

L. F. Kalinkin, Yu. I. Nagornykh.
Kosm. Issled., Tom 20, 227 - 236 (1982). In Russian.

142.513 γ-ray sources as comptonized X-ray sources.
E. E. Fenimore, R. W. Klebesadel, J. G. Laros,
R. E. Stockdale, S. R. Kane.
Nature, Vol. 297, 665 - 667 (1982).

γ-ray burst spectra have often been fit by optically thin thermal bremsstrahlung. However, at the high temperatures implied by such fits, the free-free cross-section is so much smaller than the Compton cross-section that Compton scattering might dominate the spectral formation processes. The authors have investigated the possibility that emission mechanisms based on Compton scattering can also fit the data. In particular, Monte Carlo calculations have been used to compare the γ-ray burst spectral data with black-body spectra which have undergone inverse comptonization by a much hotter, overlying plasma.

142.514 Precise source location of the anomalous 1979 March 5 gamma-ray transient.
T. L. Cline, U. D. Desai, B. J. Teegarden, W. D. Evans,
R. W. Klebesadel, J. G. Laros, C. Barat, K. Hurley, M. Niel,
G. Vedrenne, I. V. Estulin (Ehstulin), V. G. Kurt, G. A. Mersov,
V. M. Zenchenko, M. C. Weisskopf, J. Grindlay,
NASA Tech. Memo., NASA TM-83884, 14 pp. (1981).

Refinements in the source direction analysis of the observations of the unusual 1979 March 5 gamma-ray transient are presented. The final results from the interplanetary gamma-ray burst network produce a 0.1 arc-min^2 error box. It is nested inside the initially determined 2 arc-min^2 source region of Evans et al. (1980) that identified the supernova remnant N49 in the Large Magellanic Cloud as a possible source.

142.515 Hard X-rays from the gamma-ray source CG 195 + 4.
P. K. Kunte, S. V. Damle, S. Naranan,
D. Venkatesan, C. M. F. Galas, R. Lieu.
Bull. American Astron. Soc., Vol. 13, 867 (1981). – Abstract.

142.516 Gamma-ray bursts from X-ray bursts and a hot corona.
E. E. Fenimore, J. G. Laros, R. W. Klebesadel,
R. E. Stockdale.
Bull. American Astron. Soc., Vol. 13, 867 (1981). – Abstract.

142.517 Energetic phenomena observed from orbit around Venus.
J. G. Laros, R. W. Klebesadel, E. E. Fenimore, W. D. Evans.
Bull. American Astron. Soc., Vol. 13, 882 - 883 (1981).
Abstract.

142.518 Likely discovery of a gamma ray burst optical counterpart. B. E. Schaefer.
Bull. American Astron. Soc., Vol. 13, 901 (1981). – Abstract.

142.519 Cosmic gamma-ray burst spectroscopy.
E. P. Mazets, S. V. Golenetskii (Golenetskij),
V. N. Ilyinskii (Il'inskij),Yu. A. Guryan(Gur'yan),
R. L. Aptekar (Aptekar'), V. N. Panov, I. A. Sokolov,
Z. Ya. Sokolova, T. V. Kharitonova.
Astrophys. Space Sci., Vol. 82, 261 - 282 (1982).

A review is given of the gamma-ray burst energy spectrum measurements on Venera 11 and Venera 12 space probes. The gamma burst continuum approximates in shape thermal bremsstrahlung emission of a hot plasma. The radiation temperature varies over a broad range, 50–1000 keV, for different events. Spectra of many bursts contain cyclotron absorption and/or redshifted annihilation lines. Strong variability is typically observed in both continuum and line spectra. These spectral data provide convincing evidence for the gamma-ray bursts being generated by neutron stars with superstrong magnetic fields $\sim 10^{12}$–10^{13}G.

142.520 On the theory of annihilation lines in gamma-ray bursts. V. V. Zheleznyakov.
Astrophys. Space Sci., Vol. 83, 117 - 125 (1982).

The mechanism of formation of an annihilation line 0.5 MeV in gamma-ray bursts due to electron-positron pair production in strong magnetic fields of neutron stars is discussed. Bremsstrahlung from a hot polar spot is supposed to be a source of gamma-quanta which produce the pairs. It is shown that a great part of radiation with energy $E > 2mc^2$ per quantum is consumed by pair production and does not escape from the gamma-burster. This indicates a possible strong gap in continuum radiation at energies higher than 1 MeV. At the same time effective creation of pairs enables one to give a simple estimate of the expected annihilation line intensity in gamma-ray burst spectra. This estimate agrees with available observational data.

142.521 A search for high-energy gamma rays ($\geqslant 10^{15}$ eV) associated with gamma ray bursts.
R. W. Clay, P. R. Gerhardy, A. G. Gregory.
Astrophys. Space Sci., Vol. 83, 279 - 286 (1982).

A search has been made for high-energy photons detected by an extensive air shower array in coincidence with spacecraft observations of gamma ray bursts recorded between May 1978 and May 1979. No evidence has been found for the detection of coincident energetic photons above a threshold energy of $\sim 10^{15}$ eV. Using an array collecting area of ~ 30000 m^2 at $\sim 10^{16}$ eV it is found that the differential power law spectral index of the photons within a burst must exceed 1.97 ± 0.03 over the energy range from spacecraft energies to 10^{16} eV.

142.522 The 5 March 1979 event and the distinct class of short gamma bursts: are they of the same origin?
E. P. Mazets, S. V. Golenetskii (Golenetskij), Yu. A. Guryan
(Gur'yan), V. N. Ilyinskii (Il'inskij).
Astrophys. Space Sci., Vol. 84, 173 - 189 (1982).

A comparison of the 5 March 1979 event with other short gamma-ray bursts reveals considerable similarities in their features. This implies their common origin.

142.523 Cosmic gamma-ray bursts. F. Verter.
Phys. Rep., Vol. 81, 293 - 349 (1982). – Abstr. in
Phys. Abstr., Vol. 85, Abstr. 58217 (1982).

142.524 Diffuse cosmic gamma radiation with energies higher than 100 MeV at mean and high galactic latitudes. Yu. I. Nagornykh.
Kosm. Issled., Tom 20, 429 - 434 (1982). In Russian.

142.525 The log N–log S curve of gamma-ray bursts detected by the SIGNE experiments.
C. Barat, G. Chambon, K. Hurley, M. Niel, G. Vedrenne.
Astron. Astrophys., Vol. 109, L9 - L11 (1982).

The log N–log S curve for cosmic gamma-ray bursts detected by the Franco-Soviet SIGNE experiments is presented, along with a method of correcting for the different energy thresholds of these and other experiments. Selection effects due to trigger mechanisms are discussed. It is shown that log N–log S curves from different experiments are in agreement after correction for threshold and selection effects.

142.526 Extragalactic gamma radiation: use of galaxy counts as a galactic tracer.
D. J. Thompson, C. E. Fichtel.
Astron. Astrophys., Vol. 109, 352 - 354 (1982).

A derivation of the extragalactic diffuse γ radiation with energies above 35 MeV has been carried out using galaxy counts as a tracer of galactic matter. The extragalactic radiation has a differential photon number spectrum which may be expressed as a power law with index 2.35 (+0.4, −0.3) and an intensity above 35 MeV of $(5.5 \pm 1.3) \times 10^{-5}$ photons

$cm^{-2} s^{-1}$ sterad^{-1}, consistent with previous derivations. Use of a 1/sin |b| expression of the galactic component produces a poorer fit, suggesting that the high-latitude galactic γ-ray production may be dominated by cosmic ray interactions with matter rather than by Compton interactions of cosmic rays with photon fields.

142.527 Search for bursts of gamma rays of energies >1 GeV.
P. N. Bhat, N. V. Gopalakrishnan, S. K. Gupta, P. V. Ramana Murthy, B. V. Sreekantan, S. C. Tonwar, P. R. Viswanath.
Mon. Not. R. Astron. Soc., Vol. 199, 1007 - 1015 (1982).

A ground-based experiment to detect cosmic gamma-ray bursts (GRB) in which the individual gamma-rays at the top of the atmosphere have energies in the GeV range has been in operation at Ootacamund, India, for nearly 1.5 years. Not a single event was observed, although five of the GRB seen by satellite-borne detectors in the MeV energy range were in the view of and potentially observable by the experiment. Details of the experimental results and their implications to the phenomenon of GRB are presented. The experiment also enables the authors to place a 99 per cent confidence level upper limit to the rate of explosions of primordial black holes (Hawking process) in the vicinity of the solar system.

142.528 On the nature of the galactic 2CG γ-ray sources.
R. Buccheri, M. Morini, B. Sacco.
Philos. Trans. R. Soc. London, Ser. A, Vol. 301, (see 012.043), 495 - 504 (1981). – Same as 30.142.526.

The identification of two γ-ray sources of the COS-B catalogue with radio pulsars is used as an important hint for the identification of the rest of the population. The relevant distributions of γ-ray pulsars visible at the Sun within the limiting sensitivity of COS-B are derived and it is suggested that a significant fraction of the unidentified galactic γ-ray sources are pulsars.

142.529 Extended γ-ray sources and active regions in the Galaxy: the Carina and Orion complexes.
T. Montmerle.
Philos. Trans. R. Soc. London, Ser. A, Vol. 301, (see 012.043), 505 - 518 (1981). – Same as 30.142.527.

The results of γ-ray observations by the COS-B satellite lend support to the suggestion that a class of γ-ray sources comprises extended sources, associated with selected active regions in the Galaxy, and in which supernova remnants and/or strong winds from young massive stars are at work. Specifically, a case is made for the identification of the Carina complex with the γ-ray source 2CG288-00. By using a simplified model for the Carina complex, it is shown that, in this source, γ-rays may be the result of proton acceleration by stellar winds, followed by confinement by resonant Alfvén-wave scattering in the giant H II region NGC 3372. In the Orion complex, the same model leads to the conclusion that the confinement is inefficient, and that, as a consequence, ambient cosmic rays dominate.

142.530 On the search for soft X-ray and optical counterparts of selected COS-B γ-ray sources.
P. A. Caraveo.
Philos. Trans. R. Soc. London, Ser. A, Vol. 301, (see 012.043), 523 - 527 (1981).

142.531 A statistical study of the distribution of γ-ray sources.
P. A. Riley.
Philos. Trans. R. Soc. London, Ser. A, Vol. 301, (see 012.043), 529 - 531 (1981).

142.532 On the evidence for high energy γ-ray emission from the Orion nebula stemming from COS-B observations.
P. A. Caraveo.
Philos. Trans. R. Soc. London, Ser. A, Vol. 301, (see 012.043),

569 - 571 (1981). – Same as 30.142.519.

The E.S.A. COS-B satellite performed a one-month observation pointing in the direction of M42 in Orion, during July - August 1978. An excess of high energy (above 100 MeV) photons is seen in the data, well coinciding with the Orion cloud complex. Features of such an excess are discussed, such as flux value, spectral shape and possible spatial extension of the emission. Brief astrophysical implications are then derived for the physical association of the excess with the Nebula.

142.533 Extragalactic γ-rays.
A. J. Dean, D. Ramsden.
Philos. Trans. R. Soc. London, Ser. A, Vol. 301, (see 012.043), 577 - 602 (1981). – Same as 30.142.528.

Only a few extragalactic objects have been studied in the γ-ray region of the spectrum. At high energies the COS-B experiment detected emission from the quasar 3C 273 while at lower energies the results indicate that the emission from the Seyfert galaxy NGC 4151 is variable. A similar variability may also account for the conflicting reports of line emission from the radio galaxy Cen A. The implication of these and other observations in relation to the possible physical conditions in the nuclei of active galaxies, are discussed.

142.534 Low energy γ-ray observations with the MISO telescope.
R. E. Baker, L. Bassani, G. Boella, R. C. Butler, J. N. Carter, A. J. Dean, A. Della Ventura, G. Di Cocco, R. I. Hayles, F. F. Perotti, D. Ramsden, G. Villa.
Philos. Trans. R. Soc. London, Ser. A, Vol. 301, (see 012.043), 603 - 606 (1981).

The Seyfert-galaxy NGC 4151 has been observed with the MISO telescope in the energy range 20 keV to 19 MeV. A preliminary discussion of the measured spectral distribution is given.

142.535 Detection of pulsed γ-rays at energies above 300 GeV from pulsars – T.I.F.R. experiments.
B. V. Sreekantan.
Philos. Trans. R. Soc. London, Ser. A, Vol. 301, (see 012.043), 629 - 632 (1981). – Same as 30.142.529.

A brief account is given of the search for γ-rays from pulsars made by the T.I.F.R. group in India. It is concluded that for energies above 300 GeV the flux of pulsed γ-rays from the Crab and Vela pulsars is much less than would be expected by extrapolation of the spectra measured in the gigaelectronvolt range. It is still not certain whether the pulsars have been detected above 300 GeV because of variability in the apparent signals from year to year.

142.536 Very high energy γ-rays from Cygnus X-3.
Yu. I. Neshpor, Yu. L. Zyskin, J. B. (Zh. B.) Mukanov, A. A. Stepanian (Stepanyan), V. P. Fomin.
Philos. Trans. R. Soc. London, Ser. A, Vol. 301, (see 012.043), 633 - 634 (1981).

142.537 An experiment to detect γ-rays of energy above 1000 GeV from Cygnus X-3.
A. I. Gibson, A. B. Harrison, A. P. Lotts, K. J. Orford, K. E. Turver.
Philos. Trans. R. Soc. London, Ser. A, Vol. 301, (see 012.043), 635 - 636 (1981).

142.538 Observations of Cygnus X-3 at energies above 1000 GeV.
S. Danaher, D. J. Fegan, N. A. Porter, T. C. Weekes.
Philos. Trans. R. Soc. London, Ser. A, Vol. 301, (see 012.043), 637 - 639 (1981).

142.539 High energy γ-rays from the direction of the Crab pulsar.
T. Dzikowski, B. Grochalska, J. Gawin,

J. Wdowczyk.
Philos. Trans. R. Soc. London, Ser. A, Vol. 301, (see 012.043), 641 - 644 (1981).

An attempt is being made to extend the search for high energy γ-rays from point sources to much higher energies by using the method based on investigation of the muon-poor extensive air showers. A search has been made for very high energy photons from the direction of the Crab pulsar by using the Łodz extensive air shower array.

142.540 Cosmic γ-ray bursts.
G. Vedrenne.
Philos. Trans. R. Soc. London, Ser. A, Vol. 301, (see 012.043), 645 - 658 (1981). − Same as 30.142.521.

Although γ-ray bursts were discovered over ten years ago, the study of their temporal structure, their spectrum and their lg N against lg S distribution have still not enabled scientists to determine their origin. Since 1978, however, considerable progress has been made in the accuracy of locating bursts by triangulation methods, by using a large network of observations made by the Helios B solar-orbiting satellite, the interplanetary spacecraft Pioneer Venus and Venera 11 and 12, ISEE-C and Earth-orbiting satellites. In this paper an analysis of the latest observations, and results on the exact location of the arrival directions of several bursts is presented, along with the evidence they provide about the origin of this radiation.

142.541 Do γ-ray bursts contain γ-rays of energies above 1 Gev?
P. N. Bhat, N. V. Gopalakrishnan, S. K. Gupta, P. V. Ramana Murthy, B. V. Sreekantan, S. C. Tonwar.
Philos. Trans. R. Soc. London, Ser. A, Vol. 301, (see 012.043), 659 - 660 (1981). − Same as 30.142.530.

A ground-based experiment to detect gamma-ray bursts (g.r.b.) at gigaelectronvolt energies is being made at a depth of 800 g cm^{-2} in the atmosphere at Ootacamund, India. During the 1.2 years of operation of the experiment, the various satellite-borne experiments have reported observing tens of g.r.b. at megaelectronvolt energies. Of these, the source locations and times of occurrence for five g.r.b. were such that they were potentially observable in the experiment, had they contained γ-rays at gigaelectronvolt energies. None was seen. The details of the experiment and the implications of the result are presented.

142.542 Gamma-ray burst measurements at low fluxes.
K. Beurle, A. Bewick, J. S. Mills, J. J. Quenby.
Philos. Trans. R. Soc. London, Ser. A, Vol. 301, (see 012.043), 661 - 663 (1981).

142.543 Detection of γ-ray lines.
L. E. Peterson.
Philos. Trans. R. Soc. London, Ser. A, Vol. 301, (see 012.043), 669 (1981). − Abstract.

142.544 Interpretations and implications of γ-ray lines from solar flares, the galactic centre and γ-ray transients.
R. Ramaty, R. E. Lingenfelter.
Philos. Trans. R. Soc. London, Ser. A, Vol. 301, (see 012.043), 671 - 686 (1981). − Same as 30.142.522.

Observations and theories of astrophysical γ-ray line emission are reviewed and prospects for future observations by the spectroscopy experiments on the planned Gamma-Ray Observatory are discussed.

142.545 Gamma-ray line investigations with the Durham γ-ray spectrometer.
C. A. Ayre, P. N. Bhat, A. Owens, W. M. Summers, M. G. Thompson.
Philos. Trans. R. Soc. London, Ser. A, Vol. 301, (see 012.043), 687 - 691 (1981).

The second COS-B catalogue of high-energy γ-ray sources. See Abstr. 002.069.

Gamma ray astrophysics. See Abstr. 003.044.

X- and γ-ray superfast photometry. See Abstr. 031.504.

Gamma-ray-line astronomy. See Abstr. 031.553.

The atmospheric Cherenkov technique in γ-ray astronomy: the early days. See Abstr. 034.082.

Gamma-rays above 100 GeV. See Abstr. 034.083.

Future prospects for γ-ray astronomy. See Abstr. 051.027.

High energy γ-rays from a relativistic plasma. See Abstr. 062.030.

On the theory of gamma-ray amplification through stimulated annihilation radiation. See Abstr. 063.021.

The atmospheric Cherenkov technique in searches for exploding primordial black holes. See Abstr. 066.145.

Gamma ray bursts and neutron stars. See Abstr. 066.515.

On hard X-ray spectra of accreting neutron stars. See Abstr. 066.518.

Gamma ray emission from interstellar clouds: a plasma physical process capable of enhancing electron fluxes. See Abstr. 131.015.

Gamma-rays from cosmic-ray irradiated molecular clouds. See Abstr. 131.257.

High energy quasar spectra and the γ ray background. See Abstr. 141.098.

High-energy γ-ray light curve of PSR0531 + 21. See Abstr. 141.516.

The diffuse gamma-ray background and the pulsar magnetic window. See Abstr. 141.548.

On the secular variation of the γ-ray emission from PSR 0531 + 21. See Abstr. 141.551.

Observation of high energy gamma rays from Cygnus X-3. See Abstr. 142.080.

High energy gamma ray observations of Cygnus X-3. See Abstr. 142.083.

The origin of the X- and γ-ray backgrounds. See Abstr. 142.095.

X-ray observation of two gamma-ray source fields with the Einstein observatory. See Abstr. 142.096.

The temperature of thermal X-ray and γ-ray sources. See Abstr. 142.098.

The origin of the X-ray and γ-ray backgrounds. See Abstr. 142.111.

Report of IAU Commission 48: High energy astrophysics (Astrophysique de grande énergie).
See Abstr. 142.112.

The components of galactic γ-ray emission.
See Abstr. 157.001.

Gamma-ray astrophysics.
See Abstr. 157.008.

The Galaxy as the origin of gamma-ray bursts.
II. The effect of an intrinsic burst luminosity distribution on log $N(>S)$ versus log S.   See Abstr. 157.011.

The γ-ray emissivity of the local interstellar medium from correlations with gas at intermediate latitudes.
See Abstr. 157.012.

Low-latitude galactic γ-ray emission: a probe, not a proof.   See Abstr. 157.013.

Observation of Mkn 501 at gamma-ray energies.
See Abstr. 158.285.

Errata

142.901   Erratum: 'The contribution of quasars to the 2 keV − 100 MeV background radiation and the X-ray source counts at 2 keV'[Mon. Not. R. Astron. Soc., Vol. 197, 313 - 323 (1981)].
J. E. Cheney, M. Rowan-Robinson.
Mon. Not. R. Astron. Soc., Vol. 198, 767 (1982). − See Abstr. 30.142.039.

142.902   Erratum: "An X-ray active region in Orion: X-rays from a Herbig-Haro object " [Astrophys. J., Vol. 248, 591 - 595 (1981)].   S. H. Pravdo, F. E. Marshall.
Astrophys. J., Vol. 254, 826 (1982). − See Abstr. 30.142.021.

142.903   Search for high frequency pulsations in the onset of the March 5 gamma-ray burst.   M. C. Weisskopf, R. F. Elsner, P. G. Sutherland, J. E. Grindlay.
Astrophys. Lett., Vol. 22, 49 - 54 (1981). − See Abstr. 30.142.545.

142.904   Erratum: 'The high energy X-ray spectrum of 4U 0900−40 observed from OSO 8" [Astrophys. J., Vol. 250, 355 - 361 (1981)]. J. F. Dolan, D. C. Ellison, C. J. Crannell, B. R. Dennis, K. J. Frost, L. E. Orwig.
Astrophys. J., Vol. 258, 414 (1982). − See Abstr. 30.142.061.

143 Cosmic Radiation

143.001   The charge and isotopic composition of $Z = 6-14$ cosmic ray nuclei at their source.
W. R. Webber.
Astrophys. J., Vol. 252, 386 - 392 (1982).

Using data from a cosmic ray charge-isotope telescope flown on balloons, the author has determined both the charge and isotopic composition of $Z = 6-14$ cosmic ray nuclei. He observes a low abundance for the elements N and Ne in the cosmic ray source relative to solar cosmic rays. For the isotopes, he finds a cosmic ray source ratio $^{22}Ne/^{20}Ne$ that is 3.52 ± 0.67 times the solar ratio. Possible enhancements of the $^{26}Mg/^{24}Mg$ ratio which is 1.40 ± 0.24 times the solar ratio and the $^{13}C/^{12}C$ ratio which is 2.90 ± 0.93 times the solar ratio are also observed. For Ne, the underabundance of this element coupled with the overabundance of the isotope ^{22}Ne provides an important new clue to the nucleosynthesis processes producing these differences.

143.002   Calculation of cosmic ray antiproton-proton ratio.
T. K. Gaisser, B. G. Mauger.
Astrophys. J., Lett., Vol. 252, L57 - L59 (1982).

Independent calculations of the antiproton-to-proton ratio by Gaisser and Maurer and by Badhwar et al. have produced conflicting results which obscure the interpretation of recent measurements of cosmic ray antiprotons. A detailed re-examination of these calculations has been performed and the authors find that the first calculation was essentially correct and that the reported fluxes of antiprotons are significantly higher than expected for secondary antiprotons in conventional models of cosmic ray propagation.

143.003   Cosmic rays and gamma-rays from OB stars.
H. J. Völk, M. Forman.
Astrophys. J., Vol. 253, 188 - 198 (1982).

The possible acceleration of cosmic rays at the terminal shocks of OB star winds is investigated. Interest is focused on the interaction region downstream of the shock, its acceleration, and its transmission properties. Particles to be accelerated can come from the star, from the shock itself through injection out of the stellar wind plasma, from the interstellar medium, and possibly from the interface between shocked wind and ambient medium. It is shown that shock-injection of nucleons, as estimated from analogous results at the Earth's bow shock in the solar wind, will yield nonrelativistic nucleons with a power law distribution in momentum $\sim p^{-4}$ if they can be accelerated at the shock. This is believed to be possible at least intermittently on time scales limited by the stellar rotation rate. Thus, stellar winds might be sources of very low energy (nuclear) cosmic rays.

143.004   On the nature of the cosmic ray positron spectrum.
R. J. Protheroe.
Astrophys. J., Vol. 254, 391 - 397 (1982).

The author has made a new calculation of the flux of secondary positrons above 100 MeV expected for various propagation models. The models investigated are the leaky box or homogeneous model, a disk-halo diffusion model, a dynamical halo model, and the closed galaxy model. The parameters of these models have, in each case, been adjusted for agreement with the observed secondary/primary ratios and ^{10}Be abundance. The positron flux predicted for these models is compared with the available data. The possibility of a primary positron component is considered.

143.005   A numerical study of the pitch-angle scattering of cosmic rays.   J. Kóta, E. Merényi, J. R. Jokipii, D. A. Kopriva, T. I. Gombosi, A. J. Owens.
Astrophys. J., Vol. 254, 398 - 404 (1982).

The authors present the results of a careful study of finite-difference solutions to the problem of cosmic-ray transport, including pitch-angle scattering. In contrast to some recent studies, they confirm the diffusion approximation in cases where the scattering mean free path is small compared with other length scales. The reasons for the discrepancy with the conclusions of Gombosi and Owens are discussed.

143.006 The galactic cosmic-ray radial intensity gradient and large-scale modulation in the heliosphere.
R. B. McKibben, K. R. Pyle, J. A. Simpson.
Astrophys. J., Lett., Vol. 254, L23 - L27 (1982).

Measurements of fluxes from 1 AU on the IMP 8 satellite to 11 AU on the Pioneer 11 and to > 24 AU on the Pioneer 10 spacecraft show that the nucleon integral gradient is a function of the level of the ~ 11 year solar modulation cycle. The gradient decreases from ~ 4.5% to 1.5% throughout the solar minimum of 1975 - 1977, and increases to ~ 2.5 - 3.0% per AU near maximum modulation. Pioneer 10 at > 24 AU is still deep in the modulation region, leading to estimates of 50 - 70 AU for the radial distance to the heliopause at both solar minimum and solar maximum modulation.

143.007 On the reported detection of sub-GeV antiprotons in galactic cosmic rays.   D. Eichler.
Nature, Vol. 295, 391 - 393 (1982).

The author discusses scenarios in which the reported sub-GeV $\bar{p}$ excess might be produced as secondaries in high energy collisions. Such production must take place in compact, optically thick sources.

143.008 Charge abundance of cosmic rays at their source.
W. R. Webber.
Astrophys. J., Vol. 255, 329 - 340 (1982).

The relative charge abundance of galactic cosmic-ray nuclei has been measured between 600 and 1000 MeV per nucleon over a range of Z from 2 to 28. The abundances observed at Earth have been extrapolated to the cosmic-ray source yielding accurate source abundances for 15 elements. These abundances, along with recently measured isotopic abundance ratios, are compared with the average abundances observed in another sample of accelerated material – solar cosmic rays. Significant differences are found in the composition for the elements, He, C, N, and Ne. These differences seem to be explained more easily in terms of differing nucleosynthesis histories rather than preferential acceleration effects. The galactic and solar cosmic-ray abundances are also compared with a variety of compilations of unaccelerated matter.

143.009 The origin of cosmic rays.
W. Tucker, K. Tucker.
Mercury, Vol. 11, 34 - 35, 37 (1982).

143.010 Origin and anisotropy of high energy cosmic rays.
N. R. Stapley.
Thesis, Univ. Durham, England (1981). – Abstr. in Phys. Abstr., Vol. 85, Abstr. 18666 (1982).

143.011 Separation of the steady-state cosmic ray equation of transport: a generalization.   G. M. Webb.
Astrophys. Space Sci., Vol. 81, 215 - 220 (1982).

The one-dimensional, steady-state equation of transport for cosmic rays including convection, diffusion and adiabatic deceleration is separated for a spatial diffusion coefficient with an arbitrary momentum dependence and for an arbitrary spatial dependence of the convection velocity, and applies for planar, cylindrical and spherical geometries. As an application, the previously obtained spherically symmetric steady-state Green's functions, describing the propagation of cosmic rays in interplanetary space, are generalized to the case where convection velocity is a function of position.

143.012 Das Energiespektrum der kosmischen Strahlung aus atmosphärischem Cerenkovlicht.
G. Schlemmer, G. Felkel, D. Kuhn, J. Pfleiderer.
Mitt. Astron. Ges., Nr. 55, p. 180 (1982). – Abstract.

143.013 Preinjection of cosmic rays and magnetic chemically peculiar stars.   O. Havnes.
Astron. Astrophys., Vol. 110, 203 - 208 (1982).

Cosmic ray abundances and energy spectra calculated on the assumption that magnetic peculiar A stars produce low energy cosmic rays which are reaccelerated by interstellar shocks are found to be in good accordance with observations. The slowing down time of heavy elements in the low energy component is shown to be longer than previously thought mainly because they are initially accelerated in a low charge state and also to some degree because they do not become fully stripped at low energies. This increased slowing down time apparently allows reacceleration to take place. Heating of the interstellar medium by the low energy cosmic rays in the author's model is lower than limits set by observations.

143.014 Periodic and recurrent variations of cosmic rays.
A. J. Somogyi.
Report KFKI-1981-94, Hungarian Acad. Sci., Budapest. 22 pp. (1981). – Abstr. in Phys. Abstr., Vol. 85, Abstr. 32784 (1982).

143.015 An ultrahigh energy cosmic ray event with $\Sigma E_\gamma \simeq 800$ TeV.   J.-r. Ren, S.-l. Lu, H.-h. Kuang, S. Su, Y.-x. Wang, D.-c. Wang, H.-k. Fan, W.-j. Hu, Y.-g. Xue, C.-r. Wang, M. He, N.-j. Zhang, P.-y. Cao, J.-y. Li, Y.-h. Chen, S.-z. Wang, J.-g. Liu, Q.-x. Geng.
Phys. Energ. Fortis Phys. Nucl., Vol. 5, 706 - 711 (1981). In Chinese. – Abstr. in Phys. Abstr., Vol. 85, Abstr. 33873 (1982).

143.016 On a possible scheme of the 11-year modulation of cosmic rays in the light of the latest radio astronomic data.   V. I. Shishov, V. I. Vlasov, I. V. Chashej.
Geomagn. Aehron., Tom 22, 10 - 14 (1982). In Russian.

143.017 On the coefficient of diffusion of galactic cosmic rays in the interplanetary space.
V. Kh. Babayan.
Geomagn. Aehron., Tom 22, 15 - 18 (1982). In Russian.

143.018 Fluctuations and scattering of galactic cosmic rays when they are reflected from the interplanetary magnetic piston.   L. I. Dorman, V. Kh. Shogenov.
Geomagn. Aehron., Tom 22, 123 - 124 (1982). In Russian.

143.019 Fourth harmonic of the daily variation of cosmic rays during 1968-79.
S. P. Agrawal, S. P. Pathak, B. L. Mishra.
Indian J. Radio Space Phys., Vol. 10, 193 - 196 (1981). Abstr. in Phys. Abstr., Vol. 85, Abstr. 42555 (1982).

143.020 The second harmonic of the function of cosmic ray distribution in case of a strong magnetic field.
L. L. Kichatinov, Yu. G. Matyukhin.
Geomagn. Aehron., Tom 22, 192 - 196 (1982). In Russian.

143.021 Cosmic rays from Jupiter.
B. Mitra, S. K. Bose, S. R. Ganguly.
Lett. Nuovo Cimento, Vol. 33, Ser. 2, 9 - 13 (1982). – Abstr. in Phys. Abstr., Vol. 85, Abstr. 48118 (1982).

143.022 Dependence of the radiation dose aboard the Salyut 6 station on the indices of solar and geomagnetic activity.   V. A. Bondarenko, A. V. Kolomenskij, A. P. Tibanov, M. V. Tel'tsov, V. I. Shumshurov.
Kosm. Issled., Tom 20, 151 - 153 (1982). In Russian.

143.023 Reversal of the cosmic ray density gradient perpendicular to the ecliptic plane.
D. B. Swinson, H. Kananen.
J. Geophys. Res., Vol. 87, 1685 - 1687 (1982).

Annual averages of the diurnal variation in cosmic ray intensity have been determined as a function of the sense of the interplanetary magnetic field for the years 1965–1975. These data point to a cosmic ray density gradient, perpendicular to the ecliptic plane, pointing southward prior to 1969 and changing to a northward pointing gradient after the reversal of the sun's polar magnetic field in 1969–1971. This result supports numerical calculations for the prereversal and postreversal field configurations at intermediate and high cosmic ray rigidities.

143.024 Could primordial black holes be the source of the cosmic ray antiprotons?   M. S. Turner.
Nature, Vol. 297, 379 - 381 (1982).

The author explores the possibility that primordial black holes (PBHs) of mass $\sim 10^{13} - 10^{15}$ g which evaporated after decoupling ($t \gtrsim 10^{14}$ s) are the primary source of the cosmic ray $\bar{p}$s. The PBH scenario predicts a universal $\bar{p}/p$ ratio of $\sim 10^{-4} - 10^{-8}$.

143.025 Measurement of the solar diurnal anisotropy of the cosmic-ray albedo neutron flux.   S. O. Ifedili.
Sol. Phys., Vol. 76, 393 - 398 (1982).

The solar diurnal anisotropy of the cosmic-ray albedo neutron flux has been measured by a neutron detector on board the OGO-6 satellite. On the average the diurnal amplitudes and phases of the cosmic ray albedo neutron flux ($\leqslant 10$ MeV) were respectively 0.18 (± 0.02)% and 15(± 1) hr LT though there were substantial fluctuations of a few days duration which did not depend on the solar sector structure polarity and a 27-day periodicity in the diurnal amplitudes which was associated with the Sun's rotation.

143.026 The heliospheric intensity gradients of the anomalous He^4 and the galactic cosmic-ray components.
R. B. McKibben, K. R. Pyle, J. A. Simpson.
Astrophys. J., Lett., Vol. 257, L41 - L46 (1982).

Measurements with the Pioneer 10 and 11 spacecraft over the radial range from 1 to $\sim$ 25 AU during the period 1972 - 1981 show that at solar minimum (1972 - 1977) the radial gradient for the anomalous helium component [He(A)] in the energy range from 11 to 20 MeV n^{-1} is $\sim$ 14% AU^{-1}, significantly smaller than predicted for either charge state [He$^+$(A) or He^{++}(A)]. Although the He(A) disappeared at 1 AU by 1979 as a result of increasing solar modulation, it remained present in the outer solar system ($R \lesssim$ 20 AU) into 1980, when the solar polar magnetic fields were observed to reverse and solar activity was near its maximum levels. By 1981, He(A) had essentially disappeared at Pioneer 10 ($R \approx$ 24 AU) and the radial gradient had decreased to levels consistent with those expected for galactic helium with a normal modulated spectrum.

143.027 Fire-ball analysis of high energy cosmic ray interactions.   R. Hasan, M. S. Swami.
Prog. Theor. Phys., Vol. 66, 2291 - 2295 (1981). – Abstr. in Phys. Abstr., Vol. 85, Abstr. 49023 (1982).

143.028 Could the cosmic-ray knee at 10^{15} eV be due to the finite time for supernova shock acceleration?
M. A. Forman.
Bull. American Astron. Soc., Vol. 13, 796 (1981). – Abstract.

143.029 The energy spectrum of cosmic rays in the Metagalaxy.   M. M. Shapiro, R. Silberberg.
Bull. American Astron. Soc., Vol. 13, 868 (1981). – Abstract.

143.030 Detection of high energy cosmic neutrinos by means of atmospheric fluorescence.

J. Linsley.
Bull. American Astron. Soc., Vol. 13, 883 - 884 (1981). Abstract.

143.031 Proton collisions in cosmic ray sources.
S. Singh, P. N. Okeke.
Astrophys. Space Sci., Vol. 83, 75 - 79 (1982).

The authors have studied the possible proton collisions in cosmic ray sources. On the basis of the calculations of background temperature, the operational domains of various proton collisions are suggested. The energy rate loss through the photo-neutrino processes on protons is computed and the energy distribution of the secondary particles in various collisions is discussed.

143.032 Trajectory parameterization – a new approach to the study of the cosmic ray penumbra.   D. J. Cooke.
Geophys. Res. Lett., Vol. 9, 591 - 594 (1982).

A new approach to the examination of the structure of the cosmic ray penumbra has been developed, which, while utilizing the speed, efficiency and "real" geomagnetic field modeling capabilities of the digital computer, yields an analytical insight equivalent to that of the earlier and elegant approaches of Stormer, and of Lemaitre and Vallarta.

143.033 Cosmic accelerators.
T. Montmerle, C. Cesarsky.
Recherche, Vol. 13, No. 129, p. 82 - 85 (1982). In French. Abstr. in Phys. Abstr., Vol. 85, Abstr. 58209 (1982).

143.034 Limits to extragalactic cosmic rays from gamma-ray fluxes.   S. S. Said, A. W. Wolfendale, M. Giler,
J. Wdowczyk.
J. Phys. G, Vol. 8, 383 - 391 (1982). – Abstr. in Phys. Abstr., Vol. 85, Abstr. 58211 (1982).

143.035 ^{22}Ne and the nature of low energy cosmic ray sources.   S. Ramadurai.
Mon. Not. R. Astron. Soc., Vol. 200, 123 - 126 (1982).

The recent measurements on the isotopic abundances in cosmic rays of energy less than 500 MeV/n have confirmed the over-abundance of ^{22}Ne by more than a factor of 3 compared with the solar abundance. No comparable enhancements have been seen in the case of any other isotopes, especially that of ^{13}C. This is shown to be in disagreement with the models in which cosmic rays originate in the grains from novae.

143.036 Prospects for cosmic ray physics around 10^{15} eV.
T. K. Gaisser.
Cosmology and particles, (see 012.046), p. 13 - 21 (1982).

Knowledge of the chemical composition of primary cosmic ray nuclei is a prerequisite to understanding the origin, acceleration and propagation of the cosmic rays. Details of composition up to 100 GeV/nucleon have been studied extensively for many years. In the air shower energy region from 10^{14} to 10^{20} eV the subject of cosmic ray composition is still in its infancy. New air shower experiments now operating promise to give qualitatively new data on longitudinal development of individual, large showers ($10^{17} - 10^{20}$ eV).

143.037 Composition of cosmic rays at high energies.
G. B. Yodh.
Cosmology and particles, (see 012.046), p. 23 - 47 (1982).

A critical analysis of experiments pertaining to the composition of primary cosmic rays from 10^2 to 10^{11} GeV is presented. One finds that the composition of cosmic rays is continually varying from being predominantly light at 100 GeV to mostly heavy at about 10^6 GeV and reversing back to predominantly light above 10^8 GeV. The "nested leaky box" model of origin, propagation and acceleration of cosmic rays, with a galactic cut off at a rigidity of 10^5 GV/nucleon and a dominant extra-galactic component above 10^8 GeV per nucleus naturally predicts this behavior.

143.038 Cosmic ray anisotropy: $10^{12} - 10^{20}$ eV.
A. A. Watson.
Cosmology and particles, (see 012.046), p. 49 - 67 (1982).

The results of experiments designed to study the arrival direction distribution of cosmic rays of energy $10^{12} - 10^{20}$ eV are reviewed. It is shown that at all energies there is evidence for anisotropy, the amplitude of which ranges from 0.075% at the lowest energies to 90 ± 20% above $4 \cdot 10^{19}$ eV. The increase of anisotropy with energy is not smooth, showing features which occur at energies similar to those at which features are observed in the cosmic ray energy spectrum. At least up to $2 \cdot 10^{17}$ eV it seems probable that the acceleration sites lie within our Galaxy, and it is hard to escape the conclusion that particles of energy $> 10^{19}$ eV are extragalactic.

143.039 The study of air showers by the Fly's-Eye.
J. W. Elbert.
Cosmology and particles, (see 012.046), p. 69 - 84 (1982).

The Fly's-Eye is an optical-electronic system designed to detect scintillation and Cherenkov light from ultra-high-energy cosmic ray air showers. For about 9 months, 48 out of the total 67 mirrors of the system have been operated. Air showers with greater than 10^{17} eV are being detected and reconstructed by the University of Utah Cosmic Ray Group. Besides observing the spectrum and anisotropy of air showers from 10^{17} to 10^{20} eV cosmic ray nuclei, the system may also detect point sources of 10^{14} to 10^{16} eV γ-rays. Preliminary data from a search for γ-rays from the vicinity of the Crab Pulsar are presented.

143.040 The processing of nuclei by electrons in active galaxies.   R. Schaeffer.
Cosmology and particles, (see 012.046), p. 85 - 87 (1982).

The cross-sections for spallation of nuclei by electrons and γ rays are calculated and some astrophysical consequences are presented.

Catalogue of highest energy cosmic rays. Giant extensive air showers. No. 1. Volcano Ranch, Haverah Park. See Abstr. 002.075.

Spatial distribution of galactic cosmic ray density and flux.   See Abstr. 003.015.

Variability of the general magnetic field of the sun as a source of cosmic ray modulation. See Abstr. 003.027.

Cosmic radiation. Historical review. See Abstr. 003.038.

Cosmic rays and solar wind. See Abstr. 003.079.

Quasi-periodic variations of the intensity and anisotropy of cosmic radiation. See Abstr. 003.101.

Neutrino astronomy using SOCRAS, a satellite observatory for cosmic ray air showers. See Abstr. 032.518.

DUMAND (*Deep Undersea Muon and Neutrino Detection*) — an undersea neutrino telescope. See Abstr. 034.056.

Cosmic-ray-produced stable nuclides: various production rates and their implications. See Abstr. 061.037.

Does the galactic synchrotron radio background originate in old supernova remnants?   See Abstr. 066.005.

Meteorite erosion and cosmic ray variations. See Abstr. 105.012.

Origins of the low-energy relativistic interplanetary electrons.   See Abstr. 106.023.

The galactic fountain, observations of extragalactic radio sources, and the cosmic ray halo. See Abstr. 131.029.

High energy gamma rays from cosmic ray nucleons. See Abstr. 142.503.

COS-B gamma-ray measurements, cosmic rays and the local interstellar medium.   See Abstr. 142.507.

Energetic phenomena observed from orbit around Venus.   See Abstr. 142.517.

Stellar Systems, Galaxy, Extragalactic Objects, Cosmology

151 Stellar Systems (Kinematics, Dynamics, Evolution)

151.001 A model for elliptical radio galaxies with dust lanes.
T. S. van Albada, C. G. Kotanyi, M. Schwarzschild.
Mon. Not. R. Astron. Soc., Vol. 198, 303 - 310 (1982).

The authors describe the stationary states of motion of gas in a triaxial stellar system which rotates about one of its principal axes. Four dynamically different situations are possible, in two of these the gas layer is warped. The morphology of elliptical radio galaxies with dust lanes such as NGC 5128 (Cen A) and M 84 can be understood on the basis of this model; transient phenomena are not required.

151.002 Using gaseous disks to probe the geometric structure of elliptical galaxies.
J. E. Tohline, G. F. Simonson, N. Caldwell.
Astrophys. J., Vol. 252, 92 - 101, plates 1, 2 (1982).

Cold gas residing in the core of an elliptical galaxy should settle into a preferred plane of the galaxy regardless of the orientation that the gas's orbital angular momentum vector may have had when the gas entered the galaxy. The preferred plane into which the gas settles depends only on the gross geometric shape of the elliptical galaxy. A numerical model is used to show that the time scale on which this settling occurs is reasonably short if the gas disk is less than 10 kpc in size. Knowing the preferred orientation of a gaseous disk in an elliptical galaxy, the authors are able to decipher the geometric structure of 12 galaxies that possess gas in their cores. Both prolate and oblate elliptical galaxies are found to exist. The authors have also derived the intrinsic axial ratio of most of these galaxies.

151.003 Spectral stellar dynamics.
J. Binney, D. Spergel.
Astrophys. J., Vol. 252, 308 - 321 (1982).

The authors analyze orbits in two-dimensional bar like potentials in terms of the Fourier transforms of the coordinates. An orbit of typical complexity may be well represented by no more than eight sinusoidal terms per coordinate. The authors interpret the elementary frequencies of loop orbits as (1) the angular frequency of the rotating coordinate system in which the orbit most nearly closes, and (2) the orbit's fundamental frequency in that coordinate system. The elementary frequencies of a box orbit are interpreted as (1) the libration frequency of the rocking coordinate system in which the orbit is most nearly an axial orbit, and (2) the fundamental frequency of the axial orbit.

151.004 Scale-free models of galaxies. II. A complete survey of orbits.   D. O. Richstone.
Astrophys. J., Vol. 252, 496 - 507 (1982).

A complete set of orbits starting at over 400 distinct points spread out in the phase space of an oblate scale-free potential is studied. Each orbit is followed for a time corresponding to a Hubble time in a realistic galaxy potential suitable for an E5 or E6 galaxy. None of the orbits is ergodic. All of the orbits are regular – they all visit a region at least one dimension smaller than expected from the classical integrals of motion. So, for all practical purposes, they have an extra non-classical isolating integral.

151.005 Spiral arms and a supernova-dominated interstellar medium.   P. W. J. L. Brand, S. R. Heathcote.
Mon. Not. R. Astron. Soc., Vol. 198, 545 - 562 (1982).

Models of the interstellar medium (ISM) utilizing the large energy output of supernovae to determine the average kinematical properties of the gas, are subjected to an imposed (spiral) density wave. The consequent appearance of the ISM is considered. In particular the McKee–Ostriker model with cloud evaporation is used, but it is shown that the overall appearance of the galaxy model does not change significantly if a modification of Cox's mechanism, with no cloud evaporation, is incorporated. The authors find that a spiral density wave shock can only be self-sustaining if quite restrictive conditions are imposed on the values of the galactic supernova rate and the mean interstellar gas density.

151.006 Chemical evolution of galaxies – II. Variation of the heavy element yield with Z.
M. Peimbert, A. Serrano.
Mon. Not. R. Astron. Soc., Vol. 198, 563 - 572 (1982).

Recent determinations of the heavy element yield, p, are reviewed. The data indicate that p increases with metallicity or that the accretion rate is larger than the star formation rate in many galaxies. It is proposed that $p = 0.002 + 0.6\,Z$; with this relation it is possible to explain the heavy element abundances derived from irregular galaxies, blue compact galaxies, the Galaxy and M83. Within the uncertainties of the observations all the objects but one fall in the Z versus $\log M_{gas}/M_{tot}$ region limited by the simple model and an infall model with M_{gas} = const.

151.007 A comparison of simulated galaxy clustering models with observations.
S. Zięba, M. Urbanik, K. Rudnicki, S. J. Aarseth.
Astron. Astrophys., Vol. 105, 21 - 22 (1982).

The authors compare a sequence of N-body models of galaxy clustering with the actual distribution of galaxies using the method of statistical reduction. The more evolved dynamical models show some similarities with the observed sample. However, the simulated models indicate a pronounced maximum for a rather small characteristic size of clusters whereas the observed cluster size distribution is essentially flat.

151.008 Gaseous spiral arms produced by oval distortions in disc galaxies.   S.-A. Sørensen, T. Matsuda.
Mon. Not. R. Astron. Soc., Vol. 198, 865 - 872 (1982).

The response in a non-self-gravitating gaseous disc to a weak oval distortion of an axisymmetric gravitational potential is studied through numerical experiments. The response in the gas is in the form of a well-defined spiral pattern, which is well correlated with the two outer resonances. The extent and orientation of a central distorted ring is found to depend on the innermost Lindblad resonance.

151.009 Environmental effects on galaxies in clusters.
R. H. Miller, B. F. Smith.
Astrophys. J., Vol. 253, 58 - 69, plates 2 - 5 (1982).

Influences from the cluster environment build up steadily over long times so that the internal dynamics of a galaxy in a cluster change appreciably over a cluster crossing time. The authors present a first cut at studying cluster influences on the internal dynamics of galaxies by means of numerical experiments started from a rotating barlike galaxy placed in an external force field like that which a galaxy sees in a cluster. Both pattern motion and observable rotation are tidally braked by the cluster force field. The experiments verify that the braking rate scales inversely as the square of the cluster crossing time. The final shapes of tidally braked galaxies need not be spherical.

151.010 Coherent galactic oscillations.
P. E. Seiden, L. S. Schulman, J. V. Feitzinger.
Astrophys. J., Vol. 253, 91 - 100 (1982).

The stochastic self-propagating star formation model has been extended to include explicitly the interaction between stars and the interstellar gas. In addition to the constant star formation mode previously found, an oscillating mode has been discovered. In this mode, both the star formation rate and the density of active gas (gas available for star formation) are periodic functions of time. The oscillations range from almost sinusoidal behavior to regular bursting, depending on the parameters of the model. Observational evidence for oscillations and bursting is surveyed. Accompanying the oscillations is the frequent appearance of two-armed global spiral patterns. These are relatively smooth and massive structures, reminiscent of grand design patterns, in contrast to the many-armed, more filamentary patterns produced in the constant mode. One-armed patterns and ring structures are also produced.

151.011 A generating mechanism of spiral structure in barred galaxies.   K. O. Thielheim, H. Wolff.
Mon. Not. R. Astron. Soc., Vol. 199, 151 - 169 (1982).

The time-dependent response of non-interacting stars to growing oval distortions in disc galaxies is calculated by following their motion numerically and Fourier-analysing their positions. Long-lived spiral density waves are found. The linear epicyclic approximation is used to develop an analytical description of the generating mechanism.

151.012 A stationary and a slowly rotating model of a triaxial elliptical galaxy.
A. Wilkinson, R. A. James.
Mon. Not. R. Astron. Soc., Vol. 199, 171 - 196 (1982).

Evolving models of a stationary and a slowly rotating triaxial ellipsoidal system are presented. The final large-scale properties of the models are in general promisingly similar to those of observed elliptical galaxies, and there is no suggestion that the triaxial shape disappears over 4×10^9 years. The orbits of 2 per cent of the 25 000 stars have been followed in detail, and these show that essentially only two simple orbital families are involved.

151.013 Estimation of the thickness of spiral galaxies.
Y. Tong, S.-g. Wu, Q.-h. Peng.
Acta Astrophys. Sinica, Vol. 2, 30 - 34 (1982). In Chinese.

151.014 Mass segregation, relaxation, and the Coulomb logarithm in N-body systems.
R. T. Farouki, E. E. Salpeter.
Astrophys. J., Vol. 253, 512 - 519 (1982).

A series of numerical N-body calculations were carried out to determine the time scale for mass segregation in a cluster consisting of particles with a range of masses. The numerical results are used to construct a simple analytic approximate expression for this time scale which can be used for various mass distribution functions. This expression contains the so-called "Coulomb logarithm", and the upper cutoff in this logarithm was found to be close to the cluster radius. For Schechter-type mass functions, the expression for

the time scale is accurate to about 25% and can be applied to galaxy clusters for various assumed radii of galaxy halos.

151.015 Magnetic field configuration of the heliosphere and spiral galaxies.
S.-I. Akasofu, K. Hakamada.
Astrophys. J., Vol. 253, 552 - 555 (1982).

In the heliosphere, a double spiral armlike magnetic field structure results when the solar magnetic equator does not coincide with the ecliptic plane and the solar wind has a positive gradient toward higher latitudes. It is suggested that the two spiral arms of a spiral galaxy may be produced by a similar cause.

151.016 Star clusters containing massive, central black holes. IV. Galactic tidal fields.
M. J. Duncan, S. L. Shapiro.
Astrophys. J., Vol. 253, 921 - 938 (1982).

Monte Carlo simulations of star cluster evolution are presented. The Monte Carlo scheme described previously is modified to incorporate a Galactic tidal radius r_t, beyond which stars are stripped from the cluster. It is found that the late stages of cluster evolution are qualitatively unchanged by the introduction of a finite tidal cutoff. For clusters without a massive, central black hole the core collapses homologously with homology parameters independent of r_t. For clusters possessing a central black hole, core collapse is eventually reversed by the heat flux from stellar disruption by the hole. The system attains a quasi-stationary, expanding state (by which time the hole has grown to several thousand solar masses). This state appears to be (roughly) independent of (1) the initial hole mass, (2) the time during core evolution at which the hole is introduced, and (3) the value of r_t. A simple homological model for core evolution with and without black holes is presented. Numerical integrations of the resulting homological equations are in good agreement with the Monte Carlo "data".

151.017 Some axisymmetric self-similar galaxy models.
R. H. Miller.
Astrophys. J., Vol. 254, 75 - 76 (1982).

Distribution functions are presented for some axisymmetric self-similar galaxy models with flat rotation curves. In particular, velocity distribution functions for the logarithmic potentials studied by Richstone are recorded. These distribution functions depend on two integrals only. Other distribution functions (possibly involving more or other integrals) associated with the same mass model differ only by an additive function that integrates over velocities to yield zero density everywhere. However, a given distribution function can be associated with different mass models.

151.018 Transitions between epicyclic stellar orbits induced by massive gas clouds.   V. Icke.
Astrophys. J., Vol. 254, 517 - 537 (1982).

Cool gas clouds are probably the most massive objects in the plane of the galaxy. Therefore they can be expected to have considerable influence on the motion of stars. The author presents the results of extensive calculations of the gravitational scattering of stars off gas clouds. The motion of the stars is treated in the epicyclic approximation. A large series of star-cloud encounters is computed numerically; the results are used to calculate the diffusion of an ensemble of stars in phase space by assuming that the diffusion is a linear Markov process in which the transition probabilities are given by the computed encounters. It is concluded that (1) the observed dependence of the velocity dispersion on the age of stars in the solar neighborhood can be explained by repeated gravitational encounters between stars and gas clouds, (2) the mass of these clouds needs to be in the range 2×10^5 to $10^6 M_\odot$; (3) if the cloud lifetimes are short compared with the galactic year, the appearing and disappearing of clouds generates a gravitational noise that can induce the required stellar velocities if the clouds have lifetimes $\sim 10^7$ yr and peak masses of $\sim 10^6 M_\odot$.

151.019 Dynamical evolution of non-point body clusters.
V. A. Churkin.
Pis'ma Astron. Zh., Tom 8, 121 - 124 (1982). In Russian.
English translation in Soviet Astron. Lett., Vol. 8.

Peculiarities of the dynamic evolution of clusters of non-point gravitating bodies (galaxies) are considered. These clusters are shown to collapse in the course of their evolution.

151.020 On the history of the wave theory of spiral structure.
I. L. Genkin, I. I. Pasha.
Astron. Zh., Tom 59, 183 - 185 (1982). In Russian. English translation in Soviet Astron., Vol. 26, No. 1.

It is shown that B. Lindblad is not only the author of the idea of density waves in spiral galaxies but also the direct founder of the general physical and analytical principles of the modern linear wave theory of spirals.

151.021 Computer simulations of close encounters between binary and single stars: the effect of the impact velocity and the stellar masses.
L. W. Fullerton, J. G. Hills.
Astron. J., Vol. 87, 175 - 183 (1982).

A total of 45 760 simulated encounters between binary and single stars were run to study the effect of impact velocity and the masses of the three stars on the outcome of the collisions.

151.022 The ballistic particle model and the vertex deviation of young stars near the sun.
J. L. Hilton, F. Bash.
Astrophys. J., Vol. 255, 217 - 226 (1982).

The authors have examined the connection between the initial motions, at birth, of O and B stars and the long-standing problem of the vertex deviation. The ballistic particle model for spiral arm star formation is used to predict the velocities of young stars near the sun. The model is seen to be consistent with observations in that it predicts that O and B stars will have a distribution of velocities in which the direction of maximum velocity dispersion points toward $l \approx 320°$. The predicted values of the vertex deviation away from the galactic center and the velocity centroid with respect to the dynamical, local standard of rest are very near the observed values. Complications due to the effect of the second harmonic resonance in the gas and star formation along the Orion spur are discussed.

151.023 The equilibrium and bifurcation of rotating stellar systems.
R. Wiegandt.
Astron. Astrophys., Vol. 105, 326 - 328 (1982).

Using the moment equations of stellar dynamics up to the third order the conditions of equilibrium for uniformly rotating stellar systems with non-isotropic velocity dispersions can be obtained in the virial tensor form. Proceeding along the axisymmetric series of equilibrium a point can be found, where − as in the fluid case − figures with three-axial symmetry are possible as well as axisymmetric ones. This point is an upper limit for the point of bifurcation and can be described by an equation which is formally identical to the corresponding equation of hydrodynamics.

151.024 The stability of inhomogeneous axisymmetric stellar systems.
R. Wiegandt.
Astron. Astrophys., Vol. 106, 240 - 244 (1982).

In this paper the investigation of axisymmetric stellar systems with respect to the point of bifurcation and dynamical stability is pursued. Using the first variation of the second order virial equations a characteristic equation for the frequency of linear perturbations is developed, and it can be shown

that a neutral mode of oscillation occurs at the point of bifurcation, which itself lies in a region of dynamical stability. A condition for dynamical stability different from the classical one is found. Finally, the general results are applied to a class of inhomogeneous ellipsoidal models.

151.025 Excitation of warps in galaxies: fluid model of disk-halo interaction.
G. Bertin, S. Casertano.
Astron. Astrophys., Vol. 106, 274 - 286 (1982).

The authors have studied the properties of bending waves in a rotating fluid layer of finite thickness in the precence of a nonrotating thicker layer. Their analysis supports the view that a "two-stream instability" between the disk and the halo components in galaxies can be at the basis of warp excitation. Comparison with previous calculations on an infinitesimally thin disk model suggests that the authors' simple fluid scheme could be usefully applied to investigate further the dynamics of galaxy warps.

151.026 Changes in sizes and shapes of spherical galaxies in head-on collisions.
F. Ahmed, S. M. Alladin.
J. Astrophys. Astron., Vol. 2, 349 - 363 (1981).

Head-on collisions of two identical spherical galaxies are studied for two initial velocities (1) nearly equal to and (2) greater than the capture velocity. Orbits of about 500 representative stars are computed taking into account the effects of dynamical friction in the motion of the galaxies. From the computer studies the changes in the structure of the galaxies are deduced.

151.027 Vlasov simulations of stellar disks.
M. Nishida, Y. Watanabe.
Astron. Her., Vol. 74, No. 2, p. 43 - 49 (1981). In Japanese.
Abstr. in Phys. Abstr., Vol. 85, Abstr. 18782 (1982).

151.028 Monte-Carlo simulations of galaxy systems. I. The Local Supercluster.
H. T. MacGillivray,
R. J. Dodd, B. V. McNally, J. F. Lightfoot, H. G. Corwin, Jr.,
S. R. Heathcote.
Astrophys. Space Sci., Vol. 81, 231 - 250 (1982).

A computer technique is described for the Monte-Carlo simulation of the static properties (positions, magnitudes, colours, sizes, orientations and shapes) of galaxies in three-dimensional space, and the projection of those properties onto the sphere of the sky. In the present paper, the technique is used to simulate the observed geometrical properties of galaxies in the Local Supercluster. The observational data are consistent with a two-part model in which the galaxies closest to the Local Supercluster plane have random orientations while the galaxies outside the Local Supercluster plane are aligned with some dispersion about the plane.

151.029 The propagation of density waves in a galaxy.
C. K. Terzides.
Astrophys. Space Sci., Vol. 81, 345 - 355 (1982).

The author examines the behaviour of the solutions of the Lin−Shu dispersion relation in terms of the stability parameter Q and he finds that a modification of the critical values of the dispersion speeds does not change drastically this behaviour. The solutions of the Lynden Bell-Kalnajs dispersion relation behave in a more complicated way. The use of this dispersion relation imposes a different definition of the critical values of the dispersion speeds which are in general larger than the corresponding values of the Lin−Shu case.

151.030 On a model accounting for various morphological structures of galaxies.
R. Louise.
Astrophys. Space Sci., Vol. 81, 387 - 395 (1982). In French.

The density wave theory of spiral galaxies is discussed in an attempt to derive the various morphological features of real galaxies (rings, bars, bar-spirals) from a unified model of galactic dynamics. General solutions of Poisson's equation for both disk and spiral arm populations are derived.

151.031 Corrigendum: tidal disruption of a disk galaxy.
G. Som Sunder, T. K. Chaterjee.
Astrophys. Space Sci., Vol. 81, 479 - 481 (1982). – See abstracts 29.151.022, 29.151.032.

151.032 A method to solve the stellar dynamical problem.
E. Bettwieser, H. W. Yorke.
Mitt. Astron. Ges., Nr. 55, p. 120 (1982). – Abstract.

151.033 Ableitung eines Wechselwirkungsterms für inelastische Begegnungen zwischen Galaxien in Haufen.
R. Wielen, A. Goeres, P. Schwekendiek.
Mitt. Astron. Ges., Nr. 55, p. 192 - 193 (1982).

151.034 Gas ablation from a disk galaxy by dynamical pressure of intergalactic gas.
K. Toyama, S. Ikeuchi.
Astron. Her., Vol. 74, No. 4, p. 96 - 102 (1981). In Japanese. Abstr. in Phys. Abstr., Vol. 85, Abstr. 28559 (1982).

151.035 Periodic orbits in triaxial galactic models.
P. Magnenat.
Astron. Astrophys., Vol. 108, 89 - 94 (1982).
The stability properties of the main simple elliptical periodic orbits present in a triaxial logarithmic potential are numerically investigated. The effects of the potential geometry, system rotation and hamiltonian value on the stability indices are examined.

151.036 Generation of spiral density waves by a bar in differentially rotating disks.
V. I. Korchagin, Yu. G. Shevelev.
Astrofizika, Tom 17, 455 - 468 (1981). In Russian. English translation in Astrophysics, Vol. 17, No. 3.
The time dependent response of a differentially rotating gaseous disk forced by a bar for different values of its angular speed is numerically investigated. The generation of a large-scale trailing spiral pattern takes place if the bar leads to the rotation of most parts of the disk. The natural feature of large-scale patterns is the appearence of ring-like structures. The quasistationary wave pattern is established for sufficiently slow inclusion of the bar potential. The peaks of gas density appear on the leading edges of the bar.

151.037 The nonlinear theory of gravitational instability in an expanding collisionless medium.
A. M. Shukurov.
Astrofizika, Tom 17, 469 - 485 (1981). In Russian. English translation in Astrophysics, Vol. 17, No. 3.
The development of Newtonian nonlinear density perturbations in a non-collisional expanding medium possessing pressure caused by chaotic motions of particles is considered. Chaotic motions of the particles prevent appearance of density singularities which are the characteristic features of the theory of nonlinear perturbations in a noncollisional pressure-free medium. The author finds out the function describing the evolution of density perturbations in such a medium and derives relations for estimation of matter density, pressure, etc. corresponding to various initial temperatures.

151.038 On the nonlinear equations for the amplitude of spiral density waves.
V. I. Korchagin, P. I. Korchagin.
Astrofizika, Tom 17, 823 - 827 (1981). In Russian. English translation in Astrophysics, Vol. 17, No. 4.
A nonlinear equation for the amplitude of tightly wound spiral density waves is obtained without the approach of low velocity dispersion of flat subsystem of the galaxy. The values of nonlinear coefficients are calculated

for the parameters of spiral structure admitted in the Lin et. al. and Marochnik and Suchkov models, and the regions of modulation instability are determined.

151.039 Plane galactic orbits in stationary and time-dependent rotating bars.
H. Spreckels, K. O. Thielheim.
Astron. Astrophys., Vol. 108, 206 - 212 (1982).
The characteristics of plane galactic orbits in a model galaxy composed of an axially symmetric background and a weak time-dependent bar potential are investigated in the region between the centre of the galaxy and corotation. The authors studied the number and stability of the periodic orbits and the populations of non-periodic orbits following them in the case of a stationary oval distortion with varying strength and pattern velocity.

151.040 Density wave theory for spiral galaxies: effects of resonant stars at corotation.
G. Bertin, J. Haass.
Astron. Astrophys., Vol. 108, 265 - 273 (1982).
Density wave theory has already shown that galaxy disks can support discrete self-excited spiral modes independently of resonant effects at corotation. However, in the problem of excitation of spiral modes in more general galaxy models, and especially in the mechanism for selection of the prominent structures, energy and angular momentum exchange with the corotating stars might play an important role. In order to study these resonant effects and give a quantitative discussion of their importance, the authors improve the simple asymptotic differential equation that was previously used to calculate spiral modes.

151.041 Simulations of galaxy mergers. J. V. Villumsen.
Mon. Not. R. Astron. Soc., Vol. 199, 493 - 516 (1982).
The author presents a number of N-body simulations of mergers of equal and unequal galaxies. The total number of particles in the system is 1200. Two galaxies, each a spherical non-rotating system with isothermal or Hubble density profile, are put in orbit around each other where tidal effects and dynamical friction lead to merging. The final system has a Hubble profile, and in some mergers an 'isothermal' halo. Equal mass mergers are more flattened than unequal mass mergers.

151.042 Gravitational interactions and the origin of the angular momenta of galaxies. P. S. Wesson.
Vistas Astron., Vol. 25, 411 - 418 (1981).
Gravitational interactions between a "subject" galaxy which is flattened and has a quadrupole moment, and other galaxies, result in a torque which acts on the subject galaxy. It has been often suggested that this torque, acting in the early Universe, was the mechanism responsible for a major part of the angular momenta of the galaxies. This hypothesis is reviewed, and two main conclusions are drawn. Firstly, the hypothesis cannot account for the angular momentum of the Galaxy if the latter is of conventional radius (≈ 10 kpc) and had a comparable radius when it formed. Secondly, the hypothesis can account for the angular momentum of the Galaxy if it had a large radius (≈ 100 kpc) when it formed.

151.043 The effect of gravitational radiation on the secular stability of a rotating, axisymmetric galaxy.
P. O. Vandervoort, J. R. Ipser.
Astrophys. J., Vol. 256, 497 - 504 (1982).
In a post-Newtonian approximation, it is shown that the emission of gravitational radiation makes a Dedekind-like mode of second-harmonic oscillation secularly unstable in all of the rotating members of a certain Maclaurin-like sequence of axisymmetric stellar systems. The limiting, nonrotating member of the sequence is a point of neutral stability, and it

is also the point of bifurcation at which a Dedekind-like sequence of nonaxisymmetric stellar systems branches off from the Maclaurin-like sequence. The Maclaurin-like and Dedekind-like systems that are considered here belong to the family of stellar systems, of the form of elliptical disks, that has been constructed by Freeman and by Hunter.

151.044   The dynamical instability of a rotating, axisymmetric galaxy with respect to a deformation into a bar.   P. O. Vandervoort.
Astrophys. J., Lett., Vol. 256, L41 - L44 (1982).

On the basis of the tensor virial equations of the second order, a criterion has been derived for the dynamical instability of any rotating, axisymmetric galaxy with respect to a bar mode. Although the new criterion for instability differs in important ways from the well-known conjecture by Ostriker and Peebles, it does not vitiate the constraints envisaged by those authors that considerations of stability would impose on galactic dynamics and galactic structure.

151.045   Local quadratic integral in stellar dynamics.   V. A. Antonov.
Vestn. LGU, 1981, No. 19, p. 97 - 105. In Russian. – Abstr. in Ref. zh., 51. Astron., 3.51.731 (1982).

151.046   The cosmic density wave and its observable vestige.
A series of maxima in the distribution of emission-line redshifts of the observed quasars.   Y. Liu, S. Cao.
Sci. Sinica, Ser. A, Vol. 25, 80 - 88 (1982).

The authors study the cosmic density wave according to gravitational instability in the expanding universe.

151.047   The permanence of density waves of galaxies and the numerical investigation of waser mechanism.
J. Xu.
Sci. Sinica, Ser. A, Vol. 25, 191 - 202 (1982).

This paper deals with in detail the permanence of the spiral structure of galaxies and the characters of waser mechanism. A simplified model of galaxy is adopted. Various dynamical characters of density waves are studied using numerical calculation methods.

151.048   Pre-galactic constraints on galactic evolution.   J. J. Hyun.
J. Korean Astron. Soc., Vol. 14, 51 - 54 (1981).

The characteristic size and mass of galaxies as pre-galactic constraints on galactic evolution are reviewed and the general constraints for their existence in gravitationally bound systems are examined. Implications on the self-similar gravitational clustering are also discussed.

151.049   Two component model of initial mass function.   S. S. Hong.
J. Korean Astron. Soc., Vol. 14, 89 - 93 (1981).

It is of fundamental importance to know the frequency distribution of stellar masses at birth (initial mass function or IMF) for understanding the Hubble sequence and the star formation processes as well. In particular, the IMF plays an important role in the chemical evolution of a galaxy by regulating yields of heavy elements. In this paper the author analyzes the IMF deduced by Miller and Scalo, and retrieves some of the imprints left from the early period of our Galactic evolution.

151.050   Fluctuations of "instantaneous" virial masses of stationary gravitating systems.   V. Yu. Terebizh.
Astron. Zh., Tom 59, 224 - 227 (1982). In Russian. English translation in Soviet Astron., Vol. 26, No. 2.

The use of the values of the kinetic and potential energy for an arbitrary moment of time instead of the time-averaged values required by the virial theorem leads to an estimate of the mass of the system $M_{VT}(t)$ not coinciding with the real

mass M. For a system of $N \gg 1$ bodies, the standard deviation is $\sigma(V) \cong N^{-1/2}$. The difference of M_{VT} from M for a given pair or group of galaxies should not be considered as an indication to the instability of the system.

151.051   Collisions in spherical stellar systems.   V. L. Polyachenko, I. G. Shukhman.
Astron. Zh., Tom 59, 228 - 236 (1982). In Russian. English translation in Soviet Astron., Vol. 26, No. 2.

From the set of equations for the stellar distribution function and for the two-particle correlation function in action angle variables, by averaging over fast finite motions the authors obtain the general expression for the collision term of a finite stellar system with " rare " Coulomb encounters.

151.052   On the N-body problem in Dirac's cosmology.   D. Lynden-Bell.
Observatory, Vol. 102, 86 - 87 (1982).

151.053   Dynamical evolution of globular clusters: recent theoretical ideas.   M. J. Duncan, S. L. Shapiro.
Astrophysical parameters for globular clusters, (see 012.023), p. 169 - 183 (1981).

In this paper the authors briefly summarize the theoretical underpinnings of research in large N-body systems and then highlight key results which have emerged in the last two or three years. They list the key dynamical parameters which characterize globular clusters and trace the dynamical "life history" of a globular cluster from its birth in the early Universe as a bound N-body system through secular core collapse, to its ultimate, but still uncertain final fate.

151.054   Stabilization of spiral density waves in flat galaxies for a hydrodynamical model.   W. Renz.
Application of modern dynamics to celestial mechanics and astrodynamics, (see 012.026), p. 356 (1982). – Abstract.

151.055   On the possible mechanism of the recurrent activity of galactic nuclei.
G. S. Bisnovatyj-Kogan, M. M. Romanova.
Pis'ma Astron. Zh., Tom 8, 263 - 269 (1982). In Russian. English translation in Soviet Astron. Lett., Vol. 8.

A mechanism of recurrent activity is proposed in a compact star cluster on the stage of contact collisions. It is shown that cooling and compression of the gas produced during collisions in a cluster with small angular momentum may lead to recurrent activity of galactic nuclei with the characteristic time between outbursts of $\sim 10^2$ - 10^5 years.

151.056   Velocidades radiales de galaxias australes.   J. L. Sérsic, A. Araujo, J. C. Arias.
Bol. Asoc. Argentina Astron., No. 20 - 24, p. 356 (1981). Abstract.

151.057   The orientation of gas disks in tumbling prolate galaxies.   J. E. Tohline, R. H. Durisen.
Astrophys. J., Vol. 257, 94 - 102 (1982).

In an earlier paper it has been suggested that dissipative processes coupled with the differential precession of particle orbits will cause gas flowing in a static spheroidal galaxy to settle into a preferred plane of the galaxy – namely, the equatorial plane of the spheroid. In this paper, the authors discuss the existence of preferred planes for gas which flows into a tumbling, prolate spheroidal (barlike) galaxy. They argue that this type of galaxy exhibits three different dynamical regions and that the preferred plane into which the gas will settle is not the same in all three regions. The authors suggest that NGC 2685 is an example of a galaxy in which orthogonally oriented disks have been observed. A dynamical classification scheme is outlined that encompasses all prolate elliptical galaxies and, perhaps, many barred spiral galaxies as well.

151.058 **On the stability of Schwarzschild's triaxial galaxy model.** B. F. Smith, R. H. Miller.
Astrophys. J., Vol. 257, 103 - 109, plates 3, 4 (1982).

The numerical model for a triaxial galaxy constructed by Schwarzschild has been tested by means of fully self-consistent three-dimensional n-body numerical experiments that use a representation of this model to start the integration. The model is rugged and robust. No growing disturbances with growth rate in excess of 0.5 per crossing time could be detected. This limit is strong enough to indicate that the model is free of rapidly growing dynamical instabilities that could limit its value in dynamical studies.

151.059 **Galactic evolution: a survey of recent progress.** K. M. Strom, S. E. Strom.
Science, Vol. 216, 571 - 580 (1982).

Current observational knowledge bearing on the evolution of elliptical and disk galaxies is reviewed. Particular emphasis is placed on identifying the factors that appear common to all galaxies of a particular type as opposed to those that seem to depend on environmental conditions. The success of various classes of galactic formation and evolution models used to confront these data is evaluated.

151.060 **Spiral structure of a galaxy and gas motion near and beyond the corotation resonance.**
Yu. N. Mishurov.
Astron. Zh., Tom 59, 483 - 489 (1982). In Russian. English translation in Soviet Astron., Vol. 26, No. 3.

The solution for the spiral gravitational potential corresponding to "long waves" in the region before corotation was continued beyond corotation. It is shown that in the outer region an additional arm arises connected with the "short wave" mode. Solutions describing the disturbed gas motion in such a potential were obtained in terms of the linear theory.

151.061 **On the dynamics of early stages of evolution of open star clusters. II.** V. M. Danilov.
Astron. Zh., Tom 59, 490 - 499 (1982). In Russian. English translation in Soviet Astron., Vol. 26, No. 3.

The problem of disruption of a young non-stationary star cluster after "impulsive" gas loss is considered. The high-velocity stars with positive energies in the first turn leave the cluster. Impossibility of common clusters disruption is shown.

151.062 **Two dimensional hydrodynamic calculations of disc galaxies.** P. R. Woodward.
Bull. American Astron. Soc., Vol. 13, 851 - 852 (1981). Abstract.

151.063 **The stability and masses of disc galaxies.** G. Efstathiou, G. Lake, J. Negroponte.
Bull. American Astron. Soc., Vol. 13, 852 (1981). − Abstract.

151.064 **Waves of sequential star formation.** G. B. Rybicki, L. L. Cowie.
Bull. American Astron. Soc., Vol. 13, 852 (1981). − Abstract.

151.065 **The effect of gravitational radiation on the secular stability of a rotating, axisymmetric galaxy.**
P. O. Vandervoort, J. R. Ipser.
Bull. American Astron. Soc., Vol. 13, 852 (1981). − Abstract.

151.066 **The effects of galaxy collisions on the galaxy luminosity function and the background light in clusters.** G. E. Miller, J. M. Scalo.
Bull. American Astron. Soc., Vol. 13, 870 (1981). − Abstract.

151.067 **Dissipative evolution of stellar systems. I. Cooling and heating of the system by binary stars.**
V. I. Dokuchaev, L. M. Ozernoj.
Astrofiz. Issled., Izv. Spets. Astrofiz. Obs., Tom 15, 26 - 49 (1982). In Russian.

Basic processes of the formation of hard binaries with binding energy exceeding the mean kinetic energy of single stars in stellar systems are considered. It is supposed that initially there are no any binaries in the system. The rate of hard binaries formation due to tidal capture of nonbound single stars is compared with detailed calculations of the rate of hard binaries formation due to tidal hardening of soft ones. Energy dissipation processes in stellar system due to bound binaries formation as well as the heating processes due to the hardening of hard binaries are considered.

151.068 **Dissipative evolution of stellar systems. II. Influence of binary stars on the evolution of globular clusters and galactic nuclei.** V. I. Dokuchaev, L. M. Ozernoj.
Astrofiz. Issled., Izv. Spets. Astrofiz. Obs., Tom 15, 50 - 71 (1982). In Russian.

Deviations from the evaporative classical evolution of large stellar systems which result from taking into account tidal interactions of stars are considered. These dissipative interactions which is the cooling factor would lead to formation of binary stars. The last can heat the stellar system during the energy hardening as a result of interactions with single stars of the system. Hardening of the relict binaries is also considered. The influence of the heating of the system by forming binaries on globular clusters becomes essential in the course of their evolution. The influence of tidal interactions on the evolution of galactic nuclei depends on the initial velocity dispersion of the stars in the system.

151.069 **Modulation instability of spiral density waves.** V. I. Korchagin, P. I. Korchagin.
Astrofizika, Tom 16, 273 - 284 (1980). In Russian. English translation in Astrophysics, Vol. 16, No. 2.

The modulation instability of tightly wrapped spiral density waves in a differentially rotating flat subsystem of a galaxy is considered. A nonlinear equation determining the evolution of the amplitude of spiral density waves is obtained. For the typical parameters of spiral pattern instability occurs if the space scale of modulation is larger than the critical one. The time scale of instability is less or equal to the time of one revolution of the system.

151.070 **Generation of large-scale structure by mass ejection from a galactic centre.**
V. I. Korchagin, V. F. Rakhimov.
Astrofizika, Tom 17, 371 - 382 (1981). In Russian. English translation in Astrophysics, Vol. 17, No. 2.

The ejection of massive objects from a galaxy nucleus in two opposite directions in the equatorial plane leads to the generation of the two-armed trailing spiral pattern shearing by differential rotation. The polarizational energy loss of ejected bodies is obtained. It is shown that the anomalous radio arms of the galaxy NGC 4258 may be the polarizational traces of ejected bodies.

151.071 **Encounters of binaries. I. Equal energies.** S. Mikkola.
Rep. Ser., Dep. Phys. Sci., Univ. Turku, Turku-FTL-R 25, 33 pp. (1982) = Turku Univ. Obs. Informo No. 61. ISBN 951-642-150-4, ISSN 0356-9896. − Submitted to Mon. Not. R. Astron. Soc.

151.072 **Gravitationally induced spurs in the disks of nearby spiral galaxies.** G. G. Byrd.
Bull. American Astron. Soc., Vol. 14, 578 (1982). − Abstract.

151.073 **The long term evolution of rotating stellar bars.** P. Carnevali.
Bull. American Astron. Soc., Vol. 14, 578 (1982). − Abstract.

151.074 **On the origin of spiral structure in the nuclei of galaxies.**   P. Pişmiş, E. Moreno.
Bull. American Astron. Soc., Vol. 14, 578 - 579 (1982).
Abstract.

151.075 **Three dimensional stochastic self-propagating star formation (SSPSF) in spiral galaxies.**
N. F. Comins, T. Statler, B. F. Smith.
Bull. American Astron. Soc., Vol. 14, 579 (1982). – Abstract.

151.076 **A numerical experiment on a rotating galactic bar.**   R. H. Miller, P. O. Vandervoort, D. E. Welty, B. F. Smith.
Bull. American Astron. Soc., Vol. 14, 579 (1982). – Abstract.

151.077 **Bar-driven spiral density waves in galaxies.**   W. V. Schempp.
Astrophys. J., Vol. 258, 96 - 109 (1982).
A series of two-dimensional hydrodynamical models is computed to test the gravitational response of a uniform, isothermal, massless gas disk to a uniformly rotating prolate potential. The parameters describing the mass distribution and rotation properties of the prolate bar are varied from model to model. The location and strength of the region in which hydrodynamic shocks are found are dependent on the radial distribution of mass within the bar and on the axial ratio of the bar potential. The radius of corotation determines whether the galaxy has a "ring" or "spiral" morphology. Most importantly, the range from barred, through intermediate cases, to nonbarred morphologies is proposed to be the consequence of variations in the amplitude of the prolate potential. It is suggested that the full spectrum of normal two-armed trailing spiral patterns observed in galaxies may be produced by rotating barlike potentials.

151.078 **On the interpretation of rotation curves measured at large galactocentric distances.**
J. N. Bahcall, M. Schmidt, R. M. Soneira.
Astrophys. J., Lett., Vol. 258, L23 - L27 (1982).
It is shown that the inferred slope of the density law for a massive galactic halo is sensitive to the distribution of mass within the radius of interest. Illustrative rotation curves are presented that are flat to ± 1 km s^{-1} in the region from 40 to 60 kpc, but which have local logarithmic slopes for the halo density law of about 2.7 instead of the conventionally assumed logarithmic slope of 2.0. The contributions of the disk and spheroid must be taken into account in the intermediate region ($\approx$ 10 - 30 kpc) in which most current observations are being carried out. These results complicate the interpretation of galaxy rotation curves at large galactocentric distances.

151.079 **Spectrum of oscillations of the gaseous subsystem of galaxies.**   S. M. Churilov, I. G. Shukhman.
Issled. po geomagn., aehron. i fiz. Solntsa, Moskva, 1981, No. 57, p. 75 - 79. In Russian. – Abstr. in Ref. zh., 51.
Astron., 6.51.603 (1982).

151.080 **Tidal interactions of disc galaxies.**   P. L. Palmer, J. Papaloizou.
Mon. Not. R. Astron. Soc., Vol. 199, 869 - 882 (1982).
The authors develop general analytic expressions for the energy and angular momentum transferred between a cold disc galaxy and a perturbing mass, for the case when the perturber is in a parabolic, coplanar orbit about the disc. Analogous expressions are developed for the case of flat hot discs. It is shown that tidal capture of galaxies cannot take place unless the galaxies actually collide. In the case of hot discs, however, if the distribution function is isotropic, it is shown that the change in energy of the disc is always positive, in spite of the details of the encounter.

151.081 **The stability and masses of disc galaxies.**   G. Efstathiou, G. Lake, J. Negroponte.
Mon. Not. R. Astron. Soc., Vol. 199, 1069 - 1088 (1982).
Using N-body experiments the authors investigate the global stability of a series of models designed to match the observed photometric and kinematic properties of disc galaxies. The models, therefore, have an exponential surface density profile and rotation curves which are flat at large radii. The results from the models are compared with available photometric and kinematic data of disc galaxies and are used to set limits on the mass-to-light ratio of the disc component. Normalizing the stellar population models of Larson & Tinsley (1978) and Tinsley (1981) to the results for Sc galaxies and data in the solar neighbourhood, the authors find that the later-type galaxies (Sd–Sm) may have a total H I mass comparable to or exceeding the mass in disc stars.

151.082 **Merging and stripping of haloes in binary galaxy systems.**   R. G. Carlberg.
Mon. Not. R. Astron. Soc., Vol. 199, 1159 - 1168 (1982).
N-body experiments of the collision of two systems composed of an isothermal halo surrounding a central particle with 5–10 per cent of the halo's mass are performed. The presence of the core slightly suppresses the interaction on the first crossing. Subsequent orbits decay, but on a time-scale longer than dynamical friction would predict. The halo density beyond the softening radius of the core is reduced from the initial value by an average of about 30 per cent. The density distribution, in the vicinity of the cores, becomes shallower at maximum orbital separation. Integrated H I profiles used as indicators of rotation curves are consistent with the operation of this process.

151.083 **The structure and evolution of nuclear discs.**   M. E. Bailey.
Mon. Not. R. Astron. Soc., Vol. 200, 247 - 262 (1982).
The evolution of a dense cold nuclear disc of gas, formed by inflowing stellar mass loss in a spherically galaxy, is investigated as a function of galaxy parameters. In all cases the disc growth, dominated by infall, continues until the onset of local gravitational instability close to the centre in a region within which the mean mass density $\bar{\rho}_*$ is less than $10^4 - 10^5 M_\odot$ pc^{-3}. At this point massive star formation can propagate sequentially through the disc, leading to a burst of star formation. The viscosity of the disc is greatly enhanced and viscous inflow leads to nuclear activity either by accretion on to a massive black hole or by the formation of a massive central gas cloud. In the latter case the cloud is presumed to collapse to form either a black hole or a spinar. The theory imposes constraints on both the black hole and spinar models of nuclear activity.

151.084 **Propagating star formation and the structure and evolution of galaxies.**
P. E. Seiden, H. Gerola.
Fundam. Cosmic Phys., Vol. 7, 241 - 311 (1982).
The authors give a short description of the early history of propagating star formation and describe the large scale process of propagating star formation. The phenomenon itself is described and the computer simulations given. Some applications of the process to various galactic types are shown and some analytical models for propagating star formation described. The observational evidence for propagating star formation is discussed.

151.085 **On the structure of normal spiral galaxies with finite disk thickness. Part I: The modal theory.**
Z. Y. Yue.
Geophys. Astrophys. Fluid Dyn., Vol. 20, 1 - 46 (1982).
It is shown that, in a galactic disk of finite thickness, the growth rate of a spiral mode is considerably reduced compared with that in the zero-thickness disk. Thus, the effect of finite

thickness tends to limit the growth rate so that it will not be so high as to prevent the spiral mode from reaching a stationary state. On the other hand, for reasonable galactic models, the reduced growth rate is still high enough to overcome the dissipation due to turbulence and the shock wave in the interstellar medium, so that a quasi-stationary spiral structure can be maintained.

151.086 On the structure of normal spiral galaxies with finite disk thickness. Part II: Three-dimensional asymptotic density-potential relation. Z. Y. Yue.
Geophys. Astrophys. Fluid Dyn., Vol. 20, 47 - 68 (1982).

A three-dimensional asymptotic density-potential relation is established for a spiral galaxy with a disk of finite thickness. This relation is accurate to the order $|kr|^{-3}$, where k is the radial wave number and r is the radial distance from the center of the galaxy. In the case of small disk thickness, some simpler and more explicit relations are obtained with the accuracy of order $|kr|^{-2}$, which is usually sufficient for the discussion of spiral mode. A density-potential relation, accurate to the order $|kr|^{-3}$, is also obtained for the zero-thickness disk.

151.087 On the structure of normal spiral galaxies with finite disk thickness. Part III: The asymptotic solution and the quantum condition of the general eigenvalue problem. Z. Y. Yue.
Geophys. Astrophys. Fluid Dyn., Vol. 20, 69 - 110 (1982).

The asymptotic solution and the quantum condition of an eigenvalue problem which appears in the density wave theory of the spiral galaxy are re-examined from a mathematical point of view.

151.088 A rigorous solution of Poisson's equation for spiral galaxies. Y. Tong, X. Zheng. Q. Peng.
Sci. Sinica, Ser. A, Vol. 25, 285 - 294 (1982).

151.089 A model of spiral galaxies and the distribution of matter within the Galaxy.
Y. Tong, X.- t. Zheng, Q.- h. Peng.
Acta Astron. Sinica, Vol. 23, 10 - 16 (1982). In Chinese.

151.090 Binaries in dense stellar systems. S. Mikkola.
Rep. Obs. Lund, No. 18, (see 012.044), p. 57 - 62 (1982).

151.091 Structure of a disk galaxy with a massive ellipsoidal halo. N. Tajima.
Sep. print Research Institute for Theoretical Physics, Hiroshima University, Takehara, Hiroshima, Japan. 16 pp. (1982)

A model is proposed for a disk galaxy which is located in the midst of a sufficiently massive halo with the shape of a Jacobi ellipsoid. It is clarified that in a stellar trajectory there are two eigenfrequencies splitted out due to a non-axisymmetry of the gravitational potential of the halo. Using an exact solution for the stellar motion, a distribution function is constructed to derive the rotation curve in the disk. The result is comparable with the observations.

151.092 New developments in the theory of spiral galaxies. K. O. Thielheim.
Sep. print Oxford International Symposium, 14 - 19 September 1981, 17 pp. (1982).

About 30% of all galaxies exhibit spiral forms, 60% are elliptical and 10% irregular. It is the objective of galactic dynamics to explain these structural features. A first generation of self-consistent N-body simulations indicates that ellipticals are equilibrium configurations of gravitationally interacting multi-particle systems. Recent progress has been made on the modal analysis of Freeman disks. In a second generation of N-body simulations spiral density waves have been reproduced in disk configurations. The author has

considered a mechanism by which spiral density waves are produced in the surrounding disk as a consequence of the slow increase of the quadrupole moment of a central oval shaped equilibrium configuration immersed in the disk.

151.093 The phase-space distribution function of galaxies in clusters and the secondary peak.
D. Trevese, A. Vignato.
Astron. Astrophys., Vol. 110, 238 - 240 (1982).

Under the hypothesis of spherical spatial symmetry, the authors analyze the relation between the galaxy velocity distribution and the density profile of clusters. They prove that an anisotropic, and oblate, velocity distribution is necessary to obtain stationary clusters which show the observed secondary maximum in the density profile. In the light of this result the authors discuss the existing models accounting for the secondary maximum.

151.094 Gas dynamics of flow past galaxies.
G. Shaviv, E. E. Salpeter.
Astron. Astrophys., Vol. 110, 300 - 315 (1982).

Gas dynamic calculations are carried out numerically for a (spherical) galaxy with gas emission moving through intergalactic gas. The authors find: (1) For nonviscous cases the rampressure stripping is never complete and Bremsstrahlung cooling induces an instability on a relatively short timescale. (2) Viscous dissipation and heat conduction by the plasma can prevent the instability and give rise to a steady state in which all the mass produced inside the galaxy is stripped away, irrespective of the flow conditions. (3) Mass-loss occurs mostly via the tail. (4) The form and viscous drag coefficients are of the order of 0.1 and hence cannot be neglected in cluster evolution. The gravitational drag is however very small.

151.095 Trapping of orbits in barred galaxies.
G. Contopoulos.
Compendium in astronomy, (see 003.012), p. 55 - 65 (1982).
Paper presented at the workshop "Orbits in Galaxies" in Geneva on 5 - 6 May 1980.

The main factors that affect the trapping of stellar orbits around stable periodic orbits in spiral or barred galaxies are reviewed. Such trapping enhances or destroys the imposed field, therefore it is important in producing self-consistent models of galaxies. Various possible models of self-consistent bars are considered. It seems that most bars end at corotation.

151.096 Two-fluid gravitational instabilities in the Galaxy.
C. J. Jog.
Extragalactic molecules, (see 012.048), p. 199 - 210 (1982).

The author briefly describes the theoretical work on two-fluid gravitational instabilities in the Galaxy that is currently in progress.

Symposium "Origin and evolution of celestial bodies", Leningrad, 1981, May 20 - 21 on the occasion of the 100th anniversary of the Astronomical Observatory of the Leningrad University. See Abstr. 011.012.

Density scaling of the angular momentum versus mass universal relationship. See Abstr. 042.006.

The stability of n-body hierarchical dynamical systems. See Abstr. 042.053.

Exploding dynamical systems. See Abstr. 042.059.

Compatibility conditions for a non-quadratic integral of motion. See Abstr. 042.065.

A two-fluid model for Population I. See Abstr. 062.130.

On a new integral of motion in relativistic galactic dynamics.   See Abstr. 066.025.

Kinetic theory in astrophysics and cosmology. See Abstr. 066.113.

The capture of interstellar comets. See Abstr. 102.006.

Motions of the cloud medium behind large scale galactic shocks.   See Abstr. 131.010.

A relaxation oscillation model for bursts of star formation in nuclei of galaxies.   See Abstr. 131.084.

Can giant molecular clouds form in spiral arms? See Abstr. 131.265.

The effects of galaxy dynamics and mergers on the creation of extended radio emission associated with cD galaxies.   See Abstr. 141.251.

On the dynamics of early stages of evolution of open star clusters. I.   See Abstr. 153.014.

Report of IAU Commission 37: Star clusters and associations (Amas stellaires et associations). See Abstr. 153.053.

[Fe/H] and orbital parameters for globular clusters. See Abstr. 154.040.

Mass of our Galaxy as determined from tidal radii of globular clusters and dwarf spheroidal galaxies. See Abstr. 155.002.

On the evidence of a massive galactic corona. See Abstr. 155.011.

Report of IAU Commission 33: Structure and dynamics of the galactic system (Structure et dynamique du système galactique).   See Abstr. 155.056.

Velocity fields in the barred spiral galaxies NGC 2525 and NGC 7741.   See Abstr. 158.011.

Kinematics and dynamics of the barred spiral galaxy NGC 1313.   See Abstr. 158.014.

A comparison of measured spiral arm properties with model predictions.   See Abstr. 158.026.

The inhibition of star formation in barred spiral galaxies.   See Abstr. 158.075.

The dynamics of the nucleus of M31. See Abstr. 158.154.

Rotation of the bulge components of disk galaxies. See Abstr. 158.155.

Velocity and velocity dispersion profiles in NGC 3115.   See Abstr. 158.156.

Rotation of the bulge components of barred galaxies.   See Abstr. 158.193.

M/L and velocity anisotropy from observations of spherical galaxies, or must M87 have a massive black hole? See Abstr. 158.320.

H I kinematics of spiral galaxies. See Abstr. 158.355.

Systematic dynamical properties of spiral galaxies. See Abstr. 158.357.

Report of IAU Commission 28: Galaxies (Galaxies). See Abstr. 158.358.

On the proper motion of the Magellanic Clouds and the halo mass of our Galaxy.   See Abstr. 159.002.

On the mass of the Local Group and the motion of its barycentre.   See Abstr. 160.008.

A note on the dynamics of galaxy clusters. See Abstr. 160.012.

Monte-Carlo simulations of galaxy systems. II. Static properties for galaxies in rich clusters. See Abstr. 160.074.

Monte-Carlo simulations of galaxy systems. III. Static properties for galaxies in supercluster filaments. See Abstr. 160.075.

Errata

151.901   Erratum: "The dynamics of binary galaxies."
[Mon. Not. R. Astron. Soc., Vol. 195, 1037 - 1056 (1981)].   S. D. M. White.
Mon. Not. R. Astron. Soc., Vol. 198, 383 (1982). — See Abstr. 29.151.068.

151.902   Erratum: "Angular momentum orientation of spiral galaxies" [Ann. Tokyo Astron. Obs., Second Ser., Vol. 18, 164 - 174 (1981)].
T. Yamagata, M. Hamabe, M. Iye.
Ann. Tokyo Astron. Obs., Second Ser., Vol. 18, 428 (1982). See Abstr. 30.151.092.

152 Stellar Associations

152.001 R associations. VI. The reddening law in dust clouds and the nature of early-type emission stars in nebulosity from a study of five associations.
W. Herbst, D. P. Miller, J. W. Warner, A. Herzog.
Astron. J., Vol. 87, 98 - 121 (1982).
Stars illuminating reflection nebulae visible on the POSS prints have been identified in five associations. The authors give positions, identification charts, *UBVRIKLMN* photometry, and spectral types for them. The reddening law applicable to the dust clouds in which these stars are embedded is found to be steeper than normal. Eighteen early-type stars with circumstellar shells are found in the five associations.

152.002 Neutraler Wasserstoff in der Cas OB6-Assoziation.
E. Braunsfurth.
Mitt. Astron. Ges., Nr. 55, p. 128 - 129 (1982).

152.003 Sobre la correlación entre exceso de color y luminosidad. H. G. Marraco.
Bol. Asoc. Argentina Astron., No. 18, p. 20 - 22 (1980).

152.004 La asociación T de la zona de Corona Austral.
H. G. Marraco.
Bol. Asoc. Argentina Astron., No. 20 - 24, p. 155 (1981). Abstract.

152.005 La asociación Car OB1. J. C. Forte.
Bol. Asoc. Argentina Astron., No. 20 - 24, p. 247 (1981). – Abstract.

152.006 Las asociaciones Cha T1 y CrA T1 en 21 cm.
E. M. Arnal, M. Franco.
Bol. Asoc. Argentina Astron., No. 20 - 24, p. 368 (1981). Abstract.

152.007 Observaciones RI – H beta de las asociaciones Car OB 1 y Sco OB 1. J. C. Forte.
Bol. Asoc. Argentina Astron., No. 20 - 24, p. 396 (1981). Abstract.

152.008 Búsqueda de H I en la asociación Chamaleón T 1.
E. M. Arnal, M. Franco.
Bol. Asoc. Argentina Astron., No. 20 - 24, p. 435 (1981). Abstract.

152.009 Porcentaje de binarias en la asociación de Sco-Cen.
H. Levato, S. Malaroda, N. Morrell, G. Solivella.
Bol. Asoc. Argentina Astron., No. 26, p. 79 (1981). – Abstract.

152.010 Spectral classification and DDO photometry of a southern group of stars with common motions. I.
P. K. Lü.

Publ. Astron. Soc. Pacific, Vol. 94, 304 - 316 (1982).
Results of a spectroscopic and photometric study are presented for a group of southern stars with common proper motions. Many groups have been selected from the Yale Zones since their publications. A number of stars in these zones were identified previously as group members of the Hyades, 61 Cyg, Pleiades, Sirius, Scorpius-Centaurus, and Ara associations. This paper is the first report on one of four larger groups selected for a pilot study. This group contains 95 stars for which the observations consisting of image-tube spectra and DDO photometry are completed. From the analyses of the observational data (the DDO color-color diagrams, luminosity, and spectra H-R diagrams) they show a Population I main sequence with a small giant branch. The distribution of $\delta CN(0.04)$ indicates that this group of stars have CN abundances similar to a galactic cluster such as M67. The mean spectroscopic parallax yields a distance of approximately 85 pc.

On supplement 1 to the second edition of the Catalogue of Star Clusters and Associations.
See Abstr. 002.021.

Catalogue of star clusters and associations. Supplement 1, Vols. I - III. See Abstr. 002.034.

Catalogue of Star Clusters and Associations. Supplement 1. See Abstr. 002.111.

Landolt-Börnstein. See Abstr. 003.064.

OB associations and the nonuniversality of the cosmic abundances: implications for cosmic rays and meteorites. See Abstr. 107.022.

Stellar winds in the association Cyg OB2.
See Abstr. 112.017.

Spectroscopy and infrared photometry of Cyg OB 2 stars: velocity law and mass-loss rates.
See Abstr. 114.049.

On the total number of irregular variables in the T1 Mon association. See Abstr. 122.090.

Untersuchungen zur Kinematik offener Sternhaufen und OB-Assoziationen aus Radialgeschwindigkeitsmessungen.
See Abstr. 153.009.

Report of IAU Commission 37: Star clusters and associations (Amas stellaires et associations).
See Abstr. 153.053.

153 Open Clusters

153.001 **Formation rate and decay time scales for open clusters near the sun.**   S. van den Bergh.
Publ. Astron. Soc. Pacific, Vol. 93, 712 - 713 (1981/82).

Data on the ages of 63 nearby star clusters are used to study the formation and disintegration rates of open clusters. The present cluster formation rate near the sun is found to be ~0.5 × 10⁻⁶ kpc⁻² yr⁻¹.

153.002 **Spectral morphology in the open cluster Collinder 228.**   H. Levato, S. Malaroda.
Publ. Astron. Soc. Pacific, Vol. 93, 714 - 718 (1981/82).

The authors classified 58 stars in the field of the open cluster Collinder 228. Using these data and a normal reddening law the authors derived a distance for the cluster of 2.6 kpc. Among the stars observed they found two shell stars, four double-line binaries (one previously known), and three stars with evidence of a companion in their spectra. Also they found nine stars with the hydrogen lines broad in the range B0–B2.

153.003 **The angular momentum history of the Hyades K giants.**   D. F. Gray, A. S. Endal.
Astrophys. J., Vol. 254, 162 - 167 (1982).

The authors have measured the rotation velocities and macroturbulence dispersions in the four Hyades K giants. The observed rotation is ~ 40% of that expected from evolutionary models in which convective envelopes are assumed to rotate as rigid bodies. It is possible to explain this discrepancy either by postulating that a magnetic braking has occurred or that the coupling is not strong enough to maintain rigid-body rotation in convective layers. Evolutionary models having uniform specific angular momentum in the convective layers are constructed and found to agree with the observed rotation rates.

153.004 **Membership, basic parameters and luminosity function of the southern open cluster NGC 2547.**
J. J. Clariá.
Astron. Astrophys., Suppl. Ser., Vol. 47, 323 - 334 (1982).

New photoelectric UBV magnitudes and colours for 118 stars and H_β-line intensities for 23 stars in the region of the southern open cluster NGC 2547 are presented. The cluster contains only one red-giant member and two variable stars. A real gap of $\Delta(B-V) \cong 0.08$ in the main sequence distribution of stars has been detected between spectral types A3 and A7. Both photometric and spectroscopic data yield a uniform cluster reddening of $E(B-V) = 0.06 \pm 0.02$. A mean distance of 450 pc, an age of 5.7×10^7 years and other cluster parameters are derived. The luminosity function of the bright members of NGC 2547 is in good agreement with the luminosity function averaged over 62 clusters given by Taff (1974).

153.005 **UV and visible photometry of the brightest Pleiades stars.**   M. Golay, N. Mauron.
Astron. Astrophys.,Suppl. Ser., Vol. 47, 547 - 559 (1982).

This paper describes 7-colour photometric observations of the brightest stars in the Pleiades taken over the last 20 years. These data are analysed in conjunction with UV photometric and spectroscopic observations made with the OAO-II, IUE satellites and a balloon-borne photometer. Accurate determinations of the interstellar extinction are obtained and the age of the brightest Pleiades stars is given. The behaviour of the energy distribution between 2000 and 6500 Å is discussed for the case of three Be stars, among them Pleione.

153.006 **Masses of white dwarf progenitors from open cluster studies.**   B. J. Anthony-Twarog.
Astrophys. J., Vol. 255, 245 - 266 (1982).

Photographic photometry has been used to survey areas in three intermediate-age open clusters, M6, M34, and Praesepe, for white dwarfs. Photometric candidates have been found in M34 and Praesepe, clusters with turnoff masses of 3 and $2\,M_\odot$. The results of these studies when considered in the light of previous results in other cluster surveys, indicate that the upper mass limit for white dwarf progenitors is near $5\,M_\odot$.

153.007 **DDO photometry of giants in the open cluster IC 4651.**   G. H. Smith.
Astron. J., Vol. 87, 360 - 368 (1982).

DDO photometry has been obtained for 14 stars in the open cluster IC 4651. The reddening and distance modulus derived from these data are in good agreement with the results obtained from UBV photometry by Eggen. The cluster appears to have a heavy-element abundance slightly less than that of the Hyades. It is suggested that the CN-strong phenomenon among open cluster giants is the result of nonstandard stellar evolution, the occurrence of which is supported by the presence of blue stragglers in five out of the eight comparison clusters, as well as in IC 4651 itself.

153.008 **Comparisons of the HR diagrams of the youngest clusters in the Galaxy, the LMC and SMC. Evidence for a large MS widening.**   G. Meylan, A. Maeder.
Astron. Astrophys., Vol. 108, 148 - 156 (1982).

A comparative analysis of the colour magnitude diagrams of 23 well observed very young clusters is performed for four locations: clusters towards the galactic centre, the galactic anticentre, in the LMC and SMC. The relative numbers of stars outside the band of OB stars appear to increase considerably in the sequence of the four considered zones. In particular, the authors note that the relative frequency of F, G, K, and M stars increases with decreasing metallicity. Comparing models and observations of galactic clusters, the authors find that too many stars are observed outside the MS band. Extended internal mixing as well as atmospheric effects may be responsible for the observed MS widening.

153.009 **Untersuchungen zur Kinematik offener Sternhaufen und OB-Assoziationen aus Radialgeschwindigkeitsmessungen.**   F. Gieseking.
Mitt. Astron. Ges., Nr. 55, p. 117 - 118 (1982). – Abstract.

153.010 **NGC 7044, ein kaum bekannter alter galaktischer Sternhaufen.**   M. Hoffmann.
Mitt. Astron. Ges., Nr. 55, p. 166 - 167 (1982).

153.011 **Circumstellar matter in young clusters. IV. Photometric measurements of the 2200 hump.**
J. Krełowski, A. Strobel.
Acta Astron., Vol. 31, 313 - 319 (1981).

The attempt of this paper is to check if the luminosity effect described in previous papers is of local or really circumstellar origin. The ANS photometric data indicate the presence of the luminosity effect inside h and χ Per clusters. This and other arguments for the circumstellar origin of the 2200 feature and other diffuse bands are presented and discussed.

153.012 **An investigation of the Scorpius open cluster C1715−387, containing two WN7, two Of and one red supergiant members.**   P. S. Thé, M. Arens, K. A. van der Hucht.
Astrophys. Lett., Vol. 22, 109 - 118 (1982).

From Walraven photometry an additional number of 8 members of the Scorpius cluster C1715−387 were found. The determination of the cluster distance by means of the main sequence fitting procedure is inconclusive. The VRI and JHKL measurements of the two Of stars together with their Walraven WULBV and Johnson UBV observations, show that these stars do not have excess radiation up to 3.6 μm, with respect to model fluxes. The observations of the two WN7 stars (LSS 4065 and LSS 4064) in the same pass bands, compared with a model flux distribution, show infrared excess radiation caused by free-free emission in their expanding winds. Their mass-loss rates were determined. One of the newly found members is a red supergiant of $M_V \sim -5^m3$. The presence of this star in the open cluster is discussed in the light of Maeder's (1981) evolutionary scenario of mass losing massive stars in which WR stars evolve from red supergiants.

153.013 **Distribución de cúmulos galácticos observados en el sistema UBV.** A. Feinstein, J. C. Forte.
Bol. Asoc. Argentina Astron., No. 18, p. 13 - 16 (1980).

153.014 **On the dynamics of early stages of evolution of open star clusters. I.** V. M. Danilov.
Astron. Zh., Tom 59, 253 - 262 (1982). In Russian. English translation in Soviet Astron., Vol. 26, No. 2.
    The problem of disruption of young non-stationary star clusters is considered. The equations of the dynamical evolution of a star cluster, the expressions for the tidal radius of an open cluster and for the dynamical parameter of the Galaxy are obtained.

153.015 **A method for calculating galactic orbits of star clusters.** S. A. Kutuzov, L. P. Osipkov.
Dvizhenie iskusstv. i estestv. nebes. tel. Sverdlovsk, 1981, p. 46 - 62. In Russian. – Abstr. in Ref. zh., 51. Astron., 4.51.691 (1982).

153.016 **Evidence of helium abundance differences between the Hyades stars and field stars, and between Hyades stars and Coma cluster stars.**
B. Strömgren, E. H. Olsen, B. Gustafsson.
Publ. Astron. Soc. Pacific, Vol. 94, 5 - 15 (1982).
    The systematic difference in the $c_1, (b-y)$ diagram between the sequence of unevolved stars in the Hyades and the corresponding sequence for unevolved field stars with the same metal abundance is discussed. Within the framework of current models for stellar interiors and atmospheres the difference is interpreted as a helium abundance difference, such that the helium content of the Hyades is lower by 0.06 in n_{He}/n_H, as compared with the corresponding field dwarfs. Other possible explanations for this difference are discussed but found to be less probable. The corresponding difference in the $c_1, (b-y)$ diagram for the stars in the Hyades as compared with the Coma cluster stars is interpreted as a similar difference in the helium abundance.

153.017 **Collinder 228 y el complejo Eta Carinae.**
A. Feinstein, H. G. Marraco, J. C. Forte.
Bol. Asoc. Argentina Astron., No. 20 - 24, p. 157 (1981). Abstract.

153.018 **Cinemática de cúmulos abiertos.**
J. C. Forte.
Bol. Asoc. Argentina Astron., No. 20 - 24, p. 158 (1981). Abstract.

153.019 **El cúmulo abierto Trumpler 15.**
A. Feinstein, J. C. Forte.
Bol. Asoc. Argentina Astron., No. 20 - 24, p. 248 (1981). Abstract.

153.020 **Lugares de nacimiento de cúmulos abiertos.**
J. C. Forte, J. C. Muzzio.
Bol. Asoc. Argentina Astron., No. 20 - 24, p. 260 (1981). Abstract.

153.021 **Búsqueda de H neutro en los cúmulos galácticos NGC 5460, CR 394 y RU 106.**
W. G. L. Poppel, M. D. Vota.
Bol. Asoc. Argentina Astron., No. 20 - 24, p. 288 (1981). Abstract.

153.022 **Búsqueda de hidrógeno neutro en cúmulos galácticos.** E. M. Arnal.
Bol. Asoc. Argentina Astron., No. 20 - 24, p. 292 (1981). Abstract.

153.023 **H I en cúmulos galácticos.** E. M. Arnal.
Bol. Asoc. Argentina Astron., No. 20 - 24, p. 364 (1981). – Abstract.

153.024 **Evidencia observacional acerca de la pérdida de masa en gigantes rojas de NGC 7789.**
J. J. Clariá.
Bol. Asoc. Argentina Astron., No. 20 - 24, p. 405 (1981). Abstract.

153.025 **Abundancia de helio y de elementos pesados en el cúmulo de edad intermedia NGC 7789.**
J. J. Clariá.
Bol. Asoc. Argentina Astron., No. 20 - 24, p. 406 (1981). Abstract.

153.026 **El cúmulo abierto de Zeta Scl.**
H. Levato, S. Malaroda, N. Morrell.
Bol. Asoc. Argentina Astron., No. 20 - 24, p. 480 (1981). Abstract.

153.027 **Observaciones de seis cúmulos galácticos en la línea de 21 cm.**
C. A. Olano, W. Poppel, M. D. Vota.
Bol. Asoc. Argentina Astron., No. 20 - 24, p. 493 (1981). Abstract.

153.028 **Stellar content of young open clusters. I. Blue stragglers.** J.-C. Mermilliod.
Astron. Astrophys., Vol. 109, 37 - 47 (1982).
    The analysis of the colour magnitude diagrams of 75 open clusters younger than the Hyades yielded a list of 39 stars lying obviously to the left of the clusters main sequence, which have been identified as blue stragglers. The most interesting feature is the presence in this sample of numerous blue stragglers exhibiting Am, Ap, Bp (He weak, CNO), Be, and Of characteristics. The mass transfer hypothesis could explain the presence of Be and Ap stars among the blue stragglers.

153.029 **Stellar content of young open clusters. II. Be stars.**
J.-C. Mermilliod.
Astron. Astrophys., Vol. 109, 48 - 65 (1982).
    Absolute magnitudes and dereddened colours have been calculated for 94 Be stars in 34 open clusters. An analysis of the resulting composite M_V, $(U-B)_0$ diagram, based on the age group concept, has been made. Examination of the energy distribution and position of the early Be stars in the UBV plane resulted in the identification of two kinds of Be stars: the first one (70% of the early Be) contains those stars which lie within the band formed by the luminosity classes V and Ia mean loci in the $U-B$, $B-V$ plane. The second kind (30%) contains the stars which lie redder than the Ia locus. Extreme Be stars correspond in most cases to the second kind of Be stars.

153.030 **Morfología espectral en Collinder 228.**
H. Levato, S. Malaroda.
Bol. Asoc. Argentina Astron., No. 26, p. 71 (1981). – Abstract.

153.031 Analysis of UBV data in open clusters.
J.-C. Mermilliod.
Bull. Inf. Cent. Données Stellaires, No. 22, p. 70 - 74 (1982).

153.032 Hyades.   M. Šolc.
Kozmos, Vol. 13, 8 - 9 (1982). In Czech.

153.033 Bp and Ap stars in the moving cluster Scorpius-
Centaurus.
I. M. Kopylov, V. G. Klochkova, R. N. Kumajgorodskaya.
Soobshch. Spets. Astrofiz. Obs., Vyp. 32, (see 012.029),
p. 62 - 63 (1981).

153.034 Supergiants and variable stars in the χ and h Persei
cluster region.   M. Muminov.
Astron. Tsirk., No. 1186, p. 6 - 8 (1981). In Russian.

153.035 The chemical composition of the old open clusters
Melotte 66 and NGC 2243.   R. G. Gratton.
Astrophys. J., Vol. 257, 640 - 655 (1982).
The chemical composition of the old open clusters
Melotte 66 and NGC 2243 is determined by means of high-
dispersion spectra of some red giants and a model atmosphere
analysis. The two clusters are found to be quite metal poor. A
reanalysis is made of the color-magnitude diagrams of the four
old open clusters which have been studied with high dispersion
(M67, NGC 2420, NGC 2243, and Melotte 66), and new values
for the distance moduli of M67 and Melotte 66 are determined.
Age estimates are 4×10^9 years for M67 and NGC 2420 and
6×10^9 years for NGC 2243 and Melotte 66. The helium
content $\langle Y \rangle = 0.25 \pm 0.03$ is estimated from the magnitude of
the horizontal branch and the ratio of the horizontal branch
stars to red giants, and seems remarkably constant within this
sample of clusters. The results suggest that the time scale of
the galactic halo collapse is long $(5-10 \times 10^9$ years), in agreement with recent models.

153.036 An X-ray survey of the Pleiades.
J.-P. Caillault, D. J. Helfand, W. H.-M. Ku.
Bull. American Astron. Soc., Vol. 13, 811 (1981). – Abstract.

153.037 UBVR photometry of the Hyades lower main
sequence.   G. L. Grasdalen.
Bull. American Astron. Soc., Vol. 13, 871 (1981). – Abstract.

153.038 Observational isochrones determined by open
clusters and their comparisons with theoretical
results.   F. Esin.
Istanbul Üniv. Fen Fak. Mec. Seri C, Vol. 45, 137 - 150
(1980) = Publ. Istanbul Univ. Obs. No. 118 (1981).

153.039 Integrated colors of young open clusters as a func-
tion of age.   I. Tarrab.
Pré-publ. Inst. Astrophys. Paris, No. 5, 13 pp. (1982).
Submitted to Astron. Astrophys.

153.040 International Ultraviolet Explorer observations of
Hyades stars.   M.-C. S. Zolcinski, S. K. Antiochos,
R. A. Stern, A. B. C. Walker.
Astrophys. J., Vol. 258, 177 - 185 (1982).
The authors have obtained short-wavelength, low-dispersion ultraviolet spectra of seven of the brightest X-ray
emitting stars in the Hyades cluster with the IUE observatory.
Four of the stars show evidence of emission lines characteristic of transition region temperatures ($\sim 30,000-200,000$ K);
for the remaining three stars, upper limits for emission-line
fluxes were derived. In addition, the authors have observed
three of the above stars with the long-wavelength spectrograph on IUE at high dispersion and find evidence for chromospheric emission from the Mg II h and k lines for two of
these objects. They derive constraints on stellar atmospheric
parameters (chromospheric pressure, coronal temperature,

and "filling factor") and discuss the implications of their
results for models of stellar chromospheres and coronae.

153.041 Upgren 1, an old cluster near the end of its life.
A. R. Upgren, A. G. D. Philip, W. I. Beavers.
Publ. Astron. Soc. Pacific, Vol. 94, 229 - 231 (1982).
Narrow-band photometry on the four-color system and
radial velocities have been obtained for seven F-type stars near
the north galactic pole. These data indicate that five of the
seven stars are physically associated and form a small cluster.
The other two stars may also be members. The cluster distance is about 117 pc and its turnoff point is near $(B-V) = 0.5$
which suggests an age similar to that of the old open cluster
M67.

153.042 Intermediate-band photometry of the open cluster
NGC 3114.   E. G. Schmidt.
Publ. Astron. Soc. Pacific, Vol. 94, 232 - 238 (1982).
Four-color and Hβ photometry has been obtained of
33 stars in the southern galactic cluster NGC 3114. This cluster contains many interesting stars including blue stragglers, red
giants, and stars with spectral peculiarities. The distance
modulus was found to be $V_0 - M_v = 10.26$ which corresponds to
1130 pc. The color excess is $E(b-y) = 0\overset{m}{.}060$. The cluster
appears to be slightly older than the Pleiades. There are two
extreme blue stragglers and six additional stars which might be
considered milder cases of the same phenomenon.

153.043 Variable K-type stars in the Pleiades cluster.
F. Van Leeuwen, P. Alphenaar.
Messenger, No. 28, p. 15 - 18 (1982).

153.044 Open clusters in our galaxy.   G. Lyngå.
Astron. Astrophys., Vol. 109, 213 - 222 (1982).
A recently compiled computer-readable catalogue of open
cluster data has been used to study the physical nature of
open clusters. The system of open clusters has been examined
in context with questions of galactic structure. The galactic
dust layer has been studied from the extinction suffered by the
light from the clusters.

153.045 The initial mass function for young open clusters.
I. Tarrab.
Astron. Astrophys., Vol. 109, 285 - 288 (1982).
UBV data for 75 young open clusters divided into 14 age
groups, have been used to construct C-M diagrams in order to
derive the initial mass function (IMF) for the mass range 1.25
$-14 M_\odot$. This was done by counting the stars contained between theoretical isomasses. The author has found that the
average IMF for the best studied groups can be approximated
by a power law with a slope of $X = 1.7$.

153.046 The cluster NGC 5053. Catalogue of V, B–V, U–B
for 223 stars.   L. V. Zhukov.
Researches on stellar physics, 1981, (see 003.010), p. 52 - 60
(1981). In Russian. – From Ref. zh., 51. Astron., 6.51.611
(1982).

153.047 Four-color and Hβ photometry of the galactic
cluster M25.   E. G. Schmidt.
Astron. J., Vol. 87, 650 - 654 (1982).
Four-color and Hβ photometry has been obtained for
39 stars in the field of the galactic cluster M25. The membership of the individual stars is discussed. From 27 stars which
are considered likely members, the author obtains a mean
distance modulus of 8.76 and a mean color excess of
$E(b-y) = 0.341$.

153.048 The faint end of the Hyades main sequence.
J. Stauffer.
Astron. J., Vol. 87, 899 - 905 (1982).
Optical photometry and spectroscopy are presented for a

large number of low-mass proper motion members of the Hyades cluster. Infrared photometry has also been obtained for a small subset of those stars. Those data are used to help determine a cluster main sequence to $M_v \cong 13^m5$. A number of Hyades stars are proposed as probable binaries based on their positions in the V-vs-$(V-I)_K$ color-magnitude diagram. One result of that analysis is that a large fraction of the late-type stars identified as X-ray sources by Stern et al. (1981) appear to be binaries.

153.049 **Hipparcos could make the distance of the Hyades into a conventional constant.**
G. Cayrel de Strobel.
Scientific aspects of the Hipparcos space astrometry mission, (see 012.041), p. 173 - 176 (1982).

153.050 **Radial velocities of open clusters.**
S. Wramdemark.
Rep. Obs. Lund, No. 18, (see 012.044), p. 63 - 67 (1982).

153.051 **Evolutionary effects among open star clusters.**
G. Lyngå.
Rep. Obs. Lund, No. 18, (see 012.044), p. 68 - 71 (1982).

153.052 **On the radial colour variation in nine young populous clusters in the LMC.** G. Meylan.
Astron. Astrophys., Vol. 110, 348 - 350 (1982).

Differences appear between the ratios of the numbers of red to blue supergiants in the inner and outer regions of nine young populous clusters in the LMC. The ratio of the numbers of red to blue supergiants is greater in the corona than in the central part. This points towards a delayed star formation in the central region of these clusters.

153.053 **Report of IAU Commission 37: Star clusters and associations** (Amas stellaires et associations).
G. Lyngå.
Trans. IAU, Vol. XVIIIA, (see 003.013), p. 499 - 526 (1982).

On supplement 1 to the second edition of the Catalogue of Star Clusters and Associations.
See Abstr. 002.021.

Catalogue of star clusters and associations. Supplement 1, Vols. I - III. See Abstr. 002.034.

Catalogue of Star Clusters and Associations. Supplement 1. See Abstr. 002.111.

Landolt-Börnstein. See Abstr. 003.064.

All-Union conference "Star clusters and problems of stellar evolution". Sverdlovsk, 1981, April 13 - 17.
See Abst. 011.013.

Investigation of stellar motions in the field of the open cluster NGC 6913. See Abstr. 111.010.

Proper motion survey with the 48-inch Schmidt telescope. LIX. A catalogue of 929 possible candidates for Hyades membership. See Abstr. 111.033.

Mass loss rates in the open cluster IC 1805. See Abstr. 112.013.

A search for light variations of the brightest Pleiades stars. See Abstr. 113.069.

Revised RGU photometry of an anticentre field around NGC 2129. See Abstr. 113.074.

Slow flares in stellar aggregates. II. See Abstr. 113.076.

Photographic colorimetry of stellar flares in the Pleiades and Orion aggregates. See Abstr. 113.077.

The sun among the stars. V. A second search for solar spectral analogs. The Hyades' distance. See Abstr. 114.010.

Spectral classification of stars up to 15^m0 around the cluster NGC 6823. See Abstr. 114.086.

Photoelectric measures of chromospheric H and K and Hϵ in giant stars. See Abstr. 114.110.

Steps towards the abundance scale. I. The abundance of iron in the Hyades. See Abstr. 114.193.

Metallic-line stars in Praesepe open cluster: model atmosphere analysis and abundances. See Abstr. 114.197.

Linear radii of 51 red giants in galactic clusters. See Abstr. 115.004.

The absolute magnitudes of G5—M3 stars near the giant branch. See Abstr. 115.006.

Composite colour-magnitude and colour-colour diagrams for Be stars in open clusters. See Abstr. 115.011.

The bimodal distribution of rotational velocities of late B-type stars in galactic clusters. See Abstr. 116.003.

Magnetic structure in cool stars. IV. Rotation and Ca II H and K emission of main-sequence stars. See Abstr. 116.013.

The Hyades binary HD 27130 and the mass-luminosity relation and distance of the Hyades cluster. See Abstr. 120.004.

Spectroscopic orbits for three double-lined binaries in the Hyades field, 22°669, vA 771, and vB 166. See Abstr. 120.006.

Spatial distribution of flare stars in the Pleiades derived by the orthogonal expansion method. See Abstr. 122.037.

Flare stars in the Pleiades. VI. See Abstr. 122.060.

uvby β photometry of young star clusters containing β CMa stars. See Abstr. 122.204.

Beta Cephei stars in the open cluster NGC 3293. See Abstr. 122.217.

A study of the diffuse interstellar gas near the Pleiades. See Abstr. 131.035.

Radiation pressure effect on dust surrounding the Pleiades bright stars. See Abstr. 131.266.

On the distance to the giant galactic H II region NGC 3603. See Abstr. 132.021.

Formation of OB clusters. See Abstr. 132.080.

On the dynamics of early stages of evolution of open star clusters. II.   See Abstr. 151.061.

The mass (=star number) radius relations for globular and open star clusters of the Large Magellanic Cloud. See Abstr. 154.044.

Structure of the inner edge of the local spiral arm in the direction to NGC 6823.   See Abstr. 155.025.

An optical study of the magnetic field in M31. See Abstr. 158.025.

Luminosity functions of star clusters in the Small Magellanic Cloud.   See Abstr. 159.012.

The age-abundance relation for Magellanic Cloud star clusters.   See Abstr. 159.016.

Erratum

153.901   Erratum: 'Photographic photometry of the open clusters NGC 2910, NGC 2925, Ru 79 and Ru 82 in Vela II and NGC 6031 in Norma II' [Astron. Astrophys., Suppl. Ser., Vol. 45, 111 (1981)].   L. Topaktas. Astron. Astrophys., Suppl. Ser., Vol. 48, 344 (1982). – See Abstr. 30.153.001.

154 Globular Clusters

154.001 The Oosterhoff period groups and the age of globular clusters. III. The age of the globular cluster system. A. Sandage.
Astrophys. J., Vol. 252, 553 - 573 (1982).

The Oosterhoff period shifts for RR Lyrae stars in 30 galactic globular clusters are investigated. This paper reports two principal results: (i) An explanation of the RR Lyrae period shifts, their correlation with metallicity, and the progression in the period-amplitude relations towards longer periods in clusters of lower metallicity can be given from the fact that nearly all clusters have nearly equal ages, despite the large variation in [Fe/H] among the clusters. (ii) The absolute age of the globular cluster system in the galaxy is $(17 \pm 2) \times 10^9$ a, if the absolute magnitude of RR Lyrae variables in M3 is $M_v = +0.80$. This age agrees with a global value of the Hubble constant near $H_0 = 50$ km s^{-1} Mpc^{-1}.

154.002 The late-type stellar content of Magellanic Cloud clusters. J. A. Frogel, J. G. Cohen.
Astrophys. J., Vol. 253, 580 - 592 (1982).

New broad-band infrared photometric data have been obtained for 48 late-type giants in clusters in the Magellanic Clouds. Visual spectrophotometry was obtained for a subset of these stars. These observations are combined with published data for MC cluster stars and then compared with similar data for MC field giants and with predictions of various evolutionary schemes for cool, luminous, carbon and oxygen rich stars.

154.003 The anticorrelation of carbon and nitrogen on the horizontal branch of 47 Tucanae.
J. Norris, K. C. Freeman.
Astrophys. J., Vol. 254, 143 - 148 (1982).

Spectroscopic and photometric data for 14 horizontal branch stars in 47 Tuc are presented and analyzed to yield information on the systematics of the behavior of carbon and nitrogen in these objects. The observations are consistent with a bimodal distribution of cyanogen, as has been reported for the giants in this system. There is also a one-to-one anticorrelation of the behavior of cyanogen and the CH molecule. Spectrum synthesis of a selected CN-weak, CN-strong pair of stars shows that nitrogen is overabundant by ~0.9 dex, and carbon is underabundant by ~0.3 dex in the CN-strong object relative to the CN-weak one.

154.004 The cyanogen distributions in NGC 3201, M55, and M71. G. H. Smith, J. Norris.
Astrophys. J., Vol. 254, 149 - 161 (1982).

Observations have been made of the cyanogen distributions in the globular clusters NGC 3201, M55, and M71. The latter two clusters were chosen for study because they have heavy element abundances similar to those of ω Cen and 47 Tuc, respectively, but considerably smaller masses. No cyanogen enhancements were detected in M55, in marked contrast to the inhomogeneities seen in ω Cen. On the other hand, M71 was found to have a range in cyanogen strength very similar to that seen in 47 Tuc. For NGC 3201 the authors find a bimodal cyanogen distribution, though the difference between the CN-weak and CN-strong stars is not as marked as that found in NGC 6752 and M4 (which have a heavy element abundance similar to that of NGC 3201).

154.005 Evidence for extended X-ray emission from globular clusters.
F. D. A. Hartwick, A. P. Cowley, J. E. Grindlay.
Astrophys. J., Lett., Vol. 254, L11 - L13, plates L1 - L3 (1982).

Deep exposures with the Einstein Observatory show evidence for diffuse X-ray emission from three globular clusters. One possible interpretation of these observations is that the authors are observing the interaction between a cluster wind and a hot gaseous galactic halo. The one cluster for which the proper motion has been measured is consistent with this interpretation.

154.006 Comments on the origin of the carbon and nitrogen variations within NGC 6752 and 47 Tucanae.
G. H. Smith, J. Norris.
Astrophys. J., Vol. 254, 594 - 605 (1982).

This paper summarizes the available observations concerning the nitrogen and carbon abundance variations within the globular clusters NGC 6752 and 47 Tuc. Implications concerning the primordial and mixing theories for their origin are discussed. The foremost conclusion is that observations of a bimodal CN distribution coupled with the existence of a carbon, nitrogen abundance anticorrelation within a globular cluster are difficult to reconcile with a scenario in which primordial enrichment of a protocluster gas cloud has been produced by a generation of intermediate-mass stars with a composition identical to that of the present CN-weak giants.

154.007 A comment on the metal abundance of the globular cluster M71. R. A. Bell, B. Gustafsson.
Astrophys. J., Vol. 255, 122 - 127 (1982).

Cohen's analysis of echelle spectra of four red giants in the globular cluster M71 gives an iron abundance of [Fe/H] = −1.4. This disagrees strongly with the previously accepted value of [M/H] ≈ −0.3, which was mainly based on photometric data, and raises questions on the interpretation of the photometry. The authors have investigated the problem de novo by applying synthetic colors to the interpretation of the available photometry and by reexamining the echelle data. An overall metal abundance of [M/H] = −0.9 is suggested. Some suggestions for additional observations which may clarify the situation further are made.

154.008 What is the second parameter? : The anomalous globular cluster NGC 7006.
J. G. Cohen, J. A. Frogel.
Astrophys. J., Lett., Vol. 255, L39 - L43 (1982).

An infrared color-magnitude diagram for NGC 7006 and moderate dispersion digital optical spectra of eight of its members indicate a metal abundance of −1.5 dex with respect to the Sun. However, the ratio of red to blue horizontal-branch stars is quite large and is what would be expected for a cluster of much higher metallicity. The authors have determined molecular band strengths for CO in four stars, and CH and CN in five stars, and find that none of these molecular bands are anomalously strong compared to the same molecular features in other globulars of similar metallicity but varying horizontal-branch type. This is contrary to the behavior predicted if the C, N, and O abundances are the "second parameter" needed to explain anomalous horizontal-branch morphologies.

154.009 The extended giant branches of intermediate age globular clusters in the Magellanic Clouds. II.
M. Aaronson, J. Mould.
Astrophys. J., Suppl. Ser., Vol. 48, 161 - 184, plates 1 - 4 (1982).

The authors report on a study of upper asymptotic giant branch (AGB) stars in the red globular clusters of the Magellanic Clouds. A photographic near-infrared survey of the clusters is now more than half completed, and infrared (*JHK*) photometry of the stars so identified implies that the majority of the red clusters have extended giant branches and are therefore of intermediate age, unlike globular clusters in the Galaxy. With the assumption of a uniform luminosity function on the upper AGB, ages can be deduced for sufficiently well-populated cloud clusters. The intermediate age globular clusters of the Large Magellanic Cloud (LMC) are younger than those of the Small Magellanic Cloud (SMC) in the mean, and their AGB tips contain the reddest and most luminous carbon stars. However, while the LMC also contains clusters as old as galactic globular clusters, no such objects have been conclusively identified in the SMC from the present technique.

154.010   Two-dimensional classification of globular clusters.
V. Straižys.
Astrophys. Space Sci., Vol. 81, 179 - 197 (1982).

The metallicities of 75 globular clusters are determined in the Pilachowski et al. scale from the slopes of their giant branches in the $V, B - V$ diagram. In the diagram 'giant branch slope, horizontal branch morphology type', the clusters form two sequences: a vertical sequence (group I) with blue horizontal branches and [Fe/H] from -2.2 to -1.2 and a diagonal sequence (group II) with horizontal branches changing from the blue to the red parallel with an increase of metallicity from -1.8 to -1.0. The differences in horizontal branch morphology of both cluster groups are probably caused by the difference in age, which is found to be of the order 1 - 2 billion years.

154.011   Globular clusters in galaxies beyond the local group. II. The edge-on spirals NGC 891 and NGC 4565.
S. van den Bergh, W. E. Harris.
Astron. J., Vol. 87, 494 - 499 (1982).

Two nearby edge-on spiral galaxies, NGC 891 and NGC 4565, have been surveyed photographically to search for globular clusters. For NGC 4565, star counts down to a limiting magnitude J ~ 23.6 yield a total cluster population of approximately 100 ± 30. The cluster luminosity function is consistent with a distance modulus for NGC 4565 slightly smaller than that of the Virgo Cluster itself. The observed space distribution of the NGC 4565 clusters is well represented by the same radial distribution that describes the spheroidal component of the galaxy light. In NGC 891, no significant population of globular clusters was found. In both these galaxies and in other surveyed spirals, the relative numbers of globular clusters per unit galaxy luminosity are ~ 10 times smaller than in ellipticals. This result argues against the hypothesis that typical elliptical galaxies are formed by mergers of disk galaxies.

154.012   On the intrinsic widths of the subgiant and horizontal branch sequences in the globular cluster M3.   A. Sandage, B. Katem.
Astron. J., Vol. 87, 537 - 554 (1982).

Two-color photographic photometry is given for 676 stars in the globular cluster M3, obtained from measurements of 15 plates in each color of B (103a-O + WG2) and V (103a-D + GG11). The plates were taken with the Mount Wilson 2.5-m Hooker reflector, diaphragmed to 1.47 m. The main results of this study are: (1) The subgiant branch may have zero intrinsic width in color. It is almost certainly smaller than $\sigma (B-V) = 0.02$ mag. (2) The horizontal branch has an intrinsic width in magnitude of ~0.3 mag, presumably due to evolution. (3) There is no definite proof for any gap in the giant branch other than that near $V = 13.4$, similar to that found earlier in M15 but discounted then. There is a suggestion of a change in the luminosity function near $V = 14.2$-14.5.

154.013   Infrared dust emission from globular clusters.
L. Angeletti, A. Blanco, E. Bussoletti,
R. Capuzzo-Dolcetta, P. Giannone.
Mon. Not. R. Astron. Soc., Vol. 199, 441 - 449 (1982).

The implications of the presence of a central cloud in the cores of globular clusters were recently investigated. A possible mechanism of confinement of dust in the central region of the authors' cluster models was also explored. The grain temperature and infrared emission have now been computed for rather realistic grain compositions. The grain components were assumed to be graphite and/or silicates. The central clouds turned out to be roughly isothermal. An application of the theoretical results to five globular clusters showed that the predictable infrared emission for 47 Tuc, M4 and M22 should be detectable.

154.014   An ellipticity–age relation for globular clusters in the Large Magellanic Cloud – I. Measurements.
C. S. Frenk, S. M. Fall.
Mon. Not. R. Astron. Soc., Vol. 199, 565 - 580 (1982).

The authors have estimated the ellipticities of 52 globular clusters in the LMC and 93 in the Galaxy by the eye-measurement of their images on Sky Survey enlargements. These were checked against star counts in 12 clusters of the LMC sample and against star counts in 19 clusters of the galactic sample. A comparison of the two samples shows that the globular clusters in the LMC are significantly flatter than those in the Galaxy. Young clusters are flatter on average than old clusters and the shapes of the oldest clusters in the LMC are similar to those of galactic globular clusters.

154.015   Below the tip: new controversies in globular cluster research.   J. E. Hesser.
J. R. Astron. Soc. Canada, Vol. 76, 69 - 89 (1982).

154.016   Proper and space motion of the globular cluster NGC 5466.   P. Brosche, M. Geffert.
Naturwissenschaften, 69. Jahrg., 236 - 237 (1982).

154.017   The billion-year-old clusters of the Magellanic Clouds.   P. W. Hodge.
Astrophys. J., Vol. 256, 447 - 451 (1982).

Five clusters of the Magellanic Clouds in the age range of about one billion years have faint enough photometry to allow comparison with theoretical models. Semiempirical relationships are derived from various theoretical models that allow a determination of ages and chemical abundances for the clusters in question. Values for [Fe/H] ranging from -0.8 to -1.3 are found, and ages range from 0.7 to 1.1 billion years.

154.018   Radial abundance gradient in globular clusters.
M. S. Chun.
J. Korean Astron. Soc., Vol. 14, 13 - 17 (1981).

The observed radial UBV colour variations (both B–V and U–B) of some globular clusters are examined for correlations with radial variations in the integrated spectra. The results show that the presence of a radial colour gradient is correlated with the presence of a gradient of the CN (and possibly the G-band) line strength, in the sense that the CN (and possibly the G-band) is stronger in the centre (where the cluster is redder) and becomes weaker in the outer region of the cluster (where the cluster is bluer).

154.019   The natural double sequence of globular clusters.
H. Wilkens.
Bol. Asoc. Argentina Astron., No. 18, p. 73 - 75 (1980).

154.020   On the age of the globular cluster NGC 6397.
P. E. Nissen, A. Ardeberg.
Rep. Obs. Lund, No. 18, (see 012.044), p. 51 - 56 (1982).

154.021   Magellanic Cloud globular clusters and their galactic counterparts in the ultraviolet.   K. S. de Boer.
Astrophysical parameters for globular clusters, (see 012.023), p. 3 - 12 (1981).

Eleven populous clusters in the Magellanic Clouds have been observed with the International Ultraviolet Explorer between 1150 and 3200 Å. The clusters were chosen to represent the entire range of red, galactic globular cluster like clusters, to the very blue clusters, which may be still compact young clusters. Preliminary colors are presented, and a comparison is made with the UV colors of galactic globular clusters. The basic data for the understanding of the ultraviolet energy distributions (galactic globular clusters as measured by OAO-2 and by ANS, field blue horizontal branch stars and cluster BHB stars by IUE) are reviewed.

154.022 **Possible diffuse X-ray emission from globular clusters.** F. D. A. Hartwick, A. P. Cowley.
Astrophysical parameters for globular clusters, (see 012.023), p. 13 - 16 (1981).

154.023 **High resolution X-ray and optical studies of globular clusters.** J. E. Grindlay.
Astrophysical parameters for globular clusters, (see 012.023), p. 17 - 24 (1981).

Perhaps ~10% of the known globular clusters in the Galaxy contain luminous ($L_X \gtrsim 10^{35}$ erg s^{-1}) X-ray sources. The Einstein X-ray Observatory has obtained high resolution X-ray images of nine of these X-ray globulars, two of which were not known previously to be X-ray clusters. The primary motivation of this work has been to obtain precise absolute positions of the cluster X-ray sources for comparison with precise positions of the cluster centers and core radii.

154.024 **Cluster synthesis of NGC 6397 based on IUE spectra.** K. S. de Boer.
Astrophysical parameters for globular clusters, (see 012.023), p. 25 - 29 (1981).

Six known blue horizontal branch (BHB) stars in NGC 6397 have been observed with the International Ultraviolet Explorer (IUE) satellite at low spectral resolution between 1150 and 3200 Å. Together with integrated photometry of the Astronomical Netherlands Satellite (ANS), these data allow to estimate the relative number of BHB stars, main-sequence stars, and red giants contributing to the overall energy distribution. The ANS entrance aperture contained about 10 BHB stars. A sdO star, now observed with IUE, was outside the ANS aperture, but, when included in the photometric flux, would lead to reclassify NGC 6397 into the EB cluster category defined from ANS photometry of 28 galactic globular clusters.

154.025 **Methods of photometric abundance determination in globular clusters.** C. F. Trefzger.
Astrophysical parameters for globular clusters, (see 012.023), p. 33 - 44 (1981).

154.026 **Ranking globular clusters by metal abundance.** R. Zinn.
Astrophysical parameters for globular clusters, (see 012.023), p. 45 - 58 (1981).

154.027 **Radial abundance gradient in globular clusters.** M. S. Chun.
Astrophysical parameters for globular clusters, (see 012.023), p. 65 - 73 (1981).

The observed radial UBV color variations (both B−V and U−B) of some globular clusters are examined for correlations with radial variations in the integrated spectra. The results show that the presence of a radial color gradient is correlated with the presence of a gradient of the CN (and possibly the G-band) line strength, in the sense that the CN (and possibly the G-band) is stronger in the center (where the cluster is redder) and becomes weaker in the outer region of the cluster (where the cluster is bluer). This may suggest that a primordial abundance, possibly nitrogen and carbon gradient, was set up in the early stage of cluster formation.

154.028 **Further evidence for correlated cyanogen and sodium anomalies in 47 Tuc.** G. S. Da Costa.
Astrophysical parameters for globular clusters, (see 012.023), p. 75 - 84 (1981).

The observations of 22 47 Tuc giants may be summarized as follows: At any V magnitude on the giant branch within the range observed (V < 12.6 or M_V < −0.6), the CN-strong stars possess enhanced sodium abundances (by ~0.3 dex) relative to the CN-weak stars while any magnesium abundance difference is small (< 0.1 dex). Thus a correlation between the presence of sodium abundance anomalies, which are necessarily primordial in origin, and cyanogen anomalies, which could result either from mixing or primordial processes is now firmly established for 47 Tuc. Consequently, it appears most likely that all the abundance anomalies in this cluster have a primordial origin.

154.029 **Abundances in metal rich globular clusters.** D. Geisler.
Astrophysical parameters for globular clusters, (see 012.023), p. 89 - 90 (1981). − Abstract.

154.030 **The metallicity of the inner halo cluster NGC 6352.** D. Geisler, C. A. Pilachowski.
Astrophysical parameters for globular clusters, (see 012.023), p. 91 - 92 (1981). − Abstract.

154.031 **Field and globular cluster populations in our galaxy.** V. Caloi, F. Caputo, V. Castellani.
Astrophysical parameters for globular clusters, (see 012.023), p. 93 - 94 (1981). − Abstract.

154.032 **Carbon, nitrogen and heavy-element abundances on the red giant branch of Omega Centauri.** S. P. Caldwell, R. J. Dickens, R. A. Bell.
Astrophysical parameters for globular clusters, (see 012.023), p. 95 - 96 (1981). − Abstract.

154.033 **The chemical composition of globular clusters near the galactic center.** C. A. Pilachowski, C. Sneden, E. Green.
Astrophysical parameters for globular clusters, (see 012.023), p. 97 - 108 (1981).

154.034 **The cyanogen distributions of the giants in ten globular clusters and the role of mixing and primordial abundance variations.** J. Norris, G. H. Smith.
Astrophysical parameters for globular clusters, (see 012.023), p. 109 - 120 (1981).

Since Osborn (1971) first demonstrated the existence of a range in the abundance of the cyanogen radical in giants of similar effective temperature and gravity within individual globular clusters, an extensive literature has developed on the large and ubiquitous CN variations found in these systems. Most of the early investigations concerned the bands at λ4216. This feature is ideally suited for exploring variations in clusters having abundances greater than or equal to that of 47 Tuc. For lower abundances, however, it rapidly weakens and is useful only when large enhancements are involved. In the range −1.2 ≳ [Fe/H] ≳ −2.0 the intrinsically stronger violet bands at λ3883 are more useful. The present paper presents a summary of work carried out, principally at Mount Stromlo, on the systematics of the variation of the violet cyanogen bands in the giants of 8 globular clusters in this abundance range. The authors also present results on the behavior of the λ4216 bands in 47 Tuc and M71.

154.035 **The metallicity of M71.** R. C. Peterson.
Astrophysical parameters for globular clusters, (see 012.023), p. 121 - 135 (1981).

A comparison of synthetic spectra with newly-obtained

high-resolution spectra of 4500 K giants in M 5, M 71, and M 67 indicates an iron-to-hydrogen ratio that is about one-thirtieth solar in M 5, between one-tenth and one-thirtieth solar in M 71, and one-third solar in M 67. These values are found consistently over a wide range of stellar temperature in all clusters.

154.036 A comparison of blue spectra of M5 and M71 giant stars.   R. D. McClure, J. E. Hesser.
Astrophysical parameters for globular clusters, (see 012.023), p. 145 - 150 (1981).

154.037 Carbon and nitrogen abundances in giant stars of the globular clusters M 3 and M 13.   N. Suntzeff.
Astrophysical parameters for globular clusters, (see 012.023), p. 151 - 152 (1981). – Abstract.

154.038 High-dispersion spectra of BHB stars in M 4 and NGC 6397.   K. Kodaira, A. G. D. Philip.
Astrophysical parameters for globular clusters, (see 012.023), p. 153 - 160 (1981).

In the last few years, fine analyses have been carried out of high-dispersion spectra of red giants in globular clusters, yielding direct calibration of the abundance scales inferred from various abundance indices. The most fundamental abundance calibration for horizontal-branch stars has been the relation between ΔS and [Fe/H] by Butler (1975). In order to provide a direct calibration of the abundance scale for horizontal-branch stars in globular clusters and to compare this calibration with that for giants, the authors obtained, one each, 12.5 Å/mm spectrogram of three BHB stars, M4-206, M 4-553 and NGC 6397-48, together with spectrograms of several field horizontal-branch stars, with the Cassegrain image-tube spectrograph of the CTIO 4 m telescope.

154.039 Asymptotic giant branch variable stars.   R. E. White.
Astrophysical parameters for globular clusters, (see 012.023), p. 161 - 165 (1981).

It is highly possible that a specific type of cluster variable star could have been overlooked due to a variety of observational biases, to wit, a long-period (P about 100 days) and small amplitude ($\Delta m < 0.5$-mag.) variable star. If variable stars of the proposed type could be found amongst the members of the clusters' asymptotic branch (AGB), then they might provide an explanation for the observed spectroscopic disparity between many AGB stars in the same cluster. The observational sample selected was the AGB stars brighter than V = +14 in the clusters C1339+286 (NGC 5272, M 3) and C1639+365 (NGC 6205, M 13). Three observing seasons of photoelectric (V, B–V) data in M 3 and two seasons of M 13 data have been compiled into diagrams.

154.040 [Fe/H] and orbital parameters for globular clusters.   P. Seitzer, K. C. Freeman.
Astrophysical parameters for globular clusters, (see 012.023), p. 185 - 189 (1981).

To understand the formation and evolution of the galactic system one must know the relationship between dynamics and chemistry. Eggen, Lynden-Bell, and Sandage (1962) found a correlation between orbital eccentricity and metallicity for a sample of halo dwarf stars. The more metal rich stars were on less eccentric orbits than the metal poorest stars. The present authors have completed a similar analysis for globular clusters. Their findings are the exact opposite of ELS: the metal poorest clusters are on orbits of lower eccentricity than the intermediate metallicity objects. This result is found both in our galaxy and in M 31.

154.041 The determination of globular cluster axial ratio, orientation, and center.
S. J. Shawl, R. E. White, M. E. Sim.

Astrophysical parameters for globular clusters, (see 012.023), p. 191 - 192 (1981).

154.042 Image-tube radial velocities of selected globular clusters.   S. J. Shawl, J. E. Hesser, J. E. Meyer.
Astrophysical parameters for globular clusters, (see 012.023), p. 193 - 198 (1981).

154.043 Are there variable stars in the LMC star clusters NGC 2210 and NGC 1786?   J. M. Nemec.
Astrophysical parameters for globular clusters, (see 012.023), p. 215 (1981). – Abstract.

154.044 The mass (=star number) radius relations for globular and open star clusters of the Large Magellanic Cloud.   E. H. Geyer, U. Hopp.
Astrophysical parameters for globular clusters, (see 012.023), p. 235 - 238 (1981).

From star counts on B- and V-ESO Schmidt plates of 12 open clusters, 5 "blue" and 4 "red" globular clusters of the LMC the authors have derived effective cluster radii and the total numbers of member stars within these radii. There exist linear relations between the log of the star number and the log of the radius well separated for the open – and the globular clusters, but with identical slopes. There do not exist any clusters which fall in the gap between. The "blue" globular clusters populate the lower part of the globular cluster relation.

154.045 On the ellipticities of "blue" and "red" globular clusters in the Large Magellanic Cloud.
E. H. Geyer, T. Richtler.
Astrophysical parameters for globular clusters, (see 012.023), p. 239 - 242 (1981).

On ESO-Schmidt plates of the LMC the projected ellipticities of 11 "red" and 14 "blue" globular clusters have been determined by the Agfa-contourfilm technique. The comparison of the observed distribution of the ellipticities for both cluster types does not show a statistically significant difference.

154.046 Color-magnitude diagrams of the SMC clusters NGC 152 and NGC 176.
R. J. Dickson, P. W. Hodge, P. Flower.
Astrophysical parameters for globular clusters, (see 012.023), p. 243 (1981). – Abstract.

154.047 He abundance in RR-Lyrae rich clusters.   F. Caputo, V. Castellani, A. Tornambé.
Astrophysical parameters for globular clusters, (see 012.023), p. 309 - 312 (1981).

154.048 An evolutionary upper limit for the effective temperatures of horizontal-branch stars and the occurrence of gaps in their observed distribution.
F. Caputo, A. Chieffi, I. Mazzitelli, A. Tornambé , V. Castellani.
Astrophysical parameters for globular clusters, (see 012.023), p. 313 - 317 (1981).

The authors suggest that the gaps in HB distribution could be correlated with the minimum mass allowed by evolution for HB stars. If this is the case, the gap by itself could be a strong indicator of Y and/or CNO abundances.

154.049 The abundance scale of the globular clusters.   R. A. Bell.
Astrophysical parameters for globular clusters, (see 012.023), p. 325 - 336 (1981).

154.050 Theoretical isochrones and observed color-magnitude diagrams.   K. Janes.

Astrophysical parameters for globular clusters, (see 012.023), p. 349 - 356 (1981).

154.051 Ages of the oldest star clusters from new synthetic color-magnitude diagrams.
D. A. VandenBerg.
Astrophysical parameters for globular clusters, (see 012.023), p. 357 - 361 (1981).

An extensive grid of stellar evolutionary sequences and their associated isochrones have been computed for the selected metallicities Z = 0.0001, 0.001, 0.003, 0.006, 0.01 and 0.0169 with two assumed values of the helium abundance (Y = 0.2, 0.3) in each case. The following conclusions have been reached: (1) All of the globular clusters examined appear to have ages in the range of (15 - 18) $\times$ 10^9 years and may, in fact, be coeval. (2) No evidence has been found for an age − metallicity relation among the globulars. (3) If 18 billion years represents the maximum age of the globular clusters, this minimum age for the Universe requires $H_0 \lesssim 60$ km s^{-1} Mpc^{-1}. (4) An age of 10×10^9 years has been obtained from an excellent match to the observations of NGC 188 which suggests that there is an age difference of (5 - 8) billion years between the halo and the disk of the Galaxy. (5) The photometry of the metal-rich clusters, i.e. 47 Tuc, M71 and Pal 12, is best reproduced by isochrones for [Fe/H] −0.5 which contradicts recent suggestions that a value of [Fe/H] −1.2 is more appropriate for these clusters. (6) The mixing-length theory seems to be an adequate description of convection in low mass stars for a wide range of heavy element abundances.

154.052 The second parameter problem 1981.
R. T. Rood, P. O. Seitzer.
Astrophysical parameters for globular clusters, (see 012.023), p. 369 - 374 (1981).

154.053 On the evolutionary structure of globular clusters.
L. Angeletti, R. Capuzzo-Dolcetta, P. Giannone.
Astrophysical parameters for globular clusters, (see 012.023), p. 375 - 378 (1981).

Fluid-dynamical models of globular clusters were computed for various sets of initial parameters as chemical composition, total mass (in the range 10^5- 10^6 M$_\odot$), tidal radius (10-100 pc), star density profile, mass function, number (from one to eight) of star groups (each group characterized by stars with the same mass value). Mass loss from stars was considered in the framework of Population II stars. HR diagrams, synthetic colors, mass-luminosity ratios, and other structure properties were determined at various ages. In particular, the initial mass function is very important for cluster evolution.

154.054 Integrated spectrophotometric properties of globular clusters. J. P. Brodie, D. A. Hanes.
Astrophysical parameters for globular clusters, (see 012.023), p. 381 - 391 (1981).

The authors investigate the metallicity characteristics of the globular cluster system associated with Virgo's giant elliptical galaxy, M87. Using integrated spectra of 24 galactic globular clusters they have established a number of metallicity indicators, a combination of which provides an accurate [Fe/H] value for each cluster, calibrated with respect to the "old" abundance scale. These indices have been measured for four M87 globular clusters and a number of galactic nuclei.

154.055 Abundances, temperatures and reddenings of field and cluster population II giants. M. Grenon.
Astrophysical parameters for globular clusters, (see 012.023), p. 393 - 400 (1981).

In order to define the precise photometric properties of the pop. II giants in various evolutionary stages, selected bright stars have been measured in the outer coronas of the globular

clusters M 13, M 92, NGC 6397 and NGC 6752. In all cases the HR diagrams display very tight giant branches but several possible AB stars have shown themselves as non-member foreground stars according to their gravities and abundances. This paper describes the metal and CNO abundance effects and that of temperature. Methods for the derivation of physical parameters of population II stars will be presented as well as the estimation of the amount of interstellar reddening. Criteria for the detection of the C-abundance anomaly are also given.

154.056 Two-dimensional classification of globular clusters.
V. Straižys.
Astrophysical parameters for globular clusters, (see 012.023), p. 401 - 408 (1981).

154.057 Temperatures, surface gravities and masses of horizontal-branch A-stars.
D. S. Hayes, A. G. D. Philip.
Astrophysical parameters for globular clusters, (see 012.023), p. 409 - 413 (1981).

154.058 The helium content of globular clusters from their main sequences and horizontal branches.
F. Caputo, G. Cayrel de Strobel.
Astrophysical parameters for globular clusters, (see 012.023), p. 415 - 426 (1981).

The authors have applied a method allowing to determine both Y and Z for globular clusters, for which accurate UBV photometry exists. By-products of this method are: (1) A revision of the distances of the clusters, (2) A revision of the age of the globular clusters, (3) A new determination of the absolute magnitude M_V at the blue edge instability gap, and (4) A new determination of the primeval He-content of the Universe.

154.059 Differences in luminosity functions among globular clusters. E. M. Green.
Astrophysical parameters for globular clusters, (see 012.023), p. 427 - 436 (1981).

Bolometric luminosity functions (LF's) have recently been constructed for complete samples of bright giants in 17 globular clusters (Green 1981). On the assumption that M_V (HB) = 0.6 in all clusters, the observed LF's were fit to a standard theoretical LF (Y = 0.30, T = 14 billion yrs). The two main conclusions of the previous work were (1) many globulars, particularly some of the largest and best-studied clusters near the Sun, have LF's that agree quite well with canonical theoretical predictions; and (2) LF's of clusters closest to the galactic center are noticeably different. The purpose of this paper is to further examine the significance, causes, and consequences of the observed LF variations.

154.060 Search for (globular) clusters in M31: a progress report. P. Battistini, F. Bonoli, F. Fusi Pecci, R. Buonanno, C. E. Corsi.
Astrophysical parameters for globular clusters, (see 012.023), p. 439 - 440 (1981). − Abstract.

154.061 Line-strength anomalies in the integrated spectra of M31 globular clusters.
D. Burstein, S. M. Faber, C. M. Gaskell, N. Krumm.
Astrophysical parameters for globular clusters, (see 012.023), p. 441 - 450 (1981).

154.062 Integrated photometry and IIDS spectrophotometry of M33 globulars.
C. A. Christian, R. A. Schommer.
Astrophysical parameters for globular clusters (see 012.023), p. 461 - 466 (1981).

154.063 Reconnaissance of the newly discovered globular cluster system around NGC 5128 (Cen A).

J. E. Hesser, H. C. Harris, S. van den Bergh, G. L. H. Harris.
Astrophysical parameters for globular clusters, (see 012.023),
p. 467 - 472 (1981).

In this brief report the authors discuss the identification
of 12 more clusters, the overall properties of NGC 5128's
cluster system, and inferences about the NGC 5128 mass.

154.064   Globular cluster ages.   B. W. Carney.
Astrophysical parameters for globular clusters,
(see 012.023), p. 477 - 494 (1981).

**154.065   Ages of globular clusters from far ultraviolet
photometry with ANS.**
T. S. van Albada, R. J. Dickens, B. M. H. R. Wevers.
Astrophysical parameters for globular clusters, (see 012.023),
p. 495 - 499 (1981).

**154.066   Globular cluster turn-offs in the color-magnitude
diagram.   R. D. Cannon.**
Astrophysical parameters for globular clusters, (see 012.023),
p. 501 - 510 (1981).

This paper is divided into two main sections. The first
part deals with some of the observational errors and uncertainties involved in determining cluster CM diagrams, and
the second presents some of still unpublished data. The author
shows that the data on turn-off points is every bit as confusing
as the data on abundances, and that although there are
apparently sound cosmological reasons for expecting all
globular clusters to have similar ages and helium abundances,
this conclusion is not easily reconciled with the full set of CM
diagram data now available.

154.067   Uncertainties in the ages of globular clusters.
H. L. Shipman.
Astrophysical parameters for globular clusters, (see 012.023),
p. 511 - 516 (1981).

**154.068   Faint blue objects in the field of the globular cluster
M4.**
H. B. Richer, E. Chan, G. G. Fahlman, P. Hickson.
Astrophysical parameters for globular clusters, (see 012.023),
p. 519 - 525 (1981).

The authors discuss the results of a deep blink survey
centered on the globular cluster M4. Their objective was to
isolate faint blue objects which could be white dwarf members
of the cluster. The area searched lay between 4' and 24' from
the cluster center, and in this annulus 1096 faint UV bright
objects were found brighter than $U \cong 22$. The authors find
that over the area searched there should be $\cong 200$ cluster
white dwarfs brighter than $M_B \cong 10.3$. CCD photometry was
obtained of 8 candidates: 4 were clearly QSOs or faint blue
galaxies, 2 objects were of uncertain type while 2 appeared to
have colors typical of white dwarfs.

**154.069   The subgiant branch of the globular cluster
ω Centauri.   G. S. Da Costa, J. V. Villumsen.**
Astrophysical parameters for globular clusters, (see 012.023),
p. 527 - 531 (1981).

A color magnitude diagram for the subgiant branch
$(15.0 < V < 18.0)$ of the anomalous globular cluster ω Cen
has been determined. New results from this study include the
discovery of a sizeable gap at the base of the subgiant branch
which is not predicted by standard stellar evolutionary calculations. Further, the CM diagram also contains a number of "blue
stragglers" most of which are likely to be cluster members.

**154.070   Near-infrared photometry of globular clusters near
the galactic center.   M. A. Malkan.**
Astrophysical parameters for globular clusters, (see 012.023),
p. 533 - 538 (1981).

Most of the remaining undetected globular clusters in our
Galaxy are near the galactic center, and are heavily obscured

by interstellar dust. They may be found by photographic
searches at 0.8 μm. Sky-limiting exposures of hypersensitized
IV-N plates in a 1.2 m Schmidt telescope have detected
globular clusters obscured by up to 10 magnitudes of visual
extinction. This paper presents a technique for separating the
effects of reddening from those of metallicity. Both can be
determined independently with narrow-band infrared photometry of the absorptions from CO at 2.4 μm and H_2O at 2.0 μm.
The technique is applied to new observations of ten heavily
reddened globular clusters.

154.071   uvby-β photometry of turnoff stars in NGC 6397.
A. Ardeberg, P. E. Nissen.
Astrophysical parameters for globular clusters, (see 012.023),
p. 539 - 546 (1981).

**154.072   Photographic photometry of globular clusters: M5
and M15.**
R. Buonanno, C. E. Corsi, F. Fusi Pecci.
Astrophysical parameters for globular clusters, (see 012.023),
p. 551 - 553 (1981).

Using a reduction procedure based on a multicomponent
two-dimensional fit to the data derived by scanning the plates
with a PDS microphotometer an accurate photographic
photometry has been obtained for stars in the globular clusters
M5 and M15.

154.073   Superluminous giants in globular-like clusters.
P. J. Flower.
Astrophysical parameters for globular clusters, (see 012.023),
p. 555 - 559 (1981).

Superluminous giants are found near the central regions
of globular-like Magellanic Cloud clusters. They are on the
average ~2.7 mag brighter than the brightest main sequence
stars and are brighter than normal cluster giants. Their
luminosities correlate with cluster age; the older the cluster,
the fainter its SLGs. Cluster ages range from relatively young
clusters with ages ~30 million years to intermediate-age
clusters with ages <2 billion years. This implies evolutionary
masses for SLGs between ~1.5 $M_\odot$ and 7 $M_\odot$.

**154.074   Color-magnitude diagrams of M15 and M71 in the
four-color system.**
A. G. D. Philip, L. J. Relyea.
Astrophysical parameters for globular clusters, (see 012.023),
p. 561 - 569 (1981).

In this paper the authors report on the reduction of the y
and b magnitudes to obtain a four-color CM diagram of each
cluster.

**154.075   Field horizontal-branch and low metallicity stars in
the Strömgren and Geneva systems.**
B. Hauck, A. G. D. Philip.
Astrophysical parameters for globular clusters, (see 012.023),
p. 571 - 575 (1981).

Seventeen stars considered by various authors as field
horizontal-branch stars (FHB) have been found in common in
the Strömgren and Geneva photometric systems. By inspec-
tion of various photometric diagrams in both systems it can be
seen that the majority of the selected stars are located in a
specific area in each diagram. These are the FHB stars but some
stars in the list are members of other groups, i.e. normal Pop I
stars, stars with very high Balmer jumps or metal deficient stars.

**154.076   Main-sequence photometry of the metal-rich
globular cluster NGC 6352.**
J. M. Nemec, J. E. Hesser, P. Ugarte P.
Astrophysical parameters for globular clusters, (see 012.023),
p. 577 - 580 (1981).

154.077   A distant star cluster in Hydra, AM-4.
B. F. Madore, H. C. Arp.

Publ. Astron. Soc. Pacific, Vol. 94, 40 - 42 (1982).

A new apparently very sparse star cluster has been discovered on the ESO/SRC IIIa-J Southern Sky Survey. The system is probably another intergalactic globular cluster similar in distance to AM-1. The discovery of a second representative of this class of objects and especially the circumstances of its discovery raise considerable question as to the true density of these systems within the Local Group.

154.078 El cúmulo globular IC 4499.
C. R. Fourcade, J. R. Laborde, J. C. Arias.
Bol. Asoc. Argentina Astron., No. 20 - 24, p. 156 (1981).
Abstract.

154.079 El cúmulo globular NGC 1851.
C. R. Fourcade.
Bol. Asoc. Argentina Astron., No. 20 - 24, p. 159 (1981).
Abstract.

154.080 El cúmulo globular NGC 6362.
C. R. Fourcade.
Bol. Asoc. Argentina Astron., No. 20 - 24, p. 160 (1981).
Abstract.

154.081 El cúmulo globular NGC 6528.
C. R. Fourcade.
Bol. Asoc. Argentina Astron., No. 20 - 24, p. 161 (1981).
Abstract.

154.082 The globular cluster NGC 5286.
C. R. Fourcade, J. R. Laborde, J. C. Arias.
Bol. Asoc. Argentina Astron., No. 20 - 24, p. 249 - 255 (1981).

154.083 Estrellas variables en NGC 5286.
C. R. Fourcade, J. R. Laborde, A. A. Puch,
J. R. Colazo, J. C. Arias.
Bol. Asoc. Argentina Astron., No. 20 - 24, p. 256 - 257 (1981).

154.084 Estudio del cúmulo globular NGC 6752 y sus variables.
C. R. Fourcade, J. R. Laborde, A. A. Puch.
Bol. Asoc. Argentina Astron., No. 20 - 24, p. 350 (1981).
Abstract.

154.085 Estudio de estrellas variables en NGC 6362.
C. R. Fourcade, J. R. Laborde, A. A. Puch,
J. Colazo.
Bol. Asoc. Argentina Astron., No. 20 - 24, p. 351 (1981).
Abstract.

154.086 Equidensitometría en cúmulos globulares del hemisferio sur.
C. R. Fourcade, J. R. Laborde, B. Candellero.
Bol. Asoc. Argentina Astron., No. 20 - 24, p. 352 (1981).
Abstract.

154.087 Búsqueda de H I en cúmulos globulares.
E. Bajaja, E. M. Arnal.
Bol. Asoc. Argentina Astron., No. 20 - 24, p. 437 (1981).
Abstract.

154.088 Búsqueda de H I en cúmulos globulares.
E. Arnal, E. Bajaja.
Bol. Asoc. Argentina Astron., No. 20 - 24, p. 491 (1981).
Abstract.

154.089 Photometry of faint stars in globular clusters. V. The turn-off point of the cluster M10 = NGC 6254.
N. N. Samus', S. Yu. Shugarov.
Astron. Tsirk., No. 1196, p. 4 - 6 (1981). In Russian.

154.090 PAL 14: an intermediate metal abundance globular cluster in the outer galactic halo.
G. S. Da Costa, S. Ortolani, J. Mould.
Astrophys. J., Vol. 257, 633 - 639, plate 5 (1982).

The KPNO video camera system has been used to determine a color-magnitude diagram for the distant halo globular cluster Pal 14. In common with other outer halo systems, this cluster possesses a predominantly red horizontal branch. If $M_V = +0.6$ for the horizontal branch stars, then $(m-M)_0 = 19.4$, and the cluster lies 70 kpc from the galactic center and 50 kpc above the galactic plane. The giant branch stars are also red, indicating an intermediate rather than a low metal abundance for this cluster. The authors estimate $[Fe/H] = -1.55 \pm 0.25$ from the dereddened color of the giant branch at the level of the horizontal branch.

154.091 Age determinations for globular clusters.
B.C. Johnson, B. P. Flannery.
Bull. American Astron. Soc., Vol. 13, 871 - 872 (1981).
Abstract.

154.092 A comparison between the observational and theoretical H-R diagrams for the LMC star cluster 1866. S. A. Becker, G. Mathews.
Bull. American Astron. Soc., Vol. 13, 872 (1981). – Abstract.

154.093 C and N abundances in giant stars of the metal-poor globular cluster M 15.
G. E. Langer, C. F. Trefzger, N. B. Suntzeff, R. P. Kraft,
D. F. Carbon.
Bull. American Astron. Soc., Vol. 13, 872 (1981). – Abstract.

154.094 Metallicities in galactic and M 87 globular clusters.
J. P. Brodie, D. A. Hanes.
Bull. American Astron. Soc., Vol. 13, 893 (1981). – Abstract.

154.095 The system of old clusters in the Magellanic Clouds.
G. Barbaro.
Astrophys. Space Sci., Vol. 83, 143 - 162 (1982).

Integrated UBV colours have been computed for synthetic clusters older than one billion years and for two chemical composition: (a) Y = 0.30; Z = 10^{-4} and (b) Y = 0.30; Z = 10^{-2}, taking into account the contribution to the integrated light of Main Sequence, subgiant, red giant and horizontal branch stars. It has been found that integrated colours depend on Z and allow an estimate of the metal content. Old clusters in LMC and SMC have been studied in terms of colour calibrations. It was found that in the LMC clusters with Z = 10^{-2} and t >5 × 10^9 yr are lacking, clusters with relatively blue colours are similar, both in age and chemical composition, to the halo galactic globular clusters. Moreover, there is a group of clusters with $1 \times 10^9 \leqslant t \leqslant 5 \times 10^9$. In the SMC clusters with Z = 10^{-2} and t > 5 × 10^9 yr are lacking and clusters with $1 \times 10^9 \leqslant t \leqslant 5 \times 10^9$ are rare. Clusters with relatively blue colours are interpreted with the following parameters: t = 5 × 10^9 yr, $10^{-4} < Z < 10^{-3}$ and Y = 0.20. The implication of these results on the chemical history of the two galaxies is discussed.

154.096 On the structure of the 'red' globular cluster NGC 1806 in the Large Magellanic Cloud.
E. H. Geyer, U. Hopp.
Astrophys. Space Sci., Vol. 84, 133 - 141 (1982).

A structural study of the old globular cluster NGC 1806 in the LMC has been carried out by star counts on B- and V-ESO 3.6 m telescope plates with three different limiting magnitude levels. The star density distribution was obtained directly from the surface strip count function with the Plummer formalism and the generalized Schuster law according to Lohmann. The results show that the blue stellar content of NGC 1806 – the horizontal branch stars – is more concentrated towards the cluster center than the red giant and subgiant objects. Also such a dynamical mass segregation is observed for the red giants compared with the subgiants.

154.097 Globular cluster subgiant and MS properties derived from integrated light. E. M. Green.
Bull. American Astron. Soc., Vol. 14, 574 (1982). – Abstract.

154.098 The age-metallicity relationship for the clusters of the Large Magellanic Cloud. J. G. Cohen.
Astrophys. J., Vol. 143 - 153 (1982).

Moderate dispersion spectrophotometric scans with the intensified Reticon array constructed by S. Shectman have been obtained using the du Pont telescope of the Las Campanas Observatory for 38 stars expected to be members of 15 clusters in the Large Magellanic Cloud. Ages for these clusters are deduced from a transformation of their classification in the scheme of Searle, Wilkinson, and Bagnuolo. Abundances are derived from the scans using a crude analysis applied to computer-generated pseudoequivalent widths calibrated by identical observations of 42 stars in six galactic globular clusters and by several galactic supergiants. A strong age-metallicity relationship is found and the chemical history of the LMC is discussed.

154.099 Radio sources in globular clusters? M. Birkinshaw, A. J. B. Downes.
Astrophys. J., Vol. 258, 154 - 160 (1982).

Radio observations are presented of sources which have been reported in or near globular clusters. Variability of these sources at 2.7 GHz was sought on time scales from 1 day to 1 year, and high-resolution interferometric observations at 2.7 or 5.0 GHz provide positional and structural information for several objects. One long-term variable has been detected— the other sources do not vary. One source shows the aligned structure characteristic of extragalactic sources, while another has a large measured redshift. The number of sources in globular cluster fields is entirely consistent with the extragalactic source counts. The authors conclude that most of the sources are unassociated with the globular clusters toward which they lie.

154.100 TiO band strengths in metal-rich globular clusters. III. Model atmosphere calibration.
H. R. Johnson, J. R. Mould, A. P. Bernat.
Astrophys. J., Vol. 258, 161 - 164 (1982).

The authors present model atmosphere calculations of the run of TiO band strength with color temperature in globular cluster giants. This permits a more reliable calibration to be constructed of near-infrared narrow-band photometry in globulars. Published observations suggest either a substantially higher ratio of titanium to iron in the halo than in the Sun or an unrecognized problem in the recent recalibration of the globular cluster abundance scale.

154.101 Age and metal abundance of a globular cluster, as derived from Strömgren photometry.
P. E. Nissen.
Messenger, No. 28, p. 4 - 7 (1982).

154.102 Optical structure of the core of the dynamically advanced globular cluster NGC 6397.
M. Aurière.
Astron. Astrophys., Vol. 109, 301 - 304 (1982).

Long focus bi-dimensional observations of the old globular cluster NGC 6397 combined with already available data have allowed to detect a central clump of faint stars similar to the central condensation in M 15. This luminosity cusp is almost two times smaller in size and about 10 times fainter than that of M 15. Each cusp corresponds to about 6×10^{-3} times the total brightness of the whole corresponding cluster. They may be related to the "central singularity" which is expected to occur in globular cluster cores.

Catalogue of star clusters and associations. Supplement 1, Vols. I - III. See Abstr. 002.034.

Catalogue of Star Clusters and Associations. Supplement 1. See Abstr. 002.111.

Landolt-Börnstein. See Abstr. 003.064.

Model atmospheres for globular cluster stars. See Abstr. 064.046.

The helium-core flash in globular-cluster stars. See Abstr. 065.030.

Models for horizontal-branch stars with cores enriched in carbon and oxygen. See Abstr. 065.032.

Report of IAU Commission 35: Stellar constitution (Constitution des étoiles). See Abstr. 065.072.

TiO band strengths in metal-rich globular clusters IV. The solar neighborhood sequence. See Abstr. 113.083.

A carbon star in the globular cluster Lindsay 102. See Abstr. 114.038.

Lithium abundance in Population II dwarfs. See Abstr. 114.090.

Carbon and nitrogen abundances in extremely metal-deficient red giants. See Abstr. 114.097.

Six meter spectra of M 92 blue horizontal-branch stars. See Abstr. 114.186.

The Oosterhoff period groups and the age of globular clusters. IV. Field RR Lyrae stars: age of the galactic disk. See Abstr. 122.005.

Period distributions of irregularly variable RR Lyrae stars. See Abstr. 122.013.

The dwarf cepheid NJL 79 in Omega Centauri. See Abstr. 122.052.

Photometry of red variables in 47 Tucanae. See Abstr. 122.066.

RR Lyrae variable pulsations and the Oosterhoff groups. See Abstr. 122.073.

The masses and pulsations of BL Herculis variables. See Abstr. 122.074.

Period changes in BL Her stars in globular clusters. See Abstr. 122.075.

Dwarf cepheids in Omega Centauri? See Abstr. 122.076.

RR Lyrae variables in globular clusters. See Abstr. 122.142.

New double-mode RR Lyrae variables in the globular cluster M15. See Abstr. 122.160.

Masses and evolution stages of globular cluster population II cepheids. See Abstr. 122.161.

The variable stars in the globular cluster NGC 6864 (M75). See Abstr. 122.197.

An investigation of period changes in cluster BL Herculis stars.   See Abstr. 122.200.

Report of IAU Commission 27: Variable stars (Étoiles variables).   See Abstr. 122.250.

The O type subdwarf ROB 162 in the globular cluster NGC 6397.   See Abstr. 126.017.

Search for periodic pulsation in four globular cluster X-ray sources.   See Abstr. 142.078.

Star clusters containing massive, central black holes. IV. Galactic tidal fields.   See Abstr. 151.016.

Dynamical evolution of globular clusters: recent theoretical ideas.   See Abstr. 151.053.

Report of IAU Commission 37: Star clusters and associations (Amas stellaires et associations). See Abstr. 153.053.

The form of the galactic globular cluster system and the distance to the galactic centre. See Abstr. 155.001.

Search for (globular) clusters in M31. II: Photographic photometry of the candidates in a 70′ square field centered on M31.   See Abstr. 158.051.

Metallicity versus age in spiral nuclei. See Abstr. 158.194.

Luminosity functions of star clusters in the Small Magellanic Cloud.   See Abstr. 159.012.

The globular clusters of the Small Magellanic Cloud in the general diagram magnitude-diameter. See Abstr. 159.015.

The age-abundance relation for Magellanic Cloud star clusters.   See Abstr. 159.016.

Ages and abundances of Magellanic Cloud clusters. See Abstr. 159.017.

UV observations of globular clusters in the Magellanic Clouds.   See Abstr. 159.018.

New data for old Magellanic clusters. See Abstr. 159.019.

The age-metallicity relationship for the clusters of the Large Magellanic Cloud.   See Abstr. 159.020.

Early chemical and dynamical evolution of the LMC. See Abstr. 159.021.

The globular clusters in the Magellanic Clouds: theoretical considerations.   See Abstr. 159.022.

NGC 2257 and the halo of the Large Magellanic Cloud.   See Abstr. 159.025.

Observed radii and structural parameters of clusters in the SMC.   See Abstr. 159.036.

Erratum

154.901 Erratum: 'Mass segregation in globular clusters' [J. Astrophys. Astron., Vol. 2, 215 - 244 (1981)]. K. K. Scaria, M. K. V. Bappu. J. Astrophys. Astron., Vol. 2, 439 (1981). – See Abstr. 30.154.041.

155 Galaxy (Structure, Evolution)

155.001 The form of the galactic globular cluster system and the distance to the Galactic Centre.
C. S. Frenk, S. D. M. White.
Mon. Not. R. Astron. Soc., Vol. 198, 173 - 192 (1982).

The authors develop new quantitative methods for analysing the structure of the galactic globular cluster system. The cluster system is slightly flattened and there is no significant evidence for any variation in flattening as a function of metallicity. Distance modulus errors of order one magnitude are required to explain the deviation of the apparent distribution of metal-rich clusters from axial symmetry. In addition a systematic difference in distance scale of about 0.5 magnitudes is necessary to reconcile the centroid of this distribution with that of the metal-poor clusters. If the standard distance scale is adopted for metal-poor clusters, the estimated distance of the Galactic Centre is $R_\odot = 6.8 \pm 0.8$ kpc.

155.002 Mass of our Galaxy as determined from tidal radii of globular clusters and dwarf spheroidal galaxies.
K.-I. Wakamatsu.
Publ. Astron. Soc. Pacific, Vol. 93, 707 - 711 (1981/82).

The author has shown that the mean mass density of a parent galaxy within the perigalacticon distances of its companions can be expressed by the tidal radii and masses of its companions only. Applying this to the globular clusters and dwarf spheroidal systems of our Galaxy, the author has determined the mean mass density law for our Galaxy as a function of radius, and has obtained evidence against the existence of a massive halo in our Galaxy, if the boundaries of globulars and dwarf spheroidals are tidally limited.

155.003 Discussion of a deep blink survey of faint red objects toward the south galactic pole.
R. F. A. Staller, P. S. Thé, A. C. T. Bochem-Becks.
Publ. Astron. Soc. Pacific, Vol. 93, 728 - 731 (1981/82).

About 2600 red objects brighter than $R = 18^m$ were found in a field of 6 square degrees centered at R.A. = $0^h 46^m$, Dec. = $- 27°50'$(1950), close to the south galactic pole. These mostly faint objects, unbiased toward high proper motion, were detected in a blink survey on red and blue copies of Mount Palomar Sky Survey plates. A GRISM spectroscopic plate, taken with the Cerro Tololo 4-m telescope, shows that, up to the limit of this plate, the red objects are all M-type stars. If the scale height of M dwarfs $H = 260$ pc, the surface density of ~ 430 red objects per square degree is approximately twice as much as would be expected from Luyten's (1968) luminosity function.

155.004 A rotational standard of rest.
W. L. H. Shuter.
Mon. Not. R. Astron. Soc., Vol. 199, 109 - 113 (1982).

It is argued that the conventional 'local standard of rest' (LSR) is not the most appropriate standard to use for large-scale studies of galactic rotation. Three major 'deformities' in the present picture of the Galaxy are removed if a 'rotational standard of rest' is adopted, relative to which the LSR moves with a velocity of ~ 11 km s^{-1} directed towards $l \simeq 130°$.

155.005 CNO isotopes and galactic chemical evolution.
M. Tosi.
Astrophys. J., Vol. 254, 699 - 707 (1982).

In order to quantitatively reproduce the chemical abundance gradient and the age-metallicity relation observed in the disk of the Galaxy, numerical models of galactic chemical evolution have been computed under different assumptions on the star formation and infall rates. The best fit to the observed data is obtained for a model with a constant star formation rate and a uniform infall rate of $5 \times 10^{-3} M_\odot$ kpc^{-2} yr^{-1}. The general behaviour of primary and secondary elements during galactic evolution is analyzed throughout the disk by means of the computed models. The comparison between model results and observed isotopic ratios provides information about the evolution of the seven stable CNO isotopes.

155.006 Statistical modelling in the problem of deriving the spiral structure of the velocity field of stars in the Galaxy.   E. D. Pavlovskaya, A. A. Suchkov.
Astron. Zh., Tom 59, 44 - 50 (1982). In Russian. English translation in Soviet Astron., Vol. 26, No. 1.

In connection with the problem of deriving the parameters of the galactic spiral structure, the authors study statistical fluctuations in the velocity field of a given sample of stars as well as the effect of the number of stars and their distances on the parameter estimates. A sample of 100 - 150 stars within 3 - 4 kpc, their velocity dispersion being up to 20 km/s, is shown to allow the detection of the spiral structure in the case when the star distances are reliably determined and the velocity field has no additional peculiarities.

155.007 Three-colour photometry of the Milky-Way field HD 95540.   W. Becker, S. M. Hassan.
Astron. Astrophys., Suppl. Ser., Vol. 47, 247 - 256 (1982).

In a field of 0.123 square degrees near the star HD 95540 ($l = 290°$, $b = -0.3°$) RGU magnitudes of 2001 stars down to a limiting magnitude of $G = 16^m$ have been measured. Their distribution in the two-colour-diagrams suggests two extinction screens at 0.4 kpc and 2.75 kpc from the sun. The density functions for late-type giants and for main-sequence stars down to $M_G = +6^m$ are derived and discussed. The luminosity functions have been derived for two distance intervals and compared with that derived for the solar neighbourhood.

155.008 RGU three colour photometry of a field in Centaurus.   A. Spaenhauer, C. Fang.
Astron. Astrophys., Suppl. Ser., Vol. 47, 441 - 445 (1982).

2425 stars have been measured in the RGU system in a field in Centaurus down to the limiting magnitudes 16^m6, 15^m2 and 19^m0 for G, R and U respectively. Based on statistical photometric parallaxes, the density functions of main sequence stars for different intervals of absolute magnitudes as well as the density functions of late-type giants have been derived. Further the relation colour excess $E(G-R)$ vs. distance up to 2 kpc has been established as well as the luminosity functions for different intervals of distance.

155.009 Tilt of the central gas disk of the Galaxy.
A. H. Nelson, with a reply by L. Blitz and J. W.-K. Mark.
Nature, Vol. 295, 263 - 264 (1982).

155.010 A CCD image of the galactic centre.
J. W. V. Storey, J. O. Straede, P. R. Jorden, D. J. Thorne, J. V. Wall.
Nature, Vol. 296, 333 - 334 (1982).

Using a cooled CCD, the authors have obtained an image of the galactic centre at an effective wavelength of 0.9 μm. Two unresolved sources were found, separated by 3 arc s along the galactic plane. It seems likely that they are two compact H II regions seen in line emission.

155.011 On the evidence of a massive galactic corona.
K. Rohlfs.
Astron. Astrophys., Vol. 105, 296 - 299 (1982).

Flat rotation curves for galaxies can be produced either by models containing a massive, spherical corona, or by models with a Mestel-disk. It is shown that it is possible to distinguish observationally between these two alternatives by comparing the galactic force function in the radial- and in the z-direction. The dynamical evidence available in the Galaxy favours a model containing a corona.

155.012 **Bursts of star formation in the galactic centre.**
H. H. Loose, E. Krügel, A. Tutukov.
Astron. Astrophys., Vol. 105, 342 - 350 (1982).

The authors present scenarios for the possible future of the galactic centre. The results are based on one-dimensional hydrodynamic calculations which follow the evolution of the gas for several 10^8 yr. Rotation and magnetic fields are not included. The final engine of the motions of the gas is star formation. After the birth of a new generation of stars the remaining gas is churned up by supernova explosions. This terminates the star formation process and leads to an expansion of the gas.

155.013 **Predicted star counts in selected fields and photometric bands: applications to galactic structure, the disk luminosity function, and the detection of a massive halo.**
J. N. Bahcall, R. M. Soneira.
Astrophys. J., Suppl. Ser., Vol. 47, 357 - 403 (1981).

The authors present tables of predicted star counts in 17 selected directions on the sky as a function of apparent B, V, R, and I magnitudes and absolute visual magnitudes that are calculated from the Bahcall-Soneira Galaxy model. The tabulated star counts can be transformed easily into any other band using the given visual absolute and apparent magnitudes; the distribution of star colors for any two bands can also be calculated simply. The importance of I band star counts is stressed. Star counts in specified fields to $m_I = 19$ mag can reveal the faint end of the disk luminosity function down to the end of the hydrogen-burning main sequence. Star counts at high galactic latitudes to $m_I = 22$ mag should reveal the stellar constituents of a massive halo if they are massive enough to be on the hydrogen-burning main sequence.

155.014 **Surface brightness and colors of the galactic disk.**
K. Ishida, T. Mikami.
Publ. Astron. Soc. Japan, Vol. 34, 89 - 98 (1982).

Space number density distributions of the luminous stars of all spectral and luminosity classes are examined by the IRC data and are used to derive the volume emissivity in the solar neighborhood. The volume emissivity is derived to be $0.071\ L_\odot\ pc^{-3}$ in the V band. The surface brightness and colors of the galactic disk seen outside from the galactic pole are estimated to be $22.7\ (\pm .2)$ mag (V) arcsec^{-2} and $B-V =$ $= 0\overset{m}{.}62\ (\pm\ \overset{m}{.}04)$, $V-I = 1\overset{m}{.}37\ (\pm\ \overset{m}{.}07)$, $V-K = 2\overset{m}{.}71\ (\pm\ \overset{m}{.}10)$.

155.015 **Über die Altersverteilung von sonnennahen A-Sternen.** L. G. Balázs, I. Tóth.
Mitt. Astron. Ges., Nr. 55, p. 119 - 120 (1982).

155.016 **$^{12}C/^{13}C$-Verhältnisse in der galaktischen Scheibe.**
C. Henkel, T. L. Wilson, J. Martin-Pintado, T. Pauls, C. M. Walmsley.
Mitt. Astron. Ges., Nr. 55, p. 121 - 123 (1982).

155.017 **Strukturen im galaktischen Halogas in Richtung zur Großen Magellanschen Wolke.**
E. Schulz-Lüpertz, M. Grewing.
Mitt. Astron. Ges., Nr. 55, p. 123 - 128 (1982).

155.018 **Zur Winkelgeschwindigkeit des galaktischen Spiralmusters.** B. A. Balázs.
Mitt. Astron. Ges., Nr. 55, p. 199 (1982). – Abstract.

155.019 **On the three-dimensional structure of spiral arms of the Galaxy.** L. N. Kolesnik, N. G. Guseva.
Astrometr. Astrofiz., Vyp. (No.) 43, p. 67 - 74 (1981). In Russian.

2600 stars including O-B2 stars of all luminosity classes and supergiants of all spectral classes from the photoelectric catalogue of the US Naval Observatory have been used to study their distribution as a function of distance from the equatorial plane of the Galaxy. The high-luminosity stars turn out to deviate systematically from the galactic plane, lying above or below it. The thickness of the spiral arms is estimated. The bifurcation of the Sagittarius spiral arm in the direction $325°$ - $345°$ is revealed.

155.020 **The local system of stars.** W. Herbst.
Sky Telesc., Vol. 63, 574 - 577 (1982).

155.021 **A simple disk-halo model for the chemical evolution of our Galaxy.** S.-W. Lee, H. B. Ann.
J. Korean Astron. Soc., Vol. 14, 55 - 71 (1981).

On the basis of observational constraints, particularly the relationship between metal abundance and cumulative stellar mass, a simple two-zone disk-halo model for the chemical evolution of our Galaxy was investigated, assuming different chemical processes in the disk and halo and the infall rates of the halo gas defined by the halo evolution.

155.022 **Kinematical properties of the spectral group of nearby dwarfs.** S. G. Lee.
J. Korean Astron. Soc., Vol. 14, 73 - 78 (1981).

On the basis of the recently available data, the author has analysed the kinematical properties of nearby dwarfs, which are grouped by their spectral types and derived their ages from the kinematical properties. Discontinuities in the kinematical properties are found around late F stars, which appear to be caused mainly by the fact that the spectral groups earlier than late F are rather homogeneous in age while the later ones are mixed by two different age groups.

155.023 **Gould's belt: a preliminary study using the Skymap catalog.** W. E. Baggett.
Proc. Southwest Reg. Conf., Vol. 7, (see 012.019), 51 - 60 (1982).

Data from the Skymap catalog were used to study the spatial distribution of stars in Gould's Belt. Two planes were fit to B stars from the catalog, one representing the galactic belt and one Gould's Belt, and the pole position determined. The pole determined for Gould's Belt is at $l = 196°, b = 72°$, with the sun being approximately 2 pc above the plane.

155.024 **Estudio en 21 cm de los brazos interiores de la Vía Láctea mediante la aplicación de un modelo observacional de distancias.** H. G. Peña, S. Garzoli.
Bol. Asoc. Argentina Astron., No. 18, p. 36 - 46 (1980).

155.025 **Structure of the inner edge of the local spiral arm in the direction to NGC 6823.**
V. I. Kuznetsov, M. D. Metreveli.
Astrometr. Astrofiz., Vyp. (No.) 46, p. 36 - 42 (1982). In Russian.

On the basis of photometric and spectral data for faint stars up to $15\overset{m}{.}0$ an analysis was made of interstellar light absorption and space distribution of stars in the direction to NGC 6823 at distances up to 5 kpc.

155.026 **The galactic abundance gradient.**
P. A. Shaver, A. C. Danks, R. X. McGee, L. M. Newton, S. R. Pottasch.
Messenger, No. 27, p. 19 - 21 (1982).

155.027 **Cosmic chemical memory: a new astronomy.**
D. D. Clayton.
Q. J. R. Astron. Soc., Vol. 23, 174 - 212 (1982).

155.028 **The distribution of free electrons in the inner Galaxy from pulsar dispersion measures.**
D. S. Harding, A. K. Harding.
Astrophys. J., Vol. 257, 603 - 611 (1982).

The authors have statistically analyzed the dispersion measures of a sample of 149 pulsars in the inner Galaxy ($|l| < 50°$) to deduce the large-scale distribution of free thermal electrons in this region. The dispersion measure distribution of these pulsars shows significant evidence for a decrease in the electron scale height from a local value greater than the pulsar scale height to a value less than the pulsar scale height at galactocentric radii inside of ~7 kpc. An increase in the electron density (to a value around 0.15 cm^{-3} at 4–5 kpc) must accompany such a decrease in scale height. There is also evidence for a large-scale warp in the electron distribution below the $b = 0°$ plane inside the solar circle.

155.029 **The distribution of WR and supergiant stars on the galactic plane.**   G. Bertelli, C. Chiosi.
The most massive stars, (see 012.034), p. 211 - 214 (1981).

The authors suggest that the observed gradient in WR stars across the galactic plane is related to an analogous gradient in the progenitor stars, although effects of metallicity on the mean mass-loss rate cannot be excluded.

155.030 **Metallicity and the N_{RSG}/N_G ratio.**
B. Rocca-Volmerange, B. Guiderdoni.
The most massive stars, (see 012.034), p. 297 - 300 (1981).

The metallicity Z affects integrated colors of a stellar population. The authors show that an increasing metallicity and the present star formation rate induces a spectacular effect on the red supergiant/giant number ratio N_{RSG}/N_G and on the V–K color indices. A simple interpretation of the IR–excess at 2.4 μm wavelength, observed at the 5 kpc galactocentric distance of our Galaxy is also given.

155.031 **The Gould belt in linear density-wave theory.**
J. Rountree Lesh.
Bull. American Astron. Soc., Vol. 13, 786 (1981). – Abstract.

155.032 **Early-type stars in the southern galactic halo.**
J. R. Pier.
Bull. American Astron. Soc., Vol. 13, 786 (1981). – Abstract.

155.033 **A program to determine metallicity by photographic photometry of halo G and K giants at large galactocentric distances.**
R. P. Boyle, F. Smriglio, V. Straižys.
Bull. American Astron. Soc., Vol. 13, 786 (1981). – Abstract.

155.034 **Galactic chemical evolution.**
B. A. Twarog, J. C. Wheeler.
Bull. American Astron. Soc., Vol. 13, 786 (1981). – Abstract.

155.035 **Possible detection of a galactic wind.**
G. P. Garmire, J. J. Nugent.
Bull. American Astron. Soc., Vol. 13, 786 - 787 (1981). Abstract.

155.036 **The gaseous galactic halo as inferred from the line spectra of the galaxies Markarian 509 and Fairall 9.**
D. G. York, J. C. Blades, L. L. Cowie, D. C. Morton, A. Songaila, C.-C. Wu.
Bull. American Astron. Soc., Vol. 13, 787 (1981). – Abstract.

155.037 **An atlas of galactic plane 21 cm absorption.**
J. M. Dickey, S. R. Kulkarni, C. E. Heiles, J. H. Van Gorkom.
Bull. American Astron. Soc., Vol. 13, 787 (1981). – Abstract.

155.038 **An objective-prism survey of M-giant stars in the region of the South Galactic Pole.**
S. J. Schiller.
Bull. American Astron. Soc., Vol. 13, 845 (1981). – Abstract.

155.039 **A combined south-north CO survey of the galactic disk: radial distribution of molecular clouds from $l = 330°$ to 70°.**   D. B. Sanders, P. M. Solomon.
Bull. American Astron. Soc., Vol. 13, 863 (1981). – Abstract.

155.040 **[O III] imagery of the galactic plane.**
T. R. Gull, R. A. Fesen, J. Heckathorn.
Bull. American Astron. Soc., Vol. 13, 887 (1981). – Abstract.

155.041 **A corona around the Milky Way.**   J. N. Bregman.
Nature, Vol. 298, 10 - 11 (1982).

155.042 **The galactic distribution of Wolf-Rayet stars.**
B. Hidayat, K. Supelli, K. A. van der Hucht.
Contrib. Bosscha Obs. Lembang, No. 68, 19 pp. (1982).

On the basis of the most recent compilation of narrow-band photometry and absolute visual magnitudes of Wolf-Rayet stars, and adopting a normal interstellar extinction law in all directions, the galactic distribution of 132 of the 159 known galactic WR stars is presented and discussed. The spiral structure is found to be more clearly pronounced than in earlier studies. Furthermore the authors find an indication of two spiral arms at r = 4 and 6 kpc. There appears to be an asymmetry of the z-distribution of single WR stars with respect to galactic longitude. The location of the WC8.5 and WC9 stars between 4.5 and 9 kpc from the galactic center is discussed in the context of Maeder's red supergiant to WR star scenario.

155.043 **Metallicity effect and 2.4 μm excess in the galactic disk.**   B. Guiderdoni, B. Rocca-Volmerange.
Pré-publ. Inst. Astrophys. Paris, No. 2, 20 pp. (1982).
Submitted to Astron. Astrophys.

155.044 **Comparison of galactic center with other galaxies.**
G. H. Rieke, M. J. Lebofsky.
Prepr. Steward Obs., No. 365, 10 pp. (1982).

155.045 **The post-explosion shock propagation in the central region of our Galaxy.**
T. Bhattacharyya, B. Basu.
Astrophys. Space Sci., Vol. 83, 15 - 36 (1982).

Immediate consequences of nuclear explosions on the structure and physical state of a galactic disk are considered in this paper. Explosions in the nucleus of a Galaxy generate strong shock waves which, when propagating onward heat and condensing the gas, form thin dense ring-like gaseous features behind it. Such rings and dense gaseous complexes have been observed in the central region of the Galaxy. These features have been treated here as the remnants of galactic shocks generated by nuclear explosions. The authors have estimated the time elapsed since the corresponding explosion, the energy released by explosion and the initial temperature and the velocity of the shock wave thus generated. The cooling of the gas heated by strong shocks has also been considered. The high-energy radiation emitted in the cooling process is suggested here as a source for the heating of dust grains, which ultimately are radiated in the infrared spectrum.

155.046 **Kinematical confirmation of the completeness of nearby stars.**   A. R. Upgren, T. E. Armandroff.
Bull. American Astron. Soc., Vol. 14, 581- 582 (1982). Abstract.

155.047 **A hole in the Milky Way.**   M. M. Waldrop.
Science, Vol. 216, 838 - 839 (1982).

155.048 **Three-colour photometry of a field near the galactic centre (SA 133 F).** W. Becker, C. Fang.
Astron. Astrophys., Suppl. Ser., Vol. 49, 61 - 67 (1982).

Photometric observations of 1235 stars have been made in the RGU-system covering an area of 0.19 square-degrees with the centre coordinates: $l = 9°3$, $b = +12°1$, down to the limiting magnitude $G = 17^m9$. The two-colour diagrams are used for the determination of the interstellar reddening and for the identification of various stellar types. The reddening value amounts to a colour excess of $E(G-R) = 0^m45$ and is caused by interstellar material closer than 300 pc. Density functions have been determined for late-type giants of population I, for metal-poor giants, for main-sequence stars of population I and partly also for metal-poor main-sequence stars. The metal-poor giants have almost constant densities, possibly with a slight maximum at about 8 kpc. For the metal-poor main-sequence stars densities could be calculated only within the distance interval 0.5 to 1.0 kpc where they are compared with the corresponding values for the fields NGC 6171 and SA 107. The overall behaviour seems to be consistent in all three fields.

155.049 **Metallicity effect and λ2.4 μm excess in the galactic disk.** B. Guiderdoni, B. Rocca-Volmerange.
Astron. Astrophys., Vol. 109, 355 - 359 (1982).

The high metallicity ring-like region at 5 kpc from the galactic center shows an important excess of λ2.4 μm emissivity, compared to the solar neighborhood. The purpose of this work is to account for this excess by means of an evolutionary model of a synthetic stellar population.

155.050 **High order momenta of the local stellar velocity distribution.** J. Núñez, J. Torra.
Astron. Astrophys., Vol. 110, 95 - 99 (1982).

The velocity momenta through the fourth order have been calculated for different samples of the "Gliese's Catalogue of Nearby Stars" and the "512 FK 4/FK 4 Suppl. Distant Stars". In order to obtain the variances of the sample momenta the authors have derived appropriate formulae. The results agree with those found by Erickson for a sample of 869 stars extracted from the Gliese's catalogue. Differences between the obtained results for "FK 4/FK 4 Suppl." and Gliese's catalogue are interpreted through the kinematic-age relation. Wide deviations from the normal law have been found for the residual velocity distribution of nearby stars. Important variations in the vertex deviation arise, as it was foreseen, for the different samples.

155.051 **Westerbork observations of H I absorption in the direction of Sgr A.**
U. J. Schwarz, R. D. Ekers, W. M. Goss.
Astron. Astrophys., Vol. 110, 100 - 104 (1982).

The Westerbork Synthesis Radio Telescope has measured the H I 21-cm absorption in the direction of Sgr A with an angular resolution of $23 \times 143''(\alpha + \delta)$, velocity range -130 to $+130$ km s^{-1}, and a velocity resolution 4.1 km s^{-1}. Any high dispersion low optical depth component of the interstellar medium is less than one-third that proposed by Radhakrishnan and Sarma (1980) and Radhakrishnan and Srinivasan (1980). The properties of the H I absorption from the 4 kpc arm, the -31, $+23$, $+50$, and $+70$ km s^{-1} features are discussed.

155.052 **Galactic structure results derived from studies of local kinematics.** K. Lodén.
Scientific aspects of the Hipparcos space astrometry mission, (see 012.041), p. 83 - 84 (1982).

155.053 **The role of early-type stars in studies of galactic dynamics.** S. V. M. Clube.
Scientific aspects of the Hipparcos space astrometry mission, (see 012.041), p. 85 - 87 (1982).

155.054 **The decrease of stellar density perpendicular to the galactic plane – a problem?** T. Oja.
Rep. Obs. Lund, No. 18, (see 012.044), p. 30 (1982).

155.055 **The mass distribution in the galactic centre: OH/IR stars as gravitational probes.**
A. Winnberg.
Rep. Obs. Lund, No. 18, (see 012.044), p. 31 - 34 (1982).

155.056 **Report of IAU Commission 33: Structure and dynamics of the galactic system** (Structure et dynamique du système galactique). G. G. Kuzmin.
Trans. IAU, Vol. XVIIIA, (see 003.013), p. 383 - 412 (1982).

Galaxien. See Abstr. 002.102.

The great galactic centre mystery.
See Abstr. 011.014.

Flächenphotometrie der Milchstraße im Visuellen.
See Abstr. 031.535.

Application of RGU-photometry to the study of galactic structure. See Abstr. 031.571.

On the origin of the e^+ - e^- annihilation line from the galactic center. See Abstr. 061.046.

Average ^{186,187,188}Os(n, γ) cross sections and the age of the Galaxy via ^{187}Re decay to ^{187}Os. See Abstr. 061.051.

The helium to heavy element enrichment ratio, $\Delta Y/\Delta Z$. See Abstr. 065.001.

Effect of overshooting on theoretical yields. See Abstr. 065.046.

Galactic and zodiacal light surface brightness measurements with the Atmosphere Explorer satellites. See Abstr. 106.028.

Is the position of the solar system in the Galaxy exceptional? See Abstr. 107.029.

On the remarkable position of the solar system in the Galaxy. See Abstr. 107.035.

Geneva photometric boxes. II. The reddening towards the galactic poles. See Abstr. 113.008.

Three-colour photometry of a field in the galactic anticentre section near NGC 2360. See Abstr. 113.012.

Stellar luminosity functions in the R, I, J, and K bands obtained by transformation from the visual band. See Abstr. 113.014.

Fotometría UBV y H beta de estrellas débiles de tipo temprano en Norma. See Abstr. 113.028.

BV photometry of a sample of faint red stars in the direction of the south galactic pole. See Abstr. 113.086.

Photoelectric photometry of faint M-type stars in the direction of the south galactic pole. See Abstr. 113.087.

$uvby$ photometry in McCormick proper motion fields. See Abstr. 113.089.

Spectrophotometry of cool stars in the near infrared. V. Results for a region in Monoceros.
See Abstr. 114.194.

Spectrophotometry of cool stars in the near infrared. VI. Results for a region in the direction of Cepheus.
See Abstr. 114.195.

A galactic window at $l \approx 355°$, $b = -1°$ and its implications for further observations.   See Abstr. 114.202.

Survey of red giants near the galactic center direction.   See Abstr. 114.203.

Infrared photometry of Mira variables in the Baade windows and the distance to the Galactic Centre.
See Abstr. 122.001.

The Oosterhoff period groups and the age of globular clusters. IV. Field RR Lyrae stars: age of the galactic disk.   See Abstr. 122.005.

An RR Lyrae survey with the Lick astrograph. V. A survey of three fields at intermediate latitudes towards the galactic anticenter.   See Abstr. 122.029.

Metal abundances of RR Lyrae variables in selected galactic star fields. III. The Lick astrographic fields near the galactic anticenter.   See Abstr. 122.030.

On the ageing *(period gradient)* of cepheids across the spiral arm Carina-Sagittarius.
See Abstr. 122.098.

Interstellar extinction in the Perseus arm.
See Abstr. 131.018.

The galactic fountain, observations of extragalactic radio sources, and the cosmic ray halo.
See Abstr. 131.029.

The gaseous galactic halo as inferred from the line spectra of the galaxies Markarian 509 and Fairall 9.
See Abstr. 131.058.

Interstellar titanium abundances toward 19 high-latitude stars.   See Abstr. 131.079.

Neutral hydrogen towards Tycho's supernova remnant.   See Abstr. 131.096.

Interstellar absorption and distance scale in the Galaxy.   See Abstr. 131.172.

Collective formation of massive stars in galaxies.
See Abstr. 131.179.

A CO survey of the southern galactic plane.
See Abstr. 131.183.

M supergiants and star formation at the Galactic Center.   See Abstr. 131.225.

Further ($^{12}C/^{13}C$) ratios from formaldehyde: a variation with distance from the galactic center.
See Abstr. 131.242.

The large scale distribution of molecular clouds in the outer Galaxy.   See Abstr. 131.271.

Implications of collisionally-supported giant molecular clouds for spiral galactic structure and massive star formation.   See Abstr. 131.277.

High-velocity H II regions delineating a central bar in our Galaxy?   See Abstr. 132.005.

Infalling clouds with very high velocities: a collision with the Milky Way in the anticenter.   See Abstr. 132.022.

Cold H I clouds in the Galaxy.
See Abstr. 132.028.

On the nature of galactic "supershells" of neutral hydrogen.   See Abstr. 132.048.

Infalling clouds with very high velocities: a collision with the Milky Way in the anticenter.
See Abstr. 132.060.

S201 far ultraviolet imagery of Messier 8 and the Sagittarius star field.   See Abstr. 132.066.

A radio continuum survey of the northern sky at 1420 MHz – Part I.   See Abstr. 141.085.

Ionized gas clouds and the nature of apparent variability of a "point" radio source at the galactic center.
See Abstr. 141.201.

Transitions between epicyclic stellar orbits induced by massive gas clouds.   See Abstr. 151.018.

The ballistic particle model and the vertex deviation of young stars near the sun.   See Abstr. 151.022.

Two component model of initial mass function.
See Abstr. 151.049.

On the interpretation of rotation curves measured at large galactocentric distances.   See Abstr. 151.078.

A model of spiral galaxies and the distribution of matter within the Galaxy.   See Abstr. 151.089.

Two-fluid gravitational instabilities in the Galaxy.
See Abstr. 151.096.

Comparisons of the HR diagrams of the youngest clusters in the Galaxy, the LMC and SMC. Evidence for a large MS widening.   See Abstr. 153.008.

Open clusters in our galaxy.   See Abstr. 153.044.

The Oosterhoff period groups and the age of globular clusters. III. The age of the globular cluster system.
See Abstr. 154.001.

Near-infrared photometry of globular clusters near the galactic center.   See Abstr. 154.070.

Widespread galactic OH emission at 1720 MHz: a new tracer of spiral arms?   See Abstr. 156.005.

Changes of the star formation rate and the initial mass function with galactic radius.   See Abstr. 156.012.

Widespread galactic OH emission at 1720 MHz: a new tracer of spiral arms.   See Abstr. 156.021.

Large-scale distribution of galactic gamma radiation observed by COS-B.   See Abstr. 157.002.

Galaxy morphology, starclusters and the mass spectrum of star formation.   See Abstr. 158.216.

The initial mass function of massive stars and the evolution of galaxies.   See Abstr. 158.217.

Photometry of faint galaxies and the motion of the Milky Way Galaxy.   See Abstr. 158.230.

Surface photometry of edge-on spiral galaxies. III. Properties of the three-dimensional distribution of light and mass in disks of spiral galaxies.   See Abstr. 158.308.

H I kinematics of spiral galaxies. See Abstr. 158.355.

On the proper motion of the Magellanic Clouds and the halo mass of our Galaxy.   See Abstr. 159.002.

The Galaxy as a fundamental standard for extragalactic distances. I. New methods to choose between the proposed values of the Hubble constant. See Abstr. 162.118.

Erratum

155.901   Erratum: "A realistic model of the Galaxy" [Astrophys. Space Sci., Vol. 79, 289 - 319 (1981)]. K. Rohlfs, J. Kreitschmann. Astrophys. Space Sci., Vol. 82, 255 (1982). – See Abstr. 30.155.040.

156 Galaxy (Magnetic Field, Radio and Infrared Radiation)

156.001   A high resolution far-infrared survey of a section of the galactic plane. I. The nature of the sources. D. T. Jaffe, M. T. Stier, G. G. Fazio. Astrophys. J., Vol. 252, 601 - 609 (1982).
The authors have surveyed a 7.5 deg^2 portion of the galactic plane between $l^{II} = 10°$ and $l^{II} = 16°$ at 70 μm with a 1' beam. They present far-infrared, radio continuum, and ^{12}CO and ^{13}CO line observations of the 42 far-infrared sources in the survey region. The nature and energetics of these sources are analysed. Most of the sources are excited either by single early-type stars or by clusters of early-type stars. The energetics in the molecular clouds surrounding the far-infrared sources is examined and it is concluded that the sources could supply the energy necessary for the observed temperature structure and velocity field in the molecular gas.

156.002   Far IR emission of the galactic plane at high longitudes. E. Bussoletti, I. Guidi, F. Melchiorri, V. Natale. Astron. Astrophys., Vol. 105, 184 - 187 (1982).
The authors present here broad-band far IR observations (500 - 1000 μm) of the diffuse emission of the galactic plane at longitudes $\ell = 233°$, 230°, 222°, 110°. By using the method developed by Ryter and Puget (1977) they derive a normalized luminosity for two realistic grain temperatures: 10 K and 20 K. A comparison between expected and observed values indicates that dust at around 10 K is emitting the detected IR emission. The usually assumed distributions of dust and stars are sufficient to justify these results so that no extra sources of heating other than the interstellar radiation field, must be considered to interpret the observations. Dust-to-gas ratio estimations are derived for the observed regions.

156.003   Infrared mapping of the galactic plane. II. Medium-resolution maps of the Cygnus X region. S. D. Price, L. P. Marcotte, T. L. Murdock. Astron. J., Vol. 87, 131 - 140 (1982).
Medium-resolution maps of Cyg X have been generated from the AFGL mid-infrared survey data. Large-scale diffuse emission covering several square degrees is observed at 11 and 20 μm. A number of discrete sources are also found in the region. The brighter discrete sources, the emission ridges, and all the 20-μm and at least half the 11-μm large-scale features

can be explained as emission from H II regions along the line of sight.

156.004   Grain alignment in the galactic magnetic field. P. E. Johnson. Nature, Vol. 295, 371 - 375 (1982).
The linear polarization of light observed in many distant stars provides a means of mapping the magnetic field lines in our own Galaxy as well as probing the properties of the intervening dust. Processes for grain alignment in the galactic magnetic field are reviewed: evidence is examined for grains being spun up to extremely high angular frequencies by the recoil of hydrogen recombination on grains, as an essential part of the alignment process.

156.005   Widespread galactic OH emission at 1720 MHz: a new tracer of spiral arms?   B. E. Turner. Astrophys. J., Lett., Vol. 25, L33 - L37 (1982).
Widely extended, weak OH emission has been observed at 1720 MHz in 53 galactic clouds whose sizes are those of giant molecular clouds and whose distribution appears confined to two inner spiral arms. Collisional pumping of the OH at low (15−40 K) temperatures seems the most likely excitation mechanism.

156.006   The origin of the diffuse galactic far infrared and sub-millimeter emission. P. G. Mezger, J. S. Mathis, N. Panagia. Astron. Astrophys., Vol. 105, 372 - 388 (1982).
The radiation of the Galaxy at infrared and sub-millimeter wavelengths has been recently measured. It consists of contributions from compact sources (~10−20% of the total luminosity of $L_{IR} \sim 6 \times 10^9 \, L_{\odot}$ for $\lambda \gtrsim 40 \, \mu$m) and from a diffuse component. The authors explain this diffuse emission as arising from dust embedded in neutral and ionized interstellar gas. The interstellar radiation field (ISRF) is redetermined and it is found, that its intensity is nearly independent of galactocentric distance. The authors obtain good agreement between the spectrum predicted for the composite dust model heated by the ISRF and the dust emission spectrum observed for globules.

156.007   Far infrared survey of extended molecular cloud H II region complexes along the galactic plane.

R. Gispert, J. L. Puget, G. Serra.
Astron. Astrophys., Vol. 106, 293 - 306 (1982).

The authors present the complete results of their far infrared survey of the northern part of the galactic plane. The data are shown in form of brightness contour maps. A special procedure to separate extended sources from the unresolved component has been developed and a catalogue of 58 extended sources is given. The authors give a production rate of the far infrared radiation versus galactic radius and a total luminosity of the Galaxy of $5 \times 10^{10} L_\odot$ and confirm the variation of the ratio of radio continuum emission to the infrared radiation with the galactic radius. They find that this ratio correlates well with a mean dust temperature.

156.008   Where are stars formed in the Galaxy?
P. Boisse.
Recherche, Vol. 12, 873 - 875 (1981). In French. — Abstr. in Phys. Abstr., Vol. 85, Abstr. 10519 (1982).

156.009   Near-infrared multicolor observation of the diffuse galactic emission.
K. Noguchi, S. Hayakawa, T. Matsumoto, K. Uyama.
Publ. Astron. Soc. Japan, Vol. 33, 583 - 590 (1981).

Near-infrared multicolor observation of the galactic plane was carried out at 2.0 μm, 2.8 μm, and 4.5 μm in the galactic longitude range between 28° and 67° with a rocket-borne infrared telescope. The surface brightness distributions at 2.0 μm and 2.8 μm are similar to that at 2.4 μm from previous balloon observations. Near-infrared galactic emission is attributed to a common origin. Most of the near-infrared sources are considered to be late-type stars according to the color of the sources estimated from the observed spectrum corrected for interstellar extinction.

156.010   On the 2.4-μm enhancement centered at about
$l = 355°$, $b = -1°$.   K. Hamajima, T. Ichikawa,
K. Ishida, B. Hidayat, M. Raharto.
Publ. Astron. Soc. Japan, Vol. 33, 591 - 601 (1981).

The enhancement of infrared 2.4-μm diffuse radiation centered at about $l = 355°$, $b = -1°$ is examined for its possible origin making use of M-type stars detected on objective-prism plates taken with the Schmidt telescope of the Bosscha Observatory. The surface number density distribution and color excesses of these stars lead the authors to the conclusion that at least part of the enhancement is ascribed to the low interstellar extinction in that direction.

156.011   Structure of the galactic magnetic field in the solar neighborhood.   M. Inoue, H. Tabara.
Publ. Astron. Soc. Japan, Vol. 33, 603 - 615 (1981).

By means of Faraday rotation of extragalactic radio sources and pulsars, and of optical polarization of stars, the direction of the large-scale, regular magnetic field within a distance of $\cong 2$ kpc has been investigated. The results of these three independent data agree closely with one another. Therefore the authors conclude that the regular field runs in the direction $l = 100 \pm 10°$, and the magnitude of the magnetic field B is 1.6 ± 0.4 μG. The scale height of the magnetic field along z (perpendicular to the galactic plane) has been derived to be $\cong 1$ kpc, which agrees well with the scale heights of the distributions of electrons and of ions.

156.012   Changes of the star formation rate and the initial mass function with galactic radius.
J. L. Puget, R. Gispert, G. Serra.
Regions of recent star formation, (see 012.002), p. 249 - 250 (1982).

156.013   Estudio de las estructuras exteriores de la galaxia en la zona 310° ⩽ l ⩽ 33°, 7° ⩽ b ⩽ 3°.
S. Blacher, G. Dubner, S. Garzoli.

Bol. Asoc. Argentina Astron., No. 20 - 24, p. 290 (1981). Abstract.

156.014   Influencia de la turbulencia del gas y geométrica en la curva de rotación básica para la línea de 21 cm.
M. E. Zales de Caponi, S. Blacher.
Bol. Asoc. Argentina Astron., No. 20 - 24, p. 291 (1981). Abstract.

156.015   Observations of the galactic center near 1μm with a charge-couplet device (CCD).
G. R. Ricker, M. W. Bautz, D. D. DePoy, S. S. Meyer.
Bull. American Astron. Soc., Vol. 13, 786 (1981). — Abstract.

156.016   Completion of the near infrared photographic survey of the northern sky.   E. R. Craine.
Bull. American Astron. Soc., Vol. 13, 896 (1981). — Abstract.

156.017   Spectrum of the galactic magnetic fields.
A. A. Ruzmaikin (*Ruzmajkin*), A. M. Shukurov.
Astrophys. Space Sci., Vol. 82, 397 - 407 (1982).

The mean galactic magnetic field is generated and maintained due to the turbulent dynamo-action. Only the laminar and spatially-averaged characteristics of the velocity field are essential for the theory of the generation of the large-scale magnetic field; that is differential rotation, turbulent diffusion and helicity of the turbulence. On the other hand, with regard to small-scale fields the detailed properties of the turbulence such as its spectrum, the width of the inertial range, the damping scale, etc. are important. In the present paper an attempt is made to determine the statistical properties of the galactic turbulent magnetic fields and to point out some observational tests designed to verify the theory.

156.018   Cosmic radio noise at high frequencies as observed with ISS-b.   M. Kotaki, I. Kuriki, C. Katoh,
H. Sugiuchi.
J. Radio Res. Lab., Vol. 28, No. 125 - 126, p. 35 - 48 (1981). Abstr. in Phys. Abstr., Vol. 85, Abstr. 58503 (1982).

156.019   Far-infrared observations of Sagittarius A: the luminosity and dust density in the central parsec of the Galaxy.   E. E. Becklin, I. Gatley, M. W. Werner.
Astrophys. J., Vol. 258, 135 - 142 (1982).

Far-infrared observations of the central 4′ of the Galaxy with 30″ resolution made simultaneously at 30 μm, 50 μm, and 100 μm are presented. The 30 μm radiation peaks strongly at the position of the galactic center, as determined from the 2 μm surface brightness and the density of ionized gas. The 50 and 100 μm emission is much more extended along the plane and shows two emission lobes, one on either side of the 30 μm peak. At the position of the galactic center itself there is a local minimum in the 100 μm surface brightness. It is concluded that the dust density decreases inward over the central few parsecs of the Galaxy and that the dust density in the central parsec is so low that optical and ultraviolet radiation freely traverses this region (i.e., $A_v \lesssim 1$). The total luminosity of the sources heating the dust which radiates the far-infrared emission from the central few parsecs is deduced to be between 1×10^7 and $3 \times 10^7 L_\odot$.

156.020   Galactic radio emission below 16.5 MHz and the galactic emission measure.   G. R. A. Ellis.
Australian J. Phys., Vol. 35, 91 - 104 (1982).

New maps of the distribution of the galactic background radio emission are given. A map of the quantity $\int_0^b (N_e^2/T_e^{3/2}) dr$ in galactic coordinates is obtained from an analysis of the changes in the radio brightness distributions with frequency. For an assumed electron kinetic temperature of 10^4K, the emission measure is found to vary from 3.9 cm^{-6} pc near the south galactic pole to 140 cm^{-6} pc near the equator.

156.021 Widespread galactic OH emission at 1720 MHz: a new tracer of spiral arms.   B. E. Turner.
Extragalactic molecules, (see 012.048), p. 165 - 173 (1982).

Anomalous OH emission in the 1720 MHz transition has long been recognized in our Galaxy, most commonly arising in cold, dark dust clouds. The first indication that the weak, extended 1720 type of anomaly might not be confined just to small dark clouds was its observation by Haynes and Caswell (1977). These results inspired a search of the Green Bank OH survey results (Turner 1979). In the region $337° \leqslant l \leqslant 50°$, $-1° \leqslant b \leqslant 1°$, there were found at least 53 highly spatially extended 1720 MHz OH clouds. The hardest question to answer is whether these GMCs lie only in spiral arms, or whether it is merely the particular anomalous excitation mechanism that is confined to spiral arms. The question will probably be answered only by looking at other galaxies.

Parity nonconservation and the origin of cosmic magnetic fields.   See Abstr. 066.011.

CO observations around galactic longitude $l = 45°$.
See Abstr. 131.059.

A high resolution far-infrared survey of a section of the galactic plane. II. Far-infrared, CO, and radio continuum results.   See Abstr. 133.008.

Far-IR, compact sources in the galactic plane.
See Abstr. 133.017.

Radio maps of the sky.   See Abstr. 141.026.

A radio continuum survey of the northern sky at 1420 MHz — Part I.   See Abstr. 141.085.

Metallicity effect and $\lambda 2.4 \mu m$ excess in the galactic disk.   See Abstr. 155.049.

Report of IAU Commission 33: Structure and dynamics of the galactic system (Structure et dynamique du système galactique).   See Abstr. 155.056.

Comparison of the distributions of galactic γ-radiation and radio synchrotron radiation.
See Abstr. 157.014.

157 Galaxy (UV, X, Gamma Radiation)

157.001 The components of galactic γ-ray emission.
M. Salvati, E. Massaro.
Mon. Not. R. Astron. Soc., Vol. 198, 11 - 21 (1982).

The γ-ray luminosity of the galaxy is known to arise from several components. The authors distinguish three classes: a.) Strong discrete sources that are defined by the acceptance criteria of the 2CG catalogue and are mostly unidentified. Their relative contribution is estimated at $\geqslant 34\%$. b.) Normal radio pulsars should also emit γ-rays, which can account for ~ 20% of the galactic X-ray flux. c.) The remainder is attributed to cosmic-ray interactions with diffuse matter. No correlation can be established between the distribution of cosmic rays and the gas density in the galactic plane.

157.002 Large-scale distribution of galactic gamma radiation observed by COS-B.
H. A. Mayer-Hasselwander, K. Bennett, G. F. Bignami, R. Buccheri, P. A. Caraveo, W. Hermsen, G. Kanbach, F. Lebrun, G. G. Lichti, J. L. Masnou, J. A. Paul, K. Pinkau, B. Sacco, L. Scarsi, B. N. Swanenburg, R. D. Wills.
Astron. Astrophys., Vol. 105, 164 - 175 (1982).

A complete survey of the Galaxy in high-energy gamma rays has been performed with the experiment aboard the ESA satellite COS-B. Gamma-ray sky maps for energy ranges between 70 MeV and 5 GeV were derived from 4 yr of observations and are presented as contour maps and as longitude and latitude profiles for three energy intervals.

157.003 HEAO 1 measurements of the galactic ridge.
D. M. Worrall, F. E. Marshall, E. A. Boldt, J. H. Swank.
Astrophys. J., Vol. 255, 111 - 121 (1982).

The authors have systematically searched the HEAO A2 experiment data for unresolved galactic disk emission. Although there are suggestions of nonuniformities in the emission, the data are consistent with a disk of half-thickness 241 ± 22 pc and surface emissivity (2–10 keV) at galactic radius R (kpc) of $2.2 \times 10^{-7} \exp(-R/3.5)$ ergs cm^{-2} s^{-1} ($R > 7.8$ kpc), giving a luminosity of ~4.4×10^{37} ergs s^{-1}. The disk emission has a spectrum which is significantly softer than that of the high galactic latitude diffuse X-ray background, and it is most probably of discrete source origin.

157.004 Galaktische Gamma-Astronomie.
V. Schönfelder.
Naturwissenschaften, 69. Jahrg., 212 - 219 (1982).

157.005 Observaciones en contínuo de radio en 1420 MHz en la dirección de H 1538-32.
F. R. Colomb, G. M. Dubner.
Bol. Asoc. Argentina Astron., No. 26, p. 111 (1981). Abstract.

157.006 HEAO-1 observations of high-energy X-rays from the galactic center region.
J. Matteson, D. Gruber, P. Nolan, R. Proctor, L. Peterson, A. Levine, F. Primini, W. Lewin.
Bull. American Astron. Soc., Vol. 13, 867 (1981). – Abstract.

157.007 A search for 2.22 MeV gamma ray line transients with the SMM Gamma Ray Spectrometer.
J. P. Heslin, E. L. Chupp, G. Kanbach, C. Reppin, E. Rieger, G. Share, R. L. Kinzer.
Bull. American Astron. Soc., Vol. 13, 901 (1981). – Abstract.

157.008 Gamma-ray astrophysics.   C. E. Fichtel.
Bull. American Astron. Soc., Vol. 13, 908 (1981). Abstract.

157.009 Spatial and spectral characteristics of the soft X-ray diffuse background.
D. N. Burrows, W. T. Sanders, D. McCammon, W. L. Kraushaar.
Bull. American Astron. Soc., Vol. 13, 925 (1981). – Abstract.

157.010 On the origin of the 1 keV diffuse X-ray background.   J. A. Nousek, P. M. Fried, W. T. Sanders, W. L. Kraushaar.
Astrophys. J., Vol. 258, 83 - 95 (1982).

High galactic latitude data ($b < -15°$) from the Wisconsin soft X-ray sky survey are used to constrain simple geometric models for the source of the diffuse X-ray background at energies near 1 keV. Two extended enhanced features are apparent in the intensity maps. One is in Eridanus and one is in the direction of the galactic center. Away from these features, the sky is relatively uniform. The observed degree of isotropy is consistent with the model that the $0.5-1.2$ keV background consists of an isotropic extragalactic component plus a thick disk galactic component. If an $11E^{-1.4}$ photons $cm^{-2} s^{-1} sr^{-1} keV^{-1}$ spectrum is assumed for the extragalactic component, then a temperature of $2-3 \times 10^6$ K and an emission measure perpendicular to the galactic plane of 0.004 cm^{-6} pc are derived for the galactic component.

157.011 The Galaxy as the origin of gamma-ray bursts.
II. The effect of an intrinsic burst luminosity distribution on log $N(>S)$ versus log S.   M. C. Jennings. Astrophys. J., Vol. 258, 110 - 120 (1982).

The galactic monoluminosity gamma burst model of Jennings and White is generalized to include an intrinsic burst luminosity distribution of the form $n(L) \alpha L^{a-1} pc^{-3} yr^{-1} ergs^{-1}$, where L is the integrated burst luminosity. Distributions moderately peaked toward low luminosity and with a dynamic range $\lesssim 10^2$ have a profound effect on log N − log S. Such distributions permit burst sources to be spherically distributed within the Galaxy, a situation not allowed by monoluminosity bursts. Comparing models and data indicates maximum burst luminosities of $7 \times 10^{41} \lesssim L_2 ergs \lesssim 9 \times 10^{42}$ and burst rate densities of $2 \times 10^{-10} \lesssim n_0 pc^{-3} yr^{-1} \lesssim 5 \times 10^{-10}$ for halo models and $L_2 \sim 10^{40}$ ergs and $n_0 \sim 9 \times 10^{-9} pc^{-3} yr^{-1}$ for disk models.

157.012 The γ-ray emissivity of the local interstellar medium from correlations with gas at intermediate latitudes.   A. W. Strong, A. W. Wolfendale. Philos. Trans. R. Soc. London, Ser. A, Vol. 301, (see 012.043), 541 - 554 (1981). − Same as 30.157.011.

A survey of recent studies of the correlation between γ-rays from latitudes $|b| > 10°$ and gas tracers is presented. Results for the ranges 35 - 100 MeV and above 100 MeV from the SAS-2 satellite, and for energies between 70 and 5000 MeV from the COS-B satellite, are used to obtain an estimate of the γ-ray emissivity spectrum for all forms of gas. Good agreement between the two experiments is found. The presence of a component of γ-ray emission related to gas in molecular form is evident in both the SAS-2 and COS-B data. The authors discuss the correlation of the SAS-2 data with both components and show that the emissivities of each component can be independently determined. Finally, an examination of the γ-ray fluxes from specific dense clouds of molecular gas is made.

157.013 Low-latitude galactic γ-ray emission: a probe, not a proof.   G. F. Bignami. Philos. Trans. R. Soc. London, Ser. A, Vol. 301, (see 012.043), 555 - 568 (1981). − Same as 30.157.012.

The emission of high energy (above 70 MeV) γ-rays from the galactic disc has been mapped by the COS-B mission with unprecedented detail. The results for $|b| < 15°$ are seen to contain evidence of structures correlated with the Galaxy on various scales, from the 'grand design' down to granulari-

ties, showing that the diffuse interstellar medium, with its cosmic ray content, is well mapped by high energy γ-ray astronomy. Two new detailed correlations are proposed. After discussing the importance of the discrete, unresolved sources also discovered by COS-B, an astrophysical process is sketched suggesting a scenario for enhanced emission in regions where interstellar medium shocks can accelerate cosmic rays.

157.014 Comparison of the distributions of galactic γ-radiation and radio synchrotron radiation.
C. G. T. Haslam, S. Kearsey, J. L. Osborne, S. Phillipps, H. Stoffel. Philos. Trans. R. Soc. London, Ser. A, Vol. 301, (see 012.043), 573 - 575 (1981).

The authors have used the new all-sky survey of continuum radio emission at 408 MHz of Haslam et al. (1981) to compare the distribution of radio emission in a band along the galactic equator for $|b| < 20°$ with the COS-B γ-ray distribution of Mayer-Hasselwander et al. (1980).

157.015 Gamma-ray spectroscopy of the galactic center region: confirmation of the time-variability of the positron annihilation line.   W. S. Paciesas, T. L. Cline, B. J. Teegarden, J. Tueller, P. Durouchoux, J. M. Hameury. NASA Tech. Memo., NASA TM-83921, 16 pp. (1982).

The GSFC Low-Energy Gamma-Ray Spectrometer observed the region of the galactic center during a balloon flight from Alice Springs, Australia, on 1981 November 20. No significant excess over background was evident in the 511 keV annihilation line. Continuum emission was detected above 100 keV. The results confirm the recent observations of time-variability by HEAO-3 (Riegler et al. 1981, 1982). A compact source at or near the galactic center provides the most satisfactory agreement with all of the data.

Properties and performance of the MPI balloon borne Compton telescope.   See Abstr. 032.542.

Extended adiabatic blast waves and a model of the soft X-ray background.   See Abstr. 062.011.

A flux limited X-ray survey of the galactic plane. See Abstr. 142.066.

Map of the soft X-ray sky from SAS-3 observations. See Abstr. 142.067.

Gamma ray astronomy − a new window on the universe.   See Abstr. 142.504.

Diffuse cosmic gamma radiation with energies higher than 100 MeV at mean and high galactic latitudes. See Abstr. 142.524.

Report of IAU Commission 33: Structure and dynamics of the galactic system (Structure et dynamique du système galactique).   See Abstr. 155.056.

158 Single and Multiple Galaxies, Peculiar Objects

158.001   A correction: the quadratic redshift-distance law
and the observational magnitude cutoff bias.
I. E. Segal.
Astrophys. J., Vol. 252, 37 - 38 (1982).
A misleading description of the origin of the phenomenological quadratic redshift-distance law is corrected. The role of
the observational magnitude cutoff bias in the development of
the law is clarified.

158.002   Galaxy spins in the Virgo cluster.
G. Helou, E. E. Salpeter.
Astrophys. J., Vol. 252, 75 - 80 (1982).
For 20 galaxies in the Virgo cluster, the authors have
determined the sign of the inclination of the galactic plane and
the sense of the internal rotation. They thus have completely
defined the internal angular momentum vector ("spin") for
each of these galaxies. There is no evidence for a nonzero
vector sum for these 20 spins nor for any correlation between
spin and orbital motion through the cluster. The spin vectors
are not entirely random, but the nature of the nonrandomness
is not clear. The results are compatible with tidal torques as
the spin-up mechanism for galaxies and with most models for
cluster formation.

158.003   Neutral hydrogen emission and absorption in three
active Irr II galaxies.   T. X. Thuan, E. J. Wadiak.
Astrophys. J., Vol. 252, 125 - 132 (1982).
Neutral hydrogen emission-absorption is reported for
three Irr II galaxies with active nuclei: NGC 520, NGC 5363,
and NGC 5506. The peak optical depths for the H I absorption are typically about 10%. It is argued that the absorption
is due to a few "standard" clouds within several hundred
parsecs of the nucleus, in the narrow-line emission region.

158.004   The late-type stellar content of the Fornax and
Sculptor dwarf galaxies.   J. A. Frogel,
V. M. Blanco, M. F. McCarthy, J. G. Cohen.
Astrophys. J., Vol. 252, 133 - 146 (1982).
A field of area 0.13 square degrees has been surveyed for
late-type stars in each of the Fornax and Sculptor dwarf elliptical galaxies. JHK photometric data have been obtained for
most of the stars found. In Fornax, the authors have positively
identified 25 C stars and one M giant. In Sculptor, two
relatively blue C stars and a small number of possible M giants
have been identified.

158.005   Hydroxyl absorption toward galactic nuclei.
L. J. Rickard, T. M. Bania, B. E. Turner.
Astrophys. J., Vol. 252, 147 - 155 (1982).
The authors report the detection of hydroxyl absorption
toward the nuclear continuum sources in NGC 660, NGC 3227,
NGC 3504, NGC 3628, and NGC 5363. The peak apparent
optical depths are typically a few percent. The line profiles are
unpolarized, and the OH main lines are in roughly their optically thin LTE intensity ratios. The authors argue that the absorption arises mostly in molecular disks that are within a few
hundred parsecs of the nuclei.

158.006   Analysis of the Karachentsev 6 meter redshift
sample for binary galaxies.   J. W. Sulentic.
Astrophys. J., Vol. 252, 439 - 446 (1982).
An analysis of external errors in the recently published
6 m binary galaxy sample of Karachentsev has been carried out.
Emission line redshifts are found to have an uncertainty of
50 km s^{-1} when compared against accurate ($\sigma \leqslant 15$ km s^{-1})
21 cm data. Absorption line measures are found to be considerably less accurate with a mean deviation of 125 km s^{-1}.
The resultant uncertainties of redshift differentials for binaries

with separations ($s \geqslant 90''$) are about 3 times higher than
previously stated by Karachentsev. Data on 94 of the 165
mixed pairs of spiral and elliptical galaxies in the Catalogue of
Isolated Pairs of Galaxies were studied in a test of redshift-
morphology correlations previously reported in the literature.
Spiral galaxy components of mixed pairs were found to have
a statistically significant higher mean redshift (+100 km s^{-1})
than their nonspiral partners.

158.007   Colliding and merging galaxies. I. Evidence for the
recent merging of two disk galaxies in NGC 7252.
F. Schweizer.
Astrophys. J., Vol. 252, 455 - 460, plates 10, 11 (1982).
Observations of the galaxy NGC 7252 show that it has a
single nucleus, a nearly round main body marked by delicate
ripples, faint surrounding loops, and two slender tails that
project to 80 kpc and 130 kpc from the center ($H_0 = 50$). The
main body itself shows a spectrum indicative of young A stars,
and contains a small ($r \approx 4$ kpc) central disk of ionized gas.
This disk rotates with $v \sin i \approx 80-100$ km s^{-1} around a well-
defined axis, whereas the gas immediately beyond it follows a
totally different motion pattern. Five characteristics suggest a
recent merger of two similarly massive disk galaxies: the two
tails, the unusual isolation, opposite tail motions, the single
nucleus and body, and the two surviving motion systems of the
gas. The tail lengths divided by the velocities give a merger age
of ~0.5−2 × 10^9 yr, similar to that inferred from colors.

158.008   Star formation and chemical abundances in clumpy
irregular galaxies.
A. M. Boesgaard, S. Edwards, J. Heidmann.
Astrophys. J., Vol. 252, 487 - 495 (1982).
Clumpy irregular galaxies consist of several bright clumps
which are huge H II complexes and contain about 10^5 O and B
stars. Image-tube spectrograms with 1−3 Å resolution have
been obtained of the brightest emission regions of three
clumpy galaxies and one cadidate clumpy galaxy with the
Mauna Kea 2.24 m telescope. The electron temperatures were
found to be in the range 7000−9000 K and electron densities
a few hundred per cm^3 − quite typical for normal H II regions.
The galaxies appear to be normal (like Sc galaxies) in mass and
composition. Supernovae remnants are indicated by the high
[S II]/Hα ratio. Possible triggering mechanisms for the exceptional star formation activity are discussed.

158.009   On the dynamics of the broad-line gas in Seyfert 1
galaxies.   J. E. Tohline, D. E. Osterbrock.
Astrophys. J., Lett., Vol. 252, L49 - L52 (1982) = Lick Obs.
Bull., No. 903.
Recent analyses of the properties of nearby Seyfert 1 galaxies indicate that there is no correlation between the widths
of the broad permitted lines and the observed inclination of
the galaxies' spiral disks. The absence of such a correlation suggests that the lines are not rotationally broadened. However,
rapidly rotating gas in the nucleus of a spiral galaxy need not
be confined to the same plane as the galaxy's primary disk.
Therefore, the absence of a correlation between line widths
and disk inclination does not rule out rotation as the primary
broadening mechanism for lines emitted from the broad-line
gas in Seyfert 1 galaxies.

158.010   Nonthermal optical-infrared emission from
NGC 1052.
G. H. Rieke, M. J. Lebofsky, J. C. Kemp.
Astrophys. J., Lett., Vol. 252, L53 - L56 (1982).
The infrared excess of NGC 1052 is approximately 4.5%
polarized. Its spectrum resembles those of the reddest BL Lac
type sources. Despite its lack of variability, the authors con-

clude that this source is nonthermal but of exceptionally low luminosity for its type.

158.011   Velocity fields in the barred spiral galaxies NGC 2525 and NGC 7741.
C. P. Blackman, W. D. Pence.
Mon. Not. R. Astron. Soc., Vol. 198, 517 - 534 (1982).

Emission line velocity fields have been measured to an accuracy of $\sim$15 km s^{-1} in the barred spiral galaxies NGC 2525 and NGC 7741 using long slit, IPCS spectra. Within observational error, the velocity fields of both galaxies are consistent with pure circular motion. An upper limit to any systematic deviations from circular motion of the type predicted by several recent hydrodynamic models of barred spiral galaxies can be set at 20–30 km s^{-1}. The average mass-to-light ratio for the 'visible' parts of both galaxies is $\sim$1.5$\pm$0.5 in the B passband. Data from the literature on two other barred spirals, NGC 1300 and NGC 5383, which are both claimed to show deviations from circular motion of the type predicted by the hydrodynamic models, have been re-analysed. Allowing for observational error, no significant departures from circular motion are found in these galaxies either.

158.012   The optical structure of II Zw 40.
J. A. Baldwin, H. Spinrad, R. Terlevich.
Mon. Not. R. Astron. Soc., Vol. 198, 535 - 543 (1982).

New data are presented on the morphology and ionization structure and kinematics of II Zw 40, a 'detached extra-galactic H II region'. They show a core–halo structure with the ionizing stars embedded in a small nucleus and a halo composed of small optically thick filaments. Direct plates show two tails protruding from this halo. This optical structure is interpreted as an interaction between two extremely small galaxies. This encounter would then be the cause of the present burst of star formation.

158.013   The spectrum of 3C 390.3 and the hydrogen line problem.   H. Netzer.
Mon. Not. R. Astron. Soc., Vol. 198, 589 - 603 (1982).

New optical data for 3C 390.3 combined with the IUE observations of Ferland et al., enable the author to deduce some of the characteristics of this active nucleus. The ones discussed in this work can be summarized as follows: (1) Some absorbing material, which covers part of the central source, is moving away from the centre and causing double absorption features in the wings of Hβ and Hγ. (2) The separation of the Balmer lines into broad and narrow components and the calculation of the expected Lα/Hβ in a low density gas, indicate reddening of the spectrum by dust which amounts to $A_V \simeq 0.5$ mag. (3) Line transfer effects are important both in the low and high density components. The observations of Hβ and He II 4686 suggest moderate enhancement of Hβ in the broad line region.

158.014   Kinematics and dynamics of the barred spiral galaxy NGC 1313.   M. Marcelin, E. Athanassoula.
Astron. Astrophys., Vol. 105, 76 - 84 (1982).

The complete velocity field of the ionized gas in the late type barred spiral NGC 1313 has been drawn from the analysis of eight interferograms in Hα light. The velocity field shows that the rotation center of the galaxy is outside the bar. Photometry is then used to build a dynamical model where the gravitational potential is an axisymmetric "disk potential" perturbated by an excentered "bar potential". Such a model permits to find the general design of the observed velocity field in the central part of the galaxy.

158.015   Centimetre wavelengths radio studies of clumpy irregular galaxies.
J. Heidmann, U. Klein. R. Wielebinski.
Astron. Astrophys., Vol. 105, 188 - 191 (1982).

The authors made high frequency observations of six clumpy irregular galaxies, detecting 4 or possibly 5 of them. The radio spectra are steep, indicating that nonthermal emission dominates at high radio frequencies. This is so in spite of the known presence of hyperactive H II complexes in these objects. This result is similar to that obtained in a recent study of a sample of nearby spirals indicating that thermal activity must be accompanied by nonthermal events.

158.016   The distribution of thermal and nonthermal radio continuum emission of M 31.
R. Beck, R. Gräve.
Astron. Astrophys., Vol. 105, 192 - 199 (1982).

The distribution of the brightness temperature spectral index in the Andromeda Nebula was determined from radio continuum maps at 408 and 2700 MHz. The distributions of the thermal and nonthermal radio continuum emission are similar, but both are different from the distribution of the blue light which follows the total population of the disk stars. In the case of equipartition between cosmic ray and magnetic field energy densities, the cosmic ray electrons possibly originate in sources whose distribution is similar to that of the total stellar population.

158.017   IUE spectra of clumpy irregular galaxies.
P. Benvenuti, C. Casini, J. Heidmann.
Mon. Not. R. Astron. Soc., Vol. 198, 825 - 831 (1982).

The authors obtained IUE spectra at short wavelengths for two clumpy irregular galaxies, Mkn 8 and Mkn 325. Clumps could be resolved across the slit. Statistics now available on 10 clumps show that, on average, each clump radiates 100 times more at 155 nm than the giant H II region 30 Doradus and contains 0.7×10^5 early-type stars. This confirms the hypothesis that clumps may be hyperactive H II complexes and that clumpy irregulars are galaxies in which star formation occurs on an exceptional scale.

158.018   Radio emission from supernova remnants in the galaxy M 33.
S. D'Odorico, W. M. Goss, M. A. Dopita.
Mon. Not. R. Astron. Soc., Vol. 198, 1059 - 1064 (1982).

A new 21-cm map of M 33 with a resolution of 25×49 arcsec and an rms noise of 0.2 mJy per beam area has been used to search for radio emission at the positions of optically identified supernova remnants. Five well-established and three probable radio identifications are found. In the surface brightness-diameter diagram, the M 33 radio remnants agree well with galactic objects. Using the galactic relation as a reference, the authors derive an average distance to M 33 of 830 $\pm$ 100 kpc.

158.019   On the nuclear spectrum of NGC 1365.
M. G. Edmunds, B. E. J. Pagel.
Mon. Not. R. Astron. Soc., Vol. 198, 1089 - 1107 (1982).

AAT long-slit spectroscopic observations of the nucleus of the barred "hot-spot" galaxy NGC 1365 reveal a central nucleus with a spectrum resembling that of an intermediate Seyfert galaxy and a surrounding region of about 400 pc radius with a spectrum having several characteristics of a Seyfert 2. Surrounding this, in turn, are normal H II regions of low excitation suggesting an O/H abundance ratio exceeding that of the sun. The Seyfert 1 nucleus has asymmetric broad Balmer components as well as a non-thermal continuum and broad Fe II emission features. All the emission-line systems are heavily reddened. The importance of allowing for possible nonthermal excitation in the derivation of abundances from emission lines in galactic nuclei is emphasized.

158.020   Winds from dwarf galaxies and Lα absorption features in quasar spectra.
C. Fransson. R. Epstein.
Mon. Not. R. Astron. Soc., Vol. 198, 1127 - 1141 (1982).

Gas poor dwarf galaxies have necessarily evolved through a stage of extensive mass loss. If the mass shedding occurred because of a galactic wind, then the diffuse clouds of gas surrounding these dwarf galaxies could give rise to the numerous weak Lα absorption lines which are observed in the spectra of high redshift quasars. The authors present a model based on this idea which reproduces all the observed features of the Lα absorption lines.

158.021 The extreme Seyfert galaxy associated with the X-ray source 3A 0557 - 383.
A. P. Fairall, I. M. McHardy, J. P. Pye.
Mon. Not. R. Astron. Soc., Vol. 198, 13P - 17P (1982).

A Seyfert galaxy at 1950: $5^h 56^m 21^s$, $-38°20'15''$ with a velocity of recession $V_0 = 10\ 000$ km s^{-1} is identified with the X-ray source 3A 0557 - 383. The optical spectrum is very heavily reddened, so that the absolute magnitude, if corrected for extinction, reveals this to be one of the most luminous Seyferts known. It shows a very high mean X-ray luminosity of $10^{37.2}$ W (2 - 10 keV) and is believed to be variable.

158.022 Simultaneous observations of the BL Lacertae object I Zw 187.
J. N. Bregman, A. E. Glassgold, P. J. Huggins, J. T. Pollock, A. J. Pica, A. G. Smith, J. R. Webb, W. H.-M. Ku, R. J. Rudy, P. D. LeVan, P. M. Williams, P. W. J. L. Brand, G. Neugebauer, T. J. Balonek, W. A. Dent, H. D. Aller, M. F. Aller, P. E. Hodge.
Astrophys. J., Vol. 253, 19 - 27 (1982).

Two sets of simultaneous spectra separated by 10 months were obtained for the X-ray–bright BL Lac object I Zw 187. The spectra consist of data obtained with radio, infrared, optical, ultraviolet, and X-ray telescopes. After contamination by galactic light is removed from the observations, the BL Lac component has a weak 3000 Å bump superposed upon an infrared-optical-ultraviolet spectrum of slope 0.9. The X-ray data fall on or close to an extrapolation of this power law; this result is consistent with the continuum from infrared through X-ray emission arising from a single synchrotron source. Optical and X-ray fluxes are observed to vary by no more than a factor of 3, and the shortest time scales of variability in these bands are comparable, about 1 week.

158.023 Infrared polarimetry of nine Seyfert galaxies.
R. J. Rudy, P. D. LeVan, R. C. Puetter, H. E. Smith, S. P. Willner.
Astrophys. J., Vol. 253, 53 - 57 (1982).

The polarization of nine Seyfert galaxies has been measured at 2.3 μm. At a 95% confidence level no polarizations greater than 2–3% were detected. Mechanisms which could weaken or modify a strongly polarized nonthermal component are discussed. The observations are consistent with, but not conclusive evidence of, thermal emission by dust as the source of the near infrared excesses in these Seyfert galaxies.

158.024 The distribution of mass in Sc galaxies.
D. Burstein, V. C. Rubin, N. Thonnard, W. K. Ford, Jr.
Astrophys. J., Vol. 253, 70 - 85 (1982).

For a sample of 21 Sc galaxies ranging from very low to very high luminosity the authors analyze the mass distributions and show that for all galaxies the mass distribution has a single unique form, when the masses are each scaled by a mass scale length, R_m. Unlike luminosity scale lengths which are larger for more luminous (larger) galaxies, mass scale lengths are smaller for galaxies of high luminosity. From analysis of rotation curves it is concluded that the rotational velocities remain high at large nuclear distances due to the presence of significant nonluminous mass at large radii. Correlations between integral mass, luminosity, mass scale length, isophotal radius, and velocity at the isophotal radius are determined and discussed.

158.025 An optical study of the magnetic field in M31.
P. G. Martin, S. J. Shawl.
Astrophys. J., Vol. 253, 86 - 90 (1982).

The authors present optical polarization measurements for 18 globular clusters in M31. Two exceptions to the generally good agreement with previous observations are pointed out. To a first approximation, the position angles of polarization are in accord with a magnetic field direction in the galactic plane and orthogonal to the local radius vector to the nucleus. The efficiency of polarization is comparable to the maximum in our Galaxy for some lines of sight, but there are also examples of low polarization accompanying high reddening.

158.026 A comparison of measured spiral arm properties with model predictions.
R. Kennicutt, Jr., P. Hodge.
Astrophys. J., Vol. 253, 101 - 107 (1982).

The authors present measurements of the morphological properties of the spiral arms in a sample of 17 galaxies for which density wave and stochastic theory predictions are available. For both models crude correlations of predicted and observed arm properties are apparent along the Hubble sequence. In detail, however, the agreement is poor for both models. Arm pitch and arm widths are at least as well correlated with the rotational velocities of the galaxies, independent of the model. This suggests that the trends in spiral structure observed along the Hubble sequence may be largely kinematic in origin.

158.027 The galaxy components of BL Lacertae objects, N systems, and quasi-stellar objects.
J. S. Miller.
Publ. Astron. Soc. Pacific, Vol. 93, 681 - 694 (1981/82) = Lick Obs. Bull. No. 902.

Spectrophometric investigations have shown that BL Lacertae objects and N systems typically have a luminous elliptical galaxy associated with them. The galaxies detected for the BL Lac object show some range in luminosity, and the most luminous is comparable to a first-ranked cluster elliptical. The galaxies detected for N systems are generally a few magnitudes fainter than a first-ranked cluster elliptical. There are at present no convincing spectroscopic data available that show QSOs are located in galaxies of stars, though a number have been shown to have surrounding nebulosity. Upper limits to the brightness of any elliptical galaxies associated with QSOs studied in detail with spectroscopy indicate that if such galaxies are present, they are typically at least two magnitudes fainter than a first-ranked cluster elliptical.

158.028 High-resolution 1666-MHz observations of the nucleus of NGC 4151.
R. V. Booler, A. Pedlar, R. D. Davies.
Mon. Not. R. Astron. Soc., Vol. 199, 229 - 237 (1982).

The authors present a 1666-MHz radio continuum map of the nucleus of NGC 4151. The high angular resolution (0.25 arcsec) enables them to resolve the nucleus into an elongated structure approximately 3.5 arcsec × 0.5 arcsec in position angle 77°. A double source is present in the centre with a separation of 0.45 arcsec at PA 83° ± 1°. The authors interpret the observations in terms of jets pointing at PA ∼ 80° and ∼ 260° and discuss the relationship of the jets with the forbidden-line region. The optical continuum polarization of the nucleus appears to be closely aligned with the radio jet.

158.029 Galaxy infrared colour magnitude relations.
R. F. G. Wyse.
Mon. Not. R. Astron. Soc., Vol. 199, 1P - 8P (1982).

Infrared colour-absolute magnitude relations are presented for samples of spiral, and elliptical and S0 galaxies, using published data of Aaronson, Huchra and Mould and Frogel,

Persson and Aaronson. The behaviours of ellipticals and spirals are shown to be manifestly different. The spiral colour-absolute magnitude relationship is proposed as a distance indicator.

158.030 A correlation between the spectral shift and morphological type for binary galaxies.
E. Giraud, M. Moles, J.-P. Vigier.
C. R. Acad. Sci. Paris, Tome 294, Sér. II, 195 - 198 (1982). In French.

The aim of this paper is to study the possible existence of a systematic redshift difference associated with the morphological type, in a sample of pairs containing one spiral and one elliptical galaxy. This effect, already observed in the case of groups and clusters of galaxies, could be in those cases partially interpreted by a different spatial distribution of both types.

158.031 Multiaperture photometry of isolated galaxies.
N. Brosch, G. Shaviv.
Astrophys. J., Vol. 253, 526 - 538 (1982).

Multiaperture U, B, V photometry of isolated galaxies by Huchra and Thuan shows a behavior different from normal field galaxies in that the inner regions of the isolated galaxies appear to have excess blueness similar to galaxies with active nuclei. The implication of this finding is that gas, which does not escape from isolated galaxies sinks into the nucleus and gives rise to nonstellar radiation.

158.032 Mass and luminosity in spiral galaxies and the Tully-Fisher relation.   D. Burstein.
Astrophys. J., Vol. 253, 539 - 551 (1982).

The empirical relation between absolute luminosity and maximum observed rotation velocity for spiral galaxies, the Tully-Fisher (TF) relation, not only provides a promising new way to obtain the distances of galaxies, but also establishes an important link between the separate distributions of mass and luminosity in spirals. This paper explores the relationships among mass luminosity, and, if possible, Hubble type for the sample of spiral galaxies, defined by Aaronson, Huchra and Mould, in an attempt to understand the implications that the infrared TF relationship has for the structure of normal galaxies. A critical limitation of this study is that the relationship of mass to luminosity can only be examined within the confines of luminous matter. Based on the analysis of its implications the usefulness of the TF relation as a distance measuring stick is critically assessed.

158.033 Optical spectrophotometry of the nuclear region of M51.   J. A. Rose, L. Searle.
Astrophys. J., Vol. 253, 556 - 574 (1982).

The well-known spiral galaxy M51 is a typical example of a galaxy having broad emission lines and high [N II/Hα] in its nuclear region. The authors have obtained long-slit optical spectrophotometry to investigate in detail emission-line ratios and the physical conditions of the ionized gas in the nuclear region of the galaxy. The region of high [N II/Hα], covering a diameter of ~20″, is well resolved spatially in these observations. It is demonstrated that neither standard photoionization models nor collisional shock heating and ionization models are consistent with the observational results. On the other hand, a simple model incorporating a central source of nonstellar ionizing radiation with a relatively flat, perhaps power-law, spectrum supplies a good qualitative fit to the observations. It is concluded that the gas in the nuclear core of M51 is ionized by a nonstellar central source of radiation and that the galaxy contains an active nucleus similar to, but milder than, those found in the nuclei of Seyfert galaxies. The broad emission lines observed in the central core indicate turbulent velocities of up to ~200 km s^{-1} in this region. It appears likely that the central ionizing source in the nucleus of M51 is also responsible for these large turbulent velocities.

158.034 The 21 centimeter line width as an extragalactic distance indicator. II. Does the Tully-Fisher relation depend on Hubble type?   G. de Vaucouleurs, R. Buta, L. Bottinelli, L. Gouguenheim, G. Paturel.
Astrophys. J., Vol. 254, 8 - 15 (1982).

The proposal that the slope b of the Tully-Fisher (T-F) relation is the same ($b = 10$) in the blue (B) photometric band and in the infrared (H) band and that its zero point a depends on morphological type T is tested by several methods. Evidence from small samples at T = const. or in the Virgo cluster are inconclusive. However, from the sample of ~150 "best-observed" galaxies it is concluded that there is no support for the hypothesis that the slope of the T-F relation is $b = 10$ in the B band and that the zero point a is heavily dependent on morphological type. The tests confirm the validity of the T-F relation previously adopted with $b = 5$ for the B band.

158.035 The ultraviolet excess of Seyfert 1 galaxies and quasars.   M. A. Malkan, W. L. W. Sargent.
Astrophys. J., Vol. 254, 22 - 37 (1982).

The authors have combined recent spectrophotometric observations to study the spectra of eight active galactic nuclei from infrared to ultraviolet wavelengths. Power-laws are subtracted from the composite spectra for examining the ultraviolet excess. Balmer continuum emission is clearly present. The combination of a power-law with Balmer continuum flux matches the spectra of three objects very well, but cannot reproduce the flat blue and ultraviolet continuum of the other five. It is demonstrated that by adding a third component described by a blackbody at a single temperature ranging from 20 000 to 30 000 K all spectral distributions can be fitted successfully. It is noted that the ratio of thermal to power-law component correlates with total luminosity.

158.036 Photometry of resolved galaxies. I. The Pegasus dwarf irregular.   J. G. Hoessel, J. R. Mould.
Astrophys. J., Vol. 254, 38 - 49, plates 1 - 3 (1982).

The authors present color-magnitude diagrams for resolved stars in the Pegasus dwarf galaxy in the green, red, and infrared passbands of the extended Gunn photometric system. The evolved nature of the upper main sequence and the lack of luminous red supergiants indicate that recent star formation in Pegasus has been very subdued. Three star clusters are identified. Their red colors indicate they are of intermediate age or older. The authors derive a provisional distance estimate for Pegasus of 1.7 Mpc, placing it at the outer margins of the Local Group.

158.037 Abundance gradients in M31: comparison of results from supernova remnants and H II regions.
W. P. Blair, R. P. Kirshner, R. A. Chevalier.
Astrophys. J., Vol. 254, 50 - 69 (1982).

The authors have obtained spectra of 11 H II regions and additional spectra of six previously reported supernova remnants (SNRs) in M31. The SNR spectra have been used in conjunction with shock model calculations to give abundances of oxygen, nitrogen, and sulfur in the interstellar gas comprising each remnant. The authors have also determined abundances for the H II regions using the empirical method described by Pagel et al. Both nitrogen and oxygen abundances decrease by about a factor of 4 from the innermost regions studied (~ 4 kpc) to the outer regions (~ 23 kpc). These gradients are similar to those found in other intermediate and late type spiral galaxies, including our own. The mean nitrogen and sulfur abundances are similar to those of the Orion Nebula, but the mean oxygen abundance is about a factor of 2 higher, accounting for the low excitation of the M31 H II regions.

158.038 Models for far-infrared emission from normal galaxies.   M. Jura.
Astrophys. J., Vol. 254, 70 - 74 (1982).

Dust grains absorb optical and ultraviolet light and reemit

in the infrared. The author shows that the presence of grains distributed throughout either disk or spherical galaxies can lead to detectable far-infrared sources. In particular, an elliptical such as NGC 4278 with an appreciable amount of neutral gas should be easily detectable with the *Infrared Astronomical Satellite (IRAS)* if it has a dust-to-gas ratio similar to the value in the solar neighborhood. A disk should be detectable at 100 Mpc with *IRAS*. Also, disks may emit most of their energy longward of 100 μm; this could be important for estimating the mass-to-light ratios of these galaxies.

158.039 Galactic mass loss: a mild evolutionary correction to the angular size test.
D. O. Richstone, M. D. Potter.
Astrophys. J., Vol. 254, 451 - 455 (1982).

Galactic mass loss due to gas production by an evolving stellar population and removal due to supernovae or some other energy source can alter the size of a galaxy over a long time through the adiabatic invariants of the orbits. Numerical and analytic results show that if the mass profile changes homologously, the scale size of the galaxy does also. Under conventional assumptions about galaxy age and initial mass function, this bloating is smaller than the differences between open and flat cosmologies at $z = 0.5$. If, however, some elliptical galaxies maintain star formation until recent epochs and then lose a (correspondingly) larger fraction of their mass, the effect may be important. It may also be important in decreasing the size of cD galaxies as a result of gas accretion.

158.040 Galactic gas and the shapes of radio sources.
L. S. Sparke.
Astrophys. J., Vol. 254, 456 - 464 (1982).

Large double-lobed extragalactic radio sources are associated with elliptical, rather than with spiral galaxies. In an earlier paper, it was suggested that the radio morphology of elliptical galaxies results from their slow rotation. Here, more precise calculations of the flow of a slowly spinning gas are presented. The energetic beam supplying a powerful radio source is likely to propagate freely away from the galaxy, but in a weaker source it may be disrupted quickly by interactions with the interstellar medium. In the latter case, the pressure of galactic gas will confine the radiating plasma to a narrow funnel along the rotating axis, producing a thin jet of emission.

158.041 IUE observations of NGC 4649, an elliptical galaxy with a strong ultraviolet flux.
F. Bertola, M. Capaccioli, J. B. Oke.
Astrophys. J., Vol. 254, 494 - 499 (1982).

As in previously studied elliptical galaxies, the flux f_λ in NGC 4649 has a minimum at 2500 Å followed by a rapid increase towards shorter wavelengths. In NGC 4649, the flux level of this rising branch is the highest so far observed and seems not to be correlated with the luminosity or activity in the galaxy. The UV spectrum of NGC 4649 down to 2500 Å matches closely those of NGC 3379 and NGC 4472. The implications of the UV properties of elliptical galaxies on the stellar content as well as on magnitudes and colors of distant galaxies are discussed.

158.042 Carbon stars in the Carina dwarf spheroidal galaxy.
J. R. Mould, R. D. Cannon, M. Aaronson, J. A. Frogel.
Astrophys. J., Vol. 254, 500 - 506, plates 4, 5 (1982).

A $1\overset{\circ}{.}1 \times 1\overset{\circ}{.}1$ field centered on the Carina dwarf spheroidal galaxy has been scanned on objective prism plates taken with the UK 1.2 m Schmidt telescope. This search has yielded eight carbon star candidates, including two discovered previously. They are strongly concentrated toward the galaxy, so most, if not all, must be members. Near-infrared *JHK* photometry has been obtained for six of the stars. The brightest of these have derived bolometric magnitudes and colors similar to carbon stars in the Fornax dwarf galaxy. Comparison with

the intermediate-age globular clusters of the Magellanic Clouds suggests that Carina is not purely an old stellar population.

158.043 Discovery of carbon stars in the Draco dwarf spheroidal galaxy.
M. Aaronson, J. Liebert, J. Stocke.
Astrophys. J., Vol. 254, 507 - 514, plate 6 (1982).

A grating prism survey of the Draco dwarf spheroidal galaxy that is 97% areally complete has led to the discovery of three carbon stars. Membership of these stars in Draco is firmly established from luminosity, proper motion, and radial velocity considerations. Optical and preliminary infrared photometry suggest that the Draco carbon stars are more closely related to the CH stars in ω Cen than to the luminous carbon stars found in the Fornax and Carina dwarfs. Carbon stars have now been located in all four of the dwarf spheroidals that have been examined using transmission grating prism techniques. The rarity of these stars in galactic globulars, systems with which the dwarf spheroidals are often compared, indicates a fundamental population difference.

158.044 The optical warp of M31.
K. A. Innanen, K. W. Kamper, K. A. Papp, S. van den Bergh.
Astrophys. J., Vol. 254, 515 - 516, plates 7 - 9 (1982).

Digital stacking of Palomar Schmidt plates of M31 shows that the outermost isophotes of this galaxy are anti-symmetrically warped. The fact that this warp shows up in both yellow and red light suggests that it is produced by starlight rather than by emission nebulosity. This conjecture is supported by Baade's observations of the distribution of Population II giants in M31.

158.045 Kinematics of gas clouds in the nuclei of Seyfert galaxies and quasars.
Eh. A. Dibaj, K. A. Postnov, S. N. Fabrika.
Pis'ma Astron. Zh., Tom 8, 3 - 8 (1982). In Russian. English translation in Soviet Astron. Lett., Vol. 8.

The distribution of emission line Doppler widths in spectra of Seyfert 1 galaxies and quasars is investigated. It is shown that the observed velocity function can be represented neither by a model of randomly distributed discs nor by a model of one-dimensional jets. The geometry of gas clouds which are responsible for broad emission features in spectra of active nuclei seems to be more or less spherical.

158.046 A sample of faint galaxies behind the Virgo cluster.
I. D. Karachentsev.
Pis'ma Astron. Zh., Tom 8, 74 - 77 (1982). In Russian. English translation in Soviet Astron. Lett., Vol. 8.

New radial velocity measurements were carried out for faint blue galaxies in the central region of the Virgo cluster. They reveal marked excess of objects with redshifts typical for Coma members. Evidence is put forward that around the Coma and A1367 clusters there is a supercorona populated by low-luminosity emission galaxies.

158.047 Cluster galaxies with Seyfert properties.
M. A. Arakelyan, V. Yu. Terebizh.
Pis'ma Astron. Zh., Tom 8, 139 - 144 (1982). In Russian. English translation in Soviet Astron. Lett., Vol. 8.

At least four objects (NGC 1275, Mark 298, Mark 423 and III Zw 77) among the known Seyfert galaxies from Zwicky's Catalogue are members of rich clusters of galaxies from Abell's list. It is suggested that Seyfert galaxies in clusters form a specific subclass of objects which differ from the rest of Seyfert galaxies in morphological characteristics and possibly in line broadening mechanism.

158.048 On the galaxy Markarian 348.
A. R. Petrosyan.
Pis'ma Astron. Zh., Tom 8, 145 - 147 (1982). In Russian.

English translation in Soviet Astron. Lett., Vol. 8.

Results of digital processing the two-dimensional image of Markarian 348 are presented. The mass of Mark 348 together with the optical envelope is evaluated to be $10^{11} M_\odot$.

158.049 H I line studies of galaxies: I-General catalogue of 21-cm line data.
L. Bottinelli, L. Gouguenheim, G. Paturel.
Astron. Astrophys.,Suppl. Ser., Vol. 47, 171 - 192 (1982).

New observations of the 21-cm line of about 400 galaxies and a general compilation of published 21-cm line profile are combined to form a catalogue of 21-cm line widths at three levels, H I fluxes and radial velocities (1795 entries). The different corrections on line-widths, fluxes and velocities are derived and applied to produce a homogeneous catalogue of properly weighted averages of the corrected data for 1207 galaxies.

158.050 Rotation and mass of NGC 672 and IC 1727.
N. Carozzi-Meyssonnier.
Astron. Astrophys.,Suppl. Ser., Vol. 47, 237 - 246 (1982).
In French.

The two nearby interacting galaxies NGC 672 (SBS6) and IC 1727 (SBS9) have been optically studied. The author shows that IC 1727 is not a true barred spiral and that is more perturbed than NGC 672 by gravitational interaction. Masses, calculated from red spectra are similar ($8 \times 10^9 M_\odot$ for NGC 672,$10^{10} M_\odot$ for IC 1727). The spectrophotometric study, obtained from blue spectra confirms the late spectral types of the two objects and indicates homogeneous and moderate, electron temperatures and excitation degrees.

158.051 Search for (globular) clusters in M31. II: Photographic photometry of the candidates in a 70′ square field centered on M31.   R. Buonanno, C. E. Corsi, P. Battistini, F. Bònoli, F. Fusi Pecci.
Astron. Astrophys.,Suppl. Ser., Vol. 47, 451 - 461 (1982).

Accurate B and V photographic magnitudes for the globular cluster candidates in M31 down to $V = 18.0$ have been obtained. The two-dimensional fits applied to the PDS data using the formula proposed by Moffat (1969) also allow the determination of a parameter $W_{1/4}$ − defined as the half-width of the fitting surface at one fourth of the height − related to the structural propertiesof each image. The clusters within 1−2 kpc from the nucleus of M31 have a mean parameter slightly smaller than that deduced for the outer clusters possibly due to the influence of the dynamical interactions experienced close to the nuclear region of the galaxy. Many checks on the dependence of $W_{1/4}$ on background, seeing and magnitude confirm that this effect is not due to systematic bias in the photometry. The color-magnitude diagram is also briefly discussed.

158.052 Accurate positions and standard D_{25} diameters for galaxies in the central part of the Coma cluster.
G. Paturel, M. Perie, M. Rousseau.
Astron. Astrophys., Suppl. Ser., Vol. 47, 467 - 469 (1982).
In French.

Accurate positions and standard D_{25} diameters for galaxies in the central part of the Coma cluster are presented. The recently published photometric study of 61 E and S0 galaxies in the Coma cluster is now completed by the accurate determination of the positions. In addition, the previously published diameters are converted to the standard D_{25} diameter-system defined by the 25 mag arcsec^{-2} brightness level. The purpose of this work is to facilitate the insertion of these objects in future catalogues.

158.053 Influence of ellipticity on photometric profiles of elliptical galaxies.   J.-L. Nieto.
Astron. Astrophys.,Suppl. Ser., Vol. 47, 535 - 540 (1982).

The author investigates the influence of ellipticity on various measures of brightness and integrated luminosity profiles of elliptical galaxies. Under the simplifying assumptions that the equivalent profile of an elliptical galaxy follows an $r^{1/4}$law, and that isophotes are homocentric ellipses with constant ellipticity and constant position angle, he shows that: (1) a high ellipticity may lead to significant differences between the detailed profiles derived from different analyses, (2) the profile along any straight line through the center follows an $r^{1/4}$ law, (3) there exists a section whose angle with the major axis depends only on ellipticity along which the profile follows the same $r^{1/4}$ law as the equivalent profile, and (4) integrated profiles derived from photoelectric photometry through circular apertures do not significantly depart from integrated equivalent profiles except at small distances from the center.

158.054 Accurate optical positions of isolated galaxies.
N. Brosch.
Astron. Astrophys., Suppl. Ser., Vol. 48, 63 - 69 (1982).

Accurate positions of the optical centers have been measured for 12 galaxies which appear to be isolated. Coordinates for the centers to within $2''$ are given, along with those of 4 to 8 nearby stars (to within $1.''5$), which can serve to align optical and radio/IR pictures. The new coordinates are more accurate by a factor of 2 or better from those previously published.

158.055 Survey of late-type and irregular southern galaxies on plates taken with the UK 1.2-m Schmidt telescope. IV.
H. G. Corwin, Jr., A. de Vaucouleurs, G. de Vaucouleurs.
Astron. J., Vol. 87, 47 - 75 (1982).

This fourth and final installment of a survey of late-type and irregular southern galaxies is based on 148 UK 1.2-m Schmidt film copies or plates. It completes the survey from the south pole to Dec. = $-22°$ (zone $-25°$). Morphological types, luminosity classes, and inner and outer diameters are given for 500 galaxies of types Scd ($T = 6$) and later; some peculiar and irregular systems are also included.

158.056 UBV surface photometry of NGC 7479: dust and stellar content of the bar and the bar-to-arm transition region.   G. F. Benedict.
Astron. J., Vol. 87, 76 - 89 (1982).

UBV surface photometry is presented for the SB(s)c galaxy NGC 7479. These data are presented as derived aperture photometry, surface magnitude and color index maps, and luminosity and color index profiles through a number of selected features. It is found that surface photometry corroborates solid-body rotation for the bar and the bar-to-arm transition region. A bar-like region containing no new star formation is found parallel to the bright, blue bar. The dust lane associated with the southern bar-to-arm transition region is found to be the site of new star formation. Recent star formation is associated with plumes between the main spiral arms and the bar.

158.057 The nature of galactic shells.   G. Gilmore.
Nature, Vol. 295, 97 - 98 (1982).

158.058 Nature of the shells of NGC 1344.
D. Carter, D. A. Allen, D. F. Malin.
Nature, Vol. 295, 126 - 128 (1982).

Various theories have been proposed to account for the origin of the low surface brightness features resembling thin shells seen in projection around some otherwise normal elliptical galaxies. The authors present broad-band optical (B, V, R) and near-IR (J, H) photometry of one of the shells to determine their constitution and place constraints on their mode of formation. The colour indices derived suggest that the shell comprises a population of stars, perhaps bluer than the main body of the galaxy. There is no evidence for recent star formation.

158.059 Occurrence of the 3.3-μm feature in galaxies.
T. J. Lee, D. H. Beattie, I. Gatley, P. W. J. L. Brand, T. Jones, A. R. Hyland.
Nature, Vol. 295, 214 - 215 (1982).

The feature at 3.28 μm has been observed in some extragalactic objects. The authors report here new measurements which contradict some of these results. In particular, they find the feature in the late-type galaxy M83 but do not observe it in either the Seyfert galaxy NGC 4151 or the quasar 3C 273.

158.060 Asymmetric Balmer line profiles in Seyfert galaxies.
A. Lawrence.
Nature, Vol. 295, 509 - 510 (1982).

The author presents evidence that both the degree of asymmetry and the difference between Hα and Hβ profiles are correlated with the overall line width. He also argues that current evidence is most compatible with an expanding system of optically thick clouds, in which the thickest clouds are moving fastest.

158.061 The active radio galaxy 1413+135.
H. S. Murdoch.
Nature, Vol. 295, 718 (1982).

158.062 Speckle observations of the nucleus of NGC 1068.
J. Meaburn, B. L. Morgan, H. Vine, A. Pedlar, R. Spencer.
Nature, Vol. 296, 331 - 333 (1982).

The nuclei of Seyfert galaxies contain very turbulent motion and are often intense sources of radio emission. It has been proposed that very compact supermassive objects could be the source of these energetic phenomena. Speckle observations of the nucleus of NGC 1068 have revealed for the first time the presence of such an object with a diameter of $\lesssim 2.3$ pc but emitting an amount of visible light equivalent to that from 5×10^9 solar masses.

158.063 A new thermometer for external galaxies.
R. N. Martin, P. T. P. Ho, K. Ruf.
Nature, Vol. 296, 632 - 633 (1982).

The authors introduce a new spectroscopic method at radio frequencies to measure the temperature of giant cloud complexes in external galaxies. They have detected the $\lambda = 1.3$ cm J, $K = 2,2$ inversion transition of NH_3 in the nucleus of the galaxy IC 342. This result, taken together with their previous detection of the J, $K = 1,1$ transition allows to deduce a kinetic temperature of 50 ± 15 K and an equivalent size scale of at least 25 pc for the molecular gas.

158.064 Breaking the active galaxy speed record.
A. Lawrence.
Nature, Vol. 296, 706 - 707 (1982).

158.065 Gradients in the physical conditions of M101 and the pregalactic helium abundance.
J. F. Rayo, M. Peimbert, S. Torres-Peimbert.
Astrophys. J., Vol. 255, 1 - 10 (1982).

Photoelectric spectrophotometry in the 3400–7400 Å range is presented for five H II regions and the nucleus of M101. The abundances of He, N, O, Ne, S, and Ar relative to H are derived for three of the H II regions. Gradients were found in: electron temperature, He/H, N/H, O/H, S/H, and Ar/H. It was found that for a given O/H the N/O values are smaller in M101 than in the Galaxy, probably indicating that on the average the stellar population of M101 is younger than that of the Galaxy. A pregalactic helium abundance of $Y_P = 0.216 \pm 0.010$ (3 σ) was derived.

158.066 Optical polarization of the Seyfert galaxies IC 4329A and Mrk 376.
P. G. Martin, H. S. Stockman, J. R. P. Angel, J. Maza, E. A. Beaver.
Astrophys. J., Vol. 255, 65 - 69 (1982).

The optical polarization of two highly polarized Seyfert 1 galaxies, IC 4329A and Mrk 376, has been measured. In both galaxies the continuum polarization rises towards short wavelengths, and the polarization of the dominant emission lines is the same as that of the neighboring continuum. In the case of IC 4329A this, together with the orientation of the electric vector along the dark dust lane, suggests that the polarization is produced by aligned grains in the galactic disk. Modeling of this "interstellar" polarization shows the grains to be typically 3 times smaller than Galactic polarizing grains.

158.067 The Zwicky magnitude scale: how reliable is it in the estimation of blue luminosity?
G. D. Bothun, R. A. Schommer.
Astrophys. J., Lett., Vol. 255, L23 - L27 (1982).

The authors present the results of aperture photometry for approximately 250 spiral galaxies in clusters. Magnitudes have been computed on the B^T scale following the precepts of de Vaucouleurs, de Vaucouleurs, and Corwin. Comparison of the derived B^T magnitudes with the equivalent photographic magnitudes derived from Zwicky magnitudes, using the conversion formulae of Dickel and Rood and Peterson, reveals a clear bias. Converted Zwicky magnitudes overestimate the total blue luminosity by at least 0.5 mag for galaxies with $M_Z > 14.5$. The authors show that if these converted Zwicky magnitudes are used as the basis for estimating the blue luminosity, then the resulting M_H / L_B value will be systematically lower.

158.068 Giant ringlike H II regions and the distance to M101.
D. G. Lawrie, K. B. Kwitter.
Astrophys. J., Lett., Vol. 255, L29 - L31, plate L4 (1982).

The authors present a new determination of M101's distance modulus based on the diameter of the largest ringlike H II region found on Hα and [S II] λλ6717,6731 photographs. They obtain a distance modulus $\mu_0 = 28.18$ (±0.28) in excellent agreement with the value $\mu_0 = 28.29$ (±0.32) derived by de Vaucouleurs from the correlation between the velocity dispersions and diameters of H II regions. Spectrophotometry of the largest H II ring indicates an intensity ratio of [S II]/Hα ≥ 0.3. The [N II]/Hα and [S II]/Hα intensity ratios place this ring in a category intermediate between classical H II regions and typical supernova remnants.

158.069 Radio continuum observations of the spiral galaxies NGC 2841, NGC 5055, and NGC 7331.
E. Hummel, A. Bosma.
Astron. J., Vol. 87, 242 - 251 (1982).

The radio continuum emission of the nearby spiral galaxies NGC 2841, 5055, and 7331 has been mapped with the Westerbork Synthesis Radio Telescope at wavelengths of 50 and 21 cm. The radio continuum emission is distributed similarly to the light distribution but is not as flattened as the neutral hydrogen distribution. The spectral index distribution shows a general steepening of the radio spectrum toward the edge of the galaxies. This cannot be explained by the influence of the thermal emission alone but requires energy losses of the relativistic electrons which diffuse in both the radial and the z direction.

158.070 Redshifts of 31 bright galaxies.
D. Wills, B. J. Wills.
Astron. J., Vol. 87, 252 - 254 (1982).

The authors report optical spectroscopy of 31 bright galaxies, most of which are among the UGC galaxies detected at 2380 MHz by Dressel and Condon (1978).

158.071 **The luminosity dependence of spiral arm morphology and the luminosity classification.**
R. C. Kennicutt, Jr.
Astron. J., Vol. 87, 255 - 263 (1982).

Measurements of spiral arm properties are used to investigate the systematic dependence of spiral structure on galaxy luminosity and mass, and to understand the basis of van den Bergh's luminosity classification of spirals. The overall length and extent of the spiral pattern is found to correlate only loosely with luminosity, spiral arm development is dictated by additional parameters which are not coupled to galaxy mass. The relative arm width, normalized to the diameter of the galaxies, decreases slowly with luminosity, and in many cases appears to offer a more reliable luminosity estimate than arm development. Galaxy surface brightness is found to be virtually uncorrelated with luminosity. Taken together, the results suggest that luminosity classifications based on spiral arm properties may be reliably used at the $\pm 1^m$ level in normal galaxies, but no better, even if quantitative measurements are used.

158.072 **High-resolution optical surface photometry of M31.**
P. W. Hodge, R. C. Kennicutt.
Astron. J., Vol. 87, 264 - 277 (1982).

Optical surface photometry in B has been obtained for M31 with a resolution of 3 arcsec. Isophotes are given. The ellipticity curve is shown and is typical of spiral galaxies. The minor- and major-axis profiles agree well with those obtained by other means. The total mass of dust in the lane is estimated to be $2 \times 10^6 M_\odot$. Individual dust clouds are mapped and are found to be complex in structure and shallow, with average measured extinctions of only ~0.3 mag.

158.073 **The short-term radio variability of BL Lacertae objects.**   D. R. Altschüler.
Astron. J., Vol. 87, 387 - 394 (1982).

The variability properties of 37 BL Lacertae objects were monitored during a six-month interval at a frequency of 2380 MHz. Contrary to expectations, it was found that 30 of these objects showed no significant variations in this interval, and that, therefore, rapid radio variability is not a characteristic property of this class of objects. It is suggested that since these objects do not differ significantly from the subclass of high-polarization QSO's, they possibly belong to this same subclass.

158.074 **Reverberation mapping of the emission line regions of Seyfert galaxies and quasars.**
R. D. Blandford, C. F. McKee.
Astrophys. J., Vol. 255, 419 - 439 (1982).

Variations in the strengths of the central photoionizing source in a quasar or Seyfert galaxy will generate variations in the strengths and profiles of the emission lines. These "reverberations" in the emission lines will lag behind the continuum variations due to light travel time effects. A procedure is described for analyzing a time series of measurements of both the continuum and the lines. This procedure permits direct verification of the assumed causal connection of the lines to the continuum. The authors demonstrate that if the emission line region has a high degree of symmetry, then it is possible to invert the time-dependent line profiles and obtain the phase space distribution of the emission-line gas— i.e., its emissivity and the moments of its velocity distribution as functions of position.

158.075 **The inhibition of star formation in barred spiral galaxies.**   A. D. Tubbs.
Astrophys. J., Vol. 255, 458 - 466 (1982).

The lack of star formation associated with the straight dust lanes within the bar of barred spiral galaxies suggests that star formation is strongly inhibited there. The author shows that the inhibition may be an effect of the high velocities, relative to the gas in the dust lanes, of the dense clouds whose orbits intersect the dust lanes. These clouds, which presumably undergo gravitational collapse upon entering the dust lanes of normal spiral galaxies, are the precursors of star formation. In barred galaxies, the clouds in the bar enter the dust lanes at a velocity an order of magnitude higher than those in normal spirals. The high velocity can cause a quick compression of the clouds followed by a rapid expansion which disperses the majority of the cloud. In the outer parts of the barred spiral galaxy, and at the very center, the cloud velocities are lower, and star formation is not inhibited. The effectiveness of this mechanism is tested by calculating the orbits of the dense, prestellar clouds in the expected gravitational field of NGC 5383.

158.076 **Ultraviolet images of M101: observations of dust and inferences on the metallicity.**
T. P. Stecher, R. C. Bohlin, J. K. Hill, M. A. Jura.
Astrophys. J., Lett:, Vol. 255, L99 - L102 (1982).

Three ultraviolet images of the Sc I galaxy NGC 5457 (M101) were obtained by a rocket-borne telescope with an effective mean wavelength of 2250 Å, a bandpass of 970 Å, and about 10" resolution. The UV pictures reveal faint arms, which correlate well with the 21 cm H I radiograph, but which are only marginally detectable on deep visual photographs. These UV arms exhibit no individual stars or H II regions, but their surface brightness can be explained by the scattering off dust of the UV light from hot stars and bright H II regions in the disk of M101.

158.077 **The properties of AP Librae from UBV photoelectric photometry.**
B. E. Westerlund, G. Wlérick, R. Garnier.
Astron. Astrophys., Vol. 105, 284 - 292 (1982).

Photoelectric UBV observations of the BL Lacertae-type object AP Librae have been made from March, 1975, to June, 1980. During this period variations in V between 14.32 and 15.12 mag have been observed. The decrease in brightness of AP Lib is accompanied by a reddening. The measurements made it possible to identify two components in AP Lib: an extended component, possibly an E galaxy and a nucleus with "miniquasar" properties.

158.078 **The geometry of the Seyfert nucleus in NGC 4151 revisited. I. Cloudy structure from the [O III] line profile analysis.**   D. Pelat, D. Alloin.
Astron. Astrophys., Vol. 105, 335 - 341 (1982).

A detailed analysis of the [O III] 500.68 nm emission line in NGC 4151 is discussed in this paper. The cloud system in NGC 4151 involves about 6 components representing a mass of ionized gas of the order of $10^3 M_\odot$, an internal macroscopic turbulent energy of 10^{50} erg and a kinetic energy of 10^{50} erg. A broad component is revealed by the [O III] 500.68 line profile analysis: it represents 45% of the total line flux and shows an emission velocity blue-shifted by 130 km s^{-1}, with respect to the systemic velocity of the galaxy. This effect is of extreme importance for understanding the geometry and dynamics of the forbidden line region in Seyfert nuclei.

158.079 **Studies of nearly face-on spiral galaxies. I. The velocity dispersion of the H I gas in NGC 3938.**
P. C. van der Kruit, G. S. Shostak.
Astron. Astrophys., Vol. 105, 351 - 358 (1982).

The authors report H I synthesis observations and optical surface photometry of the nearly face-on ScI galaxy NGC 3938. From these observations they derive the rotation curve, and the radial variation of the H I surface density and optical surface brightness. These properties indicate that NGC 3938 is a normal system for its type. A result provides evidence that the dark mass indicated by the flat rotation curves of spiral galaxies resides in the halos of these systems.

158.080 **The photometric history of the BL Lacertae object OJ 287.**   G. Gaida, H.-J. Röser.
Astron. Astrophys., Vol. 105, 362 - 363 (1982).

The lightcurve of the BL Lacertae object OJ 287 is examined on archival plates taken in the period 1890–1930. The outburst detected in ~1912 shows a striking similarity to the one observed in 1971.

158.081 **A new bright compact galaxy in Ursa Major.**   C. Barbieri, S. Cristiani, G. Romano.
Astron. Astrophys., Vol. 105, 369 - 371 (1982).

A previously unknown compact galaxy of integrated magnitude $m_B \approx 12.4$ has been discovered as a variable "star" on Asiago Schmidt telescope plates. Large scale photographs and spectra have revealed that this remarkable object is an extremely compact blue galaxy with a strong continuum, emission and absorption lines, at a redshift $z = 0.09$ (absolute magnitude $M_B \approx -21.4$ if $H = 50$ km s^{-1} Mpc, diameter 3 kpc). The spectral characteristics resemble those of galaxies rich of early-type stars.

158.082 **Hot stars in the bulge of M31: upper limit to the star formation rate.**
J. M. Deharveng, M. Joubert, G. Monnet, J. Donas.
Astron. Astrophys., Vol. 106, 16 - 20 (1982).

UV observations in the bulge of the galaxy M31 have been obtained with the IUE satellite. The far UV surface brightness distribution is probably similar to that in the B light. A gradient in (UV–B) color is however found in the very centre of M31. Hot evolved stars are suggested as the major source of the UV flux, while no firm explanation is found for the observed UV color gradient. An upper limit to a possible additional contribution of young stars is derived; comparison with evolutionary models gives a star formation rate in the bulge of M31 at most $7.4 \times 10^{-5} M_\odot$ yr^{-1}. Examination of the case of NGC 205 leads to the opposite conclusion.

158.083 **The very large, interacting galaxy pair IC 5174/75.**   R. M. West, R. Barbier.
Astron. Astrophys., Vol. 106, 53 - 57 (1982).

The interacting galaxy pair, IC 5174/75, (R.A. = 22^{h}09^{m}8; Decl. = $-38°24'$; 1950.0), at $z = 0.036$ is very large (~320 kpc; $H_0 = 50$ km s^{-1} Mpc^{-1}). IC 5174 has two long, outer arms (total size ~300 kpc) and an active centre. It is believed that the present morphology of this system is the result of a recent encounter and that it may be qualitatively similar to the M 51 + NGC 5195 system, seen edge-on, although of quite different, absolute size.

158.084 **The optical halo around NGC 253.**   R. Beck, G. Hutschenreiter, R. Wielebinski.
Astron. Astrophys., Vol. 106, 112 - 114 (1982).

The authors investigated the blue light from an extended halo surrounding NGC 253 using both a photographic technique of image enhancement and digital smoothing of P. D. S. scans. The distribution of light can be followed out to 10 kpc from the plane of the galaxy. An arch of light is coincident with a spur of radio emission.

158.085 **The magnetic field in M31.**   R. Beck.
Astron. Astrophys., Vol. 106, 121 - 132 (1982).

The "vectors" of the linearly polarised component of the radio continuum emission at 2700 MHz (λ11.1 cm) reveal a large-scale magnetic field in M31 which is aligned along the H I spiral arm segments forming an elliptical ring-like structure in the plane of the galaxy at about 10 kpc distance from the centre. The direction of the magnetic field lines is the same as that of the rotation of M31. No large-scale magnetic field reversals on scales ⩾3 kpc could be detected between 7 and 16 kpc radius.

158.086 **The radio continuum properties of S0 galaxies.**   E. Hummel, C. G. Kotanyi.
Astron. Astrophys., Vol. 106, 183 - 189 (1982).

A large sample of S0 galaxies was observed with the Westerbork Synthesis Radio Telescope at 1415 MHz. The radio continuum properties of the sample are analyzed and compared with those of E and S galaxies. The authors suggest that the absence of a strong radio disk component and of a steep spectrum central source are due to the low gas content and that the radio power of a flat spectrum nuclear source is related to the mass and rotation of the bulge component.

158.087 **Spectroscopic observations of spheroidal systems: the bulges of M 81, NGC 4736, and the dwarf elliptical M 32.**   A. Pellet, F. Simien.
Astron. Astrophys., Vol. 106, 214 - 220 (1982).

Rotation curves and velocity dispersions have been obtained in three early type galaxies – M 81, NGC 4736, M 32 – using spectra taken with the 1.93 m telescope of the Observatoire de Haute Provence with resolutions of 1.2 and 2.5 Å over the spectral range from 4000 to 4500 Å. The authors confirm Kormendy and Illingworth's (1979) conclusion that bulges of early type spirals exhibit higher rotation velocities and smaller velocity dispersions than giant ellipticals; M 32 has a similar behaviour. No rotation appears on the minor axes of M 81 and M 32. These data and those previously obtained on M 31 are consistent with the isotropic oblate model of Binney (1978).

158.088 **Verdeling van licht, kleur en massa in sterren-stelsels.**   P. C. van der Kruit.
Zenit, 9. Jaarg., 157 - 162 (1982).

158.089 **X-ray emission from elliptical galaxies.**   V. A. Krol'.
Inst. teor. fiz. AN USSR. Prepr., 1981, No. 120, 17 pp. In Russian. – Abstr. in Ref. zh., 51. Astron., 2.51.703 (1982).

158.090 **Interaction of stars with the accretion disc around a massive black hole in the nuclei of active galaxies and quasars.**   A. S. Zentsova.
Fiz.-tekh. inst. AN SSSR. Prepr., 1981, No. 733, 16 pp. In Russian. – Abstr. in Ref. zh., 51. Astron., 2.51.746 (1982).

158.091 **Inner ring structures in galaxies as distance indicators. II. Calibration of inner ring diameters as quaternary indicators.**   R. Buta, G. de Vaucouleurs.
Astrophys. J., Suppl. Ser., Vol. 48, 219 - 237 (1982).

The linear diameters D_r of the inner ring and pseudo-ring structures in spiral galaxies of different morphological types and stage Sab to Sd in the revised Hubble system are calibrated by means of ~150 galaxies which have the best determined distance moduli previously derived from tertiary indicators via the corrected luminosity index and from the 21 cm line widths via revised versions of the Tully-Fisher relations. The logarithm of the linear diameter D_r is shown to be a linear function of the family and stage parameters, and of the corrected luminosity class. The calibration is extended to early-type spirals and lenticulars by means of ~50 objects which are probable members of groups including spirals whose distance moduli are known from radio or optical indicators. The zero point of the ring distance scale is consistent with the basic zero point defined by the primary and secondary indicators within the internal mean error of 0.2 mag or better, and has an external mean error from all sources of ~0.25 mag.

158.092 **Neutral hydrogen observations of double spiral galaxies. I. NGC 5905 and NGC 5908.**
G. A. van Moorsel.
Astron. Astrophys., Vol. 107, 66 - 74 (1982).

Westerbork 21 cm neutral hydrogen observations are presented for the pair of galaxies NGC 5905 and NGC 5908. For both galaxies a rotation curve and corresponding masses are derived. These masses are compared with the minimum mass which follows from the observed motion of the pair in its orbit.

158.093   On the sizes of rings and lenses in disk galaxies.
E. Athanassoula, A. Bosma, M. Crézé, M. P. Schwarz.
Astron. Astrophys., Vol. 107, 101 - 106 (1982).

The authors present a statistical analysis of data on sizes of rings and lenses in disk galaxies, and discuss their results in the light of current theoretical ideas. The position of rings in barred galaxies are associated with the main dynamical resonances, but for rings in non-barred galaxies the situation is less clear.

158.094   Ultraviolet photometry from the Orbiting Astronomical Observatory. XL. The energy distributions of spiral and irregular galaxies.   A. D. Code, G. A. Welch.
Astrophys. J., Vol. 256, 1 - 12 (1982).

Measurements of the total light of 40 spiral and irregular galaxies are presented. The photometry covers the wavelength range 1550–4250 Å and is calibrated on an absolute basis. On the average later-type galaxies are not only bluer at short wavelengths than ellipticals but significantly bluer than visual colors would imply. This reflects a recent history of more vigorous formation of massive stars. The shape of the upper part of the initial mass function apparently varies more among early-type galaxies, producing a wide scatter in their energy distributions. The contribution of galaxies to the observed diffuse ultraviolet background radiation is discussed.

158.095   Infrared, optical, and ultraviolet observations of hydrogen line emission from Seyfert galaxies.
J. H. Lacy, B. T. Soifer, G. Neugebauer, K. Matthews, M. Malkan, E. E. Becklin, C.-C. Wu, A. Boggess, T. R. Gull.
Astrophys. J., Vol. 256, 75 - 82 (1982).

Nearly simultaneous observations are reported of Pα, Hα, and Hβ emission from 18 Seyfert galaxies and of Lyα from eight of these galaxies. In many cases, Pα is stronger relative to the Balmer lines than is predicted by recombination calculations; reddening appears to be required. Dispersion in the Pα/Hα/Hβ ratios orthogonal to the reddening track indicates that high densities or optical depths also affect the line flux ratios. Several galaxies, notably NGC 1275, have very low Pα/Hα ratios. High densities, large optical depths, and reddening probably all contribute to the low observed Lα/Hβ ratios.

158.096   The X-ray spectrum and time variability of narrow emission line galaxies.   R. F. Mushotzky.
Astrophys. J., Vol. 256, 92 - 102 (1982).

The author reports X-ray spectral and temporal observations of six narrow emission line galaxies (NELGs) NGC 526a, NGC 2110, NGC 2992, MCG-5-23-16, NGC 5506, and NGC 7582. All of these sources are well fitted by power-law X-ray spectra of energy slope $\alpha \sim 0.8$. For three of them, a power law is a significantly better fit than an isothermal bremsstrahlung spectrum. There is evidence for Fe spectral features in three of these objects. On time scales of 6 months three of these objects, NGC 526a, NGC 2110, and MCG-5-23-16, are variable in their X-ray flux. On time scales less than one week, NGC 2110, MCG-5-23-16, and NGC 7582 showed detectable variability in at least one observation. The measured X-ray properties of the NELGs strongly resemble those for previously measured type 1 Seyferts of the same X-ray luminosity.

158.097   Spectral analysis of the asymmetric spiral pattern of NGC 4254.
M. Iye, S. Okamura, M. Hamabe, M. Watanabe.
Astrophys. J., Vol. 256, 103 - 111 (1982).

A new method is presented for measuring the strength of spiral and bar structure in galaxies. This is based on a Fourier

analysis of the intensity distribution in a polar coordinate system set up in the plane of a galaxy. Results of this technique are illustrated for the Sc I galaxy NGC 4254. It is shown that the asymmetric spiral pattern of NGC 4254 can be delineated as a superposition on the main one-armed spiral of five-armed and three-armed components. Some characteristic parameters are suggested which will quantitatively represent morphological features such as bar strength, degree of asymmetry of spiral arms, regularity of spiral winding direction, and complexity of structure.

158.098   Mid-infrared observations of Seyfert 1 and narrowline X-ray galaxies.
I. S. Glass, A. F. M. Moorwood, W. Eichendorf.
Astron. Astrophys., Vol. 107, 276 - 282 (1982).

The nuclear continua of a sample of Seyfert 1 and narrow-line X-ray galaxies are investigated by combining new broad- and intermediate-band infrared data with existing information. Power-law spectra with indices in the range $\alpha = 1.1 - 1.4$ are found to contribute significant fluxes from the infrared to the ultraviolet in both cases. The nuclear reddening estimates derived from the infrared data are generally in agreement with those derived from the visible and the UV, but are not necessarily the same as those found for the narrow-line emission regions, particularly in the case of the narrow-line X-ray galaxies. The broadband $L-N$ colour indices of the narrow-line galaxies are distributed more like the Seyfert 1s than the Seyfert 2s. The intermediateband colours show evidence for substantial silicate absorption in only three cases.

158.099   Five-channel photometry of the edge-on S0 galaxy NGC 4762.
M. Hamabe, S. Okamura, M. Iye, S. Nishimura.
Publ. Astron. Soc. Japan, Vol. 33, 643 - 651 (1981).

Photon-counting photometry of the edge-on S0 galaxy NGC 4762 is carried out in five-color bands. The galaxy is found to become bluer along the major axis as the distance from the galactic center increases. Two humps seen in the major-axis profile of NGC 4762 are most naturally interpreted as due to the local density enhancement of the disk stars without significant local variation in the stellar content.

158.100   Observational properties of the galaxy Markarian 297.
Y. Taniguchi, S. Tamura.
Publ. Astron. Soc. Japan, Vol. 33, 653 - 664 (1981).

Spectroscopic observations were made for the nonSeyfert galaxy Markarian 297. (a) Excitation conditions of the emissive gas are similar to those of H II regions in the outer arms of Sc galaxies. (b) A steep Balmer decrement indicates the presence of a large amount of dust grains in the galaxy and may explain its metal deficiency as due to the depletion of heavy elements. (c) The number of OB stars of the order of 10^5 is required as ionizing sources.

158.101   Neutral hydrogen in M31. II. Kinematical properties and density waves.   T. Sawa, Y. Sofue.
Publ. Astron. Soc. Japan, Vol. 33, 665 - 677 (1981).

The density-wave theory is applied to the neutral hydrogen (H I) data of M31 to calculate model $T(\lambda, \nu)$ and $T(\beta, \nu)$ diagrams (the brightness temperature of the 21-cm line emission plotted on the longitude-radial velocity and latitude-radial velocity planes), when the galaxy is observed with a finite antenna beam width. The calculated results reproduce many characteristic features like bifurcated distribution on observed $T(\lambda, \nu)$ and $T(\beta, \nu)$ diagrams for H I gas in M31.

158.102   Magellanic irregular galaxies and chemical evolution of galaxies.
G. Comte, J. Lequeux, G. Stasinska, L. Vigroux.
Messenger, No. 27, p. 9 - 10 (1982).

158.103 Optical positions of Seyfert galaxies.
E. D. Clements.
Abh. Hamburger Sternw., Band 10, 167 (1982).

158.104 Absolute B, V photometry of cD galaxies.
E. Valentijn.
Messenger, No. 27, p. 29 - 31 (1982).

158.105 NGC5128 – a galaxy with a recently formed disk.
M. Marcelin, J. Boulesteix, G. Courtes, B. Milliard.
Nature, Vol. 297, 38 - 42 (1982).

The authors have obtained the first complete velocity field of the ionized gas in NGC5128. It shows that the gas is rotating in a thick disk, without any evidence of the perturbations expected from the encounter of an elliptical and a spiral galaxy. The observations favour the hypothesis of the evolution of a spiral galaxy in which the formation of the disk is very late compared with the bulge.

158.106 Aktivität in den Zentren der Galaxien.
W. Kundt.
Sterne Weltraum, 21. Jahrg., 66 - 71 (1982).

158.107 Spectrophotometry and morphology of the galaxies with UV excess: III. M. A. Kazarian (*Kazaryan*), E. Ye. Khachikian (*Eh. E. Khachikyan*), A. A. Yegiazarean (*Egiazaryan*).
Astrophys. Space Sci., Vol. 82, 105 - 113 (1982).

The results of the spectrophotometry and morphology of four galaxies with ultraviolet excesses are presented. The equivalent widths and relative intensities of the emission lines are measured. The electron density, the rotation velocity of the central parts, and the masses for two of these galaxies are calculated.

158.108 The optical spectrum of the radio galaxy PKS 2152−69. G. Marenbach, I. Appenzeller.
Astron. Astrophys., Vol. 108, 95 - 98 (1982).

The authors discuss low resolution IDS and medium resolution long slit image tube spectrograms of the radio galaxy PKS 2152−69. The nucleus of this bright galaxy shows an intense emission line spectrum with LINER-like properties. The relatively weak Balmer emission lines H_α and H_β have moderately broad wings fitting the definition of the Intermediate Line Radio Galaxies. The continuum energy distribution shows a strong UV excess, but otherwise an essentially normal stellar absorption spectrum. The emission line spectrum is tentatively explained by assuming the superposition of contributions from a relatively weak photoionized central source and more extended regions of shock-heated gas.

158.109 Rotation of the dust lane in NGC 1947.
C. Möllenhoff.
Astron. Astrophys., Vol. 108, 130 - 133 (1982).

Long slit image tube spectrograms of the southern dust lane galaxy NGC 1947 were obtained in order to study the kinematics of this object. The dust lane shows fast rotation while the stellar component does (nearly) not rotate. The system shows strong similarity to NGC 5128 (Cen A), however is not a radio galaxy.

158.110 The velocity field of the ionized gas in the barred galaxy NGC 925.
M. Marcelin, J. Boulesteix, G. Courtès.
Astron. Astrophys., Vol. 108, 134 - 140 (1982).

The velocity field of the ionized gas in NGC 925 has been drawn from Perot-Fabry interferograms in $H\alpha$ light. The main perturbations of the velocity field, compared with a rotating disk, are due to the southern arm, which is the main arm of this galaxy; such perturbations, also seen in H I, are probably the signature of a density wave. The bar rotates like a solid body with an angular velocity of 55 km s^{-1} arc min^{-1}, close to

that of the disk, which indicates a rather low angular momentum for a Sc type galaxy. The systemic velocity is found to be 553 km s^{-1} ± 5 km s^{-1}. Assuming a distance of 7 Mpc the authors find a mass of 2.1 ± 0.2 X 10^{10} $M_\odot$ for NGC 925.

158.111 A survey of the distribution of λ2.8 cm radio continuum in nearby galaxies. II. NGC 6946.
U. Klein, R. Beck, U. R. Buczilowski, R. Wielebinski.
Astron. Astrophys., Vol. 108, 176 - 187 (1982).

Maps of NGC 6946 in total power and linear polarisation were made at λ2.8 cm (10.7 GHz). The total power map could be compared with maps of similar resolution made at 610 MHz. The thermal and the non-thermal components of the total emission were separated. The distribution of the radio thermal emission is in good agreement with the distribution of the optical Hα regions. The average percentage polarisation of the non-thermal emission of NGC 6946 at 2.8 cm wavelength is 10 ± 3%. The corresponding ratio between the magnetic field strength of the uniform and random field is 0.31 ± 0.06. A bisymmetric spiral model of the magnetic field can explain the observations of the polarised emission. If energy equipartition is valid, the average number density of cosmic ray electrons is ~ 5 times higher than in the solar neighbourhood while the strength of the total magnetic field is 15 ± 5 μG. The magnetic field may be of primary importance for the formation of spiral structure in NGC 6946.

158.112 Spektroskopie von Seyfert-Galaxien.
H. Schulz.
Mitt. Astron. Ges., Nr. 55, p. 94 (1982). – Abstract.

158.113 Elektronographische Beobachtungen von Galaxien mit BL Lacertae-Kernen. G. Gaida.
Mitt. Astron. Ges., Nr. 55, p. 94 - 95 (1982). – Abstract.

158.114 IC 5063: another case of intergalactic cannibalism?
I. Appenzeller, G. Gaida.
Mitt. Astron. Ges., Nr. 55, p. 95 - 96 (1982). – Abstract.

158.115 Optische Beobachtungen der pekuliaren Radiogalaxie NGC 6240. J. Fried, H. Schulz.
Mitt. Astron. Ges., Nr. 55, p. 98 - 99 (1982).

158.116 Das Magnetfeld von NGC 6946.
R. Beck, U. R. Buczilowski, U. Klein, R. Wielebinski.
Mitt. Astron. Ges., Nr. 55, p. 113 - 115 (1982).

158.117 Component analysis of characteristic parameters of elliptical and S0 galaxies.
P. Brosche, F.-T. Lentes.
Mitt. Astron. Ges., Nr. 55, p. 116 - 117 (1982).

158.118 H I-Beobachtungen stark verfärbter Galaxien und Vergleich zweier H I-Kataloge.
M. D. Gruber, G. M. Gruber, J. Pfleiderer, L. Velden.
Mitt. Astron. Ges., Nr. 55, p. 180 - 182 (1982).

158.119 Zur Beziehung zwischen optischen und Radiohelligkeiten von Galaxien.
J. Pfleiderer, P. Krommidas.
Mitt. Astron. Ges., Nr. 55, p. 188 (1982). – Abstract.

158.120 On massive-neutrino halos and galactic structures.
A. Crollalanza, J. G. Gao, R. Ruffini.
Lett. Nuovo Cimento, Vol. 32, Ser. 2, 411 - 414 (1981). Abstr. in Phys. Abstr., Vol. 85, Abstr. 28565 (1982).

158.121 Isolated triplets of galaxies. New radial velocities.
I. D. Karachentsev, V. E. Karachentseva.
Astrofizika, Tom 17, 5 - 18 (1981). In Russian. English translation in Astrophysics, Vol. 17, No. 1.

Spectral observations of the triple galaxies from the list of Karachentseva et al. (1979) have been made in 1977 - 80 with the 6 meter telescope. The values of the radial velocities for 157 galaxies from the 54 triplets are presented. The linear diameters of the galaxies and the absolute magnitudes are also given. The spectral features of some objects are noted.

158.122   Spectrophotometry of the central region of the galaxy Markarian 290.
I. I. Pronik, L. P. Metik.
Astrofizika, Tom 17, 19 - 33 (1981). In Russian. English translation in Astrophysics, Vol. 17, No. 1.

New data on the structure of the central region of the galaxy Markarian 290 are presented. Near the nucleus of the galaxy "b" towards the starlike object "a" a gaseous formation has been discovered. The hydrogen line contours and the relative intensities of emission lines have been considered. Assuming homogeneity of the physical conditions of the shell, the following parameters were obtained: T_e = 20 000°K, n_e =10^9 cm^{-3}, M = 4 $M_\odot$.

158.123   Polarization and photometric investigation of the peculiar galaxy NGC 3718.
V. A. Hagen-Thorn, I. I. Popov, V. A. Yakovleva.
Astrofizika, Tom 17, 35 - 42 (1981). In Russian. English translation in Astrophysics, Vol. 17, No. 1.

The results of polarimetric and photometric observations of the peculiar lenticular galaxy NGC 3718 are given. Appreciable polarization is found in the dark band. Reddening in this region is about $1^{m}0$. The polarization is analogous to interstellar polarization in our Galaxy.

158.124   Physical conditions in the nuclei of emission-line galaxies.
G. T. Petrov, I. M. Yankulova, V. K. Golev.
Astrofizika, Tom 17, 43 - 51 (1981). In Russian. English translation in Astrophysics, Vol. 17, No. 1.

A spectrophotometric investigation of 6 galaxies with emission lines not belonging to the Seyfert type has been carried out. The electron densities and the electron temperatures of the emitted gas are estimated. The $F_{H\beta}$ fluxes and the $L_{H\beta}$ luminosities in the H_β line are given.

158.125   Mean surface brightness of single and double galaxies.   M. A. Arakelyan, A. P. Magtesyan.
Astrofizika, Tom 17, 53 - 59 (1981). In Russian. English translation in Astrophysics, Vol. 17, No. 1.

It is shown that the mean surface brightness of spiral galaxies being the components of pairs exceeds with high statistical significance that of isolated spiral galaxies. Increase of the mean surface brightness of spirals in pairs with decrease of their mutual linear distance is stated. A small correlation between the mean surface brightness of the components of the pairs is also revealed. It decreases with the increase of the mutual linear distances.

158.126   The function of distribution of numbers of faint galaxies.   L. M. Fesenko.
Astrofizika, Tom 17, 61 - 70 (1981). In Russian. English translation in Astrophysics, Vol. 17, No. 1.

The Lick Observatory counts of galaxies are considered. The distribution of numbers of galaxies in elementary regions of $1° \times 1°$ is investigated. A mean apparent multiplicity of a galaxy was derived. The galaxy number distribution was simulated with accounting of a simple model for the number of various systems of galaxies. Based on that model an approximate expression for the galaxy number distribution was considered and was compared with observed distributions.

158.127   On the direction of coiling of galaxy spiral arms in double and multiple systems.
N. G. Kogoshvili, T. M. Borchkhadze.

Astrofizika, Tom 17, 183 - 186 (1981). In Russian. English translation in Astrophysics, Vol. 17, No. 1.

The application of the statistical method for the distribution of about 4000 spiral galaxies revealed that the number of isolated visual double systems as well as group members, in which the components have opposite directions of arm coiling, is three times as high as that of pairs, in which the components have the same direction of arm coiling.

158.128   Velocity field of the Seyfert galaxy Markarian 744.
V. L. Afanas'ev, A. I. Shapovalova.
Astrofizika, Tom 17, 403 - 420 (1981). In Russian. English translation in Astrophysics, Vol. 17, No. 3.

The results of a velocity field investigation of the Seyfert galaxy Markarian 744 which is a member of the double galaxy NGC 3786/88 are presented. From the rotation curves of the components determined up to the distance of 6 kpc for Markarian 744 and of 9 kpc for NGC 3788 the authors estimated the masses of the galaxies. It is shown that the morphological structure of the Seyfert galaxy Markarian 744 is connected with the dynamics of disks. The inner Lindblad resonance is near the disk which is 1.5 - 2 kpc in size. The inner and outer layers of the spiral arms, observed in the disk of the galaxy, have sizes of 10 and 17 kpc. They correspond to the ring position of corotation and outer resonance. There are two systems of gas with non-circular motion in the central region of Markarian 744.

158.129   The remarkable galaxy Markarian 314 with three condensations.   A. R. Petrosyan.
Astrofizika, Tom 17, 421 - 427 (1981). In Russian. English translation in Astrophysics, Vol. 17, No. 3.

On the basis of direct photographs a morphological and isodensitometrical investigation of the galaxy Markarian 314 has been carried out. The results of processing of the spectra of the central condensations of the galaxy obtained are presented. Using the values of equivalent widths and relative intensities of the observed lines for T_e = 10 000 K, the electron density, abundances of the heavy elements O, N, Ne, S, and the mass of the emitting gas are calculated. The number of O7 type stars in the condensations is evaluated. It is concluded that Markarian 314 resembles in characteristics an irregular galaxy.

158.130   Homogeneous sample of binary galaxies. II. Orbital masses.   I. D. Karachentsev.
Astrofizika, Tom 17, 429 - 444 (1981). In Russian. English translation in Astrophysics, Vol. 17, No. 3.

The distribution of 423 binary galaxies is considered according to estimates of their orbital masses. After the exclusion of false pairs and allowance for radial velocity measurement errors the mean orbital mass value appears to be in agreement with the ordinary mass estimates obtained from galactic rotation. Analyses of projection factors for different types of orbital motion of members of pairs show that the suggestion on very elongated orbits with an eccentricity $e > 0.8$ does not agree with the observational data.

158.131   Galaxies with ultraviolet continuum. XV.
B. E. Markarian, V. A. Lipovetskij, D. A. Stepanyan.
Astrofizika, Tom 17, 619 - 627 (1981). In Russian. English translation in Astrophysics, Vol. 17, No. 4.

The fifteenth list of galaxies having intense ultraviolet continuum in the spectrum is presented. The list contains data for 101 objects. The presence of emission lines either established or suspected among 82 of them is given.

158.132   Spectrophotometry of the central region of the galaxy Markarian 298=IC 1182—4.
L. P. Metik, I. I. Pronik.
Astrofizika, Tom 17, 629 - 642 (1981). In Russian. English translation in Astrophysics, Vol. 17, No. 4.

New data on the structure of the central region of the Seyfert galaxy Markarian 298=IC 1182−4 are presented.

158.133 Spectrophotometry of the Seyfert galaxy Markarian 1066.
V. L. Afanas'ev, V. A. Lipovetskij, A. I. Shapovalova.
Astrofizika, Tom 17, 643 - 660 (1981). In Russian. English translation in Astrophysics, Vol. 17, No. 4.

Spectrophotometric results are presented for the Seyfert galaxy Markarian 1066. More than 60 lines of different elements are identified in the spectrum. The continuum follows the power law $I_\nu \sim \nu^{-\alpha}$ where $\alpha = 2.5 \pm 0.5$. The electron densities Ne $\cong 700$ cm^{-3} and electron temperatures Te $\cong 10^4$ K in the [N II] region and Ne $\gtrsim 3 \times 10^5$ cm^{-3} and Te $\cong 10^4$ K in the [O III] region are deduced. The parameters of the gaseous component for the galaxies Markarian 1066 and Markarian 744 and also luminosity and mass of the nuclei are calculated. A close relation between Markarian 1066 and Markarian 744 in physical characteristics is noted.

158.134 UBVR photometry of Seyfert galaxies.
V. T. Doroshenko, V. Yu. Terebizh.
Astrofizika, Tom 17, 667 - 673 (1981). In Russian. English translation in Astrophysics, Vol. 17, No. 4.

Results of photoelectric UBVR observations of 34 galaxies including 29 galaxies with Seyfert characteristics are presented.

158.135 Homogeneous sample of binary galaxies. III. Peculiarities of kinematics and structure.
I. D. Karachentsev.
Astrofizika, Tom 17, 675 - 692 (1981). In Russian. English translation in Astrophysics, Vol. 17, No. 4.

A method is considered for the determination of the type of orbital motions in pairs of galaxies from the relation between the radial velocity difference for the components and the projection of their mutual linear separation. Application of the method to 361 pairs of galaxies has confirmed the inference made previously (1981) of the prevalence of circular orbits in binary galaxies. A comparison is made of the principal parameters of double galaxies divided according to morphological types of galaxies and also to the types of interaction between the components. The distribution of 361 pairs according to integral luminosity and mutual distance of the components is presented.

158.136 Homogeneous sample of binary galaxies. IV. Orientation, angular momentum, peculiar motions.
I. D. Karachentsev.
Astrofizika, Tom 17, 693 - 707 (1981). In Russian. English translation in Astrophysics, Vol. 17, No. 4.

Some characteristics of binary galaxies in near groups and clusters are considered. Their orbital mass-to-luminosity ratio corresponds to the normal value, obtained from the rotation of galaxies. It is found that the pairs in systems have smaller peculiar velocities than single members of the systems and are situated mainly at the periphery of groups and clusters. It can be explained by an effect of external tides on the pairs. Within the framework of the tidal hypothesis the total masses of groups and clusters are determined. For 239 isolated pairs with spiral components the distribution according to the angular momentum is obtained. The conclusion is drawn that the main angular momentum of binary galaxies is not in the internal, but in the orbital motions.

158.137 Selection effect of large importance during investigation of the velocity distribution of galaxies.
B. I. Fesenko.
Astrofizika, Tom 17, 719 - 726 (1981). In Russian. English translation in Astrophysics, Vol. 17, No. 4.

When samples of galaxies according to their apparent brightness were investigated, the groups with large apparent richness were arranged into a narrow distance interval. Therefore, in the regions of the sky with increased number of galaxies, a high maximum in the velocity distribution exists. That maximum is stimulated not by a single rich cluster but by several groups of galaxies with a similar distance from the observer. Such a system of groups imitates one rich cluster surprisingly well. In regions with a small number of galaxies a deep minimum in the velocity distribution is expected. Such a minimum is not connected with the cell structure of the universe. The correctness of these statements is confirmed by the results of the investigation of a model for the distribution of galaxies in 800 elementary regions of the sky.

158.138 Spectrophotometry of the nucleus of the galaxy Arakelyan 144.
V. A. Mineva, G. T. Petrov, V. K. Golev, Z. I. Tsvetanov.
Pis'ma Astron. Zh., Tom 8, 210 - 213 (1982). In Russian. English translation in Soviet Astron. Lett., Vol. 8.

A spectrophotometric investigation of the nucleus of the Arak 144 galaxy is carried out. The relative abundance of S$^+$ ions is by one order of magnitude higher than the mean one for emission objects of different types, while for N$^+$ ions it is equal to the typical one for various galaxies but Sy 1 galaxies.

158.139 Optical structure of the nucleus of M33.
J.-L. Nieto, M. Aurière.
Astron. Astrophys., Vol. 108, 334 - 338 (1982).

High-resolution observations of the nucleus of M33 show the structure of the nucleus: it appears stellar in its center and is surrounded by a faint nebulosity extending no further than 3″. Its profile follows quite reasonably an $r^{1/4}$ law with an effective radius of 1.″4. Fits with isothermal laws give a very small core radius. The total magnitude of the nucleus is $\mu_B = 14.50 \pm 0.10$. A velocity dispersion of $\sigma = 30$ km s^{-1}, gives a maximum mass of $6 \times 10^6 M_\odot$ and a maximum M/L ratio of 3.25.

158.140 The shape and variability of the nonthermal component of the optical spectra of active galaxies.
J. Chołoniewski.
Acta Astron., Vol. 31, 293 - 311 (1981).

The main results of this paper concern the optical spectrum of the nonthermal radiation coming from nuclei of active galaxies. The author shows, that, for type 1 Seyfert galaxies, lacertides and quasars this spectrum does not change its shape despite its strong variability of bolometric flux. This fact permits one to obtain this shape, which is, for type 1 Seyferts, concave with a mean spectral index close to 0.6, and for lacertides fullfills the power law with spectral index close to −1.

158.141 NGC 3067: additional evidence for nonluminous matter?
V. C. Rubin, N. Thonnard, W. K. Ford, Jr.
Astron. J., Vol. 87, 477 - 485 (1982).

Optical and 21-cm observations have been made of the small low-luminosity Sb III galaxy NGC 3067. Beyond the nuclear region, rotational velocities rise slowly to 151 km s^{-1} at R = 7.3 kpc, near the limit of the optical disk. The velocity gradient is 5.2 km s^{-1} kpc^{-1}. The absorption previously detected along the line of sight to the QSO 3C 232 is located near the minor axis of NGC 3067 at a projected distance of almost two galactic radii. If the absorption arises from gas in circular orbit in the plane of NGC 3067, geometrical considerations require that the rotation velocities continue to rise beyond the optical image, increasing to a velocity near 340 km s^{-1} at R = 40 kpc. Such a velocity implies that 94% of the mass is located beyond the optical image; this mass has a ratio of M/L_B greater than 100. Observations at 21 cm at Arecibo have not detected hydrogen clouds at this radial distance, but place an upper limit of 10^7 $M_\odot$ for the mass of a single cloud.

158.142 Photoelectric surface photometry of 19 spiral galaxies. B. C. Whitmore, R. P. Kirshner.
Astron. J., Vol. 87, 500 - 516 (1982).

Photoelectric surface photometry along the major and minor axes of 19 spiral galaxies has been obtained using the 1.3-m telescope at KPNO. A comparison with the standard galaxy NGC 3379 yields an average uncertainty of 0.05 μ_B (where μ_B is defined as mag/arcsec^2 in B) to a limiting surface brightness of 24.5 μ_B. A comparison of one-dimensional scans with Boroson's (1981) "global" profiles shows that most galaxies are symmetric enough to allow a determination of the total luminosity profile to better than 0.1 μ_B from scans of only the major and minor axis.

158.143 High-resolution mapping of the giant H I envelope of the Seyfert galaxy Mkn 348.
T. M. Heckman, R. Sancisi, W. T. Sullivan III, B. Balick.
Mon. Not. R. Astron. Soc., Vol. 199, 425 - 433 (1982).

The authors have mapped the giant H I envelope of Mkn 348 (NGC 262) with a spatial resolution of 1 arcmin and a velocity resolution of 17 km s^{-1} using the Westerbork Synthesis Radio Telescope. They consider hypotheses that the H I envelope (1) represents primordial material gradually being accreted by the galaxy, (2) is produced by ejecta from the active nucleus, or (3) was produced by a tidal interaction several gigayears ago between Mkn 348 and the nearby bright spiral NGC 266.

158.144 High resolution imagery of the clumpy irregular galaxy Markarian 325 = NGC 7673.
G. Coupinot, J. Hecquet, J. Heidmann.
Mon. Not. R. Astron. Soc., Vol. 199, 451 - 455 (1982).

High resolution imagery of the clumpy irregular galaxy Mkn 325 shows that some clumps have sizes $\sim$300 pc while some may still be unresolved at $\lesssim$100 pc. In spite of dimensions comparable to – or even smaller than – those of the giant H II complex 30 Doradus, one clump has a star formation rate 100 times higher.

158.145 Ring galaxies – I. Kinematics of the southern ring galaxy AM 064–741.
J. M. A. Few, B. F. Madore, H. C. Arp.
Mon. Not. R. Astron. Soc., Vol. 199, 633 - 647 (1982).

New photometric and spectroscopic data are presented for the southern ring galaxy AM 064–741. The ring nucleus has a systemic velocity of 6749 ± 67 km s^{-1} and shows rapid rotation both in the emission- and absorption-line spectrum; however, the gas appears to be rotating a factor of three times faster than the stellar component. Twenty-two radial velocity measurements of H II regions around the ring show that the ring is rotating at 311 ± 19 km s^{-1} and expanding/contracting at 128 ± 14 km s^{-1}. A companion near an extension of the minor axis of the ring is identified as the intruder responsible for the ring formation.

158.146 A high-excitation optically obscured H II region in the nucleus of NGC 5253.
D. K. Aitken, P. F. Roche, M. C. Allen, M. M. Phillips.
Mon. Not. R. Astron. Soc., Vol. 199, 31P - 35P (1982).

Spectrophotometry from 8–13 μm of the strong infrared source in the nucleus of the galaxy NGC 5253 has revealed intense emission in the [S IV] fine structure line, and a dust continuum suffering silicate absorption. These observations require either a compact cluster of several hundred early O stars or a single massive highly luminous object, within 50 pc of the nucleus and obscured in the visible, to provide both the infrared luminosity and the ionizing flux. Simple models are discussed.

158.147 Cosmic jets.
R. D. Blandford, M. C. Begelman, M. J. Rees.
Sci. American, Vol. 246, No. 5, p. 84 - 94 (1982).

The violent activity at the center of many galaxies is manifested in the production of narrow, focused streams of ionized gas. Some are a few light-years long; others are a million times longer.

158.148 The most distant known galaxies.
R. G. Kron.
Science, Vol. 216, 265 - 269 (1982).

158.149 Dwergsterrenstelsels. H. Arp.
Zenit, 9. Jaarg., 210 - 211 (1982).

158.150 The central velocity dispersion in elliptical and lenticular galaxies as an extragalactic distance indicator. G. de Vaucouleurs, D. W. Olson.
Astrophys. J., Vol. 256, 346 - 369 (1982).

The Faber-Jackson relation between absolute magnitude M_T^0 and central velocity dispersion σ_v is reexamined for a sample of 157 normal, noninteracting galaxies, 82 ellipticals and 75 lenticulars. The values of σ_v are weighted means from various sources reduced to a uniform system. Absolute magnitudes are calculated from corrected apparent magnitudes B_T^0 and distance moduli μ_0 which are weighted means of a variety of independent determinations (from optical tertiary indicators and from revised versions of the Tully-Fisher relation) for spirals associated with ellipticals or lenticulars in our sample; group distance moduli derived from redshifts were also used in some cases. The slope and zero point of the one-parameter F-J relation were derived by Jefferys's generalized least-squares technique for different subsamples. The effects of second-order parameters including color index, surface brightness, and apparent axis ratio are investigated. Small but significant effects of the first two are indicated. Generalized expressions of the F-J relation including such terms are given. Weighted mean distance moduli are given for the 157 galaxies in the sample. The mean errors of these distance moduli from all sources are of the order $\sim$0.5 mag (E) and $\sim$0.6 mag (L), depending on the precision of the input parameters.

158.151 X-ray observations of peculiar galaxies with the Einstein Observatory.
G. Fabbiano, E. Feigelson, G. Zamorani.
Astrophys. J., Vol. 256, 397 - 409 (1982).

The authors report the results of an X-ray survey with the Einstein Observatory of 33 galaxies, mostly spiral and irregular. They are chosen to have peculiar morphologies, colors, and in some cases, emission-line spectra. Thirteen galaxies were detected with (0.5–3 keV) X-ray luminosities ranging between 10^{39} and 10^{42}/ ergs s^{-1}. In some cases, extended or multiple emission regions were detected. The X-ray and optical fluxes and luminosities appear correlated. An even stronger correlation is found between X-ray and radio fluxes and luminosities. The X-ray to optical flux ratios for most of these galaxies are substantially higher than that of M31 but still significantly lower than those of the Seyfert galaxies. Several possible origins of the X-ray emission are discussed. The authors conclude that Seyfert-like nuclear emission is not the dominant mechanism of X-ray production. Comparing their results with those of the X-ray observations of our Galaxy and of the Magellanic Clouds, they find that the X-ray emission can be integrated emission from supernova remnants and Population I binary X-ray sources.

158.152 Obscuration and the various kinds of Seyfert galaxies.
A. Lawrence, M. Elvis.
Astrophys. J., Vol. 256, 410 - 426 (1982).

For a complete hard X-ray selected sample of active spiral galaxies the authors have collected X-ray luminosities, line fluxes in Hβ, [O III] λ5007, and 3.5 μm fluxes. Most of the X-ray discovered objects are "narrow emission line galaxies". Selection effects in the sample are analysed and shown to be caused by interstellar obscuration in or just around the line emitting regions. A larger sample of active galaxies of heterogeneous origin is then considered and a correlation analysis for various observational parameters is performed. Several newly found correlations are described. A key parameter is the hard X-ray luminosity. For instance, type 1 Seyferts, intermediate Seyferts, narrow emission line galaxies, and type 2 Seyferts seem to form a continuous sequence of decreasing intrinsic luminosity. It seems likely that there is only one kind of Seyfert galaxy. The observed range of qualitative appearances may be explained by a photo-ionization model if the structure of the emission line regions is a function of direction and intrinsic luminosity.

158.153   Emission-line widths in galactic nuclei.
F. R. Feldman, D. W. Weedman, V. A. Balzano, L. W. Ramsey.
Astrophys. J., Vol. 256, 427 - 434 (1982).

Line width measures for the [O III] emission lines have been obtained for 116 galactic nuclei with instrumental reso-lution of 165 km s^{-1}. Galaxies observed include 53 Seyfert 1 galaxies, 16 Seyfert 2 galaxies, and 47 "star-burst" galaxies for which the nuclear emission lines are attributed to ioniza-tion by hot, short-lived stars. The median FWHM for the [O III] lines is 375 km s^{-1} for Seyfert 1, 510 km s^{-1} for Seyfert 2, and 160 km s^{-1} for star-burst nuclei. It is suggested that an empirical criterion for dividing Seyfert galaxies from other emission line galaxies is a FWHM > 250 km s^{-1} for the [O III] lines. From dynamical arguments, the star-burst nuclei are determined to have the star formation activity restricted to a nuclear disk with radius less than a few hundred parsecs.

158.154   The dynamics of the nucleus of M31.
S. Tremaine, J. P. Ostriker.
Astrophys. J., Vol. 256, 435 - 446 (1982).

The authors construct self-consistent dynamical models of the nucleus of M31 which satisfy Stratoscope II and ground-based photometry, as well as ground-based velocity dispersion measurements. They find that the nuclear mass-to-light ratio M/L_B is very poorly determined and can take on any value between 50 and 0 in solar units. At present the only informa-tion is that the similarity of colors and line indices in the nucleus and bulk suggests that M/L_B is of order $(M/L_B)_{bulge} = 7 M_\odot/L_\odot$ (for a bulge dispersion of 180 km s^{-1}). Possible evolutionary scenarios are presented which could produce low M/L nuclei composed of stars on nearly radial orbits. The results illustrate that unless accurate velocity dispersion and photometric profiles are available, great caution must be exercised when determining mass-to-light ratios or inferring the presence of a central dark object.

158.155   Rotation of the bulge components of disk galaxies.
J. Kormendy, G. Illingworth.
Astrophys. J., Vol. 256, 460 - 480, plate 17 (1982).

Absorption-line velocities and velocity dispersions have been measured along several slit positions in the bulge compo-nents of the edge-on S0–Sb galaxies NGC 4565, 4594, 5866, and 7814. These velocity fields, together with available data on four other ocjects, confirm indications from photometric studies that there exist significant dynamical differences be-tween bulges and elliptical galaxies. The present bulges rotate more rapidly than bright ellipticals. Their kinematic properties are strikingly uniform and, in many cases, are well described by dynamical models of rotationally flattened oblate spheroids with isotropic residual velocities. The observation that bulges rotate rapidly suggests that they were formed by a dissipational process. In general, the detailed behavior of the two-dimensional velocity fields is consistent with classical dissipational collapse theories of galaxy formation.

158.156   Velocity and velocity dispersion profiles in NGC 3115.   G. Illingworth, P. L. Schechter.
Astrophys. J., Vol. 256, 481 - 496, plate 18 (1982).

An extensive series of measurements has been made of the absorption-line velocity and velocity dispersion field in the edge-on S0 galaxy NGC 3115. Major- and minor-axis velocity and dispersion profiles are presented along with profiles from spectra parallel to the principal axes but offset from the nucleus. In contrast to the case for bright elliptical galaxies, the authors find that the bulge of NGC 3115 shows sufficient rotation to be nearly consistent with being a rotationally flattened oblate spheroid with isotropic residual velocities. The kinematic properties of the bulge of NGC 3115 are similar to those of M31 and M81.

158.157   Multicolour photometry of the galaxy Markarian 141, which is a component of a double system.
L. P. Metik.
Izv. Krymskoj Astrofiz. Obs., Tom 64, 98 - 103 (1981). In Russian. English translation in Bull. Crimean Astrophys. Obs., Vol. 64.

Multicolor photometry of the galaxy Markarian 141, which is a component of a double system, has been carried out. Stellar magnitudes of the central bodies of both components of the double system have been obtained. Colour characteristics of the central part of the galaxy Markarian 141 are rather similar to those of central parts of Seyfert galaxies, while its halo and the central part of the neighbour A-galaxy are alike to that of the central parts of normal galaxies.

158.158   A surface photometry of nearby galaxies: M106, M31 and M33.   H. B. Ann, K. L. Yu.
J. Korean Astron. Soc., Vol. 14, 1 - 11 (1981).

Photoelectric drift scans of nearby galaxies, M106, M31 and M33 have been made at diurnal rate with the 61 cm Cassegrain Reflector at Sobacksan Observing Station. Luminosity profiles of M106 and M31 show the asymmetries between east and west sides of the galaxies and the near side of each galaxy exhibits a larger B−V color than the far side. B−V color distribution in the central part of M106 shows somewhat unusual feature of a blue center with red sur-rounding regions, and this is an opposite trend to the ordinary color distribution of most of external galaxies.

158.159   The first galaxies.   B. Parker.
Astronomy, Vol. 9, No. 11, p. 94 - 99 (1981).
Abstr. in Phys. Abstr., Vol. 85, Abstr. 48326 (1982).

158.160   The Ursa Minor dwarf galaxy.
R. A. Schommer, E. W. Olszewski, K. M. Cudworth.
Astrophysical parameters for globular clusters, (see 012.023), p. 453 - 459 (1981).

158.161   NGC 4258: a bent jet in a spiral galaxy.
R. H. Sanders.
Extragalactic radio sources, (see 012.025), p. 145 - 147 (1982).

158.162   Seyfert galaxies.   A. S. Wilson.
Extragalactic radio sources, (see 012.025), p. 179 - 188 (1982).

Observations of a sample of Markarian Seyferts with the VLA indicate that a large fraction possess linear radio structure on a scale of a few hundred parsecs to a few kilo-parsecs. The radio components generally straddle the optical nucleus and several sources are simple doubles. Similar struc-tures are seen in the classical Seyferts NGC 1068, 4151, and 5548. NGC 4151 is probably best interpreted as a jet. A few sources exhibit diffuse, non-aligned radio structure on a scale similar to that of the linear sources. The radio axis in linear sources is misaligned with respect to the rotation axis of the galaxy disc by a large angle. The linear sources are discussed in terms of a model of a supersonic beam or jet which is "disrupted" by interaction with interstellar gas in the inner part of the galaxy (often a spiral).

158.163   Radio continuum observations of the nuclei of nearby galaxies.
R. D. Davies, A. Pedlar, R. V. Booler.
Extragalactic radio sources, (see 012.025), p. 189 - 190 (1982).

A survey has been carried out at Jodrell Bank of the continuum radio emission from nearby galaxies. The objects include normal galaxies, Seyfert galaxies and others with nuclei active at optical and radio wavelengths. Observations of the Seyfert galaxy NGC 4151 at 1665 MHz (beam size 0.25") are discussed in some detail.

158.164   Radio emission from the Seyfert galaxy NGC 5548.
J. S. Ulvestad, A. S. Wilson, D. G. Wentzel.
Extragalactic radio sources, (see 012.025), p. 191 - 192 (1982).

Weak radio emission from the type 1.5 Seyfert galaxy NGC 5548 has been mapped with high resolution at the VLA at both 1465 and 4885 MHz. The galaxy contains the largest (5.9 kpc) triple radio source known in a Seyfert galaxy. The central component of that triple is unresolved and has a flatter spectrum than the well-resolved outer lobes.

158.165   Radio observations of Markarian 8.
D. S. Heeschen, J. Heidmann, Q. F. Yin.
Extragalactic radio sources, (see 012.025), p. 195 - 196 (1982).

Markarian 8, a clumpy irregular galaxy (Casini et al. 1979), has been observed with the VLA at 20 cm and 6 cm wavelengths.

158.166   Infrared observations of radio galaxies.
G. H. Rieke.
Extragalactic radio sources, (see 012.025), p. 239 - 246 (1982).

Recent results of infrared studies of active galaxies are reviewed. Three classes of infrared galaxies are discussed in detail: (1) galaxies undergoing a powerful burst of star formation, (2) intermediate type Seyfert galaxies, (3) an extreme species of extragalactic radio sources with strong IR radiation.

158.167   Galactic centers and twin-jets.   W. Kundt.
Extragalactic radio sources, (see 012.025), p. 265 - 268 (1982).

A schematic model is suggested to describe the various activities in the centers of massive galaxies. Basic assumptions are that (1) a uniform model can account for all the observed phenomena, such as quasars, blazars, radio galaxies, Seyferts, and the centers of normal galaxies (with disks), (2) activity in galactic centers is repetitive, and (3) in-situ-acceleration of highly relativistic electrons (other than by adiabatic compression) has an insignificant efficiency.

158.168   Detection of a broad H I absorption feature at 5300 km sec^{-1} associated with NGC 1275 (3C84).   P. C. Crane, J. M. van der Hulst, A. D. Haschick.
Extragalactic radio sources, (see 012.025), p. 307 - 308 (1982).

Observations of NGC 1275 at $\sim$1396 MHz with the NRAO line interferometer in 1974 and 1976 suggest the presence of a very broad, shallow H I absorption feature centered at $\sim$5300 km sec^{-1}. These observations were repeated in 1981 June with the Very Large Array using a greater bandwidth to determine a satisfactory baseline.

158.169   A search for H I in elliptical galaxies with nuclear radio sources.   L. L. Dressel, T. M. Bania, R. W. O'Connell.
Extragalactic radio sources, (see 012.025), p. 309 - 310

(1982).

The authors have observed twelve other elliptical galaxies with nuclear radio power $P_{2380} > 10^{22}$ WHz^{-1} at Arecibo Observatory, to determine whether a large mass of H I is a necessary auxillary to nuclear continuum emission.

158.170   Optical spectra of radio-loud and radio-quiet active galactic nuclei.   D. E. Osterbrock.
Extragalactic radio sources, (see 012.025), p. 369 - 371 (1982).

158.171   BL Lac objects and their associated galaxies.
D. Weistrop.
Extragalactic radio sources, (see 012.025), p. 377 - 382 (1982).

CCD photometry of five BL Lac objects indicates that at least three, and possibly four, are located at the centers of giant elliptical galaxies. The redshift for one of these objects, 1218+304, is estimated. A lower limit is placed on the redshift of 1219+28, for which no associated galaxy has been detected. Separation of the galaxy emission from the total observed flux makes possible comparison of the optical-far red flux from the point source alone with radio and X-ray data. This comparison suggests the emission from 1727+50 and 1218+304 can be interpreted as due to direct synchrotron emission.

158.172   Mark III VLBI observations of the nucleus of M81 at 2.3 and 8.3 GHz.   N. Bartel, B. E. Corey, I. I. Shapiro, A. E. E. Rogers, A. R. Whitney, D. A. Graham, J. D. Romney, R. A. Preston.
Extragalactic radio sources, (see 012.025), p. 387 - 388 (1982).

M81 is the nearest extragalactic object with a nucleus detectable with VLBI. The authors report here on simultaneous VLBI observations made with the Mark III system at 2.3 and 8.3 GHz.

158.173   Redshift estimates for distant radio galaxies based on broadband photometry.
J. J. Puschell, F. N. Owen, R. Laing.
Extragalactic radio sources, (see 012.025), p. 423 - 424 (1982).

The authors have begun a program of JHK photometry of elliptical galaxies with high measured redshifts and optically identified and unidentified faint steep-spectrum radio sources believed to be elliptical galaxies, using the NASA 3-m IRTF at Mauna Kea.

158.174   Colors of radio galaxies at high redshifts.
R. A. Windhorst, R. G. Kron, D. Koo, P. Katgert.
Extragalactic radio sources, (see 012.025), p. 427 - 431 (1982).

The authors present the first results of the Westerbork-Berkeley Deep Survey, the purpose of which is to derive the epoch dependence of the radio luminosity function and the optical spectral energy distribution of elliptical radio galaxies.

158.175   Observations of the Markarian galaxies 673 and 686.
A. R. Petrosyan, K. A. Saakyan, Eh. E. Khachikyan.
Pis'ma Astron. Zh., Tom 8, 270 - 274 (1982). In Russian.
English translation in Soviet Astron. Lett., Vol. 8.

Results of spectrophotometric and isodensitometric investigation of Markarian 673 and 686 having comparatively broad emission spectral lines are presented. Belonging of these galaxies to either type of objects with broad emission lines is discussed.

158.176   Redshifts of 16 Markarian galaxies.
D. M. Crenshaw, B. M. Peterson, C. B. Foltz, P. L. Byard.
Publ. Astron. Soc. Pacific, Vol. 94, 16 - 18 (1982).

New redshift measurements are given for 16 emission-line galaxies from the lists of Markarian and Lipovetsky. Data on two pairs of Markarian galaxies are presented, and estimates of the minimum mass of these systems are made.

158.177 Transformation of Zwicky magnitudes to the Holmberg and the B_T magnitude systems.
J. R. Auman, P. Hickson, G. G. Fahlman.
Publ. Astron. Soc. Pacific, Vol. 94, 19 - 25 (1982).

The photographic magnitudes given in the Zwicky Catalogue are found to have systematic errors that depend primarily on the mean surface brightness of the galaxy. Transformation formulae are given that convert the Zwicky magnitudes to the Holmberg system and to the B_T system with a significantly lower scatter than previously published transformation formulae give.

158.178 Search for redshifted CH_2O, H_2O, O_2, and NH_3 in Seyfert galaxies and quasars.
D. N. Matsakis, J. M. Bologna, P. R. Schwartz, D. L. Thacker.
Publ. Astron. Soc. Pacific, Vol. 94, 26 - 30 (1982).

A search for molecular spectral features at redshifts of optical and 21-cm lines offers some constraints on the intervening galaxy hypothesis: the CH_2O opacity toward 3C 120 and 3C 84 are less than expected from the arms of a Milky Way type galaxy and the existence of any strong H_2O masers between 0.13 and 4 kpc in front of 3C 84 is excluded. A possible detection of O_2 in 3C 84 is reported.

158.179 A note on the spectra of the nuclear "hot spots" of NGC 1097. D. L. Talent.
Publ. Astron. Soc. Pacific, Vol. 94, 36 - 39 (1982).

Spectra covering the wavelength range $\lambda\lambda 3700-4400$ are presented of the "hot spots" and amorphous nuclear region of NGC 1097. From an examination of the ratios of equivalent widths of selected absorption features it is determined that the light from the hot spots is dominated by stars of spectral types F0 to F4, while the light from the amorphous nuclear region is much redder and dominated by stars of spectral types K1 to K9.

158.180 Gas and dust in radio galaxies.
B. V. Komberg.
Zemlya Vselennaya, 1982, No. 2, p. 13. In Russian.

158.181 Análisis de la radiación proveniente de la región central de NGC 5253.
M. G. Pastoriza, H. A. Dottori.
Bol. Asoc. Argentina Astron., No. 20 - 24, p. 165 (1981). Abstract.

158.182 Estudio cinemático de NGC 1313.
G. Carranza, E. Aguero.
Bol. Asoc. Argentina Astron., No. 20 - 24, p. 166 (1981). Abstract.

158.183 Fotometría UBV de la región central de NGC 5236.
H. A. Dottori.
Bol. Asoc. Argentina Astron., No. 20 - 24, p. 167 (1981). Abstract.

158.184 Estudio interferencial de NGC 7793.
E. Aguero, G. Carranza.
Bol. Asoc. Argentina Astron., No. 20 - 24, p. 265 (1981). Abstract.

158.185 La región nuclear de NGC 5236.
G. Carranza.
Bol. Asoc. Argentina Astron., No. 20 - 24, p. 266 (1981). Abstract.

158.186 La relación gas-polvo en M31.
T. E. Gergely, E. Bajaja.

Bol. Asoc. Argentina Astron., No. 20 - 24, p. 267 (1981). Abstract.

158.187 Correlación entre las distribuciones de H I y de regiones H II en M31. E. Bajaja.
Bol. Asoc. Argentina Astron., No. 20 - 24, p. 362 (1981). Abstract.

158.188 Observación de hidrógeno neutro en galaxias.
E. Bajaja.
Bol. Asoc. Argentina Astron., No. 20 - 24, p. 366 (1981). Abstract.

158.189 Determinación de la curva de rotación de M 31 a lo largo del eje menor.
E. Bajaja, S. Blacher.
Bol. Asoc. Argentina Astron., No. 20 - 24, p. 436 (1981). Abstract.

158.190 Observación de H I en galaxias. E. Bajaja.
Bol. Asoc. Argentina Astron., No. 20 - 24, p. 438 (1981). – Abstract.

158.191 Regiones de emisión en algunas galaxias barreadas.
E. Aguero, G. Carranza.
Bol. Asoc. Argentina Astron., No. 20 - 24, p. 487 (1981). Abstract.

158.192 X-ray observations with the Einstein Observatory of emission-line galaxies.
T. Maccacaro, G. C. Perola, M. Elvis.
Astrophys. J., Vol. 257, 47 - 55 (1982).

X-ray observations of narrow-emission-line galaxies are presented and discussed. One source, NGC 1365, is found to be extended in the soft X-ray band; three others, NGC 2992, NGC 5506, and NGC 7582, have been observed to vary in intensity. The best fit spectral index and cutoff energy E_a are derived for NGC 2992, NGC 5506, and NGC 7582. The X-ray spectra of these galaxies are similar to those of type 1 Seyfert galaxies. In the case of NGC 5506 and NGC 7582, the absorbing column N_H derived is about one order of magnitude greater than predicted from the reddening of the optical continuum and of the Balmer lines. Possible explanations for the discrepancy are discussed.

158.193 Rotation of the bulge components of barred galaxies. J. Kormendy.
Astrophys. J., Vol. 257, 75 - 88 (1982).

Stellar rotation and velocity-dispersion measurements are presented for the bulge components SB0 galaxies NGC 1023, 2859, 2950, 4340, 4371, and 7743. The kinematics of nine SB bulges with data available are compared with bulges of unbarred galaxies studied by Kormendy and Illingworth. All of the SB bulges are found to rotate at least as rapidly as oblate-spheroid dynamical models which are flattened by rotation. This result confirms the conclusion of Kormendy and Illingworth that bulges rotate very rapidly. Six SB bulges found by Kormendy and Koo to be triaxial rotate even more rapidly than the oblate models. The kinematic observations imply that triaxial bulges appear to have been formed with more dissipation than ordinary bulges. These results are consistent with the hypothesis that part of the bulge in many SB galaxies consists of disk material (i.e., gas) which has been transported to the center by the bar.

158.194 Metallicity versus age in spiral nuclei.
R. W. O'Connell.
Astrophys. J., Vol. 257, 89 - 93 (1982).

It has recently been proposed that the blue nuclei of late-type spiral galaxies are old, metal-poor stellar systems in which the horizontal branch dominates the light. This is contrary to the conventional interpretation which holds that the blueness results from the presence of a substantial young (<1 Gyr) stellar population. The author shows that straight-forward comparison of narrow-band scanner colors for Sc nuclei and globular clusters in the Local Group excludes the possibility that the nuclei are old, metal-poor systems. Photometric precisions higher than 5% are in general necessary for such a distinction to be made, however.

158.195 Infrared speckle interferometry of the nucleus of NGC 1068.
D. W. McCarthy, F. J. Low, S. G. Kleinmann, F. C. Gillett. Astrophys. J., Lett., Vol. 257, L7 - L11 (1982).

At 2.26 μm nuclear emission from the Seyfert galaxy NGC 1068 is dominated by a "nuclear core" $\leqslant 0\overset{''}{.}2$ ($\leqslant 20$ pc) in diameter. Within the area scanned, $5\overset{''}{.}2 \times 10\overset{''}{.}5$, the total flux measured is 0.57 ± 0.06 Jy with 0.43 ± 0.05 Jy contained in the unresolved core.

158.196 The extended radio source in the center of M31.
R. M. Hjellming, L. L. Smarr. Astrophys. J., Lett., Vol. 257, L13 - L18 (1982).

Radio observations of the central $1° \times 1°$ region of the Andromeda Nebula (M31), made with the 1 km configuration of the NRAO VLA, show that at 1465 MHz the central radio source is nearly spherical and 2 kpc (10') in diameter, with a brightness distribution roughly proportional to $\exp(-R/1\overset{'}{.}1)$, where R is the angular distance from the radio maximum located 20" from the nucleus at a position angle of 70°. The broad structure of this object may be due to a bulge population of many weak supernova remnants or to a cosmic-ray electron-dominated galactic wind.

158.197 Variability of the galaxy Markarian 699.
K. A. Saakyan. Soobshch. Byurakan Obs., Vyp. 53, p. 99 - 101 (1982). In Russian.

The results of photometric measurements of Markarian 699 are given. The observed amplitude of brightness variation is about $0\overset{m}{.}3 - 0\overset{m}{.}4$.

158.198 High-resolution observations of M87. I. The morphology of the jet.
J.-L. Nieto, G. Lelièvre. Astron. Astrophys., Vol. 109, 95 - 100 (1982).

High-resolution (FWHM $\sim 0\overset{''}{.}5 - 1\overset{''}{.}0$) UBV data obtained with the CFH Telescope and at Pic-du-Midi Observatory allow an accurate description of the morphology of the optical M87 jet. (1) The jet consists of a large number of discrete knots which are grouped in "complexes". (2) The jet is perfectly aglined between the nucleus and its brightest part, after which it starts oscillating with an increasing amplitude. (3) The complexes, including the nucleus, are spaced regularly along the jet with an approximate period of $2\overset{''}{.}9$. (4) The optical jet and the radio jet have strikingly similar morphologies.

158.199 Cinemática de NGC 4945.
G. J. Carranza, E. L. Aguero. Bol. Asoc. Argentina Astron., No. 26, p. 119 (1981). Abstract.

158.200 H I en galaxias australes.
E. Bajaja, M. C. Martín. Bol. Asoc. Argentina Astron., No. 26, p. 120 - 122 (1981).

158.201 H I in NGC 4594.   E. Bajaja, W. W. Shane.
Bol. Asoc. Argentina Astron., No. 26, p. 123 (1981).

158.202 Resolution in normal galaxies.   G. A. Tammann.
Scientific importance of high angular resolution at infrared and optical wavelengths, (see 012.031), p. 405 - 409 (1981).

158.203 Optical observations of active galaxies and quasars at high angular resolution.   M. H. Ulrich.
Scientific importance of high angular resolution at infrared and optical wavelengths, (see 012.031), p. 411 - 421 (1981).

The author evaluates the baseline which is necessary to resolve the broad line region in the nucleus of the Seyfert galaxy NGC 4151 and discusses the conditions under which the broad line region of distant active nuclei and quasars can be resolved.

158.204 Highly compact structures in galactic nuclei and quasars.   M. J. Rees.
Scientific importance of high angular resolution at infrared and optical wavelengths, (see 012.031), p. 423 - 432 (1981).

The paper deals with three topics. First, the author mentions some ways in which our ideas about quasars can be clarified by observations with 0.1" resolution — the kind of level that can be expected from the Space Telescope. Second, if milli-arc-second resolution were attainable, there would be the prospect of studying some features of active nuclei that may correlate with VLBI radio data, where this angular precision has already been achieved. As a third and final comment, the author mentions how the apparent shapes of quasars distorted and magnified by the gravitational lens effect can offer clues to galactic structure and the nature of the "hidden mass" in the Universe.

158.205 Faint Seyfert-type galaxies. I.
B. E. Markarian, D. A. Stepanyan, V. A. Lipovetskij. Astron. Tsirk., No. 1168, p. 2 - 5 (1981). In Russian.

158.206 Some spectrophotometric data about the double galaxy NGC 3690 + IC 694.
I. M. Yankulova, G. T. Petrov, V. K. Golev. Astron. Tsirk., No. 1169, p. 1 - 3 (1981). In Russian.

158.207 On the rotation measures in the knots of the M87 jet.   V. N. Kuril'chik.
Astron. Tsirk., No. 1170, p. 2 - 4 (1981). In Russian.

158.208 Excess of L_c-radiation and hot stars in interacting galaxies and deficiency of them in isolated ones.
V. A. Dostal'. Astron. Tsirk., No. 1170, p. 4 - 6 (1981). In Russian.

158.209 Photored magnitudes of central regions of interacting galaxies.
A. S. Amirkhanyan, V. A. Dostal'. Astron. Tsirk., No. 1171, p. 4 - 7 (1981). In Russian.

158.210 Seyfert galaxies in Abell clusters.
M. A. Arakelyan, V. Yu. Terebizh. Astron. Tsirk., No. 1188, p. 1 - 3 (1981). In Russian.

158.211 Quantum effects in the redshift intervals for double galaxies.   W. G. Tifft.
Astrophys. J., Vol. 257, 442 - 449 (1982).

An improved sample of 31 double galaxies with accurate radio redshifts is defined and analyzed for the 72 km s^{-1} discrete redshift interval effect. The effect is present at the 99.8% confidence level. Some doubles show significant deviations from the principal peaks and are consistent with a population of small peaks midway between (i.e., at 36 km s^{-1} intervals). The width of the main peaks is slightly but not significantly larger than expected from observational error. Underlying natural width or structure in the main peaks cannot introduce a scatter much in excess of 6 km s^{-1}. A sample of 30 new accurate optical redshift differentials shows the same periodicity.

158.212 A color-magnitude relation for spiral galaxies.
R. B. Tully, J. R. Mould, M. Aaronson.

Astrophys. J., Vol. 257, 527 - 537 (1982).

Tight correlations are found between blue-to-infrared colors and either the H I line profile widths (masses) or the intrinsic luminosities of spiral galaxies. The correlations are derived from samples of galaxies which range in type from Sa to Im. Since colors are distance independent, the color-magnitude correlation can be used as a measuring stick. A distance modulus of 31.04±0.16 mag is found for the Virgo Cluster, if the Sandage-Tammann distance scale for nearby galaxies is adopted. The tight relationships between color and mass or luminosity, essentially independent of galaxy type, suggest that the Hubble morphological sequence is predominantly dependent on a single parameter: total mass. The color-magnitude diagram can be qualitatively understood if the initial specific star formation rate in spirals decreases with decreasing mass and, at the same time, chemical abundances decrease and/or the initial mass function flattens with decreasing mass. Spiral and lenticular galaxies inhabit quite separate regions on the color-magnitude diagram. If galaxies evolve between the two branches, the probability seems to be low of viewing them in the intermediate state.

158.213 **Echelle spectrometry of the Seyfert galaxies NGC 3783 and Markarian 509.**
B. Atwood, J. A. Baldwin, R. F. Carswell.
Astrophys. J., Vol. 257, 559 - 569 (1982).

Echelle spectrometry of two Seyfert galaxies, NGC 3783 and Markarian 509, is used to investigate the narrow and broad emission-line profiles. In NGC 3783 it is found that while the narrow emission lines have smooth profiles, they must be the summation of a number of components having different electron densities and/or temperatures. In Markarian 509 the [S II] and [N II] line profiles are similar to those of [O III], suggesting that the physical conditions as a function of velocity are more homogeneous than for NGC 3783. The broad Hα and Hβ lines in both galaxies have approximately identical profiles. For Mrk 509, a cross-correlation of the two profiles indicates that the number of optically thin unresolved clouds needed to make up the emission line region is $\gtrsim 5 \times 10^4$.

158.214 **Near-infrared spectrophotometry of four Seyfert 1 galaxies and NGC 1275.**
R. J. Rudy, B. Jones, P. D. LeVan, R. C. Puetter, H. E. Smith, S. P. Willner, A. T. Tokunaga.
Astrophys. J., Vol. 257, 570 - 577 (1982).

Low-resolution spectrophotometry from 2 to 4 μm is reported for the four Seyfert 1 galaxies Mrk 335, 3C 120, Mrk 509, NGC 7469, and the peculiar emission-line galaxy NGC 1275. The spectrum of NGC 7469 exhibits a strong 3.3 μm dust feature, indicating a thermal origin for the bulk of its considerable nonstellar infrared emission. NGC 1275 has a large stellar contribution to its infrared flux at wavelengths shortward of 3 μm. A thermal model which can explain the spectrum of NGC 1275 is discussed. Mrk 335 displays a complex spectrum suggestive of thermal dust emission. 3C 120 and Mrk 509 have nonstellar infrared emission shortward of 2 μm, but the data are ambiguous as to whether this emission is thermal or nonthermal in origin.

158.215 **Near-infrared photometry of distant radio galaxies: spectral flux distributions and redshift estimates.**
J. J. Puschell, F. N. Owen, R. A. Laing.
Astrophys. J., Lett., Vol. 257, L57 - L61 (1982).

The authors have carried out JHK photometry of optically faint radio source identifications believed to be distant elliptical galaxies in order to test the hypothesis of negligible evolution of elliptical galaxies in the red and near-infrared and to determine photometric estimates of redshifts. Whereas the spectral flux distributions are consistent with current models of galaxy evolution, photometric redshifts are systematically underestimated. Galaxies showing an anomalous flux distribu-

tion in the optical region are found to have also anomalous infrared colors. Possible explanations of this surprising result are discussed.

158.216 **Galaxy morphology, starclusters and the mass spectrum of star formation.** S. van den Bergh.
The most massive stars, (see 012.034), p. 253 - 260 (1981).

158.217 **The initial mass function of massive stars and the evolution of galaxies.** J. Lequeux.
The most massive stars, (see 012.034), p. 261 - 267 (1981).

This paper reviews the present knowledge of the initial mass function (IMF) of massive stars in the Galaxy and in external galaxies. Massive stars are considered with $M > 20 M_\odot$. Their effects on the evolution of galaxies with respect to total luminosity, heavy element production, dynamics of the interstellar matter, supernova rate, are briefly discussed.

158.218 **The heating of dust in the broad line regions of active galaxies and quasars.**
R. J. Rudy, R. C. Puetter.
Bull. American Astron. Soc., Vol. 13, 789 (1981). – Abstract.

158.219 **The appearance of broad emission lines in the optical spectrum of a BL Lac Object.**
M. H. Ulrich.
Bull. American Astron. Soc., Vol. 13, 789 (1981). – Abstract.

158.220 **Broad-emission-line profiles in Seyfert 1 galaxies.**
J. M. Shuder.
Bull. American Astron. Soc., Vol. 13, 790 (1981). – Abstract.

158.221 **Changes in the optical spectra of Seyfert 1 galaxies.**
B. M. Peterson, C. B. Foltz, E. R. Capriotti.
Bull. American Astron. Soc. Vol. 13, 790 (1981). – Abstract.

158.222 **Time-dependent models of broad emission-line profiles of Seyfert 1 galaxies.**
E. R. Capriotti, C. B. Foltz, B. M. Peterson.
Bull. American Astron. Soc., Vol. 13, 790 (1981). – Abstract.

158.223 **Prominent UV emission lines in Seyfert galaxies.**
C.-C. Wu, A. Boggess, T. R. Gull.
Bull. American Astron. Soc., Vol. 13, 790 (1981). – Abstract.

158.224 **Numerical simulations of seeing effects in the cores of elliptical galaxies.** S. Djorgovski.
Bull. American Astron. Soc., Vol. 13, 796 - 797 (1981). Abstract.

158.225 **Photometry on the giant branch of the dwarf ellipticals in the Local Group.**
J. R. Mould, J. A. Kristian, G. S. da Costa.
Bull. American Astron. Soc., Vol. 13, 797 (1981). – Abstract.

158.226 **Color gradients in low luminosity galactic spheroids.** A. Wirth, J. S. Gallagher.
Bull. American Astron. Soc., Vol. 13, 797 (1981). – Abstract.

158.227 **Orthogonal gas disks in elliptical, barred spiral, and Seyfert galaxies.**
J. E. Tohline, R. H. Durisen, D. E. Osterbrock.
Bull. American Astron. Soc., Vol. 13, 798 (1981). – Abstract.

158.228 **The distribution of mass in Sc galaxies.**
D. Burstein, V. C. Rubin, N. Thonnard, W. K. Ford, Jr.
Bull. American Astron. Soc., Vol. 13, 798 (1981). – Abstract.

158.229 **X-ray maps of peculiar galaxies.**
G. Trinchieri, G. Fabbiano.
Bull. American Astron. Soc., Vol. 13, 799 (1981). − Abstract.

158.230 **Photometry of faint galaxies and the motion of the Milky Way Galaxy.**
C. W. Baumgart, C. J. Peterson.
Bull. American Astron. Soc., Vol. 13, 800 (1981). − Abstract.

158.231 **The barred galaxy NGC 3504.** C.J.Peterson.
Bull. American Astron. Soc., Vol. 13, 801 (1981).
Abstract.

158.232 **Transform analysis of the high resolution Shane-Wirtanen catalog: the power spectrum and the bi-spectrum.**   J. N. Fry, M. Seldner.
Bull. American Astron. Soc., Vol. 13, 806 (1981). − Abstract.

158.233 **The angular correlation function at one minute of arc as a function of magnitude.**   W. L. Sebok.
Bull. American Astron. Soc., Vol. 13, 806 - 807 (1981).
Abstract.

158.234 **VLA detection of the H 110α line in M82.**
E. R. Seaquist, M. B. Bell, R. C. Bignell.
Bull. American Astron. Soc., Vol. 13, 807 (1981). − Abstract.

158.235 **On the short timescale variability of NGC 6814.**
J. H. Beall, K. S. Wood, D. J. Yentis.
Bull. American Astron. Soc., Vol. 13, 822 (1981). − Abstract.

158.236 **Multifrequency observations of 0735 + 178 and IZw 187.**
A. E. Glassgold, J. N. Bregman, P. J. Huggins.
Bull. American Astron. Soc., Vol. 13, 822 (1981). − Abstract.

158.237 **Infrared polarimetry and photometry of BL Lac objects.**
C. D. Impey, P. W. J. L. Brand, R. D. Wolstencroft,
P. M. Williams.
Bull. American Astron. Soc., Vol. 13, 823 (1981). − Abstract.

158.238 **Galaxy contributions to the optical spectra of Cyg A and Mrk 744.**   D. E. Osterbrock.
Bull. American Astron. Soc., Vol. 13, 824 (1981). − Abstract.

158.239 **Interpretation of the redshift distribution of X-ray selected active galactic nuclei.**
G. Reichert, K. O. Mason, S. Bowyer.
Bull. American Astron. Soc., Vol. 13, 847 - 848 (1981).
Abstract.

158.240 **Two X-ray-selected BL Lacertae-type objects.**
G. A. Chanan, B. Margon, D. J. Helfand,
R. A. Downes, D. Chance.
Bull. American Astron. Soc., Vol. 13, 848 (1981). − Abstract.

158.241 **X-ray spectra of active galactic nuclei with the Einstein Observatory.**
J. P. Halpern, J. E. Grindlay.
Bull. American Astron. Soc., Vol. 13, 849 (1981). − Abstract.

158.242 **X-ray observations of emission-line galaxies with the Einstein Observatory.**
R. E. Griffiths, J. Huchra, M. Davis.
Bull. American Astron. Soc., Vol. 13, 849 (1981). − Abstract.

158.243 **Observations of NGC 4151 at 2 keV - 2 MeV from HEAO-1.**
W. A. Baity, R. E. Rothschild, L. E. Peterson, F. A. Primini.
Bull. American Astron. Soc., Vol. 13, 849 - 850 (1981).
Abstract.

158.244 **Extended soft X-ray emission from NGC 4151.**
M. Elvis, U. Briel, J. P. Henry.
Bull. American Astron. Soc., Vol. 13, 850 (1981). − Abstract.

158.245 **X-ray emission from Centaurus A.**
J. Terrell.
Bull. American Astron. Soc., Vol. 13, 850 (1981). − Abstract.

158.246 **CO observations of the edge on Sb I galaxy NGC 891.**
P. M. Solomon, J. Barrett, D. B. Sanders, R. Dezafra.
Bull. American Astron. Soc., Vol. 13, 863 (1981). − Abstract.

158.247 **CO emission from the giant Sc I galaxy M101.**
P. M. Solomon, D. B. Sanders, J. Barrett,
R. Dezafra.
Bull. American Astron. Soc., Vol. 13, 863 (1981). − Abstract.

158.248 **Spectroscopy of southern galaxies from the ESO/Uppsala list.**   R. M. West.
Bull. American Astron. Soc., Vol. 13, 870 (1981). − Abstract.

158.249 **Far IR observations of NGC 4051.**
H. A. Smith, H. A. Thronson, C. J. Lada,
D. A. Harper, W. Glaccum, R. F. Loewenstein, J. Smith.
Bull. American Astron. Soc., Vol. 13, 892 (1981). − Abstract.

158.250 **The far infrared disk of M51.**   J. Smith.
Bull. American Astron. Soc., Vol. 13, 892 (1981).
Abstract.

158.251 **Models for far infrared emission from normal galaxies.**   M. Jura.
Bull. American Astron. Soc., Vol. 13, 892 (1981). − Abstract.

158.252 **IR colors of galactic nuclei.**
R. M. Price, M. Zeilik II.
Bull. American Astron. Soc., Vol. 13, 892 (1981). − Abstract.

158.253 **M supergiants and the distance to the ScI galaxy M101.**   R. M. Humphreys, S. E. Strom.
Bull. American Astron. Soc., Vol. 13, 892 (1981). − Abstract.

158.254 **A color-magnitude relationship for spiral galaxies.**
R. B. Tully, J. R. Mould, M. Aaronson.
Bull. American Astron. Soc., Vol. 13, 893 (1981). − Abstract.

158.255 **Neutral hydrogen detections in the dwarf elliptical galaxies NGC 185 and 205.**
D. W. Johnson, S. T. Gottesman.
Bull. American Astron. Soc., Vol. 13, 893 (1981). − Abstract.

158.256 **H I content of lenticular galaxies.**
C. Balkowski.
Bull. American Astron. Soc., Vol. 13, 893 (1981). − Abstract.

158.257 **Giant ring-like H II regions and the distance to M 101.**   K. B. Kwitter, D. G. Lawrie.
Bull. American Astron. Soc., Vol. 13, 894 (1981). − Abstract.

158.258 **The dust morphology of M 101 from ultraviolet images.**
T. P. Stecher, R. C. Bohlin, J. K. Hill, M. A. Jura.
Bull. American Astron. Soc., Vol. 13, 894 (1981). − Abstract.

158.259 **X-ray coronae around galaxies.**
J. N. Bregman, A. E. Glassgold, P. J. Huggins.
Bull. American Astron. Soc., Vol. 13, 894 (1981). − Abstract.

158.260 **Photoelectric UBV surface photometry of NGC 205.**   J. S. Price, G. L. Grasdalen.
Bull. American Astron. Soc., Vol. 13, 894 (1981). − Abstract.

158.261 **Mark III VLBI observations of the nucleus of M 81 at 2.3 and 8.3 GHz.**
N. Bartel, B. E. Corey, I. I. Shapiro, A. E. E. Rogers, A. R. Whitney, D. A. Graham, J. D. Romney, R. A. Preston.
Bull. American Astron. Soc., Vol. 13, 897 (1981). − Abstract.

158.262 **Further observations of the elliptical galaxy NGC 5813.**
G. Efstathiou, R. S. Ellis, D. Carter.
Anglo-Australian Obs. Prepr. No. 164, 18 pp. (1982). − Submitted to Mon. Not. R. Astron. Soc.

158.263 **Infrared colours of a complete sample of faint galaxies.** R. S. Ellis, D. A. Allen.
Anglo-Australian Obs. Prepr. No. 167, 21 pp. (1982). − Submitted to Mon. Not. R. Astron. Soc.

158.264 **Redshifts of parent galaxies of supernovae.**
R. Barbon, M. Capaccioli, R. M. West, R. Barbier.
ESO Sci. Prepr. No. 187, 8 pp. (1982). − Submitted to Astron. Astrophys. Suppl. Ser.

158.265 **Are all galactic nuclear regions sodium rich?**
M. P. Véron-Cetty, P. Véron, M. Tarenghi.
ESO Sci. Prepr. No. 190, 31 pp. (1982). − Submitted to Astron. Astrophys.

158.266 **Further examples of companion galaxies with discordant redshifts and their spectral peculiarities.**
H. Arp.
ESO Sci. Prepr. No. 196, 59 pp. (1982). − Submitted to Astrophys. J.

158.267 **Discovery of a new BL Lacertae object (1E1402.3+0416) with the Einstein Observatory.**
J. Stocke, J. Liebert, H. Stockman, J. Danziger, J. Lub, T. Maccacaro, R. Griffiths, P. Giommi.
Prepr. Steward Obs., No. 357, 18 pp. (1982).

158.268 **Observations of Paschen α in the broad-line radio galaxy 3C 445.** R. J. Rudy, A. T. Tokunaga.
Prepr. Steward Obs., No. 360, 10 pp. (1982).

158.269 **Infrared speckle interferometry of the nucleus of NGC 1068.**
D. W. McCarthy, F. J. Low, S. G. Kleinmann, F. C. Gillett.
Prepr. Steward Obs., No. 369, 19 pp. (1982).

158.270 **The noise of BL Lacertae.**
R. L. Moore, J. T. McGraw, J. R. P. Angel, R. Duerr, M. J. Lebofsky, G. H. Rieke, W. Z. Wisniewski, D. J. Axon, J. Bailey, J. M. Hough, I. Thompson, M. Breger, H. Schulz, G. C. Clayton, P. G. Martin, J. S. Miller, G. D. Schmidt, J. Africano, H. R. Miller.
Prepr. Steward Obs., No. 372, 46 pp. (1982).

158.271 **A catalog of infrared magnitudes and H I velocity widths for nearby galaxies.**
M. Aaronson, J. Huchra, J. R. Mould, R. B. Tully, J. R. Fisher, H. van Woerden, W. M. Goss, P. Chamaraux, U. Mebold, B. Siegman, G. Berriman, S. E. Persson.
Prepr. Steward Obs., No. 383, 46 pp. (1982).

158.272 **Double galaxy investigations. III. The differential redshift distribution and emission line correlations.**
W. G. Tifft.
Prepr. Steward Obs., No. 389, 37 pp. (1982). − Submitted to Astrophys. J.

158.273 **Novae as sources of nitrogen in galaxies.**
R. E. Williams.
Prepr. Steward Obs., No. 396, 17 pp. (1982).

158.274 **Galaxies with UV continuum of Seyfert type according to observations on the BTA.**
V. L. Afanas'ev, V. A. Lipovetskij, B. E. Markarian, D. A. Stepanyan.
Astrofizika, Tom 16, 193 - 206 (1980). In Russian. English translation in Astrophysics, Vol. 16, No. 2.
The results of spectroscopic observations of 36 galaxies with UV continuum accomplished on the 6-m telescope are presented. Due to different features these galaxies have been suspected to be of Seyfert type. This investigation confirmed the presence of Seyfert characteristics in 32 of them. Nine of them are related to the Sy 1 type, 20 to the Sy 2 type and three to the intermediate type, Sy 1.5.

158.275 **Emission-line distinctions in the spectra of four type 1 Seyfert galaxies.**
V. N. Popov, Eh. E. Khachikyan.
Astrofizika, Tom 16, 207 - 216 (1980). In Russian. English translation in Astrophysics, Vol. 16, No. 2.
Data on some emission line profiles in the spectra of the type 1 Seyfert galaxies Markarian 304, 335, 352 and III Zw 2 are presented. A change in the Hβ profile in the spectrum of III Zw 2 is found, represented by the appearance of a new wide blue component with −4500 to −8500 km/s doppler shift.

158.276 **Survey of mass-to-luminosity ratios for 440 binary galaxies.** I. D. Karachentsev.
Astrofizika, Tom 16, 217 - 229 (1980). In Russian. English translation in Astrophysics, Vol. 16, No. 2.
The survey of binary galaxies (Karachentsev, 1972) contains radial velocity differences for pair components, their mutual linear separation and orbital mass-to-luminosity ratio. The mean mass-to-luminosity ratio does not depend on the separation between the components and consists of the 10.8 ± 1.2. The wide volume around binary galaxies does not contain a noticeable mass excess over the scale of some hundreds of kpc.

158.277 **Distribution of apparent flattening among Seyfert galaxies.** V. T. Doroshenko, V. Yu. Terebizh.
Astrofizika, Tom 16, 393 - 396 (1980). In Russian. English translation in Astrophysics, Vol. 16, No. 2.
A comparison of apparent axial ratios of normal spirals and Seyfert galaxies witnesses a tendency of the latter to have small tilt angles on the tangent plane in relation to the celestial sphere.

158.278 **Morphology of nine galaxies with UV continua with double and multiple nuclei.**
Yu. P. Korovyakovskij, A. R. Petrosyan, K. A. Saakyan, Eh. E. Khachikyan.
Astrofizika, Tom 17, 231 - 238 (1981). In Russian. English translation in Astrophysics, Vol. 17, No. 2.
Essential criteria of external structure for the definition of the nucleus and some additional considerations concerning the multinuclei structure of the central parts of galaxies are suggested. On the basis of morphological and isodensitometrical investigations the structure of the central parts of nine galaxies with UV continua is investigated. It is shown that Markarian 19, 104, 324, 463, 823, 930 and 1027 are galaxies with double nuclei. Markarian 273 is a galaxy with three nuclei. Markarian 799 is probably a double nuclei galaxy, although it may be possible that we have a very tight pair of spirals.

158.279 **Radial velocities of 44 interacting galaxies.**
V. P. Arkhipova, V. L. Afanas'ev, V. A. Dostal', A. V. Zasov, I. D. Karachentsev, R. I. Noskova, M. V. Savel'eva.
Astrofizika, Tom 17, 239 - 244 (1981). In Russian. English translation in Astrophysics, Vol. 17, No. 2.
Radial velocities of 44 interacting systems of galaxies are determined; most of them for the first time.

158.280 Observations of galaxies of high surface brightness at 3.66 GHz.   V. G. Malumyan.
,Astrofizika, Tom 17, 245 - 248 (1981). In Russian. English translation in Astrophysics, Vol. 17, No. 2.

The results of observations of 12 galaxies of high surface brightness with the radio telescope RATAN-600 at 3.66 GHz are presented. It is shown that the objects having radio emission occur more than twice as frequently among galaxies of high surface brightness than among occasionally selected galaxies.

158.281 Homogeneous sample of binary galaxies. I. Selection and projection effects.   I. D. Karachentsev.
Astrofizika, Tom 17, 249 - 264 (1981). In Russian. English translation in Astrophysics, Vol. 17, No. 2.

For 423 pairs of galaxies of the author's catalogue (1972) the distributions according to the linear separation, radial velocity difference, absolute magnitude and other characteristics are presented. A simulation of an apparent distribution of galaxies has been provided by the Monte-Carlo method to study systematic biases of the sample distributions. Comparison of the catalogue pairs with the simulated sample allows to investigate selection and projection effects.

158.282 On the nature of jet-like formations in radio galaxies.   V. G. Gorbatskij.
Astrofizika, Tom 17, 273 - 284 (1981). In Russian. English translation in Astrophysics, Vol. 17, No. 2.

The hypothesis proposed previously to account for observed jet-like formations in some radio galaxies are briefly discussed. Many important facts cannot be explained by any of these models, in particular, the presence of bright "knots" inside a jet and sharp bends of jets. As it seems, the most plausible explanation of appearance of jets may be obtained on the assumption, that "bunches" of some small dense objects are ejected quasi periodically in the narrow cone from the nucleus of the galaxy. These objects are capable to emit relativistic electrons. On this assumption many observed features of jets may be understood.

158.283 A 15.7 min periodicity in OJ 287?
E. Valtaoja, H. Lehto, P. Teerikorpi, S. Haarala, T. Korhonen, M. Valtonen, H. Teräsranta, E. Salonen, S. Urpo, M. Tiuri, V. Piirola.
Rep. Ser., Dep. Phys. Sci., Univ. Turku, Turku-FTL-R 26, 10 pp. (1982) =Turku Univ. Obs. Informo No. 62. ISBN 951-642-151-2, ISSN 0356-9896. – Submitted to Astrophys. J.

158.284 Surface photometry of edge-on galaxies. III. Luminosity distributions in eight galaxies.
M. Hamabe, S. Okamura.
Ann. Tokyo Astron. Obs., Second Ser., Vol. 18, 191 - 204 (1982).

Detailed surface photometry is carried out for eight edge-on galaxies. An isophotal luminosity distribution is presented for each galaxy, and a luminosity profile in graphical and tabular form.

158.285 Observation of Mkn 501 at gamma-ray energies.
L. Bassani, R. C. Butler, A. J. Dean, G. Di Cocco, A. Della Ventura, F. Perotti, G. Villa.
Astrophys. Space Sci., Vol. 82, 311 - 316 (1982).

During a balloon flight in September 1979 of the MISO low-energy γ-ray telescope, the BL Lac-object Mkn 501 was studied in the hard X-ray range above 30 keV and in the low energy γ-ray range up to 19 MeV. No statistically significant X- and γ-ray fluxes were detected. The implications of the upper limits obtained are discussed in the light of the relativistic jet theories recently proposed.

158.286 Hβ luminosity of extragalactic objects and evolution of galaxies.   R. K. Thakur, R. K. Sood.
Astrophys. Space Sci., Vol. 84, 99 - 114 (1982).

In order to examine the evolutionary sequence of extragalactic objects Hβ luminosity of 27 QSOs, 26 Seyferts 1, 8 N galaxies, 11 Seyferts 2 and 2 normal galaxies has been calculated and compared with its observed value. For calculating Hβ luminosity a photoionization model has been used. It has been assumed that the $U-V$ continuum radiation is produced in the cores of the extragalactic objects by the synchrotron mechanism and as such its flux is given by a power law. This radiation ionizes hydrogen atoms in the gas clouds surrounding the cores. It is concluded that the greater the value of the observed Hβ luminosity as compared to its calculated value, the more evolved is the extragalactic object. The authors suggest that the extragalactic objects evolve along the following sequence: BL Lac objects $\rightarrow$ QSOs $\rightarrow$ Seyferts 1 $\rightarrow$ N galaxies $\rightarrow$ Seyferts 2 $\rightarrow$ normal galaxies.

158.287 A theoretical approach to the description of shapes of the galaxies.   T. Grabińska, M. Zabierowski.
Astrophys. Space Sci., Vol. 84, 251 - 253 (1982).

A theoretical study to describe shapes of the galaxies is proposed. A syntactic approach to galaxy shape description as a method for recognition of galaxies is suggested.

158.288 VLA observations of massive star formation in spiral galaxy cores.   J. Turner, P. T. P. Ho.
Bull. American Astron. Soc., Vol. 14, 571 (1982). – Abstract.

158.289 UBV observations of bright galaxies.
R. P. S. Stone.
Astrophys. J., Suppl. Ser., Vol. 48, 395 - 398 (1982) = Lick Obs. Bull., No. 900.

UBV photometry is presented of 61 bright ($V \leqslant 14.4$) galaxies of a variety of morphological types. Although an extensive series of standard stars with well-determined UBV colors was observed in order to determine the transformations to the UBV system, nevertheless a residual zero-point difference of ~ 0.035 mag in $U-B$ appears to exist between the present data and those of de Vaucouleurs and collaborators here adopted as standard. It is believed that for the particular photomultiplier-filter combination used here, the transformation determined from stellar sources may not be strictly applicable to the nonstellar energy distributions of galaxies. The discrepancy was removed through the observation of "standard galaxies", so the present results are on de Vaucouleurs's system.

158.290 Global properties of irregular galaxies.
D. A. Hunter, J. S. Gallagher, D. Rautenkranz.
Astrophys. J., Suppl. Ser., Vol. 49, 53 - 88 (1982).

The authors present optical and radio observations of global properties for a sample of noninteracting irregular galaxies and a few comparison objects. Program galaxies were chosen primarily on the basis of their blue colors and thus may be expected to represent systems with high rates of star formation. The data consist of 40" aperture photometry through intermediate-band filters, 25" aperture spectrophotometry in the region 3500–5900 Å, spectrophotometry of individual H II regions, and 21 cm H I observations. From the filter photometry the authors determine a Mg index and a reddening-free color parameter Q. An iterative population synthesis technique is applied to the large aperture spectra in order to separate Hβ emission from absorption, and a star formation rate (SFR) is estimated from the number of ionizing photons coupled with a Miller and Scalo initial mass function. Line fluxes are used to determine the abundance ratios O/H and N^+/S^+, and the radio data give information on gas masses and kinematics. No correlation is found between the SFR and global gas or abundance parameters, implying that local processes dominate.

158.291 **Gravitational mechanics of systems of galaxies.**
I. Corrections for errors in redshifts.
H. J. Rood.
Astrophys. J., Suppl. Ser., Vol. 49, 111 - 148 (1982).

Modern 21 cm hydrogen-line redshifts, accurate to
~8 km s⁻¹, are used as standards to derive zero point corrections, rms uncertainties (σ_V),and the percentage of discordant (i.e., differing by more than 3 σ_V from the zero point
redshift) optical and 21 cm redshifts from earlier catalogs.
Optical redshifts have a typical uncertainty of ~100 km s⁻¹
(independent of the epoch of the source of redshift), but the
variation in σ_V from source to source is considerable. Optical
redshifts by Kelton (σ_V = 43 km s⁻¹), the Center for Astrophysics (Davis et al.) (σ_V = 37 km s⁻¹), and Rubin, Ford, and
Thonnard (σ_V = 8 km s⁻¹) are especially noteworthy. The
rms correction factor appropriate to optical redshifts listed
in the First Reference Catalog (de Vaucouleurs and de
Vaucouleurs) is 1.33 and that for the Second Reference
Catalog (de Vaucouleurs, de Vaucouleurs, and Corwin) is
1.42. A new catalog of galaxy redshifts (CGR) containing
known redshifts for ~4000 galaxies and reliable estimates of
rms uncertainties was compiled and is available for circulation.
Results of an analysis of the CGR data pertaining to the
luminosity function and the mass-to-light ratios of binary
galaxies are described.

158.292 **Emission-line profiles in Seyfert 1 galaxies.**
D. E. Osterbrock, J. M. Shuder.
Astrophys. J., Suppl. Ser., Vol. 49, 149 - 174 (1982) = Lick
Obs. Bull., No. 905.

Observed profiles of H I, He I, and He II emission lines
in 19 Seyfert 1 galaxies are presented. The data, reduced to
energy units versus radial velocity, are given in accurate
graphical form for comparison with theoretical models. The
profiles are shown as directly observed and as corrected for
the blending effects of other lines. Some implications of
these profiles on current models of active galactic nuclei are
discussed.

158.293 **CCD photometry of Markarian 421 and 501.**
P. Hickson, G. G. Fahlman, J. R. Auman,
G. A. H. Walker, T. K. Menon, Z. Ninkov.
Astrophys. J., Vol. 258, 53 - 58 (1982).

Observations of the BL Lacertae objects Markarian 501
and 421 have been made with a CCD detector at the prime
focus of the 3.6 m Canada-France-Hawaii telescope. Each
object was observed in two bands of FWHM ≈ 700 Å centered
at 5000 Å and 6500 Å. Photometric profiles for both objects
are presented and shown to be consistent with composite
profiles from a point source and an elliptical galaxy which
obeys the de Vaucouleurs surface brightness law. The derived
photometric parameters of the underlying galaxies show them
to be apparently normal elliptical galaxies.

158.294 **Large-scale magnetic fields in spiral galaxies?**
M. Jura.
Astrophys. J., Vol. 258, 59 - 63 (1982).

In galaxies with pronounced dark lanes such as Cen A
(NGC 5128) or the Sombrero galaxy (M104 = NGC 4590),
optical polarization vectors are measured to be parallel to
the dark lane. These measurements have been interpreted as
resulting from a large-scale, systematic field parallel to the
dark lane that leads to the alignment of the grains. The author
suggests that the observed polarization might also be produced
by scattering off grains which are concentrated in the dark
lane. Further observational tests of the hypothesis that the
observed polarization indicates the magnetic field in the
galaxy are suggested.

158.295 **Observations of the mass and light distribution**
of NGC 5963, an unusual low surface brightness
spiral. W. Romanishin, S. E. Strom, K. M. Strom.
Astrophys. J., Vol. 258, 77 - 82, plate 7 (1982).

NGC 5963 is an unusual member of a class of spiral
galaxies characterized by disks of abnormally low surface
brightness. The inner region of the galaxy appears to be a
vigorous, relatively normal spiral of late type. Exterior to
this spiral is a low surface brightness disk in which low
surface brightness H II regions define a roughly spiral pattern.
From the optical surface photometry and rotation curve
reported here, the radial dependence of the local mass-to-light
ratio M/L_B is derived. In the inner high surface brightness
regions of NGC 5963, $M/L_B \approx 2$; in the outer regions,
$M/L_B \approx 18$. The observed ratio of [N II]/Hα is unusually high
in the inner disk but drops significantly just outside this
region. This result suggests that star formation, gas depletion,
and element production have proceeded much more rapidly
in the inner regions than in the outer regions of the galaxy.

158.296 **Emission-line galaxies in the direction of the**
proposed void in Bootes.
N. Sanduleak, P. Pesch.
Astrophys. J., Lett., Vol. 258, L11 - L15 (1982).

A finding list is provided for 55 galaxies that show emission lines on low-dispersion objective-prism plates and that lie
in the direction of the recently proposed void in Bootes. Seven
of these appear to have radial velocities that place them within the void. Radial velocity determinations, which are lacking
for most of the faint objects in this list, can be readily obtained
from their strong emission-line spectra and could be used for
further studies of the proposed void.

158.297 **Isophotal diameters of late-type galaxies in the**
southern hemisphere.
J. Mould, D. Ziebell.
Publ. Astron. Soc. Pacific, Vol. 94, 221 - 222 (1982).

Transformation formulae are presented for galaxy diameters and axial ratios measured by Holmberg and collaborators
to the corresponding isophotal quantities on the system of the
Second Reference Catalogue.

158.298 **Radial velocities of galaxies detected in the Arecibo**
2380 MHz survey. B. Marano, G. Vettolani.
Astron. Astrophys., Suppl. Ser., Vol. 48, 453 - 455 (1982).

Previously unknown radial velocities of 50 galaxies
detected in the Arecibo 2380 MHz survey of bright galaxies
have been obtained with the image tube spectrograph of the
152 cm Loiano telescope. A typical standard error of
100 km/s and a systematic zero error of −60±20 km/s are
derived for the velocities from a comparison with other redshift sources made on a further sample of 35 galaxies.

158.299 **Redshifts of parent galaxies of supernovae.**
R. Barbon, M. Capaccioli, R. M. West, R. Barbier.
Astron. Astrophys., Suppl. Ser., Vol. 49, 73 - 75 (1982).

Redshifts of twenty-five parent galaxies of supernovae
have been measured on spectra taken with the Shechtman
Reticon spectrograph at the 250 cm telescope of Las
Campanas. This new list raises the sample of parent galaxies
with known velocities up to 55%. A detailed discussion of the
observed features is provided for each spectrum.

158.300 **Intermediate band filter spectrophotometry of**
bright galaxies. I. Observations.
J.-E. Solheim. G. de Vaucouleurs, A. de Vaucouleurs.
Astron. Astrophys., Suppl. Ser., Vol. 49, 109 - 121 (1982).

Photoelectric photometry of galaxies in a 10-filter intermediate band system between 3400 and 5500 Ångströms is
presented. Included in the observing program are 56 galaxies
of all the main Hubble types. The observations are compared
with detailed medium resolution spectral scans and with broad
band photometry. A spectral index is defined by the
difference in received flux between the extreme bands to be
used as a type discriminant.

158.301 Dusty King spheres.   H.-E. Fröhlich.
Astron. Nachr., Band 303, 97 - 103 (1982).

In order to determine the quantitative influence of small amounts of dust on the luminosity profile of spherical galaxies, the author represents an elliptical galaxy by a King sphere filled with dust. The density of dust particles is assumed to be related to the density of stars according to a simple power law. Assuming that the observed central reddenings in NGC 4874 and NGC 4889 are really caused by dust, he estimates the optical depths in the centres of both galaxies to be about 0.75.

158.302 The amount of dust and the distance of NGC 253.
I. A. Issa.
Astron. Nachr., Band 303, 127 - 129 (1982).

The areas of 267 dark clouds in the galaxy NGC 253 were measured. It is shown that the size distribution of these clouds obeys the same exponential formula found for the galaxies NGC 3031, 5128, 5194, and 5457. From the apparent size distribution of the dark clouds the amount of dust and the distance of NGC 253 is found to $4.4 \times 10^7 M_\odot$ and 2.06 Mpc, respectively.

158.303 A new and interesting Seyfert 2 galaxy: NGC 5728.
M.-P. Véron, P. Véron, M. Tarenghi, P. Grosbøl.
Messenger, No. 28, p. 13 - 14 (1982).

158.304 De Spoelnevel.   M. Drummen.
Zenit, 9. Jaarg., 308 - 312 (1982).

158.305 Extended H I-envelopes around spiral galaxies: NGC 2655 and NGC 2715.
W. K. Huchtmeier, O.-G. Richter.
Astron. Astrophys., Vol. 109, 331 - 335 (1982).

Observations of an extended H I-envelope of an early-type spiral-galaxy are presented. The H I around the Sa-galaxy NGC 2655 extends to about 5 times its Holmberg diameter at a level of 5×10^{18} atoms cm^{-2}. Its total H I-mass is $8.6 \times 10^9 M_\odot$ or 1.8% of its total mass, which is very close to the mean value of Sa-galaxies. The companion galaxy NGC 2715 (Sc(s) II) got a total H I-mass of $4 \times 10^{10} M_\odot$. It is extremely hydrogen rich for its class. The authors suggest to identify the region where the low excitation absorption lines in quasars arise within the extended H I-envelopes around many late-type and interacting galaxies. The authors speculate that these envelopes correspond to the gas remaining from the merging of galaxies, and this should have the corresponding abundances.

158.306 Arm width as a function of absolute luminosity for bc, c spiral galaxies.   D. L. Block.
Astron. Astrophys., Vol. 109, 336 - 339 (1982).

The linear widths T of the spiral arms of 27 Shapley-Ames type bc, c galaxies contained in the author's photographic catalogue of galaxies on a uniform linear scale are presented. The widths are found to be strongly correlated to the absolute blue magnitude of the parent galaxy. Successively brighter bc, c spirals have the wider arms.

158.307 Chemical evolution of irregular galaxies.
C. Chiosi, F. Matteucci.
Astron. Astrophys., Vol. 110, 54 - 60 (1982).

The authors present a model of the chemical evolution of irregular galaxies of magellanic type, in which inflow of unprocessed gas and a novel formulation for the rate of star formation are taken into account. In particular, the rate of star formation is assumed to depend on the current total and gas mass of the galaxy. Satisfactory agreement is obtained between model results and observations.

158.308 Surface photometry of edge-on spiral galaxies.
III. Properties of the three-dimensional distribution of light and mass in disks of spiral galaxies.

P. C. van der Kruit, L. Searle.
Astron. Astrophys., Vol. 110, 61 - 78 (1982).

The authors present surface photometry of the edge-on spiral galaxies NGC 5907, 5023, 4217, and 4013, and fit the disks of these systems to the three-dimensional model light distribution proposed by van der Kruit and Searle (1981). They find good agreement and confirm that (1) all disks have a relatively sharp cut-off in the radial direction, (2) all disks have a thickness independent of radius, (3) the colours indicate that the model describes the distribution of an old disk population. The authors show that the Galaxy is similar to the systems observed only if the radial scalelength in its disk is about 5 kpc and $H \sim 75$ km s^{-1} Mpc^{-1}.

158.309 Surface photometry of edge-on spiral galaxies.
IV. The distribution of light, colour, and mass in the disk and spheroid of NGC 7814.
P. C. van der Kruit, L. Searle.
Astron. Astrophys., Vol. 110, 79 - 94 (1982).

In this final paper of their series concerning edge-on spiral galaxies the authors describe the three-dimensional light and colour distribution in a spheroid-dominated system, NGC 7814. Their previous discussions of spheroids were based on the residual light distributions that remain after subtraction of the model disk, but in NGC 7814 where only a few percent of the light comes from the disk, the authors reverse the procedure and directly model the spheroid.

158.310 Global properties of Sa-galaxies from H I-observations.   W. K. Huchtmeier.
Astron. Astrophys., Vol. 110, 121 - 137 (1982).

Observations of the neutral hydrogen content of a nearly complete sample of Sa-galaxies show clear sign of H I-deficiency within a distance of 13 Mpc of the centre of the Local Supercluster. The median of the distance-independent M_H/L_B ratio is 0.092 ± 0.011 for the whole sample and 0.074 ± 0.007 excluding the hydrogen-rich galaxies ($M_H/L_B > 0.195$). The M_H/L_B ratio changes from 0.048 ± 0.013 inside this radius to 0.108 ± 0.014 outside and is constant at greater distances within the observational errors. A population of hydrogen-rich galaxies seems to be scattered over the periphery of the Local Supercluster. The median values for the relative hydrogen content and for the mass-to-luminosity ratio for the whole sample are $M_H/M_t = 0.02 \pm 0.002$ and $M_t/L_B = 4.6 \pm 0.44$ in agreement with earlier determinations of these parameters. The absolute magnitude-intrinsic line width (Tully-Fisher-relation) depends on galaxy type.

158.311 NGC 4388: a Seyfert 2 galaxy in the Virgo cluster.
M. M. Phillips, D. F. Malin.
Mon. Not. R. Astron. Soc., Vol. 199, 905 - 913 (1982).

New optical photographic and spectroscopic observations of the edge-on spiral galaxy NGC 4388 are reported. This galaxy, which is almost certainly a member of the Virgo cluster, appears to be a barred spiral of morphological class SB(s)b pec. The nucleus emits a high-excitation, narrow emission-line spectrum of relatively low luminosity, but which is otherwise indistinguishable from that of a classical Seyfert 2 galaxy. The authors therefore group NGC 4388 with the Seyfert 2 galaxies, since its optical, X-ray and radio nuclear properties are logical extensions of the ranges observed for this class of object. NGC 4388 is the first Seyfert galaxy to be identified in the Virgo cluster − which, along with the recent discoveries of two Seyfert galaxies in the Fornax cluster, suggests that such objects are considerably more common in clusters than had previously been thought.

158.312 The near infrared properties of Seyfert and related active galaxies.   M. Ward, D. A. Allen,
A. S. Wilson, M. G. Smith, A. E. Wright.
Mon. Not. R. Astron. Soc., Vol. 199, 953 - 968 (1982).

The authors present near infrared (*J, H, K* and, in some cases, *L'* magnitudes) photometry of active galaxies (accurate to 2–3 per cent). Data on 28 galaxies, a few observed through two apertures, are given. The two-colour diagram, (*J–H*) versus (*H–K*), is used to demonstrate that for active and Seyfert galaxies a simple power law cannot account for the 1.2–2.2 μm continuum. Rather, different mixtures of a typical QSO with a 'normal' galaxy are consistent with the observed colours. Combination of the present data with earlier, mostly less accurate, measurements enables the authors to define a narrow range in the (U_0-L) (i.e. 0.36–3.5 μm) index for Seyfert 1's and low redshift QSOs. For Seyfert 2's the range in (U_0-L) is wider and the average value redder. The ratios of energies emitted at X-ray, optical and infrared bands are calculated and also found to be different in Seyfert 1's and 2's.

158.313 The near-infrared continua of BL Lacertae objects.
D. A. Allen, M. J. Ward, A. R. Hyland.
Mon. Not. R. Astron. Soc., Vol. 199, 969 - 978 (1982).

Accurate photometry at *J, H* and *K* (1.2–2.2 μm) has been secured of 53 BL Lac objects. A power-law continuum is an exceptionally good fit to the data, and the colours are distinct from those of quasars. There is no indication of the additional infrared continuum seen in quasars and which is believed to arise in circumnuclear dust. The authors argue that dust is very scarce in the nuclei of BL Lac objects, and thus one expects gas to be equally scarce. Hence one attributes the lack of optical emission lines to an absence of ionized nuclear gas. The authors further argue that BL Lac objects could underlie quasars, the latter exhibiting line and thermal continuum emission at ultraviolet, optical and near-infrared wavelengths due to the presence of circumnuclear gas and dust. The strong-lined optically violent variable quasars have colours typical of BL Lac objects rather than quasars, and may represent intermediate cases.

158.314 Infrared studies of a sample of 3C radio galaxies.
S. J. Lilly, M. S. Longair.
Mon. Not. R. Astron. Soc., Vol. 199, 1053 - 1068 (1982).

J, H and *K* photometry is presented for 35 3C radio galaxies spanning a large range of redshift. At low redshifts, $z < 0.4$, the near-infrared colours of most radio galaxies are the same as non-radio elliptical galaxies, except for the N-galaxies which show a strong infrared excess. At higher redshifts the galaxies show little evidence for spectral evolution in their infrared colours, but have optical-infrared colours considerably bluer than is predicted by a redshifted non-evolving energy distribution. The infrared Hubble diagram for the radio galaxies is constructed, and implies a high apparent value for q_0, which may indicate evolution in the infrared luminosity of these galaxies. The results are discussed in the context of current models for the evolution of the stellar population of giant elliptical galaxies.

158.315 The structure and dynamics of the nucleus of the hot spot galaxy NGC 2997.
J. Meaburn, D. L. Terrett.
Mon. Not. R. Astron. Soc., Vol. 200, 1 - 17 (1982).

Photographs of the nuclear regions of NGC 2997 in the *U, B* and *V* bands, obtained with the 1.9-m SAAO telescope, have been combined with both low and high dispersion spectra obtained with the 3.9-m Anglo-Australian telescope. Giant, radiatively ionized, H II regions, young ($\lesssim 10^7$ yr) blue OB associations and lanes of neutral dusty material form inner spiral arms which project from an amorphous nucleus of old ($\gtrsim 10^9$ yr) red stars. The rotational velocity for a radius of $\gtrsim 630$ pc is $\cong 175$ kms s^{-1} falling to $\cong 116$ km s^{-1} over the annular region of OB associations. Here it remains reasonably constant until $\lesssim 73$ pc from the centre where it drops rapidly to zero. Total masses in the annular region of $9 \times 10^8 M_\odot$ and in the central amorphous region of $2.2 \times 10^8 M_\odot$ are predicted,

whereas the mass of the ionized gas in the annulus is $1.1 \times 10^6 M_\odot$.

158.316 Infrared polarimetry and photometry of BL Lac objects.
C. D. Impey, P. W. J. L. Brand, R. D. Wolstencroft, P. M. Williams.
Mon. Not. R. Astron. Soc., Vol. 200, 19 - 40 (1982).

Infrared polarimetry and photometry have been obtained for a sample of 18 BL Lac objects. The data covers a period of one year and is part of a continuing monitoring programme; all observations were in the *J, H* and *K* wavebands. Large and variable degrees of polarization are a common property of the sample. Two BL Lac objects show wavelength-dependent polarization, with the polarization increasing towards shorter wavelengths, and two objects show evidence for position angle rotations over a five-day period. The BL Lac objects cover an enormous range of infrared luminosity; the three most luminous having $L_{IR} > 10^{46}$ erg s^{-1} and the other end of the range having infrared luminosities similar to normal elliptical galaxies.

158.317 Spectroscopic and photographic observations of the Cygnus A group and of the stellar component of the Cygnus A galaxy.
H. Spinrad, J. R. Stauffer.
Mon. Not. R. Astron. Soc., Vol. 200, 153 - 158 (1982).

Off-nuclear spectroscopic observations and photographic photometry are used to show that the non-thermal optical component of Cygnus A is situated in a large, normal-spectrum cD galaxy. The extent of the cD envelope is at least 70 kpc in radius. IDS spectra were also obtained for six nearby E/S0 galaxies. Four appear to be velocity members of the relatively poor cluster around Cygnus A. Their mean radial velocity is very close to that of the central radio galaxy. The group has a large line-of-sight velocity dispersion, $\sigma_v \cong 2000$ km s^{-1}.

158.318 Detailed observations of NGC 4151 with IUE – II. Variability of the continuum from 1978 February to 1980 May, including X-ray and optical observations.
G. C. Perola, A. Boksenberg, G. E. Bromage, J. Clavel, M. Elvis, A. Elvius, P. M. Gondhalekar, J. Lind, C. Lloyd, M. V. Penston, M. Pettini, M. A. J. Snijders, E. G. Tanzi, M. Tarenghi, M. H. Ulrich, R. S. Warwick.
Mon. Not. R. Astron. Soc., Vol. 200, 293 - 312 (1982).

NGC 4151 has been extensively monitored with IUE from 1978 February to 1980 May. The rather erratic behaviour of the ultraviolet light curve seems to be mainly due to variations which occur at rates which, if extrapolated, would produce factor two changes in time-scales between five and 30 days, implying a radius of the order of 0.01 pc for the source. The shape of the continuum can be described by a power law longward of $\lambda 2200$ and by an excess above the extrapolation of that law at shorter wavelengths, suggesting the presence of two components.

158.319 An optical and H I study of late-type low surface brightness galaxies.
A. J. Longmore, T. G. Hawarden, W. M. Goss, U. Mebold, B. L. Webster.
Mon. Not. R. Astron. Soc., Vol. 200, 325 - 346 (1982).

Neutral hydrogen and optical parameters are presented for 151 galaxies of low surface brightness selected from UK Schmidt plates. The 21-cm H I line was detected in 100 of these systems. It is found that the galaxies show the same trends of global properties with type as samples of bright galaxies, while the data are consistent with the low surface brightness (LSB) galaxies being of systematically lower mass than bright galaxies of the same type and linear dimensions. A constant value of hydrogen mass/(linear dimension)^2 is strongly suggested for LSB and bright galaxies.

158.320 M/L and velocity anisotropy from observations of spherical galaxies, or must M87 have a massive black hole?
J. Binney, G. A. Mamon.

Mon. Not. R. Astron. Soc., Vol. 200, 361 - 375 (1982).

An algorithm is developed that determines whether a given set of photometric and spectroscopic observations of a spherical galaxy are consistent with the mass-to-light ratio in the galaxy being constant. If such a model is possible, the value of the mass-to-light ratio and the values of the two independent components of velocity dispersion at each radius are determined. Tests of the algorithm demonstrate its consistency and accuracy when applied to pseudo-data generated from a variety of theoretical models. The algorithm is applied to M87. Velocity disperison profiles consistent with the observations of Sargent et al. (1978) and of Dressler (1980) combine with the photometry to yield physically plausible models that have constant mass-to-light ratio. $M/L_V = 7.6$ if $A_V = 0.14$ and $D = 16$ Mpc. The models returned by different dispersion profiles all involve a tight star cluster at $r < 100$ pc, in which the velocity dispersion is highly anisotropic towards the outside. The main body of the galaxy, which dominates the light beyond $r = 200$ pc, has only modest velocity dispersion anisotropy.

158.321 A study of some compact extragalactic objects.
C. Barbieri, S. Cristiani, G. Romano.
Astron. J., Vol. 87, 616 - 625 (1982).

This paper presents photometric data on compact extragalactic objects obtained at Asiago Observatory. The tracing of a deep photograph of BL Lac that confirms the presence of a jet-like structure in p.a. 210° is also shown.

158.322 CO observations of the SAB galaxies NGC 157, 2903, 4321, and 5248, and the Seyfert galaxy NGC 1068.   D. M. Elmegreen, B. G. Elmegreen.
Astron. J., Vol. 87, 626 - 634 (1982).

CO emission at 2.6 mm has been partially mapped in the SAB galaxies NGC 157, 2903, 4321, and 5248, and in the SA Seyfert galaxy NGC 1068. Optical velocity information has been used to estimate regions of emission within the CO beams. Upper limits and marginal detections are reported for 18 SAB and SB galaxies, including NGC 1300.

158.323 Spiral galaxies in clusters. III. Gas-rich galaxies in the Pegasus I cluster of galaxies.
G. D. Bothun, R. A. Schommer, W. T. Sullivan, III.
Astron. J., Vol. 87, 725 - 730 (1982).

The authors report the results of a 21 cm and optical survey of disk galaxies in the vicinity of the Pegasus I cluster of galaxies. The color-gas content relation for this particular cluster reveals the presence of a substantial number of blue, gas-rich galaxies. With few exceptions, the disk systems in Pegasus I retain large amounts of neutral hydrogen despite their presence in a cluster. This directly shows that environmental processes have not yet removed substantial amounts of gas from these disk galaxies. The overall properties of the Pegasus I spirals are consistent with the suggestion that this cluster is now at an early stage in its evolution.

158.324 Spiral galaxies in clusters. IV. The H I color properties of spirals in nine clusters.
G. D. Bothun, R. A. Schommer, W. T. Sullivan III.
Astron. J., Vol. 87, 731 - 738 (1982).

The authors present the results of a 21 cm H I and UBV photometric survey of 235 disk galaxies in nine nearby clusters. They find that most of their clusters do not preferentially contain gas-deficient spirals. Moreover, there is little difference between the color distributions of cluster spirals and the spirals catalogued in the Second Reference Catalog of Bright Galaxies (de Vaucouleurs et al. 1976). The Coma cluster is the only cluster in the sample exhibiting a large fraction of H I-deficient spirals and having other characteristics consistent with present-day spiral-to-S0 conversion. In general, the gas contents of cluster spirals are not well correlated with their colors.

158.325 Complete samples of active extragalactic objects. I. A 1411 MHz VLA survey centered on $\alpha = 12^h 04^m$, $\delta = +11°30'$.
J. J. Condon, M. A. Condon, C. Hazard.
Astron. J., Vol. 87, 739 - 750 (1982).

As one part of a program to detect complete samples of active extragalactic objects, the 3.36×10^{-3} sr field centered on $\alpha = 12^h 04^m$, $\delta = +11°30'$ (1950.0) was mapped with the Very Large Array at 1411 MHz. The rms source-position uncertainties are ~ 1 arcsec in each coordinate so that reliable optical identifications can be made on the basis of radio-optical position coincidence alone. The sky density of sources with flux densities $5 \leqslant S < 150$ mJy and the angular-size distribution of sources between 30 and 150 mJy were determined.

158.326 The kinematics and structure of the neutral hydrogen in the barred spiral galaxy NGC 3359.
S. T. Gottesman.
Astron. J., Vol. 87, 751 - 762 (1982) = Contrib. No. 42 Dep. of Astron., Univ. Florida.

The barred spiral galaxy NGC 3359 has been observed with the three-element line interferometer of the National Radio Astronomy Observatory. An angular resolution of $42'' \times 31''$ was achieved with a velocity resolution of 24.5 km/s. Systematic noncircular motions were found to exist which appear to be associated with features in the disk and possibly with the bar.

158.327 Results from the Orbiting Astronomical Observatory. XLI. Photometry of M31 and M33.
J. Davis, A. D. Code, J. S. Mathis, G. A. Welch.
Astron. J., Vol. 87, 849 - 858 (1982).

Filter photometry observations from the Orbiting Astronomical Observatory, in six wavelength bands between 1550 and 4250 Å, are reported for several positions along the major and minor axes in M31 and at some positions in M33. The spatial resolution is 10'. The colors along the major axis of M31 within about 20' of the nucleus are similar to the nucleus; at 45' the colors are much bluer, indicating a spiral arm population. The center of M33 is bluer than the ring of M31, and the arms of M33 even bluer. There are no colors too blue to be explained by unreddened B stars.

158.328 Hot spots in the nucleus of NGC 1365.
S. Jörsäter, P. O. Lindblad, Aa. Sandqvist.
Rep. Obs. Lund, No. 18, (see 012.044), p. 26 - 29 (1982).

158.329 Primordial helium and emission-line galaxies.
D. Kunth.
Cosmology and particles, (see 012.046), p. 241 - 251 (1982).

Recent observations of 13 low luminosity emission-line galaxies give no convincing evidence for a ΔY versus ΔZ correlation. The observed scatter is likely to result from uncontrolled parameters in the analysis rather than simple observational errors. The results taken with the best available data in the literature suggest that $Y_p = 0.240 \pm 0.015$.

158.330 A limit on the stellar population of massive halos.
D. J. Hegyi.
Cosmology and particles, (see 012.046), p. 321 - 330 (1982).

The measured rotational velocities of the edge-on spiral galaxy NGC 4565, coupled with several arguments supporting the spherical symmetry of halos, can be used to determine the space density of the halo mass. The author shows that if the halo mass surrounding NGC 4565 were contained in a population of M5 stars, the minimum expected surface brightness would exceed the measured halo surface brightness. These observations were made in the I Kron band with the annular scanning photometer.

158.331 Galaxy formation via hierarchical clustering and dissipation: the structure of disk systems.

S. M. Faber.
Astrophysical cosmology, (see 012.047), p. 191 - 217 (1982).

The White-Rees theory of galaxy formation via hierarchical clustering and core condensation is used to develop a model for the structure of disk galaxies. Disk parameters such as rotation velocity, radius and surface brightness scale directly in proportion to the corresponding halo parameters, which are in turn determined by hierarchical clustering from primordial density fluctuations. If the original fluctuation spectrum was a power law, scaling relationships among luminosity, velocity, and radius are predicted which look much like the observed Fisher-Tully and radius-luminosity laws.

158.332 Galaxy formation via hierarchical clustering and dissipation: the structure of spheroids.
S. M. Faber.
Astrophysical cosmology, (see 012.047), p. 219 - 232 (1982).

It is argued that ellipticals and spheroidal bulges are highly condensed systems in which the ordinary luminous matter falls deep into the central core of the surrounding non-luminous halo. Star formation may be triggered when the density of the luminous matter rises above a threshold level set by the halo, thus preventing further collapse. These ideas are consistent with the observed structural scaling laws for elliptical galaxies and the velocity dispersions of spheroidal bulges in spirals. This view of spheroid structure fits naturally into a picture of the Hubble sequence as a sequence of increasing dissipation and central concentration of the luminous matter relative to the surrounding dissipationless halo.

158.333 The evolution of galaxies. J. E. Gunn.
Astrophysical cosmology, (see 012.047), p. 233 - 262 (1982).

The recent observational evidence on the evolution of galaxies is reviewed and related to the framework of current ideas for galaxy formation from primordial density fluctuations. Recent strong evidence for the evolution of the stellar population in ellipticals is presented, as well as evidence that not all ellipticals behave as predicted by any simple theory. The status of counts of faint galaxies and the implications for the evolution of spirals is discussed, together with a discussion of recent work on the redshift distribution of galaxies at faint magnitudes and a spectroscopic investigation of the Butcher-Oemler blue cluster galaxies. Finally a new picture for the formation and evolution of disk galaxies which may explain most of the features of the Hubble sequence is outlined.

158.334 The colours of faint radio galaxies.
H. van der Laan, R. A. Windhorst.
Astrophysical cosmology, (see 012.047), p. 263 - 268 (1982).

The authors describe a method for forming substantially large samples of giant elliptical galaxies near the faint end of deep plate magnitude scales. It relies upon the fact, known for twenty years and established with the ever increasing numerical weight of hundreds of measured redshifts, that strong extragalactic radio sources, if they are not quasars, are uniquely and without exception associated with luminous early type galaxies, ellipticals and S0's.

158.335 Evidence for the cosmological evolution of the stellar content of radio galaxies.
S. J. Lilly, M. S. Longair.
Astrophysical cosmology, (see 012.047), p. 269 - 277 (1982).

It has long been known that normal giant elliptical galaxies observed at large redshifts cease to be optical objects but become infrared sources. The authors report on a preliminary survey of 35 galaxies from the 3 CR catalog in the redshift range from 0.03 to 1.0. The observations were made in the J, H, K wavebands at the UKIRT on Mauna Kea, Hawaii.

158.336 The radio structure of the nuclear region of NGC 1365.
A. Sandqvist, S. Jörsäter, P. O. Lindblad.
Astron. Astrophys., Vol. 110, 336 - 344 (1982).

VLA observations of the barred spiral, hot-spot nucleus, Seyfert 1 galaxy NGC 1365 in the continuum at wavelengths of 6 and 20 cm have revealed complex radio emission from the nuclear region. The radio nucleus of NGC 1365 was resolved into at least 8 nonthermal components. There is no direct correspondence between the radio sources and the H II hot spots observed with the ESO 3.6 m telescope. It is suggested that the hot spots and radio sources are signs of star burst activity in the nuclear region. An unresolved ($<1.''5 \times 6.''5$) probably background source was discovered about $1.'6$ northeast of the optical nucleus.

158.337 Active galactic nuclei and particle acceleration in accretion disks around massive black holes.
M. Kafatos, M. M. Shapiro, R. Silberberg.
Compendium in astronomy, (see 003.012), p. 323 - 345 (1982).

The nuclei of active galaxies are compact power sources, mainly nonthermal. The central power source has been interpreted using models that invoke an ultra-massive black hole or, alternatively, a spinar. The authors examine the acceleration processes near a black hole. An accretion disk around a supermassive black hole in the center of an active galactic nucleus is shown to be a likely site of particle acceleration.

158.338 CO in Seyferts. R. D. Mathieu.
Extragalactic molecules, (see 012.048), p. 13 - 16 (1982).

Because of their large infra-red excesses and luminous active nuclei, Seyfert galaxies represent prime candidates for the detection of CO emission. One of the first galaxies detected in CO was in fact the Seyfert NGC 1068 (Rickard et al. 1977). Since then, three additional detections and one possible detection have been made: NGC 4051, NGC 3227 (Bieging et al. 1980); NGC 6814, MKN 231 (?)(Blitz and Mathieu 1982). The spectra are typical of extra-galactic sources, broad (several hundred km/sec in width) and featureless. The line centers and shapes agree well with H I data, with the notable exception of NGC 3227. Correlations between CO luminosities and other gas/dust tracers have been searched for.

158.339 A search for carbon monoxide in clumpy irregular galaxies. M. A. Gordon, J. Heidmann, E. E. Epstein.
Extragalactic molecules, (see 012.048), p. 17 (1982). Abstract.

158.340 CO radial distributions in IC 342 and NGC 6946.
J. Young.
Extragalactic molecules, (see 012.048), p. 19 - 26 (1982).

The CO emission in two Scd galaxies, IC 342 and NGC 6946, has been mapped using the 14-m FCRAO telescope (HPBW = 50″) in collaboration with Nick Scoville. Spectra at 33 positions in IC 342 and 23 positions in NGC 6946 were obtained. CO rotation curves for these two galaxies have been derived assuming the same inclination, distance, systemic velocity, and position angle of the major axis as Rogstad and Shostak (1972), and good agreement with H I rotation curves has been found. Most of the optical luminosity measured in the B band probably arises from young stars and reflects recent star formation rates at different radii.

158.341 CO observations of galaxies in the Virgo Cluster.
M. Vietri.
Extragalactic molecules, (see 012.048), p. 27 - 40 (1982).

Preliminary results are reported of a project begun by A. A. Stark, J. Bally, J. Knapp and the author with the Bell Labs 7-m antenna. The aim of the project is the detection of an homogeneous sample of galaxies in the 1 → 0 transition of CO. To date, 12 galaxies have been observed. Two galaxies have definitely been detected: NGC 4254 and NGC 4303. Spectra for the (optical) central positions for these galaxies are shown.

158.342   The molecular cloud distribution in NGC 891 between 3 and 15 kpc.   P. M. Solomon.
Extragalactic molecules, (see 012.048), p. 41 - 50 (1982).

NGC 891 is an edge on (i = 88°) SbI galaxy with a large angular extent of 13!5 along the major axis. In this paper the author presents observations of CO emission at 2.6 mm along the major axis of NGC 891 carried out with the FCRAO 14 m antenna. He compares the distribution of molecular clouds with that of H I and nonthermal radio continuum within NGC 891 (Sancisi and Allen, 1979; Allen et al. 1978) and with the distribution of molecular clouds in our Galaxy.

158.343   The molecular distribution in M51.
N. Scoville.
Extragalactic molecules, (see 012.048), p. 51 - 56 (1982).

CO emission has been mapped (by Judy Young, Steve Lord, and the author) along four strips in M51 using the 14 m FCRAO telescope (HPBW = 50″) in order to sample the radial distribution of molecular gas.

158.344   CO observations of NGC 3628.
L. Tacconi, J. Young.
Extragalactic molecules, (see 012.048), p. 57 - 60 (1982).

CO emission was mapped in the edge-on Sb/Sc galaxy NGC 3628 with the 14-m antenna of the FCRAO (HPBW = 50″).

158.345   CO emission and the optical disk in the giant Sc galaxy M 101.   P. M. Solomon.
Extragalactic molecules, (see 012.048), p. 61 - 71 (1982).

In this paper the author presents observations of emission from 40 locations in M 101 and compares the radial dependence of CO surface emission with the optical light and the H I deduced from 21 cm. He finds that the CO emission and hence the interstellar molecular clouds follow the optical surface brightness of the disk in marked contrast to the atomic hydrogen. The dense star forming molecular clouds have a radial distribution which fits an exponential disk determined from optical surface brightness.

158.346   New observations of CO in central sources.
L. J. Rickard, P. Palmer, B. E. Turner.
Extragalactic molecules, (see 012.048), p. 73 - 75 (1982). Abstract.

158.347   Peculiarities of the CO emission from M 82.
A. A. Stark.
Extragalactic molecules, (see 012.048), p. 77 - 85 (1982).

The galaxy M 82 has been known to be anomalous in its CO emission ever since the first observations. Observations presented here will support this picture: ^{12}CO J = 2 → 1 and J = 1 → 0 lines which are consistent with previous observations, CO isotope lines which are consistent with optical thinness, and mapping data which suggests that molecular emission is not confined to the stellar disk of M 82 but is associated with the dusty filaments which lie above and below it. The rotation of the galaxy is apparent.

158.348   CO in the arms of Andromeda.
R. A. Linke.
Extragalactic molecules, (see 012.048), p. 87 - 92 (1982).

The author has obtained a filled CO map of a 15 × 18 arc minute field in M 31 selected to show a distinct dust lane. The integrated CO emission map shows a clear correlation with the optical obscuration. There is also an excellent correlation between CO and H I gas and therefore also with the distribution

of HII regions and OB associations as discussed by Unwin (1980). Thus it appears that all available tracers of the interstellar medium are strongly concentrated in the same regions: those of the optical spiral arms.

158.349   CO survey: comparison with morphological type.
F. Verter.
Extragalactic molecules, (see 012.048), p. 101 - 105 (1982).

The author reports on his thesis. The observational goal of the thesis is to establish the relationship between total galactic molecular gas content and morphological type. He is particularly interested in determining what effect the addition of molecular gas will have on the ratio of H I mass to optical luminosity, as a function of morphological type.

158.350   Ammonia in the spiral galaxy IC 342.
P. T. P. Ho, R. N. Martin.
Extragalactic molecules, (see 012.048), p. 111 - 113 (1982).

Using the MPI 100 m telescope, the (J,K) = (1,1) and (2,2) lines of NH_3 have been detected toward the spiral galaxy IC 342. Their nearly equal intensities imply the existence of warm (~ 50K) gas. Based on excitation arguments, the author finds that the NH_3 observations imply density $n(H_2) > 10^2 cm^{-3}$ and mass $M > 10^8 M_\odot$. These results, when combined with the infrared, radio continuum and CO observations, yield a consistent picture on star formation in the nucleus. More recent VLA observations of IC 342 in the NH_3 lines support the interpretation that there are probably many clouds rather than a single giant emission complex.

158.351   VLA observations of extragalactic ammonia.
R. N. Martin, P. T. P. Ho.
Extragalactic molecules, (see 012.048), p. 115 (1982).

Aperture synthesis observations were made of IC 342 in the (J,K) = (1,1) and (2,2) lines of NH_3. The authors detected the continuum source associated with the nucleus of IC 342. It is concluded that the NH_3 emission is in fact made up of a large number of hot molecular cores.

158.352   Observations of far-infrared emission from late-type galaxies.   L. J. Rickard, P. M. Harvey.
Extragalactic molecules, (see 012.048), p. 117 - 120 (1982).

The authors have been observing the central regions of spiral and Irr II galaxies from the Kuiper Airborne Observatory. Of 25 galaxies observed, 3 are well-known IR sources and 18 are new detections at 100 μ. The fluxes range from 3 Jy for M 81 and M 101 to 97 Jy for NGC 2146.

158.353   The far infrared disk of M 51.
J. Smith.
Extragalactic molecules, (see 012.048), p. 121 - 132 (1982).

Far infrared maps and multifrequency photometry are presented for M 51 and its companion galaxy NGC 5195. Dust reradiates about half the starlight of the M 51 + NGC 5195 system to produce the observed far infrared (80 - 200 μm) luminosity of $3 \times 10^{10} L_\odot$. Far infrared properties are given for several galactic-scale complexes of star formation in M 51. Relationships are discussed for selected maps of optical starlight and emissions from these components of M 51's interstellar medium: dust grains, CO molecules, hydrogen atoms, electrons and H II regions.

158.354   VLA detection of H 110α in M 82.
E. R. Seaquist, M. B. Bell, R. C. Bignell.
Extragalactic molecules, (see 012.048), p. 133 - 136 (1982).

The authors' purpose is to report the detection, with the VLA, of the H 110α line (4.8 GHz) in the gas rich galaxy M 82.

158.355   H I kinematics of spiral galaxies.
A. Bosma.
Extragalactic molecules, (see 012.048), p. 137 - 139 (1982).

In this contribution the author emphasizes a few points about H I studies and their relevance for the interpretation of the new CO observations of nearby spirals.

158.356 On the H I content and size of galaxies.
R. Giovanelli.
Extragalactic molecules, (see 012.048), p. 141 - 145 (1982).

The author asks, which other parameter, besides type, is important in predicting the H I content of a galaxy, or, whether it is possible to find a better one. A statistically homogeneous sample of 324 "isolated" galaxies was used to approach that question. The results allow the tentative conclusion that surface brightness, or even better, disk surface brightness, is the best distance-independent diagnostic optical tool for the H I content, expressed as M_H/L ratio, of galaxies of any morphological type. The optical diameter correlates with H I mass better than any other parameter; it is unfortunately dependent on the assumed distance scale.

158.357 Systematic dynamical properties of spiral galaxies.
N. Thonnard.
Extragalactic molecules, (see 012.048), p. 147 - 152 (1982).

The author has studied high resolution long slit spectra in the Hα region for 60 spiral galaxies of Hubble types Sa, Sb and Sc, and high signal to noise global 21-cm neutral hydrogen profiles for approximately 3/4 of these galaxies. The sample was chosen carefully to be well classified by Hubble type, to have a large spread in luminosity at each Hubble type, not to be strongly barred, and to be of high enough inclination to minimize deprojection errors. The author discusses some of the salient characteristics and systematic properties of spiral galaxies that can be inferred from the data. Distances are derived from the redshift using H = 50 km s^{-1} Mpc^{-1}.

158.358 Report of IAU Commission 28: Galaxies (Galaxies).
B. E. Westerlund.
Trans. IAU, Vol. XVIIIA, (see 003.013), p. 303 - 341 (1982).

A simple source catalogue of galaxies south of declination $-17^1/_2°$ that have been observed spectroscopically. First Version (1981).   See Abstr. 002.049.

Dumbbell galaxies and precessing radio jets. See Abstr. 002.068.

A bibliography of observations of molecular clouds in galaxies.   See Abstr. 002.096.

The ESO/Uppsala Survey of the ESO (B) Atlas. See Abstr. 002.098.

Galaxien.   See Abstr. 002.102.

Atlas of the Andromeda galaxy. See Abstr. 002.104.

Doppelgalaxien.   See Abstr. 003.041.

Extragalactic astronomy.   See Abstr. 003.114.

The Andromeda nebula. See Abstr. 003.116.

Analysis of the power spectra of galactic images and spiral structures.   See Abstr. 031.512.

On the deconvolution of brightness profiles of galaxies from seeing. Application to NGC 3379. See Abstr. 031.580.

The deconvolution of brightness profiles of galaxies from seeing: application to M32. See Abstr. 031.606.

Magnetoid interaction with surrounding stars. See Abstr. 062.020.

Hydromagnetic flows from accretion discs and the production of radio jets.   See Abstr. 062.123.

On the transport and propagation of relativistic electrons in galaxies. The effect of adiabatic deceleration in a galactic wind for the steady state case. See Abstr. 063.020.

The shower model for compact synchrotron sources and an intrinsic red-shifting mechanism. See Abstr. 063.026.

Non-thermal emission from relativistic accretion disks: a simple model for axisymmetric inhomogeneous sources.   See Abstr. 064.086.

Upper limits of a cosmic infrared background flux as determined by X- and gamma-ray observations of M 87. See Abstr. 066.034.

Self-gravitating accretion disk models for active galactic nuclei: self-consistent α-models for the broad emission-line region.   See Abstr. 066.092.

UBV sequences for South Polar Cap galaxies. See Abstr. 113.001.

Faint blue objects at high galactic latitude: summary of four Palomar Schmidt fields.   See Abstr. 113.067.

Wolf-Rayet stars in extragalactic H II regions: discovery of a peculiar WR in IC 1613/#3. See Abstr. 114.037.

Carbon abundances in giant stars of the Draco dwarf galaxy.   See Abstr. 114.093.

On the nature of the very luminous Wolf-Rayet stars in M33.   See Abstr. 114.169.

Wolf-Rayet stars in emission-line galaxies. See Abstr. 114.170.

Absolute spectrophotometry of Wolf-Rayet stars in the Local Group.   See Abstr. 114.182.

Spectrophotometry of Wolf-Rayet star candidates in M 33.   See Abstr. 114.199.

The observed upper limit to stellar luminosities in galaxies.   See Abstr. 115.018.

M supergiants and the distance to M101. See Abstr. 115.023.

Period distributions of irregularly variable RR Lyrae stars.   See Abstr. 122.013.

The Cepheid distance scale: a new application for infrared photometry.   See Abstr. 122.089.

Nova-like Hα sources in the center of M31. See Abstr. 122.159.

Detection of radio emission from optically identified supernova remnants in M31.   See Abstr. 125.001.

A search for radio emission from six historical supernovae in the galaxies NGC 5236 and NGC 5253. See Abstr. 125.051.

Extragalactic radio supernovae in NGC 4321 and NGC 6946.   See Abstr. 125.301.

Tidal stability of gas clouds in the Large Magellanic Cloud and M101.   See Abstr. 131.009.

Star formation in galactic spiral arms. See Abstr. 131.126.

The interstellar spectrum of the supernova 1980k in NGC 6946.   See Abstr. 131.128.

Collective formation of massive stars in galaxies. See Abstr. 131.179.

Massive star formation in NGC 6946. See Abstr. 131.184.

The inhibition of star formation in barred spiral galaxies.   See Abstr. 131.185.

Observations of H_2O maser emission in southern galaxies.   See Abstr. 131.253.

Review of current state of observations of extra-galactic molecules.   See Abstr. 131.268.

GMCs in M 31.   See Abstr. 131.269.

One perspective on the workshop "Extragalactic molecules".   See Abstr. 131.270.

Implications of collisionally-supported giant molecular clouds for spiral galactic structure and massive star formation.   See Abstr. 131.277.

An H II region in NGC 6744: spectrophotometry and chemical abundances.   See Abstr. 132.001.

Observations of NGC 604 over six decades in frequency.   See Abstr. 132.015.

H II regions and star formation in M 83 and M 33. See Abstr. 132.061.

A high excitation optically obscured H II region in the nucleus of NGC 5253.   See Abstr. 132.070.

A large shell nebula in NGC 55. See Abstr. 134.003.

Planetary nebulae in local group galaxies. IX. Velocity modulated photographs of the center of M31. See Abstr. 135.015.

Quasar pancakes.   See Abstr. 141.007.

Discovery of a narrow line quasar. See Abstr. 141.009.

Strong radio sources in bright spiral galaxies. II. Rapid star formation and galaxy-galaxy interactions. See Abstr. 141.010.

Milli-arcsecond jets in radio galaxies: interpretation. See Abstr. 141.029.

Integrated linear polarization of extragalactic radio sources at 10.5 GHz (λ 2.86 cm) – II.   See Abstr. 141.037.

Detection of the underlying galaxy in the QSO 3C48.   See Abstr. 141.040.

Time-dependent narrow emission-line profiles of quasars and active galactic nuclei.   See Abstr. 141.046.

Hydrogen line spectrum in quasars. II. A critical discussion of model calculations for the broad line region. See Abstr. 141.063.

Optical identification/flux density relationship for radio galaxies.   See Abstr. 141.065.

Theoretical studies of compact radio sources. I. Synchrotron radiation from relativistic flows. See Abstr. 141.066.

Theoretical studies of compact radio sources. II. Inverse-Compton radiation from anisotropic photon and electron distributions: general results and spectra from relativistic flows.   See Abstr. 141.067.

Multifrequency VLBI observations of the nucleus of NGC 1275.   See Abstr. 141.068.

Observations of Paschen α in the broad-line radio galaxy 3C 445.   See Abstr. 141.069.

Mehr-Frequenz-Spektralindexuntersuchungen an drei geschweiften Radiogalaxien.   See Abstr. 141.078.

The physical nature of the blue objects in the field of 88 Leonis.   See Abstr. 141.086.

Rotating compact radio sources and angular momentum transfer from the nucleus to outlying gas in active galaxies. See Abstr. 141.087.

The highly polarized galaxy 3C 76.1. See Abstr. 141.089.

Rotating gas and the shapes of radio sources. See Abstr. 141.103.

Extended optical line emission associated with radio galaxies.   See Abstr. 141.117.

Extended emission lines in radio galaxies. See Abstr. 141.118.

Emission lines: sign of a new energy source? See Abstr. 141.119.

The radio emission of interacting galaxies. See Abstr. 141.123.

Radio and X-ray structure of Centaurus A. See Abstr. 141.124.

Emission regions in Centaurus A. See Abstr. 141.125.

VLBI observations of the nucleus of Centaurus A. See Abstr. 141.126.

Recent WSRT and VLA observations of the jet radio galaxy NGC 6251.   See Abstr. 141.133.

Mechanisms for jets.   See Abstr. 141.148.

The nature of the energy source in radio galaxies and active galactic nuclei.   See Abstr. 141.156.

Structural evolution in the nucleus of NGC 1275. See Abstr. 141.161.

VLBI observations of M87.   See Abstr. 141.162.

Evidence for relativistic motion in the millisecond structure of BL Lac.   See Abstr. 141.187.

Symmetry in radio galaxies. See Abstr. 141.199.

A consequence of the asymmetry of jets in quasars and active nuclei of galaxies.   See Abstr. 141.200.

Compact companions to QSOs. See Abstr. 141.203.

Detection of neutral hydrogen emission and optical nebulosity in the low redshift QSO 0351 + 026. See Abstr. 141.204.

Deep CCD images of 3C 273. See Abstr. 141.207.

Rapid expansion of BL Lacertae. See Abstr. 141.208.

Radio continuum in nearby galaxies. See Abstr. 141.214.

The magnetic jet theory of the double sources in radio galaxies.   See Abstr. 141.247.

VLA scaled-array observations of the radio galaxy 3C166.   See Abstr. 141.248.

Superluminal behavior in BL Lacertae. See Abstr. 141.250.

Why is the compact radio source in M87 so stable? See Abstr. 141.254.

An extended radio source in the nuclear region of M31.   See Abstr. 141.256.

H I absorption against the nucleus and jet in Centaurus A (NGC 5128).   See Abstr. 141.263.

A search at the milliJansky level for milliarcsecond cores in a complete sample of radio galaxies. See Abstr. 141.268.

1107+036: an unusual QSO-galaxy pair. See Abstr. 141.271.

Extended radio galaxies. II. Physics of the trail region in double lobed sources.   See Abstr. 141.276.

Nuclear jets in Cygnus A. See Abstr. 141.297.

Multifrequency comparison of the total intensity and polarization distributions for 3C 31, 3C 66B, and 3C 129. See Abstr. 141.320.

PKS 2155 − 304: relativistically beamed synchrotron radiation from a BL Lacertae object.   See Abstr. 142.005.

Spectrophotometry of an X-ray source near M 33. See Abstr. 142.009.

X-ray emission from Centaurus A. See Abstr. 142.043.

X-ray emission from BL Lac objects: comparison to the synchrotron self-Compton models. See Abstr. 142.044.

The nuclear X-ray source in M81. See Abstr. 142.064.

Active galaxy contribution to the diffuse X-ray background.   See Abstr. 142.069.

Einstein observations of the confused 2A 2315 − 428 region.   See Abstr. 142.105.

Discovery of a new BL Lacertae object (1E 1402.3 + 0416) with the Einstein Observatory. See Abstr. 142.107.

Seyfert galaxies and the cosmic γ-ray diffuse background.   See Abstr. 142.509.

Extragalactic γ-rays.   See Abstr. 142.533.

Low energy γ-ray observations with the MISO telescope.   See Abstr. 142.534.

The processing of nuclei by electrons in active galaxies.   See Abstr. 143.040.

A model for elliptical radio galaxies with dust lanes. See Abstr. 151.001.

Using gaseous disks to probe the geometric structure of elliptical galaxies.   See Abstr. 151.002.

Chemical evolution of galaxies – II. Variation of the heavy element yield with Z.   See Abstr. 151.006.

The orientation of gas disks in tumbling prolate galaxies.   See Abstr. 151.057.

On the stability of Schwarzschild's triaxial galaxy model.   See Abstr. 151.058.

Galactic evolution: a survey of recent progress. See Abstr. 151.059.

Gravitationally induced spurs in the disks of nearby spiral galaxies.   See Abstr. 151.072.

Bar-driven spiral density waves in galaxies. See Abstr. 151.077.

On the interpretation of rotation curves measured at large galactocentric distances.   See Abstr. 151.078.

Tidal interactions of disc galaxies. See Abstr. 151.080.

The stability and masses of disc galaxies. See Abstr. 151.081.

Merging and stripping of haloes in binary galaxy systems.   See Abstr. 151.082.

The structure and evolution of nuclear discs.
See Abstr. 151.083.

Globular clusters in galaxies beyond the local group.
II. The edge-on spirals NGC 891 and NGC 4565.
See Abstr. 154.011.

Search for (globular) clusters in M31: a progress
report.   See Abstr. 154.060.

Line-strength anomalies in the integrated spectra of
M31 globular clusters.   See Abstr. 154.061.

Integrated photometry and IIDS spectrophotometry
of M33 globulars.   See Abstr. 154.062.

Reconnaissance of the newly discovered globular
cluster system around NGC 5128 (Cen A).
See Abstr. 154.063.

Metallicities in galactic and M87 globular clusters.
See Abstr. 154.094.

Comparison of galactic center with other galaxies.
See Abstr. 155.044.

The resolution of old red giant stars in the Sculptor
group of galaxies.   See Abstr. 160.004.

Radio and optical observations of 9 nearby Abell
clusters: A262, A347, A569, A576, A779, A1213, A1228,
A2162, A2666.   See Abstr. 160.006.

Transport processes and the stripping of cluster
galaxies.   See Abstr. 160.007.

A complete sample of Virgo cluster galaxies.
See Abstr. 160.013.

Systematic properties of compact groups of
galaxies.   See Abstr. 160.019.

Characteristics of companion galaxies.
See Abstr. 160.023.

The age of galaxies.   See Abstr. 162.117.

The Galaxy as a fundamental standard for extragalactic distances. II. A first crucial test of the short and long
distance scales through the use of the Tully-Fisher relations.
See Abstr. 162.119.

Formation of galaxies and clusters of galaxies in the
neutrino dominated Universe.   See Abstr. 162.143.

H_0, q_0 and the local velocity field.
See Abstr. 162.212.

The large scale distribution of galaxies.
See Abstr. 162.213.

Galaxy formation.   See Abstr. 162.218.

Errata

158.901   Erratum: "Classification parameters for the emis-
              sion-line spectra of extragalactic objects," [Publ.
Astron. Soc. Pacific, Vol. 93, 5 - 19 (1981).]
J. A. Baldwin, M. M. Phillips, R. Terlevich.
Publ. Astron. Soc. Pacific, Vol. 93, 817 (1981/82). — See
Abstr. 29.158.269.

158.902   Erratum: 'Is the H_2 mass of galaxies known?'
              [Comments Astrophys., Vol. 9, 117 - 125 (1981)].
J. Lequeux.
Comments Astrophys., Vol. 9, 239 (1982). — See Abstr.
29.158.135.

158.903   Dynamical or static radio halo — is there a galactic
              wind?   I. Lerche, R. Schlickeiser.
Astrophys. Lett., Vol. 22, 31 - 33 (1981). — See Abstr.
30.158.146.

158.904   Erratum: 'An extraordinary emission-line
              nebulosity associated with the Seyfert galaxy
Markarian 335' [Astrophys. J., Vol. 247, 32 - 41 (1981)].
T. M. Heckman, B. Balick.
Astrophys. J., Vol. 256, 798 (1982). — See Abstr. 29.158.276.

159 Magellanic Clouds

159.001 **The carbon abundance in the Magellanic Clouds from IUE observations of H II regions.**
R. J. Dufour, G. A. Shields, R. J. Talbot, Jr.
Astrophys. J., Vol. 252, 461 - 473 (1982).

Observations of C II], C III], and C IV lines in the ultraviolet spectra of three H II regions in the Small Magellanic Cloud (SMC) and of four H II regions in the Large Magellanic Cloud (LMC) were obtained with the IUE satellite and used to derive the carbon abundance in the Magellanic Clouds by nebular model analyses. Based on absolute emission line fluxes between the IUE and ground-based observations for the individual H II regions, it is found that $12 + \log C/H = 7.16 \pm 0.04$ for the SMC and 7.90 ± 0.15 for the LMC. The corresponding values for the average C/O ratios are $\log C/O = -0.89 \pm 0.02$ for the SMC and -0.48 ± 0.04 for the LMC. The implications of these results about the nucleosynthesis of CNO in the stars and galactic chemical evolution is discussed. The results are consistent with a scenario that nitrogen and oxygen have greater primary yields in massive stars than carbon, which may be produced predominantly in less massive $(10 \gtrsim M_\odot \gtrsim 4)$ stars.

159.002 **On the proper motion of the Magellanic Clouds and the halo mass of our Galaxy.**
D. N. C. Lin, D. Lynden-Bell.
Mon. Not. R. Astron. Soc., Vol. 198, 707 - 721 (1982).

It is shown that the SMC lies close to the spin plane of the LMC. There is a sufficient angle in the sky between the Magellanic Clouds that even if the LMC were moving purely transversely to the line of sight with the SMC moving with it, about one quarter of that transverse motion would be seen as a radial velocity of the SMC. Supplementing this idea with other arguments the authors deduce that the Magellanic stream trails 'behind' the Magellanic Clouds. Allowing for a possible halo to the Galaxy the authors compute models that last for 10^{10} yr and fit the observed velocities of the Magellanic stream.

159.003 **Electronographic observations of a field in the Small Magellanic Cloud.**
M. R. S. Hawkins, M. T. Brück.
Mon. Not. R. Astron. Soc., Vol. 198, 935 - 940 (1982).

A colour-magnitude diagram to $V \cong 21.5$ mag from electronographs of a field in the western periphery of the Small Magellanic Cloud shows the presence of a distinct stellar population of age $3 \pm 1 \times 10^9$ yr. Some older stars may also be present. The population appears to be widespread throughout the SMC.

159.004 **Faint nebulosities in the vicinity of the Magellanic H I Stream.**
P. G. Johnson, J. Meaburn, A. M. I. Osman.
Mon. Not. R. Astron. Soc., Vol. 198, 985 - 989 (1982).

Very deep Hα image tube photographs with a wide-field filter camera have been taken of the Magellanic H I Stream. A diffuse region of emission has been detected. Furthermore a mosaic of high contrast prints of IIIaJ survey plates taken with the SRC Schmidt, has been compiled over the same area. A complex region of faint, blue, filamentary nebulosity has been revealed. This appears to be reflection nebulosity either in the galactic plane or less probably, in the vicinity of the Large Magellanic Cloud. A deep Hα 1.2 m Schmidt photograph of these blue filaments reinforces the suggestion that they are reflection nebulae. The reflection and emission nebulosities in this vicinity have been compared to each other and the Magellanic H I Stream. The diffuse region of Hα emission is particularly well correlated with the Stream.

159.005 **Observations of the Magellanic Stream between declinations $-20°$ and $0°$.** R. J. Cohen.

Mon. Not. R. Astron. Soc., Vol. 199, 281 - 293 (1982).

The region of the Magellanic Stream between RA 23^h00^m and 00^h20^m and Dec $-20°$ and $0°$ (1950) has been mapped in the 21-cm line of neutral hydrogen using the Jodrell Bank Mk II telescope. The detection level of the measurements is 0.1 K. The Stream is much more extensive in this part of the sky than hitherto realized, and has a very complex filamentary structure. All the filaments follow a regular velocity pattern. In addition to the known gradient of velocity along the Stream there is a gradient transverse to the Stream. The Stream is very similar to tidal bridges and tails seen in the nearby M81 group of galaxies.

159.006 **The Ursa Minor dwarf galaxy is a member of the Magellanic Stream.** D. Lynden-Bell.
Observatory, Vol. 102, 7 - 9 (1982).

The elongation of the Ursa Minor dwarf galaxy shows that it is being disrupted in the plane defined by the Magellanic Stream. The Draco object is also elongated along the Stream. The orientations of the Carina and Sculptor systems are discussed.

159.007 **Identification of stars in the direction of the Large Magellanic Cloud (2nd series).**
C. Fehrenbach, M. Duflot.
Astron. Astrophys., Suppl. Ser., Vol. 48, 1 - 55 (1982). In French.

Stars of the Large Magellanic Cloud and galactic stars in the direction of the LMC are identified on charts published by Hodge and Wright (1967).

159.008 **A 21 cm hydrogen line survey of the Small Magellanic Cloud.** E. Bajaja, N. Loiseau.
Astron. Astrophys., Suppl. Ser., Vol. 48, 71 - 80 (1982).

Results of complete 21 cm H I line survey of the Small Magellanic Cloud, made with the IAR 30 m dish, are presented in the form of contour maps. The observations were made in the region $0° \leqslant \alpha \leqslant 30°$, $-76° \leqslant \delta \leqslant -70°$. The H I features of the eastern zone are briefly discussed.

159.009 **Ultraviolet absorption by interstellar gas near the LMC star HD 36402 in the interstellar bubble N51D.** K. S. de Boer, A. G. Nash.
Astrophys. J., Vol. 255, 447 - 457 (1982).

The authors have studied ultraviolet high-dispersion IUE spectra of the LMC star HD 36402 in the nebulosity N51D. Apart from absorption due to Milky Way gas, they find four interstellar absorption components associated with the immediate surroundings of HD 36402. The absorption at 305 km s^{-1} originates in low-density, $n(e) \approx 0.1$ cm^{-3}, 10^4 K gas. At 280 km s^{-1}, strong absorption from the excited fine-structure level of carbon points to densities $n(e) > 1$ cm^{-3} at 10^4 K. Additional weak absorption by common interstellar ions is seen near 260 km s^{-1}. From a fit to the observed Lyα profile, it is found that there is $N(H) \approx 10^{20.2}$ cm^{-2} in front of HD 36402. The fourth component near 270 km s^{-1} shows strong N V, C IV, and Si IV, more than is consistent with a wind-blown interstellar bubble; hence there is additional absorption outside the bubble. The overall pattern of absorption-line strengths suggests solar abundance ratios for the metals.

159.010 **The gas to dust ratio and the near-infrared extinction law in the Large Magellanic Cloud.**
J. Koornneef.
Astron. Astrophys., Vol. 107, 247 - 251 (1982).

The interstellar Lyman-alpha absorption-profile in the Large Magellanic Cloud (LMC) is presented and discussed.

The gas to dust ratio in the LMC derived from this profile is $N(H I)/E(B-V)= 2 \times 10^{22}$ (atoms cm^{-2} mag^{-1}) which is a factor of four higher than the galactic value. To check whether the observed dust deficiency can be explained by an anomalous large grain size distribution, the author has obtained near-infrared photometry in the J, H, and K bands of the early-type supergiants used for the derivation of the Lyman-alpha profile. It is concluded that the infrared extinction law in the LMC is very similar to its galactic counterpart. It therefore appears that the high gas to dust ratio derived would still hold if expressed as a ratio by mass.

159.011   **Carbon monoxide in the Magellanic Clouds.**
          F. P. Israel, T. de Graauw, S. Lidholm,
H. van de Stadt, C. de Vries.
Regions of recent star formation, (see 012.002), p. 433 - 438 (1982).
    The authors have detected ^{12}CO(J = 2 − 1) emission at four positions in the LMC, in the vicinity of 30 Doradus. Three other positions yielded negative results, as did four positions in the main body of the SMC. The results indicate the presence of a giant molecular cloud complex associated with the H II regions N159 and N160A, comparable to galactic complexes.

159.012   **Luminosity functions of star clusters in the Small Magellanic Cloud.**   M. Kontizas, E. Kontizas.
Astron. Astrophys., Vol. 108, 344 - 347 (1982).
    The luminosity functions (LF) of eight old star clusters of the SMC have been produced and compared with the LFs of galactic globular clusters.

159.013   **H$_2$O maser survey in the Magellanic Clouds.**
          E. Scalise, Jr., M. A. Braz.
Astron. J., Vol. 87, 528 - 531 (1982).
    An extensive search for water vapor maser emission towards the Large and Small Magellanic Clouds is reported. The 6_{16} - 5_{23} transition of water molecule was detected in two H II regions in LMC (including N 159, previously was detected) and in two H II regions in SMC. A total of 32 positions was searched in both Clouds. The strength of the sources detected is comparable with galactic masers with luminosities $L \lesssim 10^{-2} L_\odot$.

159.014   **Ofpe/WN9 circumstellar shells in the Large Magellanic Cloud.**   N. R. Walborn.
Astrophys. J., Vol. 256, 452 - 459, plates 13 - 16 (1982).
    High-resolution, 4 m echelle spectrograms of four luminous Of-like objects in the LMC with very extended envelopes are discussed. The most remarkable result is the presence of velocity-doubled nebular lines, including very strong [N II] λλ6548, 6583, in three of them, which are not in H II regions. Analysis of the nebular lines indicates enhanced nitrogen abundances relative to hydrogen and oxygen, in comparison with H II regions of the LMC. These expanding shells might be regarded as supergiant planetary nebulae, and they likely provide another example of nitrogen enrichment by massive stars. Some possibly related objects in the Galaxy are noted. The question of whether these spectra should be classified as Ofpe or WN 9is considered, with reference to spectrograms of two additional LMC stars and three in the Galaxy for which the latter type has been suggested. No identical counterparts to the six LMC objects are known in the Galaxy.

159.015   **The globular clusters of the Small Magellanic Cloud in the general diagram magnitude-diameter.**
H. Wilkens.
Bol. Asoc. Argentina Astron., No. 18, p. 72 - 73 (1980).

159.016   **The age-abundance relation for Magellanic Cloud star clusters.**   L. Searle, H. A. Smith.

Astrophysical parameters for globular clusters, (see 012.023), p. 201 - 202 (1981). − Abstract.

159.017   **Ages and abundances of Magellanic Cloud clusters.**
          P. Hodge.
Astrophysical parameters for globular clusters, (see 012.023), p. 205 - 214 (1981).
    This paper gathers age estimates, mostly based on main sequence photometry, for 48 LMC rich ("globular") clusters and 18 SMC globulars. Information on abundances in the clusters is gathered together from a variety of sources and compared for a total of 24 LMC and 5 SMC clusters. Resulting ages and abundances show a general trend of decreasing heavy elements with increasing age, though there is a surprising spread in the abundance associated with any given age.

159.018   **UV observations of globular clusters in the Magellanic Clouds.**   C. Cacciari, F. Fusi Pecci.
Astrophysical parameters for globular clusters, (see 012.023), p. 217 - 221 (1981).
    Ultraviolet observations of Galactic globular clusters have been taken with ANS, OAO 2 and IUE and, among other results, have led to the classification of galactic globulars in four classes according to the characteristics of their UV integrated spectrum. In the present paper the UV spectra of four globular clusters in the Small and Large Magellanic Cloud are shown and compared with their galactic "counterparts", in order to obtain some preliminary indication about the possible occurrence of the "second parameter" problem in the classification of MC globular clusters as well.

159.019   **New data for old Magellanic clusters.**
          S. C. B. Gascoigne, M. S. Bessell, J. Norris.
Astrophysical parameters for globular clusters, (see 012.023), p. 223 - 228 (1981).
    This paper is concerned with recent work on age and abundance determinations for a number of SMC clusters.

159.020   **The age-metallicity relationship for the clusters of the Large Magellanic Cloud.**   J. G. Cohen.
Astrophysical parameters for globular clusters, (see 012.023), p. 229 - 233 (1981).
    Spectrophotometric scans with an intensified reticon array have been obtained for 40 stars which are probable members of 15 LMC clusters. A series of calibration spectra of 42 stars in six galactic globular clusters and several galactic supergiants were obtained with the same instrumental configuration. Ages for the clusters are deduced from a transformation applied to the Searle, Wilkinson, and Bagnuolo (1980) classifications; for those clusters without SWB classes, the authors have classified them based on their published color-magnitude diagrams. Abundance is tightly correlated with age (i.e., with SWB class) for the LMC cluster system. The spread at all SWB classes except VII is within the range expected from observational and analytical errors. Thus the metallicity, except possibly during the initial formation of the LMC halo population with a mean age of about 12 billion years, has increased monotonically and smoothly with time. The data for the LMC cluster system can be fit by a simple model of an isolated well-mixed region converting gas into stars at a constant rate and in which the nuclear yield is constant.

159.021   **Early chemical and dynamical evolution of the LMC.**
          F. D. A. Hartwick, A. P. Cowley.
Astrophysical parameters for globular clusters, (see 012.023), p. 245 - 249 (1981).
    Spectroscopic observations have been made of individual red giants in nine of the oldest LMC globular clusters using the Cassegrain image-tube spectrograph on the 4-meter telescope at CTIO. In addition, similar observations were made of a number of red giants in six galactic globular clusters covering a range of metal abundances. These latter spectra were used as

standards for both radial velocity and metallicity determination. The authors give the mean cluster radial velocity together with the number of observations on which the velocity is based. All velocities have an uncertainty of ~ ±20 km s⁻¹. Metallicities for the clusters were determined by measurement of a number of different features including: the G band (CH), the continuum break at the Ca II lines, the average strength of the metallic lines, the λ3883 CN band, and the strength of Hγ relative to nearby Fe lines.

159.022 The globular clusters in the Magellanic Clouds: theoretical considerations. A. Renzini.

Astrophysical parameters for globular clusters, (see 012.023), p. 251 - 254 (1981).

159.023 A main-sequence luminosity function in the Large Magellanic Cloud. L. L. Stryker, H. R. Butcher.

Astrophysical parameters for globular clusters, (see 012.023), p. 255 - 260 (1981).

159.024 Color-magnitude diagram of a field in the SMC. M. T. Brück, M. R. S. Hawkins.

Astrophysical parameters for globular clusters, (see 012.023), p. 261 - 266 (1981).

In the Magellanic Clouds evidence accumulates that there is a widespread population of intermediate age (i.e. 1-5 × 10⁹ years). The method of deriving ages of stellar groups from the turn-off point of the main sequence color-magnitude diagrams has been widely applied in the SMC. While clusters provide individual ages, a study of the outer (halo) regions of the Cloud is important since, by analogy with our own Galaxy, it may be expected to represent the earliest generation of stars. That the SMC possesses an extensive halo in the physical meaning of a spherical envelope is known from faint star counts (Brück 1978, 1980) and the two fields the authors have studied (Brück and Marsoglu 1978; Hawkins and Brück 1981) are both situated in such peripheral regions. In this paper the authors discuss mainly the more recent of these studies.

159.025 NGC 2257 and the halo of the Large Magellanic Cloud.

L. L. Stryker, H. R. Butcher, J. L. Jewell.
Astrophysical parameters for globular clusters, (see 012.023), p. 267 - 271 (1981).

The authors study 16 regions, 9.1 degrees northeast of the bar of the Large Magellanic Cloud, surrounding the old, red globular cluster NGC 2257. They extend currently available photometry (Gascoigne 1978) down to the main-sequence turnoff and compare the cluster with standard galactic globular clusters and with the halo field, and finally the halo field to the disk field.

159.026 OB stars in the Small Magellanic Cloud. D. Crampton, J. Greasley.

Publ. Astron. Soc. Pacific, Vol. 94, 31 - 35 (1982).

Spectral classification of 19 OB stars in the SMC yields a true distance modulus of 19ᵐ1 and indicates that the foreground reddening is negligible. Through comparison of H-R diagrams, the SMC is found to be 0ᵐ5 more distant than the LMC. Equivalent-width measurements indicate that the C III, N III, and Si lines are 2 to 3 times weaker in the SMC stars than in stars in either the LMC or the Galaxy. Projected rotational velocities are also given for each star.

159.027 La estructura Hα de la Nube Menor de Magallanes. G. Carranza.

Bol. Asoc. Argentina Astron., No. 20 - 24, p. 268 (1981). Abstract.

159.028 Relevamiento de H I en las Nubes de Magallanes. E. Bajaja, N. Loiseau.

Bol. Asoc. Argentina Astron., No. 20 - 24, p. 495 (1981). Abstract.

159.029 Wolf-Rayet stars in the Magellanic Clouds. I. The WN3 binary AB 6 in the SMC. A. F. J. Moffat.

Astrophys. J., Vol. 257, 110 - 115 (1982).

Moderate dispersion spectroscopy shows AB 6 to consist of a ≳ 6 $M_\odot$ WN3 star in mutual 6ᵈ9 orbit about a ≳ 37 $M_\odot$ star of type O7. The mass ratio, WR:OB, is low compared to galactic WN binaries of high ionization, although there are no binaries with WN components as hot as this known in the Galaxy. It is possible but unlikely that the low mass ratio may be caused by the inclusion of another unresolved OB star.

159.030 Wolf-Rayet stars in the Magellanic Clouds. II. The peculiar eclipsing binary HD 5980 in the SMC.

J. Breysacher, A. F. J. Moffat, V. S. Niemela.
Astrophys. J., Vol. 257, 116 - 124 (1982).

Over seventy spectra from two observatories are combined to study this eclipsing W-R binary, which has components of type WN4 + O7 I:. The orbital period, $P = 19^d266 \pm 0^d003$, and the high eccentricity, $e = 0.49 \pm 0.10$, are confirmed. The width of the strongest optical emission line, He II 4686, decreases by at least a factor of 2 at each eclipse. Radial velocity variations of He II 4686 probably do not reflect the true W-R orbit; assumption of a mass ratio similar to those observed for other early-WN binaries, along with the more reliable absorption-line orbit of the O component, yield minimum masses of 8 $M_\odot$ for the W-R star and 27 $M_\odot$ for the O-star.

159.031 Atlas de hidrógeno neutro en la Nube Menor de Magallanes. E. Bajaja, N. Loiseau.

Bol. Asoc. Argentina Astron., No. 26, p. 124 - 128 (1981).

159.032 Is there a gaseous halo around the Large Magellanic Cloud? J. V. Feitzinger, T. Schmidt-Kaler.

Astrophys. J., Vol. 257, 587 - 591 (1982).

Arguments are given which show that the evidence for a hot gaseous halo around the Large Magellanic Cloud is not at all conclusive. The observations of interstellar lines in the UV, optical, and radio domain are well represented by a kinematic model of the LMC with a warped, two-layer structure.

159.033 Metal abundances and ages for some Magellanic Cloud variable stars.

D. Butler, P. Demarque, H. A. Smith.
Astrophys. J., Vol. 257, 592 - 602 (1982).

The authors have determined the ΔS metal abundance parameter for fifteen Magellanic Cloud field variable stars. RR Lyrae stars of ordinary luminosity in an outlying LMC field have [Fe/H] = −1.4, while those in the SMC have [Fe/H] = −1.8. Within the errors, the difference is only marginally significant. Metal abundance was also determined for two groups of variable stars which are overluminous for their periods when compared to normal Population II Cepheids. Pulsation mass and evolutionary age have also been determined. Ages range from ~2.5 × 10⁸ yr for the metal-rich stars in the central region of the SMC to ~14 × 10⁹ yr for the outlying, ordinary RR Lyrae stars. The central region of the SMC is only now reaching the degree of chemical enrichment which the Galaxy attained ten billion years ago. Both the LMC and SMC contain an old, metal-poor population of "halo" stars, just as the Galaxy does. In sharp contrast to the case for the Galaxy, however, a great deal of the total SMC chemical enrichment has taken place within the past few billion years.

159.034 Structure and distance modulus of the Small Magellanic Cloud from blue supergiants.

M. Azzopardi.
The most massive stars, (see 012.034), p. 227 - 241 (1981).

Recent determinations of the distance modulus of the

Small Magellanic Cloud (SMC) are critically assessed. A new measurement of this parameter is then described. It is based on deriving absolute visual magnitudes from equivalent $H\gamma$ widths for blue supergiants. The relation is calibrated by analysing the spectra of 103 standard stars including supergiants from spectral type O7 to F2. It is then applied to 304 SMC member stars. A mean true distance modulus of the SMC bar of $(V_0 - M_V) = 19.1 \pm 0.1$ is derived for B0 to A3 stars.

159.035 **Stellar populations in the Large Magellanic Cloud.**
L. L. Stryker.
Bull. American Astron. Soc., Vol. 13, 892 (1981). − Abstract.

159.036 **Observed radii and structural parameters of clusters in the SMC.**   M. Kontizas, E. Danezis, E. Kontizas.
Astron. Astrophys., Suppl. Ser., Vol. 49, 1 - 12 (1982).
The structural parameters of 20 star clusters of different types in the SMC have been obtained by means of star counts. Tidal and core radii of the clusters have been found and their masses are calculated. The derived masses are found to be about 10 times smaller than those in our Galaxy. The concentration parameters of the "red" clusters are comparable to values found in globular clusters of our own Galaxy, whereas the "blue" clusters do not appear similar to open clusters, showing globular cluster dynamical behaviour.

159.037 **Study of the Large Magellanic Cloud with the Fehrenbach objective-prism.**
C. Fehrenbach, M. Duflot, R. Burnage.
Messenger, No. 28, p. 8 - 9 (1982).

159.038 **Infrared interstellar extinction in the LMC.**
D. H. Morgan, K. Nandy.
Mon. Not. R. Astron. Soc., Vol. 199, 979 - 986 (1982).
$J(1.25 \, \mu m)$, $H(1.65 \, \mu m)$ and $K(2.20 \, \mu m)$ magnitudes are presented for early-type supergiants in the Large Magellanic Cloud, the observations being obtained at the 3.9-m Anglo-Australian telescope. A value of 3.21 ± 0.13 for the ratio of total to selective extinction, R, is derived for the LMC from the colour excess ratio E_{V-K}/E_{B-V}. This value is the same as the mean value of R for the Galaxy within the errors of observations. The emission line stars show infrared excesses which are consistent with emission from electrons in shells around the stars.

159.039 **The structure of the Small Magellanic Cloud.**
M. T. Brück.
Compendium in astronomy, (see 003.012), p. 297 - 313 (1982).
A survey of the distribution of various types of object in the Small Magellanic Cloud is used to build up a picture of the stellar content of its dominant features and to reconstruct in broad outline the evolutionary history of this galaxy as a whole.

159.040 **Summary of ^{12}CO $(2 \rightarrow 1)$ observations of the Magellanic Clouds.**
T. de Graauw, F. P. Israel, H. van de Stadt.
Extragalactic molecules, (see 012.048), p. 107 - 109 (1982).
During two observing runs in 1980 and 1981 the authors have made observations of ^{12}CO $(2 \rightarrow 1)$ emission in the direction of the Large and Small Magellanic Clouds. A summary of the results is given. The observations of CO(2→1) in the SMC are the first detections of CO in that galaxy.

Radial velocities from objective-prism plates in the direction of the Large Magellanic Cloud.
See Abstr. 002.063.

Der Massenverlust von SK-67/111.
See Abstr. 112.015.

Comparison of winds in the Small Magellanic Cloud and galactic early-type stars.   See Abstr. 112.022.

VBLUW **photometry of Magellanic Cloud super- and hypergiants, made in 1977 up to 1979.**
See Abstr. 113.010.

WC stars in 30 Doradus.   See Abstr. 114.013.

International Ultraviolet Explorer spectroscopy of hot stars in the LMC and SMC: the SMC extinction law, stellar flux distributions, and details of the stellar winds.
See Abstr. 114.032.

A carbon star in the globular cluster Lindsay 102.
See Abstr. 114.038.

R 136a − ein supermassiver Stern.
See Abstr. 114.060.

Der Emissionslinienstern R 50 in der Kleinen Magellanschen Wolke.   See Abstr. 114.061.

R 81: P Cygni of the LMC.   See Abstr. 114.141.

Simultaneous IUE and ground-based spectroscopic observations of the variable LMC star R 71.
See Abstr. 114.154.

H/He ratios for WN stars in the Galaxy and LMC.
See Abstr. 114.171.

Equivalent width measurements in galactic supergiant and in Small Magellanic Cloud star spectra.
See Abstr. 114.215.

R 136a in 30 Dor: an overluminous, supermassive object.   See Abstr. 115.019.

The properties of R 136, the central object in 30 Doradus.   See Abstr. 115.020.

Speckle observations of R136a.
See Abstr. 115.031.

Infrared photometry of Mira variables in the LMC and the pulsational properties of Miras.   See Abstr. 122.017.

The period-luminosity relation. IV. Intrinsic relations and reddenings for the Large Magellanic Cloud Cepheids.
See Abstr. 122.020.

Peculiarities of the cepheid distribution in the Large Magellanic Cloud.   See Abstr. 122.023.

Spectroscopic observations of the S Dor variable R71.
See Abstr. 122.053.

The Cepheid distance scale: a new application for infrared photometry.   See Abstr. 122.089.

Eruptive variable in the Large Magellanic Cloud.
See Abstr. 122.130.

New observational results on S Dor variables.
See Abstr. 122.145.

Red supergiant variables and their possible use as distance indicators.   See Abstr. 122.147.

High dispersion spectroscopy of the LMC star S Doradus during maximum light.   See Abstr. 122.247.

X-ray spectral classification of supernova remnants in the Large Magellanic Cloud.   See Abstr. 125.008.

Tidal stability of gas clouds in the Large Magellanic Cloud and M101.   See Abstr. 131.009.

Dynamics of the supergiant shell LMC 2 in the Large Magellanic Cloud.   See Abstr. 131.264.

Infalling clouds with very high velocities: a collision with the Milky Way in the anticenter.   See Abstr. 132.022.

Ring nebulae associated with Wolf-Rayet stars in the Large Magellanic Cloud. II. Kinematics of DEM 45, 137, 165, 174, and 208.   See Abstr. 134.008.

Ultraviolet spectroscopy of planetary nebulae in the Magellanic Clouds.   See Abstr. 135.004.

Radio pulsar in the Large Magellanic Cloud. See Abstr. 141.530.

A study of Sk 160/SMC X-1. See Abstr. 142.001.

A study of ultraviolet spectroscopic and light variations in the X-ray binaries LMC X-4 and SMC X-1. See Abstr. 142.028.

Ultraviolet spectra of the X-ray transient A0538−66.   See Abstr. 142.099.

Comparisons of the HR diagrams of the youngest clusters in the Galaxy, the LMC and SMC. Evidence for a large MS widening.   See Abstr. 153.008.

On the radial colour variation in nine young populous clusters in the LMC.   See Abstr. 153.052.

The late-type stellar content of Magellanic Cloud clusters.   See Abstr. 154.002.

The extended giant branches of intermediate age globular clusters in the Magellanic Clouds. II. See Abstr. 154.009.

An ellipticity–age relation for globular clusters in the Large Magellanic Cloud − I. Measurements. See Abstr. 154.014.

The billion-year-old clusters of the Magellanic Clouds.   See Abstr. 154.017.

Magellanic Cloud globular clusters and their galactic counterparts in the ultraviolet.   See Abstr. 154.021.

Are there variable stars in the LMC star clusters NGC 2210 and NGC 1786?   See Abstr. 154.043.

The mass (=star number) radius relations for globular and open star clusters of the Large Magellanic Cloud. See Abstr. 154.044.

On the ellipticities of "blue" and "red" globular clusters in the Large Magellanic Cloud. See Abstr. 154.045.

Color-magnitude diagrams of the SMC clusters NGC 152 and NGC 176.   See Abstr. 154.046.

Superluminous giants in globular-like clusters. See Abstr. 154.073.

A comparison between the observational and theoretical H-R diagrams for the LMC star cluster 1866. See Abstr. 154.092.

The system of old clusters in the Magellanic Clouds. See Abstr. 154.095.

On the structure of the 'red' globular cluster NGC 1806 in the Large Magellanic Cloud. See Abstr. 154.096.

The age-metallicity relationship for the clusters of the Large Magellanic Cloud.   See Abstr. 154.098.

Report of IAU Commission 28: Galaxies (Galaxies). See Abstr. 158.358.

160 Groups of Galaxies, Clusters of Galaxies, Superclusters

160.001 **On intracluster Faraday rotation. II. Statistical analysis.**   J. M. Lawler, B. Dennison.
Astrophys. J., Vol. 252, 81 - 91 (1982).

Faraday rotation measurements reported in an earlier paper are combined with other data, and a sample of radio sources having reliable rotation measures and seen through rich clusters of galaxies is constructed. Comparison of this sample with sources seen through the outer parts of clusters, and thus having negligible intracluster Faraday rotation, indicates that the distribution of rotation measure in the former population is broadened. The observed broadening was reproduced by Monte Carlo simulation, employing a simple physical model for the intracluster medium. A limit analysis indicates that intracluster magnetic fields must be tangled.

160.002 **Test for a richness-dependent component in the systemic redshifts of galaxy clusters.**
H. J. Rood, M. F. Struble.
Astrophys. J., Lett., Vol. 252, L7 - L10 (1982).

The sign of the redshift difference for pairs of Abell galaxy clusters is independent of the sign of the richness difference. The average redshift difference between the richer and poorer clusters in a pair is zero km s^{-1} to within 1 σ uncertainty. This result from a sample of 84 close pairs of Abell clusters and several subsamples supersedes and contradicts a previous result by Nottale based on fewer and more heterogeneous pairs.

160.003 **Determination of spatial velocity dispersion profile and stream velocity field in galaxy clusters: application to Coma.**   H. V. Capelato, D. Gerbal, G. Mathez, A. Mazure, E. Salvador-Solé.
Astrophys. J., Vol. 252, 433 - 438 (1982).

A new method, using the usual projections and inversion techniques, is given in order to recover the spatial velocity dispersion profile of clusters of galaxies. Application to the Coma cluster leads to the following description: a central steady configuration extending over 2 Mpc (H_0 = 75 km s^{-1} Mpc^{-1}), and a collapsing region up to $\sim$10 Mpc. The maximum infall velocity is $\sim$1250 km s^{-1} at $\sim$3.5 Mpc, which is fully compatible with theoretical infall velocity profiles derived for large radii and reasonable values of the cosmological density parameter Ω.

160.004 **The resolution of old red giant stars in the Sculptor group of galaxies.**   J. A. Graham.
Astrophys. J., Vol. 252, 474 - 480 (1982).

Photographs are reproduced showing the resolution of the old red giant stars in the galaxies NGC 55 and NGC 300 which belong to the nearby Sculptor group of galaxies. The red magnitude of these stars is estimated to be 22.0±0.3 mag which corresponds to a visual magnitude of 23.0 mag. A comparison with NGC 205, a companion galaxy of M31, suggests an upper limit of 1.9 Mpc (modulus 26.4 mag) for the distances to NGC 55 and NGC 300. A similar degree of resolution is not seen in NGC 253 which is probably a more distant object. It is concluded that the Sculptor galaxies are distributed over a large volume of space with NGC 55 and NGC 300 on the near side of the loose aggregate.

160.005 **Orientations of galaxies in the Local Supercluster.**   H. T. MacGillivray, R. J. Dodd, B. V. McNally, H. G. Corwin Jr.
Mon. Not. R. Astron. Soc., Vol. 198, 605 - 615 (1982).

The distribution of position angles and ellipticities for a sample of 727 spiral and irregular galaxies, selected on the basis of brightness and radial velocity from the Second Reference Catalogue of Bright Galaxies, is analysed for non-random effects. A marginally significant tendency is found for galaxies to be aligned along the plane of the Local Supercluster. This preferential alignment effect is found to exist mainly for galaxies at high supergalactic latitude and for galaxies which are seen nearly edge-on. The results are interpreted as supporting the view that superclusters formed prior to the formation of the constituent galaxies and clusters.

160.006 **Radio and optical observations of 9 nearby Abell clusters: A262, A347, A569, A576, A779, A1213, A1228, A2162, A2666.**
C. Fanti, R. Fanti, L. Feretti, A. Ficarra, I. M. Gioia, G. Giovannini, L. Gregorini, F. Mantovani, B. Marano, L. Padrielli. P. Parma, P. Tomasi, G. Vettolani.
Astron. Astrophys., Vol. 105, 200 - 218 (1982).

A survey of 61 Abell clusters included in the HEAO-2 satellite observing program was carried out at 1.4 GHz with the Westerbork Synthesis Radio Telescope. In this paper, which is the first of a series, the authors present data on the nine clusters of distance class 1 and 2 and make statistical considerations about the radio emission and structure of the cluster galaxies. The bivariate radio luminosity function of the cluster galaxies has been determined for the three morphological types, E, S0 and S+Irr, dividing these into three different classes of absolute optical magnitude. There is a marginal evidence that spiral galaxies in clusters have a lower probability of being radio sources than field ones.

160.007 **Transport processes and the stripping of cluster galaxies.**   P. E. J. Nulsen.
Mon. Not. R. Astron. Soc., Vol. 198, 1007 - 1016 (1982).

The effects of viscosity, thermal conduction and turbulence on the flow of hot gas past a galaxy are considered. Transport processes are found to cause stripping of gas from galaxies at a rate which can often exceed that due to ram pressure alone. The results are applied to M86 in the Virgo cluster and UGC 6697 in A1367. The X-ray image of M86 suggests that ion mean free paths in the Virgo intergalactic medium are much smaller than their magnetic field free values, while turbulent viscous stripping provides a natural explanation for the appearance of UGC 6697.

160.008 **On the mass of the Local Group and the motion of its barycentre.**   J. Einasto, D. Lynden-Bell.
Mon. Not. R. Astron. Soc., Vol. 199, 67 - 80 (1982).

The hypothesis that M31 and the Galaxy have almost equal and opposite momenta when seen from the barycentre of the Local Group puts a severe constraint on the component of the Sun's motions towards M31. Using this constraint and eliminating distant satellites of Andromeda from the solution, the motion of the Galaxy is found to lie 27° ± 15° from the direction of M31 and the total relative velocity of these galaxies is deduced to be 137 ± 20 km s^{-1}. Better bounds are deduced for the mass of the Local Group.

160.009 **A flower in Virgo.**   M. M. Waldrop.
Science, Vol. 215, 953 - 955 (1982).

A 9-year survey has produced the first detailed maps of the Local Supercluster, providing new evidence about how the universe evolved.

160.010 **X-ray emission from clusters of galaxies containing classical double radio sources.**
J. P. Vallée, A. H. Bridle.
Astrophys. J., Vol. 253, 479 - 484 (1982).

The authors report X-ray observations of three Abell

galaxy clusters, each containing an undeformed classical double radio source. The observations were made with the Einstein Observatory in the wavelength range 2.8–83 Å with 2' resolution. The authors derive upper limits to the X-ray emission from two of the three clusters, Abell 643 and Abell 1562. The X-ray emission of the third cluster, Abell 1763, has been mapped, showing the center and extent of its thermal gas. The classical double radio galaxy 1133+412 is near the center of the X-ray emitting region. The thermal pressure of the gas associated with the extended X-ray source can withstand the pressure of the relativistic particles and magnetic fields in the radio components. A limit of at most 1500 km s^{-1} can be placed on the velocity of separation of the radio components from the parent galaxy.

160.011 **The velocity structure of the Virgo S cloud of galaxies.** G. de Vaucouleurs.
Astrophys. J., Vol. 253, 520 - 525 (1982).

Precise distance moduli μ_0 from optical tertiary indicators and from 21 cm line widths are presented for 37 spiral galaxies of types Sab to Sdm in the Virgo S cloud. The sample is complete down to total magnitude B_T = 12.5 (27 objects) and 79% complete to B_T = 13.0 (33 objects). This material is used to: (1) derive the mean distance modulus of the S cloud, $\langle\mu_0\rangle$ = 30.88 ± 0.09 (Δ = 15.0 Mpc), in good agreement with previous determinations. The mean redshift of this sample is $\langle V_0\rangle$ = 1165 ± 133 km s^{-1} and the apparent Hubble constant H = 95 km s^{-1} Mpc^{-1} after correction for a Virgocentric velocity component of the Local Group of ~250 km s^{-1} ; (2) establish the velocity-distance relation in the direction of and within the cloud. The nonmonotonic relation is interpreted by a model consisting of a collapsing core (or shell) surrounded by an escaping halo of the kind predicted by numerical simulations of cluster evolution. The depth of the cluster derived from the distance moduli is in good agreement with the diameter ($D \approx$ 3.0 Mpc) indicated by the apparent diameter.

160.012 **A note on the dynamics of galaxy clusters.** R. G. Cooper, R. H. Miller.
Astrophys. J., Vol. 254, 16 - 21 (1982).

A single massive object forms as a result of galaxy mergers when "inelasticity" of galaxy collisions is included in a dynamical simulation of galaxy clusters. The massive object can contain as much as 80% of the cluster mass. There are very few merged objects other than the single massive object. A cluster like those seen today can form a single massive object in a burst of merging that lasts only about a crossing time; the burst can occur within the lifetime of a typical cluster (a Hubble time). The merger picture may provide a useful model for formation of poor clusters with a cD galaxy, but it does not fit the rich clusters.

160.013 **A complete sample of Virgo cluster galaxies.** R. C. Kraan-Korteweg.
Astron. Astrophys.,Suppl. Ser., Vol. 47, 505 - 512 (1982).

A catalog is presented of a complete sample of 180 Virgo cluster galaxies. The galaxies obey the conditions of lying within 6° of M 87 and of being brighter than B_T = 14.^{m}00. Positions, morphological types, magnitudes (including internal absorption corrections) as well as velocities are compiled. The few background galaxies are identified.

160.014 **A Westerbork survey of clusters of galaxies. XIV. Abell 779 and Abell 1314.**
A. S. Wilson, J. P. Vallée.
Astron. Astrophys.,Suppl. Ser., Vol. 47, 601 - 609 (1982).

The clusters of galaxies Abell 779 and Abell 1314 have been observed with the Westerbork telescope at 610 MHz (λ 49 cm) to a resolution of about 1 arc min and sensitivity (5 × r.m.s.) near 10 mJy. 111 radio sources have been detected in all. Optical identification shows that 5 or 6 radio sources are probably associated with bright cluster galaxies. The differential count of those radio sources apparently not associated with cluster galaxies has been derived in the flux density range 10–500 mJy, and is found to be similar to that obtained at 408 MHz in the 5C 5 survey.

160.015 **On the determination of velocity dispersions for cD clusters of galaxies.**
H. Quintana, D. G. Lawrie.
Astron. J., Vol. 87, 1 - 6 (1982).

The widely held belief that cD galaxies lie at the kinematical centers of their parent clusters is examined in detail. The authors find this assumption to be supported by a statistical analysis of all available velocity data. Such an assumption enables to obtain a direct measure of the true mean velocity of the cluster.

160.016 **CCD photometry of two distant clusters.**
M. Bautz, E. Loh, D. T. Wilkinson.
Astrophys. J., Vol. 255, 57 - 64 (1982).

Four color photometry of two distant clusters, Abell 370 (z = 0.37) and Cl2244−02 (z = 0.33), has been obtained with the Princeton CCD spectrometer-imager. Filters with passbands 1000 Å wide, centered at 5500, 6500, 7500, and 8500 Å, have been used. The magnitude limit of the photometry is 1.5 mag dimmer than the brightest member of Cl2244−02 and 2.0 mag dimmer than the brightest member of Abell 370. The distribution of color indices is compared with the hypothesis that all cluster members observed are elliptical galaxies.

160.017 **Abell 2069: an X-ray cluster of galaxies with multiple subcondensations.** I. M. Gioia, M. J. Geller, J. P. Huchra, T. Maccacaro, J. E. Steiner, J. Stocke.
Astrophys. J., Lett., Vol. 255, L17 - L21, plates L2, L3 (1982).

The authors present X-ray and optical observations of the cluster Abell 2069. The cluster is at a mean redshift of 0.116. The cluster shows multiple condensations in both the X-ray emission and in the galaxy surface density and, thus, does not appear to be relaxed. There is a close correspondence between the gas and galaxy distributions which indicates that the galaxies in this system do map the mass distribution, contrary to what might be expected if low-mass neutrinos dominate the cluster mass.

160.018 **Effects of different weighting procedures on the virial parameters of groups of galaxies.**
G. Giuricin, F. Mardirossian, M. Mezzetti.
Astrophys. J., Vol. 255, 361 - 372 (1982).

In view of the different procedures adopted by various authors in the estimate of the virial parameters of small groups of galaxies, the authors check whether different weighting procedures and different hypotheses on the dynamical stage of groups lead to significantly diverging results. This check was performed on the groups listed in the Turner and Gott catalog. The authors find that the values of the virial parameters are largely insensitive to different weighting procedures (including velocity and energy equipartition cases). In addition to the smallness of the crossing times, this result, together with the lack of correlation between the fraction of elliptical plus lenticular galaxy members and the linear crossing times, provides further evidence of velocity equipartition.

160.019 **Systematic properties of compact groups of galaxies.** P. Hickson.
Astrophys. J., Vol. 255, 382 - 391 (1982).

One hundred compact groups of galaxies have been identified by a systematic search of the Palomar Observatory Sky Survey red prints. Each group contains four or more galaxies within a 3 mag range, has an estimated mean surface brightness brighter than 26.0 mag arcsec^{-1}. An analysis of estimated galaxy magnitudes, morphology, and group angular size indicates: (1) There is no correlation between group density and magnitude difference between the first and second-ranked galaxies, and no preferred morphological type for the first-ranked galaxies, many of which are spiral. (2) The groups contain fewer spirals than a comparable sample of field galaxies. (3) There appears to be a deficiency of faint galaxies in comparison with rich cluster and field galaxies. Possible implications of these results for dynamical evolution are considered.

160.020   Quasar-generating superclusters: an explanation for a clumpy quasar sky?
H. R. de Ruiter, E. J. Zuiderwijk.
Astron. Astrophys., Vol. 105, 254 - 259 (1982).

The authors explore the consequences of the hypothesis that quasars are not randomly distributed in space but instead are located in superclusters. Then the sky and redshift distributions of quasars will not follow simple Poisson statistics: the quasar sky will be slightly clustered. In the framework of the hypothesis of quasar-generating superclusters, the statistics of quasars with very similar redshift, e.g. "pairs" and "triples", can provide information on quasar production and on superclusters at large redshifts.

160.021   The effect of morphological type on the spectral redshift of Perseus supercluster galaxies.
E. Giraud.
C. R. Acad. Sci. Paris, Tome 294, Sér. II, 439 - 442 (1982).
In French.

The relation between the spectral redshift of galaxies belonging to the Perseus supercluster and their morphological type on the basis of the complete sample given by Gregory et al. is studied. It is shown that in the central core the Sb and Scd galaxies have a very significant redshift excess of 288 ± 130 km sec^{-1} and that the Sbc and Sc galaxies have the same highly significant excess of $z = 0.0014 \pm 0.0007$ as a complete sample (for $m \leqslant 13$) of the Virgo I cluster.

160.022   Superclusters and voids in the distribution of galaxies.   S. A. Gregory, L. A. Thompson.
Sci. American, Vol. 246, No. 3, p. 88 - 96 (1982).

Red-shift surveys of selected regions of the sky have established the existence of at least three enormous superclusters of galaxies. The surveys also reveal that huge volumes of space are quite empty.

160.023   Characteristics of companion galaxies.   H. Arp.
Astrophys. J., Vol. 256, 54 - 74, plates 1 - 6 (1982).

From spectroscopic measures reported in a separate paper a class of objects consisting of 87 companion galaxies apparently associated with 61 larger galaxies has been selected. It is established that the spectral characteristics of the companions relative to the larger galaxy are: (1) the companions tend to have more emission and higher excitation; and (2) the companions tend to have earlier type absorption spectra. The differential redshifts of the companions relative to the larger galaxy are analyzed in two categories: (1) Where Δz is ± 800 km s^{-1}, the companions would be conventionally accepted as physically associated; nevertheless, positive Δz's outnumber negative by 36 to 15. (2) The remaining Δz's range from +4000 to +36,000 km s^{-1}. The nature of these companions indicates not many can be accidental projections of background galaxies. Some new cases of strongly interacting companions with large discordant redshifts are presented in detail. Past evidence, as well as new independent evidence that companion galaxies have some component of intrinsic redshift, is reviewed.

160.024   The shape and orientation of clusters of galaxies.
B. Binggeli.

Astron. Astrophys., Vol. 107, 338 - 349 (1982).

The paper presents evidence for the strong elongation of clusters of galaxies, based on an homogeneous and comparatively large cluster sample, as well as throws some light on the origin of the aspherical shape of clusters by investigating the relation between cluster shape and cluster environment.

160.025   Radio continuum observations at 5 GHz of Shakhbazyan's compact groups of galaxies.
Y. Sofue.
Publ. Astron. Soc. Japan, Vol. 33, 637 - 641 (1981).

Radio continuum observations at 5 GHz were made of four Shakhbazyan's compact groups of galaxies with the 100-m telescope in Bonn. Upper limits of 0.5–0.8 mJy were given as their radio flux densities, which suggest that they are not particularly radio active.

160.026   Dynamische Modelle für Galaxienhaufen.
B. Fuchs, J. Materne.
Mitt. Astron. Ges., Nr. 55, p. 108 - 112 (1982).

160.027   Mikrowellenhintergrundsabsorption in Galaxien-haufen.   D. Schallwich.
Mitt. Astron. Ges., Nr. 55, p. 112 (1982). – Abstract.

160.028   On the estimation of kinetic energies of clusters of galaxies.   M. A. Arakelyan, A. G. Kritsuk.
Astrofizika, Tom 17, 709 - 717 (1981). In Russian. English translation in Astrophysics, Vol. 17, No. 4.

The influence of a possible dependence of the velocity dispersion upon the mass of galaxies in a cluster on the estimation of kinetic energy has been considered.

160.029   Radial velocities of galaxies in the Virgo cluster.
I. D. Karachentsev, V. E. Karachentseva.
Pis'ma Astron. Zh., Tom 8, 198 - 204 (1982). In Russian.
English translation in Soviet Astron. Lett., Vol. 8.

Results of spectral observations are presented for about 100 galaxies in the Virgo cluster region. An essential part of the sample investigated consists of faint blue objects from the lists by Rubin (1967), Barbieri and Benvenuti (1974), and Börngen (1980) with apparent magnitudes $m_B = 16^m - 19^m$. The majority of them are dwarf emission members of the Virgo cluster.

160.030   A table of redshifts for Abell clusters.
C. L. Sarazin, H. J. Rood, M. F. Struble.
Astron. Astrophys., Vol. 108, L7 - L10 (1982).

A critically compiled table of redshifts for 329 Abell clusters is given. A number of photometric distance estimators were derived from the data in this table. These estimators have been used to note redshifts in the table which may be in error and to attempt to resolve discrepancies between different values of the measured redshift for a number of clusters.

160.031   Clusters of galaxies.
H. J. Rood.
Rep. Prog. Phys., Vol., 44, 1077 - 1122 (1981). – Abstr. in Phys. Abstr., Vol. 85, Abstr. 42752 (1982).

160.032   Motion of the Local Group of galaxies and isotropy of the Universe.   L. Hart, R. D. Davies.
Nature, Vol. 297, 191 - 196 (1982).

H I observations of 84 nearby Sbc spiral galaxies lying in the redshift range 1,000–5,500 km s^{-1} have been used to determine the motion of the Sun and the Local Group of galaxies relative to the backdrop of galaxies extending out to 70 Mpc. This study yields a Local Group velocity of 436 ± 55 km s^{-1} towards $l = 264° \pm 18°$, $b = 45° \pm 12°$, in close agreement with the value inferred from the dipole anisotropy of the 3 K cosmic microwave background. The component of the Local Group motion in the direction of the Virgo cluster is 410 ± 55 km s^{-1}, a value similar to that derived recently by independent methods. This implies that the local value of the density of the Universe is 0.15–0.50 of that required to close the Universe.

160.033 The CM relation at *UBV* wavelengths in the Fornax, Virgo, and Sersic 129-1 clusters.
D. Griersmith.
Astron. J., Vol. 87, 462 - 476 (1982).

New photoelectric *UBV* photometry in the range $V = 9$ to 16 mag is presented for 33 E and S0 galaxies in the southern clusters Fornax and Sersic 129-1. The raw data have been supplemented with existing data for Fornax as well as Virgo and, after correction for aperture effect, K-effect, and Galactic reddening, have been used to compare the color-absolute magnitude (CM) relations for the three clusters. The main results are: (1) The slope and intrinsic scatter of the CM relation at *UBV* wavelengths are the same within the errors in each of the Fornax, Virgo, and Sersic 129-1 clusters. At constant absolute magnitude, the colors of E and S0 galaxies are the same within each of the clusters. (2) Distance moduli relative to Virgo have been derived as Δ (m–M) = 0.23 ± 0.20 for Fornax and 3.29 ± 0.24 for Sersic 129-1. The former value agrees with previous determinations. The new relative distance modulus for Sersic 129-1 disagrees with the value predicted on the basis of H (global) = 95 km s^{-1} Mpc^{-1}. Possibilities which might explain the disagreement are discussed.

160.034 Unscrambling the Local Supercluster.
R. B. Tully.
Sky Telesc., Vol. 63, 550 - 554 (1982).

160.035 UBV photometry of nests of interacting galaxies.
V. P. Arkhipova.
Astron. Zh., Tom 59, 209 - 212 (1982). In Russian. English translation in Soviet Astron., Vol. 26, No. 2.

The results of photoelectric UBV observations of 19 nests of interacting galaxies are given. Most of the nests observed were found to be blue and to have high luminosity. The UBV and spectral data show a large amount of gas inside many nests.

160.036 Radio-optical studies of a complete sample of Abell clusters.
B. Y. Mills, R. W. Hunstead.
Extragalactic radio sources, (see 012.025), p. 85 - 86 (1982).

The majority of clusters in a complete sample of Southern Abell clusters contains at least one radio galaxy; radio galaxies are the rule rather than the exception in Abell clusters. They occur preferentially in clusters with above-average amounts of gas as shown by the Bautz-Morgan and X-ray correlations.

160.037 Studies of a complete sample of Abell clusters at 1400 MHz.
R. A. White, F. N. Owen, R. J. Hanisch.
Extragalactic radio sources, (see 012.025), p. 87 - 88 (1982).

The authors present analyses of a radio survey of Abell clusters at 1400 MHz using the NRAO 91-m telescope.

160.038 Two peculiar radio galaxies in A1367.
G. Gavazzi, W. Jaffe.
Extragalactic radio sources, (see 012.025), p. 89 - 90 (1982).

The authors observed the cluster of galaxies A1367 to map the structure of the cluster and that of the radio galaxy 3C264. They report on 3C264 and on the peculiar galaxy UGC6697.

160.039 Radio observations at 1.4 GHz of Abell clusters.
C. Fanti, R. Fanti, L. Feretti, A. Ficarra, I. M. Gioia, G. Giovannini, L. Gregorini, F. Mantovani, B. Marano, L. Padrielli, P. Parma, P. Tomasi, G. Vettolani.

Extragalactic radio sources, (see 012.025), p. 91 - 92 (1982).

The authors have observed with the Westerbork Synthesis Radio Telescope at 1.4 GHz the Abell clusters included in the HEAO-2 satellite observing program, for which radio information was not available.

160.040 Stephan's Quintet revisited.
J. M. van der Hulst, A. H. Rots.
Extragalactic radio sources, (see 012.025), p. 95 - 96 (1982).

VLA observations at 1465 MHz of the Stephan's Quintet region reveal that the arc-shaped area of emission discussed by Allen and Hartsuiker (1972) breaks up into several components. The idea that NGC 7318b is a recent interloper in the group and that the interaction resulting from this event causes the enhanced activity at the east side of NGC 7318b is adopted as still the most reasonable explanation.

160.041 A morphological classification of clusters of galaxies from Einstein images.
C. Jones, W. Forman.
Extragalactic radio sources, (see 012.025), p. 97 - 106 (1982).

It has only been with the advent of the Einstein X-ray imaging observatory that a first look at cluster X-ray morphology and classification has been possible. The efficacy of the X-ray images in cluster classification derives from their sensitivity in mapping the mass distribution within clusters. The studies of gaseous corona around individual cluster galaxies suggest the existence of massive halos. This analysis applied to entire clusters gives masses in good agreement with the virial mass. Although only about 10% of the total cluster mass is in the form of hot X-ray gas, it is a remarkably sensitive tracer of the unseen material defining the cluster potential.

160.042 Radio evolution in high redshift clusters.
W. Jaffe.
Extragalactic radio sources, (see 012.025), p. 425 - 426 (1982).

The author describes a project where he tries to follow the evolution of the Radio Luminosity Function (RLF) with redshift by observing a set of galaxy clusters with known spectroscopic redshifts. With the VLA he observed 55 clusters with redshifts of $0.25 \rightarrow 0.90$.

160.043 Momentos angulares en cúmulos de galaxias.
R. F. Sisteró, M. E. Castore de Sisteró.
Bol. Asoc. Argentina Astron., No. 20 - 24, p. 357 (1981). Abstract.

160.044 Galaxy clusters with multiple components. I. The dynamics of Abell 98.
T. C. Beers, M. J. Geller, J. P. Huchra.
Astrophys. J., Vol. 257, 23 - 32, plate 2 (1982).

The authors have measured 13 new redshifts for members in A 98, a galaxy cluster with two large enhancements in X-ray surface brightness and galaxy number density. From the 24 known redshifts, they determine line-of-sight velocity dispersions of the individual subclusters. They find the line-of-sight velocity difference to be 539 ± 361 km s^{-1}. Stable mass estimates are obtained by application of the virial theorem. The M/L_B for the subclusters are 593 (A 98 N) and 432 (A 98 S), comparable with other rich clusters. The authors develop a dynamical model for the two-body A 98 system based on the timing argument of Kahn and Woltjer. The probability that A 98 is bound is shown to be 98%. From this model and the results of previous N-body simulations, the authors show that the most likely description of the system is that it reached maximum expansion about 3.5 billion years ago and is presently in a state of collapse. The subclusters will merge in another 3 billion years.

160.045 Morphological types, colours and nuclear classes of galaxies in groups.
A. P. Magtesyan.
Soobshch. Byurakan Obs., Vyp. 53, p. 102 - 107 (1982). In

Russian.

The dependence of the morphological types, colours and Byurakan classes of galaxies in groups on their space density and relative quantity of elliptical and lenticular galaxies is considered. It is shown that the properties of galaxies in groups are similar to that in rich clusters.

160.046   Direct measurement of cluster expansion for nearby galaxy clusters.   J. S. Kaastra.
Astron. Astrophys., Vol. 109, L5 - L7 (1982).

Using the Tully-Fisher relation between the total luminosity and the H I line width of spiral galaxies as a distance indicator, a correlation between distances and redshifts of the different galaxies in nearby clusters is found. This correlation can be explained by an expansion rate of the cluster's outermost regions which is positive but less than Hubble's constant, as derived from the Tully-Fisher relation. Some dynamical implications are discussed.

160.047   H I-observations of galaxies in the Pegasus I cluster.
O.-G. Richter, W. K. Huchtmeier.
Astron. Astrophys., Vol. 109, 155 - 165 (1982).

Galaxy counts and 21 cm-H I-observations of the field of the Pegasus I cluster are presented in this paper. With the help of these observations the Pegasus I cluster is easily separated from a different clustering in the background which is identified to be part of the Perseus supercluster. No H I-deficiency has been observed for Pegasus 1 member galaxies. Together with the lack of evidence for intergalactic gas this fits well to the assumption that gas deficiency in spiral galaxies in rich clusters is caused by gas removal due to ram pressure of the motion of galaxies through the intracluster medium.

160.048   The Local Supercluster.   R. B. Tully.
Astrophys. J., Vol. 257, 389 - 422 (1982).

An attempt is made to illustrate the three-dimensional distribution of nearby galaxies. There is an evident overdensity of galaxies in the north galactic hemisphere that has been called the Local Supercluster. It is argued that this system comprises two distinct components: a disk component with 60% of the luminous galaxies, and a halo component with 40% of the luminous galaxies. With regard to the halo component, (1) almost all luminous galaxies are associated with only a small number of clouds, (2) as a consequence, most of the volume off the disk of the Local Supercluster is empty, (3) the clouds in the halo are sufficiently separated from the disk so that the two-component distinction seems warranted, and (4) at least the more prominent clouds in the halo seem to be prolate structures with their long axes directed toward the Virgo Cluster. The thinness of the disk of the supercluster, the extreme segregation of galaxies into a small fraction of the volume available, and the low local random motions are all evidence which weigh against gravitational clustering models in which galaxies formed before superclusters and in favor of the viewpoint that galaxies fragmented out of larger scale structure.

160.049   Groups of galaxies. I. Nearby groups.
J. P. Huchra, M. J. Geller.
Astrophys. J., Vol. 257, 423 - 437 (1982).

The authors present generalized techniques for finding density enhancements in the quasi–three-dimensional space known as redshift space. They use one of these techniques to examine the effects of varying selection criteria on the dynamical parameters derived for groups of galaxies. There is a broad range of selection parameters which yields well defined and well behaved groups. A whole sky catalog of nearby groups with outer number density enhancement greater than 20 is presented. The median M/L for our groups is ~170, corresponding to a cosmological density of $\Omega = 0.1$. The authors include a two-dimensional projection of several

contours near the Virgo cluster. The clustering exhibits both concentric and hierarchical structure.

160.050   X-ray observations of Abell 2218 and implications for the Sunyaev-Zel'dovich effect.
P. E. Boynton, S. J. E. Radford, R. A. Schommer, S. S. Murray.
Astrophys. J., Vol. 257, 473 - 486 (1982).

The Einstein Observatory has been used to obtain a high-resolution X-ray image and an X-ray flux measurement of the galaxy cluster Abell 2218. Aside from a minor contribution from the cD galaxy near the cluster center, the image is circularly symmetric and displays a smooth radial profile. The combined system of intracluster medium and galaxies was modeled as a gas in hydrostatic equilibrium in a common gravitational potential with an isothermal galaxy component. The gas was found to be nominally isothermal, moderately dense, and very hot – conditions consistent with the diminution in the cosmic microwave background radiation intensity reported in the direction of this cluster by some observers. In principle, the combination of such microwave measurements with X-ray and optical data allows the establishment of distances to truly remote clusters independently of the canonical chain of distance indicators and free from certain evolutionary effects.

160.051   Dynamical studies of the galaxy cluster A 115.
T. C. Beers, J. P. Huchra, M. J. Geller.
Bull. American Astron. Soc., Vol. 13, 798 (1981). – Abstract.

160.052   The velocity field in the Local Supercluster.
P. L. Schechter, M. Aaronson, J. Huchra, J. Mould, R. B. Tully.
Bull. American Astron. Soc., Vol. 13, 799 (1981). – Abstract.

160.053   A feature in the galaxy cluster Zw 0237.2 - 0146.
A. Hoag.
Bull. American Astron. Soc., Vol. 13, 799 (1981). – Abstract.

160.054   Extended X-ray emission from poor clusters with central dominant galaxies.
D. F. Cioffi, G. A. Kriss, C. R. Canizares.
Bull. American Astron. Soc., Vol. 13, 799 (1981). – Abstract.

160.055   Galaxies in the region around NGC 5846.
M. P. Haynes, R. Giovanelli.
Bull. American Astron. Soc., Vol. 13, 842 - 843 (1981). Abstract.

160.056   Effect of variable obscuration on the clustering of galaxies.   J. M. Uson, M. Seldner.
Bull. American Astron. Soc., Vol. 13, 868 (1981). – Abstract.

160.057   Lifetime constraints on massive neutrinos from ultraviolet observations of clusters of galaxies.
R. C. Henry, P. D. Feldman.
Bull. American Astron. Soc., Vol. 13, 868 (1981). – Abstract.

160.058   A dynamical study of the IC 698 group.
B. A. Williams.
Bull. American Astron. Soc., Vol. 13, 868 (1981). – Abstract.

160.059   The unusual morphology of 3 rich clusters of galaxies.
M. P. Ulmer, M. P. Kowalski, R. G. Cruddace.
Bull. American Astron. Soc., Vol. 13, 869 (1981). – Abstract.

160.060   The intrinsic shapes of cD systems.
C. Ftaclas, M. F. Struble.
Bull. American Astron. Soc., Vol. 13, 869 (1981). – Abstract.

160.061   Mass distribution in cD galaxies from Einstein observations.

D. Richstone, C. Jones, W. Forman.
Bull. American Astron. Soc., Vol. 13, 869 (1981). – Abstract.

160.062 **Low X-ray luminosity cD cluster A 2670.**
L. A. Thompson.
Bull. American Astron. Soc., Vol. 13, 869 (1981). – Abstract.

160.063 **The history of A98.**
M. J. Geller, T. C. Beers, J. P. Huchra.
Bull. American Astron. Soc., Vol. 13, 869 (1981). – Abstract.

160.064 **Limits to gas temperature variations in the central regions of the Centaurus cluster of galaxies.**
J. Schwarz, G. Stewart, A. C. Fabian.
Bull. American Astron. Soc., Vol. 13, 869 (1981). – Abstract.

160.065 **On the equilibrium distribution of the elements in the gas in the Coma cluster.**
F. Abramopoulos, G. A. Chanan, W. H.-M. Ku.
Bull. American Astron. Soc., Vol. 13, 869 - 870 (1981). Abstract.

160.066 **HEAO A-1 X-ray survey of southern clusters.**
M. P. Kowalski, M. P. Ulmer, R. G. Cruddace.
Bull. American Astron. Soc., Vol. 13, 870 (1981). – Abstract.

160.067 **Optical investigations of two X-ray clusters of galaxies: 0430.6–6133 and 0626.7–5426.**
J. Materne, G. Chincarini, M. Tarenghi, U. Hopp.
ESO Sci. Prepr. No. 181, 28 pp. (1982). – Submitted to Astron. Astrophys.

160.068 **Radio and X-ray observations of the Abell 2241 galaxy clusters.**   W. Bijleveld, E. A. Valentijn.
ESO Sci. Prepr. No. 183, 29 pp. (1982). – Submitted to Astron. Astrophys.

160.069 **Distant clusters of galaxies in the southern hemisphere – a status report.**
R. M. West, A. Kruszewski.
ESO Sci. Prepr. No. 191, 22 pp. (1982). – Submitted to Irish Astron. J.

160.070 **Gravitation screen from clusters of galaxies.**
V. K. Dubrovich.
Astrofiz. Issled., Izv. Spets. Astrofiz. Obs., Tom 15, 72 - 74 (1982). In Russian.
Radiation propagation in the universe allowing for local gravitational inhomogeneities created by protoclusters and clusters of galaxies is considered.

160.071 **Redshift quantization in compact groups of galaxies.**   W. J. Cocke, W. G. Tifft.
Prepr. Steward Obs., No. 391, 15 pp. (1982).

160.072 **A morphological study of the galaxy-rich clusters Coma and Virgo.**   A. L. Shcherbanovskij.
Soobshch. Spets. Astrofiz. Obs., Vyp. 31, p. 23 - 48 (1981). In Russian.
A morphological study of the rich clusters Coma and Virgo is carried out using the taxonomical method of hierarchical clustering described by J. Materne (1978). Determinations of radial velocities for 417 Coma galaxies and 204 Virgo ones together with their coordinates have been used and dendrograms of the internal structure of these clusters have been constructed.

160.073 **Active galaxies in groups of galaxies.**
G. M. Tovmasyan, Eh. Ts. Shakhbazyan.
Astrofizika, Tom 17, 265 - 271 (1981). In Russian. English translation in Astrophysics, Vol. 17, No. 2.
It is shown that radio emission is most often observed from first-rank galaxies. The galaxy with smaller optical luminosity but of the first rank in the group has higher probability of having radio emission than a galaxy absolutely brighter but of a lower rank in another group. For the first-rank galaxies the probability of having radio emission is about 10 times more than for the fourth-rank and fainter galaxies. This evidences in favour of a higher activity of the brightest galaxies in groups of galaxies. It is suggested that the first-rank galaxies play a definite cosmogonic role in groups of galaxies.

160.074 **Monte-Carlo simulations of galaxy systems.
II. Static properties for galaxies in rich clusters.**
H. T. MacGillivray, R. J. Dodd.
Astrophys. Space Sci., Vol. 83, 127 - 142 (1982).
The Monte-Carlo simulation technique described in Paper I is used to simulate orientations and shapes for galaxies in clusters according to 5 different schemes for true three-dimensional galaxy orientation. The projected galaxy geometric parameters are found to depend upon the type of alignment scheme present, the inclination of the parent cluster and the parent cluster shape. These results indicate that the presence and type of alignments in clusters (even if direct detection is not possible) can be deduced from analysis of the observed geometric properties of the member galaxies.

160.075 **Monte-Carlo simulations of galaxy systems.
III. Static properties for galaxies in supercluster filaments.**   H. T. MacGillivray, R. J. Dodd.
Astrophys. Space Sci., Vol. 83, 373 - 380 (1982).
Geometric properties (positions, orientations and shapes) are generated in three dimensional space for galaxies in a simulated distant supercluster filament according to the galaxy orientations and filament inclinations described in earlier papers. The distributions for projected galaxy shapes and position angles are examined. The results may be used to assist the interpretation of the observed geometrical properties for galaxies in real external superclusters.

160.076 **The velocity field in the Local Supercluster.**
M. Aaronson, J. Huchra, J. Mould, P. L. Schechter, R. B. Tully.
Astrophys. J., Vol. 258, 64 - 76 (1982).
The authors consider a model for the velocity field of the Local Supercluster which includes both the deceleration of galaxies by a spherically symmetric density enhancement and a random motion of the Local Group with respect to nearby galaxies. They adjust the free parameters in the model to minimize scatter about the infrared luminosity-H I velocity width relation (the "IR Tully-Fisher relation") yielding a best estimate for the amplitude of the deceleration pattern at the position of the Local Group of 250 ± 64 km s^{-1}, with a total velocity toward the Virgo cluster of 331 ± 41 km s^{-1}. The authors find a significant reduction in amplitude of the residuals if they test for differential rotation about the center of the Supercluster, with an estimate of the amplitude of the adopted pattern of 180 ± 58 km s^{-1} at the position of the Local Group. The derived total velocity with respect to the Supercluster shows better agreement with the velocity inferred from the 3 K background dipole anisotropy than has been previously obtained, but with some remaining differences.

160.077 **Large-scale superclusters surrounding the giant galaxy void in Bootes?**
N. A. Bahcall, R. M. Soneira.
Astrophys. J., Lett., Vol. 258, L17 - L21 (1982).
The overdensity of galaxies observed by Kirshner et al. on both redshift sides of the $z \approx 0.04$ - 0.06 galaxy void in Bootes is found to coincide in redshift space with similar overdensities of clusters and superclusters of galaxies in the northern hemisphere. The main contributors to these overdensities are the rich Hercules and Corona Borealis superclusters. It is suggested that large-scale density enhancements of galaxies associated

with the Hercules and Corona Borealis superclusters may extend to $\gtrsim 100$ - $150\,h^{-1}$ Mpc in projection and surround the giant volume of galaxy depletion in Bootes. Previous observational evidence, together with the results present here, suggests that galaxy voids may generally be associated with surrounding galaxy excesses.

160.078 Optical investigations of two X-ray clusters of galaxies: 0430.6−6133 and 0626.7−5426.
J. Materne, G. Chincarini, M. Tarenghi, U. Hopp.
Astron. Astrophys., Vol. 109, 238 - 244 (1982).

Radial velocities and density distribution in the two X-ray clusters of galaxies 0430.6−6133 and 0626.7−5426 have been determined. The cluster 0430.6−6133 is a regular cluster which has a dumbbell cD galaxy in the center. The two components will probably merge soon. The cluster 0626.7−5426 consists of two subgroups which form a supercluster. Models for the projected radial density distribution are discussed.

160.079 On the peculiar motion of the Local Group as revealed by the $B-V$ vs. HM relation for Sc I galaxies. P. Teerikorpi.
Astron. Astrophys., Vol. 109, 314 - 319 (1982).

(1) Local Sc I galaxies verify the $(B-V)^0_T$ vs. HM_c relation, previously discovered for the Rubin et al. galaxies by the present author. (2) Analysis of the $(B-V)^0_T$ vs. HM_c relation utilizing the bias-free method due to Schechter (1980) leads to significantly non-zero values for the infall velocity v_0 of the Local Group in the direction of the Virgo cluster. (3) For the closest galaxies, the dipolar correction is inadequate, and a differential expansion field may be suggested by the data. (4) With the range of values suggested elsewhere for the distance of M 101, the position of M 101 on the $(B-V)^0_T$ vs. HM_c diagram corresponds to the range of values for H_0 from about 70 to 40 km s^{-1}Mpc^{-1}.

160.080 Simultaneous study of optical and X-ray properties of the Coma cluster by Multi Mass Models.
G. Des Forêts, R. Dominguez-Tenreiro, D. Gerbal, G. Mathez, A. Mazure, E. Salvador-Solé.
Sep. print Obs. Paris-Meudon, 37 pp. (1982).

The aim of this paper is to study the Coma cluster by taking into account as numerous as possible dynamical properties of the cluster. This leads to improve previous Multi Mass Models (MMM) by removing the isothermal assumption and adding a second dynamical component accounting for the actual X-ray plasma.

160.081 The genesis of the Local Group.
D. Lynden-Bell.
Astrophysical cosmology, (see 012.047), p. 85 - 111 (1982).

The evidence for heavy halos about the Galaxy and Andromeda is discussed and possible sizes for them are estimated. On the assumption that the Local Group is gravitationally bound and that its members all came from the Big Bang, Newtonian mechanics allows one to deduce both the total mass of the Local Group and the time since its genesis in the Big Bang. Limits are placed on the mass of any possible halo of the Local Group which is not attached to the Galaxy or Andromeda. Possible roles for heavy neutrinos in Local Group dynamics are briefly considered.

Morphological classification (revised RS) of Abell clusters in $d \leqslant 4$ and an analysis of observed correlations. See Abstr. 002.004.

The mass of the neutrino from the dynamics of groups of galaxies.   See Abstr. 061.008.

Radio emission from the 2006 − 56 region. See Abstr. 141.004.

A high-resolution spectroscopic study of Q0119−046 and the nature of absorption complexes with $z_{abs} > z_{em}$.   See Abstr. 141.008.

Redshifts, first and second order clustering properties, and refined radio parameters of 4C radio galaxies in poor clusters.   See Abstr. 141.052.

A preliminary examination of the effect of cluster gas on tailed radio galaxies.   See Abstr. 141.122.

The association of QSOs with groups of galaxies. See Abstr. 141.234.

A complete X-ray sample of the high-latitude $(|b| > 20°)$ sky from HEAO 1 A-2: log N−log S and luminosity functions.   See Abstr. 142.012.

A medium sensitivity X-ray survey using the Einstein Observatory: the log N−log S relation for extragalactic X-ray sources.   See Abstr. 142.013.

Einstein observations of the confused 2A 2315−428 region.   See Abstr. 142.105.

A comparison of simulated galaxy clustering models with observations.   See Abstr. 151.007.

Environmental effects on galaxies in clusters. See Abstr. 151.009.

Mass segregation, relaxation, and the Coulomb logarithm in N-body systems.   See Abstr. 151.014.

Monte-Carlo simulations of galaxy systems. I. The Local Supercluster.   See Abstr. 151.028.

Ableitung eines Wechselwirkungsterms für inelastische Begegnungen zwischen Galaxien in Haufen. See Abstr. 151.033.

The effects of galaxy collisions on the galaxy luminosity function and the background light in clusters. See Abstr. 151.066.

The phase-space distribution function of galaxies in clusters and the secondary peak.   See Abstr. 151.093.

Galaxy spins in the Virgo cluster. See Abstr. 158.002.

Galaxy infrared colour magnitude relations. See Abstr. 158.029.

A sample of faint galaxies behind the Virgo cluster. See Abstr. 158.046.

Cluster galaxies with Seyfert properties. See Abstr. 158.047.

Accurate positions and standard D_{25} diameters for galaxies in the central part of the Coma cluster. See Abstr. 158.052.

The Zwicky magnitude scale: how reliable is it in the estimation of blue luminosity?   See Abstr. 158.067.

Global properties of Sa-galaxies from H I-observations.   See Abstr. 158.310.

NGC 4388: a Seyfert 2 galaxy in the Virgo cluster. See Abstr. 158.311.

Spectroscopic and photographic observations of the Cygnus A group and of the stellar component of the Cygnus A galaxy.   See Abstr. 158.317.

Spiral galaxies in clusters. III. Gas-rich galaxies in the Pegasus I cluster of galaxies.   See Abstr. 158.323.

Spiral galaxies in clusters. IV. The H I color properties of spirals in nine clusters. See Abstr. 158.324.

CO observations of galaxies in the Virgo Cluster. See Abstr. 158.341.

CO observations of NGC 3628. See Abstr. 158.344.

Report of IAU Commission 28: Galaxies (Galaxies). See Abstr. 158.358.

The detection of hot intergalactic gas in the NGC 3607 group of galaxies with the Einstein satellite. See Abstr. 161.004.

A detailed X-ray study of the cooling intracluster gas in A 496.   See Abstr. 161.007.

On the ionization balance of helium in the inter-galactic medium.   See Abstr. 161.008.

The velocity evolution of galaxy clustering. See Abstr. 162.015.

The cosmic distance scale: methods for determining the distance to supernovae.   See Abstr. 162.017.

A survey of galaxy redshifts. III. The density field and the induced gravity field.   See Abstr. 162.019.

Formation of galaxies and clusters of galaxies in the neutrino dominated Universe.   See Abstr. 162.143.

The peculiar velocity around a hole in the galaxy distribution.   See Abstr. 162.157.

The Galaxy as a fundamental standard for extra-galactic distances. V. Another decisive test of the long and short distance scales: the distance of the Virgo E cluster and its implications.   See Abstr. 162.182.

Formation of voids in the galaxy distribution. See Abstr. 162.184.

Dark matter.   See Abstr. 162.211.

H_0, q_0 and the local velocity field. See Abstr. 162.212.

Report of IAU Commission 47: Cosmology (Cosmologie).   See Abstr. 162.226.

Erratum

160.901 X-ray measurements and the mass of Abell 1763 out to a radius of 1.4 Mpc.   J. P. Vallée. Astrophys. Lett., Vol. 22, 63 - 68 (1981). − See Abstr. 30.160.053.

161 Intergalactic Matter

161.001 **Massive neutrino decay and the photoionization of the intergalactic medium.**   D. W. Sciama. Mon. Not. R. Astron. Soc., Vol. 198, 1P - 5P (1982).
It is pointed out that the intergalactic medium might be ionized by photons emitted by a cosmological distribution of massive neutrinos. The absence of absorption troughs in quasar spectra due to atomic hydrogen and helium, and the possible presence of a trough due to singly ionized helium, would than imply that the neutrino mass lies between 50 and 110 eV. Its radiative lifetime would have to be 10^{27} s if the critical ionization redshift ~3.5 and the present density of the IGM ~ 5×10^{-9} cm^{-3}. The apparent quasar cut-off at a red-shift ~3.5 may be due to the IGM being neutral at higher redshifts. The inferred lifetime could be compatible with particle physics expectations if the GIM mechanism is not operating.

161.002 **Cosmological constraints on hot plasma in a closed universe.**   R. D. Sherman. Astrophys. J., Vol. 256, 370 - 373 (1982).
A hot uniform collisionally heated intergalactic medium (IGM) at closure density is reconsidered. During the recombination epoch only the observational data of the soft X-ray region and the optical depths of H Lyα need be used to ascertain the viability of such a dense cosmic H and He plasma. The results indicate a present temperature in the range $1.6 \times 10^5 \lesssim T_0 \lesssim 4.2 \times 10^6$ K if the reheating of the

IGM ceased at $z \sim 3$, the latest possible time. As cessation of heating moves to earlier times, the range of T_0 allowed narrows until at the earliest possible $z \sim 12$ only $T_0 \sim 1.6 \times 10^5$ K is acceptable. This IGM avoids any hard X-ray production and thus assumes that discrete sources alone constitute the entire cosmic X-ray background.

161.003 **Intergalactic Lyman-alpha absorption lines in a close pair of high-redshift QSOs.** W. L. W. Sargent, P. Young, D. P. Schneider. Astrophys. J., Vol. 256, 374 - 385 (1982).
The authors compare high-resolution spectra of the QSOs Q1623+269 (77) and Q1623+268 (78). These objects have redshifts z_{em} = 2.518 and 2.605, respectively, and are separated by 173″ on the plane of the sky. The corresponding linear separation at $z \approx$ 2.5 lies between 1 and 2 Mpc, depending on the comological model. One strong heavy element red-shift system, z_{abs} = 2.0524, is found in the spectrum of Q1623+268, while two systems, z_{abs} = 2.0940 and 2.2405, are found in Q1623+269. A cross-correlation analysis of the Lyα absorption lines in the two QSOs shows that there is no measurable tendency for the observed wavelengths of the lines to correlate. This result contradicts a recent proposal by Oort that the Lyα lines in QSOs arise in generally distrib-uted gas in superclusters of galaxies. Both cross-correlation and autocorrelation analyses of the Lyα absorption lines show no tendency for the lines to cluster on any scale.

161.004 The detection of hot intergalactic gas in the NGC 3607 group of galaxies with the Einstein satellite.   P. Biermann, P. P. Kronberg, B. F. Madore. Astrophys. J., Lett., Vol. 256, L37 - L40 (1982).

The authors have detected extended X-ray emission around the galaxies NGC 3607, NGC 3605, and NGC 3608 in the group TG 39a = G49 using the IPC aboard the Einstein satellite. They argue that this emission is due to thermal bremsstrahlung from a hot, intergalactic gas; this gas sits in the potential well of the group, around the dominant galaxy NGC 3607. The gas has a density of about 4×10^{-3} cm^{-3}, a temperature of about 5×10^6 K, a mass of about $6 \times 10^9 M_\odot$, and a cooling time of about 5×10^9 yr. The spatial structure of the X-ray emission around NGC 3607 suggests that the dominant galaxy is being stripped at present.

161.005 The intergalactic medium.   A. C. Fabian, A. K. Kembhavi. Extragalactic radio sources, (see 012.025), p. 453 - 459 (1982).

The density of intergalactic gas may be an important parameter in the formation of extended radio sources. It may range from ~ 0.1 particle cm^{-3} in the centres of some rich clusters of galaxies down to 10^{-8} cm^{-3} or less in intercluster space. The possible influence of the intracluster gas surrounding NGC 1275 on its radio emission is discussed, and the possibility that a significant fraction of the X-ray background is due to a hot intergalactic medium is explored in some detail.

161.006 Can intergalactic suprathermal grains produce extensive air showers?   A. K. Dasgupta. Astrophys. Space Sci. Vol. 83, 51 - 61 (1982).

The events following the impact of intergalactic suprathermal grains onto the earth atmosphere are examined, and some similarity is found between the expected air shower and observations of largest cosmic ray showers. It is concluded that the largest air showers are, in any case, initiated by primaries of intergalactic origin. Whether the primaries are suprathermal dust grains or single nuclei is inconclusive.

161.007 A detailed X-ray study of the cooling intracluster gas in A496.   P. E. J. Nulsen, G. C. Stewart, A. C. Fabian, R. F. Mushotzky, S. S. Holt, W. H.-M. Ku, D. F. Malin. Mon. Not. R. Astron. Soc., Vol. 199, 1089 - 1099 (1982).

X-ray observations of the galaxy cluster A496 are presented. Imaging and spectral data, together with the presence of optical filaments, show that radiative cooling is occurring in the cluster core and indicate a mass flow rate of $\sim 200\ M_\odot$yr^{-1} on to the central galaxy. The gas temperature and density profiles within the core are similar to those in the Perseus cluster (A426) despite a six-fold difference in total X-ray luminosity. This is consistent with the model of Fabian & Nulsen for steady flows caused by cooling. The evolution of cooling flows is discussed and it is shown that thermal conduction is suppressed to well below the Spitzer rate.

161.008 On the ionization balance of helium in the intergalactic medium.   D. W. Sciama. Mon. Not. R. Astron. Soc., Vol. 200, 13P - 18P (1982).

A straightforward photoionization analysis is given of the tentative IUE observation that the intergalactic medium near the QSO Q2204−408 ($z = 3.18$) contains a He II abundance of at least 2×10^{-9} cm^{-3}. This analysis is found to be rather implausible and it is suggested that, if the IGM is indeed photoionized, the QSO may be located in an expanding supercluster of size $\gtrsim 13$ Mpc and total mean gas density $\sim 1.5 \times 10^{-5}$ cm^{-3}.

Absolute spectrophotometry of very large redshift quasars.   See Abstr. 141.042.

Theoretical models to explain the variable 21 cm absorption spectrum in AO 0235+164.   See Abstr. 141.168.

Winds from dwarf galaxies and Lα absorption features in quasar spectra.   See Abstr. 158.020.

Observations of the Magellanic Stream between declinations −20° and 0°.   See Abstr. 159.005.

On intracluster Faraday rotation. II. Statistical analysis.   See Abstr. 160.001.

Transport processes and the stripping of cluster galaxies.   See Abstr. 160.007.

X-ray observations of Abell 2218 and implications for the Sunyaev-Zel'dovich effect.   See Abstr. 160.050.

162 Universe (Structure, Evolution)

162.001 Friedmann-like singularities in Szekeres' cosmological models. S. W. Goode, J. Wainwright.
Mon. Not. R. Astron. Soc., Vol. 198, 83 - 90 (1982).

The authors establish the existence of Friedmann-like singularities in a subclass of the Szekeres cosmological models which admits no Killing vector fields, and which depends on three arbitrary functions of a single variable. They describe the asymptotic properties of the solutions as the singularity is approached, for example the behaviour of the matter density, the rate of shear of the matter, and the Weyl conformal curvature tensor.

162.002 Effects of proton decay on the cosmological future. D. A. Dicus, J. R. Letaw, D. C. Teplitz, V. L. Teplitz.
Astrophys. J., Vol. 252, 1 - 9 (1982).

The authors calculate, for an open universe, the densities of radiation and matter at large times if the proton has a lifetime τ_p of about 10^{30} years; they consider the contributions to these densities from the decay of matter in both clumps and interstellar gas. For a closed (cyclical) universe, current ideas in particle physics imply that the baryon to photon ratio will be identical for each cycle; thus the effect of entropy production will be to enlarge the cosmic scale, from cycle to cycle, by a cycle expansion factor α. The authors compute α, taking into account entropy production both by stellar nucleosynthesis and by proton decay.

162.003 G_N variability and primordial nucleosynthesis. A. Meisels.
Astrophys. J., Vol. 252, 403 - 409 (1982).

Constraints on possible G_N variability over cosmic time are derived, using the general theory of variable masses (Bekenstein's VMT). The ultimate result is that G_N can vary by no more than two or three orders of magnitude during all of cosmic time. That leaves the nonsingular cosmological models of VMT as currently viable. Possible variability is described, and specific examples of G_N behavior for some cosmological models are given.

162.004 Observer reference triad rotation, magnetic fields, and rotation in Euclidean cosmological models. A. J. Fennelly.
Astrophys. J., Vol. 252, 410 - 417 (1982).

The author shows that a cosmological model with Euclidean space sections admits fluid flows if a magnetic field is present. Further, if observer reference frames spin in their response to the universe, a vorticity may exist. Limits on any present intergalactic magnetic field of $H_R < 5.38 \times 10^{-10}$ gauss can be set.

162.005 Cosmological structure produced by a phase transition near nuclear density. C. J. Hogan.
Astrophys. J., Vol. 252, 418 - 432 (1982).

This paper shows that a cold expanding universe consisting of homogeneous low-entropy matter ($S \leqslant 1$) can create large-scale structure spontaneously. When matter near nuclear densities passes through a state of metastability during a phase transition, certain types of quantum fluctuations will trigger large-scale rupture and cracking of matter. The high-temperature phase transition could be the result of pion condensation or neutron solidification. It is demonstrated that the process will create fragments containing 10^{46} to 10^{51} nucleons. These "lumps" will generate density fluctuations which grow by the usual gravitational instability. As most important consequence it is postulated that stars with $M \geqslant 2 M_\odot$ should have been formed at high redshifts: $10^7 > z > 100$. This "pregalactic" population of stars would then be responsible for the microwave background, galaxy formation and,

probably, helium production. Their remnants could form the dark matter in clusters of galaxies and galactic halos.

162.006 Background radiation fields as a probe of the large-scale matter distribution in the Universe. N. Kaiser.
Mon. Not. R. Astron. Soc., Vol. 198, 1033 - 1052 (1982).

A "Swiss Cheese" model is used to calculate to order of magnitude the temperature fluctuation of the cosmic microwave background radiation (CMB) in a lumpy universe. The calculations are valid in a Friedmann background of aribtrary Ω provided that matter has been dominant since the photons were last scattered. The inhomogeneities may be larger than the curvature scale, as is required to deal with fluctuations on a large angular scale in a low-density universe. The author combines this model with observational limits on the fluctuations in the CMB to yield an upper limit to the present spectrum of inhomogeneities. The author also considers the limits on inhomogeneity from the isotropy of the X-ray background and finds them to be consistent with the microwave limits.

162.007 Inhomogeneous cosmology: gravitational radiation in Bianchi backgrounds. P. J. Adams, R. W. Hellings, R. L. Zimmerman, H. Farhoosh, D. I. Levine, S. Zeldich.
Astrophys. J., Vol. 253, 1 - 18 (1982).

An exact formalism is developed for describing cosmological models with strong, long wavelength gravitational waves of general polarization propagating over backgrounds corresponding to Bianchi types I through VII. The authors introduce and discuss a new metric which exhibits the appropriate symmetries of two equivalent independent polarizations of gravitational waves. The formalism is applied to an empty type I cosmology, and it is shown how the original z-dependent chaotic singularity structure transforms itself into gravitational radiation propagating along the z-axis in a Bianchi I background.

162.008 Quasars as probes of the distant and early universe. P. S. Osmer.
Sci. American, Vol. 246, No. 2, p. 96 - 104, 106 - 107 (1982).

The light from most of these enigmatic objects was emitted 15 billion years ago. Therefore they are a unique clue to how the universe looked when it was only a fourth its present age.

162.009 The Hubble diagram of quasars and the deceleration parameter q_0. F.-x. Hu.
Acta Astrophys. Sinica, Vol. 2, 1 - 7 (1982). In Chinese.

162.010 Kosmologie. 1. De uitdijing van het heelal. J. K. Katgert-Merkelijn.
Zenit, 9. Jaarg., 60 - 65 (1982).

162.011 Kosmologie. 2. Onzichtbare massa gezocht! J. K. Katgert-Merkelijn.
Zenit, 9. Jaarg., 122 - 126 (1982).

162.012 A survey of galaxy redshifts. II. The large scale space distribution. M. Davis, J. Huchra, D. W. Latham, J. Tonry.
Astrophys. J., Vol. 253, 423 - 445 (1982).

The authors have finished a redshift survey of galaxies complete to 14.5 m_B in the north and south galactic polar caps above declination = 0° and containing some 2400 galaxies. They present various projections of the resulting redshift-space maps. While different in detail, the statistical nature of the redshift-space distribuiton is very similar between the north and south. The space distribution of galaxies is frothy, characterized by large filamentary super-clusters of up to 60 Mpc in extent, and corresponding large holes devoid of galaxies. The authors also present redshift-space maps generated from n-body simulations, which very roughly match the density and amplitude of the galaxy clustering but fail to match the frothy nature of the actual distribution.

162.013 **Nonlinear effects on cosmological perturbations. I. The evolution of adiabatic perturbations.**
E. T. Vishniac.
Astrophys. J., Vol. 253, 446 - 456 (1982).

The nonlinear evolution of a power spectrum of adia-batic perturbations in the early universe is calculated under the assumptions of random phase and weak nonlinearity. The result is consistent with previous results for an ideal non-relativistic gas and is an extension of Peebles's formula. For all initial power spectra, nonlinear dissipation dominates the evolution of the power spectrum at all frequencies where the total linear dissipation since $t = 0$ has decreased the initial amplitude by more than one e-fold. This has the effect of increasing the amount of power at high frequencies where the linear theory predicts an exponential decrease. Computer simulations of the evolution of reasonable adiabatic perturba-tion spectra are presented.

162.014 **Nonlinear effects on cosmological perturbations. II. The production of isothermal perturbations by a primordial sound spectrum.**   E. T. Vishniac.
Astrophys. J., Vol. 253, 457 - 469 (1982).

In a previous paper it was claimed that dust, viscously coupled to a background gas undergoing random compres-sional motion, will cluster secularly. This process was named Purcell clustering. In this paper the exact rate of clustering is calculated for the case of dust suspended in an ideal gas and for baryons in the early universe. In addition, computer calculations of the production of isothermal perturbations in the early universe are presented. These suggest that it may be extremely difficult to produce significant isothermal perturba-tions by Purcell clustering.

162.015 **The velocity evolution of galaxy clustering.**
W. C. Saslaw, S. J. Aarseth.
Astrophys. J., Vol. 253, 470 - 478 (1982).

The authors have examined the changing velocity dis-tribution of galaxies as they cluster in computer models of the expanding universe. The models are 4000-body numerical simulations of galaxies with a large range of masses interacting gravitationally. Clustering in velocity space is measured by calculating the residual peculiar velocities around the Hubble expansion. These form "Hubble streaks" as clustering progresses. The authors distinguish isolated field galaxies from clustered galaxies. The velocity dispersion of extreme field galaxies is a good cosmological indicator of $\Omega = \rho/\rho_{crit}$. Pre-liminary comparison of several simulations with observations shows that our universe agrees better with low density models, $\Omega \lesssim 0.1$, than with high density models having $\Omega \approx 1$. The velocity dispersion of cluster centers of mass is a good cos-mological marker as well.

162.016 **Free-streaming radiation in cosmological models with spatial curvature.**   M. L. Wilson.
Astrophys. J., Lett., Vol. 253, L53 - L56 (1982).

The effects of spatial curvature on radiation anisotropy are discussed for the standard Friedmann-Robertson-Walker model universes. It is shown that the curvature is very impor-tant when considering fluctuations with wavelengths compar-able to the horizon.

162.017 **The cosmic distance scale: methods for determin-ing the distance to supernovae.**   W. D. Arnett.

Astrophys. J., Vol. 254, 1 - 7 (1982).

Several methods of determining absolute distances to supernovae are analyzed. A kinematic method which appears superior to the kinematic method of Baade is developed and illustrated with the Type I supernova 1975a in NGC 2207. A dynamical method is developed which uses the recently dis-covered analytic solutions for Type I supernovae; initial results for several clusters of galaxies are: Virgo: $D = 16 \pm 2.2$ Mpc, $H_0 = 68 \pm 11$ km s^{-1} Mpc^{-1}; Coma: $D = 100 \pm 14$ Mpc, $H_0 = 69 \pm 12$ km s^{-1} Mpc^{-1}; Pegasus I: $D = 52 \pm 8$ Mpc, $H_0 = 76 \pm 11$ km s^{-1} Mpc^{-1}; and using Type I supernovae in elliptical galaxies, $H_0 = 69 \pm 12$ km s^{-1} Mpc^{-1}.

162.018 **Bianchi type electromagnetic cosmology – Type I Hamiltonian.**
M. P. Ryan, Jr., S. M. Waller, L. C. Shepley.
Astrophys. J., Vol. 254, 425 - 436 (1982).

The dynamical effects of spatially homogeneous magnetic fields on anisotropic Bianchi Type I cosmological models are studied. Some general features to be expected of cosmic mag-netic fields which are too weak to influence cosmic dynamics are pointed out, and the potentials affecting the evolution of anisotropy for diagonal, symmetric, and general metrics are obtained by means of the Arnowitt, Deser, and Misner Hamiltonian formalism.

162.019 **A survey of galaxy redshifts. III. The density field and the induced gravity field.**
M. Davis, J. Huchra.
Astrophys. J., Vol. 254, 437 - 450 (1982).

The authors have completed a redshift survey of galaxies complete to 14.5 m_z in the north and south galactic polar caps containing some 2400 objects. They compute the number density and luminosity density of bright galaxies to a distance of 80 Mpc (H_0= 100) over the 2.7 sr of the survey. The exis-tence of a region 30 Mpc in extent beyond the Virgo cluster that is underdense by nearly a factor of 2 is demonstrated. Averaging over the full sample volume, however, the number density and luminosity density given in the north and south catalogs agree to within 7%. The mean overdensity toward the Virgo cluster is measured to be 2.0 ± 0.2. The local peculiar gravity g induced by the inhomogeneous distribution of matter at large distances is examined. In the limit of linear perturba-tion theory g is proportional to the expected peculiar motion of the local groups. From this cosmological test an estimate of the density parameter Ω can be obtained. Independently, Ω may be derived from the nonlinear, spherically symmetric Virgo infall test. Both estimates yield quite high values of Ω: $0.4 < \Omega < 0.5$ for an adopted infall velocity of 400 km s^{-1}.

162.020 **Angular fluctuations of the temperature of relic radiation in a universe with massive neutrinos.**
N. A. Zabotin, P. D. Nasel'skij.
Pis'ma Astron. Zh., Tom 8, 67 - 73 (1982). In Russian. English translation in Soviet Astron. Lett., Vol. 8.

Correlation characteristics of the microwave cosmic radiation fluctuations formed at the period of cosmological recombination are calculated in the framework of a cosmologi-cal model with massive relic neutrinos.

162.021 **"Black" regions in the universe.**
Ya. B. Zel'dovich, S. F. Shandarin.
Pis'ma Astron. Zh., Tom 8, 131 - 135 (1982). In Russian. English translation in Soviet Astron. Lett., Vol. 8.

Formation of giant regions which are nearly devoid of galaxies is explained in the frame of the fragmentation picture. This picture is associated with the scenario of large-scale struc-ture formation from primordial adiabatic density perturba-tions. Physical properties of gas inside the voids are discussed as well as its possible observational detection.

162.022 The stage of superheavy particle dominance in the universe and big bang nucleosynthesis.
A. G. Polnarev, M. Yu. Khlopov.
Astron. Zh., Tom 59, 15 - 19 (1982). In Russian. English translation in Soviet Astron., Vol. 26, No. 1.

The observed abundance of primordial He^4 is shown to provide astrophysical restrictions on possible deviations from the dominance of radiation in the big bang universe in the period of neutrons freezing out (at $t \sim 1$ s). The obtained limits together with the data on the spectrum of relic radiation practically exclude the stages of superheavy stable particle dominance at $t \geqslant 1$ s, thus putting restrictions on the modern particle theories.

162.023 How black is the Universe? R. A. Muller.
Nature, Vol. 295, 95 - 96 (1982).

162.024 Creation of open universes from de Sitter space.
J. R. Gott III.
Nature, Vol. 295, 304 - 307 (1982).

The author proposes a new cosmological model which has an early event horizon and in which the observed cosmic microwave background radiation is Hawking radiation.

162.025 Great voids in the Universe. J. Silk.
Nature, Vol. 295, 367 - 368 (1982).

162.026 A model for the cosmic creation of nuclear exergy.
K.-E. Eriksson, S. Islam, B.-S. Skagerstam.
Nature, Vol. 296, 540 - 542 (1982).

The authors have studied a model which should represent the nucleon gas of the early Universe quite well. They find that the main creation of nuclear exergy started around 10 s after the big bang, and most of the exergy was created during the first few minutes, 85% during the first hour, and that the process was essentially completed during the first 24 h. The final value of the exergy was 7.72 MeV per nucleon.

162.027 The formation of galactic haloes in the neutrino-adiabatic theory. A. L. Melott.
Nature, Vol. 296, 721 - 723 (1982).

Recent evidence of a finite rest mass for the neutrino has led to a revival of the idea that these particles may provide the unseen but dynamically indicated matter on cosmological scales of galactic halo and larger size. The author presents a numerical simulation which shows that if the initial perturbations are sufficiently anisotropic, the collapse of very large perturbations of collisionless particles leads to a condensation of particles with low velocity dispersion and high phase-space density, which may easily fragment.

162.028 Filling the void in Boötes.
V. A. Balzano, D. W. Weedman.
Astrophys. J., Lett., Vol. 255, L1 - L4 (1982).

The possible existence of a 10^6 Mpc^3 volume, empty of galaxies, suggested by Kirshner, Oemler, Schechter, and Shectman, is not confirmed with the existing sample of Markarian galaxies. Of 113 Markarian galaxies with redshifts in the region of sky surveyed by Kirshner et al., 12 have redshifts that place them in the void. The Markarian galaxy sample also shows the redshift distribution that would be expected for a homogeneous space density out to a redshift of 25,000 km s^{-1}.

162.029 An exact Bianchi type-V tilted cosmological model with matter and an electromagnetic field.
D. Lorenz.
Gen. Relativ. Gravitation , Vol. 13, 795 - 805 (1981).

The author presents a cosmological solution of the source-free Einstein-Maxwell equations with "stiff" matter and an electromagnetic null field, which is a locally rotationally symmetric tilted Bianchi type-V universe.

162.030 Effects of a positive cosmological constant on circular orbits in the Reissner-Nordström, Schwarzschild, and Kerr fields. R. J. Howes.
Gen. Relativ. Gravitation , Vol. 13, 829 - 835 (1981).

Criteria for the existence and stability of circular orbits in a general stationary axisymmetric field are established using the equation of geodesic deviation. Circular orbits in the Reissner-Nordström, Schwarzschild and Kerr fields, all with a positive cosmological constant "Λ" included, are considered and the effects of "Λ" are detailed.

162.031 The generation of isothermal perturbations in the very early universe.
J. R. Bond, E. W. Kolb, J. Silk.
Astrophys. J., Vol. 255, 341 - 360 (1982).

If the baryon asymmetry in the universe is produced at a temperature $\sim 10^{15}$ GeV as a result of baryon nonconservation, as predicted in grand unified theories, then perturbations in the baryon density should also be created at this time. Adiabatic perturbations are easily generated by energy density fluctuations. Isothermal perturbations can be created as a result of spatial variations in the local expansion rate which can result from spatial inhomogeneities in the cosmological constant, scalar curvature, or shear (and vorticity). The authors calculate numerically the entropy per baryon created in Bianchi I universes with shear, using model systems based upon the SU (5) grand unified theory. They calculate an effective viscosity in this theory, and use it to show that viscous damping of shear is not sufficiently large to smooth the universe. The survival of shear inhomogeneities results in isothermal perturbations being generated.

162.032 Baryon number creation and phase transitions in the early Universe. P. Hut, F. R. Klinkhamer.
Astron. Astrophys., Vol. 106, 245 - 253 (1982).

The authors aim at a consistent scenario for the generation of the baryon-antibaryon asymmetry in the early Universe. They discuss recent calculations using unified interactions and consider finite temperature effects, namely the existence of phase transitions. Although the expansion of the Universe can be affected significantly, the final baryon number is surprisingly insensitive to the nature of the phase transition. Problems with monopole production and the generation of the density perturbations required for galaxy formation are briefly discussed.

162.033 Quanteneffekte in der Entwicklung des frühen Universums. I. Der Einfluß der Raumzeit-Krümmung auf Felder und Teilchen. K.-H. Lotze.
Sterne, 58. Band, 22 - 29 (1982).

162.034 Anthropic-principle arguments against steady-state cosmological theories. F. J. Tipler.
Observatory, Vol. 102, 36 - 39 (1982).

162.035 To the end of time.
E. R. Harrison.
South. Stars, Vol. 29, 41 - 56 (1981).

162.036 Evolution of small perturbations of isotropic cosmological models with one-loop quantum-gravity corrections. A. A. Starobinskij.
Pis'ma v ZhEhTF, Tom 34, 460 - 463 (1981). In Russian. Abstr. in Ref. zh., 51. Astron., 2.51.778 (1982).

162.037 Quantum effects and regular models of the universe.
V. Ts. Gurovich.
Prepr. Inst. fiz. i mat. An KirgSSR. Frunze, Ilim, 1981. 106 pp. In Russian. – Abstr. in Ref. zh., 51. Astron., 2.51.779 (1982).

162.038 Analysis of the large-scale structure of the universe.
A. G. Doroshkevich, Eh. V. Kotok, Yu. S. Sigov,

S. F. Shandarin.
Inst. prikl. mat. AN SSSR. Prepr., 1981, No. 68, 28 pp. In
Russian. – Abstr. in Ref. zh., 51. Astron., 2.51.787 (1982).

**162.039 Gauge hierarchies and unusual symmetry behaviour
at high temperatures.   V. A. Kuzmin (Kuz'min),**
M. E. Shaposhnikov, I. I. Tkachev.
Phys. Lett. B, Vol. 105B, 159 - 162 (1981). – Abstr. in Phys.
Abstr., Vol. 85, Abstr. 6983 (1982).

**162.040 Cosmological constraints on masses and couplings
of leptoquarks.**
V. A. Kuzmin (Kuz'min), M. E. Shaposhnikov.
Phys. Lett. B, Vol. 105B, 163 (1981). – Abstr. in Phys. Abstr.,
Vol. 85, Abstr. 7015 (1982).

162.041 Electromagnetic field generated by cosmic vorticity.
N. A. Batakis.
Phys. Lett. A, Vol. 85A, 409 - 410 (1981). – Abstr. in Phys.
Abstr., Vol. 85, Abstr. 10539 (1982).

**162.042 Matter-antimatter domains in the Universe: a solu-
tion of the vacuum walls problem.**
V. A. Kumin (Kuz'min), M. E. Shaposhnikov, I. I. Tkachev.
Phys. Lett. B, Vol. 105B, 167 - 170 (1981). – Abstr. in Phys.
Abstr., Vol. 85, Abstr. 10540 (1982).

162.043 Perturbations of the Hubble flow.
F. Occhionero, N. Vittorio, P. Carnevali,
P. Santangelo.
Astron. Astrophys., Vol. 107, 172 - 177 (1982).
    The authors study the growth, both linear and non-linear,
of matter condensations in the Hubble flow allowing for a non-
vanishing cosmological constant.

162.044 Structure in the universe from one massive neutrino?
F. R. Klinkhamer.
Astron. Astrophys., Vol. 107, 235 - 239 (1982).
    The author discusses some severe problems confronting
theories of galaxy formation that rely on the growth of small
density perturbations and examine a scenario incorporating
massive neutrinos. He also discusses the sensitivity of such a
model to lower neutrino mass.

**162.045 Finite creation of particles in an expanding
Universe.   H. Ceccatto, A. Foussats,**
H. Giacomini, O. Zandron.
Nuovo Cimento B, Vol. 65B, Ser. 11, 233 - 247 (1981).
Abstr. in Phys. Abstr., Vol. 85, Abstr. 10955 (1982).

**162.046 Conceptual foundations of the self-similar hierarchi-
cal cosmology.   R. L. Oldershaw.**
Int. J. Gen. Syst.,Vol. 7, 151 - 157 (1981). – Abstr. in Phys.
Abstr., Vol. 85, Abstr. 14956 (1982).

**162.047 On the number of levels in the self-similar hierarchi-
cal cosmology.   R. L. Oldershaw.**
Int. J. Gen. Syst., Vol. 7, 159 - 163 (1981). – Abstr. in Phys.
Abstr., Vol. 85, Abstr. 14957 (1982).

162.048 Cosmic strings.   A. Vilenkin.
Phys. Rev. D, Vol. 24, 2082 - 2089 (1981). – Abstr.
in Phys. Abstr., Vol. 85, Abstr. 14962 (1982).

**162.049 Grand unified theories and the lepton number of
the Universe.   J. A. Harvey, E. W. Kolb.**
Phys. Rev. D, Vol. 24, 2090 - 2099 (1981). – Abstr. in Phys.
Abstr., Vol. 85, Abstr. 14963 (1982).

**162.050 Massive scalar fields. I. Generation by conformal
transformation.   M. Nagaraj.**

Vignana Bharathi, Vol. 6, No. 1, p. 18 - 26 (1980). – Abstr.
in Phys. Abstr., Vol. 85, 14967 (1982).

162.051 The empirical basis of cosmology.
J. Stock.
Fundamental physics, (see 012.008), p. 193 - 203 (1981).
Abstr. in Phys. Abstr., Vol. 85, Abstr. 14968 (1982).

**162.052 Non-singular cosmology as a result of a spontaneous
breakdown of symmetries.   S. V. Orlov.**
J. Phys. A, Vol. 14, 3039 - 3045 (1981). – Abstr. in Phys.
Abstr., Vol. 85, Abstr. 18803 (1982).

**162.053 The phase structure of the early Universe in the
minimal SU(5) grand unified theory.**
S. Parke, S.-Y. Pi.
Phys. Lett. B, Vol. 107B, 54 - 58 (1981). – Abstr. in Phys.
Abstr., Vol. 85, Abstr. 18994 (1982).

**162.054 The thermodynamics of the quark-hadron phase
transition in the early Universe.   K. A. Olive.**
Nucl. Phys. B, Vol. B190, 483 - 503 (1981). – Abstr. in Phys.
Abstr., Vol. 85, Abstr. 23356 (1982).

162.055 Quantum effects in the Bertotti-Robinson metric.
V. Sahni.
Phys. Lett. A, Vol. 86A, 87 - 90 (1981). – Abstr. in Phys.
Abstr., Vol. 85, Abstr. 23357 (1982).

162.056 Kerr metric in an expanding universe.
S. N. G. Thakurta.
Indian J. Phys. Part B, Vol. 55B, 304 - 310 (1981). – Abstr.
in Phys. Abstr., Vol. 85, Abstr. 28583 (1982).

162.057 Neutrinos of non-zero mass in Friedmann universes.
P. S. Joshi, S. M. Chitre.
Phys. Lett. A, Vol. 85A, 135 - 137 (1981). – Abstr. in Phys.
Abstr., Vol. 85, Abstr. 28585 (1982).

**162.058 Orthogonal Bianchi cosmologies: bounds on the
evolution of some physical quantities.**
J. L. Sanz.
Phys. Lett. A, Vol. 86A, 273 - 276 (1981). – Abstr. in Phys.
Abstr., Vol. 85, Abstr. 28586 (1982).

**162.059 Heating of the primordial gas by background radia-
tion if heavy elements were available.**
D. A. Varshalovich, V. K. Khersonskij, R. A. Syunyaev.
Astrofizika, Tom 17, 487 - 493 (1981). In Russian. English
translation in Astrophysics, Vol. 17, No. 3.
    The thermal evolution of the primordial H and He gas
during the cosmological expansion at the epoch $z < 150$ de-
pended on impurities of C, O, Si, S atoms. The resonance interaction of these atoms with the background radiation results in
essential heating of the matter, if the abundance of these elements were of some percent of the contemporary values. The
proper distortions of the primordial radiation spectrum are
calculated.

162.060 The missing mass – now it's a gravitino!
J. Silk.
Nature, Vol. 297, 102 - 103 (1982).

**162.061 Cosmological consequences of grand unification
beyond SU(5).   S. M. Barr.**
Nucl. Phys. B, Vol. B192, 523 - 551 (1981). – Abstr. in Phys.
Abstr., Vol. 85, Abstr. 29038 (1982).

**162.062 Are statistical notions applicable in the early
Universe?   M. Dresden.**
Unified field theories and beyond, (see 012.012), p. 175 - 200

(1981). – Abstr. in Phys. Abstr., Vol. 85, Abstr. 29055 (1982).

162.063   **Grand unification and cosmology – an environmental impact statement.**   E. W. Kolb.
Unified field theories and beyond, (see 012.012), p. 211 - 240 (1981). – Abstr. in Phys. Abstr., Vol. 85, Abstr. 29056 (1982).

162.064   **The electric dipole moment of the neutron: a cosmic seismometer?**   D. V. Nanopoulos.
Unified field theories and beyond, (see 012.012), p. 263 - 276 (1981). – Abstr. in Phys. Abstr., Vol. 85, Abstr. 29057 (1982).

162.065   **Cosmological constraints on grand unified theories.**   D. N. Schramm.
Unified field theories and beyond, (see 012.012), p. 277 - 294 (1981). – Abstr. in Phys. Abstr., Vol. 85, Abstr. 29058 (1982).

162.066   **Radiation effects in monopole pair creation.**
M. Blagojevic, S. Meljanac, I. Picek, P. Senjanovic.
Phys. Lett. B, Vol. 106B, 408 - 410 (1981). – Abstr. in Phys. Abstr., Vol. 85, Abstr. 29065 (1982).

162.067   **New cosmology. II.**   A. H. Klotz.
Acta Phys. Polonica B, Vol. B12, 921 - 933 (1981). – Abstr. in Phys. Abstr., Vol. 85, Abstr. 33191 (1982).

162.068   **Einstein-Cartan-Maxwell-Bianchi cosmologies.**
D. Lorenz.
Acta Phys. Polonica B, Vol. B12, 939 - 950 (1981). – Abstr. in Phys. Abstr., Vol. 85, Abstr. 33192 (1982).

162.069   **Primordial helium.**   Y. Andrillat.
Europhys. News, Vol. 12, No. 8 - 9, p. 7 - 9 (1981). – Abstr. in Phys. Abstr., Vol. 85, Abstr. 33193 (1982).

162.070   **Time-dependent, finite, rotating universes.**
M. J. Reboucas, J. A. S. de Lima.
J. Math. Phys., Vol. 22, 2699 - 2703 (1981). – Abstr. in Phys. Abstr., Vol. 85, Abstr. 33194 (1982).

162.071   **Spontaneous compactification, gauge symmetry and the vanishing of the cosmological constant.**
C. A. Orzalesi, M. Pauri.
Phys. Lett. B, Vol. 107B, 186 - 190 (1981). – Abstr. in Phys. Abstr., Vol. 85, Abstr. 33199 (1982).

162.072   **Effect of curvature-squared terms on cosmology.**
K. I. Macrae, R. J. Riegert.
Phys. Rev. D, Vol. 24, 2555 - 2560 (1981). – Abstr. in Phys. Abstr., Vol. 85, Abstr. 33200 (1982).

162.073   **On the influence of matter and physical fields upon the nature of cosmological singularities.**
V. A. Belinskii (*Belinskij*), I. M. Khalatnikov.
Physics reviews. Vol. 3, (see 003.002), 555 - 590 (1981). Abstr. in Phys. Abstr., Vol. 85, Abstr. 33201 (1982).

162.074   **First-order phase transitions in the early Universe.**
E. J. Weinberg.
Unified field theories and beyond, (see 012.012), p. 241 - 262 (1981). – Abstr. in Phys. Abstr., Vol. 85, Abstr. 33202 (1982).

162.075   **Formation of the large-scale structure of the Universe.**
Ya. B. Zel'dovich.
Pis'ma Astron. Zh., Tom 8, 195 - 197 (1982). In Russian.
English translation in Soviet Astron. Lett., Vol. 8.
A qualitative explanation is given for the specific large-scale structure of the Universe, according to which "dark" volumes devoid of galaxies are surrounded by rather thin "bright" layers being built up of galaxies. The explanation is based on very general ideas about the development of large-scale perturbations in a cold gas.

162.076   **The Regge-Wheeler equation with sources for both even and odd parity perturbations of the Schwarzschild geometry.**   M. Sasaki, T. Nakamura.
Phys. Lett. A, Vol. 87A, 85 - 88 (1981). – Abstr. in Phys. Abstr., Vol. 85, Abstr. 38347 (1982).

162.077   **The renormalized energy-momentum tensor in a Robertson-Walker universe.**   T. Azuma.
Prog. Theor. Phys., Vol. 66, 892 - 902 (1981). – Abstr. in Phys. Abstr., Vol. 85, Abstr. 38348 (1982).

162.078   **Cosmological constraints on elementary particle masses in Lemaître cosmological models.**
J. D. Barrow.
Phys. Lett. B, Vol. 107B, 358 - 360 (1981). – Abstr. in Phys. Abstr., Vol. 85, Abstr. 38349 (1982).

162.079   **Mass formula in a cosmogenesis model.**
P. Spindel.
Phys. Lett. B, Vol. 107B, 361 - 363 (1981). – Abstr. in Phys. Abstr., Vol. 85, Abstr. 38350 (1982).

162.080   **A possible large-scale anisotropy of the universe.**
H. H. Fliche, J. M. Souriau, R. Triay.
Astron. Astrophys., Vol. 108, 256 - 264 (1982).
The three-dimensional distribution of quasars known to date shows a singular zone, almost planar, about 200 Mpc wide, 3000 Mpc away, covering half the sky. A general stratification of the universe, parallel to this zone, is possible; this stratification seems to show up in the distribution and the kinematics of nearby galaxies (at least up to 30 Mpc). From this facts the authors get a Friedmann-Lemaitre model $q_0 = -1.1$, $\Omega_0 = 0.1$ together with a general anisotropy around the direction (5 h 46 m, +6° 50').

162.081   **Space-times with spherically symmetric hypersurfaces.**   A. Krasiński.
Gen. Relativ. Gravitation, Vol. 13, 1021 - 1035 (1981).

162.082   **Cosmological models without singularities.**
W. Petry.
Gen. Relativ. Gravitation, Vol. 13, 1057 - 1071 (1981).
A previously studied theory of gravitation in flat space-time (Petry, (1981)) is applied to homogeneous and isotropic cosmological models. There exist two different classes of models without singularities: (1) ever-expanding models, (2) oscillating models. The first class contains models with hot big bang. For these models the author gets at the beginning of the universe – in contrast to Einstein's theory – very high but finite densities of matter and radiation with a big bang of very short duration. After short time these models pass into the homogeneous and isotropic models of Einstein's theory with spatial curvature equal zero and cosmological constant $\Lambda \geqslant 0$.

162.083   **Time travel in Gödel's space.**   J. Pfarr.
Gen. Relativ. Gravitation, Vol. 13, 1073 - 1091 (1981).
The author presents an analysis of the motion of test particles in Gödel's universe. Both geodesical and nongeodesical motions are considered; the accelerations for nongeodesical motions are given. Examples for closed timelike world lines are shown and the dynamical conditions for time travel in Gödel's space-time are discussed. It is shown that these conditions alone do not suffice to exclude time travel in Gödel's space-time.

162.084 The behavior of null geodesics in a class of rotating space-time homogeneous cosmologies.
B. E. Laurent, K. Rosquist, E. Sviestins.
Gen. Relativ. Gravitation, Vol. 13, 1093 - 1115 (1981).

The null geodesics are investigated in a class of open space-time homogeneous cosmological models with rotating sheared matter (Ozsváth class III). Gödel's model is a special case. A stationary polar coordinate system is employed in which matter rotates rigidly. The geodesic equations are solved. Caustics and global rotation are discussed.

162.085 Exact solutions for the spherically symmetric vacuum field in the general scalar tensor theory.
N. Van Den Bergh.
Gen. Relativ. Gravitation, Vol. 14, 17 - 25 (1982).

162.086 Some global properties of closed spatially homogeneous space-times.   G. J. Galloway.
Gen. Relativ. Gravitation, Vol. 14, 87 - 96 (1982).

162.087 Inhomogeneous perfect fluid cosmologies.   K. C. Das.
J. Math. Phys. Sci., Vol. 15, 151 - 158 (1981). – Abstr. in Phys. Abstr., Vol. 85, Abstr. 38519 (1982).

162.088 Cosmology and broken scale invariance.   F. Cooper, G. Venturi.
Phys. Rev. D, Vol. 24, 3338 - 3340 (1981). – Abstr. in Phys. Abstr., Vol. 85, Abstr. 38545 (1982).

162.089 Two singular quantum universes.   T. Christodoulakis.
Nuovo Cimento B, Vol. 66B, Ser. 11, 81 - 95 (1981). – Abstr. in Phys. Abstr., Vol. 85, Abstr. 38549 (1982).

162.090 On the geometry of the Zel'manov-Grishchuk cosmological homogeneity criterion.
M. A. H. MacCallum, A. Spero, D. A. Szafron.
Phys. Lett. A, Vol. 87A, 157 - 158 (1982). – Abstr. in Phys. Abstr., Vol. 85, Abstr. 42761 (1982).

162.091 Finite temperature quantum fields in expanding universes.   B. L. Hu.
Phys. Lett. B, Vol. 108B, 19 - 25 (1982). – Abstr. in Phys. Abstr., Vol. 85, Abstr. 42763 (1982).

162.092 On two "self-creation" cosmologies.   G. A. Barber.
Gen. Relativ. Gravitation, Vol. 14, 117 - 136 (1982).

An attempt to produce a continuous creation theory by adapting the Brans-Dicke theory is described. The universe is seen to be created out of self-contained gravitational, scalar, and matter fields. However, the solution of the one-body problem reveals unsatisfactory characteristics of the theory, and in particular the principle of equivalence is severely violated. A second theory is described which retains the attractive features of the first theory and which does not fall foul of its objections. There do exist empirical tests for the theory which are described and which will require further examination. In the limit this theory approaches general relativity in every respect.

162.093 Primordial black holes and the cosmic baryon number – II.   D. Lindley.
Mon. Not. R. Astron. Soc., Vol. 199, 775 - 784 (1982).

Previous calculations of the generation of the cosmological baryon to photon ratio by the evaporation of primordial black holes are extended to include a closer examination of the properties of grand unified theories. The emission of magnetic monopoles is found not to present a serious problem. Using a simple 'baryosynthetic' reaction network of the kind given by Fry, Olive & Turner, the thermalization of particles emitted by black holes can be studied. This tends to reduce the influence of primordial black holes in determining the baryon to photon ratio, but they may still play an important role.

162.094 Dissipation and unification.   J. D. Barrow.
Mon. Not. R. Astron. Soc., Vol. 199, 45P - 48P (1982).

An estimate is given of the entropy production associated with anisotropy damping during the era of grand unification in the early Universe. The entropy production by collisional transport processes is negligible.

162.095 Steps toward the Hubble constant. VIII. The global value.   A. Sandage, G. A. Tammann.
Astrophys. J., Vol. 256, 339 - 345 (1982).

A new calibration is given of the mean absolute magnitude of the three brightest red and blue supergiant stars in nearby galaxies whose distances are independently known from Cepheid variables. The calibration covers 7 magnitudes in M_B of the parent galaxy and shows $\langle M_V(3) \rangle$ for red supergiants to be constant at -7.72 ± 0.06 with dispersion of $\sigma = 0.17$ mag. The brightest red supergiants have been identified and measured in the three nearby resolved galaxies IC 4182, NGC 4214, and NGC 4395 to obtain distance moduli of $(m-M)^0 = 28.21$, 29.02, and 28.82, respectively. These distances are used to calibrate the mean absolute magnitude at maximum of the two Type I supernovae (SNe I) 1937c and 1954a, after showing that $\sigma(M)$ for SNe I is small, and hence that such stars are also good distance indicators. The value $\langle M_B(\text{max}) \rangle_{\text{SNe I}} = -19.74 \pm 0.19$ is used to calibrate the velocity–apparent magnitude (Hubble) diagram for 16 SNe I, most of which have recession velocities greater than 3000 km s^{-1}, beyond the effect of any local velocity anisotropy. The result is that the global value of the Hubble constant is $H_0 = 50 \pm 7$ km s^{-1} Mpc^{-1}.

162.096 Anisotropy in nonprimordial cosmic background radiation.   C. J. Hogan.
Astrophys. J., Lett., Vol. 256, L33 - L35 (1982).

In the standard cosmological model, large-scale anisotropy in microwave background temperature is generally attributed to primordial fluctuations. However, anisotropy may also be produced by radiation emitted inhomogeneously at relatively recent epochs ($z \lesssim 200$) and thermalized by grains or molecules. This hypothesis predicts a white noise angular distribution of $\delta T/T$ on large scales, with amplitude determined by the inhomogeneity of the sources of radiation, and a smooth distribution below an angle $\theta_1 \approx 3°$-$15°$ where scattering becomes important. Nonstandard big bangs or theories of galaxy formation which have smooth initial conditions may be tested by current observations of anisotropy at $\approx 6°$ and $\approx 90°$.

162.097 Quarks and cosmology.   V. A. Kuz'min.
Usp. fiz. nauk, Tom 135, 521 - 523 (1981). In Russian. – Abstr. in Ref. zh., 51. Astron., 3.51.866 (1982).

162.098 Vacuum energy and large-scale structure of the universe.   V. F. Mukhanov, G. V. Chibisov.
Fiz. inst. AN SSSR. Prepr., 1981, No. 198, 30 pp. In Russian. Abstr. in Ref. zh., 51. Astron., 3.51.867 (1982).

162.099 Lücken im Weltraum.   M. Petroll.
Umschau, 82. Jahrg., 315 (1982).

162.100 Determination of the cosmological deceleration parameter based on statistical analysis of various quasar subsets.   F.-h. Cheng, T. Kiang, L.-z. Fang.
Acta Astron. Sinica, Vol. 22, 357 - 363 (1981). In Chinese.

162.101 Mean-energy neutrinos in the Universe.
G. S. Bisnovatyj-Kogan, Z. F. Seidov.
Astron. Zh., Tom 59, 213 - 223 (1982). In Russian. English translation in Soviet Astron., Vol. 26, No. 2.

Using recent theoretical calculations of supernovae rate data and of the chemical composition of matter the authors have calculated the density and the spectrum of neutrinos with the energy 3 - 30 MeV and their influence on the solar neutrino detector. The density of these neutrinos in the Universe at the present time is $(2 - 10) \times 10^{33}$ g/cm^3, which is larger than the density of the relict radiation. The mean-energy neutrinos have been produced after the origin of the stars at $z \lesssim 10$ and continue to be produced up to now.

162.102 A cosmological version of flatness conditions in general relativity.   M. Kihara, H. Nariai.
Prog. Theor. Phys., Vol. 66, 1498 - 1499 (1981). – Abstr. in Phys. Abstr., Vol. 85, Abstr. 42913 (1982).

162.103 The cosmological term and supersymmetry.
H. Pagels.
Gauge theories, massive neutrinos and proton decay, (see 012.021), p. 121 - 129 (1981). – Abstr. in Phys. Abstr., Vol. 85, Abstr. 43227 (1982).

162.104 The Coleman-Weinberg mechanism in early cosmology.   A. Billoire, K. Tamvakis.
Nucl. Phys. B, Vol. B200[FS4], 329 - 344 (1982). – Abstr. in Phys. Abstr., Vol. 85, Abstr. 43233 (1982).

162.105 Astrophysics and grand unification.
E. W. Kolb.
Gauge theories, massive neutrinos and proton decay, (see 012.021), p. 177 - 190 (1981). – Abstr. in Phys. Abstr., Vol. 85, Abstr. 43253 (1982).

162.106 Cosmic strings and domains in unified theories.
G. Lazarides, Q. Shafi, T. F. Walsh.
Nucl. Phys. B, Vol. B195, 157 - 172 (1982). – Abstr. in Phys. Abstr., Vol. 85, Abstr. 48371 (1982).

162.107 Interacting quantum fields in a nonsingular cosmological model.   B. L. Spokoiny (Spokoinij).
Phys. Lett. A, Vol. 87A, 211 - 214 (1982). – Abstr. in Phys. Abstr., Vol. 85, Abstr. 48372 (1982).

162.108 Creation of fermion pairs near the cosmological singularity.   A. A. Kharkov (Khar'kov).
Phys. Lett. A, Vol. 87A, 223 - 225 (1982). – Abstr. in Phys. Abstr., Vol. 85, Abstr. 48373 (1982).

162.109 Friedmann Universe in a quantum gravity model.
T. Padmanabhan.
Phys. Lett. A, Vol. 87A, 226 - 228 (1982). – Abstr. in Phys. Abstr., Vol. 85, Abstr. 48374 (1982).

162.110 Creation of Schwarzschild-de Sitter wormholes by a cosmological first-order phase transition.
K. Maeda, K. Sato, M. Sasaki, H. Kodama.
Phys. Lett. B, Vol. 108B, 98 - 102 (1982). – Abstr. in Phys. Abstr., Vol. 85, Abstr. 48376 (1982).

162.111 Multi-production of universes by first-order phase transition of a vacuum.
K. Sato, H. Kodama, M. Sasaki, K. Maeda.
Phys. Lett. B, Vol. 108B, 103 - 107 (1982). – Abstr. in Phys. Abstr., Vol. 85, Abstr. 48377 (1982).

162.112 Induced gravity, Yang Mills fields and cosmology: classical solutions.   J. Cervero.
Phys. Lett. B, Vol. 108B, 108 - 110 (1982). – Abstr. in Phys. Abstr., Vol. 85, Abstr. 48378 (1982).

162.113 On the proposed existence of an anti-gravity regime in the early Universe.   M. D. Pollock.
Phys. Lett. B, Vol. 108B, 386 - 388 (1982). – Abstr. in Phys. Abstr., Vol. 85, Abstr. 48379 (1982).

162.114 A new inflationary Universe scenario: a possible solution of the horizon, flatness, homogeneity, isotropy and primordial monopole problems.   A. D. Linde.
Phys. Lett. B, Vol. 108B, 389 - 393 (1982). – Abstr. in Phys. Abstr., Vol. 85, Abstr. 48380 (1982).

162.115 Singularity avoidance in the semi-classical Gowdy T^3 cosmological model.   B. K. Berger.
Phys. Lett. B, Vol. 108B, 394 - 398 (1982). – Abstr. in Phys. Abstr., Vol. 85, Abstr. 48381 (1982).

162.116 Models of the Universe.   P. A. M. Dirac.
Gauge theories, massive neutrinos and proton decay, (see 012.021), p. 1 - 9 (1981). – Abstr. in Phys. Abstr., Vol. 85, Abstr. 48382 (1982).

162.117 The age of galaxies.   M. G. Edmunds.
Nature, Vol. 297, 284 - 285 (1982).

162.118 The Galaxy as a fundamental standard for extragalactic distances. I. New methods to choose between the proposed values of the Hubble constant.
G. de Vaucouleurs.
C. R. Acad. Sci. Paris, Tome 294, Sér. II, 857 - 859 (1982). In French.

The traditional multi-step approach to the extragalactic distance scale can be by-passed by using the Galaxy as fundamental calibrator. The current best values of the basic scale factors of the Galaxy are tabulated. This new approach leads to several crucial tests of the long and short distance scales and corresponding values of the Hubble constant.

162.119 The Galaxy as a fundamental standard for extragalactic distances. II. A first crucial test of the short and long distance scales through the use of the Tully-Fisher relations.   G. de Vaucouleurs.
C. R. Acad. Sci. Paris, Tome 294, Sér. II, 903 - 905 (1982). In French.

The B- and H-band versions of the Tully-Fisher relation between absolute magnitude and maximum rotation velocity (or 21 cm line width) of disk galaxies are calibrated by means of the relevant parameters of the Galaxy. This leads to a first crucial test of the extragalactic distance scales and corresponding values of the Hubble constant.

162.120 Gravitational lenses and cosmological evolution.
J. A. Peacock.
Extragalactic radio sources, (see 012.025), p. 451 - 452 (1982).

162.121 Discovery of pre-galactic lithium.   B. Pagel.
Nature, Vol. 297, 456 - 457 (1982).

162.122 Lithium abundance at the formation of the Galaxy.
M. Spite, F. Spite.
Nature, Vol. 297, 483 - 485 (1982).

The authors have made new observations, aiming at a better determination of the lithium abundance at the time of formation of the Galaxy. The newly observed stars are extreme population II dwarfs (very old, very metal-poor stars). Their lithium abundance turned out to be significantly lower than the abundance in population I stars. If attributed to the big bang, this lithium abundance suggests a rather low baryonic density of the Universe, which in turn favours, under some assumptions, an open Universe. Finally, the authors suggest that the good agreement between the abundance of ^{7}Li and the deuterium ^2H abundance supports the standard model of the big bang.

162.123   Cosmology, kinetics of vacuum and neutrino rest mass.   Yu. G. Ignat'ev,
Gravitatsiya i teor. otnositel'nosti, Kazan', 1981, No. 18, p. 73 - 75. In Russian. − Abstr. in Ref. zh., 51. Astron., 4.51.753 (1982).

162.124   On estimates of the number of particles in statistical models of the Friedmann universe.
E. N. Rumyantseva.
Gravitatsiya i teor. otnositel'nosti, Kazan', 1981, No. 18, p. 104 - 109. In Russian. − Abstr. in Ref. zh., 51. Astron., 4.51.754 (1982).

162.125   Origin of initial inhomogeneities of the universe.
D. A. Kompaneets, V. N. Lukash, I. D. Novikov.
Inst. kosm. issled. AN SSSR. Prepr., 1981, No. 652, 35 pp. In Russian. − Abstr. in Ref. zh., 51. Astron., 4.51.768 (1982).

162.126   Large-scale structure of the universe. I. General properties. One- and two-dimensional models.
V. I. Arnol'd, Ya. B. Zel'dovich, S. F. Shandarin.
Inst. prikl. mat. AN SSSR. Prepr., 1981, No. 100, 31 pp. In Russian. − Abstr. in Ref. zh., 51. Astron., 4.51.769 (1982).

162.127   Cosmological Friedmann models with non-linear scalar field.   G. G. Ivanov.
Gravitatsiya i teor. otnositel'nosti, Kazan', 1981, No. 18, p. 54 - 60. In Russian. − Abstr. in Ref. zh., 51. Astron., 4.51.772 (1982).

162.128   Maximum density in clouds of heavy neutrinos.
Ya. B. Zel'dovich, S. F. Shandarin.
Pis'ma Astron. Zh., Tom 8, 259 - 262 (1982). In Russian. English translation in Soviet Astron. Lett., Vol. 8.
Formation of large-scale structure is considered in the neutrino-dominated Universe. Thermal velocities having survived since the epoch of big-bang are shown to limit significantly the maximum density of neutrinos in the firstly formed objects "pancakes".

162.129   "Black regions" in the universe.
Ya. B. Zel'dovich, S. F. Shandarin.
Zemlya Vselennaya, 1982, No. 2, p. 2 - 6. In Russian.

162.130   Problems of metagalactic astronomy.
I. S. Shklovskij.
Zemlya Vselennaya, 1982, No. 2, p. 7 - 12; No. 3, p. 17 - 23. In Russian.

162.131   Interacciones fuertes y cosmología: bootstrap a = 3″.
R. F. Sisteró.
Bol. Asoc. Argentina Astron., No. 20 - 24, p. 195 (1981). Abstract.

162.132   Las inferencias cosmológicas e históricas del Prof. Mc Crea.   C. J. Lavagnino.
Bol. Asoc. Argentina Astron., No. 20 - 24, p. 271 (1981). Abstract.

162.133   The Galaxy as a fundamental standard for extragalactic distances. III. A second crucial test of the short and long distance scales through the use of the Faber-Jackson relation.   G. de Vaucouleurs.
C. R. Acad. Sci. Paris, Tome 294, Sér. II, 981 - 983 (1982). In French.
The B- and V-band versions of the Faber-Jackson relation between absolute magnitude and central velocity dispersion of spheroidal galaxies are calibrated by means of the relevant parameters of the Galaxy. This leads to a crucial test of the extragalactic distance scales and corresponding values of the Hubble constant. Again the short scale ($H_0 = 95$ km s^{-1} Mpc^{-1}) is confirmed, the long scale ($H_0' = 50$) is rejected at highly significant levels.

162.134   Astrophysical tests for radiative decay of neutrinos and fundamental physics implications.
F. W. Stecker, R. W. Brown.
Astrophys. J., Vol. 257, 1 - 9 (1982).
The radiative lifetime τ for the decay of light neutrinos is calculated using various physical models for neutrino decay. The results are then related to the astrophysical problem of the detectability of the decay photons from cosmic neutrinos. Conversely, the astrophysical data are used to place lower limits on τ. These limits are all well below predicted values. However, an observed feature at ~ 1700 Å in the ultraviolet background radiation at high galactic latitudes may be from the decay of neutrinos with mass ~ 14 eV. This would require a decay rate much larger than the predictions of "standard" models but could be indicative of a decay rate possible in composite models or other new physics.

162.135   Relativistic hydromagnetic wave propagation and instability in an anisotropic universe.
D. Papadopoulos, F. P. Esposito.
Astrophys. J., Vol. 257, 10 - 16 (1982).
Exact equations governing finite-amplitude wave propagation in hydromagnetic media are derived within the general theory of relativity. These equations are then used to discuss the stability of a hydromagnetic cosmological model, and a general relativistic counterpart of Parker's instability is isolated.

162.136   Multipole anisotropy of the cosmic background radiation in density wave models.
R. Fabbri, I. Guidi, V. Natale.
Astrophys. J., Vol. 257, 17 - 22 (1982).
The authors study the the anisotropy of the cosmic background radiation predicted by the density-wave models at scales $\gtrsim 10°$ and establish the connection of the multipole moments of the wave pattern with the corresponding multipoles in the angular distribution of the radiation. In open spaces, multipoles of order $l > 2$ are shown to be important for long-wavelength perturbations (for a reasonable multipole content in the wave pattern) and to dominate for wavelengths much larger than the particle horizon. The interpretation of available experimental data in terms of perturbations of the particle horizon is discussed, and a sensitive test involving the gradient of the anisotropy is suggested.

162.137   Gravitational wave backgrounds and the early Universe.   B. J. Carr.
Nature, Vol. 297, 623 (1982).

162.138   Scalar particles creation rate in an expanding universe.
M. Castagnino, L. Chimento, D. Harari.
Bol. Asoc. Argentina Astron., No. 26, p. 83 - 86 (1981).

162.139   Cosmological and quantum constraint on particle masses.   C. Sivaram.
American J. Phys., Vol. 50, 279 (1982). − Abstr. in Phys. Abstr., Vol. 85, Abstr. 48462 (1982).

162.140   Two cosmological solutions of Regge calculus.
S. M. Lewis.
Phys. Rev. D, Vol. 25, 306 - 312 (1982). − Abstr. in Phys. Abstr., Vol. 85, Abstr. 48571 (1982).

162.141   Magnetic monopoles, duality and cosmological phase transitions.
C. O. Escobar, A. A. Natale, G. C. Marques.

Phys. Lett. B, Vol. 109B, 28 - 30 (1982). – Abstr. in Phys. Abstr., Vol. 85, Abstr. 48875 (1982).

162.142 **A possible solution to the problem of cosmic domain walls.** K. Tamvakis, C. E. Vayonakis.
Phys. Lett. B, Vol. 109B, 15 - 18 (1982). – Abstr. in Phys. Abstr., Vol. 85, Abstr. 48922 (1982).

162.143 **Formation of galaxies and clusters of galaxies in the neutrino dominated Universe.**
H. Sato, F. Takahara.
Prog. Theor. Phys., Vol. 66, 508 - 525 (1981). – Abstr. in Phys. Abstr., Vol. 85, Abstr. 53366 (1982).

162.144 **On the use of statistical concepts in grand unified theories.** M. Dresden.
Physica A, Vol. 110A, 1 - 40 (1982). – Abstr. in Phys. Abstr., Vol. 85, Abstr. 53394 (1982).

162.145 **On quantum theory of general-relativistic many-particle systems. II. Friedmann-Robertson-Walker universes with quantized matter.** I. Ichinose.
Prog. Theor. Phys., Vol. 66, 498 - 507 (1981). – Abstr. in Phys. Abstr., Vol. 85, Abstr. 53395 (1982).

162.146 **High-energy behaviour of the particle spectrum created by a linearly expanding universe.**
M. Castagnino.
Prog. Theor. Phys., Vol. 66, 2003 - 2010 (1981). – Abstr. in Phys. Abstr., Vol. 85, Abstr. 53396 (1982).

162.147 **Fate of wormholes created by first-order phase transition in the early universe.**
H. Kodama, M. Sasaki, K. Sato, K. Maeda.
Prog. Theor. Phys., Vol. 66, 2052 - 2072 (1981). – Abstr. in Phys. Abstr., Vol. 85, Abstr. 53397 (1982).

162.148 **Initial behavior of a quantized scalar field and the associated pair-creation in several anisotropic universes.** H. Nariai.
Prog. Theor. Phys., Vol. 66, 2073 - 2084 (1981). – Abstr. in Phys. Abstr., Vol. 85, Abstr. 53398 (1982).

162.149 **Production of magnetized black holes and wormholes by first-order phase transition in the early universe.** K. Sato.
Prog. Theor. Phys., Vol. 66, 2287 - 2290 (1981). – Abstr. in Phys. Abstr., Vol. 85, Abstr. 53399 (1982).

162.150 **Baryon asymmetry and low energy parity restoration.** A. Masiero, G. Senjanovic.
Phys. Lett. B, Vol. 108B, 191 - 195 (1982). – Abstr. in Phys. Abstr., Vol. 85, Abstr. 53400 (1982).

162.151 **Baryogenesis without the out-of-equilibrium decay of superheavy bosons.** M. S. Turner.
Phys. Rev. D, Vol. 25, 299 - 305 (1982). – Abstr. in Phys. Abstr., Vol. 85, Abstr. 53402 (1982).

162.152 **Method of determination of the relict radiation temperature in the epoch $z \geqslant 2$.**
V. K. Khersonskij.
Astron. Tsirk., No. 1168, p. 1 - 2 (1981). In Russian.

162.153 **H_0, q_0 and the local velocity field.**
A. Sandage, G. A. Tammann.
Prepr. Astron. Inst. Univ. Basel, No. 3, 72 pp. (1981).
Contents: Introduction and historical summary; H_0 from the brightest red supergiants and supernovae; The value of q_0; The very local velocity field; Prospects for the future; Appendix: remarks on various distance scales.

162.154 **Origin of the primordial inhomogeneities of the Universe.**
D. A. Kompaneets, V. N. Lukash, I. D. Novikov.
Astron. Zh., Tom 59, 424 - 433 (1982). In Russian. English translation in Soviet Astron., Vol. 26, No. 3.
The problem of the origin of primordial sound waves in the early hot Universe which gave birth to the large-scale structure of the observed Universe is considered. A general principle determining initial metric fluctuations at the moment of time close to the Planckian is proposed.

162.155 **The influence of a non-zero neutrino rest mass on the development of perturbations in an isotropic world.** A. V. Zakharov.
Astron. Zh., Tom 59, 434 - 446 (1982). In Russian. English translation in Soviet Astron., Vol. 26, No. 3.
The behaviour of long-wave perturbations in an isotropic world filled by a collisionless gas of massive particles (neutrinos) is investigated.

162.156 **The neutrino background of the universe and fluctuations of the temperature of microwave relic radiation.** N. A. Zabotin, P. D. Nasel'skij.
Astron. Zh., Tom 59, 447 - 457 (1982). In Russian. English translation in Soviet Astron., Vol. 26, No. 3.
The correlation properties of the fluctuations of microwave cosmic radiation which are formed during the period of recombination are calculated in the framework of the cosmological model with massive relic neutrinos.

162.157 **The peculiar velocity around a hole in the galaxy distribution.** P. J. E. Peebles.
Astrophys. J., Vol. 257, 438 - 441 (1982).
The development of a hole in the space distribution of galaxies is studied in a spherically symmetric zero pressure model. It is shown that the peculiar velocity around a hole of diameter D can be considerably less than the Hubble velocity HD across the hole.

162.158 **Scale-covariant gravitation and primordial nucleosynthesis.** T. Rothman, R. Matzner.
Astrophys. J., Vol. 257, 450 - 455 (1982).
The authors have developed a nucleosynthesis code which is designed to test nonstandard cosmologies and which integrates the dynamical Einstein equations for any equation of state. Using this code they have attempted to place consistent limits on any change $\dot{G}/G$ of the gravitational constant. The limit found for the best model is $\dot{G}/G \lesssim 1.7 \times 10^{-13}$ yr^{-1}. This limit is about a factor of 2 stronger than limits found in naive models where the conservation laws in the scale-covariant formalism are not taken into account.

162.159 **Relativistic collisionless particles and the evolution of cosmological perturbations.** E. T. Vishniac.
Astrophys. J., Vol. 257, 456 - 472 (1982).
At various stages in the evolution of the early universe a significant fraction of the total energy density was in the form of relativistic collisionless particles. The author considers the evolution cosmological perturbations in an Einstein–de Sitter universe filled with the usual perfect fluid plus a relativistic collisionless gas. On scales smaller than the horizon the collisionless gas undergoes rapid phase mixing and tends to a uniform distribution. However, on larger scales the ability of the collisionless gas to support a stress leads to altered growth rates for all perturbation modes except the nondecaying tensor and scalar modes. From the results it is possible to set significant limits on the amount of spatially varying shear at the time of baryon synthesis. Somewhat more stringent limits for large scales can be derived from the apparent smoothness of the microwave background. The author presents the results of a numerical integration of the evolution of adiabatic perturbations as they cross the horizon, including the effect of

massless neutrinos. The sound waves that result are lowered in amplitude by approximately 20% relative to their amplitudes in a neutrino-free cosmology.

162.160   **Pregalactic stars: precursors to galaxy formation.**
J. E. Jones.
The most massive stars, (see 012.034), p. 339 - 346 (1981).

162.161   **Horizon-free universe.**
K. Brecher, A. Frenkel.
Bull. American Astron. Soc., Vol. 13, 824 (1981). – Abstract.

162.162   **Baryon and massive neutrino normal modes in a massive neutrino dominated universe.**
F. Occhionero, N. Vittorio, M. Boccadoro, S. de Luca.
Bull. American Astron. Soc., Vol. 13, 844 (1981). – Abstract.

162.163   **Correlation methods of identification of cosmological molecular clouds.**   V. K. Dubrovich.
Astrofiz. Issled., Izv. Spets. Astrofiz. Obs., Tom 15, 21 - 25 (1982). In Russian.
Correlation properties of spatial fluctuations of the relict radiation temperature at various wavelengths in case some part of these fluctuations is created by molecular clouds of cosmological origin are considered.

162.164   **Search for initial disturbances of the universe: spectral analysis of observations at RATAN-600.**
V. K. Dubrovich, V. S. Lebedev.
Astrofiz. Issled., Izv. Spets. Astrofiz. Obs., Tom 15, 161 - 163 (1982). In Russian.
Reduction of observations with the radio telescope RATAN-600 showed the absence of quasi-periodic fluctuations of the background radiation with scales from 0.5 to 10' and relative amplitude $\Delta T/T \approx 10^{-3}$ at the level of statistical significance of 0.999.

162.165   **Three-dimensional numerical model of the formation of the large-scale structure of the universe.**
A. A. Klypin, S. F. Shandarin.
Inst. prikl. mat. AN SSSR. Prepr., 1981, No. 136, 27 pp. In Russian. – Abstr. in Ref. zh., 51. Astron., 5.51.889 (1982).

162.166   **Particle collisions in an expanding universe.**
G. S. Bisnovatyj-Kogan, I. G. Shukhman.
Zh. ehksp. i teor. fiz., Tom 82, 3 - 8 (1982). In Russian. Abstr. in Ref. zh., 51. Astron., 5.51.891 (1982).

162.167   **Black hole formation in the early universe.**
N. A. Zabotin, P. D. Nasel'skij.
Astrofizika, Tom 16, 337 - 349 (1980). In Russian. English translation in Astrophysics, Vol. 16, No. 2.
One of the possible ways of creation of primordial black holes in the early universe is considered. It has been shown that a high enough level of the primeval inhomogeneities leads to the possibility of the creation of PBHs due to clustering of black holes of minimum mass. The characteristics of the mass spectrum of developing PBHs are calculated and the restrictions of the parameters of the primeval inhomogeneity spectrum are obtained.

162.168   **Asymmetric lepton production in a universe with non-zero baryon number.**   C. Sivaram.
Astrophys. Space Sci., Vol. 82, 485 - 488 (1982).
It is pointed out that as a result of processes during different stages of stellar evolution a universe with a net baryon number develops during a Hubble age an excess of neutrinos over antineutrinos of the same order as the observed relative strength of CP violation. The same processes also favour excess positron production.

162.169   **Dynamics of a viscous universe.**
A. Woszczyna, W. Betkowski.
Astrophys. Space Sci., Vol. 82, 489 - 493 (1982).
The dynamics of the viscous Robertson-Walker universe is presented using the phase variables method. Connection between violence of the strong energy condition and the behaviour of the Universe near the singularity is discussed.

162.170   **Maxwell equations in homogeneous cosmologies.**
D. Lorenz.
Astrophys. Space Sci., Vol. 83, 63 - 67 (1982).
The author investigates the source-free Maxwell equations for all Bianchi types I–IX. Exact solutions are given for types I, II, III, IV, V, VI_0, VII_0, VIII and IX.

162.171   **The estimation of the mean density in the Universe.**
P. Flin.
Astrophys. Space Sci., Vol. 83, 437 - 439 (1982).
The mean density of matter, as estimated from deep optical samples of galaxies, is too low to close the Universe. However, some additional considerations do not exclude such a possibility.

162.172   **Cosmology of Nordström's first theory of gravitation.**   S. S. D. Willenbrock.
American J. Phys., Vol. 50, 229 - 231 (1982). – Abstr. in Phys. Abstr., Vol. 85, Abstr. 53456 (1982).

162.173   **Heavy quarks and perturbative quantum-chromodynamic calculations.**   S. Gupta, H. R. Quinn.
Phys. Rev. D, Vol. 25, 838 - 842 (1982). – Abstr. in Phys. Abstr., Vol. 85, Abstr. 54072 (1982).

162.174   **Entropy generation and phase transitions in the early Universe.**   L.-z. Fang.
Chinese Phys., Vol. 1, 617 - 618 (1981). – Abstr. in Phys. Abstr., Vol. 85, Abstr. 58513 (1982).

162.175   **On the neutrino mass, the abundance of right-handed neutrinos and the closure of the Universe.**
C.-r. Qing, Y.-s. Wu, Z.-x. He, Z.-x. Zhang, Z.-l. Zou.
Chinese Phys., Vol. 1, 619 - 623 (1981). – Abstr. in Phys. Abstr., Vol. 85, Abstr. 58514 (1982).

162.176   **Cosmological evolution of monopoles connected by strings.**   A. Vilenkin.
Nucl. Phys. B, Vol. B196, 240 - 258 (1982). – Abstr. in Phys. Abstr., Vol. 85, Abstr. 58516 (1982).

162.177   **Supercooled phase transitions in the very early Universe.**   S. W. Hawking, I. G. Moss.
Phys. Lett. B, Vol. 110B, 35 - 38 (1982). – Abstr. in Phys. Abstr., Vol. 85, Abstr. 58517 (1982).

162.178   **Geodesic instability and internal time in relativistic cosmology.**   C. M. Lockhart, B. Misra, I. Prigogine.
Phys. Rev. D, Vol. 25, 921 - 929 (1982). – Abstr. in Phys. Abstr., Vol. 85, Abstr. 58518 (1982).

162.179   **Colliding gravitational waves in expanding cosmologies.**   J. Centrella, R. A. Matzner.
Phys. Rev. D, Vol. 25, 930 - 941 (1982). – Abstr. in Phys. Abstr., Vol. 85, Abstr. 58519 (1982).

162.180   **Supercooling in SU(5) and temperature dependence of the gauge coupling constant.**
G. P. Cook, K. T. Mahanthappa.
Phys. Rev. D, Vol. 25, 1154 - 1156 (1982). – Abstr. in Phys. Abstr., Vol. 85, Abstr. 58520 (1982).

162.181 **The Galaxy as a fundamental standard for extra-**
galactic distances. IV. New checks of the short
distance scale and application of the "sosies" method.
G. de Vaucouleurs.
C. R. Acad. Sci. Paris, Tome 294, Sér. II, 1087 - 1089 (1982).
In French.
Two additional tests of the extragalactic distance scales
and corresponding values of the Hubble constant
are presented: (a) verification of the zero point of the luminosity
index scale, (b) comparison of the Galaxy to its "sosies". Both
confirm closely the short distance scale ($H_0 = 95 \, km\,s^{-1}Mpc^{-1}$)
and reject the long scale ($H_0' = 50$) at highly significant levels.

162.182 **The Galaxy as a fundamental standard for extra-**
galactic distances. V. Another decisive test of the
long and short distance scales: the distance of the Virgo E
cluster and its implications. G. de Vaucouleurs.
C. R. Acad. Sci. Paris, Tome 294, Sér. II, 1131 - 1133 (1982).
In French.
The validity of the short scale and the impossibility of
the long scale are again confirmed by comparison of the
implications of the alternative estimates of the distance of
the Virgo E cluster for the absolute magnitudes of the globular
clusters, the RR Lyrae and the Mira-type variables.

162.183 **Isotopic abundance ratios and Dirac's Large**
Numbers Hypothesis. E. B. Norman.
Astrophys. J., Vol. 258, 41 - 42 (1982).
Some effects of the spontaneous creation of nucleons
within atomic nuclei are examined. It is shown that certain
observed isotopic abundance ratios are in sharp disagreement
with those expected from a recently proposed model of
multiplicative creation based upon Dirac's Large Numbers
Hypothesis.

162.184 **Formation of voids in the galaxy distribution.**
S. J. Aarseth, W. C. Saslaw.
Astrophys. J., Lett., Vol. 258, L7 - L10 (1982).
Voids form naturally as an initially homogeneous system
of galaxies clusters gravitationally. The size distribution of
voids has a characteristic shape. It does not depend strongly on
initial conditions or cosmology. Therefore, comparison with
observations provides a powerful general test of gravitational
clustering theories.

162.185 **The effect of the electromagnetic fields on the**
evolution of homogeneous cosmological models near
the singularity. B. L. Spokoinij.
Gen. Relativ. Gravitation, Vol. 14, 279 - 291 (1982).
Homogeneous cosmological models of all Bianchi types
with a general (nondiagonal) metric are considered near the
singularity. The spaces are filled with an arbitrary sourceless
six-component electromagnetic field (EMF) and a perfect fluid
at rest. It is shown that in the general case the models of
types VI_0, VII_0, VIII, and IX have an oscillatory regime. The
models of all the other types have Kasner asymptotics. The
main result of the present paper is the derivation of the law of
rotation of the Kasner axes.

162.186 **Fondamenti del big-bang.** A. Masani.
G. Astron., Vol. 7, 253 - 276 (1981).

162.187 **Quanteneffekte in der Evolution des frühen**
Universums. II. Kosmologische Konsequenzen der
Teilchenerzeugung. K.-H. Lotze.
Sterne, 58. Band, 86 - 92 (1982).

162.188 **Perturbation of the magnitude-redshift relation in**
an inhomogeneous relativistic model: the redshift
equations. L. Nottale.
Astron. Astrophys., Vol. 110, 9 - 17 (1982).
The question of the possible perturbation of the magnitude-redshift relation by clusters of galaxies is dealt within the
framework of a locally non homogeneous cosmological model.

162.189 **Cosmological density fluctuations produced by a**
Goldstone field. A. Vilenkin.
Phys. Rev. Lett., Vol. 48, 59 - 61 (1982).
It is shown that a massless Goldstone field produced at a
cosmological phase transition near $T \sim 10^{17}$ GeV can generate
density fluctuations sufficient to explain the galaxy formation.

162.190 **Gravitational entropy in a self-consistent thermo-**
dynamic cosmological solution.
G. Horwitz, D. Weil.
Phys. Rev. Lett., Vol. 48, 219 - 222, with a correction p. 1136
(1982).
The entropy of a Robertson-Walker universe in statistical
thermodynamic stable equilibrium with conformal bosons is
calculated with use of a new self-consistent method. Its value
is similar to but higher than the entropy of an alternative
empty de Sitter solution, which may help in interpretation of
Hawking thermal states.

162.191 **Simulation of gravitational superclustering of**
massive neutrinos. A. L. Melott.
Phys. Rev. Lett., Vol. 48, 894 - 896 (1982).
Results of a numerical simulation of large-scale, plane,
symmetric gravitational clustering of massive neutrinos are
presented. It is shown that neutrinos may cluster in galactic
halos in the adiabatic scenario of galaxy formation. The
population of test particles exhibits very little phase mixing,
suggesting that the scale of galactic halos may be set by the
neutrino mass.

162.192 **Cosmology for grand unified theories with radiatively**
induced symmetry breaking.
A. Albrecht, P. J. Steinhardt.
Phys. Rev. Lett., Vol. 48, 1220 - 1223 (1982).

162.193 **Reheating an inflationary universe.**
A. Albrecht, P. J. Steinhardt, M. S. Turner,
F. Wilczek.
Phys. Rev. Lett., Vol. 48, 1437 - 1440 (1982).

162.194 **Effects of anisotropy and dissipation on the**
primordial light-isotope abundances.
T. Rothman, R. Matzner.
Phys. Rev. Lett., Vol. 48, 1565 - 1568 (1982).
Account is taken of the dissipation of anisotropy due to
neutrino viscosity and nearly collisionless radiation in a
Bianchi type-I cosmology during primordial nucleosynthesis.
For experimentally allowed cross sections, and for moderate
anisotropy, the final ^4He mass fraction is significantly less than
in the standard model, while the D fraction increases much less
than would be expected in a low-helium isotropic cosmology.

162.195 **Formation of galaxies in a gravitino-dominated**
universe.
J. R. Bond, A. S. Szalay, M. S. Turner.
Phys. Rev. Lett., Vol. 48, 1636 - 1640 (1982).
If gravitinos of mass 1 keV (or similar particles) dominate
the mass of the universe, a critical scale of galactic size arises
due to their collisionless phase mixing. It is shown that density
perturbation spectrum of gravitinos is relatively flat between
galactic and cluster scales, unlike the massive neutrino case. If
gravitinos form the dark matter, initial adiabatic fluctuations
lead to a hierarchical picture of clustering. Galaxies form first,
but dissipation is necessary for their survival.

162.196 **Gravitational lenses and cosmological evolution.**
J. A. Peacock.
Mon. Not. R. Astron. Soc., Vol. 199, 987 - 1006 (1982).
The effect of gravitational lensing on the apparent

cosmological evolution of extragalactic radio sources is investigated. Models for a lens population consisting of galaxies and clusters of galaxies are constructed and used to calculate the distribution of amplification factors caused by lensing. Although many objects at high redshifts are predicted to have flux densities altered by 10–20 per cent relative to a homogeneous universe, flux conservation implies that de-amplification is as common as amplification. The effects on cosmological evolution as inferred from source counts and redshift data are thus relatively small; the slope of the counts is not large enough for intrinsically rare lensing events of high amplitude to corrupt observed samples. Lensing effects may be of greater importance for optically selected quasars, where lenses of mass as low as $\sim 10^{-4} M_\odot$ can cause large amplifications.

162.197   Particle creation by the expansion of the universe.
L. Parker.
Fundam. Cosmic Phys., Vol. 7, 201 - 239 (1982).

The purpose here is to acquaint the reader with some of the main features and consequences of particle creation by the gravitational field of the expanding universe. That process is a natural consequence of quantum field theory and general relativity (or analogous curved spacetime theories of gravity).

162.198   Kerr-Newman metric in cosmological background.
L. K. Patel, H. B. Trivedi.
J. Astrophys. Astron., Vol. 3, 63 - 67 (1982).

A new solution of Einstein-Maxwell field equations is presented. The material content of the field described by this solution is a perfect fluid plus sourceless electromagnetic fields. The metric of the solution is explicitly written. This metric is examined as a possible representation of Kerr-Newman metric embedded in Einstein static universe. The Kerr-Newman metric in the background of Robertson-Walker universe is also briefly described.

162.199   The large scale structure of the Universe I. General properties. One- and two-dimensional models.
V. I. Arnold *(Arnol'd)*, S. F. Shandarin, Ya. B. Zeldovich *(Zel'dovich)*.
Geophys. Astrophys. Fluid Dyn., Vol. 20, 111 - 130 (1982).

Evolution of initially smooth perturbations in a cold self-gravitating medium in a Friedmann Universe gives rise to the formation of singularities in the distribution of density in a manner similar to that of catastrophe theory. The authors present the full list of singularities for the one- and two-dimensional cases. They discuss the geometrical and some dynamical properties of each kind of singularity. They give also asymptotic laws for the growth of the density near each kind of singularity. This list of singularities gives the elements from which the large scale structure of the Universe is constructed.

162.200   A GUT-ed tour through the early universe.
D. V. Nanopoulos.
Cosmology and particles, (see 012.046), p. 89 - 102 (1982).

A simplified view of the evolution of the universe is presented. The implications of Grand Unified Theories at each characteristic period in the history of the universe are pointed out. A new mechanism for cosmological baryon production, through the decays of superheavy fermions, is discussed in some detail.

162.201   Cosmology and the neutron electric dipole moment.
J. Ellis.
Cosmology and particles, (see 012.046), p. 103 - 116 (1982).

There is a contribution to the neutron electric dipole moment d_n from the CP violating θ vacuum parameter of QCD. Diagrams analogous to those responsible for the baryon number of the universe also contribute to θ, providing an order of magnitude lower bound on d_n in terms of the baryon-to-photon ratio n_B/n_γ. The comparison between d_n and n_B/n_γ

gives us information about entropy generation after the epoch of baryon generation in the very early Universe.

162.202   Elementary particle phase transitions in the very early Universe.   S. A. Bludman.
Cosmology and particles, (see 012.046), p. 117 - 136 (1982).

The restoration of elementary particle symmetry at high temperatures induces a huge cosmological constant (vacuum energy density) which exceeds the thermal energy density if the Higgs meson mass is small enough. If the Universe began at low entropy (Tepid Universe), this prevents any initial cosmological singularity and prevents massive monopole production in the initial Grand Unified Theories phase transition.

162.203   Galaxies may be single particle fluctuations from an early, false-vacuum era.   W. H. Press.
Cosmology and particles, (see 012.046), p. 137 - 156 (1982).

Evolving from an early, hot Friedmann phase, the universe may go over to a "Guth era" of exponential expansion. The horizon size at the end of the Guth era (which can be as large as today's comoving scale of galaxies or cluster) is shown to correspond to a scale not much larger than the Compton wavelength of the matter fields at the beginning of the Guth era. Means of "freezing in" these initial fluctuations are suggested.

162.204   Constraints on neutrinos and axions from cosmology.
D. N. Schramm.
Cosmology and particles, (see 012.046), p. 189 - 202 (1982).

A review is made of the astrophysical arguments with regard to neutrino properties. It is shown that the best fit to the present baryon density and ^{4}He abundance is obtained with three neutrino species. The possible role of massive neutrinos in the dark mass of galaxies is discussed.

162.205   Why and how to detect the cosmological neutrino background.   J. Schneider.
Cosmology and particles, (see 012.046), p. 203 - 210 (1982).

The Standard Big Bang theory predicts, parallel to the cosmic radiation background at 2.7K, a cosmological neutrino background. The author discusses the cosmological significance and the feasibility of the detection of this background.

162.206   Primordial nucleosynthesis.
J. Audouze.
Cosmology and particles, (see 012.046), p. 231 - 240 (1982).

A summary of the primordial nucleosynthesis occurring during the early phases of the universe is provided. The observed abundances of the light elements D, ^{3}He, ^{4}He and ^{7}Li such as the processes responsible for their formation are recalled. D and ^{7}Li can be used to probe the present density of the universe and its dynamical properties on very large scales while the ^{4}He abundance is related to the lepton density and the rate of expansion of the universe. Moreover the influence of the very hypothetical mass of the neutrinos on the early evolution of the universe is mentioned.

162.207   Anisotropy of the cosmic microwave background radiation.   J. Silk.
Cosmology and particles, (see 012.046), p. 253 - 271 (1982).

Theoretical predictions of the angular anisotropy in the cosmic microwave background radiation on both small and large angular scales are described. The role of massive neutrinos is reviewed with regard to their effect on the background radiation anisotropy and on the galaxy correlation function over very large scales. A brief comparison is made with recent observational data on the background radiation, ranging from angular scales of a few arc-minutes to the dipole and quadrupole components.

162.208   The evolution of structure in the Universe: observational considerations.   R. B. Partridge.
Cosmology and particles, (see 012.046), p. 273 - 295 (1982).

Large scale structure in the Universe – galaxies and clusters, for instance – is believed to arise from small density perturbations generated at an early epoch in the expanding Universe. This paper treats the observational evidence which bears on the nature, spectrum and initial amplitude of such density perturbations. Special attention is devoted to studies of the microwave background radiation. The observational results, while not conclusive, favor isothermal over adiabatic perturbations.

162.209 **Population III objects and the shape of the cosmological background radiation.**
J. Heyvaerts, J.-L. Puget.
Cosmology and particles, (see 012.046), p. 297 - 316 (1982).

The spectrum of the cosmological radiation may keep track of non thermal processes having followed the decoupling era, in the form of departures from a strictly Planckian spectrum. In this paper the authors examine the consequences of energy and metals release by a population of pregalactic objects. The formation of the universal spectrum under these conditions is described in a self-consistent manner. It is concluded that the observed spectrum can be explained if the star burst occurred before the epoch $z \simeq 30$ and after $z \simeq 300$.

162.210 **Distortion of the microwave background by dust.**
M. Rowan-Robinson.
Cosmology and particles, (see 012.046), p. 317 - 320 (1982).

The Woody and Richards distortion of the microwave background has a natural explanation within the framework of the isothermal density fluctuation picture. A pregalactic generation of "stars" makes light and metals. The latter are able to condense into dust grains at a redshift ~ 150 - 225, which then absorb the starlight and reradiate it in the infrared. At the present epoch we see this emission redshifted into the millimetre range of the spectrum.

162.211 **Dark matter.**   G. Lake.
Cosmology and particles, (see 012.046), p. 331 - 341 (1982).

The author reviews what is known about the form and distribution of dark matter as deduced from the internal dynamics and clustering of galaxies. From their internal dynamics, there are indications that later type galaxies have relatively more dark mass. It is shown that all available evidence argues for the continuity of the galaxian two-point covariance function over a factor of roughly 10^3 in radius and a factor of 100 in luminosity. Using velocity data on scales from $50h^{-1}$ kpc to $10h^{-1}$ Mpc and the Cosmic Virial Equation, the deduced value of Ω is found to consistently lie in the range of 0.08 - 0.12. It is further argued that massive neutrinos are not consistent with this result.

162.212 **H_0, q_0 and the local velocity field.**
A. Sandage, G. A. Tammann.
Astrophysical cosmology, (see 012.047), p. 23 - 83 (1982).

162.213 **The large scale distribution of galaxies.**
M. Davis.
Astrophysical cosmology, (see 012.047), p. 113 - 143 (1982).

The author gives a review of the recently completed Harvard-Smithsonian Center for Astrophysics (CfA) galaxy survey, which is a magnitude limited redshift survey complete to a Zwicky magnitude of 14.5. The catalog contains some 2400 galaxies in a solid angle of 2.7 steradian. This sample is a compromise in that it has large angular coverage and permits a detailed study of the galaxy distribution to a distance of roughly 80 h^{-1} Mpc, twice the distance of the Shapley-Ames sample. The overall distribution of galaxies is described and compared with existing N-body simulations. The problems and challenges posed by the CfA sample for theories of galaxy and cluster formation are outlined.

162.214 **The nature of the largest structures in the universe.**
J. H. Oort.
Astrophysical cosmology, (see 012.047), p. 145 - 163 (1982).

162.215 **The nature and origin of large-scale density fluctuations.**   P. J. E. Peebles.
Astrophysical cosmology, (see 012.047), p. 165 - 187 (1982).

An important recent development in cosmology has been the discovery of large angular scale fluctuations $\delta T/T \sim 1 \times 10^{-4}$ in the brightness of the microwave background. The author reviews here the constraints this new measure may provide on models for the nature and origin of the departures from a homogeneous universe.

162.216 **Primordial nucleosynthesis and its consequences.**   J. Audouze.
Astrophysical cosmology, (see 012.047), p. 395 - 425 (1982).

The purpose of this review is to show that the light element (D, ^{3}He, ^{4}He and ^{7}Li) abundances are among the most powerful tools to decipher the physical conditions of the early universe. The most recent data especially those concerning the D, ^{4}He and ^{7}Li abundances are first reviewed. Then a brief summary of the nucleosynthetic properties of the Big Bang models, especially the most classical one called also the "canonical" Big Bang model, is presented.

162.217 **Fundamental tests of galaxy formation theory.**   J. Silk.
Astrophysical cosmology, (see 012.047), p. 427 - 472 (1982).

Galaxy formation is a highly complex process, involving both gravitational and dissipational interactions. Three tests are reviewed here that utilize the large-scale structure of the universe as an environment where traces of the seed fluctuations from which galaxies formed may be sought. Out to a scale of about 20 Mpc one can study the dynamics and structure of the Local Supercluster of galaxies, the density contrast of which is barely into the non-linear regime. On larger scales, out to about 100 Mpc, one can examine the large-scale matter distribution, which appears to contain numerous filamentary and shell-like structures and large holes. Finally, the large-scale angular anisotropy of the cosmic microwave background radiation can probe the spectrum of density fluctuations in the matter distribution both now and at a very early epoch in the universe.

162.218 **Galaxy formation.**   J. P. Ostriker.
Astrophysical cosmology, (see 012.047), p. 473 - 493 (1982).

162.219 **Remarks on a possible pregalactic "population III".**   M. J. Rees.
Astrophysical cosmology, (see 012.047), p. 495 - 500 (1982).

162.220 **Elementary particle physics in the very early universe.**   S. Weinberg.
Astrophysical cosmology, (see 012.047), p. 503 - 528 (1982).

162.221 **Massive neutrinos in cosmology and galactic astronomy.**   D. W. Sciama.
Astrophysical cosmology, (see 012.047), p. 529 - 556 (1982).

This article gives a brief qualitative discussion of the consequences for cosmology and galactic astronomy of a neutrino type possessing a restmass of order tens of electron volts. Emphasis is laid on the possible detectability of ultraviolet photons which may be emitted by massive neutrinos dominating both the universe and individual galaxies including the Milky Way. One speculative possibility is that ionisation observed both in the intergalactic medium and in the halo of our Galaxy may be produced by such photons.

162.222 Some remarks on phase-density constraints on the masses of massive neutrinos.   J. E. Gunn.
Astrophysical cosmology, (see 012.047), p. 557 - 562 (1982).

162.223 The boundary conditions of the universe.   S. W. Hawking.
Astrophysical cosmology, (see 012.047), p. 563 - 574 (1982).

This paper considers the questions of what are the boundary conditions of the universe and where should they be imposed. It is difficult to define boundary conditions at the initial singularity and, even if one could, they would be insufficient to determine the evolution of the universe. In order to overcome this problem it is suggested that one should adopt the Euclidean approach and evaluate the path integral for quantum gravity over positive definite metrics. If one took these metrics to be compact, one would avoid the need to specify any boundary conditions for the universe. This approach might explain why the apparent cosmological constant is zero, why the universe is spatially flat, and why it was in thermal equilibrium at early times.

162.224 Spontaneous birth of the closed universe and the anthropic principle.   Ya. B. Zeldovich (*Zel'dovich*).
Astrophysical cosmology, (see 012.047), p. 575 - 579 (1982).

162.225 Cosmology and fundamental physics.
M. S. Longair.
Astrophysical cosmology, (see 012.047), p. 583 - 598 (1982).

162.226 Report of IAU Commission 47: Cosmology (Cosmologie).   G. O. Abell.
Trans. IAU, Vol. XVIIIA, (see 003.013), p. 635 - 648 (1982).

Universe.   See Abstr. 003.036.

Cosmology, physics, and philosophy.
See Abstr. 003.047.

Extragalactic adventure: our strange universe.
See Abstr. 003.061.

Modern cosmology.   See Abstr. 003.112.

Extragalactic astronomy.   See Abstr. 003.114.

The unified model of the Universe. The geometrically unified field solution.   See Abstr. 003.117.

The expanding Universe.   See Abstr. 003.121.

The first three minutes. Modern outlook on the origin of the Universe.   See Abstr. 003.137.

Aufbau und Entwicklung des Weltalls. I. Historische Wurzeln der Kosmologie.   See Abstr. 004.071.

Neutrino 81. Proceedings of a conference held at Maui, Hawaii, July 1981.   See Abstr. 012.050.

Coming attractions in SUMs and cosmology.
See Abstr. 022.095.

Exploding dynamical systems.
See Abstr. 042.059.

Mass of a neutrino in physics of elementary particles and in big-bang cosmology.   See Abstr. 061.009.

Dirac neutrinos in early cosmology.
See Abstr. 061.015.

Right-handed and left-handed neutrinos and the two galactic populations of the Universe. Additional evidence for the neutrino mass.   See Abstr. 061.016.

Majorana masses, photon gas heating and cosmological constraints on neutrinos.   See Abstr. 061.018.

Right-handed neutrino interactions in the Early Universe.   See Abstr. 061.019.

Galactic halos, globular clusters and massive neutrinos.   See Abstr. 061.025.

Neutrino mass and cosmological baryon excess in left-right symmetric grand unified theories.
See Abstr. 061.026.

The rest mass of neutrinos and clustering in the early Universe.   See Abstr. 061.030.

Radiative decay lifetime of neutrinos and the evolution of the Universe after the recombination era.
See Abstr. 061.031.

Neutrinos of finite rest mass in astrophysics and cosmology.   See Abstr. 061.040.

Kinetics of molecular hydrogen formation, thermochemical evolution of primordial matter of protogalaxies and characteristics of collapsing protostars of the first generation.
See Abstr. 061.041.

Cosmological bounds on the masses of stable, right-handed neutrinos.   See Abstr. 061.044.

Does the standard hot-big-bang model explain the primordial abundances of helium and deuterium?
See Abstr. 061.055.

Comment on "Does the standard hot-big-bang model explain the primordial abundances of helium and deuterium? ".   See Abstr. 061.056.

Large scale distribution of elements and gravity.
See Abstr. 061.057.

Evolution and nucleosynthesis of primordial massive stars.   See Abstr. 065.047.

Properties and cosmological consequences of very massive objects.   See Abstr. 065.048.

The evolution of the physical characteristics of theoretical stellar models with variable G (Brans-Dicke cosmological theory) in isochrones of one, three and five billion years.   See Abstr. 065.070.

Quantum conformal fluctuations in a singular space-time.   See Abstr. 066.016.

Thermalization of starlight by elongated grains: could the microwave background have been produced by stars?   See Abstr. 066.023.

Shear-free collapse with heat flow.
See Abstr. 066.062.

All nontwisting N's with cosmological constant.
See Abstr. 066.063.

Quantum field theory in curved space-time. Massive and massless vector fields.   See Abstr. 066.064.

Gravitational vacuum hypothesis and cosmology with variable particle number.   See Abstr. 066.080.

Uniqueness of the propagator in spacetime with cosmological singularity.   See Abstr. 066.081.

Stability of gravity with a cosmological constant. See Abstr. 066.101.

H-space with a cosmological constant. See Abstr. 066.122.

Twisted symmetry breaking on the projective hypersphere: a model of the small cosmological constant. See Abstr. 066.131.

A new torsion balance for studies in gravitation and cosmology.   See Abstr. 066.132.

Superposition of Planckian spectra and the distortions of the cosmic microwave background radiation. See Abstr. 066.135.

Hydrodynamics in the O_4 gravity. See Abstr. 066.137.

Solitonic gravitational waves in Bianchi II cosmologies. 1. The general framework.   See Abstr. 066.139.

Is quantum gravity deterministic and/or time symmetric?   See Abstr. 066.141.

The large-number hypothesis and the Earth's expansion.   See Abstr. 081.052.

The Hubble diagram for type I supernovae. See Abstr. 125.052.

A revised statistical estimate of the counts of faint radio sources at 5 GHz.   See Abstr. 141.003.

Do quasars have cosmologically long lifetimes? See Abstr. 141.006.

Evidence for a decrease in the space density of quasars at $z \gtrsim 3.5$.   See Abstr. 141.020.

The periodicity in the distribution of quasar redshifts and the density perturbations in the early Universe. See Abstr. 141.062.

New study on quasars and isotropy of H_0. See Abstr. 141.076.

Superluminal motion in NRAO 140 and a possible future method for constraining H_0 and q_0. See Abstr. 141.181.

Cosmological evolution of QSOs and radio galaxies from radio-selected samples.   See Abstr. 141.196.

Quasars in the universe.   See Abstr. 141.272.

C IV absorption in an unbiased sample of 33 QSOs: evidence for the intervening galaxy hypothesis. See Abstr. 141.289.

Cosmological implications of the redshift distribution of QSO absorption systems.   See Abstr. 141.294.

An investigation of the properties of double radio sources using the Spearman partial rank correlation coefficient. See Abstr. 141.296.

The space distribution of quasars. See Abstr. 141.313.

Quasars in the universe.   See Abstr. 141.314.

Evidence from deep radio surveys for cosmological evolution.   See Abstr. 141.316.

Some aspects of the cosmological evolution of extragalactic radio sources.   See Abstr. 141.317.

On evolutionary models of radio sources. See Abstr. 141.318.

The origin of the X-ray and γ-ray backgrounds. See Abstr. 142.111.

Cosmic density waves and its observable vestige. See Abstr. 151.046.

On the N-body problem in Dirac's cosmology. See Abstr. 151.052.

Mass and luminosity in spiral galaxies and the Tully-Fisher relation.   See Abstr. 158.032.

The 21 centimeter line width as an extragalactic distance indicator. II. Does the Tully-Fisher relation depend on Hubble type?   See Abstr. 158.034.

Galactic mass loss: a mild evolutionary correction to the angular size test.   See Abstr. 158.039.

The central velocity dispersion in elliptical and lenticular galaxies as an extragalactic distance indicator. See Abstr. 158.150.

The first galaxies.   See Abstr. 158.159.

Quantum effects in the redshift intervals for double galaxies.   See Abstr. 158.211.

Emission-line galaxies in the direction of the proposed void in Bootes.   See Abstr. 158.296.

Infrared studies of a sample of 3C radio galaxies. See Abstr. 158.314.

Primordial helium and emission-line galaxies. See Abstr. 158.329.

Galaxy formation via hierarchical clustering and dissipation: the structure of disk systems. See Abstr. 158.331.

Galaxy formation via hierarchical clustering and dissipation: the structure of spheroids. See Abstr. 158.332.

The evolution of galaxies. See Abstr. 158.333.

Evidence for the cosmological evolution of the stellar content of radio galaxies.   See Abstr. 158.335.

Motion of the Local Group of galaxies and isotropy of the Universe.   See Abstr. 160.032.

The Local Supercluster.   See Abstr. 160.048.

Groups of galaxies. I. Nearby groups.
See Abstr. 160.049.

Large-scale superclusters surrounding the giant galaxy void in Bootes?   See Abstr. 160.077.

The genesis of the Local Group.
See Abstr. 160.081.

Cosmological constraints on hot plasma in a closed universe.   See Abstr. 161.002.

Author Index

The authors are listed in alphabetical order according to the initial letter following the first names.

Subject Index

Starting with Volume 18 of *Astronomy and Astrophysics Abstracts,* some alterations concerning formation, arrangement, and versatility of the key words have been made. In order to provide an adequate description of a paper, specific key words are used as frequently as possible. References to a whole subject category are suppressed now. The user, therefore, has to refer to the contents at the beginning of each volume.

Whenever possible, the key words are formed in such a way that there are two different supplementary terms, e.g. the pair

interstellar matter

molecules.

An effort is made to choose preferably terms which can be inverted in order to increase the usefulness of this index. In the example given there are the two entries

interstellar matter

molecules

and

molecules

interstellar matter.

Exceptions to the rule of inversion of terms are given in all cases where the second key word is either a very specific one (e.g. Urca processes) or a general one (e.g. history). The use of substantives is preferred. In order to obtain the possibility to extend a one-term key word in a two-term one, combinations as

Mars   or   sun

atmosphere   active regions

are changed into

Mars atmosphere   and   solar active regions,

respectively.

Starting with Volume 30 of *Astronomy and Astrophysics Abstracts,* this Index is given in a four-column arrangement. The overall size further has been optimized by suppression of a repetition of the first key word in a two-term combination. Thus, the primary term is printed only once for all the following secondary key words which belong to this header.

The user is – in any case – requested to look for synonymous entries, because further references to this topic might exist elsewhere in the Index under another current astronomical term.

ASTRONOMY AND ASTROPHYSICS ABSTRACTS

A Publication of the Astronomisches Rechen-Institut Heidelberg

Member of the Abstracting Board
of the International Council of Scientific Unions

Editors: S. Böhme, W. Fricke, H. Hefele, I. Heinrich, W. Hofmann,
D. Krahn, V. R. Matas, L. D. Schmadel, G. Zech

Volume 1	Literature 1969, Part 1	X + 435 pp. (1969)
Volume 2	Literature 1969, Part 2	X + 516 pp. (1970)
Volume 3	Literature 1970, Part 1	X + 490 pp. (1970)
Volume 4	Literature 1970, Part 2	X + 562 pp. (1971)
Volume 5	Literature 1971, Part 1	X + 505 pp. (1971)
Volume 6	Literature 1971, Part 2	X + 560 pp. (1972)
Volume 7	Literature 1972, Part 1	X + 526 pp. (1972)
Volume 8	Literature 1972, Part 2	X + 594 pp. (1973)
Volume 9	Literature 1973, Part 1	X + 610 pp. (1973)
Volume 10	Literature 1973, Part 2	X + 661 pp. (1974)
Volume 11	Literature 1974, Part 1	X + 579 pp. (1974)
Volume 12	Literature 1974, Part 2	X + 699 pp. (1975)
Volume 13	Literature 1975, Part 1	X + 632 pp. (1975)
Volume 14	Literature 1975, Part 2	X + 747 pp. (1976)
Volume 15/16	Author and Subject Indexes to Volumes 1–10	
	Literature 1969–1973	VII + 655 pp. (1976)
Volume 17	Literature 1976, Part 1	XII + 645 pp. (1976)
Volume 18	Literature 1976, Part 2	X + 859 pp. (1977)
Volume 19	Literature 1977, Part 1	X + 732 pp. (1977)
Volume 20	Literature 1977, Part 2	X + 786 pp. (1978)
Volume 21	Literature 1978, Part 1	X + 834 pp. (1978)
Volume 22	Literature 1978, Part 2	X + 849 pp. (1979)
Volume 23/24	Author and Subject Indexes to Volumes 11–14 and 17–22	
	Literature 1974–1978	VIII +1127 pp. (1979)
Volume 25	Literature 1979, Part 1	X + 872 pp. (1979)
Volume 26	Literature 1979, Part 2	X + 794 pp. (1980)
Volume 27	Literature 1980, Part 1	X + 939 pp. (1980)
Volume 28	Literature 1980, Part 2	X + 841 pp. (1981)
Volume 29	Literature 1981, Part 1	X + 853 pp. (1981)
Volume 30	Literature 1981, Part 2	X + 792 pp. (1982)
Volume 31	Literature 1982, Part 1	X + 776 pp. (1982)

Published for Astronomisches Rechen-Institut by
Springer-Verlag Berlin Heidelberg New York

W. Högner, N. Richter

Isophotometric Atlas of Comets

This beautiful atlas contains a carefully selected collection of material needed for the study of the physics of comets. The authors scrutinized more than 300 photographs taken in the years 1902–1967. They applied photographic equidensities as quasiisophotes according to the method of Lau and Krug by using the Sabattier effect. Reproductions of

- the original photographs,
- their isophote diagrams,
- enlarged isophote diagrams of the cometary heads and nuclei

are presented. The most important documentary and astronomical data of the objects represented in the atlas are compiled in a special set of tables. The IAU considers the compilation of this atlas to be of "extreme value" to the astronomical community.

Part 1
1980. 90 plates, comments and tables
ISBN 3-540-09171-8

Part 2
1980. 55 plates, comments and tables
ISBN 3-540-09172-6

K.R. Lang

Astrophysical Formulae
A Compendium for the Physicist and Astrophysicist

2nd corrected and enlarged edition. 1980. 46 figures, 69 tables. XXIX, 783 pages
ISBN 3-540-09933-6

"... For astronomers, teachers and students it represents an important reference source for fundamental formulae used in astrophysics. For a student it may serve as a compact review of a familiar field or a handy aid to gain a rapid insight into the techniques of new fields of astrophysics. For a teacher the **Astrophysical Formulae** will be a useful guide in the very broad field of modern astrophysics... It is a work that everybody from us needs and that will spare much of our time and effort."
Journal of the British Interplanetary Society

Springer-Verlag
Berlin
Heidelberg
New York